Geography in America at the Dawn of the 21st Century

Edited by

GARY L. GAILE

and

CORT J. WILLMOTT

OXFORD
UNIVERSITY PRESS

OXFORD
UNIVERSITY PRESS

Great Clarendon Street, Oxford OX2 6DP

Oxford University Press is a department of the University of Oxford.
It furthers the University's objective of excellence in research, scholarship,
and education by publishing worldwide in

Oxford New York

Auckland Cape Town Dar es Salaam Hong Kong Karachi
Kula Lumpur Madrid Melbourne Mexico City Nairobi
New Delhi Shanghai Taiper Toronto

With offices in

Argentina Austria Brazil Chile Czech Republic France Greece
Guatemala Hungary Italy Japan South Korea Poland Portugal
Singapore Switzerland Thailand Turkey Ukraine Vietnam

Published in the United States
by Oxford University Press Inc., New York

First published 2003

British Library Cataloguing in Publication Data
Data available

Library of Congress Cataloging in Publication Data
Data available

ISBN 0-19-823392-2

10 9 8 7 6 5 4 3 2

Typeset by Graphicraft Limited, Hong Kong
Printed in Great Britain
on acid-free paper by
Antony Rowe Limited,
Chippenham, Wiltshire

To Susan and Pat

Foreword

In a world in which both natural environment and human society are changing rapidly and profoundly this review of the past decade is a unique and basic appraisal of recently unfolding knowledge. Drawing upon observations from diverse geographic analysis, it reviews significant developments that are important for at least four groups of readers.

The observations and interpretation are of unique value for beginning or prospective students who are exploring the dimensions of the geographic discipline and its challenges to their future careers. It identifies up-to-date and fruitful ideas, personnel, and institutions.

For researchers and teachers who are committed to geography as a discipline the book reviews much with which they are familiar but certainly will give them new insights into subfields other than their own, and assure their recognition of new advances in methods and findings. The typical scholar attending an annual disciplinary meeting may expect to learn the latest findings in her or his own specialty but cannot attend discussions of other specialty areas, and will find them thoughtfully summarized in this unique volume.

Scientists and teachers from other disciplines can find discerning reviews of recent geographic findings and methods of possible interest without being obliged to search through a large number of publications to identify relevant reports.

Finally, readers with broad concerns for social and environmental change will find guidance to recent geographic contributions to understanding a wide range of scientific questions from dimensions of global climatic change to local land use. For example, there are lessons for national environmental policy as well as local city planning.

In facilitating these various uses of the volume the editors have avoided undue emphasis upon any one philosophical orientation or any one methodology. Diversity and lack of overall bias are evident while allowance is made for a solid representation of different approaches to a rapidly changing world. The resulting appraisal is thoughtful, creative, and comprehensive.

Gilbert White

Acknowledgements

Geography in America at the Dawn of the 21st Century required the participation of numerous people, and of the Association of American Geographers (AAG) and its specialty groups. The AAG specialty groups and chapter authors, in particular, deserve most of the credit for their enthusiastic support and participation. Long hours were spent by specialty-group authors and other group members to produce high-quality, state-of-the-research chapters that are the core of this book.

Oxford University Press was both encouraging and patient while, at the same time, nudging us to bring this book to fruition. Anne Ashby was a magnificent force. Not only did she encourage us with dining in OUP's exquisite facilities, but she also said "Now, Gary" just when it was needed. Sarah Holmes was our bulwark, she was unflappable when most folks would have flapped. Finally, Sylvia Jaffrey wielded a copy-editing effort never matched by even seasoned veterans. You all have our deepest thanks. We are most grateful.

Presentations of draft chapters at the 1999 Annual AAG Meeting in Hawaii were ably guided by the following session chairs, who have our heartfelt thanks: Ed Aguado, Reg Golledge, Mike Goodchild, Will Graf, John Paul Jones III, Paul Knox, Sallie Marston, Alec Murphy, Billie Lee Turner II, Tom Wilbanks.

External reviewers also worked diligently, and they have our gratitude. They provided us with the indispensable insights that were required to ensure the quality and comprehensiveness of the chapters, and thereby the volume itself. They are: Ed Aguado, Dan Arreola, Roger Barry, Bernie Bauer, Dan Bedford, D. Gordon Bennett, Bill Berentsen, Brian Berry, Bill Bowen, Kathleen Braden, Tony Brazel, Ray Bromley, Lyn Brown, Babs Buttenfield, Anne Buttimer, Karl Butzer, Nel Caine, Barbara Carmichael, Bill Clark, Keith Clarke, Vic Conrad, Frank Davis, George Demko, Mona Domosh, James Eflin, Ken Foote, Melissa Gilbert, Pat Gilmartin, Jim Goodman, Peter Gould, Lisa Graumlich, Eve Gruntfest, Susan Hanson, Peter Heffington, Dave Hill, Jim Huff, Richard Jackson, Bob Kates, Cindi Katz, William Koelsch, Helga Leitner, Gordon Lewthwaite, David Ley, C. P. Lo, Ed Malecki, Dick Marston, Russ Mather, Steve Matthews, Jonathan Mayer, Mike McNulty, Judy Meyers, Julian Minghi, Lisle Mitchell, Hal Moellering, Barbara Morehouse, Alec Murphy, Darrell Napton, Sam Natoli, Duane Nellis, Tim Oakes, Val Preston, Marie Price, Paul Robbins, Eric Shepard, Doug Sherman, Ira Sheskin, Neil Smith, Christoph Stadel, Phil Suckling, Robert Stock, Graham Tobin, Bret Wallach, Marv Waterstone, Barry Weller, Tom Wilbanks.

It is important to acknowledge those places that influenced either the inspiration for or production of this book. It was on a deck at the University of Wisconsin, overlooking Lake Mendota, that the "Geography in America" concept was developed, nearly seventeen years ago. *Geography in America at the Dawn of the 21st Century* grew out of the first (1989) *Geography in America* volume, which was conceived at the Lake Mendota venue. In addition, we would like to acknowledge the inspirational contributions of Boulder, Colorado; Cambridge, UK; London, UK; Cambria, California; Honolulu, Hawaii; and Maui, Hawaii. Newark, Delaware also came into play, but not necessarily in an inspirational way. The Continental Divide and Pacific Ocean played an important editorial role—chapter manuscripts edited on a deck overlooking the mountains or Pacific Ocean were required to maintain the editor's undivided attention. We would also like to acknowledge Trios, Mateo, the Deer Park, Rhumbas, Creekside Gardens Café, Nepenthe, and the Boulderado for providing much needed ambience.

The Geography Departments at the Universities of Colorado and Delaware are places within which much of the production of this book took place. At Colorado, special thanks go to Marcia Signer and Brian King for keeping things moving when Gary was moving. At Delaware, Janice Spry, Linda Parrish, and Karen Stabley helped keep "our ship" on course when administrative demands stalled Cort. Michelle Johnson, Elsa Nickl, Kenji Matsuura, Tony Seraphin, and Pat Willmott also helped Cort with the compilation of the name index.

Our friends assisted, tolerated, or simply distracted us, and helped to keep us sane, and we are grateful. These folks include: the Fergusons, Ed Aguado, Mike McNulty, Bill and Irene Clark, Jim and Anne Huff, Wes and Jan, Elayne, Buzz, Melissa and the kids, Perfect, Lefty, Cuddles, Smokie, Mike and Deb Pagano, Sam Fitch, Leslie Durgen, Russ Mather, Dick and Beth Svee, Kenji Matsuura, Fritz Nelson, Tower of Power, Junior Wells, Christian McBride, and the people of Cambria.

Our families have been most supportive, and we are especially thankful for their understanding of our unremitting, unsocial behavior caused by the many hours that *Geography in America at the Dawn of the 21st Century* consumed. Susan, Pat, Jeff, Abby, and Mike, and Julia—we love you immensely and we'll be home more—soon.

Boulder and Newark G. L. G. and C. J. W.

Contents

List of Figures

List of Tables

Abbreviations

AAG	Association of American Geographers
ACS	American Community Survey
ACSM	American Congress on Surveying and Mapping
ACSUS	Association for Canadian Studies in the United States
AEGSG	American Ethnic Geography Specialty Group
AGILE	Association for Geographic Information Laboratories in Europe
AGS	American Geographical Society
AP	Advanced Placement
ARGUS	Activities and Readings in the Geography of the United States
ARGWorld	Activities and Resources in the Geography of the World
ASCUS	Association for Canadian Studies in the United States
ASEAN	Association of South East Asian Nations
ASPG	American Society of Professional Geographers
ASPRS	American Society for Photogrammetry and Remote Sensing
CAAA	Clean Air Act Amendments (US)
CAD	Computer Assisted Design
CARLU	Contemporary Agriculture and Rural Land Use
CCM	Community Climate Model
CESG	Cultural Ecology Specialty Group
CGP	*Current Geographical Publications*
CLAG	Conference of Latin Americanist Geographers
CoMa	Coastal and Marine Geography Specialty Group
CRP	Conservation Reserve Program
CSG	Cryosphere Specialty Group
CSM	Climate System Model
CSSG	Canadian Studies Specialty Group
DAWN	Development Alternatives with Women for a New Era
DCM	Digital cartographic model
DEM	Digital elevation model
DLM	Digital landscape model
DOD	Department of Defense
DSM	Demand-side management
EDA	Economic Development Administration
EDRA	Environmental Design Research Association
ENSO	El Niño southern oscillation
EPBG	Environmental Perception and Behavioral Geography
EROI	Energy return on investment
ESG	European Specialty Group
ESRI	Environmental Science Research Institute
ESS	Earth system science
ETM+	Enhanced Thematic Mapper Plus
EWS	Early Warning System
FCCC	Framework Convention on Climate Change
FDI	Foreign direct investment

FGDC	Federal Geographic Data Committee
FIPS	Federal Information Processing Standard
FTA	Free Trade Agreement
GA	Geographical Association
GAD	Gender and Development
GCLP	The Global Change and Local Places Project
GCM	General circulation model
GENIP	Geography Education National Implementation Project
GESG	Geography Education Specialty Group
GIGI	Geographic Inquiry into Global Issues
GIS	Geographical Information Systems
GISci	Geographical Information Science
GISP	Greenland Ice Sheet Projects
GISSG	GIS Specialty Group
GISHE	Geographical Information Systems in Higher Education
GITA	Geospatial Information and Technology Association
GORABS	Geography of Religions and Belief Systems
GPOW	Geographic Perspectives on Women
GPS	Global Positioning System
GSG	Geomorphology Specialty Group
HDCD	Historical Data Climate Dataset
HDGC	Human Dimensions of Global Change
IAG	International Association of Geomorphologists
IAPS	International Association for People–Environment Studies
IBG	Institute of British Geographers
ICA	International Cartographic Association
ICT	Information and Communication Technologies
IDNDR	International Decade for Natural Disaster Reduction
IFSAR	Interferometric synthetic aperture Radar
IGBP	International Geosphere-Biosphere Program
IGU	International Geographic Union
IJGIS	*International Journal of Geographical Information Systems*
IJPG	*International Journal of Population Geography*
IFOV	Instantaneous Field of View
IMF	International Monetary Fund
IPCC	Intergovernmental Panel on Climate Change
IRGEE	*International Research in Geographical and Environmental Education*
ISO	International Standards Organization (International Organization for Standardization)
IVHS	Intelligent Vehicle Highway System
IWMI	International Water Management Institute
IWRA	International Water Resources Association
JANIS	Joint Army and Navy Intelligence Studies
JGHE	*Journal of Geography in Higher Education*
JOG	*Journal of Geography*
LA	Location Allocation
LIDAR	Light Detection And Ranging
LRS	Linear Referencing Systems
LULC	Land use and land cover
LULCC	Land-use and land-cover change
MAUP	Modifiable areal unit problem
MGSG	Medical Geography Specialty Group
MODIS	Moderate Resolution Imaging Spectroradiometer

MOOTW	Military operations other than war
MSC	Mapping Science Committee
MSS	Multispectral Scanner
NAEP	National Assessment of Educational Progress
NAFTA	North American/Atlantic Free Trade Area
NAO	North Atlantic Oscillation
NASA	National Aeronautics and Space Administration
NCAR	National Center for Atmospheric Research
NCES	National Center for Education Statistics
NCGE	National Council for Geographic Education
NCGIA	National Center for Geographic Information and Analysis
NCSS	National Council for Social Studies
NDVI	Normalized Difference Vegetation Index
NGDS	National Geographic Data System
NGO	Non-Governmental Organization
NGS	National Geographic Society
NOAA	National Oceanic and Atmospheric Administration
NSDI	National Spatial Data Infrastructure
NSF	National Science Foundation
NSIDC	National Snow and Ice Data Center
OGC	Open GIS Consortium
PDSI	Palmer Drought Severity Index
PNA	Pacific-North American
RSSG	Remote Sensing Specialty Group
RTS	Recreation, Tourism, and Sport
SAP	Structural adjustment program
SAR	Synthetic aperture Radar
SD	Sustainable development
SDSS	Spatial Decision Support Systems
SDTS	Spatial Data Transfer Standard
SIU	Southern Illinois University
SOI	Southern Oscillation Index
SGSG	Socialist Geography Specialty Group
SPOT	Système pour l'observation de la Terre
SRTM	Shuttle Radar Topography Mission
SSSG	Sexuality and Space Specialty Group
STDS	Spatial Data Transfer Standard
TGSG	Transportation Geography Specialty Group
TIGER	Topographically Integrated Geographic Encoding and Referencing
TRI	Toxic Release Inventory
UCGIS	University Consortium for Geographic Information Science
UCOWR	University Council on Water Resources
UNCED	United Nations Conference on Environment and Development
UNFAO	United Nations Food and Agriculture Organization
URISA	Urban and Regional Information Systems Association
USACE	United States Army Corps of Engineers
USDA	United States Department of Agriculture
USDOE	United States Department of Energy
USEPA	United States Environmental Protection Agency
UWIN	University Water Information Network
VJESG	Values, Justice and Ethics Specialty Group
WID	Women in Development
WMO	World Meteorological Organization

List of Contributors

Stuart Aitken, San Diego State University, saitken@mail.sdsu.edu
James P. Allen, California State University-Northridge, james.allen@csun.edu
Douglas M. Amadeo, University of Nebraska-Lincoln, damadeo@unlserve.unl.edu
J. Clark Archer, University of Nebraska-Lincoln, jarcher@unlibfo.unl.edu
Trevor Barnes, University of British Columbia, tbarnes@pop.geog.ubc.ca
Thomas J. Bassett, University of Illinois-Urbana, bassett@uiuc.edu
Sarah W. Bednarz, Texas A&M University, s-bednarz@tamu.edu
James E. Bell, US Dept. of State, jbell@pd.state.gov
Kate Berry, University of Nevada-Reno, kberry@scs.unr.edu
Mark A. Blumler, State University of New York-Binghamton, mablum@binghamton.edu
Daniel G. Brown, Michigan State University, brownda@pilot.msu.edu
David R. Butler, Southwest Texas State University, db25@swt.edu
George Carney, Oklahoma State University, cgeorge@okway.okstate.edu
Cesar Caviedes, University of Florida, caviedes@geog.ufl.edu
Shaul Cohen, University of Oregon, scohen@oregon.uoregon.edu
Craig E. Colten, Louisiana State University, ccolten@lsu.edu
John A. Cross, University of Wisconsin-Oshkosh, cross@uwosh.edu
Susan Cutter, University of South Carolina, scutter@gwm.sc.edu
Lori Daniels, University of British Columbia, daniels@geog.ubc.ca
Karen de Bres, Kansas State University, karendb@ksu.edu
Douglas Deur, Louisiana State University, ddeur@unix1.sncc.lsu.edu
Roger M. Downs, Pennsylvania State University, rd7@psu.edu
Leslie A. Duram, Southern Illinois University-Carbondale, duram@siu.edu
Glen Elder, University of Vermont, gelder@zoo.uvm.edu
Andrew W. Ellis, Arizona State University, andrew.w.ellis@asu.edu
Gregory Elmes, West Virginia University, elmes@wvgeog.wvnet.edu
Kurt E. Englemann, University of Washington, kengel@u.washington.edu
Lawrence E. Estaville, Southwest Texas State University, le02@swt.edu
C. Cindy Fan, University of California-Los Angeles, fan@geog.ucla.edu
William Forbes, North Texas State University, wforbes@unt.edu
Donald A. Friend, Minnesota State University, friend@mnsu.edu
Gary L. Gaile, University of Colorado-Boulder, gaile@spot.colorado.edu
Wil Gesler, University of North Carolina-Chapel Hill, wgesler@email.unc.edu
Amy Glasmeier, Pennsylvania State University, akg1@ems.psu.edu
Patricia Gober, Arizona State University, gober@asu.edu

Andrew R. Goetz, University of Denver, agoetz@du.edu

Reginald G. Golledge, University of California-Santa Barbara, golledge@geog.ucsb.edu

Sucharita Gopal, Boston University, suchi@crsa.bu.edu

Anton Gosar, University of Ljubljana, anton.gosar@guest.arnes.si

Dean Hanink, University of Connecticut, hanink@uconnvm.uconn.edu

Susan Hardwick, University of Oregon, susanh@oregon.uoregon.edu

James W. Harrington, University of Washington, jwh@u.washington.edu

Andrew Herod, University of Georgia, aherod@arches.uga.edu

Kenneth M. Hinkel, University of Cincinnati, ken_hinkle@compuserve.com

Rex Honey, University of Iowa, rex-honey@uiowa.edu

Peter J. Hugill, Texas A&M University, pjhugill@tamu.edu

John R. Jensen, University of South Carolina, jrjensen@sc.edu

Ezekiel Kalipeni, University of Illinois-Urbana, kalipeni@uiuc.edu

Sylvia-Linda Kaktins, University of Toledo, skaktin@UTNet.UToledo.edu

David Keeling, Western Kentucky University, david.keeling@wku.edu

Karen Kemp, University of Redlands, karen_kemp@redlands.edu

Judith Kenny, University of Wisconsin-Milwaukee, jkenny@csd.uwm.edu

Lawrence Knopp, University of Minnesota-Duluth, lknopp@umn.edu

Daniel Knudsen, Indiana University, knudsen@indiana.edu

Audrey Kobayashi, Queens University, kobayasi@post.queensu.ca

Boian Koulov, George Washington University, bkoulov@gwu.edu

David Legates, University of Delaware, legates@udel.edu

Thomas R. Leinbach, University of Kentucky, leinbach@pop.uky.edu

Alan A. Lew, Northern Arizona University, Alan.lew@nau.edu

Martin Lewis, Stanford University, mwlewis@stanford.edu

Diana Liverman, University of Oxford, diana.liverman@environmental-change.oxford.ac.uk

Deborah Anne Luchsinger, National Renewable Energy Laboratory, deborah_luchsinger@nrel.gov

Laurence J. G. Ma, University of Akron, larryma@uakron.edu

Susan Macey, Southwest Texas State University, sm07@swt.edu

Susan Mains, University of the West Indies, Mona, susan.mains@uwimona.edu.jm

David M. Mark, University of Buffalo, dmark@geog.buffalo.edu

Geoffrey J. Martin, independent scholar, none

Linda McCarthy, University of Wisconsin-Milwaukee, lmccarth@uwm.edu

Mary G. McDonald, University of Hawaii-Manoa, mcdonald@hawaii.edu

Patrick McGreevy, Clarion University, mcgreevey@vaxa.clarion.edu

Robert McMaster, University of Minnesota, mcmaster@atlas.socsci.umn.edu

Linda O. Mearns, National Center for Atmospheric Research, lindam@ncar.ucar.edu

Christopher D. Merrett, Western Illinois University, cd-merrett@wiu.edu

Klaus J. Meyer-Arendt, University of West Florida, kjma@uwf.edu

Don Mitchell, Syracuse University, dmmitc01@maxwell.syr.edu
Beth Mitchneck, University of Arizona, bethm@u.arizona.edu
Ines M. Miyares, Hunter College, imiyares@geo.hunter.cuny.edu
Mark S. Monmonier, Syracuse University, mon2ier@syr.edu
Burrell Montz, University of Binghamton, montz@binghamton.edu
Stanley Morain, University of New Mexico, smorain@edac.unm.edu
Karen N. Morin, Bucknell University, morin@bucknell.edu
Ellen Mosley-Thompson, The Ohio State University, thompson.4@osu.edu
Garth Myers, University of Kansas, gmyers@ukans.edu
Heidi Nast, DePaul University, hnast@depaul.edu
Oberhauser Ann, West Virginia University, aoberhau@wvu.edu
Benjamin Ofori-Amoah, University of Wisconsin-Stevens Point, bofori@uwsp.edu
Joesph Oppong, University of North Texas, oppong@unt.edu
Eugene J. Palka, United States Military Academy-West Point, be4546@usma.edu
Clifton W. Pannell, University of Georgia, cpannell@franklin.uga.edu
Martin J. Pasqualetti, Arizona State University, pasqualetti@asu.edu
Cynthia Pope, Central Connecticut State University, cynthia_pope@yahoo.com
James Proctor, University of California-Santa Barbara, jproctor@geog.ucsb.edu
Carolyn Prorok, Slippery Rock State University, carolyn.prorok@sru.edu
Norbert Psuty, Rutgers University, psuty@imcs.rutgers.edu
Dale A. Quattrochi, National Atmospheric and Sciences Administration, dale.quattrochi@msfc.nasa.gov
Bruce A. Ralston, University of Tennessee, bralston@utk.edu
Michael R. Ratcliffe, US Government Department of the Census, michael.r.ratcliffe@census.gov
Merrill K. Ridd, University of Utah, merrill.ridd@geog.utah.edu
David Rigby, University of California-Los Angeles, rigby@geog.ucla.edu
David J. Robinson, Syracuse University, drobins@maxwell.syr.edu
Jeffrey C. Rogers, Ohio State University, jrogers@geography.ohio-state.edu
Peter Rogerson, University of Buffalo, rogerson@acsu.buffalo.edu
Thomas A. Rumney, Plattsburgh State University, thomas.rumney@plattsburgh.edu
Donna Rubinoff, University of Colorado-Boulder, rubinoff@colorado.edu
Robert Rundstrom, University of Oklahoma, rrundstrom@ou.edu
Scott Salmon, Miami University, salmonsc@muohio.edu
Fred M. Shelley, Southwest Texas State University, fs03@swt.edu
Nanda Shrestha, Florida A&M University, nanda.shrestha@famu.edu
Geoffrey C. Smith, University of Manitoba, smithgc@ms.umanitoba.ca
Barry D. Solomon, Michigan Technological University, bdsolomo@mtu.edu
Lynn Staeheli, University of Colorado-Boulder, lynner@spot.colorado.edu
Philip E. Steinberg, Florida State University, psteinbe@coss.fsu.edu
Robert H. Stoddard, University of Nebraska-Lincoln, rstoddar@unlinfo.unl.edu

Frederick Stutz, San Diego State University, stutz@mail.sdsu.edu

Kok-Chiang Tan, University of Guelph, kctan@uoguelph.ca

Gerard Toal, Virginia Technological University, toalg@vt.edu

Nancy Torrieri, US Government Department of the Census, nancy.k.torrieri@census.gov

Billie Lee Turner II, Clark University, bturner@vax.clarku.edu

James A. Tyner, Kent State University, jtyner@kent.edu

Thomas T. Veblen, University of Colorado-Boulder, veblen@spot.colorado.edu

JoAnn Vender, Pennsylvania State University, jodivender@yahoo.com

Steven J. Walsh, University of North Carolina-Chapel Hill, swalsh@email.unc.edu

John F. Watkins, University of Kentucky, geg173@uky.edu

James L. Wescoat, Jr., University of Illinois-Urbana, wescoat@staff.uiuc.edu

Gilbert White, University of Colorado-Boulder, gilbert.white@colorado.edu

Cort J. Willmott, University of Delaware, willmott@udel.edu

Dick Winchell, Eastern Washington University, dick.winchell@mail.ewu.edu

Julie A. Winkler, Michigan State University, winkler@pilot.msu.edu

Dawn Wright, Oregon State University, dawn@dusk.geo.orst.edu

Brent Yarnal, Pennsylvania State University, alibar@essc.psu.edu

Kenneth R. Young, University of Texas-Austin, kryoung@mail.utexas.edu

Terence Young, California State Polytechnic University-Pomona, tgyoung@mindspring.com

Susy Svatek Ziegler, University of Minnesota, ziegler@geog.umn.edu

Karl S. Zimmerer, University of Wisconsin-Madison, zimmerer@facstaff.wisc.edu

CHAPTER 1

Introduction

Geography in America has become more robust, more recognized, more marketable, more unified, and more diversified since the first publication of *Geography in America* (Gaile and Willmott 1989*a*). American geographers have built on geography's traditional strengths, while simultaneously embracing valuable new ideas and evaluating important new perspectives that have challenged the established theory and knowledge base of the discipline (National Research Council 1997). The robustness of American geography is well illustrated within the chapters in this book. Across the discipline from Geographic Information Science to the regional geography of Africa, American geographers have been able to respond constructively to new challenges and criticism, including the clear need to understand and evaluate the causes and effects of the events of September 11, 2001.

Defining and Characterizing Geography

American geography at the dawn of the twenty-first century can be characterized by its *unity amidst diversity*. While our traditional focus on place—and on spatial relationships within and among places—continues to provide unity, a growing variety of research problems, methods, subfields, and epistemologies is increasing our diversity. While we well recognize the difficulty in defining "geography" satisfactorily (Gaile and Willmott 1989*b*), we also are persuaded that an understanding of our shared perspectives, principles, and goals holds the greatest promise for effectively integrating diversity into our discipline. For this reason, we offer a synopsis of the nature and practice of geography, which draws from earlier work and especially from the above-mentioned National Research Council (NRC) report.

Several years ago, Gilbert White asked us personally to define "geography," and we give a slightly revised version of that definition and characterization here. We continue to believe that geography "is not bounded," but now feel that a meaningful definition and characterization of the nature and practice of geography is both possible and useful.

Definition

Geography is the study and science of environmental and societal dynamics and society–environment interactions as they occur in and are conditioned by the real world. Geographic investigations into these are influenced by the character of specific places, as well as by spatial relationships among places and processes at work over a hierarchy of geographic scales.

Characterization

Reciprocal influences, i.e. of environmental and societal dynamics on geographic places and regions, are of commensurate importance within geography. Appreciation for and understanding of the interplay between societal and environmental dynamics within and across the myriad of geographic contexts is a recurring theme, as are field research and efforts to improve the quality of the human experience and the environment through informed intervention. Geographers, at an increasing rate, are investigating how processes and resultant patterns vary over the range of geographic scales from the local to the global. They occasionally work to improve geographic theory but, more commonly, geographers' interests lie in solving real-world problems that have a significant geographic dimension. Geographic approaches to problem-solving (methods) are quite varied, but typically include visualization and digital analyses (often using maps or geographic information systems—GIS), as well as verbal, mathematical, and cognitive assessments.

Through the combined use of geographic theory, knowledge, and methods, geographers endeavor to describe, evaluate, explain, ameliorate, and forecast important changes taking place on the surface of the Earth. Most geographers also have a deep aesthetic appreciation for landscape and the web of interacting societal and environmental processes that produce it. Geography is a discipline dedicated to the understanding and appreciation of environmental and societal processes and their interactions.

Categorizing the Work of American Geographers

Categories of geographic research can be identified as much by their distinctive perspectives as they can by their subject matter (NRC 1997). A geographic work, as a consequence, can be categorized according to its perspective, its subject matter, or both, or perhaps by its modes of representation (visual, verbal, mathematical, digital, or cognitive). The NRC (ibid.) defined these three dimensions of geography as its "domains of synthesis" (subject matter), ". . . ways of looking at the world" (perspectives), and "spatial representation" (ways of representing geographic phenomena or processes). According to the NRC, three main "domains of [geographic] synthesis" can described as "environmental dynamics," "environmental/societal dynamics," and "human/societal dynamics," while geographers' "ways of looking at the world" are designed to reveal "integration in place," "interdependencies between places," and "interdependencies among [spatial] scales." The elements of the third NRC dimension, "ways of representing . . ." are listed above. Process and change are at the heart of modern geography, and the NRC report underscored this by using the term "dynamics" in the name of each domain of synthesis. The NRC depiction of the three dimensions of contemporary geography was well conceived; thus, we use elements of it to help us identify main sections within this book.

Growth and Change over the Last Half-Century

Geography, as an academic discipline, changed fundamentally during the second half of the twentieth century. The number of geographers in the academy swelled substantially, as did the number of schools and colleges that teach modern geography.

The traditional, intellectual corpus of geography remains strong, despite significant change brought by at least three "revolutions" during the last half-century. The first of these was the Quantitative Revolution of the 1960s and 1970s, which was an effort to replace the descriptive "exceptionalism," which dominated geography for decades into the 1950s, with normative and empirical approaches to analysis and inference. The debate between Hartshorne (1955) and Schaefer (1953) is a classic that defined these positions (see Billinge *et al.* (1984) for a set of "recollections" of this revolution). The second revolution was Marxist in its orientation (see Harvey 1973, and early issues of the "radical" journal *Antipode*), and it was critical of certain practices and viewpoints, including reductionism associated with applied statistics, inequities inherent in the capitalist system, and, by extension, the Vietnam War. For the current state of this research see Ch. 15, from the Socialist specialty group. The concern for inequalities of power also precipitated the somewhat simultaneous evolution (as opposed to revolution) of a gendered geography (Hayford 1974). This evolution has reached "establishment" status (see Ch. 47, from the Geographic Perspectives on Women specialty group). Gender has established itself as an important approach to understanding, especially in human geography when differential power relations occur. More recently, the Postmodern Revolution (Harvey 1989; Soja 1989) was a wide-ranging critique of the academic system, and especially of its traditional modernist approaches to knowledge production. These three intellectual revolutions bear some striking similarities. All were reactions to weaknesses in the mainstream practice of geography at the time, and all acquired converts from the pool of successful mainstream practitioners, a number of whom became the champions of the revolution. All also spawned active cadres of "true believers" who, at times, denigrated the work of other geographers as irrelevant, wrong, or counter-productive. The three revolutions and challenges to orthodoxy have all waned, but each has made an indelible imprint on American geography.

There is no question that the intellectual growing-pains experienced during these revolutions were sometimes unpleasant; none the less, we believe that their net influences on geography have made it a much more robust discipline. Consider, for example, that "empirical verifiability"—a maxim of the Quantitative Revolution—most certainly cannot answer all important questions, but it frequently can augment or strengthen our knowledge of a subject. It also is true that the dialectical approach offered during the Marxist Revolution—as well as the questioning of the system within which we operate—cannot always provide practical solutions, but it often sheds light on very important relationships between economics and power. And, while the Postmodern Revolution often left us without an adequate way of moving forward, it taught us to examine carefully the deeper meanings in our text, problems, and the very way that we go about producing knowledge. If there is one thing that we have learned, it is to be tolerant of alternative or even revolutionary thought as, in the end, it may be good for us.

Continuing Self-Examination

American geographers' introspection, with respect to the nature and practice of their discipline, has continued since the publication of *Geography in America* in 1989. Among these works was a rather different self-examination of the discipline, edited by Ronald Abler, Melvin Marcus, and Judith Olson, and entitled *Geography's Inner World: Pervasive Themes in Contemporary American Geography.* This innovative volume appeared in 1992, and it explored cross-cutting themes rather than specialty area or topical interests. Among the themes considered were geographers' modes of communication, analysis and modeling, and visualization amongst others. *Geography in America* and *Geography's Inner World* made fine companion volumes, since they both attempted to characterize geography in the 1980s or early 1990s, but from different perspectives.

Five years after *Geography's Inner World* was published, the National Research Council issued its report on *Rediscovering Geography: New Relevance for Science and Society* (NRC 1997). It reaffirmed that "location matters," and that understanding spatial relationships and place remain fundamental to understanding "the evolving character and organization of the Earth's surface." Its primary purpose, however, was to identify issues and constraints for the discipline, largely within the United States, as well as to clarify research and teaching priorities. Although written mainly for leaders and decision-makers from government, education, and the private sector, rather than for geographers, a sizable portion of the report is devoted to characterizing the nature and practice of geography. This report's re-sorting of geographers' research and teaching interests and their "ways of looking at the world" into a three-dimensional "matrix of geographic perspectives" is particularly intriguing, and it helped guide the organization of the sections within this book. And, at this writing, a new assessment is being compiled by Donald Dahmann. It will be called the *Geography in America Timeline* (no relation to either this or the former *Geography in America* volume), and will chronicle the history of geography in the United States.

Of critical interest are new positions and debates ongoing in the discipline. Golledge's (2002) spatial vision of the nature of geographic knowledge is quite different from Turner's (2002) environment/society-based view. The debate in response to Turner (Butzer 2002; Kates 2002; Wescoat 2002) and the fact that Cutter, Golledge, and Graf (2002) had the courage to tackle the challenge of addressing what are the "big questions" in geography provide convincing evidence of the robustness of geography in America.

It is clear that American geographers continue to have an interest in examining their place in the world of academia, as well as the state of their discipline. *Geography in America at the Dawn of the 21st Century* is an effort to assess the latter, through the varied lenses of the specialty groups of the AAG. It is a reference work primarily. It attempts to present the most significant work done by American geographers in the last decade or so, since these contributions lay the foundation for geography in the twenty-first century. A reading of the following chapters will give you a relatively comprehensive understanding of what American geographers have been thinking about and been doing for the last decade or so.

A Community of Diverse Thought

Tradition and Stability

Each of the three revolutions of the last half-century tried to rewrite the essentials of geography. None succeeded in making revolutionary changes, but all contributed to the evolution and expansion of geography. It is safe to say that the traditions of geography have endured and are robust (NRC 1997), but now more in-depth thought and analysis—much of which was introduced by our revolutionaries—make the discipline a more rigorous and meaningful academic pursuit.

While current work tends to be more sophisticated than earlier contributions, many past contributions from geographers were seminal. Traditional geographic work was not characterized by multiple forms of understanding; none the less, it typically provided detailed and thoughtful findings, often obtained from hard-won field observations and experiences hitherto unknown. Reading Carl Sauer, Richard Hartshorne, Joseph Spencer, William Morris Davis, or C. Warren Thornthwaite inspires both a respect for their contributions, and an understanding of how newfound complexities and alternative perspectives can improve upon traditional approaches.

New Ways of Thinking

When the last *Geography in America* was published in 1989, social theoretical changes were rocking the foundations of the social sciences. Compared to the relatively modest shocks of applications of structuration (Palm 1986) and realism (Lawson and Staeheli (1991), new social theory was leaning towards the third revolution of Post-modernism and Post-structuralism.

The Post-Modern Moment

Several geographers, critical of contemporary geographic thought, looked outside the bounds of American academia to find inspiration in the signal post-modern works of Derrida, Foucault, and others. Post-modernism gained a cache in geography as it did in other social sciences. New journals were founded, "critical" became an icon, the "cultural turn" appeared, and "new" became an established euphemism for rejecting the past. Post-modernism not only did not accept the body of former knowledge, it clearly rejected it. Post-modernism was incredibly appealing from an academic standpoint by being intellectually hypercritical of all knowledge. While this heightened level of criticism did serve to expose problematic areas in geographic research, it also led to an intellectual cul-de-sac where nothing but criticism was acceptable. This criticism of the focus on criticism has led to its near-demise. None the less, the post-modern critique has left us all with better ways of inspecting our work.

Specialization Matures

Since the publication of *Geography in America* in 1989, specialization has matured in the discipline. As indicated by a comparison of specialty groups now and then, specialization is both growing and evolving.

Specialization has always been a concern amongst the leaders of the AAG. They have witnessed other disciplines torn asunder, losing their unity to subdisciplinary schisms. The specialty group framework, initially advocated by the AAG in 1978, has allowed the discipline of geography to maintain its unity and celebrate its diversity. Of the fifty-three specialty groups existent in 2002, five have emerged since the inception of this book, four have merged to form two combined specialty groups, and three others have changed their titles. These changes are indicative of a healthy dynamism within the specialty group framework of the AAG.

Vital Signs

Geography's vital signs are strong. In the academic year 1999–2000, the discipline in the United States awarded 200 Ph.D. degrees and one-third of these were earned by women (US Dept. of Education Survey). This is the highest number of Ph.Ds awarded in the US since the mid-1970s. One key measure of the vitality of a discipline—jobs available for its Ph.Ds—is particularly encouraging for geography. Indeed, there are more jobs available currently for Ph.D. geographers than there are Ph.Ds to fill them. A recent *AAG Newsletter* (2002), for instance, reported that there were 1.3 geography jobs per Ph.D. produced by the American post-secondary education system in the academic year 2000–2001. Many of these jobs, of course, owe their genesis to the considerable and growing demand for education and training in GISci. The strength of geography departments within the American collegiate system has increased concomitantly over the last decade; and, now, there are more Departments of Geography than at any time in the past.

Geography curricula within the K-12 system also have expanded dramatically, owing in large measure to the establishment of state Geographic Alliances in the 1980s (Hill and LaPrairie 1989). The emerging importance of geography to the nation was summarized within the 1997 NRC report, *Rediscovering Geography: New Relevance for Science and Society*, and further recognized by the establishment of a permanent Geography Committee at the NRC. During the 1990s American geography enjoyed growth and increased recognition, and its prospects for the twenty-first century appear excellent.

Over the last decade American geographers have increasingly gained credibility. Geographers are frequently called to Congressional hearings, scientific forums, and discussions of global organizations. The

AAG Newsletter does a fine job of informing us of these many kudos and accomplishments. A classic example of such a geographer is Gilbert White who in 1999 was presented with the National Science Medal by President William Clinton. The mere facts that Nobel Peace Prize winner Kofi Annan spoke to the 2001 AAG Meetings in New York and that Nobel Peace Prize winner Nelson Mandela opened the 2002 International Geographical Congress in Durban speaks strongly to the highly credible image geography has forged for itself.

The AAG publication *Guide to Programs in Geography in the United States and Canada* appears annually and offers data to help assess our discipline's vital signs. As noted in the first *Geography in America*, there was considerable concern in the discipline when several university-level Geography departments were eliminated or lost their department status. This included such historically notable departments as Northwestern, Chicago, Michigan, Columbia, and Pittsburgh. Despite this, that book noted grounds for optimism, including the creation of twenty-first new degree programs since the 1970s (Gaile and Willmott 1989*b*: p. xxxiv). The current picture is much more optimistic. Between our first publication in 1989 and 2002, Geography programs listed in the *Guide* have increased notably. We have gone from 210 listed departments to 232 (a rise of 10%). Ph.D. programs have risen from 51 to 62 (a rise of over 20%)—for data see the *Guide* in respective years.

Membership of AAG peaked in the mid-1990s (at 7,381 in 1995) and is again on an upswing at 6,731 in 2001 (*AAG Newsletter* 2002: 8), about what it was in 1990. It is speculated that membership in all academic societies is being adversely affected by the greatly heightened access to free information on the World Wide Web. Geographers' main professional meeting, the annual meeting of the Association of American Geographers (AAG) in Los Angeles in 2001, attracted 3,741 geographers (ibid. 9).

In addition, new journals are appearing, research funds continue to flow, and geographers continue to play an important role in applied work around the globe. We have been able to maintain a loyal and optimistic view of our discipline. In sum, our vital signs of productivity, marketability, and institutional presence remain quite strong.

Overview of *Geography in America at the Dawn of the 21st Century*

Each specialty-group chapter has been placed within one of seven main topical sections. Parts I–III (Environmental Dynamics, Human/Society Dynamics, and Environment/Society Dynamics) were those defined in *Rediscovering Geography* (NRC 1997). Placement was made according to whether the chapter deals mainly with natural processes, human/social systems, or the interplay between people and their environment. Chapters with a primary focus on approaches to representation (methods) were assigned to Part IV, Geographic Methods, while chapters concerned mainly with solving practical problems appear within Part V, Geographers at Work, which also includes the History of Geography specialty-group chapter. Part VI, Regional Geography, contains chapters with an overarching regional focus or orientation. Chapters with a strong ethical position or religious interest appear in Part VII, Values, Rights, and Justice. The main problem in assigning specialty-group chapters to Parts was, of course, that each chapter contained elements of all three NRC dimensions, as well as overlaps with other dimensions. None the less, we feel that we were able to assign the vast majority of chapters unambiguously to the most appropriate Part, based upon its primary affinity with that Part.

Environmental Dynamics

Prospects for global change—and by extension, regional and local change—has energized the traditional subfields of physical geography, which include climatology, geomorphology, and biogeography. Physical geographers continue to investigate those physical, biological, and chemical systems that influence the land surface, with a keen interest in the effects of human intervention. All types of environments have been examined, but cold regions have drawn special attention recently because global-change effects may be more dramatic there. American geographers' heightened interest in cold environments also spawned two new AAG specialty groups in the late 1990s, the Cryosphere Specialty Group and the Mountain Specialty Group. Issues of scale and nonlinearity, including sub-grid-scale biases and the determination of "appropriate scale," have become of concern across the spectrum of physical geography. Rapidly developing technologies, such as GIS and remote sensing, have allowed physical geographers to make more in-depth as well as more spatially extensive analyses of our "natural" world, while the processes at work there have been increasingly revealed through numerical modeling on the newest generation of very-fast computers. Within Part I, Environmental Dynamics, chapters written by members of the AAG's Climate, Geomorphology, Biogeography, Cryosphere, and Mountain Geography Specialty Groups document the contributions of American physical geographers during the last decade of the twentieth century.

Biogeographers examine the distribution of organisms and the ecosystems within which species of interest live. Research often is approached from either an ecological or evolutionary standpoint, although applied work may involve blends of perspectives. Analyses of species and ecosystem responses to disturbance, of human-induced gaps within an ecosystem for instance, are increasingly important applications of biogeographic principles and methods. Topical interests included: plant and animal distributions; vegetation–environment relations; vegetation dynamics and disturbance ecology; influence of climate variation on vegetation; paleobiogeography; cultural biogeography; and nature conservation.

Geographer/climatologists have been working on a wide variety of climate and climate-related problems, although potential climate change is a frequently recurring theme. Research into the variability of atmospheric circulation—especially on interannual time scales—received a great deal of attention. Establishing teleconnections was of special interest, because they may foretell modes of regional or larger-scale variability in climate. Particular patterns of synoptic-scale variability were of interest as well, as they can be linked to human health-risk factors and mortality. Land-surface/climate interactions was another major theme, as was the interrelationships between climate and the hydrologic cycle. Other studies focused on the detection of climate change in observations or on the modeling and simulation of potential climate change. Communications among the members of the Climate Specialty Group—and with the larger community of other climatologists—was greatly enhanced by the development of a well-thought-out list server (CLIMLIST), developed by John Arnfield at The Ohio State University.

Cryospheric studies have come to the fore recently as the effects of global warming on snow, ice, and high-latitude ecosystems may be nontrivial. It has been conjectured, for instance, that increased melting of organic-rich permafrost may increase the rate at which CO_2 and CH_4 are emitted into the atmosphere. It also is possible that increased melting of glaciers, sea ice, or snow cover may further increase air temperature and raise sea level. Although many of the geographers who work in the cryosphere were trained as climatologists, geomorphologists, or biogeographers, their research tends to integrate approaches from all three subfields. Some work is theoretical, but much is empirical—based upon field or remote measurements in efforts to monitor and document variability and change. American geographers also have been studying mountain environments for quite some time

Geographer/geomorphologists, in their investigations of landforms and landforming processes often concerned themselves with local- or regional-scale systems. They also worked to transform their subdiscipline into a more theoretically based science. The traditional concept of geomorphic equilibrium, for example, was re-examined, and chaos theory and nonlinear dynamics were offered as more comprehensive, organizing frameworks. Field observation remained a cornerstone of geomorphology, although efforts to integrate it with computational methodologies were explored. Topical areas of active interest were: fluvial geomorphology; eolian and coastal geomorphology; weathering; mass wasting, periglacial, and glacial geomorphology; Quaternary geomorphology; biogeomorphology, environmental geomorphology; geoarcheology; and planetary geomorphology.

However, interest increased to a level sufficient to form the Mountain Geography Specialty Group in 1999. Most research in this subfield has been physically oriented (hence the inclusion of this chapter in Part I, Environmental Dynamics), although there is an increasing interest in human–environmental relationships and sustainability within mountainous settings.

Human/Society Dynamics

Human/Society Dynamics retains its position of dominance in terms of the activity of American geographers, accounting for the greatest levels of membership in any of our six aggregated groups. Given the trends towards globalization and political transition, this has been an especially exciting area of research enquiry. It is within this area that the greatest level of theoretical development has taken place, including the Postmodern Revolution. It is also in this area that the events of September 11, 2001 will be of the greatest interest (see Clarke *et al.* 2002; Sorkin and Zukin 2002).

Cultural Geography is a topical area that is situated at the heart of the theoretical debate. This chapter very fairly illustrates the views of both "traditional" materialist cultural geographers and postmodern, non-materialist "new" cultural geographers. Cultural geographers explore culture, space, and landscape using humanist, structuralist, and post-structuralist approaches. They also study everyday life and popular and folk culture. This expansive group engages in research over a broad list of fascinating topics from the American street to cemeteries to rock and roll to landscapes of resistance.

Cultural Ecology is, on the one hand, one of the most traditionally rooted specialty groups in geography, and on the other one of the most highly active and dynamically changing specialty groups. Based on the works of Carl Sauer, Karl Butzer, and their followers, traditions flow deep. However, the evolution of "political ecology" has significantly transformed this tradition-based specialty group. Indeed, political ecology is one of the more exciting developments in geography in the past decade. As we go to press, the specialty group is actively discussing a name change to more formally incorporate political ecology, and this new work is detailed in Chapter 8.

Economic Geography also benefits intellectually from dynamism. In keeping with the rapid changes of the global economy, what was formerly the Industrial Geography specialty group renamed itself the Economic Geography specialty group. Indeed, simple former categorizations of industrial, mercantile, and service became increasingly irrelevant as the US economy became globalized and went through major transitions. In the US, there was transition from high volume to high value-added production, from industrial to knowledge-based production, from a focus on physical capital to a focus on human capital. The new blurred boundaries in the economic world provide a fertile ground for research.

Environmental Perception and Behavioral Geography (EPBG) has its strongest roots in the Quantitative Revolution and its behavioral sciences corollary. EPBG geographers study "human activities, human experiences, and all forms of empirical surroundings." A great debate between quantitative and qualitative work is going on among ESPG geographers. Also, this specialty is in the forefront of research dealing with the virtual world.

Historical Geography has also benefited from heady debates between modernists and postmodernists. Historical geographers' scholarship has thrived from working with other specialties with a variety of perspectives, notably, gender-based, GIS, and applied scholarship. Among the foci of historical geographers are world systems analysis, migration in the North American context, capitalist development from a historical point of view, human/environment interactions concomitant with land modification, and the examination of the geography of Native Americans.

The Berlin Wall came down in 1989 when the initial *Geography in America* was published (no causal inference intended), and political geographers have been working fervently ever since. The rapid demise of the Second World was clearly the singular most important event of the end of the last century. Coupled with the terrorist attack of September 11, 2001, political geographers found themselves with much to understand and explain. The fertile new field of the "War against Terror" coupled with the traditional bailiwick of understanding elections (brought home by the Florida election debacle) keep these geographers very, very busy. Toss in the debates about the North American Free Trade Agreement (NAFTA), the World Trade Organization (WTO), and the World Bank and this group of geographers have a very rich current and potential research agenda.

Demographic dynamics challenge today's population geographers. The First International Population Geographies Conference was held in July 2002 in St Andrews, Scotland, and Ch. 13 from the Population Geography specialty group clearly adheres to the focus of international work. The lines of research identified are dominated by migration and mobility. Fertility and mortality play a minor role. Issues of ethnicity, social context, and public policy strongly influence this research. A call to more strongly involve issues of gender, racism, agism, and class conflicts is being heeded.

Geography has finally come out of the closet to acknowledge that Sexuality and Space are important topics for research and understanding. This insightful and provocative chapter goes so far as to question geography's origins. Scholars in this field also are willing to explore admittedly *dangerous* work to provide important insight into our social fabric. A reading of this chapter will likely raise more new questions for geographers than any other in this volume.

One might initially think that the Socialist Geography specialty group was dismissed in the earlier part of this Introduction as part of the Marxist Revolution that has now passed. Yet socialist geographers effectively argue that it is important to challenge those economic and political structures that perpetuate inequalities (see the David Harvey quote near the end of this Introduction). These clever folk look at the production of knowledge in a challenging way. They continue to question orthodoxy and established systems, often focusing on issues of social justice.

Transportation geographers have clearly been on the move. The research of this group investigates societal change, sustainable transport, information and communications technology, globalization, and institutional issues. Whether "distance matters" is a topic that has featured on the cover of *The Economist.* It still does.

Consideration of most things urban changed after September 11, 2001. Urban geography will be faced with the strongest challenge to explore and explain these issues (see Clarke *et al.* 2002). Chapter 17, Urban Geography, clearly leans towards a post-structuralist interpretation of the city before September 11. Contested spaces, spatial constructions of social life, and explorations of urban landscapes of resistance are all intellectually provocative. Whether urban geographers of all stripes can transform their existing research agenda and rise to the intellectual challenge of post-September 11 urban scholarship is a heady question.

Environment/Society Dynamics

Improving our understanding of relationships between environmental processes and human activities has been a goal of geographers for much of the twentieth century (Thomas *et al.* 1956). Recently, however, global-change research and "Human Dimensions of Global Change" initiatives have made us more keenly aware of the importance of understanding these relationships. The deleterious effects of misguided, human modification of our environment have become abundantly clear. In recognition of the many issues associated with human use of the environment, a growing number of geographers have begun to bridge the gaps between science and social-science approaches in order to study the links (and feedbacks) between society and the environment. Although systems of interest include significant, environmental, and social components, among American geographers social aspects received the most attention. Specialty group chapters within which the need to understand environment/society relationships is central in Part III.

A relatively new AAG specialty group (founded in 1995), the Human Dimensions of Global Change Specialty Group was established to help foster the "study of societal causes and consequences of changes in the global environment, as well as individual and institutional responses to these changes." Many group members shared research interests that cut across the science/social-science or physical-geography/human-geography divide. Typical areas of interest were human vulnerability and adaptation, impacts of climate change, and the reasons for and consequences of land-use and land-cover change. In addition to their own research, a number of geographers in this area have been actively involved in crafting national and international research agendas in the global change and human dimensions of global changes arenas.

A comprehensive understanding of the production, disposition, and human use of water also requires expertise that intersects subfields of physical and human geography. Solutions to contemporary water problems—such as floods, droughts, wetland losses, and groundwater depletion—require a broad understanding of the hydrologic cycle, and especially of the impacts of human activities on the cycle. Informed water-use policies, of course, also depend fundamentally on this understanding as well as on legal and ethical considerations. Much of the water-resources work by American geographers, however, was conducted from human geography perspectives. Questions of how best to manage the resources (policy and legal questions) are often central. Water-resources geographers examined problems across a broad range of scales, from individual and household to issues of national and international importance.

Energy production and use in the environment raises an extremely wide array of questions, as there are many forms of energy and energy-related resources. Geographers who study energy and environment must cut across geography's subfields to investigate adequately and understand "earth-energy associations." They have looked into virtually all types and phases of energy, including energies derived from fossil fuels, nuclear fission, and running water. Among geographers, a traditional focus has been on economics and availability, although geopolitical issues have increased in importance recently. An interest in efficiency and conservation, as well as in the development of sustainable energy resources and systems, has begun to develop.

Research in coastal and marine geography also expanded in response to the broader issues of global change and environmental degradation. Coastal and marine systems are intertwined; however, coastal and marine research tended to be segregated into three sub-areas: coastal physical geography, marine physical geography, and coastal-marine human geography. Coastal settings were of considerable interest because of their growing populations and economic importance, and their increasing vulnerability to sea-level rise and extreme weather. Marine research by geographers also expanded, in part, because of global-change interests in monitoring the ocean surfaces. Many aspects of the human geography of coastal and marine regions were investigated, including culture, economics, politics, resource management, and environmental and development planning. Assessments of risk and hazards were made as well.

It is significant that the Contemporary Agriculture and Rural Land Use (CARLU) and Rural Development groups are merging. Both shared the "rural" identity, but CARLU was largely domestic and Rural Development had a broader scope. Indeed, this merger points to the grass-roots origins of most specialty groups and indicates that the dynamism of the specialty group structure allows for reasonable change. CARLU attempts to understand the change in the American rural landscape.

The Rural Development specialty group has a more global perspective, looking at extractive industries, sustainability, and social capital.

Geographic Methods

Development and use of geographic methods grew rapidly over the 1990s. Technological advances in computing and satellite observation, in particular, laid the foundation for important contributions in geographic information science (GISci), remote sensing (RS), cartography, and mathematical modeling and quantitative methods (MMQM). Dramatic improvements in computational speed, visualization technology, and data-storage media not only made modern GIS possible, but they facilitated geographic modeling as well as the quantitative analysis of large spatial, especially remotely sensed, data sets.[1] Cartographic visualization and animation also benefited. Exploration of the innumerable possibilities made available by the Internet has only just begun. The 1990s saw unprecedented advances in geographic methods for analysis and modeling.

The field of GIS was quite young when the first *Geography in America* was published in 1989. The AAG GIS Specialty Group (SG) was youthful as well. Over the course of the 1990s, across virtually the whole of America, interest in GIS exploded, as it also did within the AAG. The AAG GIS SG grew rapidly, and became the largest SG within the AAG9 reaching a peak of 1,949 members in 2000 (*AAG Newsletter* 2002). Interests of its members expanded from primarily technical to a wide range of theoretical and practical issues inherent in geographic data and their analysis. The SG was evolving into a more broadly based Geographic Information Science organization. Among the emerging areas of interest were representational issues (e.g. how to represent "fields" and "objects"), analytical issues (e.g. how to incorporate spatial statistics and other forms of models), data quality and error propagation issues, integrating GIS with other media (e.g. decision-making tools, remote sensing, and environmental modeling), and GIS and Society. The more traditional, technical issues (such as interoperability and parallel processing) continued to attract active investigation, however. Ethical, legal, and educational considerations also grew in importance. Perhaps the most fundamental weakness within GIS has been its limited ability to evaluate change over time or model (Willmott and Gaile 1992) the dynamics of process. None the less, there is no question that GIS is a most exciting subfield. Its far-reaching popularity also is imbuing a wide cross-section of Americans with an understanding of why "geography matters."

Remote sensing (RS) is a means of observing and measuring aspects of the earth's surface from a distance. Photographic and non-photographic instruments (usually aboard air- or space-borne platforms) have been used. As instrumentation continues to improve, the spatial and temporal resolution of the observations, as well as their accuracy, should do so also. Some of these improvements should be dramatic (e.g. spatial resolutions of < 30 m should soon be commonplace), and should allow for much more reliable analyses of land-surface patterns, processes, and change. Better observations of land-surface and environmental change are of particular interest in this age of global change, as are better approaches to obtaining social and population data from satellite-based observations. Other challenges facing the remote-sensing community include: dealing with staggering increases in available data; improving currently underdeveloped theories (models) of the mechanisms of change and then meaningfully observing those mechanisms remotely; and integrating RS observations with other types of geographic data. Remotely sensed data are fast becoming the major source of information about geographic change.

The subfield of cartography has been in transition since 1989, when the first *Geography in America* was published. Over the course of the 1990s, an almost complete automation of the cartographic process occurred, and significant portions of traditional cartographic work (e.g. terrain modeling, creation and refinement of geographic data structures, generalization, and spatial interpolation) became increasingly conducted under the guise of GIS. What was dynamic and interactive cartography also began to be referred to as "geographic visualization." None the less, representational issues remained important (for mapping in general, including via GIS) and so cartographers continued to make important contributions to map design, symbolization, and generalization. Improvements were made, for instance, in the use of color, in automated type placement, and in the selection of the "best" map projection. Cartographers also introduced better models of generalization, feature

[1] A Geographic Information System (GIS) is a computer-based system for collecting, storing, manipulating, analyzing, and visualizing geographic and other spatial information. Emphasis on the technical aspects of GIS, however, has raised concerns that important issues were being overlooked. In an effort to insure that relevant intellectual and scientific issues—as well as the technical ones—are integral to the field, the more inclusive name of Geographic Information Science (GIScience or GISci) was proposed by Michael Goodchild. GISsience is more comprehensive than GIS, and is increasingly the preferred designation.

representation, and spatial interpolation, as well as approaches to interactive visualization and animation. Cartographic research into these and other areas of map design, communication, and generalization remains strong.

American geographers in the mathematical models and quantitative methods (MMQM) subfield produced a considerable amount of innovative research, even though the specialty group was relatively small. Advances in computational technology played a key role in the development and application of sophisticated (sometimes nonlinear) models, as well as in the analysis of very large data sets. A growing number of these approaches to data analysis, such as exploratory techniques, were integrated into GIS. Methods or areas of interest to both physical and human geographers included: chaos theory, cluster analysis, exploratory data analysis, Fourier analysis, fractal evaluations, linear programming, analysis and optimization of sample networks, neural networks, and spatial scale and aggregation. Within human geography, problem-specific models were developed to evaluate spatial behavior, choice, decision, and process, as well as the evolution of complex spatial systems. Within physical geography, problem-specific models were used to evaluate sub-grid-scale biases, spatial downscaling, and non-linear dynamics. Many specific models and methods are described in Ch. 27. The 1990s were a very productive decade for geographers who contributed in the MMQM subfield.

Geographers at Work

The late Peter Gould's *The Geographer at Work* (1985) is an impressive volume that addresses the interaction of the intellectual and pragmatic work of geographers. Today's geographers do not "assume away" the real world in order to achieve some theoretical purity. Many geographers are actively involved in using a high level of scholarship to address real-world issues.

In 1989, *Geography in America* began with a chapter on Geography Education. The reasoning then was simple. This disciplinary area had been rejuvenated by the efforts of Dave Hill, Nick Helburn, and Bob and Sarah Bednarz, amongst others. The continuing efforts of this group of scholars/educators have made a significant impact on the official presence of geography in the national educational arena. The evolution of the Geographic Alliance network is one of the best things that has happened to geography in the last two decades. The inclusion of geography into the five educational standards advocated by the Clinton Administration is a clear signal that geographers' messages are being heard.

Things that could get you!—these could well define the topics of hazards research in geography. Stemming from the seminal works of Gilbert White, geographers have made impressive strides in informing about the risk of hurricanes, floods, tornadoes, earthquakes, and other natural disasters. They have also been very active in exploring the area of human-made disasters such as nuclear-risk.

Spatial diffusion studies stemming from the Quantitative Revolution of the 1960s have strongly influenced studies of epidemiology. Medical geographers have built on this core of expertise. This group's research not only identifies possible interventions, but also identifies secondary factors that play a strong part in disease transmission. Contemporary medical geographers also uncover contextual explanations for the prevalence of disease.

The initial *Geography in America* did not include military geography, since there was no specialty group at that time. The important reality of geographers' work in military intelligence in and after World War II need not be denied. Ranging from simple tasks of map-reading to complex tasks of geopolitical understanding, military geographers work in a very applied fashion based on the needs of national security. Geography is a major part of any conflict. Until we live in a utopian conflict-free world, military geography should be a major part of our research effort.

Old folks rule! The Aging and Aged specialty group clearly focuses on the demographic shift where elderly people have a much more predominant position in American society. How our retiring populations migrate, what their needs are, and how they impact local governments is a truly rewarding field of research.

Let's have fun! Let's understand having fun! The easily disparaged Recreation, Tourism, and Sport specialty group in fact attempts to understand the soul of society; it has gained respectability, especially since tourism is now regarded as a major factor in local economic development. Geographers have engaged in research that explains human behavior in new areas. It is admitted that these geographers studying fun deserve serious respect.

Applied geographers most often find themselves in a client-driven, problem-solving work mode. They work for the Census, US Agency for International Development, the World Bank, the State Department, the Department of Housing and Urban Development, and a wide variety of Non-Governmental Organizations. These geographers are on the front line of policy initiatives.

A new specialty group in geography focuses on the history of geography. It is the work of geographers to chronicle their research advances. This volume especially prizes such documentation and hopes to find a place in its critical chronology.

Regional Geography

The study of geographic regions has always been in the mainstream of the work of geographers. Whether they be geomorphologists, cultural geographers, or remote sensors, geographers have often specialized in a region in addition to their topical or methodological specialties. Indeed, regional geography, academically strongest during the "area studies" era of the first half of the twentieth century, has weathered the storms of the various "isms" and continues to play a central role in geographic research. It is important to note that not all regions one might find in a World Regional textbook are represented by specialty groups. Further, there is clearly some overlap (e.g. specialty groups exist for both Asia and China). Much of the work in the regional subfields will be cross-referenced in other topical chapters in this volume.

Africa has witnessed signal changes in the last decade. Apartheid ended in South Africa, democracies grew where dictatorships previously prevailed elsewhere in Sub-Saharan Africa, and social development in health and education made major strides. None the less, the media focused attention on the crises that gripped the continent versus the successes that were achieved.

Africa remains a favored playground of development theorists who explore post-colonial, post-structural, and postmodern approaches to the study of its geography. Political ecology finds fertile fields for development here. The Boserupian argument received major attention by geographers arguing about the relationships between agricultural intensification and population change, including "More People, Less Erosion" (Tiffin *et al.* 1994). Additional studies of migration, refugees, environmental change, and globalization make the work of this specialty group such that it has a broad impact on many of our substantive subfields.

American Ethnic Geography focuses on the experiences of ethnic groups in the United States and Canada. This group traces its roots to the quantitative revolution and the Civil Rights movements of the 1960s. Over the years it has blended traditional cultural geography with historical and population geographies to explore both past and current geographical and spatial trends in ethnic studies, including revitalization studies.

The diverse study of American Indian issues by American geographers spans the gamut from physical sciences to the humanities. Its long history in geography dates back through the Berkeley School. A primary focus on land and legal issues is clear. Yet revolutionary research is occurring in this subfield, including Shari Fox's (yet unpublished) research on Inuit perceptions of climate change. The Canadian Studies Specialty Group studies some similar issues, but also focuses on new cross-border studies, including the effects of the North American Free Trade Agreement (NAFTA).

The Asian and China Specialty Groups study an amazing array of dynamic changes, including the very important political and economic transitions of recent years. Understanding the Middle East is more critical than ever in the wake of September 11, and a small band of geographers is exploring this difficult-to-research region in the tradition of geographer Sir Richard Burton. Making Asia less enigmatic is a goal of American geographers. Whether it is the complexity of the transitions in India, Indonesia, or Vietnam, our scholars are applying themselves to understand contemporary changes happening against a context of deep histories. In China, the transition to a market economy and the incorporation of Hong Kong stand juxtaposed against Western models of transition (e.g. the World Bank Structural Adjustment model of change). China is more open to study than before and American geographers have clearly risen to the opportunity.

It is important to note that during the production of the last *Geography in America* no European specialty group existed. This glaring omission has been rectified, and is especially fitting given the coincidence with Europe's bold experiment at economic and political integration—the European Union. Shedding some of their parochialism, American geographers have become engaged in a variety of collaborative research efforts with European geographers in a wide array of research areas. The fall of the Berlin Wall in 1989 increased the interaction and unification of the former East and West European territories (most notably in Germany). It also led to major political and economic changes in the former Soviet Union which are well documented by research from the Russia, Central Eurasian and East European specialty group.

Geographic studies in Latin America continue to maintain an important place in scholarly research on the continent. This very solid specialty group provides a comprehensive documentation on the wide array of research on path-breaking themes. Stemming from tradition cultural ecological research, new and important political ecology studies are emerging. Development studies include the increasingly important issues of social capital and civic society, and sustainable development has established strong roots in Latin America, signaled by the Rio Convention. Global change studies of forests, oceans, and other biospheres inform our understanding of major climatic events.

None of these brief summaries provides a satisfactory introduction to the wealth and breadth of research that occurs in the regional chapter, research that spans the full range of geographic enquiry. What is important is that geographers remain committed to regional research and to solving problems in specific places.

Values, Rights, and Justice

Perhaps one of the most prominent and laudable changes made by geographers since the last publication of the initial *Geography in America* in 1989 is signaled by the advent of specialty groups focused on issues of values, rights, and justice. Three new specialty groups emerged since 1989 with foci on ethical issues. To this assemblage we have added the Geographical Perspectives on Women specialty group, acknowledging that its focus on unequal gender-based power relationships strikes at the heart of questions of values, rights, and justice. It should be noted that, since this publication began production, the Values, Ethics, and Justice specialty group has merged with the human rights group founded in 1997. The Values, Ethics, and Justice specialty group immediately began to ask important questions. The single query, "How far do we care?" calls into question a wide range of foreign policy initiatives and responses. Ethical questions cut across the entire discipline, and the group has taken up the challenge to address and analyze them. Certainly issues of feminism, sexuality, socialism, and native identity come strongly into play.

Geographers responding to the human rights agenda have addressed issues of oppression from gay rights to women's rights. Chapter 46 lauds the diverse efforts of Richard Hartshorne, Harold Rose, Gilbert White, and Richard Morrill for advancing human rights understanding. The study of human rights is extremely complicated, given the importance of the problems and the weakness of the data involved. Combining qualitative, quantitative, and ethical perspectives, geographers in this area contribute both ethical and pragmatic research to inform policy-makers spanning the local-to-global continuum. Rex Honey deserves special commendation for championing this group's effort.

The authors of Ch. 47, Geographic Perspectives on Women, notes that gender and feminist perspectives are now commonly found in textbooks, course offerings, and recent publications in geography. This was not the case when the first *Geography in America* was published. The maturation of this specialty group is evident in feminist analyses of methodology and in approaches to pedagogy. In addition, the progress in geography on studies of identity and difference has been led by female geographers. Contributions to research on gender and work, Third World development, and cultural geography round out the prevalent work of members of this specialty group. As this chapter indicates, there is every reason to believe that gender and feminist research will continue to make major advances.

The study of the geography of religion and belief systems concerns another new specialty group that explores issues of values and ethics. In today's world of religion-influenced crises, it is more incumbent on geographers than ever to understand other religions and belief systems. In addition to traditional empirical studies, Ch. 48 includes both humanistic and critical approaches to this important topic.

Challenges Facing American Geography

How times have changed! Geographers *do* need to become better forecasters. Since the publication of *Geography in America* in 1989, geographers face a daunting research agenda. Global warming is not mere speculation. The digital era has changed computer, remote sensing, and all locational prospects. Globalization has caused a serious rethinking of both political and economic geographies. Human rights now have a voice. The environment has global representation as witnessed by the succession of summits, most recently in 2002 in South Africa. South Africa itself is now led by a democratic majority and is exploring significant changes to its geographic landscape, e.g. in terms of urban demarcation. The Cold War is over, the Berlin Wall has been breached, but geopolitics still plays a heady roll. Global conflicts take on a very new set of dilemmas. Human rights matter. The debate over HIV/AIDS and prescription drugs raises huge ethical versus capitalist questions. The supra-national organizations of the World Bank, the International Monetary Fund (IMF), and the Global Agreement on Trade and Tariffs (GATT) both "rule the world" and raise protests around the globe. Indeed, recent events charge American geographers to better explain our world.

The events of September 11, 2001 affected everyone's views and many research agendas (see Clarke *et al.* 2002; Sorkin and Zukin (2002)). Geographers, with their charge of understanding the world, were more affected than most. As Neil Smith, a resident of south Manhattan who experienced the event at close hand said, "All terrorism is local" (Smith 2001). David Harvey (2002: 61) states the left position cogently:

> For those who have a more jaundiced view of what neoliberal globalization and market freedom have really been about in these last few years, the towers therefore symbolized something far more sinister. They represented the callous disregard of U.S. financial and commercial interests for global poverty and suffering; the militarism that backs authoritarian regimes wherever convenient (like the Mujahadeen and the Taliban in their early years); the insensitivity of the U.S.-led globalization practices to local cultures, interests, and traditions; the disregard for environmental degradation and resource depletion (all those SUV's powered by Saudi oil generating greenhouse gases and now, in New York City, adorned with plastic U.S. flags made in China); irresponsibly selfish behavior with respect to a wide range of international issues such as global warming, AIDS and labor rights; the use of international institutions such as the International Monetary Fund and the World Bank for partisan political processes; the shallow and often hypocritical stances with respect to human rights and terrorism; and the fierce protection of patent rights of multinationals (a principle that the U.S. enforced with respect to the AIDS epidemic in Africa but then cynically overthrew when it needed Cipro drugs to combat the anthrax menace at home).

David Harvey is probably our most insightful geographer. Faced with these chains of clear blame, many might well surrender—but to where? We do live in an imperfect world. Pragmatists would argue that we get past this in the best way, both ethically and pragmatically. Indeed pragmatism is revitalized in geography (Wescoat 1992). The debate between the theoretical and the practical should always take place in academia. In geography it continues, and that is very healthy.

References

Abler, R. F., Marcus, M. G., and Olson, J. M. (eds.) (1992). *Geography's Inner Worlds: Pervasive Themes in Contemporary American Geography*. New Brunswick: Rutgers University Press.

Association of American Geographers Newsletter (2002). "2001 Membership Statistics and Annual Meeting Report," 37/6: 8–9.

Billinge, M., Gregory, D., and Martin, R. (eds.) (1984). *Recollections of a Revolution: Geography as a Spatial Science*. London: Macmillan.

Butzer, K. W. (2002). "The Rising Costs of Contestation." *Annals of the Association of American Geographers*, 92/1: 75–8.

Clarke, S. E., Gaile, G. L., and Pagano, M. A. (2002). "Urban Scholarship After September 11, 2001." *Urban Affairs Review*, 37/3: 60–7.

Cutter, S. L., Golledge, R. G., and Graf, W. L. (2002). "The Big Questions in Geography." *The Professional Geographer*, 54/3: 305–17.

Gaile, G. L., and Willmott, C. J. (eds.) (1989*a*). *Geography in America*. Columbus, Ohio: Merrill Publishing Co.

Gaile, G. L., and Willmott, C. J. (1989*b*). "Foundations of Modern American Geography," in Gaile and Willmott (eds.) (1989*a*: pp. xxiv–xliv).

Golledge, R. G. (2002). "The Nature of Geographic Knowledge." *Annals of the Association of American Geographers*, 92/1: 1–14.

Gould, P. R. (1985). *The Geographer at Work*. Boston: Routledge, Kegan & Paul.

Hartshorne, R. (1955). "'Exceptionalism in Geography' Re-examined." *Annals of the Association of American Geographers*, 45: 205–44.

Harvey, D. (1973). *Social Justice and the City*. London: Edward Arnold.

—— (1989). *The Condition of Postmodernity*. Oxford: Blackwell.

—— (2002). "Cracks in the Edifice of the Empire State," in Sorkin and Zukin (2002).

Hayford, A. M. (1974). "The Geography of Women: An Historical Introduction." *Antipode*, 6/2: 1–19.

Hill, A. D. and LaPrairie, L. A. (1989). "Geography in American Education," in G. L. Gaile and C. J. Willmott (eds.) (1989*a*).

Kates, R. W. (2002). "Humboldt's Dream, Beyond Disciplines and Sustainability Science: Contested Identities in a Restructuring Academy." *Annals of the Association of American Geographers*, 92/1: 79–81.

Lawson, V. A., and Staeheli, L. A. (1991). "On Critical Realism, Geography, and Arcane Sects!" *The Professional Geographer*, 43/2: 231–3.

National Research Council (NRC) (1997). *Rediscovering Geography: New Relevance for Science and Society*. Washington, DC: National Academy Press.

Palm, Risa (1986). "Coming Home." *Annals of the Association of American Geographers*, 76: 469–79.

Schaeffer, F. (1953). "Exceptionalism in Geography: A Methodological Examination." *Annals of the Association of American Geographers*, 43: 226–49.

Smith, N. (2001). "The Geography of Terror." Invited Presentation to the Developing Area and Research (DART) Program at the University of Colorado. November.

Soja, E. W. (1989). *Postmodern Geographies*. New York: Verso.

Sorkin, M. and Zukin, S. (eds.) (2002). *After the World Trade Center: Rethinking New York City*. New York: Routledge.

Thomas, W. L., Sauer, C. O., Bates, M., and Mumford, L. (1956). *Man's Role in Changing the Face of the Earth*. Chicago: University of Chicago Press.

Tiffin, M., Mortimore, M., and Gichuki, F. (1994). *More People, Less Erosion: Environmental Recovery in Kenya*. Chichester, NY: J. Wiley.

Turner, B. L., II (2002). "Contested Identities: Human-Environment Geography and Disciplinary Implications in a Restructuring Academy." *Annals of the Association of American Geographers*, 92/1: 52–74.

Wescoat, J. L., Jr. (1992). "Common Themes in the Work of Gilbert White and John Dewey: A Pragmatic Appraisal." *Annals of the Association of American Geographers*, 82: 587–607.

—— (2002). "Environmental Geography—History and Prospect." *Annals of the Association of American Geographers*, 92/1: 81–3.

Willmott, C. J., and Gaile, G. L. (1992). "Modeling," in R. F. Abler, M. G. Marcus, and J. M. Olson (eds.), *Geography's Inner Worlds: Pervasive Themes in Contemporary American Geography*. New Brunswick: Rutgers University Press, 163–86.

PART I

Environmental Dynamics

CHAPTER 2

Biogeography

Kenneth R. Young, Mark A. Blumler, Lori D. Daniels, Thomas T. Veblen, and Susy S. Ziegler

Biogeographers study the distributions of organisms and the systems those species inhabit. Biogeography can be viewed both as a form of geographical enquiry applied to plants and animals, and also as a biological science concerned with geography. Thus, biogeography is interdisciplinary, like other "composite" sciences such as geomorphology (Bauer 1996; Osterkamp and Hupp 1996). Veblen (1989) provided an overview of biogeography in the late 1980s. He commented on the nature of biogeography as practiced in academic geography programs, finding most similarity in approach and subject matter with ecologists and ecology.

Three broad research orientations can be identified (K. R. Young 1995): ecological, evolutionary, and applied. Each orientation includes both theoretical frameworks and empirical foundations. Ecological approaches relate plant and animal distributions to current biological and physical processes, including interactions among species, precipitation and temperature regimes, and soil nutrient dynamics. Evolutionary approaches accommodate genetic and population changes in species over long time-periods, in addition to historical processes as affected by Earth history, plate tectonics, and climate change; these approaches have been labeled as "classical biogeography" (Veblen 1989). Complete biogeographical explanations often require detailed information on both ecological conditions and historical changes over centuries or millennia or even millions of years. Biogeographical approaches also are applied to the evaluation of important societal issues, for example through the study of nature reserves. Of practical and theoretical concern are situations where species or their distributions and abundances are modified by human influences. This is the part of biogeography closest to geography's mainstream research interests in human–nature interrelations, and is called "cultural biogeography." Some people characterize geography as the study of the Earth as modified by humans; in this case, biological geography (biogeography) would include the study of how species and living land cover have been altered by people.

Research by Biogeographers

Plant and Animal Distributions

Species distributions can change over short and long time-scales (Hengeveld 1990; Dingle 1996). Biogeographers who study the shifting spatial distribution patterns of specific species of plants or animals often focus their research on biophysical factors that determine the range limits of the species and how those factors change through time. These controls include the effects of other organisms, the physical conditions of the environment, and disturbance. Physical controls on plant distributions that change include climate (K. C. Parker 1993), hydrogeomorphic conditions (Shankman and Kortright 1994), soil moisture (Terwilliger and Zeroni 1994), and availability of other kinds of resources (Brown 1989; Terwilliger 1997). Many studies focus on the dynamic range margins of a single species in order to explain current distributions and to predict future extent. As examples, Knapp and Soulé (1998) studied the expansion of western juniper in central Oregon, Conkey *et al.* (1995) analyzed disjunct stands at the southeastern edge of the jack pine's range in Maine, and Ziegler (1995, 1997) sought to explain the persistence of eastern white pine at its southern limit in Wisconsin.

Some biogeographers analyze the spatial distribution of abundance or the genetic characteristics of a species. Pérez's (1991, 1998) comprehensive analysis of high elevation paramos in Venezuela illustrated that the distribution and abundance of plant species of contrasting life forms are determined by the interactions of several environmental factors. Parker and Hamrick (1992) examined the genetic diversity and clonal structure of a columnar cactus. Genetic variation within and between populations could be related both to the history of range expansion and retraction, and to ecological traits of a pine species (K. C. Parker *et al.* 1997*a*). Jelinski and Fisher (1991) found that the spatial organization of genetic diversity in poplar trees had implications for the herbivores that consumed their leaves. A likely direction for future studies is the wider use of genetic markers to shed light on past and present distributions.

Although *Dynamic Zoogeography* (Udvardy 1969) was a seminal work in the development of biogeography, relatively few geographers focus exclusively on the distributions of animals. Examples of zoogeography, however, illustrate the breadth of subjects studied. Gurnell (1998) described the changing distributions of the European beaver, linking the animal's hydrogeomorphological impact to particular environmental conditions. In contrast, DeMers (1993) observed the range of the western harvester ant in North Dakota expanding via roadside ditches and stimulated by regional climate. As a third example, Yaukey (1994) combined computer simulations and field observations to understand patterns in social dominance among five races of dark-eyed junco.

Vegetation–Environment Relations

Biogeographic studies of environmental influences on vegetation include analyses of individual plants (discussed in the previous section), landscape analyses of vegetation patterns, and global scale modeling of climatic influences on vegetation distribution. A spatial relationship can be further examined for causality. For example, local soil characteristics have been correlated with patterns in coniferous forests (Taylor 1990; Barrett *et al.* 1995), at altitudinal treeline (Malanson and Butler 1994), on limestone-derived substrates (Franklin and Merlin 1992), and on arid alluvial fans (K. C. Parker 1995). Elevation, latitude, and edaphic effects explained local to subregional vegetation patterns in the Sonoran Desert (K. C. Parker 1991), vegetation gradients in the Sierra-Cascade mountains (A. J. Parker 1995), and those along a topographic gradient in Georgia (Hoover and Parker 1991).

Traditional multivariate analyses of vegetation–environment relations (Gauch 1982) are now being used along with geographic information systems (GIS) applied to satellite and aerial photography images, and with approaches derived from landscape ecology (M. G. Turner and Gardner 1991; M. G. Turner *et al.* 2001). As an example, Callaway and Davis (1998) used historical aerial photographs, vegetation maps, and field data to analyze regeneration of coastal live oak. Similarly, aerial photography, GIS, and digital elevation models, combined with field observations, have been used to map and analyze treeline vegetation in Colorado (Baker and Weisberg 1995) and Montana (D. G. Brown 1994).

At a regional scale, species distributions and vegetation types are being predicted from bioclimatic and other environmental variables (Franklin 1995, 1998). In the Andes, elevation influences vegetation composition and structure (K. R. Young 1993; Keating 1999). Similarly, tree species richness in montane forests of northwestern Argentina corresponds to latitude, elevation, climatic stress, and the influence of mountain ranges (Grau and Brown 1998). At the global scale, Box's (1996) work has been important in defining plant functional types. Neilson (1993) has constructed a model for predicting vegetation from spatially defined climatic constraints, which when combined with soil data, makes refined predictions of vegetation and hydrological balance (Bachelet *et al.* 1998).

Vegetation Dynamics and Disturbance Ecology

Because natural systems are always changing, a focus on the causes and consequences of those dynamics is an important area of research. The study of vegetation dynamics tends to be differentiated by vegetation type, spatial scale, and disturbance type (Veblen 1992). For instance, D. A. Brown (1993) found that disturbance was a major factor affecting the local and landscape patterns of Great Plains grasses. For forests, which have received the major share of recent research efforts, studies of within-stand dynamics focus mainly on endogenous disturbance in the form of fine-scale tree-fall gaps. However, forest dynamics are also studied in relation to coarse-scale exogenous disturbances, environmental fluctuations, and demographic changes of the tree species related to whole stand (cohort) replacement patterns. Data on tree size and age structures and tree spatial distributions have facilitated interpretations of ecological processes and stand history in a range of forest types across North America (Taylor and Halpern 1991; Frelich and Graumlich 1994; Daniels *et al.* 1995; K. C. Parker *et al.* 1997*b*; Arabas 2000; Ziegler 2000), and elsewhere (Grau *et al.* 1997; Enright and Goldblum 1998; Rigg *et al.* 1998). Occasionally, such approaches have been combined with monitored permanent plots (Kupfer and Runkle 1996). Kellman and Tackaberry (1993) evaluated differential mortality due to fire and tree-falls in tropical riparian forest fragments in Belize.

An important research theme in the study of gap dynamics has been species partitioning of gap environments that might be explained by differences in life history traits. For example, tree-ring studies in Argentina and New Zealand show species-specific differences in responses to gap creation (Rebertus and Veblen 1993; Runkle *et al.* 1995). Demographic parameters such as mortality and recruitment rates have been measured for tree species in tropical rain forests of Australia (Herwitz and Young 1994), temperate forests of western North America (Daniels and Klinka 1996), subalpine forests in China (Taylor *et al.* 1996), and timberline forests in Peru (K. R. Young 1991). Malanson (1996) used models to assess the effects of demographic parameters on stand-level dynamics. The composition of recovering fragmented and human-altered forests can also be affected by edge effects and by the available species pools (Beatty 1991; Goldblum 1997; Kupfer *et al.* 1997).

Research on the effect of coarse-scale disturbances on vegetation dynamics includes detailed analyses of how stands develop following fires, windstorms, forest dieback, insect outbreaks, and geomorphic events, as well as studies of the spatial and temporal variations in the disturbance *per se*. An important theme has been the heterogeneity of burns and of post-fire patterns across a wide range of ecosystem types from tropical ecosystems (Horn 1997; Keating 1998) to eucalypt woodlands (Enright *et al.* 1997), the prairie-forest ecotone (Leitner *et al.* 1991), chaparral and sage scrub (O'Leary 1990*b*; Minnich and Bahre 1995), oak woodlands (Callaway and Davis 1993), and pine forests (Parker and Parker 1994).

A common landscape pattern in the western US is a shift from park-like stands to denser forests of ponderosa pine, which coincides with changes in livestock grazing, decreased ignitions by Native Americans, fire exclusion, and climate changes (Savage 1991; Mast *et al.* 1998). Changes in fire regimes related to settlement have been reported throughout North America (Goldblum and Veblen 1994; Grissino-Mayer *et al.* 1995; Taylor and Skinner 1998; Wolf and Mast 1998), for temperate latitudes of South America (Szeicz *et al.* 1998; Veblen *et al.* 1999), and for the neotropics (Horn 1998). Wind disturbance may preferentially damage larger, faster-growing early successional species and accelerate succession towards later stages (Veblen *et al.* 1989*a*; Dyer and Baird 1997*b*). Similarly, Hadley and Savage (1996) suggest that wind disturbance can hasten the development of forest-interior characteristics by creating gaps and heterogeneity. Repeated wind-caused damage was the main determinant of forest structure on Tierra del Fuego (Rebertus *et al.* 1997), as were the hurricanes that affect Puerto Rico (Scatena and Lugo 1995).

In a southern California montane forest, Savage (1994) attributed widespread tree mortality to fire suppression, air pollution, drought, competition, and insect infestation. Tree-ring reconstructions have proven useful for understanding interactions among fire regimes, forest health (disease and insect outbreaks), and topographic controls (Hadley and Veblen 1993; Hadley 1994; Veblen *et al.* 1994).

Reviews by Malanson (1993) and K. C. Parker and Bendix (1996) evaluated interactions between landform and vegetation processes, identifying riparian environments as a major focus of research. This approach addresses the influences both of the physical environment on vegetation (Bendix 1994), and of vegetation on flooding and sedimentation (Malanson and Kupfer 1993). Vegetation composition, structure, and succession have been attributed to characteristics of gravel bars and flood plains in Montana (Malanson and Butler 1991), Iowa (Craig and Malanson 1993), Wyoming (Miller *et al.* 1995), Colorado (Baker 1990), and Kenya (Medley 1992). Floods have a dominant influence on many types of riparian vegetation (Kupfer and Malanson 1993; Birkeland 1996), including that of abandoned channels (Shankman and Drake 1990). Flooding due to beavers also has important impacts on vegetation (Butler 1995).

Mass movements can have major influences on vegetation (Hunter and Parker 1993; K. C. Parker and Bendix 1996). Impacts of avalanches have been modeled

(Walsh *et al.* 1994). Mass movements also can have more subtle influences on tree growth (K. R. Young and León 1990) and post-disturbance forest development (Veblen *et al.* 1989*b*). Conversely, the effects of vegetation on slope failure have been examined by Terwilliger (1990).

Influences of Climatic Variation on Vegetation

There has been a substantial increase in research on climate fluctuations and the responses of vegetation to these fluctuations across a wide range of spatial and temporal scales (Daniels and Veblen 2000). This research assists both understanding of how past climatic variation has shaped present landscapes and assessment of how landscapes may change in response to future variation (Graumlich 1991; MacDonald *et al.* 1993; Malanson and O'Leary 1995; Savage *et al.* 1996; Larsen and MacDonald 1998; Schwartz 1998). Three dominant research approaches are evident: (1) direct effects of climatic variation on growth and survivorship of individual species, (2) indirect effects mediated by climatically altered disturbance regimes, and (3) modeling of climatic influences on vegetation dynamics and landscape changes. Potential feedbacks of vegetation on atmospheric processes have been examined by Klinger *et al.* (1994) and Klink (1995).

Much attention has been directed to treeline ecotones considered sensitive to climate variation (Graumlich and Brubaker 1995; Kupfer and Cairns 1996; Burwell 1998). Dendroecological examinations have revealed the sensitivity of treeline to changes in temperature and precipitation (Butler *et al.* 1994; Szeicz and MacDonald 1995; Hessl and Baker 1997). At sites in North America, increases in regional temperature after 1880 are associated with augmented tree growth and recruitment (Taylor 1995; MacDonald *et al.* 1998*b*), whereas severe droughts cause mortality in North and South America (Lloyd and Graumlich 1997; Villalba and Veblen 1998). A series of ecosystem and forest process models has been developed to investigate causal mechanisms for treeline location and dynamics (Scuderi *et al.* 1993; D. G. Brown 1994; Malanson 1997; Cairns 1998).

Overpeck *et al.* (1990) predicted increased rates of disturbance in forest ecosystems under global warming scenarios, which stimulated research of climatic influences on disturbance regimes, particularly fire. Several studies in boreal Canada have used instrumental, sedimentary, and tree-ring records to show changes in fire regimes that correspond with climatic variation over the past several centuries (Larsen and MacDonald 1995; Larsen 1997). In northern Patagonia, years of widespread fire are strongly linked to interannual variations in atmospheric circulation features and tree demography (Villalba and Veblen 1998; Veblen *et al.* 1999). Changes in frequencies of widespread fire track multi-decadal variations in the amplitude of the El Niño-Southern Oscillation (Kitzberger and Veblen 1997).

Biogeographers are developing and applying a wide variety of modeling approaches to the task of understanding climatic influences on vegetation dynamics. In addition to treeline research, biogeographers have used models to investigate responses to increased cloud cover (Malanson and Cairns 1995) and the influences of disturbance, land use, and climate change (Neilson and Marks 1994; Baker 1995; Dyer 1995) at a range of spatial scales. Such simulations can be used to test scenarios impossible to evaluate with empirical studies.

Paleobiogeography

The study of the biogeography of the past, paleobiogeography, is a vast topic, potentially as deep in time as the origin of life on Earth and as wide in subject matter as any organism and environment that has existed. Traditionally, geographers have concentrated on changes in the last 15,000 years, contributing to multidisciplinary attempts to document the magnitude and rate of changes in vegetation, climate, disturbances, and human impact. Recent research approaches also have involved the use of museum and herbarium collections, morphological and molecular techniques, new fossil finds, and computerized programs to determine likely lineages of specific taxonomic groups (Myers and Giller 1988; Forey *et al.* 1992). This research avenue should not be overlooked, especially as new information on plate tectonics and past climate change can be incorporated (Hallam 1994; Brenchley and Harper 1998). This integration of recent discoveries has been done for pines (*Pinus*, Kremenetski *et al.* 1998; MacDonald *et al.* 1998*a*) and southern beeches (*Nothofagus*, Veblen *et al.* 1996).

Many biogeographers assess past vegetation change by studying pollen and macrofossils extracted from lake sediments or other substrates. For example, Liu (1990) sampled lake sediments, measured the importance and influx rates of pollen taxa and of fossilized leaves, seeds, invertebrates, and vertebrates, and provided a chronology of biotic communities following deglaciation in Ontario. Other examples of paleobiogeographic research include Horn (1993) for highlands in Costa Rica, Graumlich and Davis (1993) for the Great Lakes region,

Hall and Valastro (1995) for the southern Great Plains, Brubaker and McLachlan (1996) for the Pacific Northwest, and MacDonald *et al.* (1991) for boreal forest in Canada. Additional sources of data are contained in packrat middens (Jennings and Elliot-Fisk 1993) and the rings of living and dead trees (Lloyd and Graumlich 1997).

Commonly the goals of paleobiogeographic studies are to elucidate past climate change inferred from known vegetation–environmental relationships. This allows empirical tests of climate scenarios derived from computer modeling (R. S. Thompson *et al.* 1993; Benson *et al.* 1997; Whitlock and Bartlein 1997). Pollen from ice caps (Liu *et al.* 1998, L. G. Thompson *et al.* 1995), tree-rings and wood anatomy (Graumlich 1993; Woodcock 1992), charcoal (Horn and Sanford 1992; Millspaugh and Whitlock 1995), and sediments (Fletcher *et al.* 1993; Liu and Fearn 1993) are important sources of information on past climate regimes, hurricane frequency, and sea-level rise.

These techniques also can be used to date and evaluate human impact on vegetation and physical environments. Northrop and Horn (1996) used pollen and charcoal analyses to document human settlement and cultivation in Costa Rica. Similar data have been used to reassess "pristine" tropical forests (Kennedy and Horn 1997), and the nature of ancient landscapes in California (Mensing 1998), Mexico (Goman and Byrne 1998), and China (Liu and Qiu 1994). The ability to extract information from fossilized pollen and other indicators depends on the refinement of techniques (Gajewski 1993; Larsen and MacDonald 1993; MacDonald 1993; Moser *et al.* 1996; Whitlock and Millspaugh 1996; Horn *et al.* 1998). Also important is documentation of modern pollen rain in order to permit comparisons with the past (Rodgers and Horn 1996; Orvis 1999).

Cultural Biogeography

Culturally modified landscapes include towns and cities, agricultural regions, tree plantations, and natural areas affected by human-set fires or alien species. The study of these cultural landscapes has roots in early cultural geography, with its attention to the history of human–environment relationships (C. O. Sauer 1956; Glacken 1967). Cultural biogeographers also aim to understand natural landscapes before and during human modification. As an example, the lowland grasslands of California were thoroughly transformed by invasion of Mediterranean species as the Spanish arrived (Blumler 1995). In addition, advances in ecological and geographical theory have reshaped the traditional nature/human dichotomy (Zimmerer 1994; Proctor 1998; Wolch and Emel 1998) and perceptions of natural versus human-caused disturbances (Blumler 1998*a*; Vale 1998).

Vegetation can be highly influenced by air pollution (O'Leary 1990*a*; Chang and Terwilliger 2000). A striking contrast exists in the study of pollutants between the experimental emphasis of the late Walt Westman (1979) and his students (Malanson and Westman 1991; Preston 1993), and other approaches that emphasize cartographic analyses and careful attention to the historical record (Savage 1997).

Biotic invasion not only is a temporal and historical process, but also has a spatial dynamic (Brothers 1992; Medley 1997; Mensing and Byrne 1998). These exotic, invading species can alter landscapes, at times with unforeseen consequences (Westman 1990; Veblen *et al.* 1992; Knapp 1996; McCay 2000), including the modification of disturbance regimes. Some biogeographers carry out experimental research on non-native species (Beatty and Licari 1992), while others have scrutinized the historical record to determine the dates of arrival and spread (Blumler 1995).

Cytogenetic evidence on the relationships between early crops and their wild ancestors has made it possible to compare the relative importance of diffusion and independent invention (Blumler 1992), first evaluated by C. O. Sauer (1952). Global analysis demonstrates that diffusion is far more important than invention. New evidence that climate change can often be rapid, including during the time that agriculture began, has caused some archaeologists to accept such change as a major player in the Neolithic transition; biogeographers are among those contributing to the new hypotheses (Blumler 1991, 1996; Blumler and Byrne 1991). Blumler (1998*b*) also has investigated evolutionary patterns in cultivated and wild wheat, while J. D. Sauer (1993) summed up these topics in a wide-ranging volume. Finally, paleobiogeographers continue to make discoveries that shed light on the timing and location of agricultural dispersals (Fearn and Liu 1995).

Concern about the sustainability of land-use systems, especially in developing countries, has increased recently. Biogeographers studying sustainability employ diverse approaches, such as remote sensing, to examine crop productivity (Lambin *et al.* 1993); estimating potential for disease outbreaks (Cairns 1994); researching the ethnobotany of tropical forests (Medley 1993*a*; Voeks 1996, 1997); evaluating the consequences of the spread of eucalypts in the world (Doughty 2000); studying the effects of livestock grazing (Sluyter 1996; Blumler 1998*a*); and investigating crop genetic diversity (Zimmerer 1996, 1998).

Biogeographers interested in cultural landscapes also study the ecology of densely populated areas, particularly city parks and urban or suburban forests (Welch 1994; Yaukey 1996). Schiller and Horn (1997) evaluated the use of urban greenway habitats by fox and deer in the southeastern United States. Medley *et al.* (1995*a*) identified differences in landscape structure of forest vegetation along an urban to suburban gradient.

Nature Conservation

Increasingly, biogeographers apply their research to nature conservation issues: biodiversity (biological diversity), global change, and the management of ecosystems and protected areas. A unifying theme of conservation studies is the importance of understanding the role of people in changing patterns in nature.

The causes of biodiversity and its conservation have been important foci of research across a range of spatial scales and habitat types (Savage *et al.* 1992; O'Leary 1993; S. S. Young and Herwitz 1995). As an example, Kellman *et al.* (1998) showed that fire-sensitive forest plants in Belize survive in the forest interior if there is a protective edge community, and that immigration rates must offset local extinction rates in order to preserve plant diversity. Threats to biological diversity were analyzed in California by Walter (1998) and in eastern Peru by K. R. Young (1996). Impacts of large herbivores, both native and introduced, have been examined in North America (Hansen *et al.* 1995; Baker *et al.* 1997), southern South America (Relva and Veblen 1998), and Africa (M. D. Turner 1998). Anthropogenic disturbances receiving recent research attention include logging and road-building on forested landscapes of Wyoming (Reed *et al.* 1996; Tinker *et al.* 1997), and urban and agricultural land use in the eastern US (Medley *et al.* (1995*b*). Biogeographers also study the impacts of mines on the natural environment (Brothers 1990; Knapp 1991) and of the channelizing of streams (Shankman 1996).

Biogeographic research often has implications for ecological restoration (Cowell 1993). Knapp (1992) documented vegetation recovery in two Great Basin ghost towns, concluding that complete recovery is improbable. Mast *et al.* (1999) reconstructed the age structure of an Arizona ponderosa pine forest in 1876 (prior to Euro-American settlement), and used the results to guide restoration efforts. Similar reconstructions are proving useful for restoration where forest fires have been suppressed (Baker 1994). Analysis of nineteenth-century land-survey records has revealed changes in vegetation and, by implication, disturbance patterns over the past two centuries (Cowell 1995; Dyer and Baird 1997*a*; Hansen *et al.* 1995). Similarly, repeat photography has proven effective for documenting and depicting landscape changes (Bahre 1991; Veblen and Lorenz 1991; Vale and Vale 1994). Conifer forests in California are good examples of how fire exclusion has resulted in fuel accumulations and exceptional fire hazards (Minnich *et al.* 1995). As changes in fire regimes and their ecological consequences vary, management prescriptions become contentious issues (Shinneman and Baker 1997).

Some research has management implications for wildlife, as shown by the following examples: Smith *et al.* (1991) used an assessment of the range of Rocky Mountain bighorn sheep to select reintroduction sites. K. R. Young (1994, 1997) described the effects of roads and deforestation on the specialized biota of tropical montane areas. Naughton-Treves (1998) discussed the tensions created as large primates from a national park in Uganda raided adjacent cropland. Medley (1993*b*) found that monkey populations in Kenya declined with habitat loss, showing that forest preservation or restoration should be the focus of conservation efforts (Medley 1998). Taylor and Qin Zisheng (1998) addressed issues of giant panda conservation related to forest and bamboo habitats.

Biogeographers recognize the importance of scale and context in nature conservation. Baker (1992) and Savage (1993) argued that land managers who work toward successful conservation must consider landscape processes, such as disturbance and vegetation dynamics. Similarly, protection of natural resources is more successful if the social context is understood (Metz 1990; Paulson 1994; K. R. Young and León 1995). Vale (1989) addressed the issue of recreation in protected areas, concluding that a balance between nature protection and recreational use is desirable.

Technology can be used to aid conservation efforts. As examples, Homer *et al.* (1993) used satellite data to model attributes of sage grouse winter habitat in Utah, Franklin and Steadman (1991) found that GIS-based habitat mapping could be an important step in successful conservation of Polynesian land bird species, and Ortega-Huerta and Medley (1999) similarly evaluated jaguar habitat in Mexico. Modeling is also used increasingly: DeMers *et al.* (1995) examined animal colonization success in relation to habitat connectivity in two Ohio landscapes. Models of climate change indicate that vegetation patterns in Yellowstone National Park could shift dramatically and counter-intuitively, with important management implications (Bartlein *et al.* 1997). Current conservation objectives might not accommodate the types of change in species distributions that are predicted to occur. Modeling and the incorporation of remotely sensed data and GIS undoubtedly will be even more important in future nature conservation studies.

The Future of Biogeography

Membership in the Biogeography Specialty Group of the Association of American Geographers (AAG) has stayed relatively constant, with about 300 members in both the 1980s and 1990s. At the same time, there has been a substantial quantitative and qualitative advance in research on vegetation dynamics and related topics. Following a mid-century hiatus in ecological research by geographers (Veblen 1989), the rapid development that began in the 1970s has accelerated through the 1980s and 1990s. Progress is also notable in the leading roles of biogeographers in the production of major interdisciplinary syntheses on vegetation dynamics, biogeography, landscape ecology, and conservation (Glenn-Lewin *et al.* 1992; Malanson 1993; Butler 1995; G. A. J. Scott 1995; Veblen *et al.* 1996; Enright and Hill 1995; Kellman and Tackaberry 1997; Overpeck *et al.* 1997; Whittaker 1998; Zimmerer and Young 1998; Mladenoff and Baker 1999; Knight *et al.* 2000). Increased interdisciplinary collaboration and involvement in management issues also characterize the maturation of biogeographic research.

Some of the recent major thematic trends in the study of vegetation dynamics that promise to be fertile research directions include: (1) impacts of climatic variation, both directly on plant demography and indirectly through altered disturbance regimes, (2) multi-scale approaches from individual plants to landscapes or regions, (3) interactions among different kinds of disturbances in spatially heterogeneous landscapes, and (4) integration of field studies with modeling approaches. Paleobiogeography will continue to prosper by adapting approaches for studying vegetation to the challenge of acquiring information on past time-periods.

Biogeographers will also pursue issues of cultural biogeography and nature conservation in a wide variety of geographical settings, using a range of research approaches. There is an important need to link that research with the management and restoration of protected areas. Biogeographical expertise is behind such biodiversity planning efforts as described by J. M. Scott *et al.* (1993) and Beatley and Manning (1997), and for the delimitation and description of biological and ecological regions (Bailey 1996; D. E. Brown *et al.* 1998). These are all traditional strengths in geography, to which biogeographers can contribute or which they may critique (McKendry and Machlis 1993; Kupfer 1995; Bush 1996).

The traditional subdivision of geography into "human" and "physical" can be unhelpful to biogeographers who study biological processes that may have no satisfactory analogs in the social or physical sciences (Mayr 1997; Weingart *et al.* 1997). Another problem in defining biogeography originates in the teaching programs of most geography departments where the biogeography discussed in introductory courses and textbooks is almost unrecognizable compared to the research topics that define the field today. For example, an emphasis on descriptive narratives and unsophisticated ecological truisms can only discourage or mislead students (Rogers 1983). Conspicuously absent are evolutionary and genetic theories, which make some topics inaccessible to students. The textbooks used most often in upper level courses, J. H. Brown and Lomolino (1998) and Cox and Moore (2000), successfully describe ecological and evolutionary/historical approaches to biogeography, although much work by geographers is overlooked. The balanced presentation in MacDonald's (2003) new textbook largely corrects that oversight. A recent volume by Spellerberg and Sawyer (1999) introduces applied biogeography.

The most exciting developments in science take place between and among established specialties where innovation is necessary and new insights are likely. Areas of cross-disciplinary interaction are likely to be found in the transfer of information, techniques, and research approaches between geographers that study disease and agriculture, and those interested in the changing distributions of species and the human utilization and manipulation of plants and animals. Finally, it is noteworthy that the study of change in geomorphology, hydrology, and climatology provides a vocabulary and research agenda similar to those utilized by biogeographers (Hupp *et al.* 1995). It is likely that these paths connecting biogeography to other parts of geography will also be particularly rewarding at the beginning of the new millennium.

Acknowledgements

For comments, we thank Bill Baker, Sally Horn, Phil Keating, Blanca León, George Malanson, Kim Medley, Alan Taylor, and Joy Wolf.

References

Arabas, K. B. (2000). "Spatial and Temporal Relationships among Fire Frequency, Vegetation, and Soil Depth in an Eastern North American Serpentine Barren." *Journal of the Torrey Botanical Society*, 127: 51–65.

Bachelet, D., Brugnach, M., and Neilson, R. P. (1998). "Sensitivity of a Biogeography Model to Soil Properties." *Ecological Modelling*, 109: 77–98.

Bahre, C. J. (1991). *A Legacy of Change: Historic Human Impact on Vegetation of the Arizona Borderlands*. Tucson, Ariz.: University of Arizona Press.

Bailey, R. B. (1996). *Ecosystem Geography*. Heidelberg: Springer-Verlag.

Baker, W. L. (1990). "Species Richness of Colorado Riparian Vegetation." *Journal of Vegetation Science*, 1: 119–24.

— (1992). "The Landscape Ecology of Large Disturbances in the Design and Management of Nature Reserves." *Landscape Ecology*, 7: 181–94.

— (1994). "Restoration of Landscape Structure Altered by Fire Suppression." *Conservation Biology*, 8: 763–69.

— (1995). "Longterm Response of Disturbance Landscapes to Human Intervention and Global Change." *Landscape Ecology*, 10: 143–59.

Baker, W. L., and Weisberg, P. J. (1995). "Landscape Analysis of the Forest-Tundra Ecotone in Rocky Mountain Park, Colorado." *Professional Geographer*, 47: 361–75.

Baker, W. L., Munroe, J. A., and Hessl, A. E. (1997). "The Effects of Elk on Aspen in the Winter Range in Rocky Mountain National Park." *Ecography*, 20: 1–11.

Barrett, L. R., Liebens, J., Brown, D. G., Schaetzl, R. J., Zuwerink, P., Cate, T. W., and Nolan, D. S. (1995). "Relationships between Soils and Presettlement Forests in Baraga County, Michigan." *American Midland Naturalist*, 134: 264–85.

Bartlein, P. J., Whitlock, C., and Shafer, S. L. (1997). "Future Climate in the Yellowstone National Park Region and its Potential Impact on Vegetation." *Conservation Biology*, 11: 782–92.

Bauer, B. O. (1996). "Geomorphology, Geography, and Science," in R. L. Rhoads and C. E. Thorn (eds.), *The Scientific Nature of Geomorphology*. Chichester, NY: John Wiley & Sons, 381–413.

Beatley, T., and Manning, K. (1997). *The Ecology of Place: Planning for Environment, Economy, and Community*. Washington, DC: Island Press.

Beatty, S. W. (1991). "Colonization Dynamics in a Mosaic Landscape: The Buried Seed Pool." *Journal of Biogeography*, 18: 553–63.

Beatty, S. W., and Licari, D. L. (1992). "Invasion of Fennel (*Foeniculum vulgare*) into Shrub Communities on Santa Cruz Island, California." *Madroño*, 39: 54–66.

Bendix, J. (1994). "Scale, Direction, and Pattern in Riparian Vegetation–Environment Relationships." *Annals of the Association of American Geographers*, 84: 652–65.

Benson, L., Burdett, J., Lund, S., Kashgarian, M., and Mensing, S. (1997). "Nearly Synchronous Climate Change in the Northern Hemisphere during the Last Glacial Termination." *Nature*, 388: 263–5.

Birkeland, G. H. (1996). "Riparian Vegetation and Sandbar Morphology along the Lower Little Colorado River, Arizona." *Physical Geography*, 17: 534–53.

Blumler, M. A. (1991). "Modelling the Origins of Legume Domestication and Cultivation." *Economic Botany*, 45: 243–50.

— (1992). "Independent Inventionism and Recent Genetic Evidence on Plant Domestication." *Economic Botany*, 46: 98–111.

— (1995). "Invasion and Transformation of California's Valley Grassland, a Mediterranean Analogue Ecosystem," in R. Butlin and N. Roberts (eds.), *Human Impact and Adaptation: Ecological Relations in Historical Times*. Oxford: Blackwell, 308–32.

— (1996). "Ecology, Evolutionary Theory, and Agricultural Origins," in D. R. Harris (ed.), *The Origins and Spread of Agriculture and Pastoralism in Eurasia*. London: UCL Press, 25–50.

— (1998*a*). "Biogeography of Land Use Impacts in the Near East," in Zimmerer and Young (1998: 215–36).

— (1998*b*). "Introgression of Durum into Wild Emmer and the Agricultural Origin Question," in A. B. Damania and J. Valkoun (eds.), *The Origins of Agriculture and the Domestication of Crop Plants in the Near East*. Aleppo: ICARDA, 247–62.

Blumler, M. A., and Byrne, R. (1991). "The Ecological Genetics of Domestication and the Origins of Agriculture." *Current Anthropology*, 32: 23–54.

Box, E. O. (1996). "Plant Functional Types and Climate at the Global Scale." *Journal of Vegetation Science*, 7: 309–20.

Brenchley, P. J., and Harper, D. A. T. (1998). *Palaeoecology: Ecosystems, Environments and Evolution*. London: Chapman & Hall.

Brothers, T. S. (1990). "Surface-Mine Grasslands." *Geographical Review*, 80: 209–25.

— (1992). "Postsettlement Plant Migrations in Northeastern North America." *American Midland Naturalist*, 128: 72–82.

Brown, D. A. (1989). "Physiological and Morphological Problems for Dispersal and Survival of Grasses in a Changing Environment," in G. P. Malanson (ed.), *Natural Areas Facing Climate Change*. The Hague: SPB Academic Publishing, 25–38.

— (1993). "Early Nineteenth-Century Grasslands of the Midcontinent Plains." *Annals of the Association of American Geographers*, 83: 589–612.

Brown, D. E., Reichenbacher, F., and Franson, S. E. (1998). *A Classification of North American Biotic Communities*. Salt Lake City: University of Utah Press.

Brown, D. G. (1994). "Predicting Vegetation Types at Treeline Using Topography and Biophysical Disturbance Variables." *Journal of Vegetation Science*, 5: 641–56.

Brown, J. H., and Lomolino, M. V. (1998). *Biogeography* (2nd edn.). Sunderland, Mass: Sinauer Associates.

Brubaker, L. B., and McLachlan, J. S. (1996). "Landscape Diversity and Vegetation Response to Long-Term Climate Change in the Eastern Olympic Peninsula, Pacific Northwest, USA," in B. Walker and W. Steffen (eds.), *Global Change and Terrestrial Ecosystems*. Cambridge: Cambridge University Press, 184–203.

Burwell, T. (1998). "Successional Patterns of the Lower Montane Treeline, Eastern California." *Madroño*, 45: 12–16.

Bush, M. B. (1996). "Amazonian Conservation in a Changing World." *Biological Conservation*, 76: 219–28.

Butler, D. R. (1995). *Zoogeomorphology: Animals as Geomorphic Agents*. Cambridge: Cambridge University Press.

Butler, D. R., Malanson, G. P., and Cairns, D. M. (1994). "Stability of Alpine Treeline in Glacier National Park, Montana, U.S.A." *Phytocoenologia*, 22: 485–500.

Cairns, D. M. (1994). "Spatial Pattern Analysis of Witches' Broom Disease of Cacao at a Landscape in Rondonia, Brazil." *Tropical Agriculture*, 71: 31–5.

— (1998). "Modeling Controls on Pattern at Alpine Treeline." *Geographical and Environmental Modelling*, 2: 43–63.

Callaway, R. M., and Davis, F. W. (1993). "Vegetation Dynamics, Fire, and the Physical Environment in Coastal Central California." *Ecology*, 74: 1567–78.

— (1998). "Recruitment of *Quercus agrifolia* in Central California: The Importance of Shrub-Dominated Patches." *Journal of Vegetation Science*, 9: 647–56.

Chang, E., and Terwilliger, V. J. (2000). "The Effects of Air Pollution on Vegetation from a Geographic Perspective." *Progress in Physical Geography*, 24: 53–74.

Conkey, L. E., Keifer, M., and Lloyd, A. H. (1995). "Disjunct Jack Pine (*Pinus banksiana* Lamb.) Structure and Dynamics, Acadia National Park, Maine." *Ecoscience*, 2: 168–76.

Cowell, C. M. (1993). "Ecological Restoration and Environmental Ethics." *Environmental Ethics*, 15: 19–32.

— (1995). "Presettlement Piedmont Forests: Patterns of Composition and Disturbance in Central Georgia." *Annals of the Association of American Geographers*, 85: 65–83.

Cox, C. B., and Moore, P. D. (2000). *Biogeography: An Ecological and Evolutionary Approach* (6th edn.). Oxford: Blackwell Scientific Publications.

Craig, M. R., and Malanson, G. P. (1993). "River Flow Events and Vegetation Colonization of Point Bars in Iowa." *Physical Geography*, 14: 436–48.

Daniels, L. D., and Klinka, K. (1996). "The Dynamics of Old-Growth *Thuja-Tsuga* Forests near Vancouver, British Columbia," in J. S. Dean, D. M. Meko, and T. W. Swetnam (eds.), *Tree Rings, Environment and Humanity: Radiocarbon*. Tucson, Ariz.: University of Arizona Press, 379–92.

Daniels, L. D., and Veblen, T. T. (2000). "ENSO Effects on Temperature and Precipitation of the Patagonian-Andean Region: Implications for Biogeography." *Physical Geography*, 21: 223–43.

Daniels, L. D., Marshall, P. L., Carter, R. E., and Klinka, K. (1995). "Age Structure of *Thuja plicata* in the Tree Layer of Old-Growth Stands near Vancouver, British Columbia." *Northwest Science*, 69: 175–83.

DeMers, M. N. (1993). "Roadside Ditches as Corridors for Range Expansion of the Western Harvester Ant (*Pogonomyrmex occidentalis* Cresson)." *Landscape Ecology*, 8: 93–102.

DeMers, M. N., Simpson, J. W., Boerner, R. E. J., Silva, A., Berns, L., and Artigas, F. (1995). "Fencerows, Edges, and Implications of Changing Connectivity Illustrated by Two Contiguous Ohio Landscapes." *Conservation Biology*, 9: 1159–68.

Dingle, H. (1996). *Migration: The Biology of Life on the Move*. New York: Oxford University Press.

Doughty, R. W. (2000). *The Eucalyptus: A Natural and Commercial History of the Gum Tree*. Baltimore: Johns Hopkins University Press.

Dyer, J. M. (1995). "Assessment of Climatic Warming Using a Model of Forest Species Migration." *Ecological Modelling*, 79: 199–219.

Dyer, J. M., and Baird, P. R. (1997*a*). "Remnant Forest Stands at a Prairie Ecotone Site: Presettlement History and Comparison with Other Maple-Basswood Stands." *Physical Geography*, 18: 146–59.

— (1997*b*). "Wind Disturbance in Remnant Forest Stands along the Prairie-Forest Ecotone, Minessota, USA." *Plant Ecology*, 129: 121–34.

Enright, N. J., and Goldblum, D. (1998). "Stand Structure of the Emergent Conifer *Agathis ovata* in Forest and Maquis, Province Sud, New Caledonia." *Journal of Biogeography*, 25: 641–8.

Enright, N. J., and Hill, R. S. (eds.) (1995). *Ecology of the Southern Conifers*. Melbourne: Melbourne University Press.

Enright, N. J., Goldblum, D., Ata, P., and Ashton, D. H. (1997). "The Independent Effects of Heat, Smoke and Ash on Emergence of Seedlings from the Soil Seed Bank of a Healthy *Eucalyptus* Woodland in Grampians (Gariwerd) National Park, Western Victoria." *Australian Journal of Ecology*, 22: 81–8.

Fearn, M. L., and Liu, K. B. (1995). "Maize Pollen of 3,500 B.P. from Southern Alabama." *American Antiquity*, 60: 109–17.

Fletcher, C. H., III, Van Pelt, J. E., Brush, G. S., and Sherman, J. (1993). "Tidal Wetland Record of Holocene Sea-Level Movements and Climate History." *Palaeogeography, Palaeoclimatology, Palaeoecology*, 102: 177–213.

Forey, P. L., Humphries, C. J., Kitching, I. J., Scotland, R. W., Siebert, D. J., and Williams, D. M. (1992). *Cladistics: A Practical Course in Systematics*. Oxford: Clarendon Press.

Franklin, J. (1995). "Predictive Vegetation Mapping: Geographic Modelling of Biospatial Patterns in Relation to Environmental Gradients." *Progress in Physical Geography*, 19: 474–99.

— (1998). "Predicting the Distribution of Shrub Species in Southern California from Climate and Terrain-Driven Variables." *Journal of Vegetation Science*, 9: 733–48.

Franklin, J., and Merlin, M. (1992). "Species-Environment Patterns of Forest Vegetation on the Uplifted Reef Limestone of Atiu, Mangaia, Ma'uke and Miti'aro, Cook Islands, West Central Pacific Ocean." *Journal of Vegetation Science*, 3: 3–14.

Franklin, J., and Steadman, D. W. (1991). "The Potential for Conservation of Polynesian Birds through Habitat Mapping and Species Translocation." *Conservation Biology*, 5: 506–21.

Frelich, L. E., and Graumlich, L. J. (1994). "Age-Class Distribution and Spatial Patterns in an Old-Growth Hemlock-Hardwood Forest." *Canadian Journal of Forest Research*, 24: 1939–47.

Gajewski, K. (1993). "The Role of Paleoecology in the Study of Global Climatic Change." *Review of Palaeobotany and Palynology*, 79: 141–51.

Gauch, H. G. (1982). *Multivariate Analysis in Community Ecology*. New York: Cambridge University Press.

Glacken, C. (1967). *Traces on the Rhodian Shore: Nature and Culture in Western Thought from Ancient Times to the End of the Eighteenth Century*. Berkeley: University of California Press.

Glenn-Lewin, D. C., Peet, P. K., and Veblen, T. T. (eds.) (1992). *Plant Succession: Theory and Prediction*. London: Chapman & Hall.

Goldblum, D. (1997). "The Effects of Treefall Gaps on Understory Vegetation in New York, USA." *Journal of Vegetation Science*, 8: 125–32.

Goldblum, D., and Veblen, T. T. (1994). "Fire History of a Ponderosa Pine/Douglas Fir Forest in the Colorado Front Range." *Physical Geography*, 13: 133–48.

Goman, M., and Byrne, R. (1998). "A 5,000-Year Record of Agriculture and Tropical Forest Clearance in the Tuxtlas, Veracruz, Mexico." *The Holocene*, 8: 83–9.

Grau, H. R., Arturi, M. F., Brown, A. D., and Aceñolaza, P. G. (1997). "Floristic and Structural Patterns along a Chronosequence of Secondary Forest Succession in Argentinean Subtropical Montane Forests." *Forest Ecology and Management*, 95: 161–71.

Grau, H. R., and Brown, A. D. (1998). "Structure, Composition, and Inferred Dynamics of a Subtropical Montane Forest of Northwest Argentina," in F. Dallmeier and J. A. Comiskey (eds.), *Forest Biodiversity in North, Central and South America and the Caribbean: Research and Monitoring*. Pearl River, NY: UNESCO and Parthenon Publishing Group, 715–26.

Graumlich, L. J. (1991). "Subalpine Tree Growth, Climate, and Increasing CO2: An Assessment of Recent Growth Trends." *Ecology*, 72: 1–11.

— (1993). "A 1000-Year Record of Temperature and Precipitation in the Sierra Nevada." *Quaternary Research*, 39: 249–55.

Graumlich, L. J., and Brubaker, L. B. (1995). "Long-Term Records of Growth and Distribution of Conifers: Integration of Paleoecology and Physiological Ecology," in T. M. Hinckley and W. Smith (eds.), *Ecophysiology of Coniferous Forests*. San Diego: Academic Press, 37–62.

Graumlich, L. J., and Davis, M. B. (1993). "Holocene Variation in Spatial Scales of Vegetation Pattern in the Upper Great Lakes." *Ecology*, 74: 826–39.

Grissino-Mayer, H. D., Baisin, C. H., and Swetnam, T. W. (1995). "Fire History in the Pinaleño Mountains of Southeastern Arizona: Effects of Human-Related Disturbances," in *Biodiversity and Management of the Madrean Archipelago*. USDA Forest Service General Technical Report RM-GTR-264, 399–407.

Gurnell, A. M. (1998). "The Hydrogeomorphological Effects of Beaver Dam-Building Activity." *Progress in Physical Geography*, 22: 167–89.

Hadley, K. S. (1994). "The Role of Disturbance, Topography and Forest Structure in the Development of a Montane Forest Landscape." *Bulletin of the Torrey Botanical Club*, 12: 47–61.

Hadley, K. S., and Savage, M. (1996). "Wind Disturbance and Development of a Near-Edge Forest Interior, Mary's Peak, Oregon Coast Range." *Physical Geography*, 17: 47–61.

Hadley, K. S., and Veblen, T. T. (1993). "Stand Response to Western Spruce Budworm and Douglas-Fir Bark Beetle Outbreaks, Colorado Front Range." *Canadian Journal of Forest Research*, 23: 479–91.

Hall, S. A., and Valastro, S. (1995). "Grassland Vegetation in the Southern Great Plains during the Last Glacial Maximum." *Quaternary Research*, 44: 237–45.

Hallam, A. (1994). *An Outline of Phanerozoic Biogeography*. Oxford Biogeography, 10. Oxford: Oxford University Press.

Hansen, K. J., Wyckoff, W., and Banfield, J. (1995). "Shifting Forests: Historical Grazing and Forest Invasion in Southwestern Montana." *Forest and Conservation History*, 39: 66–76.

Hengeveld, R. (1990). *Dynamic Biogeography*. Cambridge: Cambridge University Press.

Herwitz, S. R., and Young, S. S. (1994). "Mortality, Recruitment, and Growth Rates of Montane Tropical Rain Forest Canopy Trees on Mount Bellenden-Ker, Northeast Queensland, Australia." *Biotropica*, 26: 350–61.

Hessl, A. E., and Baker, W. L. (1997). "Spruce-Fir Growth Form Changes in the Forest-Tundra Ecotone of Rocky Mountain National Park, Colorado, USA." *Ecography*, 20: 356–67.

Homer, C. G., Edwards, T. C., Ramsey, R. D., and Price, K. P. (1993). "Use of Remote-Sensing Methods in Modeling Sage Grouse Winter Habitat." *Journal of Wildlife Management*, 57: 78–84.

Hoover, S. R., and Parker, A. J. (1991). "Spatial Components of Biotic Diversity in Landscapes of Georgia, USA." *Landscape Ecology*, 5: 125–36.

Horn, S. P. (1993). "Postglacial Vegetation and Fire History in the Chirripó Páramo of Costa Rica." *Quaternary Research*, 40: 107–16.

— (1997). "Postfire Resprouting of *Hypericum irazuense* in the Costa Rican páramos: Cerro Asunción revisited." *Biotropica*, 29: 529–31.

— (1998). "Fire Management and Natural Landscapes in the Chirripó páramo, Chirripó National Park, Costa Rica," in Zimmerer and Young (1998: 125–46).

Horn, S. P., and Sanford R. L., Jr. (1992). "Holocene Fires in Costa Rica." *Biotropica*, 24: 354–61.

Horn, S. P., Rodgers J. C., III, Orvis, K. H., and Northrop, L. A. (1998) "Recent Land Use and Vegetation History from Soil Pollen Analysis: Testing the Potential in the Lowland Humid Tropics." *Palynology*, 22: 167–80.

Hunter, J. C., and Parker, V. T. (1993). "The Disturbance Regime of an Old-Growth Forest in Coastal California." *Journal of Vegetation Science*, 4: 19–24.

Hupp, C. R., Osterkamp, W. R., and Howard, A. D. (1995). *Biogeomorphology, Terrestrial and Freshwater Systems*. Amsterdam: Elsevier.

Jelinski, D. E., and Fisher, L. J. (1991). "Spatial Variability in the Nutrient Composition of *Populus tremuloides*: Clone-to-Clone Differences and Implications for Cervids." *Oecologia*, 88: 116–24.

Jennings, S. A., and Elliott-Fisk, D. L. (1993). "Packrat Midden Evidence of Late Quaternary Vegetation Change in the White Mountains, California-Nevada." *Quaternary Research*, 39: 214–21.

Keating, P. L. (1998). "Effects of Anthropogenic Disturbances on Paramo Vegetation in Podocarpus National Park, Ecuador." *Physical Geography*, 19: 221–38.

— (1999). "Changes in Paramo Vegetation along an Elevation Gradient in Southern Ecuador." *Journal of the Torrey Botanical Society*, 126: 159–75.

Kellman, M., and Tackaberry, R. (1993). "Disturbance and Tree Species Coexistence in Tropical Riparian Forest Fragments." *Global Ecology and Biogeography Letters*, 3: 1–9.

— (1997). *Tropical Environments: The Functioning and Management of Tropical Ecosystems*. London: Routledge.

Kellman, M., Tackaberry, R., and Rigg, L. (1998). "Structure and Function in Two Tropical Gallery Forest Communities: Implications for Forest Conservation in Fragmented Systems." *Journal of Applied Ecology*, 35: 195–206.

Kennedy, L. M., and Horn, S. P. (1997). "Prehistoric Maize Cultivation at the La Selva Biological Station, Costa Rica." *Biotropica*, 29: 368–70.

Kitzberger, T., and Veblen, T. T. (1997). "Influences of Humans and ENSO on Fire History of *Austrocedrus chilensis* Woodlands in Northern Patagonia, Argentina." *Ecoscience*, 4: 508–20.

Klinger, L. F., Zimmerman, P. R., Greenberg, J. P., Heidt, L. E., and Guenther, A. B. (1994). "Carbon Trace Gas Fluxes along a Successional Gradient in the Hudson Bay Lowland." *Journal of Geophysical Research*, 99: 1469–94.

Klink, K. (1995). "Temporal Sensitivity of Regional Climate to Land-Surface Heterogeneity." *Physical Geography*, 16: 289–314.

Knapp, P. A. (1991). "Long-Term Soil and Vegetation Recovery in Five Semiarid Montana Ghost Towns." *Professional Geographer*, 43: 486–99.

—— (1992). "Secondary Plant Succession and Vegetation Recovery in Two Western Great Basin Desert Ghost Towns." *Biological Conservation*, 60: 81–9.

—— (1996). "Cheatgrass (*Bromus tectorum* L.) Dominance in the Great Basin Desert—History, Persistency, and Influence to Human Activities." *Global Environmental Change—Human and Policy Dimensions*, 6: 37–52.

Knapp, P. A., and Soulé, P. T. (1998). "Recent *Juniperus occidentalis* (Western Juniper) Expansion on a Protected Site in Central Oregon." *Global Change Biology*, 4: 347–57.

Knight, R. L., Smith, F. W., Buskirk, S. W., Romme, W. H., and Baker, W. L. (eds.) (2000). *Forest Fragmentation in the Southern Rocky Mountains*. Boulder: University Press of Colorado.

Kremenetski, C., Liu, K. B., and MacDonald, G. M. (1998). "The Late Quaternary Dynamics of Pines in Northern Asia," in D. M. Richardson (ed.), *Ecology and Biogeography of Pinus*. Cambridge: Cambridge University Press, 95–106.

Kupfer, J. A. (1995). "Landscape Ecology and Biogeography." *Progress in Physical Geography*, 19: 18–34.

Kupfer, J. A., and Cairns, D. M. (1996). "The Suitability of Montane Ecotones as Indicators of Global Climatic Change." *Progress in Physical Geography*, 20: 253–72.

Kupfer, J. A., and Malanson, G. P. (1993). "Observed and Modeled Directional Change in Riparian Forest Composition at a Cutbank Edge." *Landscape Ecology*, 8: 185–99.

Kupfer, J. A., and Runkle, J. R. (1996). "Early Gap Successional Pathways in a Beech-Maple Forest Preserve: Patterns and Determinants." *Journal of Vegetation Science*, 7: 247–56.

Kupfer, J. A., Malanson, G. P., and Runkle, J. R. (1997). "Factors Influencing Species Composition in Canopy Gaps: The Importance of Edge Proximity in Hueston Woods, Ohio." *Professional Geographer*, 49: 165–78.

Lambin, E. F., Cashman, P., Moody, A., Parkhurst, B. H., Pax, M. H., and Schaaf, C. B. (1993). "Agricultural Production Monitoring in the Sahel Using Remote Sensing: Present Possibilities and Research Needs." *Journal of Environmental Management*, 38: 301–22.

Larsen, C. P. S. (1997). "Spatial and Temporal Variations in Boreal Forest Fire Frequency in Northern Alberta." *Journal of Biogeography*, 24: 663–73.

Larsen, C. P. S., and MacDonald, G. M. (1993). "Lake Morphometry, Sediment Mixing and the Selection of Sites for Fine Resolution Palaeoecological Studies." *Quaternary Science Reviews*, 12: 781–92.

—— (1995). "Relations between Tree-Ring Widths, Climate, and Annual Area Burned in the Boreal Forest of Alberta." *Canadian Journal of Forest Research*, 25: 1746–55.

—— (1998). "An 840-year Record of Fire and Vegetation in a Boreal White Spruce Forest." *Ecology*, 79: 106–18.

Leitner, L. A., Dunn, C. P., Guntenspergen, G. R., Stearns, F., and Sharpe, D. M. (1991). "Effects of Site, Landscape Features, and Fire Regime on Vegetation Patterns in Presettlement Southern Wisconsin (USA)." *Landscape Ecology*, 5: 203–18.

Liu, K. B. (1990). "Holocene Paleoecology of the Boreal Forest and Great Lakes-St. Lawrence Forest in Northern Ontario." *Ecological Monographs*, 60: 179–212.

Liu, K. B., and Fearn, M. L. (1993). "Lake-Sediment Record of Late Holocene Hurricane Activities from Coastal Alabama." *Geology*, 21: 793–6.

Liu, K. B., and Qiu, H. L. (1994). "Late-Holocene Pollen Records of Vegetational Changes in China: Climate or Human Disturbances?" *Terrestrial, Atmospheric and Oceanic Sciences*, 5: 393–410.

Liu, K. B., Yao, Z., and Thompson, L. G. (1998). "A Pollen Record of Holocene Climatic Changes from the Dunde Ice Cap, Qinghai-Tibetan Plateau." *Geology*, 26: 135–8.

Lloyd, A. H., and Graumlich, L. J. (1997). "Holocene Dynamics of Treeline Forests in the Sierra Nevada." *Ecology*, 78: 1199–210.

McCay, D. H. (2000). "Effects of Chronic Human Activities on Invasion of Longleaf Pine Forests by Sand Pine." *Ecosystems*, 3: 283–92.

MacDonald, G. M. (1993). "Reconstructing Plant Invasions Using Fossil Pollen Analysis." *Advances in Ecological Research*, 24: 67–110.

—— (2003). *Biogeography: Introduction to Space, Time and Life*. New York: John Wiley & Sons.

MacDonald, G. M., Cwynar, L. C., and Whitlock, C. (1998*a*). "The Late Quaternary Dynamics of Pines in Northern North America," in D. M. Richards (ed.), *Ecology and Biogeography of Pinus*. Cambridge: Cambridge University Press, 122–36.

MacDonald, G. M., Larsen, C. P. S., Szeicz, J. M., and Moser, K. A. (1991). "The Reconstruction of Boreal Forest Fire History from Lake Sediments: A Comparison of Charcoal, Pollen, Sedimentological and Geochemical Indices." *Quaternary Science Reviews*, 10: 53–71.

MacDonald, G. M., Szeicz, J. M., Claricoates, J., and Dale, K. A. (1998*b*). "Response of the Central Canadian Treeline to Recent Climatic Changes." *Annals of the Association of American Geographers*, 88: 183–208.

MacDonald, G. M., Edwards, T. W. D., Moser, K. A., Pienitz, R., and Smol, J. P. (1993). "Rapid Response of Treeline Vegetation and Lakes to Past Climate Warming." *Nature*, 361: 243–6.

McKendry, J. E., and Machlis, G. E. (1993). "The Role of Geography in Extending Biodiversity Gap Analysis." *Applied Geography*, 13: 135–52.

Malanson, G. P. (1993). *Riparian Landscapes*. Cambridge: Cambridge University Press.

—— (1996). "Effects of Dispersal and Mortality on Diversity in a Forest Stand Model." *Ecological Modelling*, 87: 103–10.

—— (1997). "Effects of Feedbacks and Seed Rain on Ecotone Patterns." *Landscape Ecology*, 12: 27–38.

Malanson, G. P., and Butler, D. R. (1991). "Floristic Variation among Gravel Bars in a Subalpine River, Montana, USA." *Arctic and Alpine Research*, 23: 273–8.

Malanson, G. P., and Butler, D. R. (1994). "Tree-Tundra Competitive Hierarchies, Soil Fertility Gradients and Treeline Elevation in Glacier National Park, Montana." *Physical Geography*, 15: 166–80.

Malanson, G. P., and Cairns, D. M. (1995). "Effects of Increased Cloud-Cover on a Montane Forest Landscape." *Ecoscience*, 2: 75–82.

Malanson, G. P., and Kupfer, J. A. (1993). "Simulated Fate of Leaf Litter and Large Woody Debris at a Riparian Cutbank." *Canadian Journal of Forest Research*, 23: 582–90.

Malanson, G. P., and O'Leary, J. F. (1995). "The Coastal Sage Scrub-Chaparral Boundary and Response to Global Climatic Change," in J. M. Moreno and W. C. Oechel (eds.), *Global Change and Mediterranean-type Ecosystems*. New York: Springer-Verlag, 203–24.

Malanson, G. P., and Westman, W. E. (1991). "Modeling Interactive Effects of Climate Change, Air Pollution, and Fire on a California (USA) Shrubland." *Climatic Change*, 18: 363–76.

Mast, J. N., Fulé, P. Z., Moore, M. M., Covington, W. W., and Waltz, A. E. M. (1999). "Restoration of Presettlement Age Structure of an Arizona Ponderosa Pine Forest." *Ecological Applications*, 9: 228–39.

Mast, J. N., Veblen, T. T., and Linhart, Y. B. (1998). "Disturbance and Climatic Influences on Age Structure of Ponderosa Pine at the Pine/Grassland Ecotone, Colorado Front Range." *Journal of Biogeography*, 25: 743–55.

Mayr, E. (1997). *This is Biology: The Science of the Living World*. Cambridge, Mass.: Belknap Press.

Medley, K. E. (1992). "Patterns of Forest Diversity along the Tana River, Kenya." *Journal of Tropical Ecology*, 8: 353–71.

—— (1993*a*). "Extractive Forest Resources of the Tana River National Primate Reserve, Kenya." *Economic Botany*, 47: 171–83.

—— (1993*b*). "Primate Conservation along the Tana River, Kenya: An Examination of the Forest Habitat." *Conservation Biology*, 7: 109–21.

—— (1997). "Distribution of the Non-Native Shrub *Lonicera maackii* in Kramer Woods, Ohio." *Physical Geography*, 18: 18–36.

—— (1998). "Landscape Change and Resource Conservation along the Tana River, Kenya," in Zimmerer and Young (1998: 39–55).

Medley, K. E., McDonnell, M. J., and Pickett, S. T. A. (1995*a*). "Forest-Landscape Structure along an Urban-to-Rural Gradient." *Professional Geographer*, 47: 159–68.

Medley, K. E., Okey, B. W., Barrett, G. W., Lucas, M. F., and Renwick, W. H. (1995*b*). "Landscape Change with Agricultural Intensification in a Rural Watershed, Southwestern Ohio, U.S.A." *Landscape Ecology*, 10: 161–76.

Mensing, S. A. (1998). "560 Years of Vegetation Change in the Region of Santa Barbara, California." *Madroño*, 45: 1–11.

Mensing, S. A., and Byrne, R. (1998). "Pre-Mission Invasion of *Erodium cicutarium* in California." *Journal of Biogeography*, 24: 757–62.

Metz, J. J. (1990). "Conservation Practices at an Upper-Elevation Village of West Nepal." *Mountain Research and Development*, 10: 7–15.

Miller, J. R., Schulz, T. T., Hobbs, N. T., Wilson, K. R., Schrupp, D. L., and Baker, W. L. (1995). "Changes in the Landscape Structure of a Southeastern Wyoming Riparian Zone Following Shifts in Stream Dynamics." *Biological Conservation*, 72: 371–9.

Millspaugh, S. H., and Whitlock, C. (1995). "A 750-Year Fire History Based on Lake Sediment Records in Central Yellowstone National Park, USA." *The Holocene*, 5: 283–92.

Minnich, R. A., and Bahre, C. J. (1995). "Wildland Fire and Chaparral Succession along the California-Baja California Boundary." *International Journal of Wildland Fire*, 5: 13–24.

Minnich, R. A., Barbour, M. G., Burk, J. H., and Fernau, R. F. (1995). "Sixty Years of Change in Californian Conifer Forests of the San Bernardino Mountains." *Conservation Biology*, 9: 902–14.

Mladenoff, D. J., and Baker, W. L. (eds.) (1999). *Spatial Modeling of Forest landscapes: Approaches and Limitations*. Cambridge: Cambridge University Press.

Moser, K. A., MacDonald, G. M., and Smol, J. P. (1996). "Applications of Freshwater Diatoms to Geographical Research." *Progress in Physical Geography*, 20: 21–52.

Myers, A. A., and Giller, P. S. (1988). *Analytical Biogeography: An Integrated Approach to the Study of Animal and Plant Distributions*. London: Chapman & Hall.

Naughton-Treves, L. (1998). "Predicting Patterns of Crop Damage by Wildlife around Kibale National Park, Uganda." *Conservation Biology*, 12: 156–68.

Neilson, R. P. (1993). "Transient Ecotone Response to Climatic Change: Some Conceptual and Modelling Approaches." *Ecological Applications*, 3: 385–95.

Neilson, R. P., and Marks, D. (1994). "A Global Perspective of Regional Vegetation and Hydrologic Sensitivities from Climatic Change." *Journal of Vegetation Science*, 5: 715–30.

Northrop, L. A., and Horn, S. P. (1996). "PreColumbian Agriculture and Forest Disturbance in Costa Rica: Palaeoecological Evidence from Two Lowland Rainforest Lakes." *The Holocene*, 6: 289–99.

O'Leary, J. F. (1990*a*). "Californian Coastal Sage Scrub: General Characteristics and Considerations for Biological Conservation," in A. A. Schoenherr (ed.), *Endangered Plant Communities of Southern California*. Southern California Botanists Special Publication No. 3, 24–41.

—— (1990*b*). "Post-Fire Diversity Patterns in Two Subassociations of Californian Coastal Sage Scrub." *Journal of Vegetation Science*, 1: 173–80.

—— (1993). "Towards Greater Uniformity of Species Diversity Studies in Mediterranean-type Ecosystems." *Landscape and Urban Planning*, 24: 185–90.

Ortega-Huerta, M. A., and Medley, K. E. (1999). "Landscape Analysis of Jaguar (*Panthera onca*) Habitat Using Sighting Records in the Sierra de Tamaulipas, Mexico." *Environmental Conservation*, 26: 257–69.

Orvis, K. H. (1999). "Modern Surface Pollen from Three Transects across the Southern Sonoran Desert Margin, Northwestern Mexico." *Palynology*, 22: 197–211.

Osterkamp, W. R., and Hupp, C. R. (1996). "The Evolution of Geomorphology, Ecology, and other Composite Sciences," in R. L. Rhoads and C. E. Thorn (eds.), *The Scientific Nature of Geomorphology*. Chichester, NY: John Wiley & Sons, 415–41.

Overpeck, J. T., Hughen, K., Hardy, D., Bradley, R., Case, R., Douglas, M., Finney, B., Gajewski, K., Jacoby, G., Jennings, A., Lamoureux, S., Lasca, A., MacDonald, G., Moore, J., Retelle, M., Smith, S., Wolfe, A., and Zielinski, G. (1997). "Arctic Environmental Change of the Last Four Centuries." *Science*, 278: 1251–6.

Overpeck, J. T., Rind, D., and Goldberg, R. (1990). "Climate-Induced Changes in Forest Disturbance and Vegetation." *Nature*, 343: 51–3.

Parker, A. J. (1995). "Comparative Gradient Structure and Forest Cover Types in Lassen Volcanic and Yosemite National Parks, California." *Bulletin of the Torrey Botanical Club*, 122: 58–68.

Parker, A. J., and Parker, K. C. (1994). "Structural Variability of Mature Lodgepole Pine Stands on Gently Sloping Terrain in Taylor Park Basin, Colorado." *Canadian Journal of Forest Research*, 24: 2020–9.

Parker, K. C. (1991). "Topography, Substrate, and Vegetation Patterns in the Northern Sonoran Desert." *Journal of Biogeography*, 18: 151–63.

— (1993). "Climatic Effects on Regeneration Trends for Two Columnar Cacti in the Northern Sonoran Desert." *Annals of the Association of American Geographers*, 83: 452–74.

— (1995). "Effects of Complex Geomorphic History on Soil and Vegetation Patterns on Arid Alluvial Fans." *Journal of Arid Environments*, 30: 19–39.

Parker, K. C., and Bendix, J. (1996). "Landscape-Scale Geomorphic Influences on Vegetation Patterns in Four Environments." *Physical Geography*, 17: 113–41.

Parker, K. C., and Hamrick, J. L. (1992). "Genetic Diversity and Clonal Structure in a Columnar Cactus, *Lophocereus schottii*." *American Journal of Botany*, 79: 86–96.

Parker, K. C., Hamrick, J. L., Parker, A. J., and Stacy, E. A. (1997*a*). "Allozyme Diversity in *Pinus virginiana* (Pinaceae): Intraspecific and Interspecific Comparisons." *American Journal of Botany*, 84: 1372–82.

Parker, K. C., Parker, A. J., Beaty, R. M., Fuller, M. M., and Faust, T. D. (1997*b*). "Population Structure and Spatial Pattern of Two Coastal Populations of Ocala Sand Pine (*Pinus clausa* (Chapm. Ex Engelm.) Vasey ex Sarg. Var. *clausa* D. B. Ward)." *Journal of the Torrey Botanical Society*, 124: 22–33.

Paulson, D. D. (1994). "Understanding Tropical Deforestation: The Case of Western Samoa." *Environmental Conservation*, 21: 326–32.

Pérez, F. L. (1991). "Soil Moisture and the Distribution of Giant Andean Rosettes on Talus Slopes of a Desert Paramo." *Climate Research*, 1: 217–31.

— (1998). "Human Impact on the High Páramo Landscape of the Venezuelan Andes," in Zimmerer and Young (1998: 147–83).

Preston, K. P. (1993). "Selection for Sulfur Dioxide and Ozone Tolerance in *Bromus rubens* along the South Central Coast of California." *Annals of the Association of American Geographers*, 83: 141–55.

Proctor, J. D. (1998). "The Social Construction of Nature: Relativist Accusations, Pragmatist and Critical Realist Responses." *Annals of the Association of American Geographers*, 88: 352–76.

Rebertus, A. J., and Veblen, T. T. (1993). "Structure and Tree-Fall Gap Dynamics of Old-Growth *Nothofagus* Forests in Tierra del Fuego, Argentina." *Journal of Vegetation Science*, 4: 641–54.

Rebertus, A. J., Kitzberger, T., Veblen, T. T., and Roovers, L. M. (1997). "Blowdown History and Landscape Patterns in the Andes of Tierra del Fuego, Argentina." *Ecology*, 78: 678–92.

Reed, R. A., Johnson-Barnard, J., and Baker, W. L. (1996). "Contribution of Roads to Forest Fragmentation in the Rocky Mountains." *Conservation Biology*, 10: 1098–106.

Relva, M. A., and Veblen, T. T. (1998). "Impacts of Introduced Large Herbivores on *Austrocedrus chilensis* Forests in Northern Patagonia, Argentina." *Forest Ecology and Management*, 108: 27–40.

Rigg, L. S., Enright, N. J., and Jaffré, T. (1998). "Stand Structure of the Emergent Conifer *Araucaria laubenfelsii*, in Maquis and Rainforest, Mont Do, New Caledonia." *Australian Journal of Ecology*, 23: 528–38.

Rodgers, J. C., III, and Horn, S. P. (1996). "Modern Pollen Spectra from Costa Rica." *Palaeogeography, Palaeoclimatology, Palaeoecology*, 124: 53–71.

Rogers, G. F. (1983). "Growth of Biogeography in Canadian and U.S. Geography Departments." *Professional Geographer*, 35: 219–26.

Runkle, J. R., Stewart, G. H., and Veblen, T. T. (1995). "Sapling Diameter Growth in Gaps for Two *Nothofagus* Species in New Zealand." *Ecology*, 76: 2107–17.

Sauer, C. O. (1952). *Agricultural Origins and Dispersals*. New York: American Geographic Society.

— (1956). "The Agency of Man on the Earth," in W. L. Thomas (ed.), *Man's Role in Changing the Face of the Earth*. Chicago: University of Chicago Press, i. 49–69.

Sauer, J. D. (1993). *Historical Geography of Crop Plants: A Select Roster*. Boca Raton, Fla.: CRC Press.

Savage, M. (1991). "Structural Dynamics of a Southwestern Pine Forest under Chronic Human Influence." *Annals of the Association of American Geographers*, 81: 271–89.

— (1993). "Ecological Disturbance and Nature Tourism." *Geographical Review*, 83: 290–300.

— (1994). "Anthropogenic and Natural Disturbance and Patterns of Mortality in a Mixed Conifer Forest in California." *Canadian Journal of Forest Research*, 24: 1149–59.

— (1997). "The Role of Anthropogenic Influences in a Mixed-Conifer Forest Mortality Episode." *Journal of Vegetation Science*, 8: 95–104.

Savage, M., Brown, P. M., and Feddema, J. (1996). "The Role of Climate in a Pine Forest Regeneration Pulse in the Southwestern United States." *Ecoscience*, 3: 310–18.

Savage, M., Reid, M., and Veblen, T. T. (1992). "Diversity and Disturbance in a Colorado Subalpine Forest." *Physical Geography*, 13: 240–9.

Scatena, F. N., and Lugo, A. E. (1995). "Geomorphology, Disturbance, and the Soil and Vegetation of Two Subtropical Wet Steepland Watersheds of Puerto Rico." *Geomorphology*, 13: 199–213.

Schiller, A., and Horn, S. P. (1997). "Wildlife Conservation in Urban Greenways of the Mid-Southeastern United States." *Urban Ecosystems*, 1: 103–16.

Schwartz, M. D. (1998). "Green-Wave Phenology." *Nature*, 394: 839–40.

Scott, G. A. J. (1995). *Canada's Vegetation: A World Perspective*. Montreal: McGill-Queen's University Press.

Scott, J. M., Anderson, H., Davis, F., Caicoo, S., Csuti, B., Edwards, T. C., Noss, R., Ulliman, J., Groves, C., and Wright, R. G. (1993). "Gap Analysis: A Geographic Approach to Protection of Biological Diversity." *Wildlife Monographs*, 123: 1–41.

Scuderi, L. A., Schaaf, C. B., Orth, K. U., and Band, L. E. (1993). "Alpine Treeline Growth Variability: Simulation Using an Ecosystem Process Model." *Arctic and Alpine Research*, 25: 175–82.

Shankman, D. (1996). "Stream Channelization and Changing Vegetation Patterns in the U.S. Coastal Plain." *Geographical Review*, 86: 216–32.

Shankman, D., and Drake, L. G. (1990). "Channel Migration and Regeneration of Bald Cypress in Western Tennessee." *Physical Geography*, 11: 343–52.

Shankman, D., and Kortright, R. M. (1994). "Hydrogeomorphic Conditions Limiting the Distribution of Bald Cypress in the Southeastern United States." *Physical Geography*, 15: 282–95.

Shinneman, D. J., and Baker, W. L. (1997). "Nonequilibrium Dynamics between Catastrophic Disturbances and Old-Growth Forests in Ponderosa Pine Landscapes of the Black Hills." *Conservation Biology*, 11: 1276–88.

Sluyter, A. (1996). "The Ecological Origins and Consequences of Cattle Ranching in Sixteenth-Century New Spain." *Geographical Review*, 86: 161–77.

Smith, T. S., Flinders, J. T., and Winn, D. S. (1991). "A Habitat Evaluation Procedure for Rocky Mountain Bighorn Sheep in the Intermountain West." *Great Basin Naturalist*, 51: 205–25.

Spellerberg, I. F., and Sawyer, J. W. D. (1999). *An Introduction to Applied Biogeography*. Cambridge: Cambridge University Press.

Szeicz, J. M., and MacDonald, G. M. (1995). "Recent White Spruce Dynamics at the Subarctic Alpine Treeline of North-Western Canada." *Journal of Ecology*, 83: 873–85.

Szeicz, J., Zeeb, B. A., Bennett, K. D., and Smol, J. P. (1998). "High Resolution Paleoecological Analysis of Recent Disturbance in a Southern Chilean *Nothofagus* Forest." *Journal of Paleolimnology*, 20: 235–52.

Taylor, A. H. (1990). "Habitat Segregation and Regeneration Patterns of Red Fir and Mountain Hemlock in Ecotonal Forests, Lassen Volcanic National Park, California." *Physical Geography*, 11: 36–48.

—— (1995). "Forest Expansion and Climate Change in the Mountain Hemlock (*Tsuga mertensiana*) Zone, Lassen Volcanic National Park, California, U.S.A." *Arctic and Alpine Research*, 27: 207–16.

Taylor, A. H., and Halpern, C. B. (1991). "The Structure and Dynamics of *Abies magnifica* Forests in the Southern Cascade Range, USA." *Journal of Vegetation Science*, 2: 189–200.

Taylor, A. H., and Qin Zisheng (1998). "Forest Landscape Dynamics and Panda Conservation in Southwestern China," in Zimmerer and Young (1998: 56–74).

Taylor, A. H., and Skinner, C. N. (1998). "Fire History and Landscape Dynamics in a Late-Successional Reserve, Klamath Mountains, California, USA." *Forest Ecology and Management*, 111: 285–301.

Taylor, A. H., Qin Zisheng, and Jie, L. (1996). "Structure and Dynamics of Subalpine Forests in the Wang Lang Natural Reserve, Sichuan, China." *Vegetatio*, 124: 25–38.

Terwilliger, V. J. (1990). "Effects of Vegetation on Soil Slippage by Pore Pressure Modification." *Earth Surface Processes and Landforms*, 15: 553–70.

—— (1997). "Changes in the $\delta^{13}C$ Values of Trees during a Tropical Rainy Season: Some Effects in Addition to Diffusion and Carboxylation by Rubisco?" *American Journal of Botany*, 84: 1693–700.

Terwilliger, V. J., and Zeroni, M. (1994). "Gas Exchange of a Desert Shrub (*Zygophyllum dumosum* Boiss.) under Different Soil Moisture Regimes during Summer Drought." *Vegetatio*, 115: 133–44.

Thompson, L. G., Mosley-Thompson, E., Davis, M. E., Lin, P.-N., Henderson, K. A., Cole-Dai, J., Bolzan, J. F., and Liu, K. B. (1995). "Late Glacial Stage and Holocene Tropical Ice Core Records from Huascarán, Peru." *Science*, 269: 46–50.

Thompson, R. S., Whitlock, C., Bartlein, P. J., Harrison, S. P., and Spaulding, W. G. (1993). "Climatic Changes in the Western United States since 18,000 yr B.P.," in H. E. Wright, Jr. *et al.* (eds.), *Global Climates since the Last Glacial Maximum*. Minneapolis: University of Minnesota Press, 468–513.

Tinker, D. B., Resor, C. A. C., Beauvais, G. P., Kipfmueller, K. F., Fernandes, C. I., and Baker, W. L. (1997). "Watershed Analysis of Forest Fragmentation by Clearcuts and Roads in a Wyoming Forest." *Landscape Ecology*, 12: 1–17.

Turner, M. D. (1998). "Long-Term Effects of Daily Grazing Orbits on Nutrient Availability in Sahelian West Africa: 1. Gradients in the Chemical Composition of Rangeland Soils and Vegetation." *Journal of Biogeography*, 25: 669–82.

Turner, M. G., and R. H. Gardner (eds.) (1991). *Quantitative Methods in Landscape Ecology: the Analysis and Interpretation of Landscape Heterogeneity*. New York: Springer-Verlag.

Turner, M. G., Gardner, R. H., and O'Neill, R. V. (2001). *Landscape Ecology in Theory and Practice: Pattern and Process*. New York: Springer.

Udvardy, M. D. F. (1969). *Dynamic Zoogeography*. New York: Von Nostrand Reinhold.

Vale, T. R. (1989). "Vegetation Management and Nature Protection," in G. P. Malanson (ed.), *Natural Areas Facing Climate Change*. The Hague: SPB Academic Publishing, 75–86.

—— (1998). "The Myth of the Humanized Landscape: An Example from Yosemite National Park." *Natural Areas Journal*, 18: 231–6.

Vale, T. R., and Vale, G. R. (1994). *Time and the Tuolumne Landscape: Continuity and Change in the Yosemite High Country*. Salt Lake City: University of Utah Press.

Veblen, T. T. (1989). "Biogeography," in C. Wilmott and G. Gaile (eds.), *Geography in America*. Columbus, Ohio: Merrill Publishing, 28–46.

—— (1992). "Regeneration Dynamics," in D. C. Glenn-Lewin, R. K. Peet, and T. T. Veblen, *Plant Succession: Theory and Prediction*. London: Chapman & Hall, 152–87.

Veblen, T. T., and Lorenz, D. C. (1991). *The Colorado Front Range: A Century of Ecological Change*. Salt Lake City: University of Utah Press.

Veblen, T. T., Hill, R. S., and Read, J. (eds.) (1996). *The Ecology and Biogeography of Nothofagus Forests*. New Haven: Yale University Press.

Veblen, T. T., Hadley, K. S., Reid, M. S., and Rebertus, A. J. (1989*a*). "Blowdown and Stand Development in a Colorado Subalpine Forest." *Canadian Journal of Forest Research*, 19: 1218–25.

Veblen, T. T., Kitzberger, T., Villalba, R., and Donnegan, J. (1999). "Fire History in Northern Patagonia: The Roles of Humans and Climatic Variation." *Ecological Monographs*, 69: 47–67.

Veblen, T. T., Mermoz, M., Martin, C., and Kitzberger, T. (1992). "Ecological Impacts of Introduced Animals in Nahuel Huapi National Park, Argentina." *Conservation Biology*, 6: 71–83.

Veblen, T. T., Ashton, D. H., Rubulis, S., Lorenz, D. C., and Cortés, M. (1989*b*). "*Nothofagus* Stand Development on In-Transit Moraines, Casa Pangue Glacier, Chile." *Arctic and Alpine Research*, 21: 144–55.

Veblen, T. T., Hadley, K. S., Nel, E. M., Kitzberger, T., Reid, M., and Villalba, R. (1994). "Disturbance Regime and Disturbance Interactions in a Rocky Mountain Subalpine Forest." *Journal of Ecology*, 82: 125–35.

Villalba, R., and Veblen, T. T. (1998). "Influences of Large-Scale Climatic Variability on Episodic Tree Mortality at the Forest-Steppe Ecotone in Northern Patagonia." *Ecology*, 79: 2624–40.

Voeks, R. A. (1996). "Tropical Forest Healers and Habitat Preferences." *Economic Botany*, 50: 354–73.

— (1997). *Sacred Leaves of Candomble: African Magic, Medicine, and Religion in Brazil.* Austin: University of Texas Press.

Walsh, S. J., Butler, D. R., Allen, T. R., and Malanson, G. P. (1994). "Influence of Snow Patterns and Snow Avalanches on the Alpine Treeline Ecotone." *Journal of Vegetation Science*, 5: 657–72.

Walter, H. S. (1998). "Land Use Conflicts in California," in P. W. Rundel, G. Montenegro, and F. M. Jaksic (eds.), *Landscape Degradation and Biodiversity in Mediterranean-Type Ecosystems*. Berlin: Springer-Verlag, 107–26.

Weingart, P., Mitchell, S. D., Richerson, P. J., and Maasen, S. (eds.) (1997). *Human by Nature: Between Biology and the Social Sciences*. Mahwah, NJ: Lawrence Erlbaum Associates.

Welch, J. M. (1994). "Street and Park Trees of Boston: A Comparison of Urban Forest Structure." *Landscape and Urban Planning*, 29: 131–43.

Westman, W. E. (1979). "Oxidant Effects on Californian Coastal Sage Scrub." *Science*, 205: 1001–3.

— (1990). "Park Management of Exotic Plant Species: Problems and Issues." *Conservation Biology*, 3: 251–60.

Whitlock, C., and Bartlein, P. J. (1997). "Vegetation and Climate Change in Northwest America during the Past 125 k yr." *Nature*, 388: 57–61.

Whitlock, C., and Millspaugh, S. H. (1996). "Testing Assumptions of Fire History Studies: An Examination of Modern Charcoal Accumulation in Yellowstone National Park." *The Holocene*, 6: 7–15.

Whittaker, R. J. (1998). *Island Biogeography: Ecology, Evolution, and Conservation*. Oxford: Oxford University Press.

Wolch, J., and Emel, J. (1998). *Animal Geographies: Place, Politics, and Identity in the Nature–Culture Borderlands*. London: Verso.

Wolf, J. J., and Mast, J. N. (1998). "Fire History of Mixed Conifer Forests of the North Rim, Grand Canyon National Park, Arizona." *Physical Geography*, 19: 1–14.

Woodcock, D. W. (1992). "Climate Reconstruction Based on Biological Indicators." *Quarterly Review of Biology*, 67: 457–77.

Yaukey, P. (1994). "Variation in Racial Dominance Within the Winter Range of the Dark-Eyed Junco (*Junco hyemalis* L.)." *Journal of Biogeography*, 21: 359–68.

— (1996). "Patterns of Avian Population Density, Habitat Use, and Flocking Behavior in Urban and Rural Habitats during Winter." *Professional Geographer*, 48: 70–81.

Young, K. R. (1991). "Natural History of an Understory Bamboo (*Chusquea* sp.) in a Tropical Timberline Forest." *Biotropica*, 23: 542–54.

— (1993). "Tropical Timberlines: Changes in Forest Structure and Regeneration between Two Peruvian Timberline Margins." *Arctic and Alpine Research*, 25: 167–74.

— (1994). "Roads and the Environmental Degradation of Tropical Montane Forests." *Conservation Biology*, 8: 972–6.

— (1995). "Biogeographical Paradigms Useful for the Study of Tropical Montane Forests and their Biota," in S. P. Churchill, H. Balslev, E. Forero, and J. L. Luteyn (eds.), *Biodiversity and Conservation of Neotropical Montane Forests*. New York: New York Botanical Garden, 79–87.

— (1996). "Threats to Biological Diversity Caused by Coca/Cocaine Deforestation in Peru." *Environmental Conservation*, 23: 7–15.

— (1997). "Wildlife Conservation in the Cultural Landscapes of the Central Andes." *Landscape and Urban Planning*, 38: 137–47.

Young, K. R., and León, B. (1990). "Curvature of Woody Plants on the Slopes of a Timberline Montane Forest in Peru." *Physical Geography*, 11: 66–74.

— (1995). "Connectivity, Social Actors, and Conservation Policies in the Central Andes: The Case of Peru's Montane Forests," in S. P. Churchill, H. Balslev, E. Forero, and J. L. Luteyn (eds.), *Biodiversity and Conservation of Neotropical Montane Forests*. New York: New York Botanical Garden, 653–61.

Young, S. S., and Herwitz, S. R. (1995). "Floristic Diversity and Co-occurrences in a Subtropical Broad-leaved Forest and Two Contrasting Regrowth Stands in Central-West Yunnan Province, China." *Vegetatio*, 119: 1–13.

Ziegler, S. S. (1995). "Relict Eastern White Pine (*Pinus strobus* L.) Stands in Southwestern Wisconsin." *American Midland Naturalist*, 133: 88–100.

— (1997). "White Pine in Southwestern Wisconsin: Stability and Change at Different Scales," in R. Ostergren and T. Vale (eds.), *Wisconsin Land and Life: Geographic Portraits of the State*. Madison: University of Wisconsin Press, 81–94.

— (2000). "A Comparison of Structural Characteristics between Old-Growth and Second-Growth Hemlock-Hardwood Forests in Adirondack Park, New York." *Global Ecology and Biogeography*, 9: 373–89.

Zimmerer, K. S. (1994). "Integrating the 'New Ecology' and Human Geography: Promise and Prospects." *Annals of the Association of American Geographers*, 84: 108–25.

— (1996). *Changing Fortunes: Biodiversity and Peasant Livelihood in the Peruvian Andes*. Berkeley: University of California Press.

— (1998). "The Ecogeography of Andean Potatoes: Versatility in Farm Regions and Fields can Aid Sustainable Development." *BioScience*, 48: 445–55.

Zimmerer, K. S., and Young, K. R. (eds.) (1998). *Nature's Geography: New Lessons for Conservation in Developing Countries*. Madison: University of Wisconsin Press.

CHAPTER 3

Climate

Jeffrey C. Rogers, Julie A. Winkler, David R. Legates, and Linda O. Mearns

Public awareness of climate and its societal impact has substantially increased in the recent decade. An extraordinarily persistent El Niño from 1992–5 followed by another strong event in 1997–8 have been particularly newsworthy. Severe storms, both on the mesoscale and synoptic scale, and extreme events such as cold waves, heat waves, flooding, and regional droughts regularly continue to draw attention. Concern about anthropogenic climate warming has engendered considerable public debate over its detection, potential impacts, and public policy issues. In keeping with the importance of climatic issues, American geographer-climatologists are contributing extensively to the climate research literature and the understanding of both the physical aspects and impacts of the climate system.

We have elected to use this forum to demonstrate the remarkable breadth of climatic research interests among American geographers during the past decade, rather than to focus on a few key contributions. A thematic approach is used to organize the large body of literature. This approach eschews the more traditional definitions of the subfields of climatology (e.g. physical, dynamic, synoptic, and applied climatology), which we felt did not accurately reflect the integrative and innovative nature of much of the current research by geographer-climatologists. We found that a large portion of the recent contributions of geographer-climatologists can be organized around the thematic areas of atmospheric circulation, surface–atmosphere interactions, hydroclimatology, and climatic change. In addition, we identified a substantial research effort focused on the evaluation of climatological observations and the development of analytical techniques appropriate for climatological research. A small number of papers by geographer-climatologists whose explicit intention was the formulation of policy for the private and public sectors was also identified.

Atmospheric Circulation

Atmospheric circulation has been a major research focus in the last decade. Geographer-climatologists have been concerned with the frequency, spatial and temporal variability, and physical causes of extratropical and tropical circulation systems. Climatological analysis of atmospheric circulation has traditionally fallen within the subfields of dynamic and synoptic climatology. Dynamic climatology encompasses the largest scales of atmospheric circulation with emphases on theoretical and modeling approaches to climate analysis (Rayner *et al.* 1991). Synoptic climatology incorporates more regional scales, often with application of circulation and synoptic weather types to explain local climate variability (Harman and Winkler 1991; Yarnal 1993). An abundance of statistical and analytical approaches to circulation analysis, numerical modeling methods and the frequent use of oceanic and land surface datasets has also led to one suggestion that both fields fall under the broader umbrella of climate dynamics (Oliver *et al.* 1989).

An important focus of circulation research has been the contribution of atmospheric teleconnections to climate variability. Three teleconnections of particular interest are: (1) the Southern Oscillation, an atmospheric mass oscillation varying longitudinally across the Pacific basin with modes of variability closely linked to El Niño and La Niña; (2) the Pacific–North American (PNA) pattern, a mid-tropospheric circulation configuration, whose modes appear to be at least partly driven by the phase of the Southern Oscillation; and (3) the North Atlantic Oscillation (NAO), a sea-level pressure oscillation in atmospheric mass occurring over the North Atlantic Ocean.

Teleconnection patterns are strongly correlated with regional-scale climate variability in some parts of the world. For example, Palecki and Leathers (1993) found that over 70 per cent of the variance in the January land-surface temperature record for the Northern Hemisphere can be related to the phase and strength of teleconnection patterns. Other examples include the correspondence in China between variations in the Southern Oscillation Index (SOI) and precipitation and temperature anomalies (Song 1998), the association between positive values of the SOI (i.e. La Niña conditions) and negative precipitation anomalies across the southern United States (Vega *et al.* 1998), and the opposing precipitation patterns along the Pacific and Caribbean coasts of Costa Rica during El Niño events (Waylen *et al.* 1996). Northwestern North America and the southeastern United States are continental areas where PNA "centres of action" develop and strongly influence regional climate (Leathers *et al.* 1991; Vega *et al.* 1995; Yarnal and Leathers 1988). In particular, the positive mode of the PNA pattern sets the stage for cold waves over North America (Konrad 1996). Upper-level airflow steers large polar anticyclones from their source region in the Arctic along cross-continental trajectories, leading to cold waves and citrus freezes in Florida (Rogers and Rohli 1991) and other parts of the South (Rohli and Rogers 1993; Rohli and Henderson 1998). The comparative roles of stationary and eddy transports of sensible heat flux also vary with the mode of the PNA (Rogers and Raphael 1992) and NAO (Carleton 1988). Cyclonic activity associated with the Icelandic Low (Serreze *et al.* 1997) and the North Atlantic storm track (Rogers 1997) appear to be strongly linked to the NAO. The NAO also significantly affects circulation and climate variability toward the equator and into southern Africa (McHugh and Rogers 2001). Other investigations of large-scale circulation patterns, although not directly tied to teleconnection variations, include analyses of the spatial and temporal variability of the North American circumpolar vortex (Burnett 1993; Davis and Benkovic 1994), the Atlantic subtropical anticyclone (Davis *et al.* 1997), the quasi-stationary waves of the Northern (Harman 1991) and Southern (Raphael 1998; Burnett and McNicoll 2000) Hemispheres and the standing and transient eddies associated with sensible heat transport (Raphael and Rogers 1992; Raphael 1997).

Geographer-climatologists have also explored linkages between synoptic-scale weather features and climate variability. Circulation typing and composite analysis are two popular means for investigating these linkages. Brinkmann (2000) examined methods for improving the internal performance of correlation-based typing schemes. Circulation typing, whether objective or subjective, has proven invaluable for assessing the relative frequency of different synoptic-scale weather patterns (Mock *et al.* 1998; Kalkstein *et al.* 1990), delineating the synoptic settings associated with mesoscale derecho events (Bentley and Mote 2000; Bentley *et al.* 2000), and understanding the causes of large-scale weather systems such as east coast cyclones (Davis *et al.* 1992). Synoptic typing schemes have also been employed in applied studies to evaluate the characteristics of air masses (Kalkstein *et al.* 1996, 1998; Schwartz 1995), identify the synoptic controls of pollutant transport and concentrations (Comrie and Yarnal 1992; Comrie 1996; Pryor *et al.* 1995; Pryor 1998), and understand the synoptic-scale weather patterns contributing to enhanced health risk and human mortality (Greene and Kalkstein 1996; Kalkstein 1991; Kalkstein and Greene 1997; Jamason *et al.* 1997; Smoyer *et al.* 2000). Jones and Davis (2000) furthermore apply these methods to understanding variations in grape productivity. Composite analysis has been used to explore the linkages between synoptic-scale circulation and regional heavy precipitation events (Mote *et al.* 1997; Konrad and Meentemeyer 1994; Keim 1996; Winkler 1988) and the spatial and temporal variability of lower tropospheric (e.g. 850 hPa) observed and geostrophic winds over central and eastern North America (Winkler *et al.* 1996).

Considerable recent effort has focused on direct examination of synoptic-scale weather systems, emphasizing their development and behavior (Angel and Isard 1997; Bierly 1997; Bierly and Harrington 1995; Bierly *et al.* 2000; Serreze and Barry 1988) as well as their effects on coastal climates (Raphael and Mills 1996; Rohli and Keim 1994) and the Eurasian continental interior

(Rogers and Mosley-Thompson 1995). A suite of papers explored the temporal and spatial variability of significant airflow features including the North American monsoon (Carleton *et al.* 1990; Adams and Comrie 1997; Comrie and Glenn 1998), the Great Plains low-level jet (Walters 2001; Walters and Winkler 2001), and airstreams within mid-latitude cyclones (Bierly and Winkler 2001). Furthermore, the causes of tropical systems and prediction of hurricane wind strength were evaluated by Whitney and Hobgood (1997). Geographers have also been concerned with diurnal variations of extratropical weather phenomena and have identified strong diurnal signals in the frequency and characteristics of derechos (Bentley and Mote 1998), intense precipitation (Winkler 1992), and cloud-to-ground lightning flashes (Hanuta and LaDochy 1989; King and Balling 1994; Walters and Winkler 1999).

Surface–Atmosphere Interactions

Although study of surface–atmosphere interactions has always been an important component of climatology, a marked increase in research in this area has been influenced by: (1) the impact of human activities on the atmosphere and vice versa; (2) the necessity to separate natural climate variability from anthropogenic influences; and (3) demand for improved surface–atmosphere algorithms for numerical simulations. Often collectively referred to as physical climatology or boundary-layer climatology, surface–atmosphere interactions incorporate the subfields of agricultural climatology, bioclimatology, energy-balance climatology, microclimatology, mountain and alpine climatology, and urban climatology. Analysis of surface–atmosphere interactions increasingly involves a mix of spatial scales ranging from the micro to the macro and incorporates a spectrum of analytical methods, including numerical simulations and statistical analysis of *in situ* and remotely sensed observations. Summarized below are a few of the areas within the broad arena of surface–atmosphere interactions in which geographer-climatologists have recently contributed.

The influence of topographic variations on surface–atmosphere interactions continues to garner attention among geographer-climatologists. Isard (1989) found that the local- to regional-scale spatial pattern of daily energy and moisture fluxes at Niwot Ridge in the Colorado Front Range is more influenced by cloud cover than by topographic position and that the daily radiation load is usually higher on east-facing slopes where wind speed and sensible heat flux are small. In contrast, microscale spatial variations in the surface heat budget at Niwot Ridge appear to be due to different subclasses of tundra vegetation (Greenland 1993). Interannual variations in the energy balance at this locale can be attributed to yearly differences in precipitation amount (Greenland 1991). Basist *et al.* (1994) found that exposure to the prevailing winds is the single most important feature relating topography to the spatial distribution of precipitation in mountainous regimes.

The influence of vegetation cover on the overlying atmosphere has also been investigated. For example, suburban neighborhoods with a greater amount of tree and shrub cover were found to have lower albedos, lower surface temperatures, and enhanced latent, sensible, and storage heat fluxes compared to neighborhoods with little or no tree cover (Grimmond *et al.* 1996). Also, remotely-sensed observations were used to show that the frequency of convective cloud days in the rural Midwest is higher for surfaces having a high relative density of forest vegetation and lower for those with a high density of crops (Carleton *et al.* 1994). Cutrim *et al.* (1995) found that in the Amazonian state of Rondonia shallow cumulus clouds are more frequent where the rainforest has been cleared. In addition, Giambelluca *et al.* (1997) found that, in spite of similar albedos, primary and secondary tropical forests have very different energy and mass exchanges. O'Brien (1996) points out that the relationship between tropical deforestation and climate is complex and remains weakly supported by empirical data. She argues that few comprehensive empirical studies have been conducted at the local to regional scales. Results from micro- and global-scale studies are instead often extrapolated to encompass these scales.

A number of investigators have used numerical simulations to better understand the impact of vegetation heterogeneities on local/regional climate. Klink and Willmott (1994) found that increasing the area of bare soil downwind of irrigated maize produces nearly linear changes in daily average surface temperature and average heat flux, whereas upwind bare soil forces nonlinear responses. Klink (1995*a*, *b*) subsequently showed that roughness and canopy resistance discontinuities play a larger role in the regional average energy balance than does albedo heterogeneity. Rowe (1991) argued that large-scale characterizations of albedo, such as those employed in regional and global climate models, need to incorporate within grid-cell heterogeneity of plant canopy structure. Song *et al.* (1997) found that the use of heterogeneous versus homogeneous (i.e. spatially averaged) soil moisture fields in model simulations had a large influence on calculated latent heat flux, air temperature, planetary boundary layer height, and turbulent exchanges. Finally, Tsvetsinskaya *et al.* (2001*a*, *b*) coupled a dynamic (i.e. interactive) crop model into a regional climate model. Simulations for dry and wet years over the United States suggested that the interactive vegetation strongly influenced the climate simulated by the regional model, particularly in dry years, compared to the static (i.e. non-growing) vegetation case.

Investigations of the interaction between urban surfaces and the atmosphere have incorporated a wide range of spatial scales. Analyses at the micro- to local scales have evaluated the aerodynamics of urban areas (Grimmond and Oke 1998; Grimmond *et al.* 1998), estimated the effect of urban canyon geometry on nocturnal cooling rates (Arnfield 1990), linked the strength of the urban heat island to variations in urban system symmetry/asymmetry and orientation (Todhunter 1990), modeled the storage heat term of an urban canyon (Arnfield and Grimmond 1998), detected upward-directed vertical velocities near the top of deep urban canyons (Arnfield and Mills 1994), estimated the nocturnal release of heat from buildings and substrate of a city core (Oke *et al.* 1999), and compared the winter and springtime energy balances at a suburban location (Grimmond 1992). Complex interactions between the micro- and local scales in urban/suburban environments are illustrated by Schmid *et al.* (1991) who identified spatial variations of the order of 10^2–10^3 meters in the energy-balance heat-flux terms in suburban Vancouver. They attributed these horizontal variations to microadvective interactions between surface types at small scales. At the local to regional scale, Stoll and Brazel (1992) found that surface/air temperature relationships in urban areas can be significantly influenced by advection from adjacent land uses and Nasrallah *et al.* (1990) speculated that low rates of urban temperature change in Kuwait City, when compared to arid North American cities, may be explained by lack of greenbelt development and wider use of locally derived building materials having similar thermal properties to the surrounding desert terrain. Gallo *et al.* (1996) found that stations with surrounding rural land use/cover have relatively large diurnal temperature ranges compared to stations with urban-related land use/cover. Evoking still broader scales of analysis, Grimmond and Oke (1995) compared the urban energy balances of four United States cities characterized by a range of synoptic regimes and surface morphologies.

Surface heterogeneity resulting from snow cover also has a substantial effect on the overlying atmosphere. Observations suggest that temperatures are typically 6–8 °C lower on days with snow cover than those without (Baker *et al.* 1992; Leathers *et al.* 1995), a tendency confirmed with a one-dimensional snowpack model (Ellis and Leathers 1998). In addition, Leathers and Robinson (1993) found that extensive snow cover could modify air masses and affect air temperature far south of the snowpack. A number of recent studies have investigated the dramatic effect of snow melt on the surface albedo of polar regions (Anderson 1997; Barry 1996; Barry *et al.* 1993; Robinson *et al.* 1992). On a larger scale, Ye (2000, 2001) shows that Eurasian snow cover is itself linked to sea surface temperature conditions in the Atlantic and Pacific Oceans.

One other important example of surface–atmosphere interaction is the "green wave." The onset of spring is a time when increasing solar radiation melts snow cover, forces changes in synoptic storm tracks, modifies the surface energy balance and allows vegetation to resume growth (Schwartz 1992). Rapid albedo and transpiration increases are triggered by the appearance of spring foliage, resulting in further modifications to sensible-latent heat exchange and other lower tropospheric parameters such as surface maximum air temperature, diurnal temperature range, lapse rate, vapor pressure, visibility and wind (Schwartz 1992, 1996*b*; Schwartz and Karl 1990).

Hydroclimatology

Hydroclimatology is a subdiscipline for which no strict definition exists. Langbein (cited in Mather 1991) suggested that hydroclimatology is the "study of the influence of climate upon the waters of the land" while Hirschboeck (1988) offered that it is "an approach to studying hydrologic events within their climatological context." A more holistic, encompassing definition is used here, whereby hydroclimatology is simply considered the study of atmospheric moisture and surface water. Although it can be argued that a narrow definition better sets hydroclimatology apart from other subfields, a more holistic definition recognizes the important integrative role and multidisciplinary significance of hydroclimatic research.

In the past decade, hydroclimatic research undertaken by geographer-climatologists has focused on precipitation variability, precipitation estimation, floods and droughts, and the hydrologic cycle. Research on precipitation variability has largely been dominated by studies of the spatial and temporal patterns of precipitation. Not surprisingly, the southeastern United States, a region that experiences frequent extreme hydroclimatic events, has been the focus of a number of studies that have explored regional and seasonal variations in overall precipitation frequency (Robinson and Henderson 1992; Changnon 1994; Henderson and Vega 1996), thunderstorm activity (Easterling 1991), heavy rain events (Keim 1996; Konrad 1994, 2001), the occurrence of freezing rain and sleet (Gay and Davis 1993), and the frequency of snowstorms (Suckling 1991). Elsewhere, regional studies of temporal trends in snowfall and snow cover have identified increases in the Great Lakes, High Plains, and the polar regions (Robinson and Dewey 1990; Hughes and Robinson 1996; Ye and Mather 1997), have examined changes in North American mountain snowpacks resulting from El Niño and La Niña variations (Clark *et al.* 2001), and have investigated precipitation and river discharge variability over the Amazon River Basin (Vörosmarty *et al.* 1996; DeLiberty 2000). Legates (2000) also has been instrumental in providing real-time estimates of precipitation using surface radar estimates that have been calibrated using rain gauges. On a much broader spatial scale, geographer-climatologists have developed and evaluated global precipitation climatologies (Legates and Willmott 1990; Legates 1995), which have been used to initiate and validate global climate models.

Recent research has also focused on accurate estimation of snowfall and snow cover (cf. Hughes and Robinson 1996). The motivation for these studies is the significant gage-induced biases resulting from the deleterious effects of wind on snowfall measurement by traditional can-type gages (Legates and DeLiberty 1993; Groisman and Legates 1994, 1995). Other problems with gage-based precipitation measurements, in general, include spatial fidelity, missing or incomplete metadata, improper siting, instrument changes and relocations, and gage measurement biases. In response to these concerns, remote-sensing techniques of precipitation estimation have increasingly become an important research topic in hydroclimatology. For example, techniques for estimating precipitation from satellite observations have been validated for the Pacific Atolls (Morrissey and Greene 1993) and the open ocean (Greene *et al.* 1997).

Floods and droughts also have garnered attention over the past decade. Muller *et al.* (1990), for example, found that in Louisiana the often-used Palmer Drought Severity Index (PDSI) is not well correlated with river stage. By contrast, Soulé and Shankman (1990) concluded that the PDSI is a good indicator of river stage in western Tennessee because of its slow response time. Other regional studies have linked winter atmospheric circulation patterns to annual streamflow in the western United States (McCabe 1995), evaluated the relative influences of temperature and precipitation anomalies on the timing of streamflow in the Sierra Nevada region of California (Aguado *et al.* 1992), provided a flood climatology for North Carolina for the explicit purpose of examining the relative effects of climate change and human influences (Konrad 1998), and described the spatial and temporal variability of droughts in the Midwest (Changnon 1996) and the contiguous United States (Soulé 1992). Todhunter (2001) also examined the northern Red River snowmelt flood from a hydroclimatological perspective.

Geographer-climatologists have continued their long-term interest in the hydrologic cycle and in the evaluation and use of hydrologic models (Mahmood 1998*a*, *b*; Frakes and Yu 1999). Over the past decade, the water balance and its spatial/temporal variability have been studied at the continental (Legates and Mather 1992; Feddema 1998), regional (Wolock and McCabe 1999*b*; Grundstein and Bentley 2001), and basin (McCabe and Wolock 1992; Shelton 1998) scales. To facilitate these investigations, studies of the spatial distributions of plant-usable soil water potential (Dunne and Willmott 1996), the relationship between surface roughness and soil moisture (Klink and Willmott 1994), and the use of satellite data to estimate watershed evapotranspiration (Song *et al.* 2000*a*, *b*) also have been initiated.

Climate Change

Given that climate change has been a major focus of climatological research over the past decade, it is no surprise that geographical-climatologists have been involved in many aspects of climate change research. The subsections below focus on paleoclimatic reconstruction, the detection of climate variability and change from the observational record, and climate modeling and scenario development.

Paleoclimatology

Paleoclimatic analysis plays a key role in helping climatologists reconstruct historic climate records and in understanding the extent to which the climatic environment changes with time. Climate reconstruction makes use of: (1) pollen analysis (e.g. Bartlein *et al.* 1995); (2) regional tree-ring dendrochronologies (e.g. Stahle and Cleaveland 1992); (3) lake and river sediments

(Liu *et al.* 1992); and (4) stratigraphic variations in the chemical and dust content of cores obtained from ice caps (Mosley-Thompson 1996).

Proxy data have been used to reconstruct paleoclimatic time series for the United States. An 800-year tree-ring reconstruction of growing season drought (Stahle *et al.* 1998) shows that the early American Roanoke colony was established at the outset of the worst drought in the record. Woodhouse and Overpeck's (1998) comprehensive review of the paleoclimatic literature suggests that twentieth-century United States droughts were eclipsed several times by droughts over the last 2,000 years. Tree rings have also been used to reconstruct time series of the number of wintertime precipitation days in the southwestern United States (Woodhouse and Meko 1997) and to evaluate precipitation variability along central coastal California (Haston and Michaelsen 1994). Sandy lake sediment records have permitted estimates of historic occurrences of intense hurricanes in Alabama (Liu and Fearn 1993).

Efforts are increasingly being made to link proxy climate records to atmospheric circulation data and to output from numerical models. For example, Mock and Bartlein (1995) developed twentieth-century atmospheric circulation analogs for late Quaternary climates in order to demonstrate how climatic heterogeneity in the western United States is the rule rather than the exception. Hirschboeck *et al.* (1996) identified anomalous synoptic-scale circulation patterns associated with frost-ring formation in North American trees and constructed circulation response patterns for tree-ring sites in Oregon and New Mexico. Paleo-circulation output from the NCAR (National Center for Atmospheric Research) Community Climate Model (CCM) was reviewed by Bartlein *et al.* (1998) in order to evaluate its implications for historic distributions of three plant taxas, broadly representing North American vegetation cover, and results were subsequently compared to proxy-based records of vegetation distributions. Oxygen isotopic records from shallow ice cores taken from the Greenland ice-cap are a proxy for air temperature records and Rogers *et al.* (1998) demonstrate their link to sea-level pressure variations and the NAO.

Detection Studies

Climate change detection studies have attempted to identify trends in the observed time series of several climate variables. For example, the possible link between air temperature and anthropogenic effects prompted analyses of the fluctuations in average air temperature at annual (Balling *et al.* 1998), seasonal (Balling *et al.* 1990; Hartley and Robinson 2000; Skaggs *et al.* 1995), and daily (Henderson and Muller 1997; Michaels *et al.* 1998) temporal scales. Long-term variability in snow cover extent (Robinson and Dewey 1990; Ye *et al.* 1998; Clark *et al.* 1999; Frei *et al.* 1999), the linkages between variations in snow cover and air temperature (Leathers *et al.* 1993; Groisman and Easterling 1994; Hughes and Robinson 1996), and the possible influence of snow cover on trends in diurnal temperature range (Cerveny and Balling 1992) also have been investigated. In addition, variations through time in the frequency of atmospheric circulation patterns have been the focus of several studies (Kalkstein *et al.* 1990; Rohli and Henderson 1997). Other authors have identified trends in precipitation, streamflow, and run-off (Keim *et al.* 1995; McCabe and Wolock 1997), and in pollution concentrations (Cerveny and Balling 1998).

Climate Modeling and Scenario Development

The past decade has seen tremendous advances in the modeling of the earth-atmosphere-ocean system, although geographer-climatologist involvement in this research continues to be limited. The scale of the work and required computer resources limit development of general circulation models (GCMs) to large research laboratories. Climate modeling in the past decade has developed from the use of atmosphere-only climate models with very simple oceans, to highly complex models that include complete interactions between the earth, atmosphere, and oceans (Meehl 1998; Mearns 1999). Since the late 1980s atmospheric models have been coupled with three-dimensional dynamical ocean models, permitting much more realistic modeling of interannual and longer-term variability of the coupled system. Ocean models permit detailed modeling of internal horizontal and vertical heat transport (Washington and Meehl 1989).

As the climate modeling field has grown, opportunities for collaborations with modelers, or running and evaluating model experiments, has increased. Raphael (1998) recently evaluated NCAR Climate System Model (CSM) runs to examine the simulation of quasi-stationary waves in the southern hemisphere. Schwartz (1996*a*) and Brinkmann (1993) evaluated atmospheric GCM control runs to compare modeled air mass frequencies with those observed in Midwestern climate data. Marshall *et al.* (1997) conducted experiments with the NCAR CCM to determine the effect of model resolution and different precipitation parameterizations on how well the model reproduces precipitation. Mearns *et al.* (1990) evaluated how well a version of the CCM and several other climate models reproduced higher-order moments of temperature and precipitation variability. McGinnis and Crane (1994) compared observed Arctic climate variability to that simulated in four GCMs and found that strong model-prescribed coupling between summertime climate variability and the overlying atmosphere was not occurring in the observational data. More recently, Kothavala (1997) examined the reproduction of precipitation extremes in several different climate models and investigated their changes in frequency in perturbed climate runs. Geographer-climatologists have increasingly used GCM results to estimate possible climatic changes in other environmental variables at the watershed scale (Wolock and McCabe 1999*a*; Shelton 2001).

Regional models examine only a portion of the earth's surface, usually at a spatial resolution higher than that of GCMs (on the order of tens of kilometers). The models are driven at their boundaries either by reanalyses from weather prediction models or from GCM output. The basic strategy is to rely on the GCM (or reanalyses) to reproduce the large-scale circulation of the atmosphere while the regional model simulates sub-GCM-scale distributions of climate, such as precipitation, temperature, and winds over the small area of interest (Giorgi and Mearns 1999). The GCM provides the initial and lateral boundary conditions for driving the regional climate model. Numerous experiments with regional models, driven by control and doubled CO_2 output from GCMs, have been performed for domains such as the continental United States and Europe (Giorgi *et al.* 1994, 1998). Mearns *et al.* (1995) evaluated experiments over the United States to determine how well regional models reproduce high-frequency climatic variability and how such variability could change under perturbed climate conditions.

High-resolution climate-change scenarios have been created in recent years using regional climate models or statistical downscaling techniques (Giorgi *et al.* 2001). Statistical downscaling involves use of GCM results and statistical relationships between large-scale and location-specific variables such as temperature or precipitation. The relationships between the large-scale circulation and local variables are assumed to apply in the climate model output, with large-scale circulation variables driving the analysis. The method takes advantage of the fact that GCMs generally simulate larger-scale circulation features better than regional or local climates. Hence changes in the circulation are accepted from the model, but the relationship between the circulation and the local climate is taken from observations. A good overview of statistical downscaling is provided in Hewitson and Crane (1996). Geographers active in this area of research have employed a wide variety of techniques for developing the statistical relationships including neural networking (e.g. McGinnis 1997; Crane and Hewitson 1998) as well as more traditional regression techniques (Easterling 1999).

Interest has grown in comparing results of regional climate model experiments and statistical downscaling as means of producing high-resolution climate-change scenarios. For example, Mearns *et al.* (1999*a*) compared the results of a stochastically generated statistical downscaling technique with those of regional modeling experiments for $1 \times CO_2$ and $2 \times CO_2$ conditions in the Great Plains. They found that climate changes generated by the two methods were different, even though they were developed from output from the same GCM experiments. High-resolution scenarios have been little used in climate impact assessments, but such applications are beginning to appear (e.g. Mearns *et al.* 1999*b*, 2001*a*, *b*). Climate change scenario development often involves assessments of the impacts of climatic change on resource systems (e.g. agriculture, water resources). In this regard, such work has a human component and is covered in Ch. 18 on Human Dimensions.

Climate Methodologies: Data and Data Manipulation

Multivariate statistical analyses are now routinely employed in climatological analyses. However, over the past decade a number of novel developments, refinements, and applications have been made for a variety of statistical methods, including basic correlation (Legates and Davis 1997; Legates and McCabe 1999), vector correlation (Hanson *et al.* 1992), and predictive logistic regression (Travis *et al.* 1997). In synoptic climatology, discriminant function analysis has been used to differentiate between weather types (Greene and Kalkstein 1996; Kalkstein *et al.* 1996) and factor analytical techniques have been extended to vector data (Klink and Willmott 1989). The usefulness of rotation procedures in principal components analysis continues to be debated (White *et al.* 1991; Legates 1991*b*). Climatologists also have applied a variety of time-series analysis tools (Harrington and Cerveny 1988; Legates 1991*a*;

Faiers *et al.* 1994; and Keim and Cruise 1998). Balling (1997), Robeson and Shein (1997), and Robeson and Janis (1998) provide useful assessments of the issues and concerns related to evaluating temporal autocorrelation and variability in spatial patterns.

Procedures for evaluating data quality have become an important research focus. Notably, Peterson and Easterling (1994) and Easterling and Peterson (1995) examined data homogeneity issues, whereas Willmott *et al.* (1994) focused on network adequacy and accuracy. The development of large climatic databases also has focused attention on spatial interpolation methods and the assessment of regional/global-scale spatial and temporal variability (Willmott and Legates 1991; Robeson 1994, 1995; Willmott and Matsuura 1995; Willmott *et al.* 1996). Most of this research has moved away from the development of spatial interpolation algorithms to the analysis of how variations in station distributions over time affect areal and temporal estimates of climate variables.

For some specific applications, additional spatial statistical methods have been developed and/or extended. Comrie (1992), for example, developed a procedure to detrend or "de-climatize" climatological data, while Robeson and Shein (1997) have examined the spatial coherence of wind data using mean absolute deviations. Robeson (1997) also provides an evaluation of spherical methods used for spatial interpolation of climate data.

Policy-Oriented Research

Much of the climatological research conducted within the discipline of geography during the past decade has an applied aspect. This tendency toward applied research by geographer-climatologists in part reflects geography's overall focus on society and environmental problems. Obvious applications of the research include, but are not limited to, improved weather and climate forecasting, the development and enhancement of observational networks, policy formulation for the amelioration of anthropogenic influences on climate, improved agricultural practices, better exchange of climatological information, reduction of air pollution, strategies to reduce weather-related deaths, and improved educational materials. Even research that tends more heavily toward the basic research side of the basic–applied continuum provides the necessary building blocks for future applications of climatological knowledge to the solution of social, economic, and environmental problems.

A small number of articles during the past decade, however, have had as their explicit intent the formulation of policy for the public and private sectors. For example, Schmidlin *et al.* (1998) have examined risk factors for death during tornadic storms. This research has contributed to improved information on safety procedures. Schmidlin *et al.* (1992) also have calculated updated design ground snow loads used by engineers and planners for roof design. Policy-related agricultural applications include the analysis of the climatological factors affecting the spread of western corn rootworm along with the implications for modifying established crop rotation practices in the American Midwest (Spencer *et al.* 1999) and the use of remotely sensed observations and Geographic Information Systems (GIS) to evaluate the influence of spatial variations in climate, soil, and land-cover variables on soybean yields in order to determine vulnerable areas during drought (Carbone *et al.* 1996). Suckling (1997) analyzed the meso-scale spatial coherence of solar radiation, which is an important consideration for effective solar-energy applications. One final example is Robinson's (1997) examination of how water availability and water demand vary with time and climate and the consequent effects on energy production.

Conclusion

This review demonstrates the remarkable breadth and expansion of climatological research within the discipline of geography during the past ten years, focusing on key themes of common interest among climatologists. Climatologists are actively engaged in understanding the physical processes at the core of the climate system through data gathering and statistical analysis, the formation of physically based qualitative models, and the use of numerical models. They are also applying climatological information to the solution of societal and environmental problems and slowly are becoming more involved in policy formulation. The upcoming decade will bring challenges to geographer-climatologists as the public becomes increasingly aware of the importance of climate to society and climate research remains a major focus in the scientific community. Some of the major challenges lie ahead in climate-change detection; effort will be needed to help detect statistically significant changes in global and regional temperatures and precipitation as well as in the frequency and magnitude of extreme events. Climate fluctuation is often very regionalized and it will become increasingly necessary to understand its environmental impact on local and regional scales, including potential impacts within the urban environment. Output from GCMs will increasingly be directed toward application to regional-scale climate analyses. The climate community is also dramatically pushing the outer envelope of long-range predictions of atmospheric circulation and climate and there will be a need not only for better models but also improved understanding of atmospheric circulation and climate interrelationships. Finally, growing concern nationally about water rights and usage, as well as water quality issues, make it imperative that geographer-climatologists join efforts to improve our understanding of how the atmosphere interacts with the hydrologic cycle across various spatial and temporal scales.

References

Adams, D. K., and Comrie, A. C. (1997). "The North American Monsoon." *Bulletin of the American Meteorological Society*, 10: 2197–213.

Aguado, E., Cayan, D., Riddle, L., and Roos, M. (1992). "Climatic Fluctuations and the Timing of West-Coast Streamflow." *Journal of Climate*, 5: 1468–83.

Anderson, M. R. (1997). "Determination of a Melt Onset Date for Arctic Sea Ice Regions Using Passive Microwave data." *Annals of Glaciology*, 25: 382–7.

Angel, J. R., and Isard, S. A. (1997). "An Observational Study of the Influence of the Great Lakes on the Speed and Intensity of Passing Cyclones." *Monthly Weather Review*, 125: 2228–37.

Arnfield, A. J. (1990). "Canyon Geometry, the Urban Fabric and Nocturnal Cooling: A Simulation Approach." *Physical Geography*, 11: 220–39.

Arnfield, A. J., and Grimmond, C. S. B. (1998). "An Urban Canyon Energy Budget Model and its Application to Urban Storage Heat Flux Modeling." *Energy and Buildings*, 27: 61–8.

Arnfield, A. J., and Mills, G. M. (1994). "An Analysis of the Circulation Characteristics and Energy Budget of a Dry, Asymmetric, East–West Urban Canyon. 1. Circulation Characteristics." *International Journal of Climatology*, 14: 119–34.

Baker, D. G., Ruschy, D. L., Skaggs, R. H., and Wall, D. B. (1992). "Air Temperature and Radiation Depression Associated with a Snow Cover." *Journal of Applied Meteorology*, 31: 247–54.

Balling, R. C., Jr. (1997). "Analysis of Daily and Monthly Spatial Variance Components in Historical Temperature Records." *Physical Geography*, 18: 544–52.

Balling, R. C., Jr., Skindlov, J. A., and Phillips, D. H. (1990). "The Impact of Increasing Summer Mean Temperatures on Extreme Maximum and Minimum Temperatures in Phoenix, Arizona." *Journal of Climate*, 3: 1491–4.

Balling, R. C., Jr., Vose, R. S., and Weber, G.-R. (1998). "Analysis of Long-Term European Temperature Records: 1751–1995." *Climate Research*, 10: 193–200.

Barry, R. G. (1996). "The Parameterization of Surface Albedo for Sea Ice and Its Snow Cover." *Progress in Physical Geography*, 20: 63–79.

Barry, R. G., Serreze, M. C., Maslanik, J. A., and Preller, R. H. (1993). "The Arctic Sea-Ice Climate System—Observations and Modeling." *Reviews of Geophysics*, 31: 397–422.

Bartlein, P. J., Edwards, M. E., Shafer, S. L., and Barker, E. D., Jr. (1995). "Calibration of Radiocarbon Ages and the Interpretation of Paleoenvironmental Records." *Quaternary Research*, 44: 417–24.

Bartlein, P. J., Anderson, K. H., Anderson, P. M., Edwards, M. E., Mock, C. J., Thompson, R. S., Webb, R. S., Webb, T., III, and Whitlock, C. (1998). "Paleoclimate Simulations for North America over the Past 21,000 Years: Features of the Simulated Climate and Comparisons with Paleoenvironmental Data." *Quaternary Science Reviews*, 17: 549–85.

Basist, A., Bell, G. D., and Meentemeyer, V. (1994). "Statistical Relationships between Topography and Precipitation Patterns." *Journal of Climate*, 7: 1305–15.

Bentley, M. L., and Mote, T. L. (1998). "A Climatology of Derecho-Producing Mesoscale Convective Systems in the Central and Eastern United States, 1986–1995. Part I: Temporal and Spatial Distribution." *Bulletin of the American Meteorological Society*, 79: 2527–40.

—— (2000). "A Synoptic Climatology of Cool-Season Derecho Events." *Physical Geography*, 21: 21–37.

Bentley, M. L., Mote, T. L., and Byrd, S. F. (2000). "A Synoptic Climatology of Derecho Producing Mesoscale Convective Systems in the North-Central Plains." *International Journal of Climatology*, 20: 1329–49.

Bierly, G. D. (1997). "The Role of Stratospheric Intrusions in Colorado Cyclogenesis." *Physical Geography*, 51: 340–8.

Bierly, G. D., and Harrington, J. A., Jr. (1995). "A Climatology of Transition Season Colorado Cyclones: 1961–1990." *Journal of Climate*, 8: 853–63.

Bierly, G. D., and Winkler, J. A. (2001). "A Composite Analysis of Airstreams within Cold-Season Colorado Cyclones." *Weather and Forecasting*, 16: 57–60.

Bierly, G. D., Harrington, J. A., Jr., and Wilhelm, D. F. (2000). "Climatology of Surface Cyclone Trajectory and Intensity for Heavy-Snow Events at Three Midwestern Stations." *Physical Geography*, 21: 522–37.

Brinkmann, W. A. R. (1993). "Development of an Airmass-Based Regional Climate Change Scenario." *Theoretical and Applied Climatology*, 47: 129–36.

—— (2000). "Modification of a Correlation-Based Circulation Pattern Classification to Reduce Within-Type Variability of Temperature and Precipitation." *International Journal of Climatology*, 20: 839–52.

Burnett, A. W. (1993). "Size Variations and Long-Wave Circulation within the January Northern Hemisphere Circumpolar Vortex: 1946–89." *Journal of Climate*, 6: 1914–20.

Burnett, A. W., and McNicoll, A. R. (2000). "Interannual Variations in the Southern Hemisphere Winter Circumpolar Vortex: Relationships with the Semiannual Oscillation." *Journal of Climate*, 13: 991–9.

Carbone, G. J., Narumalani, S., and King, M. (1996). "Application of Remote Sensing and GIS Technologies with Physiological Crop Models." *Photogrammetric Engineering and Remote Sensing*, 62: 171–9.

Carleton, A. M. (1988). "Meridional Transport of Eddy Sensible Heat in Winters Marked by Extremes of the North Atlantic Oscillation, 1948/49–1979/80." *Journal of Climate*, 1: 212–23.

Carleton, A. M., Carpenter, D. A., and Weber, P. J. (1990). "Mechanisms of Interannual Variability of the Southwest United States Summer Rainfall Maximum." *Journal of Climate*, 3: 999–1015.

Carleton, A. M., Travis, D., Arnold, D., Brinegar, R., Jelinski, D. E., and Easterling, D. R. (1994). "Climate-Scale Vegetation-Cloud Interactions during Drought Using Satellite Data." *International Journal of Climatology*, 14: 593–624.

Cerveny, R. S., and Balling, R. C., Jr. (1992). "The Impact of Snow Cover on Diurnal Temperature Range." *Geophysical Research Letters*, 19: 797–800.

—— (1998). "Weekly Cycles of Air Pollutants, Precipitation and Tropical Cyclones in the Coastal NW Atlantic Region." *Nature*, 394: 561–3.

Changnon, D. (1994). "Regional and Temporal Variations in Heavy Precipitation in South Carolina." *International Journal of Climatology*, 14: 165–77.

—— (1996). "Changing Temporal and Spatial Characteristics of Midwestern Hydrologic Droughts." *Physical Geography*, 17: 29–46.

Clark, M. P., Serreze, M. C., and McCabe, G. J. (2001). "Historical Effects of El Niño and La Niña Events on the Seasonal Evolution of the Montane Snowpack in the Columbia and Colorado River Basins." *Water Resources Research*, 37: 741–57.

Clark, M. P., Serreze, M. C., and Robinson, D. A. (1999). "Atmospheric Controls on Eurasian Snow Extent." *International Journal of Climatology*, 19: 27–40.

Comrie, A. C. (1992). "A Procedure for Removing the Synoptic Climate Signal from Environmental Data." *International Journal of Climatology*, 12: 177–83.

—— (1996). "An All-Season Synoptic Climatology of Air Pollution in the U.S.-Mexico Border Region." *The Professional Geographer*, 48: 237–51.

Comrie, A. C., and Glenn, E. C. (1998). "Principal Components-Based Regionalization of Precipitation Regimes across the Southwest United States and Northern Mexico, with an Application to Monsoon Precipitation Variability." *Climate Research*, 10: 201–15.

Comrie, A. C., and Yarnal, B. (1992). "Relationships between Synoptic-Scale Atmospheric Circulation and Surface Ozone Concentrations in Metropolitan Pittsburgh, Pennsylvania." *Atmospheric Environment*, 26B: 301–12.

Crane, R. G., and Hewitson, B. C. (1998). "Doubled CO_2 Precipitation Changes for the Susquehanna Basin: Down-scaling from the GENESIS General Circulation Model." *International Journal of Climatology*, 18: 65–76.

Cutrim, E., Martin, D. W., and Rabin, R. (1995). "Enhancement of Cumulus Clouds over Deforested Lands in Amazonia." *Bulletin of the American Meteorological Society*, 76: 1801–5.

Davis, R. E., and Benkovic, S. R. (1994). "Spatial and Temporal Variations of the January Circumpolar Vortex over the Northern Hemisphere." *International Journal of Climatology*, 14: 415–28.

Davis, R. E., Dolan, R., and Demme, G. (1992). "Synoptic Climatology of Atlantic Coast North-Easters." *International Journal of Climatology*, 13: 171–89.

Davis, R. E., Hayden, B. P., Gay, D. A., Phillips, W. L., and Jones, G. V. (1997). "The North Atlantic Subtropical Anticyclone." *Journal of Climate*, 10: 728–44.

DeLiberty, T. L. (2000). "A Regional Scale Investigation of Climatological Tropical Convection and Precipitation in the Amazon Basin." *The Professional Geographer*, 52: 258–71.

Dunne, K. A., and Willmott, C. J. (1996). "Global Distribution of Plant-Extractable Water Capacity of Soil." *International Journal of Climatology*, 16: 841–59.

Easterling, D. R. (1991). "Climatological Patterns of Thunder-storm Activity in South-Eastern USA." *International Journal of Climatology*, 11: 213–21.

—— (1999). "Development of Regional Climate Scenarios Using a Downscaling Approach." *Climatic Change*, 41: 615–34.

Easterling, D. R., and Peterson, T. C. (1995). "A New Method for Detecting Undocumented Discontinuities in Climatological Time Series." *International Journal of Climatology*, 15: 369–77.

Ellis, A. W., and Leathers, D. J. (1998). "A Quantitative Approach to Evaluating the Effects of Snow Cover on Cold Airmass Temperatures across the U.S. Great Plains." *Weather and Forecasting*, 13: 688–701.

Faiers, G. E., Grymes, J. M., III, Keim, B. D., and Muller, R. A. (1994). "A Reexamination of Extreme 24-hour Rainfall in Louisiana, USA." *Climate Research*, 4: 25–31.

Feddema, J. J. (1998). "Estimated Impacts of Soil Degradation on the African Water Balance and Climate." *Climate Research*, 10: 127–41.

Frakes, B., and Yu, Z. (1999). "An Evaluation of Two Hydrologic Models for Climate Change Scenarios." *Journal of the American Water Resources Association*, 35: 1351–63.

Frei, A., Robinson, D. A., and Hughes, M. G. (1999). "North American Snow Extent: 1900–1994." *International Journal of Climatology*, 19: 1517–34.

Gallo, K. P., Easterling, D. R., and Peterson, T. C. (1996). "The Influence of Land Use/Land Cover on Climatological Values of the Diurnal Temperature Range." *Journal of Climate*, 9: 2941–4.

Gay, D. A., and Davis, R. E. (1993). "Freezing Rain and Sleet Climatology of the Southeastern USA." *Climate Research*, 3: 209–20.

Giambelluca, T. W., Holscher, D. Bastos, T. X., Frazao, R. R., Nullet, M. A., and Ziegler, A. D. (1997). "Observations of Albedo and Radiation Balance over Post Forest Land Surfaces in the Eastern Amazon Basin." *Journal of Climate*, 10: 919–28.

Giorgi, F., and Mearns, L. O. (1999). "Regional Climate Modeling Revisited: An Introduction to the Special Issue." *Journal of Geophysical Research*. 104: 6335–52.

Giorgi, F., Shields-Brodeur, C., and Bates, G. T. (1994). "Regional Climate Change Scenarios over the United States Produced with a Nested Regional Climate Model." *Journal of Climate*, 7: 375–99.

Giorgi, F., Mearns, L., Shields, S., and McDaniel, L. (1998). "Regional Nested Model Simulations of Present Day and

$2 \times CO_2$ Climate over the Central Great Plains of the United States." *Climatic Change*, 40: 457–93.

Giorgi, F., Hewitson, B., Christensen, J., Hulme, M., Von Storch, H., Whetton, P., Jones, R., Mearns, L., and Fu, C. (2001). "Regional Climate Information: Evaluation and Projections" (ch. 10), in J. T. Houghton *et al.* (eds.), *Climate Change 2001: The Scientific Basis, Contribution of Working Group I to the Third Assessment Report of the IPCC*. Cambridge University Press: Cambridge, 739–68.

Greene, J. S., and Kalkstein, L. S. (1996). "Quantitative Analysis of Summer Air Masses in the Eastern United States and an Application to Human Mortality." *Climate Research*, 7: 43–53.

Greene, J. S., Morrissey, M. L., and Ferraro, R. R. (1997). "Verification of a Scattering-Based Algorithm for Estimating Rainfall over the Open Ocean." *Theoretical and Applied Climatology*, 56: 33–44.

Greenland, D. E. (1991). "Surface-Energy Budgets over Alpine Tundra in Summer, Niwot Ridge, Colorado Front Range." *Mountain Research and Development*, 11: 339–51.

— (1993). "Spatial Energy Budgets in Alpine Tundra." *Theoretical and Applied Climatology*, 46: 229–39.

Grimmond, C. S. B. (1992). "The Suburban Energy Balance: Methodological Considerations and Results for a Mid-Latitude West Coast City under Winter and Spring Conditions." *International Journal of Climatology*, 12: 481–97.

Grimmond, C. S. B., and Oke, T. R. (1995). "Comparison of Heat Fluxes from Summertime Observations in the Suburbs of Four North American Cities." *Journal of Applied Meteorology*, 14: 873–89.

— (1998). "Aerodynamic Properties of Urban Areas Derived from Analysis of Surface Form." *Journal of Applied Meteorology*, 38: 1262–92.

Grimmond, C. S. B., Souch, C., and Hubble, M. D. (1996). "Influence of Tree Cover on Summertime Surface Energy Balance Fluxes, San Gabriel Valley, Los Angeles." *Climate Research*, 6: 45–57.

Grimmond, C. S. B., King, T. S., Roth, M., and Oke, T. R. (1998). "Surface Aerodynamic Characteristics of Urban Areas: Anemometric Analyses." *Boundary Layer Meteorology*, 89: 1–24.

Groisman, P. Ya., and Easterling, D. R. (1994). "Variability and Trends of Total Precipitation and Snowfall over the United States and Canada." *Journal of Climate*, 7: 184–205.

Groisman, P. Ya., and Legates, D. R. (1994). "Accuracy of Historical United States Precipitation Data." *Bulletin of the American Meteorological Society*, 75: 215–27.

— (1995). "Documenting and Detecting Long-Term Precipitation Trends: Where We Are and What Should be Done." *Climatic Change*, 31: 601–22.

Grundstein, A. J., and Bentley, M. L. (2001). "A Growing-Season Hydroclimatology, Focusing on Soil Moisture Deficits, for the Ohio Valley Region." *Journal of Hydrometeorology*, 2: 345–55.

Hanson, B., Klink, K., Matsuura, K., Robeson, S. M., and Willmott, C. J. (1992). "Vector Correlation: Review, Exposition, and Geographic Application." *Annals of the Association of American Geographers*, 82: 103–16.

Hanuta, I., and LaDochy, S. (1989). "Thunderstorm Climatology Based on Lightning Detector Data, Manitoba, Canada." *Physical Geography*, 10: 101–10.

Harman, J. R. (1991). *Synoptic Climatology of the Westerlies: Process and Patterns*. Washington, DC: Association of American Geographers.

Harman, J. R., and Winkler, J. A. (1991). "Synoptic Climatology: Themes, Applications, and Prospects." *Physical Geography*, 12: 220–30.

Harrington, J. A., Jr., and Cerveny, R. S. (1988). "Temporal Statistics: An Application in Snowfall Climatology." *Physical Geography*, 9: 337–53.

Hartley, S., and Robinson, D. A. (2000). "A Shift in Winter Season Timing in the Northern Plains of the USA as Indicated by Temporal Analysis of Heating Degree Days." *International Journal of Climatology*, 20: 365–79.

Haston, L., and Michaelsen, J. (1994). "Long-Term Central Coastal California Precipitation Variability and Relationships to El-Niño-Southern Oscillation." *Journal of Climate*, 7: 1373–87.

Henderson, K. G., and Muller, R. A. (1997). "Extreme Temperature Days in the South-Central United States." *Climate Research*, 8: 151–62.

Henderson, K. G., and Vega, A. J. (1996). "Regional Precipitation Variability in the Southern United States." *Physical Geography*, 17: 93–112.

Hewitson, B. C., and Crane, R. G. (1996). "Climate Downscaling: Techniques and Application." *Climate Research*, 7: 85–95.

Hirschboeck, K. K. (1988). "Flood Climatology", in V. R. Baker *et al.* (eds.), *Flood Geomorphology*. New York: John Wiley & Sons, 27–49.

Hirschboeck, K. K., Ni, F., Wood, M. L., and Woodhouse, C. A. (1996). "Synoptic Dendroclimatology: Overview and Outlook," in J. S. Dean, D. M. Meko, and T. W. Swetnam (eds.), *Tree Rings, Environment and Humanity*. Tucson, Ariz.: Radiocarbon Publishers, 205–23.

Hughes, M. G., and Robinson, D. A. (1996). "Historical Snow Cover Variability in the Great Plains Region of the United States: 1910 through 1993." *International Journal of Climatology*, 16: 1005–18.

Isard, S. A. (1989). "Topographic Controls in an Alpine Fellfield and their Ecological Significance." *Physical Geography*, 10: 13–31.

Jamason, P. F., Kalkstein, L. S., and Gergen, P. J. (1997). "A Synoptic Evaluation of Asthma Hospital Admissions in New York City." *American Journal of Respiratory and Critical Care Medicine*, 156: 1781–8.

Jones, G. V., and Davis, R. E. (2000). "Using a Synoptic Climatological Approach to Understand Climate-Viticulture Relationships." *International Journal of Climatology*, 20: 813–37.

Kalkstein, L. S. (1991). "A New Approach to Evaluate the Impact of Climate on Human Mortality." *Environmental Health Perspectives*, 96: 145–50.

Kalkstein, L. S., and Greene, J. S. (1997). "An Evaluation of Climate/Mortality Relationships in Large US Cities and the Possible Impacts of a Climate Change." *Environmental Health Perspectives*, 103: 84–93.

Kalkstein, L. S., Dunne, P. C., and Vose, R. S. (1990). "Detection of Climatic Change in the Western North American Arctic Using a Synoptic Climatological Approach." *Journal of Climate*, 3: 1153–67.

Kalkstein, L. S., Sheridan, S. C., and Graybeal, D. Y. (1998). "A Determination of Character and Frequency Changes in Air Masses Using a Spatial Synoptic Classification." *International Journal of Climatology*, 18: 1223–36.

Kalkstein, L. S., Nichols, M. C., Barthel, C. D., and Greene, J. S. (1996). "A New Spatial Synoptic Classification: Application to

Air-Mass Analysis." *International Journal of Climatology*, 16: 983–1004.

Keim, B. D. (1996). "Spatial, Synoptic, and Seasonal Patterns of Heavy Rainfall in the Southeastern United States." *Physical Geography*, 17: 313–28.

Keim, B. D., and Cruise, J. F. (1998). "A Technique to Measure Trends in the Frequency of Discrete Random Events." *Journal of Climate*, 11: 848–55.

Keim, B. D., Faiers, G. E., Muller, R. A., Grymes, J. M. III, and Rohli, R. V. (1995). "Long-Term Trends of Precipitation and Runoff in Louisiana, USA." *International Journal of Climatology*, 15: 531–41.

King, T. S., and Balling, R. C., Jr. (1994). "Diurnal Variations in Arizona Monsoon Lightning Data." *Monthly Weather Review*, 122: 1659–64.

Klink, K. (1995*a*). "Surface Aggregation and Subgrid-Scale Climate." *International Journal of Climatology*, 15: 1219–40.

— (1995*b*). "Temporal Sensitivity of Regional Climate to Land-Surface Heterogeneity." *Physical Geography*, 16: 289–314.

Klink, K., and Willmott, C. J. (1989). "Principal Components of the Surface Wind Field in the United States: A Comparison of Analyses Based upon Wind Velocity, Direction, and Speed." *International Journal of Climatology*, 9: 293–308.

— (1994). "Influence of Soil Moisture and Surface Roughness Heterogeneity on Modeled Climate." *Climate Research*, 4: 105–18.

Konrad, C. E., II (1994). "Moisture Trajectories Associated with Heavy Rainfall in the Appalachian Region of the United States." *Physical Geography*, 15: 227–48.

— (1996). "Relationships between the Intensity of Cold Air Outbreaks and the Evolution of Synoptic and Planetary Scale Features over North America." *Monthly Weather Review*, 124: 1067–83.

— (1998). "A Flood Climatology of the Lower Roanoke River Basin in North Carolina." *Physical Geography*, 19: 15–34.

— (2001). "The Most Extreme Precipitation Events over the Eastern United States from 1950 to 1996: Considerations of Scale." *Journal of Hydrometeorology*, 2: 309–25.

Konrad, C. E., II, and Meentenmeyer, V. (1994). "Lower Tropospheric Warm Air Advection Patterns Associated with Heavy Rainfall over the Appalachian Region." *The Professional Geographer*, 46: 143–55.

Kothavala, Z. (1997). "Extreme Precipitation Events and the Applicability of Global Climate Models to Study Floods and Droughts." *Mathematics and Computers in Simulation*, 43: 261–8.

Leathers, D. J., and Robinson, D. A. (1993). "The Association between Extremes in North American Snow Cover Extent and United States Temperatures." *Journal of Climate*, 6: 1345–55.

Leathers, D. J., Ellis, A. W., and Robinson, D. A. (1995). "Characteristics of Temperature Depressions Associated with Snow Cover across the Northeast United States." *Journal of Applied Meteorology*, 34: 381–90.

Leathers, D. J., Yarnal, B., and Palecki, M. A. (1991). "The Pacific/North American Teleconnection Pattern and United States Climate. Part I: Regional Temperature and Precipitation Associations." *Journal of Climate*, 4: 517–28.

Leathers, D. J., Mote, T. L., Kuivinen, K. C., McFeeters, S., and Kluck, D. R. (1993). "Temporal Characteristics of USA Snowfall 1945–1946 through to 1984–1985." *International Journal of Climatology*, 13: 65–76.

Legates, D. R. (1991*a*). "An Evaluation of Procedures to Estimate Monthly Precipitation Probabilities." *Journal of Hydrology*, 122: 129–40.

— (1991*b*). "The Effect of Domain Shape on Principal Components Analyses." *International Journal of Climatology*, 11: 135–46.

— (1995). "Global and Terrestrial Precipitation: A Comparative Assessment of Existing Climatologies." *International Journal of Climatology*, 15: 237–58.

— (2000). "Real-Time of Radar Precipitation Estimates." *The Professional Geographer*, 52: 235–46.

Legates, D. R., and Davis, R. E. (1997). "The Continuing Search for an Anthropogenic Climate Change Signal: Limitations of Correlation-Based Approaches." *Geophysical Research Letters*, 24: 2319–22.

Legates, D. R., and DeLiberty, T. L. (1993). "Precipitation Measurement Biases in the United States." *Water Resources Bulletin*, 29: 855–61.

Legates, D. R., and McCabe, G. J., Jr. (1999). "Evaluating the Use of "Goodness-of-Fit" Measures in Hydrologic and Hydroclimatic Model Validation." *Water Resources Research*, 35: 233–41.

Legates, D. R., and Mather, J. R. (1992). "An Evaluation of the Average Annual Global Water Balance." *Geographical Review*, 82: 253–67.

Legates, D. R., and Willmott, C. J. (1990). "Mean Seasonal and Spatial Variability in Gauge-Corrected, Global Precipitation." *International Journal of Climatology*, 10: 111–27.

Liu, K.-B., and Fearn, M. L. (1993). "Lake-Sediment Record of Late Holocene Hurricane Activities from Coastal Alabama." *Geology*, 21: 793–6.

Liu, K.-B., Sun, S., and Jiang, X. (1992). "Environmental Change in the Yangtze River Delta since 12,000 Years B.P." *Quaternary Research*, 38: 32–45.

McCabe, G. J., Jr. (1995). "Relations between Winter Atmospheric Circulation and Annual Streamflow in the Western United States." *Climate Research*, 5: 139–48.

McCabe, G. J., Jr., and Wolock, D. M. (1992). "Effects of Climatic Change and Climatic Variability on the Thornthwaite Moisture Index in the Delaware River Basin." *Climatic Change*, 20: 143–53.

— (1997). "Climate Change and the Detection of Trends in Annual Runoff." *Climate Research*, 8: 129–34.

McGinnis, D. L. (1997). "Estimating Climate-Change Impacts on Colorado Plateau Snowpack Using Downscaling Methods." *The Professional Geographer*, 49: 117–25.

McGinnis, D. L., and Crane, R. G. (1994). "A Multivariate Analysis of Arctic Climate in GCMs." *Journal of Climate*, 7: 1240–50.

McHugh, M. J., and Rogers, J. C. (2001). "North Atlantic Oscillation Influence on Precipitation Variability around the Southeast African Convergence Zone." *Journal of Climate*, 14: 3631–42.

Mahmood, R. (1998*a*). "Air Temperature Variations and Rice Productivity in Bangladesh: A Comparative Study of the Performance of the YIELD and the CERES-Rice Models." *Ecological Modelling*, 106: 201–12.

— (1998*b*). "Thermal Climate Variations and Potential Modification of the Cropping Pattern in Bangladesh." *Theoretical and Applied Climatology*, 61: 231–43.

Marshall, S., Roads, J. O., and Oglesby, R. J. (1997). "Effects of Resolution and Physics on Precipitation in NCAR's CCM." *Journal of Geophysical Research*, 102: 19529–41.

Mather, J. R. (1991). "A History of Hydroclimatology." *Physical Geography*, 12: 260–73.

Mearns, L. O., (1999). "Climatic Change and Variability," in K. R. Reddy and H. Hodges (eds.), *Climate Change and Global Crop Productivity*, Melbourne: CAB.

Mearns, L. O., Easterling, W., Hays, C., and Marx, D. (2001*a*). "Comparison of Agricultural Impacts of Climate Change Calculated from High and Low Resolution Climate Model Scenarios: Part I. The Uncertainty Due to Spatial Scale." *Climatic Change*, 51: 131–72.

Mearns, L. O., Giorgi, F., McDaniel, L., and Shields, C. (1995). "Analysis of the Variability of Daily Precipitation in a Nested Modeling Experiment: Comparison with Observations and 2 Times CO_2 results." *Global and Planetary Change*, 10: 55–78.

Mearns, L. O., Schneider, S. H., Thompson, S. L., and McDaniel, L. R. (1990). "Analysis of Climate Variability in General Circulation Models: Comparison with Observations and Changes in Variability in 2 times CO_2 experiments." *Journal of Geophysical Research*, 95: 20469–90.

Mearns, L. O., Bogardi, I., Giorgi, F., Matyasovszky, I., and Palecki, M. (1999*a*). "Comparison of Climate Change Scenarios Generated from Regional Climate Model Experiments and Empirical Downscaling." *Journal of Geophysical Research*, 104: 6603–21.

Mearns, L. O., Mavromatis, T., Tsvetsinskaya, E., Hays, C., and Easterling, W. (1999*b*). "Comparative Responses of EPIC and CERES Crop Models to High and Low Resolution Climate Change Scenarios." *Journal of Geophysical Research*, 104: 6623–46.

Mearns, L. O., Hulme, M., Carter, T. R., Leemans, R., Lal, M., and Whetton, P. (2001*b*). "Climate Scenario Development" (ch. 13), in J. T. Houghton *et al.* (eds.), *Climate Change 2001: The Scientific Basis, Contribution of Working Group I to the Third Assessment Report of the IPCC*. Cambridge: Cambridge University Press, 583–638.

Meehl, G. A. (1998). "Climate Modeling," in D. J. Karoly and D. Vincent (eds.), *Meteorology of the Southern Hemisphere*, Boston: American Meteorological Society, 365–410.

Michaels, P. J., Balling, R. C., Jr., Vose, R. S., and Knappenberger, P. C. (1998). "Analysis of Trends in the Variability of Daily and Monthly Historical Temperature Measurements." *Climate Research*, 10: 27–33.

Mock, C. J., and Bartlein P. J. (1995). "Spatial Variability of Late-Quaternary Paleoclimates in the Western United States." *Quaternary Research*, 44: 425–33.

Mock, C. J., Bartlein, P. J., and Anderson, P. M. (1998). "Atmospheric Circulation Patterns and Spatial Climatic Variations in Beringia." *International Journal of Climatology*, 18: 1085–104.

Morrissey, M. L., and Greene, J. S. (1993). "Comparison of Two Satellite-Based Rainfall Algorithms using Pacific Atoll Rain Gage Data." *Journal of Applied Meteorology*, 32: 411–25.

Mosley-Thompson, E. (1996). "Holocene Climate Changes Recorded in an East Antarctic Ice Core," in P. D. Jones, R. Bradley, and J. Jouzel (eds.), *Climatic Variations and Forcing Mechanisms of the Last 2000 Years*, NATO Advanced Research Series I, 41: 263–79.

Mote, T. L., Gamble, D. W., Underwood, S. J., and Bentley, M. L. (1997). "Synoptic-Scale Features Common to Heavy Snowstorms in the Southeast United States." *Weather and Forecasting*, 12: 5–23.

Muller, R. A., Keim, B. D., and Hoff, J. L. (1990). "Application of Climatic Divisional Data to Flood Interpretations: An Example from Louisiana." *Physical Geography*, 11: 353–62.

Nasrallah, H. A., Brazel, A. J., and Balling, R. C., Jr. (1990). "Analysis of the Kuwait City Urban Heat Island." *International Journal of Climatology*, 10: 401–6.

O'Brien, K. L. (1996). "Tropical Deforestation and Climate Change." *Progress in Physical Geography*, 20: 311–35.

Oke, T. R., Sproken-Smith, A., Jauregui, E., and Grimmond, C. S. B. (1999). "The Energy Balance of Central Mexico City during the Dry Season." *Atmospheric Environment*, 33: 3919–30.

Oliver, J. E., Barry, R. G., Brinkmann, W. A. R., and Rayner, J. N. (1989). "Climatology", in G. L. Gaile and C. J. Wilmott (eds.), *Geography In America*. Columbus, Ohio: Merrill, 47–69.

Palecki, M. A., and Leathers, D. J. (1993). "Northern Hemisphere Extratropical Circulation Anomalies and Recent January Land Surface Temperature Trends." *Geophysical Research Letters*, 20: 819–22.

Peterson, T. C., and Easterling, D. R. (1994). "Creation of Homogeneous Composite Climatological Reference Series." *International Journal of Climatology*, 14: 671–9.

Pryor, S. C. (1998). "A Case Study of Emission Changes and Ozone Responses." *Atmospheric Environment*, 32: 123–31.

Pryor, S. C., McKendry, I. G., and Steyn, D. G. (1995). "Synoptic-Scale Meteorological Variability and Ozone Concentrations in Vancouver, B. C." *Journal of Applied Meteorology*, 34: 1824–33.

Raphael, M. N. (1997). "The Relationship between the Transient, Meridional Eddy Sensible and Latent Heat Flux." *Journal of Geophysical Research*, 102: 13487–94.

—— (1998). "Quasistationary Waves in the Southern Hemisphere: An Examination of their Simulation by the NCAR Climate System Model, With and Without an Interactive Ocean." *Journal of Climate*, 11: 1405–18.

Raphael, M. N., and Mills, G. M. (1996). "The Role of Mid-Latitude Pacific Cyclones in the Winter Precipitation of California." *The Professional Geographer*, 48: 251–62.

Raphael, M. N., and Rogers, J. C. (1992). "The Meridional Flux of Eddy Sensible Heat at 700 mb in the Northern Hemisphere Winter." *Physical Geography*, 13: 1–13.

Rayner, J. N., Hobgood, J. S., and Howarth, D. A. (1991). "Dynamic Climatology: Its History and Future." *Physical Geography*, 12: 207–19.

Robeson, S. M. (1994). "Influence of Spatial Sampling and Interpolation on Estimates of Air Temperature Change." *Climate Research*, 4: 119–26.

—— (1995). "Resampling of Network-Induced Variability in Estimates of Terrestrial Air Temperature Change." *Climatic Change*, 29: 213–29.

—— (1997). "Spherical Methods for Spatial Interpolation: Review and Evaluation." *Cartography and Geographic Information Systems*, 24: 3–20.

Robeson, S. M., and Janis, M. J. (1998). "Comparison of Temporal and Unresolved Spatial Variability in Multiyear Time-Averages of Air Temperature." *Climate Research*, 10: 15–26.

Robeson, S. M., and Shein, K. A. (1997). "Spatial Coherence and Decay of Wind Speed and Power in the North-Central United States." *Physical Geography*, 18: 479–95.

Robinson, D. A., and Dewey, K. F. (1990). "Recent Secular Variations in the Extent of Northern Hemisphere Snow Cover." *Geophysical Research Letters*, 17: 1557–60.

Robinson, D. A., Serreze, M. C., Barry, R. G., Scharfen, G., and Kukla, G. (1992). "Large-Scale Patterns and Variability of Snowmelt and Parameterized Surface Albedo in the Arctic Basin." *Journal of Climate*, 5: 1109–19.

Robinson, P. J. (1997). "Climate Change and Hydropower Generation." *International Journal of Climatology*, 17: 983–96.

Robinson, P. J., and Henderson, K. G. (1992). "Precipitation Events in the South-East United States of America." *International Journal of Climatology*, 12: 701–20.

Rogers, J. C. (1997). "North Atlantic Storm Track Variability and its Association to the North Atlantic Oscillation and Climate Variability of Northern Europe." *Journal of Climate*, 10: 1635–47.

Rogers, J. C., and Mosley-Thompson, E. (1995). "Atlantic Arctic Wave Cyclones and the Mild Siberian Winters of the 1980s." *Geophysical Research Letters*, 22: 799–802.

Rogers, J. C., and Raphael, M. N. (1992). "Meridional Eddy Sensible Heat Fluxes in the Extremes of the Pacific/North American Teleconnection Pattern." *Journal of Climate*, 5: 127–39.

Rogers, J. C., and Rohli, R. V. (1991). "Florida Citrus Freezes and Polar Anticyclones in the Great Plains." *Journal of Climate*, 4: 1103–13.

Rogers, J. C., Bolzan, J. F., and Pohjola, V. A. (1998): "Atmospheric Circulation Variability Associated with Shallow-Core Seasonal Isotopic Extremes near Summit Greenland." *Journal of Geophysical Research-Atmospheres*, 103: 11205–19.

Rohli, R. V., and Henderson, K. G. (1997). "Winter Anticyclone Changes on the Central Gulf Coast of the USA." *International Journal of Climatology*, 17: 1183–93.

— (1998). "Upper-Level Steering Flow and Continental Anticyclones on the Central Gulf Coast of the United States." *International Journal of Climatology*, 18: 935–54.

Rohli, R. V., and Keim, B. D. (1994). "Spatial and Temporal Characteristics of Extreme-High-Temperature-Events in the South-Central United States." *Physical Geography*, 14: 1–15.

Rohli, R. V., and Rogers, J. C. (1993). "Atmospheric Teleconnections and Citrus Freezes in the Southern United States." *Physical Geography*, 14: 1–17.

Rowe, C. M. (1991). "Modeling Land-Surface Albedos from Vegetation Canopy Architecture." *Physical Geography*, 12: 93–114.

Schmid, H. P., Cleugh, H. A., Grimmond, C. S. B., and Oke, T. R. (1991). "Spatial Variability of Energy Fluxes in Suburban Terrain." *Boundary-Layer Meteorology*, 54: 249–76.

Schmidlin, T. W., Edgell, D. J., and Delaney, M. A. (1992). "Design Ground Snow Loads for Ohio." *Journal of Applied Meteorology*, 31: 622–7.

Schmidlin, T. W., King, P. S., Hammer, B. O., and Ono, Y. (1998). "Behavior of Vehicles during Tornado Winds." *Journal of Safety Research*, 29: 181–6.

Schwartz, M. D. (1992). "Phenology and Springtime Surface-Layer Change." *Monthly Weather Review*, 120: 2570–8.

— (1995). "Detecting Structural Climate Change: An Air Mass-Based Approach in the North Central United States, 1958–1992." *Annals of the Association of American Geographers*, 85: 553–68.

— (1996*a*). "An Air Mass-Based Approach to Regional GCM Validation." *Climate Research*, 6: 227–35.

— (1996*b*). "Examining the Spring Discontinuity in Daily Temperature Ranges." *Journal of Climate*, 9: 803–8.

Schwartz, M. D., and Karl, T. R. (1990). "Spring Phenology: Nature's Experiment to Detect the Effect of 'Green-up' on Surface Maximum Temperature." *Monthly Weather Review*, 118: 883–90.

Serreze, M. C., and Barry, R. G. (1988). "Synoptic Activity in the Arctic Basin, 1979–85." *Journal of Climate*, 1: 1276–95.

Serreze, M. C., Carse, F., Barry, R. G., and Rogers, J. C. (1997). "Icelandic Low Cyclone Activity: Climatological Features, Linkages with the NAO and Relationships with Recent Changes Elsewhere in the Northern Hemisphere Circulation." *Journal of Climate*, 10: 453–64.

Shelton, M. L. (1998). "Seasonal Hydroclimate Change in the Sacramento River Basin, California." *Physical Geography*, 19: 239–55.

— (2001). "Mesoscale Atmospheric 2 × CO_2 Climate Change Simulation Applied to an Oregon Watershed." *Journal of the American Water Resources Association*, 37: 1041–52.

Skaggs, R. H., Baker, D. G., and Ruschy, D. L. (1995). "Interannual Variability Characteristics of the Eastern Minnesota (USA) Temperature Record: Implications for Climate Change Studies." *Climate Research*, 5: 223–7.

Smoyer, K. E., Kalkstein, L. S., Greene, J. S., and Ye, H. (2000). "The Impacts of Weather and Pollution on Human Mortality in Birmingham, Alabama and Philadelphia, Pennsylvania." *International Journal of Climatology*, 20: 881–97.

Song, J. (1998). "Reconstruction of the Southern Oscillation from Dryness/Wetness in China for the Last 500 Years." *International Journal of Climatology*, 18: 1345–55.

Song, J., Willmott, C. J., and Hanson, B. (1997*a*). "Influence of Heterogeneous Land Surfaces on Surface Energy and Mass Fluxes." *Theoretical and Applied Climatology*, 58: 175–88.

Song, J., Wesely, M. L., Coulter, R. L., and Brandes, E. A. (2000*a*). "Estimating Watershed Evapotranspiration with PASS. Part I: Inferring Root-Zone Moisture Conditions Using Satellite Data." *Journal of Hydrometeorology*, 1: 447–61.

Song, J., Wesely, M. L., LeMone, M. A., and Grossman, R. L. (2000*b*). "Estimating Watershed Evapotranspiration with PASS. Part II: Moisture Budgets during Drydown Periods." *Journal of Hydrometeorology*, 1: 462–73.

Soulé, P. T. (1992). "Spatial Patterns of Drought Frequency and Duration in the Contiguous USA Based on Multiple Drought Event Definitions." *International Journal of Climatology*, 12: 11–24.

Soulé, P. T., and Shankman, D. (1990). "The Relationship of Palmer's Drought Indices to River Stage in Western Tennessee." *Physical Geography*, 11: 206–19.

Spencer, J. L., Isard, S. A., Levine, E. (1999). "Free Flight of Western Corn Rootworm (Coleoptera: Chrysomelidae) to Corn and Soybean Plants in a Walk-in Wind Tunnel." *Journal of Economic Entomology*, 92: 146–55.

Stahle, D. W., and Cleaveland, M. K. (1992). "Reconstruction and Analysis of Spring Rainfall over the Southeastern U.S. for the Past 1000 Years." *Bulletin of the American Meteorological Society*, 73: 1947–61.

Stahle, D. W., Cleaveland, M. K., Blanton, D. B., Therrell, M. D., and Gay, D. A. (1998). "The Jamestown and Lost Colony Droughts." *Science*, 280: 564–7.

Stoll, M. J., and Brazel, A. J. (1992). "Surface-Air Temperature Relationships in the Urban Environment of Phoenix, Arizona." *Physical Geography*, 13: 160–79.

Suckling, P. W. (1991). "Spatial and Temporal Climatology of Snowstorms in the Deep South." *Physical Geography*, 12: 124–39.

— (1997). "Spatial Coherence of Solar Radiation for Regions in the Central and Eastern United States." *Physical Geography*, 18: 53–62.

Todhunter, P. E. (1990). "Microclimatic Variations Attributable to Urban-Canyon Asymmetry and Orientation." *Physical Geography*, 11: 131–41.

— (2001). "A Hydroclimatological Analysis of the Red River of the North Snowmelt Flood Catastrophe of 1997." *Journal of the American Water Resources Association*, 37: 1263–78.

Travis, D. J., Carleton, A. M., and Changnon, S. A. (1997). "An Empirical Model to Predict Widespread Occurrences of Contrails." *Journal of Applied Meteorology*, 36: 1211–20.

Tsvetsinskaya, E., Mearns, L. O., and Easterling, W. (2001*a*). "Investigating the Effect of Seasonal Plant Growth and Development in 3-Dimensional Atmospheric Simulations. Part I: Simulation of Surface Fluxes over the Growing Season." *Journal of* Climate, 14: 692–709.

— (2001*b*). "Investigating the Effect of Seasonal Plant Growth and Development in 3-Dimensional Atmospheric Simulations. Part II: Atmospheric Response to Crop Growth and Development." *Journal of Climate*, 14: 711–29.

Vega, A. J., Henderson, K. G., and Rohli, R. V. (1995). "Comparison of Monthly and Intermonthly Indices for the Pacific-North American Teleconnection Pattern." *Journal of Climate*, 8: 2097–103.

Vega, A. J., Rohli, R. V., and Henderson, K. G. (1998). "The Gulf of Mexico Mid-Tropospheric Response to El Niño and La Niña Forcing." *Climate Research*, 10: 115–25.

Vörosmarty, C. J., Willmott, C. J., Choudhury, B. J., Schloss, A. J., Stearns, T. K., Robeson, S. M., and Dorman, T. J. (1996). "Analyzing the Discharge Regime of a Large Tropical River through Remote Sensing, Ground-Based Climatic Data, and Modeling." *Water Resources Research*, 32: 3137–50.

Walters, C. K. (2001). "Airflow Configurations of Warm Season Southerly Low-Level Wind Maxima in the Great Plains. Part II: Spatial and Temporal Characteristics and Relationship to Convection." *Weather and Forecasting*, 16: 513–30.

Walters, C. K., and Winkler, J. A. (1999). "Diurnal Variations in the Characteristics of Cloud-to-Ground Lightning Activity in the Great Lakes Region of the United States." *The Professional Geographer*, 51: 349–66.

Walters, C. K., and Winkler, J. A. (2001). "Airflow Configurations of Warm Season Southerly Low-Level Wind Maxima in the Great Plains. Part I: The Synoptic and Subsynoptic-Scale Environment." *Weather and Forecasting*, 16: 531–51.

Washington, W. M., and Meehl, G. A. (1989). "Climate Sensitivity due to Increased CO_2: Experiments with a Coupled Atmosphere and Ocean General Circulation Model." *Climate Dynamics*, 4: 1–38.

Waylen, P. R., Caviedes, C. N., and Quesada, M. E. (1996). "Interannual Variability of Monthly Precipitation in Costa Rica." *Journal of Climate*, 9: 2606–13.

White, D., Richman, M., and Yarnal, B. (1991). "Climate Regionalization and Rotation of Principal Components." *International Journal of Climatology*, 11: 1–25.

Whitney, L. D., and Hobgood, J. S. (1997). "The Relationship between Sea Surface Temperatures and Maximum Intensities of Tropical Cyclones in the Eastern North Pacific Ocean." *Journal of Climate*, 10: 2921–30.

Willmott, C. J., and Legates, D. R. (1991). "Rising Estimates of Terrestrial and Global Precipitation." *Climate Research*, 1: 179–86.

Willmott, C. J., and Matsuura, K. (1995). "Smart Interpolation of Annually Averaged Air Temperature in the United States." *Journal of Applied Meteorology*, 34: 2577–86.

Willmott, C. J., Robeson, S. M., and Feddema, J. J. (1994). "Estimating Continental and Terrestrial Precipitation Averages from Rain-Gauge Networks." *International Journal of Climatology*, 14: 403–14.

Willmott, C. J., Robeson, S. M., and Janis, M. J. (1996). "Comparison of Approaches for Estimating Time-Averaged Precipitation Using Data from the USA." *International Journal of Climatology*, 16: 1103–15.

Winkler, J. A. (1988). "Climatological Characteristics of Summertime Extreme Rainstorms in Minnesota." *Annals of the Association of American Geographers*, 78: 57–73.

— (1992). "Regional Patterns of the Diurnal Properties of Heavy Hourly Precipitation." *The Professional Geographer* 44: 127–46.

Winkler, J. A., Harman, J. R., Waller, E. A., and Brown, J. T. (1996). "Climatological Characteristics of Springtime Lower Tropospheric Airflow over Central and Eastern North America." *International Journal of Climatology*, 16: 739–55.

Wolock, D. M., and McCabe, G. J., Jr. (1999*a*). "Estimates of Runoff Using Water-Balance and Atmospheric General Circulation Models." *Journal of the American Water Resources Association*, 35: 1341–50.

— (1999*b*). "Explaining Spatial Variability in Mean Annual Runoff in the Conterminous United States." *Climate Research*, 11: 149–59.

Woodhouse, C. A., and Meko, D. (1997). "Number of Precipitation Days Reconstructed from Southwestern Tree Rings." *Journal of Climate*, 10: 2663–9.

Woodhouse, C. A., and Overpeck, J. T. (1998). "2000 Years of Drought Variability in the Central United States." *Bulletin of the American Meteorological Society*, 79: 2693–714.

Yarnal, B. (1993). *Synoptic Climatology in Environmental Analysis: A Primer*. London: Belhaven Press.

Yarnal, B., and Leathers, D. J. (1988). "Relationships between Interdecadal and Interannual Climatic Variations and their Effect on Pennsylvania Climate." *Annals of the Association of American Geographers*, 78: 624–41.

Ye, H. (2000). "Decadal Variability of Russian Winter Snow Accumulation and its Associations with Atlantic Sea Surface Temperature Anomalies." *International Journal of Climatology*, 20: 1709–28.

— (2001). "Characteristics of Winter Precipitation Variation over Northern Central Eurasia and their Connections to Sea Surface Temperatures over the Atlantic and Pacific Oceans." *Journal of Climate*, 14: 3140–55.

Ye, H., and Mather, J. R. (1997). "Polar Snow Cover Changes and Global Warming." *International Journal of Climatology*, 17: 155–62.

Ye, H., Cho, H., and Gustafson, P. (1998). "The Changes of Russian Winter Snow Accumulation during 1936–1983 and Its Spatial Pattern." *Journal of Climate*, 11: 856–63.

CHAPTER 4

Cryosphere

Kenneth M. Hinkel, Andrew W. Ellis, and Ellen Mosley-Thompson

Introduction

The cryosphere refers to the Earth's frozen realm. As such, it includes the 10 percent of the terrestrial surface covered by ice sheets and glaciers, an additional 14 percent characterized by permafrost and/or periglacial processes, and those regions affected by ephemeral and permanent snow cover and sea ice. Although glaciers and permafrost are confined to high latitudes or altitudes, areas seasonally affected by snow cover and sea ice occupy a large portion of Earth's surface area and have strong spatiotemporal characteristics.

Considerable scientific attention has focused on the cryosphere in the past decade. Results from 2 × CO_2 General Circulation Models (GCMs) consistently predict enhanced warming at high latitudes, especially over land (Fitzharris 1996). Since a large volume of ground and surface ice is currently within several degrees of its melting temperature, the cryospheric system is particularly vulnerable to the effects of regional warming. The Third Assessment Report of the Intergovernmental Panel on Climate Change (IPCC) states that there is strong evidence of Arctic air temperature warming over land by as much as 5 °C during the past century (Anisimov *et al.* 2001). Further, sea-ice extent and thickness has recently decreased, permafrost has generally warmed, spring snow extent over Eurasia has been reduced, and there has been a general warming trend in the Antarctic (e.g. Serreze *et al.* 2000).

Most climate models project a sustained warming and increase in precipitation in these regions over the twenty-first century. Projected impacts include melting of ice sheets and glaciers with consequent increase in sea level, possible collapse of the Antarctic ice shelves, substantial loss of Arctic Ocean sea ice, and thawing of permafrost terrain. Such rapid responses would likely have a substantial impact on marine and terrestrial biota, with attendant disruption of indigenous human communities and infrastructure. Further, such changes can trigger positive feedback effects that influence global climate. For example, melting of organic-rich permafrost and widespread decomposition of peatlands might enhance CO_2 and CH_4 efflux to the atmosphere.

Cryospheric researchers are therefore involved in monitoring and documenting changes in an effort to separate the natural variability from that induced or enhanced by human activity. This entails, by extension, understanding how cryogenic processes may be affected under a warming scenario; e.g. enhanced coastal thermoerosion, changes in precipitation patterns, surface run-off and glacier mass balance, assessment of avalanche risk, and understanding the increased potential for detachment slides or thermokarst.

Cryosphere specialists in the field of geography generally integrate elements of climatology, geomorphology, and hydrology. Although differentiated by the specific subfield, methods of data collection and analysis, and diverse backgrounds and training, they are united in several respects: (1) a shared interest in near-surface water at and below the freezing point; (2) a reliance on primary data sources derived from field sampling or remotely-sensed imagery; (3) an interest in extrapolating results spatially and/or temporally; and (4) a desire to understand the synergistic dynamics between the Earth's cryosphere and current, past, and future climate.

Snow Cover and Sea Ice

The spatial and temporal interrelationships among Earth's troposphere, snow cover, and sea ice have been the subject of much research among geographers over the past several decades. At winter peaks in spatial extent, snow covers approximately 46 million km² of the Northern Hemisphere landmass, while sea ice covers 14–16 million km² of the Arctic Ocean and 17–20 km² of Antarctica's Southern Ocean. Recent interest in snow cover and sea ice has stemmed from the realization of their significance in climate diagnostics and as potential monitors and instruments of global climate change. While studies of large-scale snow cover and sea ice dominate research activities, traditional small-scale studies of snow as a freshwater source, flood threat, and avalanche danger has also been maintained within geography. Sea-ice cover is a critical component of the climate system as it reduces solar radiation receipt at the Earth's surface by increasing the albedo. Equally important, it reduces the flux of heat, moisture, and momentum between the atmosphere and ocean. There is significant concern regarding the effect that potential future warming may have upon the extent and thickness of polar sea ice (Shapiro-Ledley 1993).

Quality snow-cover and sea-ice data have become increasingly valuable to the geographer. Traditional ground-based observations of snow depth provide for very long data records, although the data often require intensive quality control. The Historical Daily Climate Dataset (HDCD; Robinson 1993) includes rigorously quality-controlled daily snow-depth data for approximately 1,000 stations within the United States. The advancement of remote sensing technology, and its application to the spatial problems facing a geographer, have promoted the use of visible satellite imagery in the monitoring of snow-cover extent on regional-to-global scales. Taken together, ground-based and remotely sensed snow-cover products have been at the center of numerous geographic studies of snow-cover extent variability, reflecting climate variations on various temporal and spatial scales (e.g. Barry *et al.* 1995; Hall *et al.* 2000; Robinson *et al.* 1995; Walsh 1995). During the past decade, the evolution of microwave remote sensing systems has allowed for improved estimation of the physical properties of snow cover, including snow-water equivalence, snow-cover extent, snow depth, and the onset of snow melt (e.g. Hall *et al.* 1996; Sturm *et al.* 1993; Tait *et al.* 1999). When combined with the technology of geographic information systems (GIS), passive microwave data are beginning to be used for climate-change studies within which the amount of available water, the date of peak accumulation, and the associated spatial distribution can be monitored (Goodison and Walker 1993). Frequent cloud cover and long periods of wintertime darkness have made passive microwave instruments essential in monitoring sea ice to produce databases that extend back through the late 1970s. This has yielded numerous studies of interannual variation and trends in sea-ice coverage, thickness, and concentration (e.g. Cavalieri *et al.* 1997; Maslanik *et al.* 1996; Vinnikov *et al.* 1999).

In the area of process-based snow-cover research within geography, the most basic principle continually studied is the radiational effect of snow cover (e.g. Baker *et al.* 1991; Ellis and Leathers 1998). Over the past decade, snow-cover researchers in geography have come fully to appreciate the significant effects of snow cover on large-scale atmospheric temperature patterns. Much climatological research has focused on the synergistic relationship between lower-atmospheric temperature patterns and snow cover on regional to continental scales (e.g. Baker *et al.* 1992; Leathers and Robinson 1993; Leathers *et al.* 1995). As a result, climatologists have become increasingly concerned with the role of snow cover in climate diagnostics and climate change. Research in this area has included study of the influence of snow cover and sea ice on mid-latitude cyclone intensities and trajectories, and the co-variability between snow cover and geopotential height fields and atmospheric teleconnections (e.g. Changnon *et al.* 1993; Clark *et al.* 1999; Robinson and Dewey 1990).

Over the past decade, much of the geographic research associated with the hydrological and mechanical characteristics of snow cover has centered on improving simulation of the operative physical processes. Recent geographic research in the area of snow melt has been dominated by studies designed to improve the quantitative methods behind water-yield forecasting on temporal scales of days to months (e.g. Davis *et al.* 1995; Grundstein and Leathers 1998; Rowe *et al.* 1995). In working toward the goal of improved forecasts, a strong emphasis has been placed on accurately representing and modeling snow-water equivalence (e.g. Marshall *et al.* 1994; Mote and Rowe 1996; Schmidlin *et al.* 1995). Geographic research associated with the mechanics of snow cover has focused on increasing the base of knowledge surrounding the snow avalanche phenomena. Over the past decade, geographers have researched the climatology of avalanches (e.g. Mock and Kay 1992), the characteristics and geological controls of avalanche paths (e.g. Butler and Walsh 1990), the variability of snow and snowpack strength in relation to terrain (e.g. Birkeland *et al.* 1995; Elder 1995), and avalanche forecasting (e.g. McClung and Tweedy 1994).

Clearly, the nature of snow-cover research conducted by geographers across North America is very diverse. Many of the significant relations between snow cover and Earth's climate have been identified in recent years. The association between snow cover and glacier growth and decay is dictated by the mass balance equation, as discussed in the next section.

Glaciers

The term "glacier" refers to perennial alpine glaciers, ice caps, the major ice sheets of Greenland and Antarctica, and the extensive continental glaciers that repeatedly expanded over the northern parts of North America, Scandinavia, and Europe during recurrent glacial stages. Glaciers respond to regional and global changes in both ambient temperature and the balance between the mass received by snow accumulation and that lost by ablation processes. The observed twentieth-century warming, and the anticipated future warming, raise concerns about future mass balance changes to the Antarctic and Greenland ice sheets that collectively contain about 70 meters of equivalent sea-level rise. Due to their size and slow response times, it is difficult to quantify their current mass balances, although Antarctica's balance is thought to be near zero or slightly positive while Greenland's net balance may be slightly negative (IPCC 2001: ch. 11). Current estimates suggest that Greenland is close to balance at elevations above 2,000m, but ice in many coastal areas has thinned in the last decade, particularly in the southeast (Thomas and PARCA Investigators 2001). Smaller ice caps, glaciers, and rock glaciers may serve as critical harbingers of current climatic change because they respond more quickly to environmental changes (Dyurgerov and Meier 1999). Recent observations indicate that most, if not all, ice fields in the tropics and subtropics are currently experiencing rapid retreat (Hastenrath and Greischer 1997; Thompson *et al.* 2000, and references therein).

The Antarctic and Greenland ice sheets, as well as carefully selected tropical and subtropical ice caps, continuously preserve the annual snowfall and its chemical constituents over many millennia. These frozen archives provide critical information about the Earth's past climatic conditions from areas where few paleoclimatic or meteorological records exist. Ice-core paleoclimate histories fill a critical temporal gap between the shorter, high-resolution records available from corals, tree-rings and lake sediments and the longer, lower-temporal resolution histories from deep ocean cores. Ice-core histories also fill spatial gaps by providing climatic information from the polar regions and from high elevation, remote sites in the tropics and the mid-latitudes. Ice cores from the central part of the Greenland ice sheet reveal large and rapid changes in the North Atlantic climate regime during the Late Glacial Stage (Dansgaard *et al.* 1989; Taylor *et al.* 1993). Ice cores from the Guliya ice cap on the Tibetan Plateau (Thompson *et al.* 1997) confirm that these rapid changes occurred well beyond the confines of the North Atlantic. The recognition that the Earth's climate system is capable of large and abrupt changes is relatively new. The more parochial view tends to consider climate change as a gradual process, proceeding slowly due to the large inertia in the climate system. Ice cores from the South American Andes (Thompson *et al.* 1995, 1998) confirm that the Late Glacial Stage cooling in the tropics was concomitant and comparable in magnitude to that in the mid- and higher latitudes. Additionally, these ice-core histories reveal that the Younger Dryas cold event, centered at 12,600 yr BP, lasting roughly one millennium and first recognized in the North Atlantic region, was also characteristic of the South American climate regime. The Younger Dryas ended concurrently over both South America and the North Atlantic sector.

Delmas *et al.* (1992) provides a comprehensive overview of the spectrum of paleoclimatic information available from ice cores. Due to the low concentrations of many atmospheric constituents in the polar atmosphere, polar ice cores provide information unavailable elsewhere. For example, the dustiness of the global atmosphere during the Late Glacial Stage is revealed by comparison of dust histories from Antarctica and Greenland (Mosley-Thompson and Thompson 1994). Similarly, bipolar comparisons of the excess sulfate histories

(Mosley-Thompson *et al.* 1993) reveal volcanic eruptions capable of perturbing the stratosphere and thereby temporarily reducing global surface temperatures. Finally, the gases trapped within the bubbles of polar ice cores reveal the pre-anthropogenic concentrations of greenhouse gases such as carbon dioxide and methane, and conclusively demonstrate human modification of atmosphere chemistry by the burning of fossil fuels (Barnola *et al.* 1991).

Permafrost and Periglacial Geomorphology

Several significant events have occurred in the past decade in the field of geocryology. First, two topic-specific books were published that supplement Washburn's (1980) classic *Geocryology: A Survey of Periglacial Processes and Environments*. In 1989, Peter Williams and Michael Smith of Carleton University published *The Frozen Earth: Fundamentals of Geocryology*. The productive partnership between soil physics (Williams) and geography (Smith) is reflected in their approach, which is soundly based on thermodynamic principles and supported by extensive field and laboratory research. French's (1996) second edition of *The Periglacial Environment* is a survey text covering modern and past (Pleistocene) processes and landforms.

A second significant event took place in summer 1998 when Canada was host to the Seventh International Conference on Permafrost. Sponsored by the International Permafrost Association, the conference proceedings have historically provided an important outlet for geocryological research. North American geographers were well represented, and the plenary talk was given by Chris Burn, professor of geography at Carleton University.

A third event of note was the publication in 1992 of the Proceedings of the 22nd Annual Binghamton Symposium in Geomorphology. Edited by geographers John Dixon (Arkansas) and Athol Abrahams (SUNY-Buffalo), *Periglacial Geomorphology* contains fourteen chapters, of which the senior author of eleven is a geographer. Indeed, fifteen of the twenty-three contributors are geographers and two-thirds of these are from American universities.

A final significant event relates to the establishment of a quarterly journal devoted to geocryology, *Permafrost and Periglacial Processes*. Although it is specialized and has a worldwide subscription of only 200, it constitutes the single most concentrated outlet for geocryological research in North America and Eurasia. An occasional special issue covers such topics as cryosols and, most recently, periglacial cryostratigraphy, paleoenvironments, and processes.

Process-based geomorphological and surficial studies constitute an important research component in the community (Thorn 1992). The purpose is to understand and quantify fundamental physical and chemical processes in the periglacial environment. As such, these include evaluating the impact of aeolian sediment transport (Lewkowicz 1998), coastal and deltaic processes (Walker 1991, 1998), and the potential impact of rising sea levels on coastal zones (Walker 1992). Hydrologic studies address sediment transport (Lewkowicz and Wolfe 1994), discharge (Caine 1996), nivation and snowbank hydrology (Lewkowicz and Harry 1991) and solute transport in alpine streams (Caine and Thurman 1990). Although most of the field sites are from higher latitudes including the Antarctic (Hall 1993), other studies have been conducted in alpine regions of Canada (Harris 1994) and the Tibetan Plateau (Wang and French 1994). Clark and Schmidlin (1992) and Marsh (1998) provide reviews of relic periglacial landforms in the eastern United States.

A major emphasis in process-based geomorphology is associated with mass movement in periglacial regions. This includes creep (Harris *et al.* 1993), solifluction (Smith 1992), active-layer detachment slides and rapid mass movement (Lewkowicz and Hartshorn 1998).

The unique set of landforms characteristic of periglacial environments has been the focus of a number of process-based field studies. Perhaps the best known are the publications of J. R. Mackay addressing the formation and characteristics of pingos in the Tuktoyaktuk region of northwestern Canada. These studies cover a period of nearly forty years, and recent summaries of unique long-term observations and field experiments are now available (Mackay 1998). The growth mechanisms, internal structure, and chemical properties of palsas and related frost mounds have been reviewed by Nelson *et al.* (1992). Pediments (French and Harry 1992) and cryoplanation terraces (Nelson 1998) have been analyzed for formative process and climatic significance, as have thaw lakes (Burn 1992).

A large number of studies have focused on processes in the permafrost and active layer. Mackay (1992) evaluated the frequency and patterns of ground cracking and the development of ice-wedge polygons in tundra (Mackay 1997). Burn (1997) has examined the near-surface cryostratigraphy for paleoenvironmental reconstruction in the western Arctic coastal region of

Canada. The importance of cryostructures in evaluating the regional history and cryoprocesses has been demonstrated by Murton and French (1994) in the same region.

Considerable effort has been devoted to the study of heat-transfer processes in the active layer and upper permafrost (Nelson *et al.* 1993). The primary goal is to identify the factors that determine active layer thaw and permafrost stability, so as to make more realistic predictions given a regional warming scenario. These include site-specific studies to model conductive and nonconductive heat-transfer processes (Hinkel *et al.* 1997; Outcalt *et al.* 1990) and to estimate soil thermal properties (Allard and Fortier 1990). Field studies have quantified the impact of surface disruption by fire (Burn 1998) and forest clearing (Nicholas and Hinkel 1996) on permafrost degradation.

Studies of the seasonal development of the active layer at the scale of regional watersheds (Nelson *et al.* 1997) and estimates of scale-dependent thaw depth variability (Nelson *et al.* 1998) reflect an effort to extrapolate plot results to the regional scale. At a more extensive areal and longer temporal scale, a concerted effort has been made to understand the impact of global climatic change in periglacial environments (Smith and Riseborough 1996; Woo *et al.* 1992) and to model the potential impact of climatic warming on permafrost stability (Anisimov *et al.* 1997).

Toward the New Millenium

The Cryosphere Specialty Group (CSG) was formally organized at the 1997 meeting of the AAG in Fort Worth, Texas. The initiative was largely the result of efforts by H. Jesse Walker to "foster communication between practitioners dealing with the various elements of the cryosphere, to establish linkages with related organizations, and to enhance research on and teaching of cryospheric topics" (Bylaws of the CSG of the AAG 1998). As such, it is a topical and regional specialty group that includes those geographers who might otherwise refer to themselves as climatologists, hydrologists, or geomorphologists.

Researchers focusing on Earth's cryosphere have a promising future within geography. The field is still relatively young and practitioners are favorably positioned to provide needed information on the unique processes that characterize this large and varied portion of the earth. These insights are also essential to improve understanding of the global climate system, its past variations, and potential future changes. Therefore, as a subdiscipline, cryospheric research will continue to be of increasing importance to its parent disciplines of hydrology, climatology, and geomorphology.

For snow-cover researchers in geography, the beginning of the twenty-first century will likely see a continuation of the increasing demand for knowledge of the synergistic relationship between snow cover and Earth's atmosphere on the larger spatial scales. The use of satellite-derived snow-cover products is expected to increase during the next several decades as data records lengthen and additional sensing platforms are launched. Improvements in satellite technology should promote regular monitoring of the physical properties of snow cover, particularly the water content, and representation of snow cover via quantitative modeling. From one-dimensional snow-melt models to three-dimensional GCMs, the recognition of snow cover as a significant global climate component, water source, flood threat, and avalanche danger should be enhanced through geographic research in the early years of the twenty-first century.

In the field of glacier research, the future will likely continue to focus on obtaining and interpreting high-resolution ice core records. The results from the Greenland Ice Sheet Projects (GISP) indicate that significant temperature oscillations occur at decadal to sub-decadal scales, although no forcing mechanism has been unambiguously identified. Future projects will likely include further sampling of high-latitude ice sheets and glaciers, and temperate glaciers in both hemispheres to determine interhemispheric synchronicity. Since high-quality ice-core records serve as repositories of atmospheric gases and particulates, they can be used to model the evolution of atmospheric chemistry and atmosphere–ocean dynamics, and to validate GCM models.

Research in periglacial geomorphology will continue to emphasize field-based projects to quantify mass and energy fluxes. Given that many periglacial processes operate at relatively slow rates or, in some cases, at high rates over very short time-periods, long-term efforts to monitor and model these processes in remote areas are required. In addition to collecting baseline data, much of the effort will concentrate on understanding the impact of climate change and human activity on process rates. Further research will be directed toward extrapolating site-specific results to larger regions and across longer time-frames. In this effort, digital databases including digital elevation models (DEMs), vegetation atlases, and ground ice maps will ultimately prove invaluable to developing and validating coupled terrestrial-atmospheric-hydrologic-vegetation models.

Several unifying trends in cryogenic research can be identified. First, a concerted effort is underway to develop cryospheric databases. The intent is to collect and organize relevant historical data, and to provide a repository for the wealth of digital data currently being collected by automated sensors. This information is then organized and made available to the community through a website. To a large degree, this effort has been spearheaded at the national and international levels by Roger Barry, Director of the World Data Center-A for Glaciology/National Snow and Ice Data Center (NSIDC), the archive for cryospheric data (Clark and Barry 1998).

Second, there is an increased emphasis on numerical modeling of temporal and spatial patterns. These models are often run on a GIS platform and utilize remotely sensed imagery or algorithms to spatially extrapolate site-specific measurements. For this reason, the issues of scaling and scale-dependant variability are likely to have high research priority in the near future. Data collections such as those discussed above will be instrumental in these efforts.

Finally, following the general trend in science, geocryologists are becoming more interdisciplinary in their outlook. This creates an opportunity for practitioners to promote and demonstrate the utility of cryospheric research as it relates to regional and global issues. Although the importance of glaciers as recorders and harbingers of climate change is well known to the international community of earth scientists, the role of permafrost has been notably neglected. This oversight also extends to the GCM modeling efforts, which often lack a realistic permafrost component. The situation can only be rectified by participating in national and international workshops and organizations (e.g. IPCC and the WMO), and by targeting prominent journals for dissemination of important findings. Similarly, cryosphere geographers must advocate and promote the importance and utility of discipline-specific perspectives in addressing the issues as we enter the twenty-first century.

References

Allard, M., and Fortier, R. (1990). "The Thermal Regime of a Permafrost Body at Mont du Lac des Cygenes, Quebec." *Canadian Journal of Earth Sciences*, 27: 694–7.

Anisimov, O. A., Shiklomanov, N. I., and Nelson, F. E. (1997). "Effects of Global Warming on Permafrost and Active-Layer Thickness: Results from Transient General Circulation Models." *Global and Planetary Change*, 15: 61–77.

Anisimov, O., Fitzharris, B., Hagen, J. O., Jeffries, R., Marchant, H., Nelson, F. E., Prowse, T., and Vaughan, D. G. (2001). "Polar Regions (Arctic and Antarctic)," in O. Anisimov and B. Fitzharris (eds.), *Climate Change 2001: Impacts, Adaptation, and Vulnerability, the Contribution of Working Group II to the Third Assessment Report of the Intergovernmental Panel on Climate Change*. Cambridge: Cambridge University Press, 801–41.

Baker, D. G., Ruschy, D. L., and Skaggs, R. H. (1992). "Air Temperature and Radiation Depressions Associated with a Snow Cover." *Journal of Applied Meteorology*, 30: 247–54.

Baker, D. G., Skaggs, R. H., and Ruschy, D. L. (1991). "Snow Depth Required to Mask the Underlying Surface." *Journal of Applied Meteorology*, 30: 387–92.

Barnola, J., Pimenta, M. P., Raynaud, D., and Korotkevich, T. S. (1991). "CO_2–Climate Relationships as Deduced from the Vostok Ice Core: A Re-examination Based on New Measurement and on a Re-evaluation of the Air Dating." *Tellus*, 43B: 83–90.

Barry, R. G., Fallot, J. M., and Armstrong, R. L. (1995). "Twentieth-Century Variability in Snowcover Conditions and Approaches to Detecting and Monitoring Changes: Status and Prospects." *Progress in Physical Geography*, 19: 520–32.

Birkeland, K. W., Hansen, K. J., and Brown, R. L. (1995). "The Spatial Variability of Snow Resistance on Potential Avalanche Slopes." *Journal of Glaciology*, 41: 183–90.

Burn, C. R. (1992). "Thermokarst Lakes." *Canadian Geographer*, 36: 81–5.

—— (1997). "Cryostratigraphy, Paleogeography, and Climate Change during the Early Holocene Warm Interval, Western Arctic Coast, Canada." *Canadian Journal of Earth Sciences*, 34: 912–25.

—— (1998). "The Response (1958–1997) of Permafrost and Near-Surface Ground Temperatures to Forest Fire, Takhini River Valley, Southern Yukon Territory." *Canadian Journal of Earth Sciences*, 35: 184–99.

Butler, D. R., and Walsh, S. J. (1990). "Lithologic, Structural, and Topographic Influences on Snow Avalanche Path Location, Eastern Glacier National Park, Montana." *Annals of the Association of American Geographers*, 80: 362–78.

Caine, N. (1996). "Streamflow Patterns in the Alpine Environment of North Boulder Creek, Colorado Front Range." *Zeitschrift für Geomorphologie*, Suppl., 104: 27–42.

Caine, N., and Thurman, E. M. (1990). "Temporal and Spatial Variations in the Solute Content of an Alpine Stream, Colorado Front Range." *Geomorphology*, 4: 55–72.

Cavalieri, D., Gloersen, J. P., Parkinson, C. L., Comiso, J. C., and Zwally, H. J. (1997). "Observed Hemispheric Asymmetry in Global Sea Ice Changes." *Science*, 278: 1104–6.

Changnon, D., McKee, T. B., and Doesken, N. J. (1993). "Annual Snowpack Patterns across the Rockies: Long-Term Trends and Associated 500-mb Synoptic Patterns." *Monthly Weather Review*, 121: 633–47.

Clark, G. M., and Schmidlin, T. W. (1992). "Alpine Periglacial Landforms of Eastern North America: A Review." *Permafrost and Periglacial Processes*, 3: 225–30.

Clark, M. J., and Barry, R. G. (1998). "Permafrost Data and Information: Advances since the Fifth International Conference on Permafrost," in Lewkowicz and Allard (1998: 181–8).

Clark, M. P., Serreze, M. C., and Robinson, D. A. (1999). "Atmospheric Controls on Eurasian Snow Extent." *International Journal of Climate*, 19: 27–40.

Dansgaard, W., White, J. W. C., and Johnsen, S. J. (1989). "The Abrupt Termination of the Younger Dryas Climate Event." *Nature*, 339: 532–3.

Davis, R. E., McKenzie, J. C., and Jordan, R. (1995). "Distributed Snow Process Modeling: An Image Processing Approach." *Hydrological Processes*, 9: 865–75.

Delmas, R. J. (1992). "Environmental Information from Ice Cores." *Reviews of Geophysics*, 30: 1–21.

Dixon, J. C., and Abrahams, A. D. (eds.) (1992). *Periglacial Geomorphology*. New York: Wiley.

Dyurgerov, M. B., and Meier, M. F. (1999). "Twentieth Century Climate Change: Evidence from Small Glaciers." *Proceedings of the National Academy of Sciences*, 97: 1406–11.

Elder, K. J. (1995). "Snow Distribution in Alpine Watersheds." Ph.D. dissertation, University of California, Santa Barbara.

Ellis, A. W., and Leathers, D. J. (1998). "The Effects of a Discontinuous Snow Cover on Lower Atmospheric Temperature and Energy Flux Patterns." *Geophysical Research Letters*, 25: 2161–4.

Fitzharris, B. B. (1996). "The Cryosphere: Changes and their Impacts," in R. T. Watson, M. C. Zinyowera, and R. H. Moss (eds.), *Climate Change 1995: Impacts, Adaptation, and Migration of Climate Change: Scientific-Technical Analyses. Contribution of Working Group II to the Second Assessment Report of the Intergovernmental Panel on Climate Change*. Cambridge: Cambridge University Press.

French, H. M. (1996). *The Periglacial Environment*. Harlow: Addison Wesley Longman.

French, H. M., and Harry, D. G. (1992). "Pediments and Cold-Climate Conditions, Barn Mountains, Unglaciated Northern Yukon, Canada." *Geografiska Annaler*, 74A: 145–57.

Goodison, B. E., and Walker, A. E. (1993). "Use of Snow Cover Derived from Satellite Passive Microwave Data as an Indicator of Climate Change." *Annals of Glaciology*, 17: 137–42.

Grundstein, A. J., and Leathers, D. J. (1998). "Factors Influencing the Variability of Midwinter Snow Depth Decrease in the Northern Great Plains of the United States." *Physical Geography*, 18: 408–23.

Hall, D. K., Tait, A. B., Foster, J. L., Chang, A. T. C., and Allen, M. (2000). "Intercomparison of Satellite-derived Snow-cover Maps." *Annals of Glaciology*, 31: 369–76.

Hall, D. K., Foster, J. L., Chang, A. T. C., Cavalieri, D. J., Wang J. R., and Benson, C. S. (1996). "Analysis of Melting Snow Cover in Alaska Using Aircraft Microwave Data (April 1995)." *Proc. IGARSS '96*, Lincoln, Nebraska.

Hall, K. (1993). "Enhanced Bedrock Weathering in Association with Late-Lying Snowpatches: Evidence from Livingston Island, Antarctica." *Earth Surface Processes and Landforms*, 19/2: 121–9.

Harris, C., Gallop, M., and Coutard, J.-P. (1993). "Physical Modelling of Gelifluction and Frost Creep: Some Results of a Large-Scale Laboratory Experiment." *Earth Surface Processes and Landforms*, 18: 383–98.

Harris, S. A. (1994). "Climatic Zonality of Periglacial Landforms in Mountain Areas." *Arctic*, 47: 184–92.

Hastenrath, S., and Greischer, L. (1997). "Glacier Recession on Mount Kilimanjaro, East Africa." *Journal of Glaciology*, 43: 455–9.

Hinkel, K. M., Outcalt, S. I., and Taylor, A. E. (1997). "Seasonal Patterns of Coupled Flow in the Active Layer at Three Sites in Northwest North America." *Canadian Journal of Earth Sciences*, 34: 667–78.

Intergovernmental Panel on Climate Change (2001). *Climate Change 2001: The Scientific Basis. Contribution of Working Group I to the Third Assessment Report of the Intergovernmental Panel on Climate Change*. Cambridge: Cambridge University Press.

Leathers, D. J., and Robinson, D. A. (1993). "The Association between Extremes in North American Snow Cover Extent and United States Temperatures." *Journal of Climate*, 6: 1345–55.

Leathers, D. J., Ellis, A. W., and Robinson, D. A. (1995). "Characteristics of Temperature Depressions Associated with Snow Cover across the Northeast United States." *Journal of Applied Meteorology*, 34: 381–90.

Lewkowicz, A. G. (1998). "Aeolian Sediment Transport during Winter, Black Top Creek, Fosheim Peninsula, Ellesmere Island, Canadian Arctic." *Permafrost and Periglacial Processes*, 9: 35–46.

Lewkowicz, A. G., and Allard, M. (eds.) (1998). *Permafrost—Seventh International Conference*. Quebec: Centre d'études nordiques.

Lewkowicz, A. G., and Harry, D. G. (1991). "Internal Structure and Environmental Significance of a Perennial Snowbank, Melville Island, NWT." *Arctic*, 44: 74–82.

Lewkowicz, A. G., and Hartshorn, J. (1998). "Terrestrial Record of Rapid Mass Movements in the Sawtooth Range, Ellesmere Island, Northwest Territories, Canada." *Canadian Journal of Earth Sciences*, 35: 55–64.

Lewkowicz, A. G., and Wolfe, P. M. (1994). "Sediment Transport in Hot Weather Creek, Ellesmere Island, N.W.T., Canada, 1990–1991." *Arctic and Alpine Research*, 26: 213–26.

McClung, D. M., and Tweedy, J. (1994). "Numerical Avalanche Prediction: Kootenay Pass, British Columbia, Canada." *Journal of Glaciology*, 40: 350–8.

Mackay, J. R. (1992). "The Frequency of Ice-Wedge Cracking (1967–1987) at Garry Island, Western Arctic Coast, Canada." *Canadian Journal of Earth Sciences*, 29: 236–48.

—— (1997). "A Full-Scale Field Experiment (1978–1995) on the Growth of Permafrost by Means of Lake Drainage, Western Arctic Coast: A Discussion of the Method and Some Results." *Canadian Journal of Earth Sciences*, 34: 17–33.

—— (1998). "Pingo Growth and Collapse, Tuktoyaktuk Peninsula Area, Western Arctic Coast, Canada: A Long-Term Field Study." *Geographie physique et Quaternaire*, 52: 271–323.

Marsh, B. (1998). "Wind-Traverse Corrugations in Pleistocene Periglacial Landscapes of Central Pennsylvania." *Quaternary Research*, 49: 149–56.

Marshall, S., Roads, J. O., and Glatzmaier, G. (1994). "Snow Hydrology in a General Circulation Model." *Journal of Climate*, 7: 1251–69.

Maslanik, J. A., Serreze, M. C., and Barry, R. G. (1996). "Recent Decreases in Arctic Summer Ice Cover and Linkages to Atmospheric Circulation Anomalies." *Geophysical Research Letters*, 23: 1677–80.

Mock, C. J., and Kay, P. A. (1992). "Avalanche Climatology of the Western United States, with an Emphasis on Alta, Utah." *Professional Geographer*, 44: 307–18.

Mosley-Thompson, E., and Thompson, L. G. (1994). "Dust in Polar Ice Sheets." *Analusis*, 22: 44–6.

Mosley-Thompson, E., Thompson, L. G., Dai, J., Davis, M. E., and Lin, P. N. (1993). "Climate of the Last 500 Years: High Resolution Ice Core Records." *Quaternary Science Reviews*, 12: 419–30.

Mote, T. L., and Rowe, C. M. (1996). "A Comparison of Microwave Radiometric Data and Modeled Snowpack Conditions for Dye2, Greenland." *Meteorology and Atmospheric Physics*, 59: 245–55.

Murton, J. B., and French, H. M. (1994). "Cryostructures in Permafrost, Tuktoyaktuk Coastlands, Western Arctic Canada." *Canadian Journal of Earth Sciences*, 31: 737–47.

Nelson, F. E. (1998). "Cryoplanation Terrace Orientation in Alaska." *Geografiska Annaler*, 80A: 135–52.

Nelson, F. E., Hinkel, K. M., and Outcalt, S. I. (1992). "Palsa-Scale Frost Mounds," in Dixon and Abrahams (1992: 305–25).

Nelson, F. E., Hinkel, K. M., Shiklomanov, N. I., Mueller, G. R., Miller. L. L., and Walker, D. A. (1998). "Active-Layer Thickness in North-Central Alaska: Systematic Sampling, Scale, and Spatial Autocorrelation." *Journal of Geophysical Research—Atmospheres*, 103: 28963–73.

Nelson, F. E., Shiklomanov, N. I., Mueller, G. R., Hinkel, K. M., Walker, D. A., and Bockheim, J. G. (1997). "Estimating Active-Layer Thickness over a Large Region: Kuparuk River Basin, Alaska, U.S.A." *Arctic and Alpine Research*, 29: 367–78.

Nelson, F. E., Lachenbruch, A. H., Woo, M.-k., Koster, E. A., Osterkamp, T. E., Gavrilova, M. K., and Cheng, G. D. (1993). "Permafrost and Changing Climate," in *Proceedings of the Sixth International Conference on Permafrost*. Wushan, Guangzhou: South China University of Technology Press, 987–1005.

Nicholas, J. R. J., and Hinkel, K. M. (1996). "Concurrent Permafrost Aggradation and Degradation Induced by Forest Clearing, Central Alaska, U.S.A." *Arctic and Alpine Research*, 28: 294–9.

Outcalt, S. I., Nelson, F. E., and Hinkel, K. M. (1990). "The Zero-Curtain Effect: Heat and Mass Transfer across an Isothermal Region in Freezing Soil." *Water Resources Research*, 26: 1509–16.

Robinson, D. A. (1993). "Historical Daily Climate Data for the United States," in *Preprints, Eighth Conference on Applied Climatology*. Anaheim, Calif.: American Meteorological Society, 264–9.

Robinson, D. A., and Dewey, K. F. (1990). "Recent Secular Variations in the Extent of Northern Hemisphere Snow Cover." *Geophysical Research Letters*, 17: 1557–60.

Robinson, D. A., Frei, A., and Serreze, M. C. (1995). "Recent Variations and Regional Relationships in Northern Hemisphere Snow Cover." *Annals of Glaciology*, 21: 71–6.

Rowe, C. M., Kuivinen, K. C., and Jordan, R. (1995). "Simulation of Summer Snowmelt on the Greenland Ice Sheet Using a One-Dimensional Model." *Journal of Geophysical Research*, 100: 16265–73.

Schmidlin, T. W., McKay, M., and Cemper, R. P. (1995). "Automated Quality Control Procedure for the 'Water Equivalent on the Ground' Measurement." *Journal of Applied Meteorology*, 29: 1136–41.

Serreze, M. C., Walsh, J. E., Chapin, F. S., III, Osterkamp, T., Dyurgerov, M., Romanovsky, V., Oechel, W. C., Morison, J., Zhang, T., and Barry, R. G. (2000). "Observational Evidence of Recent Change in the Northern High-Latitude Environment." *Climatic Change*, 46: 159–207.

Shapiro-Ledley, T. (1993). "Sea Ice: A Factor in Influencing Climate on Short and Long Time Scales." *NATO ASI-Series*, 1: 533–56.

Smith, D. J. (1992). "Long-Term Rates of Contemporary Solifluction in the Canadian Rocky Mountains," in Dixon and Abrahams (1992: 203–21).

Smith, M. W., and Riseborough, D. W. (1996). "Permafrost Monitoring and Detection of Climate Change." *Permafrost and Periglacial Processes*, 7: 301–9.

Sturm, M., Grenfell, T. C., and Perovich, D. K. (1993). "Passive Microwave Measurements of Tundra and Taiga Snow Covers in Alaska, USA." *Annals of Glaciology*, 17: 125–30.

Tait, A. B., Hall, K. K., Foster, J. L., and Chang, A. T. C. (1999). "High Frequency Passive Microwave Radiometry over a Snow-Covered Surface in Alaska." *Photogrammetric Engineering and Remote Sensing*, 65: 689–95.

Taylor, K. C., Lamorey, G. W., Doyle, G. A., and Alley, R. B. (1993). "The 'Flickering Switch' of Late Pleistocene Climate Change." *Nature*, 361: 432–6.

Thomas, R. H., and PARCA Investigators (2001). "Program for Arctic Regional Climate Assessment (PARCA): Goals, Key Findings, and Future Directions." *Journal of Geophyiscal Research*, 106 (D24): 33691–705.

Thompson, L. G., Mosley-Thompson, E., and Henderson, K. A. (2000). "Ice-Core Paleoclimate Records in Tropical South America Since the Last Glacial Maximum." *Journal of Quaternary Science*, 15/4: 377–94.

Thompson, L. G., Mosley-Thompson, E., Davis, M. E., Lin, P.-N., Henderson, K. A., Cole-Dai, J., Bolzan, J. F., and Liu, K.-b. (1995). "Late Glacial Stage and Holocene Tropical Ice Core Records from Huascarán, Peru." *Science*, 269: 46–50.

Thompson, L. G., Yao, T., Davis, M. E., Henderson, K. A., Mosley-Thompson, E., Lin, P.-N., Beer, J., Synal, H.-A., Cole-Dai, J., and Bolzan, J. F. (1997). "Tropical Climate Instability: The Last Glacial Cycle from a Qinghai-Tibetan Ice Core." *Science*, 276: 1821–5.

Thompson, L. G., Davis, M. E., Mosley-Thompson, E., Sowers, T. A., Henderson, K. A., Zagorodnov, V. S., Lin, P.-N., Mikhalenko, V. N., Campen, R. K., Bolzan, J. F., Cole-Dai, J., and Franco, B. (1998). "A 25,000-Year Tropical Climate History from Bolivian Ice Cores." *Science*, 282: 1858–64.

Thorn, C. E. (1992). "Periglacial Geomorphology: What, Where, When?" in Dixon and Abrahams (1992: 1–30).

Vinnikov, K. Y., Robock, A., Stouffer, R. J., Walsh, J. E., Parkinson, C. L., Cavalieri, D. J., Mitchell, J. F. B., Garrett, D., and Zakharov, V. G. (1999). "Global Warming and Northern Hemisphere Sea Ice Extent." *Science*, 286: 1934–7.

Walker, H. J. (1991). "Bluff Erosion at Barrow and Wainwright, Arctic Alaska." *Zeitschrift für Geomorphologie*, Suppl., 81: 53–61.

Walker, H. J. (1992). "Sea Level Change: Environmental and Socio-economic Impacts." *GeoJournal*, 26: 511–20.

—— (1998). “Arctic Deltas.” *Journal of Coastal Research*, 14: 718–38.

Walsh, J. E. (1995). “Long-Term Observations for Monitoring of the Cryosphere.” *Climatic Change*, 31: 369–94.

Wang, B., and French, H. M. (1994). “Climate Controls and High-Altitude Permafrost, Qinghai-Xizang (Tibet) Plateau, China.” *Permafrost and Periglacial Processes*, 5: 87–100.

Washburn, A. L. (1980). *Geocryology: A Survey of Periglacial Processes and Environments*. New York: Halsted Press.

Williams, P. J., and Smith, M. W. (1989). *The Frozen Earth: Fundamentals of Geocryology*. New York: Cambridge University Press.

Woo, M.-k., Lewkowicz, A. G., and Rouse, W. R. (1992). “Response of the Canadian Permafrost Environment to Climatic Change.” *Physical Geography*, 13: 287–317.

CHAPTER 5

Geomorphology

David R. Butler

Introduction

Geomorphology is the science that studies landforms and landforming processes. Topics of research in geomorphology during the 1990s represent the diversity of the discipline, as practiced by both academics and nonacademic applied geographers in government and private positions. Discussions on the role and importance of scientific theory and social relevance in geomorphology have become increasingly common, although agreement has not been forthcoming. Issues of scale, both spatial and temporal, appear at the forefront of many current papers in the discipline, but little consensus has been reached as to what constitutes the appropriate scale for studies in geomorphology. The use of a broad diversity of research tools also characterizes American geomorphology, including fieldwork, computer and/or laboratory modeling, surface exposure dating, historical archival work, remote sensing, global positioning systems, and geographic information systems. Problems arise, however, when attempting to integrate the results of fine-scale fieldwork with coarser-scaled simulation models.

Key Themes in Geomorphology During the 1990s

The 1990s saw a renewed debate in the role of scientific theory in geomorphology. Prominent in that debate were issues of temporal and spatial scale. Significant discussions, culminating in the 2000 Binghamton Symposium on the integration of computer modeling and fieldwork in geomorphology, were also engendered by perceived clashes between the roles of fieldwork and the "new technology" in geomorphology.

Scientific Theory and Geomorphology

The 1990s have seen continuing interest in defining the role of geomorphology as a science. Most geomorphologists have accepted applied geomorphology (in the sense of Sherman 1989) as a logical extension of the environmental linkages of the science. The social relevancy of geomorphological research can be established without sacrificing the intellectual core of the discipline (Sherman 1994).

However, questions continue as to what actually constitutes "the scientific nature of geomorphology". Rhoads and Thorn (1993) raised the question of the role of theory in geomorphology in an essay that ultimately led to the 1996 Binghamton Symposium on the scientific nature of the discipline (Rhoads and Thorn 1996). Their goals in hosting the symposium were to "initiate a broad examination of contemporary perspectives on the scientific nature of geomorphology. This initial exploration of methodological and philosophical diversity within geomorphology is viewed as a necessary first step in the search for common ground among the diverse group of scientists who consider themselves geomorphologists" (ibid. p. x). Virtually all geomorphologists would agree that indeed they are scientists, and that geomorphology is a science. Whether exploration of the diversity within geomorphology is a "necessary first step," or whether it is even necessary to search for common ground, remains open to debate.

Changing paradigms (in the sense of Sherman 1996) have characterized the history of twentieth-century geomorphology, and the 1990s have been no exception. The 23rd Binghamton Symposium (Phillips and Renwick 1992) examined the topic of geomorphic systems. The traditional concept of equilibrium was broadly questioned at the meeting and subsequent papers, but the issue of the role of equilibrium as a central paradigm in geomorphology was reopened by Thorn and Welford (1994*a*). They advocated a revised version of G. K. Gilbert's dynamic equilibrium, a revision based on sediment transfer, which they termed "mass flux equilibrium." This advocacy met with considerable discussion (Phillips and Gomez 1994) if not outright scorn (Kennedy 1994), but was vigorously defended (Thorn and Welford 1994*b*). The Mississippi River flood of 1993 provided a case study for continuing reassessments of the role of process frequency versus magnitude in determining the amount of geomorphic work accomplished on the landscape (Magilligan *et al.* 1998).

Issues of Scale

Chaos theory (Malanson *et al.* 1990, 1992; Phillips 1992), and the related concept of non-linear dynamical systems (Phillips and Renwick 1992; Phillips 1993, 1999) has provided an integrative framework within which many geomorphologists have examined issues of both temporal and spatial scale. What is ordered and regular at one scale (whether temporal or spatial) may be disordered and irregular, if not downright unpredictable, at another scale. Phillips (1995, 1997*a*) addressed the issues of scale as they apply to biogeomorphology and to humans as geomorphic agents, and Pope *et al.* (1995) advocated the use of multiple scales of investigation in order more clearly to understand spatial variations in weathering. Sherman (1995) struck a similar tone in his discussion of the problems of scale in the modeling and interpretation of coastal dunes. Hudson and Mossa (1997) suggested that the role of scale is involved in increases of the duration of effective discharge on three Gulf Coast rivers. Walsh *et al.* (1998) provided examples of several geomorphic processes that appear to operate similarly regardless of the spatial scale, but also summarized processes that may operate differently at different scales. Chaos theory and non-linear dynamical systems may provide room for those who advocate the continued utility of some form of equilibrium theory, while at the same time providing room for those whose work illustrates that equilibrium simply cannot be applied at all spatio-temporal scales to all geomorphic processes.

One of the strengths of geomorphology has been its ability to transfer the components of theory between different subfields of geomorphology. For example, Bauer and Schmidt (1993) illustrated the application of coastal theory to a fluvial system, in an examination of waves and sandbar erosion in the Grand Canyon. Advances in biogeomorphology have been adopted in fluvial geomorphology, and the converse is also true. Several additional examples illustrating the synergy of intellectual concepts utilized across the subfields of geomorphology are presented in the topical summaries provided in the following section.

Techniques in Geomorphology and the Technological Revolution

Unlike the 1970s and 1980s when many geomorphologists resisted the technological advances of the period, the 1990s have seen remote sensing and Geographic Information Systems (GIS) become common tools in the field. Vitek *et al.* (1996) provided a useful review of the journey from paper maps to GIS and virtual reality, and Walsh *et al.* (1998) demonstrated how issues of scale, pattern, and process in geomorphology can be usefully examined using both remote sensing and GIS. A special issue of *Geomorphology* (Butler and Walsh 1998) was devoted to the application of remote sensing and GIS in the study of geomorphology, and Harden (1992) illustrated the use of a GIS to incorporate and quantify the effects of roads and footpaths on soil erosion and sediment yield in an Andean watershed. The 2000 International Binghamton Geomorphology Symposium focused on the integration of computer modeling and fieldwork. Questions of accuracy of computer modeling of geomorphic processes exist across scales, from sub-meter to grid study units covering hundreds or thousands of square kilometers. Simulation models may provide widely varying estimates of erosion, depending upon the unit of spatial aggregation utilized. Fieldwork would seem to be in no danger of being replaced or deemed unnecessary, but instead clearly complements such modeling efforts.

Whereas GIS and remote sensing provide a macro-view of geomorphological environments, questions of geomorphic interest have also been examined through the application of microscopic and laboratory techniques. Dorn (1995) used digital processing of backscatter electron imagery more precisely to quantify chemical weathering. The utility of radiocarbon dating is well established in Quaternary geomorphology, but recently controversy has arisen as to the efficacy of radiocarbon measurements of rock varnish age (Krinsley *et al.* 1990; Dorn and Phillips 1991; Dorn 1996, 1998*a*; Beck *et al.* 1998; Dalton 1998). The development of dating techniques utilizing cosmogenic nuclides, and radionuclides released into the environment by the testing of thermonuclear weapons, has revolutionized studies of landscape evolution over long time-scales (Harbor 1999).

Geomorphometry continues to occupy a small but useful place among the techniques of geomorphology (Woldenberg 1997). Pike (1995) reviewed the practice and progress in geomorphometry and provided a useful bibliography (1996) dealing with the quantitative representation of topography.

Although the revolution in technological advances has dominated much of geography as well as geomorphology during the 1990s, it is useful to be reminded of the utility of historical data sources and the refinement and extension of familiar field techniques as well. Trimble and Cooke (1991) provided a thorough review of the diversity of historical data sources available to geomorphologists, and Trimble (1998) illustrated how those data sources can be utilized in dating fluvial processes. Balling and Wells (1990) utilized historical rainfall data to examine the question of climate change versus human land-use patterns as the driving force behind arroyo activity in New Mexico. Garcia and Brook (1996) accomplished historical reconstructions of the position of the channel of Georgia's Ocmulgee River using early land-survey maps and aerial photographs. Marcus *et al.* (1992) examined the methods used for estimating Manning's *n* in small mountain streams, and James (1997) reconstructed channel incision on the American River, Califonia, using streamflow gage records. Hupp and Carey (1990) extended principles of dendrogeomorphology to estimate slope retreat in Kentucky.

Advances in Topical Specialties in Geomorphology

Topical advances and significant publications typify every subfield within geomorphology since the publication of Marston's (1989) review. Geomorphologists are active in illustrating the applied aspects of many components of their work, and integrate their work with other subfields of physical geography including climatology, biogeography, hydrology, glaciology, and pedology. Geomorphologists are also active in working with K–12 teachers and students through state geographic alliances, and with various branches of federal, state, and local government agencies. The following sections describe some of the primary research foci with the various subfields of geomorphology during the 1990s. Because of space limitations, each section can provide only a brief overview of the diversity of topics examined.

Fluvial Geomorphology

Research in fluvial geomorphology has advanced during the 1990s on a variety of fronts. Several damaging floods, including the famous 1993 Mississippi River flood (Magilligan *et al.* 1998), provided opportunities for assessment of the geomorphic effects of, and forms resulting from, high-magnitude events (Woltemade 1994; Myers and Swanson 1996). The interaction of high-water events with sediment deposition and patterns of riparian vegetation (Hupp and Simon 1991; Marston *et al.* 1995; Birkeland 1996; Hupp and Osterkamp 1996; Bendix 1997, 1998) illustrates a major linkage between fluvial and biogeomorphology (discussed below), as did several studies examining channel migration and vegetative responses (e.g. Malanson and Butler 1990; Shankman and Drake 1990). The removal of riparian vegetation as a result of expanding urbanization, or conversely the reoccupation of stream channel margins by riparian vegetation, profoundly alters local sediment budgets (Trimble 1990, 1995, 1997*a*, *b*). Forest removal for fuel was shown to exacerbate problems with monsoonal flooding in central Nepal (Marston *et al.* 1996).

Overbank sedimentation is also intimately tied together with overall floodplain evolution. Hupp and Bazemore (1993) described both temporal and spatial patterns of sedimentation in wetlands of western Tennessee. Students of Jim Knox continued to examine historical alluviation, channel incision, spatial variation in stream power along a channel reach, and floodplain evolution in portions of west-central Wisconsin (Lecce

1997*a*, *b*; Faulkner 1998). Hudson and Kesel (2000) examined channel migration and meander-bend morphology on the lower Mississippi River for the period 1877 and 1924, prior to channel cutoffs, revetments, and changes in sediment regime. They found that the heterogeneity of floodplain deposits strongly influenced meander-bend migration, suggesting that rivers with complex floodplain deposits exhibit spatial patterns and relationships that deviate from models that are based on homogeneity of floodplain deposits.

Since the publication of Chin's (1989) excellent review of step-pools in streams, the origin, stability, and processes of step-pool sequences and associated patterns of pools and riffles have been increasingly scrutinized (Abrahams *et al.* 1995; Thompson *et al.* 1996; Robert 1997; Chin 1998, 1999*a*, *b*). Step-pool sequences have been hypothesized to originate under conditions of high flow that induces particle sorting and the formation of antidunes. Flume experiments were not consistent with this theory (Abrahams *et al.* 1995), but morphologic data from hundreds of step-pools in the Santa Monica Mountains of California (Chin 1998) suggest that the antidune model is appropriate (Chin 1999*b*).

Surface roughness, both at bed surfaces and on land surfaces subject to overland flow, has been a major area of research at the smaller scale of hillslopes. Abrahams and colleagues have extensively monitored the effects of bed form and land surface roughness on overland flow and rill hydraulics on an instrumented hillslope in southern Arizona (Abrahams and Parsons 1994*b*; Abrahams and Li 1998; Abrahams *et al.* 1996, 1998; Li and Abrahams 1999), as well as through laboratory flume analyses (Abrahams *et al.* 2000).

Eolian and Coastal Geomorphology

Several major books encapsulating the latest research trends in eolian geomorphology were published in the 1990s (Abrahams and Parsons 1994*a*; Lancaster 1995; Tchakerian 1995). Research topics illustrate that eolian geomorphologists are cognizant of the work of biogeomorphologists (see below), and consider the significant role of vegetation in influencing eolian processes (Lee 1991*a*, *b*). Eolian geomorphology is studied in coastal dune (Bauer and Schmidt 1993), arid (Tchakerian 1991; Williams and Lee 1995), and semi-arid settings (Bach 1998). The Dust Bowl region of the southern Great Plains continues to be the focus of research on blowing dust (Lee *et al.* 1994; Lee and Tchakerian 1995).

Research in coastal geomorphology transcends AAG specialty group boundaries, as exemplified by the frequent co-sponsoring of paper sessions at the annual meeting by the Geomorphology Specialty Group (GSG) in concert with the Coastal and Marine Specialty Group. Coastal geomorphology has seen progression in the understanding of dune formation and sandy beach environments (Sherman and Bauer 1993), deltaic environments (Walker 1998), and sediment dynamics (Mossa 1996). Marcus and Kearney (1991) examined historically rapid sedimentation in tributary estuaries of Chesapeake Bay, and compared the amount of sediment deriving from upland versus coastal sources in a Chesapeake Bay estuary. They found that coastal contributions to estuarine sediment were four to twelve times higher than fluvial inputs, and that coastal erosion is the dominant process associated with sediment inputs along many tributary estuaries during the past several centuries. Sherman and Bauer's (1993) paper provides one possible, and realistic, roadmap for the direction of coastal studies during the next several decades.

Weathering Geomorphology

Weathering is "the breakdown and decay of earth materials *in situ*" (Pope *et al.* 1995). Weathering geomorphology, including karst geomorphology, does not claim a large number of practitioners in American geomorphology, but those involved with the topic are fervent in their devotion and productive in their research (see, for example, the recent body of work on geochemical processes and pedogenesis in Kärkevagge, Sweden, including Dixon *et al.* 1995; Darmody *et al.* 2000*a*, *b*; and Allen *et al.* 2001). One of the primary topics in American geomorphology in the field of weathering is the dating of weathered rock surfaces. An entire issue of *Physical Geography* (12/4, 1991) was devoted to the topic. Numerous studies, described in the techniques section above, have been carried out on dating of rock varnish on exposed surfaces.

Geomorphic aspect can be an important factor in determining spatial variation in rates of weathering and subsequent soil development (Hunckler and Schaetzl 1997). Lithology is clearly important in determining the significance of aspect on weathering rates. Meierding (1993*a*) illustrated that aspect differences did not affect weathering recession rates of marble pillar surfaces, but significantly accelerate rates of weathering on north-facing sandstone cliffs in New Mexico in comparison to southeast-facing cliffs. Rates of weathering have also been affected by air pollution in North America (Meierding 1993*b*), although the effects of air pollution on weathering rates are clearly correlated with climate as well as levels of pollution. Recently, Pope and

Rubenstein (1999) have provided a theoretical framework, as well as a representative case study, for examining human-impacted weathering.

Karst geomorphology as practiced by American geomorphologists has taken on a strong tropical component in the 1990s, in concert with an appreciation for the environmental impact of humans on karst landscapes. Brook and Hanson (1991) utilized sophisticated statistical analyses including double fourier series analysis to examine the cockpit and doline karst in Jamaica. Day (1993*a*, *b*) has illustrated the profound impacts of human activities on diverse karst sites in Central America and the Caribbean. Principal human impacts include forest clearance, conversion of land to agricultural use, urbanization and industrialization, and quarrying and mining. Resulting impacts on the region's karst processes include increased runoff, decreased groundwater discharge, and increased siltation.

Mass Wasting, Periglacial, and Glacial Geomorphology

American geomorphologists involved in the study of mass wasting paid particular attention in the 1990s to the processes and hazards produced by debris flows and debris avalanches (Orme 1990; Kull and Magilligan 1994; Butler and Walsh 1994; Butler and Malanson 1996; Vaughn 1997; Marston *et al.* 1998). The application of remote sensing and GIS technologies for the mapping, and understanding the distribution, of forms of landslides was reviewed by Brunsden (1993), and illustrated by Walsh and Butler (1997). Mass movement in the Himalayas was the topic of a special issue of *Geomorphology* (Shroder 1998).

Periglacial processes and forms of mass wasting continue to fascinate geomorphologists. Beyer (1997) examined the distribution of particle sizes within nonsorted stony earth circles in Colorado, and Wilkerson (1995) described the rates of heave and resulting surface rotation of particles in periglacial frost boils in California's White Mountains. Nicholas and Garcia (1997) examined the origin of fossil rock glaciers in the La Sal Mountains of Utah, and found both mass movement and periglacial creep to be causal agents of movement. Pérez (1990) described how particles move downslope on mountain snowpatches, with implications for the development of protalus ramparts. He also examined the role of stone size on talus slopes as it affects conservation of soil moisture (Pérez 1998). Caine (1992) illustrated longer-term (up to ten years) sediment fluxes across the Martinelli snowpatch on Niwot Ridge, Colorado. He also showed (Caine 1995) that the indirect presence of snow patterns on erosion is much greater than the direct effects caused by processes such as wet-snow avalanches. Butler and Walsh (1990) illustrated that the spatial distribution of snow-avalanche paths in northwestern Montana was influenced by the spatial patterns of lithologic outcrops, faults, and pre-existing topography. In the same area, Butler and Malanson (1990) described how current process rates on avalanche paths could not account for their size, and suggested that rates of incision were vastly greater during Pleistocene deglaciation.

Process and form development was the theme of a special issue of *Geomorphology* devoted to glacial geomorphology (Harbor 1995). In that issue, a variety of topics were examined by American geomorphologists including the development of glacial-valley cross-sections, modeling of ice-cap glaciation, and the effects of Quaternary glacial erosion on river diversion. The effects of late nineteenth- and twentieth-century climatic change on recent glacier fluctuations in the Wind River Range of Wyoming was described by Marston *et al.* (1991). Glacial recession during this period has produced many moraine-dammed lakes in mountain ranges around the world, and several papers have examined the natural hazards associated with those lakes (summarized in Cenderelli 2000).

Quaternary Geomorphology

Quaternary studies continue to occupy a significant niche in American geomorphology. Increasingly, Quaternary studies are multidisciplinary in nature, such that a geographically trained geomorphologist may work with scientists specializing in climate modeling, pedology, techniques of paleoenvironmental reconstruction, paleontology, and/or geoarcheology. The use of cosmogenic isotopes has in many cases revolutionized dating control in studies of long-term landscape evolution.

Paleoenvironmental reconstructions in Quaternary geomorphology utilize a variety of soil properties (Dixon 1991), relict landforms (Orme and Orme 1991; Mossa and Miller 1995), and relative-age dating techniques (Nicholas and Butler 1996; Liebens and Schaetzl 1997) to place landform development into a temporal sequence. Microlaminations within rock varnish deposits can be used to record fluctuations in the Quaternary alkalinity of volcanic rocks (Dorn 1990).

In an extension of his work on the fluvial history of the Driftless Area of Wisconsin, Jim Knox and his students (Leigh and Knox 1994; Mason and Knox 1997) examined the loess and colluvium of the region. Colluvial age indicated accelerated late Wisconsinan hillslope erosion in the area associated with moist climatic conditions. Leigh (1994*b*) specifically examined the Roxana silt of the region, and described its lithology, source, and paleoenvironmental implications. Elsewhere, Arbogast and Johnson (1998) examined the effects of environmental change on the Quaternary landscapes of south-central Kansas. In a paper that combined Quaternary science and biogeomorphology, Johnson and Balek (1991) illustrated the significance of soils microfauna in the genesis of Quaternary stonelines.

Biogeomorphology

Biogeomorphology is a relatively recent arrival on the scene in geomorphology, but a great deal of research has been conducted by American geomorphologists in the field. These contributions can be subdivided into those that examine the interaction of geomorphic processes with plants, and those that examine the interaction of animals with geomorphic processes.

Examples of the interaction between floods and riparian vegetation have been described in the section on fluvial geomorphology. On a hillslope environment, Abrahams *et al.* (1994, 1995) examined the role that vegetation plays in resisting overland flow and rill erosion in the arid environment of southern Arizona. Also in southern Arizona, Parker (1995) described how the complex geomorphic histories of arid alluvial fans produce profound effects on the spatial patterns of vegetation and soils on those fans. Parker and Bendix (1996) provided additional examples of the influences of landscape-scale geomorphic processes on vegetation patterns.

The interrelationships between tree uprooting (treethrow), mass wasting, and pedogenesis were described in a series of review papers by Schaetzl and others (Schaetzl and Follmer 1990; Schaetzl *et al.* 1990). Treethrow results in a pit-and-mound microtopography that has long-term influences on mass wasting (Small 1997). Slope angle was shown to be a significant variable in leading to mass movement by tree uprooting (Norman *et al.* 1995). Schaetzl (1990) illustrated the effects of treethrow microtopography on the characteristics and genesis of spodosols in Michigan.

Prior to the 1990s, only a few examples of research existed on the interaction of geomorphic processes and animals. During this decade, the role of cattle as geomorphic agents has been scrutinized in a number of studies. Trimble (1994) described specific examples of the effects of cattle on streambeds in Tennessee, and Trimble and Mendel (1995) provided a thorough review of the cow as a geomorphic agent. Magilligan and McDowell (1997) examined what occurs when the geomorphic influences of cattle are removed from a stream channel through the elimination of cattle grazing, and McDowell and Magilligan (1997) provided a valuable overview of the response of stream channels to the removal of cattle grazing disturbance.

The geomorphic role of naturally occurring wild animal populations was shown to be widespread and significant by Butler (1995). Animals examined ranged from insects and other invertebrates, through all forms of vertebrates and with special emphasis on mammals. Specific species examined for their geomorphic roles included grizzly bears (*Ursus arctos horribilis*) (Butler 1992; Baer and Butler 2000), kangaroo rats (Neave and Abrahams 2001) and North American beavers (*Castor canadensis*). Marston (1994) illustrated how the removal of beaver in small mountain valleys of the western United States had led to river entrenchment. Butler and Malanson (1995) illustrated the significant role that beaver dams and their attendant ponds have on sediment retention in mountain valleys of Montana. Hillman (1998) described the influence of beaver-dam failure on flood-wave attenuation along a second-order boreal stream, and Meentemeyer *et al.* (1998) described landforms produced by the burrowing actions of beavers.

Even less research has been done on the effects of geomorphic processes and habitat alteration on wild animal populations. Morris (1992) illustrated the effects of stream-flow diversion on the habitats and spawning habits of fish in Washington state, but more work needs to be done to understand how human alteration of surface geomorphic processes can influence the health and distribution of wild animal populations.

Environmental Geomorphology

One of the most pronounced trends in recent years has been the return of geomorphology to a strong relationship with environmental science and management. A number of research topics characterize this trend, including the general effects of humans as geomorphic agents, the impacts of human land-use changes on surface run-off and erosion, and the examination of anthropogenically introduced trace metals in streams.

The general impact of humans as agents of geomorphic change has been examined by Goudie (1993) and Phillips (1991, 1997*a*). Environmental consequences of footpath and road construction include soil trampling and accelerated surface run-off and attendant erosion (Harden 1992; Vogler and Butler 1996; Wallin and Harden 1996). Agricultural development and shifting patterns of cultivation and land abandonment produce temporally and spatially variable rates of sediment erosion and deposition that can profoundly alter landscapes (Harden 1993, 1996; Beach 1994; Magilligan and Stamp 1997; Phillips 1997*b*). Recently, Trimble and Crosson (2000) questioned the uncritical use of models of soil erosion in the United States as the basis for science or for national policy, and called for a comprehensive national system of monitoring soil erosion and consequent downstream sediment movement.

The mining of floodplain sediments disrupts sediment dynamics of fluvial systems and brings attendant environmental impacts (Walker 1994), including channel planform and land-cover change (Mossa and McLean 1997) and channel incision (James 1991). Mining adjacent to floodplains leads to the introduction of trace metals and radionuclides into stream systems. Such elements pollute stream systems, but provide useful markers for examining the rates and spatial patterns of dispersion and diffusion. Lecce and Pavlowsky (1997) have examined the storage of mining-related zinc in floodplain sediments of Wisconsin's Blue River, and Leigh (1994*a*) described mercury contamination and floodplain sedimentation associated with former gold mines in the Appalachian Mountains of north Georgia. Hupp *et al.* (1993) examined the trapping of trace elements in wetlands of the Chickahominy River in Virginia, and Andrew Marcus and associates have examined the distributions of copper trace metal concentrations in streambed sediments in Alaska (Marcus 1996; Marcus *et al.* 1996). One of the most significant pollution events in a fluvial system was the introduction of thorium-230, via the failure of a holding pond, into the Rio Puerco of New Mexico (Graf 1990). The broader conceptual issues and methodologies for examining nuclear contamination in that fluvial system were also subsequently summarized by Graf (1994).

Geoarcheology

Geomorphologists have been at the forefront of recent advances in geoarcheology (Butzer 1997), frequently as part of a multidisciplinary team of palynologists and paleoecologists, archeologists, historians, and pedologists (cf. Beach and Dunning 1997; Dahlin *et al.* 1998). Geomorphologists attempt to reconstruct the interplay of edaphic variation, climate and climate changes, and cumulative or changing land-use practices in understanding past erosion and sedimentation as well as in understanding contemporary vegetation (Butzer and Butzer 1997). The influence of Karl Butzer in geoarcheology, through his own research and through his influence on a generation of students, was honored in a special issue of *Geoarchaeology* (12/4, 1997).

Planetary Geomorphology

One of the most welcome pieces of news associated with the technological advances of the 1990s is the reissuing of the out-of-print, color-rich *Geomorphology from Space—A Global Overview of Regional Landforms* (Short and Blair 1986) as a CD-ROM by NASA's Goddard Space Flight Center. The Goddard Center has also placed the entire volume on a convenient web-site: (<http://daac.gsfc.nasa.gov/DAAC_DOCS/geomorphology/GEO_HOME_PAGE.html>, last accessed 12 November 2001), where every image is downloadable for research and teaching purposes alike. This volume serves as an excellent resource for the study of the planetary geomorphology of planet Earth. The publication of the US Geological Survey's digital shaded-relief map of the United States (Thelin and Pike 1991) serves as an additional outstanding resource for teachers of Earth geomorphology.

Exploration of the planetary and lunar bodies of the solar system provided rich data sources for geomorphologists during the 1990s. Perhaps most notable were the images from NASA's Mars Pathfinder mission that reached Earth in 1997. Thousands of digital images of the surface of Mars allowed planetary geomorphologists to compare and contrast surface processes on Mars and Earth. Frankel's (1996) book on volcanoes of the solar system extended this trend to several other bodies as well, including Earth's moon, Venus, Mars, and the moons of Jupiter, Saturn, Uranus, and Neptune. The study of long-runout landslides on extraterrestrial bodies also continues (Dade and Huppert 1998), with emphasis on identification of the mechanisms that contribute to the mobility of rockfalls on Earth as well as on other planetary bodies. As Baker (1993) pointed out in his review of extraterrestrial geomorphology, recent discoveries illustrate the role of extraterrestrial studies for understanding the science not only of Earthlike planets but particularly of Earth itself.

Geomorphology and "The Outside World"

As the discussion of topical specialties has made clear, much of modern research in geomorphology has applications to the "real" or "outside" world that exists beyond the hallowed halls of academia (although it is worth noting that a significant part of that research previously described has been conducted by geomorphologists other than academics). Many geomorphologists extend the applications of their research into the real world through active consulting roles. However, it is not only in the realm of applied geomorphology where geomorphologists, academic and non-academic, can influence the real world. Several senior-level academic geomorphologists have moved beyond the professorial ranks and entered the realm of upper administration at major universities. In those positions, such as vice-president and provost, or associate vice-president for academic affairs, or director of a major research center, geomorphologists influence the progress of university-level education and help shape the direction of future geomorphological research.

The role of geomorphologists in helping to shape the future of American schools is not restricted to the university level. The necessity to know and understand geography, including geomorphology, in our increasingly networked world has been supported by the National Academy of Sciences and the National Research Council through the Rediscovering Geography Committee (1997). Several physical geographers, including 1998–9 AAG President Will Graf, comprised part of that committee and spoke eloquently for the need for more geography at all grade levels in the public schools. Geomorphologists also work directly with K–12 teachers and students through the network of state geographic alliances. Several geomorphologists were team members of Mission Geography, an ambitious partnership between NASA and the major national geography organizations to produce curriculum supplements for grades K–4, 5–8, and 9–12 (Bednarz and Butler 1999). These curriculum supplements utilize NASA imagery and products to illustrate both visual examples and conceptual issues in geography, with numerous cases directly illustrating modern issues in geomorphology.

Meetings and Organizations

The International Association of Geomorphologists (IAG) was created in the mid-1980s in response to a perceived absence of international leadership and coordination of research in geomorphology. The IAG has successfully established a quadrennial meeting calendar, with meetings having been held in Frankfurt, Germany, in 1989; in Hamilton, Ontario, in 1993; in Bologna, Italy, in 1997; and in Tokyo, Japan, in 2001. Regional conferences in locations such as Turkey are now held in the mid-year between two successive IAG meetings. The IAG has established a publication series, issuing volumes of importance associated in some cases with the international conferences. Papers from the Second IAG meeting in Frankfurt were published in the early 1990s in a series of Supplements of *Zeitschrift für Geomorphologie*. Papers from the Third IAG in Hamilton were published by Wiley as a series of IAG-sponsored books (e.g. Hickin 1995; Slaymaker 1995, 1996) or as special issues of major journals such as *Physical Geography* (Abrahams and Marston 1993). The nation-by-nation review of national efforts in geomorphology (Walker and Grabau 1993) was a milestone in examining the international status of the science of geomorphology. At the time this review is being written, papers from the Fourth IAG meeting have not been published in any consistent format, although the papers from the Binghamton Geomorphology Symposium on Engineering Geomorphology, held in concert with the Fourth IAG meeting, have been published in a special issue of *Geomorphology* (Giardino *et al.* 1999). The IAG issues several newsletters each year, published in English in *Earth Surface Processes and Landforms*, *Geomorphology*, and *Zeitschrift für Geomorphologie*. The newsletters are also published electronically on the Internet's GEOMORPHLIST.

GEOMORPHLIST, begun modestly in the mid-1980s as the Internet voice of the GSG, has grown to become an electronic organization that encompasses more than 500 subscribers from countries around the world. It serves as a major forum for information gathering for academics pursuing research questions, for dissemination of job openings in geomorphology, and as a news outlet. The Geomorphology Specialty Group has for several years been publishing its semi-annual newsletter on GEOMORPHLIST, and recently the newsletter has become strictly electronic.

Within the United States and Canada, the Binghamton Symposia in Geomorphology has continued to be a major annual event, with meetings held on a wide diversity of topics in geomorphology (Table 5.1). Begun under the direction of Drs Marie Morisawa and Donald Coates of SUNY-Binghamton, the Binghamton Symposia have always been considered a premier event in North American geomorphology. Publication of the papers of the annual meetings, whether as a special issue of *Geomorphology* subsequently also published as a book volume by Elsevier, or as a stand-alone edited book (e.g. Dixon and Abrahams 1992; Rhoads and Thorn 1996), has always been eagerly awaited.

The annual meetings of the Association of American Geographers (AAG) continue to be primary outlets for the presentation of research findings of American geomorphologists trained in geography, especially through extremely successful Special Sessions sponsored or co-sponsored by the GSG. Special sessions in fluvial, coastal, and Quaternary geomorphology have been particularly

Table 5.1 *The Binghamton Symposia in Geomorphology, 1989–2001*

Conference	Year	Topic	Organizer(s)/Editor(s)
20	1989	Appalachian geomorphology	T. W. Gardner and W. D. Sevon
21	1990	Soils and landscape evolution	P. L. K. Knuepfer and L. D. McFadden
22	1991	Periglacial geomorphology	J. C. Dixon and A. D. Abrahams
23	1992	Geomorphic systems	J. D. Phillips and W. H. Renwick
24	1993	Geomorphology: the research frontier and beyond	J. D. Vitek and J. R. Giardino
25	1994	Geomorphology and natural hazards	M. Morisawa
26	1995	Biogeomorphology, terrestrial and freshwater systems	C. R. Hupp, W. R. Osterkamp, and A. D. Howard
27	1996	The scientific nature of geomorphology	B. L. Rhoads and C. E. Thorn
28	1997	Engineering geomorphology	J. R. Giardino and R. A. Marston
29	1998	Coastal geomorphology	P. E. Gares and D. Sherman
30	1999	Geomorphology in the public eye	P. Knuepfer and J. F. Petersen
31	2000	Integration of computer modeling and field observations in geomorphology	J. F. Shroder Jr. and M. Bishop
32	2001	Mountain geomorphology	D. R. Butler, G. P. Malanson, and S. J. Walsh

notable on the annual meeting programs during the 1990s, as have been the recent series of special sessions organized in the field of weathering geomorphology. Continued presences of GSG members at annual meetings of the Geological Society of America and the American Geophysical Union, and the biannual meetings of the American Quaternary Association, attest, however, to the multidisciplinary feelings of many American geographic geomorphologists.

The Geomorphology Specialty Group has continued its tradition of awarding the G. K. Gilbert Award for Excellence in Geomorphological Research at its Business Meeting during the AAG Annual Meetings, although the award has not been given every year (Table 5.2). The variety of topics studied by award recipients speaks to the richness and breadth of geomorphological research during the past decade. The specific titles of the publications for which the Gilbert Award was bestowed are accessible on the GSG website: www.cla.sc.edu/GEOG/gsgdocs/Awards/AwardsHistory.html, last accessed 12 November 2001.

The Distinguished Career Award of the GSG has been bestowed upon several geomorphologists whose outstanding career contributions have earned the lasting respect of their colleagues. The first awardee was Jesse Walker in 1989, and subsequent recipients have included Ross Mackay (1990), Neil Salisbury (1992), M. Gordon "Reds" Wolman (1993), Theodore M. Oberlander (1994), Harold "Duke" Winters (1995), Derek Ford (1996), and Nicholas Lancaster in 1997. At the AAG Annual Meeting in Fort Worth in 1997, the GSG unanimously voted to name the Distinguished Career Award in honor of a true giant of geomorphology, both literally and figuratively, the late Mel Marcus. The first Mel Marcus Distinguished Career Award was awarded in 1999 to Dick Reeves. Jack Ives received the award for 2000, and also received the career achievement award from the Mountain Geography Specialty Group. The 2001 recipient was Jim Knox.

Publication Outlets for Geomorphologists

Marston's (1989) review illustrated the broad diversity of journal outlets in which American geomorphologists publish, and those trends in general continued through

Table 5.2 *Recipients of the G. K. Gilbert Award, presented by the AAG Geomorphology Specialty Group*

Year	Recipient	Contribution to geomorphology
1989	No Award Given	
1990	Don Johnson and Donna Watson-Stegner	Evolution model of pedogenesis
1991	Alan Howard	Optimal Drainage Networks
1992	Don Currey	Quaternary palaeolakes in the evolution of semidesert basins, with special emphasis on Lake Bonneville and the Great Basin, USA
1993	William C. Mahaney	*Ice on the Equator*
1994	T. Nelson Caine	Sediment transfer on the floor of the Martinelli Snowpatch, Colorado Front Range
1995	No award given	
1996	James C. Knox	Large increases in flood magnitudes in response to modest changes in climate
1997	Jonathan D. Phillips	Deterministic uncertainty in landscapes
1998	David R. Butler	*Zoogeomorphology: Animals as Geomorphic Agents*
1999	T. R. Paton, G. S. Humphreys, and P. B. Mitchell	*Soils: A New Global View*
2000	Ellen Wohl, Doug Thompson, and Andy Miller	Canyons with undulating walls
2001	Karl F. Nordstrom	*Beaches and Dunes of Developed Coasts*

the 1990s. Only a few new journals (e.g. *Permafrost and Periglacial Processes*, and the Association of Polish Geomorphologists' *Landform Analysis*) have appeared during the 1990s as new potential outlets for geomorphic research results.

Book chapters and conference proceedings continue to be a primary publication outlet for geomorphologists as well, but particularly notable during the 1990s was the large number of influential research monographs and books published. These books cover the wide spectrum of topics that represents modern geomorphology. Major books were published on fluvial geomorphology and riparian landscapes (Malanson 1993; Leopold 1994), eolian and desert geomorphology (Abrahams and Parsons 1994*a*; Lancaster 1995; Tchakerian 1995), animals as geomorphic agents (Butler 1995), geomorphic responses to climate change (Bull 1991), anthropogenic influences in geomorphology (Costa *et al.* 1995; Graf 1994), rock coatings (Dorn 1998*b*), and earth surface systems and non-linear dynamical systems (Phillips 1999). Special issues of major journals, which essentially produce book-length compendiums, have examined specific topics including drainage basin sediment budgets (Abrahams and Marston 1993), glacial processes and form development (Harbor 1995), eolian environments (Hesp 1996; Lancaster 1996); recent developments in geoscience education and Quaternary geomorphology (Tormey 1996), the application of remote sensing and geographic information systems in geomorphology (Butler and Walsh 1998), and mass movement in the Himalayas (Shroder 1998). A memorial issue of *Mountain Research and Development*, dedicated to the late Barry Bishop of the National Geographic Society, also contained numerous pieces on mountain geomorphology (Marcus and Marcus 1996).

Conclusions

As the twenty-first century begins in earnest, American geomorphologists should be proud of their accomplishments. Geomorphology is a vibrant and significant component of American geography. The annual business meetings of the Geomorphology Specialty Group are eagerly anticipated by its members, not just for the sake of accomplishing business tasks and elections, but because American geomorphology is exciting and alive, and American geomorphologists enjoy getting together and sharing stories of their latest accomplishments.

Nevertheless, questions remain as to where geomorphology is headed in the twenty-first century. Smith (1993: 251) expressed concern that fluvial geomorphology is "dismally organized, without focus or direction, and is practiced by individualists who rarely collaborate in numbers significant enough to generate major research initiatives." Is such a statement, for fluvial geomorphology in particular but also for geomorphology in general, an accurate assessment of the state of geomorphology at the beginning of the new millennium? I believe that the answer is a resounding "No." The research described throughout this chapter provides numerous counter-points to that view. Collaborative efforts exist across the board, whether from among a group of geomorphologists, such as the group who examined the geomorphic effects of the 1993 Mississippi River flood (Magilligan *et al.* 1998); a group of physical geographers working together on issues of scale and technological applications (Walsh *et al.* 1998); or in a multidisciplinary group (Dahlin *et al.* 1998). I have shown how geomorphologists have become active in integrating our discipline into education initiatives across grade levels, so that future generations will, it is hoped, have a greater appreciation of geography as a discipline, and of geomorphology in particular. Geomorphologists are active in environmental issues, and in such roles are highly visible to the general public.

Questions remain, of course, as to what will be the hot topics of geomorphological research in years to come. Will issues of scale continue to be relevant as better technology allows for the creation of more accurate simulation models? Which branches of geomorphology will prosper, and which may wither and die? I make no claim to be clairvoyant, and quite honestly, your guesses would be as good as mine. I do believe that room for improvement exists in applying the research concepts of geomorphology to the real world. As pointed out by Giardino and Marston (1999) in their assessment of the future of engineering geomorphology, becoming involved in policy formulation needs to be a major new arena in the new millennium. Trimble and Crosson (2000) echo the importance of using the principles of our discipline to reveal where glaring policy changes are needed. Geomorphology continues to provide the answers to stimulating scientific questions. Those answers, whether developed individually (and I believe, in contrast to Smith (1993), that individual advancements of the discipline are quite possible) or collectively, now must be put forward into the public and policy arenas, for the development of scientifically wise, data-grounded policy. Wherever geomorphology goes topically, it will enjoy its greatest success when it is clear that the discipline is vibrant, exciting, and involved in the betterment of humankind.

Acknowledgements

The geomorphology community owes a sincere debt of gratitude to Dr Jeff Lee of Texas Tech University in the United States for his outstanding job as moderator of GEOMORPH-LIST from its inception in the 1980s until late 1998, when he passed the baton on to Dr William Locke of Montana State University. I thank the many geomorphologists who sent reprints or suggestions for papers and issues to include in this chapter. I also thank Cort Willmott, Karl Butzer, and two anonymous referees for their valuable comments on a previous draft of this chapter. Responsibility for its contents, however, remains mine.

References

Abrahams, A. D., and Li, G. (1998). "Effect of Saltating Sediment on Flow Resistance and Bed Roughness in Overland Flow." *Earth Surface Processes and Landforms*, 23: 953–60.

Abrahams, A. D., and Marston, R. A. (eds.) (1993). "Drainage Basin Sediment Budgets—A Collection of Papers Presented at the Third International Geomorphology Conference, Hamilton, Ontario, August 1993." *Physical Geography*, 14/3: 221–320.

Abrahams, A. D., and Parsons, A. (eds.) (1994*a*). *Geomorphology of Desert Environments*. London: Chapman & Hall.

____ (1994*b*). "Hydraulics of Interrill Overland Flow on Stone-Covered Desert Surfaces." *Catena*, 23: 111–40.

Abrahams, A. D., Gao, P., and Aebly, F. A. (2000). "Relation of Sediment Transport Capacity to Stone Cover and Size in Rain-Impacted Interrill Flow." *Earth Surface Processes and Landforms*, 25: 497–504.

Abrahams, A. D., Li, G., and Atkinson, J. F. (1995). "Step-Pool Streams: Adjustment to Maximum Flow Resistance." *Water Resources Research*, 31/10: 2593–602.

Abrahams, A. D., Li, G., and Parson, A. J. (1996). "Rill Hydraulics on a Semiarid Hillslope, Southern Arizona." *Earth Surface Processes and Landforms*, 21/1: 35–47.

Abrahams, A. D., Parsons, A. J., and Wainwright, J. (1994). "Resistance to Overland Flow on Semiarid Grassland and

Shrubland Hillslopes, Walnut Gulch, Southern Arizona." *Journal of Hydrology*, 156: 431–46.

—— (1995). "Effects of Vegetation Change on Interrill Runoff and Erosion, Walnut Gulch, Southern Arizona." *Geomorphology*, 13/1–4: 37–48.

Abrahams, A. D., Li, G., Krishnan, C., and Atkinson, J. F. (1998). "Predicting Sediment Transport by Interrill Overland Flow on Rough Surfaces." *Earth Surface Processes and Landforms*, 23: 1087–99.

Allen, C. E., Darmody, R. G., Thorn, C. E., Dixon, J. C., and Schlyter, P. (2001). "Clay Mineralogy, Chemical Weathering and Landscape Evolution in Arctic-Alpine Sweden." *Geoderma*, 99: 277–94.

Arbogast, A. F., and Johnson, W. C. (1998). "Late-Quaternary Landscape Response to Environmental Change in South-Central Kansas." *Annals of the Association of American Geographers*, 88/1: 126–45.

Bach, A. J. (1998). "Assessing Conditions Leading to Severe Wind Erosion in the Antelope Valley, California, 1990–1991." *The Professional Geographer*, 50/1: 87–97.

Baer, L. D., and Butler, D. R. (2000). "Space-Time Modeling of Grizzly Bears." *The Geographical Review*, 90/2: 206–21.

Baker, V. R. (1993). "Extraterrestrial Geomorphology: Science and Philosophy of Earthlike Planetary Landscapes." *Geomorphology*, 7/1–3: 9–35.

Balling, R. C., and Wells, S. G. (1990). "Historical Rainfall Patterns and Arroyo Activity within the Zuni River Drainage Basin, New Mexico." *Annals of the Association of American Geographers*, 80/4: 603–17.

Bauer, B. O., and Schmidt, J. C. (1993). "Waves and Sandbar Erosion in the Grand Canyon: Applying Coastal Theory to a Fluvial System." *Annals of the Association of American Geographers*, 83/3: 475–97.

Beach, T. (1994). "The Fate of Eroded Soil: Sediment Sinks and Sediment Budgets of Agrarian Landscapes in Southern Minnesota, 1851–1988." *Annals of the Association of American Geographers*, 84/1: 5–28.

Beach, T., and Dunning, N. (1997). "An Ancient Maya Reservoir and Dam at Tamarindito, El Peten, Guatemala." *Latin American Antiquity*, 8/1: 20–9.

Beck, W., Donahue, D. J., Jull, A. J. T., Burr, G., Broecker, W. S., Bonani, G., Hajdas, I., and Malotki, E. (1998). "Ambiguities in Direct Dating of Rock Surfaces Using Radiocarbon Measurements." *Science*, 280: 2132–5.

Bednarz, S. W., and Butler, D. R. (1999). " 'Mission Geography' and the Use of Satellite Imagery in K–12 Geographic Education—a NASA-GENIP Partnership." *Geocarto International*, 14/4: 85–90.

Bendix, J. (1997). "Flood Disturbance and the Distribution of Riparian Species Diversity." *The Geographical Review*, 87/4: 468–83.

—— (1998). "Impact of a Flood on Southern California Riparian Vegetation." *Physical Geography*, 19/2: 162–74.

Beyer, P. J. (1997). "Particle Size Distribution within Nonsorted Stony Earth Circles, Colorado." *Physical Geography*, 18/2: 176–94.

Birkeland, G. H. (1996). "Riparian Vegetation and Sandbar Morphology along the Lower Little Colorado River, Arizona." *Physical Geography*, 17/6: 534–53.

Brook, G. A., and Hanson, M. (1991). "Double Fourier Series Analysis of Cockpit and Doline Karst near Browns Town, Jamaica." *Physical Geography*, 12/1: 37–54.

Brunsden, D. (1993). "Mass Movement; the Research Frontier and Beyond: A Geomorphological Approach." *Geomorphology*, 7/1–3: 85–128.

Bull, W. B. (1991). *Geomorphic Responses to Climatic Change*. New York: Oxford University Press.

Butler, D. R. (1992). "The Grizzly Bear as an Erosional Agent in Mountainous Terrain." *Zeitschrift für Geomorphologie*, 36/2: 179–89.

—— (1995). *Zoogeomorphology: Animals as Geomorphic Agents*. Cambridge: Cambridge University Press.

Butler, D. R., and Malanson, G. P. (1990). "Non-equilibrium Geomorphic Processes and Patterns on Avalanche Paths in the Northern Rocky Mountains, U.S.A." *Zeitschrift für Geomorphologie*, 34/3: 257–70.

—— (1995). "Sedimentation Rates and Patterns in Beaver Ponds in a Mountain Environment." *Geomorphology*, 13/1–4: 255–69.

—— (1996). "A Major Sediment Pulse in a Subalpine River Caused by Debris Flows in Montana, USA." *Zeitschrift für Geomorphologie*, 40/4: 525–35.

Butler, D. R., and Walsh, S. J. (1990). "Lithologic, Structural, and Topographic Influences on Snow-Avalanche Path Location, Eastern Glacier National Park, Montana." *Annals of the Association of American Geographers*, 80/3: 362–78.

—— (1994). "Site Characteristics of Debris Flows and Their Relationship to Alpine Treeline." *Physical Geography*, 15/2: 181–99.

—— (eds.) (1998). "Special Issue—The Application of Remote Sensing and Geographic Information Systems in the Study of Geomorphology." *Geomorphology*, 21/3–4: 179–349.

Butzer, K. W. (1997). "Late Quaternary Problems of the Egyptian Nile: Stratigraphy, Environments, Prehistory." *Paléorient*, 23/2: 151–73.

Butzer, K. W., and Butzer, E. K. (1997). "The 'Natural' Vegetation of the Mexican Bajío: Archival Documentation of a 16th-Century Savanna Environment." *Quaternary International*, 43/44: 161–72.

Caine, N. (1992). "Sediment Transfer on the Floor of the Martinelli Snowpatch, Colorado Front Range, U.S.A." *Geografiska Annaler*, 74A/2–3: 133–44.

—— (1995). "Snowpack Influences on Geomorphic Processes in Green Lakes Valley, Colorado Front Range." *The Geographical Journal*, 161/1: 55–68.

Cenderelli, D. A. (2000). "Floods from Natural and Artificial Dam Failures," in E. E. Wohl (ed.), *Inland Flood Hazards*. Cambridge: Cambridge University Press, 73–103.

Chin, A. (1989). "Step-Pools in Stream Channels." *Progress in Physical Geography*, 13: 391–408.

—— (1998). "On the Stability of Step-Pool Mountain Streams." *Journal of Geology*, 106/1: 59–69.

—— (1999*a*). "The Morphologic Structure of Step-Pools in Mountain Streams." *Geomorphology*, 27: 191–204.

—— (1999*b*). "On the Origin of Step-Pool Sequences in Mountain Streams." *Geophysical Research Letters*, 26/2: 231–8.

Costa, J. E., Miller, A. J., Potter, K. W., and Wilcock, P. R. (eds.) (1995). *Natural and Anthropogenic Influences in Fluvial Geomorphology*. Washington, DC: American Geophysical Union.

Dade, W. B., and Huppert, H. E. (1998). "Long-Runout Rockfalls." *Geology*, 26/9: 803–6.

Dahlin, B. H., Andrews, A. P., Beach, T., Bezanilla, C., Farrell, P., Luzzadder-Beach, S., and McCormick, V. (1998). "Punta

Canbalam in Context—a Peripatetic Coastal Site in Northwest Campeche, Mexico." *Ancient Mesoamerica*, 9/1: 1–15.

Dalton, R. (1998). "Dating Expert Comes Under Scrutiny." *Nature*, 392: 218–19.

Darmody, R. G., Thorn, C. E., Dixon, J. C., and Schlyter, P. (2000). "Soils and Landscapes of Kärkevagge, Swedish Lappland." *Soil Science Society of America Journal*, 64: 1455–66.

Darmody, R. G., Thorn, C. E., Harder, R. L., Schlyter, J. P. L., and Dixon, J. C. (2000). "Weathering Implications of Water Chemistry in an Arctic-Alpine Environment, Northern Sweden." *Geomorphology*, 34: 89–100.

Day, M. J. (1993*a*). "Human Impacts on Caribbean and Central American Karst." *Catena*, Suppl. 25: 109–25.

— (1993*b*). "Resource Use in the Tropical Karstlands of Central Belize." *Environmental Geology*, 21: 122–8.

Dixon, J. C. (1991). "Alpine and Subalpine Soil Properties as Paleoenvironmental Indicators." *Physical Geography*, 12/4: 370–84.

Dixon, J. C., and Abrahams, A. D. (eds.) (1992). *Periglacial Geomorphology*. Chichester, NY: Wiley.

Dixon, J. C., Darmody, R. G., Schlyter, P., and Thorn, C. E. (1995). "Preliminary Investigation of Geochemical Process Responses to Potential Environmental Change in Kärkevagge, Northern Scandinavia." *Geografiska Annaler*, 77A/4: 259–67.

Dorn, R. I. (1990). "Quaternary Alkalinity Fluctuations Recorded in Rock Varnish Microlaminations on Western U.S.A. Volcanics." *Palaeogeography, Palaeoclimatology, Palaeoecology*, 76: 291–310.

— (1995). "Digital Processing of Back-Scatter Electron Imagery: A Microscopic Approach to Quantifying Chemical Weathering." *Geological Society of America Bulletin*, 107: 725–41.

— (1996). "Uncertainties in the Radiocarbon Dating of Organics Associated with Rock Varnish: A Plea for Caution." *Physical Geography*, 17/6: 585–91.

— (1998*a*). "Ambiguities in Direct Dating of Rock Surfaces Using Radiocarbon Measurements—Response." *Science*, 280: 2135–9.

— (1998*b*). *Rock Coatings*. Amsterdam: Elsevier.

Dorn, R. I., and Phillips, F. M. (1991). "Surface Exposure Dating: Review and Critical Evaluation." *Physical Geography*, 12/4: 303–33.

Faulkner, D. J. (1998). "Spatially Variable Historical Alluviation and Channel Incision in West-Central Wisconsin." *Annals of the Association of American Geographers*, 88/4: 666–85.

Frankel, C. (1996). *Volcanoes of the Solar System*. Cambridge: Cambridge University Press.

Garcia, J. E., and Brook, G. A. (1996). "Reconstruction of the Ocmulgee River, Georgia: 1807–1971." *Southeastern Geographer*, 36/2: 192–206.

Giardino, J. R., and Marston, R. A. (1999). "Introduction—Engineering Geomorphology: An Overview of Changing the Face of the Earth." *Geomorphology*, 31/1–4: 1–11.

Giardino, J. R., Marston, R. A., and Morisawa, M. (eds.) (1999). "28th Annual Binghamton Symposium: Changing the Face of the Earth—Engineering Geomorphology." *Geomorphology*, 31/1–4: 1–439.

Goudie, A. (1993). "Human Influence in Geomorphology." *Geomorphology*, 7/1–3: 37–59.

Graf, W. L. (1990). "Fluvial Dynamics of Thorium-230 in the Church Rock Event, Puerco River, New Mexico." *Annals of the Association of American Geographers*, 80/3: 327–42.

— (1994). *Plutonium and the Rio Grande: Environmental Change and Contamination in the Nuclear Age*. New York: Oxford University Press.

Harbor, J. M. (ed.) (1995). "Special Issue—Glacial Geomorphology: Process and Form Development." *Geomorphology*, 14/2: 85–196.

— (1999). "Special Issue—Cosmogenic Isotopes in Geomorphology." *Geomorphology*, 27/1–2: 1–172.

Harden, C. P. (1992). "Incorporating Roads and Footpaths in Watershed-Scale Hydrologic and Soil Erosion Models." *Physical Geography*, 13/4: 368–85.

— (1993). "Upland Erosion and Sediment Yield in a Large Andean Drainage Basin." *Physical Geography*, 14/3: 254–71.

— (1996). "Interrelationships between Land Abandonment and Land Degradation: A Case from the Ecuadorian Andes." *Mountain Research and Development*, 16/3: 274–80.

Hesp, P. A. (ed.) (1996). "Special Issue—Aeolian Environments." *Geomorphology*, 22/2: 111–204.

Hickin, E. J. (ed.) (1995). *River Geomorphology*. Chichester, NY: Wiley.

Hillman, G. R. (1998). "Flood Wave Attenuation by a Wetland Following a Beaver Dam Failure on a Second Order Boreal Stream." *Wetlands*, 18/1: 21–34.

Hudson, P. F., and Kesel, R. H. (2000). "Channel Migration and Meander-Bend Curvature in the Lower Mississippi River Prior to Major Human Modification." *Geology*, 28/6: 531–4.

Hudson, P. F., and Mossa, J. (1997). "Suspended Sediment Transport Effectiveness of Three Large Impounded Rivers, U.S. Gulf Coastal Plain." *Environmental Geology*, 32/4: 263–73.

Hunckler, R. V., and Schaetzl, R. J. (1997). "Spodosol Development as Affected by Geomorphic Aspect, Baraga County, Michigan." *Soil Science Society of America Journal*, 61: 1105–15.

Hupp, C. R., and Bazemore, D. E. (1993). "Temporal and Spatial Patterns of Wetland Sedimentation, West Tennessee." *Journal of Hydrology*, 141: 179–96.

Hupp, C. R., and Carey, W. P. (1990). "Dendrogeomorphic Approach to Estimating Slope Retreat, Maxey Flats, Kentucky." *Geology*, 18: 658–61.

Hupp, C. R., and Osterkamp, W. R. (1996). "Riparian Vegetation and Fluvial Geomorphic Processes." *Geomorphology*, 14: 277–95.

Hupp, C. R., and Simon, A. (1991). "Bank Accretion and the Development of Vegetated Depositional Surfaces along Modified Alluvial Channels." *Geomorphology*, 4: 111–24.

Hupp, C. R., Woodside, M. D., and Yanosky, T. M. (1993). "Sediment and Trace Element Trapping in a Forested Wetland, Chickahominy River, Virginia." *Wetlands*, 13/2: 95–104.

James, L. A. (1991). "Incision and Morphologic Evolution of an Alluvial Channel Recovering from Hydraulic Mining Sediment." *Geological Society of America Bulletin*, 103/6: 723–36.

— (1997). "Channel Incision on the Lower American River, California, from Streamflow Gage Records." *Water Resources Research*, 33/3: 485–90.

Johnson, D. L., and Balek, C. L. (1991). "The Genesis of Quaternary Landscapes with Stone-Lines." *Physical Geography*, 12/4: 385–95.

Kennedy, B. A. (1994). "Requiem for a Dead Concept." *Annals of the Association of American Geographers*, 84/4: 702–5.

Krinsley, D. H., Dorn, R. I., and Anderson, S. W. (1990). "Factors that Interfere with the Age Determination of Rock Varnish." *Physical Geography*, 11/2: 97–119.

Kull, C. A., and Magilligan, F. J. (1994). "Controls over Landslide Distribution in the White Mountains of New Hampshire." *Physical Geography*, 15/4: 325–41.

Lancaster, N. (1995). *Geomorphology of Desert Dunes*. London: Routledge.

— (ed.) (1996). "Special Issue—Response of Aeolian Processes to Global Change." *Geomorphology*, 17/1–3: 1–271.

Lecce, S. A. (1997*a*). "Nonlinear Downstream Changes in Stream Power on Wisconsin's Blue River." *Annals of the Association of American Geographers*, 87/3: 471–86.

— (1997*b*). "Spatial Patterns of Historical Overbank Sedimentation and Floodplain Evolution, Blue River, Wisconsin." *Geomorphology*, 18/3–4: 265–77.

Lecce, S. A., and Pavlowsky, R. T. (1997). "Storage of Mining-Related Zinc in Floodplain Sediments, Blue River, Wisconsin." *Physical Geography*, 18/5: 424–39.

Lee, J. A. (1991*a*). "Near-Surface Wind Flow around Desert Shrubs." *Physical Geography*, 12/2: 140–6.

— (1991*b*). "The Role of Desert Shrub Size and Spacing on Wind Profile Parameters." *Physical Geography*, 12/1: 72–89.

Lee, J. A., and Tchakerian, V. P. (1995). "Magnitude and Frequency of Blowing Dust on the Southern High Plains of the United States, 1947–1989." *Annals of the Association of American Geographers*, 85/4: 684–93.

Lee, J. A., Allen, B. L., Peterson, R. E., Gregory, J. M., and Moffett, K. E. (1994). "Environmental Controls on Blowing Dust Direction at Lubbock, Texas, U.S.A." *Earth Surface Processes and Landforms*, 19: 437–49.

Leigh, D. S. (1994*a*). "Mercury Contamination and Floodplain Sedimentation from Former Gold Mines in North Georgia." *Water Resources Bulletin*, 30/4: 739–49.

— (1994*b*). "Roxana Silt of the Upper Mississippi Valley: Lithology, Source, and Paleoenvironment." *Geological Society of America Bulletin*, 106/3, 430–40.

Leigh, D. S., and Knox, J. C. (1994). "Loess of the Upper Mississippi Valley Driftless Area." *Quaternary Research*, 42/1: 30–40.

Leopold, L. B. (1994). *A View of the River*. Cambridge, Mass.: Harvard University Press.

Li, G., and Abrahams, A. D. (1999). "Controls on Sediment Transport Capacity in Laminar Interrill Flow on Stone-Covered Surfaces." *Water Resources Research*, 35/1: 305–10.

Liebens, J., and Schaetzl, R. J. (1997). "Relative-Age Relationships of Debris Flow Deposits in the Southern Blue Ridge, North Carolina." *Geomorphology*, 21/1: 53–67.

McDowell, P. F., and Magilligan, F. J. (1997). "Response of Stream Channels to Removal of Cattle Grazing Disturbance: Overview of Western U.S. Exclosure Studies," in S. S. Y. Wang, E. J. Langendoen, and F. D. Shields, Jr. (eds.), *Management of Landscapes Disturbed by Channel Incision: Stabilization, Rehabilitation, Restoration*. Oxford, Miss.: University of Mississippi, 469–75.

Magilligan, F. J., and McDowell, P. F. (1997). "Stream Channel Adjustments Following Elimination of Cattle Grazing." *Journal of the American Water Resources Association*, 33/4: 867–78.

Magilligan, F. J., and Stamp, M. L. (1997). "Historical Land-Cover Changes and Hydrogeomorphic Adjustment in a Small Georgia Watershed." *Annals of the Association of American Geographers*, 87/4: 614–35.

Magilligan, F. J., Phillips, J. D., James, L. A., and Gomez, B. (1998). "Geomorphic and Sedimentological Controls on the Effectiveness of an Extreme Flood." *Journal of Geology*, 106/1: 87–95.

Malanson, G. P. (1993). *Riparian Landscapes*. Cambridge: Cambridge University Press.

Malanson, G. P., and Butler, D. R. (1990). "Woody Debris, Sediment, and Riparian Vegetation of a Subalpine River, Montana, USA." *Arctic and Alpine Research*, 22/2: 183–94.

Malanson, G. P., Butler, D. R., and Georgakakos, K. P. (1992). "Nonequilibrium Geomorphic Processes and Deterministic Chaos." *Geomorphology*, 5/3–5: 311–22.

Malanson, G. P., Butler, D. R., and Walsh, S. J. (1990). "Chaos Theory in Physical Geography." *Physical Geography*, 11/4: 293–304.

Marcus, M. G., and Marcus, W. A. (eds.) (1996). "In Memoriam: Barry Chapman Bishop, 1932–1994." *Mountain Research and Development*, 16/3: 185–333.

Marcus, W. A. (1996). "Segment-Scale Patterns and Hydraulics of Trace Metal Concentrations in Fine Grain Sediments of a Cobble and Boulder Bed Mountain Stream, Southeast Alaska." *Mountain Research and Development*, 16/3: 211–20.

Marcus, W. A., and Kearney, M. S. (1991). "Upland and Coastal Sediment Sources in a Chesapeake Bay Estuary." *Annals of the Association of American Geographers*, 81/3: 408–24.

Marcus, W. A., Ladd, S. C., and Crotteau, M. (1996). "Channel Morphology and Copper Concentrations in Stream Bed Sediments," in *Tailings and Mine Waste '96* (Proceedings of the Third International Conference on Tailings and Mine Waste). Rotterdam: A. A. Balkema, 421–30.

Marcus, W., Roberts, K., Harvey, L., and Tackman, G. (1992). "An Evaluation of Methods for Estimating Manning's *n* in Small Mountain Streams." *Mountain Research and Development*, 12/3: 227–39.

Marston, R. A. (1989). "Geomorphology," in G. Gaile and C. Willmott (eds.), *Geography in America*. Columbus, Ohio: Merrill Publishing Co., 70–94.

— (1994). "River Entrenchment in Small Mountain Valleys of the Western USA: Influence of Beaver, Grazing and Clearcut Logging." *Revue de Geographie de Lyon*, 69/1: 11–15.

Marston, R. A., Kleinman, J., and Miller, M. (1996). "Geomorphic and Forest Cover Controls on Monsoon Flooding, Central Nepal Himalaya." *Mountain Research and Development*, 16/3: 257–64.

Marston, R. A., Miller, M. M., and Devkota, L. P. (1998). "Geoecology and Mass Movement in the Manaslu-Ganesh and Langtang-Jugal Himals, Nepal." *Geomorphology*, 16/1–3: 139–50.

Marston, R. A., Pochop, L. O., Kerr, G. L., Varuska, M. L., and Veryzer, D. J. (1991). "Recent Glacier Changes in the Wind River Range, Wyoming." *Physical Geography*, 12/2: 115–23.

Marston, R. A., Girel, J., Pautou, G., Piegay, H., Bravard, J.-P., and Arneson, C. (1995). "Channel Metamorphosis, Floodplain Disturbance, and Vegetation Development: Ain River, France." *Geomorphology*, 13/1–4: 121–31.

Mason, J. A., and Knox, J. C. (1997). "Age of Colluvium Indicates Accelerated Late Wisconsinan Hillslope Erosion in the Upper Mississippi Valley." *Geology*, 25/3: 267–70.

Meentemeyer, R. K., Vogler, J. B., and Butler, D. R. (1998). "The Geomorphic Influences of Burrowing Beavers on Streambanks,

Bolin Creek, North Carolina." *Zeitschrift für Geomorphologie*, 42/4: 453–68.

Meierding, T. C. (1993*a*). "Inscription Legibility Method for Estimating Rock Weathering Rates." *Geomorphology*, 6: 273–86.

— (1993*b*). "Marble Tombstone Weathering and Air Pollution in North America." *Annals of the Association of American Geographers*, 83/4: 568–88.

Morris, S. E. (1992). "Geomorphic Assessment of the Effects of Flow Diversion on Anadromous Fish Spawning Habitat: Newhalem Creek, Washington." *Professional Geographer*, 44/4: 444–52.

Mossa, J. (1996). "Sediment Dynamics in the Lowermost Mississippi River." *Engineering Geology*, 45: 457–79.

Mossa, J., and McLean, M. (1997). "Channel Planform and Land Cover Changes on a Mined River Floodplain." *Applied Geography*, 17/1: 43–54.

Mossa, J., and Miller, B. J. (1995). "Geomorphic Development and Paleoenvironments of Late Pleistocene Sand Hills, Southeastern Louisiana." *Southeastern Geology*, 35/2: 79–92.

Myers, T., and Swanson, S. (1996). "Stream Morphologic Impact of and Recovery from Major Flooding in North-central Nevada." *Physical Geography*, 17/5: 431–45.

Neave, M., and Abrahams, A. D. (2001). "Impact of Small Mammal Disturbances on Sediment Yield from Grassland and Shrubland Ecosystems in the Chihuahuan Desert." *Catena*, 44: 285–303.

Nicholas, J. W., and Butler, D. R. (1996). "Application of Relative Age-Dating Techniques on Rock Glaciers of the La Sal Mountains, Utah: An Interpretation of Holocene Paleoclimates." *Geografiska Annaler*, 78A/1: 1–18.

Nicholas, J. W., and Garcia, J. E. (1997). "Origin of Fossil Rock Glaciers, La Sal Mountains, Utah." *Physical Geography*, 18/2: 160–75.

Norman, S. A., Schaetzl, R. J., and Small, T. W. (1995). "Effects of Slope Angle on Mass Movement by Tree Uprooting." *Geomorphology*, 14/1: 19–27.

Orme, A. J., and Orme, A. R. (1991). "Relict Barrier Beaches as Paleoenvironmental Indicators in the California Desert." *Physical Geography*, 12/4: 334–46.

Orme, A. R. (1990). "Recurrence of Debris Production under Coniferous Forest, Cascade Foothills, Northwest United States," in J. B. Thornes (ed.), *Vegetation and Erosion*. Chichester, NY: Wiley, 67–84.

Parker, K. C. (1995). "Effects of Complex Geomorphic History on Soil and Vegetation Patterns on Arid Alluvial Fans." *Journal of Arid Environments*, 30/1: 19–39.

Parker, K. C., and Bendix, J. (1996). "Landscape-Scale Geomorphic Influences on Vegetation Patterns in Four Environments." *Physical Geography*, 17/2: 113–41.

Pérez, F. L. (1990). "Surficial Talus Fabric and Particle Gliding Over Snow on Lassen Peak, California." *Physical Geography*, 11/2: 142–53.

— (1998). "Conservation of Soil Moisture by Different Stone Covers on Alpine Talus Slopes (Lassen, California)." *Catena*, 33: 155–77.

Phillips, J. D. (1991). "The Human Role in Earth Surface Systems: Some Theoretical Considerations." *Geographical Analysis*, 23: 316–31.

— (1992). "Qualitative Chaos in Geomorphic Systems, with an Example from Wetland Response to Sea Level Rise." *Journal of Geology*, 100: 365–74.

— (1993). "Instability and Chaos in Hillslope Evolution." *American Journal of Science*, 293/1: 25–48.

— (1995). "Biogeomorphology and Landscape Evolution: The Problem of Scale." *Geomorphology*, 13/1–4: 337–47.

— (1997*a*). "Humans as Geological Agents and the Question of Scale." *American Journal of Science*, 298: 98–115.

— (1997*b*). "A Short History of a Flat Place: Three Centuries of Geomorphic Change in the Croatan National Forest." *Annals of the Association of American Geographers*, 87/2: 197–216.

— (1999). *Earth Surface Systems*. Oxford: Blackwell.

Phillips, J. D., and Gomez, B. (1994). "In Defense of Logical Sloth." *Annals of the Association of American Geographers*, 84/4: 697–701.

Phillips, J. D., and Renwick, W. H. (eds.) (1992). *Geomorphic Systems*. Amsterdam: Elsevier.

Pike, R. J. (1995). "Geomorphometry—Progress, Practice, and Prospect." *Zeitschrift für Geomorphologie*, Suppl., 101: 221–38.

— (1996). "A Bibliography of Geomorphometry, the Quantitative Representation of Topography—Supplement 2.0." *U.S. Geological Survey Open-File Report 96–726*. Menlo Park, Calif.: US Geological Survey.

Pope, G. A. (1995). "Newly Discovered Submicron-Scale Weathering in Quartz: Geographical Implications." *Professional Geographer*, 47: 375–87.

Pope, G. A., and Rubenstein, R. (1999). "Anthroweathering—Theoretical Framework and Case Study for Human-Impacted Weathering." *Geoarchaeology: An International Journal*, 14/3: 247–64.

Pope, G. A., Dorn, R. I., and Dixon, J. C. (1995). "A New Conceptual Model for Understanding Geographical Variations in Weathering." *Annals of the Association of American Geographers*, 85/1: 38–64.

Rediscovering Geography Committee (1997). *Rediscovering Geography*. Washington, DC: National Academy Press.

Rhoads, B. L., and Thorn, C. E. (1993). "Geomorphology as Science: The Role of Theory." *Geomorphology*, 6: 287–307.

— (eds.) (1996). *The Scientific Nature of Geomorphology*. Chichester, NY: Wiley.

Robert, A. (1997). "Characteristics of Velocity Profiles along Riffle-Pool Sequences and Estimates of Bed Shear Stress." *Geomorphology*, 19/1–2: 89–98.

Schaetzl, R. J. (1990). "Effects of Treethrow Microtopography on the Characteristics and Genesis of Spodosols, Michigan, USA." *Catena*, 17: 111–26.

Schaetzl, R. J., and Follmer, L. R. (1990). "Longevity of Treethrow Microtopography: Implications for Mass Wasting." *Geomorphology*, 3: 113–23.

Schaetzl, R. J., Burns, S. E., Small, T. W., and Johnson, D. L. (1990). "Tree Uprooting: Review of Types and Patterns of Soil Disturbance." *Physical Geography*, 11/3: 277–91.

Shankman, D., and Drake, L. G. (1990). "Channel Migration and Regeneration of Bald Cypress in Western Tennessee." *Physical Geography*, 11/4: 343–52.

Sherman, D. J. (1989). "Geomorphology: Praxis and Theory," in M. S. Kenzer (ed.), *Applied Geography: Issues, Questions, and Concerns*. Dordrecht: Kluwer, 115–31.

— (1994). "Social Relevance and Geographical Research." *Geographical Review*, 84/3: 336–41.

— (1995). "Problems of Scale in the Modeling and Interpretation of Coastal Dunes." *Marine Geology*, 124: 339–49.

— (1996). "Fashion in Geomorphology," in Rhoads and Thorn (1996: 87–114).

Sherman, D. J., and Bauer, B. O. (1993). "Coastal Geomorphology through the Looking Glass." *Geomorphology*, 7/1–3: 225–49.

Short, N. M., and Blair Jr., R. W. (eds.) (1986). *Geomorphology from Space: A Global Overview of Regional Landforms*. Washington, DC: NASA.

Shroder, J. F., Jr. (ed.) (1998). "Mass Movement in the Himalaya." *Geomorphology*, 26/1–3: 1–222.

Slaymaker, O. (ed.) (1995). *Steepland Geomorphology*. Chichester, NY: Wiley.

— (ed.) (1996). *Geomorphic Hazards*. Chichester, NY: Wiley.

Small, T. W. (1997). "The Goodlett-Denny Mound: A Glimpse at 45 Years of Pennsylvania Treethrow Mound Evolution with Implications for Mass Wasting." *Geomorphology*, 18/3–4: 305–13.

Smith, D. G. (1993). "Fluvial Geomorphology: Where Do We Go from Here?" *Geomorphology*, 7/1–3: 251–62.

Tchakerian, V. P. (1991). "Late Quaternary Aeolian Geomorphology of the Dale Lake Sand Sheet, Southern Mojave Desert, California." *Physical Geography*, 12/4: 347–69.

— (ed.) (1995). *Desert Eolian Geomorphology*. London: Chapman & Hall.

Thelin, G. P., and Pike, R. J. (1991). "Landforms of the Coterminous United States—A Digital Shaded-Relief Portrayal." Essay to accompany *Map 1-2206*. Washington, DC: US Geological Survey.

Thompson, D. M., Wohl, E. E., and Jarrett, R. D. (1996). "A Revised Velocity-Reversal and Sediment-Sorting Model for a High-Gradient, Pool-Riffle Stream." *Physical Geography*, 17/2: 142–56.

Thorn, C. E., and Welford, M. R. (1994*a*). "The Equilibrium Concept in Geomorphology." *Annals of the Association of American Geographers*, 84/4: 666–96.

— (1994*b*). "No Dirge, No Philosophy, Just Practicality." *Annals of the Association of American Geographers*, 84/4: 706–9.

Tormey, B. B. (ed.) (1996). "Special Issue—Recent Developments in Quaternary Geology: Implications for Geoscience Education and Research." *Geomorphology*, 16/3: 195–276.

Trimble, S. W. (1990). "Geomorphic Effects of Vegetation Cover and Management: Some Time and Space Considerations in Prediction of Erosion and Sediment Yield," in J. B. Thornes (ed.), *Vegetation and Erosion*. Chichester, NY: Wiley, 55–65.

— (1994). "Erosional Effects of Cattle on Streambanks in Tennessee, U.S.A." *Earth Surface Processes and Landforms*, 19: 451–64.

— (1995). "Catchment Sediment Budgets and Change," in A. Gurnell and G. Petts (eds.), *Changing River Channels*. Chichester, NY: Wiley, 201–15.

— (1997*a*). "Contribution of Stream Channel Erosion to Sediment Yield from an Urbanizing Watershed." *Science*, 278: 1442–4.

— (1997*b*). "Stream Channel Erosion and Change Resulting from Riparian Forests." *Geology*, 25/5: 467–9.

— (1998). "Dating Fluvial Processes from Historical Data and Artifacts." *Catena*, 31: 283–304.

Trimble, S. W., and Cooke, R. U. (1991). "Historical Sources for Geomorphological Research in the United States." *Professional Geographer*, 43/2: 212–28.

Trimble, S. W., and Crosson, P. (2000). "U.S. Soil Erosion Rates—Myth and Reality." *Science*, 289: 248–50.

Trimble, S. W., and Mendel, A. C. (1995). "The Cow as a Geomorphic Agent—a Critical Review." *Geomorphology*, 13/1–4: 233–53.

Vaughn, D. M. (1997). "A Major Debris Flow along the Wasatch Front in Northern Utah, USA." *Physical Geography*, 18/3: 246–62.

Vitek, J. D., Giardino, J. R., and Fitzgerald, J. W. (1996). "Mapping Geomorphology: A Journey from Paper Maps through Computer Mapping to GIS and Virtual Reality." *Geomorphology*, 16/3: 233–49.

Vogler, J. B., and Butler, D. R. (1996). "Pedestrian- and Bicycle-Induced Path Erosion on a University Campus." *Physical Geography*, 17/5: 485–94.

Walker, H. J. (1994). "Environmental Impact of River Dredging in Arctic Alaska (1981–1989)." *Arctic*, 47/2: 176–83.

— (1998). "Arctic Deltas." *Journal of Coastal Research*, 14/3: 718–38.

Walker, H. J., and Grabau, W. E. (1993). *The Evolution of Geomorphology: A Nation-by-nation Summary*. Chichester, NY: Wiley.

Wallin, T. R., and Harden, C. P. (1996). "Estimating Trail-Related Soil Erosion in the Humid Tropics: Jatun Sacha, Ecuador, and La Selva, Costa Rica." *Ambio*, 25/8: 517–22.

Walsh, S. J., and Butler, D. R. (1997). "Morphometric and Multispectral Image Analysis of Debris Flows for Natural Hazard Assessment." *Geocarto International*, 12/1: 59–70.

Walsh, S. J., Butler, D. R., and Malanson, G. P. (1998). "An Overview of Scale, Pattern, Process Relationships in Geomorphology: A Remote Sensing and GIS Perspective." *Geomorphology*, 21/3–4: 183–205.

Wilkerson, F. D. (1995). "Rates of Heave and Surface Rotation of Periglacial Frost Boils in the White Mountains, California." *Physical Geography*, 16/6: 487–502.

Williams, S. H., and Lee, J. A. (1995). "Aeolian Saltation Transport Rate: An Example of the Effect of Sediment Supply." *Journal of Arid Environments*, 30: 153–60.

Woldenberg, M. J. (1997). "James Keill (1708) and the Morphometry of the Microcosm," in D. R. Stoddart (ed.), *Process and Form in Geomorphology*. London: Routledge, 243–64.

Woltemade, C. J. (1994). "Form and Process: Fluvial Geomorphology and Flood-Flow Interaction, Grant River, Wisconsin." *Annals of the Association of American Geographers*, 84/3: 462–79.

CHAPTER 6

Mountain Geography

Donald A. Friend

Why Mountain Geography?

The raw facts alone make mountains worthy of geographic interest: mountains constitute 25 per cent of the earth's surface; they are home to 26 per cent of the world's populace; and generate 32 per cent of global surface run-off (Meybeck *et al.* 2001). More than half the global population depends directly on mountain environments for the natural resources of water, food, power, wood, and minerals; and mountains contain high biological diversity; hence they are important in crop diversity and crop stability (Ives 1992; Smethurst 2000; UNFAO 2000). Elevation, relief, and differences in aspect make mountains excellent places to study all processes, human and physical: high energy systems make mountains some of the most inhospitable of environments for people and their livelihoods, and strikingly distinct changes in environment over short distances make mountains ideally suited to the study of earth surface processes. Mountains are often political and cultural borders, or in some cases, political, cultural, and biological islands. With ever-increasing populations placing ever-increasing environmental pressure on mountains, mountain environments are heavily impacted and are therefore quickly changing. Moreover, they are more susceptible to adverse impacts than lowlands and are degrading accordingly. Whatever environmental change or damage happens to mountain peoples and environments then moves to lower elevations, thus affecting all. Three seminal texts indicate an ongoing interest in mountain geography: the oldest, Peattie (1936), is still in print; the newest, Messerli and Ives (1997) is contemporary; and Price (1981) is now being rewritten. Indeed, mountain geography as a field in its own right has led to the recent formation of the Mountain Geography Specialty Group of the Association of American Geographers (Friend 1999).

With increasing importance placed on *sustainability science* (Kates *et al.* 2001), mountain geography is at the cutting edge of inter- and multidisciplinary research that serves to unify rather than further specialize scholarly geography (Friend 1999). The United Nations proclaimed 2002 the International Year of Mountains and has devoted an entire chapter (13) of its Agenda 21 from the Rio Earth Summit to mountain sustainable development (Friend 1999; Ives and Messerli 1997; Ives *et al.* 1997*a*, *b*; Messerli and Ives 1997; Sène and McGuire 1997; UNFAO 1999, 2000). The current themes in mountain geographical research occupy all parts of the people–environment continuum, very often meeting in the middle (Messerli and Ives 1997; Price 1981; Stone 1992; Thompson *et al.* 1986).

Research Themes in Mountain Geography

Western researchers began studying mountains in the nineteenth century. Alexander von Humboldt, Albrecht Penck, and Charles Darwin were some of the earliest scholars interested in mountains (Sarmiento 1999). In general, they studied the relationships between elevation and soil, plants, climate, and landforms, thus creating the first understanding of mountain geography. In the mid- and late twentieth century, Carl Troll articulated the study of relationships between geology, geography, and ecology in mountains and coined the term "geoecology," focusing on altitudinal zonation and verticality, where mountains are comprised of rings of altitudinal zones, each unique in terms of ecology and human activity (Gade 1996; Troll 1968, 1971).

Mountain studies are often distinctly physical: The classic text, *Mountain Weather and Climate* (Barry 1992), is now in its second edition and the "bible" for those interested in the topic. A new text, *Mountain Meteorology* (Whiteman 2000) attests to the continued interest and importance of mountain weather. Snow avalanches and their mechanics, prediction, and relationship to climate and people are of continuing interest and are being addressed in new ways integrating technology, people, and climate to predict avalanches (Birkeland 1998, 2001; Birkeland *et al.* 2001; Hardy *et al.* 2001; Mock and Birkeland 2000). Studies and texts addressing mountain geomorphology are of course found in the literature as mountains are some of the most geomorphically active landscapes: a special issue of the journal *Geomorphology* devoted exclusively to "Mass Movement in the Himalaya" is an excellent example of the level of study devoted to only one aspect of mountain geomorphology (Shroder 1998). Specialized work on other aspects of mountain environments is also common: rivers (Marston *et al.* 1997; Wohl 2000), environmental change (Price 1999; Williams *et al.* 1996), arid slopes and lands (Friend 2000; Friend *et al.* 2000; Marston and Dolan 2000), and rock glaciers (Barsch 1996) are among the many geomorphic topics addressed. Biogeographic studies of various basic (Butler 1995; Hadley 1994; Sarmiento 2000; Young 1996) and applied (Allen and Hansen 1999; Byers 1991; Zimmerer 1998) topics appear regularly in both the geographic and ecologic/biotic literature.

Some mountain studies are purely human or cultural: the spiritual and historical significance of mountains is of both popular and academic interest, as are historical accounts of mountain activities (Bernbaum 1990, 1997; Blake 1999*a*, *b*; Rowan and Rowan 1995). Studies of human and people–environment issues in mountains are particularly important (Allan *et al.* 1988; Denniston 1995; Halvorson 2000; Ives *et al.* 1997*a*, *b*; Messerli and Ives 1997).

Integrative research on mountain agriculture is common (Harden 2001; Jodha 1997; Rhoades 1997; Zimmerer 1998); much work focuses on hazards and their human dimensions, i.e. drought, flooding, avalanches, and other slope failures, which also integrates human and physical geography (Bachman 1999; Hewitt 1997; Marston *et al.* 1996; Messerli and Ives 1997; Owen *et al.* 1995).

In the past two decades or so, studies integrating policy and mountain peoples and environments have emerged (Bishop 1990; Blaikie and Brookfield 1987; Brower 1990; Byers 1996, 2000; Halvorson 2000; Inyan and Williams 2001; Ives and Messerli 1989; Ives *et al.* 1997*a*; Stevens 1993, 1997; Thompson *et al.* 1986; Young 1996, 1997; Zimmerer 1993), with more recent studies due in part to the United Nations Earth Summit:

> In 1992 mountains were restored to the map of world concern at the U.N. Conference on Environment and Development in Rio de Janeiro with the publication of *An Appeal for the Mountains* (UNCED (United Nations Conference on Environment and Development-Mountain Agenda) 1992). Just as biologists broadened their attention from mere studies of animal biology to include larger concerns for managing and protecting wildlife in the 1970s, so did researchers studying mountains realize that to continue working on them they had best learn to conserve and protect them (TMI 1995).
>
> Much of the mountain geography literature of today is driven by interest in conserving environments. Scholars have identified nine areas of particular concern: cultural diversity; sustainable development; production systems and alternative livelihoods; local energy demand and supply in mountains; tourism; sacred, spiritual, and symbolic significance of mountains; mountains as sources of water; mountain biodiversity; and climate change and natural hazards (TMI 1995). (Smethurst 2000)

Thus, all parts of the people–environment continuum are now being addressed in Mountain Geography with particular attention paid to sustainable development (Denniston 1995; Friend 1999; Inyan and Williams 2001; Ives *et al.* 1997*a*; Messerli and Ives 1997; Sène and McGuire 1997).

Development of the Mountain Geography Specialty Group

In 1998 a group was formed to bridge the subspecialties and bring together all geographers working in mountain environments and on mountain issues; it was recognized that mountain environments are most sensitive to natural or human-induced change, a fact that calls for the attention of geographers uniquely trained in identifying linkages between earth systems and social science (Friend 1999). The group gained enough support to be officially recognized at the next Annual Meeting of the AAG in 1999, where mission and founding statements were adopted along with by-laws. The Mission Statement reads,

> The Mountain Geography Specialty Group serves to foster communication, promote basic and applied research, enhance education, and encourage service related to mountain peoples and mountain environments, and their interactions.

The group has seen exceptional growth in its three short years in existence and at the time of writing has approximately 150 members, sponsoring special sessions each year at the Annual Meeting of the AAG that have created a niche where broad-based research covering the people–environment continuum can be presented in one place. Indeed, the mission of the group is being honored by bringing together many individuals who work in mountains or on mountain issues. The founding committee are: Karl Birkeland (US Forest Service and Montana State University); Kevin S. Blake (Kansas State University); Barbara Brower (Portland State University); Alton C. Byers (The Mountain Institute); Leland R. Dexter (Northern Arizona University); Donald A. Friend (Chair) (Minnesota State University); Katherine J. Hansen (Montana State University); Richard A. Marston (Oklahoma State University).

Conclusion

The challenges that mountain geography faces are indeed opportunities, especially in 2002, which was the International Year of Mountains. The work of mountain geographers is in high demand as it is deemed critical to global sustainability efforts, and was showcased during 2002 and will be for several years afterward at many special events including the Rio + 10 conference: The World Summit on Sustainable Development. As mountain environments have been recognized by the international community as among the most crucial to long-term global sustainability, mountain peoples and issues must then also be included in any discussion or study of mountains (Rhoades 1997; TMI 1995). Kates *et al.* (2001), progenitors of sustainability science, pose several core questions that will serve as the chief challenges and opportunities for mountain geographers in the coming years:

Core Questions of Sustainability Science

1. How can the dynamic interactions between nature and society—including lags and inertia—be better incorporated in emerging models and conceptualizations that integrate the Earth system, human development, and sustainability?
2. How are long-term trends in environment and development, including consumption and population, reshaping nature–society interactions in ways relevant to sustainability?
3. What determines the vulnerability or resilience of the nature–society system in particular kinds of places and for particular types of ecosystems and human livelihoods?
4. Can scientifically meaningful "limits" or "boundaries" be defined that would provide effective warning of conditions beyond which the nature–society systems incur a significantly increased risk of serious degradation?
5. What systems of incentive structures—including markets, rules, norms and scientific information—can most effectively improve social capacity to guide interactions between nature and society toward more sustainable trajectories?
6. How can today's operational systems for monitoring and reporting on environmental and social conditions be integrated or extended to provide more useful guidance for efforts to navigate a transition toward sustainability?
7. How can today's relatively independent activities of research planning, monitoring, assessment, and decision support be better integrated into systems for adaptive management and societal learning?

Moreover, sustainable mountain development, as called for by ch. 13 of Agenda 21 of the United Nations Conference on Environment and Development (UNCED), puts mountain geography at the forefront of sustainability science, which has been the focus of much work on mountains for several years (Ives *et al.* 1997a; Kates *et al.* 2001; Sène and McGuire 1997; UNCED 1992).

Many American geographers are already working in mountain environments or on issues related to mountain peoples and policy, but are often unaware of their colleagues' efforts. As is often the case in the discipline of geography, human and physical geographers do not interact much, but the field of mountain geography cuts across those boundaries bringing together various specialties (Friend 1999). Beginning with von Humboldt and Darwin, mountain peoples and environments have long been of interest to geographers: in the mid-twentieth century Peattie (1936, 1942–52) wrote and edited extensively on mountain peoples and environments; several recent texts are available on the subject (Allan *et al.* 1988; Bernbaum 1990; Funnell and Parish 2001; Gerrard 1990; Messerli and Ives 1997; Parish 2001; Price 1981); there are at least two major research journals devoted exclusively to things of the mountains: *Mountain Research and Development* and the *Himalayan*

Research Bulletin, with several others partially devoted, e.g. *Arctic, Antarctic and Alpine Research*, and *Permafrost and Periglacial Processes*; and, according to *GeoBase/ Geographical Abstracts*, since 1974 there have been over 26,000 scholarly articles published using "mountain/s" in the title or as keywords.

Mountain geography as a distinct field of study is thriving and growing and is on the cutting edges of geography and the emerging field of sustainability science. The proposed, new field of study, montology (Ives *et al.* 1997*a*), is catching on, with publications (Haslett 1998; Sarmiento 2000) and conferences (Montology 2001, 2002). Montology proposes the same interdisciplinary multispatial-scale approach as sustainability science but with a focus purely on mountain peoples and environments. Montology is "part science, part humanities, part social science and part folk science" (Ives *et al.* 1997*a*). Indeed, we have much work to look forward to as mountains are increasingly recognized as critical to global environmental health, which, of course, involves the good work of mountain geographers.

References

Allan, N. J. R., Knapp, G., and Stadel, C. (eds.) (1988). *Human Impact on Mountains*. Lanham, Md.: Rowman & Littlefield.

Allen, K., and Hansen, K. (1999). "Geography of Exotic Plants Adjacent to Campgrounds, Yellowstone National Park, USA." *Great Basin Naturalist*, 59/4: 315–22.

Bachman, D. (1999). "European Avalanches of 1998–99." *The Avalanche Review*, 17/6: 11.

Barry, R. G. (1992). *Mountain Weather and Climate*. 2nd edn. London: Routledge.

Barsch, D. (1996). *Rockglaciers: Indicators for the Present and Former Geoecology in High Mountain Environments*, Springer Series in Physical Environment. Berlin: Springer.

Bernbaum, E. (1990). *Sacred Mountains of the World*. San Francisco: Sierra Club Books.

— (1997). "The Spiritual and Cultural Significance of Mountains," in Messerli and Ives (1997: 39–60).

Birkeland, K. W. (1998). "Terminology and Predominant Processes Associated with the Formation of Weak Layers of Near-Surface Faceted Crystals in the Mountain Snowpack." *Arctic and Alpine Research*, 30/2: 193–9.

— (2001). "Spatial Patterns of Snow Stability throughout a Small Mountain Range." *Journal of Glaciology*, 47/157: 176–86.

Birkeland, K. W., Mock, C. J., and Shinker, J. J. (2001). "Avalanche Extremes and Atmospheric Circulation Patterns." *Annals of Glaciology*, 32: 135–40.

Bishop, B. C. (1990). *Karnali Under Stress: Livelihood Strategies and Seasonal Rhythms in a Changing Nepal Himalaya, Geography Research Paper Nos. 228–229*. Chicago: Committee on Geographical Studies, University of Chicago.

Blaikie, P., and Brookfield, H. (1987). *Land Degradation and Society*. London: Methuen.

Blake, K. (1999*a*). "Peaks of Identity in Colorado's San Juan Mountains." *Journal of Cultural Geography*, 18/2: 29–55.

— (1999*b*). "Sacred and Secular Landscape Symbolism at Mount Taylor, New Mexico." *Journal of the Southwest*, 41/4: 487–509.

Brower, B. (1990). "Crisis and Conservation in Sagarmatha National Park, Nepal." *Society & Natural Resources*, 4/2: 15–163.

Butler, D. R. (1995). *Zoogeomorphology: Animals as Geomorphic Agents*. Cambridge: Cambridge University Press.

Byers, A. (1991). "Mountain Gorilla Mortality and Climatic Factors in the Parc National des Volcans, Ruhengeri Prefecture, Rwanda, 1988." *Mountain Research and Development*, 11/2: 145–51.

— (1996). "Historical and Contemporary Human Disturbance in the Upper Barun Valley, Makalu-Barun National Park and Conservation Area, East Nepal." *Mountain Research and Development*, 16/3: 235–47.

— (2000). "Contemporary Landscape Change in the Huascaran National Park and Buffer Zone, Cordillera Blanca, Peru." *Mountain Research and Development*, 20/1: 52–63.

Denniston, D. (1995). "High Priorities: Conserving Mountain Ecosystems and Cultures." *World Watch Paper*, 123.

Friend, D. A. (1999). "Mountain Chronicle: Formation of a Mountain Geography Specialty Group within the Association of American Geographers (AAG)." *Mountain Research and Development*, 19/2: 167–8.

— (2000). "Revisiting William Morris Davis and Walther Penck to Propose a General Model of Slope 'Evolution' in Deserts." *The Professional Geographer*, 52/2: 164–78.

Friend, D. A., Phillips, F. M., Campbell, S. W., Liu, T., and Sharma, P. (2000). "Evolution of Desert Colluvial Boulder Slopes." *Geomorphology*, 36/1–2: 19–45.

Funnell, D., and Parish, R. (2001). *Mountain Environments and Communities*. London: Routledge.

Gade, D. W. (1996). "Carl Troll on Nature and Culture in the Andes." *Erdkunde*, 50/4: 301–16.

Gerrard, A. J. (1990). *Mountain Environments: An Examination of the Physical Geography of Mountains*. Cambridge, Mass.: MIT Press.

Hadley, K. S. (1994). "The Role of Disturbance, Topography, and Forest Structure in the Development of a Montane Forest Landscape." *Bulletin of the Torrey Botanical Club*, 121: 47–61.

Halvorson, S. J. (2000). "Geographies of Children's Vulnerability: Households and Water-Related Disease Hazard in the Karakoram Mountains, Northern Pakistan." Ph.D. Dissertation, University of Colorado, Boulder.

Harden, C. (2001). "Soil Erosion and Sustainable Mountain Development: Experiments, Observations and Recommendations from the Ecuadorian Andes." *Mountain Research and Development*, 21/1: 77–83.

Hardy, D., Williams, M. W., and Escobar, C. (2001). "Near-Surface Faceted Crystals, Avalanches and Climate in High-Elevation, Tropical Mountains of Bolivia." *Cold Regions Science and Technology*, 33: 291–302.

Haslett, J. R. (1998). "A New Science: Montology" *Global Ecology and Biogeography Letters*, 7/3: 228–9.

—— (1997). "Risk and Disasters in Mountain Lands," in Messerli and Ives (1997: 371–408).

Inyan, B. J., and Williams, M. W. (2001). "Protection of H eadwater Catchments from Future Degradation: San Miguel River Basin, Colorado." *Mountain Research and Development*, 21/1: 54–60.

Ives, J. D. (1992). Preface to P. B. Stone (ed.), *The State of the World's Mountains: A Global Report*. London: Zed Press, pp. xiii–xvi.

Ives, J. D., and Messerli, B. (1989). *The Himalayan Dilemma: Reconciling Development and Conservation*. New York: Routledge.

—— (1997). Preface to Messerli and Ives (1997).

Ives, J. D., Messerli, B., and Rhoades, R. E. (1997*a*). "Agenda for Sustainable Mountain Development," in Messerli and Ives (1997).

Ives, J. D., Messerli, B., and Spiess, E. (1997*b*). "Mountains of the World—A Global Priority," in Messerli and Ives (1997: 1–16).

Jodha, N. S. (1997). "Mountain Agriculture," in Messerli and Ives (1997: 313–35).

Kates, R. W., Clark, W. C., Correll, R., Hall, J. M., Jaeger, C. C., Lowe, I., McCarthy, J. J., Schellnhuber, H. J., Bolin, B., Dickson, N. M., Faucheux, S., Gallopin, G. C., Grübler, A., Huntley, B., Jäger, J., Jodha, N. S., Kasperson, R. E., Mabogunje, A., Matson, P., Mooney, H., Moore, B., III, M., O'Riordan, T., and Svedin, U. (2001). "Sustainability Science." *Science*, 292: 641–2.

Marston, R. A., and Dolan, L. S. (2000). "Effectiveness of Sediment Control Structures Relative to Spatial Patterns of Upland Soil Loss in an Arid Watershed, Wyoming." *Geomorphology*, 31/1–4: 313–23.

Marston, R. A., Fritz, D. E., and Nordberg, V. (1997). "The Impact of Debris Torrents on Substrates of Mountain Streams." *Geomorphologie: Relief, Processus, Environment*, 1: 21–32.

Marston, R. A., Kleinman, J., and Miller, M. M. (1996). "Geomorphic and Forest Cover Controls on Flooding: Central Nepal Himalaya." *Mountain Research and Development*, 16/3: 257–64.

Messerli, B., and Ives, J. D. (eds.) (1997). *Mountains of the World: A Global Priority*. New York: Parthenon.

Meybeck, M., Green, P., and Vörösmarty, C. (2001). "A New Typology for Mountains and Other Relief Classes." *Mountain Research & Development*, 21/1: 34–5.

Mock, C. J., and Birkeland, K. W. (2000). "Snow Avalanche Climatology of the Western United States Mountain Ranges." *Bulletin of the American Meteorological Association*, 81/10: 2367–92.

Montology (2001). Conference Name: Applied Montology: Comparative Geographies of the Andes and the Appalachians. University of Georgia (‹http://www.uga.edu/clacs/Montology.html›, accessed 29 May 2002).

—— (2002). Conference Title: International Montology: The State and Development Issues of Mountain Systems St. Petersburg, Russian Federation (‹http://www.icimod.org.sg/iym2002/calendar/highsummit/global.htm›, accessed 29 May 2002).

Owen, L. A., Sharma, M. C., and Bigwood, R. (1995). "Mass Movement Hazard in the Garhwal Himalaya: The Effects of the 20 October 1991 Garhwal Earthquake and the July–August 1992 Monsoon Season," in D. F. M. McGregor and D. A. Thompson (eds.), *Geomorphology and Land Management in a Changing Environment*. London: Wiley & Sons, 69–88.

Parish, R. (2001). *Mountain Environments*. Harlow: Pearson Education.

Peattie, R. (1936). *Mountain Geography: A Critique and Field Study*. Cambridge, Mass.: Harvard University Press; rpt. 1969, New York: Greenwood Press.

—— (ed.) (1942–52). 9 vols. i. The Friendly Mountains: Green, White, and Adirondacks (1942); ii. The Great Smokies and the Blue Ridge (1943); iii. The Rocky Mountains (1945); iv. The Pacific Coast Ranges (1946); v. The Sierra Nevada (1947); vi. The Berkshires: The Purple Hills (1948); vii. The Inverted Mountains: Canyons of the West (1948); viii. The Cascades: Mountains of the Pacific Northwest (1949); ix. The Black Hills (1952). American Mountain Series. New York: Vanguard Press.

Price, L. W. (1981). *Mountains and Man*. Berkeley: University of California Press.

Price, M. (ed.) (1999). *Global Change in the Mountains*. New York: Parthenon.

Rhoades, R. E. (1997). *Pathways Toward a Sustainable Mountain Agriculture for the 21st Century: The Hindu Kush-Himalayan Experience*. Kathmandu: International Centre for Integrated Mountain Development (ICIMOD).

Rowan, P., and Rowan, J. H. (eds.) (1995). *Mountain Summers*. Gorham, NH: Gulfside Press.

Sarmiento, F. O. (1999). "To Mount Chimborazo in the Steps of Alexander von Humboldt." *Mountain Research and Development*, 19/2: 77–8.

—— (2000). "Breaking Mountain Paradigms: Ecological Effects on Human Impacts in Man-aged Tropandean Landscapes." *Ambio*, 29/7: 423–31.

Sène, E. H., and McGuire, D. (1997). "Sustainable Mountain Development—Chapter 13 In Action," in Messerli and Ives (1997: 447–53).

Shroder, J. F., Jr. (ed.) (1998). Special Issue: Mass Movement in the Himalaya. *Geomorphology*, 26/1–3: 1–222.

Smethurst, D. (2000). "Mountain Geography." *The Geographical Review*, 90/1: 35–56.

Stevens, S. (1993). *Claiming the High Ground: Sherpas, Subsistence, and Environmental Change in the Highest Himalaya*. Berkeley: University of California Press.

—— (ed.) (1997). *Conservation Through Cultural Survival: Indigenous Peoples and Protected Areas*. Washington: Island Press.

Stone, P. B. (ed.) (1992). *The State of the World's Mountains: A Global Report*. London: Zed Books.

Thompson, M., Warburton, M., and Hatley, T. (1986). *Uncertainty on a Himalayan Scale*. London: Ethnographica.

TMI (The Mountain Institute) (1995). *International NGO Consultation on the Mountain Agenda: Summary Report and Recommendations to the U.N. Commission on Sustainable Development, April 1995*. Washington, DC: The Mountain Institute.

Troll, C. (1968). "The Cordilleras of the Tropical Americas: Aspects of Climatic, Phytogeographical and Agrarian Ecology." *Colloquium Geographicum*, 9 (August): 15–56.

—— (1971). "Landscape Ecology (Geoecology) and Biogeoecology. A Terminological Study." *Geoforum*, 8: 43–6.

UNCED (United Nations Conference on Environment and Development–Mountain Agenda) (1992). *An Appeal for the Mountains*. New York: United Nations.

UNFAO (United Nations Food and Agriculture Organization) (1999). *Managing Fragile Ecosystems: Sustainable Mountain Development*. Fifth ad hoc Inter-Agency Meeting on Follow-up to Agenda 21, ch. 13. FAO Headquarters.

—— (2000). *International Year of the Mountains: Concept Paper*. Rome.

Whiteman, C. D. (2000). *Mountain Meteorology: Fundamentals and Applications*. New York: Oxford University Press.

Williams, M. W., Baron, J. S., Caine, N., Sommerfeld, R., and Sanford, R., Jr. (1996). "Nitrogen Saturation in the Colorado Front Range." *Environmental Science and Technology*, 30: 640–6.

Wohl, E. E. (2000). *Mountain Rivers*. Water Resources Monograph Ser., 14. Washington, DC: American Geophysical Union.

Young, K. R. (1996). "Threats to Biological Diversity Caused by Coca/Cocaine Deforestation in Peru." *Environmental Conservation*, 23/1: 7–15.

—— (1997). "Wildlife Conservation in the Cultural Landscapes of the Central Andes." *Landscape and Urban Planning*, 38: 137–47.

Zimmerer, K. (1993). "Soil Erosion and Social Dis(courses) in Cochabamba, Bolivia." *Economic Geography*, 69/3: 312–27.

—— (1998). Ecogeography of the Cultivated Andean Potatoes. *Bioscience*, 48/6: 445–54.

Part II

Human/Society Dynamics

CHAPTER 7

Cultural Geography

Garth A. Myers, Patrick McGreevy, George O. Carney, and Judith Kenny

Introduction

> We have not really prescribed limitations of inquiry, method, or thought upon our associates. From time to time there are attempts to the contrary, but we shake them off after a while and go about doing what we most want to do. . . . We thrive on cross-fertilization and diversity.
>
> Sauer (1956)

> You can't go wrong when you call something cultural, for it is the one term that, without necessarily specifying anything, carries the full weight of all possible forms of specificity.
>
> Gallagher (1995: 307)

Both these quotations, one recent and one nearly a half-century old, point to the monumental task before us in attempting to report on the progress of cultural geography over the past dozen years. Many things get called cultural geography, for many different reasons, with varying purposes in mind. Different people who consider themselves cultural geographers often have wildly different ideas of what this label means, as well as radically different approaches to what they do. We cannot pretend to encompass the whole of this body of work, and we must admit as much at the outset. Instead, let us begin with the specialty group itself, since it provides some focus and continuity for taking stock of the subfield.

The Cultural Geography Specialty Group's membership has increased slowly but steadily since the group's inception in the late 1980s. With 465 members, the CGSG was, as of 2000, the Association of American Geographers' fourth-largest specialty group out of fifty-seven, behind the GIS, Urban Geography, and Remote Sensing groups. In terms of topical proficiency among AAG members, cultural geography looms even larger. Cultural geography is the third most frequently claimed area of proficiency, behind only GIS and Urban Geography, with 848 practicing professionals, or 13 per cent of the AAG membership. And, given Gallagher's and Sauer's points, the number of people who might be claimed by someone as cultural geographers would be much larger than this.

Reflecting on these numbers, it appears that, far from being a moribund subfield dying out in the face of a technological revolution in the discipline, cultural geography, however it may be defined, is actually flourishing on the eve of a new millennium. A quiet groundswell of interest in the diverse array of matters cultural investigated by the specialty group's members and well-wishers is evident in the "cultural turn" across the social sciences during the 1980s and 1990s (Chaney 1994). The past decade has seen a number of best-selling books and important scholarly texts which, if not typically written by cultural geographers, directly address the meaning of places, regions, or landscapes, or the importance of cultural geography to world history (Zukin 1991;

Cronon 1991; Crosby 1986; Diamond 1997; Schama 1997). Cultural geography, in the form of attention to the meaning of landscape, place, and space to society in general, artists, and identity politics, has taken center-stage in many areas of the humanities, such as in literary criticism, philosophy, art history, and history (Yeager 1996; Appadurai 1992; Casey 1997; R. Young 1995; McClintock 1995; Scott 1998). There are many other signs of strength for cultural enquiry in geography—new journals, new specialty groups with close linkages, and good enrollments in cultural classes on North American campuses, to name but a few.

This has been an exceedingly productive decade or so for cultural geographers from a variety of perspectives within the subfield. For instance, Rowntree, Foote, and Domosh make the point of saying in the first edition of *Geography in America* that, as of 1988, what they termed the "new" (generally, post-structuralist) cultural geography was more talked about than done. Many people have set about doing it in the past dozen years, and the output is a decidedly varied lot. The diversity within such nominally "new-cultural" edited volumes as Duncan and Ley (1993), Barnes and Duncan (1992), and Anderson and Gale (1992) is testimonial to this. "Diverse" and "extensive" are words that would characterize the output from more traditional cultural geography in recent years, too (see the broad array of approaches in Earle *et al.* (1996), Foote *et al.* (1994), or Carney (1998*a*), for instance).

Actually, Rowntree, Foote, and Domosh (1988: 209) went to great pains to reject the stereotyped dichotomy of traditional versus new cultural geography that emerged in the 1980s. While we tend to concur with their conclusion that "constructing such a dichotomy is an ill-founded strategy that privileges and reifies one segment over the other without the necessary critique and interactive discourse," the fact is that such dichotomies continue to be constructed. In his contribution to Foote *et al.*'s reader in cultural geography, a piece entitled "After the Civil War: Reconstructing Cultural Geography as Heterotopia," James Duncan (1994) makes the claim that the differences cannot be reconciled into one single subfield with a unified theory or method. Duncan argues that cultural geographers must "celebrate difference" (in his paper, difference means the different approaches of cultural geographers, but implies difference according to race, gender, sexual orientation, and the like). While we join Duncan in this celebration, in this chapter we must add the following caveat. In the interests of avoiding overlap, we do not deal extensively here with works explicitly announcing themselves as "cultural ecology" or "historical geography" even though most scholars in the two specialty groups by these names would in all likelihood consider themselves cultural geographers and indeed may even pay dues to our specialty group. All the same, it is our goal to set out a broad-minded appreciation of the vast subfield's recent scholarship regardless of the ideological, methodological, or theoretical divisions within our ranks.

Approaching Cultural Geography

According to the authors of *Geography in America*'s Cultural Geography chapter, this subfield "has been treated as an intellectual ambient or background out of which have come more focused subfields." Rowntree *et al.* (1988: 209) used that observation as a counterpoint for trying to organize what was then a very new specialty group. The express purpose of the specialty group was, and is, to serve as a forum for method and theory linked to increased interest in "culture, space, and landscape" throughout the social sciences. It has served this role well through panel and speaker sponsorship, and award activities at the annual meetings. The theory and practice of teaching cultural geography has increasingly become part of this forum alongside method and theory. While we concur that cultural geography cannot and should not be boiled down to a single formula, we do seek to contest the notion that this ambience is all there is. We argue that the study of the many cultural meanings of landscapes and places occupies something quite close to a unifying theme in cultural geography, in spite of the wide variety of approaches to questions of meaning. A second common theme linking most cultural geographers is an emphasis on examining popular, folk, and vernacular cultures, and even "high culture," in their geographical dimensions. Hence in a later section of the chapter we examine landscape and place studies, as well as cultural studies more generally, from this variety of perspectives. But first, it is important to understand a little more about these different perspectives.

The first edition's chapter (1988: 210) begins with a discussion of the "epistemological spectrum embraced by cultural geographers." Their spectrum included humanism, positivism, structuralism, and post-structuralism. Little, if any, cultural geography written in the 1990s could be said to originate from a positivist perspective, and one of the few solid strands of agreement among the subfield's practitioners would probably be on the limitations of positivist analysis for cultural enquiry in geography. Humanistic and post-structuralist approaches have expanded apace in the years since 1988, as have structuralist approaches. However, in the case of the latter, a more appropriate term would probably be materialist approaches, since practitioners of this type of cultural geography appear to have learned much lately from humanistic and post-structuralist critiques of the pitfalls of structuralism, even as they retain a political-economy perspective on cultural questions. Moreover, all these approaches have been influenced substantially by feminist theory, perhaps most substantively evidenced by the strong interconnections with cultural geography of the *Feminist Glossary of Human Geography* (McDowell and Sharp 1999).

In the successive sections below, we examine humanistic, post-structuralist, and materialist cultural geography of the past decade or so. In this review, we take account of both methodological and theoretical or conceptual distinctions. In some cases the differentiation between these approaches is stark, but in other cases differences are minor. Several authors cited in one camp can be said comfortably to cohabit the intellectual terrain of other camps, and they are cited as such, across the divides of this heuristic device. Without being overly sanguine, it can be said that, if we indeed have a civil war in the subfield, it is generally a civil one. Nowhere is this civility more in evidence than in the first several volumes of the annual, *Philosophy and Geography*, edited by Andrew Light and Jonathan M. Smith (1997, 1998*a*, *b*). Besides the co-editors, geographers, and philosophers as far afield ideologically from one another as Neil Smith, Edward Casey, Baird Caldicott and Henri Lefebvre appear in the volumes, and respectfully agree to disagree. It is a healthy sign of what we hope "celebrating difference" comes to mean in the subfield of cultural geography.

Barnett (1998: 31–4) has recently editorialized that surprisingly little theorizing has gone on within the seemingly endless discussions of culture theory in geography during the last ten years or so. Barnett's claim is that culture theorists take for granted far too many of their terms of reference and basic assumptions about what culture is and does. Many of the most taken-for-granted assumptions center on uses of language. Smith (1996) and Curry (1996) have taken on the immense challenge of problematizing how and why geographers write as they do. One of Curry's conclusions is that geographers' written output is itself a product of particular places, and particular political and ideological constellations of authority. The division of cultural geography's "work in the world" below probably oversimplifies these constellations into three, when the real "geography of geography" is myriad and legion. We none the less see most of the written output of cultural geographers as being related to at least one of these three meta-constellations, tied to the distinct concepts of culture employed in the approaches.

Humanistic Approaches

Humanistic geography emerged in the 1970s as a reaction to the geometric determinism of logical positivism and spatial science during the quantitative revolution in geography. The humanistic approach to cultural geography is concerned with questions related to the human meanings and values associated with the interpretation of cultural landscapes and places. While scientific geographers characterize their own approach as nomothetic in contrast to a traditional ideographic approach, humanistic geographers seek to emphasize a third dimension: meaning. So in addition to describing and explaining a landscape or place, they want to ask, what does it mean to be human beings? To the extent that meanings play a role in the creation of cultural environments where people live, it is also concerned with explanation. Humanistic geography focuses on human creativity, human consciousness, and understanding the human condition, with understandable ties to the traditional humanities disciplines of history, philosophy, and literature (Buttimer 1993; Conzen 1990; Zelinsky 1994; Jordan *et al.* 1997; Francaviglia 1991; Entrikin 1991; Tuan 1996). Those who would consider themselves humanistic geographers engage and accept a wide range of humanistic philosophies, including phenomenology, idealism, materialism, pragmatism, and realism.

From the humanistic perspective, cultural geography is more of an art than a science. Anne Buttimer (1993) declares that "there must be more to human geography than the danse macabre of materially motivated robots." Landscape study has been central to the intellectual maturation of humanistic geography because it has provided an explicit vehicle for description and analysis of the interaction between humans and their culturally constructed systems. Two methods of interpreting the landscape predominate. The first emphasizes the tangible elements of the cultural landscape (including within it much of the work of cultural ecologists), while the second stresses the cultural perception of human surroundings.

In the first approach, landscapes are viewed as visible expressions of material culture by documenting patterns of houses, barns, fences, land-use systems, and other settlement characteristics. These artefacts are then placed within a larger cultural context to yield insights into social processes, such as the diffusion of technologies and ideas or distinct cultural groups. The past decade has seen an impressive array of studies expanding our understanding of patterns and processes in the tangible landscape from a humanistic perspective. This vein of enquiry among American and Canadian geographers probably still has its center of strength in studies of North America (Conzen 1993; Hart 1998; Zelinsky 1992; Jordan and Kaups 1997; Jakle and Sculle 1994; Jakle *et al.* 1996; Noble and Wilhelm 1995). One of the premier authorities on the cultural geography of the United States, Terry G. Jordan-Bychkov has authored numerous works on phenomena ranging from cemeteries to cattle ranching (Jordan 1993). His research has examined the diffusion of a variety of Old World traits and their impact on the cultural landscape of the United States with particular emphasis on material culture in a folk-culture context (Jordan and Kaups 1997; and Jordan *et al.* 1997). Citing the "fertile bicontinental traditions of landscape study" fostered by William G. Hoskins and John Brinckerhoff Jackson as its inspiration, Michael Conzen introduced the edited collection *The Making of the American Landscape* (1990: p. vii). This volume represents an effort to address the major cultural and historical themes in the construction of America's regional landscapes. Wilbur Zelinsky has used his creativity and inventiveness to search for more and better methods of measuring the American cultural system and how its major components have varied through time and over space (Zelinsky 1992). One of his strongest suits is the ability to examine phenomena, particularly those belonging to popular or vernacular culture, that others have overlooked (Zelinsky 1994). Donald Meinig is another humanistic landscape scholar whose work reaches a broad audience without sacrificing its distinct geographical perspective. Volumes ii and iii of *The Shaping of America* (1993 and 1998) will remain standards for generations of Americanists. The fourth volume in Meinig's *The Shaping of America* series is in preparation, to be entitled, *Global America, 1915–1992.* In recent years, studies have appeared that focus on non-North American contexts, to expand this approach beyond its roots (Newman 1995; Silberfein 1998; Butzer 1992).

The goal of perception studies, the second approach, is to understand how people perceive and respond to their cultural environments. Some earlier studies were intuitive and interpretative, others empirical and behavioral. An expanded engagement with environmental historians, folklorists, anthropologists, ethnomusicologists, and architects has generally left the humanistic geographers less caught up with issues of the mind and psychology. It is more common to find a melding of the material culture concerns with those of environmental perception. This is true, for instance, of Shortridge's (1990) enquiry into the meaning of the Midwest to American culture. Still, several important volumes have drawn explicit attention to geography's inner worlds in a humanistic manner (Porteous 1990; Feld and Basso 1996; Sibley 1996). Perhaps the strongest direction of perception studies, albeit here heavily influenced by post-structuralist thought, involves the study of memory (Schama 1997; Tuan 1996; Sidorov 2000; Till 1999) or psychoanalysis (Pile 1993, 1996). The continuing work of Yi-Fu Tuan (1989, 1993, 1996, 2000) represents a rich vein in cultural geography's humanistic tradition. Always beginning near the heart of geography's central concerns, Tuan's curiosity reaches out to embrace the breadth of human experience in a way that few geographers have approached. A new edited volume (Adams *et al.* 2001) celebrates Tuan's influence at the time of his formal retirement. Tuan's tremendous productivity has been part of a wave of work that has deepened and broadened the philosophical sophistication of humanistic geography (see for instance Sack 1997).

Re-Reading Cultural Geography (Foote *et al.* 1994) attempts to survey the entire range of cultural geographic work, as an update to the Wagner and Mikesell (1962) classic. It contains a number of papers that take a broadly post-structuralist approach (such as those by Cosgrove and Duncan). However, it is a work largely absent of the "new" cultural geography and it sticks fairly close to a humanistic line in its effort to define the subfield (Zelinsky 1995).

The humanistic perspective, applied constructively in joining the divergent viewpoints of the sciences and the humanities, enhances the holistic nature of the discipline. The eagerness of many post-structuralist and materialist cultural geographers (for example Schein 1997) to engage and at times embrace elements of a humanistic perspective evidences its enduring strengths.

Post-Structuralist Approaches

Although relatively few cultural geographers explicitly label themselves as post-structuralists, we use the term to characterize the growing number who are distrustful of the totalizing theoretical claims of structuralist and positivist approaches and equally uncomfortable with the unproblematic empiricism they see in many traditional approaches (Doel 1999). One root of post-structuralism in cultural geography is the aforementioned humanistic tradition (authors such as Lowenthal, Wright, Tuan, or Buttimer) and especially its critiques of positivism's claim of authority (Entrikin 1991). Both Bouman (2001) and Foote *et al.* (2000) are examples of a blending of humanistic and poststructuralist approaches, in that they attempt to show the variety of meanings that can be attributed to events and places, and how complex it is to understand and mediate between them, let alone to inform public action. A similar blurring or blending of approaches is increasingly evident in *The Journal of Cultural Geography*'s latest issues as well. There is a more recent influence of cultural Marxism (see below under materialist approaches), and at least one reader that comfortably combines materialist, feminist, post-structuralist, and humanistic approaches in the analysis of place (McDowell 1997). But most of these geographers share a wariness of certain structuralist approaches that also relegate the cultural to the status of epiphenomenon (Entrikin 1990; Duncan 1990).

Significant influences on these practitioners come from beyond the discipline as well. Scholars in literary theory, anthropology, and cultural studies introduced work that served as a post-structuralist influence on cultural geographers as early as the 1970s. Anthropologist Clifford Geertz, historian and philosopher Michel Foucault, and literary theorist Edward Said, for instance, provided inspiration for many to reconceptualize their tasks. Geertz (1973) phrased his methodological concerns in terms of "thick description," citing the need to find meaning in the context of an array of cultural "texts." Foucault (1972) developed a post-structuralist interpretation of intellectual history in which he established the significance of discourse analysis. Discourses can be defined as social frameworks that enable and limit ways of thinking and acting. Thus, inherent in the concept of discourse are relations between discourses, knowledges, representations, and power. Following Foucault, Said (1978) employed discourse analysis to examine historical European representations of the "Orient." His work on Orientalism is perhaps the best-known analysis of imperial practices and discourses of the Other. By insisting on the systematic nature of European representations of colonized regions—and the power relationship associated with these imperial geographies—Said drew the attention of cultural geographers (as well as many non-geographers, of course).

Stimulating debates across disciplinary boundaries, these three scholars offered new perspectives as they situated place and culture at the center of analysis. Feminist theory further extended our conceptualization of discourses of gender and identity politics (Haraway 1991; Domosh 1996). In Nast's work (1996), for example, Foucault's more literary and discursive ideas are grounded in the spatial realities of gender and power relations in northern Nigeria.

Such influences reflect the theoretical invigoration surrounding the concepts of culture, place, and landscape in the expanding interdisciplinary (and largely post-structuralist) field of cultural studies. Rather than viewing culture as an unproblematic series of traits, or the unified possession of a group, these geographers describe culture metaphorically as an arena, a contested terrain. Furthermore, culture is considered in terms of process. Their conceptualizations of place and landscape, traditionally central concerns of cultural geography, are also problematized. Various metaphors can be employed. Place, for example, can be a spectacle, a text, a drama, a dialogue, or discourse of many voices. It is the site where a recursive process involving human agency, structure, landscape, and environment unfolds. Post-structuralist cultural geographers offer interpretations that are tentative, dynamic, situated in time and place, and open-ended (Duncan 1994; Blaikie 1995).

In addition to numerous articles and monographs, several important edited volumes have appeared in the last decade. Each of these volumes combines theoretical assessments with interpretations of actual landscapes or concrete geographical issues. Among the earliest of these collections, *The Iconography of Landscape* (Cosgrove and Daniels 1988), highlighted a central theme of the new cultural geography by focusing on the interrelationships among ideology, power, place, and landscape. In an introductory essay to the volume, Cosgrove and Daniels point out that while "every culture weaves its world out of image and symbol" (p. 8), we can only understand the text those images and symbols comprise by recognizing that their meanings are often unstable and opaque. Mark Harrison's investigation of crowd behavior in nineteenth-century English towns, for example, shows how working-class demonstrations attempted to redefine the meaning of certain symbolic urban sites.

In *Writing Worlds: Discourse, Text and Metaphor in the Representation of Landscape*, Barnes and Duncan (1992: 3) suggest that landscape and writing about landscape are complexly intertextual in such a way that writing does more than simply reflect the world: it helps to constitute it. The papers in this volume emphasize not only the instability of landscape meaning, but also the lack of authorial control, polyvocality and "irresolvable social contradictions" that render landscapes analogous to literary texts (1992: 7).

Stating an explicit commitment to post-structuralist analysis in cultural geography, Kay Anderson and Faye Gale (1992) introduced a series of essays in *Inventing Places: Studies in Cultural Geography* with a consideration of the recent turn toward culture in popular commentaries and academic study. Raising theoretical and methodological concerns, this collection delves into "the everyday knowledge of ordinary and elevated folk" (p. 10). This is accomplished in many of the case studies by a post-structuralist reading of the landscape. One contributor, however, argues for the appropriateness of a cultural geography without landscape by focusing on the representations of a particular place and people. Peter Jackson's (1992) analysis of visual representations of race and constructions of culture takes Edward Curtis's late nineteenth- and early twentieth-century photographs of North American Indians as its focus. Relating Said's discussion of imaginative geographies to America's construction of Native Americans, he examines visual images and their links to historically and culturally specific forms of domination. Joan Schwartz (1995) and James Ryan (1997) further demonstrate in their analyses the radical insights provided by photographs. Such images serve as artefacts for interpretation just as they problematize visual representation.

Duncan and Ley, in *Place/Culture/Representation* (1993), identify a distinction between traditional approaches to cultural geography, which attempt to represent the world through mimesis, and postmodern and interpretative approaches, which deny the possibility of an objective point from which to make a perfect representation. Our way of representing the world may seem "natural," but others with a different point of view can see our discrepancies. A representation is therefore always a partial truth, "the outcome of a relation between an empirical world and a historical subject" (1993: 4). James Duncan's (1993*b*) article, on "Sites of Representation," focuses on these issues in his attention to the making of cultural geographies. The tropes (or rhetorical devices) of mimesis—the claim to representational accuracy and authority, and spatialized time, the representation of a foreign place as a previous historical period—are illustrated by examining nineteenth-century European representations of Africa. The influence of Said and other scholars of post-colonial studies can be seen in this essay, and in other recent volumes dealing with colonial travel writing (Blunt 1994; Blunt and Rose 1994; Duncan and Gregory 1999; Phillips 1997). As cultural geographers have taken up such analyses of colonial geographies, many maintain an interest in examining colonial discourses as a means of addressing Western discourses, thus revealing the centrality of imperialism in the cultural representations of Britain to the British (Kenny 1995).

Geography's complicity in the development of imperialism and colonialism occupies the attention of many post-structuralists, under the rubric of post-colonial theory (Godlewska and Smith 1994; Heffernan and Dixon 1991). Post-colonial critiques have broadened this opening perspective, however, to accommodate the ambivalence of colonial discourse in social relations between imperial and indigenous elites (Crush 1996; Chatterjee and Kenny 1999). Even the very methods of and approaches to research, in the manner that they may replicate colonialist or imperialist categories, come under scrutiny, as in a challenging piece by Jennifer Robinson (1994). From literary theory (R. Young 1990, 1995; Bhabha 1984), geographers have taken up the concepts of mimicry and cultural hybridity to evaluate the significance of cultural hegemony in the imperialist project. By grounding research in historical and geographical specificity, increasingly cultural geographers argue situations of complicity (if not consent) as well as resistance (Myers 1999; K. Mitchell 1997).

Jane Jacobs and Ruth Fincher introduce their edited volume, *Cities of Difference* (1998), by asking how contemporary theories of difference might enhance our understanding of traditional urban studies. They advocate an approach described as a cultural political economy approach to explore the relationship between urban space and identity politics (see the discussion of materialist approaches below). The fragmented nature of contemporary urban areas reflects multiple axes of difference, including race, ethnicity, class, gender, sexuality, and measures of able-bodiedness, and requires an appreciation of the nuanced cultural politics of cities. This collection demonstrates applications for feminist and post-colonial theory and the need for reconceptualization of cultural geographies of postmodernity.

The new cultural geography has become increasingly post-structuralist and increasingly visible. There are several new journals (*Ecumene*—recently relaunched and retitled as *Cultural Geographies*—as well as *Social and Cultural Geography*, and *Gender, Place, and Culture*) and introductory textbooks (M. Crang 1998; Cloke *et al.* 1999; Massey *et al.* 1999) that present cultural geography in the light of contemporary theoretical issues. The dialogue within cultural geography has served to problematize the distinction between the traditional and the new. Price and Lewis (1993) criticize the new cultural geography for simplifying the Sauerian tradition and advocating an exclusionary theoretical expertise. Olwig (1996) more obliquely suggests the new cultural geographers have taken a wrong turn by "dematerializing" the study of landscape. Replies by Cosgrove (1993), Duncan (1993*a*) and Jackson (1993) to the Price and Lewis piece in particular suggest that critics overemphasize the role of cultural ecology in the Berkeley School and ignore the diversity of approaches that have been characterized as "new cultural geography." A small but growing contingent of cultural geographers appear to utilize elements of the traditional empirical cultural geography in pursuit of new cultural geography theoretical aims. Timothy Oakes (1997), in an example of this new type of empirically grounded yet theoretically rigorous analysis, reinterprets not only the tradition of cultural geographic conceptualizations of place, but also those of literary modernism. He sees in both the recognition that places of modernity are inherently paradoxical sites that resist "theoretical closure" (Oakes 1997: 523). By our definition, of course this is a post-structuralist view of traditional approaches to place.

In an age of "blurred genres" (Geertz 1983) when the boundaries between disciplines and subdisciplines seem to be dissolving, what distinguishes cultural geography is partly the dialogue with its own tradition. While post-structuralists such as John Paul Jones III (1998) find themselves in dialogue with the tradition of spatial analysis, cultural geography's post-structuralists engage a pluralistic tradition which is arguably more hospitable to their varied approaches.

Post-structuralism's grip on cultural geography is quite strong. The skepticism about totalizing visions emanating from post-structuralists has taken root in Lewis (the same Lewis of the much-cited critique of new cultural geography!) and Wigen's (1997) intriguing "critique of metageography," wherein various myths about how the world is represented in the Western imagination are debunked. Even a more traditional book such as the edited volume, *Fast Food, Stock Cars and Rock-n-Roll* (Carney 1996) contains doses of social theory in some of its contributions. Moreover, post-structuralism's influences have permeated Marxist thought in geography, to which we now turn.

Materialist Approaches

Materialist approaches to geography generally revolve around Marxism of one sort or another. Much of the Marxist tradition in geography is typically classed with urban or economic rather than cultural, subfields. However, the past decade has seen something of a cultural turn in Marxist geography. The Marxist-inspired analyses of Massey (1993), Harvey (1989, 1996), and Soja (1989, 1997) float in and around cultural issues so enticingly and passionately that many "cultural studies" edited volumes, critics, and syllabi outside geography take these three (largely economic) geographers as geography's main contributors to cultural questions (see also Thrift 2000). The cultural Marxism of Raymond Williams (1977, 1980, 1982) and Antonio Gramsci (1971) is embedded in the very language of much of the new cultural geography discussed above as post-structuralist cultural geography. For instance, Cosgrove's (1989) influential piece, with its engagingly written claim that "geography is everywhere," poses the suggestion that the study of landscape ought to proceed according to a schema differentiating "dominant, residual, emergent, and excluded" landscapes. The first three terms are a direct development from Williams's (1977) schema for differentiating cultural movements in a social formation. Peter Jackson (1989) and James Duncan (1990) similarly lean heavily on Williams, who was himself deeply influenced by Gramsci's ideas.

Other, more explicitly Marxist cultural geographers have taken an avowedly materialist approach to questions of culture in the subfield, even while learning from the post-structuralists to take issues of representation, symbolism, language, and discourse more seriously. These materialist cultural geographers, however, take issue with what they see as post-structuralists' over-reliance on metaphorical and representational analysis and idealism, at the expense of realist and grounded analyses of actual landscapes and places (see the debates between Walton (1995, 1996), Peet (1996*a*), and D. Mitchell (1996*b*) in the *Professional Geographer*). Dick Peet's (1996*b*) essay on the mainstream and alternative memorializations of Daniel Shays in western Massachusetts grounds issues of representation in their political and material context. As with post-structuralists in cultural geography, materialist researchers have turned cultural geography toward more urban and industrial settings in comparison to the cultural geography of earlier years. The emphasis here is more often on realist contexts of struggle, such as the port docks and union halls of Andrew Herod's (2000) work, or the city streets of the American South in Alderman's (1996) study. Don Mitchell's growing body of work examines immigrant labor, public space, and housing rights in various California and Eastern US urban settings (1993, 1995, 1996*a*), and he has extended his claims about the "work of landscape" into a materialist's version of a critical introduction to cultural geography (Mitchell 2000).

Places—their iconography, their representation, their soul—appear to matter far more to materialist and Marxist cultural geographers in the 1990s than they did to earlier Marxist-influenced geographers. The role of place in resistance to the penetration of capitalism and colonialism is central to such richly cultural works as those of Brenda Yeoh (1997), Mike Davis (1990, 1998), or the anthropologist Donald Moore (1993, 1997). Moore, like a small but growing contingent of left-leaning anthropologists, historians, or literary theorists, is refreshingly attentive to what cultural geographers (including non-Marxist thinkers) do and say, enlivening his theorization of place with rich empirical narration and Gramscian concepts of power. Place-consciousness and Gramscian analysis also merge in the works of such historians as Jonathan Glassman (1995) or literary theorists such as Said (1993, 1995). Gramsci's influence within cultural geography itself is often more subtle, as in the works cited above by Cosgrove, P. Jackson, or Duncan. Withers (1988) utilizes Gramsci's idea of hegemony to frame his study of the transformation of Gaelic Scotland's cultural geography. Johnson (1992) takes Gramsci's ideas on the role of intellectuals in cultural hegemony into the analyses of the geography of educators in Ireland, and Myers (1994, 1998) assesses the applicability of Gramscian hegemony theory in colonial British Africa, for instance.

Ultimately, the challenge for cultural geographers inspired by materialist and Marxist frameworks of understanding is to give genuine and urgent attention to questions of individual agency and consciousness so often subsumed under the weight of economistic thinking in Marxism. The need to balance materialist concerns with the constraints imposed by economic structures with Marxism's new-found interest in cultural matters is, perhaps in a different form, the same struggle for balance encountered by cultural geographers across the ideological spectrum. This may have to do with what cultural geographers ultimately must depend upon in their works, and that is narrative structure, and the struggle to make respectable facts out of subjective realities and gathered lore. As Marxism makes claims to be a science, cultural studies geographers who embrace it have much to balance in their analyses. As Entrikin (1991: 58) put it:

> One of the goals of the modern cultural geographer is to interpret the meaning of places. The geographer becomes a translator, translating the story of places in such a way that the subjective and objective realities that compose our understanding of place remain interconnected. The geographer as narrator . . . constructs a narrative aimed at the different concerns of objective representation and truth. In this way the geographer strives to be scientific. However, the goal of scientific objectivity is only one among several possible goals. Another concern is to gain insight into the experience of place as context. . . . This goal is not always compatible with the scientific viewpoint.

No matter their theoretical underpinnings, cultural geographers struggle with the "betweenness" of their approaches to the subject matter—between science and art, between objectivity and subjective experience. This struggle comes to the fore in studies that explicitly aim at interpreting places and landscapes, and this body of work is our next focus.

Studying Cultural Geography: Places, Landscapes, Everyday Life, Popular and Folk Culture

At least a half-dozen different ways of studying landscape find their home in cultural geography from the theoretical perspectives outlined above. Each of these includes, in some way, the general effort to uncover the landscape's cultural meanings. Some will view the landscape as an ecological artefact. Environmental historians, for instance, use landscape as the organizing theme for work of this type, generated primarily from archival sources, as in the work of "new western historians" (Worster 1993; Cronon 1991; Limerick *et al.* 1991). At the same time that cultural geographers are starting to come to grips with the ongoing and future currents of globalization or global cultural flows as they impact places and landscapes (Appadurai 1990), interestingly enough, history has re-emerged as a central concern in the subfield. The authors of the corresponding chapter in *Geography in America* (Rowntree *et al.* 1988) admit to cultural geography as preferring "diachronic depth as central to its methodology." Cultural geography's relationship to environmental history has indeed expanded in the past decade. The popular and academic growth of this sort of cultural geography can also be seen in the "new historicism" where literary scholars are now deeply concerned with context, the world in which a work emerged.

In cultural ecology and historical geography, particularly in the tradition of humanistic enquiry, landscapes become evidence for culture origins and the diffusion of ideas, as well as the data banks of material culture (Meinig 1993; Wishart 1994; Francaviglia 1991; Pasqualetti 1997). The visual or material artefacts concerning human occupation and settlement take precedence in the work of these material culture scholars and those in such allied fields as architectural history, folklore, and historical archaeology. Together with J. B. Jackson's (1994) emphasis on vernacular cultural landscapes, the new western history movement has heavily influenced the output of cultural geographers who work on landscape and culture questions in the Midwest, Great Plains, Mountain West, and Pacific Coast (Alanen 2000, 1997;

Blake 1995; Kearns 1998; Gumprecht 1998; Starrs 1998; John 2001; Sluyter 2001). The enduring legacies of the Progressives in America's landscape also occupy the attentions of cultural geographers (T. Young 1993, 1995, 1996). Heritage tourism, and the representation of the histories of landscapes and places in it, is often the place "where geography and history meet" for many cultural geographers (Johnson 1996: 551; Chang 1999).

The past decade, however, has seen a growth not only in these areas of cultural geographic enquiry into landscapes (many of which date back at least four decades to the work of such scholars as Sauer and Kniffen), but also in research which seeks landscape meaning in the arts. Art, literature, music, and film analyses of landscape meaning are increasingly common in the works of cultural geographers. McGreevy's (1994) study of the meaning and making of Niagara Falls is one example. Geographers have recognized that much can be learned by the way people depict the landscape in art, photography, literature, music, and film (Turner 1989; Carney 1994; Leyshon *et al.* 1995, 1998; Zonn 1990; Shortridge 1991; Sternberg 1998). Much of this recent research emphasizes the aesthetic or scenic component of a landscape's heritage. The visual aspects of landscapes are no longer taken simply as fact in this work, but instead are scrutinized. The ideological qualities of vision are interrogated (Schwartz 1995; Rose 1992; Daniels 1987; Atkinson and Cosgrove 1998). Finally, post-structuralist approaches also study the iconography of landscape, seeing in its texts, symbols, and signs the moving targets and refractions of cultural identities. Here, landscapes have been treated metaphorically as texts that were authored and could be read by insightful observers, as in "reading the cultural landscape" (P. Lewis 1979). Others utilize iconography as a vehicle for landscape analysis, or semiotics that conceptualize landscape into sign and symbol systems (Duncan 1990).

In 1988, Rowntree *et al.* felt it remained to be seen whether the British sort of social geography would influence North American cultural geographers' study of North America. It is clear a decade later that it has, alongside the continued interest in J. B. Jackson-style studies. Tim Cresswell's (1996) study of derelicts, hippies, radical women, and other "heretics" transgressing urban order in both US and European settings, McGreevy's (1990) article on the social meaning of the American celebration of the Christmas ritual, and Steve Herbert's (1996) ethnography of normative orders in the conceptions of space and place among police officers in Los Angeles, in different ways attest to the "British" influences on landscape and place studies. Even more traditional pieces aimed at reading the landscape (see Blake and Arreola 1996) now take on questions of power relations and social structure, which Rowntree, Foote, and Demosh found to be only a beginning trend.

Place occupies a niche as a subject of enquiry distinct from landscape. Perhaps it was Carl Sauer who was the first American geographer concerned with the concept of place. In his dissertation on the Missouri Ozarks, he calls it his "home geography." Later, Yi-Fu Tuan put place in the context of humanistic geography. He declares that it is the feeling and emotional attachment we have to a place that gives it human dimension and meaning (Tuan 1976: 269). Abler (1987: 513) states that geographers should "speak first and foremost of places and regions." He goes on to say these should be "real places, especially their internal workings, what they look like, what they smell like, what it feels like to be there" (ibid.). Pierce Lewis (1985: 468) posited that we should maintain "a passion for the earth (topophilia, or a love of places), more especially some beloved part of the earth. It is a passion that equates geography with particular places at particular times and does it at a gut level, without any attempt to analyze or dissect that place, or subject it to scientific scrutiny." Wilbanks says that places are what folks "out there" want us geographers to talk and write about. The American public, according to Wilbanks (1994), demands that geographers write about places because people are fascinated and curious about them. In understanding the "mysteries" of places, cultural geographers operate under the obligation to provide the "clues" for unraveling the "mysteries" of places. Continuing the metaphor, cultural geographers are often like "place detectives" in uncovering the knowledge and bonds that exist in a people/place relationship. Among these are the look, feel, smell, taste, and sound of a place; human senses that help us discern the "sense of place." We also want to know about the main players in this place, or the "characters" of a place who add the human element to it. Moreover, we are concerned with the imprints that people make on a place. Thus, cultural geographers are in the business of decoding the character of places generated by both folk and popular culture. Increasingly, this brings cultural geographers into the study of politics as well (Keith and Pile 1993; Agnew and Duncan 1989).

Cultural geographers interpret all levels of culture (high–low, elitist–populist, crude versus the fine arts, and folk–popular) and the manner they are manifested spatially. From the 1970s to the present, several cultural geographers have examined various popular culture phenomena, especially literature, foodways, music, architecture, sports, and film, from a humanistic viewpoint (de Wit 1992; Flack 1997; Kong 1995; Nash and

Carney 1996; Adams 1992; Aitken 1991; Aitken and Zonn 1993; Bale 1989; Carney 1998*b*). By and large, these studies focus on an analysis of distribution patterns, delineation of culture regions, or identifying the origins and charting the diffusion paths of the trait and its associated characteristics. Recently, their work has been joined by post-structuralist and materialist studies of everyday life, popular culture, and folk culture, from within cultural geography (Moss 1992).

Future Directions in Cultural Geography

As the title of two panels at the AAG Annual Meeting in Hawaii in 1999 suggested, "making cultural geography work" has joined the more abstract discussions of theory that dominated the 1980s. This doesn't mean that debates over theory have been eclipsed in the subfield—far from it—but the terms of discussion in the 1990s shifted toward getting on and actually *doing cultural geography*—this latter phrase in fact is the title of a new edited volume on practical methodological concerns (Shurmer-Smith 2002). One aspect of "making cultural geography work" has involved cultural geographers' engagement with issues of ethics in the expansion of Geographic Information Systems (GIS). John Pickles's (1995) edited volume and a series of pieces by Michael Curry have initiated what might be called a "cultural studies of science" approach to GIS from cultural geographers. Indeed, their cultural geographic sensitivity to the complexity of places and landscapes allows them to recognize the reduction and systematic bias that inevitably accompanies GIS representations. This practice of employing the central strengths of cultural geography to investigate issues beyond our traditional ken is growing. Pragmatic questions unrelated to geotechnology are present, for instance, in Kong, Yeoh, and Teo's (1996) study of what it is like to be old in Singapore. Cultural geography, to Kong and her co-authors, offers the tools for understanding the experience of place among Singapore's elderly population, in an effort to enhance their quality of life. Foote (1997) takes readers from the battle of Gettysburg to the bombing site in Oklahoma City for a highly pertinent interrogation of the ways that places of violence and tragedy, like the former World Trade Center, have left indelible marks on the American landscape.

Cultural geographers have also begun to examine the implications of globalization for cultural questions. Joe Wood's (1997) piece on Vietnamese place-making in Northern Virginia and several recent foodways volumes (Pillsbury 1990, 1998; Shortridge and Shortridge 1998) open the door to this type of analysis for North American studies. Michael Watts's (1991) sweeping analyses of a religious movement in northern Nigeria, Bale and Sang's (1996) fascinating excavation of *Kenyan Running* in this period of immense global change, and Chang's (1999) study of heritage tourism in Singapore are but three examples of valuable contributions from non-North American contexts. Even if, as Zelinsky proclaimed, "we have begun flashing light into hitherto shadowy corners of the cultural cosmos" (1992: 144), there remain many unexplored frontiers of cultural geography. The transnationalization of culture—the recent large-scale sharing among the peoples of many lands of cultural items previously restricted to individual countries, e. g. music, clothing, technology, and foodways (Zelinsky 1992)—is one of those unexplored frontiers.

Other areas of enquiry are still in need of attention. Not much has been accomplished since the 1980s in the field of religious studies in cultural geography, in spite of some intriguing openings into new ideas (Nagar and Leitner 1998; Kong 1993; Heatwole 1989; Hopkins 1990; Stump 2000). There have been many studies on folk architecture, but we need more analyses on popular and academic architecture (e.g. Kenny 1997; Hubka and Kenny 1999; Domosh 1988, 1989; Lees 2001; Till 1995). Metaphorically speaking, the geographic body is bare when it comes to clothing, adornment, and attire. Metaphorically speaking, the geographic cupboard is partially filled—but more studies are needed—on foodways, particularly those with innovative methodological and conceptual approaches (Law 2001; Alexander 2000).

The publication of the National Geography Standards in 1994, although such standards are by their very nature just the sort of template we suggest most cultural geographers shy away from, is a milestone suggestive of another key aspect of the more practical application of insights from cultural geography. That aspect is, simply put, that teaching matters a great deal to cultural geographers since it is indeed a large part of what we do. Like investigating places and landscapes, teaching is a central practice of our subfield, and theory may be just as relevant to the latter as the former. If knowledge is socially constructed, how do we justify lecturing to students and asking them to memorize the facts we have distilled? The specialty group's annual business meetings routinely discuss the teaching experiences of members. Cultural geographers regularly publish in geography education journals or edited volumes (see for instance Fredrich and Fuller 1998; Fredrich 1998). Kit Salter's (1994) pithy "put aside your books" article in Foote *et al.* (1994) captures the desire among many cultural geographers to roll up their sleeves and get to work in the classroom training people in how to appreciate the landscapes and places around them. Geographers across the ideological spectrum have long expressed exasperation with the discipline's inability to capitalize on the implicit popularity of geography (Wright 1926; Kropotkin 1885; Harvey 1984), and many cultural geographers seem intent on declaring that now is the time to capitalize, through teaching and through work as public intellectuals.

Conclusion

Much like the discipline as a whole, the subfield of cultural geography is diverse and eclectic. Cultural geographers study a myriad of phenomena. As such, diversity and eclecticism have become traditions.

All of this rolling up of sleeves in practical research and teaching does not mean that theoretical perspectives and insights cannot help us do the traditional tasks of cultural geography better, but simply that cultural geographers are among those in our discipline who still see the importance of those tasks. Partly because of experiences during the so-called quantitative revolution, many cultural geographers remain wary of a theoretical template that would reduce or constrict the latitude of their research. Some are skeptical of the assumption of progress that seems to pervade some theoretical prescriptions, evident in such terms as the new cultural geography, or post-structuralism, or post-modernism. Many cultural geographers are driven by the urge to conduct substantive research involving in-depth fieldwork or small-scale community studies—especially the hands-on experience of interviewing local informants and investigating local sources. They are impatient with philosophical debate. Yet theory, at its best, leads us to expand our view, to be aware of our commitments, to see the larger context. For instance, we can see in Duncan and Duncan (2001) clear ways in which highly theoretical cultural geography can offer insight into current public policy issues. In some ways, all the theoretical debates in literary studies have come down to paying close attention to the full context: ideological, intellectual, and material. Theory is important, because it keeps expanding our understanding of these contexts.

Our traditional appreciation of the multidimensional complexity of the human world warns us against erecting false disciplinary walls. Many cultural geographers are naturals at cross-disciplinary work. It is well past time for us vigorously to engage other researchers both within and outside the discipline, especially folklorists, historians of all breeds (environmental, architectural, and social), anthropologists, cultural studies, and popular culture enthusiasts (Dogan and Pahre 1990). Cultural geographers could begin by increasing their involvement with multidisciplinary groups.

In conclusion we wish to affirm that the vitality of cultural geography rests on two principles. First, we must celebrate the diversity of approaches to understanding the meaning of place and landscape; and second, we must recognize that the future lies in a less parochial cultural geography, more open to crossing disciplinary boundaries in every direction. It is the strength of the cultural geographic perspective—rather than its absence—that makes this expansiveness possible. In the other direction is suffocation.

References

Abler, R. (1987). "What Shall We Say? To Whom Shall We Speak?" *Annals of the Association of American Geographers*, 77/4: 511–24.

Adams, P. (1992). "Television as Gathering Place." *Annals of the Association of American Geographers*, 82/1: 117–35.

Adams, P., Hoelscher, S., and Till, K. (eds.) (2001). *Textures of Place: Exploring Humanist Geographies.* Minneapolis: University of Minnesota Press.

Agnew, J., and Duncan, J. (eds.) (1989). *The Power of Place.* Boston: Unwin Hyman.

Aitken, S. (1991). "A Transactional Geography of the Image-Event: The Films of Scottish Director, Bill Forsyth." *Transactions of the Institute of British Geographers* 16: 105–18.

Aitken, S., and Zonn, L. (eds.) (1993). *Place, Power, Situation, and Spectacle: A Geography of Film.* Savage, Md.: Rowman & Littlefield.

Alderman, D. (1996). "Creating a New Geography of Memory in the South: (Re)naming of Streets in Honor of Martin Luther King, Jr." *Southeastern Geographer*, 36/1: 51–69.

Alanen, A. (1997). "Homes on the Range: Settling the Penokee-Gogebic Iron Ore District of Northern Wisconsin and Michigan," in R. Ostergren and T. Vale (eds.), *Wisconsin Land and Life*. Madison: University of Wisconsin Press.

—— (2000). *Preserving Cultural Landscapes in America*. Baltimore: Johns Hopkins University Press.

Alexander, D. (2000). "The Geography of Italian Pasta." *The Professional Geographer*, 52/3: 553–66.

Anderson, K., and Gale, F. (eds.) (1992). *Inventing Places: Studies in Cultural Geography*. Melbourne: Longman Chesire.

Appadurai, A. (1990). "Disjuncture and Difference in the Global Cultural Economy," in M. Featherstone (ed.), *Global Culture*. London: Sage.

Atkinson, D., and Cosgrove, D. (1998). "Urban Rhetoric and Embodied Identities: City, Nation, and Empire at the Vittorio Emanuele II Monument in Rome, 1870–1945." *Annals of the Association of American Geographers*, 88/1: 28–49.

Bale, J. (1989). *Sports Geography*. New York: Spon.

Bale, J., and Sang, J. (1996). *Kenyan Running*. London: Frank Cass.

Barnes, T., and Duncan, J. (1992). *Writing Worlds: Discourse, Text and Metaphor in the Representation of Landscape*. London: Routledge.

Barnett, C. (1998). "Cultural Twists and Turns." *Environment and Planning D: Society and Space*, 16/6: 631–4.

Bhabha, H. (1984). "Of Mimicry and Man: The Ambivalence of Colonial Discourse." *October*, 28: 125–33.

Blaikie, P. (1995). "Changing Environments or Changing Views? A Political Ecology for Developing Countries." *Geography*, 80/3: 203–14.

Blake, K. (1995). "Zane Grey and Images of the American West." *Geographical Review*, 85/2: 202–16.

Blake, K., and Arreola, D. (1996). "Residential Subdivision Identity in Metropolitan Phoenix." *Landscape Journal*, 15/1: 23–35.

Blunt, A. (1994). *Travel, Gender, and Imperialism: Mary Kingsley and West Africa*. New York: Guilford Press.

Blunt, A., and Rose, G. (eds.) (1994). *Writing Women and Space: Colonial and Postcolonial Geographies*. New York: Guilford.

Bouman, M. (2001). "A Mirror Cracked: Ten Keys to the Lake Calumet Region." *Journal of Geography*, 100/3: 104–9.

Buttimer, A. (1993). *Geography and the Human Spirit*. Baltimore: Johns Hopkins University Press.

Butzer, K. (1992). "The Americas Before and After 1492," *Annals of the Association of American Geographers*, Special Edition, 82/3.

Carney, G. (1994). *The Sounds of People and Places: A Geography of American Folk and Popular Music*. Lanham, Md.: Rowman & Littlefield.

—— (1996). *Fast Food, Stock Cars, and Rock-and-Roll: Place and Space in American Pop Culture*. Lanham, Md.: Rowman & Littlefield.

—— (1998*a*). *Baseball, Barns, and Bluegrass: A Geography of American Folklife*. Lanham, Md.: Rowman & Littlefield.

—— (1998*b*). "Geography of Music." *Journal of Cultural Geography*, 18: 1–97 (Special Issue).

Casey, E. (1997). *The Fate of Place: A Philosophical History*. Berkeley: University of California Press.

Chaney, D. (1994). *The Cultural Turn*. London: Routledge.

Chang. T. (1999). "Local Uniqueness in the Global Village: Heritage Tourism in Singapore." *The Professional Geographer*, 51/1: 91–103.

Chatterjee, S., and Kenny, J. (1999) "Creating a New Capital: Colonial Discourse and the Decolonization of Delhi." *Historical Geography*, 27: 73–98.

Cloke, P., Crang, P., and Goodwin, M. (eds.) (1999). *Introducing Human Geographies*. New York: Arnold.

Conzen, M. (1990). *The Making of the American Landscape*. Boston: Unwin Hyman.

Conzen, M., Rumney, T., and Wynn, G. (eds.) (1993). *A Scholar's Guide to Geographical Writing on the American and Canadian Past*. Chicago: University of Chicago Press.

Cosgrove, D. (1989). "Geography is Everywhere: Culture and Symbolism in Human Landscapes," in D. Gregory and R. Walford (eds.), *Horizons in Human Geography*. London: Macmillan.

—— (1993). "On 'The Reinvention of Cultural Geography' by Price and Lewis: Commentary." *Annals of the Association of American Geographers*, 83/3: 515–17.

Cosgrove, D., and Daniels, S. (1988). *The Iconography of Landscape*. Cambridge: Cambridge University Press.

Crang, M. (1998). *Cultural Geography*. London: Routledge.

Cresswell, T. (1996). *In Place/Out of Place: Geography, Ideology and Transgression*. Minnesota: University of Minnesota Press.

Cronon, W. (1991). *Nature's Metropolis*. New York: Norton.

Crosby, A. (1986). *Ecological Imperialism*. New York: Cambridge University Press.

Crush, J. (1996). "The Culture of Failure: Racism, Violence and White Farming in Colonial Swaziland." *Journal of Historical Geography*, 22: 177–97.

Curry, M. (1996). *The Work in the World*. Minneapolis: University of Minnesota Press.

Daniels, S. (1987). "Marxism, Culture and the Duplicity of Landscape," in R. Peet and N. Thrift (eds.), *New Models in Geography*, ii. London: Unwin Hyman.

Davis, M. (1990). *City of Quartz*. London: Verso.

—— (1998). *The Ecology of Fear*. London: Verso.

deWit, C. (1992). "Food-Place Associations on American Product Labels." *Geographical Review*, 82: 323–30.

Diamond, J. (1997). *Guns, Germs, and Steel: The Fates of Human Society*. New York: Norton.

Doel, M. (1999). *Poststructuralist Geographies: The Diabolical Art of Spatial Science*. Lanham, Md.: Rowman & Littlefield.

Dogan, M., and Pahre, R. (1990). *Creative Marginality: Innovation at the Intersections of Social Sciences*. Boulder: Westview Press.

Domosh, M. (1988). "The Symbolism of the Skyscraper: Case Studies of New York's First Tall Buildings." *Journal of Urban History*, 14: 321–45.

—— (1989). "A Method for Interpreting Landscape: A Case Study of the New York World Building." *Area*, 21: 347–55.

—— (1996). "Feminism and Human Geography," in Earle, Mathewson, and Kenzer (1996).

Duncan, J. (1990). *The City as Text*. Cambridge: Cambridge University Press.

—— (1993*a*). "On 'The Reinvention of Cultural Geography' by Price and Lewis: Commentary." *Annals of the Association of American Geographers*, 83/3: 515–17.

—— (1993*b*). "Sites of Representation: Place, Time, and Discourse of the Other," in Duncan and Ley (1993).

—— (1994). "After the Civil War: Reconstructing Cultural Geography as Heterotopia," in Foote, Hugill, Mathewson, and Smith (1994).

Duncan, J., and Duncan, N. (2001). "The Aestheticization of the Politics of Landscape Preservation." *Annals of the Association of American Geographers* 91/2: 387–409.

Duncan, J., and Gregory, D. (eds.) (1999). *Writes of Passage: Reading Travel Writing*. London: Routledge.

Duncan, J., and Ley, D. (eds.) (1993). *Place/Culture/Representation*. London: Routledge.

Earle, C., Mathewson, K., and Kenzer, M. (eds.) (1996). *Concepts in Human Geography*. Lanham, Md.: Rowman & Littlefield.

Entrikin, J. N. (1991). *The Betweenness of Place*. Baltimore: Johns Hopkins University Press.

Feld, S., and Basso, K. (1996). *Senses of Place*. Santa Fe: School of American Research.

Flack, W. (1997). "American Microbreweries and Neolocalism: 'Ale-ing' for a Sense of Place." *Journal of Cultural Geography*, 16: 37–53.

Foote, K. (1997). *Shadowed Ground: America's Landscapes of Violence and Tragedy*. Austin: University of Texas Press.

Foote, K., Toth, A., and Arvay, A. (2000). "Hungary After 1989: Inscribing a New Past on Place." *Geographical Review*, 90/3: 301–34.

Foote, K., Hugill, P., Mathewson, K., and Smith, J. (eds.) (1994). *Re-Reading Cultural Geography*. Austin: University of Texas Press.

Foucault, M. (1972). *The Archeology of Knowledge and the Discourse on Language*. New York: Random House.

Francaviglia, R. (1991). *Hard Places: Reading the Landscape of America's Historic Mining Districts*. Iowa City: University of Iowa Press.

Fredrich, B. (1991). "Food and Culture: Using Ethnic Recipes to Demonstrate the Post-Columbian Exchange of Plants and Animals." *Journal of Geography*, January–February: 11–15.

Fredrich, B., and Fuller, K. (1998). "Linking Geography and Art: Inness' 'The Lackawanna Valley.'" *Journal of Geography*, November–December: 254–62.

Gallagher, C. (1995). "Raymond Williams and Cultural Studies," in C. Prendergast (ed.), *Cultural Materialism: On Raymond Williams*. Minneapolis: University of Minnesota Press.

Geertz, C. (1973). *The Interpretation of Cultures*. New York: Basic Books.

—— (1983). *Local Knowledge*. New York: Basic Books.

Glassman, J. (1995). *Feasts and Riot*. Portsmouth, NH: Heinemann.

Godlewska, A., and Smith, N. (eds.) (1994). *Geography and Empire*. Oxford: Blackwell.

Gramsci, A. (1971). *Selections from the Prison Notebooks*. New York: International Publishers.

Gregory, D. (1994). *Geographical Imaginations*. Oxford: Blackwell.

Gumprecht, B. (1998). *The Los Angeles River: Its Life, Death, and Possible Rebirth*. Baltimore: Johns Hopkins University Press.

Haraway, D. (1991). *Symians, Cyborgs and Women*. London: Routledge.

Hart, J. F. (1998). *The Rural Landscape*. Baltimore: Johns Hopkins University Press.

Harvey, D. (1984). "On the History and Present Condition of Geography: An Historical Materialist Manifesto." *Professional Geographer*, 3: 1–11.

—— (1989). *The Condition of Postmodernity*. Oxford: Blackwell.

—— (1996). *Justice, Nature, and the Geography of Difference*. Oxford: Blackwell.

Heatwole, C. (1989). "Sectarian Ideology and Church Architecture." *Geographical Review*, 79: 63–78.

Heffernan, M., and Dixon, C. (eds.) (1991). *Colonialism and Development in the Contemporary World*. London: Mansell.

Herbert, S. (1996). *Policing Space*. Minneapolis: University of Minnesota Press.

Herod, A. (2000). "Implications of Just-in-Time Production for Union Strategy: Lessons from the 1998 General Motors-United Auto Workers Dispute." *Annals of the Association of American Geographers*, 90/3: 521–47.

Hopkins, J. (1990). "West Edmonton Mall: Landscape of Myths and Elsewhereness." *The Canadian Geographer*, 34: 2–17.

Hubka, T., and Kenny, J. (1999). "The Transformation of the Worker's Cottage in Milwaukee's Polish Community," *Perspectives in Vernacular Architecture VIII*, Spring.

Jackson, J. B. (1994). *A Sense of Place, A Sense of Time*. New Haven: Yale University Press.

Jackson, P. (1989). *Maps of Meaning: An Introduction to Cultural Geography*. London: Unwin Hyman.

—— (1992). "Constructions of Culture, Representations of Race: Edward Curtis' Way of Seeing," in Anderson and Gale (1992).

—— (1993). "On 'The Reinvention of Cultural Geography' by Price and Lewis: Commentary." *Annals of the Association of American Geographers*, 83/3: 515–17.

Jacobs, J., and Fincher, R. (eds.) (1998). *Cities of Difference*. New York: Guilford.

Jakle, J., and Sculle, K. (1994). *The Gas Station in America*. Baltimore: Johns Hopkins University Press.

Jakle, J., Sculle, K., and Rogers, J. (1996). *The Motel in America*. Baltimore: Johns Hopkins University Press.

John, G. (2001). "Cultural Nationalism, Westward Expansion, and the Production of the Imperial Landscape: George Catlin's Native American West." *Ecumene*, 8/2: 175–203.

Johnson, N. (1992). "Nation-Building, Language, and Education: The Geography of Teacher Recruitment in Ireland, 1925–1955." *Political Geography*, 11: 170–89.

—— (1996). "Where Geography and History Meet: Heritage Tourism and the Big House in Ireland." *Annals of the Association of American Geographers*, 86/3: 551–66.

Jones, J. P. (1998). "My Dinner with Derrida, or Spatial Analysis and Post-Structuralism Do Lunch." *Environment and Planning A*, 30/2: 247–60.

Jordan, T. (1993). *North American Cattle-Ranching Frontier: Origins, Diffusion, and Differentiation*. Albuquerque: University of New Mexico Press.

Jordan, T., and Kaups, M. (1997). *The American Backwoods Frontier: An Ethnic and Ecological Interpretation*. Baltimore: Johns Hopkins University Press.

Jordan, T., Kilpinen, J., and Gritzner, C. (1997). *The Mountain West: Interpreting the Folk Landscape*. Baltimore: Johns Hopkins University Press.

Kearns, G. (1998). "The Virtuous Circle of Facts and Values in the New Western History." *Annals of the Association of American Geographers*, 88/3: 377–409.

Keith, M., and Pile, S. (eds.) (1993). *Place and the Politics of Identity*. New York: Routledge.

Kenny, J. (1995). Climate, Race, and Imperial Authority: The Symbolic Landscape of the British Hill Station in India. *Annals of the Association of American Geographers*, 85: 694–714.

Kenny, J. (1997). "Polish Routes to Americanization: House Form and Landscape on Milwaukee's Polish South Side," in R. Ostergren and T. Vale (eds.), *Wisconsin Land and Life*. Madison: University of Wisconsin Press.

Kong, L. (1993). "Ideological Hegemony and the Political Symbolism of Religious Buildings in Singapore." *Environment and Planning D: Society and Space*, 11: 23–45.

— (1995). "Popular Music in Geographical Analyses." *Progress in Human Geography*, 19: 183–98.

Kong, L., Yeoh, B., and Teo, Y. (1996). "Singapore and the Experience of Place in Old Age." *Geographical Review*, 86/4: 529–49.

Kropotkin, P. (1885). "What Geography Ought to Be." *The Nineteenth Century*, 18: 940–56.

Law, L. (2001). "Home Cooking: Filipino Women and Geographies of the Senses in Hong Kong." *Ecumene*, 8/3: 264–83.

Lees, L. (2001). "Towards a Critical Geography of Architecture: The Case of an Ersatz Colosseum." *Ecumene*, 8/1: 51–86.

Lewis, M., and Wigen, K. (1997). *The Myth of Continents*. Berkeley: University of California Press.

Lewis, P. (1979). "Axioms for Reading the Landscape," in D. Meinig (ed.), *The Interpretation of Ordinary Landscapes*. New York: Oxford University Press.

— (1985). "Beyond Description." *Annals of the Association of American Geographers*, 75/4: 465–78.

Leyshon, A., Matless, D., and Revill, G. (1995). "The Place of Music." *Transactions of the Institute of British Geographers*, NS 20: 423–85 (Special Issue).

— (1998). *The Place of Music*. New York: Guilford Press.

Light, A., and Smith, J. (eds.) (1997). *Philosophy and Geography I: Space, Place and Environmental Ethics*. Lanham, Md.: Rowman & Littlefield.

— (1998*a*). *Philosophy and Geography II: The Production of Public Space*. Lanham, Md.: Rowman & Littlefield.

— (1998*b*). *Philosophy and Geography III: Philosophies of Place*. Lanham, Md.: Rowman & Littlefield.

Limerick, P., Milner, C., and Rankin, C. (1991). *Trails: Toward a New Western History*. Lawrence, Kan.: University of Kansas Press.

McClintock, A. (1995). *Imperial Leather: Race, Gender and Sexuality in the Colonial Contest*. New York: Routledge.

McDowell, L. (ed.) (1997). *Undoing Place? A Geographical Reader*. New York: Arnold.

McDowell, L., and Sharp, J. (eds.) (1999). *A Feminist Glossary of Human Geography*. New York: Arnold.

McGreevy, P. (1990). "Place in the American Christmas." *Geographical Review*, 80/1: 32–41.

— (1994). *Imagining Niagara: Meaning and the Making of Niagara Falls*. Amherst: University of Massachusetts Press.

Massey, D. (1991). "A Global Sense of Place." *Marxism Today*: June: 24–9.

Massey, D., Allen, J., and Sarre, P. (eds.) (1999). *Human Geography Today*. London: Blackwell.

Meinig, D. (1993). *The Shaping of America: A Geographical Perspective on 500 Years of History*, ii. *Continental America, 1800–1867*. New Haven: Yale University Press.

Meinig, D. W. (1998). *The Shaping of America: A Geographical Perspective on 500 Years of History*, iii. *Transcontinental America, 1850–1915*. New Haven: Yale University Press.

Mitchell, D. (1993). "Public Housing in Single-Industry Towns: Changing Landscapes of Paternalism," in Duncan and Ley (1993).

— (1995). "There's No Such Thing as Culture: Towards a Reconceptualization of the Idea of Culture in Geography." *Transactions of the Institute of British Geographers, New Series*, 20: 102–16.

— (1996*a*). *The Lie of the Land*. Minneapolis: University of Minnesota Press.

— (1996*b*). "Sticks and Stones: the Work of Landscape." *The Professional Geographer*, 48/1: 94–6.

— (2000). *Cultural Geography: a Critical Introduction*. Oxford: Blackwell.

Mitchell, K. (1997). "Different Diasporas and the Hype of Hybridity." *Environment and Planning D: Society and Space*, 15.

Moore, D. (1993). "Contesting Terrain in Zimbabwe's Eastern Highlands: Political Ecology, Ethnography, and Peasant Resource Struggles." *Economic Geography*, 69/4: 380–401.

— (1997). "Remapping Resistance: 'Ground for Struggle' and the Politics of Place," in S. Pile and M. Keith (eds.), *Geographies of Resistance*. London: Routledge.

Moss, P. (1992). "Where is the 'Promised Land'?: Class and Gender in Bruce Springsteen's Rock Lyrics." *Geografiska Annaler*, 74B: 167–87.

Myers, G. (1994). "From 'Stinkibar' to 'The Island Metropolis': The Geography of British Hegemony in Zanzibar," in Godlewska and Smith (1994).

— (1998). "Intellectual of Empire: Eric Dutton and Hegemony in British Africa. *Annals of the Association of American Geographers*, 88/1: 1–27.

— (1999). "Colonial Discourse and Africa's Colonized Middle: Ajit Singh's Architecture." *Historical Geography*, 27.

Nagar, R., and Leitner, H. (1998). "Contesting Social Relations in Communal Places," in Jacobs and Fincher (1998).

Nash, P., and Carney, G. (1996). "The Seven Themes of Music Geography." *The Canadian Geographer*, 40: 69–74.

Nast, H. (1994). "The Impact of British Imperialism on the Landscape of Female Slavery in the Kano Palace, Northern Nigeria." *Africa*, 64: 34–73.

— (1996). "Islam, Gender, and Slavery in West Africa circa 1500: A Spatial Archeology of the Kano Palace, Northern Nigeria." *Annals of the Association of American Geographers*, 86/1: 44–77.

Newman, J. (1995). *The Peopling of Africa*. New Haven: Yale University Press.

Noble, A., and Wilhelm, H. (1995). *Barns of the Midwest*. Athens, Ohio: Ohio University Press.

Oakes, T. (1997). "Place and the Paradox of Modernity." *Annals of the Association of American Geographers*, 87/3: 509–31.

Olwig, K. (1996). "Recovering the Substantive Nature of Landscape." *Annals of the Association of American Geographers*, 86/4: 630–53.

Pasqualetti, M. (1997). *The Evolving Landscape: Homer Aschmann's Geography*. Baltimore: Johns Hopkins University Press.

Peet, R. (1996*a*). "Discursive Idealism in the 'Landscape as Text' School." *The Professional Geographer*, 48/1: 96–8.

— (1996*b*). "A Sign Taken for History: Daniel Shays Memorial in Petersham, Massachusetts." *Annals of the Association of American Geographers*, 86/1: 21–43.

Phillips, R. (1997). *Mapping Men and Empire: A Geography of Adventure*. London: Routledge.
Pickles, J. (ed.) (1995). *Ground Truth*. New York: Guilford.
Pile, S. (1993). "Human Agency and Human Geography Revisited: a Critique of 'New Models' of the Self." *Transactions of the Institute of British Geographers,* NS, 18: 122–39.
— (1996). *The Body and the City: Psychoanalysis, Space, and Subjectivity*. London: Routledge.
Pillsbury, R. (1990). *From Boarding House to Bistro: The American Restaurant Then and Now*. Boston: Unwin Hyman.
— (1998). *No Foreign Food: The American Diet in Time and Place*. Boulder: Westview Press.
Porteous, J. D. (1990). *Landscapes of the Mind: Worlds of Sense and Metaphor*. Toronto: University of Toronto Press.
Price, M., and Lewis, M. (1993). "The Reinvention of Cultural Geography." *Annals of the Association of American Geographers*, 83: 1–17.
Robinson, J. (1994). "White Women Researching/Representing 'Others': From Antiapartheid to Postcolonialism?" in Blunt and Rose (1994).
Rose, G. (1993). *Feminism and Geography*. Cambridge: Polity Press.
Rowntree, L., Foote, K., and Domosh, M. (1988). "Cultural Geography," in G. Gaile and C. Willmott (eds.), *Geography in America*. Columbus, Ohio: Merrill.
Ryan, J. (1997). *Picturing Empire*. London: Reaktion Books.
Sack, R. (1997). *Homo Geographicus: A Framework for Action, Awareness and Moral Concern*. Baltimore: Johns Hopkins University Press.
Said, E. (1978). *Orientalism*. New York: Vintage.
— (1993). *Culture and Imperialism*. New York: Harper & Row.
— (1995). "Secular Interpretation, the Geographical Element, and the Methodology of Imperialism," in G. Prakash (ed.), *After Colonialism*. Princeton: Princeton University Press.
Salter, K. (1994). "Cultural Geography as Discovery," in Foote, Hugill, Mathewson, and Smith (1994).
Sauer, C. (1956). "The Education of a Geographer." *Annals of the Association of American Geographers*, 46: 287–99.
Schama, S. (1997). *Landscape and Memory*. New York: Columbia University Press.
Schein, R. (1997). "The Place of Landscape: A Conceptual Framework for Interpreting an American Scene." *Annals of the Association of American Geographers*, 87: 660–80.
Schwartz, J. (1995). "We Make Our Tools and Our Tools Make Us: Lessons from Photographs for the Practice, Politics, and Poetics of Diplomatics." *Archivaria: The Journal of the Association of Canadian Archivists*, 40: 40–74.
Scott, J. (1998). *Seeing Like a State: How Certain Schemes to Improve the Human Condition Have Failed*. New Haven: Yale University Press.
Shortridge, B. G., and Shortridge, J. R. (eds.) (1998). *The Taste of American Place: A Reader on Regional and Ethnic Foods*. Lanham, Md.: Rowman & Littlefield.
Shortridge, J. R. (1990). *The Middle West*. Lawrence: University of Kansas Press.
— (1991). "The Concept of the Place-Defining Novel in American Popular Culture." *Professional Geographer*, 43: 280–91.
Shurmer-Smith, P. (2002). *Doing Cultural Geography*. Thousand Oaks, Calif.: Sage.
Sibley, D. (1996). *Geographies of Exclusion*. New York: Routledge.
Sidorov, D. (2000). "National Monumentalization and the Politics of Scale: The Resurrections of the Cathedral of Christ the Savior in Moscow." *Annals of the Association of American Geographers*, 90/3: 548–72.
Silberfein, M. (ed.) (1998). *Rural Settlement Structure and African Development*. Boulder: Westview.
Sluyter, A. (2001). "Colonialism and Landscape in the Americas: Material/Conceptual Transformations on Continuing Consequences." *Annals of the Association of American Geographers*, 91/2: 410–28.
Smith, J. (1996). "Geographic Rhetoric: Modes and Tropes of Appeal." *Annals of the Association of American Geographers*, 86/1: 1–20.
Soja, E. (1989). *Postmodern Geographies*. London: Verso.
— (1997). *Thirdspace*. Oxford: Blackwell.
Spain, D. (1992). *Gendered Spaces*. Chapel Hill, NC: University of North Carolina Press.
Starrs, P. (1998). *Let the Cowboy Ride: Cattle Ranching in the American West*. Baltimore: Johns Hopkins University Press.
Sternberg, R. (1998). "Fantasy, Geography, Wagner, and Opera." *Geographical Review*, 88: 327–48.
Stump, R. W. (2000). *Boundaries of Faith: Geographical Perspectives on Religious Fundamentalism*. Lanham, Md.: Rowman & Littlefield.
Thrift, N. (2000). "Performing Cultures in the New Economy." *Annals of the Association of American Geographers*, 90/4: 674–92.
Till, K. (1995). "Neotraditional Towns and Urban Villages: The Cultural Production of a Geography of 'Otherness.'" *Evironment & Planning D: Society & Space*, 11: 709–32.
— (1999). "Staging the Past: Landscape Designs, Cultural Identity and Erinnerungspolitik at Berlin's Neue Wache." *Ecumene*, 6/3.
Tuan, Y.-F. (1976). "Humanistic Geography." *Annals of the Association of American Geographers*, 66: 266–76.
— (1989). *Morality and Imagination: Paradoxes of Progress*. Madison: University of Wisconsin Press.
— (1991). "Language and the Making of Place." *Annals of the Association of American Geographers*, 81: 684–96.
— (1993). *Passing Strange and Wonderful: Aesthetics, Nature, and Culture*. Washington, DC: Island Press.
— (1996). *Cosmos and Hearth: A Cosmopolite's Viewpoint*. Minneapolis: University of Minnesota Press.
— (2000). *Escapism*. Baltimore: Johns Hopkins University Press.
Turner, F. (1989). *Spirit of Place: The Making of an American Literary Landscape*. San Francisco: Sierra Club Books.
Wagner, P., and Mikesell, M. (1962). *Readings in Cultural Geography*. Chicago: University of Chicago Press.
Walton, J. (1995). "How Real(ist) Can You Get?" *The Professional Geographer*, 47/1: 61–5.
Watts, M. (1991). "Mapping Meaning, Denoting Difference, Imagining Identity: Dialectical Images and Postmodern Geographies." *Geografiska Annaler*, 73: 7–16.
Wilbanks, T. (1994). "Sustainable Development in Geographic Perspective." *Annals of the Association of American Geographers*, 84/4: 541–56.
Williams, R. (1973). *The Country and The City*. London: Chatto & Windus.

Williams, R. (1977). *Marxism and Literature.* Oxford: Oxford University Press.
— (1980). *Problems in Materialism and Culture.* London: New Left Books.
— (1982). *The Sociology of Culture.* New York: Schocken.
Wishart, D. (1994). *An Unspeakable Sadness.* Lincoln, Nebr.: University of Nebraska Press.
Withers, C. (1988). *Gaelic Scotland.* London: Routledge.
Wood, J. (1997). "Vietnamese American Place Making in Northern Virginia." *Geographical Review*, 87/1: 58–72.
Worster, D. (1993). *The Wealth of Nature.* Oxford: Oxford University Press.
Wright, J. (1926). "A Plea for the History of Geography." *Isis*, 8: 477–91.
Yeager, P. (ed.) (1996). *The Geography of Identity.* Ann Arbor: University of Michigan Press.
Yeoh, B. (1996). *Contesting Space: Power Relations and the Urban Built Environment in Colonial Singapore*. Oxford: Oxford University Press.
Young, R. (1990). *White Mythologies*. London: Routledge.
— (1995). *Colonial Desire*. London: Routledge.
Young, T. (1993). "San Francisco's Golden Gate Park and the Search for a Good Society, 1865–1880." *Environmental History*, 37: 4–13.
— (1995). "Modern Urban Parks." *Geographical Review*, 85/4: 535–51.
— (1996). "Social Reform Through Urban Parks: the American Civic Association's Program for a Better America." *Journal of Historical Geography*, 22/4: 460–72.
Zelinsky, W. (1992). *The Cultural Geography of the United States: A Revised Edition.* Englewood Cliffs, NJ: Prentice Hall.
— (1994). *Exploring the Beloved Country: Geographic Forays into American Society and Culture.* Iowa City: University of Iowa Press.
— (1995). "Review of Re-Reading Cultural Geography." *Annals of the Association of American Geographers*, 85/4: 750–3.
Zonn, L. (1990). *Place Images in Media: Portrayal, Experience, and Meaning*. Savage, Md.: Rowman & Littlefield.
Zukin, S. (1991). *Landscapes of Power.* Berkeley: University of California Press.

CHAPTER 8

Cultural Ecology

Thomas J. Bassett and Karl S. Zimmerer

Cultural ecology in the 1990s was a highly productive and rapidly growing specialty group within geography. The group's scholarship has contributed to a number of core themes and concepts in geography and in related fields within the social and biogeophysical sciences and humanities (Butzer 1989, 1990*a*; Porter 1991; B. L. Turner 1997*a*; Zimmerer 1996*c*). This review evaluates the central research contributions—findings, themes, concepts, methods—of North American geographical cultural ecology over this decade (1990–9). The evaluation is based on the clustering of the contributions of the 1990s into eight main areas: long-term cultural ecology; resource management; local knowledge; pastoralism; environmental politics; protected areas; gender ecology; and environmental discourses (Figs 8.1 and 8.2). Notable accomplishments and characteristic approaches are reviewed in each area. Emphasis is placed on the continued evolution of the common ground of cultural ecology and its most prominent offshoot, political ecology.

A nature-culture or nature-society core is central to advances of the 1990s. This core is made up of interacting dialectical processes of culture-and-consciousness and domestic-and-political economy, on the one hand, and non-human nature, on the other hand (Zimmerer and Young 1998: 5). Increased awareness of this recursive interaction has led to a historical perspective that is common to much work in cultural and political ecology during the past decade (Figs 8.1 and 8.2). Culture and society in environmental interactions are considered with new importance granted to the multiple forms and contingencies of spatial scale, from the local to the global, as well as varied temporal frames. Culture and society are conceptualized in new ways while, at the same time, the biogeophysical environments themselves are thought of as increasingly complex and less spatially and temporally predictable than was previously presumed. The nature-culture core has placed cultural and political ecology at the center of the new millennium's concerns about environmental degradation and planning, conservation, biodiversity, indigenous knowledge, and

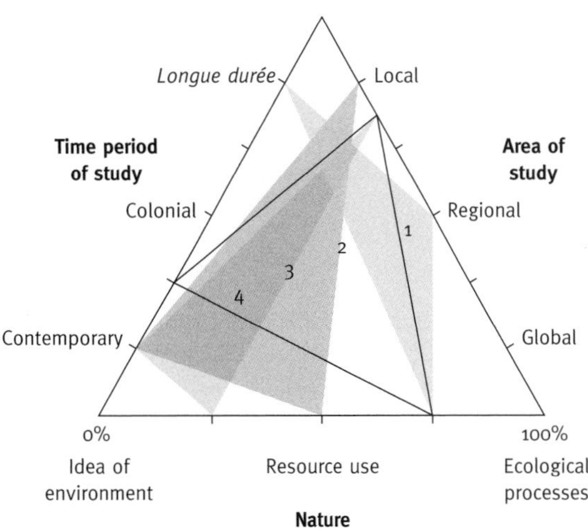

Fig. 8.1 The types of emphasis on historical time, spatial scale, and environments that are associated with studies of: (1) long-term historical cultural ecology; (2) natural resource management; (3) local knowledge; and (4) environmental politics

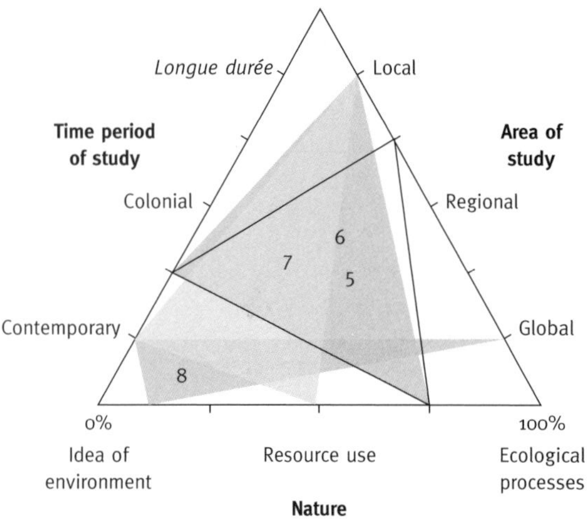

Fig. 8.2 The types of emphasis on historical time, spatial scale, and environments that are associated with studies of: (5) pastoralism; (6) protected areas; (7) gender ecology; and (8) environmental discourse

the multiple ways that various groups manage natural resources, shape landscapes, and struggle over resource access and control (Peet and Watts 1993).

Globalization, which refers to awareness of the multi-scale, inter-connections of environmental and human processes across large areas, has served as an umbrella for much cultural and political ecology. Major themes of globalization that expanded during the 1990s include protected areas and conservation strategies; environmental politics; gender ecology; and environmental discourses. Cultural and political ecology's focus on development and environmental concerns is invigorated by new theoretical and methodological approaches to the topics of sustainable and alternative development, environmental and biological conservation, and social empowerment and human rights. The new emphasis on globalization has shown sizable overlap with cognate fields. Works on these expanded themes led to a notable diversification of cultural and political ecological research in geography in the period 1990–9.

Historical Cultural Ecology: *Longue Durée* Human Environmental Change

Cultural ecology has contributed to major debates on environmental changes through its perspective on the long-term historical changes of human–environmental relations (Fig. 8.1—area 1). The quincentennial of the Colombian encounter with the New World was a main focal point (Butzer 1990*b*, 1992). A special issue of the Annals of the AAG, titled "The Americas Before and After 1492: Current Geographical Research", edited by Butzer (1992), produced a series of state-of-the-art studies on contact landscapes (Doolittle 1992; Gade 1992; Whitmore and Turner 1992); demographic change (Lovell 1992); and ideas of nature prior to and following European contact (Denevan 1992; Sluyter 1999). Overseeing these portraits is Butzer's (1990*b*, 1990*c*, 1992, 1996*a*) command of long-term historical cultural ecology, somewhat akin to Ferdinand Braudel's longue durée. Butzer marshals the longue durée of cultural historical ecology to challenge the so-called "Black" and "Green" legends of previous environments that were purportedly pristine in both the New and Old Worlds. Adopting a spatial emphasis, Doolittle Wnds that southwestern and eastern woodland agricultural types were as complex as the mosaic environments in which they were sited. The synthesis of Whitmore and Turner similarly shows that Mesoamerican cultivated landscapes were "patchwork-like microsystems, Wne-tuned to small-scale environmental variations, while others were dominated by zonal patterns keyed to the broad environmental zones created by elevation, aspect, and slope" (Whitmore and Turner 1992: 403).

The Colombian quincentennial also generated much interest in long-term and episodic historical cultural ecology. Horticulture-style intensity, scale, and management of agriculture were prominent, shaped by the local conditions as much as by rule of the Amazon chiefdoms, Aztec and Inca empires, early civilizations such as the Tiwanaku and Maya, and various other complex pre-European societies (Butzer 1992; Denevan 1992; Doolittle 1992; Dunning 1995; Knapp 1991; Zimmerer 1995). Pre-colonial farming of staple food plants shaped lifeways and landscapes that were found to be more culturally and historically inscribed, ecologically versatile, and sociopolitically constructed than previously thought (Zimmerer 1996*a*). The centrality of agriculture to the environmental history of the Americas recommended the adoption of an agrocentric perspective and corresponding methods in field research on long-term change (Butzer 1990*c*; Gade 1992, 1999; Whitmore and Turner 1992).

Cultural and ecological landscapes of the Americas during pre-colonial epochs were often regional mosaics in which farmlands for local food supply and extra-local demands were adjoined by forests, wetlands, savannas and prairies, and desert scrub (Butzer 1992; Denevan 1992; Doolittle 1992; Gade 1992, 1999; Whitmore and Turner 1992; see Fig. 8.1—area 1). Forest clearing for farming often entailed the use of fire and, in certain cases, led to significant soil erosion and sediment deposition in the tropical environments of Mesoamerica and South America. To investigate these histories, cultural ecology has improved its multimethod repertoires of such techniques as the analysis of pollen micro-fossils, sediment cores, gastropod populations, tree rings, geo-archaeological features and other artefacts, as well as the critical appraisal of archival documents that refer to landscapes (Butzer and Butzer 1993, 1997; Doolittle, Neely, and Pool 1993; Dunning *et al.* 1998; Horn 1998; Knapp 1998; Sluyter 1996, 1997*a*, *b*; Williams 1992; Zimmerer 1993*a*). Further refining and application of these multimethod techniques will help to investigate the range of spatial and temporal scales needed to guide understandings of human environmental change and land use and environmental planning.

Artefacts such as roadways, house and burial mound sites, farm terraces, and dams and canals, that today are often mistaken as virgin nature, were abundant in lowland tropics; this realization is owed largely to cultural ecology (Butzer 1993; Denevan 1991, 1996). Complex remains in the Mayan lowlands were uncovered and illuminated by Dunning and his collaborators (Beach and Dunning 1995; Dunning 1995; Dunning and Beach 1994; Dunning *et al.* 1997, 1998). Their findings on Mayan field terracing and erosion control during the Late Classic elucidated adaptations that were non-deterministic but that redefined environmental surroundings. Amerindian modifications—such as soil and water works of the Quichua and their predecessors in highland Ecuador, pebble-mulch terracing by Anasazi of the Colorado Plateau, and widespread wetland agriculture and field terracing—amounted to one of the world's major ecological revolutions (Gartner 1997; Knapp 1991; Mathewson 1990; Lightfoot 1993, 1994; Sluyter 1994). Irrigation was a common element. Crucial ties existed between local innovation and state controls of irrigation during political transitions of pre-European as well as the European colonial periods (Butzer 1996*b*; Doolittle 1993, 1995; Knapp 1992; Zimmerer 1995).

In some quarters, the Spanish conquest is believed to have severely disrupted the harmony of Amerindians and nature prior to contact, thus devastating the environment. Denevan (1992), Butzer (1993), Sluyter (1997*b*, 1999), and others have challenged these "myths" by showing the *longue durée* of land-use changes. They demonstrate that various New World landscapes were modified by human action prior to the Conquest and argue that negative consequences of Spanish colonialism have been exaggerated. Butzer (1996*a*: 145) argues that sustainable Mediterranean agroecosystems were superimposed on indigenous systems and did not result in environmental degradation, at least during the first century of colonization. Whitmore and Turner (1992: 420) suggest that the "Mestizo" landscapes were "more or less ecologically sustainable." In subsequent research the public, political, and polemical use of ecomyths, both historical and contemporary, were examined as part of the research process (Sluyter 1999; Zimmerer 1996*a*). Other works showed that modifications, such as early colonial livestock-grazing and pre-European fire-setting, were highly varied in the spatial context, thus leaving some environments moderately intact (Sluyter 1998).

Debunking a "Pristine Myth" (the *Leyenda Verde* in Latin America) and a corollary "Myth of the Ecologically Noble Savage" fueled some highly successful overviews of the cultural ecology of the Americas (Butzer 1990*b*, 1992, 1993; Denevan 1992; Doolittle 1992; Gartner 1997; Sluyter 1994; Turner and Butzer 1992; Whitmore and Turner 1992). This perspective highlighted the modifications—major in some cases—that were incurred in many places across the Americas by non-European peoples. Turner and Butzer (1992) provide a balanced cultural ecological discussion of the Colombian encounter.

Diasporic or cross-regional cultural ecologies, forged within the power relations of European colonialism, were etched in landscape changes that are of great potential relevance to present-day interests in globalization and global human environmental change. Mixing of both cultural and environmental lifeways under Spanish, Portuguese, and British colonialism produced hybrid cultural ecologies of indigenous, mestizo, creole, and African-American slave peoples. Associated elements included biota (crops, weeds, livestock) to tools, technologies, and land-use institutions (communal crop and livestock arrangements) (Carney 1993*b*, 1996*a*, *b*, 1998; Butzer 1996*a*; Gade 1992; Voeks 1997). The cultural ecological roles of African-American slaves and their descendants—including gendered relations—were crucial to the make-up of various colonial and later post-colonial environments, resource management techniques, and related knowledge as shown by findings on American rice farming and the Condomblé use of medicinal and ritual plants.

Complexity in Natural Resource Management: History, Culture, and Scale

Studies of human–environment interaction (soils and water; forests and wildlife; agroecology, food plants and consumption; rangelands; and mountains) have offered important insights into the complexity that arises from historical, cultural, and scale-related processes. The attention given to human impacts on, and responses to these resource environments has contributed to our understanding of the multi-scalar complexities of the cultural ecology dialectic with an emphasis on local and regional settings and recent historical periods (Fig. 8.1—area 2). Human ecological thresholds, non-linear relations, coupling effects, and path dependencies are common to these people–environment interactions.

Changes in farm and range soils have played important roles as *both* a consequence *and* an environmental conditioner of human livelihoods and experience (Carney 1991; Gray 1999; Grossman 1997; Rocheleau *et al.* 1995; Zimmerer 1994). Complex, multifaceted cause–effect relations have been elucidated by uncovering the earlier histories of soil erosion and conservation that took shape prior to the 1990s. The cause–effect relations in complex natural resource management have become better understood by placing emphasis on geographical scale as a primary factor in human perceptions, social policies, and environmental change processes themselves.

Contributions on the people–soils nexus offer a number of fresh insights. Relations of gender, ethnicity, and political economic factors (markets, state policies on food, agriculture, and conservation) have strongly shaped, and been molded by, the changes that occur at the people–soils interface (Brannstrom and Oliveira 2000; Carney 1991; Gray 1999; Grossman 1997; Moran 1995; Rocheleau *et al.* 1995; Zimmerer 1993*b*). During the 1990s the worsening shortfall of labor-time for soil conservation works, which is often gender-related, has induced widespread soil degradation, with this effect locally and regionally varied. Conservation prospects, including those based on "traditional" or "indigenous" techniques must be evaluated in conjunction with such limiting factors. No longer does it suffice to ask simply whether a technique was sustainable in the past, for it is clear that favorable social conditions are as crucial as the soundness of methods (Beach and Dunning 1995; Brookfield and Humphreys 1994). Soils knowledge of local land users too is seen as shaped by social relations (Carney 1993*a*; WinklerPrins 1999; Zimmerer 1991*b*). While offering key insights, their knowledge is not made easily commensurate with soil science, thus furthering the need for research on bridging local and scientific styles of soils and environmental management for the purpose of combining conservation with development.

New analytical frameworks for studying the cultural and political ecology of people–forests interaction have been developed and applied in a variety of settings (Aageson 1998*a*, *b*; Hecht 1994; Hecht and Cockburn 1990; Metz 1990; Moran 1993; Rocheleau and Ross 1995; Stevens 1993*a*). Political economy, land-use factors (such as markets and especially international or "global" market integration, tree tenure), and cultural beliefs are found to shape deforestation and conservation, including forest restoration in a diversity of situations (Paulson 1994; Steinberg 1998*a*; Stevens 1993*a*; Walters 1997). Greater attention is being paid to the human use and changing ecology of non-timber forest products, both in developing countries or the "South" as well as places in urban industrial societies (close even to the urban centers of California) (Hansis 1998; Metz 1994, 1998). While some studies show that forests managed as components of long-term agricultural fallows still maintain useful and relatively intact ecosystems, other research points to more degraded forests associated with socio-economic and cultural changes (Steinberg 1998*a*; Voeks 1996).

Use and conservation of wildlife are a central theme in cultural ecology scholarship of the 1990s. Adoption of the people–wildlife framework furnishes new insights by linking wildlife populations to human-disturbed habitats (commonly forests and agricultural clearings), with special reference to the use, degradation, and conservation management/restoration of the plant habitats (Coggins 1999; Medley 1998; Naughton-Treves 1997; Young 1997; Zimmerer and Young 1998). People–wildlife studies were typically set in protected areas and focused on one or a few wildlife species of special relevance to conservation. Examples included the Meihuashan Reserve in southeastern China, where prey-base ungulates rely on forest types that are being altered by local Han villagers (Coggins 1999) and a pair of endangered primates in the Tana River National Primate Reserve in Kenya where key riverine forest habitats are being altered (Medley 1998).

Cultural ecological studies of agroecology and food plants have supplied new understandings of the multifaceted links among the transitions in production and consumption practices, agricultural change, livelihood quality, and conservation. Keys to sound agroecologies and food security, such as the adaptive diversity of food plants and the availability of high-quality dietary items, are shown to be shaped within dense, multi-scale networks of land users, customary resource rights, market and consumption practices, and policy-making (Carney 1993*a*; Cleveland *et al.* 1995; Paulson and Rogers 1997; Zimmerer 1996*a*). Links among *in situ* conservation of diverse food plants, household labor supplies, and political economic conditions has been examined by a number of scholars. In the Andean countries, government policies, markets for products and off-farm labor, and local labor recruiting have historically influenced the geography and utilization of diverse and nutritionally important food plants, whose cultivation is kept active by the locally better-off who increasingly grow them as luxury-like items (Zimmerer 1991*a*). Such findings at the people–agriculture interface are leading to calls for programs that support the capacities of farmers to refashion elements of "traditional" and "modern" agriculture into their own cultural ecological amalgams of *in situ* conservation (Cleveland *et al.* 1995; Paulson and Rogers 1997; Zimmerer 1996*a*).

Research at the people–agriculture interface also has explored the effects of changes in cultural values and practices on cuisine, diet, and dooryard gardening. Ethnographic-style study of diet quality and its cultural ecology showed complex relations to development transitions in Papua New Guinea and St Vincent in the Eastern Caribbean (Grossman 1991, 1993, 1998*a*). Cultural and ethnic diversity's relation to diverse food plants and cuisine—from the restaurant scene of San Diego to the kitchen clusters of Quechua women in the Andes and Somoan women's committees—was often a matter of innovative reinventions and outcomes that were highly varied in terms of environmental and dietary quality (Fredrich 1991; Gade 1999; Paulson and Rogers 1997; Steinberg 1998b; Zimmerer 1996*a*). Gardens were particularly dynamic sites of cultural ecological change. Factors such as women's control over resources and off-farm migration had important implications for the role of gardening in resource conservation or decline.

People-mountain studies continue to furnish a useful forum for the rethinking of certain core concerns in cultural ecology. Many of these studies have been sited in the mountains of South Asia. In Khumbu, Stevens (1993*a*) produced a major cultural ecological chronicle of Sherpa subsistence and human–environmental relations. Much-debated evidence, ideas, and theories of human-environmental changes in the region were elucidated from the perspectives of political economy and ecology, discourse analysis, field-based assessments of resource use, and contemporary environmental studies (Allan 1991, 1995; Brower and Dennis 1998; Guthman 1997; Metz 1991, 1998; Stevens 1993*a*). Biogeophysical limits and constraints, including the role of "natural" hazards, were woven into interpretations that put primary emphasis on interconnections with place-based, cultural, and social factors (Allan 1991, 1995). The people–mountain interface was widely promoted and publicized as a top-level priority for global environmental institutions and organizations including the follow-up to Agenda 21 of the United Nations Conference on Environment and Development that was held in Rio de Janeiro in 1992 (Ives and Messerli 1990; Messerli and Ives 1997; Mountain Agenda 1992).

Local Knowledge: Identity, Social Movements, and New Ecological Models

Cultural ecologists enlarged their interest and made important contributions to local knowledge studies in the 1990s. Earlier works on indigenous agricultural knowledge and on land degradation and society around the world were a major catalyst. Cultural and political ecologists extended local knowledge research by examining its influence on the construction of ethnic identities and new social movements (Bebbington 1996; Zimmerer 1992; see Fig. 8.1—area 3). A second extension was focused on the integration and complex interrelations of so-called local knowledge with scientific forms of knowledge as well as, in some cases, their differences, limits, and possible incongruency. Peasant agroecological knowledge is seen as one of the most important but most neglected resources in Africa, Latin America, and Asia. Cultural ecological research in the 1990s amplified a call for an alternative agricultural development approach that is at once location-specific and ecologically particularistic and which builds upon community agricultural knowledge. It sought, in particular, to situate local knowledge and practices within political, socioeconomic, cultural, and historically changing contexts (Bassett 1994; Batterbury 1996; Bebbington 1996; Peet and Watts 1996; Voeks 1998). Some of this research distinguishes between "community ecological knowledge" (Richards 1985: 141) and knowledge and practices that are differentiated by gender, ethnicity, and race, and economic status (Carney 1991, 1993*a*, *b*, 1998; Gray 1999; Rocheleau and Ross 1995; Rocheleau *et al.* 1996; Schroeder 1993; Schroeder and Suryanata 1996). Such differences are highlighted in contestations over resource access and control, a theme that runs through much of this recent literature (see Environmental Politics below).

Cultural ecologists demonstrated the multifaceted historical rationality of indigenous technologies (irrigation, raised fields, and foodplant diversity) with respect to environmental variability (Knapp 1991; Whitmore and Turner 1992). Cultural ecologists show how these strategies changed under different economic and demographic regimes, and how their impact on the environment was also varied (Butzer 1990*a*, *c*; Grossman 1998*b*; B. L. Turner *et al.* 1990). These studies indicate that maintaining "traditional" technology may or may not be a priority of land users whose visions and politics for improving their livelihoods often involve adopting "modern" technologies (Agrawal 1995). The symbolic importance of indigenous technology in terms of ethnic identity and cultural politics is a recurring theme of 1990s research (Batterbury 1996; Bebbington 1991; Zimmerer 1996*a*). Work by Bebbington on indigenous organizations, NGOs, and social capital emphasizes the role of rural development institutions in mediating rural resource management and technological change (Bebbington 1991, 1993, 1997; Bebbington *et al.* 1996).

A critical look at the ecological models underlying classic cultural ecological research is leading to fresh insights into landscape patterns and transformations (Metz 1998; Zimmerer and Young 1998). The adaptation model of vertical, layer-cake zonation in mountain environments is being rethought (Zimmerer 1999). The desertification model of advancing desert-like conditions along a linear front due to "overcultivation, overgrazing, deforestation" is challenged by Bassett and Koli Bi (1999) who point to an increase in tree cover in northern Côte d'Ivoire under heightened grazing pressure and changing fire regimes. Land use in the Peruvian and Bolivian Andes shows "an imbrication into irregular mosaics." Non-equilibrium ecological theories emerging in such cognate fields as agroecology and landscape ecology are providing cultural ecologists with new conceptual frameworks with which to study the interplay of local knowledge and land-use/cover change (Zimmerer 1994*a*). For example, the influence of non-equilibrium dynamics (such as rainfall variability) on the spatial and temporal dynamics of rangeland production suggests that the opportunistic herding practices characteristic of transhumant pastoralism are rational resource management strategies (M. Turner 1998, 1999*c*). Indeed, herding practices are often far from optimal due to the varied effects of labor relations between herd owners and hired herders, land-use conflicts, and agricultural commitments (Bassett 1993*b*; 1994; M. Turner 1999*b*, *c*, *d*; Heasley and Delehanty 1996).

Pastoralism: Non-Equilibrium Dynamics and Conservation

The spatial and temporal distribution of livestock on rangelands is a classic cultural-ecological theme (Fig. 8.2—area 5). Porter and Sheppard provide a lucid description of how livestock raising among the Pokot of northern Kenya is a "carefully crafted yet flexible 'dance' that is sensitive to time, place, distance, stock-grazing habits, stock endurance, and the happenstance of rain" (Porter and Sheppard 1998: 266). Varied cultural and political-geographical institutions such as cattle exchanges (*tilia*) and territorial units (e.g. *karok*) across a range of environments are critical to Pokot pastoralism (ibid. 260–303). Research conducted in both East and West Africa show environmental change taking place as the state and global economic forces intrude into rural communities and indigenous institutions of herd management lose their flexibility (Heasley and Delehanty 1995; M. Turner 1999*c*, *d*; Unruh 1995*a*). These studies are notable for their focus on social as well as ecological processes in contrast to conventional approaches that link grazing pressure to imbalances between stocking rates and carrying capacity.

For example, research in the Sudano-Sahelian region of West Africa questions the utility of equilibrium-based ecological concepts such as carrying capacity and plant succession by showing how irregular rainfall and fire create non-equilibrium range conditions (M. Turner 1993, 1998). Thus, the timing and intensity of grazing is linked, in part, to shifting spatial and temporal patterns of range resources. Grazing pressure is also linked to the quality of herd management (Bassett 1994; M. Turner 1999*b*, *c*), opportunities and constraints affecting herd mobility (M. Turner 1999*e*), particularly herder access to key resources such as high quality dry season pastures. Access to range resources is constrained by state policies biased towards sedentary agriculture (Unruh 1990), private property regimes (Bassett 1993*a*), and the exclusion of grazing from national parks (Naughton-Treves 1997; Neumann 1995; 1997; M. Turner 1999*d*).

The appeal of community-based natural resource management initiatives such as the Village Lands Management approach (*Gestion de Terroirs Villageois*) is that they ostensibly valorize indigenous ecological knowledge and offer greater flexibility and autonomy to local peoples. However, the spatial emphasis of these approaches on bounded village territories and their neglect of non-local resources such as transhumant routes, dry season pastures, and other geographically dispersed key resources, points to the limitations of this approach to pastoral livelihoods (M. Turner 1999*a*). Since resource access and control rights are negotiated within and between communities differentiated by gender, ethnicity, and economic status, it is not surprising that so-called participatory planning projects are used by some community fractions to restrict the rights of other community members (Carney 1993; Gray 1999; Ribot 1996; Schroeder 1993, 1999; Schroeder and Suryanata (1996).

Environmental Politics: Multi-Scale Struggles over Natural Resource Control

The emphasis given to conflicts over natural resource access, control, and management, has led some cultural and political ecologists to place politics front and center of their discussions. The focus on power relations and political processes that influence the dynamics of environmental change occurs at multiple scales, with emphasis on local and regional settings (Fig. 8.1—area 4). From the micro-politics of the household and community to macro processes originating at the national and international levels, explicit connections are made between poverty and power relations and between environmental degradation and political-economic processes—what Bryant and Bailey (1997: 27–47) call "the politicized environment". Edited collections by Schroeder and Neumann (1995) and Peet and Watts (1996), along with Bryant and Bailey's synthetic work, *Third World Political Ecology*, represent a concerted effort "to refine and deepen the political" in examining human–environmental interactions (Peet and Watts 1996: 39).

At the local level, several studies focus on gender politics and resource control conflicts linked to interventions by the state in agricultural development and environmental "stabilization" projects (Carney 1993*a*; Schroeder 1995, 1999). In the case of a state-directed irrigated rice development scheme in the Gambia, Carney and Watts (1990) show how intra-household conflicts over land and labor control were fought out in the naming of fields as either individual (*kamanyango*) or family (*maruo*). These designations held important implications for control over farm labor and output. Their analysis shows how production politics are simultaneously material, cultural, and symbolic processes that in some circumstances adversely influence agricultural performance.

Moving back and forth between local and national levels, Rangan (1995, 1996) and Watts (1998*b*) situate their studies of deforestation in the Garwhal Himalayas and the ecological devastation of petroleum extraction in Ogoniland (Nigeria), respectively, in relation to local-national politics over resource control. In contrast to the populist interpretations of these struggles as exemplars of grassroots environmentalism in which "the people" are characterized by a set of common interests, Rangan and Watts emphasize the conflicting goals and politics and the heterogeneous groups associated with these movements. Their findings challenge the overly simplistic populist interpretations that have become common in conjunction with neoliberal and "newly democratic" states.

Bryant and Bailey similarly privilege the political dimensions of environmental change and resource conflicts. "Putting politics first," they define political ecology as "an inquiry into the political sources, conditions, and ramifications of environmental change . . . [thus] the role of power in the mediation of relations between actors over environmental matters becomes of paramount importance (Bryant and Bailey, 1997: 188, 191). Ironically, their actor-oriented approach, which systematically addresses roles played by the state, multilateral institutions, business, environmental non-governmental organizations, and grassroots actors, may leave little room for the environment itself. Vayda and Walters (1999) critique this "politics without ecology" approach for claiming to explain environmental change without demonstrating the environmental effects of resource struggles.

An explicit attempt to forge analytical links between social dimensions of resource management, environmental change, and policy-making is addressed by a number of authors in a special issue of *Land Degradation and Development* edited by Batterbury and Bebbington (1999). Blaikie (1994) discusses epistemological and methodological issues involved in conducting what Batterbury *et al.* (1997) call the "hybrid research" agenda. Zimmerer's study of crop plant biodiversity in the Peruvian Andes integrates cultural, historical, economic, and environmental history with ecological analysis (Zimmerer 1996*a*).

Protected Areas: Conflicts, Markets, and Conservation Strategies

The creation of national parks and other types of protected areas has historically entailed the expulsion of indigenous peoples from ancestral lands and denied them access to resources critical to their livelihoods. Cultural ecological research on protected areas highlights conflicts over natural resource access, control and management among park authorities, conservation organizations, and local peoples. These studies examine the historical origins of what Neumann (1998) calls "the national park ideal," document land- and sea-user resistance to protected area policies, and propose alternative strategies that privilege local customary tenure and contemporary community-based resource management (Fig. 8.2—area 6).

Stevens (1997*a*) chronicles the history of protected area ideas with emphasis on the international diffusion of the Yellowstone National Park model to countries around the world. He shows how the Yellowstone model, which dramatically reduces access to and use of natural resources by local populations, is being modified by a variety of "participatory local management" approaches (Stevens 1997*b*). His fieldwork in Nepal suggests that local resource management practice can be the basis of alternative conservation models (Stevens 1993*a*, *b*).

Research in Africa and Central America demonstrates that peoples across the globe have resisted the erosion of resource control through non-compliance, trespass, and even armed struggle. Neumann argues that the non-cooperation of Meru peasants expelled from their former lands now enclosed by Arusha National Park in Tanzania, is morally justified by them on the basis of ancestral occupation and customary claims to resources. He notes that this "everyday resistance" is widespread and is based on "a unity of social identity, local history, and landscape" notwithstanding important social and ethnic differences within Meru communities (Neumann 1998: 175–6). Longitudinal field research among the Miskito, Kuna, and Suma indigenous peoples of Central America shows the wisdom of linking natural resource conservation to the empowerment of local resource users (Herlihy 1992, 1993, 1997; Nietschmann 1997). Nietschmann views the coastal Miskito struggle to defend its coral reefs from the depredations of industrial fishing fleets, "drug-trafficking lobster pirates," and "predatory colonial conservation" organizations as a unified resistance. He argues that the Miskito possess a conservation ethic and profound knowledge of the sea, and have historically used a panoply of cultural, social, and political institutions to manage their marine resources. He argues against the establishment of a biosphere reserve as a way to preserve local fisheries because it would transfer control over coral reef management from local peoples to central governments and international conservation organizations. He is highly critical of the latter for being "highly regulatory, legalistic, centralized, top-down, based on imposed universal notions, non participatory and anti fisherman" (Nietschmann 1997: 223–4). Nietschmann forcefully argues for community-based resource management schemes "in which outsiders may be invited to participate" to be authored and owned by local communities (Nietschmann 1997: 223).

Elsewhere in Latin America protected areas have likewise sprung up across the landscape and serve as sites for studies of the multi-scale fashioning of conservation and its relations to human rights and environmental justice (Herlihy 1993; Whitesell 1997). Deforestation and settlement activities at forest frontiers near the boundaries of parks such as Costa Rica's Corcovado were documented and used to make the call for the local integration of forest management and tourism activities into the local economy (Naughton 1993). Integration of local people into the Vizcaíno Biosphere Reserve in Mexico exacerbated the marginalization of those people, although recent signs point to new hope that comes out of expanded collaborations among these people, reserve staff, and NGOs (Young 1999*a*, *b*). Still other studies show how conflicts and the redefinition of resource and conservation claims takes place at multiple scales in these complex cultural ecological transitions (Aageson 1998*a*; Horn 1998; Sundberg 1998).

Co-management of conservation projects involving communities and government agencies also became a focus of cultural and political ecology research (Paulson 1998; Stevens 1997*b*). In numerous cases, such as forestry projects in The Gambia, community-based conservation was utilized by multi-scale NGOs, donors, and government sponsors and superiors for devolving the work responsibilities of environmental stabilization while at the same time expanding management activities and minimizing costs (Schroeder 1999).

Market-based conservation is promoted as a viable alternative to preserving tropical forests from the highly destructive practices of commercial logging, large-scale livestock raising, and pioneer agriculture. Despite the interest in market-oriented conservation, there is surprisingly little data on peasant farmer incomes in rain forest areas. Coomes's research on market production and revenues among Amazonian peasants in northeastern Peru seeks to fill this gap (Coomes 1996*b*; Coomes and Barham 1997; Coomes and Burt 1997). His findings point to considerable specialization and income inequalities in an area of land abundance and low population density. In terms of economic returns per hectare, the most productive land-use options are agriculture, livestock, and fishing, while hunting and forest extraction offer the lowest returns (Coomes 1996*b*: 55). His studies indicate that agroforestry practices are more profitable and less damaging to the environment for land-rich than for poorer households (Coomes and Barham 1997, 39). Coomes's findings on the diversity of household income and resource management practices in agroforestry systems complicate the notion of "traditional" agroforestry as an inherently stable, egalitarian, and "sustainable" conservation model.

Common themes running through protected area studies in cultural and political ecology are: the imposition of Western conservation ideals on foreign lands; critiques of colonial-style views of indigenous resource management as inherently destructive; links between cultural identity and landscapes; the importance of local knowledge in resource conservation; new models of participatory planning and management in protected areas that respect indigenous rights to resources; and the recognition that "indigenous peoples" and social movements are characterized as much by their heterogeneity as their (uneasy) alliance around certain issues (Coomes and Barham 1997; Young 1999*a*; Rangan 1996).

Gender Ecology: The Sites of Resource Rights and Territories

Gender ecology has expanded into a major area in the cultural and political ecology research of the 1990s. Rocheleau and her colleagues proposed a global–local perspective of feminist political ecology that is built around the three themes of gendered knowledge, gendered environmental rights, resource use, and responsibilities, and gendered environmental politics and activism (Rocheleau 1995; Rocheleau and Ross 1995; Rocheleau *et al.* 1996). Such research revealed the cultural ecological importance of local-scale gendering and corresponding territorialization of ownership and use rights over land, trees, water, wildlife, and other rural resources (see Fig. 8.2—area 7). Gendered resource rights and territories were rife with implications for "global" projects on development, commodity production, conservation, and environmental stabilization and sustainability. Indeed, while intrahousehold struggles between men and women often took place at a specific site, they could lead to major region-wide shifts in economic power and key human environmental processes such as the size and composition of herds being grazed in Sahelian West Africa (M. Turner 1999*a*, *b*, *d*).

A noteworthy series of gender ecology studies clustered around the topic of irrigated rice development, vegetable gardening, and orchard production in The Gambia (Carney 1993). Agricultural development involving irrigation and mechanization were part of a 150-year social environmental history that was propelled by colonial and national governments and international aid donors and the women and men of farm households (Carney and Watts 1990). Beginning in 1984 the Jahaly Pachar irrigation scheme and others in The Gambia, including projects that fit within the "women in development" rubric, transformed wetland agriculture by intensifying the work regimes and reducing the food-producing autonomy of women and farm households (Carney 1991, 1993*a*). Intra-household gender conflicts were intensified along with farm production and, concomitantly, the meanings of gendered power relations within agriculture (such as land tenure rights) were also sites of cultural ecological struggles. A garden boom also placed cultural ecological change in the context of intra-household conflicts. Women vegetable growers, aided by horticultural development projects and booming international export markets, were often at odds with the tree-planting and tenure claims of their male spouses who were aided by commodity-based environmental stabilization and conservation schemes (Schroeder 1993, 1995, 1996, 1999).

Environmental Discourses: Post-Structuralism and Policy

Analytical perspectives associated with post-structural critiques have widely influenced cultural ecology in the 1990s. Discourse theory, in particular, has had great appeal to scholars long interested in ethnoscientific knowledge and the relationship between culture, science, resource management, and policy-making among international and increasingly global institutions (see Fig. 8.2—area 8).[1] Cultural and political ecologists frequently contest Western environment and development policy discourses that represent indigenous resource management as destructive, inefficient, and "traditional" (i.e. non-modern) (Nietschmann 1997: 216). The dominant environment and development discourse at the end of the twentieth century is "sustainable development," a fuzzy green concept that means different things to different people (Adams 1995; Escobar 1996: 48–54; B. L. Turner 1997*b*; Goldman 1995; Peet and Watts 1996: 14–18). Adams traces the roots of the sustainable development discourse to Western environmentalism and its characteristic "tension between reformism and radicalism, between technocentrism and ecocentrism" (Adams 1995: 88).

[1] Stuart Hall defines discourses as "ways of talking, thinking, or representing a particular subject or topic. They produce meaningful knowledge about that subject. This knowledge influences social practices, and so has real consequences and effects. Discourses are not reducible to class-interests, but always operate in relation to power—they are part of the way power circulates and is contested. The question of whether a discourse is true or false is less important than whether it is effective in practice. When it is effective—organizing and regulating relations of power (say, between the West and the Rest)—it is called a 'regime of truth'" (Hall 1995: 205).

Mainstream sustainable development discourses are technocentrist and reformist with their emphasis on regulation, better planning, and rational land use.

The environmental degradation narrative most forcefully challenged by cultural and political ecologists is the neo-Malthusian model of land degradation in which population pressure on resources is considered to be the most important causal factor behind environmental problems. Kummer and Turner's (1994) case study of deforestation in the Philippines points to large-scale logging, not population, as the principal factor driving deforestation in that country. B. L. Turner *et al.* (1993) present a number of compelling case studies demonstrating that population growth does not invariably lead to environmental degradation. When accompanied by economic diversification, population growth can be an important factor behind agricultural intensification (Hyden *et al.* 1993). Goldman (1992, 1993*a*) shows how farmers in a high population density area of southeastern Nigeria successfully cope with declining soil fertility by managing fallow fields in innovative ways.

Zimmerer (1996*b*) compares different discourses on soil erosion in Bolivia as articulated by peasant peoples, rural trade unions (also known as peasant leagues or farmer organizations), and development institutions. The latter group, which included NGOs, international aid donors and government agencies, invariably linked soil erosion to the inadequate technical knowledge of land users. In contrast, the trade unions or peasant leagues blamed unfavorable economic policies for forcing peasants to mine soils while farmers themselves framed the erosion problem with reference to religious beliefs. They explained the more frequent occurrence of the highly erosive "crazy rains" as being precipitated by a breakdown in civility and a neglect of ritual obligations. Grossman (1997) argues that colonial discourses on soil erosion differed between Africa and the Caribbean in large part because of differences in state policies and political economies. Rocheleau *et al.* (1995) critique crisis narratives that extrapolate local problems such as soil erosion to regional scales because of the tendency to homogenize the diversity of actors and landscapes to a single scale.

Conclusion: New Directions

Research by cultural ecologists in the 1990s represents both significant continuity and substantial change. Emphasis on cultivated landscapes, indigenous technical knowledge, and population, land use, and environmental change have been advanced significantly. Many of the themes and regional specializations discussed by Butzer (1989) in his review of the 1980s scholarship endured into the 1990s, including early irrigation (Knapp 1991; Zimmerer 1995), population (Brookfield 1995; Newman 1995; Whitmore 1991, 1996), food systems (Downing *et al.* 1996; Watts 1994, 1996), agricultural intensification (Goldman 1993*b*; Turner, *et al.* 1993; Turner and Shajaat Ali 1996), and agricultural history (Carney 1993*b*, 1996*a*, *b*, 1998; Doolittle 1992; Zimmerer 1996*a*). New research directions are also noteworthy, the most salient being in cultural ecology's burgeoning offshoot of political ecology. The number of articles published by CESG members that take an explicit political ecological focus is remarkable in itself (see Bryant 1992, 1998; Blaikie 1994, 1995, 1998; and Bryant and Bailey 1997; Forsyth 1996). A question of increased relevance to this literature is whether political ecology is exclusively an interdisciplinary perspective or whether it can or should contain transdisciplinary or disciplinary contributions (such as a geographical political ecology). Other new directions that were notable during the 1990–1999 period include:

1. The incorporation of insights from the "new" ecology into cultural and political ecological research (Bassett 1994; M. Turner 1993; Zimmerer 1994*a*, 1996*a*, 1999).
2. The infusion of new cultural and human–ecological theory into cultural and political ecology, specifically discourse theory (Watts 1993; Peet and Watts 1996) and theories of anthropogenic nature and land use (Neumann 1997, 1998; Zimmerer and Young 1998).
3. The focus on protected areas (Stevens 1997b; Neumann 1996), property rights regimes (Bassett and Crummey 1993; Watts 1998*a*), market- and community-based natural resource management (Coomes 1996*b*; Coomes and Barham 1997; Coomes and Burt 1997), and issues of human rights and environmental justice (Herlihy 1993; Nietschmann 1997; Neumann 1995).
4. The contributions of cultural ecology to environmental history (Butzer 1990, 1992; Butzer and Butzer 1997; Sluyter 1997*a*, 1998).
5. The use of new methodologies such as GIS/RS to examine land-use/cover patterns (Reenberg 1995; Reenberg and Paarup-Laursen 1997; Kull 1998; Bassett and Koli Bi 1999; B. L. Turner 1997*a*).
6. The prominence of cultural ecological studies in interdisciplinary research on the human dimensions of global environmental change (Turner *et al.* 1990; Meyer and Turner 1992).

7. The expanding emphasis on urban environmental issues and the incorporation of biogeophysical or landscape analysis of factors such as risk and scientific management concepts (Myers 1999; Pelling 1997: Swyngedouw 1997, 1999).
8. The growing focus on advanced industrial societies and heightened interest in whether it supports the interpretation of "ecological modernization" due to globalization (Blaikie 1998; Bryant and Bailey 1997).

If the vigor of a scholarly field can be measured by the productivity, range, and quality of its members' research, then geographical cultural ecology was particularly robust in the 1990s. Cultural ecological research offers fresh insights into the cultural, historical, political, and biophysical dimensions of human–environmental interactions. As the developing world becomes increasingly integrated into the tumultuous orbit of the global economy, the works of cultural and political ecology are certain to enhance our understanding of the nature and direction of nature–society dynamics at a variety of spatial and temporal scales.

Acknowledgements

This chapter was written equally by the co-authors. It is dedicated to the memory of Bernard Q. Nietschmann, extraordinary academic innovator, activist, and teacher who is a lasting influence in the fields of cultural and political ecology. We would like to thank Rob Daniels (Illinois) and Morgan Robertson (Wisconsin) for their research assistance and the members and Advisory Panel of the Cultural Ecology Specialty Group for their comments and advice.

References

Aageson, D. L. (1998*a*). "Indigenous Resource Rights and Conservation of the Monkey-Puzzle Tree (*Araucaria araucana*, Araucariaceae): A Case Study from Southern Chile." *Economic Botany*, 52/2: 146–60.

— (1998*b*). "On the Northern Fringe of the South American Temperate Forest: The History and Conservation of the Monkey-Puzzle Tree," *Environmental History*, 3/1: 64–85.

Adams, W. M. (1995). "Green Development Theory," in J. Crush (ed.), *Power of Development*. London: Routledge, 87–99.

Agrawal, A. (1995). "Dismantling the Divide between Indigenous and Scientific Knowledge." *Development and Change*, 26: 413–39.

Allan, N. J. R. (1991). "From Autarky to Dependency: Society and Habitat Relations in the South Asian Mountain Rimland." *Mountain Research and Development*, 11/1: 65–74.

— (ed.) (1995). *Mountains at Risk: Current Issues in Environmental Studies*. New Delhi: Manohar.

Bassett, T. (1993*a*). "Introduction: The Land Question and Agricultural Transformation in Africa," in Bassett and Crummey (1993: 3–31).

— (1993*b*). "Land Use Conflicts in Pastoral Development in Northern Côte d'Ivoire," in Bassett and Crummey (1993: 131–54).

— (1994). "Hired Herders and Herd Management in Fulani Pastoralism (Northern Côte d'Ivoire)." *Cahiers d'Études Africaines*, 133–5: 147–73.

Bassett, T., and Crummey, D. (eds.) (1993). *Land in African Agrarian Systems*. Madison: University of Wisconsin Press.

Bassett, T., and Koli Bi, Zuéli (1999). "Fulbe Pastoralism and Environmental Change in Northern Côte d'Ivoire," in M. DeBruijn and H. van Dijk (eds.), *Pastoralism under Pressure*. Amsterdam: Brill, 139–60.

Batterbury, S. (1996). "Planners or Performers? Reflections on Indigenous Dryland Farming in Northern Burkina Faso." *Agriculture and Human Values*, 13/3: 12–22.

Batterbury, S., and Bebbington, A. (1999). "Environmental Histories, Access to Resources and Landscape Change: An Introduction," *Land Degradation and Development*, 10/4: 279–89.

Batterbury, S., Forsyth, T., and Thomson, K. (1997). "Environmental Transformations in Developing Countries: Hybrid Research and Democratic Policy," *The Geographical Journal*, 163/2: 126–32.

Beach, T., and Dunning, N. P. (1995). "Ancient Maya Terracing and Modern Conservation in the Petén Rain Forest of Guatemala," *Journal of Soil and Water Conservation*, 50/2: 138–45.

Bebbington, A. (1991). "Indigenous Agricultural Knowledge Systems, Human Interests, and Critical Analysis: Reflections on Farmer Organization in Ecuador," *Agriculture and Human Values*, VIII (1/2): 14–24.

— (1993). "Modernization from Below: An Alternative Indigenous Development?" *Economic Geography*, 69/3: 274–92.

— (1996). "Movements, Modernizations, and Markets: Indigenous Organizations and Agrarian Strategies in Ecuador," in Peet and Watts (1996: 86–109).

— (1997). "Social Capital and Islands of Sustainability in the Rural Andes," *Geographical Journal*, 163/2: 189–97.

Bebbington, A., Quisbert, J., and Trukillo, G. (1996). "Technology and Rural Development Strategies in a Small Farmer Organization: Lessons from Bolivia for Rural Policy and Practice," *Public Administration and Development*, 16: 195–213.

Blaikie, P. (1994). "Political Ecology in the 1990s: An Evolving View of Nature and Society," *CASID Distinguished Speaker Series, No. 13*. E. Lausing: Michigan State University.

Blaikie, P. (1995). "Changing Environments or Changing Views? A Political Ecology for Developing Countries." *Geography*, 80/3: 203–14.

— (1998). "A Review of Political Ecology: Issues, Epistemology, and Analytical Narratives." *Zeitschrift für Wirtschaftsgeographie*, 3–4: 131–47.

Blaikie, P., and Brookfield, H. (eds.) (1987). *Land Degradation and Society*. London: Metheun.

Brannstrom, C., and Oliveira, M. S. (2000). "Human Modification of Stream Valleys in the Western Plateau of São Paulo, Brazil: Implications for Environmental Narratives and Management" *Land Degradation and Development*, 11: 535–48.

Brookfield, H. (1995). "The 'Population–Environment Nexus' and PLEC," *Global Environmental Change*, 5/4: 381–93.

Brookfield, H., and Humphreys, G. S. (1994). "Evaluating Sustainable Land Management: Are We on the Right Track?" in J. D. Etchevers (ed.), *Transactions of the 15th World Congress of Soil Science*, Acapulco, Mexico.

Brower, B., and Dennis, A. (1998). "Grazing the Forest, Shaping the Landscape? Continuing the Debate about Forest Dynamics in Sagarmatha National Park, Nepal," in Zimmerer and Young (1998: 184–208).

Bryant, R. (1992). "Political Ecology: An Emerging Research Agenda in Third World Studies," *Political Geography Quarterly*, 11/1: 108–25.

Bryant, R. (1998). "Power, Knowledge, and Political Ecology in the Third World: A Review." *Progress in Human Geography*, 22/1: 79–94.

Bryant, R., and Bailey, S. (1997). *Third World Political Ecology*. London: Routledge.

Butzer, K. W. (1989). "Cultural ecology," in G. Gaile and C. Willmott (eds.), *Geography in America*. Columbus, Ohio: Merrill, 192–208.

Butzer, K. W. (1990*a*). "Ethno-Agriculture and Cultural Ecology in Mexico: Historical Vistas and Modern Implications," *Yearbook, Conference of Latin Americanist Geographers*, 17/18: 139–52.

— (1990*b*). "The Indian Legacy in the American Landscape," in M. Conzen (ed.), *The Making of the American Landscape*. Boston: Unwin Hyman, 27–50.

— (1990*c*). "The Realm of Cultural-Human Ecology: Adaptation and Change in Historical Perspective," in B. L. Turner *et al.* (1990: 685–701).

— (ed.) (1992). *The Americas before and after 1492: Current Geographical Research*, Special Issue of the *Annals of the Association of American Geographers*, 82/3.

— (1993). "No Eden in the New World," *Nature*, 362: 15–17.

— (1996*a*). "Ecology in the Long View: Settlement Histories, Agroecosystemic Strategies, and Ecological Performance," *Journal of Field Archaeology*, 23: 141–50.

— (1996*b*). "Irrigation, Raised Fields and State Management: Wittfogel Redux?" *Antiquity*, 70: 200–4.

Butzer, K. W., and Butzer, E. K. (1993). "The Sixteenth-Century Environment of the Central Mexican Bajío: Archival Reconstruction from Colonial Land Grants and the Question of Spanish Ecological Impact," in K. Mathewson (ed.), *Culture, Form, and Place: Essays in Cultural and Historical Geography*, Geoscience and Man, 32: 89–124.

— (1997). "The 'Natural' Vegetation of the Mexican Bajío: Archival Documentation for a 16th-Century Savanna Environment," *Quaternary International*, 43/44: 161–72.

Carney, J. (1991). "Indigenous Soil and Water Management in Senegambian Rice Farming Systems," *Agriculture and Human Values*, 8/1–2: 37–48.

— (1993*a*). "Converting the Wetlands, Engendering the Environment: The Intersection of Gender with Agrarian Change in The Gambia," *Economic Geography*, 69: 329–49.

— (1993*b*). "From Hands to Tutors: African Expertise in the South Carolina Economy." *Agricultural History*, 67/3: 1–30.

— (1996*a*). "Landscapes of Technology Transfer: Rice Cultivation and African Continuities." *Technology and Culture*, 37/1: 5–35.

— (1996*b*). "Rice Milling, Gender, and Slave Labor in Colonial South Carolina," *Past and Present*, 153: 108–34.

— (1998). "The Role of African Rice and Slaves in the History of Rice Cultivation in the Americas," *Human Ecology*, 26: 525–45.

Carney, J., and Watts, M. (1990). "Manufacturing Dissent: Work, Gender and the Politics of Meaning in a Peasant Society," *Africa*, 60/2: 207–41.

Cleveland, D. A., Bowannie, F., Jr., Eriacho, D. F., Laahty, A., and Perramond, E. (1995). "Zuni Farming and United States Government Policy: The Politics of Biological Diversity in Agriculture," *Agriculture and Human Values*, 12/3: 2–18.

Coggins, C. (1999). "Wildlife Conservation and Bamboo Management in China's Southeast Uplands," *Geographical Review*, 90/1: 83–111.

Coomes, O. (1996*a*). "Income Formation among Amazonian Peasant Households in Northeastern Peru: Empirical Observations and Implications for Market-Based Conservation,"*Yearbook of the Conference on Latin Americanist Geographers*, 22: 51–64.

— (1996*b*). "State Credit Programs and the Peasantry under Populist Regimes: Lessons from the APRA Experience in Peruvian Amazonia," *World Development*, 24/8: 1333–46.

Coomes, O., and Barham, B. (1997). "Rain Forest Extraction and Conservation in Amazonia," *Geographical Journal*, 163/2: 180–8.

Coomes, O., and Burt, G. (1997). "Indigenous Market-Oriented Agroforestry: Dissecting Local Diversity in Western Amazonia." *Agroforestry Systems*, 37: 27–44.

Denevan, W. M. (1992). "The Pristine Myth: The Landscape of the Americas in 1492," *Annals of the Association of American Geographers*, 82/3: 369–85.

— (1996). "A Bluff Model of Riverine Settlement in Prehistoric Amazonia," *Annals of the Association of American Geographers*, 86/4: 654–81.

Doolittle, W. (1992). "Agriculture in North America on the Eve of Contact: A Reassessment," *Annals of the Associaton of American Geographers*, 82/3: 386–401.

— (1993). "Canal Irrigation at Casas Grandes: A Technological and Developmental Assessment of its Origins," in A. I. Woosley and J. C. Ravesloot (eds.), *Culture and Contact: Charles C. DiPeso's Gran Chichimeca*. Albuquerque: University of New Mexico Press, 133–50.

— (1995). "Indigenous Development of Mesoamerican Irrigation," *Geographical Review*, 85: 301–23.

Doolittle, W. E., Neely, J. A., and Pool, M. D. (1993). "A Method for Distinguishing between Prehistoric and Recent Water and Soil Control Features," *Kiva*, 59/1: 7–25.

Downing, T. E., Watts, M., and Bohle, H. (1996). "Climate Change and Food Insecurity: Toward a Sociology and Geography of Vulnerability," in T. Downing (ed.), *Climate Change and World Food Security*. Berlin: Springer.

Dunning, N. P. (1995). "Coming Together at the Temple Mountain: Environment, Subsistence and the Emergence of Lowland Maya Segementary State," in N. Grube (ed.), *The Emergence of Lowland Maya Civilization*. Mockmühl: Verlag von Flemming.

Dunning, N. P., and Beach, T. (1994). "Soil Erosion, Slope Management, and Ancient Terracing in the Maya Lowlands," *Latin American Antiquity*, 5/1: 51–69.

Dunning, N. P., Beach, T., and Rue, D. (1997). "The Paleoecology and Ancient Settlement of the Petexbatun Region, Guatemala," *Ancient Mesoamerica*, 8: 255–66.

Dunning, N. P., Rue, D. J., Beach, T., Covich, A., and Traverse, A. (1998). "Human-Environment Interactions in a Tropical Watershed: The Paleoecology of Laguna Tamarindito, El Petén, Guatemala," *Journal of Field Archaeology*, 25: 139–51.

Escobar, A. (1996). "Constructing Nature: Elements for a Post-structural Political Ecology." in Peet and Watts (1996: 46–85).

Forsyth T. (1996). "Science, Myth and Knowledge: Testing Himalayan Environmental Degradation in Thailand." *Geoforum*, 27/3: 375–92.

Fredrich, B. E. (1991). "Food and Culture: Using Ethnic Recipes to Demonstrate the Post-Columbian Exchange of Plants and Animals," *Journal of Geography*, 90/1: 11–15.

Gade, D. W. (1992). "Landscape, System, and Identity in the Post-Conquest Andes," *Annals of the Association of American Geographers*, 82/3: 460–77.

—— (1999). *Nature and Culture Convergent: Geography, Historical Ecology, and Ethnobiology in the Andes*. Madison: University of Wisconsin Press.

Gartner, W. G. (1997). "Four Worlds without an Eden: Pre-Columbian Peoples and the Wisconsin Landscape," in R. C. Ostergren and T. R. Vale (eds.), *Wisconsin Land and Life*. Madison: University of Wisconsin Press, 331–50.

Goldman A. (1992). "Resource Degradation, Agriculural Change, and Sustainability in Farming Systems of Southeast Nigeria," in J. L. Moock and R. Rhoades (eds.), *Diversity, Farmer Knowledge, and Sustainability*. Ithaca, NY: Cornell University Press.

—— (1993*a*). "Population Growth and Agricultural Change in Imo State, Southeastern Nigeria," in B. L. Turner *et al.* (1993: 250–301).

—— (1993*b*). "Agricultural Innovation in Three Areas of Kenya: Neo-Boserupian Theories and Regional Characterization," *Economic Geography*, 69/1: 44–71.

—— (1995). "Threats to Sustainability in African Agriculture: Searching for Appropriate Paradigms," *Human Ecology*, 23/3: 291–334.

Gray, L. (1999). "Is Land being Degraded? A Multi-Scale Perspective on Landscape Change in Southwestern Burkina Faso," *Land Degradation and Development*, 10/4: 329–43.

Grossman, L. S. (1991). "Diet, Income, and Subsistence in an Eastern Highland Village, Papua New Guinea," *Ecology of Food and Nutrition*, 26: 235–53.

—— (1993). "The Political Ecology of Banana Exports and Local Food Production in St. Vincent, Eastern Caribbean." *Annals of the Association of American Geographers*, 83/2: 347–67.

—— (1997). "Soil Conservation, Political Ecology, and Technological Change on Saint Vincent," *Geographical Review*, 87/3: 353–74.

—— (1998*a*). "Diet, Income, and Agriculture in an Eastern Caribbean Village," *Human Ecology*, 26/1: 21–42.

—— (1998*b*). *The Political Ecology of Bananas: Contract Farming, Peasants, and Agrarian Change in the Eastern Caribbean*. Chapel Hill, NC: University of North Carolina Press.

Guthman, J. (1997). "Representing Crisis: The Theory of Himalayan Environmental Degradation and the Project of Development in Post-Rana Nepal," *Development and Change*, 28/1: 45–69.

Hall, S. (1995). "The West and the Rest: Discourse and Power," in S. Hall *et al.* (eds.), *Modernity: An Introduction to Modern Societies*. Oxford: Blackwell.

Hansis, R. (1998). "A Political Ecology of Picking: Non-Timber Forest Products in the Pacific Northwest," *Human Ecology*, 26: 49–68.

Heasley, L., and Delehanty, J. (1996). "The Politics of Manure: Resource Tenure and the Agropastoral Economy in Southwestern Niger," *Society and Natural Resources*, 9: 31–46.

Hecht, S. B. (1994). "The Logic of Livestock and Deforestation in Amazonia," *Bioscience*, 43: 687–95.

Hecht, S. B., and Cockburn, A. (1990). *Fate of the Forest*. 2nd edn., New York: Harper Collins.

Herlihy, P. H. (1992). "Wildlands Conservation in Central America during the 1980s: A Geographical Perspective," *Yearbook, Conference of Latin Americanist Geographers*, 17/18: 31–43.

—— (1993). "Securing a Homeland: The Tawakha Sumu of Mosquitia's Rain Forest," in J. Dow and R. V. Kemper (eds.), *State of the Peoples: A Global Human Rights Report on Societies in Danger*. Boston: Beacon Press.

—— (1997). "Indigenous Peoples and Biosphere Reserve Conservation in the Mosquitia Rain Forest Corridor, Honduras," in S. Stevens (ed.), *Conservation through Cultural Survival: Indigenous Peoples and Protected Areas*. Washington, DC: Island Press, 99–129.

Horn, S. P. (1998). "Fire Management and Natural Landscapes in the Chirripo Paramo, Chirripo National Park, Costa Rica," in Zimmerer and Young (1998: 125–46).

Hyden, G., Kates, R., and Turner, B. L., II (1993). "Beyond Intensification," in B. L. Turner *et al.* (1993: 401–39).

Ives, J. D., and Messerli, B. (1990). "Progress in Theoretical and Applied Mountain Research, 1973–1989, and Major Future Needs," *Mountain Research and Development*, 10/2: 101–27.

Knapp, G. (1991). *Andean Ecology: Adaptive Dynamics in Ecuador*. Boulder: Westview Press.

—— (1992). *Riego precolonial y tradicional en la sierra norte del Ecuador*. Quito: Abya-Yala.

—— (1998). "Quilotoa Ash and Human Settlements in the Equatorial Andes," in P. Mothes (ed.), *Actividad volcánica y pueblos precolombinos en el Ecuador*. Quito: Abya-Yala, 139–55.

Kull, C. (1998). "Leimavo Revisited: Agrarian Land-Use Change in the Highlands of Madagascar," *Professional Geographer*, 50/2: 163–76.

Kummer, D., and Turner, B. L., II (1994). "The Human Causes of Deforestation in Southeast Asia," *Bioscience*, 44/5: 323–8.

Lambin, E. F., and Guyer, J. I. (1994). "The Complementarity of Remote Sensing and Anthropology in the Study of Complex Human Ecology." *Working Paper* 175. African Studies Center, Boston University.

Lightfoot, D. R. (1993). "The Cultural Ecology of Puebloan Pebble-Mulch Gardens," *Human Ecology*, 21/2: 115–43.

—— (1994). "Morphology and Ecology of Lithic-Mulch Agriculture," *The Geographical Review*, 84/2: 172–85.

Liverman, D., Moran, E., Rindfuss, R. R., and Stern, P. C. (eds.) (1998). *People and Pixels: Linking Remote Sensing and Social Science*. Washington, DC: National Academy Press.

Lovell, W. G. (1992). "Heavy Shadows and Black Night: Disease and Depopulation in Colonial Spanish America," *Annals of the Association of American Geographers*, 82/3: 426–43.

Mathewson, K. (1990). "Rio Hondo Reflections: Notes on Puleston's Place and the Archaeology of Maya Landscapes," in M. D. Pohl (ed.), *Ancient Maya Wetland Agriculture: Excavations on Albion Island, Northern Belize*. Boulder: Westview Press, 21–51.

Medley, K. E. (1998). "Landscape Change and Resource Conservation along the Tana River, Kenya," in Zimmerer and Young (1998: 39–55).

Messerli, B., and Ives. J. D. (1997). *Mountains of the World: A Global Priority*. New York: Parthenon.

Metz, J. J. (1990). "Conservation Practices at an Upper-Elevation Village of West Nepal," *Mountain Research and Development*, 10/1: 7–15.

—— (1991). "A Reassessment of the Causes and Severity of Nepal's Environmental Crisis," *World Development*, 19/7: 805–20.

—— (1994). "Forest Product Use at an Upper Elevation Village in Nepal," *Environmental Management*, 18/3: 371–90.

—— (1998). "The Ecology, Use, and Conservation of Temperate and Subalpine Forest Landscapes of West Central Nepal," in Zimmerer and Young (1998: 287–306).

Meyer, W., and Turner, B. (1992). "Human Population Growth and Global Land-Use/Cover Change," *Annual Review of Ecological Systems*, 23: 39–61.

Moran, E. (1993). "Deforestation and Land Use in the Brazilian Amazon," *Human Ecology*, 21: 1–21.

—— (1995). "Rich and Poor Ecosystems of Amazonia: An Approach to Management," in T. Nishizawa and J. I. Uitto (eds.), *The Fragile Tropics of Latin America: Sustainable Management of Changing Environments*. Tokyo: United Nations Press.

Moran, E., and Brondizio, E. (1998). "Land-Use Change after Deforestation in Amazonia," in D. Liverman *et al.* (eds.) *People and Pixels: Linking Remote Sensing and Social Science*. Washington, D.C. National Academy Press. 94–120.

Mountain Agenda (1992). *An Appeal for the Mountains*. Berne: Institute of Geography, University of Berne.

Myers, G. A. (1999). "Political Ecology and Urbanisation: Zanzibar's Construction Materials Industry." *Journal of Modern African Studies*, 37/1: 83–108.

Naughton, L. (1993). "Conservation versus Artisanal Gold Mining in Corcovado National Park, Costa Rica: Land Use Conflicts at Neotropical Wilderness Frontiers," *Yearbook, Conference of Latin Americanist Geographers*, 19: 47–55.

Naughton-Treves, L. (1997). "Farming the Forest Edge: Vulnerable Places and People around Kibale National Park, Uganda," *Geographical Review*, 87/1: 27–46.

Neumann, R. (1995). "Local Challenges to Global Agendas: Conservation, Economic Liberalization, and the Pastoralists' Rights Movement in Tanzania," *Antipode*, 27/4: 363–82.

—— (1997). "Primitive Ideas: Protected Buffer Zones and the Politics of Land in Africa," *Development and Change*, 28: 559–82.

—— (1998). *Imposing Wilderness: Struggles over Livelihood and Nature Preservation in Africa*. Berkeley and Los Angeles: University of California Press.

Newman, J. L. (1995). *The Peopling of Africa: A Geographic Interpretation*. New Haven: Yale University Press.

Nietschmann, B. (1997). "Protecting Indigenous Coral Reefs and Sea Territories, Miskito Coast, RAAN, Nicaragua," in S. Stevens (ed.), *Conservation through Survival: Indigenous Peoples and Protected Areas*. Washington, DC: Island Press, 193–224.

Paulson, D. D. (1994). "Understanding Tropical Deforestation: The Case of Western Samoa," *Environmental Conservation*, 21/4: 326–32.

—— (1998). "Collaborative Management of Public Rangeland in Wyoming: Lessons in Co-management," *Professional Geographer*, 50/3: 301–15.

Paulson, D. D., and Rogers, S. (1997). "Maintaining Subsistence Security in Western Samoa," *Geoforum*, 28/2: 173–87.

Peet, R., and Watts, M. (1993). "Introducton: Development Theory and Environment in the Age of Market Triumphalism," *Economic Geography*, 69/3: 227–53.

—— (1996). *Liberation Ecologies: Environment, Development, Social Movements*. London: Routledge.

Pelling, M. (1997). "What Determines Vulnerability to Floods: A Case Study in Guyana." *Environmental and Urbanisation*, 10: 469–86.

Porter, P. (1991). "Cultural Ecology," in G. Dunbar (ed.), *Modern Geography: An Encyclopedic Survey*. New York: Garland, 38–9.

Porter, P., and Sheppard, E. (1998). *A World of Difference: Society, Nature, Development*. New York: Guilford.

Rangan, H. (1995). "Contested Boundaries: State Policies, Forest Classifications, and Deforestation in the Garhwal Himalayas," *Antipode*, 27/4: 343–62.

—— (1996). "From Chipko to Uttaranchal: Development, Environment, and Social Protest in the Garhwal Himalayas, India," in Peet and Watts (1996: 205–26).

Reenberg, A. (1995). "The Spatial Pattern and Dynamics of a Sahelian Agro-Ecosystem: Land Use Systems Analysis Combining Household Survey with Georelated Information," *GeoJournal*, 37/4: 489–99.

Reenberg, A., and Paarup-Laursen, B. (1997). "Determinants for Land Use Strategies in a Sahelian Agro-Ecosystem—Anthropological and Ecological Geographical Aspects of Natural Resource Management," *Agricultural Systems*, 53: 209–29.

Ribot, J. (1996). "Participation without Representation: Chiefs, Councils and Forestry Law in the West African Sahel," *Cultural Survival Quarterly*, 20: 40–4.

Richards, P. (1985). *Indigenous Agricultural Revolution*. London: Hutchinson.

Rocheleau, D. (1995). "Gender and Biodiversity: A Feminist Political Ecology Perspective," *IDS Bulletin*, 26/1: 9–16.

Rocheleau, D., and Ross, L. (1995). "Trees as Tools, Trees as Text: Struggles over Resources in Zambrana-Chacuey, Dominican Republic," *Antipode*, 27: 407–28.

Rocheleau D., Steinberg, P., and Benjamin, P. (1995). "Environment, Development, Crisis and Crusade: Ukambani, Kenya, 1890–1990," *World Development*, 23/6: 1037–51.

Rocheleau, D., Thomas-Slayter, B., and Wangari, E. (1996). *Feminist Political Ecology: Global Issues and Local Experience*. London: Routledge.

Schroeder, R. (1993). "Shady Practices: Gender and the Political Ecology of Resource Stabilization in Gambian Garden/Orchards," *Economic Geography*, 69/4: 349–65.

—— (1995). "Contradictions along the Commodity Road to Environmental Stabilization: Foresting Gambian Gardens," *Antipode*, 27/4: 325–42.

—— (1999). *Shady Practices: Agroforestry and Gender Politics in the Gambia*. Berkeley: University of California Press.

Schroeder, R., and Neumann, R. (1995). "Manifest Ecological Destinies: Local Rights and Global Environmental Agendas," *Antipode*, 27/4: 321–4.

Schroeder, R., and Suryanata, K. (1996). "Gender and Class Power in Agroforestry Systems," in Peet and Watts (1996: 188–204).

Sluyter, A. (1994). "Intensive Wetland Agriculture in Mesoamerica: Space, Time, and Form," *Annals of the Association of American Geographers*, 84/4: 557–84.

—— (1996). "The Ecological Origins and Consequences of Cattle Ranching in Sixteenth-Century New Spain," *Geographical Review*, 86/2: 161–77.

—— (1997*a*). "Landscape Change and Livestock in Sixteenth-Century New Spain: The Archival Data Base," *Yearbook of the Conference of Latin Americanist Geographers*, 23: 27–39.

—— (1997*b*). "Regional, Holocene Records of the Human Dimension of Global Change: Sea-Level and Land-Use Change in Prehistoric Mexico," *Global and Planetary Change*, 14: 127–46.

—— (1998). "From Archive to Map to Pastoral Landscape: A Spatial Perspective on the Livestock Ecology of Sixteenth-Century New Spain," *Environmental History*, 3/4: 508–28.

—— (1999). "The Making of the Myth in Postcolonial Development: Material-Conceptual Landscape Transformation in Sixteenth-Century Veracruz," *Annals of the Association of American Geographers*, 89/3: 377–401.

Steinberg, M. K. (1998*a*). "Political Ecology and Cultural Change: Impacts on Swidden-Fallow Agroforestry Practices among the Mopan Maya in Southern Belize," *Professional Geographer*, 50/4: 407–17.

—— (1998*b*). "Neotropical Kitchen Gardens as a Potential Research Landscape for Conservation Biologists," *Conservation Biology*, 12/5: 1–3.

Stevens, S. (1993*a*). *Claiming the High Ground: Sherpas, Subsistence, and Environmental Change in the Highest Himalaya*. Berkeley: University of California Press.

—— (1993*b*). "Indigenous Peoples and Protected Areas: New Approaches to Conservation in Highland Nepal," in L. Hamilton, D. Bauer, and H. Takeuchi (eds.), *Parks, Peaks, and People*. Honolulu: East-West Center, 73–88.

—— (1997*a*). "The Legacy of Yellowstone," in S. Stevens (ed.), *Conservation through Survival: Indigenous Peoples and Protected Areas*. Washington, DC: Island Press, 13–32.

—— (1997*b*). "New Alliances for Conservation," in S. Stevens (ed.), *Conservation through Survival: Indigenous Peoples and Protected Areas*. Washington, DC: Island Press, 33–62.

Sundberg, J. (1998). "Strategies for Authenticity, Space, and Place in the Maya Biosphere Reserve, Petén, Guatemala," *Yearbook, Conference of Latin Americanist Geographers*, 24: 85–96.

Swyngedouw, E. (1997). "Power, Nature and City: Water and the Political Ecology of Urbanization in Guayaquil, Ecuador: 1880–1990." *Environment and Planning* A, 29: 311–32.

Turner, B. L., II (1997*a*). "Spirals, Bridges, and Tunnels: Engaging Human-Environment Perspectives in Geography," *Ecumene*, 4/2: 196–217.

—— (1997*b*). "The Sustainability Principle in Global Agendas: Implications for Understanding Land-Use/Cover Change," *The Geographical Journal*, 163/2: 133–40.

Turner, B. L., II, and Butzer, K. (1992). "The Columbian Encounter and Land-Use Change," *Environment*, 34/8: 16–20, 38–44.

Turner, B. L., II, and Shajaat Ali, A. M. (1996). "Induced Intensification: Agricultural Change in Bangladesh with Implications for Malthus and Boserup," *Proceedings of the National Academy of Sciences*, 93: 14984–91.

Turner, B. L., II, Hyden, G., and Kates, R. (1993). *Population Growth and Agricultural Change in Africa*. Gainesville, Fla.: University Press of Florida.

Turner, II, B. L., *et al.* (1990). *The Earth as Transformed by Human Action: Global and Regional Changes in the Biosphere over the Past 300 Years*. Cambridge: Cambridge University Press.

Turner, M. D. (1993). "Overstocking the Range: A Critical Analysis of the Environmental Science of Sahelian Pastoralism," *Economic Geography*, 69/4: 402–21.

—— (1998). "The Interaction of Grazing History with Rainfall and its Influence on Annual Rangeland Dynamics in the Sahel," in Zimmerer and Young (1998: 237–261).

—— (1999*a*). "Conflict, Environmental Change, and Social Institutions in Dryland Africa: Limitations of the Community Resource Management Approach," *Society and Natural Resources*, 12: 643–57.

—— (1999*b*). "Labor Process and the Environment: The Effects of Labor Availability and Compensation on the Quality of Herding in the Sahel," *Human Ecology*, 27: 267–96.

—— (1999*c*). "Merging Local and Regional Analyses of Land-Use Change: The Case of Livestock in the Sahel," *Annals of the Association of American Geographers*, 89/2: 191–219.

—— (1999*d*). "No Space for Participation: Pastoralist Narratives and the Etiology of Park-Herder Conflict in Southeastern Niger," *Land Degradation and Development*, 10: 343–61.

—— (1999*e*). "The Role of Social Networks, Indefinite Boundaries and Political Bargaining in Maintaining the Ecological and Economic Resiliency of the Transhumance Systems of Sudano-Sahelian West Africa," In M. Niair-Fuller (ed.), *Managing Mobility in African Rangelands: The Legitimization of Transhumance*. London: Intermediate Technology Publications.

Unruh, J. (1990). "Integration of Transhumant Pastoralism and Irrigated Agriculture in Semi-Arid East Africa," *Human Ecology*, 18/3: 223–46.

—— (1995*a*). "Post-Conflict Recovery of African Agriculture: The Role of 'Critical Resource' Tenure," *Ambio*, 24/6: 343–8.

—— (1995*b*). "The Relationship between Indigenous Pastoralist Resource Tenure and State Tenure in Somalia," *Geojournal*, 36/1: 19–26.

Vayda, A. P., and Walters, B. B. (1999). "Against Political Ecology," *Human Ecology*, 27/1: 167–79.

Voeks, R. A. (1996). "Extraction and Tropical Rain Forest Conservation in Eastern Brazil," *Tropical Rainforest Research Current Issues*, 477–87.

—— (1997). *Sacred Leaves of Candomblé: African Magic, Medicine, and Religion in Brazil*. Austin: University of Texas Press.

Voeks, R. A. (1998). "Ethnobotanical Knowledge and Environmental Risk: Foragers and Farmers in Northern Borneo," in Zimmerer and Young (1998: 307–26).

Walters, B. B. (1997). "Human Ecological Questions for Tropical Restoration: Experiences from Planting Native Upland Trees and Mangroves in the Philippines," *Forest Ecology and Management*, 99: 275–90.

Watts, M. (1993). "Development I: Power, Knowledge, Discursive Practice," *Progress in Human Geography*, 17/2: 257–72.

— (1994). "Living under Contract: Agro-Food Complexes in the Third World," in P. Little and M. Watts (eds.), *Living Under Contract: Contract Farming and Agrarian Transformation in Africa*. Madison: University of Wisconsin Press, 21–77.

— (1996). "Development III: The Global Agrofood System and Late Twentieth-Century Development (or Kautsky *redux*)," *Progress in Human Geography*, 20/2: 230–45.

— (1998*a*). "Agrarian Thermidor: State, Decollectivization, and the Peasant Question in Vietnam," in I. Szelényi (ed.), *Privatizing the Land: Rural Political Economy in Post-Communist Societies*. London: Routledge, 149–88.

— (1998*b*). "Nature as Artifice and Artifact," in B. Braun and N. Castree (eds.) *Remaking Reality: Nature at the Millennium*. London: Routledge, 243–68.

Whitesell, E. A. (1997). "Accomodating Multiple Human Rights and Non-Human Rights: The Dilemmas of Conservation in Brazilian Amazonia," Paper presented at the Meeting of the Association of American Geographers (Fort Worth, Texas).

Whitmore, T. (1991). "A Simulation of the Sixteenth-Century Population Collapse in the Basin of Mexico," *Annals of the Association of American Geographers*, 81/3: 464–87.

— (1996). "Population Geography of Calamity: The Sixteenth and Seventeenth Century Yucatan," *International Journal of Population Geography*, 2: 291–311.

Whitmore, T., and Turner, B. (1992). "Landscapes of Cultivation in Mesoamerica on the Eve of the Conquest." *Annals of the Association of American Geographers*, 82/3: 402–25.

Williams, B. J. (1992). "Tepetate in 16th Century and Contemporary Folk Terminology, Valley of Mexico," *Terra* 10: 483–93.

WinklerPrins, A. M. G. A. (1999). "Local Soil Knowledge: A Tool for Sustainable Land Management," *Society and Natural Resources*, 12: 151–61.

Young, E. (1999*a*). "Local People and Conservation in Mexico's El Vizcaíno Biosphere Reserve," *Geographical Review*, 89: 364–90.

— (1999*b*). "Balancing Conservation with Development in Small-Scale Fisheries: Is Ecotourism an Empty Promise?" *Human Ecology*, 27/4: 581–620.

Young, K. R. (1997). "Wildlife Conservation in the Cultural Landscapes of the Central Andes," *Landscape and Urban Planning*, 38: 137–47.

Zimmerer, K. S. (1991*a*). "Labor Shortages and Crop Diversity in the Southern Peruvian Sierra," *Geographical Review*, 81/4: 414–32.

— (1991*b*). "Wetland Production and Smallholder Persistence: Agricultural Change in a Highland Peruvian Region," *Annals of the Association of American Geographers*, 81/3: 414–32.

— (1992). "Biological Diversity and Local Development: 'Popping Beans' in the Central Andes," *Mountain Research and Development*, 12/1: 47–61.

— (1993*a*). "Agricultural Biodiversity and Peasant Rights to Subsistence in the Central Andes during Inca Rule," *Journal of Historical Geography*, 19: 15–32.

— (1993*b*). "Soil Erosion and Labor Shortages in the Andes with Special Reference to Bolivia, 1953–91," *World Development*, 21/10: 1659–75.

— (1994*a*). "Human Geography and the 'New Ecology': The Prospect and Promise of Integration," *Annals of the Association of American Geographers*, 84/1: 108–25.

— (1994*b*). "Local Soil Knowledge: Answering Basic Questions in Highland Bolivia," *Journal of Soil and Water Conservation*, January/February: 29–34.

— (1995). "The Origins of Andean Irrigation," *Nature*, 378: 481–3.

— (1996*a*). *Changing Fortunes: Biodiversity and Peasant Livelihood in the Peruvian Andes*. Los Angeles and Berkeley: University of California Press.

— (1996*b*). "Discourses on Soil Loss in Bolivia: Sustainability and the Search for Socioenvironmental 'Middle Ground'," in Peet and Watts (1996: 110–24).

— (1996*c*). "Ecology as Cornerstone and Chimera in Human Geography," in C. Earle, K. Mathewson, and M. Kenzer (eds.), *Concepts in Human Geography*. London: Rowman & Littlefield, 161–88.

— (1999). "The Overlapping Patchworks of Mountain Agriculture in Peru and Bolivia: Toward a Regional-Global Landscape Model," *Human Ecology*, 27/1: 135–65.

Zimmerer, K. S., and Young, K. R. (1998). *Nature's Geography: New Lessons for Conservation in Developing Countries*. Madison: University of Wisconsin Press.

CHAPTER 9

Economic Geography

James W. Harrington, Trevor J. Barnes, Amy K. Glasmeier, Dean M. Hanink, and David L. Rigby

Introduction

To read the comparable chapter on economic geography in *Geography in America* is to recall a world, and a way of viewing that world, that seems remote. For one thing, that chapter was called Industrial Geography. There were good reasons why industrial geography was so prominent in the last report. The 1970s and 1980s were a period of fundamental industrial change in Western economies involving deindustrialization and lay-offs, restructuring of methods of production, the emergence of new manufacturing and service sectors, and new forms of international economic organization supported by innovations in telecommunications, transportation, and corporate organization and management. All those substantive issues remain important, and in some cases central, to present economic geographical research. Changed, though, is the conceptualization of those issues.

In particular, newer approaches tend to blur the boundary between the economic part of economic geography, and other social, cultural, and political geographical practices. Some have labeled this move "the cultural turn" (Crang 1997; Thrift and Olds 1996; Barnes 1996*b*), but this description is too narrow because more than just the cultural is at stake. Rather, the very idea of the economic is being reconceived. The economic is no longer conceptualized as sovereign, isolated, and an entity unto itself, but porous and dependent, bleeding into other spheres as they bleed into it. To use Karl Polyani's (1944) term, which is often deployed in this literature, the economy is "embedded" within broader processes.

There are at least two reasons for the reconceptualization of the economic by economic geographers. One is internal to the academy, and is bound up with a broader intellectual shift in the social sciences and humanities that is increasingly suspicious of essentialized entities such as "the economy" (Barnes 1996*a*; Gibson-Graham 1996; Lee and Wills 1997). A second source of change is the actual geography of economic activities. The economic geographical landscape of the 1990s seems quite different from the one written about in the last report, and thereby demands a new theoretical vocabulary in which to be represented. In the last report, for example, there was no mention of Fordism or post-Fordism, flexibility or economies of scope, localities or local modes of regulation, growth coalitions or territorial complexes, or glocalization or even globalization. Since then, however, these terms, and others like them, have become central to the discipline, defining a new theoretical lexicon and a new set of problems to investigate. These same two changes have affected the conduct of most of the human-geographic research presented in this volume.

Nowhere has this been more evident than in the recent examination by economic geographers of the changing and increasingly complex set of relationships that operate within and between the two spatial scales of the local and the global. Both scales have long been the foci of economic geography. The difference now, though, is the explicit theorization of both scales, and their relationship. On the one hand, the local and the regional have been rediscovered theoretically as pivotal sites for labor markets, consumption, the formation of entrepreneurial growth coalitions, and technological innovation and growth. The local and the regional serve not just as empirical background for wider study, but provide a necessary conceptual component in understanding why any economic geographical activity occurs at all. On the other hand, much is made theoretically of globalization especially with respect to the movement of financial capital, multinational corporation investment and strategy, and the effects of new information technology. A key question that remains is the interrelationship between the two scales, and it now motivates some of the most innovative empirical and theoretical work in the discipline (Peck 1996; Cox 1997; Storper 1997; Gertler and Barnes 1999).

Our chapter reviews the attempts to reconceptualize both the economic and the geographic. The chapter is organized into three main sections. The first reviews the work undertaken by geographers and others that attempt to embed the economy within the social, the political, the cultural, and the environment. The second section reviews some of the recent writings that reconceive the boundaries of regions, and in particular the complex relationships that exist among different geographical scales. The last major section addresses the wider policy relevance of economic geographical research and its impact on the public sphere, as well as the limitations of that impact.

Reconceiving Economy

Much innovative work carried out in economic geography since the mid-1980s attempted to explore the relationship between the economic and the non-economic. Up until that time, the two dominant theoretical approaches in economic geography—orthodox neoclassical economics and Marxism—strove to keep the economic as a distinct sphere, to make it conceptually and analytically separate from everything else. In contrast, recent work in the discipline attempts to show how economic-geographic processes are mutually dependent with other processes. In discussing the various reconceptualizations of the economic by economic geographers we divide the review into different broad approaches that have been used, and then by substantive topic. This makes for a messy narrative. But the project of reconceptualizing the economic continues on the general level of approaches and in the context of specific topics.

Approaches

1. *Spatial Divisions of Labor and Localities.* Massey's book *Spatial Divisions of Labor* (1984) was a key early text to question the conceptual isolation of economic processes. While the economy remained vitally important to her, Massey stressed the role of place-bound social and cultural variables such as gender, religious organization, and class politics in determining industrial location. Using what Warde (1985) called a geological metaphor, Massey argued that the industrial investment found at a given place is determined in large part by the social and cultural conditions, which, in turn, were partly a response to yet earlier rounds of investment. The social, the economic, and place, in Massey's view, are recursively connected, determining the conditions of any given locale. In turn, her insight became the basis of the British "Locality Project" headed by P. Cooke (1989, 1990) and which, following Massey, attempted a series of local studies connecting place, the social, and the economic.

A North American version of the same project was carried out by Cox and Mair (1988, 1991). Their thesis hinged on what they termed "local dependency," defined as "the dependence of various actors—capitalist firms, politicians, people—on the reproduction of certain social relations within a particular territory." As in Massey's work, Cox and Mair recognized that the economic is dependent upon a set of non-economic, place-situated institutions and relations. However, certain factions of capital can become "hypermobile," able to escape local constraints. The consequence is the disruption of existing local dependency relations and intercommunity competition, as local communities, primarily represented by business interests, attempt to lure investment funds towards themselves and away from other communities (see also Harvey's (1989) related idea of urban entrepreneurialism).

The British and American project focused on the generation and impact of place characteristics. "Place matters!" as Massey declared. It matters, in part, because abstract notions of the economy found in neoclassical and Marxist economics cannot be sustained when brought down to earth; they became complicated by a range of diverse social practices.

2. *Regulationism, Flexible Specialization, and Post-Fordism.* A different way to conceive the economic in

economic geography emerged shortly after Massey's book was published, and later came to dominate certain parts of the discipline: regulation theory. A response to the Western industrial economic collapse of the 1970s, regulation theory was first offered by Aglietta (1979). It represented a new historical vision of capitalist accumulation separated into distinct regimes, each characterized by a series of dominant industries, by different forms of industrial organization, and by specific sets of regulatory relationships managing a capital–labor accord and effective demand. Aglietta's arguments were extended by Boyer (1992) and Lipietz (1986), including the separation of the regime of accumulation from the mode of regulation.

Prototype US versions of regulation theory were presented by Bowles *et al.* (1986) in the form of social structures of accumulation, but the main US riposte to the French regulationists was provided by Piore and Sabel (1984) who claimed that particular forms of industrial organization, such as Fordism, dominated periods of capitalist production and that these periods were punctuated by industrial divides. The economic crisis of the 1970s was seen as the second industrial divide, marking the end of Fordism and the emergence of a new, hegemonic form of industrial organization they labeled flexible specialization.

The consistency of some of these approaches, and the simple notion that capitalist history can be readily divided into distinct periods of growth and decline were questioned by Webber (1991), and by Webber and Rigby (1996, 1999). While Hirst and Zeitlin (1991) were proponents in the UK, the flexible specialization model of the post-Fordist future generated skepticism (Amin and Robbins 1990), and at times outright hostility (Lovering 1990, 1991; Scott 1991). Donaghu and Barff (1990) used a study of athletic footwear to show how flexibility can be maintained with worldwide production. In a detailed case study, Ó hUallacháin and Matthews (1996) illustrated that vertical integration and scale economies provide powerful advantages in primary sectors; Ó hUallacháin and Wasserman (1999) showed how flexibility can be managed with vertical integration. Others disputed the flexible "foundations" of new industrial districts (Amin and Robbins 1991; Glasmeier 1994; Harrison 1994; Malecki 1995; Markusen 1996; Park and Markusen 1995; Sayer and Walker 1992). Gertler (1988, 1992) offered a more balanced and more keenly detailed critique of the concept of flexibility.

In British economic geography, especially, French regulationist theory became almost *de rigeur*. An important conceptual development was the idea of a local mode of regulation (Tickell and Peck 1992; Peck and Tickell 1995) which can operate in contradiction to a national mode. Peck and Tickell stressed, as did regulation theory more generally, that the economic does not stand aloof, but is always bound and tied to local regulatory institutions, from government training schemes to household-level practices.

The wider point is that regulationists attempted to move away from a narrowly conceived economism that had dominated Marxist economics, and its geographical counterpart. Both radical economists such as Lipietz, and radical economic geographers such as Peck, suggested that one cannot understand the economy without relating it to a broader set of social, political, and even cultural regulatory institutions and norms.

3. *Institutionalism and Evolutionary Economics.* Institutions and norms are the very stuff of institutionalism, which has made a recent resurgence in economics, and which finally arrived in economic geography during the 1990s. The institutionalist perspective originated with the maverick, late nineteenth- and early twentieth-century American economist, Thorstein Veblen. For Veblen, institutions (defined as "settled habits of thought" (Veblen 1919: 239)) influenced every aspect of human behavior and were central to even the most economic of phenomena such as the market.

Martin (1994) and Sunley (1996) have explicitly called for an economic geography modeled along Veblenesque lines. The work of Amin and Thrift (1994*b*, 1997) has clear resonances with Veblen's concern for institutions. In fact, they placed Veblen's interests within the wider field of what they call "socioeconomics" that brings together theories of institutional change, sociological economics, social networks and embeddedness, and political economy. In particular, Amin and Thrift made use of the idea of "institutional thickness," place-bound institutional structures that are open, interlinked, and reflexive, that can give a voice to a place, as well as allowing transfers of appropriate knowledge and information to enhance economic performance. In a related vein, P. Cooke and Morgan (1993, 1998) used a network model of institutions by which to understand what they called "the co-operative advantage of regions"—that is, different institutions coming together for the benefit of the region as a whole. Storper (1997) developed the idea of place-specific "conventions" used by "reflexive" firms to negotiate and overcome risk (Storper and Salais 1997). In all these cases, the economy does not innocently exist, but it is the result of active institutional construction in particular places.

Veblen made direct use of biological analogies in his writings; indeed, he talked about his work as a form of "post-Darwinian economics" in which institutions continually need to adapt to the changing external environment. Nelson and Winter (1982) developed these ideas into a framework of "evolutionary economics." The idea of evolution became much more mathematically recondite as well as spatially germane with the recent writings by the economists Romer (1986), Arthur (1989), and Krugman (1991*a*). Two particularly important concepts that emerge from their work are those of path dependency (the idea that past decisions about technology will influence future ones), and technological "lock in" (the notion that once a technological choice is made, however inefficient, it cannot be reversed). Webber and Rigby (1996) provided mathematically sophisticated economic geographical versions of these ideas. Storper (1997), and earlier Storper and Walker (1989), made special use of path dependency in understanding regional economic specialization. The gist of the argument is that new technology development requires a set of self-reinforcing, place-based "relational assets" or "untraded interdependencies". Those "assets" or "interdependencies" operate cumulatively over time, continually bolstering the initial technology selected, making it increasingly competitive. As a result, such places become locked into a particular specialization, with their future trajectory determined by a set of past decisions. Page and Walker (1991) provided an exposition of the linked historical development of agriculture, agro-industry, and Fordist manufacturing in the American midwest.

Storper's work has implications for the long stream of research on technology-intensive production. Many of the relational assets and untraded interdependencies that maintain a particular economic specialization in place are possible because of spatial proximity. Some kinds of industrial activities require more geographical embedding than others, where geographical embedding means spatial proximity among firms; or, as Gertler (1995) phrased it, the importance of "being there." Specifically, for those industrial sectors relying on specialized information or skill or rapidly changing innovations, spatial proximity among firms facilitates the frequent interaction, both formal and informal, that engenders the social virtues of trust, cooperation, and exchange of information (tacit and explicit) necessary for success. Embeddedness of these industrial sectors is necessarily geographical; place and space enter into the very constitution of the industry (Saxenian 1994).

4. *Spatial Science, Analytical Political Economy, and the New Economic Geography.* Alongside the introduction of a "softer" social and cultural approach to understanding the economy, progress has been made in analytic economic geography (and its closely aligned field of regional science). This has taken two very different forms over the last decade and a half.

First, there has emerged from economics the so-called "new economic geography," put forward by Paul Krugman (1995) and others (e.g. Henderson 1996; Kilkenny 1998). Krugman's (1991*a*) earlier work in economics had been instrumental in revolutionizing international trade theory in its recognition of the importance of increasing returns to scale of production as the microeconomic foundation of much of the world's contemporary pattern of trade. The "new growth theorists" provided crucial conceptual and analytic progress in endogenizing technical innovation as part of the explanation of economic growth (Romer 1986 and 1990; Lucas 1988). Together, these breakthroughs have allowed the conceptual and modeling tools of neoclassical microeconomics to be brought to bear on issues of agglomeration and regional economic differentiation (Krugman 1995).

While the economists' version of economic geography has brought greater use of the phrase, economic geographers have for the most part been critical of its narrow conception of both the economy, composed only of rational decision-makers, and geography, conceived as a Euclidean space (Martin 1999). Even on its own terms, economic geography critics point to the unabashed neoclassical origins including its patently unrealistic assumption of spatial equilibrium (Clark 1998; Martin and Sunley 1996; Plummer *et al.* 1998; Rigby 1991; Webber 1996). In addition, Krugman's work is highly reductionist in its approach, and even he acknowledges that his aggregate analyses simply cannot account for a number of salient issues in real-world economic geography (Krugman 1998; Martin and Sunley 1998). In contrast, the much more empirically grounded and geographically complex work of Porter (1990) addressed some of these same issues of national (or regional) economic performance, and has been more widely used by economic geographers (Ettlinger and Patton 1996; Lindahl and Beyers 1999).

A second group, also critical of the new economic geography, are analytical political economic geographers, keen to provide an alternative to the equilibrium arguments of Krugman *et al.* using formal mathematical logic (Plummer *et al.* 1998; Sheppard 2000). The inspiration for this group was Harvey's (1982) *Limits to Capital*, a geographical reworking of Marx. In a series of earlier papers, Harvey (1975) and Massey (1973) had begun to use Marx's (1970) model of capitalist accumulation to rethink the character of industrial and regional development. Much of the work that has followed formalizes and extends these links, in part to provide a more secure foundation for political economic research occurring in geography.

One of the earliest examples of such work was Sheppard's (1984) construction of a capitalist space economy based upon an interregional input–output system and a transportation commodity coupled with a Marxian model of the labor value of commodities. Liossatos (1988) offered another model with similar intent. Sheppard together with Barnes, who had himself been revisiting theories of agricultural rent from a Sraffan perspective (Barnes 1984, 1988), also attacked the inconsistencies of neoclassical models of equilibrium prices in settings where geography was treated as more than simply another subscript (Barnes and Sheppard 1984; Sheppard and Barnes 1986). Their *Capitalist Space Economy* (Sheppard and Barnes 1990) combined and extended the earlier pieces offering a fundamental remapping of economic geography in the political economy tradition.

Webber (1982) outlined a model of uneven regional development based on the political economic frameworks of dependency theory and unequal exchange. These ideas were elaborated with Foot in empirical analyses of trade between the Canadian and Philippines economies (Foot and Webber 1983; Webber and Foot 1982). Webber and Rigby (1986) showed how to measure key variables in a Marxist accounting scheme and used those techniques to examine the determinants of the rate of profit in the Canadian manufacturing sector. Webber advanced these arguments and engaged in a series of decompositions of profitability and technical change across a series of Canadian industries (Webber 1987*a*; Webber and Tonkin 1987). He also developed an interregional model of capital flows that showed how dynamic equilibrium profit rates may be unequal and that capital may flow from high to low profit rate regions, posing a serious challenge to orthodox accounts (Webber 1987*b*). Equilibrium models of capitalist dynamics were also questioned by Rigby (1990) who sought to shore-up Marx's model of the falling rate of profit. Webber (1996) and Webber and Rigby (1996, 1999) later pushed this analysis further to reveal the irrelevance of equilibrium models of prices, profits, and capital flows and offered a consistent theoretical alternative.

Taking a different tack, a recent series of papers relaxed familiar assumptions of homogeneous space and perfect information, showing that output, prices, and profits may all vary between firms in equilibrium (Sheppard *et al.* 1992). In addition, firms may benefit from pursuing rate-of-profit maximizing strategies rather than total-profit maximizing strategies, a result that upsets much conventional microeconomic theory (Plummer *et al.* 1998; Sheppard *et al.* 1998).

Of the four approaches discussed in this first section, the last represents the most continuity in the theorization of the economy and its relationship to other spheres. The analytical political economy approach, however, certainly recognizes the importance of social class in setting some critical economic variables such as prices and profit rates, and Sheppard and Barnes (1990) at least recognized the role of place-specific, cultural factors in theorizing collective action (see also Wills's 1996 complementary approach).

Topics

Our review so far has focused on four broad approaches to the discipline that have emerged since the mid-1980s. But much of the work within economic geography since that time cannot be fitted neatly within this grid, and is driven by subject matter as much as by methodological approach. That said, even here the theme of reconceiving the economic remains strong.

1. *Labor, Gender, and Ethnicity*. In many ways, geographical studies of the labor market were made for the new, catholic approach to economic geography. Labor markets are alive with a whole series of cross-cutting social influences including gender, ethnicity, forms of regulation, cultural norms and expectations, and issues of identity. Sayer and Walker (1992) suggested that the organization of work, and the divisions thereby created among social groups, sexes, and places, are a preeminent lens through which to comprehend the integration of economic and social processes. Economists have brought analytic models to bear on this complexity, such as Tesfatsion's (1998) computational experiments with individual job seekers operating within networks. But much of the recent work in economic geography deals with that complexity by using non-analytical methods. Three traditions are developing: an established segmentation approach that derives from political economy and emphasizes particularly the uneven operation of the labor market; an approach that makes use of institutional concepts presented above; and an embryonic approach that draws from post-structural thought and emphasizes issues of identity and power.

The segmentation approach stems from work carried out by US radical economists in the 1970s that found significant empirical differences in wages and conditions of work between two sets of equally trained workers: whites and African-Americans. The conclusion was that there was not a single labor market in operation, but two segmented ones that were labeled primary and secondary. Furthermore, principally determining who ended up in which one were the socially defined characteristics of the workers. To simplify: women and visible ethnic minorities tended to be found in the secondary labor market characterized by poor wages and working condition, while white men predominated in the primary labor market characterized by good wages and working conditions.

Labor markets appear to be segmented in many dimensions, and the spatial dimension itself interacts with the others (Hanson and Pratt 1992). Hanson and Pratt (1991, 1995) described how geographic and activity separation reinforce the process of occupational segmentation and "typing." Geographically circumscribed employment opportunities seem to be a result of the circumscribed employer searches for labor within certain occupations, as well as the circumscribed social and geographic networks of some potential workers. These findings result from the investigations of "spatial mismatch" of employment opportunities and residential locations of low- to moderately-skilled and educated workers. (A very readable updating of the original spatial mismatch focus on African-American men in manufacturing sectors appeared in *The Professional Geographer*, 48 (T. J. Cooke 1996; Hodge 1996; Holloway 1996; McLafferty and Preston 1996; Wyly 1996).)

Peck (1996) made the argument that labor market segmentation is always the consequence of *institutional forces* operating at the local scale which determine the demand, supply, and regulation of the labor force. Herod (1996, 1997, 1998) pointed to the utility of studying the active role played by labor and labor's institutions in shaping the geography of economic activity ("a labor geography"), rather than treating labor passively as the dupe of capital ("a geography of labor"). These institutions are not necessarily self-regulating: matching labor reproduction and labor demand do not necessarily develop in ways that allow continued regional economic growth and distribution (Peck 1994). The role of *institutionalist* analysis of local labor processes, then, is to identify the components and overall nature of these institutional ensembles, to understand their congruence or incongruence, to relate a localized ensemble to perceived problems in regional labor processes, and to identify potential institutional interventions (Harrington and Ferguson 2001).

The third, still emerging, approach is couched in terms of identity and household roles. Research by Hanson and Pratt (1995), Gibson-Graham (1996), and McDowell (1996) recognizes that jobs and the institutions associated with them are often race- and gender-specific (Matthaei 1995). Unless one fits those characteristics, or can adapt to fit them, then obtaining a job so labeled is very difficult. This operates at all levels of the job market from the merchant banking to sewing machine operators in clothing factories. In each case a myriad of powerful social forces construct, maintain, and reproduce different worker identities, funneling them into particular industry-occupation slots. Oberhauser (1993, 1995) studied the increased importance of home-based production by women in Appalachia, in response to the decline in formal, largely male employment. The form, sector, and gender-specificity of home production affects divisions of labor within the household, in the formal labor force, and provides new possibilities for economic development.

2. *Economic Geography of the Environment.* The "environment" is at once a basis for, result of, and object of production and consumption. However, environmental analysis in the context of economic geography remains a relatively undeveloped subfield (Hanink 1995). To some degree, the relative neglect of the environment by economic geographers is due to the legacy of environmental determinism, an approach that is still contentious within the discipline (Bassin 1992; Peet 1993). There is no doubt that economic change and environmental conditions are related (Straussfogel 1997; Wallner *et al.* 1996; Wilbanks 1994), and some analysis of the environment's role as an agent of economic development is being conducted by economists (Gallup *et al.* 1999). Economic geographers, however, have been focusing on the reverse relationship: the problems of environmental degradation induced by economic change (see the two special issues of *Economic Geography* edited and introduced by Peet and Watts (1993)). Another important environment–economy link investigated by geographers is defined in the issue of environmental justice (Bowen *et al.* 1995; Cutter 1995; Jerret *et al.* 1997).

Additional work that can be categorized as economic-environmental geography includes Benton's (1996) examination of the impact of environmental policy on trade, and Robinson's (1995) review of the impact of air pollution controls on industrial location in the US. Becker and Henderson (1999) examined the "greening of industry," focusing on the impacts of environmental regulations on the probability of plant exit. Perhaps the most promising research agenda for economic geographers in this area consists of analyses of the relationship between a region's economy and its environment treated as an asset (Hanink 1997). Reed (1995), for example, provided a case study of the local tensions that arise when a natural resource-based economy declines and new options for economic use of the environment arise. The integration of the theory of uneven development into resource–environment analysis by Roberts and Emel (1992) provided a strong conceptual basis for further related economic-environment analyses at the regional scale. This work also hinted at the possibility of the kind of blurring of economics that has been the theme of this section. Their argument was that inclusion of the environment necessarily stretches the definition of the economy: the economy cannot be a closed, self-contained entity because nature continually seeps in and disrupts. The environment has an agency, or causative power, that puts into question conventional economic theorizing.

3. *Consumption and the New Retailing Geography.* Retail geography was one of the principal vehicles during the 1960s for prosecuting a spatial science approach. By the early 1970s it was in decline, and by the 1980s it was effectively in abeyance. In recent years, however, studies of retailing have gained new vigor, with work conducted within each of the major approaches of economic geography.

Within the spatial analysis paradigm, O'Kelly and Miller (1989), Brown (1992), and Parr (1997) have generalized the widely used Reilly's law of retail gravitation, tying it more closely to spatial interaction models in general and to the economic delineation of market areas specifically. Thill (1992) analytically compared the spatial results of competition among individual establishments and among chains of establishments, finding similarity between the two systems under most configurations. Mulligan and Fik (1989) modeled and tested the joint influence of competitive structures and geographic configuration on retail pricing. Plummer *et al.* (1998) performed similar analyses, from the theoretical basis of political economy.

Some geographers currently working on retailing shun statistical methods, and are much more likely to draw upon social theory and cultural studies than economics (Wrigley and Lowe 1996; Bell and Valentine 1997; Crewe and Gregson 1998; Miller *et al.* 1998). The argument, in line with the general thesis proposed in this section, is that the economic act of shopping is also simultaneously cultural and social. We are what we shop; that is, our consumer choices are irrevocably bound up with our identities, which are themselves both fluid and a consequence of forces that lie outside ourselves. For this group of researchers, explanations of shopping behavior based upon rational-choice models are inadequate. We don't make decisions that way, and moreover, the work that presumes that we do erases place-specific factors that are important to understanding how we do make choices. A number of geographical case studies exist now at very different scales in showing how place matters in determining retail habits: at one end of the scale is Crewe and Gregson's work on car boot sales in Britain, and at the other, Canadian writing on the opening of the West Edmonton Mall, then the largest suburban shopping mall in North America (E. L. Jackson and Johnson 1991; Jones 1991; Simmons 1991).

Another subject of the new retail geography is the structural, technological, and regulatory change that has swept retailing in most parts of the world. Retail outlets have become larger, more consolidated into chains, and less tightly regulated by governments. In particular, rapid corporate consolidation and outlet reconfiguration in UK food retailing led to empirical and then increasingly theoretical research into food distribution, corporate strategy, and regulation within British geography (Wrigley 1994; Hallsworth and Taylor 1996; Clark and Wrigley 1997; Hallsworth 1997; Marsden *et al.* 1998). Pollard (1996) studied competitive and regulatory change in the US banking industry from an analogous, retail-service, intra-urban viewpoint. Even more extreme regulatory changes have been occasioned by the removal of trade restrictions between parts of an integrating border region (Slowe 1991), the entry of market competition into eastern German retailing (Coles 1997), and the recognition and financing of micro-enterprises as a bridge between the formal and informal economies (Simon 1998). Finally, the phenomenon of electronic retailing, in which the shopping function and (for software, information, and entertainment) the delivery function are distributed at no marginal transport cost, has not yet appeared in the economic geographic literature (see Wyckoff 1997).

A final body of work on retailing and consumption has its feet in both the old and the new retail geography. Called geodemographics, it examines the use of economic and demographic data about households, aggregated to fine geographic scales (the neighborhood, block group, or postal code), by both advertisers and retail location analysts. Developed in the early 1970s, the approach combines the concepts of urban ecology (competition for urban residential space by different social groups), the techniques of factorial ecology (distinguishing the central tendencies of specified groups), Census data on household incomes, ethnicity, sex, and age mixes, and the geography of US postal codes. However, when these data are manipulated in a robust geographic information system (GIS), users can relate the characteristics of different geographies (postal codes, Census designation, municipal boundaries), can create their own, proprietary geographies, can engage in (simple) spatial analysis, and can devise network or routing configurations. This combination of geodemographic data and GIS has become the fastest-growing use of economic-geographic principles outside the academy, and the fastest-growing employment opportunity for graduates trained in GIS and rudimentary economic geography concepts. English-language research and scholarly writing on these subjects are largely confined to British geographers (Longley and Clarke 1995; Birkin *et al.* 1996), and a broader critique of the consequent, explicit commodification of people and consumption (Goss 1995*a*; Leslie 1995). American geographers have been more active in the applied literature, helping to disseminate the use of these tools (Warden 1993; Klosterman and Xie 1997; Thrall 1999).

Computer-based technologies—GIS, spatial analysis, visualization, and tools for collaboration (including across places and at different times)—are being developed into decision-support systems useful for retailing, public-service provision, infrastructure planning, and industrial location (Mennecke 1997). Economic geography's concern for the social and institutional bases of economic activity is sorely needed in the analysis of these tools and their use (Crewe and Lowe 1995; Goss 1995*b*).

Reconceiving Regions: Defining "Local" in a Global Economy

The advent of electronic communications and computing, and the transport and logistics advances they engendered, have rendered key factors of regional development—capital, resources, and technology—increasingly mobile. In a world where financial capital is highly mobile, resources are easily transported, and much technical capability is readily diffused, what remains local? For one thing, the market for much production remains local or national, especially the provision and delivery of services. More importantly, even "global" corporations and intercorporate networks seek out, not just *any* places, but the best places in the world for particular functions, based on some combination of transitory localized characteristics and long-term, hard-to-replicate localized characteristics.

The transitory characteristics are the simple cost factors of low wages, low taxes, and limited government regulation. These hardly *anchor* productive activities, and unless they are matched with developmental policies of infrastructure and educational investment, they represent a dead end of competition for the lowest of the low, a competition that provinces, regions, or localities in wealthy countries are not likely to win. The longer-term, hard-to-replicate characteristics are the supply factors of communications and transport infrastructures, educated and innovative workforces, and environmental amenities, and the demand factors of growing household, commercial, and government markets for high-quality goods and services. These characteristics are not only hard to replicate, and thus earn high economic rents, but are hard to sustain, and require large economic investments (Storper 1995, 1997). Labor processes and labor-related institutions are important components of these truly localized characteristics.

Storper (1995: 209) suggested that production systems gain locational specificity "in three ways—the labor market, the input–output systems, and the knowledge system." Labor reproduction, allocation, training, and mobility form important routes of information transfer in localized production systems and are sources of much of the localized economic and social impact of economic activity. Economic geography has long emphasized the localization of input–output linkages, including recent work that decomposes inter-industry structures into regional subsystems (Sonis and Hewings 1998) and that allows use of national data to drive regional commodity-industry models (R. W. Jackson 1998). Warf (1998) presented the ways in which geographic scale (i.e. the definition of "region") affects the use and interpretation of input–output analysis. Studies of regional information flows underline the complex ways in which localization does and does not come about through the "knowledge system."

Defining Regions by Defining Processes: Agglomeration and Technology-Based Industrial Districts

From the mid-1980s onwards Scott (1988*a*, *b*) developed what he called a neo-Weberian theory of industrial location that drew heavily upon Williamson's (1975) firm-level transactional approach. Making a distinction between vertically integrated and disintegrated production, Scott later effectively equated the former with Fordism and the latter with flexible specialization. Furthermore, there were strong geographical effects. The process of vertical disintegration or externalization, in particular, produced geographically compact and distinct industrial districts (first recognized by Alfred Marshall) or territorial complexes (Scott and Storper 1992). Such districts or complexes were recursively reinforced by various kinds of agglomeration and external economies as well as specific institutions and local patterns of activity. That is, to use the term already employed, vertically disintegrated production was locally embedded. But how local? Winder (1999) used historical records to argue that the entire manufacturing belt of the late nineteenth-century US operated as a fairly integrated "industrial district," with inter-firm and inter-industry linkages carrying critical technological change throughout the region.

By the end of the 1980s it was clear that a simple transactions cost approach was insufficient to account for the varied relationships that bind individual firms and workers to one another and to particular industrial districts. Thus, relations among firms increasingly were seen as governed by various forms of what Storper (1995) called untraded interdependencies (see also Amin and Thrift 1994*a*; Camagni 1991; Camerer and Vepsalainen 1988; Grabher 1993). In large part these interdependencies were understood, after Granovetter (1985), as broader sets of social relations that over time coalesce to form regional "cultures," or tacitly understood conventions/institutions that encourage trust, reduce uncertainty, and guide behavior. The individual firm became less significant as the critical locus of competitive advantage. Case studies revealed the varied institutional foundations of industrial and regional performance (Saxenian 1994; Storper 1993; Todtling 1992; Ettlinger 1994). Storper's work complemented Scott's by its emphasis on the importance of technological innovation and change, thereby adding dynamism to Scott's model. Storper's important geographical point, and made with Richard Walker (Storper and Walker 1989), was that technologically innovative, propulsive industries are initially footloose, but once they locate, agglomerate, and become embedded within a territorial complex, the potential for further locational change is restricted.

In an attempt to understand the shared "technological capital" of industrial districts as the motor of agglomeration, researchers have conceived a regional variant of the national system of innovation (DeBresson and Amesse 1991; Lundvall 1992; Nelson 1993; Freeman 1991, 1995; Braczyk *et al.* 1998). The advantages of agglomeration are seen to emerge from a shared knowledge base, from enhanced local information exchange and learning (Lundvall and Johnson 1994; Malmberg and Maskell 1997; Scott 1995), especially that of a tacit variety, from multiple sources of innovation (von Hippel 1988), from the collective sharing of knowledge spillovers (Anselin *et al.* 1997; Jaffe *et al.* 1993), and from varied social, cultural, and institutional factors that support regional milieux (Maillat 1995).

These theoretical claims suggest marked variation in the innovative capacity of different regions. Regional differences in research and development expenditures, considered as an input to innovation, were confirmed by Malecki (1991) and by Feldman and Florida (1994). Regional differences in patents, one measure of innovative output, are reported by Malecki (1991) and by Feldman (1994). While von Hippel (1988) outlined the various sources of innovation, Webber *et al.* (1992) emphasized that innovation is only one of a number of processes that influence the evolution of technology in space. More recent empirical work has shown that regional differences in technology are significant and that they tend to persist over time (Rigby and Essletzbichler 1997; Essletzbichler *et al.* 1998). Gertler (1993, 1995) accounted for some of these differences, noting the institutional barriers to technology transfer.

Defining Regions by Defining Processes: The Example of Labor

Labor figures prominently in all explanations of and prescriptions for regional growth and development, traditionally as a key "factor" of production and industrial location. The heterogeneity of labor has been recognized, and increasing the "quality" of labor through education and training has become an important tool of national and regional development policy. However, most regional-development writing lacks explicit attention to the reproduction of labor qualities, the structuring of labor markets, the process of employment search, labor control, and the design of work—referred to collectively as "labor processes." In a world whose local and national economies are becoming globally integrated, these labor processes define and distinguish places. Labor processes are central to the relationship between global economic change and local development. The development of dominant institutions for labor regulation, based on the size and power of particular sectors in the "high-technology" economy (including high-value-added services) has been a recurrent theme in recent studies of regional economies (P. Cooke and Morgan 1994; Saxenian 1994; Massey 1995; DiGiovanna 1996). In addition, explicit conceptualization of labor should allow better-informed interregional comparisons, leading eventually to structural assessment of economic development policy with respect to region-specific labor processes.

The concept of the local labor market (or "commuting shed") is central to economic geography. Unfortunately, it is a very difficult concept. The definition of local labor markets is subject to the difficulty of all delineations based on potential interaction, though practicalities of data availability generally drive the implementation of the concept (Schubert *et al.* 1987). The geographic scale of the job search and/or personal network of the searcher, and the scale of the recruitment mechanisms (formal, work-network, or personal network) of the employer varies tremendously by gender, race, occupation, and residential location. Recognizing the interdependencies that develop among institutional arrangements in a local labor market, Jonas (1996: 328) introduced the concept of a "local labor-control regime . . . the gamut of practices, norms, behaviors, cultures and institutions within a locality through which labor is integrated into production." His use of the term "regime" reflects an attempt to link local labor practices to national and international regulation of production. Referring to the institutional approach presented earlier in the chapter, labor markets are localized by the ways in which institutional development and interaction are unique to individual regions, even though many of the social institutions and (corporate and labor) organizations exist at a spatial scale larger than the region (Jonas 1992; Saxenian 1994; Herod 1995; Peck 1996: ch. 4; Rutherford 1998).

Producer Services

Since the mid-1980s, research in economic geography has included an active focus on intermediate, or producer services. This body of research presents a set of scale questions: what is the scale of producer-service locational needs and market areas? Is there a parallel hierarchy of economic and geographic scale for service provision, marketing, and economic impact?

These services include the most rapidly growing sectors of services, as well as some of the most highly paid and prevalent in the central districts of the largest metropolitan areas. Observers have noted the high growth rates of producer-service employment, establishments, and self-employment in the suburbs of major metropolitan areas. This trend has been ascribed to a need for proximity to increasingly suburbanized clients in all sectors (Harrington and Campbell 1997), a search for suburban, female workers whose commuting ranges are constrained by household responsibilities (Nelson 1986), and the availability of large-scale, speculative, suburban commercial real-estate development (Daniels 1991). However, producer-service activity suburbanizes selectively, with some sectors concentrating in certain suburbs. Market access has been shown to be key to producer-service location and suburbanization, though its influence varies by sector and establishment size (Coffey *et al.* 1996). The optimum location is that which minimizes the total distance to clients, up to the number of clients where the cost of servicing the next-most-distant client equals the revenue that would be gained from that client (Lentnek *et al.* 1992, 1995).

At the regional scale, the growth of producer-service employment has served to diversify local industry structures. While only a few producer-service sectors (such as software development and capital-asset management) are the source of major inventions, their roles in local economies include formalizing and circulating best practice among their clients. In this way, producer services can be seen as agents who endogenize (local and non-local) external economies of information flow. Economies of scale in the generation and dissemination of useful technical and market information are an important explanation of the trend toward increased use of specialized producer services. Thus, small-scale productive activities within market areas well served by such services may benefit and perform accordingly. This logic sees producer services as key to regional economic development (MacPherson 1997; Mackun and MacPherson 1997).

The ways in which producer services firms and internalized producer services activities are organized geographically and how they affect local markets for labor and information, are worthy topics for investigation. Esparza and Krmenec (1994, 1996) used survey data from producer-service establishments in the midwestern US to uncover a significant and striking break between the largely national and international markets of Chicago establishments and the regionalized market areas of establishments in second-, third-, and fourth-tier cities. Eberts and Randall (1998) found the distribution of gender and of part-time status among the producer-service employment in Saskatchewan to be more related to location within the regional hierarchy of cities than to the specific producer-service sector. Beyers and Lindahl (1997, 1999) used organizational-growth-strategy typologies to relate organizational forms of producer services providers to specific sectors, labor practice, and age of the enterprises. Wood (1991), Perry (1992), and Coffey and Bailly (1993) have postulated how producer services support localized flexibilities of scale and scope.

International Economy and Globalization

Geographical analysis of the international economy and of globalization as an economic process was considerably broadened in scope during the 1990s. The negotiation of Canada–US and US–Mexico–Canada trade agreements motivated many studies of the subnational, regional impact of trade, of changes in trade regimes, and of foreign direct investment (Erickson and Hayward 1991; Conroy and Glasmeier 1992; Ó hUallacháin 1993; Hanink 1994; Warf and Randall 1994; Hayward and Erickson 1995). The trade studies generally employed commodity-flow and input–output analyses, applied creatively in light of the dearth of subnational data on international trade. Grant (1994) called for more politically informed analysis of trade regimes and trade flows. Political factors, as well as economic ones, were incorporated into trade analyses by Glasmeier *et al.* (1993), O'Loughlin and Anselin (1996), and Poon (1997). While the early emphasis was on trade in goods, more recent work has focused on the tertiary and quaternary sectors in the international economy (Bagchi-Sen and Sen 1997; Mitchelson and Wheeler 1994), with a particular focus on international finance (Daly and Stimson 1994; Leyshon 1995; Ó hUallacháin 1994; Roberts 1995). The global context of labor-market organization also has received particular attention (Ball 1997; Herod 1995; Zabin 1997).

The regional impact of FDI, especially investment in developing countries, has been a recurrent research theme (Eng and Lin 1996; Leung 1996; South 1990). A significant portion of FDI originating in richer countries can be traced to cost-cutting, but a number of investigations have considered FDI in the context of more complex business strategies motivating (the much greater) capital flows between rich countries (Angel and Savage 1996; Florida and Kenney 1994; O'Farrell *et al.* 1995; Schoenberger 1990). In a related approach, Ó hUallacháin and Matthews (1996) have described the strategic implications for a domestic industry of corporate restructuring by foreign competitors.

An especially interesting line of research in economic geography on globalization has followed a hierarchical model that considers the interaction between a larger entity and a smaller one. Dicken (1994), for example, examined the tensions that arise when states attempt to exercise control over transnational flows, while Florida (1996) examined regional response to the stress of economic globalization. In contrast to the typical hierarchical approach, Storper (1997) and Scott (1998) suggested the primacy of the regional economy over the global one. In effect, they developed a model of dynamic comparative advantage—encompassing economic/technological rejuvenation—in which certain regions are able to counteract the tendency toward spatial economic leveling which globalization would accomplish under neoclassical assumptions.

Economic Geography within Public Policy

The popular press abounds with discussions of the globalization of the economy, international competitiveness, and the presumed cultural bases for economic differentials (though which cultural characteristics are deemed to be virtuous depends on which national or regional economies are ascendant at the moment). These popular concerns are indeed the concerns of economic geography. Together, they provide an opportunity and a challenge to those who study it. Can the academic excitement over embedded economies, analytically tractable treatments of uneven development, formulation of local processes within global movements, and the formation of localized advantage be translated into usable analyses for governments, NGOs, and citizens?

In terms of volume of published work that reflects potentially useful analyses, the answer is "yes." Twenty years ago it was uncommon to find policy-relevant articles in economic geography-related journals. Over the last two decades this absence has lessened rather dramatically. Today, significant shares of the submissions to major journals in this area are articles about problems directly relevant to policy-makers (Peck 1997). These articles span a wide range of topics. Economic geographers have written about such current issues as the spatial mismatch between poor inner-city residents and suburban job markets (see above); the effectiveness of local training programs (Peck and Jones 1995); the effect of development processes on land-use decisions (Hall 1997); the formulation of the nation's urban policy (Gaffikin and Warf 1993; Berry 1994; Glasmeier and Harrison 1997); the health and functionality of American cities (Clarke and Gaile 1998; Glickman *et al.* 1998); the geography of metropolitan structural change (Hewings *et al.* 1998); the regional economic impact of industry relocation (Markusen 1988; Knudsen 1992; Erickson 1994; Markusen and Oden 1996); the relationships among corporate behavior and technology adoption, and the technological innovativeness of local regions (Malecki and Tootle 1996; 1997); the formation of new industrial complexes (Scott 1998; Storper 1995; Saxenian 1994; Glasmeier 1999); the causes and consequences of US–Mexico border development (Garcia de Fuentes and Medina 1996); economic policy and regional development in the PRC (Xie and Dutt 1990; Wei 1996, 1999; Ying 1999); the effectiveness of regional development policies (Isserman 1996*a*, *b*; Glasmeier *et al.* 1996; Glasmeier and Leichenko 1996; Glasmeier

1998); the roles of producer services in regional development (Harrington *et al.* 1992; Beyers and Lindahl 1996*a*, *b*); trade-related economic development (Hayward and Erickson 1995; Leichenko and Erickson 1997); and national economic development policies (Harrison *et al.* 1995). These inquiries range in style from the highly empirical to the highly theoretical.

Do those with a need for analysis and insight connect with these inquiries? According to Peck (1997) and others, economic geographic inquiry abounds with policy-relevant questions, but those who conduct it are far less linked to the policy process than are neoclassical economists and neoclassically trained public policy analysts. This finding was noted ten years ago when the first edition of this volume was produced and appears to hold, perhaps to a lesser degree, today. The finding is disappointing in light of the rich and growing tradition of economic geographic inquiry and its focus on policy-relevant problems.

One explanation for the limited impact of economic geography on policy-making is the need for policy-makers to see the relationship between the causes and consequences of policies. Economic geographers (and social scientists generally) are often unsure of the causal chain linking problems and processes found in the real world. Moreover, many are unwilling to ascribe some phenomenon to a single cause. A second difficulty is establishing links with a policy audience, though economic geographers have done this through appointments and consultation with government agencies, NGOs, private-sector economic development organizations, and private foundations. Third, the limited value that the academic hierarchy places on policy analysis and advising reduces incentives for academic geographers to establish such links. Junior scholars may find it hard to justify putting time and effort into activities that inform teaching and other research, but do not count toward promotion. A fourth difficulty is the need to translate academic frameworks and research findings into accessible policy prose. Academics often ask questions in ways that are initially difficult to decipher and to link to policy concerns.

Even with these limitations, however, it is clear that economic geographers are both conducting and disseminating relevant research to broad policy audiences, and with higher frequency. One example of the role of geographers in policy analysis can be seen at the US Federal level in the published compilation of research titles and authors funded by the Economic Development Administration (EDA)'s Research and National Technical Assistance office (US Department of Commerce 1994, 1996, 1997). More than thirty authors who have written on a wide variety of subjects received funding from the EDA during the last decade. We also find economic geographers prominently situated on the boards and editorial panels of federal agencies such as the Housing and Urban Development Administration, the Appalachian Regional Commission, and the United States Department of Agriculture's Economic Research Service.

Despite the need to make better links with policy-makers, there is growing interest in and credibility given to economic geographic inquiry in the policy process. Economic geographers have many opportunities to contribute to policy debates, not only to help fathom the nature of policy questions, but also to answer questions about *where* change is likely to occur. The ability to contribute to both very concrete and very abstract discussions positions members of the subdiscipline to engage the policy process at its most academic and activist levels.

Conclusion

Reviewing this review, and comparing its contents with the nature of the American economy in the year 2000, one is struck by the relative absence of consumption and consumer-oriented sectors (retailing, health, and entertainment) in this review, compared to the proportion of US economy involved in household consumption and in the provision of consumer services. This reflects, in part, the limited space for this chapter, and the authors' attempt to emphasize two particular themes in the subdiscipline. However, it also reflects omissions in the attention of economic geographers, which need to be addressed in the next ten years. The "economic" surely includes these processes.

Comparing this review to the concerns of popular and business writing suggests another set of omissions: the ways in which information and communications technologies are changing distribution systems. Economic geographers recognize the development and operation of production systems that span sectors, firms (as well as home workers and public establishments), and locations. However, communication, distribution, and logistics within production systems have taken on central importance in the satisfaction (or creation) of demand and the extraction of profit. One particularly visible aspect of this revolution, internet retailing, is only a very small part of the changes in the means and the geography of distribution. These changes affect the location and development of employment at intra-urban, regional, and international scales and, thus, affect the organization of "the region."

One final omission, at least in this review, is work by American economic geographers on the negative impacts of circulation and accumulation. The economic and social processes of interest are inherently uneven across and within places. While the seeming success stories of growing and restructuring industrial districts have been told, less attention has been paid to the obvious and not-so-obvious failures (Walker (1997) provides an exception). Empirical analysis and theoretical developments are needed to comprehend the relationships among improvement and decline at all scales.

However, this is not a call for attention to be diverted from issues of creation of regional well-being amidst internationalization, operation of labor markets within and across regions, or the environmental bases and impacts of localized economic activity. Rather, the expansion of the scope and substance of economic geography requires expansion of the number of researchers and users of economic geography. Such expansion may come from heightened recruitment of and support for graduate students, invigorated research by those currently in the field, and even greater cooperation with those working on these issues in related subdisciplines and other disciplines.

Note

The authors appreciate useful comments from many colleagues, though the errors of commission and especially of omission are their own.

References

Aglietta, M. (1979). *A Theory of Capitalist Regulation*. London: New Left Books.

Amin, A., and Robbins, K. (1990). "The Re-Emergence of Regional Economies? The Mythical Geography of Flexible Accumulation." *Environment and Planning D: Society and Space*, 8: 7–34.

— (1991). "These Are Not Marshallian Times," in Camagni (1991: 105–18).

Amin, A., and Thrift, N. (eds.) (1994*a*). *Globalization, Institutions, and Regional Development in Europe*. Oxford: Oxford University Press.

— (1994*b*). "Institutional Issues for the European Regions: From Markets and Plans to Socioeconomics and Powers of Association". *Economy and Society*, 24, 41–66.

— (1997). "Globalization, Socioeconomics, Territoriality," in Lee and Wills (1997: 133–46).

Angel, D., and Savage, L. (1996). "Global Localization? Japanese Research and Development Laboratories in the USA." *Environment and Planning A*, 28: 819–33.

Anselin, L., Varga, A., and Acs, Z. (1997). "Local Geographic Spillovers between University Research and High Technology Innovations". *Journal of Urban Economics*, 42: 422–48.

Arthur, W. B. (1989). "Competing Technologies, Increasing Returns, and Lock-in by Historical Events." *Economic Journal*, 99: 116–31.

Bagchi-Sen, S., And Sen, J. (1997). "The Current State of Knowledge in International Business And Producer Services." *Environment and Planning A*, 29: 1153–74.

Ball, R. (1997). "The Role of the State in the Globalisation of Labour Markets: The Case of the Philippines. *Environment and Planning A*, 29: 1603–28.

Barnes, T. (1984). "Theories of Agricultural Rent Within the Surplus Approach". *International Review of Regional Science*, 9: 125–40.

— (1988). "Scarcity and Agricultural Land Rent in Light of the Capital Controversies: Three Views." *Antipode*, 2: 207–38.

— (1996*a*). *Logics of Dislocation: Models, Metaphors and Meanings of Economic Space*. New York: Guilford.

— (1996*b*). "Political Economy II: Compliments of the Year." *Progress in Human Geography*, 20/4: 521–8.

Barnes, T., and Sheppard, E. (1984). "Technical Choice and Reswitching in Space Economies." *Regional Science and Urban Economics*, 14: 345–62.

Bassin, M. (1992). "Geographical Determinism in *Fin-De-Siècle* Marxism: Georgii Plekhanov and the Environmental Basis of Russian History." *Annals of the Association of American Geographers*, 82: 3–22.

Becker, R. A., and Henderson, J. V. (1999). "Effects of Air Quality Regulation on Polluting Industries." Paper Presented At the 95th Meeting of the Association of American Geographers, Honolulu, Hawaii, March 23–7.

Bell, D., and Valentine, G. (1997). *Consuming Geographies*. London: Routledge.

Benton, L. (1996). "The Greening of Free Trade? The Debate About the North American Free Trade Agreement (NAFTA) and the Environment." *Environment and Planning A*, 28: 2155–77.

Berry, B. J. L. (1994). "Editorial: Let's Have More Policy Analysis". *Urban Geography*, 15: 315–17.

Beyers, W. B., and Lindahl, D. P. (1996*a*). "Explaining the Demand For Producer Services: Is Cost-Driven Externalization the Major Factor?" *Papers in Regional Science*, 75/3: 351–74.

— (1996*b*). "Lone Eagles and High Fliers in Rural Producer Services." *Rural Development Perspectives*, 11/3: 2–10.

— (1999). "Workplace Flexibilities in the Producer Services." *Service Industries Journal*, 19/1: 35–61.

Birkin, M., Clarke, G., Clarke, M., and Wilson, A. (1996). *Intelligent GIS: Location Decisions and Strategic Planning*. Cambridge: Geoinformation International.

Bowen, W., Salling, M., Haynes, K., and Cyran, E. (1995). "Toward Environmental Justice: Spatial Equity in Ohio and Cleveland." *Annals of the Association of American Geographers*, 82: 3–22.

Bowles, S., Gordon, D., and Weisskopf, T. (1986). "Power and Profits: the Social Structure of Accumulation and the Profitability of the Postwar US Economy." *Review of Radical Political Economics*, 18: 132–67.

Boyer, R. (1992). *La Théorie De La Régulation*. Paris: Economica.

Braczyk, H.-J., Cooke, P., and Heidenreich, M. (eds.) (1998). *Regional Innovation Systems*. London: UCL Press.

Brown, S. (1992). "The Wheel of Retail Gravitation." *Environment and Planning A*, 24/1: 1409–29.

Camagni, R. (ed.) (1991). *Innovation Networks: Spatial Perspectives*. New York: Belhaven Press.

Camerer, C., and Vepsalainen, A. (1988). "The Economic Efficiency of Corporate Culture." *Strategic Management Journal*, 9: 115–26.

Clark, G. L. (1998). "Stylized Facts and Close Dialogue: Methodology in Economic Geography." *Annals of the Association of American Geographers*, 88: 73–87.

Clark, G. L. and Wrigley, N. (1997). "The Spatial Configuration of the Firm and the Management of Sunk Costs." *Economic Geography*, 73/3: 285–304.

Clarke, S. E., and Gaile, G. L. (1998). *The Work of Cities*. Minneapolis: University of Minnesota Press.

Coffey, W. J. and Bailly, A. S. (1993). "Producer Services and Systems of Flexible Production." *Urban Studies*, 29: 857–68.

Coffey, W. J., Drolet, R., and Polese, M. (1996). "The Intrametropolitan Location of High Order Services: Patterns, Factors and Mobility in Montreal." *Papers in Regional Science*, 75/3: 293–323.

Conroy, M. E., and Glasmeier, A. K. (1992). "Winners and Losers on the NAFTA 'Fast Track.'" *The Aspen Institute Quarterly* (Autumn): 1–20.

Coles, T. (1997). "Trading Places: The Evolution of the Retail Sector in the New German Lander Since Unification." *Applied Geography*, 17/4: 315–33.

Cooke, P. (ed.) (1989). *Localities: the Changing Face of Urban Britain*. London: Unwin Hyman.

—— (1990). *Back To the Future: Modernity, Postmodernity and the Locality*. London: Unwin Hyman.

Cooke, P., and Morgan, K. "The Network Paradigm: New Departures in Corporate and Regional Development." *Environment and Planning D*, 11/5: 543–64.

—— (1993). (1994). "Growth Regions under Duress: Renewal Strategies in Baden Württemberg and Emilia-Romagna," in Amin and Thrift. (1994: 91–117).

—— (1998). *The Associational Economy: Firms, Regions and Innovations*. Oxford: Oxford University Press.

Cooke, T. J. (1996). "City-Suburb Differences in African-American Male Labor Market Achievement." *The Professional Geographer*, 48/4: 458–67.

Cox, K. (ed.) (1997). *Spaces of Globalization: Reasserting the Power of the Local*. New York: Guilford.

Cox, K., and Mair, A. (1988). "Locality and Community in the Politics of Local Development." *Annals, Association of American Geographers*, 78: 307–25.

—— (1991). "From Localized Social Structures to Localities as Agents." *Environment and Planning A*, 23: 197–213.

Crang, P. (1997). "Cultural Turns and the (Re)Constitution of Economic Geography: Introduction to Section One," in Lee and Wills (1997: 3–15).

Crewe, L., and Gregson, N. (1998). "Tales of the Unexpected: Exploring Car Boot Sales as Marginal Spaces of Contemporary Consumption." *Transactions, Institute of British Geographers*, 23: 39–53.

Crewe, L., and Lowe, M. (1995). "Gap On the Map? Towards a Geography of Consumption and Identity." *Environment and Planning A*, 27/12: 1877–98.

Cutter, S. (1995). "Race, Class and Environmental Justice." *Progress in Human Geography*, 19: 111–22.

Daly, M., and Stimson, R. (1994). "Dependency in the Modern Global Economy: Australia and the Changing Face of Asian Finance." *Environment and Planning A*, 26: 415–34.

Daniels, P. W. (1991). "Service Sector Restructuring and Metropolitan Development: Processes and Prospects," in P. W. Daniels (ed.), *Services and Metropolitan Development: International Perspectives*. London: Routledge, ch. 1.

Debresson, C., and Amesse, F. (1991). "Networks of Innovators: A Review and Introduction to the Issue." *Research Policy*, 20: 363–79.

Dicken, P. (1994). "Global–Local Tensions: Firms and States in the Local Space-Economy." *Economic Geography*, 70: 101–28.

Digiovanna, S. (1996). "Industrial Districts and Regional Economic Development: A Regulation Approach." *Regional Studies*, 30/4: 373–86.

Donaghu, M. T., and Barff, R. (1990). "Nike Just Did It: International Subcontracting and Flexibility in Athletic Footwear Production". *Regional Studies*, 24/6: 537–52.

Eberts, D., and Randall, J. E. (1998). "Producer Services, Labor Market Segmentation and Peripheral Regions: The Case of Saskatchewan." *Growth and Change*, 29/4: 401–22.

Eng, I., and Lin, Y. (1996). "Seeking Competitive Advantage in an Emergent Open Economy: Foreign Direct Investment in Chinese Industry." *Environment and Planning A*, 28: 1113–38.

Erickson, R. A. (1989). "The Influence of Economics on Geographic Inquiry." *Progress in Human Geography*, 13: 223–50.

—— (1994). "Technology, Industrial Restructuring, and Regional Development." *Growth and Change*, 25/3: 353–79.

Erickson, R., and Hayward, D. (1991). "The International Flows of Industrial Exports From US Regions." *Annals of the Association of American Geographers*, 81: 371–90.

Esparza, A. X., and Krmenec, A. J. (1994). "Producer Services in the Space Economy: A Model of Spatial Interaction." *Papers in Regional Science*, 73: 55–72.

—— (1996). "The Spatial Markets of Cities Organized in A Hierarchical System." *Professional Geographer*, 48/4: 367–78.

Essletzbichler, J., Haydamack, B. W., and Rigby, D. L. (1998). "Regional Dynamics of Technical Change in the US Structural Fabricated Metals Industry." *Geoforum*, 29: 23–35.

Ettlinger, N. (1994). "The Localization of Development in Comparative Perspective." *Economic Geography*, 70: 144–66.

Ettlinger, N., and Patton, W. (1996). "Shared Performance: The Proactive Diffusion of Competitiveness and Industrial and Local Development." *Annals of the Association of American Geographers*, 86/2: 286–305.

Feldman, M. (1994). *The Geography of Innovation*. Dordrecht: Kluwer.
Feldman, M., and Florida, R. (1994). "The Geographic Sources of Innovation: Technological Infrastructure and Product Innovation in the United States." *Annals of the Association of American Geographers*, 84: 210–29.
Florida, R. (1996). "Regional Creative Destruction: Product Organization, Globalization, and the Economic Transformation of the Midwest." *Economic Geography*, 72: 314–34.
Florida, R., and Kenney, M. (1994). "The Globalization of Japanese R&D: The Economic Geography of Japanese R&D Investment in the United States." *Economic Geography*, 70: 344–69.
Foot, S., and Webber, M. (1983). "Unequal Exchange and Uneven Development." *Environment and Planning D: Society and Space*, 1: 281–304.
Freeman, C. (1991). "Networks of Innovators: A Synthesis of Research Issues." *Research Policy*, 20: 499–514.
—— (1995). "The 'National System of Innovation' in Historical Perspective." *Cambridge Journal of Economics*, 19: 5–24.
Gaffikin, F., and Warf, B. (1993). "Urban Policy and the Post-Keynesian State in the United Kingdom and the United States." *International Journal of Urban and Regional Research*, 17/1: 67–84.
Gallup, J. L., Sachs, J. D., and Mellinger, A. D. (1999). "Geography and Economic Development," in B. Pleskoric and J. E. Stiglitz (eds.), *World Bank Annual Conference on Development Economics 1998*. Washington DC: The World Bank.
Garcia de Fuentes, A., and Medina, S. P. (1996). "Factores de Localización de la Industria Maquiladora: El Caso de Yucatán, Mexico." *Mexico Yearbook*, Conference of Latin Americanist Geographers.
Gertler, M. S. (1988). "The Limits to Flexibility: Comments on the Post-Fordist Vision of Production and its Geography." *Transactions, Institute of British Geographers*, 13: 419–32.
—— (1992). "Flexibility Revisited: Districts, Nation-States, and the Forces of Production." *Transactions, Institute of British Geographers*, 17: 259–78.
—— (1993). "Implementing Advanced Manufacturing Technologies in Mature Industrial Regions: Towards a Social Model of Technology Production." *Regional Studies*, 27: 665–80.
—— (1995). "'Being There': Proximity, Organization and Culture in the Development and Adoption of Advanced Manufacturing Technologies." *Economic Geography*, 71: 1–26.
Gertler, M. S., and Barnes, T. (eds.) (1999). *Towards a New Industrial Geography: Regions, Regulation, and Institutions*. London: Routledge.
Gibson-Graham, J. K. (1996). *The End of Capitalism (As We Knew It)*. Oxford: Blackwell.
Glasmeier, A. (1994). "Flexible Districts, Flexible Regions? The Institutional and Cultural Limits to Districts in an Era of Globalization and Technological Paradigm Shifts," in Amin and Thrift (1994: 28–38).
—— (1998). "The Relevance of Firm Learning to the Design of Manufacturing Modernization Programs". *Economic Development Quarterly*, 12/2: 107–24.
—— (1999). "Territory-Based Regional Development Policy and Planning in a Learning Economy: The Case of 'Real Service Centers' in Industrial Districts." *European Planning Studies* (Spring).
Glasmeier, A., and Harrison, B. (1997). "Why Business Alone Won't Redevelop the Inner City: A Friendly Critique of Michael Porter's Approach to Urban Revitalization." *Economic Development Quarterly*, 11/1: 28–38.
Glasmeier, A. K., and Leichenko, R. M. (1996). "From Free Market Rhetoric to Free Market Reality: The Future of the US South in an Era of Globalization." *International Journal of Urban and Regional Research*, 20/4: 601–24.
Glasmeier, A., Feller, I., and Mark, M. M. (1996). "Evaluating Manufacturing Modernization Programs." *Research Policy*, 25: 309–19.
Glasmeier, A. K., Tompson, J. W., and Kays, A. J. (1993). "The Geography of Trade Policy: Trade Regimes and Location Decisions in the Textile and Apparel Complex." *Transactions of the Institute of British Geographers*, 18: 19–35.
Glickman, N., Lahr, M., and Wyly, E. (1998). "A Top-10 List of Things to Know about American Cities." *Cityscape*, 3/3: 7–32.
Goss, J. (1995*a*). "Marketing the New Marketing: The Strategic Discourse of Geodemographic Information Systems," in J. Pickles (ed.), *Ground Truth: The Social Implications of Geographic Information Systems*. New York: Guilford, 130–70.
—— (1995*b*). "'We Know Who You Are and We Know Where You Live': The Instrumental Rationality of Geodemographic Systems." *Economic Geography*, 71: 171–98.
Grabher, G. (1993). "The Weakness of Strong Ties: The Lock-in of Regional Development in the Ruhr Area," in G. Grabher (ed.), *The Embedded Firm: On the Socioeconomics of Industrial Networks*. New York: Routledge, 255–77.
Granovetter, M. (1985). "Economic Action and Social Structure: The Problem of Embeddedness." *American Journal of Sociology*, 91: 481–510.
Grant, R. (1994). "The Geography of International Trade." *Progress in Human Geography*, 18: 298–312.
Hall, P. (1997). "The Future of the Metropolis and Its Form." *Regional Studies*, 31/3: 211–20.
Hallsworth, A. G. (1997). "Rethinking Retail Theory: Circuits of Power as an Integrative Paradigm." *Geographical Analysis*, 29/4: 329–38.
Hallsworth, A. G., and Taylor, M. (1996). "'Buying Power': Interpreting Retail Change in a Circuits of Power Framework." *Environment and Planning A*, 28/12: 2125–37.
Hanink, D. (1994). *The International Economy: A Geographical Perspective*. New York: John Wiley & Sons.
—— (1995). "The Economic Geography in Environmental Issues: A Spatial-Analytic Approach." *Progress in Human Geography*, 19: 372–87.
—— (1997). "Economic Geography and the Environment." Paper presented at the Workshop on the Future of Economic Geography, September 26–8, Washington, DC (‹http://www.geog.uconn.edu/aag-econ/glaswksp.pdf›, last accessed 11 October 2001).
Hanson, S., and Pratt, G. (1991). "Job Search and the Occupational Segregation of Women." *Annals of the Association of American Geog*raphers, 81/2: 229–53.
—— (1992). "Dynamic Dependencies: A Geographic Investigation of Local Labor Markets." *Economic Geography*, 68/4: 373–405.
—— (1995). *Gender, Work, and Space*. London: Routledge.
Harrington, J. W., and Campbell, H. S. (1997). "The Suburbanization of Producer-Service Employment." *Growth and Change*, 28/4: 335–59.

Harrington, J. W., and Ferguson, D. 2001. "Social Processes and Regional Economic Development," in B. Johansson, C. Karlsson, and R. Stough (eds.), *Theories of Endogenous Regional Growth*. Berlin: Springer-Verlag, 49–76.

Harrington, J. W., MacPherson, A. D., and Lombard, J. R. (1992). "Interregional Trade in Producer Services: Review and Synthesis." *Growth and Change*, 24/4: 75–94.

Harrison, B. (1994). *Lean and Mean: The Changing Landscape of Corporate Power in the Age of Flexibility*. New York: Basic Books.

Harrison, B., Weiss, M., and Grant, J. (1995). *Building Bridges: Community Development Corporations and the World of Employment Training*. New York: Ford Foundation.

Harvey, D. (1975). "The Geography of Capitalist Accumulation: A Reconstruction of Marxian Theory." *Antipode*, 7: 9–21.

—— (1982). *The Limits To Capital*. Oxford: Blackwell.

—— (1989). "From Managerialism to Entrepreneurialism: The Transformation in Urban Governance in Late Capitalism". *Geographiska Annaler* B, 71: 3–18.

Hayward, D. J., and Erickson, R. A. (1995). "The North American Trade of US States: A Comparative Analysis of Industrial Shipments (1983–91)." *International Regional Science Review*, 18/1: 1–31.

Henderson, J. (1996). "Ways to Think About Urban Concentration: Neoclassical Urban Systems Versus the New Economic Geography." *International Regional Science Review* 19: 31–6.

Herod, A. (1995). "The Practice of International Labor Solidarity and the Geography of the Global Economy." *Economic Geography*, 71/4: 341–63.

—— (1996). "Labor's Spatial Praxis and the Geography of Contract Bargaining in the US East Coast Longshore Industry (1953–89)." *Political Geography*, 16: 145–70.

—— (1997). "From a Geography of Labor to a Labor Geography: Labor's Spatial Fix in the Geography of Capitalism". *Antipode*, 29: 1–31.

—— (ed.) (1998). *Organizing the Landscape: Geographical Perspectives on Labor Unionism*. Minneapolis: University of Minnesota Press.

Hewings, G. J. D., Sonis, M., Guo, J., Israilevich, P. R., and Schindler, G. R. (1998). "The Hollowing-Out Process in the Chicago Economy (1975–2011)." *Geographical Analysis*, 30/3: 217–33.

Hirst, P. Q., and Zeitlin, J. (1991). "Flexible Specialization Versus Post-Fordism: Theory, Evidence, and Policy Implications." *Economy and Society*, 20: 1–56.

Hodge, D. C. (1996). "And in Conclusion: 'It Depends.'" *The Professional Geographer*, 48/4: 417–19.

Holloway, S. R. (1996). "Job Accessibility and Male Teenage Employment (1980–1990.) *The Professional Geographer*, 48/4: 445–58.

Isserman, A. (1996*a*). *Socioeconomic Review of Appalachia: The Evolving Appalachian Regional Economy*. Paper Prepared For the ARC in Partial Fulfillment of ARC Contract No. 95–13.

—— (1996*b*). *Socioeconomic Review of Appalachia: Appalachia Then and Now. An Update of "The Realities of Deprivation" Reported to the President in 1964*. Paper Prepared For the Appalachian Regional Commission in Partial Fulfillment of ARC Contract No. 95–13.

Jackson, E. L., and Johnson, D. B. (1991). "Geographic Implications of Mega-Malls." *Canadian Geographer* 35/3: 226–32.

Jackson, R. W. (1998). "Regionalizing National Commodity-by-Industry Accounts". *Economic Systems Research*, 10/3: 223–38.

Jaffe, A., Trajtenberg, M., and Henderson, R. (1993). "Geographic Localization of Knowledge Spillovers as Evidenced by Patent Citations". *Quarterly Journal of Economics*, 108: 577–98.

Jerret, M., Eyles, J., Cole, D., and Reader, S. (1997). "Environmental Equity in Canada: An Empirical Investigation into the Income Distribution of Pollution in Ontario." *Environment and Planning A*, 29: 1777–800.

Jonas, A. E. G. (1992). "Corporate Takeover and the Politics of Community: The Case of Norton Company in Worcester." *Economic Geography*, 68/4: 348–72.

—— (1996). "Local Labor Control Regimes: Uneven Development and the Social Regulation of Production." *Regional Studies*, 30/4: 323–38.

Jones, K. (1991). " Mega-Chaining, Corporate Concentration, and the Mega-Malls." *Canadian Geographer*, 35/3: 241–9.

Kilkenny, M. (1998). "Transport Costs, the New Economic Geography, and Rural Development". *Growth and Change*, 29: 259–80.

Klosterman, R. E., and Xie, Y. (1997). "Retail Impact Analysis with Loosely Coupled GIS and a Spreadsheet." *International Planning Studies*, 2/2: 175–92.

Knudsen, D. C. (1992). "Manufacturing Employment Change in the American Midwest (1977–86)." *Environment & Planning A*, 24/9: 1303–16.

Krugman, P. (1991). *Geography and Trade*. Cambridge, Mass.: MIT Press.

—— (1995). *Development, Geography, and Economic Theory*. Cambridge, Mass.: MIT Press.

—— (1998). "Space: The Final Frontier." *Journal of Economic Perspectives*, 12: 161–74.

Lee, R., and Wills, J. (eds.) (1997). *Geographies of Economies*. London: Arnold.

Leichenko, R. M., and Erickson, R. A. (1997). "Foreign Direct Investment and State Export Performance." *Journal of Regional Science*, 37/2: 307–29.

Lentnek, B., Macpherson, A., and Phillips, D. (1992). "Optimum Producer Service Location." *Environment and Planning A*, 24: 467–79.

—— (1995). "A Market Coverage Model for Producer Services." *Papers in Regional Science*, 74/4: 389–99.

Leslie, D. A. (1995). "Global Scan: The Globalization of Advertising Agencies, Concepts, and Campaigns." *Economic Geography*, 71: 402–26.

Leung, C. (1996). "Foreign Manufacturing Investment and Regional Industrial Growth in Guangdong Province, China." *Environment and Planning A*, 28: 513–36.

Leyshon, A. (1995). "Geographies of Money and Finance I." *Progress in Human Geography*, 19: 531–43.

Lindahl, D. P., and Beyers, W. B. (1999). "The Creation of Competitive Advantage by Producer Service Firms." *Economic Geography*, 75/1: 1–20.

Liossatos, P. (1988). "Value and Competition in a Spatial Context: A Marxian Model." *Papers of the Regional Science Association*, 45: 87–103.

Lipietz, A. (1986). "New Tendencies in the International Division of Labor: Regimes of Accumulation and Modes of Regulation," in A. J. Scott and M. Storper (eds.), *Production Work, Territory: The Geographical Anatomy of Industrial Capitalism*. Winchester, Mass.: Allen & Unwin, 16–40.

Longley, P., and Clarke, G. (eds.) (1995). *GIS For Business and Service Planning*. Cambridge, UK: Geoinformation International.

Lovering, J. (1990). "Fordism's Unknown Sucessor: A Comment on Scott's Theory of Flexible Accumulation and the Re-Emergence of Regional Economies." *International Journal of Urban and Regional Research*, 14: 159–74.

Lovering, J. (1991). "Theorizing Postfordism: Why Contingency Matters (A Further Response to Scott)." *International Journal of Urban and Regional Research*, 15: 298–301.

Lucas, R. (1988). "On the Mechanics of Economic Development." *Journal of Monetary Economics*, 22: 3–42.

Lundvall, B. A. (ed.) (1992). *National Systems of Innovation: Towards a Theory of Innovation and Interactive Learning*. London: Pinter.

Lundvall, B. A., and Johnson, B. (1994). "The Learning Economy." *Journal of Industry Studies*, 1: 23–42.

McDowell, L. (1996). *Capital Cultures*. Oxford: Blackwell.

Mackun, P., and MacPherson, A. (1997). "Externally-Assisted Product Innovation in the Manufacturing Sector: The Role of Location, In-House R&D and Outside Technical Support." *Regional Studies*, 31/7: 659–68.

McLafferty, S., and Preston, V. (1996). "Spatial Mismatch and Employment in A Decade of Restructuring." *The Professional Geographer*, 48/4: 420–31.

MacPherson, A. (1997). "The Role of Producer Service Outsourcing in the Innovation Performance of New York State Manufacturing Firms." *Annals of the Association of American Geographers*, 87: 52–71.

Maillat, D. (1995). "Territorial Dynamics, Innovative Milieus and Regional Policy." *Entrepreneurship and Regional Development*, 7: 157–65.

Malecki, E. J. (1991). *Technology and Economic Development*. Harlow: Longman.

—— (1995). "Guest Editorial: Flexibility and Industrial Districts." *Environment and Planning A*, 27: 11–14.

Malecki, E. J., and Tootle, D. M. (1996). "The Role of Networks in Small Firm Competitiveness." *International Journal of Technology Management*, 11: 43–57.

—— (1997). "Networks of Small Manufacturers in the USA: Creating Embeddedness," In M. Taylor and S. Conti (eds.), *Interdependent and Uneven Development: Global-Local Perspectives* Aldershot: Ashgate, 195–221.

Malmberg, A. (1997). "Industrial Geography: Location and Learning." *Progress in Human Geography*, 21: 573–82.

Malmberg, A., and Maskell, P. (1997). "Toward an Explanation of Regional Specialization and Industry Agglomeration." *European Planning Studies*, 5: 25–41.

Markusen, A. (1988). "Planning For Industrial Decline: Lessons From Steel Communities." *Journal of Planning Education and Research*, 7: 173–84.

—— (1996). "Sticky Places in Slippery Space: A Typology of Industrial Districts." *Economic Geography*, 72: 293–313.

Markusen, A., and Oden, M. (1996). "A Defense Conversion: Investing in the Peace Dividend," in T. Schafer and J. Faux (eds.), *Reclaiming Prosperity: A Blueprint for Progressive Economic Reform*. Washington, DC: Economic Policy Institute.

Marsden, T., Harrison, M., and Flynn, A. (1998). "Creating Competitive Space: Exploring the Social and Political Maintenance of Retail Power." *Environment and Planning A*, 30/3: 481–98.

Martin, R. (1994). "Economic Theory and Human Geography," in D. Gregory, R. Martin, and G. Smith, (eds.), *Human Geography: Society, Space and Social Science*. Basingstoke: Macmillan, 21–53.

Martin, R. (1999). "The New 'Geographical Turn' in Economics: Some Critical Reflections." *Cambridge Journal of Economics*, 23: 65–91.

Martin, R., and Sunley, P. (1998). "Slow Convergence: The New Endogenous Growth Theory and Regional Development." *Economic Geography*, 74: 201–27.

—— (1996). "Paul Krugman's Geographical Economics and its Implications for Regional Development Theory: A Critical Assessment." *Economic Geography*, 72: 259–92.

Marx, K. (1970). *Capital* (3 vols.). London: International.

Massey, D. (1973). "Towards a Critique of Industrial Location Theory." *Antipode*, 5: 33–9.

—— (1984). *Spatial Divisions of Labor: Social Structures and the Geography of Production*. London: Macmillan.

—— (1995). "Masculinity, Dualisms and High Technology." *Transactions of the IBG*, 20: 487–99.

Matthaei, J. (1995). "The Sexual Division of Labor, Sexuality and Lesbian/Gay Liberation: Towards a Marxist-Feminist Analysis of Sexuality in US Capitalism." *Review of Radical Political Economy*, 27/2: 1–37.

Mennecke, B. E. (1997). "Understanding the Role of Geographic Information Technologies in Business." *Journal of Geographic Information and Decision Analysis*, 1/1: 44–68.

Miller, D., Jackson, P., Thrift, N., Holbrook, B., and Rowlands, M. (1998). *Shopping, Places and Identity*. London: Routledge.

Mitchelson, R., and Wheeler, J. (1994). "The Flow of Information in a Global Economy: The Role of the American Urban System in 1990." *Annals of the Association of American Geographers*, 84: 87–107.

Mulligan, G., and Fik, T. (1989). "Asymmetrical Price Conjectural Variation in Spatial Models." *Economic Geography*, 65: 19–32.

Nelson, K. (1986). "Labor Demand, Labor Supply, and the Suburbanization of Low-Wage Office Work," in A. J. Scott and M. Storper (eds.), *Production, Work, Territory*. Boston: Unwin Hyman, ch. 8.

Nelson, R. (ed.) (1993). *National Innovation Systems: A Comparative Analysis*. Oxford: Oxford University Press.

Nelson, R., and Winter, S. (1982). *An Evolutionary Theory of Economic Change*. Cambridge, Mass.: Harvard University Press.

Oberhauser, A. M. (1993). "Industrial Restructuring and Women's Homework in Appalachia: Lessons From West Virginia." *Southeastern Geographer*, 33/1: 23–43.

Oberhauser, A. M. (1995). "Gender and Household Economic Strategies in Rural Appalachia." *Gender, Place and Culture*, 2/1: 51–70.

O'Farrell, P., Moffat, L., and Wood, P. (1995). "Internationalisation by Business Services: A Methodological Critique of Foreign-Market Entry-Mode Choice," *Environment and Planning A*, 27: 683–97.

Ó hUallacháin, B. (1993). "Industrial Geography." *Progress in Human Geography*, 17: 548–55.
—— (1994). "Foreign Banking in the American Urban System of Financial Organization." *Economic Geography*, 70: 206–28.
Ó hUallacháin, B., and Matthews, R. A. (1996). "Restructuring of Primary Industries: Technology, Labor, and Corporate Strategy and Control in the Arizona Copper Industry." *Economic Geography*, 72/2: 196–215.
Ó hUallacháin, B., and Wasserman, D. (1999). "Vertical Integration in a Lean Supply Chain: Brazilian Automobile Component Parts." *Economic Geography*, 75/1: 21–42.
O'Kelly, M. E., and Miller, H. J. (1989). "A Synthesis of Some Market Area Delimitation Models." *Growth and Change*, 20/3: 14–33.
O'Loughlin, J., and Anselin, L. (1996). "Geo-Economic Competition and Trade Bloc Formation: United States, German, and Japanese Exports (1968–1992)." *Economic Geography*, 72: 131–60.
Page, B., and Walker, R. (1991). "From Settlement to Fordism: The Agro-Industrial Revolution in the American Midwest." *Economic Geography*, 67/4: 281–15.
Park, S., and Markusen, A. (1995). "Generalizing New Industrial Districts: A Theoretical Agenda and an Application from a Non-Western Economy." *Environment and Planning A*, 27: 81–104.
Parr, J. B. (1997). "The Law of Retail Gravitation: Insights from Another Law." *Environment and Planning A*, 29/8: 1477–92.
Peck, J. (1994). "Regulating Labour: The Social Regulation and Reproduction of Local Labour-Markets." A. Amin and N. Thrift (1994: 147–76).
—— (1996). *Work-Place: The Social Regulation of Labor Markets.* New York: Guilford Press.
—— (1997). "Economic Geography in Government Policy." Paper presented at the Workshop on the Future of Economic Geography, September 26–8, Washington, DC (‹http://www.geog.uconn.edu/aag-econ/glaswksp.pdf›, last accessed 11 October 2001).
Peck, J., and Jones, M. (1995). "Training and Enterprise Councils: Schumpeterian Workfare State, Or What?" *Environment and Planning A*, 27: 1361–96.
Peck, J., and Tickell, A. (1995). "The Social Regulation of Uneven Development: 'Regulatory Deficit,' England's South-East and the Collapse of Thatcherism." *Environment and Planning A*, 27: 15–40.
Peet, R. (1993). "Reinventing Marxist Geography: A Critique of Bassin". *Annals of the Association of American Geographers*, 83: 156–60.
Perry, M. (1992). "Flexible Production, Externalization and the Interpretation of Business Service Growth." *Service Industries Journal*, 12: 1–16.
Piore, M. J., and Sabel, C. F. (1984). *The Second Industrial Divide: Possibilities For Prosperity.* New York: Basic Books.
Plummer, P., Haining, R., and Sheppard, E. (1998). "Spatial Pricing in Interdependent Markets." *Environment and Planning A*, 30/1: 67–84.
Pollard, J. S. (1996). "Banking at the Margins: A Geography of Financial Exclusion in Los Angeles." *Environment and Planning A*, 28/7: 1209–32.
Polyani, K. (1944). *The Great Transformation.* New York: Rinehart.
Poon, J. (1997). "The Cosmopolitanization of Trade Regions: Global Trends and Implications (1965–1990)." *Economic Geography*, 73: 390–404.
Porter, M. (1990). *The Competitive Advantage of Nations.* New York: Free Press.
Reed, M. (1995). "Cooperative Management of Environmental Resources: A Case Study from Northern Ontario, Canada." *Economic Geography*, 71: 132–49.
Rigby, D. (1990). "Technical Change and the Rate of Profit: An Obituary For Okishio's Theorem." *Environment and Planning A*, 22: 1029–50.
—— (1991). "The Existence, Significance, and Persistence of Profit Rate Differentials." *Economic Geography*, 67: 210–22.
Rigby, D., and Essletzbichler, J. (1997). "Evolution, Process Variety, and Regional Trajectories of Technological Change in US Manufacturing" *Economic Geography*, 73: 269–84.
Roberts, R., and Emel, J. (1992). "Uneven Development and the Tragedy of the Commons: Competing Images for Nature-Society Analysis." *Economic Geography*, 68: 249–71.
Roberts, S. (1995). "Small Place, Big Money: The Cayman Islands and the International Financial System." *Economic Geography*, 71: 237–56.
Robinson, K. (1995). "Industrial Location and Air Pollution Controls: A Review of Evidence from the USA." *Progress in Human Geography*, 19: 222–44.
Romer, P. (1986). "Increasing Returns and Long-Run Growth." *Journal of Political Economy*, 94: 1002–37.
Romer, P. (1990). "Endogenous Technological Change." *Journal of Political Economy* 98: 71–102.
Rutherford, T. D. (1998). "'Still in Training?' Labor Unions and the Restructuring of Canadian Labor Market Policy." *Economic Geography*, 74/2: 131–48.
Saxenian, A. (1994). *Regional Advantage: Culture and Computing in Silicon Valley and Route 128.* Cambridge. Mass.: Harvard University Press.
Sayer, A., and Walker, R. (1992). *The New Social Economy.* Oxford: Blackwell.
Schoenberger, E. (1990). "US Manufacturing Investments in Western Europe: Markets, Corporate Strategy, and the Competitive Environment." *Annals of the Association of American Geographers*, 80: 379–93.
Schubert, U., Gerking, S., Isserman, I., and Taylor, C. (1987). "Regional Labor Market Modeling: A State of the Art Review," in Fischer, M., and Nijkamp, P. (eds.), *Regional Labour Markets.* Amsterdam: North-Holland.
Scott, A. J. (1988*a*). *Metropolis: From the Division of Labor to Urban Form.* Berkeley: University of California Press.
—— (1988*b*). *New Industrial Spaces.* London: Pion.
—— (1991). "Flexible Production Systems: Analytical Tasks and Theoretical Horizons—a Reply To Lovering." *International Journal of Urban and Regional Research*, 15: 130–4.
—— (1995). "The Geographic Foundations of Industrial Performance." *Competition and Change: The Journal of Global Business and Political Economy*, 1: 51–66.
—— (1998). *Regions and the World Economy: The Coming Shape of Global Production, Competition and Political Order.* Oxford: Oxford University Press.
Scott, A. J., and Storper, M. (1992). "Regional Development Reconsidered," in H. Ernste and V. Meier (eds.), *Regional*

Development and Contemporary Industrial Responses: Extending Flexible Specialization. London: Belhaven, 3–24.

Sheppard, E. (1984). "Value and Exploitation in the Capitalist Space Economy." *International Regional Science Review*, 97–108.

— (2000). "Geography or Economics? Conceptions of Space, Time, Interdependence, and Agency," in G. Clark, M. Gertler, and M. A. Feldman (eds.), *A Handbook of Economic Geography*. Oxford: Oxford University Press, 99–124.

Sheppard, E., and Barnes, T. (1986). "Instabilities in the Geography of Capitalist Production: Collective vs. Individual Profit Maximization." *Annals of the Association of American Geographers*, 76: 493–507.

— (1990). *The Capitalist Space Economy*. London: Unwin Hyman.

Sheppard, E., Haining, R., and Plummer, P. (1992). "Spatial Pricing in Interdependent Markets." *Journal of Regional Science*, 32: 55–75.

Sheppard, E., Plummer, P., and Haining, R. (1998). "Profit Rate Maximization in Interdependent Markets." *Journal of Regional Science*, 38: 659–67.

Simmons, J. (1991). "The Regional Mall in Canada." *Canadian Geographer*, 35/3: 232–40.

Simon, P. B. (1998). "Informal Responses to Crises of Urban Employment: An Investigation into the Structure and Relevance of Small-Scale Informal Retailing in Kaduna, Nigeria." *Regional Studies*, 32/6: 547–57.

Slowe, P. M. (1991). "The Geography of Borderlands: The Case of the Quebec–US Borderlands." *Geographical Journal*, 157/2: 191–8.

Sonis, M., and Hewings, G. J. D. (1998). "Economic Complexity as Network Compilation: Multiregional Input–Output Structural Path Analysis." *Annals of Regional Science*, 32/3: 407–36.

South, R. (1990). "Transnational 'Maquiladora' Location." *Annals of the Association of American Geographers*, 80: 549–70.

Storper, M. (1993). "Regional 'Worlds' of Production: Learning and Innovation in the Technology Districts of France, Italy, and the USA." *Regional Studies*, 27: 433–55.

— (1995). "The Resurgence of Regional Economies, Ten Years Later: The Region as a Nexus of Untraded Interdependencies." *European Urban and Regional Studies*, 2/3: 191–221.

— (1997). *The Regional World: Territorial Development in a Global Economy*. New York: Guilford Press.

Storper, M., and Salais, R. (1997). *Worlds of Production: The Action Frameworks of the Economy*. Cambridge, Mass.: Harvard University Press.

Storper, M., and Walker, R. (1989). *The Capitalist Imperative: Technology, Territory and Industrial Growth*. Oxford: Blackwell.

Straussfogel, D. (1997). "Redefining Development as Humane and Sustainable." *Annals of the Association of American Geographers*, 87: 280–305.

Sunley, P. (1996). "Context in Economic Geography: The Relevance of Pragmatism." *Progress in Human Geography*, 20: 338–55.

Tesfatsion, L. (1998). "Ex Ante Capacity Effects in an Evolutionary Labor Market with Adaptive Search." Economic Report, Iowa State University (October).

Thill, J. C. (1992). "Competitive Strategies for Multi-Establishment Firms." *Economic Geography*, 68/3: 290–309.

Thrall, G. I. (1999). "Retail Location Analysis, Strategic Step 7: Judgment." *Geo Info Systems*, 9/2: 36–7.

Thrift, N., and Olds, K. (1996). "Refiguring the Economic in Economic Geography." *Progress in Human Geography*, 20: 21–37.

Tickell, A., and Peck, J. (1992). "Accumulation, Regulation and the Geographies of Post-Fordism: Missing Links in Regulationist Research." *Progress in Human Geography*,16: 190–218.

Todtling, F. (1992). "Technological Change at the Regional Level: The Role of Location, Firm Structure, and Strategy." *Environment and Planning A*, 24: 1565–84.

US Department of Commerce. Economic Development Administration. (1994). *Economic Studies of the Economic Development Administration: An Annotated Bibliography 1986–1994*. Washington, DC: EDA.

US Department of Commerce. Economic Development Administration. (1996). *Recent Economic Studies of the Economic Development Administration: An Annotated Bibliography Update 1995*. Washington, DC: EDA.

Veblen, T. (1919). *The Place of Science in Modern Civilization and Other Essays*. New York: B. W. Huebsch.

von Hippel, E. (1988). *The Sources of Innovation*. New York: Oxford University Press.

Walker, R. (1997). "California Rages," In R. Lee and J. Wills (eds.), *Geographies of Economies*. London: Arnold.

Wallner, H., Narodoslawsky, M., and Moser, F. (1996). "Islands of Sustainability: A Bottom-Up Approach Towards Sustainable Development." *Environment and Planning A*, 28: 1763–78.

Warde, A. (1985). "Spatial Change, Politics and the Division of Labour," in D. Gregory and J. Urry (eds.), *Social Relations and Spatial Structures*. London: Macmillan.

Warden, J. T. (1993). "Industrial Site Selection: A GIS Case Study." *Geo Info Systems*, 3/6: 36–45.

Warf, B. (1998). "Spatial Scale, Backward Linkages, and Economic Multipliers." *Applied Geographic Studies*, 2/1: 17–27.

Warf, B., and Randall, J. (1994). "The US-Canada Free Trade Agreement: Impacts on US States and Canadian Provinces." *International Regional Science Review*, 17/1: 99–119.

Webber, M. (1982). "Agglomeration and the Regional Question." *Antipode*, 14: 1–11.

— (1987*a*). "Quantitative Measurement of Some Marxist Categories." *Environment and Planning A*, 19: 1303–22.

— (1987*b*). "Rates of Profit and Interregional Flows of Capital." *Annals of the Association of American Geographers*, 77: 63–75.

— (1991). "The Contemporary Transition." *Environment and Planning D: Society and Space*, 9: 165–82.

— (1996). "Profitability and Growth in Multiregion Systems: Theory and A Model." *Economic Geography*, 72/3: 335–52.

Webber, M., and Foot, S. (1982). "The Measurement of Unequal Exchange." *Environment and Planning A*, 16: 927–47.

Webber, M., and Rigby, D. (1996). *The Golden Age Illusion*. New York: Guilford.

— (1999). "Accumulation and the Rate of Profit: Regulating the Macroeconomy." *Environment and Planning A*, 31: 141–64.

— (1986). "The Rate of Profit in Canadian Manufacturing." *Review of Radical Political Economics*, 18: 33–55.

Webber, M., Sheppard, E., and Rigby, D. (1992). "Forms of Technical Change." *Environment and Planning A*, 24: 1679–709.

Webber, M., and Tonkin, S. (1987). "Technical Changes and the Rate of Profit in the Canadian Food Industry, 1950–1981." *Environment and Planning A*, 19: 1579–96.

Wei, Y. (1996). "Fiscal Systems and Uneven Regional Development in China, 1978–1991." *Geoforum*, 27/3: 329–44.

—— (1999). "Regional Inequality in China." *Progress in Human Geography*, 23/1: 48–58.

Wilbanks, T. (1994). "'Sustainable Development' in Geographical Development." *Annals of the Association of American Geographers*, 84: 541–56.

Williamson, O. (1975). *Markets and Hierarchies*. New York: Free Press.

Wills, J. (1996). "Geographies of Trade Unionism: Translating Traditions across Space and Time." *Antipode*, 28: 352–78.

Winder, G. M. (1999). "The North American Manufacturing Belt in 1880: A Cluster of Regional Industrial Systems or One Large Industrial District?" *Economic Geography*, 75/1: 71–92.

Wood, P. (1991). "Flexible Accumulation and the Rise of the Business Services." *Transactions of the Institute of British Geographers*, 16: 160–72.

Wrigley, N. (1994). "After the Store Wars: Towards a New Era of Competition in UK Food Retailing?" *Journal of Retailing and Consumer Services*, 1/1: 5–20.

Wrigley, N., and Lowe, M. (eds.) (1996). *Retailing, Consumption and Capital: Towards the New Retail Geography*. Harlow: Longmans.

Wyckoff, A. (1997). "Imagining the Impact of Electronic Commerce." *OECD Observer*, 208: 5–8.

Wyly, E. K. (1996). "Race, Gender, and Spatial Segmentation in the Twin Cities." *The Professional Geographer*, 48/4: 431–44.

Xie, Y., and Dutt, A. K. (199). "Regional Investment Effectiveness and Development Levels in China." *Geojournal*, 20/4: 393–407.

Ying, L. G. (1999). "China's Changing Regional Disparities during the Reform Period." *Economic Geography*, 75/1: 59–70.

Zabin, C. (1997). "US-Mexico Economic Integration: Labor Relations and the Organization of Work in California and Baja California Agriculture." *Economic Geography*, 73: 337–55.

CHAPTER 10

Environmental Perception and Behavioral Geography

Douglas M. Amedeo and Reginald G. Golledge

Three foci and combinations of them constitute the basis for nearly all of EPBG's concerns: human activities, human experiences, and all forms of empirical surroundings. With these as fundamentals, this specialization is viewed as a working framework for exploring situations of people engaged in activities and having experiences in ordinary spatial and/or environmental contexts (Goody and Gold 1985, 1987; Aitken *et al.* 1989; Golledge and Timmermans 1990*a*, *b*; Aitken 1991, 1992; Kitchin 1996).

The Emergence of the EPBG Specialization

For at least the 1960s and 1970s, common ways of exploring human geographic issues entailed framing them as spatial-like representations for observation and study. The main focus was on structural features of spatial patterns such as density differences, dispersions, clusters, arrangements, shapes, configurations, connectivities, and spatial hierarchies, among others, and the typical research goal was to describe and account for those features over time (Amedeo and Golledge 1986; Abler *et al.* 1971; Haggett 1966; Haggett and Chorley 1969). The reasoning employed in works such as these followed the ways in which the issues themselves were represented. It included examinations and evaluations of spatial co-variances, distance-decay regularities, contagious and competitive effects in spatial diffusions, spatial clustering in regionalizing and in regional ecologies, and more general applications of process-form type arguments. These approaches were largely structural in perspective and had few, if any, provisions for consideration of individual behavior and experience.

Concurrent with this spatial-structural perspective, however, efforts were also being devoted to understanding human decision-making in spatial contexts. Consumer choices in market places, industrial and retail location decisions, trip determinations, competitive decision-making in space, and spatial-allocation determinations were some of the main topics investigated (Golledge and Stimson 1997). This emphasis on spatial decision-making no doubt generated the impulse for further behavioral research in geography. It did so largely through the effects of its successes and limitations, both of which provided many opportunities for thoughtful criticisms and explorations into additional behavioral issues.

Its successes demonstrated that knowledge about people acting in spatial contexts could be gained by focusing on *individuals* as spatial decision-makers and studying their concomitant spatial search and learning processes (Golledge 1969; Golledge and Brown 1967; Gould 1965). On the other hand, it was evident from these studies that prevailing ways of looking at individual decision-making in space needed to be qualified and expanded conceptually to obtain more comprehensive understanding of human activities in environmental contexts.

This was because, in initial attempts to understand spatial decision-making, many ordinary human characteristics of the actors were formally disallowed to meet the needs and capabilities of the models used. Individuals were assumed to have, when engaged in their activities, limited but prominent objectives or goals (e.g. maximize utility, profits, or satisfaction, minimize cost, effort, disutility, or friction, etc.), and did not emote, contemplate, or reflect on the circumstances confronted. Little attention was paid to how individuals processed the external information confronted in the course of their activities, how they rationalized this information with information from past events, and how their past experiences entered into their understanding of current circumstances, and influenced their intended actions. The employment of simplifying assumptions in exploratory investigations such as these was and, indeed, still is a common research strategy in attempts to comprehend complex human–environment issues. But, as others, including those utilizing this strategy began to point out, the use of model-type individuals tended greatly to inhibit opportunities for further explorations.

Missing from such model individuals were greater considerations of the facets of what it means to be human, and this was soon to be recognized by members of the evolving EPBG specialization. An abundance of new works gradually surfaced that would attempt to move closer to the actual person . . . the person who, in addition to being a biological entity and thus susceptible like all living things to the adversities and benefits generated by the physical fundamentals of our surroundings; perceives, cognizes, remembers, reflects, has thoughts, adjusts, adapts, emotes; is particularly oriented by notions of *self*; is a member of a cultural/social system; has goals, objectives, purposes; possesses beliefs, values, motives, preferences, and, presumably, idiosyncrasies; and, not least of all, often relates in a very strong ego-involvement sense to surroundings such as places, landscapes, environments, and/or settings (Lowenthal 1967; Downs and Stea 1973, 1977; Tuan 1974; Golledge and Rushton 1976; Saarinen 1976; Moore and Golledge 1976; Jakle *et al.* 1976; Porteous 1977; Gold 1980; Golledge and Stimson 1997). Greater reflections on these aspects of this more human person by many in the field, may be said, because of their conceptually far-reaching implications, to have *greatly changed* and broadened the research agendas and foci of many geographers concerned with human issues. Indeed, a distinctive way of looking at human–environment relations, Environmental Perception and Behavioral Geography, emerged in the discipline.

A Specialization That Extends the Discipline's Basic Concerns

But, given its distinctive emphasis, to what extent does this specialization diverge from the basic premisses of geography itself? For EPBG, interests in behavioral issues extend far beyond the issues themselves. Similar to the wider interest of geographers in general, members of EPBG are fundamentally concerned with all those behavioral issues that ultimately, through their implications, contribute to long-term knowledge about durable human–environment relations. York (1987: 1–2) states that "In a fundamental sense, most of geography's interests can be reduced to an 'ultimate' concern with comprehending person–environment relations . . . The expression *person–environment relations* refers to the great variety of relatively enduring mutual connections that evolve over time as a 'natural' result of the inevitable necessity for individuals to carry out their behaviors and have their experiences *in* and *with* reference to environmental contexts," (our italics replace York's underline). As examples of durable relations, York (ibid.) points to the familiar "sustaining 'dependencies' between environments and individuals that are gradually established over time by those working directly with their surroundings in activities related to subsistence and/or production." She also refers to person–environment relations that are "especially evident in the strong affections for, attachments to, apprehensions about, or persistent preferences for different surroundings or places." She writes about relations "entrenched in the self-and-social category of life," which are often "manifested in the direct and indirect links between the use, ownership, and/or control of space and environments and the management of self–other relationships." Still other examples are her references to the spiritual and philosophical long-term connections that are established between people and places (as in sacred places) and the types of relations generated over time in such domains as health, aesthetics, and geriatrics. York (1987: 2) summarizes her observations on this larger interest in the discipline itself by indicating that, "in all of these examples, the important point is this: people, because they must behave and experience *in* and *with reference to* environments, establish long-term relationships with their surroundings that are, above all, durable, that influence and are influenced by ongoing behavior and experiences, and that represent an essential part of existence."

From geography's perspective, then, little is new when noting the foundations of the EPBG specialization. It still focuses on the wide variety of built and non-built environments encountered by individuals; continues to subscribe to the long-standing implication in the discipline of general ecological constraints in human functioning; holds as a basic tenet the condition that information about the external world is structured spatially; and, not least of all, presumes that humans must function in a world that is fundamentally environmental. All of these positions that ground the specialization empirically, at the very least, parallel the larger discipline's tacit but widely understood premiss that understanding human activities and experiences must be in terms of the environmental contexts in which they occur.

There is, however, a position in this specialization that receives less prominence in the larger discipline. This is that integral and essential to the enactment of all activities and the occurrences of experiences in environmental contexts is the human processing of information. It is a position that follows plausibly from the specialization's greater emphasis on the more human-faceted individual and that inextricably links the components, *environmental perception* and *behavioral geography*, conceptually.

Why Environmental Perception *and* Behavioral Geography?

This is an important question, for the answer dictates the conceptual coherency of the specialization itself. Environmental perception is a human knowing-process that frames what is apprehended externally within an environmental context or setting relevant for activity and experience. It is a process that cognitively coordinates external with internal sources of information, so as to rationalize both with respect to the other in terms of agreement, completion, and meaning. It entails a variety of internal and external activations, sensory receptor information-acquisition, attention and memory activities, application of experience-based representations such as orientation, place, and other cognitive structures, encoding activities, and the exercising of cognitive idiosyncrasies.

A number of points illustrate why a knowing process such as *environmental* perception is needed when dealing with behavioral fundamentals such as activities and experiences in environments. One is the implication that follows from the foundation of geography itself, namely that environments or surrounds, because they are, in effect, the settings for activities, constitute external information sources for their execution and completion. Another point follows from the beliefs of cognitive scholars themselves regarding the information *relevance* of actual surroundings for activities. Neisser (1976), for example, uses the term "reality" in his text on cognition. For him, "reality" refers to environments actually confronted by humans in their daily activities. He states,

> This trend [over-reliance of laboratory studies in research on cognition] can only be reversed, I think, if the study of cognition takes a more "realistic" turn . . . First, cognitive psychologists must make a greater effort to understand cognition as it occurs in the ordinary environment and in the context of natural purposeful activity . . . Second, it will be necessary to pay more attention to the details of the real world in which perceivers and thinkers [note: action and thought] live, and the fine structure of information which that world makes available to them. (1976: 7)

Likewise, Blumenthal (1977: 147), in his work on Cognition, points out that "The ultimate expression of the development of a living system is to maintain its spatial and temporal integrity and to maintain itself against the flux of the environment."

It is clear that these cognitive scholars think much like geographers have always thought, namely, that consideration of actual environments or surroundings is essential to understanding how humans deal with and comprehend their world. Neisser's expression, "as it occurs in the ordinary environment and in the context of natural purposeful activity" not only reinforces the significance of geography's perspective about environments but also its general curiosity about human activities in those settings. It follows, then, that all environments, essentially by their presence, constitute external sources of information for human activities. Hence, the need to consider a relevant knowing process like *environmental* perception in this specialization is compelling.

External Information of Environments and Its Relevance to Human Activity

It is noted that Neisser, in his remarks about everyday environments as reflective of reality, makes reference not to hypothetical or esoteric behavior, but "natural purposeful activity." So do S. Hanson and P. Hanson (1993) in their reference to "activities in everyday life." Their work reminds us that a rather noticeable emphasis in EPBG is on the everyday or ordinary activities and experiences of humans.

But if that is so, what, then, is the common relevance of external information for "natural purposeful" or "everyday life" activity? The answer to that relates to the major categories of external information needed for the enactment, continuance, and completion of activities in general. For example, activities of individuals can be viewed in two ways: those whose enactment and completion require considerable locomotion and orientation throughout a setting, and those whose definition and effectiveness depend heavily on the nature of and interrelationships among things within the setting. In practice, of course, everyday activities depend exclusively neither on one nor the other of these types; their separation here is used to point out that individuals must have two basic categories of external information to begin and successfully complete practically all activity in a setting: knowledge about the setting's spatial-structural aspects and knowledge about its social-cultural-physical makeup (see also Hanson 1999). The need for this external information reflects the fact that individuals ordinarily engage in activities that correspond to or fit cultural-social-physical *contexts* and that relate to specific *arenas*.

Then, too, it is clear that, throughout any activity episode, it is necessary that an individual be able continuously to evaluate and monitor effectiveness of activity enactment in terms of its initiation, continuance, and completion. The individual accomplishes this by assessing how successfully the intended activity develops within the setting's physical and structural confines, the degree to which the activity fulfills immediate purposes, whether it conforms to the setting's cultural-social-physical demands, and, if relevant, how well it satisfies personal aspirations about the presentation of self in the setting. Thus, the context-arena information that all immediate environments manifest must have relevance to the effective execution of intended activities. What is significant, however, is that this information made available by the presence of an environment is neither ordered nor made relevant for a *particular* encounter. Its relevance for activity is neither given, immediately obvious, or indubitable. Its amount is apt to be enormous, and not all of it is likely to be relevant for the enactment of an intended activity at the moment. So, for these and other reasons, directed processing of such information must occur in order to ascertain its usefulness for activity enactment. Hence, it is evident that, given its fundamental interests, a compelling need exists in behavioral geography, for the provision of a knowing process, namely, environmental perception and cognition, that, by the way it assesses external information and relates it to stored information, acknowledges its environmental nature.

It is obvious that, for the execution, continuance, and completion of everyday activity, external information relevant to its enactment must be known or perceived. Such processing comes about from the ways in which external and internal sources of information are mutually appraised *cognitively*, one in terms of the other, to produce percepts of surroundings immediately useful for an individual engaged in the enactment of activity in those surroundings. Thus, individuals engage in a "knowing" process through which they acquire, synthesize, and integrate external or environmental information with internal sources of knowledge to form, in their perceptions, a *contextual-arena* basis for immediate ongoing activity. Internal, environmentally related knowledge, in the sense above, directs what information is acquired during the environmental knowing process and organizes its elaboration to render it informative. The terms "mental representation," and "cognitive map" are commonly used to refer to this experience-driven internal knowledge. As to the importance of "*representation*" in environmental perceptual-cognitive processing, G. Mandler (1985: 31) indicates that "all actions and thoughts require some underlying representation."

Of no small interest to individuals in the execution of their activities is the coherency and interdependency of external information as it is manifested in immediate surroundings. Hence, in the conduct of our activity episodes, the usual way of reflecting on external information is in terms of interrelated or organized information whose coherency is exemplified in real-world environmental settings such as farm fields, barns, machine-sheds, hiking trails, playgrounds, grocery stores, roads, classrooms, gas-stations, funeral parlors, tennis courts, theaters, ski-slopes, among many others. Indeed, this is the common way we relate our experiences when we converse about our activities to others. Surroundings such as these are referred to as places, settings, scenes, landscapes, environments, and so on for the plausible reason that information of activity-relevance occurs in those forms. Each of these settings can be viewed either—or both—from their arena or their context characteristics. In general, because of this inherent information-coherency characteristic of environmental presentations and the need to reflect that coherency in cognitive processes designed to apprehend or know such information, representations utilized in environmental perception are probably schema in structure. G. Mandler (1985: 36), in his comments about "schemas as representational systems," indicates that "Schemas are cognitive structures, which is the more general term used for underlying representations in cognitive systems . . . Schemas are used primarily *to organize experience*, and in that role they overlap with some aspects of 'plans' and 'images' . . . Schemas are built up in the course of *interaction with the environment*" (italics ours). (See also Neisser (1976) and J. M. Mandler (1984).)

Incentives for the Emergence and Development of EPBG

The more specific reasons for the emergence and the development of a Behavioral Geography Specialization were actually many and interrelated (Golledge and Timmermans 1990). The specialization's beginning was certainly influenced by the gradual shift from interests in aggregations of human events over extensive spaces to greater concerns about *individuals* in relatively more immediate spatial and/or environmental contexts. This increasing focus on individuals placed greater emphasis on the significance of environments in human activity and experience, generated wider considerations of human information-processing, and prompted more extensive explorations into those human attributes related to functioning in and experiencing environments.

Then, too, emphasis on the perfectly rational individual, equipped with full and relevant information, and acting strictly in a way to satisfy self-interest was being reduced and replaced by individuals closer to the more human-faceted person described earlier. Shifts like this did not make research in general much easier; they added enormous conceptual and methodological complexity to the things behavioral geographers were inclined to study. But the shift had its clear advantages as well. It encouraged and accommodated a number of alternative perspectives for exploring human–environment relations (e.g. ethnomethology, phenomenology, gender studies, biographies, case studies, and grounded-theory approaches) emerging at the time in geography and earlier elsewhere. These perspectives required greater considerations of a variety of human facets frequently held constant by the modeling requirements in more deductively structured approaches (see, for example, Aitken 1991, 1992).

But the shift in the unit of focus from aggregates to individuals and the emphasis on the multifaceted individual reinforced an emerging desire among behavioral geographers to explore not only biological and physical constraints on human functioning in environments but also how humans might relate to their surroundings in other ways (e.g. aesthetically, affectively, spiritually, philosophically, etc.). Inevitably, with this desire, came the important question of what constitutes an environment. It was clear, for example, that the world was becoming significantly more urban (World Resources 1996: 150), so that the surrounds confronted by many individuals throughout much of their existence were largely of the built type. At least two broad categories of environments, then, emerged with the development of behavioral geography: built and non-built types. And, as expected, what was meant by the classical rendition of human–environment relations widened considerably.

From this wider look at the role of a variety of surroundings in human activity and experience, many traditional issues were revisited for, perhaps, their potential additional implications. These included the significance of place in human experience, the role of the spatial facet of environments in human activity, and the apprehension of environments in both experiences and activities. The latter, in particular, together with this greater emphasis on various aspects of the more human-faceted individual, exemplified a growing interest in human processing of information in behavior and experience. The result, of course, was the rapid development of the other component of this specialization, environmental perception.

The emergence of this behavioral specialization was also a consequence of geographers cultivating curiosities about behavioral issues while working with individuals in other fields, becoming familiar with related research published in such outlets as *Environment and Behavior*, *Journal of Environmental Psychology*, *Environmental Systems*, and *Journal of Architectural and Planning Research*, and participating in conferences held by interdisciplinary organizations such as the Environmental Design Research Association (EDRA) and the International Association for People–Environment Studies (IAPS). Throughout, of course, there was a need for geographers to expand their own conceptualizations about human–environment relations, and, thus, to pay attention to what like-minded individuals from these other fields were developing (Ittelson *et al.* 1974; Stokols 1987; Kuipers 1978; Kaplan and Kaplan 1982; Rapoport 1982; Pick and Acredolo 1983; Spencer *et al.* 1989; Loomis *et al.* 1998; and Garling and Golledge 1993).

From responses to specific incentives such as these, the number and variety of behaviorally related problems being explored in the specialization virtually exploded and has significantly influenced the work in other emphases in the larger discipline (see e.g. works in cartography such as Lloyd and Steinke 1985; Eastman 1985*a*, *b*; Amedeo and Kramer 1991; and, certainly, MacEachren 1995).

What emerged from all this, then, was a specialization with a distinctive concern about how individuals relate to or experience actual environments. Individuals were to be the focus of analysis and many of their attributes were to be given more serious consideration than in the past. Environments were to be treated more like physical-sociocultural systems, and, as part of their behavioral episodes, individuals were to be viewed as acquiring situational information by transacting with those environmental systems both in a functional and, particularly, in a cognitive process sense.

Thus, what is communicated or implied by the expression, behavioral *geography* is the basic thought that there is something significant about a geographic context that bears on and/or has an effect upon many behavioral issues in general. That significance is fundamentally found in the fact that the world, as it is encountered by humans, is geographical . . . which is to say, is spatial and/or environmental. *External information* related to human activity and experience, then, occurs in *that kind* of world. When information is presented, framed, or contexted in that way, it effects the way it is known and the way it is related to. Quite simply, and in general, how information is presented or available is important in how that information is understood.

Some Topical Areas Reflecting the Conceptual Nature of EPBG

In no particular order of importance, then, some of the more specific topics reflecting this specialization's interests that have received considerable attention over the last decades include *environments in the experiences of elderly people* (Golant 1986; Golant *et al.* 1988; Callahan 1992; Golant 1992; Plane 1992; Laws 1993; McHugh *et al.* 1995; Gilderbloom and Markham 1996; McHugh and Mings 1996; Pandit 1997); *gender issues in spatial and/or environmental contexts* (Self *et al.* 1992; Jones *et al.* 1997; Hanson and Pratt 1995; Self and Golledge 1995; Bondi 1990; England 1993; Cope 1996; Golledge *et al.* 1995; Pratt and Hanson 1988; McLafferty and Preston 1991; Gilbert 1997, 1998; Reprint Series Bibliography 1996); *place experiences, sense of place, place attachments, and self and environment* (Rowles 1990, 1993; Altman and Low 1992; Altman and Churchman 1994; Chawla 1992; Massey 1993; McDowell 1993; Aitken 1994); *affective responses in and to environments* (Ulrich 1983; Amedeo and York 1984, 1988; Amedeo 1993; Oakes 1997); *responses to environmental hazards* (Palm and Hodgson 1992, 1993); *spatial and/or environmental perception and cognition* (see major reviews listed previously, and Amedeo and York 1990; Golledge and Stimson 1997; Montello 1991, 1998); *environmental aesthetics* (Porteous 1996; Amedeo *et al.* 1989; Amedeo, 1999); *spatial and environmental essentials with regard to special populations* (Golledge 1991, 1993, 1995; Amedeo and Speicher 1995); *spatial decision making and choice behaviors specifically with regard to their relationships to transport issues such as modal choice, characteristics of Intelligent Transportation Systems, and Advanced Traveler Information systems* (Kwan 1998; Golledge 1998; Gärling *et al.* 1989; Summers and Southworth 1998; Hanson and Huff 1988); *information processing and the diffusion of both information and innovations* (e.g. Gould 1993); *environment-behavior issues associated with particular groups such as the poor, ethnic minorities, the homeless, and disabled people* (Kobayashi and Peake 1994; Kodras and Jones 1991; Dear 1987; Dear and Gleeson 1990; Preston *et al.* 1993; Pratt and Hanson 1993; Katz and Monk 1993; Rowe and Wolch 1990); *visualization, cognitive cartography, and new representational formats and the worlds of children* (e.g. Freundschuh 1990; Blaut 1997; Liben and Downs 1997; Aitken 1994; Hermon 1999), *the language of spatial relations* (Freundschuh *et al.* 1989); *cognitive maps* (Tolman 1948; Lynch 1960; Downs and Stea, 1977; Liben 1982; Kitchin 1994); *external reflections, indicators, or representations of cognitive maps* (Montello *et al.* 1999; Aitkin 1994; Richardson *et al.* 1999); *learning and spatial knowledge (*Couclelis *et al.* 1987; McNamara 1992; Stevens and Coupe 1978; Thorndyke and Hayes-Roth 1982; Golledge *et al.* 1993; Lloyd 1993; Portugali 1992; Gärling 1994; MacEachren 1992b; Montello 1998); *spatial familiarity* (Gale *et al.* 1990); *cognitive mapping without sight* (Klatzky *et al.* 1990; Kitchin *et al.* 1998); *characteristics of cognitive maps* (Gale 1985; Buttenfield 1986; Kitchin 1995); *spatial knowledge acquisition* (Piaget and Inhelder 1967; Hart and Moore 1973; MacEachren 1992*a*; Lloyd and Cammack 1996; Montello 1998; Gallistel 1990*a*, *b*; Golledge 1999); and *naive geography and the nature of everday knowledge* (Egenhofer and Mark 1995; Montello 1993, 1999; Montello and Golledge 1998).

Observations about Some Wider Interests in EPBG

Some topics more than others in EPBG seem, at first glance, to exemplify a "behavioral" sense. The negotiation of space in trip-making studies is an example where such a sense appears to be immediately apparent. For other topics in the specialization, particularly sense of place, place attachment, affective responses to surroundings, and landscape aesthetics, the behavioral dimension is often less obvious. This explicit–implicit distinction should not, however, be taken as indicative of relevance and significance for the long-term conceptual development of this specialization.

For example, *place* research, especially when many of its conceptual implications are clearly understood, may eventually be designated a *fundamental* area of concern for the comprehension of human–environment relations. This is mainly because *affect* and *self* are inherently rooted in place concepts such as "place experience," "sense of place," and "place attachment" and both represent *primary* influences in cognitive processing related to thought, perception, and behavior (Altman and Low 1992; Tuan 1974, 1977; Relph 1976; Buttimer and Seamon 1980).

With regard to the potential primacy of *affect*, for example, Blumenthal (1977: 102) indicates that "the emotional augmentation of experience links enduring needs and dispositions to the psychological present. It can direct the course of cognition, the retrieval of memories, the structuring of thoughts, or the formation of perceptions. In this way, emotion contributes to the larger continuity of human experience." Actually, much research has established that spaces and places, as well as people and situations, can evoke significant affective responses (Strongman 1987; Amedeo and York 1984, 1988; Amedeo 1993; Ulrich 1983). Ittelson *et al.* (1974: 88) remark that "spaces and places, no less than people, can evoke intense emotional responses. Rooms, neighborhoods, and cities can be 'friendly,' 'threatening,' 'frustrating,' or 'loathsome;' they can induce hate, love, fear, desire and other affective states." In his discussion of emotions in person-environment-behavior episodes, Amedeo (1993) argues that emotions experienced in a particular environment can change an individual's intended activity in that setting or alter its quality and tone significantly.

As to the primacy of *self* in cognitive processing associated with environmental perception, Blumenthal (1977: 147) connects this concept to concerns about human–environment relating in this way: "The ultimate expression of the development of a living system is to *maintain its spatial and temporal integrity and to maintain itself against the flux of the environment. Likely more than anything else, the self concept contributes to this maintenance* and gives continuity to our experience throughout our waking hours" (italics ours).

Geographers have, of course, argued for years about the significance of place, particularly in the cultural and historical branches of the discipline. There has, at times, been a bit of impatience with the progress in studies dealing with the place concept; but now it is becoming increasingly clear that the concept *place* is enormously complicated. In fact, much literature demonstrates that assertion. Relph (1976, 1984); Altman and Low (1992), Buttimer and Seamon (1980) have made the importance of that message vivid, but so too has Johnston (1991) in his allusions to the often cursory and even misinterpretations of this concept. In all likelihood, *place*, as it is understood by those who find it significant in their life's experiences, is multidimensional in nature. As more and more place studies are completed in this specialization, particularly in areas related to sense of place (Tuan 1971), place attachment (Relph 1984), place and identity (Williams and Roggenbuck 1989), place and well-being (Roggenbuck and Ham 1986), it will become increasingly clear that *self* and *affect* are heavily involved and at the foundations of many place issues (Scherl 1989, 1991).

But what of studies investigating landscape *aesthetics*; how do they relate to the behavioral concerns of EPBG? Much of the research in landscape aesthetics has been devoted to the study of landscape appeal or attractiveness, or, in short, scenicness (see at least Nasar 1988; Porteous 1996; Amedeo *et al.* 1989; Ulrich 1984). Amedeo (1999: 329) points out,

> the notion scenic is an absorbing one. Scenicness is often the basis for expressing *environmental preferences* and, in tourism, recreational, and residential circumstances, for example, may have significant economic exchange-value. Certainly few individuals are ever dispassionate when contemplating scenicness. Reactions to scenicness frequently have *strong emotional and aesthetic undertones*, and discussions involving this notion are nearly always intense and rarely apathetic. [italics ours]

But in addition to *affect* and *preferences* highlighted by Amedeo, which, in any event, are both critical topics in behavioral issues, the study of landscape aesthetics has also begun to find its way into health issues, particularly from the perspective of the alleged therapeutic benefits of natural environments in cases of stress and healing (see e.g. Ulrich 1979, 1984; Ulrich and Simons 1986).

Landscape aesthetics is, in any event, certainly one of those research areas that has effectively illuminated the cognitive rationalization of accumulated experience (i.e. internal information) with empirical circumstances (i.e. external information) in the apprehension of surroundings and their appeal. For example, outcomes from this research frequently support the idea that perceiving scenic quality is an attribution process guided by some internal "rule" (e.g. aesthetic schema) which, itself, is activated when evoked by specific interpretations of external information. In this research, it appears that it is more plausible to argue for a process involving mutual interactions of empirical circumstances with cognitive activity in the attribution of scenicness than it is to insist that either one information source or the other (i.e. either external or internal) prevails in influencing that assessment (Amedeo 1999). So again we have an area of research that has significant implications for the behavioral specialization, particularly in its contributions to understanding environmental perception and its continuous broader emphasis that two fundamental sources of information, empirical circumstances and accumulated experiences, must be rationalized in any activity episode or experience with regard to their enactment or happening in environment.

Understanding Activities Independent of Environments

Since the EPBG specialization emphasizes the *fundamental importance* of environment in activity episodes (a basic premiss of geography, itself), one may wonder whether it is not possible to *fully comprehend* such episodes without reference to environments. That is to say, why be concerned with environments at all? Why not explore human activities independent of environments? The response to questions such as these is an old but compelling one: activities are more fully understood when observed in the setting or the environment of which they are a part and in terms of which they generally are conceived. Despite their apparent substantial social and psychological meaning, it would be incomplete to evaluate activities independent of the setting in which they occur. An important reason for this is that too many stimulus-effects potentially important to their clarification may be overlooked in the assessment of their nature. The same point, of course, can be made about the occurrences of experiences in environmental contexts. In general, external information necessary for both the enactment of activities and the onset of experiences ordinarily appears as, and is encountered in, an environmental configuration. From a definition-of-the-situation perspective, such a configuration is a rather complex gestalt containing information about content and relations, environmental patterning-effects on both, and information about properties unique to the patterning itself. Facets of environmental information-displays such as *arena*, *context*, and perhaps even *ambiance*, when viewed interdependently, tend to exemplify this gestalt-like character of environments. Hence, since information external to individuals is generally manifested as part of environmental arrays and *environmental-type* schemata are believed to guide and/or direct apprehension of such arrays it is reasonable to expect that, in general, the process of perceiving the external world involves apprehending its information both ecologically and componently. In other words, because information external to individuals is usually an inextricable part of a physical setting, comprehension of it is influenced by that mode of its appearance. This suggests that configuration properties of environmental arrays such as details about spacing, position, connection, orientation, organization, temporality, and ambiance (Golledge 1992) may, to some extent, qualify how content information such as physical, social, and psychological details become known in any environmental encounter. Thus, contemplating an activity from the perspective of it being at least in part a response to an environment should not only take into account the way in which external information necessary to it normally appears in surroundings but also any qualifying implications those appearances might have on all aspects of such information. What this amounts to for cognitively oriented theory about activity is that different external informational circumstances are encountered when the activity is pictured as happening in environments than when it is not conceived of in that way.

Whim or Plausible Extension of the Geography Discipline?

What we have in this specialization, then, is a number of important concepts, the linking of which describes fundamentals in our interest in behavioral issues. These include activity and experiences in environments, external information manifested spatially in the content of environments, internal information based on experiences, perceptual-cognitive processing as manifested in environmental perception, and representations or mental structures, such as environmental schemata, that are integrations of previous experiences.

The emphasis in this specialization on human processing for rationalizing external information as it is manifested in empirical environments with internal information developed from previous experiences during activity episodes reinforces a sobering but fundamental point made by the cognitive scientist George Mandler (1985: 49). He states,

> No biological or environmental constraints fully determine human thought and action, but neither does any schema or cognitive structure. To say that a particular set of actions is contextually constrained is to say that we are able to develop mental structures that respond, when necessary, to the specific demands and conditions of certain contexts and environments. To say that perceptual schemas determine what is seeable is to say that we have developed structures that constrain our analyses of certain physical events.

Specializations that arise in disciplines because of whims differ from those that surface because the circumstances in their discipline's way of looking at things compel their eventual emergence. The undisputable aspects *of our world*, though seemingly obvious and mundane, make this specialization a rather compelling one, in the sense that some work of this nature must be undertaken in the discipline—otherwise, it seems, many interesting questions in the field go unexplored. The wonder is that, in the long history of the discipline, environmental perception and behavioral geography did not surface sooner.

Outlook for the EPBG Specialization

Substantial interest in environmental perception and behavioral geography can now be found not only in the United States, Canada, and Mexico, but also in Japan, Hong Kong, Malaysia, Australia, New Zealand, India, Nigeria, Ghana, South Africa, Switzerland, Israel, France, Italy, Spain, Belgium, the Netherlands, Denmark, Sweden, Finland, Norway, Lithuania, Estonia, Russia, Brazil, Venezuela, Chile, Ecuador, and the United Kingdom as well. This emphasis has become robust enough in some countries to warrant nationalistic reviews (e.g. Wakabayashi 1996; Portugali 1996; Aragones and Arredondo 1985). A variety of textbooks dealing with various behavioral issues have been produced since the early 1970s (e.g. Gold 1980; Golledge and Stimson 1997; Matthews 1992; Walmsley and Lewis 1984; Porteous 1976; Golledge and Timmermans 1988; Portugali 1996; Gärling 1994; Bovy and Stern, 1992). Research about cognitive maps now appears on a regular basis in the major journals of many countries, and Masters and Ph.D. dissertations exploring new facets of such issues have slowly increased. A cursory review of AAG Directories since 1991 indicates that an average of twenty MA and Ph.D.s are completed each year, and that the discipline's two professional journals have published twenty-four papers (*Annals*) and thirty-five papers (*Professional Geographer*) whose themes are in the general area of environmental perception and behavior. Many members of this specialty group also publish widely in allied journals (e.g. *Risk and Hazard Research*; *Leisure Research*; *Environment and Behavior*; *Journal of Environmental Psychology*; *Journal of Architectural and Planning Research*; *Urban Studies*; *Journal of Marketing*; *Urban History*; *Transportation*; *Landscape Journal*; *Journal of Planning, Education, and Research*; and the refereed *Proceedings of the Annual Conference of the Environmental Design Research Association*) and have written invited chapters in specialized topic books within this wider general area (e.g. Cohen 1985; Stokols and Altman 1987; Gärling 1994).

In 1990 Golledge and Timmermans wrote two reviews of research in environmental perception and behavior focusing *only* on spatial cognition and preference and choice. They found over 400 relevant publications in geography and related disciplines that were thematically linked to geographic concepts, models, and theories dealing with spatial cognition, preference, and choice. If one looks through the relevant chapter of the first *Geography in America* (Gaile and Willmott 1989) and all the more relatively current reviews cited throughout this chapter, a sense of a growing, enthusiastic specialty group is evident. Its broader interest in human–environment relations is quite central to contemporary concerns about global and local problems of resource management and use, to behavior, actions, and experiences of individuals in a wide variety of environmental contexts, to the issues associated with historical meaning and preservation of environment, to studies of children's environments, and to investigations into the significance of place and space, among many others.

EPBG concerns are also appearing in current transportation research. Since the mid-1980s, a paradigm shift from supply to demand considerations has taken place in this area. In particular, disaggregate household-based studies have focused on the derived demand for different modes, different time of travel and work schedules, human responses to real-time traffic conditions, and stated and revealed preferences for alternative forms of travel behavior (Gärling *et al.* 1994; Golledge *et al.* 1994; Hanson and Hanson 1993; Bovy and Stern 1990; Leiser and Zilberschatz 1989; Stern and Leiser 1988; Kwan 1995; Halperin 1988; Huff 1986; and T. Bell 1999). One of the most active areas of research in EPBG continues to be in spatial knowledge acquisition (Stea 1997; Blaut 1991; Downs and Liben 1986, 1987; Downs 1994; Golledge 1992, 1993; MacEachren 1991, 1992*a*, *b*; Golledge *et al.* 1995; Golledge 1999), and cognitive maps (Saarinen 1988, 1998; Golledge 1999; Kitchin 1995; Lloyd 1989, 1993; Lloyd and Cammack 1996; Lloyd *et al.* 1995; Lloyd and Heivly 1987). Relatively recent emphases attracting attention from EPBG members are geographic education (Hardwick 1997; Boehm and Petersen 1997), relationships between virtual and real worlds (Dow 1999; Foote 1997), and tourism issues such as destination image, perceived dimensions of behavior, and eco-tourism (Desbarats 1983; Fly 1986; Katz and Kirby 1991; Pigram 1993).

EPBG: A Specialization Open to Multiple Perspectives

A salient and, perhaps, the most intellectually stimulating dimension in this specialization is the presence of multiple perspectives underlying its research approaches. This is a noticeable change in the way behavioral research in geography was approached three decades ago; but it is unquestionably a welcome one. It largely stems from a number of factors. For example, earlier in this chapter we remarked on the trend surfacing in this specialization in which more human facets of an individual would receive attention in research. We referred to that kind of individual as "a person who, in addition to being a biological entity and, thus, susceptible like all living things to the adversities and benefits generated by the physical fundamentals of our surroundings, perceives, cognizes, remembers, reflects; has thoughts, adjusts, adapts, emotes; is particularly oriented by notions of *self*; is a member of a cultural/social system; has goals, objectives, purposes; possesses beliefs, values, motives, preferences, and, presumably, idiosyncracies; and, not least of all, often relates in a very strong ego-involvement orienting sense to his or her surroundings, whether those surroundings are labeled as places, landscapes, environments, settings and the like." We also observed that viewing individuals in this manner complicated research enormously. Why this is so is not only because more facets and their potential interactions are to be considered, but also because research, in general, becomes, as a result, more susceptible to critical analysis when this is viewed as the state of a person. Such criticisms typically question existing and well-established approaches such as those of statistical inferential frameworks and deductive model-building and, when systematized, gradually take form as additional and recognizable perspectives.

For example, criticisms leveled against what were mostly deductively structured arguments to investigation included such things as the lack of ties to context in the analysis of processes and the events they generate, gender biases in findings and investigations, researcher-value contamination effects in assessments, artificiality in the simplicity of model individuals, absence of significant appreciation for subject-definitions of relevant categories, constrained structures in the elicitation of "data," inadequate safeguards for separating out researcher experiences from research interpretations of subject experiences, aloofness of researcher from subjects and their experiences, ignorance of general demand requirements generated in many research designs, attempts to measure concepts for which full understanding of their underlying characteristics was lacking, ignorance of cultural and historical influences in the events being analyzed, unsubstantiated claims about "random samples" and the nature of "populations," divorce of theory formation from actual empirical circumstances, among many others.

Thus, a number of alternative approaches or research perspectives have emerged in response to these criticisms in the social and behavioral sciences in general and in this specialization in particular. Collectively, they have been referred to as "qualitative inquiries." Under that designation, Patton (1990) and Creswell (1994, 1998) commonly include ethnography, phenomenology, grounded theory, biography, and case study as well-recognized *traditions* in qualitative research approaches. These perspectives are oriented around the idea of responding to these criticisms by designing approaches to research that directly confront and deal with, to some extent, the prominence of their presence. To some, the descriptor "qualitative" suggests the lack of quantification in these approaches. Creswell in both his 1994 and 1998 books on both qualitative and quantitative approaches demonstrates that meaning is faulty. It is clear that needs and occasions for the application of numbers, operations on numbers, collection of data, and analyses are not the distinguishing features between "qualitative inquiries" and so-called "quantitative approaches." What qualitative implies is a focused and deliberate attention on the *quality* of processes, experiences, behavior, and individuals and how that *quality*, because of its essentialness to how these things are known, in effect, qualifies them.

References

Abler, R., Adams, J. S., and Gould, P. (1971). *Spatial Organization: The Geographer's View of the World*. Englewood Cliffs, NJ: Prentice-Hall.

Aitken, S. C. (1991). "Person-Environment Theories in Contemporary Perceptual and Behavioral Geography I: Personality, Attitudinal and Spatial Choice Theories." *Progress in Human Geography*, 15/2: 179–93.

—— (1992). "Person-Environment Theories in Contemporary Perceptual and Behavioral Geography II: The Influence of Ecological, Environmental Learning, Societal/Structural, Transactional and Transformational Theories." *Progress in Human Geography*, 16/4: 553–62.

—— (1994). *Putting Children in Their Place*. Washington, DC: Association of American Geographers.

Aitken, S. C., Cutter, S. L., Foote, K. E., and Sell, J. L. (1989). "Environmental Perception and Behavioral Geography," in G. L. Gaile and C. J. Wilmott (1989: 218–38).

Altman, I., and Low, S. M. (eds.) (1992). *Place Attachment*. New York: Plenum Press.

Altman, I., and Churchman, A. (eds.) (1994). *Women and the Environment*. New York: Plenum Press.

Amedeo, D. (1993). "Emotions in Person-Environment-Behavior Episodes," in R. G. Golledge and T. Gärling (1993: 83–116).

—— (1999). "External and Internal Information in Versions of Scenic-Quality Perceptions." *Journal of Architectural and Planning Research*, 16/4: 328–52.

Amedeo, D., and Golledge, R. G. (1986). *An Introduction to Scientific Reasoning in Geography*. Robert E. Krieger.

Amedeo, D., and Kramer, P. (1991). "User Perceptions of Bi-Symbol Maps." *Cartographica*, 28/1: 28–53.

Amedeo, D., and Speicher, K. (1995). "Essential Environmental and Spatial Concerns for the Congenitally Visually Impaired." *Journal of Planning Education and Research*, 14/2: 101–10.

Amedeo, D., and York, R. A. (1984). "Grouping in Affective Responses to Environments: Indications of Emotional Norm Influence in Person-Environment Relations." *EDRA 15: Proceedings of the Fifteenth International Conference of the Research Association*, 193–205.

—— (1988). "Affective States in Cognitively-Oriented Person-Environment-Behavior Frameworks." *EDRA 19: Proceedings of the Nineteenth International Conference of the Environmental Design Research Association*, 203–11.

—— (1990). "Indications of Environmental Schemata from Thoughts about Environments." *Journal of Environmental Psychology*, 10/3: 219–53.

Amedeo, D., Pitt, D., and Zube, E. (1989). "Landscape Feature Classification as a Determinant of Perceived Scenic Value." *Landscape Journal: Design, Planning, and Management of the Land*, 8/1: 36–50.

Aragones, J., and Arredondo, J. (1985). "Structure of Urban Cognitive Maps." *Journal of Environmental Psychology*, 5: 197–212.

Bechtel, R. B. (1997). *Environment & Behavior: an Introduction*. Thousand Oaks, Calif: Sage.

Bell, T. L. (1999). "Attitudes Towards Big-Box Retailers: Enthusiastic Support or Begrudging Acceptance?" Paper presented at the 95th Annual Meeting of the Association of American Geographers, Honolulu, Hawaii, 23–7 March.

Bell, S., and Gärling, T. (1999). "Time Estimation of Trip Chains." *Journal of Applied Geographic Studies*, 3/2: submitted.

Blaut, J. M. (1991). "Natural Mapping." *Transactions of the Institute of British Geographers*, 16: 55–74.

—— (1997). "The Mapping Abilities of Young Children: Children Can." *Annals of the Association of American Geographers*, 87 1: 152–8.

Blumenthal, A. L. (1977). *The Process of Cognition*. Englewood Cliffs, NJ: Prentice-Hall.

Boehm, R. G., and Petersen, J. F. (eds.) (1997). *The First Assessment: Research in Geographic Education*. San Marcos, Tex.: Gilbert M. Grosvenor Center for Geographic Education, Southwest Texas State University.

Bondi, L. (1990) "Progress in Geography and Gender: Feminism and Difference." *Progress in Human Geography*, 14: 438–45.

Bovy, P. H. L., and Stern, E. (1990) *Route Choice: Wayfinding in Transport Networks*. Dordrecht: Kluwer.

Buttenfield, B. P. (1986). "Comparing Distortion on Sketch Maps and Mds Configurations." *Professional Geographer*, 38/3: 238–46.

Buttimer, A., and Seamon, D. (eds.) (1980). *The Human Experience of Space and Place*. New York: St. Martin's Press.

Callahan, J. J. (1992). "Aging in Place." *Generations*, 16: 5–6.

Chawla, L. (1992). "Childhood Place Attachments," in Altman and Low (1992: 63–86).

Cohen, R. (1985). *The Development of Spatial Cognition*. Hillsdale, NJ: Lawrence Erlbaum.

Cope, M. (1996). "Weaving the Everyday: Identity, Space, and Power in Lawrence, Massachusetts, 1920–1939." *Urban Geography*, 17: 179–204.

Couclelis, H., Golledge, R. G., Gale, N. D., and Tobler, W. R. (1987). "Exploring the Anchor-point Hypothesis of Spatial Cognition." *Journal of Environmental Psychology*, 7: 99–122.

Creswell, J. W. (1994). *Research Design: Qualitative and Quantitative Approaches*. London: Sage.

—— (1998). *Qualitative Inquiry and Research Design; Choosing among Five Traditions*. London: Sage.

Dear, M. J. (1987). "Community Solutions to the Homelessness Problem." Paper presented at the Association of American Geographers Annual Meeting, Portland, Oregon.

Dear, M. J., and Gleeson, B. (1990). *Community Attitudes Toward the Homeless*. Los Angeles Homelessness Project. Reprint 10. Department of Geography, University of Southern California, Los Angeles.

Desbarats, J. (1983). "Spatial Choice and Constraints on Behavior." *Annals of the Association of American Geographers*, 73/3: 340–57.

Dow, D. (1999). "The Effects of Symbolic Iconicity on the Integration of Spatial Knowledge Acquired from the Fly-through Navigation of Simulated Environments." Ph.D. dissertation proposal. San Diego, Calif.: Geography Department, San Diego State University.

Downs, R. M. (1994). "The Need for Research in Geography Education: IT Would Be Nice to Have Some Data." *Journal of Geography*, 93: 57–60.

Downs, R. M., and Liben, L. S. (1986). *Children's Understanding of Maps*. Scholarly Report Series No. 8. Center for the Study of Child and Adolescent Development. The Pennsylvania State University.

—— (1987). "Children's Understanding of Maps," in P. Ellen and C. Thinus-Blanc (eds.), *Cognitive Processes and Spatial Orientation in Animal and Man*. Dordrecht: Martinus Nijhoff, 201–19.

Downs, R. M., and Stea, D. (1973). *Image and Environment: Cognitive Mapping and Spatial Behavior*. Chicago: Aldine.

—— (1977). *Maps in Minds: Reflections on Cognitive Mapping*. New York: Harper & Row.

Eastman, J. R. (1985*a*). "Cognitive Models and Cartographic Design Research." *Cartographic Journal*, 22/2: 95–101.

—— (1985*b*). "Graphic Organization and Memory Structures for Map Learning." *American Cartographer*, 22: 1–20.

Egenhofer, M. J., and Mark, D. M. (1995). *Naive Geography*. Technical Report 95–8. National Center for Geographic Information and Analysis (NCGIA), Orono, Me.: University of Maine, and Buffalo: State University of New York (SUNY).

England, K. V. L. (1993). "Suburban Pink Collar Ghettos: The Spatial Entrapment of Women?" *Annals of the Association of American Geographers*, 83/2: 225–42.

Fly, J. (1986). "Nature, Outdoor Recreation and Tourism: The Basis for Regional Population Growth in Northern Lower Michigan." Unpublished doctoral dissertation, University of Michigan, Ann Arbor.

Foote, K. E. (1997). "The Virtual Geography Department: An Overview." Paper presented at the 93rd Annual Meeting of the Association of American Geographers, Fort Worth, Texas, 1–5 April.

Freundschuh, S. M. (1990). "Can Young Children Use Maps to Navigate?" *Cartographica*, 27/1: 54–66.

Freundschuh, S. M., and Sharma, M. (1996). "Spatial Image Schemata, Locative Terms, and Geographic Spaces in Children's Narrative: Fostering Spatial Skills in Children." *Cartographica*, 32/2: 38–49.

Freundschuh, S. M., Gould, M. D., Mark, D. M., Egenhofer, M. J., Kuhn, W., McGranaghan, M., and Svorou, S. (1989). *Working Bibliography on "Languages of Spatial Relations."* (1st edn.) Technical Paper 89–10, January. National Center for Geographic Information and Analysis, State University of New York (SUNY), Buffalo.

Gaile, G. L., and Willmott, C. J. (eds.) (1989). *Geography in America*. Columbus, Ohio: Merrill.

Gale, N. D. (1985). "Route Learning by Children in Real and Simulated Environments." Ph.D. Dissertation, University of California, Santa Barbara.

Gale, N. D., Golledge, R. G., Halperin, W. C., Couclelis, H. (1990). "Exploring Spatial Familiarity." *The Professional Geographer*, 42/3: 299–313.

Gallistel, C. R. (1990*a*). "Representations in Animal Cognition: An Introduction." *Cognition*, 37/1–2: 1–22.

—— (1990*b*). *The Organization of Learning*. Cambridge, Mass: MIT Press.

Gärling, T. (1994). "How Do Urban Residents Acquire, Mentally Represent, and Use Knowledge of Spatial Layout?" in T. Gärling (ed.), *Readings in Environmental Psychology*, X. *Urban Cognition*. London: Academic Press, 1–12.

Gärling, T., and Golledge, R. G. (1993). *Behavior and Environment: Psychological and Geographical Approaches*. Amsterdam: Elsevier.

Gärling, T., Kwan, M.-P., and Golledge, R. G. (1994). "Computational-Process Modeling of Household Activity Scheduling." *Transportation Research B*, 28B/5: 355–64.

Gärling, T., Brännäs, K., Garvill, J., Golledge, R. G., Gopal, S., Holm, E., and Lindberg, E. (1989). "Household Activity Scheduling," in *Transport Policy, Management and Technology Towards 2001: Selected Proceedings of the Fifth World Conference on Transport Research*, 4. Ventura, Calif.: Western Periodicals, 235–48.

Gersmehl, P. H. (1992). "Themes and Counterpoints in Geographic Education." *Journal of Geography*, 91/3: 119–23.

Gilbert, M. R. (1997). "Identity, Space, and Politics: A Critique of the Poverty Debates," in Jones, Nast, and Roberts (1997: 29–45).

—— (1998) "Race, Space, and Power: The Survival Strategies of Working Poor Women." *Annals of the Association of American Geographers*, 88/4: 595–621.

Gilderbloom, J. I., and Markham, J. P. (1996). "Housing Modification Needs of the Disabled Elderly: What Really Matters?" *Environment and Behavior*, 28/4: 494–511.

Golant, S. M. (1986). "Subjective Housing Assessment by the Elderly: A Critical Information Source for Planning and Program Evaluation." *The Gerontologist*, 26: 122–7.

—— (1992). *Housing America's Elderly: Many Possibilities/Few Choices*. Newbury Park, Calif.: Sage.

Golant, S. M., Rowles, G. D., and Meyer, J. W. (1988). "Aging and the Aged," in Gaile and Willmott (1989: 451–66).

Gold, J. R. (1980). *An Introduction to Behavioural Geography*. Oxford: Oxford University Press.

Golledge, R. G. (1969). "The Geographical Relevance of Some Learning Theories," in K. R. Cox and R. G. Golledge (eds.), *Behavioral Problems in Geography: A Symposium*. Northwestern University, Studies in Geography, No. 17. Evanston, Ill.: Department of Geography, Northwestern University, 101–45.

—— (1991). "Tactual Strip Maps as Navigational Aids." *Journal of Visual Impairment and Blindness*, 85/7: 296–301.

—— (1992). "Do People Understand Spatial Concepts?: The Case of First-Order Primitives," in A. U. Frank, I. Campari, and U. Formentini (eds.), *Theories and Methods of Spatio-Temporal Reasoning in Geographic Space*. International Conference

GIS—from Space to Territory: Theories and Methods of Spatio-Temporal Reasoning. Pisa, Italy, 21–3 September, Proceedings. New York: Springer-Verlag, 1–21.

—— (1993). "Geography and the Disabled: A Survey with Special Reference to Vision Impaired and Blind Population." *Transactions of the Institute of British Geographers*, 18: 63–85.

—— (1995). "Primitives of Spatial Knowledge," in T. L. Nyerges, D. M. Mark, R. Laurini, and M. J. Egenhofer (eds.), *Cognitive Aspects of Human-Computer Interaction for Geographic Information Systems*. Dordrecht: Kluwer, 29–44.

—— (1998). "The Relationship Between Geographic Information Systems and Disaggregate Behavioral Travel Modeling." *Geographical Systems*, 5: 9–17.

—— (ed.) (1999). *Wayfinding Behavior: Cognitive Mapping and Other Spatial Processes*. New York: Johns Hopkins University Press.

Golledge, R. G., and Brown, L. A. (1967). "Search, Learning, and the Market Decision Process." *Geografiska Annaler B: Human Geography*, 49: 116–24.

Golledge, R. G., and Rushton, G. (eds.) (1976). *Spatial Choice and Spatial Behavior: Geographic Essays on the Analysis of Preferences and Perceptions*. Columbus, Ohio: Ohio State University Press.

Golledge, R. G., and Stimson, R. J. (1997). *Spatial Behavior: A Geographic Perspective*. New York: Guilford Press.

Golledge, R. G., and Timmermans, H. (eds.) (1988). *Behavioural Modelling in Geography and Planning*. London: Croom Helm.

—— (1990*a*). "Application of Behavioural Research on Spatial Problems I: Cognition." *Progress in Human Geography*, 14/1: 57–99.

—— (1990*b*). "Application of Behavioural Research on Spatial Problems II: Preference and Choice." *Progress in Human Geography*, 14/3: 311–54.

Golledge, R. G., and Zhou, J. (1999). "GPS Monitored Travel Behavior." Unpublished manuscript, Geography Department, University of California at Santa Barbara.

Golledge, R. G., Dougherty, V., and Bell, S. (1995). "Acquiring Spatial Knowledge: Survey Versus Route-Based Knowledge in Unfamiliar Environments." *Annals of the Association of American Geographers*, 85/1: 134–58.

Golledge, R. G., Kwan, M.-P., and Gärling, T. (1994). "Computational-Process Modeling of Household Travel Decisions Using a Geographical Information System." *Papers in Regional Science*, 73/2: 99–117.

Golledge, R. G., Ruggles, A. J., Pellegrino, J. W., and Gale, N. D. (1993). "Integrating Route Knowledge in an Unfamiliar Neighborhood: Along and Across Route Experiments." *Journal of Environmental Psychology*, 13/4: 293–307.

Goodey, B., and Gold, J. R. (1985). "Behavioral and Perceptual Geography: From Retrospect to Prospect." *Progress in Human Geography*, 9/4: 585–95.

—— (1987). "Environmental Perception: The Relationship with Urban Design." *Progress in Human Geography*, 11/1: 126–33.

Gould, P. (1965). "Wheat on Kilimanjaro: The Perception of Choice within Game and Learning Model Frameworks," in L. von Bertalanffy and A. Rapoport (eds.), *Yearbook of the Society for General Systems Research*, lO: 157–66.

—— (1993). *The Slow Plague: A Geography of the Aids Pandemic*. Cambridge: Blackwell.

Haggett, P. (1966). *Locational Analysis in Human Geography*. New York: St. Martin's Press.

Haggett, P., and Chorley, R. J. (1969). *Network Analysis in Geography*. New York: St. Martin's Press.

Halperin, W. C. (1988). "Current Topics in Behavioral Modelling of Consumer Choice," in R. G. Golledge and H. Timmermans (eds.), *Behavioral Modelling in Geography and Planning*. London: Croom Helm, 1–26.

Hanson, S. (1999). "Isms and Schisms: Healing the Rift Between the Nature-Society and Space-Society Traditions in Human Geography." *Annals of the Association of American Geographers*, 89/1: 133–43.

Hanson, S., and Hanson, P. (1993). "The Geography of Everyday Life," in Gärling and Golledge (1993: 249–69).

Hanson, S., and Huff, J. O. (1982). "Assessing Day-to-Day Variability in Complex Travel Patterns." *Transportation Research Record*, 891: 18–24.

—— (1988). "Repetition and Day-to-Day Variability in Individual Travel Patterns: Implications for Classification," in Golledge and Timmermans (1988: 368–98).

Hanson, S., and Pratt, G. (1995). *Gender, Work, and Space*. London: Routledge.

Hardwick, S. W. (1997). "Meeting the Needs of Special Populations in Geography: Key Issues and Research Questions." Paper prepared for presentation at the Center for Geographic Education sponsored conference "The First Assessment: Research in Geographic Education" held at Southwest Texas State University, San Marcos, Texas, 22–5 May.

Hart, R. A., and Moore, G. T. (1973). "The Development of Spatial Cognition: A Review," in Downs and Stea (1973: 246–88).

Hermon, T. (1999). "Children's Production of Space and Self: Play and the Personal Geographies of Urban Fourth and Fifth-Graders." Ph.D. dissertation proposal. San Diego, Calif.: Geography Department, San Diego State University.

Huff, J. O. (1986). "Geographic Regularities in Residential Search Behavior." *Annals of the Association of American Geographers*, 76: 208–27.

Huff, J. O., and Hanson, S. (1989). "Measurement of Habitual Behaviour: Examining Systematic Variability in Repetitive Travel," in P. Jones (ed.), *New Approaches in Dynamic and Activity Based Approaches to Travel Analysis*. Brookfield, Vt.: Gower, 229–49.

Ittelson, W. H., Proshansky, H. M., Rivlin, L. G., and Winkel, G. H. (1974). *An Introduction to Environmental Psychology*. New York: Holt, Rinehart & Winston.

Jakle, J., Brunn, S., and Roseman, C. (1976). *Human Spatial Behavior*. North Sciatute, Mass.: Duxberry Press.

Johnston, R. J. (1991). *A Question of Place: Exploring the Practice of Human Geography*. Cambridge, Mass.: Blackwell.

Jones, J. P. III, Nast, H. J., and Roberts, S. M. (eds.) (1997). *Thresholds in Feminist Geography: Difference, Methodology, and Representation*. Lanham, Md.: Rowman & Littlefield.

Kaplan, S., and Kaplan, R. (1982). *Cognition and Environment: Functioning in an Uncertain World*. New York: Praeger.

Katz, C., and Kirby, A. (1991). "In the Nature of Things: The Environment and Everyday Life." *Transactions of the Institute of British Geographers*, 16: 259–71.

Katz, C., and Monk, J. (eds.) (1993). *Full Circles: Geographies of Women over the Life Course*. London: Routledge.

Kitchin, R. M. (1994). "Cognitive Maps: What Are They and Why Study Them?" *Journal of Environmental Psychology*, 14/1: 1–19.

— (1995). *C-Map: A Software Package*. Swansea: Department of Geography, University of Wales.

— (1996). "Increasing the Integrity of Cognitive Mapping Research: Appraising Conceptual Schemata of Environment-Behavior Interaction." *Progress in Human Geography*, 20/1: 56–84.

Kitchin, R. M., Jacobson, D. R., Golledge, R. G., and Blades, M. (1998). "Belfast Without Sight: Exploring Geographies of Blindness." *Irish Geographer*, 31/1: 34–46.

Klatzky, R. L., Loomis, J. M., Golledge, R. G., Cicinelli, J. G., Doherty, S., and Pellegrino, J. W. (1990). "Acquisition of Route and Survey Knowledge in the Absence of Vision." *Journal of Motor Behavior*, 22/1: 19–43.

Kobayashi, A., and Peake, L. (1994). "Unnatural Discourse: 'Race' and Gender in Geography." *Gender, Place and Culture*, 1: 225–43.

Kodras, J., and Jones, J. P. III (1991). "A Contextual Examination of the Feminization of Poverty." *Geoforum*, 22: 159–71.

Kuipers, B. J. (1978). "Modeling Spatial Knowledge." *Cognitive Science*, 2: 129–53.

Kwan, M.-P. (1995). "GISICAS: an Activity-Based Spatial Decision Support System for ATIS." Paper presented at the International Conference on Activity-based Approaches: Activity Scheduling and the Analysis of Activity Patterns, 25–8 May. Eindhoven, The Netherlands: Eindhoven University of Technology.

— (1998). "Space-Time and Integral Measures of Individual Accessibility: A Comparative Analysis Using a Point-Based Framework." *Geographical Analysis*, 30/3: 191–216.

Laws, G. (1993). "The Land of Old Age: Society's Changing Attitudes Toward Urban Built Environments for Elderly People." *Annals of the Association of American Geographers*, 83: 672–93.

Leiser, D., and Zilberschatz, A. (1989). "The TRAVELLER: A Computational Model of Spatial Network Learning." *Environment and Behavior*, 21/4: 435–63.

Liben, L. S. (1982). "Children's Large-Scale Spatial Cognition: Is the Measure the Message?" in R. Cohen (ed.), *New Directions for Child Development: Children's Conceptions of Spatial Relationships*. San Francisco: Jossey-Bass.

Liben, L. S., and Downs, R. M. (1997). "Can-ism and Can'tianism: A Straw Child." *Annals of the Association of American Geographers*, 87/1: 159–67.

Lloyd, R. E. (1989). "Cognitive Maps: Encoding and Decoding Information." *Annals of the American Association of Geographers*, 79/1: 101–24.

— (1993) "Cognitive Processes and Cartographic Maps," in Gärling and Golledge (1993: 141–69).

Lloyd, R. E., and Cammack, R. (1996). "Constructing Cognitive Maps with Orientation Biases," in Portugali (1996: 187–213).

Lloyd, R. E., and Heivly, C. (1987). "Systematic Distortion in Urban Cognitive Maps." *Annals of the Association of American Geographers*, 77: 191–207.

Lloyd, R. E., and Steinke, T. (1985). "Comparison of Quantitative Point Symbols: The Cognitive Process." *Cartographica*, 22/1: 59–77.

Lloyd, R. E., Cammack, R., and Holliday, W. (1995). "Learning Environments and Switching Perspectives." *Cartographica*, 32/2: 5–17.

Loomis, J. M., Golledge, R. G., and Klatzky, R. L. (1998). "Navigation System for the Blind: Auditory Display Modes and Guidance." *Presence: Teleoperators and Virtual Environments*, 7/2: 193–203.

Lowenthal, D. (ed.) (1967). "Environmental Perception and Behavior." Department of Geography Research Paper, 109. Chicago: University of Chicago.

Lynch, K. (1960). *The Image of the City*. Cambridge, Mass.: MIT Press.

McDowell, L. (1993). "Space, Place, and Gender Relations: Part 2. Identity, Difference, Feminist Geometries and Geographies." *Progress in Human Geography*, 17: 305–18.

MacEachren, A. M. (1991). "The Role of Maps in Spatial Knowledge Acquisition." *Cartographic Journal*, 28/2: 152–62.

— (1992*a*). "Application of Environmental Learning Theory to Spatial Knowledge Acquisition from Maps." *Annals of the Association of American Geographers*, 82/2: 245–74.

— (1992*b*) "Learning Spatial Information from Maps: Can Orientation-Specificity be Overcome?" *Professional Geographer*, 44/4: 431–43.

— (1995). *How Maps Work; Representation, Visualization, and Design*. New York: Guilford.

McHugh, K. E., and Mings, R. C. (1996). "The Circle of Migration: Attachment to Place in Aging." *Annals of the Association of American Geographers*, 86/3: 530–50.

McHugh, K. E., Hogan, T. D., and Happel, S. K. (1995). "Multiple Residence and Cyclical Migration: A Life Course Perspective." *Professional Geographer*, 47: 251–67.

McLafferty, S., and Preston, V. (1991). "Gender, Race and Commuting among Service Sector Workers." *The Professional Geographer*, 43: 1–15.

McNamara, T. P. (1992). "Spatial Representation." *Geoforum*, 23/2: 139–50.

Mandler, G. (1985). *Cognitive Psychology: An Essay in Cognitive Science*. Hillsdale, NJ: Lawrence Erlbaum.

Mandler, J. M. (1984). *Stories, Scripts, and Scenes: Aspects of Schema Theory*. Hillsdale, NJ: Lawrence Erlbaum.

Massey, D. (1993). "Power-Geometry and Progressive Sense of Place," in J. Bird *et al.* (eds.), *Mapping the Futures: Local Cultures*. London: Routledge, 59–69.

Matthews, M. H. (1992). *Making Sense of Place: Children's Understanding of Large-Scale Environments*. Savage, Md.: Barnes & Noble.

Monmonier, M. S., and Schnell, G. A. (1988). *Map Appreciation*. Englewood Cliffs, NJ: Prentice Hall.

Montello, D. R. (1991). "The Measurement of Cognitive Distance; Methods and Construct Validity." *Journal of Environmental Psychology*. 11/2: 101–22.

— (1993) "Scale and Multiple Psychologies of Space," in A. U. Frank and I. Campari (eds.), *Spatial Information Theory: a Theoretical Basis for GIS. Lecture Notes in Computer Science 716*. Proceedings, European Conference, COSIT '93. Marciana Marina, Elba Island, Italy, September. New York: Springer-Verlag, 312–21.

— (1998). "A New Framework for Understanding the Acquisition of Spatial Knowledge in Large-Scale Environments," in M. Egenhofer and R. G. Golledge (eds.), *Spatial and Temporal Reasoning in Geographic Information Systems*. New York: Oxford University Press, 143–54.

—— (1999). *Thinking of Scale: The Scale of Thought*. Technical Report, NCGIA Varenius Project, UCSB, 11–12.

Montello, D. R., and Golledge, R. G. (1998). *Scale and Detail in Geography*. NCGIA Varenius Project Initiative, May.

Montello, D. R., Lovelace, K. L., Golledge, R. G., and Self, C. M. (1999). "Sex-Related Differences and Similarities in Geographic and Environmental Spatial Abilities" (submitted).

Moore, G. T., and Golledge, R. G. (eds.). (1976). *Environmental Knowing: Theories, Research, and Methods*. Stroudsburg, Pa.: Dowden, Hutchinson, & Ross.

Nasar, J. L. (ed.) (1988). *Environmental Aesthetics: Theory, Research, and Applications*. New York: Cambridge University Press.

Neisser, U. (1976). *Cognition and Reality: Principles and Implications of Cognitive Psychology*. San Francisco: W. H. Freeman.

Oakes, T. (1997). "Place and the Paradox of Modernity." *Annals of the Association of American Geographers*, 87/3: 509–31.

Palm, R. I., and Hodgson, M. E. (1992). *After a California Earthquake: Attitude and Behavior Change*. Chicago: University of Chicago Press.

—— (1993). *Natural Hazards in Puerto Rico; Attitudes, Experiences, and Behavior of Homeowners*. Program on Environment and Behavior, Monograph 55, Institute of Behavioral Science, University of Colorado.

Pandit, K. (1997). "Cohort and Period Effects in U.S. Migration: How Demographic and Economic Cycles Influence the Migration Schedule." *Annals of the Association of American Geographers*, 38/3: 439–50.

Patton, M. Q. (1990). *Qualitative Evaluation and Research Methods*. San Diego: Sage.

Piaget, J., and Inhelder, B. (1967). *The Child's Conception of Space*. New York: Norton.

Pick, H. L. Jr., and Acredolo, L. P. (eds.) (1983). *Spatial Orientation: Theory, Research and Application*. New York: Plenum.

Pigram, J. J. (1993). "Human-Nature Relationships: Leisure Environmnents and Natural Settings," in Gärling and Golledge (1993: 400–26).

Plane, D. A. (1992). "Age-Composition Change and the Geographical Dynamics of Interregional Migration." *Annals of the Association of American Geographers*, 82: 64–85.

Porteous, J. D. (1976). "Home, the Territorial Core." *Geographic Review*, 66: 383–90.

—— (1977). *Environment and Behavior: Planning and Everyday Urban Life*. Reading, Mass.: Addison-Wesley.

—— (1996). *Environmental Aesthetics: Ideas, Politics, and Planning*. New York: Routledge.

Portugali, J. (ed.) (1992). "Geography, Environment and Cognition." *Geoforum*, 23/2: 107–247. (12 articles, various authors.)

—— (ed.) (1996). *The Construction of Cognitive Maps*. Dordrecht: Kluwer.

Pratt, G., and Hanson, S. (1988). "Gender, Class, and Space." *Environment and Planning D: Society & Space*, 6: 15–35.

—— (1993). "Women and Work Across the Life Course: Moving Beyond Essentialism," in Katz and Monk (1993: 27–54).

Preston, V., McLafferty, S., and Hamilton, E. (1993). "The Impact of Family Status on Black, White, and Hispanic Women's Commuting." *Urban Geography*, 14: 228–50.

Rapoport, A. (1982). *The Meaning of the Built Environment: A Nonverbal Communication Approach*. Beverly Hills, Calif.: Sage.

Relph, E. C. (1976). *Place and Placelessness. Research in Planning and Design Series*, 1. 1st edn. London: Pion.

—— (1984). *Place and Placelessness. Research in Planning and Design Series*, 1. 2nd edn. London: Pion.

Reprint Series (1996). *Gender and Spatial Behavior Bibliography*. Goleta, Calif.: Santa Barbara Geographical Press.

Richardson, A. E., Montello, D. R., and Hegarty, M. (1999). "Spatial Knowledge Acquisition from Maps, and from Navigation in Real and Virtual Environments." *Memory & Cognition* (forthcoming).

Rogennbuck, J., and Ham, S. (1986). "Use of Information and Education in Recreation Management," in *A Literature Review—the President's Commission on American Outdoors*. Washington, DC: US Government Printing Office, 59–71.

Rowe, S., and Wolch, J. (1990). "Social Networks in Time and Space: Homeless Women in Skid Row, Los Angeles." *Annals of the Association of American Geographers*, 80: 184–204.

Rowles, G. D. (1990). "Place Attachment Among the Small Town Elderly." *Journal of Rural Community Psychology*, 1: 103–20.

—— (1993). "Evolving Images of Place in Aging and 'Aging in Place'." *Generations*, 17: 65–70.

Saarinen, T. F. (1976). *Environmental Planning: Perception and Behavior*. Boston: Houghton-Mifflin.

—— (1988). "Centering of Mental Maps of the World." *National Geographic Research*, 4/1: 112–27.

—— (1998). "World Sketch Maps: Drawing Skills or Knowledge." Discussion Paper 98–7, Department of Geography and Regional Development, University of Arizona, Tucson.

Scherl, L. (1989). "Self in Wilderness: Understanding the Psychological Benefits of Individual–Wilderness Interaction through Self-Control." *Leisure Sciences*, 11: 123–35.

—— (1991). "Developing an Understanding of Recreation Experiences in the Marine Park: Implications for Management." Paper presented at the World Leisure Research Congress (July), Sydney, Australia.

Self, C. M., and Golledge, R. G. (1995). "Sex-Related Differences in Spatial Ability: What Every Geography Educator Should Know." *Journal of Geography*, 93/5: 234–43.

Self, C. M., Gopal, S., Golledge, R. G., and Fenstermaker, S. (1992). Gender-Related Differences in Spatial Abilities. *Progress in Human Geography*, 16/3: 315–42.

Spencer, C., Blades, M., and Morsley, K. (1989). *The Child in the Physical Environment*. New York: John Wiley.

Stea, D. (1997). "A Diachronic Perspective on Universal Mapping." Paper presented at the 93rd Annual Meeting of The Association of American Geographers, Ft. Worth, Texas, April 1–5.

Stern, E., and Leiser, D. (1988). "Levels of Spatial Knowledge and Urban Travel Modeling." *Geographical Analysis*, 20: 140–56.

Stevens, A., and Coupe, P. (1978). "Distortions in Judged Spatial Relations." *Cognitive Psychology*, 10: 422–37.

Stokols, D., and Altman, I. (1987). *Handbook of Environmental Psychology*. New York: Wiley, i and ii.

Strongman, K. T. (1987). *The Psychology of Emotion*. New York: Wiley.

Summers, M., and Southworth, F. (1998). "Design of a Testbed to Assess Alternative Traveler Behavior Models within an Intelligent Transportation System Architecture." *Geographic Systems*, 5/1–2: 91–116.

Thorndyke, P. W., and Hayes-Roth, B. (1982). "Differences in Spatial Knowledge Acquired from Maps and Navigation." *Cognitive Psychology*, 14: 560–89.

Tolman, E. C. (1948). "Cognitive Maps in Rats and Men." *Psychological Review*, 55: 189–208.

Tuan, Y.-F. (1971). "Geography, Phenomenology, and the Study of Human Nature." *Canadian Geographer*, 15: 181–92.

— (1974). *Topophilia: A Study of Environmental Perception, Attitudes, and Values*. Englewood Cliffs, NJ: Prentice-Hall.

— (1977). *Space and Place: the Perspective of Experience*. Minneapolis: University of Minnesota Press.

Ulrich, R. S. (1979). "Visual Landscapes and Psychological Well-Being." *Landscape Research*, 4: 17–23.

— (1983). "Aesthetic and Affective Responses to Natural Environments," in I. Altman and J. F. Wohlwill (eds.), *Behavior and the Natural Environment*. New York: Plenum, 88–125.

— (1984). "View Through a Window May Influence Recovery from Surgery." *Science*, 224: 420–1.

Ulrich, R. S., and Simons, R. (1986). "Recovery from Stress during Exposure to Everyday Outdoor Environments." *Proceedings of the 17th Annual Environmental Design Research Association (EDRA) Conference*, Atlanta.

Wakabayashi, Y. (1996). "Behavioral Studies on Environmental Perception by Japanese Geographers." *Geographical Review of Japan*, 69B/1: 83–94.

Walmsley, D. J., and Lewis, G. (1984). *Human Geography: Behavioural Approaches*. New York: Longman.

Williams, D., and Roggenbuck, J. (1989). "Measuring Place Attachment." Paper presented at the NRPA Symposium on Leisure Research, San Antonio, Texas.

World Resources (1996–7). New York: Oxford University Press, 150.

York, R. A. (1987). "Examining Verbal Responses to Physical Settings for Reflections of Environmental Schemata." Ph.D. dissertation, Geography Department, University of Nebraska, Lincoln.

CHAPTER 11

Historical Geography

Craig E. Colten, Peter J. Hugill,
Terence Young, and Karen M. Morin

Introduction

Gazing down on the field of historical geography from a lofty vantage point, the most obvious conclusion one can draw is that it is alive and well. Despite gloomy forecasts in the 1980s (Wyckoff and Hausladen 1985), the number of significant titles published in recent years and the consistency of historical geographic scholarship testifies to the vitality of this subdiscipline. Johns Hopkins, along with Texas, California, Chicago, and other university presses have released handsome and important contributions. Recently, the second and third volumes of the highly regarded *Historical Atlas of Canada* (Harris and Mathews 1987–93) have appeared; and Thomas McIlwraith and Edward Muller (2001) have revised the standard 1980s text on North American historical geography. *The Journal of Historical Geography* has a healthy backlog of manuscripts; *The Geographical Review* regularly features work from specialty group members; and *Historical Geography* has grown in size and substance. Although the number of academic job listings for historical geography may never challenge the opportunities in GIS, a sizable and energetic corps of practitioners is hard at work, whatever their individual job titles.

The decade that has elapsed since Earle *et al.*'s (1989) review of the field (see also Conzen, Rumney, and Wynn 1993) has been particularly productive for historical geographers in terms of theory and approach. Studies framed by colonialism, capitalist development, postmodernism, feminism, and environmental history are all inherently interdisciplinary and add to the complex intellectual current in which historical geography finds itself. This diversity poses a particular problem for the authors of a chapter with panoramic intent. Like a bird's-eye view of a nineteenth-century city, the most prominent structures, or themes, stand out in the foreground. Common dwellings, or the vast body of supporting literature, blend into a less distinct background pattern. Outstanding singular efforts rise like spires above the cluttered landscape. This chapter hopes to call attention to the scholarship found both along the main thoroughfares and the back streets in the bird's-eye view, while also pointing out unique contributions.

Anne Mosher's (1999) outline of several major trends in historical geography scholarship provides the framework for this chapter. She first identified a substantive thrust in the use of world-systems analysis. Second comes the concentrated study of migration to and within North America, and third is an examination of capitalist development from our historical geographic viewpoint. Human use and modification of the environment is a fourth focus, and the examination of Native Americans a fifth. This essay will compress the discussion of Native Americans in a section on landscape analysis, regional approaches, and other significant, but more singular works. It will also offer a section on feminist themes. Applied historical geography and geographic information systems complete the listing.

Macro-Scale Historical Geography

Since 1989 large-scale analysis of historical geographic problems has come a long way. The previous volume of *Geography in America* noted Immanuel Wallerstein's (1974) influence on geographers. Although Wallerstein is a historical sociologist, his analysis is essentially historico-geographic in nature, with an explicitly geographic model of a world economy broken into core, semiperiphery, and periphery that owes a great deal to theories of economic development. Wallerstein's model is also implicitly geopolitical. Since 1989, sociologists, Christopher Chase-Dunn (Chase-Dunn and Hall 1997), Michael Mann (1993), and Charles Tilly (1990), the political scientist George Modelski (1988), and the anthropologist Tom Hall (Chase-Dunn and Hall 1997) have made useful contributions to a revised view of the world-system development.

Within large-scale geography three major pathways have been trod, not always by scholars normally regarded as historical geographers in a strict sense. The first path has been a search for historical patterns in past geographies over the 500-year development of the capitalist world economy. This takes the form of the application of world-system theory and Kondratiev long-cycle theory. The second path has been a renewed concern with geopolitics, this time as a matter of the historical development of the world-system. This second path shows signs of subsuming the first. It contains both explicitly global and regional views. The third pathway has been an attack upon traditional American cultural geography for various flaws in its use of historical analysis, but one that also slips over into the question of how the capitalist world-system originated.

The most significant pathway has been a renewed search for pattern in past geographies. A renewed interest in political geography, notably in a more sophisticated geopolitics, has become a major part of this search. World-system theory has been embraced by some historical geographers, but it has also been criticized because it is a nomothetic analysis of a past that does not lend itself to prediction (Hugill 1997). Some adherents of world-system theory argue that the last 500 years represent simply an aberrant phase in world history now coming to an end (Taylor 1996). In 1991 Brian Berry, returning to his roots as a student of H. C. Darby, argued in *Long-Wave Rhythms in Economic Development and Political Behavior* that the history of the capitalist world economy shows clear regularities and that its future development is subject to prediction. The pioneering work on this was done in the 1920s by the Russian economist Nikolai Kondratiev who identified waves of development of approximately fifty years' duration. Berry, however, was working with economistic data gathered at the level of the nation-state and rejected the possibility of pushing Kondratiev waves further back than the late 1700s.

In 1988 George Modelski and William Thompson argued in *Seapower in Global Politics, 1494–1993* that two Kondratiev cycles combined to form a century-long cycle of world leadership and that, on the basis of their empirical analysis of naval power, there had been five such cycles since the late 1400s. In 1993, in *World Trade Since 1431: Geography, Technology, and Capitalism*, Peter Hugill similarly argued for such world leadership cycles; that geographers should accept the possibility of a longer, less economistic, view; and that they should pay particular attention to the technology that seems to drive Kondratiev upswings. Hugill showed how the three main eras of technics identified by Lewis Mumford could be expanded to form the basis for these five cycles by including software as well as hardware technologies. Hugill introduced a more explicitly geopolitical model in this work by accepting the navalist view of a world history first analyzed by the American Admiral Alfred Thayer Mahan in the late nineteenth century.

Peter Taylor's *The Way the Modern World Works: World Hegemony to World Impasse* (1996) draws heavily on Wallerstein's work to argue that hegemony is a condition often sought by core states in the capitalist world-system but achieved only three times since the Treaty of Westphalia, by Holland in the early 1600s, Britain in the early 1800s, and the United States in the mid-1900s. In each case a new hegemonic landscape of production and consumption was introduced. Taylor argues that the end of hegemony is at hand because of the internal contradictions of the high-consumption phase of the capitalist world economy currently dominated by the United States.

In *Geopolitics: Re-visioning World Politics*, John Agnew (1998) defines three eras of geopolitical thinking. Civilizational geopolitics was practiced by European and European origin states in the aftermath of the Treaty of Westphalia. The rest of the world was described as a field for the civilizing influence of a vitalized European culture considered to have direct links back to Rome and Greece and that was easily diffused by Europeans. Sovereignty was defined as vested first in monarchs, then in monarchs controlled by some representative assembly, and then in peoples. The French Revolution redefined sovereignty as vested in territory. The geographic boundaries of the state became intensely important. The organic theory of the state propounded by the pioneering German geopolitician Friedrich Ratzel in the late 1800s defined a new, naturalized geopolitics. This geopolitics was readily subverted into the belief that some peoples were better fitted by nature to settle and use a specific territory than others. The result was belief in a master race, environmentalism, and other extreme positions. Agnew argues that modern America's geopolitics is ideological, and as such, it demands an ideological other. With the demise of the USSR that "other" is in short supply.

Two volumes of Don Meinig's *The Shaping of America* have appeared since 1989, vol. ii, *Continental America, 1880–1867* in 1993, and vol. iii, *Transcontinental America, 1850–1915*, in 1998. In both volumes Meinig argues the historical geographer's case for a rather different view of American history than the nationalist one favored by American historians. Meinig writes at the beginning of vol. ii how the need to incorporate a subject people with an alien culture brought under American jurisdiction by the Louisiana Purchase ensured that America became an empire before it became a nation. From that point on America's imperial tendencies were never far below the surface. Imperial America expanded aggressively into formerly Spanish territory in the 1830s and 1840s, disputed British claims to the Oregon territory, and forced Mormon dissidents to remain in the Union. The idea of an American nation developed in the North before any other region. An increasingly nationalistic North expressed its imperial tendencies in the subjugation of the South in the Civil War and in renewed imperial expansion as part of the New Imperialism of the late 1800s. Both the Civil War and the New Imperialism were geopolitical endeavors. Although Meinig never makes the link explicitly, it is clear that America was behaving as an essentially organic state through much of the 1800s, expanding on the lines defined in the geopolitical writings of Friedrich Ratzel.

In 1999 Hugill argued in *Global Communications Since 1844: Geopolitics and Technology* for an explicit link between world leadership cycles, geopolitics, and technology. He noted that, early this century, an explicitly navalist and Mahanian geopolitics gave way to continental thinking, our understanding of which was driven by the writings of the British geographer, Halford Mackinder. Air power seemed at first to end Britain's insular geography and required direct British involvement in a war for the heartland of Europe. Shortly before World War II, technical change, especially in the aspect of telecommunications that led to radar, altered the nature of air power and made possible the beginnings of a return to a navalist geopolitics wherein the destruction of cities by strategic air power replaced blockade as the main weapon to force civilian populations to surrender. America's rise to hegemony was delayed by its investment in the 1910s and 1920s in less effective forms of telecommunications than Britain, notably in radio.

The third and distinct path for recent scholarship in large-scale historical geography has been a critical assessment of the cultural concept of diffusion. Jim Blaut (1993) has noticed the tendency of diffusion to become "diffusionism." He argues that all components of the world economy were roughly equal in wealth and power before the early 1400s and that diffusion is a long, ongoing process of the exchange of ideas and material technologies. Diffusionism contends that Europe alone has displayed special inventiveness and that European ideas have been regarded by European peoples as naturally superior and have been widely spread by the European colonizations. This is essentially the same model that Agnew develops as "civilizational geopolitics" but with undertones of political correctness. Blaut makes common cause with Wallerstein in arguing that Europe can only be understood in a world-system context. In particular, the appropriation of the surplus production of non-European regions was the principle cause of Europe's rapidly increasing wealth after 1400. Hugill and Foote (1994) have followed Sauer in arguing that scholars need to separate the uncommonly occurring innovations from the commonly occurring diffusion of those innovations.

Future research in macro-scale historical geography will need to be sensitive to a wide variety of issues that tend to revolve around whether historical geographers want to be seen as historians or as social scientists who prefer to work with historical data. Chase-Dunn and Hall (1997) have argued, *contra* Wallerstein, that a hierarchy of world-systems exists and that the world-system should be seen as a construction by a particular anthropological or social group rather than as an overarching economic or political unit. Such units tend to be particularistic, best understood by descriptive techniques. Geographers are divided about whether there is order and predictability to the world-system, thus whether it is properly a theory. If it can be looked at only as a sequence of unpredictable historic events is it properly a subject for social-science geography? The issue of hegemonic succession will dominate a revived geopolitics written increasingly by historical geographers and in which a search for applicable theory must be uppermost. Finally, since long cycles in the world economy seem driven by technical change, concentrations of innovation in time and space and the diffusion of those innovations are critical issues that historical geographers must address more carefully.

Migration

To say that the migration of peoples to North America produced a host of new societies in frontier and urban environments is an understatement, and the study of this migration and settlement remains a strong component of historical geography. Explanation of spatial patterns and process involves theories of cultural ecology, colonialism, and economic development.

In an important reassessment of the backwoods frontier, Terry Jordan and Matti Kaups (1989) employ the theory of cultural preadaptation to argue that the particular techniques of the Finns were particularly suited to forest settlement, and the Finns enhanced their fitness with the adoption of traits from the Delaware Indians. This pioneer syncretism largely established the successful toolkit used by subsequent backwoods settlers.

The transition from dispersed colonial settlement to an urban-based society is the subject of Robert Mitchell and Warren Hofstra's (1995) collaboration. Examining several theories, they offer an explanation for the development of rural settlement in coastal Virginia in contrast to the rise of a town-based pattern in the Shenandoah Valley. Staple theory, they suggest, explains these regional differences and long-distance trade theory accounts for the emerging pattern of commercial ties between backcountry towns and seaboard mercantile cities.

Cole Harris (1997) provides another important theoretical analysis of migration and settlement in the Americas. Considering British Columbia's "resettlement" through a series of essays, Harris seeks explanation in social theory used "suggestively" rather than deductively to catch "the opportunity for historical geographical synthesis" (Harris 1997: p. xiv). He persuasively presents the local resettlement as part of a larger colonial process and considers how this ongoing action contributed to the displacement of native peoples and shaped the relationships of immigrant groups. Additional discussions of the theoretical power of colonial geographies appeared in a special issue of *Historical Geography* (Kenny 1999).

Migration is also tied to the successful planting of society in particular and sometimes adverse environments. Robert Sauder (1989*b*) analyzes land selection in the arid Owens Valley of California which indicates a pattern comparable to more humid lands. Bradley Baltensperger (1993) considers the availability of marginal lands, technological change, economic viability of farms, and climatic variation in explaining farm enlargement on the Great Plains. National institutions, such as church, corporations, and the Crown, also influenced migration and the stability of immigrant groups according to John Lerh and Yossi Katz (1995). In a societal variant of preadaptation, Lehr and Katz argue that the long-term stability was determined in part by the degree to which immigrant and host institutions in western Canada were "congruent or dissonant."

David Ward's (1989) analysis of changing conceptions of the slum and ghetto in American cities is a vehicle for exploring public policy. He argues that "inner-city slums . . . were part of a more complex and contingent set of environmental restraints on economic advancement and assimilation" and they were a consequence of the uneven pattern of industrialization (Ward 1989: 8). Others have examined the geography of immigrant communities in cities as a function of ethnic, labor, and economic forces (Schreuder 1990; Hiebert 1991, 1993).

Capitalist Development

The study of industrial restructuring, which produced a changing distribution of workplaces, and upheavals in residential location as well, has become an increasingly popular subject for historical geographers. On the impact of capital at the scale of individual cities, the work of Richard Harris stands out. He has offered an insightful analysis of the remaking of New York in the first half of the twentieth century, paying particular attention to the relationship between work and housing (Harris 1993). As industry moved from Manhattan, it lured blue-collar workers, but left behind the lowest-income laborers who turned to the service trades. Additionally, Harris (1994 and 1996) exposes the diverse nature of suburban development in Chicago and Toronto. He traces the creation of both elite and working-class suburbs that traditional models of suburbanization inadequately consider.

Robert Lewis (1994 and 2000) argues that the formation of industrial districts at the urban periphery also included diverse forms. His analysis of Montreal's East End illustrates that development of industrial suburbs involved "specific nucleations of productive spaces that were aligned with the search for new cost structures, the development of new forms of labour power, and the transformation of urban space, all of which were fashioned by waves of industrial growth" (Lewis 1994: 154). These productive spaces allowed manufacturers to implement new productive strategies and to promote the clustering of industries of mixed sizes. William Wyckoff's (1995) examination of capital withdrawal and its impacts in Butte, Montana, places the closing bracket on the cycle of twentieth-century industrialization.

Both the presence and absence of capital have proven to be powerful forces in shaping the residential geography of the city. Anne Mosher and Deryck Holdsworth (1992) point out the significance of alley dwellings as an "organic" response to the need for residential housing in highly controlled and hierarchical settings. In rapidly industrializing locations, alley houses allowed immigrant workers to find shelter that was otherwise unavailable. Mosher (1995) also considers the means employed by manufacturers to control their workforce by shaping the landscape they lived in. By constructing a planned town, capitalists, drawing on an environmentalist logic, sought to make their workers both content and compliant.

Taking a different view of the influences on residential geography, Laura Pulido *et al.* (1996) suggest race and racial attitudes are important factors in evolving patterns of environmental inequity. They argue that restrictions on mobility and the imposition of environmental disamenities result from shifting attitudes about certain racial groups and the power relationships embodied in urban growth. Pulido (2000) challenges us to rethink our notion of "environmental racism" by adopting the notion of "white privilege" as a powerful force in shaping urban geography. David Delaney (1998) seeks explanatory power in the legal restrictions placed on racial groups in their search for housing. These approaches expand the discussion of capital's impact on industrial location, suburbanization, and ethnic patterning to incorporate social attitudes as well and the public recognition of environmental conditions.

At a grander scale of analysis, James Lemon (1996) argues that North American cities were the beneficiaries of nature's largesse, that city growth was rooted in the agricultural lands, minerals, and forest resources of their associated hinterlands. His critique holds that the limits to the dream of unchecked growth lay in the countryside and that Americans squandered their opportunity through environmental degradation and resource depletion.

The transformation of rural economies also constitutes part of the assessment of capitalist development. Anne Knowles (1997) considers the response of Welsh immigrants to American capitalism and, in doing so, examines the role of ethnic groups in establishing a capitalist economy. Don Mitchell (1996) presents a critical commentary on the power of capital to shape the agricultural landscape of California's migrant workers while ensuring the survival of capitalist agriculture.

The analytical framework offered by theories of capital development offer expanded opportunities to explore the development of industrial, agricultural, and residential landscapes. At a fundamental level, they illuminate power relationships ignored by traditional explanations.

Environmental Historical Geography

The historical geography chapter in the last *Geography in America* noted a "significant" upsurge of interest in environmental relations during the 1970s (Earle *et al.* 1989). In addition, it identified a sharp distinction between the efforts of the physical, that is scientific, geographers and the historical geographers relying on humanistic and social-scientific approaches. This methodological distinction generally continued through the 1990s but with a greater convergence between the two approaches and a tremendously increased volume of research. Scientific inquiries frequently became more sensitive to historic contingencies even as humanistic investigations became better informed by science. This Geertzian "blurring of genres" suggests geographers will play key roles in environmental studies and public policy during the first decade of the twenty-first century.

Without a doubt, the 1990s was a surging, successful decade for scholars with environmental interests. Among the many approaches to the topic, historical studies in particular have blossomed and historical geographers have published many well-regarded books, including edited volumes organized around diverse methods, themes, and regions (Colten and Skinner 1996; Dilsaver and Colten 1992; Murphy and Johnson 2000; Wyckoff and Dilsaver 1995) as well as single-topic works (Benton 1998; Colten 2000; Daniels 1999; Dilsaver and Tweed 1990; Gumprecht 1999; Palka 2000; Starrs 1998; Williams 1989; Wyckoff 1999). Historical geographers also found themselves embraced in scholarly journals published by such disciplines as archeology (Pope and Rubenstein 1999), ecology (Marston and Anderson 1991), geology (Tinkler and Parish 1998), history (Young 1993), and landscape architecture (Wood 1992).

The writings of most environmental historical geographers revolve primarily about three centers. The first and by far the largest is the relationship between people and material changes in the natural world—with studies of ecosystem change a part of this group. Most inquiries in this vein are strongly positivistic, but where the previous *Geography in America* noted only a few science-based studies, now there are many. Most of the past decade's studies have been geomorphological, with a particular interest in the impact of human activities on weathering and erosion, especially in streams (Beach 1994; Brown *et al.* 1998; Marcus, Nielson, and Cornwell 1993; Marcus and Kearney 1991; Marston and Wick 1994; Mossa and McLean 1997; Phillips 1997; Tinkler and Parish 1998). The remainder are more biogeographic (Savage 1991; Everitt 1998) or synthetic (Buckley 1993; Gade 1991; Meierding 1993; "Regional Perspectives on Twentieth Century Environmental Change" 1998). These reductive approaches were complemented by a larger, two-track group of holistic, wide-ranging studies with frequent links to social and economic change. Like environmental historians, one set of scholars explores the transformation of rural and wild environments ("The Americas Before and After 1492" 1992; Buckley 1998; Hansen, Wyckoff, and Banfield 1995; Palka 2000; Hatvany 1997; Lewis 1989; Offen 1998; Prince 1995; Sauder 1989*b*; Sluyter 1996; Starrs 1998; Williams 1989; Williams 2000), while a smaller but growing set focuses on urban settings (Boone 1996; "City and the Environment" 1997; Colten 1990; Colten 1994*a*; Colten 1998; Colten 2000; Colten and Skinner 1996; Gumprecht 1999; Lawrence 1993*a*; Lawrence 1993*b*; J. D. Wood 1991).

The second center of environmental interest emphasizes the roles of attitudes, values, and other ideas associated with material change in the natural environment. An important avenue within environmental studies, this approach is underexplored by historical geographers. Once again, the majority consider rural and wild areas (Allen 1992; Baltensperger 1992; Bertolas 1998; Bowden 1992; Frenkel 1992; Jackson 1992; Logan 1992; Lowenthal 2000; Matless 2001; Prince 1997; Shultis 1995). Only a few authors have focused on urban environments (D. Wood 1992; Young 1993, 1995).

In the third and final center, historical geographers turn their attention to the changing politics of the environment. A small yet significant batch of critical scholars focus on social conflicts surrounding the use of the environment (Fitsimmons and Gottlieb 1996; Proctor 1995) while the remainder, like many environmental historians, explore their interest in policy development. Unlike historians, however, the geographers often bring to bear a richer, more scientific understanding of the natural world. Physical geographers in particular have taken their understanding of historic, systemic changes and projected them forward as policy proposals (Marcus 1994). Historical geographers, by contrast, have explored past politics to expose mismanaged environments (Colten and Skinner 1996; Sauder 1989*a*) and the development of natural resource agencies (Benton 1998; Dearden and Berg 1993; Dilsaver 1994; Dilsaver and Colten 1992; Dilsaver and Tweed 1990; Dilsaver and Wyckoff 1999; Shrubsole 1992; Teisch 1999; Wescoat, Halvorson, and Mustafa 2000).

Landscapes and other Themes

Landscapes, always a vital part of historical geographical scholarship, have been showcased recently by two major collections. Michael Conzen (1990) and his contributors offer a national portrait that places landscape development in historical and regional contexts. Karl Raitz (1996*b*) assembled a group of authors who examine the processes underlying the creation of the National Road and the landscapes it produced. Among the individual book efforts, James Vance's (1995) treatment of railroad evolution rescues the study of this transportation network from the hands of railroad buffs and adds great insight. John Jakle and Keith Sculle (1994, 1996, 1999) have delivered a trio of works on gas stations, motels, and fast food joints. These works follow the traditional and highly productive approach of treating landscapes as assemblages of artefacts that present culturally significant historical records.

A host of postmodern views of landscape have emerged, some of which also include the more traditional perspectives (Schein 1997; Domosh 1996). The representation of particular views of past landscapes in museums, historic structures, and ritual has also drawn attention (DeLyser 1999; Johnson 1996; Osborne 1998; Peet 1996; Hoelscher 1998). These latter works present the landscape as a reflection of past discourses and/or our current and sometimes fuzzy social memory. Another vigorous discussion questions whether public space has lost or retained its social significance (Goheen 1994, 1998; Domosh 1998).

Studies with a regional focus continue to provide shining examples of historical geography at its finest. John Hudson (1994), in a work he describes as "geographical history," traces the evolution of the agricultural system that defines the corn belt. Others explore the development of idealized townscapes that represent a particular region (J. S. Wood 1997), the creation of a regional designation (Shortridge 1989), and the conflicts over land-use policies that define the ranching West (Starrs 1998). Charles Aiken updates discussions about the plantation South with his analysis of the changing form of the agricultural institution and its landscapes. In addition, Wyckoff (1999) presents the power of historical geography to illuminate locational patterns, characteristics of place, and the creation of landscapes in Colorado; and Richard Nostrand (1992) examines the evolution of the Hispanic Homeland. A special edition of the *Journal of Historical Geography* presents essays about regional environmental myths (Bowden 1992), while a volume on the mountainous West explores important themes in that region's development (Wyckoff and Dilsaver 1995).

Native Americans provide a focus for still further historical geographies, although much remains to be done in this arena. Klaus Frantz (1993 and 1999) and Malcolm Comeaux (1991) discuss the efforts to create separate spaces for American Indians, while David Wishart exposes the dislocation of Nebraska Indians (Wishart 1994). Matthew Hannah (1993) draws on the "panoptic" logic of Michel Foucault to examine the spatial conditions of the Oglala Lakota Sioux. Mapmaking and map use among native peoples were the subjects of a collection edited by Malcolm Lewis (1998).

Carville Earle's (1992) reexamination of historical questions from the perspective of geographical history is another significant contribution from the past decade. As he argues, geographical history enables a locational and ecological reinterpretation of historical problems in such fields as colonial settlement, Southern agriculture, and Southern urbanization.

Feminist Historical Geography

A decade ago Jeanne Kay noted the continued "male orientation and near-absence of material on women in North American regional historical geography, despite nearly 20 years of scholarly publications in women's history" (1991: 435; also Kay 1990). Recently, many geographers have responded to Kay's challenge. The "women's issue" of simply writing women into historical geography remains a concern to many scholars, while other demonstrate a more fundamental interest in the production of gender differences themselves, and how they work within and through economic, political, cultural, and sexual differences in the creation of past geographies.

Feminist intervention into landscape interpretation studies were influenced by Vera Norwood and Janice Monk's path-breaking collection *The Desert is No Lady* (Norwood and Monk 1987). Recent works informed by this perspective include Jeanne Kay's (1997) study of Utah Mormon pioneer women's concepts of land and nature tied to biblical metaphors. Reading historical landscapes as narratives that create inclusionary and exclusionary concepts of nation and citizenship has long been a concern of feminist geography (Monk 1992; Gulley 1993).

New ways of discussing ethnic migration, settlement patterns, and labor relations and movements are emerging. Historians provide good models for how to incorporate migratory women into early agricultural and mining-camp work (Ruiz 1998). Minority women's experiences discussed by literary critics and historians (Deutsch 1987; Anzaldúa 1987) help shed new light on American and European imperial processes. Nadine Schuurman (1998) analyzes First Nation women's mobility through various communities in British Columbia during the second half of the nineteenth century.

The spatialization and politics of identity formation are a major emphasis of North American feminist historical geography. Many studies tie the mutual constitution of gender and ethnicity with its impact on employment patterns, access to public spaces, and the practice of politics (Deutsch 1994, 1998; Estrada 1998; Cope 1998*a*). Such works have been aided by the development of different understanding of the relationship between race and gender. Audrey Kobayashi and Linda Peake (1994), for example, have attempted to "unnaturalize" the discourses of race and gender common to many geographic narratives.

Sarah Deutsch (1998) demonstrates differences between the experiences of immigrant Italian and Jewish garment workers and Irish telephone operators in early twentieth-century Boston. These laboring women encountered quite uneven access to public protest and public space based on their alliances with unions, police, elite women's organizations, and the municipal political machine. Similarly, Silvia Estrada's (1998) examination of prostitutes, factory workers, and street vendors in Tijuana, Mexico, shows that the spatial regulation of women's work in public spaces was linked to economic changes in the city since the nineteenth century.

Drawing on Doreen Massey's (1994) ideas about place as "constellations of relations," Meghan Cope (1998*a*) examines the ways that relations in and between home and work in the woolen mills in Lawrence, Massachusetts, contributed to the social construction of place. Her work illustrates how specific social relations of gender and ethnicity were (re)produced through intersecting divisions of labor and multiple axes of social division. And Kate Boyer (1998), in her study of clerical workers in late nineteenth- and early twentieth-century Montreal, argues that ideas of respectability mediated female use of public space. These workers challenged meanings of respectability by maintaining professions in the public financial sector, spaces in which all but "fallen" women were formerly "out of place."

Gendered notions of citizenship, community, and historical contextualizations of social and spatial constraints in the practice of public politics has received some attention in feminist historical geography (Cope 1998*b*; Mattingly 1998). Meghan Cope (1998*b*), for example, argues that white settler women in nineteenth-century Colorado enacted citizenship in everyday, extra-institutional ways, by building multiple reciprocal networks of home, family, and community. Such studies are implicitly informed by historian Joan Scott's (1989) argument that public, institutionalized forms of politics and government are limited in the extent to which they can reflect women's status historically. Scott argues for moving beyond a notion of politics as formal operations of government, to a definition that more broadly assesses all contests for power.

Issues of community building and citizenship parallel a large body of works that focus on the gendering of social space more generally, most especially in their challenge to the supposed public–private dichotomy. Women's "home extended outward" in urban social work has received particular attention, including Eileen McGurty's (1998) study of settlement house workers in turn-of-the-century Chicago, and their efforts at reform and neighborhood organization.

Several authors illustrate links between cultural or legal practices and the production of public space at the turn of the twentieth century, providing historical grounding for contemporary issues. These include downtown shopping areas in Eastern cities (Domosh 1996, 1998), a park in San Francisco (Schenker 1996), and Vancouver's streets after midnight (Boyer 1996). Mona Domosh (1998) argues that New York City public spaces were the scenes of slight, everyday "tactical" transgressions, such as women performing bourgeois respectability after 4 p.m. when they should have been at home.

Tourism, as a gendered, classed, racialized, and sexualized process, has taken on special significance in historical works, especially as many tourist destinations were established within the context of Euro-American colonialism or imperialism. Recent critiques have focused on the ways in which social forces positioned women as consumers of historical tourist sites or producers of cultural knowledge about them, and situated women in feminized job categories in historical places (Squire 1993, 1995; Smith 1989; Morin 1999). Sheilagh Squire (1995), for instance, documents women's contributions to regional development of tourism in the Canadian Rockies from 1885–1939, as explorers, scientists, alpinists, and genteel tourists.

In this vein scholars have also conceptualized intersections among British and American imperialisms and Victorian gender relations in women's travel narratives of North America (George-Findlay 1996; Morin 1998, 1999; Morin and Kay Guelke 1998). These works examine concepts of difference as European women negotiated encounters with local people. They also link historical geography with the insights of post-colonial critiques of subaltern subjectivity, agency, and resistance to colonialism and imperialism. Morin and Kay Guelke (1998), for instance, examine the efforts of Mormon wives to counteract their negative public images in nineteenth-century Utah, by presenting a positive view of polygamy to British women travelers.

Applied Historical Geography

Applied historical geography uses the speciality's techniques to solve practical problems and to present scholarly findings to audiences beyond the academy. This volume of work is significant, but it is often under-acknowledged since many applied contributions do not find an outlet in traditional scholarly publications. They have been most obvious in the arenas of (1) cultural resource management and preservation, (2) tourism and museum interpretation, (3) litigation support, (4) natural resources management, and (5) hazards.

During the past decade or so, the National Park Service has moved from recognizing unique architectural treasures to considering regional folk housing, vernacular landscapes, neighborhoods, and urban plans. Geographers who have held these landscape features to be their domain for decades are helping to create a rationale for geographically based preservation efforts (Datel and Dingemans 1988; Sauder and Wilkinson 1989).

Geographers have also examined the function of preservation efforts. Richard Francaviglia (1996) evaluates how preservation of main-street features evokes certain responses, for example, and Ary Lamme (1989) argues that if preservation is done without a focus or without a simplification of the history, the message becomes confusing. Geographers face the challenge of seeing many layers in the landscape and finding it hard to reduce their analysis to a period of great achievement (Jakle and Wilson 1992). Still it is in the inclusiveness of the geographic approach that can contribute to the preservation effort. The development of cultural-resource inventories is one important avenue (Carney 1984, 1991).

With state and federal programs seeking to preserve and interpret scenic byways and historic highways, there is a place for historical geographers to document the landscapes along these routes (Krimm 1990; McIlwraith 1995; Raitz 1996*a*). Other route-oriented projects included development of interpretative material for the first National Heritage Corridor, the National Illinois and Michigan Canal (Conzen and Lim 1991; Conzen and Carr 1988).

Museum exhibits also provide a means to showcase historical geographical scholarship in an accessible format. Two major traveling, waterborne exhibits that focused on inland waterways drew on geographic interpretations (Jakle 1991; Wilhelm 1991; Colten 1994*b*). The Map Division of the Library of Congress has also presented exhibits that offer a historical geographic perspective on the ethnic migration and settlement, particularly the Portuguese and German communities. Others contributed to the development of exhibits on Oklahoma folklife and urban parks. The diversity of these efforts indicates the role geographers can play in these public forums for scholarly work. In addition to actual involvement in public programs, there is an emerging literature that evaluates the "art and science" of landscape interpretation (Francaviglia 1991) and critiques the historical narrative in heritage landscapes (Johnson 1996).

Historical geographers have contributed their expertise in legal matters that begin with the interpretation of fragile old maps, but then go far beyond. Ary Lamme (1990), for example, discusses the role for geographers in preservation-related lawsuits. Based on geographer's interest in land-use regulation, he argues that expertise in "sense of place" and "site and situation" can help clarify the value of historical landscape elements. Others offer testimony and provide research on cases that consider historical water issues (Kindquist 1994, 1997) and hazardous waste management practices (Colten 1996, 1998).

Several recent works on natural-resource management illustrate the value of long-term analysis of management procedures to inform current policy. John Wright (1993), through a series of case studies traces the evolving public policies that allow communities to set aside and preserve open space. Historical discussions of national forests (Geores 1996) and national parks (Dilsaver and Tweed 1990; Dilsaver 1994; Meyer 1996) highlight the changing nature of resource management and provide guidance for those currently in charge of federal properties.

Geographic Information Systems and Historical Geography

As government agencies and private-sector businesses scramble to digitize sets of geographic data, they often forget about the past. This presents another challenge for historical geographers who must advocate the inclusion of past land uses and land covers. Illinois has created coverages of initial land sales by the General Land Office (Schroeder 1995) and historical industries depicted in Sanborn Fire Insurance Maps (Colten 1995). The historic structures and past vegetation that have been digitized for Louisiana enables the analysis of their relationship to colonial land claims (Mires 1993). In Washington, historical land-use information is now seen as vital to resource management (Vrana 1989).

Use of GIS as a cultural resource-management tool is also expanding. Many states are developing coverages of historic sites, archeological resources, and cemeteries. These coverages facilitate the standard environmental impact statements and also provide a means for analysis of landscape features.

Beyond recognition of the need to compile and maintain historical coverages, there is increasing use of GIS tools to analyze past activities on the land. One impressive project being done in conjunction with the US Geological Survey involves the reconstruction of past urbanized areas to use in the projection of future impacts on urban areas (Kirtland *et al.* 1994; Foresman *et al.* 1997). Known as the temporal urban mapping project (<http://rockyweb.cr.usgs.gov/html/growth/lulcmap.html>, last accessed 14 September 2001) it contains integrated coverages for the Washington-Baltimore and San Francisco areas (Ratcliffe and Foresman 1999).

Numerous projects now underway bode well for the incorporation of GIS analysis in historical geographic analysis. Richard Healey (1999) is analyzing the regional economic growth of the northeast US and Anne Knowles (1999) is "visualizing" the US iron frontier. Roger Miller (1995) recently has offered observations on the possibilities for linking social theory and GIS analysis.

With an emphasis on current or "real-time" data, there will always be a tendency to discard "outdated" files. Historical geographers must see to it that such layers of electronic data are archived, just as librarians and archivists do with old city directories and manuscript census records.

Conclusion

As other disciplines, such as historical sociology and environmental history, look to historical geographic scholarship they see a solidly rooted academic specialty that provides a relatively small, but insightful set of analyses and interpretations. When historical geographers interact with interdisciplinary scholars, in such fields as gender studies, they significantly add to those discussions. By lending their talents and perspectives to applied and GIS projects, historical geographers expand their contributions and increase the value of those endeavors. External recognition of historical geography's contribution may be the specialty's greatest asset. By contrast, the greatest challenge to historical geography may be acknowledging that the diverse theoretical approaches and topical interests strengthen, rather than weaken, the field. Critical self-appraisal is valuable, as is the infusion of new techniques and approaches. The past decade has shown that historical geography can embrace new methods and theoretical positions, but at the same time, long-established modes of inquiry substantially contribute to our understanding of past geographies and geographical change.

Acknowledgements

The authors are deeply grateful to James Shortridge and Joe Wood for offering comments on a previous draft of this chapter. Craig Colten is from Louisiana State University; Peter Hugill from Texas A & M University; Terence Young from California State Polytechnic University-Pomona, and Karen Morin from Bucknell University.

Portions of "Feminist Historical Geography" appeared as Karen M. Morin and Lawrence D. Berg, "Emplacing Current Trends in Feminist Historical Geography," *Gender, Place and Culture*, 6/4 (1999), 311–30. It appears here with permission.

References

Agnew, John (1998). *Geopolitics: Re-visioning World Politics.* London: Routledge.

Aiken, Charles (1998). *The Cotton Plantation South since the Civil War.* Baltimore: Johns Hopkins University Press.

Allen, J. L. (1992). "Horizons of the Sublime: The Invention of the Romantic West." *Journal of Historical Geography*, 18: 27–40.

"The Americas Before and After 1492: Current Geographical Research." (1992). *Annals of the Association of American Geographers*, Special Issue, 82/3.

Anzaldúa, Gloria (1987). *Borderlands/La Frontera: The New Mestiza.* San Francisco: Spinsters.

Baltensperger, B. H. (1992). "Plains Boomers and the Creation of the Great American Desert Myth." *Journal of Historical Geography*, 18: 59–73.

—— (1993). "Larger and Fewer Farms: Patterns and Causes of Farm Enlargement on the Central Great Plains, 1930–1978." *Journal of Historical Geography*, 19/3: 299–313.

Beach, T. (1994). "The Fate of Eroded Soil: Sediment Sinks and Sediment Budgets of Agrarian Landscapes in Southern Minnesota, 1851–1988." *Annals of the Association of American Geographers*, 84: 5–28.

Benton, Lisa M. (1998). *The Presidio: From Army Post to National Park.* Boston: Northeastern University Press.

Berry, Brian J. L. (1991). *Long-wave Rhythms in Economic Development and Political Behavior.* Baltimore: Johns Hopkins University Press.

Bertolas, Randy James (1998). "Cross-Cultural Environmental Perception of Wilderness." *Professional Geographer*, 50: 98–111.

Blaut, James M. (1993). *The Colonizer's Model of the World: Geographical Diffusionism and Eurocentric History.* New York: Guilford.

Boone, Christopher G. (1996). "Language Politics and Flood Control in Nineteenth-Century Montreal." *Environmental History*, 1: 70–85.

Bowden, M. J. (1992). "The Invention of American Tradition." *Journal of Historical Geography*, 18: 3–26.

Boyer, Kate (1996). "What's a Girl Like You Doing in a Place Like This? A Geography of Sexual Violence in Early Twentieth Century Vancouver." *Urban Geography*, 17: 286–93.

—— (1998). "Anxiety, and Changing Meanings of Public Womanhood in Early Twentieth-Century Montreal." *Gender, Place and Culture*, 5: 261–76.

Brown, Erik Thorson, Stallard, Robert F., Larsen, Matthew C., Bourlès, Didier L., Raisbeck, Grant M., and Yiou, Françoise (1998). "Determination of Predevelopment Denudation Rates of an Agricultural Watershed (Cayaguás River, Puerto Rico) Using In-situ-produced ^{10}Be in River-borne Quartz." *Earth and Planetary Science Letters*, 160: 723–8.

Buckley, Geoffrey L. (1993). "Desertification of the Camp Creek Drainage in Central Oregon." *Yearbook of the Association of Pacific Coast Geographers*, 55: 97–126.

—— (1998). "The Environmental Transformation of an Appalachian Valley, 1850–1906." *Geographical Review*, 88: 175–98.

Carney, George (1984). "The Shotgun House in Oklahoma." *Journal of Cultural Geography*, 4: 51–71.

— (1991). "Historic Resources of Oklahoma's All-Black Towns: A Preservation Profile." *Chronicles of Oklahoma*, 59/Summer: 116–33.

Chase-Dunn, Christopher, and Hall, Thomas D. (1997). *Rise and Demise: Comparing World-Systems*. Boulder: Westview.

"City and the Environment" (1997). *Historical Geography*, Special Issue, 25.

Colten, Craig E. (1990). "Historical Hazards: The Geography of Relict Industrial Wastes." *Professional Geographer*, 42: 143–56.

— (1994*a*). "Chicago's Waste Lands: Refuse and Disposal and Urban Growth." *Journal of Historical Geography*, 19: 124–42.

— (1994*b*). "Harvesting the River: Extending a Museum's Outreach," in J. D. and D. F. Britton (eds.), *History Outreach: Programs for Museums, Historical Organizations, and Academic History Departments*. New York: Krieger, 101–16.

— (1995). "Derelict Properties: An Urban Environmental Problem." *GIS and Mapnotes*, 13/Fall: 18–21.

— (1998). "Industrial Topography, Groundwater, and the Contours of Environmental Knowledge." *Geographical Review*, 88: 199–218.

— (ed.) (2000). *Transforming New Orleans and Its Environs: Centuries of Change*. Pittsburgh: University of Pittsburgh Press.

Colten, Craig E., and Skinner, Peter N. (1996). *The Road to Love Canal: Managing Industrial Waste before EPA*. Austin: University of Texas Press.

Comeaux, Malcolm L. (1991). "Creating Indian Lands: The Boundary of the Salt River Indian Community." *Journal of Historical Geography*, 17/3: 241–56.

Conzen, Michael P. (ed.) (1990). *The Making of the American Landscape*. Boston: Unwin.

Conzen, Michael P., and Carr, Kay J. (1988). *The Illinois and Michigan Canal National Heritage Corridor: A Guide to its History and Sources*. De Kalb: Northern Illinois University Press.

Conzen, Michael P., and Lim, Linda S. (1991). *Illinois Canal Country: The Early Years in Comparative Perspective*. Chicago: University of Chicago, Committee on Geographical Studies, Studies on the Illinois and Michigan Canal Corridor, 5.

Conzen, Michael P., Rumney, Thomas, and Wynn, Graeme (1993). *A Scholar's Guide to Geographical Writing on the American and Canadian Past*. Chicago: University of Chicago Department of Geography, Research Paper 235.

Cope, Meghan (1998*a*). "Home–Work Links, Labor Markets, and the Construction of Place in Lawrence, Massachusetts, 1920–1939." *Professional Geographer*, 50: 126–40.

— (1998*b*). "She Hath Done What She Could: Community, Citizenship, and Place among Women in Late Nineteenth-Century Colorado." *Historical Geography*, 26: 45–64.

Daniels, Stephen (1999). *Humphry Repton: Landscape Gardening and the Geography of Georgian England*. New York: Yale University Press.

Datel, Robin E., and Dingemans, Dennis J. (1988). "Why Places Are Preserved: Historic Districts in American and European Cities." *Urban Geography*, 9/1: 37–52.

Dearden, Philip, and Berg, Lawrence D. (1993). "Canada's National Parks: A Model of Administrative Penetration." *The Canadian Geographer*, 37: 194–211.

Delaney, David (1998). *Race, Place and the Law: 1836–1948*. Austin: University of Texas Press.

DeLyser, Dydia (1999). "Authenticity on the Ground: Engaging the Past in a California Ghost Town." *Annals of the Association of American Geographers*, 89/4: 602–32.

Deutsch, Sarah (1987). *No Separate Refuge: Culture, Class, and Gender on an Anglo-Hispanic Frontier in the American Southwest, 1880–1940*. New York: Oxford University Press.

— (1994). "Recovering the City: Women, Space, and Power in Boston, 1870–1910." *Gender and History*, 6: 202–23.

— (1998). "Commentary: Women, Difference, and the Public Terrain." *Historical Geography*, 26: 83–91.

Dilsaver, Lary M. (1994). "Preservation Choices at Muir Woods." *Geographical Review*, 84: 290–305.

Dilsaver, Lary M., and Colten, Craig E. (1992). *The American Environment: Interpretations of Past Geographies*. Lanham, Md.: Rowman & Littlefield.

Dilsaver, Lary M., and Tweed, William C. (1990). *Challenge of the Big Trees: A Resource History of Sequoia and Kings Canyon National Parks*. Three Rivers, Calif.: Sequoia Natural History Association.

Dilsaver, Lary M., and Wyckoff, William. (1999). "Agency Culture, Cumulative Causation, and Development in Glacier National Park." *Journal of Historical Geography*, 25: 75–92.

Domosh, Mona (1996). *Invented Cities: The Creation of Landscape in Nineteenth Century New York and Boston*. New Haven: Yale University Press.

— (1997). "With 'Stout Boots and a Stout Heart:' Historical Methodology and Feminist Geography," in J. P. Jones, III, H. L. Nast, and S. M. Roberts (eds.), *Thresholds in Feminist Geography: Difference, Methodology, Representation*, Lanham, Md.: Rowman & Littlefield, 225–37.

— (1998). "Gorgeous Incongruities: Polite Politics and Public Space in Nineteenth Century New York City." *Annals of the Association of American Geographers*, 88/2: 209–26.

Earle, Carville (1992). *Geographical Inquiry and American Historical Problems*. Stanford: Stanford University Press.

Earle, Carville, *et al.* (1989). "Historical Geography", in Gary L. Gaile and Court J. Willmott (eds.), *Geography in America*. Columbus: Merrill, 156–91.

Estrada, Silvia Lopez (1998). "Women, Urban Life, and City Images in Tijuana, Mexico." *Historical Geography*, 26: 5–25.

Everitt, Benjamin L. (1998). "Chronology of the Spread of Tamarisk in the Central Rio Grande." *Wetlands*, 18: 658–68.

Fitzsimmons, Margaret, and Gottlieb, Robert (1996). "Bound and Binding the Metropolitan Space: The Ambiguous Politics of Nature in Los Angeles," in A. J. Scott, E. Soja, and R. Weinstein (eds.), *The City: Los Angeles and Urban Theory at the End of the Twentieth Century*. Berkeley: University of California Press.

Foresman, T. W., Pickett, S. T. A., and Zipperer, W. C. (1997). "Methods for Spatial and Temporal Land Use and Land Cover Assessment for Urban Ecosystems and Application in the Greater Baltimore-Chesapeake Region." *Urban Ecosystems*, 1/4: 201–16.

Francaviglia, Richard V. (1991). *Hard Places: Reading the Landscape of America's Historic Mining Districts*. Iowa City: University of Iowa Press.

— (1996). *Main Street Revisited: Time, Space, and Image Building in Small-town America*. Iowa City: University of Iowa Press.

Frantz, Klaus (1993). *Indian Reservations in the United States*. Geography Research Paper, 242. Chicago: University of Chicago Press.

Frantz, Klaus (1999). *Indian Reservations in the United States: Territory, Sovereignty, and Socioeconomic Change.* Chicago: University of Chicago Press.

Frenkel, Stephen (1992). "Geography, Empire, and Environmental Determinism." *Geographical Review*, 82: 143–53.

Gade, Daniel W. (1991). "Weeds in Vermont as Tokens of Socioeconomic Change." *Geographical Review*, 81: 153–69.

Geores, Martha E. (1996). *Common Ground: The Struggle for Ownership of the Black Hills National Forest.* Lanham, Md.: Rowman & Littlefield.

George-Findlay, Brigette (1996). *The Frontiers of Women's Writing: Women's Narratives and the Rhetoric of Westward Expansion.* Tucson: University of Arizona Press.

Goheen, Peter G. (1994). "Negotiating Access to Public Space in Mid-Nineteenth Century Toronto." *Journal of Historical Geography*, 20/4: 430–49.

—— (1998). "Public Space and the Geography of the Modern City." *Progress in Human Geography*, 22/4: 479–96.

Gulley, Hugh E. (1993). "Women and the Lost Cause: Preserving a Confederate Identity in the American Deep South." *Journal of Historical Geography*, 19: 125–41.

Gumprecht, Blake (1999). *Los Angeles River: Its Life, Death, and Possible Rebirth.* Baltimore: Johns Hopkins University Press.

Hannah, Matthew G. (1993). "Space and Social Control in the Administration of the Oglala Lakoth ("Sioux"), 1871–1879." *Journal of Historical Geography*, 19/4: 412–32.

Hansen, Kathy, Wyckoff, William, and Banfield, Jeff (1995). "Shifting Forests: Historical Grazing and Forest Invasion in Southwestern Montana." *Forest and Conservation History*, 39: 66–76.

Harris, R. Cole (1997). *The Resettlement of British Columbia: Essays on Colonialism and Geographical Change.* Vancouver: UBC Press.

Harris, R. Cole (ed.), and Mathews, Geoffrey J. (cartographer) (1987–1993). *Historical Atlas of Canada, V. i–iii.* Toronto: University of Toronto Press.

Harris, Richard (1993). "Industry and Residence: The Decentralization of New York City, 1900–1940." *Journal of Historical Geography*, 19/2: 169–90.

—— (1994). "Chicago's Other Suburbs." *Geographical Review*, 84/4: 394–410.

—— (1996). *Unplanned Suburbs: Toronto's American Tragedy, 1900 to 1950.* Baltimore: Johns Hopkins University Press.

Harris, Richard, and Lewis, Robert (1998). "Constructing a Fault(y) Zone: Misrepresentations of American Cities and Suburbs, 1900–1950." *Annals of the Association of American Geographers*, 88/4: 622–39.

Hatvany, Matthew G. (1997). "Un Paysage Agraire Original: Les Aboiteaux de Kamouraska," in C. Boudreau, S. Courville, and N. Séguin (eds.), *Atlas Historique du Québec: Le Territoire.* Sainte-Foy: Les Presses de l'Université Laval, 64–5.

Healey, Richard G. (1999). "Implementing an Internet-based Historical GIS of Regional Industrial Development." Presented to the annual meeting of the Association of American Geographers, Honolulu, Hawaii.

Hiebert, Daniel (1991). "Class, Ethnicity and Residential Structure: The Social Geography of Winnepeg, 1901–1921." *Journal of Historical Geography*, 17/1: 56–86.

—— (1993). "Jewish Immigrants and the Garment Industry of Toronto, 1901–1931." *Annals of the Association of American Geography*, 83/2: 243–71.

Hoelscher, Steve (1998). *Heritage on Stage: The Invention of Ethnic Place in America's Little Switzerland.* Madison: University of Wisconsin Press.

Hudson, John C. (1994). *Making the Corn Belt: A Geographical History of Middle-Western Agriculture.* Bloomington: University of Indiana Press.

Hugill, Peter J. (1993). *World Trade Since 1431: Geography, Technology, and Capitalism.* Baltimore: Johns Hopkins University Press.

—— (1997). "World-System Theory: Where's the Theory?" *Journal of Historical Geography*, 23: 344–9.

—— (1999). *Global Communications Since 1844: Geopolitics and Technology.* Baltimore: Johns Hopkins University Press.

Hugill, Peter J., and Foote, Kenneth (1994). "Re-Reading Cultural Geography," in Kenneth Foote, Peter Hugill, Kent Mathewson, and Jonathan Smith (eds.), *Re-Reading Cultural Geography.* Austin: University of Texas Press, 9–23.

Jackson, R. H. (1992). "The Mormon Experience: The Plains as Sinai, the Great Salt Lake as the Dead Sea, and the Great Basin as Desert-cum-Promised Land." *Journal of Historical Geography*, 18: 41–58.

Jakle, John A. (1991). "The Ohio Valley Revisited: Images from Nicholas Cresswell and Reuben Gold Thwaites," in Robert Reid (ed.), *Always a River.* Bloomington: University of Indiana Press, 32–66.

Jakle, John A., and Sculle, Keith A. (1994). *The Gas Station in America.* Baltimore: Johns Hopkins University Press.

—— (1999). *Fast Food: Roadside Restaurants in the Automobile Age.* Baltimore: Johns Hopkins University Press.

Jakle, John A., and Wilson, David (1992). *Derelict Landscapes: The Wasting of America's Built Landscape.* Lanham, Md.: Rowman & Littlefield.

Jakle, John A., Sculle, Keith A., and Rogers, Jefferson (1996). *The Motel in America.* Baltimore: Johns Hopkins University Press.

Johnson, Nuala, C. (1996). "Where Geography and History Meet: Heritage Tourism and the Big House in Ireland." *Annals of the Association of American Geographers*, 86/3: 551–66.

Jordan, Terry G., and Kaups, Matti (1989). *The American Backwoods Frontier: An Ethnic and Ecological Interpretation.* Baltimore: Johns Hopkins University Press.

Kay, Jeanne (1990). "The Future of Historical Geography in the United States." *Annals of the Association of American Geographers*, 80: 618–21.

—— (1991). "Landscapes of Women and Men: Rethinking the Regional Historical Geography of the United States and Canada." *Journal of Historical Geography*, 17: 435–52.

—— (1997). "Sweet Surrender but What's the Gender: Nature and the Body in the Writings of Nineteenth-century Mormon Women," in J. P. Jones, III, H. L. Nast, and S. M. Roberts (eds.), *Thresholds in Feminist Geography: Difference, Methodology, Representation.* Lanham, Md.: Rowman & Littlefield, 361–82.

Kenny, Judith (1999). "Colonial Geographies: Accommodation and Resistance—An Introduction." *Historical Geography*, 27: 1–4.

Kindquist, Cathy (1994). "The South Park Water Transfers: Socio-Economic and Environmental Impacts on the Area of Origin," in *Symposium Proceedings, Effects of Human Induced Change on Hydrologic Systems*, Huntsville, Ala: American Water Resources Association, 343–52.

— (1997). "Political Feasibility and Local Resistance to Conjunctive Use Projects: A Colorado Test Case," in *Symposium Proceedings, Conjunctive Use of Water Resources: Aquifer Storage and Recovery*, Herndon, Va.: American Water Resources Association, 191–200.

Kirtland, D., Gaydos, L., DeCola, K. C., Acevedo, L., and Bell, C. (1994). "An Analysis of Human Induced Land Transformation in the San Francisco/Sacramento Area." *World Research Review*, 6/2: 206–17.

Knowles, Anne K. (1997). *Calvinists Incorporated: Welsh Immigrants on Ohio's Industrial Frontier*. Geography Research Paper 240. Chicago: University of Chicago.

— (1999). "Visualizing the US Iron Frontier." Presented to the annual meeting of the Association of American Geographers, Honolulu, Hawaii.

Kobayashi, Audrey, and Peake, Linda (1994). "Unnatural Discourse: 'Race' and Gender in Geography." *Gender, Place and Culture*, 1: 225–43.

Krimm, Arthur (1990). "Mapping Route 66," in Jan Jennings (ed.), *Roadside America: The Automobile in Design and Culture*. Ames: Iowa State University Press, 198–208.

Lamme, Ary J. III (1989). *America's Historic Landscapes: Community Power and the Preservation of Four National Historic Sites*. Knoxville: University of Tennessee Press.

— (1990). "Spatial Criteria in Supreme Court Decisions on Preservation." *Geographical Review*, 80/4: 343–54.

Lawrence, Henry W. (1993*a*). "The Greening of the Squares of London: Transformation of Urban Landscapes and Ideals." *Annals of the Association of American Geographers*, 83: 90–118.

— (1993*b*). "The Neoclassical Origins of Modern Urban Forests." *Forest and Conservation History*, 37: 26–36.

Lehr, John C., and Katz, Yossi (1995). "Crown, Corporation and Church: The Role of Institutions in the Stability of Pioneer Settlements in the Canadian West, 1870–1914." *Journal of Historical Geography*, 21/4: 413–29.

Lemon, James (1996). *Liberal Dreams and Nature's Limits: Great Cities of North America since 1600*. New York: Oxford University Press.

Lewis, G. Malcolm (ed.) (1998). *Cartographic Encounters: Perspectives on Native American Mapmaking and Map Use*. Chicago: University of Chicago Press.

Lewis, Michael E. (1989). "National Grasslands in the Dustbowl." *Geographical Review*, 79: 161–71.

Lewis, Robert D. (1994). "Restructuring and the Formation of an Industrial District in Montreal's East End, 1850–1914." *Journal of Historical Geography*, 20/2: 143–57.

— (2000). *The Making of an Industrial Landscape, 1850 to 1930*. Baltimore: Johns Hopkins University Press.

Logan, L. (1992). "The Geographical Imagination of Frederick Remington: The Invention of the Cowboy West." *Journal of Historical Geography*, 18: 75–90.

Lowenthal, David (2000). "Nature and Morality from George Perkins Marsh to the Millennium." *Journal of Historical Geography*, 26/1: 3–23.

McGurty, Eileen M. (1998). "Trashy Women: Gender and the Politics of Garbage in Chicago, 1890–1917." *Historical Geography*, 26: 27–43.

McIlwraith, Thomas F. (1995). "The Ontario Country Road as a Cultural Resource." *Canadian Geographer*, 39/4: 323–35.

McIlwraith, Thomas F., and Muller, Edward K. (2001). *North America: The Historical Geography of a Changing Continent*. Lanham, Md.: Rowman & Littlefield.

Mann, Michael (1993). *The Sources of Social Power*, ii. *The Rise of Classes and Nation-states, 1760–1914*. New York: Cambridge University Press.

Marcus, W. Andrew (1994). "The Lower Potomac Estuary—From Tidewater to the Chesapeake Bay: Past Trends and Future Needs." *Potomac Valley Chronicle*, 1: 36–47.

Marcus, W. Andrew, and Kearney, Michael S. (1991). "Upland and Coastal Sediment Sources in a Chesapeake Bay Estuary." *Annals of the Association of American Geographers*, 81: 408–24.

Marcus, W. Andrew, Nielsen, C. C., and Cornwell, J. C. (1993). "Sediment Budget-Based Estimates of Trace Metal Inputs to a Chesapeake Estuary." *Environmental Geology*, 22: 1–9.

Marston, Richard A., and Anderson, Jay E. (1991). "Watersheds and Vegetation of the Greater Yellowstone Ecosystem." *Conservation Biology*, 5: 338–46.

Marston, Richard A., and Wick, David A. (1994). "Natural Variability and Forestry Impacts in Mountain Streams of Southeastern Wyoming," in R. A. Marston and V. A. Hasfurther (eds.), *Effects of Human-Induced Changes on Hydrologic Systems, Proceedings of 1994 Annual Symposium, American Water Resources Association*. Jackson Hole, WY.: American Water Resources Association, 225–34.

Massey, Doreen J. (1994). *Space, Place and Gender*. Minneapolis: University of Minnesota Press.

Matless, David (2000). "Bodies Made of Grass Made of Earth Made of Bodies: Organicism, Diet, and National Health in Mid-Twentieth Century England." *Journal of Historical Geography*, 27/3: 355–76.

Mattingly, Doreen J. (1998). "Gender and the City in Historical Perspective: An Introduction." *Historical Geography*, 26: 1–4.

Meierding, Thomas C. (1993). "Marble Tombstone Weathering and Air Pollution in North America." *Annals of the Association of American Geographers*, 83: 568–88.

Meinig, Donald W. (1993). *The Shaping of America*, ii. *Continental America, 1800–1867*. New Haven: Yale University Press.

— (1998). *The Shaping of America*, iii. *Transcontinental America, 1850–1915*. New Haven: Yale University Press.

Meyer, Judith (1996). *The Spirit of Yellowstone: The Cultural Evolution of a National Park*. Lanham, Md.: Rowman & Littlefield.

Miller, Roger P. (1995). "Beyond Method, Beyond Ethics: Integrating Social Theory into GIS and GIS into Social Theory." *Cartographic and Geographic Information Systems*, 22/1: 98–103.

Mires, P. B. (1993). "Relationships of Louisiana Colonial Land Claims with Potential Natural Vegetation and Historic Standing Structures." *Professional Geographer*, 45/3: 342–50.

Mitchell, Don (1996). *The Lie of the Land*. Minneapolis: University of Minnesota Press.

Mitchell, Robert, and Hofstra, Warren (1995). "How Do Settlement Systems Evolve? The Virginia Backcountry during the Eighteenth Century." *Journal of Historical Geography*, 21/2: 123–47.

Modelski, George, and Thompson, William (1988). *Seapower in Global Politics, 1494–1993*. Seattle: University of Washington Press.

Monk, Janice (1992). "Gender in the Landscape: Expressions of Power and Meaning," in K. Anderson and F. Gale (eds.), *Inventing Places: Studies in Cultural Geography*. Melbourne: Longman, 123–38.

Morin, Karen M. (1998). "British Women Travellers and Constructions of Racial Difference across the Nineteenth-Century American West." *Transactions of the Institute of British Geographers*, 23: 311–30.

—— (1999). "Peak Practices: Englishwomen's 'Heroic' Adventures in the Nineteenth-Century American West." *Annals of the Association of American Geographers*, 89: 489–514.

Morin, Karen M., and Guelke, Jeanne Kay (1998). "Strategies of Representation, Relationship, and Resistance: British Women Travelers and Mormon Plural Wives, ca. 1870–1890." *Annals of the Association of American Geographers*, 88: 436–62.

Mosher, Anne E. (1995). "Something Better than the Best: Industrial Restructuring, George McMurtry and the Creation of the Model Industrial Town of Vandergrift, Pennsylvania, 1883–1901." *Annals of the Association of American Geographers*, 85/1: 84–107.

—— (1999). "Pre-millennial Musings on the State of Historical Geography." Paper presented to annual meeting of the Association of American Geographers, Honolulu, Hawaii.

Mosher, Anne E., and Holdsworth, Deryck W. (1992). "The Meaning of Alley Housing in Industrial Towns: Examples from Late-Nineteenth and Early-Twentieth Century Pennsylvania." *Journal of Historical Geography*, 18/2: 174–89.

Mossa, J., and McLean, M. B. (1997). "Channel Planform and Land Cover Changes on a Mined River Floodplain: Amite River, Louisiana, USA." *Applied Geography*, 17: 43–54.

Murphy, Alexander B., and Johnson, Douglas L. (eds.) (2000). *Cultural Encounters with the Environment: Enduring and Evolving Geographic Themes*. Lanham, Md.: Rowman & Littlefield.

Norwood, Vera, and Monk, Janice (eds.) (1987). *The Desert is No Lady: Southwestern Landscapes in Women's Writing and Art*. New Haven: Yale University Press.

Nostrand, Richard (1992). *The Hispano Homeland*. Norman: University of Oklahoma Press.

Offen, Karl H. (1998). "An Historical Geography of Chicle and Tunu Gum Production in Northeastern Nicaragua." *Conference of Latin Americanist Geographers*, 24: 57–74.

Osborne, Brian S. (1998). "Constructing the Landscapes of Power: The George Etienne Cartier Monument, Montreal." *Journal of Historical Geography*, 24/4: 431–58.

Palka, Eugene J. (2000). *Valued Landscapes in the Far North: A Geographical Journey through Denali National Park*. Lanham, Md.: Rowman & Littlefield.

Peet, Richard (1996). "A Sign Taken for History: Daniel Shays' Memorial in Petersham, Massachusetts." *Annals of the Association of American Geographers*, 86/1: 21–43.

Phillips, Jonathan (1997). "A Short History of a Flat Place: Three Centuries of Geomorphic Change in the Croatan." *Annals of the Association of American Geographers*, 87: 197–216.

Pope, Gregory A., and Rubenstein, Ruth (1999). "Anthroweathering: Theoretical Framework and Case Study for Human-impacted Weathering." *Geoarchaeology: An International Journal*, 14: 247–64.

Prince, Hugh C. (1995). "A Marshland Chronicle, 1830–1860: From Artificial Drainage to Outdoor Recreation in Central Wisconsin." *Journal of Historical Geography*, 21: 3–22.

—— (1997). *Wetlands of the American Midwest: A Historical Geography of Changing Attitudes*. Chicago: University of Chicago Press.

Proctor, James D. (1995). "Whose Nature? The Contested Moral Terrain of Ancient Forests," in William Cronon (ed.), *Uncommon Ground: Toward Reinventing Nature*. New York: W. W. Norton.

Pulido, Laura (2000). "Rethinking Environmental Racism: White Privilege and Urban Development in Southern California." *Annals of the Association of American Geographers*, 90/1: 12–40.

Pulido, Laura, Sidawi, Steve, and Vos, Robert O. (1996). "An Archeology of Environmental Racism in Los Angeles." *Urban Geography*, 17/2: 419–39.

Raitz, Karl (ed.) (1996*a*). *A Guide to the National Road*. Baltimore: Johns Hopkins University Press.

—— (ed.) (1996*b*). *The National Road*. Baltimore: Johns Hopkins University Press.

Ratcliffe, Michael R., and Foresman, T. W. (1999). "Historical Geography, Deep Ecology, and GIS: The View from Baltimore." Paper presented to the annual meeting of the Association of American Geographers, Honolulu, Hawaii.

"Regional Perspectives on Twentieth-Century Environmental Change" (1998). *The Canadian Geographer*, Special Issue, 42/2.

Ruiz, Vicky (1998). *From Out of the Shadows: Mexican Women in Twentieth-Century America*. New York: Oxford University Press.

Sauder, Robert A. (1989*a*). "Patenting an Arid Frontier: Use and Abuse of the Public Land Laws in Owens Valley, California." *Annals of the Association of American Geographers*, 79: 544–69.

—— (1989*b*). "Sod Land Versus Sagebrush: Early Land Appraisal and Pioneer Settlement in an Arid Intermountain Frontier." *Journal of Historical Geography*, 15: 402–19.

Sauder, Robert A., and Wilkinson, Teresa (1989). "Preservation Planning and Geographic Change in New Orleans' Vieux Carre." *Urban Geography*, 10/1: 41–61.

Savage, Melissa (1991). "Structural Dynamics of a Southwestern Pine Forest under Chronic Human Influence." *Annals of the Association of American Geographers*, 81: 271–89.

Schein, Richard H. (1997). "The Place of Landscape: A Conceptual Framework for an American Scene." *Annals of the Association of American Geographers*, 87/4: 660–80.

Schenker, Heather M. (1996). "Women's and Children's Quarters in Golden Gate Park, San Francisco." *Gender, Place and Culture*, 3: 293–308.

Schreuder, Yda (1990). "The Impact of Labor Segmentation on the Ethnic Division of Labor and the Immigrant Residential Community: Polish Leather Workers in Wilmington, Delaware, in the Early-Twentieth Century." *Journal of Historical Geography*, 16/4: 402–24.

Schroeder, Erich K. (1995). "Integrating 19th-Century Land Office Records on a Geographic Information System." *Illinois GIS and Mapnotes*, 13/Fall: 12–8.

Schuurman, Nadine (1998). "Contesting Patriarchy: Nlha7pamus and Stl'atl'imx Women and Colonialism in Nineteenth-century British Columbia." *Gender, Place and Culture*, 5: 141–58.

Scott, Joan W. (1989). "History in Crisis? The Others' Side of the Story." *American History Review*, 94: 680–1.

Shortridge, James (1989). *The Middle West: Its Meaning in American Culture*. Lawrence: University of Kansas Press.

Shrubsole, Dan (1992). "The Grand River Conservation Commission: History, Activities, and Implications for Water Management." *The Canadian Geographer*, 36: 221–36.

Shultis, John (1995) "Improving the Wilderness: Common Factors in Creating National Parks and Equivalent Reserves during the Nineteenth Century." *Forest and Conservation History*, 39: 121–9.

Sluyter, Andrew (1996). "The Ecological Origins and Consequences of Cattle Ranching in Sixteenth-Century New Spain." *Geographical Review* 86: 161–77.

Smith, Cyndi (1989). *Off the Beaten Track: Women Adventurers and Mountaineers in Western Canada*. Jasper, Ala.: Coyote Books.

Squire, Sheilagh J. (1993). "Valuing Countryside: Reflections on Beatrix Potter Tourism." *Area*, 25: 5–10.

— (1995). "In the Steps of 'Genteel Ladies': Women Tourists in the Canadian Rockies, 1885–1939." *The Canadian Geographer*, 39: 2–15.

Starrs, Paul F. (1998). *Let the Cowboy Ride: Cattle Ranching in the American West*. Baltimore: Johns Hopkins University Press.

Taylor, Peter J. (1996). *The Way the Modern World Works: World Hegemony to World Impasse*. New York: Wiley.

Teisch, Jessica (1999). "The Drowning of Big Meadows: Nature's Managers in Progressive-Era California." *Environmental History*, 4: 32–53.

Tilly, Charles (1990). *Coercion, Capital, and European States, AD 990–1990*. Oxford: Blackwell.

Tinkler, Keith J., and Parish, John (1998). "Recent Adjustments to the Long Profile of Cooksville Creek, An Urbanized Bedrock Channel in Mississauga, Ontario." *Geophysical Monograph*, 107: 167–87.

Vance, James E., Jr. (1995). *The North American Railroad: Its Origin, Evolution, and Geography*. Baltimore: Johns Hopkins University Press.

Vrana, Ric (1989). "Historical Data as an Explicit Component of Land Information Systems." *International Journal Geographical Information Systems*, 3/1: 33–49.

Wallerstein, Immanual (1974). *The Modern World-System*. New York: Academic Press.

Ward, David (1989). *Poverty, Ethnicity, and the American City, 1840–1925*. Cambridge: Cambridge University Press.

Wescoat, James, Jr., Halvorson, Sarah, and Mustafa, Daanish (2000). "Water Management in the Indus Basin of Pakistan: A Half-Century Perspective." *International Journal of Water Resources Development*, 16/3: 391–406.

Wilhelm, H. G. H. (1991). "Settlement and Selected Landscape Imprints in the Ohio Valley," in Robert Reid (ed.), *Always a River*. Bloomington: University of Indiana Press, 67–104.

Williams, Michael (1989). *Americans and Their Forests: A Historical Geography*. New York: Cambridge University Press.

— (2000). "'Dark Ages and Dark Areas': Global Deforestation in The Deep Past." *Journal of Historical Geography* 26/1: 28–46.

Wishart, David (1994). *Unspeakable Sadness: The Dispossession of the Nebraska Indians*. Lincoln: University of Nebraska Press.

Wood, Denis (1992). "*Culture Naturale*: Some Words about Gardening." *Landscape Journal*, 11: 58–65.

Wood, J. David (1991). "Moraine and Metropolis: The Oak Ridges and the Greater Toronto Area." *The International Journal of Environmental Studies*, 39: 45–53.

Wood, Joseph S. (1997). *The New England Village*. Baltimore: Johns Hopkins University Press.

Wright, John B. (1993). *Rocky Mountain Divide: Selling and Saving the West*. Austin: University of Texas Press.

Wyckoff, William (1999). *Creating Colorado: The Making of a Western American Landscape, 1860–1940*. New Haven: Yale University Press.

— (1995). "Postindustrial Butte." *Geographical Review*, 85/4: 478–96.

Wyckoff, William, and Dilsaver, Lary M. (eds.) (1995). *The Mountainous West: Explorations in Historical Geography*. Lincoln: University of Nebraska Press.

Wyckoff, William, and Hausladen, Gary (1985). "Our Discipline's Demographic Futures: Retirements, Vacancies, and Appointment Priorities." *Professional Geographer*, 37/3: 339–43.

Young, Terence. (1993). "San Francisco's Golden Gate Park and the Search for a Good Society, 1865–1880." *Forest and Conservation History*, 37: 4–13.

— (1995). "Modern Urban Parks." *Geographical Review*, 85: 544–60.

CHAPTER 12

Political Geography

Gerard Toal (Gearóid Ó Tuathail) and Fred M. Shelley

Introduction

The decade and a half since the last review article on political geography by Reynolds and Knight (1989) in *Geography In America* has been one of extraordinary geopolitical transformation and change. Not only did the Cold War come to an end with the fall of the Berlin Wall and the dissolution of the Soviet Union but the spectacular terrorist attacks of September 2001 brought the "post-Cold War peace" to an end also. In the early 1990s the threat of superpower nuclear war faded as an omnipresent nightmare in international relations. Yet new threats and dangers quickly emerged to take the place of those imagined during the Cold War. Concern grew about "rogue states," genocidal ethnonationalism, global warming, and the dangers of nuclear proliferation (Halberstam 2001; Klare 1995; Odom 1998). Fears about terrorism also grew with a series of bombings, from Paris, London, and Moscow to Oklahoma City, New York, and Atlanta. United States troops and embassies in Saudi Arabia, Tanzania, Kenya, and Yemen were the targets of terrorist attacks. But it was only after the disruption, shock, and panic of the devastating terrorist attacks of 11 September 2001 and subsequent incidents of bioterrorism that world politics was given new definition and clarity by the world's most powerful state. The new metanarrative of geopolitics is the "war against terror."

Beyond the high dramas of geopolitics, already existing trends in everyday economic and political life deepened in the last decade and a half. New social movements have forced questions concerning the politics of identity and lifestyles onto the political agenda. The globalization of financial markets, telecommunication systems, and the Internet further rearranged governing notions of "here" and "there," "inside" and "outside," "near" and "far." With global media networks broadcasting news twenty-four hours a day and the Internet spreading a world wide web, the "real" geographies of everyday life were becoming strikingly virtual as well as actual (Wark 1994; Mulgan 1997). Informationalization, and the relentless pace of techno-scientific modernity were transforming everyday life and education in the United States' colleges and universities. Celebrated by the culture of transnational corporate capitalism, these tendencies brought enormous wealth to some, further polarizing income inequalities across the planet while also introducing unprecedented vulnerabilities and uncertainties into what was becoming "global everyday life."

Political geographic research during this intense decade and a half of transformation has been triangulated between these multi-scalar geopolitical transformations, the emergence of new intellectual discourses within academia, and the legacy of political-geographic research traditions. Many of the problematics defining the late twentieth and early twenty-first century are inescapably political-geographic questions, from murderous spatial practices such as "ethnic cleansing" or hyperbolic spatial narratives about "borderlessness" and "the end of distance" to generalized concern about the changing status of key human geographic notions such as "territory," "community," "scale," "place," and "democracy." The trends first identified by Reynolds and Knight (1989) have deepened. As they discerned, political geography is increasingly defined and dominated by critical post-positivist approaches and perspectives, though traditional regionalist and positivist legacies persist.

Contemporary research on political-geographic questions is located within the mainstream of contemporary social science. It spans the study of global economic transformations, geopolitical restructuring, the politics of identity, technological change and territoriality, politics of the household and interpersonal relations, and the politics of the environment. With its inherited discourses on place and politics, technological transformations and geopolitical space, nature, and the contested politics of human–environment relations, political geographers are well positioned to contribute to the larger social science conversation about the human condition in the twenty-first century.

The intellectual domain called "political geography" is a convenient fiction around which some scholars identify themselves while others do not. Rather than perpetuating the notion of a neatly delimited disciplinary landscape of well-defined subfields (an inherent danger in a volume of this kind), we have chosen to address "untidy political geographies," plural problematics of political and politicized geographies. We do so in order to do justice to the many clusters of political and geographic research traversing and transgressing English-speaking Anglo-American geography today. Whether scholarship is called "political geography" or not is less significant than how it creatively reworks understandings of the political and the geographic. Thus this chapter represents a retrospective and prospective consideration of the global geopolitical processes, contemporary intellectual movements, and current political geographic scholarship (re)making "political geography" as a thinking space within academia at the dawn of the twenty-first century.

Political Geographic Problematics from the Fall of the Berlin Wall

World-systems theorists have often noted that history can be interpreted in terms of long and relatively stable geopolitical orders punctuated by short, rapid periods of transition. The late 1980s and early 1990s was such a transition period. The "long 1989" of popular revolts against Communism began with the unsuccessful student protest in Tianamen Square, continued with the more successful "people's revolutions" in Eastern Europe, and eventually led to the disintegration of the Soviet Union in 1991. The "long 1989" brought a close to what Hobsbawm has termed the "Short Twentieth Century," which began with the outbreak of World War I in the Balkans and the Bolshevik revolution of 1917, was punctuated by World War II and the emergence of the Cold War (Taylor 1990), and ended with renewed warfare in the Balkans and the collapse of the Soviet Union.

By Hobsbawm's logic, the twentieth century ended with the "long 1989," and we have been living in its wake ever since. This experience needs to be conceptualized not only within the historical imagination offered to us by Hobsbawm but also within a geographical imagination that stresses the particularly spatial and geopolitical dimensions of this epochal change (Taylor 1993). From 1989 onwards, political geographers actively engaged these geopolitical transformations producing accessible prescient studies on a "world in crisis" (Johnston and Taylor 1989), the significance of 1989 (Nijman 1992), the political geography of the "new world order" (Williams 1993), and "geographies of global change" (Johnston *et al.* 1995, 2002).

The fall of the Berlin Wall in November 1989 stimulated new ways of thinking about borders and boundaries. Liberated from the authoritarian Communist structures that had stultified democracy and bureaucratized repression, the peoples of the Eastern bloc were free to "return to geography" and reinvent the meanings and landscapes of a Central Europe. Western academic and research institutions responded to the challenge of democratization by sponsoring research in Central and Eastern Europe that yielded important results (Murphy 1995).

Political geographers were active in researching the new political geographies taking shape in Central Europe (O'Loughlin and van Der Wusten 1993). The collapse of the one-party Communist system in Eastern Europe and the dissolution of the Soviet Union raised three fundamental political-geographic problematics. The first concerned the search for new ideologies of legitimization in multicultural, multiethnic, and multinational states. The Soviet Union was the third multinational empire to collapse in Europe in the twentieth century. The earlier collapses of the Ottoman and Habsburg empires led to horrific crimes of genocide as power elites tried to form territorial nation-states amidst populations and peoples of diverse cultures, traditions, and identities (Cigar 1995). As earlier in the century, the battle would be between more exclusivist ethnic versions of "the nation" and more inclusivist multi-ethnic and multicultural versions of the state as a national community. Could a new civil nationalism be created on more democratic principles to replace the illegitimized and anti-democratic "civil nationalism" permitted under Communism?

A second political-geographic problematic concerned the borders of the newly independent states and their relationship to the new Russian federation. The legacy of Stalin's brutal deportations and the idiosyncratic redrawing of the map by Soviet leaders was a political geographic landscape seething with injustice and grievance, revanchism, and resurgent national romanticism. The political frontiers and status of historically distinctive regions within the Russian federation and within the newly declared independent states were also part of this problematic (Smith 1996).

The third problematic concerned the future form of the state in the former Communist lands. What type of state would replace the authoritarian command-and-control Soviet state? Would Western-style liberal democracy or something else prevail in this region? As all three problematics worked themselves out in the 1990s they gave rise to dramas that helped define the decade: pogroms and war in Azerbaijan and Armenia over the region of Nagorno Karabakh; the horror of ethnic cleansing in the former Yugoslavia as a Serbian Communist elite shifted to ethnic nationalism to relegitimate itself; violent secessionist movements in Georgia, Moldova, and Chechnya; and the emergence of illiberal democracy in Russia and corrupt "gangster states" across the regions as rising mafia elites got rich plundering natural resources while the majority of peoples saw their living standards plummet (Luke and Ó Tuathail 1998*b*; O'Lear 2001, 2002). Political geographers studying the region focused on many different aspects of these problematics: the dynamics of national self-determination in the former Soviet Union (Smith 1994, 1998), the diffusion of democracy (Bell and Staeheli 2001; O'Loughlin *et al.* 1998*b*), state formation in southeastern Europe (White 2000), the new electoral geography of the former Soviet Union (O'Loughlin *et al.* 1996), the political economy of post-Communist transition (Pickles and Smith 1998), and the meaning of the new territorial order in Eurasia (Kolossov and O'Loughlin 1998, 1999).

As the end of the Cold War was rearranging political geographies across Eurasia, economic globalization was rearranging the conditions within which political geographies operated across the developed and developing world (Short 2001). "Globalization" was one of the buzzwords of the 1990s endlessly asserted to be the defining process of the late twentieth century, a seemingly inevitable transformation from the era of nationally structured capitalism to a new era of global capitalism. The term, however, was embedded within the hegemonic neoliberal worldview that dominated elite thinking in developed regions. This loose and poorly conceptualized description was also part of this ideology's push to naturalize the transcendence of the state and the borderless world of commerce it proclaimed as inevitable (Escobar 2001; Herod *et al.* 1998).

The rhetoric of globalization marked a moment of transition beyond existing territorial organizations of capitalism, beyond the techno-territorial complexes, national bargain, capital–labor, and capital–capital relationships that defined it in the post-war period. Displacing and replacing the spaces of nationally organized capitalism were a series of supranational territorialities of capitalism, emergent networks of institutions and actors that are connected by technological systems and binding flows (Castells 1996). Most significant was the interconnected domain of "global financial space" headquartered in global cities and wired to major world markets and crucial offshore sites beyond international financial regulations (Leyshon and Thrift 1997). In addition, globalization referred to capitalism's latest spatial division of labor with its international technopoles, its front office/back office divisions, its subcontracting and flexible manufacturing global webs, its keiretsu and branch plant networks, its export processing zones, and its "just-in-time" production and distribution systems (Cox 1997; Daniels and Lever 1996). Frequently described as global, these economic and techno-territorial complexes are in actuality highly concentrated in certain locations, bypassing and ignoring large portions of the globe.

The development of new techno-territorial complexes associated with finance and manufacturing profoundly changed the conditions of geopolitical power in the late twentieth century. In broadly tilting the relationship of power between states and markets towards the latter, globalization raised a series of questions which political geographers sought to engage. What is the future of the state in a world of powerful transnational capital flows? What are the political implications of the creation of free trade areas such as NAFTA and the European Union? What are the political-geographic implications of states being at the mercy of global financial turbulence and the unregulated movement of "hot money'? What are the prospects for deliberative democracy and global governance in a world shaped by the speed of financial markets (Kofman and Youngs 1996; M. Low 1998; Merrett 1996; Murphy 1996; Warf and Purcell 2001)?

Also rearranging political geographic problematics at the opening of the twenty-first century are revolutions in the mode of information. Two developments in the 1990s were of particular significance. The first was the rise to prominence of 24-hour news networks with global telecommunication systems, which enabled them to report live from theaters of conflict and drama across the globe. Pioneered by Ted Turner's Cable News Network (CNN), this planetary coverage capacity was first made possible by INTELSAT satellites in 1981. By the mid-1990s, over a hundred 24-hours-a-day long-term television channels and many other short-term 24-hour services were operating using INTELSAT satellite systems. This ability to project real-time images of political crises involving what were nominally strangers in faraway lands not only rearranged traditional geographic notions of "proximity" and "distance", but it also held the potential to rearrange established geographies of community, responsibility, and identity (Morley and Robins 1995). Telecommunicational bonds of sympathy were established between Western viewing audiences and Chinese students in Tianamen Square, Kurdish refugees fleeing the Iraqi army after the Gulf War, and starving Somalians (Adams 1996). Pictures of massive population displacement and genocide in Rwanda forced Western powers to react to the crisis there, though their actions were too little too late. Telecommunications transformed, together with many other factors, the geopolitical significance of ostensibly marginal strategic places such as Bosnia and Kosovo as a consequence of disturbing pictures of victims of ethnic cleansing (Ó Tuathail 1999*a*).

The second transformation in the mode of information was the explosive growth of the Internet. More than a tool of information dissemination and display, the Internet quickly became a new medium for the visualization of previously marginal and/or repressed identities, a new forum for the conduct of politics, and a powerful news network with the potential of undermining the state control of information systems. The expansion of the Internet and the development of cyberspace as a new domain of geopolitics inspired and galvanized a new generation of researchers with interests in techno-political geographic questions (O'Lear 1996, 1999; Purcell and Kodras 2001; Spiegel 2000 and responses). Not only may cyberspace alter the political-geographic landscape as communities join across space to foster political change, but the problematic of cyberspace governance challenges global civil society to manage a domain of communication that transcends state boundaries and democratic state laws (Klein 2001).

A final process reconstituting political geographic problematics in the 1990s is the continuing dialectic of modernization, the deepening of techno-scientific modernity, and the contradictions this has provoked. Advances in informational systems, medicine, bioengineering, and chemistry together with expansionism in unsustainable ways of living continue to transform everyday life across the planet. The growing environmental contradictions of the West's "megamachinic" and technologically dependent systems of advanced production, consumption and pollution forced the environment onto the political agenda decades ago, but it was not until the 1990s that a concerted effort was made to address these contradictions in a systematic way at the global scale (Mumford 1964; Solecki and Shelley 1996). The Earth Summit conference in Rio in 1992 and the follow-up summits that eventually produced the Kyoto Accords were important moments in the attempt to address deepening global environmental problems, but they are likely to be perceived as failures in the years to come. Short-term national interests dominated long-term environmental governance aspirations in the United States and elsewhere. The earth's ozone layer continues to disintegrate at alarming rates while the earth's climate continues to warm. Fifteen of the hottest years on record have occurred since 1980, with the last years of the century characterized by some of the highest temperatures and most violent storms on record.

The pattern of previously latent side-effects becoming more manifestly central to the fate of modernization is not confined to the environment. Broader side-effects are evident in the emergence of new diseases such as BSE/CJD (Bovine Spongiform Encephalopathy/Creutzfeldt-Jacob Disease) and the explosion of concern with the manufactured risks, dangers, and vulnerabilities brought into being by our reliance on the science and technology that huge corporations and oligopolistic markets, with diminishing government oversight, determine are the "better things for better living." Contemporary anxieties about catastrophic terrorism are emblematic of a "risk society" that is increasingly going to have to face the dangers and vulnerabilities built into its own (mal)functioning. While not yet a significant site of political geographic theorization, problematization, and research, the challenge of negotiating and managing our civilization of deep technology will inevitably touch most political geographic research in the twenty-first century (Luke 1998).

Intellectual Trends and Political Geographic Research

Research within the international discipline of geography in the 1990s has been conditioned by its encounter with a variety of transnational intellectual currents and tendencies. Of these, four are particularly worthy of note, not because they have necessarily recast political geography but because their still unfolding effects are pluralizing the places where researchers find political-geographic problematics and liberalizing the perspectives used to study political geographies. If forced to summarize this tendency, it might be argued that political geography was and is becoming decentered in positive ways as postmodernism, Foucaultian problematics of power/knowledge, environmental discourse, and risk society studies are adapted and worked into research. Not all view these tendencies as positive. Debate on the so-called "cultural turn" clarifies what geographical knowledge is to many—objective data or resultant patterns and forms separate from questions of interpretation and meaning—and what it can become when pushed beyond its unreflexive and anti-hermeneutic assumptions and methodologies.

In *The Postmodern Condition: A Report on Knowledge*, Lyotard (1984: p. xxiv) simplified his understanding of the "postmodern" by defining it as "incredulity towards metanarratives." The "postmodern condition" for Lyotard was one where the grand myths of humanism (human emancipation and liberation) and big science (progress and freedom) were coming undone and being replaced by a proliferation of local discourses and pragmatic languages. In the debate that followed, "postmodernism" became a floating sign for a series of intellectual trends and tendencies that, depending on one's point of view, either threatened to undermine all that was coherent and rational about academic knowledge or offered the possibility of a radical academia that was open to the heterogeneity of voices and experiences that constitute humanity (Dear 2001 and responses). The specter of postmodernism haunted academic conversations and in the guise of post-structuralist theory and "an insurrection of subjugated knowledges" (Foucault 1980) opened up a range of new political-geographic problematics. These included studies of the spatial dimensions of the politics of identity, what might be termed the geopolitics of identity, which ranged from considerations of the politics of cultural formations, national identity, memorialization and heritage to investigations of the geopolitics constructing boundaries between selves and others (Dear and Flusty 2001; Keith and Pile 1993; Pile and Thrift 1995; Sibley 1995).

One vital aspect of the geopolitics of identity is the geographical politics of sexual identity (Bell and Valentine 1995). Discourses of feminism and problematizations of the body contributed to the emergence of a plethora of political geographic problematics revolving around marginality and location, with research on the margins revealing the heretofore invisible and unproblematized center (Duncan 1996; Nast and Pile 1998). Other aspects of the geopolitics of identity included post-colonial theory (McClintock *et al.* 1997) and emergent theorizations of racial identity and "whiteness" (Jackson 1998). "Postmodernism" was also a sign used to describe specific methodologies which problematized meaning such as semiotics, deconstruction, and discourse analysis.

A second intellectual movement reshaping political geographic problematics, sometimes encompassed within postmodernism or post-structuralism, is that associated with the writings of Foucault. Soja (1989) argued that Foucault's engagements with the histories of institutions of power, his effort to develop a genealogy of concepts and institutions disciplining the human body, and his concern with the strategies and tactics of power, what he casually referred to as the "geopolitics" of power, made him a "postmodern geographer." The questions Foucault asked and the way he went about answering them have reverberated across many different academic disciplines. Foucault's work historicizes disciplinary knowledge and problematizes its operation as a "technology of power" that opens up certain possibilities for human liberation while also closing off others. His work has in part inspired a wave of reflective studies of the history of geographical knowledge and its relationship to states, empires, intellectual institutions, and identity regimes (Driver 2001; Gregory 1995; Ryan 1997).

A third intellectual tendency reinventing political geographic problematics is the proliferation of discourses on the environment within contemporary social science. Once a confined domain, environmental geography has diffused into all aspects of geographical research. This growing environmentalizing of geographical imaginations at the end of the twentieth century has also ushered in a certain politicization of human–environmental relations (Harvey 1996). New clusters of knowledge are forming across the social sciences around such issues as environmental discourse and practice (Benton and Short 1999), the ecological politics of development practices (political ecology, social ecology, and anti-development; Blaikie *et al.* 1994; Escobar 1995), social movements and the environment (ecological politics/liberation ecologies; Peet and Watts 1996; Steinberg and Clark 1999), the role of the environment in conflict (environmental security; Dalby 1996, 2002), feminist theory and ecological struggles (feminist political ecology; Rocheleau *et al.* 1996), and environmental justice (Heiman, 1996; Low and Gleeson 1998). All are marked by vital intersections of politics and geography, which are now getting the attention they deserve from political geographers.

A final intellectual tendency within the academy in the last decade is the effort to engage and theorize the nature of contemporary modernity. This tendency encompasses the "new sociology" of Beck *et al.* (1994) on "risk society" and "reflexive modernization." It also encompasses the ambitious project of Castells (1996, 1997, 1998) to elaborate the subjectivities, organizational forms, and practices characterizing "the information age." A second set of literatures informing this general subject are those coming out of science studies where the work of Haraway (1997), Latour (1993), and others is forcing a reconceptualization of fundamental ontological boundaries between humans and machines, individuals and network, researcher and researched, the organic and the mechanical (Luke 1996*b*). Already making an impact on the fringes of established political geographic research, this intellectual current is likely only to strengthen in the twenty-first century.

Political Geographic Research Clusters

As suggested at the outset, contemporary political geographic research can be understood as triangulated between a context of global processes and tendencies, an intellectual environment characterized by certain discourses, and its own inherited conceptual traditions and vocabularies. The field at the end of the twentieth century is a decentered one with much contemporary research taking place at interfaces with other fields: history, culture, international relations, ecology, sociology, and science studies. Good regional studies continue to be produced (Barton 1997; Chaturvedi 1996; Graham 1998; Heffernan 1998), along with local political geographies of world cities (Agnew 1995*a*; Cybriwsky 1995; A. Jones 1998; Nijman 1999; Taylor 2000; Ward 1995) as well as suburbs (J. Wood 1997) and peripheral localities (Hanna 1996). For the sake of convenience we have identified eight research clusters that capture the variety of contemporary political geographic research on the eve of the twenty-first century (for other overviews see Agnew 1997; Waterman 1998; Agnew *et al.* 2002). Discussion is inevitably brief, but the general theme is one of the theoretical renewal and reinvention of the inherited discursive formation of political geography. We discuss the clusters in no particular order of importance.

Political Geographies of Territorial Nation States

The study of the manifest jurisdictional geographies created by states has always been a concern of political geographers. Historically, this has given rise to a research tradition devoted to the analysis of the boundaries of states. Of particular interest are legal disputes between states over territorial frontiers and resources. This research appeared in the International Boundaries Research Unit at Durham University publication *Boundary and Territory Bulletin* (terminated in 2003) and in the journal *Geopolitics and International Boundaries*. Studies include general considerations of international boundary disputes, and attempts to rearrange the political map after conflicts (Corson and Minghi 1998).

Growing in importance over the decade with the "cultural turn" has been a concern with the relationship between conceptual and material borders and boundaries. Three sets of studies in particular have led to a deepening theorization of the meaning of borders and boundaries in political geography. The first was a notable set of articles born out of a path-breaking cooperative relationship between the Palestinian Ghazi Falah and the Israeli David Newman which explored the Israeli-Palestinian peace process and the attempt by both sides to find a "good border" (Falah and Newman 1995, 1996; Newman and Falah 1995, 1997). Falah himself outlined the "de-signification" of Palestine during the 1947 war and possible territorial scenarios for Israel/Palestine (Falah 1997, 1998). Falah and Newman's work reinforced how border disputes are never a technical matter of cartography but at the very heart of constellations of power, identity, and geography that comprise states. These and other themes specific to the Arab world are pursued in *The Arab World Geographer*, a journal founded and edited by Falah while Newman has gone on to edit *Geopolitics* (Newman 1998; Newman and

Kliot 1999). The second notable study was by the Finnish geographer Anssi Paasi (1996), whose richly theorized work on the Finnish-Russian border is unlikely to be surpassed. Recent cooperative work between Newman and Paasi has sought explicitly to rethink boundary studies and political geography (Newman and Paasi 1999). The third set of studies, by Matt Sparke and colleagues, engages the renegotiation of boundaries, identities, and civil society in transnational regions (Sparke 1998*a*, 2000*a*, 2000*b*, 2002, 2003; Sparke and Lawson 2002).

Newman's ascendancy to the editorship of the journal *Geopolitics and International Boundaries* and the renaming of the journal in 1999 as simply *Geopolitics* signaled a move away from traditional boundary disputes political geography to a more theoretically informed study of the geography-identity-power problematic that underpins the myth of territorial nation states and writes global space as borders, orders, and identities. While there are no true territorial nation-states or entities where geography, collective identity, and the governing political unit perfectly correspond, this has not stopped states and certain nationalist forces from trying to rearrange real geographies to correspond to the idealized and essentialist geographies imagined by the territory-nation-state myth. Both explicit drives and implicit cultural tendencies to create such political geographies from above have provoked sometimes equally essentialist and violent political geographies from below in the form of regionally based secessionist movements (Williams 1994). Even if direct violence is largely absent from these struggles as in the case of Canada/Quebec (Kaplan 1994; Knight 1998) and Italy/Padania (Agnew 1995*b*; Giordano 2000), perceptions of the violence of controlling identity regimes—complexes of gender, class, race, religion, language, and nationality—and state structures are always present. That nationalism is a complex of many different cross-cutting and nested identities is now widely recognized and documented in a series of recent studies (Kaplan and Herb 1999; Shin 2001; Yiftachel 1999, 2000). How globalization and post-Cold War geopolitics impact nationalist movements are important research questions (Bradnock 1998; special issue of *Geopolitics* 3/3 Winter 1998). This literature affirms a growing sophistication in how political geographers research and understand the boundaries peoples, places, and power create (Radcliffe 1998). No longer are such key spatial concepts as "borders," "territories," and "scale" treated as self-evident and stable notions separate from struggles over identity, power, "history," and meaning (Cox 1998; Kearns 2001; Sparke *et al.* 2002). International Relations is turning to political geographers and explicit politic-geographic questions in recent studies of these concepts and geographies of movements and flows (*Millennium* 1999).

Cultural Political Geographies: Geographies of Power

One striking consequence of the academic trends discussed earlier has been the politicization of space and spatial studies. Nowhere has this had more dramatic effects than in "cultural geography," which has been at the center of a creative rethinking of the politics of place and the place of politics in contemporary knowledge. Politicized cultural geographies have dissolved notional borders between political geography and cultural geography, unleashing creative forms of knowledge that have problematized previously neglected subjects and objects of analysis. Dear (1997: 221) notes the significance of Keith and Pile's (1993) collection in helping redefine the intersection between political and cultural geography around notions of cultural politics, identity, difference, and the politics of spatiality. Cultural geography meets political geography as both trace their common concern with power geographies of various kinds, from the politics of community to the embodied geopolitics of identity.

Another significant volume in this general intellectual movement is that of Painter (1995), which seeks to renew political geography by infusing it with the more dynamic theoretical literatures of new cultural geography. Painter's text, a second edition of which is in development, outlines an equivalent "new political geography" that contributes to contemporary debates on state formation, liberal democracy, post-colonialism, geopolitics, and social movements. This general reinvigoration of the political/cultural conversation was consolidated by the appearance of a series of excellent collections in the mid-1990s such as Duncan (1996) and Pile and Keith (1997). More recent works have deepened this conversation (Agnew *et al.* 2002; Anderson *et al.* 2002; Jackson 2002; D. Mitchell 2000).

From the many themes introduced or given new life by creativity on the cultural-political frontier, five stand out. The first concerns the relationships of states to their citizens and how these relationships are mediated by regimes of identity and structures of privilege and place. Painter and Philo (1995) identify a certain subdisciplinary convergence around the theme of "spaces of citizenship" which involves study of the exclusionary and inclusionary practices found in the political and socio-cultural spaces of and across states. Marston (1995) traces how the shifting contours of the private/public divide condition citizenship. Mitchell (2000, 2002*a*, *b*, *c*) explores the dilemmas of identity in transnational networks and communities. Related to this is a growing literature on the relationship between citizenship, hegemonic national identity, and "race" (Bonnett 1998, 2000; Jackson 1998). Given the crucial importance of white supremacist myths in so many instances of violence in contemporary America, this research has tremendous potential to provide critical insights into identity assemblage processes in the United States (Kirby 1997; Flint 2001; Gallaher 1997, 2000). Related also is a literature addressing how racially based practices of social exclusion work spatially (Sibley 1995; MacLaughlin 1998).

A second theme considers how power operates and how it is contested in multiple ways by different actors. A new concern with "geopolicing"—the geography of police practices—has emerged, drawing upon Foucaultian themes of governmentality and population management. Herbert (1997) traces how the Los Angeles Police Department establishes territoriality and police space, an activity that is becoming more dependent upon electronic surveillance systems and databases (rendering the police merely another group of knowledge workers in risk society: M. Davis 1990; Ericson and Haggerty 1997). Blomley (1995), Delaney (1993), and Blomley *et al.* (2000) explore the legal faces of the geographies of power. Proctor and Smith (1999) explore geographies of ethics and morality. The geopolitics of migration and the construction of migrants as threats are documented by Leitner (1997) and Tesfahuney (1998). The various essays in *Geographies of Resistance* and *Entanglements of Power* provide a complex picture of resistance not simply as heroic rebellion against the state but as a varied and diverse set of practices replete with tensions and contradictions (Pile and Keith 1997; Sharp *et al.* 2000). Watts's (1997) study of Nigeria is particularly remarkable. These studies are also part of a growing literature on social movements within geography (Cresswell 1996; Miller 1999; Routledge 1993; Staeheli 1994).

The third theme concerns public space as a domain of power (Staeheli and Thompson 1997). Studies vary from analyses of public parks (Mitchell 1995) and highways (Rollins 1995) to the mythic identities written in stone in public memorials and museums (Atkinson and Cosgrove 1998; Charlesworth 1994; Johnson 1997). Webster and Leib (2001) and Alderman (2000) have examined the iconography of the American South, pointing out that the Confederate battle flag and other symbols of the South during the Civil War have different significations to groups of people within the region. A fourth theme, inspired by a sensitivity to imperialism and north–south multicultural politics, concerns hybridity and (post-)colonial identity (Jacobs 1996; Slater 1997). A fifth theme deserves separate consideration because of the challenge it represents to traditional modes of thought: sexual politics.

Sexual Politics and the Body: On the Margins that Reveal the Center

In the last two decades "feminist geography" has a significant transformative impact on many different geographical subfields. Within a self-conscious political geography, however, that impact has been marginal (J. P. Jones *et al.* 1997). As feminist geography in the 1990s deepened and broadened into a series of problematics concerned with the social construction of gender, the social organization of desire, and the politics of location and situated knowledges, it has slowly begun to have an impact on how existing political-geographic problematics are studied, while also revealing previously unacknowledged questions for study.

Masculinity and its multiple hegemonic forms has become a serious object of research (Bonnett 1996). Gay and lesbian geographers have asserted the heterogeneity of gender experiences as research has shifted beyond locating gay communities to a concern with sexual identities and the body as a site of politics (Bell and Valentine 1995). The "idea of knowledge as embodied, engendered and embedded in the material context of place and space" is one researchers have pursued across a variety of domains, identities, and locations (Duncan 1996: 1; Gibson-Graham 1996; McDowell and Sharp 1997; Pile and Thrift 1995). Previously closeted questions concerning "sexual identities" and "body politics" gained more visibility in the 1990s but are still marginalized by disciplinary institutions and attacked by "heteronormative" (i.e. heterosexually normalizing) culture (Brown 2000; Nast and Pile 1998; O'Reilly and Webster 1998; Valentine 1999).

Marginality, however, can be made powerful as a position of situated knowledge and critique. Insurgent knowledge from the margins reveals the unproblematized identities, epistemological assumptions, and power politics of the center. Historically the center of knowledge in political geography was in keeping with Western intellectual cultural norms, namely, disembodied, masculinist, and heteronormative. That the existence and power of this gendered grid of intelligibility is now revealed and in question is a tribute to the pioneering intellectual efforts of certain academics

(Brown 1997*b*; Knopp 1992; Seager 1993). Established research on growth coalition politics, citizenship, social movements, and nationalism have been supplemented and also reconceptualized by gender-problematizing research on gay neighborhood politics (Knopp 1995), sexual citizenship (Bell 1995), AIDS activism (Brown 1997*a*, *b*), and a range of studies on the sexing of "the nation," from research on the embodied public performativity of "national identity" (T. Davis 1995; Marston and Mulligan 1998), masculinity, memorials, and national myths (Johnson 1997), to the everyday securing of gendered symbolizations of "the nation" (Radcliffe and Westwood 1996).

Further studies exploring the "sexing" of geopolitics, the state, nations, and citizens remain to be pursued (Nast 1998). This avenue of research is hardly homogeneous and important tensions exist between differing strategies of research and theorization. Resisted by some for heteronormative reasons and haunted by possible essentialist standpoint politics, "marginal knowledges" are nevertheless remarkably central to even the most traditional and dominant political geographic problematics of our time. Any study of borders has to confront their overdetermined symbolic and imaginary significance. Any study of state and social movement violence must inevitably engage how violence is a means of asserting and performing certain idealized subjectivities (Dalby 1994; Jeffords 1989, 1994; Sparke 1994, 1998*b*). Any study of contemporary nationalism and genocide must confront the "male fantasies" underwriting "final solutions" on the one hand, and rape warfare on the other hand (Allen 1996). Political geographic problematics have always been embodied and sexed; it was only in the late twentieth century that this was being acknowledged.

Political Economic Geographies: Charting Global Change

Research at the intersection of political and economic geography over the last decade has been active as geographers have sought to grapple with globalization, risk society, and the creation of a new geopolitical world order. Four distinct literatures can be identified in this area. The first can be described as "geopolitical economy," and found its most noteworthy expression in Agnew and Corbridge's (1995) *Mastering Space*. This volume outlined and developed a well-argued synthesis of geopolitics and international political economy, providing strong empirical and theoretical arguments about geopolitical order, geopolitical discourse, territoriality, hegemony, and neoliberalism. The book served to introduce many outside the geographical community to the importance of a materialist geographical perspective on questions of geopolitical change and international political economy. Steinberg (2001) considers similar issues, while decentering them through his historical study of a space that typically is marginal to political-economic analysis: the world-ocean. While distinct, given its focus on post-Communist transformations, Pickles and Smith's (1998) *Theorizing Transition* shares a similar commitment to a "geopolitical economy" analysis with an emphasis on institutions, state formations, and modes of regulation (see also Lee and Wills 1997).

A somewhat distinct second literature is more eclectic in its treatment of geopolitical questions and global economic change. Represented by Demko and Wood's *Reordering the World* (1994, 1999), this literature does in part overlap with the "geopolitical economy" perspective of Corbridge and Agnew, but is generally more traditional in its focus on classic geopolitical dilemmas—boundaries, sovereignty, and the evolution of the geopolitical system, and on current policy issues such as international migration, refugees, and humanitarian crises (Cohen 1991; Glassner 1996; Hyndman 2000; Rumley *et al.* 1996; W. Wood 1994, 1996). All of these issues are of considerable significance today and are likely to become more significant as international institutions and military alliances struggle to contain the consequences of "failed states," genocidal practices, and proliferating techno-scientific risks (Ó Tuathail 1998). As Chief Geographer at the US Department of State, Wood has pioneered and championed the use of GIS as spatial database management systems in "applied political geography" challenges such as complex emergencies and war crimes investigations (W. Wood 2000; Dziedzic and Wood 1999; W. Wood and Smith 1997).

The work of Nijman and Grant constitutes a third perspective. Strongly empirical and institutional rather than cultural and discursive, both have charted various aspects of globalization and geopolitical change. Nijman (1993) offers a quantitatively based analysis of the political geography of superpower conflict. Grant (1993) studied the institutional politics and political economy of the US–Japan trade dispute. Working together, they have traced the evolution of the political geography of foreign aid, with particular emphasis on Japan and the United States (Grant and Nijman 1997, 1998; see also Grant 1995; Fielden 1998; Nijman 1995). Their recent work documents and traces the impact of transnational corporations on urbanization and development in Africa and India (Grant and Nijman 2002; also Grant and Short 2002).

Finally, the continuing development of world-systems theory represents another literature on global change. Taylor's (1996) book is a remarkable creative development of Wallerstein's initial ideas on the three hegemonies of the United Provinces, the United Kingdom, and the United States, introducing such new concepts as "ordinary modernity" and "world impasse" (also Taylor and Flint 1999). Chase-Dunn and Hall (1997) have extended the theoretical underpinnings of world-systems theory to a more global perspective, examining interactions between pre-industrial, non-European societies. Straussfogel (1997*a*), Flint and Shelley (1996), and Shelley and Flint (2000) review the development of the world-systems literature from a geographical perspective, linking this literature to both larger trends in social science and the increasingly important role of place, scale, and representation in political geography.

Critical Geopolitics: Problematizing Geopolitical Practices

Within political geography there has always been a tradition of skepticism towards orthodox geopolitics, the intellectual and political practice of interpreting the earth and global political transformations for the benefit of one's own state and its leaders. Orthodox geopolitics is problem-solving geopolitics for state strategy and foreign policy practice. It takes the existing power structures for granted and works within these to provide conceptualization and advice to foreign policy decision-makers. Critical geopolitics, by contrast, is a problematizing theoretical enterprise that places the existing structures of power and knowledge in question. A convenient name for a disparate set of literatures and tendencies, it congealed in the 1990s into a developed critique of orthodox geopolitics and the non-reflective, simplistic nostrums associated with it (Ó Tuathail and Dalby 1994; Dalby and Ó Tuathail 1996). Critical geopolitics seek to recover the complexities of global political life and expose the power relationships that characterize knowledge about geopolitics concealed by orthodox geopolitics. It deconstructs the self-interested ways in which orthodox geopolitics reads the world political map by projecting cultural and political assumptions upon it while concealing these very assumptions. Geopolitics, critical geopoliticians argue, operates with a "view from nowhere," a seeing that refuses to see itself and the power relationships that make it possible (Dalby 1990; Ó Tuathail 1996; Ó Tuathail, Dalby, and Routledge 1998).

Critical geopolitics can be divided into four different types of research: formal, practical, popular, and structural geopolitics (Ó Tuathail 1999*b*). Formal geopolitics is the Foucaultian-inspired genealogies of "geopolitical thought" and geopolitical traditions (Atkinson and Dodds 1999; O'Loughlin 1994). Broader revisionist histories of geographical knowledge have affirmed how it is all, in certain ways, geopolitical (Blunt and Rose 1994; Livingstone 1993; Gregory 1994). In tracing geographical knowledge, these histories have documented the often close relationship between geographical knowledge and such political practices as state formation, colonialism, racism, and nationalism (Bell, Butlin, and Heffernan 1995; Godlewska and Smith 1994; Kearns 2002; Hooson 1994). Formal geopolitics is merely a local variant of this general revisionist literature with studies of German geopolitics by Bassin (1996), Murphy (1997), Herb (1997) and Natter (2002) being particularly noteworthy.

Practical geopolitics is concerned with the geographical politics involved in the everyday practice of foreign policy. It addresses how common geographical understandings and perceptions enframe foreign policy conceptualization and decision-making. A good recent example of the significance of inherited geographical understandings is how the geographical notion of "the Balkans" helped condition how US foreign policy makers approached, conceptualized, and responded to the Bosnian civil war, with damaging results for the region and for European security (Todorova, 1997; Ó Tuathail 1999*a*, 2002). Much of the most creative work on practical geopolitical issues is outside political geography (Campbell 1998; Krishna 1994; Shapiro 1997) but not all (Dodds 1997).

Popular geopolitics refers to the geographical politics created and debated by the various media shaping popular culture. It addresses the social construction and perpetuation of certain collective national and transnational understandings of places and peoples beyond one's own borders, what Dijkink (1996) refers to as "national identity and geopolitical visions" (see Dodds 1998; Sharp 1998). Finally, structural geopolitics involves the study of the structural processes and tendencies that condition how all states practice foreign policy (Agnew 1998). Today, these processes include, as already noted, globalization, informationalization, and the proliferating risks unleashed by the successes of our techno-scientific civilization across the planet (see Newman 1998). Studies of the geopolitical effects of informationalization and the media are also appearing (Luke and Ó Tuathail 1997; Robins 1996; Myers *et al.* 1996). What remains to be addressed and theorized in detailed ways is the impact of informationalization on the practice of international relations. Critical studies of techno-scientific risks and proliferating weapons of mass destruction are only just beginning but hold much promise, especially when theorized through emergent literatures on critical security studies and techno-scientific risk society (Dalby 1997).

Structures and Outcomes of Governance

The geopolitical upheavals of the late 1980s and early 1990s reinvigorated a long-established research cluster on the structure and outcomes of governance. This cluster includes the implementation of democratic processes in previously non-democratic countries and the outcomes of elections and referenda. The processes of globalization, informationalization, and techno-scientific modernity have stimulated considerable rethinking of the primacy of the state as the unit of analysis in social science (Flint and Shelley 1996; Taylor 1994, 1996). In addition, the end of the Cold War created a window of opportunity to revitalize the promise of collective global governance by interstate organizations, institutions, and regimes. The United Nations Charter authorizes the Security Council to establish peacekeeping missions, but Cold War antagonism sharply curtailed peacekeeping operations. In 1992 UN Secretary-General Boutros-Boutros Ghali published *An Agenda for Peace* that outlined an ambitious vision of the UN's potential for multilateral conflict management (Weiss *et al.* 2001). Ghali's vision and the Clinton administration's brief enthusiasm for UN-driven multilateralism, however, soon floundered as right-wing forces in the US Senate blocked American payments to the world body while UN forces suffered humiliating setbacks in Cambodia, Somalia, Bosnia, and Rwanda. Still, more than two-thirds of all peacekeeping missions authorized since the UN was founded in 1945 occurred after the fall of the Berlin Wall.

The end of the Cold War also revitalized the activities of regional international organizations such as the European Union (EU) and NATO. Member states have ceded many functions to the EU and sought political stability through the expansion of NATO. At the same time, the end of the Cold War eliminated one of the important rationales underlying unification. Some Europeans became skeptical about the deepening and widening of the EU, and the expansion of NATO, once the Soviet threat disappeared. Whether further unification under the auspices of the EU will continue to occur is an open question. Anxiety over what might constitute political identity and community in the future has encouraged many ordinary Europeans to question the desirability of continued unification (Morley and Robins 1995). The geopolitical consequences of the expansion of NATO are also unclear as the organization attempts to move from being a military alliance to a "security community organized around shared values." The resurgence of xenophobic and racist politics in Austria, Hungary, the Czech Republic, and elsewhere demonstrate that Europe's struggle with racism and multiculturalism is still ongoing.

In North America, the signing of the North American Free Trade Agreement (NAFTA) generated controversy. In general, research by political geographers has borne out the prediction that NAFTA's impacts would reinforce gaps between cores and peripheries. Merrett (1996) documented that actual blue-collar job loss in Canada exceeded even the most pessimistic predictions by NAFTA's opponents during debate over ratification. This loss may have contributed to increasing nationalist sentiment in Quebec during the 1990s (Kaplan 1994). The small but media-savvy Zapatista uprising in Chiapas, Mexico, symbolically challenged the perceived imperialism of free trade.

The perceived ceding of sovereignty by states to organizations such as the UN and the EU has fueled the ambition of regional secessionist movements in Scotland (Davidson 1996) and northern Italy (Agnew 1995) to reimagine themselves within a "Europe of regions" not states. Former colonial states are often the strongest proponents of existing state boundaries, a good illustration of the cultural imposition of core upon periphery described by Straussfogel (1997*b*) and Taylor (1996). Even though the boundaries between present-day states in Africa were delineated with no reference to regional identities and cultures, the Organization of African Unity steadfastly opposed efforts to create new states, a policy that was not always successful as Eritrea secured independence in 1993. Independence and secessionist movements not only in Africa but also in Kashmir, Chiapas, Nicaragua, East Timor, and other culturally distinctive portions of former European colonies in the less-developed world have, for the most part, been unsuccessful (Bradnock 1998).

The apparent cession of sovereignty by states to international authorities and organizations has been paralleled by devolution and abdication of state authority over various services and activities. Transnational networks engaged in smuggling, drug-running, prostitution, and the transportation of illegal migrants pose daily challenges to the control of borders by states (Brunn 1998; Luke and Ó Tuathail 1998*b*). The imposition of Western cultural values on indigenous societies has also led to the creation of interesting new forms of political organization with quasi-state authority, such as the hometown associations of Nigeria. Services once considered the exclusive responsibility of the state have been delegated to lower levels of government or have been privatized (Luke 1996*a*; Murphy 1996). The increasing authority of cross-national enterprise may have reinforced a view, often articulated by leaders of transnational corporations with a vested interest in promoting globalization, that non-state institutions in the private and non-profit sectors are better able to reinvigorate a sense of community in an increasingly atomistic, individualistic civil society (Staeheli 1994; Staeheli and Thomson 1997). Political geographers have devoted much attention to understanding the causes and consequences of devolution and privatization in the United States and elsewhere. Restructuring of the role of government has been investigated with respect to a variety of services and functions, including agriculture, education (Shelley 1997), welfare reform (Cope 1997), poverty (Kodras 1997), and environmental policy. As the twenty-first century progresses, political geographers will have much to say about the effects of devolution and privatization on the state and its evolving structures of governance.

Electoral Geography and Representation

The end of the Cold War, the impacts of globalization and enhanced telecommunications, and the global diffusion of democracy have revitalized the realm of electoral geography, which for decades has been a major thrust of research in political geography. Over the past several decades, in fact, it was through electoral geography that many of the major intellectual trends of twentieth-century social science became infused into political geography, from positivism and statistical analysis to world-systems theory (Archer and Taylor 1981). During the 1990s electoral geography was criticized as excessively mechanistic and overly reliant on rational choice and economic interpretations of voter behavior, ignoring social and cultural factors that also have influenced voter decisions (Painter 1995). In large measure, electoral geographers have responded to this challenge, and today's electoral geography has successfully incorporated social and cultural perspectives that complement long-standing research traditions focused on economic considerations. Contemporary electoral geography involves much greater recognition of the role of local context in electoral outcomes (Eagles 1995) and an explicit treatment of cultural as well as economic influences on local and regional voting outcomes.

The spread of democracy associated with the end of the Cold War has meant that many formerly Communist countries are holding elections for the first time in recent memory. Political geographers have been active in identifying and interpreting electoral patterns in the former Soviet Union (O'Loughlin and Bell 1999; O'Loughlin *et al.* 1997), Slovakia (Brunn and Vlckova 1994), Hungary (Martis *et al.* 1992), Moldova (O'Loughlin *et al.* 1998*a*), Ukraine (O'Loughlin and van Der Wusten 2001), Turkey (Secor 2001), and Mexico (Frohling *et al.* 2001). Discourses of ethnicity and national identity play crucial roles in political mobilization and the geography of elections in these countries.

The end of the Cold War also altered electoral politics in the West. Gender, cultural factors, and nationalism have to varying degrees influenced electoral geography in the post-Cold War United States (Shelley and Archer 1994; Shelley *et al.* 1996; Archer *et al.* 2001) and Europe (Davidson 1996; Agnew 1995). Integration of social and cultural considerations into electoral geography has also given political geographers the opportunity to deepen understanding of the historical relationships between elections and social, economic, and political change, for example the rise of the Nazi Party in Germany (Flint 1995) and in the United States the restructuring of the Rustbelt (Shelley and Archer 1989) and the South (Shelley and Archer 1995). Election outcomes are influenced by electoral systems, as the very close and controversial United States presidential election of 2000 illustrated. In the United States, as in many countries, the geographic structure of the electoral system and its interface with the judicial system can influence the outcome of an election, even if the losing candidate gets a plurality of popular votes as Al Gore did in his loss to George W. Bush in 2000 (Webster and Leib 2002).

Direct democracy remains an important component of democratic governance in many states, as well as in regions and localities. Analysis of the geographic distribution of initiatives, referenda, and other direct democratic processes has often proven a particularly valuable source of information to political geographers, because under direct democracy voters are expressing opinions on individual policy issues. Direct democracy provides especially valuable information about cultural and identity politics, and is therefore critical to the understanding of social and cultural linkages to political processes. O'Loughlin *et al.* examined a historical sequence of national referenda in Ireland on controversial religious-oriented questions such as the legalization of divorce and abortion. The effects of cultural and economic forces on gay rights referenda have been examined in Colorado and in Oregon (O'Reilly and Webster 1998).

Although direct democracy is a valuable source of information because voters express their views on public policy issues directly, in today's complex world the large majority of public policy decisions are made by elected representatives. In most countries, representatives to local and national legislative bodies are elected from territorially defined districts. Geographers have long recognized that the process of boundary delineation can have profound effects on public policy outcomes. Since the civil rights movement of the 1960s, the United States federal government has actively intervened to ensure the rights of African-Americans and other minority groups to vote. Recognizing that racially motivated gerrymandering could render minority votes meaningless, the government has worked to ensure that minorities are fairly represented in the districting process (Grofman *et al.* 1992). Following the 1990 census, several states interpreted this responsibility as a charge to create "majority–minority" districts, in which a majority of the population were members of minority groups and which were typically expected to elect minority legislators. In order to do so, states such as North Carolina, Louisiana, and Texas delineated oddly shaped majority–minority districts. The constitutionality of districts such as these was promptly challenged in the courts, and the United States Supreme Court ruled that race alone could not justify oddly shaped districts, which had to be justified on other grounds. During litigation, the expertise of many political geographers has been tapped by the American judiciary. In analyzing such cases, however, political geographers began to grapple with the often profound and troubling implications of the legal questions (Leib 1998). How do ethnic divisions in American society influence representation? Can single-member districts adequately ensure African-Americans and other minority groups reasonable access to political decision-making? Or should some alternative method of representation replace the American tradition of territorially based representation (Guinier 1994)?

Techno-Political Geographies, Development, and the Environment in Risk Society

One research constellation that is likely to flourish in the twenty-first century is the study of the techno-political geographies thrown into relief by the deepening and sometimes catastrophic (mal)functioning of reflexive modernization (Beck 1997, 1998). Research on techno-political systems and risks already have a long and under-appreciated history in political geography, in the work of Brunn (1999) on telecommunications and futurism, Flynn *et al.* (1995) on nuclear waste disposal hazards, Morrill (1999) on land-use conflicts, and Seager (1993) on environmental politics. This literature is being augmented by a new generation of scholars interested in socio-technical networks such as the Internet and virtual reality (Crang 1999) and in the political geographies generated by crises in the managing of "nature," "resources," and the manufactured risks our civilization has chosen to live with (Solecki 1996). Williams (1999), N. Low and Gleeson (1998), and the studies in Peet and Watts (1996) reveal complex political-geographic problematics involving "development," "the environment," and social movements that require more explicit theorization by political geographers. Exemplary of the ongoing theoretical (re)invention required to address twenty-first century problematics is the critical literature now being produced on water and marine political geography (Dow 1999; Steinberg 2001). Future research will emerge out of the creative imbrication of literatures in, for example, political ecology, post-structuralist feminism, critical geopolitics, and science studies. Such creative syntheses will be required to understand the complex techno-political geographies that will set the parameters for life and political struggles in the twenty-first century.

Conclusion: Contradictions of the Post September 11 Age

Amidst all the everyday structural violence of world politics and the wars of the 1990s, it is undoubtedly ethnocentric to proclaim the end of the "post-Cold War peace" and a new era in world politics in the wake of the September 11, 2001 terrorist attacks against the World Trade Center in New York and the Pentagon in Arlington, Virginia. Yet, these attacks were unprecedented global events that struck at symbolic heartlands in the affluent world, at buildings representing transnational corporate capitalism and the military might of the most powerful state in the world. The spectacle of destruction, death, and suffering the terrorist attacks left in their wake were projected to the world and became the justification for a "new war" against "terrorists with global reach" (i.e. with an ability to target the US "homeland"). The United States government quickly built an international coalition to overthrow the Taliban regime in Afghanistan as the first phase of a global military campaign against terrorism named "Enduring Freedom." In boosting defense spending back to Cold War levels, making common cause with Russia against "international terrorism," establishing military bases in Central Asia, sending military advisers to fight terrorists across the globe (the Philippines, Georgia, and Yemen), and providing new infusions of aid to the one-time pariah state of Pakistan (while maintaining levels to Israel and Egypt), the United States set down the revised geopolitical parameters of the post-September 11 world. In identifying global terrorism and an "axis of evil" states (Iran, Iraq, and North Korea) bent on acquiring weapons of mass destruction as the new enemies of humanity, the US state returned to its familiar role as a power with a world historic mission as leader of the "civilized world" against evil otherness. Manichean geopolitics was back—"if you are not with us, you are against us"—and so was the patriotic pleasure of a heroic United States, "still standing tall."

The geopolitics of the post-September 11 era, however, has many paradoxes and contradictions. First, the vulnerability of advanced technoscientific systems—financial markets, nuclear power plants, modern transportation systems, power grids, contemporary metropolitan regions—to "asymmetrical threats" (Pentagon-speak for threats posed by small non-state terrorist actors) reveal a powerful weakness within the ostensible strength and military might of techno-scientific risk society. Second, an obsession with absolute invulnerability—the logic of military securitizations—sits uneasily with the dynamics of capitalist globalization where conditions of borderlessness, deregulation, and insecurity are the norm. Third, the United States, a state built around faith in techno-scientific progress, has declared war against "technological progress" in the form of the diffusion of weapons of mass destruction to an "axis of evil" states. Missing from the new geopolitics of the post-September 11 world is self-examining reflection on the threats that do not fit the comfortable theological geopolitics of "evil otherness," threats produced by our own techno-scientific risk society as a matter of its routine functioning (from the weaponized anthrax produced by US military labs to the nuclear waste lying around in "temporary" storage facilities throughout the country). Ultimately, there is no single state or actor that is able to control the unfolding dynamics and dramas of the new geopolitical age. All of humanity is hurtling along an uncertain trajectory towards an unknowable future. Politico-geographic questions will inevitably be part of our unsure and insecure future. We can only hope they are less bellicose and bloody than those that scarred the twentieth century.

Acknowledgements

This chapter is dedicated to the memory of Graham Smith, Department of Geography at the University of Cambridge, who died at the age of 46 in 1999. He was the foremost political geographer of the peoples of the former Soviet Union. We wish to acknowledge the helpful comments of John O'Loughlin and Phil Steinberg on earlier drafts of this chapter, and to thank the many members of the Political Geography Specialty Group who provided us with suggestions and citations.

References

Adams, P. (1996). "Protest and the Scale Politics of Telecommunications." *Political Geography*, 15: 419–41.

Agnew, J. (1995*a*). *Rome*. Chichester, NY: Wiley.

— (1995*b*). "The Rhetoric of Regionalism: The Northern League in Italian Politics, 1983–94." *Transactions, Institute of British Geographers*, NS 20: 156–72.

— (1997). *Political Geography: A Reader*. London: Arnold.

— (1998). *Geopolitics: Re-Visioning World Politics*. London: Routledge.

Agnew, J., and Corbridge, S. (1995). *Mastering Space: Hegemony, Territory and International Political Economy*. London: Routledge.

Agnew, J., Mitchell, K., and Toal, G. (2002). *A Companion to Political Geography*. Oxford: Blackwell.

Allen, B. (1996). *Rape Warfare: The Hidden Genocide in Bosnia-Herzegovina and Croatia*. Minneapolis: University of Minnesota.

Alderman, D. H. (2000). A Street Fit for a King: Naming Places and Commemoration in the American South. *Professional Geographer*, 52/4: 672–84.

Anderson, K., Domosh, M., Pile, S., and Thrift, N. (2002). *Handbook of Cultural Geography*. London: Sage.

Archer, J. C., and Taylor, P. J. (1981). *Section and Party*. Chichester, NY: Wiley.

Archer, J. C., Lavin, S. J., Martis, K. C., and Shelley, F. M. (2001). *Atlas of American Politics, 1960–2000*. Washington, DC: CQ Press.

Atkinson, D., and Cosgrove, D. (1998). "Urban Rhetoric and Embodied Identities: City, Nation, and Empire at the Vittorio Emanuele II Monument in Rome, 1870–1945." *Annals of the Association of American Geographers*, 88: 28–49.

Barton, J. (1997). *A Political Geography of Latin America*. London: Routledge.

Bassin, M. (1996). Nature, Geopolitics and Marxism: Ecological Contestations in Weimar Germany. *Transactions, Institute of British Geographers*, NS 21: 315–41.

Beck, U. (1997). *The Reinvention of Politics*. Cambridge: Polity.

— (1998). *Democracy Without Enemies*. Oxford: Blackwell.

Beck, U., Giddens, A., and Lash, S. (1994). *Reflexive Modernization*. Stanford: Stanford University Press.

Bell, D. (1995). "Pleasure and Danger: The Paradoxical Spaces of Sexual Citizenship." *Political Geography*, 14/2: 139–53.

Bell, D., and Valentine, G. (1995). *Mapping Desire: Geographies of Sexualities*. London: Routledge.

Bell, J., and Staeheli, L. (2001). "Discourses of Diffusion and Democratization." *Political Geography*, 20: 175–95.

Bell, M., Butlin, R., and Heffernan, M. (eds.) (1995). *Geography and Imperialism, 1820–1940*. Manchester: Manchester University Press.

Benton, L., and Short, J. (1999). *Environmental Discourse and Practice*. Oxford: Blackwell.

Blaikie, P., Cannon, T., Davis, I., and Wisner, B. (1994). *At Risk: Natural Hazards, People's Vulnerability, and Disasters*. London: Routledge.

Blomley, N. (1995). *Law, Space and the Geographies of Power*. New York: Guilford.

Blomley, N., Delaney, D., and Ford, R. T. (eds.) (2000). *The Legal Geographies Reader*. Oxford: Blackwell.

Blunt, A., and Rose, G. (eds.) (1994). *Writing Women and Space*. New York: Guilford.

Bonnett, A. (1996). "The New Primitives: Identity, Landscape and Cultural Appropriation in the Mythopoetic Men's Movement." *Antipode*, 28: 273–291.

— (1998). "Who was White? The Disappearance of Non-European White Identities and the Formation of European Racial Whiteness." *Ethnic and Racial Studies*, 21: 1029–55.

— (2000). *Anti-Racism*. London: Taylor & Francis.

Bradnock, R. (1998). Regional Geopolitics in a Globalizing World: Kashmir in Geopolitical Perspective. *Geopolitics*, 3/2: 1–29.

Brown, M. (1997*a*). "Radical Politics out of Place? The Curious Case of ACT UP Vancouver," in Pile and Keith (1997: 152–67).

— (1997*b*). *Replacing Citizenship: AIDS Activism and Radical Democracy*. New York: Guilford.

— (2000). *Closet Space: Geographies of Metaphor from the Body to the Globe*. New York: Routledge.

Brunn, S. D., and Vlckova, V. (1994). "Parties, Candidates and Competitive Regions in the 1992 Slovak National Council Elections," *Geograficky Casopis*, 46: 231–46.

— (1998). "When Walls Come Down and Borders Open: New Geopolitical Worlds at the Grassroots in Eastern Europe, in F. M. Davidson, J. I. Leib, F. M. Shelley, and G. R. Webster (eds.), *Teaching Political Geography*. Washington, DC: National Council for Geographic Education, 31–7.

— (1999). "A Treaty of Silicon for the Treaty of Westphalia? New Territorial Dimensions of Modern Statehood." *Geopolitics*, 3/1, 106–32.

Campbell, D. (1998). *National Deconstruction: Violence, Identity and Justice in Bosnia*. Minneapolis: University of Minnesota.

Castells, M. (1996). *The Rise of the Network Society*. Oxford: Blackwell.

— (1997). *The Power of Identity*. Oxford: Blackwell.

— (1998). *End of Millennium*. Oxford: Blackwell.

Charlesworth, A. (1994). "Contesting Places of Memory: The Case of Auschwitz." *Society and Space*, 12: 579–94.

Chase-Dunn, C., and Hall, T. (1997). *Rise and Demise: Comparing World Systems*. Boulder: Westview.

Chaturvedi, S. (1996). *The Polar Regions: A Political Geography*. Chichester, NY: Wiley.

Cigar, N. (1995). *Genocide in Bosnia: The Policy of "Ethnic Cleansing."* College Station, Tex.: A & M Press.

Cohen, S. (1991). "Global Geopolitical Change in the Post-Cold War Era." *Annals of the Association of American Geographers*, 81: 551–80.

Cope, M. (1997), "Responsibility, Regulation, and Retrenchment: The End of Welfare?" in L. A. Staeheli, J. E. Kodras, and C. Flint (eds.), *State Devolution in America: Implications for a Diverse Society*. Thousand Oaks, Calif.: Sage, 181–205.

Corson, M., and Minghi, J. (1998). "Prediction and Reality: Two Years after Dayton." *Geopolitics and International Boundaries*, 2: 14–27.

Cox, K. (ed.) (1997). *Space of Globalization: Reasserting the Power of the Local*. New York: Guilford.

— (1998). "Spaces of Dependence, Spaces of Engagement and the Politics of Scale, or, Looking for Local Politics." *Political Geography*, 17: 1–24.

Crang, P. (ed.) (1999). *Virtual Geographies*. London: Routledge.

Cresswell, T. (1996). *In Place, Out of Place: Geography, Ideology and Transgression*. Minneapolis: University of Minnesota.

Cybriwsky, R. (1995). *Tokyo*. Chichester, NY: Wiley.

Dalby, S. (1990). *Creating the Second Cold War: The Discourse of Politics*. London: Pinter.

— (1994). "Gender and Critical Geopolitics: Reading Geopolitical Discourse in the New World Disorder." *Society and Space*, 12: 595–612.

— (1996). "The Environment as Geopolitical Threat: Reading Robert Kaplan's 'Coming Anarchy'." *Ecumune*, 3: 472–96.

— (1997). "Contesting an Essential Concept: Reading the Dilemmas in Contemporary Security Discourse," in K. Krause and M. Williams (eds.) *Critical Security Studies*. Minneapolis: University of Minnesota Press, 3–31.

— (2002). *Environmental Security*. Minneapolis: University of Minnesota.

Dalby, S., and Ó Tuathail, G. (1996). "Editorial Introduction: The Critical Geopolitics Constellation: Problematizing Fusions of Geographical Knowledge and Power." *Political Geography*, 15: 451–6.

Daniels, P. W., and Lever, W. F. (eds.) (1996). *The Global Economy in Transition*. Harlow: Longman.

Davidson, F. M. (1996). "The Fall and Rise of the SNP since 1983: Analysis of a Regional Party." *Scottish Geographical Magazine*, 112: 11–19.

Davis, M. (1990). *City of Quartz*. London: Verso.

Davis, T. (1995). "The Diversity of Queer Politics and the Redefinition of Sexual Identity and Community in Urban Spaces," in Bell and Valentine (1995: 284–303). London: Routledge.

Dear, M. (1997). "Identity, Authenticity and Memory in Place-Time," in Pile and Keith (1997: 219–35).

— (1999). "Telecommunications, Gangster Nations and the Crisis of Representative Democracy: An Editorial Comment." *Political Geography*, 18: 81–4.

—— (2001). "The Politics of Geography: Hate Mail, Rabid Referees, and Culture Wars." *Political Geography*, 20: 1–12.

Dear, M., and Flusty, S. (eds.) (2001). *Spaces of Postmodernity: Readings in Human Geography*. Oxford: Blackwell.

Delaney, D. (1993). "Geographies of Judgment: The Doctrine of Changed Conditions and the Geopolitics of Race." *Annals of the Association of American Geographers*, 83: 48–65.

Delaney, D., and Leitner, H. (1997). "The Political Construction of Scale." *Political Geography*, 16: 93–7.

Demko, G., and Wood, W. (1994). *Reordering the World: Geopolitical Perspectives on the Twenty First Century*. Boulder: Westview.

—— (1999). *Reordering the World: Geopolitical Perspectives on the Twenty First Century*. 2nd edn. Boulder: Westview.

Dijkink, G. (1996). *National Identity and Geopolitical Visions*. London: Routledge.

Dodds, K. (1997). *Geopolitics in Antarctica*. Chichester, NY: Wiley.

—— (1998). "Enframing Bosnia," in G. Ó Tuathail and S. Dalby, (eds.), *Rethinking Geopolitics*. London: Routledge, 170–97.

Dodds, K., and Atkinson, D. (eds.) (1999). *Geopolitical Traditions: Critical Histories of a Century of Geopolitical Thought*. London, Routledge.

Dow, K. (1999). "Caught in the Currents: Pollution, Risk and Environmental Change in Marine Space." *Professional Geographer*, 51/3: 414–25.

Driver, F. (2001). *Geography Militant: Cultures of Exploration and Empire*. Oxford: Blackwell.

Duncan, N. (ed.) (1996). *BodySpace: Destabilization Geographies of Gender and Sexuality*. London: Routledge.

Dziedzic, M., and Wood, W. (2000). "Kosovo Brief—Information Management: A New Opportunity for Cooperation between Civilian and Military Entities." *Virtual Diplomacy Series*, 9. Washington DC: United States Institute of Peace.

Eagles, M. (ed.) (1995). *Spatial and Contextual Models in Political Research*. London: Taylor & Francis.

Ericson, R., and Haggerty, K. (1997). *Policing the Risk Society*. Toronto: University of Toronto Press.

Escobar, A. (1995). *Encountering Development*. Princeton: Princeton University Press.

Escobar, A. (2001). "Culture Sits in Places: Reflections on Globalism and Subaltern Strategies of Localization." *Political Geography*, 20: 139–74.

Falah, G. (1997). "Re-envisioning Current Discourse: Alternative Territorial Configurations of Palestinian Statehood." *The Canadian Geographer*, 41: 307–30.

—— (1998). "The 1948 Israeli-Palestinian War and Its Aftermath: The Transformation and De-signification of Palestine's Cultural Landscape." *Annals of the Association of American Geographers*, 86: 256–85.

Falah, G., and Newman, D. (1995). "The Spatial Manifestation of Threat: Israelis and Palestinians See a 'Good' Border." *Political Geography*, 14: 689–706.

—— (1996). "State Formation and the Geography of Palestinian Self-Determination." *Tijdschrift voor Economische en Sociale Geografie*, 87: 60–72.

Fielden, M. (1998). "The Geopolitics of Aid: The Provision and Termination of Aid to Afghan Refugees in North West Frontier Province, Pakistan." *Political Geography*, 17: 459–88.

Flint, C. (1995). "A TimeSpace for Electoral Geography: Economic Restructuring, Political Agency and the Rise of the Nazi Party." *Political Geography*, 20: 301–29.

—— (2001). "Right-Wing Resistance to the Process of American Hegemony: The Changing Political Geography of Nativism in Pennsylvania, 1920–1998." *Political Geography*, 20: 763–86.

Flint, C., and Shelley, F. M. (1996). "Structure, Agency and Context: The Contributions of Geography to World-Systems Theory." *Sociological Inquiry*, 66: 496–508.

Flynn, J., Chalmers, J., Easterling, D., Kasperson, R., Kunreuther, H., Mertz, C. K., Mushkatel, A., Pijawkal, K. D., and Slovic, P., with Dotto, L. K. (1995). *One Hundred Centuries of Solitude: Redirecting America's High-Level Nuclear Waste Policy*. Boulder: Westview Press.

Foucault, M. (1980). *Power/Knowledge*. New York: Pantheon.

Frohling, O., Gallaher, C., and Jones, J. P. (2001). "Imagining the Mexican Election." *Antipode*, 33: 1–16.

Gallaher, C. (1997). "Identity Politics and the Religious Right: Hiding Hate in the Landscape." *Antipode*, 29: 256–277.

—— (2000). "Global Change, Local Angst: Class and the American Patriot Movement." *Environment and Planning D: Society and Space*, 18: 667–91.

Gibson-Graham, J. K. (1996). *The End of Capitalism (As We Knew It)*. Oxford: Blackwell.

Giordano, B. (2000). "Italian Regionalism or 'Padanian' Nationalism—the Political Project of the Lega Nord in Italian Politics." *Political Geography*, 19: 445–71.

Glassner, M. I. (1996). "Political Geography and the United Nations." *Political Geography*, 15: 227–30.

Godlewska, A., and Smith, N. (1994). *Geography and Empire*. Oxford: Blackwell.

Graham, B. (ed.) (1998). *Modern Europe: Place, Culture, Identity*. London: Arnold.

Grant, R. (1993). "Trading Blocs or Trading Blows? The Macroeconomic Geography of US and Japanese Trade Policies." *Environment and Planning A*, 25: 273–91.

—— (1995). "Reshaping Japanese Aid for the Post-Cold War Era." *Tijdschrift voor Economische en Sociale Geografie*, 86: 235–48.

Grant, R., and Nijman, J. (1997). "Historical Changes in U.S. and Japanese Foreign Aid to the Asia-Pacific Region." *Annals of the Association of American Geographers*, 87: 32–51.

—— (1998). *The Global Crisis in Foreign Aid*. Syracuse, NY: Syracuse University Press.

—— (2002). "Globalization and the Corporate Geography of Cities in the Less-Developed World." *Annals of the Association of American Geographers* (forthcoming).

Grant, R., and Short, J. (2002). *Globalization and the Margins*. London: Palgrave.

Gregory, D. (1994). *Geographical Imaginations*. Cambridge, Mass.: Blackwell.

Grofman, B., Handley, L., and Niemi, R. (1992). *Minority Representation and the Quest for Voting Equality*. New York: Cambridge University Press.

Guinier, L. (1994). *The Tyranny of the Majority*. New York: Free Press.

Halberstam, D. (2001). *War in a Time of Peace*. New York: Scribner.

Hall, D., and Danta, D. (1996). *Reconstructing the Balkans: A Geography of the New Southeast Europe*. Chichester, NY: Wiley.

Hanna, S. P. (1995). "Finding a Place in the World-Economy: Core–Periphery Relations, the Nation-State and the Underdevelopment of Garrett County, Maryland." *Political Geography*, 14: 451–72.

Haraway, D. (1997). *Modest_Witness@Second_Millennium*. London: Routledge.
Harvey, D. (1996). *Justice, Nature and the Geography of Difference*. Oxford: Blackwell.
Heffernan, M. (1998). *The Idea of Europe: A Political Geography of Europe*. London: Arnold.
Heiman, M. (1996). "Race, Waste and Class: New Perspectives on Environmental Justice." *Antipode*, 28: 111–21.
Held, D. (1995). *Democracy and Global Order*. Stanford: Stanford University Press.
Herb, G. (1997). *Under the Map of Germany: Nationalism and Propaganda 1918–1945*. London: Routledge.
Herbert, S. (1996). *Policing Space*. Minneapolis: University of Minnesota.
—— (1999). "The End of the Territorially-Sovereign State? The Case of Crime Control in the United States." *Political Geography*, 18: 149–72.
Herod, A., Ó Tuathail, G., and Roberts, S. (1998). *An Unruly World? Globalization, Governance and Geography*. London: Routledge.
Hooson, D. (ed.) (1994): *Geography and National Identity*. Oxford: Blackwell.
Hyndman, J. (2000). *Managing Displacement: Refugees and the Politics of Humanitarianism*. Minneapolis: University of Minnesota.
Jackson, P. (1998). "Constructions of 'Whiteness' in the Geographical Imagination." *Area*, 30: 90–106.
—— (ed.) (2002). *Transnational Spaces*. New York: Routledge.
Jacobs, J. (1996). *Edge of Empire: Postcolonialism and the City*. London: Routledge.
Jaggers, K., and Gurr, T. R. (1995). "Tracking Democracy's Third Wave with the Polity III Data." *Journal of Peace Research*, 32: 469–82.
Jeffords, S. (1989). *The Remasculinization of America: Gender and the Vietnam War*. Bloomington: Indiana University Press.
—— (1994). *Hard Bodies: Hollywood Masculinity in the Reagan Era*. New Brunswick: Rutgers University Press.
Johnson, N. (1997). "Cast in Stone: Monuments, Geography and Nationalism," in Agnew (1997: 347–64).
Johnston, R. J., and Taylor, P. (eds.) (1989). *A World in Crisis? Geographical Perspectives*. Oxford: Blackwell.
Johnston, R. J., Taylor, P., and Watts, M. (eds.) (1995). *Geographies of Global Change*. Oxford: Blackwell.
—— (eds.) (2002). *Geographies of Global Change*. 2nd edn. Oxford: Blackwell.
Jones, A. (1998). "Re-Theorizing the Core: A 'Globalized' Business Elite in Santiago, Chile." *Political Geography*, 17: 295–318.
Jones, J. P., Nast, H., and Roberts, S. (eds.) (1997). *Thresholds in Feminist Geography: Difference, Methodology, Representation*. New York: Rowman & Littlefield.
Kaplan, D. (1994). "Two Nations in Search of a State: Canada's Ambivalent Spatial Identities." *Annals of the Association of American Geographers*, 84/4: 584–606.
Kaplan, D., and Herb, G. H. (1999). *Nested Identities*. New York: Rowan & Littlefield.
Kearns, G. (2001). "'Educate that Holy Hatred': Place, Trauma and Identity in the Irish Nationalism of John Mitchel." *Political Geography*, 20: 885–911.
—— (2002). "Imperial Geopolitics: Geopolitical Visions at the Dawn of the American Century," in Agnew, Mitchell, and Toal (2002).
Keith, M. and Pile, S. (eds.) (1993). *Place and the Politics of Identity*. London: Routledge.
Keohane, and Nye, (1998). "Power and Interdependence in the Information Age." *Foreign Affairs*, 77/5, 81–9.
Kirby, A. (1997). "Is the State our Enemy?" *Political Geography*, 16: 1–13.
Klare, M. (1995). *Rogue States and Nuclear Outlaws*. New York: Hill & Wang.
Klein, H. (ed.) (2001). "Global Democracy and the ICANN Elections", *Journal of Policy, Regulation, and Strategy for Telecommunications Information and the Media*, Special Issue, 3/4: 255–358.
Knight, D. (1998). "Canada and its Political Fault-Lines: Reconstitution or Dis-integration?" in D. G. Bennett (ed.), *Tensions Areas of the World*. 2nd edn. Dubuque, Iowa: Kendall Hunt, 207–28.
Knopp, L. (1992). "Sexuality and the Spatial Dynamics of Capitalism." *Environment and Planning D: Society and Space*, 10: 651–69.
—— (1995). "Sexuality and Urban Space: A Framework for Analysis," in Bell and Valentine (1995: 149–61).
Kodras, J. E. (1997). "Globalization and Social Restructuring of the American Population: Geographies of Exclusion and Vulnerability," in L. A. Staeheli, J. E. Kodras, and C. Flint (eds.), *State Devolution in America: Implications for a Diverse Society*. Thousand Oaks, Calif.: Sage, 41–59.
Kofman, E., and Peake, L. (1990). "Into the 1990s: A Gendered Analysis of Political Geography." Political Geography 9: 313–36.
Kofman, E., Peake, L., and Youngs, G. (eds.) (1996). *Globalization: Theory and Practice*. London: Pinter.
Kolossov, V., and O'Loughlin, J. (1998). "New Borders for New World Orders: Territorialities at the Fin-de-siècle." *Geojournal*, 44: 259–73.
—— (1999). "Pseudo-States as Harbingers of a Post-Modern Geopolitics: The Example of the Trans-Dniester Moldovan Republic (TMR)." *Geopolitics*, 3: 151–76.
Krishna, S. (1994). "Cartographic Anxiety: Mapping the Body Politic in India." *Alternatives*, 19: 507–21.
Kuehls, T. (1996). *Beyond Sovereign Territory: The Space of Ecopolitics*. Minneapolis: University of Minnesota.
Latour, B. (1993). *We Have Never Been Modern*. Cambridge, Mass.: Harvard University Press.
Lee, R. and Wills, J. (eds.) (1997). *Geographies of Economies*. London: Arnold.
Leib, J. (1998). "Political Geography and Voting Rights in the United States," in F. M. Davidson, J. I. Leib, F. M. Shelley, and G. R. Webster (eds.), *Teaching Political Geography*. Indiana, Pa.: National Council for Geographic Education, 59–68.
Leitner, H. (1997). "Reconfiguring the Spatiality of Power: The Construction of a Supranational Migration Framework for the European Union." *Political Geography*, 16: 123–44.
Leyshon, A., and Thrift, N. (1997). *Money/Space: Geographies of Monetary Transformation*. London: Routledge.
Livingstone, D. (1993). *The Geographical Tradition*. Cambridge, Mass.: Blackwell.

Low, M. (1998). "Representation Unbound: Globalization and Democracy," in Cox (1997: 240–80).

Low, N., and Gleeson, B. (1998). *Justice, Society and Nature*. London Routledge.

Luke, T. (1996*a*). "Governmentality and Contragovernmentality: Rethinking Sovereignty and Territoriality after the Cold War." *Political Geography*, 15: 491–508.

— (1996*b*). "Liberal Society and Cyborg Subjectivity: The Politics of Environments, Bodies and Nature." *Alternatives*, 21/1: 1–30.

— (1998). *Ecocritique*. Minneapolis: University of Minnesota.

Luke, T., and Ó Tuathail, G. (1997). "On Videocameralistics: The Geopolitics of Failed States, the CNN International and (UN)governmentality. *Review of International Political Economy*, 4: 709–33.

— (1998*a*). "Flowmations, Fundamentalism and Fast Geopolitics: 'America' in an Accelerating World," in Herod, Ó Tuathail, and Roberts (1998).

— (1998*b*). "The Fraying Modern Map: Failed States and Contraband Capitalism." *Geopolitics*. 3/3: 14–33.

Lyotard, J. F. (1984). *The Postmodern Condition: A Report on Knowledge*. Minneapolis: University of Minnesota.

MacLaughlin, J. (1998). "The Political Geography of Anti-Traveler Racism in Ireland: The Politics of Exclusion and the Geography of Closure." *Political Geography*, 17: 417–36.

Marston, S. (1995). "The Private Goes Public: Citizenship and the New Spaces of Civil Society." *Political Geography*, 14: 194–8.

Marston, S., and Mulligan, A. (1998). "Ethnicities, Nationalisms and Sexualities: Identity Politics in the St. Patrick's Day Parades in New York City in the 1990s." Paper presented at the "Nationalisms and Identities in a Globalized World Conference, Maynooth, Ireland, 17–23 August 1998.

Martis, K. C., Kovacs, Z., Kovacs, D., and Peter, S. (1992). "The Geography of the 1990 Hungarian Parliamentary Elections." *Political Geography*, 11: 293–305.

Mattelart, A. (1994). *Mapping World Communication: War, Progress, Culture*. Minneapolis: University of Minnesota Press.

Merrett, C. (1996). *Free Trade: Neither Free Nor About Trade*, Montreal: Black Rose.

McClintock, A., Mufti, A., and Shohat, E. (1997). *Dangerous Liaisons: Gender, Nation and Postcolonial Perspectives*. Minneapolis: University of Minnesota.

McDowell, L., and Sharp, J. (1997). *Space, Gender, Knowledge*. London: Arnold.

Millennium (1999). "Territorialities, Identities and Movement in International Relations." *Millennium: Journal of International Studies*, Special Issue 28: 3.

Miller, B. (2000). *Geography and Social Movements*. Minneapolis: University of Minnesota.

Minghi, J. (1998). "Border Disputes: From Argentina to Somalia," in D. G. Bennett (ed.), *Tensions Areas of the World*. 2nd edn. Dubuque, Iowa: Kendall Hunt, 71–88.

Mitchell, D. (1992). "Iconography and Locational Conflict from the Underside: Free Speech, People's Park, and the Politics of Homelessness in Berkeley, California." *Political Geography*, 11: 152–69.

— (1995). "The End of Public Space? People's Park, Definitions of the Public, and Democracy." *Annals of the Association of American Geographers*, 85: 108–33.

— (2000). *Cultural Geography: A Critical Introduction*. Oxford: Blackwell.

Mitchell, K. (2000). "Networks of Ethnicity," in Trevor Barnes and Eric Sheppard (eds.), *The Blackwell Companion to Economic Geography*. Oxford: Blackwell, 392–407.

— (2002*a*). "Geographies of Transnationality," in K. Anderson, M. Domosh, S. Pile, and N. Thrift (eds.), *The Cultural Geography Handbook*. London: Blackwell.

— (2002*b*). "Global Bodies in Homely Spaces," in A. Blunt and A. Varsley (eds.), *Geographies of Home*. New York: Routledge.

— (2002*c*). "Transnationalism in the Margins: Hegemony and the Shadow State," in Jackson (2002).

Mittelman, J. (ed.) (1996). *Globalization: Critical Reflections*. Boulder: Lynne Rienner.

Morley, D., and Robins, K. (1995). *Spaces of Identity: Global Media, Electronic Landscapes and Cultural Boundaries*. London: Routledge.

Morrill, R. (1999). "Inequalities of Power, Costs and Benefits across Geographic Scales: The Future Uses of the Hanford Reservation." *Political Geography*, 18: 1–24.

Mulgan, G. (1997). *Connexity: How to Live in a Connected World*. Boston: Harvard Business School Press.

Mumford, L. (1964). *The Pentagon of Power*. San Diego: Harcourt Brace Jovanovich.

Murphy, A. (1995). *Geographic Approaches to Democratization*. A report to the National Science Foundation.

— (1996). "The Sovereign State System as Political-Territorial Ideal: Historical and Contemporary Considerations," in T. Biersteker and C. Weber (eds.), *State Sovereignty as Social Construct*. Cambridge: Cambridge University Press.

Murphy, D. (1997). *The Heroic Earth: Geopolitical Thought in Weimar Germany, 1918–1933*. Kent, Ohio: Kent State University Press.

Myers, G., Klak, T., and Koehl, T. (1996). "The Inscription of Difference: News Coverage of the Conflicts in Rwanda and Bosnia." *Political Geography*, 15: 21–46.

Natter, W. (2002). "Nazi Geopolitics," in Agnew, Mitchell, and Toal (2002).

Nast, H. (1998). "Unsexy Geographies." *Gender, Place and Culture*, 5/2: 191–206.

Nast, H., and Pile, S. (eds.) (1998). *Places Through the Body*. London: Routledge.

Newman, D. (ed.) (1998). "Boundaries, Territory and Postmodernity." Special Issue, *Geopolitics*.

Newman, D., and Kliot, N. (eds.) (1999). "Geopolitics at the End of the Twentieth Century." Special Issues, *Geopolitics*, 4/1–2.

Newman, D., and Falah, G. (1995). "Small State Behaviour: On the Formation of a Palestinian State in the West Bank and Gaza Strip." *The Canadian Geographer*, 39: 219–34.

— (1997). "Bridging the Gap: Palestinian and Israeli Discourses on Autonomy and Statehood." *Transactions of the Institute of British Geographers*, NS 22: 111–29.

Newman, D., and Paasi, A. (1999). "Fences and Neighbours in the Postmodern World: Boundary Narratives in Political Geography." *Progress in Human Geography*, 22: 186–207.

Nijman, J. (1992). "The Political Geography of the Post-Cold War World: Introduction and Closing Remarks." *The Professional Geographer*, 44: 1–3, 28–9.

— (1993). *The Geopolitics of Power and Conflict*. London: Belhaven.

Nijman, J. (1995). "Reshaping US Foreign Aid Policy: Continuity and Change." *Tijdschrift voor Economische en Sociale Geografie*, 86: 219–34.

— (1999). "Cultural Globalization and the Identity of Place: The Reconstruction of Amsterdam." *Ecumene*, 6/2.

Odom, W. (1998). *The Collapse of the Soviet Military*. New Haven: Yale University Press.

O'Lear, S. (1996). "Using Electronic Mail (E-Mail) Surveys for Geographic Research: Lessons from a Survey of Russian Environmentalists." *Professional Geographer*, 48: 209–17.

— (1999). "Networks of Engagement: Electronic Communication and Grassroots Environmental Activism in Kaliningrad," *Geografiska Annaler*, 81: 165–78.

— (2001). "Azerbaijan: Territorial Issues and Internal Challenges in Mid-2001," *Post Soviet Geography and Economics*, 42: 305–12.

— (2002). "Armenian Energy: Establishing or Eroding Sovereignty?" *Journal of Central Asian Studies* (forthcoming).

O'Loughlin, J. (1993). "Geo-economic Competition in the Pacific Rim: The Political Geography of Japanese and US Exports, 1966–1988." *Transactions, Institute of British Geographers*, NS 18: 438–59.

O'Loughlin, J., and van Der Wusten, H. (eds.) (1993). *The New Political Geography of Eastern Europe*. Chichester, NY: Wiley.

— (ed.) (1994). *Dictionary of Geopolitics*. Westport, Conn.: Greenwood Press.

— (2001). "The Regional Factor in Contemporary Ukrainian Politics: Scale, Place, Space or Bogus Effect." *Post-Soviet Geography and Economics*, 42/1: 1–33.

O'Loughlin, J., and Bell, J. E. (1999). "The Political Geography of Civic Engagement in Ukraine, 1994–1998." *Post-Soviet Geography and Economics*, 39.

O'Loughlin, J., Kolossov, V., and Tchepalyga, A. (1998*a*). "National Construction, Territorial Separatism and Post-Soviet Geopolitics: The Example of the Transdniester Moldovan Republic," *Post Soviet Geography and Economics*, 38: 332–58.

O'Loughlin, J., Kolossov, V., and O. Vendina (1997). "The Electoral Geographies of a Polarizing City: Moscow, 1993–1996." *Post-Soviet Geography and Economics*, 38: 567–601.

O'Loughlin, J., Mayer, T., and Greenberg, E. (eds.) (1995). *War and Its Consequences: Lessons from the Persian Gulf Conflict*. New York: HarperCollins.

O'Loughlin, J., Shin, M., Talbot, P. (1996). "Political Geographies and Cleavages in the Russian Parliamentary Elections." *Post-Soviet Geography*, 37: 355–85.

O'Loughlin, J., Ward, M., Lofdahl, C., Cohen, J., Brown, D., Reilly, D., Gleditsch, K., and Shin, M. (1998*b*). "The Diffusion of Democracy, 1946–1994." *Annals of the Association of American Geographers* 88 4: 545–74.

Oppenheimer, A. (1996). *Bordering on Chaos: Guerillas, Stockbrokers, Politicians, and Mexico's Road to Prosperity*. Boston: Little Brown.

O'Reilly, K., and Webster, G. (1998). "A Sociodemographic and Partisan Analysis of Voting in Three Anti-Gay Rights Referenda in Oregon." *Professional Geographer*, 50: 498–515.

Ó Tuathail, G. (1996). *Critical Geopolitics: The Politics of the Writing of Global Space*. Minneapolis: University of Minnesota.

— (1997*a*). "At the End of Geopolitics? Reflection on a Plural Problematic at the Century's End." *Alternatives*, 22: 35–56.

— (1997*b*): "Emerging Markets and Other Simulations: Mexico, Chiapas and the Geo-financial Panopticon." *Ecumune*, 4: 300–17.

— (1998). "Deterritorialized Threats and Global Dangers: Geopolitics, Risk Society and Reflexive Modernization." *Geopolitics*, 3/1: 17–31.

— (1999*a*). "A Strategic Sign: The Geopolitical Significance of 'Bosnia' in U.S. Foreign Policy." *Environment and Planning D: Society and Space*, 17.

— (1999*b*). "Understanding Critical Geopolitics: Geopolitics and Risk Society." *Journal of Strategic Studies*, 22/2–3: 107–24.

— (2000). "The Postmodern Geopolitical Condition: States, Statecraft, and Security at the Millennium." *Annals of the Association of American Geographers*, 90/1: 166–78.

— (2001). Geopolitics @ Millennium: Paranoid Fantasies and Technological Fundamentalism Amidst the Contradictions of Contemporary Modernity," in A. Gosar (ed.), "Political Geography in the 21st Century: Understanding the Place—Looking Ahead." *Geographica Slovenica*, 34.

— (2002). "Theorizing Practical Geopolitical Reasoning: The Case of U.S. Bosnia Policy in 1992." *Political Geography*, 21.

Ó Tuathail, G., and Dalby, S. (1994). "Critical Geopolitics: Unfolding Spaces for Thought in Geography and Global Politics." Special issue on Critical Geopolitics. *Environment and Planning D: Society and Space*, 12: 513–14.

— (1998). *Rethinking Geopolitics*. London: Routledge.

Ó Tuathail, Dalby, S., and Routledge, P. (1998). *The Geopolitics Reader*, London: Routledge.

Paasi, A. (1996). *Territories, Boundaries and Consciousness: The Changing Geographies of the Finnish-Russian Border*. Chichester, NY: Wiley.

Painter, J. (1995). *Politics, Geography and "Political Geography."* London: Edward Arnold.

Painter, J., and Philo, C. (1995). "Spaces of Citizenship: An Introduction." *Political Geography*, 14/2: 107–20.

Peet, R., and Watts, M. (eds.) (1996). *Liberation Ecologies: Environment, Development, Social Movements*. London: Routledge.

Pickles, J., and Smith, A. (eds.) (1998). *Theorizing Transition: The Political Economy of Post-Communist Transformations*. London: Routledge.

Pile, S., and Keith, M. (eds.) (1997). *Geographies of Resistance*. London: Routledge.

Pile, S., and Thrift, N. (eds.) (1995). *Mapping the Subject: Geographies of Cultural Transformation*. London: Routledge.

Radcliffe, S. (1998). "Frontiers and Popular Nationhood: Geographies of Identity in the 1995 Ecuador-Peru Border Dispute." *Political Geography*, 17: 273–94.

Radcliffe, S., and Westwood, S. (1996). *Remaking the Nation: Place, Identity and Politics in Latin America*. London: Routledge.

Proctor, J., and Smith, D. (eds.) (1999). *Geography and Ethics: Journeys in a Moral Terrain*. London: Routledge.

Purcell, D., and Kodras, J. E. (2001). "Information Technologies, Representational Spaces, and the Marginal State: Redrawing the Balkan Image of Slovenia." *Information, Communications, and Society*. 4/3: 1–29.

Reynolds, D., and Knight, D. (1989). "Political Geography," in Gary Gaile and Cort Willmott (eds.), *Geography in America*. Columbus: Merrill, 582–618.

Robins, K. (1996). *Into the Image: Culture and Politics in the Field of Vision*, London: Routledge.

Rocheleau, D., Thomas-Slayter, B., and Wangari, E. (eds.) (1996). *Feminist Political Ecology*. London: Routledge.

Rollins, W. (1995). "Whose Landscape? Technology, Fascism, and Environmentalism on the National Socialist Autobahn." *Annals of the Association of American Geographers*, 83/3: 494–520.

Routledge, P. (1993). *Terrains of Resistance: Nonviolent Social Movements and the Contestation of Place in India*. New York: Praeger.

Rumley, D., Chiba, T., Takagi, A., and Fukushima, Y. (eds.) (1996). *Global Geopolitical Change in the Asia-Pacific: A Regional Perspective*. Aldershot: Avebury.

Ryan, J. (1997). *Picturing Empire*. Chicago: University of Chicago Press.

Seager, J. (1993). *Earth Follies: Feminism, Politics and the Environment*. London: Earthscan.

Secor, A. (2001). "Ideologies in Crisis: Political Cleavages and Electoral Politics in Turkey in the 1990s." *Political Geography*, 20: 539–60.

Shapiro, M. (1997). *Violent Cartographies: Mapping Cultures of War*. Minneapolis: University of Minnesota.

Sharp, J. (1998). "Reel Geographies of the New World Order: Patriotism, Masculinity, and Geopolitics in Post-Cold War American Movies," in Ó Tuathail and Dalby (1998: 152–69).

—— (2000). *Condensing the Cold War: Reader's Digest and American Identity*. Minneapolis: University of Minnesota.

Sharp, J., Routledge, P., Philo, C., and Paddison, R. (2000). *Entanglements of Power: Geographies of Domination/Resistance*. London: Routledge.

Shelley, F. M. (1997). "Education Policy and the 104th Congress," in L. A. Staeheli, J. E. Kodras, and C. Flint (eds.), *State Devolution in America: Implications for a Diverse Society*. Thousand Oaks, Calif.: Sage, 221–32.

Shelley, F. M., and Archer, J. C. (1989). "Sectionalism and Presidential Politics in America: A Twentieth-Century Perspective from Illinois, Indiana and Ohio," *Journal of Interdisciplinary History*, 20: 227–55.

—— (1994). "Some Geographical Aspects of the 1992 American Presidential Election," *Political Geography*, 13: 137–59.

—— (1995). "The Volatile South: A Historical Geography of Presidential Elections in the South," *Southeastern Geographer*, 35: 22–36.

Shelley, F. M., and Flint, C. (2000). "Geography, Place, and World-Systems Analysis," in T. D. Hall (ed.), *The World-Systems Reader*. Austin: University of Texas Press, 69–82.

Shelley, F. M., Archer, J. C., Davidson, F. M., and Brunn, S. D. (1996). *The Political Geography of the United States*. New York: Guilford.

Shin, M. (2001). "The Politicization of Place in Italy." *Political Geography*, 20: 331–52.

Short, J. (2001). *Global Dimensions: Space, Place and the Contemporary World*. London: Reaktion.

Sibley, D. (1995). *Geographies of Exclusion: Society and Difference in the West*. London: Routledge.

Slater, D. (1997). Geopolitical Imaginations across the North-South Divide: Issues of Difference, Development and Power. *Political Geography*, 16: 631–54.

Smith, G. (ed.) (1994). *The Baltic States*. New York: St. Martin's Press.

—— (ed.) (1996). *The Nationalities Question in the Post-Soviet States*. 2nd edn. London: Longman.

Smith, G., Law, V., Wilson, A., Bohr, A., Allworth, E. (1998). *Nation-Building in the Post-Soviet Borderlands*. Cambridge: Cambridge University Press.

Soja, E. (1989). *Postmodern Geographies*. London: Verso.

Solecki, W. (1996). "Paternalism, Pollution and Protest in a Company Town." *Political Geography*, 15: 5–20.

Solecki, W. D., and Shelley, F. M. (1996). "The Environment as a Developing Political Issue in the 1950s." *Government and Policy*, 14: 215–30.

Sparke, M. (1994). "Writing on Patriarchal Missiles: The Chauvinism of the 'Gulf War' and the Limits of Critique." *Environment and Planning A*, 26: 1061–89.

—— (1998*a*). "From Geopolitics to Geoeconomics: Transnational State Effects in the Borderlands." *Geopolitics*, 3/2: 62–98.

—— (1998*b*). "Outsides inside Patriotism: The Oklahoma Bombing and the Displacement of Heartland Geopolitics," in Ó Tuathail and Dalby (1998: 198–223).

—— (2000*a*). "Chunnel Visions: Unpacking the Anticipatory Geographies of an Anglo-European Borderland." *Journal of Borderland Studies*, 15: 2–34.

—— (2000*b*). "Excavating the Future in Cascadia: Geoeconomics and the Imagined Geographies of a Cross-Border Region," *BC Studies*, 127/Autumn: 5–44.

—— (2002). "Troubling Borderlands: Cascadia and the Neoliberal Natural History of a Cross-Border Region," in Andrew Holman (ed.), *Into the Borderlands: New Essays in the History of Canadian-American Relations*. Toronto: Canadian Scholars' Press.

—— (2003). *Hyphen-Nation-States: Critical Geographies of Displacement and Disjuncture*. Minneapolis: University of Minnesota Press.

Sparke, M., and Lawson, V. (2002). "Geoeconomics: Entrepreneurial Political Geographies of the Global-Local Nexus," in Agnew, Mitchell, and Toal (2002).

Sparke, M., Katz, C., and Newstead, C. (2002). "The Cultural Geography of Scale-Jumping: Landscapes of Reterritorialization in the Americas," in *The Cultural Geography Handbook*, Oxford: Blackwell.

Spiegel, S. (2000). "Traditional Space versus Cyberspace: The Changing Role of Geography in Current International Politics." *Geopolitics*, 5/3: 114–25.

Staeheli, L. (ed.) (1994). "Special Issue: Empowering Political Struggle." *Political Geography*, 13: 387–476.

Staeheli, L., and Thompson, A. (1997). "Citizenship, Community and Struggles for Public Space." *Professional Geographer*, 49: 28–38.

Steinberg, P. (2002). *The Social Construction of the Ocean*. Cambridge: Cambridge University Press.

Steinberg, P., and Clark, G. (1999). "Troubled Water? Acquiescence, Conflict, and the Politics of Place in Watershed Management." *Political Geography*, 18: 477–508.

Straussfogel, D. (1997*a*). "A Systems Perspective on World-Systems Theory." *Journal of Geography*, 96: 119–26.

Straussfogel, D. (1997*b*). "World-System Theory: Toward a Heuristic and Conceptual Tool, *Economic Geography*, 73.

Taylor, P. J. (1990). *Britain and the Cold War: 1945 as Geopolitical Transition*. London: Pinter.

Taylor, P. J. (ed.) (1993). *Political Geography of the Twentieth Century*. London: Belhaven.
— (1994). "The State as Container: Territoriality in the Modern World-System." *Progress in Human Geography*, 18: 151–62.
— (1995). "Beyond Containers: Inter-nationality, Inter-stateness, Inter-territoriality." *Progress in Human Geography*, 19: 1–14.
— (1996). *The Way the Modern World Works: World Hegemony to World Impasse*. Chichester, NY: Wiley.
— (2000). "World Cities and Territorial States under Conditions of Contemporary Globalization." *Political Geography*, 19: 5–32.
Taylor, P. J., and Flint, C. (1999). *Political Geography*. 4th edn. New York: Wiley.
Tesfahuney, M. (1998). "Mobility, Racism and Geopolitics." *Political Geography*, 17: 499–515.
Thrift, N. (1998). "The Rise of Soft Capitalism," in Herod, Ó Tuathail, and Roberts (1998: 25–71).
Todorova, M. (1997). *Imagining the Balkans*. New York: Oxford University Press.
Valentine, G. (1999). "'Sticks and Stones May Break My Bones': A personal geography of harassment." *Antipode*, 30: 305–32.
Ward, P. (1995). *Mexico City*. Chichester, NY: Wiley.
Warf, B. (1997). "The Geopolitics/Geoeconomic of Military Base Closures in the USA." *Political Geography*, 16: 541–64.
Warf, B., and Purcell, D. (2001). "The Currency of Currency: Speed, Sovereignty, and Electronic Finance," in T. R. Leinbach and S. D. Brunn (eds.), *Worlds of E-Commerce: Economic, Geographical and Social Dimensions*. Chichester, NY: John Wiley, 223–40.
Wark, McK. (1994). *Virtual Geography: Living with Global Media Events*. Bloomington: Indiana University Press.
Waterman, S. (1998). "Political Geography as a Mirror of Political Geography." *Political Geography* 17: 373–88.
Watts, M. (1997). "Black Gold, White Heat: State Violence, Local Resistance and the National Question in Nigeria," in Pile and Keith (1997: 33–67).
Webster, G. R., and Leib, J. (2001). "Whose South Is It Anyway? Race and the Confederate Battle Flag in South Carolina." *Political Geography*, 271–99.
— (2002). "Introduction to Forum: The 2000 Presidential Election and the Florida Debacle in Geographic Context." *Political Geography*, 21: 67–70.
Weiss, T., Forsythe, D., and Coate, R. (2001). *The United Nations and Changing World Politics*. 3rd edn. Boulder: Westview.
White, G. (2000). *Nationalism and Territory: Constructing Group Identity in Southeastern Europe*. Lanham, Md.: Rowman & Littlefield.
Williams, C. (ed.) (1993). *The Political Geography of the New World Order*. London: Belhaven.
— (1994). *Called Unto Liberty!* Cleavedon, Avon: Multilingual Matters.
Williams, R. (1999). "Environmental Injustice in America and Its Politics of Scale." *Political Geography*, 18: 49–74.
Wood, J. (1997). "Vietnamese American Place Making in Northern Virginia," *Geographical Review*, 87: 58–72.
Wood, W. (1994). "Forced Migration: Local Conflicts and International Dilemmas." *Annals of the Association of American Geographers*, 84: 607–34.
— (1996). "From Humanitarian Relief to Humanitarian Intervention: Victims, Interveners, and Pillars." *Political Geography*, 15: 671–96.
— (2000). "Complex Emergency Response Planning and Coordination: Potential GIS Applications." *Geopolitics*, 5/1: 19–36.
Wood, W., and Smith, D. (1997). "Mapping War Crimes: GIS Analyses Ethnic Cleansing Practices in Bosnia." *GIS World*, September: 56–8.
Yiftachel, O. (1999). "Between Nation and State: Fractured Regionalism among Palestinian-Arabs in Israel." *Political Geography*, 18: 285–307.
— (2000). "Ethnocracy" and its Discontents: Minorities, Protests, and the Israeli Polity." *Critical Inquiry*, 26: 725–56.

CHAPTER 13

Population Geography

Patricia Gober and James A. Tyner

Geographic issues loom large as the American population begins the new millennium. Regional fertility differentials are growing, social networks focus new immigrants on a small number of port-of-entry metropolitan areas and states, highly channelized migration streams redistribute population in response to economic and social restructuring, and a highly variegated landscape of aging has emerged. Perhaps at no other time in its history has the field of population geography been confronted with a more intellectually important and socially relevant research agenda. Building upon its strong tradition in spatial demography and incorporating an increasingly diverse set of quantitative and qualitative methodologies, population geography today seeks a more complete understanding of human movement, regional demographic variability, and the social context within which these population processes occur. In addition, population geographers increasingly tackle issues of policy significance.

After a brief review of the history of population geography and an empirical analysis of its presence in geography's major journals, we summarize six lines of contemporary research including studies of: (1) internal migration and residential mobility; (2) international migration, transnationalism, and the nexus of internal and international migration systems; (3) immigrant assimilation, acculturation, and the emergence of ethnic enclaves; (4) regional demographic variability; (5) the social context for population processes; and (6) public policy research. We conclude by identifying major challenges facing the field today and fruitful new directions for research including the need for greater emphasis on environmental issues, integration with geography's new technologies, and more social relevance.

Background

Although geographers long had integrated population characteristics into their broader regional studies, population geography emerged as a distinct field of study only in the early 1950s. It, like urban geography, surfaced from a discipline that was strongly rooted in the study of rural cultural landscapes and regional inventories. Its birth was marked by the 1953 AAG presidential address of Glenn Trewartha, a noted climatologist and population geographer. Trewartha lamented the neglect of population in the discipline of geography, which was at that time organized into the subdivisions of physical and cultural geography. He argued for a new threefold structure organized around population, the physical earth, and the cultural landscape. According to Trewartha (1953: 87), population would never be included adequately in regional studies and regional courses of instruction "until the field of population geography is developed as a specialized systematic branch of our discipline."

Trewartha's call to arms was answered in the mid-1960s with Zelinsky's (1966: 5–6) book on population geography in which he identified three "distinct and ascending levels of discourse: (1) the simple description of the location of population numbers and characteristics, (2) the explanation of the spatial configuration of these numbers and characteristics, and (3) the geographic analysis of population phenomena." In keeping with broader intellectual trends in social science and human geography, Demko *et al.* (1970) refined this view to focus on spatial analysis, logical positivism, and quantitative methods. Subsequent interest in population exploded during the 1970s and 1980s. When the previous *Geography in America* volume was published in 1989, an impressive 10 per cent of the articles in the *Annals*, *Geographical Review*, and *Professional Geographer* dealt with population-related topics (White *et al.* 1989).

Emphasis on spatial analysis and logical positivism was reinforced by population geography's close ties with demography. Formal demography is concerned with the collection, adjustment, presentation, and projection of population data and has a strong empirical, statistical, and mathematical bent. Social demography is broader and seeks explanation for population patterns based on the theories and subject matter of various disciplines, including sociology, economics, political science, and geography. Most population geographers are trained in the rigorous methods of formal demography, but align themselves more closely with the social aspects of the discipline. Both social and formal demography retain an almost exclusive commitment to positivism, empiricism, and quantification, even as the social sciences have moved to more multifaceted approaches to the study of human behavior.

Only recently has the subdiscipline of population geography been seriously challenged to question its traditional methods of inquiry, its assumptions about population processes, and the legitimacy of its data. P. White and Jackson (1995) argue that population geography overemphasizes population events at the expense of the longer biographical history and the wider political economy, is preoccupied with data at the expense of wider social theory, accepts the constraints of data rather than questioning categories and probing their social meaning, is far too attached to essentialist categories such as gender and age, and is reluctant to delve into the larger social world in which population processes take place. Fielding (1992), McHugh (2000*b*), and Watkins (1999) call for a more ethnographic approach to studying human migration in which movement is seen as an outgrowth of people's life histories, their current circumstances, and future expectations.

Visibility of Population Geography

Following the S. E. White *et al.* (1989) tradition of analyzing population geography's contribution to disciplinary discourse, we took a census of the *Annals* and *Professional Geographer* between the first issue of 1990 and the third issue of 1998. Of the 226 articles in the *Annals* during this period, 17, or 7.5 per cent, dealt with population themes. Comparable figures for the *Professional Geographer* were 33, or 13.6 per cent, of a total of 242 articles. Given that population geographers comprise only 4.8 per cent (329 out of 6,910 in 1998) of members of the Association of American Geographers, it is clear that Trewartha's dream of a prominent place for population in the discipline of geography has been achieved.

A important, coming-of-age event in the history of population geography was the establishment in 1994 of the *International Journal of Population Geography* (*IJPG*), published in the United Kingdom by John Wiley & Sons. Although the subdiscipline long used interdisciplinary outlets for its work, most notably, *Demography*, the *International Migration Review, Environment and Planning*, and the *Journal of Regional Science*, the *IJPG* was the first journal to carry population geography research exclusively, to bring together the work of population planners and practitioners and social scientists interested in population, to provide a forum for debate of methodological and theoretical issues relevant to population geography, and to facilitate cross-national comparisons of population processes.

Research Themes

This essay cannot begin to cover exhaustively the rich variety of subject matter pursued under the rubric of population geography. Instead, we provide a representative sample of contemporary research themes and issues of debate. We identify the large questions that are being asked by population geographers today, the methods they are using, key areas of controversy, and how findings relate to matters of wider societal significance.

Internal Migration and Residential Mobility

From its beginnings, the study of human movement has formed the intellectual core of population geography because migration is, by its very nature, both a demographic event and a geographic process. Although migration specialists have pursued a wide range of methodological issues including fuzzy-set migration regions (Plane 1998), migration drift (Plane 1999), and methods of representing structural change in migration transition patterns over time (Rogers and Wilson 1996), four topical themes embody current migration research: (1) the effects of economic restructuring on migration patterns and processes; (2), the effects of demographic cycles on migration rates and timing; (3) the integration of migration and residential mobility into a life-course perspective; and (4) ethnographic approaches to migration.

The relationship between migration and the economic system is of long-standing concern in population geography (Brown 1991). Today it is manifest in studies linking the powerful forces of regional restructuring to the size and direction of migration flows. An edited volume entitled *Migration and Restructuring in the US: A Geographic Perspective* by Pandit and Withers (1999) summarized the nature of restructuring forces and outlined effects on migration systems at both the national and regional levels. In that volume, Brown *et al.* (1999) observed that our basic conceptualization of migration—as linked to regional wage, job opportunity, and information differentials—changed little since the 1960s despite fundamental reordering of relationships among labor, capital, and economic growth. They asked whether downsizing has disconnected economic growth from the demand for more workers, and hence in-migration; whether the shift from manufacturing to services could lead to job change without migration or out-migration by some population segments and in-migration by others; and whether de-linking of labor from place of employment, facilitated by the communications revolution, undermines the traditional connection between the generation of wealth, employment growth, and in-migration.

The way individual regions respond to restructuring captured the attention of population geographers during the 1990s. Cushing (1999) confirmed that migration is not performing its normative role of relocating unemployed labor in Appalachia where many middle-aged and older workers have low educational attainment and extraordinarily strong ties to place; White (1994) demonstrated that isolated sections of the Great Plains continued to lose population to nearby cities that are supported by groundwater exploitation; and Brown *et al.* (1999) found extremely high levels of intra-regional differentiation in the Ohio River Valley making it difficult to generalize about the effects of restructuring at a broad regional scale. In high-amenity, high-growth counties of the non-metropolitan West, von Reichert and Rudzitis (1994) established that retirees are attracted to low-wage destinations where amenities are captured in the labor market whereas labor-force migrants are drawn to high-wage destinations where amenities are not reflected in the labor market. In a separate study, they found older migrants are more likely than younger ones to accept amenity-driven reductions in wages (von Reichert and Rudzitis 1992).

Similar questions are being asked about connections between economic restructuring and migration systems abroad. Economic reform and the relaxation of migration restrictions in China redirected migration flows in favor of regions with high per capita income growth and foreign investment (Fan 1996). Also in China, the introduction of market forces in urban land provision led to residential mobility patterns that favored greater residential segmentation by class and age. In Guangzhou, households operating in the open market sector were more likely to purchase housing in the inner city while those in social housing made more outward moves (Li and Sui 2001). In Ecuador, land reform in the 1960s and 1970s led to the disintegration of the semi-feudal hacienda system which, in turn, increased the number of small agricultural landholders, diversified farm labor, and increased temporary labor migration (circulation) initially to the construction and service sectors in Guayaquil and Quito, but more recently, to the New York City metropolitan area (Jokisch 1997). And in Germany, reunification resulted in a dramatic increase in east-to-west migration furthering the processes of deconcentration in the West and concentration in the East (Kontuly 1997).

In a series of articles, Kontuly and others explored the counterurbanization hypothesis using the relationship between net migration and population size as an indicator of whether national settlement systems are concentrating or deconcentrating (Kontuly and Bierens 1990; Geyer and Kontuly 1993; Kontuly and Schon 1994). Results point to cycles of net migration that correspond to city size and age of development. In addition to economic restructuring, reasons for counterurbanization include economic cyclical forces, environmental factors, residential preferences, government policy, and technological innovation (Kontuly 1998).

Stimulated by the aging of the baby-boom generation, internal migration in the US has been linked to demographic cycles, or in other words, to the aging of generations of varying sizes. This line of research is informed by the "Easterlin effect," which states that individuals from large cohorts (people born at the same time) face greater competition for jobs and housing than their counterparts in small cohorts (Easterlin 1980). These unfavorable conditions lead, in turn, to depressed mobility and migration rates. Rogerson (1987), Long (1988), Plane (1992), and Pandit (1997*a*) established that migration rates for young adults do, in fact, rise and fall with generations of different sizes. Depressed migration rates of the 1970s corresponded to the entry of the large baby-boom generation into the labor force and into age categories where the propensity to migrate is high. An elaboration on this theme is Pandit's (1997*b*) examination of the effects of generation size on the timing of migration. Results revealed that members of small cohorts move earlier in their life cycles than members of large cohorts. Reconstructed age schedules of migration supported the notion of delayed mobility among baby boomers. In yet another variation on this theme, Pandit (1997*a*) simultaneously evaluated the effects of demographic and economic cycles on the migration schedule and found both to be important, although generation size was the more influential.

Plane (1992) added geography to this theme by attributing accelerated population deconcentration during the 1970s to demographic cycles. Many baby boomers, according to Plane, left the Northeast and Midwest in the face of stagnant job growth and a labor market crowded with contemporaries. At the same time, new employment opportunities allowed the South and West better to retain their young adults and attract in-migrants from the North. Although tight labor market conditions did not by themselves explain the migration changes of the 1970s, they added weight and intensity to patterns that were established earlier. Flows among the working-age population were reinforced by growing cohorts of persons 60 to 70 years of age, beginning around 1960 and extending until 1985, many of whom formed highly efficient migration streams directed southward. More recently, Pandit (2000) showed that both the level and timing of the migration schedule vary regionally, and that states with high mobility generally display older mobility distributions.

During the past 10 to 15 years, many migration and mobility studies adopted a life-course perspective. Frustrated with the inability of the traditional family life cycle to capture changes in household organization, housing careers, and geographic mobility, sociologists, demographers, and population geographers embraced the notion of the life course as an organizing framework for socio-demographic change (Elder 1977; Clausen 1986). The life course refers to "pathways which individuals follow through life and incorporates the multitude of roles that individuals experience with respect to education and work, marriage and parenthood, and residence and community life" (Gober 1992: 174). Life-course analysis examines the timing and sequences of demographic events and their relationships to other events (Withers 1997).

In the realm of residential mobility, the life-course approach stimulates investigations into the triggers, or stimuli, that create changes in the residential environment and influence the likelihood of moving. Examples of triggers include having a child (Clark and Dieleman 1996; Deileman *et al.* 1995), getting married (Odland and Shumway 1993; Clark *et al.* 1994), and obtaining a divorce (Dieleman and Schouw 1989). Clark and Withers (1999) found that a household that experiences an employment transition is 2.4 times more likely to move than a household that experiences no change in job. This ratio increases to 3.0 for married households with one worker but drops to 2.0 for married households with two workers suggesting that, because they must account for the commuting costs and the location preferences of two members, two-worker households are less responsive to job changes than one-worker households.

Growing use of the life-course perspective and longitudinal methods of analysis prompted debate about the efficacy of traditional cross-sectional approaches (comparing characteristics of people at the same point in time) to studying migration and mobility. Davies and Pickles (1985, 1991) argued that the results of cross-sectional analysis are biased and conclusions drawn from them about migration are misleading. Further, they concluded that longitudinal models, in which panels of individuals are followed through time and their life events recorded, are a conceptually and methodologically superior way of studying such an inherently dynamic process as migration. Longitudinal approaches allow the mobility process to be viewed in context, as an outgrowth of a sequence of other events such as marriage, birth, income change, and fluctuations in the housing and labor markets. In an empirical comparison of cross-sectional and longitudinal approaches to studying the relationship between mobility and room stress, Clark (1992) found that the two approaches yielded similar results. Further, Dieleman (1995: 676) described the Herculean efforts involved in using longitudinal approaches and concluded that, "while the longitudinal approach substantiates and enriches the results of cross-sectional analyses of mobility and tenure choice; it does not invalidate these results."

Ironically, the life-course perspective is embraced by the most quantitative and the most qualitative approaches to studying human migration. Watkins (1999) cautioned that unquestioned use of census data and large surveys leads to an outsider's view of migration as a static process related to other indicators of a person's *current* circumstances such as age, marital status, family composition, and economic welfare when, in fact, migration is a dynamic event inexorably linked to an accumulated life history. This argument is quite similar to that made by the life-course modelers who use longitudinal analysis to tease out the relationships between migration and past life events. Ethnographers part company with this perspective in their belief that human lives are less a series of life events and more an accumulated set of thoughts, perceptions, feelings, aspirations, and experiences—information not gleaned easily from census data and panel surveys. Migration evolves out of this complicated life history, and in-depth personal narratives are needed to elucidate the meaning of migration for people, their families, and their communities.

While we know intuitively that migration is a cultural event, there has been relatively little ethnographic research about migration (Fielding 1992). Important exceptions are McHugh and Mings' (1996) and McHugh's (2000*b*) examination of elderly seasonal migrants, Stack's (1996) ethnography of African-American migration, and Watkins (1999) description of elderly in rural northern Minnesota. McHugh and Mings' study of seasonal migrants belied the notion of migration as a permanent, one-way move. Biographical portraits of five couples moving between summer and winter homes demonstrated the circularity rather than linearity of migration, attachments to multiple places, the sense of belonging and collective identity that arises in winter communities, and the importance of a migratory lifestyle in maintaining a sense of independence in older age. Watkins (1999) narrated the story of four elderly persons whose lives were joined by place and migration. Although their moves were a predictable outgrowth of regular life-course transitions, they were intertwined with cumulative life experiences filled with feelings, memories, and perceptions.

Stack has been collecting life histories of African-Americans for more than twenty years. Her recent book, *Call to Home*, depicted the return migration of African Americans to the rural Carolinas as an outgrowth of a long-established north–south system of circulation (Stack 1996). Waves of earlier migrants from the rural South to the urban North maintained ties to the South through regular visits, ownership of property, remittances, and the sending of children home for summer vacation. Return migrants are now remaking the local culture and politics of the places they left behind twenty and thirty years ago.

International Migration

The magnitude of international migration increased recently due to the globalization of the international economy, widening regional economic and demographic differences, expanding social networks connecting countries and communities, the fall of the Soviet Union, and increasing ethnic strife and territorial conflict. In 1992, more than 100 million persons lived outside the country of their birth, representing almost 2 per cent of the world's population (Castles and Miller 1993: 4). A major conclusion to be gleaned from the research of population geographers on international migration is the deep dissatisfaction with the stereotypical view of immigration as a voluntary, complete, and permanent process. The notion of international immigration, therefore, is being broadened to capture the varied experiences of refugees; so-called non-immigrants who reside in the US and other countries for significant periods of time such as students, temporary workers, circular migrants, and expatriates; and undocumented workers (Kraly 1997).

Research reveals the ambiguity of separating refugees, strictly defined as those people living outside the country of their nationality and unwilling to return because of a "well-founded fear of persecution," from economic migrants (Jones 1989; Bascomb 1993; Wood 1994). Refugees are, in fact, motivated by a set of forces similar to those that influence other migrants such as regional disparities in income and welfare, the presence of kinship networks that provide information and support, insecurities associated with growing ethnic tensions, and the weakening of traditional values in throes of modernization. Wood (1994) argued that the view of refugees as humanitarian problems separated from their economic and political context leads to policies geared toward relieving short-term crises rather than addressing the longer-term and larger-scale causes of dislocation. Empirical support for the importance of economic factors for refugee movements comes from Jones (1989) who found that economic setbacks were more important than political violence in explaining the spatial distribution of origin areas for Salvadoran refugees to the United States. Political deaths were related to internal displacement but not to migration to the United States. Bascomb (1993) linked the Eritrean refugee resettlement process to agrarian transformation in Sudan in the 1980s. The shift from subsistence to a market economy and a growing shortage of land, exacerbated by drought in Sudan, led to growing economic marginalization and social differentiation of Eritrean refugees.

US immigration statistics are weakened by their failure to incorporate emigration, or movement from the country. Between 1900 and 1980, approximately 30 million immigrants came to the US, of whom nearly 10 million returned or moved on to another country (Warren and Kraly 1985). Approximately 195,000 foreign-born residents emigrate from the United States each year (Kraly 1997). In Canada, among every 100 immigrants, 30 to 45 eventually emigrate (Beaujot and Rappak 1989). Kraly and Warren (1991) tracked 10 million non-immigrant aliens to the US in 1983 and found that more than 100,000 departed after one year of stay. Although these individuals meet the United Nation's definition of long-term immigrants (a person who crosses an international boundary and lives in the country or intends to live in the country for more than one year), they do not appear in official records either as immigrants to or emigrants from the US. Kraly and Warren (1992) revised US immigration data to reflect the UN demographic concept of long-term immigration by considering new immigrants (permanent resident aliens) arriving in the year, temporary migrant arrivals (non-immigrants) who subsequently adjust to permanent resident status, arrivals of asylees and refugees, and non-immigrants who arrive during the year and stay more than 12 months before departing. Revised estimates are 12 per cent higher than Immigration and Naturalization Service's published estimates of immigration for 1983.

Traditional, data-driven conceptualizations of immigration also assume that an immigrant cuts his or her ties to home and maintains a single residence and single focus of activity in the new host country. Evidence mounts that many international migrants do not emigrate permanently but spend sojourns in North America and return to their homes in Mexico, the Caribbean, and other parts of Latin America and Asia, repeating this cycle a number of times during their lifetimes (E. Conway *et al.* 1990; Bailey and Ellis 1993). D. Conway (1997) recounted the long history of Barbadian circulation, emigration, and return migration which has created a complex network of friends, family, and community across the Caribbean and in Britain, Canada, and the US. Barbadian identity today is defined both by this tradition of mobility, the vast diaspora it has created, and a deep attachment to the Caribbean home.

Mountz and Wright (1996) used ethnographic techniques to describe the seamless web of interconnections between Mexican workers in Poughkeepsie, New York, and family and friends in the rural Zapotech community of San Agustin. Migrants are mainly males who work in Poughkeepsie and maintain regular, indeed daily, contact with wives, children, and extended families in San Agustin. In addition to returning for fiestas, funerals, and other village events and sending remittances to family members and in support of community activities, migrants telephone frequently, receive and send videotapes of community activities, and share gossip about daily life in the transnationized community. Migration to the US does not lead to a break with life in Mexico but establishes a new transnational scope to social and economic life.

An important sub-theme to emerge from studies of the transmigration process deals with remittances sent home by temporary and permanent migrants. Jones (1998) asked whether remittances increase or decrease income inequalities in origin regions and found that the answer depended upon the temporal and spatial scale of analysis. Inter-family inequalities first decrease then increase as a place's migration experience deepens, but throughout this process, rural incomes improve relative to urban ones. In relating remittances to migration, Jones found that migrants often invest their remittances in a way that makes them less dependent upon future migration. D. Conway and Cohen (1998) also linked migration and remittances in an ethnographic study of households in the Oaxacan village of Santa Ana del Valle. Migration, circulation, and the remittances they produce are rites of passage for young men, a way to finance expenses for young families, and a means of meeting festival and political expenses for older villagers. These processes affect village gender relations as women left behind gain economic independence, invest remittances, and assume more important roles in local government.

An important controversy centers on whether immigration to the US is linked to internal migration patterns of the native born. There is, in fact, a net out-migration of native born, especially the poor and unskilled, from high-immigration states and metropolitan areas (Frey 1995*a*, *b*, 1996*a*, *b*). Frey interpreted this empirical fact as evidence of "demographic balkanization," the geographic fragmentation of the nation into zones where immigration is the major force of demographic change and regions where immigrants are largely absent. In his view, poorly educated and low-income natives are pushed from high-immigration regions by increasing competition for jobs, social problems stemming from rapid demographic change, and prejudice associated with greater racial and ethnic diversity. Wright *et al.* (1997) disagreed with this interpretation and contended that high-immigration cities lose native born because of their size, not because of large immigrant flows to them. The economies of large cities are restructuring in ways that put severe pressure on the wages of the unskilled, leading to the out-migration of low-wage, unskilled labor. Further, Ellis and Wright (1999) cautioned against the use of the term "balkanization" to describe the immigration patterns, asserting that it invokes an image of racial disharmony and territorial conflict in the former Yugoslavia.

Wright and Ellis (1996, 1997) explored the effects of immigration on the intra-urban sectoral division of labor in Los Angeles and New York, the nation's two leading immigrant destinations. In New York, immigrants entered the labor market by filling vacancies left by retiring or out-migrating whites, particularly during the 1970s. Immigrants to Los Angeles were incorporated into the metropolitan economy through job growth rather than demographic succession. None the less, significant sectoral shifts did occur among the native born and immigrants to Los Angeles. Native-born whites moved from employment in manufacturing, services, and public administration to emerging sectors in the restructured economy; native-born blacks lost relative position in almost all sectors; and immigrants gained a comparative advantage in manufacturing, business services, hospitals, and social services.

Immigrant Assimilation and Adjustment and the Emergence of Immigrant Enclaves

New immigrant source areas and changing patterns of immigration led to the redistribution of ethnicity at various geographic scales. In a study of residential concentrations of twelve groups of new immigrants to Los Angeles, Allen and Turner (1996) found that US-born members of ethnic groups are more residentially dispersed than immigrants, recent immigrants are more likely to live in ethnic concentrations than those who arrived earlier, and immigrants who reside in ethnic concentrations are less fluent in English, have lower levels of educational attainment and lower incomes than those living outside such concentrations. But, contrary to expectations, a majority of recent immigrants do not live in concentrated ethnic zones, raising questions about the importance of ethnic concentration in the adjustment process of new immigrants. At the larger regional scale, Wong (1998) examined patterns of intermarriage as a surrogate for ethnic integration or segregation and found that western states, including Alaska and Hawaii, have high levels of integration, whereas Appalachia is least integrated.

Following on the pioneering work of Desbarets (1985) with Southeast Asian refugees, and in keeping with population geography's core interest in mobility and migration is research on the secondary migration of immigrants. Questions center on the interrelationships among movement behavior, immigrant adjustment, and immigrant concentrations. More specifically, research has addressed whether living in an immigrant enclave deters mobility (Kritz and Nogle 1994; Neuman and Tienda 1994; Newbold 1996); whether the foreign born move toward immigrant concentrations in their secondary migrations (Belanger and Rogers 1992; Frey 1995*a*; Nogle 1997; Gober 1999); whether immigrants with low levels of human capital, i.e. low educational attainment and poor English-speaking skills, tend to locate in immigrant concentrations for their ethnic support networks (Nogle 1997; McHugh *et al.* 1997), and how immigrant settlement systems evolve through time in response to changing opportunities and pressures for the immigrant group and for society at large (Newbold 1999). McHugh *et al.* (1997) used the concept of segmented assimilation to explain the multi-dimensional nature of recent Cuban migration in the US. Segmented assimilation says that the one-size-fits-all approach to immigrant adjustment grossly oversimplifies the way new immigrants, as individuals and as communities, engage their host society. The adjustment process is influenced by personal characteristics and cultural background, reasons for immigration (forced versus voluntary), the availability of local support services, labor market opportunities, and the nature of the local urban environment. Cuban migration streams consist of substreams, made of up immigrants differentiated by age, class, time of arrival, and tendencies to settle in the Miami area. The lure of Miami is stronger for poor than for affluent Cubans.

Population geographers also examined whether immigrant concentrations facilitate or deter economic and social assimilation. In a study of Indo-Chinese refugees in St Paul, Minnesota, Kaplan (1997) found geographic concentration to be a positive factor in the assimilation process. In a series of articles about ethnicity in Miami, Boswell and others concluded that "whether segregation is beneficial or harmful depends upon whether it is voluntary" (Boswell 1993; Boswell and Cruz-Báez 1997: 491; Boswell *et al.* 1998). Outside the US, Glavac and Waldorf (1998) studied Vietnamese immigrants in Australia, Selya (1992) looked at illegal migration in Taiwan, and in the edited volume, *EthniCity*, Roseman *et al.* (1996) provided cases studies of immigrant assimilation throughout Germany, France, Italy, the Netherlands, Spain, South Africa, Singapore, and Austria.

Regional Demographic Variability

Despite its overriding concern with mobility and migration, population geography continues to describe and explain regional patterns of demographic diversity, including studies of birth and death rates, sex ratios, population density, abortion rates, and age structure. In a study of geographic patterns of birth and death rates, fertility rates, and proportion of births to young mothers, Morrill (1993) found persistently wide regional disparities in demographic characteristics even in the face of converging levels of educational achievement, median income, and occupational opportunity and national systems of communication and advertising. Despite these homogenizing processes, immigration and innovation renew regional demographic diversity. Supporting evidence is provided by Pandit and Bagchi-sen (1993) who established growing divergence in regional fertility rates between 1970 and 1990. Traditional North–South distinctions were replaced by East–West differences. Regional ethnic composition was increasingly an important factor in accounting for the spatial variation in US fertility. Pandit (1992*a*, *b*) also examined the effects of migration on inter-regional fertility differentials in the US.

Challenging conventional wisdom about population redistribution in the US, Fonseca and Wong (2000) found no significant relationship between population growth and increases in population density at the state level between 1980 and 1990. Despite the now familiar patterns of population growth favoring the South and West, states of the northeastern megalopolis and older large metropolitan areas of the North experienced increases in density on a par with southern and western states. Densification of the nation thus is occurring in a great range of places—all of which need to be attentive to issues of urban sprawl, environmental degradation, traffic congestion, and open space preservation.

Gober (1994, 1997) investigated the relative roles of demographic demand factors versus supply considerations in explaining wide state-to-state disparities in abortion rates in the US. Although the right to an abortion is theoretically the law of the land as outlined in the Supreme Court's 1973 *Roe* v. *Wade* decision, states exert widely differing levels of control on access to abortion services. Results show that demand factors, such as state income levels, the percent Catholic, and the mobility status of the population influence abortion rates indirectly through effects on state laws restricting abortion access and on the availability of abortion services. In a cross-national comparison of abortion rates, rights, and access in the US and Canada, Gober and Rosenberg (2001) found growing regional differences in the incidence of and access to abortion in the two countries, despite different abortion histories, cultural attitudes, and health-care systems.

Arguably no aspect of demographic diversity has attracted more attention in recent years than the geographical aspects of aging, with obvious implications for health care, social security and pensions, the provision of local services, and prospects for future growth. In 2000, the population over 65 varied from 5.7 per cent of Alaska's population to 17.6 per cent of Florida's (US Bureau of the Census 2000*b*). During the past several decades, aging-in-place has been an especially strong component of elderly change because large birth and immigrant cohorts of the early part of the twentieth century passed into seniorhood. The US elderly population grew by 75 per cent between 1970 and 2000 compared to national population growth of only 39 per cent (US Bureau of the Census 1970, 2000*a*).

One issue involving the distribution of elderly is the relative importance of elderly migration versus aging-in-place in shaping regional processes of aging in the US (Rogers 1993*a*; Frey 1995*a*), Canada (Moore and McGuinness 1999), and other industrialized nations (Rogers 1993*b*). Frey (1995*a*) identified four types of US states: (1) elderly in-migration states; (2) elderly out-migration states; (3) high aging-in-place states; and (4) low aging-in-place states. In Canada, Moore and McGuinness (1999) found that almost 43 per cent of census divisions experienced significant aging-in-place ameliorated by migration, 28 per cent experienced both aging-in-place and increased aging due to migration, fewer than 10 per cent of areas were either stable or experienced a decline in aging, and the remaining 20 per cent were subject to migration-dominated aging. Population aging is an economic disadvantage in Canada because communities with limited local resources often shoulder a disproportionate burden from growth in elderly populations.

Rogerson (1999) found that the baby boom generation displays a bicoastal pattern. Areas such as the upper Midwest and Deep South, where the baby boom cohort initially was concentrated, experienced net declines over the past three decades, suggesting that there has been a reduction in the spatial concentration of baby boomers. Tying together the distributions of baby boomers and their elderly parents, Rogerson *et al.* (1997) demonstrated that functional limitations and the need for support closes the distance between family members. Parents who are not well off also are more likely to live with adult children than more affluent elders. In another study, Lin and Rogerson (1995) found that rural parents have less access to children than urban ones.

Social Theory and Population Processes

Throughout the 1990s population geography has been criticized for not addressing broader debates of human geography, namely social theory and gender issues (Findlay and Graham 1991; Fincher 1993; P. White and Jackson 1995). Other population geographers, however, suggest that many of these critiques have been misplaced, and that population geography does not warrant its position as straw dog. Ogden (1998: 105), for example, contends that "population geography as a field is not in crisis." To Ogden and others (Skeldon 1995), population geography has a long-standing engagement with broader theoretical, philosophical, and methodological debates.

It is true, however, that social theory has permeated the subfield of population geography more thoroughly in the United Kingdom than in North America (Findlay and Graham 1991; Halfacree and Boyle 1993; McHugh 2000*b*). Also, a review of the literature indicates that "much population geography is written by those who do not think of themselves" as population geographers (Ogden 1998: 105).

Incorporation of social theory into population geography involves the search for alternative approaches to the study of migration as well as integration of new subject matter into the field, most notably a confrontation with race and gender. With respect to methods, population geographers in the 1990s advocated more ethnographic field work (Li and Findlay 1996; McHugh and Mings 1996; Findlay and Li 1997; McHugh 2000*b*). As McHugh (2000*a*, *b*) suggested, ethnography opens up thorny issues of authority, positionality, and representation in research. Moreover, ethnographic approaches allow researchers to "enliven migration studies in geography and foster linkages with other branches of the discipline, opening up new vistas in migration, culture and society" (McHugh 2000*b*: 85–6).

Population geographers also integrated structuration theory into their work. Structuration offers the potential to combine elements of both structural and behavioral approaches, which are otherwise largely disparate. In their widely cited article, Goss and Lindquist (1995) incorporated elements of structuration in a case study of the highly institutionalized Philippine labor migration industry. They suggested that international migration is "best examined as the articulation of agents with particular interests and playing specific roles within an institutional environment, drawing knowledgeably upon sets of rules in order to increase access to resources" (ibid. 345). Halfacree (1995) likewise used structuration to examine the gendered character of migration within the United States.

Population geographers provided more humanistic understandings of migration. Miles and Crush (1993) and Vandsemb (1995), for example, utilized personal narratives to understand the personal dimensions of migration. Miles and Crush (1993: 92) argued, for example, that "life-history collection opens a window on the struggles of ordinary people." Vandsemb (1995: 415) concurred, writing that "narratives make it possible to look at actual decisions and actions and to perceive behind these practices the network of social relations that allowed them to take place."

The diversity of methodologies employed by population geographers is perhaps best illustrated within the realm of gender and migration. Whereas scholars such Bailey, Cooke, Shumway, and Waldorf maintained a strong link with positivism, they provided empirically grounded research documenting significant gender differences in the migration experience of women and men (Waldorf 1995; Cooke and Bailey 1996; Bailey and Cooke 1998; Shumway and Cooke 1998). Most importantly, they provided concrete evidence that women, and especially married women, are most often negatively affected in their earning potential owing to patriarchal structures within households. These findings were supported by a series of articles examining the gendered migration experiences of Puerto Rican women (Conway *et al.* 1990; Bailey and Ellis 1993; Ellis *et al.* 1996). Arguing for a merging of structural and behavioral approaches, these researchers identified that "duration of mainland sojourns is determined not so much by job or economic-structural factors as by the gendered nature of women's responsibilities" (Ellis *et al.* 1996: 46). Similarly, Fan and Huang (1998) interpreted Chinese female marriage migration from a combination of structural and individual perspectives. They suggested that "peasant women in disadvantaged positions are motivated to interpret migration as not simply a life event, but as an alternative to their limited social and economic mobility" (Fan and Huang 1998: 246).

One of the most significant contributions on the study of gender and migration is Chant's (1992) edited volume entitled *Gender and Migration in Developing Countries.*

Momsen and Radcliffe advocated a household strategies approach. In this way, researchers are better able to examine the confluence of intra-household resource and decision-making structures, hierarchies of power within households, and socially determined, gender-segregated labor markets (Chant and Radcliffe 1992: 23).

Drawing on the aforementioned structuration approaches advocated by Halfacree (1995) and Goss and Lindquist (1995), Tyner (1994, 1996*b*, 1999*a*, *b*) forwarded an institutional approach to examine the manifestation of patriarchal structures within systems of government-sponsored international labor migration. Using the Philippines as a case study, Tyner qualitatively examined how the recruitment and deployment of labor is organized around specific gendered assumptions of male versus female occupations, and how this contributes to an increased vulnerability of female migrant workers. In addition, Tyner (1996*a*, 1997) examined how gendered representations influence the construction of migration policies designed to protect migrant workers.

Lastly, Tyner (1998, 1999*b*) examined the social construction of race and nation via the formation of immigration legislation. Through examinations of the incarceration of Japanese-Americans and also on the exclusions of Philippine immigrants from the US, Tyner detailed how the scientific study and control of population has been central to the construction of race and nation. He concluded that "Jim Crow laws, zoning restrictions, and anti-miscegenation laws have been employed to maintain a separation of races. Immigration legislation, likewise, has historically been employed as a means of restricting the unwanted in the construction of the state" (Tyner 1999*a*: 71).

Population Geography and Public Policy

Inspired by Morrill's (1981) seminal work on legislative redistricting, population geographers continued to pursue research questions that inform public policy at the local, regional, and national levels. Much of this work, however, is not in scholarly books or professional journals but takes the form of informal reports about population change for local and regional planning boards; population projections and other demographic analyses for communities and school districts; participation on national boards that tackle population-related issues such as urban growth, regional redistribution, immigration, and the census; and legal consulting in matters of legislative redistricting and school-district boundaries.

In a series of articles using California cities as examples, Clark and Morrison demonstrated the power of population geography and demographic analysis to inform the process of drawing voting districts (Clark and Morrison 1991, 1995; Morrison and Clark 1992). The Voting Rights Act of 1965 prohibits the drawing of district boundaries in a way that lessens the voting strength of a legally protected group. This seemingly straightforward mandate is subject to myriad legal challenges, often on the basis of the demographic circumstances of a particular community. Important to these circumstances are structural factors that include age and citizenship characteristics of minority group members, socio-economic status and the propensity to vote, and geographic considerations such as the degree of spatial concentration of a particular group. Forming a district that empowers blacks is often easier because they are typically concentrated in cities, in contrast to Asians who are more scattered in their residential patterns. Hispanics fall between these two extremes (Clark and Morrison 1995).

Motivated by the legal question of whether local school boards, by manipulating the boundaries of school attendance areas, are liable for maintaining and increasing segregation, Clark (1987, 1995) evaluated the relative importance of demographic change versus changing boundaries on the racial composition of local schools. He found that racial change in schools is more strongly related to local demographic change than to school district boundaries and concluded that "geography and demography have modified the effect of *Brown* and the potential for future judicial intervention in schools systems" (Clark 1995: 664).

At the regional scale, White (1994) related population redistribution in western Kansas to a proposed policy to convert the region into a "Buffalo Commons." He noted the fallacy in assuming that the region is depopulating and is thus ripe for abandonment. In fact, population is redistributing itself away from rural areas and small towns in favor of places of more than 500 inhabitants that are sustained by groundwater exploitation from the Ogallala aquifer. Rather than recreate a Buffalo Commons in the High Plains, regional policy should, according to White, favor the concentration of redevelopment efforts on regional centers with groundwater access—places that are large enough and have resources enough to take advantage of new economic opportunities in the region.

The recent rise in immigration has reactivated national concern with overcrowded housing (officially defined as more than one person per room). During most of this century, the incidence of overcrowding steadily declined, and scholarly and policy interest in the subject waned. Since 1980, however, overcrowding is on the rise, especially among renters. Immigration is one explanation. New immigrants often share accommodations with family and friends until they are able to establish independent households. Myers and Lee (1996) showed that immigrants experience significant reductions in overcrowding with increased duration of residence owing to rising incomes. Controlling for income, Hispanics experience markedly higher rates of overcrowding than Asians or non-Hispanic whites.

Myers *et al.* (1996) explored the policy ramifications of maintaining a national standard of overcrowdedness in an increasingly multicultural society. Policies to increase supply and ease affordability constraints may not achieve the policy goal of reducing overcrowding because the incidence of overcrowding is more closely linked to the distribution of Hispanics and immigrants than to housing supply and affordability variables. They also show that, controlling for both household size and income, Asian and Hispanic households are much more likely to live in overcrowded households than either white or black households. That overcrowding still remains in Asian and Hispanic households with incomes more than twice the average suggests that overcrowding may be more a matter of personal preference than affordability or supply constraints. The implication is that national crowding standards should be relaxed to reflect the growing cultural diversity in our society. Rather than judge households by a single middle-class majority standard, we should allow for greater social expression from place to place.

Conclusions

Now, almost fifty years since Trewartha's call for the study of population as a distinct branch of geography, the subdiscipline of population geography faces both challenges and opportunities related to its distinctive status. Arguably, its most significant challenge lies in how to respond to P. White and Jackson's (1995) call for a more critical, qualitative, and socially engaged research agenda without abandoning altogether the field's core strengths in spatial demography and statistical analysis. With a few notable exceptions, population geographers have ignored rather than engaged social theorists in debates about subject matter, context, data, and methods. As a result, the field—at least as practiced by people who call themselves population geographers—is somewhat disconnected from the epistemological debates that have engaged human geography in the past ten years.

A second, and to some extent, related challenge is the growing fragmentation of the field. Although population geographers studied a wide variety of topics twenty-five years ago, it was possible to identify prevailing paradigms around which research was organized, including the neoclassical model of migration, the concept of place utility, the process of residential mobility, and migration decision-making. Today, it is far more difficult to discern the big ideas that drive work in the field. The breadth of this review is symptomatic of the stunning variety of research that fits within the rubric of population geography. Carried to an extreme, however, few, if any, topics reach the point of critical mass, and the potential for cumulative generalization is limited.

Strategic opportunities for population geography lie in reaching out to rapidly growing new branches of the discipline, in other words, in looking outward rather than inward. The field of environmental geography is expanding rapidly as the discipline responds to national calls for greater collaboration between science and social science and for more mission-oriented, policy-relevant research. Despite the fact that population is at the heart of most environmental problems, population geographers have paid relatively little attention to environmental issues and have engaged infrequently with colleagues in physical geography. The process of human migration, traditionally the core interest of population geography, is largely responsible for concentrating population in urban areas and in coastal locations with ramifications for the build-up of carbon dioxide in the atmosphere, the declining health of coastlines, reductions in biodiversity, and degradation of land. Fruitful new avenues of research lie in articulating the relationships between population dynamics and environmental problems at a range of geographic scales from the local to the global.

Considerable potential exists in building stronger ties to geography's vibrant GIS community. Population geographers have been slow to integrate geographic information analysis into studies of migration, demographic diversity, residential segregation, and residential mobility even as spatial analysis, the mainstay of population geography, is being revolutionized by GIS. All too often, aspatial statistical methods are used to study the inherently spatial processes of population geography. One important exception is Wong's (1997) use of GIS to incorporate interactions between adjacent areal units and the geometric characteristics of areal units in measures of residential segregation. Given the statistical and technical bent of many population geographers and the widespread availability of georeferenced population data, population geography should be a leader in the use of GIS to solve substantive research questions.

And finally, population geography should confront, rather than ignore, geography's growing concern with gender relations, racism, agism, and class conflicts. Population geographers have yet, for example, to respond to Jackson's call for a more critical assessment of race and segregation. Over a decade ago Jackson (1989: 176) argued that "Geographical work has been concerned mainly with measuring patterns of minority-group concentration and dispersal generally, failing to ask questions about the *social* significance of *spatial* segregation." Population events and processes occur in a wider social world, demanding greater diversity of ontological, epistemological, and methodological tools. We see the new millennium as a time of great potential and promise for a vigorous expansion of the field of population geography while, concurrently, not losing sight of its strong foundation.

References

Allen, J. P., and Turner, E. (1996). "Spatial Patterns of Immigrant Assimilation." *The Professional Geographer*, 48/2: 140–55.

Bailey, A. J., and Cooke, T. J. (1998). "Family Migration and Employment: The Importance of Migration History and Gender." *International Regional Science Review*, 21/2: 99–118.

Bailey, A. J., and Ellis, M. (1993). "Going Home: The Migration of Puerto Rican-Born Women from the United States to Puerto Rico." *The Professional Geographer*, 45/2: 148–58.

Bascomb, J. (1993). "The Peasant Economy of Refugee Resettlement in Eastern Sudan." *Annals of the Association of American Geographers*, 83/2: 320–46.

Beaujot, R., and Rappak, J. P. (1989). "The Link between Immigration and Emigration in Canada, 1945–1986." *Canadian Studies in Population*, 16: 201–16.

Belanger, A., and Rogers, A. (1993). "The Internal Migration and Spatial Redistribution of the Foreign-Born Population in the United States: 1965–70 and 1975–80." *International Migration Review*, 26/4: 1342–69.

Boswell, T. D. (1993). "Racial and Ethnic Segregation Patterns in Metropolitan Miami, Florida, 1980–1990." *Southeastern Geographer*, 33/1: 82–109.

Boswell, T. D., and Cruz-Báez, A. D. (1997). "Residential Segregation by Socioeconomic Class in Metropolitan Miami: 1990." *Urban Geography*, 18/6: 474–96.

Boswell, T. D., Cruz-Báez, A. D., and Zijlstra, P. (1998). "Housing Preferences and Attitudes of Blacks toward Housing Discrimination in Metropolitan Miami." *Urban Geography*, 19/3: 189–210.

Brown, L. A. (1991). *Place, Migration and Development in the Third World: An Alternative View*. London: Routledge.

Brown, L. A., Lobao, L., and Digiacinto, S. (1999). "Economic Restructuring and Migration in an Old Industrial Region," in Pandit and Withers (1999: 37–58).

Castles, S., and Miller, M. J. (1993). *The Age of Migration: International Population Movements in the Modern World*. New York: Guilford Press.

Chant, S. (ed.) (1992). *Gender and Migration in Developing Countries*. New York: Belhaven Press.

Chant, S., and Radcliffe, S. A. (1992). "Migration and Development: the Importance of Gender," in Chant (1992: 1–29).

Clark, W. A. V. (1987). "Demographic Change, Attendance Area Adjustment and School District impacts." *Population Research and Policy Review*, 6: 199–222.

—— (1992). "Comparing Cross-Sectional and Longitudinal Analyses of Residential Mobility and Migration." *Environment and Planning A*, 24: 1291–302.

—— (1995). "The Expert Witness in Unitary Hearings: The Six Green Factors and Spatial-Demographic Change." *Urban Geography*, 16/8: 644–79.

Clark, W. A. V., and Dieleman, F. M. (1996). *Households and Housing: Choices and Outcomes in the Housing Market*. New Brunswick, NJ: Rutgers University Press, Center for Urban Policy Research.

Clark, W. A. V., and Morrison, P. A. (1991). "Demographic Paradoxes in the Los Angeles Voting Rights Case." *Evaluation Review*, 15/6: 712–26.

—— (1995). "Demographic Foundations of Political Empowerment in Multiminority Cities." *Demography*, 32/2: 183–201.

Clark, W. A. V., and Withers, S. D. (1999). "Changing Jobs and Changing Houses: Mobility Outcomes of Employment Transitions." *Journal of Regional Science*, 39/4: 653–73.

Clark, W. A. V., Duerloo, M. C., and Dieleman, F. M. (1994). "Tenure Changes in the Context of Micro-level Family and Macro-level Economic Shifts." *Urban Studies*, 31: 137–54.

Clausen, J. A. (1986). *The Life Course: A Sociological Perspective*. Englewood Cliffs, NJ: Prentice-Hall.

Conway, D. (1997) "Why Barbados Has Exported People: International Mobility as a Fundamental Force in the Creation of Small Island Society," in J. M. Carrion (ed.), *Ethnicity, Race and Nationality in the Caribbean*. San Juan: Institute of Caribbean Studies, University of Puerto Rico, 274–308.

Conway, D., and Cohen, J. H. (1998). "Consequences of Migration and Remittances for Mexican Transnational Communities." *Economic Geography*, 74/1: 26–44.

Conway, E., Ellis, M., and Shiwdhan, N. (1990). "Caribbean International Circulation: Are Puerto Rican Women Tied-Circulators?" *Geoforum*, 2/1: 51–66.

Cooke, T. J., and Bailey, A. J. (1996). "Family Migration and the Employment of Married Women and Men." *Economic Geography*, 72: 38–48.

Cushing, B. (1999). "Migration and Persistent Poverty in Rural America," in Pandit and Withers (1999: 15–36).

Davies, R. B., and Pickles, A. R. (1985) "Longitudinal Versus Cross-Sectional Models for Behavioral Research: A First-Round Knock Out." *Environment and Planning A*, 17: 1315–29.

— (1991). "An Analysis of Housing Careers in Cardiff." *Environment and Planning A*, 23: 629–50.

Demko, G. J., Rose, H. M., and Schnell, G. A. (1970). "The Geographic Study of Population," in G. J. Demko, H. M. Rose, and G. A. Schnell (eds.), *Population Geography: A Reader*. New York: McGraw-Hill, 1–5.

Desbarets, J. (1985). "Indochinese Resettlement in the United States." *Annals of the Association of American Geographers*, 75: 522–38.

Dieleman, F. M. (1995). "Using Panel Data: Much Effort, Little Reward?" *Environment and Planning A*, 27: 676–81.

Dieleman, F. M., and Shouw, R. J. (1989). "Divorce, Mobility and Housing Demand." *European Journal of Population*, 5: 235–52.

Dieleman, F. M., Clark, W. A. V., and Duerloo, M. C. (1995). "Falling Out of the Homeowner Market." *Housing Studies*, 10: 3–15.

Duerloo, M. C., Clark, W. A. V., and Dieleman, F. M. (1994). "The Move to Housing Ownership in Temporal and Regional Context." *Environment and Planning A*, 26: 1659–70.

Easterlin, R. (1980). *Birth and Fortune: The Impact of Numbers on Personal Welfare*. New York: Basic Books.

Elder, G. H. (1977). "Family History and the Life Course." *Journal of Family History*, 2: 278–304.

Ellis, M., and Wright, R. (1999). "The Balkanization Metaphor in the Analysis of US Immigration." *Annals of the Association of American Geographers*, 88: 686–98.

Ellis, M., Conway, D., and Bailey, A. J. (1996). "The Circular Migration of Puerto Rican Women: Towards a Gendered Explanation." *International Migration*, 34/1: 31–62.

Fan, C. C. (1996). "Economic Opportunities and Internal Migration: A Case Study of Guangdong Province, China." *The Professional Geographer*, 48/1: 28–45.

Fan, C. C., and Huang, Y. (1998). "Waves of Rural Brides: Female Marriage Migration in China." *Annals of the Association of American Geographers*, 88/2: 227–51.

Fielding, T. (1992). "Migration and Culture," in T. Champion and T. Fielding (eds.), *Migration Processes and Patterns*, i. *Research Progress and Prospects*. London: Belhaven Press, 201–12.

Fincher, R. (1993). "Commentary: Gender Relations and the Geography of Migration." *Environment and Planning A*, 25: 1703–5.

Findlay, A., and Graham, E. (1991). "The Challenge Facing Population Geography." *Progress in Human Geography*, 15: 149–62.

Findlay, A. M., and Li, L. N. (1997). "The Meaning of Migration: A Biographical Approach to Understanding Hong Kong Emigration," *Area*, 29: 33–44.

Fonseca, J. W., and Wong, D. W. (2000). "Changing Patterns of Population Density in the United States." *Professional Geographer*, 52/3: 504–17.

Frey, W. H. (1995*a*). "Elderly Demographic Profiles of US States: Aging-in-Place, Migration and Immigration Impacts." *Research Report No. 95–325*. Ann Arbor: Population Studies Center, University of Michigan.

— (1995*b*). "Immigration, Domestic Migration, and Demographic Balkanization in America: New Evidence for the 1990s." *Population and Development Review*, 22: 741–63.

— (1995*c*). "Immigration Impacts on Internal Migration of the Poor: 1990 Census Evidence for US States." *International Journal of Population Geography*, 1/1: 51–67.

— (1996*a*). "Immigration and Internal Migration 'Flight' from US Metropolitan Areas: Toward a New Demographic Balkanization." *Urban Studies*, 32: 733–57.

— (1996*b*). "Immigrant and Native Migrant Magnets." *American Demographics*, November: 1–5.

Geyer, H. S., and Kontuly, T. (1993). "A Theoretical Foundation for the Concept of Differential Urbanization." *International Regional Science Review*, 15/2: 157–77.

Glavac. S. M., and Waldorf, B. (1998). "Segregation and Residential Mobility of Vietnamese Immigrations in Brisbane, Australia." *Professional Geographer*, 50/3: 344–57.

Gober, P. (1992). "Urban Housing Demography." *Progress in Human Geography*, 16/2: 171–89.

— (1994). "Why Abortion Rates Vary: A Geographical Examination of the Supply of and Demand for Abortion Services in the United States in 1988." *Annals of the Association of American Geographers*, 84/2: 230–50.

— (1997). "The Role of Access in Explaining State Abortion Rates." *Social Science and Medicine*, 44/7: 1003–16.

— (1999). "Settlement Dynamics and Internal Migration of the U.S. Foreign-Born Population," in Pandit and Withers (1999: 231–49).

Gober, P., and Rosenberg, M. W. (2001). "Looking Back, Looking Around, Looking Forward: A Woman's Right to Choose," in I. Dyck, N. Lewis, and S. McLafferty (eds.), *Geographies of Women's Health*. London: Routledge, 88–104.

Goss, J., and Lindquist, B. (1995). "Conceptualizing International Labor Migration: A Structuration Perspective." *International Migration Review*, 29/2: 317–51.

Halfacree, K. H. (1995). "Household Migration and the Structuration of Patriarchy: Evidence from the USA." *Progress in Human Geography*, 19/2: 159–82.

Halfacree, K. H., and Boyle, P. J. (1993). "The Challenge Facing Migration Research: The Case for a Biographical Approach." *Progress in Human Geography*, 17: 333–48.

Jackson, P. (1989). "Geography, Race, and Racism," in R. Peet and N. Thrift (eds.), *New Models in Geography: The Political-Economy Perspective*. London: Unwin Hyman, 176–95.

Jokisch, B. D. (1997). "From Labor Circulation to International Migration: The Case of South-Central Ecuador." *Yearbook, Conference of Latin Americanist Geographers*, 23: 63–75.

Jones, R. C. (1989). "Causes of Salvadoran Migration to the Untied States." *Geographical Review*, 79/2: 183–94.

— (1998). "Remittances and Inequality: A Question of Migration Stage and Geographic Scale." *Economic Geography*, 74/1: 8–25.

Kaplan, D. H. (1997). "The Creation of an Ethnic Economy: Indochinese Business Expansion in Saint Paul." *Economic Geography*, 73/2: 214–33.

Kontuly, T. (1997). "Political Unification and Regional Consequences of German East–West Migration." *International Journal of Population Geography*, 3/3: 31–47.

Kontuly, T. (1998) "Contrasting the Counterurbanisation Experience in European Nations," in P. J. Boyle and K. H. Halfacree (eds.), *Migration into Rural Areas: Theories and Issues*. London: Wiley, 61–78.

Kontuly, T., and Bierens, H. J. (1990). "Testing the Recession Theory as an Explanation for the Migration Turnaround." *Environment and Planning A*, 22: 253–70.

Kontuly, T., and Schon, K. P. (1994). "Changing Western German Internal Migration Systems during the Second Half of the 1980s." *Environmental and Planning A*, 26: 1521–43.

Kraly, E. P. (1997). "Emigration: Implications for U.S. Immigration Policy Research," in E. Loaza and S. Martin (eds.), *Mexico/U.S. Migration Patterns, Research Papers*. Washington, DC: Commission on Immigration Reform.

Kraly, E. P., and Warren, R. (1991). "Long-Term Immigration to the United States: New Approaches to Measurement." *International Migration Review*, 25/1: 60–92.

— (1992). "Estimates of Long-Term Immigration to the United States: Moving US Statistics toward United Nations Concepts." *Demography* 29/4: 613–28.

Kritz, M. M., and Nogle, J. M. (1994). "Nativity Concentration and Internal Migration among the Foreign-Born." *Demography*, 31: 509–24.

Li, L. N., and Findlay, A. M. (1996). "Placing Identity," *International Journal of Population Geography*, 2: 361–78.

Li, S., and Sui, Y. (2001). "Residential Mobility and Urban Restructuring under Market Transition: A Study of Guangzhou, China." *Professional Geographer*, 53/2: 219–29.

Lin, G., and Rogerson, P. A. (1995). "Elderly Parents and the Geographic Availability of their Adult Children." *Research on Aging*, 17: 303–31.

Long, L. (1988). *Migration and Residential Mobility in the United States*. New York: Russell Sage.

McHugh, K. E. (2000*a*). "The 'Ageless Self'? Emplacement of Identities in Sun Belt Retirement Communities." *Journal of Aging Studies*, 14/1: 103–15.

— (2000*b*). "Inside, Outside, Upside Down, Backward, Forward, Round and Round: A Case for Ethnographic Studies in Migration." *Progress in Human Geography*, 24/1: 71–89.

McHugh, K. E., and Mings, R. (1996). "The Circle of Migration: Attachment to Place in Aging." *Annals of the Association of American Geographers*, 86: 530–50.

McHugh, K .E., Skop, E. H., and Miyares, I. M. (1997). "The Magnetism of Miami: Segmented Paths in Cuban Migration." *Geographical Review*, 87: 504–19.

Miles, M., and Crush, J. (1993). "Personal Narratives as Interactive Texts: Collecting and Interpreting Migration Life-Histories." *Professional Geographer*, 45/1: 84–94.

Moore, E. G., and McGuinness, D. (1999). "Geographical Dimensions of Aging," in Pandit and Withers (1999: 139–73).

Morrill, R. L. (1981). *Political Redistricting and Geographic Theory*. Resource Publications in Geography. Washington, DC: Association of American Geographers.

— (1993). "Development, Diversity, and Regional Demographic Variability in the U.S." *Annals of the Association of American Geographers*, 83/3: 406–33.

Morrison, P. A., and Clark, W. A. V. (1992). "Local Redistricting: The Demographic Context of Boundary Drawing." *National Civic Review*, Winter: 57–63.

Mountz, A., and Wright, R. A. (1996). "Daily Life in the Transnational Migrant Community of San Agustin, Oaxaca, and Poughkeepsie, New York." *Diaspora*, 5/3: 403–25.

Myers, D., and Lee, S. W. (1996). "Immigration Cohorts and Residential Overcrowding in Southern California." *Demography* 33/1: 51–65.

Myers, D., Baer, W. C., and Choi, S. (1996). "The Changing Problem of Overcrowded Housing." *American Planning Association Journal*, 62/1: 66–84.

Neuman, K. E., and Tienda, M. (1994). "The Settlement and Secondary Migration Patterns of Legalized Immigrants: Insight from Administrative Records," in B. Edmonston and J. Passel (eds.), *Immigration and Ethnicity: The Integration of America's Newest Immigrants*. Lanham, Md.: Urban Institute Press, 187–226.

Newbold, K. B. (1996). "Internal Migration of the Foreign-born in Canada." *International Migration Review*, 30: 728–47.

— (1999). "Evolutionary Immigrant Settlement Patterns: Concepts and Evidence," in Pandit and Withers (1999: 250–72).

Nogle, J. M. (1997). "Internal Migration Patterns for US Foreign-Born, 1985–1990." *International Journal of Population Geography*, 3/1: 1–13.

Odland, J., and Shumway, J. M. (1993). "Interdependencies in the Timing of Migration and Mobility Events." *Papers of the Regional Science Association*, 72: 221–37.

Ogden, P. E. (1998). "Population Geography." *Progress in Human Geography*, 22/1: 105–14.

Pandit, K. (1992*a*). "Regional Fertility Differentials and the Effect of Migration: An Analysis of U.S. State Level Data." *Geographical Analysis*, 24/4: 352–64.

— (1992*b*). "Snowbelt to Sunbelt Migration and the North–South Fertility Convergence: Exploring the Theoretical Links." *Southeastern Geographer*, 32: 138–47.

— (1997*a*). "Cohort and Period Effects in US Migration: How Demographic and Economic Cycles Influence the Migration Schedule." *Annals of the Association of American Geographers*, 87/3: 439–50.

— (1997*b*). "Demographic Cycle Effects on Migration Timing and the Delayed Mobility Phenomenon." *Geographical Analysis*, 29/3: 187–99.

— (2000). "Regional Variation in Mobility Levels and Timing in the United States." *Professional Geographer*, 53/3: 483–93.

Pandit, K., and Bagchi-sen, S. (1993). "The Spatial Dynamics of U.S. Fertility, 1970–1990." *Growth and Change*, 24: 229–46.

Pandit, K. and Withers, S. D. (1999). *Migration and Restructuring in the U.S.: A Geographic Perspective*. Boulder: Rowman & Littlefield.

Plane, D. A. (1992). "Age-Composition Change and the Geographical Dynamics of Interregional Migration in the U.S." *Annals of the Association of American Geographers*, 82: 64–85.

— (1998). "Fuzzy-Set Migration Regions." *Geographical & Environmental Modeling*, 2/2: 141–62.

— (1999). "Migration Drift." *The Professional Geographer*, 51/1: 1–11.

Rogers, A. (1992*a*). "Elderly Migration and Population Redistribution in the United States," in A. Rogers (ed.), *Elderly Migration and Population Redistribution*. London: Belhaven, 226–48.

— (1992*b*). *Elderly Migration and Population Redistribution*. London: Belhaven.

Rogers, A., and Wilson, R. T. (1996). "Representing Structural Change in U.S. Migration Patterns." *Geographical Analysis*, 28/1: 1–17.

Rogerson, P. A. (1987). "Changes in U.S. National Mobility Levels." *The Professional Geographer*, 39: 344–51.

Rogerson, P. A., Burr, J. A., and Lin, G. (1997). "Changes in Geographic Proximity between Parents and their Adult Children." *International Journal of Population Geography*, 3/2: 121–36.

Roseman, C. C., Laux, H. D., and Thieme, G. (eds.) (1996). *EthniCity*. London: Rowman & Littlefield.

Selya, R. M. (1992). "Illegal Migration in Taiwan: A Preliminary Overview. *International Migration Review*, 26/3: 787–805.

Shumway, J. M., and Cooke, T. J. (1998). "Gender and Ethnic Concentration and Employment Prospects for Mexican-American Migrants." *Growth and Change*, 29: 23–43.

Skeldon, R. (1995). "The Challenge Facing Migration Research: A Case for Greater Awareness." *Progress in Human Geography*, 19/1: 91–6.

Stack, C. (1996). *A Call to Home: African Americans Reclaim the Rural South*. New York: Basic Books.

Trewartha, G. (1953). "A Case for Population Geography." *Annals of the Association of American Geography*, 43: 71–97.

Tyner, J. A. (1994). "The Social Construction of Gendered Migration from the Philippines." *Asian and Pacific Migration Journal*, 3/4: 589–617.

—— (1996*a*). "Constructions of Filipina Migrant Entertainers." *Gender, Place and Culture: A Journal of Feminist Geography*, 3/1: 77–93.

—— (1996*b*). "The Gendering of Philippine International Labor Migration." *The Professional Geographer*, 48/4: 405–16.

—— (1997). "Constructing Images, Constructing Policy: The Case of Filipina Migrant Performing Artists." *Gender, Place and Culture: A Journal of Feminist Geography*, 4/1: 19–35.

—— (1999*a*). "The Geopolitics of Eugenics and the Exclusion of Philippine Immigrants from the United States." *Geographical Review*, 89/a: 54–73.

—— (1999*b*). "The Global Context of Gendered Labor Migration from the Philippines to the United States." *American Behavioral Scientist*, 42/4: 665–83.

US Bureau of the Census (1970). Census of Population, 1970, i.

US Bureau of the Census (2000*a*). 2000 Census of Population and Housing, Profiles of General Demographic Characteristics, <http://quickfacts.census.gov/qfd/states/12000.html>, last accessed 4 October 2001.

US Bureau of the Census. (2000*b*). 2000 Census of Population and Housing, State and County Quickfacts. http://quickfacts.census.gov/qfd/states/12000.html, last accessed 4 October 2001.

Vandsemb, B. H. (1995). "The Place of Narrative in the Study of Third World Migration: The Case of Spontaneous Rural Migration in Sri Lanka." *The Professional Geographer*, 47/4: 411–25.

von Reichert, C., and Rudzitis, G. (1992). "Multinomial Logistic Models Explaining Income Changes of Migrants to High-Amenity Counties." *Review of Regional Studies*, 2/1: 25–42.

—— (1994). "Rent and Wage Effects on the Choice of Amenity Destinations of Labor Force and Nonlabor Force Migrants: A Note." *Journal of Regional Science*, 34/3: 444–55.

Waldorf, B. (1995). "Determinants of International Return Migration Intentions." *The Professional Geographer*, 47/2: 125–36.

Warren, R., and Kraly, E. P. (1985). "The Elusive Exodus: Emigration from the United States." *Population Trends in Public Policy*, 8. Washington, DC: Population Reference Bureau.

Watkins, J. F. (1999). "Life Course and Spatial Experience: A Personal Narrative Approach in Migration Studies," in Pandit and Withers (1999: 294–312).

White, P., and Jackson, P. (1995). "(Re)theorizing Population Geography." *International Journal of Population Geography*, 1/2: 111–21.

White, S. E. (1994). "Ogallala Oases: Water Use, Population Redistribution, and Policy Implications in the High Plains of Western Kansas, 1980–1990." *Annals of the Association of American Geographers*, 84/1: 29–45.

White, S. E., Brown, L. A., Clark, W. A. V., Gober, P., Jones, R., McHugh, K. E., and Morrill, R. L. (1989). "Population Geography," in G. L. Gaile and C. J. Willmott (eds.), *Geography in America*. Columbus: Merrill, 258–89.

Withers, S. D. (1997). "Methodological Considerations in the Analysis of Residential Mobility: A Test of Duration, State Dependence, and Associated Events." *Geographical Analysis*, 29/4: 354–74.

Wong, D. W. S. (1997). "Spatial Dependency of Segregation Indices." *Canadian Geographer*, 40: 128–36.

—— (1998). "Spatial Patterns of Ethnic Integration in the United States." *The Professional Geographer*, 50/1: 13–30.

Wood, W. B. (1994). "Forced Migration: Local Conflicts and International Dilemmas." *Annals of the Association of American Geographers*, 84/4: 607–34.

Wright, R., and Ellis, M. (1996). "Immigrants and the Changing Racial/Ethnic Division of Labor in New York City: 1970–1990." *Urban Geography*, 17/4: 317–53.

—— (1997). "Nativity, Ethnicity, and the Evolution of the Intraurban Division of Labor in Metropolitan Los Angeles." *Urban Geography*, 18: 243–63.

Wright, R. A., Ellis, M., and Reibel, M. (1997). "The Linkage between Immigration and Internal Migration in Large Metropolitan Areas." *Economic Geography*, 73/2: 234–54.

Zelinsky, W. (1966). *A Prologue to Population Geography*. Englewood Cliffs, NJ: Prentice-Hall.

CHAPTER 14

Sexuality and Space

Glen Elder, Lawrence Knopp, and Heidi Nast

Introduction

> I have noticed with some dismay in recent years the appearance of tables representing various strange groups attending meetings of the Association of American Geographers. Marxist Geographers and Gay Geographers come to mind, and I wonder what next? Are we going to have a table of Whores in Geography, and Russian Communist Geography? ... As for special tables, rooms and meeting times for such groups as Gay Geographers, we should flatly refuse any such groups the right to such representation. When engaging in their gay behavior they are not acting as geographers. . . . Our exclusion of such groups cannot be taken as a moralistic stand on the part of the Association, but simply as a professional one. It is not our business to support the Gay or the Street Walkers, or the Democrats or the Republicans. None of these groups, though they may have members or practitioners in geography, can be said to be geographers, *per se*. They should then not be permitted official or even associative status at our meetings. We have plenty to do in geography, and room for greater diversity of professional interest than almost any other society. There are, however, limits. We should confine our meetings to geography by geographers and for geographers. All others keep out.
>
> Carter (1977: 101–2)[1]

[1] Thanks to William Koelsch for alerting us to the existence of this letter.

In 1996, the Sexuality and Space Specialty Group (SSSG) came into being as a forum for addressing the sorts of sentiments expressed in the letter above, and for exploring the unquestioned heterosexuality of the geographical enterprise. While the sentiments expressed may seem extreme, they point to disciplinary resistances to certain lines of inquiry. The comments and the subsequent creation of the SSSG reveal how the topical contours of geography are, and always have been, politically negotiated. Until recently, sexuality research in geography had been considered especially out of place (see Valentine 1998; Chouinard and Grant 1995).

Organized collectively under the aegis of the AAG, the SSSG represents considerable political will and work. Its presence underscores how marginalized groups can never take for granted their place in society, including the academy. Inclusion of the SSSG in the AAG (and in this volume) attests both to the skills and persistent efforts of geographers, who found their research topic choices sidelined or frowned upon in the past, and to larger shifts in socio-political contexts, concerns and opportunities. Today, gay men, lesbians, and Marxist geographers hold positions of governance in the national association. We not only meet openly at conferences, we publish, edit journals, and otherwise assert our presence in the discipline.

The SSSG is one of the AAG's newest specialty groups. While it is not made up only (or even primarily) of gay and lesbian geographers, its existence is a result of mainly lesbian, bisexual, and gay male and feminist geographers working, despite overt and subtle hostility, rejection, and indifference, to analyze sexuality using geographical tools and frameworks. The SSSG does not limit its focus to the study of gay, bisexual, or lesbian lives, but fosters and encourages interrogation of interrelationships between all sorts of sexualities and the spatiality of everyday life.

The governing by-laws of the specialty group include the following objectives:

1. To encourage geographic research and scholarship on topics related to sexuality.
2. To promote educational ways for communicating geographic perspectives on sexuality that will inform both curricular and pedagogical needs.
3. To promote interest in geography on issues related to sexuality.
4. To promote the exchange of ideas and information about intersections of geography and sexuality.

For the SSSG to achieve these goals, it has become clear that the discipline needs to understand how heterosexual norms and expectations reside in geography's cultural practices and theorizings. In other words, our task is to chip away at our discipline's deeply flawed heteronormative representations of the world.

This chapter introduces the reader to how geographers have studied geography, and how we can do so in the future. We first point to recent analyses and trends in sexuality research in parts of Anglo-Western and non-Anglo-Western worlds, respectively, particularly those related to non-heterosexualities. Analyses of non-heterosexualities importantly indicate the socially constructed, oppressive, and often violent ways in which contemporary heterosexual practices work, and foreground alternative sexuality experiences that shape how sex is imagined, practiced, and theorized spatially. Lastly, we review and provide examples of geography's heterosexism and recent theoretical trends in sexuality research in geography.

As is typical of much academic practice, nuanced, reflexive theorizing has not always been matched by or applied in the context of equally nuanced empirical work. In part this is due to the rather vexing methodological and ethical problems implied by such theorizing. But at least as important is the fact that empirical work in the area has always been, and remains, *dangerous*—for the powerful as well as the powerless, within and outside academia (see e.g. Valentine 1998; England 1999; Nast 1999). Furthermore, a dialectic involving (theoretically underdeveloped) inductive and (empirically underdeveloped) deductive reasonings operates in our subfield just as it does in most others. It should come as no great surprise, then, that some of the earliest empirical work concerning sexuality and geography was not necessarily strongly informed theoretically. Nor should this be viewed as particularly troubling. Simply raising sexuality as an issue was itself an enormously courageous act; doing so in a theoretically critical way entailed profound professional, if not personal, consequences.

Non-Heterosexualities in the Anglo-Western World

In the world of English-language geography, conscious and explicit engagements with non-heterosexualities can probably be said to have begun in the late 1970s and early 1980s, with the works of Ketteringham (1979, 1983), Weightman (1980, 1981), and McNee (1984, 1985). McNee's work in particular was important, for at least two reasons. First, it was inspired (if not explicitly informed) by a broader social and geographic theory, anarchism, which had a respectable (though, in the context of the Cold War, marginal) niche in the canon of geographic thought at the time (largely through the work of Russian geographer Pietr Kropotkin). Second, and at least as important, was the fact that it was accompanied by various forms of activism at AAG meetings and elsewhere, including the public advertising of informal "gay caucus" meetings and bold, iconoclastic gender/sexuality performances in paper sessions. In one session, for example, McNee presented in drag. More typically at others, he affected a hypermasculine cowboy persona. He also sponsored and publicized a tour of one AAG site's red light district, including significantly its gay and lesbian spaces. The result was that the discipline's living corpus—its members—was forced to confront the interrelated issues of gender and sexuality in ways most members had heretofore been able to avoid. McNee himself, who had been an economic geographer of some renown, became quite marginalized (if somewhat as an endearing curiosity) in the discipline as well as in his own department and the local community of business and government in Cincinnati, where he lived and worked.[2]

The scholarly and activist interventions of McNee *et al.* were informed by the emerging lesbian and gay rights movements in the US and elsewhere in the West. To the extent that this movement tended to take for

[2] Fortunately for him, he found a new place of belonging in the Cincinatti gay and lesbian community.

granted modern Western notions of sexuality, identity, and social process, this activist scholarship did not question gender and sexual categories as much as it might have.[3] Nor did it question the urban, Western, and "diffusionist" bias (cf. Blaut 1977, 1987, 1992) inherent in geographical studies of social change. The bulk of the early work inspired by and/or coeval with it, therefore, tended to reproduce (and even harden) these notions. Levine (1979), Murray (1979), Lee (1980), Murphy (1980), Castells and Murphy (1982), Lauria and Knopp (1985), and Knopp (1986, 1987), for example, all focused on non-heterosexualities in urban, Western contexts. Throughout the 1980s, studies of gay and/or lesbian neighborhoods, urban spatial concentrations of gays and lesbians, and inner-city urban change such as gentrification predominated.

This first wave of scholarly work on relationships between geography and sexuality was cast theoretically largely in terms of some variant of urban political economy, usually Marxist. But because traditional Marxism offered little in the way of an explicit theorization of sexualities, scholars turned to various feminists (e.g. Snitow *et al.* 1983; Mackenzie and Rose 1983), Marxist theories of the family (e.g. Zaretsky 1976; Stone 1977), and the emerging field of gay and lesbian studies (e.g. Adam 1978; Wolfe 1978; Weeks 1977, 1985; D'Emilio 1981, 1983*a*, *b*; Escoffier 1985) for help. The result was an at times awkward and uneasy blending of masculinist and heteronormative Marxist social theory with a much more critical and contextually sensitive, but still mostly underdeveloped, collection of other critical approaches. Some British and Canadian as well as US geographers were among the first to undertake a theoretical and political critique of this union (see Adler and Brenner 1992; Bell 1991, 1992, 1993, 1995; Binnie 1992, 1993; Davis 1991, 1992; Ingram 1993; Rothenberg and Almgren 1992; Rothenberg 1995; Valentine 1993*a*, *b*). Even this work, however, has tended to focus on urban Western contexts as the presumed center of cultural innovation and resistance to heterosexual hegemony (for notable exceptions, see Kramer 1995; for a recent effort to examine sexuality outside the urban, if still Western, metropolis see Phillips *et al.* 2000).

Most recently, there has been a proliferation of English-language geographical scholarship on sexualities cast explicitly in terms of queer theory, including the beginnings of a focus on non-urban, if still largely Western, contexts and processes (e.g. Brown 1997, 2000; Ingram *et al.* 1997; Bell *et al.* 1994; Callard 1996; Peake 1993; Elder 1999; England 1999; Gibson-Graham 1999; Knopp 1999, 2000; Nast 1999). These and other new works are cast in terms of intellectual and political frameworks that seek to deconstruct binarisms and other categorizations that defy the complexity and fluidity of human social life and experience. In so doing, most newer geographical studies seek to disclose and challenge a range of hierarchical and unjust power relations characterizing contemporary Western and, presumably, non-Western societies.[4] These approaches are as yet still quite underdeveloped and will need ultimately to be joined to efforts emerging from non-Western contexts (see below). They will also need to confront the dilemma of how to define a politics and world-view that is at once flexible yet clear in its sympathies and commitments. The challenges are significant, if not daunting. Yet they are clearly welcome, as the proliferation of new work, new conference sessions, and new activisms within and outside the academy make clear.

Non-Heterosexualities in Non-Western Contexts

In an effort to provide evidence of the different ways in which sexuality is inscribed and encoded across space, geographers have extended their gaze beyond the confines of Western urban contexts. While a great deal of this work has also focused on non-heterosexuality, Nast (forthcoming) has theorized the foundational geographical importance of state concubinage in an early Nigerian city-state and the immanent place of "race" and heteronormativity in various nationalist and colonial projects (Nast 1998, 2000). Besides this work and a recent call by Seager (1997) for geographers to examine the heterosexual and global trafficking in women, work on sexuality in non-Western contexts and in geography specifically, has tended strongly to focus on the non-heterosexual. This work has sought to reveal how the

[3] At the same time, it was probably somewhat more conscious of the constructedness and fluidity of these categories than some of the early work inspired by it (e.g. Lauria and Knopp 1985). What it lacked was a vocabulary (like queer theory) that would allow the discussion and extension of this into other realms such as economic geography, for example.

[4] Some such work, however, is deliberately ambiguous about the implied political commitments and orientations of authors; this in turn, has precipitated some critical discussions about the politics of "queer" (cf. e.g. Bell, Binnie, Cream, and Valentine 1994, and Knopp 1995).

strict and interconnected policing of gender and sex as experienced in Western contexts is not universal.

Work in this regard has convincingly revealed that gender bimorphism does not underpin sexuality in all contexts. Put another way, sexuality is not universally tied to the Western heteropatriarachal model of masculinity and femininity. Empirical case studies from South Africa (Elder 1995, 1998) and India (Balachandran 1996), for example, demonstrate how the politics of desire do not follow a universal pattern. Instead, in both cases, such politics reveal that the histories and geographies of places shape the terms of sexual political engagement in those places. In fact, the terms "gay" and "lesbian" become vacuous categories when abstracted from their Western settings. Indeed, Mayer (2000) reveals that all sexual identities make little sense when removed from their local, regional, or national settings. In short, the politics of sex are always contextual.

More recently, and informed by post-colonial critiques, work on sexuality in geography has shown that the situated experiences of groups and individuals who do not observe context-specific heterosexuality, which some have called queer (see Elder 1999), can and should inform analyses of the Western gay, lesbian, and/or queer landscapes. Queerness as a term does not universalize the experience of those who do not practice heterosex, but highlights the *contextual* nature of that oppositional (and usually non-heteronormative, non-monogamous) desire. "Queer" by this definition is a term of political engagement and not necessarily an identity.

From this perspective, and by contrast, the notion of a gay and/or lesbian identity, under the conditions of postmodernity, are free-floating but nevertheless powerful signifiers of sexual struggle. The political struggles of North American and Western European urban queers who are known or identify as gay and lesbian is now ubiquitous. Gay Pride festivals, for example, are now annual summer events in both hemispheres and in cities as diverse as Harare, Zimbabwe, and Montreal, Canada. In fact, the international visibility of gay and lesbian life has led to an unproblematized celebration of global community (e.g. Miller 1993), something recently called into question (*Antipode* 2002). By using the term "queer", in contrast, geographers (e.g. Gibson-Graham 1996) celebrate the local albeit connected, oppositional, and contested, politics of sexuality. By locating and localizing the politics of desire in this way, such local, national, and regional politics are not held up to the now internationally recognized standards of a gay and lesbian political identity model emanating from Western urban contexts.

How Sex Works in Space

Thus far, this chapter has sought to describe previous works. Moving on from that overview of the field, we seek to ask three interrelated questions: First, how might sex be theorized sociospatially? Second, what sorts of spatial, cultural, political, and economic work does sexuality do? Third, how and why has it been so effective? These are questions that an increasing number of geographers are beginning to explore, though they often do so within a heterosexualized context, without explicitly naming or recognizing the limited qualities of this form of sexuality *per se.* Tyner (1996), for example, describes the spatiality of heterosex work in various locales, but do not comment or elaborate on the specificity of the sexuality (*hetero*sex) that they address, an indication that their work is located within (or at least does not explicitly prioritize the undermining of) a heteronormative imaginary. Nor do they examine the dynamics of desire that propel the heterosex they describe—something that is essential if we are to go beyond the nomothetic.

Geographers exploring *non*-heterosexualities also leave desire untheorized, dwelling on descriptive qualities such as: *where* non-heterosexuality takes place; *who* is involved, what kinds of problems "they" have, and either how they are oppressed and/or working in emancipatory ways (see Nast 1998 for a relevant literature review; also, see below). In so doing, these works leave the domain of sex and desire stranded from political, cultural, historical, economic, and social-geographic reasonings.[5]

All sexualities, however, have logics produced through specific histories and geographies. Nuclear versions of the family (codified by Freud as "oedipal") for example, emerged coevally with eighteenth- and nineteenth-century industrial capitalisms, nationalisms, trans-Atlantic slavery, and colonization. The productive forces of each of these institutions commingled with, overlapped, reinforced, dissipated, contradicted, and generally overdetermined one another. Nature at this time was fetishized and eroticized as maternal land in need of direction and thus colonization and conquest, with women simultaneously being described as biologically incapable of participating in newly emergent secular nation-states' civic domains. Moreover, ideals of maternity itself were representationally tied to chastity and passivity, *vis-à-vis* men, with such maternal

[5] In this sense, the title of the 1995 text, *Mapping Desire* is misleading; none of the chapters theorize what constitutes desire.

qualities being extolled in heteropatriarchal images of many nation states (Nast 1998; see Mayer 2000 for international empirical evidence of the relationship between gender, nation, and sex).

Racism too was integral to nineteenth-century constructions of the oedipal or nuclear family. Ideals of motherhood in the United States, for example, were racialized and racistly split. On the one hand, "good" daughters symbolically embodied an idyllic future motherhood and were represented in terms of a purity colored white. Good daughters/mothers passively followed white male wisdom and guidance. On the other hand, black women's bodies were disparaged through representations strategically dovetailing into white racist exploitation of black bodies (Nast 2000). Images of manhood were similarly racistly driven: On the one hand, white sons were touted as the embodiment of a superior fatherhood inherently able to create and sustain moral law; on the other hand, black men were infantilized as immoral and shiftless "sons" incapable of attaining real paternal status. Moreover, black sons were constructed as sexually desirous of white daughters and mothers, the symbolic embodiment of the worst filial crime—incest. These strategically racist depictions of exploited labor help account for why many white settler communities across colonial and neocolonial time and place are separated from colored "others" through spatial forms of containment such as native towns, bantusans, reservations, and segregated black areas. In the United States, black men are still infantilized as "boys" or "sons" to the extent that they are cast as dangerously desirous of white women. The latter, in turn, continue to epitomize idylls of good motherhood. The black male body is hence in many instances feared, unconsciously symbolized as an abomination deserving of spatial and/or bodily castration and death (Nast 2000). This brief discussion draws attention to a deficit of theorizing sexuality and sex in geography, discussed below.

(Hetero)sex in Theory

The sociospatial effects of heterosexist language and practices in geography have been paid scant attention, though several scholars outside geography have explicitly theorized heterosex's spatiality. We argue that the earliest, and most detailed *theoretical* introspection into the inherent sexualization of the discipline and its theories is found in Rose (1993). She centers her arguments on the masculinism of time geography, cultural geography, and humanistic geographers. Though she does not say so explicitly, it is clear from her discussion that the masculinity she identifies as embedded in geographical theories is ensconced in Western ideals and practices of heterosexuality, especially nuclear familial settings wherein women are identified as nature and called to submit to a heteropatriarchal order. For this reason, in the discussion that follows, we use the word "heteromasculinity" in place of her word, masculinity.

Rose (1993) locates time geography, in particular, in the repressive regime of "objective" social scientific inquiry, which does not consider or validate non-white, non-male otherness, belying structural racism and heteromasculinism (below). Rose argues that the unconscious maleness of time geography reflects systematic fears about the maternal. At the same time, she argues that humanistic and cultural geography have a special fascination with the maternal, encoded theoretically through an emphasis on subjective experience and on romanticized aesthetics of maternal nature, feeling, and place. This combination of subjectivity and maternal love is epitomized, perhaps, in the notion of topophilia.

Rose claims that objective and subjective modalities of masculinity feed off one another. Her work suggests that many human geographers, especially those invested in field work, phenomenology, and the cultural, are embedded in a language of maternal nostalgia, something Porteous's (1990: 68–85) less theoretically nuanced work makes abundantly clear. This language, unconsciously or consciously, implicitly or explicitly, betrays infantile fantasies of union with a pure and plenitudinous maternal earth. Such fantasies of maternal union between man and nature are similarly displaced onto geographical renderings of landscape and nature. Time geographers, in contrast, draw upon universal images of the white adult man, implicitly shunning the infantile desires theorized as inherent in topophilia. In so doing, time geography effaces the maternal body and embodied difference altogether.

We would argue that two additional geographical domains gather force from, and contribute to shoring up, heteromasculinity: physical geography and Marxist geography. Physical geography is grounded historically in heterosexualized practices of voyeurism and looking, tied historically to the language and exercises of colonization and mapmaking. In this sense, it implicitly conceives of nature and earth as maternal, both of the former having been called to order through imperializing conquests of might and science. The sexual politics of conquest (earth and mother metaphorically equated) are carried into the geographical present through veiled heterosexual allusions that inhere in current technological practices and rhetorics involving, for example, field probes and Peeping Tom satellites. In this instance, the "global" view replaces the national one in "science," a new science that expresses desires through developing means for colonizing, controlling, and mastering phenomena that are global in reach (global warming, global informational systems, etc.). The objective, in both senses of the word, involves a forceful positioning and reading of the world as a playground for male probes. Probing can be infantile (the child's finger exploring or cataloguing mother/earth's body), adolescent (racing to know mother/earth's secrets before others do), or paternal (reducing mother/earth to manipulable raw data and statistics). In any case, the raw material of mother/earth/landscape/nature is rendered as incapable of speech and therefore carefully (scientifically) spoken for. In this sense, "objective science" is artifice, a sham intended to shield the hetero-masculine subject from view.

Marxist geography similarly engages the hetero-masculine in theoretical and empirical ways. On the one hand, Marxist geographers provide an analytical structure for entertaining embodied geographical difference: production is envisaged as male and reproduction as female. On the other hand, this structure places two different versions of the heteromasculine (discussed above) next to one another: the infantile (or topophilic and subjective) and the paternal (or scientific and objective). The feminine, when positioned alongside these two, forms a third term that makes for a more complex framework that strikingly resembles the nuclear family: mother–father–son (Kobayashi and Nast 1996: 84). In this case, brawny sons (laborers) and capitalist patriarchs struggle over who will control an otherwise alienated mother/nature. Early Marxian discussions of the "emasculinization" of labor speak volumes about how theorized tensions between labor and capital encapsulate and reproduce heteromasculine, familial anxieties over who controls the maternal, encoded once again as nature, resources, or earth. Marxist geographical theorizing thus not only excluded and suppressed maternal agency (reproduction is passive), it inflated the explanatory potential, and thereby the value of, the two synergistic modes of the masculine: capital (as paternal cunning and logic) and labor, within which lies the culturally loaded notion of "struggle." Masculinized Marxist identification with labor's muscular struggles helps explain historical antagonisms between Marxist and humanist endeavors. For Marxists, humanists are "soft," which might be read in terms of being infantile—or tied to the mother, whereas for humanistic geographers, Marxists are "hard"—or adolescent. In either case, both sets of theorists are positioned as sons, albeit at different stages in life. In contrast, theorists of "science" occupy, like "capital," a position of hetero-paternal authority.

How the two modalities of the heteromasculine (father and son) have been used in geographical theory mirrors how the two have been used practically in the world at large. Western colonization and industrial capitalisms, for example, were embodied by many different renditions of competing father–son desires. Witness the celebratory popular images of brave explorers and the desires of colonizing sons to have and to hold maternal/landscapes not theirs to own. Moreover, witness the misogynistic paternal desires of industrialists and financiers to exploit mother/nature so as to pursue profit for profit's sake. And then there are the infantile fascinations and nostalgia for nature, registered in detailed scientific cataloguings of flora and fauna.

Heteromasculine subordination of the maternal-earth-nature-environment (as passively open to scrutiny) is of course resisted most emphatically by feminist geographers. But until the racism and heteromasculinity of geography is systematically interrogated and undone in the context of the larger sociospatial, historical, and cultural structures alluded to above, "other" voices will continue to be muffled in the main circuits of geographical knowledge production.

Conclusion

Efforts to place questions of sexuality firmly on geographical agenda in the US culminated in 1996 with the establishment of the SSSG, its creation part of a larger shift in societal relations. Geographers have brought special attention to the spatiality of sexuality, that is, how sexuality is organized and given meaning spatially through praxis. The SSSG has also sought to foster interdisciplinary debate between sexuality scholars because along with advances that Anglo geography has made in studying sexuality, a parallel debate about the spatialities of sexualities has been unfolding outside geography and sometimes with very little input from geographers (e.g. see the 1997 edited collection *Queers in Space* by Ingram, Bouthillette, and Retter; Colomina 1992, *Sexuality and Space*). Fortunately, and as a result of the highly geographical nature of the sexuality debate, leading authors on sexuality have willingly participated in the annual geography meetings. The outcomes of these interdisciplinary exchanges have been fruitful. Presentations at AAG annual conferences by George Chauncey, author of

Gay New York (1995) in 1995,[6] Eve Kosofsky Segwick, author of several texts including the ground-breaking work *Epistemology of the Closet* (1992) in 1996, and in 1998 Frank Browning and Will Fellows, authors of *A Queer Geography* (1996) and *Farm Boys* (1998) respectively, stimulated and extended interdisciplinary engagements.[7]

By problematizing the geographical imagination as (amongst other things) a heterosexual one, we bring into question geography's origins, epistemologies, languages, experiences, and paradigms. By building on the insights of feminist geographers, we have argued that by accepting maleness and femaleness as analytical categories in and of themselves, most geographers sidestep the procreational norms through which the world is sexually structured and known. By procreational, we mean the many practical and symbolic ways in which notions of modern motherhood, fatherhood, and (nuclear, heterosexed) family life insinuate their ways into cultural bodies, places, and imaginings: from constructions of normative nuclear familial life and goals, to heteropatriarchal framings of the nation-states, to the sexualized language with which many of us write or explain the world (Nast 1998).

One effect of sidestepping procreational biases in our thinking is that sex in general, and heterosex in particular seems innocuous or invisible in geography. Yet, heterosexuality presents itself in the discipline and practice of geography in many ways. Most problematically, heterosex is unconsciously buried in our epistemologies and concerns. Part of the reason for this burial is that sex (any sort) has until recently been empirically disregarded. Unlike in anthropology, in geography sexual practices and spaces have been tittered at as embarrassingly unutterable (McNee 1984). If anything, sex is imagined to be located somewhere "out there," in societally marginal places or on bodies deemed heterosexually deviant or deviantly non-heterosexual. In this sense, geography clearly operates within and reproduces what Butler (1990: 151 n. 6) calls the "heterosexual matrix":

> that grid of cultural intelligibility through which bodies, genders, and desires are naturalized . . . a hegemonic discursive/epistemic model of gender intelligibility that assumes that for bodies to cohere and make sense there must be a stable sex expressed through a stable gender (masculine expresses male, feminine expresses female) that is oppositionally and hierarchically defined through the compulsory practice of heterosexuality.

Simply put, the SSSG seeks to chart a navigable path through geography's heterosexual matrix for all geographers. The SSSG also seeks to create an intellectual home for geographers interested in the relationships between sexualities and space. The intellectual engagement of geographers with sexuality research holds tremendous potential for theorizing *and* practicing emancipatory geographic futures. One of the principles that the SSSG incorporated into the group's by-laws, for example, is the guarantee that there is sexed diversity in its operations. In particular, we mandated a co-chairship to be assumed by those "who occupy different sexual subject positions," an attempt to validate the fluidity of sexuality and the different experiences different sexualities bring to bear in geographic research and teaching.

[6] For a written account of this exchange, see Elder (1996).

[7] For a written account of this exchange, see Elder (2000).

References

Nine of the listed references are papers given at conferences and may be difficult to locate. However, it is a conscious strategy by us to include these papers for two reasons. First, as a political strategy we argue that drawing on different kinds of sources strengthens our chapter. Second, several of the authors were graduate students or junior faculty at the time who risked (and some lost) careers by asking questions about geography's sexual biases. Our reliance on these difficult-to-locate references thereby acknowledges and attempts to reclaim those voices.

Adam, B. (1978). *The Survival of Domination*. New York: Elsevier North-Holland.

Adler, S., and Brenner, J. (1992). "Gender and Space: Lesbians and Gay Men in the City." *International Journal of Urban and Regional Research*, 16: 24–34.

Balachandran, C. (1996). "Geographies of Gay India at Home and the Diaspora: Some Thoughts." Presented at Annual Meeting of the Association of American Geographers.

Bell, D. (1991). "Insignificant Others: Lesbian and Gay Geographies." *Area*, 23: 323–9.

Antipode (2002). "Queer Patriarchies, Queer Racisms," International. Special Issue, 34/5: 835–998.

— (1992). "What We Talk About When We Talk About Love." *Area*, 24: 409–10.

— (1993). "The Politics of Sex: Queer as Fuck?" Presented at New Theoretical Directions in Political Geography conference, University of Birmingham.

— (1995). "Pleasure and Danger: The Paradoxical Spaces of Sexual Citizenship." *Political Geography*, 14: 139–53.

Bell, D., and Valentine, G. (1995). *Mapping Desire*. London: Routledge.

Bell, D., Binnie, J., Cream, J., and Valentine, G. (1994). "All Hyped Up and No Place to Go." *Gender, Place and Culture*, 1: 31–47.

Binnie, J. (1992). "Fucking Among the Ruins: Postmodern Sex in Postindustrial Places." Presented at Sexuality and Space Network conference on Lesbian and Gay Geographies, University College London.

— (1993). "Invisible Cities/Hidden Geographies: Sexuality and the City." Presented at Social Policy and the City conference, Liverpool.

Blaut, J. (1977). "Two Views of Diffusion." *Annals of the Association of American Geographers*, 67: 343–9.

— (1987). "Diffusionism: A Uniformitarian Critique." *Annals of the Association of American Geographers*, 77: 30–47.

— (1992). *The Colonizer's Model of the World: Geographical Diffusionism and Eurocentric History*. New York: Guilford.

Brown, M. (1997). *Replacing Citizenship: AIDS Activism and Radical Democracy*. New York: Guilford.

— (2000). *Closet Space: Geographies of Metaphor from the Body to the Globe*. New York: Routledge.

Browning, F. (1996). *A Queer Geography: Journeys toward a Sexual Self*. New York: Crown Publishers.

Butler, J. (1990). *Gender Trouble*. New York: Routledge.

Callard, F. (1996). "The Body in Theory." Presented at the Annual Meeting of the Association of American Geographers.

Carter, G. (1977). "A Geographical Society Should Be a Geographical Society." *Professional Geographer*, 29: 101–2.

Castells, M., and Murphy, K. (1982). "Cultural Identity and Urban Structure: The Spatial Organization of San Francisco's Gay Community," in N. Fainstein and S. Fainstein (eds.), *Urban Policy under Capitalism*. Beverly Hills: Sage, 237–59.

Chauncey, G. (1995). *Gay New York*. New York: Basic Books.

Chouinard, V., and Grant, A. (1995). "On Not Being Anywhere Near 'the Project': Ways of Putting Ourselves in the Picture." *Antipode* 27: 137–66.

Colomina, B. (1992). *Sexuality and Space*. New York: Princeton Architectural Press.

Davis, T. (1991). "'Success' and the Gay Community: Reconceptualizations of Space and Urban Social Movements." Presented at First Annual National Graduate Student Conference on Lesbian and Gay Studies.

— (1992). "Where Should We Go from Here? Towards an Understanding of Gay and Lesbian Communities." Presented at the 27th International Geographical Congress.

D'Emilio, J. (1981). "Gay Politics, Gay Communities: the San Francisco Experience". *Socialist Review*, 55: 77–104.

— (1983*a*). "Capitalism and Gay Identity," in A. Snitow, C. Stansell, and S. Thompson (eds.), *Powers of Desire: The Politics of Sexuality*. New York: Monthly Review Press, 100–13.

— (1983*b*). *Sexual Politics, Sexual Communities: The Making of a Homosexual Minority in the United States, 1940–1970*. Chicago: University of Chicago Press.

Elder, G. (1995). "Of Moffies, Kaffirs, and Perverts: Male Homosexuality and the Discourse of Moral Order in the Apartheid State," in Bell and Valentine (1995: 56–65).

— (1996). "Reading between the Spaces in George Chauncey's Gay New York." *Society and Space*, 14: 755–8.

— (1998). "The South African Body Politic: Space, Race and Heterosexuality," in H. Nast and S. Pile (eds.), *Places Through the Body*. New York: Routledge, 153–64.

— (1999). "Queerying Boundaries in the Geography Classroom." *Journal of Geography in Higher Education*, 86–93.

— (2000). "Queers, Space, Queer-Space and Environment: A Review Essay of *A Queer Geography* by Frank Browning and *Farm Boys* by Will Fellows." *Professional Geographer*, 155–8.

England, K. (1999). "Sexing Geography, Teaching Sexualities." *Journal of Geography in Higher Education*, 23: 94–101.

Escoffier, J. (1985). "Sexual Revolution and the Politics of Gay Identity." *Socialist Review*, 15: 119–53.

Fellows, W. (1996). *Farm Boys: Lives of Gay Men from the Rural Midwest*. Wisconsin: University of Wisconsin Press.

Gibson-Graham, J. K. (1996). "Querying Globalization," in J. K. Gibson-Graham (ed.), *The End of Capitalism (As We Knew It)*. Oxford: Blackwell, 120–47.

— (1999). "Queer(y)ing Capitalism in and out of the Classroom." *Journal of Geography in Higher Education*, 23: 80–5.

Ingram, G., (1993). "Queers in Space: Towards a Theory of Landscape, Gender and Sexual Orientation." Presented at Queer Sites conference, University of Toronto.

Ingram, G., Bouthillette, A., and Retter, Y. (1997). *Queers in Space: Communities/Public Places/Sites of Resistance*. Seattle: Bay Press.

Ketteringham, W. (1979). "Gay Public Space and the Urban Landscape: A Preliminary Assessment." Paper delivered to the Association of American Geographers.

— (1983). "The Broadway Corridor: Gay Businesses as Agents of Revitalization in Long Beach, California." Paper delivered to the Annual Meeting of the Association of American Geographers.

Knopp, L. (1986). "Gentrification and Gay Community Development: A Case Study of Minneapolis." Paper delivered to the Annual Meeting of the Association of American Geographers.

— (1987). "Social Theory, Social Movements and Public Policy: Recent Accomplishments of the Gay and Lesbian Movements in Minneapolis, Minnesota." *International Journal of Urban and Regional Research*, 11: 243–61.

— (1995). "If You're Going to Get All Hyped Up You'd Better Go *Somewhere!*" *Gender, Place and Culture*, 2: 85–8.

— (1999). "Out in Academia: The Queer Politics of one Geographer's Sexualisation." *Journal of Geography in Higher Education*, 23: 116–23.

— (2000). "A Queer Journey to Queer Geography," in P. Moss (ed.), *Engaging Autobiography: Geographers Writing Lives*. Syracuse, NY: Syracuse University Press, 78–98.

Kobayashi, A., and Nast, H. (1996). "(Re)corporealizing Vision," in N. Duncan (ed.), *BodySpace*. New York: Routledge, 75–96.

Kramer, J. (1995). "Bachelor Farmers and Spinsters: Gay and Lesbian Identities and Communities in Rural North Dakota," in Bell and Valentine (1995: 182–99).

Lauria, M., and Knopp, L. (1985). "Towards an Analysis of the Role of Gay Communities in the Urban Renaissance." *Urban Geography*, 6: 152–69.

Lee, D. (1980). "The Gay Community and Improvements in the Quality of Life of San Francisco." MCP thesis (University of California).

Levine, M. (1979). "Gay Ghetto." *Journal of Homosexuality*, 4: 363–77.

Mackenzie, S., and Rose, D. (1983). "Industrial Change, the Domestic Economy and Home Life," in J. Anderson, J. Duncan, and R. Hudson (eds.), *Redundant Spaces in Cities and Regions?* London: Academic Press, 155–200.

McNee, B. (1984). "If You Are Squeamish . . ." *East Lakes Geographer*, 19: 16–27.

— (1985). "Will Gays Find Justice in the Queen City?" *Urban Resources*, 2: C1–C5.

Mayer, T. (2000). *Gender Ironies of Nationalism: Sexing the Nation*. New York: Routledge.

Miller, N. (1993). *Out in the World: Gay and Lesbian Life from Buenos Aires to Bangkok*. New York: Vintage Books.

Murphy, K. (1980). "Urban Transformations: A Case Study of the Gay Community in San Francisco." Berkeley: MCP thesis (University of California).

Murray, S. (1979). "The Institutional Elaboration of a Quasi-Ethnic Community." *International Review of Modern Sociology*, 9: 165–77.

Nast, H. (1998) "Unsexy Geographies," *Gender, Place and Culture*, 5: 191–206.

— (1999). "'Sex', 'Race' and Multiculturalism: Critical Consumption and the Politics of Course Evaluations." *Journal of Geography in Higher Education*, 23: 102–15.

— (2000). "Mapping the 'Unconscious.'" *Annals of the Association of American Geographers*, 90/2: 215–55.

— (forthcoming). *Concubines and Power*. Minneapolis: University of Minnesota Press.

Pile, S. (1996) *The Body and the City: Psychoanalysis, Space and Subjectivity*. London: Routledge.

Peake, L. (1993). "'Race' and Sexuality: Challenging the Patriarchal Structuring of Urban Social Space." *Environment and Planning D: Society and Space*, 11: 415–32.

Phillips, R., Watt, D., and Shuttleton, D. (2000). *De-centering Sexualities: Politics and Representations beyond the Metropolis*. London: Routledge.

Porteous, J. (1990). "Bodyscape," in J. Porteous (ed.), *Landscapes of the Mind*. Toronto: University of Toronto Press, 68–85.

Rose, G. (1993). *Feminism and Geography: The Limits of Geographical Knowledge*. Minneapolis: University of Minnesota Press.

Rothenberg, T. (1995). "'And She Told Two Friends': Lesbians Creating Urban Social Space," in Bell and Valentine (1995: 165–81).

Rothenberg, T., and Almgren, H. (1992). "Social Politics of Space and Place in New York City's Lesbian and Gay Communities." Presented at 27th International Geographical Congress, Washington, DC, August.

Seager, J. (1997). "Commentary." *Environment and Planning A*, 29: 1521–3.

Segwick, E. (1992). *Epistemology of the Closet*. Berkeley: University of California Press.

Skelton, T. (1995). "Boom, Bye, Bye: Jamaican Ragga and Gay Resistance," in Bell and Valentine (1995: 264–83).

Snitow, A., Stansell, C., and Thompson, S. (1983). *Powers of Desire: The Politics of Sexuality*. New York: Monthly Review Press.

Stone, L. (1977). *The Family, Sex and Marriage in England, 1500–1800*. New York: Harper & Row.

Tyner, J. (1996). "Constructions of Filipina Migrant Entertainers." *Gender, Place and Culture*, 3: 77–93.

Valentine, G. (1993*a*). "(Hetero)sexing Space: Lesbian Perceptions and Experiences of Everyday Spaces." *Environment and Planning D: Society and Space*, 11: 395–413.

— (1993*b*). "Negotiating and Managing Multiple Sexual Identities: Lesbian Time-Space Strategies." *Transactions of the Institute of British Geographers*, 18: 237–48.

— (1998). "Sticks and Stones may Break My Bones: A personal geography of harassment.". *Antipode* 30: 305–32.

Weeks, J. (1977). *Coming Out: Homosexual Politics in Britain from the Nineteenth Century to the Present*. London: Quartet.

— (1985). *Sexuality and its Discontents: Myths, Meanings and Modern Sexualities*. London: Routledge & Kegan Paul.

Weightman, B. (1980). "Gay Bars as Private Places." *Landscape*, 24: 9–17.

— (1981). "Commentary: Towards a Geography of the Gay Community." *Journal of Cultural Geography*, 1: 106–12.

Wolfe, D. (1978). *The Lesbian Community*. Berkeley: University of California Press.

Zaretsky, E. (1976). *Capitalism, the Family and Personal Life*. New York: Harper Colophon Books.

CHAPTER 15

Socialist Geography

Scott Salmon and Andrew Herod

The production of knowledge is a political act.[1] As such, geographical knowledge reflects and embodies the material conditions and social relations existing at the time of its production. This recognition serves as our point of intellectual engagement with *Geography in America at the Dawn of the Twenty-First Century* and provides the framework within which we interpret a number of changes within "socialist" geography during the 1990s. Thus, in this chapter we do not subscribe to a *progressivist* account of intellectual practice, one that proposes a model of social progress towards an ultimate "truth" through the teleology of reason, technology, production, and so on. Rather, our review of socialist geography in this chapter is a *problematizing* and *contextualizing* one, a treatment that seeks to remain open to both historical transformation and geographical particularity, and to the recognition that knowledge production is a discursive act that is inherently reflective of power relations.

Believing that the production of knowledge and the creation of a more just society are inextricably linked processes, leftist geographers have historically sought to challenge those bodies of knowledge that maintain (implicitly or explicitly) the current economic and political structures of society that favor the haves over the have-nots, encourage environmental destruction in the pursuit of profit, foster racism and patriarchal systems of living, and generally reinforce social inequality and hinder the pursuit of social justice. Drawing precisely upon this notion that the production of knowledge is a political act, socialist geographers in the late 1960s came together to "promote critical analysis of geographic phenomena, cognizant of geographic research on the well-being of social classes; to investigate the issue of radical change toward a more collective society; and to discover the impact of economic growth upon environmental quality and upon social equity" (Socialist Geography Specialty Group 1999). Although the broad political goals of the Socialist Geography Specialty Group (SGSG) have not changed since the 1960s, the fact that the world has been transformed dramatically in the interceding years means that the focus and approaches adopted by leftist geographers within the AAG (and elsewhere) have, of necessity, evolved to meet these challenges and new realities. The changing material geographies of contemporary capitalism have required new tools of analysis and new foci of intellectual inquiry on the part of socialist geographers.

There have always been, as will become obvious in the pages that follow, multiple currents within geography's leftist discourse that, while not necessarily sharing a common epistemological foundation, reflect in their various ways a common concern with social justice. Obviously, the act of definition is inherently political and, as such, is both contested and changing. Indeed the label applied to the left wing of geographical practice has

The Socialist Geography Specialty Group has been renamed as the Socialist and Critical Geography Specialty Group

[1] For more on this idea in the context of the discipline of geography see e.g. Driver (1992); Harvey (1984); Hudson (1977); Livingstone (1992*b*); Peet (1985); Thrift (1996).

changed over time in response to the constantly evolving nature of work produced by leftist geographers, an evolution itself shaped by the changing economic and political context within which this process of knowledge production has been taking place. Hence, the ensemble of leftist geography over time has been variously referred to as "radical," "socialist," "socialist-feminist," "Marxist," "political economy," and, more recently, "critical." While we use many of these terms at different points in this narrative, we recognize that no single one can accurately capture the diversity of knowledge production on the geographic left. Consequently, this chapter is an engagement with an intellectual terrain, rather than an argument for a particular type of theory or version of "socialist geography."[2]

At this point it is also, perhaps, appropriate to talk a little about what this chapter is not. Specifically, we have deliberately chosen not to try to present an exhaustive or taxonomic account of all the work produced during the 1990s by leftist geographers and/or members of the SGSG. We have decided not to do so for several reasons, not least of which is the fact that such taxonomies rarely provide much more than can readily be obtained by a thorough library or Internet search. Indeed, technological advancements and the ever-quickening pace of contemporary academic knowledge production mean that such a glorified bibliography of this sort would rapidly become obsolete. Moreover, given the constraints of this venue, the issue of inclusivity is important to us. Singling out particular authors for attention or attempting to identify key texts while excluding the work of a multitude of others who, almost by definition, would be considered to be "on the margins" smacks of elitism and cronyism, and may simply serve to further the personality cults to which academia is often prone. Instead, what we try to do here is to provide some broad observations concerning some of the significant trends that have either emerged or solidified in socialist geography during the 1990s, and to discuss some of the challenges that confront the field now that we have passed into the new millennium. The approach we take is thus designed to provide a starting point for students and others to get a feel for where socialist geography has been and where it might go in the future. Finally, given our own geographic locations, this is largely a narrative concerning developments within Anglo-American socialist geography during the decade since *Geography in America* was first published. We do not claim that what we present below is anything but a partial account. However, given that the purpose of the present book is to provide an outline of scholarly work in geography in *America* (which, rather ethnocentrically, seems to refer primarily to the United States!) at the dawn of the twenty-first century, we hope that readers will forgive the North Atlantic focus of the chronicle that appears below.

[2] For this reason—and following Walker (1989)—we have adopted the somewhat vague, and occasionally clumsy, descriptor of "leftist" to describe the entirety of this terrain, although we have used more precise labels, such as "feminist" or "socialist" when they are appropriate.

Socialist Geography and (the Demise of) the "Old" World Order

Radical geography emerged within the Anglo-American arena in the late 1960s, at once the product of, and response to, the social and intellectual context of the time. Initially, radicalism within geography was inspired by the political uprisings of the 1960s. About such things the conventional geography of the day had little to say. Thus, the early efforts of radicals within the discipline were directed toward addressing pressing social concerns by shifting the topical focus of their research into new arenas. This concern with relevance led to the introduction of work on poverty, hunger, health, and crime to human geographers who had, to that point, largely ignored them. The result, as Peet (1977) notes, was a geography that was more relevant to social issues but that was nevertheless still tied to a philosophy of science, theoretical discourse, and methodology developed within the framework of a fairly politically conservative discipline.

Frustration with the apparent inability of conventional geographic theory to provide a meaningful foundation for a more relevant and more radical geography led a number of geographers toward an engagement with theories of social justice and, ultimately, Marxism. Indeed, to some extent the widespread adoption of Marxism within human geography can be attributed to the dearth of alternatives within the discipline and the fact that prior to the 1960s—especially in comparison with other social sciences—there was little in the way of social theory in geography beyond positivist idealism (Urry 1989; Smith 1989). Peet (1977: 17), for example, has suggested that radical geography developed largely as "a negative reaction to the established discipline." The particular circumstances of geography's initial interlocution with Marxism have been well documented elsewhere and need not be recounted here (see e.g. Harvey 1973; Peet 1977, 1998; Peet and Thrift 1989; Walker 1989). Rather, it suffices to say that during the 1970s Marxism inspired a powerful critique of the existing state of "establishment geography" (Elliot-Hurst 1973; Anderson 1973; Slater 1975; Massey 1973). Much of this early leftist scholarship within geography was devoted to developing a Marxist framework with which to approach the traditional concerns of geography. At the same time, geographers were critically examining almost every aspect of life in modern capitalism, attempting to demonstrate the ways in which capitalism structured urban landscapes and regional geographies and fueled exploitation of the natural environment and processes of underdevelopment at the global scale.

This is not to suggest that the radical discourse of the 1970s and early 1980s was unified or without dissension. As leftist geographers became more versed in social theory and began to read not just Marx but, among others, Weber, Durkheim, Kropotkin, Luxemburg, Sartre, Freud, Foucault, and Habermas, the radical project was both expanded and increasingly contested. An early and influential challenge to the Althusserian-inspired Marxism that dominated leftist geography in the early 1970s was the humanist critique. Drawing on debates beyond geography, this critique attacked extant Marxist work in geography for its alleged reification of structures to the extent that the resulting explanation denied people social purposefulness (cf. Duncan and Ley 1982; Chouinard and Fincher 1983). This humanist critique prompted a wide-ranging debate on the left (and beyond) regarding the relative importance of structures and human agency in determining the making of geography and history, the role of the individual in social change, and, particularly, the constitutive role of space in structuring social relations (Thrift and Peet 1989). While these concerns led some to Hägerstrand's time geography (e.g. Pred 1981), many more gravitated towards Giddens's structuration theory as a framework that was broadly sympathetic to historical materialism and which demonstrated how social structures were "instantiated" in particular spatial structures (Giddens 1979, 1981).

Riding the wave of this structure–agency debate, the introduction of the realist philosophy of science to human geography was heralded by many leftist geographers as a means to reinvigorate Marxist methodology, one which enabled structural explanation while avoiding the "excesses" of Althusserian Marxism (see Sayer 1984). In particular, proponents of realism argued that Marxism had overestimated the range of structurally determined (necessary) relationships in contemporary capitalism, suggesting that contingent relationships accounted for a much greater proportion of contemporary geographic forms and processes. This emphasis on contingency led to a growing number of "locality studies," which sought to collect detailed empirical evidence to assist in the identification of the nature, causes, and consequences of spatial differentiation in processes of change (e.g. Dickens 1988; Cooke 1989; Bagguley *et al.* 1990). These developments were not uniformly welcomed, however. Many who had previously worked within the Marxist tradition embraced locality studies as a viable means by which to provide detailed information about the "place" of the local economy within the global and so to come to grips, conceptually and methodologically, with a world rapidly being transformed by global forces. In contrast, those less enamored interpreted this turn as little more than a convenient cover for a retreat from traditional Marxist scholarship into atheoretical empiricism (Smith 1987; see the special issue of *Environment and Planning A*, 1991, and the brief summary of the debate in Walker 1989 for more on this).

Rooted more squarely within the Marxist tradition, the French Regulation School approach also seemed to offer a means to move beyond the impasse of structural determinism.[3] Inspired by the work of French political economists (e.g. Aglietta 1979; Lipietz 1987), many economic geographers were attracted by the apparent potential of this framework to offer a historically and geographically grounded account of the dynamics of capitalist development in different national contexts. Arising largely from an internal critique, the Regulation School sought to explain the apparent paradox within capitalism between an inherent tendency (long identified by orthodox Marxist theory) toward instability and crisis and capitalism's evident ability to coalesce and stabilize around a particular set of institutions, norms, and regulatory frameworks that served to secure relatively long periods of economic stability. This theoretical project was underpinned by a recognition that the stagnation of the world economy represented not a cyclical lull but a generalized and sustained crisis of the institutional forms that had guided most advanced capitalist economies through the post-war boom. Given that regulationist accounts posited societal reproduction as the driving force of capitalist dynamics, the goal of those who drew on the Regulationist School was therefore to identify the structures and mechanisms that

[3] While Jessop (1992), for example, identifies seven "regulationist schools," the French variant was by far the most influential within geography. There is a large body of literature on the French Regulation School (for reviews see Boyer 1990; Dunford 1990; and Jessop 1990, 1992) and only the briefest summary is presented here.

enabled any particular "regime of accumulation" to maintain itself in specific historical and geographical contexts. Within geography, this was manifest in something of a preoccupation with the historical periodization of national economies and, in particular, with the demise of the Fordist regime of accumulation, the apparent emergence of a "post-Fordist" successor, and the rise of "flexible" forms of economic organization and "new industrial spaces" (e.g. Amin and Robbins 1990; Gertler 1988; Sayer 1989; Schoenberger 1988; Scott 1988).

Changing Times: Changing Geographies

By the 1990s then, two major changes—both shaped by the changing world beyond the academy—had occurred within leftist geography. First, whereas during the 1970s Marxist geographers were principally concerned with introducing Marx to geography, the fundamental geographical restructuring of advanced capitalist economies brought about by deindustrialization, gentrification, the growing consolidation of a new international division of labor, and other such processes meant that during the 1980s the goal largely shifted to introducing geography to Marx to show how the spatial organization of society made a difference to how capitalism worked (e.g. Soja 1980; Harvey 1982; Massey 1984; Smith 1984). Second, at a fairly fundamental level, the rapidly changing course of history and geography had begun to strain the capacity of traditional Marxist explanations. Far-reaching transformations were bringing about epochal shifts in regimes of capital accumulation—from Fordist mass production based in mass consumption to post-Fordist flexible production systems and segmented markets—that were accompanied by the rise of service employment and information-based industries, and the increasing interdependence of the world economy which was dramatically restructuring the role of cities, regions, and nations. Similarly, driven by austerity and ideologically inspired programs, many nation-states were reinventing themselves as they began to dismantle Keynesian social welfare programs oriented toward managing the "reserve army of labor" (i.e. the poor) and increasingly spent fiscal resources upon military and police functions while seeking to implement Schumpterian "workfare" programs that would discipline labor markets (Jessop 1993; Peck 1996). These political-economic changes were accompanied by a raft of social transformations in the "Global North," including the growing participation of women in the paid workforce and the emergence of a (largely non-white) "underclass" subsisting in the growing informal economy in most advanced industrial economies.

The "new times" of late twentieth-century capitalism were also manifest in a changing political climate, triggered in part by events with global ramifications. Among other things, the demise of an oppressive "state-socialism" in the Soviet Union and Eastern Europe presented some serious challenges, together with some liberating opportunities, to socialist praxis in the West. While those on the political right such as Fukuyama (1992)—and even some on the political left—heralded the "final victory of capitalism," others saw the creative possibilities in disconnecting Marxist ideas from the authoritarianism of the state-socialism of the Soviet system (Sayer and Folke 1991). The continued decline of social-democratic politics signaled also a fundamental shift to the political right that was increasingly mirrored by a growing conservatism on college and university campuses amongst students, faculty, and administrators alike. The radical individualist tenor of the times evidently had a broad-based appeal, even amongst those who did not subscribe to the tenets of neo-liberalism. Furthermore, new political movements emerged that confounded many of the existing political alignments—based on capital and class—by cutting across them. Thus, the growth of feminism, environmentalism, the lesbian and gay rights movements, anti-racist, anti-ablist, and homeless movements all presented divergent views of oppression and conflict that at once both informed and transformed the leftist discourse within geography.

For Marxist geography, these were simultaneously the best of times and the harbinger of worse times to come. The success of Marxist, or Marxian-inspired, political economy within the discipline (Peet and Thrift 1989; Walker 1989) was accompanied by a markedly less combative tone and, in some cases, something of a tactical retreat from the political component of the Marxist project. By the 1990s many of geography's first-generation Marxists had achieved a degree of seniority within the institutional structures of academia. Indeed, many of the early radicals were now respected members of the establishment they had once attacked. Increasingly called upon to justify their scholarship by budget-conscious administrators, many Marxist scholars embraced realist methodologies as a means to turn their attention to "practical" research. While the early years of leftist geography had been productively devoted to a critique of the existing geographical frameworks and, subsequently, to applying the insights of Marxist theory to a range of different subfields within the discipline, the identity of leftist geography, and the hegemony of Marxist geography in particular, was now itself in dispute. Changing times within the academy and the world at large were being mirrored in a number of theoretical challenges to the centrality of the Marxist canon and a growing—some would say necessary—self-criticism about the nature of leftist geography. It is to some of those issues that we now turn.

A New World Order? Leftist Geography During the 1990s

In the conclusion to the chapter "Geography from the Left" that appeared in *Geography in America*, Richard Walker (1989: 638) suggested that "Marxism has for long provided the fulcrum of opposition to conventional theory in geography, but there has been a movement away from Marxism in the 1980s for political and intellectual reasons." Looking back on the 1990s from the dawn of the new millennium, it is readily apparent that the trend identified by Walker has only become more pronounced in the years since *Geography in America* was published. While Marxism is still a central strand of leftist geography, other theoretical approaches and focal points of inquiry have emerged to challenge its once-hegemonic position, a situation that has either expanded the ambit of leftist geography or fragmented it inexorably, depending upon one's point of view.

Cultural Politics and Non-Essentialism

Perhaps the single most significant transformation in leftist geography during the 1990s was the emergence of what has been termed by some as the "cultural turn" (Thrift and Olds 1996; for an assessment, see Barnett 1998) and a growing call for the development of "non-essentialist" analysis and ways of knowing. Whereas research in the 1970s and 1980s typically attempted to answer "economic" questions—explaining patterns of poverty, industrial restructuring, urban development, and the like—research agendas of the 1990s were more concerned with the cultural aspects of life under capitalism and a questioning of the analytical categories by which such life has been interpreted and understood. Informed to a large degree by post-structuralist, post-colonial, and feminist theory, together with a reinvigorated "cultural Marxism," such work sought to examine how theory is culturally situated, how culture shapes the production of the categories that we use to make sense of the world, and how cultural contestations shape the landscape of capitalism (e.g. Pile 1994). In particular, many feminist geographers argued against "essentialism" in theorizing (i.e. against approaches that assume all women share essential common interests regardless of race, religion, class, and geographic location). Instead, they suggested that the complexities of social existence required "non-essentialist" approaches, that is to say approaches that recognize that white, wealthy, North American women may have such different sets of interests than do black, poor, third-world women that they have little in common with each other. In related vein, some feminist work in the 1990s even challenged categories such as "the economic," "the cultural," and "the social" as they had traditionally been applied to geography. Hanson and Pratt (1995), for example, suggested that such categories have often reflected the sexist assumptions that shaped the early years of the discipline—"economic geography," they argued, was taken to refer to the world of work beyond the home and, implicitly, to the world of men, whereas "social geography" frequently connoted those things related to the "realm of reproduction," both biological and social, which was understood to be the realm of women.

While much of the research around "culture" and anti-essentialism focused upon issues of gender politics, multiculturalism (K. Mitchell 1993), and diversity, this work also raised new questions about "race" and geography. Certainly, the issue of race had been broached in the socialist geography of the 1960s and 1970s. However, its treatment then had been largely in terms either of how geography as a discipline had historically been complicit in the suppression of non-Western peoples through its links to imperialism (cf. Hudson 1977; Livingstone 1992*a*; Godlewska and Smith 1994) or as part of a fairly superficial analysis of the geography of race and its intersection with the geography of urban poverty. By the 1990s, though, leftist geographers had come to think of race in much deeper terms. First, there was a new interest in how racial categories were themselves manifest spatially. This approach argued that if racial categories were socially constructed as part of various racial projects at different historical time periods in different places, then such racial categories were clearly also geographically informed and constituted (Omi and Winant 1994; Jackson 1989, 1994; Jackson and Penrose 1993). Second, there was a concern that the dominant models of social behavior assumed social actors that were either not racialized or, alternatively, implicitly treated as "white." Such approaches effectively erased race from social analysis. A third theme related to the ways in which concepts developed in the context of Western political and economic development (such as "class") were frequently applied to non-Western situations without any apparent recognition that they contained culturally specific assumptions (e.g. Myers 1994).

Criticism of the use of meta-categories (i.e. categories that were assumed to explain all, regardless of the specific cultural or geographic context within which they were applied) were also seen in a renewed interrogation of the rather monolithic category "labor," which had been so central to early Marxist analysis within geography. Specifically, a number of Marxist writers had become disenchanted with the theoretical approaches to understanding the uneven development of capitalism—and labor's role in this—that had prevailed during the 1970s and 1980s. Many of these writers felt the extant theorizing of uneven development and capitalist geographies paid too much attention to the actions of capital, relegating workers to residual status by treating them either as simple "factors" in the location decisions of capital (important only in terms of wage rates or levels of unionization, for example) or as the passive victims of transformations in the economic landscape wrought by capital (Herod 1998, 2001; Martin *et al.* 1996; D. Mitchell 1996; Peck 1996; Wills 1996). By attempting to produce less capital-centric accounts of the making of the geography of capitalism this growing interest in labor geography sought to do two things.

First, it tried to show how workers' different geographic locations and positions within industrial sectors might lead them to adopt very different political and organizing strategies, often putting them in conflict with workers located elsewhere. Rather than assuming such political differences were simply part of a false consciousness on the part of workers unified within the singular and totalizing category of labor, this work attempted to understand why workers in different regions might have real differences in their sets of interests which lead to quite varied political stances on issues such as unemployment policies, wage rates, and the like. Second, it tried to show how workers might often play active roles in shaping the geography of capitalism as they search for a "spatial fix" (Harvey 1982) which they believe to be useful for furthering their own political and economic agendas. For example, workers may actively engage in local place-based boosterism to encourage investment in their communities as part of what they see as a way of ensuring their own social reproduction, even if this is at the expense of workers elsewhere. Likewise, efforts to build solidarity between workers in different places requires coming to grips with geographical differences, constraints, and opportunities, so that workers' political praxis must also be seen as spatial praxis. In turn, the choices workers make in their political and spatial praxis influences the types of landscapes that are subsequently constructed.

Issues of Geographic Scale

Another central element in theorizing during the 1990s about the production of landscapes under capitalism, or any other social system for that matter, involved the question of geographic scale. Historically, scale had been regarded by geographers as a relatively straightforward term. Typically, it has been conceived of as either a handy mental device for delineating the landscape—an approach characterized, for example, in the Hartshornian regional geography of the 1930s–1950s (see Hartshorne 1939; Pudup 1988; Smith 1989; Herod 1991) in which one geographer's region was as good as the next—or as somehow natural and fixed divisions of space. During the 1990s, however, there was a proliferation of writing about the "social production of scale" that made clear that scales of social organization are not simply premade waiting to be used but are actively created by the participants involved. Much of this early writing revolved around Smith's (1984) arguments about how scale was produced out of tendencies within capital both to fix itself in place during the production process—thereby differentiating the landscape between developed and underdeveloped places—but also to try to equalize conditions across the landscape through the process of competition. However, later writing suggested that this was too capital-centric an approach to understanding the production of scale and that a more catholic approach would be one that examined how various social groups produced and used scales as part of their political praxis (e.g. women's groups (Staeheli 1994); gay activists (Brown 1995); environmentalists (Williams 1999); unions (Herod 1997, 2001); the homeless (Smith 1993); and others).

This focus on issues of scale was also related to a growing interest in the nature of economic and political globalization. Specifically, some authors saw globalization as the ultimate playing out of capital's expansionary trends that had been identified nearly a century before by Lenin (1939) in which the "annihilation of space by time" (Marx 1973) eviscerated geographical differences between places and thus made geography unimportant. Others argued that—perhaps paradoxically—in an increasingly interconnected global economy where flows of people, goods, and capital across national borders appeared to be occurring at ever-greater levels, local geographies were becoming *more* important because very minute differences in the economic, political, or cultural attributes of places may lead global capital to choose one location over another. Seeking to capture this tension between the global and the local, several writers began to examine what they called the "glocalization" of economic activity; that is to say, how corporations that operate across the planet nevertheless frequently attempt to tailor their marketing and production strategies to very local conditions (e.g. Kanter 1995; Mair 1997; Swyngedouw 1997).

Elsewhere, this attempt to theorize the connection between local social actors and their wider institutional context led others to an engagement with questions concerning the connection between globalization and local politics. Recognizing that the local state and local politics were part of the spatially differentiated regulation of particular accumulation regimes, several writers advanced the notion of a "local mode of regulation" (Peck and Tickell 1992, 1994). In this way, apparently global processes of economic restructuring and the transition from a Fordist to a post-Fordist accumulation regime were used to explain the rise of "entrepreneurial" political regimes in particular localities (e.g. Goodwin *et al.* 1993; Hall and Hubbard 1998; Lauria 1997). Subsequently, debates over the relative significance of "the local" and "the global" raised issues concerning political strategy, how social actors might seek to "jump scales" either to broaden their base of activities (e.g. to expand activity from local to regional or national spheres) or to limit them (e.g. in the case of unionized workers who withdraw from national contracts to negotiate locally), and the ways in which "the local" and "the global" are represented discursively (for more on this latter, see Gibson-Graham 1996).

The Politics of Narrative

Matters of the discursive representation of globalization and global capital became particularly significant towards the end of the 1990s. For some the issue was over just how global the global economy had become and whether the processes of internationalization to which we were apparently bearing witness were new either in form or in impact (Hirst and Thompson 1996; Dicken *et al.* 1997; Leyshon 1997). For others, such as Gibson-Graham (1996), the central concern was the way in which global capital had been represented—by those not just on the political right but also in much Marxist economic geography—as hegemonic and seemingly infinitely flexible and adaptive. By representing capital in such terms, Gibson-Graham argued, it became increasingly difficult to imagine how confronting it in any meaningful way could transform social relations. In turn, this made it very hard either to develop policy designed to challenge the logic of neoliberalism, which suggested that capital should be allowed to flow across the globe in as unhindered a manner as possible (cf. Ohmae 1990, 1995; Bryan and Farrell 1996), or to present those opposed to capital as having any type of agency or capacity for action, particularly at the global scale—workers and others were portrayed as capable of operating only at the sub-global level, a situation which resulted in a theoretical and political concession of "the global" to capital and the neoliberals.

Such issues of language and narrative were related both to changing material conditions in contemporary global capitalism—the emergence of a new economic and political world order—and also changing intellectual currents, particularly the growing influence during the 1980s and 1990s of postmodern thought among leftist geographers. Indeed, the influence of postmodernism within leftist geography has perhaps been one of the most contentious issues that those on the left have faced during the past decade, largely due to the fact that the term itself has been used flexibly to encompass many things by proponents and foes alike. While some embraced postmodernism as a progressive and liberating means of challenging Marxism as an essentialist, modernist meta-narrative that was too focused on class, others interpreted the rise of postmodernism in geography as a neoconservative move designed to counter the dominance of Marxism, particularly in the context of "localities research." Still others adopted a position that seemed to suggest that Marxism itself could be "postmodernized" through an attempt to develop non-essentialist theory and by providing greater attention to the role of space in the reproduction and operation of capitalism. Adherents of this approach argued that it is orthodox Marxism's focus upon time and historical change that have marked it as a modernist project, not its focus on class (see Soja 1989, 1996).

While the ascension of postmodernism on the left generated some political controversy, it also led to a diversification of topical focus and approach amongst geographers. For some, a geographic engagement with postmodernism drew attention to changes in the material environment itself, giving rise to new currents within urban geography such as examinations of how postmodern ideas have impacted architectural styles and urban form during the past twenty years or so (e.g. Knox 1991, 1993).

These ideas also inspired new approaches to the "natural" environment, which, in the wake of the massive political-economic, technological, and environmental changes associated with globalization, had become the focus of growing political conflict. Amidst apocalyptic predictions of impending global ecological doom, geographers sought to deconstruct taken-for-granted discourses concerning the "environment" and "development" and emphasized the centrality of the idea of a "politicized environment" to new social movements in many parts of the world (e.g. Peet and Watts 1996; Bryant and Bailey 1997). Others developed this logic further, arguing that "nature" itself has to be understood as a socially constructed (and manipulated) artefact—a source of social power as well as the font of a potentially liberatory politics (e.g. Braun and Castree 1998).

For many, then, an engagement with postmodernism prompted a deeper questioning of the "accepted ways of thinking about the subject matter of geography, the ways of knowing that subject matter, and the ways of communicating the results of geographical inquiry" (Curry 1991: 210). Whereas modernist ways of knowing (such as orthodox Marxism) tended to see the world in terms of an absolute knowledge that could be revealed through the application of a "scientific" method of investigation, those who advocated a postmodern approach tended to see knowledge as "more relative and variable" (Curry 1991: 222). Such divergent views led to a long-running debate concerning whether postmodernism was really a separate entity from modernism or whether the transient, the fleeting, the contingent, the diverse, and the paradoxical which postmodernism seemed to celebrate were merely the underbelly of modernism itself. This latter point of view was forcefully argued by Harvey (1989), amongst others, who saw postmodernism as little more than what Jameson (1984) has refferred to as the "cultural logic of late capitalism."

In many ways these oppositional viewpoints were typical of the polarizing influence of postmodernism among many on the left. Indeed, the "postmodern challenge" served to inflame a number of lingering resentments associated with the Marxist project. Certainly, for some, postmodernism entailed a liberating celebration of difference in both personal identity and theoretical approach precisely because it did not involve recourse to explanation grounded in totalizing meta-narratives such as "class." In this way it represented a clear break from the dogma of much Marxist theorizing of the 1960s and 1970s which tended to view all struggle outside the workplace as simply "displaced class struggle" and, therefore, relatively unimportant. For proponents, then, postmodernism validated a focus on broader social dimensions such as gender, race, and sexual orientation, and legitimized voices other than that of the working-class male as offering a relevant view of the world. For detractors, however, the influence of postmodernism was undeniably pernicious, representing a neoconservatism appropriate for the feel-good times of the Reagan and post-Reagan era in which ethics gave way to aesthetics, and linguistic dexterity too often replaced concrete political action (Palmer 1990). From this perspective, "diversity" was often regarded as a pseudonym for a radical individualism that, taken to its logical extreme, implied that organizing toward collective action was difficult, if not impossible.

This latter debate was, perhaps, most rancorous in terms of the perceived relative places of class and gender, Marxism and feminism, in understanding the world. Although there is a long history of scholarship that sees class and gender relations as inextricably linked (e.g. Hartmann 1981), the issue of postmodernism in leftist geography ignited a number of smoldering debates between some Marxists and some feminists. Much of this debate swirled around David Harvey's 1989 book *The Condition of Postmodernity*. Whereas Harvey argued that postmodernism as a cultural and economic condition could be usefully analyzed through the class lens of Marxism, some feminists suggested that such an interpretation—whether consciously or not—reinforced a political position which "assumed that the only enemy is capitalism" and neglected other forms of oppression (Massey 1991: 31). Following McDowell's (1991) admonition that the baby of Marxist insight and class analysis need not be discarded with the bathwater of the male-centric vision with which it had become associated, a number of feminists actively sought to integrate the categories of class and gender (e.g. Bondi 1991; MacKenzie 1989). However, the concern remains that this has not been a wholly reciprocal rapprochement and that, unfortunately, the Marxist mainstream has been slower to incorporate the insights of feminist scholarship than feminism has been to engage traditional Marxist categories such as class (McDowell 1992).

Conclusion

The 1990s, then, were tumultuous years on the geographic left. Although interpretations vary widely, during the space of a decade a dramatically new world order rapidly unfolded, bringing with it an entirely new set of challenges and concerns for those within geography working for progressive social change. At the same time, intellectual and ideological trends first identified in the 1980s continued to impact academia, the discipline of geography, and the left within it. Not all these changes could be considered benign. Nevertheless, they have been reflected in the diversity of knowledge produced by geographers on the left during this period.

In assessing the contemporary state of leftist geography, it is clear that the impetus for change on the geographic left during the 1990s emerged simultaneously from "external" events of global change and ongoing "internal" debates within the traditional academic forums and, as the decade drew to a close, the digital venues of cyberspace. The (continuing) ramifications of the collapse of the former Soviet Union, the subsequent rise of ethnic and national conflicts, and the Western military aggression that they sometimes elicited caused many to rethink traditional political alignments and policy orientations. This aspect of the new world order was accompanied by a significant shift to the political right within social democracies of the advanced capitalist core, a shift that was eventually mirrored within much of the academy among both faculty and students. Within geography, epistemological challenges grounded in postmodern perspectives also prompted many leftist geographers to rethink the way they view and understand the world and has led to a variety of new perspectives and topical foci. This has most clearly been reflected in the left's marked "cultural turn" and the associated shift away from traditional class-based explanations in favor of accounts emphasizing alternative axes of oppression, such as gender, "race," ethnicity, sexuality, physical ability, and environmental inequity. At the same time, attempts to come to grips with the rapid pace of global change, both empirically and theoretically, prompted many to rethink the way change is theorized and explained, leading to the development of new non-essentialist frameworks and fresh interpretations of traditional geographic concepts, such as spatial scale. While this chapter has, of necessity, focused only on selected elements of these shifts, we hope it has captured some of the emerging contours of this new intellectual landscape.

In his chronicle of the consolidation of the left within geography, Peet (1977) suggested there is a dialectical relationship between the development of a radical consciousness and the material context in which ideas develop. This would seem to be as true today as it was then, although perhaps not in the way that Peet initially envisioned. In responding to the challenges of the 1990s, the left has transformed itself. The flow of leftist thought continues unabated but the topical focus of this work has broadened considerably and the theoretical wellsprings that propel these discursive currents have multiplied dramatically. In the course of this transformation numerous voices have emerged to rearticulate and reinvent the leftist political project. As a result, the traditional concerns of socialist geography no longer dominate the leftist agenda and Marxism has been displaced as the hegemonic theoretical paradigm by a more diverse, if less coherent, set of voices. While some might lament the declining centrality of the Marxist paradigm and the loss of traditional certainties this entails, few would deny that the transformations that have propelled leftist geography into the new millennium testify to its continuing intellectual vibrancy and vitality. Perhaps, after all, the only constant is change, and the left has certainly redefined itself in concert with the emergence of a new world order. As we enter the new millennium the discourse of the left is no longer dominated by a single paradigm, but it remains to be seen whether diversity is a permanent condition or whether a single paradigm will (re)emerge to unify the geographic left once more.

Finally, in line with the broader theme of this chapter—that knowledge production is shaped by the material contexts within which it occurs—we want to end by highlighting what we see as a disturbing attack on public education during the 1990s that affects not just leftist geography but academic geography and academia in general. This attack has been manifested through a growing "corporatization" of education and, in the United States at least, an assault on secular education, not just at the university level but throughout. During the past decade we have seen two groups of conservatives—whose goals in practice have frequently overlapped—lay siege to public education at all levels as part of a broader *Kulturkampf* on multiculturalism, secularism, the welfare state, and the values of "liberalism."[4] On the one

[4] We use the term *Kulturkampf* here precisely because this is the term that has been used by many conservatives in an attempt to characterize the "cultural war" taking place in the United States between the values of conservatism and those of liberalism. It has been used perhaps most controversially by United States Supreme Court Chief Justice Antonin Scalia, in an apparent literal translation of the English term into German. However, it should be pointed out that in German this phrase has very different connotations and refers to an anti-Catholic campaign waged by the Prussian and Imperial German Government in the 1870s through appointing priests, regulating the administration of parochial schools, and confiscating Church property, and in which Chancellor Otto von Bismarck referred to Catholics as *Reichsfinde* ("enemies of the nation"). The *Kulturkampf* was thus both a cultural and a political war against a class of persons on the basis of their religious beliefs, a point that seems to have been lost on the Catholic Scalia. In German, then, the term *Kulturkampf* implies something even stronger than it does in English.

hand, religious and social conservatives have struggled to wrest education away from its perceived control by secularists whose "left-wing agenda" includes teaching about multiculturalism, gay rights, "revisionist history" (such as teaching US history from the perspective of non-whites), and other "non-traditional" values. Fiscal conservatives, on the other hand, have increasingly tried to encourage private funding of education as a means to reduce public expenditures and to make academia more entrepreneurial. These latter efforts have often been furthered by university administrators, who increasingly are not academics but individuals brought in from the business world to run institutions of higher education according to "the principles of the marketplace" and whose goal is, by deft financial management, to augment university endowments (K. Mitchell 1999). This latter trend is evidenced by the increasing amount of paid, subcontracted work universities conduct for corporations (in the case of geography this has often focused on the application of GIS, something which has been examined in Pickles 1995), the growing "casualization" of teaching through the use of graduate students and temporary instructors who cost universities less to employ, increased workloads, and attacks on tenure and job security which can be read as a means to erase labor market rigidities and bring about more flexible academic workforces (cf. Nelson 1997; Wills 1996). It is our worry that such developments threaten academic freedom—particularly that research or teaching which challenges the corporate interests on whom schools and universities increasingly rely—and universities' traditional roles as sources of social critique. This is something that should give all progressive academic geographers—socialist and non-socialist alike—cause for concern.

References

Aglietta, M. (1979). *A Theory of Capitalist Regulation.* London: New Left Books.

Amin, A. (1994). *Post-Fordism: A Reader.* Oxford: Blackwell.

Amin, A., and Robbins, K. (1990). "The Re-emergence of Regional Economies? The Mythical Geography of Flexible Accumulation." *Environment and Planning D: Society and Space*, 8/1: 7–34.

Anderson, J. (1973). "Ideology in Geography: An Introduction." *Antipode*, 5: 1–6.

Bagguley, P., Mark-Lawson, J., Shapiro, D., Urry J., Walby, S., and Warde A. (1990). *Restructuring: Place, Class and Gender.* London: Sage.

Barnett, C. (1998). "The Cultural Turn: Fashion or Progress in Human Geography?" *Antipode*, 30/4: 379–94.

Blaut, J. M. (1994). "Robert Brenner in the Tunnel of Time." *Antipode*, 26/4: 351–74.

Bondi, L. (1991). "Gender Divisions and Gentrification: A Critique." *Transactions of the Institute of British Geographers*, NS 16: 190–8.

Boyer, R. (1990). *The Theory of Regulation: A Critical Introduction.* New York: Columbia University Press.

Braun, B., and Castree, N. (eds.) (1998). *Remaking Reality: Nature at the Millennium.* New York: Routledge.

Brown, M. P. (1995). "Sex, Scale, and the 'New Urban Politics': HIV-Prevention Strategies from Yaletown, Vancouver," in D. Bell and G. Valentine (eds.), *Mapping Desire: Geographies of Sexuality.* London: Routledge, 245–63.

Bryan, L., and Farrell, D. (1996). *Market Unbound: Unleashing Global Capitalism.* New York: John Wiley.

Bryant, R., and Bailey, S. (1997). *Third World Political Ecology.* New York: Routledge.

Chouinard, V., and Fincher, R. (1983). "A Critique of 'Structural Marxism and Human Geography.'" *Annals of the Association of American Geographers*, 73: 137–46.

Cooke, P. (ed.) (1989). *Localities.* London: Unwin Hyman.

Cox, K., and Mair, A. (1991). "From Localised Social Structures to Localities as Agents." *Environment and Planning A*, 23: 197–213.

Curry, M. (1991). "Postmodernism, Language, and the Strains of Modernism." *Annals of the Association of American Geographers*, 81/2: 210–28.

Dicken, P., Peck, J., and Tickell, A. (1997). "Unpacking the Global," in R. Lee and J. Wills (eds.), *Geographies of Economies.* London: Arnold, 158–66.

Dickens, P. (1988). *One Nation? Social Change and the Politics of Locality.* London: Pluto Press.

Driver, F. (1992). "Geography's Empire: Histories of Geographical Knowledge." *Environment and Planning D: Society and Space*, 10: 23–40.

Duncan, J., and Ley, D. (1982). "Structural Marxism in Human Geography: A Critical Assessment." *Annals of the Association of American Geographers*, 72: 30–59.

Dunford, M. (1990). "Theories of Regulation." *Environment and Planning D: Society and Space*, 8: 297–321.

Elliot-Hurst, M. (1973). "Establishment Geography: How to be Irrelevant in Three Easy Lessons." *Antipode*, 5/2: 40–59.

Environment and Planning A (1991). Special Issue on New Perspectives on the Locality Debate. *Environment and Planning A*, 232.

Fukuyama, F. (1992). *The End of History and the Last Man.* New York: Free Press.

Gertler, M. (1988). "The Limits to Flexibility: Comments on the Post-Fordist Vision of Production." *Transactions, Institute of British Geographers*, 13: 419–32.

Gibson-Graham, J. K. (1996). *The End of Capitalism (As We Knew It): A Feminist Critique of Political Economy.* Cambridge, Mass.: Blackwell.

Giddens, A. (1979). *Central Problems in Social Theory*. London: Macmillan.
— (1981). *A Contemporary Critique of Historical Materialism*. London: Macmillan.
Godlewska, A., and Smith, N. (eds.) (1994). *Geography and Empire*. Cambridge, Mass.: Blackwell.
Goodwin, M., Duncan, S., and Halford, S. (1993). "Regulation Theory, the Local State and the Transition of Urban Politics." *Environment and Planning D: Society and Space*, 11/1: 67–88.
Hall, T., and Hubbard, P. (eds.) (1998). *The Entrepreneurial City: Geographies of Politics, Regime, and Representation*. New York: Wiley.
Hanson, S., and Pratt, G. (1995). *Gender, Work, and Space*. New York: Routledge.
Hartmann, H. (1981). "The Unhappy Marriage of Marxism and Feminism: Towards a More Progressive Union," in R. Dale, G. Esland, R. Fergusson, and M. MacDonald (eds.), *Education and the State*, ii. *Politics, Patriarchy and Practice*. Barcombe: Falmer Press, 191–210.
Hartshorne, R. (1939). *The Nature of Geography: A Critical Survey of Current Thought in Light of the Past*. Lancaster, Pa.: Association of American Geographers.
Harvey, D. (1973). *Social Justice and the City*. London: Edward Arnold.
— (1982). *The Limits to Capital*. Oxford: Basil Blackwell.
— (1984). "On the History and Present Condition of Geography: An Historical Materialist Manifesto." *Professional Geographer*, 36/1: 1–11.
— (1989). *The Condition of Postmodernity*. Oxford: Basil Blackwell.
Herod, A. (1991). "The Production of Scale in United States Labour Relations." *Area*, 23/1: 82–8.
— (ed.) (1998). *Organizing the Landscape: Geographical Perspectives on Labor Unionism*. Minneapolis: University of Minnesota Press.
— (2001). *Labor Geographies: Workers and the Landscapes of Capitalism*. New York: Guilford.
Hirst, P., and Thompson, G. (1996). *Globalization in Question*. Cambridge: Polity Press.
Hudson, B. (1977). "The New Geography and the New Imperialism: 1870–1918." *Antipode*, 9/1: 12–19.
Jackson, P. (1989). *Maps of Meaning: An Introduction to Cultural Geography*. London: Unwin Hyman.
— (1994). "Constructions of Criminality: Police–Community Relations in Toronto." *Antipode*, 26/3: 216–35.
Jackson, P., and Penrose, J. (eds.) (1993). *Constructions of Race, Place and Nation*. Minneapolis: University of Minnesota Press.
Jameson, F. (1984). "Postmodernism, or the Cultural Logic of Late Capitalism." *New Left Review*, 146: 53–92.
Jessop, B. (1990). "Regulation Theories in Retrospect and Prospect." *Economy and Society*, 19/2: 153–216.
— (1992). "Fordism and Post-Fordism: A Critical Reformulation," in M. Storper and A. Scott (eds.), *Pathways to Industrialization and Regional Development*. London: Routledge, 43–65.
— (1993). "Towards a Schumpterian Workfare State? Preliminary Remarks on Post-Fordist Political Economy." *Studies in Political Economy*, 40: 7–39.
Kanter, R. M. (1995). *World Class: Thriving Locally in the Global Economy*. New York: Simon & Schuster.
Knox, P. (1991). "The Restless Urban Landscape: Economic and Sociocultural Change and the Transformation of Metropolitan Washington, DC." *Annals of the Association of American Geographers*, 81/2: 181–209.
— (ed.) (1993). *The Restless Urban Landscape*. Englewood Cliffs, NJ: Prentice Hall.
Lauria, M. (ed.) (1997). *Reconstructing Urban Regime Theory: Regulating Urban Politics in a Global Economy*. London: Sage.
Lenin, V. (1939 [1900]). *Imperialism: The Highest Stage of Capitalism*. New York: International Publishers.
Leyshon, A. (1997). "True Stories? Global Dreams, Global Nightmares, and Writing Globalization," in R. Lee and J. Wills (eds.), *Geographies of Economies*. London: Arnold, 133–46.
Lipietz, A. (1987). *Mirages and Miracles: The Crisis of Global Fordism*. London: Verso.
Livingstone, D. N. (1992*b*). *The Geographical Tradition: Episodes in the History of a Contested Enterprise*. Oxford: Blackwell.
— (1992*a*). "A 'Sternly Practical' Pursuit: Geography, Race and Empire," in Livingstone (1992*a*: 216–59).
McDowell, L. (1991). "The Baby and the Bathwater: Diversity, Deconstruction and Feminist Theory in Geography." *Geoforum*, 22: 123–33.
— (1992). "Multiple Voices: Speaking from Inside and Outside the Project." *Antipode*, 24: 56–72.
McDowell, L., and Court, G. (1994). "Missing Subjects: Gender, Power, and Sexuality in Merchant Banking." *Economic Geography*, 70/3: 229–51.
MacKenzie, S. (1989). "Women in the City," in Peet and Thrift (1989: ii. 109–26).
Mair, A. (1997). "Strategic Localization: The Myth of the Postnational Enterprise," in K. Cox (ed.), *Spaces of Globalization: Reasserting the Power of the Local*. New York: Guilford, 64–88.
Martin, R., Sunley, P., and Wills, J. (1996). *Union Retreat and the Regions: The Shrinking Landscape of Organised Labour*. London: Jessica Kingsley.
Marx, K. (1973). *Grundrisse: Foundations of the Critique of Political Economy*. New York: Random House.
Massey, D. (1973). "Towards a Critique of Industrial Location Theory," *Antipode*, 5: 33–9.
— (1984). *Spatial Divisions of Labour*. London: Methuen.
— (1991). "Flexible Sexism." *Environment and Planning D: Society and Space*, 9: 31–57.
Merrett, C. (1999). "Culture Wars and National Education Standards: Scale and the Struggle over Social Reproduction." *Professional Geographer*, 51/4: 598–609.
Mitchell, D. (1996). *The Lie of the Land: Migrant Workers and the California Landscape*. Minneapolis: University of Minnesota Press.
Mitchell, K. (1993). "Multiculturalism, or the United Colors of Capitalism?" *Antipode*, 25/4: 263–94.
— (1999). "Scholarship Means Dollarship, or, Money in the Bank is the Best Tenure." *Environment and Planning A*, 31/3: 381–8.
Myers, G. A. (1994). "Eurocentrism and African Urbanization: The Case of Zanzibar's Other Side." *Antipode*, 26/3: 195–215.
Nelson, C. (ed.) (1997). *Will Teach for Food: Academic Labor in Crisis*. Minneapolis: University of Minnesota Press.
Ohmae, K. (1990). *The Borderless World: Power and Strategy in the Interlinked Economy*. New York: HarperBusiness.

Ohmae, K. (1995). *The End of the Nation State: The Rise of Regional Economies*. New York: Free Press.

Omi, M., and Winant, H. (1994). *Racial Formation in the United States: From the 1960s to the 1990s*. New York: Routledge.

Palmer, B. D. (1990). *Descent into Discourse: The Reification of Language and the Writing of Social History*. Philadelphia: Temple University Press.

Peck, J. (1996). *Work-Place: The Social Regulation of Labor Markets*. New York: Guilford.

Peck, J., and Tickell, A. (1992). "Local Modes of Social Regulation? Regulation Theory, Thatcherism and Uneven Development." *Geoforum*, 23: 347–64.

— (1994). "Searching for a New Institutional Fix: The After-Fordist Crisis and the Global-Local Disorder," in Amin (1994: 280–315).

Peet. R. (1977). "The Development of Radical Geography in the United States", in R. Peet (ed.), *Radical Geography*. London: Meuthen.

— (1985). "The Social Origins of Environmental Determinism." *Annals of the Association of American Geographers*, 75/3: 309–33.

— (1998). *Modern Geographical Thought*. Oxford: Blackwell.

Peet, R., and Thrift, N. (1989). "Political Economy and Human Geography," in R. Peet and N. Thrift (eds.), *New Models in Geography*, ii. *The Political Economy Perspective*. London: Unwin Hyman, 3–29.

Peet, R., and Watts, M. (eds.) (1996). *Liberation Ecologies: Environment, Development, Social Movements*. New York: Routledge.

Pickles, J. (ed.) (1995). *Ground Truth: The Social Implications of Geographic Information Systems*. New York: Guilford Press.

Pile, S. (1994). "Masculinism, the Use of Dualistic Epistemologies and Third Spaces." *Antipode*, 26/3: 255–77.

Pred, A. (1981). "Social Reproduction and the Time-Geography of Everyday Life." *Geografiska Annaler, Series B*, 63: 5–22.

Pudup, M. B. (1988). "Arguments Within Regional Geography." *Progress in Human Geography*, 12/3: 369–90.

Sayer, A. (1984). *Method in Social Science: A Realist Approach*. London: Hutchinson.

— (1989). "Post-Fordism in Question." *International Journal of Urban and Regional Research*, 13: 666–93.

Sayer, A., and Folke, S. (1991). "What's Left to Do? Two Views from Europe." *Antipode*, 23: 240–8.

Schoenberger, E. (1988). "From Fordism to Flexible Accumulation: Technology, Competitive Strategies, and International Location." *Environment and Planning D: Society and Space*, 6/3: 245–62.

Scott, A. (1988). *New Industrial Spaces*. Berkeley: University of California Press.

Slater, D. (1975). "The Poverty of Modern Geographic Inquiry," *Pacific Viewpoint*, 16: 159–76.

Smith, N. (1984). *Uneven Development: Nature, Capital and the Production of Space*. Oxford: Blackwell.

— (1989). "Geography as Museum: Private History and Conservative Idealism in *The Nature of Geography*," in J. N. Entrikin and S. D. Brunn (eds.), *Reflections on Richard Hartshorne's* The Nature of Geography. Washington, DC: Association of American Geographers, 91–120.

— (1993). "Homeless/Global: Scaling Places," in J. Bird, B. Curtis, T. Putnam, G. Robertson, and L. Tucker (eds.), *Mapping the Futures: Local Culture, Global Change*. London: Routledge, 87–119.

Socialist Geography Specialty Group (1999). *Declaration of Principles*. Available at ‹http://www.aag.org›, last accessed 10 October 2001.

Soja, E. (1980). "The Socio-Spatial Dialectic." *Annals of the Association of American Geographers*, 70: 207–25.

— (1989). *Postmodern Geographies: The Reassertion of Space in Critical Social Theory*. London: Verso.

— (1996). *Thirdspace: Journeys to Los Angeles and other Real-and-Imagined Places*. Oxford: Blackwell.

Staeheli, L. (1994). "Empowering Political Struggle: Spaces and Scales of Resistance." *Political Geography*, 13: 387–91.

Swyngedouw, E. (1997). "Neither Global nor Local: 'Glocalization' and the Politics of Scale," in K. Cox (ed.), *Spaces of Globalization: Reasserting the Power of the Local*. New York: Guilford, 137–66.

Thrift, N. (1996). "Flies and Germs: A Geography of Knowledge," in N. Thrift, *Spatial Formations*. London: Sage, 96–124.

Thrift, N., and Olds, K. (1996). "Refiguring the Economic in Economic Geography." *Progress in Human Geography*, 20/3: 311–37.

Urry, J. (1989). "Sociology and Geography," in R. Peet and N. Thrift (eds.), *New Models in Geography*, ii. *The Political Economy Perspective*. London: Unwin Hyman, 295–317.

Walker, R. (1989). "Geography from the Left," in G. L. Gaile and C. J. Willmott (eds.), *Geography in America*. Columbus, Ohio: Merrill, 619–50.

Williams, R. W. (1999). "Environmental Injustice in America and its Politics of Scale." *Political Geography*, 18/1: 49–73.

Wills, J. (1996). "Uneven Reserves: Geographies of Banking Trade Unionism." *Regional Studies*, 30: 359–72.

— (1996). "Labouring for Love? A Comment on Academics and their Hours of Work." *Antipode*, 28/3: 292–303.

CHAPTER 16

Transportation Geography

Andrew R. Goetz, Bruce A. Ralston, Frederick P. Stutz, and Thomas R. Leinbach

Introduction

Transportation geography is the study of the spatial aspects of transportation. It includes the location, structure, environment, and development of networks as well as the analysis and explanation of the interaction or movement of goods and people (Black 1989). In addition it encompasses the role and impacts—both spatial and aspatial—of transport in a broad sense including facilities, institutions, policies and operations in domestic and international contexts. It also provides an explicitly spatial perspective, or point of view, within the interdisciplinary study of transportation.

There has been substantial progress in the development of the transportation geography subfield over the last ten years. In 1993, the *Journal of Transport Geography* was started in the UK, providing the subfield with its own eponymous journal. Several second editions of key textbooks were published, including *The Geography of Transportation* (Taaffe *et al.* 1996), *The Geography of Urban Transportation* (Hanson 1995), and *Modern Transport Geography* (Hoyle and Knowles 1998). The Transportation Geography Specialty Group (TGSG) instituted the Edward L. Ullman Award for scholarly contributions to the subfield; recipients have included Edward Taaffe, Harold Mayer, Howard Gauthier, William Garrison, William Black, James Vance, Susan Hanson, Morton O'Kelly, Bruce Ralston, Donald Janelle, Thomas Leinbach, Brian Slack, and Kingsley Haynes. The specialty group also began honoring students who have written the best doctoral dissertations and masters theses each year, and a TGSG web page was created. The University of Washington Department of Geography instituted the Douglas K. Fleming lecture series in transportation geography at AAG annual meetings. Finally, transport geographers have played prominent roles in a Geography and Regional Science Program organized joint National Science Foundation/European Science Foundation initiative on Social Change and Sustainable Transport (SCAST) (Leinbach and Smith 1997; Button and Nijkamp 1997). This initiative led to the development of the North American-based Sustainable Transportation Analysis and Research (STAR) network led by geographer William Black as a counterpart to the European-based Sustainable Transport in Europe and Links and Liaisons with America (STELLA) network. Together, these initiatives and research networks offer significant opportunities for geographers to contribute to a growing body of literature on the environmental, economic, and equity implications of transportation systems.

Transportation geography has natural linkages with many other geographic subfields. Co-sponsored sessions at AAG meetings have been organized with specialty groups including Applied Geography, Economic (formerly Industrial), Geographic Information Systems (GIS), Regional Development and Planning, Recreation Tourism and Sport, Spatial Analysis and Modeling (formerly Mathematical Models and Quantitative Methods), and Urban (see related chapters in this volume). There remains, however, a much greater opportunity to develop additional and deeper linkages. Transportation is at the heart of many topics central to geography, and more interaction with other geographers would benefit the study of transportation and geography.

This chapter will first provide a brief historical overview of transportation geography, followed by a discussion of major topics in transport geography over the last decade divided into three major sections: (1) modeling, network analysis, and GIS; (2) government policy, industrial change, international development, and historical studies; and (3) information technology, environmental, behavioral, and social issues. The conclusion suggests major topics for research in transportation geography for the twenty-first century.

Transportation Geography and History of Geographic Thought

In a seminal article, Taaffe and Gauthier (1994) analyzed the paradigmatic development of transportation geography within the context of the ecological, area study, and spatial organization traditions of geography. They classified the post-1970 pluralistic body of transport geography literature into six subcategories: *model-building, GIS, analytical-empirical, behavioral, historical-cultural, and Marxist-social theory*, and discussed each as applied to the three human geographic traditions, plus policy studies. Even though each of the subcategories represents complex philosophical bases, there is nevertheless a generalized continuum across the subcategories from positivist to non-positivist orientations.

The majority of research in transport geography historically has been more positivist in orientation within the spatial organization framework. The *spatial interaction and network analysis* research traditions of Ullman (1956), Taaffe (1956), Garrison *et al.* (1959), Garrison (1960), Garrison and Marble (1961), and Taaffe *et al.* (1963) provided early foundations for more contemporary work in modeling, GIS, and analytical-empirical studies. The *behavioral* research tradition of Golledge (1980), Burnett (1980), Gauthier and Mitchelson (1981), Louviere (1981), Pipkin (1981), Southworth (1981), Hanson (1982), and Horowitz (1985) emerged from studies on urban travel that developed economic and psychological behavioral choice models using disaggregate data. Much of the behavioral emphasis in transport geography has been positivist in orientation, but important influences have emerged from non-positivist, humanistic approaches (Hagerstrand 1982; Ley 1983; Seamon 1979; Tuan 1971) that involve perception, cognition, and more subjective assessments of concepts such as time, space, place, and movement. Non-positivist approaches were also found in the *historical-cultural* tradition of Borchert (1967, 1987), Mayer and Wade (1969), Meinig (1986), and Vance (1986).

Most interestingly, there has been very little transport geography research within the *Marxist-social theory* arena (the work of Eliot Hurst 1973, 1974 notwithstanding) even though it represents a major thrust in human geography, particularly in urban, economic, and political geography (Peet 1998). As stated earlier, the study of transportation is interdisciplinary, but has traditionally been dominated by civil engineers on the "hard" physical science side, and by neoclassical economists on the "soft" social science side. The study of transportation suffers from this bias, and thus any serious transport researchers are subject to having to operate within this milieu. The need for geographers to be well-versed in complementary disciplines that represent systematic subfields is long-standing. For example, economic geographers must be able to communicate with at least some groups of economists, while urban geographers become well-versed in social theory through interaction with sociologists, political scientists, and other social scientists. To some extent, these processes have served to fragment geography, as linkages between interdisciplinary specialists have in many cases become stronger than linkages with other geographers (Johnston 1998). This is a critical concern for geography, as various groups continue to develop separate languages that serve to exclude and, in some cases, vilify rather than to establish and develop common ground.

In this context, some have referred to a "ghettoization" of transport geography, whereby an underclass stigmatization has been assigned simply because some geographers like to study transportation, and find no other support networks for their views. In a realm where even the questioning of transport policy can lead to pariah status among the dominant neoclassical transport economists (see e.g. Goetz and Dempsey 1989; Dempsey and Goetz 1992), it is not at all surprising that so little transport work is being conducted within the Marxist-social theory framework. This is not to suggest that there is no need for more work coming from this perspective. To the contrary, transport studies should be very amenable to meaningful social-theory perspectives, and it can be geographers who lead the way in this regard. Several recent contributions are indicative of the potential (Castells 1996, 1999; England 1993; Herod 1998; Hodge 1990; Sheppard 1995; Warf 1988). The NSF-ESF SCAST initiative and development of the STAR and STELLA networks, intended to stimulate cooperative and collaborative research on transport between European and North American scholars, also represent opportunities to develop stronger linkages and create new synergies among geographers and other researchers in regards to a wide range of current issues that are relevant to transportation (van Geenhuizen *et al.* 1999).

With this broad overview of research traditions, the chapter now embarks on a more thorough discussion of major research conducted in transportation geography over the last decade.

Contemporary Transportation Geography

Modeling, Network Analysis, and GIS

The research over the last 10 years in these areas has been influenced by three major factors: developments in information technology, industry restructuring as a result of deregulation and transportation technology changes, and federal policy and funding. These factors have combined to spur advances in the timeliness and amount of data available, new research questions, and improved theoretical models in transportation geography. While many of these advances have been impressive, there remain many transportation geography questions unanswered as we head into the next century.

1. *Information Technology*. Advances in information technology have led to improvements in information exchange, increased information capture, and better tools for the maintenance and display of transport information. An important part of these developments has been the rise of Geographic Information Systems technology applied to transportation issues, known as GIS-T. Using GIS as a platform upon which to build transportation models and to build and maintain transportation databases seems only natural. In addition to better GIS software and data models for transportation, transport geographers have benefited from the increased availability of digital transportation data. There have been many GIS-T publications, including texts on the subject by Miller and Shaw (2001) and Thill (2000) (see GIS chapter for additional related information).

2. *The Impact of GIS-T*. GIS-T has become a major area of research. Evidence of the activities in GIS-T include the following: The Bureau of Transportation Statistics (BTS), a branch of the US Department of Transportation, provides numerous sources of data on transportation, including network databases, surveys of person travel behavior, and studies of commodity flows; there is an annual GIS-T conference where researchers from federal, state, and local agencies, GIS vendors, consultants, and academicians meet to present their research in this area; and several GIS vendors, such as Caliper, ESRI, and Intergraph market transportation-specific GIS software. During the past decade we have seen advances in software, hardware, and operating systems that allow researchers to use and maintain large databases and perform sophisticated analyses on relatively inexpensive computing equipment. The improvement in GIS and personal computer capabilities over the last decade has allowed many standard transportation models to be incorporated into GIS software, or the standard models to become more visual.

One of the problems in using GIS for transportation is that the data structures for cartographic display and relational table operations are inefficient for exploring transportation data or for optimal use of transport algorithms. As a result, much of the GIS-T research has focused on developing proper transport-oriented data models, such as Linear Referencing Systems (LRS). This approach allows users to access transportation data in more natural ways, such as by mile-posting or by routes (National Cooperative Highway Research Program (NCHRP) 1997). Another difficult issue in GIS-T is conflation—the task of merging coordinate and attribute data from two different geo-spatial databases (Federal Geographic Data Committee (FGDC) 1998; Sutton 1997). Much of the work in LRS and conflation and other transport database-building problems has taken place outside the mainstream academic journals—done by consultants, university transportation centers, and software developers.

It is fair to say that we have not developed new transportation algorithms that exploit the spatial analysis skills, such as spatial search, of many GIS systems. Instead, we have grafted existing models onto GIS structures. Nearly all these grafting efforts have focused on aggregate models of transportation, although there is interest in using GIS to facilitate disaggregate transportation models (Goodchild 1998). The recent work of Kwan (1999*a*, *b*, 2000) has illuminated temporal human spatial behavior by using a GIS environment.

Most use of GIS in transportation is in building and maintaining databases. These databases are often used for facilities management, but they also reflect the need to study the effects of proposed policies. At the federal level, the need for good transportation information is focused through the BTS. At the close of this decade, the amount, quality, timeliness, and availability of transportation information far exceeds that of a decade ago. Currently, there are multiple GIS databases with transportation information. The FGDC (1998) lists numerous transportation network data sources just for roads. In addition, the BTS distributes transportation GIS databases for the US and North America. Finally, many state and local governments maintain similar databases for their spatial domains.

The existence of these networks allows us to test modeling procedures (traditionally tested on randomly generated networks) on real actual networks to see which ones work best in the "real world." For example, Zhan and Noon (1998) have used several road networks to test the efficiency of various shortest-path algorithms. While there has been an increase in GIS transport network availability, with much of it free and available on the Web, not all networks have the same structure or contain the same attribute information. Thus, there are many organizations that are trying to determine what should be included in transportation databases and how they should be distributed. These include the National Imagery and Mapping Agency (formerly the Defense Mapping Agency), the Intelligent Transport Systems working group, and the FGDC. In addition to these GIS databases, there are many efforts to collect and distribute transportation data that can be used to calibrate transport models. This information includes the Commodity Flow Survey, the American Travel Survey, and the National Personal Travel Survey, to name a few. These various GIS databases and transport flow databases have made national-level transportation modeling much easier over the past decade. For more information, visit the BTS website at <http://www.bts.gov>.

By combining GIS network databases, flow data, and models, transport geographers have been able to develop Spatial Decision Support Systems (SDSS) to analyze various policy scenarios. In the United States, much of this work is government funded and is closely tied to the data collection efforts of the BTS. As a result, programs such as the Highway Performance Modeling System that are often seen as GIS database-building exercises also can be viewed as necessary steps in analytical modeling. Indeed, the boundaries between database building, applications of models, and theory construction are more blurred than ever. For the US and North America, several studies have been done on modeling the impacts of landbridge (Southworth *et al.* 1998), intermodal transport modeling (Southworth *et al.* 1997), and the impacts of the North American Free Trade Agreement (NAFTA). Wong and Meyer (1993) evaluated Meals on Wheels program efficiency, while Aitken *et al.* (1993) used 1,600 community surveys to build 3-D GIS plots to locate a suburban beltway for Caltrans, mitigating neighborhood impact and inequality of access. Nyerges *et al.* (1997) developed an SDSS for group decision-making about the location of transport facilities. Transport-related SDSS also have been developed for areas outside the United States, including south Asia (Ralston *et al.* 1994), Latin America (Louriero and Ralston 1996), and Africa (Liu *et al.* 1993). Leinbach (1995) has suggested that more research on third-world transport and development needs to be done, and GIS-T has a role to play in such studies, including spurring better data-collection efforts in areas where we poorly understand the relationship between transportation and development (World Bank 1994).

Another area of interest in information technology (IT) deals with the impact of improved information flow on transportation efficiency and equity (see also *Communications and Information Technology* below). The journal *Geoinformatica* (2000) devoted a special issue to GIS and Intelligent Transportation Systems (ITS). As with traditional transport, IT may be yet another area where the "rich get richer and the poor get poorer." In fact, there is some evidence of this in the shipper-carrier bidding processes in the private sector. The larger companies are better able to take advantage of opportunities made possible by better supply and demand information. As we look ahead, it is not clear how increased information availability can be incorporated into our traditional transportation models, whether they are models of private-sector goods movement, urban transportation, or the distribution of transport opportunities throughout society. The empirical questions are many and obvious, but they have yet to have a strong impact on the theoretical structures used by transport geographers.

In the private sector, the rise of Just-In-Time (JIT) delivery, express logistics carriers, and the move to supply chain optimization all reflect the importance of information on logistics. We need to see how classic models in transportation geography can change to accommodate such issues as inventory carrying costs, the interactions of shippers and carriers, the rise of third-party logistics, and the like. While some authors have considered this problem (Osleeb and Ratick 1990; DeWitt *et al.* 1997) the current major transportation geography texts published in America do not focus on the supply-chain optimization approach, with its emphasis on total logistics costs and transparent information flows. Yet this approach is quite prevalent in the business literature.

In addition, the transportation functions found in commercial GIS packages do not address many logistics issues. Indeed, the GIS vendors do a fine job on node-based routing models with their emphasis on space-based transportation costs. This "route-centric" approach addresses short-term planning but ignores issues of time trade-offs, the shipper-carrier load tender/bidding process (GIS software assumes that the shipper is the carrier, or there is at most one carrier), and the existence of third-party logistics firms. It appears that by focusing solely on measures based on topology and geometry, the current state-of-the-art in GIS-T may be leading us away from asking the correct questions about location, allocation, routing, and other aspects of spatial organization. These areas clearly fall within the domain of transportation geography and they reflect the tactical and strategic decisions firms must make. Until we develop models that address information-rich environments, the relevance of transport geography and GIS-T to both private- and public-sector planning will be limited. Rodrigue (2002) has been active in rallying transportation geographers to consider supply-chain approaches in geography. In all likelihood, models that address information-rich transport planning processes also will have implications for transport equity issues raised by Hanson (1998).

3. *Network Restructuring as a Result of Deregulation and Technology Changes.* The deregulation of transport markets in the US and other countries, combined with changes in transport technology, such as containerization, have led to changes in how transport systems are organized. It was widely believed that deregulation would result in more efficient transport, but that is not always the case. A major result of these changes has been the rise of hub-and-spoke networks, most notably in the airline and ocean shipping industries. The importance of such systems has caught the attention of theoretical and empirical transport geographers alike (see "Government Policy, Industrial Change, International Development, and Historical Studies" below for more on the empirical work).

Studies have focused heavily on the structure of networks (Bowen 2002; Buckwalter 2001; Fleming and Hayuth 1994; Lee *et al.* 1994; Reynolds-Feighan 1992; Shaw 1993; Shaw and Ivy 1994), while on the more theoretical side there has been much interest in hub-and-spoke networks as an outcome of a network design problem (Bryan 1998; Horner and O'Kelly 2001; Kuby and Gray 1993; O'Kelly 1998; O'Kelly and Bryan 1998; O'Kelly and Miller 1994; O'Kelly *et al.* 1996). In these works the goal is finding the theoretical reasons for the rise of hub-and-spoke networks, the locations of hubs, and their possible evolution. This body of work looks at the hub location problem as an optimization process. Like all theoretical models, this research is based on assumptions, some of which, such as symmetry of flows, are more questionable than others. These problems are mathematically difficult in their own right. As we try to relax some of the more restrictive assumptions, it is not clear that they will become any more tractable. None the less, this approach to network design is important if we are to simulate the effects of economic and policy changes on the structure of hub-and-spoke networks. As O'Kelly (1998) points out in his Fleming Lecture, such networks come in a variety of sizes and functions.

4. *Policy and Funding.* As described above, the work done in GIS-T has allowed transport geographers to assess the impacts of policies such as NAFTA. As the political barriers to trade come down in the EU and North America, this type of work will continue. In addition to the freight modeling work mentioned above, urban transport modeling efforts have been heavily influenced by federal funding.

The ITS initiative grew out of the 1991 Intermodal Surface Transportation Efficiency Act (ISTEA), which has been replaced by the 1998 Transportation Equity Act for the Twenty-first Century (TEA21). The names of these programs stress the importance that the concepts of intermodalism, efficiency, and equity have had on transportation policy and practice (Goetz and Vowles 2000; Hanson 2000; Hodge 1988, 1990, 1995). In the urban transportation modeling arena, the early 1980s saw the move from all-or-nothing or incremental assignments to more elegant theoretical models such as network equilibrium models. But these models had some simplifying assumptions that often were questionable. For example, equilibrium models assume the demand for transport is even (there is constant pressure on the transport system) over the modeling period. In recent years, spurred on by the demands of ITS and the data-collection efforts in support of it, dynamic traffic assignment models have become the focus of much research. Not surprisingly, the theoretical aspects of dynamic traffic assignment have been the focus of the operations research community. However, ITS calls for more than theoretical models. A major goal of ITS is to develop methods, technology, and organizations to support real-time traffic-management systems (Chin *et al.* 1999). The functions real-time dynamic traffic management should support are (Maiou and Summers 1997):

- Estimating and predicting network status.
- Providing information to travelers on modes, travel times, and routes. This information should support traffic management objectives.
- Supporting other ITS functions and goals.

These functions clearly require the use of some standard transportation models, such as trip generation, mode choice, and route choice, along with newer models of real-time updating of route suggestions. These will be dependent, in part, on GIS-T databases and modeling capability (Xiong and Gordon 1995). In fact, it is difficult to imagine a dynamic traffic analysis (DTA) system that does not use GIS. There are a host of technical issues, such as algorithmic design and implementation, and how to transmit information to drivers, along with more societal issues, such as access to information for disadvantaged groups and the societal goals of DTA versus the individual goals of travelers.

Urban transport is just one part of the ITS effort. More efficient freight flows, reduced accident rates, better traveler information, maximizing use of transport infrastructure, and more rapid emergency response are just some of the envisioned benefits of ITS (Garrison and Ward 2000). The considerable federal funds allocated to these efforts insure that much research on transportation systems and regional trade will continue. Transport geographers already are playing an important role in the collection and analysis of transport data and the development of modeling and information systems. Whether the promises of ITS are fully realized or the distribution of benefits does enhance transport equity no doubt will be the focus of more research in the coming years (Haynes *et al.* 2000).

5. *Traditional Issues.* There are several other research trends that do not fit neatly into the structure outlined above, but which are of interest. These include the continued interest in transportation and energy (Greene 1997; Greene and Fan 1995; Greene and Han 1996), modeling and calibration of transportation models (Agyemang-Duah and Hall 1997; Aljarad and Black 1995; Black 1992, 1995; Black and Thomas 1998; Fotheringham and O'Kelly 1989; Knudsen 1990; Kuby *et al.* 1991; O'Kelly *et al.* 1995), transportation and land use relationships (Giuliano 1995; Giuliano and Small 1999; Miller 1999; O'Kelly 1988; O'Kelly and Bryan 1996; Ralston and Liu 1989; Southworth 1995; Sutton 1999; Warren 1993), and visualization of transportation activities (Black 1997a; Marble *et al.* 1995). Another of these seminal topics is the relationship between aging and mobility. Much more research needs to be carried out on this theme (Stamatiadis *et al.* 1996).

Government Policy, Industrial Change, International Development, and Historical Studies

1. *Deregulation, Globalization, and Industrial Structure.* Government policies, particularly deregulation, liberalization, and privatization, have had a profound effect on transportation, fundamentally altering the structure, organization, and operation of the airline, railroad, trucking, bus, maritime, and intermodal industries. These policies have also served as catalysts for industry globalization, which in turn has also greatly affected transportation. Transportation is a cause of globalization, in terms of shrinking the planet through space-time convergence and making possible increased global interaction, but is also affected by globalization processes external to transportation, such as telecommunications technology development, increased world trade, and global business alliances (Janelle and Beuthe 1997). Industrial restructuring is a result of numerous macro-scale forces, but the principal impetus for triggering many of the major changes in transportation over the last two decades has been policies aimed at deregulating or liberalizing government regimes.

Since the US domestic airline industry was one of the first transportation industries to be significantly deregulated (in 1978), and because of its higher public visibility, it has attracted a great deal of research attention. Fleming (1991), Sorenson (1991), Debbage (1993), and Shaw and Ivy (1994) each addressed issues of airline competition in the wake of deregulation by examining levels of industry and market concentration, and the decline of the first wave of low-cost airlines. Accessibility for many airline passengers increased (Chou 1993; Maraffa and Finnerty 1993) as hub airports accommodated more flights within expanded hub-and-spoke networks (Ivy 1993; Reynolds-Feighan 1992, 1998; Shaw 1993). Studies of airport planning in Denver have shown how aviation authorities responded to capacity problems caused by increasing air traffic at hub airports (Dempsey *et al.* 1997; Goetz and Szyliowicz 1997).

Hub cities benefited from increased air transportation employment and related economic activity, frequency of service, and lower fares, except in cases where hubs were dominated by one or two airlines such as in the concentrated hubs of the US Southeast (e.g. Atlanta, Charlotte, Cincinnati, and Memphis) that led to a geographic pattern of higher average fares in that region (Goetz and Sutton 1997). Successful lower-cost air carriers, particularly Southwest Airlines, have had a dramatic effect in lowering average airfares and increasing traffic in the US

Southwest and in other selected metropolitan areas (e.g. Baltimore-Washington) where this service was located (Vowles 2000*a*). Smaller communities have fared less well as commuter air service has been sporadic (Vowles 1999). The effects of proposed airline alliances were also studied by Vowles (2000*b*), while in a recent Fleming lecture, Goetz (2002) examined competitive and antitrust implications raised by alleged predatory behavior and proposed (United–USAirways) and actual (American–TWA) mergers.

The most recent and devastating concerns in the airline industry revolve around the terrorist attacks of September 11, 2001 when two United and two American Airlines planes were hijacked and deliberately crashed into the World Trade Center building in New York City, the Pentagon in Washington, DC, and an open field in southwestern Pennsylvania, resulting in thousands of deaths, a war in Afghanistan, and an international counteroffensive against terrorism. The aftershocks have dealt a serious blow to the entire US airline industry even though the federal government agreed to a $15 billion airline aid package after the unprecedented four-day commercial airline shutdown and subsequent calamitous decline in passenger traffic. In particular, United Airlines, the second largest airline in the world, has been devastated, losing $2.1 billion in 2001 and another $3.2 billion in 2002, due not only to the attacks but also previous mismanagement and ongoing labor disputes, leading the carrier to declare Chapter 11 bankruptcy in December 2002. The structure of the industry itself may change dramatically as a result of these and other financial losses, the federal government takeover of aviation security, and the distinct possibility of future mergers, acquisitions, bankruptcies, and/or additional government intervention. The US air transportation system has been profoundly affected by the attacks of September 11, a date that will undoubtedly become a major historical watershed in aviation research.

The terrorist attacks of September 11 also affected the entire global aviation industry, as worldwide travel demand declined dramatically in the wake of the attacks. One clear lesson learned from these events is the profound impact that the airline industry specifically and transportation generally have on national and regional economies, itself a manifestation of an increasingly interconnected global economic system. Much of this increased economic integration in aviation occurred over the last twenty years as domestic deregulation policies were extended to international liberalization of air service through "open skies" bilateral agreements largely involving the US, several EU states (particularly Netherlands and Germany), and Canada, thus promoting the globalization of international aviation (Debbage 1994; Ivy 1995). East Asian NICs have adopted a more pragmatic liberalization policy whereby the state has retained more control over airlines (Bowen and Leinbach 1995; Leinbach and Bowen 1996), even though liberal bilateral agreements involving countries such as Singapore, Korea, and Thailand were expanded. The international airline industry finds itself buffeted not only by the constant threat of terrorism, but also by periodic instabilities in regional economies, such as the 1998 financial crisis in Asia (Rimmer 2000).

The maritime industry and port system have undergone tremendous changes as a result of rationalization, globalization, and technological developments in containerization, with geographers making significant research contributions in this context. Many of the same issues affecting the airline industry are also affecting maritime shipping, such as industry consolidation through alliances and acquisitions (Alix *et al.* 1999) and the effects of globalization on local transport terminal operations (McCalla 1999; Rodrigue 1999). Debate has centered on port concentration issues (Hayuth 1988; Kuby and Reid 1992); the development of inland load centers, satellite terminals, and the transhipment function (Fleming 2000; Slack 1990, 1999); the role of port authorities in the urban political economy (Slack 1993; Warf 1988); labor disputes in US ports (Herod 1998); regional politics in public works waterway projects (Bierman and Rydzkowski 1991); the effects of containerization on port morphology in Asia (Airriess 1989, 1991, 1993, 2001; Wang and Slack 2000); and the effects of natural disasters on transport systems, such as impact of the 1995 earthquake on the port of Kobe, Japan (Chang 2000).

2. *International Development.* Studies on transportation and development include those by Leinbach (1989, 1992, 1995, 2001*a*) on the role that transport and related policies play in influencing third-world development, especially in Southeast Asia (Leinbach and Chia Lin Sien 1989). In a recent Fleming lecture, Leinbach (2000) has suggested that new research must be undertaken on transport's role in development that captures the thrust of newer alternative frameworks, critical issues, and individual and family mobility needs. Other development-related research focused on the Asian realm includes the role of transportation in Taiwan's regional development (Shaw and Williams 1991), studies on the rapid transit system in Calcutta (Dutt and Mukhopadhyay 1992), and in an ongoing study of air cargo services and the electronics industry in Southeast Asia, Bowen *et al.* (2002) show the complex interrelations between the state, rent-seeking behavior, industrialization strategies,

and air transport services. The important role that air transportation plays in urban systems and national economic development was illustrated in the cases of Southeast Asia (Bowen 2000) and the former Soviet Union (Sagers and Maraffa 1990), as well as the United States (Goetz 1992; Irwin and Kasarda 1991; Ivy *et al.* 1995; Minshall *et al.* 1995).

3. *Historical/Cultural Studies.* Finally, several excellent historical transportation geography studies were published during the 1990s. In one of the best studies ever produced that illustrates the important role of transportation in the evolution of a gateway city's nodal region, Cronon (1991) analyzed urban system development through an exploration of how railroads created the trade hinterlands of Chicago leading to the remarkable growth of "nature's metropolis." Likewise, Vance (1995) added a final volume to his stellar tradition of historical transportation geography research by delving into the origin, evolution, and geography of North American railroads, focusing on the critical role they played in fostering economic and urban development. In what is proving to be an intriguing topic for historical transport researchers (see e.g. Lewis 1997), Moon (1994) examined the development of the US Interstate Highway System, particularly the dramatic geographic and land-use impacts that it has engendered, while Rollins (1995) interpreted the symbolic landscape of the German Autobahn highway system.

Information Technology, Environment, Travel Behavior, and Social Issues

1. *Communications and Information Technology.* Communications and IT are transforming the world economy and the world of transportation at rates never before thought possible (Brunn and Leinbach 1991; Hepworth 1990). Profound implications, even many that cannot be measured, accompany this IT explosion. At the center of this information boom are the microprocessor, networked computers, and the Internet. The impact of the IT explosion on transportation and communications is just starting to be included in studies by transport geographers (Stutz and de Souza 1998). Central among questions of how IT will affect transportation of the future might be how IT can help mitigate inequality of access and lead us toward more sustainable transportation (Hanson 1998; Hanson 2000).

Intelligent Vehicle Highway System (IVHS) technologies are aimed at accommodating the tremendous increase in travel with present smart roadways (Hau 1990; Hodge *et al.* 1996; Stutz and de Souza 1998). The Automated Highway System (AS) program is one component of the US Department of Transportation's research initiative in ITS, the expenditures for which have averaged about $200 million annually for the last five years. The AS research was authorized in ISTEA, which called on USDOT to develop an automated highway and vehicle prototype from which fully automated IVHS can be developed.

Telecommuting is an IT work option that is starting to be explored by transportation geographers. Telecommuting may actually help clear choked suburban freeways. It is just one of a set of transportation control measures to manage demand for urban transportation and reduce the need for additional infrastructure. Used in conjunction with other planning applications such as transit-oriented development (TOD), congestion pricing, and IVHS, telecommuting may be an effective tool for reorienting the social, built, and physical environments and their attendant links (Gillespie and Richardson 2000; Janelle 1995; Saxena and Mokhtarian 1997). New developments in flow systems include spatially separated computer networks linking facilities of multinational corporations. Computer innovations now tie the home environment to remote information sources, including business and banking services, travel and commercial services, and library and telephone directory services.

Yet Janelle (1997) has argued that the substitution hypothesis (that communication can be substituted for transportation) is in doubt. He notes many questions that communication technologies raise for transport geographers, including the possibility for greater volatility in transport demand, vulnerability of systems dependent on just-in-time logistics, and, more generally, the role information technology can play in transport efficiency. A recent collection of essays focuses accessibility in the context of information and cyberspace (Janelle and Hodge 2000). Still another discusses the structure and role of information technology and mobility in the perspective of electronic commerce (Leinbach and Brunn 2001). In a lead essay Leinbach (2001*b*) points up the nature of this development and areas for research.

2. *Environment, Congestion Management, and Sustainability.* Along with technical advances and rapidly growing applications of IT has come better understanding of the true societal cost paid for a polluted environment and the awareness of the fragility of the human–earth ecosystem (Greene and Wegener 1997; Stutz 1995). The growth of populations and economies has sharply intensified the weight we place on that ecosystem, while

simultaneously emphasizing the relatively new prospect that natural resources are not unlimited (Stutz and de Souza 1998). In past years, mobility was the dominant issue. Today, however, environmental considerations are receiving a higher priority. Transportation's organizations are being required to show an increasing environmental consciousness as a basis for their programs. The 1991 ISTEA legislation altered the practice of transportation planning by explicitly linking transportation funding to air quality compliance. Transportation priorities are changing as a result, with transportation geographers and other professionals trying to demonstrate how transportation programs will contribute to the attainment of air quality and other environmental objectives (Bae 1993; Stutz 1995).

Demand for transportation services are typically increasing faster than population, with congestion an increasing factor across many modes and regions. The period required for the expansion of transportation infrastructure is usually twenty to thirty years, and in some cases, such as that of airport expansion, the project may take longer, be much more expensive than originally projected, or never come to pass at all (Cidell and Adams 2001; Dempsey *et al.* 1997). When combined with the enormous costs that can be associated with infrastructure expansion in a period of metropolitan financial constraints, the result is often an extended lag between awareness of a need and effective action in response. Since the period for the improvement is inherently lengthy, this dynamic virtually guarantees that supply will be many years behind demand, often resulting in congestion (Hodge 1992; Schintler 1997).

As a result of ISTEA and the Clean Air Act Amendments (CAAA), metropolitan planning organizations (which are now playing a more important role in metropolitan transportation planning (Goetz *et al.* 2002)) and states have included more transportation control measures (TCMs) and trip-making reductions in their transportation and clean air plans. Two of ISTEA's funding provisions—the Congestion Mitigation and Air Quality (CMAQ) Improvement Program and the flexible use of Surface Transportation Funds—have particularly encouraged the planning and implementation of TCMs. Transportation geographers and journals are just now starting to address such issues (Giuliano *et al.* 1993; Modarres 1993; Stutz 1995).

The concept of sustainability is increasingly being applied to transportation. It is generally argued that the current fossil-fueled motor vehicle/highway scenario is not sustainable due to the limited nature of petroleum reserves, worsening air quality problems, space demands for automobiles in cities, global warming, congestion on roadways and the incipient loss of time and productivity, continued urban sprawl, and finally, rising fatalities on highways (Black 1996, 1997*b*, 2001; Button and Nijkamp 1997; Greene and Wegener 1997; Janelle 1997). Promising areas for research of sustainable transport include improvement of technology, congestion pricing, and integrating land-use planning and transport, e.g. transit-oriented development (Bernick and Cervero 1997; Cervero 1998; Fielding 1995; Gassaway 1992; Giuliano 1995; Giuliano and Golob 1990; Hodge 1992; Jones 1995; Moon 1990; Nijkamp and Pepping 1998; Stutz 1995; Warren 1993).

3. *Urban Travel Behavior and Social Issues.* The strategy of balancing jobs and housing throughout metropolitan areas gained credibility in the early 1990s as a potential solution to the problem of increased suburban traffic congestion caused by explosive employment growth. The problem resulted from increased commuting times to work centers from low-density residential subdivisions that shifted further into the suburban fringe. Some observers have noted that if nearby affordable housing development had accompanied job center growth, less auto traffic would have resulted. This led to a groundswell of support for the jobs/housing balance concept. A related issue is the possible spatial mismatch between where the jobs are located and where the workers live, exacerbated by differences in race as well as income. Local and regional planners increasingly support the notion of expanding the supply of housing in job-rich areas and the quantity of jobs in housing-rich areas. Geographers have repeatedly shown that residential location is based on many factors in addition to the home/work separation, such as quality of neighborhoods, schools, amenities, and perceived safety (Giuliano 1998; Giuliano and Small 1993; Levinson 1998; Peng 1997; Taylor and Ong 1995).

Although women continue to increase in importance in the American workforce, salaried women, more often than their male co-workers, have the bulk of the responsibility for day-to-day maintenance of households. Thus they make shorter and more numerous trips, and often need the flexibility of driving alone (Rosenbloom and Burns 1993). The National Personal Transportation Survey of 1990 and 1995 show that women in metropolitan households with children make 21 per cent more trips per day than men in the same households and that 45 per cent of all trips occur as a part of trip chains, comprised primarily of servicing passenger and personal business (running errands) types of trips. Because trip chains may be more dependent on single-occupancy vehicles than work trips, the travel behavior of women has important implications for urban transport planning and management (Blumen 1994; Burns 2000; England 1993; Harris and Bloomfield 1997; Johnston-Anumonwo 1992, 1995, 1997; Law 1999; McLafferty and Preston 1991; Sermons and Koppelman 2001). Voluntary mobility of homeless persons seems to be related to location of resources and social support (Wolch *et al.* 1993); while the travel behavior of blind or vision impaired people, especially for bus travel, can be time-consuming, frustrating, and difficult (Marston *et al.* 1997).

Activity-based travel demand modeling was proposed twenty-five years ago as an alternative to the trip-based modeling framework and the discrete choice, utility-maximizing models that were being incorporated into the traditional urban transportation modeling system in the 1980s (Agyemang-Duah and Hall 1997; Bacon 1992; Fotheringham and Trew 1993; Horowitz 1991; Lo 1990; Pipkin 1995; Thill 1992; Thill and Horowitz 1997). Substantial progress has been made recently in advancing from activity-based travel analysis (with emphasis on descriptive analysis and understanding) to activity-based travel forecasting models that can be used effectively for addressing contemporary planning and policy issues (Baker 1994, 1996; Golledge and Timmermans 1990; Goodchild *et al.* 1993; Stutz *et al.* 1992; Timmermans and Golledge 1990). A recent intriguing paper by Nijkamp and Baaijens (1999) points up the notion of *time pioneers* and the existence of reduced travel behaviors.

Transportation Geography for the Twenty-First Century

The NSF/ESF SCAST Initiative provides a framework for charting future developments in transportation geography (Leinbach and Smith 1997; Button and Nijkamp 1997) in both North America (Black 1997*b*) and Europe. Research issues in five major areas were identified: (1) societal change (Giuliano and Gillespie 1997); (2) sustainable transport (Greene and Wegener 1997); (3) information and communications technology (Hodge and Koski 1997); (4) globalization (Janelle and Beuthe 1997); and (5) institutional issues (Stough and Rietveld 1997). As we embark further into the twenty-first century, these issues are likely to dominate the study of transportation, with geographers well-positioned by inclination and orientation to make major research contributions.

References

Agyemang-Duah, K., and Hall, F. (1997). "Spatial Transferability of an Ordered Response Model of Trip Generation." *Transportation Research A*, 31/5: 389–402.

Airriess, C. A. (1989). "The Spatial Spread of Container Transport in a Developing Regional Economy: North Sumatra, Indonesia." *Transportation Research A*, 23/6: 453–61.

— (1991). "Global Economy and Port Morphology in Belawan, Indonesia." *Geographical Review*, 81/2: 183–96.

— (1993). "Export-Oriented Manufacturing and Container Transport in ASEAN." *Geography*, 78/1: 31–42.

— (2001) "The Regionalization of Hutchison Port Holdings in Mainland China," *Journal of Transport Geography*, 9: 267–78.

Aitken, S., Stutz, F., Prosser, R. and Chandler, R. (1993). "Neighborhood Integrity and Residents' Familiarity: Using a GIS to Investigate Peace Identity." *Tijdschrift Voor Economische en Sociale Geografie*, 34: 1–12.

Alix, Y., Slack, B., and Comtois, C. (1999). "Alliance or Acquisition? Strategies for Growth in the Container Shipping Industry, The Case of CP Ships." *Journal of Transport Geography* 7: 203–8.

Aljarad, S. N., and Black, W. (1995). "Modeling Saudi Arabia–Bahrain Corridor Mode Choice." *Journal of Transport Geography*, 3: 257–68.

Bacon, R. W. (1992). "Working, Shopping and House Rents." *Geographical Analysis*, 24: 268–80.

Bae, C. H. C. (1993). "Air Quality and Travel Behavior: Untying the Knot." *Journal of the American Planning Association*, 59/3: 65–75.

Baker, R. G. V. (1994). "On Travel Behavior Relative to a General Place Utility Field." *Environment and Planning A*, 26: 1455–74.

— (1996). "Multipurpose Shopping Behaviour at Planned Suburban Shopping Centres: A Space-Time Analysis." *Environment and Planning A*, 28: 611–30.

Bernick, M., and Cervero, R. (1997). *Transit Villages in the 21st Century*. New York: McGraw Hill.

Bierman, D. E., and Rydzkowski, W. (1991). "Regional Politics in Public Works Projects: The Tennessee–Tombigbee Waterway." *Transportation Quarterly*, 45/2: 169–80.

Black, W. R. (1989). "Transportation Geography," in G. Gaile and C. Willmott (eds.), *Geography in America*. Columbus, Ohio: Merrill, 316–32.

—— (1992). "Network Autocorrelation in Transport Network and Flow Systems." *Geographical Analysis*, 24: 207–22.

—— (1995). "Spatial Interaction Modeling Using Artificial Neural Networks." *Journal of Transport Geography*, 3: 159–66.

—— (1996). "Sustainable Transportation: A US Perspective." *Journal of Transport Geography*, 4/3: 151–9.

—— (1997*a*). "Virtual Reality and Three-Dimensional Visualization." *Journal of Transport Geography*, 5: 47.

—— (1997*b*). "North American Transportation: Perspectives on Research Needs and Sustainable Transportation." *Journal of Transport Geography*, 5/1: 12–19.

—— (2001). "An Unpopular Essay on Transportation." *Journal of Transport Geography*, 9: 1–11.

Black, W. R., and Thomas, I. (1998). "Accidents on Belgium's Motorways: A Network Autocorrelation Analysis." *Journal of Transport Geography*, 6: 23–31.

Blumen, O. (1994). "Gender Differences in the Journey to Work." *Urban Geography*, 15/3: 223–45.

Borchert, J. R. (1967). "American Metropolitan Evolution." *The Geographical Review*, 57: 301–32.

—— (1987). *America's Northern Heartland*. Minneapolis: University of Minnesota Press.

Bowen, J. T., Jr. (2000). "Airline Hubs in Southeast Asia: National Economic Development and Nodal Accessibility." *Journal of Transport Geography*, 8: 25–41.

—— (2002). "Network Change, Deregulation, and Access in the Global Airline Industry." *Economic Geography*, 78/4: 425–40.

Bowen, J. T., Jr., and Leinbach, T. R. (1995). "The State and Liberalization: The Airline Industry in the East Asian NICs." *Annals of the Association of American Geographers*, 85: 468–93.

Bowen, J. T., Jr., Leinbach, T. R., and Mabazza, D. (2002). "Air Cargo Services, The State, and Industrialization Strategies: The Redevelopment of Subic Bay, The Philippines." *Regional Studies*, 36/4.

Brunn, S. D., and Leinbach, T. R. (eds.) (1991). *Collapsing Space and Time: Geographic Aspects of Communications and Information*. London: HarperCollins.

Bryan, D. (1998). "Extensions to the Hub Location Problem: Formulations and Numerical Examples." *Geographical Analysis*, 30/4: 315–30.

Buckwalter, D. W. (2001). "Complex Typology in the Highway Network of Hungary, 1990 and 1998." *Journal of Transport Geography*, 9: 125–35.

Burnett, K. P. (1980). "Spatial Constraints Oriented Modeling: Empirical Analysis." *Urban Geography*, 1: 153–66.

Burns, E. (2000). "Travel, Gender and Work: Emerging Commuting Choices in Inner-City Phoenix," in J. Wheeler, *et al. Cities in the Telecommunications Age*. New York: Routledge, 267–82.

Button, K., and Nijkamp, P. (1997). "Social Change and Sustainable Transport." *Journal of Transport Geography*, 5/3: 215–18.

Castells, M. (1996). *The Rise of the Network Society*. Cambridge, Mass: Blackwell.

—— (1999). "Theories of Social Change," Panel Presentation in Plenary Session, Conference on Social Change and Sustainable Transport, National Science Foundation and European Science Foundation, Berkeley, California, 10 March.

Cervero, R. (1998). *The Transit Metropolis: A Global Inquiry*. Washington, DC: Island Press.

Chang, S. E. (2000). "Disasters and Transport Systems: Loss, Recovery, and Competition at the Port of Kobe after the 1995 Earthquake." *Journal of Transport Geography*, 8: 53–65.

Chin, S. M, Greene, D. L., Hopson, J., Hwang, H. L., and Thompson, B. (1999). "Towards National Indicators of VMT and Congestion Based on Real-Time Traffic Data." Paper 990966, presented at the Transportation Research Board Meetings, Washington, DC.

Chou, Y.-H. (1993). "Airline Deregulation and Nodal Accessibility." *Journal of Transport Geography*, 1: 36–46.

Cidell, J., and Adams, J. (2001). "The Groundside Effects of Air Transportation." Research Report CTS-01-02, University of Minnesota Center for Transportation Studies.

Cronon, W. G. (1991). *Nature's Metropolis: Chicago and the Great West*. New York: W. W. Norton.

Debbage, K. G. (1993). "U.S. Airport Market Concentration and Deconcentration." *Transportation Quarterly*, 47: 115–36.

—— (1994). "The International Airline Industry: Globalization, Regulation, and Strategic Alliances." *Journal of Transport Geography*, 3: 190–203.

Dempsey, P. S., and Goetz, A. R. (1992). *Airline Deregulation and Laissez-Faire Mythology*. Westport, Conn.: Quorum Books.

—— Goetz, A. R., and Szyliowicz, J. S. (1997). *Denver International Airport: Lessons Learned*. New York: McGraw Hill.

DeWitt, W. J., Langley, C. J., and Ralston, B. A. (1997). "The Impact of GIS on Supply Chains." *Logistics Technology International*, 34–7.

Dutt, A. K., and Mukhopadhyay, A. (1992). "The Mass Rapid Transit System in Calcutta—A Clean Efficient & Well Maintained Metro." *Public Transport International*, 1: 70–81.

Eliot-Hurst, M. E. (1973). "Transportation and the Societal Framework." *Economic Geography*, 49: 163–80.

—— (1974). "The Geographic Study of Transportation, its Definition, Growth, and Scope," in M. E. Eliot-Hurst (ed.), *Transportation Geography: Comments and Readings*. New York: McGraw Hill.

England, K. (1993). "Suburban Pink Collar Ghettos: The Spatial Entrapment of Women?" *Annals of the Association of American Geographers*, 83/2: 225–42.

Federal Geographic Data Committee (FGDC) (1998). "Road Data Model—Content Standard & Implementation Guide."

Fielding, G. (1995). "Congestion Pricing and the Future of Transit." *Journal of Transport Geography*, 3/4: 239–46.

Fleming, D. K. (1991). "Competition in the U.S. Airline Industry." *Transportation Quarterly*, 45/2: 181–210.

—— (2000). "A Geographical Perspective of the Transhipment Function." *International Journal of Maritime Economics*, 11/3: 163–176.

Fleming, D. K., and Hayuth, Y. (1994). "Spatial Characteristics of Transportation Hubs: Centrality and Intermediacy." *Journal of Transport Geography*, 2: 3–18.

Fotheringham, A. S., and O'Kelly, M. E. (1989). *Spatial Interaction Models: Formulations and Applications*. Dordrecht: Kluwer Academic.

Fotheringham, A. S., and Trew, R. (1993). "Chain Image and Store-Choice Modeling: The Effects of Income and Race." *Environment and Planning A*, 25: 179–96.

Garrison, W. L. (1960). "Connectivity of the Interstate Highway System." *Papers of the Regional Science Association*, 6: 121–38.

Garrison, W. L., and Marble, D. F. (1961). *The Structure of Transportation Networks*. Washington, DC: U.S. Department of Commerce, Office of Technical Services.

Garrison, W. L., and Ward, J. D. (2000). *Tomorrow's Transportation: Changing Cities, Economies, and Lives*. Boston: Artech House.

Garrison, W. L., Berry, B. J. L., Marble, D. F., Nystuen, J. D., and Morrill, R. L. (1959). *Studies of Highway Development and Geographic Change*. Seattle: University of Washington Press.

Gassaway, A. (1992). "The Adequacy of Walkways for Pedestrian Movement along Public Roadways in the Suburbs of an American City." *Transportation Research A*, 26/5: 361–79.

Gauthier, H. L., and Mitchelson, R. L. (1981). "Attribute Importance and Mode Satisfaction in Travel Mode Choice Research." *Economic Geography*, 57: 348–61.

Geoinformatica (2000). Special Issue on Geographic Information Systems and Intelligent Transportation Systems, 4/2.

Gillespie, A., and Richardson, R. (2000). "Teleworking and the City: Myths of Workplace Transcedence and Travel Reduction," in J. O. Wheeler *et al.*, *Cities in the Telecommunications Age*. New York: Routledge, 228–48.

Giuliano, G. (1995). "Land Use Impacts of Transportation Investments: Highway and Transit," in S. Hanson (1995: 305–41).

—— (1998). "Information Technology, Work Patterns and Intrametropolitan Location: A Case Study." *Urban Studies*, 35/7: 1077–95.

Giuliano, G., and Gillespie, A. (1997). "Research Issues Regarding Societal Change and Transport." *Journal of Transport Geography*, 5/3: 165–76.

Giuliano, G., and Golob, T. F. (1990). "Using Longitudinal Methods for Analysis of a Short-Term Transportation Demonstration Project." *Transportation*, 17: 1–28.

Giuliano, G., and Small, K. (1993). "Is the Journey to Work Explained by Urban Structure?" *Urban Studies*, 30/9: 1485–500.

—— (1999). "The Determinants of Growth of Employment Subcenters." *Journal of Transport Geography*, 7: 189–201.

Giuliano, G., Hwang, K., and Wachs, M. (1993). "Employee Trip Reduction in Southern California: First Year Results." *Transportation Research A*, 27: 125–87.

Goetz, A. R. (1992). "Air Passenger Transportation and Growth in the U.S. Urban System, 1950–1987." *Growth and Change*, 23/2: 218–42.

—— (2002). "Deregulation, Competition, and Antitrust Implications in the U.S. Airline Industry." *Journal of Transport Geography*, 10: 1–19.

Goetz, A. R., and Dempsey, P. S. (1989). "Airline Deregulation Ten Years After: Something Foul in the Air." *Journal of Air Law and Commerce*, 54: 927–63.

Goetz, A. R., and Sutton, C. J. (1997). "The Geography of Deregulation in the U.S. Airline Industry." *Annals of the Association of American Geographers*, 87/2: 238–63.

Goetz, A. R., and Szyliowicz, J. S. (1997). "Revisiting Transportation Planning and Decision-Making Theory: The Case of Denver International Airport." *Transportation Research Part A*, 31/4: 263–80.

Goetz, A. R., and Vowles, T. M. (2000). "Progress in Intermodal Passenger Transportation: Private Sector Initiatives." *Transportation Law Journal*, 27: 475–97.

Goetz, A. R., Dempsey, P. S., and Larson, C. (2002). "Metropolitan Planning Organizations: Findings and Recommendations for Improving Transportation Planning." *Publius: The Journal of Federalism*, 32/1: 87–105.

Golledge, R. G. (1980). "A Behavioral View of Mobility and Migration Research." *The Professional Geographer*, 32: 14–21.

Golledge, R. C., and Timmermans, H. (1990). "Applications of Behavioral Research on Spatial Problems I: Cognition." *Progress in Human Geography*, 14: 57–99.

Goodchild, M. F. (1998). "Geographic Information Systems and Disaggregate Transportation Modeling." *Geographical Systems*, 5: 19–44.

Goodchild, M. F., Klinkenberg, B., and Janelle, D. G. (1993). "A Factorial Model of Aggregate Spatiotemporal Behavior: Application to the Diurnal Cycle." *Geographical Analysis*, 25: 277–94.

Greene, D. L. (1997). "Economic Scarcity: Forget Geology, Beware Monopoly." *Harvard International Review*, 19/3: 16–22.

Greene, D. L., and Fan, Y. (1995). "Transportation Energy Intensity Trends, 1972–1992." *Transportation Research Record*, 1475: 10–19.

Greene, D. L., and Han, X. (1996). "The Unintended Consequences of Transportation." Transportation Statistics Annual Report 1995, Bureau of Transport Statistics, Washington, DC.

Greene, D. L., and Wegener, M. (1997). "Sustainable Transport." *Journal of Transport Geography*, 5/3: 177–90.

Hagerstrand, T. (1982). "Diorama, Path, and Project." *Tijdschrift voor Economische en Sociale Geografie*, 73: 323–39.

Hanson, S. (1982). "The Determinants of Daily Travel-Activity Patterns: Relative Location and Socio-Demographic Factors." *Urban Geography*, 3: 179–202.

—— (ed.) (1995). *The Geography of Urban Transportation*, 2nd edn. New York: Guilford.

—— (1998). "Off the Road? Reflections on Transportation Geography in the Information Age." *Journal of Transport Geography*, 6/4: 241–9.

—— (2000). "Reconceptualizing Accessibility," in Janelle and Hodge (2000: 267–78).

Harris, R., and Bloomfield, A. (1997). "The Impact of Industrial Decentralization on the Gendered Journey to Work, 1900–1940." *Economic Geography*, 73/1: 94–117.

Hau, T. (1990). "Electronic Road Pricing: Developments in Hong Kong, 1983–1989." *Journal of Transport Economics and Policy*, 24/2: 203–14.

Haynes, K., Bowen, W. M., Arieria, C. R., Burhans, S., Salem, P. L. and Shafie, H. (2000). "Intelligent Transportation Systems Benefit Priorities: An Application to the Woodrow Wilson Bridge." *Journal of Transport Geography*, 8: 129–39.

Hayuth, Y. (1988). "Rationalization and Deconcentration of the U.S. Container Port System." *Professional Geographer*, 40/3: 279–88.

Hepworth, M. E. (1990). *Geography of the Information Economy*. New York: Guilford Press.

Herod, A. (1998). "Discourse on the Docks: Containerization and Inter-Union Work Disputes in U.S. Ports, 1955–1985." *Transactions of the Institute of British Geographers*, 23: 177–91.

Hodge, D. C. (1988). "Fiscal Equity in Urban Mass Transit Systems: A Geographic Analysis." *Annals of the Association of American Geographers*, 78/2: 288–306.

— (1990). "Geography and the Political Economy of Urban Transportation." *Urban Geography*, 11/1: 87–100.

— (1992). "Urban Congestion: Reshaping Urban Life." *Urban Geography*, 13/6: 577–88.

— (1995). "My Fair Share: Equity Issues in Urban Transportation," in Hanson (1995: 359–75).

Hodge, D. C., and Koski, H. (1997). "Information and Communication Technologies and Transportation: European—US Collaborative and Comparative Research Possibilities." *Journal of Transport Geography*, 5/3: 191–7.

Hodge, D. C., Morrill, R. L., and Stanilov, K. (1996). "Implications of Intelligent Transportation Systems for Metropolitan Form." *Urban Geography*, 17/8: 714–39.

Horner, M. W., and O'Kelly, M. E. (2001). "Embedding Economies of Scale Concepts for Hub Network Design." *Journal of Transport Geography*, 9: 255–65.

Horowitz, J. L. (1985). "Travel and Location Behavior: State of the Art and Research Opportunities." *Transportation Research*, 19A: 441–53.

— (1991). "Modeling the Choice of Choice Set in Discrete-Choice Random-Utility Models." *Environment and Planning A*, 23: 1237–46.

Hoyle, B., and Knowles, R. (eds.) (1998). *Modern Transport Geography*, 2nd edn. Chichester: John Wiley & Sons.

Irwin, M. D., and Kasarda, J. D. (1991). "Air Passenger Linkages and Employment Growth in U.S. Metropolitan Areas." *American Sociological Review*, 56: 524–37.

Ivy, R. L. (1993). "Variations in Hub Service in the U.S. Domestic Air Transportation Network." *Journal of Transport Geography*, 1: 211–18.

— (1995). "The Restructuring of Air Transport Linkages in the New Europe." *Professional Geographer*, 47/3: 280–8.

Ivy, R. L., Fik, T. J., and Malecki, E. J. (1995). "Changes in Air Service Connectivity and Employment." *Environment and Planning A*, 27: 165–79.

Janelle, D. G. (1995). "Metropolitan Expansion, Telecommuting, and Transportation," in Hanson (1995: 407–34).

— (1997). "Sustainable Transportation and Information Technology: Suggested Research Issues." *Journal of Transport Geography*, 5/1: 39–40.

Janelle, D. G., and Beuthe, M. (1997). "Globalization and Research Issues in Transportation." *Journal of Transport Geography*, 5/3: 199–206.

Janelle, D. G., and Hodge, D. (eds.) (2000). *Information, Place and Cyberspace: Issues in Accessibility*. Berlin: Springer.

Johnston, R. J. (1998). "Fragmentation around a Defended Core: The Territoriality of Geography." *The Geographical Journal*, 164/2: 139–47.

Johnston-Anumonwo, I. (1992). "The Influence of Household Type on Gender Differences in Work Trip Distance." *Professional Geographer*, 44/2: 161–9.

— (1995). "Racial Differences in the Commuting Behavior of Women in Buffalo, 1980–1990." *Urban Geography*, 16/1: 23–45.

— (1997). "Race, Gender, and Constrained Work Trips in Buffalo, NY, 1990." *Professional Geographer*, 49/3: 306–17.

Jones, M. C. (1995). "Street Barriers in American Cities." *Urban Geography*, 16/2: 112–22.

Knudsen, D. C. (1990). "A Dynamic Analysis of Interaction Data: U.S. Rail Freight Flow, 1972–81." *Geographical Analysis*, 22: 259–69.

Kuby, M. J., and Gray, R. G. (1993). "The Hub Network Design Problem with Stopovers and Feeders: The Case of Federal Express." *Transportation Research Part A*, 27: 1–12.

Kuby, M. J., and Reid, N. (1992). "Technological Change and the Concentration of the U.S. General Cargo Port System: 1970–88." *Economic Geography*, 68/3: 272–89.

Kuby, M. J., Ratick, S., and Osleeb, J. (1991). "Modeling U.S. Coal Export Planning Decisions." *Annals of the Association of American Geographers*, 81/4: 627–49.

Kwan, M.-P. (1999*a*). "Gender and Individual Access to Urban Opportunities: A Study Using Space Time Measures," *Professional Geographer*, 51/2: 210–27.

— (1999*b*). "Gender, the Home Work Link, and Space-Time Patterns of Nonemployment Activities," *Economic Geography*, 75/4: 370–94.

— (2000). "Analysis of Human Spatial Behavior in a GIS Environment: Recent Developments and Future Prospects," *Journal of Geographical Systems*, 2: 85–90.

Law, R. (1999). "Beyond 'Women and Transport': Towards New Geographies of Gender and Daily Mobility." *Progress in Human Geography*, 23/4: 567–88.

Lee, J., Chen, L., and Shaw, S. L. (1994). "A Method for the Exploratory Analysis of Airline Networks." *The Professional Geographer*, 46: 468–77.

Leinbach. T. R. (1989). "Transport Policies in Conflict: Deregulation, Subsidies, and Regional Development in Indonesia." *Transportation Research A*, 23/6: 467–75.

— (1992). "Small Towns, Rural Linkages, and Employment." *International Regional Science Review*, 14/3: 317–23.

— (1995). "Transport and Third World Development: Review, Issues, and Prescription." *Transportation Research A*, 29/5: 337–44.

— (2000). "Mobility in Development Context: Changing Perspectives, New Interpretations, and the Real Issues." *Journal of Transport Geography*, 8/1: 1–9.

— (2001*a*). "Transport Infrastructure Challenges in Urban Southeast Asia: Financial Crises, Liberalization, and Sustainability," in Carla Chifos and Ruth Yabes (eds.), *Southeast Asian Urban Environments: Structured and Spontaneous*. Tempe: Arizona State University Press, 115–132.

— (2001*b*). "The Emergence of the Digital Economy and Electronic Commerce," in Leinbach and Brunn (2001: 3–26).

Leinbach, T. R., and Bowen, John T. (1996). "Development and Liberalization: The Airline Industry in ASEAN," in G. Hufbauer and C. Findlay (eds.), *Flying High: Liberalizing Civil Aviation in the Asia Pacific*. Washington, DC: Institute for International Economics, 79–97.

Leinbach, T. R., and Brunn, S. D. (eds.) (2001). *The Worlds of Electronic Commerce*. Chichester: John Wiley.

Leinbach, T. R., and Chia Lin Sien (1989). *Southeast Asian Transport: Issues in Development*. Singapore: Oxford University Press.

Leinbach, T. R., and Smith, J. H. (1997). "Development of a Cooperative International Interdisciplinary Program on Social Change and Sustainable Transport." *Journal of Transport Geography*, 5/1: 1–3.

Levinson, D. (1998). "Accessibility and the Journey to Work." *Journal of Transport Geography*, 6/1: 11–21.

Lewis, T. (1997). *Divided Highways: Building the Interstate Highways, Transforming American Life*. New York: Viking Penguin.

Ley, D. (1983). *A Social Geography of the City*. New York: Harper & Row.

Liu, C., Zhang, M., and Ralston, B. A. (1993). "TRAILMAN: The Transportation and Inland Logistics Manager." US Agency for International Development.

Lo, L. (1990). "A Translog Approach to Consumer Spatial Behavior." *Journal of Regional Science*, 30/3: 393–413.

Louriero, C. F. G., and Ralston, B. A. (1996). "Investment Selection Model for Multicommodity Multimodal Transportation Networks." *Transportation Research Record*, 1522: 38–46.

Louviere, J. J. (1981). "A Conceptual and Analytical Framework for Understanding Spatial and Temporal Choices." *Economic Geography*, 57: 304–14.

McCalla, R. J. (1999). "Global Change, Local Pain: Intermodal Seaport Terminals and their Service Areas." *Journal of Transport Geography*, 7: 247–54.

McLafferty, S., and Preston, V. (1991). "Gender, Race, and Commuting among Service Sector Workers." *Professional Geographer*, 43/1: 1–15.

Maiou, S. P., and Summers, M. (1997). "Real-Time Dynamic Traffic Assignment Systems." FHWA Program Review Meeting, Oak Ridge, Tennessee.

Maraffa, T., and Finnerty, T., Jr. (1993). "Changes in the Interurban Accessibility of Ohio Cities during the Era of Airline Deregulation." *Professional Geographer*, 45: 389–98.

Marble, D., Gou, Z., and Liu, L. (1995). "Visualization and Exploratory Data Analysis of Interregional Flows." *Proceedings of Geographic Information Systems for Transportation Symposium*, 1: 128–36.

Marston, J., Golledge, R., and Costanzo, C. (1997). "Investigating Travel Behavior of Nondriving Blind and Vision Impaired People: The Role of Public Transit." *Professional Geographer*, 49/2: 235–45.

Mayer, H. A., and Wade, R. C. (1969). *Chicago: Growth of a Metropolis*. Chicago: University of Chicago Press.

Meinig, D. W. (1986). *The Shaping of America: A Geographical Perspective on 500 Years of History*. New Haven: Yale University Press.

Miller, H. J. (1999). "Measuring Space-Time Accessibility Benefits within Transportation Networks: Basic Theory and Computation Procedures." *Geographical Analysis*, 31/2: 187–213.

Miller, H. J., and Shaw, S.-L. (2001). *Geographic Information Systems for Transportation: Principles and Applications*. Oxford: Oxford University Press.

Minshall, C. W., Buxbaum, R. W., and Wright, C. J. (1995). "The Economics of Aerospace in Ohio." *Transportation Quarterly*, 49/3: 101–26.

Modarres, A. (1993). "Evaluating Employer-Based Transportation Demand Management Programs." *Transportation Research A*, 27/4: 291–97.

Moon, H. (1990). "Land Use around Suburban Transit Stations." *Transportation*, 17: 67–88.

—— (1994). *The Interstate Highway System*. Washington, DC: Association of American Geographers.

National Cooperative Highway Research Program (NCHRP) (1997). "A Generic Data Model for Linear Referencing Systems," NCHRP Research Results Digest, 218.

Nijkamp, P., and Baaijens, S. (1999). "Time Pioneers and Travel Behaviour: An Investigation into the Viability of 'Slow Motion.'" *Growth and Change*, 30/2: 239–63.

Nijkamp, P., and Pepping, G. (1998). "A Meta-analytical Evaluation of Sustainable City Initiatives." *Urban Studies*, 35/9: 1481–500.

Nyerges, T. L., Montejano, R., Oshiro, C., and Dadswell, M. (1997). "Group-Based Geographic Information Systems for Transportation Improvement Site Selection." *Transportation Research C*, 5: 349–69.

O'Kelly, M. E. (1988). "Aggregate Rent and Surplus Measurement in a von Thunen Model." *Geographical Analysis*, 20: 187–97.

—— (1998). "A Geographer's Analysis of Hub-and-Spoke Networks." *Journal of Transport Geography*, 6: 171–86.

O'Kelly, M. E., and Bryan, D. (1996). "Agricultural Location Theory: von Thunen's Contribution to Economic Geography." *Progress in Human Geography*, 20: 457–75.

—— (1998). "Hub Location with Flow Economies of Scale." *Transportation Research Part B*, 32/8: 605–16.

O'Kelly, M. E., and Miller, H. J. (1994). "The Hub Network Design Problem: A Review and Synthesis." *Journal of Transport Geography*, 2: 31–40.

O'Kelly, M. E., Song, W., and Shen, G. (1995). "New Estimates of Gravitational Attraction by Linear Programming." *Geographical Analysis*, 27: 271–85.

O'Kelly, M. E., Bryan, D., Skorin-Kapov, D., and Skorin-Kapov, J. (1996). "Hub Network Design with Single and Multiple Allocation: A Computational Study." *Location Science*, 4: 125–38.

Osleeb, J. P., and Ratick, S. J. (1990). "A Dynamic Location-Allocation Model for Evaluating the Spatial Impacts for Just-in-Time Planning." *Geographical Analysis*, 22: 50–69.

Peet, R. (1998). *Modern Geographical Thought*. Oxford: Blackwell.

Peng, Z. (1997). "The Jobs-Housing Balance and Urban Commuting." *Urban Studies*, 34/8: 1215–35.

Pipkin, J. S. (1981). "The Concept of Choice and Cognitive Explanations of Spatial Behavior." *Economic Geography*, 57: 315–31.

—— (1995). "Disaggregate Models of Travel Behavior," in Hanson (1995: 188–218).

Ralston, B. A., and Liu, C. (1989). "Accessibility Maximizing on Congested Networks." *Geographical Analysis*, 2: 236–50.

Ralston, B. A., Tharakan, G., and Liu, C. (1994). "A Spatial Decision Support System for Transportation Policy Analysis." *Journal of Transport Geography*, 2: 101–10.

Reynolds-Feighan, A. J. (1992). *The Effects of Deregulation on U.S. Air Networks*. Berlin: Springer-Verlag.

—— (1998). "The Impact of U.S. Airline Deregulation on Airport Traffic Patterns." *Geographical Analysis*, 30/3: 234–53.

Rimmer, P. J. (2000). "Effects of the Asian Crisis on the Geography of Southeast Asia's Air Traffic." *Journal of Transport Geography*, 8: 83–97.

Rodrigue, J. P. (1999). "Globalization and the Synchronization of Transport Terminals." *Journal of Transport Geography*, 7: 255–61.

—— (2002). "Globalization and the Geography of Logistics." Paper presented at the 98th Annual Meeting of the Association of American Geographers, Los Angeles, Calif.

Rollins, W. H. (1995). "Whose Landscape? Technology, Fascism, and Environmentalism on the National Socialist *Autobahn*." *Annals of the Association of American Geographers*, 85/3: 494–520.

Rosenbloom, S., and Burns, E. (1993). "Gender Differences in Commuter Travel in Tucson: Implications for Travel Demand Management Programs." *Transportation Research Record*, 1404: 82–90.

Sagers, M., and Maraffa, T. (1990). "Soviet Air-Passenger Transportation Network." *Geographical Review*, 80/3: 266–78.

Saxena, S., and Mokhtarian, P. L. (1997). "The Impact of Telecommuting on the Activity Space of Participants." *Geographical Analysis*, 29: 124–44.

Schintler, L. (1997). "Congestion." *Journal of Transport Geography*, 5/1: 27.

Seamon, D. (1979). *A Geography of the Lifeworld: Movement, Rest, and Encounter*. New York: St. Martin's Press.

Sermons, M. W., and Koppelman, F. S. (2001). "Representing the Differences between Female and Male Commute Behavior in Residential Location Choice Models." *Journal of Transport Geography*, 9: 101–10.

Shaw, S. L. (1993). "Hub Structures of Major U.S. Passenger Airlines." *Journal of Transport Geography*, 1: 47–58.

Shaw, S. L., and Ivy, R. L. (1994). "Airline Mergers and their Effect on Network Structure." *Journal of Transport Geography*, 2/4: 234–46.

Shaw, S. L., and Williams, J. F. (1991). "Role of Transportation in Taiwan's Regional Development." *Transportation Quarterly*, 45/2: 271–96.

Sheppard, E. (1995). "Modeling and Predicting Aggregate Flows," in Hanson (1995: 100–28).

Slack, B. (1990). "Intermodal Transportation in North America and the Development of Inland Load Centers." *Professional Geographer*, 42/1: 72–83.

—— (1993). "Pawns in the Game: Ports in a Global Transportation System." *Growth and Change*, 24: 579–88.

—— (1999). "Satellite Terminals: A Local Solution to Hub Congestion." *Journal of Transport Geography*, 7: 241–6.

Sorenson, N. (1991). "The Impact of Geographic Scale and Traffic Density on Airline Production Costs: The Decline of the No-Frills Airlines." *Economic Geography*, 67/4: 333–45.

Southworth, F. (1981). "Calibration of Multinomial Logit Models of Mode and Destination Choice." *Transportation Research B*, 15: 315–25.

—— (1995). "A Technical Review of Urban Land Use—Transportation Models as Tools for Evaluating Vehicle Travel Reduction Strategies." ORNL–6881, Oak Ridge, Tenn.

Southworth, F., Peterson, B. E., and Chin, S. M. (1998). "Methodology for Estimating Freight Shipment Distances for the 1997 Commodity Flow Survey." Report prepared for the Bureau of Transportation Statistics, US Department of Transportation.

Southworth, F., Xiong, D., and Middendorf, D. (1997). "Development of Analytic Intermodal Freight Networks for Use within a GIS." *Proceedings of the GIS-T 97 Symposium*. American Association of State Highway and Transportation Officials Conference, 201–18.

Stamatiadis, N., Leinbach, T. R., and Watkins, J. (1996). "Travel Among Non-Urban Elderly," *Transportation Quarterly*, 50/3: 113–21.

Stough, R. R., and Rietveld, P. (1997). "Institutional Issues in Transport Systems." *Journal of Transport Geography*, 5/3: 207–14.

Stutz, F. P. (1995). "Environmental Impacts', in Hanson (1995: 376–406).

Stutz, F. P., and de Souza, A. R. (1998). "Transportation and Communications in World Economy," in F. P. Stutz and A. R. de Souza, *The World Economy: Resources, Location, Trade, and Development*, 3rd edn. Upper Saddle River, NJ: Prentice Hall, 163–227.

Stutz, F. P., Parrott, R., and Kavanagh, P. (1992). "Charting Urban Space-time Population Shifts with Trip Generation Models." *Urban Geography*, 13: 468–74.

Sutton, C. J. (1999). "Land Use Change along Denver's I-225 Beltway." *Journal of Transport Geography*, 7/1: 31–41.

Sutton, J. (1997). "Data Attribution and Network Representation Issues in GIS and Transportation." *Transportation Planning and Technology*, 21: 25–44.

Taaffe, E. J. (1956). "Air Transportation and United States Urban Distribution." *The Geographical Review*, 46: 219–38.

Taaffe, E. J., and Gauthier, H. L. (1994). "Transportation Geography and Geographic Thought in the United States: An Overview." *Journal of Transport Geography*, 2/3: 155–68.

Taaffe, E. J., Gauthier, H. L., and O'Kelly, M. E. (1996). *The Geography of Transportation*, 2nd edn. Upper Saddle River, NJ: Prentice Hall.

Taaffe, E. J., Morrill, R. L., and Gould, P. R. (1963). "Transport Expansion in Underdeveloped Countries: A Comparative Analysis." *The Geographical Review*, 53: 503–29.

Taylor, B., and Ong, P. (1995). "Spatial Mismatch or Automobile Mismatch? An Examination of Race, Residence and Commuting in US Metropolitan Areas." *Urban Studies*, 32/9: 1453–73.

Thill, J. C. (1992). "Choice Set Formation for Destination Choice Modeling." *Progress in Human Geography*, 16: 361–83.

—— (2000). *Geographic Information Systems in Transportation Research*. London: Pergamon Press.

Thill, J. C., and Horowitz, J. (1997). "Travel-Time Constraints on Destination-Choice Sets." *Geographical Analysis*, 29/2: 108–23.

Timmermans, H. J. P., and Golledge, R. G. (1990). "Applications of Behavioral Research on Spatial Problems II: Preference and Choice." *Progress in Human Geography*, 14: 311–54.

Tuan, Y. F. (1971). "Geography, Phenomenology, and the Study of Human Nature." *The Canadian Geographer*, 15: 181–92.

Ullman, E. L. (1956). "The Role of Transportation and the Bases for Interaction," in W. L. Thomas (ed.), *Man's Role in Changing the Face of the Earth*. Chicago: University of Chicago Press, 862–80.

van Geenhuizen, M., Nijkamp, P., and Black, W. R. (1999). "Social Change and Sustainable Transport: A Manifesto on Transatlantic Research Opportunities." Keynote Lecture for the NSF-ESF Conference on Social Change and Sustainable Transport, 10–13 March, University of California, Berkeley.

Vance, J. E., Jr. (1986). *Capturing the Horizon: The Historical Geography of Transportation Since the Transport Revolution of the Sixteenth Century*. New York: Harper & Row.

—— (1995). *The North American Railroad*. Baltimore: Johns Hopkins University Press.

Vowles, T. M. (1999). "Predicting the Loss of Commuter Air Service in the United States." *Journal of Air Transport Management*, 5: 13–20.

Vowles, T. M. (2000*a*). "The Effect of Low Fare Air Carriers on Airfares in the US." *Journal of Transport Geography*, 8: 121–8.

—— (2000*b*). "The Geographic Effects of U.S. Airline Alliances." *Journal of Transport Geography*, 8: 277–85.

Wang, J., and Slack, B. (2000). "The Evolution of a Regional Container Port System: The Pearl River Delta." *Journal of Transport Geography*, 8: 263–75.

Warf, B. (1988). "The Port Authority of New York-New Jersey." *Professional Geographer*, 40/3: 288–97.

Warren, W. (1993). "A Transportation View of the Morphology of Cities." *Transportation Quarterly*, 47/3: 367–77.

Wolch, J., Rahimian, A., and Koegel, P. (1993). "Daily and Periodic Mobility Patterns of the Urban Homeless." *Professional Geographer*, 45/2: 159–69.

Wong, D. W. S., and Meyer, J. W. (1993). "A Spatial Decision Support System Approach to Evaluate the Efficiency of a Meals-on-Wheels Program." *Professional Geographer*, 45: 332–41.

World Bank (1994). Bangladesh Transport Sector Study.

Xiong, D., and Gordon, S. (1995). "Data Structure and Information Coding for ITS Location Referencing Messages." *Proceedings of the GIS-T Symposium*, 1: 165–77.

Zhan, F. B., and Noon, C. (1998). "Shortest Path Algorithms: An Evaluation using Real Road Networks." *Transportation Science*: 32: 65–73.

CHAPTER 17

Urban Geography

Stuart Aitken, Don Mitchell and Lynn Staeheli

Introduction

The study of urbanization processes and urban spaces is contentious and problematic. Different disciplines focus on different processes and ways of knowing, and urban life—its contexts and problems—is tugged and twisted in so many directions that it is difficult to know the appropriate questions to ask, let alone to articulate future research directions. Mayors and other city leaders are concerned about civic boosterism and the quality of life in their cities, planners try to manage competing claims on space and movement, and environmentalists grapple with degradation and equity, while economists conjure up more appropriate models of development and growth. The urban arena is a context for competing intellectual claims and traditions that at times converge on consensus but more often than not garner dissent. We forefront our appraisal of the subfield with a contention that guides most of what is to follow. The contention is important because it necessarily limits the kinds of research we talk about. We argue that with the emergence of a more sophisticated articulation of spatial theory in the last decade, geographers are now well positioned to say something important about the urban issues that are shaping the new millennium.[1] This sea change occurred in the 1990s and now places many aspects of geographic research at the forefront of urban analysis. The articulation of spatial theory comes in large part from two sources: first, critical geography with its focus on the spatial construction of social life and, second, from emerging ideas about technology and space. It is not our intention to dismiss the importance of empirical and interpretative studies, which are discussed tangentially in relation to the central theoretical themes of the chapter. In this review, however, we emphasize the articulation of spatial theory as a significant development in urban geography as we enter the twenty-first century.

We begin in the first main section by picking up where "The Urban Problematic" left off in *Geography in America* (1989). We describe the ways in which American geography is rising to the challenge of understanding the tremendous changes that are underway in cities and argue that the work of urban geographers focusing on the roles of space and scale is critical to understanding these changes. Whereas Sallie Marston and her colleagues (1989: 667) note that "geographic contributions to [the] basic discourse on the nature of the urban question have been, for the most part, imitative of those already evolving in other disciplines," we argue that geography now is at the forefront of many of the important questions about urban life. Empirically,

[1] This chapter was originally drafted in 1998 and presented at the Annual Meeting of the Association of American Geographers in 1999. The paper was submitted to the editors in late spring, 1999. Comments from editors and reviewers were received in mid-summer of 2000 and the paper was redrafted and returned to the editors in the late summer 2000 as a final draft. Urban geography is a dynamic field. Owing to the long period between composition and publication, the chapter does not include the most recent scholarship in the field.

urban geography is about describing, interpreting and analyzing a set of events, meanings, experiences, institutions, and artefacts as they are understood socially and spatially. Theoretically, the "urban problematic" is about the social fragmentation that accompanies differential access to power and the control of space; a study of this problematic highlights urban processes as they relate to lived experiences and the production of cultural and spatial forms. That these positions are the central theoretical and empirical underpinnings of urban geography may be debated, but there is little controversy about the importance of "things urban" in the popular social imaginary. For most people, "urban" holds important meanings and connotations because it suggests certain kinds of spaces and places. Today most social, economic, and urban theorists, irrespective of their philosophical and ideological perspectives, agree with the traditional geographic claim that both space and place are profoundly important in how social life is constituted (Knopp 1995: 151).

Urban geography as a subfield, then, serves as an important arena for debate on various topics and areas of tension pertaining to processes of urbanization, urban economics, social differentiation of urban space, and the ways that everyday life is contextualized by that space. Of course, urban geographers in North America comprise an eclectic group of scholars, practitioners, policy-makers, teachers, and lay persons whose work covers a breadth that cannot be represented in a set of monographs let alone a single chapter. In what follows we do not attempt a comprehensive assessment of the entire breadth of the subfield but, rather, we isolate what we consider to be a set of interdependent themes that summarize some of the major contributions of American urban geographers in the 1990s. Because our synthesis coalesces around a set of interdependent themes, some very important aspects of the subfield that do not relate specifically to these themes may not gain the coverage to which they would otherwise be entitled. Indeed, there are important aspects of urban geographic research such as those pertaining to transportation and policy (Hanson 1995; Vuchic 2000), or ecology and urban sustainability (Bromley 1990; Fitzpatrick and LaGory 2000) that we either omit entirely or refer to only in passing, as these topics are dealt with in other chapters in this volume. We note the omissions in this chapter and identify some insightful recent monographs in part as an apology to those geographers who do not find their work, or the arena in which their work resides, acknowledged here. We are committed to an assessment of urban geography that coheres around a set of theoretical constructs rather than to a disconnected and fragmented overview of what is perhaps one of geography's most eclectic subfields.

After revisiting the challenges set forth by Sallie Marston, George Towers, Martin Cadwallader, and Andrew Kirby in the first volume of *Geography in America* (1989), we then highlight in the second section the ways in which theories of space, of the spatiality of urban life, and of scale and technology have been used to meet some of these challenges. In particular, we focus on geographic research that highlights difference, that explores the privatization of urban space, and that shows how urban space is visualized, monitored, "raced," and represented. We end this second section by focusing on scale and globalization, and thereby examining the work of North American geographers in the global south along with contributions that add to our understanding of how the local/global dialectic plays out in urban space.

In an attempt to ground rather than fetishize the theories of space, spatiality, and scale elaborated upon in the second section, the third section discusses work by urban geographers that explores the social processes behind the segmentation of, and segregation in, contemporary cities. Here we examine four topics—the fragmentation and segregation of urban areas, spatial politics within localities, public space, and new spaces of technology—as examples of the ways in which the theories raised in the second section help unravel specific urban problems and produce important perspectives on the transformation of urban spaces. Each of these topics, in turn, sheds light on other topics, such as housing practices, labor markets, consumption practices, and globalization. We conclude, therefore, with a call for urban geographers to build on the strengths of the theories and methods they developed in the past decade in what will surely be an era of keen urban development, contestation, and transformation. For it is in the realm of the urban that social life is more and more being lived for the majority of the world's population.

Urban Problematics

In "The Urban Problematic," Marston and her colleagues set out to establish the roots of urban geography in America and geographers' contemporary contributions to "the urban question." The problematic broadly encompassed an analytic perspective on systems of cities, explanatory models, and empirical research that focused on inter-urban and intra-urban contexts. The urban question was broadly defined through the Chicago school of urban ecology and the more contemporary work of sociologists such as Manuel Castells (1977) and Peter Saunders (1981). Marston *et al.* (1989: 667–8) argued that urban geography should evolve through a historical consciousness that recognizes how "the massive baggage [of a] collective past" shapes both city form and urban theory, while also remaining alive to contemporary processes and events that themselves may be radically transformative. In this section, we review the argument presented by Marston *et al.*, and describe some of the ways in which theoretical work has responded to the challenges of the urban problematic they describe, as well as to new problematics raised by changes in contemporary cities.

The Urban Problematic Revisited

For Marston and her colleagues, the urban problematic revolved around considerations of what constituted the existence of "the urban category" as a distinct social and spatial entity with explicit functions and underlying processes. They note that in pursuing this question, urban geography has had great difficulty overcoming the major problems faced by the legacy of the Chicago School's ecological theories (1989: 666; see also Dear and Flusty 1998: 51). In the 1970s and 1980s, quantitative urban geographers posed models that attempted to capture the morphology of the city through factor ecology, density functions, trend-surface analysis, and bid-rent curves. "The Urban Problematic" notes that despite technical advances in spatial statistics and computing power, such models failed to do justice to the different institutional structures that molded the context and development of cities (Marston *et al.* 1989: 657). Even aspects of spatial segregation—the central focus of urban factorial ecology—are elided by urban ecology. Edward Soja (1989: 241–2) notes that the notion of an "ecological order" is highly ironic when considering Los Angeles, "arguably the most segregated city in the country." He argues that urban factorial ecology "components are so numerous that they operate statistically to obscure the spatiality of social class relations deeply embedded in the zones and wedges of the urban landscape, as if they needed to be obscured any further." In the 1990s, many urban geographers turned their attention to what seemed to be hidden by factorial methods: the spatial practices and processes that shape material and social relations within cities.

Scholars from other disciplines were also concerned to move beyond ecological approaches and, in particular, beyond the determinism that seemed to characterize some variants of ecology. Many of these scholars, however, did so in ways that appeared to argue that space was unimportant to the operation of processes driving urbanization. One of the most influential of such scholars was Peter Saunders (1981: 278) who argued that "the problem of space . . . can and must be severed from the concern with specific social processes . . . [with the understanding that] all social processes occur in a spatial and temporal context." Marston *et al.* (1989: 666) emphasize that the importance of Saunders' thesis was threefold. First, it moved away from the spatial determinism of the ecological models. Second, and perhaps more importantly, his approach involved a theoretical perspective that highlighted cities as arenas in which more generalized processes are resolved.[2] Third, and more controversially within geography, Saunders seemed to argue that space-based theorizing was futile.

Many urban geographers welcomed and used Saunders' theoretical claim that urban areas were sites in which broader processes were played out and made more concrete. Research on gentrification is one such example. In the late 1980s and 1990s, such research had become a basis for understanding much larger social and economic issues. For example, it highlighted key aspects of economic and social restructuring in Western society and suggested how the role of local government was being redefined (Hamnett 1991; N. Smith 1986, 1992*b*, 1993; van Weesep 1994). The complexity of gentrification is elucidated in descriptions that focus interdependently on human agency (Duncan 1993), meanings and representations (Mills 1993), consumption (Warde 1991), and the general restructuring of the political economy (B. Wilson 1992; N. Smith 1996; Scott 1998). Issues of political identity are indelibly linked with gentrification as women and minorities struggle to appropriate affordable housing and accessible facilities (Bondi 1991; Rose 1988; Breitbart 1990; Warde 1991; Ruddick 1996*b*). Rose and Villeneuve (1998) broaden the discussion to speculate on housing formation through class and labor segmentation with a particular emphasis on disparities amongst the growing number of dual income earners in North American metropolitan areas. It may be argued that gentrification forefronts a spatial expression of fundamental social change because it is tied to so many other urban processes (Hamnett 1991: 76).

[2] One of the problems of the ecological models is that the city was an a priori object of analysis with its own dynamics and social relations. Saunders' thesis suggests that cities need to be dealt with as arenas of socio-spatial relations rather than "real" objects. Marston and her colleagues (1989: 667) use this thesis to raise the efficacy of an "urban question," arguing that perhaps urban geography will be eclipsed by locality studies.

Research on gentrification in conjunction with work on immigration and residential segregation (W. Clark 1998; Ellis and Wright 1998), labor market segmentation (Hanson and Pratt 1995), and structural changes in urban retail (Hopkins 1991; Goss 1993) highlights the importance of understanding the relations between theoretical abstractions and concrete outcomes and a commitment to the geographical and the local as key issues in the urban problematic (Cox and Mair 1989; Harvey 1996; N. Smith 1996; Peet 1998). In this approach, the theorization of space and of *spatiality* is not futile, but central. It is no longer enough to uncover the ways that the spatial is a construction of social forces. It is now important that we understand the ways that the social is a construction of spatial forces. As a definition of spatiality, this is more than a play on words because it focuses attention on the centrality of space in helping unravel complex urban social questions.

Space and Urban Problematics

Soja (1996: 2) notes that a growing community of scholars and citizens is beginning to think about the *spatiality* of urban life in much the same way that past wisdom persistently focused on its intrinsically revealing historical and social qualities. This spatial turn in scholarship and popular culture (cf. de Certeau 1984; Gottdiener 1985; Logan and Molotch 1987; Soja 1989; Harvey 1989; Lefebvre 1991) suggests a way of thinking that was under-theorized in urban geography prior to the 1990s. Today, the legitimacy of geographic perspectives in urban studies stems in part from the reassertion of space in social discourse (Harvey 1989, 1993, 1996, 2000; Soja 1989, 1996, 2000; Dear 2000; Dear and Flusty 1998). There are several ways in which this reassertion has been important. We focus on five interrelated topics that are by no means exclusive and exhaustive: the importance of spatiality and difference; the privatization and surveillance of urban space; the meaning of new technologies to understanding and ordering urban geographies; the representation of urban environmental racism; and the importance of "representation" to understanding urban life in general. In the following pages, we attempt to recreate some of the debates about each.

1. *Spatiality and Difference.* Marston *et al.* (1989: 658) note that during the 1980s urban geographers paid little attention to such aspects of difference as race, gender, ethnicity, and segregation. In the years since, the "reassertion of space in social theory" (Soja 1989) has led geographers to begin the project of theorizing new expressions of identity politics and spatial differentiation in the city. Empirical studies focusing on everyday lives, articulated especially but not exclusively by feminist geographers, suggested new ways of understanding the varied and multi-layered contexts of urban experience (Hanson 1992). As a consequence, research on difference and diversity is perhaps the most enduring hallmark of urban geography in the 1990s, as evidenced by *Cities of Difference*, a remarkable collection of essays edited by Ruth Fincher and Jane M. Jacobs (1998).

Understanding social and spatial difference is important today because of increasing concern over the multiple realities, complex daily lives, and varied experiences of people living and working in cities. It may be argued that contemporary social theorizing in urban geography is a reaction to the universal rational modeling of the 1970s where averages and norms represented individual lives. Although this may be so, current concerns also derive from the writings of those "others" who understand intuitively the importance of identity politics and the power of space and scale to constrain those politics. This is not to say that commentators and researchers of urban life have overlooked the diversity within cities or that urban geographers have a monopoly on the study with their focus on urban inequality and social polarization. Katherine Gibson (1998: 302) points out that these concerns date from at least the writings of Friedrich Engels and other nineteeth-century reformers, and are also present in the writing of Chicago School sociologists and the more recent welfare geographers.[3] Rather than suggesting concern over difference is new we want to note a sea change in the acknowledgement and appreciation of difference, and a move away from attempting to compartmentalize diversity and universalize our understanding of it. Much of this change resonates with an intensification of political struggles by women, minorities, and other oppressed groups, and some of it comes from recent developments in social theory. Feminist, queer studies, anti-racist scholars, and scholars concerned with ability and disability have raised academic consciousness about the ways in which difference structures cities and urban life (Hanson and Pratt 1995; Valentine 1992; Bell and Valentine 1995; Chouinard and Grant 1995; Knopp 1995, 1998; Brown 1997; Butler and Bowlby 1997; S. Kirby and Hay 1997).

Within urban geography, the study of difference has generally taken three forms: (1) representations of difference (e.g. Harvey 1989; Soja 1996); (2) the role

[3] This latter work on urban welfare is exemplified by David Smith's (1997) attempt to enlarge the concept of social justice at the interface of geography and ethics (see also Herman and Mattingly's (1999) focus on ethics and fieldwork in inner city communities).

of geography in creating difference (e.g. Hanson and Pratt 1990; Valentine 1989, 1997; Holloway 1998); and (3) social polarization and its effects (e.g. Pinch 1994; Gibson-Graham 1996; O'Loughlin and Friedrichs 1996). Katherine Gibson (1998: 302) argues that perspectives on difference are diverging at the moment, whereas in the 1970s and 1980s they found contact in questions of relevance, comparison, measurement, and mapping. She argues that the urban literature on social polarization is now largely a discourse about class and the transformation of class relations. The contemporary literature on social polarization, for the most part, approaches difference from economic standpoints and recognizes that equity issues are now systemic and global, and largely beyond the control of any one city government. The role of geography in creating social difference is no longer focused on issues of urban access, equity, and social welfare. Researchers are increasingly interested in how space elaborates what Pierre Bourdieu (1984) refers to as cultural capital and social capital, and how these forms of capital contextualize space (cf. Fernández Kelly 1994; Holloway 1998). From the perspective of representation, theories of difference highlight gender, race, ability, sexuality, ethnicity, and age as valued aspects of multi-layered political identities. The urban context of representations of difference revolve around issues of display (Rendell 1998), identity (Daly 1998), and a panoptic institutionalized gaze that segregates and controls (Graham and Marvin 1996). It is also about different spatial practices and ways of knowing, however. Following from this, Gibson (1998: 304) argues that a politics of difference does not focus solely on the emergence of new class structures, but is enacted around a multiplicity of identities, knowledges, and activism. The problem with these three discourses is that the last two are embedded in psychoanalytic, post-structural, and feminist theories that are, for the most part, ahistorical, whereas the first draws from Marxist and post-Marxist structural theories that seek economic and class-based interpretations. Creating a plausible synthesis at the level of theory is exceedingly difficult. Gibson (1998: 308) notes that Iris Marion Young's (1990*b*) exposé of the politics of urban difference, however, accomplishes such a synthesis with a novel appeal to an explicit and public political imaginary that is emancipatory at the day-to-day and individual level. That is, theory attentive to social practice in the city can show clearly the linkages between—and importance of—both political-economic structure and the formation and contestations of social identities as these linkages are forged in people's everyday lives and political activism.

The concern for difference and the sites in which it is constructed and expressed takes on increasing importance in the context of political debates in many industrialized countries about the fragmentation of urban society; these debates are typically framed in terms of the role of community (Staeheli 1997). The search for identity in community, however, often leads to concerns that difference is homogenized and that those who cannot conform are either excluded or marginalized (Young 1990*b*). In the 1970s, Gerald Suttles (1972) argued that the search for community was based on a mythical desire for unity and wholeness. As an extension of the Chicago school of urban ecology, Suttles' work appears dated today, but the problem of myth construction that he identified is none the less important. It is evident, for example, in the work of contemporary geographers who understand the power of myth in giving form and meaning to urban lives, but who also position their research around an examination of how such myth creates borders, exclusions, and dichotomies (Sibley 1995; Aitken 1998). In the case of neighborhood planning, for example, "neo-traditional" and "new urbanist" schools of architecture and planning have furthered mythic constructions with their rhetoric of "community" and "urbanity," while at the same time establishing a narrowly drawn, quite exclusive, definition of who is counted among the legitimate population of "urbanites." Likewise, new urbanism's privatization of spaces and social functions historically deemed public (such as parks, neighborhood centers, shopping districts) has raised geographers' concerns about uncovering the ideologies and political economies that drive the new urbanist movement (Till 1994; McCann 1994; Falconer Al-Hindi and Staddon 1997; Veregge 1997). Playing on a nostalgia for a "simpler time" when neighbors waved from the front porch and kids played in the streets, neo-traditional design has become perhaps the preeminent means for securing these amenities at a time when public investment in the spaces of social reproduction is declining (Katz 1998*a*, *b*). Indeed, Robyn Dowling (1998*a*) argues further that this nostalgia is part of a larger cultural geography of exclusion that is contained within contemporary suburban ethics that uphold neo-traditional values even when those values are not designed into the community. But in communities designed specifically with neo-traditionalism in mind, developers such as Andres Duany in Florida and Peter Calthorpe in California promote their developments as a return to old-fashioned family and community values. These architects and planners use the ideas of Kevin Lynch, Jane Jacobs, and Herbert Gans to provide a foundation for creating mythic landscapes that are ingratiating for those who can afford them and exclusionary for those who cannot.

Evan McKenzie (1994) notes that, neo-traditional or not, residential "privatopias" are created through the property associations that increasingly dominate urban areas, and they are predicated upon developers contriving the semblance of community life that residents seek. This community life is enclosed, insular, secure, and surrounded by "people like us." Property associations often create and maintain "public" spaces (parks, recreation sites, clubhouses, etc.), access to which is restricted to members of the association and their guests. Moreover, such spaces, which like malls are legally private property, are immune from the public forum rules that govern state-held public space. Gated suburbs, the most extreme spatial form of property associations, additionally invert the traditional relationship in the bourgeois city between public and private, making the public *interior* to the private by putting formerly public amenities behind the fences, guardhouses, and gates that mark the perimeter of such neighborhoods. While often associated with urban life in the US, these communities are found in cities around the world (cf. Caldeira 1999; Marcuse 1995; King 1999; Saff 2001).

For Young (1990*a*), local autonomy is problematic because it involves claims about sovereignty. With decentralized autonomy, McKenzie's privatopias and other small communities exercise control to the extent that citizens in each municipality decide their own form of government, rules and laws, how their land and economic resources are used, and so forth. Such local autonomy can only foster inequality among communities and thereby the oppression of individuals who do not live in more privileged and powerful communities. Like many of the Chicago ecologists, Young believes that large cities liberate people from conformist pressures, but she is skeptical of any kind of political value for difference at the scale of the community or local government. Rather, difference is constituted, experienced, and politicized through the infinitely unique spatial and temporal distinctions that exist throughout all cities. Young (1990*a*) wants to promote the promiscuous mingling of different peoples that Mumford (1961) long ago identified as the essence of urbanity. Mumford, of course, influenced the ideas of Gans, Jacobs, Suttles, and Lynch whose work is foundational for the new urbanism school. Young's solution is different because it highlights the "inexhaustible" experience of difference in cities by contriving larger-scale regional authorities that can ensure power is not based solely upon local sovereignty: "Where there are diverse and unequal neighborhoods, towns and cities, whose residents move in and out of one another's locales and interact in complex webs of exchange, only sovereign authority whose jurisdiction controls them all can mediate relations justly" (Young 1990*a*: 250). Aitken (1998: 187) argues that although Young's solution may appear utopian, it is important for urban geography precisely because it points to the problematizing of sites—and spaces—within the city as social practices as well as a political acts. In most respects, Young's solution is the antithesis of new urbanist and neo-traditional values.

2. *Spatiality and Privatization.* McKenzie's (1994) arguments about "privatopia," and Young's about the kinds of justice available in different kinds of spaces, are central to debates about the privatization of urban space (and hence urban spatiality) that animate contemporary urban geography. Malls, festival marketplaces, Business Improvement Districts, the rise of private security policing, and new forms of surveillance and control of urban spaces indicate to many an important transformation in the structure, societal understanding, and use of urban space. Mike Davis (1990) has diagnosed the development of what he calls "fortress cities" as a response to perceived urban disorder and decay. Such fortress cities have the effect, according to Davis, of separating out undesirable populations and containing them (if imperfectly) in "control zones" isolated from other segments of the population (Davis 1998). Similarly, the contributors to Sorkin's (1992*b*) *Variations on a Theme Park* argue that many of the functions traditionally conducted in urban public spaces are being moved "indoors" to the privatized space of the mall (Crawford 1992), the pedestrian skyway (Boddy 1992), or the self-contained theme park (Winner 1992; Sorkin 1992*a*). The underlying assumption of this work is that urban life—urban spatiality—is more and more being parsed on the basis of private-property rights (Blomley 1999).

Furthermore, analysts such as Nijman (1999) and Zukin (1991, 1995) see the model of the theme park and the mall re-radiating out into the urban landscape, as whole districts are made over in the image of the pleasurable spectacle (see also Crilley 1993; Knox 1993). But as Jon Goss (1996) argues, the privatization of urban space does not necessarily suggest homogenization of values, nor does it mean the end of difference. Focusing on the Aloha Towers Marketplace in Honolulu, Goss elaborates this, suggesting that private space may actually provide *new* opportunities for the performance and expression of subjugated identities. Goss (1993, 1996, 1999) finds the privatization of urban space to be a highly ambiguous affair, encapsulating on the one hand an incredible, intricately planned technology of social and spatial control *and* on the other hand new opportunities for social interaction and new forms of conviviality. Neil Smith (1996) sees no such ambivalence, finding in the rush to privatization (coupled with all manner of punitive laws governing the use of public space) the development of a "revanchist city" built on the model of total domination by the relations of property and the laws designed to protect that property.

A number of geographers are concerned that the move towards residential surveillance systems and "armed response teams" (private police forces) have become a means of "fortressing" private neighborhoods (Davis 1990; Dillan 1994; Graham and Marvin 1996). As urban areas become more privatized and home-centered, many neighborhoods are equipping themselves electronically against incursions and perceived threats to property values. The drive towards social exclusion suggested in the previous section is bolstered by fortressing which is as much a technological as a physical process. The fortress metaphor describes a landscape that is demarcated by physical borders such as gates and walls as well as surveillance devices such as remotely controlled cameras (Fyfe and Bannister 1996, 1998). We will discuss theoretical interpretations of technological space in a moment, but for now it is important to note that although privatization may be suggested by postmodern architecture, the walling of neighborhoods and control through private security guards (as Davis 1990 argues), it is also supported by a sophisticated array of surveillance technologies including GIS to integrate infrared sensors, motion detectors, and closed-circuit televisions. Dear and Flusty (1998: 61) suggest that this emerging urban landscape is characterized by "commudities" (centers of command and control) and resistance is discouraged through "praedatorianism," which they define as a forceful interdiction by praedatorian guards with varying degrees of legitimacy.

Urban geography seems ambivalent about the meaning or status of privatization and urban fortressing. Some suggest fortressing is a logical extension of modernism (Graham and Marvin 1996; Sui 1997); others argue that it is a component of a radically different form of urbanism (Dear and Flusty 1998). Whatever the status of these debates within urban geography, the important point is that research has been directed towards the ways that privatization leads to a structural transformation of urban space and spatiality. One important indication of the scale of this transformation can be found in the ready acceptance of new forms of surveillance over everyday life in cities. Fyfe and Bannister (1996, 1998) detail the growing use of, and the widespread popular support for, closed-circuit television surveillance of city streets, parks, and pedestrian districts in Britain. In the United States, technological surveillance of city streets extends, for example, to the use of sensitive sound-pinpointing devices that guide police officers to the scene of gunshots or other disturbances (Graham and Marvin 1996). Police more and more use sophisticated computer-based "profiling" techniques to detain or remove potentially "dangerous" persons from parks and city streets (Davis 1990; Herbert 1996; R. Saunders 1999). While such technological innovations in the field of surveillance are not complete, some critics argue that their panoptic reach, coupled with the increased privatization of traditional publicly supported functions, adds up to a vision of cities as places that can be perfectly controlled, perfectly knowable, perfectly planned.[4] But not all analyses of the role of technology, or even surveillance, in urban spaces and urban life, are quite so dystopian. Indeed, there is a growing body of spatial scientific theory that emanates from, and empowers, technological advances, burgeoning data sets, and visualization methods that are increasingly sensitive to spatial nuances.

3. *Spatiality, Technology, and Visualization.* The rapid proliferation of computer-based communication technologies such as local access networks and the World Wide Web, coupled with desk-top Computer Assisted Design (CAD) and Geographic Information System (GIS) packages, suggests important changes in the ways in which urban problems are dealt with and theorized. Some geographers suggest that these technologies are now the most important determinants of contemporary planning practice and urban theory (Batty 1993; Nijkamp and Scholten 1993). Others are more cautious about the technologies' potential (Innes and Simpson 1993), or raise concerns about mechanistic solutions taking the place of the face-to-face daily practices of planning (Healey 1992; Hillier 1993; Aitken and Michel 1995; Obermeyer 1998). Still others suggest that communications technologies and GIS may be used to empower otherwise marginalized communities (Harris and Weiner 1998; Elwood and Leitner 1998). In attempting to understand the impacts of technology on the spatiality of the city, we examine the extent to which GIS involves a revitalization of old explanatory models which, to some critics, implies a problematic reassertion of rational discourses in urban geography. Alternatively, some adherents of GIS argue that new spatial platforms not only enable old models to be used in new ways, but also herald the advent of new theoretical perspectives. In addition to the exhumation of old models, greater

[4] Zukin (1995) points out that these new forms of control extend from garbage collecting, to homelessness intervention, to policing itself.

computing power and large databases have enabled new models from the physical sciences to be tested and their relevance to urban space assessed.

Daniel Sui (1994: 260) points out that the integration of GIS with spatial analysis and modeling encourages a variety of sophisticated urban research ranging from studies of urban spatial structure to urban crime analyses and transportation planning (see also Shaw 1993). Foody (1995), for example, merges artificial neural network techniques with ancillary data from GIS to classify urban land cover. Pond and Yeates (1994) use GIS visualization technologies to identify land in transition to urban uses. Perhaps the most influential new types of urban modeling are derived from fractal geometries and cellular automata. Sui (1997) notes that cities modeled through fractals move well beyond the Euclidean logic of the 1970s because they describe spaces and patterns that are asserted by some to be "natural." Sui (ibid. 76) and others argue that fractal geometry "has proved to be instrumental in visualizing dynamic systems" because urban areas are fractal by nature. Fractals describe patterns that remain constant with changes in scale. These patterns are persistent in "nature" (e.g. coastlines and leaves), suggesting to some urban researchers that they may be used (heuristically at least) to describe city morphology. Some extend this argument to speculate on ways that urban fractal patterns suggest social processes. For the most part, these processes are reduced to what earlier urban models describe as population density functions, rank-size distributions, and spatial dependence (e.g. Anselin *et al.* 1993; Batty and Xie 1994*b*).

Another model-based approach that theoretically links physical systems to attempts at understanding urban growth is that prescribed by cellular automata (tessellation automation or self–replicating entities in proximate space) (K. Clark *et al.* 1997: 249). The motivation for modeling cities as cellular automata is to understand dynamics and linkages from a perspective that is independent of scale. The underlying argument is that local processes may be linked to global patterns. Helen Couclelis (1997) notes that GIS has facilitated the ability to use cellular automaton-based urban models as forecasting and policy evaluation tools. These models have been used to simulate growth in Savannah (Batty and Xie 1994*a*), Cincinnati (White *et al.* 1997), Washington/Baltimore and the San Francisco Bay Area (Kirtland *et al.* 1994; K. Clark *et al.* 1997; K. Clark and Gaydos 1998). The latter studies incorporate remotely sensed spatial data integrated with other geographic data in a GIS to model and predict human-induced land-cover change. These models are useful from the standpoint of policy regarding how urban growth might be channeled and managed.

As Sui (1997: 75) notes these attempts to use digital representation and modeling are common to urban management efforts. While interest in the applicability of GIS and other technology to urban problems continues to grow, there has simultaneously been a general decline of interest amongst geographers in the study of so-called city systems or urban fields (cf. Wang 2000; Markusen *et al.* 2000). As a result, urban modeling and representation in the 1990s has been focused at the local scale or on specific populations. For example, although there is sustained speculation on the importance of commuting fields to larger urban structures and networks of cities (W. Clark and Kuijpers-Linde 1994), a large part of this interest focuses on equity issues as they relate to gender and racial diversity in the labor force (McLafferty and Preston 1991; Pratt and Hanson 1991; England 1993; Hanson and Pratt 1995). One novel area in the study of commuting fields is the application of GIS and space-time geography to model urban accessibility (H. Miller 1991) and issues of gender equity (Kwan 1999).

Although urban modelers are cautious about making claims that GIS is a panacea for planning problems, most agree that the management of information and the management of urban systems "have in the last decade become twins" (Nijkamp and Scholten 1993: 85) and that this is an appropriate coupling. Consistent with this perspective, many GIS researchers favor rational instrumentalist perspectives. Rational instrumentalism is based on a modernist discourse that adheres to the premiss that through the application of rational-scientific methods and technology it is possible to build better cities (Cosgrove 1990; Lake 1992). The main assumption is that human spatial patterns and behaviors (at least in the aggregate) are predictable. Instrumentalism is allied to rational-choice theory which portrays society as a set of goal-oriented individuals who seek to maximize wealth and efficiency through fixed choices from amongst alternative courses of action (Barnes and Shepherd 1992). Nigel Thrift (1996) cautions that much of this work is driven by "technological determinism" rather than any real concerns for urban theory. Robert Lake (1993: 404) adds to this critique by suggesting that the "rational model" used in planning and applied urban geography "has been actively resurrected and rehabilitated by the ascendancy of GIS to a position near or at the core of both planning and geography," even though "the post positivist assumptions embraced by GIS have long since been jettisoned by academic theorists." The question of how GIS is appropriated and used to "solve" urban problems is critically assessed in literature that focuses on how day-to-day planning processes relate to communicative rather than instrumental perspectives. Within rational instrumentalism, scant attention is paid to alternative planning strategies such as the recent "communicative and contextual turn" (Healey 1992). Communicative rationality differs from instrumental rationality because the former recognizes the importance of dialogue and the day-to-day processes through which understandings are reached and collective identities constructed (B. Miller 1992: 24). To the extent that communicative rationality better accounts for the significance of dialogue between affected groups, there is often greater ownership in knowledge production whether through GIS or other information technologies (Aitken and Michel 1995: 26).

The problem, of course, is that societal and cultural contexts, existing technology, software logic, and spatial theories shape the kind of geographic information systems available today which, for the most part, remain inaccessible to many groups and individuals (Obermeyer 1998: 65). Even so, GIS packages do not necessarily serve one specific and narrowly science-based epistemology. Recent work on community participation with GIS suggests a wide array of epistemological positions emanating from notions of democracy and citizenship rather than rationality and science (Craig *et al.* 2001). There has been growing concern in the literature for public participation GIS whereby community groups may engage proactively with the planning decisions and policy-making that shapes their lives. Sarah Elwood (2001; Elwood and Leitner 1998), for example, shows how the introduction of technology at the local planning level actually changes the way some residents think about the planning process in ways that are empowering. But she also notes that such technology may also disenfranchise certain sectors of a community. Work such as this that embraces the complexity of the interface between technology and users is advancing the way urban planning contexts (and democracy) are theorized. Elwood and Leitner (1998) contend that it is necessary to ascertain community perspectives through participatory research and sharing information in order to know how GIS might appropriately be used in any particular setting. Part of this process may result in a transformation of how communities imagine their place in a city. Communities rely on particular—sometimes unexamined—*representations* of what the city is and what it means. In the next subsection, we focus on one particular context of GIS visualization that has been quite fruitful of late: urban environmental racism. This is followed by a more general discussion of urban representation as it relates to work in critical geography.

4. *Representing Environmental Racism and Empowering Marginalized Communities.* David Wilson (1998: 254) notes that how "people, places and processes are represented determines who are a city's villains, victims, and salvationists amid urban change." He argues that "representations are the conduit through which planning and policy interventions are understood and advanced, laying out city 'truths' and 'facts' in a seamless everyday." "Representation" has thus emerged as a keyword in urban geography in the 1990s. As computer-assisted design [CAD] has become more important in city planning (Levy 1998) and as GIS-based analyses of urban social, community and environmental problems are developed (Thrall 1993; Thrall and Ruiz 1994; Thrall *et al.* 1994; Sui 1994, 1997; Obermeyer 1998; Griffith *et al.* 1998), the question of how to best visualize the city remains paramount.

Spatial technologies are increasingly used to visualize and conceptualize environmental racism, a process whereby waste and pollution facilities are located disproportionately in poor and minority urban neighborhoods. Sui (1994) cites the ground-breaking work of Burke (1993) who uses socioeconomic data from the Census's TIGER files and Toxic Release Inventory (TRI) data to determine where toxic release facilities are located in Los Angeles County. Statistical analysis suggested a strong association between income, minority status, and the location of toxic release sites. In general, the poorer the area and the higher its minority percentage, the greater the number of toxic waste facilities in the area. Burke's work was followed by other studies by geographers on urban environmental health issues (Cole and Eyles 1997; McMaster *et al.* 1997; Jerrett *et al.* 1998).

Scale is a problematic issue in much of this work. A study in Ohio by Bowen *et al.* (1995), for example, suggests that although there are high correlations between racial variables and toxic releases at the county level, a census-track examination of the most heavily urbanized counties revealed no statistical relation between race and toxicity. The importance of this study is that it raises issues of spatial scale, but it also foregrounds several methodological advisories for future research. Laura Pulido and her colleagues (Pulido 1996; Pulido *et al.* 1996) caution that works such as those of Bowen and his colleagues focus solely on problematic census variables (such as race) and often miss the importance of evaluating social processes, including class formation and the conceptualization of racism.

Urban geographers and planners have been increasingly concerned about marginalized groups' lack of access to GIS technology, databases, and decision-making

(Curry 1995; Weiner *et al.* 1995; Pickles 1995). Although community groups need spatial information to help them uncover environmental inequities and "local resources" (Brandt and Craig 1994), Yapa (1991) points out that GIS is usually too expensive and requires an expertise that often goes beyond that of community members. The way GIS is constituted can make it difficult for lay-persons to participate in ongoing policy and planning debates. In addition, GIS-based graphics and visualization techniques lend an aura of persuasiveness to policy reports and arguments made by city planners despite academic pleas for skepticism (Aitken and Michel 1995; Curry 1995). Work by Bob McMaster and his colleagues (McMaster *et al.* 1997) in Minneapolis has sought to introduce the capabilities of GIS and MapInfo to community groups so that they may access publicly available information on local toxic hazards through TRI, Petrofund, and Superfund sites, and also resource databases on schools, community centers, senior care, daycare centers, and local parks. Elwood and Leitner (1998: 87) suggest that while the development of local databases may contribute to the empowerment of community groups, it might also make them vulnerable to greater surveillance and control. The point here is that researchers can share their knowledge in a participatory setting that might enable appropriate and ethical kinds of collaboration with community groups. Although this kind of knowledge is clearly important, perhaps some of the most significant advances on understanding urban representations and local spaces is non-technical, deriving from post-structural urban theory.

5. *Spatiality, Representation, and Meaning.* If one angle on the question of representation has been through the ways new technology change the way we "see" the city, then another has been through the way the city itself is imagined. Larry Knopp (1998: 150–1) suggests that for urban geographers today, urbanization may be viewed as a representation of social relations and meanings in space, at densities and scales that are at once sufficiently large and complex as to feel overwhelming and almost incomprehensible, while at the same time remaining navigable and meaningful from the vantage point of people's daily lives. Knopp argues further that the same holds true for various sub-areas of the city. As such, one of the more recent challenges to urban geography is to understand the contradictions and power dynamics that exist for local places and communities in the contexts of larger city and regional spaces (cf. Young 1990*a*; de Certeau *et al.* 1998).

A metaphor for urban representation in the 1990s has been to conceptualize the city as a text (Duncan 1990; Donald 1992). Drawing on sources ranging from de Certeau's (1984) discussion of the little tactics through which the city is read and interpreted against the grid of power to the literary theories of Barthes and LaCapra, the textual metaphor asks analysts to focus on "the discourses, symbols, metaphors and fantasies through which we ascribe meaning to the modern experience of urban living" (Donald 1992: 6). For de Certeau (1984), in fact, the very spatiality of urban life needs to be seen as a certain kind of (perhaps illegible) text. And a recent collection of essays on "planning, identity, and control in public space," called *Images of the Street* (Fyfe 1998*a*), is introduced by an essay called "reading the street" (Fyfe 1998*b*). Anthony King (1996: 4), however, suggests that "focusing solely on *discursive* representations, the way the city is read or written, takes our analytical attention from other representational levels." In particular, we need to focus on the built environment as itself "a representation of specific ideologies, of social, political, economic, and cultural relations and practices" and the field of visual representation "where visual signifiers refer to some other signified."

In more general terms, Lefebvre (1991) sees urban space as produced through a dialectic interaction between different kinds of representational practices. Lefebvre's theory of the production of space rests upon a tripartite scheme that envisages space as being constructed through "representations of space" (the space of planners and rationality); "representational spaces" (the spaces of everyday life); and spatial practices (the practices that link and transform space, practices that are negotiated through representations of space and representational spaces). Lefebvre's focus on representation as key to his theory of the production of space is established in *opposition* to the argument that urban spaces are in any simple ways "texts." His point is that representational practices are of the utmost importance, but only if connected to the materiality or physicality of space itself. Lefebvre's ideas have reverberated throughout urban geography, even if his cautionary notes about reading the city as a text have not been heeded.

Perhaps one of the reasons they have not been heeded stems from the fact that more and more cities are establishing themselves precisely *as* representations meant to be read, particularly by tourists or suburban visitors. Cities now often sell themselves on the basis of particular representational strategies (Kearns and Philo 1994). Nijman (1999) and Crump (1999) have shown, for quite different settings, that the construction of a particular image for a city—a brand name, as it were—is crucial to capturing tourists and footloose capital. And mindful of their status as unstable representations, cities have frequently turned to just the sorts of privatization strategies noted above as a means of assuring *continued* inward investment. Urban geographers have thus been increasingly concerned with the status of cities as contested sites of tourism in which highly sanitized urban histories are put on display to be consumed by (presumably undiscriminating) tourists (Boyer 1992; Schein 1997). This is not just an issue for the developed world, but is perhaps even more acute in the developing one (Oakes 1999). Work by urban geographers on the post-colonial tourist gaze in the global south, for example, suggests an interesting focus on the homogenization of symbols (Chang 1999). Several geographers suggest ways that urban development can be re-presented using the context of cities in the global south that does not rely on post-colonial tourism (Bromley 1990; Lawson and Klak 1990).

The point of this latter work is that a city representing itself for tourists is perhaps a different thing than one built on an image of heavy industry. Whether that is the case or not, many analysts have argued that the city serves as a "stage" (Daniels and Cosgrove 1993) upon which social life is enacted. Still other representations see the city as a playground for capital (Harvey 1989), or indeed, as "a city," in an abstract, asocial sense. For Shields (1996: 227), "the notion of 'the city,' *the city itself, is a representation*. It is a gloss on an environment which designates by fiat, resting only on an assertion of the self-evidence that a given environment *is* 'a city.' " Such a position, as Shields recognizes, often leads to precisely the wrong questions. Rather than asking the "useful question of *whose* city it is," scholars tend to adopt what Zukin (1996: 44) calls "a connoisseur's view" often forgetting that "one person's 'text' [or stage or playground] is another person's shopping center or office building" (Zukin 1996: 43). The important point is that "the city" can no longer be taken for granted as a self-evident "thing" (Donald 1992; Shields 1996), nor as an a priori scale of analysis. Rather, it is a site of interpretation, a set of social relations that take on differing meanings depending on how they are looked at—and for what purpose. For example, in a discussion of heritage tourism in Singapore, Chang (1999: 92) notes that "Cities exemplify local places in a global village because they are tied to global networks of capital flows and movements of people and technology while also serving as nodes where global processes converge." He describes "imaging strategies" whereby tourism development in Singapore raises local issues of race and identity. This is especially clear when the "global city" (a representation in and of itself) is a focus of attention against the counterpoint of local agency.

Scaling the City: Connecting the Global to the Local

The impacts of globalization on the city have been pronounced and, in the 1990s, the examination of "global cities," together with a thoroughgoing reconceptualization of geographical scale itself, has formed a significant problematic in urban research. Doreen Massey (1994, 1997) has argued that transformations in technology, political economy, and geopolitics constitute a new geography or a new spatiality that connects and reconceptualizes places. Some recent discussion in geography suggests that a critical theory of scale must speak not only to the connection between the local and the global, but also to the social construction of spaces that are responsible for social fragmentation (Herod 1991, 1997; Agnew 1993; Smith 1992*a*, *b*, 1993; Delaney and Leitner 1997). Taken together, this work augments Massey's (1994) concerns over the importance of understanding the *power geometry* of the flows and interconnections that link places and the differential positioning of individuals and social groups within them. To put that another way, "time-space compression" has led both to a reconfiguration of geographical scale itself *and* the emergence of particular "command and control centers" in the global economy; these centers go by the shorthand "global cities." We examine debates about these aspects of the urban problematic—scale and global cities—in turn.

1. *The Production of Scale.* Massey's notion of "power geometry" is aligned with an emerging focus on the "production of geographical scale" (N. Smith 1992*a*; Herod 1991; Jonas 1994; B. Miller 1994; Cox 1998). For Neil Smith (1992*a*: 66), "scale demarcates the site of social contest, the object as well as the resolution of contest. . . . It is geographical scale that defines the boundaries and binds the identities around which control is executed *and* contested." In this view, "the urban" emerges as a particular constellation of social relations formed through contestation over the boundaries and borders that "contain" or give form to social processes. "Scale" is therefore not seen as something that independently exists and has (a priori) causal power, but rather as something that is actively constructed, maintained, and reproduced. The "local" as a scale of social relations, political activity, or even "community," has to be constructed, just as the "global" has been *made* through trade agreements, the development of institutions like the World Trade Organization that oversee global economic practices, and specific actions and policies (Sassen 1991, 1998; Piven 1998, 1999). But once made, and even though continually contested, scales like the global and

the local clearly have important effects. These effects can be seen in everything from the way cities respond to sequential rounds of investment and disinvestment to the fashioning of new "diasporic" social identities (King 1990, 1993). And they provide the frame within which urban politics are now conducted.

Specifically, Cox (1998) notes that the spaces and scales of dependence and the spaces and scales of engagement may not be congruent. He argues that the processes and politics that create dependence do not necessarily correspond with the spaces and scales in which political agents can press their cases. In other words, the processes that create issues to be addressed through political action may be formed at different scales than those in which political agents can operate. This may be particularly important for political activism around issues of globalization in which activists may not have access to the spaces or scales at which decisions are made and processes shaped. The "urban," like the "state" or the "global," thus organizes different *kinds* of power, and urban politics cannot be understood as isolated from these other scales (see also B. Miller 1994). In other words, there is a central irony at work in the current restructuring of geographical scale. Though economic relations are in many important ways becoming more global, political *responses* to these relations are becoming increasingly localized (Preteceille 1990; Swyngedouw 1997). For Smith (1993), this insight has indicated the need to better understand how social movements—such as those engaged by homeless people, anti-racist activists, and so forth—can "jump scale" to begin to contest structures of power and dominance at the very scales at which they exist, while at the same time finding means not just to control place, but to command space (Harvey 1989: 234–5). For some, the disenfranchisement of local political responses has led to a certain pessimism over the future, as locally tied business people, politicians, and news commentators express an increasing lack of control over the fate of their places. As Beauregard (1993) reminds us, however, such "voices of decline" have a long lineage that suggests an ongoing tension between global politics and local representations. The situation of particular places *vis-à-vis* the "geometry of power" has long been an animating force for urban politics. In one sense, then, the contemporary conflation of the global and the local ("glocalization," as Swyngedouw (1989) calls it) speaks to a host of issues that have always connected urban areas to larger political issues. In another sense, however, we are only beginning to develop the theoretical tool that enables understanding of the emergence of global cities.

2. *Global Cities.* The role of places within a larger geometry of power is crucial for political agents, but it is also crucial for the functioning of the globalizing economy. While the globalization thesis itself is very much open to debate, there is no question that cities are differently situated in relationship to the global economy. Cities have become primary nodes in a global *spatial* division of labor that has seen a sharp concentration of economic functions in some cities (Sassen 1991; Knox and Taylor 1995), the stripping of productive capacity from others, and a massive economic reorganization in still others (Warf and Erickson 1996). While it would be hard to imagine *any* city that is not to some degree "global"—that is, not in some way continuing to be transformed by inflows and outflows of capital operating at a global scale, migrations of new peoples from distant lands, or access to global media, entertainment, and other commodities—in the literature the terms "global cities" and "world cities" are usually limited to those "nodal points that function as control centers for the interdependent skein of material, financial and cultural flows which, together, support and sustain globalization" (Knox 1995: 236).

Following Rodriguez and Feagin (1986), global hierarchies of cities are often posited. A first tier, consisting of London, New York, and Tokyo, houses key financial and information nodal points for the global economy as a whole. A second tier of cities functions as nodes for particular transnational regions (Los Angeles, Frankfurt, etc.) (Markusen *et al.* 2000). A third tier is comprised of international cities of lesser importance, such as Seoul, Madrid, and Sydney. And a fourth level exists of nationally important cities with some strong international links (e.g. San Francisco, Osaka, Milan). Knox (1995: 239) adds a fifth tier of places such as Rochester, NY, Columbus, Ohio, or the "technopolises" of Japan, where "an imaginative and aggressive leadership has sought to carve out distinctive niches in the global market place." Together, such places have helped to create a "transnational producer-services class" (Sklair 1991), a class whose specific cultural traits have drawn increasing attention from geographers (McDowell 1995). For urban geographers, much of the interest in the global cities literature has been in understanding how particular places—such as Tokyo (Cybriwsky 1997), Los Angeles (Scott and Soja 1996), Manaus (Diniz and Santos 2000), Amsterdam (Nijman 1999), London (Thrift 1994), Singapore (Chang 1999), or New York City (N. Smith 1996)—internalize and spatially configure global-scale flows of capital, people, media, and technology (cf. Appadurai 1990). This work is concerned with the importance of specific places in the geography of global flows.

As important as *economic* globalization has been for cities, the transformation wrought by *cultural* globalization (whether that is understood in terms of the globalization of media images and patterns of consumption, or new migrations) is at least as important. For example, Maoz Azaryahu (1999) argues that controversy over the siting of a McDonald's close to an Israeli war memorial was as much about the endorsement and adoption of American values as it was about the commodification of a sacred space. It is also about appropriate symbolic distances and barriers (the McDonald's was more acceptable when its "golden arches" were de-emphasized and the smell from the deep-fryers was filtered). Marcuse (1995) articulates this theoretically by noting that the "globalization of the city" has led to new spatial configurations of urban space. This space is not at all marked by openness (as the image of a borderless world the very notion of "globalization" plays upon) but instead by walls. Specifically, Marcuse argues that in contemporary technologically advanced cities, five distinctive types of residential area are emerging: elite quarters built as enclaves and isolated buildings which also serve as a site for "command and control" functions for the city and economy (see also Dear and Flusty 1998; King 1999); gentrified districts occupied largely by the "transnational producer-services class" and the cultural workers of the global city; a suburban city for skilled workers, civil servants, and middle-level professionals; a tenement district occupied by immigrants and other low-paid workers; and "an abandoned city, the end result of trickle-down, left for the poor, the unemployed, the excluded" (Marcuse 1995: 246). The irony, Marcuse suggests, is that in such a city it is the *walls* (both real and metaphorical) rather than the connections between the quarters of the city that are of most importance. The city, he argues, is at its most divided at least since the medieval period (see also Davis 1990).

The "global city"—whether seen in terms of economics, culture, or urban structure—and the reconfiguration of geographical scale—whether seen as an economic process, a point of political struggle, or a form of social relations—indicate a profound transformation in the way urban life is lived. The important point that needs stressing here is that urban representations are not just about meaning but also about the material conditions of day-to-day life in cities and how these material conditions change. The "urban problematic" is every bit as important to understand now as it was when Marston and her colleagues tried to come to terms with the eclecticism that categorized urban geography a decade ago. While changes in urban theory have created new avenues for research, even more important have been changes in the spaces of cities themselves—the rapid urban restructuring that marked the last quarter of the century. The balance of this chapter attempts to chart just a few of the transformations in urban space and urban theory—and lived experience.

Urban Transformations

Tiananmen, Leipzig, Berlin, Prague, Budapest, Bucharest. As protestors in 1989 took to parks, streets, squares, and town centers across Asia and Eastern Europe, events clearly indicated the vital importance of urban space to political control—its structure and representation, its use and occupation, its power and control. The revolutions of 1989 were not only revolutions in the halls of government; they were popular uprisings filling the very urban spaces that had to this point provided the setting for the reproduction of the society the people were protesting against. By transforming space, society itself was transformed (Lefebvre 1991).

In the decades surrounding 1989, American and Western European urban streets and parks were caught up in the urban restructuring that accompanied what Harvey (1989) identified as a thorough sea change in the global political economy and were transformed by the explosive growth of homelessness (Dear and Wolch 1987; Takahashi 1996; Wolch and Dear 1996). Women and children, men both young and old, people of color as well as whites, found themselves thrown onto the streets, there to make whatever life they could in the putatively public spaces of the city. Meanwhile, the urban and suburban middle classes more and more found their entertainment in the spaces of the mall and festival marketplace, spaces themselves that seemed to portend a new relationship between property, citizenship, and consumption. By the beginning of the 1990s, the continued rapid pace of urban restructuring contributed to a growing sense of what some analysts describe as a "crisis" of the city: an end to urbanity and all that has implied for the construction of a vibrant public sphere (Berman 1984; 1986; Davis 1990, 1998; Goheen 1994, 1998; Zukin 1995). Some have argued that this transformation has led to cities characterized by fragmented social and spatial relations in which the construction of a "public" is impossible. At the same time, others note that the processes that have led to new forms of fragmentation may also open the possibility of forming new connections and linkages between people.

Urban space, then, is not only a barometer *for* but is also a creator *of* urban social transformation. This section builds on the debates and problematics examined in the previous section and explores how they have shaped research in four specific areas: the sorting of urban spaces, the spaces and scales of politics, the analysis of urban public and private space; and the development of new spaces such as those that involve consumption and the Web.

Sorting Urban Social Space

The long history of research in urban geography on power and segregation in cities continued in the 1990s with protracted discussions of the finer sorting of the urban social fabric through residential segregation (Young 1990*a*), urban "balkanization" through immigration (Ellis and Wright 1998) and intra-urban migration (W. Clark 1992), the notion of a segmented and differentiated labor force (Hanson and Pratt 1995; Ellis and Wright 1999), and the creation of designer shopping malls (Hopkins 1991; Goss 1993). Segregation indices of the 1970s and 1980s have been combined with contemporary visualization techniques to help urban geographers focus on changing patterns of population segmentation (Plewe and Bagchi-Sen 2001). These techniques suggest that there has been an astonishing persistence of very high levels of racial residential segregation in large cities. In considering explanations for this in American cities, Douglas Massey and Nancy Denton (1993) note that this is not simply a question of the over-representation of African-Americans in the poorer classes, and a coincidence of class and ethnicity. Neither can we look simply at the workings of an anonymous capitalist market, for the processes that lead to increasing residential segregation are more complex than this. In addition to the machinations of the capitalist economy, geographers have studied the practices of developers, landlords, and financial and government institutions from around the globe (Kaplan 1996; Dowling 1998*b*; Saff 1998; Takahashi 1998).

There is some equivocation about the extent to which the segregation manifest in urban space is the result of distinctive economic and social polarization processes. In a study of Los Angeles, for example, W. Clark and McNicholas (1995) argue that segregation may not necessarily be a function of polarization: African-American households exhibit less economic polarization, for example, than the population as a whole. Studies of the cultural and political processes at work in both (second-tier) global cities such as Los Angeles (Ellis and Wright 1999) and in the global south suggest that the dynamic relationship of social polarization and racial segregation is indeed complex. John Western (1996), for example, interviewed people in the 1970s who had been removed by the South African government from racially mixed to "purified" and polarized areas established for racially defined groups. The South African government's policy of apartheid was aimed at creating residential segregation, but one of the chief results of the removals was the break-up of family and friendship networks that had built up over decades (see also Robinson 1996). More recently, Grant Saff (1994, 1998, 2001) has focused on institutionally created polarization in post-apartheid South Africa with a particular concern for the power inherent in maintaining real-estate values. He has shown the ways in which a certain "deracialization" of space has led to resegregation on the basis of class (and hence, often, still of race), even as economic polarization continues apace in the "new South Africa." In her work in Dar es Salaam, Richa Nagar (1997; Nagar and Leitner 1998) focuses on the complexity of cultural, ethnic, and gender webs that segregate contemporary residential space. Of special importance in advancing our understanding of social sorting, Nagar outlines the effects of community politics and activism as part of the process of residential segregation and, by so doing, points to ways that conceptualizations of, and activism in, urban spaces are themselves political projects of great complexity.

Spaces of Politics

As protesters took to the streets in Tiananmen, Leipzig, Berlin, Prague, Budapest, and Bucharest, they *made* new spaces of politics. These spaces pose questions and challenges to theoretical explanations of politics that centered either on the state or the economy. The protests typify many other attempts to create alternative political spaces in which processes related to cultural transformation, identity, and place-making are central (Magnusson 1996; Keith and Pile 1993; A. Kirby 1993; Brown 1997; Fincher and Jacobs 1998).

Economic transformations that reordered markets and challenged the primacy of the nation state have connected global and local scales in ways that have profound implications for localities and urban politics (S. Smith 1989; N. Smith 1996; Massey 1994; Cox 1997). While many researchers have demonstrated the debilitating effects of these changes on marginalized groups and in creating finer social segmentation within the city, others have shown that the processes of restructuring and globalization have provided new resources for community activism. Geographers in particular have argued that these processes have created new linkages between places and opened scales of politics that are differentially accessible to political agents.

Globalization clearly and directly affects the political opportunity structure and contexts in which activists operate. Susan Fainstein (1997), however, notes an equally significant impact in that globalization shapes the issues around which activists organize. More specifically, Margit Mayer (1991) argues that grassroots urban politics may profit from economic restructuring in the sense that new representations and mobilizations are possible (witness the local mobilizations against the World Trade Organization in Seattle and Washington DC). Mobilization may be around attempts to capture the benefits of globalization, about mitigating the impacts of globalization, or about fighting the oppressions and injustices blamed on globalization. In some cases, globalization may itself be an object of mobilization. Each of the above have seen significant organizational efforts, and have been the subject of research on community activism. Complicating the picture further, Massey (1997) has drawn attention to the ways in which processes of globalization highlight the different interests in, and interpretation of, places held by the residents of a given place. As she notes, places may be interpreted by some "as locales in which to construct some kind of life, maybe even some kind of resistance; by others as places of deadly entrapment" (1997: 111). Community activism and political practice is not necessarily—or even primarily—based on spatial communities that are internally coherent or spatially contiguous (1997: 112). The connections and flows that join community activists in spaces of engagement, then, may not necessarily be place-based or uniquely represent the interests of residents in a local community or neighborhood. The combination of the heightened fragmentation of interests within a place, and processes that create "community without propinquity" (Webber 1964), leads to a new spatiality of community activism.

1. *Capturing Benefits: Local Economic Development Strategies and Community Activism.* Susan Clarke and Gary Gaile begin their book *The Work of Cities* with the following observation:

> To many of us, local economic development is an arcane world of revenue bonds, roads, dubious revitalization projects, and, even worse, corrupt deals over tax breaks for private investors. But in the 1990s, this agenda includes world trade centers in Durham and Lubbock, "internationally friendly" infrastructure and thirty sister cities in Portland, training software specialists in Ann Arbor and Stamford, and setting up public-private partnerships for telecommunications in cities like Milpitas. What is going on?
>
> It's simple—cities are adapting to restructuring and globalization trends with a range of policy choices unanticipated by scholars and unheard of just a few years ago. (1998: 1)

Similarly, Cox and his contributors (1997) note the power (however limited) of local agents to intervene in global economic processes. In addition, A. Kirby *et al.* (1995) note the increased attention to "municipal foreign policy" as representatives of localities attempt to create economic development opportunities that link their locality with localities in other countries. The "glocalization" (Swyngedouw 1989) of the political economy of which these actions are a part is significant for many reasons. The first is the obvious presence of localities in economic development efforts at the international and global scales. From the perspective of this section, the more important reasons have to do with the ways in which economic development has become the object of community activism and the ways in which this has reworked the boundaries and relations between the state, market, and civil society.

Local economic development in industrialized countries has been marked by a blurring of boundaries between the state, market, and civil society. Public-private partnerships—the hallmark of contemporary economic development—typically involve local business and government leaders, but also community leaders who represent educational and civic organizations. While the tendency may be to dismiss these partnerships as representing only elite interests, examples of involvement by non-elite organizations and groups have also been highlighted. Examples include organizations representing people of color (Martin 1998; Pincetl 1994; Pulido 1994), the working class or labor (Steinberg 1994; Herod 1991), neighborhood organizations (Clarke and Gaile 1998; Fainstein 1997), among others. The attention to these groups highlights efforts to refocus the study of urban political economy and urban politics away from "big politics" to the ways in which economic development intersects with the political strategies of marginalized—and accordingly overlooked—groups (Staeheli and Clarke 1995; Brown 1999). These efforts are especially notable in communities that are attempting to stave off disinvestment (e.g. Fitzgerald 1991; Clarke and Gaile 1998). Through their involvement in these efforts, the marginalized have also contributed to the redrawing of boundaries between state and market; to what effect is unclear at this point.

2. *Resisting the Burdens of Globalization.* The efforts described above represent attempts to capture the benefits of globalization by repositioning localities within the new economy. A significant amount of community work, however, involves attempts to mitigate—or in some cases resist—the effects of globalization on localities and on specific groups within them. These efforts are varied in nature, but most involve attempts to create new political spaces in which marginalized groups are empowered and in which the social polarization heightened by globalization (O'Loughlin and Friedrichs 1996) can be combated. They represent attempts to counteract the effects of power derived from money (Harvey 1997) by creating spaces of engagement in which money is not the only—or most significant—resource.

Not surprisingly, geographers have paid particular attention to activism that involves the built environment. As Glenda Laws (1994) argued, the built environment can be oppressive for marginalized groups, but also can provide the basis for resisting oppression. In keeping with this belief, geographers have examined housing associations and movements (H. Clark 1994) and community and neighborhood redevelopment (e.g. Steinberg 1994; N. Smith 1996; Keith 1997). This work has highlighted the attempts by social groups to create spaces in which the effects of globalization are mitigated. Some analysts conceptualize these attempts as relying on stores of "social capital" within groups that enable group members to work cooperatively in resisting the effects of social, political, and economic change.

Equally significant, however, have been analyses of groups who are themselves part of the process of globalization: immigrants. This research has highlighted the ways in which immigrants confront political, economic, and social structures that construct immigrants as "outsiders" within the political spaces of the city (Pincetl 1994; K. Anderson 1991; K. Mitchell 1997; Nagel 1999). In a sense, this work explores the ways in which immigrants create a space to simply *be* in the political spaces of the city. Once again, the value of "social capital" as a political resource has become a focus of research. Such research seeks to explicate the ways that "outsiders" can gain access to the "social capital" that may operate within a place and the ways in which various forms of "social capital" allow access to political spaces (Bourdieu 1984; Putnam 1993). "Social capital" is seen to be the "thing" that allows immigrants and other marginalized groups to combat the effects of material power and the decisions made in spaces and scales to which they do not have access (cf. Fernández Kelly 1994). In social struggle, marginalized groups are not without allies—or even sometimes access to powerful tools for analyzing and combating the forces of marginalization. Of particular importance in the structuring of their lives and struggles are the reformulated public spaces of the city.

Public Space and the "New City"

Urban geographic research on public space has first and foremost been marked by its interest in the particular events—protests, occupations, riots, encampments, violence against women, homelessness—that have erupted in and over public space in the last generation (Valentine 1989, 1992; Pain 1991; N. Smith 1989, 1996; Adams 1992; D. Mitchell 1992, 1995, 1996, 1997; Hershkovitz 1993; Blomley 1994; Killian 1998; Goheen 1994, 1998; Cresswell 1996; Ruddick 1996*a*; Staeheli and Thompson 1997). But such events are typically studied as indicators of larger issues such as: the changing social meaning and location of public space; the social, political, and economic forces that produce public space; strategies of control and use of public space; privatization; the relationship between public space and the more general social-political public sphere; "deviance" and marginalization of particular public space users; gender, sexuality, and normative public spaces; political protest in public space; and "the end of public space" in American and other cities. These topics are rarely treated independently of each other. Rather, studies usually examine, empirically and theoretically, the interrelationship between different aspects of the role of public and private in the construction of the urban problematic.

Public space can be defined as "not only a region of social life located apart from the realm of family and close friends, but also . . . [the] realm of acquaintances and strangers" (Sennett 1992: 17, quoted in Goheen 1998: 479). According to D. Mitchell (1995), public space is important to social change because it provides a space *for* representation. This is true both in terms of *political* representation—public space is where members of the public come together to make demands against the state, to agitate for changes in civil society, or to contest the power of the economy—and in terms of constructing and representing social identity. Accordingly, the status of public space can be understood as the outcome of struggle around two opposed ideals about public space, one that sees public space as an ordered, rational, retreat for leisure, recreation, and the performance of urban spectacles, and the other that envisions public space as a site of struggle for political inclusion by marginalized groups (D. Mitchell 1995). When social actions guided by these two ideals clash, not only is the status and nature of public space created, but so too are social understandings about what constitutes "the public" and democracy in given places.

This argument has been criticized (Domosh 1998; Staeheli 1996; Lees 1998; Killian 1998) on a number of grounds, including its promotion of outright struggle as the moving force in public space production. Critics argue that public spaces, and their relationship to urban publics, are produced in less overt ways. Domosh (1998), for example, argues that what she terms as "polite politics" (the quotidian interaction between men and women on city streets, for example) can be every bit as important in giving public space its structure and meaning as riots and overt policing. For Domosh, as for other critics, public space is *never*—never even *ideally*—a space of complete openness and inclusiveness. It makes little sense, therefore, to focus on ideals of openness and inclusiveness. Instead, public space should be seen as always riven through with structures and relations of power. These structures may be based upon sexism, racism, ablism, or agism and the relations they exude are predisposed toward constraining the power, movement, and accessibility of those who are designated as not part of the hegemonic norm. The space of the public may be construed as the space of hegemonic normativity. But space is constantly made "public," according to Domosh, through the everyday small actions—the "micro-politics"—of those who must move through it.

If the means by which public space is socially produced marks one of the important arenas for urban theory in the 1990s, then another is the question of what *function* public space serves in contemporary and historical cities. Urban geographers have developed numerous, and often contradictory, answers to this question over the past decade, answers that are themselves dependent at least in part on how analysts understand more general processes of urban restructuring. To understand the function of public space, in other words, it is important to examine just who *controls* it (and to what ends) and who has *access* to it (and for what purposes).

To meet this imperative, geographers have turned to debates over the status of the "public sphere." Jurgen Habermas (1989), arguably the leading philosopher of the public sphere, suggests that it was historically constituted through the association of *private* individuals who come together to discuss and question the nature of state authority. And urban space—coffee houses, clubhouses, and the like—were the spaces *par excellence* for the formation of the public sphere. Yet Howell (1993) showed that Habermas's notion of the public sphere suffered from an impoverished theory of space and spatiality. Howell thus encouraged geographers to better theorize the production of spaces that made the public sphere possible, gave it its shape, and reproduced its social inclusions and exclusions (see Goheen 1998). Similarly, feminists (Hanson 1992; Fraser 1992) have shown that the construction of an urban public sphere depended on a highly gendered spatial division of labor. "Public" space was constructed as a male domain, and thus its very constitution *as* public was first and foremost an act of exclusion—a limiting of access (E. Wilson 1991; Domosh 1998). What Habermas identified as *potentially* a "universal" public sphere has been shown to be enormously constricted (cf. Howell 1993; Marston 1990). The relationship between changing structures of citizenship and the structure of the public sphere suggests that a key focus of research should be how the boundaries between the public and the private, the legitimate and the illegitimate, are maintained, challenged, and transformed (Brown 1997).

Crucially, such research needs clearer theories about the relationships between public and private spaces on the one hand, and particular social actions on the other. Staeheli (1996) has argued that geographers (among others) have the tendency to confuse the public and private status of actions (political organizing, child-rearing, sexual activity) with the public and private status of the spaces in which they occur (bedrooms, city streets, council chambers). Analysts tend to equate public actions with public spaces, and private with private. Instead, Staeheli suggests that geographers need to focus more clearly on the relationship between activity and space, to see each as existing on a continuum, and to understand that these continua might describe a multi-dimensional space in which the questions that develop are ones about the transgression of socially coded spaces and activities.[5] In other words, it is the constitution and transformation of public space—and hence the public sphere—that is of crucial importance. That is why "end of public space" arguments have had the force that they have (Sorkin 1992*b*; but see Lees 1998).

Whatever the status of these debates within urban geography, the important point is that research has been directed towards the nature, politics, and structural transformation of public space. This transformation of public space, in turn, has been linked to the explosion of the homeless and other marginal peoples in cities. Building on work concerned with the location of

[5] The Queer Nation "kiss-ins" of the early 1990s gained their power from conducting an ostensibly "private" activity (same-sex kissing) in an ostensibly "public" space (the steps of a statehouse). It was the transgressive quality of the action—made transgressive by the social coding of space—that gave it its power (see also Cresswell 1996; Goss 1996; Sibley 1995; Bell and Valentine 1995).

service-facilities for the homeless and mentally ill (Dear and Wolch 1987; Wolch and Dear 1996), research on homelessness in the 1990s has expanded to include studies of the time-geography of homeless women (Rowe and Wolch 1990); the spaces of homeless youth (Ruddick 1996*b*); laws governing the behavior of homeless people and regulating the spaces in which they live (D. Mitchell 1998*a*, *b*); and the changing structure of old skid-row districts (Rahimian *et al.* 1992; Wolch *et al.* 1993). Homeless people destabilize normalized notions of space by performing actions socially deemed "private" (sleeping, eating, defecating, and urinating) in spaces deemed "public." In this way they transgress the sanctioned coding of public space and are thus subject to increasing social and spatial control (N. Smith 1996; D. Mitchell 1997).

New Spaces of Technology and Consumption

Connected to the production and transformation of public space, and the development of new political forms, has been the contradictory role of new technologies in urban settings. As noted above, GIS, closed-circuit television, radar imaging, and other technologies have been promoted and used both as tools of control and as means of resistance and liberation. In this subsection we explore some of the complex ways in which technology has been used and theorized in urban geography.

The 1990s witnessed the replacement of mass-produced products and media services in urban areas with carefully tailored products and services that were designed for small market niches. GIS and communication technologies are now used to geo-market products to specific areas with particular income and social characteristics. This shift requires much more specific information on shopping behaviors, debts, and habits as well as reading and viewing tastes, some of which is provided by data compiled from debit cards and credit referencing. The largest private information disseminators in the US such as Equifax and Trans Union sell information based on detailed birth, family, migration, address, telephone, social security, salary and medical histories, credit transactions, mortgages, bankruptcies, and tax and legal records of US residents (Graham and Marvin 1996: 216). The field of geodemographics is now burgeoning because GIS facilitates access to, and spatial rendering of, databases that heretofore would have been too cumbersome to integrate. Curry (1995) points out that the technology promotes a widespread use of unregulated data. He is further concerned that the marketing strategies employed through geodemographics employ visual representations which can attribute actions and beliefs to residents of particular urban areas for the purposes of marketing particular products. Not only does this raise significant issues concerning privacy and confidentiality, but technologies combining GIS and spatial analyses along with increasing consumer reflexivity also drive a large part of the growing segmentation of urban markets. The segmentation of cities as described throughout this chapter, when combined with political shifts to the right, fuels an increasing use of technology as a surveillance tool for improving social control.

If Paul Adams (1992) can describe television as a "gathering place" then it seems even more probable that interactive computer networks have the same kind of place-like characteristics. While TV's appearance is one of one-way communication, in reality, as McLuhan long ago observed, it demands an actively participating *audience* (Adams 1992; Stevenson 1998; Hartley 1992; Carpignano *et al.* 1990). That the audience may be spatially fragmented is less important than the fact that the broadcast draws them together to create what McLuhan called a "fictive we" in a way similar to how print media helped construct the "imagined community" that Benedict Anderson (1983) sees as so important to the development of modern nationalism. "Cyberspace" likewise constitutes a new type of "public" space. To the degree that television and computer technologies create a "fictive we," they also thereby create a public space in which to situate that audience. But here, obviously, the material structure of that space is far different from the structure of a local urban public space such as a square or sidewalk.

David Gelernter was one of the first writers to expand on the idea of a virtual cyber-city. He describes "Mirror World" as "an eye on the world from your computer terminal through virtual technology" (Gelernter 1991: 1), seeing free Internet access coupled with new Web technologies as liberatory, enabling private citizens to know more fully the workings of their society. The way Gelernter conceptualizes it, Mirror World is an attempt to "capture the whole country . . . a bottomless cascade tracing the state of the government, the economy, the polity downwards level by level from the big trends on the top to a billion details far below. When you wander backwards through time in the Mirror World, you have raw history, the past complete and unedited in your grasp." Gelernter's utopian vision is of a scientific viewing tool that combines the power of the microscope, the telescope, and the camera obscura so that the human-scale social world may be exhibited and viewed at any resolution. There are important new urban geographies and histories created in these mirror worlds that have inspired writers to speculate about Virtual Communities (Rheingold 1994), Cities of Bits (W. Mitchell 1995), CyberCities (Boyer 1996), and so forth (cf. Sui 1997: 79). According to W. Mitchell (1995), these cities are anti-spatial. Many geographers beg to differ, suggesting a quintessential spatial configuration that may imply new geographies or may simply manifest old geographies in new way. To take just one concrete example of the latter, Wheeler and O'Kelly (1999) point out that congested and unreliable aspects of today's Internet suggest certain advantages that result from being atop the already existing hierarchy of cities: Cyberspace favors those places with large numbers of high speed connections—typically "global" cities.

Urban geographers, then, are keen to understand electronic space and its relation to the city (Graham and Marvin 1996). What sort of space is it? What is its relationship to the constitution and transformation of the public sphere? Who controls it? Who does and does not have access to it? The questions are the same as they are for the traditional public spaces of the city described earlier. Stevenson (1998: 198–9) argues that the bourgeois public sphere was itself in part a product of a certain type of media—print—and thus the "irreversible" ascendance of new forms of communication—radio, television, the Internet—require a "reconceptualization of the public sphere, one that involves reconfiguring social communication around its new forms of mediation." Derek Gregory (1994) is skeptical of the benefits that may accrue from communications technologies because of inherent issues related to surveillance and voyeurism. Gregory suggests that with the World Wide Web we are no longer *gazing at* the world but rather, with the prospect of movement and rapid scale changes, we are *traveling through* a technologically created world of information. The point Gregory makes is that the consequences of systematic surveillance of this kind might problematically dissolve the distinction between what is real and what is represented. David Harvey (1996: 279–80) takes this point a little further, cautioning that some landscapes created and perpetuated on the Web are disarmingly utopian and hide the ultimate annihilation of space, time, and bodies by capitalist technology. These landscapes are modeled, coded, and represented in large part by the engineers of the system. Moreover, the people at the computer keyboards are real people and Web representations originate in, and must return to, corporeal and physical space. The Web, and the apostles of its liberatory potential, often neglect this material foundation of lived experience.

The questions that arise around the production of "new" public spaces such as television and cyberspace are thus the same as those that need investigating for "old" public spaces—they are questions about by whom and how "the public" is constituted, what the relationship is between the public sphere and other aspects of society, the nature of the structural transformations underway in public spaces and spheres, and the role that public space (and its privatization) plays in structuring social struggle and contestation, reinforcing relations of power, and providing a space *for* representation.

Conclusion

We are intensely aware of dramatic changes in the urban problematic over the last ten years. The decade witnessed an important growth in our understanding of urban space with specific concerns revolving around the social constructions of space and scale, or spatiality. In this chapter, we focused on the spatiality of difference, the urban geography of social control, and the transformation of urban places and spaces at the end of the twentieth century. We suggested that a "new city" is emerging out of transformations of public and private spheres, and that urban geographers have been at the forefront of those studying it. Geographers have shown how the spatiality of technology demands a broadened understanding of the ways urban spaces are controlled and segmented. An important frontier for urban theory and practice in the twenty-first century will be understanding the *relationship* between electronic and technological spaces to a reconstituted and mutated urban space. Information technology and the varied processes of globalization signaled unparalleled time-space compression in the last half of this century. The work of urban geographers has been essential for understanding these changes not only in terms of applying the technology, but also in understanding its assumptions and limitations, its potential benefits and its deep costs. Finally, urban geographers' research into the spatiality of representation has suggested some new ways that we imagine our cities and how these imaginings continually transforming the politics and culture of urban life. In this arena, geographic work highlights social meanings and political identities, at scales that are complex and perhaps overwhelming, while at the same time remaining meaningful in terms of understanding people's day-to-day lives. Commentators note more community activism and greater efforts of local agents to intervene in global political and economic processes. Attempts to understand agency, difference, and representation set important agendas of research and practice throughout the decade but it is perhaps the profound changes in cities and urban life that has inspired some of the decade's most noteworthy work.

Urban geographers must build on the advances made over the past decade ("advances" only possible because of often contentious debate within the discipline), for urbanization has been an exceptionally important global issue of the end of this century and it promises to be an even more important aspect of the next. Cities now account for half the world's population. By the year 2025, about 5 billion people will live in cities. There is global shift to technology-, industry-, and service-based economies in many areas of the world that were once categorized as rural. If the beginnings of the new millennium herald the kind of changes urban geographic research has grappled with in the 1990s, and if urban geographers rise to the challenge in the coming decade as creatively as they did in the past one, then the subfield will continue its strong contribution to our understanding of the multi-layered complexities of the world. It is clear that the complexities associated with these global changes will spawn new urban mutations the like of which we can only imagine. It is our firm belief that geographers are—and should remain—at the forefront not only of understanding contemporary urban space, but also of imagining and mapping its futures.

References

Adams, Paul (1992). "Television as Gathering Place." *Annals of the Association of American Geographers*, 82: 117–35.

Agnew, John (1993). "Representing Space: Space, Scale and Culture in Social Science," in James Duncan and David Ley (eds.), *Place/Culture/Representation*. London: Routledge, 251–71.

Aitken, Stuart (1998). *Family Fantasies and Community Space*. New Brunswick: Rutgers University Press.

Aitken, Stuart, and Michel, Suzanne (1995). "Who Contrives the 'Real' in GIS? Geographic Information, Planning and Critical Theory." *Cartography and Geographical Information Systems*, 22: 17–29.

Anderson, Benedict (1983). *Imagined Communities*. London: Verso.

Anderson, Kay (1991). *Vancouver's Chinatown: Racial Discourse in Canada, 1875–1980*. Montreal: McGill-Queen's University Press.

Anselin, Luc, Dobson, Roger F., and Hudak, S. (1993). "Linking GIS and Spatial Data Analysis in Practice." *Geographical Systems*, 1: 2–23.

Appadurai, Arjun (1990). "Disjuncture and Difference in the Global Cultural Economy." *Theory, Culture and Society*, 7: 295–310.

Azaryahu, Maoz (1999). "McDonald's or Golani Junction? A Case of a Contested Place in Israel." *The Professional Geographer*, 51: 481–92.

Barnes, Trevor, and Shepherd, Eric (1992). "Is There a Place for the Rational Actor? A Geographical Critique of the Rational Choice Paradigm." *Economic Geography*, 68: 1–21.

Batty, Michael (1993). "Using Geographical Information Systems in Urban Planning and Policy-making," in Manfred Fischer and Peter Nijkamp (eds.), *Geographical Information Systems, Spatial Modeling and Policy Evaluation*. New York: Springer-Verlag, 51–69.

Batty, Michael, and Xie, Y. C. (1994*a*). "From Cells to Cities." *Environment and Planning B*, 21: 31–48.

— (1994*b*). "Urban Analysis in a GIS Environment: Population Density Modeling Using ARC/INFO." *International Journal of Geographical Information Systems*, 8: 451–70.

Beauregard, Robert (1993). *Voices of Decline: The Postwar Fate of US Cities*. Oxford: Blackwell.

Bell, David, and Valentine, Gill (1995). *Mapping Desire*. New York: Routledge.

Berman, M. (1984). *All That is Solid Melts into Air*. New York: Vintage.

— (1986). "Taking It to the Streets: Conflict and Community in Public Space." *Dissent*, Fall: 476–85.

Blomley, Nicholas (1994). *Law, Space, and the Geographies of Power*. New York: Guilford.

— (1999). "Violent Meanings and the Enactment of Property in Early Vancouver." Unpublished Paper, Department of Geography, Simon Frasier University.

Boddy, Trevor (1992). "Underground and Overhead: Building the Analogous City," in Michael Sorkin (ed.), *Variations on a Theme Park: The New American City and the End of Public Space*. New York: Hill & Wang, 123–53.

Bondi, Liz (1991). "Gender Divisions and Gentrification: A Critique." *Transactions of the Institute of British Geographers*, 16: 157–70.

Bourdieu, Pierre (1984). *Distinction: A Social Critique of the Judgment of Taste*. London: Routledge.

Boyer, M. C. (1992). "Cities for Sale: Merchandizing History at South Street Seaport," in Michael Sorkin (ed.), *Variations on a Theme Park: The New American City and the End of Public Space*. New York: Hill & Wang, 181–204.

— (1996). *CyberCities: Visual Perception in the Age of Electronic Communication*. New York: Princeton Architectural Press.

Bowen, William, Salling, Mark, Hayes, Kingsley, and Cryan, Ellen (1995). "Toward Environmental Justice: Spatial Equity in Ohio and Cleveland." *Annals of the Association of American Geographers*, 85: 641–63.

Brandt, M., and Craig, William (1994). "Data Providers Empower Community GIS Efforts." *GIS World*, 7: 49–51.

Breitbart, Myrna M. (1990). "Quality Housing for Women and Children." *Canadian Women's Studies*, 11: 19–24.

Bromley, Ray (1990). "A New Path to Development? The Significance and Impact of Hernando De Soto's Ideas on Underdevelopment, Production and Reproduction." *Economic Geography*, 66: 328–43.

Brown, Michael (1997). *RePlacing Citizenship: AIDS Activism and Radical Democracy*. New York: Guilford.

—— (1999). "Reconceptualizing Public and Private in Urban Regime Theory: Governance in AIDS Politics". *International Journal of Urban and Regional Research* 23: 70–87.

Burke, L. M. (1993). "Race and Environmental Equity: A Geographical Analysis in Los Angeles." *Geo Info Systems*, 9: 44–50.

Butler, Ruth, and Bowlby, Sophie (1997). "Bodies and Spaces: An Exploration of Disabled People's Experiences of Public Space." *Environment and Planning D: Society and Space*, 15: 411–33.

Caldeira, Teresa (1999). "Fortified Enclaves: The New Urban Segregation," in J. Holston (ed.), *Cities and Citizenship*. Durham, NC: Duke University Press, 114–38.

Carpignano, P., Andersen, R., Aronowitz, S., and Difazio, W. (1990). "Chatter in the Age of Electronic Reproduction: Talk Television and the 'Public Mind'." *Social Text*, 25/26: 33–55.

Castells, Manuel (1977). *The Urban Question*. London: Arnold.

Chang, T. C. (1999). "Local Uniqueness in the Global Village: Tourism in Singapore." *Professional Geographer*, 51: 91–104.

Chouinard, Vera, and Grant, Ali (1995). "On Not Being Anywhere Near the 'Project': Revolutionary Ways of Putting Ourselves in the Picture." *Antipode*, 27: 137–66.

Clark, Helene (1994). "Taking up Space: Redefining Political Legitimacy in New York City." *Environment and Planning A*, 26: 937–56.

Clark, Keith, and Gaydos, Leonard (1998). "Loose-Coupling a Cellular Automaton and GIS: Long-Term Urban Growth Prediction for San Francisco and Washington/Baltimore." *International Journal of Geographical Information Science*, 12: 699–714.

Clark, Keith, Hoppen, Stacey, and Gaydos, Leonard (1997). "A Self-modifying Cellular Automation Model of Historical Urbanization in the San Francisco Bay Area." *Environment and Planning B: Planning and Design*, 24: 247–61.

Clark, W. A. V. (1992). "Comparing Cross-sectional and Longitudinal Analyses of Residential Mobility and Migration." *Environment and Planning A*, 24: 1291–307.

—— (1998). *The California Cauldron: Immigration and the Fortunes of Local Communities*. New York: Guilford.

Clark, W. A. V., and Kuijpers-Linde, Marianne (1994). "Commuting in Restructuring Urban Regions." *Urban Studies*, 31: 465–84.

Clark, W. A. V., and McNicholas, M. (1995). "Re-examining Economic and Social Polarisation in a Multi-ethnic Metropolitan Area: The Case of Los Angeles." *Area*, 28: 56–63.

Clarke, Susan, and Gaile, Gary (1998). *The Work of Cities*. Minneapolis: University of Minnesota Press.

Cole, Donald C., and Eyles, John (1997). "Environments on Human Health and Well-Being in Local Community Studies." *Toxicology and Industrial Health*, 13: 259–67.

Cosgrove, Dennis (1990). "Environmental Thought and Action: Pre-modern and Post-modern." *Transactions of the Institute of British Geographers*, 15: 344–58.

Couclelis, Helen (1997). "From Cellular Automata to Urban Models: New Principles for Model Development and Implementation." *Environment and Planning B: Planning and Design*, 17: 585–96.

Cox, Kevin (ed.) (1997). *Spaces of Globalization*. New York: Guilford.

—— (1998). "Spaces of Dependence, Spaces of Engagement and the Politics of Scale, or: Looking for Local Politics." *Political Geography*, 17: 1–24.

Cox, Kevin, and Mair, Andrew (1989). "Levels of Abstraction in Locality Studies." *Antipode*, 21: 121–32.

Craig, William, Harris, Trevor, and Weiner, Daniel (eds.) (2001). *Community Participation and Geographic Information Systems*. London: Taylor & Francis.

Crawford, Margaret (1992). "The World in a Shopping Mall," in Sorkin (1992*b*: 3–30).

Cresswell, Tim, (1996). *In Place/Out of Place: Geography, Ideology and Transgression*. Minneapolis: University of Minnesota.

Crilley, Darrell (1993). "Megastructures and Urban Change: Aesthetics, Ideology and Design," in Knox (1993: 127–64).

Crump, Jeffrey (1999). "What Cannot be Seen Will Not be Heard: The Production of Landscape in Moline, Illinois." *Ecumene*, 6: 295–317.

Curry, Michael (1995). "GIS and the Inevitability of Ethical Inconsistency," in Pickles (1995: 68–87).

Cybriwsky, Roman (1997). *Historical Dictionary of Tokyo*. Lanham, Md.: Scarecrow Press.

Daly, Gerald (1998). "Homelessness and the Street: Observations from Britain, Canada and the United States," in Fyfe (1998*a*: 111–28).

Daniels, Stephen, and Cosgrove, Denis (1993). "Spectacle and Text: Landscape Metaphors in Cultural Geography," in James Duncan and David Ley (eds.), *Place/Culture/ Representation*. London: Routledge, 57–77.

Davis, Mike (1990). *City of Quartz: Excavating the Future in Los Angeles*. London: Verso.

—— (1998). *Ecologies of Fear: Los Angeles and the Imagination of Disaster*. New York: Metropolitan Books.

Dear, Michael (2000). *The Postmodern Urban Condition*. Oxford: Blackwell.

Dear, Michael, and Flusty, Steven (1998). "Postmodern Urbanism." *Annals of the Association of American Geographers*, 88: 50–72.

Dear, Michael, and Wolch, Jennifer (1987). *Landscapes of Despair*. Princeton: Princeton University Press.

de Certeau, Michel (1984). *The Practice of Everyday Life*. Berkeley: University of California Press.

de Certeau, Michel, Giard, Luce, and Mayol, Pierre (1998). *The Practice of Everyday Life*, ii. *Living and Cooking*. Minneapolis: University of Minnesota Press.

Delaney, David, and Leitner, Helga (1997). "The Political Construction of Scale." *Political Geography*, 16: 93–7.

Dillan, D. (1994). "Fortress America." *Planning*, 60/4: 8–12.

Diniz, Clélio Campolina, and Santos, Fabiana Borges T. (2000). "Manaus: Vulnerability in a Satellite Platform," in Markusen, Lee, and DoGiovanna (2000: 125–46).

Domosh, Mona (1998). "Those 'Gorgeous Incongruities': Polite Politics and Public Space on the Streets in Nineteenth-century New York City." *Annals of the Association of American Geographers*, 88: 209–26.

Donald, J. (1992). "Metropolis: The City as Text," in R. Bocock and K. Thompson (eds.), *Social and Cultural Forms of Modernity*. Cambridge: Polity Press, 417–61.

Dowling, Robyn (1998*a*). "Neotraditionalism in the Suburban Landscape: Cultural Geographies of Exclusion in Vancouver, Canada." *Urban Geography*, 19: 105–22.

— (1998*b*). "Suburban Stories, Gendered Lives: Thinking through Difference," in Fincher and Jacobs (1998: 49–68).

Duncan, James (1990). *The City as Text: The Politics of Landscape Interpretation in the Kandyan Kingdom*. Cambridge: Cambridge University Press.

— (1993). "Elite Landscapes as Cultural (Re)Productions: The Case of Shaughnessy Heights", in Kay Anderson and Faye Gale (eds.), *Inventing Places: Studies in Cultural Geography*. Melbourne: Longman Cheshire, 37–51.

Elwood, Sarah (2001). "Powderhorn Park," in Craig, Harris, and Weiner (2001: 77–89).

Elwood, Sarah, and Leitner, Helga (1998). "GIS and Community-Based Planning: Exploring the Diversity of Neighborhood Perspectives and Needs." *Cartography and Geographic Information Systems*, 25: 77–88.

Ellis, Mark, and Wright, Richard (1998). "The Balkanization Metaphor in the Analysis of U.S. Immigration." *Annals of the Association of American Geographers*, 88: 686–98.

— (1999). "The Industrial Division of Labor among Immigrants and Internal Migrants to the Los Angeles Economy." *International Migration Review*, 33: 26–37.

England, Kim (1993). "Suburban Pink Collar Ghettos: The Spatial Entrapment of Women?" *Annals of the Association of American Geographers*, 83: 225–42.

Fainstein, Susan (1997). "Justice, Politics and the Creation of Urban Space," in A. Merrifield and E. Swyngedouw (eds.), *The Urbanization of Injustice*. New York: New York University Press, 18–44.

Falconer Al-Hindi, Karen, and Staddon, Chad (1997). "The Hidden Histories and Geographies of Neo-traditional Town Planning: The Case of Seaside, Florida." *Environment and Planning D: Society and Space*, 15: 349–72.

Fernández Kelly, Maria Patricia (1994). "Towanda's Triumph: Social and Cultural Capital in the Transition to Adulthood in the Urban Ghetto." *International Journal of Urban and Regional Research*, 18: 88–111.

Fincher, Ruth, and Jacobs, Jane (1998). *Cities of Difference*. New York: Guilford.

Fitzgerald, Joan (1991). "Class as Community: The New Dynamics of Social Change." *Environment and Planning D: Society and Space*, 9: 117–28.

Fitzpatrick, Kevin, and LaGory, Mark (2000). *Unhealthy Places: The Ecology of Risk in the Urban Landscape*. New York: Routledge.

Foody, G. M. (1995). "Land Cover Classification by an Artificial Neural Network with Ancillary Information." *International Journal of Geographic Information Systems*, 9: 527–42.

Fraser, Nancy (1992). "Rethinking the Public Sphere: A Contribution to the Critique of Actually-Existing Democracy," in Craig Calhoun (ed.), *Habermas and the Public Sphere*. Cambridge, Mass.: MIT Press, 109–42.

Fyfe, Nicholas (ed.) (1998*a*). *Images of the Street: Planning, Identity, and Control in Public Space*. London: Routledge.

— (1998*b*). "Introduction: Reading the Street," in Fyfe (1998*a*: 1–10).

Fyfe, Nicholas, and Bannister, Jon (1996). "City Watching: Closed Circuit Television Surveillance in Public Spaces," *Area*, 28: 37–46.

— (1998). " 'The Eyes Upon the Street': Closed-Circuit Television Surveillance and the City," in Fyfe (1998*a*: 254–67).

Gelernter, David (1991). *Mirror Worlds or the Day Software Puts the Universe in a Shoebox . . . How It Will Happen and What It Will Mean*. Oxford: Oxford University Press.

Gibson, Katherine (1998). "Social Polarization and the Politics of Difference: Discourses in Collision or Collusion?" in Fincher and Jacobs (1998: 301–16).

Gibson-Graham, J. K. (1996). *The End of Capitalism (As We Knew It)*. Oxford: Blackwell.

Goheen, Peter (1994). "Negotiating Access to Public Space in Mid-Nineteenth Century Toronto." *Journal of Historical Geography*, 20: 430–49.

— (1998). "Public Space and the Geography of the Modern City." *Progress in Human Geography*, 22: 479–96.

Goss, Jon (1993). "The 'Magic of the Mall': An Analysis of Form, Function, and Meaning in the Contemporary Retail Built Environment." *Annals of the Association of American Geographers*, 83: 18–47.

— (1996). "Disquiet on the Waterfront: Reflections on Nostalgia and Utopia, the Urban Archetypes of Festival Marketplaces." *Urban Geography*, 17: 221–47.

— (1999). "Once Upon a Time in the Commodity World: An Unofficial Guide to Mall of America." *Annals of the Association of American Geographers*, 89: 45–75.

Gottdiener, Mark (1985). *The Social Production of Urban Space*. Austin: University of Texas Press.

Graham, Stephen, and Marvin, Simon (1996). *Telecommunications and the City: Electronic Space, Urban Places*. London: Routledge.

Gregory, Derek (1994). *Geographical Imaginations*. Cambridge, Mass.: Blackwell.

Griffith, Daniel, Doyle, Philip, Wheeler, David, and Johnson, David (1999). "A Tale of Two Swaths: Urban Childhood Blood-Levels across Syracuse, NY." *Association of American Geographers*, 88: 640–55.

Habermas, Jurgen (1989). *The Structural Transformation of the Public Sphere*. Cambridge, Mass.: MIT Press.

Hamnett, C. (1991). "The Blind Men and the Elephant: The Explanation of Gentrification." *Transactions of the Institute of British Geographers*, 16: 173–89.

Hanson, Susan (1992). "Geogaphy and Feminism: Worlds in Collision?" *Annals of the Association of American Geographers*, 82: 569–86.

— (1995). *The Geography of Urban Transportation*. New York: Guilford Press.

Hanson, Susan, and Pratt, Geraldine (1990). "Geographic Perspectives on Occupational Segregation of Women." *National Geographic Research*, 6: 376–99.

—— (1995). *Gender, Work and Space*. New York: Routledge.

Harris, Trevor, and Weiner, Daniel (1998). "Empowerment, Marginalization and 'Community-Integrated' GIS." *Cartography and Geographic Information Systems*, 25: 67–76.

Hartley, John (1992). *The Politics of Pictures: The Creation of the Public in the Age of Popular Media*. London: Routledge.

Harvey, David (1989). *The Condition of Postmodernity*. Oxford: Blackwell.

—— (1993). "From Space to Place and Back Again: Reflections on the Condition of Postmodernity," in Jon Bird, Barry Curtis, Tim Putnam, George Robertson, and Lisa Tickner (eds.), *Mapping the Futures: Local Cultures, Global Change*. London: Routledge, 3–29.

—— (1996). *Justice, Nature and the Geography of Difference*. Cambridge, Mass.: Blackwell.

—— (1997). "The Environment of Justice," in A. Merrifield and E. Swyngedouw (eds.), *The Urbanization of Injustice*. New York: New York University Press, 65–99.

—— (2000). *Spaces of Hope*. Berkeley: University of California Press.

Healey, Patsy (1992). "Planning through Debate: The Communicative Turn in Planning Theory." *Transportation Planning Research*, 63: 143–62.

Herbert, Steve (1996). "The Normative Ordering of Police Territoriality." *Annals of the Association of American Geographers*, 86: 567–82.

Herman, Tom, and Mattingly, Doreen (1999). "Community, Justice, and the Ethics of Research: Negotiating Reciprocal Research Relations," in James Proctor and David Smith (eds.), *Geography and Ethics: Journeys in a Moral Terrain*. London: Routledge, 209–22.

Herod, Andrew (1991). "The Production of Scale in the United States Labour Relations," *Area*, 23: 82–8.

—— (1997). "Labor's Spatial Praxis and the Geography of Contract Bargaining in the US East Coast Longshore Industry, 1953–1989." *Political Geography*, 16: 154–69.

Herschkovitz, Linda (1993). "Tiananmen Square and the Politics of Place." *Political Geography*, 12: 395–420.

Hillier, John (1993). "To Boldly Go where No Planners Have Ever . . ." *Environment and Planning D: Society and Space*, 11: 89–113.

Holloway, Sarah (1998). "Local Childcare Cultures: Moral Geographies of Mothering and the Social Organization of Preschool Education." *Gender, Place and Culture*, 5: 29–53.

Hopkins, Jeff (1991) "West Edmonton Mall as a Centre for Social Interaction." *The Canadian Geographer*, 35: 268–79.

Howell, Phillip (1993). "Public Space and the Public Sphere: Political Theory and the Historical Geography of Modernity." *Environment and Planning D: Society and Space*, 11: 303–22.

Innes, J. E., and Simpson, D. M. (1993). "Implementing GIS for Planning: Lessons from the History of Technological Innovations." *Journal of the American Planning Association*, 59: 230–6.

Jerrett, Michael, Eyles, John, and Cole, Donald (1998). "Socioeconomic and Environmental Covariates of Premature Mortality in Ontario." *Social Science & Medicine*, 47: 33–50.

Jonas, Andrew (1994). "The Scale Politics of Spatiality." *Environment and Planning D: Society and Space*, 12: 257–64.

Kaplan, David (1996). "What Is Measured in Measuring the Mortgage Market." *The Professional Geographer*, 48: 356–66.

Katz, Cindi (1998*a*). "Disintegrating Developments: Global Economic Restructuring and the Eroding Ecologies of Youth," in T. Skelton and G. Valentine (eds.), *Cool Places: Geographies of Youth Cultures*. London: Routledge, 130–44.

—— (1998*b*). "Power, Space and Terror: Social Reproduction and the Public Environment." Unpublished paper, Graduate Center, CUNY.

Kearns, Gerry, and Philo, Chris (eds.) (1994). *Selling Places: Culture and Capital in the Contemporary City*. Oxford: Pergamon.

Keith, Michael (1997). "Street Sensibility? Negotiating the Political by Articulating the Spatial," in A. Merrifield and E. Swyngedouw (eds.), *The Urbanization of Injustice*. New York: New York University Press, 137–60.

Keith, Michael, and Pile, Steven (1993). *Place and the Politics of Identity*. London: Routledge.

Killian, Ted (1998). "Public and Private, Power and Space," in Andrew Light and Jonathan Smith (eds.), *The Production of Public Space*. Philosophy and Geography, 2. Lanham, Md.: Rowman & Littlefield 115–34.

King, Anthony (1990). *Global Cities: Post-Imperialism and the Internationalization of London*. London: Routledge.

—— (1993). "Identity and Difference: The Internationalization of Capital and the Globalization of Culture," in Knox (1993: 83–110).

—— (ed.) (1996). *Re-Presenting the City: Ethnicity, Capital and Culture in the 21st-Century City*. New York: New York University Press.

—— (1999). "New Thinking about the Globalization Thesis and Cities in the Periphery." Paper presented to the Global Affairs Institute Colloquium on Globalization at the Margins, Syracuse University, 29 January.

Kirby, Andrew (1993). *Power/Resistance: Local Politics and the Chaotic State*. Bloomington: Indiana University Press.

Kirby, Andrew, Marston, Sallie, and Seasholes, Kenneth (1995). "World Cities and Global Communities: The Municipal Foreign Policy Movement and New Roles for Cities," in Knox and Taylor (1995: 267–79).

Kirby, Stewart, and Hay, Iain (1997). "(Hetero)sexing Space: Gay Men and 'Straight' Space in Adelaide, South Australia," *Professional Geographer*, 42: 295–305.

Kirtland, D., Gaydos, L., Clark, K., De Cola, L., Acevedo, W., and Bell, C. (1994). "An Analysis of Transformations in the San Francisco Bay/Sacramento Area." *World Resource Review*, 6: 206–17.

Knopp, Larry (1995). "Sexuality and Urban Space: A Framework for Analysis," in Bell and Valentine (1995: 149–61).

—— (1998). "Sexuality and Urban Space: Gay Male Identity Politics in the United States, the United Kingdom, and Australia," in Fincher and Jacobs (1998: 149–75).

Knox, Paul (ed.) (1993). *The Restless Urban Landscape*. Englewood Cliffs, NJ: Prentice Hall.

—— (1995). "World Cities and the Organization of Global Space," in R. J. Johnston, Peter Taylor, and Michael Watts (eds.), *Geographies of Global Change: Remapping the World in the Late Twentieth Century*. Oxford: Blackwell, 232–47.

Knox, Paul, and Taylor, Peter (eds.) (1995). *World Cities in a World System*. Cambridge: Cambridge University Press.

Kwan, Mei-Po (1999). "Gender and Individual Access to Urban Opportunities: A Study Using Space-time Measures." *The Professional Geographer*, 51: 210–26.

Lake, Robert W. (1992). "Planning and Applied Geography." *Progress in Human Geography*, 16: 414–21.

—— (1993). "Planning and Applied Geography: Positivism, Ethics and Geographical Information Systems." *Progress in Human Geography*, 17: 404–13.

Laws, Glenda (1994). "Oppression, Knowledge and the Built Environment." *Political Geography*, 13: 7–32.

Lawson, Victoria, and Klak, Thomas (1990). "Conceptual Linkages in the Study of Production and Reproduction in Latin American Cities." *Economic Geography*, 66: 310–27.

Lees, Loretta (1998). "Urban Renaissance and the Street: Spaces of Control and Contestation," in Fyfe (1998*a*: 236–53).

Lefebvre, Henri (1991). *The Production of Space*. Cambridge: Blackwell.

Levy, Richard M. (1998). "The Visualization of the Street: Computer Modeling and Urban Design," in Fyfe (1998*a*: 58–74).

Logan, John R., and Molotch, Harvey (1987). *Urban Fortunes: The Political Economy of Place*. Berkeley: University of California Press.

McCann, Eugene (1994). "Neotraditional Developments: The Anatomy of a New Urban Form." *Urban Geography*, 13: 210–33.

McDowell, Linda (1995). "Body Work: Heterosexual Gender Performances in City Workplaces," in Bell and Valentine (1995: 75–95).

McKenzie, Evan (1994). *Privatopia: Homeowners' Associations and the Rise of Residential Private Government*. New Haven: Yale University Press.

McLafferty, Sara, and Preston, Valerie (1991). "Gender, Race and Commuting Among Service Sector Workers." *The Professional Geographer*, 43: 1–15.

McMaster, Robert, Leitner, Helga, and Shepherd, Eric (1997). "GIS-based Environmental Equity and Risk Assessment: Methodological Problems and Prospects." *Cartography and Geographic Information Systems*, 24: 172–89.

Magnusson, Warren (1996). *The Search for Political Space*. Toronto: University of Toronto Press.

Marcuse, Peter (1995). "Not Chaos, But Walls: Postmodernism and the Partitioned City," in S. Watson and K. Gibson (eds.), *Postmodern Cities and Spaces*. Oxford: Blackwell, 224–53.

Markusen, Ann, Lee, Yong-Sook, and DiGiovanna, Sean (eds.) (2000). *Second Tier Cities*. Minneapolis: University of Minnesota Press.

Marston, Sallie (1990). "Who Are 'the People'? Gender, Citizenship and the Making of the American Nation," *Environment and Planning D: Society and Space*, 6: 449–58.

Marston, Sallie A., Towers, George, Cadwallader, Martin, and Kirby, Andrew (1989). "The Urban Problematic," in Gary L. Gaile and Cort J. Willmott (eds.), *Geography in America*. Columbus: Merrill, 651–72.

Martin, Patricia (1998). "On the Frontier of Globalization: Development and Discourse along the Rio Grande." *Geoforum*, 29: 217–36.

Massey, Doreen (1994). *Space, Place, and Gender*. Minneapolis: University of Minnesota Press.

—— (1997). "Space/Power, Identity/Difference: Tensions in the City," in A. Merrifield and E. Swyngedouw (eds.), *The Urbanization of Injustice*. New York: New York University Press, 100–16.

Massey, Douglas, and Denton, Nancy (1993). *American Apartheid*. Cambridge, Mass.: Harvard University Press.

Mayer, Margit (1991). "Politics in the Post-Fordist City." *Socialist Review*, 21: 105–11.

Miller, Byron (1992). "Collective Action and Rational Choice: Place, Community and the Limits to Individual Interest." *Economic Geography*, 68: 22–42.

—— (1994). "Political Empowerment, Local-Central State Relations, and Geographically Shifting Political Opportunity Structures: Strategies of the Cambridge, Massachusetts Peace Movement." *Political Geography*, 13: 393–406.

Miller, Harvey J. (1991). "Modelling Accessibility Using Space-Time Prisms within Geographic Information Systems." *International Journal of Geographical Information Systems*, 5: 287–301.

Mills, Caroline (1993). "Myths and Meanings of Gentrification," in James Duncan and David Ley (eds.), *Place/Culture/Representation*. London: Routledge, 149–70.

Mitchell, Don (1992). "Iconography and Locational Conflict from the Underside: Free Speech, People's Park, and the Politics of Homelessness in Berkeley, California." *Political Geography Quarterly*, 11: 152–69.

—— (1995). "The End of Public Space? People's Park, Definitions of the Public, and Democracy." *Annals of the Association of American Geographers*, 85: 108–33.

—— (1996). "Political Violence, Order, and the Legal Construction of Public Space: Power and the Public Forum Doctrine." *Urban Geography*, 17: 158–78.

—— (1997). "The Annihilation of Space By Law: The Roots and Implications of Anti-Homeless Laws in the United States." *Antipode*, 29: 303–35.

—— (1998*a*). "Anti-Homeless Laws and Public Space I: Begging and the First Amendment." *Urban Geography*, 19: 6–11.

—— (1998*b*). "Anti-Homeless Laws and Public Space II: Further Constitutional Issues." *Urban Geography*, 19: 98–104.

Mitchell, Katharyne (1997). "Different Diasporas and the Hype of Hybridity." *Environment and Planning D: Society and Space*, 15: 533–54.

Mitchell, W. J. (1995). *City of Bits: Space, Place, and the InfoBahn*. Cambridge, Mass.: MIT Press.

Mumford, L. (1961). *The City in History: Its Origins, Its Transformations, and Its Prospects*. New York: Harcourt, Brace & World.

Nagar, Richa (1997). "The Making of Hindu Communal Organizations, Places, and Identities in Postcolonial Dar es Salaam." *Environment and planning D: Society and Space*, 15: 707–21.

Nagar, Richa, and Leitner, Helga (1998). "Contesting Social Relations in Communal Places: Identity Politics among Asian Communities in Dar es Salaam," in Fincher and Jacobs (1998: 226–51).

Nagel, Caroline (1999). "Social Justice, Self-Interest, and Salman Rushdie: Re-assessing Identity Politics in Multicultural Britain," in James Proctor and David M. Smith (eds.), *Geography and Ethics: Journeys in a Moral Terrain*. London: Routledge, 132–46.

Nijkamp, Peter, and Scholten, H. J. (1993). "Spatial Information Systems: Design, Modeling, and Use in Planning." *International Journal of Geographical Information Systems*, 7: 85–96.

Nijman, Jan (1999). "Cultural Globalization and the Identity of Place: The Reconstruction of Amsterdam." *Ecumene*, 6: 146–64.

Oakes, Timothy. (1999). "Eating the Food of the Ancestors: Place, Tradition, and Tourism in a Chinese Frontier Town." *Ecumene*, 6: 123–45.

Obermeyer, Nancy (1998). "The Evolution of Public Participation GIS." *Cartography and Geographic Information Systems*, 25: 65–6.

O'Loughlin, John, and Friedrichs, Jürgen (1996). *Social Polarization in Post-Industrial Metropolises*. Berlin: Walter de Gruyter.

Pain, Rachel (1991). "Space, Sexual Violence, and Social Control: Integrating Geographical and Feminist Analyses of Women's Fear of Crime." *Progress in Human Geography*, 15: 379–94.

Peet, Richard (1998). *Modern Geographical Thought*. Oxford: Blackwell.

Pickles, John (ed.) (1995). *Ground Truth: The Social Implications of Geographic Information Systems*. New York: Guilford.

Pincetl, Stephanie (1994). "Challenges to Citizenship: Latino Immigrants and Political Organizing in the Los Angeles Area." *Environment and Planning A*, 26: 915–36.

Pinch, Steven (1994). "Social Polarization: A Comparison of Evidence from Britain and the United States." *Environment and Planning A*, 25: 779–95.

Piven, Frances Fox (1998). "Organizing the Poor for the Year 2000." Paper presented to the Syracuse Social Movements Initiative, Syracuse University, 29 October.

—— (1999). "Welfare Reform and the Economic and Cultural Reconstruction of Low Wage Labor Markets." *City and Society*, Annual Review, 21–36.

Plewe, Brandon, and Bagchi-Sen, Sharmitstha (2001). "The Use of Weighted Ternary Histograms for the Visualization of Residential Segregation." *The Professional Geographer*, 53: 347–60.

Pond, B., and Yeates, M. (1994). "Rural/Urban Land Conversion III: A Technical Note on Leading Indicators of Urban Land Development." *Urban Geography*, 15: 207–22.

Pratt, Geraldine, and Hanson, Susan (1991). "On the Links between Home and Work: Family-Household Strategies in a Buoyant Labour Market." *International Journal of Urban and Regional Research*, 15: 55–71.

Preteceille, E. (1990). "Political Paradoxes of Urban Restructuring: Globalization of the Economy and Localization of Politics," in John Logan and Todd Swanstrom (eds.), *Beyond the City Limits: Urban Policy and Economic Restructuring in Comparative Perspective*. Philadelphia: Temple University Press, 27–59.

Pulido, Laura (1994). "Restructuring and the Contraction and Expansion of Environmental Rights in the United States." *Environment and Planning A*, 26: 915–36.

—— (1996). "A Critical Review of the Methodology of Environmental Racism Research." *Antipode*, 28:142–59.

Pulido, Laura, Sidawi, Steve, and Vos, Robert O. (1996). "An Archaeology of Environmental Racism in Los Angeles." *Urban Geography*, 17: 419–55.

Putnam, Robert (1993). "The Prosperous Community: Social Capital and Public Life." *The American Prospect*, 13: 37.

Rahimian, R., Wolch, J., and Koegel, P. (1992). "A Model of Homeless Migration: Homeless Men on Skid Row, Los Angeles." *Environment and Planning A*, 24: 1317–36.

Rendell, Jane (1998). "Displaying Sexuality: Gendered Identities and the Early Nineteenth-Century Street," in Fyfe (1998*a*: 75–91).

Rheingold, Howard (1994). *The Virtual Community: Homesteading on the Electronic Frontier*. Reading, Mass.: Addison-Wesley.

Robinson, Jennifer (1996). *The Power of Apartheid: State, Power and Space in South African Cities*. Oxford: Butterworth-Heinemann.

Rodriguez, N. P., and Feagin, J. (1986). "Urban Specialization and the World System." *Urban Affairs Quarterly*, 22: 187–219.

Rose, Damaris (1988). "A Feminist Perspective on Employment Restructuring and Gentrification: The Case of Montreal," in Jennifer Wolch and Michael Dear (eds.), *The Power of Geography: How Territory Shapes Social Life*. Boston: Unwin Hyman, 118–35.

Rose, Damaris, and Villeneuve, Paul (1998). "Engendering Class in the Metropolitan City: Occupational Pairings and Income Disparities among Two-Earner Couples." *Urban Geography*, 19: 123–37.

Rowe, Stacy, and Wolch, Jennifer (1990). "Social Networks in Time and Space: Homeless Women on Skid Row, Los Angeles." *Annals of the Association of American Geographers*, 80: 184–204.

Ruddick, Susan M. (1996*a*). "Constructing Difference in Public Spaces: Race, Class, and Gender as Interlocking Systems." *Urban Geography*, 17: 132–51.

—— (1996*b*). *Young and Homeless in Hollywood*. London: Routledge.

Saff, Grant (1994). "The Changing Face of the South African City: From Urban Apartheid to the Deracialization of Space," *International Journal of Urban and Regional Restructuring*, 18: 377–91.

—— (1998). "The Effects of Informal Settlement on Suburban Property Values in Cape Town." *The Professional Geographer*, 50: 449–64.

—— (2001). "Exclusionary Discourse in Suburban Cape Town," *Ecumene* 8/1: 87–107.

Sassen, Saskia (1991). *The Global City: New York, London, Tokyo*. Princeton: Princeton University Press.

—— (1998). "Globalization and Its Discontents." Paper presented to the Global Affairs Colloquium on Globalization at the Margins, Syracuse University, 6 November.

Saunders, Peter (1981). *Social Theory and the Urban Question*. London: Hutchinson.

Saunders, Ralph (1999). "The Space Community Policing Makes and the Body That Makes It." *The Professional Geographer*, 51: 135–46.

Schein, Richard (1997). "The Place of Landscape: A Conceptual Framework for an American Scene." *Annals of the Association of American Geographers*, 87: 660–80.

Scott, Allen (1998). *Regions and the World Economy: The Coming Shape of Global Production, Competition, and Political Order*. Oxford: Oxford University Press.

Scott, Allen, and Soja, Edward (eds.) (1996). *The City: Los Angeles and Urban Theory and the End of the Twentieth Century*. Berkeley: University of California Press.

Sennett, Richard (1992). *The Fall of Public Man*. New York: Norton.

Shaw, S. L. (1993). "GIS for Urban Travel Demand Analysis: Requirements and Alternatives." *Computers, Environment and Urban Systems*, 17: 15–29.

Shields, Rob (1996). "A Guide to Urban Representation and What To Do About It: Alternative Traditions of Urban Theory," in A. King (ed.), *Re-Presenting the City: Ethnicity, Capital and Culture in the 21st-Century Metropolis*. New York: New York University Press, 227–52.

Sibley, David (1995). *Geographies of Exclusion*. London: Routledge.

Sklair, L. (1991). *Sociology of the Global System*. Baltimore: Johns Hopkins University Press.

Smith, David M. (1997). "Back to the Good Life: Towards an Enlarged Conception of Social Justice." *Environment and Planning D: Society and Space*, 15: 19–35.

— (1986). "Gentrification, the Frontier, and the Restructuring of Urban Space," in N. Smith and P. Williams (eds.), *Gentrification and the City* Boston: Allen & Unwin.

— (1989). "Tompkins Square Park." *The Portable Lower East Side*, 6: 1–28.

— (1992*a*). "Contours of a Spatialized Politics: Homeless Vehicles and the Production of Geographical Scale." *Social Text*, 33: 55–81.

— (1992*b*). "New City, New Frontier: The Lower East Side as Wild West," in Michael Sorkin (ed.), *Variations on a Theme Park: Scenes from the New American City*. New York: Hill & Wang, 61–93.

— (1993). "Homeless/Global: Scaling Places," in Jon Bird, Barry Curtis, Tim Putnam, George Robertson, and Lisa Tickner (eds.), *Mapping the Futures: Local Cultures, Global Change*. London: Routledge, 87–119.

— (1996). *The New Urban Frontier: Gentrification and the Revanchist City*. London: Routledge.

Smith, Susan (1989). "Society, Space, and Citizenship: A Human Geography for the 'New Times.'" *Transactions of the Institute of British Geographers*, 14: 144–56.

Soja, Edward W. (1989). *Postmodern Geographies: The Reassertion of Space in Critical Social Theory*. London: Verso.

— (1996). *Thirdspace: Journeys to Los Angeles and Other Real-and-Imagined Places*. Cambridge: Blackwell.

— (2000). *Postmetropolis: Critical Studies of Cities and Regions*. Oxford: Blackwell.

Sorkin, Michael (1992*a*). "See you in Disneyland," in Michael Sorkin (1992*b*: 205–32).

— (ed.) (1992*b*). *Variations on a Theme Park: The New American City and the End of Public Space*. New York: Hill & Wang.

Staeheli, Lynn A. (1996). "Publicity, Privacy, and Women's Political Action." *Environment and Planning D: Society and Space*, 14: 601–19.

— (1997). "Citizenship and the Search for Community," in L. Staeheli, J. Kodras, and C. Flint (eds.), *State Devolution in America: Implications for a Diverse Society*. Thousand Oaks, Calif.: Sage Publications, 60–75.

Staeheli, Lynn A., and Clarke, Susan E. (1995). "Gender, Place, and Citizenship," in J. Garber and R. Turner (eds.), *Gender and Urban Research*. Thousand Oaks, Calif.: Sage Publications, 3–23.

Staeheli, Lynn A., and Thompson, Albert (1997). "Citizenship, Community, and Struggles for Public Space." *The Professional Geographer*, 49: 28–38.

Steinberg, Philip (1994). "Territorial Formation on the Margin: Urban Anti-planning in Brooklyn." *Political Geography*, 13: 461–76.

Stevenson, John (1998). "The Mediation of the Public Sphere: Ideological Origins, Practical Possibilities," in A. Light and J. Smith (eds.), *The Production of Public Space*. Lanham, Md.: Rowman & Littlefield, 189–206.

Sui, Daniel (1994). "GIS and Urban Studies: Postivism, Post-positivism, and Beyond." *Urban Geography*, 15: 258–78.

— (1997). "Reconstructing Urban Reality: From GIS to Electropolis." *Urban Geography*, 18: 74–89.

Suttles, Gerald (1972). *The Social Construction of Communities*. Chicago, University of Chicago Press.

Swyngedouw, Erik (1989). "The Heart of the Place: The Resurrection of Locality in an Age of Hyperspace." *Geografiska Annaler B*, 71: 31–42.

— (1997). "Neither Global nor Local: 'Globalization' and the Politics of Scale," in Kevin Cox (ed.), *Spaces of Globalization*. New York: Guilford Press, 137–66.

Takahashi, Lois (1996). "A Decade of Understanding Homelessness in the USA: From Characterization to Representation." *Progress in Human Geography*, 20: 291–310.

— (1998). "Community Responses to Human Service Delivery in United States Cities," in Fincher and Jacobs (1998: 120–48).

Thrall, Grant (1993). "Basic GIS Software Can Help Make Urban Policy." *Geo Info Systems*, 3: 58–64.

Thrall, Grant, and Ruiz, M. (1994). "A History of Implementing an Urban GIS, Part One: Design, Tribulations, and Failure." *Geo Info Systems*, 4: 50–8.

Thrall, Grant, Bates, B., and Ruiz, M. (1994). "A History of Implementing an Urban GIS, Part Two: Two Solutions towards a Working GIS." *Geo Info Systems*, 4: 46–51.

Thrift, Nigel (1994). "On the Social and Cultural Determinants of International Financial Centers: The Case of the City of London," in S. Corbridge, R. Martin, and N. Thrift (eds.), *Money, Space and Power*. Oxford: Basil Blackwell, 327–55.

— (1996). "New Urban Eras and Old Technological Fears: Reconfiguring the Goodwill of Electronic Things." *Urban Studies*, 33: 1463–94.

Till, Karen (1994). "Neotraditional Towns and Urban Villages: The Cultural Production of a Geography of 'Otherness'." *Environment and Planning D: Society and Space*, 11: 709–32.

Valentine, Gill (1989). "The Geography of Women's Fear," *Area*, 21: 385–90.

— (1992). "Images of Danger: Women's Sources of Information about the Spatial Distribution of Male Violence." *Area*, 24: 22–9.

— (1997). "'My Son's a Bit Dizzy,' 'My Wife's a Bit Soft': Gender, Children and Cultures of Parenting." *Gender, Place and Culture*, 4: 63–88.

van Weesep, Jan (1994). "Gentrification as a Research Frontier." *Progress in Human Geography*, 18: 74–83.

Veregge, Nina (1997). "Traditional Environments and the New Urbanism: A Regional and Historical Critique." *Traditional Dwellings and Settlement Review*, 3: 49–62.

Vuchic, Vukan (2000). *Transportation for Livable Cities*. New Brunswick: Center for Urban Policy Research Press.

Wang, Fahui (2000). "Modeling Commuting Patterns in Chicago in a GIS Environment: A Job Accessibility Model." *The Professional Geographer*, 52: 120–33.

Warde, A. (1991). "Gentrification as Consumption: Issues of Class and Gender." *Environment and Planning D: Society and Space*, 9: 223–32.

Warf, Barney, and Erickson, Rodney (1996). "Globalization and the U.S. City System." *Urban Geography*, 17: 1–4.

Webber, Melvin (1964). "The Urban Place and the Nonplace Urban Realm," in M. Webber (ed.), *Explorations into Urban Structure*. Philadelphia: University of Pennsylvania Press, 79–153.

Weiner, Daniel, Warner, Timothy, Harris, Trevor, and Levin, Richard (1995). "Apartheid Representations in a Digital Landscape: GIS, Remote Sensing and Local Knowledge in Kiepersol, South Africa." *Cartography and Geographic Information Systems*, 22: 58–69.

Western, John (1996[1981]). *Outcast Cape Town*. Berkeley: University of California Press.

Wheeler David C., and O'Kelly, Morton E. (1999). "Network Typology and City Accessibility of the Commercial Internet." *The Professional Geographer*, 51: 327–39.

White, R. G., Engelen. G., and Uljee, I. (1997). "The Use of Constrained Cellular Automata for High-Resolution Modelling of Urban Land-use Dynamics." *Environment and Planning B: Planning and Design*, 24: 323–43.

Wilson, B. M. (1992). "Structural Imperatives Behind Racial Change in Birmingham, Alabama." *Antipode*, 24: 171–202.

Wilson, David (1998). "Urban Representation as Politics," *Urban Geography*, 19: 254–61.

Wilson, Elizabeth (1991). *The Sphinx in the City: Urban Life, the Control of Disorder, and Women*. Berkeley: University of California Press.

Winner, Langdon, (1992). "Silicon Valley Mystery House," in Sorkin (1992*b*: 31–60).

Wolch, J., and Dear, M. (1996). *Malign Neglect: Homelessness in an American City*. San Francisco: Jossey-Bass.

Wolch, J., Rahimian, A., and Koegel, P. (1993). "Daily and Periodic Mobility Routines of the Urban Homeless." *Professional Geographer*, 45: 159–69.

Yapa, L. S. (1991). "Is GIS Appropriate Technology?" *International Journal of Geographical Information Systems*, 5: 41–58.

Young, Iris Marion (1990*a*). *Justice and the Politics of Difference*. Princeton: Princeton University Press.

—— (1990*b*). "The Ideal of Community and the Politics of Difference," in L. Nicholson (ed.), *Feminism/Postmodernism*. New York: Routledge, 300–23.

Zukin, Sharon (1991). *Landscapes of Power: From Detroit to Disney World*. Berkeley: University of California Press.

—— (1995). *The Cultures of Cities*. Oxford: Blackwell.

—— (1996). "Space and Symbols in an Age of Decline," in A. King (ed.), *Re-Presenting the City: Ethnicity, Capital and Culture in the 21st-Century Metropolis*. New York: New York University Press, 43–59.

Part III

Environment/Society Dynamics

CHAPTER 18

The Human Dimensions of Global Change

Diana Liverman, Brent Yarnal, and Billie Lee Turner II

The human–environment condition has emerged as one of the central issues of the new millennium, especially as it has become apparent that human activity is transforming nature at a global scale in both systemic and cumulative ways. Originating with concerns about potential climate warming, the global environmental change agenda rapidly enlarged to include changes in structure and function of the earth's natural systems, notably those systems critical for life, and the policy implications of these changes, especially focused on the coupled human–environment system. Recognition of the unprecedented pace, magnitude, and spatial scale of global change, and of the pivotal role of humankind in creating and responding to it, has led to the emergence of a worldwide, interdisciplinary effort to understand the human dimensions of global change. The term "global change" now encompasses a range of research issues including those relating to economic, political, and cultural globalization, but in this chapter we limit our focus to global environmental change and to the field that has become formally known as the human dimensions of global (or global environmental) change. We also focus mainly on the work of geographers rather than attempting to review the whole human dimensions research community.

Intellectually, geography is well positioned to contribute to global environmental change research (Liverman 1999). The large-scale human transformation of the planet through activities such as agriculture, deforestation, water diversion, fossil fuel use, and urbanization, and the impacts of these on living conditions through changes in, for example, climate and biodiversity, has highlighted the importance of scholarship that analyzes the human–environmental relationship and can inform policy. Geography is one of the few disciplines that has historically claimed human–environment relationships as a definitional component of itself (Glacken 1967; Marsh 1864) and has fostered a belief in and reward system for engaging integrative approaches to problem solving (Golledge 2002; Turner 2002). Moreover, global environmental change is intimately spatial and draws upon geography-led remote sensing and geographic information science (Liverman *et al.* 1998).

Geographers anticipated the emergence of current global change concerns (Thomas *et al.* 1956; Burton *et al.* 1978) and were seminal in the development of the multidisciplinary programs of study into the human dimensions of global change. They helped to galvanize a large but dispersed research community through such efforts as "The Earth as Transformed by Human Action" symposium and volume (Turner *et al.* 1990) and the International Human Dimensions Programme (IHDP[1])

[1] For more information on the International Human Dimensions Programme see <http://www.ihdp.org/>, last accessed 10 November 2002.

and work closely with international and national scientific organizations to create local-to-global research (Kates 2001; National Research Council 1999*a*; National Research Council 2001; Stern *et al.* 1992; White 1982).

Human dimensions of global change research is broadly defined as the study of the societal causes and consequences of changes in the global environment, as well as the analysis of individual and institutional responses to these changes (National Research Council 1999*a*). From an initial focus on climate change, human dimensions research has expanded to address changes in biodiversity, land, health, and water, to examine a broad set of driving forces including energy, consumption, and institutions, and to link global with local scales of human–environment interactions (Brookfield 1995; Wilbanks and Kates 1999). Geographers have made major contributions to understanding the human dimensions of these issues, most notably to scholarship on the impacts of climate change and on the social causes of land-use and land-cover change.

The Human Dimensions of Global Change Specialty Group of the Association of American Geographers (AAG) was organized by Robert Ford, Jimmie Eflin, and Brent Yarnal in 1995 following a successful NSF-funded workshop at the 1993 AAG meetings in Atlanta. Its goal is to promote the varied interests of geographers who are united by research, teaching, or service that in one way or another involve the human dimensions of global-scale processes that affect or are affected by environmental changes (Brunn and O'Lear 1999). Many of the members of this specialty group have contributed to the Human Dimensions of Global Change Hands-On project, coordinated by Susan Hanson, that has created a series of educational modules dealing with important global change issues.[2] Another example of the institutionalization of human dimensions research was the establishment of a major journal, *Global Environmental Change*, which has been edited by geographers (Martin Parry, Timothy O'Riordan, and J. Kenneth Mitchell), and publishes a wide range of articles by geographers.

In this chapter we describe the emergence of interest in the human dimensions of global change and the ways in which geographers have contributed to this field, and suggest some future directions.

[2] For more information on the Hands-On project, including several on-line modules, see <http://www.aag.org/HDGC/Hands_On.html>, last accessed 10 November 2002.

Origins and Institutional Setting of Human Dimensions of Global Change Research

Interest in global change grew rapidly in the 1980s because of increasing concern about issues such as global warming, ozone depletion, and the loss of tropical forests. Geographers played a major role in program development. Formal identification of the subfield occurred in the late 1980s through the work of Robert Kates, Gilbert White, "Reds" Wolman, and others working through the National Academy of Sciences and international bodies, such as the International Council of Scientific Unions. A white paper prepared by Roger Kasperson was the springboard for a workshop that led to the creation of committees on human dimensions research by the Social Science Research Council in 1988 and the National Research Council in 1989. The white paper also prompted the formation of an interdisciplinary review panel and a call for proposals for research on the human dimensions of global change by the National Science Foundation in 1989. Geographers were central to these efforts, as members of the panels and committees (Roger Barry, Roger Kasperson, Diana Liverman, Billie Lee Turner II, and Thomas Wilbanks), as program officers (Tom Baerwald at NSF), and as guides to the overall enterprise through the National Research Council Board on Sustainable Development (Robert Kates).

Geographers helped make the human dimensions theme visible through the production of National Research Council (NRC) reports such Global Environmental Change: Understanding the Human Dimensions (Stern *et al.* 1992). The report identified five major driving forces for global change—population growth, economic growth, technological change, institutions, and attitudes and beliefs—and recommended a program of Federal support for research, data, fellowships, and centers for the study of human dimensions of global change (Stern *et al.* 1992). Geographers continued to shape national research agendas through their coordinating roles in other NRC reports, such as People and Pixels: Linking Remote Sensing and Social Science (Liverman *et al.* 1998), Making Climate Forecasts Matter (Stern and Easterling 1999), Human Dimensions of Global Environmental Change (National Research Council 1999*a*), and Grand Challenges in the Environmental Sciences (National Research Council 2001).[3] Geographers

[3] Many of the NRC reports are available to read on-line at <http://www.nas.edu>.

have also been closely involved in the management and research projects of CIESIN (Center for International Earth Science Information Network) that has a particular focus on the development of datasets that combine social and environmental data.[4]

An emerging international commitment paralleled these United States-led efforts to the study of the human dimensions of global change. International endeavors include the Intergovernmental Panel on Climate Change (IPCC) and International Human Dimensions Programme of the International Geosphere-Biosphere Programme (IHDP-IGBP), as well as regional initiatives such as the Inter American Institute (IAI) and the Global Change System for Analysis, Research and Training (START).[5] Geographers have played a leading role through their authorship in the Intergovernmental Panel on Climate Change (IPCC), their organization of the Land-Use and Land-Cover Change (LUCC), Global Environmental Change and Human Security, and Industrial Transformation projects of IHDP, and their considerable presence at the biennial Open Science Meetings of the Human Dimensions of Global Change Community, and the 2001 Open Science Conference on Global Environmental Change in Amsterdam (sponsored by IGBP, IHDP, and World Climate Research Programme).

In the remainder of this chapter, we assess geographic contributions to understanding the human dimensions of global change by discussing the specific research themes of climate and society, land-use change, industrial ecology, integrated assessment, and international environmental policy. We conclude by considering future directions for the field and the role of geography in that future.

The Impacts of Global and Regional Climate Change on Society

Geographers have made major contributions to the international understanding of climate impacts on society including an emphasis on vulnerability and adaptation, the development of innovative methods, and broadening the range of sectoral and regional case studies. These contributions build on a much longer tradition of natural hazards research and climate impact assessment in geography in which a range of methods have been used to understand climate impacts on society (Kates *et al.* 1985; White 1974).

Vulnerability and Adaptation

Geographers played an important role in the reconceptualization of climate impact assessment to account for vulnerability, rather than to assume that the most important determinant of climate impacts is the physical environment. Case studies and theoretical analyses demonstrated that the impacts of climatic variations and changes depend as much, if not more, on social, political, and economic conditions as they do on the magnitude of the climatic events, a position increasingly labeled vulnerability assessment (Cutter *et al.* 2000; Dow 1992; Liverman 1994; Watts and Bohle 1993). Geographers' work on vulnerability expanded the worldwide reach of climate impact studies at a time when most studies of global warming focused on the United States and Europe (Dilley 2000; Downing 1991; Vogel 1998). International collaborative projects involved geographers from Latin America, Asia, and Africa in studies of how climate change might affect agriculture, water resources, and ecosystems and highlighted the great variations in vulnerability within and between countries (Parry *et al.* 1988; Vogel 1995; Watson *et al.* 1996, 1998). Analyses have been influenced by the work of feminist and other geographers who have argued that women, children, the elderly, people of color, and others are often more vulnerable than other groups to environmental change because of the ways in which racism, sexism, and other prejudices and social structures marginalize and disempower these people (Cutter 1996; Vogel 1998). In short, from an original climate impacts focus, global change research now addresses vulnerability through the full array of environmental changes underway (Kasperson and Kasperson 2001).

Geographers have also argued strongly for focusing on the possibility of adaptation in climate impact studies because farmers and other people are likely to adjust to climate changes and predictions (Burton 1997; Easterling 1996; Smit *et al.* 1996). William Easterling and colleagues used historical analogues from the 1930s drought to investigate climate change impacts on agriculture in the Midwest/Great Plains region and showed how farmer adjustments reduce the severity of impacts

[4] Center for International Earth Science Information Network <http://www.ciesin.org/>, last accessed 10 November 2002.

[5] For more information on IPCC (<http://www.ipcc.ch>), IHDP (<http://www.ihdp.org/>), IAI (<http://www.iai.int>), START (<http://www.start.org/>), all last accessed 10 November 2002.

and permit adaptation, especially under irrigated conditions (Easterling *et al.* 1993). This study is one of several recognizing that higher carbon dioxide may enhance plant productivity and that farmers adapt to new conditions. Barry Smit and colleagues demonstrated the significance of adaptation and presented an analytical framework in several studies of climate variability and agriculture in Canada (Smithers and Smit 1997).

Sectoral Studies

Geographers' expertise in human–environment interactions across a range of sectors and in specific locales provided the basis for their contributions in assessing climate impacts on a number of key sectors including agriculture, water resources, ecosystems, human health, and coastal zones.

Geographers have been particularly effective in their studies of climate impacts on agriculture. They contributed to the shift from regression studies based on monthly equilibrium scenarios to crop modeling efforts that incorporate the direct physiological effects of higher carbon dioxide levels. They also employed transient climate scenarios and daily data, examined spatial shifts in crop potential, expanded analyses in developing countries, and replaced the concept of the "dumb" farmer with that of skilled managers capable of flexible adaptation to climate change. For example, Linda Mearns and colleagues used the CERES crop simulation model to show that it is important to account for potential changes in climate variability when assessing climate change impacts on wheat yields in the United States because extreme events are critical to yield variability and crop failures (Mearns *et al.* 1996, 1997). A comparative study employed transient scenarios, a variety of adaptations, and modeling the physiological benefits of higher carbon dioxide to investigate potential climate change impacts on crop yields internationally (Rosenzweig and Parry 1994). The authors concluded that although global warming might produce only a small decrease in overall global crop production, developing countries would experience production declines and would be less able to adapt through technology.

Other geographers focused on the potential impacts of climate change on water resources in different regions (Cohen 1992; Morehouse 2000; Riebsame 1995). For example, Stewart Cohen suggested that the potential impacts of climate change combined with projected population growth on water resources in Canada include decreased supplies in the Great Lakes and increased municipal water use in cities such as Toronto (Cohen 1986, 1992). In contrast, Neff *et al.* (2000) demonstrated that some areas, such as the Mid-Atlantic Region of the United States, might receive increased precipitation with climate change and, therefore, may have to adapt to wetter rather than drier conditions.

Biogeographers joined ecologists in the concern that climate change may be so severe or rapid that ecosystems will be unable to survive or adapt to the new climate conditions (Malanson 1989; Peters and Lovejoy 1992). For example, William Baker (1995) suggested that global warming might shorten intervals between fires and that this altered disturbance regime may create perpetual disequilibrium in some landscapes. Pat Bartlein *et al.* (1997) used relationships between climate and vegetation in Yellowstone National Park to show that a doubling of atmospheric carbon dioxide could lead to reduced range or extinction of high elevation species and to argue that the projected climate changes are much greater than those in the paleoecological record.

Larry Kalkstein has written extensively on climate change and human health by establishing statistical relationships between climate and mortality in various cities (e.g. Kalkstein and Davis 1989). In a study of forty-four large American cities, he found that mortality can increase by up to fifty deaths per day for synoptic conditions associated with warm dry or moist air masses in the eastern and Midwestern United States. He predicted a dramatic increase in weather-related mortality if climate changes to warmer conditions by 2050 (Kalkstein and Greene 1997) and suggested similar effects for cities in China and Egypt, where he also examined the potential impacts of climate change on air pollution and vector-borne diseases (Kalkstein and Smoyer 1993). Geographers also promoted the idea that sea-level rise poses important challenges to coastal development, management, and planning (Warrick and Farmer 1990) and conducted local scenarios and policy options for regions such as the eastern United States (Nicholls and Leatherman 1996; Vellinga and Leatherman 1989) and southern England (Bray *et al.* 1997; Warrick *et al.* 1993).

Although many geographers have worked under the assumption that the model projections of global warming are reasonable, others believe that the threat of global warming may have been exaggerated, that there are explanations of warming trends other than greenhouse gas emissions, and that humans will be able to adapt easily to climate change. Prominent among these is Robert Balling, who has written many papers and a book on the climate change issue showing other possible explanations for warming, including land-use change and urbanization (Balling and Idso 1990; Balling 1992).

Regional Assessments

The coarse spatial and temporal resolution of the climate model results used to construct climate change scenarios hampered attempts to undertake regional case studies of climate impacts. Geographers contributed to the regional and temporal specificity of impact assessments by analyzing daily and synoptic data to show the significance of climate variability and extremes and by using mesoscale climate models and statistical downscaling techniques to provide more regional and temporal detail for climate scenarios (Giorgi *et al.* 1998; Hewitson and Crane 1996). Regional climate impacts studies with heavy involvement of geographers include studies of Africa (Justice and Desanker 2001), Canada (Brklacich *et al.* 1998; Cohen *et al.* 1998), and Mexico (Liverman and O'Brien 1991) as well as a number of the regional and cross-cutting reports that were written for the United States National Assessment (National Assessment Synthesis Team, 2001).

Intergovernmental Program on Climate Change

The overall influence of geographic research is clear in the key roles played by many geographers in international assessments of climate change. Every five years, beginning in 1990, the international scientific community has committed to produce reports on the state of understanding of climate change, impacts, and mitigation through IPCC. These comprehensive reports provide a basis for international negotiations and agreements, such as the Framework Convention on Climate Change, that seek to manage the human impact on Earth's climate.

Both physical and human geographers contributed to the 1990 IPCC report, especially to the second volume on the potential impacts of global warming on sectors such as agriculture, water resources, ecosystems, urban settlements, and sea-level rise (Tegart *et al.* 1990) with contributions from geographers such as Stewart Cohen, Stephen Lonergan, Martin Parry, and Richard Warrick. The international study of climate impacts on agriculture led by Martin Parry with contributions from geographers such as Thomas Downing, Masatoshi Yoshino, and Timothy Carter provided critical results for the IPCC agriculture chapter showing how climate variability and change might affect agriculture across the world (Parry 1990).

By 1995, geographers were playing an even more significant role in IPCC. The lack of consensus about how climate may change at the regional level, and the recognition that changes in social systems may be more important than changes in natural systems in determining climate impacts, reoriented Working Group 2 of the second IPCC assessment to pay more attention to vulnerability and regional impacts (Watson *et al.* 1996, 1998). Parry and Carter (1998) developed the overall technical guidelines for assessing impacts and adaptations, arguing for a focus on vulnerability. Geographers were co-authors or were cited in many chapters from the Working Group 2 report on impacts, including: Adrian Aguilar, Ian Douglas, Diana Liverman, and Masatoshi Yoshino on urban settlements; Martin Parry, Michael Brklacich, Linda Mearns, and Thomas Downing on agriculture; Larry Kalkstein on human health; Roger Barry, Blair Fitzharris, and Lisa Graumlich on mountain and cryosphere ecosystems; and Stephen Leatherman and Richard Warrick on sea-level rise.

The IPCC Third Assessment Report in 2000 continued to involve geographers. In Working Group 1, which was dedicated to the physical science of climate change, geographers Roger Barry, Timothy Carter, Robert Crane, David Easterling, Bruce Hewitson, Linda Mearns, and Robert Wilby served as either lead or contributing authors (Houghton *et al.* 2001). There was even greater representation by geographers in Working Group 2, which focused on impacts, adaptation, and vulnerability (McCarthy *et al.* 2001). Nigell Arnell, Ian Burton, Stewart Cohen, Thomas Downing, Bill Easterling, Blair Fitzharris, Larry Kalkstein, Paul Kay, Linda Mortsch, Robert Kates, Susanne Moser, Linda Mearns, Roger Pulwarty, Dipo Odejuwon, Barry Smit, Richard Warrick, Thomas Wilbanks, and others participated in the process of putting together this document contributing chapters on methods and key sectors such as agriculture, health, and human settlements. Unfortunately, geographers had little input to Working Group 3 on mitigation (Metz *et al.* 2001).

Climate Variability and El Niño

Climate impact research has much to contribute beyond the global warming issue, especially in understanding and reducing the impacts of extreme events and year-to-year climate variability. While hazards research continues to provide insights into the impacts of drought, floods, and severe storms, the 1990s saw a growth in interest in seasonal to decadal climate fluctuations. Climate variability associated with frequent El Niño and La Niña events that alter Pacific sea surface temperatures, and improvements in scientific ability to make seasonal forecasts based on ocean conditions have resulted in increased attention to the impacts and responses to large-scale interannual climate variability. Whilst Cesar Caviedes, Vera Markgraf, and Henry Diaz have sustained a long-term interest in the impacts of El Niño (Caviedes 1984; Markgraf and Diaz 1992, 2000), other geographers have only recently started to focus on the human and biological effects of this phenomenon (Eakin 1999; Nkemdirim 2001). Because the El Niño-Southern Oscillation (ENSO) index correlates strongly with extreme climate events around the world, research on impacts and the use of forecasts is needed in regions as diverse as the Andes, Northeast Brazil, Southern Africa, India, and Indonesia. A review and an agenda for research on the human dimensions of climate variability and seasonal forecasting was developed by a National Research Council panel chaired by William Easterling (Stern and Easterling 1999).

There are still many unresolved research questions in climate-society research and a continuing role for geographers in sectoral and regional studies. For example, the spatial and temporal resolution of many climate change scenarios is still inadequate for regional policymaking. Vulnerabilities and abilities to adapt are continually changing with new socioeconomic conditions and institutions (Kasperson *et al.* 1995). Moreover, as seasonal forecasts improve, there are tremendous challenges in communicating uncertainty, ensuring access to the predictions, and estimating the costs and benefits of the forecasts, errors, and adjustments.

The Social Causes and Consequences of Land-Use and Land-Cover Change

Interest in the causes of local and regional land-use changes is long-standing in geography. Geography maintains a rich tradition of study of landscape and ecosystem, documenting and explaining the history of human transformation of land cover and interpreting the major forces that drive land use (Humboldt 1845–62; Thomas *et al.* 1956). The significant role of land-use and land-cover change in many facets of the human–environment condition vaulted the subject (and with it, geography) into the center-stage of global change research (IGBP-IHDP 1995; National Research Council 2001). Land-use and land-cover change research contributes to the overall scientific understanding of biogeochemical cycles (especially the carbon cycle), regional climate modification, and alterations in natural ecosystems. It provides insights into the drivers giving rise to these changes (Geist and Lambin 2002; Lambin *et al.* 2001) and into the vulnerability of places and people in the face of actual and potential change (IGBP-IHDP 1999). It also provides a critical basis for policies to mitigate and adapt to climate change, to conserve biodiversity, and to reduce land degradation. Geist and Lambin (2002) constructed a meta-analysis of human dimensions research on tropical deforestation and showed that the causes are complex, varying from region to region and involving many biophysical and socio-economic factors.

The inclusion of land-use studies on the overall human dimensions research agenda has attracted geographers to the subfield and created new opportunities for funding in various national and international venues. Interdisciplinary collaboration to study land use as a human dimensions issue has been fostered by geographers through the organization of several conferences and workshops and through the development of national and international research agendas (Meyer and Turner 1994). Many geographers have become involved in the International Human Dimensions Programme/International Geosphere Biosphere Programme (IHDP/IGBP) core project on Land-Use and Land-Cover Change (LUCC), with its coordinated, comparative, multilevel strategy for understanding, monitoring, and modeling land use (IGBP-IHDP 1995, 1999; Turner *et al.* 1995*b*) and for managing core projects internationally. Indeed, the international office of LUCC sits in the Department of Geography, University of Louvain, Belgium.[6] In developing frameworks, case studies, and models of how social forces drive changes in land use and land cover, this type of comparative research program has the potential to explain and predict land-use change, but also to assist in identifying strategies for managing land use and protecting ecosystems (Eastman and Anyamba 1996; Turner 2001).

The 1992 quincentennial of the "Columbian Encounter" served as springboard for geography to demonstrate to the global change community the scale and magnitude of past land-use and land-cover change in the Americas and elsewhere (Turner and Butzer 1992). In a special issue of the *Annals of the Association of American Geographers* (1992) and a volume from a 1992 SCOPE conference (Turner *et al.* 1995*a*), with contributions by Karl Butzer, Elisabeth Butzer, Ian Douglas,

[6] See <http://www.geo.ucl.ac.be/LUCC/>, last accessed 10 November 2002.

Alfred Siemens, Neil Smith, and others), geographers showed that—before the European conquest—many regions of the Americas were radically transformed from their pre-use state. Additional studies also disputed the view that significant land degradation in the Americas began with the colonial experience (Butzer 1993; O'Hara *et al.* 1993; Sluyter 1996), challenging the "pristine myth" of indigenous land use in the Americas (Denevan 2001; Doolittle 2001). Such findings demand recalibration of various global change models.

Considerable attention has also been given to current land-use and land-cover change, from the global to local scales, with much of it focused on Amazonia and the causes of tropical deforestation (Turner 2001). Various remote sensing techniques have been used to address the amount of deforestation and some of its dynamics (Lambin 1997). Intensive work by David Skole and others, for example, has documented the pace and magnitude of Amazonian deforestation (Rignot *et al.* 1997; Skole and Tucker 1993), demonstrating the important but often overlooked role of regrowth in the carbon cycle. Land-use modeling (Riebsame *et al.* 1995) is becoming an important area of research, as demonstrated by the special journal issue on predicting land-use change (Veldkamp and Lambin 2001), which featured geographers such as Steven Walsh and Robert Pontius.

Such remote sensing work in combination with ground studies has increased understanding of the social processes driving land-cover change (Guyer and Lambin 1993; Lambin *et al.* 2001; Skole *et al.* 1994; Sierra and Stallings 1998). Again, work focused on tropical deforestation tempers the simple view that population is causing land-cover change through more subtle, but complex interpretations showing the connections among economy, politics, and policy. For instance, Susanna Hecht has shown the importance of Brazilian state policies supporting cattle and frontier settlement in the conversion of forest to pasture (Hecht 1985; Hecht and Cockburn 1989). Nigel Smith has demonstrated the importance of new transportation corridors and the value of Amazonian biological and cultural diversity (Smith 1982, 1999). Harold Brookfield and David Kummer have established the role of logging for export and the weak control of logging concessions in the deforestation of much of southeastern Asia and Indonesia (Brookfield and Byron 1990; Brookfield *et al.* 1995; Kummer 1991). Similar findings have been made for other land covers in Asia (Apan and Peterson 1998; Bensel and Kummer 1996; Bernard and De Koninck 1996; Schreier *et al.* 1994), in Africa (Ite and Adams 1998; Jarosz 1993; Lambin and Ehrlich 1997; Sussman *et al.* 1994), and in Latin America beyond the Amazon (Brothers 1997; Klooster and Masera 2000; O'Brien 1998).

Increasingly, political ecology, cultural ecology, and institutional approaches are fused with risk, hazards, and vulnerability concepts to address land-use and land-cover change (Batterbury and Bebbington 1999; Blaikie and Brookfield 1987). The importance of institutions, such as land tenure, and the ways in which local people respond to them has been demonstrated, for example, by Metz (1994) in Nepal, and Bassett (1988) and Batterbury and Warren (2000) in western Africa. Food vulnerability has increasingly been tied to land-cover change as land-use conversion to intensive commercial agriculture, ranching, or private ownership may alter the entitlements of the poor to the use of common land, forests, or employment (Kasperson and Kasperson 2001; Vogel 1995; Yarnal 1994). This fusion, including the use of remote sensing and geographic information systems (GIS), may lead to a new generation of geographers interested in land transformation with genuine appreciation for varied approaches of study.

Industrial Ecology and Emissions

Fewer geographers have been involved in the study of industrial ecology[7] and greenhouse gas emissions, although those involved, such as Cutler Cleveland, David Angel, and Tom Wilbanks, have made significant contributions, especially those practitioners from the energy field (Cleveland and Matthias 1998). The Industrial Transformation project of the IHDP has engaged geographers (such as David Angel) in international comparative studies of relationships between economic restructuring, technological change, environmental impacts, and mitigation policies (Angel and Rock 2002).[8] Another area in which geographers can contribute to the understanding of the environmental impacts of energy use and industrial development is to relate economic and industrial geography to regional resource use and pollution patterns and to link global economic and global environmental change (Angel

[7] Defined as the study of the local and global uses, flows, and environmental impacts of materials and energy in economic and industrial development.

[8] See <http://130.37.129.100/ivm/research/ihdp-it>, accessed 10 November 2002 (Industrial Transformation: A Science Project of the International Human Dimensions Programme on Global Environmental Change IHDP-IT).

and Rock 2002; Clark 1993; Gibbs 1997, 2002; Hall *et al.* 1986; O'Brien and Leichenko 2000). The increasing mobility of capital has facilitated the expansion of transnational corporations and massively restructured the geography of manufacturing, agriculture, human settlements, and the associated environmental impacts (Bridge 2002).

Other important research questions include the impact of environmental policy and regulations on industrial geographies and the competitive potential of different technologies. Regional geographers can provide detailed insights into critical patterns of energy consumption and materials flows, particularly in rapidly changing regions such as the former Soviet Union and Eastern Europe, China, or Southern Africa.

Geographers can make useful contributions to the study of longer-term climate change and atmospheric chemistry particularly through understanding of regional economic restructuring, regional geographies, and consumption patterns. In these studies, geographers need to be aware of new developments in atmospheric chemistry, which, for example, point to the significance of regional sulfur emissions and nitrogen cycling in global change. Vaclav Smil has written extensively on energy use, pollution, and the impact on key biogeochemical cycles as well as providing assessments of regional conditions in China (Smil 1985, 1994).

Industrial ecology has tended to focus on material and energy flows through time and space, which, as a result of the concern with global warming, has led to numerous studies linking fossil-fuel burning and other industrial activities to emissions of carbon dioxide and other radiatively active gases. One major focus of the emissions studies has been the prediction of overall greenhouse gas trends and the allocation of national and regional responsibilities (e.g. Hall, Cleveland, and Kaufmann 1986; Stern and Common 2001).

Danny Harvey has written on the highly controversial estimates of the combined greenhouse gas indices called Global Warming Potentials (GWPs) and has been joined by others in discussing the politics of responsibility for emissions (Waterstone 1985; Harvey 1993; Redclift and Sage 1998). He has also discussed Canada's carbon dioxide emissions and possible mitigation options (Harvey *et al.* 1997). Other national studies include an analysis of links between socioeconomic restructuring and greenhouse gas emissions in Bulgaria (Yarnal 1997).

As discussion has shifted to mitigation and the implementation of the Kyoto protocols on reducing greenhouse emissions, public and academic interest has been generated in the emissions and mitigation policies of more localized regions, cities, and communities. Geographers have taken a leadership role in local greenhouse gas emissions studies (Gibbs 2002). The Global Change and Local Places (GCLP) project, organized under the umbrella of the Association of American Geographers, has focused attention on the local human activities emitting greenhouse gases, the socioeconomic forces driving local emissions, and the potential for local mitigation of these emissions (Kates and Torrie 1998; Wilbanks and Kates 1999). This project involves case studies in the northwestern Ohio (University of Toledo), western North Carolina (Appalachian State University), southwestern Kansas (Kansas State University), and central Pennsylvania (Pennsylvania State University). The project has refined methods for estimating local emissions (Kates *et al.* 1998) and it has produced essential insights into the local geographies of greenhouse gas emissions and socioeconomic drivers, including the importance of industrial restructuring in Ohio, natural gas production and cattle feedlots in Kansas, biomass energy use in North Carolina, and coal use in Pennsylvania (Angel *et al.* 1998; Easterling *et al.* 1998). An important part of GCLP has been research on local perceptions of greenhouse gas emissions and the potential for taking local actions to reduce emissions (Harrington 2001; O'Connor *et al.* 2002).

Integrated and Regional Assessment

The need to understand the complex relationships among the social causes, consequences, and responses to global change has led some policy-makers and researchers to seek integrated analytical approaches to the problem. Such integrated assessment methods promote the development of models that help develop projections for global warming and subsequent policy scenarios. The concept of integration has been very broadly and varyingly defined. It can mean the end-to-end connection of a causal chain from the change and spatial pattern of fossil fuel emissions or of land use, through biophysical and socioeconomic impacts, to social consequences, including consideration of adaptation and mitigation options. Integration can also involve expanding each link of this chain to consider more source activities and emissions, other atmospheric and biotic processes, a larger number of sectors, involvement of stakeholders in the modeling effort, more spatial detail or heterogeneity, and links among global change issues such as ozone depletion and global warming. Although most integrated assessments so far have been based on formal (computer) models, integration can also include structured, cross-disciplinary discourse; judgemental or qualitative integration of data, theory, and different types of models; or structured heuristic processes such as simulations, scenario exercises, and policy exercises.

Geographers have been drawn into both quantitative and qualitative integrated assessment efforts through collaborations with integrated assessment centers such as Carnegie-Mellon University and various European groups. Geographers have also participated in the regular meetings of the Energy Modeling Forum that has provided the basis for many of the emissions scenarios for the IPCC.[9] Other geographers have expressed some skepticism about the coarse regional, sectoral, and social aggregations of the mostly global models, the degree and representation of uncertainties, and the difficulties in using economic measures as the integrator of impacts (Demeritt and Rothman 1999). Geographers have tended to emphasize the interdisciplinary and integrated analysis of global changes at the regional level with a particular focus on the integrated regional assessment of climate change and variability.

Like the more general integrated assessment, integrated regional assessment is an interdisciplinary, iterative process that often involves scientific, policy, and societal stakeholders. Its aim is to promote a better understanding of—and more informed decisions on—how locales and regions contribute to, and are affected by global change (Knight 2001; Yarnal 1998). It is a rapidly growing and important element of HDGC research (Cash and Moser 2000; Easterling 1997).

Geographers have taken a leading role in several large, integrated regional assessments of climate change impacts on river basins. These assessments include the Mackenzie Basin Impacts Study (Cohen 1997*a*, *b*; Lonergan *et al.* 1993), the Great Lakes-St Lawrence Basin Project (Mortsch and Mills 1996), and the Susquehanna River Basin Integrated Assessment (Yarnal 1998). Riebsame *et al.* (1995) studied how climate change might influence the development trajectories of five international river basins: the Nile, Zambezi, Indus, Mekong, and Uruguay. Easterling *et al.* (1993) was involved in the MINK (Missouri, Iowa, Nebraska, Kansas) integrated assessment of climate change impacts on agriculture.

A shift to regional studies within IPCC and the desire to demonstrate the local relevance of global change research funding resulted in the recent establishment of several regional climate assessment centers in the United States and in the US National Assessment of the Potential Consequences of Climate Variability and Change for the Nation. Geographers such as Greg Knight, Brent Yarnal, and Bill Easterling at Penn State's CIRA (Center for Integrated Regional Assessment) and Diana Liverman, Andrew Comrie, and Barbara Morehouse at the University of Arizona's CLIMAS (Climate Assessment for the Southwest) have leading roles in some of the established centers and are involved in planning new ones in other regions (Climate Research 2000, 2002). The US National Assessment, the first phase of which ran from 1997 to 2001, involved many geographers from the assessment centers and elsewhere. Their task was to determine the local, regional, sectoral, and national implications of climate change and climate variability within the United States in the context of other existing and potential future environmental, economic, and social stresses (National Assessment Synthesis Team 2001).[10] Most of these new regional assessment projects incorporate extensive interaction with stakeholders to understand their perceptions, information needs, and ability to adapt to climate variability. These projects also are identifying new research challenges in the comprehensive understanding of regional vulnerabilities, the communication of information and forecasts, and the limitations of current climate research in providing data of relevance to local communities.

Among other international efforts, the Human–Environment Regional Observatory (HERO) project aims at developing infrastructure to monitor and study the local and regional dimensions of global change. It blends place-based fieldwork with advanced information technology to foster collaboration across research sites. The five-year project features geographers from Clark University, Pennsylvania State University, Kansas State University, University of Arizona, and the United States Geological Survey.[11]

International Environmental Policy

The response to global change has been examined within a broader geographic literature on international environmental policy. Political geographers have paid little attention to international environmental issues with

[9] Energy Modeling Forum <http://www.stanford.edu/group/EMF/home/index.htm>.

[10] US National Assessment <http://www.usgcrp.gov/usgcrp/nacc/default.htm>, last accessed 10 November 2002.

[11] HERO <http://hero.geog.psu.edu/>, last accessed 25 February 2003.

the exceptions of Simon Dalby (1996, 2000). The most consistent analysts of international environmental policy from a geographic perspective have been Gilbert White in the United States (White 1982, 1991, 1993) and Timothy O'Riordan in Britain, especially through his regular editorials and reviews of environmental institutions in the journal *Environment* (O'Riordan 1991, 2001, 2002; O'Riordan and Jordan 2002).

Geographers have studied specific international policy issues. For example, Peter Morrisette (1991) dissected international agreements on ozone depletion to show how the parties achieved cooperation and to identify how the lessons might apply to other issues such as global warming. Danny Harvey has evaluated and criticized the policy of joint implementation, a system by which a company or country can claim credit for reducing greenhouse gas emissions in another company or country through investments in reforestation or energy efficiency (Harvey 1995; Harvey *et al.* 1997). Steve Lonergan (1999) focuses on issues of environmental security, such as the relation between environmental degradation, migration, and conflict, and also led in the development of the IHDP core project on Global Environmental Change and Human Security.[12]

Critical Perspectives

A set of challenges to global change research has emerged from social theory and science studies. Various authors argue that quantitative modeling is inappropriate or unsuccessful when it comes to predicting social trends and human futures, that global change is essentially about power relations and equity, and that international environmental policy and global change research are constructed as discourses to serve particular power groups or to spread global capitalism through economic integration (Dalby 2000; Proctor 1998; Redclift and Benton 1994; Waterstone 1993; Wescoat 1993) Some geographers believe that social problems are more urgent—poverty, homelessness, trade—or that local environmental issues are more important and salient (Peet and Watts 1996). Others believe that geographers are being coopted to serve the agenda and to maintain the funding of the earth scientists, or that it is important to remain skeptical about the meta-narratives of global change and the debates over scientific "truth" (Blaikie 1996; Buttel and Taylor 1994; Demeritt 2001).

Future Directions

Although there has been considerable progress in understanding the human dimensions of global change, there are still many unresolved questions and several important new areas for research. Of the many possible arenas for geographic contributions, some of the most significant include:

- Promoting regional approaches to global change research, including the regional dynamics of changes in land use and land cover, greenhouse gas emissions, and vulnerabilities;
- Advancing the downscaling of climate impacts and forecasts from global, to regional, to local scales;
- Understanding the interactions between global change and socioeconomic globalization, including the responses of states, local communities, ecosystems, and individuals to these changes through actions such as consumption, technological shifts, adaptation, social movements, and institutional change;
- Integrating remote sensing, geographic information systems, and other information technologies with social and biophysical data to monitor, explain, and develop policy to cope with global change;
- Anticipating and responding to emerging and new driving forces and global change issues, to sudden or surprising change, and to more certain environmental forecasts in ways that are sensitive to heterogeneous human and physical geographies.

Geographers are in a unique position to contribute to the overall human dimensions of global change research challenge because geography, by tradition, has bridged the physical and social sciences and worked between global and local scales. Their ability to translate across disciplines and scales and to work with a variety of quantitative and qualitative methods allows them to provide intellectual leadership and to develop collaborative projects within and beyond the discipline.[13] The critical stance promoted by social theorists demands self-conscious participation in policy-making and agenda setting, attention to the broader political and social

[12] IHDP Global Environmental Change and Human Security <http://www.gechs.org/>, last accessed 10 November 2002.

[13] See, for example, the large number of entries by geographers in Oxford University Press's *Encyclopedia of Global Change* (Goudie 2002).

discourse on the environment, and sensitivity to issues of power and marginalized groups.

As such, a human dimension of global change research is part of a larger scientific enterprise called sustainability science (Kates *et al.* 2001). Sustainable development and sustainability has engaged geographers, who have provided significant input into reports such as the United Nations' Agenda 21 and the United States National Research Council's vision, *Our Common Journey: A Transition Toward Sustainability* (National Research Council 1999*b*). The emergence of sustainability science focuses human dimensions researchers on the essential larger question: how can humans live with each other and with nature in ways that provide for social, economic, and ecological diversity and balance over the long term? Defining the research agenda to answer this question provides great challenges and opportunities for geographers that will occupy them for the next decade and beyond.

References

Alves, D. S., and Skole, D. L. (1996). "Characterizing Land Cover Dynamics Using Multi-temporal Imagery." *International Journal of Remote Sensing*, 17: 835–9.

Angel, D. P., and Rock, M. (2000). *Asia's Clean Revolution: Industry, Growth and the Environment*. Sheffield: Greenleaf Publishing.

Angel, D. P., Attoh, S., Kromm, D., DeHart, J., Slocum, R., and White, S. (1998). "The Drivers of Greenhouse Gas Emissions: What Do We Learn from Local Case Studies?" *Local Environment*, 3: 263–77.

Annals of the Association of American Geographers (1992). Special Issue on the Columbian Encounter, 82 /1.

Apan, A. A., and Peterson, J. A. (1998). "Probing Tropical Deforestation—the Use of GIS and Statistical Analysis of Georeferenced Data." Applied Geography, 18: 137–52.

Baker, W. L. (1995). "Long-Term Response of Disturbance Landscapes to Human Intervention and Global Change." *Landscape Ecology*, 10: 143–59.

Balling, R. C., Jr. (1992). *The Heated Debate: Greenhouse Predictions Versus Climate Reality*. San Francisco: Pacific Research Institute for Public Policy.

Balling, R. C., Jr., and Idso, S. B. (1990). "Confusing Signals in the Climatic Record." *Atmospheric Environment*, 24A: 1975–7.

Bartlein, P. J., Whitlock, C., and Shafer, S. L. (1997). "Future Climate in the Yellowstone National Park Region and its Potential Impact on Vegetation." *Conservation Biology*, 11: 782–92.

Bassett, T. J. (1988). "The Political Economy of Peasant-Herder Conflicts in the Ivory Coast." *Annals of the Association of American Geographers*, 73: 453–72.

Batterbury, S. P. J., and Bebbington, A. J. (1999). "Environmental Histories, Access to Resources and Landscape Change: An Introduction." *Land Degradation & Development*, 10/4: 279–88.

Batterbury, S. P. J., and Warren, A. (2001). "The African Sahel 25 Years after the Great Drought: Assessing Progress and Moving towards New Agendas and Approaches." *Global Environmental Change*, 11/1: 1–8.

Bensel, T. G., and Kummer, D. M. (1996). "Fuelwood Consumption and Deforestation in the Philippines." *Human Organization*, 55: 498–550.

Bernard, S., and De Koninck, R. (1996). "The Retreat of the Forest in Southeast Asia: A Cartographic Assessment." *Singapore Journal of Tropical Geography*, 17: 1–14.

Blaikie, P. M. (1996). "Post-modernism and Global Environmental Change." *Global Environmental Change*, 6: 81–5.

Blaikie, P. M., and Brookfield, H. C. (1987). *Land Degradation and Society*. London: Methuen.

Bray M., Hooke, J., and Carter, D. (1997). "Planning for Sea-Level Rise on the South Coast of England: Advising the Decision-Makers." *Transactions of the Institute of British Geographers*, 22: 13–30.

Bridge, G. (2002). "Grounding Globalization: The Prospects and Perils of Linking Economic Processes of Globalization to Environmental Outcomes." *Economic Geography*, 78/3.

Brklacich, M., Bryant, C., Veenhof, B., and Beauchesne, A. (1998). "Implications of Global Climate Change for Canadian Agriculture: A Review and Appraisal of Research from 1984 to 1997, 1: Synthesis and Research Needs," in G. Koshida and W. Avis (eds.) *Canada Country Study: Climate Impacts and Adaptation*, vii. *National Sectoral Volume*. Downsview: Environment Canada, 219–56.

Brookfield, H. B. (1995). "Population, Land Management and Environmental Change (PLEC)." *Global Environmental Change*, Special Issue, 5: 4.

Brookfield, H., and Byron, B. (1990). "Deforestation and Timber Extraction in Borneo and the Malay Peninsula—the Record Since 1965." *Global Environmental Change*, 1: 42–56.

Brookfield, H. B. Y., Potter, L., and Byron, Y. (1995). *In Place of the Forest: Environmental and Socio-Economic Transformation in Borneo and the Eastern Malay Peninsula*. Tokyo: United Nations University Press.

Brothers, T. S. (1997). "Deforestation in the Dominican Republic: A Village-Level View." *Environmental Conservation*, 24: 213–23.

Brunn S., and O'Lear S. (1999). "Research and Communication in the 'Invisible College' of the Human Dimensions of Global Change Research Community." *Global Environmental Change*, 9/4: 285–301.

Burton, I. (1997). "Vulnerability and Adaptive Response in the Context of Climate and Climate Change." *Climatic Change*, 36: 185–96.

Burton, I., Kates, R. W., and White, G. F. (1978). *The Environment as Hazard*. New York: Oxford University Press.

Buttel, F., and Taylor, P. (1994). "Environmental Sociology and Global Environmental Change: A Critical Assessment," in Redclift and Benton (1994: 228–55).

Butzer, K. W. (1992). "The Americas Before and After 1492—an Introduction to Current Geographical Research." *Annals of the Association of American Geographers*, 82: 345–68.

—— (1993). "Archaeology—No Eden in the New-World." *Nature*, 362: 15.

Cash, D. W., and Moser, S. C. (2000). "Linking Global and Local Scales: Designing Dynamic Assessment and Management Processes." *Global Environmental Change*, 10/2: 109–20.

Caviedes, C. N. (1984). "El Niño 1982–83." *Geographical Review*, 74: 267–90.

Clark, G. (1993). "Global Competition and the Environmental Performance of Resource Firms: Is the 'Race to the Bottom' Inevitable?" *International Environmental Affairs*, 5/3:147–72.

Cleveland, C. J., and Matthias, R. (1998). "Indicators of Dematerialization and the Materials Intensity of Use." *Industrial Ecology*, 2: 13–49.

Climate Research (2000). "Mid-Atlantic Regional Assessment of Climate Change Impacts," *Climate Research*, Special Issue 7, 14/3.

Climate Research (2002). "Climate Assessment for the Southwest" *Climate Research*, Special Issue 12, 21/3.

Cohen, S. J. (1986). "Climatic Change, Population Growth, and their Effects on Great Lakes Water Supplies." *Professional Geographer*, 38: 317–23.

Cohen, S. J. (1992). "Possible Impacts on Water-Resources in the Great Lakes and the Saskatchewan River Subbasin." *Canadian Geographer*, 36: 69–72.

—— (ed.) (1997*a*). *Mackenzie Basin Impact Study: Final Report*. Downsview: Canadian Climate Center, Atmospheric Environment Service, Environment Canada.

—— (1997*b*). "Scientist–Stakeholder Collaboration in Integrated Assessment of Climate Change: Lessons from a Case Study of Northwest Canada." *Environmental Modeling and Assessment*, 2: 281–93.

Cohen, S., Demeritt D., Robinson J., and Rothman J. (1998). "Climate Change and Sustainable Development: Research Highways in Need of a Merging Lane," in M. Mayer and W. Avis (eds.), *Canada Country Study: Climate Impacts and Adaptation*, viii. *National Cross-Cutting Issues*, in Downsview: Environment Canada, Environmental Adaptation Research Group, 167–203.

Cutter, S. L. (1996). "Vulnerability to Environmental Hazards." *Progress in Human Geography*, 20: 529–39.

Cutter, S. L., Mitchell J. T., and Scott M. S. (2000). "Revealing the Vulnerability of People and Places: A Case Study of Georgetown County, South Carolina." *Annals of the Association of American Geographers*, 90/4: 713–37.

Dalby, S. (1996). "Reading Rio, Writing the World: The New York Times and the Earth Summit." *Political Geography*, 15: 593–614.

—— (2000). "Environmental Security," in Miriam Lowi and Brian Shaw (eds.), *Environment and Security: Discourses and Practices*. London: Macmillan, 84–100.

Demeritt, D., and Rothman, D. S. (1999). "Counting the Costs of Climate Change: An Assessment and Critique." *Environment & Planning A*, 31: 389–408.

Demeritt, D. (2001). "The Construction of Global Warming and the Politics of Science." *Annals of the Association of American Geographers*, 91/2: 307–37.

Denevan, W. M. (2001). *Cultivated Landscapes of Native Amazonia and the Andes*. Oxford Geographical and Environmental Studies. Oxford: Oxford University Press.

Dilley, M. (2000). "Reducing Vulnerability to Climate Variability in Southern Africa: The Growing Role of Climate Information." *Climatic Change*, 45/1: 63–73.

Doolittle, W. M. (2001). *Cultivated Landscapes of Native North America*. Oxford Geographical and Environmental Studies. Oxford: Oxford University Press.

Dow, K. (1992). "Exploring Differences in Our Common Future(s): The Meaning of Vulnerability to Global Environmental Change." *Geoforum*, 23: 417–36.

Downing, T. E. (1991). "Vulnerability to Hunger in Africa—a Climate Change Perspective." *Global Environmental Change*, 1: 365–80.

Eakin, H. (1999). "Seasonal Climate Forecasting and the Relevance of Local Knowledge." *Physical Geography*, 20/6: 447–60.

Eastman, J. R., and Anyamba, A. (1996). "Prototypical Patterns of ENSO-Drought and Drought Precursors in Southern Africa." *The 13th PECORA Symposium Proceedings*. Sioux Falls, S. Dak., 20–2 August 1996.

Easterling, W. E. (1996). "Adapting North American Agriculture to Climate Change." *Agricultural and Forest Meteorology*, 63: 1–53.

—— (1997). "Why Regional Studies are Needed in the Development of Full-Scale Integrated Assessment Modeling of Global Change Processes." *Global Environmental Change*, 7: 337–56.

Easterling, W. E., Crosson, P. R., Rosenberg, N. J., McKenney, M., Katz L. A., and Lemon, K. (1993). "Agricultural Impacts of and Responses to Climate Change in the Missouri-Iowa-Nebraska-Kansas (Mink) Region." *Climatic Change*, 24: 23–61.

Easterling, W., Polsky, C., Muraco, W. A., Goodin, D., Mayfield, M., and Yarnal, B. (1998). "Changing Places, Changing Emissions: The Effect of Spatial Scale on the Measurement of Greenhouse Gas Emissions." *Local Environment*, 3: 247–62.

Geist, H. J., and Lambin, E. F. (2002). "Proximate Causes and Underlying Driving Forces of Tropical Deforestation." *BioScience*, 52: 143–50.

Gibbs, David (2002). *Local Economic Development and the Environment*. New York: Routledge.

Gibbs, D., and Healey, M. (1997). "Industrial Geography and the Environment." *Applied Geography*, 17/3: 193–201.

Giorgi, F., Mearns, L. O., Shields, C., and McDaniel L. (1998). "Regional Nested Model Simulations of Present Day and 2 × CO2 Climate over the Central Plains of the US." *Climatic Change*, 40: 457–93.

Glacken C. J. (1967). *Traces on the Rhodian Shore: Nature and Culture in Western Thought from Ancient Times to the End of the Eighteenth Century*. Berkeley and Los Angeles: University of California Press.

Golledge, R. G. (2002). "The Nature of Geographic Knowledge." *Annals of the Association of American Geographers*, 92/1: 1–14.

Goudie, A. S. (ed.) (2002). *Encyclopedia of Global Change: Environmental Change and Human Society*. Oxford: Oxford University Press.

Guyer, J., and Lambin, E. F. (1993). "Land-Use in an Urban Hinterland—Ethnography and Remote-Sensing in the Study of African Intensification." *American Anthropologist*, 95: 839–59.

Hall, C. A. S., Cleveland, C. J., and Kaufmann, R. K. (1986). *Energy and Resource Quality: The Ecology of the Economic Process*. New York: Wiley Interscience. Repr. Niwot, Colo: University of Colorado Press, 1992.

Harrington, L. M. B. (2001). "Attitudes toward Climate Change: Major Emitters in Southwestern Kansas." *Climate Research*, 16: 113–22.

Harvey, L. D. D. (1993). "A Guide to Global Warming Potentials (GWPs)." *Energy Policy*, 21: 24–34.

—— (1995). "Creating a Global Warming Implementation Regime." *Global Environmental Change*, 5: 415–32.

Harvey, L. D. D., Torrie, R., and Skinner, R. (1997). "Achieving Ecologically-Motivated Reductions of Canadian CO2 Emissions." *Energy*, 22: 705–24.

Hecht, S. B. (1985). "Environmental, Development and Politics: Capital Accumulation and the Livestock Sector in Eastern Amazonia." *World Development*, 13: 663–84.

Hecht, S. B., and Cockburn, A. (1989). *The Fate of the Forest: Developers, Destroyers and Defenders of the Amazon*. London: Verso.

Hewitson, B. C., and Crane, R. G. (1996). "Climate Downscaling: Techniques and Application." *Climate Research*, 7: 85–95.

Houghton, J. T., Ding, Y., Griggs, D. J., Noguer, M., van der Linden, P. J., Dai, X., Maskell, K., and Johnson, C. A. (eds.) (2001). *Climate Change 2001: The Scientific Basis*. Cambridge: Cambridge University Press.

Humboldt, A. von (1845–62). *Kosmos, Entwurf einer Physischen Weltbeschreibung*, 5 vols. Stuttgart: Cotta Verlag.

IGBP-IHDP (1995). *Land-Use and Land-Cover Change: Science/Research Plan*. Stockholm and Geneva: International Geosphere-Biosphere Programme, IGBP Report 35, HDP Report 7.

—— (1999). *Land-Use and Land-Cover Change, Implementation Strategy*. IGBP Report 46, IHDP Report 10. Prepared by Scientific Steering Committee and International Project Office of LUCC. Stockholm and Bonn.

IHDP (2001). *Industrial Transformation: Science Plan*, IHDP Report 12. Bonn: International Human Dimensions of Global Environmental Change Programme.

Ite, U., and Adams, W. M. (1998). "Forest Conversion, Conservation and Forestry in Cross River State, Nigeria." *Applied Geography*, 18: 301–14.

Jarosz, L. (1993). "Defining and Explaining Tropical Deforestation—Shifting Cultivation and Population-Growth in Colonial Madagascar (1896–1940)." *Economic Geography*, 69: 366–80.

Justice, C. O., and Desanker, P. V. (2001). "Africa and Global Climate Change: Critical Issues and Suggestions for Further Research and Integrated Assessment Modeling." *Climate Research*, 17: 93–103

Kalkstein, L. S., and Davis, R. E. (1989). "Weather and Human Mortality—an Evaluation of Demographic and Interregional Responses in the United States." *Annals of the Association of American Geographers*, 79: 44–64.

Kalkstein, L. S., and Greene, J. S. (1997). "An Evaluation of Climate/Mortality Relationships in Large US Cities and the Possible Impacts of a Climate Change." *Environmental Health Perspectives*, 105: 84–93.

Kalkstein, L. S., and Smoyer, K. E. (1993). "The Impact of Climate Change on Human Health—Some International Implications." *Experientia*, 49: 969–79.

Kasperson, J. X., and Kasperson, R. E. (2001). *International Workshop on Vulnerability and Global Environmental Change*. Stockholm Environment Institute. SEI Risk and Vulnerability Programme Report 2001–01.

Kasperson, J. X., Kasperson, R. E., and Turner, B. L., II (eds.) (1995). *Regions at Risk: Comparisons of Threatened Environments*. Tokyo: United Nations University.

Kates, R. W., Ausubel, J., and Berberian, M. (eds.) (1985). *Climate Impact Assessment: Studies of the Interaction of Climate and Society*. New York: John Wiley.

Kates, R. W., Clark, W. C., Corell, R., Hall, J. M., Jaeger, C. C., Lowe, I., McCarthy, J. J., Schellnhuber, H. J., Bolin, B., Dickson, N., Faucheux, M., Gallopin, S., Gilberto, C., Gruebler, A., Huntley, B., Jäger, J., Jodha, N. S., Kasperson, R. E., Mabogunje, A., Matson, P., Mooney, H., Moore, B., III, O'Riordan, T., and Svedin, U. (2001). "Sustainability Science." *Science*, 292: 641–2.

Kates, R. W., and Torrie, R. (1998). "Global Change in Local Places." *Environment*, 40/2: 5, 39–41.

Kates, R., Mayfield, M., Torrie, R., and Witcher, B. (1998). "Methods for Estimating Greenhouse Gases from Local Places." *Local Environment*, 3: 279–97.

Klooster, D., and Masera, O. (2000). "Community Forest Management in Mexico: Carbon Mitigation and Biodiversity Conservation through Rural Development." *Global Environmental Change*, 10/4: 259–72.

Knight C. G. (2001). "Regional Assessment," in *Encyclopedia of Global Change*. New York: Oxford University Press.

Kummer, D. M. (1991). *Deforestation in the Postwar Philippines*. Geography Research Paper. 234. Chicago: University of Chicago.

Lambin, E. F. (1997). "Modelling and Monitoring Land-Cover Change Processes in Tropical Regions." *Progress in Physical Geography*, 21: 375–93.

Lambin, E., and Ehrlich, D. (1997). "Land-Cover Changes in Sub-Saharan Africa (1982–1991): Application of a Change Index Based on Remotely Sensed Surface Temperature and Vegetation Indices at a Continental Scale." *Remote Sensing of Environment*, 61: 181–200.

Lambin, E. F., B. L., Turner, II, Geist, H., Agbola, S., Angelsen, A., Bruce, J. W., Coomes, O., Dirzo, R., Fischer, G., Folke, C., George, P. S., Homewood, K., Imbernon, J., Leemans, R., Li, X., Moran, E. F., Mortimore, M., Ramakrishnan, P. S., Richards, J. F., Skånes, H., Steffen, W., Stone, G. D., Svedin, U., Veldkamp, T., Vogel, C., and Xu, J. (2001). "The Causes of Land-Use and Land-Cover Change—Moving Beyond the Myths." *Global Environmental Change: Human and Policy Dimensions*, 11: 5–13.

Liverman, D. M. (1994). "Vulnerability to Global Environmental Change," in S. Cutter (ed.), *Environmental Risks and Hazards*. New York: Prentice Hall, 326–42.

—— (1999). "Geography and the Global Environment." *Annals of the Association of American Geographers*, 89/1: 107–24.

Liverman, D. M., and O'Brien, K. (1991). "Global Warming and Climate Change in Mexico," *Global Environmental Change*, 1/4: 351–64.

Liverman, D. M., Moran, E. F., Rindfuss, R. R., and Stern, P. C. (eds.) (1998). *People and Pixels: Linking Remote Sensing and Social Science*. Washington: National Academy Press.

Lonergan, S. C. (ed.) (1999). *Environmental Change, Adaptation and Security*. Dordecht: Kluwer.

Lonergan, S., Difrancesco, R., and Woo, M. K. (1993). "Climate Change and Transportation in Northern Canada: an Integrated Impact Assessment." *Climatic Change*, 24: 331–51.

McCarthy, J. J., Canziani, O. F., Leary, N. A., Dokken, D. J., and White, K. S. (eds.) (2001). *Climate Change 2001: Impacts, Adaptation, and Vulnerability*. Cambridge: Cambridge University Press.

Malanson, G. P. (ed.) (1989). *Natural Areas Facing Climate Change*. The Hague: SPB Academic.

Markgraf, V., and Diaz, H. F. (1992). *El Niño: Historical and Paleoclimatic Aspects of the Southern Oscillation*. New York: Cambridge University Press.

—— (2000). *El Niño and the Southern Oscillation: Multiscale Variability and Global and Regional Impacts*. New York: Cambridge University Press.

Marsh, G. P. (1864). *Man and Nature: or the Earth as Modified by Human Action*. New York: Scribners.

Mearns, L. O., Rosenzweig, C., and Goldberg, R. (1996). "The Effect of Changes in Daily and Interannual Climatic Variability on CERES-Wheat: A Sensitivity Study." *Climatic Change*, 32: 257–92.

—— (1997). "Mean and Variance Change in Climate Scenarios: Methods, Agricultural Applications, and Measures of Uncertainty." *Climatic Change*, 35: 367–96.

Metz, J. J. (1994). "Forest Product Use at an Upper Elevation Village in Nepal." *Environmental Management*, 18: 371–90.

Metz, B., Davidson, O., Swart, R., and Pan, J. (eds.) (2001). *Climate Change 2001: Mitigation*. Cambridge: Cambridge University Press.

Meyer, W. B., and Turner, B. L., II. (eds.) (1994). *Changes in Land Use and Land Cover: A Global Perspective*. New York: Cambridge University Press.

Morehouse, B. (2000). "Climate Impacts on Urban Water Resources in the Southwest: The Importance of Context." *Journal of the American Water Resources Association*, 36/2: 265–327.

Morrisette, P. M. (1991). "The Montreal Protocol: Lessons for Formulating Policies for Global Warming." *Policy Studies Journal*, 19: 152–61.

Mortsch, L., and Mills, B. (eds.) (1996). *Great Lakes-St. Lawrence Basin Project Progress Report 1: Adapting to the Impacts of Climate Change and Variability*. Downsview: Canadian Climate Center, Atmospheric Environment Service, Environment Canada.

National Assessment Synthesis Team (2001). *Climate Change Impacts on the United States: The Potential Consequences of Climate Variability and Change*. Report for the US Global Change Research Program. Cambridge: Cambridge University Press.

National Research Council (1999*a*). *Human Dimensions of Global Environmental Change: Research Pathways for the Next Decade*. Washington, DC: National Academy Press.

—— (1999*b*). *Our Common Journey: A Transition to Sustainability*. Washington, DC: National Academy Press.

—— (2001). *Grand Challenges in the Environmental Sciences*. Committee on the Grand Challenges in the Environmental Sciences. Washington, DC: National Academy Press.

Neff, R., Chang, H., Knight, C. G., Najjar, R. G., Yarnal, B. and Walker H. A. (2000). "Impacts of Climate Variation and Change on Mid-Atlantic Region Hydrology and Water Resources." Climate Research, 11: 207–18.

Nicholls, R. J., and Leatherman, S. P. (1996). "Adapting to Sea-Level Rise: Relative Sea-Level Trends to 2100 for the United States." *Coastal Management*, 24: 301–24.

Nkemdirim, L. C. (ed.) (2001). *El Niño-Southern Oscillation and Its Global Impacts*. Bonn: Commission on Climatology, International Geographical Union.

O'Brien, K. (1998). *Sacrificing the Forest: Environmental and Social Struggles in Chiapas*. Boulder: Westview Press.

O'Brien, K., and Leichenko, R. (2000). "Double Exposure: Assessing the Impacts of Climate Change within the Context of Economic Globalization." *Global Environmental Change*, 10: 221–32.

O'Connor, R. E., Bord, R. J., Yarnal, B., and Wiefek, N. (2002). "Who Wants to Reduce Greenhouse Gas Emissions?" *Social Science Quarterly*, 83:1–17.

O'Connor, R. E., Yarnal, B., Neff, R., Bord, R., Wiefek, N., Reenock, C., Shudak, R., Jocoy, C. L., Pascale, P., and Knight, C. G. (1999). "Community Water Systems and Climate Variation and Change: Current Sensitivity and Planning in Pennsylvania's Susquehanna River Basin." *Journal of the American Water Resources Association*, 35:1411–19.

O'Hara, S., Street-Perrott, F. A., and Burt, T. P. (1993). "Accelerated Soil-Erosion around a Mexican Highland Lake Caused by Pre-Hispanic Agriculture." *Nature*, 362: 48–51.

O'Riordan, T. (1991). "Stability and Transformation in Environmental Government." *Political Quarterly*, 62: 167–78.

—— (ed.) (2001). *Globalism, Localism, and Identity: Fresh Perspectives on the Transition to Sustainability*. London: Earthscan Publications.

—— (2002). "Green Politics and Sustainability Politics." *Environment*, 44/3: 1.

O'Riordan, T., and Jordan, A. (2002). "Institutions for Global Environment Change." *Global Environmental Change*, 21/2:143.

Parry, M. L. (1990). *Climate Change and World Agriculture*. London: Earthscan.

Parry, M. L., and Carter, T. R. (1998). *Climate Impact and Adaptation Assessment: A Guide to the IPCC Approach*. London: Earthscan.

Parry, M. L., Carter, T. R., and Konijn, N. T. (eds.) (1988). *The Impact of Climatic Variations on Agriculture*. Dordrecht: Kluwer, i and ii.

Peet, R., and Watts, M. (eds.) (1996). *Liberation Ecologies: Environment, Development, Social Movements*. New York: Routledge.

Peters, R. L., and Lovejoy, T. E. (eds.) (1992). *Global Warming and Biological Diversity*. New Haven: Yale University Press.

Proctor, J. D. (1998). "The Meaning of Global Environmental Change—Rethinking Culture in Human Dimensions Research." *Global Environmental Change*, 8: 227–48.

Raskin, P., Banuri T., Gallopín G., Gutman P., Hammond A., Kates R. W., and Swart R. (2002). *Great Transition: The Promise and Lure of the Times Ahead*. Boston: Stockholm Environment Institute-Tellus Institute.

Redclift, M., and Benton, T. (eds.) (1994). *Social Theory and the Global Environment*. New York: Routledge.

Redclift, M., and Sage, C. (1998). "Global Environmental Change and Global Inequality—North/South Perspectives." *International Sociology*, 13: 499–516.

Riebsame W. E., Strzepek, K. M., Wescoat, J., Perritt, R., Gaile, G. L., Jacobs, J., Leichenko, R., Magadza, C., Phien, H., Urbiztondo, B. J., Restrepo, P., Rose, W. R., Saleh, M., Ti, L. H., Tucci, C., and Yates D. (1995). "Complex River Basins," in K. M. Strzepek and J. B. Smith (eds.), *As Climate Changes: International Impacts and Implications*. Cambridge: Cambridge University Press, 57–91.

Riebsame, W. E., Meyer, W. B., and Turner, B. L., II (1994). "Modeling Land-Use and Cover as Part of Global Environmental Change." *Climatic Change*, 28: 45–64.

Rignot, E., Salas, W. A., and Skole, D. L. (1997). "Mapping Deforestation and Secondary Growth in Rondonia, Brazil, Using Imaging Radar and Thematic Mapper Data." *Remote Sensing of Environment*, 59: 167–79.

Rosenzweig, C., and Parry, M. L. (1994). "Potential Impact of Climate Change on World Food Supply." *Nature*, 367: 133–8.

Schreier, H. S., Brown, S., Schmidt, M. G., Shah, P. B., Shrestha, B., Nakarmi, G., and Wymann, S. 1994. "Gaining Forests but Losing Ground—a GIS Evaluation in a Himalayan Watershed." *Environmental Management*, 18: 139–50.

Seto, K. C., Kaufmann, R. K., and Woodcock C. E. (2000). "Agricultural Land Conversion in Southern China." *Nature*, 406: 121.

Sierra, R., and Stallings, J. (1998). "The Dynamics and Social Organization of Tropical Deforestation in Northwest Ecuador 1983–1995." *Human Ecology*, 26: 135–61.

Skole, D., and Tucker, C. (1993). "Tropical Deforestation and Habitat Fragmentation in the Amazon—Satellite Data from 1978 to 1988." *Science*, 260: 1905–10.

Skole, D., Chomentowski, W. H., Salas, W. A., and Nobre, A. D. (1994). "Physical and Human Dimensions of Deforestation in Amazonia." *Bioscience*, 44: 314–22.

Sluyter, A. (1996). "The Ecological Origins and Consequences of Cattle Ranching in Sixteenth-Century New Spain." *Geographical Review*, 86: 161–77.

Smil, V. (1985). *Carbon Nitrogen Sulfur: Human Interference in Grand Biospheric Cycles*. New York: Plenum Press.

—— (1994). *Energy in World History*. Boulder: Westview Press.

Smit, B., McNabb, D., and Smithers, J. (1996). "Agricultural Adaptation to Climatic Variation." *Climatic Change*, 33: 7–29.

Smith, N. J. H. (1982). *Rainforest Corridors: The Transamazon Colonization Scheme*. Berkeley: University of California Press.

—— (1999). *The Amazon River Forest: A Natural History of Plants, Animals, and People*. New York: Oxford University Press.

Smithers, J., and Smit, B. (1997). "Human Adaptation to Climatic Variability and change." *Global Environmental Change*, 7: 129–46.

Stern, D., and Common, M. S. (2001). "Is There an Environmental Kuznets Curve for Sulfur?" *Journal of Environmental Economics and Management*, 41: 162–78.

Stern, P. C., and Easterling, W. E. (eds.) (1999). *Making Climate Forecasts Matter*. Washington, DC: National Academy Press.

Stern, P. C., Young, O. R., and Druckman, D. (eds.) (1992). *Global Environmental Change: Understanding the Human Dimensions*. Washington, DC: National Academy Press.

Sussman, R. G., Green, G. M., and Sussmen, L. K. (1994). "Satellite Imagery, Human Ecology, Anthropology, and Deforestation in Madagascar." *Human Ecology*, 22: 333–54.

Tegart, W. J. M., Sheldon, G., and Griffiths, D. C. (eds.) (1990). *Climate Change: The IPCC Impacts Assessments*. Intergovernmental Panel On Climate Change, Working Group 2. Canberra: Australian Government Publication Service.

Thomas, W. L., Sauer, C. O., Bates, M., and Mumford, L. (1956). *Man's Role in Changing the Face of the Earth*. Chicago: University of Chicago Press.

Turner, B. L., II (1997). "The Sustainability Principle in Global Agendas: Implications for Understanding Land-Use/Cover Change." *Geographical Journal*, 163/2: 133–40.

Turner, B. L., II, (2001). "Land-Use and Land-Cover Change: Advances in 1.5 Decades of Sustained International Research." *GAIA-Ecological Perspectives in Science, Humanities, and Economics*, 10/4: 269–72.

—— (2002). "Contested Identities: Human–Environment Geography and Disciplinary Implications in a Restructuring Academy." *Annals of the Association of American Geographers*, 92/1: 52–74.

Turner, B. L., II, and Butzer, K. W. (1992). "The Columbian Encounter and Land-Use Change." *Environment*, 43/8: 16–20, 37–44.

Turner, B. L., II, Gómez Sal, A., González Bernáldez, F., and Di Castri, F. (eds.) (1995*a*). *Global Land-Use Change: A Perspective from the Columbian Encounter*. Madrid: Consejo Superior de Investigaciones Científicas.

Turner, B. L., II, Clark, W. C., Kates, R. W., Richards, J. F., Mathews, J. T., and Meyer, W. B. (eds.) (1990). *The Earth as Transformed by Human Action: Global and Regional Changes in the Biosphere over the Past 300 Years*. Cambridge: Cambridge University Press.

Turner, B. L., II, Skole, D., Sanderson, S., Fischer, G., Fresco, L., and Leemans, R. (1995*b*). *Land-Use and Land-Cover Change: Science/Research Plan*. Stockholm and Geneva: International Geosphere-Biosphere Programme, IGBP Report 35, HDP Report 7.

Veldkamp, A., and Lambin, E. (2001). "Predicting Land-Use Change." *Agriculture, Ecosystems & Environment*. Special Issue, 85/1-3.

Vellinga, P., and Leatherman, S. P. (1989). "Sea-Level Rise, Consequences and Policies." *Climatic Change*, 15: 175–89.

Vogel, C. (1995). "People and Drought in South Africa: Reaction and Mitigation," in T. Binns (ed.), *People and the Environment in Africa*. London: Wiley, 249–56.

—— (1998). *Vulnerability and global environmental change*, LUCC Newsletter 3 (March): 15–19.

Warrick, R. A., and Farmer, G. (1990). "The Greenhouse Effect, Climatic Change and Rising Sea Level—Implications for Development." *Transactions of the Institute of British Geographers*, 15: 5–20.

Warrick, R. A., Barrow, E. M., and Wigley, T. M. L. (eds.) (1993). *Climate and Sea Level Change: Observations, Projections, and Implications*. Cambridge: Cambridge University Press.

Waterstone, M. (1985). "The Equity Aspects of Carbon Dioxide-Induced Climate Change." *Geoforum*, 16: 301–6.

—— (1993). "Adrift on a Sea of Platitudes—Why We Will Not Resolve the Greenhouse Issue." *Environmental Management*, 17: 141–52.

Watson R. T., Zinyowera, M. C., and Moss, R. T. (eds.) (1996). *Climate Change 1995: Impacts, Adaptations and Mitigation of Climate Change*. New York: Cambridge University Press.

Watson, R. T., Zinyowera, M. C., Moss, R. T., and Dokken, D. J. (eds.) (1998). *The Regional Impacts of Climate Change: An Assessment of Vulnerability*. New York: Cambridge University Press.

Watts, M. J., and Bohle, H. G. (1993). "The Space of Vulnerability: The Causal Structure of Hunger." *Progress in Human Geography*, 17: 43–67.

Wescoat, J. L., Jr. (1993). "Resource Management: UNCED, GATT, and Global Change." *Progress in Human Geography*, 17: 232–40.

White, G. F. (ed.) (1974). *Natural Hazards: Local, National, Global*. New York: Oxford University Press.

— (1982). "Ten Years after Stockholm." *Science*, 216/4576: 569.

— (1991). "UNCED and Uncertainty." *Environment*, 33/5: p. i.

— (1993). "The Global Environment: What Can We Do?" *GeoJournal*, 30: 367–78.

Wilbanks, T. J., and Kates R. W. (1999). "Global Change in Local Places." *Climatic Change*, 43/3: 601–28.

Yarnal, B. (1994). "Agricultural Decollectivization and Vulnerability to Environmental Change—a Bulgarian Case Study." *Global Environmental Change*, 4: 229–43.

— (1997). "Socioeconomic Transition and Greenhouse-Gas Emissions from Bulgaria." *Geographical Review*, 86: 42–58.

— (1998). "Integrated Regional Assessment and Climate Change Impacts in River Basins." *Climate Research*, 11: 65–74.

Yarnal, B., Kalkstein, L. S., Scheraga, J. D. (eds.) (2000). "Mid-Atlantic Regional Assessment of Climate Change Impacts." *Climate Research*, Special Issue 7, 14/3.

CHAPTER 19

Water Resources

James L. Wescoat, Jr.

Perspectives

Water resources geography expanded its spatial, regional, and intellectual horizons during the 1990s. Tobin *et al.* (1989) reviewed earlier US geographers' contributions to the hydrologic sciences, water management, water quality, law, and hazards; and they identified three emerging topics: (1) theory development and model formulation; (2) applied problem-solving and policy recommendations; and (3) international water problems. This chapter assesses progress along those and other fronts, beginning with historical and disciplinary perspectives.

Charting the progress of a field requires a sense of its history, and Platt's (1993) review of geographic contributions to water resource administration in the US offers a useful perspective on policy-related research, beginning with George Perkins Marsh and John Wesley Powell. Doolittle (2000) reaches back to Native American antecedents in water resource management in North America (cf. chapters in this volume on cultural ecology, historical geography, and Native American geography). Carney (1998) sheds light on African influences on rice cultivation in the southeastern US. Research on European antecedents ranges from seventeenth-century "hydrologic" theories in England (Tuan 1968) to hydraulic engineering at the École des Ponts et Chausées in France, water courts in Spain, and more distant Muslim and Asian contacts (e.g. Beach and Luzzader-Beach 2000; Bonine 1996; Butzer 1994; Lightfoot 1997; Swyngedouw 1999; Wescoat 2000).

In the field of water law and institutions, Templer (1997) has linked recent geographic work on Western water laws with earlier research in political geography. A historical geographic study of water rights transfers from irrigated ranches in the South Platte River headwaters to Denver, Colorado, has shed new light on how urban economic and political power employ and reshape water law (Kindquist 1996). The battle between Owen's Valley and Los Angeles continues to stimulate historical geographic research on relations among facts, laws, and their social meanings (Sauder 1994). Although a geographic perspective of international water laws has yet to be written, databases on transboundary conflicts and agreements shed light on the evolution of international water law (Wolf 1997, 1999*a*, *b*; and <http://www.transboundarywaters.orst.edu>, last accessed 10 February 2003). Historical or contemporary, the pragmatic spirit of water resources geography remains strong (Wescoat 1992). It combines a commitment to addressing social and environmental issues with innovative and imaginative approaches.

Writing at the end of the 1980s, however, Platt (1993: 48) argued that, "geographical research on water issues today is less holistic, less prescriptive and less visionary" than the river basin studies of the 1930s and 1950s. One geographer who contributed to those early river basin studies and who has consistently asked what difference research makes for the well-being of people and places—Gilbert F. White (1997)—has reflected back on four key scientific debates ("watersheds") of that period: (1) soil taxonomy versus landscape classification; (2) upstream versus downstream approaches in watershed management; (3) separation of water and sediment pollution from a more comprehensive view of water quality; and (4) emphasis on individual versus community-based resource management. He wonders how water resources management might have been developed if researchers and organizations had concentrated on landscape ecology, upstream land-use, water quality, and community-based approaches a half century ago.

G. F. White (1998) has also offered a personal, half-century, perspective on the quest for "integrated water development." He drew three major lessons from United Nations water programs: (1) the difficulties water managers face in examining the full range of choices open to them; (2) the lack of consistent criteria for evaluating proposed or completed projects; and (3) a continuing failure to analyse in detail, through post audits, the actual consequences of water decisions for the environments and peoples affected.

Interestingly, the genre of water resources review articles seems to have originated in mid-twentieth century papers by Ackerman *et al.* (1974). Prior to that time, reviews focused on broader fields such as natural resources or more specific topics (e.g. geographers' contributions to regional river basin planning). Christopher Lant (1998 and pers. comm.) has reflected upon trends in water resources research and the coherence of a field as wide-ranging as water resource geography. The American Geographical Society *Research Catalogue* indicates how far-flung water resources geography was through the 1960s (Table 19.1).

In1974, the AGS *Research Catalogue, First Supplement* added category number 5310012 on "Water Supply," which included 35 entries on the US (14), Japan (7), and a sprinkling of other places. Although the number of water resources articles in *Current Geographical Publications* increased, classification problems continued (Table 19.2).

This broad distribution of topics reflects stimulating and diverse connections between water resources geography and related fields of geography and other disciplines. Coherence might have been more difficult were it not for the development of key integrative concepts (e.g. water budget analysis, watershed management, and river basin planning); integrative institutions (e.g. AAG Water Resource Specialty Group, American Water Resources Association, and University Council on Water Resources Research); and integrative intellectuals (e.g. Ackerman, Barrows, Mather, and White).

There have been few biographical studies of these early pioneers of water resource geography. Exceptions include obituaries, Mather's (1997) biography of Warren Thornthwaite, Reuss's (1993) oral history with White, and the *Geographers on Film* series. The history of geographic thought about water resources, including debates over Wittfogel's "hydraulic hypothesis" for the origin of complex social organization and despotism, has barely been scratched (Cosgrove and Petts 1990; Tuan 1964; Wescoat 2000).

The greater part of water resource geography thrives on "live" contemporary water problems. Recent floods, wetlands losses, groundwater depletion, and policy debates capture the imagination and effort of most water resource geographers. They seek to understand and help rectify these important water problems. This presentist emphasis reinforces the need for regular reviews of the

Table 19.1 *American Geographical Society topics relevant to water resources (1964)*

Section	Topic
4	**Physical Geography**
401	Distribution of land and water
43	*Geomorphology*
433	Work of flowing water
434	Work of subterranean water
435	Work of (or in) standing water
436	Ice sheets, glaciers, ground ice and their work
44	*The Waters of the Land. Hydrography*
441	Rivers—potamology
4415	Floods (including flood control and water pollution)
4419	Animal and vegetal life in rivers
442	Lakes; limnology
443	Subterranean waters, including ground water
444	Springs, hot springs, geysers

Table 19.1 *continued*

Section	Topic
445	Water pollution
45	*Oceanography*
46	*Meteorology and Climatology*
461	Meteorology
4614	Atmospheric circulation
46144	Secondary circulation and its effects (cyclones, thunderstorms, tornadoes, etc.)
4615	Moisture in the atmosphere
46159	Droughts
462	Changes of Climate
5	**Human Geography**
501	*Relation of Men and the Geographical Environment*
5011	Adjustment of man to the geographical environment
50115	Geography of natural calamities
5012	Influence of man on the geographic environment
50125	"New" or "made" land
51	*Physiological geography*
52	*Geography of Population*
52513	Rural water supply
5252	City geography
52523	Urban water supply
52529	Port and harbor facilities
53	*Economic Geography*
531	Natural resources and industries
531001	Conservation of natural resources
53101	Land utilization
531012	Comprehensive regional surveys; regional planning
5312	Agriculture
53122	Agricultural methods
531223	Irrigation
531224	Drainage
5313	Forests and forestry
53137	Functional value of the forest cover (as protection against soil erosion, excessive run-off, etc.)
5315	Products of oceans, rivers, and lakes
53152	Conservation of fishing resources
53155	Fisheries
53156	Whaling
53157	Minerals from sea water
532	Geography from manufacturing industries: industrial geography
5321	Power in general
53212	Water power
53215	Tidal power
533	Transportation and communication
5332	Water transportation
53321	Rivers and lakes
53322	Canals
53323	Marine transportation

Note: There were no water entries under Political, Social and Cultural, or Military Geography in 1964.

Table 19.2 *Current geographical publications—catalogue topics*

Hydrology
Water run-off
Floods
Flood control
Estuaries
River biology
Lake water movement
Groundwater
Hot springs
Water pollution
Water resources
Economic Geography (57 categories, including)
Water supply
Reservoirs
Irrigation
Fish
Ships
River transportation
Canals
Waste disposal

Source: <http://leardo.lib.uwm.edu/cgp>, last accessed 10 February 2003.

field. If research of the 1990s has dealt largely with problems of the 1990s, there is a continuing need to ask how the field has evolved in relation to its changing academic, social, and environmental contexts.

Scope and Method of Review

This chapter surveys US contributions to water resource geography with an emphasis on the 1990s. It asks how geographers have responded to the challenges presented in earlier reviews. It gives special but not exclusive emphasis to research by members of the Association of American Geographers Water Resources Specialty Group including recent Ph.Ds. Historically close relations between water resource geography, climatology, hazards research, and especially hydrology are briefly updated. Previous reviews gave less attention to linkages with historical, political, economic, urban, and cultural geography, so they receive closer attention here.

The chapter extends beyond academic institutions to identify centers of water resources geography in government (e.g. US Army Corps of Engineers), research organizations (e.g. the National Research Council), and consulting firms. Geographers' contributions in these organizations are less well catalogued and thus less widely employed in education than academic publications. Water resources education flourished in the 1990s (Halvorson and Wescoat 2001; Rense 1996; Waterstone 1991; Woltemade 1997). The lag in textbook publishing following Mather's *Water Resources: Distribution, Use, and Management* (1984) ended with S. A. Thompson's (1998, 1999) volumes on water resources management and hydrology.

Although broad in analytic scope, few texts extend far beyond North America (Waterstone 1991). This volume, being titled *Geography in America*, also restricts the scope of international water resource geography relevant for students and scholars. Canadian universities such as Waterloo and Wilfrid Laurier have produced a significant number of the water resource geography Ph.Ds in North America. Geographers contribute to renowned water research centers in every region of the world, which calls for a broader review than is attempted here. Pioneering contributions on international water problems range from A. Michel (1967) on partition and negotiation in the Indus River basin; White (1963) on social aspects of international water development in the Mekong basin; and White *et al.* (1972) on domestic water supply in East African countries. In a rare longitudinal study, J. Thompson *et al.* (2001) resurveyed water supply behaviors in the same villages studied thirty years earlier in *Drawers of Water*. In a rare comparative study, Cosgrove *et al.* (1995) critique modern water engineering and landscape change in Europe and the Mediterranean. Inspired by these examples, this chapter adopts a broad definition of "Geography in America." It includes work by geographers who have studied in the US at some point, or have written in US journals or about US water resources.

To identify these works, the Water Resources Specialty Group called for papers and comments via its newsletter, electronic mail, web page, and annual meetings. Electronic library searches were conducted with Water Resources Abstracts, GEOBASE, Worldcat, Article1st, Dissertation Abstracts, and Uncover, using the key-words "water" and "geograph*." Snowball sampling of references in the above works identified additional publications. These methods helped locate works by geographers who do not belong to the Water Resources Specialty Group. These searches yielded a 36-page working bibliography, only a small portion of which could be reviewed here. Much more effort is needed to compile and review research in public water agencies and private companies (Marston, pers comm. 1999).

Previous reviews have employed six main approaches: (1) water sectors (e.g. irrigation, urban water, flood hazards, etc.); (2) water systems (e.g. water supply, demand, and distribution, and wastewater); (3) hydrologic processes (e.g. watershed, riparian, groundwater); (4) major research trends (e.g. from river basin surveys to behavioral approaches and institutional analysis); (5) water problems (e.g. scarcity, pollution, and hazards); or (6) a mixture of water sectors, topics, problems, and trends. This chapter cuts across these approaches by focusing on geographic scales of inquiry. It asks: which of the cells in Table 19.3 have, and have not, received attention?

As with other approaches, geographic scales involve theoretical, methodological, and substantive challenges. But just as physical geographers study the scale and "scaling" of precipitation and other hydrologic processes (Hirschboeck 1999), human geographers struggle with scales of water resources management, inquiry, and understanding (Swyngedouw 2000). S. Michel (2000) links struggles over geographic scale with those over the "political scope" of transborder wastewater management. Indian water rights claims on reservations, which have a semi-sovereign political status, involve adjudication problems between state and federal courts (Berry 1998). Votteler (1998) describes an ongoing "collision" between federal environmental law, state groundwater law, and private water rights over the Edwards Aquifer in Texas.

Although water issues involve multiple contested scales of environmental and social interaction, scale provides a useful way to review major emphases and omissions in water resource geography. It facilitates joint consideration of physical, human, and environment–society relations in water resource geography. Although contested, water management does differ across scales of community-based water organizations, state-administered water rights, interstate agreements, river basin organizations, national policies, and international treaties.

This review proceeds from research on household to global scales. Most water resource geographers work at multiple scales between these two ends of the spectrum, e.g. at watershed, river basin, national, and international scales. Focusing on each scale, however, some geographic topics and regions are well covered and others neglected, which raises interesting questions about priorities in water resource geography. Why has there been more geographic research on water problems in Colorado than California? More geographic research on the US–Mexico border than the US–Canada border? More US research on Central Asia than Russia? Why are there so few systematically comparative and multi-scalar studies? After surveying work at different scales, the concluding section of the chapter returns to the implications of these patterns for future research.

Table 19.3 *Approaches to water resources research*

Scales of Inquiry: local, regional, global

Problems
- Scarcity
- Pollution
- Floods
- Droughts

Sectors
- Irrigation
- Hydropower
- Municipal
- Industrial
- Flood Hazards

Systems
- Supply
- Demand
- Distribution
- Waste Disposal

Hydrology
- Hydroclimatology
- Fluvial geomorphology
- Watershed processes
- Riparian/aquatic ecosystems
- Hydrogeology

Research Trends
- River basin surveys
- Environmental perception and behavior
- Institutions (law, policy, organizations)
- Culture, power and governance

Geographic Regions
- North America
- Latin America
- S and SE Asia
- Middle East
- Africa
- Oceania
- Europe/Russia

Biomes
- Polar
- Taiga/tundra
- Mid-latitude temperate
- Steppe/prairie
- Desert
- Humid tropics
- Island
- Coastal
- Montane

Individuals and Households

Small-scale research in water resource geography ranges from individuals to households and points to sites. Hydrologists employ point data to construct water budgets and hydrographs (Keim 1997; Rense 1995). Geomorphologists employ plots and transects to develop process models of erosion, sedimentation, and run-off (Harden 1992). Geographers increasingly seek to link in situ measurements with remote sensing and GIS methods to model and visualize larger-scale atmospheric and hydrologic processes, including storm systems, snowmelt, run-off, and non-point source pollution (e.g. Williams *et al.* 1996).

During the second half of the twentieth century many water resource geographers focused on individual perceptions, attitudes, and values toward water supply, floods, droughts, water quality, recreation, and aesthetic experience. Some of this research employed behavioral scientific methods and pragmatic or utilitarian philosophies aimed at improving public policy, while others strove for a phenomenological understanding of water experience. Emphasis on individual experience and methodologies was criticized for neglecting broader structural processes and inequalities, which led to new lines of multi-scalar, structural, and gender analysis (Mustafa 1998; Ollenburger and Tobin 1997; Paul 1998; cf. also natural hazards chapter).

But the 1990s also witnessed new lines of local water inquiry at the household and site scales. Household research underscored the importance of sanitation behavior *vis-à-vis* domestic water supply in Egypt (el-Katsha and White 1989). Participant observation and ethnographic methods have extended research on women's roles in managing domestic water, sanitation, and health (Halvorson 2000). Again, when linked with larger community and regional sampling, these household studies have shed new light on regional water-related disease hazards and water quality policies from Puerto Rico to East Africa and Pakistan (Arbona 1992; Mitchell 1998; J. Thompson *et al.* 2001).

Less research has focused on waters at the site scale (e.g. springs, pools, fords, fountains, and irrigated fields), notwithstanding their importance for environmental and social well-being, and the attention they receive in landscape research, environmental design, and urban water conservation. For example, Tuan (1984) associates the "play of fountains" with other forms of human affection for, and domination over, nature. Others strive to understand changing patterns of irrigated suburban landscapes in arid environments (Baumann *et al.* 1998; Planning and Management Consultants 1993).

Points, plots, individuals, households, and sites constitute relatively small units of water resource observation and experience. They provide proxy data for larger phenomena, and when linked with larger scales of inquiry, offer insights into some of the most immediately practical, aesthetic, and spiritual dimensions of water use.

Communities and Watersheds

Water resource geography has continuing vitality at the community and watershed scales. Studies of community water management have examined how water knowledge of county commissioners, ranching communities, and local officials affects local decision-making in regions of economic growth and resource policy tension (Berry *et al.* 1996; Root *et al.* 1998). Kromm and White (1992) have produced a major corpus of work on irrigation management and groundwater conservation in the High Plains of Kansas and adjoining states (S. E. White and Kromm 1995). Templer (1996) has pioneered research on conjunctive water management in the southern High Plains of Texas.

Outside the Great Plains region (and some individual studies), research on community-based irrigation management has focused on developing countries. Geographers associated with the International Water Management Institute (IWMI, formerly IIMI) have conducted detailed field-based research on irrigation efficiency and equity in Pakistan, India, and Sri Lanka (e.g. Vander Velde and Svendsen 1994; Vander Velde and Johnson 1992). In the same region, Allan (1986) has documented community water management in the upper Indus while Rahman (1993) has written about irrigation development in the lower Indus basin of Pakistan and organized comparative studies. Sukhwal (1993) has examined agricultural and environmental effects of the Indira Gandhi Canal near the border in Rajasthan, India. The Andean region has stimulated research on irrigation communities using cultural and political ecology approaches (e.g. Denevan 2001; Perrault *et al.* 1998). These studies have provided a partial foundation for comparative irrigation research in different cultural ecological contexts.

Research on community vulnerability and response to natural hazards, especially floods, has continuing vitality (cf. Natural Hazards chapter for a more complete treatment). Montz and Tobin (1996; Tobin and Montz 1997) have probed the environmental and social effects of local floods in many states, linking flood effects with key federal, state, and local policy variables. In a rare long-term *ex post* evaluation of flash flood hazards, Gruntfest (1997) organized a conference on the 1976 Big Thompson flood. Clark (1998) has applied concepts of vulnerability to community-scale coastal storm hazards in New England.

The 1990s witnessed increasing attention to water quality problems, especially groundwater contamination from agricultural and industrial chemicals. Rajagopal and Tobin (1992) have developed a sustained program of research on groundwater contamination and remediation alternatives in Iowa (Tobin and Rajagopal 1993). Giambelluca *et al.* (1996) focus on groundwater contamination in the vulnerable island environment of Hawaii. Luzzader-Beach (1997) traces groundwater plumes in California while Mack (1995) models infiltration and run-off at landfills. In light of their increasing importance for water supply, and advances in hydrogeology and GIS visualization, groundwater quality problems deserve even greater attention from geographers.

Research on riparian and wetland environments links water use, land use, water quality, and water policy. In the Midwest, Woltemade (1994) studied the relations between flood flows and fluvial geomorphology in Wisconsin, and went on to posit links with aquatic ecology (Woltemade 1997). Lant and Mullens (1991) use contingent valuation methods to estimate the economic value of wetlands and riparian protection in the Midwest (cf. Lant *et al.* 1995). Duram (1995) examines how protection of the Cheyenne Bottoms wetlands fits within the context of Kansas water policy.

Further West, Pitlick and Van Steeter (1998) have studied how fluvial geomorphology experiments affect aquatic habitat management in the emerging literature on regulated rivers and stream restoration. Schmidt (1998, 1999) has contributed to the literatures on regulated rivers and river restoration in studies of sediment dynamics beneath Flaming Gorge Dam and Glen Canyon Dam in the Colorado River basin. James (1999) has analyzed the physical linkages between river channel dynamics and the historical record of hydraulic mining in California. Working in several regions—from the Southeast and upper Midwest to southern California—Trimble (1995, 1997*a*, *b*) has shed light on reservoir sedimentation and accelerated stream channel erosion. Trimble (1993) has also linked soil erosion research with larger scales of watershed analysis, the renewal of which was an exciting development of the 1990s.

Watershed research has emphasized the effects of land-use and land-cover change on run-off, sedimentation, stream habitat, and water supply (Wescoat 1999). Marston (1991, 1995) examines vegetation change and watershed processes in the Rocky Mountains, southern France, and central Himalayas (Marston *et al.*). Marston (1994) and Trimble and Mendel (1995) examine the effects of animals (e.g. beaver and cattle) on fluvial geomorphology at the watershed scale. Upper watershed restoration in agricultural regions links fluvial geomorphology, restoration ecology, and agronomic science (Rhoads and Herricks, 1996). Working in the Philippines, Dubois (1990) linked grazing with land-use change, watershed sedimentation, and coastal environmental impacts—one of the few studies to integrate freshwater and marine resources (see also Buddemeier 1996; Mullens 1995; and Pulwarty and Redmond 1997, on climate variability, water policy, and anadromous fisheries in the Pacific Northwest).

Watershed councils, stakeholder groups, and initiatives proliferated during the 1990s. Geographic research by Michaels (2001) examines the origins and efficacy of these groups in the Northeastern US. Steinberg and Clark (1999) focus on the political geography of conflict over watershed management in metropolitan Boston, which bears comparison with Platt's (1995) analysis of Boston water supply planning. The National Research Council's (1999) *New Strategies for America's Watersheds*, chaired by geographer William Graf with contributions from Stanley Trimble and others, constitutes an important synthesis of physical, social, and environmental policy dimensions of watershed management in the US.

Watersheds continue to be a strong focus of geographic research in the US. They link physical and ecological concerns with late twentieth-century innovations in locally based, but often state and federally supported, governance initiatives. At the same time, small rural communities face severe financial and technical constraints on water supply and quality improvements, and they have received relatively little attention (cf. Berry *et al.* 1996). Geographers have given more attention to acute water problems faced by the rural poor in developing countries than in the US.

From Urban to Metropolitan Water Systems

With a few important exceptions, urban water resource geography remains underdeveloped relative to watershed and regional scales of research. In 1998, Baumann *et al.* published *Urban Water Demand Management and Planning*, drawing together lessons from three decades of research. Dziegielewski *et al.* (1994) and colleagues at Planning and Management Consultants, Inc. (1993) have extended urban water demand forecasting into the closely related fields of urban water conservation and drought management (Opitz 1998).

Building upon the watershed-scale research described above, Wolman *et al.* (1995) have analyzed effects of urbanization on land-use change, hydrology, and water quality. Riebsame (1997) documents population growth, land-use change, and potential water resources implications in the Western US. The 1990s witnessed a convergence of riparian and wetlands protection with urban planning (Schmid 1994). Platt (1994) situates these urban water problems in relation to a broader program of research and planning on "Ecological Cities." Sheaffer and Stevens (1983) anticipated these developments in projects that link urban and rural wastewater reuse in northern Illinois, Michigan, and, recently, the Susquehanna River basin. Gumprecht (1999) wrote an award-winning book on the history and emerging restoration efforts on the Los Angeles River, which reflects back upon urban historical expropriation of regional water supplies (Kindquist 1996).

The dramatic pace of urbanization in Latin America, Africa, and Asia has stimulated several studies of water problems in megacities, secondary cities, peri-urban areas, and squatter settlements (for an early study see Ridgley 1989). In contrast with above-mentioned research on water pricing and systems analysis, Swyngedouw (1997) stakes out a political-economic analysis of urban water crises in Ecuador. Moran (2000) focuses on urban water privatization in Eastern Europe, a trend that brings water resource geography within the orbit of regional restructuring and governance struggles. These water problems in urbanizing regions stand out as a broad class of challenges to be addressed in the twenty-first century.

State and Interstate Relations

Emphasis on state water management comes as little surprise in a country with a federal system of government. Constitutional authority for water law and policy rests principally at the state level of government. Many single-state studies do not apply to the entire state but rather to smaller areas. Conducting research in a single state controls for many legal and policy variables, but it can limit the generalizability of the study. Fields of state water resources research in the 1990s included: (1) Western water law; (2) state hydroclimatology; (3) state water policy; and (4) interactions among adjacent states.

Work by lawyer-geographers such as Bulman, Matthews, Platt, Templer, and Votteler has established water law as a research topic in water resource geography (for a review see Templer 1997). Matthews's (1984) AAG monograph on *Water Resources, Geography, and Law*, which deserves a sequel, laid a strong foundation. In several dozen papers, Templer has charted developments in Texas groundwater law, Edwards Aquifer management, river basin adjudication, and conjunctive water management. Votteler (1998) extends research on Texas water law to include interactions with federal water and environmental laws. Bulman (1994) addresses water and environmental law in northern California and the Pacific Northwest, while Matthews (1994) has examined efforts to frame a model state water code.

Again, geographers have focused on the Great Plains where Brooks and Emel (1995) critique the effects of different state water laws on groundwater depletion in Nebraska (cf. also Roberts and Emel 1992). Emel and Roberts (1995) focus on the difference that state institutions make for groundwater management in the High Plains, especially distinctions between Texas and New Mexico. Kromm and White (1992) examine the difference state water policies make for groundwater use and conservation in Kansas. White's (1994) research on population change in the Great Plains parallels that of Riebsame (1997) in the West. Research on state water use and policies in Plains states has surprisingly few counterparts in other states. Hordon *et al.* (1998) analyze state water demand, and it seems likely that a larger body of state water resources research exists in unpublished theses, government documents, and consulting reports by geographers.

Similarly, we find a large number of reports prepared by geographers on state water resources and hydroclimatic issues. Notable examples include Mather's (1991) analysis of the potential impacts of climate change on Delaware; Hodny's (1998) water budget analysis of Delaware; and Luzzader-Beach (1995) on groundwater quality in California.

A deficiency in US water resource geography involves interstate water issues. This review identified few studies of interstate compacts (McCormick 1994; Waterstone 1994). Even if more studies exist, they seem incommensurate with the conflicts among the states, not to mention the emerging innovations in interstate adaptive water management (e.g. in the Platte River and

Columbia River basins). The pace of interstate negotiations and adjudication extends from years to decades. The conditions under which states choose to cooperate, and the geographic alternatives they consider, pose key research problems for trans-boundary river, aquifer, and atmospheric resources.

Given the primacy of state governments in water policy, it comes as something of a surprise that more research has not focused on the performance of those policies. Funding for state water resources research institutes has declined in recent years, although geographers have a long history of federal water policy analysis dating back to New Deal times. However, state constitutional authority over water remains, and state water management has been an important employment opportunity for geography students, which argues for greater research at state and interstate levels.

River Basins, Regions, and Regulated Rivers

The neglect of interstate water relations may be explained in part by emphasis on interstate river basins and regulated rivers. River basin surveys were one of the earliest forms of water resource geography, dating back to the work of Philippe Buache in eighteenth-century France, developing slowly in the nineteenth century with surveys of the upper Mississippi River by Claude Nicollet and Western basins by John Wesley Powell, and then expanding rapidly in the 1930s. Contributions by Barrows, White, Hudson, Colby, and others to the Tennessee Valley Authority and New Deal river basin development were followed by a wave of river basin planning in the 1970s that collapsed again during the Reagan administration.

The 1990s witnessed important regional-scale research on several topics: (1) drought in the Southeastern and Western US; (2) river basin studies; and (3) regulated rivers research. Droughts in the 1980s stimulated regional research in the US (Dziegielewski *et al.* 1993; Wilhite *et al.* 2000) and Mexico (Liverman 1990, 1999). Notwithstanding the high economic costs of the 1980s droughts, lessons were learned from previous major droughts of the 1930s, 1950s, and 1970s. Implementation of conservation practices exceeded expectations, and the drought in California initiated significant water banking and reclamation reform initiatives. Brooks and Emel (1995) took a regional approach to Great Plains water management as part of a comparative international study of critical regional environmental problems.

Although river basin studies had a formative, and continuing, role in water resource geography, the revitalization of watershed management and devolution of federal governance have led to a quiescent period of river basin planning, *per se* (Muckleston 1990). Highlights of the 1990s included research on conflicts among environmental, hydropower, and tribal water claims in the Columbia River basin (this chapter went to press as energy conflicts in California and the Middle East were fundamentally altering the political and economic context of hydro power and dam decommissioning in the West) (Bulman 1994; Mullens 1995). For the Rio Grande, Graf (1994) tracked the sources and sinks of plutonium contamination in the upper basin. Geographers examined the effects of climate variability on run-off in river basins such as the Colorado and Sacramento (Shelton 1998).

The National Oceanic and Atmospheric Administration has supported regional water and climate assessments. Pennsylvania State University geographers have developed a collaborative research program on the Susquehanna River basin, linking climate variability with water supply, water quality, and economic change in eastern Pennsylvania (O'Connor *et al.* 1999; Yarnal 1992; Yarnal *et al.* 2000). In addition to its collaborative scope, this project advances earlier work on hydrologic impacts of regional climate variability by giving greater attention to flood hazards (Yarnal *et al.* 1997). Geographers at the University of Arizona have organized a comparably broad program of regional research of hydroclimatic variability, groundwater depletion, and social vulnerability (Liverman 1999; Morehouse 2000*a*, *b*). A team at the University of Colorado at Boulder has begun regional water resources assessments in the Rocky Mountain region. Although common in physical geography, long-term collaborative water resources research has proven challenging to organize and sustain.

The Colorado River has played a key role in emerging research on "regulated rivers." Schmidt (1999) and Pitlick and Van Steeter (1998) have studied relations between hydrologic flows, reservoir releases, fluvial geomorphology, and aquatic ecosystem effects in the Colorado River basin. These physical geographic studies contribute to broader programs of "adaptive management," which involve stakeholder-directed processes of water management experiments, scientific monitoring, and policy adjustment (National Research Council 1999). Pulwarty and Redmond (1997) connect these river basin experiments with climate variability. At the same time, Gosnell questions the extent to which the search for "reasonable and prudent alternatives" excludes difficult choices, such as "jeopardy findings" under the Endangered Species Act, especially where they come into conflict with vested or as yet unfulfilled Indian water rights.

Indian Reservations and Autonomous Territories

The 1990s saw significant new research on Indian water rights in the US, and those of indigenous peoples in other regions. Jacobsen (1992) documented the disproportionately slow pace and quality of development in the Navajo Indian Irrigation Project (NIIP). Berry (1990) analyzed implications of the McCarren Amendment for adjudicating Indian water rights claims in state rather than federal courts. McNally and Matthews (1995) examined the trend toward settlement of Indian reserved water rights cases with an emphasis on Montana case studies (e.g. of the Fort Peck reservation).

In contrast with the rapidly growing literature on environmental justice and racism in urban environments, the acute water problems on Indian reservations are relatively neglected and rarely framed as issues of racism (though see Berry 1998; and Wescoat *et al.* 2002). Research on indigenous irrigation systems in the Andes was noted earlier (Denevan 2001). But this review did not identify comparative geographic research on first-people's water claims. In the Western US and perhaps elsewhere it seems likely that the unmet claims of tribes will alter the logic and outcomes of water policy debates (cf. Ch. 38 in this volume by the Native American Specialty Group).

National Water Studies

Geographic research at the national scale (i.e. on national policies and projects) presents a mixed picture. Devolution of federal water programs included dismantling of the US Water Resources Council. At the same time, geographers have made important contributions to major federal water agencies such as the US Army Corps of Engineers (e.g. Gerald Galloway, who chaired the post-1993 Upper Mississippi River flood study that came to be known as the Galloway Report, and James Johnson, chief of planning, USACE); and the Bureau of Reclamation (e.g. Dan Beard, former Commissioner; and Shannon Cunniff, director of research).

For reasons discussed earlier, US scientific and policy research at the national scale deals increasingly with water quality and water-related hazards (Rajagopal and Tobin 1992; Lant and Roberts 1990). In an evocatively titled but critical article, Tobin (1995) laments the continuing emphasis of federal flood policy on engineering structures. Duram and Brown (1999) shed light on links between federal agricultural policies and non-point source pollution.

A growing number of geographers contribute to studies by the National Research Council, Water Science and Technology Board (on which geographer Jeffrey Jacobs serves as senior staff scientist). Geographers' contributions to recent NRC committee reports include:

- *River Resource Management in the Grand Canyon (1996)*
- *A New Era for Irrigation (1996)*
- *Sustaining Our Water Resources (1999)*
- *Water for the Future: The West Bank and Gaza Strip, Israel, and Jordan (1999)*
- *New Strategies for America's Watersheds (1999)*
- *Downstream: Adaptive Management for the Grand Canyon Ecosystem* (1999)
- *Watershed Management for Potable Water Supply: Assessing the New York City Strategy (2000)*

Graf (1992, 2001) analyzed emerging linkages between water science and public policy in the Western US, and Mitchell offers one of the few comparative frameworks for analysis of national water policies.

Some geographers have written about national water policies in other countries (e.g. Knight and Staneva 1996 on water problems in Bulgaria). Sparked in part by the Three Gorges Dam controversy, geographers have given increasing attention to water issues and approaches in China (Boxer 1998; Edmonds 1992; Jhaveri 1988; Luk and Whitney 1993). Brush *et al.* (1987) organized a joint US–China conference on water and sediment problems in the Yellow River (Huang He) basin.

Far more attention could and should be paid to regional effects of national water legislation, such as the Clean Water Act and Safe Drinking Water Act. Geographers have given relatively little attention to Federal and Supreme Court case law on water issues. The substantial restructuring of national water codes in countries from South Africa to New Zealand argues for renewed emphasis on national comparisons—of the sort that are now occurring at the international scale.

International Waters

Tobin *et al.* (1989) identified international water resources as a research priority. Early contributions included White (1957, 1963) in the Mekong, A. Michel (1967) in the Indus, and Day on the lower Rio Grande basin. The 1990s witnessed a dramatic burst of research at international scale on: (1) transboundary river basin planning; (2) international water conflict; and (3) comparative international water management. The Middle East and US–Mexico borderlands have drawn the greatest attention, followed by the Aral Sea, Mekong, Indus, Nile, and La Plata river basins. The limited review possible here allows only a fraction of this recent work to be cited.

The largest body of recent international research has focused on the Jordan, Litani, and Tigris-Euphrates basins. In addition to a monograph analyzing Jordan basin conflicts, negotiations, and prospects, Wolf (1995) has compiled a transboundary river inventory (n = 261) and water treaty database (http://www.transboundary-waters.orst.edu,/last accessed18August2003) as a basis for identifying decision-making alternatives and negotiation strategies. He and others argue persuasively that while water issues have generated many serious conflicts, and sometimes physical violence, they have not led to the sorts of wars anticipated by some international affairs and security analysts. Kolars (1994, 1997) has produced a substantial body of research on development of the Euphrates river basin and the controversial Southeastern Anatolia Development Project (GAP) (Kolars and Mitchell 1991). Amery (1993) has focused on disputes between Lebanon and Israel over the Litani River basin, and Amery and Wolf (2000) have compiled an inspiring set of essays on water and the "geography of peace" in the Middle East. An unprecedented report of four national academies of science—the US National Academy of Sciences, Royal Scientific Society of Jordan, Israel Academy of Sciences and Humanities, and Palestine Academy for Science and Technology—identified scientific gaps and technical opportunities for cooperation on water and environmental management in the Jordan River basin.

Compared to the political geographic thread in much Middle East water research, work on the US–Mexico borderlands has a political-economic and, secondarily, environmental emphasis. Brown and Mumme assess the formation of binational watershed councils in the Rio Grande/Rio Bravo and Tijuana River basins. S. Michel (2001) includes an analysis of bilateral institutions in her study of wastewater politics in the Tijuana basin. Varady *et al.* examine transboundary water institutions in the San Pedro River basin of southern Arizona. Eaton (1992) has modeled international water development problems in the lower Rio Grande basin. Although less geographic research deals with the US–Canada border, important studies have focused on St Lawrence-Lake Champlain (Kujawa 1995) and the International Joint Commission, on which geographer Gerald Galloway served as US representative.

Research on the Aral Sea basin has focused most directly on environmental effects of interstate, and now international, water development. Micklin (1988, 1992) has produced two decades of sustained research in the region, ranging from documentation of sea-level decline and desertification to water development scenarios and their implications (Micklin and Williams 1996). Smith (1994, 1995) focuses on impacts of irrigation development and climate variability in the Amu and Syr Darya basins, with special emphasis on their deltaic environments. Bedford (1996) links scenarios of climate variability in the upper Amu Darya basin with hydrologic effects and potential international ramifications.

Studies in other international river basins and aquifers have expanded the scope of international water resource geography. Jacobs (1995) has studied development of the Mekong River basin, the evolving roles of the Mekong Commission, and its resilience under scenarios of hydroclimatic variability. Perritt (1989) has undertaken comparable studies in international river basins of South America and Africa. Flint (1995) has asked how the UN Convention on Non-Navigable Uses of International Watercourses might affect negotiations among riparian nations in the Nile River basin, a process that is now underway. In South Asia, research has reflected upon the past 50 and 500 years of water development in the Indus basin (Wescoat *et al.* 2000) and prospects for collaboration in the Ganges-Brahmaputra basin (Eaton 1992).

In addition to single basin or aquifer studies, comparative research on international river basins has advanced in recent years. Elhance (2000) compares patterns of conflict and cooperation in six international river basins, concluding that institutional apparatus, precedents, and prospects are emerging for increased cooperation and conflict management. Jacobs (1999) compares modern development of the Mekong and Mississippi rivers, both of which had strong involvement of the US Army Corps of Engineers. Several papers compare water laws and treaties in different regions and basins (Wolf 1997; Wescoat 1996). International water resource geography thus seems to have flourished in the post-Cold War era, at a time of decreasing bilateral water programs, and expanding multilateral and global initiatives.

Global and Multi-Scale Research

Research on the water resources dimensions of global environmental change has stimulated research in water resource geography, but not as much as might be expected (Liverman *et al.* 1997; Riebsame *et al.* 1995). The main type of global change research estimates the regional impacts and adjustments associated with general circulation model (GCM) scenarios of global climate change. Studies of this sort focus on the Aral Sea region, US–Mexico borderlands, and complex river basins of Asia, Africa, South America, and North America (Bedford 1996; Feddema 1992, 1996; Riebsame 1995; cf. Ch. 18, Human Dimensions of Global Change).

An important conclusion of GCM-based scenario analyses of the early 1990s was the need to give greater emphasis to regional and meso-scale research on hydroclimatic scenarios, impacts, and adaptations. Geographers have thus focused less in recent years on estimating global hydrology and water supply parameters and more on regional vulnerability to climatic variability derived from historical and paleoclimate records. Similarly, only a few water resources geographers have participated actively in emerging global forums (e.g. the International Conference on Water and Environment, Dublin 1992; the Second World Water Forum at The Hague, 2000; and annual meetings of the International Water Resources Association, Global Water Partnership, Stockholm International Water Institute, and World Water Council). Differences between global and regional-scale inquiry are evident in *The Earth as Transformed by Human Action* chapters on global water use (L'vovich and White 1990) and regional water quality.

Reflecting back upon this review, most studies involve multiple scales of inquiry, inference, and action. Scale and scaling are growing theoretical and methodological concerns in the hydrologic sciences (Cayan *et al.* 1993; Keim 1997; Hirschboeck 1999; Marston *et al.* 1996; Pitlick 1997). And they receive increasing attention in water resource geography. For example, flood hazards in the Himalayas and Ganges-Brahmaputra basin have stimulated research on the relative importance of upstream environmental degradation and downstream precipitation events, revealing that while upstream processes did not produce large flood events and losses in Bengal, they have caused severe flood-related damages in the upper and middle reaches of the basin (Gamble and Meentemeyer 1996; Marston *et al.* 1996). Comparable research following the 1993 flood in the upper Mississippi and lower Missouri basins has shed light on damages ranging in extent from regional river basin run-off and innundation to local levee failures (Pitlick 1997; Knox 1993; Woltemade 1997).

Political ecology constitutes another emerging link between global and multi-scale water research (see Ch. 8, Cultural Ecology, and 11, Historical Geography). Few studies focus primarily on water, which makes sense within conceptual frameworks focused on social access, power, and economic livelihood (see Swyngedouw 2000, for a critique of rescaling and "globalization"). However, they do show how multiple scales of political-economic and cultural change have influenced irrigation agriculture in Europe, Africa, and the Americas (e.g. Butzer 1990; Denevan 2001; Doolittle 2000; Perrault *et al.* 1998). Other works address multiple scales of water management, but few focus on relationships among scales, which is crucial for interregional comparisons.

The Role of Individuals and Organizations

In addition to the substantive themes discussed above, three institutional conclusions emerge from this review. First, every *major* research trend mentioned above involves contributions from a small number of geographers—often less than five and never more than ten. The establishment of a trend is often driven by one or two individuals (e.g. White in flood hazards; Templer and Matthews in Western water law). When one or two collaborate closely, especially at a single institution, the results over a decade stand out for the discipline at large (e.g. Kromm and White; Montz and Tobin).

A second observation is that only a few institutions employ more than one or two water resource geographers, which may diversify but also dilute our research contributions. Southern Illinois University (SIU) stands out as the major exception. By focused hiring, professional service, individual initiative, and university support, SIU became headquarters for the University Water Information Network (UWIN), the University Council on Water Resources (UCOWR), the International Water Resources Association (IWRA), and editorial offices for the *Journal of the American Water Resources Association*. Geographers at SIU established a leading private water consulting firm, Planning and Management Consultants, Ltd. (e.g. see 1993), which employs graduates of its department.

Third, private and government sector contributions to water resource geography still constitute major gaps in our knowledge. Geographers in private consulting and water companies have few close links with the Water Resources Specialty Group (though the electronic efforts of Douglas Gamble, Faye Anderson, and others are changing that). Geographers in public service have included the director of planning for the US Army Corps of Engineers (James Johnson); senior environmental planner for the UN Food and Agriculture Organization's Investment Centre (Random Dubois); executive director of the Metropolitan Water District of Southern California (John Wodraska); planner for the Governor's Commission for a Sustainable South Florida (Bonnie Kranzer); and several state climatologists. But probably scores of our graduates work with local, state, and national water agencies; and closer ties with them are vital for the discipline, our graduates, and environmental inquiry. As noted above, the National Research Council (NRC) involves a growing number of geographers on committees and staff.

Summary and Implications

At the beginning of the twenty-first century, water resource geographers have developed a specialty group and rich array of contributions. At the same time, a small number of individuals are involved in any particular region and sector. The largest number of water resource geographers work in the US, particularly the Western states. Interestingly, more work focuses on water issues in the interior West (Rocky Mountains/Great Plains) than the Pacific Coast. More work on US–Mexico than US–Canada transboundary water issues. The 1990s witnessed exciting development of international water resources research, especially North America, the Middle East, and Central Asia. Polar regions and Oceania have received the least attention to date. As in past decades, more geographers work on rural and regional than urban or site-scale water problems, and more on environmental quality than social well-being. All these patterns reflect commitments of small numbers of water resources geographers working in interdisciplinary groups but maintaining contact with one another through the specialty group.

These academic and institutional trends have several implications. First, the Water Resources Specialty Group has created a community that, if nurtured and sustained, could extend geographers' contributions to a broader array of issues, organizations, environments, and peoples. Second, much intellectually creative research on water resources occurs outside the specialty group, which argues for building and strengthening bridges with physical geographers in hydrology, climatology, and biogeography; human geographers working on social and cultural theory, political economy, ethics and values, and justice and law; environmental geographers in fields of natural hazards, cultural ecology, and the human dimensions of global change; and geographic information and mapping scientists. During the 1990s graduate students took a leading role in forging such bridges through joint AAG sessions with other specialty groups and their dissertations, and doubtless that will continue. Perhaps the greatest gap between recent research and student interests involves the use of GIS and remote sensing technologies in water resources management (Reitsma 1996).

Reviewing the water resource geography literature by scale, although problematic in some theoretical and methodological ways, cut across sectoral, systemic, and regional interests. It also identifies research needed to conduct multiple-scale, interregional comparative research. In the post-Cold Water era, *ex-post* evaluation of US involvement in water development in Asia, Africa, and Latin America is a pressing, but as yet undeveloped, research field. Research on privatization of water and wastewater systems, and sociospatial networks of resistance to that and other trends, has been limited (Emel 1990; Moran 2000). Advances in social scientific research on irrigation outside the US suggest that the twenty-first century should become a period of comparative international research and adoption of innovations from other regions of the world.

New networks of scholars and water stakeholders are pioneering methods of adaptive management of large-scale regional water systems, such as the Columbia, Upper Mississippi, Everglades, and Grand Canyon ecosystems. These innovative science-policy experiments are potentially relevant for other complex river systems around the world, but few dissertations or comparative case studies are underway. Finally, as physical geographers focus analytically on scaling, water resource geographers could deepen our understanding of how interacting scales of human–environment relations shape our use and experience of this precious resource.

Acknowledgements

I wish to thank the many WRSG members who sent reprints, survey responses, and suggestions for this chapter, only a small portion of which could be cited here. The University of Colorado Water Resource Geography Reading Group offered creative suggestions on organization, priorities, and recent citations. I extend thanks, in particular, to Hannah Gosnell, Mark Haggerty, Sarah Halvorson, Lisa Headington, Paul Lander, Tamara Laninga, Klara Mezgolits, Daanish Mustafa, and Andrea Ray. Chris Lant, the editors, and anonymous reviewers offered constructive criticism for this account of our field.

References

Allan, N. J. (1986). "Communal and Independent Mountain Irrigation Systems in North Pakistan," in M. Rahman (ed.), *Proceedings of the IGU Irrigation Conference*. Lahore: International Geographical Union.

Amery, H. A. (1993). "The Litani River of Lebanon." *Geographical Review*, 83: 229–37.

Amery, H. A., and Wolf, A. T. (2000). *Water in the Middle East: A Geography of Peace*. Austin: University of Texas Press.

Arbona, S. (1992). "Land Use and Water Pollution in Puerto Rico." Unpublished Ph.D. dissertation. East Lansing: Michigan State University.

Baumann, D. D., Boland, J. J., and Haneman, W. M. (eds.) (1998). *Urban Water Demand Management and Planning*. New York: McGraw-Hill.

Beach, T., and Luzzader-Beach, S. (2000). "The Physical Geography and Environmental Archaeology of the Alanya Region," in *Landscape and the State in Medieval Anatolia: Seljuk Gardens and Pavilions of Alanya, Turkey*. BAR International Series, 893. Oxford: Archaeopress, 115–38.

Bedford, D. P. (1996). "International Water Management in the Aral Sea Basin." *Water International*, 21/2: 63–9.

Berry, C. A. (1998). "Race for Water? American Indians, Eurocentrism and Western Water Policy," in David Camacho (ed.), *Environmental Justices, Political Struggles: Race, Class and the Environment*. Durham, NC: Duke University Press.

Berry, C. A., Markee, N. L., Stewart, M. J., and Giewat, G. (1996). "County Commissioners' Water Knowledge." *Journal of the American Water Resources Association*, 32/5: 1089–99.

Bonine, M. E. (1996). "Qanats and Rural Societies: Sustainable Agriculture and Irrigation Cultures in Contemporary Iran," in Jonathan B. Mabry (ed.), *Canals and Communities: Small-Scale Irrigation Systems*. Tucson: University of Arizona Press, 183–209.

Boxer, B. (1998). *China's Water Problems in the Context of U.S.–China Relations*, Woodrow Wilson Center, Environmental Change and Security Project, China Environment Series, 2: 20–7.

Brooks, E., and Emel, J. (1995). "The Llano Estacado of the American Southern High Plains," in J. X. Kasperson, R. E. Kasperson, and B. L. Turner II (eds.), *Regions at Risk: Comparisons of Threatened Environments*. Tokyo: United Nations University Press, 255–303.

Brush, L. M., Wolman, M. G., and Huang, B. W. (1987). *Taming the Yellow River: Silt and Floods*. Dordrecht: Kluwer Academic.

Buddemeier, R. W. (ed.) (1996). *Groundwater Discharge in the Coastal Zone: Proceedings of an International Symposium*. Texel: LOICZ (Land–Ocean Interactions in the Coastal Zone, International Geosphere-Biosphere Programme).

Bulman, T. L. (1994). "When the River Runs Dry: Management Responses to Water Diversions," in *Effects of Human-Induced Changes on Hydrologic Systems*. Proceedings of the American Water Resources Association, 707–16.

Butzer, K. W. (1990). "The Realm of Cultural-Human Ecology: Adaptation and Change in Historical Perspective," in *The Earth as Transformed by Human Action: Global and Regional Changes in the Biosphere over the Past 300 Years*. Cambridge: Cambridge University Press, 685–701.

—— (1994). "The Islamic Traditions of Agroecology: Crosscultural Experience, Ideas, and Innovations," *ECUMENE: Journal of Environment, Culture, Meaning*, 1: 1–50.

Carney, J. (1998). "The Role of African Rice and Slaves in the History of Rice Cultivation in the Americas," *Human Ecology*, 24/4: 525.

Cayan, D. R., Riddle, L. G., and Aguado, E. (1993). "The Influence of Precipitation and Temperature on Seasonal Streamflow in California." *Water Resources Research*, 29/4: 1127–40.

Clark, George E., *et al.* (1998). "Assessing the Vulnerability of Coastal Communities to Extreme Storms: The Case of Revere, MA, USA." *Mitigation and Adaptation Strategies for Global Change*, 3: 59–82

Cosgrove, D., and Petts, G. (eds.) (1990). *Water, Engineering and Landscape*. London: Belhaven.

Denevan, W. M. (2001). *Cultivated Landscapes of Native Amazonia and the Andes*. Oxford: Oxford University Press.

Doolittle, W. E. (1990). *Canal Irrigation in Prehistoric Mexico: The Sequence of Technological Change*. Austin: University of Texas Press.

—— (2000). *Cultivated Landscapes of Native North America*. Oxford: Oxford University Press.

Dubois, R. (1990). *Soil Erosion in a Coastal River Basin: A Case Study from the Philippines*. Geography Research Paper, 232. Chicago: Department of Geography, University of Chicago.

Duram, L. A. (1995). "Water Regulation Decisions in Central Kansas Affecting Cheyenne Bottoms Wetland and Neighboring Farmers." *Great Plains Research*, 95: 5–19.

Duram, L. A., and Brown, K. G. (1999). "An Assessment of Public Participation in Federally Funded Watershed Planning Initiatives." *Society and Natural Resources*, 12: 455–67.

Dziegielewski, B., Garbharran, H., and Langowski, J. F. (1993). *Lessons Learned from the California Drought of 1987–1992*. IWR Report 93-NDS-5. Fort Belvoir, Va.: US Army Corps of Engineers Water Resources Support Center, Institute for Water Resources.

Dziegielewski, B., Opitz, E. M., Kiefer, J. C., and Baumann, D. D. (1993). "Evaluation of Urban Water Conservation Programs: A Procedures Manual." Prepared for California Urban Water Agencies, Sacramento, California. Denver: American Water Works Association.

Eaton, D. (ed.) (1992). *Challenges in the Binational Management of Water Resources in the Rio Grande/Rio Bravo*. Austin: LBJ School of Public Affairs.

—— (ed.) (1992). *The Ganges-Brahmaputra Basin: Water Resource Cooperation Between Nepal, India, and Bangladesh*. Austin: LBJ School of Public Affairs.

Edmonds, R. (1992). "The Sanxia (Three Gorges) Project: the Environmental Arguments Surrounding China's Super Dam." *Global Ecology and Biogeography Letters*, 2/4: 105–25.

Elhance, A. P. (2001). *Hydropolitics in the Third World: Conflict and Cooperation in International River Basins*. Washington: US Institute of Peace Press.

el-Katsha, S., and White, A. U. (1989). "Women, Water and Sanitation: Household Behavioral Patterns in Two Egyptian Villages." *Water International*, 14: 103–11.

Emel, J. (1990). "Resource Instrumentalism, Privatization, and Commodification." *Urban Geography*, 11: 527–47.

Emel, J., and Roberts, R. (1995). "Institutional Form and Its Effect on Environmental Change: The Case of Groundwater in the Southern High Plains." *Annals of the Association of American Geographers,* 85: 664–83.

Feddema, J. J. (1999). "Future African Water Resources: Interactions between Soil Degradation and Global Warming." *Climatic Change,* 42/3: 561–96.

Feddema, J. J., and Mather, J. R. (1992). "Hydrologic Impacts of Global Warming over the United States," in S. K. Majumdar *et al.* (eds.), *Global Climate Change: Implications, Challenges and Mitigation Measures*. Easton, Pa.: Pennsylvania Academy of Science, 50–62.

Flint, C. G. (1995). "Recent Developments of the International Law Commission Regarding International Watercourses and their Implications for the Nile River." *Water International*, 20/4: 197–206.

Gamble, D., and Meentemeyer, V. (1996). "The Role of Scale in Research on the Himalaya–Ganges–Brahmaputra Interaction." *Mountain Research and Development*, 16: 149–55.

Giambelluca, T. W., Loague, K., Green, R. E., and Nullet, M. A. (1996). "Uncertainty in Recharge Estimation: Impact on Groundwater Vulnerability Assessments for the Pearl Harbor Basin, O'ahu, Hawai'i, U.S.A." *Journal of Contaminant Hydrology*, 23/1–2: 85 ff.

Gottlieb, R., and Fitzsimmons, M. (1991). *Thirst for Growth: Water Agencies as Hidden Government in California*. Tucson: University of Arizona Press.

Graf, W. L. (1992). "Science, Public Policy, and Western American Rivers." *Transactions of the Institute of British Geographers*, NS 17: 5–19.

—— (1994). *Plutonium and the Rio Grande: Environmental Change and Contamination in the Nuclear Age*. New York: Oxford University Press.

—— (2001). "Damage Control: Dams and the Physical Integrity of America's Rivers." *Annals of the Association of American Geographers*, 91: 1–27.

Gruntfest, E. 1997. *Twenty Years Later: What We Have Learned Since the Big Thompson Flood*. Proceedings. Boulder: Natural Hazards Research Applications and Information Center.

Gumprecht, B. (1999). *The Los Angeles River: Its Life, Death and Possible Rebirth*. Baltimore: The Johns Hopkins University Press.

Halvorson, S. J. (2000). "Geographies of Children's Vulnerability: Households and Water-Related Disease Hazard in the Karakoram Mountains, Northern Pakistan." Unpublished Ph.D. dissertation, University of Colorado at Boulder.

Harden, C. (1993). "Land-Use, Soil Erosion and Reservoir Sedimentation in an Andean Drainage Basin in Ecuador." *Mountain Research and Development*, 13/2: 177–84.

Harden, C., and Wallin, T. (1996). "Quantifying Trail-Related Erosion at Two Sites in the Humid Tropics: Jatun Sacha, Ecuador, and La Selva, Costa Rica." *Ambio*, 25/7: 517–22.

Hirschboeck, K. K. (1999). "A Room with a View; Some Geographic Perspectives on Dilettantism, Cross-training, and Scale in Hydrology." *Annals of the Association of American Geographers*, 894: 69 ff.

Hordon, R. M., Lawrence, K. D., and Lawrence, S. M. (1998). "Water Demand Estimation on a State-by-State Basis." *Advances in Business and Managment Forecasting*, 2: 101–12.

Hodny, J. W. (1998). "Water Resources of Delaware: A Climatic Water Budget Microanalysis." Unpublished Ph.D. dissertation, University of Delaware.

Hunter, J. M., and Arbona, S. I. (1995). "Paradise Lost: An Introduction to the Geography of Water Pollution in Puerto Rico." *Social Science and Medicine*, 16: 1127–45.

Jacobs, J. W. (1995). "Mekong Committee History and Lessons for River Basin Development," *The Geographical Journal*, 161/2: 135–48.

—— (1999). "Comparing River Basin Development Experiences in the Mississippi and the Mekong." *Water International*, 24/3: 196–203.

Jacobsen, J. E. (1992). "The Navajo Indian Irrigation Project and Quantification of Navajo Winters Rights." *Natural Resources Journal*, 32/4: 825–54.

James, L. A. (1999). "Time and the Persistence of Alluvium: River Engineering, Fluvial Geomorphology, and Mining Sediment in California." *Geomorphology*, 31: 265–90.

Jhaveri, N. (1988). "The Three Gorges Debacle", *The Ecologist,* 18, 56–63.

Keim, B. D. (1997). "Preliminary Analysis of the Temporal Patterns of Heavy Rainfall in the Southeastern United States." *Professional Geographer*, 49/1: 202–11.

Kindquist, C. E. (1996). "The South Park Water Transfers: The Geography of Resource Expropriation in Colorado, 1859–1994." Unpublished Ph.D. dissertation, University of British Columbia.

Knight, C. G., and Staneva, M. P. (1996). "Water Resources of Bulgaria: An Overview." *Geojournal*, 40/4: 347–62.

Knox, J. C. (1993). "Large Increases in Flood Magnitude in Response to Modest Changes in Climate." *Nature*, 361: 430–2.

Kolars, J. F. (1994). "Problems of International River Management: The Case of the Euphrates," in A. Biswas (ed.),

International Waters of the Middle East from Euphrates-Tigris to Nile, New Delhi: Oxford University Press, 44–94.

—— (1997). "River Advocacy and Return Flow Management on the Euphrates/Firat River." *Water International*, 22/1: 49–53.

Kolars, J. F., and Mitchell, W. A. (1991). *The Euphrates River and the Southeast Anatolia Development Project*. Carbondale: Southern Illinois University.

Kromm, D. E., and White, S. (eds.) (1992). *Groundwater Exploitation in the High Plains*. Lawrence: University of Kansas Press.

Kujawa, B. R. (1995). "Conceptualizing Geographies of Environmentalism: Planning for the Future of Lake Champlain." *Proceedings of the New England-St. Lawrence Valley Geographical Society*, 25: 95–108.

Lant, C. L. (1998). "The Changing Nature of Water Management and Its Reflection in the Academic Literature." *Water Resources Update*, 110: 18–22.

Lant, C. L., and Mullens, J. B. (1991). "Lake and River Quality for Recreation Management and Contingent Valuation." *Water Resources Bulletin*, 27/3: 453–60.

Lant, C. L., and Roberts, R. S. (1990). "Greenbelts in the Cornbelt: Riparian Wetlands, Intrinsic Values, and Market Failure." *Environment and Planning A*, 22: 1375–88.

Lant, C. L., Kraft, S. E., and Gillman, K. R. (1995). "Enrollment of Filter Strips and Recharge Areas in the CRP and USDA Easement Programs." *Journal of Soil and Water Conservation*, 50/1: 193–200.

Leichenko, R., and Wescoat, J. L., Jr. (1993). "Environmental Impacts of Climate Change and Water Development in the Indus Delta Region." *Water Resources Development*, 9: 247–61.

Lightfoot, D. R. (1997). "Qanats in the Levant: Hydraulic Technology at the Periphery of Early Empires." *Technology and Culture*, 38/2: 432 ff.

Lightfoot, D. R., and Eddy, F. W. (1995). "The Construction and Configuration of Anasazi Pebble-Mulch Gardens in the Northern Rio Grande." *American Antiquity*, 60/3: 459 ff.

Liverman D. M. (1999). "Adaptation to drought in Mexico," in D. Wilhite (ed.), *Drought*. New York: Routledge, ii. 35–45.

—— (1990). "Drought and Agriculture in Mexico: The Case of Sonora and Puebla in 1970." *Annals of the Association of American Geographers*, 80/1: 49–72.

Liverman, D. M., Merideth, R. W., and Holdsworth, A. (1997). "Climate Variability and Social Vulnerability in the U.S.–Mexico Border Region: An Integrated Assessment of the Water Resources of the San Pedro River and Santa Cruz River Basins." United States Geological Survey (USGS). *The Impact of Climate Change and Land Use in the Southwestern United States*<http://geochange.er.usgs.gov/sw/>.

Luk, S. H., and Whitney, J. (eds.) (1993). *Megaproject: A Case Study of China's Three Gorges Project*. Armory, NY: M. E. Sharpe.

Luzzadder-Beach, S. (1997). "Spatial Sampling Strategies for Detecting a Volcanic Ground Water Plume in Sierra Valley, California." *Professional Geographer*, 49/2: 179–92.

L'vovich, M., and White, G. F. (1990). "Use and Transformation of Terrestrial Water Systems," in B. L. Turner II *et al.* (eds.) *The Earth as Transformed by Human Action*. New York: Cambridge University Press, 235–52.

McCormick, Z. L. (1994). "Interstate Water Allocation Compacts in the Western United States—Some Suggestions." *Water Resources Bulletin*, 30/3: 385 ff.

McGinnis, David L. (1994). "Climate Change and GCM Simulation of Water Resources from Mountain Snowpack." Unpublished Ph.D. dissertation, State College, Pennsylvania State University.

Mack, M. J. (1995). "HER—Hydrologic Evaluation of Runoff: The Soil Conservation Service Curve Number Technique as an Interactive Computer Model." *Computers & Geosciences*, 21/8: 929–37.

McNally, M., and Matthews, O. P. (1995). "Changing the Balance in Western Water Law? Montana's Reservation System." *Natural Resources Journal*, 35/3: 671 ff.

Marston, R. (1991). "Watersheds and Vegetation of the Greater Yellowstone Ecosystem." *Conservation Biology*, 5/3: 338–46.

—— (1994). "River Entrenchment in Small Mountain Valleys of the Western USA: Influence of Beaver, Grazing and Clearcut Logging." *Geographie de Lyon*, 69/1: 11–15.

Marston, R., Kleinman, J., and Miller, M. (1996). "Geomorphic and Forest Cover Controls on Monsoon Flooding, Central Nepal Himalaya." *Mountain Research and Development*, 16/3: 257–64.

Marston, R., Girel, J., Pautou, G., Piegay, H., Bravard, J. P., and Arreson, C. (1995). "Channel Metamorphosis, Floodplain Disturbance, and Vegetation, Development: Ain River, France." *Geomorphology*, 13: 121–31.

Mather, J. R. (1984). *Water Resources: Distribution, Use and Management*. New York: John Wiley.

—— (1991). *Climate Change and Delaware's Water Resources Future: First Year Completion Report, Project Number 03*. Wilmington, Del.: Center for Climatic Research, Dept. of Geography, University of Delaware.

—— (1996). *The Genius of C. Warren Thornthwaite, Climatologist-Geographer*. Norman: University of Oklahoma Press.

Matthews, O. P. (1984). *Water Resources, Geography, and Law*. Resource Publications in Geography. Washington, DC: Association of American Geographers.

—— (1994). "Changing the Appropriation Doctrine under the Model State Water Code." *Water Resources Bulletin*, 30/2: 189–96.

Michaels, S. (2001). "Making Collaborative Watershed Management Work: The Confluence of State and Regional Initiatives." *Environmental Management* 27: 27–35.

Michel, A. A. (1967). *The Indus Rivers: A Study of the Effects of Partition*. New Haven: Yale University Press.

Michel, S. (2000). "Place, Politics, and Water Pollution in the Californias: A Geographical Analysis of Water Quality Politics in the Tijuana-San Diego Metropolitan Region." Unpublished Ph.D. dissertation, University of Colorado, Department of Geography at Boulder.

Micklin, P. P. (1992). "The Aral Crisis: Introduction to the Special Issue." *Post-Soviet Geography*, 33/5: 269–83.

Micklin, P. P., and Williams, W. D. (eds.) (1996). *The Aral Sea Basin*. NATO Workshop, Tashkent, 1994. New York: Springer Verlag.

Mitchell, J. E. (1998). "Early Childhood Diarrhea and Primary School Performance in the Northern Areas of Pakistan." Unpublished Ph.D. dissertation, University of Colorado at Boulder.

Montz, B. E., and Tobin, G. A. (1996). "The Environmental Impacts of Flooding: A Case Study of St. Maries, Idaho."

Papers and Proceedings of the Applied Geography Conferences, 19: 15–23.

Moran, S. D. (2000). "Fluid Categories: Water System Management in Post-Communist Poland." Unpublished Ph.D. dissertation, Clark University, Worcester, Mass.

Morehouse, B. J., Carter, R. H., and Spouse, T. W. (2000*a*). "Section 1: Dimensions of Borderland Water Conflicts: Past, Present, and Future—the Implications of Sustained Drought for Transboundary Water Management in Nogales, Arizona, and Nogales, Sonora." *Natural Resources Journal*, 40/4: 783–808.

—— (2000*b*). "Water Resources and Climate Change—Climate Impacts on Urban Water Resources in Southwest: The Importance of Context." *Journal of the American Water Resources Association*, 36/2: 265–79.

Mullens, J. B. (1995). "Implementation of Regional Plans in the Pacific Northwest: An Analysis of the Northwest Power Planning Council's Water Budget and Model Conservation Standards, 1984–1993." Unpublished Ph.D. dissertation, Oregon State University, Corvallis.

Muckleston, K. 1990. "Integrated Water Management in the United States." in B. Mitchell (ed.), *Integrated Water Management*. New York: Belhaven, 22–44.

Mustafa, D. (1998). "Structural Causes of Vulnerability to Flood Hazard in Pakistan." *Economic Geography*, 74/3: 289–306.

National Research Council: Water Science and Technology Board (1992). *Sustaining our Water Resources*. Washington, DC: National Academy Press.

—— (1996*a*). *River Resource Management in the Grand Canyon*. Washington, DC: National Academy Press.

—— (1996*b*). *A New Era for Irrigation*. Washington, DC: National Academy Press.

—— (1999*a*). *Water for the Future: The West Bank and Gaza Strip, Israel, and Jordan*. Washington, DC: National Academy Press.

—— (1999*b*). *New Strategies for America's Watersheds*. Washington, DC: National Academy Press.

—— (1999*c*). *Downstream: Adaptive Management for the Grand Canyon Ecosystem*. Washington, DC: National Academy Press.

—— (2000). *Watershed Management for Potable Water Supply: Assessing the New York City Strategy*. Washington, DC: National Academy Press.

O'Connor, R. E. (1999). "Water Resources and Climate Change —Weather and Climate Extremes, Climate Change, and Planning: Views of Community Water System Managers in Pennsylvania's Susquehanna River Basin and Others." *Journal of the American Water Resources Association*, 35/6: 1411–20.

Ollenburger, J. C., Tobin, G. A. (1997). "Water and Postdisaster Stress," in E. Enarson and B. Hearn Morrow (eds.), *The Gendered Terrain of Disaster: Through Women's Eyes* Westport: Praeger, 95–107.

Paul, Bimal Kanti (1998). "Coping with the 1996 Tornado in Tangail, Bangladesh: An Analysis of Field Data." *The Professional Geographer*, 50/3: 287 ff.

Perrault, T., Bebbington, A., and Carroll, T. (1998). "Indigenous Irrigation Organizations and the Formation of Social Capital in Northern Highland Ecuador." *Yearbook of the Conference of Latin American Geographers*, 24: 1–16.

Perritt, R. (1989). "African River Basin Development: Achievements, the Role of Institutions, and Strategies for the Future." *Natural Resources Forum*, 13/3: 204 ff.

Pitlick, J., and Van Steeter, M. (1998). "Geomorphology and Endangered Fish Habitats of the Upper Colorado River, 2. Linking Sediment Transport to Habitat Maintenance." *Water Resources Research*, 34/2: 303 ff.

Pitlick, J. (1997). "A Regional Perspective of the Hydrology of the 1993 Mississippi River Basin Floods." *Annals of the Association of American Geographers*, 87/1: 135 ff.

Planning and Management Consultants, Ltd. (1993). *Evaluating Urban Water Conservation Programs: A Procedures Manual*. Denver: American Water Works Association.

Platt, R. (1993). "Geographers and Water Resource Policy," in M. Reuss (ed.), *Water Resources Administration in the United States*. East Lansing: Michigan State University and American Water Resources Association, 36–54.

—— (1995). "The 2020 Water Supply Study for Metropolitan Boston: The Demise of Diversion." *Journal of the American Planning Association*, 61: 185–99.

Pulwarty, R. S., and Redmond, K. T. (1997). "Climate and Salmon Restoration in the Columbia River Basin: The Role and Usability of Seasonal Forecasts." *Bulletin of the American Meteorological Society*, 78: 381–97.

Rahman, M. (1993). "Irrigation and Farm Water Management in Pakistan." *Geojournal*, 31: 363–71.

Rajagopal, R., and Tobin, G. (1992). "Economics of Ground-Water Quality Monitoring: A Survey of Experts." *Environmental Monitoring and Assessment*, 22: 39–56.

Reitsma, R. F. (1996). "Structure and Support of Water-Resources Management and Decision-making." *Journal of Hydrology*, 177/3–4: 253 ff.

Rense, W. C. (1995). "Applications of the Thornthwaite Water Budget to Mountain Geosystem Analysis," in R. B. Singh (ed.), *Global Environmental Change: Perspectives of Remote Sensing and Geographic Information System*. New Delhi: Oxford-IBH, 57–70.

—— (1996). "The Role of Hydrology Education in the Environmental Sciences," in R. B. Singh (ed.), *Disasters, Environment and Development*. Proceedings of the IGU seminar, New Delhi, December 9–12, 1994. New Delhi: Oxford-IBH, 567–74.

Reuss, M. (1993). *Water Resources People and Issues: Interview with Gilbert F. White*. Fort Belvoir: Institute of Water Resources, Office of History, US Army Corps of Engineers.

Rhoads, B. L., and Herricks, E. E. (1996). "Naturalization of Headwater Agricultural Streams in Illinois: Challenges and Possibilities" in A. Brookes and Shields, D. (eds.), *River Channel Restoration*. Wiley, Chichester, 331–67.

Ridgley, M. A. (1989). "Evaluation of Water-Supply and Sanitation Options in Third World Cities: An Example from Cali, Columbia." *Geojournal*, 18/2: 199–211.

Riebsame, W. E., with Wescoat, J. L., Jr., and Morrisette, P. (1997). "Western Land Use Trends and Policy: Implications for Water Resources." Denver: Western Water Policy Review Advisory Commission.

Riebsame, W. E., *et al.* (1995). "Complex River Basins," in K. Strzepek and J. Smith, *As Climate Changes: International Impacts and Implications*. Cambridge: Cambridge University Press, 57–91.

Roberts, R., and Emel, J. (1992). "Groundwater Management in the Southern High Plains: Questioning the Tragedy of the Commons." *Economic Geography*, 68/3: 249–71.

Root, A. L., Berry, K. A., Markee, N. L., and Giewat, G. (1998). "Representing Ranching Communities: Comparing the Water Policy-Making Environment of County Commissioners with and without Ranching Constituencies." *Proceedings of the American Water Resources Association*. Reno, NY: American Water Resources Association. 317–24.

Sauder, R. A. (1994). *The Lost Frontier: Water Diversion in the Growth and Destruction of Owens Valley Agriculture*. Tucson: University of Arizona Press.

Schmid, J. A. (1994). "Wetlands in the Urban Landscape of the United States," in R. H. Platt, R. A. Rowntree, and P. C. Muick (eds.), *The Ecological City: Preserving and Restoring Urban Biodiversity*. Amherst: University of Massachusetts, 106–33.

Schmidt, J. C., Webb, R. H., Valdez, R. A., Marzolf, G. R., and Stevens, L. E. (1998). "Science and Values in River Restoration in the Grand Canyon." *BioScience*, 48/9: 735–47.

Schmidt, J. C., Andrews, E. D., Wegner, D. L., Patten, D. T., Marzolf, G. R., and Moody, T. O. (1999). "Origins of the 1996 Controlled Flood in Grand Canyon," in R. H. Webb *et al.* (eds.), *The Controlled Flood in Grand Canyon*. Monograph 110. Washington, DC: American Geophysical Union.

Sheaffer, J. R. and Stevens, L. (1983). *Future Water: An Exciting Solution to America's Most Serious Resource Crisis*. New York: W. Morrow.

Shelton, M. L. (1998). "Seasonal Hydroclimate Change in the Sacramento River Basin, California." *Physical Geography*, 19/3: 239–55.

Smith, D. (1994). "Change and Variability in Climate and Ecosystem Decline in Aral Sea Basin Deltas." *Post-Soviet Geography*, 35: 142–65.

—— (1995). "Environmental Security and Shared Water Resources in Post-Soviet Central Asia." *Post-Soviet Geography*, 36: 351–70.

Steinberg, P. E., and Clark, G. E. (1999). "Troubled Water? Acquiescence, Conflict and the Politics of Place in Watershed Management." *Political Geography*, 18/4: 83–114.

Sukhwal, B. L. (1993). "Water Resources and their Management in the Dry Region: A Case Study of the Thar Desert, Rajasthan, India." *Geographical Review of India* 55/3: 5–17.

Swyngedouw, E. (1997). "Power, Nature, and the City: The Conquest of Water and the Political Ecology of Urbanization in Guayaquil, Ecuador: 1880–1990." *Environment & Planning A*, 29/2: 311 ff.

—— (1999). "Modernity and Hybridity: Nature, Regeneracionismo, and the Production of the Spanish Waterscape (1890–1930)." *Annals of the Association of American Geographers*, 89: 443–65.

—— (2000). "Authoritarian Governance, Power, and the Politics of Rescaling." *Environment and Planning D: Society and Space*, 18: 63–76.

Templer, O. W. (1997). "Water Law and Geography: A Geographic Perspective," in Fred M. Shelley, Gary Thompson, and Chand Wije (eds.), *Geography, Environment, and American Law*. Boulder: University Press of Colorado, 61–78.

—— (1997). "Surface Water Development and Salinity on the High Plains and Rolling Plains of West Texas," in *Forum of the Association for Arid Lands Studies*, 13: 19–27.

Templer, O. W., and Urban, L. V. (1997). "Integrated Use of Surface and Groundwater on the High Plains of West Texas," in *Papers and Proceedings of the Applied Geography Conferences*, 22: 119–27.

Thompson, J., Porras, I. T., Tumwine, J. K., Mujwahuzi, M. R., Katui-Katua, M., Johnstone, N., and Wood, L. (2001). *Drawers of Water II: 30 Years of Change in Domestic Water Use and Environmental Health in East Africa*. London: International Institute for Environment and Development.

Thompson, S. A. (1998). *Hydrology for Water Managment*. Rotterdam: A. A. Balkema.

—— (1999). *Water Use, Management and Planning in the United States*. New York: Academic Press.

Tobin, G. A. (1995). "The Levee Love Affair: A Stormy Relationship." *Water Resources Bulletin*, 31/3: 359–67.

Tobin, G., and Montz, B. (1997). "Impacts of a Second 'Once in a Lifetime' Flood on Property Values: Linda and Olivehurst, California Revisited." *Papers and Proceedings of the Applied Geography Conferences*, 20: 83–90.

Tobin, G., and Rajagopal, R. (1993). "Groundwater Contamination and Protection Problems in a Small Rural Community." *The Social Science Journal*, 30/1: 113–28.

Tobin, G. A., Baumann, D. D., Damron, J. E., Emel, J. L., Hirschboeck, K. K, Matthews, O. P., and Montz, B. E. (1989). "Water Resources," in Gary Gaile and Cort Wilmott (eds.), *Geography in America*. Columbus: Merrill, 113–40.

Trimble, S. W. (1993). "The Distributed Sediment Budget Model and Watershed Management in the Paleozoic Plateau of the Upper Midwestern United States." *Physical Geography*, 14/3: 285–303.

—— (1997*a*). "Contribution of Stream Channel Erosion to Sediment Yield from an Urbanizing Watershed." *Science*, 278/21: 1442–4.

—— (1997*b*). "Streambank Fish-Shelter Structures Help Stabilize Tributary Streams in Wisconsin." *Environmental Geology*, 32/3: 230–4.

Trimble, S. W., and Mendel, A. C. (1995). "The Cow as a Geomorphic Agent—A Critical Review." *Geomorphology*, 13: 233–53.

Tuan, Yi-Fu (1968). *The Hydrologic Cycle and the Wisdom of God*. Toronto: University of Toronto, Department of Geography.

—— (1984). *Dominance and Affection: The Making of Pets*. New Haven: Yale University Press.

Vander Velde, E. J., Svendsen, M. (1994). "Goals and Objectives of Irrigation in Pakistan—A Prelude to Assessing Irrigation Performance." *Quarterly Journal of International Agriculture [Zeitschrift für ausländische Landwirtschaft]*, 33/3: 222 ff.

Vander Velde, E. J., and Johnson, R. J. (1992). "Tubewells in Pakistan Distributary Canal Commands: Lagar Distributary, Lower Chenab Canal System, Punjab." *IIMI Working Paper*, 21. Colombo: International Irrigation Management Institute.

Votteler, T. H. (1998). "The Little Fish that Roared: Endangered Species Act, State Groundwater Law, and Private Property Rights Collide over the Texas Edwards Aquifer." *Environmental Law*, 28/4: 845–79.

Wallach, B. (1996). "Irrigation," in *Losing Asia: Modernization and the Culture of Development*. Baltimore: Johns Hopkins University Press, 55–102.

Waterstone, Marvin (1992). "Water in the Global Environment." *Pathways in Geography Series*, 3. Indiana, Pa.: National Council for Geographic Education.

Wescoat, James L., Jr. (1992). "Common Themes in the Work of Gilbert White and John Dewey: A Pragmatic Appraisal." *Annals of the Association of American Geographers*, 82: 587–607.

—— (1996). "Main Currents in Multilateral Water Agreements: A Historical-Geographic Perspective, 1648–1948." *Colorado Journal of International Environmental Law and Policy*, 7: 39–74.

—— (2000). "Wittfogel East and West: Changing Perspectives on Water Development in South Asia and the United States, 1670–2000," in A. B. Murphy and D. L. Johnson (eds.), *Cultural Encounters with the Environment: Enduring and Evolving Geographic Themes*, Lanham: Rowman & Littlefield, 109–32.

Wescoat, J. L., Jr., Halvorson, S. J., and Mustafa, D. (2000). "Water Management in the Indus Basin of Pakistan: A Half-Century Perspective," *International Journal of Water Resources Development*, 16: 391–406.

Wescoat, J. L., Jr., Halvorson, S. J., Headington, L., and Replogle, J. (2002). "Water, Poverty, Equity and Justice in Colorado: A Pragmatic Approach," in K. Mutz and G. Bryner (ed.), *Justice and Natural Resources*. Covello, Calif.: Island Press, 57–85.

White, G. F. (1957). "A Perspective of River Basin Development." *Law and Contemporary Problems*, 22: 157–88.

—— (1963). "Contributions of Geographical Analysis to River Basin Development." *Geographical Journal*. CXXIX: 412–36.

—— (1998). "Reflections on the 50-year International Search for Integrated Water Management." *Water Policy*, 1: 21–7.

—— (1997). "Watersheds and Streams of Thought," in H. N. Barakat and A. K. Hegazy (eds.), *Reviews in Ecology: Desert Conservation and Development*. Cairo: UNESCO, 89–98.

White, Stephen E. (1994). "Ogallala Oases: Water Use, Population Redistribution, and Policy Implications in the High Plains of Western Kansas, 1980–1990." *Annals of the Association of American Geographers*, 84/1: 29–45.

White, Stephen E., and Kromm, D. (1995). "Local Groundwater Management Effectivenes in the Colorado and Kansas Ogallala Region." *Natural Resources Journal*, 35: 275–307.

Wilhite, D. A., Hayes, M. J., Knutson, C., Smith, K. H. (2000). "Dialogue on Water Issues—Planning for Drought: Moving from Crisis to Risk Management." *Journal of the American Water Resources Association*, 36/4: 697–710.

Williams, M., Losleben, M., Caine, N., and Greenland, D. (1996). "Changes in Climate and Hydrochemical Responses in a High-Elevation Catchment in the Rocky Mountains, USA. *Limnology and Oceanography*, 41/5: 939–46.

Wolf, A. T. (1995). *Hydropolitics along the Jordan River: The Impact of Scarce Water Resources on the Arab-Israeli Conflict*. Tokyo: United Nations University Press.

—— (1997). "International Water Conflict Resolution: Lessons from Comparative Analysis." *International Journal of Water Resources Development*, 13/3: 333–66.

—— (1999*a*). "Criteria For Equitable Allocations: The Heart of International Water Conflict." *Natural Resources Forum*, 23/1: 3–30.

—— (1999*b*). "The Transboundary Freshwater Dispute Database Project." *Water International*, 24/2: 160–3.

Wolman, M. G. (1995). *Natural and Anthropogenic Influences in Fluvial Geomorphology: The Wolman Volume*. Geophysical Monograph, 89. Washington, DC: American Geophysical Union.

Woltemade, C. J. (1994). "Form and Process: Fluvial Geomophology and Flood-Flow Interaction, Grant River Wisconsin." *Annals of the Association of American Geographers*, 84: 462–79.

—— (1997). "Integration of Hydrology, Geomorphology and Aquatic Ecology in Management of the Upper Mississippi River Basin." *Water Resources Education, Training, and Practice: Opportunities for the Next Century*. Proceedings of the AWRA. 29 June–3 July, 455–64.

Yarnal, B. (1992). "Human Dimensions of Global Environmental Changes in the Susquehanna River Basin: A Call for Research." *Pennsylvania Geographer*, 30/2: 19–34.

Yarnal, B., Lakhtakia, M. N., Yu, Z., White, R. A., Pollard, D., Miller, D. A., and Lapenta, W. M. (2000). "A Linked Meteorological and Hydrological Model System: The Susquehanna River Basin Experiment (SRBEX)." *Global and Planetary Change*, 25/1: 149 ff.

Yarnal, B., Johnson, D. L., Frakes, B. J., Bowles, G. I., and Pascale, P. (1997). "The Flood of '96 and its Socioeconomic Impacts in the Susquehanna River Basin." *Journal of the American Water Resources Association*, 33: 1299–312.

CHAPTER 20

Energy Geography

Barry D. Solomon, Martin J. Pasqualetti, and Deborah A. Luchsinger

Introduction

Fossil fuels powered the Industrial Revolution and they continue to dominate our lives as we enter the twenty-first century. Yet there are clear signs that the grip they have on every sector of society must soon relax in favor of other energy sources. Such a transition will not come because we are running out of fossil fuels, but rather because the environmental and social costs of their rapid use threaten our very existence on the planet.

This is an expected development. From the time when fossil fuels first enabled and magnified humans' dominion over the earth, the costs they brought—as any good economist would argue—have been inseparable from their benefits. Although the benefits were explicit and the local costs were experienced by many, it was not until skilled writers such as Zola, Orwell, Llewellyn, and Dickens vividly portrayed them that their widespread and pernicious nature was broadcast to those outside their immediate reach. Nowadays the problems we are grappling with have spread to the global scale, including atmospheric warming, thinning ozone, and rising exposure to above-background radioactivity.

Understanding earth–energy associations is a task well matched to the varied skills of geographers. The worth of such study is increasingly apparent as the world's human population continues to rise, as fossil fuels become more difficult to wrest from the earth, and as we continue to realize that there will be no risk-free, cost-free, or impact-free rabbits coming out of the alternative energy hat. In this chapter, we review developments in energy geography in the US and Canada as posted to the literature since the first edition of *Geography in America*, including a sprinkling from overseas to provide context.

Owing to the fundamental nature of energy, we have accordingly cast a wide net in our background research, albeit with some boundaries. For example, while we discuss several important contributions to energy research by physical and environmental geographers, we excluded consideration of such themes as energy budgets, most climate change research, and mine-land reclamation and radioactive waste transport studies by hydrologists and geomorphologists. Since most of these topics are covered by other specialties within North American geography, we focus on contributions by human geographers.

Energy Studies and Specialists

Core topics of energy geography have traditionally included resource development, power-plant siting, land use, environmental impact assessment, energy distribution, and transport, spatial patterns of consumption, and diffusion of conservation technologies. During the 1980s the radius of this core expanded to include risk perception and emergency behavior. While some of this latter work has been continued by Pijawka and Mushkatel (1991), Kasperson and Kasperson (1996),

Metz (1996), and Greenberg *et al.* (1998), among others, its emphasis has diminished. In part this is in response to changes in technology, economics, policy, and international activities; a confidence that the essential characteristics of these issues have been identified; and the absence of any major accidents or incidents in the last ten years. As a result, some geographers who figured prominently in the energy chapter of the first edition of *Geography in America* have redirected their time and attention away from energy studies. Being part of a rather small specialty, North American energy geographers are a scattered tribe. Nevertheless, a few institutions stand out—Boston University, Arizona State University, University of Toronto, and especially Oak Ridge National Laboratory.

Within the published record, mostly non-geography, interdisciplinary journals have been the dominant outlet for work in this field, but the maturity that has come to the subdiscipline is demonstrated by an increasing number of outstanding books. Particularly substantive of the recent offerings have been several books by Smil, including one on energy in the biosphere (1999) and one on the history of energy use (1994*b*). Also impressive has been Odell's classic study of OPEC oil producers (Odell 1986); Kuby's 1996 learning module on population growth, energy use, and pollution for the human dimensions of global change component of the Commission on College Geography; and a review and retrospective overview of energy by Wilbanks (1988).

The most dominant theme of books by energy geographers has been nuclear power, not a surprising development, given public and governmental attention and research sponsorship. The most comprehensive book on this topic is a global review by Mounfield (1991), a British geographer. Others include several on radioactive wastes (e.g. Jacob 1990; Blowers *et al.* 1991; Openshaw *et al.* 1989); nuclear accidents (Gould 1990); and nuclear power-plant decommissioning (Pasqualetti 1990; Pasqualetti and Pijawka 1996). That several of these authors are from the UK suggests that geographers there continue to make at least as strong a contribution to our understanding of energy issues as their American counterparts, despite the smaller number of professional geographers who make their home there.

Finite Energy Resources

Conventional energy resources in the US include the non-renewable fossil fuels of petroleum, natural gas, and coal, as well as falling water and that which derives from the binding energy in atoms. Together these sources supply 95 per cent of domestic energy consumption.

Much of the attention to these resources has traditionally focused on economics and availability. Catalysts expanded in more recent years to include geopolitical issues more dominantly, especially those focused on the instabilities in the Middle East and the worries about nuclear power after the reactor explosion at Chernobyl. In response, more attention has focused on the newly-discovered "resources" of energy efficiency and conservation. As the 1990s drew to a close, however, public interest in energy matters was modest, perhaps reflecting the low cost of oil.

Despite such changes, a core of geographers continued contributing in all phases of energy studies including fossil fuels. The most visible work by geographers on petroleum, focusing on the scarcity of the resource base, is centered at Boston University. This line of research was inspired by the seminal work of M. King Hubbert, a petroleum geologist who predicted with startling accuracy as early as 1956 the ultimate size and production peak of oil resources for the US and the world. The historical production pattern he suggested would follow a classic bell-shaped curve, reflecting the interplay over time of resource depletion, real oil prices, technical change, and political decisions on long-run production costs (Kaufmann 1991; Cleveland 1991). These authors have long forecast that the yield (amount of oil added to proven reserves) per effort from additional domestic well drilling will cease to be a net source of energy early in the twenty-first century. Thus, domestic oil discovery, production, and proven oil reserves are on an inexorable downward path (Cleveland and Kaufmann 1991; Cleveland 1993; Ruth and Cleveland 1993).

Given the increasingly dismal domestic supply outlook, major US petroleum companies have increasingly turned to foreign sources of crude oil (Solomon 1989). While over half of all US oil is now imported and import dependence is forecast to reach 70 per cent by 2010, US suppliers rely much less on the volatile Middle East than they did in recent decades. Currently, a major source of oil imports is Venezuela, the largest OPEC producer in the Western Hemisphere. The trend toward globalization of the oil industry is likely to continue, and policies to reverse it are likely to have significant economic and environmental costs (Kaufmann and Cleveland 1991). Such trends beg the question of when the national security costs of increased foreign dependence will force changes in US energy policy. After all, Middle Eastern OPEC nations still hold most of the world's oil reserves, and these nations have been regaining lost market share (Odell 1986, 1992; Solomon 1989; Greene 1997; Greene *et al.* 1998). This is a concern not only for the US, but also for Western Europe and Japan (Lakshmanan and Han 1995; Lakshmanan and Andersson 1991–2). Many of these OPEC countries are well poised to expand oil production, and only with great difficulty have they been adjusting to slack oil market conditions (Auty 1991).

A search for foreign oil supplies outside OPEC has led us to Mexico, Canada, and the former Soviet Union. Economic and political risks have discouraged foreign investors in Mexico. Canadian oil development has been focusing on their northern hinterlands, much as the US has been exploring the resource potential of Alaska's Northern Slope. Two recent studies (DiFrancesco 1998; DiFrancesco and Anderson 1999) examined the interregional economic impacts of oil resource development nearby, in the Beaufort Sea. Using multiregional input–output analysis, the authors showed that little or none of the net economic benefits may accrue to Canada's Northwest Territories. Sagers (1993*b*) examined the oil, gas, and coal industries in the former Soviet republics, which are struggling to sustain production levels in the post-socialist era, especially in Russia where most of the huge fossil-fuel reserves are concentrated. The dilapidated state of Russian and Ukrainian oil refineries is common, with capital lacking for much-needed modernization (Sagers 1993*a*). Such conditions are especially troubling given the well-known inefficiencies of the region's energy transportation system (Sagers and Green 1986).

Fortunately, many energy alternatives to the world's continued oil addiction are justifying their increasing attention. For example, natural gas, although largely ignored until the 1950s, enjoyed a boost in the US and Britain in the 1990s (Spooner 1995). As exploration for gas not associated with oilfields increased proved reserves have been revised steadily upward. Yet despite a modest upswing in US gas production and use in the 1990s, it too is subject to depletion, and yield per effort is declining for reasons similar to that for oil (Cleveland and Kaufmann 1997). Russia, with the largest gas reserves in the world, has been regrettably unable to overcome the variety of economic modernization problems that beset its other energy industries (Sagers 1995).

Coal, as the fossil fuel of greatest abundance in the US, commands attention. In contrast synthetic fuels from coal, touted as substitutes for foreign imports in the late 1970s and early 1980s, have failed to materialize, probably victim to the low energy prices of the past fifteen years. The strong research tradition of American geographers on the topics of coal and electric power has focused on operations research methods to model optimal production, networks, and shipment decisions. Thus, many geographers have modeled such problems. Kuby *et al.* (1991) for example, demonstrated the utility of mathematical programming models in capturing the complexities of the coal transportation energy system. Later, Kuby *et al.* (1993) developed a large mixed-integer programming model for China's coal and electricity distribution system. The latter model has been used by the World Bank and Chinese government officials, and was later extended to consider energy efficiency investments (Xie and Kuby 1997). In the US, a model for optimal electric utility planning under environmental constraints in the US was developed by Hillsman *et al.* (1994), with alternative configurations reviewed by Hobbs (1995).

Since coal supplies the majority of our electricity, it is tightly interwoven with geographic issues of the power industry, especially since the industry began to deregulate and restructure in the mid- to late 1990s (Elmes 1996). Restructuring the US coal and electric power industries has been a gradual process in response to free trade agreements (Calzonetti 1990), changing environmental and economic regulations, as well as market forces (Elmes and Harris 1996). These forces accelerated in the 1990s. In their factor analysis of the coal industry, Elmes and Harris (1996) demonstrated the importance of such national, economic, and political forces on regional markets. The deregulation of the US electric utility industry promises to bring even greater spatial and economic impacts. As shown by several geographers (e.g. Elmes 1996; Solomon and Heiman 2001), this industry currently can be characterized by instability and uncertainty as competitive forces increase in power generation and transmission access, if not transmission services. Also unclear (e.g. Hwang *et al.* 1995), is whether or not inefficient spatial pricing practices or demand-side management conservation can survive in a deregulated environment. Early results have not been encouraging, as California, the first major state to deregulate, experienced serious market disruptions and apparent electricity shortages during 2000–1 (Solomon and Heiman 2001).

Environmental externalities are those problems not accounted for in the market price of energy, such as reduced visibility resulting from power plant emissions. Once these social costs are correctly internalized into market prices, producers and consumers presumably will adjust their decisions to reflect the true costs of energy resources. Geographers have joined economists in examining such issues, most prominently in the context of energy policy issues raised by global climate change, and sulfur dioxide control under the US Clean Air Act Amendments (CAAA) of 1990. In a detailed multi-year study of fuel-cycle externalities sponsored by the US Government and the European Community, Lee (1996, 1998) demonstrated the feasibility of the damage-function approach. His work showed that these costs could range from 3 to 30 per cent of the cost of electricity generation, depending on the proximity of the power plants to human populations, and the technology used. Such values are conservative, however, since they admittedly underestimated the cost of CO_2 emissions on the environment and human health. Wilbanks (1990) and Das and Wilbanks (1997) have examined the issues and options for environmentally sound power development in Pakistan and India. Their focus on institution building, institutional reform, and privatization provides insightful lessons for developing nations, as well as economies in transition such as those of the former Soviet Union (e.g. Pryde 1991). One recently successful approach used in the US has been to market on a voluntary basis electricity generated from clean, renewable energy resources to consumers at slight price premiums. Systems to certify so-called green power have been adopted in California, Pennsylvania, New Jersey, and Connecticut.

Hobbs (1994) showed that in light of the CAAA cap on total nationwide SO_2 emissions the appropriate environmental cost adder for these emissions may be either positive or negative, depending on the location. This results because the market now sets a price for the constrained SO_2 emissions, though it does not necessarily account adequately for the external costs and is not a spatial price. The emission price is for an allowance, equal to the right to emit one ton of SO_2, the trading of which is a prominent feature of the CAAA. As Solomon (1998) has shown, trading has been heaviest in the northeast states, Ohio, Illinois, California, and Arizona. This trading program is a major geographic experiment, and serves as an institutional model for the reduction of greenhouse warming gases such as CO_2 and CH_4 (Solomon 1995; Solomon and Lee 2000). In an international context, Kaufmann *et al.* (1998) have shown that the spatial intensity of economic activity, rather than income, provides the impetus for policies and technologies that reduce SO_2 emissions. The authors conducted econometric analysis using panel data for twenty-three nations over fifteen years.

Environmental geographers have been heavily involved in an international dialogue of identifying appropriate energy policy responses to global warming and climatic changes (Smil 1990; Wilbanks 1992*a*; Harvey 1992), including some climatologists who have argued that the warming concern is exaggerated (Balling 1992). The basis for the policy concern is a substantial scientific research record that includes much work of climatologists on different sides of the issue (cf. Jain and Bach 1994; Balling 1994). Geographers (and others) have recognized that many preferred policy options in a greenhouse-constrained world also reduce conventional air pollutants and have a variety of other ancillary benefits (Wilbanks 1992*b*). These options include increased natural gas use, energy conservation, and the further development of renewable and possibly nuclear energy sources (Solomon 1992; Schwengels and Solomon 1992). Despite this fairly obvious strategy, the implementation of the Framework Convention on Climate Change of 1992 (FCCC) has been slowed by the US recalcitrance and difficulty in crafting a fair and efficient protocol for international reductions in greenhouse gas emissions (Solomon and Ahuja 1991). Nevertheless, agreement was reached at a conference in Marrakesh, Morocco, in late 2001.

Three major types of greenhouse-related energy studies have dominated the work of geographers: detailed technology assessments at the US national level, sectoral-specific studies, and foreign, national-level case studies. The leading example of the first study type is Brown *et al.* (1998), which provided an assessment of the potential impact of advanced energy technologies on greenhouse gas reductions. The authors found that large carbon reductions are possible at incremental costs below the value of energy saved, especially efficiency measures. As former President Clinton acknowledged, this work provided a foundation for the US commitment to reduce emissions made at the 1997 Kyoto, Japan, summit on global climate change. Sectoral-specific studies have further highlighted the role of energy efficiency technologies (Wilbanks 1994*a*) and the buildings, industrial, and transportation sectors in particular (Schwengels and Solomon 1992; Hillsman and Southworth 1990; Hillsman 1995*a*; Lakshmanan and Han 1997). In his comparative assessment of advanced fossil-fueled, photovoltaic, and solar-hydrogen technologies for electricity generation, Harvey (1995) focused on the conditions under which these technologies would be cost-competitive and thus used to cut emissions.

Since global warming and climatic changes do not respect political boundaries, case studies can be instructive wherever they are conducted. For example, Bach (1995), Harvey *et al.* (1997), and Smil (1994*a*) sketch the broad range of policies that would be required significantly to reduce emissions in Germany, Canada, and China, respectively. These studies not only underscore the great difficulty of significant emission reductions, but they also emphasize the necessity to implement policy reforms outside the energy sectors, including such matters as urban design and transportation policies. To accelerate the process of emissions reduction in developing countries and economies in transition, the FCCC approved a pilot program in joint implementation of the treaty, whereby two or more countries cooperatively reduce or sequester emissions in their countries. Implementation of this pilot program has been slowed, however, by concerns among developing countries that national sovereignty would be violated and that Western nations or Japan would gain a competitive economic advantage. Geographers have generally been sympathetic to these concerns, questioning the policy's fairness and cost-effectiveness (Harvey and Bush 1997; Lofsted *et al.* 1996), while identifying constructive solutions that involve compromises by the applicable countries (Lee *et al.* 1997).

Among conventional energy resources, the only renewable source that is well established in North America is hydroelectricity, accounting for 7 per cent of US power supply and 61 per cent in Canada. Hydroelectric installations, however, are often not sustainable because gradual siltation of their reservoirs may limit facility lifetime to 50–200 years. Since the 1980s, hydroelectric power has attracted increasingly negative attention in the US and elsewhere, where the large part of its potential already has been tapped. Identified shortcomings include forced migration of people, displacement and destruction of salmon stocks, and downstream ecological changes.

Human geographers have written little on hydropower facility construction in the last decade, with the major exception of some mammoth dam projects in developing countries (e.g. Sternberg 1996–7; Elhance 1999). Instead, emphasis in the US has shifted from dam construction to dam removal and downstream restoration (Kellner 1994). Several dams have been suggested for demolition, although attention has been focused on small structures such as the Elwha Dam in Washington (Pohl 1999). Of the large dams, the most prolonged attention has focused on the removal or neutralization of Glen Canyon Dam in Arizona. Such opposition was fomented by the flooding of spectacular Glen Canyon upstream and by the significant habitat changes in Grand Canyon downstream, the latter the subject of recent research by physical geographers at Arizona State University, such as Will Graf.

Of all present-day energy resources, nuclear fission power has put more geographers into the mainstream of energy studies than any other subject, and this includes numerous contributions by our British colleagues such as Mounfield, Openshaw, O'Riordan, Blowers, and Fernie. While nuclear power was once the hottest topic among energy geographers' themes, it has cooled off as many of the issues came to be conceptually understood and were edged more into the political arena. In part, this move reflected the passing peak of interest attracted first by the accident at Three Mile Island (TMI) in 1979 and Chernobyl in 1986. In contrast to the well-regarded contributions by US geographers, especially following TMI (e.g. Johnson and Zeigler 1989), little has been published recently. The fall-off in geographic studies of nuclear power seems to signal, regrettably, that geographers are leaving to others the still-fertile geographic themes of land contamination, atmospheric dispersal of radioactive emissions, site remediation, shipment and disposal of contaminated materials, nuclear policy decisions, human health and safety, and power-plant siting. Much of the recent work on nuclear power has focused outside the US (e.g. Pryde and Bradley 1994, for the former Soviet Union; Kuhn 1998, for Canada; and Lofstedt 2001 for Sweden). Marples (1996), a historian who occasionally publishes in geography journals, has done some of the best work on Chernobyl.

Part of the waning interest in nuclear power is undoubtedly a result of the cancellation of over a hundred nuclear reactors since TMI (see e.g. Obermeyer 1990), and the success of demand-side management in postponing the need for many new energy facilities. These factors have also moved power-plant siting out of the geographic spotlight (cf. O'Riordan *et al.* 1988).

The only nuclear issues continuing to receive attention are those associated with the back-end of the fuel cycle, including the persistent problem of radioactive waste disposal. This latter attention has been a response to the urgency that has been placed on this matter by the nuclear industry and various governments, and because the topic embraces such a wide spectrum of spatial considerations. These factors include repository site selection (Blowers *et al.* 1991; Jacob 1990; Openshaw *et al.* 1989; Cook *et al.* 1990; Flynn *et al.* 1997; Kuhn 1998); waste transport (Brainard *et al.* 1996); risk perception (Kasperson and Kasperson 1996; Greenberg and Gerrard 1995; Metz 1996; Pijawka and Mushkatel 1991), changes in property values (Metz and Clark 1997); and long-term stewardship (Pasqualetti 1997).

Several new nuclear topics have also gained stock, a reflection of society's more complete sense of understanding of the entire nuclear fuel chain. Two topics predominate here. The first is power-plant decommissioning, which considers not just the disposal of the wastes produced by power plants but the safe disposal of the reactors themselves. The most complete consideration of this topic has been by Pasqualetti, especially on the socioeconomics (1989, 1990) and land use (Pasqualetti and Pijawka 1996).

Nuclear site remediation is presently the most visible topic of interest. This interest largely concerns the clean-up and restoration of former nuclear weapons sites (Greenberg *et al.* 1998). It is the defense-industry equivalent of power-plant decommissioning, and includes a great deal of site decontamination. Geographers are particularly well suited to lend their expertise in this arena, but as yet such contributions have been few and mostly consultative.

Development of Sustainable Energy Resources

Renewable energy technologies have substantially different spatial characteristics than fossil and nuclear fuels, characteristics that are not yet fully explored. Geographers have made some contributions in this regard, largely in the 1980s, highlighted by Pryde's fine book on the subject (1983), but there is much more to do. Some of that work has been by physical geographers, i.e. specialists in climatology, geomorphology, hydrology, and biogeography. Indeed, concerns about global warming and nuclear hazards have prompted renewed attention to renewable resources.

Recent work by North American geographers on alternative energy sources has focused on the changing geographies of power generation (Elmes 1996; Kuhn 1992), and the process of commercializing of new energy technologies (Brown 1997). Brown notes that the movement of new technological innovations from government-sponsored research programs into the private sector has several obstacles to overcome, most notably the lack of commercialization opportunities (Macey and Brown 1990).

Of the many forms of renewable energy, biomass has received the most attention by human geographers, in part because it is the major energy resource *currently* used in many developing countries and is common in a few Western nations such as Scandinavia (Kasanen and Lakshmanan 1989). Along with its relative abundance, biomass also has challenging limitations, owing to its wide variety of forms ranging from wood to crops to garbage. Alternative pricing for conventional fuels is creating a growing opportunity for biomass fuel penetration in agricultural markets. Woodfuel use and sustainable development has been addressed most broadly by Jones (1994); by Hosier (1992), and Osei (1996) for Africa; and by Hosier and Bernstein (1992) for Haiti. In the Western context of Sweden, Lofstedt (1998) addressed the controversy regarding the use of biomass technology, and appropriate methods for communicating its risks to the public.

Wind energy has been for several years the fastest growing energy source to be utilized in the world, and in the late 1990s Germany overtook the US as global leader. A comparison and critique of the varying policy approaches to renewable energy development in Germany and the US was recently provided by Heiman (2002). Wind energy use and public perception in California, the location of its most extensive US development, has occasionally been examined by geographers, most recently in articles and a new book that examine the issue of aesthetic intrusion on the land (Pasqualetti 2000, 2001; Pasqualetti *et al.* 2002). Extensive study of wind energy also has been made by climatologists, for example by Robeson and Shein (1997) and Barthelmie (2002) who looked at the influence of climatic variability on wind resources in the US Great Plains, and offshore Europe areas of great wind-power potential.

Like wind and hydropower, immobility presents geographic challenges for the development of geothermal energy, particularly the pressure it puts on natural resources. Work of geographers on this topic has slowed, however, most of it being summarized in the first edition of this volume. Geographers have contributed only two recent studies, one which looked at the problems of air pollution and chemical-waste contamination (Pasqualetti and Dellinger 1989), and the other which examined the challenges of energy densities for direct (non-electric) applications (Somer 1999).

Net energy analysis techniques were used by Cleveland and Herendeen (1989) to calculate the energy return on investment (EROI) for solar parabolic collector systems over a range of output temperatures and geographic areas. They found that the EROI over three successive generations has significantly improved. Despite many inviting and urgent geographic questions about solar energy development since then, however, geographers have been concentrating their attention elsewhere. This may well change in coming years, especially as solar energy is combined with hydrogen, itself a viable future energy carrier (Harvey 1995). As with wind, solar (especially photovoltaics) is becoming increasingly cost-effective for consumers. For now, however, the three most prominent emerging energy topics are efficiency, sustainability, and industrial ecology.

Sustainable Development Theory and Industrial Ecology

Sustainable development (SD) is a concept that should be embraced by all people of the Earth, not just academic geographers. Roughly speaking, SD is concerned with meeting economic and social needs of all people in an ecologically sustainable manner. As such, it is at least in part a geographic concept involving energy and resource use. In his comprehensive review, Wilbanks (1994*b*) offered eight postulates toward a geographic system theory of SD: it must be pursued in open systems; it is a process, not a product; feedback mechanisms should be developed to improve system quality; it will exhibit considerable geographic differentiation; major decision points occur during times of stress; such stresses are more a function of rates of change in the parameters of a locality than of magnitudes of change; rates of change that can be assimilated vary among systems according to their resilience to rapid change; and in the long run SD is probably unrealizable in most localities until it is also approached in most others.

A physical view of sustainability has been proposed by Hannon *et al.* (1993), who suggested that the entropy rate of a climax ecosystem should be the reference limit for the sustainability of economic systems. Taking this one step further, Kates *et al.* (2001) call for more extensive research focus on place-based, sustainability science, which should be reconnected to the political agenda for SD.

Sustainable development thus requires that, by necessity, the energy and materials flows in economic systems must be examined in great detail (Ruth 1993; cf. Wilbanks 1992*b*; Osei 1996). An increasing number of such studies have been completed in the new and related field of industrial ecology. The first major sector study of energy and materials flow was the input–output analysis by Hannon and Brodrick (1982) of the iron and steel industry. They examined the increasing role and potential for recycling and energy conservation in this industry. An update by Ruth (1995) using dynamic econometric methods was pessimistic on the prospects for substantial further savings. Other dynamic modeling case studies have been completed for the copper, aluminum, lead, zinc, glass, and pulp and paper sectors, with an increasing emphasis on de-materialization and its technological limits (Cleveland and Ruth 1998).

We have briefly noted the potential roles of energy efficiency (a technology and innovation response) and conservation (a behavioral response) in the contexts of global warming, climate change, and sustainable development. Such strategies can be justified on other grounds, such as for their economic, pollution prevention, and even comfort merits. Indeed, conservation and energy efficiency have been among the major sources of energy services in the US since the 1970s oil shocks, though this is often not reflected in conventional energy statistics. Geographers have contributed to our understanding of the potential for and effectiveness of various options for energy efficiency and conservation, mostly focusing on evaluation of assistance programs.

The early important work of Hannon (1975) used input–output analysis to analyze the direct and indirect energy requirements of various personal consumption activities, and the vexing problem of how to promote conservation with rising incomes. The energy activities under the control of consumers, and thus most amenable to personal energy conservation, are found in the home and in transportation. A case study in the Pacific Northwest region (Brown 1993) found that marketing was more effective than building codes in stimulating the purchase of more energy-efficient new homes by wealthier consumers. Another long-term study evaluated the cost-effectiveness of the US Department of Energy's (USDOE) low-income home weatherization assistance program. While overall the program was found effective, the value of using alternative evaluation methods and different stakeholder perspectives was shown (Brown and Berry 1995; cf. Hobbs and Horn 1997).

A series of studies has considered the effectiveness of utility-sponsored demand-side management programs (DSM), focusing on energy-efficiency technologies. A summary of these extensive programs in California showed that overall they performed very well in the early 1990s (Brown and Mihlmester 1995), before deregulation diminished such efforts. Such changes underscore the difficulty of evaluating persistence of program impacts (Brown *et al.* 1996), though regulatory guidance is available (Solomon and Meier 1995). With less utility funds available for DSM programs it may be necessary to leverage them with public conservation funds. If so, special difficulties arise in evaluating coordinated DSM programs (Hill and Brown 1995).

The continued effectiveness of energy efficiency in a growing economy depends on technological innovation, which is poorly understood. Brown (1997) has evaluated the USDOE Energy-Related Inventions Program, a long-term commercialization assistance program. She found that this program has had an 8 : 1 return in terms of market sales to program costs, among other economic benefits. Others have considered the technological challenges in the case of industry (e.g. Hannon and Brodrick 1982; Ruth 1995) and transportation (Hillsman 1995*b*), though adequate spatial data are often lacking (Southworth and Peterson 1990). Others have pointed out the importance of examining not just technology, but consumers (Pasqualetti *et al.* 1995). Finally, international case studies of energy conservation have not been ignored. Such studies of the former Soviet Union (Tretyakova and Sagers 1990) and China (Xie and Kuby 1997) underscore the difficulty of achieving significant savings in centrally-planned economies.

Challenges and Future Progress

Many rich veins wend through the energy landscape, and the tools and perspectives of geographers are particularly appropriate to the task of finding and mining them. Among the veins emerging recently are those related to the restructuring of the electric utility industry through privatization and competition. The sea change of deregulation already has encouraged substantial spatial competition in the marketing of energy services, including green power (i.e. power supplied by renewable resources) in an increasing number of states. A related trend that may accelerate is the movement away from large central power plants to distributed power generation. In addition, the deregulation of this industry raises vexing questions of power reliability and regional supply vulnerability. These are among the fine new ores for geographic assay, particularly the encouragement of sustainability and the development of green power. With a little imagination we can envision a downsized and much more distributed power-generation base and a more diverse resource-development portfolio (Wilbanks 1988), and the enlightening contributions geographers can make. Along with deregulation has come increased attention to cross-border energy issues, especially between the US and Mexico. Whereas Canada supplies the US with substantial amounts of natural gas and electricity (Calzonetti 1990), Mexico so far has been only a small trading partner in these commodities. This is now changing as the US looks toward Mexico with the thought of tapping more substantially into that country's growing reserves of natural gas and crude oil. More immediately, however, power plants are being constructed just south of the border that will be delivering all their electricity to California, thereby speeding up the permitting process and avoiding the stricter environmental requirements to the north (Pasqualetti 2002).

The perspectives and skills of geography are especially suited to addressing many questions that accompany increased attention to solar energy, geothermal energy, and wind power. Each resource presents different problems than those now supplying us, solar being spatially ubiquitous and the other two being spatially concentrated and site-specific. The geographers' art and science, including the use of geographic information systems, is especially applicable to such research.

Four other themes deserve special mention. In the recent past, while some energy issues such as oil depletion, global warming, and radioactive waste disposal have received substantial attention from geographers, other timely topics such as energy and urban form have been seriously neglected. The latter topic has stimulated only a few recent government-sponsored studies (e.g. Southworth 1995) and one book by a geographer, herself British (Owens 1986). Energy geographers and land-use planners should join forces to address the many relationships that arise.

A second important topic that has gone largely unstudied by contemporary geographers is the vulnerability of different types of energy systems to natural and anthropogenic disasters. This subject is an increasing concern to the insurance industry, in light of the increase in extreme weather events that has been accompanying suspected global warming and climate changes.

A third attractive energy theme for geographers is less a resource than an organizational framework. It falls under the rubric "landscapes of power." Such landscapes are created during the development, distribution, and use of energy. Naturally, they include mines, but there are many other examples as well, including land-use changes, visual intrusions, transmission corridors and port facilities, site decommissioning and remediation, and the influence of energy demand on city form and function. Landscapes of power have the breadth and adhesion to become the lingua franca of energy geography.

A fourth and most general theme worth more attention by geographers is energy education. Although several energy courses are offered around the US they have become fewer in number since the first edition of *Geography in America*. In print, only Kuby (1996) and Pasqualetti *et al.* (e.g. 1995) have recently contributed on this topic. The latter article developed from several years of work in the private sector developing a computer-oriented education program that used the promise of environmental benefits to encourage energy conservation and efficiency.

This chapter has suggested throughout that despite the many contributions that geographers have made to energy studies, there is still much left to do. Indeed, there are so many bright and attractive research topics crisscrossing the energy landscape that the number of geographers working on them (or even available to work on them) is insufficient to the task. The importance of energy to everyone makes this unending and challenging field of inquiry especially appropriate and rewarding for all those geographers we can recruit to our ranks.

Acknowledgements

We thank Marilyn Brown and an anonymous referee for helpful comments on a previous version of this chapter.

References

Auty, R. M. (1991). "Third World Response to Global Processes: The Mineral Economies." *The Professional Geographer*, 43: 68–76.

Bach, W. (1995). "Coal Policy and Climate Protection: Can the Tough German CO_2 Reduction Targets be Met by 2005?" *Energy Policy*, 23: 85–91.

Balling, R. C., Jr. (1992). *The Heated Debate: Greenhouse Predictions Versus Climate Reality*. San Francisco: Pacific Research Institute for Public Policy.

—— (1994). "Greenhouse Gas and Sulfate Aerosol Experiments Using a Simple Global Energy-Balance Model." *Physical Geography*, 15: 299–309.

Barthelmie, R. J. (2002). "Modelling and Measurements of Coastal Wind Speeds," in S. K. Majumdar, E. W. Miller, and A. I. Panah (eds.), *Renewable Energy: Trends and Prospects*. Easton, Pa.: Pennsylvania Academy of Science, 961–75.

Blowers, A., Lowry, D., and Solomon, B. D. (1991). *The International Politics of Nuclear Waste*. London: Macmillan.

Brainard, J., Lovett, A., and Parfitt, J. (1996). "Assessing Hazardous Waste Transport Risks Using GIS." *International Journal of Geographical Information Systems*, 10: 831–49.

Brown, M. A. (1993). "The Effectiveness of Codes and Marketing in Promoting Energy-Efficient Home Construction." *Energy Policy*, 21: 391–402.

—— (1997). "The Energy-Related Inventions Program: Evaluation Challenges and Solutions," in S. A. Rood, (ed.), *Proceedings, Technology Transfer Metrics Summit*, Santa Fe, NM, 28 April–2 May, 1996. Chicago: Technology Transfer Society, 171–85.

Brown, M. A., and Berry, L. G. (1995). "Determinants of Program Effectiveness: Results of the National Weatherization Evaluation." *Energy*, 20: 729–43.

Brown, M. A., and Mihlmester, P. E. (1995). "Actual v. Anticipated Savings from DSM Programs: An Assessment of the California Experience," in *Proceedings, 1995 International Energy Program Evaluation Conference*, Conf-950817, Chicago, 295–301.

Brown, M. A., Kreitler, V., and Wolfe, A. K. (1996). "The Persistence of Program Impacts: Methods, Applications, and Selected Findings." *Energy Services Journal*, 2: 167–92.

Brown, M. A., Levine, M. D., Romm, J. P., Rosenfeld, A. H., and Koomey, J. G. (1998). "Engineering-Economic Studies of Energy Technologies to Reduce Greenhouse Gas Emissions: Opportunities and Challenges." *Annual Review of Energy and the Environment*, 23: 287–385.

Calzonetti, F. J. (1990). "Canadian–U.S. Electricity Trade and the Free Trade Agreement: Perspectives from Appalachia." *Canadian Journal of Regional Science*, 13: 171–7.

Cleveland, C. J. (1991). "Physical and Economic Aspects of Resource Quality: The Cost of Oil Supply in the Lower 48 United States, 1936–1988." *Resources and Energy*, 13: 163–88.

—— (1993). "An Exploration of Alternative Measures of Natural Resource Scarcity: The Case of Petroleum Resources in the US." *Ecological Economics*, 7: 123–57.

Cleveland, C. J., and Herendeen, R. (1989). "Solar Parabolic Collectors: Successive Generations are Better Net Energy and Exergy Producers." *Energy Systems and Policy*, 13/1: 63–77.

Cleveland, C. J., and Kaufmann, R. K. (1991). "Forecasting Ultimate Oil Recovery and its Rate of Production: Incorporating Economic Forces into the Models of M. King Hubbert." *The Energy Journal*, 12/2: 17–46.

—— (1997). "Natural Gas in the U.S.: How Far can Technology Stretch the Resource Base?" *The Energy Journal*, 18/2: 89–108.

Cleveland, C. J., and Ruth, M. (1998). "Indicators of Dematerialization and the Materials Intensity of Use." *Journal of Industrial Ecology*, 2/3: 15–50.

Cook, B. J., Emel, J. L., and Kasperson, R. E. (1990). "Organizing and Managing Radioactive Waste Disposal as an Experiment." *Journal of Policy Analysis and Management*, 9: 339–66.

Das, S., and Wilbanks, T. J. (1997). *Private Power Development and Environmental Protection in India*. Oak Ridge, Tenn.: Oak Ridge National Laboratory, ORNL/TM-13454, prepared for the U.S. Agency for International Development.

DiFrancesco, R. J. (1998). "Large Projects in Hinterland Regions: A Dynamic Multiregional Input–Output Model for Assessing the Economic Impacts." *Geographical Analysis*, 30: 15–34.

DiFrancesco, R. J., and Anderson, W. P. (1999). "Developing Canada's Arctic Oil Resources: An Assessment of the Interregional Economic Impacts." *Environment and Planning A*, 31: 459–76.

Elhance, A. (1999). *Hydropolitics in the 3rd World: Conflict and Cooperation in International River Basins*. Herndon, Va.: United States Institute of Peace Press.

Elmes, G. A. (1996). "The Changing Geography of Electric Energy in the United States: Retrospect and Prospect." *Geography*, 81: 347–60.

Elmes, G. A., and Harris, T. M. (1996). "Industrial Restructuring and the United States Coal Energy System, 1972–1990: Regulatory Change, Technological Fixes, and Corporate Control." *Annals of the Association of American Geographers*, 86: 507–29.

Flynn, J., Kasperson, R. E., Kunreuther, H., Slovic, P. (1997). "Overcoming Tunnel Vision: Redirecting the US High-Level Nuclear Waste Program." *Environment*, 39/3: 6–11, 25–30.

Gould, P. R. (1990). *Fire in the Rain: The Democratic Consequences of Chernobyl*. Cambridge: Polity Press.

Greenberg, M. R., and Gerrard, M. B. (1995). "Whose Backyard, Whose Risk: Fear and Fairness in Nuclear Waste Siting." *Journal of Urban Technology*, 3: 113–15.

Greenberg, M. R., Krueckenberg, D., Lowrie, H. M., Simon, D., Isserman, A., and Sorensen, D. (1998). "Socio-Economic Impacts of U.S. Nuclear Weapons Facilities: A Local Scale Analysis of Savannah River, 1950–1993." *Applied Geography*, 18: 101–16.

Greene, D. L. (1997). "Economic Scarcity: Forget Geology, Beware Monopoly." *Harvard International Review*, 19/3: 16–19, 65–6.

Greene, D. L., Jones, D. W., and Leiby, P. N. (1998). "The Outlook for U.S. Oil Dependence." *Energy Policy*, 26: 55–69.

Hannon, B. M. (1975). "Energy Conservation and the Consumer." *Science*, 189: 95–102.

Hannon, B. M., and Brodrick, J. (1982). "Steel Recycling and Energy Conservation." *Science*, 216: 485–91.

Hannon, B., Ruth, M., and Delucia, E. (1993). "A Physical View of Sustainability." *Ecological Economics*, 8: 253–68.

Harvey, L. D. D. (1992). "Implementation of Mitigation at the Local Level: The Role of Municipalities," in S. K. Majumdar, L. S. Kalkstein, B. M. Yarnal, E. W. Miller, and L. W. Rosenfeld (eds.), *Global Climate Change: Implications, Challenges & Mitigation Measures*. Easton, Pa.: Pennsylvania Academy of Science, 423–38.

—— (1995). "Solar-Hydrogen Electricity Generation in the Context of Global CO_2 Emission Reduction." *Climatic Change*, 29: 53–89.

Harvey, L. D. D., and Bush, E. J. (1997). "Joint Implementation: A Strategy for Combating Global Warming?" *Environment*, 39/8: 14–20, 36–43.

Harvey, L. D. D., Torrie, R., and Skinner, R. (1997). "Achieving Ecologically-Motivated Reductions of Canadian CO_2 Emissions." *Energy*, 22: 705–24.

Heiman, M. K. (2002). "Power for the People: A Comparison of the US and German Commitments to Renewable Energy," in S. K. Majumdar, E. W. Miller, and A. I. Panah (eds.), *Renewable Energy: Trends and Prospects*. Easton, Pa.: Pennsylvania Academy of Science, 403–11.

Hill, L. J., and Brown, M. A. (1995). "Issues in Estimating the Cost-Effectiveness of Coordinated DSM Programs." *Utilities Policy*, 5: 47–53.

Hillsman, E. L. (1995*a*). "Transportation Sector", in J. Sathaye and S. Meyers (eds.), *Greenhouse Gas Mitigation Assessment: A Guidebook*. Dordrecht: Kluwer Academic Publishers, 6–1 – 6–15.

—— (1995*b*). "Transportation DSM: Building on Electric Utility Experience." *Utilities Policy*, 5: 237–49.

Hillsman, E. L., and Southworth, F. (1990). "Factors that May Influence Responses of the US Transportation Sector to Policies for Reducing Greenhouse Gas Emissions." *Transportation Research Record*, 1267: 1–11.

Hillsman, E. L., Alvic, D. R., and Bennett, J. B. (1994). "The Bureau of Mines Electric Utility Model." *Operations Research*, 42: 998–1009.

Hobbs, B. F. (1994). "What do SO_2 Emissions Cost? Allowance Prices and Externality Adders." *Journal of Energy Engineering*, 120/3: 122–32.

—— (1995). "Optimization Methods for Electric Utility Resource Planning." *European Journal of Operational Research*, 83: 1–20.

Hobbs, B. F., and Horn, G. T. F. (1997). "Building Public Confidence in Energy Planning: A Multimethod MCDM Approach to Demand-Side Planning at BC Gas." *Energy Policy*, 25: 357–75.

Hosier, R. H. (1992). "Energy and Environmental Management in Eastern African Cities." *Environment and Planning A*, 24: 1231–54.

Hosier, R. H., and Bernstein, M. A. (1992). "Woodfuel Use and Sustainable Development in Haiti." *The Energy Journal*, 13/2: 129–56.

Hwang, M., Calzonetti, F. J., and Mann, P. C. (1995). "Regional Disparities in Pricing Strategies of Electric Utilities." *The Annals of Regional Science*, 29: 351–62.

Jacob, G. (1990). *Site Unseen: The Politics of Siting a Nuclear Waste Repository*. Pittsburgh: University of Pittsburgh Press.

Jain, A. K., and Bach, W. (1994). "The Effectiveness of Measures to Reduce the Man-Made Greenhouse Effect: The Applicability of a Climate Policy Model." *Theoretical and Applied Climatology*, 49: 103–18.

Johnson, J. H., Jr., and Zeigler, D. J. (1989). "Information Planning and the Geographical Context of Radiological Emergency Management." *Transactions of the Institute of British Geographers*, NS 14: 350–63.

Jones, D. W. (1994). "Energy Use and Fuel Substitution: Lessons Learned and Applications to Developing Countries," in W. Bentley and M. Gowen (eds.), *Forest Resources and Wood-Based Biomass Energy as Rural Development Assets*. New Delhi: Oxford-IBH, 69–104.

Kasanen, P., and Lakshmanan, T. R. (1989). "Residential Heating Choices of Finnish Households." *Economic Geography*, 65: 130–45.

Kasperson. R. E., and Kasperson, J. X. (1996). "The Social Amplification and Attenuation of Risk." *The Annals of the American Academy of Political and Social Science*, 545: 95–105.

Kates, R. W., Clark, W. C., Corell, R., Hall, J. M., Jaeger, C. C., Lowe, I., McCarthy, J. J., Schellnhuber, H. J., Bolin, B., Dickson, N. M., Faucheux, S., Gallopin, G. C., Gruebler, A., Huntley, B., Jager, J., Jodha, N. S., Kasperson, R. E., Mabogunje, A., Matson, P., Mooney, H., Moore, B., III, O'Riordan, T., and Svedin, U. (2001). "Sustainability Science." *Science*, 292: 641–2.

Kaufmann, R. K. (1991). "Oil Production in the Lower 48 US: Reconciling Curve Fitting and Econometric Models." *Resources and Energy*, 13: 111–27.

Kaufmann, R. K., and Cleveland, C. J. (1991). "Policies to Increase US Oil Production: Likely to Fail, Damage the Economy, and Damage the Environment." *Annual Review of Energy and the Environment*, 16: 379–400.

Kaufmann, R. K., Davidsdottir, B., Garnham, S., and Paulie, P. (1998). "The Determinants of Atmospheric SO_2 Concentrations: Reconsidering the Environmental Kuznets Curve." *Ecological Economics*, 25: 209–20.

Kellner, P. (1994). "From the Pastoral to the Promethean: Electric Power Regimes in the Pacific Northwest and the Rhetoric of the Electric Sublime." Ph.D. dissertation, University of California at Los Angeles.

Kuby, M. (1996). "Population Growth, Energy Use, and Pollution: Understanding the Driving Forces of Global Change." *Commission on College Geography*, ii. *Developing Active Learning Modules on the Human Dimensions of Global Change*. Washington, DC: Association of American Geographers.

Kuby, M., Ratick, S. J., and Osleeb, J. P. (1991). "Modeling US Coal Export Planning Decisions." *Annals of the Association of American Geographers*, 81: 627–49.

Kuby, M., Neuman, S., Zhang, C., Cook, P., Zhou, D., Friesz, T., Shi, Q., Gao, S., Watanatada, T., Cao, W., Sun, X., and Xie, Z. (1993). "A Strategic Investment Planning Model for China's Coal and Electricity Delivery System." *Energy*, 18: 1–24.

Kuhn, R. G. (1992). "Canadian Energy Futures: Policy Scenarios and Public Preferences." *The Canadian Geographer*, 36: 350–65.

—— (1998). "Social and Political Issues in Siting a Nuclear Fuel Waste Disposal Facility in Ontario, Canada." *Canadian Geographer*, 42: 14–28.

Lakshmanan, T. R., and Andersson, A. (1991–2). "Western European Energy Policy in Turmoil and Transition." *Harvard International Review*, 14/2: 17–21.

Lakshmanan, T. R., and Han, X. (1995). "Changes in the Energy Cost of Goods and Services of the Japanese Economy." *The Annals of Regional Science*, 29: 277–302.

—— (1997). "Factors Underlying Transportation CO_2 Emissions in the USA: A Decomposition Analysis." *Transportation Research*, 2: 1–15.

Lee, R. (1996). "The U.S.-EC Fuel Cycle Externalities Study: The U.S. Research Team's Methodology, Results and Conclusions," in *The External Costs of Energy, First EC/OECD/IEA Workshop on Energy Externalities*. Brussels: European Commission, Directorate General XII. Oak Ridge National Laboratory, 109–38.

—— (1998). "The External Costs of Electricity Generation: An 'Insider's' Perspective on the Lessons from the US Experience." *Energy Studies Review*, 8: 177–86.

Lee, R., Kahn, J. R., Marland, G., Russell, M., Shallcross, K., and Wilbanks, T. J. (1997). "Understanding Concerns about Joint Implementation." Knoxville, Tenn.: University of Tennessee, Joint Institute for Energy & Environment.

Lofstedt, R. E. (1998). "Sweden's Biomass Controversy: A Case Study of Communicating Policy Issues." *Environment*, 40/4: 16–20, 42–5.

—— (2001). "Playing Politics with Energy Policy: The Phase-Out of Nuclear Power in Sweden." *Environment*, 43/4: 20–33.

Lofstedt, R. E., Sepp, K., and Kelly, R. (1996). "Partnerships to Reduce Greenhouse Emissions in the Baltic." *Environment*, 38/6: 16–20, 40–2.

Macey, S. M., and Brown, M. A. (1990). "Demonstration as a Policy Instrument with Energy Technology Examples." *Knowledge: Creation, Diffusion, Utilization*, 11: 219–36.

Marples, D. R. (1996). *Belarus: From Soviet Rule to Nuclear Catastrophe*. London: Macmillan.

Metz, W. C. (1996). "Historical Application of a Social Amplification of Risk Model: Economic Impacts of Risk Events at Nuclear Weapons Facilities." *Risk Analysis*, 16: 185–93.

Metz, W. C., and Clark, D. E. (1997). "The Effect of Decisions about Spent Nuclear Fuel Storage on Residential Property Values." *Risk Analysis*, 17: 571–82.

Mounfield, P. (1991). *World Nuclear Power*. London: Routledge.

Obermeyer, N. J. (1990). "Bureaucrats, Clients, and Geography: The Bailly Nuclear Power Plant in Northern Indiana." Research Paper 216, Department of Geography, University of Chicago.

Odell, P. R. (1986). *Oil and World Power: Background of the Oil Crisis*. 8th edn. New York: Viking Penguin.

—— (1992). "Prospects for Non-OPEC Oil Supply." *Energy Policy*, 20: 931–41.

Openshaw, S., Carver, S., and Fernie, J. (1989). *Britain's Nuclear Waste: Safety and Siting*. London: Pinter.

O'Riordan, T., Kemp, R., and Purdue, M. (1988). *Sizewell B: An Anatomy of the Inquiry*. London: Macmillan.

Osei, W. Y. (1996). "Rural Energy Technology: Issues and Options for Sustainable Development." *Geoforum*, 27: 63–74.

Owens, S. (1986). *Energy, Planning and Urban Form*. London: Pion.

Pasqualetti, M. J. (1989). "Introducing the Geosocial Context of Nuclear Decommissioning: Policy Implications in the U.S. and Great Britain." *Geoforum*, 20: 391–6.

—— (ed.) (1990). *Nuclear Decommissioning and Society: Public Links to a New Technology*. London: Routledge.

—— (1997). "Landscape Permanence and Nuclear Warnings." *The Geographical Review*, 87: 73–91.

—— (2000). "Morality, Space, and the Power of Wind-Energy Landscapes." *The Geographical Review*, 90: 381–94.

—— (2001). "Wind Energy Landscapes: Society and Technology in the California Desert." *Society and Natural Resources*, 14/8: 689–99.

—— (2002). "The US–Mexico Border Energy Zone," in L. Fernandez and R. Carson (eds.), *Both Sides of the Border: Transboundary Environmental Management Issues Facing Mexico and the United States*. Dordrecht: Kluwer Academic Publishers, 323–46.

Pasqualetti, M. J., and Dellinger, M. (1989). "Hazardous Waste from Geothermal Energy: A Case Study." *The Journal of Energy and Development*, 13: 275–95.

Pasqualetti, M. J., and Pijawka, K. D. (1996). "Unsiting Nuclear Power Plants: Decommissioning Risks and their Land Use Context." *The Professional Geographer*, 48: 57–69.

Pasqualetti, M. J., Gipe, P., and Righter, R. (eds.). (2002). *Wind Energy in View: Landscapes of Power in a Crowded World*. San Diego: Academic Press.

Pasqualetti, M. J., Jennings, R., Harrington, M., and Boscamp, R. (1995). "DSM Programs Must Target Consumers, Not Just Technology." *Public Utilities Fortnightly*, 133/2: 23–6.

Pijawka, K. D., and Mushkatel, A. H. (1991). "Public Opposition to the Siting of the High-Level Nuclear Waste Repository: The Importance of Trust." *Policy Studies Review*, 10: 180–94.

Pohl, M. (1999). "The Elwha Dam, Washington: Downstream Impacts and Policy Implications." Ph.D. dissertation, Arizona State University.

Pryde, P. R. (1983). *Nonconventional Energy Resources*. New York: Wiley.

—— (1991). "The Environmental Costs of Eastern Resources: Air Pollution in Siberian Cities." *Soviet Geography*, 32: 403–11.

Pryde, P. R., and Bradley, D. J. (1994). "The Geography of Radioactive Contamination in the Former USSR." *Post-Soviet Geography*, 35: 557–93.

Robeson, S. M., and Shein, K. A. (1997). "Spatial Coherence and Decay of Wind Speed and Power in the North-Central United States." *Physical Geography*, 18: 479–95.

Ruth, M. (1993). *Integrating Economics, Ecology and Thermodynamics*. Dordrecht: Kluwer.

—— (1995). "Technology Change in US Iron and Steel Production: Implications for Material and Energy Use, and CO_2 Emissions." *Resources Policy*, 21: 199–214.

Ruth, M., and Cleveland, C. J. (1993). "Nonlinear Dynamic Simulation of Optimal Depletion of Crude Oil in the Lower 48 United States." *Computers, Environment, and Urban Systems*, 17: 425–35.

Sagers, M. J. (1993*a*). "Former Soviet Refineries Face Modernization, Restructuring." *Oil and Gas Journal*, 91/48: 23–7.

—— (1993*b*). "The Energy Industries of the Former USSR: A Mid-Year Survey." *Post-Soviet Geography*, 34: 341–418.

—— (1995). "The Russian Natural Gas Industry in the Mid-1990s." *Post-Soviet Geography*, 36: 521–64.

Sagers, M. J., and Green, M. B. (1986). *The Transportation of Soviet Energy Resources*. Totowa, NJ: Rowman & Littlefield.

Schwengels, P. F., and Solomon, B. D. (1992). "Energy Technologies for Reducing Greenhouse Gas Emissions in Developing Countries and Eastern Europe," in *Environmentally Sound Technology for Sustainable Development*, United Nations, *ATAS Bulletin*, 7: 38–47.

Smil, V. (1990). "Planetary Warming: Realities and Responses." *Population and Development Review*, 16: 1–29.

— (1994*a*). "China's Greenhouse Gas Emissions." *Global Environmental Change*, 4: 279–86.

— (1994*b*). *Energy in World History*. Boulder: Westview Press.

— (1999). *Energies: An Illustrated Guide to the Biosphere and Civilization*. Cambridge, Mass.: MIT Press.

Solomon, B. D. (1989). "The Search for Oil: Factors Influencing US Investment in Foreign Petroleum Exploration and Development." *The Professional Geographer*, 41: 39–50.

— (1992). "Energy Supply Technologies for Reducing Greenhouse Gas Emissions," in S. K. Majumdar, L. S. Kalkstein, B. M. Yarnal, E. W. Miller and L. W. Rosenfeld (eds.), *Global Climate Change: Implications, Challenges & Mitigation Measures*. Easton, Pa.: Pennsylvania Academy of Science, 410–22.

— (1995). "Global CO_2 Emissions Trading: Early Lessons from the U.S. Acid Rain Program." *Climatic Change*, 30: 75–96.

— (1998). "Five Years of Interstate SO_2 Allowance Trading: Geographic Patterns and Potential Cost Savings." *The Electricity Journal*, 11/4: 58–70.

Solomon, B. D., and Ahuja, D. R. (1991). "International Reductions of Greenhouse-Gas Emissions: An Equitable and Efficient Approach." *Global Environmental Change*, 1: 343–50.

Solomon, B. D., and Heiman, M. H. (2001). "The California Electricity Crisis: Lessons for Other States." *The Professional Geographer*, 53: 463–8.

Solomon, B. D., and Lee, R. (2000). "Emissions Trading Systems and Environmental Justice." *Environment*, 42/8: 32–45.

Solomon, B. D., and Meier, A. K. (1995). *Conservation Verification Protocols, Version 2.0*. Washington, DC: US Environmental Protection Agency, EPA 430/B-95-012.

Somer, C. (1999). "The Economic Geography of Geothermal District Energy Development in a Small, Western Community." M.A. thesis, Arizona State University.

Southworth, F. (1995). *A Technical Review of Urban Land Use-Transportation Models as Tools for Evaluating Vehicle Travel Reduction Strategies*. Oak Ridge, Tenn.: Oak Ridge National Laboratory, ORNL-6881, prepared for the US Department of Energy.

Southworth, F., and Peterson, B. E. (1990). "Disaggregation within National Vehicle Miles of Travel and Fuel Use Forecasts in the United States", in L. Lundqvist, L. G. Mattsson, and E. A. Eriksson (eds.), *Spatial Energy Analysis*. Aldershot: Gower, 199–224.

Spooner, D. J. (1995). "The 'Dash for Gas' in Electricity Generation in the UK." *Geography*, 80: 393–406.

Sternberg, R. (1996–7). "The Spatial Parameters of Hydroelectric Energy in the Tropical World." *Revista Geografica*, 123: 163–86.

Tretyakova, A., and Sagers, M. J. (1990). "Trends in Fuel and Energy Use and Programs for Energy Conservation by Economic Sector in the USSR." *Energy Policy*, 18: 726–39.

Wilbanks, T. J. (1988). "Impacts of Energy Development and Use, 1888–2088", in *Earth '88: Changing Geographic Perspectives*. Washington: National Geographic Society, 96–114.

— (1990). "Implementing Environmentally Sound Power Sector Strategies in Developing Countries." *Annual Review of Energy*, 15: 255–76.

— (1992*a*). "Energy Policy Responses: Concerns About Global Climate Change", in S. K. Majumdar, L. S. Kalkstein, B. M. Yarnal, E. W. Miller, and L. W. Rosenfeld (eds.), *Global Climate Change: Implications, Challenges & Mitigation Measures*. Easton, Pa.: Pennsylvania Academy of Science, 452–70.

— (1992*b*). "The Case for Energy Efficiency Improvement as a Global Strategy," in M. Kuliasha, A. Zucker, and K. J. Ballew (eds.), *Technologies for a Greenhouse-Constrained Society*. Boca Raton: Lewis Publishers, 587–615.

— (1994*a*). "Energy Efficiency Improvement: Making a No Regrets Option Work." *Environment*, 36/9: 16–20, 36–44.

— (1994*b*). " 'Sustainable Development' in Geographic Perspective." *Annals of the Association of American Geographers*, 84: 541–56.

Xie, Z., and Kuby, M. (1997). "Supply-Side–Demand-Side Optimization and Cost–Environment Tradeoffs for China's Coal and Electricity System." *Energy Policy*, 25: 313–26.

CHAPTER 21

Coastal and Marine Geography

Norbert P. Psuty, Philip E. Steinberg, and Dawn J. Wright

Introduction

The 1990s witnessed a significant increase in popular interest in the US regarding the geography of the world's coastal and marine spaces. Factors motivating this renewed interest included growing public environmental awareness, a decade of unusually severe coastal storms, more frequent reporting of marine pollution hazards, greater knowledge about (and technology for) depleting fishstocks, domestic legislation on coastal zone management and offshore fisheries policies, new opportunities for marine mineral extraction, heightened understanding of the role of marine life in maintaining the global ecosystem, new techniques for undertaking marine exploration, the 1994 activation of the United Nations Convention on the Law of the Sea, reauthorization of the US Coastal Zone Management Act in 1996, and designation of 1998 as the International Year of the Ocean.

Responding to this situation, the breadth of perspectives from which coastal and marine issues are being encountered by geographers, the range of subjects investigated, and the number of geographers engaging in coastal-marine research have all increased during the 1990s. As West (1989*a*) reported in the original *Geography in America*, North American coastal-marine geography during the 1980s was focused toward fields such as coastal geomorphology, ports and shipping, coastal zone management, and tourism and recreation. Research in these areas has continued, but in the 1990s, with increased awareness of the importance of coastal and marine areas to physical and human systems, geographers from a range of subdisciplines beyond those usually associated with coastal-marine geography have begun turning to coastal and marine areas as fruitful sites for conducting their research. Climatologists are investigating the sea in order to understand processes such as El Niño, remote-sensing experts are studying how sonic imagery can be used for understanding species distribution in three-dimensional environments, political ecologists are investigating the ocean as a common property resource in which multiple users' agendas portend conflict and cooperation, and cultural geographers are examining how the ocean is constructed as a distinct space with its own social meanings and "seascapes."

Despite (or perhaps because of) this expansion in coastal-marine geography, the subdiscipline remains fragmented into what we here call "Coastal Physical Geography," "Marine Physical Geography," and "Coastal-Marine Human Geography." Clearly, to understand coastal-marine spaces fully one must integrate both the human and the physical and the coastal and the marine, but few are achieving this integration. Even within the three sub-subdisciplinary labels that we use to organize this chapter, there are divisions among groups of scholars who could benefit from each others' work and prosper through collaboration. Coastal geomorphologists tend to focus either on applied issues surrounding the instantaneous impact of human manipulation or issues in coastal dynamics, but few bridge these literatures to examine human/physical process interaction in coastal systems. Coastal-zone political ecologists have little overlap with those who are involved in designing tourism promotion schemes for coastal areas. The AAG's Coastal and Marine Geography Specialty Group (CoMa) can play a role in facilitating this cross-fertilization within the subdiscipline as well as promoting outreach to non-geographers who research related topics.

Reflecting the current state of coastal-marine geography, the next three sections of this chapter review recent trends in coastal physical, marine physical, and human geography. This is followed by a brief discussion of CoMa, its history, and the role that it has undertaken in bridging some aspects of the subdiscipline. The chapter concludes with a discussion of future directions for coastal-marine research in geography.

Coastal Physical Geography

Physical geographers have contributed to various aspects of coastal geomorphology during the recent decade, employing a multitude of techniques and methodologies on a wide range of morphologies (Mossa *et al.* 1992; Morang *et al.* 1993). Philosophically, there are the traditional dichotomies between basic and applied inquiry, between intensive empirical observations and broad regional explanatory description, between modern process-response studies and historical approaches examining Holocene (or older) evolutionary systems, between studying an altered site or a natural system, between refinements of technology and instrumentation and the application of analytical tools. There is no single best way to contribute to the body of knowledge, and there is no reason for every scientist to parrot the same approach or technique. Diversity of approach and contribution is an attribute in itself and it fosters progress. As ably described by Sherman and Bauer (1993*a*), there are a variety of scales and approaches in coastal geomorphology and it is more important that the inquiry have a basis in theory and be cognizant of the existing literature than whether it is on one side or the other in a dichotomy.

Very basic inquiry regarding the mechanics of wave and sediment interaction to drive sediment transport in the nearshore zone has slowly revealed the processes that control the temporal and spatial scales of beach change. Whereas much of this research is within the realm of geophysics and engineering, the highly instrumented inquiry of Greenwood and his collaborators and students (Greenwood *et al.* 1991; Osborne and Greenwood 1992; Aagaard and Greenwood 1994) has produced insights into the significance of wave groupiness and infragravity wave frequencies as causative factors in suspended sediment transport. Others have pursued the subharmonic and/or infragravity wave frequency theme in relating components of beach morphologies to higher energy densities at these longer wave periods (Jagger *et al.* 1991; Bauer and Greenwood 1992; Allen *et al.* 1996). The important issue of mobilization of sediment on the beach face has been a particular focus of Horn (1997). Summary statements on basic processes and responses in the beach and nearshore zone were included in a special issue on Coastal Geomorphology in *Geographical Review* (Sherman 1988).

The process-response paradigm in coastal geomorphology is that larger storms beget larger waves and currents which, in turn, drive more sediment transport and create permanent changes at the coast. Although the eastern seaboard of the US is often battered by subtropical cyclones (hurricanes), the mid-latitude cyclones accompanied by strong frontal winds (northeasters) are more frequent bearers of high waves and storm surge along the Mid-Atlantic and North Atlantic shores. Through a series of papers that began more than a decade ago, Dolan and Davis (1992, 1994) have developed a well-recognized northeaster-storm intensity scale that has a foundation in weather and storm wave characteristics, and expands into a measure of storm-induced damage. Importantly, this effort synthesized a broad range of data into the domain of coastal change and used an inherently geographical approach to spatially organize and partition coastal storm characteristics. A further extension of the interest in coastal storms was generated by Engstrom (1994, 1996) who brought a geographical and geomorphological perspective to descriptions of storms from a previous century.

Coastal dunes of all shapes and sizes are highly valued for their role as an ecological niche in the midst of dense coastal development. The Coastal Zone Management Act specifically calls for the creation and enhancement of dunes in the coastal zone. There is increasing recognition of the variety of dune features that exist at the coast (Nordstrom *et al.* 1990) and the interplay between the beach and coastal dunes (Psuty 1988). The measurement of sand gains and losses in the dunes and the exchanges of sediment from the beach to the dune have been conducted along many shorelines (Davidson-Arnott and Law 1996; Gares *et al.* 1996; N. Jackson and Nordstrom 1997). The research has led to a series of site-specific descriptions and categorizations of dune types and their interaction with local dynamics (McCann and Byrne 1989; Gares 1992). In addition, there has been growth of a developmental model of the spatial/temporal evolution of the coastal foredune and the coastal dune system within the framework of beach/dune interaction. This follows the concept of a developmental sequence produced by Short and Hesp (1982) and Hesp and Thom (1990) and elaborated on by Psuty (1988) and Sherman and Bauer (1993*b*), and recognizes sediment budget as a formational variable (Psuty 1992*b*).

Further contributions involve the very difficult task of elucidating and improving aeolian transport equations in the multi-varied coastal foredune system where it will take development of improved instrumentation before measurements appropriate to model testing can be accomplished. Bauer and Namikas (1998) have constructed a rapid response saltation trap that could greatly improve our understanding of the application of transport equations to dune development.

Interest in regional coastal geomorphology is directed primarily toward barrier beach systems and their spatial/temporal evolution and displacements. Stone and McBride developed empirical data sets and models that relate barrier island shifts on the Gulf Coast to regional sediment budget and to alongshore/cross-shore transfers (Stone *et al.* 1992; McBride and Byrnes 1997; Stone and McBride 1998). They point to stepwise non-periodic shifts accompanied by short-term oscillations. A similar approach has been used to study changes on more-localized barrier spits in other locales (Davidson-Arnott and Fisher 1992; Ollerhead and Davidson-Arnott 1993). A blending of sea-level rise effects, beach displacement, and stages of barrier island transgression is presented by Dubois (1995) in a multi-scaled integration of the coherence of beach morphology within a spatially mobile and transgressing barrier island. Further, the opportunity for three-dimensional analysis of the barrier with ground-penetrating radar (Jol *et al.* 1996) provides another perspective on morphological development. Whereas most of the geomorphological research is in the beach, dune, and nearshore region, McBride and Moslow (1991) treat processes and characteristics of offshore sand ridges which are spatially related to the presence of inlets in the barrier island system along the East Coast and to sediment leakage from the ebb-tidal deltas as the barrier island system transgresses inland.

Coastal change caused by human interaction with natural processes adds another level of investigation (Walker 1988; Nordstrom 1994, 2000). Many studies provide examples of the localized human modifications that drive further human manipulative responses (Walker 1990*a*, *b*), often in association with structures in the water (Nakashima and Mossa 1991; Psuty and Namikas 1991) or with buildings (N. Jackson *et al.* 2000). Cultural modifications of the coastal environment have been instrumental in reworking the form and sediment distribution to such an extent that it would be perilous to ignore the limitations established by human-induced topography or human manipulations of the sediment budget in developed areas.

Most coastal geomorphological studies have been directed toward sandy shorelines rather than bedrock or cliffed coasts despite the worldwide presence of bedrock coasts. Trenhaile (1987) is a leading researcher on rocky coasts and has produced an excellent synthesis of bedrock coastal processes and geomorphology. At the local scale, Lawrence and Davidson-Arnott (1997) examined erosion of a bluff and the adjacent submarine platform and several other studies examined shingle beaches (McKay and Terich 1992; Sherman *et al.* 1993) to determine changes in spatial accumulations of coarse materials.

Despite the strong emphasis on modern-day processes, there remains a thread of inquiry that harkens back to the roots of American coastal geomorphology and the identification of coastal features associated with paleolakes in the tradition of Gilbert and his classic study of Glacial Lake Bonneville (Gilbert 1890). Currey (1990) has revisited much of Gilbert's Great Salt Lake study area and he continues to develop the Quaternary sequences of paleoshorelines in the interior basins at a variety of scales. Sack (1994) likewise carries on this field of inquiry.

Research into bays and estuaries has generated several major thrusts. Characteristics of beach/dune features of the lower energy estuarine environments in the northeastern US have been a focus of N. Jackson and Nordstrom (1992), whereas Armbruster *et al.* (1995) described the responses of beaches on the inland margins of Gulf Coast barrier islands in association with the passage of hurricanes and cold fronts. Nordstrom (1992) has developed a comprehensive framework of the processes and responses appropriate to beaches on shorelines of estuaries. He has also expanded into broader ecological issues of estuarine systems (Nordstrom and Roman 1996).

Estuarine-based research has considered issues of sedimentation and wetland development as part of the mix of sea-level rise and changing sediment availability. Reed (1990, 1995*a*) focused on wetland characteristics and composition in coastal Louisiana, leading to models of wetland changes and adaptations on decadal time-scales in association with varying sediment supply and relative sea-level rise rates. Kearney (1996) expanded the concepts of wetland deterioration associated with sediment deficits into decadal and centurial time. Psuty (1992*a*) generated a model that relates vertical and horizontal displacements of wetlands to rates of sea-level rise and sediment delivery on centurial and longer time-scales. In a different climatic context but in a similar sedimentation/sea-level relationship, Ellison (1993) and Ellison and Stoddart (1991) identify responses in mangrove communities. Important concepts that are emerging from these and similar investigations (Phillips 1992, 1997) concern the non-linearity of the changes and the relaxation times inherent in any modification to a system that is exchanging sediment spatially and temporally.

Although most shorelines are eroding in the presence of sea-level rise and a negative sediment budget, coastal zone population continues to increase. The inevitable result is increased concern for the manifestations of human development, for the economic value, and for the amenities of the coast. This has led to heightened interest in coastal dynamics and to improved knowledge of shoreline change, rates of change, and forecasts of future shoreline positions. A common form of establishing shoreline erosion rates is to secure the oldest surveyed shoreline position (usually mapped in the mid-nineteenth century) and compare it with shorelines from aerial photos and recent surveys. Some of the modern GIS registration techniques have improved comparisons of shorelines from historic maps and aerial photos. The additional incorporation of kinematic GPS shoreline determination and LIDAR imagery (Daniels *et al.* 1999) into the spatial matrix has resulted in the analysis of shoreline changes that together cover more than a century.

Whereas establishment of the displacement of the shoreline by this comparative means has value, there are hazards in extending the past trend into the future. Dolan *et al.* (1991) and Fenster *et al.* (1993) called attention to the episodic variation in shoreline position in many of the data sets based on relatively few points derived from historic data, and suggested that the sampling period will strongly influence the derived trend. Crowell *et al.* (1997) agreed with the problems associated with the use of a few data points and suggested that sea-level rise curves can be used as a surrogate for site-specific shoreline change rates. Obviously, there are complicating factors related to human manipulations of shoreline position that will also modify future trends, but that is a variable that could be woven into the fabric of analysis and could be another component of the application of this approach.

In the midst of a multitude of empirical observations of barrier island and estuarine change is the inescapable conclusion that the annual and decadal scales are non-linear, and that whereas many of the centurial time domains may have a trend there is a lot of scatter about any sort of trend line. Evidence continues to come forth regarding the difficulty of applying a narrow cause-and-effect relationship when the system itself is dynamic and sediment budgets are anything but constant.

In the original *Geography in America* volume, West (1989*a*) indicated that coastal geomorphologists were strong in empiricism but needed to strengthen their contributions to the theory side of the subdiscipline. Coastal geomorphologists are still strong in data gathering and observational science, and this is inherent to geomorphology. However, there has been a broadening perspective of conceptual themes, theoretical frameworks, and methodological approaches that are providing integrative vehicles for our inquiry. Importantly, the geographical geomorphologists are incorporating human activities as a process in landform evolution and are contributors in the emerging holistic approach to the very dynamic coastal zone.

Marine Physical Geography

The Coastal Zone Management Act of 1972 defines the coastal zone as a transition from land to the US territorial sea, consisting mainly of the swash zone, bays, dunes, estuaries, intra-coastal developments and waterways, coastal wetlands, marshes, and the like. But what of the open sea, often beyond sight of land? This is the domain of marine geography, which involves the understanding and characterization of space, place, and pattern of the open water and ice found seaward of the coast. American geographers have contributed little to marine research until recent decades, although the first textbook of modern marine science, written by Lt. Matthew Fontaine Maury of the US Navy in 1855, was entitled *The Physical Geography of the Sea* (Maury 1855). It was the post-World War II exploitation of offshore resources, as well as the environmental movements of the 1960s arising from coastal population and industrial growth, that directed some geographers to open water (West 1989*a*). But even today there are very few geographers working in this domain.

The study of marine physical geography received a major boost in the 1990s with the rise of earth system science (ESS) (Williamson 1994). The goal of the US Government-sponsored ESS initiative is to obtain a scientific understanding of the *entire* earth system (atmosphere, oceans, ice cover, biosphere, crust, and interior) on a global scale. ESS seeks to describe how component parts of Earth and their interactions have evolved, how they function, and how they may be expected to continue evolving at all time-scales (Nierenberg 1992). The recent emphasis on ESS, particularly with regard to the oceans, stems from the realization that many of Earth's resources are diminishing rapidly. A further factor is the growing awareness that an environmentally secure future requires a more integrated and coordinated approach toward understanding the consequences of global change, both for humanity and for managing global resources. Geographers have responded to these issues by broadening their focus beyond traditional boundaries.

Important emphases of ESS during the 1990s have been the study of synoptic weather patterns over the oceans, tracking and modeling of El Niño, mapping of water quality and pollution, and determination of various biophysical properties of the oceans, including temperature, chlorophyll pigments, suspended sediment, and salinity. Geographers involved in these studies have relied mainly on remote sensing techniques that are often ground-truthed with vessels at sea. For example, Siegel and Michaels (1996) have evaluated the role of light in the cycling of carbon, nitrogen, silica, phosphorus, and sulfur in the upper ocean. Their shipboard data have provided an "optical link" to global ocean color imagery derived from the SeaWifs satellite sensor (Garver *et al.* 1994). Lubin *et al.* (1994) and Ricchiazzi and Gautier (1998) have assessed the impact of seasonal ozone depletion on the intensity of surface radiation in the Antarctic and how this affects the ecology of the Southern Ocean. Geographers have participated in numerous field campaigns to Palmer Station, Antarctica, to determine the ecological processes linking annual pack-ice extent to biological dynamics of different trophic levels (R. Smith *et al.* 1998). Washburn *et al.* (1998) have used high-frequency radio radar to map ocean surface currents off the California coast to interpret changes in the populations of various marine species. Schweizer and Gautier (1997) launched an ambitious series of multimedia educational materials and workshops on El Niño, replete with both multispectral satellite imagery and shipboard sea-surface temperature maps.

ESS has also played a role in the creation of the Ridge Interdisciplinary Global Experiments (RIDGE) program, a successful research initiative of the US National Science Foundation in the 1990s that will be continued into the twenty-first century. RIDGE was launched in response to the growing realization that knowledge of the global mid-ocean ridge (seafloor-spreading centers) is fundamental to the understanding of key processes in a multitude of disciplines, including marine biology, geochemistry, physical oceanography, geophysics, and marine geology (National Research Council 1988). This has prompted several major coordinated experiments on the seafloor, involving multiple arrays of instruments (Wright 1999) for the study of geological, physical, chemical, and biological processes within and above the seafloor (Detrick and Humphris 1994). The resulting data range from measurements of temperature and chemistry of hydrothermal vent fluids and plumes, to the microtopography of underwater volcanoes, to the magnitudes and depths of earthquakes beneath the seafloor, to the biodiversity of hydrothermal vent fauna. Geographers have been involved in the first implementations of GIS to support these investigations both at sea and onshore (Wright 1996), as well as in the development of a long-term scientific information management infrastructure for the data (Wright *et al.* 1997). The current state of the art in marine (and coastal) applications of GIS is summarized in Wright and Bartlett (2000), an international collaborative effort between geographers, oceanographers, geodetic scientists, computer scientists, and coastal managers.

The International Year of the Ocean (1998), sponsored by the United Nations, called attention to an increasing need for investigations into deep ocean, island, and coastal management, all in the context of ESS. Specifically, ch. 17 of the 1992 UN Conference on Environment and Development's *Agenda 21* report calls for the assessment and management of fisheries, a de facto guarantee of biodiversity protection (Vallega 1999). Kracker (1999) has quantified aquatic landscapes via a traditional landscape ecology approach, incorporating underwater acoustic remote sensing techniques in determining abundance and distribution patterns for regions of intensive fish production.

Human Coastal-Marine Geography

The 1990s were a period in which marine and coastal areas became an increasingly significant object of study for human geographers interested in environmental planning, resource management, and development policy, as well as related topics in cultural, political, and economic geography. Coastal areas, in particular, have presented a growing range of issues of concern for human geographers. Although the coastal zone comprises just 17 per cent of the contiguous US land area, it is home to 56 per cent of the country's population. Daily, 3,600 people are added to the coastal zone, increasing population density in US coastal areas from 187 people per square mile in 1960, to 273 in 1994, and to a projected 327 in 2015 (NOAA 1998). Growth rates of coastal-zone populations are similarly dramatic around the world, and a host of research topics is associated with this increased population density. Marine areas also present numerous topics for human geographic research. During the 1990s, the rate of extraction of living resources from marine areas has remained at (or, for many species, above) maximum sustainable yields, extraction of non-living resources (especially petroleum) from marine areas has continued to play an important role in the world economy, and global shipping, which had plateaued during the recession of the 1980s, increased again during the 1990s with a commensurate increase in world trade.

This increased importance of coastal and marine areas to society has been matched by increased attention from human geographers (H. Smith and Vallega 1991). Complementing the extensive work on coastal hazards conducted by physical geographers, a number of coastal-marine human geographers have turned their attention to the human aspects of hazard creation, risk assessment, environmental perception, mitigation policies, and evacuation procedures. While most of this literature has focused on storm-related coastal hazards (Meyer-Arendt 1992; Baker 1995; Platt 1995; Clark *et al.* 1998; Dow and Cutter 1998), a smaller body of research has been produced on marine hazards associated with shipping and resource extraction (Argent and O'Riordan 1995; Dow 1999*a*, *b*).

The ever-increasing size of ships and the tightness of their schedules has led to an interest in the attendant changes in the shipping industry, the viability of individual ports, and the implications of transportation space's transformation into one seamless surface of intermodal transportation flows. A number of geographers have researched the impacts of containerization and shipping industry organization on port location and related industries (Slack 1990; Slack *et al.* 1996). Other geographers have placed changes in shipping technology and shipping regulations within the overall history of change in the global political-economic system (Hugill 1993; Steinberg 1998).

A secondary effect of containerization has been the abandonment of downtown and small-city ports in favor of a small number of very large, capital-intensive terminals. This has led to urban decay in old port areas and to opportunities for urban waterfront renewal, a subject that has attracted attention from scholars whose approaches range from studying the potential of waterfront renewal projects for stimulating economic development (Meyer-Arendt 1995; West 1989*b*), to focusing on the political-economic forces that drive renewal programs (Kilian and Dodson 1995; DeFilippis 1997), to researching and critiquing the representations of maritime life in the marketplaces and maritime festivals that often are the centerpieces of waterfront renewal projects (Goss 1996; Kilian and Dodson 1996; Atkinson and Laurier 1998; Laurier 1998; Steinberg 1999*b*).

Whereas tourism promotion is a pressing issue for the nation's decaying urban waterfronts, it is also a concern in other coastal and marine spaces. With tourism's rise as a global industry, the development and marketing of coastal and marine recreation spaces has taken a leading role in many countries' development strategies (Orams 1999). In some instances, tourists are encouraged to enjoy the ocean and its resources from the vantage point of a beach, in other instances from a cruise ship, and in other instances from the underwater perspective of the scuba diver. Geographic research on coastal and marine tourism typically goes beyond an investigation of its potential for economic development to include environmental, cultural, and political issues as well. Contributors to Wong's (1993) volume discuss how coastal recreation both reflects and impacts local environments (see also Meyer-Arendt 1991), Trist (1999) uses political ecology to analyze the images of the Caribbean Sea promoted by the marine tourism industry and the various demands of yachting, cruise ships, and diving on the Caribbean island of St Lucia; Young (1999*a*) studies political and cultural conflicts concerning whale-watching in Mexico; while Laurier (1999) focuses on the perceptions of the ocean held by recreational yachters.

This cultural-political turn in the study of coastal and marine tourism is part of a larger trend wherein the sea is becoming an increasingly popular topic for scholars who utilize a combination of cultural geography, cultural ecology, political economy, political ecology, and/or discourse analysis to interpret the ways in which various cultures perceive the sea and allocate access to its diverse resources (Nichols 1999; Young 1999*b*; Glaesel 2000). Recently, this perspective has been joined with one that emphasizes the ocean as a "socially constructed" space that is discursively and materially shaped by societies as they use the ocean. Proponents of this constructivist view stress that the ever-changing social construction of ocean-space serves to limit and enable further social uses of the ocean (Steinberg 1999*b*) and that in many social systems the ocean is a space that unites, rather than divides, land-based societies (Lewis and Wigen 1999).

Along with this fusing of political geography and cultural geography, there has been a continuation of research in the "classical" political geographic tradition, centering primarily on marine boundaries and international conventions that regulate exploitation of the ocean's resources (Earney 1990; Glassner 1990; Blake 1992) as well as issues in ocean management policy. In this area also, the field of inquiry has expanded recently, as scholars have integrated the study of marine boundaries with research on marine tenure systems, property rights, and territoriality in their efforts to investigate the legal norms that underlie marine boundaries between and within societies (S. Jackson 1995; Schug 1996; Scott and Mulrennan 1999; Steinberg 1999*c*).

The AAG's Coastal and Marine Geography Specialty Group (CoMa)

Recognition of the importance of the global ocean came early to the AAG. The first organized meeting of the Marine Geography Committee (MGC) of the AAG was held in 1970 in San Francisco, where it sponsored a session of six papers covering coastal geomorphology, fisheries, marine law, coastal research in Europe, the urban-maritime interface, and developing federal coastal interests and research funding. The first Chair was Evelyn Pruitt. Although Committee membership was limited to a handful of geographers who were appointed, participation in the MGC-sponsored sessions at the Annual Meetings of the AAG gradually increased, and by 1978 a Marine Geography Directory listed eighty-four persons. When specialty groups were created by the AAG in 1979, the MGC structure was dissolved and the broad membership was reconstituted as the Marine Geography Specialty Group, which in 1981 was rechristened the Coastal and Marine Specialty Group (CoMa). In comparison to the first meeting in 1970, at the 1999 annual meeting in Honolulu CoMa (with a membership of 170) sponsored five special sessions, featuring twenty-four paper presentations.

During the 1990s, about 85 per cent of presentations in CoMa-sponsored sessions concerned coastal topics. However, presentations on non-coastal marine topics roughly tripled over the course of the decade. This shift accompanied a dramatic increase in research into global earth systems, along with rising interest in global environmental concerns, global change, and the effects of human-induced change. Further, the ocean and coastal areas have drawn increasing attention from human geographers interested in policy, resource management, and development issues. Coastal-marine geographers are broadening the range and depth of physical, cultural, and economic issues initiated by the interest group, expanding upon initiatives begun decades earlier.

Starting in 1991, CoMa has recognized outstanding professional contributions by a coastal/marine geographer by conferring the Richard Joel Russell Award (President of the Association of American Geographers (1948), President of the Geological Society of America (1957), and member of the National Academy of Sciences (inducted in 1959)). Six members have received the honor to date (Table 21.1).

While much of the work performed by US coastal and marine geographers has been directed at the CoMa community, there have been sustained efforts to reach beyond CoMa, beyond the subdiscipline, and beyond the United States. There is a natural affinity between coastal and marine geographers and the Coastal Commission of the International Geographical Union. Whereas the IGU Coastal Commission has a broad topical range, most of the American involvement has

Table 21.1 *Richard J. Russell Award recipients*

Year	Recipient	Institution
1991	H. Jesse Walker	Louisiana State University
1992	Filmore Earney	Northern Michigan University
1993	Norbert P. Psuty	Rutgers University
1996	Karl F. Nordstrom	Rutgers University
1997	Douglas J. Sherman	University of Southern California
1999	Bernard O. Bauer	University of Southern California

been in the realm of coastal geomorphology and in Commission leadership. Norbert Psuty was the Vice-Chair (1984–92) and Chair (1992–4) of the Coastal Commission, and he was editor of its semi-annual newsletter (1984–96). Douglas Sherman is presently on the Board of the IGU Coastal Commission. Among Commission products contributed by American geographers were the Coastal Geomorphology Bibliography, 1986–1900 (Sherman 1992), the *Journal of Coastal Research* Special Issue on Dune/Beach Interaction (Psuty 1988), the special section in the *JCR* on wetlands (Reed 1995*b*), and the Special Issue of the *Zeitschrift für Geomorphologie* on Rapid Coastal Changes (Kelletat and Psuty 1996). American geographers also contributed to IGU Coastal Commission publications on coastal recreation and tourism (Fabbri 1990; Wong 1993). The IGU/American collaborative effort should get a boost in the future as a result of the OCEANS program, an IGU initiative dedicated to cooperation with UNESCO's International Oceanographic Commission (Vallega 1999).

American coastal geomorphologists are active in the quadrennial Conference of the International Association of Geomorphologists and have contributed to several follow-up publications (Paskoff and Kelletat 1991; Sherman and Bauer 1993*a*). They are represented in publications from two major coastal geomorphological symposia: *Coastal Sediments '91* (Kraus *et al.* 1991) and *Large Scale Coastal Behavior '93* (List 1993). An international coastal geomorphology memorial symposium honoring Bill Carter was held in association with the 1994 San Francisco AAG meeting, and it resulted in a volume dedicated to his memory (*Journal of Coastal Research*, Summer 1996). Recently, Paul Gares and Douglas Sherman organized the 1998 Binghamton Symposium in Geomorphology with the theme of Coastal Geomorphology. Of the presentations, slightly under half were by American geographers. Other major initiatives undertaken by coastal-marine geographers include a 1999 focus section of *The Professional Geographer* on ocean-space (Steinberg 1999*a*), a 1999 issue of *Geographical Review* devoted to the concept of world-regions defined by ocean basins (Wigen and Harland-Jacobs 1999), and a volume on the marine and coastal applications of GIS (Wright and Bartlett 2000).

These edited volumes and special issues bring together geographers using a variety of methodologies and perspectives, and may form the basis for a more unified and coherent subdiscipline of coastal and marine geography, expanding upon the beginnings chronicled by West (1989*a*). The contributions of the 1990s detailed in this chapter demonstrate that much has been accomplished, but that much also remains to be done.

Future Opportunities

There are numerous research agendas remaining in coastal and marine geography. In coastal physical topics, human manipulation of coastal topography and sediment budget are probably under-appreciated and subsumed as a small perturbation in either the instantaneous time-scale or that of the Holocene or longer. However, many of the contemporary issues in applied coastal geomorphology are part of the decadal up to centurial time-scale. This is the time-scale of interest to humans and that they influence. Recognition of the changes that are possible within this range and the influence of humans, therefore, is a task with strong feedback relationships.

Much physical geography research is addressing the non-linear nature of change in the marine and coastal zone: sea-level rise, sediment delivery, storminess, human intervention, nutrient flux, biomass production, etc. This is in recognition of the need for an improved understanding of the importance of scale in any inquiry and the bringing together of instantaneous models with developmental history models. Consideration of the aperiodic nature of natural processes contributes to the understanding of the oscillations of resources in a management context. Increasingly, there must be more recognition of the spatial/temporal role of humans in affecting aspects of the coastal and marine system.

There are many unsolved issues in the effective management, visualization, and analysis of marine and coastal data, particularly with regard to GIS. Marine physical geography is relatively youthful, and thus there are huge opportunities for geographers, particularly in mapping the parts of the ocean that are out of reach of satellite sensors (e.g. the water column and the ocean floor). To realize these opportunities, geographers must continue to collaborate with those working in corollary subdisciplines (e.g. remote sensing, GIS, geomorphology, etc.) and also with classically trained oceanographers, ocean engineers, and marine policy specialists.

Human geographers are expanding their productivity in marine and coastal issues in many of the traditional areas, while also testing their skills in uncharted waters. The areas of hazards, tourism, and trade remain major research domains, but they have been joined by an increasing emphasis on issues of culture, representation, and resource-competition. The challenge for human geographers is to merge the study of conceptual issues in the human–ocean relationship with practical problem-solving in ocean management.

In many ways, the aforementioned divisions of "coastal physical," "marine physical," and "human" are arbitrary, reflective of the current state of affairs, but definitely on a continuum toward total integration, particularly in light of increasing human-induced environmental threats to the health of the oceans. Solving these problems will require interdisciplinary, collaborative efforts across the social and natural sciences. These new directions do not so much supplant the more traditional lines of coastal and marine geographical research as they complement them, and there should be a body of literature developing in the next decade that fuses traditional with innovative perspectives into an improved analytical understanding of the complex interactions that transpire in coastal and marine systems.

Acknowledgements

This chapter would not have been possible without members of CoMa who contributed suggestions; the anonymous reviewers who provided excellent criticism; and Regan Fawley who assisted in manuscript preparation.

References

For a more extensive bibliography, including works not cited here, please see <http://dusk.geo.orst.edu/gia>, last accessed 13 November 2002.

Aagaard, T., and Greenwood, B. (1994). "Suspended Sediment Transport and the Role of Infragravity Waves in a Barred Surf Zone." *Marine Geology*, 118: 23–48.

Allen, J., Psuty, N., Bauer, B., and Carter, R. (1996). "A Field Assessment of Contemporary Models of Beach Cusp Formation." *Journal of Coastal Research*, 12: 622–9.

Argent, J., and O'Riordan, T. (1995). "The North Sea," in J. X. Kasperson, R. E. Kasperson, and B. L. Turner II (eds.), *Regions at Risk: Comparisons of Threatened Environments*. Tokyo: United Nations University Press, 367–419.

Armbruster, C., Stone, G., and Xu, J. (1995). "Episodic Atmospheric Forcing and Bayside Foreshore Erosion: Santa Rosa Island Florida." *Gulf Coast Association of Geological Societies, Transactions*, 45: 31–8.

Atkinson, D., and Laurier, E. (1998). "A Sanitised City? Social Exclusion at Bristol's 1996 International Festival of the Sea." *Geoforum*, 29: 199–206.

Baker, E. (1995). "Public Response to Hurricane Probability Forecasts." *The Professional Geographer*, 47: 137–47.

Bauer, B., and Greenwood, B. (1992). "Modifications of a Linear Bar-Trough System by a Standing Edge Wave." *Marine Geology*, 92: 177–204.

Bauer, B., and Namikas, S. (1998). "Design and Field Test of a Continuously Weighing, Tipping-Bucket Assembly for Aeolian Sand Traps." *Earth Surface Processes and Landforms*, 23: 1171–84.

Blake, G. (1992). *Maritime Boundaries*. London: Routledge.

Clark, G., Moser, S., Ratick, S., Dow, K., Meyer, W., Emani, S., Jin, W., Kasperson, J., Kasperson, R., and Schwartz, H. (1998). "Assessing the Vulnerability of Coastal Communities to Extreme Storms: The Case of Revere, MA, USA." *Adaptation and Mitigation Strategies for Global Change*, 3: 59–82.

Crowell, M., Douglas, B., and Leatherman, S. (1997). "On Forecasting Future U.S. Shoreline Positions: A Test of Algorithms." *Journal of Coastal Research*, 13: 1245–55.

Currey, D. (1990). "Quaternary Paleolakes in the Evolution of Semidesert Basins, with Special Emphasis on Lake Bonneville and the Great Basin, USA." *Palaeogeography, Palaeoclimatology, Palaeoecology*, 76: 189–214.

Daniels, R. C., McCandless, D., and Huxford, R. H. (1999). "Coastal Change Rates for Southwest Washington and Northwest Oregon," in G. Gelfenbaum and G. Kaminsky (eds.), *Southwest Washington Coastal Erosion Study Workshop Report 1998*, Open-File Report 99–524. Menlo Park, Calif.: US Geological Survey.

Davidson-Arnott, R., and Fisher, J. (1992). "Spatial and Temporal Controls on Overwash Occurrence on a Great Lakes Barrier Spit." *Canadian Journal of Earth Sciences*, 29: 102–17.

Davidson-Arnott, R., and Law, M. (1996). "Measurement and Prediction of Long-Term Sediment Supply to Coastal Foredunes." *Journal of Coastal Research*, 12: 654–63.

DeFilippis, J. (1997). "From a Public Re-Creation to Private Recreation: The Transformation of Public Space at South Street Seaport." *Journal of Urban Affairs*, 19: 405–17.

Detrick, R., and Humphris, S. (1994). "Exploration of Global Oceanic Ridge System Unfolds." *EOS, Transactions, American Geophysical Union*, 75: 325–6.

Dolan, R., and Davis, R. (1992). "An Intensity Scale for Atlantic Coast Northeast Storms." *Journal of Coastal Research*, 8: 840–53.

—— (1994). "Coastal Storm Hazards," in C. W. Finkl (ed.), *Coastal Hazards, Journal of Coastal Research*, Special Issue, 12: 103–14.

Dolan R., Fenster, M., and Holme, S. (1991). "Temporal Analysis of Shoreline Recession and Accretion." *Journal of Coastal Research*, 7: 723–44.

Dow, K. (1999*a*). "Caught in the Currents: Pollution, Risk, and Environmental Change in Marine Space." *The Professional Geographer*, 51: 414–26.

—— (1999*b*). "The *Extra*ordinary and the Everyday in Explanations of Vulnerability to an Oil Spill." *Geographical Review*, 89: 74–93.

Dow, K., and Cutter, S. (1998). "Crying Wolf: Repeat Responses to Hurricane Evacuation Orders." *Coastal Management*, 26: 237–52.

Dubois, R. (1995). "The Transgressive Barrier Model: An Alternative to Two-Dimensional Volume Balance Models." *Journal of Coastal Research*, 11: 1272–86.

Earney, F. (1990). *Marine Mineral Resources: Ocean Management and Policy*. London: Routledge.

Ellison, J. (1993). "Mangrove Retreat with Rising Sea Level, Bermuda." *Estuarine, Coastal and Shelf Science*, 37: 75–87.

Ellison, J., and Stoddart, D. (1991). "Mangrove Ecosystem Collapse During Predicted Sea-Level Rise: Holocene Analogues and Implications." *Journal of Coastal Research*, 7: 151–65.

Engstrom, W. (1994). "Nineteenth-Century Coastal Gales of Southern California." *Geographical Review*, 84: 306–15.

— (1996). "The California Storm of January 1862." *Quaternary Research*, 46: 141–8.

Fabbri, P. (1990). *Recreational Uses of Coastal Areas*. Dordrecht: Kluwer.

Fenster, M., Dolan, R., and Elder, J. (1993). "A New Method for Predicting Shoreline Positions From Historical Data." *Journal of Coastal Research*, 9: 147–71.

Gares, P. (1992). "Topographic Changes Associated with Coastal Dune Blowouts at Island Beach State Park, New Jersey." *Earth Surface Processes and Landforms*, 17: 589–604.

Gares, P., Davidson-Arnott, R., Bauer, B., Sherman, D., Carter, R., Jackson, D., and Nordstrom, K. (1996). "Alongshore Variations in Aeolian Transport: Carrick Finn Strand, Republic of Ireland." *Journal of Coastal Research*, 12: 673–82.

Garver, S., Siegel, D., and Mitchell, B. (1994). "Variability in Near Surface Particulate Absorption Spectra: What Can a Satellite Ocean Color Imager See?" *Limnology and Oceanography*, 39: 1349–67.

Gilbert, G. (1890). *Lake Bonneville*, Monograph 1. Washington, DC: US Geological Survey.

Glaesel, H. (2000). "State and Local Resistance to the Expansion of Two Environmentally Harmful Marine Fishing Techniques in Kenya." *Society & Natural Resources*, 13: 321–38.

Glassner, M. (1990). *Neptune's Domain: A Political Geography of the Sea*. Boston: Unwin Hyman.

Goss, J. (1996). "Disquiet on the Waterfront: Reflections on Nostalgia and Utopia in the Urban Archetypes of Festival Marketplaces." *Urban Geography*, 17: 221–47.

Greenwood, B., and Osborne, P. (1990). "Vertical and Horizontal Structure in Cross-Shore Flows: An Example of Undertow and Wave Set-Up on a Barred Beach." *Coastal Engineering*, 14: 543–80.

Greenwood, B., Osborne, P., and Bowen, A. (1991). "Measurements of Suspended Sediment Transport: Prototype Shorefaces," in N. Kraus, K. Ginerich, and D. Keiebel (eds.), *Coastal Sediments '91*. New York: American Society of Civil Engineers, 284–99.

Hesp, P., and Thom, B. (1990). "Geomorphology and Evolution of Erosional Dunefields," in K. Nordstrom, N. Psuty, and B. Carter (eds.), *Coastal Dunes: Form and Process*. Chichester: John Wiley & Sons, 253–88.

Horn, D. P. (1997). "Beach Research in the 1990s." *Progress in Physical Geography*, 21: 454–70.

Hugill, P. (1993). *World Trade Since 1431: Geography, Technology, Capitalism*. Baltimore: Johns Hopkins University Press.

Jackson, N., and Nordstrom, K. (1992). "Site-Specific Controls on Wind and Wave Processes and Beach Mobility on Estuarine Beaches." *Journal of Coastal Research*, 8: 88–98.

— (1997). "Effects of Time-Dependent Moisture Content of Surface Sediments on Aeolian Transport Rates Across a Beach, Wildwood, New Jersey, USA." *Earth Surface Processes and Landforms*, 22: 611–21.

Jackson, N., Nordstrom, K., Bruno, M., and Spalding, V. (2000). "Classification of Spatial and Temporal Changes to a Developed Barrier Island, Seven Mile Beach, New Jersey, USA," in O. Slaymaker (ed.), *Global Change*. London: John Wiley & Sons.

Jackson, S. (1995). "The Water is Not Empty: Cross-Cultural Issues in Conceptualising Sea Space." *Australian Geographer*, 26: 87–96.

Jagger, K., Psuty, N., and Allen, J. (1991). "Caleta Morphodynamics, Perdido Key, Florida, USA." *Zeitschrift für Geomorphologie*, Suppl. 81: 99–113.

Jol, H., Smith, D., and Meyers, R. (1996). "Digital Ground Penetrating Radar (GPR): A New Geophysical Tool for Coastal Barrier Research Examples from the Atlantic, Gulf, and Pacific Coasts, USA." *Journal of Coastal Research*, 12: 960–8.

Kearney, M. (1996). "Sea-Level Change During the Last Thousand Years in Chesapeake Bay." *Journal of Coastal Research*, 12: 977–83.

Kelletat, D., and Psuty, N. (eds.) (1996). *Field Methods and Models to Quantify Rapid Coastal Changes. Zeitschrift für Geomorphologie*, Suppl. 102. Berlin: Gebrüder Borntraeger.

Kilian, D., and Dodson, B. (1995). "The Capital See-Saw: Understanding the Rationale for the Victoria and Alfred Redevelopment." *South African Geographical Journal*, 77: 12–20.

— (1996). "Forging a Postmodern Waterfront: Urban Form and Spectacle at the Victoria and Alfred Docklands." *South African Geographical Journal*, 78: 29–40.

Kracker, L. (1999). "The Geography of Fish: The Use of Remote Sensing and Spatial Analysis Tools in Fisheries Research." *The Professional Geographer*, 51: 440–50.

Kraus, N., Ginerich, K., and Kriebel, D. (eds.) (1991). *Coastal Sediments '91, Proceedings*. New York: American Society of Civil Engineers.

Laurier, E. (1998). "Replication and Restoration: Ways of Making Maritime Heritage." *Journal of Material Culture*, 3: 21–50.

— (1999). "That Sinking Feeling: Elitism, Working, Leisure, and Yachting," in D. Crouch (ed.), *Leisure Practices and Geographical Knowledge*. London: Routledge, 195–213.

Lawrence, P., and Davidson-Arnott, R. (1997). "Alongshore Wave Energy and Sediment Transport on Southeastern Lake Huron, Ontario, Canada." *Journal of Coastal Research*, 13: 1004–15.

Lewis, M., and Wigen, K. (1999). "A Maritime Response to the Crisis in Area Studies." *Geographical Review*, 89: 161–8.

Li, R., Qian, L., and Blais, J. (1995). "A Hypergraph-Based Conceptual Model for Bathymetric and Related Data Management." *Marine Geodesy*, 18: 173–82.

List, J. (ed.) (1993). *Large Scale Coastal Behavior '93*, USGS Open File Report 93–381.

Lubin, D., Ricchiazzi, P., Gautier, C., and Whritner, R. (1994). "A Method for Mapping Antarctic Surface UV Radiation Using Multispectral Satellite Imagery," in C. Weiler and P. Penhale (eds.), *UV Radiation and Biological Research in Antarctica*, 62. Washington, DC: American Geophysical Union, 53–81.

McBride, R., and Byrnes, M. (1997). "Regional Variations in Shore Response Along Barrier Island Systems of the Mississippi River

Delta Plain: Historical Change and Future Prediction." *Journal of Coastal Research*, 13: 628–55.

McBride, R., and Moslow, T. (1991). "Origin, Evolution, and Distribution of Shoreface Sand Ridges, Atlantic Inner Shelf, USA." *Marine Geology*, 97: 57–85.

McCann, S., and Byrne, M. (1989). "Stratification Models for Vegetated Coastal Dunes in Atlantic Canada," in C. Gimingham, W. Ritchie, B. Willetts, and A. Willis (eds.), *Coastal Sand Dunes*. Edinburgh: Royal Society of Edinburgh, Scotland, Section B (Biological Sciences), 203–15.

McKay, P., and Terich, T. (1992). "Gravel Barrier Morphology: Olympic National Park, Washington State, USA." *Journal of Coastal Research*, 8: 813–29.

Maury, M. (1855). *The Physical Geography of the Sea*. New York: Harper.

Meyer-Arendt, K. (1991). "Tourism Development on the North Yucatan Coast: Human Response to Shoreline Erosion and Hurricanes." *GeoJournal*, 23: 327–36.

— (1992). "Historical Coastal Environmental Changes: Human Response to Shoreline Erosion," in L. Dilsaver and C. Colten (eds.), *The American Environment: Interpretations of Past Geographies*. Savage, Md.: Rowman & Littlefield, 217–33.

— (1995). "Casino Gaming in Mississippi: Location, Location, Location." *Economic Development Review*, 13: 27–33.

Morang, A., Mossa, J., and Larson, R. (1993). *Technologies for Assessing the Geologic and Geomorphic History of Coasts*, Technical Report CERC-93-5. Vicksburg: US Army Waterways Experiment Station, Coastal Engineering Research Center.

Mossa, J., Meisburger, E., and Morang, A. (1992). *Geomorphic Variability in the Coastal Zone*, Technical Report CERC-92-4. Vicksburg: U.S. Army Corps of Engineers Waterways Experiment Station, Coastal Engineering Research Center.

Nakashima. L., and Mossa, J. (1991). "Responses of Natural and Seawall-Backed Beaches to Recent Hurricanes on the Bayou Lafourche Headland, Louisiana." *Zeitschrift für Geomorphologie*, 35: 239–56.

National Oceanic and Atmospheric Administration (NOAA) (1998). Thomas J. Culliton, "Population: Distribution, Density and Growth." *State of the Coast Report*. Silver Spring, Md.: NOAA. URL: ‹http://state-of-coast.noaa.gov/bulletins/html/pop_01/pop.html›, last accessed 13 November 2002.

National Research Council (1988). *The Mid-Ocean Ridge: A Dynamic Global System*. Washington, DC: National Academy Press.

Nichols, K. (1999). "Coming to Terms with Integrated Coastal Management." *The Professional Geographer*, 51: 388–99.

Nierenberg, W. (ed.) (1992). *Encyclopedia of Earth System Science*. San Diego, Calif.: Academic Press.

Nordstrom, K. (1992). *Estuarine Beaches*. London: Elsevier Science Publishers.

— (1994). "Beaches and Dunes of Developed Coasts." *Progress in Physical Geography*, 18: 497–516.

— (2000). *Beaches and Dunes of Developed Coasts*. Cambridge: Cambridge University Press.

Nordstrom, K., and Roman, C. (eds.) (1996). *Estuarine Shores: Evolution, Environments and Human Alterations*. London: John Wiley & Sons.

Nordstrom, K., Psuty, N., and Carter, B. (eds.) (1990). *Coastal Dunes: Form and Process*. Chichester: John Wiley & Sons.

Ollerhead, J., and Davidson-Arnott, R. (1993). "Controls on Barrier Spit Evolution: A Comparison of Buctouche Spit, New Brunswick and Long Point spit, Ontario, Canada," in J. List (ed.), *Large Scale Coastal Behavior '93*, USGS Open File Report 93–381: 151–3.

Orams, M. (1999). *Marine Tourism: Development, Impacts and Management*. London: Routledge.

Osborne, P., and Greenwood, B. (1992*a*). "Frequency Dependent Cross-Shore Suspended Sediment Transport 1: A Non-Barred Shoreface, Queensland Beach, Nova Scotia, Canada." *Marine Geology*, 106: 1–24.

Paskoff, R., and Kelletat, D. (1991). *Geomorphology and Geoecology: Coastal Dynamics and Environments. Zeitschrift für Geomorphologie*, Suppl. 81. Berlin: Gebrüder Borntraeger.

Phillips, J. (1992). "Qualitative Chaos in Geomorphic Systems, with an Example From Wetland Response to Sea Level Rise." *Journal of Geology*, 100: 365–74.

— (1997). "Human Agency, Holocene Sea Level, and Floodplain Accretion in Coastal Plain Rivers." *Journal of Coastal Research*, 13: 854–66.

Platt, R. (1995). "Evolution of Coastal Hazards Policies in the United States." *Coastal Management*, 22: 265–84.

Psuty, N. (ed.) (1988). *Dune/Beach Interaction, Journal of Coastal Research*, Special Issue 3.

— (1992*a*). "Estuaries: Challenges for Coastal Management," in P. Fabbri (ed.), *Ocean Management in Global Change*. London: Elsevier Applied Science, 502–18.

— (1992*b*). "Spatial Variation in Coastal Foredune Development," in R. Carter, T. Curtis, and M. Sheehy-Skeffington (eds.), *Coastal Dunes: Geomorphology, Ecology and Management for Conservation*. The Hague: Balkema, 3–13.

Psuty, N., and Namikas, S. (1991). "Beach Nourishment Episodes at the Sandy Hook Unit, Gateway National Recreation Area, New Jersey, USA: A Preliminary Comparison," in Kraus, Ginerich, and Kriebel (1991: 2116–29).

Reed, D. (1990). "The Impact of Sea Level Rise on Coastal Salt Marshes." *Progress in Physical Geography*, 14: 24–40.

— (1995*a*). "The Response of Coastal Marshes to Sea Level Rise: Survival or Submergence." *Earth Surface Processes and Landforms*, 20: 39–48.

— (ed.) (1995*b*). "Special Thematic Section on Wetlands." *Journal of Coastal Research*, 20: 295–380.

Ricchiazzi, P., and Gautier, C. (1998). "Investigation of the Effect of Surface Heterogeneity and Topography on the Radiation Environment of Palmer Station, Antarctica, with a Hybrid 3-D Radiative Transfer Model." *Journal of Geophysical Research*, 103: 6161–76.

Sack, D. (1994). "Geomorphic Evidence of Climate Change from Desert-Basin Paleolakes," in A. Abrahams and A. Parsons (eds.), *Geomorphology of Desert Environments*. London: Chapman & Hall, 617–30.

Schug, D. (1996). "International Maritime Boundaries and Indigenous People: A Case Study of the Torres Strait." *Marine Policy*, 20: 209–22.

Schweizer, D., and Gautier, C. (1997). "Ocean Expeditions: El Niño," in *6th Symposium on Education, American Meteorological Society*. Long Beach, Calif.: American Meteorological Society, 1–10.

Scott, C., and Mulrennan, M. (1999). "Land and Sea Tenure at Erub, Torres Strait: Property, Sovereignty and the Adjudication of Cultural Continuity." *Oceania*, 70: 146–76.

Sherman, D. (ed.) (1988). "Theme Issue on Coastal Geomorphology." *Geographical Review*, 78: 115–240.

—— (ed.) (1992). *International Bibliography on Coastal Geomorphology. Journal of Coastal Research*, Special Issue 16.

Sherman, D., and Bauer, B. (1993*a*). "Coastal Geomorphology Through a Looking Glass." *Geomorphology*, 7: 225–49.

—— (1993*b*). "Dynamics of Beach-Dune Systems." *Progress in Physical Geography*, 17: 413–47.

Sherman, D., Orford, J., and Carter, R. (1993). "Development of Cusp-Related Gravel Size and Shape Facies at Malin Head, Ireland." *Sedimentology*, 40: 1139–52.

Short, A., and Hesp, P. (1982). "Wave, Beach and Dune Interactions in S.E. Australia." *Marine Geology*, 48: 259–84.

Siegel, D., and Michaels, A. (1996). "On Non-Chlorophyll Light Attenuation in the Open Ocean: Implications for Biogeochemistry and Remote Sensing." *Deep-Sea Research*, 43: 321–45.

Slack, B. (1990). "Intermodal Transportation in North America and the Development of Inland Load Center." *The Professional Geographer*, 42: 72–83.

Slack, B., Comtois, C., and Sletmo, G. (1996). "Shipping Lines as Agents of Change in the Port Industry." *Maritime Policy and Management*, 23: 289–300.

Smith, H., and Vallega, A. (1991). *The Development of Integrated Sea-Use Management*. London: Routledge.

Smith, R., Baker, K., Byers, M., and Stammerjohn, S. (1998). "Primary Productivity of the Palmer Long Term Ecological Research Area and the Southern Ocean." *Journal of Marine Systems*, 17: 245–59.

Stammerjohn, S., and Smith, R. (1996). "Spatial and Temporal Variability of Western Antarctic Peninsula Sea Ice Coverage," in R. Ross, E. Hofmann, and L. Quetin (eds.), *Foundations for Ecological Research West of the Antarctic Peninsula, 70*. Washington, DC: American Geophysical Union, 81–104.

Steinberg, P. (1998). "Transportation Space: A Fourth Spatial Category for the World-Systems Perspective?" in P. Ciccantell and S. Bunker (eds.), *Space and Transport in the World System*. Westport, Conn.: Greenwood, 17–34.

—— (ed.) (1999*a*). "Focus Section: Geography of Ocean-Space." *The Professional Geographer*, 51: 366–450.

—— (1999*b*). "The Maritime Mystique: Sustainable Development, Capital Mobility, and Nostalgia in the World-Ocean." *Environment and Planning D: Society & Space*, 17: 403–26.

—— (1999*c*). "Lines of Division, Lines of Connection: Stewardship in the World Ocean." *Geographical Review*, 89: 254–64.

Stone, G., and McBride, R. (1998). "Louisiana Barrier Islands and their Importance in Wetland Protection: Forecasting Shoreline Change and Subsequent Response of Wave Climate." *Journal of Coastal Research*, 14: 900–15.

Stone, G., Stapor, F., Jr., May, J., and Morgan, J. (1992). "Multiple Sediment Sources and a Cellular, Non-Integrated Longshore Drift System: Northwest Florida and Southeast Alabama Coast, USA." *Marine Geology*, 105: 141–54.

Trenhaile, Alan S. (1987). *The Geomorphology of Rock Coasts*. New York: Oxford University Press.

Trist, C. (1999). "Recreating Ocean Space: Recreational Consumption and Representation of the Caribbean Marine Environment." *The Professional Geographer*, 51: 376–87.

Vallega, A. (1999). "Ocean Geography vis-à-vis Global Change and Sustainable Development." *The Professional Geographer*, 51: 400–14.

Walker, H. (ed.) (1988). *Artificial Structures and Shorelines*. Dordrecht: Kluwer.

—— (1990*a*). "The Coastal Zone," in B. Turner (ed.), *The Earth as Transformed by Human Action*. Cambridge: Cambridge University Press, 271–94.

—— (1990*b*). "Nature, Humans, and the Coastal Zone." *The International Journal of Social Education*, 5: 50–62.

Washburn, L., Emery, B., and Paduan, J. (1998). "Preliminary Results from an Array of HF Radars for Mapping Surface Currents in the Santa Barbara Channel." *Eos, Transactions of the American Geophysical Union*, 79: F393.

West, N. (1989*a*). "Coastal and Marine Geography," in G. Gaile and C. Willmott (eds.), *Geography in America*. Columbus: Merrill, 141–54.

—— (1989*b*). "Urban-Waterfront Developments: A Geographic Problem in Search of a Model." *Geoforum*, 20: 459–68.

Williamson, P. (1994). "Integrating Earth System Science." *Ambio*, 23: 3.

Wigen, K., and Harland-Jacobs, J. (eds.) (1999). "Special Issue: Oceans Connect." *Geographical Review*, 89: 161–313.

Wong, P. (ed.) (1993). *Tourism vs. Environment: The Case for Coastal Areas*. Dordrecht: Kluwer.

Wright, D. (1996). "Rumblings on the Ocean Floor: GIS Supports Deep-Sea Research." *Geo Info Systems*, 6: 22–9.

—— (1999). "Getting to the Bottom of It: Tools, Techniques, and Discoveries of Deep Ocean Geography." *The Professional Geographer*, 51: 426–34.

Wright, D., and Bartlett, D. (eds.) (2000). *Marine and Coastal Geographical Information Systems*. London: Taylor & Francis.

Wright, D., Fox, C., and Bobbitt, A. (1997). "A Scientific Information Model for Deep-Sea Mapping and Sampling." *Marine Geodesy*, 20: 367–79.

Young, E. (1999*a*). "Balancing Conservation with Development in Small-Scale Fisheries: Is Ecotourism an Empty Promise?" *Human Ecology*, 27: 581–620.

—— (1999*b*). "Local People and Conservation in Mexico's El Vizcaino Biosphere Reserve." *Geographical Review*, 89: 364–90.

CHAPTER 22

Contemporary Agriculture and Rural Land Use

Leslie Aileen Duram and J. Clark Archer

The Contemporary Agriculture and Rural Land Use (CARLU) Specialty Group was organized in 1985 (Napton 1989) to provide a forum for researchers who identify, describe, and explain the geographical patterns of agricultural activity and rural land use. Indeed, rural and agricultural geographers study many aspects of rural land use, including rural settlement, rural environmental management, the globalization of primary industries (i.e. agriculture, forestry, and mining), and also utilize spatial technologies for rural systems analysis. The various dimensions, consequences and policy implications of long-term sustainability of rural landscapes in industrialized, capitalist countries and particularly in North America, have been matters of special attention (Pierce 1994; Troughton 1995; Ilbery 1998).

The early Jeffersonian ideal of a nation populated predominately by rural freeholders remains a popular and persistent theme in American culture. The country craft motifs of cows, chickens, and apples adorn many urban kitchens. Nearly all children know Laura Ingalls Wilder's popular stories about a *Farmer Boy* (Wilder 1933) or a *Little House on the Prairie* (Wilder 1935). But the agrarian conditions Wilder describes in these stories near the start of the twentieth century bear little resemblance to the conditions faced by farmers in rural areas at the start of the twenty-first century due to social and agricultural change (Bell 1989; Baltensperger 1991; Roberts 1996; Lang *et al.* 1997; Lawrence 1997). Likewise, the quaint scenes of chickens and pigs printed on paper towels do not hint at current environmental and social concerns with large-scale livestock production in the US (Furuseth 1997; Hart and Mayda 1997). In many ways these historically imbedded ideals clash with the current reality of rural areas.

Rural and agricultural researchers provide insight into how rural North America evolved to look like it does today. Their research helps describe the cultural, economic, environmental, political, and social forces that influenced and continue to influence rural places. This research often suggests what alternatives are available for rural areas in the future. Following the introduction, this chapter is organized according to four main research themes: rural regions, agricultural location theory, rural land-use change, and agricultural sustainability.

Introduction

Current geographic research stresses the study of contemporary agriculture and related aspects of rural land use and rural settlement in the United States, but these patterns can be better understood within the context of other times or settings. For example, the initial biotechnical foundations of much of modern agriculture can be traced to neolithic era plant and animal domestication efforts that were geographically concentrated within particular culture hearths of the Old and New worlds (Sauer 1969; Mannion 1997). Except in Mexico, which usually is deemed part of Latin America, North America did not contain major neolithic cultural hearths of plant or animal domestication. Consequently, practically all the crops and livestock now important to agriculturalists in North America were introduced from elsewhere. Only a tiny fraction (1%) of cultivated land in North America is devoted to crops that are actually native to this part of the world (Grigg 1984: 162–4); instead, most cultivated land in North America is devoted to crops first domesticated in Southwest Asia (41%), Latin America (26%), East Asia (14%), or Northern Europe (9%).

Globally, the leading crops are wheat, rice, and corn, which each individually outrank the production of all remaining cereals combined (World Resources Institute 1996: 233). Although modern varieties of wheat, rice, and corn have extended the ranges of these crops well beyond their original settings, their contemporary patterns of cultivation continue to echo the environmental conditions under which their wild and domestic ancestors evolved (Fredrich 1991; Doolittle 1992; Huke *et al.* 1993; Woodhead *et al.* 1994; Slafer 1994; Smit *et al.* 1997).

At the broadest conceptual level there have been two main geographical research streams that focused on locational patterns of agricultural and rural land use. Although increasingly interrelated in practice, these two streams have rather distinctive intellectual origins, with studies of agricultural regionalization usually more closely linked to cultural and historical-settlement geography, and studies of agricultural location theory usually more closely linked to economic geography.

Rural Regions and Landscapes

Agricultural regionalization has been a prominent interest of rural geographers. An early study by O. E. Baker (1921) examined regional patterns of agriculture in the United States using US Census data and stressing the "physical factors" of topography, soils, moisture, and temperature. Grigg (1969: 100) reviewed and reprinted maps from several efforts to regionalize agriculture globally to suggest that "when the world as a whole is being dealt with, climatic criteria are probably the best indicators of agricultural potentiality. But in smaller areas," he went on, "soil type and morphology, drainage conditions, and slope assume a far greater significance." Yet, he argued the limitations of classifications based strictly on physical environmental factors, because social conditions often cause spatial variations in agriculture.

Probably the most influential effort to depict agricultural regions on a global scale was undertaken more than a half-century ago by Whittlesey (1936). Whittlesey's map is still used widely in textbooks and atlases. For instance, the world map entitled "Major Agricultural Regions" in a recent edition of *Goode's World Atlas* (Hudson and Espenshade (eds.) 2000: 38) is a "Revision of *Agricultural Regions* by Whittlesey, Annals Association of American Geographers 1936." Whittlesey delimited agricultural regions based on both environmental conditions that set "the limits of the range for any crop or domestic animal," and social conditions including "density of population, stage of technology, and inherited tradition" (Whittlesey 1936: 208). He differentiated thirteen types of agriculture at the global scale, including such types as livestock ranching, shifting cultivation, commercial crop farming, and specialized horticulture (ibid. 213–40). Many discussions of global agriculture still use his terminology, but agricultural production has changed substantially over the past century. Clearly, an important research opportunity exists to update Whittlesey's global agricultural regionalization as the twenty-first century dawns.

Regional variations have been a recurrent focus of rural geographers. For example, Hudson (1994) traced the historical evolution of one of the world's preeminent agricultural regions in *The Making of the Corn Belt.* Hudson weaves a complex tapestry of commercial and cultural linkages with distant settings and assesses their impacts on the decision-makers and workers whose hopes and efforts forged the region's rural landscape. In the last chapter, pointedly titled "The Corn Business," Hudson (ibid. 208) reports that, due to changes in technology and marketing, "The 'typical' Corn Belt farm has vanished and has been replaced by a small cluster of metal buildings surrounding a suburban-type tract house" while "steam hisses and aromas pour constantly forth from the huge factories" that process a continuing flood of Corn Belt grain. Other authors describe the current interplay between structural and agency factors in farmer decision-making in the Corn Belt region (Lighthall 1995; Duram 2000).

In *The Land that Feeds Us*, Hart (1991) examines the agricultural settlement patterns of America's primary farming regions in the eastern half of the country. Case-study descriptions of representative farms help to provide scope and depth to the study. Hart expresses what many rural geographers realize about the importance of fieldwork: "The best way to understand farms is to talk to the people who are trying to make a living by farming" (ibid. 13). *The Rural Landscape* (Hart 1998) is a compilation of fieldwork, archival research, and cartographic investigation in pursuit of broad geographical patterns. The author describes a portrait of the landforms, vegetation, and settlements that together comprise the rural landscape. Again, Hart argues that fieldwork is necessary for regional understanding: "The only proper way to learn about and understand the landscape is to live in it, think about it, explore it, ask questions about it, contemplate it, and speculate about it" (ibid. 1).

Providing a historical context, Aiken's (1998) book on *The Cotton Plantation South Since the Civil War* focuses on a commercial agricultural system that came to dominate and even to define an entire region. The book presents the rise, transformation, and implications of the southern plantation system, including the linkages to the institution of slavery during the antebellum era. This explains the scope and intensity of the pervasive linkages among diverse aspects of land tenure, land use, social status, political power, and economic well-being in the rural South where the plantation system has predominated. Other researchers investigate current issues in the rural South including population change (Bascom and Gordon 1999), economic restructuring (Moore 1999), and agricultural sustainability (Furuseth 1997).

In *Let the Cowboy Ride*, Starrs (1998) rounds up field observations together with wide-ranging archival research to present a gritty portrait of cattle ranching in the American West. Insights are built on field observations and interviews with ranchers, elected officials, federal land managers, conservationists, and others in five case-study sites situated in Nebraska, Nevada, New Mexico, Texas, and Wyoming. Each site involves ranching in its own distinctive mixture of environment, economy, settlement history, and local culture. But throughout, there is attention to the recurring themes of "how cattle ranching evolved and why its history is rife with conflict that reflects an essential geographical ignorance and prejudice" (ibid. 4). The roots of the conflict, according to Starrs (ibid. 237) can be traced to faulty land tenure laws ill suited to the drier western third of the United States enacted by a federal "Congress [that] was strongly prejudiced against more appropriate land systems, such as the large grants used throughout the Hispanic New World or the large leases that gave Canadian ranchers secure tenure on provincial rangelands." In addition to explaining why the cowboys and the feds cannot ever get along, the book also depicts ranch life as "something transcendent, beyond the demands of 'mere money' " (ibid. 78).

Drawing from regional examples, the book *At Odds with Progress* (Wallach 1991) illustrates the cultural geography of rural America through specific cases of our simultaneous mistreatment and careful stewardship of the environment. Wallach takes the reader from Maine and Tennessee to California and Washington, with North Dakota and Texas in between. He depicts real people on the landscape and the complexity of their natural resource use: agriculture, logging, oil drilling, water projects, aquifer use, and grazing lands. The author claims that Americans are fatalists, resigned to the fact that "progress" will consume our rural surroundings. Yet, at the same time, we conceal this pessimism in three disguises: efficiency, social welfare, and ecology. Thus Americans do not openly protect the environment simply because it is the right thing to do. Instead, we shroud our true environmentalist concerns and actions in such socially and politically acceptable terms as economic efficiency, ecosystem integrity, and mitigating risks. The outcome is that many of our actions are those of conservationists, but we never confront the "progress" of development. The author notes that Americans should be proud of their conservation accomplishments, although these are "things we've done in spite of ourselves" (ibid. 63).

Research at the regional scale often allows rural geographers to show how environmental, agricultural, social, cultural, economic, political, and settlement features combine and interact in distinctive ways. In addition, many researchers have been successful in identifying common underlying patterns so as to establish general principles perhaps applicable in diverse settings. Indeed, a global view of country-level food "self-sufficiency" offers interesting patterns (Grigg 2001). The search for common patterns has sometimes emphasized cultural factors such as settlement history or land ownership (Emel and Roberts 1995; Hewes 1996; Williams 1996; Berardi 1998*b*; Heasley and Guries 1998) or physical factors such as topography, soil, or climate (Gersmehl and Brown 1986; B. D. Baker and Gersmehl 1991; Riebsame 1991; Cross 1994; Kromm 1994; Loveland *et al.* 1995; Nellis *et al.* 1997*b*; Rosenzweig and Hillel 1998). Many of the most broadly directed efforts to identify common underlying patterns of rural land use in diverse settings have focused on economic considerations, including such matters as agricultural market demand, supply competition, production costs, transportation charges, and economic restructuring (Hudson 1991; Bowler 1992; Grigg 1995; Ilbery and Bowler 1996; Schroeder 1997; Wells 2000).

Agricultural Location Theory

The foundations of economically oriented research on rural land-use patterns were set by German economic theoretician Johann Heinrich von Thünen early in the nineteenth century. Studying costs and revenues to be expected from selling agricultural commodities at a central market in Prussia, von Thünen deduced that "fairly sharply differentiated concentric rings or belts will form around the Town, each with its own particular staple product" (Hall 1966: 8). Many efforts have been undertaken to test or extend von Thünen's principles in other settings or times (Grigg 1984; Kellerman 1989*a*, *b*; Munton *et al.* 1994; Archer and Lonsdale 1997; Hite 1997; Visser 1999). Research inspired by von Thünen has been surveyed recently (O'Kelly and Bryan 1996), but it can be noted that spatial patterns similar to those deduced by von Thünen have been identified at various geographical scales, from that of the farm enterprise (Aitchison 1992) to the world as a whole (Chisholm 1962; Peet 1969).

Indeed, it could be argued from von Thünen's agricultural location theory that many of the recent local dislocations caused by the transnational globalization of industrialized commercial agriculture reflect yet another round of readjustments to the effects of progressively diminished transport costs at very large geographical scales (Bonanno *et al.* 1994; McMichael 1994; Ilbery and Bowler 1996) although trade liberalization (Ufkes 1993) deregulation (Archer and Lonsdale 2001), and environmental conflicts (Hollander 1995) also play roles in agricultural globalization. In the late nineteenth and early twentieth centuries similar localized disruptions caused by declining transportation costs and increasing competition from yet more distant suppliers adversely affected agriculturalists in Europe (O'Rourke 1997), New England (Glade 1991), and parts of the Midwest (Gregson 1993; Hudson 1994; Page 1996).

Market information about supply and demand balances in markets for agricultural commodities now moves with extraordinary rapidity. For example, a recent study of spatial price adjustments in daily transactions at twenty-eight beef packing plants at various locations in the United States showed that price differences adjusted for transport cost differentials tended to diminish at a rate of about one-third each day (Schroeder 1997). Because of the higher costs of shipping live cattle or carcasses longer distances, prices at plants at greater distances from one another tended to become coordinated more slowly, which "implies a distance-decay in strength of spatial price linkages" (ibid. 361). But the finding that plants under the same ownership exhibited close price coordination regardless of separating distances could be cause for concern in light of other adverse consequences of recent market concentration in the beef packing industry (Broadway 1990; Stull *et al.* 1995; Ufkes 1995; Broadway 2000).

In the modern era, some of the effects of lowered transport costs can be observed in the wide supermarket choices available to urban consumers who can purchase numerous food products from many locational sources at relatively low prices. An inspection of food labels in a typical American supermarket is apt to turn up places of origin throughout the United States, elsewhere in North and South America, and also in Europe, Africa, and Asia (de Wit 1992; Grigg 1995; Jarosz and Qazi 2000), although there are regional variations in food preferences in different parts of the United States (Shortridge and Shortridge 1989; R. B. Larson 1998).

Within the agricultural sector, declining transportation costs are likely to increase the locational role of environmental conditions. While typically neglected in abbreviated discussions of von Thünen's work, the implication of an inverse relationship between the cost and speed of transportation on the one hand, and the relative influence of natural environmental factors on the other is one of the most consequential aspects of his work from a modern perspective. Evidence for such a relationship was found by Gregson (1993: 332), who studied manuscript census data on crop selection for a sample of Missouri farms in 1860, 1870, and 1880 to find that cropping patterns increasingly conformed with variations in soil type once railroads had reached the study area. In her words (ibid. 334), "As the constraints imposed by high transport costs were ameliorated, the constraints imposed by nature became more binding." Similarly, Leaman and Conkling (1975: 430–1) found that farming patterns in central New York were in "closer conformity to the physical environment" in 1860 than they had been in 1840 when "land quality did not appear to be an important locational determinant." And Archer and Lonsdale (1997) found that between 1978 and 1992 county-level agricultural land values across the United States became more closely aligned with both distances to metropolitan centers and environmental regions as delimited by the US Department of Agriculture.

Indeed, geographical variations in topography, soil, and climate seem even more significant today than they were during the heyday of "environmental determinism," when O. E. Baker (1921: 17) asserted in the *Annals of the Association of American Geographers* that "the physical factors or conditions determine in large degree the utilization of land in a region; and these physical factors become more important as . . . agriculture and forestry become more highly organized and commercialized." Ironically, "regional rural geography" and "agricultural location theory" have become more tightly intertwined as advances in transportation and production technologies have reduced economic distances and expanded efficient scales of production. Lower transportation costs have greatly enhanced the comparative economic advantages enjoyed by farmers in more fertile settings, even at the expense of farmers who might be closer to markets but who must till lands that are naturally less agriculturally productive. Land-use decision-making and management of the natural environment have become key to long-term sustainability of agriculture and rural regions.

Rural Land-Use Change

In 1890 the rural North American frontier closed, according to the noted historian Frederick Jackson Turner (1920). Maps based on the 1890 US Census of Population still showed that portions of the arid West remained below the critical population thresholds of 2 and 6 persons per square mile that Turner deemed indicative of frontier and fully settled areas, but these patches were regarded as small interruptions in an otherwise continuous expanse of settlement from the Atlantic Coast to the Pacific Coast. Cities and towns were becoming more and more important as economic and cultural nodes in the settlement structure of the United States (Berry and Horton 1970), but farms and ranches still provided the basic livelihoods of 9.1 million workers out of a total national labor force of 23.3 million workers in 1890 (Kurian 1994: 74). Agricultural labor continued to increase up to the time of the 1920 Census, which established a maximum of 12.4 million agricultural workers out of a total workforce of 38.2 million. Since 1920—which also was the Census year in which more than half the US population lived in urban places for the first time—the size of the US agricultural workforce has declined steadily. By 1990, the total US labor force had expanded to 126.4 million, but the agricultural component of the US labor force had shrunk to barely 3.2 million workers. Hence, fewer than 3 out of 100 workers (2.5%) were employed on US farms or ranches in 1990. These population declines took place in Canada as well, where farm population decreased by 66 per cent and the number of farms went down 50 per cent from 1941 to 1991 (Troughton 1995: 296).

Although population distribution has shifted, rural areas continue to provide food, minerals, fuel, and recreation for the entire continent, and to some extent, the world. Indeed one could argue that 'the single most important resource of America is its rural lands' (Lapping *et al.* 1989: 103). Rural geographers investigate the interrelationships between society and the rural environment and specifically the linkages between nature, technology, and production and consumption of rural resources (FitzSimmons 1990; Roberts 1995; Marsden *et al.* 1998).

Rural natural resource management incorporates environmental policy, socioeconomic factors, and ecological variables to address a variety of topics including land use, sustainable development, pollution, and mapping. One component of rural land use is played out in the management of public lands, which are utilized for recreation, resource extraction, and research (Harrington and Roberts 1988; Curry-Roper 1989; Duram 1995; Harrington 1996). Recreational access to these public resources varies because of topography, roads, and land cover (Millward 1996). The multipurpose goals of the use of such lands, which belong to the illusive public "at large," often lead to complex management strategies (Paulson 1998).

Sustainable rural development and the notion of uneven development provide a framework through which to address rural social and ecological decline (Roberts and Emel 1992; Wilbanks 1994; Napton 1997; Furuseth and Thomas 1997; Wilson 1999). The emergence of rural and small-town businesses play an important role in economic development (Aspaas 1997). The various dimensions of rural public policy can greatly impact rural land use (Frederic 1991; Platt 1996).

Increasingly, issues in rural development actually address rural–urban interactions. As North American cities expand, often in the form of low-density suburban land uses, farmland protection becomes necessary (Napton 1990; G. E. Walker 1997) and the use of rural lands at the boundary of urban areas comes into question (Beesley and Walker 1990; Halseth and Rosenberg 1995; Feldman and Jonas 2000). In addition, population dynamics in rural regions can impact rural recreation activities (Mason 1989; Reed and Gill 1997), and cause rural landscape change (Riebsame *et al.* 1996). Other population issues such as migration and changing demographic characteristics also impact rural development (White 1992; Everitt and Gfellner 1996; Lonsdale and Archer 1998, 1999; Gill 1999; Sommers *et al.* 1999; Beyers and Nelson 2000). See Ch. 23, Rural Development, for additional information on these topics.

Local people increasingly join together to voice their concerns about rural resource issues. The NIMBY (Not in My Back Yard) movement has influenced rural environmental dialog and actions (Lant and Sherill 1995). Mistrust of government agencies (Binney *et al.* 1996) and the complexities of rural pollution are best studied from a contextual approach that acknowledges the social, economic, and political factors involved (Solecki 1992; Solecki and Shelley 1996). Local resistance to land development is most apparent on the rural–urban fringe and in areas of rapid population growth (Halseth 1996; Gosnell *et al.* 1996; Shumway 1997). Research methods that encourage community participation can influence environmental policy outcomes (Berardi 1998*a*; Berardi and Donnelly 1999).

With rapid growth of spatial data availability, remote sensing and GIS have become important tools in rural and agricultural analyses. Application of these technologies can assist in the provision of rural services (Gribb *et al.* 1990), in rural resource planning (Nellis 1992), and in investigating rural local knowledge (Weiner *et al.* 1995). GIS and remote sensing can be used to study land-use and land-cover change (R. Walker and Solecki 1999; Cihlar and Jansen 2001) and evaluate land uses to make recommendations regarding policies such as the Conservation Reserve Program (CRP) (Nellis *et al.* 1997*a*; Wu *et al.* 1997). Geographic technologies are increasingly important in agricultural production (Heimlich 1998) and specifically helpful in field-scale modeling (Goense *et al.* 1996) and production forecast modeling (Corbett *et al.* 2002). Computer cartography allows researchers to analyze and display vast amounts of data in a concise format in order to understand rural change in a given region (Beazley and Beesley 1996).

Agricultural Sustainability

Agricultural sustainability is an important concern in rural North America, and agricultural policies have increasingly addressed environmental problems within the changing structure of agricultural production (Napton 1994; Roberts and Dean 1994; Pierce 1994). Researchers and the public increasingly realize the potential ecological impacts of our current industrial farming system. According to a congressional study, rural regions in the US should be targeted for multiple "agro-environmental priorities" to address water quality, wildlife habitat, and soil quality problems (US Congress 1995: 10). Environmental concerns with agriculture have increasingly focused on the impacts of agrichemicals on the nation's water supply (both ground water and surface water). By the late 1980s, seventeen types of pesticides were detected in groundwater in twenty-three states, and while concentrations may not be high enough to be acutely toxic, there is concern about long-term chronic health effects (Hallberg 1987). According to the *Yearbook of Agriculture* (USDA 1991: 82): "About 10% of community water systems and 4% of rural domestic wells contained detectable concentrations of pesticides, while over 50% of wells surveyed contained detectable concentrations of nitrates." A decade later, "prevalent nitrate contamination was detected in shallow ground water (less than 100 feet below land surface) beneath agricultural and urban areas, where about 15 per cent of all samples exceeded the USEPA drinking-water standard" (US Geological Survey 1999). In surface water, the USDA (1991: 83) reported that of the approximately 165,000 miles of rivers in the US that are contaminated by non-point source (run-off) pollution, agriculture is the source of 64 per cent of this pollution. According to a 1999 US Geological Survey study, "nationally, 11 herbicides, 1 herbicide degradation product, and 3 insecticides were detected in more than 10 percent of [stream] samples" (S. J. Larson *et al.* 1999).

Knowledge of this widespread environmental degradation has encouraged an increase in alternative methods of production that require fewer agrichemicals. According to a major study by the National Research Council (1989: 6): "Wider adoption of proven alternative systems would result in greater economic benefits to farmers and environmental gains to the nation." Another National Research Council study (1993: 11) recommends policies that encourage "economically viable cropping systems that incorporate cover crops, multiple crops, and other innovations . . . to meet long-term soil and water quality goals." Indeed alternative agriculture, especially organic farming methods, has gained increasing acceptance in North America (Duram 1998). There are notable local variations in the types of crops produced and the degree to which these farms fit into the conventional industrial model of production (Buck *et al.* 1997; Duram 1997; Guthman 1998; Coombes and Campbell 1998). Certainly, there is great potential for further investigation into the impacts of the industrialization of agriculture (Page 1996). Various research approaches (i.e. postmodern, Marxist, political economy) are employed by rural social geographers (Phillips 1998) and rural researchers to explain these structural changes in agricultural production (Marsden and Little 1990; Friedland *et al.* 1991; Bonanno *et al.* 1994; McMichael 1994; Marsden *et al.* 1998). Indeed

Whatmore (1995) concisely describes how the new political economy approach addresses our current global agro-food system. She ponders the structural shift in agricultural production: "how does farming, the anchor of commonsense understandings of food production, fit into the creation of oven-ready meals; genetically engineered plants and animals; or synthetic foodstuffs?" (Whatmore 1995: 36). This question emphasizes that important research opportunities exist in the context of current agricultural restructuring.

Research shows that environmental policies can have long-term impacts on agricultural regions. In the US, the Conservation Reserve Program (CRP) has taken land out of production and has many potential impacts, such as decreased erosion, increased farm income, and reduced residential development (Johnson and Maxwell 2001; Lant *et al.* 2001). However, this and other agro-environmental programs may be less effective than expected, as when "slippage" of land into production occurs while other land is being removed through the CRP (Leathers and Harrington 2000). Continued research into the impacts of agri-environmental policies will yield practical findings and applications.

Water use and availability is another prominent rural resource concern. In the Great Plains, for example, agriculture in eight states depends on water from the Ogallala Aquifer. In some places withdrawal rates have exceeded natural renewal rates, leading to dire predictions for the region's agricultural future. Research suggests that population dynamics (Popper and Popper 1999) and individual water use behavior are key components to a long-term management solution (Kromm and White 1991; White 1994). In addition to individuals acting within their socio-economic setting, water institutions also play a role in determining resource use through policy action (Gottlieb and FitzSimmons 1991; Roberts 1992; Templer 1992; Emel and Roberts 1995; White and Kromm 1995; Kromm and White 1996; Hogan and Fouberg 1998).

While the public commonly equates farming with crop production, in fact the impacts of livestock production are also an important issue in rural areas. Large-scale hog production has brought social change and ecological decline to rural areas (Furuseth 1997; Hart and Mayda 1997). Ecological degradation is a specific concern as livestock grazing can cause streambank erosion (Trimble and Mendel 1995).

From the individual decision-maker's point of view, addressing and mediating environmental degradation involves several steps. People must be aware of a problem, have knowledge of alternative actions, have a motivation for change, and have the resources to enact that change (Padgitt and Petrzelka 1994). Each of these stages, from perception of ecological degradation to sources of information and economic status, provide avenues of investigation. Diverse factors influence agricultural decision-making, which in turn impacts sustainability at the local level. Variations in communities and their worldview influence agricultural methods in a given place (Bellows 1997; Curry-Roper 1997; Curry 2000). Local agricultural land uses develop through a combination of individual decision-making and farmers' perceptions of ecological conditions (Cross 1994; Chiotti and Johnston 1995; Dagel 1997). Adoption of alternative methods is linked to sources of information and the perceived accessibility of alternative information from agricultural extension and research (Paulson 1995; Duram 1999; Duram and Larson 2001). Further, perceptions of agricultural chemical risk varies among farmers (Tucker and Napier 2001) and adoption of sustainable methods is also influenced by broader policy and economic concerns (Lowe *et al.* 1994; Lighthall and Roberts 1995). Current research includes investigations into the role of women in agricultural production, in the context of agricultural restructuring and decision-making (Everitt 1988; Leckie 1993; Whatmore *et al.* 1994; Sachs 1996). There is clearly a need for continued research into agricultural decision-making within the ever-changing social, economic, and political structures of modern agriculture.

Conclusion

This chapter provides an overview of current research themes among rural and agricultural geographers, within the context of key historical works in rural geography. In this conclusion, we forecast trends in future rural issues and research concerns. Based on the four substantive sections of this chapter, the following questions will likely echo into the new century. First, how can we describe the functional and formal attributes of dynamic rural regions? Second, will locational theory and economic factors provide similar rural patterns in the future? Third, what management actions will individuals and society take to ensure sustainable rural land use? Fourth, what alternatives will we select to ensure long-term agricultural sustainability?

These questions, however, will be asked with pressing urgency. In 1950, world population was 2.5 billion, in 2000 it was about 6 billion, and by 2050 we could easily reach 9.3 billion (World Resources Institute 1998: 244).

This places ever-increasing demands on rural resources: competition for agricultural land, recreational opportunities, and natural resource extraction must increase substantially to keep pace with human population growth. From a global perspective, per capita cropland was .25 hectares in 1984, and down to .22 hectares by 1994 (World Resources Institute 1998: 286). Certainly, technological advances have interceded to mediate this decline, just as striving for political stability and social justice would also aid in maintaining or improving people's quality of life. If population and resource trends continue, however, we must take decisive action to sustain our rural landscapes.

The importance of rural lands is obvious at all geographic scales. For example, rural North America is a significant contributor to global agricultural production, national public land preservation, regional ecosystemic processes, and local social stability. Thus future rural land-use management must balance the social, economic, political, and ecological concerns of a landscape that is under pressure of increasing competition. In the future, rural and agricultural geographers will likely continue to investigate these important issues through both theoretical and applied analyses to describe and understand rural landscapes in the twenty-first century.

References

Aiken, C. S. (1998). *The Cotton Plantation South since the Civil War*. Baltimore: Johns Hopkins University Press.

Aitchison, J. (1992). "Farm Types and Agricultural Regions," in Bowler (1992: 109–33).

Archer, J. C., and Lonsdale, R. E. (1997). "Geographical Aspects of U.S. Farmland Values and Changes During the 1978–1992 Period." *Journal of Rural Studies*, 13/4: 399–413.

—— (2001). "Great Plains Settlement: Globalization and Deregulation," in Heikki Jussia, Roser Majoral, and Fernanda Delgado-Cravidao (eds.), *Globalization and Marginality in Geographical Space: Political, Economic, and Social Issues of Development in the New Millennium*. Aldershot: Ashgate, 37–52.

Aspaas, H. R. (1997). "Women's Home-Based Businesses." *Small Town*, Jan.–Feb.: 4–11.

Baker, B. D., and Gersmehl, P. J. (1991). "Temporal Trends in Soil Productivity Evaluations." *Professional Geographer*, 43/3: 304–18.

Baker, O. E. (1921). "The Increasing Importance of the Physical Conditions in Determining the Utilization of Land for Agricultural and Forest Production in the United States." *Annals of the Association of American Geographers*, 11: 17–46.

Baltensperger, B. H. (1991). "A County that Has Gone Downhill." *Geographical Review*, 81/4: 433–42.

Bascom, J., and Gordon, R. (1999). "Country Living: Rural Non-Farm Population Growth in the Coastal Plain Region of North Carolina," in N. Walford, J. C. Everitt, and D. E. Napton (eds.), *Reshaping the Countryside: Perceptions and Processes of Rural Change*. Wallingford: CAB International, 77–90.

Beazley, S. L., and Beesley, K. B. (1996). *Monitoring Change in Agriculture Using Computer-Assisted Cartography I: Land Use*. Rural Research Centre. Truro, Nova Scotia: Nova Scotia Agriculture College.

Beesley, K. B., and Walker, G. (1990). "Residence Paths and Community Perception: A Case Study from the Toronto Urban Field." *Canadian Geographer*, 34/4: 318–30.

Bell, M. M. (1989). "Did New England Go Downhill?" *Geographical Review*, 79/4: 450–66.

Bellows, A. (1997). "The Invisible Food and Agricultural Revolution." IFOAM: *Ecology and Farming*, September: 19–20.

Berardi, G. (1998*a*). "Application of Participatory Rural Appraisal in Alaska." *Human Organization*, 57/4: 438–46.

—— (1998*b*). "Natural Resource Policy, Unforgiving Geographies, and Persistent Poverty in Alaska Native Villages." *Natural Resources Journal*, 38/1: 85–108.

Berardi, G., and Donnelly, S. (1999). "Rural Participatory Research in Alaska: The Case of Tanakon Village." *Journal of Rural Studies*, 15/2: 171–8.

Berry, B. J., and Horton, F. E. (1970). *Geographic Perspectives on Urban Systems*. Englewood Cliffs, NJ: Prentice Hall.

Beyers, W., and Nelson, P. (2000). "Contemporary Development Forces in the Nonmetropolitan West: New Insights from Rapidly Growing Communities." *Journal of Rural Studies*, 16/4: 459–74.

Binney, S. E., Mason, R., Martsolf, S. W., and Detweiler, J. H. (1996). "Credibility, Public Trust and the Transportation of Radioactive Waste Through Local Communities." *Environment and Behavior*, 28/3: 283–301.

Bonanno, A., Busch, L., Friedland, W. H., Gouveia, L., and Mingione, E. (eds.) (1994). *From Columbus to ConAgra: The Globalization of Agriculture and Food*. Lawrence, Kan.: University Press of Kansas.

Bowler, I. R. (ed.) (1992). *The Geography of Agriculture in Developed Market Economies*. New York: Wiley.

Broadway, M. (1990). "Meatpacking and its Social and Economic Consequences for Garden City, Kansas in the 1980s." *Urban Anthropology*, 19/4: 321–44.

—— (2000). "Planning for Social Change in Small Towns or Trying to Avoid the Slaughterhouse Blues." *Journal of Rural Studies*, 16/1: 37–46.

Buck, D., Getz, C., and Guthman, C. (1997). "From Farm to Table: The Organic Vegetable Commodity Chain of Northern California." *Sociologia Ruralis*, 37/1: 3–20.

Chiotti, Q. P., and Johnston, T. (1995). "Extending the Boundaries of Climate Change Research: A Discussion of Agriculture." *Journal of Rural Studies*, 11/3: 335–50.

Chisholm, M. (1962). *Rural Settlement and Land Use*. London: Hutchinson.

Cihlar, J., and Jansen, L. (2001). "From Land Cover to Land Use: A Methodology for Efficient Land Use Mapping over Large Areas." *The Professional Geographer* 53/2: 275–89.

Coombes, B., and Campbell, H. (1998). "Dependent Reproduction of Alternative Modes of Agriculture: Organic Farming in New Zealand." *Sociologia Ruralis*, 38/2: 127–45.

Corbett, J., White, J., and Collins, S. (2002). "Spatial Decision Support Systems and Environmental Modeling: An Application Approach," in Keith C. Clarke, B. Parks, and M. Crane (eds.), *Geographic Information Systems and Environmental Modeling*. Upper Saddle River, NJ: Prentice-Hall, 51–66.

Cross, J. A. (1994). "Agroclimatic Hazards and Dairy Farming in Wisconsin." *Geographical Review*, 84/3: 277–89.

Curry, J. (2000). "Community Worldview and Rural Systems: A Study of Five Communities in Iowa." *Annals of the Association of American Geographers*, 90/4: 693–712.

Curry-Roper, J. (1989). "The Impact of the Timber and Stone Act on Public Land Ownership in Northern Minnesota." *Journal of Forest History*, 33/2: 70–9.

—— (1997). "Community-Level Worldviews and the Sustainability of Agriculture," in B. Ilbery, Q. Chiotti and T. Rickard (eds.), *Agricultural Restructuring and Sustainability: A Geographical Perspective*. Wallingford: CAB International, 101–15.

Dagel, K. C. (1997). "Defining Drought in Marginal Areas: The Role of Perception." *Professional Geographer*, 49/2: 191–202.

de Wit, C. (1992). "Food-Place Associations on American Product Labels." *Geographical Review*, 82/3: 323–30.

Doolittle, W. E. (1992). "Agriculture in North America on the Eve of Contact: A Reassessment." *Annals of the Association of American Geographers*, 82/3: 386–401.

Duram, L. A. (1995). "Public Land Management: The National Grasslands. Part One: Historical Development; Part Two: Case Study of Pawnee National Grassland; Part Three: Future Management Scenarios." *Rangelands*, 17/2: 36–42.

—— (1997). "A Pragmatic Study of Conventional and Alternative Farmers in Colorado." *The Professional Geographer*, 49/2: 202–13.

—— (1998). "Organic Agriculture in the United States: Current Status and Future Regulation." *Choices: Food, Farm and Resource Issues*, 2nd quarter: 34–38.

—— (1999). "Factors in Organic Farmers' Decisionmaking: Diversity, Challenge, and Obstacles." *American Journal of Alternative Agriculture*, 14/1: 2–10.

—— (2000). "Agents Perceptions of Structure: How Illinois Organic Farmers View Political, Economic, and Social Factors." *Agriculture and Human Values*, 17/1: 35–47.

Duram, L. A., and Larson, K. L. (2001). "Agricultural Research and Alternative Farmers' Information Needs." *The Professional Geographer*, 53/1: 84–96.

Emel, J., and Roberts, R. S. (1995). "Institutional Form and Its Effect on Environmental Change: The Case of Groundwater in the Southern High Plains." *Annals of the Association of American Geographers*, 85/4: 664–83.

Everitt, J. C. (1988). "Husband–Wife Role Variation as a Factor in the Social Space Definition of Farmers." *Social Science Quarterly*, 69/1: 154–76.

Everitt, J. C., and Gfellner, B. (1996). "Elderly Mobility in a Rural Area: The Example of Southwest Manitoba." *Canadian Geographer*, 40/4: 338–51.

Feldman, T., and Jonas, A. (2000). "Sage Scrub Revolution? Property Rights, Political Fragmentation, and Conservation Planning in Southern California under the Federal Endangered Species Act." *Annals of the Association of American Geographers*, 90/2: 256–92.

FitzSimmons, M. (1990). "The Social and Environmental Relations of US Agricultural Regions," in P. Lowe, T. Marsden, and S. Whatmore (eds.), *Technological Change and the Rural Environment*. London: David Fulton, 8–32.

Frederic, P. (1991). "Public Policy and Land Development: The Maine Land Use Regulation Commission." *Land Use Policy*, 8/1: 50–62.

Fredrich, B. E. (1991). "Food and Culture: Using Ethnic Recipes to Demonstrate the Post-Columbian Exchange of Plants and Animals." *Journal of Geography*, 90/1: 11–15.

Friedland, W. H., Busch, L, Buttel, F. H., and Rudy, A. P. (eds.) (1991). *Towards a New Political Economy of Agriculture*. Boulder, Cl.: Westview.

Furuseth, O. J. (1997). "Restructuring of Hog Farming in North Carolina: Explosion and Implosion." *Professional Geographer*, 49/4: 391–403.

Furuseth, O. J., and Thomas, D. S. K. (1997). "Moving from Principles to Policy: A Framework for Rural Sustainable Community Development in the United States," in Ivonne Audirac (ed.), *Rural Sustainable Development in America*. New York: John Wiley & Sons, 131–45.

Gersmehl, P. J., and Brown, D. A. (1986). "The Diminishing Advantage of Muscatine Silt Loam and the Dixification of the Midwest." *Journal of Geography*, 85/5: 212–17.

Gill, A. M. (1999) "Competitive Tensions in Community Planning", in N. Walford, J. C. Everitt, and D. E. Napton (eds.), *Reshaping the Countryside: Perceptions and Processes of Rural Change*. Wallingford: CAB International, 157–68.

Glade, D. W. (1991). "Weeds in Vermont as Tokens of Socioeconomic Change." *Geographical Review*, 81/2: 153–69.

Goense, D., Hofstee, J. W., and van Bergeijk, J. (1996). "An Information Model to Describe Systems for Spatially Variable Field Operations." *Computers and Electronics in Agriculture*, 14/2–3: 197–214.

Gosnell, H., Riebsame, W., and Theobald, D. (1996). "Land Use and Landscape Change in the Colorado Mountains: A Case Study of the East River Valley." *Mountain Research and Development*, 16/4: 407–18.

Gottlieb, R., and FitzSimmons, M. (1991). *Thirst for Growth: Water Agencies as Hidden Government in California*. Tucson, Ariz.: University of Arizona Press.

Gregson, M. E. (1993). "Rural Response to Increased Demand: Crop Choice in the Midwest, 1860–1880." *Journal of Economic History*, 53/2: 332–45.

Gribb, W., Czerniak, R., and Harrington, J. (1990). "Rural Addressing and Computer Mapping in New Mexico." *Professional Geographer*, 42/4: 471–80.

Grigg, D. (1969). "The Agricultural Regions of the World: Review and Reflections." *Economic Geography*, 45/2: 95–132.

—— (1984). *An Introduction to Agricultural Geography*. London: Hutchinson.

—— (1995). "The Geography of Food Consumption: A Review." *Progress in Human Geography*, 19/3: 338–54.

—— (2001). "Food Imports, Food Exports and their Role in National Food Consumptions." *Geography*, 86/2: 171–6.

Guthman, J. (1998). "Regulating Meaning, Appropriating Nature: The Codification of California Organic Agriculture." *Antipode*, 30/2: 135–54.

Hall, P. (ed.) (1966). *Von Thünen's Isolated State*. Trans. C. M. Wartenberg London: Pergamon.

Hallberg, G. (1987). "Agricultural Chemicals in Ground Water: Extent and Implications." *American Journal of Alternative Agriculture*, 2/1: 3–15.

Halseth, G. (1996). "Community and Land-Use Planning Debate: An Example from Rural British Columbia." *Environment and Planning A*, 28/7: 1279–98.

Halseth, G., and Rosenberg, M. (1995). "Complexity in the Rural Canadian Housing Landscape." *Canadian Geographer*, 39/4: 336–52.

Harrington, L. M. B. (1996). "Regarding Research as a Land Use." *Applied Geography*, 16/4: 265–77.

Harrington, L. M. B., and Roberts, R. S. (1988). "Wilderness Research: Effects of Administering Agency." *Society and Natural Resources*, 1/3: 215–25.

Hart, J. F. (1991). *The Land that Feeds Us*. New York: Norton.

—— (1998). *The Rural Landscape*. Baltimore, Md.: Johns Hopkins University Press.

Hart, J. F., and Mayda, C. (1997). "Pork Palaces on the Panhandle." *Geographical Review*, 87/3: 396–400.

Heasley, L., and Guries, R. P. (1998). "Forest Tenure and Cultural Landscapes: Environmental Histories in the Kickapoo Valley," in H. M. Jacobs (ed.), *Who Owns America? Social Conflict over Property Rights*. Madison, Wis.: University of Wisconsin Press, 182–207.

Heimlich, R. (1998). "Precision Agriculture: Information Technology for Improved Resource Use." *Agricultural Outlook*, 250: 19–23.

Hewes, L. (1996). "Making a Pioneer Landscape in the Oklahoma Territory." *Geographical Review*, 86/4: 588–603.

Hite, J. (1997). "The Thünen Model and the New Economic Geography as a Paradigm for Rural Development Policy." *Review of Agricultural Economics*, 19/2: 230–40.

Hogan, E., and Fouberg, E. H. (1998). "*The Geography of South Dakota: Revised Edition*." Sioux Falls, S. Dak.: Augustana College Center for Western Studies.

Hollander, G. M. (1995). "Agroenvironmental Conflict and World Food System Theory: Sugarcane in the Everglades Agricultural Area." *Journal of Rural Studies*, 11/3: 309–18.

Hudson, J. C. (1991). "New Grain Networks in an Old Urban System," in John Fraser Hart (ed.), *Our Changing Cities*. Baltimore, Md.: Johns Hopkins University Press, 86–107.

—— (1994). *Making the Corn Belt: A Geographical History of Middle-Western Agriculture*. Bloomington, Ind.: Indiana University Press.

Hudson, J. C., and Espenshade, E. B. (eds.) (2000). *Goode's World Atlas*. Chicago: Rand McNally.

Huke, R., Huke, E., Woodhead, T., and Huang, J. (1993). *Rice-Wheat Atlas of China*. Manila: International Rice Research Institute.

Jarosz, L., and Qazi, J. (2000). "The Geography of Washington's World Apple: Global Expressions in an Local Landscape." *Journal of Rural Studies* 16/1: 1–14.

Ilbery, B. (1998). *The Geography of Rural Change*. Harlow: Longman.

Ilbery, B. W., and Bowler, I. R. (1996). "Industrialization and World Agriculture", in I. Douglas, R. Huggett, and M. Robinson (eds.), *Companion Encyclopedia of Geography, the Environment, and Humankind*. New York: Routledge, 228–48.

Johnson, J., and Maxwell, B. (2001). "The Role of the Conservation Reserve Program in Controlling Rural Residential Development." *Journal of Rural Studies*, 17/3: 323–32.

Kellerman, A. (1989*a*). "Agricultural Location Theory, 1. Basic Models." *Environment and Planning A*, 21/10: 1381–96.

—— (1989*b*). "Agricultural Location Theory, 2. Relaxation of Assumptions and Applications." *Environment and Planning A*, 21/11: 1427–46.

Kromm, D. E. (1994). "Water Conservation in the Irrigated Prairies of Canada and the United States." *Canadian Water Resources Journal*, 18/4: 451–8.

Kromm, D. E., and White, S. E. (1991). "Reliance on Sources of Information for Water-Saving Practices by Irrigators in the High Plains of the U.S.A." *Journal of Rural Studies*, 7/4: 411–21.

—— (1996). "Changing Patterns of Irrigation in the American High Plains." *Arid Ecosystems*, 2/4: 88–94.

Kurian, G. T. (1994). *Datapedia of the United States 1790–2000, America Year by Year*. Lanham, Md.: Bernan Press.

Lang, R. E., Popper, D. E., and Popper, F. J. (1997). "Is There Still a Frontier? The 1890 Census and the Modern American West." *Journal of Rural Studies*, 13/4: 377–86.

Lant, C., Loftus, T., Kraft, S., and Bennett, D. (2001). "Land-Use Dynamics in a Southern Illinois (USA) Watershed." *Environmental Management*, 28/3: 325–40.

Lant, C., and Sherrill, J. (1995). "The Role of NIMBY Groups in 22 Sanitary Landfill Sitings in Illinois." *The Environmental Professional*, 17: 243–50.

Lapping, M., Daniels, T., and Keller, J. (1989). *Rural Planning and Development in the United States*. New York: Guilford.

Larson, R. B. (1998). "Regionality of Food Consumption." *Agribusiness*, 14/3: 213–26.

Larson, S. J., Gilliom, R. J., and Capel, P. D. (1999). Pesticides in Streams of the United States—Initial Results from the National Water-Quality Assessment Program U.S. Geological Survey Water-Resources Investigations Report 98-4222, 92 pp. At URL: ‹http://ca.water.wr.usgs.gov/pnsp/rep/wrir984222/›, last accessed 28 January 2003.

Lawrence, M. (1997). "Heartlands or Neglected Geographies? Liminality, Power, and the Hyperreal Rural." *Journal of Rural Studies*, 13/1: 1–17.

Leaman, J. H., and Conkling, E. C. (1975). "Transport Change and Agricultural Specialization." *Annals of the Association of American Geographers*, 65/3: 425–32.

Leathers, N., and L. M. B. Harrington (2000). "Effectiveness of Conservation Reserves: 'Slippage' in Southwestern Kansas." *The Professional Geographer*, 52/1: 83–93.

Leckie, G. (1993). "Female Farmers in Canada and the Gender Relations of a Restructuring Agricultural System." *The Canadian Geographer*, 37/3: 212–30.

Lighthall, D. R. (1995). "Farm Structure and Chemical Use in the Corn Belt." *Rural Sociology*, 60: 505–20.

Lighthall, D. R., and Roberts, R. S. (1995). "Towards an Alternative Logic of Technological Change: Insights from Corn Belt Agriculture." *Journal of Rural Studies*, 11/3: 319–34.

Lonsdale, R. E., and Archer, J. C. (1998). "Emptying Areas of the United States, 1990–1995." *Journal of Geography*, 97/3: 108–22.

Lonsdale, R. E., and Archer, J. C. (1999). "Demographic Factors in Characterizing and Delimiting Marginal Lands," in H. Jussila, R. Majoral, and C. C. Mutambirwa (eds.), *Marginality in Space—Past, Present and Future*. Aldershot: Ashgate, 135–49.

Loveland, T. R., Merchant, J. W., Brown, J. F., Ohlen, D. O., Reed, B. C., Ohlen, P., and Hutchinson, J. (1995). "Map Supplement: Seasonal Landcover Regions of the United States." *Annals of the Association of American Geographers*, 85/2: 339–55.

Lowe, P., Marsden, T., and Whatmore, S. (eds.) (1994). *Regulating Agriculture*. London: David Fulton.

McMichael, P. (ed.) (1994). *The Global Restructuring of Agro-Food Systems*. Ithaca, NY: Cornell University Press.

Mannion, A. M. (1997). *Global Environmental Change: A Natural and Cultural Environmental History*. 2nd edn. New York: Addison Wesley Longman.

Marsden, T., and Little, J. (eds.) (1990). *Political, Social, and Economic Perspectives on the International Food System*. Brookfield, Vt.: Avebury.

Marsden, T., Munton, R., Ward, N., and Whatmore, S. (1998). "Agricultural Geography and the Political Economy Approach: A Review." *Economic Geography*, 72/4: 361–75.

Mason, D. S. (1989). "Private Deer Hunting on the Coastal Plain of North Carolina." *Southeastern Geographer*, 29/1: 1–16.

Millward, H. (1996). "Countryside Recreational Access in the United States: A Statistical Comparison of Rural Districts." *Annals of the Association of American Geographers*, 86/1: 102–22.

Moore, T. (1999). "Deindustrialisation and Rural Economic Restructuring in Southern West Virginia," in N. Walford, J. C. Everitt, and D. E. Napton (eds.), *Reshaping the Countryside: Perceptions and Processes of Rural Change*. Wallingford: CAB International, 123–34.

Munton, R., Moran, W., and Chisholm, M. (1994). "Classics in Human Geography Revisited: Chisholm, Rural Settlement and Land Use." *Progress in Human Geography*, 18/1: 61–4.

Napton, D. E. (1989). "Contemporary Agriculture and Rural Land Use", in G. L. Gaile and C. J. Willmott (eds.), *Geography in America*. Columbus, NY: Merrill, 333–50.

—— (1990). "Regional Farmland Protection: The Twin Cities Experience." *Journal of Soil and Water Conservation*, 45/4: 446–9.

—— (1994). "The Future of U.S. Agricultural Conservation Policy." *Progress in Rural Planning and Policy*, 4: 55–64.

—— (1997). "Restructuring for Rural Sustainability: Overcoming Scale Conflicts and Cultural Biases," in B. Ilbery, Q. Chiotti, and T. Rickard (eds.), *Agricultural Restructuring and Sustainability: A Geographical Perspective*. Wallingford: CAB International, 329–40.

National Research Council (1989). *Alternative Agriculture*. Washington, DC: National Academy Press.

—— (1993). *Soil and Water Quality: An Agenda for Agriculture*. Washington, DC: National Academy Press.

Nellis, M. D. (1992). "Geographic Information Systems for Rural Resource Planning in Kansas." *Progress in Rural Policy and Planning*, 2: 30–3.

Nellis, M. D., Harrington, L. M. B., and Sheeley, J. (1997*a*). "Policy, Sustainability and Scale: The U.S. Conservation Reserve Program," in B. Ilbery, Q. Chiotti, and T. Rickard (eds.), *Agricultural Restructuring and Sustainability: A Geographical Perspective*. Wallingford: CAB International, 219–32.

Nellis, M. D., Ransom, M., Price K., and Egbert S. (1997*b*) "Evaluating Soil Properties of CRP Land Using Remote Sensing and GIS in Finney County, Kansas." *Journal of Soil and Water Conservation*, 52/5: 352–58.

O'Kelly, M., and Bryan, D. (1996). "Agricultural Location Theory: von Thünen's Contribution to Economic Geography." *Progress in Human Geography*, 20/4: 457–75.

O'Rourke, K. H. (1997). "The European Grain Invasion, 1870–1913." *The Journal of Economic History*, 57/4: 775–801.

Padgitt, S., and Petrzelka, P. (1994). "Making Sustainable Agriculture the New Conventional Agriculture: Social Change and Sustainability," in J. Hatfield and D. Karlen (eds.), *Sustainable Agriculture Systems*. Boca Raton, Fla.: Lewis, 262–85.

Page, B. K. (1996). "Across the Great Divide: Agriculture and Industrial Geography." *Economic Geography*, 72/4: 376–97.

Paulson, D. D. (1995). "Minnesota Extension Agents' Knowledge and Views of Alternative Agriculture." *American Journal of Alternative Agriculture*, 10/3: 122–8.

—— (1998). "Collaborative Management of Public Rangeland in Wyoming: Lessons in Co-Management." *The Professional Geographer*, 50/3: 301–15.

Peet, R. (1969). "The Spatial Expansion of Commercial Agriculture in the Nineteenth Century: A Von Thünen Interpretation." *Economic Geography*, 45/4: 283–301.

Phillips, M. (1998). "The Restructuring of Social Imaginations in Rural Geography," *Journal of Rural Studies*, 14/2: 121–53.

Pierce, J. (1994). "Towards the Reconstruction of Agriculture: Paths of Change and Adjustment," *Professional Geographer*, 46/2: 178–90.

Platt, R. (1996). *Land Use and Society: Geography, Law, and Public Policy*. Washington, DC: Island.

Popper, D. E., and Popper, F. J. (1999). "The Buffalo Commons: Metaphor as Methods." *Geographical Review*, 90/3: 491–510.

Reed, M. G., and Gill, A. M. (1997). "Tourism, Recreational and Amenity Values in Land Allocation: An Analysis of Institutional Arrangements in the Post-Productivist Era." *Environment and Planning A*, 29: 2019–40.

Riebsame, W. (1991). "Sustainability of the Great Plains in an Uncertain Climate," *Great Plains Research*, 1/1: 133–51.

Riebsame, W., Gosnell, H., and Theobald, D. M. (1996). "Land Use and Landscape Change in the Colorado Moutains: Theory, Scale and Pattern," *Mountain Research and Development*, 16/4: 395–405.

Roberts, R. S. (1992). "Groundwater Management Institutions," in D. E. Kromm and S. E. White (eds.), *Groundwater Exploitation in the High Plains*. Lawrence, Kan.: University of Kansas Press, 88–109.

—— (1995). "Agency, Regional Differentiation and Environment in Rural Conflict and Change." *Journal of Rural Studies*, 11/3: 239–42.

—— (1996). "Recasting the 'Agrarian Question': The Reproduction of Family Farming in the Southern High Plains." *Economic Geography*, 72/4: 398–415.

Roberts, R. S., and Dean, L. E. (1994). "An Inquiry into Lowi's Policy Typology: The Conservation Coalition and the 1985 and 1990 Farm Bills." *Environment and Planning C: Government and Policy*, 12/1: 71–86.

Roberts, R. S., and Emel, J. L. (1992). "Uneven Development and the Tragedy of the Commons: Competing Images for Nature-Society Analysis." *Economic Geography*, 68/3: 249–71.

Rosenzweig, C., and Hillel, D. (1998). *Climate Change and the Global Harvest: Potential Impacts of the Greenhouse Effect on Agriculture*. New York: Oxford University Press.

Sachs, C. (1996). *Gendered Fields: Rural Women, Agriculture, and the Environment*. Boulder, Col.: Westview.

Sauer, C. O. (1969). *Agricultural Origins and Dispersals: The Domestication of Animals and Foodstuffs*. 2nd edn. Cambridge, Mass.: MIT Press.

Schroeder, T. C. (1997). "Fed Cattle Spatial Transactions Price Relationships." *Journal of Agricultural and Applied Economics*, 29/2: 347–62.

Shortridge, B. G., and Shortridge, J. R. (1989). "Consumption of Fresh Produce in the Metropolitan United States." *Geographical Review*, 79/1: 79–98.

Shumway, J. M. (1997). "Hot, Medium, and Cold: The Geography of Nonmetro Population Growth and Change in the Mountain West," *Small Town*, 28/2:16–23.

Slafer, G. A. (ed.) (1994). *Genetic Improvement of Field Crops*. New York: M. Deckker.

Smit, B., Blain, R., and Keddie, P. (1997). "Corn Hybrid Selection and Climatic Variability: Gambling With Nature?" *Canadian Geographer*, Fall, 41/4: 429–38.

Solecki, W. D. (1992). "Rural Places and Circumstances of Acute Chemical Disasters." *Journal of Rural Studies*, 8/1: 1–13.

Solecki, W. D., and Shelley, F. M. (1996). "Pollution, Political Agendas, and Policy Windows—Environmental Policy on the Eve of Silent Spring." *Environment and Planning* C, 14/4: 451–68.

Sommers, L. M., Mehretu, A., and Pigozzi, B. W. (1999). "Towards Typologies of Socio-Economic Marginality: North/South Comparisons," in H. Jussila, R. Majoral, and C. C. Mutambirwa (eds.) *Marginality in Space—Past, Present and Future*. Aldershot: Ashgate, 7–24.

Starrs, P. F. (1998). *Let the Cowboy Ride: Cattle Ranching in the American West*. Baltimore, Md.: Johns Hopkins University Press.

Stull, D., Broadway, M., and Griffith, D. (eds.) (1995). *Any Way You Cut It: Meat Processing and Small Town America*. Lawrence, Kan.: University Press of Kansas.

Templer, O. (1992). "Hydrology and Texas Water Law: A Logician's Nightmare." *Great Plains Research*, 2/1: 27–50.

Trimble, S. W., and Mendel, A. C. (1995). "The Cow as a Geomorphic Agent: A Critical Review." *Geomorphology*, 13/1–4: 233–53.

Troughton, M. (1995). "Presidential Address: Rural Canada and Canadian Rural Geography—An Appraisal." *The Canadian Geographer*, 39/4: 290–305.

Tucker, M., and Napier, T. L. (2001). "Determinants of Perceived Agricultural Chemical Risk in Three Watersheds in the Midwestern United States." *Geographical Review*, 90/1: 57–82.

Turner, F. J. (1920). *The Frontier in American History*. New York: Holt.

Ufkes, F. (1993). "Trade Liberalization, Agro-food Politics and the Globalization of Agriculture," *Political Geography*, 12/3: 215–31.

—— (1995). "Lean and Mean: US Meat-Packing in an Era of Agro-Industrial Restructuring," *Environment and Planning D: Society and Space*, 13/6: 683–705.

US Congress (1995). "Targeting Environmental Priorities in Agriculture: Reforming Program Strategies." Office of Technology Assessment OTA-ENV-640. Washington DC: US Government Printing Office.

USDA (United States Department of Agriculture) (1991). "Agriculture, Agricultural Chemicals, and Water Quality," in Agriculture and the Environment: Yearbook of Agriculture. Washington DC: US Government Printing Office, 78–85.

U.S. Geological Survey (1999). The Quality of Our Nation's Waters—Nutrients and Pesticides: US Geological Survey Circular 1225, 82 pp. At URL: ‹http://water.usgs.gov/pubs/circ/circ1225/›, last accessed 28 January 2003.

Visser, S. (1999). "Scale-Economy Conditions for Spatial Variations in Farm Size." *Geographical Analysis*, 31/1: 27–44.

Walker, G. E. (1997). "Sustainable Agriculture and Its Social Geographic Context in Ontario," in B. Ilbery, Q. Choitti, and T. Rickard (eds.), *Agricultural Restructuring and Sustainability: A Geographical Perspective*. Walling ford: CAB International, 313–28.

Walker, R., and Solecki, W. (1999). "Managing Land Use and Land-Cover Change: The New Jersey Pinelands Biosphere Reserve." *Annals of the Association of American Geographers*, 89/2: 220–37.

Wallach, B. (1991). *At Odds with Progress: Americans and Conservation*. Tuscan, Ariz.: University of Arizona.

Weiner, D., Warner, T. A., Harris, T. M., and Levin, L. M. (1995). "Apartheid Representations in a Digital Landscape: GIS, Remote Sensing and Local Knowledge in Kiepersol, South Africa." *Cartography and GIS*, 22/1: 30–44.

Wells, M. J. (2000). "Politics, Locality, and Economic Restructuring: California's Central Coast Strawberry Industry in the Post-World War II Period." *Economic Geography*, 76/1: 28–49.

Whatmore, S. (1995). "From Farming to Agribusiness: the Global Agro-Food System," in R.J. Johnston, P. Taylor, and M. Watts (eds.), *Geographies of Global Change: Remapping the World in the Late Twentieth Century*. Oxford: Blackwell, 36–49.

Whatmore, S., Marsden, T., and Lowe, P. (1994). *Gender and Rurality*. London : D. Fulton.

White, S. E. (1992). "Population Change in the High Plains Ogallala Region: 1980–1990." *Great Plains Research* 2/2: 179–97.

—— (1994). "Ogallala Oases: Water Use, Population Redistribution, and Policy Implications in the High Plains of Western Kansas, 1980–1990." *Annals of the Association of American Geographers*, 84/1: 29–45.

White, S. E., and Kromm, D. (1995). "Local Groundwater Management Effectiveness in the Colorado and Kansas Ogallala Region." *Natural Resources Journal*, 35/2: 275–307.

Whittlesey, D. (1936). "Major Agricultural Regions of the Earth." *Annals of the Association of American Geographers*, 26/4: 199–240.

Wilbanks, T. J. (1994). Presidential Address: "Sustainable Development in Geographic Perspective." *Annals of the Association of American Geographers*, 84/4: 541–56.

Wilder, L. I. (1933). *Farmer Boy*. New York: Harper.

—— (1935). *Little House on the Prairie*. New York: Harper.

Williams, M. (1996). "European Expansion and Land Cover Transformation", in I. Douglas, R. Huggett, and M. Robinson (eds.), *Companion Encyclopedia of Geography, The Environment, and Humankind*. New York: Routledge, 182–205.

Wilson, R. (1999). "'Placing Nature': The Politics of Collaboration and Representation in the Struggle for La Sierra in San Luis, Colorado." *Ecumene*, 6/1: 1–28.

World Resources Institute, with the United Nations Environment Program, the United Nations Development Program, and the World Bank (1996). *World Resources, 1996–97*. New York: Oxford University Press.

—— (1998). *World Resources, 1998–99*. New York: Oxford University Press.

Woodhead, T., Huke, R., Huke, E., and Balaba, L. (1994). *Rice-Wheat Atlas of India*. Manila: International Rice Research Institute.

Wu, J., Nellis, M. D., Ransom, M. D., Price, K. P., and Egbert, S. L. (1997). "Evaluating Soil Properties of CRP Land Using Remote Sensing and GIS in Finney County, Kansas." *Journal of Soil and Water Conservation*, 52/5: 352–8.

CHAPTER 23

Rural Development

William Forbes and Sylvia-Linda Kaktins

Overview of the Relationship Between Geography and Rural Development

Rural development could be defined simply as economic development in rural areas. However, practitioners and researchers find rural development involves more than mere economic strategies. Many rural communities struggle with changes from resource extractive to service-based economies, along with cultural impacts of globalization (Harrington 1995; Ewert 1997). Rural development in response is becoming integrative like geography, considering class structure, community values, natural resources, social capital, sustainability, and regional and global forces (World Commission on Environment and Development 1987; Straussfogel 1997; Heartland Center for Leadership Development 1998).

Rural development has represented an explicit research perspective within geography since 1982. Geographers, through their ability to integrate human and physical aspects of *place*, can help communities assess complex change and devise strategies to meet their goals (Stoddart 1986; Turner 1989; Abler *et al.* 1992). Integrated descriptions of human and physical aspects of *place* can benefit relationships with undergraduate students (Marshall 1991), other geographers (Bowler *et al.* 1992), rural development researchers in other fields, and rural development practitioners (Kenzer 1989). Geographers may be especially useful in the interdisciplinary world of sustainable development (Wilbanks 1994).

Rural Development Specialty Group

The Rural Development Specialty Group began in 1982 as the result of an International Geographic Union (IGU) working group meeting in Fresno, California. The group was formed "to promote sharing of ideas and information among geographers interested in the many facets of rural development." Richard Lonsdale (University of Nebraska) and Donald Q. Innis (State University of New York at Geneseo) were co-founders. Subsequent leaders included Vincent Miller (Indiana University of Pennsylvania), John Dietz (University of Northern Colorado), Al Larson (University of Illinois at Chicago), Paul Frederic (University of Maine at Farmington), Henry Moon (University of Toledo), Brad Baltensperger (Michigan Technological University), Karen Nichols (State University of New York at Geneseo), William Forbes (University of North Texas), and Peter Nelson (Middlebury College). The group may soon merge with the Contemporary Agriculture and Rural Land Use Specialty Group, forming a larger Rural Geography Specialty Group that will continue to provide a forum for rural development research in geography.

Key Careers in Rural Development Geography

Donald Q. Innis (1924–88) was trained by Griffith Taylor (University of Toronto) and Ph.D. adviser Carl O. Sauer. Following Sauer's example, Innis strongly emphasized fieldwork. His 1959 Berkeley dissertation, *The Human Ecology of Jamaica*, was based on living with and becoming accepted by Jamaican peasant farmers (Merrill 1990). Innis expanded his interests in tropical peasant agriculture during his career at Geneseo State University. Contrasting with promoters of modern agricultural technology, he argued that long-established patterns of peasant agriculture represented a stable option for development. Work in Latin America, Asia, and Africa strengthened his position.

In addition to producing numerous publications, Innis conducted workshops on sustainable intercropping systems at AAG meetings. The Rural Development Specialty Group dedicated a special session to Donald Q. Innis at the 1990 Toronto meeting. Today the Group recognizes individuals who have made an outstanding contribution in the area of rural development through its Donald Q. Innis Award. The first winner of this award was Richard Lonsdale of the University of Nebraska.

Richard Lonsdale received his Ph.D. in 1960 from Syracuse University. Lonsdale joined the staff at the University of Nebraska-Lincoln in 1971, where he is now a Professor Emeritus in Geography. Lonsdale has conducted significant research on special problems of marginal regions and rural industrialization that continues to this day (Lonsdale 1996, 1998; Lonsdale and Archer 1997). He is an active participant in annual meetings of the IGU Commission on the Dynamics of Marginal Regions (University of Nebraska 2001).

Contributions by Geographers to Rural Development Theory

Peet and Hartwick (1999) provide a valuable, chronological overview of development theories. Their critique calls for democratically realigning development to better meet the needs of the poor. Much recent contribution by geographers to rural development theory has been to recommend *inclusiveness*, suggesting methodology to address facets of development typically not at the forefront of economic policy. Key topics in this area include: poverty/uneven development, women, psychology, humanism, post-colonialism, political ecology, the environment, and sustainability.

Recent geographic studies in poverty and uneven development included a proposal for a new index of development status (Tata and Schultz 1988). Subsequent criticism suggests that world human welfare is too complex to measure with a single number (Yapa and Zelinsky 1988) and any index used must account for sustainability (Daly 1988). A follow-up effort was made to analyze more explicitly the links and disparity between economic development and human welfare (Holloway and Pandit 1992). A postmodern view questions the value of all three major development paradigms: neoclassical economics, Marxism, and sustainable development (Yapa 1996). Modernism is defended, through literature and definition of place, as a valuable approach that helps interpret the cultural dynamics of socioeconomic transformation and restructuring (Oakes 1997).

Women play important roles in rural development in terms of rural community leadership (Heartland Center for Leadership Development 1998). There is potential common ground between geography and feminism in finding significance in everyday life, appreciating the importance of context, and thinking about cultural difference (Hanson 1992). Feminist geographic research can be more inclusive and help transform power relations (Staehli and Lawson 1995). Women's studies can be particularly relevant in the developing world (Momsen and Townsend 1987; review by Hapke 1999).

Psychology and geography were linked to examine the spatial marginalization of individuals and social groups (Sibley 1996). Humanism fosters better communication between cultures—an important aspect in today's rural development arena (Buttimer 1990). Post-colonialism and migration are key themes of Crush (1995). Political ecology (see Ch. 12, Political Geography) has offered valuable critique through Smith (1991), Harvey (1996), Peet and Watts (1996), Klooster (2000*a*), Zimmerer (2000), and others.

Concern for the environment has heightened through sustainable development, popularized with the Brundtland Commission's publication of *Our Common Future* (World Commission on Environment and Development 1987), and by the 1992 United Nations Conference on Environment and Development in Rio de Janeiro. Geographers' contributions to sustainable development theory address international agricultural research (Bebbington and Carney 1990), geographic perspectives (Wilbanks 1994), present turmoil in development studies (Peet 1994), and broad-based measurements of development (Straussfogel 1997; Durrant 1998).

Due to the increasingly varied nature of rural development, standard positivist science may not always be the most appropriate methodology for geographical rural development research. Even when researching economic aspects of development, geographers tend to rely more on empirical observation than on economic theory (Clark 1998). Realist approaches can be uniquely valuable in identifying contextual social structures, processes, causes, and effects (Lawson and Staehli 1990). Porter and Sheppard (1998) also emphasize site difference, along with local and global relationships, in development theory. New paradigms are important, but artificial divisions can also occur and then disappear—linking theory with the legacy of on-the-ground geographic research is encouraged (Brown 1999). Curry (2000) finds linkages between past methods, material economic systems, and cultural meaning in rural Iowa.

Geographic research has challenged widely held myths about cultural ecology and human–nature relationships, which in turn influence development approaches. Grossman (1993) challenged the notion that developing world export agriculture undermines local food production, leading to local food shortages and increased food imports. Geographers in cultural ecology (see Ch. 8) highlighted severe environmental impacts in the pre-Columbian Americas, challenging the myth of Native Americans as ideal ecologists (Butzer 1992; Denevan 1992). Recent research in ecology supports the notion of a dynamic, non-pristine nature, leading to proposals to align human geography with this new research (Zimmerer 1994).

Other myth-challenging studies have been done on development and the human–nature relationship. Hecht and Cockburn (1989) illuminated causes of deforestation in the Amazon. Sluyter (1999) searched in Veracruz for origins of today's man versus wilderness post-colonial development myth. Rudel *et al.* (2000) noted how Puerto Rican development related to the highest reforestation rates in the Western Hemisphere. British geographers Fairhead and Leach (1996) outlined development myths concerning the West African savanna.

The ability of geographers to challenge development myths using non-geographers' research is also important. Geomorphologists Messerli and Ives found forest regeneration rates so different from World Bank reports that they were forced to reevaluate a widely held concept, that severe Himalayan deforestation directly influenced catastrophic flooding in Bangladesh (Allan 1990). Award-winning journalist Salopek (2000) questions the intrusiveness of development, given its potential cultural impacts in Mexico's Sierra Madre Occidental.

Topics in Rural Development Geography

Overviews of rural development have been provided in recent years from both geographic (Hoggart and Buller 1987) and nongeographic (Galston and Baehler 1995) perspectives. Geographers have attempted to clarify what is rural (Cleland 1995; Berry *et al.* 2000). Non-geographers have contributed valuable overviews of third-world rural development (Dixon 1990) and rural policy (Gilg *et al.* 1994). Valuable geographic research has also been done under special topics in rural development.

Recent geographic research addressed economic issues in rural development through: general overviews (Auty 1995); case studies (Schaeffer and Loveridge 2000); local policy (Glasmeier 2000); branch plants and globalization (Glasmeier *et al.* 1995); enterprise zones (Haley 1994; Theirl 1994); business retention and expansion programs (Loveridge *et al.* 1995); transport costs (Kilkenny 1998); relationships between foreign aid and economic growth (Bowen 1995); and ability of rural communities to deal with change (Bowler *et al.* 1992; Neil *et al.* 1992; Marsden *et al.* 1993; Bernstein *et al.* 1992; Beesley *et al.* 2001).

Economic class structure was examined through studies on uneven development (Cole 1987); labor (Marsden *et al.* 1992); rural homelessness (Lawrence 1995); and historical relationships of migration to poverty (Rain 1999). Indigenous culture economies have been addressed through studies on development (Young 1995); Indian gaming (Winchell *et al.* 1998); land rights (Cullen 1997); and planning (Marchand and Winchell 1992; Winchell 1992). Readers may be interested in the economic geography chapter (Ch. 9), and a special issue of *Growth and Change* (1998: 29/3) addressing rural development, including economic analysis methods.

Sustainable rural development is a rich topic for geographers, with overviews addressing the role of geography (Wilbanks 1994), definition and maintenance of sustainable communities (Conroy 1997), and "ecogeography" and rural development (Tricart and Kiewietdejonge 1992). Nongeographers' case studies can also be valuable (Ghai 1994). Agricultural and land-use issues are addressed primarily by the Contemporary Agriculture and Rural Land Use Specialty Group (see corresponding chapter).

Tourism is a growing field in rural development geography as many communities look to a service-oriented economy to diversify livelihoods. Careful planning and implementation is needed to avoid impacts to cultural and ecological integrity. Geographers have addressed:

overviews (Butler *et al.* 1998); economics (Harris and Nelson 1993; Hanink and White 1999); community misrepresentation (Belsky 1999); hidden impacts (Place 1997); sustainable tourism development (Hall and Lew 1998); and regional studies of Appalachia (Sirk 1994), Texas (Kimmel 1997), and China (Oakes 1998).

Rural telecommunications is another growing aspect of rural development. Recent studies have been done by geographers (Malecki 2001) and nongeographers (Allen and Dillman 1994). Technology is also assisting geographers through rural addressing and computer mapping (Gribb *et al.* 1990; Short 1998). Glasmeier and Howland (1995) addressed the transition of rural economies through service-based technology.

Transportation was a major focus of the Rural Development Specialty Group from 1985 to 1995. Moon (1987, 1988, 1990, 1994) was a primary researcher of the effects of the interstate highway system on small towns.

Rural health services figure importantly in quality of life issues. Geographers have investigated: health care services in rural North America (Gesler and Ricketts 1992); traditional medicine in Africa (Good 1987); successful rural health service delivery in Nepal (Shrestha 1988); relationship of place and geographic methods to health services (Kearns 1993; Ricketts *et al.* 1994); rural health in Kansas (Paul and Nellis 1996); definitions of rural (Johnson-Webb *et al.* 1997); and rural physician shortages (Cutchin 1997; Baer *et al.*1998). Tom Ricketts (University of North Carolina) and Gary Hart (University of Washington) are among several key figures applying geography at rural health research centers.

Rural Development Geography in the US and Canada

Geographers have applied research related to rural development in specific regions of the world. Studies of Appalachia and the Southeast have recently addressed overviews (Glasmeier *et al.* 2000); economic effects of development planning (Isserman and Rephann 1995); women and the informal sector (Oberhauser 1999); the Community Reinvestment Act (Zhou and Shaw 2000); and change in small town central business districts (Lentz 1996). Midwest/Great Plains issues include: groundwater (Kromm and White 1992; White 1994; Emel and Roberts 1995); uneven development (McQuillan 1996); farm size and ownership (Hart 1991); regional overviews (Borchert 1987; Shortridge 1989); a proposal for a "buffalo commons" (Popper and Popper 1987); defining heartlands (Lawrence 1997) and the Great Plains (Rossum and Lavin 2000); and small-town tourism (Gariglietti 1998).

Geographers addressed recent rural changes in the western US by analyzing: conservation and development near Yellowstone (Rasker 1993); footloose entrepreneurs (Rasker and Glick 1994); uneven development (Lorah 1994); economic base models (Nelson and Beyers 1998); input–output models (Robison 1997); Southwest employment trends (Mulligan 1997); population and employment growth (Vias 1999); migration (Rudzitis 1993); Pacific fisheries (Mansfield 2001); transitions from "Old" to "New" West in rural Utah (Shumway and Durrant 2000); and mapping the "New" West (Riesbame 1997). Readers may be interested in following this topic further through a book (Rudzitis 1996) or special issue of *Rural Development Perspectives* (1999: 14/2).

Northeast rural development issues addressed include: highways and sprawl (Ford 1995) and forest regulation (Frederic 1997). Canadian rural development geography studies have included Ontario (Bailey and Hooey 2000; Hooey 2000); Eastern Canada (Beesley and Daborn 1997); British Columbia (Booth and Halseth 1998); and Manitoba (Kaktins 1998). The 1997 Rural Resources Rural Development Conference brought together many Canadian researchers that addressed rural development issues (Beesley and Daborn 1997).

Rural Development Geography and the Developing World

Latin American studies have a long tradition in geography (see Ch. 43, Latin America) that contributes to rural development analysis. Examples of topics are: agricultural change and biodiversity in highland Peru (Zimmerer 1991, 1996); government policy, adaptive agriculture, and regional economic change in Ecuador (Lawson 1988; Knapp 1991); relationship of drought vulnerability, new agricultural technologies, and land tenure in Mexico (Liverman 1990); comparative development dates and species declines in the US Southwest and Chihuahua (Forbes 2001); political ecology of forestry in Oaxaca (Klooster 2000a); maquiladoras and their relationship to exports, local development (South 1990, 1993), and women's labor (Cravey 1998); relationship of population pressure and deforestation in the Dominican

Republic (Zweifler *et al.* 1994; Sambrook *et al.* 1999); and World Bank projects and squatters in NE Brazil (Muller 1996; Muller and Lages 1995; Muller and Palhano Siba 1998).

Rural development is perhaps most challenging in Africa. Geographers have contributed to an improved understanding of African rural systems (see Ch. 36, Africa). Studies have addressed: overviews of development (Crush 1995); peasant-herder conflicts in the northern Ivory Coast (Bassett 1988); disruption of traditional roles of children in economic development in the Sudan (Katz 1991); central place theory in Kenya (Fox 1991); Kenyan women and the informal economy (Thomas-Slayter and Rocheleau 1995; Aspaas 1998); gender and agrarian change in The Gambia (Carney 1993); gender and fisheries in Ghana (Endemano Walker 2001); deforestation in Ghana enhanced by structural readjustment (Owusu 1998); investment patterns of cocoa farmers in Ghana (Awanyo 1998); causes of deforestation and land-use change in Madagascar (Kull 1998); politics of manure in Niger (Heasley and Delehanty 1996) and fuelwood in Malawi (Kalipeni and Feder 1999); development from within (Fraser-Taylor and Mackenzie 1992); rural–urban interface (review by Logan 1993); relationships between rural settlement structure and development processes (Silberfein 1998); and sub-Saharan regional disparity (review by Gaile 1990).

Geographers have contributed to understanding of rural development issues in Asia (see Ch. 39, Asian Geography) by addressing: Philippine market exchange (M. W. Lewis 1989); socialist regional development in China (Lo 1989); rural economic restructuring in China (Veeck and Pannell 1989); environmental degradation and agrarian policy reform in China (Zurick 1988; Powell 1992; Muldavin 1997); uneven regional development in China (Fan 1995); urban–rural interaction in China (Lin 2001); the economic geography of China (Sun 1988); adventure travel in Nepal (Zurick 1992); development and conservation in Nepal (reviewed by Allan 1990); frontier settlement by landless migrants in Nepal (Shrestha 1989); rapid rural appraisal in Thailand (Lovelace 1988); common property in India (Jodha 1990); gender and fishing in India (Hapke 2001); rural and urban infant mortality rates in Bangladesh (Paul 2000); Soviet Central Asia (R. A. Lewis 1992); and modernization (Wallach 1996).

Directions for Rural Development Geography in the Twenty-First Century

Regional and global changes affecting rural communities and their human–nature relationships are considerable (Bowler *et al.* 1992; Beesley *et al.* 2001). Geographers have a tradition of interpreting identities of place and the forces that change them (Abler *et al.* 1992), and can play a valuable role in rural development that attempts to understand and direct this change (Kenzer 1989) through efforts in rural development theory, practice, and education (Peet and Hartwick 1999; Demko 1988; Glasmeier *et al.* 1997).

Geographers should continue to provide critiques of development theory through modern and postmodern perspectives (Oakes 1997; Yapa 1996; Bebbington 2000; Klooster 2000*b*). Moral consideration is also seen as important in geographers' contributions to sustainable development (Wilbanks 1994). Efforts have been made to outline morality in a diversity of cultural contexts (Tuan 1989). Sack (1999) notes the importance of place in adjusting intrinsic moral truths. Consideration of diverse value systems (see Ch. 45, Values, Ethics, and Justice) is important as we help to direct change of rural communities.

Specialization (Goodchild and Janelle 1988) yet integration (Turner 1989) is required in the practice of rural development geography. Geography should focus on providing scientific grounding for integration of the human and physical subsystems of our world (Rigby and Willmott 1998; Zimmerer and Young 1998). Rural communities lie at the direct interface of the human–nature relationship. Their small scale offers an opportunity for applied geographers to make a difference.

Rural development and geography are both integrative fields. Future work may be done more by interdisciplinary teams than by individual geographers. Geographers are encouraged to look for linkages with both other geographers and other disciplines (Brown 1999; Hanson 1999; Gober 2000). Freshwater (2000) and Drabenstott and Sheaff (2001) note that, in looking to the future of rural America, increasingly diverse rural issues can no longer be adequately addressed by broad agricultural policies—more site-specific local and regional partnerships are needed. The diverse research interests of geographers, whether or not they are explicitly titled "rural development," enable them to be effective partners, educating society about the dynamic, integrative nature of rural communities in the twenty-first century.

References

Abler, R. F., Marcus, M. G., Olson, J. M. (eds.) (1992). *Geography's Inner Worlds*. New Brunswick, NJ: Rutgers University Press.

Allan, N. J. R. (1990). Review of J. D. Ives, B. Messerli, "The Himalayan Dilemma: Reconciling Development and Conservation," 1989. *Annals of the Association of American Geographers*, 317–8.

Allen, J. C., and Dillman, D. A. (1994). *Against All Odds: Rural Community in the Information Age*. Boulder, Col.: Westview Press.

Aspaas, H. R. (1998). "Heading Households and Heading Businesses: Rural Kenyan Women in the Informal Sector." *Professional Geographer*, 50/2: 192–204.

Auty, R. M. (1995). *Patterns of Development: Resources, Policy and Economic Growth*. New York: Edward Arnold.

Awanyo, L. (1998). "Culture, Markets, and Agricultural Production: A Comparative Study of the Investment Patterns of Migrant and Citizen Cocoa Farmers in the Western Region of Ghana." *Professional Geographer*, 50/4: 516–30.

Baer L. D., Ricketts, T. C., Konrad, T. R., and Mick, S. S. (1998). "Do International Medical Graduates Reduce Rural Physician Shortages?" *Medical Care*, 36/11: 1534–44.

Bailey, T. J., and Hooey, C. A. (2000). "Perceptions of Development and Planning in a Small Town: The Case of Napanee, Ontario." *Papers and Proceedings of the Applied Geography Conferences*, 23: 45–54. Denton, Tex.: Applied Geography Conferences, Inc.

Bassett, T. J. (1988). "The Political Ecology of Peasant-Herder Conflicts in the Northern Ivory Coast." *Annals of the Association of American Geographers*, 78/3: 453–72.

Bebbington, A. (2000). "Reencountering Development: Livelihood Transitions and Place Transformations in the Andes." *Annals of the Association of American Geographers*, 90/3: 495–520.

Bebbington, A., and Carney, J. (1990). "Geography in the International Agricultural Research Centers: Theoretical and Practical Concerns." *Annals of the Association of American Geographers*, 80/1: 34–48.

Bernstein, H., Crow, B., and H. Johnson (eds.) (1992). *Rural Livelihoods: Crises and Responses*. Oxford: Oxford University Press

Beesley, K., and Daborn, L. A. (eds.) (1997). *Proceedings: Rural Resources, Rural Development Conference*. Truro, Nova Scotia, Canada: Rural Research Centre.

Beesley, K., Millward, H., Ilbery, B., and Harrington, L. (eds.) (2001). *The New Countryside: Geographic Perspectives on Rural Change*. Waterloo: Wilfrid Laurier University Press.

Belsky, J. M. (1999). "Misrepresenting Communities: The Politics of Community-Based Rural Ecotourism in Gales Point Manatee, Belize." *Rural Sociology*, 64/4: 641–66.

Berry, K. A., Markee, N. L., Fowler, N., and Giewat, G. R. (2000). "Interpreting What is Rural and Urban for Western U.S. Counties." *Professional Geographer*, 52/1: 93–105.

Booth, A., and Halseth, G. (1998). *Community Participation and the New Forest Economy: British Columbia Models of Community Participation and Examples of Management—an Annotated Bibliography*. 2nd edn. A Report for the BC Resources Communities Project. Prince George: University of Northern British Columbia.

Borchert, J. (1987). *America's Northern Heartland. An Economic and Historical Geography of the Upper Midwest*. Minneapolis: University of Minnesota Press.

Bowen, J. (1995). "Foreign Aid and Economic Growth: An Empirical Analysis." *Geographical Analysis*, 27/3: 249–61.

Bowler, I. R., Bryant, C. R., and Nellis, M. D. (eds.) (1992). *Contemporary Rural Systems in Transition*. Wallingord, UK: CAB International, i and ii.

Brown, L. A. (1999). "Change, Continuity, and the Pursuit of Geographic Understanding." Presidential Address. *Annals of the Association of American Geographers*, 89/1: 1–25.

Butler, R. W., Hall, C. M., and Jenkins, J. (eds.) (1998). *Tourism and Recreation in Rural Areas*. Chichester: John Wiley.

Buttimer, A. (1990). "Geography, Humanism, and Global Concern." *Annals of the Association of American Geographers*, 80/1: 1–33.

Butzer, K. W. (1992). "The Americas Before and After 1492: An Introduction to Current Geographical Research." *Annals of the Association of American Geographers*, 82/3: 345–68.

Carney, J. (1993). "Converting the Wetlands, Engendering the Environment: The Intersection of Gender with Agrarian Change in The Gambia." *Economic Geography*, 69/4: 329–49.

Clark, G. L. (1998). "Stylized Facts and Close Dialogue: Methodology in Economic Geography." *Annals of the Association of American Geographers*, 88/1: 73–87.

Cleland, C. (1995). "A Measure of Rurality. *Papers and Proceedings of the Applied Geography Conferences*, 18: 183–9. Denton, Tex.: Applied Geography Conferences.

Cole, J. (1987). *Development and Underdevelopment*. New York: Methuen.

Conroy, M. E. (1997). "The Challenges of Economic Geography in Defining, Creating, and Defending Sustainable Communities," in Glasmeier, Leichenko, and Fuellhart (1997).

Cravey, A. J. (1998). *Women and Work in Mexico's Maquiladoras*. Lanham, Md: Rowman & Littlefield.

Crush, J. S. (1995). *Power of Development*. London: Routledge.

Cullen, B. T. (1997). "What Rights Do Indigenous People Have to Land? Conflict Over Native Title in Australia." *Papers and Proceedings of the Applied Geography Conferences*, 20: 163–9. Denton, Tex.: Applied Geography Conferences.

Curry, J. M. (2000). "Community Worldview and Rural Systems: A Study of Five Communities in Iowa." *Annals of the Association of American Geographers*, 90/4: 693–712.

Cutchin, M. P. (1997). "Physician Retention in Rural Communities: The Perspective of Experiential Place Integration." *Health and Place*, 3: 25–41.

Daly, H. (1988). "Comment on Tata and Schultz." *Annals of the Association of American Geographers*, 78/4: 612–13.

Demko, G. J. (1988). "Geography Beyond the Ivory Tower." *Annals of the Association of American Geographers*, 78/4: 575–9.

Denevan, W. M. (1992). "The Pristine Myth: The Landscapes of the Americas in 1492." *Annals of the Association of American Geographers*, 82/3: 369–85.

Dixon, C. (1990). *Rural Development in the Third World*. New York: Routledge.

Drabenstott, M., and Sheaff, K. H. (2001). "Exploring Policy Options for a New Rural America: A Conference Summary," in,

Exploring Policy Options for Rural America, Conference Proceedings. Kansas City, Mo.: Center for the Study of Rural America, Federal Reserve Bank of Kansas City. ‹http://www.kc.frb.org/RuralCenter/conference/ruralconf-main.htm›, last accessed 12 December 2001.

Durrant, J. O. (1998). "On 'Redefining Development as Humane and Sustainable.'" *Annals of the Association of American Geographers*, 88/2: 311–14.

Emel, J., and Roberts, R. (1995). "Institutional Reform and its Effect on Environmental Change: The Case of Groundwater in the Southern High Plains." *Annals of the Association of American Geographers*, 85/4: 664–83.

Endemano Walker, B. L. (2001). "Sisterhood and Seine-Nets: Engendering Development and Conservation in Ghana's Marine Fishery." *Professional Geographer*, 53/2: 160–77.

Ewert, E. C. (1997). "From Extractive to Attractive: Demographic and Economic Boom and Bust in the Non-Metropolitan American West." *Papers and Proceedings of the Applied Geography Conferences*, 20: 354. Denton, Tex.: Applied Geography Conferences.

Fairhead, J., and Leach, M. (1996). *Misreading the African Landscape: Society and Ecology in the Forest Savanna Mosaic*. New York: Cambridge University Press.

Fan, C. C. (1995). "Of Belts and Ladders: State Policy and Uneven Regional Development in Post-Mao China." *Annals of the Association of American Geographers*, 85/3: 421–49.

Forbes, W. (2001). "Great Extirpations: A Tale of Two Diversities in the US Southwest and Northwestern Chihuahua." *Papers and Proceedings of the Applied Geography Conferences*, 24: 41–9. Denton, Tex.: Applied Geography Conferences.

Ford, J. J. (1995). "Impact of State Highway Access Policy on Local Land Uses: A Case Study of Semi-Rural Sprawl in Pennsylvania." *Papers and Proceedings of the Applied Geography Conferences*, 18: 125–32. Denton, Tex.: Applied Geography Conferences.

Fox, R. (1991). "The Use of Central Place Theory in Kenya's Development Strategies." *Tijdschrift voor Economishe en Sociale Geografie*, 82/2: 106–27.

Fraser-Taylor, D. R., and Mackenzie, F. (eds.) (1992). *Development from Within: Survival in Rural Africa*. New York: Routledge, Chapman & Hall.

Frederic, P. (1997). Northern New England and New York Forests: Management Options. *Proceedings: New England-St. Lawrence Valley Geographical Society*, 20: 1–6.

Freshwater, D. (2000). "Rural America at the Turn of the Century: One Analyst's Perspective." *Rural America*, 15/3: 2–7. USDA Economic Research Service, ‹http://www.ers.usda.gov/publications/ruralamerica/sep2000/›, last accessed 3 December 2001.

Gaile, G. L. (1991). Review of A. Mehretu, "Regional Disparity in Sub-Saharan Africa: Structural Readjustment of Uneven Development," 1989. *Annals of the Association of American Geographers*, 81/1: 178–80.

Galston, W. A., and Baehler, K. J. (1995). *Rural Development in the United States: Connecting Theory, Practice and Possibilities*. Washington, DC: Island Press.

Gariglietti, R. (1998). "A Tourism Development Technique Applicable to Small Midwestern Towns." *Papers and Proceedings of the Applied Geography Conferences*, 21: 462. Denton, Tex: Applied Geography Conferences.

Gesler, W. M., and Ricketts, T. C. (eds.) (1992). *Health in Rural North America: The Geography of Health Care Services and Delivery*. New Brunswick, NJ: Rutgers University Press.

Ghai, D. (ed.) (1994). *Development and Environment: Sustaining People and Nature*. Oxford: Blackwell.

Gilg, A. W., Dilley, R. S., Furuseth, O., McDonald, G., and Murdoch, J. (1994). *Progress in Rural Policy and Planning*. New York: John Wiley & Sons.

Glasmeier, A. K. (2000). "Economic Geography in Practice: Local Economic Development Policy," in M. Gertler, G. Clark, and M. A. Feldman, (eds.), *Handbook of Economic Geography*. Oxford: Oxford University Press.

Glasmeier, A. K., and Howland, M. (1995). *From Combines to Computers: Rural Services and Development in the Age of Information Technology*. Albany, NY: State University of New York Press.

Glasmeier, A. K., Leichenko, R. M., and Fuellhart, K. (1997). *Global and Local Challenges to Theory, Practice, and Teaching in Economic Geography—Final Report of the 1997 NSF Workshop on the Future of Economic Geography*. Washington, DC: National Science Foundation.

Glasmeier, A. K., Kays, A., Thompson, J. W., and Gurwitt, R. (1995). *Branch Plants and Rural Development in the Age of Globalization*. Monograph, State Overview Series, Rural Economic Policy Program. Aspen, Col.: The Aspen Institute.

Glasmeier, A. K., Witson, R., Wood, L., and Fuellhart, K. (2000). "Appalachia: Rich in Natural Resource Wealth, Poor in Human Opportunity," in *A Geographic View of Pittsburgh and the Alleghenies: Precambrian to Post-Industrial*. Washington, DC: Association of American Geographers.

Gober, P. (2000). "In Search of Synthesis." Presidential Address, *Annals of the Association of American Geographers*, 90/1: 1–11.

Good, C. M. (1987). *Ethnomedical Systems in Africa: Patterns of Traditional Medicine in Rural and Urban Kenya*. New York: Guilford.

Goodchild, M. F., and Janelle, D. G. (1988). "Specialization in the Structure and Organization of Geography." *Annals of the Association of American Geographers*, 78/1: 1–28.

Gribb, W. J., Czerniak, R. J., and Harrington, J. A., Jr. (1990). "Rural Addressing and Computer Mapping in New Mexico." *Professional Geographer*, 42/4: 471–80.

Grossman, L. S. (1993). *The Political Ecology of Bananas: Contract Farming, Peasants, and Agrarian Change in the Eastern Caribbean*. Chapel Hill: University of North Carolina Press.

Haley, M. A. U. (1994). "Enterprise Zones as a Strategy for Community Development." *Papers and Proceedings of the Applied Geography Conferences*, 17: 219. Denton, Tex.: Applied Geography Conferences.

Hall, C. M. (1994). *Tourism and Politics: Policy, Power and Place*. New York: John Wiley & Sons.

Hall, C. M., and Lew, A. (1998). *Sustainable Tourism: A Geographical Perspective*. Harlow: Allison, Wesley, & Longman.

Halseth, G. (1998). *Cottage Country in Transition: A Social Geography of Change and Contention in the Rural-Recreational Countryside*. Montreal: McGill-Queen's University Press.

Hanink, D. M., and White, K. (1999). "Distance-Effects in the Demand for Wild Land Recreation Services: The Case of

National Parks in the United States." *Environment and Planning A*, 31: 477–92.

Hanson, S. (1992). "Geography and Feminism: Worlds in Collision?" *Annals of the Association of American Geographers*, 82/4: 569–86.

— (1999). "Isms and Schisms: Healing the Rift between the Nature-Society and Space-Society Traditions in Human Geography." *Annals of the Association of American Geographers*, 89/1: 133–43.

Hapke, H. M. (1999). Review of C. Sachs, "Gendered Fields: Rural Women, Agriculture, and Environment," 1996. *Annals of the Association of American Geographers*, 89/2: 369–71.

— (2001). "Gender, Work, and Household Survival in South Indian Fishing Communities: A Preliminary Analysis." *Professional Geographer*, 53/3: 313–31.

Harrington, J. W., Jr. (1995). "Producer Services Research in U.S. Regional Studies." *Professional Geographer*, 47/1: 87–96.

Harris, J., and Nelson, J. G. (1993). "A Total Economy View of Tourism and Recreation Development." *Papers and Proceedings of the Applied Geography Conferences*, 16: 157. Denton, Tex.: Applied Geography Conferences.

Hart, J. F. (1991). "Part-Ownership and Farm Enlargement in the Midwest." *Annals of the Association of American Geographers*, 81/1: 66–79.

Harvey, D. (1996). *Justice, Nature, and the Geography of Difference*. Cambridge, Mass.: Blackwell.

Heartland Center for Leadership Development (1998). Research Report: *Twenty Clues to Rural Community Survival*. Lincoln, Nebr.: Heartland Center for Leadership Development.

Heasley, L., and Delehanty, J. (1996). "The Politics of Manure: Resource Tenure and the Agropastoral Economy in Southwestern Niger." *Society and Natural Resources*, 9: 31–46.

Hecht, S. B., and Cockburn, A. (1989). *The Fate of the Forest: Developers, Destroyers, and Defenders of the Amazon*. New York: Verso.

Hoggart, K., and Buller, H. (1987). *Rural Development: A Geographical Perspective*. NY: Croom Helm & Methuen.

Holloway, S. R., and Pandit, K. (1992). "The Disparity Between the Level of Economic Development and Human Welfare." *Professional Geographer*, 44/1: 57–71.

Hooey, C. A. (2000). "Impacts of a Large Auto Plant on a Small Town: An Examination of Short and Long Term Change." *Papers and Proceedings of the Applied Geography Conferences*, 23: 37–44. Denton, Tex.: Applied Geography Conferences.

Isserman, A., and Rephann, T. (1995). "The Economic Effects of the Appalachian Regional Commission—An Empirical Assessment of 26 Years of Regional Development Planning." *Journal of the American Planning Association*, 61/3: 345–64.

Jodha, N. S. (1990). "Depletion of Common Property Resources in India: Micro-Level Evidence," in G. McNicoll, and M. Cain (eds.), *Rural Development and Population: Institutions and Policy*. New York: Oxford University Press.

Johnson-Webb, K. D., Baer, L. D., and Gesler, W. M. (1997). "What is Rural? Issues and Considerations." *Journal of Rural Health*, 13/3: 253–6.

Kaktins, S. (1998). "Brandon's Hog Processing Plant and Quality of Life in Brandon, Manitoba, Canada." *Papers and Proceedings of the Applied Geography Conferences*, 21: 18–24. Denton, Tex.: Applied Geography Conferences.

Kalipeni, E., and Feder, D. (1999). "A Political Ecology Perspective on Environmental Change in Malawi with the Blantyre Fuelwood Project Area as a Case Study." *Politics and the Life Sciences*, 18/1: 37–54.

Katz, C. (1991). "Sow What You Know: The Struggle for Social Reproduction in Rural Sudan." *Annals of the Association of American Geographers*, 81/3: 488–514.

Kearns, R. A. (1993). "Place and Health: Towards a Reformed Medical Geography." *Professional Geographer*, 45/2: 139–47.

Kenzer, M. S. (1989). *Applied Geography: Issues, Questions, and Concerns*. Dordrecht: Kluwer, 141–4.

Kilkenny, M. (1998). "Transport Costs, the New Economic Geography, and Rural Development." *Growth & Change*, 29: 259–80.

Kimmel, J. R. (1997). "Rural Tourism in Texas: Geographic Potentials and Problems." *Papers and Proceedings of the Applied Geography Conferences*, 20: 245–52. Denton, Tex.: Applied Geography Conferences.

Klooster, D. (2000*a*). "Community Forestry and Tree Theft in Mexico: Resistance or Complicity in Conservation?" *Development and Change*, 31: 281–306.

— (2000*b*). Review of R. Peet, E. Hartwick, "Theories of Development," 1999. *Professional Geographer*, 52/2: 364–5.

Knapp, G. (1991). *Andean Ecology: Adaptive Dynamics in Ecuador*. Boulder, Col.: Westview Press.

Kromm, D. E., and White, S. E. (eds.) (1992). *Groundwater Exploitation in the High Plains*. Lawrence, Kan.: University Press of Kansas.

Kull, C. A. (1998). "Leimavo Revisited: Agrarian Land-Use Change in the Highlands of Madagascar." *Professional Geographer*, 50/2: 163–76.

Lawrence, M. (1995). "Rural Homelessness: A Geography without a Geography." *Journal of Rural Studies*, 11/3: 297–307.

— (1997). "Heartlands or Neglected Geographies? Liminality, Power, and the Hyperreal Rural." *Journal of Rural Studies*, 13/1: 1–17.

Lawson, V. A. (1988). "Government Policy Biases and Ecuadorian Agricultural Change." *Annals of the Association of American Geographers*, 78/3: 433–52.

Lawson, V. A., and Staehli, L. A. (1990). "Realism and the Practice of Geography." *Professional Geographer*, 42/1: 13–20.

Lentz, R. H. (1996). "What's Happened to the Small Town Central Business District: A Case Study of Monroe, N.C." *Papers and Proceedings of the Applied Geography Conferences*, 19: 338. Denton, Tex.: Applied Geography Conferences.

Lewis, M. W. (1989). "Commercialization and Community Life: The Geography of Market Exchange in a Small-Scale Phillippine Society." *Annals of the Association of American Geographers*, 79/3: 390–410.

Lewis, R. A. (ed.) (1992). *Geographic Perspectives on Soviet Central Asia*. London: Routledge.

Lin, G. C. S. (2001). "Evolving Spatial Form of Urban-Rural Interaction in the Pearl River Delta, China." *Professional Geographer*, 53/1: 56–70.

Liverman, D. M. (1990). "Drought Impacts in Mexico: Climate, Agriculture, Technology, and Land Tenure in Sonora and Puebla." *Annals of the Association of American Geographers*, 80/1: 49–72.

Lo, C. P. (1989). "Recent Spatial Restructuring in Zhujiang Delta, South China: A Study of Socialist Regional Development

Strategy." *Annals of the Association of American Geographers*, 79/2: 293–308.

Logan, I. (1993). Review of J. Baker, P. O. Pedersen (eds.), "The Rural-Urban Interface in Africa: Expansion and Adaptation," 1992. *Annals of the Association of American Geographers*, 84/1: 148–51.

Lonsdale, R. E. (1996). "Some Questionable Economic Development Strategies Practiced at the Community Level," in E. Furlani (ed.), *Development Issues in Marginal Regions*. Mendoza: Universidad Nacional de Cuyo.

— (1998). "A Century of Shifting Perceptions and Development Issues of Marginality," in H. Jussila, W. Leimgruber, and R. Majoral (eds.), *Perceptions of Marginality*. Aldershot: Ashgate.

Lonsdale, R. E., and Archer, J. C. (1997). "Geographical Aspects of U.S. Farmland Values and Changes During the 1978–1992 Period." *Journal of Rural Studies*, 13/4: 399–413.

Lorah, P. (1997). "Divergent Paths: The Environment and Uneven Development in the Rural Rocky Mountain West." *Papers and Proceedings of the Applied Geography Conferences*, 20: 371. Denton, Tex.: Applied Geography Conferences.

Lovelace, G. W., Subhadhira, S., and Simaraks, S. (eds.) (1988). *Rapid Rural Appraisal in Northeast Thailand: Case Studies*. Khon Kaen University, Thailand: KKU-Ford Rural Systems Research.

Loveridge, S., Smith, T. R., and Morse, G. W. (1995). "Volunteer Visitor Business Retention and Expansion Programs," in D. W. Sears and J. N. Reid (eds.), *Rural Development Strategies*. Chicago: Nelson-Hall.

McQuillan, D. A. (1996). Review of C. S. Tauxe, "Farms, Mines, and Main Streets. Uneven Development in a Dakota County," 1993. *Annals of the Association of American Geographers*, 86/3: 589–90.

Malecki, E. J. (2001). "Going Digital in Rural America," in *Exploring Policy Options for Rural America*, Conference Proceedings. Kansas City, Mo.: Center for the Study of Rural America, Federal Reserve Bank of Kansas City. ‹http://www.kc.frb.org/RuralCenter/conference/ruralconfmain.htm›, last accessed 12 December 2001.

Mansfield, B. (2001). "Property Regime or Development Policy? Explaining Growth in the US Pacific Groundfish Fishery." *Professional Geographer*, 53/3: 384–97.

Marchand, M. M., and Winchell, D. G. (1992). "Tribal Implementation of GIS: A Case Study of Planning Applications with the Colville Confederated Tribes." *American Indian Culture and Research Journal*, 16/4: 175–83.

Marsden, T., Lowe, P., and Whatmore, S. (eds.) (1992). *Labour and Locality: Uneven Development and the Rural Labour Process*. London: David Fulton.

Marsden, T., Murdoch, J., Lowe, P., Munton, R., and Flynn, A. (1993). *Constructing the Countryside*. Boulder, Col.: Westview Press.

Marshall, B. (ed.) (1991). *The Real World: Understanding the Modern World through the New Geography*. Boston: Houghton Mifflin.

Merrill, G. (1990). "Obituary: Donald Q. Innis (1924–1988)." *Canadian Geographer*, 34/1: 77–8.

Momsen, J. H., and Townsend, J. G. (eds.) (1987). *Geography of Gender in the Third World*. London: Hutchinson Education; Albany: State University of New York.

Moon, H. (1987). "Interstate Highway Interchanges Reshape Rural Communities." *Rural Development Perspectives*, 4/1: 35–8.

— (1988). "Modelling Land Use Change Around Nonurban Interstate Highway Interchanges." *Land Use Policy*, 5/4: 397–407.

— (1990). "Interstate Villages as Urban Places." *Small Town*, 19/4: 4–14.

— (1994). *The Interstate Highway System*. Resource Publications in Geography Series. Washington, DC: Association of American Geographers.

Muldavin, J. S. S. (1997). "Environmental Degradation in Heilongjiang: Policy Reform and Agrarian Dynamics in China's New Hybrid Economy." *Annals of the Association of American Geographers*, 87/4: 579–613.

Muller, K. D. (1996). "The World Bank and Rural Development: Northeast Brazil." *Papers and Proceedings of the Applied Geography Conferences*, 19: 343. Denton, Tex.: Applied Geography Conferences.

Muller, K. D., and Lages, A. M. G. (1995). "Emerging Rural Settlement Patterns in the Sugar Cane Regions of Alagoas, Northeast Brazil." *Papers and Proceedings of the Applied Geography Conferences*, 18: 151–3. Denton, Tex.: Applied Geography Conferences.

Muller, K. D., and Palhano Siba, P. R. (1998). "Survival of Rural Squatter Settlements: Sem Terra (Without Land) Movement in NE Brazil." *Papers and Proceedings of the Applied Geography Conferences*, 21: 485. Denton, Tex.: Applied Geography Conferences.

Mulligan, G. F. (1997). "Interindustry Employment Requirements in Southwestern Communities." *Papers and Proceedings of the Applied Geography Conferences*, 20: 377. Denton, Tex.: Applied Geography Conferences.

Neil, C., Tykklainen, M., and Bradbury, J. (eds.) (1992). *Coping with Closure: An International Comparison of Mine Town Experiences*. New York: Routledge, Chapman & Hall.

Nelson, P. and Beyers, W. (1998). "Using Economic Base Models to Explain New Trends in Rural Income." *Growth and Change* 29: 321–44.

Oakes, T. (1997). "Place and the Paradox of Modernity." *Annals of the Association of American Geographers*, 87/3: 509–31.

— (1998). *Tourism and Modernity in China*. London: Routledge.

Oberhauser, A. M. (1999). *Locating Gender in Rural Economic Networks*. Working Paper Series, 9905. Morgantown, W.Va.: Regional Research Institute, West Virginia University.

Owusu, J. H. (1998). "Current Convenience, Desperate Deforestation: Ghana's Adjustment Program and the Forestry Sector." *Professional Geographer*, 50/4: 418–36.

Paul, B. K. (2000). "Rural–Urban Differences in Infant Mortality in Bangladesh: A Longitudinal Study." *Papers and Proceedings of the Applied Geography Conferences*, 23: 318–22. Denton, Tex.: Applied Geography Conferences.

Paul, B. K., and Nellis, D. (1996). "Health Care 'By-Passers' in Rural Kansas." *Papers and Proceedings of the Applied Geography Conferences*, 19: 195–200. Denton, Tex.: Applied Geography Conferences.

Peet, R. (1994). Review of F. J. Schuurman (ed.), "Beyond the Impasse: New Directions in Development Theory." *Annals of the Association of American Geographers*, 84/2: 339–41.

Peet, R., with Hartwick, E. (1999). *Theories of Development*. New York: Guilford Press.

Peet, R., and Watts, M. (eds.) (1996). *Liberation Ecologies: Environment, Development, Social Movements*. London: Routledge.

Place, S. E. (1997). "Ecotourism: The Contradictions of Sustainable Development on the Caribbean Shore of Costa Rica." *Papers and Proceedings of the Applied Geography Conferences*, 20: 253–9. Denton, Tex.: Applied Geography Conferences.

Popper, D. E., and Popper, F. J. (1987). "The Great Plains: From Dust to Dust: A Daring Proposal for Dealing with an Inevitable Disaster." *Planning*, 53/6: 12–18.

Porter, P. W., and Sheppard, E. S. (1998). *A World of Difference: Society, Nature, Development*. New York: Guilford Press.

Powell, S. G. (1992). *Agricultural Reform in China: From Communes to Commodity Economy 1978–1990*. Manchester: Manchester University Press.

Rain, D. (1999). *Eaters of the Dry Season*. Boulder, Col.: Westview Press.

Rasker, R. (1993). "Rural Development, Conservation, and Public Policy in the Greater Yellowstone Ecosystem." *Society and Natural Resources*, 6: 109–26.

Rasker, R., and Glick, D. (1994). "Footloose Entrepreneurs: Pioneers of the New West?" *Illahee*, 10: 34–43.

Ricketts T. C., Savitz, L. A., Gesler, W. M., and Osborn, D. N. (eds.) (1994). *Geographic Methods for Health Services Research: A Focus on the Rural–Urban Continuum*. Lanham, Md.: University Press of America.

Riesbame, W. (ed.) (1997). *Atlas of the New West: Portrait of a Changing Region*. New York: W. W. Norton.

Rigby, D. L., and Willmott, C. (1998). *Infrastructure Needs in Geography and Regional Science. A Report to the National Science Foundation*. Washington, DC: National Science Foundation.

Robison, H. (1997). "Community Input–Output Models for Rural Areas Analysis with an Example from Idaho." *The Annals of Regional Science*, 31: 325–51.

Rossum, S., and Lavin, S. (2000). "Where Are the Great Plains? A Cartographic Analysis." *Professional Geographer*, 52/3: 543–52.

Rudel, T. K., Perez-Lugo, M., and Zichal, H. (2000). "When Fields Revert to Forest: Development and Spontaneous Reforestation in Post-war Puerto Rico." *Professional Geographer*, 52/3: 386–97.

Rudzitis, G. (1993). "Nonmetropolitan Geography: Migration, Sense of Place, and the American West." *Urban Geography*, 14: 574–85.

—— (1996). *Wilderness and the Changing American West*. New York: John Wiley & Sons.

Sack, R. D. (1999). "A Sketch of a Geographic Theory of Morality." *Annals of the Association of American Geographers*, 89/1: 26–44.

Salopek, P. (2000). "Pilgrimage Through the Sierra Madre." *National Geographic*, 116/6: 56–81.

Sambrook, R. A., Pigozzi, B. W., and Thomas, R. N. (1999). "Population Pressure, Deforestation, and Land Degradation: A Case Study from the Dominican Republic." *Professional Geographer*, 51/1: 25–40.

Schaeffer, P. V., and Loveridge, S. (eds.) (2000). *Small Town and Rural Economic Development: A Case Studies Approach*. Westport, Conn.: Praeger.

Short, R. L. (1998). "Implementing an Automated Process for Delineating Urban and Rural Settlements." *Papers and Proceedings of the Applied Geography Conferences*, 21: 502. Denton, Tex.: Applied Geography Conferences.

Shortridge, J. R. (1989). *The Middle West: Its Meaning in American Culture*. Lawrence, Kan.: University Press of Kansas.

Shrestha, N. R. (1988). "Human Relations and Primary Care Delivery in Rural Nepal: The Case of Deurali." *Professional Geographer*, 40/2: 202–13.

—— (1989). "Frontier Settlements and Landlessness among Hill Migrants in Nepal Tarai." *Annals of the Association of American Geographers*, 79/3: 370–89.

Shumway, J. M., and Durrant, J. O. (2000). "In Pursuit of Understanding Place: The Transformation of the Rural West From Old to New in Emery County, Utah." *Papers and Proceedings of the Applied Geography Conferences*, 23: 81–8. Denton, Tex.: Applied Geography Conferences.

Sibley, D. (1996). *Geographies of Exclusion: Society and Difference in the West*. New York: Routledge.

Silberfein, M. (ed.) (1998). *Rural Settlement Structure and African Development*. Boulder, Col.: Westview Press.

Sirk, R. A. (1994). "Patchwork in the Glen and the Four Rivers Region: Evolving Organizations for Marketing Rural Appalachian Tourism and Recreation." *Papers and Proceedings of the Applied Geography Conferences*, 17: 148–55. Denton, Tex.: Applied Geography Conferences.

Sluyter, A. (1999). "The Making of Myth in Postcolonial Development: Material-Conceptual Landscape Transformation in Sixteenth-Century Veracruz." *Annals of the Association of American Geographers*, 89/3: 377–401.

Smith, N. (1991). *Uneven Development: Nature, Capital and the Production of Space*. Cambridge, Mass.: Blackwell.

South, R. B. (1990). "Transnational 'Maquiladora' Location." *Annals of the Association of American Geographers*, 80/4: 549–70.

—— (1993). Review of P. A. Wilson, "Exports and Local Development: Mexico's New Maquiladoras." *Annals of the Association of American Geographers*, 83/4: 748–51.

Staeheli, L. A., and Lawson, V. A. (1995). "Feminism, Praxis, and Human Geography." *Geographical Analysis*, 27/4: 321–38.

Stoddart, D. R. (1986). *On Geography and Its History*. Oxford: Basil Blackwell.

Straussfogel, D. (1997). "Redefining Development as Humane and Sustainable." *Annals of the Association of American Geographers*, 87/2: 280–305.

Sun, J. (ed.) (1988). *The Economic Geography of China*. Hong Kong: Oxford University Press.

Tata, R. J., and Schultz, R. R. (1988). "World Variation in Human Welfare: A New Index of Development Status." *Annals of the Association of American Geographers*, 78/4: 580–93.

Theirl, S. R. (1994). "A Strategy for Community Investment: The Western Reserve Enterprise Zone." *Papers and Proceedings of the Applied Geography Conferences*, 17: 240. Denton, Tex.: Applied Geography Conferences.

Thomas-Slayter, B., and Rocheleau, D. (1995). *Gender, Environment and Development in Kenya: A Grassroots Perspective*. Boulder, Col.: Lynne Rienner.

Tricart, J., and Kiewietdejonge, C. (1992). *Ecogeography and Rural Management: A Contribution to the International Geosphere-Biosphere Programme*. New York: Longman Scientific & Technical.

Tuan, Y. (1989). *Morality and Imagination: Paradoxes of Progress*. Madison: University of Wisconsin Press.

Turner, B. L., II (1989). "The Specialist-Synthesis Approach to the Revival of Geography: The Case of Cultural Ecology." *Annals of the Association of American Geographers*, 79/1: 88–100.

University of Nebraska (2001). Department of Geography website ‹http://www.unl.edu/unlgeog/home.htm›, last accessed 4 November 2001.

Veeck, G., and Pannell, C. W. (1989). "Rural Economic Restructuring and Farm Household Income in Jiangsu, People's Republic of China." *Annals of the Association of American Geographers*, 79/2: 275–92.

Vias, A. C. (1999). "An Analysis of Population and Employment Growth in the Nonmetropolitan Rocky Mountain West, 1970–1995." Dissertation; University of Arizona, Tucson: Department of Geography and Regional Development.

Wallach, B. (1996). *Losing Asia: Modernization and the Culture of Development*. Baltimore: Johns Hopkins University Press.

Walters, B. B., Cadelina, A., Cardano, A., and Visitacion, E (1999). "Community History and Rural Development: Why Some Farmers Participate more Readily than Others." *Agricultural Systems*, 59: 1–22.

White, S. E. (1994). "Ogallala Oases: Water Use, Population Redistribution, and Policy Implications in the High Plains of Western Kansas, 1980–1990." *Annals of the Association of American Geographers*, 84/1: 29–45.

Wilbanks, T. (1994). "'Sustainable Development' in Geographic Perspective." *Annals of the Association of American Geographers*, 84/4: 541–56.

Winchell, D. G. (1992). "Tribal Sovereignty as the Basis for Tribal Planning." *The Western Planner*, 13/3: 9–11.

Winchell, D. G., Lounsbury, J. F., and Sommers, L. M. (1998). "Indian Gaming in the U.S.: Distribution, Significance and Trends." *Focus*, 44/4: 1–10.

World Commission on Environment and Development (1987). *Our Common Future*. New York: Oxford University Press.

Yapa, L. (1996). "What Causes Poverty?: A Postmodern View." *Annals of the Association of American Geographers*, 86/4: 707–28.

Yapa, L., and Zelinsky, W. (1988). "How Not to Study the Geography of Human Welfare." *Annals of the Association of American Geographers*, 78/4: 609–11.

Young, E. (1995). *Third World in the First: Development and Indigenous Peoples*. New York: Routledge.

Zhou, B., Shaw, W. (2000). "The Community Reinvestment Act as a Policy Option for Enhancing Regional Money Supply: Evidence from Kentucky." *Papers and Proceedings of the Applied Geography Conferences*, 23: 64–71. Denton, Tex.: Applied Geography Conferences.

Zimmerer, K. S. (1991). "Wetland Production and Smallholder Persistence: Agricultural Change in a Highland Peruvian Region." *Annals of the Association of American Geographers*, 81/3: 443–63.

—— (1994). "Human Geography and the 'New Ecology'," *Annals of the Association of American Geographers* 84/1: 108–25.

—— (1996). *Changing Fortunes: Biodiversity and Peasant Livelihood in the Peruvian Andes*. Berkeley: University of California Press.

—— (2000). "The Reworking of Conservation Geographies: Nonequilibrium Landscapes and Nature-Society Hybrids." Millenium Essay, *Annals of the Association of American Geographers*, 90/2: 356–69.

Zimmerer, K. S., and Young, K. A. (eds.) (1998). *Nature's Geography: New Lessons for Conservation in Developing Countries*. Madison: University of Wisconsin Press.

Zurick, D. N. (1988). "Resource Needs and Land Stress in Raptia Zone, Nepal." *Professional Geographer*, 40/4: 428–43.

—— (1992). "Adventure Travel and Sustainable Tourism in the Peripheral Economy of Nepal." *Annals of the Association of American Geographers*, 82/4: 608–28.

Zweifler, M. O., Gold, M. A., and Thomas, R. N. (1994). *Land Use Evolution in Hill Regions of the Dominican Republic*. Lansing: Michigan State University.

Part IV

Geographic Methods

CHAPTER 24

Geographic Information Systems

Daniel G. Brown, Gregory Elmes, Karen K. Kemp, Susan Macey, and David Mark

Introduction

The role of GIS within the discipline of geography, not to mention its role within the daily operation of a very large range of human enterprises in the developed world, has undergone major changes in the decade-plus since the first edition of *Geography in America* (Gaile and Willmott 1989) was published. Not the least of these major changes is an important redefinition of the acronym. In 1989, GIS meant only "geographic information systems" and referred to an immature but rapidly developing technology. Today, many geographers make an emphatic distinction between the technology (GISystems or GIS) and the science behind the technology (GIScience or GISci). This important transition from a focus on the technology to a focus on the far-ranging theoretical underpinnings of the technology and its use are clearly reflected in the research progress made in this field in the past decade. This chapter highlights some of the significant aspects of this diverse research and its related impacts on education and institutions. The chapter focuses on the work of North American geographers, though reference to the work of others is unavoidable. We recognize the many and increasing contributions of our colleagues in other disciplines and overseas to the development of GISci, but focus our attention to the scope defined by the present volume. The chapter closes with speculations on the future of GIS in geography in America in the coming decade.

The Emergence of Geographic Information Science

In the late 1980s, geographic information systems (GIS) were large stand-alone software and information systems being applied to a growing range of application areas. Today GIS are well integrated into the normal operations of a large range of industries as diverse as forestry, health care delivery, retail marketing, and city planning. Developments in the capabilities of and access to GIS technology during the past decade have paralleled developments in the computer industry as a whole. Similarly, academic research into the fundamental concepts and theories that underlie GIS has matured and become better connected across multiple disciplines. Drawing on fields as diverse as computer science, cognitive science, statistics, decision science, surveying, remote sensing, and social theory, "geographic information science" (Goodchild 1992*b*) has emerged as an important synthesizing influence during the 1990s. The term "geographic information science" (GISci) was coined and came into widespread use in the 1990s, and suggests a shift in focus of many in the community away from the technology for its own sake and toward more challenging intellectual issues.

In a much referenced paper in which he argued for the identification and fostering of a science of geographic information, Goodchild (1992*b*: 41) defined GISci research as "research on the generic issues that surround the use of GIS technology, impede its successful implementation, or emerge from an understanding of its potential capabilities." In justifying a change in name from the *International Journal of Geographical Information Systems* to the *International Journal of Geographical Information Science*, Fisher (1997*a*: 1) relates that the journal has published articles that "have contributed not to the understanding or development of systems as such, but to the science which underpins and exploits the systems." Pickles (1997: 369) offered this definition for a science of GIS: "the scholarly investigation of its origins, logics, systems, new capacities, and new uses." These definitions are construed broadly enough to include both "research about GIS" and "research using GIS," but it seems obvious that the former would be more central to a geographic information science than the latter. However, it should be noted that the very large size of the GIS Specialty Group (GISSG), which in the late 1990s was the largest of the AAG's SGs, likely reflects the fact that many geographers use or hope to use GIS in their research or teaching. Core GIS and GISci researchers in American geography would be a smaller group.

GIS and Geography

Although geographers have remained central within the emerging GISci community, and although the GISSG has become the largest specialty group within the AAG, there are signs of discomfort in the relationship between academic geography and GIS as one of its children, albeit jointly parented with other disciplines. An early indication of this discomfort came in the form of Terry Jordan's (1988) presidential address to the AAG, in which he argues that GIS is merely a tool constituting "non-intellectual expertise" (Jordan 1988). We, on the other hand, believe that GIS can be thought to pose enough intellectual challenge to comprise a science.

One indicator of the relationship between GIS and geography is the nature of publications in the discipline. A survey of articles published in the *Annals of the Association of American Geographers*, an admittedly narrow gage of the discipline, turns up eight research articles published between 1990 and 1998 that included "GIS" or "geographic information systems" in the title, abstract, or keyword list. Of the eight articles that specifically mention GIS, three (Egbert and Slocum 1992; Peuquet 1994; Curry 1997) discuss "research about GIS," whereas the remaining five present "research using GIS," mostly in physical geography (Butler and Walsh 1990; Dobson *et al.* 1990; Savage 1991; Palm and Hodgson 1992; and Sabin and Holliday 1995). The list is a bit longer when articles that fall within areas that often qualify as GISci but did not mention GIS are added, including spatial cognition (Golledge *et al.* 1992; MacEachren 1992*a*) and cartography (Olson and Brewer 1997). In a summary of research articles submitted for publication in the *Annals* between 1987 and 1993, Brunn (1995) listed manuscripts in twenty-one topics. GIS was not included in the list explicitly, but was lumped with the topic titled "Models and Techniques." A little over 2 per cent of the articles submitted to the *Annals* during this period fell into this category, and since other techniques are included, only a fraction of those would be considered GIS. This category would clearly not include articles on research that uses GIS. Based on the research articles published in the AAG's flagship journal, one might be excused for concluding that academic geographers view GIS largely as a tool worthy of only limited basic research. Brunn's (1995) work seems to indicate that the problem is related to a low number of submissions to the *Annals*, rather than to any editorial bias against GIS. Geographers in GISci are publishing their work in other outlets.

The pages of the academic journals of the AAG also saw a continuation of the debate about the nature of GIS and its relationship with geography (Dobson 1993; Pickles 1997; Wright *et al.* 1997). Wright *et al.* (1997) summarize a discussion on this topic that occurred spontaneously in a computer-based newsgroup. Explicit in their evaluation was a consideration of three disparate positions on the role of GIS: as a tool, as toolmaking, and as a science. This discussion raises basic questions about the epistemological and ontological foundations of GIS, which had not been asked earlier in its development. In his comments on the work by Wright *et al.* Pickles (1997) lauds them for raising these issues, but remains critical of the limited engagement that they and others in the GIS community have made with theories and critiques of science within and outside geography.

In the United States, John Pickles has been instrumental in encouraging and facilitating a rethinking of the place of GIS within the discipline and within society in general. His 1995 edited volume, *Ground Truth: The Social Implications of Geographic Information Systems* (Pickles 1995), is essentially a call to action. Some of the criticisms aimed at quantification in geography in the 1970s have been redirected at GIS, partly in response to the boosterism that has surrounded the technology (e.g. Openshaw 1991) and partly as a general indictment of the positivist theoretical environment within which many GIS practitioners work.

An earlier call to reexamine the role of GIS and the computer within geography came from Dobson (1993), in a paper which is a revision of his earlier paper on "Automated Geography" (Dobson 1983). Dobson's optimism about the potential for a revolution in the way geographers work, and the kind of work they do, lies in what he sees as the facilitation by computer technology generally, and GIS in particular, of integrated analysis, in the vein of the "landscape" approach of the late nineteenth and early twentieth centuries. Although the enthusiasm for integration and wholism may have been shared by many, others felt this optimism tempered by the observation that there remained fundamental computational (Armstrong 1993) and theoretical (Marble and Peuquet 1993) limits to what GIS can allow geographers to accomplish. Further, some commentators questioned the sorts of analyses GIS facilitates and whether these were the ones needed (Pickles 1993; Sheppard 1993).

Table 24.1 *Research challenges identified by the University Consortium for Geographic Information Science (UCGIS) in 1996 and 2000*

Spatial data acquisition and integration
Distributed computing
Extensions of geographic representation
Cognition of geographic information
Interoperability of geographic information
Scale
Spatial analysis in a GIS environment
The future of the spatial information infrastructure
Uncertainty in spatial data and GIS-based analyses
GIS and society
Geospatial data mining and knowledge discovery
Ontological foundations for geographic information science
Geographic visualization
Analytical cartography
Remotely-acquired data and information in GIScience

Source: UCGIS (1996, 2000).

GIS-Related Research in Geography in America

The diversity of geography's traditional areas of study is strongly reflected in the diversity of themes pursued by geographers doing research in GISci and in the application of related technologies. While spatial analysis underlies all practical applications of GIS, it also provides a solid foundation for the development of new tools and the exploration of new ways of examining the geographic world through this digital medium. The traditions of cartography provide the foundation for the map interface commonly used in GIS while cartographic research themes have new pertinence when applied to GIS. Human and cultural geography themes have recently risen in prominence as interest in issues surrounding the use and value of GIS has been expressed.

This section summarizes several research themes within GISci to which geographers contribute. These themes are grouped into the following categories: computational and technological issues, representational issues, analytical issues, data quality and error propagation, integrating GIS, and GIS and society. These categories have some, but not perfect, resemblance to those presented in Goodchild's (1992*b*) discussion of GISci and those generated by the University Consortium for Geographic Information Science (UCGIS, more on this organization later in this article) in 1996 and 2000 (Table 24.1). We cite a number of papers as examples of particular research areas only, rather than as a listing of all work in a given area. Furthermore, we do not cite the hundreds of applications of the numerous advances described, many of which are made in disciplines other than geography.

Computational and Technological Issues

While many of the computational and technological research issues have been undertaken by computer and information scientists, geographers and other geographic information scientists have made some significant contributions by capitalizing on generic advances in computing technology.

Recent special issues of the *International Journal of Geographical Information Science* highlight two of these themes: parallel processing and interoperability. In the special issue on parallel processing, Mower (1996) lists a number of spatial data handling problems that have been addressed with parallel processing. These include image processing, drainage basin analysis, network analysis, cartographic name placement, line intersection detection, and viewshed analysis. Healey (1996) points to a number of advances that have facilitated the adoption of parallel processing approaches for GIS. Significant investment in parallel technology by large hardware firms (such as Cray), development of standards for parallel software (MPI), and increased availability and need for large data sets in GIS all provided impetus for further research and development in parallel GIS algorithms. Armstrong and Densham (1992) presented a conceptual framework for developing spatial algorithms within a parallel processing environment (e.g. spatial interpolation; Armstrong and Marciano 1996). Xiong and Marble (1996) illustrated approaches to decomposing spatial analysis problems using the empirical example of a traffic flow analysis in a parallel environment.

Research in interoperability for GIS addresses "incompatibilities in data formats, software products, spatial conceptions, quality standards, models of the world and whatever else make 'GIS interoperability a dream for users and a nightmare for systems developers' " (Vckovski 1998: 297). While interoperability as it is defined by the Open GIS Consortium (described below in the section on institutions), is in large part a type of standards effort, it is a fundamental foundation for the much-anticipated revolution of object-component software. In this scenario, which is rapidly unfolding, the huge, complex software packages typical of many commercial GIS are unbundled to allow small, inexpensive components to be served across the Internet on a pay-per-use basis. This has lead to the emergence of yet another meaning for the S in GIS—geographic information services. Such low-cost services might facilitate democratization of GIS and its use as a means of empowering the less advantaged members of society, a theme we will revisit in more detail below.

Much of the work in interoperability addresses technical methods and mechanisms for overcoming the barriers to full integration of disparate and heterogeneous data, but some focus on the development and refinement of data standards, in particular, the spatial data transfer standard (SDTS) (e.g., Arctur *et al.* 1998; Moellering 1992). The Internet has provided an important vehicle for development of data-sharing environments (Nebert 1999). However, barriers to technical interoperability have been easier to challenge successfully than those related to semantic interoperability—incompatibilities of languages, symbolic representations, and syntax (UCGIS 1996). For systems to be interoperable there must be a consistent set of interpretations of the encoded information—one system must be capable of understanding the meaning of another system's data.

Other technologies are changing the way GIS practitioners do their work. One of the more important of these is most certainly the Internet. An estimated 540 million people used the Internet daily in early 2002. A number of articles discussing both the opportunities and technologies involved with GIS on the Internet have appeared in the industry's trade magazines, including *GIS World* (now *GEOWorld*) and *Geo Info Systems* (now *Geospatial Solutions*). Although there has been much boosterism (McKee 1999), the capabilities, impacts, and benefits of delivering on-demand maps and GIS functionality are now becoming a very active area of research and development (see for example Carver *et al.* 1997; Lin and Zhang 1998; Peng 1999; Harrower *et al.* 2000; MacEachren 1998). Another related development is that of mobile computing, which integrates wireless communications technology (cellular and satellite) with global positioning systems (GPS) and GIS for greatly enhanced field data collection and real-time mapping (Novak 1995), and which has led to the emergence of new commercial enterprises in location-based services. Although these technologies are clearly changing GIS practice (Blennow and Persson 1998), education and research activities related to mobile GIS have largely been focused in the engineering disciplines.

Representational Issues

The ways in which GIS represent real-world geographic phenomena limit current GIS. Chrisman *et al.* (1989) discussed the problematic representational issue of raster versus vector in the previous volume of *Geography in America*. Although this topic remains with us yet, the necessity of a functional distinction between these two fundamentally different representations has been reduced. Some GIS vendors now provide functionality that can integrate operations on raster and vector format data using automatic data conversion. In the 1990s, focus shifted to understanding the conceptual underpinnings of these different representations. Couclelis (1992) and Goodchild (1992*a*) outline the underlying perceptions of the world that give rise to the raster and vector data models, namely the concepts of fields and objects. Others have attempted to develop a stronger philosophical foundation for GIS through the consideration of ontology for both objects (B. Smith and Mark 1998, 2001) and fields—continuous surfaces of measured attributes (Peuquet *et al.* 1999). Work in this area has helped to better define the relationships between real-world phenomena, human understanding of those phenomena, and their representation within digital databases. Better definition of these relationships holds the promise of giving rise to alternative and/or better representational structures, more intelligent use of those structures, and more intuitive interfaces for human–computer interaction.

Meanwhile, in continued examinations of the object model, the observation that some objects suffer from indeterminacy came to the fore. This observation manifested itself in a concern both for the nature of boundaries between objects in an uncertain world (Mark and Csillag 1989) and for the representation of so-called fuzzy objects (Fisher 1994) and their relationships with boundaries (Brown 1998). In Europe, a research initiative on "Geographic Objects with Indeterminate Boundaries" was undertaken to explore alternative representations (Burrough and Frank 1996). As in most such European meetings, American geographers made significant contributions, including the identification of problems associated with indeterminate objects (Couclelis 1996) and potential GIS-based representational solutions (Usery 1996). Object-orientation has become an important technical vehicle through which object representations can be implemented and has given rise to GIS data representations that better reflect spatial semantics (Leung *et al.* 1999), encapsulate semantic and topological relationships among objects (Tang *et al.* 1996), and facilitate modeling of geographical movement processes (Westervelt and Hopkins 1999).

Work has begun to extend traditional representations in other ways as well. Fundamental to these extensions are the incorporation of full three-dimensional data structures into existing GIS and the development of robust spatio-temporal representations (often called four-dimensional). Concepts of how to represent time within a spatial data model were developed by Langran (1992) and Peuquet (1994), but practical successes in implementing temporal data structures were rare. E. J. Miller (1997) describes an approach to interpolation in four dimensions.

Related to these issues, research into spatial cognition, a tradition in geography that extends well before the 1990s (cf. Blaut and Stea 1971; Downs and Stea 1973; Golledge and Zannaras 1973) has recently attracted a larger audience as the GISci research community becomes increasingly aware that the ways in which humans view geographic space can be used to develop formal models of that space. Such formal models can subsequently be used as strong theoretical bases for data models, data structures, and approaches to data visualization (Mark and Frank 1991). The work has examined how people acquire spatial knowledge (e.g. Golledge *et al.* 1992) and process visual information from maps (Lloyd 1997), and informs our GIS-based approaches to communicating knowledge. Issues of spatio-temporal cognition and their implications for GIS design were addressed by Egenhofer and Golledge (1998). Major research efforts have focused on cognition in order to contribute to the development of better representations of geographic space through understanding how people perceive such issues as scale and dynamics (Mark *et al.* 1999). Cognitive and computational research agendas in GISci have been connected through work on formalizing spatial relations (Egenhofer and Franzosa 1991; Mark and Egenhofer 1994), on spatial databases and query languages (Egenhofer 1992, 1997), and on human–computer interaction for GIS (Mark and Gould 1991; Medyckyj-Scott and Hearnshaw 1993).

Scale, one of geography's fundamental concepts, is intimately related to how we represent the real world in the digital medium. Scale is fundamental in our cognition, measurement, representation, and presentation of geographic information. Early in the decade, McMaster and Buttenfield compiled an edited volume that considered the issues associated with map scale and how they translate into the digital environment. Lam and Quattrochi (1992) summarized the multiple definitions of scale and placed it in the context of the fractal model that engaged geographic information scientists for many years (see also Lam and DeCola 1993). Work on the dependence of geographical representations on scale produced important information about the degree of spatial dependence and its effects on representations of geographic phenomena and processes (Bian and Walsh 1993; Stoms 1994; Walsh *et al.* 1999). Issues surrounding scale in GIS, remote sensing, and environmental modeling were addressed in an edited volume by Quattrochi and Goodchild (1997) in which authors discuss empirical observations of the effects of scale change, fractal models, and data structures and tools for analysis at multiple scales.

Visualization of geographic information is another fundamental geographic theme that has emerged as a critical representation issue for GIS. MacEachren *et al.* (1992: 101) defined geographic visualization as "the use of concrete visual representations whether on paper or through computer displays or other media to make spatial contexts and problems visible, so as to engage the most powerful of human information-processing abilities, those associated with vision." Hearnshaw and Unwin (1994) edited a volume devoted to exploring the linkages between new visualization approaches and data stored in a GIS, and a paper in the book by MacEachren and Taylor (1994) illustrates the importance of visualization within modern cartography. Weibel and Buttenfield (1992) produced specific recommendations for software designers and GIS users to improve the graphic output from GIS for use in analysis and decision-making. There has been a substantial amount of work on developing approaches to graphically portraying information about uncertainty in geographic information (MacEachren 1992*b*; Beard and Mackaness 1993), including that presented in a special issue of *Cartographica* (Beard *et al.* 1991). Hunter and Goodchild (1995) reviewed a number of techniques for handling and displaying uncertainty information about digital elevation models (DEMs). These approaches included calculating epsilon bands, mapping probabilities, and managing the error. Special issues of *Cartography and Geographic Information Systems* (MacEachren and Monmonier 1992) and *Computers and Geosciences* (MacEachren and Kraak 1997) were devoted to the topic of geographic visualization in general.

Analytical Issues

Interestingly, the foundations created by early work in spatial analysis during geography's quantitative revolution continue to provide the basis for many of the analytical tools in today's commercially available GIS. A special issue in the *International Journal of Geographical Information Systems* highlighted the range of spatial analysis methods used within GIS (Burrough 1990), as did books by Fischer *et al.* (1996), Fotheringham and Rogerson (1994) and, more recently, Fischer (1999).

When coupled with statistical and computer modeling software, GIS has been advanced by the implementation of a broad array of new tools, many of which have been borrowed liberally from developments in other fields and adapted for spatial analytical applications. Geostatistics, an advanced toolkit developed for geological applications is now becoming commonly used in physical geography applications (Oliver *et al.* 1989) and is integrated into the functionality of many GIS packages or made available through direct linkages with standalone software (Cook *et al.* 1994). Human and urban geography have benefited through better inferential statistical capabilities from the development of functional links between GISystems and various spatial statistical and modeling packages (Sui 1998). Anselin *et al.* (1993), Griffith *et al.* (1990), and Can (1992) have created substantial stand-alone and/or add-on modules that provide a range of spatial analysis capabilities for GIS, mostly related to characterizing spatial structure and incorporation of spatial autocorrelation into standard statistical models. In some cases, researchers have extended the conceptual underpinnings of an existing spatial analytical approach to facilitate its implementation within a GIS/spatial context, for example H. J. Miller's (1991) use of the space-time prism concept.

Although not specific to GIS and spatial analysis, other non-parametric approaches have been implemented in association with GIS applications to provide predictive capabilities where non-linear relationships exist. Artificial neural networks provide such non-parametric approaches for image classification and pattern recognition and thus have some application within geography and GIS (Fischer and Gopal 1993; Hewitson and Crane 1994; Brown *et al.* 1998). Classification and regression trees (CART) and generalized additive models (GAM) are examples of other such tools (Franklin 1995 provides a good summary of predictive mapping approaches for vegetation). Spatial modeling has also benefited from developments in agent-based approaches to modeling. For example, Roy and Snickars (1996) and Clarke and Gaydos (1998) use cellular automata and Westervelt and Hopkins (1999) use agent-based simulation to model spatial processes within a GIS context.

As data volumes have increased, exploratory spatial data analysis has become increasingly used and accepted. Bailey and Gatrell (1995) take an explicitly exploratory approach to spatial data analysis, and the volume edited by Fischer *et al.* (1996) contains a substantial amount of work devoted to exploratory analysis. Work on exploring the spatial autocorrelation structure of geographic data (Griffith 1993) has contributed significantly to both exploratory and inferential statistical applications within GIS. The book by H. J. Miller and Han (2001) on geographic data mining and knowledge discovery describes more recent advances in exploratory spatial data analysis. As might be expected, however, old foundations cannot be quickly unseated and this increasing emphasis on exploratory analysis and data mining still troubles those who believe that theory ought to drive data analysis, rather than the reverse (Taylor and Johnston 1995).

Uncertainty and Error Propagation

As the field has matured, methods for identifying, correcting, tracking, and visualizing errors and uncertainty in spatial data with GIS-based analyses have commanded the attention of many GIS researchers. While not clearly an extension of traditional geographic research, this topic has allowed many geographers to extend their knowledge of the behavior of geographic phenomena. At the end of the 1980s a research initiative by the National Center for Geographic Information and Analysis (NCGIA, more on this organization later) resulted in an important early book on the topic (Goodchild and Gopal 1989). Intensive research by geographers and others continued on these themes and results are collected in the proceedings of three conferences in a series on spatial accuracy assessment in natural resources held in the 1990s (Congalton 1994; Mowrer *et al.* 1996; Lowell and Jaton 1999).

Much of the early work was focused on developing adequate descriptions of the error in spatial data (Goodchild 1989). This led to work on error modeling that involves describing both the overall magnitude of error and how the error is thematically or spatial distributed (Goodchild *et al.* 1992; Lanter and Veregin 1992). In some cases, such descriptions are obtained analytically (Thapa and Bossler 1992), but in many cases richer descriptions of error can be obtained through simulation (Fisher 1990).

Once modeled, it is possible to examine how error propagates through sequences of analytical processes. "Error propagation modeling refers to the attempt to emulate the processes of source error modification and transference, with the goal of estimating the error characteristics of derived data products" (Veregin 1996: 419). Although the changes in error levels through simple GIS-based analyses can often be quantified analytically (Lanter and Veregin 1992), simulation and probability approaches have proven more successful for even slightly complex analyses (Fisher 1992; Heuvelink 1993; Veregin 1996). Monte Carlo methods are now commonly employed to evaluate the uncertainty in the output of a GIS analysis based on some knowledge about and a model of the error in the input (Emmi and Horton 1995). Obtaining estimates of error is important for understanding the degree to which the results of analyses should be used to support a conclusion or decision. Propagation of error pattern through calculation of derivatives of DEMs was used to diagnose and treat a common systematic error pattern (Brown and Bara 1994).

Integrating GIS with Other Tools

In many applications, GIS is one tool among many used to address a particular question or problem. In applying GIS to these problems, researchers or practitioners have had to develop approaches to combining the capabilities of GIS and other tools. Here we discuss three additional areas in which GIS is used in a supporting role and the input that geographers have made in support of their integration: environmental modeling, decision support systems, and remote sensing.

The integration of GIS and environmental models and the use of such tools to understand a broad range of environmental processes were the topics of a series of conferences held during the 1990s (Goodchild *et al.* 1993, 1996; and NCGIA 1996). The proceedings volumes from these conferences provide excellent compendia of information on the range of solutions that have been developed for addressing environmental issues ranging from atmospheric science through hydrology to ecology. Solutions range from developing efficient computer codes for transforming and passing data between different software products, to methods for handling fundamental differences in the conceptual models upon which different software tools are based. Indeed, many of these issues are now being addressed generically within the interoperability context.

Many of the analytical developments in GIS, including integration with models, have been specifically aimed at developing a basis for spatial decision support systems (SDSS; Armstrong 1994). In addition to integrating GIS with models (Armstrong *et al.* 1991) and knowledge-based approaches (Kirkby 1996), SDSS developments draw on the decision sciences (such as multicriteria evaluation) to improve the utility of GIS in decision-making contexts (Carver 1991; Jankowski and Richard 1994; Eastman *et al.* 1995).

Early in the 1990s, the challenges of integrating GIS and remote sensing were seen to include: error analysis, data structures and access to data; data processing flow and methodology; the future computing environment; and institutional issues (Star 1991). Many of these challenges have been addressed through research in other areas of GISci and are noted elsewhere. The influence of resolution of both images and other geospatial data, and the effect that observation scale has on analytical power were addressed at length in Quattrochi and Goodchild (1997). However, much remains to be done, as Jensen *et al.* (1998) suggest: do the current data collection and integration strategies fulfill our needs? Is it possible to integrate accurately the increasingly more abundant and precise spatial data with other current and historical datasets to solve complex problems? Are there significant gaps in the remotely sensed data currently being collected? If data required by user communities are not available, how can the data be obtained and who should collect it?

GIS and Society

The introduction to this chapter alluded to the debate on the engagement of GIS and society within academic geography, a debate that has emerged from acrimony and division to constructive engagement and synthesis during the last decade. The discourse between GISci researchers and critical social theorists has focused on what has been characterized as the reconstitution of logical positivism, reductionism, and nomothetic methods through GIS. A strong polarization of ideas and scholarship was apparent at the outset of the 1990s (e.g. Taylor 1990; Taylor and Overton 1991; Openshaw 1991), but the intellectual atmosphere of the 1990s was characterized by increasing communication and debate between scholars of different perspectives. Early critiques of GIS from the social-theoretic perspective were posited by John Pickles' "GIS and the surveillant society" (Pickles 1991) and Neil Smith's "History and Philosophy of Geography: Real Wars, Theory Wars" (Smith 1992). Sui (1994) considered the influence of GIS and postmodern theories in urban geography. If at the end of the decade there was not complete accord, there were signs that the philosophical tensions were furnishing constructive syntheses.

The progress from antipathy to constructive engagement may be traced through a sequence of meetings and subsequent publications. Foremost among these publications is the edited book *Ground Truth* (Pickles 1995), mentioned earlier. It offers a seminal critique of underlying epistemological and methodological differences within the academic community. A meeting held in November 1993 in Friday Harbor, Washington, spurred by the propositions that later appeared in *Ground Truth*, resulted in the publication of a special issue of *Cartography and Geographic Information Systems* focusing on GIS and society (Poiker and Sheppard 1995). A meeting held in Minnesota in 1996 on "GIS and Society: The Social Implications of How People, Space and Environment Are Represented in GIS," (Harris and Weiner 1996) was a direct outcome from Friday Harbor. Other outgrowths of the debate were a special issue of *Geographical Review* (Adams and Warf 1997), examining cyberspace and geographies of the information society, and a special issue of *Cartography and Geographic Information Science* focusing on public participation GIS (Obermeyer 1998).

At the end of the 1990s, Pickles (1999: 58–9) concluded that the GIS community continues to be "reticent to acknowledge the conditions of its own construction . . . and that has by and large failed to develop . . . critical reflection upon its own practice," but that the absence of "GIS and society" questions has been rectified. Several essays in the 2nd edition of the volume, titled *Geographical Information Systems*, i. *Principles and Technical Issues*; ii. *Management Issues and Applications* (Longley *et al.* 1999), grounded GIS in different viewpoints and provide a deeper background in cognitive studies and management perspectives, including three chapters on the impact of societal issues on GIS with respect to education (Forer and Unwin 1999), privacy (Curry 1999), and geospatial data policies (Rhind 1999). Many of these perspectives were absent from the first edition of the same volume (Maguire *et al.* 1991).

As part of the discussions of the UCGIS's research priority in GIS and society, academics outlined five distinct research directions, which may be summarized as the institutional approach; the legal and ethical perspective; the intellectual history perspective; the critical social theory perspective; and the public participation GIS perspective. There is necessarily a degree of overlap between these areas of study.

The institutional approach identifies a set of questions directed at uncovering the status and magnitude of GIS implementation by public and private institutions and their rates of adoption of GIS. Institutional issues around GIS use have to do with both the role of GIS within existing institutions and the role of new GIS-related institutions, such as those concerned with sharing spatial data and the National Spatial Data Infrastructure (Weatherbe and Calkins 1991). The National Research Council's Mapping Science Committee (MSC) has published regularly on institutional issues (NRC 1993, 1994, 1998, 2001). The MSC reports have advanced policy dialog of data production, partnerships with government, state–local interaction with federal agencies and distributed GIS. MSC has brought research contributions of federal agencies, such as the National Mapping Division of the USGS and the Federal Geographic Data Committee to the attention of Congress as well as the wider academic community. The committee is identifying data needs for place-based decision-making, research priorities in geography at the USGS, and incorporating GIS in sustainable development and K-12 education.

Proponents of the institutional perspective are concerned with issues of efficiency and effectiveness, measuring the costs and benefits of GIS and evaluating the degree of equity in their distribution among individuals and social groups (Obermeyer and Calkins 1991). Theories and methods have been developed to evaluate the outcomes of the use of GIS decision-making and policy (Epstein *et al.* 1996; Ventura 1995). The effect of GIS on interagency communication is evaluated, as are GIS's effects on citizens' relationships with government as reflected in people's beliefs and actions regarding the management of land and resources (D. A. Smith and Tomlinson 1992; Tulloch *et al.* 1997).

The legal and ethical perspective is concerned with the legal and ethical setting of GIS, including the various mechanisms governing access to spatial data, and the consequences of the proliferation of proprietary spatial databases. How these changes are rooted in governmental and legal regulations (Onsrud and Rushton 1995), the ethical implications of these changes (Curry 1997), issues of liability (Epstein *et al.* 1998), and possible legal remedies (Onsrud 1995) have all been subjects of investigation.

Important within the legal and ethical perspective is the question whether GIS dissolves traditional protections of privacy. Curry (1997, 1999) argued that the use of geographic information increases our ability to extract characteristics about individuals from aggregated data and reconstruct "virtual people." This raises important questions about the role of private life and the control that individuals can or should have over information about them. Within this context, it can be argued that a technology such as GIS, though in many aspects driven by needs in the military, has been accepted as largely uncontrollable; societal norms, such as those about what constitutes a private life, are adjusted in the face of the technology, rather than the reverse.

Drawing on existing literature on ethics and science or professional behavior, the issue of ethical behavior within the GIS profession has been addressed by a number of authors. Fisher (1997*b*) outlines professional ethical issues as they relate to six different actors in the GIS profession: vendors, data providers, researchers, appliers, educators, and the public. Onsrud (1993) discussed the elements that constitute unethical conduct. In a more critical account, Curry (1995: 69) argues that by their very nature "the creation and maintenance of GIS involve ethical inconsistencies." This argument relates back to the issue of privacy and the problem of storing data on individuals without their knowledge or consent.

The intellectual history perspective addresses the evolution of geographic information technologies and the dynamics by which dominant technologies are selected from a variety of potential geographic information technologies at critical points in time. This perspective attempts to reveal the societal, institutional, and personal influences governing these choices and their consequences. It questions whether and why productive alternative technologies have been overlooked. Coppock and Rhind set out the historical mainstreams of GIS development. Foresman's edited volume, *History of GIS* (1998), is an important historical record of technical, institutional, and intellectual progress, while other studies focus on software innovation. Chrisman *et al.* (1992) appraise the development and influence of the ODYSSEY system. Mark (1997) examines the convoluted road to "discovery" of the Triangulated Irregular Network (TIN) data model.

The critical social theory perspective examines limitations in the ways that populations, location conflicts, and natural resources are represented within GIS and the extent to which these limits can be overcome by extending geographic information representations (Pickles 1995, 1997, 1999; Poiker and Sheppard 1995). The critique emphasizes the ways in which the nature of and access to GIS simultaneously marginalize and empower groups with overlapping or opposing interests. Furthermore, questions are raised of how the evolution of geographic information technologies reflects societal structures and priorities as well as the practices of those who develop and utilize the technologies (Harris and Weiner 1996). The implications of differential access to information and technology is a major thread to emerge from these inquiries.

The public participation GIS perspective centers on the effective and widespread exploitation of GIS by the general public and by community and grassroots groups (Harris and Weiner 1998; Craig *et al.* 1999). Public participation has implications for empowerment within community groups using GIS and the consequent restructuring of power relations. The implications of GIS for the public quickly move into the territory of political and social theory. The term "GIS2," referring to an alternative, grassroots GIS, emerges from an activist agenda directed at democratization of information and society (Harris and Weiner 1996). Interest in Community-Integrated GIS (CiGIS) decision-making has developed as a case in point of GIS2. One research theme stems from the integration of both cultural and resource data in multipurpose land information systems (MPLIS) and the involvement of the public through the land planning process. An outcome of this approach is the "Shaping Dane" demonstration, a community-based land-use project in Wisconsin. Citizen planning is enabled through web-based dissemination of data, digital library information, GIS analysis, and display tools and feedback mechanisms (Land Information and Computer Graphics Facility 2001). New geographic representations and technologies are needed to address problems arising from the use of current GIS technologies in these contexts (Obermeyer 1998; Schroeder 1997).

A detailed research agenda for GIS and society, in part driven by momentum from within the GISci community, in part directed at issues raised by critical social theory, has emerged at the end of the twentieth century. Technological progress and data availability may be said to drive some aspects of GIS and society research—for example interoperability, open-GIS, mobility, and location-based services. Basic research continues to address the improvement of data models to reflect richer social, cultural, and political landscapes. GIS researchers have benefited from methods derived from critical social theory, while simultaneously many human geographers have recognized the value of GISci, and find its theory, tools, and methods to be a valuable mode of inquiry alongside qualitative methods. The increasing impact of geographic information in daily life promises to be a rich field for the future development of geographic inquiry and knowledge for both the physical and human sides of the discipline.

GIS Education in Geography in America

There has been substantial growth and maturation of GIS education since 1989. In the early 1980s, techniques education in geography was predominantly focused on cartography and remote sensing, with only passing mention of GIS (Dahlberg and Estes 1982). Though Chrisman *et al.* (1989) made little mention of GIS education, the conclusions did lament the lack of textbooks and an inadequate supply of faculty with sufficient training to teach GIS. This scant attention stemmed partly from the attitude of the period. From the 1960s through the 1980s, GIS education was largely confined to colleges and universities and predominantly seen as an adjunct to research needs (Chrisman 1998).

In the 1990s, several factors created an unprecedented growth in demand for GIS education. Its growing use in a broad array of application areas coupled with the greater affordability, availability and ease-of-use of GIS hardware and software have generated a strong demand for GIS education and training. This demand has provided a huge increase in teaching opportunities but, unfortunately, much of the early growth focused on technical training rather than on building strong intellectual foundations (Sui 1995; Warren 1995).

Clearly, geography does not have a monopoly on GIS education. GIS courses may now be found in health, environmental studies, biology, forestry, computer science and many other disciplines. The growth of business applications of GIS technology is being paralleled by its penetration into top business schools such as Wharton, Clemson, and Harvard (M. L. Johnson 1998; Murphy 1997). Many geographers now recognize that GIS has become imbedded in the activities of a diversity of professional occupations. At the same time, they have learned of the key role GIS can play in the teaching of important spatial concepts (DeCola and El Haimus 1995; Svingen 1994), a "trojan horse" model. Down's article "The Geographic Eye: Seeing through GIS?" (1997) eloquently illustrates how various geographic skills are enhanced through the use of GIS.

Since the early 1990s, a number of new instructional resources have become available. The first widely used text for GIS, *Principles of Geographical Information Systems for Land Resources Assessment* (Burrough 1986), had no competition for several years. Another signal that the field is maturing can be seen by the large number of undergraduate level texts that have appeared in the last few years, many of them by geographers (Burrough and McDonnell 1998; Chrisman 1997; Clarke 1997; Davis 1996; DeMers 1997; Heywood *et al.* 1998). While most of those identified above are conceptual and general in focus, many GIS books on the market have an application orientation intended to serve both the demands of a specific market segment and the broader level of delivery now needed in GIS education. However, 2001 saw the first incorporation of the word "science" into the title of a GIS textbook (Longley *et al.* 2001).

Perhaps the most significant early contributor of curricular materials in the 1990s was the NCGIA. The *NCGIA Core Curriculum in GIS* (Goodchild and Kemp 1992), first distributed in draft form in 1990 when GIS textbooks were very scarce, allowed faculty with little GIS background but with knowledge of quantitative geography or cartography to develop and teach adequate GIS courses without undue amounts of preparation time. Distributed around the world in paper format and in several languages beginning in 1992, it became an important foundation for many GIS courses during the first half of the decade (Kemp 1997) The NCGIA later participated in the development of three additional curricula distributed on-line, including a Core Curriculum in GISci <www.ncgia.ucsb.edu/giscc>, a Core Curriculum for Technical Programs <www.ncgia.ucsb.edu/cctp>, and the Remote Sensing Core Curriculum, now managed by the American Society for Photogrammetry and Remote Sensing <www.asprs.org> (websites last accessed 14 November 2002).

Added to this base is the increasing availability through the Internet of a broad range of GIS education resources. Databases, online demonstrations, course modules and tutorials, and even older versions of GIS software are now freely available. In addition, GIS software vendors are supporting the educational community by providing GIS software and curriculum materials. A sampling includes Environmental Systems Research Institute's Virtual Campus <campus.esri.com>, Intergraph's downloadable GIS demos <www.intergraph.com/gis/demos>, and MapInfo's series of demonstration modules <www.mapinfo.com> (websites last accessed 14 November 2002). Many of these programs have been aimed at the high school and community college levels (A. Johnson 1998; Kavanaugh-Brown 1998; Nellis 1993; Padgett *et al.* 2000). While much of the educational material is now being published on the Web, we are also seeing regular pieces on GIS education in such popular journals as *Converge* and *Geo Info Systems/Geospatial Solutions* (Ramirez 1995, 1996; Phoenix 1996; Kemp and Wright 1997; Macey 1997). More frequent articles, special sections and special issues on GIS education have appeared in academic journals such as the *Journal of Geography* and *Transactions in GIS*.

Funding from the NSF has been instrumental in fostering GIS education expansion on several levels. The Instrumentation and Laboratory Improvement (ILI) and Advanced Technological Education (ATE) programs have supported new initiatives at the university and community college levels. The ILI program has assisted many schools in integrating technology with innovative coursework in GIS. The ATE program is aimed at developing new initiatives in technical education for community colleges, where surveying and computer-aided design (CAD) faculty have often taken the lead in teaching GIS.

While many programs and materials exist, there continues to be a need for more mechanisms to support program development and information sharing (Kemp and Unwin 1997). Academic organizations and conferences have provided a continuing forum for GIS educators. The semi-annual series of conferences on GIS in Higher Education beginning in 1991, but with roots back to the mid-1980s, provided GIS educators with a much-needed forum for interaction. As GIS education evolved, the series metamorphosed into various conferences on GIS and Education and GIS in Education and the focus expanded from a concern for course content and technology to pedagogical discussion and delivery mechanisms. In 1995, Woronov wrote of GIS as being in its infancy in pre-college education, and lamented the fact that there were still many logistical, pedagogical, and technical issues to be resolved. In 1997, the GISci community through the UCGIS identified a broad range of challenges that GIS education continued to face (Table 24.2; Kemp and Wright 1997). A summit on GISci education at the last GIS/LIS conference in 1998 brought together representatives from several academic, professional, government, and private organizations. The issues discussed—the appropriate curriculum content for different constituencies, accreditation and certification, methods of delivering GIS education, the roles of universities and the private sector in GIS training, educational partnerships, and distance learning in GIS—reflect the concerns of the academic community (Wright and DiBiase 2000). Many challenges remain, but the widespread extension of GIS education down into K-12 and community colleges and into professional Masters programs is a demonstration that pressures from the workplace for properly trained *and* educated personnel are mounting. Across America, high demand for young geography faculty who can teach GIS and/or GISci demonstrates how broadly the technology is taught within the discipline.

Table 24.2 *Education challenges identified by the University Consortium for Geographic Information Science in 1997*

Emerging technologies for delivering GIS education
Supporting infrastructure
Access and equity
Alternative designs for curriculum content and evaluation
Professional education
Research-based graduate GIS education
Learning with GIS
Accreditation and certification

Source: Kemp and Wright (1997).

The maturing in the perception of GIS from the 1980s view of GIS as simply a technique to the current emphasis on GIS as a field of inquiry with substantial conceptual underpinnings has raised the question of how GIS could be introduced without teaching the geographic concepts that underlie the system. Walsh (1992: 55) argued that "to know GIS one must first know geography." There has been continuing debate over the concern that "we may be failing to communicate the importance of the geography within the automation" (Posey 1993: 456). Indeed, there remains a dichotomy in GIS education about the two primary avenues through which GIS is incorporated into the curriculum: teaching about GIS, and teaching with GIS—a dichotomy that echoes completely the distinction between research with GIS and research about GIS. While there remains a need to train students in the basic principles and techniques of GIS, the strictly "technique" approach is waning in favor of embedding GIS within an applied or concept-based curriculum, ensuring that fundamental spatial literacy is a component of this general program.

It is relevant to consider four aspects of the relationship between GIS and higher education, especially in geography: geography as the home discipline of GIS; GIS as a collection of marketable skills; GIS as enabling technology for science; and GIS as an intellectual theme within geography (Kemp *et al.* 1992). GIS should no longer be seen as just a technique subject. At the tertiary level, we have come full circle with renewed interest in graduate and research education in GIS, as evidenced by its incorporation within the richer fabric of GISci (Mark 1998). At the high-school level, in addition to providing skills relevant to the workforce, GIS technology is being integrated into the curriculum as a tool to simulate real-life situations and foster skills crucial to developing higher-level thinking and problem-solving (McGarigle 1998; Ramirez 1995). Thus, by emphasizing the applied side of GIS as a means to explore substantive themes in various contexts, the concern that GIS would become a purely technical subject (King 1991), isolated from real world, societal issues has been alleviated.

Given the lack of geography programs in US high schools, the most common home for GIS education at this level remains in science and geoscience programs (Bednarz and Ludwig 1997; DeCola and El Haimus 1995; McGarigle 1997). The visualization and problem-solving abilities afforded by GIS have led to its incorporation particularly in the earth science and environmental studies areas of the curriculum. GIS is also often used as an interdisciplinary modeling tool in real-life applications (Kelly 1998; McGarigle 1997, 1998). GIS has successfully been used to heighten environmental literacy and awareness among high-school students. "Students who have an understanding of the GIS structure have excelled in the cognitive and affective domains concerning environmental issues which is an important step in achieving environmental literacy" (Ramirez 1996: 38).

These achievements augur well for an increased understanding of the value of the spatial perspective. However, the lack of preparation in geographic literacy among students and teachers remains a handicap that we are only starting to address with special programs (Bednarz and Ludwig 1997; Palladino 1998). While concerns voiced at the First National Conference on the Educational Applications of GIS (Woronov 1995), namely that GIS use in schools implies changes in classroom practice and requires high levels of technical expertise, are fading, GIS often remains a "hard sell" to busy teachers and the larger community (McGarigle 1997). While Fitzpatrick (1991: 159) noted that GIS had reached "a level where elementary students, using commonplace but powerful hardware, can easily conduct sophisticated analyses after only brief instruction," concerns about the necessarily technical nature of the subject and its place in the curriculum continue to be difficult to overcome. Indeed, the introduction of GIS into pre-college classrooms is progressing at a "frustrating pace" (Audet and Paris 1997: 293).

At all levels of education, barriers such as the lack of resources, administrative support, teacher training, and rewards, have been noted by several authors (Audet and Paris 1997; Bednarz and Ludwig 1997; Macey 1998; Murphy 1997; Posey 1993). In the 1980s, the primary barriers were technological in nature resulting from the need for high-end hardware and generally expensive software programs. In the 1990s, hardware and software improvements, plus the availability of educational pricing have greatly reduced, though not entirely eliminated, these obstacles. The provision and maintenance of the laboratory infrastructure for all education contexts remains a concern (Macey 1998; Palladino and Kemp 1991). In 1997, the UCGIS identified "Supporting Infrastructure" as one of its GIS education challenges. This challenge identified not only the obvious technical infrastructure issues, but it noted the lack of academic recognition and support that is given to those teaching this highly technical, rapidly evolving subject. At the primary- and high-school levels, the relatively low availability of pre- and post-service teacher training and of acceptance of GIS into the curriculum continue to hamper its adoption (Bednarz and Ludwig 1997). Efforts by several organizations including the NCGIA and the National Council for Geographic Education (NCGE) in concert with vendor support are making inroads into the provision of teacher training. Other initiatives, including the development of pedagogical tools such as *Urban World* (Thompson *et al.* 1997) and *The Geographer's Craft* (Foote 1997), are easing the heavy development burden that still characterizes the teaching of GIS. In addition, the World Wide Web is opening up new opportunities in GIS education, as distance learning initiatives, such as those at Pennsylvania State University <www.worldcampus.psu.edu/pub/programs/gis> and at the University of Southern California <www.usc.edu/dept/geography/learngis> come on line (websites last accessed 14 November 2002).

In the middle of the decade, Nellis (1994) set out three agendas of reform to enable the integration of spatial technologies into geographic education: building consensus about the learning and teaching of geography, training well-integrated users of technology within geographic curricula, and restructuring the discipline to bring about real change. It is now recognized that in geography, the role of technology such as GIS "can lead to more active learning and adventurous teaching" (Nellis 1994: 36). It can also be a valuable integrator. However, while Goodchild (1992*b*) called for an education system that responds rapidly to new research and is able to build new concepts quickly into new programs, he notes that "unfortunately, the higher education sector is too often characterized by conservatism, and it may take many years for new ideas to work themselves into the curriculum" (ibid. 42). As Macey (1997) notes, "the greatest impact in GIS education will be felt at the bottom, grass-roots level in the coming decade." Much remains to be done to achieve the much-heralded promise of GIS through widespread education that involves not only the technology but also the critical fundamental spatial concepts and the value of the spatial perspective.

Table 24.3 *GIS institutions established in the United States, 1988–1999*

1988	National Center for Geographic Information and Analysis established by the National Science Foundation
1990	FGDC (Federal Geographic Data Committee), an interagency committee, organized under OMB Circular A-16
1991	NSGIC (National States Geographic Information Council)
1993	Open GIS Project began
1994	NSDI (National Spatial Data Infrastructure) established by Executive Order 12906
1994	Open GIS Consortium founded
1995	University Consortium for Geographic Information Science officially established

GIS Institutions and Technological Change

The decade of the 1990s has been a period of significant change in the institutions and technologies surrounding GIS throughout the world, and American geographers have played significant roles in some of these changes. From the viewpoint of the first years of the twenty-first century, it is difficult to recall what it was like without "geographic information science," FGDC, OGC, UCGIS, and other institutions and technologies. None of the above institutions existed in 1989! Nor did the World Wide Web, whose development also has had a major influence on academia and society in general, and on the GIS community and GIS practice in particular. Although academics and other researchers had access to the Internet for email and file transfer in the 1980s, the Web was a major change in usability and extent of the Internet. The Virtual Geography Department, the NCGIA on-line Core Curricula, and the ESRI Virtual Campus mentioned above are examples of the importance of the Web in GIS and geographic education.

GIS Institutions

One of the most notable developments in GIS in the 1990s was the establishment of a large number of institutions, such as organizations, journals, conference series, etc. In 1989, there were very few formal institutions specifically focused on GIS, in the United States or elsewhere. Academic GIS activities were mainly coordinated by geographic, cartographic, or surveying organizations, and these same older fields also provided journals and conferences for publication and dissemination of results. In the government sector, mapping agencies often led GIS activities, but there was little formal coordination. In the private sector, the situation was similar. By 2002, all this had changed (Table 24.3).

The GIS Specialty Group (GISSG) of the Association of American Geographers (AAG) was itself very young when the first edition of *Geography in America* (Gaile and Willmott 1989) was written. The GISSG came into existence in 1986/7, and elected its first officers at the 1987 AAG meeting in Portland, Oregon. Though the organization neither conducts nor publishes research, its membership organizes sessions and occasionally offers workshops within the AAG Annual Meetings. It is probably the only GIS-related institution whose membership consists exclusively of (self-identified) geographers. Membership in the SG grew steadily until 1995, doubling in seven years from its inaugural membership. Absolute numbers have been roughly constant since then, and about 20 per cent of all AAG members are members of the GISSG, a testimony to the broad impacts of, and interest in, GIS within the discipline.

The National Center for Geographic Information and Analysis (NCGIA) was a very young organization in

Table 24.4 *Research initiatives of the National Center for Geographic Information and Analysis (NCGIA), 1988–1997*

1. Accuracy of spatial databases
2. Languages of spatial relations
3. Multiple representations
4. Use and value of geographic information
5. Large spatial databases
6. Spatial decision support systems
7. Visualization of spatial data quality
8. Formalizing cartographic knowledge
9. Institutions sharing geographic information
10. Spatio-temporal reasoning in GIS
(11. cancelled)
12. Integration of remote sensing and GIS
13. User interfaces for GIS
14. GIS and spatial analysis
15. Multiple roles for GIS in US global change research
16. Law, information policy, and spatial databases
17. Collaborative spatial decision-making
18. (Replaced by Conference) Spatial technologies, geographic information, and the city
19. GIS and society: the social implications of how people, space, and environment are represented in GIS
20. Interoperating GISs
21. Formal models of common-sense geographic worlds

Table 24.5 *Research initiatives of Project Varenius, 1997–1999*

Panel on Cognitive Models of Geographic Space
- Scale and detail in the cognition of geographic information
- Cognitive models of dynamic geographic phenomena and representations
- Multiple modalities and multiple frames of reference for spatial knowledge

Panel on Computational Implementations of Geographic Concepts
- Interoperating geographic information systems
- The ontology of fields
- Discovering geographic knowledge in data-rich environments

Panel on Geographies of the Information Society
- Measuring and representing accessibility in the information age
- Place and identity in an age of technologically regulated movement
- Empowerment, marginalization and public participation GIS

1989, and is mentioned just briefly by Chrisman *et al.* (1989). It was formally established in December 1988 by a research grant from the geography program of the U.S. National Science Foundation to a three-site consortium of the University of California, Santa Barbara, the State University of New York at Buffalo, and the University of Maine. From the perspective of this review, it is important to note that two of the three consortium centers are located in geography departments (at Santa Barbara and Buffalo). The original award was for five years and was subsequently extended for three additional years, ending as an NSF grant in December 1996. Since then, the NCGIA has continued as an independent research consortium.

During this initial eight-year period, the NCGIA identified twenty-one Research Initiatives intended to focus research attention on significant problem areas that impeded the growing use of GIS (see Table 24.4). Eighteen of these Initiatives included Specialist Meetings, workshops that brought together researchers in diverse but related fields of study to examine the current situation and to lay out an agenda for research. Many geographers have been important participants in these meetings and the subsequent research activities.

During this same period, the NCGIA contributed to GIS education and supported several conferences and symposia. Additional NSF funding was obtained in 1997 to initiate the Varenius Project which explicitly seeks to "advance the science of geographic information" (Goodchild *et al.* 1999). Building upon the success of their original activities, an additional set of nine Research Initiatives grouped within three thematic areas was identified (see Table 24.5). Many of these expanded upon earlier topics and all have sponsored Specialist Meetings attended by geographers.

Spurred on by the activities of the NCGIA, a series of informal meetings at conferences in 1992 and 1993 led to a workshop held in Boulder, Colorado, in December 1994 (Mark and Bossler 1995). At this workshop, forty-two participants representing thirty-three universities, one national laboratory, and one professional society (the AAG), met to create formally the University Consortium for Geographic Information Science to represent the GISci research community in the United States. Of the forty-two people who attended the Boulder workshop, thirty-two were American geographers.

The UCGIS was incorporated in 1995. A major goal of the UCGIS is to increase funding for GISci research, and it has conducted several Congressional events aimed at influencing US policy on the funding of GISci. In addition, the UCGIS has established and promoted a set of national "challenges" for GISci research and education aimed at encouraging research and financial attention to these themes (see Tables 24.1 and 2).

UCGIS had sixty-five full members in 2002, of which sixty-one were universities or university consortia. Six private firms are affiliate members. Geographers form the largest single group among more that 1,000 GISci researchers listed in UCGIS member portfolios. Three of the first five presidents of UCGIS were former chairs of the AAG GISSG.

In 1990, the US National Academy of Sciences established the Mapping Science Committee to provide independent advice to government agencies and others on geographic data and information issues. The Mapping Science Committee was active throughout the decade, and by 2001 had produced ten reports, some of which have been highly influential, including the monographs *Toward a Coordinated Spatial Data Infrastructure for the Nation* and *Distributed Geolibraries: Spatial Information Resources* (NRC 1993, 1999). More of the work of the MSC is describe in the section on GIS and Society (above).

During the 1990s, several key developments occurred within the US Federal government in the area of geographic data standards and infrastructure coordination. The Federal Geographic Data Committee (FGDC), an interagency committee organized under OMB Circular A-16, was formed in 1990 to coordinate geospatial information activities within the US Federal government. This committee is chaired by the Secretary of the Interior, and gave high visibility to GIS and geospatial data throughout the 1990s. The main purpose of FGDC is coordination across Federal agencies, and between the Federal government and governments at other levels. Thus far, FGDC has put an emphasis on standards for geographic data and for metadata. In particular, the FGDC is responsible for coordinating the National Spatial Data Infrastructure (NSDI), which was established on 11 April 1994 by Executive Order 12906. The purpose of this was to establish "a coordinated National Spatial Data Infrastructure to support public and private sector applications of geospatial data in such areas as transportation, community development, agriculture, emergency response, environmental management, and information technology" (Clinton 1994). The Executive Order charged the FGDC with the responsibility to implement NSDI by coordinating activities within the federal government and encouraging participation by state, local, and tribal governments.

Standards are a key element in such an infrastructure project. In the early 1980s, the US Geological Survey became the lead agency for geographic data standards within the Federal government. In 1992, after twelve years of development, the Spatial Data Transfer Standard (SDTS) was approved as a Federal Information Processing Standard (FIPS) and designated as FIPS 173. Compliance with SDTS is mandatory for federal agencies exchanging geospatial data outside their agencies. Responsibility for maintaining and extending SDTS also lies with the FGDC.

Other levels of government in the United States also formed organizations during the 1990s. The National States Geographic Information Council (NSGIC) held its first meeting 27–9 October 1991 in Atlanta. The majority of US states are now members of NSGIC, which has an annual meeting and is an official stakeholder organization of the FGDC. In many states there is a close relationship between a land-grant university with state extension functions and GIS within state agencies, and thus many US academic geographers play an active role in NSGIC. The National Association of Counties also coordinates some county-level GIS activities across the nation and several states now have formed statewide geographic information councils or agencies.

The private sector also became organized in the 1990s. The Open GIS Consortium (OGC) was founded in August 1994 to provide a formal structure and process for developing an interoperability specification for GIS. Primarily focused in the private sector in the United States, OGC currently has members from many countries, and from the public, private, and academic sectors. OGC is particularly active in ensuring compliance with international geospatial standards activities through the International Organization for Standardization (ISO).

Two important GIS-related organizations were formed in Europe during the same period. In 1993, the European Science Foundation (ESF) established the GISDATA Scientific Programme, which continued in operation through 1997 (Masser and Salgé 1996). Like the US NCGIA, this program identified a number of important research topics and held several research-oriented workshops. US researchers were active participants in all these workshops. As well, the Association for Geographic Information Laboratories in Europe (AGILE) was established in 1998 shortly after the formation of the UCGIS. AGILE's mandate is to promote academic teaching and research on GIS at the European level and to ensure the continuation of the networking activities that emerged through the GISDATA Programme, and through the European Union's Fifth Framework and the Directorate General XIII.

Table 24.6 *GIS journals established, 1979–1999*

1979	*Geo-Processing* (closed mid-1980s)
1987	*International Journal of Geographical Information Systems*
1990	*Cartography and Geographic Information Systems**
	*Surveying and Land Information Systems**
1994	*Geographical Systems*
1995	*Journal of Geographic Information Sciences*
1996	*Transactions in GIS*
1997	*GeoInformatica*
	*International Journal of Geographical Information Science**
1999	*Spatial Cognition and Computation*

* Name change of pre-existing journal.

Journals and Conferences

In 1989 the *International Journal of Geographical Information Systems* (*IJGIS*) was the only scholarly journal explicitly devoted to GIS, though a number of special issues devoted to GIS had appeared in journals up to that time, notably a series in *Photogrammetric Engineering and Remote Sensing*. By the end of the 1990s there were at least eight such journals, two of which were formed by name changes of cartographic or surveying journals, and five that are new (see Table 24.6). Another shift of note was the 1997 name change of *IJGIS*, with the "S" changing from "systems" to "science" at that time. Several trade magazines for GIS and related areas, such as *GIS World* (now *GEOWorld*), *Geo Info Systems* (now *Geospatial Solutions*), and *Business Geographics* (which ceased publication in 2001), also were established in the 1990s.

The 1990s also saw the establishment of a variety of conference series in GIS. One such series, the GIS/LIS meeting, went through an entire life cycle in the period covered by this book. The first GIS/LIS meeting co-sponsored by the AAG, the Urban and Regional Information Systems Association (URISA), the American Congress on Surveying and Mapping (ACSM), and the American Society for Photogrammetry and Remote Sensing (ASPRS) was held in San Antonio, Texas, in November 1988. AM/FM International, later known as Geospatial Information and Technology Association (GITA), also joined the GIS/LIS consortium, and a total of eleven meetings were held, the last in November 1998 in Fort Worth, Texas. The GIS/LIS meetings provided a single venue for GIS researchers and practitioners to exchange ideas. In 1998, some of the sponsoring organizations decided to withdraw to concentrate on their own national meetings. The field has changed a great deal since the late 1980s, and GIS/LIS, which played a critical role in these developments, was not as clearly needed in 1999.

Also in the 1990s, several specialized GIS-related international meeting series became established, while others seemed to be in decline. The Auto Carto meetings were the premier outlet for automated or analytical cartography in the 1970s and early 1980s, but the series apparently ended with Auto Carto 13 in Seattle in 1997. The Spatial Data Handling (SDH) meetings, founded under the aegis of the International Geographical Union's Commission on Spatial Data Handling, began in 1984 and have been held in alternate years, covering all technical and theoretical aspects of GIS. A meeting called GIScience 2000 was held in Savannah, Georgia, at the end of October 2000, and it can be expected to become a meeting series if the follow-up meeting in Boulder, Colorado, in September 2002 is judged to be a success.

The decline of generic or broadly based GIS-related meetings may be due in part to the emergence of strong new meetings of a more specialized nature. The first international symposium on large spatial databases was held in 1989, and subsequently has met in odd-numbered years under the acronym SSD, producing each time a fully-refereed volume of papers published in Springer-Verlag's Lecture Notes in Computer Science series (cf. Buchmann *et al.* 1989). This conference brings together more computational GIS researchers on the one hand and the spatial and spatio-temporal database communities from computer science and engineering on the other. Similarly, another research community, focused on spatial cognition and geographic reasoning, crystallized with a meeting held in Spain in 1990 (Mark and Frank 1991). This was followed by another stand-alone meeting on cognition, computation, and GIS in Italy in 1992 (Frank *et al.* 1992), and the next year the first Conference on Spatial Information Theory (COSIT '93; Frank and Campari 1993) was held, also in Italy. This also became a series, with meetings in Austria (1995), Pennsylvania (1997), Germany (1999), and California (2001). Other important conferences to appear in this decade were in the Integrating GIS and Environmental Modeling series (1991, 1993, 1996, and 2000) and the international GeoComputation series (held annually since 1996). A series of international meetings on spatial data accuracy in natural resource applications began in 1994 with a meeting in Williamsburg, Va. This is a biennial series that includes geographers in collaboration with spatial statisticians and scientists in natural resources. The first International Symposium on Geographic Information and Society, held in Minneapolis in June 1999, may turn into a parallel international conference series for this growing research area.

Summary and Prospects

The transition from a society based on paper maps to one based on digital geographic information has not been carefully planned. It is impossible to predict scientific discoveries, and in a field where the technology exhibits obsolescence within months and where human skills may be mismatched with technology within a year or two, the dangers of forecasting are especially great. Nevertheless, we provide some speculations about the future of this important area of geography.

We can be confident that the price-performance of computing will continue to improve. It seems equally likely, however, that the sizes of datasets will grow to use or exceed the new capacity of the improved systems. The exponential growth of digital geographic information will continue, and will generate concomitant demand for professionals at all levels, people who understand its uses and foundations. Nevertheless, use of geographic information and associated technology by untrained, or even unaware, individuals will increase even more dramatically. Geographic technology and services will be embedded in the artefacts of everyday life—automobiles, personal digital assistants, websites, and communication devices. The growth of wireless communication technologies will allow every appliance to have access to the Web, and to have its position in geographic space known to the networks, creating an entirely new geography of technology and users. Academic geography and geographers will be competing with other spatially enabled disciplines to explore and explain the actual and virtual worlds we will be creating. Unless students of geography maintain very high levels of technical ability and instrumentation, it is likely that the contribution of geography to GISci will diminish over the next decade. The transition from analog to digital, working itself out in so many arenas, will inevitably generate more geographies of the information society, providing the discipline with numerous, new intellectual challenges.

In the technical area, it is clear that our representations of geographic phenomena must be extended and enhanced. The ability to handle intensively sampled spatio-temporal information, including motion, change, and dynamic fields, will be one focus, and true three-dimensional GIS will be another. Data mining and knowledge discovery techniques will be needed to help scientists, decision-makers, and the public in general make sense of the exceptional quantities of relatively raw or uninterpreted data that will be available on the Internet within the next decade. The impact of interoperability, object-component systems, and open systems will continue to expand, with increasing focus developing on semantic interoperability.

Complex, two-way interactions between geographic information technologies and society will continue to require basic and applied research, which will feed into policy decisions and new laws. The GIS and society debate will broaden as greater and more diverse segments of society come in contact with digital geographic information. Additionally, the GIS and society debate will deepen as new models of representation are introduced, especially those that are able to capture the dynamic and temporal behavior of individuals. The debate on the pros and cons of surveillance and the meaning of privacy in a spatially enabled society will crescendo, but will not be resolved within a decade. Geographers will be well placed to lead in research on this critical dimension of GISci. We will be challenged on the one hand to make positive contributions to systems and services for the citizen, for example in public administration, health, and for special-needs groups such as the elderly and disabled. On the other hand, we will be tasked to provide sound theory and epistemology for a science that changes almost as quickly as it is recorded.

For geographers who study GIS and who use it in their research, the challenge will be to maintain a focus and to provide understanding and developments based on geography's long experience with exploring changing worlds. Multidisciplinary explanation will become the norm, yet geographers should strive to continue to provide a distinctive contribution. The entrance of GIS into primary and secondary schools promises the development of a generation of spatially literate students. Similarly, spatial skills are being acquired rapidly in business and government. A wider proportion of the American public is beginning to be aware that "Geography Matters." The wider use of GIS across many segments of society coupled with rigorous geographic education could portend a richer appreciation of the discipline in 2010 than today.

References

Adams, P. C., and Warf, B. (eds.) (1997) *The Geographical Review*, Special Issue, 87/2.

Anselin, L., Dodson, R. F., and Hudak, S. (1993). "Linking GIS and Spatial Data Analysis in Practice." *Geographical Systems*, 1: 2–23.

Arctur, D., Hair, D., Timson, G., Martin, E. P., and Fegeas, R. (1998). "Issues and Prospects for the Next Generation of the Spatial Data Transfer Standard (SDTS)." *International Journal of Geographical Information Science*, 12/4: 403–26.

Armstrong, M. P. (1993). "On Automated Geography!" *Professional Geographer*, 45/4: 440–2.

— (1994). "Requirements for the Development of GIS-Based Group Decision Support Systems." *Journal of the American Society for Information Science*, 45/9: 669–77.

Armstrong, M. P., and Densham, P. J. (1992). "Domain Decomposition for Parallel Processing of Spatial Problems." *Computers, Environment, and Urban Systems*, 16: 497–513.

Armstrong, M. P., and Marciano, R. J. (1996). "Local Interpolation Using Distributed Parallel Supercomputer." *International Journal of Geographical Information Systems*, 10/6: 713–29.

Armstrong, M. P., Rushton, G., and Honey, R. (1991). "Decision Support for Regionalization: A Spatial Decision Support System for Regionalizing Service Delivery Systems." *Computers, Environment and Urban Systems*, 15/1–2: 37.

Audet, R. H., and Paris, J. (1997). "GIS Implementation Model for Schools: Assessing the Critical Concerns." *Journal of Geography*, 96/4: 293–300.

Bailey, T. C., and Gatrell, A. C. (1995). *Interactive Spatial Data Analysis*. New York: John Wiley & Sons.

Beard, M. K., and Mackaness, W. A. (1993). "Visual Access to Spatial Data Quality." *Cartographica*, 30/3: 37–45.

Beard, K. M., Clapham, S., and Buttenfield, B. P. (1991). *Initiative 7 Specialist Meeting: Visualization of Spatial Data Quality*. Technical Report 91–26. Santa Barbara, Calif.: National Center for Geographic Information Analysis.

Bednarz, S. W., and Ludwig, G. (1997). "Ten Things Higher Education Needs to Know about GIS in Primary and Secondary Education." *Transactions in GIS*, 2/2: 123–33.

Bian, L., and Walsh, S. J. (1993). "Scale Dependencies of Vegetation and Topography in a Mountainous Environment of Montana." *Professional Geographer*, 45/1: 1–11.

Blaut, J. M., and Stea, D. (1971). "Studies in Geographic Learning." *Annals, Association of American Geographers*, 61: 387–93.

Blennow, K., and Persson, P. (1998). "Modelling Local-Scale Frost Variations Using Mobile Temperature Measurements with a GIS." *Agricultural and Forest Meteorology*, 89/1: 59.

Brown, D. G. (1998). "Classification and Boundary Vagueness in Mapping Presettlement Forest Types." *International Journal of Geographical Information Science*, 12/2: 105–29.

Brown, D. G., and Bara, T. J. (1994). "Recognition and Reduction of Systematic Error in Elevation and Derivative Surfaces from 7-1/2 Minute DEMs." *Photogrammetric Engineering and Remote Sensing*, 60/2: 189–94.

Brown, D. G., Lusch, D. P., and Duda, K. A. (1998). "Supervised Classification of Glaciated Landscape Types Using Digital Elevation Data." *Geomorphology*, 21/3–4: 233–50.

Brunn, S. D. (1995). "Geography's Research Performance Based on Annals Manuscripts, 1987–1993." *Professional Geographer*, 47/2: 204–15.

Buchmann, A., Gunther, O., Smith, T., and Wang, Y. (eds.) (1989). *Symposium on the Design and Implementation of Large Spatial Databases*. Lecture Notes in Computer Science, 409. Berlin: Springer-Verlag, 271–86.

Burrough, P. A. (1986). *Principles of Geographical Information Systems for Land Resources Assessment*. Oxford: Clarendon Press.

— (ed.) (1990). "Methods of Spatial Analysis in GIS." *International Journal of Geographical Information Systems*, Special Issue, 4/3.

Burrough, P. A., and Frank, A. U. (eds.) (1996). *Geographic Objects with Indeterminate Boundaries. GISDATA II*. London: Taylor & Francis.

Burrough, P. A., and McDonnell, R. A. (1998). *Principles of Geographic Information Systems*. 2nd edn. New York: Oxford University Press.

Butler, D. R., and Walsh, S. J. (1990). "Lithologic, Structural, and Topographic Influences on Snow Avalanche Patch Location in Eastern Glacier National Park, Montana." *Annals, Association of American Geographers*, 80/3: 362–78.

Calkins, H. W., and Weatherbe, R. (1995). "Taxonomy of Spatial Data Sharing," in H. Onsrud and G. Rushton, (eds.), *Sharing Geographic Information*. New Brunswick: Center for Urban Policy Research, 65–75.

Can, A. (1992). "Residential Quality Assessment: Alternative Approaches Using GIS." *Annals of Regional Science*, 26: 97–110.

Carver, S. J. (1991). "Integrating Multi-Criteria Evaluation with Geographical Information Systems." *International Journal of Geographical Information Systems*, 5/3: 321–39.

Carver, S., Blake, M., Turton, I., and Duke-Williams, O. (1997). "Open Decision-Making: Evaluating the Potential of the World Wide Web," in Z. Kemp (ed.), *Innovations in GIS*, 4. New York: Taylor & Francis, 267–77.

Chrisman, N. (1997). *Exploring Geographic Information Systems*. New York: John Wiley & Sons.

— (1998). "Academic Origins of GIS," in Foresman (1998: 3–43).

Chrisman, N. R., Dougenik, J. A., and White, D. (1992). "Lessons from the Design of Polygon Overlay Processing from the ODYSSEY Whirlpool Algorithm." *Proceedings 5th International Symposium on Spatial Data Handling*, 2: 401–10.

Chrisman, N. R., Cowen, D. J., Fisher, P. F., Goodchild, M. F., and Mark, D. M. (1989). "Geographic Information Systems," in Gaile and Willmott (1989: 76–96).

Clarke, K. C. (1997). *Getting Started with Geographic Information Systems*. Upper Saddle River, NJ: Prentice Hall.

Clarke, K. C., and Gaydos, L. J. (1998). "Loose-Coupling a Cellular Automaton Model and GIS: Long-Term Urban Growth Prediction for San Francisco and Washington/Baltimore." *International Journal of Geographical Information Science*, 12/7: 699–714.

Clinton, W. J. (1994). *Coordinating Geographic Data Acquisition and Access: The National Spatial Data Infrastructure*. Office of the President, Executive Order 12906.

Congalton, R. G. (ed.) (1994). *International Symposium on Spatial Accuracy of Natural Resource Data Bases: "Unlocking the Puzzle."* Conference Proceedings, 16–20 May 1994. Washington, DC: American Society for Photogrammetry and Remote Sensing.

Cook, D., Cressie, N., Majure, J., and Symanzik, J. (1994). "Some Dynamic Graphics for Spatial Data (with Multiple Attributes) in a GIS." *Proceedings in Computational Statistics, 11th Symposium*, Heidelberg: Physica-Verlag, 105–19.

Couclelis, H. (1992). "People Manipulate Objects (but Cultivate Fields): Beyond the Raster-Vector Debate in GIS," in Frank, Campari, and Formentini (1992: 65–77).

—— (1996). "Towards an Operational Typology of Geographic Entities with Ill-Defined Boundaries," in Burrough and Frank (1996: 45–55).

Craig, W. J., Harris, T. M., and Weiner, D. (eds.) (1999). *Empowerment, Marginalization and Public Participation GIS*, Project Varenius Report. Santa Barbara, Calif.: National Center for Geographic Information and Analysis.

Curry, M. R. (1995). "Geographic Information Systems and the Inevitability of Ethical Inconsistency," in J. Pickles (1995: 68–87).

—— (1997). "The Digital Individual and the Private Realm." *Annals, Association of American Geographers*, 87/4: 681–99.

—— (1999). "Rethinking Privacy in a Geocoded World," in Longley, Maguire, Goodchild, and Rhind (1999: 757–66).

Dahlberg, R. E., and Estes, J. E. (1982). "Education in Cartography and Remote Sensing for Applied Geography," in J. W. Frazier (ed.), *Applied Geography. Selected Perspectives*. Englewood Cliffs, NJ: Prentice Hall, 235–62.

Davis, B. E. (1996). *GIS: A Visual Approach*. Santa Fe, NM: OnWord Press.

De Cola, L., and El Haimus, A. (1995). "Teaching Geographic Concepts with GIS." *GIS World*, 8/5: 68–72.

DeMers, M. N. (1997). *Fundamentals of Geographic Information Systems*. New York: John Wiley & Sons.

Dobson, J. E. (1983). "Automated Geography." *Professional Geographer*, 35: 135–43.

—— (1993). "The Geographic Revolution: A Retrospective on the Age of Automated Geography." *Professional Geographer*, 45/4: 431–9.

Dobson, J. E., Rush, R. M., and Peplies, R. W. (1990). "Forest Blowdown and Lake Acidification." *Annals, Association of American Geographers*, 80/3: 343–61.

Downs, R. M. (1997). "The Geographic Eye: Seeing Through GIS?" *Transactions in GIS*, 2/2: 111–22.

Downs, R. M., and Stea, D. (1973). *Image and the Environment: Cognitive Mapping and Spatial Behavior*. Chicago: Aldine.

Eastman, J. R., Jin, W., Kyem, P. A. K., and Toledano, J. (1995). "Raster Procedures for Multi-Criteria/Multi-Objective Decisions." *Photogrammetric Engineering and Remote Sensing*, 61/5: 539–47.

Egbert, S. L., and Slocum, T. A. (1992). "Exploremap: An Exploration System for Choropleth Maps." *Annals, Association of American Geographers*, 82/2: 275–88.

Egenhofer, M. (1992). "Why not SQL!" *International Journal of Geographical Information Systems*, 6/2: 71–85.

—— (1997). "Query Processing in Spatial-Query-by-Sketch." *Journal of Visual Languages and Computing*, 8/4: 403–24.

Egenhofer, M. J., and Golledge, R. G. (eds.) (1998). *Spatial and Temporal Reasoning in Geographic Information Systems*. Oxford: Oxford University Press.

Egenhofer, M., and Franzosa, R. (1991). "Point-Set Topological Spatial Relations." *International Journal of Geographical Information Systems*, 5/2: 161–74.

Egenhofer, M., Glasgow, J., Gunther, O., Herring, J., and Peuquet, D. (1999). "Progress in Computational Methods for Representing Geographic Concepts.' *International Journal of Geographical Information Science*, 13/8: 775–96.

Emmi, P. C., and Horton, C. A. (1995). "A Monte Carlo Simulation of Error Propagation in a GIS-Based Assessment of Seismic Risk." *International Journal of Geographical Information Systems*, 9/4: 447–61.

Epstein, E. F., Hunter, G. J., and Agumya, A. (1998). "Liability Insurance and the Use of Geographical Information." *International Journal of Geographical Information Science*, 12/3: 203.

Epstein, E. F., Tulloch, D. L., Niemann, B. J., Ventura, S. J., and Limp, F. W. (1996). "Comparative Study of Land Records Modernization in Multiple States." *Proceedings GIS/LIS '96*. Bethesda, Md.: American Society for Photogrammetry and Remote Sensing.

Fischer, M. M. (1999). "Spatial Analysis: Retrospective and Prospect," in Longley, Maguire, Goodchild, and Rhind (1999: 283–92).

Fischer, M. M., and S. Gopal. (1993). "Neurocomputing—a New Paradigm for Geographic Information Processing." *Environment & Planning A*, 25/N6: 757–60.

Fischer, M., Scholten, H. J., and Unwin, D. (eds.) (1996). *Spatial Analytical Perspectives on GIS*. New York: Taylor & Francis.

Fisher, P. F. (1990). "Simulation of Error in Digital Elevation Models." *Papers and Proceedings of Applied Geography Conferences*, 13: 37–43.

—— (1992). "First Experiments in Viewshed Uncertainty: Simulating Fuzzy Viewsheds." *Photogrammetric Engineering and Remote Sensing*, 58: 345–52.

—— (1994). "Probable and Fuzzy Models of the Viewshed Operation," in M. Worboys (ed.), *Innovations in GIS*. London: Taylor & Francis, 161–75.

—— (1997*a*). "Editorial: Welcome to the International Journal of Geographical Information Science." *International Journal of Geographical Information Science*, 11/1: 1–3.

—— (1997*b*). "The Ethics of Six Actors in the Geographical Information Systems Arena," in Z. Kemp (ed.), *Innovations in GIS*, 4. New York: Taylor & Francis, 227–33.

Fitzpatrick, C. (1991). "Teaching Geography with Computers." *Journal of Geography*, 92/4: 156–9.

Foote, K. E. (1997). "The Geographer's Craft: Teaching GIS in the Web." *Transactions in GIS*, 2/2: 137–50.

Forer, P., and Unwin, D. (1999). "Enabling Progress in GIS and Education," in Longley, Maguire, Goodchild, and Rhind (1999: 747–56).

Foresman, T. W. (ed.) (1998). *The History of Geographic Information Systems: Perspectives from the Pioneers*. Upper Saddle River, NJ: Prentice Hall.

Fotheringham, S., and Rogerson, P. (eds.) (1994). *Spatial Analysis and GIS*. New York: Taylor & Francis.

Frank, A. U., and Campari, I. (eds.) (1993). *Spatial Information Theory: A Theoretical Basis for GIS*. Lecture Notes in Computer Sciences, 716. Berlin: Springer-Verlag.

Frank, A. U., Campari, I., and Formentini, U. (eds.) (1992). *Theories and Methods of Spatio-Temporal Reasoning in Geographic Space*. Lecture Notes in Computer Science, 639. Berlin: Springer-Verlag.

Franklin, J. (1995). "Predictive Vegetation Mapping: Geographic Modelling of Biospatial Patterns in Relation to Environmental Gradients." *Progress in Physical Geography*, 19/4: 474–99.

Gaile, G. L., and Willmott. C. J. (1989). *Geography in America*. Columbus, Ohio: Merrill.

Golledge, R. G., Gale, N., Pellegrino, J. W., and Doherty, S. (1992). "Spatial Knowledge Acquisition by Children: Route Learning and Relational Distances." *Annals, Association of American Geographers*, 82/2: 223–44.

Golledge, R., and Zannaras, G. (1973). "Cognitive Approaches to the Analysis of Human Spatial Behavior," in W. Ittleson (ed.), *Environment and Cognition*. New York: Academic Press.

Goodchild, M. F. (1989). "Modeling Error in Objects and Fields," in Goodchild and Gopal (1989: 107–13).

—— (1992*a*). "Geographical Data Modeling." *Computers and Geosciences*, 18: 401–18.

—— (1992*b*). "Geographical Information Science." *International Journal of Geographical Information Systems*, 6/1: 31–45.

Goodchild, M., and Gopal, S. (1989). *The Accuracy of Spatial Databases*. New York: Taylor & Francis.

Goodchild, M. F., and Kemp, K. K. (eds.) (1992). *NCGIA Core Curriculum in GIS*. Santa Barbara, Calif.: National Center for Geographic Information and Analysis.

Goodchild, M. F., Egenhofer, M., and Fegeas, R. (eds.) (1999). *Interoperating Geographic Information Systems*. Dordrecht: Kluwer.

Goodchild, M. F., Guoqing, S., and Shiren, Y. (1992). "Development and Test of an Error Model for Categorical Data." *International Journal of Geographical Information Systems*, 6/2: 87–104.

Goodchild, M. F., Parks, B. O., and Steyaert, L. T. (eds) (1993). *Environmental Modeling with GIS*. New York: Oxford University Press.

Goodchild., M. F., Egenhofer, M. J., Kemp, K. K., Mark, D. M., and Sheppard, E. (1999). "Introduction to the Varenius Project." *International Journal of Geographic Information Science*, 13/8: 731–45.

Goodchild, M. F., Steyaert, L. T., Parks, B. O., Johnston, C., Maidment, D., Crane, M., and Glendinning, S. (eds) (1996). *GIS and Environmental Modeling: Progress and Research Issues*. Fort Collins, Col.: GIS World.

Griffith, D. A. (1993). "Advanced Spatial Statistics for Analyizing and Visualizing Geo-referenced Data." *International Journal of Geographical Information Systems*, 7/2: 107–23.

Griffith, D. A., Lewis, R., Li, B., Vasiliev, I., Knight, S., and Yang, X. (1990). "Developing Minitab Software for Spatial Statistical Analysis: A Tool for Education and Research." *Operational Geographer*, 8: 28–33.

Harris, T. M., and Weiner, D. (1996). *GIS and Society: The Social Implications of How People, Space and Environment are Represented in GIS*. Technical Report 96–7. Santa Barbara, Calif.: National Center for Geographic Information and Analysis.

—— (1998). "Empowerment, Marginalization, and 'community-integrated' GIS." *Cartography and Geographic Information Systems*, 25: 67–76.

Harrower, M., MacEachren, A. M., and Griffin, A. (2000). "Developing a Geographic Visualization Tool to Support Earth Science Learning." *Cartography and Geographic Information Science*, 27/4: 279–93.

Healey, R. G. (1996). "Editorial: Special Issue on Parallel Processing in GIS." *International Journal of Geographical Information Systems*, 10/6: 667–8.

Hearnshaw, H. M., and Unwin, D. J. (eds.) (1994). *Visualization and Geographic Information Systems*. New York: John Wiley & Sons.

Heuvelink, G. B. M. (1993). *Error Propagation in Quantitative Spatial Modelling: Applications in Geographical Information Systems*. Utrecht: Netherlands Geographical Studies, University of Utrecht.

Hewitson, B. C., and Crane, R. G. (eds.) (1994). *Neural Nets: Applications in Geography*. Dordrecht: Kluwer.

Heywood, I., Cornelius, S., and Carver, S. (1998). *An Introduction to Geographic Information Systems*. New York: Addison Wesley Longman.

Hunter, G. J., and Goodchild, M. F. (1995). "Dealing with Error in Spatial Databases: A Simple Case Study." *Photogrammetric Engineering and Remote Sensing*, 61/5: 529–37.

Jankowski, P., and Richard, L. (1994). "Integration of GIS-based Suitability Analysis and Multicriteria Evaluation in a Spatial Decision Support System for the Route Selection." *Environment and Planning* B, 21/3: 323.

Jensen, J., Saalfeld, A., Broome, F., Cowen, D., Price, K., Ramsey, D., and Lapine, L. (1998). *Spatial Data Acquisition and Integration*. UCGIS White Paper ‹www.cla.sc.edu/geog/ucgis/›, last accessed 14 November 2002.

Johnson, A. (1998). "Filling a Vital Niche in GIS Professional Education GIS Programs at Community Colleges." *ESRI ARC News*, 20/2: 32.

Johnson, M. L. (1998). "GIS in Business: Issues to Consider in Curriculum." *Journal of Geography*, 97/3: 98–105.

Jordan, T. (1988). "The Intellectual Core: President's Column." *AAG Newsletter*, 23: 1.

Kavanaugh-Brown, J. (1998). "Students Contribute to Communities." *Government Technology*, 11/10: 48.

Kelly, D. (1998). "Have a Problem? Ask Greenbrier High School Students to Solve It." *Converge*, 1/4: 14–18.

Kemp. K. K. (1997). "The NCGIA Core Curriculum in GIS and Remote Sensing." *Transactions in GIS*, 2/2: 181–90.

Kemp K. K., and Unwin, D. J. (1997). "From Geographic Information Systems to Geographic Information Studies: An Agenda for Educators." *Transactions in GIS*, 2/2: 103–9.

Kemp, K. K., and Wright, R. (1997). "UCGIS identifies GIScience Education Priorities." *Geo Info Systems*, 7/9: 16–20.

Kemp, K., Goodchild, M., and Dodson, R. (1992). "Teaching GIS in Geography." *The Professional Geographer*, 44: 11–191.

King, G. Q. (1991). "Geography and GIS Technology." *Journal of Geography*, 90/2: 66–72.

Kirkby, S. D. (1996). "Integrating a GIS with an Expert System to Identify and Manage Dryland Salinization." *Applied Geography*, 16/4: 289.

Lam, N., and DeCola, L. (eds.) (1993). *Fractals in Geography*. Englewood Cliffs, NJ: Prentice Hall.

Lam, N. and Quattrochi, D. A. (1992). "On the Issues of Scale, Resolution, and Fractal Analysis in the Mapping Sciences." *Professional Geographer*, 44: 88.

Land Information and Computer Graphics Facility (2001). "Web-Based Community Planning for the Citizen Planner," *Land Information Bulletin*, National Consortium for Rural Geospatial Innovations. Madison: University of Wisconsin.

Langran, G. (1992). *Time in Geographic Information Systems*. London: Taylor & Francis.

Lanter, D. P., and Veregin, H. (1992). "A Research Paradigm for Propagating Error in Layer-Based GIS." *Photogrammetric Engineering and Remote Sensing*, 58: 825–33.

Leung, Y., Leung, K. S., He, J. Z. (1999). "A Generic Concept-Based Object-Oriented Geographical Information System." *International Journal of Geographical Information Systems*, 13/5: 475–98.

Lin, H., and Zhang, L. (1998). "Internet-Based Investment Environment Information System: A Case Study on BKR of China." *International Journal of Geographical Information Science*, 12/7: 715–25.

Lloyd, R. E. (1997). *Spatial Cognition: Geographic Environments*. Dordrecht: Kluwer.

Longley, P. A., Maguire, D. J., Goodchild, M. F., and Rhind, D. W. (eds.) (1999). *Geographical Information Systems*, i. *Principles and Technical Issues*; ii. *Management Issues and Applications*. 2nd edn. New York: John Wiley & Sons.

Longley, P. A., Goodchild, M. F., Maguire, D. J., and Rhind, D. W. (2001). *Geographic Information Systems and Science*. New York: John Wiley & Sons.

Lowell, K., and Jaton, A. (eds.) (1999). *Spatial Accuracy Assessment: Land Information Uncertainty in Natural Resources*. Chelsea, Mich.: Ann Arbor Press.

MacEachren, A. M. (1992*a*). "Application of Environmental Learning Theory to Spatial Knowledge Acquisition from Maps." *Annals, Association of American Geographers*, 82/2: 245–74.

—— (1992*b*). "Visualizing Uncertain Information." *Cartographic Perspectives*, 13: 10–19.

MacEachren, A. M., and Kraak, M.-J. (eds.) (1997). "Special Electronic Issue on Exploratory Cartographic Visualization." *Computers & Geosciences*, 23/4.

MacEachren, A. M., and Monmonier, M. (1992). "Introduction: Special Issue on Geographic Visualization". *Cartography & GIS*, 19/4: 197–200.

MacEachren, A. M., and Taylor, D. R. F. (eds.) (1994). *Visualization in Modern Cartography*. Oxford: Pergamon.

MacEachren, A. M., Boscoe, F. P., Haug, D., Pickle, L. W. (1998). "Geographic Visualization: Designing Manipulable Maps for Exploring Temporally Varying Georeferenced Statistics." *Proceedings of the IEEE Information Visualization Symposium*, Research Triangle Park, NC, 19–20 Oct. 1998, 87–94.

MacEachren, A. M., Buttenfield, B., Campbell, J., DiBiase, D., and Monmonier, M. (1992). "Visualization," in R. Abler, M. Marcus, and J. Olson (eds.), *Geography's Inner Worlds*. Rutgers, NJ: Rutgers University Press, 99–137.

Macey, S. M. (1997). "GIS Education: From the Top Down and the Bottom Up." *Geo Info Systems*, 7/4: 16–17.

—— (1998). "Managing a Computer Teaching Laboratory." *Journal of Geography*, 97/1: 13–18.

McGarigle, B. (1997). "GIS Goes to School." *Government Technology*, 10/9: 28–9.

—— (1998). "HAZMAT HIGH: High School Students Learn GIS and Help with Emergency Planning." *Converge*, 1/3: 16–20.

McKee, L. (1999). "Catch the Integrated Geoprocessing Trend." *GEOWorld*, 12/2: 34.

Maguire, D. J., Goodchild, M. F., and Rhind, D. W. (eds.) (1991). *Geographical Information Systems: Principles and Applications*. Harlow: Longman Scientific and Technical.

Marble, D. F., and Peuquet, D. J. (1993). "The Computer and Geography: Ten Years Later." *Professional Geographer*, 45/4: 446–8.

Mark, D. M. (1997). "The History of Geographic Information Systems: Invention and Re-invention of Triangulated Irregular Networks (TINS)." *Proceedings, GIS/LIS'97*. Bethesda, Md.: American Society for Photogrammetry and Remote Sensing.

—— (1998). "Integrative Graduate Education and Research Training in Geographic Information Science." *Proceedings of the GIS/LIS*. Fort Worth, Tex.: 618–24. CD-ROM.

Mark, D. M., and Bossler, J. (1995). "The University Consortium for Geographic Information Science." *Geo Info Systems*, 5/4: 38–9.

Mark, D. M., and Csillag, F. (1989). "The Nature of Boundaries on "Area-Class" Maps." *Cartographica*, 26/1: 65–78.

Mark, D. M., and Egenhofer, M. J. (1994). "Modeling Spatial Relations between Lines and Regions: Combining Formal Mathematical Models and Human Subjects Testing." *Cartography and Geographic Information Systems*, 21/4: 195–212.

Mark, D. M., and Frank, A. U. (1991). *Cognitive and Linguistic Aspects of Geographic Space*. Dordrecht: Kluwer.

Mark, D. M., and Gould, M. D. (1991). "Interacting with Geographic Information: A Commentary." *Photogrammetric Engineering and Remote Sensing*, 57/11: 1427–30.

Mark, D. M., Freksa, C., Hirtle, S. C., Lloyd, R., and Tversky, B. (1999). "Cognitive Models of Geographic Space." *International Journal of Geographic Information Science*, 13/8: 747–74.

Masser, I., and Salgé, F. (1996). "The GIS Data Series: Editors' Preface," in Burrough and Frank (1996: pp. ix–x).

Medyckyj-Scott, D., and Hearnshaw, H. M. (eds.) (1993). *Human Factors in Geographical Information Systems*. London: Belhaven Press.

Miller, E. J. (1997). "Towards a 4D GIS: Four-Dimensional Interpolation Utilizing Kriging," in Z. Kemp (ed.), *Innovations in GIS*, 4. New York: Taylor & Francis, 181–197.

Miller, H. J. (1991). "Modeling Accessibility Using Space-Time Prism Concepts Within a GIS." *International Journal of Geographic Information Systems*, 5/3: 287–301.

Miller, H. J., and Han, J. (eds.) (2001). *Geographic Data Mining and Knowledge Discovery*. London: Taylor & Francis.

Moellering, H. (1992). "Opportunities for Use of the Spatial Data Transfer Standard at the State and Local Level." *Cartography and Geographic Information Systems*, 19/5: 332–4.

Mower, J. E. (1996). "Developing Parallel Procedures for Line Simplification." *International Journal of Geographical Information Science*, 10/6: 699–712.

Mowrer, H. T., Czaplewski, R. L., and Hamre, R. H. (eds.) (1996). *Spatial Accuracy Assessment in Natural Resources and Environmental Sciences: Second International Symposium*. Conference Proceedings, 21–3 May 1996. General Technical Report RM-GTR-277. Fort Collins, Col.: Rocky Mountain Forest and Range Experiment Station.

Murphy, L. D. (1997). "Why Aren't Business Schools Teaching Business Geographics?" *Business Geographics*, 5/2: 24–6.

NCGIA (1996). *Proceedings, Third International Conference/ Workshop on Integrating GIS and Environmental Modeling.*

Santa Fe, NM, 21–6 January 1996. Santa Barbara, Calif.: National Center for Geographic Information and Analysis.

Nebert, D. (1999). "Interoperable Spatial Data Catalogs." *Photogrammetric Engineering and Remote Sensing*, 65/5: 573–5.

Nellis, M. D. (1993). "The New Geography: ARCVIEW for Teaching Geography." *Geographic Insights*, 3: 6.

— (1994). "Technology in Geographic Education: Reflections and Future Directions." *Journal of Geography*, 93/1: 36–9.

NRC (1993). *Towards a Coordinated Spatial Data Infrastructure for the Nation*. Mapping Science Committee. Washington, DC: National Research Council.

— (1994). *Promoting the National Spatial Data Infrastructure Through Partnerships*. Mapping Science Committee. Washington, DC: National Research Council.

— (1999). *Distributed Geolibraries: Spatial Information Resources, Summary of a Workshop*. Washington, DC: National Academy Press.

— (2001). *National Spatial Data Infrastructure Partnership Programs: Rethinking the Focus*. Washington, DC: National Academy Press.

Novak, K. (1995). "Mobile Mapping Technology for GIS Data Collection." *Photogrammetric Engineering and Remote Sensing*, 61/5: 493.

Obermeyer, N. J. (ed.) (1998). "Public Participation GIS." *Cartography and Geographic Information Systems*, Special Issue, 25/2.

Oliver, M., Webster, R., and Gerrard, J. (1989). "Geostatistics in Physical Geography. Part II: Applications." *Transactions of the Institute of British Geographers*, 14: 270–86.

Olson, J. M., and Brewer, C. A. (1997). "An Evaluation of Color Selections to Accommodate Map Users with Color Vision Impairments." *Annals of the Association of American Geographers*, 87/1: 103–34.

Onsrud, H. J. (1993). "Identifying Unethical Conduct in the Use of GIS." *Cartography and Information Systems*, 22/1: 90.

— (1995). "The Role of Law in Impeding and Facilitating the Sharing of Geographic Information," in Onsrud and Rushton (1995: 292–306).

Onsrud, H. J., and Rushton, G. (eds.) (1995). *Sharing Geographic Information*. New Brunswick, NJ: Center for Urban Policy Research Press (CUPR).

Openshaw, S. (1991). "A View on the GIS Crisis in Geography, or Using GIS to Put Humpty Dumpty Back Together Again." *Environment and Planning* A, 23: 621–8.

Padgett, D. A., Li, L., and Mignolo, A. (2000). "Developing a GIS Curriculum. The Small Liberal Arts College Experience." *Geo Info Systems*, 10/4: 36–9.

Palladino, S. D. (1998). *K-14: A Review of Eight Years of NCGIA Efforts*. ‹www.ncgia.ucsb.edu/~spalladi/documents/ESRI.html›, last accessed 14 November 2002.

Palladino, S. D., and Kemp, K. K. (eds.) (1991). *GIS Teaching Facilities: Six Case Studies on the Acquisition and Management of Laboratories*. Technical Paper 91–21. Santa Barbara, Calif.: National Center for Geographic Information and Analysis.

Palm, R., and Hodgson, M. (1992). "Earthquake Insurance: Mandated Disclosure and Homeowner Response in California." *Annals, Association of American Geographers*, 82/2: 207–22.

Peng, Z.-R. (1999). "An Assessment Framework for the Development of Internet GIS," *Environment and Planning B*, 26/1: 117.

Peuquet, D. J. (1994). "It's About Time: A Conceptual Framework for the Representation of Temporal Dynamics in Geographic Information Systems." *Annals, Association of American Geographers*, 84/3: 441–61.

Peuquet, D. J., Smith, B., and Brogaard, B. (1999). *The Ontology of Fields*. Santa Barbara, Calif.: National Center for Geographic Information and Analysis.

Phoenix, M. (1996). "GIS's Role in Higher Education." *Geo Info Systems*, 6/5: 50–1.

Pickles, J. (1991). "Geography, G.I.S., and the Surveillant Society." *Papers and Proceedings of the Applied Geography Conference*, 14: 80–91.

— (1993). "Discourse on Method and the History of Discipline: Reflections on Dobson's 1983 Automated Geography." *Professional Geographer*, 45/4: 451–5.

— (ed.) (1995). *Ground Truth: The Social Implications of Geographic Information Systems*. New York: Guilford Press.

— (1997). "Tool or Science? GIS, Technoscience, and the Theoretical Turn." *Annals, Association of American Geographers*, 87/2: 363–72.

— (1999). "Arguments, Debates and Dialogues: The GIS – Social Theory Debate and the Concern for Alternatives," in Longley, Maguire, Goodchild, and Rhind (1999: 49–60).

Poiker, T., and Sheppard, E. (eds.) (1995). "GIS and Society." *Cartography and Geographic Information Systems*, Special Issue, 22.

Posey, A. S. (1993). "Automated Geography and the Next Generation." *The Professional Geographer*, 45/4: 455–6.

Quattrochi, D. A., and Goodchild, M. F. (eds.) (1997). *Scale in Remote Sensing and GIS*. New York: CRC Publishers.

Ramirez, M. (1995). "Closing the Gap: GIS in the High School Classroom." *Geo Info Systems*, 5/4: 52–5.

— (1996). "Florida Schools Study GIS for Environmental Education." *Geo Info Systems*, 6/5: 35–8.

Rhind, D. W. (1999). "National and International Geospatial Data Policies," in Longley, Maguire, Goodchild, and Rhind (1999: 767–87).

Roy, G. G., and Snickars, F. (1996). "City Life: A Study of Cellular Automata in Urban Dynamics," in Fischer, Scholten, and Unwin (1996: 213–28).

Sabin, T. J., and Holliday, V. (1995). "Playas and Lunettes on the Southern High Plains: Morphometric and Spatial Relationships." *Annals, Association of American Geographers*, 85/2: 286–305.

Savage, M. (1991). "Structural Dynamics of a Southwestern Pine Forest under Chronic Human Influence." *Annals, Association of American Geographers*, 81/2: 271–89.

Schroeder, P. C. (1997). "GIS in Public Participation Settings." Paper presented at University Consortium for Geographic Information Science, Annual Assembly and Summer Retreat, 15–21 June. Bar Harbor, Me.: UCGIS.

Sheppard, E. (1993). "Automated Geography: What Kind of Geography for What Kind of Society?" *Professional Geographer*, 45/4: 457–60.

— (1998). "Working Bibliography on Geographies of the Information Society." ‹www.ncgia.ucsb.edu/varenius/info_soc_panel/bibliog2.html›, last accessed 15 November 2002.

Sheppard, E., Couclelis, H., Graham, S., Harrington, J. W., and Onsrud, H. (1999). "Geographies of the Information Society." *International Journal of Geographic Information Science*, 13/8: 797–823.

Smith, B., and Mark, D. M. (1998). "Ontology and Geographic Kinds," in T. K. Poiker and N. Chrisman (eds.), *Proceedings, 8th International Symposium on Spatial Data Handling (SDH '98)*. Vancouver: International Geographical Union, 308–20.

—— (2001). "Geographic Categories: An Ontological Investigation." *International Journal of Geographical Information Science*, 15/7: 591–612.

Smith, D. A., and Tomlinson, R. F. (1992). "Assessing Costs and Benefits of Geographical Information Systems: Methodological and Implementation Issues." *International Journal of Geographical Information Systems*, 6: 257–71.

Smith, N. (1992). "History and Philosophy of Geography: Real Wars, Theory Wars." *Progress in Human Geography*, 16/2: 257–71.

Star, J. L. (ed.) (1991). *Proceedings: The Integration of Remote Sensing and Geographic Information Systems*. Special Session of ACSM-ASPRS Annual Convention, Baltimore, 1991. Washington, DC: American Society for Photogrammetry and Remote Sensing.

Stoms, D. M. (1994). "Scale Dependence of Species Richness Maps." *Professional Geographer*, 46/3: 346–58.

Sui, D. Z. (1994). "GIS and Urban Studies: Positivism, Post-Positivism, and Beyond." *Urban Geography*, 14: 258–78.

—— (1995). "A New Pedagogic Framework to Link GIS to Geography's Intellectual Core." *Journal of Geography*, 94/6: 578–91.

—— (1998). "GIS-Based Urban Modelling: Practices, Problems, and Prospects." *International Journal of Geographical Information Science*, 12/7: 649–50.

Svingen, B. E. (1994). "New Technologies in the Geography Classroom." *Journal of Geography*, 93/4: 180–5.

Tang, A. Y., Adams, T. M., Usery, E. L. (1996). "A Spatial Data Model Design for Feature-Based Geographical Information Systems." *International Journal of Geographical Information Systems*, 10/5: 643–59.

Taylor, P. J. (1990). "Editorial Comment: GKS." *Political Geography Quarterly*, 9: 211–12.

Taylor, P. J., and Johnston, R. J. (1995). "Geographic Information Systems and Geography," in Pickles (1995: 51–67).

Taylor, P. J., and Overton, M. (1991). "Further Thoughts on Geography and GIS: A Preemptive Strike?" *Environment and Planning A*, 24: 463–6.

Texas, University of (1998). *The Virtual Geography Department Home Page*. ‹www.utexas.edu/depts/grg/virtdept/main.html›, last accessed 15 November 2002.

Thapa, K., and Bossler, J. (1992). "Accuracy of Spatial Data Used in Geographic Information Systems." *Photogrammetric Engineering and Remote Sensing*, 58: 835–41.

Thompson, D., F. Lindsay, E., Davis, P. E., and Wong, D. W. S. (1997). "Towards a Framework for Learning with GIS: The Case of *Urban World*, a Hypermap Learning Environment Based on GIS." *Transactions in GIS*, 2/2: 151–67.

Tulloch, D. L., Epstein, E. F., Niemann, B. J., and Ventura, S. J. (1997). *Multipurpose Land Information Systems Development Bibliography: A Community-Wide Commitment to the Technology and Its Ultimate Applications*. Technical Report 97-1. Santa Barbara, Calif.: National Center for Geographic Information Analysis.

UCGIS (1996). "Research Priorities for Geographic Information Science." *Cartography and Geographic Information Systems*, 23/3 ‹www.ncgia.ucsb.edu/other/ucgis/CAGIS.html›, last accessed 15 November 2002.

—— (1998). *White Paper on Research in GIS and Society* ‹www.ucgis.org›, last accessed 15 November 2002.

—— (2000). *Emerging Research Themes*. White papers accessible at ‹www.ucgis.org/emerging/›, last accessed 15 November 2002.

Usery, L. (1996). "A Conceptual Framework and Fuzzy Set Implementation for Geographic Features," in Burrough and Frank (1996: 71–85).

Vckovski, A. (1998). "Guest Editorial. Interoperability in GIS." *International Journal of Geographical Information Science*. Special Issue, 12/4: 297–8.

Ventura, S. J. (1995). "The Use of Geographic Information Systems in Local Government." *Public Administrative Review*, 55: 461–7.

Veregin, H. (1996). "Error Propagation through the Buffer Operation for Probability Surfaces." *Photogrammetric Engineering and Remote Sensing*, 62/4: 419–28.

Walsh, S. J. (1992). "Spatial Education and Integrated Hands-on Training: Essential Foundations of GIS." *Journal of Geography*, 91/2: 54–61.

Walsh, S. J., Evans, T. P., Welsh, W. F., Entwisle, B., Rindfuss, R. R. (1999). "Scale-Dependent Relationships between Population and Environment in Northeastern Thailand." *Photogrammetric Engineering and Remote Sensing*, 65/1: 97–105.

Warren, S. (1995). "Teaching GIS as a Socially Constructed Technology." *Cartography and Geographic Information Systems*, 22/1: 70–7.

Weibel, R., and Buttenfield, B. P. (1992). "Improvement of GIS Graphics for Analysis and Decision-Making." *International Journal of Geographical Information Systems*, 6/3: 223–45.

Westervelt, J. D., and Hopkins, L. D. (1999). "Modeling Mobile Individuals in Dynamic Landscapes." *International Journal of Geographical Information Systems*, 13/3: 191–208.

Woronov, T. (1995). "Issues in GIS in Education," in D. Barstow and M. D. Gerrard (eds.), *Conference Report: First National Conference on the Educational Applications of Geographic Information Systems (EdGIS)*. Washington, DC: National Science Foundation, 110–13.

Wright, D., and DiBiase, D. (2000). "Distance Education in GIScience." *UCGIS White Paper*, ‹dusk.geo.orst.edu/disted›, last accessed 15 November 2002.

Wright, D. J., Goodchild, M. F., and Proctor, J. D. (1997). "GIS: Tool or Science? Demystifying the Persistent Ambiguity of GIS as 'Tool' versus 'Science'." *Annals, Association of American Geographers*, 87/2: 346–62.

Xiong, D., and Marble, D. F. (1996). "Strategies for Real-Time Spatial Analysis Using Massively Parallel SIMD Computers: An Application to Urban Traffic Flow Analysis." *International Journal of Geographical Information Systems*, 10/6: 769–89.

CHAPTER 25

Remote Sensing*

Dale A. Quattrochi, Stephen J. Walsh, John R. Jensen, and Merrill K. Ridd

Remote Sensing and Its Relationship to Geography

As noted in the first edition of *Geography in America*, the term "remote sensing" was coined in the early 1960s by geographers to describe the process of obtaining data by use of both photographic and nonphotographic instruments (Gaile and Wilmot 1989: 46). Although this is still a working definition today, a more explicit and updated definition as it relates to geography can be phrased as: "remote sensing is the *science*, *art*, and *technology* of identifying, characterizing, measuring, and mapping of Earth surface, and near Earth surface phenomena from some position above using photographic or nonphotographic instruments." Both patterns and processes may be the object of investigation using remote sensing data. The science dimension of geographic remote sensing is rooted in the fact that: (1) it is dealing with primary data, wherein the investigator must have an understanding of the environmental phenomena under scrutiny, and (2) the investigator must understand something of the physics of the energy involved in the sensing instrument and the atmospheric pathway through which the energy passes from the energy source, to the Earth object, to the sensor. The art dimension of geographic remote sensing has to do with the creative ways that the scientific interpretations are presented for visualization and measurement. The technological dimension of geographic remote sensing has to do with the constantly evolving hardware, software, and algorithmic manipulation and modeling involved in the collection, processing, and interpreting of data regarding the Earth phenomena under investigation. It is the rapidly advancing combination of these three dimensions over recent decades that has brought remote sensing to be a vibrant and dynamic part of the discipline of geography today.

We wish not to dwell at length on the historical aspects of remote sensing as it relates to geography. This has been done quite adequately in the first edition of *Geography in America* as well as in other publications, such as the American Society of Photogrammetry and Remote Sensing (ASPRS) *Manual of Remote Sensing* series (e.g. Colwell 1983), that is now going through a third edition and complete update, and is being presented as a compendium of individual volumes that deal with specific aspects of remote sensing science. Instead, after a brief reflection on key developments, we wish to provide some semblance of the state of the art of remote sensing today, and to look into the proverbial crystal ball to assess what prospects, challenges, and opportunities lie ahead at the dawn of the new millennium for geographic remote sensing in the future. This, however, is not an easy task. Given the rapid changes that are occurring in the number of current and planned remote sensing data acquisition platforms and sensing instruments, as well as the computational technologies for processing remote sensing data, any discussion of what remote sensing will

* Direct inquiries about the use of material in the chapter to Dale Quattrochi, National Aeronautics and Space Administration, Earth Science Department, Marshall Space Flight Center, Huntsville, Alabama 35812.

be like in the future is subject to complete revision within a very short time frame. Still, it is useful as a benchmark to lay out a "roadmap" (albeit, the roads on the map are evolving into super highways in real time) to achieve a better understanding of not only how remote sensing will be transformed in the future, but also how these changes will affect geography as a discipline. As will be seen via the following discussion, the prospects for geographical remote sensing in the future are bright, and the opportunities for advancing both the science of remote sensing and the overall recognition of geography as a leader in using remote sensing techniques for analysis of the world around us, are truly remarkable.

The Presence of Remote Sensing in Geography

It is evident from the vast body of literature published by geographers, that remote sensing is now well ensconced as a fundamental geospatial analysis tool and an integral component of geography as a discipline. This is illustrated by the plethora of articles by geographers in remote sensing journals, such as *Photogrammetric Engineering and Remote Sensing, Remote Sensing of Environment, International Journal of Remote Sensing*, and *Geocarto International*. Additionally, remote sensing as a presence within geography is exemplified by the fact that the Association of American Geographer's (AAG), Remote Sensing Specialty Group (RSSG) is one of the largest specialty groups within the Association, with approximately 8 per cent of AAG specialty group members belonging to it in 2000 (AAG 2001*a*). There are at least 135 departments in the United States and Canada that list remote sensing as either a specialty area or as a course taught within the department (AAG 2001*b*). Thus, remote sensing has become a solid, respected scholarly pursuit as well as having a large teaching presence within geography departments in America.

In addition to its role as a geographic technique, remote sensing has in conjunction with its integration into geographic information systems (GIS), become an extremely powerful geospatial analysis tool. This is evident from a perusal of both the remote sensing and GIS literature, where remote sensing is a key data source used in quantitative spatial models within a GIS purview. For example, under the aegis of the National Center for Geographic Information and Analysis (NCGIA) <http://www.ncgia.ucsb.edu> and the University Consortium for Geographic Information Science (UCGIS) <http://www.ucgis.org> (both last accessed 6 March 2002), the partnership between remote sensing and GIS has been strengthened and enhanced, which ultimately will synergistically benefit both of these techniques individually as geospatial analysis entities, as well as geography as a whole.

Future Role of Remote Sensing in Geography

From the perspective of looking into the proverbial crystal ball, the advent of a host of new sensing systems that have either just recently been launched, or will be put into orbit within the next five years, will improve the ability to make more accurate measurements of the Earth and perform more robust spatial analysis of land surface patterns and processes. With the enhanced spatial, temporal, and radiometric resolutions of these systems, in concert with improved data processing algorithms and techniques, many new geographical applications of remote sensing data will emerge to strengthen the existing bond between geography and remote sensing. The prospects for using the data from these new remote sensing systems, however, also presents some new and difficult—even daunting—challenges to geographical analysis, particularly in the integration of these data with other types of geospatial information. In a like manner, the prospects for new educational outreach opportunities of using remote sensing data are bright. Given the wide range of real and potential applications of the data from remote sensing systems currently in operation or to be launched in the near future, it is obvious that these data can, and will, be employed to extend the depth and breadth of geographical remote sensing in education at the K-12, community college, undergraduate, graduate, and general public levels. The diversity and varying characteristics of these new remote sensing datasets, however, pose a significant challenge for their optimal use in education; e.g. in designing courses and materials that illustrate how these different remote sensing data can be used from a geographical perspective, for obtaining a synoptic view of the physical, biological, and social interactions that have shaped the Earth's landscapes.

To examine what the prospects are for remote sensing in geography in the future, we focus our discussion on four broad areas:

- new sensors and improved spatial, temporal, and radiometric resolutions;
- new geographical applications of remote sensing;

- new challenges for geographical analysis in using and integrating remote sensing data with other types of geospatial information, particularly as related to assessment of the human environment; and
- new educational outreach opportunities for extending geographic remote sensing analysis.

For each of these areas, we wish to provide some semblance of what the accrued benefits to geography will be, as well as give an idea of what challenges will be faced in working with remote sensing data in the future. Moreover, we hope that an examination of these four areas will whet the appetite for both the continued and expanded usage of remote sensing data in geography at the dawn of the new millennium and beyond.

New and Forthcoming Remote Sensing Systems and Attributes

As of late 2000, there were approximately a dozen governmental or commercial land imaging satellite systems in orbit with spatial resolutions from 1 to 30m. The number of these systems is expected roughly to double within the decade, with these advanced systems having an impressive array of spatial, temporal, and radiometric resolutions that will significantly improve the amount and types of data that can be used for geographical applications. Most of these new sensors will have geospatial measurement capabilities that significantly enhance those available today. Figure 25.1 gives a synopsis of the past, present, and future land remote sensing satellites with spatial resolutions from 1m or less to 30m. It is evident from the figure that the amount of data available now, or expected to be available soon, is truly staggering. To illustrate how rapidly the number of land remote sensing satellites has grown recently, Fig. 25.2 presents a graphical perspective on the number of satellite systems that are currently operational, or will become so between 1996 and 2006, with spatial resolutions from 1 to 30m. The figure also illustrates the proportions between government- and commercially owned and operated satellite systems. It is evident from an inspection of Fig. 25.2 that although government-operated systems dominated up to 1999, the number of commercial satellite systems with spatial resolutions of 1 to 30m will exceed those operated or launched by government entities.

Overall, the current or planned satellites and sensor systems shown in Fig. 25.1 can be grouped into four general categories:

- The Landsat or "Landsat-like" group
- The high-resolution group
- The hyperspectral group
- The radar group

Note in Fig. 25.1 that most of the satellites in the first group (Landsat-like) have already been launched, with a few yet to come. Most of them have both a panchromatic band (with spatial resolutions of 2.5–20m) and multispectral bands (with resolutions of 6–30m). The number of spectral bands ranges from three to seven within this first group. The next two sections indicate the specifications of the high spatial resolution group, first those with both a panchromatic and multispectral bands, and then those with only a panchromatic band, and their respective resolutions, swath width, and global coverage. Note that three of these sensors in the high-resolution group have spatial resolutions less than 1m.

The remote sensing systems in the hyperspectral group sacrifice spatial resolution in favor of greater spectral power better to discriminate and characterize environmental features on the surface of the Earth. These satellite sensors either focus, or will focus, on measuring the thermal properties of objects, or other properties such as variations in moisture or brightness among the great variety of materials and objects on the land. The final group in the table shows five radar satellite systems. Figure 25.3 provides further information about selected instruments and the nations or private companies engaged in satellite-based sensor system development. At least ten countries and four major companies are engaged in remote sensing enterprises, as they look towards an expanding market for specified products. Geography, in its many dimensions, benefits mightily from every one of these data development innovations.

This impressive array of satellite sensor systems provides geographers with an expanding opportunity, if not an obligation, in research, application, and education. Additionally there are numerous airborne systems in service and more to come, refining still further the spatial and spectral resolutions of data that are, and will be, of great value to geographic investigators.

The Landsat or "Landsat-like" Group

The satellites, sensor systems, and data that are part of the Landsat-like group closely follow in the footsteps of the Landsat series of land remote sensing spacecraft that have been in orbit since 1972. The data from the Landsat satellites have been used extensively in geographical remote sensing research since they were first acquired, and may fundamentally be thought of as being the real

Satellite	Launch year	Band resolution in meters						No. of spectral bands				Swath		Global	
		PAN band			MS bands							Kilometers		Cover/yr	
		<=1	1.8–3	>2.5	>10	5.–10	<=4	VNIR	SWIR	MWIR	TIR	>60	<30	>7×	<6×
High global repeat (Landsat-like)															
Landsat 5	1984.250				30			4	2		1	185		23	
SPOT 1	1986.016			10	20			3				120		15	
SPOT 2	1990.100			10	20			3				120		15	
IRS-1C	1995.930			5	23			3	1			142		18	
IRS-1D	1997.790			5	23			3	1			142		18	
SPOT 4	1998.230			10	20			3	1			120		15	
Landsat 7	1999.288			15	30			4	2		1	185		23	
CBERS-1	1999.790			20	20			4	2		1	120		15	
IRS-P6	2001.400				23	6		4,3	1			142	24	18	3
CBERS-2	2001.790			20	20			4	2		1	120		15	
SPOT 5	2002.120		2.5			10		3	1			120		15	
ALOS-1	2002.500		2.5			10		4				70	35	9	4
Resource 21-A	2004.800					10		4	1			320		40	
Resource 21-B	2004.950					10		4	1			320		40	
High res, PAN & MS															
SIE IKONOS-2	1999.731	1					4	4					12		1.5
OrbView-4	2001.496	1					4	4					8		1
EW QuickBird-2	2001.624	0.6					2.5	4					16		2
OrbView-3	2002.047	1					4	4					8		1
ISI Eros-B1	2003.470	1					4	4					16		2
EW QuickBird-3	2003.500	0.6					2.5	4					16		2
SIE IKONOS-BIK-	2004.000	0.5					2	4					12		1.5
ISI Eros-B2	2003.915	1					4	4					16		2
ISI Eros-B3	2004.496	1					4	4					16		2
ISI Eros-B4	2004.916	1					4	4					16		2
ISI Eros-B5	2005.414	1					4	4					16		2
ISI EROS-B6	2005.915	1					4	4					16		2
High res, PAN only															
SPIN-2	1998.242		2									180			
ISI Eros-A1	2000.951		1.8										13		1.6
Cartosat-1	2002.000		2.5										30		3.7
Cartosat-2	2003.000	1											30		3.7
Multi-hyperspectral															
Terra (ASTER)	1999.956				15			3	6		5	60		7.5	
MTI	2000.170				20	5		7	3	2	3		13		1.6
EO-1	2000.931				30			115	115				15		1.9
OV-4 Warfighter	2001.247					8		100	100				5		0.6
NEMO	2002.500			5				105	105				30		3.7
ARIES	2003.748			10				32	32				15		1.9
Radar															
ERS-2	1995.300						10					100		12.3	
Radarsat 1	1995.750						8.5					50			6
Envisat	2001.500						10					100		12.3	
ALOS-1	2002.500						10					70		8.7	
Radarsat 2	2003.329					3						50			6

Fig. 25.1 Data characteristics of current and planned land imaging satellites

Source: W. E. Stoney, Mitretek Systems, Corp. – Summary of Land Imaging Satellites Planned to go Operational by 2006

driving force behind the widespread use of these data in geographical pursuits. Landsat satellites incorporate wide swath widths (e.g. 185km² or 100 nautical mi²), which when divided into individual images or "scenes" represent a standard "footprint" of Landsat data. The pixel spatial resolutions of Landsat data have varied from 80m for the early satellites in the series (Landsat's 1–3) to 30m for later platforms (i.e. Landsat's 4, 5, and 7). The sensor design of the Landsat series has been characterized by broad spectral coverage. (e.g. from visible to thermal infrared wavelengths). The primary sensor on Landsat's 1–3 was the Multispectral Scanner (MSS) which had four broad spectral bands from 0.5 to 1.1μm and a spatial resolution, or pixel size of 80m. Landsat's 4 and 5 employed the Thematic Mapper (TM) that had 7 spectral channels over a broad band of wavelengths from 0.45 to 12.50μm. This encompasses the visible, reflective infrared, middle reflective infrared, and thermal infrared regions of the electromagnetic spectrum. TM data have a pixel resolution of 30m in all channels except for the thermal infrared (TIR), which has a pixel resolution of 120m. The newest satellite in this series, Landsat 7, was launched on 15 April 1999 and employs an updated version of the TM sensor called the Enhanced Thematic Mapper Plus (ETM+) (NASA 1999). The ETM+ has the same general bandwidths and pixel spatial resolutions of the TM for the visible and reflective infrared bands, but the thermal infrared channel has a resolution of 60m as opposed to 120m. Additionally, the ETM+ has a co-registered panchromatic band with a spatial resolution of 15m. A detailed history of the Landsat program can be found via a number of references; refer to *Landsat Data User Notes* published by the EROS Data Center (NOAA 1975–1984), and the NASA Landsat 7 home page on the Internet at <http://landsat7.usgs.gov> (6 March 2002). A good overview of the characteristics of the Landsat and Landsat-like satellites and the data from the MSS, TM, and ETM+ sensors is provided in Jensen (2000: 184–201).

Moreover, the French SPOT (*Système pour l'observation de la Terre*) series of satellites, initially launched in 1986, are a complement to Landsat. SPOT is principally a three-band sensor (with bands in the visible and near infrared portions of the electromagnetic spectrum) with a 20m pixel resolution. These bands are co-registered with a panchromatic (black and white) band that has a 10m pixel spatial resolution. More information on the SPOT satellite system and data products can be accessed from the SPOT Image website at <http://www.spotimage.fr/home/> (6 March 2002).

Because of the relatively broad spectral and spatial resolution of the data from these satellites, as well as the temporal coverage (e.g. 18-day repeat cycle for Landsat 1, 2, and 3 and a 16-day cycle for Landsat 4, 5, and 7), Landsat and SPOT data have been primarily used for regional-scale applications. Since the launch of the Landsat and SPOT series of satellites, data from these sensors have been used for a plethora of geographical applications ranging from land-use and land-cover change analysis, vegetation and soil assessment, agricultural applications, urban analysis, and geomorphology (Jensen 2000).

The High Spatial Resolution Group

One of the most exciting advances that has occurred in civilian remote sensing has been the advent of satellite-based sensor systems capable of collecting data at very high spatial resolutions. These satellites provide high spatial resolution data in either panchromatic or visible multispectral (~.45–.65μm) formats. The High Resolution Group of satellite sensors are essentially focused on commercial applications and are designed to provide mapping and GIS products at spatial resolution scales currently provided by airborne sensors. These systems have been developed to address "human scale" applications at local levels (e.g. identification of buildings, individual land-cover surfaces or land uses). As seen in Fig. 25.1, there are twelve land imaging satellites with high spatial resolution panchromatic or multispectral bands, and four satellite sensor systems with a high resolution panchromatic band, either currently in orbit or planned to be launched by 2006. These systems have spatial resolutions of 1m or less in the panchromatic bands and resolutions of 4m or less in the multispectral bandwidths. These sensor systems will offer a tremendous leap in our capabilities to better discern very fine-scale targets for use in a variety remote sensing-based surveys, such as socioeconomic characterization (e.g. counts of individual dwellings), more robust and accurate land-cover/land-use analyses (e.g. down to Anderson land-cover classification level 4 (Anderson *et al.* 1976), improved cadastral and transportation infrastructure mapping (e.g. building height and volume, street and road conditions), and utility infrastructure assessment and planning (e.g. power line right-of-way mapping). An example of high resolution panchromatic data from the Space Imaging Systems IKONOS satellite is given in Fig. 25.4 (for more information on IKONOS data, see the Space Imaging website at <www.spaceimaging.com> (6 March 2002). This figure shows an IKONOS pan image of the Washington, DC area at a 1m spatial resolution. It is evident in Fig. 25.2 that significant detail regarding the size, shape, and overall spatial relationships of buildings, streets, parking lots, and other land cover and land uses within the image can be discerned.

There are trade-offs in using these high resolution data, however, in that the footprint of the data is quite small in comparison with the that from the sensors in the Landsat-like group (e.g. IKONOS data have a footprint size of 13km^2), they are not usually routinely obtained like Landsat, and the cost for these images is substantially higher than data from the Landsat-like group of sensors. Still, these high spatial resolution

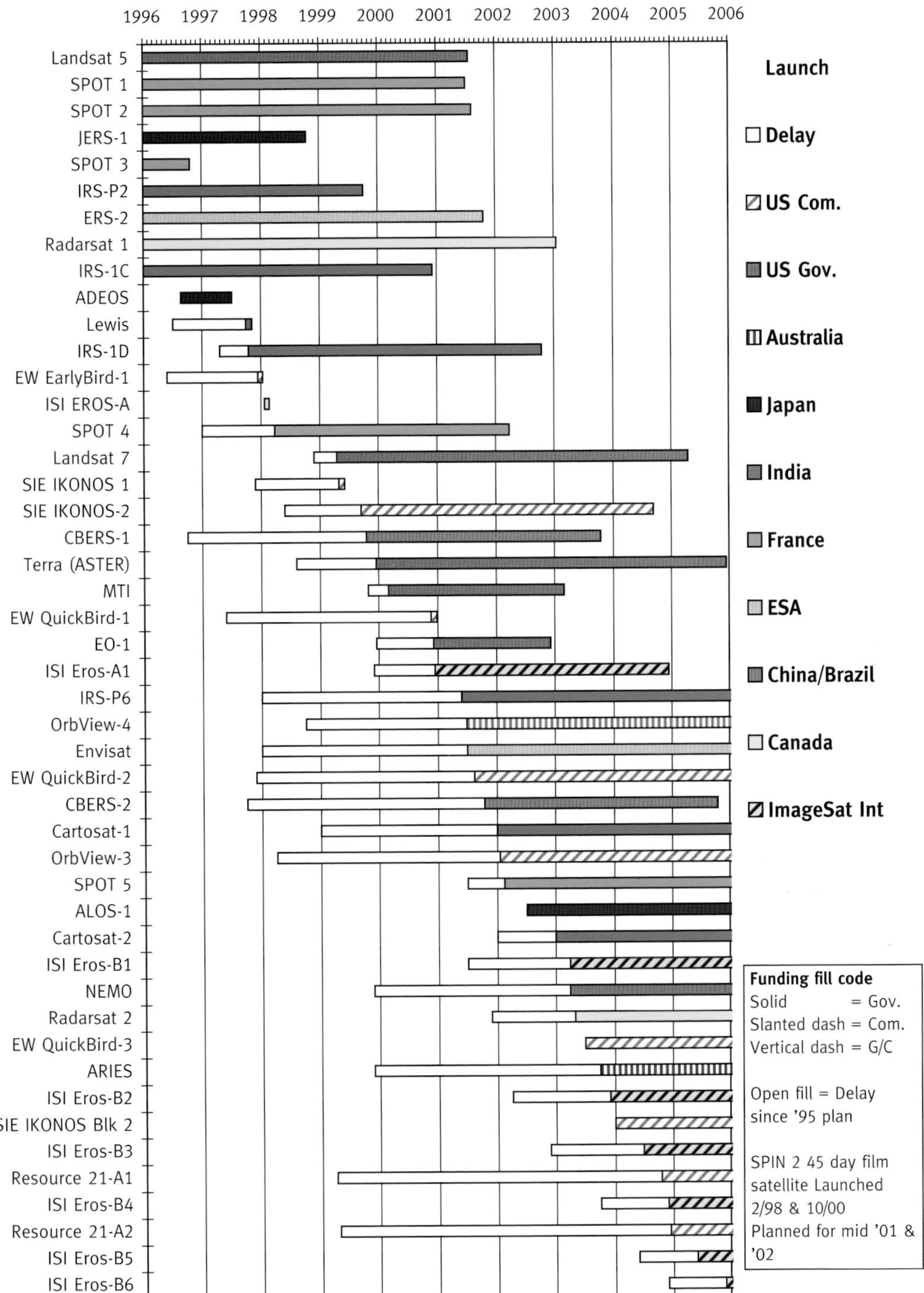

Fig. 25.2 Land imaging satellites, 1–30m resolution, past, present, and planned

Source: W. E. Stoney, Mitretek Systems, Corp.—Summary of Land Imaging Satellites

Satellite funder	Satellite	Launch	Sensor types	Resolution in meters											Stereo type	SW KM	Global cover repeat days
					Thematic mapper bands												
				PAN	VNIR				SWIR		MWIR	TIR	R/L				
					1	2	3	4	5	7		6	res	band			
Landsat like, frequent global coverage																	
INDIA (Gov.)	IRS-1 C,D	'95,'97	M&P	5		23	23	23	70						C/T	70,142	48,24
INDIA (Gov)	IRS-P6	'01	M		23	6,23	6,23	6,23	70						C/T	24/142	125,24
FRANCE (Gov)	SPOT 4	'98	M&P	10		20	20	20	20						C/T	120**	26
FRANCE (Gov)	SPOT 5	'02	M&P	5,2.5		10	10	10	20						TBD	120**	26
CHINA-BRAZIL (Gov)	CBERS-1,2	'99,'01	M&P	20,80	20	20	20	20	80	80		160			C/T	120	26
U.S. (Gov)	Landsat 7	'99	M&P	15	30	30	30	30	30	30		60				185	16
JAPAN	ALOS-1	'02	M&P&R	2.5	10	10	10	10							F/A-C	70&35,70	48
R21, A1,2	Resource 21	'04	M		10	10	10	10	20							320	4***
High resolution, small area coverage																	
Space Imaging	Ikonus 2	'99	M&P	1	4	4	4	4							F/A	12	247
EarthWatch	QuickBird 2,3	'01,'03	M&P	0.6	2.5	2.5	2.5	2.5							F/A	16	185
Orbimage	OrbView-4,3	'01,'02	M&P	1	4	4	4	4							F/A	8	Non SS
ImageSat	Eros-B1	'02	M&P	1	TBD										F/A	16	185
ImageSat	Eros-B2,3,4	'02–'03	M&P	1	TBD										F/A	16	185
Space Imaging	Ikonus Blk 2	'04	M&P	0.5	2	2	2	2							TBD	TBD	TBD
Russia (Gov)	SPIN-2	'98–'02	P(f)	2,10											F/A	180,200	Non SS
ImageSat	Eros-A1	'00	P	1*											F/A	13	228
India (Gov)	Cartosat 1(IRS-P5)	'02	P	2.5											F/A-C	30	99
India (Gov)	Cartosat 2	'03	P	1											TBD	30	99
Multi & hyperspectral experimental																	
U.S./JAPAN (Gov)	Terra	'99	M			15	15	15	30	5 @ 30		5@90				60	49
OrbImage (Gov Inst)	OrbView-4	'00	H		200 bands @ 8											5	600
US (NASA)	EO-1	'00	H&M		233 bands @ 30											15	200
US (DoE)	MTI	'00	M		4 bands @ 5, 3 @ 20				3 @ 20		2 @ 20	3 @ 20				13	228
US (NAVY)	NEMO	'01	H	5	210 bands @ 30											30	100
AUSTRALIA	ARIES	'03	H	10	32 bands @ 30					32 @ 30						15	200
Radar																	
CANADA (Gov)	Radarsat 1	'95	R										8.5	C		50	
CANADA (Gov)	Radarsat 2	'01	R										3	C		50	
ESA (Gov)	ERS-2, ENVISAT	'95,'00	R										10,30	C		100	
JAPAN (Gov)	ALOS	'02	M&P&R										10	L		70	

C — C-band radar
(f) — Film
H — Hyperspectral
IR — Infrared
L — L-band radar
M — Multi-spectral
MIR — Mid-wave IR
P — Panchromatic
R — Radar
SWIR — Short-wave IR
TBD — To be determined
TIR — Thermal IR
VNIR — Visible and near IR

Non-SS: SPIN-2 flies in a 66° non-sun-synchronous orbit
F/A = fore/aft stereo, F/A-C = continuous fore/aft stereo, C/T = side to side stereo
All stereo satellites have 2–3-day site repeat capabilities
Some 1-meter systems note that the resolution is .82 meters at nadir center
Where there are two values, they are in the order of the listed sensors
* System is capable of 1 meter but Israeli policy may initially restrict data to 1.8
** Swath is achieved by two side-by-side instruments
*** 3.5–4-day global repeat coverage will be provided by 2 or 4 satellites

Companion low res wide area sensors

Sat	Inst.	No. of bands	Res meters	Swath (km)
IRS-1C,D	WIFS	3	188	810
IRS-P6	AWIFS	3	70	720
SPOT 4,5	Veg.	4	1000	2200
CBERS	WFI	2	260	900
Terra	MODIS	2	250	2330
		5	500	2330
		29	1000	2330

Fig. 25.3 Band resolutions of land imaging satellite systems

Source: W. E. Stoney, Mitretek Systems, Corp. – Summary of Land Imaging Satellites Planned to go Operational by 2006

Fig. 25.4 KONOS panchromatic image of Washington, DC.
Courtesy of Space Imaging, 1999

systems will have a revolutionary impact on geographical applications in remote sensing by providing imagery that offers great enough spatial resolutions for accurately identifying, observing, and mapping features that heretofore was available only from aircraft remote sensing platforms.

One other feature of the high spatial resolution group of satellites that makes data from them appealing for geographical analysis is that they have stereoscopic capabilities. The satellites can be pointed fore and aft to produce stereo pairs of imaged data. This provides the capability for analysis of topographic characteristics that are of utility for a host of geographic research applications, such as topographic mapping, slope analyses, or land-use planning. The ability to provide a stereoscopic perspective when coupled with a digital image format makes geographic research that must include a vertical dimension a reality for any part of the world, even where accurate topographic maps are not currently available.

Hyperspectral/Experimental Systems

Overall, the hyperspectral class of Earth remote sensing satellites may generally be classified as "experimental systems." Hyperspectral systems are characterized by very high spectral resolution (i.e. 30–300 spectral bands). These hyperspectral instruments or "imaging spectrometers" have the capability to acquire simultaneously images in many relatively narrow, contiguous, or non-contiguous spectral bands throughout the ultraviolet, visible, and mid-infrared portions of the electromagnetic spectrum from 0.4 to 2.5μm (400–2,500nm). This combination of narrow spectral bands across the full range of the reflective spectrum provides the diagnostic power to detect subtle variations in moisture, chemical composition, color, and brightness of organic and inorganic substances in vegetation, soil, rocks, and building materials. The value of hyperspectral imaging lies in its ability to provide high spectral resolution data for each image pixel. Information of this type cannot be identified with data from either high spatial resolution/broadband or low spectral resolution/broadband systems, such as IKONOS or Landsat (Jensen 2000: 227). For example for biogeographic studies, imaging spectrometers can provide information on species type, chemistry, absorbing and scattering constituents, etc. that cannot be discriminated using non-hyperspectral remote sensing.

A number of hyperspectral remote sensing satellites are planned to be launched in the near future (see Fig. 25.3). One airborne imaging spectrometer developed by the NASA Jet Propulsion Laboratory in Pasadena, California called the Airborne Visible Infrared Imaging Spectrometer (AVIRIS), has been used since 1992 for remote sensing research in geology, forestry, agriculture, arid land, rangeland, and urban investigations. The AVIRIS sensor has 224 spectral channels with spectral resolutions from 0.4 to 2.5μm, and is normally flown onboard the NASA ER-2 research aircraft at an altitude of 20km above mean terrain, which provides a ground spatial resolution of 20m. (See the AVIRIS website at <http://makalu.jpl.nasa.gov/aviris.html> (6 March 2002) for more information on the characteristics of this hyperspectral airborne sensor as well as numerous examples of data collected by the AVIRIS.)

In December 1999, the NASA Terra spacecraft was launched which has a number of Earth remote sensing instruments on board. One of these, the Moderate-resolution Imaging Spectroradiometer or MODIS, is a hyperspectral system with thirty-six spectral bands that image data in the visible to thermal infrared portions of the electromagnetic spectrum (i.e. from 0.620 to 14.385 μm). Spatial resolutions of MODIS range from 250m to 1km in different spectral bands. The thirty-six bands on MODIS allow for a wide array of Earth remote sensing uses including land-cover classification, land/cloud properties, ocean color analysis, and atmospheric temperature. Extensive information on the MODIS sensor is available on the MODIS website at <http://modis.gsfc.nasa.gov> (6 March 2002).

The Radar Group

The sensors associated with the Landsat-like, High Resolution, and Hyperspectral/Experimental groups are referred to as *passive* remote sensing systems because they record electromagnetic energy that is reflected (e.g. blue, green, red, near-infrared, and mid-infrared light) or emitted (e.g. thermal infrared) energy from the surface of the Earth (Jensen 2000). *Active* remote sensing systems are not dependent upon the sun's electromagnetic energy or Earth surface thermal properties. Active sensors create their own electromagnetic energy that: (1) is transmitted from the sensor toward the terrain; (2) interacts with the terrain producing a backscatter of energy; and (3) is recorded by the remote sensing system's receiving apparatus (ibid.). Active systems as opposed to passive remote sensing systems are largely unaffected by the atmosphere (e.g. clouds, water vapor), which means they can be used to collect data in areas of the world where clouds are persistent, thereby presenting considerable challenges to collecting data from passive remote sensors. The most widely used active remote sensing systems for Earth science research are *active microwave*; i.e. *Radar* (Radio detection and ranging) and *LIDAR* (Light Detection and Ranging). Radar is based on the transmission of long-wavelength microwaves (e.g. 3–2.5cm) through the atmosphere. The Radar system then records the amount of energy backscattered from the terrain. LIDAR is based on the transmission of relatively short-wavelength laser light (e.g. 0.90μm) and then recording the amount of light backscattered from the target on the ground (ibid.).

Although airborne Radar and LIDAR systems have been in operation for some time, satellite-based civilian remote sensing Radar systems are relatively recent phenomena. The European Space Agency (ESA) launched the European Remote Sensing Satellite ERS-1 and ERS-2 in 1991 and 1995, respectively. The Japanese launched the Japanese Earth Resources Satellite JERS-1 in 1992, and the Canadian government launched the RADARSAT-1 satellite in 1995 (ibid.). Other Radar Earth remote sensing satellites are scheduled to be launched in the near future (Fig. 25.1).

The Radar group of satellite remote sensing systems supply image resolutions from 5 to 100m at swath widths of 20 to 500km and at different wavelengths (i.e. C, L, and X band). (See ibid.: ch. 9, 285–331, for a good overview of the theory and mechanics of active microwave systems.) Figure 25.5 provides an example of a Radar image from RADARSAT. The primary advantages of active microwave satellite systems is: (1) they are all-weather sensors and can penetrate clouds; (2) they have synoptic views of large areas for mapping at 1:25,000 to 1:400,000 scales; (3) imagery from these sensors can be collected even at night; (4) they permit imaging at shallow look angles, resulting in different

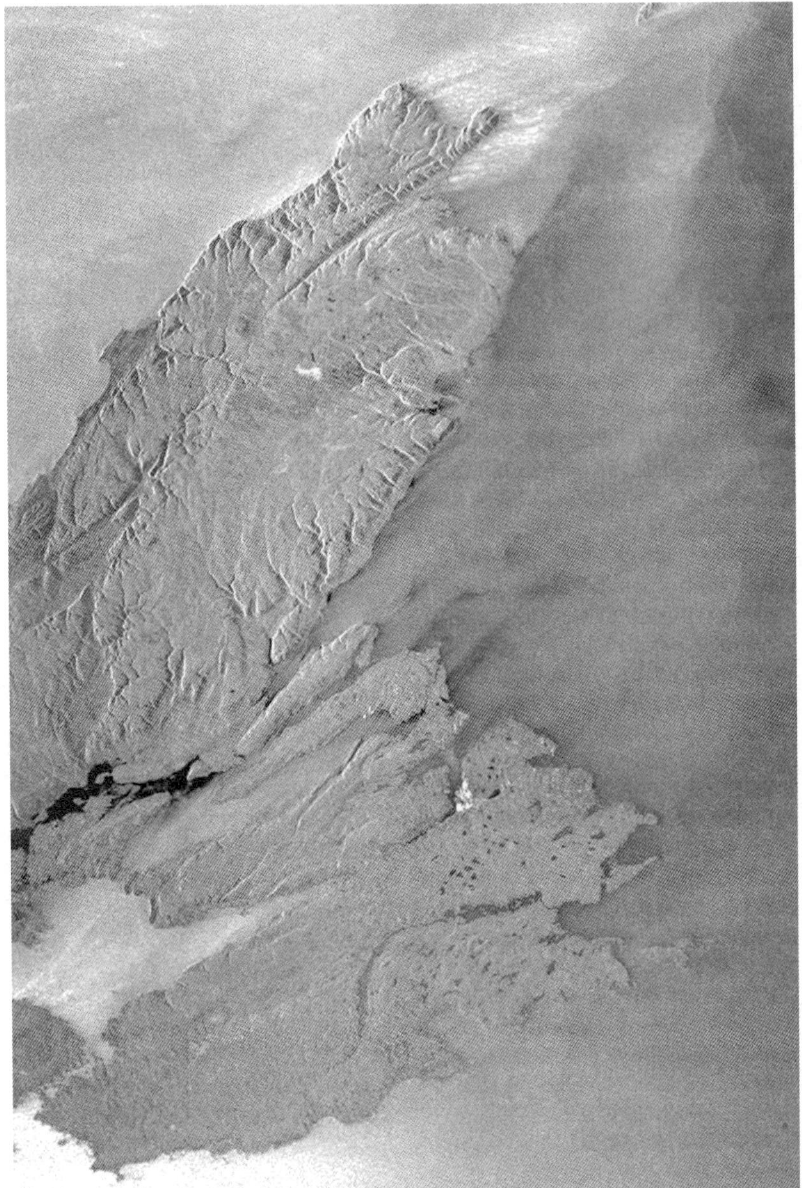

Fig. 25.5 RADARSAT image of Cape Breton Highlands Region of Nova Scotia
Courtesy of RADRSAT International, 1995

perspectives that cannot always be obtained from aircraft-based remote sensing systems; and (5) they provide information on surface roughness, dielectric properties, and moisture content in ways which passive remote sensing systems cannot provide. These advantages make active microwave systems useful tools for obtaining remote sensing data and can be used in an array of biophysical geographic research studies, particularly over wide areas or where clouds impede observation using passive systems.

New Applications of Remote Sensing Data: Human–Environment Interactions, and Assessment of Cultural and Infrastructure Resources

As we sit at the beginning of the new millennium with bright prospects for a wealth of new Earth remote sensing data that are either currently available, or will be forthcoming in the very near future (Figs 25.1–3), it appears that the multitude of applications of these data to geographical research initiatives are limited only by the imagination of the user. To illustrate how remote sensing data will continue to become an even more vital and essential aspect of geography, we focus attention on what the applications of remote sensing are, and will be, for developing a better understanding of human– and population–environment interactions, and assessment of cultural and infrastructure resources. Remote sensing systems are being used to characterize associated landscape features and patterns (e.g. population-related structures as well as the composition, spatial organization, and dynamics of land cover/land use) as signatures of human–environment interactions. These interactions operate at a host of space-time scales, and can be sensed through remote sensing systems that support a variety of spatial and temporal resolutions. Because of the enhanced (and diverse) spatial and spectral resolutions that are, and will become available, particularly as part of the High Resolution and Hyperspectral/Experimental groups of satellite sensing systems, it is anticipated that vast strides will be made in the uses of remote sensing data for identifying, observing, and modeling socioeconomic and human-scale attributes or characteristics of people, place, and environment, increasingly examined within a policy-relevant context. As a result of these new systems and data, new inroads will be made in extracting selected socioeconomic characteristics directly from remotely sensed data. For example, population estimations may be performed via counts of individual dwelling units, or quality of life indicators may be obtained using data from these new sensors, as supported by extensive ancillary information such as census estimates, local infrastructure, and related information. The integration of multiple remote sensing systems achieved through an assembled image time-series including, for example, Landsat TM, ETM+, and IKONOS data, will afford views of the Earth that can describe characteristics relating to the morphology of cities and urban places, the source or destination of migrant populations, and the human imprint on the landscape told in the form of land-cover and land-use fragmentation patterns. Other land conversion scenarios, possibly involving deforestation and agricultural extensification and intensification, land degradation through use and misuse of the land, and an expansion of the human "reach" across the landscape through an enhanced transportation infrastructure, will also be more readily identified and measured via the combined analyses of data from different sensors. These types of studies that seek to understand the interactions between humans, the environment, and their social and biophysical systems are central to assessing quality of life, for considering government's role in land management through public and environmental policy, and in examining concerns for land, its resources, and its dynamism. The science of remote sensing has much to offer to the discourse involving human–environment interactions. Defining the spatial and temporal linkages between people and pixels and people and the environment through image characterizations of the landscape, image change-detections, feature classifications, calibration and validation of models, and the support of land-cover/land-use simulations are part of the current and anticipated integration of remote sensing in studies involving human–environment interactions (Messina and Walsh 2001). Challenges abound, but so do the possibilities!

"Socialization of the pixel" suggests that remotely sensed data transformed into information and represented at the pixel-level should include the human dimension as well as the more conventional biophysical dimension. It is this linking across thematic domains, often within a scale, pattern, and processes paradigm (Walsh *et al.* 1999, 2001), that enriches the remote sensing data and increases its utility through direct observation and through and indirect association of people, pixels, and the environment (Walsh and Crews-Meyer 2002). Organizations such as the Land Cover/Land Use Change Program and the Large-Scale Biosphere Atmosphere Experiment in Amazonia Project within the National Aeronautics and Space Administration (NASA), the BioComplexity Program within the National Science Foundation (NSF), the Open Meeting of the Human Dimensions of Global Environmental Change Research Community, and the Global Change Open Science Conference are seeking to understanding the human dimensions of land-use and land-cover change, and the science of remote sensing is central to that characterization.

Remote Sensing of Human Environment and Urban-Suburban Cultural Resources

Geographers and other scientists require detailed information about the spatial characteristics of the social-cultural landscape (Liverman *et al.* 1998; Millington *et al.* 2001; Walsh and Crews-Meyer, 2002). Qualitative and quantitative cultural information may be obtained on the ground using direct observation. In addition, significant spatially distributed information about the cultural landscape may be obtained by analyzing remotely sensed data using analog or digital image processing techniques. This section summarizes some of the more important types of cultural information that are required for studies that integrate population and the environment and the general state-of-the-art of how such data can be collected using remote sensing technology.

1. *Human–Environment Interactions: Remote Sensing Practices, Opportunities, and Challenges for Landscape Characterization and Assessing the Human Dimensions of Land-Use and Land-Cover Change.* Since at least the mid-1990s, special sessions of the Association of American Geographers (AAG) have been regularly convened to investigate human–environment interactions occurring in a host of local and regional settings. Digital spatial technologies including remote sensing and geographic information systems (GIS) have been typically used in such studies—remote sensing for characterizing the biophysical landscape, and GIS for linking social and geographical variables to landscape states and conditions, often within a raster data model and through a relational database structure. These studies utilize satellite images and aerial photography to characterize land-use and land-cover (LULC) composition and structure through single date satellite images and aerial photography, as well as through an image and photographic time-series that represents historical views, landscape dynamics, and space-time scales of observation. Analyses are generally organized within a spatially explicit framework that rely upon: (1) the spatial, temporal, spectral, and radiometric resolutions of remote sensing systems for characterizing the landscape through scale-dependent LULC classification schemes; (2) GIS for assessing geographic site and situation of local and regional study sites and for defining topologic relationships between landscape elements, and (3) population surveys for conducting socioeconomic and demographic analyses at various units of observation. Moreover, these studies have applied ecological pattern metrics to satellite-based classifications as an approach for quantitatively defining the composition and spatial organization of LULC and changes that occurred across the landscape over both time and space scales. Empirical models have also been derived that integrate with social, biophysical, and geographical variables to highlight hypothesized relationships between population and the environment, and the concept of scale dependence used to explore ranges of spatial and temporal scales in which variables exhibited autocorrelation or randomness.

Geographers continue to strive to integrate thematic domains, space-time scales, and theory relevant to the natural and social sciences, all within a spatially explicit context. Such an approach provides a link to pattern and process at defined scales to examine questions related to land degradation, deforestation and agricultural expansion, loss of biodiversity, and trajectories of LULC change. Fundamental to the emerging discourse on linking people and the land through studies of human–environment interactions is the question of landscape characterization that involves elements of biophysical remote sensing, where LULC and key attributes of landscape state (e.g. cover types, change classes, and pixel histories of LULC dynamics) and condition variables (e.g. plant productivity, leaf area, and senescence) are mapped, analyzed, and linked to human activities.

Recent research continues the legacy of LULC monitoring, yet pushes the research focus forward by exploring the complexities of social and environmental drivers of land-cover/land-use change (LULCC) and the differential strengths of these drivers and their interactions across multiple scales of analysis (e.g. Walsh *et al.* 1999, 2001). North American remote sensing specialists from departments of geography are joining scholars from other disciplines and countries at the forefront of this increasingly interdisciplinary and international field. Rindfuss and Stern (1998), speaking to a broad interdisciplinary audience, addressed the needs and challenges of linking remote sensing to social science in the introductory chapter of an edited volume, *People and Pixels* (Liverman *et al.* 1998). The overarching theme of this volume, as well as much of the LULCC research in general, revolved around the theoretical and methodological difficulties related to linking social theoretical paradigms and socioeconomic data with natural scientific paradigms and remotely sensed data.

Scale dependence is examined as part of the LULC and LULCC research and cast within a human dimensions context by characterizing the landscape from a host of aircraft and satellite systems, developing image time-series for tracking the trajectories of change across time, and documenting landscape characteristics through social surveys and remote sensing systems. Historical aerial photography in analog or digital form continues to play a central role in assessing LULC and LULCC either (1) as a validation data set for comparison to processed satellite digital data, or (2) for landscape characterization, generally as a consequence of their historical coverage or spatial resolution through its defined minimum mapping unit. Researchers also often integrate aerial photography, optical satellite systems, and social surveys in a synergistic system where the social survey validates the remote sensing classifications of LULC by comparing, for instance, total land under cultivation from the survey questionnaire to the percent area of a region in agriculture from the remotely sensed classification. Collectively, social surveys and remote sensing systems describe the biophysical and social condition for lands and people associated with human settlements and their associated territories through the integration of longitudinal survey data and a satellite time-series. Together, they report on current or historical demographic characteristics of a specific social unit (e.g. household, community, district, country) and define landscape patterns and trends through image "snapshots." Figure 25.6 shows how LULC data from both satellites and GIS analyses are considered over time and through spatial, social, and biophysical concatenations of space to yield LULC outputs that may be represented through multivariate models, data visualizations, image change-detections, probability analyses associated with Markov transition probabilities, and LULC simulations based on cellular automata.

The human dimensions community is examining the causes of LULCC defined through remote sensing systems, both directly and indirectly. Explicitly stated is the commitment to, and necessity of, integrating approaches that are both multi-thematic and multi-scale to explicate the interactions between humans and the environment. The theoretical context for an interdisciplinary and scale-explicit analysis of human–environment interactions draws primarily upon landscape ecology, which states that spatial patterns observed on the landscape (e.g. via remotely sensed data) are indicative of the processes shaping that landscape. Landscape ecology argues that all landscapes contain structure (spatial organization), function (dynamic interaction among the landscape components), and change (in structure and function) (Foreman and Godron 1986). In addition, human and political ecology describe the importance of the social dimension on landscape form and function. Human ecology places humans as important actors on the landscape that shape and are shaped by it, whereas political ecology emphasizes the importance of exogenous factors (e.g. market conditions and political institutions) and nested effects on human and biophysical landscapes. In remote sensing, the spatial and temporal resolutions become central factors in landscape characterization that reflect the interrelationships of scale, pattern, and process on LULC and LULCC. Biophysical, social, and

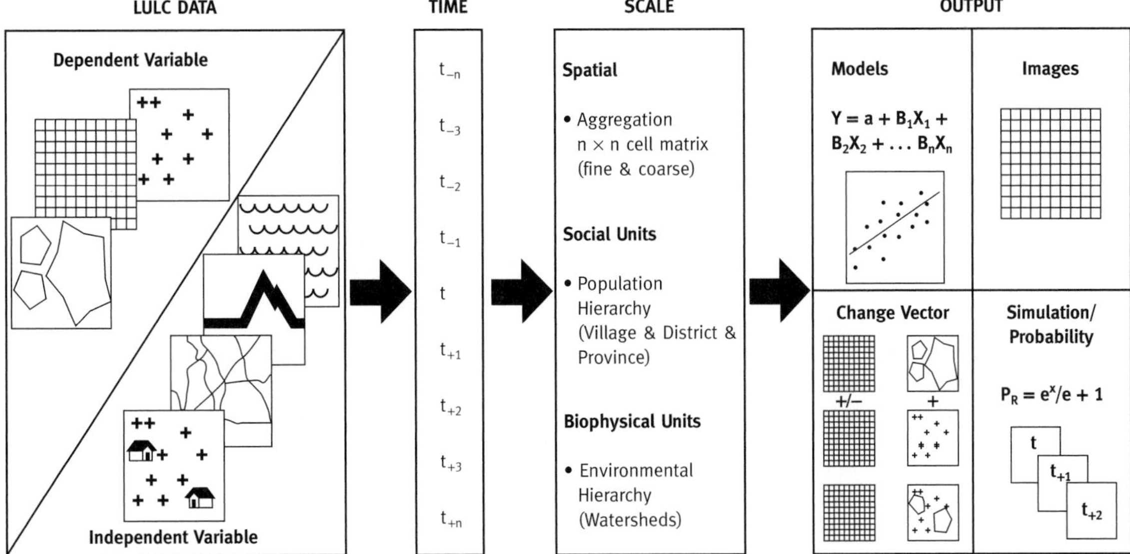

Fig. 25.6 Illustration of how land-use and land-cover data from both satellites and GIS analyses can be integrated over time and through spatial, social, and biophysical concatenations of space, to yield land-use and land-cover change outputs

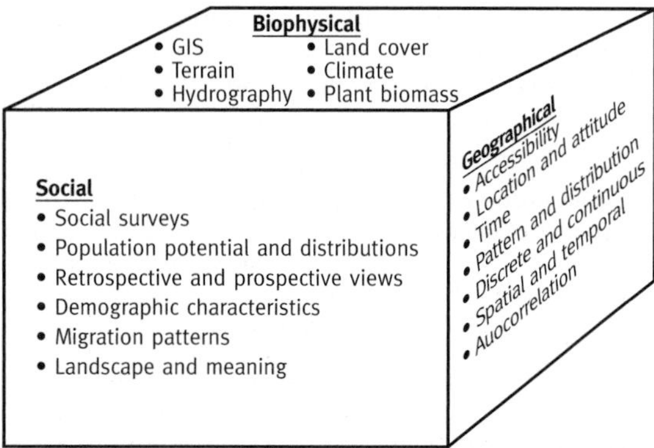

Fig. 25.7 Data "cube" showing biophysical, social, and geographical factors that are associated with explanations of land-use and land-cover change output

geographical factors are associated with explanations of LULCC, and each thematic domain has a number of approaches and variables associated that must be brought to bear on questions integrating population and the environment (Fig. 25.7).

Appropriate (spatial and temporal) scales of observation, including both grain and extent, are fundamental to scale-dependent research that examines human–environment interactions (Walsh *et al.* 1999). In the natural sciences, spatial scale is often thought of as a set of nested hierarchies that involve natural units, such as drainage basins or terrain facets. "Natural units" in the social sciences are more commonly thought of as aggregations of populations defined through political boundaries that indicate nested reporting units (e.g. block, block group, tract) or social hierarchies (e.g. household, community, district). In the social sciences, data collection is often achieved through the use of social surveys that examine the human dimension (and occasionally the environmental dimension) through questions examining current conditions as well as retrospective and prospective issues, whereas remote sensing data collection involves the IFOV (Instantaneous Field of View) of sensors, the periodicity of satellite orbits and aircraft overflights, and the generation of multiple "snapshots" in time through the generation of an image time-series. In addition, the remote sensing perspective provides an arbitrary partitioning of space into pixels ranging in dimensionality from the fine to the coarse, and are generally insensitive to the geometry or linkages of the scale of social entities as well as the landscape matrix. Therefore, data transformations between the social-biophysical and the spectral-spatial domains are often implemented for improving data compatibility for integrated analyses, and for enhancing the ability to define social meaning from the biophysical landscape through bi-directional studies that either emanate from the land and move towards people, or that emanate from people and move towards the land. Depending upon such analytical directions, the pixel, patch, class, or landscape may be the unit of observation or the household, community, district, or province may be more appropriate.

2. *Linking Remotely Sensed Data to Socio-Economic Data For Integrated Land Use and Land Cover Analysis.* Since the inception of the Landsat program in the early 1970s, digital remote sensing data and analytical approaches have played a pivotal role for geographers in quantifying and mapping LULC. Quantifying and mapping LULC will continue to be of significant importance as improvements in technology and methods (e.g. new and improved sensors, advanced classification techniques, increased processing speed and storage capacity) promise to enhance our abilities to monitor changes on the Earth's surface.

To examine connections between social processes and LULC, it is necessary to link two very different types of data. The most commonly available information about LULC is from remotely sensed data, and the unit of observation is the pixel, which may be aggregated into plots, patches, landscapes, regions, or other areal units. On the social side, data may come from household surveys and censuses, qualitative interviews and observations, focus groups, key informants, oral histories, and administrative monitoring systems. The unit of observation may be the person, household, village, subdistrict, province, or country. Units of observation and analysis for the LULC do not match those on the social side, thus requiring approaches to make these different data types compatible.

To elucidate the above relationships more completely, it is necessary to describe: (1) the nature of the linkage between population and environmental variables; (2) how remote sensing has been applied to characterize LULC and LULCC within this context; (3) approaches that facilitate the integration of discrete versus continuous variables representing place and space; and (4) emergent areas of inquiry that provide challenges to the remote sensing professional and the scientific community to apply appropriate theory and to develop effective methods that can characterize LULC in highly variable environments (e.g. dense crown closure of rainforests, and highly dissected agricultural uplands). These population and environment linkages may also have peculiar mapping requirements (e.g. in phase with population migration patterns, and out of phase with monsoonal precipitation) and environmental constraints (e.g. field sizes and spatial resolutions, and field logistics and geographic access). Mapping LULC in these situations provides challenges and opportunities for linking derived information to socioeconomic systems and for characterizing landscape form and function through optical and non-optical systems matched to the system dynamics being considered, and the scale dependence of the human–environment interactions under study.

The human–environment research community posits that LULC and LULCC are fingerprints of social, biophysical, and geographical drivers of landscape change. LULC is assessed across time and space to discern patterns, trends, and trajectories. Integrative methods have been developed to characterize the population as well as the biophysical landscape and spatially to represent nodes of economic opportunity and places of varying resource endowments. The methods used and products derived need to inform policy-makers, who further influence the systems through feedbacks and linear and non-linear responses (Fig. 25.8). Therefore, the evaluation of the landscape through remote sensing approaches is iterative, sensitive to short- and long-term forces of change, and influenced by the scales of observation.

Table 25.1 is a representative example of the type and scale of socioeconomic data that has been advocated for LULC research by the International Geosphere-Biosphere Program (IGBP) <http://www.igbp.kva.se/cgi-bin/php/frameset.php> (6 March 2002) and the International Human Dimensions of Environmental Change Programme (IHDP) <http://www.uni-bonn.de/ihdp> (6 March 2002). While many social datasets contain fine-scale data at the household level, it is uncommon to find work that explicitly links household socioeconomic data with satellite-based LULC data. More often, work that links land-cover data to social science data tends to favor high levels of aggregation.

There has been relatively little work that links specific households units to satellite-based measures of LULC occurring on household plots despite the fact that many decisions about land use affecting land cover are likely made at the household level. A comprehensive understanding of the forces that drive LULCC requires studies of the decisions households make about land use, in addition to policy and environmental processes at work at higher levels. The few studies that have operated at the household level have focused mainly on Amazonia, where pieces of land extend back from roads in a fishbone or piano key pattern, and the household responsible for that land lives in a dwelling unit located within the parcel's boundaries (Moran *et al.* 1994). In Amazonia, by locating the parcel of land on a remotely sensed image, defining its location via digitization and overlay of existing cadastral maps, and finding and interviewing members of the household, researchers have been able to link household-level social data with LULC data (McCracken *et al.* 1999). Unfortunately, in many developing countries, cadastral maps may not be available. In the absence of cadastral maps, it can be very difficult to define household plots, thereby complicating

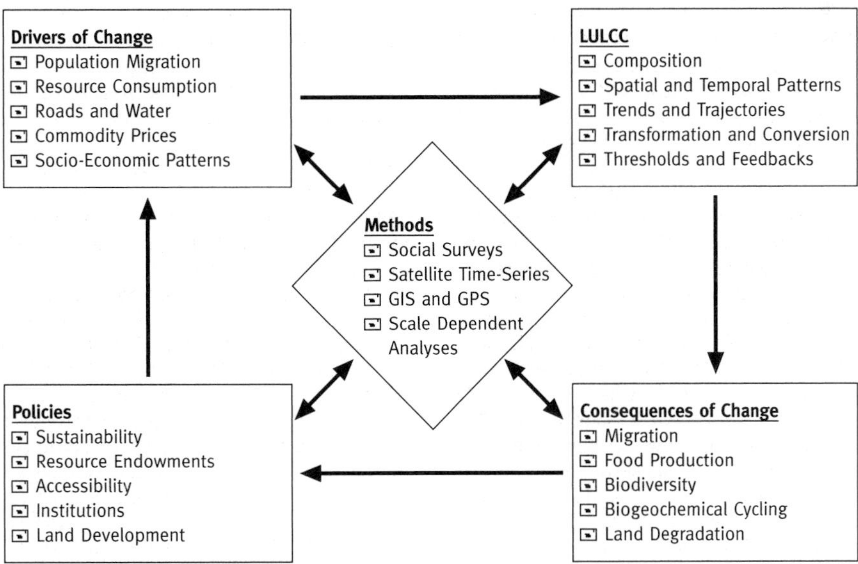

Fig. 25.8 Flow diagram illustrating the methods used, products derived, and overall interactions between them for input to policy-making decisions

Table 25.1 *Selected socioeconomic data advocated for land-use and land-cover research by the International Geosphere-Biosphere Programme*

Data	Spatial Scale			
Food and energy consumption	Household	District	National	Global
Income and savings	Household			
Social cohesion	Household			
Farming techniques	Household			
Attitudes to risk	Household		National	
Attitudes to the environment	Household	District	National	
Employment opportunities		District	National	
Literacy and education	Household	District	National	
Gender rules	Household			
Resource access and use	Household	District		
Prices (inputs and outputs)		District	National	Global
Markets		District	National	Global
Fertility rates	Household	District	National	
Migration	Household	District	National	
Land tenure and ownership	Household			
Communication		District	National	
Technology	Household	District	National	Global
Urbanization		District	National	
Trade	Household	District	National	Global
Population indicators	Household	District	National	Global
Health	Household	District	National	
Pests and diseases		District	National	Global

Source: IGBP/IHDP-LUCC and IGBP-DIS (1997).

the linkage of satellite data to household plots and requiring that other methods be used, such as field-based GPS measurements of plot locations. With locational information about lands used or owned by a household, plot histories can be assembled through social surveys and a satellite time-series used to define LULC and LULCC patterns depending upon the spatial-temporal resolution of sensor systems. Linear mixture modeling has been used to define LULC proportions within a defined pixel in an attempt to decompose pixel response patterns to their component parts relative to the concept of an integrated pixel. In a number of environments, the size of LULC plots associated with a single household may be below the spatial resolution of customary LULC mapping systems (e.g. Landsat TM). In such situations, using field-based approaches such as a spectroradiometer to define spectral responses for specific landscape elements, and the integration of data from higher spatial resolution systems (e.g. IKONOS data) for rendering detailed LULC information for scaling to regional study area dimensions, show promise for extending the usability of remote sensing systems where spatial resolution is a constraining factor.

The examples described above all require the use of a variable and socially defined, yet spatially explicit, unit of observation. For example, national boundaries and household plot boundaries come in many different shapes and sizes. The linkage method basically involves overlaying a defined boundary on either preprocessed satellite data or a satellite-derived product (e.g. NDVI (Normalized Difference Vegetation Index) or LULC). In these cases, pixels are aggregated and linked to the appropriate social entity (e.g. a country, district, village, or household). LULC is reported as the percent area of LULC types occurring within the social entity, with spatial reference to a discrete point location specifying, for example, the centroid of a nuclear village settlement pattern or the geographic center of a district or sub-district. This linkage direction has been referred to as "linking pixels to people." Another approach works in the opposite direction by linking people to pixels using distribution models or population potential surfaces. For example, Walsh *et al.* (1999) used a distribution model to spread village population data, represented at discrete points, to individual pixels continuously covering the study area. Statistical models were then run using a sample of pixels to assess both human and environmental associations in conjunctions with an NDVI. Walsh *et al.* (2001) performed a similar type of analysis using a population potential model to place people on the landscape relative to friction coefficients that indicated the ease of movement across the landscape and distance decay exponents for discerning the geographic reach of the population across the landscape.

3. *Emergent Approaches for Linking Population and the Environment.* The distinction between land use and land cover and the exploration of LULC dynamics is emerging as the dominant theme in applied remote sensing. The advantage of remote sensing in LULC exploration is that the remotely sensed data are easily interpretable with minimal inference. The scales, patterns, and processes of LULCC vary in response to spatial, social, and environmental forces. While the majority of land-use studies are executed at regional to global levels producing conceptual models of LULCC, the need to address actual local conditions and policy ramifications requires a multi-scale dynamic modeling scheme (Turner *et al.* 1995). To combine these social and physical variables in a local framework at multiple scales of aggregation, the use of hierarchical modeling methods has come to the forefront in conjunction with multi-scalar analysis of remotely sensed data.

Hierarchy theory, a subset of general systems theory, refers to levels of organization and the spatio-temporal relationships among those levels. In landscape ecology, hierarchy theory has been used to describe the structure of ecological systems. In complex systems, different organizational levels can be distinguished; hierarchy theory emphasizes that a proper understanding of a phenomenon requires that references to the next higher and lower scales of resolution be made (Odum 1994). Essentially, phenomena can be dissected from the complex spatio-temporal context (O'Neill 1988). The complexity of data and interactions defining an environmental matrix are, in part, remotely observable. One goal of remote sensing activities, and LULCC modeling in particular, is the extraction of the whole system by identifying and defining the relevant parts.

While hierarchy and scale are related concepts, they have different implications with respect to remote sensing. Hierarchy refers to levels of organization, while scale refers to the respective spatial or temporal dimension (Turner and Gardner 1991). The landscape ecology concepts most closely aligned with scale are grain and extent. Grain refers to the size of the smallest units that can be distinguished (resolution), while extent refers to the size of the study area (loosely related to synoptic coverage or image footprint). It is not possible to identify the whole entity unless the extent is large enough to include all the parts. Historically, it was considered impossible to detect phenomena finer than the grain of observations. However, as sensor designs have changed and, more importantly, as changing research directions have targeted such sub-pixel analyses, the fuzzy or partial classification has appeared. As different ecological processes are dominant at different spatial and temporal scales, the interpretation of ecosystems is highly dependent upon the scale of observation. This can, for example, be examined by applying different grains for the units of analysis and comparing the results (Walsh *et al.* 1997, 1999). These methods may lead to a better understanding of which processes dominate at different spatial scales. However, no a priori fundamental hierarchical levels are imposed by nature, and so the applied scales depend on the problem and study area. Furthermore, hierarchies can change with time, due to the resilience and adaptability of systems. To account for scale effects, a hierarchical approach offers insights through concept and practice in the study of complex systems (Kolasa and Pickett 1991; Messina and Walsh 2001).

As hierarchical and multi-scalar methods of landscape evaluation have emerged, the methods for classifying both the physical environment and the representative images have changed. Classification is the process of assigning class membership to all pixels within an image. The relationship between the spectral classes and the physical environment is determined to create informational classes. Once the informational classes are derived, comparisons are possible among image products derived from different years. Historically, lack of direct correspondence between the spectral classes and the informational classes was cause for serious concerns regarding the utility of the data.

Change detection processes have typically followed one of a series of methods. Image algebra (e.g. differencing and ratioing) applies arithmetic operations to pixels in each image, and then forms the change image from the resulting values. Post-classification change detection requires two or more independent classifications created using similar methods. Multi-date composites are assembled from multiple images into a single unit using statistical or traditional classifiers. Vector analysis compares both the spectral position of each pixel in spectral space and locations among image time-series data (Allen and Kupfer 2001). Texture-based classifiers attempt to categorize the image data based upon spectral and spatial relationships among neighboring pixels (Jensen 2000).

When considering LULC systems, geographers usually think of autonomous systems proceeding with or without human interference. However, processes of change reveal a succession of states. Those states function to varying degrees as ecological units. By modifying or regulating human activity, it is possible to control the manifestation of change upon the landscape and consequently, the processes of change are linked to those controls that affect the unfolding LULC sequence. To build a dynamic model of an LULC system, it is important to select the level of detail or scale that is useful and then capture the laws of change at that level of detail (Messina *et al.* 1999; Messina and Walsh 2001). The art of model-building revolves around selecting the level of detail that permits the development of simple and parsimonious processes that lead to theory.

The development of stochastic or deterministic models characterizing the environment requires the use of surrogate landscape variables. Remotely sensed data are by their nature classified according to a prescribed variety of characteristics, either ones of interest to science or more commonly ones that are technically feasible. In this artificial classification, characterization takes the form of pixels further modeled through some classification or surrogate data analysis scheme. These derived data then become inputs and validations for all manner of environmental LULCC models. The difficulty in modeling human–environment interactions and LULC dynamical systems has historically been the necessary dual simulation mode of model construction. Human systems are typically modeled using stochastic methods, while many natural systems are adequately modeled deterministically. Combining the two methodologically polar components into an effective approximation of reality requires the use of alternative modeling techniques.

Cellular automata modeling, used throughout the 1990s in the urban and social simulation community, has surfaced in geography for its ability to represent systems using minimal empirical evidence (Clarke *et al.* 1997). Quite often it is the inability to collect in situ data effectively and functionally that leads to the use of remotely sensed data. While it is possible to represent the elements and interactions of a system without explicitly trying to explain their interdependencies and distributions, a more complete model of LULCC should also contain one or more functional explanations for the interactions observed. Remotely sensed data serve as one of the best sources for all manner of LULC dynamical system models. The new sensors providing higher spatial, spectral, and temporal resolution feed this growing interest in general systems theory and dynamical systems modeling.

4. *Emergent Opportunities for Remote Sensing in the Future.* The policy relevance of both scale and human–environment interactions as evidenced above is a growing issue for geographers. At a recent NASA LULC workshop, a program director from the National Science Foundation stressed the need for environmental research addressing questions of how humans are impacted and in turn impact the environment, and challenged the LULC community to conduct research relevant to decision-makers to effect sound environmental policy. Remote sensing scientists hold a unique advantage in addressing this stated need in that the synoptic nature of their data lends itself to decision support systems; the challenge remains to integrate the human and biophysical components, and to characterize LULC composition and spatial and temporal dynamics through multiple techniques operating within a spatially explicit context. The application of a case study approach to detect emergent principles and relationships that can then be applied to other areas and other times would seem useful. For results to be generalizable to other studies, they need to be spatially explicit and of commensurate resolutions (especially spatial resolution). Scale-dependent studies have also begun on the human side: that is, in settings where the human imprint is pronounced and where proximate and direct causes for understanding human–environment interactions are attributed to society or the linkage among social, biophysical, and geographical domains, manifested as exogenous or endogenous variables. At present, scale dependence research is being pursued by replicating approaches that have been successful with biophysical variables, processes, and systems. The nature and confidentiality of demographic data, mobility of the human population, importance of social versus agglomeration units for scaling, and the object–field dichotomy in human–environment representation, however, each point to the need for theoretical as well methodological advances in linking people and the land. This is because characterization of the land through remote sensing systems, in conjunction with the feedbacks and thresholds between population and the environment, adds complexity to the social-biophysical milieu.

The synthesis of synoptic coverage (to model state or condition) and local parameters (to model process) generates not only a multi-scale understanding of landscape change, but more importantly, a spatially explicit path for generalization and interpretation of findings across scales and across research projects. Spatially explicit data allow the union of disparate data sources such as satellite imagery and social surveys. Geographers know the importance of a spatial perspective—that location and organization matter to both ecological and anthropogenic processes. A spatial perspective is especially important for understanding human–environment interactions, where layers of spatial dependence in "natural" and "human" ecosystems intertwine into a complex web of feedbacks and possible non-linear responses. Setting both the physical (via remote sensing) and the human (via social surveys) elements into the appropriate geographic and spatial context is what allows for the assessment and possible prediction of both human population dynamics (e.g. migration) and socially significant events (e.g. droughts). These models of human–environment interactions are increasingly represented through LULCC, and are built through both direct or observable human impacts on LULC (e.g. deforestation, road building, and water works), as well as through proximate and non-observable human forces (e.g. political, social, economic, and cultural institutions).

One of the challenges to research of human–environment interactions, remote sensing analysis, and geographic thought, is to generally define questions that are relevant to policy-makers and the larger population. Not coincidentally, a number of recent Request for Proposals (RFPs) aimed at the remote sensing community have included statements directing researchers to state explicitly how their findings are policy-relevant and how results will contribute to the greater good of society and assist decision-makers in formulating more effective policies. Assessing policy and institutional change is inherently difficult, in that such changes are rarely directly observable. Furthermore, the policy record (should it exist) reflects changes that are intended, and may differ greatly from actual change observed due to political feasibility, intervening events, time lags, or implementation gaps. While it is not the role of remote sensing analysts to judge the merit of policies aimed at environmental management, they can assist in measuring the outcomes of such management to infer effectiveness. For example, the myriad of policies that affect global carbon cycling (including but not limited to those affecting inflation rates, economic development, commodity pricing, logging bans, and land management) interact in a variety of ways, and while their relative influences are open to debate, their impact in the landscape is less so. Forest cover may increase, decrease, reorganize spatially, or remain stable, but such measures are observable, quantifiable, and therefore able to be compared across time and space through remote sensing technologies. While such measures may or may not encompass the totality of a causal explanation for observed conditions, they do provide the conditions themselves. Environmental monitoring programs must first start with an inventory, and then move to repeated inventories to monitor change. Because of the benefits associated with remotely sensed data including synoptic coverage, targeted spectral range, and repeatability, the remote sensing community is in a unique position to contribute to answering some of the most pressing human–environment issues of our time—global carbon cycling, deforestation, habitat loss, and loss of biodiversity.

Assessing Cultural Information from Remote Sensing-Derived Data

The general process of obtaining information from remote sensing Earth observation is shown in Fig. 25.9

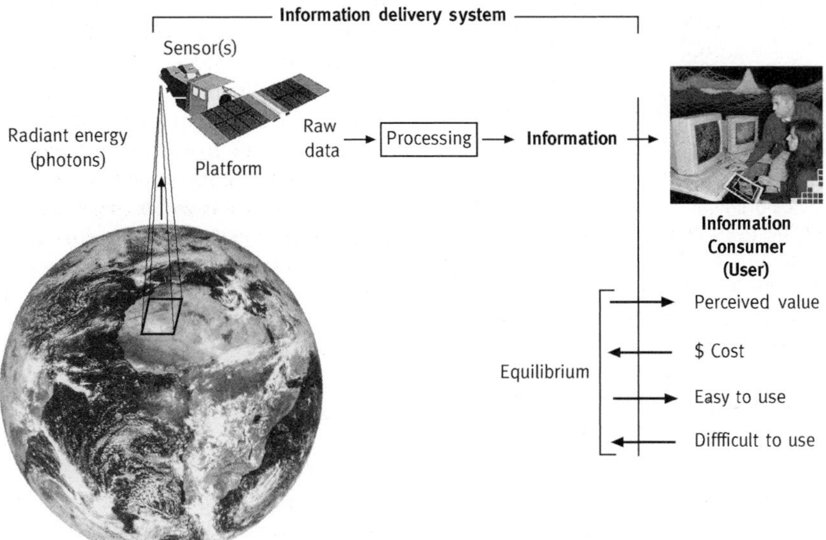

Fig. 25.9 Remote sensing Earth observation economics

Table 25.2 *Urban/suburban attributes and the minimum remote sensing resolutions required to provide such information*

	Minimum Resolution Requirements		
Attributes	Temporal	Spatial	Spectral
Land Use/Land Cover			
L1: USGS level 1	5–10 years	20–100m	V-NIR-MIR-Radar
L2: USGS level 2	5–10 years	5–20m	V-NIR-MIR-Radar
L3: USGS level 3	3–5 years	1–5m	Pan-V-NIR-MIR
L4: USGS level 4	1–3 years	0.25–1m	Panchromatic
Building and property infrastructure			
B1: building perimeter, area, height, and cadastral information (property lines)	1–5 years	0.25–0.5m	Pan-Visible
Transportation infrastructure			
T1: general road centerline	1–5 years	1–30m	Pan-V-NIR
T2: precise road width	1–2 years	0.25–0.5m	Pan-V
T3: traffic count studies (cars, airplanes, etc.)	5–10 min	0.25–0.5m	Pan-V
T4: parking studies	10–60 min	0.25–0.5m	Pan-V
Utility infrastructure			
U1: general utility line mapping and routing	1–5 years	1–30m	Pan-V-NIR
U2: precise utility line width, right-of-way	1–2 years	0.25–0.6m	Pan-Visible
U3: location of poles, manholes, substations	1–2 years	0.25–0.6m	Panchromatic
Digital elevation model (DEM) creation			
D1: large scale DEM	5–10 years	0.25–0.5m	Pan-Visible
D2: large scale slope map	5–10 years	0.25–0.5m	Pan-Visible
Socioeconomic characteristics			
S1: local population estimation	5–7 years	0.25–5m	Pan-V-NIR
S2: regional/national population estimation	5–15 years	5–20m	Pan-V-NIR
S3: quality of life indicators	5–10 years	0.25–30m	Pan-V-NIR
Energy demand and conservation			
E1: energy demand and production potential	1–5 years	0.25–1m	Pan-V-NIR
E2: building insulation surveys	1–5 years	1–5m	TIR
Critical environmental area assessment			
C1: stable sensitive environments	1–2 years	1–10m	V-NIR-MIR
C2: dynamic sensitive environments	1–6 months	0.25–2m	V-NIR-MIR-TIR
Disaster emergency response			
DE1: pre-emergency imagery	1–5 years	1–5m	Pan-V-NIR
DE2: post-emergency imagery	12 hr–2 days	0.25–2m	Pan-V-NIR-Radar
DE3: damaged housing stock	1–2 days	0.25–1m	Pan-V-NIR
DE4: damaged transportation	1–2 days	0.25–1m	Pan-V-NIR
DE5: damaged utilities, services	1–2 days	0.25–1m	Pan-V-NIR
Meteorological data			
M1: weather prediction	3–25 min	1–8km	V-NIR-TIR
M2: current temperature	3–25 min	1–8km	TIR
M3: clear air and precipitation mode	6–10 min	1km	WSR-88D Radar
M4: severe weather mode	5 min	1km	WSR-88D Radar
M5: monitoring urban heat island effect	12–24 hr	5–30m	TIR

Source: Jensen and Cowan (1999).

(Jensen 1996). The first step involves relatively sophisticated digital (or analog) image processing to collect and convert the raw remote sensor data into calibrated, geocoded (accurate x, y, z location), reflectance, emittance, or backscattered data. The knowledge base associated with pre-processing remotely sensed data is well developed with many robust algorithms that actually work (e.g. Jensen 1996; Schott 1998). It is then possible to perform specialized types of analyses on the pre-processed radiometrically and geometrically calibrated remote sensor *data* and convert it into *information* that is of value to users such as geographers, planners, and/or engineers. Good examples might be (1) the creation of an accurate digital elevation model of an urban landscape using LIDAR data to locate a cellular phone tower, or (2) the accurate mapping of residential building perimeter and height, sidewalks, and rights-of-way using stereoscopic vertical aerial photography and photogrammetric techniques to perform a population estimation. In both cases, the information derived must be of sufficient value and be easy enough to use that it justifies the added expense of conducting a remote sensing investigation.

1. *Innovations in Remote Sensing Systems and Information Extraction Methods Used to Assess Cultural Information: Satellite and Airborne Sensors and Analysis Techniques.* Jensen and Cowen (1999) reviewed many of the urban/suburban data collection requirements of geographers, planners, and engineers as they manage cities, counties, or councils of government. How often the data or information must be collected (i.e. its temporal resolution) and the range of remote sensing spatial resolutions (e.g. 1 × 1m; 1 × 1ft ground resolved distance) necessary to extract the desired information are summarized in Table 25.2 and depicted as ellipses in Fig. 25.10. The temporal and spatial resolution of existing and proposed remote sensing systems that can provide the desired information are shown as shaded rectangles in Fig. 25.10. It is apparent from this and Table 25.2 that many urban/suburban applications require very high spatial resolution remote sensor data, often ≤ 1 × 1m.

2. *Urban/Suburban Land Use and Land Cover.* Geographers and other scientists constantly monitor land use change (conversion) within the city and at the urban fringe to understand the developmental processes that are taking place (Cullingworth 1997). They use a variety of remote sensing instruments and image analysis methods to extract the desired urban/suburban land-use or land-cover information (American Planning Association 1998). At one end of the continuum is large-scale aerial photography which continues to yield most of the high resolution USGS Level 3 and 4 land-use or land-cover information (Anderson *et al.* 1976; Colwell 1997). The characteristics of a typical aerial analog frame camera are shown in Fig. 25.11a. Recently, digital frame cameras based on area-array charge-coupled-device (CCD) technology have become available (Light 1999; Fulghum 2000). Digital frame cameras (Fig. 25.11b) contain multiple arrays (matrices) of detectors that are sensitive to specific regions of the electromagnetic spectrum. Numerous private commercial firms such as Emerge & Landcare Aviation, Inc. and Image America, Inc. have developed specialized digital camera remote sensing systems (Light 1999). On-board kinematic Ground Positioning Systems (GPS) and Inertial Measurement Units (IMU) obtain positional x, y, z information at the exact instant that each pixel is recorded, making it possible to arrange many flight lines of these digital images into accurate, seamless mosaics of orthophotography using very few ground control points and a digital elevation model. For example, a 30.5cm (1 × 1ft) digital metric camera image of Harbor Town, Hilton Head, South Carolina obtained using an area-array sensor with 3,072 by 2,048 pixels per frame is shown in Fig. 25.12. A church in Missouri recorded at 7.62 × 7.62cm (3 × 3in) spatial resolution using a digital frame camera with 32,000 by 8,000 pixels per frame is shown in Fig. 25.13.

Levels 1–3 nominal scale land-use and/or land-cover information may be obtained using a variety of orbital and sub-orbital remote sensing systems (Green *et al.* 1994; Skole 1994). Satellite multispectral scanning remote sensing systems (Fig. 25.11c) such as the Landsat Multispectral Scanner (MSS) or Landsat 7 Enhanced Thematic Mapper Plus (ETM+) having a spatial resolution of approximately 20–100m can provide levels 1–2 land-use/cover information. For example, a simple level 1 land-cover map of urban information of Charleston, South Carolina, obtained using Landsat MSS 80m data is shown in Fig. 25.14.

Digital linear-array remote sensing technology (Fig. 25.11d) has now been placed in space and is of significant benefit for relatively high resolution land-use and land-cover mapping. For example, Space Imaging, Inc. launched IKONOS-2 in 1999. The IKONOS-2 sensor system obtains panchromatic imagery at 1 × 1m and four bands of multispectral data at 4 × 4m. An IKONOS-2 panchromatic image of the Columbia, South Carolina airport is shown in Fig. 25.15. The aforementioned soft-copy photogrammetric techniques can be applied to the IKONOS-2 stereoscopic 1 × 1m panchromatic data to extract many of the types of information listed in Table 25.2.

Imaging spectrometry is the simultaneous acquisition of images in many relatively narrow, spectral bands

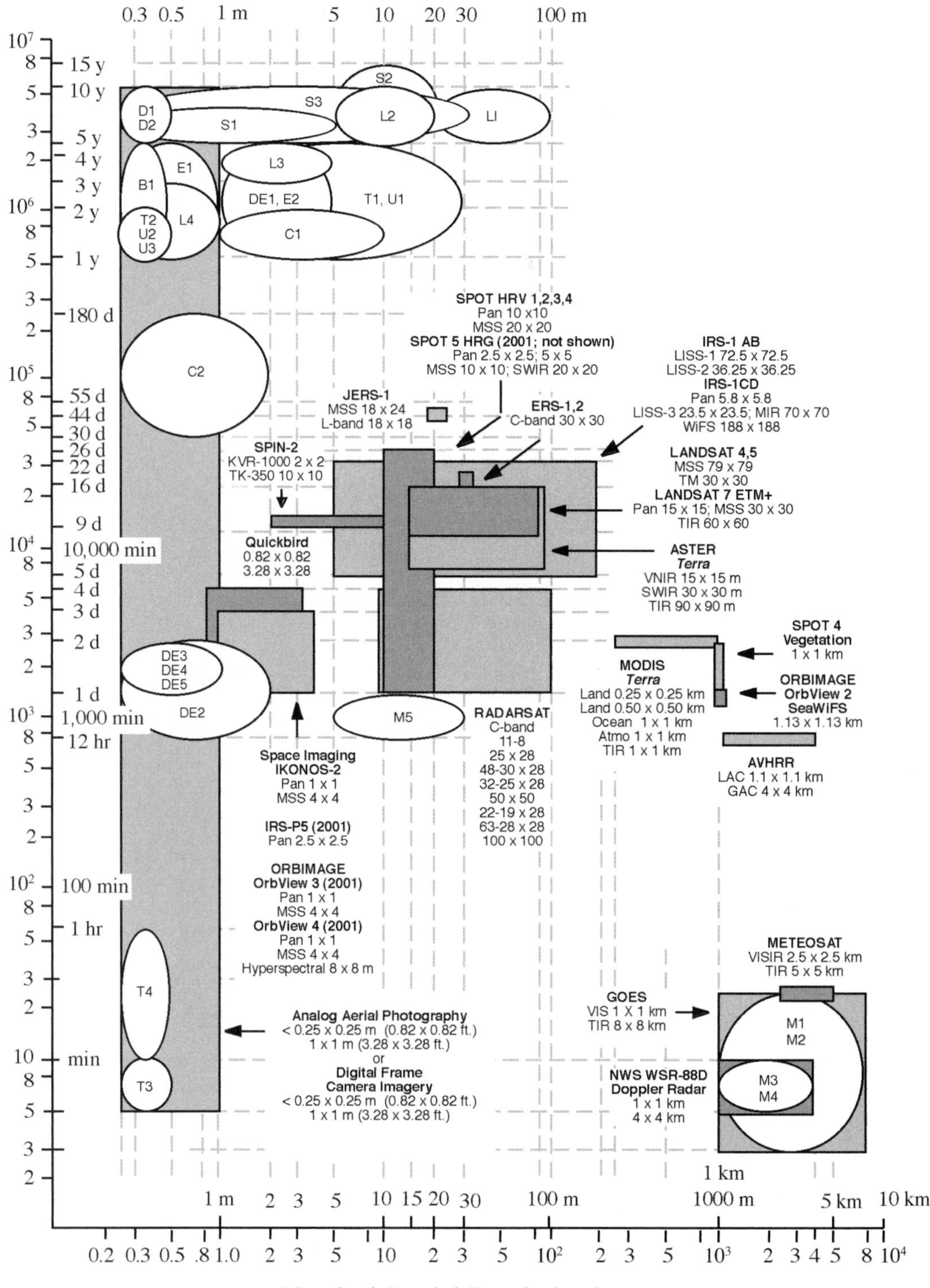

Fig. 25.10 Nominal spatial resolution in meters for remote sensing satellite data

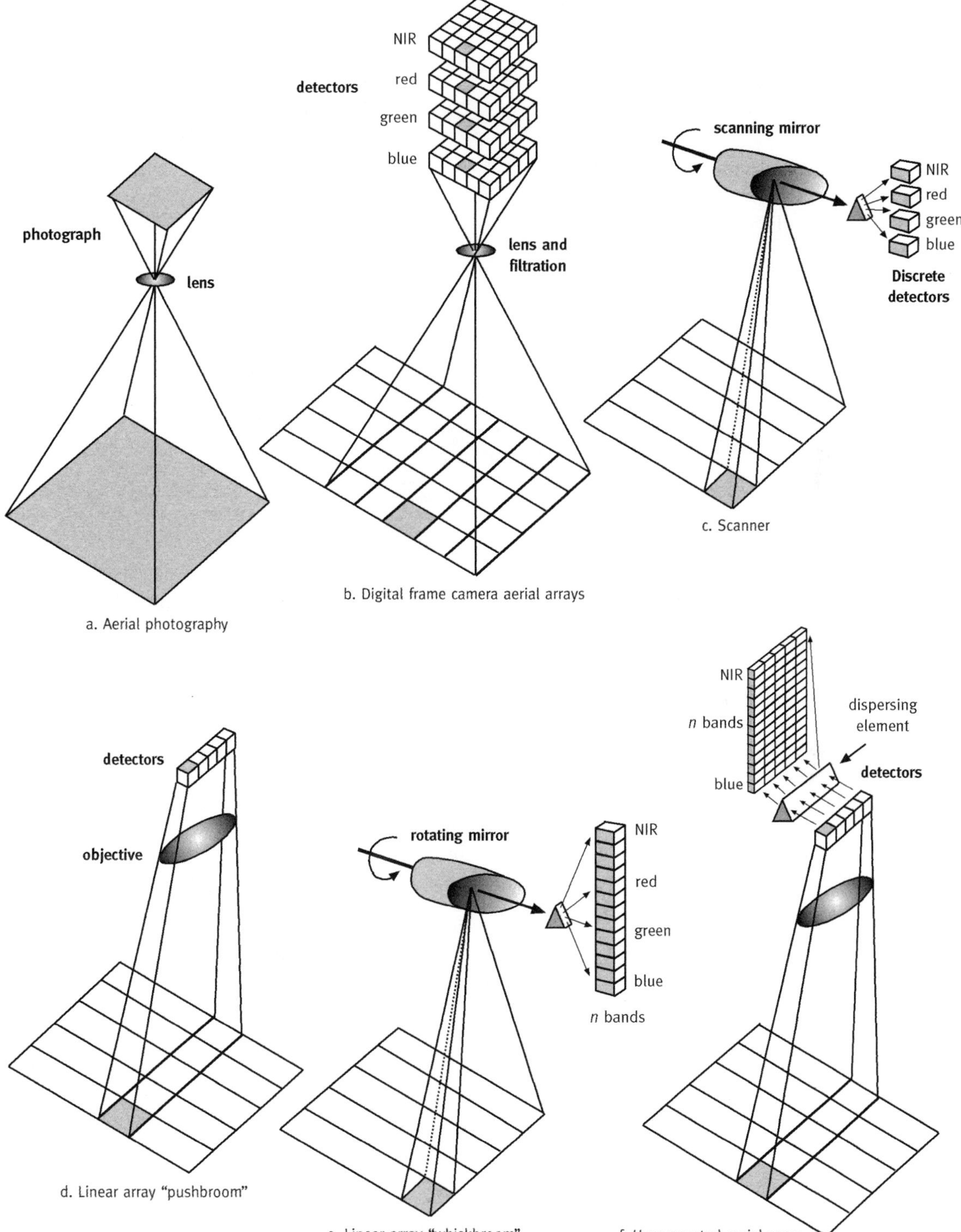

Fig. 25.11 Operational characteristics of imaging devices

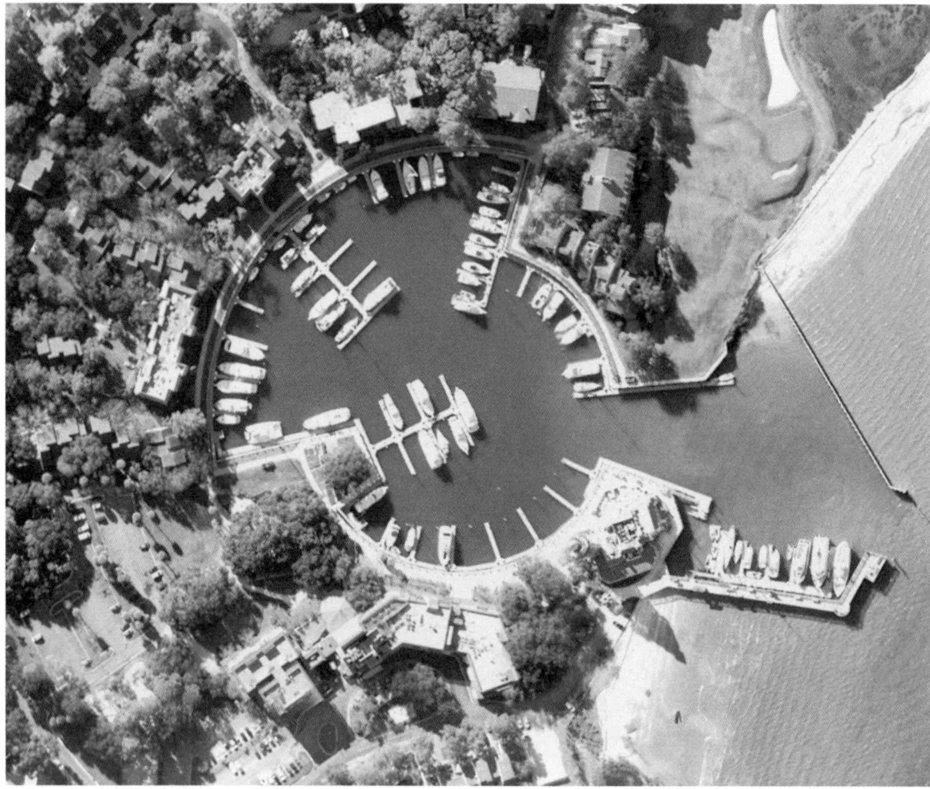

Fig. 25.12 Digital metric camera image of Hilton Head, South Carolina

throughout the ultraviolet, visible, and infrared portions of the electromagnetic spectrum. Many surface materials (e.g. asphalt and concrete road surface materials and right-of-way land cover such as vegetation, soil, and rock) have diagnostic absorption features that are often only 20–40nm wide. Therefore, hyperspectral remote sensing systems that acquire data in contiguous 10nm bands may sometimes capture spectral data with sufficient resolution for the direct identification of those materials with diagnostic spectral features. Two practical approaches to imaging spectrometry are shown in Figs 25.11e and f. The "whiskbroom" scanner linear array is analogous to the scanner approach used for the ETM+, except that radiant flux from within the IFOV is passed onto a spectrometer, where it is dispersed and focused onto a linear array of detectors (Fig. 25.11e). Thus, each pixel is simultaneously sensed in as many spectral bands as there are detector elements in the linear array. This is how NASA's AVIRIS sensor collects data. In 1998 and 1999, AVIRIS was flown on an NOAA Twin Otter aircraft acquiring data at an altitude of approximately 3,810m (12,500ft) above ground level (Fig. 25.16). This resulted in hyperspectral data with a spatial resolution of approximately 3.5 × 3.5m. For example, Fig. 25.17 depicts the spectral signature of healthy and potentially stressed bahia grass on clay-capped hazardous waste sites on the Savannah River Site, South Carolina, derived from analysis of AVIRIS data. The only hyperspectral satellite remote sensing system is the Moderate Resolution Imaging Spectrometer (MODIS) on board the NASA Terra satellite. It collects data in 36 co-registered spectral bands. MODIS's coarse spatial resolution ranges from 250 × 250m (bands 1–2), to 500 × 500m (bands 3–7) and 1 × 1km (bands 8–36). MODIS has one of the most comprehensive radiometric calibration systems and can be used to obtain level 1 land-cover information.

Significant advances in the processing of multispectral remote sensing have occurred. For example, there are improved multispectral classification algorithms based on (1) more robust statistical pattern recognition (e.g. Lillesand and Kiefer 2000); (2) expert systems (e.g. ERDAS's Knowledge Classifier); (3) fuzzy logic (e.g. Ji and Jensen 1996); and (4) neural networks (e.g. Jensen *et al.* 1999; Research Systems, Inc., 2000). A graphical

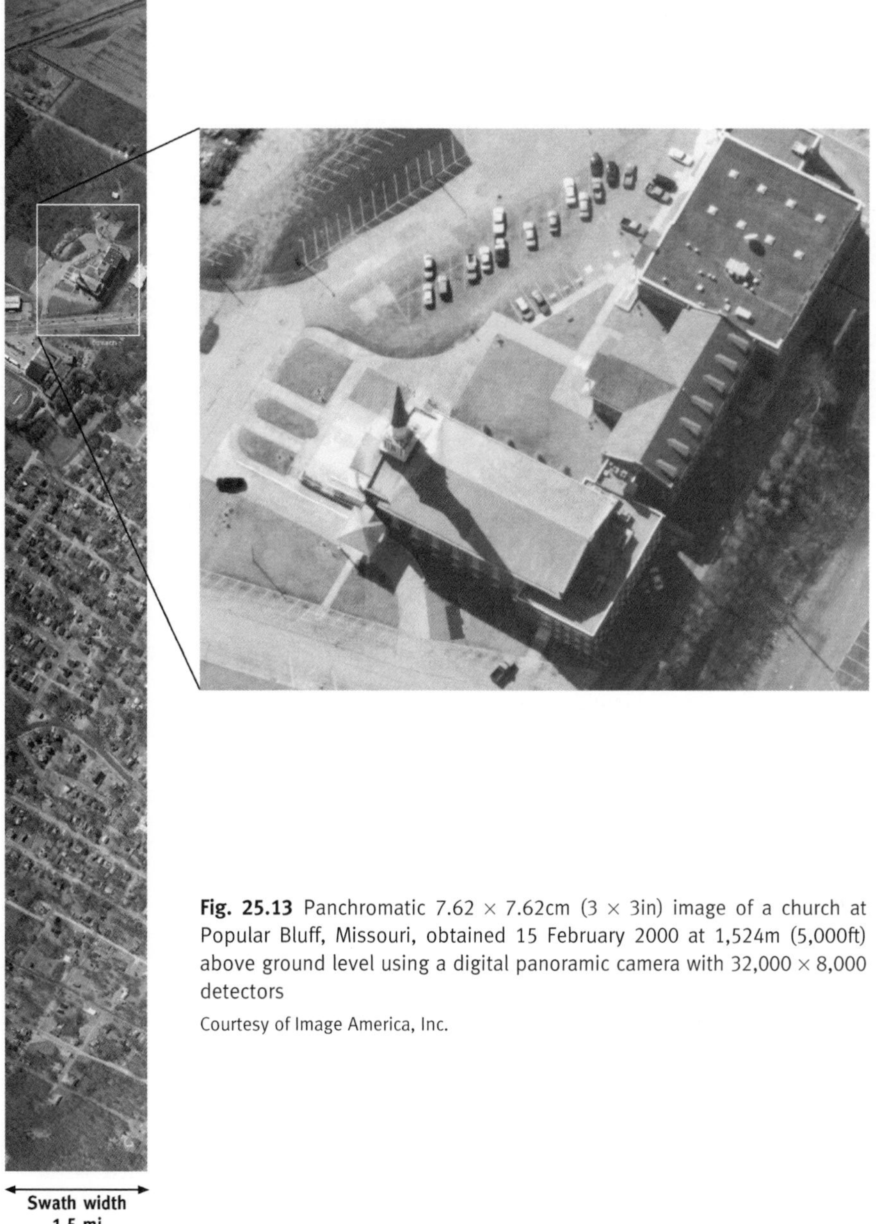

Fig. 25.13 Panchromatic 7.62 × 7.62cm (3 × 3in) image of a church at Popular Bluff, Missouri, obtained 15 February 2000 at 1,524m (5,000ft) above ground level using a digital panoramic camera with 32,000 × 8,000 detectors

Courtesy of Image America, Inc.

user interface associated with a neural network image classification system is shown in Fig. 25.18.

3. *Building and Property Infrastructure Information.* While land-use and land-cover information is of value for many urban applications, it is not as useful as information about the actual building infrastructure. The extraction of building infrastructure information today is based primarily on soft-copy photogrammetry (Greve 1996). This means that the individual frames of digital imagery are processed on personal computers using specialized photogrammetric software. The user works interactively with the stereoscopic three-dimensional models viewed on the CRT screen or conducts batch processing to achieve the desired result. While the

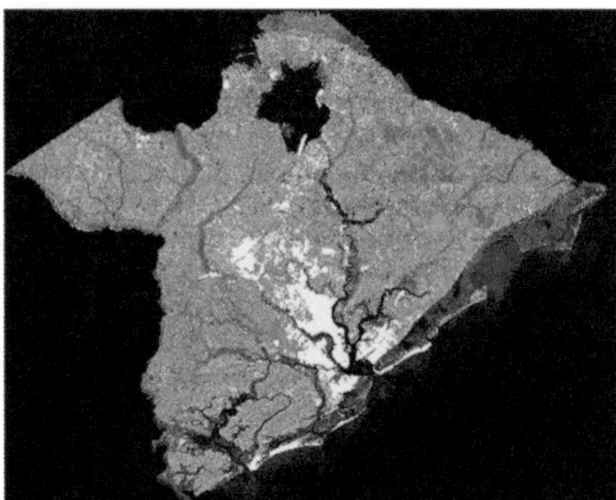

Fig. 25.14 Simple level-1 land-cover map of urban information for Charleston, South Carolina, obtained using Landsat MSS data

majority of the functions are complex, soft-copy photogrammetry visual analysis software is relatively easy to use by novice interpreters. For example, Fig. 25.19a depicts fundamental two-dimensional planimetric building and street infrastructure detail. Figure 25.19b displays three-dimensional capitol building infrastructure information derived from stereoscopic digital imagery. Digital imagery may also be overlayed onto digital elevation models to create unique three-dimensional landscapes that can be navigated using virtual reality techniques. For example, Fig. 25.19c depicts a three-dimensional image of a portion of Rosslyn, Virginia, draped over a digital elevation model and viewed from an oblique perspective. This three-dimensional soft-copy photogrammetry visualization environment is becoming very important for (1) extracting useful x, y, z three-dimensional urban/suburban infrastructure information, and (2) communicating concepts and proposed changes to groups of people in a three-dimensional environment.

4. *Transportation Infrastructure.* Geographers and engineers routinely analyze remote sensor data to: (1) update transportation network maps; (2) evaluate road, railroad, and airport runway and tarmac conditions; (3) study urban traffic patterns at congestion points such as tunnels, bridges, shopping malls, and airports; and (4) conduct parking studies (Haack *et al.* 1997; Jensen 2000). Street network centerline updating in rapidly developing areas may be performed every 1–5 years and in areas with minimum tree density (or during the leaf-off season) using imagery with a spatial resolution of 1–30m (Lacy 1992). If more precise road dimensions are required, such as the exact center of the road or the width of the road and sidewalks, then a spatial resolution of ≤ 0.2–0.5m is required (Jensen *et al.* 1994). Currently, only analog or digital aerial photography can provide such planimetric information.

Fig. 25.15 IKONOS panchromatic image of the Columbia, South Carolina, airport

Courtesy of Space Imaging

Road, railroad, and bridge conditions (cracks, potholes, etc.) are routinely monitored both in situ and using high spatial resolution remote sensor data (Fig. 25.20). Stereoscopic analysis of high spatial resolution imagery (≤ 0.25–0.5m) by a trained analyst can provide significant information about the condition of the road and railroad (Haack *et al.* 1997).

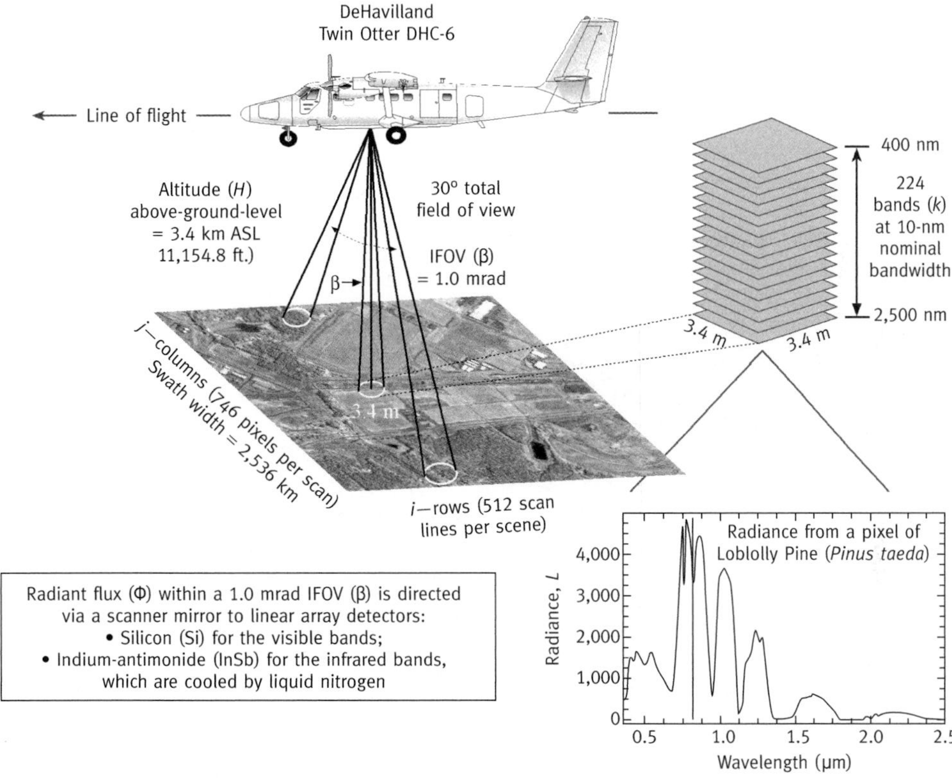

Fig. 25.16 Illustration of airborne hyperspectral data acquisition characteristics and spectral responses

Traffic count studies of automobiles, airplanes, boats, pedestrians, and people in groups require very high temporal resolution data ranging from 5 to 10 minutes. Even when such timely data are available, it is difficult to resolve a car or boat using even 1 × 1m data. This requires high spatial resolution imagery (≤ 0.25–0.5m). Such information can only be acquired using aerial photography, digital cameras, or video sensors that are either located on the top edges of buildings looking obliquely at the terrain, or placed in aircraft or helicopters and flown repetitively over the study areas. When such information is collected at an optimum time of day, future parking and traffic-movement decisions can be made. Parking studies require the same high spatial resolution (≤ 0.25–0.5m) but slightly lower temporal resolution (10–60 min).

5. *Socioeconomic Information.* Some socioeconomic information may be directly or indirectly extracted from remotely sensed data. Two of the most important types of socioeconomic information are population estimation and quality of life assessment. The most accurate remote sensing method of population estimation is to count individual dwelling units (Lo 1995; Holz 1988; Haack *et al.* 1997). The number (and type) of individual dwelling units are then multiplied by the average number of persons that are known to inhabit such dwellings in a given culture. For example, Fig. 25.21 depicts individual huts in a village in Africa. When required, this is performed every 5–7 years using high spatial resolution remotely sensed data (0.25–5m). The dwelling-unit approach is not suitable for a regional or national census of population because it is too time-consuming and costly (Sutton *et al.* 1997). Fortunately, there is relationship between population and the general urbanized built-up area (settlement size) extracted from a remotely sensed imagery (Olorunfemi 1984). Urbanized area (hectares) is multiplied by an empirically derived population value for the culture being examined. Another population estimation technique is based on more detailed levels 1–3 land-use information. Geographers establish an empirical value for the population density for residential land-use class. Summing the estimated totals for each land-use category provides the total population projection (Lo 1995). The urban built-up area and land-use data method can be based on coarser spatial resolution multispectral remote sensor

Fig. 25.17 AVIRIS data illustrating the spectral signature of healthy and potentially stressed bahia grass on clay

a. Color composite of the Mixed Waste Management Facility clay caps at the Savannah River Site, South Carolina (RGB = 50, 30, 20)

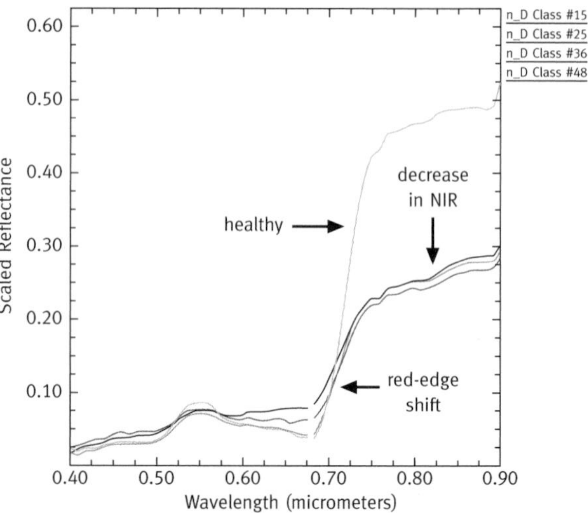

b. Spectral reflectance curves of four bahia grass endmembers (one healthy and three stressed)

c. Classification map showing those areas with the most potential bahia grass stress in yellow, red, and white

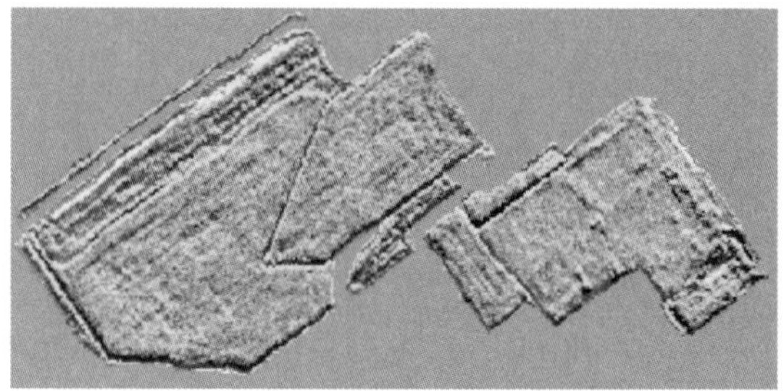

d. Second derivative image (RGB = bands x, x, x)

data (5–20m) every 5–15 years. Henderson and Xia (1998) discuss how Radar imagery can be of value when conducting population estimates using these same techniques.

Quality of life indicators such as house value, median family income, average number of rooms, average rent, and education can be estimated by extracting some of the urban attributes found in Table 25.3 using high spatial resolution (0.25–30m) imagery (e.g. Haack *et al.* 1997; Lo and Faber 1998; Jensen and Cowen 1999). Note that the attributes are arranged by *site* (building and lot) and *situation*. The remote sensing-derived attributes are correlated with in situ observations to compute the quality of life indicators.

6. *Utility Infrastructure*. Commercial, industrial, and residential land uses consume enormous amounts of electrical power, natural gas, and potable water, and require extensive telephone and cable infrastructure (Haack *et al.* 1997). These same land uses generate great quantities of refuse, waste water, and sewage. The most fundamental cartographic problem is to map the general centerline of the utility of interest such as a power line right-of-way (Philipson 1997). The majority of utility rights-of-way may be observed well on imagery with 1–30m spatial resolution obtained once every 1–5 years if the utility is not buried or obscured by trees (Feldman *et al.* 1995). When it is necessary to inventory the precise location of transmission towers, utility poles, manhole covers, the true centerline of the utility, the width of the utility right-of-way, and the dimensions of buildings, pumphouses, and substations then it is necessary to have imagery with a spatial resolution of ≤ 0.25–0.6m (Jadkowski *et al.* 1994). Ideally, new facilities are inventoried every 1–2 years.

7. *Urban Digital Elevation Model Creation*. Many urban/suburban applications require very accurate x, y, z elevation data as well as slope and aspect information. Photogrammetric methods of deriving digital elevation models have been with us since the 1930s. Therefore, this section focuses only on new developments associated

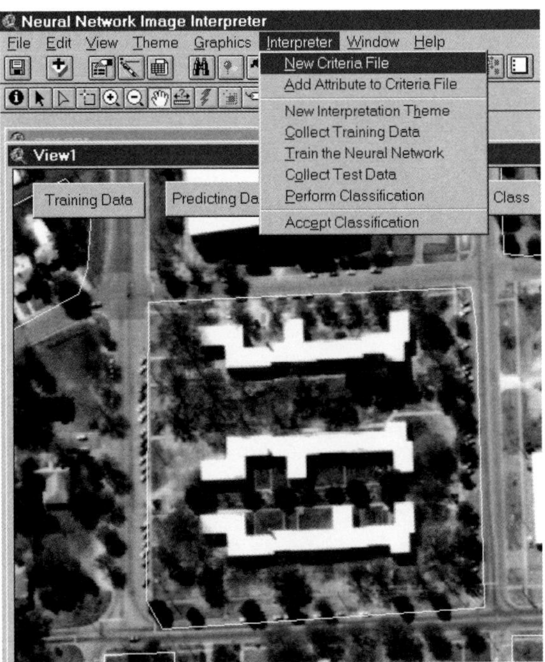

Fig. 25.18 Illustration of a neural network graphical interface with a neural network image classification system

with LIDAR and interferometric SAR remote sensing to extract accurate digital elevation models.

Many cities, counties, and even entire states are having private commercial firms (many of them photogrammetric engineering firms) collect LIDAR data to obtain accurate digital elevation models. LIDAR sensors measure active laser pulse travel time from the transmitter on board the aircraft to the target and back to the receiver (Cowen *et al.* 2000). As the aircraft moves forward, a scanning mirror directs the laser pulses back and forth across-track. This results in a series of data points arranged across the flight line. For example, Fig. 25.22a is a vertical aerial photograph of an area near Princeville, NC. LIDAR elevation postings obtained in a portion of the region are shown in Fig. 25.22b. Multiple flight lines can be combined to cover the desired area. Data-point density is dependent on the number of pulses transmitted per unit time, the scan angle of the instrument, the elevation of the aircraft above-ground-level, and the forward speed of the aircraft. The greater the scan angle off-nadir, the more vegetation that must be penetrated to receive a pulse from the ground assuming a uniform canopy (Jensen 2000).

LIDAR data avoids the problems of aerial triangulation and orthorectification because *each* LIDAR measurement is individually georeferenced (Flood and Gutellius 1997). Information about the scanning mirror and aircraft attitude allow precise determination of where the LIDAR instrument was pointed at the time of an individual laser pulse. Exact aircraft position is determined from on-board GPS and inertial navigation

Table 25.3 *Urban/suburban attributes that could be extracted from remote sensor data using the fundamental elements of image interpretation and used to assess housing quality and quality of life*

	Attributes
Site	*Building*
	single- or multiple-family
	size (sq. ft.)
	height (ft.)
	carport or garage (attached, detached)
	age (derived by convergence of evidence)
	Lot
	size (sq. ft.)
	front yard (sq. ft.)
	back yard (sq. ft.)
	street frontage (ft.)
	driveway (paved, unpaved)
	fenced
	pool (in-ground, above-ground)
	patio, deck
	outbuildings (sheds)
	density of buildings per lot
	percent landscaped
	health of vegetation (e.g. NDVI greenness)
	property fronts paved or unpaved road
	abandoned autos
	refuse
Situation	*Adjacency to Community Amenities*
	schools
	churches
	hospitals
	fire station
	library
	shopping
	open space, parks, golf courses
	Adjacency to Nuisances and Hazards
	heavy street traffic
	railroad or switchyard
	airports and/or flightpath
	freeway
	located on a floodplain
	sewage treatment plant
	industrial area
	power plant or substation
	overhead utility lines
	swamps and marsh
	steep terrain

Fig. 25.19 Illustration of soft-copy photogrammetry for assessment of building infrastructure

a. Building perimeter information extracted from vertical aerial photography using soft-copy photogrammetric techniques

b. Perimeter, height, and volume information extracted from stereoscopic aerial photography of the capitol using soft-copy photogrammetric techniques

c. Orthophoto of Rosslyn, Virginia, overlayed on a digital elevation model and viewed from an oblique perspective

Fig. 25.20 High spatial resolution image of railroad and road bridges

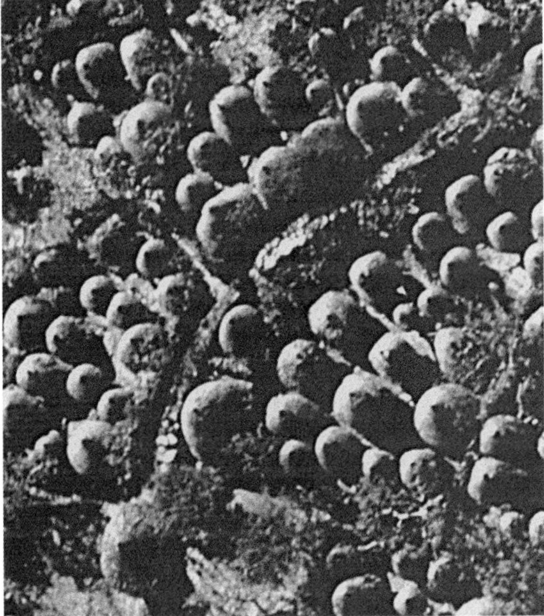

Fig. 25.21 High spatial resolution image of individual huts in Africa

equipment. The combination of all these factors allows three-dimensional georeferenced coordinates to be determined for each laser pulse.

To a certain degree, LIDAR can penetrate vegetation canopy and map the surface below. Penetrating the canopy is more difficult using aerial photography and Interferometric SAR. Although some of the LIDAR laser energy will be backscattered by vegetation above the ground surface, only a portion of the laser needs to reach the ground to produce a surface measurement. Both of these partial returns (vegetation and ground) can be recorded by the LIDAR instrument, allowing measurements of both vegetation canopy height and ground-surface elevation. When LIDAR imagery is obtained during leaf-off periods in the early winter, the effects of the tree canopy on the extraction of accurate elevation values is minimal. However, when LIDAR imagery is obtained during leaf-on periods the LIDAR-derived DEM may be usable, but less accurate. Digital elevation models for an area in North Carolina derived using LIDAR, photogrammetry, and interferometric SAR data are shown in Fig. 25.22c, d, and e, respectively. LIDAR is best acquired during leaf-off conditions and the algorithm used to reduce the effects of canopy penetration should be as accurate as possible. Figure 25.23 summarizes the relationship between LIDAR elevation data and geodetically surveyed elevation data for 1468 points in North Carolina obtained in 2000.

An exciting event of the past decade has been the development of Radar interferometry whereby Radar images of the same location on the ground are recorded by (1) two different antennas on the same platform (aircraft or satellite) at the same time (called single-pass interferometry), or (2) by a single antenna on an aircraft or spacecraft at different times. Analysis of the resulting two interferograms allows precise measurement of the range (distance) to any specific x, y, z point found in each image of the interferometric pair. Private commercial companies acquire interferometric synthetic-aperture Radar (IFSAR) data. For example, the Intermap X-band Star 3*i* system generates high quality 3 × 3m X-band microwave imagery plus a detailed digital elevation model of the terrain. Interferometric SAR data can provide extremely high precision topographic information (x, y, z) that can be just as accurate as digital elevation models derived using traditional optical photogrammetric techniques and LIDAR if certain conditions are met (e.g. the data are acquired when the leaves are off). Figure 25.22c depicts topographic information for an area in North Carolina extracted from interferometric SAR data.

The first satellite single-pass interferometric SAR was the Shuttle Radar Topography Mission (SRTM). It had a C-band and an X-band antenna separated by 60m (200ft) (Fig. 25.24a, b). SRTM data are being used to produce a digital elevation model of 80 per cent of the Earth at 20 × 20m postings. Such information is valuable for regional elevation, slope, and aspect information particularly in developing countries where such data have never before been available. Two examples of DEMs produced from SRTM data and overlayed with Landsat TM data (red, green, and blue bands or TM bands 1, 2, 3) are found in Fig. 25.24c and d. The first

a. Rectified color-infrared airphoto

b. Individual LIDAR postings overlayed on rectified aerial photography

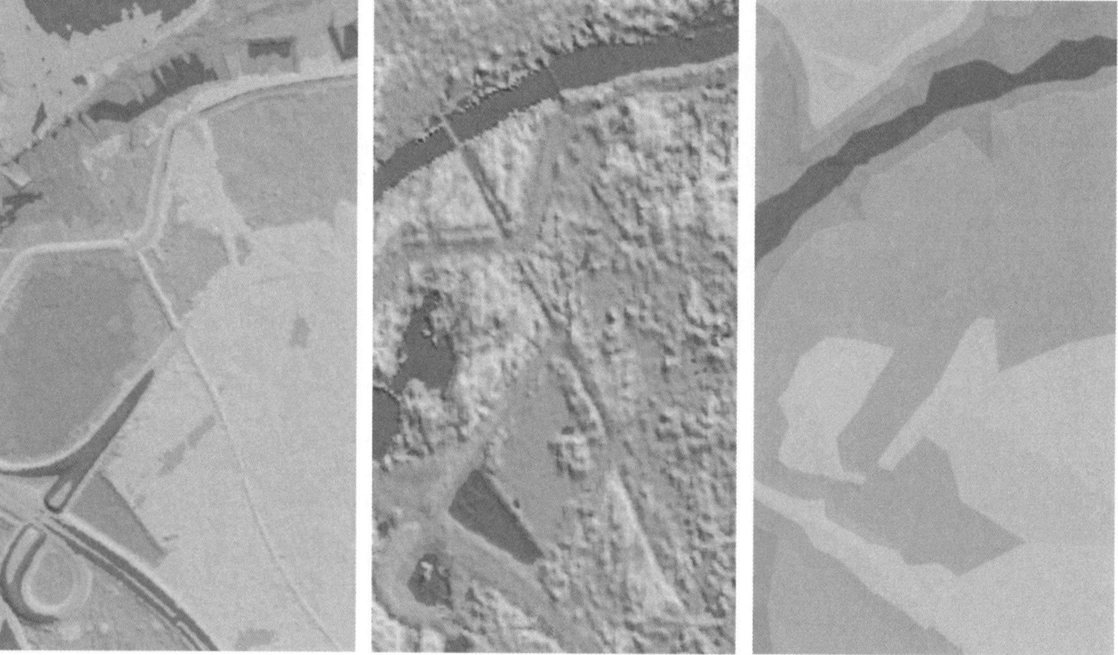

c. Digital elevation model derived from LIDAR data near Princeville, North Carolina

d. DEM derived from interferometric SAR data

e. USGS DEM obtained using photogrammetric techniques

Fig. 25.22 Illustration of LIDAR imagery for obtaining accurate digital elevation models

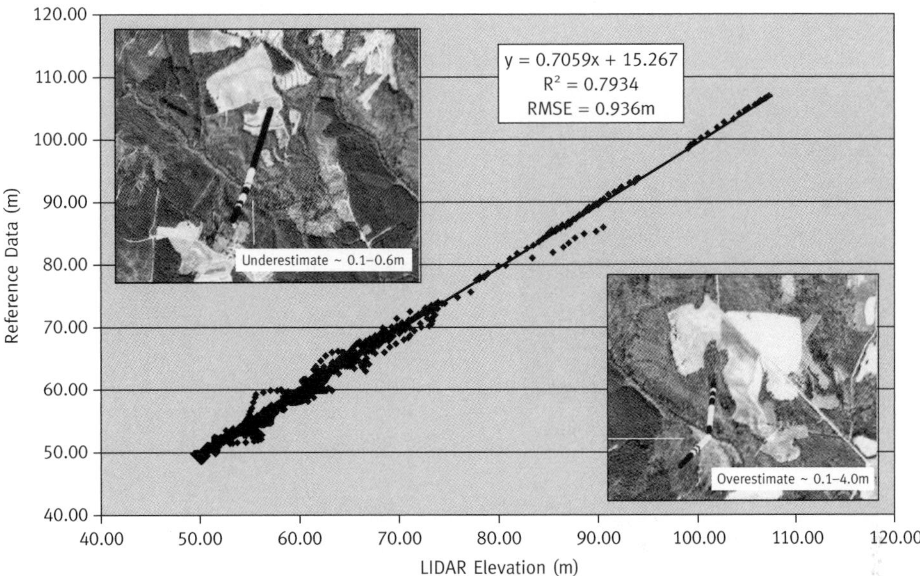

Fig. 25.23 Description of the relationship between LIDAR-derived elevation data and geodetically surveyed elevation data

DEM depicts the San Fernando Valley, California, that has a population of more than 1 million people. It has a view toward the northeast with a 3× vertical exaggeration. The second DEM is a 2× vertically exaggerated image of a portion of the San Andreas Fault near Bakersfield, California.

8. *Energy Demand and Conservation.* Power outages are becoming increasingly common. Energy demand may be estimated using remotely sensed data, particularly in developing countries. The square footage of individual buildings is determined from high spatial resolution imagery. Local information about energy consumption is obtained for a representative sample of single and multiple family homes in the area. Regression relationships are then derived to predict the anticipated energy consumption for the region. It is also possible to predict how much solar photovoltaic energy potential a geographic region has by modeling the individual rooftop square footage and orientation with known photovoltaic generation constraints. Both these applications require high spatial resolution imagery (0.25–0.5m) (Angelici *et al.* 1980; Curran and Hobson 1987). Finally, high spatial resolution (1–5m) predawn thermal infrared imagery (8–14mm) can be used to inventory the relative quality of housing insulation if (1) the rooftop material is known (e.g. metal, asphalt or wood shingles), (2) precipitation is not present on the roof, and (3) the orientation and slope of the roof are known (Eliasson 1992). If energy conservation or the generation of solar photovoltaic power were important, these variables would probably be collected every 1–5 years.

9. *Disaster Emergency Response.* A rectified, *pre-disaster* remote sensing image database is essential if important humanitarian and economic decisions are to be made rapidly after a disaster occurs. The pre-disaster remote sensor data should be updated every 1–5 years to keep it current. It should be high spatial resolution (1–5m) multispectral data if possible (Jensen *et al.* 1998). High resolution (< 0.25–2m) panchromatic and/or near-infrared data should be acquired within 12 hours to 2 days after a disaster (Fig. 25.25; Davis 1993; Schweitzer and McLeod 1997). Post-disaster images are registered to the pre-disaster images and change detection takes place (Jensen 1996). If precise, quantitative information about damaged housing stock, disrupted transportation arteries, the flow of spilled materials, and damage to above-ground utilities is required, it is advisable to acquire post-disaster 0.25–1m panchromatic and near-infrared data within 1–2 days.

10. *Critical Environmental Area Assessment.* Urban/suburban environments often include very sensitive areas such as wetlands, endangered species habitat, parks, land surrounding treatment plants, and the land in urbanized watersheds that provides the run-off for potable water. Relatively stable sensitive environments need to be monitored only every 1–2 years using a multispectral remote sensor collecting 1–10m data. For

a. Shuttle Endeavor with 60m boom retracted

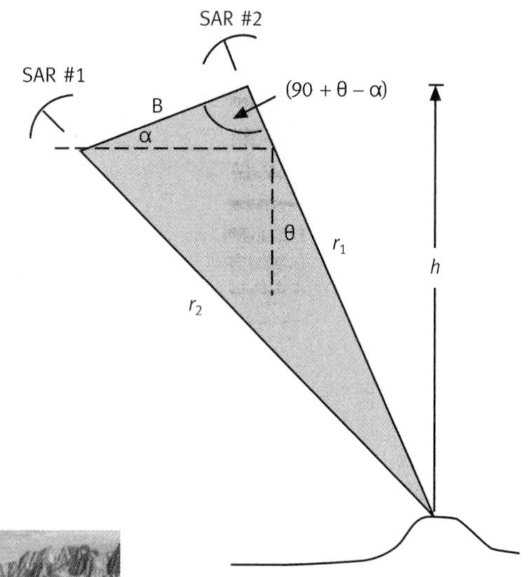

b. Interferometry geometry

c. SRTM digital elevation model of the San Fernando Valley, California

d. SRTM digital elevation model of the Temblor Range and the San Andreas fault near Bakersfield, California

Fig. 25.24 Illustration of image collection and images from the Space Shuttle Radar Topography Mission
Courtesy of NASA Jet Propulsion Laboratory

Fig. 25.25 Example of post-disaster high spatial resolution image of tornado damage

extremely critical areas that could change rapidly, multispectral remote sensors (including a thermal infrared band) should obtain < 0.25–2m spatial resolution data every 1–6 months.

Educational Initiatives for Remote Sensing

The major advances in geographic remote sensing have occurred principally at research universities. At those institutions remote sensing education has kept pace with the frontier, and, in fact, helped to drive it forward. As graduate students and professors address research problems they have proposed to funding agencies, results and findings from the lab are quickly introduced into advanced coursework, extending the advancements in geographic remote sensing science, art, and technology to more students. Through publication of research results, the literature becomes enriched with advancing concepts, techniques, and applications. Professors and students at other institutions pick up a new idea and push it further. Thus, at major research institutions geographic education is alive, well, and thriving.

Gradually, as these well-trained graduate students move on to other institutions as faculty members, seeking funding of their own, geographic remote sensing research and education expand outward. It is beyond this upper level of self-perpetuating research and education that geographic remote sensing education needs a more vigorous and systematic promotion.

Educational opportunities for geographic remote sensing are unlimited. Just as maps and photographs of the environment have become universal in all forms of media, remote sensing is emerging in like manner to reach the public at large. But now, with the digital domain upon us, remotely sensed imagery of every place on Earth can reach even the most remote site as the new century emerges. Desktop computers and supporting peripheral hardware have pervaded the American academic landscape. Sophisticated and friendly software are available at affordable prices. Spatial digital data are showering down on us in increasing quantities from dozens of sources, carrying information on an expanding array of earth phenomena ranging from cloud-top and oceanic temperatures to terrain configuration to agricultural crops to housing. Opportunities for infusing geographic remote sensing education into society are at an all-time high, and rising. This section addresses educational outreach in two dimensions: formal and informal education.

Formal Education

Through the established educational infrastructure there is much to be done to develop and infuse remote sensing into research universities, four-year colleges, junior colleges, and K-12 pre-college education.

1. *Research Universities.* The development of theoretical constructs, algorithms, and technical applications to various geographic problems is in its own right a mission of great importance for geography and other environmental and human sciences. Beyond the scientific and technological research and applications to myriad environment/human problems, remote sensing specialists in geography have a great responsibility in formal education. That responsibility clearly applies to higher education, and especially to research institutions to lead the way.

As remote sensing technology and data sources continue to expand and improve, the obligation for remote sensing specialists in geography expands with it. This is where the new trends and applications are developed for promulgation to the educational community and public at large.

But advanced development of tools for environmental applications is not enough. At research institutions, there needs to be collaboration between those advancing the technology and those colleagues in the systematic geographic sciences, both physical and human geography, whose scientific endeavors will be enriched by the acquisition and application of new remote sensing data and tools. We emphasize that this initiative applies to physical as well as human geography applications and courses. It is likely that most systematic courses in geography could be enriched by creative integration of advancing remote sensing data as spatial, spectral, and radiometric resolution improves. To select the right data source, resolution, and application for the particular project and coursework is a challenge and an opportunity. This initiative applies to, for example, biogeography, hydrogeography, climatology, geomorphology, agrogeography, urban geography and planning, population geography, transportation geography, and others.

Consider how this will improve and advance geographic science in general. The pace that is set at major universities will readily transmit to the scores of colleges that have introduced, and are introducing, remote sensing into the curriculum. Consider also the numerous ways such systematic innovations provide a stage and mechanism for outreach to resource and management agencies and community service opportunities at local, state, and national levels. Additionally, consider the opportunities for students as interns and employees. Ultimately, consider the enhanced image of geography as not only a fundamental research discipline, but also a practical and applied science serving the community directly.

And now, for regional geography. Can remote sensing add anything that systematic geographies cannot to the study and coursework regarding regions? Many in remote sensing have felt that every regional geography course, at every scale, could be substantially enriched by skillful application of remote sensing, both digital and manual. Especially if integrated with GIS, building datasets including remote imagery and classifications that could be queried and combined in various layers to address specific problems or issues could improve course content as well as pedagogy, emphasizing inquiry and inductive reasoning to the benefit of students as at all levels of education

Methods of integrating remote sensing with GIS, GPS, field techniques, quantification, and modeling can improve both systematic and regional geography. The more effectively these techniques, along with field investigation, are integrated, the more geographic education is improved.

In our major universities remote sensing specialists in geography must remain at the cutting edge of research in respective science applications for the health and vigor of the discipline. Remaining competitive for extramural funding requires an aggressive research program in scientific application and technological innovation. American geographic scientists in remote sensing are among the most advanced in the world. Collaborative research with allied scientists also invigorates geography. Many university professors have created tutorials and course materials that can be placed on the Internet or otherwise made available. The discipline will be well served across universities and colleges as this initiative continues.

2. *Four-year and Community Colleges.* Considerable work is done in our baccalaureate institutions in terms of applied remote sensing. This work can be expanded and improved as "first-tier schools" advance and publish application techniques. Such activities provide vital training for students in preparation for either advanced academic training or employment opportunities. Coupling remote sensing procedures with GIS data and strategies further prepares the student, and promotes outreach to the community. Increasing numbers of geography professors are engaged in extension to local communities to serve special needs and provide opportunities for students.

It should be recognized that remote sensing, whether digital or manual, always deals with primary data, as distinct from GIS operations which are basically dealing with secondary data. This is to say that remote sensing scholars are looking "directly" at the environment with no intervening interpreter. Remote sensing, therefore, becomes a primary tool for environmental analysis, totally in the control of the observer. Environmental characterization, mensuration, monitoring, and mapping are in the hands of the remote sensing scholar. This is a significant capability accessible to for teachers and students everywhere.

Because costs of robust hardware and sophisticated software are increasingly favorable and friendly, along with widely available data of multiple scales and resolutions, remote sensing is within reach of college teachers even in remote places of modest financial means. Students and teachers have access to global resources with increasing availability.

Curriculum development using remote sensing at four-year colleges and junior colleges is an open-ended opportunity. Not only technical courses in remote sensing itself, but remote sensing infusion into physical geography and human geography coursework increase the vigor, realism, and currency of the course. The environment becomes more alive and meaningful through the visualization of remote imagery. With digital multispectral image analysis, substantial insights and understanding of the biophysical environment as well as the physical and human interface emerge for the student and professor. In as much as many datasets can be obtained, often at modest cost, for any locality, coursework is readily enhanced. Better yet, such imaging promotes field exploration and analysis to further enlighten the student and create environmental understanding, encouraging further investigation. Such activity promotes increased powers of observation, inquiry, and experiential education.

Regional and global education are likewise increased, as data at many scales, resolutions, and thematic substance are available. Much of this is available at no cost, on-line, or in CD or other formats. Coursework on any theme and any part of the world can be enriched. Windows technology provides for comparing several scenes concurrently. One of the more attractive advantages of such initiatives is that in the digital domain, up-scaling and down-scaling is so efficient. Through roam, zoom, and scroll capabilities, the student can explore at will.

3. *K-12 Schools.* Student scholars from kindergarten through high school can enjoy similar opportunities at the hands of an energetic teacher. With data from satellite-based sensors becoming widely available, curricular and instructional opportunities are limitless. Nearly every discipline taught from kindergarten to high school can find a valuable ally and resource in remote sensing.

Tutorials and curricular modules are slowly coming, and much remains to be done to bring volumes of remotely sensed data to the classroom. These great and expanding volumes of data are like a huge library being filled with information (or potential information). While pretty pictures can be accessed for visual impression, pedagogic links to the data library need to be prepared to serve the K-12 teacher and student. Collegiate remote sensing specialists would do well to provide thematic modules and other pedagogic links to serve in this area.

Even image processing software is available at no cost for pre-college schools. For example, the image analysis software produced by LARS (Laboratory for Applied Remote Sensing) at Purdue University (see the LARS website at <www.lars.purdue.edu> (6 March 2002) for more information on this software) provides for multispectral image analysis and classification. It has been made available to schools at no cost through the GLOBE (Global Learning and Observation to Benefit the Environment: see the website at <www.globe.gov> (6 March 2002) for more information) environmental science program. Protocols for remotely sensed image interpretation are built into the program. Hundreds of K-12 schools in the US and dozens of countries abroad are engaged in the program.

The future is very bright and pedagogically exciting for remote sensing in the pre-college classroom and laboratory. As for geography education, satellite remote sensing imagery (digital and manual) is the most visual and student-enriching medium since conventional aerial photography. Local aerial photos have long been a staple for exciting students in learning about their neighborhoods. Teaching local and regional geography will be improved with manual image interpretation alone. But in this era of computer-literate beginning students, the field is ripe and ready to harvest. With access to data around the world, every pre-college geography classroom ought to jump to the opportunity. College specialists will need to help K-12 teachers rise to the challenge.

Finally, geography educators ought to take the lead in environmental education. The only rivals are earth science and biological science. In the typical K-12 setting, neither is adequately dealing with environmental science. Both, especially earth science, are focusing on "vertical" relationships and materials at the micro-level, with very little attention to "horizontal" or spatial relationships. Little is done with textbooks or state-established guidelines to develop the all-important dimension of spatial patterns and analysis of spatial dynamics. In the 40-plus years since the Sputnik scare stimulated science education in America, this condition remains the case, leaving not only a void in American environmental science education, but also a great opportunity, indeed obligation, for academic geography to fill the gap. It remains for geography, armed with remote sensing, to rise up and fill this great need in geographic and environmental education for the American school system, students, and the public at large.

Informal Education

Remote sensing has an important role outside institutionalized education systems. As data volumes and access increase and hardware/software become more widespread, cheaper, and friendlier, remote sensing information will become common throughout the cyber world and into every corner of society. The academic remote sensing community should seize the opportunity to promote and guide the process.

1. *Outreach to Agencies.* Many geography students, undergraduate and graduate, find employment with agencies that use or could use remote sensing effectively in their respective roles. Well-trained students can have a significant influence on developing the application within the agency. This is especially true with federal and state resource management agencies. This is so because

remotely sensed data is primary data, and "sees" the environment and landscape as it is. The temporal frequency of multispectral satellite data makes it particularly useful to forestry, range management, agriculture, watershed, wildlife habitat management, etc. A student well trained in remote sensing image analysis can introduce and influence proper use of the technology within the organization, especially in view of increasing computer-based data handling throughout the infrastructure of organizations. To link this technology with GIS is natural as so many agencies are building digital maps and databases. Alert professors will seek out opportunities to extend their remote sensing expertise into the community, further strengthening academic ties and service (with or without funding support) to the community, and strengthening student internships and employment avenues.

Among local agencies such as city and county planning, remote sensing can be invaluable. Most planning agencies still rely on analog aerial photographs, typically out-of-date, to establish land use and patterns of growth. This is typically more time-consuming, expensive, and inconsistent in interpretation, as well as out-of-date. Well-chosen satellite imagery (airborne imagery if it can be afforded) can provide a vital assessment of patterns of growth, monitoring of built vs. non-built environments, change in land cover, and environmental change and impact, to the substantial benefit of agencies dealing with areas of rapid urbanization and industrial expansion. Many planning agencies are advancing into GIS technology, building map and file databases, unaware of the great value of remote sensing application to the planning process. Again, alert professors with remote sensing skills will find specific opportunities to enter the community agencies and promote this valuable dimension of primary data analysis as an integral part of their digital domain in fulfilling their planning roles more effectively. At the very least, rasterized remote imagery makes a realistic visual and effective background to digital maps for analysis and public display. But much more than this, the environmental information inherent in imagery opens many possibilities the typical planner is unaware of. Students in training at colleges and universities, those who see city and regional planning as a future direction for employment, would be well advised to add remote sensing training to their academic toolkit while in school. It would behoove professors and students to set up demonstrations at school or at the agency to promote this important dimension to planning.

2. *Outreach to the Business World.* Many businesses, including engineering and environmental firms, are beginning to utilize digital database development and GIS. For reasons similar to those stated above for agencies, they would benefit significantly from remote sensing applications. For example, engineering firms dealing with highway and corridor development build data layers for which imagery coupled with DEM (Digital elevation models) and other coverages could provide important environmental information. Agribusinesses concerned with crop conditions and yield forecasting benefit from multi-temporal image analysis. Thermal imaging of buildings in the built environment provides significant opportunities to assist in light-reflecting roofing materials selection for air-conditioning cost reduction. Many public buildings are among the worst violators of energy loss decisions. Local and national weather broadcast companies have effectively used satellite data since the early days of TIROS in the 1960s. Certainly more could be done with terrestrial image presentation.

3. *Public at Large.* This is perhaps our largest and certainly most diffuse target audience. Just as photography has become a widely utilized and highly informative medium for public education, remote imagery from space and aircraft can have a similar impact on public education. Massive quantities of imagery are being compiled in archives available on the Internet. Schoolrooms, households, broadcast and print media, and others have unlimited access to images of the world. Each home becomes a virtual domain classroom with the world available to parents and children. While digital image analysis and classification are not likely to be accessible in most households, unenhanced imagery itself will enlighten viewers to the geographic environmental and spatial characteristics of places on the planet. Perhaps this will further stimulate interest in academic pursuits, and vice versa.

Remote Sensing Geographic Education: Epilogue for the Future

The effort here has been to broaden and sharpen the view of opportunities and obligations for remote sensing specialists in geography regarding a heightened sense of mission. From basic research in science and technology, to applied research in unending thematic areas, to classroom and laboratory institutions, to exportable course materials and module development, to outreach in formal, informal, and public education, geographers have a more significant mission than ever before in the history of our discipline. Remote sensing makes geography more visual and more vital than ever before, from local to global dimensions. We cannot afford to shrink or shirk this responsibility.

The Interface Between, and the Integration of, Remote Sensing and Geography in the New Millennium

Given the dazzling leaps that remote sensing technology and data analysis techniques have made since the first edition of *Geography in America* was published, it is safe to say that remote sensing has now attained new heights of importance and distinction within geography. Geographers have not only contributed to the widespread uses of remote sensing data for a wide array of environmental analyses, but geography as a discipline has been significantly advanced by its long-standing embrace of remote sensing. It also appears that with the new types of sensors to be launched in the near future and the data to be collected by these remote sensing platforms, the interface between geography and remote sensing will become an even stronger bond. The exploitation of data from this multitude of current and future remote sensing systems will provide for new and innovative geographical research possibilities for many years to come. In the same way that geographic information systems have invigorated the whole discipline of geography, the overall abundance, spatial characteristics, and fidelity of data from current and future remote sensing systems will contribute to keeping geography fresh, vibrant—and relevant—as a field of study. There are, however, several key issues that offer considerable challenges to geographic research in remote sensing in the future. The answers to the questions surrounding these issues are not by any means trivial, but they will be fecund topics for intensive research in the future as we assess in depth the relevance and appropriateness of data from these new remote sensing systems to geography.

Multi-scale Challenges

Although the interface between remote sensing and geography remains strong, it is the integration of the multitude of kinds of remote sensing data collected at varying space, time, and radiometric resolutions, that is perhaps one of the greatest challenges that geography faces now, as well as in the future. We know that spatial heterogeneity constrains our ability to transform information from one scale to another. This is a paramount element to consider in our attempts to integrate and analyze remote sensing data obtained at different spatial scales from different sensors. Thus, a fundamental issue of extreme importance to geographical research is to understand how to combine data with different spatial and temporal resolutions in a meaningful way to intercompare data from these sensors successfully. Some of the more important remote sensing data integration questions that will need to be addressed by geographers in the new millennium are:

1. *Invariants of Scale.* What properties of physical and human systems are invariant with respect to scale and how can these be analyzed using multi-scaled remote sensing data?
2. *Ability to Change Scale.* Can we develop a generic set of methods for disaggregating coarse-scale data and aggregating fine-scale data, in ways that are compatible with our understanding of geographical and Earth system processes?
3. *Measures of the Impact of Scale.* How is the observation of processes affected by changes of scale and how can we measure the degree to which processes are manifested at different scales using multi-scaled remote sensing data?
4. *Implementation of Multi-scale Approaches.* What is the potential for integrated tools to support multi-scale databases and associate modeling analysis within an integrated GIS (IGIS) structure? What are the problems that must be overcome in integrating remote sensing data obtained at different spatial, temporal, or spectral scales and resolutions?

Generic Integration (Conflation) Issues

Corresponding with questions related to how to integrate multi-scaled remote sensing data, are the general issues associated with the integration of many different types of geographic data within a multi-scaled remote sensing data analysis. Data integration strategies and methods have not kept pace with advances in data collection. Scientists and the public must be able to *conflate* data from many different remote sensing and non-remote sensing sources (Saalfeld 1988). A general theoretical and conceptual framework is needed to be able to accommodate at a minimum, 5 distinct forms of data integration:

1. *In situ measurement to in situ measurement* (calibration, adjustment, variance, etc.)
2. *In situ measurement to foundation map* (point-to-map; registration, verification)
3. *Vector to foundation map* (map-to-map; vector segmentation scheme integration; different scales, different geographic coverage, etc.)
4. Image to foundation map (image-to-map; for elevation mapping, map revision)
5. Image to foundation image (image-to-image; involving different spatial, spectral, temporal, and radiometric resolutions).

A single common framework is needed that will integrate diverse types of spatial data and allow:

- horizontal integration (merging adjacent data sets)
- vertical data integration (operations involving the overlaying of maps), and
- temporal data integration.

Spatial data integration must incorporate differences in spatial data content, scales, data acquisition methods, standards, definitions, practices, manage uncertainty and representation differences, and detect and deal with redundancy and ambiguity of representation. These and related integration issues are being undertaken as part of the active and future research focus of the UCGIS.

Metadata Issues

Data about data or *metadata* are becoming increasingly important with the tremendous amounts of multiple space, time, and spectral resolution remote sensing data that are currently available, or will become so in the near future. Metadata allow us to understand the origin of the data, its geometric characteristics, its attributes, and what type of cartographic, digital image processing or modeling has already been applied to the data. A user must have a complete understanding of the content and quality of a digital spatial dataset (inclusive of remote sensing data) in order to make maximum use of its information content. Research on metadata as related to both current and new remote sensing data must focus on:

1. Organization, storage, and provision of metadata to the geographical remote sensing data user community;
2. Development of improved web-based interfaces for efficiently browsing and downloading metadata;
3. Documentation of the genealogy (i.e. lineage) of *all* the *operations* that have been performed or applied to a dataset.

Again, active research on these topics is being conducted under the auspices of the UCGIS as well as a multitude of federal and state government organizations. With the acquisition of the copious amounts of data to be obtained from remote sensing systems that will be launched in the next 5–10 years (see Fig. 25.1), however, metadata issues will become even more significant within the overall pursuit of geographical research for many years to come.

Geography and Remote Sensing in the New Millennium: Building Upon the Past—Advancing Into the Future

At the end of the chapter on remote sensing in the first edition of *Geography in America*, a paragraph was quoted from an article entitled "Impacts of Remote Sensing on U.S. Geography" (Estes *et al.* 1980). The first sentence of this paragraph states that "The exploitation of the improved or unique information available to the geographer via the application of remote sensing techniques has barely begun." It may be perceived that, given the tremendous amount of research performed by geographers who have used remote sensing data to advance the overall width, breadth, and depth of our discipline, this quotation is now out of date and inoperative. As we look to the future of remote sensing and geography from where we sit at the outset of the new millennium, however, we can construe that this statement is as true today as it was then. We *are* on the brink of a virtual remote sensing data explosion; the types and kinds of data that will be provided by sensing systems in the very near future will potentially drive a new revolution of geographic research—in development of data analysis techniques, in applications of these data to geographical problems, and in integrating these data with many other geographical datasets. In essence, we *have* barely begun to exploit the improved or unique information available to geographers via the applications of the remote sensing data to come in the very near future. An old adage that has been used (perhaps with disdain) to define geography is that "geography is what geographers do." A more modern and more succinct statement that can be articulated on what the role of remote sensing and geography has become, and will continue to be, in the new millennium is—"Remote Sensing and Geography: JUST DO IT!"

References

Association of American Geographers (2001*a*). *Newsletter*, 26/6 (June).

Association of American Geographers (2001*b*). *Guide to Programs in Geography in the United States and Canada 2000–2001*. Washington, DC. Association of American Geographers.

Allen, J. C., and Barnes, D. F. (1985). "The Causes of Deforestation in Developing Countries." *Annals of the Association of American Geographers*, 75/2: 163–84.

Allen, T. R., and Kupfer, J. A. (2001). "Spectral Response and Spatial Pattern of Fraser Fir Mortality and Regeneration, Great Smoky Mountains, USA." *Plant Ecology*, 156/1: 59–74.

American Planning Association (1998). *Land-Based Classification Standards*, American Planning Association Research Department, ‹http://www.planning.org/lbcs›, last accessed 6 March 2002.

Anderson, J. R., Hardy, E., Roach, J., and Witmer, R. (1976). *A Land Use and Land Cover Classification System for Use with Remote Sensor Data*. Professional Paper, 964. Washington, DC: USGS.

Angelici, G. L., Bryant, N. A., Fretz, R. K., and Friedman, S. Z. (1980). *Urban Solar Photovoltaics Potential: An Inventory and Modeling Study Applied to the San Fernando Valley Region of Los Angeles*. Report 80-43. Pasadena: JPL.

Clarke, K. C., Gaydos, L., and Hoppen, S. (1997). "A Self-Modifying Cellular Automaton Model of Historical Urbanization in the San Francisco Bay Area." *Environment and Planning B*, 23: 247–61.

Colwell, R. N. (1997), "History and Place of Photographic Interpretation." *Manual of Photographic Interpretation*. Bethesda: American Society for Photogrammetry and Remote Sensing, 33–48.

—— (ed. in chief) (1983). *Manual of Remote Sensing*. Falls Church, Va.: American Society for Photogrammetry and Remote Sensing.

Cowen, D. J., Jensen, J. R., Hendrix, C., Hodgson, M. E., and S. R. Schill (2000). "A GIS-Assisted Rail Construction Econometric Model that Incorporates LIDAR Data." *Photogrammetric Engineering and Remote Sensing*, 66/11: 1323–8.

Curran, P. J., and Hobson, T. A. (1987). "Landsat MSS Imagery to Estimate Residential Heat-Load Density." *Environment and Planning*, 19: 1597–610.

Cullingworth, B. (1997), *Planning in the USA: Policies, Issues and Processes*. London: Routledge.

Davis, B. A. (1993). "Mission Accomplished." *NASA Tech Briefs*, 17/1: 14–16.

Eliasson, I. (1992). "Infrared Thermography and Urban Temperature Patterns." *International Journal of Remote Sensing*, 13: 869–79.

Estes, J. E., Jensen, J. R., and Simonett, D. S. (1980). "Impacts of Remote Sensing on U.S. Geography." *Remote Sensing of Environment*, 10: 43–80.

Feldman, S. C., Pelletier, R. E., Walser, E., Smoot, J. R., and Ahl, D. (1995). "A Prototype for Pipeline Routing Using Remotely Sensed Data and Geographic Information System Analysis." *Remote Sensing of Environment*, 53: 123–31.

Flood, M., and Gutellius, B. (1997). "Commercial Implications of Topographic Terrain Mapping Using Scanning Airborne Laser Radar." *Photogrammetric Engineering and Remote Sensing*, 63: 327.

Foreman, R. T. T., and Godron, M. (1986). *Landscape Ecology*. New York: John Wiley.

Fulghum, D. A. (2000), "Digital Aircraft Camera Offers Cheap, Fast Images." *Aviation Week and Space Technology*, 31 July.

Gaile, G. E., and Wilmott, C. J. (eds.) (1989). *Geography in America*. Columbus, Ohio: Merrill.

Green, K., Kempka, D., and Lackey, L. (1994). "Using Remote Sensing to Detect and Monitor Land-Cover and Land-Use Change." *Photogrammetric Engineering and Remote Sensing*, 60: 331–7.

Greve, C. W. (1996). *Digital Photogrammetry: An Addendum to the Manual of Photogrammetry*, Bethesda: American Society for Photogrammetry and Remote Sensing.

Haack, B., Guptill, S., Holz, R., Jampoler, S., Jensen, J., and Welch, R. (1997). Urban Analysis and Planning. *Manual of Photographic Interpretation*, Bethesda: American Society for Photogrammetry and Remote Sensing, 517–53.

Henderson, F. M., and Utano, J. J. (1975). "Assessing General Urban Socioeconomic Conditions with Conventional Air Photography." *Photogrammetria*, 31: 81–9.

Henderson, F. M., and Xia, Z.-G. (1998). "Radar Applications in Urban Analysis, Settlement Detection, and Population Estimation," in F. M. Henderson and A. J. Lewis (eds.), *Manual of Remote Sensing*, ii. *Principles and Applications of Imaging Radar*. Bethesda, Md.: American Society for Photogrammetry and Remote Sensing, 733–68.

Holz, R. K. (1988). Population Estimation of Colonias in the Lower Rio Grande Valley Using Remote Sensing Techniques." Annual Meeting of the Association of American Geographers, Phoenix, Ariz.

IGBP/IHDP-LUCC and IGBP-DIS (1997). International Geosphere-Biosphere Programme/International Human Dimensions of Global Change Program—Land Use and Land Cover Change Project, and International Geosphere-Biosphere Programme-Data Information System, Stockholm and Paris.

Jadkowski, M. A., Convery, P., Birk, R. J., and Kuo, S. (1994). "Aerial Image Databases for Pipeline Rights-of-Way Management." *Photogrammetric Engineering and Remote Sensing*, 60: 347–53.

Jensen, J. R. (1996). *Introductory Digital Image Processing: A Remote Sensing Perspective*. Saddle River, NJ: Prentice-Hall.

—— 2000. *Remote Sensing of the Environment: An Earth Resource Perspective*, Saddle River, NJ: Prentice-Hall.

Jensen, J. R., and Cowen, D. J. (1999). "Remote Sensing of Urban/Suburban Infrastructure and Socio-Economic Attributes." *Photogrammetric Engineering and Remote Sensing*, 65: 611–22.

Jensen, J. R., Halls, J., and Michel, J. (1998), "A Systems Approach to Environmental Sensitivity Index (ESI) Mapping for Oil Spill Contingency Planning and Response," *Photogrammetric Engineering and Remote Sensing*, 64/12: 1003–14.

Jensen, J. R., Qiu, F., and Ji, M. (1999). "Predictive Modeling of Coniferous Forest Age Using Statistical and Artificial Neural Network Approaches Applied to Remote Sensing Data." *International Journal of Remote Sensing*, 20/14: 2805–22.

Jensen, J. R., Cowen, D. C., Halls, J., Narumalani, S., Schmidt, N., Davis, B. A., and Burgess, B. (1994). "Improved Urban Infrastructure Mapping and Forecasting for BellSouth Using

Remote Sensing and GIS Technology." *Photogrammetric Engineering and Remote Sensing*, 60: 339–46.

Ji, M., and Jensen, J. R. (1996). "Fuzzy Training in Supervised Image Classification," *Journal of Geographic Information Sciences*, 2/2: 1–12.

Kolasa, J., and Pickett, S. T. A. (1991). *Ecological Heterogeneity*. New York: Springer-Verlag.

Lacy, R. (1992). "South Carolina Finds Economical Way to Update Digital Road Data," *GIS World*, 5: 58–60.

Light, D. L. (1999). "An Airborne Digital Imaging System," *Proceedings of the International Symposium on Spectral Sensing Research*, Las Vegas, 31 October.

Lillesand, T., and Kiefer, R. (2000). *Elements of Image Interpretation*. 4th edn. New York: John Wiley.

Liverman, D., Moran, E. F., Rindfuss, R. R., and Stern, P. C. (1998). *People and Pixels: Linking Remote Sensing and Social Science*. Washington: National Academy Press.

Lo, C. P. (1986). "The Human Population." *Applied Remote Sensing*. New York: Longman, 40–70.

—— (1995). "Automated Population and Dwelling Unit Estimation from High-Resolution Satellite Images: A GIS Approach." *International Journal of Remote Sensing*, 16: 17–34.

Lo, C. P., and Faber, B. J. (1997). "Integration of Landsat Thematic Mapper and Census Data for Quality of Life Assessment." *Remote Sensing of Environment*, 66: 143–57.

McCracken, S. D., Brondizio, E. S., Nelson, D., Moran, E. F., Siqueira, A. D., and Rodriguez-Pedraza, C. (1999). "Remote Sensing and GIS at the Farm Property Level: Demography and Deforestation in the Brazilian Amazon." *Photogrammetric Engineering and Remote Sensing* 65/11: 97–105.

Messina, J. P., and Walsh, S. J. (2001). "2.5D Morphogenesis: Modeling Landuse and Landcover Dynamics in the Ecuadorian Amazon." *Plant Ecology*, 156/1: 75–88.

Messina, J. P., Walsh, S. J., Taff, G., and Valdivia, G. (1999). "The Application of Cellular Automata Modeling for Enhanced Land Cover Classification in the Ecuadorian Amazon." *Proceedings of 4th International Conference on GeoComputation*. GeoComputation CD-ROM. Fredericksburg, Va.: Mary Washington College, 81.

Millington, A. C., Walsh, S. J., and Osborne, P. E. (eds.) (2001). *GIS and Remote Sensing Applications in Biogeography and Ecology*. Boston: Kluwer.

Moran, E. F., Brondizio, E., and Mausel, P. (1994). "Integrating Amazonian Vegetation, Land-Use, and Satellite Data." *BioScience*, 44: 329–38.

NASA (1999). *EOS Data Products Handbook*, 2. Washington, DC: National Aeronautics and Space Administration.

NOAA (1975–84). *Landsat Data Users Notes*. Washington, DC: National Oceanic and Atmospheric Administration:

Odum, H. T. (1994). *Ecological and General Systems: An Introduction to Systems Ecology*. Niwot, Colo.: University Press of Colorado.

Olorunfemi, J. F. (1984). "Land Use and Population: a Linking Model." *Photogrammetric Engineering and Remote Sensing*, 50: 221–7.

O'Neill, R. (1988). "Hierarchy Theory and Global Change," in T. R. Rosswall, R. G. Woodmansee, and P. G. Risser (eds.), *Scales and Global Change: Spatial and Temporal Variability in Biospheric and Geospheric Processes*. York: Wiley, 29–46.

Philipson, W. (1997). *Manual of Photographic Interpretation*. Bethesda, Md.: American Society for Photogrammetry and Remote Sensing.

Research Systems, Inc. (2000). "Applying Neural Network Classification." *The Environment for Visualizing Images — ENVI, Version 3.4*. Boulder: Research Systems, Inc., 514–16.

Rindfuss, R. R., and Stern, P. C. (1998). "Linking Remote Sensing and Social Science: The Needs and the Challenges." in D. Liverman, E. R. Moran, R. R. Rindfuss, and P.C. Stern (eds.), *People and Pixels*. Washington, DC: National Academy Press, 1–27.

Saalfield, A. (1998). "Conflation: Automated Map Compilation." *International Journal of Geographical Information systems*, 2/3: 217–28.

Schott, J. R. (1989). "Image Processing of Thermal Infrared Images." *Photogrammetric Engineering and Remote Sensing*, 44/9: 1311–21.

Schweitzer, B., and McLeod, B. (1997), "Marketing Technology that is Changing at the Speed of Light," *Earth Observation Magazine*, 6: 22–4.

Skole, D. L. (1994). "Data on Global Land-Cover Change: Acquisition, Assessment, and Analysis," in W. B. Meyer and B. L. Turner, II, (eds.), *Changes in Land Use and Land Cover: A Global Perspective*. Cambridge: Cambridge University Press, 437–72.

Sutton, P., Roberts, D., Elvidge, C., and Meij, H. (1997). "A Comparison of Nighttime Satellite Imagery and Population Density for the Continental United States." *Photogrammetric Engineering and Remote Sensing*, 63/11: 1303–13.

Turner, B. L., II, *et al.* (1995). "Land-Use and Land-Cover Change," *Science/Research Plan, International Geosphere-Biosphere Program and International Human Dimensions of Global Change Program, Stockholm and Paris, Report*, 35.

Turner, M. G., and Gardner, R. H. (1991). "Quantitative Methods in Landscape Ecology: An Introduction," in Turner and Gardner (eds.), *Quantitative Methods in Landscape Ecology*. Ecological Studies. New York: Springer-Verlag, 3–14.

Wagman, D. (1997). "Fires, Hurricanes Prove No Match for GIS." *Earth Observation Magazine*, 6: 27–9.

Walsh, S. J., and Crews-Meyer, K. A. (eds.) (2002). *Linking People, Place, and Policy: A GIScience Approach*. Boston: Kluwer.

Walsh, S. J., Crawford, T. W., Welsh, W. F., and Crews-Meyer, K. A. (2001). "A Multiscale Analysis of LULC and NDVI Variation in Nang Rong District, Northeast Thailand." *Agriculture, Ecosystems, and Environment*, 85: 47–64.

Walsh, S. J., Moody, A., Allen, T. R., and Brown, G. (1997). "Scale Dependence of NDVI and Its Relationship to Mountainous Terrain," in D. A. Quattrochi, and M. F. Goodchild (eds.), *Scale in Remote Sensing and GIS*. Boca Raton, Fla.: Lewis 27–55.

Walsh, S. J., Evans, T. P., Welsh, W. F., Entwisle, B., and Rindfuss, R. R. (1999). "Scale Dependent Relationships Between Population and Environment in Northeast Thailand." *Photogrammetric Engineering and Remote Sensing*, 65/1: 97–105.

Walsh, S. J., Evans, T. P., Welsh, W. F., Rindfuss, R. R., and Entwisle, B. (1998). "Population and Environmental Characteristics Associated with Village Boundaries and Landuse/Landcover Patterns in Nang Rong District, Thailand," in *Proceedings of the 1996 Pecora Symposium on Human Interactions with the Environment: Perspectives From Space*, 302–11.

CHAPTER 26

Cartography

Mark Monmonier and Robert B. McMaster

Introduction

Summarizing a decade of cartographic research in a short chapter is difficult: bias is inevitable, randomness is indefensible, breadth is tricky, and coherence is essential. Rather than attempt a broad, shallow survey, we chose to focus on some of the period's significant conceptual frameworks, and relate each model to one or more related research papers published since A. Jon Kimerling (1989) summarized cartographic research for the first volume of *Geography in America*. This has been a transition period in which the discipline has witnessed several significant changes, including: (1) the nearly complete automation of the cartographic process and a proliferation of maps produced by desktop mapping systems and GISs; (2) the inclusion of significant amounts of core cartographic research—such as terrain modeling, geographic data structures, generalization, and interpolation—within the growing discipline of GIS; and (3) the wide adoption of the term "geographic visualization" to describe the dynamic, interactive component of cartography. These developments and the migration of more and more cartographic interests into the newly created discipline of GIS have raised concern about whether our discipline would survive. These doubts are offset by growing recognition that research and education on representational issues in GIS is critical, and that research in map design, symbolization, and generalization cannot be neglected.

Cartography remains an independent discipline. Our two journals, *Cartography and Geographic Information Science* (recently renamed with *Science* replacing *Systems*) and *Cartographic Perspectives*, are thriving. American cartographic researchers also publish their work in *Cartographica*, *GeoInfo Systems*, *GIS World*, and the *International Journal of Geographic Information Science*. The Mapping Science Committee of the National Academy of Sciences and the recently formed Committee on Geography represent our interests at the national level, as do the Cartography and Geographic Information Society (a member organization of the American Congress on Surveying and Mapping), the North American Cartographic Information Society, the University Consortium for Geographic Information Science, and the AAG's Cartography Specialty Group.

During the decade our educators, researchers, and essayists have published many textbooks and monographs, including the sixth edition of *Elements of Cartography* (Robinson *et al.* 1995); several new editions of Borden Dent's *Cartography: Thematic Map Design* (most recently 1999); Terry Slocum's *Thematic Cartography and Visualization* (1999); John Snyder's (1993) seminal work on projections, *Flattening the Earth: Two Thousand Years of Map Projections*; Alan MacEachren's *How Maps*

Work (1995); Denis Wood's (1992) social critique of cartography, *The Power of Maps*; and a series of books by Mark Monmonier, including *Maps with the News: The Development of American Journalistic Cartography* (1989*b*), *How to Lie with Maps* (1991, rev. 1996), *Mapping it Out: Expository Cartography for the Humanities and Social Sciences* (1993), *Drawing the Line: Tales of Maps and Cartocontroversy* (1995), *Cartographies of Danger: Mapping Hazards in America* (1997), and *Air Apparent: How Meteorologists Learned to Map, Predict, and Dramatize the Weather* (1999). Five more Auto-Carto conferences (9 through 13) met, respectively, in Baltimore (1989 and 1991), Minneapolis (1993), Charlotte (1995), and Seattle (1997). American researchers also contributed to the International Cartographic Association's (ICA) biennial conferences and proceedings, including those held in Budapest (1989), Bournemouth (1991), Cologne (1993), Barcelona (1995), and Stockholm (1997). Two US National Reports to the International Cartographic Association have been completed, including the 1991 special issue of *Cartography and Geographic Information Systems* on the "History and Development of United States Academic Cartography." And Woodward and Harley's monumental history of cartography project published three new tomes: vol. ii, bk. 1, *Cartography in the Traditional Islamic and South Asian Societies* (Harley and Woodward 1992); vol. ii, bk. 2, *Cartography in the Traditional East and Southeast Asian Societies* (Harley and Woodward 1994); and vol. ii, bk. 3, *Cartography in the Traditional African, American, Arctic, Australian, and Pacific Societies* (Woodward and Lewis 1998). Despite the untimely death of Brain Harley in 1991, the project moves steadily towards its goal of six volumes, the last of which will address the twentieth century.

Map Design and Communication

Modeling Cartographic Communication

Throughout the 1970s and early 1980s, cartographers developed and critiqued a plethora of models of the cartographic communication process. Some of the more influential models were those developed by Muehrcke (1983) and Monmonier (1977). Although communication models commanded less attention in the 1990s, Robert Lloyd (1996) and his colleagues recently published a model offering fresh insight on the interaction of the map, the map reader, and the map designer. As Fig. 26.1 indicates, the relationship between maps and the map reader is based on both unsupervised and reinforcement learning. Unsupervised learning, as Lloyd (1996: 211) explained, "results in knowledge being acquired without the map reader knowing if the processing has produced a proper result," and reinforced learning "involves indirect feedback from the consequences of decisions made by the map reader." In a similar way, the model addresses connections between maps and map designer, and map reader and map designer. Although not as prominent as in the 1970s and 1980s, when work in the psychophysical aspects of cartography reached its peak, research in experimental methods continues at several centers.

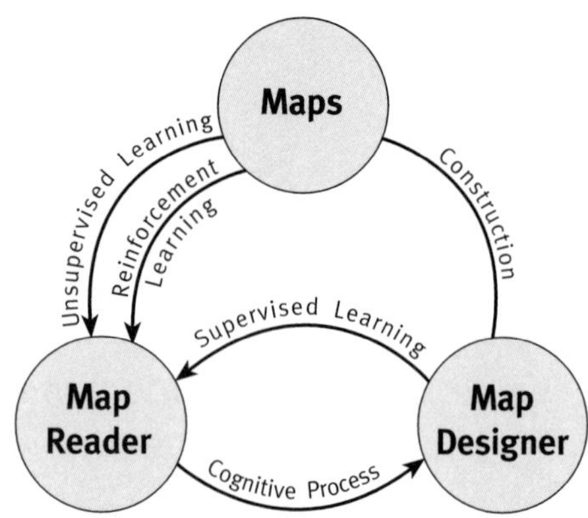

Fig. 26.1 Relationships among maps, map readers, and map designers illustrating the knowledge needed for map design

We can report on several studies in this area of research. For instance, work in the experimental cognitive domain addressed the effects of sequencing in choropleth mapping (Patton and Cammack 1996), the evaluation of four types of multivariate, quantitative point symbols (Nelson and Gilmartin 1996), the use of LCD technology to display choropleth maps and comparison of design options (McGranaghan 1996), the effectiveness of selecting landmarks for use as navigational aids (Deakin 1996), and the importance of brightness difference and its relationship to figural development in cartographic design (C. Wood 1994). Mark and Egenhofer (1994) describe the results of two human-subjects experiments to test how people think about spatial relations between lines and regions. Their results indicated that the 9-intersection model (a formal model of topological spatial relations) forms a sound basis for characterizing line/region relations in the specific case of roads and parks. Other studies have looked at gender differences in map reading (Kumler and Buttenfield 1996).

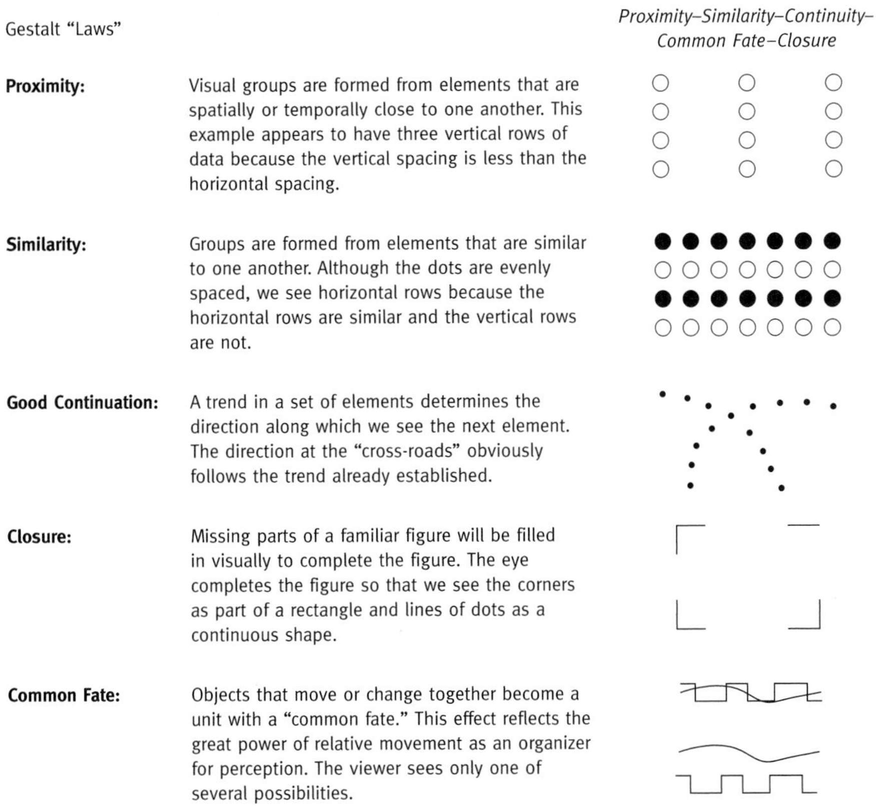

Gestalt "Laws"

Proximity–Similarity–Continuity–Common Fate–Closure

Proximity: Visual groups are formed from elements that are spatially or temporally close to one another. This example appears to have three vertical rows of data because the vertical spacing is less than the horizontal spacing.

Similarity: Groups are formed from elements that are similar to one another. Although the dots are evenly spaced, we see horizontal rows because the horizontal rows are similar and the vertical rows are not.

Good Continuation: A trend in a set of elements determines the direction along which we see the next element. The direction at the "cross-roads" obviously follows the trend already established.

Closure: Missing parts of a familiar figure will be filled in visually to complete the figure. The eye completes the figure so that we see the corners as part of a rectangle and lines of dots as a continuous shape.

Common Fate: Objects that move or change together become a unit with a "common fate." This effect reflects the great power of relative movement as an organizer for perception. The viewer sees only one of several possibilities.

Fig. 26.2 Gestalt "laws" of proximity, similarity, continuity, common fate, and closure

Gestalt Principles

Cartographic educator John Belbin's model in Fig. 26.2 provides a concise verbal and graphic summary of five gestalt "laws" useful in positioning labels on maps and within map keys. As the diagrams illustrate, the laws of similarity and proximity promote efficient and unambiguous classifications and symbol–label associations, whereas the laws of good continuation and common fate explain why labels can cross symbols or other labels and remain intelligible. Among other applications, the law of closure explains why separated portions of a overlapping graduated circles and other partly hidden symbols are seen as a coherent whole.

Developed in the first two decades of the twentieth century by psychological theorists interested in visual perception, cartographic adaptations of gestalt principles are not new. Indeed, gestalt postulates informed Eduard Imhof's (1962, 1974) classic paper on label placement; Arthur Robinson and Barbara Petchenik (1976) mentioned gestaltists in *The Nature of Maps*; and the third edition of Robinson's textbook, *Elements of Cartography* (Robinson and Sale 1969), wove gestalt-like concepts into its treatment of figure–ground relationships. More recently, MacEachren (1995: 70) noted renewed interest in gestalt concepts as an approach to understanding grouping in human vision.

Belbin's (1996) contribution is a concise, richly illustrated pedagogic essay that demonstrates the principles' relevance to cartographic design. In particular, he examines a set of rules (not previously published) devised by the late Newman Bumstead, a cartographic editor at the National Geographic Society. Knowingly or not, Bumstead incorporated gestalt principles into five rules for spacing letters, words, lines, and blocks of text. Bumstead's rules are a key concept in Belbin's map design course, in which students master ten single-spaced pages of similar guidelines, and those who violate them find the offending text stamped "B V," for "Bumstead violated." However heavy-handed, these rules and "laws" are a useful prescription for readability and aesthetics as well as an exemplar of cartographic theory with practical applications.

In a summary chapter to *Cartographic Design*, the book in which Belbin's essay appears, Philip Muehrcke (1996) lays out four fundamental questions on cartographic design:

1. What do we mean by cartographic design?
2. What are the most basic principles of map design?
3. How do we distinguish well-designed maps from poorly designed ones?
4. Is it possible to teach a map design sense, or is it a talent some innately have and others do not?

Muehrcke also identifies a series of important current themes/topics in cartographic design, including:

1. The long-standing split between practitioners and researchers persists today.
2. There is a downward trend in the volume of map design literature, but an upward trend in mapping.
3. We now have the tools to move away from the practice of attempting to provide a single best map.
4. We now have the capacity to incorporate more design dimensions or facets into our work. Effective integration of sound, animation, and interactions raises many new design challenges.
5. The gender issue is touchy. Rather than focus on potential gender differences, we might better expose map users to the range of useful strategies people are known to employ with respect to performing different tasks.
6. With the aid of GIS technology, we should expect to see dasymetric mapping displace choropleth mapping.
7. The changes in the ways maps are used in the electronic age are probably far more significant than changes in how they are made.
8. There is a growing literature documenting dissatisfaction with cartographers and their maps.

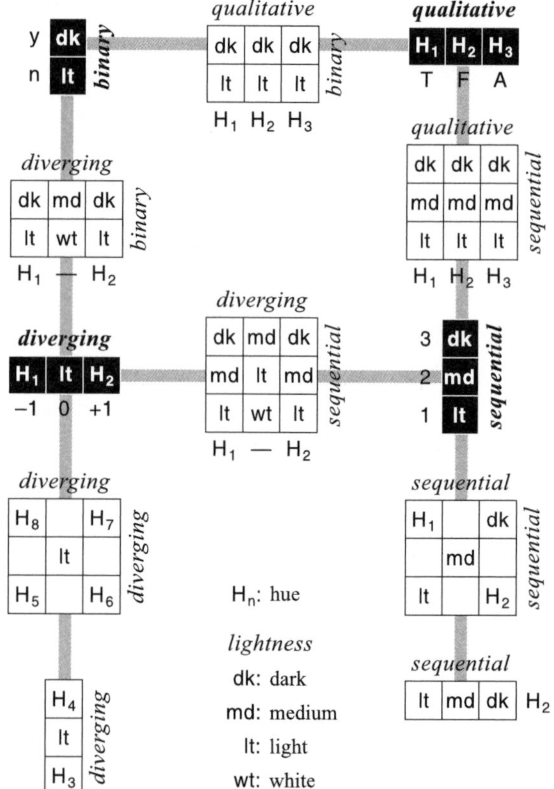

Fig. 26.3 Color guidelines as represented by schematic legends of the more basic one- and two-variable color maps

Color Guidelines for One- and Two-Variable Maps

Cynthia Brewer's (1994) model of color symbolization in Fig. 26.3 consists of twelve schematic map legends that describe color schemes varying in hue or degree of lightness. (In referring to a color's intensity, brightness, darkness, or luminance, Brewer uses "lightness" instead of "value" to avoid confusion between "color value" and "data value.") Her model is based on four distinct conceptualizations of data: "qualitative" for a variable with three or more categories differing in kind; "binary" for a two-category distinction between presence and absence; "sequential" for ordinal, interval, or ratio measurements with the low-to-high sequence typical of most choropleth maps; and "diverging" for a quantitative variable such as a rate of change diverging in two directions (positive and negative) from low, near-zero rates near the middle of its range. At the upper right in the diagram a schematic legend describes the recommended strategy for a one-variable qualitative map on which different hues (H_1, H_2, H_3) represent nominal differences (T, F, A), and two steps below another model legend describes a typical one-variable sequential choropleth map on which increasing degrees of lightness (lt, md, dk) reflect increasing higher data values (1, 2, 3).

These two legends form the right side of a rectangle the corners of which represent applications on one-variable maps of the four basic data types. On the left side of the rectangle, in the upper corner, dark and light (dk, lt) colors portray presence (y) and absence (n) on a binary map, and two steps below, in the lower-left corner, a pair of different hues (H_1, H_2) represents the more extreme values (−1, +1) on a data scale diverging from comparatively neutral values (0) portrayed by a light (lt), appropriately subdued color.

As the corners represent univariate maps, the sides depict bivariate displays. Along each side, a two-way legend describes the color schema for a two-variable map based on the data types in the adjoining corners, and extending downward from the rectangle's lower corners similar constructions describe color schemes for two-variable maps on which both variables are either diverging or sequential. Color plates in Brewer's article explain the application of this model as well as a more complex version that incorporates pattern variation and "balance" schemes and two-way color schemes for binary–binary, qualitative–qualitative, qualitative–diverging, and binary–sequential maps.

Despite its complexity, the model in Fig. 26.3 encourages more visually effective and aesthetically pleasing quantitative maps than the one-hue-at-most caveats found in most cartographic texts. Brewer, who acknowledges earlier work by Judy Olson (1987) and Janet Mersey (1990), concedes that many aspects of the schema have yet to be validated empirically. Even so, this conceptual framework underlies several recent empirical studies, including a recent exploration of design adjustments for color-impaired map viewers by Olson and Brewer (1997). In addition to influencing the design of an innovative mortality atlas published by the National Center for Health Statistics (Pickle *et al.* 1996; also see Brewer *et al.* 1997), Brewer's model is discussed prominently in Alan MacEachren's (1995: 295–7) *How Maps Work* as well as recent map design textbooks by Borden Dent (1999) and Terry Slocum (1999).

Automated Type Placement

Among the decade's more innovative approaches to automating the placement of type on maps is James Mower's (1993) parallel computing approach. His procedure used a Connection Machine 2 (CM-2)—a massively parallel computer developed by Thinking Machines, Inc., which has several advantages:

1. The number of available processors is large enough to allow each name or feature in the project database to be assigned to a unique processor.
2. Competition for map space can be managed through the interconnection of conflicting processors.
3. The project can be implemented in a parallel variant of a common high-level computer language.

Mower's study indicates that parallel computing environments offer fast and flexible alternatives to serial models of computation. His test results included: features and labels can be selected from a single database over varying windows and map scales; the execution performance of the procedure degrades slowly as scale decreases; if the number of names in the database window is less than the number of processors, increases in execution time are related primarily to increases in feature density, not to increases in the overall number of names in the map database window; feature selection is limited to the most important places as scale decreases; and the selection of unimportant features is suppressed in areas of high feature density.

Figure 26.4a and b illustrates some of the advantages of the parallel approach where, as one zooms in, the density of labels changes (26.4a to b). As Mower (1993: 78) points out, "as the scale increases and the field of view decreases, the percentages of features selected from each population decile increase steadily." Note in particular the density of labels in the triangle created by Syracuse, Binghamton, and Amsterdam increases significantly as more space becomes available with increasing scale. More recently, Steven Zoraster (1997) detailed a specific approach for label placement: the application of a stochastic optimization algorithm called "simulated annealing" to find both feasible and near-optimal solutions for point feature label placement. He applied this strategy in labeling petroleum-industry base maps.

Selecting an Appropriate Map Projection

John Snyder, the decade's leading authority on map projections, has created a formal model of the map projection selection process (Fig. 26.5). According to Snyder (1987: 33), the following characteristics are crucial in projection selection: (1) the size of the area to be mapped (global, hemisphere, or smaller); (2) the special properties that are required (equal-area, conformal, etc.); (3) the extent of the region (along great circle, along a small circle, radial); and (4) the general location of the region (polar, equatorial, mid-latitude). The generalized model presented here involves determining, from the initial map application and region size, the (1) distortion property, (2) special characteristics, and (3) extent of the projection. The framework allows for each of these three to be broken down further where, for instance, the extent is determined by the distortion property of the projection as well as the actual location being mapped. Snyder also created a more detailed model in which the actual size of the geographical area to be mapped—world,

Fig. 26.4 Examples of Mower's parallel computer type-placement methodology

Fig. 26.4 *Continued*

hemisphere, or continent, ocean, and smaller region—and type of property to be retained—conformal, equal-area, equidistant, global-look, and straight rhumb lines—are used as a first set of criteria for selection. If selecting a world projection that needs to be conformal, for instance, options include: constant scale along equator, constant scale along meridian, constant scale along oblique great circle, and entire earth depicted. Further levels in the conceptual decision tree involve scaling and centering of the projection.

Nyerges and Jankowski (1989) used Snyder's conceptual framework in the development of an expert system

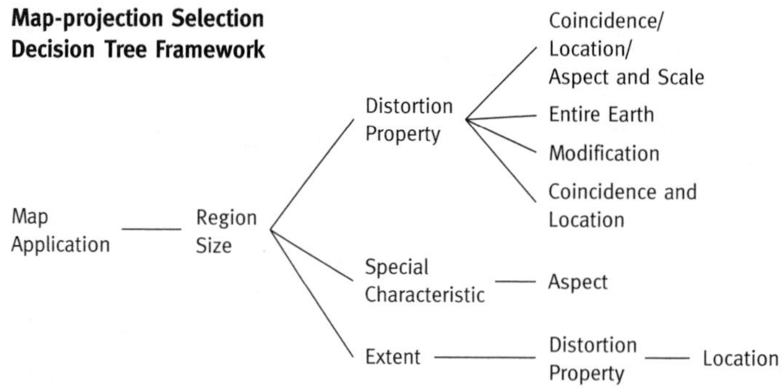

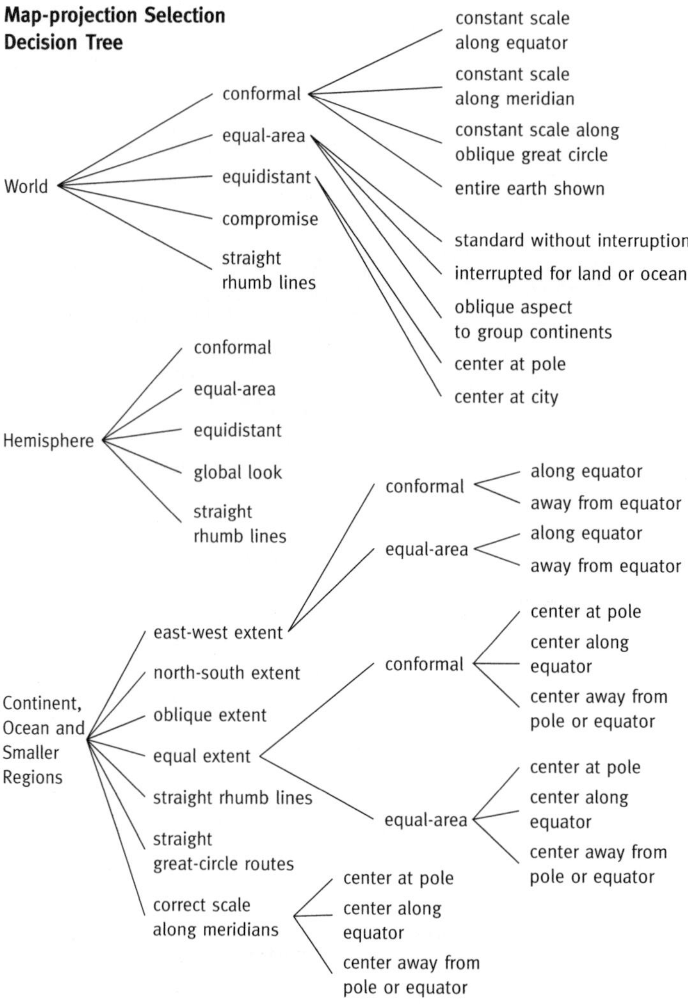

Fig. 26.5 Snyder's Map Projection Decision Tree

for projection selection (Fig. 26.5), which depicts the hierarchy of information needed. The procedure is based on a hierarchy of information that begins with the identification of the "Geographic Area Category." Other information needed by the system includes "Category Object," "Geographic Attributes of Object," "Map Function," "Geometric Properties," "Type of Display," and finally the "Map Scale." In a related study, DeGenst and Canters (1996) present a procedure for the application-dependent choice of a map projection suitable for implementation in a digital environment. To guarantee that the selection process leads to an unambiguous choice, they severely limit the number of map projections available.

Other recent work on map projections includes a collaborative design by Athelstan Spilhaus and John Snyder (1991). The authors assert that the graticule of meridians and parallels is a largely artificial type of map boundary that can detract from the display of irregular features such as oceans. Such natural boundaries as shorelines may be used instead as the boundary of world maps. The principle of natural boundaries has been applied to several examples of equal-area or conformal ocean maps with single or multiple lobes. The authors argue that by careful selection of the poles and centers, these maps can show both oceans and continents in their entirety on a single map. Other work by Snyder (1994) assesses the Hill Eucyclic Projection.

Other researchers have also developed new map projections. Albert Christensen (1992) presents a computer solution to the Chamberlin Trimetric projection with a numerical method that circumvents the need for closed formulas to analyze distortions. The existence and consequence of singularities in the computer algorithms are discussed in detail, with suggestions to minimize their adverse affect. Linear, area, and angular distortions are estimated for the Chamberlin projection, and compared with values analytically computed for the Transverse Mercator and Albers Equal-Area Conic projections. In addition, distance distortions for the three projections are computed, listed, and compared. Christensen concludes that the Chamberlin projection is an excellent compromise between the other two, provided the discontinuities are resolved. In a similar vein, Daniel Strebe (1994) presents a series of new projections, including the patchwork conic, an amalgamation of Equidistant Conic projections, and the Generalized Equidistant Cylindric, which has a conformal path of correct scale that is definable by the projection designer, and the Generalized Equidistant Conic.

For the purposes of global environmental monitoring, Denis White and his colleagues created a sampling design called "global tessellation" based on a systematic grid that can adequately assess the condition of many types of resources, and also retain flexibility for addressing new issues as they arise. After a review of existing approaches to constructing regular subdivisions of the Earth's surface, White *et al.* (1992) proposed the development of the sampling grid on the Lambert azimuthal equal-area projection of the Earth's surface to the face of a truncated icosahedron fit to the globe. The benefits of this geometric model include less deviation in area when subdivided as a spherical tessellation than any of the spherical platonic solids, and less distortion in shape over the extent of a face when used for a projection surface by the Lambert azimuthal projection. Additionally, one hexagon face of the truncated icosahedron covers the entire conterminous United States, and can be decomposed into a triangular grid at an appropriate density for sampling. The geometry of the triangular grid provides for varying the density, and points on the grid can be addressed in several ways.

For a similar environmental project, Daniel Carr and his colleagues (1992) present concepts that motivate the use of hexagon mosaic maps and hexagon-based ray-glyph maps. In this context, the term "hexagon mosaic map" refers to maps that use hexagons to tessellate major areas of a map, such as land masses. The approach applied here uses kriging to produce estimates for a hexagonal grid. The estimates are then transformed into a hexagon grid-cell choropleth map or hexagon mosaic map. A hexagon mosaic map can have interpretation advantages over a contour map, because the regular tessellation suggests the use of an estimation process and facilitates thought about confidence intervals. The hexagon edges at the class boundaries imply the estimation lattice that has been used since, in contrast, smooth contour lines give little clue to this underlying estimation step. The value for each hexagon typically is presumed to be an estimate that represents the whole hexagon region. The value for the hexagon does not have to match the value at any particular sampling site within the hexagon. In contrast, many people interpret contour lines as precise, and knowledgeable local experts argue about the placement of the lines.

Cartographic Generalization

One indication of the amount of research completed in cartographic generalization is the number of books and monographs published on this topic. Works published in the late 1980s, or in the early part of the decade, include a special issue of *Cartographica* on "Numerical Map Generalization" (McMaster 1989), *Map Generalization: Making Rules for Knowledge Representation* (Buttenfield and McMaster 1991), and *Generalization in Digital Cartography* (McMaster and Shea 1992). More recent works include Esa Maria Joao's *Causes and Consequences of Map Generalization* (Joao 1998) and *GIS and Generalization: Methodology and Practice*, an anthology edited by Jean-Claude Muller *et al.* (1995). In addition, *Geographic Information Research: Bridging the Atlantic* (Craglia and Couclelis 1997), based on a 1995 conference sponsored by the National Science Foundation and European Science Foundation, contains papers on diverse aspects of cartographic generalization.

The McMaster and Shea Model

Among the various generalization models proposed in recent years, the theoretical framework created by McMaster and Shea (1992) is among the most general. Published in the AAG resource publication, *Generalization in Digital Cartography*, the model (Fig. 26.6) accounts for three significant components of the process: the philosophical objectives, or "why" we generalize; a cartometric evaluation, or "when" to generalize; and the fundamental spatial and attribute transformations, which describe "how" to generalize. Philosophical objectives include several theoretical elements, including reducing the complexity, maintaining attribute accuracy, aesthetic quality, and a logical hierarchy, and consistently applying the rules. Overall, these relate to both application-specific goals of mapping: the map purpose and intended audience, the appropriateness of scale, the retention of clarity, and in a digital domain, the computational objectives, including cost effectiveness of the algorithms, maximum data reduction, and minimum memory/disk requirements.

An area of increasing interest in map generalization is cartometric evaluation, described in the central portion of Fig. 26.6. A series of geometric conditions, such as congestion, coalescence, conflict, complication, inconsistency, and imperceptibility, must be detected through a series of spatial and holistic measures, including density, distribution, length and sinuosity, shape, distance, gestalt, and abstract techniques. The third component, the issue of "when to generalize," involves the transformation controls of operational generalization systems. These include operator, algorithm, and parameter selection. The third part of the model identifies the necessary spatial and attribute transformations for generalization, whereas other models have divided the operators into those applied to raster and vector data.

The domain of line simplification, smoothing, displacement, and amalgamation has been a focus of considerable interest. Li and Openshaw (1993), for instance, developed a "natural principle" approach for line simplification while Mackaness (1994) discussed an algorithm that identifies and resolves spatial conflicts for feature displacement. Monmonier (1996*c*) presented six distinct generalizations useful for dynamic cartography. More recently, an issue of the journal *Cartography and Geographic Information Science* provides additional perspectives on map generalization, including work by Saalfeld on improvements to the Douglas algorithm (Saalfeld 1999), on scale-specificity and the concept of characteristic points (Dutton 1999), a new method for the displacement of cartographic features (Harrie 1999), and object-oriented strategies (Ormsby and Mackaness 1999).

Kilpelainen's Model

Although the European literature contains a number of conceptual frameworks, few have had as significant an influence on American workers as the models of Kilpelainen and Weibel. As Fig. 26.7 illustrates, Kilpelainen (1995) developed alternative frameworks for the representation of multi-scale databases. Assuming a master cartographic database, called the Digital Landscape Model (DLM), she proposed a series of methods for generating smaller-scale Digital Cartographic Models (DCMs). The master DLM is the largest-scale most accurate database possible, whereas secondary DLMs are generated for smaller-scale applications. Digital Cartographic Models, on the other hand, are the actual graphical representations, derived through a generalization/symbolization of the DLM. In frame A of her model, each DCM, labeled Generalized version 1, 2, . . . n, is generated directly from the initial master database. In frame B, however, a separate DLM is created for each scale/resolution, and the DCM is directly generated from each DLM. Alternatively, in frames C and D the master DLM is used to generate smaller-scale DLMs, which are then used to generate a DCM at that level. In certain instances, "secondary" DLMs are used to generate "tertiary" DLMs (note level 2 to level 3 in frame D). The assumption is that DCMs are generated on an as-needed basis.

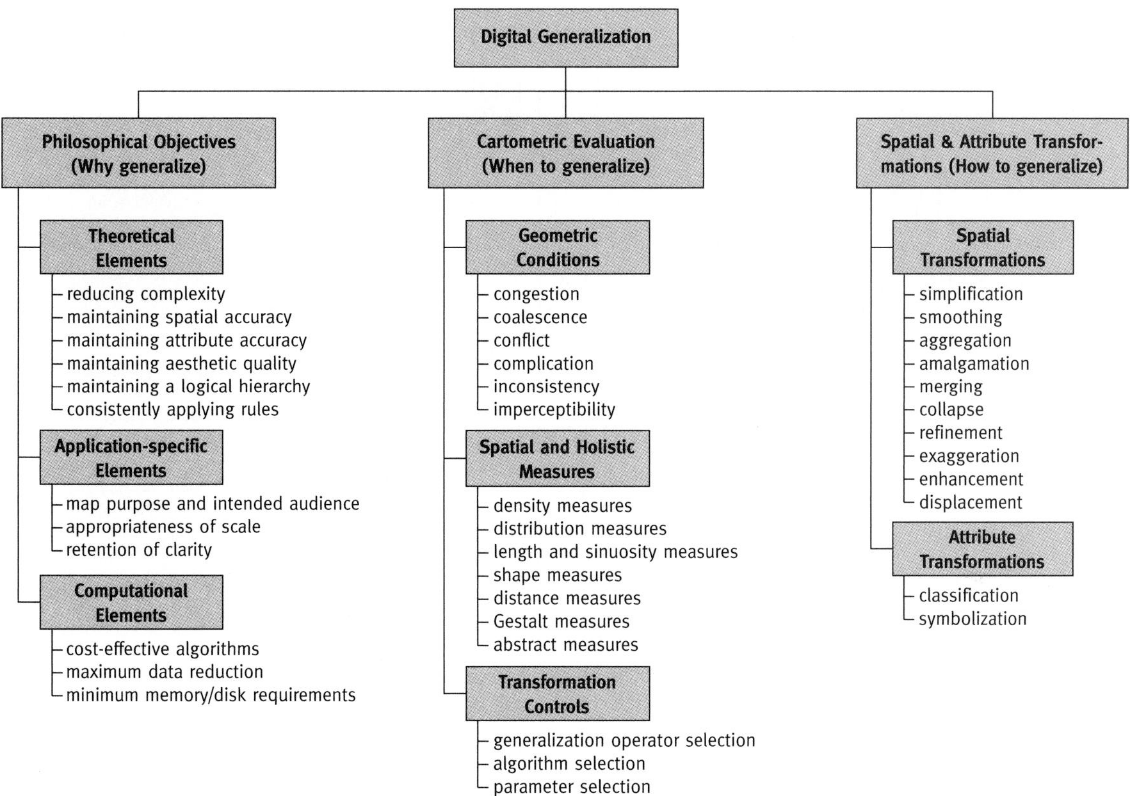

Fig. 26.6 McMaster and Shea model of generalization

Weibel's Model

Robert Weibel at the University of Zurich has worked extensively in the area of methods for terrain generalization. The research has two primary objectives: to design a strategy for terrain generalization that is adaptive to different terrain types, scales, and map purposes, and to implement and evaluate some components of this approach and its potential. Toward these ends, he developed the model of terrain generalization in Fig. 26.8. The model itself consists of five major stages for the generalization of terrain: structure recognition, process recognition, process modeling, process execution, and data display and evaluation of results. In structure recognition, the specific cartographic objects —as well as their spatial relations and measures of importance—are selected from the source data. Next, process recognition identifies the necessary generalization operators and parameters. Process recognition, "determines specifically how the source data are to be transformed, which types of conflicts have to be identified and resolved, and which types of objects and structures have to be carried into the target database" (Weibel *et al.* 1992: 134). Process modeling then compiles the rules and procedures from a process library.

The final stages involve process execution, in which the rules and procedures are applied to create the generalization, and data display. Specifically, the approach includes three different generalization methods: a global filtering procedure, a selective (iterative) filtering method, and a heuristic approach based on the generalization of the terrain's structure lines. For a given generalization problem that is constrained by the terrain character, map objective, scale, graphic limits, and data quality, the appropriate technique is selected through structure and process recognition procedures. A major contribution of the work is the distinction (top of Fig. 26.8) between global structure recognition and its local counterpart that serves as the basis for heuristic generalization. The author also depicts the application of specific generalization operations, including selection, simplification, combination, and

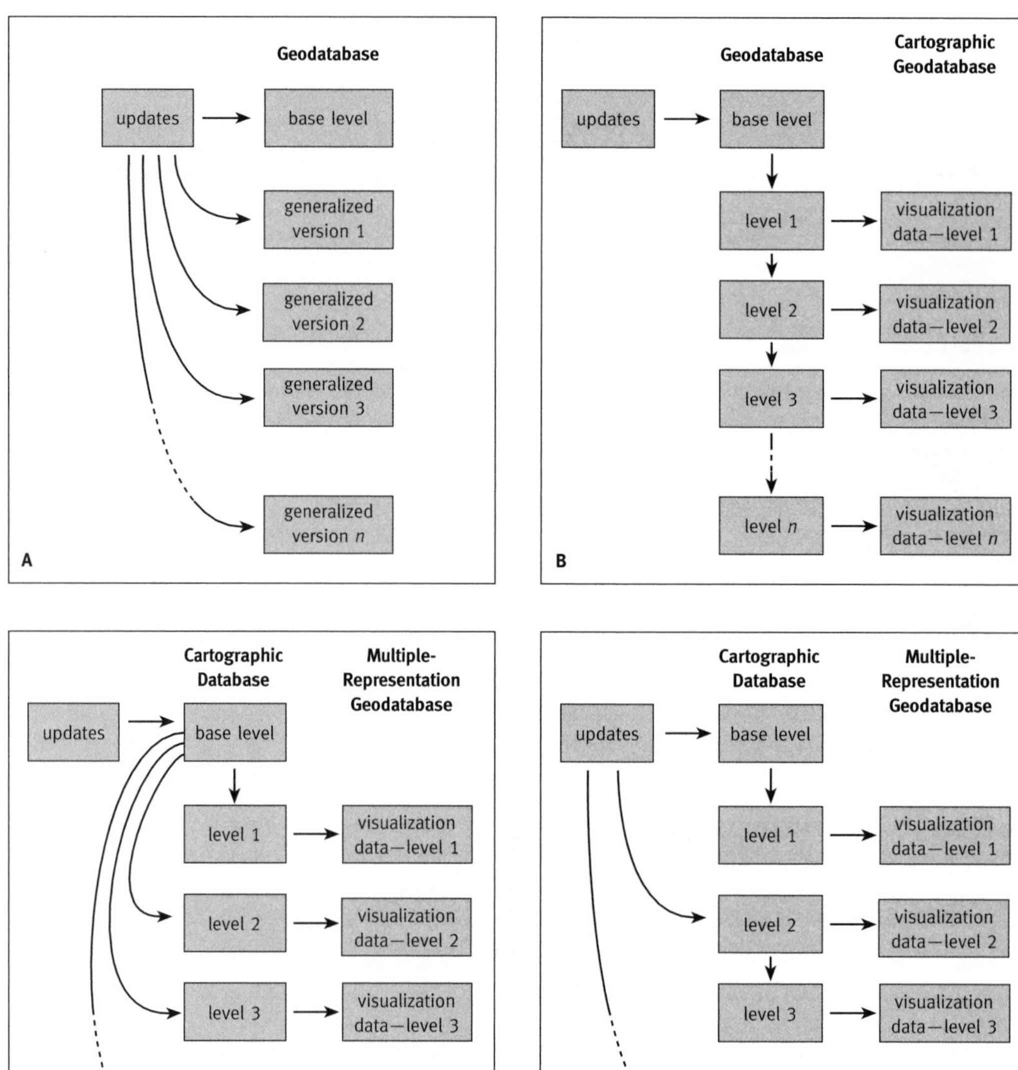

Fig. 26.7 Different update propagations in a geodata database

displacement. Figure 26.8 also shows the heuristic generalization processes of simplification, combination, and displacement.

Related to Weibel's model, further work on terrain modeling has addressed a variety of issues, including sensitivity analysis and algorithm development. The research by Chang and Tsai (1991) examines the effect of the spatial resolution of digital elevation models (DEMs) on slope and aspect data. After a review of computing methods for slope and aspect and such factors as DEM resolution, topographic complexity, and quality of DEM data, their research presents two experiments using DEMs from 8 to 80m intervals. Results of the experiments show that the accuracy of slope and aspect data, as well as the mean and standard deviation of slope values, decrease with lower DEM resolutions. Comparison of

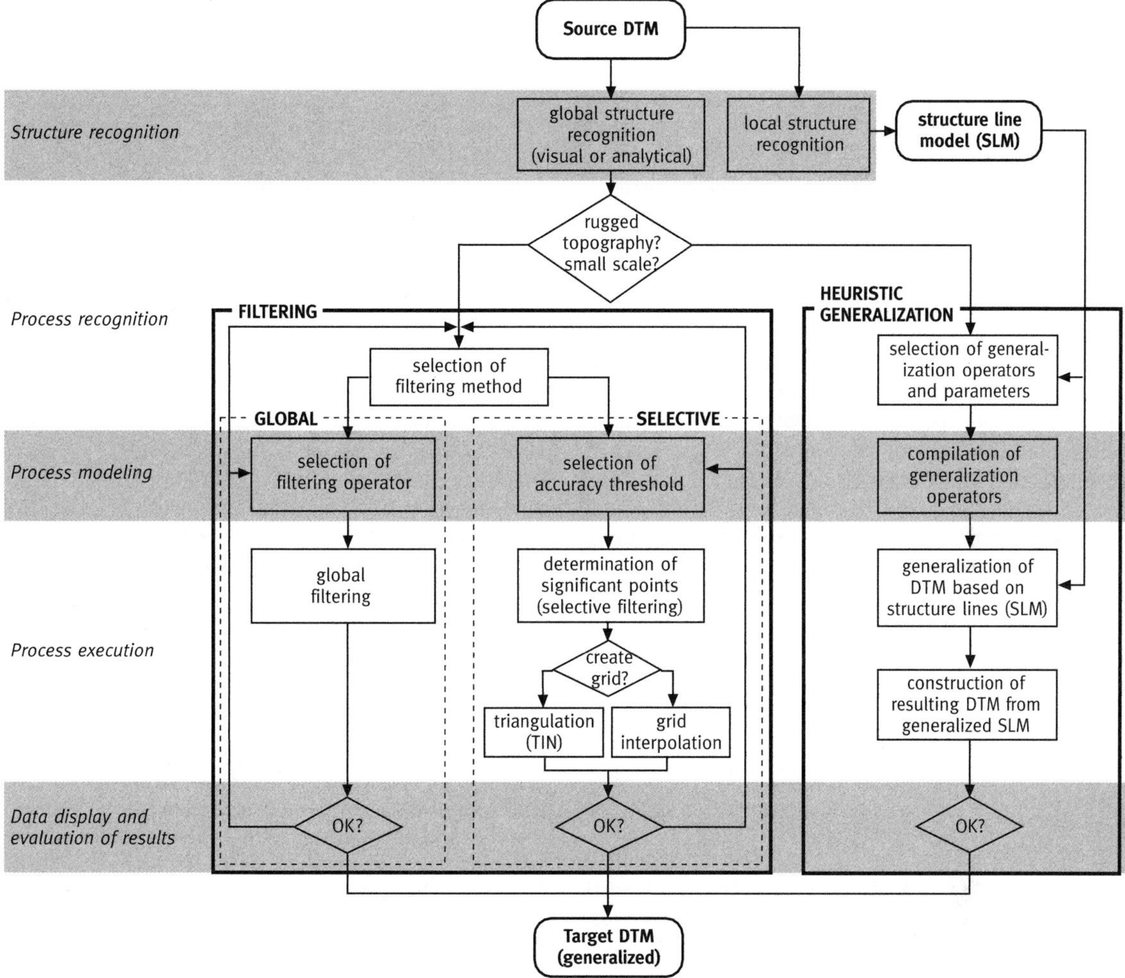

Fig. 26.8 Schematic flow of the strategy for computer-assisted terrain generalization

slope and aspect maps generated from different resolutions reveals that slope differences concentrate in areas of steep slopes, whereas aspect differences occur largely in comparatively flat areas with minor landform features. Slope differences can be explained statistically by relative relief, and aspect differences by relative relief and standard deviation of elevation. Using inductive procedures and domain-specific knowledge to identify and correct errors, Bennett and Armstrong (1996) developed a new approach for extracting basin morphology from a DEM. Specifically, a bit-mapped classification scheme is used to place each point in a DEM into one of six possible hydrologic categories. Using parallel computing technologies, Ding and Densham (1994) devised an algorithm that reduced solution times for hill shading on DEMs by over 70 per cent. Also noteworthy is Menno-Jan Kraak's (1993) "cartographic terrain modeling" (CTM) project, which allows for the manipulation and visualization of TINs. As part of the project, Kraak developed a conditioned chain split algorithm that allows ancillary polygonal information, such as forest stands, lakes, or urban boundaries, to be integrated into the TIN.

Geographic Data Interpolation

Although new conceptual models of the geographic data interpolation process have not been published in the past ten years, several important research projects have been completed. Hodgson (1992), in developing new methods for spatial interpolation, identified spatial searching, such as the identification of k-nearest neighbors to a point, as one of the most time-intensive tasks in vector-based geographic information systems transformational or analytical operations. His work examined the assumptions and resulting limitations of k-neighbor searching algorithms applied to spherical coordinates. Most existing k-neighbor searching algorithms reported in the literature are inappropriate for searching with spherical coordinates because they assume that the data are monotonic in each direction. On the sphere the monotonic coordinate system breaks down at the poles and at 180 degrees east and west. Hodgson identified one of the simplest, yet most efficient, k-neighbor searching algorithms that might be applied to spherical coordinates if constrained. His results indicated that, in comparing the processing efficiency among the brute-force searching method, a constrained heuristic k-neighbor searching algorithm, and a modified k-neighbor algorithm, processing times could be decreased by as much as 99 per cent using rapid searching methods.

Other researchers have recognized the importance of spherical interpolation. Robeson (1997) provides both a review and an assessment of spherical methods for interpolation. Declercq (1996) tested a series of interpolation routines, including polynomials, splines, linear triangulation, proximation, distance weighting, and kriging, on their efficiency at visualizing spatial patterns. His conclusions, based on measures of (1) accuracy of predicted values at unvisited points, (2) preservation of distinct spatial patterns, and (3) processing time, indicate that highly accurate interpolations do not always produce realistic spatial patterns, that the effectiveness of distance weighting and kriging methods was found to be largely dependent on the number of neighbors used, and, for both gradually and abruptly changing data, geographic reality was viewed best with the squared inverse distance weighting method.

Data Standards, Cartographic Features, and Geographic Reality

After nearly ten years of development and testing, the Department of Commerce approved the Spatial Data Transfer Standard (SDTS) as Federal Information Processing Standard (FIPS) 173 on 29 July 1992. The success of SDTS and FIPS was the result of many years of hard work by Harold Moellering of Ohio State University, who headed the US effort. FIPS 173 will transfer digital spatial datasets between different computer systems, making data sharing practicable. This standard is of significant interest to users and producers of digital spatial data because of the potential for increased access to and sharing of spatial data, the reduction of information loss in data exchange, the elimination of the duplication of data acquisition, and the increase in the quality and integrity of spatial data. The FIPS 173 implementation was effective from 15 February 1993, and is now mandatory for all federal agencies. FIPS 173 serves as the national spatial data transfer mechanism for all federal agencies, and will be available for use by state and local governments, the private sector, and research and academic organizations. In terms of data standards, we have now moved well beyond the planning stages to implementation, as detailed in the 1992 special issue of *Cartography and Geographic Information Systems* on "Implementing the Spatial Data Transfer Standard," with papers on "The Spatial Data Transfer Standard: A Management Perspective," "the SDTS Topological Vector Profile," "Developing a Raster Profile for SDTS," "SDTS Software Support: The FIPS 123 Function Library," and "The Design of a Spatial Data Transfer Processor." For instance, FIPS 173 provides standardized concepts for dealing with the notions of the spatial objects of points, lines, and areas, and the varying degrees of topological relationships between and among the objects. Fundamental to this work is the concept of features, as identified by the National Committee on Digital Cartographic Data Standards and the US Geological Survey, and the definitions of "entities," "objects," and "features."

Stephen Guptill's model of feature categories (1990), as reported in Usery (1993) and outlined in Fig. 26.9, models geographic reality using the fundamental views—cover, geoposition, ecosystem, division, and morphology—and subviews based on the United States Geological Survey's topographic and land-use/land-cover maps classified into overlapping world views. The major cover type subviews, for instance, include barren land, built-up land, cultivated cropland, vegetation, and water. Such object modeling has become more common as object-oriented approaches for spatial data structure are explored.

Usery (1993) explored feature-based approaches to GISs and developed a conceptual framework that is used for structuring geographic entities as features. In describing the hierarchical levels of cover sets (Fig. 26.10), he explained that, "while at one level roads, streams, counties, and urban places form members of different sets of geographical entities, at a higher level of abstraction cover sets such as transportation systems include both roads and streams; political systems include urban places and counties" (ibid. 7).

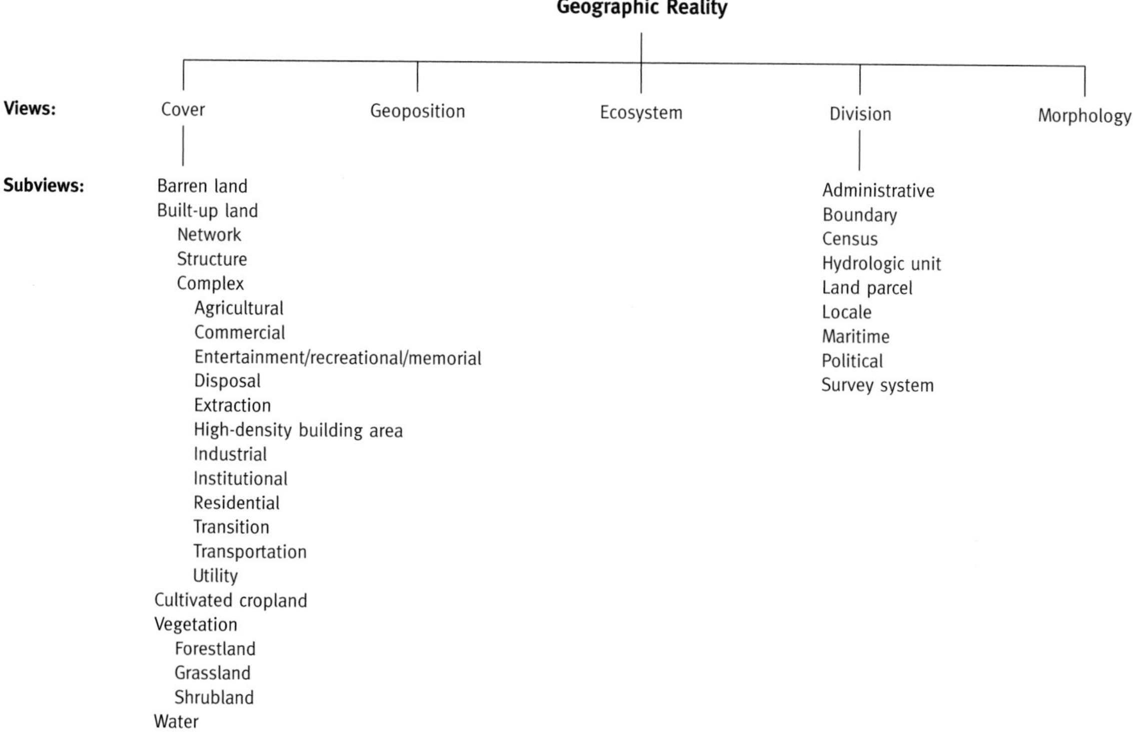

Fig. 26.9 US Geological Survey geographic reality based on topographic and land-use/land-cover maps classed into overlapping world views and subviews for DLG-E

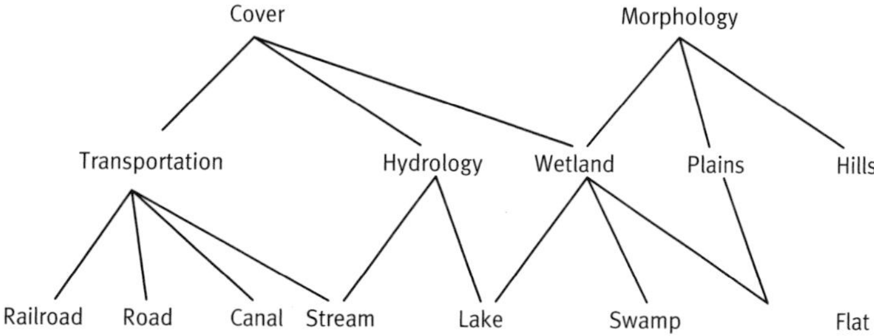

Fig. 26.10 A geographical example of hierarchical levels of cover sets

Dynamic Cartography

Cartography Cubed

Alan MacEachren's (1994) conceptual model of map-use space in Fig. 26.11 treats communication and visualization as opposite ends of a continuum defined by three orthogonal axes. A dimension labeled "human-map interaction" differentiates maps intended for communication, which usually permit little or no interaction, from maps used for visualization, for which interaction is high. A second axis addresses the user's objectives by contrasting communication's role in "presenting knowns" with visualization's mission of "revealing unknowns." A third dimension contrasting public and private use recognizes that communication is usually a public process in which a map author prepares a specific design for a mass audience of readers, CD buyers, or web surfers, whereas visualization typically allows individual users, typically working alone, to satisfy individual interests.

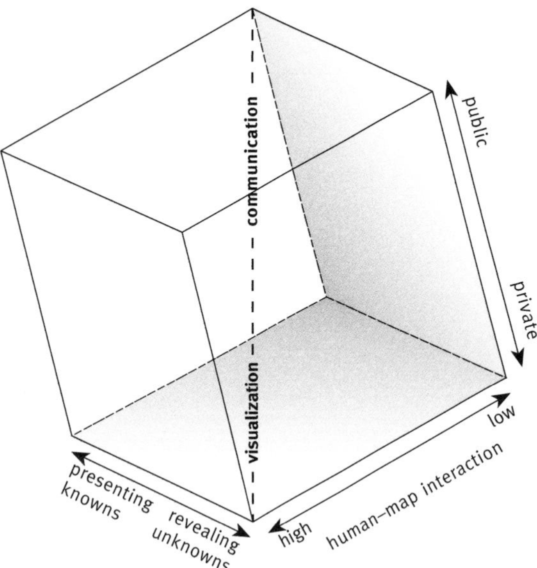

Fig. 26.11 Cartography cubed

Named "cartography cubed" because of its shape, MacEachren's model is a logical extension of an earlier model devised by his colleague, David DiBiase (1990), who distinguished visual thinking, in which exploration and confirmation occur largely in private, from visual communication, which involves synthesis and public presentation. Eight exemplars, one at each of the cube's corners, justify the three separate axes. Along the public–private axis, for instance, the "you-are-here" locator map in a shopping mall is a low-interaction device for presenting known information in public, whereas the equally low-interaction plat map at the county courthouse serves a conceptually similar objective in private. MacEachren's model also recognizes the discovery value for private users of dynamic low-interaction "closed" graphic scripts in which the viewer's "user profile" guides a customized exploratory analysis (Monmonier 1994) as well as the high-interaction equivalent of a "density dial" that facilitates experimentation with a choropleth map's category breaks (Ferreira and Wiggins 1990).

Each axis is a continuum, not a dichotomous scale. Degree of interactivity varies depending upon what the software allows the user to change and how rapidly the computer responds. What's more, maps of well-known relationships can be useful as starting points or pump primers in a visualization support system designed to reveal unknowns. And the use of email or the Internet to point out interesting maps or share opinions about the reliability or implications of geographic data blurs the distinction between public and private use. In identifying three key sources of variation in map use, MacEachren's model underscores the absence of sharp boundaries between cartographic communication and geographic visualization.

Mapping Time

Irina Vasiliev's (1996) conceptual model in Fig. 26.12 groups cartographic treatments of spatial-temporal information into five categories, each subdivided according to the dimensionality (point, line, area) of its symbols. Her first two categories address the representation of time as either moment or duration. Individual symbols can represent unique moments in time, as when the year-date 1893 is attached to the city label "Auburn," whereas all symbols on the map collectively represent the same moment when the date is attached to the title. Similarly, a map showing the duration of a continuous event or sequence of events relies upon labels indicating either an interval of time or a coherent series of instants in time. As several of Vasiliev's examples point out, arrows or lines can usefully reinforce descriptions of sequence and continuity. Her schema's third category, "structured time," recognizes the need for temporal labels or special symbols to represent standard time, as on a time-zone map with clock-face symbols, and periodicity, as on maps describing periodic markets or the frequency of service along delivery routes. Because market days are qualitatively different, colored point symbols and type can usefully reinforce day-of-the-week labels such as "Mondays" and "Tuesdays & Thursdays." The fourth category, which treats time as a measure of distance and vice versa, recognizes the common goals of isochronic lines, labels describing travel time, and point symbols spaced evenly in time but not space. Her final category, "space as clock," recognizes the representational roles of globe clocks, "chronosphere" animations, and Meso-American calendar maps, all based on the viewer's understanding of basic interrelationships of the Earth, the sun, and time.

An extension and elaboration of earlier models by Mei-Ling Hsu (1979) and John K. Wright (1944), Fig. 26.12 summarizes Vasiliev's (1997) comprehensive historical examination of cartographic treatments of time. A catalog of conventional approaches, her structural framework underscores the importance of spatial-temporal maps in the discourse of geography and other disciplines concerned with movement, change, and process (Monmonier 1993). In particular, it emphasizes the complementarity of graphic symbols and temporal labels, each of which brings different kinds of meaning

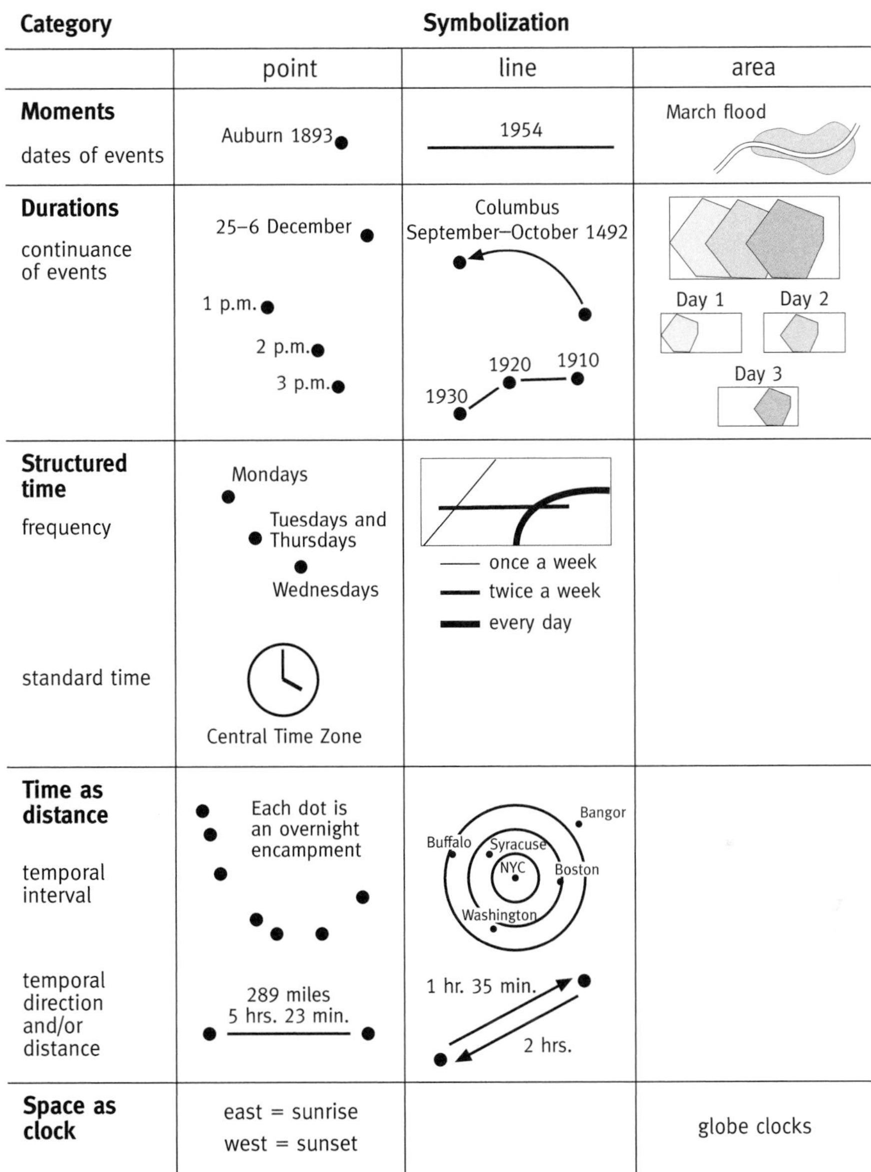

Category
Symbolization
point
line
area
Moments
dates of events
Auburn 1893
1954
March flood
Durations
continuance of events
25–6 December
1 p.m.
2 p.m.
3 p.m.
Columbus
September–October 1492
1920
1910
1930
Day 1
Day 2
Day 3
Structured time
frequency
Mondays
Tuesdays and Thursdays
Wednesdays
once a week
twice a week
every day
standard time
Central Time Zone
Time as distance
temporal interval
Each dot is an overnight encampment
Bangor
Buffalo
Syracuse
NYC
Boston
Washington
temporal direction and/or distance
289 miles
5 hrs. 23 min.
1 hr. 35 min.
2 hrs.
Space as clock
east = sunrise
west = sunset
globe clocks

Fig. 26.12 Catalog of strategies for representing spatial-temporal information on maps

to the map (MacEachren 1995: 245–69). Useful as a pedagogic device, the catalog can also inform the content and layout of symbol menus in GIS and online mapping.

Animation's New Visual Variables

Until recently map authors have had to rely largely on the eight visual variables identified by Jacques Bertin (1983): a pair of locational coordinates and the six retinal variables of size, shape, hue, value, texture/pattern, and orientation. Dynamic cartography enriched the cartographic toolkit by allowing map authors to expand or compress time in much the same way they manipulate the map's locational coordinates. And as DiBiase *et al.* (1992) pointed out, animation offers further flexibility with three dynamic variables: duration, rate of change, and order.

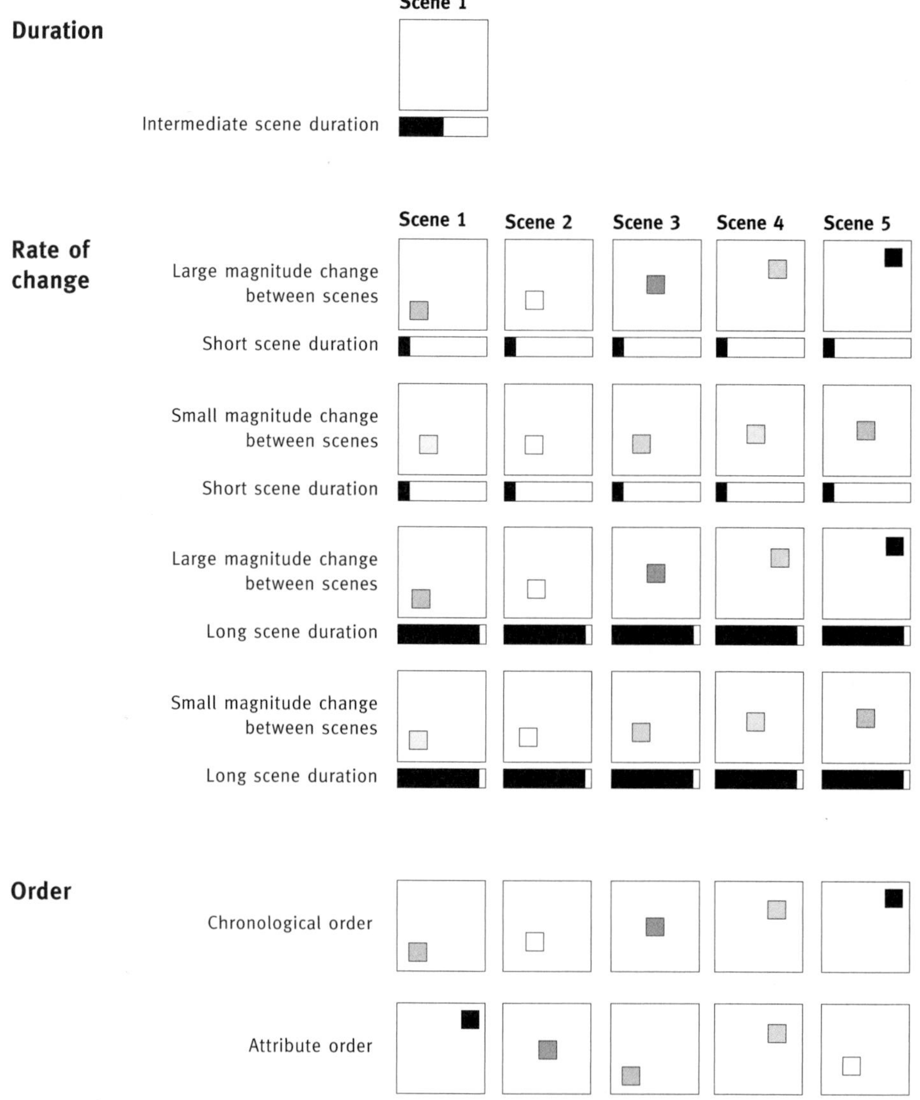

Fig. 26.13 Cartographic animation affords three new visual variables: duration, rate of change, and order

The conceptual model in Fig. 26.13 defines the dynamic variables and explains their uses. Duration, described graphically with a duration bar, refers simply to the length of time a particular view, or "scene," occupies the screen: a map on the screen for a half-second thus has a shorter duration than if it were shown for five seconds. The second dynamic variable, rate of change, links duration to the magnitude of change (in position or graytone) between scenes. The diagram's second row demonstrates that combining a large incremental change with a short duration produces a high rate of change, whereas the fifth row indicates that linking a small incremental movement with a long scene duration yields a comparatively small rate of change. Rows 3 and 4 describe the trade-off between duration and the magnitude, with the short duration and small magnitude in row 3 producing a smoother, more continuous sequence than the longer duration and larger magnitude in row 4.

A map author can apply the third dynamic variable, order, to either time or the attribute itself. For example, a chronologically ordered animation describing the spread of a disease will generally produce a comparatively

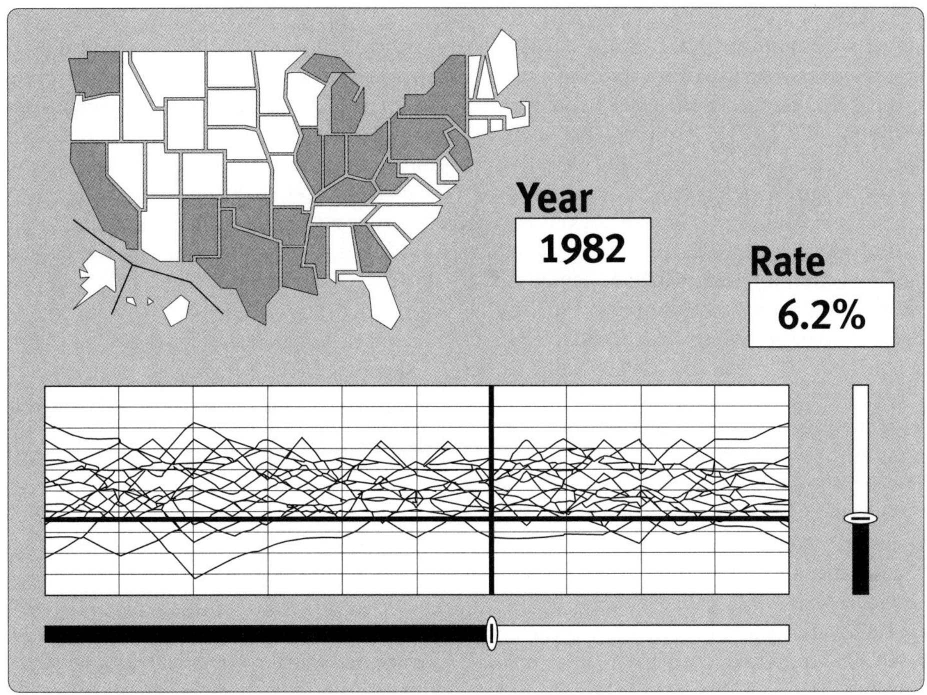

Fig. 26.14 Model interface for the interactive display of complementary cartographic and statistical representations of a spatial-temporal variable

continuous sequence of scenes, especially if the epidemic is a contagious process, without pronounced hierarchical influences. By contrast, a sequenced choropleth map (Slocum *et al.* 1990) showing mortality rates grouped by category will probably entail jerky transitions, especially if areas with high rates adjoin areas with low rates. Both orderings are potentially valuable, the former to illustrate the process of contagious diffusion and the latter to underscore salient differences in impact. In addition to orderings based on class intervals, Peterson (1993) identified several other potentially useful non-temporal cartographic animations, including the fly-through, the graphic zoom (without any change in level of detail), and the cartographic zoom, which varies both scale and the level of generalization.

Cartographic Complementarity and Interactive Displays

Monmonier's (1996*a*) juxtaposition of the map and time-series graph in Fig. 26.14 exemplifies the cartographic complementary possible with high-interaction graphics. The data in this illustration are average annual unemployment rates for the fifty states. Zigzag lines on the graph depict temporal trends for the nation's nine census divisions for the 21-year period 1970 through 1990, and the pair of bold vertical and horizontal lines are movable temporal and statistical cut-points called "combs," which in this view identify states with unemployment rates above 6.2 per cent for 1982. The comb aligned with the graph's temporal axis allows the viewer to vary the year, while the comb parallel to the graph's value axis separates areal units into two categories. Few labels are needed because the user can quickly determine the range for both scales by moving the combs. The movable cut-point for the value scale obviates an elaborate key as well as the inherent uncertainty of the traditional choropleth map, on which fixed, often arbitrary category breaks obscure within-category variation. And the temporal comb frees the viewer from conventional displays that typically offer only one or two years of data, or dilute meaningful temporal variation with multi-year averages. Although the display in Fig. 26.14 presents annual averages, a more flexible interface with the temporal equivalent of a "zoom" command could encourage interactive exploration of seasonal, monthly, or even weekly variations in unemployment.

Interactive linking of a time-series graph and a map is one of many cartographic applications of "dynamic multiple linked windows," a concept in exploratory data analysis pioneered by statisticians Richard Becker and William Cleveland (1987). To promote interactive analysis of three or more variables, Becker and Cleveland displayed their data pairwise as an array of scatterplots, each describing a single bivariate relationship. Every case (that is, each row in the user's data table) is represented by a dot in each scatterplot, one of which contains a movable rectangle, or "brush," used to select for highlighting all dots within its perimeter. In addition to highlighting cases within the brush, the system helps the viewer search for patterns involving other variables by highlighting the corresponding dots in all the other scatterplots. Monmonier (1989*a*) extended this treatment to spatial data by adding a cartographic window that highlights the locations of places within the scatterplot brush and provides a "geographic brush" for selecting places individually or by region. Other applications of geographic brushing include interactive cluster analysis (MacDougall 1992) and dynamic statistical summaries based on micromaps linked to frequency and time-series graphs (Carr *et al.* 1998).

Multiple linked windows have additional applications in geographic visualization. Slider bars with which the user manipulates scale as well as time afford a dynamic centrographic analysis of bivariate distributions in which cross-sectional density profiles aligned with cardinal directions complement a plan view of each distribution's center of mass (Monmonier 1992). And a programmed "graphic narrative" can interrogate a spatial-temporal distribution with a dynamic rate-of-change map in which the time interval's starting point and duration vary systematically (Monmonier 1996*b*: 180–3).

Information Flow, Public Policy, and the History of Cartography

Data Sharing

The models in Fig. 26.15 recognize the importance of public policy in cartographic communication and map use, particularly the role of government agencies at federal, state, and local levels in determining map content as well as the costs and benefits to specific users and the general public. Because geographic data are costly to

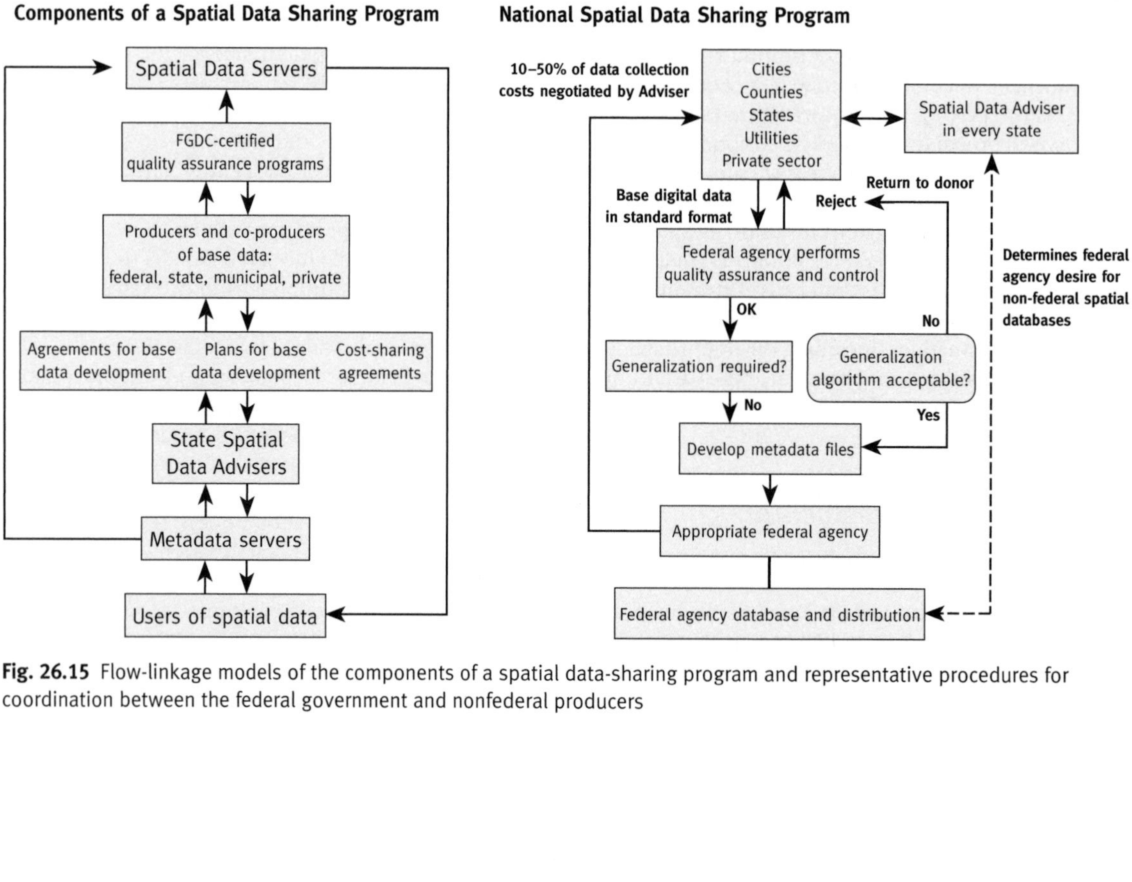

Fig. 26.15 Flow-linkage models of the components of a spatial data-sharing program and representative procedures for coordination between the federal government and nonfederal producers

collect and many users have similar needs, data sharing can yield substantial savings, especially if the private sector participates as both producer and user. But as this pair of diagrams indicate, data sharing is a complex process requiring the identification of a lead agency at the federal level, conscientious consultation with potential users, carefully planning for collection and distribution, and formal cost-sharing agreements. Cooperators also need to agree on specific data products, methods and levels of generalization, and procedures for quality control and quality assurance.

The two diagrams are a conceptual model of the spatial data sharing program designed by the National Research Council's Mapping Science Committee (1993). Appointed to investigate policies and methods for improving the accuracy of geographic information and controlling costs, the committee proposed a national spatial data infrastructure (NSDI), which would integrate data collected by states, municipalities, and private organizations with data collected by federal agencies such as the Geological Survey, the Bureau of the Census, and the Natural Resources Conservation Service (formerly the Soil Conservation Service). Although government agencies and private-sector producers might negotiate with each other directly, they depend at least indirectly upon the Federal Geographic Data Committee (FGDC), which coordinates and disseminates standards for quality assurance. Equally essential are information servers that support network distribution and metadata that "describe the content, ancestry and source, quality, data base schema, and accuracy of the data" (Mapping Science Committee 1993: 97). Effective and efficient data sharing requires data catalogs with a complete and systematic description of the data and their limitations.

The left-hand diagram identifies key components in the spatial data sharing program, and the right-hand diagram provides further details, including principal responsibilities, decisions, and avenues of communication. Every state would have a spatial data adviser, who would determine which data products developed within the state are suitable for the National Geographic Data System (NGDS) and would initiate negotiations between the producer and the appropriate federal agency. If the proposed product merits integration with the NGDS, the federal agency would support production through cost-sharing or work-sharing. Indeed, as John Bossler (1995) notes, the age-old system of barter can be an efficient strategy for acquiring needed geographic data. Moreover, as Coleman and Nebert (1998) point out, expanding data sharing beyond national borders holds distinct advantages for map users in Europe and North America.

Networks and Cartographic Centralization

The developmental model in Fig. 26.16 addresses another, more specifically cartographic aspect of cost-sharing: the dissemination of centrally produced news maps. Based on a study of the development of American journalistic cartography in the nineteenth and twentieth centuries (Monmonier 1989*b*), the diagram identifies three phases in the transmission of cartographic artwork from producer to publisher. Before the Associated Press established its Wirephoto network in 1935, three sources accounted for most news maps: the newspaper's own artists (including local freelancers), government agencies (including the city planning department and the Weather Bureau), and news/feature syndicates, which supplied stereomats by mail or rail express. Because few newspapers had a staff artist responsible for information

Pre-photowire

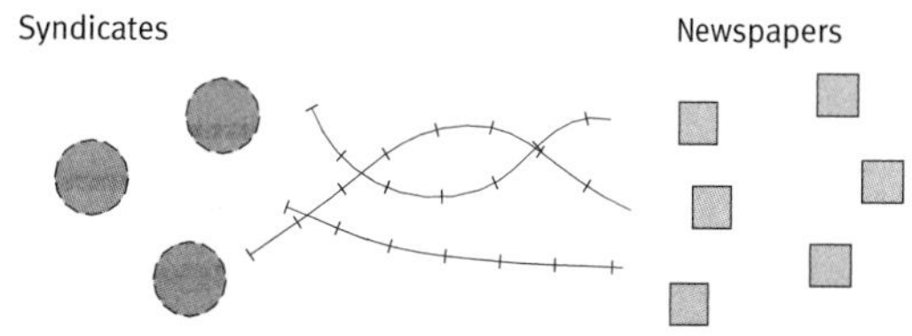

Photowire

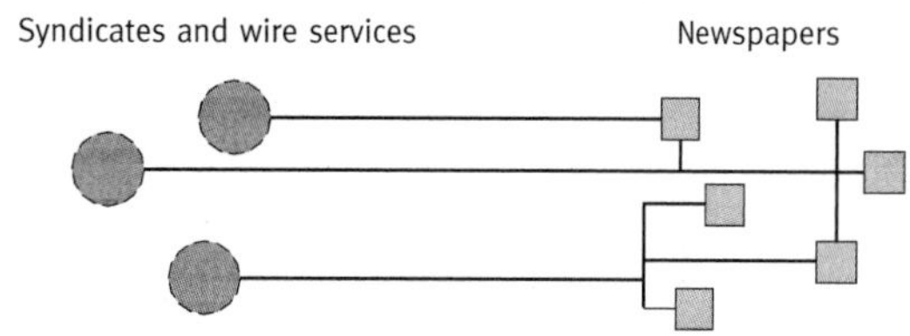

Satellite and microcomputer

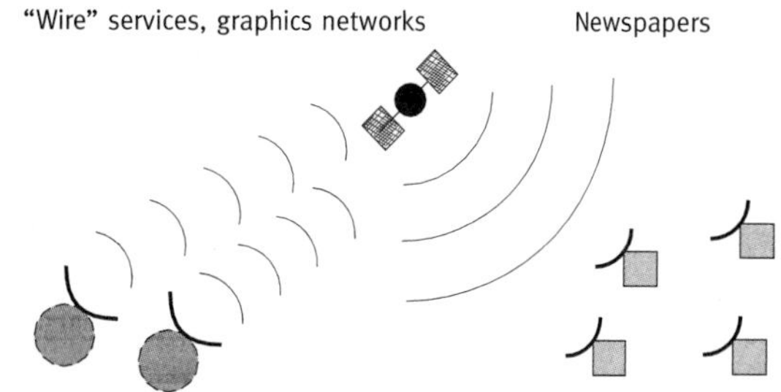

Fig. 26.16 Stages in the development of centrally produced journalistic cartography

graphics, maps rarely addressed the geographic aspects of breaking news. Expansion of photowire service in the 1940s and 1950s gave small- and medium-circulation newspapers access to news maps produced by the AP and United Press International. But because facsimile transmission equipment was designed to "move" halftone photographs, not line art, wire-service maps often had a somber appearance, with rough-edged labels and linework on a blotchy white background.

In the mid-1980s satellite telecommunications and microcomputers not only made news maps more widely available but improved their appearance. Because the new technology reduced delivery time from several minutes for a rasterized photowire map to less than a minute for a simple MacDraw file, maps were no longer in fierce competition with photos for transmission time. And before printing a crisp image on its Macintosh and Laserwriter, a subscribing newspaper could adapt an AP or KNT (Knight-Ridder-Tribune) map to its preferred typeface and column width. What's more, with a computer in place for customizing wire-service graphics, news editors soon recognized the ease with which an enterprising nonartist could craft maps for local news stories. Threatened by increased competition from television and *USA Today*, newspapers greatly increased their use of maps and other graphics.

The AP Wirephoto network also pioneered the regular distribution of centrally produced weather maps, which are now widely distributed by telephone facsimile, satellite, TV cable networks (Carter 1998), and the World Wide Web (Monmonier 1999). Underscoring the historic importance of distribution channels as a crucial element in cartographic communication and map use (Woodward 1974), the Internet threatens to displace both electronic and paper atlases as the dominant reference tool for cartographic information (Peterson 1997). But as Jeremy Crampton (1998) points out, conceptual flaws in online mapping and GIS tools challenge professional cartographers and geographic educators to shape map content and educate consumers. Like other innovative technologies, online cartography creates new problems as it solves old ones.

Conclusions

The discipline, or subdiscipline, of cartography is indeed at a crossroads as we enter into the twenty-first century. After the adjustments based on the rapid automation of mapping in the 1970s and 1980s, the discipline has made further significant intellectual changes due to the meteoric rise of geographic information systems and science in the 1990s. There has often been debate, as typified in Judy Olson's 1996 AAG Presidential Plenary address on the relationship between cartography and GIS. Cartographers might argue that theirs is an actual discipline with hundreds, or thousands, of years of history, while GIS is nothing more than a sophisticated set of "tools" or methods useful in spatial analysis. On the other hand, pure GIS specialists would argue that cartography is simply the visual output of a GIS process. Others have developed and nurtured the term "visualization" to represent what traditionally has been called cartography. One issue is certain: the proliferation of maps produced by GISs, mostly poorly designed, has resulted in a call for better education on the design of maps, including symbolization, generalization, classification, and use of color. In summarizing the discipline's accomplishments over the past ten years several trends emerge, including the strong continued research focus on analytical cartography, the emphasis on an animated—four-dimensional—cartography, a growing emphasis on geographic visualization, and a popularization of maps and mapping.

Research in analytical cartography represents intellectual material at the intersection of cartography and geographic information science. Specific research areas, including geographic data structures, data standards, digital terrain modeling, cartographic transformations, spatial interpolation, and generalization are often claimed by both disciplines. The disciplinary home for this research is irrelevant, however; the important point is that cutting-edge research continues, particularly in analytical generalization for which a series of books, workshops, and special issues of journals have appeared over the past ten years. One can also see a rapid growth in the development of four-dimensional cartographies. Although much of the theoretical work—including the development of a set of dynamic visual variables—has come out of the Pennsylvania State University program, many others have provided interesting applications as applied to both human-social and biophysical problems. This is one area of mapping in which the theoretical ideas are far ahead of the actual GIS and mapping software—successful animations must be composed in non-GIS packages such as Macromedia Director or programmed more or less from scratch in C++.

Related to the efforts in four-dimensional cartography is a new area of research in geographic visualization, a term not found in Kimerling's (1989) research summary. This same summary from 1989 did not mention the term. Today, an increasing amount of research that, in the past, has been within the purview of cartography is now called visualization, scientific visualization, or geographic visualization. For instance, one can point to the chapter (MacEachren *et al.* 1992) on "Visualization" in *Geography's Inner Worlds*, which was an attempt to provide a horizontal synthesis of geography across the subdisciplines. There has also been a growing number of special issues of journals and publications on this topic. From the standpoint of the "pure" cartographer, most disturbing is the metamorphosis at some institutions from what is left of the cartography curriculum into one of visualization. Finally, maps have become extremely popular in our society—in the news, on TV, in transportation systems, and even on T-shirts. The social critique of cartography, which started with the early essays by Brian Harley in *Cartographica* (Harley 1989), and further efforts by Denis Wood, Mark Monmonier, and research in the GIS and society debates, has addressed the influence both of social and cultural practices on the form of the map itself, and of maps and spatial data on the society in which they exist. The intense debates in the early part of the decade have evolved—fortunately—into a civil and productive level of discourse. What is acknowledged by all is the powerful and persuasive capability of the many existing cartographies that are proliferating in our society. Thus this area of research, *cartography and society*, will see significant attention over the next decade.

References

Becker, R. A., and Cleveland, W. S. (1987). "Brushing Scatterplots," *Technometrics*, 29: 127–42.

Belbin, J. A. (1996). "Gestalt Theory Applied to Cartographic Text," in C. H. Wood and C. P. Keller (eds.), *Cartographic Design: Theoretical and Practical Perspectives*. Chichester: John Wiley, 253–69.

Bennett, D. A., and Armstrong, M. P. (1996). "An Inductive Knowledge-Based Approach to Terrain Feature Extraction," *Cartography and Geographic Information Systems*, 23: 3–19.

Bertin, J. (1983). *Semiology of Graphics: Diagrams, Networks, Maps*, trans. W. J. Berg. Madison: University of Wisconsin Press, 41–97.

Bossler, J. D. (1995). "Facilitating the Sharing of Spatial Data: Perspectives from the Mapping Sciences Committee," in H. J. Onsrud and G. Rushton (eds.), *Sharing Geographic Information*. New Brunswick, NJ: Center for Urban Policy Research, 347–54.

Brewer, C. A. (1994). "Color Use Guidelines for Mapping and Visualization," in A. M. MacEachren and D. R. F. Taylor (eds.), *Visualization in Modern Cartography*. London: Pergamon, 123–47.

Brewer, C. A., MacEachren, A. M., Pickle, L. W., and Herrmann, D. (1997). "Mapping Mortality: Evaluating Color Schemes for Choropleth Maps," *Annals of the Association of American Geographers*, 87: 411–38.

Buttenfield, B. P., and McMaster, R. B. (eds.) (1991). *Map Generalization: Making Rules for Knowledge Representation*. Harlow: Longman.

Carr, D. B., Olsen, A. R., and White, D. (1992). "Hexagon Mosaic Maps for Display of Univariate and Bivariate Geographical Data," *Cartography and Geographic Information Systems*, 19: 228–36.

Carr, D. B., Olsen, A. R., Courbois, J. P., Pierson, S. M., and Carr, D. A. (1998). "Linked Micromap Plots: Named and Described," *Statistical Computing and Graphics Newsletter*, 9: 24–32.

Carter, J. R. (1998). "Uses, Users, and Use Environments of Television Maps," *Cartographic Perspectives*, 30: 18–37.

Chang, K., and Tsai B. (1991). "The Effect of DEM Resolution on Slope and Aspect Mapping," *Cartography and Geographic Information Systems*, 18: 69–77.

Christensen, A. H. J. (1992). "The Chamberlin Trimetric Projection," *Cartography and Geographic Information Systems*, 19: 88–100.

Coleman, D. J., and Nebert, D. D. (1998). "Building a North American Spatial Data Infrastructure," *Cartography and Geographic Information Systems*, 25: 151–60.

Craglia, M., and Couclelis, H. (eds.) (1997). *Geographic Information Research: Bridging the Atlantic*. London: Taylor & Francis.

Crampton, J. (1998). "The Convergence of Spatial Technologies," *Cartographic Perspectives*, 30: 3–5.

Deakin, A. K. (1996). "Landmarks as Navigational Aids on Street Maps," *Cartography and Geographic Information Systems*, 23: 21–36.

Declercq, F. A. N. (1996). "Interpolation Methods for Scattered Sample Data: Accuracy, Spatial Patterns, Processing Time," *Cartography and Geographic Information Systems*, 23: 128–44.

DeGenst, W., and Canters, F. (1996). "Development and Implementation of a Procedure for Automated Map Projection Selection," *Cartography and Geographic Information Systems*, 23: 145–64.

Dent, B. D. (1999). *Cartography: Thematic Map Design*. 5th edn. Boston: McGraw-Hill, 303–4.

DiBiase, D. (1990). "Visualization in the Earth Sciences." *Earth and Mineral Sciences*, Bulletin of the College of Earth and Mineral Sciences, Pennsylvania State University, 59/2: 13–18.

DiBiase, D., MacEachren, A. M., Krygier, J. B., and Reeves, C. (1992). "Animation and the Role of Map Design in Scientific Visualization," *Cartography and Geographic Information Systems*, 19: 201–14, 265–6.

Ding, Y., and Densham, P. J. (1994). "A Loosely Synchronous, Parallel Algorithm for Hill Shading Digital Elevation Models," *Cartography and Geographic Information Systems*, 21: 5–14.

Dutton, G. (1999). "Scale, Sinuosity, and Point Selection in Digital Line Generalization," *Cartography and Geographic Information Science*, 26: 33–53.

Ferreira, J., and Wiggins, L. (1990). "The Density Dial: A Visualization Tool for Thematic Mapping," *Geo Info Systems*, 1/6: 69–71.

Guptill, S. C., Boyko, K. J., Domaratz, M. A., Fegeas, R. G., Rossmeissl, H. J., and Usery, E. L. (1990). "An Enhanced Digital Line Graph Design," *USGS Circular 1048*. Reston, Va.: USGS.

Harley, J. B. (1989). "Deconstructing the Map," *Cartographica*, 26/2: 1–20.

Harley, J. B., and Woodward, D. (1992). *The History of Cartography: Cartography in the Traditional Islamic and South Asian Societies*. Chicago: University of Chicago Press, vol. ii bk. 1.

——— (1994). *The History of Cartography: Cartography in the Traditional East and Southeast Asian Societies*. Chicago: University of Chicago Press, vol. ii bk. 2.

Harrie, L. (1999). "The Constraint Method for Solving Spatial Conflicts in Cartographic Generalization," *Cartography and Geographic Information Science* 26: 55–69.

Hodgson, M. E. (1992). "Rapid K-Neighbor Identification Algorithm for the Sphere," *Cartography and Geographic Information Systems*, 19: 228–36.

Hsu, M-L. (1979). "The Cartographer's Conceptual Process and Thematic Symbolization," *The American Cartographer*, 6: 117–27.

Imhof, E. (1962). "Die Anordnung der Names in der Karte," *International Yearbook of Cartography*, 2: 93–129.

—— (1974). "Positioning Names on Maps," trans. G. F. McCleary. *The American Cartographer*, 2: 128–44.

Joao, E. M. (1998). *Map Generalization: Causes and Consequences*. London: Taylor & Francis.

Kilpelainen, T. (1995). "Requirements of a Multiple Representation Database for Topographical Data with Emphasis on Incremental Generalization," *Proceedings of the 17th International Cartographic Conference*, 2 (Barcelona, Spain): 1815–25.

Kimerling, A. J. (1989). "Cartography," in G. L. Gaile and C. J. Willmott (eds.), *Geography in America*. Columbus, Ohio: Merrill.

Kraak, M.-J. (1993). "Cartographic Terrain Modeling in a Three-Dimensional GIS Environment," *Cartography and Geographic Information Systems*, 20: 13–18.

Kumler, M. P., and Buttenfield, B. P. (1996). "Gender Differences in Map Reading Abilities: What Do We Know? What can We Do?" in C. H. Wood and C. P. Keller (eds.), *Cartographic Design: Theoretical and Practical Perspectives*. Chichester: John Wiley, 125–36.

Li, Z., and Openshaw, S. (1993). "A Natural Principle for the Objective Generalization of Digital Maps," *Cartography and Geographic Information Systems*, 20: 19–30.

Lloyd, R., Rostkowska-Covington, E., and Steinke, T. (1996). "Feature Matching and the Similarity of Maps," in C. H. Wood and C. P. Keller (eds.), *Cartographic Design: Theoretical and Practical Perspectives*. Chichester: John Wiley, 211–35.

MacDougall, E. B. (1992). "Exploratory Analysis, Dynamic Statistical Visualization, and Geographic Information Systems," *Cartography and Geographic Information Systems*, 19: 237–46.

MacEachren, A. M. (1994). "Visualization in Modern Cartography: Setting the Agenda," in A. M. MacEachren and D. R. F. Taylor (eds.), *Visualization in Modern Cartography*. London: Pergamon, 1–12.

—— (1995). *How Maps Work*. New York: Guilford Press.

MacEachren, A. M., Buttenfield, B. P., Campbell, J. B., DiBiase, D. W., and Monmonier, M. (1990). "Visualization," in R. F. Abler, M. G. Marcus, and J. M. Olson (eds.), *Geography's Inner Worlds: Pervasive Themes in Contemporary American Geography*. New Brunswick, NJ: Rutgers University Press, 99–137.

McGranaghan, M. (1996). "An Experiment with Choropleth Maps on a Monochrome LCD Panel," in C. H. Wood and C. P. Keller (eds.), *Cartographic Design: Theoretical and Practical Perspectives*. Chichester: John Wiley, 177–90.

Mackaness, W. A. (1994). "An Algorithm for Conflict Identification and Feature Displacement in Automated Map Generalization," *Cartography and Geographic Information Systems*, 21: 219–32.

McMaster, R. B. (ed.) (1989). "Numerical Generalization in Cartography," *Cartographica*, 26/1.

McMaster, R. B., and Shea, K. S. (1992). *Generalization in Digital Cartography*. Washington, DC: Association of American Geographers.

Mapping Science Committee, National Research Council (1993). *Toward a Coordinated Spatial Data Infrastructure for the Nation*. Washington, DC: National Academy Press.

Mark, D. M., and Egenhofer, M. J. (1994). "Modeling Spatial Relations Between Lines and Regions: Combining Formal Mathematical Models and Human Subjects Testing," *Cartography and Geographic Information Systems*, 21: 195–212.

Mersey, J. E. (1990). "Colour and Thematic Map Design: The Role of Colour Scheme and Map Complexity in Choropleth Map Communication," *Cartographica*, 27/3 [monograph 41].

Monmonier, M. (1977). *Maps, Distortion, and Meaning*. Washington, DC: Association of American Geographers.

—— (1989*a*). "Geographic Brushing: Enhancing Exploratory Analysis of the Scatterplot Matrix," *Geographical Analysis*, 21: 81–4.

—— (1989*b*). *Maps with the News: The Development of American Journalistic Cartography*. Chicago: University of Chicago Press.

—— (1992). "Directional Profiles and Rose Diagrams to Complement Centrographic Cartography," *Journal of the Pennsylvania Academy of Science*, 66: 29–34.

—— (1993). *Mapping It Out: Expository Cartography for the Humanities and Social Sciences*. Chicago: University of Chicago Press, 189–205.

—— (1994). "Graphic Narratives for Analyzing Environmental Data," in A. M. MacEachren and D. R. F. Taylor (eds.), *Visualization in Modern Cartography*. London: Pergamon, 201–13.

—— (1995). *Drawing the Line: Tales of Maps and Cartocontroversy*. New York: Henry Holt.

—— (1996*a*). "Cartographic Complementarity: Objectives, Strategies, and Examples," in C. H. Wood and C. P. Keller (eds.), *Cartographic Design: Theoretical and Practical Perspectives*. Chichester: John Wiley, 77–90.

—— (1996*b*). *How to Lie with Maps*. 2nd edn. Chicago: University of Chicago Press.

—— (1996*c*). "Temporal Generalization for Dynamic Maps," *Cartography and Geographic Information Systems*, 23: 96–8.

—— (1997). *Cartographies of Danger: Mapping Hazards in America*. Chicago: University of Chicago Press.

—— (1999). *Air Apparent: How Meteorologists Learned to Map, Predict, and Dramatize Weather*. Chicago: University of Chicago Press.

Mower, J. E. (1993). "Automated Feature and Name Placement on Parallel Computers," *Cartography and Geographic Information Systems*, 20: 69–82.

Muehrcke, P. C. (1983). *Map Use: Reading, Analysis, and Interpretation*. Madison, Wis.: JP Publications.

—— (1996). "The Logic of Map Design," in C. H. Wood and C. P. Keller (eds.), *Cartographic Design: Theoretical and Practical Perspectives*. Chichester: John Wiley, 271–8.

Muller, J.-C., Lagrange, J.-P., and Weibel, R. (1995). *GIS and Generalization: Methodology and Practice*. London: Taylor & Francis.

Nelson, E., and Gilmartin, P. (1996). "An Evaluation of Multivariate Quantitative Point Symbols for Maps," in C. H. Wood and C. P. Keller (eds.), *Cartographic Design: Theoretical and Practical Perspectives*. Chichester: John Wiley, 191–210.

Nyerges, T., and Jankowski, P. (1989). "A Knowledge Base for Map Projection Selection," *The American Cartographer*, 16: 29–38.

Olson, J. M. (1987). "Color and the Computer in Cartography," in H. J. Durrett (ed.), *Color and the Computer*. Orlando, Fla.: Academic Press, 205–20.

Olson, J. M., and Brewer, C. A. (1997). "An Evaluation of Color Selections to Accommodate Map Users with Color-Vision Impairments," *Annals of the Association of American Geographers*, 87: 103–34.

Ormsby, D., and Mackaness, W. (1999). "The Development of Phenomenological Generalization Within an Object-Oriented Paradigm," *Cartography and Geographic Information Science*, 26: 70–80.

Patton, K., and Cammack, R. G. (1996). "An Examination of the Effects of Task Type and Map Complexity on Sequenced and Static Choropleth Maps," in Wood and C. P. Keller (eds.), *Cartographic Design: Theoretical and Practical Perspectives*. Chichester: John Wiley, 237–52.

Peterson, M. P. (1993). "Interactive Cartographic Animation," *Cartography and Geographic Information Systems*, 20: 40–4.

—— (1997). "Trends in Internet Map Use," *Proceedings of the International Cartographic Conference, Stockholm*, 1635–42.

Pickle, L. W., Mungiole, M., Jones, G. K., and White, A. A. (1996). *Atlas of United States Mortality*. Rockville, Md.: National Center for Health Statistics (U.S. Department of Health and Human Services publication no. (PHS) 97–1015).

Robeson, S. M. (1997). "Spherical Methods for Spatial Interpolation: Review and Evaluation," *Cartography and Geographic Information Systems*, 24: 3–20.

Robinson, A. H., and Petchenik, B. B. (1976). *The Nature of Maps: Essays Toward Understanding Maps and Mapping*. Chicago: University of Chicago Press, 74–5, 82–5.

Robinson, A., and Sale, R. (1969). *Elements of Cartography*. 3rd edn. New York: John Wiley, 257–60.

Robinson, A. H., Morrison, J. L., Muehrcke, P. C., Kimerling, A. J., and Guptill, S. C. (1995). *Elements of Cartography*. 6th edn. New York: John Wiley.

Saalfeld, A. (1999). "Topologically Consistent Line Simplification with the Douglas-Peucker Algorithm," *Cartography and Geographic Information Science*, 26: 7–18.

Slocum, T. A. (1999). *Thematic Cartography and Visualization*. Upper Saddle River, NJ: Prentice-Hall, 106–7.

Slocum, T. A., Roberson, S. H., and Egbert, S. L. (1990). "Traditional Versus Sequenced Choropleth Maps: An Experimental Investigation," *Cartographica*, 27/1: 67–88.

Snyder, J. P. (1987). *Map Projections—A Working Manual*, US Geological Survey Professional Paper 1395. Washington, DC: US Government Printing Office.

—— (1993). *Flattening the Earth: Two Thousand Years of Map Projections*. Chicago: University of Chicago Press.

—— (1994). "The Hill Eucyclic Projection," *Cartography and Geographic Information Systems*, 21: 213–18.

Spilhaus, A., and Snyder, J. P. (1991). "World Maps With Natural Boundaries," *Cartography and Geographic Information Systems*, 18: 246–54.

Strebe, D. R. (1994). "The Generalized Conic Projection," *Cartography and Geographic Information Systems*, 21: 233–41.

Usery, E. L. (1993). "Category Theory and the Structure of Features in Geographic Information," *Cartography and Geographic Information Systems*, 20: 5–12.

Vasiliev, I. (1996). "Design Issues to be Considered When Mapping Time," in C. H. Wood and C. P. Keller (eds.), *Cartographic Design: Theoretical and Practical Perspectives*. Chichester: John Wiley, 137–46.

—— (1997). *Mapping Time*, Cartographica monograph 49. Toronto: University of Toronto Press.

Weibel, R. (1992). "Model and Experiments for Adaptive Computer-Assisted Terrain Generalization," *Cartography and Geographic Information Systems*, 19: 133–53.

White, D., Kimerling, A. J., and Overton, W. S. (1992). "Components of a Global Sampling Design for Environmental Monitoring," *Cartography and Geographic Information Systems*, 19: 5–22.

Wood, C. (1994). "Effects of Brightness Difference on Specific Map-Analysis Tasks: An Empirical Analysis," *Cartography and Geographic Information Systems*, 21: 15–31.

Wood, D. (1992). *The Power of Maps*. New York: Guilford Press.

Woodward, D. (1974). "The Study of the History of Cartography: A Suggested Framework," *The American Cartographer*, 1: 101–15.

Woodward, D., and Lewis, G. M. (1998). *The History of Cartography: Cartography in the Traditional African, American, Arctic, Australian, and Pacific Societies*. Chicago: University of Chicago Press, vol. ii bk. 3.

Wortman, K., and J. L. Morrison (eds.) (1992). "Implementing the Spatial Data Transfer Standard," *Cartography and Geographic Information Systems*, 19/5.

Wright, J. K. (1944). "A Proposed Atlas of Diseases: Appendix 1—Cartographic Considerations," *Geographical Review*, 34: 649–52.

Zoraster, S. (1997). "Practical Results Using Simulated Annealing for Point Feature Label Placement," *Cartography and Geographic Information Systems*, 24: 228–38.

CHAPTER 27

Mathematical Models and Quantitative Methods

David R. Legates, Sucharita Gopal, and Peter Rogerson

Although the use of mathematical models and quantitative methods in geography accelerated in earnest with the development of quantitative geography and regional science in the late 1950s, such techniques had already made their way into the mainstream of physical geography much earlier. Today, mathematical models and quantitative methods are used in a number of subfields in geography with their proliferation being aided, in part, by the widespread use of remote sensing, geographic information systems (GIS), and computer-based technology. As a consequence, geography as a whole has witnessed a new growth in the development of models and quantitative methods over the last decade, and it is this growth that we seek to elucidate here.

Highlighting the advances in the use of models and methods in geography is a difficult undertaking. Such techniques are so widely used in GIS and remote sensing that many developments in these areas also could be considered in this chapter. Moreover, modeling and quantitative techniques are so strongly integrated within some geographic subfields (e.g. climatology and geomorphology, economic and urban geography, regional science) that it is often difficult to separate technique development from application. This is illustrated by the fact that many members of the Association of American Geographers who frequently use and develop quantitative techniques and models are not active participants in the Mathematical Models and Quantitative Methods Specialty Group, choosing instead to favor specialty groups with a more topical, rather than methodological, focus. In a very real sense, the quantitative revolution has been completed in many subfields of geography, with the goals and aims of the revolutionaries having long since passed into the mainstream.

Furthermore, geographers who are involved with quantitative methods and mathematical models are extremely diverse in their interests and applications—they contribute to an extremely wide variety of disciplines. While they excel at spreading the geographic word to other disciplines, summarizing their multifarious contributions is nearly impossible. The rather trite statement, "Geography is what geographers do," seems to apply strongly here. Geographers are largely a collection of individuals who, although united by their interest in spatial models and methods, are unique in the ways that they make contributions to various fields.

Thus, here the focus will be on the innovation and methodological development associated with mathematical modeling and quantitative methods that has occurred over the last decade of the twentieth century. It will be taken for granted that in many subfields of geography, rote application of existing techniques and models is commonplace. Instead of dwelling on pure application, therefore, this chapter will focus on the new and innovative solutions to geographic problems that have been developed, and on the pioneering work that has occurred in some subfields where the use of such methods has been lacking. Particular focus will be placed on the development of models and methods that address the spatial nature of geographic phenomena. After all, it is this spatial characteristic that sets geography apart from its sister disciplines and is an issue often overlooked by them.

It is clear that the old adage "necessity is the mother of invention" rings loudly throughout this chapter. Few geographers create models or develop new quantitative applications from a purely theoretical standpoint. Rather, virtually all these methods have arisen from the need to solve certain geographical problems and we salute those who have chosen to refine their techniques rather than simply employ existing models and methods in what may be an inappropriate application.

Quantitative Methods and Mathematical Models in Human Geography

The 1990s have seen considerable development in the mathematical models and quantitative methods used by human geographers. In particular, significant developments have occurred in three major areas: spatial analysis, behavioral modeling, and computational intelligence.

Spatial Analysis and Geographic Information Systems

Over the last decade, considerable developments in spatial analysis and GIS techniques have occurred. Spatial analysis incorporates research into spatial statistics and data analysis, exploratory and confirmatory spatial data analysis, and spatial modeling. From its initial emphasis on the use of quantitative procedures and techniques to analyse patterns of points, lines, areas, and surfaces, developments in spatial analysis have become increasingly associated with modeling spatial choice and spatial processes, as well as determining their implications for the spatio-temporal evolution of complex spatial systems (Thill and Horowitz 1991). In particular, recent emphases have focused on areas of spatial modeling that encompass a diverse array of models that transcend both the environmental and social sciences. Much of this growth has been spurned by the rapidly advancing computer technology (both hardware and software) and, in particular, that related to GIS (see Fischer *et al.* 1996; Zhang and Griffith 2000).

The origin of spatial analysis is associated with the growth of quantitative geography and regional science in the 1950s. In previous decades, growth was limited by scarce and expensive computing technology, limited datasets with sometimes questionable reliability and little metadata information, and an over-reliance on linear statistical models. The 1990s, however, have seen many of these limitations overcome. For example, spatial analysis has become strongly linked with GIS as evidenced by the books published on the subject (Cho 1996; Fischer and Nijkamp 1993; Fischer and Getis 1997; Fotheringham and Rogerson 1994; Longley *et al.* 1997). Spatial analysis, therefore, has developed as a direct result of major breakthroughs in cost, speed, and data storage capacity of computer hardware as well as the advent of stand-alone workstations and PCs and readily available package software. Moreover, the establishment of academic centres in GIS, the proliferation of professional and academic conferences (e.g. GIS/LIS, USIRA, UCGIS), and the creation of new journals all have resulted in significant developments in spatial analysis.

The last decade has also witnessed the beginnings of an incorporation of powerful spatial analytical and modeling capabilities into GIS (see Anselin 1995 and Fischer and Getis 1997 for examples). Anselin and Getis (1992) and Fischer (1998) summarize spatial analysis research in this decade as being either exploratory spatial data analysis (a data-driven approach) or confirmatory spatial modeling (a model-driven approach). In many applications, the distinction between the two is blurred.

Exploratory spatial data analysis can be considered as the core or foundation of modern spatial analysis involving the analysis of patterns of points, lines, and surfaces defined in two or three-dimensional space. Studies in this area have focused on techniques to describe spatial distributions, identify spatial outliers, discover spatial clustering, and examine spatial non-stationarity (Anselin 1990). For example, Openshaw (1991) improved his computationally intensive GAM (Geographical Analysis Machine) to elicit patterns in data as well as identify outliers. Fotheringham and Zhan (1996) and Rogerson (2001*a*, *b*) have successfully applied statistical analysis techniques to detect geographic clusters. Anselin and Bao (1997) have attempted to integrate spatial analysis fully within a GIS framework. Their program can estimate descriptive statistics as well more advanced spatial statistics related to spatial autocorrelation. Thus, successful attempts have been made to provide spatial data exploration tools as well as to visualize patterns.

Spatial analysis in the 1990s has highlighted the importance and relevance of spatial modeling that encompasses both physical and social science models. Deterministic process models and stochastic process models have been developed that range across a variety of spatial scales. In the social sciences, a wide range of spatial models have been developed for the purposes of describing, analyzing, forecasting, and policy appraising economic developments within a set of localities or regions. In particular, the spatial analysis tradition has been strongly oriented toward location-allocation problems in space (i.e. siting of retail stores) and spatial interaction models dealing with flows of people, commodities, and resources between regions (e.g. Fortney 1996). Both of these represent powerful techniques to the spatial analyst interested in the location of activities.

Spatial analysis lies at the heart of spatial interaction modeling, broadly defined as a movement or communication over space that results from decision-making processes. Spatial interaction modeling encompasses such diverse behavior as migration, commercial trade, technological diffusion, travel-to-work, telephone calls, and airline passenger traffic (Fotheringham and O'Kelly 1989; Flowerdew and Boyle 1995; Plane 1999). The expansion method of generating models that embed temporal or spatial shifts in key parameters provides a method to uncover greater specificity of spatial relationships. This method, first introduced by Casetti (1972), continues to be applicable in a number of applications (see Casetti 1992; Jones and Casetti 1992; Anselin 2000). In traditional, constrained spatial interaction models, the number of predicted movers leaving origins and entering destinations is constrained to match exactly the observed number (Roy and Flood 1992; Pooler 1994). Relaxing the constraints to vary over a range of values provides greater flexibility in calibration and leads to relaxed spatial interaction models (Pooler 1994). More recent contributions to this area have involved computationally intensive methods including neural networks for estimation of flows (Fischer and Gopal 1994) and genetic algorithms for the estimation of globally optimal parameters (Diplock and Openshaw 1996).

Location and allocation models are other important tools available in spatial analysis that allows decision-makers to optimize the location of private and public consumer-oriented facilities, including stores, schools, and hospitals (Densham 1994; Birkin *et al.* 1996; Arentze *et al.* 1996). The use of these models has become commonplace and its incorporation into a GIS has led to the development of spatial decision support systems (Janssen and Rietveld 1990; Frank *et al.* 2000). Issues related to database structures of spatial decision support systems also have been addressed from the perspective of spatial modeling (Armstong and Densham 1990). Often these models contain a spatial choice function specified using one or more criteria that are important to decision-makers. For example, multicriteria decision-making considers the impacts of choice alternatives along multiple dimensions in order to choose the best alternative (Carver 1991; Jankowski 1995; Eastman *et al.* 1995). Spatial decision support systems have spun off other useful frameworks that incorporate more societal and public participation as in the *Spatial Understanding Support System* (Couclelis and Monmonier 1995) to resolve the "not in my backyard" syndrome.

Over the last decade, spatial statistics has been a major focus of geographic research involving mathematical models and quantitative methods. Early studies by Anselin (1990, 1992) and Griffith (1990) addressed the question of what is special about spatial data. Other studies have clarified the statistical and econometric issues involved in the use of spatial data (Anselin and Florax 1995; Haining 1990; Cressie 1993; Griffith 1998). In particular, spatial dependence and heterogeneity are the two effects that have received the most attention (Anselin and Getis 1992), with a particular resurgence in research devoted to the modifiable areal unit problem (MAUP) in census geography. MAUP research has focused on the well-known scale problem that occurs when areal units are progressively aggregated into fewer and larger units for analysis. For example, Tobler (1991) has shown that a spatial process being examined at one scale might not be the same spatial process examined at a different scale. Scale studies have been carried out using techniques such as Fourier transforms (Jensen 1996), variograms (Xia and Clarke 1997), fractals (Benguigui 1995; Appleby 1996), chaos theory (Nijkamp and Reggiani 1990, 1992; White 1990; Thill and Wheeler 1995), linear programming (O'Kelly and Lao 1991; O'Kelly *et al.* 1995), and spatial interpolation (Armstrong 2000). In addition, MAUP research has focused on the aggregation or zonal problem, which results when alternative aggregation schemes are employed at equal or similar spatial scales or resolutions. This aggregation problem arises due to uncertainty about how the data are to be aggregated to form a given number of zones.

Thus, research has tended to document the statistics that are affected or not affected by aggregation effects of the MAUP (see Fotheringham and Wong 1991; Amrhein 1995; Amrhein and Reynolds 1996; Wong and Amrhein 1996; Wrigley 1995). In practice, the scale and aggregation problems interact with each other and with spatial autocorrelation in the variables of the data set. The statistical analysis of dependence and heterogeneity in spatial patterns involves finding the degree of spatial association or autocorrelation. Research in this area has focused on local versus global patterns of spatial association. Noteworthy are the development of *G* statistics introduced by Getis and Ord (1992) and Ord and Getis (1995) and Anselin's LISA (1995) for the study of local pattern in spatial data. Techniques to measure and incorporate spatial non-stationarity have included the expansion method (Jones and Casetti 1992; Casetti 1997) and geographically weighted regression (Fotheringham *et al.* 1997). The increasing availability of spatial statistics software has resulted in greater awareness and use of spatial statistics in many geographical applications. A recent textbook that focuses on statistical methods in geography by Rogerson includes a chapter that provides a "gentle" introduction to Casetti's expansion method and to spatial regression; a recent book by Fotheringham *et al.* (2000) summarizes many of the recent developments in exploratory spatial analysis.

As mentioned earlier, modern computer technology has enabled researchers to apply existing spatial analysis methods in a number of applications including transportation, planning, and the allocation and provision of services such as schools, hospitals, and policing. Business organizations, market companies, major retailers, and supermarket chains also use spatial analysis, and consulting companies use it to locate their customers and analyze market potential. For example, transportation models have aided micro market area analysis. The main objectives of marketing geography are market area delineation and market potential. In the past, Euclidean distance had been used, and this became problematic when a model was applied to a small region. Now, the network analysis capabilities of spatial analysis packages have made it possible to estimate route distances, assuming stores and consumers are distributed over the network. Moreover, availability of TIGER data has meant that street-level information now is readily available. Thus, research in this context tends to stress computational or algorithmic implementations. Okabe and Sadahiro (1994) and Okabe *et al.* (1995) have developed practical computational methods of spatial analysis for a GIS environment. More recently, Miller and Wu have used three complementary approaches to derive a concept of space-time accessibility that is consistent with existing understanding of accessibility measures (e.g. Miller 1998, 1999; Miller and Wu 2000).

As an additional example, research in public health has focused particularly on the spatial progression of infectious diseases. With respect to AIDS, Gould (1993) and Golub *et al.* (1993) theorized about the spread of the disease and present evidence of a spatial hierarchical and expansionary trend. Thomas (1994, 1996) has provided a modeling framework for the space-time structure of AIDS while Ord and Getis (1995) used their local spatial autocorrelation statistic to examine the development of the spatial process of AIDS over time and at a local spatial scale. In addition, local statistics are useful in identifying the existence of regions in which the incidence of AIDS is greater than elsewhere. Diffusion models for describing its spread at an aggregate scale have been proposed (Lam and DeCola 1993; Lam *et al.* Liu 1996) while structural time-series approaches have been applied to forecast epidemics, including measles (Logan and Cliff 1997) and typhoid fever (Smallman-Raynor and Cliff 2001). Such approaches offer an advantage over standard autoregressive/moving average models in that they can model the impact of interventions (e.g. mass vaccination) in the time-series process. In light of this, Wise *et al.* (1997) have developed a suite of classification and regionalization tools suitable for the interactive spatial analysis of health data.

Behavioral Modeling

Spatial decisions and processes are fundamental to the understanding of spatial structure. Behavioral modeling has been strongly influenced by efforts in several disciplines, including economics, psychology, marketing, decision theory, transportation, and regional science. Trends that began in the 1980s and continued to grow through the 1990s include transitions from the aggregate to the disaggregate level, from deterministic to probabilistic frameworks, from static to dynamic modeling, and from analytical to computational perspectives. In particular, a variety of computational process models have been developed to show how individuals make activity scheduling and travel choices (Ettema *et al.* 1995; Kwan and Golledge 1997; Kwan 1998).

As an example of behavior modeling, Dieleman (1992) and Odland (1997) have used longitudinal approaches to study migration patterns. Longitudinal approaches treat incidents of migration and relocation as transitions in the personal locational histories of the individual. This general framework allows for the analysis of interdependencies between location histories and the development of other histories for the same individual. Timing of events in two or more histories can provide useful information on migration processes.

Spatial shopping choice modeling is another example of behavioral modeling that is important in urban planning, marketing, and transportation. Recent progress in this area includes improving the validity and predictive performance of spatial interaction models by relaxing some of the underlying assumptions. A more recent extension is to link multipurpose shopping and space-time shopping behavior (Baker 1994). An alternative, the stated preference modeling approach, popular in the 1980s, has witnessed further improvements and extensions. For example, Timmermans (1996) extended this approach to the case of sequential choice behavior. Horowitz (1997) and Thill and Horowitz (1997*a*, *b*) went further and proposed a new model—the Approximate Nested Choice-Set Destination Choice—that assumes that the choice set is constrained within a travel perimeter defined by the individual's time budget. In general, spatial modelers have paid most attention to models that are computationally tractable and they have tried to incorporate time to make models more realistic.

Computational Intelligence

The 1990s witnessed an increasing use of computational intelligence technology that was relevant for spatial analysis. Computational intelligence technologies provide a way to improve both spatial data analysis and models to meet the large-scale data processing needs of GIS and remote sensing systems. As with most other aspects of technology-driven research, much of the growth in computational intelligence was spurred by the availability of large amounts of spatial data, and the advent of high-speed computers and parallel computer architectures for real-time analysis and modeling. Moreover, the relevance of spatial analysis and modeling in both applied and empirical contexts has provided an impetus for using non-linear methods and more intelligent techniques. Computational intelligence methods include neural networks, genetic programming, simulation using cellular automata, and fuzzy logic modeling. Note that computational intelligence can be distinguished from the artificial intelligence expert systems approach that was popular in the 1980s in that computational intelligence denotes the lowest forms of intelligence that can be characterized by some form of computational adaptivity and fault tolerance instead of explicitly representing knowledge in the true artificial intelligence sense. Thus, this type of spatial analysis and modeling has a number of practical applications including census geography, global change research, land-surface characterization, strategic land-use and transportation planning, environmental analysis and planning, and the allocation of education, health, and other services.

The growing realization of the limitations of conventional tools and models as vehicles for exploring spatial patterns and relationships in GIS and remote sensing has led researchers to explore neural network modeling for spatial analysis. In short, neural networks are based on human information processing methods. They (1) offer greater flexibility and freedom in model design and variable representation; (2) are robust and fault tolerant and can therefore deal with noisy or missing data and fuzzy information; (3) can deal efficiently with large datasets and can be implemented on parallel computers (see Birkin *et al.* 1995); (4) are amenable to real-time processing; and (5) can implicitly incorporate the special nature of spatial data. Neural networks can be viewed as non-linear extensions of conventional statistical models and are applicable for pattern classification and function approximation. They have been used to model spatial interaction (Gopal and Fischer 1996; Fischer and Gopal 1994) and forecast trip distributions (Mozolin *et al.* 2000) as well as to classify census regions (Openshaw and Rao 1995) and land use/land cover using remotely sensed and other data (Leung 1997; Wilkinson 1997). Fischer (1997) provides a comprehensive review of the many types of neural networks and selection criteria.

Evolutionary algorithms are computer-based problem-solving systems that use some of the known mechanisms of evolution as key elements in their design and implementation. A variety of evolutionary algorithms have been proposed, including genetic algorithms, evolutionary programming, and genetic programming. All share a common conceptual base of simulating the evolution of individual structures via processes of selection, mutation, and reproduction. These processes depend on the perceived performance of the individual structures as defined by an environment. Genetic algorithms are a family of computational models that can be used in search, optimization, and machine learning, where the character string of the individual structure (i.e. chromosome) can be used to encode the values for the different parameters being optimized. As an example, Diplock (1998) generated models of spatial interaction using genetic algorithms where the individual genes were replaced by components of equations.

Introduced in 1968, fuzzy sets now have a large following in various scientific and engineering disciplines. Much of their popularity arises from the development of computer chips that can effectively represent fuzzy logic, their ability to handle imprecise data and model natural language, and the growing integration of neural networks and other models with fuzzy logic. Fuzzy logic can be used to model the imprecision that is characteristic of human reasoning and real-world phenomena and it offers an alternative paradigm for building more intelligent modeling systems. In the use of fuzzy logic in spatial analysis, Leung (1988) and Burrough (1989) were early pioneers. More recent work has involved fuzzy spatial interaction models (Openshaw 1996), fuzzy neural network classifiers (Gopal and Fischer 1997), and decision support systems based on fuzzy rules (Leung 1988).

Cellular automata are defined as discrete dynamical systems that possess a simple structure; that is, they are an array of identically programmed cells that interact with one another by a finite set of prescribed rules. In particular, they are capable of exhibiting complex self-organizing behaviour (Xie 1996). For example, Batty *et al.* (1997) and Batty (1998) used cellular automata to model urban evolution over time. Through use of these methodologies, they found it possible to simulate changes in different urban activities and gain greater insights into the underlying structure and process. Other work in this area by R. White and Engelen (1993), Portugali *et al.* (1994), and Xie (1994) demonstrates the utility of this approach in simulating urban micro–macro dynamics.

Quantitative Methods and Mathematical Models in Physical Geography

Research in physical geography—biogeography, climatology, geomorphology, and hydrology—has been largely quantitative since the dawn of the quantitative revolution in geography (Burton 1963). Almost half a century later, most of the research in these subfields continues to utilize analytic and quantitative methodology to some degree. Correlation and regression, simple tests of hypotheses, discriminant analysis, principal components and factor analysis, and other univariate and multivariate statistical procedures are prevalent in physical geography research. Over the past decade, however, physical geographers have continued to develop new methods and expand upon existing techniques for data analysis and theory construction. Moreover, mathematical models are becoming increasingly employed to examine and test geographical hypotheses. This section will focus on new and innovative methods and modeling applications that have been developed in physical geography over the last decade.

New Applications of Quantitative Methods

Multivariate statistical analyses often are the staple of many sophisticated statistical applications in physical geography. Although many such procedures have been available for more than fifty years, geographers continue to develop and adapt multivariate analyses to solve spatial problems. The unique relationships that spatial analyses and spatial autocorrelation present pose a particular challenge for physical geographers. For example, novel developments and applications in correlation/regression analysis include research in basic correlation and alternative measures (Kessler and Neas 1994; Legates and Davis 1997; Legates and McCabe 1999), directional correlograms (Nelson *et al.* 1998), vector correlation (Hanson *et al.* 1992), nonparametric and predictive logistic regression models (Brown 1994; Ridenour and Giardino 1995*b*; Travis *et al.* 1997) and the use of mean absolute differences in regression analyses (Robeson and Shein 1997). Discriminant function analysis also has been applied in novel ways to examine, for example, fluvial geomorphology and hydraulic geometry (Ridenour and Giardino 1995*a*) and synoptic climatology (Kalkstein *et al.* 1996). Principal components analysis (PCA) too has been increasingly applied to geographic problems with research into vector PCA (Klink 1986; Klink and Willmott 1989) and component rotation procedures (D. White *et al.* 1991) although significant criticisms of the latter have been identified (Legates 1991*a*; 1993). Davis and Kalkstein (1990) have combined PCA and cluster analysis to automate a spatial synoptic climatological classification while Rhoads (1991*b*, 1992) discusses a variety of advanced statistical models for use in fluvial geomorphology, including multivariate functions, lag relationships, multicollinearity issues, and continuously varying parameter models.

Time-series analysis has always been an important tool in physical geography research. Recent innovative trends in time-series analyses have focused on developing spectral analysis methods (Outcalt *et al.* 1992), issues associated with temporal smoothing and filtering (Howarth and Rogers 1992), the use of linear programming (Kuby *et al.* 1997) and data transformation methods (Outcalt *et al.* 1997), applications for vector data (Klink 1999), and the use of time-series analysis (Townsend and Walsh 1998) and harmonic analysis (Jakubauskas *et al.* 2001) with remotely sensed data. Research also has focused on examining and developing methods to assess frequency distributions (Harrington and Cerveny 1988; Legates 1991*b*) with particular emphasis on trends in extreme events (Knox and Kundzewicz 1997; Keim and Cruise 1998). Issues in temporal autocorrelation (Robeson and Shein 1997) and temporal variations in spatial patterns and variance (Balling 1997; Robeson and Janis 1998) also have seen considerable advances and applications.

One of the emerging areas of quantitative analysis is the use of fractals in geography. The fractal dimension of geographic phenomena has become an important topic in physical geography, with most notable applications in geomorphology. Use of fractals in geomorphology goes back to the late 1960s but their utility has only been recognized more recently. Fractals can be defined as a non-differentiable function exhibiting self-similarity that is independent of both sampling interval and level of resolution (Mark 1984). Clearly, the nature of geomorphologic data often lends itself to a representation by a fractal dimension that is less than the true dimensionality. Klinkenberg and Goodchild (1992), for example, used fractals to characterize digital elevation model topography. Similarly, fractals have been employed by Andrle and Abrahams (1989) to describe the roughness of talus slopes, by Outcalt and Melton (1992) to describe natural terrain texture, by Outcalt *et al.* (1994) to characterize physiography, by Wilson and Dominic (1998) and Colby (2001) to estimate surface topography and structure, and by Wilson (2001) to examine fracture and faulting networks. Andrle (1992), Goodchild and Klinkenberg (1993), and Phillips (1993*a*) also applied fractal methodology to river channel morphology with generally good results. In biogeography, canopy heights have been modeled using fractals (Drake and Weishampel 2001). Overall, the use of fractal methodology in physical geography is very promising, although nearly all studies stress the need to evaluate the goodness-of-fit of fractals in any given application prior to their use.

Chaos theory and non-linear dynamics also are increasingly becoming the focus of geographical research. Chaotic behavior is used to describe processes that are unstable and have the characteristic that small initial perturbations can rapidly intensify and lead to significant and substantial differences between the perturbed and unperturbed solutions. Non-linear dynamical systems theory (NDS) includes chaotic behavior as well as fractals, dissipative structures, bifurcation, and catastrophe theory (Phillips 1992). Such non-linearities probably represent much of the processes that dominate physical geography; hence the potential for such methods (see Malanson *et al.* 1990 for a more complete discussion). Landscape representation and evolution, for example, have been modeled using chaos and NDS theory by Malanson *et al.* (1992), Phillips (1993*b*, 1994, 1995*a*, *b*, 1999*a*), and Gomez and Phillips (1999). Although chaos and non-linear dynamics was first recognized and developed in meteorology, their widest applications to date in physical geography appear to be in geomorphology (see Phillips 1996, 1999*b*).

Cutting-edge research in physical geography also has focused on the use of artificial neural networks to solve modeling and spatial interpolation problems. In biogeography in particular, artificial neural networks have been used to map vegetation (Franklin 1995) and model land-cover change (Gopal and Woodcock 1996) as well as to determine canopy directional reflectance (Abuelgasim *et al.* 1996; Abuelgasim *et al.* 1998), land surface albedo (Liang *et al.* 1999), vegetation age (Jensen *et al.* 1999) and type (Cairns 2001). Gopal *et al.* (1999) used artificial neural networks to classify global land cover from AVHRR (Advanced Very High Resolution Radiometer) data. Artificial neural networks have been used in climatology to forecast ozone (Comrie 1997) and to spatially downscale climate data (Snell *et al.* 2000; Schoof and Pryor 2001). Frakes and Yu (1999) also have used artificial neural networks to evaluate climate change scenarios.

In climatology, issues of station network adequacy, accuracy, and spatial interpolation have been frequently addressed. Such concerns have become exceedingly important with the development of large datasets and the need to evaluate spatial and temporal trends in climate variables. Research has focused on network adequacy and accuracy (Willmott *et al.* 1994), data homogeneity issues (Peterson and Easterling 1994), and the effect of network design on climate maps (Kutiel and Kay 1996). In general, however, the major focus of large climate dataset analysis has been on spatial interpolation methodology and the assessment of regional- and global-scale spatial and temporal variability. Willmott and Legates (1991), Willmott *et al.* (1991, 1994), Robeson (1994, 1995), Willmott *et al.* (1996), Willmott and Robeson (1995), and Willmott and Matsuura (1995), Robeson and Willmott (1996), and Robeson and Janis (1998) have conducted research on spatial interpolation and time averaging. Most of this research, however, has moved from the development of spatial interpolation algorithms to the analysis of how variations in station distributions over time affect areal and temporal estimates of climate variables.

Other spatial statistical methods also have been developed for specific applications. These include methods to model karst landscape topography using double fourier series analysis (Brook and Hanson 1991), to detrend or declimatize climatological data (Comrie 1992), to characterize geomorphic line complexity (Andrle 1994), to describe the two-dimensional shape of glacial valley cross-sections (James 1996), and to represent the spatial coherence of wind data using mean absolute deviations (Robeson and Shein 1997). Legates and Davis (1997) and Legates and McCabe (1999) also further examined and developed methods for model evaluation.

Models in Physical Geography

With the advent of faster computers and more efficient hardware and software techniques, modeling in physical geography has developed significantly over the last decade. More complex models have been developed that have allowed physical geographers to model processes at a more physically based level. Through comparisons with older models, research has demonstrated that this new generation of models is better able to simulate reality and provide more accurate answers to existing questions. Thus, physical geographers have focused more on model refinements and enhancements that have opened areas of research that before simply were glossed over.

Issues of spatial scaling have become one of the emerging areas of modeling applications in physical geography. Although climate, landforms, vegetation, and soils data are measured at a variety of spatial scales, these processes often are modeled at a much coarser resolution than their real natural variability. Consequently, research into aggregation and representation of sub-grid processes as well as spatial downscaling from coarser resolution models have been the focus of research in all aspects of physical geography, including climatology (Klink 1995; Hewitson and Crane 1996; Palutikof *et al.* 1997; Winkler *et al.* 1997; Crane and Hewitson 1998), geomorphology (Walsh *et al.* 1997), biogeography (Phillips 1995*a*; Malanson and Armstrong 1997), and hydrology (Shelton 1989; Band *et al.* 1993; Mahmood 1996; Humes *et al.* 1997).

At the global/continental scale, modeling research has focused largely on general circulation model (GCM) simulations. Such models have been used to analyze the potential for climate change under a number of assumptions about the future climate. Research includes modeling of the active soil layer thickness (Anisimov *et al.* 1997) and permafrost zonation (Anisimov and Nelson 1996, 1997) as well as issues involving aggregation to a lower resolution (Klink 1995) and downscaling to a higher resolution (Hewitson and Crane 1996; Palutikof *et al.* 1997; Winkler *et al.* 1997; Crane and Hewitson 1998). Simpler models also have been employed to examine the effects of global climate change (e.g. Balling 1994).

Larger-scale climatological modeling has focused on urban climatology, ice sheet modeling, and agricultural crop modeling applications. Energy-balance models of urban (Arnfield 1988, 1990; Todhunter 1990; Arnfield and Mills 1994*a*, *b*) and suburban (Grimmond 1992) canyons has been the focus of most urban climatologists. In particular, net radiation and sensible heat fluxes have been evaluated with respect to building orientation, albedo, diurnal and seasonal variability, and circulation characteristics. Modeling the errors associated with ground-based radar estimates of precipitation was a major focus of Legates (2000). Todhunter (1989) has also examined the urban energy balance with respect to stratification by synoptic weather types while Loewenherz and Wendland (1988) have developed a statistically based predictive model to simulate seasonal ground frost occurrence. Rowe *et al.* (1995) and Mote and Rowe (1996) have examined and developed models of snowpack depth and melt for the Greenland Ice Sheet. Agricultural applications of climatological models include research on rice yields (Mahmood and Hayes 1995; Mahmood 1998) and the crop yield response to climate changes (Easterling *et al.* 1996). Particular crop models used by climatologists include the CERES, EPIC, and YIELD crop models.

Geomorphologic modeling has focused on eolian and fluvial processes. Eolian research by geographers includes modeling of arid environments and coastal dunes. Eolian processes in arid environments have been examined in a number of studies, including modeling of alluvial fan entrenchment and segmentation (Hooke and Dorn 1992) and analysis of sediment fluxes with the incorporation of unsteady state processes (Bauer *et al.* 1998). In the coastal zone, models have been developed to describe sediment transportation and eolian saltation as a function of wind shear velocities, humidity, and other environmental variables (e.g. Sherman and Hotta 1990; Sherman 1992; Sherman and Lyons 1994; Namikas and Sherman 1994, 1998). Sherman *et al.* (1996, 1998) also provide an evaluation and comparison of wind-blown sand models while Gomez and Troutman (1997) examine errors in bed load sampling. Modeling efforts in fluvial geomorphology have been concentrated, for example, on determining the sediment transport by overland flow (Abrahams *et al.* 1988), estimating the sediment budget of river deltas (Kesel *et al.* 1992), evaluating the formation of desert pavements by raindrop erosion (Wainwright *et al.* 1995), examining the structure of stream confluence morphology formation (Rhoads and Kenworthy 1995, 1998), and modeling sediment sorting processes in pool-riffle streams (Thompson *et al.* 1996). Trofimov and Phillips (1992) also examined the impacts of instability of geomorphic systems in geomorphological modeling and forecasting.

Modeling in biogeography has developed considerably over the past decade. In addition to the research on climate impacts on crop development discussed above, research in this area has focused on ecotone modeling, phenological development and maturity, and estimation of vegetation canopy parameters. A physiologically mechanistic model of the alpine treeline location has been developed by Cairns (1994) and used to examine the alpine carbon balance by Cairns and Malanson (1997). Malanson and Cairns (1997) have modeled tree migration rates as they relate to a variety of vegetative parameters. Schwartz (1994) and Schwartz *et al.* (1997) have examined vegetation phenology and maturity as it relates to climate variability. Rowe (1991) developed a model to estimate canopy albedo from vegetation canopy architecture and Grimmond *et al.* (1996) have modeled the effect of vegetation cover on the surface energy balance.

Developing methodological and modeling advances in hydrological research also has been the focus of physical geographers. For example, hydrologic modeling has been conducted to evaluate the water balance of lower coastal plains watersheds (Sun and Brook 1988), examine spatial variability of hydrologic parameters for large watersheds (Shelton 1989) and an urban area (Anderson 1991), and monitor soil water and meteorological conditions using radar estimates and station-based observations (Legates *et al.* 1996, 1998). Hydrological research has been quite varied and intertwines with applications in biogeography, climatology, and fluvial geomorphology. As examples of research in this area, Klink and Willmott (1994) determined the influence of soil moisture and surface roughness heterogeneity on the modeled climate, Humes *et al.* (1997) evaluated sensible heat fluxes over a semi-arid landscape, Rhoads (1991*a*) evaluated the use of a downstream hydraulic geometry model for fluvial analyses, Li *et al.* (1996) examined the determination of overland flow hydraulics for use in fluvial process modeling, and Townsend and Walsh (1996, 1998) developed a GIS approach to model soil wetness potential for a forest ecosystem. Vandiver and Kirsch (1997) also have developed a neutron-attenuation method to measure soil-water content.

Models and Methods in Physical and Human Geography—Similar Paths?

As evidenced in this chapter, a number of research trends characterize the decade of the 1990s. First, there has been a marked incorporation of advanced technology that has led to better and more sophisticated models in both physical and human geography. This trend is unlikely to be abated in the near future. Second, geographers have now been able statistically to analyze and integrate large databases. Advances in computer technology have enhanced our ability to handle large quantities of data, and an increasing trend to use and model large amounts of spatial data now is evident. Computational intelligence methods provide our first glimpses into attempts to incorporate large datasets into mathematical models and exploratory data analysis (Getis 1999). In particular, "spatial data mining"—exploring high-level spatial information and knowledge in large spatial databases—has become an important issue over the last few years. For example, Han *et al.* (1997) are developing a prototype system to automate tools for discovery of interesting and useful knowledge obtained from the large spatial databases that result from data collected by satellite telemetry systems, remote sensing systems, regional sales systems, and other data collection tools.

Is the research on mathematical models and quantitative methods being conducted by physical and human geographers following a similar path? Evidence here tends to indicate that the answer is a resounding affirmative. Research on chaos theory, cluster analysis, exploratory data analysis, fourier analysis, fractals, linear programming, network optimization and analysis, neural networks, scale and aggregation issues, spatial interpolation and spatial variograms, spatial modeling, and the development of spatial decision support systems have parallels in both areas of geographical inquiry. Even the development of cellular automata has a parallel with many of the grid-based models used in climatology. Similarly, neural networks have been applied for land-cover mapping and classification problems in remote sensing and spatial interaction modeling as well as downscaling in climatology with much disciplinary cross-fertilization. Thus, both human and physical geographers must become increasingly aware of each other's research so that innovations in one discipline can be readily transferred to the other. It is to be hoped that the twenty-first century will yield an era of enhanced integration of human and physical geography in the development of mathematical models and quantitative methods. Both geographies will play an increasing role in our understanding and modeling of the myriad of processes that occur on planet Earth.

References

Abrahams, A. D., Luk, S.-H., and Parsons, A. J. (1988). "Threshold Relations for the Transport of Sediment by Overland Flow on Desert Hillslopes." *Earth Surface Processes and Landforms*, 13: 407–19.

Abuelgasim, A. A., Gopal, S., Irons, J. R., and Strahler, A. H. (1996). "Classification of ASAS Multiangle and Multispectral Measurements Using Artificial Neural Networks." *Remote Sensing of Environment*, 57: 79–87.

Abuelgasim, A. A., Gopal, S., and Strahler, A. H. (1998). "Forward and Inverse Modelling of Canopy Directional Reflectance using a Neural Network." *International Journal of Remote Sensing*, 19: 453–71.

Amrhein, C. G. (1995). "Searching for the Elusive Aggregation Effect: Evidence from Statistical Simulations." *Environment & Planning A*, 27: 105–19.

Amrhein, C. G., and Reynolds, H. (1996). "Using Spatial Statistics to Assess Aggregation Effects." *Geographical Systems*, 3: 143–58.

Anderson, K. L. (1991). "Spatial Changes in the Hydrology of Portions of the Twin Cities, Minnesota: A Simulation Study." *Physical Geography*, 12: 147–66.

Andrle, R. (1992). "Estimating Fractal Dimension with the Divider Method in Geomorphology." *Geomorphology*, 5: 131–41.

—— (1994). "The Angle Measure Technique: A New Method for Characterizing the Complexity of Geomorphic Lines." *Mathematical Geology*, 26: 83–97.

Andrle, R., and Abrahams, A. D. (1989). "Fractal Techniques and the Surface Roughness of Talus Slopes." *Earth Surface Processes and Landforms*, 14: 197–209.

Anisimov, O. A., and Nelson, F. E. (1996). "Permafrost Distribution in the Northern Hemisphere under Scenarios of Climatic Change." *Global and Planetary Change*, 14: 59–72.

—— (1997). "Permafrost Zonation and Climate Change in the Northern Hemisphere: Results from Transient General Circulation Models." *Climatic Change*, 35: 241–58.

Anisimov, O. A., Shiklomanov, N. I., and Nelson, F. E. (1997). "Global Warming and Active-Layer Thickness: Results from Transient General Circulation Models." *Global and Planetary Change*, 15: 61–77.

Anselin, L. (1990). "What is Special about Spatial Data?" in D. A. Griffith (ed.), *Spatial Statistics, Past Present and Future*. Ann Arbor, Mich.: Institute of Mathematical Geography, 63–77.

—— (1992). "Space and Applied Econometrics—Introduction." *Regional Science and Urban Economics*, 22: 307–16.

—— (1995). "Local Indicators of Spatial Association-LISA." *Geographical Analysis*, 27: 93–115.

—— (2000). "The Alchemy of Statistics, or Creating Data Where No Data Exist." *Annals of the Association of American Geographers*, 90: 586–92.

Anselin, L., and Bao, S. (1997). "Exploratory Spatial Data Analysis Linking *SpaceStat* and *ArcView*," in Fischer and Getis (1997: 35–59).

Anselin, L., and Florax, R. J. G. M. (1995). *New Directions in Spatial Econometrics*. Springer-Verlag: Berlin.

Anselin, L., and Getis, A. (1992). "Spatial Statistical Analysis and Geographical Information Systems." *The Annals of Regional Science*, 26: 19–23.

Appleby, S. (1996). "Multifractal Characterization of the Distribution Pattern of the Human Population." *Geographical Analysis*, 28: 147–60.

Armstrong, M. P. (2000). "Geography and Computational Science." *Annals of the Association of American Geographers*, 90: 146–56.

Armstrong, M. P., and Densham, P. J. (1990). "Database Organization Strategies for Spatial Decision Support System." *International Journal of GIS*, 4: 2–18.

Arnfield, A. J. (1988). "Validation of an Estimation Model for Urban Surface Albedo." *Physical Geography*, 9: 361–72.

—— (1990). "Canyon Geometry, the Urban Fabric and Nocturnal Cooling: A Simulation Approach." *Physical Geography*, 11: 220–39.

Arentze, T. A., Borgers, A. W. J., and Timmermans, H. J. P. (1996). "Design of a View-Based DSS for Location Planning." *International Journal of GIS*, 10: 219–36.

Arnfield, A. J., and Mills, G. M. (1994*a*). "An Analysis of the Circulation Characteristics and Energy Budget of a Dry, Asymmetric, East–West Urban Canyon, I. Circulation Characteristics." *International Journal of Climatology*, 14: 119–34.

—— (1994*b*). "An Analysis of the Circulation Characteristics and Energy Budget of a Dry, Asymmetric, East–West Urban Canyon, II. Energy Budget." *International Journal of Climatology*, 14: 239–61.

Baker, R. (1994). "An Assessment of the Space-Time Differential Model for Aggregate Trip Behavior to Planned Suburban Shopping Centers." *Geographical Analysis*, 26: 341–63.

Balling, R. C., Jr. (1994). "Greenhouse Gas and Sulfate Aerosol Experiments Using a Simple Global-Energy-Balance Model." *Physical Geography*, 15: 299–309.

—— (1997). "Analysis of Daily and Monthly Spatial Variance Components in Historical Temperature Records." *Physical Geography*, 18: 544–52.

Band, L. E., Patterson, P., Nemani, R., and Running, S. W. (1993). "Forest Ecosystem Processes at the Watershed Scale: Incorporating Hillslope Hydrology." *Agricultural and Forest Meteorology*, 63: 93–126.

Batty, M. (1998). "Urban Evolution on the Desktop: Simulation with the Use of Extended Celluar Automata." *Environment and Planning A*, 309: 1943–67.

Batty, M., Couclelis, H., and Eichen, M. (1997). "Urban Systems as Celluar Automata." *Environment and Planning B*, 24: 159–64.

Bauer, B. O., Yi., J., Namikas, S. L., and Sherman, D. J. (1998). "Event Detection and Conditional Averaging in Unsteady Aeolian Systems." *Journal of Arid Environments*, 39: 345–75.

Benguigui, L. (1995). "Fractal Analysis of the Public Transportation System of Paris." *Environment and Planning A*, 27: 1147–61.

Birkin, M., Clarke, M., and George, F. (1995). "The Use of Parallel Computers to Solve Nonlinear Spatial Optimisation Problems: An Application to Network Planning." *Environment and Planning A*, 27: 1049–68.

Birkin, M., Clarke, G., Clarke, M., and Wilson, A. (1996). *Intelligent GIS: Location Decisions and Strategic Planning*. Cambridge: Geoinformation International.

Brook, G. A., and Hanson, M. (1991). "Double Fourier Series Analysis of Cockpit and Doline Karst near Browns Town, Jamaica." *Physical Geography*, 12: 37–54.

Brown, D. G. (1994). "Comparison of Vegetation-Topography Relationships at the Alpine Treeline Ecotone." *Physical Geography*, 15: 125–45.

Burrough, P. A. (1989). "Fuzzy mathematical Methods for Soil Survey and Land Evaluation." *Journal of Soil Science*, 40: 477–92.

Burton, I. (1963). "The Quantitative Revolution and Theoretical Geography." *The Canadian Geographer*, 7: 151–62.

Cairns, D. M. (1994). "Development of a Physiologically Mechanistic Model for Use at the Alpine Treeline Ecotone." *Physical Geography*, 15: 104–24.

—— (2001). "A Comparison of Methods for Predicting Vegetation Type." *Plant Ecology*, 156: 3–18.

Cairns, D. M., and Malanson, G. P. (1997). "Examination of the Carbon Balance Hypothesis of Alpine Treeline Location in Glacier National Park, Montana." *Physical Geography*, 18: 125–45.

Carver, S. J. (1991). "Integrating Multi-criteria Evaluation with Geographical Information Systems." *International Journal of GIS*, 5: 321–39.

Casetti, E. (1972). "Generating Models by the Expansion Method: Applications to Geographic Research." *Geographical Analysis*, 4: 81–91.

—— (1992). "Bayesian Regression and the Expansion Method." *Geographical Analysis*, 24: 58–74.

—— (1997). "Mixed Estimation and Expansion Method: An Application to the Spatial Modelling of the Aids Epidemic," in Fischer and Getis (1997: 15–34).

Cho, Y.-H. (1996). *Exploring Spatial Analysis in GIS*. Santa Fe, NM: OnWord Press.

Colby, J. D. (2001). "Simulation of a Costa Rican Watershed: Resolution Effects and Fractals." *Journal of Water Resources Planning and Management-ASCE*, 127: 261–70.

Comrie, A. C. (1992). "A Procedure for Removing the Synoptic Climate Signal from Environmental Data." *International Journal of Climatology*, 12: 177–83.

—— (1997). "Comparing Neural Networks and Regression Models for Ozone Forecasting." *Journal of the Air & Waste Management Association*, 47: 653–63.

Couclelis, H., and Monmonier, M. (1995). "Using SUSS to Resolve NIMBY: How Spatial Understanding Support Systems Can Help with the 'Not in My Back Yard' Syndrome." *Geographical Systems*, 2: 83–102.

Crane, R. G., and Hewitson, B. C. (1998). "Doubled CO_2 Precipitation Changes for the Susquehanna Basin: Down-scaling from the Genesis General Circulation Model." *International Journal of Climatology*, 18: 65–76.

Cressie, N. (1993). *Statistics for Spatial Data*. New York: John Wiley.

Davis, R. E., and Kalkstein, L. S. (1990). "Development of an Automated Spatial Synoptic Climatological Classification." *International Journal of Climatology*, 10: 769–94.

Densham, P. (1994). "Integrating GIS and Spatial Modeling: Visual Interactive Modeling and Location Selection." *Geographical Systems*, 1: 203–21.

Dieleman, F. M. (1992). "Struggling with Longitudinal Data and Modeling in the Analysis of Residential Mobility." *Environment and Planning A*, 26: 1659–70.

Diplock, G. (1998). "Building New Spatial Interaction Models by Using Genetic Programming and a Supercomputer." *Environment and Planning A*, 30: 1893–904.

Diplock, G., and Openshaw, S. (1996). "Using Simple Genetic Algorithms to Calibrate Spatial Interaction Models." *Geographical Analysis*, 28: 262–79.

Drake, J. B., and Weishampel, J. F. (2001). "Multifractal Analysis of Canopy Height Measures in a Longleaf Pine Savanna." *Forest Ecology and Management*, 128: 121–7.

Easterling, W. E., Chen, X., Hays, C., Brandle, J. R., and Zhang, H. (1996). "Improving the Validation of Model-Simulated Crop Yield Response to Climate Change: An Application to the EPIC Model." *Climate Research*, 6: 263–73.

Eastman, J. R., Weigen, J., Kyem, P., and Toledano, J. (1995). "Raster Procedures for Multi-criteria/Multi-objective Decisions." *Photogrammetric Engineering and Remote Sensing*, 61: 539–47.

Ettema, D., Borgers, A., and Timmermans, H. P. J. (1995). "SMASH (Simuation Model of Activity Scheduling Heuristics): Empirical Test and Simulations Issues." Paper presented at the International Conference on Activity Based Approaches: Activity Scheduling and Analysis of Activity Patterns, 25–8 May. Eindhoven University of Technology, The Netherlands.

Fischer, M. M. (1997). "Computational Neural Networks: A New Paradigm for Spatial Analysis." *Environment and Planning A*, 29: 1873–91.

—— (1998). "Computational Neural Networks: A New Paradigm for Spatial Analysis." *Environment and Planning A*, 30: 1873–91.

Fischer, M. M., and Getis, A. (1997). *Recent Developments in Spatial Analysis*. Berlin: Springer Verlag.

Fischer, M. M., and Gopal, S. (1994). "Neural Network Models and Interregional Telephone Traffic: Comparative Performances between Multilayer Feedforward Networks and the Conventional Spatial Interaction Model." *Journal of Regional Science*, 34: 503–27.

Fischer, M. M., and Nijkamp, P. (1993). *Geographic Information Systems, Spatial Modelling and Policy Evaluation*. Berlin: Springer Verlag.

Fischer, M. M., Scholten, H., and Unwin, D. (1996). *Spatial Analytical Perspectives in GIS in Environmental and Socio-Economic Sciences*. London: Taylor & Francis.

Flowerdew, R., and Boyle, P. J. (1995). "Migration Models Incorporating Interdependence of Movers." *Environment and Planning A*, 27: 1493–502.

Fortney, J. (1996). "A Cost-Benefit Location-Allocation Model for Public Facilities: An Econometric Approach." *Geographical Analysis*, 28: 67–92.

Fotheringham, A. S., and O'Kelly, M. E. (1989). *Spatial Interaction Models: Formulations and Applications*. Dordecht: Kluwer.

Fotheringham, A. S., and Rogerson, P. (1994). *Spatial Analysis and GIS*. London: Taylor & Francis.

Fotheringham, A. S., and Wong, D. W. (1991). "The Modifiable Areal Unit Problem in Multivariate Statistical Analysis." *Environment and Planning A*, 23: 1025–44.

Fotheringham, A. S., and Zhan, F. B. (1996). "A Comparison of Three Exploratory Methods for Cluster Detection in Spatial Point Patterns." *Geographical Analysis*, 28: 200–18.

Fotheringham, A. S., Brunsdon, and Charlton, M. (2000). *Quantitative Geography: Perspectives on Spatial Analysis*. London: Sage.

Fotheringham, A. S., Charlton, M., and Brunsdon, C. (1997). "Measuring Spatial Variations in Relationship with Geographically Weighted Regression," in Fischer and Getis (1997: 60–82).

Frakes, B., and Yu, Z. B. (1999). "An Evaluation of Two Hydrologic Models for Climate Change Scenarios." *Journal of the American Water Resources Association*, 35: 1351–63.

Frank, W. C., Thill, J.-C., and Batta, R. (2000). "Spatial Decision Support System for Hazardous Material Truck Routing." *Transportation Research Part C—Emerging Technologies*, 8: 337–59.

Franklin, J. (1995). "Predictive Vegetation Mapping: Geographic Modelling of Biospatial Patterns in Relation to Environmental Gradients." *Progress in Physical Geography*, 19: 474–99.

Getis, A. (1999). "Some Thoughts on the Impact of Large Data Sets on Regional Science." *Annals of Regional Science*, 33: 145–50.

Getis, A., and Ord, J. K. (1992). "The Analysis of Spatial Association by Use of Distance Statistics." *Geographical Analysis*, 24: 189–206.

Golub, A., Gorr, W. L., Gould, P. R. (1993). "Spatial Diffusion of the HIV/AIDS Epidemic: Modeling Implications and Case Study of AIDS Incidence in Ohio." *Geographical Analysis*, 25: 85–100.

Gomez, B., and Phillips, J. D. (1999). "Deterministic Uncertainty in Bed Load Transport." *Journal of Hydraulic Engineering-ASCE*, 125: 305–8.

Gomez, B., and Troutman, B. M. (1997). "Evaluation of Process Errors in Bed Load Sampling Using a Dune Model." *Water Resources Research*, 33: 2387–98.

Goodchild, M. F., and Klinkenberg, B. (1993). "Statistics of Channel Networks on Fractional Brownian Surfaces," in Lam and DeCola (1993: 122–41).

Gopal, S., and Fischer, M. M. (1996). "Learning in Single Hidden-Layer Feedforward Network Models." *Geographical Analysis*, 28: 38–55.

—— (1997). "Fuzzy ARTMAP—a Neural Classifier for Multispectral Image Classification," in Fischer and Getis (1997: 306–35).

Gopal, S., and Woodcock, C. (1996). "Remote Sensing of Forest Change Using Artificial Neural Networks." *IEEE Transactions on Geoscience and Remote Sensing*, 34: 398–404.

Gopal, S., Woodcock, C. E., and Strahler, A. H. (1999). "Fuzzy Neural Network Classification of Global Land Cover from a 1 Degrees AVHRR Data Set." *Remote Sensing of Environment*, 67: 230–43.

Gould, P. R. (1993). "The Geography of AIDS." *New York Times Book Review*, 29 August, 23.

Griffith, D. A. (1990). "Supercomputing and Spatial Statistics—a Reconnaissance." *The Professional Geographer*, 42: 481–92.

—— (1998). "Advanced Spatial Statistics." *Special Topics in the Exploration of Quantitative Spatial Data Series*, Berlin: Martinus Nijhoff.

Grimmond, C. S. B. (1992). "The Suburban Energy Balance: Methodological Considerations and Results for a Mid-latitude West Coast City under Winter and Spring Conditions." *International Journal of Climatology*, 12: 481–97.

Grimmond, C. S. B., Souch, C., and Hubble, M. D. (1996). "Influence of Tree Cover on Summertime Surface Energy Balance Fluxes, San Gabriel Valley, Los Angeles." *Climate Research*, 6: 45–57.

Haining, R. (1990). *Spatial Data Analysis in the Social and Environmental Sciences*. Cambridge: Cambridge University Press.

Han, J., Koperski, K., and Stefanovic, N. (1997). "GeoMiner: A System Prototype for Spatial Data Mining." *Proceedings of the 1997 ACM-SIGMOD International Conference on Management of Data (SIGMOD'97)*, Tucson, Ari.

Hanson, B., Klink, K., Matsuura, K., Robeson, S. M., and Willmott, C. J. (1992). "Vector Correlation: Review, Exposition, and Geographic Application." *Annals of the Association of American Geographers*, 82: 103–16.

Harrington, J. A., Jr., and Cerveny, R. S. (1988). "Temporal Statistics: An Application in Snowfall Climatology." *Physical Geography*, 9: 337–53.

Hewitson, B. C., and Crane, R. G. (1996). "Climate Downscaling: Techniques and Application." *Climate Research*, 7: 85–95.

Hooke, R. Le B., and Dorn, R. I. (1992). "Segmentation of Alluvial Fans in Death Valley, California: New Insights from Surface Exposure Dating and Laboratory Modeling." *Earth Surface Processes and Landforms*, 17: 557–74.

Howarth, D. A., and Rogers, J. C. (1992). "Problems Associated with Smoothing and Filtering of Geophysical Time-Series Data." *Physical Geography*, 13: 81–99.

Humes, K. S., Kustas, W. P., and Goodrich, D. C. (1997). "Spatially Distributed Sensible Heat Flux over a Semiarid Watershed. Part I: Use of Radiometric Surface Temperature and a Spatially Uniform Resistance." *Journal of Applied Meteorology*, 36: 281–92.

Jakubauskas, M. E., Legates, D. R., and Kastens, J. H. (2001). "Harmonic Analysis of Time-Series AVHRR NDVI Data." *Photogrammetric Engineering and Remote Sensing*, 67: 461–70.

James, L. A. (1996). "Polynomial and Power Functions for Glacial Valley Cross-Section Morphology." *Earth Surface Processes and Landforms*, 21: 413–32.

Jankowski, P. (1995). "Integrating GIS and Multi-criteria Decision Making Methods." *International Journal of GIS*, 9: 251–73.

Janssen, R., and Rietveld, P. (1990). "Multi-criteria Analysis and GIS: An Application of Agricultural Landuse in The Netherlands," in H. J. Scholten and J. C. Stilwell (eds.), *Geographical Information Systems and Regional Planning*. Dordrecht: Kluwer.

Jensen, J. R. (1996). *Introductory Digital Image Processing: A Remote Sensing Perspective*. Upper Saddle River, NJ: Prentice-Hall.

Jensen, J. R., Qiu, F., and Ji, M. H. (1999). "Predictive Modelling of Coniferous Forest Age Using Statistical and Artificial Neural Network Approaches Applied to Remote Sensor Data." *International Journal of Remote Sensing*, 20: 2805–22.

Jones, J.-P., III, and Casetti, E. (1992). *Applications of the Expansion Method*. London: Routledge.

Kalkstein, L. S., Nichols, M. C., Barthel, C. D., and Greene, J. S. (1996). "A New Spatial Synoptic Classification: Application to Air-mass Analysis." *International Journal of Climatology*, 16: 983–1004.

Keim, B. D., and Cruise, J. F. (1998). "A Technique to Measure Trends in the Frequency of Discrete Random Events." *Journal of Climate*, 11: 848–55.

Kesel, R. H., Yodis, E. G., and McCraw, D. J. (1992). "An Approximation of the Sediment Budget of the Lower

Mississippi River Prior to Major Human Modification." *Earth Surface Processes and Landforms*, 17: 711–22.

Kessler, E., and Neas, B. (1994). "On Correlation, with Applications to the Radar and Raingage Measurement of Rainfall." *Atmospheric Research*, 34: 217–29.

Klink, K. (1986). "Space-Time Analysis of the Surface Wind Field Using Vector-Based Principal Components Analysis." *Publications in Climatology*, 39/2.

— (1995). "Surface Aggregation and Subgrid-Scale Climate." *International Journal of Climatology*, 15: 1219–40.

— (1999). "Climatological Mean and Interannual Variance of United States Surface Wind Speed, Direction and Velocity." *International Journal of Climatology*, 19: 471–88.

Klink, K., and Willmott, C. J. (1989). "Principal Components of the Surface Wind Field in the United States: A Comparison of Analyses Based upon Wind Velocity, Direction, and Speed." *International Journal of Climatology*, 9: 293–308.

— (1994). "Influence of Soil Moisture and Surface Roughness Heterogeneity on Modeled Climate." *Climate Research*, 4: 105–18.

Klinkenberg, B., and Goodchild, M. F. (1992). "The Fractal Properties of Topography: A Comparison of Methods." *Earth Surface Processes and Landforms*, 17: 217–34.

Knox, J. C., and Kundzewicz, Z. W. (1997). "Extreme Hydrological Events, Palaeo-Information and Climate Change." *Hydrological Sciences Journal-Journal des Sciences Hydrologiques*, 42: 765–79.

Kuby, M. J., Cerveny, R. S., and Dorn, R. I. (1997). "A New Approach to Paleoclimatic Research Using Linear Programming." *Palaeogeography, Palaeoclimatology, Palaeoecology*, 129: 251–67.

Kutiel, H., and Kay, P. A. (1996). "Effects of Network Design on Climatic Maps of Precipitation." *Climate Research*, 7: 1–10.

Kwan, M. (1998). "Space-Time and Integral Measures of Individual Accessibility: A Comparative Analysis Using a Point-Based Framework." *Geographical Analysis*, 30: 191–216.

Kwan, M., and Golledge, R. (1997) "Computational Process Modeling of Disaggregated Travel Behavior," in Fischer and Getis (1997: 171–85).

Lam, N. S.-N., and DeCola, L. (1993). *Fractals in Geography*. Englewood Cliffs, NJ: Prentice-Hall.

Lam, N. S.-N., Fan, M., and Liu, K.-B. (1996). "Spatial-Temporal Spread of the AIDS Epidemic, 1982–1900: A Correlogram Analysis of Four Regions of the United States." *Geographical Analysis*, 28: 93–107.

Legates, D. R. (1991*a*). "The Effect of Domain Shape on Principal Components Analyses." *International Journal of Climatology*, 11: 135–46.

— (1991*b*). "An Evaluation of Procedures to Estimate Monthly Precipitation Probabilities." *Journal of Hydrology*, 122: 129–40.

— (1993). "The Effect of Domain Shape on Principal Components Analyses: A Reply." *International Journal of Climatology*, 13: 219–28.

— (2000). "Real-Time Calibration of Radar Precipitation Estimates." *The Professional Geographer*, 52: 235–46.

Legates, D. R., and Davis, R. E. (1997). "The Continuing Search for an Anthropogenic Climate Change Signal: Limitations of Correlation-Based Approaches." *Geophysical Research Letters*, 24: 2319–22.

Legates, D. R., and McCabe, G. J., Jr. (1999). "Evaluating the Use of 'Goodness-of-Fit' Measures in Hydrologic and Hydroclimatic Model Validation." *Water Resources Research*, 35: 233–41.

Legates, D. R., Nixon, K. R. Stockdale, T. D., and Quelch, G. E. (1996). "Soil Water Management Using a Water Resource Decision Support System and Calibrated WSR-88D Precipitation Estimates." *Symposium on GIS and Water Resources*, American Water Resources Association, 427–35.

— (1998). "Use of the WSR-88D Weather Radars in Rangeland Management." *Specialty Conference on Rangeland Management and Water Resources*, American Water Resources Association, 55–64.

Leung, Y. (1988). *Spatial Analysis and Planning Under Imprecision*. Amsterdam: North-Holland.

— (1997). "Feedforward Neural Network Models for Spatial Data Classification and Rule Learning," in Fischer and Getis (1997: 336–59).

Li, G., Abrahams, A. D., and Atkinson, J. F. (1996). "Correction Factors in the Determination of Mean Velocity of Overland Flow." *Earth Surface Processes and Landforms*, 21: 509–15.

Liang, S. L., Strahler, A. H., and Walthall, C. (1999). "Retrieval of Land Surface Albedo from Satellite Observations: A Simulation Study." *Journal of Applied Meteorology*, 38: 712–25.

Loewenherz, D. S., and Wendland, W. M. (1988). "Seasonal Ground Frost Occurrence: Developing a Statistically-Based Predictive Model." *Physical Geography*, 9: 186–98.

Logan, J. D., and Cliff, A. D. (1997). "A Structural Time Series Approach to Forecasting the Space-Time Incidence of Infectious Diseases: Post-war Measles Elimination Programmes in the United States and Iceland," in Fischer and Getis (1997: 101–27).

Longley, P., Goodchild, M., Maguire, D., and Rhind, D. (1997). *Geographical Information Systems: Techniques, Management and Applications*. Cambridge: Geoinformation International.

Mahmood, R. (1996). "Scale Issues in Soil Moisture Modelling: Problems and Prospects." *Progress in Physical Geography*, 20: 273–91.

— (1998). "Air Temperature Variations and Rice Productivity in Bangladesh: A Comparative Study of the Performance of the YIELD and the CERES-Rice models." *Ecological Modelling*, 106: 201–12.

Mahmood, R., and Hayes, J. T. (1995). "A Model-Based Assessment of Impacts of Climate Change on Boro Rice Yield in Bangladesh." *Physical Geography*, 16: 463–86.

Malanson, G. P., and Armstrong, M. P. (1997). "Issues in Spatial Representation: Effects of Number of Cells and Between-Cell Step Size on Models of Environmental Processes." *Geographical and Environmental Modelling*, 1: 47–64.

Malanson, G. P., and Cairns, D. M. (1997). "Effects of Dispersal, Population Delays, and Forest Fragmentation on Tree Migration Rates." *Plant Ecology*, 131: 67–79.

Malanson, G. P., Butler, D. R., and Georgakakos, K. P. (1992). "Nonequilibrium Geomorphic Processes and Deterministic Chaos." *Geomorphology*, 5: 311–22.

Malanson, G. P., Butler, D. R., and Walsh, S. J. (1990). "Chaos Theory in Physical Geography." *Physical Geography*, 11: 293–304.

Mark, D. M. (1984). "Fractal Dimension of a Coral Reef at Ecological Scales: A Discussion." *Marine Ecology Progress Series*, 14: 293–4.

Miller, H. J. (1998). "Emerging Themes and Research Frontiers in GIS and Activity-Based Travel Demand Forecasting." *Geographical Systems*, 5: 189–98.

—— (1999). "Measuring Space-Time Accessibility Benefits within Transportation Networks: Basic Theory and Computational Procedures." *Geographical Analysis*, 31: 187–212.

Miller, H. J., and Wu, Y.-H. (2000). "GIS Software for Measuring Space-Time Accessibility in Transportation Planning and Analysis." *GeoInformatica*, 4: 141–59.

Mote, T. L., and Rowe, C. M. (1996). "A Comparison of Microwave Radiometric Data and Modeled Snowpack Conditions for Dye 2, Greenland." *Meteorology and Atmospheric Physics*, 59: 245–56.

Mozolin, M., Thill, J.-C., and Usery, E. L. (2000). "Trip Distribution Forecasting with Multiplayer Perceptron Neural Networks: A Critical Evaluation." *Transportation Research Part B—Methodological*, 34: 53–73.

Namikas, L. M., and Sherman, D. J. (1994). "A Review of the Effects of Surface Moisture Content on Aeolian Land Transport," in V. Tchakerian (ed.), *Desert Aeolian Processes*. New York: Chapman & Hall.

—— (1998). "AEOLUS II: An Interactive Program for the Simulation of Aeolian Sedimentation." *Geomorphology*, 22: 135–49.

Nelson, F. E., Hinkel, K. M., Shiklomanov, N. I., Mueller, G. R., Miller, L. L., and Walker, D. A. (1998). "Active-Layer Thickness in North Central Alaska: Systematic Sampling, Scale, and Spatial Autocorrelation." *Journal of Geophysical Research*, 103: 28, 963–73.

Nijkamp, P., and Reggiani, A. (1990). "Theory of Chaos: Relevance for Analysing Spatial Processes," in M. M. Fischer, P. Nijkamp, and Y. Y. Papageorgiou (eds.), *Spatial Choices and Processes*. Amsterdam: North-Holland, 49–79.

—— (1992). *Interaction, Evolution and Chaos in Space*. Berlin: Springer-Verlag.

O'Kelly, M. E., and Lao, Y. (1991). "Mode Choice in a Hub-and-Spoke Network: A Zero-One Linear Programming Approach." *Geographical Analysis*, 23: 283–97.

O'Kelly, M. E., Song, W., and Shen, G. (1995). "New Estimates of Gravitational Attraction by Linear Programming." *Geographical Analysis*, 27: 271–85.

Odland, J. (1997). "Longitudinal Approaches to Analyzing Migration Behavior in the Context of Personal Histories," in Fischer and Getis (1997: 149–70).

Okabe, A., Sadahiro, Y. (1994). "A Statistical Method for Analyzing the Spatial Relationship between the Distribution of Activity Points and the Distribution of Activity Continuously Distributed over a Region." *Geographical Analysis*, 26: 152–67.

Okabe, A. Y., Yomono, H., and Kitamura, M. (1995). "Statistical Analysis of the Distribution of Points on a Network." *Geographical Analysis*, 27: 152–75.

Openshaw, S. (1991). "Developing Appropriate Spatial Analysis Methods for GIS," in D. Maguire, M. F. Goodchild, and D. Rhind (eds.), *Geographical Systems: Principles and Applications*. London: Longman. i. 389–402.

—— (1996). "Fuzzy Logic as a New Scientific Paradigm for Doing Geography." *Environment and Planning A*, 28: 761–8.

Openshaw, S., and Rao, L. (1995). "Algorithms for Reengineering 1991 Census Geography." *Environment and Planning A*, 27: 425–46.

Ord, J. K., and Getis, A. (1995). "Local Spatial Autocorrelation Statistics: Distributional Issues and an Application." *Geographical Analysis*, 27: 286–306.

Outcalt, S. I., and Melton, M. A. (1992). "Geomorphic Application of the Hausdorff-Besicovich Dimension." *Earth Surface Processes and Landforms*, 17: 775–87.

Outcalt, S. I., Hinkel, K. M., and Nelson, F. E. (1992). "Spectral Signature of Coupled Flow in the Refreezing Active Layer, Northern Alaska." *Physical Geography*, 13: 273–84.

—— (1994). "Fractal physiography?" *Geomorphology*, 11: 91–106.

Outcalt, S. I., Hinkel, K. M., Meyer, E., and Brazel, A. J. (1997). "Application of Hurst Rescaling to Geophysical Serial Data." *Geographical Analysis*, 29: 72–87.

Palutikof, J. P., Winkler, J. A., Goodess, C. M., and Andresen, J. A. (1997). "The Simulation of Daily Temperature Time Series from GCM Output. 1. Comparison of Model Data with Observations." *Journal of Climate*, 10: 2497–513.

Peterson, T. C., and Easterling, D. R. (1994). "Creation of Homogeneous Composite Climatological Reference Series." *International Journal of Climatology*, 14: 671–9.

Phillips, J. D. (1992). "Nonlinear Dynamical Systems in Geomorphology: Revolution or Evolution?" *Geomorphology*, 5: 219–29.

—— (1993*a*). "Interpreting the Fractal Dimension of River Networks," in Lam and DeCola (1993: 142–57).

—— (1993*b*). "Spatial Domain Chaos in Landscapes." *Geographical Analysis*, 25: 101–17.

—— (1994). "Deterministic Uncertainty in Landscapes." *Earth Surface Processes and Landforms*, 19: 389–401.

—— (1995*a*). "Biogeomorphology and Landscape Evolution: The Problem of Scale." *Geomorphology*, 13: 337–47.

—— (1995*b*). "Nonlinear Dynamics and the Evolution of Relief." *Geomorphology*, 14: 57–64.

—— (1996). "Nonlinear Dynamics and Predictability in Geomorphology," in B. Rhoads and C. Thorn (eds.), *The Scientific Nature of Geomorphology*. New York: John Wiley, 315–36.

—— (1999*a*). "Divergence, Convergence, and Self-Organization in Landscapes." *Annals of the Association of American Geographers*, 89: 466–88.

—— (1999*b*). "Spatial Analysis in Physical Geography and the Challenge of Deterministic Uncertainty." *Geographical Analysis*, 31: 359–72.

Plane, D. A. (1999). "Migration Drift." *The Professional Geographer*, 51: 1–11.

Pooler, J. (1994). "A Family of Relaxed Spatial Interaction Models." *The Professional Geographer*, 46: 210–17.

Portugali, J., Beenson, I., and Omer, I. (1994). "Socio-spatial Residential Dynamics: Stability and Instability within the Self-Organizing City." *Geographical Analysis*, 26: 321–40.

Rhoads, B. L. (1991*a*). "A Continuously-Varying Parameter Model of Downstream Hydraulic Geometry." *Water Resources Research*, 27: 1865–72.

—— (1991*b*). "Multicollinearity and Parameter Estimation in Simultaneous-equation Models of Fluvial Systems." *Geographical Analysis*, 23: 346–61.

—— (1992). "Statistical Models of Fluvial Systems." *Geomorphology*, 5: 433–55.

Rhoads, B. L., and Kenworthy, S. T. (1995). "Flow Structure at an Asymmetrical Stream Confluence." *Geomorphology*, 11: 273–93.

Rhoads, B. L., and Kenworthy, S. T. (1998). "Time-Averaged Flow Structure in the Central Region of a Stream Confluence." *Earth Surface Processes and Landforms*, 23: 171–91.

Ridenour, G. S., and Giardino, J. R. (1995*a*). "Discriminant Function Analysis of Compositional Data: An Example from Hydraulic Geometry." *Physical Geography*, 15: 481–92.

—— (1995*b*). "Log-Ratio Linear Modelling of Hydraulic Geometry Using Indices of Flow Resistance as Covariates." *Geomorphology*, 14: 65–72.

Robeson, S. M. (1994). "Influence of Spatial Sampling and Interpolation on Estimates of Air Temperature Change." *Climate Research*, 4: 119–26.

—— (1995). "Resampling of Network-Induced Variability in Estimates of Terrestrial Air Temperature Change." *Climatic Change*, 29: 213–29.

Robeson, S. M., and Janis, M. J. (1998). "Comparison of Temporal and Unresolved Spatial Variability in Multiyear Time-Averages of Air Temperature." *Climate Research*, 10: 15–26.

Robeson, S. M., and Shein, K. A. (1997). "Spatial Coherence and Decay of Wind Speed and Power in the North-Central United States." *Physical Geography*, 18: 479–95.

Robeson, S. M., and Willmott, C. J. (1996). "Spherical Spatial Interpolation and Terrestrial Air Temperature Variability," in M. F. Goodchild *et al.* (eds.), *GIS and Environmental Modeling: Progress and Research Issues*. Fort Collins, Colo.: GIS World Books, 111–15.

Rogerson, P. A. (2001*a*). "A Statistical Method for the Detection of Geographic Clustering." *Geographical Analysis*, 33: 215–27.

—— (2001*b*). "Monitoring Point Patterns for the Development of Space-Time Clusters." *Journal of the Royal Statistical Society Series A—Statistics in Society*, 164: 87–96.

Rowe, C. M. (1991). "Modeling Land-surface Albedos from Vegetation Canopy Architecture." *Physical Geography*, 12: 93–114.

Rowe, C. M., Kuivinen, K. C., and Jordan, R. (1995). "Simulation of Summer Snowmelt on the Greenland Ice Sheet Using a One-Dimensional Model." *Journal of Geophysical Research*, 100: 16, 265–73.

Roy, J., and Flood, J. (1992). "Interregional Migration Modeling via Entropy and Information Theory." *Geographical Analysis*, 24: 16–34.

Schoof, J. T., and Pryor, S. C. (2001). "Downscaling Temperature and Precipitation: A Comparison of Regression-Based Methods and Artificial Neural Networks." *International Journal of Climatology*, 21: 773–90.

Schwartz, M. D. (1994). "Monitoring Global Change with Phenology—the Case of the Spring Green Wave." *International Journal of Biometeorology*, 38: 18–22.

Schwartz, M. D., Carbone, G. J., Reighard, G. L., and Okie, W. R. (1997). "A Model to Predict Peach Phenology and Maturity Using Meteorological Variables." *HortScience*, 32: 213–16.

Shelton, M. L. (1989). "Spatial Scale Influences on Modeled Runoff for Large Watersheds." *Physical Geography*, 10: 368–83.

Sherman, D. J. (1992). "An Equilibrium Relationship for Shear Velocity and Apparent Roughness Length in Aeolian Saltation." *Geomorphology*, 5: 419–31.

Sherman, D. J., and Hotta, S. (1990). "Aeolian Sediment Transport: Theory and Measurement," in K. F. Nordstrom *et al.* (eds.), *Coastal Dunes: Form and Process*. Chichester: John Wiley, 17–37.

Sherman, D. J., and Lyons, W. (1994). "Beach-State Controls on Aeolian Sand Delivery to Coastal Dunes." *Physical Geography*, 15: 381–95.

Sherman, D. J., Bauer, B. O., Gares, P. A., and Jackson, D. W. T. (1996). "Wind-blown Sand at Castroville, California." *Coastal Engineering 1996. Proceedings, 25th International Conference*, Coastal Engineering Research Council, ASCE, Orlando, Florida, 4214–27.

Sherman D. J., Jackson, D. W. T., Namikas, S. L., and Wang, J. (1998). "Wind-blown Sand on Beaches: An Evaluation of Models." *Geomorphology*, 22: 113–33.

Smallman-Raynor, M., and Cliff, A. D. (2001). "Epidemic Diffusion Processes in a System of US Military Camps: Transfer Diffusion and the Spread of Typhoid Fever in the Spanish–American War, 1898." *Annals of the Association of American Geographers*, 91: 71–91.

Snell, S. E., Gopal, S., and Kaufmann, R. K. (2000). "Spatial Interpolation of Surface Air Temperatures Using Artificial Neural Networks: Evaluating their Use for Downscaling GCMs." *Journal of Climate*, 13: 886–95.

Sun, C.-H., and Brook, G. A. (1988). "A Hydrologic Model for Lower Coastal Plain Watersheds, Southeast United States." *Physical Geography*, 9: 15–34.

Thill, J.-C., and Horowitz, J. L. (1991). "Estimating a Destination-Choice Model from a Choice-Based Sample with Limited Information." *Geographical Analysis*, 23: 298–315.

—— (1997*a*). "Modeling Non-work Destination Choices with Choice Sets Defined by Travel-time Constraints," in Fischer and Getis (1997: 186–208).

—— (1997*b*). "Travel-time Constraints on Destination-choice Sets." *Geographical Analysis*, 29: 108–23.

Thill, J.-C., and Wheeler, A. K. (1995). "On Chaos, Continuous-Time and Discrete-Time Models of Spatial System Dynamics." *Geographical Systems*, 2: 121–30.

Thomas, R. W. (1994). "Forecasting Global HIV-AIDS Dynamics: Modelling Strategies and Preliminary Simulations." *Environment and Planning A*, 26: 1147–66.

—— (1996). "Alternative Population Dynamics in Selected HIV/AIDS Modeling Systems: Some Cross-national Comparisons." *Geographical Analysis*, 28: 108–25.

Thompson, D. M., Wohl, E. E., and Jarrett, R. D. (1996). "A Revised Velocity-Reversal and Sediment-Sorting Model for a High-Gradient, Pool-Riffle Stream." *Physical Geography*, 17: 142–56.

Timmermans, H. J. P. (1996). "A Stated Choice Model of Sequential Mode and Destination Choice Behaviour for Shopping Trips." *Environment and Planning A*, 28: 173–84.

Tobler, W. R. (1991). "Frame Independent Analysis," in M. Goodchild and S. Gopal (eds.), *Accuracy of Spatial Data Bases*. New York: Taylor & Francis, 115–22.

Todhunter, P. E. (1989). "An Approach to the Variability of Urban Surface Energy Budgets under Stratified Synoptic Weather Types." *International Journal of Climatology*, 9: 191–201.

—— (1990). "Microclimatic Variations Attributable to Urban-Canyon Asymmetry and Orientation." *Physical Geography*, 11: 131–41.

Townsend, P. A., and Walsh, S. J. (1996). "Estimation of Soil Parameters for Assessing Potential Wetness: Comparison of Model Responses through GIS." *Earth Surface Processes and Landforms*, 21: 307–26.

—— (1998). "Modeling Floodplain Inundation Using an Integrated GIS with Radar and Optical Remote Sensing." *Geomorphology*, 21: 295–312.

Travis, D. J., Carleton, A. M., and Changnon, S. A. (1997). "An Empirical Model to Predict Widespread Occurrences of Contrails." *Journal of Applied Meteorology*, 36: 1211–20.

Trofimov, A. M., and Phillips, J. D. (1992). "Theoretical and Methodological Premises of Geomorphological Forecasting." *Geomorphology*, 5: 203–11.

Vandiver, D. W., and Kirsch, S. W. (1997). "A Field Test of an Electrical-Capacitance Method for Measuring Soil-water Content." *Physical Geography*, 18: 80–7.

Wainwright, J., Parsons, A. J., and Abrahams, A. D. (1995). "A Simulation Study of the Role of Raindrop Erosion in the Formation of Desert Pavements." *Earth Surface Processes and Landforms*, 20: 277–91.

Walsh, S. J., Butler, D. R., and Malanson, G. P. (1997). "An Overview of Scale, Pattern, and Process Relationships in Geomorphology: A Remote Sensing and GIS Perspective." *Geomorphology*, 21: 183–205.

White, D., Richman, M., and Yarnal, B. (1991). "Climate Regionalization and Rotation of Principal Components." *International Journal of Climatology*, 11: 1–25.

White, R. (1990). "Transient Chaotic Behavior in a Hierarchical Economic System." *Environment and Planning A*, 22: 1309–21.

White, R., and Engelen, G. (1993). "Cellular Automata and Fractal Urban Form: A Cellular Modeling Approach to the Evolution of Urban Land Use Patterns." *Environment and Planning A*, 25: 1775–99.

Wilkinson, G. (1997). "Neurocomputing for Earth Observation—Recent Developments and Future Challenges," in Fischer and Getis (1997: 289–305).

Willmott, C. J., and Legates, D. R. (1991). "Rising Estimates of Terrestrial and Global Precipitation." *Climate Research*, 1: 179–86.

Willmott, C. J., and Matsuura, K. (1995). "Smart Interpolation of Annually Averaged Air Temperature in the United States." *Journal of Applied Meteorology*, 34: 2577–86.

Willmott, C. J., and Robeson, S. M. (1995). "Climatologically Aided Interpolation (CAI) of Terrestrial Air Temperature." *International Journal of Climatology*, 15: 221–9.

Willmott, C. J., Robeson, S. M., and Feddema, J. J. (1991). "Influence of Spatially Variable Instrument Networks on Climatic Averages." *Geophysical Research Letters*, 18: 2249–51.

—— (1994). "Estimating Continental and Terrestrial Precipitation Averages from Rain-gauge Networks." *International Journal of Climatology*, 14: 403–14.

Willmott, C. J., Robeson, S. M., and Janis, M. J. (1996). "Comparison of Approaches for Estimating Time-Averaged Precipitation Using Data from the USA." *International Journal of Climatology*, 16: 1103–15.

Wilson, T. H. (2001). "Scale Transitions in Fracture and Active Fault Networks." *Mathematical Geology*, 33: 591–613.

Wilson, T. H., and Dominic, J. (1998). "Fractal Interrelationships between Topography and Structure." *Earth Surface Processes and Landforms*, 23: 509–25.

Winkler, J. A., Palutikof, J. P., Andresen, J. A., and Goodess, C. M. (1997). "The Simulation of Daily Temperature Time Series from GCM Output. 2. Sensitivity Analysis of an Empirical Transfer Function Methodology." *Journal of Climate*, 10: 2514–32.

Wise, S., Haining, R., and Ma, J. (1997). "Regionalization Tools for the Exploratory Spatial Analysis of Health Data," in Fischer and Getis (1997: 83–100).

Wong, D., and Amrhein, C. G. (1996). "Research on the MAUP: Old Wine in a New Bottle or Real Breakthrough?" *Geographical Systems*, 3: 73–6.

Wrigley, N. (1995). "Revisiting the Modifiable Areal Unit Problem and the Ecological Fallacy," in A. D. Cliff, P. R. Gould, A. G. Hoare, and N. J. Thrift (eds.), *Diffusing Geography: Essays for Peter Haggett*. Cambridge, Mass.: Blackwell.

Xia, Z.-G., and Clarke, K. C. (1997). "Approches to Scaling of Geospatial Data," in D. A. Quattrochi and M. F. Goodchild (eds.), *Scale in Remote Sensing and GIS*. Boca Roton, Fla.: Lewis Publishers, 309–60.

Xie, Y. (1994). "Analytical Models and Algorithms for Cellular Urban Dynamics." Ph. D. Dissertation, Department of Geography, State University of New York, Buffalo.

—— (1996). "A Generalized Model for Cellular Urban Dynamics." *Geographical Analysis*, 28: 350–73.

Zhang, Z. Q., and Griffith, D. A. (2000). "Integrating GIS Components and Spatial Statistical Analysis in DBMSs." *International Journal of Geographical Information Science*, 14: 543–66.

PART V

Geographers at Work

CHAPTER 28

Geography Education

Sarah W. Bednarz, Roger M. Downs, and JoAnn C. Vender

The Changing Role of Geography Education in American Geography

The evolution of the profession of geography as an academic discipline has been intertwined with the teaching of geography in schools and colleges (Warntz 1964; Blouet 1981; Cormack 1997; Douglas 1998). Even today, the largest proportion of members of the Association of American Geographers (AAG) is employed in higher education and is charged with teaching high school graduates. The short-term fortunes of academic Departments of Geography are a direct function of student credit hours generated. Therefore, the long-term viability of Departments is a function of significant numbers of students being willing—or required—to take college geography courses.

Motivation for optional or mandatory participation in geographic learning at all levels of instruction is a cause of and a response to society's valuation of geographic knowledge. Over the past two decades, American society has placed an increasing value on geographic literacy, although what it means to be geographically literate remains subject to debate. In this chapter, we use the definition from the National Geography Standards (hereafter referred to as the Standards) (Geography Education Standards Project 1994: 34), "The outcome of *Geography for Life* is a geographically informed person (1) who sees meaning in the arrangement of things in space; (2) who see relations between people, places, and environments; (3) who uses geographic skills; and (4) who applies spatial and ecological perspectives to life situations."

The increasing valuation of geographic knowledge has been facilitated by an infrastructure ranging from the National Geographic Society's (NGS) state alliance network to the AAG's Commission on College Geography, publicized through activities such as the National Geographic Bee, and Environmental Science Research Institute (ESRI)'s "Geography Matters" campaign, and codified through public commitment to programs such as the Standards and the National Assessment of Educational Progress (NAEP).

In light of the changing role of geography in American society, the Geography Education Specialty Group (GESG) adopted a revised mission statement in 1999. Its goal is, "To promote research on the lifelong development of knowledge about the world through geography; to develop the theory and foster the practice of teaching and learning geography in formal and informal educational contexts; and to be an advocate for geographic literacy" (AAG GESG 1999: 1).

This statement, while forward-looking, also captures the evolution of geography education over the past decade. The key recognition is that geography education is *not* part of, and therefore not a special form of, education in general. It involves education in and about the domain of geographic knowledge (its content, skills and perspectives) by people who are geographically trained. Education is the fundamental process of reproduction of a discipline. It is *geography* education, not *geographic* education. The process involves the teaching and learning of geography in formal and informal contexts. The process is lifelong in the context of an individual and continuous in the context of the field and domain of geography. The process of education requires an informed interplay between theory and practice.

Inasmuch as the production of new geographic knowledge depends on the reproduction of geographically informed people *and* on the willingness of society to support that production process in colleges and universities, then geography education is central to the future of geography. That position has not been, and still is not, valued or widely appreciated by academic geographers. The adjectival distancing achieved by the phrase "geographic education" has bracketed the fundamental mechanism of geography's reproduction and apparently made it the concern, at best, of only a small part of the profession and at worst, of someone else. Although GESG members explicitly identify themselves as geography educators, we believe that all geography professors are educators.

In this chapter we will show how—and why—geography education has become increasingly central to geography and to American society. We will describe the development of the infrastructure on which this central position rests. Wilbanks (1997: p. ix) argues that, "[There] is a well-documented growing perception (external to geography as a discipline) that geography is useful, perhaps even necessary, in meeting certain societal needs. As a result, many parties . . . have been asking more from the information, techniques, and perspectives associated with geography than the nation's scientific and educational systems are delivering; and the gap between demand and supply may be widening." Wilbanks (ibid. pp. ix–x) follows with the case statement for geography education, "The most salient aspect of this demand is in education reform, especially in grade K-12, where geography education is indeed expanding rapidly."

Expansion over the past decade has been characterized by the development of new mechanisms and the improvement of existing mechanisms for supplying society with people who are geographically literate. It has involved public awareness campaigns; the development of new curriculum materials; the revision of teacher training programs; experiments in pedagogy; the incorporation of new technologies; the creation and implementation of new standards, frameworks, and curricula at national, state, and local levels; expansion of graduate courses and degree-granting programs focused on geography education; the evolution and growth of journals; the establishment of university research centers focusing on geography education; the coordination of grassroots organizations within geography and the building of links to national educational organizations. In short, it has been a process of infrastructure development to meet the demand for the reproduction of geographically informed people.

We begin analysis of infrastructure development with the GESG. In microcosm, this history of the interplay between missions and products reflects an evolving understanding of the role of geography education, as seen from within the discipline. The result is a growing emphasis on a service function for GESG in particular and for geography education in general.

The main part of the chapter will describe the development of the infrastructure supporting geography education, with a focus on the K-12 level. We account for the foundations of the geography reform movement in national-level activities (the National Education Goals, NAEP, and the Standards). Those foundations, though *national* in scope, were *voluntary* in terms of adoption and could only be implemented through state and grassroots structures. The activities required new materials and technologies to exploit their potential. They required new curricular structures, pedagogical practices, and teacher training programs. Above all, they required the development of a cadre of committed geography educators in schools, state Departments of Education, colleges, and universities.

Wilbanks (1997: p. x) offers an important caveat to this infrastructure construction program, "An expansion in classroom instruction without a strong foundation in knowledge and skills, however, does not serve the demands of education well; and it is important to balance this one kind of response to external demands with an assessment of the knowledge base that undergirds it." Although Wilbanks was referring to geography as whole, we apply his message to efforts to develop a theoretical base for geography education. Practice has exceeded theory, and in the conclusion, we assess challenges for the next decade.

The Evolution of the Geography Education Specialty Group

The activities and membership of GESG reflect its evolution along three lines: first, from a focus on teaching geography in higher education to a concern with geography education in all educational contexts and at all ages; second, from a focus on pedagogy to a balance between practice and theory; and third, from a focus on dissemination and facilitation of good practice to a balance between service and research. We discuss the origin and changing emphases from GESG's inception in 1979 to the present.

A petition to form a specialty group on Geography in Higher Education was submitted by A. David Hill. The group's working title, "College Geography," was changed at its first meeting during the AAG's 1979 annual meeting in Philadelphia. The mission statement in the 1982 by-laws reads, "This Group is broadly concerned with research, development, and practice in the learning and teaching of geography and with examining and strengthening the role of geography in higher education and in all of education by focusing on the development of learners, teachers, curricula, and programs (AAG GHESG 1982: 1)." Significantly, the statement expanded the scope to include pre-collegiate education, and perhaps informal education, by virtue of the vagueness of the phrase "in all of education." However, the focus remained exclusively on higher education throughout its first decade. It is also significant that "geography education" has, in the United States, been taken to exclude what in the United Kingdom is referred to as the geography of education (see Bondi and Matthews 1988). The latter has a focus on spatial patterns in the provision of inputs to the educational system: schools, buses, equipment, etc.

From the start, the group was service-oriented. It was intent on defining the role of geography in higher education, assessing its condition, making connections among subdisciplines and geography departments, and exploring methods for teaching geography at the collegiate level. The first annual report (1980) noted:

> [the] purpose of this group is to strengthen the role of geography in higher education and to pull together the other twenty-four specialty groups in order to design curriculum which will include the findings and materials presented each year at the AAG annual meeting. Also, the higher education specialty group will promote collaborative relationships between geography departments at institutions of higher education and between AAG specialty groups. (Wallace 1980: 2)

The group helped to augment an infrastructure for the professional development of collegiate faculty and graduate students in the 1980s. It offered sessions on: leadership for department heads, teaching of geography for graduate students and teaching assistants, integrating women into the curriculum, teaching the non-traditional student, introductory geography as general education, and computer-assisted instruction.

The late 1980s' renaissance in geography education increased attention to geography as a discipline and to curriculum reform at all levels (see "The Changing Practice of Geography Education" section of this chapter). As a result, in 1991 a name change was proposed, "to reflect the interest in, and importance of, geographic education at all levels of the education system. Recent activities, especially the development of more than 40 state-wide Geographic Alliances, and major collaborative efforts between the AAG and the National Council for Geographic Education (NCGE) on research and instructional issues relating to geographic education, prompted the Group to consider a name change" (Smith 1991: 2). The old and new guard debated:

> It must be stressed, however, that this change does not mean a reorientation of the Group *away* from issues of higher education . . . many of the . . . senior-level professors who believe strongly that issues of geography in higher education must be represented fully in this Group, and who have worked diligently since the Group's inception to do so[,] would not support a move toward a total precollegiate focus. (ibid. 2–3)

After a ballot in 1992, the group called itself the Geography Education Specialty Group. The mission statement was modified: "geography in higher education and in all of education" became "geography at all levels of education," thus de-emphasizing higher education and re-emphasizing the sequence of kindergarten through graduate school.

Contemporaneous with the name change was a dramatic increase in membership. Annual reports indicate that membership doubled from 111 in 1980 to 234 in 1990. Membership rose dramatically (37 percent) between 1992 and 1993. Although membership peaked during the years 1993–5, coinciding with the heyday of Standards development (see Table 28.1), both raw numbers and the percentage of AAG members joining GESG are higher in the late 1990s than in the early part of the decade.

GESG chairs have noted that members identify primarily with disciplinary subfields other than geography education. This is substantiated by the significant difference between the number of GESG members and those AAG members identifying "educational geography" as one of their three topical proficiencies. The topical proficiency is about half of the Specialty Group membership.[1]

[1] Members are allowed to list only 3 topical proficiencies on their membership/renewal forms; there are 6 blanks for specialty groups, although the web page states that members may join as many specialty groups as they wish.

Table 28.1 *Growth of Geography Education Specialty Group*

Year	Educational Geography	GESG	AAG total	GESG (%)
1989	165	217	6,321	3.43
1990	169	234	6,271	3.73
1991	169	225	6,290	3.58
1992	198	272	6,647	4.09
1993	217	429	6,997	6.13
1994	170	477	7,204	6.62
1995	181	411	7,381	5.57
1996	190	374	7,271	5.14
1997	197	388	7,026	5.52
1998	202	348	6,910	5.03
1999	194	653	6,527	10.
2000	200	297	6,523	4.55

Source: Association of American Geographers

During the mid-1990s, GESG focused on facilitating national projects designed to enhance the practice of geography at pre-collegiate and higher education levels: National Geography Standards, curriculum projects such as the Activities and Readings in the Geography of the United States (ARGUS) and Geographic Inquiry into Global Issues (GIGI), Advanced Placement geography, and pre-service teacher education. Connections with cognate organizations—notably NCGE and NGS—continued to develop. The group sponsored sessions dealing with structural and pedagogical issues of higher education.

Attention remains focused on large-scale initiatives to develop infrastructure. Initiatives include issues related to teaching geography at the collegiate level; the assessment of student learning; the need for research on geography learning and teaching (encouraged by the 1998 establishment of a graduate student paper competition); possibilities for learning, teaching, networking, and collaboration via the Internet and information technologies; and addressing lifelong learning within and outside formal education.

In microcosm, debates within GESG reflect concerns for the discipline as a whole. Thus, despite movements toward an encompassing view of the educational process, and to linkages among theory, practice, service, and research, there remains a fundamental ambiguity about the status of GESG and geography education within the larger discipline. Is its role purely that of service to the profession? Is there new theoretical knowledge to be developed within geography education or is it purely an applied science (or craft)? Is geography education, as far as the AAG is concerned, simply an issue of higher education or does the mandate extend to pre-school education? Is its infrastructure equivalent to the urban infrastructure: necessary, indispensable, but out of sight and out of mind unless it breaks down? The infrastructure had broken down by the late 1970s, as evidenced by laments about geographic illiteracy. We argue that geography cannot afford to disregard the infrastructure that reproduces a geographically literate or informed population: people who will take geography courses, support geography in the future, and perhaps become geographers.

The Changing Practice of Geography Education

Documenting Progress in Geography Education

The literature of geography education is suffused with the progressivist rhetoric of renaissance and reform, couched within a decadal time-frame. Hill and LaPrairie (1989) noted a nascent renaissance in geography education. R. S. Bednarz and Petersen (1994) edited articles on geography education under the title "A Decade of Reform." The preface to *Geography for Life* (Geography Education Standards Project 1994: 9) refers to the Standards as the culmination of a decade of reform. Grosvenor's (1995) retrospective on the campaign to enhance geography education was titled "In Sight of the Tunnel: The Renaissance of Geography Education." In the introduction to *The First Assessment: Research in Geographic Education*, Boehm (1997: 1) characterized the last decade as "a period of change in the manner in which geography is taught," and as a "renaissance in geographic education." R. S. Bednarz (1997), reflecting on his tenure as editor of the *Journal of Geography*, called it a "decade of progress."

Renaissance is defined as a revival, a period of marked improvement. Since the publication of *Geography in America*, geography education has seen marked improvement in three respects: (1) in its quantitative presence in schools, colleges, and universities across the country;

(2) in its qualitative status as a school subject, and (3) in its functional character: that is, its curriculum, instructional, and assessment practices. The revival of geography education has been the topic of reviews and retrospectives. Taken together they present the story of a period in which geography education grew and matured as a school topic, as a player in national and local educational policy circles, and as a subfield of the discipline.

The foundation for the renaissance was laid in the preceding decade (1979–88). Hill and LaPrairie (1989) identified four precipitating factors: (1) reports that alerted the public to the state of geographic illiteracy (e.g. National Assessment of Education Progress 1979; National Commission on Excellence in Education 1983; National Governor's Association 1986); (2) the publication of the *Guidelines for Geographic Education: Elementary and Secondary Schools* (Joint Committee on Geographic Education 1984), which presented geography as a unified scientific discipline that teachers and the public could understand via five themes and which laid the foundation for subsequent efforts; (3) the 1985 formation of the Geography Education National Implementation Project (GENIP) to unite efforts of the AAG, the American Geographical Society (AGS), NCGE, and NGS (Natoli 1994); and (4) the establishment in 1986 of state Alliances for geography education under the auspices of NGS (Dulli 1994; Grosvenor 1995).

In this section, we concentrate on the development of the infrastructure supporting geography education, especially developments in curriculum, instruction, and assessment at the K-12 level. We begin with evidence of the presence of geography in K-16 education in terms of numbers of students enrolled in courses. Part of the explanation for this growth lies in the national-level cornerstones established in the first half of the decade, and so we discuss three key events: (1) the inclusion of geography in the National Education Goals (1990); (2) the development and results of the National Assessment for Educational Progress (NAEP) test (1994); and (3) the development of the Standards, *Geography for Life* (1994).

In each case, the key events, though couched in national (and often nationalistic) tones, were merely opportunities for change. Systemic change depends on implementation, and therefore we document how one cornerstone—the National Geography Standards—has been implemented. This illustrates how the microstructure of the infrastructure works in practice and leads to a discussion of the effects and implications of systemic educational change, in which we consider: (1) the role of professional associations and universities in engineering change; (2) changes in instruction and curriculum; and (3) efforts to prepare geography teachers through pre-service teacher preparation programs and in-service staff development.

Reproducing Geographically Literate Americans

The scope of geography education is expanding, as *Rediscovering Geography* (National Research Council 1997) indicated. Although much of the evidence is anecdotal, key facts confirm the impression. In 1992–3, 155,800 students in Texas took a year-long high-school course in world geography. That same year, 328,367 students took world history. In 2000–1, 306,413 students took world geography whereas 264,402 students took world history (Texas Education Agency 2001). In Tennessee, in 1985–6, 10,829 students completed a high school geography course. That number climbed to 29,000 in 1992–3 (R. S. Bednarz 1996). Similar trends are evident elsewhere, particularly in states in which the practice of geography education has changed dramatically such as Colorado, Oregon, Delaware, and Florida.

At the same time, enrollments in college geography classes have grown and the number of geography majors expanded. In Texas, geography majors at Texas A&M University rose from 70 in 1985 to 190 in 2000–1. Florida State University saw an increase, from 48 majors in 1989 to 64 in 1994, to 110 in 2000–1. Institutions as disparate as Temple University, the University of Nebraska, and Indiana State University reported growth in enrollments and majors (ibid.). Gober *et al.* (1995) present data supporting the growth thesis. For example, 57.6 per cent of Geography Department chairs reported modest or significant increases in geography graduates over the preceding five years whereas only 6 per cent reported a modest or significant decrease (ibid.: 197, table 4). Even a conservative estimate suggests that "The numbers of (geography) undergraduate degrees are expected to rise until the mid 1990s, hold steady until 2000, and then increase again early in the next decade" (ibid. 194).

The link between school and college geography is clear. Expansion at the college level is connected to the development of an enhanced infrastructure supporting K-12 geography. For example, a survey of Texas A&M geography majors in 1996 indicated that 47 per cent had taken a high school geography course, and a fifth of majors said they found geography via their high school class. Students taking geography in high school started their majors earlier. Of students who did not take geography in high school, 68 per cent began majoring during their junior year whereas 73 per cent who took high school geography selected a geography major in the freshman or sophomore year (R. S. Bednarz 1996). South Dakota witnessed a similar effect after it instituted a high-school geography requirement (Gritzner and Hogan 1992). Many states are establishing high school graduation requirements: Texas; Colorado (for admission to the University of Colorado-Boulder); Minnesota; and Tennessee.

Advocacy of geographic literacy remains the driving force behind systemic change. The concerted efforts of the professional geography associations have driven the renaissance. For example, the AAG supported geography education through three National Science Foundation (NSF)-sponsored projects: ARGUS, ARGWorld, and Human Dimensions of Global Change (HDGC). NGS expanded the Alliance network, coordinated public awareness initiatives, and lobbied state and federal agencies and policy-makers successfully on behalf of geography. NCGE developed teaching initiatives, technology training initiatives, and K-12 curriculum materials. AGS used its newsletter *Ubique*, to provide information to members, many of whom are in business, not academia.

These events, coupled with the advocacy of the professional associations, laid the groundwork on which the successes of this past decade were built. They enabled geography's inclusion in the National Education Goals, the first key event in the renaissance.

The Three National Cornerstones

1. The National Education Goals, adopted by the nation's governors and President George Bush in 1989, established six goals for educational improvement. Five core academic subjects were identified. Goal 3 stated that "American students will leave grades four, eight, and twelve having demonstrated competency in challenging subject matter, including English, mathematics, science, history, and geography." To achieve this goal, "world class" national content and student performance standards were to be developed in core subject areas and a national voluntary assessment system (NAEP) established to measure achievement.

Many of these ideas became lost in ensuing partisan political battles that pitted federal versus state authority in education (Ravitch 1995; Hershberg 1997; Nash *et al.* 1997). Nevertheless, the presence of geography in the National Goals (passed into Public Law 103–227 as *Goals 2000: The Educate America Act*) provided essential validation for the grassroots efforts by state Alliances to improve student learning in geography. Goal 3 led to the next key events: NAEP and the Standards.

2. NAEP Geography Test is directed by the US Department of Education's National Center for Education Statistics (NCES). NAEP measures student performance in areas such as mathematics, science, and reading, usually at grades 4, 8, and 12. Its findings are reported as "The Nation's Report Card."

In 1994, a stratified random sample of 18,600 students took the first comprehensive examination in geography conducted by NAEP. In an ideal world, the NAEP assessment would have used the Standards as its test framework in order to provide baseline data measuring student attainment of the Standards (Bettis 1997). However, the NAEP process began before work on the Standards. The content assessed by NAEP is congruent with expectations in *Geography for Life*, but it is organized differently.

A framework was developed around three content dimensions: (1) Space and Place: knowledge of geography as it relates to particular places on Earth, to spatial patterns on Earth's surface, and to physical and human processes that shape such spatial patterns; (2) Environment and Society: knowledge of geography as it relates to interactions between environment and society; and (3) Spatial Dynamics and Connections: knowledge of geography as it relates to spatial connections among people, places, and regions (NAEP Geography Consensus Project 1992). The five-themes framework was not used as the basis of the test framework despite calls to do so (Boehm 1997). The themes worked poorly for assessment purposes because they are interrelated, making it difficult to develop assessment items that measure knowledge in just one content dimension. In addition, the test development team perceived them to be overly simple and not representative of the breadth, depth, and perhaps, most importantly, the language of contemporary geography.

The NAEP results were not surprising given the geographic illiteracy of the American population (Persky *et al.* 1996). They revealed that: (1) little geography (less than one hour per week) was taught at grade 4; (2) ethnic and racial subgroups performed differently. At all grades, White and Asian students scored, on average, higher than Black and Hispanic students; (3) male students outperformed female students; (4) location played a role in geographic understanding. Student scores varied with type of community (central city, urban fringe/large town, and rural/small town) and region of the nation; and (5) at each grade level, about 70 per cent of students achieved at the basic level (partial mastery of knowledge and skills needed) and approximately 25 percent of the students (22 percent grade 4, 28 percent grade 8, 27 percent grade 12) reached the proficient achievement level (solid academic performance).

The 2001 administration of the NAEP assessment revealed small—but statistically significant—improvements in results at grades 4 and 8 and no change at grade 12. Average scores of grade 4 and 8 students improved in comparison to 1994; in both grades, score increases occurred among the lower-performing students. At grade 4 the average score of Black students was higher in 2001 than 1994. While there were few changes between 1994 and 2001 in the time spent on geography instruction in grade 4, at grade 8 students in 2001 reported more time on geography than in 1994. The 2001 assessment showed that 21 per cent of grade 4, 30 per cent of grade 8, and 25 per cent of grade 12 students attained the proficient achievement level; at grades 4 and 8, 74 per cent of students attained the basic level while 71 per cent did so at grade 12 (Weiss *et al.* 2002).

Other than the official reports issued by NCES, no systematic research has explored the implications of NAEP for geography education. This can be explained, in part, by the failure of the NCES to offer the NAEP assessment on a meaningful schedule. The NAEP assessment was to have been repeated in 1998, which would have allowed educators to measure change in student understanding based on grade cohorts moving up. For example, the grade cohort assessed in 1994 at grade 4 would be assessed again at grade 8. Unfortunately, NCES did not put geography on such a schedule. The second NAEP assessment was conducted in 2001.

For geography education this is a missed opportunity. The third section of this chapter focuses on research in geography education, but we note that the lack of baseline research assessing the current level of geography knowledge and skills is a significant factor hampering the efforts to improve geography education.

3. The National Geography Standards were developed as part of the Goals 2000 initiative under the auspices of the US Department of Education and the four professional geography associations. *Geography for Life*, specifying "what every young American should know and be able to do in geography," summarized the "challenging subject matter" required to meet the National Education Goals. It was developed through a broad national consensus process as a voluntary set of content standards (Geography Education Standards Project 1994: 243–53; de Souza and Munroe 1995). The framework consists of eighteen ideas, designated as "Standards," that are organized under six "essential elements." Statements related to each Standard and benchmarked at grades 4, 8, and 12 clarify what students should know, understand, and be able to do as a result of their knowledge and understanding.

Although released in 1994, geographers have attempted little discussion or critical evaluation of the quality and utility of the Standards. They were developed at the same time as the controversial National History Standards, which deflected the attention of the politicized education community away from geography's efforts. Unfortunately it may also have dampened initial efforts to disseminate the Standards. Some educationists have criticized the Standards as developmentally inappropriate (Seefeldt 1995) and confusing to individuals charged with using them to develop state-level curriculum standards (Munroe and Smith 1998). Despite such scattered comments, the National Geography Standards have been widely accepted and used successfully in curriculum development projects, for example in New York, Florida, Arizona, Indiana, and Texas, and in national initiatives such as *Path Toward World Literacy*, a Standards-based guide to K-12 geography published in 2001 (Boehm 2001).

The Standards initiated significant reforms in the practice of geography education. They brought school geography closer to contemporary academic geography in terms of content, perspectives, and skills. The Standards recognized that geography is a physical and human science, and featured Physical Systems as an essential element. Systems-theory concepts organize the human and physical content. Five Standards feature content focused on environment–society links, providing depth and breadth to the familiar theme of human–environment interaction. The assumption that students need to understand (1) the nature of change in places as well as over time, (2) that everything takes place in a geographic context, and (3) that how humans perceive their geographic context plays a role in history and strengthens the partnership between history and geography. The Standards introduce geographic information systems (GIS) into geography education.

The Standards align geography education with other academic disciplines in terms of pedagogy and its role in K-12 education. Knowledge and skills are integrated in statements that specify student performance expectations —what students should know and be able to do, thus making a clear connection between *knowing* and *doing* geography. The Standards' geography skills highlight the importance of requiring students to think and solve problems through the application of the spatial and the ecological perspective. Finally, the Standards argue that geography is for life, throughout the school year, throughout all subjects, K-12 and beyond, reflecting geography's increasing role in society.

Standards Implementation: An Example of the Microstructure of the Infrastructure

The US education system is a dispersed decision-making structure composed of 100,000 schools in 15,000 local districts in 50 states, Washington DC, Puerto Rico, Guam, and the Virgin Islands. Private, state, and national groups mediate connections between and among these elements. The result is a complex, spatially fragmented system reflecting the fundamental belief that education is best controlled at the local level.

Implementing policy reform in such an educational system means taking a multi-front approach to all levels: state Standards-setting initiatives and teacher certification requirements; district-wide curriculum development; and changing the practice of teachers, classroom by classroom. It means working with publishers to provide materials that support new curricula and methods of instruction. Despite efforts by the geography education community to implement the Standards, progress has been slow.

Research in Standards implementation has concentrated at the state level and focused on social studies (Glidden 1998, Education Week 1999). Three studies (S. W. Bednarz 1998; Munroe and Smith 1998, 2000) found state-level geography standards documents to be uneven in quality and connection with the Standards. An examination of middle- and high-school social studies textbooks found no evidence of Standards-based student expectations (Martin 1997). A cursory glance at the four leading elementary social studies textbook series published in 1998 indicated that only one (Houghton Mifflin's *We the People*) uses the Standards as its organizing framework for geography while the bestseller (McGraw-Hill/Macmillan's *Adventures in Time and Place*) used the five themes. Social studies textbooks issued in 2002 appear to be more closely aligned with the Standards than previous editions.

At the classroom level, implementation can be measured using a variety of criteria. Variability in implementation is a result of teachers' experience, knowledge of geography, teaching skills, access to information about Standards, sense of self as a change agent, institutional and personal incentives to implement Standards, and teaching context (S. W. Bednarz 1996). State-based Alliances have supported implementation through institutes, workshops, and the development of materials correlated to the Standards. However, despite wide praise and Alliance-based efforts, implementation remains uneven. A system of high-quality instructional materials and staff development must support Standards. Necessary materials include (1) "concept" books, texts needed for teachers and students to study the discipline's conceptual underpinnings; (2) extended lesson books with examples of student work that meet the Standards; and (3) courses of study, the layout of a program that will enable students to reach the Standards for a given year (Tucker and Codding 1998). Geographers are gradually producing these materials (e.g. Boehm 2001; Vender 2003) but more is needed.

Change and Continuity in the Practice of Geography Education

Geography education has changed significantly in response to infrastructure changes during the last decade. Marran (1994) compared the "old" geography with the "new" geography (see Table 28.2). Old geography was textbook-dependent, teacher-directed, and focused on memorizing facts about the principal products of world regions. This approach drove the subject into near extinction in American schools by the 1970s. New geography, emphasizes *doing* geography. It embraces constructivist ideas about "best learning and teaching practice" particularly the value of hands-on, active learning and inquiry approaches that had failed to catch on during the 1960s curriculum projects such as the High School Geography Project (Slater 1982; Hill 1992; Klein 1995; Stoltman 1998). The conditions that scuttled the 1960s projects may still exist today: perceiving projects to be nationalizing the curriculum; wariness of non-traditional learning environments and non-textbook materials; open-ended assessments; but most importantly, the dearth of well-prepared geography teachers.

The practice of geography education has changed in its nature and content. States have revised curricula to move to Standards-based education. Three curricular models operate in the US. In the most common model, geography is a component of the "social studies," sharing time in a crowded curriculum with history, economics, political science, and other social sciences. New York, Florida, Illinois, Michigan, and Texas follow this format, based loosely on the National Council for the Social Studies Standards (NCSS 1994). Geography exists as a separate subject, more or less linked to the Standards depending upon the state, but it is taught and assessed within the realm of social studies.

Table 28.2 *Old and new geography*

Old geography	New geography
Oriented to specific place/location	Emphasis on spatial relationships
Structured on the recall of information	Problem-solving encouraged
Fact-based objective testing	Connected to critical thinking skills
Limited skill development	Depth replaces breadth
Teacher-directed and teacher-shaped	Collaborative learning strategies
Textbook-driven	Research-based
Student as segregated learner	Adaptable to new technology
Minimal problem-solving	Observation through field work
Hooray! It's field trip day!	Human–environment interaction emphasis
Regional emphasis	Framework/Standards driven
Ethnocentric/nationalistic bias	

Source: J. F. Marran (1994). "Discovering Innovative Curricular Models for School Geography," in R. S. Bednarz and J. F. Petersen (1994: 23–30).

The second model, characterized as the histocentric model because of that subject's primacy in the curriculum, is found in Virginia, California, and Massachusetts. Geography is relegated to secondary status. While not all geographers see this as pernicious (Entrikin 1998), the model makes it difficult to teach geography as a discipline in its own right (Munroe and Smith 1998: p. viii). The third model, with geography as a stand-alone subject, is found only in Colorado, which has separate geography standards.

One consequence of the changing curriculum, particularly in states that have developed approaches congruent with the Standards, has been a shift in the middle- or high-school course that features geography. "Old geography" organized content as a survey of world regions. The course, while about geography, did not engage students in geography *per se*. New state standards encourage teachers to follow a topical, systematic approach *and* to focus on geographic inquiry about places. With the advent of the Advanced Placement Human Geography program in 2001, more high-school teachers will become knowledgeable about college-level geography (typically organized by systematic topic) in order to help students pass the credit-granting examination (College Board 1999). Academically motivated students will arrive at college ready for a higher-level, more rigorous geography education.

Another innovation with ramifications at the university level has been the integration of geographic information systems (GIS) technology with the practice of geography education. The rate of adoption by instructors has been slow. Relevant factors relate to (1) hardware and software requirements, the need for data, and other technical obstacles; (2) teacher training and a paucity of curriculum materials; and (3) motivation, reward, and broader systemic issues. Incentives and disincentives for any innovation influence the adoption process. Teachers are reluctant to learn a new technology without assurance of institutional support or career advantages (Audet and Paris 1997; S. W. Bednarz and Ludwig 1997; Kerski 2000). None the less, teaching with GIS and other visualization techniques is becoming more common (Sui 1995; Audet and Paris 1997; S. W. Bednarz and Ludwig 1997). Geography educators have embraced CD-ROMs, and web-based and web-assisted instruction (Carstensen *et al.* 1993; Nellis 1994; Proctor 1995; Krygier *et al.* 1997; Foote 1997; Solem 1999).

Geography educators have sought new venues to teach about geography, making real the idea of "geography for life." The Smithsonian Institution project, Earth2U, Exploring Geography, encouraged children and parents to think spatially. Museums, particularly children's museums, have adopted geography themes to focus activities and exhibits. As another indicator of societal interest in geography, agencies such as United States Geological Survey, Federal Emergency Management Agency, the State Department, the National Aeronautics and Space Administration (NASA), and the

Central Intelligence Agency have launched programs to teach geography through curriculum materials, websites, and other formal and informal means.

As geography has become a more important part of the curriculum, the need for teaching materials has increased. Few materials have been produced for young learners, a reflection of the minor role geography plays in the elementary curriculum, although NGS has produced "teacher toolkits" with lessons appropriate for grades K-12 and has recently developed elementary-level "Reading Expeditions" and "Map Essentials" curriculum extensions for geography and social studies. Most curriculum projects have been aimed at middle- and high-school students, producing materials focused on the geography of the United States (ARGUS and ARGUS CD), the world (GIGI), the special needs of underrepresented students (Finding a Way) and general geography (GeoLinks). Recently completed are two projects, ARGWorld (Activities and Resouces in the Geography of the World)—an NSF-funded project to develop materials to teach world geography in grades 6 through 9, and Mission Geography—a NASA/GENIP collaboration to produce materials linking the Standards with NASA's missions and results.

The development of materials is not limited to pre-collegiate levels. Three projects have developed teaching and resource materials for higher education: (1) Project GeoSim (Carstensen *et al.* 1992); (2) the Human Dimensions of Global Change project; and (3) the Virtual Geography Department project (Foote 1999).

Teacher education has played a crucial role in infrastructure development. As the practice of geography education has changed, a richer geography, more congruent with the aims, perspectives, and methodology of the academic discipline, has evolved. Translating the "new geography" into classrooms requires teacher preparation (pre-service education in colleges and universities) and staff development (in-service education in schools). NGS has led the effort to reach in-service teachers through its Alliance network. The Alliances, established as collaborative arrangements between university geographers and classroom teachers, have instituted summer workshops serving more than 11,000 teachers, developed curriculum materials, and organized political action on behalf of geography (Salter 1991; Bockenhauer 1993; Dulli 1994; Grosvenor 1995). In academic year 1997/8 at least 38,500 educators participated in Alliance workshops and institutes (Vender 1999). Alliance members have worked to implement the Standards by serving on state and local standards- and assessment-writing committees (Vender 1999). The Alliance network has changed how teachers teach geography, their understanding of the discipline, and the status of geography (Cole and Ormrod 1995; Grosvenor 1995; Lockyear 1997). The Alliance network is envied by other disciplines (Giese 1999) and has been recognized as a model for professional development (Bockenhauer 1993; Vender 1999).

Unfortunately, pre-service geography education has not changed in as positive or rapid fashion as in-service education. Boehm *et al.* (1994: 89–98) commented, "geographic education faces serious shortcomings based on its failure to create and maintain strategies for effective pre-service teacher education. It is axiomatic that if all we did is provide in-service training in geography for teachers then we institutionalize the continual need for further in-service teacher training in geography." GENIP released recommendations based on an analysis of the geography education system shown in Fig. 28.1 (Bednarz and Bednarz 1995: 482–6; Goodman 1994). However, the wider geography community has shown little interest in the issue. It is difficult to change a political process such as teacher certification; both geographers and professors of education are sometimes unwilling to work cooperatively to change degree programs in order to raise geography course requirements for pre-service educators (Libbee 1995). Thus, with notable exceptions,[2] a major shortage of well-prepared geography teachers in US classrooms remains.

The State of Educational Practice in Geography

Infrastructure implies system and coordination. Unfortunately, the system of American education is anything but coordinated, and geography education is no exception. Built around local control, American education is highly fragmented, and again, geography education is no exception. Key structural constraints exist in the current practice of geography education: the lack of geographically informed teachers; weak pre-service programs for would-be geography teachers; a subject that has been eviscerated by the loss of physical geography to Earth Science and diluted by an incorporation of human geography under the rubric of social studies; limited access to the classroom computing technology that is essential for GIS, remote sensing, and cartography; and the continual fight for curriculum space.

[2] A notable example of such an effort is a model pre-service initiative in Tennessee that has been adopted by at least nine higher-education institutions in Tennessee (Jumper 1998).

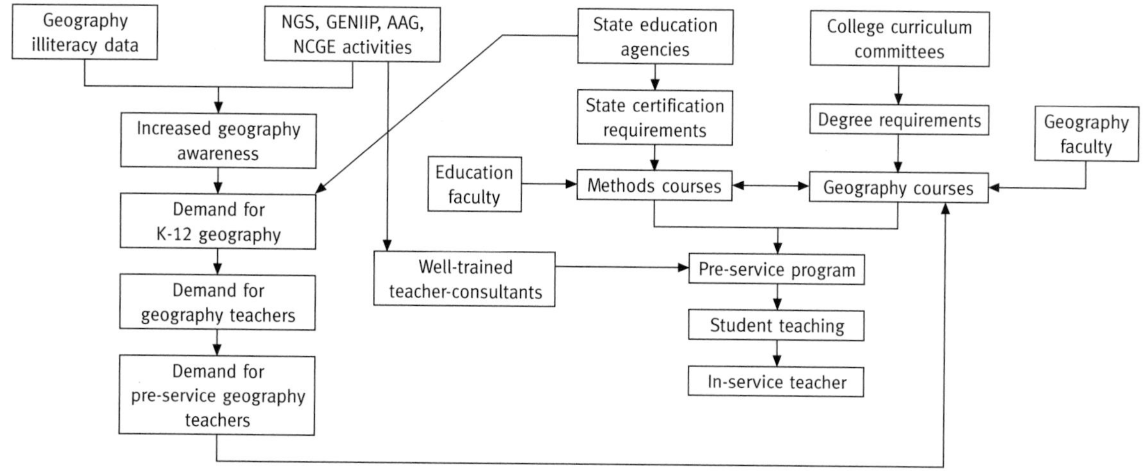

Fig. 28.1 Geography education system

Source: Bednarz and Bednarz (1995: 483)

Despite these constraints, in states such as Colorado, Minnesota, Tennessee, and Texas, evidence exists of an increase of geographically literate students entering college geography programs. Despite the general aversion to national control of education, geography is a key player in national debates about educational policy and practice. But as Wilbanks (1997: p. x) argues, educational practice without either a strong foundation in knowledge and skills or an assessment of the supporting knowledge base is not well grounded. In the next section, we turn to the research support for the theoretical basis of geography education.

The Emerging Theory of Geography Education

The Nature of Knowledge in Geography Education

Of necessity, geography education is an interdisciplinary enterprise. While grounded in the content of geography, the process must acknowledge work in areas ranging from education to cognitive development and cognitive science to technology development. It is an applied field in which there is a mutual relationship between theory and praxis. It is a field in which the normative and prescriptive are confounded with the positive and descriptive, and in which it is impossible to avoid ideology and teleology. In effect, education establishes desirable goals and develops a system for achieving those goals. In describing the state of the knowledge base for geography education, we must keep these characteristics in mind.

From the perspective of geography education, we need to answer questions about content and pedagogy. What should be taught to whom? Why should it be taught? When should it be taught? How should it be taught? How can we measure the success of what is taught? These questions can be organized into categories based on the interaction between two ideas; first, the use of a lifespan developmental framework and second, a novice-expert mastery framework (Downs 1994*a*). The first category focuses on properties of the domain of geographic expertise: (1) *character*: what is the essential character of geographic expertise? (2) *genesis*: what is the origin of geographic expertise? (3) *nature*: what are the essential components of geographic expertise? The second category explores the fostering of expertise in the domain of geography: (4) *ontogenesis*: how does expertise in geography develop? (5) *identification*: what are the procedures for identifying geographic aptitude and assessing geographic performance? and (6) *nurturing*: what models of teaching might work and why?

To be successful, we need theoretically motivated answers to these questions. Of necessity, such a theory must have a prescriptive element, a goal to which the educational system aspires. In the case of geography education, the goal is geographic literacy, a goal captured by the idea of the *geographically informed person*. Given this goal and the need to articulate the relationship between content and pedagogy, we can outline the parts of the infrastructure that support the production of knowledge about the process of reproducing people who are geographically literate.

The Basic Sources of Information in Geography Education

Two bibliographies provide access to the literature. The first, with an American focus, is Forsyth's (1995) *Learning Geography: An Annotated Bibliography of Research Paths.* The second, with a European (in general) and British (in particular) focus is Foskett and Marsden's (1998) *A Bibliography of Geographical Education 1970–1997.* These sources provide a wide-ranging survey of items, 5,708 in the case of Foskett and Marsden and more than 500 in the case of Forsyth. Despite substantive overlap, they use different organizing headers and therefore provide complementary and comprehensive first-level mappings of the structure and coverage of the literature.

In addition, there are topically focused reviews such as Stoltman (1997) on curriculum and instruction research, Bettis (1997) on assessment and reform in geographic education, Graves (1997) on geographical textbooks, and Wise (1997) on school atlases. Hill has provided updates on geography education in North America (see Hill 1989, 1992, 1994).

As in all fields, journals constitute a vital part of the infrastructure. In the United States the *Journal of Geography* (*JOG*), flagship publication of NCGE, is the lead outlet for geography education. *JOG* publishes research ranging from teaching strategies and methods to articles of general interest to geographers. Significant changes in the types of articles published in *JOG* are evident in the last decade. Geography teaching strategies and methods articles constituted over two-thirds (69%) of the articles in the period 1988–93 but declined to 38 per cent in 1994–7. Articles on research in geography learning and thinking doubled, from 7.5 per cent during 1988–93 to 14.5 per cent during 1994–7. Articles related to geography education as an institution (e.g. policy development, program description, and analyses of the role of geography within the educational system) increased dramatically. In the period 1988–93, an average of 6 per cent of articles focused on such research; this increased to 28 per cent in the period 1994–7, mirroring changes in the capacity and nature of the infrastructure of geography education in the United States.

Competing with *JOG* is the *Journal of Geography in Higher Education* (*JGHE*), published in the United Kingdom. It focuses on issues related to the scholarship of learning and teaching geography in higher education, posing three questions about articles for publication: (1) Is it about educational issues? (2) Is it connected with higher education? and (3) Is it relevant to geographers? Articles in *JGHE* have become more focused on research and less on practical matters in response to the introduction of research assessment exercises in the UK.

Also in the United Kingdom, the Geographical Association (GA) caters to British geography instructors through publications aimed at different educational levels: *Primary Geography* (early elementary students), *Teaching Geography* (late elementary to high school), and *Geography* (advanced high school, teachers, undergraduates, and interested lay people). *Geography* is the most scholarly journal, publishing review and research articles of use to instructors developing lectures; *Teaching Geography* features articles on teaching methods and resources as well as issues related to changes in educational policies.

The Commission on Geographical Education of the International Geographical Union inaugurated the journal *International Research in Geographical and Environmental Education* (*IRGEE*) in 1991. *IRGEE* has become a significant outlet for fundamental research in geography. It began publishing two issues per year, and in 2000 went to four issues per year and initiated an additional monograph series.

Other journals publishing work in geography education in the United States include *Social Education* (aimed at classroom practitioners) and *Theory and Research in Social Education* (for college and university faculty) published by the National Council for the Social Studies (NCSS). A new journal, *Research in Geography Education*, published by the Grosvenor Center at Southwest Texas State University, commenced in 2000.

Essential parts of the infrastructure for generating research are mechanisms associated with colleges and universities: teaching and research program specialties, the production of Masters theses and doctoral dissertations, and conferences at which research can be presented. As Table 28.3 indicates, the number of programs listing Geographic Education as a specialty increased from forty-nine in 1989 to sixty-three in 2000. University-based centers in geography education have been organized at the University of Colorado, San Jose State University, Southwest Texas State University, Hunter College (CUNY), and the University of South Carolina. Table 28.4 shows that the number of theses and dissertations directly connected with geography education has increased from at least eight in 1989 to twenty-one in 1997, but ten in 2000. In addition to

Table 28.3 *US collegiate programs with a specialty in geographic education*

Year	Programs
1989	49
1990	51
1991	51
1992	41
1993	61
1994	60
1995	64
1996	58
1997	64
1998	64
1999	65
2000	63

Source: *AAG Guide to Departments of Geography*, 1989–90 through 2000–1.

Table 28.4 *Theses and dissertations related to geography education*

Year	Masters	Ph.Ds	Total
1989	4	4	8
1990	5	3	8
1991	6	3	9
1992	9	12	21
1993	9	7	16
1994	10	6	16
1995	3	1	4
1996	10	5	15
1997	13	8	21
1998	4	6	10
1999	n.r.	4	4
2000	6	4	10
TOTAL	79	63	142

Note: n.r. = none reported

Sources: *AAG Guide to Departments of Geography*, 1989–90 through 1998–9; Dissertation Abstracts Online; personal communication.

traditional outlets for research presentations, new opportunities for sharing research have been established. Southwest Texas State University sponsored conferences (biannually since 1993) to showcase research in geography education. The GIS in Higher Education (GISHE) conferences (1997, 1998, 1999) highlighted research in teaching GIS and remote sensing for K-12 and higher education audiences. In 2000 ESRI inaugurated a GIS Education pre-conference as part of its annual Users Conference.

Evidence exists, therefore, of increases in the rate of production of new knowledge and the opportunity to participate in formal instructional programs. More journals publish work on geography education; the nature of what is being published is changing, and institutional support for geography education is increasing. What is the character of the resultant work and what type of knowledge does it provide?

The Knowledge Base in Geography Education

Over the past decade, research foci have included: (1) studies of the role of map and globe skills, especially in the elementary school (e.g. Castner 1990, 1995; Gregg 1997; Johnson and Gondesen 1991; Leinhardt *et al.* 1998; Muir and Cheek 1991); (2) debates about the appropriate role of place location instruction and knowledge (e.g. S. W. Bednarz 1995; Boehm and Petersen 1987; Gregg and Leinhardt 1993; S. W. Bednarz 1995; Gregg *et al.* 1997); (3) the role of spatial ability in geographic understanding (e.g. Self and Golledge 1994); (4) the nature of the geographic learning process (e.g. Hickey and Bein 1996); (5) explorations of the influences of teaching methods on student understanding (e.g. Hertzog and Lieble 1996; Hill 1996; Klein 1995; Lyman and Foyle 1991; Stoltman 1991); (6) alternative models for and approaches to curriculum design (e.g. S. W. Bednarz 1996; Farrell and Cirrincione 1989; Marran 1994; Stoltman 1997; Thornton and Wenger 1990); (7) the development of programs of instruction for alternative populations (e.g. Andrews *et al.* 1991; LeVasseur 1993); (8) the links of geography education to other knowledge domains such as environmental education (e.g. Corcoran 1996), art (e.g. Fredrich *et al.* 1996), literature (e.g. Oden 1989), and music (e.g. Byklum 1994); and (9) the nature of teacher education in the United States (e.g. Blackwell 1995; Boehm *et al.* 1994). While not exhaustive, this list of topics is illustrative of the range of research into geography education.

Taken in totality, this work can be characterized as small-scale (in terms of numbers of participants); largely asynchronous (few longitudinal studies); rarely controlled (re: formal experimental design); and often descriptive and anecdotal. Theory can be built only from and grounded in empirical data and there is a dearth of data in geography education (see Downs 1994*b*). Of the work that does it exist, it can be said that:

> much of the existing work in geography education fails to meet generally accepted research standards in terms of design, execution, and reporting. There are too many one-of-a-kind, ad hoc studies that do not lead to a cumulative understanding of essential phenomena. Thus we lack a range of valid and reliable instruments for assessment. Therefore, we need to pay attention to the basics of the empirical method: sample selection, hypothesis formulation, data quality, statistical analysis, reporting requirements, research ethics, etc. (ibid. 129)

How can we overcome the lack of data and theoretical grounding on which we can foster geographic literacy and develop a generation of geographically informed people?

Structural reasons exist for this situation, most of which find their roots in the long-standing split within the geography community between researchers, who identify with the AAG, and practitioners, who identify with the NCGE. To AAG researchers, apparently there were no issues of importance to be studied in geography education. To NCGE practitioners, little material of value in teaching geography came from the researchers. There was a reciprocal lack of interest—and sometimes lack of respect—between the groups and traditions.

Although these attitudes have resulted in a persistent split we see signs that it is being overcome. Both researchers and practitioners have failed to develop strong links with theory developed in cognate domains of knowledge, especially theory in cognitive development and education. Many ideas appeared to be generated from "within" geography, although this has begun to change significantly over the past decade. Now, in some areas (e.g. spatial cognition and spatial ability), the connections between geography and cognate disciplines are genuinely reciprocal.

Over the past twenty years, the attention of the geography education community has focused on a fight for survival and a fight for space in the curriculum. In the case of the former, the renaissance of geography education in the 1980s consumed the attention of scholars and practitioners. In the case of the latter fight, the need to reestablish a role for geography in the American educational establishment has consumed the bulk of attention in the 1990s. In both cases, it fostered an imperative to do things now, to get things going; this has militated a more detached study of what we—as geography educators—are doing, and how and why.

The awkward relationship between geography and the pre-service educational establishment has posed problems for the theory development. Two structural constraints have eviscerated geography in K-12 education in America. Physical geography has been seen as the purview of Earth Science, while the remainder of geography has been seen as falling under the umbrella of social studies, a curricular domain dominated by history. To the extent that social studies is not a traditional discipline (and therefore is a-theoretical) and to the extent that it is dominated by history, geography has played a relatively small—and subservient—role in the preservice instructional program. In that sense, there has been little need for theory.

The Future of Geography Education in the United States

In their chapter in *Geography in America* (ch. 1), Hill and LaPrairie developed a scenario for the future of geography in education. They built it on a belief that geography in higher education was enjoying—and would continue to enjoy—a higher status. They saw the situation for school geography as being predicated on the school reform program overcoming significant obstacles. If there were success in the school and the higher education arenas, then, "we may reach a time when *each* student will learn *some* geography, and the subject will become prominent and respected in the schools. Then, college geography will face its greatest challenge: many more students with solid geographic educations will be looking to expand their geographic studies in college" (Hill and LaPrairie 1989: 20). They continued:

> In this future scenario, the current level of sophistication of college geography courses would become inadequate, because material taught in college introductory courses would be taught at the precollege level. The need for geography-teacher training would place new demands on college faculty. The translation of research into educational materials would constitute a large task. Certainly the need would prevail to clarify the nature and purpose of geography in American education.

We would like to report that Hill and LaPrairie were not only prescient (which they were) but that their scenario has been played out, successfully. As *Rediscovering*

Geography (NRC 1997) made clear, geography education is crucial to the future of the discipline. As the empirical studies on market trends indicate, there is growth in enrollment in courses and majors at the college level and there are good prospects for employment. As data from states such as Texas demonstrate, increasing numbers of K-12 students are taking geography courses. As this chapter makes clear, geography has built an elaborate infrastructure for the reproduction of people who are geographically informed.

As one result, geography is garnering respect in terms of social valuation. One indication is the decision by National Aeronautic and Space Administration (NASA) to list geography alongside mathematics, science, engineering, and technology as the five educational foci of NASA Education programs. The investment in a curriculum project, Mission Geography, is a significant step in what promises to be a long-term joint venture. The introduction of Advanced Placement (AP) Human Geography marks another milestone in geography's fight for recognition. The AP course offers a challenging capstone experience for high-school students, thus encouraging more students to begin as geography majors in college (rather than transfer in during their programs as is the case at present). The second NAEP test in 2001 allowed us to see areas of progress and stasis since 1994 (Weiss *et al.* 2002).

Despite these achievements, we are a long way from reaching the goal of universal geographic literacy. Despite significant progress, the Standards have not yet been translated into state standards and school district curricula, and classroom practice *nationwide*. The old obstacles remain. Pre-service training remains woefully weak and largely inaccessible to reform. Up-to-date materials are not available to teachers. Geography lags in access to computing technology and training that would enable us to take advantage of GIS. Research in geography education remains insufficient to ground the necessary development of curricula and materials. We have not yet clarified the nature and purpose of geography in American education for ourselves, let alone for society at large.

We have reached the point where *some* students are learning *some* geography and *some* of that geography reflects the *best* of what our discipline can offer to students and society. Geography is prominent in some states and in some schools in those states. Glimmerings of evidence indicate that more freshmen are electing geography as a college major. The support infrastructure remains strong and committed. GIS is becoming increasingly important in society and education and offers a major opportunity for geography to respond to Hill and LaPrairie's (1989: 19) scenario, "Americans, consummate pragmatists, will judge geography by what it proves it can do to help them improve their lives and their worlds, as they define them." Understanding the potential links among GIS, education, and a way of thinking that is central to geography, spatial thinking, is the focus of a three-year study under the aegis of the National Academy of Sciences. The intent of the study is to make recommendations about software, hardware, GIS design, and curriculum guidelines to ensure that American students have the opportunity to learn to think spatially.

Our challenge as geography educators is proving the ESRI slogan correct, that geography matters. We cannot improve on Hill and LaPrairie's scenario. We want to reiterate their concluding sentence, "If geographers can unite to marshal their scarce resources, if they are able to invent and support devices that systematically encourage cumulative improvement, and if they can convince non-geographers that geography is essential, the status of their field will improve" (1989: 20).

References

AAG GESP (Association of American Geographers Geography Education Specialty Group) (1999). "Geography Education Specialty Group By-Laws, April 1999." Document on file with AAG office in Washington, DC.

AAG GHESP (Association of American Geographers Geography in Higher Education Specialty Group) (1999). "Geography in Higher Education, April 1982." Document on file with AAG office in Washington, DC.

AAG (Association of American Geographers) (1989/90–2000/1). *AAG Guide to Departments of Geography*. Washington, DC.: Association of American Geographers.

Andrews, S. K., Otis-Witborn, A., and Young, T. M. (1991). *Beyond Seeing and Hearing: Teaching Geography to Sensory Impaired Children*. Indiana, Pa.: NCGE.

Audet, R. H., and Paris, J. (1997). "GIS Implementation Model for Schools." *Journal of Geography*, 96: 14–21.

Bednarz, R. S. (1996). "Factors Affecting the Choice of Geography as a Major." Unpublished paper, National Council for Geographic Education, Santa Barbara, Calif.

—— (1997). "A Decade of Progress in Geographic Education." *Journal of Geography*, 96/6: 278–9, 324.

Bednarz, R. S., and Petersen, J. F. (1994). *A Decade of Reform in Geographic Education: Inventory and Prospect*. Indiana, Pa.: National Council for Geographic Education.

Bednarz, S. W. (1995). "Using Mnemonics to Teach Place Geography." *Journal of Geography*, 94/1: 330–8.

Bednarz, S. W. (1996). "The Effects of the National Geography Standards on High School Geography," in T. van der Zijp, J. van der Schee, and H. Trimp (eds.), *Innovation in Geographical Education: Proceedings, Commission on Geographical Education*. Amsterdam: Free University Center for Geographic Education, 252–6.

—— (1998). "State Standards: Implementing Geography for Life." *Journal of Geography*, 97/2: 83–9.

Bednarz, S. W., and Bednarz, R. S. (1995). "Preservice Geography Education." *Journal of Geography*, 94/5: 482–6.

Bednarz, S. W., and Ludwig, G. M. (1997). "Ten Things Higher Education Needs to Know about GIS in Primary and Secondary Education." *Transactions in GIS*, 2/2: 123–33.

Bettis, N. C. (1997). "Assessment and Reform in Geographic Education: 1930–1997," in R. G. Boehm and J. F. Petersen (eds.), *The First Assessment*. San Marcos, Tex.: Grosvenor Center for Geographic Education, 252–72.

Blackwell, P. J. (1995). "Reform of Teacher Education: An Opportunity for Geographers." *Journal of Geography*, 94/5: 495–8.

Blouet, B. W. (ed.) (1981). *The Origins of Academic Geography in the United States*. Hamden, Conn.: Shoestring Press.

Bockenhauer, M. H. (1993). "The National Geographic Society's Teaching Geography Project." *Journal of Geography*, 92/3: 121–6.

Boehm, R. G. (1997). "The First Assessment: A Contextual Statement," in R. G. Boehm and J. F. Petersen (eds.), *The First Assessment*. San Marcos, Tex.: Grosvenor Center for Geographic Education, 1–17.

—— (ed.) (2001). *Path Toward World Literacy: A Scope and Sequence in Geographic Education K-12*. Washington, DC: National Geographic Society.

Boehm, R. G., and Petersen, J. F. (1987). "Teaching Place Names and Locations in Grades 4–8: Map of Errors." *Journal of Geography*, 86/4: 167–70.

Boehm, R. G., Brierley, J., and Sharma, M. (1994). "The Bête Noire of Geographic Education: Teacher Training Programs," in R. G. Bednarz and J. F. Petersen (eds.), *A Decade of Reform in Geographic Education: Inventory and Prospect*. Indiana, Pa.: National Council for Geographic Education, 89–98.

Bondi, L., and Matthews, M. H. (1988). *Education and Society: Studies in the Politics, Sociology and Geography of Education*. London: Routledge.

Byklum, D. (1994). "Geography and Music: Making the Connection." *Journal of Geography*, 93/6: 274–80.

Carstensen, L. W., Shaffer, C. A., Morrill, R. W., and Fox, E. A. (1993). "Geo-Sim: A GIS-Based Simulation Laboratory for Introductory Geography." *Journal of Geography*, 92/5: 218–26.

Castner, H. W. (1990). *Seeking New Horizons: A Perceptual Approach to Geographic Education*. Montreal/Kingston, Canada: McGill/Queen's University Press.

—— (1995). *Discerning New Horizons: A Perceptual Approach to Geographic Education*. Indiana, Pa.: NCGE.

Cole, D. B., and Ormrod, J. E. (1995). "Effectiveness of Teaching Pedagogical Content Knowledge through Summer Geography Institutes." *Journal of Geography*, 94/3: 427–33.

College Board (1999). *Advanced Placement Course Description, Human Geography*. New York: The College Board.

Corcoran, P. B. (1996). "Environmental Education in North America: Alternative Perspectives." *IRGEE*, 5/2: 117–31.

Cormack, L. B. (1997). *Charting an Empire: Geography at the English Universities, 1580–1620*. Chicago: University of Chicago Press.

de Souza, A. R., and Munroe, S. S. (1994). "Implementation of Geography Standards: Strategies and Initiatives," in R. S. Bednarz and J. F. Petersen (1994: 107–16).

Douglas, M. P. (1998). *The History, Psychology, and Pedagogy of Geographic Literacy*. Westport, Conn.: Praeger.

Downs, R. M. (1994*a*). "Being and Becoming a Geographer: An Agenda for Geography Education." *Annals of the Association of American Geographers*, 84/2: 175–91.

—— (1994*b*). "The Need for Research in Geography Education: It Would Be Nice to Have Some Data," in R. S. Bednarz and J. F. Petersen (1994: 127–33).

Dulli, R. E. (1994). "Improving Geography Learning in the Schools: Efforts by the National Geographic Society," in R. S. Bednarz and J. F. Petersen (1994: 121–4).

Education Week (1999). *Quality Counts '99: Rewarding Results, Punishing Failure*. Washington, DC: Education Week, 18/17.

Entrikin, J. N. (1998). "Blurred Boundaries: Humanism and Social Science in Historical Geography," *Historical Geography*, 26: 93–9.

Farrell R. T., and Cirrincione, J. M. (1989). "The Content of the Geography Curriculum: A Teacher's Perspective." *Social Education*, 53/2: 225–7.

Foote, K. E. (1997). "The Geographer's Craft: Teaching GIS in the Web." *Transactions in GIS*, 2/2: 137–50.

—— (1999). "Editorial: Bringing Faculty Online: Inspiring and Sustaining Innovation in Information and Computer Technologies (ICT)," *Journal of Geography in Higher Education*, 23/1: 5–7.

Forsyth, A. S. (1995). *Learning Geography: An Annotated Bibliography of Research Paths*. Indiana, Pa.: National Council for Geographic Education.

Foskett, N., and Marsden, B. (1998). *A Bibliography of Geographical Education 1970–1997*. Sheffield: The Geographical Association.

Fredrich, B., Fuller, K., and Fuerst, A. (1996). "Linking Geography and Art: Inness 'The Lackawanna Valley'." *Journal of Geography*, 95/6: 254–61.

Geography Education Standards Project (1994). *Geography for Life: National Geography Standards*. Washington, DC: National Geographic Society.

Giese, J. (1999). "Social Science on the Frontier: What's New in the Social Science Disciplines and Social Science Education." *Bulletin, Social Science Education Consortium*.

Glidden, H. (1998). *Making Standards Matter 1998*. Washington, DC: American Federation of Teachers.

Gober, P., Glasmeier, A., Goodman, J. M., Plane, D. A., Stafford, H. A., and Wood, J. S. (1995). "Employment Trends in Geography, Part I: Enrollment and Degree Patterns." *Professional Geographer*, 47/3: 317–28.

Goodman, J. (1994). "Findings of the GENIP Preservice Task Force." Unpublished report submitted to GENIP.

Graves, N. (1997). "Textbooks and Textbook Research in Geographical Education: Some International Views." *International Research in Geographical and Environmental Education*, 6/1: 60–2.

Gregg, M. (1997). "Problem Posing for Maps: Utilizing Understanding." *Journal of Geography*, 96/5: 250–7.

Gregg, M., and Leinhardt, G. (1993). "Geography in History: What is the Where?" *Journal of Geography*, 92/2: 56–63.

Gregg, M., Stainton, C., and Leinhardt, G. (1997). "Strategies for Geographic Memory." *IRGEE*, 6/1: 41–58.

—— (1997). "Strategies for Geographic Memory." *International Research in Geographical and Environmental Education*, 6/1: 41–58.

Gritzner, C. F., and Hogan, E. P. (1991). "Indirect Benefits of South Dakota's High School Geography Requirement." Paper presented at the Annual Meeting of the Association of American Geographers, Miami, Fla.

Grosvenor, G. M. (1995). "In Sight of the Tunnel: The Renaissance of Geography Education." *Annals*, 85/3: 409–20.

Hershberg, T. "Explaining Standards." *Education Week*, 10 December 1997: 33.

Hertzog, C. J., and Lieble, C. (1996). "A Study of Two Techniques for Teaching Introductory Geography: Traditional Approach versus Cooperative Learning in the University Classroom." *Journal of Geography*, 95/6: 274–80.

Hickey, G. M., and Bein, F. L. (1996). "Students' Learning Difficulties in Geography and Teachers' Interventions." *Journal of Geography*, 95/3: 118–22.

Hill, A. D. (1989). "Geography and Education: North America," *Progress in Human Geography*, 13/4: 589–98.

—— (1992). "Geography and Education: North America." *Progress in Human Geography*, 16/2: 232–42.

—— (1994). "Geography and Education: North America." *Progress in Human Geography*, 18/1: 65–73.

—— (1996). "Geographic Inquiry into Global Issues (GIGI)," in T. van der Zijpp, J. van der Schee, and H. Trimp (eds.) Innovation in Geographical Education: Proceedings, Commission on Geographical Education. Amsterdam: Free University Center for Geographic Education, 270–2.

Hill, A. D., and LaPrairie, L. A. (1989). "Geography in American Education," in G. L. Gaile and C. J. Willmott (eds.), *Geography in America*. Columbus: Merrill, 1–26.

Johnson, P. E., and Gondesen, M. E. (1991). "Teaching Longitude and Latitude in the Upper Elementary Grades." *Journal of Geography*, 90/2: 73–8.

Joint Committee on Geographic Education (1984). *Guidelines for Geographic Education*. Washington, DC: Association of American Geographers and National Council for Geographic Education.

Jumper, S. R. (1998). Personal communication.

Kerski, Joseph J. (2000). "The Implementation and Effectiveness of Geographic Information Systems Technology and Methods in Secondary Education." Ph.D. Dissertation, University of Colorado.

Klein, P. (1995). "Using Inquiry to Enhance Learning and Appreciation of Geography." *Journal of Geography*, 94/2: 358–67.

Krygier, J. B., Reeves, C., DiBiase, D., and Cupp, J. (1997). "Design, Implementation and Evaluation of Multimedia Resources for Geography and Earth Science Education." *Journal of Geography in Higher Education*, 21/1: 17–39.

Leinhaardt, G., Stainton, C., and Bausmith, J. M. (1998). "Constructing Maps Collaboratively." *Journal of Geography*, 97/1: 19–30.

LeVasseur, M. (1993). *Finding a Way: Encouraging Underrepresented Groups in Geography: An Annotated Bibliography*. Indiana, Pa.: NCGE.

Libbee, M. (1995). "Strengthening Certification Guidelines." *Journal of Geography*, 94/5: 501–4.

Lockyear, E. J. (1997). "Assessing a Professional Development Program for Geography Teachers in Colorado." MA thesis, University of Colorado.

Lyman, L., and Foyle, H. (1991). "Teaching Geography Using Cooperative Learning." *Journal of Geography*, 90/5: 223–6.

Marran, J. F. (1994). "Discovering Innovative Curricular Models for School Geography," in R. S. Bednarz and J. F. Petersen (1994: 23–30).

Martin, C. (1997). *Textbook Analysis*. Indiana, Pa.: National Council for Geographic Education and GENIP.

Muir, S. P., and Cheek, H. N. (1991). "Assessing Spatial Development: Implications for Map Skill Instruction." *Social Education*, 55/5: 316–19.

Munroe, S. S., and Smith, T. (1998). *State Geography Standards*. Washington, DC: Thomas B. Fordham Foundation.

—— (2000). *Update, State Geography Standards*. Washington, DC: Thomas B. Fordham Foundation.

NAEP Geography Consensus Project (1992). *Geography Framework for the 1994 National Assessment of Educational Progress*. Washington, DC: National Assessment Governing Board.

Nash, G. B., Crabtree, C., and Dunn, R. E. (1997). *History on Trial: Culture Wars and the Teaching of the Past*. New York: Knopf.

National Assessment of Educational Progress (NAEP) (1979). *Summaries and Technical Documentation for Performance Changes in Citizenship and Social Studies Assessment, 1969–1976*. Denver, Colo.: NAEP.

National Commission on Excellence in Education (1983). *A Nation at Risk: The Imperative for Educational Reform*. Washington, DC: US Government Printing Office.

National Council for the Social Studies (1994). *Curriculum Standards for Social Studies*. Washington, DC: NCSS.

National Governor's Association (1986). *Time for Results: The Governor's 1991 Report on Education*. Washington, DC: National Governor's Association.

National Research Council (1997). *Rediscovering Geography*. Washington, DC: National Research Council.

Natoli, S. J. (1994). "Guidelines for Geographic Education and the Fundamental Themes of Geography," in R. S. Bednarz and J. F. Petersen (1994: 13–22).

Nellis, M. D. (1994). "Technology in Geographic Education: Reflections and Future Directions," in R. S. Bednarz and J. F. Petersen (1994: 51–8).

Oden, P. (1989). "Geography is Everywhere in Children's Literature." *Journal of Geography*, 88/3: 151–8.

Persky, H. R., Reese, C. M., O'Sullivan, C. Y., Lazer, S., and Shakrani, S. (1996). *NAEP 1994 Geography Report Card*. Washington, DC: National Center for Education Statistics.

Proctor, J. D. (1995). "Multimedia Guided Writing Modules for Introductory Human Geography." *Journal of Geography*, 94/6: 571–82.

Ravitch, D. (1995). *Debating the Future of American Education: Do We Need National Standards and Assessments?* Washington, DC: Brookings Institution Brown Center of Education.

Salter, C. L. (1991). "Geographic Alliances: Be Moved but Not Alarmed." *Professional Geographer*, 43/1: 102–4.

Seefeldt, C. (1995). "Transforming Curriculum in Social Studies," in S. Bredekamp and T. Rosegrant (eds.), *Reaching Potentials: Transforming Early Childhood Curriculum and Assessment.* Washington, DC: NAEYC, 109–24.

Self, C. M., and Golledge, R. G. (1994). "Sex-Related Differences in Spatial Ability: What Every Geography Educator Should Know." *Journal of Geography*, 93/5: 234–43.

Slater, F. (1982). *Learning Through Geography*. London: Heinemann Educational Books.

Smith, R. W. (1991). "Geography in Higher Education Specialty Group Annual Report 1990–1991." Report submitted to AAG office, 26 September. Document on file with AAG office in Washington, DC.

Solem, M. N. (1999). "The Diffusion, Adoption, and Effects of Instruction with the Internet in U.S. College Geography." Ph.D. Dissertation, Department of Geography, University of Colorado, Boulder.

Stoltman, J. P. (1991). "Research on Geography Teaching," in J. P. Shaver (ed.), *Handbook of Research in Social Studies Teaching and Learning*. New York: Macmillan, 437–47.

Stoltman, J. P. (1997). "Geography Curriculum and Instruction Research Since 1950 in the United States," in R. G. Boehm and J. F. Petersen (eds.), *The First Assessment*. San Marcos, Tex.: Grosvenor Center for Geographic Education, 131–70.

Sui, D. Z. (1995). "A Pedagogic Framework to Link GIS to the Intellectual Core of Geography." *Journal of Geography*, 94/6: 578–99.

Texas Education Agency (2001). Unpublished data on student enrollment. Austin: TEA.

Thornton, S. J., and Wenger, R. N. (1990). "Geography Curriculum and Instruction in Three Fourth Grade Classrooms." *Elementary School Journal*, 90/5: 515–31.

Tucker, M. S., and Codding, J. B. (1998). *Standards for Our Schools*. San Francisco: Jossey-Bass.

United States Congress House Committee on Education and the Workforce (2001). *No Child Left Behind Act of 2001: Report of the Committee on Education and the Workforce, House of Representatives, on H.R. 1, together with Additional and Dissenting Views (Including Cost Estimate of the Congressional Budget Office)*. Washington, DC: US Government Printing Office.

United States Department of Education (2002). *No Child Left Behind*. Website available at URL: ‹http://www.nochildleftbehind.gov/›, accessed 23 February 2003.

Vender, J. C. (1999). "Summary and Review of 1998 Alliance Annual Reports." Unpublished report prepared for the National Geographic Society Education Foundation, Washington, DC.

—— (2003). *Teaching to the Standards: A K-12 Scope and Sequence in Geography*. Washington, DC: Geographic Education National Implementation Project.

Wallace, E. (1980). "Specialty Group Report—Geography in Higher Education." Memoraandum submitted to AAG Gouncil, 1 May. Document on file with AAG office in Washington, DC.

Warntz, W. (1964). *Geography Then and Now: Some Notes on the History of Academic Geography in the United States*. New York: American Geographical Society, Research Series, 25.

Weiss, A. R., Lutkus, A. D., Hildebrant, B. S., and Johnson, M. S. (2002). *The Nation's Report Card: Geography 2001*. Washington DC: National Center for Education Statistics.

Wilbanks, T. J. (1997). "Preface," in National Research Council, *Rediscovering Geography*. Washington, DC: National Research Council, pp. ix–xii.

Wise, M. (1997). "The School Atlas, 1885–1915." *Paradigm*, 23: 1–11.

CHAPTER 29

Hazards

Burrell E. Montz, John A. Cross, and Susan L. Cutter

Introduction

In August of 1992, Hurricane Andrew battered southeastern Florida, causing fifty-eight deaths, and more than $27 billion in property losses (National Climatic Data Center 1999). The following year, widespread flooding occurred within the Upper Mississippi River basin, inundating 5.3 million hectares during the worst flood to affect much of the region in this century. The Northridge earthquake (magnitude 6.7) led to sixty-one deaths and more than $20 billion in property damage and loss in 1994. A year later, Kobe, Japan, experienced a magnitude 6.9 earthquake. Despite massive efforts to prepare for such events, more than 6,000 lives were lost, and $150–200 billion in property damage was experienced. In 1998, Hurricane Mitch devastated Honduras, Nicaragua, and other parts of Central America. More than 5,600 people died in Honduras alone and approximately 70,000 homes were damaged. In Nicaragua, more than 850,000 people were affected, with approximately 2,860 deaths. Estimates of losses in agriculture, housing, transportation and other infrastructure are in excess of $1.3 billion dollars (United Nations Office for the Coordination of Humanitarian Affairs 1998). These are just a few, albeit particularly devastating, events that continued to focus our attention in the 1990s on hazards and disasters.

The widespread news media coverage of these disaster events provided a backdrop for fictional portrayals as Hollywood rediscovered the disaster movie genre. With enhanced special effects and big-named stars, popular films such as *Twister*, *Volcano*, *Dante's Peak*, *Armageddon*, *Deep Impact*, *Titanic*, and *A Civil Action* added a different slant to the media coverage of disasters and the public's perception of hazards throughout the decade. The public's interest and fascination in actual disasters also propelled several books to the bestseller list (Barry 1997; Junger 1997; Larson 1999).

Both the fictional representations and the consequences of real disasters illustrate the shift in our understanding of the forces at work in such events. Some of the damage in Hurricane Andrew, for example, is attributed to inadequate enforcement of building standards. In Kobe, structures engineered to withstand seismic activity failed, prompting concern about just how safe infrastructure is in tectonically active areas. And Hurricane Mitch's devastating toll cannot be explained solely by the storm. Decades of land abuse and a combination of social, political, and economic factors combined with the storm to cause the severe losses.

In order to capture both the evolution of concerns in hazards research as well as the conceptual and methodological approaches used in this research, the chapter addresses three broad questions:

1. What has been the focus of hazards research in geography in the last decade?
2. Have interpretations of hazards changed and, if so, how?
3. What does the future portend in terms of significant hazard issues and developments?

Geographic research on hazards reflects a creative tension between theory and practice. The core focus is on the relationship between the physical/environment and human use systems. During the last decade, we have seen significant changes in how we evaluate hazards and hazardousness. These range from overt recognition of the diversity of conditions that contribute to vulnerability and the hazardousness of place to the increasingly complex interaction among technological, human, and physical/environmental systems such as those found in the world's mega-cities. Many of the basic research questions remain much the same as they were ten or even twenty years ago (Mitchell 1989). What has changed, however, is our broadened understanding of the nature of hazards themselves and the range of causal or contributing factors. For example, hazardousness and vulnerability have increased in many places, even as our understanding of physical processes and the environment has become more sophisticated. Increasingly, we look to various mixes of social, political, and economic factors that contribute to hazardousness and vulnerability, as they interact with one another and with the physical environment.

We have seen a shift in the classification of hazards and the types of events/phenomena we study. Instead of neatly dividing the field into natural hazards and technological hazards as was done a decade earlier, hazards are now viewed as a continuum of interactions among physical/environmental, social, and technological systems ranging from extreme natural events to technological failures to social disruptions to terrorism. Adding complexity is the recognition that social and political factors can increase vulnerability of populations to a multitude of hazards. Increasingly, the study of multi-hazards is essential, inasmuch as the occurrence of one event often triggers other events, crossing the artificial boundaries that earlier hazards scholars had drawn. All these constitute hazards geography in the 1990s and beyond.

Before we examine some of the scholarly trends during the past decade, it is important to understand the larger context in which much of this research has taken place. The following discussion helps illuminate some of the trends in research as well as providing some understanding of constraints that have thwarted advancements in the field.

Societal and Institutional Contexts

One of the most significant shifts in hazards research in the last two decades has been the increased recognition of the need to consider the contexts (social, political, etc.) within which our research takes place. This is as salient to our evaluation of the status of hazards research at the end of the century as it was a decade ago. The previous review of hazards research in geography (Mitchell 1989) came on the eve of the United Nations' designation of the 1990s as the International Decade for Natural Disaster Reduction (IDNDR), which focused worldwide attention on natural hazards and disasters. The United Nations resolution establishing the IDNDR had as its objective to "reduce through concerted international actions, especially in developing countries, loss of life, property damage and social and economic disruption caused by natural disasters" (UN 96th Plenary Meeting, reprinted in Housner 1989: 74). Disaster prevention and mitigation would be achieved through technical assistance, technology transfer, demonstration projects, education, and training, goals specifically mentioned in the UN resolution (Housner 1989). Domestically, the US National Committee for the Decade of Natural Disaster Reduction urged a "fundamental shift in public perceptions of natural disasters" in an effort to stop the escalating costs of disasters, and a multidisciplinary program for natural hazard reduction was proposed (US National Committee 1991: 11).

Numerous activities to enhance risk assessment, mitigation, and warning systems, both domestically and internationally, were undertaken as Decade projects (US National Committee 1994), including the intensive study of Mt. Rainier as a "Decade Volcano," the provision of next generation radar (NEXRAD) coverage of the nation, the modernization of the US Earthquake Monitoring System, and the enhancement of NOAA Weather Radio. A tremendous number of reports, maps, and videos were published during the Decade, increasing both the scientific literature and products aimed at heightening public awareness. At the same time, the IDNDR focused attention on global hazards along with the continuing work of the Intergovernmental Panel on Climate Change (Houghton *et al.* 1990, 1996; Watson *et al.* 1995). Geographers had prominent roles in these efforts, insuring that social scientific issues were prominent. Still, some argued at the beginning of the Decade (Mitchell 1988) and midway through (White 1996) that social science was largely ignored and that the Decade would have little long-term impact (BOND 1999).

Similarly in 1994, the Natural Hazards Center at Boulder, Colorado, began a three-year study of the state-of-the-art knowledge in hazards research. The first assessment of natural hazards provided an appraisal of research needs and the social utility of allocating funds for research on natural hazards (White and Haas 1975).

Among the contributions of the first assessment was the establishment of the Natural Hazards Research and Applications Information Center at the University of Colorado, Boulder (<http://www.colorado.edu/hazards>, last accessed 19 November 2002), which serves as a national/international clearinghouse for information on natural hazards and human adjustments to them. Twenty years after the first assessment, the National Science Foundation funded a Second Assessment to gauge the contributions to theory and practice from the research community during the period 1974–94. It is not surprising that geographers had a prominent role in that review, with approximately sixteen out of the more than 100 contributors. Similar to its predecessor, the Second Assessment (Mileti 1999) concluded that hazards problems no longer can be viewed (or solved) in isolation but are symptomatic of larger, more complex societal issues. The idea that people can use technology to control nature to ensure safety is no longer viable and has escalated losses from hazards. This Assessment calls for a policy of "sustainable hazards mitigation" which links local environmental management (including hazard mitigation) with social resilience, community adaptation, and economic development. Improving our ability to achieve sustainable hazards mitigation requires researchers and practitioners to shift strategies to include the following (ibid. 12):

- Adopt a global systems perspective
- Accept responsibility for hazards and disasters
- Anticipate ambiguity and change
- Reject short-term thinking
- Account for social forces
- Embrace sustainable development principles.

Unlike many subfields in geography, hazards is situated within an interdisciplinary group of hazard/disaster researchers and practitioners. More often than not, the science and application of hazards geography have more in common (both intellectually and pragmatically) with this group than with the geographic discipline itself. This is fostered by the annual workshop sponsored by the Natural Hazards Center which brings together researchers and practitioners to share and exchange ideas. Hazards geographers routinely provide expert testimony and advice to many hazard/disaster-oriented agencies such as FEMA, EPA, DOE, and USAID as well as to state and local governments. The outward-looking view of many hazard geographers—seen in publication content and outlets, student placements, and student topics for theses and dissertations—is reflective of the broader hazards community within which we operate.

Finally, the role of the National Science Foundation in targeting opportunities for hazards research is important in understanding the development and current orientation of the field. Most of the funded hazards-oriented research by geographers was not supported by the Geography and Regional Science Program at NSF, but rather through the Engineering Directorate (via the Hazard Reduction Program within Civil and Mechanical Systems, which is part of NSF's support for the National Earthquake Hazards Reduction Program). The clear orientation toward physical and engineering sciences, a criterion to make it relevant to seismic risks, and an emphasis on human adjustments using a positivist research methodology, have fostered a particular line of sponsored research in the hazards field. What this has meant is that many of the more important hazards questions or topics (e.g. an assessment of the effectiveness of the National Flood Insurance Program, NFIP, in reducing flood losses, equity in the distribution of disaster assistance, factors influencing the social construction of risks) have not been conducted because of lack of funding. Narrowly focused sponsored research may be driving the field more than the pursuit of intellectually important or pragmatic questions, especially during the last two decades.

The Focus of Hazards Research

In his 1989 review for *Geography in America*, Mitchell noted the expansion of the hazards research agenda in geography (Mitchell 1989). Among other things, he traced the evolution of hazards research from early work that focused on rational theories of individual decision-making through analyses of cognitive and behavioral factors influencing decisions, to research examining constraints on responses to hazards. This evolution continued into the 1990s. While the themes have not necessarily undergone significant change, the approaches, methodologies, and constructs have. Thus, researchers seek to solve real-world problems, continuing to build on the work of Gilbert White (White 1945, 1964; Platt 1986) and have made a difference in influencing public policy (Changnon 1996; Interagency Floodplain Management Review Committee 1994). At the same time, there is expanding work on theoretical frameworks (Alexander 1997) and integrating conceptualizations (Palm 1990; Cutter 1993; Tobin and Montz 1997; Hewitt 1997), including research on the impacts of climate change (Changnon 1993; Rosensweig and Parry 1994;

Warrick *et al.* 1993) and human adaptation to global change (Kasperson *et al.* 1990; Stern and Easterling 1999).

Geographers continue to approach hazards research from a variety of perspectives. This diversity is viewed as a traditional strength of hazards research, yet some would argue that it is also its greatest weakness. One indicator of the focus of hazards research in geography is dissertations, which often represent cutting-edge developments in the field. A cursory review of dissertation titles and abstracts, using keywords relating to concepts (such as hazard, risk, and vulnerability), types of hazard (tornado, thunderstorm), or event (Three Mile Island, Hurricane Andrew) found approximately 100 works by geographers during the 1990s, nearly two-thirds more than the number found during the 1980s.

During the 1980s, dissertation research on elusive hazards such as slow-onset events (drought and famine) was minuscule compared to the more familiar rapid-onset hazards and disasters (earthquakes and floods). The popularity of natural hazard topics continued into the 1990s, although dissertations examining technological issues increased from the previous decade. Likewise, studies addressing issues associated with multi-hazards and previously ignored hazards, such as global warming, are increasing in number. Some of this doctoral research represents a continuation of paradigms and trends from the previous decade, but major shifts in the direction and location of the research are apparent. New topics are being explored and previous topics are being reworked from different perspectives. Although still dominant, the proportion of dissertations emphasizing human use systems continued to decline, as has the share coming from a physical geographic perspective. Conversely, studies that develop predictive models of hazards, evaluate the utility of Geographic Information Systems in hazards analysis, or explore methods of mapping hazards became more commonplace during the 1990s. Another major trend in hazards dissertation research was the movement away from studying specific types of hazards toward the study of total risk, environmental risk and equity, generic hazard risk perceptions, or disaster preparedness—all topics that incorporate a variety of hazard threats, natural and technological, physical and social.

In keeping with the interdisciplinary and dynamic nature of the field, changes in centers of hazards research have been significant. The University of Colorado at Boulder and Clark University continued as prominent centers for such study at the beginning of the decade. By the end of the decade, with the directorship of Colorado's Hazards Center shifting from a geographer to a sociologist, it appears that other institutions may claim leadership roles in training hazards geographers (Cross 1998). New centers of graduate study of hazards were established during the 1990s, most notably at the University of South Carolina and Southwest Texas State University. Likewise, Rutgers University and the University of Waterloo, both of which began producing hazards dissertations during the 1980s, continued to be active.

Theorizing Hazards and Risks

The dynamic tension between theory and practice became more apparent during the 1990s. A number of publications lamented the lack of theory underpinning hazards research (Alexander 1991, 1997; Hewitt 1997; Lindell *et al.* 1997, among others), resulting in large part from the historical emphasis in hazards research on solving practical problems. During the last decade, however, there was a flurry of theoretically and conceptually based research.

The traditional natural hazards paradigm views the relationship between people and their environment as a series of adjustments in both the physical and human use systems. In updating their earlier work, Burton *et al.* (1993) explored those theories that explained empirically observed hazard response or addressed distinctive elements in the intersection of society, technology, and the environment, what they term theories of the middle range. They also commented on the movement within the field from a focus on extreme events, in the 1970s, to more complex conditions, in the 1990s, and from the individual to social groups.

The need to understand the physical forces and dynamics of the hazards events themselves has been a fruitful area of research, but one that has not been fully recognized by other subfields. For example, Gares *et al.* (1994: 2) argued that geomorphologists have underestimated their contributions to hazards research because they have worked outside "the broader natural hazards research paradigm." Within the hazards community, however, such work is routinely used and acknowledged. The research by physical geographers continues to inform us about the nature of hazards and events such as work on the hydrology of the 1993 Mississippi floods (Pitlick 1997), severe storm frequencies and intensities (Faiers *et al.* 1994; Engstrom 1994, Paulson 1993; Konrad and Meentemeyer 1994; R. E. Davis and Rogers 1992), and avalanche hazards (Butler and Walsh 1990; Mock and Kay 1992). Some of the textbooks written during the decade used the traditional hazards approach, focusing on various hazards (mostly extreme natural events) including their physical basis and measurement, as well as human perception, response, and adjustment mechanisms (Alexander 1993; Smith 1996).

Acknowledging the critiques of the natural hazards paradigm especially from the political economy and ecology schools and social theorists (Beck 1992, 1999), hazards in context provides an expansion of the human–ecological approach to include the social, political, and economic settings within which the hazard takes place. Mitchell *et al.* (1989) introduced the conceptual model and demonstrated the significance of context in understanding the British response to a severe windstorm. Expanding on the theme, Palm (1990) proposed her integrative framework that examined the relationship between the physical setting, political and economic context, and the role and influence of individuals as intervening factors that both constrain and enable responses to hazards. She then applied this framework using a case study of seismic hazards and public responses to them in California. Other works have addressed cultural context as it influences response (Palm and Carroll 1998), geographic context that serves to define risk (Hewitt 1997), socioeconomic contexts in which responses are made and frequently limited (Blaikie *et al.* 1994; Cross 1994), and institutional contexts that set the framework in which administrative and legal decisions are made (Palm and Hodgson 1992; Penning-Roswell 1996; Platt 1994*a*, *b*; Tobin 1999).

The interaction among nature, society, and technology creates a mosaic of risks and hazards that affect people and the places where they live and work. In 1971, Hewitt and Burton proposed a regional ecology of hazards that spatially delineated the hazards affecting a particular locale. In the 1990s this concept was expanded into the hazards of place (Cutter and Solecki 1989) or hazardscape (also riskscape) (Cutter 1993) which examined the distributive patterns of hazards and underlying processes that gave rise to them. A hazardscape can be viewed as a landscape of many hazards affecting a region (similar to a regional ecology), or the comparisons of one type of hazard across a broader spatial area such as a state or nation. Originally formulated using one type of hazard (airborne toxic releases), the hazards of place conceptualization has since been expanded to include other examples, most notably pesticide drift (Tiefenbacher 1998), toxic releases (Cutter and Solecki 1996; Colten 1990), tornado activity (Schmidlin and Schmidlin 1996), and multiple hazards assessments of specific places (Cross 1992; Ahmad 1992; Palm and Hodgson 1993; Montz and Tobin 1998*a*; Cutter *et al.* 1999).

Working within the risk community, Kasperson *et al.* (1988) suggested that the social construction of risk ultimately places expert judgement at odds with general public perceptions of risk. They proposed the social amplification of risk model that helps explain why minor risks (as defined by technical experts) are amplified and thus generate public concern, while more major risks (according to technical experts again) are attenuated and not seen as problematic by the general public (Kasperson and Kasperson 1996). The social amplification of risks has generated considerable interest and controversy outside the geographical community and has helped stimulate interdisciplinary research in social theories of risk (Krimsky and Golding 1992).

In the latter half of the decade, most conceptually driven research focused on vulnerability (Cutter 1996; Dow and Downing 1995): vulnerability as a biophysical condition or source of exposure (Rosenfeld 1994); vulnerability as socially constructed or social responses (Blaikie *et al.* 1994; Bohle *et al.* 1994; Cannon 1994; Watts and Bohle 1993; Chen 1994; Yarnal 1994); and vulnerability as both a biophysical condition and a social response (Liverman 1990*a*; Dow, 1992, 1999; Cutter *et al.* 2000). Just as we have recognized the complex conditions of which hazards and disasters are an integral part, we also acknowledge that vulnerability is equally multifaceted. Places and people are vulnerable because they are situated in areas that are at particularly high risk to the occurrence of an event (biophysical vulnerability). Places and people are vulnerable because of settlement patterns that have ignored hazards or because wealth and access to resources are unevenly distributed throughout a society (social vulnerability). And, places and people are vulnerable because each of these factors works separately and in various combinations to make locations and groups more or less vulnerable. Regardless of the causal mechanism, vulnerability research continues to dominate the conceptual and theoretical domain of hazards research.

Changing Interpretations of Hazards

The nature of hazard continues to change, as have our interpretations of them. While evident in the 1980s, the recognition and acceptance that hazards and disasters are not just physical events but rather are socially constructed situations have influenced much of theoretical and empirical work. Emphases have shifted toward analyses of vulnerability and hazardousness from many perspectives including feminist theory (Fordham and Ketteridge 1998), environmental equity issues (Heiman 1996; Lansana-Margai 1995; Kasperson and Dow 1991; Cutter 1995*a*, *b*; Pulido *et al.* 1996) and what might be termed the democratization of knowledge about hazards and the politicization of risk (O'Riordan 1998; Cutter 1993; Beck 1999). This research suggests that hazards are now part of everyday life requiring a proactive approach by those at risk. There is a convergence of scientific and less scientific understandings of hazards and risks and public views of risk are acknowledged in societal debates about risks and their management. Along with this is an acceptance of the role of uncertainty in both our understanding of the physical systems and in gauging human responses. Methodologically, the employment of both quantitative and qualitative approaches has helped enrich the subfield.

While the very nature of hazards is changing, much remains to be learned about the hazards with which we have traditionally dealt, such as floods, earthquakes, hurricanes, tornadoes, and the like. Indeed, both the dynamics of the events and our responses to them remain important lines of inquiry as evidenced by the work following the 1993 Mississippi River floods (Changnon 1996; Interagency Floodplain Management Review Committee 1994; Myers and White 1993), the 1997 Red River of the North floods (Todhunter 1998), and the Kobe earthquake (Chang 2000). At the same time, we are beginning to recognize the significance of slow-onset, more pervasive hazards as they contribute to hazardousness and vulnerability. Thus hazard geographers are addressing such larger-scale issues as drought, climate change, and ENSO from physically based perspectives, while others are interested in the challenges these new hazards present in terms of how societies might cope with these complex physical and social conditions (Liverman 1990*b*; Mitchell and Ericksen 1992; Dow 1992). Particularly salient examples of this expansion of hazards research are the works addressing the drought-hunger-food security nexus (Bohle *et al.* 1994; Chen 1994; Downing 1996).

Technological hazards are receiving increased attention by geographers. This trend is attributable to a number of factors including the widespread occurrence of such hazards, the increased attention given to them in public policy, and their close association with the complex conditions noted above. Some of this work focuses on specific types of hazards (Colten 1991; Tiefenbacher 1998; Cutter and Ji 1997, Flynn *et al.* 1997), events in particular places such as urban areas (Cutter and Tiefenbacher 1991) or specific regions (Cutter and Solecki 1996), long-term recovery from technological disasters (Mitchell 1996), the relationship between natural and technological hazards (Showalter and Myers 1994), and the integration of natural and technological hazards in regions (Hewitt 1997; Kasperson *et al.* 1995).

The relationship between hazards and environmental quality has long been recognized, but it increased in interest with specific events of the 1990s, such as flood waters coming in contact with sewage and agricultural chemicals in the 1993 Mississippi River floods. This has led to a marriage of sorts between hazards research and environmental impact assessment (Montz and Tobin 1998*b*; Dixon and Montz 1996; Montz and Tobin 1997) and between natural and technological hazards as they interact in policy and legal contexts (Gruntfest and Pollack 1994).

Disasters and Development

The decade has seen increasing interest in the relationship between disasters and development issues. As countries and regions strive to develop economically, the influence of disastrous events can only be negative, as evidenced by the 5 per cent reduction in GNP some countries have experienced during disaster years (Burton *et al.* 1993). With an estimated annual average cost to the global economy of more than $50 billion (Glickman *et al.* 1992), natural hazards put tremendous pressure on the resources of affected nations. An important part of the relationship between disasters and development centers on variations in vulnerability because the negative impacts of events are not shared equally among or between populations. Geographers and others have recognized that vulnerability varies with differences in wealth, power, and control over resources (Cannon 1994; Hewitt 1995, 1997; Rosenfeld 1994). This focus on vulnerability is not new, as Hewitt (1983) illustrates. However, it has been more closely associated with sustainability and development issues in recent work, as the perpetual disaster-damage-repair-disaster cycle diverts resources away from development.

Some have argued that the focus of hazards research must be on vulnerability of populations because, without such a focus, progress, as measured through reduced losses and disruption, cannot be made. For instance, Blaikie *et al.* (1994) argue that mitigation directed to specific sectors of society (i.e. marginalized populations who are most vulnerable) will protect the greatest number of people and thus should be emphasized. They and others (Hewitt 1997; Varley 1994; Wisner 1998) suggest that only through modifying social structures can vulnerability be appropriately and successfully addressed.

Forces that fuel population growth in urban areas, generated both within and outside the cities, interact with other socioeconomic forces that marginalize some groups and relegate them to unsafe structures in very hazardous areas. The increasing vulnerability and exposure to hazards in mega-cities has resulted in research that has dealt with the dynamic nature of cities and the resulting disaster vulnerability (Mitchell 1995, 1998, 1999; Uitto 1998), the social geography of cities and its relationship to vulnerability (Takahashi 1998), and how risk varies over time in mega-cities (Kakhandiki and Shah 1998). As cities grow in size, so too will hazardousness and vulnerability grow. The underlying conditions that promote growth of urban areas will, at the same time, serve to increase vulnerability, and this increase in vulnerability will not be equally spread throughout the population. As a result, issues of environmental equity come to the fore, though their salience varies from place to place. Different groups face dissimilar levels of vulnerability, and these differences have significant implications for economic development.

Technological Innovations

In addition to some of the theoretical and conceptual developments, technological advancements have also stimulated hazards research in the 1990s. The use of GIS and other computer-assisted decision support systems has proliferated in both the hazards and risk areas (Beroggi and Wallace 1995; Carrara and Guzzetti 1996; Gattrell and Vincent 1991). For example, the use of remote sensing in assessing fire hazards (Helfert and Lulla 1990), calculating the flood inundation area after the 1993 Mississippi floods (Lougeay *et al.* 1994), and in sensitivity mapping for oil spills (Jensen *et al.* 1990, 1993) has demonstrated their utility to the field. Similarly, GIS has contributed to the evaluation of chemical hazards (Sorenson *et al.* 1992), toxic releases (Chakraborty and Armstrong 1997; McMaster *et al.* 1997; Scott and Cutter 1997), modeling airborne exposure pathways (Hepner and Finco 1995; Scott 1999), earthquake hazards (Hodgson and Palm 1992; Palm and Hodgson 1992), and comprehensive analyses of hazardousness and vulnerability (Cutter *et al.* 1999; Montz 1994). Hazard mapping, which has a long tradition in the field, has benefited most from the improvements in technology (Monmonier 1997).

The rapid and easy dissemination of real-time hazard and disaster information via the World Wide Web and the analytical and cartographic capabilities of GIS exemplify several of the technological changes that have revolutionized hazards research and mitigation efforts in the 1990s. The Doppler radar images that we see displayed on our computer monitors obtained through the World Wide Web attest to these technological changes in weather monitoring and forecasting. Similarly, improved satellite-based remote sensing provides enhanced images of developing hurricanes, detection of thermal anomalies related to volcanic eruptions, sea surface temperatures, and wildfires. These are just a few ways in which technology has dramatically altered our ability to detect threatening events and provide warnings. Indeed, a proliferation of websites, offering a wide diversity of real-time hazard and disaster event information, has greatly facilitated the study of hazards at all levels (Cross 1997; Gruntfest and Weber 1998). The use of the World Wide Web has transformed research and teaching in hazards as well as emergency management. Similarly, the use of these technologies for detection, warning, and mitigation has led to evaluations of effectiveness (Gruntfest and Waterincks 1998; Gruntfest and Carsell 2000), further illustrating the applied nature of much hazards work.

Hazards Education

The 1990s saw tremendous growth in hazards education at multiple levels as evidenced by the proportion of North American doctoral dissertations in geography centering on hazards and the number of courses devoted to hazards in North American colleges and universities (Cross 2000). Instruction has been aided by a wide choice of new or revised textbooks written by geographers (Alexander 1993, 2000; Blaikie *et al.* 1994; Bryant 1991; Burton *et al.* 1993; Cutter 1993, 1994; Ebert 1997; Hewitt 1997; Smith 1996; Tobin and Montz 1997), plus several prepared by geologists. Although the majority of first-year college courses dealing with hazards concentrates on the physical dimensions of these events, social response to hazards and mitigation is prominently featured in most upper-level undergraduate hazards courses. In addition, many introductory geography textbooks published during the 1990s, whether physical, regional, or social, included discussions about hazards.

Pedagogical approaches to hazards education at both the college and pre-collegiate levels have also been studied by Montz *et al.* (1989), Alexander (1991, 1997), and Lidstone (1996), among others. Lidstone (1996: 8) promotes, "geographical education in our schools [to] provide the medium for creating a citizenship able to come to terms with living in dynamic social and physical environments." Nevertheless, Alexander (1997: 298) warns about a "general lack of holistic analyses that treat hazard, risk and disaster as integrated phenomena."

A variety of publications by geographers during the decade conveyed hazards information to the general public, including DeBlij *et al.* (1994), Monnomier (1997), and M. Davis (1998). Articles highlighting hazards issues appeared in both the *National Geographic* and the *Canadian Geographic*, which included detailed map inserts showing spatial patterns of hazard risks in North America and Canada, respectively (Parfit 1998; Lanken 1996). The US Geological Survey published a variety of hazard overview maps, including *This Dynamic Planet*, which provided a global view of seismic and volcanic hazards.

Regional overviews of hazards were also published overseas, including the highly detailed *Atlas of Natural Disasters in China* (Suishan and Shengchao 1992). Other detailed English-language profiles of hazards appeared, including Europe (Embleton and Embleton-Hamann 1997) South Asia (Haque 1997), Africa (Freeth *et al.* 1992), the West Indies (Ahmad 1992), and high mountain areas (Kalvoda and Rosenfeld 1998). The International Decade for Natural Disaster Reduction highlighted hazards education and fostered the attention given to regional surveys of hazards, though the long-term impacts of this remain unknown.

Disasters in the Twenty-First Century: Commonplace or Extreme Events?

The focus of hazards research in geography evolved significantly, in part building on initiatives from earlier research and in part reflecting the shifts in hazard types and complexities that came to be recognized during the decade. Much of the recent hazards work is hybrid in nature, usually linking hazards with some other field such as global change or GIS. Research continues to analyze the physical and human use systems in the face of extreme events as we seek to understand better each component as well as their interactions. Similarly, hazard researchers address real-world problems requiring practical solutions, but continue working to develop conceptual frameworks that provide foundations for evaluating hazards as they change over time and space. Thus, the traditional foci of hazards geography remain central to the subfield, but the need to adapt our research is clear. For example, hazards researchers are examining the impacts of climate change, the human dimensions of global change, hazards associated with exposure to environmental pollution, and the relationship between hazards and sustainability. We have moved beyond the notion that hazards are only extreme events, and now note that everyday risks are just as important. Yet, for all our research accomplishments during this International Decade, we must admit that property losses and the human costs of disasters are still rising, which leaves us with much more work to do.

The events of September 11, 2001 and instances of bioterrorism in different locations have had important impacts on hazards research. While terrorism was a topic recognized by hazards geographers (Mitchell 1979; Hewitt 1987, 1997), its importance will increase. Immediately in the aftermath of September 11, hazards researchers, with geographers among them, brought their expertise to bear on both quick response research and longer-term efforts. These events will also likely influence thinking about and definitions of such concepts as risk, vulnerability, preparedness, and warning. The events are different in nature than those hazards researchers have traditionally addressed. However, the theories, conceptual frameworks, and methodologies that guide hazards research have direct application here.

The interdisciplinary nature of hazards research in geography and the cross-disciplinary collaborations that were prominent in the past will serve us well in the future as we cope with the ever-increasing complexity in our social, physical, and technological systems. Hazards geographers have a well-established track record of working on applied problems with research that makes a material difference in improving society. It is difficult, however, to relate such successes as the implementation of mitigation measures that saved lives or reduced property damage directly to geographic research. Still, the evolution of hazards research during the 1990s toward emerging hazards and different views of hazardousness is illustrative of the dynamic nature of the field and of its concern with addressing real issues. The release, while this chapter was in final editing, of Kasperson and Kasperson's (2001) comprehensive overview of global environmental risk and Cutter's (2001) assessment of the regional variability in hazard events and losses in the United States, is indicative of how interrelated the study of hazards, risk, vulnerability, and global change have become. Our interpretations of hazards are changing, and the hazards themselves are changing. Thus, the need to search for practical answers becomes more important. Perhaps the same factors that promote applied hazards research will also require changes in the way we conceptualize hazards, vulnerable populations, and societal forces and their interactions. Our knowledge and practice can help us cope with existing hazards and better respond to unprecedented environmental surprises in the future. This is the challenge facing hazards geographers in the twenty-first century.

References

Ahmad, R. (1992). "Natural Hazards in the Caribbean." *Journal of the Geological Society of Jamaica*, Special Issue, 12.

Alexander, D. (1991). "Natural Disasters: A Framework for Research and Teaching." *Disasters*, 15: 209–26.

— (1993). *Natural Disasters*. New York: Chapman & Hall.

— (1997). "The Study of Natural Disasters, 1977–1997: Some Reflections on a Changing Field of Knowledge." *Disasters*, 21: 284–304.

— (2000). *Confronting Catastrophe: New Perspectives on Natural Disasters*. Oxford: Oxford University Press.

Barry, J. M. (1997). *Rising Tide: The Great Mississippi Flood of 1927 and How it Changed America*. New York: Simon & Schuster.

Beck, U. (1992). *Risk Society: Towards a New Modernity*. Los Angeles: Sage Publications.

— (1999). *World Risk Society*. New York: Blackwell.

Beroggi, G. E. G., and Wallace, W. A. (eds.) (1995). *Computer Supported Risk Management*. Dordrecht: Kluwer.

Blaikie, P., Cannon, T., Davis, I., and Wisner, B. (1994). *At Risk: Natural Hazards, People's Vulnerability, and Disasters*. London: Routledge.

Bohle, H. G., Downing, T. E., and Watts, M. J. (1994). "Climatic Change and Social Vulnerability: Toward a Sociology and Geography of Food Insecurity." *Global Environmental Change*, 4: 37–48.

BOND (Board on Natural Disasters), US National Research Council (1999). "Mitigation Emerges as Major Strategy for Reducing Losses Caused by Natural Disasters." *Science*, 284: 1943–7.

Bryant, E. (1991). *Natural Hazards*. Cambridge: Cambridge University Press.

Burton, I., Kates, R. W., and White, G. F. (1993). *The Environment as Hazard*. 2nd edn. New York: Guilford Press.

Butler, D. R., and Walsh, S. J. (1990). "Lithologic, Structural, and Topographic Influences on Snow Avalanche Path Location, Eastern Glacier National Park, Montana" *Annals of the Association of American Geographers*, 80/3: 362–78.

Cannon, T. (1994). "Vulnerability Analysis and Explanation of 'Natural' Disasters," in Varley (1994: 13–30).

Carrara, A., and Guzzetti, F. (eds.) (1995). *Geographical Information Systems in Assessing Natural Hazards*. Dordrecht: Kluwer Academic.

Chakraborty, J., and Armstrong, M. P. (1997). "Exploring the Use of Buffer Analysis for the Identification of Impacted Areas in Environmental Equity Assessment." *Cartography and Geographic Information Systems*, 24/3: 145–57.

Chang, S. E. (2000). "Disasters and Transport Systems: Loss, Recovery, and Competition at the Port of Kobe after the 1995 Earthquake." *Journal of Transport Geography*, 8/1: 53–65.

Changnon, S. A. (1993). "Changes in Climate and Levels of Lake Michigan: Shoreline Impacts at Chicago." *Climate Change*, 23: 213–30.

— (ed.) (1996). *The Great Flood of 1993: Causes, Impacts, and Responses*. Boulder: Westview Press.

Chen, R. S. (1994). "The Human Dimension of Vulnerability," in R. Socolow, C. Andrews, F. Berkhout, and V. Thomas (eds.), *Industrial Ecology and Global Change*. Cambridge: Cambridge University Press, 85–105.

Colten, C. E. (1990). "Historical Hazards: The Geography of Relict Industrial Wastes." *The Professional Geographer*, 42/2: 143–56.

— (1991). "A Historical Perspective on Industrial Wastes and Groundwater Contamination." *The Geographical Review*, 81/2: 215–28.

Cross, J. A. (1992). "Natural Hazards within the West Indies." *Journal of Geography*, 91/5:190–9.

— (1994). "Agroclimate Hazards and Dairy Farming in Wisconsin." *The Geographical Review*, 84: 277–89.

— (1997). "Natural Hazards and Disaster Information on the Internet." *Journal of Geography*, 96: 307–14.

— (1998). "A Half Century of Hazards Dissertation Research in Geography." *International Journal of Mass Emergencies and Disasters*, 16: 199–212.

— (2000). "Hazards Courses in North American Geography Programs." *Environmental Hazards*, 2/2: 77–86.

Cutter, S. L. (1993). *Living with Risk: The Geography of Technological Hazards*. New York: Edward Arnold.

— (1995*a*). "Race, Class and Environmental Justice." *Progress in Human Geography*, 19/1: 107–18.

— (1995*b*). "The Forgotten Casualties: Women, Children, and Environmental Change." *Global Environmental Change*, 5/3: 181–94.

— (1996). "Vulnerability to Environmental Hazards." *Progress in Human Geography*, 20/4: 529–39.

— (ed.) (2001). *American Hazardscapes: The Regionalization of Hazards and Disasters*. Washington DC: Joseph Henry Press.

Cutter, S. L., and Ji, M. (1997). "Trends in US Hazardous Materials Transportation Spills." *The Professional Geographer*, 49/3: 318–31.

Cutter, S. L., and Solecki, W. D. (1989). "The National Pattern of Airborne Toxic Releases." *Professional Geographer*, 41/2: 149–61.

Cutter, S. L., and Solecki, W. D. (1996). "Setting Environmental Justice in Space and Place: Acute and Chronic Airborne Toxic Releases in the Southeastern United States." *Urban Geography*, 17/5: 380–99.

Cutter, S. L., and J. Tiefenbacher (1991). "Chemical Hazards in Urban America." *Urban Geography*, 12/5: 417–30.

Cutter, S. L., J. T. Mitchell, and M. S. Scott. 2000. "Revealing the Vulnerability of People and Places: A Case Study of Georgetown County, South Carolina." *Annals of the Association of American Geographers*, 90/4: 713–37.

Cutter, S. L., Thomas, D. S. K., Cutler, M. E., Mitchell, J. T., and Scott, M. S. (1999). *South Carolina Atlas of Environmental Risks and Hazards*. Columbia: University of South Carolina Press.

Davis, M. (1998). *Ecology of Fear: Los Angeles and the Imagination of Disaster*. New York: Metropolitan Books.

Davis, R. E., and Rogers, R. F. (1992). "A Synoptic Climatology of Severe Storms in Virginia." *The Professional Geographer*, 44/3: 319–32.

DeBlij, H. *et al.* (1994). *Nature on the Rampage*. Washington: Smithsonian Books.

Dixon, J. E., and Montz, B. E. (1996). "Strategic Environmental Assessment in New Zealand: Lessons from Natural Hazards," in M. DoRosario Partidario (ed.), *Improving Environmental Assessment Effectiveness: Research, Practice, and Training*. Fargo, N. Dak.: International Association of Impact Assessment, 79–84.

Dow, K. (1992). "Exploring Differences in Our Common Future(s): The Meaning of Vulnerability to Global Environmental Change." *Geoforum*, 23/3: 417–36.

—— (1999). "The Extraordinary and the Everyday in Explanations of Vulnerability to an Oil Spill." *The Geographical Review*, 89/1: 74–93.

Dow, K., and Downing, T. E. (1995). "Vulnerability Research: Where Things Stand." *Human Dimensions Quarterly*, 1: 3–5.

Downing, T. E. (ed.) (1996). *Climate Change and World Food Security*. Berlin: Springer-Verlag.

Ebert, C. H. V. (1997). *Disasters: Violence of Nature and Threats by Man*. 3rd edn. Dubuque: Kendall Hunt.

Embleton, C., and C. Embleton-Hamann (eds.) (1997). *Geomorphological Hazards of Europe*. Amsterdam: Elsevier.

Engstrom, W. N. (1994). "Nineteenth-Century Coastal Gales of Southern California." *The Geographical Review*, 84/3: 306–15.

Faiers, G. E., Keim, B. D., and Hirschboeck, K. K. (1994). "A Synoptic Evaluation of Frequencies and Intensities of Extreme Three- and 24-hour Rainfall in Louisiana." *The Professional Geographer*, 46/2: 156–63.

Flynn, J., Kasperson, R. E., Kunreuther, H., and Slovic, P. (1997). "Overcoming Tunnel Vision: Redirecting the US High-Level Nuclear Waste Program." *Environment*, 39/3: 6–11, 25–30.

Fordham, M., and Ketteridge, A. M. (1998). "Men Must Work and Women Must Weep: Examining Gender Stereotypes in Disasters," in E. Enarson and B. H. Morrow (eds.), *The Gendered Terrain of Disaster: Through Women's Eyes*. Westport, Conn.: Praeger, 81–94.

Freeth, S. J., Ofoegbu, C.O., and Mosto Onuoha, K. (eds.) (1992). *Natural Hazards in West and Central Africa*. Braunschweig: Vieweg.

Gares, P.A., Sherman, D. J., and Nordstrom, K. F. (1994). "Geomorphology and Natural Hazards." *Geomorphology*, 10: 1–18.

Gattrell, A. C., and Vincent, P. (1991). "Managing Natural and Technological Hazards," in I. Maser and M. Blakemore (eds.), *Handling Geographical Information: Methodology and Potential Applications*. New York: Longman Scientific, 148–80.

Glickman, T. S., Golding, D., and Silverman, E. D. (1992). *Acts of God and Acts of Man: Recent Trends in Natural Disasters and Major Industrial Accidents*. Center for Risk Management, Discussion Paper 92-02, Washington, DC: Resources for the Future.

Gruntfest, E., and Carsell, K. (2000). *The Warning Process: Toward an Understanding of False Alarms*. Washington, DC: United States Bureau of Reclamation.

Gruntfest, E., and Pollack, D. (1994). "Warnings, Mitigation, and Litigation: Lessons for Research from the 1993 Floods." *Water Resources Update*, 95: 40–4.

Gruntfest, E., and Waterincks, P. (1998). *Beyond Flood Detection: Alternative Applications of Real-Time Data*. Washington, DC: United States Bureau of Reclamation.

Gruntfest, E., and Weber, M. (1998). "Internet and Emergency Management: Prospects for the Future." *International Journal of Mass Emergencies and Disasters*, 16: 55–72.

Haque, C. E. (1997). *Hazards in a Fickle Environment: Bangladesh*. Dordrecht: Kluwer Academic.

Heiman, M. K. (1996). "Waste Management and Risk Assessment: Environmental Discrimination through Regulation." *Urban Geography*, 17: 400–18.

Helfert, M. R., and Lulla, K. P. (1990). "Mapping Continental-Scale Biomass Burning and Smoke Pails over the Amazon Basin as Observed from the Space Shuttle." *Photogrammetric Engineering and Remote Sensing*, 56: 1367–73.

Hepner, G. F., and Finco, M. V. (1995). "Modeling Dense Gaseous Contaminant Pathways over Complex Terrain Using a Geographic Information System." *Journal of Hazardous Materials*, 42: 187–99.

Hewitt, K. (ed.) (1983). *Interpretations of Calamity*. Winchester, Mass.: Allen & Unwin.

—— (1987). "The Social Space of Terror: Towards a Civil Interpretation of Total War." *Society and Space, Environment and Planning*, D5: 445–74.

—— (1995). "Excluded Perspectives in the Social Construction of Disaster." *International Journal of Mass Emergencies and Disasters*, 13: 317–39.

—— (1997). *Regions of Risk: A Geographical Introduction to Disasters*. Harlow: Longman.

Hewitt, K., and Burton, I. (1971). *The Hazardousness of Place: A Regional Ecology of Damaging Events*. Research Publication, 6. Toronto: University of Toronto, Department of Geography.

Hodgson, M. E., and Palm, R. (1992). "Attitude and Response to Earthquake Hazards: A GIS Design for Analyzing Risk Assessment." *Geo Info Systems*, 2/7: 40–51.

Houghton, J. T., Jenkins, G. J., and Ephraums, J. J. (eds.) (1990). *Climate Change: The IPCC Scientific Assessment*. Cambridge: Cambridge University Press.

Houghton, J. T. *et al.* (eds.) (1996). *Climate Change 1995: The Science of Climate Change: Contribution of Working Group I to the Second Assessment Report of the Intergovernmental Panel on Climate Change*. Cambridge: Cambridge University Press.

Housner, G. W. (1989). "An International Decade of Natural Disaster Reduction: 1990–2000." *Natural Hazards*, 2/1: 45–75.

Interagency Floodplain Management Review Committee (1994). *Sharing the Challenge: Floodplain Management into the 21st Century (Galloway Report)*. Washington DC: US Government Printing Office.

Jensen, J. R., Ramsay, E. W., Holmes, J. M., Michel, J. E., Savitsky, B., and Davis, B. A. (1990). "Environmental Sensitivity Index (ESI) Mapping for Oil Spills Using Remote Sensing and Geographic Information System Technology." *International Journal of Geographical Information Systems*, 4/2: 181–201.

Jensen, J. R., Narumalani, S., Weatherbee, O., Murday, M., Sexton, W. J., and Green, C. J. (1993). "Coastal Environmental Sensitivity Mapping for Oil Spills in the United Arab Emirates Using Remote Sensing and GIS Technology." *Geocarto International*, 8/2: 5–13.

Junger, S. (1997). *The Perfect Storm*. New York: W. W. Norton.

Kakhandiki, A., and Shah, H. (1998). "Understanding Time Variation of Risk—Crucial Implications for Megacities Worldwide." *Applied Geography*, 18/1: 47–53.

Kalvoda, J., and Rosenfeld, C. L. (eds.) (1998). *Geomorphological Hazards in High Mountain Areas*. Dordrecht: Kluwer Academic.

Kasperson, J. X., and Kasperson, R. E. (eds.) (2001). *Global Environmental Risk*. Tokyo: United Nations University Press.

Kasperson, J. X., Kasperson, R. E., and Turner, B. L., II (1995). *Regions at Risk: Comparisons of Threatened Environments*. Tokyo: United Nations University Press.

Kasperson, R. E., and Dow, K. (1991). "Developmental and Geographical Equity in Global Environmental Change: A Framework for Analysis." *Evaluation Review*, 15: 149–61.

Kasperson, R. E., and Kasperson, J. X. (1996). "The Social Amplification and Attenuation of Risk." *Annals of the American Academy of Political and Social Science*, 545 (May): 95–105.

Kasperson, R. E., Dow, K., Golding, D., and Kasperson, J. X. (eds.) (1990). *Understanding Global Environmental Change: The Contributions of Risk Analysis and Management*. Worcester, Mass.: Clark University Press.

Kasperson, R. E., Renn, O., Slovic, P., Brown, H. S., Emel, J., Goble, R., Kasperson, J. X., and Ratick, S. (1988). "The Social Amplification of Risk: A Conceptual Framework." *Risk Analysis*, 8/2: 177–87.

Konrad, C.F., II, and Meentemeyer, V. (1994). "Lower Tropospheric Warm Air Advection Patterns Associated with Heavy Rainfall over the Appalachian Region." *The Professional Geographer*, 46/2:143–55.

Krimsky. S., and Golding, D. (eds.) (1992). *Social Theories of Risk*. Westport, Conn.: Praeger.

Lanken, D. (1996). "Funnel Fury." *Canadian Geographic*, 116 4 (July/August): 24–31. ("Natural Hazard" poster map supplement.)

Lansana-Margai, F. (1995). "Evaluating the Potential for Environmental Quality Improvement in a Community Distressed by Manmade Hazards." *Journal of Environmental Management*, 44: 181–90.

Larson, E. (1999). *Isaac's Storm: A Man, A Time, and the Deadliest Hurricane in History*. New York: Crown Publishers.

Lidstone, J. (1996). "Disaster Education: Where We Are and Where We Should Be," in Lidstone, J. (ed.), *International Perspectives on Teaching about Hazards and Disasters*. Clevedon: Channel View Publications, 7–18.

Lindell, M. K. *et al.* (1997). "Adoption and Implementation of Hazard Adjustments." *International Journal of Mass Emergencies and Disasters*, 15/3: 323–453.

Liverman, D. M. (1990*a*). "Vulnerability, Resilience, and the Collapse of Society," in R. E. Kasperson, K. Dow, D. Golding, and J. X. Kasperson (eds.), *Understanding Global Environmental Change: The Contributions of Risk Analysis and Management*. Worcester, Mass.: Clark University, 27–44.

—— (1990*b*). "Drought Impacts in Mexico: Climate, Agriculture, Technology, and Land Tenure in Sonora and Puebla." *Annals of the Association of American Geographers*, 80/1: 49–72.

Lougeay, R., Baumann, P., and Nellis, M. D. (1994). "Two Digital Approaches for Calculating the Area of Regions Affected by the Great American Flood of 1993." *Geocarto International*, 9/4: 53–9.

McMaster, R. B., Leitner, H., and Sheppard, E. (1997). "GIS-Based Environmental Equity and Risk Assessment: Methodological Problems and Prospects." *Cartography and Geographic Information Systems*, 24/3: 172–89.

Mitchell, J. K. (1979). "Social Violence in Northern Ireland," *Geographical Review*, 69/2: 179–201.

—— (1988). "Confronting Natural Disasters: An International Decade for Natural Hazard Reduction." *Environment*, 30/2: 25–9.

—— (1989). "Hazards Research," in G. L. Gaile and C. J. Wilmott (eds.), *Geography in America*. Columbus: Merrill, 410–24.

—— (1995). "Coping with Natural Hazards and Disasters in Megacities: Perspectives on the Twenty-First Century." *GeoJournal*, 37/3: 303–11.

—— (ed.) (1996). *The Long Road to Recovery: Community Responses to Industrial Disaster*. Tokyo: United Nations University.

—— (1998). "Hazards in Changing Cities: Introduction." *Applied Geography*, 18/1: 1–6.

—— (ed.) (1999). *Crucibles of Hazard: Mega-cities and Disasters in Transition*. Tokyo: United Nations University.

Mitchell, J. K., and Ericksen, N. J. (1992). "Effects of Climate Change on Weather-Related Disasters," in Irving M. Mintzer (ed.), *Confronting Climate Change: Risks, Implications and Responses*. Cambridge: Cambridge University Press, 141–51.

Mitchell, J. K., Devine, N., and Jagger, K. (1989). "A Contextual Model of Natural Hazard." *The Geographical Review*, 79: 391–409.

Mileti, D. S. (1999). *Disasters by Design: A Reassessment of Natural Hazards in the United States*. Washington DC: Joseph Henry Press.

Mock, C. J., and Kay, P. A. (1992). "Avalanche Climatology of the Western United States, with an Emphasis on Alta, Utah." *The Professional Geographer*, 44/3: 307–18.

Monmonier, M. (1997). *Cartographies of Danger: Mapping Hazards in America*. Chicago: University of Chicago Press.

Montz, B. E. (1994). "Methodologies for Analysis of Multiple Hazard Probabilities: An Application in Rotorua, New Zealand." Unpublished research report, Center for Environmental and Resource Studies, University of Waikato, Hamilton, New Zealand.

Montz, B. E., and Tobin, G. A. (1997). "The Environmental Impacts of Flooding in St. Maries, Idaho." Quick Response Report, 93. Boulder, Colo.: Natural Hazards Research and Applications Information Center.

Montz, B. E., and Tobin, G. A. (1998*a*). "Hazardousness of Location: Critical Facilities in the Tampa Bay Region." *Papers and Proceedings of the Applied Geography Conferences*, 21: 315–22.

—— (1998*b*). "Flooding: A Strategic Assessment of Environmental Impacts." *Applied Geographic Studies*, 2: 43–57.

Montz, Burrell E., Tobin, G. A., and Medford, O. E. (1989). "The Urban Floodplain as a Focus for Geographic Education." *Journal of Geography*, 88: 214–19.

Myers, M. F., and White, G. F. (1993). "The Challenge of the Mississippi Flood." *Environment*, 35: 6–9, 25–35.

National Climatic Data Center (1999). "Billion Dollar U.S. Weather Disasters 1980–1999." ‹http://www.ncdc.noaa.gov/oa/reports/billionz.html›, last accessed 3 February 2003.

O'Riordan, T. (1998). "Environmental Management and Governance: Intergovernmental Approaches to Hazards and Sustainability." *Disasters*, 22: 176–84.

Palm, R. (1990). *Natural Hazards: An Integrative Framework for Research and Planning*. Baltimore: Johns Hopkins University Press.

Palm, R., and Carroll, J. (1998). *Illusion of Safety: Culture and Earthquake Response in California and Japan*. Boulder: Westview Press.

Palm, R., and Hodgson, M. E. (1992). "Earthquake Insurance: Mandated Disclosure and Homeowners Response in California." *Annals of the Association of American Geographers*, 82/2: 207–22.

—— (1993). *Natural Hazards in Puerto Rico: Experience, Attitudes and Behavior of Homeowners*. Boulder, Colo.: Institute of Behavioral Science, University of Colorado.

Parfit, M. (1998). "Living with Natural Hazards." *National Geographic*, 1941 (July): 1–39. ("Natural Hazards of North America" double map supplement.)

Paulson, D. D. (1993). "Hurricane Hazard in Western Samoa." *The Geographical Review*, 83/1: 43–53.

Penning-Roswell, E. C. (1996). "Flood-Hazard Response in Argentina." *The Geographical Review*, 86: 72–90.

Pitlick, J. (1997). "A Regional Perspective of the Hydrology of the 1993 Mississippi River Basin Floods." *Annals of the Association of American Geographers*, 87: 135–51.

Platt, R. H. (1986). "Floods and Man: A Geographer's Agenda," in R. W. Kates and I. Burton (eds.), *Geography, Resources and Environment, ii. Themes from the Work of Gilbert White*. Chicago: University of Chicago Press, 28–68.

—— (1994*a*). "Evolution of Coastal Hazard Policies in the United States." *Coastal Management*, 22: 265–84.

—— (1994*b*). "Parsing Dolan." *Environment*, 36: 4–5, 43.

Pulido, L., Sidawi, S., and Vos, R. O. (1996). "An Archeology of Environmental Racism in Los Angeles." *Urban Geography*, 17: 419–39.

Rosenfeld, C. L. (1994). "The Geographical Dimensions of Natural Disaster." *International Geographical Union Bulletin*, 44: 5–11.

Rosensweig, C., and Parry, M. L. (1994). "Potential Impact of Climate Change on World Food Supply." *Nature*, 367: 133–8.

Schmidlin, T. W., and Schmidlin, J. A. (1996). *Thunder in the Heartland: A Chronicle of Outstanding Weather Events in Ohio*. Kent, Ohio: Kent State University Press.

Scott, M. S. 1999. "The Exploration of an Air Pollution Hazard Scenario using Dispersion Modeling and a Volumetric Geographic Information System." Ph.D. Dissertation, University of South Carolina. Ann Arbor, Mich.: Dissertation microfiche.

Scott, M. S., and Cutter, S. L. (1997). "Using Relative Risk Indicators to Disclose Toxic Hazard Information to Communities." *Cartography and Geographic Information Systems*, 24/3: 158–71.

Showalter, P., and Myers, M. F. (1994). "Natural Disasters in the United States as Release Agents of Oils, Chemical, or Radiological Materials between 1980–1989: Analysis and Recommendations." *Risk Analysis*, 14: 169–82.

Smith, K. (1996). *Environmental Hazards: Assessing Risk and Reducing Disaster*. 2nd edn. London: Routledge.

Sorenson, J. H., Carnes, S. A., and Rogers, G. O. (1992). "An Approach for Deriving Emergency Planning Zones for Chemical Munitions Emergencies." *Journal of Hazardous Materials*, 30: 223–42.

Stern, P., and Easterling, W. E. (eds.) (1999). *Making Climate Forecasts Matter, Panel on the Human Dimensions of Seasonal-to-Interannual Climate Variability*. Washington, DC: National Academy Press.

Suishan, Y., and Shengchao, P. (eds.) (1992). *Atlas of Natural Disasters in China*. Beijing: Science Press.

Takahashi, S. (1998). "Social Geography and Disaster Vulnerability in Tokyo." *Applied Geography*, 18/1: 17–24.

Tiefenbacher, J. (1998). "Mapping the Pesticide Driftscape: Theoretical Patterns of the Drift Hazard." *Geographical and Environmental Modeling*, 27: 75–93.

Tobin, G. A. (1999). "Sustainability and Community Resilience: The Holy Grail of Hazards Planning?" *Environmental Hazards*, 1/1: 13–25.

Tobin, G. A., and Montz, B. E. (1997). *Natural Hazards: Explanation and Integration*. New York: Guilford.

Todhunter, P. E. (1998). "Flood Hazard in the Red River Valley: A Case Study of the Grand Forks Flood of 1997." *North Dakota Quarterly*, 65: 254–75.

Uitto, J. I. (1998). "The Geography of Disaster Vulnerability in Megacities: A Theoretical Framework." *Applied Geography*, 18/1: 7–16.

United Nations Office for the Coordination of Humanitarian Affairs (1998). "Central America Hurricane Mitch Fact Sheet, 33." ‹http://wwwnotes.reliefweb.int›, last accessed 19 November 2002.

US National Committee for the Decade of Natural Disaster Reduction (1991). *A Safer Future: Reducing the Impacts of Natural Disasters*. Washington, DC: National Academy Press.

—— (1994). *Facing the Challenge: The US National Report to the IDNDR World Conference on Natural Disaster Reduction, Yokohama, Japan*. Washington, DC: National Academy Press.

Varley, A. (ed.) (1994). *Disasters, Development, and Environment*. Chichester: John Wiley.

Warrick, R. A., Barrow, E. M., and Wigley, T. M. L. (eds.) (1993). *Climate and Sea-level Change: Observations, Projections, and Implications*. Cambridge: Cambridge University Press.

Watson, R. T., Zinyowera, M. C., and Moss, R. H. (1995). *Climate Change 1995—Impacts, Adaptations and Mitigation of Climate Change: Scientific-Technical Analyses: Contribution of Working Group 11 to the Second Assessment Report of the Intergovernmental Panel on Climate Change*. Cambridge: Cambridge University Press.

Watts, M. J., and Bohle, H. G. (1993). "The Space of Vulnerability: The Causal Structure of Hunger and Famine." *Progress in Human Geography*, 17: 43–67.

White, G. F. (1945). *Human Adjustments to Floods*. Department of Geography Research Paper, 29. Chicago: University of Chicago Press.
—— (1964). *Choice of Adjustments to Floods*. Department of Geography Research Paper, 93. Chicago: University of Chicago Press.
—— (1996). "Emerging Issues in Global Environmental Policy." *Ambio*, 25/1: 58–60.
White, G. F., and Eugene Haas, J. (1975). *Assessment of Research on Natural Hazards*. Cambridge: MIT Press.
Wisner, B. (1998). "Marginality and Vulnerability: Why the Homeless of Tokyo Don't 'Count' in Disaster Preparations." *Applied Geography*, 18/1: 25–33.
Yarnal, B. (1994). "Socioeconomic Restructuring and Vulnerability to Environmental Hazards in Bulgaria." *Disasters*, 18: 95–106.

CHAPTER 30

Medical Geography

Wil Gesler

Medical geographers employ geographical concepts and techniques to study issues related to disease and health. In its early stages of development as a distinct geographic subdiscipline, from the 1950s and into the 1980s, medical geography focused on disease ecology and health-care delivery as topics and spatial analysis as technique. These three areas have maintained their importance and research productivity within them has increased over the last decade. At the same time, since the 1980s, medical geography has evolved into new areas of concern. Both those who continue to call themselves medical geographers and those who do not identify closely with the subdiscipline have moved toward a geography of health that is less concerned with disease and the medical world and more with well-being and social models of health and health care (Rosenberg 1998). Health geography is characterized by an emphasis on place and place meaning, grounding in socio-cultural theory, and a critical perspective on health issues (Kearns and Moon 2000). The evolution of medical geography led to lively debates in the mid-1990s (Kearns 1993; Mayer and Meade 1994; Litva and Eyles 1995; Philo 1996) that have been put into historical perspective by Del Casino and Dorn (1998). By the end of the 1990s, the dichotomy between old and new medical geographers constructed during the debate was giving way to complementarity and synthesis. As examples, disease ecology was opened out to include political economic concerns (Mayer 1996) and multi-level modeling combined aspects of spatial analysis with a focus on place (Duncan *et al.* 1996; Verheij 1999).

The structure of this chapter results from a decision made by the Medical Geography Specialty Group (MGSG) to base its contribution to this volume on papers presented at two special sessions on "Retrospect and Prospect" during the 1998 Association of American Geographers meetings in Boston. The six presenters were Michael Greenberg on disease ecology, Ellen Cromley on health services, Gerard Rushton on spatial analysis, Susan Elliott on women's health, Jennifer Wolch on mental health, and Joseph Scarpaci on the developing world. The first three topics represent the more established themes of medical geography; topics four and five represent some of the new directions the field has taken in recent years; and the sixth topic combines research from both older and newer traditions. The work of the six authors has been supplemented by papers and abstracts received in response to an appeal to the specialty group to submit items of their own or ones that they found to be especially important or useful, as well as suggestions made by two anonymous reviewers.

Disease Ecology

Disease ecologists study how human populations, physical and built environments, and human behavior interact either to prevent or to produce disease. This leads them into many fascinating avenues of research such as examining the impact of dam building in Africa on the spread of schistosomiasis or how nuclear radiation may be related to cancer incidence. Greenberg (1998) reported that a bibliographic search for disease ecology studies for the period 1990–7 produced 190 papers from more than 100 journals, including thirteen geography journals. Disease ecology is thus very much alive and well (Meade 1999).

Work in disease ecology covered a wide spectrum of topics over the last decade. As examples, Hunter (1990) carried out a meticulous analysis of the bot fly maggot in Latin America and Hunter and Arbona (1995) looked at the health effects of water pollution in Puerto Rico. Haggett (2000) summarized his work over three decades on modeling epidemics. Not surprisingly, the 1990s produced a large number of papers on HIV/AIDS. Examples include Loytonen's (1991) work on the diffusion of the virus in Finland; Cliff and Smallman-Raynor's (1992) analysis of global patterns and local spatial processes; Barnett and Blaikie's (1992) work on the sociopolitical aspects of AIDS in Africa; Brown's (1995) ethnographic investigation of local responses to AIDS in Vancouver; and the work of Wilton (1996) who traced a small group of people through five stages of reaction to their HIV/AIDS diagnosis. The topic of emerging diseases was seized upon by many health professionals, geographers included, who were very alarmed about the potential for the rapid spread of such diseases as Legionnaires' Disease, Lyme Disease, and several African tropical diseases (e.g. Lassa, Marburg, and Ebola viruses) (Haggett 1994).

In the related areas of migration and comparative studies, medical geographers and others continued to examine the health impacts of worldwide population movements and such culture traits as diet, smoking, and violence among different groups. Migration studies are represented by Kliewer (1992) who looked at the influence of population movements on regional variations of stomach and colon cancer mortality in the Western US; comparative studies include an international comparison of mortality and morbidity patterns by Walter and Birnie (1991).

A topic receiving increasing attention from medical geographers, epidemiologists, and others is environmental health in neighborhoods where there are high levels of poverty and unemployment. This topic tests the geographer's ability to gain a holistic view of the effects of multiple hazards such as ozone, genetics, segregation, and poor housing, to name but a few. Some studies along these lines look at socioeconomic impacts on health; these include the implications of economic recession and structural adjustment on AIDS in Africa (Sanders and Sambo 1991) and socio-demographic determinants of death among blacks and whites in the US (Rogers 1992). More localized stressors are represented by studies of the health effects of air pollution in several specific sites around the globe, including one in Mexico City (Romieu *et al.* 1993) and another in Beijing (Schwartz 1995).

Both natural and technological hazards have the potential to create nightmare high-risk environments. Floods, earthquakes, tornadoes, leaks from chemical weapons sites, radiation from nuclear power plants, and many other hazards all need to be examined for their health impacts. Over the past decade, researchers have carried out several studies on hazardous waste sites and their effects on various populations, for example on the children of homeless families (Roth and Fox 1990). Environmental justice becomes an issue when these sites are located so that people with certain demographic and socioeconomic characteristics are more exposed to them. Other hazards studies recently include two on the effects of radon on health, one around the Sellafield nuclear power plant in England (Gardner 1991) and one in Belarus (Marples 1993).

Health Services

The domain of medical geographers who study the delivery of health care includes studies of the distribution of resources such as clinics, nurse practitioners, and CT scanners; access to and utilization of health care; regionalization of resources; and location/allocation modeling. Ellen Cromley (1998) noted several important research trends in this field over the last decade, a period in which there have been very significant changes in health-service systems.

One of the trends is to examine proposed national reforms and ask how the changes actually work on the ground. In the case of the health promotion approach embodied in Canada's Achieving Health for All plan, medical geographers pointed to the need for better understanding of health behavior within communities (Taylor 1990; Eyles 1990). In a paper by Kronick *et al.* (1993), some very straightforward concepts of economic geography were used to answer a simple question: where are the places in the US where managed competition as a market-driven approach to health-care reform can work? The results indicated that reform of the US health care system by managed competition would be feasible only in major metropolitan areas.

A second set of studies investigated the geography of health care considered as an industry. Providers of acute care hospital services and ancillary services such as housekeeping are now organized into national and, in some cases, multinational corporations (Mohan 1991). In the United States, the development of multiple-site physician practices has been investigated by Cromley and Albertsen (1993) using survey research for one type of physician specialist in a single medical market and by Albert and Gesler (1997) for different types of physicians in North Carolina. Health insurance companies (McManus 1993) and hospitals and their networks have been examined using GIS technology to evaluate utilization, describe geographical coverage, and make location decisions (Love and Lindquist 1995).

A third trend has been the continued development of research approaches to traditional concerns of the geography of health services, namely the geographical distribution of physicians and geographical variations in access and utilization. New efforts are being made to model physician locations longitudinally (Seifer *et al.* 1995; Konrad and Li 1995; Ricketts *et al.* 1996) by looking at physician migration into and out of regions rather than describing physician/population ratios cross-sectionally or considering physician retention alone.

Another area of interest has been help-seeking behavior. Senior *et al.* (1993) developed statistical methods for analyzing disaggregated data obtained from interviews with mothers to investigate the effects of transportation and time-space and gender-role constraints on the utilization of infant immunization services. Investigations of informal caregiving to the elderly (Joseph and Hallman 1998) illustrate the help-seeking theme by pointing out geographical complexities in the delivery of this type of care. In addition to the journey to receive medical care or provide assistance as a part of one's daily travel and activity pattern, residential relocation for health care is receiving attention. For example, studies of migration for medical care have investigated the high mobility and migration paths of individuals infected with HIV (Ellis 1996).

Spatial Analysis

Analysis of point, line, area, and surface patterns of disease and health-care phenomena is often of direct benefit to human populations. As examples, medical geographers may use spatial analysis to examine such issues as the spread of malaria or the distribution of physicians. Cliff and Haggett's (1988) atlas of disease is an excellent example of work in this area. The last decade has seen both an overall expansion in research productivity in this area and moves toward more sophisticated statistical techniques (e.g. use of the K function) and a vastly increased use of computers and GIS technology (e.g. to map disease-agent habitats). The computer and GIS revolutions that were first making an impact on medical geography a decade ago are now beginning to be fully realized. Programs for mapping and spatial analysis are much more sophisticated and user-friendly than they were only a few years ago. Gatrell and Bailey (1996) are a good source for the variety of GIS and statistical software medical geographers can use to visualize, explore, and model disease and health-care datasets. The paper by Gatrell and Senior (1999) and the volume edited by Albert *et al.* (2000) review recent applications of GIS in medical geography.

Spatial analytic techniques, enhanced by GIS capabilities, have been used in recent years to investigate a number of intriguing questions. A perennial problem for medical geographers, as well as epidemiologists and others, is whether the reported locations of disease cases (e.g. stomach cancer) truly form a cluster (Waller and Jacquez 1995). Thus Gatrell *et al.* (1996) constructed simulation envelopes to assess evidence for childhood leukemia clustering in west-central Lancashire; and Rushton *et al.* (1996) used simulations in their studies of birth defects in Des Moines, Iowa.

Spatial analysis can be applied to help answer many puzzling problems in disease ecology. Along these lines, the geostatistical technique of kriging was used to examine the spatial and temporal distributions of anopheline mosquitoes in an Ethiopian village (Ribeiro *et al.* 1996). A great amount of attention has been paid recently to the spread of AIDS, an obviously important health as well as social and economic problem. Among many studies, Lam *et al.* (1996) made innovative use of spatial correlograms to examine the diffusion of AIDS in four US regions. How are environmental changes linked to disease occurrence? This question led to a study by Aase and Bentham (1994) of the connection between ozone depletion and the geography of malignant melanoma in Nordic countries. Geographers who deal with the delivery of health care have always been concerned with spatial inequalities in the provision of practitioners and facilities. To shed some light on this question, Lowell-Smith (1993) employed location quotients, Gini indices, and Lorenz curves in a study of the distribution of freestanding ambulatory surgery centers in the US at different geographic scales.

Women's Health

The next two areas of medical geographic focus discussed in this chapter—women's health and mental health—tend, in very general terms, to have come within the purview of medical geographers somewhat later than disease ecology, health services, and spatial analysis. They also borrow heavily from other geographic subdisciplines such as social and feminist geography, as well as other social sciences. This has meant that, in many cases, scholars who may not think of themselves as medical geographers have written geographies of health and disease. This is certainly the case for the geographic study of the health of women. As Susan Elliott (1998) noted, the medical geography literature in this area is quite thin. However, the literature expands considerably when work by non-medical geographers is also taken into account. For example, although women's health may not be the primary focus, work by feminist geographers certainly deals with the topic (Jones *et al.* 1997).

A good place to begin a review of women's health research is the edited collection by Matthews (1995) that contains a wide range of papers whether categorized by subject matter, theoretical approach, methods used, or geographic scale. Elliott (1998) suggested that one way to describe recent work in women's health is in terms of three major components of the Population Health Framework (PHF) that guides Canadian health and health promotion policy (Evans and Stoddard 1994). Work in this area consistently demonstrates a relationship between health status and economic inequality or relative deprivation through the use of national and international comparisons. Unfortunately, the vast majority of the data used to support these relationships were collected solely from men.

The first component of the PHF is the socioeconomic determinants of health. Here we note that the literature around gender differences in health is built upon a number of "conventions" which are currently being debunked. For example, one of the most oft-quoted "facts" is that "women get sick, men die." However, MacIntyre *et al.* (1996), using data from two large-scale epidemiologic surveys in Scotland and the UK, indicated no significant gender differences in reporting any long-standing (chronic) illness at any age. Rather, it would appear that sex differences in health vary according to the particular condition in question as well as life-cycle stage. A second myth is the assumption that all household members have relatively equal access to resources. This is certainly not always the case; many women have restricted access to household resources and/or little or no say in how household income is spent.

In contrast to the research on men's health inequalities, most research on women's health has taken a "social roles" perspective, the second PHF component. In order fully to understand women's health experience, we need to look at both women's social roles (e.g. paid worker, partner, household manager, and mother) and the material circumstances within which those roles are enacted (Macran *et al.* 1994). Women with multiple roles end up with more duties, time pressures, and life stresses than their male counterparts, thus potentially increasing their risk of acute and chronic illness. While this "multiple burden hypothesis" (a third myth) makes intuitive sense, the evidence for it is often equivocal.

Very little has been done within medical geography on the third PHF component, issues of environment and women's health. Indeed, much of what exists pertains solely to the realm of reproductive health (Lewis and Kieffer 1994). There is recognition within the literature, however, of the importance of investigating the impacts of the biophysical environment on the health of women (Blocker and Egberg 1989). Kettel (1996) offers a useful conceptual model for gender-sensitive research and policy analysis that centers on women's interaction with the biophysical environment. In essence, her framework underscores the fact that women's interaction with the biophysical environment occurs within their own life spaces.

Mental Health

Medical geographers have been studying mental health since the 1960s, but this topic has not always received the attention it deserves. Those who study people with mental illnesses or their care need to know something about the psychology, sociology, and anthropology of the subject, which means that geographers who become involved in this subspecialty tend to think in interdisciplinary terms.

Jennifer Wolch (1998) summarized the work that medical geographers have accomplished in mental health over the last few decades in terms of waves. During the first wave, which took place in the 1970s and 1980s, studies that used such techniques as locational analysis looked at the shift of mentally disabled people from large asylums to community-based treatment and then onto the streets. This work informed patient advocates and mental health practitioners and influenced public policies and programs. The end of this wave in the late 1980s and early 1990s coincides with the decade under review in this chapter. This period witnessed a focus on the urban homeless (Laws 1992; Wolch *et al.* 1993; Wolch and Dear 1993) and an increasing use of social theory, feminism, qualitative methods, and the "interpretative turn."

A second wave of research began to turn away from large-scale distributions of mental health phenomena and considerations of space and to focus instead on two new directions: mental disability as difference and the analysis of place. However, work continued on distributions of mentally ill people, historical geographies of the sites of asylums and poorhouses, and homelessness (Dear *et al.* 1997; Joseph and Kearns 1996; Parr 1997). As work shifted from deviance to difference, positionality, and identity (the first new trend), newer researchers tended to shun direct engagement with policy issues. "Difference" studies were informed by ideas from Foucault, post-structural psychoanalytic theory, post-modernism, and feminism. Researchers, including Dyck (1995), Moss and Dyck (1996), and Dorn (1998) focused on the body, interpersonal relationships, representational practices, and micro-geographies of power in everyday life.

The second trend within the second wave entailed a rediscovery of place. Place studies of mental health as well as in other areas of medical geography rejected the traditional concerns of the subdiscipline with medicine and disease and advocated a "geography of health" (Kearns and Gesler 1998). As examples, Kearns and Dyck (1996) discussed nursing education and biculturalism in New Zealand; and Ruddick (1996) showed how the use of particular local places by homeless youths in Hollywood influenced the location of facilities.

The Developing World

Over the last decade, medical geographers maintained a long-standing interest in disease and health issues in countries in Latin America, Africa, and Asia. This work serves as a fitting end to discussion of the six sub-themes as it combines research from both the older and newer traditions of medical geography. Some of the best examples of this work can be found in the edited volume on health and development issues by Verhasselt and Phillips (1994) and a series of eighteen papers in a special edition of *Social Science and Medicine*, with an introduction by Verhasselt and Pyle (1993).

Over the last decade, work continued on the mapping of and explanations for spatial patterns of mortality and morbidity. In Malawi, Kalipeni (1993) examined the spatial variation of infant mortality from 1977 to 1987 and found associations with a number of demographic and socioeconomic variables. Paul (1993) found that demographic, nutritional, environmental, cultural, and behavioral factors were related to spatial patterns of maternal mortality ratios in the countries of Africa. In Visakhapatnam, India, Asthana (1995) discovered that, even though there were considerable differences in infrastructural and socioeconomic development, morbidity rates did not vary among five slum settlements.

Research in disease ecology also continued in developing country settings. Watts (1998) showed how a shift from Imperial Medicine to Primary Health Care in Africa over the last century aided a largely successful disease eradication effort in the 1980s. Focusing on the Upper Region of Ghana, Hunter (1997) showed how guinea worm disease declined in most of his study areas from 1960 to 1990, due mainly to improved water security provided by hand pump tube wells. Papers by Hayes *et al.* (1990) and Poland *et al.* (1990) questioned the utility of household surveys as a reliable data source in less-developed countries, and examined the roles of various home environments in transmitting disease and teaching health behaviors.

Turning next to health-care delivery, we find studies that concentrated on the spatial distribution of health-care resources and access and utilization issues. Smith (1998) examined the state of health care following the widespread implementation of market socialism in China and found that, although status and access to care have both improved overall, pre-existing gender and geographic inequalities have not been eliminated. Good (1991) reminded us that a substantial portion of health care in Africa is delivered by Protestant and Roman Catholic missions and set out a research agenda for studies of this phenomenon. Bailey and Phillips (1990) looked at the influence of distance, transport availability, and access on utilization of health care by the residents of three pairs of study sites with contrasting social status in Kingston, Jamaica.

To round out this section, several illustrations are provided of a very diverse body of work that emphasized the political, social, and economic contexts of disease and health and the role of place and subjective experience. Along these lines, Kalipeni and Oppong (1998) discovered that political factors were of utmost importance for the health of refugees fleeing conflicts in Africa, and Miranda *et al.* (1995) looked at the effect of creating greater free-market conditions for health care in Chile by the military government. The "third wave" of the AIDS pandemic was the economic, social, political, and cultural response to the disease. This response was discussed by Earickson (1990) as an economic challenge to the Third World. In a study of south Sulawesi, Indonesia, Ford *et al.* (1997) looked at the sexual culture, AIDS awareness, and public health response. Lewis (1998), using literature on gender relations, domestic violence, aging, and socioeconomic development, discussed key issues for women's health in the Pacific Islands. Finally, Scarpaci (1998) showed how long-term and short-term urban policy goals in poor countries have affected the delivery of social services such as health.

What Lies Ahead?

What are the challenges and opportunities for medical geographers in the years to come? This question will be answered, first, in terms of the six sub-areas reviewed above and then in a more general way.

One of the greatest challenges facing disease ecology studies is finding and training researchers to investigate a seemingly endless number of diseases. This is especially true in developing countries where infectious and contagious diseases continue to create havoc and where resources for training researchers and facilitating studies continue to be minimal. On the positive side, medical geographers have made very substantial contributions to the concepts and methods of disease ecology. Their holistic approach to investigations of population, environment, and behavior is invaluable.

Three important factors set the context for research in disease ecology, offering both opportunities and constraints: (1) world events such as the recent conflict in Afghanistan that cause rapid movements of people and health risks; (2) political systems that exert pressure for answers to health questions; and (3) technology and data systems that allow us to study what once we could not (Greenberg 1998). Many imperatives for useful research exist with this context. We need, as examples, to find out much more about how cultural traits affect the incidence of such diseases as stomach cancer, the differential effects of indoor and outdoor lead on populations with different ethnic backgrounds and income levels, the psychological impacts of natural hazards, and what such health reforms as managed care mean for levels of morbidity.

Where does the geography of health services stand today and what are some future directions? An insightful review of the medical geographer's work on access to health services by Powell (1995) points out that much of the work is aspatial and studies by geographers are not widely cited, even by other medical geographers. One of the greatest problems for this subfield is creating a meaningful view of what is being studied across a range of relevant geographic scales. Shannon and Pyle's (1993) medical atlas of the twentieth century and the second edition of *The Dartmouth Atlas of Health Care* (Center for Evaluative Clinical Sciences 1998) provide us with some valuable overviews. However, we still do not have clear pictures of where the 560,000 physicians, 6,000 hospitals, and other health services sites are located within the US and how they are connected. It would be of great value to use our rapidly developing spatial database technologies to create a digital spatial database of the US health-care system that could be accessed over the Internet to display data at a variety of geographic scales. We very much need comprehensive research on geographical variation in state health services delivery policy and regulations in the US. Another neglected topic is the impact of communication technology on health services. Use of the Internet as a source of medical care information (Rosenfeld and Kolk 1995), telemedicine, and the role of global positioning systems (GPS) in emergency services (Riordan 1997) are just a few developments of interest.

For spatial analyists in all areas of geography, an ongoing problem is finding datasets related to research questions that are accurate and as complete as possible. This is especially true in developing countries. Another difficulty is ironing out statistical problems such as the effect of spatial autocorrelation. Also, users of spatial analytic techniques need to be careful that an overemphasis on methods does not override the practical relevance of their work. Despite these challenges, the future looks very bright indeed for spatial analysis. Multi-level modeling has made a significant contribution to our understanding of health and disease phenomena at different geographic scales. Opportunities to investigate the components of disease ecology (e.g. disease agent habitats, vector movement, and population distributions) are boundless. We now have the capability to establish disease surveillance systems that can track, for example, the outbreak of cholera in an area. There is an increasing awareness by the Centers for Disease Control, the Environmental Protection Agency, the National Cancer Institute, and many public health departments that GIS and spatial analysis are relevant for investigating many of the questions of interest to them. Rushton (1998) is very optimistic about the increasing use of GIS-H (GIS for Health) systems for integrating the acquisition, storage, manipulation, and analysis of health data.

It is well known that studies involving women and their concerns have come relatively late into the geographic literature. Medical geographers have some catching up to do as well. Despite the paucity of studies in women's health, however, there are some signs of progress. There is a small but growing literature on women's differential access to health care. For example, it has been found that, despite the fact that women share equally in the risk of heart attack (especially after menopause), they are much more likely to die of an attack (Wenger *et al.* 1993). A solid foundation of research in the area of health and functioning has been built by health geographers who investigate how women cope with chronic illness using qualitative/interpretative approaches (Dyck 1995; Moss and Dyck 1996). Inertia in women's health studies has been overcome and the momentum gained in recent years promises to continue and will probably increase.

To move forward with answering the question, "Why are some women healthy and others not?" Elliott (1998) suggests that (1) research that takes place within a population health perspective must be theoretically informed (e.g. by using Lewis's (1998) tripartite foundation of medical geography, feminist geography, and social theory); (2) the framework must be situated in relations of power within society and with respect to imbalances in access to resources; and (3) baseline data on women's health must be available.

In the mental health field, Wolch (1998) believes that a new model of community mental health care emerged from second wave research. This third wave has for its context globalization of the economy, welfare reform and devolution, and postmodern urbanism. However, Wolch also feels that much of the new work is not carried out with policy relevance in mind. Researchers in mental health largely abandoned large-scale or longitudinal projects and turned to small-scale studies using qualitative methods. Furthermore, she detects a "deep-seated cynicism about the ability of academic research to do anything except support a corrupt political and policy process" (ibid. 4). She urges mental health geographers to return to policy-relevant research, without necessarily abandoning disability and in-place studies. New studies are desperately needed to examine the current "displacement model" of service delivery that continually recycles poor, homeless, and mentally disabled people from place to place. Basic questions for mental health geographers to address include how new local landscapes of madness are created, where new sites of mental health treatment are located and how this is linked to welfare state policy, and where are the sites of resistance to the displacement model. As with many other aspects of medical or health geography, the times call for political awareness and for creative blending of small- and large-scale studies, as well as mixtures of theoretical perspectives and methodological tools. Although much remains to be done, geographic mental health research shows signs of great vitality. Recent evidence of this is the edited volume on "Post-asylum Geographies" in *Health and Place* (Philo 2000).

There is no question that work in developing countries, where most of the world's population resides, will continue to offer a tremendous challenge to medical geographers in the future. The will, the funds, and the time for researchers to study health and disease in these countries can present almost insurmountable difficulties. It would be highly desirable to have this research carried out by scholars who are native to the places being studied, but they face even higher barriers in terms of training and resources than do scholars from the developed world. What medical geographers have going for them here is a very strong tradition of field studies, carried out by a handful of pioneers, in non-Western settings.

All of the five subtopics discussed previously cry out for study in developing areas. We need to answer, as examples, such questions as why malaria is showing a resurgence in some countries, how the dumping of industrial wastes affects populations, how GIS can be used to monitor emerging diseases, how impoverished governments can deliver health care most effectively, how mental health is dealt with in both the private and public sectors, and how cultural beliefs affect women's health in different places.

What does the future hold for other aspects of medical geography besides the six sub-areas already discussed? Teaching students the concepts of the field should be a priority. An important step in this direction was taken by Matthews and Rosenberg (1995) who edited a set of journal papers on "Teaching Medical Geography" that came out of an MGSG workshop. Four general texts that are available currently are Curtis and Taket (1996), Ricketts *et al.* (1994), Meade and Earickson (2000), and Gesler and Kearns (2002). Given current trends, work in medical geography will increasingly be informed by social theory as many medical geographers and those who call themselves health geographers evolve the sub-discipline in new directions. Medical geographers find themselves discussing such ideas as therapeutic landscapes (Gesler 1992; Williams 1999), political ecology (Mayer 1996), the medically inscribed body (Dorn and

Laws 1994), neoliberalism and health (Kearns and Moon 2000), commitment to social justice and transformative politics (Earickson 2000), and the ascription of pathology to ethnic groups and places (Craddock 2000). At the same time, we can expect further studies that combine both traditional and newer ideas. Another trend, which also saw its beginnings in the 1990s, is toward more qualitative research (e.g. Cutchin 1997). Medical geographers have increasingly used such techniques as snowball sampling and in-depth interviews. A landmark publication here was a set of papers collected by Susan Elliott (1999) that explored qualitative approaches in health geography.

Acknowledgements

I would like to acknowledge the input of two anonymous reviewers who suggested very useful structural changes to the chapter as well as specific topics and research studies for inclusion. I would also like to thank Dr Melinda Meade for presenting the paper at the 1999 AAG meetings in Hawaii.

References

Aase, A., and Bentham, G. (1994). "The Geography of Malignant Melanoma in the Nordic Countries: The Implications of Stratospheric Ozone Depletion." *Geografiska Annaler*, 76B: 129–39.

Albert, D. P, and Gesler, W. M. (1997). "Multiple Locations of Practice in North Carolina: Findings and Health Care Policy Implications." *Carolina Health Services And Policy Review*, 4: 55–75.

Albert, D. P., Levergood, B. and Gesler, W. M. (eds.) (2000). *Spatial Analysis, Geographic Information Systems, and Remote Sensing Applications to Health- and Disease-Related Issues*. Ann Arbor: Ann Arbor Press.

Asthana, S., "Variations in Poverty and Health Between Slum Settlements; Contradictory Findings from Viskhapatnam, India." *Social Science & Medicine*, 40/2: 177–88.

Bailey, W., and Phillips, D. R. (1990). "Spatial Patterns of Use of Health Services in the Kingston Metropolitan Area, Jamaica." *Social Science & Medicine*, 30/1: 1–12.

Barnett, T., and Blaikie, P. (1992). *AIDS in Africa: Its Present and Future Impacts* London: Belhaven.

Blocker, T. J., and Egberg, D. L. (1989). "Environmental Issues as Women's Issues." *Social Science Quarterly*, 70: 586–93.

Brown, M. (1995). "Ironies of Distance: An Ongoing Critique of the Geographies of AIDS." *Environment and Planning D: Society and Space*, 13: 159–83.

Center for Evaluative Clinical Sciences (1998). *The Dartmouth Atlas of Health Care 1998*. New York: Oxford University Press.

Cliff, A., and Smallman-Raynor, M. (1992). "The AIDS Pandemic: Global Geographic Patterns and Local Spatial Processes." *Geographic Journal*, 158/2: 182–98.

Cliff, A., and Haggett, P. (1988). *Atlas of Disease Distributions: Analytical Approaches to Epidemiological Data*. Oxford: Blackwell.

Craddock, S. (2000). *City of Plagues: Disease, Poverty and Deviance in San Francisco*. Minneapolis: University of Minnesota Press.

Cromley, E. K. (1998). "The Geography of Health Services Then and Now." Paper presented at the annual meetings of the Association of American Geographers, Boston.

Cromley, E. K., and Albertsen, P. C. (1993). "Multiple-Site Physician Practices and Their Effect on Service Distribution." *Health Services Research*, 28/4: 503–22.

Curtis, S., and Taket, A. (1996). *Health and Societies: Changing Perspectives*. London: Arnold.

Cutchin, M. (1997). "Physician Retention in Rural Communities: The Perspective of Experiential Place Integration." *Health and Place*, 3/1: 25–41.

Dear, M., Wilton, R., Gaber, S. L., and Takahashi, L. (1997). "Seeing People Differently: The Sociospatial Construction of Disability." *Environment and Planning D: Society & Space*, 15/4: 455–80.

Del Casino, V. J., Jr., and Dorn, M. L. (1998). "Doubting Dualism: A Genealogical Reading of Medical Geographies." Unpublished manuscript.

Dorn, M. (1998). "Beyond Nomadism: The Travel Narratives of a Cripple," in H. J. Nast and S. Pile (eds.), *Places Through the Body*. London: Routledge, 183–206.

Dorn, M., and Laws, G. (1994). "Social Theory, Medical Geography and Body Politics: Extending Kearns' Invitation." *Professional Geographer*, 46: 106–10.

Duncan, C., Jones, G., and Moon, G. (1996). "Health-Related Behavior in Context: A Multilevel Modeling Approach." *Social Science and Medicine*, 42: 817–30.

Dyck, I. (1995). "Hidden Geographies: The Changing Lifeworlds of Women with Multiple Sclerosis." *Social Science & Medicine*, 40/3: 307–20.

Earickson, R. J. (1990). "International Behavioral Responses to a Health Hazard: AIDS." *Social Science & Medicine*, 31/9: 951–62.

—— (2000). "Health Geography: Style and Paradigms." *Social Science & Medicine*, 50: 457–8.

Elliott. S. J. (1998). "Why are Some Women Healthy and Others Not?: A Framework for Understanding Geographies of Women's Health." Paper presented at the annual meetings of the Association of American Geographers, Boston.

Elliott, S. J., *et al.* (1999). "Focus: Qualitative Approaches in Health Geography." *Professional Geographer*, 51/2: 240–320.

Ellis, M. (1996). "The Postdiagnosis Mobility of People with AIDS." *Environment and Planning A*, 28/6: 999–1017.

Evans, R., and Stoddard, G. (1994). "Producing Health, Consuming Health Care," in R. Evans, Barer, M., and Marmor, T. (eds.), *Why Are Some People Healthy and Others Not?* New York: Aldine.

Eyles, J. (1990). "The Problem of Marketing Health Promotion Strategies." *The Canadian Geographer*, 34/4: 341–6.

Ford, N., and Koetsawang, S. (1991). "The Socio-cultural Context of the Transmission of HIV in Thailand." *Social Science & Medicine*, 33/4: 405–14.

Gardner, M. (1991). "Fathers' Occupational Exposure to Radiation and the Raised Level of Childhood Leukemia Near the Sellafield Nuclear Plant." *Environmental Health Perspectives*, 94: 5–8.

Gatrell, A. C., and Bailey, T. C. (1996). "Interactive Spatial Data Analysis in Medical Geography." *Social Science & Medicine*, 46/6: 843–55.

Gatrell, A. C., and Senior, M. (1999). "Health and Health Care Applications," in P. A. Longley, M. F. Goodchild, D. J. Maguire, and D. W. Rhind (eds.), *Geographic Information Systems*, ii. *Management Issues and Applications*. 2nd edn. New York: John Wiley, 925–38.

Gatrell, A. C., Bailey, T. C., Diggle, P. J., and Rowlingson, B. S. (1996). "Spatial Point Pattern Analysis and Its Application in Geographical Epidemiology." *Transactions of the Institute of British Geographers*, NS 21: 256–74.

Gesler, W. M. (1992). "Therapeutic Landscapes: Medical Issues in Light of the New Cultural Geography." *Social Science & Medicine*, 34: 735–46.

Gesler, W. M., and Kearns, R. A. (2002). *Culture/Place/Health*. London: Routledge.

Good, C. M. (1991). "Pioneer Medical Missions in Colonial Africa." *Social Science & Medicine*, 32/1: 1–10.

Greenberg, M. (1998). "Disease Ecology in Medical Geography, State of the Art." Paper presented at the annual meetings of the Association of American Geographers, Boston.

Haggett, P. (1994). "Geographical Aspects of the Emergence of Infectious Disease." *Geografiska Annaler*, 76B/2: 71–89.

—— (2000). *The Geographical Structure of Epidemics*. Oxford: Clarendon Press.

Hayes, M. V., Taylor, S. M., Bayne, L. R., and Poland, B. D. (1990). "Reported Versus Recorded Health Service Utilization in Grenada, West Indies." *Social Science & Medicine*, 31/4: 455–60.

Hunter, J. M. (1990). "Bot Fly Maggot Infestation in Latin America." *Geographical Review*, 80/4: 382–98.

—— (1997). "Bore Holes and the Vanishing of Guinea Worm Disease in Ghana's Upper Region." *Social Science & Medicine*, 45/1: 71–89.

Hunter, J. M., and Arbona, S. I. (1995). "Paradise Lost: An Introduction to the Geography of Water Pollution in Puerto Rico." *Social Science & Medicine*, 40/10: 1331–55.

Jones, J. P., Nast, H. J., and Roberts, S. M. (1997). *Thresholds in Feminist Geography: Difference, Methodology, Reproduction*. Lanham, Md.: Rowman & Littlefield.

Joseph, A. E., and Hallman, B. C. (1998). "Over the Hill and Far Away: Distance as a Barrier to the Provision of Assistance to Elderly Relatives." *Social Science & Medicine*, 46/6: 631–9.

Joseph, A. E., and Kearns, R. A. (1996). "Deinstitutionalization Meets Restructuring: The Closure of a Psychiatric Hospital in New Zealand." *Health & Place*, 2/3: 179–89.

Kalipeni, E. (1993). "Determinants of Infant Mortality in Malawi: A Spatial Perspective." *Social Science & Medicine*, 37/2: 183–98.

Kalipeni, E., and Oppong, J. (1998). "The Refugee Crisis in Africa and Implications for Health and Disease: A Political Ecology Approach." *Social Science & Medicine*, 46/12: 1637–53.

Kearns, R. A. (1993). "Place and Health: Toward a Reformed Medical Geography." *The Professional Geographer*, 45: 139–47.

Kearns, R. A., and Dyck, I. (1996). "Cultural Safety, Biculturalism and Nursing Education in Aotearoa, New Zealand." *Health & Social Care in the Community*, 4/6: 371–4.

Kearns, R. A., and Gesler, W. M. (eds.) (1998). *Putting Health into Place: Landscape, Identity and Well-Being*. Syracuse: Syracuse University Press.

Kearns, R. A., and Moon, G. (2000). "Abstracting Health from Medical Geography: Novelty, Place and Theory After a Decade of Change". Paper presented at the Medical Geography Symposium, Montreal.

Kettel, B. 1996. "Women, Health and Environment." *Social Science & Medicine*, 42: 1367–79.

Kliewer, E. (1992). "Influence of Migrants on Regional Variations of Stomach and Colon Cancer Mortality in the Western US." *International Journal of Epidemiology*, 21/3: 442–9.

Konrad, T. R., and Li, H. (1995). "Migrating Docs: Studying Physician Practice Location." *Journal of the American Medical Association*, 274/24: 1914.

Kronick, R., Goodman, D. C., Wennberg, J., and Wagner, E. (1993). "The Demographic Limitations of Managed Competition." *The New England Journal of Medicine*, 328/2: 148–52.

Lam, N. S-N., Fan, M., and Liu, K.-B. (1996). "Spatial-Temporal Spread of the AIDS Epidemic, 1982–1990: A Correlogram Analysis of Four Regions of the United States." *Geographical Analysis*, 28/2: 93–107.

Laws, G. (1992). "Emergency Shelter Networks in an Urban Area: Serving the Homeless in Metropolitan Toronto." *Urban Geography*, 13/2: 99–126.

Lewis, N. D. (1998). "Intellectual Intersections: Gender and Health in the Pacific." *Social Science & Medicine*, 46/6: 641–59.

Lewis, N. D., and Kieffer, E. (1994), "The Health of Women: Beyond Maternal and Child Health," in D. R. Phillips and Y. Verhasselt (eds.), *Health and Development*. London: Routledge, 122–37.

Litva, A., and Eyles, J. (1995). "Coming Out: Exposing Social Theory in Medical Geography." *Health & Place*, 1/1: 5–14.

Love, D., and Lindquist, P. (1995). "The Geographical Accessibility of Hospitals to the Aged: A Geographic Information Systems Analysis Within Illinois." *Health Services Research*, 29/6: 629–51.

Lowell-Smith, E. G. (1993). "Regional and Intrametropolitan Differences in the Location of Freestanding Ambulatory Surgery Centers." *Professional Geographer*, 45/4: 398–407.

Loytonen, M. (1991). "The Spatial Diffusion of Human Immunodeficiency Virus Type I in Finland 1982–1997." *Annals of the Association of American Geographers*, 81/1: 127–51.

MacIntyre, S., Hunt, K., and Sweeting, H. (1996). "Gender Differences in Health: Are Things Really as Simple as They Seem?" *Social Science and Medicine*, 42: 617–24.

McManus, J. M. (1993). "Mapping Helps Met Track Its Health Care Networks." *National Underwriter*, 7 June: 38.
Macran, S., Clarke, L, Sloggett, A., and Bethune, A. (1994). "Women's Socioeconomic Status and Self-assessed Health: Identifying Some Disadvantaged Groups." *Sociology of Health and Illness*, 16: 182–91.
Marples, D. (1993). "A Correlation Between Radiation and Health Problems in Belarus." *Post-Soviet Geography*, 34/5: 281–92.
Matthews, S. (ed.). (1995). Papers on "Geographies of Women's Health." *Geoforum*, 26/3.
Matthews, S., and Rosenberg, M. (eds.) (1995). "Teaching Medical Geography." *Journal of Geography in Higher Education*, 19: 317–34.
Mayer, J. (1996). "The Political Ecology of Disease as One New Focus for Medical Geography." *Progress in Human Geography*, 20/4: 441–56.
Mayer, J., and Meade, M. S. (1994). "A Reformed Medical Geography Reconsidered." *Professional Geographer*, 46: 103–5.
Meade, M. S. (1999). "Disease Ecology: Backwards and Forwards." Unpublished manuscript.
Meade, M. S., and Earickson, R. J. (2000). *Medical Geography* (2nd edition). New York: Guilford.
Miranda, E., Scarpaci, J. L., and Irarrazaval, I. (1995). "A Decade of HMOs in Chile: Market Behavior, Consumer Choice and the State." *Health & Place*, 1/1: 51–9.
Mohan, J. (1991). "The Internationalisation and Commercialisation of Health Care in Britain." *Environment and Planning A*, 23/6: 853–67.
Moss, P., and Dyck, I. (1996). "Inquiry Into Environment and Body: Women, Work, and Chronic Illness." *Environment & Planning D: Society & Space*, 14/6: 737–53.
Parr, H. (1997). "Mental Health, Public Space, and the City: Questions of Individual and Collective Access." *Environment and Planning D: Society & Space*, 15/4: 435–54.
Paul, B. K. (1993). "Maternal Mortality in Africa: 1980–87." *Social Science & Medicine*, 37/6: 745–52.
Philo, C. (1996). "Staying In? Invited Comments on 'Coming Out: Exposing Social Theory in Medical Geography.'" *Health & Place*, 2: 35–40.
— (ed.) (2000). "Post-Asylum Geographies." *Health & Place*, 6/3: 135–259.
Poland, B. D., Taylor, S. M., and Hayes, M. V. (1990). "The Ecology of Health Services Utilization in Grenada, West Indies." *Social Science & Medicine*, 30/1: 13–24.
Powell, M. (1995). "On the Outside Looking In: Medical Geography, Medical Geographers, and Access to Health Care." *Health & Place*, 1/1: 41–50.
Ribeiro, J. M. C., Seulua, F., Abose, T., Kidane, G., and Teklehaimanot, A. (1996). "Temporal and Spatial Distribution of Anopheline Mosquitoes in an Ethiopian Village: Implications for Malaria Control Strategies." *Bulletin of the World Health Organization*, 74/3: 299–305.
Ricketts, T. C., Savitz, L. A., Gesler, W. M., and Osborne, D. N. (eds.) (1994). *Geographic Methods for Health Services Research: A Focus on the Rural–Urban Continuum*. Lanham, Md.: University Press of America.
Ricketts, T. C., Tropman, E., Slifkin, R. T., and Konrad, T. R. (1996). "Migration of Obstetrician-Gynecologists into and out of Rural Areas, 1985 to 1990." *Medical Care*, 34/5: 428–38.
Riordan, T. (1997). "A New Device Helps Emergency Workers Find People Who Use a Mobile Phone to Call for Help." *The New York Times*, 10 November: D2.
Rogers, R. (1992). "Living and Dying in the USA: Sociodemographic Determinants of Death Among Blacks and Whites." *Demography*, 29/2: 287–304.
Romieu, I., Lugo, M. C., Velasco, S. R., Sanchez, S., Meneses, F., and Hernadez, M. (1993). "Air Pollution and School Absenteeism Among Children in N. Mexico City." *American Journal of Epidemiology*, 136/12: 1524–31.
Rosenberg, M. W. (1998). "Research Review 5: Medical or Health Geography? Populations, People and Places." *International Journal of Population Geography*, 4: 211–26.
Rosenfeld. J. J., and Kolk, M. V. (1995). *The Internet Compendium Subject Guides to Health and Science Resources*. New York: Neal-Schuman Publishers.
Roth, L., and Fox, E. R. (1990). "Children of Homeless Families: Health Status and Access to Health Care Racial Inequity in Hazardous Waste Distribution." *Journal of Community Health*, 15/4: 275–84.
Ruddick, S. M. (1996). *Young and Homeless in Hollywood: Mapping Social Identities*. New York: Routledge.
Rushton, G., (1998). "Spatial Analysis in Geography: Progress and Prospect." Paper presented at the annual meetings of the Association of American Geographers, Boston.
Rushton, G., Krishnamurthy, R., Krishnamurti, D., Lolonis, P., and Song, H. (1996). "The Spatial Relationship Between Infant Mortality and Birth Defect Rates in a U.S. City." *Statistics in Medicine*, 15: 1907–19.
Sanders, D., and Sambo, A. (1991). "AIDS in Africa—the Implications of Economic Recession and Structural Adjustment." *Health Policy and Planning*, 6/2: 157–66.
Scarpaci, J. L. (1998). "Urban Social Policy in Poor Countries: Theoretical and Methodological Directions." *Urban Geography*, 19/3: 262–82.
Schwartz, J. (1995). "Air Pollution and Mortality in Residential Areas of Beijing, China." *American Journal of Respiratory and Critical Care Medicine*, 49/4: 216–22.
Seifer, S. D., *et al.* (1995). "Graduate Medical Education and Physician Practice Location." *Journal of the American Medical Association*, 274/9: 685–91.
Senior, M. L., New, S. J., Gatrell, A. C., and Francis, J. (1993). "Geographic Influences on the Uptake of Infant Immunisations: 1. Concepts, Models, and Aggregate Analyses." *Environment and Planning A*, 25/3: 425–36.
Shannon, G. W., and Pyle, G. F. (1993). *Disease and Medical Care in the United States*. New York: Macmillan.
Smith, C. J. (1998). "Modernization and Health Care in Contemporary China." *Health & Place*, 4/2: 125–39.
Taylor, S. M. (1990). "Geographic Perspectives on National Health Challenges." *The Canadian Geographer*, 34/4: 334–8.
Verhasselt, Y., and Pyle, G. (Introducers) (1993). Eighteen Papers on "Geographical Inequalities of Mortality in Developing Countries." *Social Science and Medicine*, 36/10.
Verhasselt, Y., and Phillips, D. R. (1994). *Health and Development*. London: Routledge.
Verheij, R. (ed.) (1999). *Urban-Rural Variations in Health Care*. Utrecht: NIVEL, Netherlands Institute of Primary Health Care.

Waller, L. A., and Jacquez, G. M. (1995). "Disease Models Implicit in Statistical Tests of Disease Clustering." *Epidemiology*, 6/6: 584–90.

Walter, S. D., and Birnie, S. (1991). "Mapping of Mortality and Morbidity Patterns: An International Comparison." *International Journal of Epidemiology*, 20/3: 678–89.

Watts, S. (1998). "Perceptions and Priorities in Disease Eradication: Dracunculiasis Eradication in Africa." *Social Science & Medicine*, 46/7: 799–810.

Wenger, N. K., Speroff, L., and Packard, B. (1993). "Cardiovascular Health and Disease in Women." *New England Journal of Medicine*, 329: 247–63.

Williams, A. (ed.) (1999). *Therapeutic Landscapes: The Dynamic Between Wellness and Place*. Lanham, Md.: University Press of America.

Wilton, R. D. (1996). "Diminished Worlds? The Geography of Everyday Life with HIV/AIDS." *Health & Place*, 2/2: 69–83.

Wolch, J. (1998). "From Distributions of Deviance to Definitions of Difference: Past and Future Mental Health Geographies." Paper presented at the annual meetings of the Association of American Geographers, Boston.

Wolch, J., and Dear, M. (1993). *Malign Neglect: Homelessness in an American City*. San Francisco: Jossey-Bass.

Wolch, J., Rahimian, A., and Koegel, P. (1993). "Daily and Periodic Mobility Patterns of the Urban Homeless." *Professional Geographer*, 45/2: 159–69.

CHAPTER 31

Military Geography

Eugene J. Palka

Introduction

In the benchmark publication *American Geography: Inventory and Prospect* (1954), Joseph Russell reported that military geography had long been recognized as a legitimate subfield in American geography. Despite the occasional controversy surrounding the subfield since his assessment (Association of American Geographers 1972; Lacoste 1973), and the general period of drought it experienced within American academic geography during the Vietnam era, military geography displays unquestionable resilience at the dawn of the twenty-first century. The subfield links geography and military science, and in one respect is a type of applied geography, employing the knowledge, methods, techniques, and concepts of the discipline to military affairs, places, and regions. In another sense, military geography can be approached from an historical perspective (Davies 1946; Meigs 1961; Winters 1998), with emphasis on the impact of physical or human geographic conditions on the outcomes of decisive battles, campaigns, or wars. In either case, military geography continues to keep pace with technological developments and seeks to apply geographic information, principles, and tools to military situations or problems during peacetime or war.

Throughout the twentieth century, professional and academic geographers made enormous contributions to the US Military's understanding of distant places and cultures. The vast collection of Area Handbooks found in most university libraries, serves as testament to the significant effort by geographers during wartime. Although some of the work remains hidden by security classification, a casual glance at Munn's (1980) summary of the roles of geographers within the Department of Defense (DOD) enables one to appreciate the discipline's far-reaching impact on military affairs.

The value of military geography within a theater of war can hardly be disputed. The subfield has also been important during peacetime, however, providing an important forum for the continuing discourse among geographers, military planners, political officials, and government agencies, as each relies upon geographic tools and information to address a wide range of problems within the national security and defense arenas.

Despite the subdiscipline's well-established tenure, the Military Geography Specialty Group is in its infancy. The time-lag is attributable to the subfield's tumultuous experience during the Vietnam era and the associated demise that ensued. The recent formation of a specialty group is noteworthy because it indicates that military geography has weathered the storm and its roots are still firmly implanted within American geography. The timing of this forum is most fortuitous because it provides an unparalleled opportunity to piece together the annals of the subfield, offer a current appraisal, and provide a clear vision necessary to forge ahead into the next century. To these ends, this chapter traces the history of military geography as a conscious field of study, chronicles its origin and development within the United States, provides a current assessment, proposes an agenda for the future, and illuminates some of the inherent challenges that will invariably surface along the journey.

Compared to others, this chapter contains a greater emphasis on the historical development of the subfield since a military geography chapter was not included in the first edition of *Geography in America* (1989).

The Military Geographical Legacy

It is probable that the use of geographic knowledge in military decision-making predates written history. Thompson (1962*b*) traced the first use of military geography to Megiddo (near present day Haifa) in 1479 BC. Later historical writings are replete with examples of famous military leaders whose actions were influenced by their interpretation of geographic factors. Thucydides' history of the Peloponnesian War from 431–404 BC (Meigs 1961), Xenophon's account of the march of 10,000 Greek mercenaries across Asia Minor to the shores of the Black Sea in 400 BC (Rouse 1959), and Ceasar's geographic insights during the Gallic Wars from 60–55 BC (Edwards 1939), constitute a small sample of the hundreds of early writings that include a military geographic perspective.

As a formal field of study, military geography was primarily a European innovation, with the French, Germans, and British pioneering most of the work during the nineteenth and early twentieth centuries. Theophile Lavallée's (1836) publication of *Géographie Physique, Historique et Militaire*, is generally regarded as the first publication exclusively devoted to military geography. The field gained added credibility a year later when Albrecht von Roon (1837), a captain on the Prussian General Staff, published a work containing detailed physiographic descriptions of military regions of Europe.

During the late nineteenth and early twentieth centuries, military geography was employed to meet national objectives under the rubric of "grand strategy" (Peltier and Pearcy 1966). Brown's (1885) military geography of the United States and Canada was the first American publication devoted exclusively to the subfield, but his effort went largely unnoticed because it was written for use as a textbook at the Infantry and Cavalry Schools. Mahan (1890) provided the first widely recognized American contribution related to the field and laid the foundation for what was later to become strategic geography.

Although American publications were sparse at the turn of the twentieth century, the British continued to develop a significant body of literature and provided landmark works during the era by Maguire (1899), Mackinder (1902), May (1909), and MacDonnell (1911). These comprehensive works were clearly indicative of the maturity of the subfield in Great Britain at that time, and proved to be influential on American perspectives in later decades.

Twentieth-Century Military Geography in America

At the turn of the century, several unpublished papers and lectures devoted to military topography were used for instructional purposes at the US Military Academy at West Point, the Command and General Staff College at Fort Leavenworth, and the Army War College at Carlisle (Thompson 1962*b*). Unfortunately, none of the papers reached a wider audience nor entered into the geographic literature. Semple (1903) and Brigham (1903), however, provided examples of military geography, if only indirectly, within the context of other publications. In *American History and Its Geographic Conditions*, Semple included chapters on the geography of the Civil War, and sea and land operations during the war of 1812, viewed from her classic environmental deterministic perspective. Brigham took a similar approach in his chapter on the Civil War, in *Geographic Influences in American History*. Aside from these two publications, there were no published works on military geography at the time.

World War I

The first formal demand for military geography in the United States surfaced during World War I. American geographers initially assisted the war effort by providing written descriptions of the physical landscapes surrounding major training camps throughout the country. Descriptions were subsequently printed on the backs of training maps and were used to teach basic terrain analysis skills to leaders and soldiers. The military geographic emphasis at that time, and throughout the war, focused on the physical geographic aspects of the battlefield, envisioned primarily as the domain of ground forces. A total of fifty-one members of the AAG participated in World War I and/or the Peace Conference that followed (Martin and James 1993). Regardless of whether they served in the Military or Civilian Service, or conducted supporting research within academia, the emphasis was on applied geographic research, specifically the application of geographic principles and knowledge to solving the Military's wartime problems. D. W. Johnson (1917) published *Topography and Strategy in the War*, but his later work, *Battlefields of the World War, Western and Southern Fronts: A Study in Military Geography* (Johnson 1921), became the most widely recognized military geographic publication stemming from the war years. Perhaps the most lingering effect of World War I on military geography was the notion that the subfield focused exclusively on wartime military problems.

World War II

When World War II arrived, geographers again provided widespread support to the war effort. By 1943, over 300 geographers were working in Washington, in the Office of Strategic Services, War Department, Intelligence Division, and in the Army Map Service (Martin and James 1993). During the war, military geography progressed beyond the basic practice of collecting and compiling data, to providing continual assessments of both the physical and human geography of specific regions. These efforts culminated in the Joint Army and Navy Intelligence Studies (JANIS), which were essentially the regional geographies of selected theaters. Regional studies were helpful not only to commanders and military planners, but also to the Army quartermaster and the research and development teams who were responsible for designing uniforms, vehicles, equipment, weapons, and materials.

Within academia, training programs specifically designed for military personnel spurred widespread interest in military geography. Poole (1944) proposed a formal agenda for training military geographers and aspects of his syllabus were instituted at universities and military schools alike. Throughout World War II, a plethora of articles and studies were published on various aspects of military geography (Ackerman 1945; Palka and Lake 1988). Arthur Davies' (1946) "Geographical Factors in the Invasion and Battle of Normandy" evolved as one of the best-known American publications during the period and continues to be regarded as a classic analysis.

Military geography continued to thrive after the war. Many geographers returned to academia and shared their wartime experiences and perspectives via publications and lectures (Committee on Training and Standards in the Geographic Profession 1946; Mason 1948*a*, *b*). The subdiscipline continued to be an integral part of the geography curriculum at numerous universities throughout the country and was well represented by the Committee on Military Geography within the AAG. The Committee was established to advise professional military schools on the design and implementation of military geography courses within their respective programs (Renner 1951). The subfield continued to flourish during the initial stages of the Cold War era, and the Korean War served as a catalyst, just as World War II had a few years prior. Russell's (1954) treatment of military geography in *American Geography: Inventory and Prospect* confirmed the healthy status of the subfield, but reinforced the perception that the scope was restricted to wartime concerns.

The Vietnam Era and the Aftermath

On the eve of the Vietnam era, several noteworthy publications emerged. "The Potential of Military Geography" (Peltier 1961) was an emphatic statement of the legitimacy and utility of the subfield. Meigs (1961) examined geographical factors that influenced the outcome of the Peloponnesian War, and drew parallels with World War II operations in North Africa and Sicily. Jackman's (1962) article, entitled "The Nature of Military Geography," provided a clear, concise, theoretical perspective and became the most frequently cited paper on military geography. Although unpublished, Thompson (1962*b*), under the mentorship of Preston James, provided exceptional coverage of the history of military geographic thought and superbly developed the theoretical underpinnings. *Military Geography* (Peltier and Pearcy 1966) was published a few years later, and until recently, was the most comprehensive and focused publication on the topic, essentially serving as the guide-post for the subfield between the end of World War II and the mid-1990s. The legitimacy of the subfield was reinforced in the AAG's (1966) special bulletin, *Geography as a Professional Field*, which featured a section entitled "Careers in Military Geography."

During the Vietnam War and its aftermath, military geography lost its appeal among university geography programs, and by the early 1980s, the subdiscipline had been relegated to a position of obscurity within the AAG's Applied Geography specialty group. The declining popularity of the subfield was decidedly apparent during the Vietnam era. The trend was underscored by: the scarcity of contemporary publications, the absence of a specialty group within the AAG, the reduced numbers of AAG members claiming military geography as a topical proficiency, the scant number of course offerings at universities, and the lack of any dissertations in the subfield from 1969 to 1982 (Browning 1983).

The demise of military geography coincided with the widespread social and political unrest that occurred in America during the mid-1960s and early 1970s. During that era, anti-war sentiments and a general mistrust of the federal government prompted geographers to become increasingly concerned with being socially, morally, and ecologically responsible in their research efforts and professional affiliations with government agencies. Contributing to the war effort in Vietnam came to be regarded as irresponsible by many members of the AAG. The controversy surrounding the Vietnam War cast a persistent shadow on military geography throughout the 1970s. In retrospect, the adverse effect of the unpopular war was understandable, if not predictable. The applied wartime focus of military geography, perceived for many years as its *raison d'être*, came under intense scrutiny, and without alternative applications, the subfield lost its appeal within academia.

The Waning Years

Despite the success of American military operations in Grenada and Panama during the early and mid-1980s, the justification for those incursions was controversial. Consequently, there was little renewed interest in military geography. Nevertheless, several publications managed to surface among the few practitioners. Some employed military geographic studies as a means to analyze regional conflicts (Soffer 1982), or to assess the security implications of land-use patterns (O'Sullivan 1980; Soffer and Minghi 1986). Others such as Munn (1980) reinforced the Military's continued interest in the subfield and outlined the role of geographers within DOD. Among military publications, Palka (1987, 1988) examined the routine requirement for geographic perspectives and tools in military planning, and Garver (1981, 1984) and Galloway (1984, 1990) compiled an excellent melange of contemporary readings for use in their military geography course at West Point. More recently, Winters (1991) discussed the influence of geomorphologic and climatologic factors in his historical account of a Civil War battle, and O'Sullivan (1991) provided extensive treatment of the relationships between terrain and small unit engagements.

A notable trend that occurred during the latter stages of the Cold War era was the gradual amalgamation of boundaries between political geography, military geography, and military intelligence, an almost inevitable condition foreseen decades earlier by Jackman (1962). Military geography frequently meshed with political geography at one end of the spectrum (O'Sullivan and Miller 1983; Soffer and Minghi 1986), and overlapped with military intelligence at the other end (O'Sullivan 1991). A slight revival even occurred as the use of applied geography towards "peace" (as opposed to "war") initiatives gained popularity (Pepper and Jenkins 1985). Peace geography appeared destined to overtake military geography, as the former was perceived as being more socially responsible and politically correct than the latter. Thus, by the end of the 1980s, military geography struggled with an image problem, ambiguity, and legitimacy.

War in the Persian Gulf

It would be virtually impossible to overstate the importance of military geography during the Gulf War. The reliance on geographical information was fundamental to training, equipping, and deploying forces to Southwest Asia. Satellite imagery, aerial photographs, computer-assisted cartography, the global positioning system, geographic information systems, and a wide range of map types and scales were routinely used to develop plans and conduct operations at all echelons. Additionally, pertinent regional and systematic geographies provided fundamental informational considerations to support decision-making processes from the outset of Desert Shield to the end of Desert Storm. Ironically, the Gulf War did *not* contribute substantially to military geography's resurgence, but the end of the Cold War *has* proven to be a major catalyst.

End of the Cold War: Beginning of a New Era?

The fall of the Berlin Wall and the demise of the former USSR have prompted significant changes in the National Security Strategy of the United States (Shalikashvili 1995). The dramatic shift in strategic orientation has warranted substantial changes in the size, force structure, and disposition of the US Military. Moreover, although the numbers of "forward deployed" units have decreased considerably, the country's involvement in "military operations other than war" (MOOTW) has increased at an unprecedented rate. Between 1989 and 1997, the Military participated in forty-five operations other than war, more than triple the total number conducted during the entire Cold War era from 1947 to 1989 (Binnendijk 1998).

The end of the Cold War has provided an opportune time to reexamine the scope of military geography (Anderson 1993; Palka 1995; Palka and Galgano 2000; O'Sullivan 2001). For Russell (1954) and Jackman (1962), military geography included the whole range of geographic research as it applied to military problems. Both, however, conceptualized military problems within a wartime context. "Contemporary" military problems are significantly different from those that were envisioned by either Russell or Jackman (Palka and Galgano 2000; O'Sullivan 2001). To be sure, security remains the fundamental concern of the US Military, but since 1991, there has been a considerable expansion of activities in the humanitarian sphere (Anderson 1994; Palka 1995) and environmental security realm (Butts 1993, 1994).

Current Appraisal

There is ample reason to be enthusiastic about the future prospects for military geography. A number of indicators suggest that the subfield is currently healthier than at any time since World War II. Recent developments have not only provided a tremendous boost to the subdiscipline, but have paved the way for even greater success in the future.

First, the end of the Cold War and the Military's subsequent involvement with MOOTW have served to fuel the fire and to generate wider-ranging possibilities for military geographic studies (Goure 1995; Gutmanis 1995; Palka 1995; Palka and Galgano 2000). Geographic applications to support warfighting, although invaluable, have not always been popular. Indeed, such applications encountered significant resistance and may have actually led to the demise of the subfield during controversial wars. Humanitarian and peacekeeping operations, however, have been much less disputable and appear to be extremely attractive to a larger audience and new generation of geographers.

Second, the landmark works by Collins (1998), Winters (1998), and Palka and Galgano (2000) essentially fill a void that has existed for more than a quarter of a century. For the first time since Peltier and Pearcy's (1966) publication, a comprehensive work on military geography is available for use as a college textbook. In *Military Geography for Professionals and the Public*, Collins (1998), considers a wide range of environmental conditions, employs historical vignettes, and addresses warfare at the tactical, operational, and strategic levels. His work will undoubtedly appeal to a diverse group of users and will surely find its way into curricula at military service schools.

In *Battling the Elements*, Harold Winters (1998) highlights the impacts of weather, climate, terrain, soils, and vegetation on the outcomes of important military operations. He clearly illustrates from an historical perspective that the physical settings in which battles are fought are neither passive nor presumable. Rather, environmental components are active agents that have the potential to shape conflict and decisively influence the outcome. Winters demonstrates in a compelling fashion that the relationships between the environment and combat are highly variable, often unpredictable, and always formidable. *Battling the Elements* is fundamentally geographic, but will also be of interest to historians and military personnel.

More recently in *The Scope of Military Geography*, Palka and Galgano (2000) retain the wartime focus of traditional military geography; yet broaden the scope of the subfield to incorporate a wide range of MOOTW as well as peacetime endeavors. Their book emphasizes the synergy between geography and military operations across a spectrum from peacetime to war and takes into account the current nature of military activities.

Third, the recent establishment of a military geography specialty group indicates renewed interest and enthusiasm. Since the 1996 meeting of the AAG in Charlotte, multiple sessions have been scheduled during the annual conference to accommodate the plethora of papers within the subfield. Balanced participation from academics, planners, military personnel, and government professionals is perceived as a major strength of the organization. Moreover, the range of experience among members, from graduate students to professors emeritus ensures continuity of key issues and themes, yet promotes a thorough mixing of ideas and perspectives.

Fourth, recent technological advancements continue to highlight the capabilities of geographic tools to facilitate problem-solving at multiple scales. Corson and Minghi (1996*a*) showcased the applications of *Powerscene* (a computer-based terrain visualization system) to revolutionize making boundaries. They further elaborated on the utility of *Powerscene* during the negotiations of the Dayton Peace Accords (Corson and Minghi 1996*b*), and in so doing, rekindled a traditional military geographic concern with boundaries, a topic that dates back to World War I (Russell 1954; James and Martin 1979).

Finally, military geography as a worthwhile endeavor is resurfacing in professional military schools. A recent article in *Parameters* by the former chairman of the Department of National Security and Strategy at the Army War College, entitled "The Immutable Importance of Geography," provides a case in point (Hansen 1997).

New Directions and an Agenda for the Future

In order to maintain momentum well into the twenty-first century, a few paradigm shifts are necessary. The first shift involves the fundamental requirement to move beyond the traditional, narrowly construed focus on wartime military problems, and take into account peacetime concerns. Ironically, peacetime concerns have always consumed a far greater percentage of the Military's time and effort, yet military geographers have generally ignored these areas. Notable exceptions include McColl's (1993) detailed examination of international refugees and their political and military implications. More recently, Shaw *et al.* (2000) employed an ecological classification scheme to better manage the US Army's training lands, and Malinowski and Brockhaus (1999) applied spatial analytical techniques to address the Army's "recruiting" problem, currently one of the Military's most perplexing issues. These examples epitomize the diverse opportunities that exist in the peacetime arena.

A second necessary adjustment concerns the traditional, but outdated practice of placing military geography within either a tactical or strategic context (Russell 1954; Jackman 1962; Peltier and Pearcy 1966). Since 1982, the "operational" level of war has been incorporated into US Military doctrine as an intermediate echelon, a change that has gone relatively unnoticed among academics. The tactical level is focused on engagements and battles; whereas the strategic level is geared towards accomplishing national objectives. The operational level is essentially the critical link between the capabilities of tactics and the goals of strategy, and involves campaigns and the accomplishment of strategic goals by military forces at the theater level. Thus, military geographers must be cognizant of doctrinal developments that render previous paradigms obsolete.

A third transition involves the need to embrace two contemporary military realms, MOOTW and environmental security. These areas represent uncharted waters within an expanded military geography. Both spheres have evolved during the post-Cold War era and appear destined to demand the lion's share of emphasis within the subfield in the future. Expansion into new realms will *not* dilute the subdiscipline, nor will it challenge the age-old definition of military geography. It *will* maximize the subfield's potential and enable practitioners to address military "problems" across the entire spectrum of contemporary military employment scenarios during peacetime and in war.

Military operations other than war include: nation assistance; security assistance; humanitarian assistance and disaster relief; support to counter-drug operations; peacekeeping; arms control; combating terrorism; shows of force; noncombatant evacuation; and support to domestic civil authority (Center for Army Lessons Learned (CALL) 1993*a*, *b*, *c*, *d*). Recent examples are too many to list, but have been well publicized. Trends suggest that the Military will continue to be called upon to respond to refugee problems, floods, earthquakes, hurricanes, and forest fires, domestically and abroad. Additionally, the Military continues to be actively involved in operations to improve infrastructure and health conditions throughout Latin America, Africa, Asia, and the Pacific (Shannon and Sullivan 1993; Binnendijk 1998). These types of missions are innately geographic and afford ideal opportunities for military geographers to apply their expertise to worthwhile causes and showcase military geography as a versatile problem-solving subdiscipline.

Environmental security is the other arena that offers the greatest possibilities for the future of military geography. Environmental security has been an integral part of the country's National Security Strategy since 1991. Annual updates have expanded the concept based on input from national and international opinions which recognize that environmental issues have major impacts on health and economies, and are increasingly viewed as threats to development and political stability (Butts 1993). The environment is not a new security concept, but in the past it has generally been regarded as the victim rather than the cause of conflict (Butts 1994). Today, the environment plays an unquestionable role in regional stability, and it has become a fundamental military concern during peacetime training.

The US Military is linked to environmental security in several respects. First, it is often employed to resolve conflicts resulting from regional instability prompted by environmental problems. Second, in support of the country's National Security Strategy, the Military is involved in a wide range of security assistance scenarios to help host countries resolve their internal environmental problems. In such instances, the US Military has helped host governments to improve fisheries, construct water-supply systems, implement flood control, develop irrigation, construct sanitary landfills, administer forest management programs, and perform wildlife protection and management (Butts 1993). Third, the Department of Defense is one of the most important environmental resource managers in the world, overseeing 25 million acres of domestic holdings and more than 2 million acres of land abroad. Thus, environmental stewardship is a fundamental military concern during peacetime (Shaw *et al.* 2000).

Domestically, DOD's formal environmental restoration programs have been underway since 1984. A wide range of multifaceted programs have been administered to remedy various age-old pollution problems, and specific measures have been implemented to better manage natural resources, prevent contamination, protect endangered species, and comply with environmental laws and legislation. Complex environmental problems frequently demand multidisciplinary approaches, and so it is not surprising that many non-geographers have addressed the Military's peacetime environmental concerns. (Diersing *et al.* 1988; Varren *et al.* 1989). Geographers, however, have been conspicuously absent from the environmental security arena, a surprising trend, given the discipline's traditional interest in land-use planning.

Expansion into the MOOTW or environmental security arenas does not necessitate abandoning traditional themes in military geography. What is called for is a broadening of the military geographic perspective to keep pace with contemporary military concerns. *Not* to expand the scope of military geography accordingly would severely constrain the continued growth of the subfield. Moreover, the fate of the subdiscipline would continue to be tied to public opinion, a condition that has resulted in a "roller-coaster ride" for the subfield throughout the twentieth century.

A fourth paradigm shift targets the generic classification schemes previously employed to distinguish or categorize work in military geography. One scheme differentiated between topical, systematic, or regional endeavors, based on the nature of geography (Peltier and Pearcy 1966). Another common practice classified work as either tactical or strategic military geography, based on the level of warfare. The former scheme has limited utility, and the latter has long been obsolete.

I am convinced of the need to organize research themes and clarify the boundaries of the subfield in order to reduce traditional ambiguities and enhance coherence. As such, I propose a model based on context, scale, approach, and perspective (see Fig. 31.1). This design is capable of addressing military problems across a full spectrum of employment scenarios, from peacetime to war. The scheme also identifies the magnitude of the operation (scale), and the nature of the geography involved (approach). By also indicating the perspective (applied or historical), this model provides a comprehensive, yet useful method for organizing and classifying research in military geography. Perhaps most important, the model conceptualizes the scope of the subfield.

A Peek over the Horizon and Foreseeable Challenges

The question of *scope* has been historically troublesome for military geography. The broadened scope for the subfield (envisioned in Fig. 31.1) serves as an enabler rather than a constraint. Military geography requires the additional latitude to pursue complex problems at various scales across multiple disciplinary and subdisciplinary boundaries, from either an applied or historical perspective. Recent works by Corson and Minghi (1996*a*, *b*, 1997) clearly illustrate the positive benefits of this mindset.

The *social acceptance* of military geographic research themes and applications has been a significant concern since the Vietnam era. The term "military" has long rendered the subfield more vulnerable than most to public opinion (Coniglio 1981). Indeed, the mistrust of the federal government and of the Military during the Vietnam era nearly drove military geography to extinction. In light of the ever-changing nature of public opinion, questions regarding the social and political correctness of geographic applications to military affairs are almost inevitable. By expanding its scope, however, to include humanitarian and environmental concerns that occur during peacetime or MOOTW contexts (refer to Fig. 31.1), military geography will be able to offer additional research possibilities that support noble causes from virtually anyone's perspective. It is postulated that the corresponding benefits to the subfield would be increased popularity and stability.

Continual *academic interest* in military geography represents another anticipated challenge and is closely related to the concern with social acceptance. Indeed, public needs and opinions often influence academic pursuits. In the case of military geography, interest has also been somewhat proportional to the number of geographers in academia with military experience.

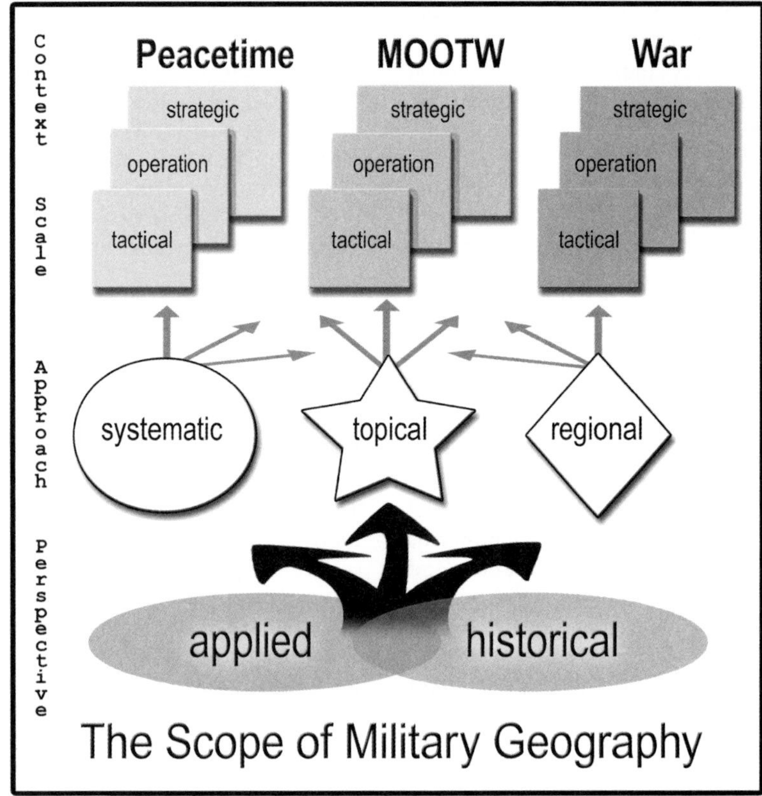

Fig. 31.1 The scope of military geography

Academic interest in military geography peaked during and after World Wars I and II. Since the end of compulsory service, however, military experience within the geography ranks has continuously declined. Thus, it has become much more of a challenge to generate interest in military problems among geographers. Yet, a fundamental pitfall that must be avoided is the unfounded notion that military geography is principally the domain of geographers who have military experience (Poole 1944; Russell 1954). On the contrary, the most productive military geographers in recent years have been *non-military* personnel. While military service can surely enhance understanding of military science, the former is neither a prerequisite for, nor a guarantee of comprehending the latter. To be sure, scholarship in the subfield demands a thorough grasp of both geography and military science, but *how* the knowledge is acquired is irrelevant. The recently organized specialty group provides an excellent forum for alleviating such fallacies and promoting a continuous dialog and free exchange of ideas among members of diverse backgrounds.

Finally, the *dynamic nature* of military affairs poses a considerable, if not, continuous challenge to military geography. Although the primary mission of the US Military has remained virtually unchanged over the past two hundred years, the Military's doctrine is always in the process of evolving in response to national objectives, new technology, the variable nature of the threat, and the world's ever-changing balance of power. Military geographers must therefore stay appraised of recent developments in both geography and military science and should avoid complacency within existing paradigms.

Conclusion

Military geography in America has evolved from infancy during the course of the twentieth century. The subfield gained definition and distinction during the course of the World Wars and the Korean Conflict, only to have its legitimacy questioned during the Vietnam era. After a period of stagnation during the 1970s and early 1980s, military geography has rebounded and accelerated to new heights during the 1990s.

On any given day, the US Military is represented in nearly half the countries of the world, actively engaged in training, peacekeeping, humanitarian assistance missions, environmental security, and other military operations other than war (Binnendijk 1998). Moreover, the military is charged with continually preparing to respond to crises across the full range of military operations from humanitarian assistance to fighting and winning major theater wars. The problems encountered within each scenario are varied, complicated, and often unexpected. Many of the problems, however, are innately geographic. The tools, techniques, and knowledge of the geographer can be fundamentally relevant and unquestionably useful in solving military problems in peacetime or war. Military geographers, be they systematic, regional, or technical specialists, can enhance the success of military operations through professional writing, guest lectures in military units, seminars with the senior leadership of the military, and formal teaching at the military's institutions of higher learning. Indeed, historical precedents have long been established in each of these cases (Poole 1944; Forde 1949; Warman 1954; Jackman 1971).

The current emphasis on MOOTW is indicative of America's global commitment, yet also reflective of the increased flexibility (since the end of the Cold War) to employ military forces in a wide range of humanitarian scenarios. Similarly, the commitment to environmental security has been expressed in our National Security Strategy since 1991. These relatively recent changes in the US Military's orientation demand a military geography that is broader in scope and capable of focusing on the full range of contemporary military problems in peacetime and in war. The core of "traditional" military geography is still of timeless value to the Military, and will continue to be an integral aspect of all plans and operations during armed conflicts. Yet, now more than ever before, geographers specializing in various subfields or regions can make substantial contributions to the Military's humanitarian relief efforts and environmental programs by recognizing the previously untapped potential of military geography.

The synergistic effect of recent publications, an active specialty group within the AAG, and the endless possibilities afforded by MOOTW and environmental security, will undoubtedly propel military geography into the twenty-first century. The groundwork has been laid and the conditions are right for the subfield to thrive within American geography in the foreseeable future.

References

Ackerman, Edward A. (1945). "Geographic Training, Wartime Research, and Immediate Professional Objectives." *Annals*, 35: 121–43.

Anderson, E. (1993). "The Scope of Military Geography." Editorial. *GeoJournal*, 31/2: 115–17.

— (1994). "The Changing Role of the Military." Editorial. *GeoJournal*, 34/2: 131–32.

Association of American Geographers (1966). *Geography as a Professional Field*. Bulletin 1966, no. 10. Edited by Preston E. James and Lorrin Kennamer. Washington, DC: US Department of Health, Education and Welfare.

— (1972). Minutes of the Annual Business Meeting, Boston, 18 April 1971. *The Professional Geographer*, 24/1: 36–9.

Binnendijk, Hans (ed. in chief) (1998). *Strategic Assessment*. Washington, DC: Institute for National Strategic Studies, National Defense University.

Brinkerhoff, John Richmond (1964). "The Nature of Modern Military Geography." Masters Thesis. Falls Church, Va.

Brigham, Albert Perry (1903). *Geographic Influences in American History*. Boston: Ginn & Co.

Brown, Lieutenant W. C. (1885). *Military Geography*. Fort Benning, Ga.: United States Infantry and Cavalry School.

Browning, Clyde E. (1983). *A Bibliography of Dissertations in Geography: 1969 to 1982*. Studies in Geography, 18. Chapel Hill, NC: University of North Carolina, Department of Geography.

Butts, Kent Hughs (ed.) (1993). "Environmental Security: What's DOD's Role?" Special Report of the Strategic Studies Institute. Carlisle, Pa.: US Army War College.

— (1994). Environmental Security: A DOD Partnership for Peace. Special Report of the Strategic Studies Institute. Carlisle, Pa.: US Army War College.

Center for Army Lessons Learned (CALL). (1993*a*). "Somalia." Newsletter 93-1, January 1993. Fort Leavenworth, Kan.: US Army Combined Arms Command.

— (1993*b*). "Counterdrug Operations II." Newsletter 93-5, September 1993. Fort Leavenworth, Kan.: US Army Combined Arms Command.

— (1993*c*). "Operations Other than War, II: Disaster Assistance." Newsletter 93-6, October 1993. Fort Leavenworth, Kan.: US Army Combined Arms Command.

— (1993*d*). "Operations Other than War, III: Civil Disturbance." Newsletter 93-7, November 1993. Fort Leavenworth, Kan.: US Army Combined Arms Command.

Cole, Brigadier D. H. (1950). *Imperial Military Geography*, 10th edn. London: Sifton Praed & Co.

Collins, John M. (1998). *Military Geography for Professionals and the Public*. Washington, DC: National Defense University Press.

Committee on Training and Standards in the Geographic Profession (1946). "Lessons from the War-Time Experience for Improving Graduate Training for Geographic Research." *Annals*, 36: 195–214.

Coniglio, James V. (1981). "Military Geography: Legacy of the Past and New Directions." in *Readings in Military Geography*, compiled by Colonel John B. Garver, Jr. West Point, NY: Department of Geography & Computer Science, US Military Academy Garver (1981).

Corson, Mark W., and Minghi, Julian V. (1996*a*). "*Powerscene*: Application of New Geographic Technology to Revolutionise Boundary Making." *IBRU Boundary and Security Bulletin*. Summer 1996.

—— (1996*b*). "The Political Geography of the Dayton Accords." *Geopolitics and International Boundaries*, 1/1: 77–97.

—— (1997). "Prediction and Reality in Bosnia: Two Years after Dayton." *Geopolitics and International Boundaries*, 2/3: 14–27.

Davies, A. (1946). "Geographical Factors in the Invasion and Battle of Normandy." *Geographical Review*, 36/4: 613–31.

Diersing, V. E., Shaw, R. B., Warren, S. D., and Novak, E. W. (1988). "A User's Guide For Estimating Allowable Use of Tracked Vehicles in Nonwooded Military Training Lands." *Journal of Soil and Water Conservation*, 43/2: 191–5.

Dupuy, Richarad Ernest (1939). *World in Arms; A Study in Military Geography*. Harrisburg, PA: Military Service Publishing Company.

Edwards, H. J. (trans.) (1939). Ceasar, *Gallic Wars*, Loeb Classical Library. London: William Heinemann.

Forde, H. M. (1949). "An Introduction to Military Geography." *Military Review*, 28: 30–6, 55–62.

Frederick the Great. (1747). *Instructions for His Generals*, trans. by T. R. Phillips. Harrisburgh, PA.

Gaile, G. L., and Willmott, C. J. (eds.) (1989). *Geography in America*. Columbus, Ohio: Merrill Publishing Company.

Galloway, Colonel Gerald E., Jr. (compiler) (1990). *Readings in Military Geography*. West Point, NY: Department of Geography, US Military Academy.

Garver, Colonel John B., Jr. (compiler) (1981). *Readings in Military Geography*. West Point, NY: Department of Geography & Computer Science, US Military Academy.

Garver, John B., Jr., and Galloway, Gerald E., Jr. (compilers) (1984). *Readings in Military Geography*. West Point, NY: Department of Geography & Computer Science, US Military Academy.

Goure, Daniel (1995). "Non-Lethal Force and Peace Operations." *GeoJournal*, 37/2: 267–75.

Gutmanis, Ivars (1995). "United States International Policies and Military Strategies in the Era of Defunct Aggressor." *GeoJournal*, 37/2: 257–66.

Hansen, David G. (1997). "The Immutable Importance of Geography." *Parameters* (Spring): 55–64.

Jackman, Albert (1962). "The Nature of Military Geography." *The Professional Geographer*, 14: 7–12.

—— (1971). "Military Geography." in Roger A. Leestma (ed.), *Research Institute Lectures on Geography, Special Report ETL-SR-71*, Fort Belvoir, Va.: US Army Engineer Topographic Laboratories.

James, P. E., and Jones, Clarence F. (1954). *American Geography: Inventory and Prospect*. Syracuse, NY: Syracuse University Press.

James, P. E., and Martin, G. J. (1979). *The Association of American Geographers: The First Seventy-Five Years, 1904–1979*. Washington, DC: Association of American Geographers.

Johnson, D. W. (1917). *Topography and Strategy in the War*. New York: Henry Holt.

—— (1921). *Battlefields of the World War, Western and Southern Fronts: A Study in Military Geography*. Research Series No. 3. New York: American Geographical Society,.

Lacoste, Yves (1973). "An Illustration of Geographical Warfare: Bombing of the Dikes on the Red River, North Vietnam." *Antipode*, 5/2: 1–13.

Lavallée, T. S. (1836). *Géographie Physique, Historique et Militaire*. Paris.

McColl, Robert W. (1993). "The Creation and Consequences of International Refugees: Politics, Military, and Geography." *GeoJournal*, 31/2: 169–77.

MacDonnell, Colonel A. C. (1911). *The Outlines of Military Geography*. London: Hugh Rees, LTD.

Mackinder, Halford J. (1902). *Britain and the British Seas*. New York: D. Appleton and Co.

Maguire, T. Miller (1899). *Outlines of Military Geography*. Cambridge: Cambridge University Press.

Mahan, Alfred Thayer (1890). *The Influence of Sea Power upon History, 1660–1783*. Boston: Little, Brown & Co.

Malinowski, Jon, and Brockhaus, John (1999). "Correlates of U.S. Army Recruiting Success in Texas." Paper presented at the annual conference of the AAG, 23–8 March 1999. Honolulu, HI.

Martin, Geoffrey J., and James, Preston E. (1993). *All Possible Worlds: A History of Geographical Ideas*. New York: John Wiley & Sons.

Mason, Charles H. (1948*a*). "The Role of the Geographer in Military Training." *Bulletin of the American Society for Professional Geographers* (April): 7.

—— (1948*b*). "Role of the Geographer in Military Planning." *Annals*, 38/1: 104–5.

May, Colonel Edward Sinclair (1909). *An Introduction to Military Geography*. London: Hugh Rees.

Meigs, Peveril (1961). "Some Geographical Factors in the Peloponnesian War." *Geographical Review*, 51/3: 370–80.

Munn, Alvin A. (1980). "The Role of Geographers in the Department of Defense." *The Professional Geographer*, 32/3: 361–4.

O'Loughlin, J., and van der Wusten, H. (1986). "Geography of War and Peace: Notes for a Contribution to a Revived Political Geography." *Progress in Human Geography*, 10/3: 434–510.

—— (1993). "Political Geography of War and Peace," in P. J. Taylor (ed.), *Political Geography of the Twentieth Century: A Global Analysis*. London: Belhaven Press.

O'Sullivan, Patrick (1980). "Warfare in Suburbia." *Professional Geographer*, 32/3: 355–60.

—— (1991). *Terrain and Tactics*. New York: Greenwood Press.

—— (2001). *The Geography of War in the Post Cold War World*. Lewiston, NY: Edwin Mellen Press.

O'Sullivan, Patrick, and Miller, Jesse W., Jr. (1983). *The Geography of Warfare*. New York: St. Martin's Press.

Palka, Eugene J. (1987). "Aerial Photography." *Infantry Journal* (May–June): 12–14.

— (1988). "Geographic Information in Military Planning." *Military Review*, 68/3: 52–61.

— (1995). "The US Army in Operations other than War: A Time to Revive Military Geography." *GeoJournal*, 37/2: 201–8.

Palka, Eugene J., and Galgano, Francis A. (eds.) (2000). *The Scope of Military Geography*. New York: McGraw-Hill.

Palka, Eugene J., and Lake, Dawn (1988). *A Bibliography of Military Geography*. 4 vols. West Point, NY: US Military Academy Press.

Peltier, Louis C. (1961). "The Potential of Military Geography." *Professional Geographer*, 13/6: 3–4.

Peltier, Louis C., and Pearcy, G. Etzel (1966). *Military Geography*. Princeton, NJ: D. Van Nostrand.

Pepper, David, and Jenkins, Alan (eds.) (1985). *The Geography of Peace and War*. Oxford: Basil Blackwell.

Poole, Sidman P. (1944). "The Training of Military Geographers." *Annals*, 34/4: 202–6.

Renner, George (1951). "Instructor's Manual for Political and Military Geography." New York: US Air Force Reserve Officer Training Program, Headquarters Continental Air Command Directorate, Mitchell Air Force Base.

Roon, Capt. Albrecht Theodor Emil Graf von (1837). *Militarische Landerbeschreiburg von Europa*, i. *Mittel-und Sud-Europa*. Berlin.

Rouse, W. H. D. (1964). *The March Up Country*. A translation of Xenophon's *Anabasis*. Ann Arbor, Mich.: University of Michigan Press.

Russell, J. A. (1954). "Military Geography," in P. E. James, and C. F. Jones (eds.) *American Geography: Inventory and Prospect*. Syracuse, NY: Syracuse University Press (1954).

Semple, Ellen Churchill (1903). *American History and Its Geographic Conditions*. Boston and New York: Houghton, Mifflin.

Shalikashvili, John M. (1995). "National Military Strategy of the United States of America." Washington, DC: US Government Printing Office.

— (1997). "National Military Strategy of the United States of America." Washington, DC: US Government Printing Office.

Shannon, Hon. John W., and Sullivan, General Gordon R. (1993). "Strategic Force—Decisive Victory: A Statement on the Posture of the United States Army, Fiscal Year 1994." Presented to the Committees and Subcommittees of the United States Senate and the House of Representatives, First Session, 103rd Congress.

Shaw, Robert B., Doe, William W. III, Palka, Colonel Eugene J., and Macia, Thomas E. (2000). "Where Does the U.S. Army Train to Fight? Sustaining Army Lands for Readiness in the 21st Century." *Military Review*, 80/5, (September–October 2000): 68–77.

Soffer, Arnon (1982). "Topography Conquered: The Wars of Israel in Sinai." *Military Review*, 62/4: 60–72.

Soffer, Arnon, and Minghi, Julian V. (1986). "Israel's Security Landscapes: The Impact of Military Considerations on Land Uses." *Professional Geographer*, 38/1: 28–41.

Taylor, E. R. G. (1948). "Geography in War and Peace." *Geographical Review*, 38/1: 132.

Taylor, T. G. (1951). *Geography in the Twentieth Century*. New York: Philosophical Library.

Thompson, Edmund R. (1962*a*). "Military Geography in the Nineteenth Century." Paper delivered at the meeting of the AAG, 24 April.

— (1962*b*). "The Nature of Military Geography: A Preliminary Survey." Ph.D. Dissertation. Syracuse University.

Warman, Henry J. (1954). *Geography-Backgrounds, Techniques and Prospects (For Teachers)*. Worcester, Mass.: Clark University Print Shop.

Varren, S. D., Diersing, V. E., Thompson, P. J., and Goran, W. D. (1989). "An Erosion-Based Land Classification System for Military Installations." *Environmental Management*, 13/2: 251–7.

Whitman, J. (1994). "A Cautionary Note on Humanitarian Intervention." *GeoJournal*, 34/2: 167–75.

Winters, Harold A. (1991). "The Battle That Was Never Fought: Weather and the Union Mud March of January 1863." *Southeastern Geographer*, 31/1: 31–8.

— (1998). *Battling the Elements: Weather and Terrain in the Conduct of War*. Baltimore, Md.: The Johns Hopkins University Press.

Yake, Daniel, and Hanson, Tim. (1995). "When Disaster Strikes." *Off Duty*. America Edition, February/March 1995: 6–8.

CHAPTER 32

Aging and the Aged

Susan M. Macey, Geoffrey C. Smith, and John F. Watkins

Introduction

Elders are the fastest growing segment of the American population. In 1900, average life expectancy was 47 years. In the 1990s this figure stood at 78 years (Satcher 1996). Thus, not only are there both a higher percentage and a greater number of elderly individuals, but they are living longer, thus presenting a unique opportunity and challenge for geographic research. An earlier summary of the geographic literature on aging details several well-developed themes (Golant *et al.* 1989). These include residential location and migration patterns, activity patterns, and environmental relationships. The same themes persist and have been joined by work in health and service provision. Newer issues appearing in the literature concern the implications of a spatial shift in the elderly population for personal and environmental outcomes, with both the natural and built environments being the objects of study. A notable characteristic of these geographic studies is their broad range of scales from the macro-level (migration) to the micro-level (daily living space).

This chapter seeks to highlight representative and influential contributions made by geographers to our understanding of how these demographic and spatial shifts affect the North American scene and how they will continue to impact America in the twenty-first century. It takes a broad, but selective view of current aging research as geographers are just one group of social scientists studying the elderly population and much collaboration and overlap in interests occur. The objective is not to discuss the findings of individual research, but rather to explore the breadth of issues examined by geographers. Several themes will be explored, including demographics and the components of population change, migration patterns, residential location and housing, service and health delivery, and environmental issues of particular relevance to the elderly population.

The Demographics and Dynamics of our Aging Population

The aging of populations, and the growing number of older individuals, implies that the spatial mobility of the aged will be a growing force shaping societies. The migration patterns of older persons have attracted considerable attention among scholars in a host of disciplines and have continued to be a research focus in geography. Taken together, migration and "aging in place" are two major components of demographic change producing spatial distributions of older people. Studies have been undertaken at different spatial scales, ranging from the living space to regional patterns.

Regional Patterns of Migration

Where do elderly migrants come from, where do they go, and what shapes their migration choices? These are questions that for decades have been posed by geographers, and they remain the basis of recent and ongoing research. Mobility studies within the United States have evolved in their spatial scale, moving from the state-level and regional examinations that dominated the 1980s (Golant 1990*b*; Watkins 1989) to county-level studies (Rowles and Watkins 1993; Watkins 1990). These latter studies have confirmed that, besides the maintenance of south and westward shifts of aged persons through retirement migration—Florida and Arizona retain their lead positions as favored destinations—there is an emergence of county clusters within many states of the country that serve as destination islands. Although these clusters may at times represent areas of high natural amenities (Rogers 1992), there appears to be a rapid growth within metropolitan counties, especially in the south (Frey 1993; Golant 1990*a*).

Elderly migrant selectivity and choices have been addressed through several distinct lines of research. D. E. Clark *et al.* (1996), for example, took a "traditional" approach with an application of push-pull models and a difference model, using US Census data on personal and place characteristics for several elderly migrant age groups. Newbold (1996) also looked at this mix of characteristics in the US, while Everitt and Gfellner (1996) focused on Canadian elders. Several authors have emphasized housing choices and values and economic factors in their examinations (W. A. V. Clark and Davies 1990; W. A. V. Clark and White 1990; Steinnes and Hogan 1992; Warnes and Ford 1995) while others have examined return migration, with a general assumption that returnees are inspired by the location of family and friends who reside at or near the migrant's place of birth (Rogers 1990; Newbold 1996; Stoller and Longino 2001). More directly, Carlson *et al.* (1998) looked specifically at the "family and friends" element of the migration decision in Idaho. Cuba and Longino (1991) underscored the origin-destination proximity in moves to Cape Cod, with the maintenance of social ties being a principal factor in the migrant decision. Cuba (1991) goes on to examine in more detail the retirement migration decision-making strategy, finding that the destination choice was commonly made well in advance of the actual decision to move.

Components of Population Change

It has been well documented that migration is a dominant component of population change at sub-national levels, increasing in importance at progressively finer spatial scales. Several geographers have examined the relative importance of migration and "aging in place." Pynoos (1990) provided an inventory of more than ninety federal programs that he suggests are related to aging in place and argued for more explicit concern with the implications of adopting aging in place as a policy priority. From a demographic perspective, Ahmed and Smith (1992) explored the comparative importance of fertility, mortality, and net migration as factors in US state-level aging. Rogers and Woodward (1992) used these components, along with distinct migration flow data, in projection models to identify the variety of tempos of aging in the country. Frey (1995) also looked at the notion of aging in place, extending his analysis to determine the differential characteristics of elders who either migrate or age in place. This extension allowed for more clarity in determining the impacts of migration and aging in place on both origin and destination areas. Rowles (1993) explicitly discussed the "place of aging" as venue for those who either age in place or migrate, paying special attention to the development of place ties and the ways in which individuals form identifications with place, with enhanced life quality being an important outcome of more firmly established ties. In a later work, Rowles (1994) summarized the emerging body of writing on how the elderly experience and develop affinity for place, tracing the development of the "aging in place" philosophical rationale, and suggesting that personal and societal images of the relationship between aging and place are evolving and assuming new forms.

At the regional level, Rogers and Raymer (1999) used data from the US Census Bureau and other sources to examine the influence of birthplace on the internal migration and spatial distribution patterns of elderly foreign- and native-born populations in the US from 1950 to 1990. Their study focused on regional population stocks and their sources of growth and change. The results of a later report by the same authors (Rogers and Raymer 2001) reveal the interregional migration patterns exhibited considerable stability between 1950 and 1990, contributing less than aging in place to the shaping of regional elderly population growth and distribution patterns. A study of the components of change (mortality and estimated migration) in Pennsylvania by Schnell (1994) is one of the few to include institutional as well as residential populations.

The Spatial-Temporal Continuum

Mobility as a human action embraces a broad range of spatial and temporal scales. Defined as a permanent move across an administrative boundary, migration has often been distinguished by the distance moved. Are elders moving locally, say, within a county, migrating to contiguous county locations, or crossing one or many state lines to reach their destinations? Several of the studies already mentioned have explicitly categorized and examined elderly migrants along this spatial continuum (e.g. Rowles and Watkins 1993; Warnes and Ford 1995; Watkins 1990). Also recognized is the temporal continuum in mobility, including the prevalence of seasonal movers (e.g. snowbirds), and the accumulation of spatial experience as prelude to a permanent move. Craig (1992) and Hogan and Steinnes (1994) looked specifically at the levels, characteristics, and determinants of snowbird moves in the United States. Other studies explored the connections between seasonal mobility and permanent moves, testing the hypothesis that greater seasonal experience in place might increase the likelihood of a permanent move to that place (McHugh 1990; Hogan and Steinnes 1993). Both studies found that seasonal moves are not a precursor to migration but form a distinct lifestyle.

A final element of the temporal continuum in mobility relates to age. Contemporary research on elderly mobility is moving away from a singular definition of the elderly population that uses a minimum age criterion of 60 or 65 years for inclusion in the group. The works by Rogers and his colleagues are perhaps most indicative of the variations to be found within the older age groups (Rogers 1990, 1992; Rogers and Raymer 1999; Rogers and Woodward 1992). These scholars commonly use five-year age groups in their modeling exercises, and have clearly justified the need to distinguish, for example, between younger retiree migrants and "old and oldest" elderly migrants. Speare and Meyer (1988) used Annual Housing Survey data to associate reasons for moving with age, and consequently with destination choice. Similarly, Clark and Davies (1990) examined specific elderly age groups in their housing study and, in their conclusions, called for the treatment of age as a continuum rather than a set of some predetermined age groups. This concept of a continuum reaches beyond the labeling of individuals according to increasingly finer age groupings. "Age" is rich with meanings and connotations that differ from person to person, from culture to culture, and from time to time. The whole concept of age forms the basis for much debate in gerontology, and the debate has moved only slowly into the field of aging studies in geography (see e.g. Laws 1995; McHugh 2000).

The People and Places of Mobility

Research efforts in the geography of aging and the aged have moved firmly toward promoting a clearer understanding of the people and places associated with migration. One topic of inquiry has been that of family and friends. Besides studies that recognize social ties as one of many determinants of mobility (Carlson *et al.* 1998; Cuba and Longino 1991; Walters 2000), several works have focused almost solely on these ties. Rogerson *et al.* (1993), Lin and Rogerson (1995), and Lin (1997) are central among these focused works. They examine the actual distances between elders and their adult children, seek explanations for these distances, and explore how distances and household structure are used in mobility decisions. The importance of this line of research lies in the provision of support across generations. Although increases in the prevalence of institutionalization have been touted as a trend that may carry us through the twenty-first century, the reality is that most elders live independently or with children. Indeed, cross-generational caregiving is a vital component of local services for the aged population (Morrison 1990) and will be necessary for promoting life quality in the face of lengthening life expectancies resulting from medical advances and recent behavioral changes.

A second line of research has emerged that deals with the impacts of elderly migration on place. Rowles and Watkins (1993), in a study of selected Appalachian communities, touched on the breadth of impacts that might be experienced. These potential impacts range from savings and investment transfers and increased local spending on housing and luxury items, to increased service demands and the polarization of and conflict between socially and culturally disparate groups. Other studies have focused on specific impacts. Bennett (1993, 1996) explored local expenditure patterns in southeastern coastal destination counties and also addressed the issue of enhanced local volunteerism associated with retirement in-migration. Service demands and provision have been examined by several scholars, including Joseph and Cloutier (1991) in a study of rural Ontario and McHugh and Mings (1994) as part of their ongoing study of seasonal migrants in Arizona. Both of these works pointed up the impact of the age progression of functional status and the importance of prior experience as a precondition of dependence and service use. In a later study, McHugh and Mings (1996) used several case studies to explore the evolving meaning of home, place, and migration among seasonal migrants over their life course. Further research (Cuba 1992) has probed the notion of migrant perspectives of place. Cuba's study, conducted on Cape Cod, examined how migrants across a spectrum of ages perceive change at their destination, and how these perceptions influence assimilation into local society and culture. Finally, other work (Warnes *et al.* 1999) has investigated the personal outcomes of British expatriate retirees in southern Europe.

Residential Location

In addition to the study of the changing spatial distributions of older people at different scales of analysis, knowledge of where older people live is of vital importance for providers of a variety of services and resources targeted to the elderly population. Since 1988, research investigating residential patterns of the elderly population has been mainly focused on three spatial scales: national, regional (areas within a nation), and urban.

At the national scale, Rogerson (1996) conducted a state-level study of elderly population growth in the United States during the period 1985–90, while Moore *et al.* (1997) examined shifts in Canada's elderly population by census division from 1986 to 1991. Both studies revealed a marked tendency for the spatial variation of the elderly population to decline during the study period. In a further study, Rogerson (1998) used a geographic information system to provide a geography of African American elderly populations. Other authors (Frey 1992; Smith 1998*a*) have revealed that older people in both the United States and Canada have become increasingly metropolitanized in recent years, with a clear majority of the elderly persons in each national population now living in metropolitan areas. The increasing attraction of metropolitan areas and resort communities for elderly individuals has likewise been noted in several regional studies of the elderly population of Ontario, Canada (Dahms 1996; Rosenberg and Moore 1990; Rosenberg *et al.* 1989). However, it is noteworthy that Rosenberg and Moore (1990) also drew attention to the vulnerability of rapidly aging Ontario counties to public policy initiatives concerning the structure of transfer payments. In other work at the regional scale, Sheskin and his colleagues provide an ethnic perspective in their studies of the Jewish elderly in Florida (Sheskin 1999; Sheskin *et al.* 1993).

Recent research on elderly residential patterns in North American cities has also highlighted change in the degree of spatial segregation of the aged from their younger counterparts. Golant (1990*a*) found that in terms of a simple central city versus suburbs geographical division, the non-institutionalized elderly population (age 65 and over) in US metropolitan areas became progressively less segregated from younger age cohorts during the period 1970–88. This trend has been partly attributed to the aging in place of suburban residents. Similar recent trends toward elderly suburbanization have been disclosed in the context of Canadian metropolitan areas (Moore *et al.* 1997; Smith 1998*a*). Overall, age-segregation in North American metropolitan areas is typically moderate, with higher levels generally associated with "younger" communities in terms of the ages of their physical structures and the percentages of their populations comprised of elderly residents.

The Environment and the Aged

How we age is as vital a question as where we age. These attributes may be two sides of the same coin, for example, when we examine housing type and its impact on lifestyle, or access to health care and other services. We now appreciate that elderly individuals are not necessarily passive in their interaction with the environment. The situation is often complex, involving the interplay of the elderly person's socioeconomic and demographic attributes, as well as the circumstances presented by the environment.

Housing for the Elderly Population

While the overwhelming majority of the elderly population live in conventional dwellings in the community, a wide range of alternative age-integrated and age-segregated housing environments are available to meet the diverse needs of older persons with varying degrees of dependency. These alternatives include various forms of congregate housing, government-subsidized senior citizen apartments, single-room occupancy hotels (SROs), continuing care retirement communities (CCRCs), shared housing, assisted care facilities, rooming houses, residential care facilities, nursing homes, and many other forms of housing. It is notable that a comprehensive integrated account of this bewildering array of housing options, and their appropriateness for the elderly population of the United States, is provided in a book authored by a geographer (Golant 1992).

In general terms, geographic work concerning aspects of elderly housing is distinguished by a relative emphasis on various location and environmental issues. Using the US 1987 Annual Housing Survey, Golant and La Greca (1994*a*) discovered spatial variation in the quality of housing occupied by elderly homeowners and renters in terms of city–suburban, metropolitan–non-metropolitan, and regional differences. In contrast, the same authors found that, after controlling for socio-demographic variables, chronological age and length of residence were poor predictors of housing quality (Golant and La Greca 1995). However, differences were disclosed among elderly subgroups defined on the basis of racial or ethnic criteria (Golant and La Greca 1994*b*). Theoretical perspectives have also been offered concerning the residential adjustments of elderly people who require more supportive shelter and care settings than are provided by conventional dwellings and neighborhoods (Golant 1998). This work has particularly focused on the circumstances under which characteristics of various group residential settings will be congruent with the needs of elderly occupants. Reflecting society's maturing attitude towards elderly people, many current studies examine the active role of seniors in shaping their circumstances and environment. For example, Golant's conceptual model (1998) focused on how changes in personal and environmental outcomes are experienced differently by older persons following a residential change. This model attempted to portray the complex array of individual qualities that influence how older people evaluate their shelter and care setting.

Yet other work has investigated characteristics of the local environments of various types of elderly housing. The service infrastructures of the neighborhoods occupied by specialized seniors' housing projects have been assessed (Everitt and Gfellner 1995; Smith and Gauthier 1995), while Rollinson (1990, 1991) examined the physical characteristics of the local environments of single-room occupancy hotels that accommodate the elderly poor. From a historical perspective, Laws (1993, 1994, 1995) proposed that urban built environments for elderly individuals are reciprocally related to changing social perspectives on the aging process. She suggested that certain forms of age-segregation in the city, such as the prevalence of inner city elderly concentrations after World War II, might have been territorial expressions of intergenerational conflict. It has also been suggested that some current age-segregated built environments, like social housing, might be incongruent with the "postmodern life course," which both rejects rigid age scheduling and celebrates diverse lifestyles (Holdsworth and Laws 1994).

Service Delivery and Utilization

More than a decade ago it was noted that, despite the focal importance assumed by public-service equity issues in locational analysis, research focused on explicitly geographical aspects of elderly service delivery and utilization was fragmentary and lacking a clear paradigm (Golant *et al.* 1989). While these observations remain equally valid, it is none the less possible to identify some recurrent themes in recent research on elderly services conducted by geographers and scholars in related disciplines.

Considerable attention has been directed towards geographic aspects of informal support services provided for elderly people by adult children (DeWit and Frankel 1988; Rogerson *et al.* 1993; Smith 1998*b*). This interest partly reflects recent social policies of the United States and other Western nations that have tended to favor the development of the informal support system over costly formal support services that are subject to increasing public accountability (McCaslin and Golant 1990). The results of some recent work, however, suggest that trends toward economic instability and family disruption associated with post-industrial social change are increasingly limiting the potential for intergenerational spatial propinquity and interaction, particularly in lower-income families (Greenwell and Bengtson 1997; Joseph and Hallman 1996). On the other hand, the results of an analysis of three nationally representative British datasets produced between 1996 and 1999 (Grundy and Shelton 2001) do not disclose a trend toward reduced contact between adult children and their non-coresident parent. In related work, Rogerson (1998) stressed the need for the availability of informal supports to be taken into account in attempts to target services to African-American elderly populations.

Meyer and her associates have employed formal spatial modeling procedures to evaluate the delivery and utilization of congregate meal programs and home-care services in Connecticut (Meyer and Cromley 1991; Meyer *et al.* 1991; Wong and Meyer 1993). An interaction potential model was formulated and applied to determine spatial variations in the availability of nutrition services in the service region of the Western Connecticut Area Agency on Aging (Meyer and Cromley 1991). Alternative iterations of the model (based on different personal mobility assumptions) each revealed considerable variation in availability that, in turn, was strongly related to patterns of use of the services.

Behavioral modeling approaches have been employed to investigate the elderly individual's cognition, evaluation, and utilization of various service resources. They included shopping centers (Smith 1992), health and social services (Joseph and Cloutier 1990), and key categories of service outlets (Smith and Gauthier 1995; Smith and Sylvestre 2001). Overall, the findings of this work indicate that behaviors related to service utilization vary among elderly segments defined according to gender, residential location, housing type, health status, and access to transportation resources. Access to alternative types of transportation among the elderly population has been found to vary according to a range of factors including gender, urban–rural residence, age, and car ownership (Cutler and Coward 1992).

A substantial body of work has focused on elderly service issues in rural areas. Particular problems related to the dispersed nature of elderly populations, geographic isolation, limited public-sector capacity, frequent unavailability of public transportation, and the relatively high incidence of poverty (Joseph and Cloutier 1990, 1991; Joseph and Fuller 1991; Rowles and Johansson 1993). The rural elderly population is frequently disadvantaged in terms of service accessibility relative to their urban counterparts. Moreover, demographic and economic changes are threatening informal support systems and local community resources that act as traditional buffers to severe service deprivation among the rural elderly population. Rowles and Johansson (1993) examined persistent elderly poverty in rural Appalachia. They expressed a need for a "new vision" that would provide a culturally appropriate, locally based response to the structural conditions that perpetuate poverty in the elderly population. They spoke of the physical environment as a "hostile terrain" with disadvantages of inaccessibility and high travel and service costs accentuated by the continuing lack of basic services in many parts of the region. Rowles *et al.* (1996) addressed the issue of long-term care for the rural elderly population. The authors identified several guiding principles for rural long-term care that focus on the reorientation of attitudes away from the traditional adoption of urban models toward promoting a more flexible, client-centered philosophy that better reflects rural needs and culture.

Health and the Quality of Life

Several studies have identified the need to address demand issues for health care and support. Everitt and Gfellner's series of surveys and workshops on determining seniors' independence in the Westman region of Manitoba, Canada, provide insight into elderly people's health, satisfaction with life, physical activity level, availability of formal and informal support in communities, use of public transportation, and knowledge of and ability to use available services (Everitt and Gfellner 1996, 1997, 1998). While nursing home relocation has received considerable attention, the little-studied phenomenon of room changes within nursing homes was the subject of a study by Everard *et al.* (1994). They monitored room changes in four nursing homes over a one-year period. Through participant observation, repeated interviews, and event analyses, they developed detailed case studies and a model of the decision-making process involved in room changes.

Quality of life has been examined at the micro- as well as the macro-level. Elderly tenants in Florida's rent-subsidized apartment buildings have been subjects for Golant (1999) who documented the unmet needs of those elderly tenants whose physical impairments and psychological problems threaten their ability to live independently. His study described the difficulties housing owners, sponsors, and administrators face when trying to help their tenants age in place and the different helping strategies they adopted. At the macro-scale, Golant (1992) identified places with larger than average residential concentrations of elderly individuals. He considered how this larger elderly presence is linked to a locality's overall quality of life, population characteristics, economic well-being, political orientation, government expenditures on the services and facilities needed by the elderly population, and the incidence of elderly housing problems. He also provided an overview of community leaders' beliefs concerning whether a large population of elderly individuals was a political asset or liability.

Given that most elderly Americans are women and that the imbalance becomes more pronounced with age (Satcher 1996), gender issues are also underrepresented in the literature. Harper and Laws (1995) particularly noted the "feminization" of our aging society. Macey and Schneider (1995) addressed the common image of frailty in old age that is generally associated with gender. Their study analyzed the demographic and location characteristics of elderly deaths classified as due to neglect, lack of food or water, exposure, and privation. Moss (1997) examined the importance of micro-space to older women living with arthritis while Moss and Dyck (1996) inquired into the relationship between women, work, and chronic illness. However, in general, there has been scant attention given to the impact of crime, abuse, or gender issues on the elderly population in the geographic literature.

A small but rapidly growing segment of the elderly population, the oldest old, has received attention in other disciplines, but little special emphasis in geographic studies to date. Those aged 85 and over are projected to be the fastest-growing sector of the elderly population into the next century (US Bureau of the Census 1989). Suzman *et al.* (1992) set a benchmark for our knowledge of this sector of the elderly population. A common theme is the differential effects of socio-economic status, race, and rural versus urban residence on health of the oldest-old (Clayton *et al.* 1994; Johnson 1994).

Health promotion is also a common issue for social scientists whose concerns often parallel those of geographers (Cowart *et al.* 1995; Hickey and Stilwell 1991). Cowart *et al.* (1995) examined the implications of health promotion for older rural African Americans through the examination of a program aimed at empowering the local population to assess and prioritize their health needs to improve their own level of health. The influence of age and race on elderly mortality and morbidity were also examined more generally by Gibson (1994) who suggested that a geographic perspective could aid in the explanation of the unusual phenomenon of blacks being less disadvantaged in the older age groups for mortality and morbidity.

Environmental Concerns

The specific impact of environmental issues on elderly individuals was a relatively new concern appearing in the late 1980s. Although Bennett (1996) included an assessment of the impact of retirement development on the environment in rural coastal areas, environmental issues are generally addressed as just one of many categories of location attributes. Not since the late 1980s have explicit energy issues relating to the elderly population been addressed. Two studies by Macey (1988, 1989) examined the special needs and concerns of the elderly population related to energy consumption and health. Further work by Berry and Brown (1988) reviewed participation by elderly individuals in residential energy conservation programs. While these authors state that few conservation programs have been successful at reaching the elderly population and coverage of elderly individuals by utility, federal, and state programs is limited, little follow-up work has been done to address needs in this field.

Another area that has received considerable study by geographers at large, natural hazards, has seen little specific attention given to the elderly population in such circumstances. Some studies have singled out elderly people as a special subgroup in the population. In examining forty-eight cities across the US (Kalkstein and Davis 1989), and in a more detailed empirical analysis of St Louis (Kalkstein 1991), these authors noted that elderly persons appear to be disproportionately stressed by weather extremes. However, these studies examined the elderly population (those 65 years or older) as just one of six age categories. Macey and Schneider (1993, 1994) extended our knowledge by focusing specifically on those 60 years of age and older. Their studies of deaths from excessive heat and excessive cold detailed differences in mortality according to age, gender, race, and location attributes.

Pathways to the Twenty-First Century

Aging is an unavoidable dynamic. Very few countries in today's world are actively getting "younger" in terms of their age profiles; the dominant world trend is towards an aging society in which elderly populations are not only growing in size but are also growing as a share of the total population. As a component of demographic change, elderly mobility will remain a focal area of research as geographers attempt to deepen their understanding of evolving economic and social conditions, primarily as a consequence of the American Baby Boomer's movement through retirement. Migrant "winners and losers" and the determinants of elderly migration will likely remain at the forefront of research efforts. As health and service sectors are likely to become strained, research in these areas will also be a priority. But geography has much more to offer in studies of the elderly population besides the traditional measurement and mapping of the spatial manifestations of aging and the aged. How have elders gathered and used spatial information throughout their lives, and how do their experiences inform their current decisions (Watkins 1999)? What are the structural elements of age as a characteristic that transcend the simple measure of chronological existence (McHugh 2000)? Can understanding of mobility as a spatial transition be extended to other issues of aging (Uhlenberg 1995)? Can geographers effectively adopt "different" approaches in their studies better to apply our unique geographic perspectives (McHugh 1999)? Much work remains to be done in furthering understanding of the specific interaction of elderly people with their environment, be it at the macro- or micro-scale. Geographers on both sides of the Atlantic (Harper and Laws 1995; Warnes 1990) have advocated wider exchanges between geographers and social gerontologists. The growing need to prepare for an aging society provides both an opportunity and a challenge for geographic research to expand its horizons.

References

Ahmed, B., and Smith, S. K. (1992). "How Changes in Components of Growth Affect the Population Aging of States." *Journals of Gerontology: Social Sciences*, 47/1: S27–S37.

Bennett, D. G. (1993). "Retirement Migration and Economic Development in High-Amenity, Nonmetropolitan Areas." *The Journal of Applied Gerontology*, 12/4: 466–81.

— (1996). "Implications of Retirement Development in High-Amenity Nonmetropolitan Coastal Areas." *Journal of Applied Gerontology*, 15: 345–60.

Berry, L. G., and Brown, M. A. (1988). "Participation of the Elderly in Residential Conservation Programmes." *Energy Policy*, 16/2: 152–63.

Carlson, J. E., Junk, V. W., Fox, L. K., Rudzitis, G., and Cann, S. E. (1998). "Factors Affecting Retirement Migration to Idaho: An Adaptation of the Amenity Retirement Migration Model." *The Gerontologist*, 38/1: 18–24.

Clark, D. E., Knapp, T. A., and White, N. E. (1996). "Personal and Location-Specific Characteristics and Elderly Interstate Migration." *Growth and Change*, 27 (Summer): 327–51.

Clark, W. A. V., and Davies, S. (1990). "Elderly Mobility and Mobility Outcomes: Households in the Later Stages of the Life Course." *Research on Aging*, 12/4: 430–62.

Clark, W. A. V., and White, K. (1990). "Modeling Elderly Mobility." *Environment and Planning A*, 22/7: 909–24.

Clayton, G. M., Dudley, W. N., Patterson, W. D., Lawhorn, L. A., Poon, L. W., Johnson, M. A., and Martin, P. (1994). "The Influence of Rural/Urban Residence on Health in the Oldest-Old." *International Journal of Aging and Human Development*, 38/1: 65–89.

Cowart, M. E., Sutherland, M., and Harris, G. J. (1995). "Health Promotion for Older Rural African Americans: Implications for Social and Public Policy." *Journal of Applied Gerontology*, 14/1: 33–46.

Craig, W. J. (1992). "Seasonal Migration of the Elderly: Minnesota Snowbirds." *Southeastern Geographer*, 32/1: 38–50.

Cuba, L. (1991). "Models of Migration Decision Making Reexamined: The Destination Search of Older Migrants to Cape Cod." *The Gerontologist*, 31/2: 204–9.

— (1992). "Aging Places: Perspectives on Change in a Cape Cod Community." *Journal of Applied Gerontology*, 11/1: 64–83.

Cuba, L., and Longino, C. F. J. (1991). "Regional Retirement Migration: The Case of Cape Cod." *Journals of Gerontology: Social Sciences*, 46/1: S33–S42.

Cutler, S. J., and Coward, R. T. (1992). "Availability of Personal Transportation in Households of Elders: Age, Gender, and Residence Differences." *The Gerontologist*, 32: 77–82.

Dahms, F. (1996). "The Greying of the South Georgian Bay." *The Canadian Geographer*, 40: 148–63.

DeWit, D. J., and Frankel, B. G. (1988). "Geographic Distance and Intergenerational Contact: A Critical Assessment and Review of the Literature." *Journal of Aging*, 2: 25–43.

Everard, K., Rowles, G. D., and High, D. M. (1994). "Nursing Room Changes: Toward a Decision-Making Model." *The Gerontologist*, 34/4: 520–7.

Everitt, J. C., and Gfellner, B. M. (1995). *Elderly Persons' Housing in Brandon, Manitoba: Status, Challenges, and Prospects.* A Report prepared for the Health Policy Division, Policy and Consultation Branch, Health Canada. Faculty of Science, Brandon University, Brandon, Manitoba.

— (1996). "Elderly Mobility in a Rural Area: The Example of Southwest Manitoba." *The Canadian Geographer*, 40/4: 338–51.

— (eds.) (1997). *Determining Seniors' Independence: A Participatory Workshop*. Paper No. 4. Brandon: Brandon University Press.

— (eds.) (1998). *Determining Seniors' Independence: A Profile of Brandon and the Westman Region*. Working Paper No. 3. Brandon: Brandon University Press.

Frey, W. H. (1992). "Metropolitan Redistribution of the US Elderly: 1960–70, 1970–80, 1980–90." in A. Rogers (ed.), *Elderly Migration and Population Redistribution*. London: Belhaven, 123–42.

— (1993). "U.S. Elderly Population Becoming More Concentrated." *Population Today*, (April): 3 P.

— (1995). "Elderly Demographic Profiles of U.S. States: Impacts of 'New Elderly Births,' Migration, and Immigration." *The Gerontologist*, 35/6: 761–70.

Gibson, R. (1994). "The Age-by-Race Gap in Health and Mortality in the Older Population: A Social Science Research Agenda." *The Gerontologist*, 34: 454–62.

Golant, S. M. (1990*a*). "Metropolitanization and Suburbanization of the U.S. Elderly Population: 1970–1988." *The Gerontologist*, 30/1: 80–5.

— (1990*b*). "Post-1980 Regional Migration Patterns of the U.S. Elderly Population." *Journals of Gerontology: Social Sciences*, 45/4: S135–S140.

— (1992). *Housing America's Elderly: Many Possibilities/Few Choices.* Newbury Park, Calif.: Sage Publications.

— (1998). "Changing an Older Person's Shelter and Care Setting: A Model to Explain Personal and Environmental Outcomes," in R. J. Scheidt and P. G. Windley (eds.), *Environment and Aging Theory*, Westport, Conn.: Greenwood Press, 33–60.

— (1999). *The Casera Project.* Tallahassee, Fla.: Stephen M. Golant and Margaret Lynn Duggar and Associates.

Golant, S. M., and La Greca, A. J. (1994*a*). "City-Suburban, Metro-Nonmetro, and Regional Differences in the Housing Quality of U.S. Elderly Households." *Research on Aging*, 16: 322–46.

— (1994*b*). "Differences in the Housing Quality of White, Black, and Hispanic U.S. Elderly Households." *Journal of Applied Gerontology*, 13: 413–37.

— (1995). "The Relative Deprivation of U.S. Elderly Households as Judged by Their Housing Problems." *Journal of Gerontology: Social Sciences*, 50B: S13–S23.

Golant, S. M., Rowles, G. D., and Meyer, J. W. (1989). "Aging and the Aged," in G. L. Gaile and C. J. Willmott, (eds.), *Geography in America*. Columbus, Ohio: Merrill, 451–66.

Greenwell, L., and Bengston, V. L. (1997). "Geographic Distance and Contact between Middle-Aged Children and their Parents." *Journal of Gerontology: Social Sciences*, 52B: S13–S26.

Grundy, E., and Shelton, N. (2001). "Contact between Adult Children and their Parents in Great Britain 1986–99." *Environment and Planning A*, 33/4: 685–97.

Harper, S., and Laws, G. (1995). "Rethinking the Geography of Aging." *Progress in Human Geography*, 19: 199–221.

Hickey, T., and Stilwell, D. L. (1991). "Health Promotion for Older People: All is Not Well." *The Gerontologist*, 31: 822–9.

Hogan, T. D., and Steinnes, D. N. (1993). "Elderly Migration to the Sunbelt: Seasonal versus Permanent." *The Journal of Applied Gerontology*, 12/2: 246–60.

— (1994). "Toward an Understanding of Elderly Seasonal Migration Using Origin-Based Household Data." *Research on Aging*, 16/4: 463–75.

Holdsworth, D. W., and Laws, G. (1994). "Landscapes of Old Age in British Columbia." *The Canadian Geographer*, 38: 174–81.

Johnson, C. L. (1994). "Differential Expectations and Realities: Race, Socioeconomic Status and Health of the Oldest-Old." *International Journal of Aging and Human Development*, 38: 13–27.

Joseph, A. E., and Cloutier, D. S. (1990). "A Framework for Modeling the Consumption of Health Services by the Rural Elderly." *Social Science and Medicine*, 30: 45–52.

— (1991). "Elderly Migration and Its Implications for Service Provision in Rural Communities: An Ontario Perspective." *Journal of Rural Studies*, 7/4: 433–44.

Joseph, A. E., and Fuller, A. M. (1991). "Towards an Integrative Perspective on Housing Services and Transportation Implications of Rural Aging." *Canadian Journal on Aging*, 10: 127–48.

Joseph, A. E., and Hallman, B. C. (1996). "Caught in the Triangle: The Influence of Home, Work, and Elder-Location on Work-Family Balance." *Canadian Journal on Aging*, 15: 393–412.

Kalkstein, L. S. (1991). "A New Approach to Evaluate the Impact of Climate on Human Mortality." *Environmental Health Perspectives*, 96: 145–50.

Kalkstein, L. S., and Davis, R. E. (1989). "Weather and Human Mortality: An Evaluation of Demographic and Interregional Responses in the United States." *Annals of the Association of American Geographers*, 79/1: 44–64.

Laws, G. (1993). "'The Land of Old Age': Society's Changing Attitudes toward Urban Built Environments for Elderly People." *Annals of the Association of American Geographers*, 83: 672–93.

— (1994). "Aging, Contested Meanings, and the Built Environment." *Environment and Planning A*, 26: 1787–802.

— (1995). "Understanding Ageism: Lessons from Feminism and Postmodernism." *The Gerontologist*, 35/1: 112–18.

Lin, G. (1997). "Elderly Migration: Household versus Individual Approaches." *Papers in Regional Science*, 76/3: 285–300.

Lin, G., and Rogerson, P. A. (1995). "Elderly Parents and the Geographic Availability of their Adult Children." *Research on Aging*, 17/3: 303–31.

Macey, S. M. (1988). "Energy Consumption and the Elderly." *The Canadian Geographer*, 32: 167–9.

— (1989). "Hypothermia and Energy Conservation: A Tradeoff for the Aged?" *International Journal of Aging and Human Development*, 29/2: 149–59.

Macey, S. M., and Schneider, D. F. (1993). "Deaths from Excessive Heat and Excessive Cold among the Elderly." *The Gerontologist*, 33/4: 497–500.

— (1994). "Dying of Cold in the Sunbelt: Elderly Hypothermia Mortality Rates". *OMEGA Journal of Death and Dying*, 30/1: 13–21.

— (1995). "Frailty and Mortality among the Elderly." *The Journal of Applied Gerontology*, 14/1: 22–32.

McCaslin, R., and Golant, S. M. (1990). "Assessing Social Welfare Programs for the Elderly: The Specification of Functional Goals." *Journal of Applied Gerontology*, 9: 4–19.

McHugh, K. E. (1990). "Seasonal Migration as a Substitute for, or Precursor to, Permanent Migration." *Research on Aging*, 12/2: 229–45.

— (1999). "Inside, Outside, Upside Down, Backward, Forward, Round and Round: A Case for Ethnographic Studies in Migration." *Progress in Human Geography*, 23/4: 589–607.

— (2000). "The 'Ageless Self'? Emplacement of Identities in Sunbelt Retirement Communities." *Journal of Aging Studies*, 14/1: 103–15.

McHugh, K. E., and Mings, R. C. (1994). "Seasonal Migration and Health Care." *Journal of Aging and Health*, 6/1: 111–32.

— (1996). "The Circle of Migration: Attachment to Place in Aging." *Annals of the Association of American Geographers*, 86/3: 530–50.

Meyer, J. W., and Cromley, E. K. (1991). "Nutrition Service Availability: An Interaction Potential Model." *Journal of Applied Gerontology*, 10: 431–43.

Meyer, J. W., Lusky, R., and Wright, A. (1991). "Title III Services: Variation of Use within a State." *Journal of Applied Gerontology*, 10: 140–56.

Moore, E. G., Rosenberg, M., and McGuinness, D. (1997). *Growing Old in Canada: Demographic and Geographic Perspectives*. Toronto: Statistics Canada.

Morrison, P. A. (1990). "Demographic Factors Reshaping Ties to Family and Place." *Research on Aging*, 12/4: 399–408.

Moss, P. (1997). "Negotiating Spaces in Home Environments: Older Women Living with Arthritis." *Social Science and Medicine*, 45/1: 23–33.

Moss, P., and Dyck, I. (1996). "Inquiry into Environment and Body: Women, Work and Chronic Illness." *Environment and Planning D: Society and Space*, 14/6: 737–53.

Newbold, B. K. (1996). "Determinants of Elderly Interstate Migration in the United States: 1985–1990." *Research on Aging*, 18/4: 451–76.

Pynoos, J. (1990). "Public Policy and Aging in Place: Identifying the Problems and Potential Solutions," in D. Tilson (ed.), *Aging in Place: Supporting the Frail Elderly in Residential Environments*. Glenview, Ill.: Scott, Foresman, 167–208.

Rogers, A. (1990). "Return Migration to Region of Birth Among Retirement-Age Persons in the United States." *Journals of Gerontology: Social Sciences*, 45/3: S128–S134.

— (ed.) (1992). *Elderly Migration and Population Redistribution: A Comparative Study*. London: Belhaven Press.

Rogers, A., and Raymer, J. (1999). "The Regional Demographics of the Elderly Foreign-Born and Native-Born Populations in the United States Since 1950." *Research on Aging*, 21: 3–35.

— (2001). "Immigration and the Regional Demographics of the Elderly Population in the United States." *Journal of Gerontology: Social Sciences*, 56B/1: S44–S55.

Rogers, A., and Woodward, J. A. (1992). "Tempos of Elderly Age and Geographic Concentration." *The Professional Geographer*, 44/1: 72–83.

Rogerson, P. A. (1996). "Geographic Perspectives on Elderly Population Growth." *Growth and Change*, 27: 75–95.
— (1998). "The Geography of Elderly Minority Populations in the United States." *International Journal of Geographic Information Science*, 12: 687–98.
Rogerson, P. A., Weng, R. H., and Lin, G. (1993). "The Spatial Separation of Parents and their Adult Children." *Annals of the Association of American Geographers*, 83/4: 656–71.
Rollinson, P. A. (1990). "The Everyday Geography of Poor Elderly Hotel Tenants in Chicago." *Geografiska Annaler*, 72B: 47–57.
— (1991). "Spatial Isolation of Single-Room-Occupancy Hotel Tenants." *The Professional Geographer*, 43: 456–64.
Rosenberg, M. W., and Moore, E. G. (1990). "The Elderly, Economic Dependency, and Local Government Revenues and Expenditures." *Environment and Planning C*, 8: 149–65.
Rosenberg, M. W., Moore, E. G., and Ball, S. B. (1989). "Components of Change in the Elderly Population of Canada." *The Canadian Geographer*, 33: 218–29.
Rowles, G. D. (1993). "Evolving Images of Place in Aging and 'Aging in Place'." *Generations*, 17/2: 65–70.
— (1994). "Evolving Images of Place in Aging and Aging-in-Place," in D. Shenk and W. A. Achenbaum (eds.), *Changing Perceptions of Aging and the Aged*. New York: Springer, 115–25.
Rowles, G. D., Beaulieu, J. E., and Myers, W. W. (eds.) (1996). *Long-Term Care for the Rural Elderly. New Directions, Services and Policy*. New York: Springer.
Rowles, G. D., and Johansson, H. K. (1993). "Persistent Elderly Poverty in Rural Appalachia." *Journal of Applied Gerontology*, 12: 349–67.
Rowles, G. D., and Watkins, J. F. (1993). "Elderly Migration and Development in Small Communities." *Growth and Change*, 24/4: 509–38.
Satcher, D. (1996). "The Aging of America." *Journal of Health Care for the Poor and Underserved*, 7/3: 179–82.
Schnell, G. A. (1994). "The Aged in Pennsylvania: The Components of Population Change, 1980–1990." *Journal of the Pennsylvania Academy of Science*, 68: 80–5.
Sheskin, I. M. (1999). "Florida's Jewish Elderly." *Responses to an Aging Florida*. (Spring 1999): 18–21.
Sheskin, I. M., Zadka, P., and Henry Green, H. (1993). "A Comparative Profile of Jewish Elderly in South Florida and Israel," in U. Schmelz and S. DellaPergola (eds.), *Papers in Jewry Demography, 1989*, Jewish Population Studies, 25. Jerusalem: The Avraham Harman Institute of Contemporary Jewry, Hebrew University of Jerusalem, 154–64.
Smith, G. C. (1992). "The Cognition of Shopping Centers by the Central Area and Suburban Elderly: An Analysis of Consumer Information Fields and Evaluation Criteria." *Urban Geography*, 13: 142–63.
— (1998*a*). "Change in Elderly Residential Segregation in Canadian Metropolitan Areas, 1981–91." *Canadian Journal on Aging*, 17: 59–82.
— (1998*b*). "Residential Separation and Patterns of Interaction between Elderly Parents and their Adult Children." *Progress in Human Geography*, 22: 368–84.
Smith, G. C., and Gauthier, J. J. (1995). "Evaluation and Utilization of Local Service Environments by Residents of Low-Rent Senior Citizen Apartments." *Canadian Journal of Urban Research*, 4: 305–23.
Smith, G. C., and Sylvestre, G. M. (2001). "Determinants of the Travel Behavior of the Suburban Elderly." *Growth and Change*, 32/3: 395–412.
Speare, A., Jr., and Meyer, J. W. (1988). "Types of Elderly Residential Mobility and their Determinants." *Journals of Gerontology: Social Sciences*, 43/3: S574–S581.
Steinnes, D. N., and Hogan, T. D. (1992). "Take the Money and Sun: Migration as a Consequence of Gains in Unaffordable Housing Markets." *Journals of Gerontology: Social Sciences*, 47/4: S197–S203.
Stoller, E. P., and Longino, C. F. (2001). "'Going Home' or 'Leaving Home'? The Impact of Person and Place Ties on Anticipated Counterstream Migration." *The Gerontologist*, 41/1: 96–102.
Suzman, R. M., Manton, K. G., and Willis, D. P. (1992). "Introducing the Oldest Old," in Suzman, Manton, and Willis (eds.), *The Oldest Old*. New York: Oxford University Press, 3–14.
Uhlenberg, P. (1995). "A Note on Viewing Functional Change in Later Life as Migration." *The Gerontologist*, 35/4: 549–52.
US Bureau of the Census (1989). "Projections of the Population of the United States by Age, Sex, and Race; 1988 to 2080." *Current Population Reports*, series P-25, No. 1018. Washington, DC.
Walters. W. H. (2000). "Types and Patterns of Later-Life Migration." *Geografiska Annaler*, 82B/3: 129–47.
Warnes, A. M. (1990). "Geographical Questions in Gerontology: Needed Directions for Research." *Progress in Human Geography*, 14: 24–56.
Warnes, A. M., and Ford, R. (1995). "Housing Aspirations and Migration in Later Life: Developments During the 1980s." *Papers in Regional Science*, 74/4: 361–87.
Warnes, A. M., King, R., Williams, A. M., and Patterson, G. (1999). "The Well-being of Expatriate Retirees in Southern Europe." *Ageing and Society*, 19/6: 717–40.
Watkins, J. F. (1989). "Gender and Race Differentials in Elderly Migration." *Research on Aging*, 11/1: 33–52.
— (1990). "Appalachian Elderly Migration: Patterns and Implications." *Research on Aging*, 12/4: 409–29.
— (1999) "Life Course and Spatial Experience: A Personal Narrative Approach in Migration Studies," in K. Pandit and S. Davies-Withers (eds.), *Migration and Restructuring in the United States*. Boulder: Rowman & Littlefield, 294–312.
Wong, D. W. S., and Meyer, J. W. (1993). "A Spatial Decision Support System Approach to Evaluate the Efficiency of a Meals-On-Wheels Program." *The Professional Geographer*, 45: 332-41.

CHAPTER 33

Recreation, Tourism, and Sport

Klaus J. Meyer-Arendt and Alan A. Lew

In North America, the subdisciplines of recreation geography, tourism geography, and sport geography (RTS) are alive and well. From their beginnings as serious research topics in the 1930s (cf. Mitchell and Smith 1989), the RTS subfields have gradually emerged as legitimate and significant areas of study within North American geography. Among the three subdisciplines, tourism geography has experienced the greatest growth in recent years, and in view of the role of tourism in the world economy, that growth trend is expected to continue.

This chapter presents an overview of research in recreation geography, tourism geography, and sport geography by North American geographers since 1988. Research conducted prior to that date was summarized in Mitchell and Smith's (1989) chapter in the first volume of *Geography of America* (Gaile and Willmott 1989), and readers are urged to consult that reference. An excellent summary of themes in RTS research from a global perspective is provided by Hall and Page (1999).

Definitions

According to a traditional, dualistic Western definition, all time can be divided into two categories: work and leisure. Leisure, or non-work time, is filled with various activities (or "non-activities") such as watching television, playing games, and socializing. Whereas the study of many leisure activities falls within the domains of psychology, physical education, and sociology, most leisure activities also lend themselves to geographic analysis. This is where the origins of RTS geography lie. Tourism and recreation activities exhibit distinct place, time, distance, and activity patterns. For example, tourism is typically more passive and entails more distant and extended travel than does most recreation. Sport is a form of recreation that includes both active participation and passive spectator activities.

Leisure studies is a broad and multidisciplinary research area that encompasses most of the RTS literature, and that has engendered its own body of literature that geographers have contributed to. However, the terms "leisure geography" or "geography of leisure" never came into common use among North American geographers. Mitchell and Smith (1989) noted that the term "recreation geography" was coined in 1954, and up through the 1970s it seemed best to reflect the predominant interests of North American geographers studying leisure activities. During this period, geographers focused on recreation resource inventories, spatial patterns of recreational area usage, recreation participation, recreation perception, recreation planning and management, and urban recreation. An emphasis on outdoor recreation research was part of a broader societal concern with recreational resources in the public domain. Hence, many of the studies focused on public-sector concerns such as management, carrying capacity, and the valuation of wilderness (Hall and Page 1999). Tourism emerged as an area of geographic research in the 1960s and 1970s along with a growth in private-sector interests in travel patterns, tourist demands, economic impacts, and marketing. About this same time, researchers in the small but distinct sports geography arena sought their own identity, creating the trifold split in North American leisure geography and resulting in the "Recreation, Tourism, and Sport" name being applied to the newly formed specialty group within the Association of American Geographers (AAG) in 1974.

Table 33.2 *Recipients of the John Rooney Applied RTS Award for outstanding contributions to Applied Recreation, Tourism, and Sport Geography*

1997	Robert A. Britton, American Airlines (Fort Worth, Texas)
1998	Mark J. Okrant, Plymouth (NH) State College
1999	Lloyd E. Hudman, Brigham Young University
2000	Geoffrey Wall, University of Waterloo

The RTS Specialty Group of the AAG

In 2000 there were about 280 RTS members, including many non-North American geographers, an increase of 40 per cent over twelve years. To encourage development of the RTS field, in 1988 the specialty group instituted the Roy Wolfe Award for RTS geographers with distinguished research and service records (Table 33.1). In 1997, a second award, the John Rooney Applied RTS Award, was created for meritorious contributions to applied RTS geography (Table 33.2). And since the early 1990s, the specialty group has held a student paper competition to encourage growth in RTS research among younger scholars (Table 33.3). The group organizes and sponsors many sessions at the annual Association of American Geographers meetings, and although the number of recreation and sport presentations have remained fairly stable since 1988, tourism presentations have increased significantly (Fig. 33.1).

Table 33.1 *Recipients of the Roy Wolfe Award for outstanding research and service contributions to the Recreation, Tourism, and Sport specialty group*

1988	Lisle S. Mitchell, University of South Carolina
1989	John F. Rooney, Jr., Oklahoma State University
1990	Richard V. Smith, Miami University of Ohio
1991	Geoffrey Wall, University of Waterloo
1992	Peter E. Murphy, University of Victoria
1993	Richard W. Butler, University of Western Ontario
1994	Carlton S. Van Doren, Texas A&M University
1995	Charles A. Stansfield, Rowan College of New Jersey
1996	Robert L. Janiskee, University of South Carolina
1997	None
1998	Klaus J. Meyer-Arendt, University of West Florida
1999	Rudi Hartmann, University of Colorado at Denver
2000	Alan A. Lew, Northern Arizona University
2001	Alison Gill, Simon Fraser University
2002	Keith Debbage, University of North Carolina, Greensboro
2003	Dimitri Ioannides, Southwest Missouri State University

In recent years, the RTS specialty group has collaborated with the International Geographical Union's Study Group on the Geography of Sustainable Tourism in organizing paper sessions at the AAG annual meetings. North American RTS members have also been actively involved in other organizations, including holding two recent chair positions of the International Academy for the Study of Tourism (Richard Butler and Geoffrey Wall). Since 1996, RTS has had presence on the World Wide Web to promote group activities (cf. <www.geog.nau.edu/rts/>, accessed 21 Sept 2001).

Methods of Assessing Recent RTS Research

To summarize research on RTS topics both by North American geographers and about North American geography from 1988 to 2000, efforts were made to be consistent by replicating and expanding upon the methodology used in the previous assessment (Mitchell and Smith 1989). To build the bibliography, solicitations were sent out on the 150-member RTS e-mail listserv, and *Current Geographical Publications, Books in Print*, and the content of leading journals in the field were surveyed for entries starting in 1988. Several 1987 references that were not in Mitchell and Smith's (1989) bibliography were also included in the final tabulation. RTS research published in monographs, book chapters, proceedings, and various forms of *gray* literature was,

Table 33.3 *Recipients of RTS Student Research Paper Awards for papers presented at the annual Association of American Geographers meetings*

1994	Ali A. Abusalih, Mississippi State University ("Biloxi and Dockside Casino Gambling: Impacts on the Coastal Landscape") Thomas Herman, San Diego State University ("Taming a Remote Wilderness: Recording and Evaluating Resource Conditions")
1995	Tou C. Chang, McGill University ("The United Colours of Heritage Tourism: Local Uniqueness in a Global Village") Michael Hawkins, Louisiana State University ("Guides to Paradise: Tourist Images of Jamaica from Victorian Times to the Present") Jennifer R. Beltz, Northern Arizona University ("Ecotourism in Brazil")
1996	Tou C. Chang, McGill University ("Singapore's Little India: A Tourist Attraction as a Contested Landscape") Paul Mackun, SUNY at Buffalo ("Tourism in the Third Italy")
1997	Barbara J. McNichol, University of Calgary ("Determining Group Images of Proposed Tourism Resort Developments") Jonathan Wessell, Western Michigan University ("Recreation Quality and Usage: A Comparative Study of Three West Michigan Rail-Trails")
1998	Lisa De Chano, Southwest Texas State University ("Geographical Analysis of NHL Player Origins")
1999	Margaret Pawlick, University of Western Ontario ("Ethical Considerations in Recreation Service Provision for Specific Populations")
2000	Charles Brian Copp, Florida Atlantic University ("Small Business Networks in Slovakia's Tourism Industry")
2001	Hunter Shobe, University of Oregon ("Territorial Identity, Ethnic Identity, and Sport: The Case of Catalonia and FC Barcelona") Travis Wampler, Southwest Missouri State University ("Route 66: The Forgotten Modes of Transportation, Urbanity, and Culture in America")

unfortunately, excluded from the bibliographic search. Only those references with at least one resident North American (US or Canada) geographer were included.

The primary source of entries were the fifteen journals in which RTS academics are most likely to publish. Ten of these were reviewed by Mitchell and Smith (1989), including: *Annals of Tourism Research; Annals of the Association of American Geographers; Canadian Geographer; Economic Geography; Geographical Review; Journal of Cultural Geography; Journal of Geography; Journal of Travel Research; Leisure Sciences; The Professional Geographer*. New to this edition were *Tourism Management*, *Focus* (a non-refereed magazine published by the American Geographical Society), *Journal of Travel Research* (Tourism and Travel Research Association), the *Yearbook of the Conference of Latin Americanist Geographers*, and *Sport Place. Sport Place* was founded by John F. Rooney (Oklahoma) in 1987 and was the first RTS-focused journal. Reflecting growing academic interest, the 1990s has witnessed an explosion in new tourism journals, including *Tourism Geographies*, founded in 1999 by Alan A. Lew (Northern Arizona). Although virtually all these new journals include geographers on their editorial boards and contain research articles by North American RTS geographers, they were too recent to include in this review.

Sport geography, the smallest of the RTS subdisciplines, is the least represented in the mainstream journal literature selected for this review. To remedy this situation, Thomas Rumney (SUNY-Plattsburgh) was asked to update his thorough bibliography of sport geography (Rumney 1995) so that North American references could be included for this chapter.

Once the references were assembled, they needed to be organized in a logical and systematic manner. This was no easy task, as Hall and Page (1999) discovered in their review of the "main approaches" to the geography of tourism and recreation. Despite some overlap, the following "themes and approaches" were identified for each of the RTS subdisciplines.

Themes and Approaches in Geographic Research on Tourism

General Books and Articles

Since 1988, there have been scores of books and general articles on the geography of tourism. Many books, both authored and edited, covered the entire field of tourism geography, usually in an introductory manner (e.g. Hall and Page 1999; Hudman and Jackson 1998; Ringer 1998; Seaton *et al.* 1994; Williams 1998). Other books addressed specific themes, such as coastal tourism (Fabbri 1990), urban tourism (Murphy 1997), sustainable tourism (Hall and Lew 1998), and economic geography (Ioannides and Debbage 1998). In addition to books, a few journal articles reviewed the overall scope of tourism geography (cf. Mitchell 1994), while Smith and Godbey (1991) examined the linkages among leisure, recreation, and tourism. Links between tourism and economic geography were addressed by Ioannides (1995*b*), between tourism and international understanding by Mings (1988), and between ecology and tourism by Farrell and Runyan (1991). A focus on tourist attractions was provided by Lew (1994) and Wall (1997*c*). Methodological approaches were discussed by Hartmann (1988), Murphy and Carmichael (1991), and Smith (1989, 1994), among others.

Travel

The study of travel, or the movement of people over space, continues to be a significant research theme in tourism as well as recreation, and researchers employ a wide variety of methodologies. Approaches include the spatial analysis of travel patterns (Hudman and Davis 1994; Ioannides and Debbage 1997; Mings and McHugh 1992), economic impacts of travel (Roehl *et al.* 1993), application of descriptive frameworks and models (Loban 1997; Oppermann 1994, 1995*b*), travel planning (Fesenmaier and Johnson 1989), the role of welcome centers (Stewart *et al.* 1993), motivational aspects (Mansfield 1992*b*), risk assessment (Roehl and Fesenmaier 1992), industry contributions (Marti 1991; Van Doren 1993), pilgrimage tourism (Hudman and Jackson 1992), and descriptive approaches (Lanegran and St. Peter 1993; Zurick 1995*a*).

Historical Tourism

Historical tourism is important for understanding both a slice of the past and the evolutionary background of modern touristic phenomena, and several geographers have embraced this research approach. The theme of historical tourism in national parks and wilderness areas appeared in articles on railroads and national parks (Hall and Shultis 1991), Bar Harbor in Maine (Hornsby 1993), the Catskills (Johnson 1990), and romanticized wildernesses (Squire 1994*b*, 1995).

Perception

Perceptual studies in tourism geography are less common than they were in the 1970s and 1980s, though they still encompassed a wide variety of topics in the literature. Most of this research was devoted to perceptions of destinations (Allen 1988; Crang 1997; Lew 1992), although perceptions of tourism (Husbands 1989) and social impacts (Mansfield 1992*a*) were also addressed.

Environmental Aspects

Studies on the environmental aspects of tourism today often fall within the relatively new category of "ecotourism." However, most studies of ecotourism focus upon the tourist and tourism development, rather than the physical environment as a physical geographer or an ecologist might understand the term. The environmental issues in tourism geography are very broad (well summarized by Mieczkowski 1995), and therefore the more policy-oriented ecotourism studies have been moved to the Planning and Management discussion, below. When this was done, the remaining books and articles more clearly emphasized the physical environment aspects of tourism (Savage 1993; Wilkinson 1992; Wong 1993). The impacts of environmental changes upon the pattern of tourism were addressed by Meyer-Arendt (1987, 1991) and Place (1988); while tourism and disaster planning was examined by Murphy and Bayley (1989).

Destination Studies

Tourism, by its nature, is place-specific, and therefore lends itself to geographical analysis. Not surprisingly, tourism geography research had emphasized destinations —both as specific sites and broader physical or political regions. Typically, such studies assess tourism development patterns or impacts within given areas, and many are descriptive rather than statistically or methodologically framed. This latter issue was a major criticism of tourism research in its early years, and while the situation has improved since then, a theoretical basis is still often lacking in these studies.

Examples of regional destination studies of North America included research on regional tourism variations across the entire US (Lollar and Van Doren 1991) and individual states and regions, such as Virginia (Andrews 1990), and the Northwest Territories (Hamley 1991). Physical regions were a focus in studies of places such as the Great Lakes (Chubb 1989), Vancouver Island

(Murphy and Keller 1990), and the Colorado Rockies (Hartmann 1992). More dispersed territories, sharing culture more than physical cohesion, such as Western US Indian reservations (Browne and Nolan 1989; Lew 1996) and Route 66 (Mariolle 1996), have also been studied, though are less common. Other North American regional studies are more focused, such as the application of GIS to tourism in Wisconsin (Foust and Botts 1991), regional travel flows in New England (Hamilton-Jones 1991), and Pacific Northwest shopping districts (Lew 1988).

Analyses of foreign tourist regions by North American geographers included general descriptions and impact studies of the Alps (Kariel 1989; Price 1987), China (Lew and Yu 1995; Oakes 1998; Toops 1992), Dominica (Sharkey and Momsen 1995), Dominican Republic (Sambrook *et al.* 1994), Estonia (Jaakson 1996), Laos (Harnapp and Harnapp 1998), Nepal (Rense 1988; Stevens 1991, 1993), Vietnam (Lenz 1993), and Zambia (Teye 1988). More specific destination studies included analyses of Jamaica's hotel industry (Taylor 1988, 1993), as well as tourism and civil unrest in Sri Lanka (Scott 1988), mineral tourism in Japan (Cybriwski and Shimizo 1993), village tourism in Fiji (Vlack 1993) and Bali (Hussey 1989, Long and Wall 1996), urban tourism in the Caribbean (Weaver 1993*b*), northern Mexican border cities (Curtis and Arreola 1991) and general "border tourism" (Timothy 1995), the impacts of religious tourism to Mt. Sinai (Hobbs 1992), and site-specific tourism impacts resulting from popular movies (Riley *et al.* 1998).

Specialized Tourism

Narrowly focused theme destinations have become an important part of the tourism landscape and tourism marketing; perhaps the fastest-growing segment of the industry today. Specialized tourism includes a broad range of industry segments such as cultural tourism, heritage tourism, resort tourism, retirement tourism, farm tourism, festival and event tourism, gourmet tourism, and casino gambling, among others.

Cultural tourism typically involves visits to ethnic culture groups, often indigenous and in their native environments, and many anthropologists and sociologists have focused their tourism research on this issue. Geographers have studied indigenous cultural tourism, often under the heading of "ecotourism," and their research has included a book on tourism and indigenous peoples (Butler and Hinch 1996) and articles on native culture and tourism in Zambia (Husbands and Thompson 1990), on the Navajo Indian Reservation (Jett 1990), in Belize (King 1997; Steinberg 1994), and in the Canadian Arctic (Milne *et al.* 1995; Hinch 1998). Cultural tourism that is more broadly defined to incorporate dominant Western traditions include studies on Amish and Mennonite communities in Pennsylvania (Hovinen 1995; Wall and Oswald 1990), and religious tourism in Europe and elsewhere (Nolan and Nolan 1992; Rinschede 1992). This latter category overlaps with heritage tourism, which encompasses using cultural artefacts from the past to promote tourism. A few examples of heritage tourism research include studies of historic preservation in Singapore (Chang 1999), Ireland (Johnson 1996), and coastal Cape May, New Jersey (Stansfield 1990).

The study of resorts, particularly seaside resorts, is one of the oldest research thrusts in tourism geography. Popular themes have included resort evolution, resort morphology, typologic analyses, and locational analyses. While some resort studies were more general in scope, such as Roehl and Van Doren's (1990) survey of American resort hotels, most could easily be categorized under regional and place studies due to their more general and descriptive nature. Examples include studies of Revere Beach, Massachusetts (Berman 1989), resort typology studies of Britain (DeBres 1991) and the Dominican Republic (Meyer-Arendt *et al.* 1992), and impact studies of "Spring Break" students on Padre Island, Texas (Gerlach 1989), resorts in Tecolutla, Mexico (Meyer-Arendt 1990*b*), and Disneyland Paris (d'Hauteserre 1999). Wall (1996*a*) and (Getz 1993), summarized conclusions from their extensive work on resort communities and tourism business districts, respectively. Others focused more specifically on the resort development cycle in such places as the Bahamas (Debbage 1990) and Cyprus (Ioannides 1992), while various types of resort morphology studies were done on the Gulf of Mexico (Meyer-Arendt 1990*a*), Hawaii's Waikiki (Mitchell 1996), and American amusement parks (LaPolla 1988). Retirement tourism, resembling resorts in some ways, was discussed in articles on changing patterns of retirement counties (Graff and Wiseman 1990), the RV resort landscape (Mings and McHugh 1989; Parsons 1992), and retirement resort cyclicity (Foster and Murphy 1991).

Farm tourism and rural tourism are increasing popular forms of specialized tourism, yet few geographers have conducted research in this arena. Exceptions include Oppermann's (1995*a*) study of German farm tourism and Weaver and Fennell's (1997) profile of Saskatchewan vacation farms. Similarly *festival and event tourism* received limited geographic attention, exceptions being two books by Getz (1991, 1997) and articles on historic houses and events (Janiskee 1996), history-themed festivals (Janiskee 1990), convention tourism (Zelinsky 1994), and the housing evictions caused by mega-events (Olds 1998).

Casino gambling is another form of specialized tourism that has drawn significant interest since it has become legalized in many parts of the US in recent years. Two edited books, one covering North American trends (Meyer-Arendt and Hartmann 1998) and one covering Indian gaming and tourism (Lew and Van Otten 1998), offered well-informed overviews. Additional research articles on gaming included works on Connecticut's Foxwood Casino (Carmichael *et al.* 1995; d'Hauteserre 1998), Nevada (Roehl 1995; Sommers and Lounsbury 1991), and Indian gaming in general (Winchell *et al.* 1997/8).

Marketing and Economic Aspects of Tourism

Marketing is one of the largest areas of tourism research, but one to which geographers have contributed only a small portion of this literature (most significantly, Oppermann 1997 and Wall and Heath 1992). Geographers wrote much more on *economic development* aspects of tourism, several of which were cited under the general topic of tourism development, above. More narrowly focused economic works included studies of regional economic impacts (Gribb 1990; Jeffrey and Hubbard 1988), tourism employment (Cukier-Snow and Wall 1993), cross-border shopping (Timothy and Butler 1995), and tourism multipliers (Wall 1997*b*).

Planning and Management

Much of the recent literature on tourism management and planning has centered on the concept of "sustainable tourism" (i.e. the application of sustainable development principles to tourism). From a business perspective, this has often been in the form of "ecotourism," while non-governmental organizations have also been known to use the term "alternative tourism." One major book on sustainable tourism with a significant North American emphasis has appeared (Hall and Lew 1998), while several articles have debated whether such forms of tourism are truly sustainable, or alternative, or ecologically friendly (e.g. Butler 1990, 1991*a*; Ross and Wall 1999). Despite this, these concepts have led to a greater concern for resource management and community-oriented planning (Boyd and Butler 1996; Farrell 1999). More narrowly focused planning articles included those on the destination life cycle and planning (Getz 1992), convention tourism planning (Getz *et al.* 1998), managing mountain tourism (Gill and Williams 1994; Price 1992), community-driven planning (Murphy 1995), tourism and urban regeneration (Owen 1990), and tourism planning in developing countries (Ioannides 1995*a*). Ecotourism literature was frequently destination-focused, such as articles on Thailand (Dearden 1991; Hvenegaard and Dearden 1998), Costa Rica (Place 1991), and Nepal (Zurick 1992). Because of their small size and vulnerability, islands have been major venues for tourism management research, including ecotourism. Significant contributions include books by Briguglio *et al.* (1996) and Conlin and Baum (1995), and articles on Caribbean islands (Albers 1991; Weaver 1993*a*; Wilkinson 1989) and the South Pacific (Zurick 1995*b*).

Themes and Approaches in Geographic Research on Recreation

Leisure

Leisure studies, as discussed earlier in the chapter, is a broad field in which active and passive human use of "free," non-work time is analyzed. Only a few geographers have been involved in this avenue of research, which is dominated by psychologists, sociologists, and kinesiologists. Among geographers, Ed Jackson of the University of Alberta, along with his colleagues and former students, has addressed various aspects of patterns of leisure participation (e.g. Jackson 1988, 1991, 1994; Jackson and Henderson 1995). Others have examined recreational boating usage (Richter 1992), recreation marketing (Peterson 1991), recreation travel (Kim and Fesenmaier 1990), leisure and the environment (Glyptis 1993), recreational preferences (Raitz and Dakhil 1989), behaviorial modeling (Fesenmaier 1990), neighborhood community centers (Winder 1998), and climatic influences upon recreation (Konrad 1995). Outdoor recreation in Canada was the subject of a book by Wall (1989).

Management and Impact Assessment

The dual themes of park management and impact assessment are rooted in some of the oldest geographic studies of outdoor recreation, and remain important components of the subdiscipline today. Avoidance of potential user conflicts is a chief goal in many of these studies, though the range of research on recreation management issues is considerable. General literature was found to focus on topics such as US national park values (Lemons 1987), rivers (Wikle 1991), upland areas (Sidaway and Thompson 1991), and volunteers in recreation land management (Bristow 1998). More work, however, consisted of site-specific experiences and lessons, including recreation resource compatibility problems in Illinois (Bristow *et al.* 1995), and park management controversies in New Mexico (Harvey 1998) and at the Grand Canyon (Morehouse 1996). Impact assessments were mostly site-specific, such as estimating greyhound track attendance (Foust and Botts 1989), environmental impacts of recreation in California (Goodenough 1992) and at US national parks (Flint 1998), modeling wildlife boundaries in Canada's Pacific Rim National Park (Dearden 1988), park protection from human settlements in Peru (K. Young 1993), and mountain environments (Kariel and Drapier 1992; Thorsell and Harrison 1992).

Resource Inventory and Valuation

Another long-term research thrust in recreation geography has been in the inventory and valuation of recreational resources. This research has focused both on the resources themselves and also the recreational user (i.e. the supply and demand of recreation). In terms of the inventory of supply, articles have appeared on the mapping of recreation resources (Clarke 1988), including the use of GIS in the Everglades National Park (Welch *et al.* 1995). Other articles have covered urban-area outdoor resources (McPherson 1992; Rogers and Rowntree 1988; Talarchek 1990), mountain park resources (Saremba and Gill 1991), and water recreation (Smardon 1988; Willis and Garrod 1991). Inventory and valuation studies based upon demand included studies of urban recreationists (Halseth and Rosenberg 1995; Mandel 1998; Strapp 1988), recreation expenditures and opportunity theory (Lieber *et al.* 1989), ethnicity and recreation behavior (Pfister 1993), and recreational access (Millward 1996).

Parks

Other than the management, impacts and resource inventory literature cited above, literature on parks has been mostly descriptive and historical in nature, and contributions by RTS specialty group geographers have been few. Focusing first on non-urban parks, historical studies reviewed included articles on Sequoia National Park (Dilsaver 1987) and US national parks in general (Johnson 1994; Sellars 1997). A small number of articles focused on physical geography, such as climate change impacts on the Nahanni National Park (Staple and Wall 1996). Examples of the many general overview articles included introductions to national parks in the southern US (Boorstein 1992), Latin American and Caribbean national parks (Eyre 1990), Bruce National Park in Canada (Gorrie 1987), Egypt's St Katherine National Park (Hobbs 1996), and the administrative structure of Canada's national parks (Dearden 1993). Research on urban parks included articles on greenway park management (Little 1990), equity and park access issues (Talen 1997), a historical study of San Francisco's Golden Gate Park (T. Young 1993), and the changing role of modern urban parks (Mitchell 1995; T. Young 1995).

Themes and Approaches in Geographic Research on Sport

General Sport Geography and Landscape

General sport geography references included books that provided an overview of the subdiscipline, as well as articles that were general or otherwise transcended the analysis of individual sports. Important books included those written by the prolific English sport geographer John Bale on broad aspects of sport geography from an introduction to sports geography and foreign athletes at American universities (Bale 1989, 1991) to sport landscape studies (Bale 1992, 1994, 1998). Other significant overview books were *The Theater of Sport* (Raitz 1995), and *Atlas of American Sport* (Rooney and Pillsbury 1992). General research articles included an overview of North American sports (Louder 1991), college athletic conferences (Abbott 1990), high-school sports (Ashley 1988), sports territory (Augustin 1997), sports and television (Clay 1988*b*), athletic participation (Goudge 1993), an analysis of the football field as a landscape feature (Clay 1988*a*), historical development of soccer (Bowden 1990), and a bibliography of sports geography (Rumney 1995).

Specific Sports

By far the greatest volume of literature in sport geography was dedicated to individual sports. In terms of the total number of articles reviewed, American football was the most researched sporting activity, followed in order by baseball, golf, and snow skiing. A sampling of topics discussed in the football articles include college recruiting (Ferrett and Ojala 1992; Schnell 1990), American football in Europe (Little 1990; Maguire 1990), regional differences in press coverage (Shelley and McConnell 1993) and radio broadcasting of college football (Roseman and Shelley 1988), and the impact of domed football stadiums (Zeller and Jurkovac 1989). Baseball articles covered topics such as geographic names in baseball nomenclature (Elliott 1988), meteorological impacts on the trajectory of hit baseballs (Kraft and Skeeter 1995) and climate impacts of the success of college and professional baseball teams (McConnell 1994; Skeeter 1988), as well as topics similar to those in football (McEachern and Russell 1992; Ojala and Gadwood 1989; Shelley and Shelley 1993).

Golf articles often described the current situation in specific places, such as Pennsylvania (Miller 1990). Other works, however, were more analytical, including research on golf supply and demand (Rooney 1989, 1993) and an analysis of golf course microenvironments (Stadler and Simone 1988). Snow skiing, the last of the four dominant topics of research, included articles on ski area development in Quebec (Archenbault 1993) and the impact of global warming on ski areas (Brottan and Wall 1993; Lipski and McBoyle 1991). The literature review found a few articles on other topics, including fishing (Dargitz 1988; Wall 1988), ice hockey (Genest 1996), ice bowling (Griepentrog 1992), soccer (Handley *et al.* 1994), whitewater rafting (Mayfield and DeHart 1989), stock car racing (Pillsbury 1989), rowing (Rumney 1988), American cricket (Tooley 1988), motorcycle racing (Van Zuyk 1988), endurance sports (Wiggins and Soulé 1993), paddlesports (Wikle 1993), and bridge (Brown 1988). Surprisingly, only one article (Rooney 1990) touched upon the geography of basketball.

RTS Geography in the New Millennium

Recreation, tourism, and sport geography are each alive, well and growing in the first decade of the third millennium. The number of RTS specialty group members is at an all-time high, publications are numerous, new journals and books are burgeoning, and—for the first time in many decades—in 1999 the Association of American Geographers met at a resort conference facility (Hilton Hawaiian Village in Waikiki). Although the outlook for RTS geography seems rosy, there are certain trends that require elaboration.

Research in RTS Geography

In the US, RTS topics have traditionally been considered frivolous and not worthy of serious academic study, especially within the discipline of geography. This has been reflected in the paucity of geography departments that specifically advertise for RTS-trained academics—even when the position being filled was formerly held by an RTS scholar. This situation has been less true in Canada and elsewhere, which has resulted in some of the world's leading RTS geographers being non-Americans. Because of the slow rate at which American geography has come to accept the rising role of recreation and tourism in contemporary society, many RTS geographers have allied themselves with non-geography colleagues in multidisciplinary centers of tourism and recreation research. Hall and Page (1999) noted a trend for tourism programs—and tourism geographers—to be housed in business schools where recognition of the applied value of tourism studies is apparently greater than in departments of geography. This situation has been similar for recreation geographers as departments of leisure studies, tourism studies, and recreation and parks have grown faster than RTS geography programs. (Salem State College, with one of the largest undergraduate geography programs in the US, in large part because of its tourism track, is a significant exception to this.) Because of the conservative nature of academic geography, it is anticipated that this trend toward employment in interdisciplinary tourism programs will continue into the near future.

One of the difficulties in RTS geography's lack of status in academe is that much RTS research is applied in nature and is, therefore, less prominent in the discipline's journals. RTS geographers often work as consultants to assist local communities, regional governments, and the private sector, reflecting the importance of leisure in contemporary American society. As such, RTS geographers have been less involved in the more theoretical debates that guide much of the literature in the leading geography journals (for exceptions see Britton 1991 and Squire 1994*a*). More active participation by RTS geographers in these discussions would help to raise the subdiscipline's profile. RTS research could benefit by expanding its theoretical basis to include new ideas from closely aligned areas of economic and cultural geography where social theory and the geography of consumption offer considerable potential for informing RTS research.

For RTS geographers, the tensions between employment in geography and more applied leisure fields (including recreation, leisure studies, and hospitality and tourism) has also presented a challenge as to just where their research and publication "home" is situated. Because of the synthesis that geographers naturally bring to their work, RTS scholars tend easily to place their research within the literature of the larger bodies of tourism and leisure studies. For reasons cited above, they have tended to avoid publishing within the mainstream geography journals, and as a result have often fallen away from the major disciplinary debates and writings within geography. This has further contributed to an isolation of the subdiscipline. The growth of RTS geography as a field of study, and probably more importantly as an economic activity, has resulted in more articles (especially on tourism topics) appearing in leading North American geography journals in the 1990s. We expect that this presence will continue to grow in the coming years, though RTS geographers will also need to make conscious efforts to reach out and embrace their own discipline.

Teaching RTS Geography

The *Guide to Programs in Geography in the United States and Canada* (2000–1 edition, Association of American Geographers) listed "Recreation/Tourism" program specialties in sixty-two US and Canadian departments of geography (25% of the total). At the same time, *Schwendeman's Directory of Collegiate Geography of the US* (2000 edn., Eastern Kentucky University), listed thirty-four programs that taught a "travel/tourism" class in either 1998 or 1999 (up from eighteen two years earlier). While RTS appears to be a major teaching subject in geography, few departments actually employ geographers trained in RTS subdisciplines, and even fewer produce Ph.Ds specializing in RTS topics. This is especially true of larger universities than smaller, regional universities where responses to rapidly changing economic and social conditions demand greater flexibility.

Of the geography programs that claimed recreation/tourism as a program specialty, forty-four responded to a survey of RTS course offerings by Bristow and Carmichael (1999). The survey results (Table 33.4) indicated that the overall number of courses offered was quite small and were not offered on an annual basis. Canadian offerings were proportionally much greater than American offerings. In addition, the authors found a decline in recreation geography courses and an increase in tourism geography course offerings since the early 1970s, reflecting the non-replacement of retiring recreation geographers. Many of the recreation courses that were formerly offered in geography have been subsequently absorbed by expanding leisure and recreation studies programs.

Table 33.4 *Course offerings in RTS geography in North America, 1998*

Course Contents	USA	Canada	North America
Recreation Geography	2	5	7
Tourism Geography	15	4	19
Combined	7	1	8
TOTAL	24	10	34

Source: Bristow and Carmichael (1999).

One way that RTS geographers are making their presence better known in the teaching realm is by posting their courses on the World Wide Web (Bristow and Carmichael 1999). With the recent proliferation of distance-learning courses, the trend toward web-based teaching is sure to continue. The applied nature of RTS geography makes it particularly well suited to the Web, and more courses taking advantage of this new technology will likely increase the visibility and viability of the RTS subdisciplines.

Future Outlook

In the post-industrial world that has come to characterize North America, leisure, recreation, travel, and sporting activities have taken center stage in the lives of many, if not most, people. In particular, travel and exploration (arguably a form of tourism) form the basis of the geographer's fascination with places, and are often viewed as a shared characteristic among those who have heeded geography's calling. Today, tourism has become the world's largest industry, and the demand for academic and applied tourism research is greater than ever. Recreation and sport research, although not growing as fast, will none the less also remain important as demand for outdoor recreation continues to grow and tickets to popular sporting events continue to be scalped at high prices.

Recreation, tourism, and sport are multidisciplinary areas of study, and geographers have been at the forefront of research in these arenas. The ever-growing body of research in these fields is shifting, however, to multidisciplinary departments, multi-disciplinary journals, and the private sector (where applied research is more lucratively rewarded). Geographers will continue to play important roles, especially in tourism research, because so many of the themes and problems are inherently geographic. Many RTS geographers may shift their academic linkages to nongeography programs and interdisciplinary academic conferences, thus contributing to the centrifugal spin-off of the many subdisciplines of geography.

In North American, the RTS Specialty Group of the Association of American Geographers is the leading organization bringing together professional RTS geographers and students. While the RTS Newsletter and the RTSNET-L e-mail listserv connects RTS members and affiliated colleagues, the annual meeting of the AAG continues to be the single most important venue for sharing research and ideas. A respectable number of Canadian RTS geographers also attend the AAG meetings, although GEOPARETO (GEOgraphy of PARks, REcreation and TOurism) provides a much smaller venue for members of the Canadian Association of Geographers. In spite of centrifugal forces sending RTS geographers to multidisciplinary departments and multidisciplinary conventions, the AAG meetings will continue to play an important role in RTS geography.

Research on the geography of recreation, tourism, and sport historically has been fragmented rather than integrative. It has also been more descriptive than conceptual, or even applied. The trends of the 1990s indicate that more rigorous methodologies and integrative approaches are being employed, holding promise for higher standards in future RTS geography research. Mitchell and Smith (1989) identified eighteen research agenda needs for RTS geography, most of which are as relevant today as they were a decade ago. Key areas that we see as most important to the future development of RTS geography in the coming decade include:

- 1. RTS geography needs to be better integrated into the mainstream of North American geographic literature by incorporating recent developments in cultural and economic geography theory and method into RTS research, and publishing research in more general geography journals. This may occur naturally as RTS issues increasingly become central to North American life, economy, and environment.
- 2. The applicability of RTS geography to major contemporary issues could be better demonstrated. These include environmental and cultural resource management, environmental justice, elderly and handicap access, urban and regional development, and other key areas where geography has traditionally played a leading role. The relatively few RTS geographers have already made significant contributions, but much more can and should be done.
- 3. The three subfields of recreation, tourism, and sport should be better integrated with each other and with the broader topic of leisure geography, so that each may better benefit from theoretical and methodological advances made by the others. Despite cohabitation within the AAG's RTS specialty group, more could be done to foster cross-fertilization of ideas and research. The three subdisciplines, as well as the broader discipline of geography, would all benefit from the synergy that further collaboration would engender.

Acknowledgements

Assistance provided by Rob Bristow, Barbara Carmichael, Lisle Mitchell, Wes Roehl, and Tom Rumney is greatly appreciated.

References

General

Bristow, R. and Carmichael, B. (1999). "The State of Teaching Recreation and Tourism Geography in North America," in R. Donnell (ed.), *Proceedings of the New England St Lawrence Valley Geographical Society Annual Meeting*, XXVIII. Framingham State College.

Gaile, G. L., and Willmott, C. J. (eds.) (1989). *Geography in America*. Columbus, Ohio: Merrill.

Mitchell, L. S., and Smith, R. V. (1989). "The Geography of Recreation, Tourism, and Sport," in Gaile and Willmott (1989: 387–408).

Tourism

Albers, B. (1991). "Planning in Paradise, Developing an Environmentally Sound Resort in the Caribbean." *Geo Infosystems*, 1/10: 26–38.

Allen, R. F. (1988). "Image and Reality: The Critical Use of Tourist Brochures in World Geography Classes." *Journal of Geography*, 88/1: 6–10.

Andrews, A. (1990). "Regional Aspects of Tourism in Virginia." *The Virginia Geographer*, 22/2: 20–30.

Berman, M. (1989). "Revere Beach: A Peculiarly American Seaside Resort." *Focus*, 39/3: 4–8.

Boyd, S. W., and Butler, R. W. (1996). "Managing Ecotourism: an Opportunity Spectrum Approach." *Tourism Management*, 17/8: 557–66.

Briguglio, L., Archer, B., Jafari, J., and Wall, G. (eds.) (1996). *Sustainable Tourism in Islands and Small States: Issues and Policies*. London: Cassell.

Britton, S. (1991). "Tourism, Capital, and Place: Towards a Critical Geography of Tourism." *Environment and Planning D: Society and Space*, 9: 451–78.

Browne, R., and Nolan, M. L. (1989). "Western Indian Reservation Tourism Development." *Annals of Tourism of Research*, 16/3: 360–76.

Butler, R. W. (1990). "Alternative Tourism: Pious Hope or Trojan Horse?" *Journal of Travel Research*, 23/3: 91–6.

— (1991*a*). "Tourism, Environment, and Sustainable Development." *Environmental Conservation*, 18/3: 201–8.

— (1991*b*). "West Edmonton Mall as a Tourist Attraction." *Canadian Geographer*, 35/3: 287–94.

Butler, R. W., and Hinch, T. (eds.) (1996). *Tourism and Indigenous Peoples*. London: International Thomson Business Press.

Carmichael, B., Peppard, D. M., Jr., and Boudreau, F. (1995). "Megaresort on my Doorstep: Local Resident Attitudes toward Foxwood Casino and Casino Gambling on Nearby Indian Reservation Land." *Journal of Travel Research*, 34/3: 9–16.

Chang, T. C. (1999). "Local Uniqueness in the Global Village: Heritage Tourism in Singapore." *The Professional Geographer*, 51/1: 91–103.

Chubb, M. (1989). "Tourism Patterns and Determinants in the Great Lakes Region: Populations, Resources, Roads and Perceptions." *Geojournal*, 19/3: 297–302.

Conlin, M., and Baum, T. (eds.) (1995). *Island Tourism: Management Principles and Practice*. Chichester: Wiley.

Crang, M. (1997). "Picturing Practices: Research Through the Tourist Gaze." *Progress in Human Geography*, 21/3: 359–73.

Cukier-Snow, J., and Wall, G. (1993). "Tourism Employment: Perspectives from Bali." *Tourism Management*, 14/3: 195–201.

Curtis, J. R., and Arreola, D. D. (1991). "Zonas de Tolerancia on the Mexican Border." *The Geographical Review*, 81/3: 333–46.

Cybriwski, R., and Shimizo A. (1993). "Ashio: Rocks, Rust and Tourism on Japan's Copper Mountain." *Focus*, 43/1: 22–8.

Dearden, P. (1991). "Tourism and Sustainable Development in Northern Thailand." *Geographical Review*, 81/4: 400–13.

Debbage, K. G. (1990). "Oligopoly and the Resort Cycle in the Bahamas." *Annals of Tourism of Research*, 17: 513–27.

— (1991). "Spatial Behavior in a Bahamian Resort." *Annals of Tourism of Research*, 18: 251–68.

DeBres, K. (1991). "Seaside Resorts, Working, Museums, and Factory Shops: Three Vignettes as British Tourism Enters the Nineties." *Focus*, 41/2: 10–16.

d'Hauteserre, A. M. (1998). "Foxwoods Casino Resort: An Unusual Experiment in Economic Development." *Economic Geography*, Special Issue.

— (1999). "The French Mode of Social Regulation and Sustainable Tourism Development: The Case of Disneyland Paris." *Tourism Geographies*, 1/1: 86–107.

Fabbri, P. (ed.) (1990). *Recreational Uses of Coastal Areas*. Dordrecht: Kluwer.

Farrell, B. H. (1999). "Conventional or Sustainable Tourism? No Room for Choice." *Tourism Management*, 20/2: 189–91.

Farrell, B. H., and Runyan, D. (1991). "Ecology and Tourism." *Annals of Tourism Research*, 18: 26–40.

Fesenmaier, D. R., and Johnson, B. (1989). "Involvement-Based Segmentation." *Tourism Management*, 10/4: 293–300.

Foster, D. M., and Murphy, P. (1991). "Resort Cycle Revisited: The Retirement Connection." *Annals of Tourism of Research*, 18: 553–67.

Foust, B., and Botts, H. (1991). "Tourism Gets a Boost from Wisconsin's Recreational Resource GIS." *Geo Infosystems*, 1/8: 43–5.

Gerlach, J. (1989). "Spring Break at Padre Island: A New Kind of Tourism." *Focus*, 39/1: 13–16.

Getz, D. (1991). *Festivals, Special Events, and Tourism*. New York: Van Nostrand Reinhold.

— (1992). "Tourism Planning and Destination Life Cycle." *Annals of Tourism of Research*, 19: 752–70.

— (1993). "Planning for Tourism Business Districts." *Annals of Tourism of Research*, 20: 583–600.

— (1997). *Event Management and Event Tourism*. Elmsford, NY: Cognizant Communications Corporation.

Getz, D., Anderson, D., and Sheehan, L. (1998). "Roles, Issues and Strategies for Convention and Visitors Bureaux in Destination Planning and Product Development: A Survey of Canadian Bureaux." *Tourism Management*, 19/4: 331–40.

Gill, A., and Williams, P. (1994). "Managing Growth in Mountain Communities." *Tourism Management*, 15/3: 212–20.

Goss, J. (1993). "Placing the Market and Marketing the Place: Tourist Advertising of the Hawaiian Islands, 1972–1992." *Environment and Planning D: Society and Space*, 11: 663–88.

— (1999). "'From Here to Eternity': Voyages of Re(dis)covery in Tourist Landscapes of Hawai'i," in D. W. Woodcock (ed.), *Hawai'i: New Geographies*. Honolulu: University of Hawai'i, Department of Geography, 153–77.

Graff, T. O., and Wiseman, R. F. (1990). "Changing Pattern of Retirement Counties Since 1965." *Geographical Review*, 80/3: 239–51.

Gribb, W. J. (1990). "Tourism as a Form of Regional Economic Development." *Papers and Proceedings of Applied Geography Conferences*, 13: 96–104.

Hall, C. M., and Page, S. J. (1999). "The Geography of Tourism and Recreation: Environment, Place and Space." New York: Routledge.

Hall, C. M., and Shultis, J. (1991). "Railways, Tourism and Worthless Lands: The Establishment of National Parks in Australia, Canada, New Zealand and the United States." *Australian-Canadian Studies—A Journal for the Humanities and the Social Sciences*, 8/2: 57–74.

Hall, C. M., and Lew, A. A. (eds.) (1998). *Sustainable Tourism: A Geographical Perspective*. London: Addison Wesley Longman.

Hamilton-Jones, D. R. W. (1991). "Future Trends in North-South Tourism from New England." *St Lawrence Valley Geographical Society Proceedings*, 21: 94–104.

Hamley, W. (1991). "Tourism in the Northwest Territories." Geographical Review, 81/4: 389–99.

Harnapp, V., and Harnapp, L. (1998). "Laos: Now Open for Tourist Business." *Focus*, 45/2: 6–12.

Hartmann, R. (1988). "Combining Field Methods in Tourism Research." *Annals of Tourism Research*, 15: 88–105.

— (1992). "Ski Resort Development in Colorado: Mining the 'White Gold'," in A. David Hill (ed.), *Interdependence in Geographic Education*. Boulder: University of Colorado Center for Geographic Education, 53–61.

Hinch, T. (1998). "Ecotourist and Indigenous Hosts: Diverging Views on their Relationship." *Current Issues in Tourism*, 1/1: 120–4.

Hobbs, J. J. (1992). "Sacred Space and Touristic Development at Jebel Musa (Mt. Sinai), Egypt." *Journal of Cultural Geography*, 12/2: 99–114.

Hornsby, S. J. (1993). "The Gilded Age and the Making of Bar Harbor." *Geographical Review*, 83/4: 455–68.

Hovinen, G. (1995). "Heritage Issues in Urban Tourism: An Assessment of New Trends in Lancaster County." *Tourism Management*, 16/5: 381–8.

Hudman, L. E., and Davis, J. E. (1994). "Changes and Patterns of Origin Regions of International Tourism." *Geojournal*, 34/4: 481–90.

Hudman, L. E., and Jackson, R. H. (1992). "Mormon Pilgrimage and Tourism." *Annals of Tourism of Research*, 19/1: 107–21.

— (1998). *The Geography of Travel and Tourism*. Albany, NY: Delmar Publishers.

Husbands, W. (1989). "Social Status and Perception of Tourism in Zambia." *Annals of Tourism of Research*, 16/2: 237–53.

Husbands, W., and Thompson, S. (1990). "The Host Society and the Consequences of Tourism in Livingstone, Zambia." *International Journal of Urban and Regional Research*, 14/3: 490–513.

Hussey, A. (1989). "Tourism in a Balinese Village." *Geographical Review*, 79/3: 311–25.

Hvenegaard, G. T., and Dearden, P. (1998). "Ecotourism Versus Tourism in a Thai National Park." *Annals of Tourism Research*, 25: 700–20.

Ioannides, D. (1992). "Tourism Development Agents: The Cypriot Resort Cycle." *Annals of Tourism of Research*, 19/4: 711–31.

— (1995*a*). "Planning for International Tourism in Less Developed Countries: Toward Sustainability?" *Journal of Planning and Literature*, 9/3: 25–7.

— (1995*b*). "Strengthening the Ties Between Tourism and Economic Geography: A Theoretical Agenda." *Professional Geographer*, 47/1: 49–60.

Ioannides, D., and Debbage, K. (1997). "Post-Fordism and Flexibility: The Travel Industry Polyglot." *Tourism Management*, 18/4: 229–41.

Ioannides, D., and Debbage, K. (eds.) (1998). *The Economic Geogaphy of the Tourism Industry: A Supply-Side Analysis*. New York: Routledge.

Jaakson, R. (1996). "Tourism in Transition in Post-Soviet Estonia." *Annals of Tourism Research*, 23: 617–34.

Janiskee, R. L. (1990). "History-Themed Festivals: A Special Events Approach to Rural Recreation and Tourism." *Papers and Proceedings of Applied Geography Conferences*, 13: 111–17.

— (1996). "Historic Houses and Special Events." *Annals of Tourism Research*, 23: 398–414.

Jeffrey, D., and Hubbard, N. J. (1988). "Foreign Tourism, the Hotel Industry, and Regional Economic Performance." *Regional Studies*, 14/4: 329–40.

Jett, S. C. (1990). "Culture and Tourism in the Navajo Country." *Journal of Cultural Geography*, 11/1: 85–107.

Johnson, K. A. (1990). "Origins of Tourism in the Catskill Mountains." *Journal of Cultural Geography*, 11/1: 5–16.

Johnson, N. C. (1996). "Where Geography and History Meet: Heritage Tourism and the Big House in Ireland." *Annals of the Association of American Geographers*, 86/3: 551–66.

Kariel, H. G. (1989). "Socio-Cultural Impacts of Tourism in the Austrian Alps." *Mountain Research and Development*, 9/1: 59–70.

King, T. D. (1997). "Folk Management and Local Knowledge: Lobster Fishing and Tourism at Caye Caulker, Belize." *Coastal Management*, 25/4: 455–69.

Lanegran, D. D., and St. Peter, P. H. (1993). "Travel, Tourism and Geographic Fieldwork: Project Marco Polo 1992." *Journal of Geography*, 92/4: 160–5.

LaPolla, F. (1988). "The Psychogeography of American Amusement Parks." *Revue Française d'Études Americaines*, 13/36: 235–40.

Lenz, R. (1993). "On Resurrecting Tourism in Vietnam." *Focus*, 43/3: 1–6.

Lew, A. A. (1988). "Tourism and Place Studies: An Example of Older Retail Districts in Oregon." *Journal of Geography*, 87/4: 122–6.

— (1992). "Perceptions of Tourists and Tour Guides in Singapore." *Journal of Cultural Geography*, 12/2: 45–52.

— (1994). "A Framework of Tourist Attraction Research," in B. J. R. Ritchie and C. R. Goeldner (eds.), *Travel, Tourism, and Hospitality Research: A Handbook for Managers and Researchers*. New York: John Wiley, 291–304.

— (1996). "Tourism Management on American Indian Lands in the US." *Tourism Management*, 17/5: 355–65.

Lew, A. A., and Van Otten, G. (eds.) (1998). *Tourism and Gaming on American Indian Lands*. Elmsford, NY: Cognizant Communications Corporation.

Lew, A. A., and Yu, L. (eds.) (1995). *Tourism in China: Geographical, Political, and Economic Perspectives*. Boulder, Colo.: Westview Press.

Loban, S. (1997). "A Framework for Computer-Assisted Travel Counseling." *Annals of Tourism Research*, 24: 813–34.

Lollar, S. A., and Van Doren, C. (1991). "U.S. Tourist Destinations: A History of Desirability." *Annals of Tourism of Research*, 18: 622–38.

Long, V., and Wall, G. (1996). "Successful Tourism in Nusa Lembongan, Indonesia?" *Tourism Management*, 17/1: 43–50.

Mansfield, Y. (1992*a*). "Group-Differentiated Perceptions of Social Impacts Related to Tourism Development." *Professional Geographer*, 44: 377–92.

— (1992*b*). "From Motivation to Actual Travel." *Annals of Tourism Research*, 19: 399–419.

Mariolle, E. M. (1996). "On the Road Again: Route 66 as a Tourist Attraction." *Illinois Geographical Society Bulletin*, 38/2: 24–33.

Marti, B. (1991). "The Cruise Ship Vessel Sanitation Program." *Journal of Travel Research*, 33: 29–38.

Meyer-Arendt, K. J. (1987). "Recreational Landscape Evolution along the North Yucatán Coast," *Yearbook, Conference of Latin Americanist Geographers*, 13: 45–50.

Meyer-Arendt, K. J. (1990*a*). "Recreational Business Districts in Gulf of Mexico Seaside Resorts." *Journal of Cultural Geography*, 11/1: 39–56.

— (1990*b*). "Tecolutla: Mexico's By-Passed 'Acapulco East,'" *Focus*, 40/3: 1–6.

— (1991). "Tourism Development on the North Yucatan Coast: Human Response to Shoreline Erosion and Hurricanes." *Geojournal*, 23: 327–36.

Meyer-Arendt, K. J., and Hartmann, R. (eds.) (1998). *Casino Gambling in America: Origins, Patterns, and Impacts*. Elmsford, NY: Cognizant Communication Corp.

— Sambrook, R., and Kermath, B. (1992). "Seaside Resorts in the Dominican Republic: A Typology." *Journal of Geography*, 91/5: 219–25.

Mieczkowski, Z. (1995). *Environmental Issues of Tourism and Recreation*. Lanham, Md.: University Press of America.

Milne, S., Ward, S., and Wenzel, G. (1995). "Linking Tourism and Art in Canada's Eastern Arctic: The Case of Cape Dorset." *Polar Record*, 31/176: 25–36.

Mings, R. C. (1988). "Assessing the Contribution of Tourism to International Understanding." *Journal of Travel Research*, 27/2: 33–8.

Mings, R. C., and McHugh, K. E. (1989). "The RV Resort Landscape." *Journal of Cultural Geography*, 10/1: 35–50.

— (1992). "The Spatial Configuration of Travel to Yellowstone National Park." *Journal of Travel Research*, 30/4: 38–46.

Mitchell, L. S. (1994). "Research on the Geography of Tourism," in J. R. B. Ritchie and C. Goeldner (eds.), *Travel, Tourism and Hospitality Research*. New York: John Wiley, ch. 17.

— (1996). "Morphology of Seaside Resorts: Myth and Reality of Waikiki," *Les Cahiers du Tourisme*, Series C, No. 199. Aix-en-Provence: Centre des Hautes Études Touristique.

Murphy, P. E. (1995). "Quality Management in Urban Tourism: Balancing Business and Environment." *Tourism Management*, 16/5: 345–6.

— (ed.) (1997). *Quality Management in Urban Tourism*. Chichester: Wiley.

Murphy, P. E., and Bayley, R. (1989). "Tourism and Disaster Planning." *Geographical Review*, 79: 36–46.

Murphy, P. E., and Carmichael, B. A. (1991). "Assessing the Tourism Benefits of an Open Access Sports Tournament: the 1989 B.C. Winter Games." *Journal of Travel Research*, 29/3: 32–6.

Murphy, P. E., and Keller, C. P. (1990). "Destination Travel Patterns: An Examination and Modeling of Tourist Patterns on Vancouver Island, B.C.," *Leisure Sciences*, 12/1: 49–66.

Nolan, M. L., and Nolan, S. (1992). "Religious Sites as Tourism Attractions in Europe." *Annals of Tourism of Research*, 19: 68–78.

Oakes, T. S. (1998). *Tourism and Modernity in China*. New York: Routledge.

— (1992). "Cultural Geography and Chinese Ethnic Tourism." *Journal of Cultural Geography*, 12/2: 3–17.

Olds, K. (1998). "Urban Mega-Events, Evictions and Housing Rights: The Canadian Case." *Current Issues in Tourism*, 1/1: 2–46.

Oppermann, M. (1994). "A Model of Travel Itineraries." *Journal of Travel Research*, 33/4: 57–61.

— (1995*a*). "Holidays on the Farm: A Case Study of German Hosts and Guests." *Journal of Travel Research*, 34/1: 63–7.

— (1995*b*). "Travel Life Cycle." *Annals of Tourism Research*, 22: 535–52.

— (ed.) (1997). *Geography and Tourism Marketing*. Albany, NY: Haworth Press.

Owen, C. (1990). "Tourism and Urban Regeneration." *Cities*, 7/3: 194–201.

Parsons, J. J. (1992). "Quartzsite, Arizona: A Woodstock for Rivers." *Focus*, 42/3: 1–3.

Place, S. E. (1988). "The Impact of National Park Development on Tortuguero, Costa Rica." *Journal of Cultural Geography*, 9/1: 37–52.

— (1991). "Nature Tourism and Rural Development in Tortuguero." *Annals of Tourism of Research*, 18: 186–201.

Price, M. F. (1987). "Tourism and Forestry in the Swiss Alps: Parasitism or Symbiosis?" *Mountain Research and Development*, 7/1: 1–12.

— (1992). "Patterns of the Development of Tourism in Mountain Environments." *Geojournal*, 27/1: 87–96.

Rafferty, M. D. (1993). *A Geography of World Tourism*. Englewood Cliffs, NJ: Prentice Hall.

Rense, W. C. (1988). "Tourism in Nepal," *The Pennsylvania Geographer*, 26/1, 2: 11–18.

Riley, R. W., Baker, D., and Van Doren, C. S. (1998). "Movie Induced Tourism." *Annals of Tourism Research*, 25: 919–35.

Ringer, G. (ed.) (1998). *Destinations: Cultural Landscapes of Tourism*. London: Routledge.

Rinschede, G. (1992). "Forms of Religious Tourism." *Annals of Tourism of Research*, 19: 51–67.

Roehl, W. S. (1990). "Travel Agent Attitudes Toward China after Tiananmen Square." *Journal of Travel Research*, 29/2: 16–22.

— (1995). "Competition, Casino Spending, and Use of Casino Amenities." *Journal of Travel Research*, 34: 57–62.

Roehl, W. S., and Fesenmaier, D. R. (1992). "Risk Perceptions and Pleasure Travel: An Exploratory Analysis." *Journal of Travel Research*, 30/4: 17–26.

Roehl, W. S., and Van Doren, C. S. (1990). "Locational Characteristics of American Resort Hotels." *Journal of Cultural Geography*, 11/1: 71–83.

Roehl, W. S., Fesenmaier, J., and Fesenmaier, D. R. (1993). "Highway Accessibility and Regional Tourist Expenditures." *Journal of Travel Research*, 31/3: 58–63.

Ross, S., and Wall, G. (1999). "Ecotourism: Towards Congruence Between Theory and Practice." *Tourism Management*, 20/1: 123–32.

Sambrook, R. A., Kermath, B. M., and Thomas, R. N. (1994). "Tourism Growth Poles Revisited: A Strategy for Regional Economic Development in the Dominican Republic," *Yearbook, Conference of Latin Americanist Geographers*, 20: 87–96.

Savage, M. (1993). "Ecological Disturbance and Native Tourism." *Geographical Review*, 83/3: 290–300.

Scott, C. R. (1988). "The Shattered Dream: Tourism and Civil Disorder in Sri Lanka." *The Pennsylvania Geographer*, 26/1, 2: 6–10.

Seaton, A. V., Jenkins, C. L., Wood, R. C., Dieke, P. U. C., Bennett, M. M., MacLellan, L. R., and Smith, R. (1994). *Tourism: The State of the Art*. New York: Wiley.

Sharkey, D. A., and Momsen, J. H. (1995). "Tourism in Dominica: Problems and Prospects." *Caribbean Geography*, 6/1: 40–51.

Shaw, G., and Williams, A. M. (1994). *Critical Issues in Tourism: A Geographical Perspective*. Cambridge: Blackwell.

Smith, S. L. J. (1989). *Tourism Analysis: A Handbook*. Harlow: Longman Scientific and Technical.

—— (1994). "A Guide to the Use of Quantitative Method Applicable to Tourism and Recreation." *Annals of Tourism Research*, 21: 582–95.

Smith, S. L. J., and Godbey, O. C. (1991). "Leisure, Recreation and Tourism." *Annals of Tourism of Research*, 18: 85–100.

Sommers, L. M., and Lounsbury, J. P. (1991). "Border Boom Towns of Nevada." *Focus*, 41/4: 12–18.

Squire, S. J. (1994*a*). "Accounting for Cultural Meanings: The Interface Between Geography and Tourism Studies Re-Examined." *Progress in Human Geography*, 18/1: 1–16.

—— (1994*b*). "The Cultural Values of Literary Tourism." *Annals of Tourism Research*, 21: 103–20.

—— (1995). "In the Steps of Genteel Ladies: Women Tourists in the Canadian Rockies 1885–1939." *Canadian Geographer*, 39/1: 2–15.

Stansfield, C. A. (1990). "Cape May: Selling History By the Sea." *Journal of Cultural Geography*, 11/1: 25–38.

Steinberg, M. K. (1994). "Tourism Development and Indigenous People: The Maya Experience in Southern Belize." *Focus*, 44/2: 17–20.

Stevens, S. F. (1991). "Sherpas, Tourism, and Cultural Change in Nepal's Mount Everest Region." *Journal of Cultural Geography*, 12/1: 39–58.

—— (1993). "Tourism Change and Continuity in the Mount Everest Region, Nepal." *Geographical Review*, 83: 410–27.

Stewart, W. P., Anderson, B. S., Fesenmaier, D. R., and Lue, C. (1993). "Highway Welcome Center Surveys: Problems With Nonresponse Bias." *Journal of Travel Research*, 31/3: 53–7.

Taylor, F. F. (1988). "The Ordeal of the Infant Hotel Industry in Jamaica." *The Journal of Imperial and Commonwealth History*, 16/2: 201–17.

—— (1993). *To Hell with Paradise: A Journal of the Jamaican Tourist Industry*. Pittsburgh: University of Pittsburgh Press.

Teye, V. (1988). "Geographic Factors Affecting Tourism in Zambia." *Annals of Tourism of Research*, 15: 487–503.

Timothy, D. J. (1995). "Political Boundaries and Tourism: Borders as Tourist Attractions." *Tourism Management*, 16/7: 525–32.

Timothy, D. J., and Butler, R. (1995). "Cross-Border Shopping: A North American Perspective." *Annals of Tourism Research*, 22: 16–34.

Toops, S. W. (1992). "Tourism in Xinjiang, China." *Journal of Cultural Geography*, 12/2: 19–34.

Van Doren, C. S. (1993). "Pan Am's Legacy to World Tourism." *Journal of Travel Research*, 32/1: 3–12.

Vlack, R. (1993). "The Impact of Tourism in Fiji: A Comparison of Two Villages." *Focus*, 43/2: 1–7.

Wall, G. (1996*a*). "Integrating Integrated Resorts." *Annals of Tourism Research*, 23: 713–17.

—— (1996*b*). "Scale Effects on Tourism Multipliers." *Annals of Tourism Research*, 24/2: 446–50.

—— (1997). "Tourist Attractions: Points, Lines and Areas." *Annals of Tourism Research*, 24/1: 240–3.

Wall, G., and Heath, E. (1992). *Marketing Tourism Destinations: A Strategic Planning Approach*. New York: Wiley.

Wall, G., and Oswald, B. (1990). "Cultural Groups as Tourist Attractions: A Comparative Study," *Les Cahiers du Tourisme*, Series C, No. 138. Aix-en-Provence: Centre de Hautes Études Touristiques.

Weaver, D. B. (1993*a*). "Ecotourism in the Small Island Carribbean." *Geojournal*, 31/4: 325–40.

—— (1993*b*). "Model of Urban Tourism for Small Caribbean Islands." *Geographical Review*, 83/2: 134–40.

—— (1997). "A Regional Framework for Planning Ecotourism in Saskatchewan." *The Canadian Geographer*, 41/3: 281–93.

Wilkinson, P. F. (1989). "Strategies for Tourism in Island Microstates." *Annals of Tourism Research*, 16/2: 153–77.

—— (1992). "Tourism—the Curse of the Nineties: Belize—an Experiment to Integrate Tourism and the Environment." *Community Development Journal*, 27/4: 386–95.

Williams, S. (1998). *Tourism Geography*. New York: Routledge.

Wilson, V. R. (1995). "Expected Impact of Tourism on the Business Community of Windberg, Pennsylvania." *The Pennsylvania Geographer*, 33/1: 71–89.

Winchell, D. G., Lounsbury, J. F., and Sommers, L. M. (1997/8). "Indian Gaming in the U.S.: Distribution, Significance, and Trends." *Focus*, 44/4: 1–10.

Wong, P. P. (ed.) (1993). *Tourism vs. Environment: The Case for Coastal Areas*. Dordrecht: Kluwer.

Zelinsky, W. (1994). "Conventionland, USA: The Geography of a Latterday Phenomenon." *Annals of the Association of American Geographers*, 84: 68–86.

Zurick, D. N. (1992). "Adventure Travel and Sustainable Tourism in the Peripheral Economy of Nepal." *Annals of the Association of American Geographers*, 82/4: 608–28.

—— (1995*a*). *Errant Journeys: Adventure Travel in a Modern Age*. Austin: University of Texas Press.

—— (1995*b*). "Preserving Paradise." *Geographical Review*, 85: 157–72.

Recreation and Parks

Boorstein, M. F. (1992). "The Wonders and Origins of Four National Parks of the Southern United States." *Focus*, 42/4: 26–31.

Bristow, R. S. (1998). "Volunteer-Based Recreation Land Management." *Parks and Recreation*, 33/8: 70–6.

Bristow, R. S., Lieber S. R., and Fesenmaier, D. R. (1995). "The Compatibility of Recreation Activities in Illinois." *Geografiska Annaler*, Series B, 77B: 3–15.

Clarke, L. M. (1988). "Recreational Mapping at a Tourist Attraction: An Experimental Map Evaluation Involving User Feedback." *The Society of University Cartographers*, 22/2: 13–18.

Dearden, P. (1988). "Protected Areas and the Boundary Model: Meares Island and the Pacific Rim National Park." *Canadian Geographer*, 32/3: 256–65.

—— (1993). "Canada's National Parks: A Model of Administrative Penetration." *Canadian Geographer*, 37/3: 194–211.

Dilsaver, L. M. (1987). "The Evolution of Land Use in Giant Forest, Sequoia National Park." *Yearbook, Association of Pacific Coast Geographers*, 49: 35–50.

Eyre, C. A. (1990). "The Tropical National Parks of Latin America and the Caribbean: Present Problems and Future Potential," *Yearbook, Conference of Latin Americanist Geographers*, 16: 15–33.

Fesenmaier, D. R. (1990). "Theoretical and Methodological in Behavioral Modeling: Introductory Comments." *Leisure Sciences*, 12/1: 1–8.

Flint, D. (1998). U.S. National Parks: Visitor Experience and Resource Protection." *Geography Review*, 11/5: 20–4.

Foust, B., and Botts, H. (1989). "Estimating Attendance and Handle for Greyhound Parimutuel Facilities: The Case of Wisconsin License Applications." *Papers and Proceedings of Applied Geography Conferences*, 12: 16–24.

Glyptis, S. (ed.) (1993). *Leisure and the Environment*. Belhaven, London.

Goodenough, R. (1992). "The Use of Environmental Impact Assessment in the Management of Open Space and Recreational Land in California." *International Journal of Environmental Studies*, 40/2–3: 171–84.

Gorrie, P. (1987). "The Bruce: Our Newest National Park." *Canadian Geographic*, 107/5: 62–71.

Halseth, G., and Rosenberg, M. W. (1995). "Cottagers in an Urban Field." *Professional Geographer*, 47: 148–59.

Harvey, M. W. T. (1998). "Defending the Park System: The Controversy Over Rainbow Bridge." *New Mexico Historical Review*, 75/1: 45–67.

Hobbs, J. J. (1996). "Speaking with People in Egypt's St. Katherine National Park." *Geographical Review*, 86/1: 1–21.

Jackson, E. L. (1988). "Integrating Ceasing Participation with Other Aspects of Leisure Behavior." *Journal of Leisure Research*, 20/1: 31–45.

—— (1991). "Shopping and Leisure: Implications of West Edmonton Mall for Leisure and for Leisure Research." *Canadian Geographer*, 35/3: 280–6.

—— (1994). "Geographical Aspects of Constraints on Leisure and Recreation." *Canadian Geographer*, 38/2: 110–21.

Jackson, E. L., and Henderson, K. (1995). "Gender-Based Analysis of Leisure Constraints." *Leisure Sciences*, 17/1: 31–51.

Johnson, J. M. (ed.) (1994). *Exploration and Mapping of the National Parks*. Winnetka: Speculum Orbis Press.

Kariel, H. G., and Drapier, D. L. (1992). "Outdoor Recreation in Mountains." *Geojournal*, 27/1: 97–101.

Kim, S., and Fesenmaier, D. R. (1990). "Evaluating Spatial Structure Effects in Recreation Travel." *Leisure Sciences*, 12/1: 367–81.

Konrad, C. E., II (1995). "When to Plan Outdoor Activities: The Daytime and Seasonal Patterns of Wetness in the Southeastern United States." *Southeastern Geographer*, 35/2: 150–67.

Lemons, J. (1987). "United States' National Park Management: Values, Policy, and Possible Hints for Others." *Environmental Conservation*, 14/4: 329–40.

Lieber, S. R., Fesenmaier, D. R., and Bristow, R. S. (1989). "Recreation Expenditures and Opportunity Theory: The Case of Illinois." *Journal of Leisure Research*, 21/2: 106–23.

Little, C. E. (1990). *Greenways for America*. Baltimore: Johns Hopkins University Press.

McPherson, E. G. (1992). "Accounting for Benefits and Costs of Urban Greenspace." *Landscape and Urban Planning*, 22/1: 41–51.

Mandel, C. (1998). "Soothing the Urban Soul." *Canadian Geographic*, 118/4: 30–40.

Millward, H. (1996). "Countryside Recreational Access in the United States: A Statistical Comparison of Rural Districts." *Annals of the Association of American Geographers*, 86/1: 102–22.

Mitchell, D. (1995). "The End of Public Space?: People's Park, Definitions of the Public, and Democracy." *Annals of the Association of American Geographers*, 85/1: 108–33.

Morehouse, B. (1996). "Conflict, Space, and Resource Management at Grand Canyon." *Professional Geographer*, 48: 46–57.

Peterson, K. (1991). "The Delineation of Recreational Boating Markets: Multivariate Approach." *Journal of Leisure Research*, 23/3: 209–24.

Pfister, R. E. (1993) "The Concept of Ethnic Identity: A New Avenue to Understanding Leisure Preferences and Behaviors," in D. Chavez, A. Ewert, and A. Magill (eds.), *Culture, Conflict, and Communication in the Wildland/Urban Interface*. Boulder, Colo.: Westview Press, 53–68.

Raitz, K., and Dakhil, M. (1989). "A Note about Information Sources for Preferred Recreational Environments." *Journal of Travel Research*, 27/4: 45–9.

Richter, D. M. (1992). "Transient Boat Usage on Geneva Lake, Wisconsin." *The Wisconsin Geographer*, 8: 48–55.

Rogers, G. F., and Rowntree, R. A. (1988). "Intensive Surveys of Structure and Change in Urban National Areas." *Landscape and Urban Planning*, 15/1–2: 11–22.

Saremba, J., and Gill, A. (1991). "Value Conflicts in Mountain Park Settings." *Annals of Tourism of Research*, 18: 455–72.

Sellars, R. W. (1997). *Preserving Nature in the National Parks: A History*. New Haven, Conn.: Yale University Press.

Sidaway, R., and Thompson, D. (1991). "Upland Recreation: The Limits of Acceptable Change." *Ecos*, 12/1: 31–9.

Smardon, R. C. (1988). "Water Recreation in North America." *Landscape and Urban Planning*, 16/1–2: 127–43.

Staple, T., and Wall, G. (1996). "Climate Change and Recreation in Nahanni National Park Reserve." *Canadian Geographer*, 40/2: 109–20.

Strapp, J. D. (1988). "The Resort Cycle and Second Homes." *Annals of Tourism of Research*, 15: 504–16.

Talarchek, G. M. (1990). "The Urban Forest of New Orleans: An Explanatory Analysis of Relationships." *Urban Geography*, 11/1: 65–86.

Talen, E. (1997). "The Social Equity of Urban Service Distribution: An Exploration of Park Access in Pueblo, Colorado and Macon, Georgia." *Urban Geography*, 18/6: 521–41.

Thorsell, J. W., and Harrison, J. (1992). "National Parks and Nature Reserves in Mountain Environments and Development." *Geojournal*, 27/1: 113–26.

Wall, G. (ed.) (1989). *Outdoor Recreation in Canada*. Toronto: Wiley.

Welch, R., Remillard, M., and Doren, R. F. (1995). "GIS Database Development for South Florida's National Parks and Preserves." *Photogrammetric Engineering and Remote Sensing*, 61/11: 1371–81.

Wikle, T. A. (1991). "Evaluating the Acceptability of Recreation Rationing Policies Used on Rivers." *Environmental Management*, 15/3: 389–94.

Willis, K., and Garrod, G. (1991). "Valuing Open Access Recreation on Inland Waterways: On Site Recreation Surveys and Selection Effects." *Regional Studies*, 25/6: 511–24.

Winder, G. M. (1998). "Neighborhood Community Centers and the Changing Leisure Culture in Brandon." *The Canadian Geographer*, 42/1: 86–93.

Young, K. R. (1993). "National Park Protection in Relation to Ecological Zonation of a Neighboring Human Community: An Example from Northern Peru." *Mountain Research and Development*, 13/3: 267–80.

Young, T. (1993). "San Francisco's Golden Gate Park and the Search for a Good Society 1865–1880." *Forest and Conservation History*, 37/1: 4–13.
— (1995). "Modern Urban Parks." *Geographical Review*, 85/4: 535–51.

Sport

Abbott, C. (1990). "College Athletic Conferences and American Regions." *Journal of American Studies*, 24/2: 211–21.
Archenbault, M. (1993). "Le Ski Alpin au Quèbec: Approche Économique et Aid a la Droision." *Revue de Geographie Alpin*, 81/1: 47–60
Ashley, R. M. (1988). "Using Interest in High School Sports to Develop Geography's Five Themes." *The Illinois Geographical Society*, 30/2: 37–40.
Augustin, J. (1997). "Les Territoires Incertains du Sport." *Les Cahiers de Géographie du Quebec*, 41/114: 405–12.
Bale, J. R. (1989). *Sports Geography*. London: E. and F. N. Spon.
— (1991). *The Brawn Drain: Foreign Student-Athletes in American Universities*. Urbana: University of Illinois Press.
— (1992). *Sport, Space and the City*. London: Routledge.
— (1994). *The Landscapes of Sport*. Leicester: Leicester University Press.
— (1998). "The Place of 'Place' in Cultural Studies in Sports." *Progress in Human Geography*, 12/4: 507–24.
Bowden, M. J. (1990). *Hamlets: A History and Geography of Sutton Fuller Hamlets Soccer Club 1969–1989*. Sutton, Mass.: Putnam House.
Brottan, J., and Wall, G. (1993). "Prospects for Downhill Skiing in a Warmer World," in Marie Sanderson (ed.), *The Impact of Climate Change on Water in the Grand River Basin, Ontario*. Waterloo: University of Waterloo, Geography Publication, 40: 93–104.
Brown, R. (1988). "Duplicate Bridge: An Indoor Sport." *Sport Place*, 2/3: 24–31.
Clay, G. (1988*a*). "The Roving Eye, Run It Down the Field Again, Fellows: The Football Field as a Generic Unit of Territory." *Sport Place*, 2/2: 23–5.
— (1988*b*). "The Roving Eye, the Arena Effects: Television Über Alles." *Sport Place*, 2/2: 20–33.
Dargitz, R. E. (1988). "Angling Activity of Urban Youth: Factors Associated with Fishing in a Metropolitan Context." *Journal of Leisure Research*, 20/3: 192–207.
Elliott, H. M. (1988). "Cardinal Directions, South Paws, and Geographic Names." *Sport Place*, 2/2: 40–1.
Ferrett, R. L., and Ojala, C. F. (1992). "Geographic Shifts in Divisional College Football Recruiting Since the Inception of Proposition 48." *Sport Place*, 6/1: 29–38.
Genest, S. (1996). "Le Canada et l'internationalisation du hockey sur glace," in J. Augustin and C. Sorbetz (eds.), *La Culture du Sport au Quèbec*. Talence: Éditions de la Maison des Sciences de l'Homme l'Aquitaine, 177–186.
Goudge, T. L. (1983). "Interscholastic Athletic Participation: A Geographical Analysis of Opportunity and Development." *Proceedings, U.S. Olympic Academy*, 7: 165–202.
Griepentrog, C. (1992). "Brothertown Fishing Club Ice Bowling Tournament." *Sport Place*, 6/2: 21–6.
Handley, A. C., *et al.* (1994). "Youth Soccer in the United States." *Geographical Bulletin*, 36/1: 3–20.
Kraft, M. D., and Skeeter, B. R. (1995). "The Effect of Meteorological Conditions on Fly Ball Distance in North American Major League Baseball Games." *Geographical Bulletin*, 37/1: 40–8.
Lipski, S., and McBoyle, G. (1991). "The Impact of Global Warming on Downhill Skiing in Michigan." *East Lakes Geographer*, 26: 37–51.
Little, A. (1990). "The Gridiron Invasion: American Football in Britain." *Sport Place*, 4/2: 20–5.
Louder, D. (1991). "Étude Géographique du Sport en Amerique du Nord: Surval et Critique," in D. Errais, D. Mathieu, and J. Praicheux (eds.). Besançon: UFR et STAPS, Géopolitique du Sport.
Maguire, J. (1990). "American Football: An Emergent Sport Form in English Society." *Sport Place*, 4/2: 13–14.
Mayfield, M. W., and DeHart, J. L. (1989). "Whitewater Rafting in the Southeastern United States: Current Status and Constraints on Future Growth." *Sport Place International*, 3/3: 14–19.
McConnell, H. (1994). "Baseball is a Warm-Weather Game: A Geographer's Look at the College World Series, 1947–1993." *Sport Place*, 8/1: 5–37.
McEachern, P. D., and Russell, P. I. (1992). "I Heard It on the Radio: Distribution of Radio Networks in Major League Baseball," *Sport Place International*, 6/3: 3–17.
Miller, E. W. (1990). "Golf in Pennsylvania." *The Pennsylvania Geographer*, 28/1: 42–50.
Ojala, C. F., and Gadwood, M. T. (1989). "The Geography of Major League Baseball Player Production, 1876–1988." *Sport Place*, 3/3: 24–35.
Pillsbury, R. (1989). "A Mythology at the Brink: Stock Car Racing in the American South." *Sport Place International*, 3/3: 2–12.
Pleumaron, A. (1992). "Course and Effect: Golf Tourism in Thailand." *The Ecologist*, 22/3: 104–10.
Raitz, K. B. (1995). *The Theater of Sport*. Baltimore: Johns Hopkins University Press.
Rooney, J. F., Jr. (1989). "American Golf Courses: A Regional Analysis of Supply." *Sport Place International*, 3/1, 2: 2–17.
— (1990). "The Time for Minor League Football and Basketball is Now." *Sport Place International*, 4/1: 17–21.
— (1993). "The Golf Construction Boom, 1987–1993." *Sport Place*, 7/1: 15–22.
Rooney, J. R., Jr., and Pillsbury, R. (1992). *Atlas of American Sport*. New York: Macmillan.
Roseman, C. C., and Shelley, F. M. (1988). "The Geography of Collegiate Football Radio Broadcasting." *Sport Place*, 2/2: 42–50.
Rumney, T. A. (1988). "Rowing in America: A Look from the Sliding Seat." *Sport Place*, 2/3: 32–9.
— (1995). "Bibliography of Sport's Geography." *Sport Place*, 9/1: 4–30.
Schnell, G. A. (1990). "Geographic Origins of Football Players at Pennsylvania's Colleges." *Pennsylvania Geographer*, 28/1: 15–29.
Shelley, A. M., and Shelley, F. M. (1993). "Changing Migration Patterns of Major League Baseball Players." *Sport Place*, 7/1: 23–35.
Shelley, F. M., and McConnell, H. (1993). "So Who's Number One? An Investigation of Geographical Bias in the Ranking of College Football Teams." *Sport Place*, 7/2: 19–29.

Skeeter, B. R. (1988). "The Climatically Optimal Major League Baseball Season in North America." *The Geographical Bulletin*, 30/2: 97–104.

Stadler, S. J., and Simone, M. A. (1988). "Applied Macroanalysis of Golf Course Environments." *Papers and Proceedings of Applied Geography Conferences*, 11: 34–42.

Tooley, C. (1988). "American Cricket." *Sport Place*, 2/2: 17–18.

Van Zuyk, P. (1988). "Battleline in the Desert: A Report on the Barstow-to-Vegas Motorcycle Race and the California Desert." *Sport Place*, 2/1: 15–20.

Wall, G. (1988). "Demand and Supply of Sport Fishing." *Annals of Tourism of Research*, 15: 561–2.

Wiggins, G. A., and Soulé, P. T. (1993). "Environmental and Topographical Factors Affecting Male 'Middle-of-the Pack' Dualthlon Performance." *Geographical Bulletin*, 35/2: 86–97.

Wikle, T. A. (1993). "Spatial Aspects of Paddlesports in the United States." *Sport Place*, 7/1: 3–14.

Zeller, R. A., and Jurkovac, T. (1989). "A Dome Stadium: Does It Help the Home Team in the National Football League?" *Sport Place International*, 3/3: 36–9.

CHAPTER 34

Applied Geography*

Nancy K. Torrieri and Michael R. Ratcliffe

Introduction

Applied geographers solve problems that inform decision-making and policy. The problems most often arise in government or the private sector, require practical, rapid, and cost-effective solutions, and are usually client-driven. Often, government or the private sector funds the research of applied geographers, and the research results have specific and usually current implications for government policies and programs or business strategies. Applied geographers use techniques, tools, methods, and concepts from the discipline of geography, but borrow liberally from other disciplines as necessary. This approach is well suited to solving problems that have multiple, not merely spatial, dimensions.

As a subdiscipline of geography, applied geography has an ill-defined niche. Is applied geography a subdiscipline similar to biogeography, geomorphology, or economic geography? Or does it occupy an intermediary position within the discipline, synthesizing concepts from various subdisciplines? A certain ambiguity exists as to the point of origin of applied geography as a subdiscipline as well as whether it is even a distinct subdiscipline within geography (Hart 1989). The nature of applied geography has been a staple topic of discussion in professional papers and applied geography texts. Such discussions tend to focus on applied geography's role within the discipline and what differentiates applied geographers from other geographers. Kenzer (1989) provides an excellent and lively account, in part noting that aspects of the discussion are bound up in the historic differences between academic and non-academic (professional) geographers. Kenzer and the contributors to the volume he edited provide plenty of grist for discussion. Over the past twenty years considerable interest has been attracted toward this discussion. In particular, we draw attention to work by Frazier (1982), Kenzer (1989), Hart (1989), Palm and Brazel (1992), Johnston (1993); Frazier *et al.* (1995), Harvey (1997), and Golledge *et al.* (1997) for discussions of the origins, purpose, and focus of applied geography.

For the purposes of this review, applied geography is treated as one of many subdisciplines or specialties. Notwithstanding attaining the status of a subdiscipline, however, there is no single body of research that applied geographers can point to as clearly their own. There is no set of techniques, concepts, methodologies, and theories that is used only by applied geographers. So, what distinguishes applied geography from other subdisciplines within geography? We would argue that the orientation toward using geographic methods, concepts, and theories to solve specific problems and questions, whether brought by a potential client or found within society, differentiates applied geography from other specialties within geography (Palm and Brazel 1992).

Applied Geography in the Nineteenth and Early Twentieth Centuries

Although some point to the "quantitative revolution" as the point of origin for applied geography, others find its roots in the exploration and writing of eighteenth- and nineteenth-century geographers (often coinciding with the extension of national hegemonies) that fostered the development of geography as an independent discipline (Palm and Brazel 1992; Turner 2002). In the US, federal government policy goals in the late nineteenth century provided the impetus for exploration and data collection and analysis (Frazier *et al.* 1995). The US government's topographic surveys in the 1870s and 1880s played an important role in the development and application of geographic techniques. Although focused primarily on mapping the intermountain West to identify locations of mineral resources, these surveys also gathered considerable amounts of information about the physical environment and the peoples and cultures of the region. John Wesley Powell's work on the arid regions provides a particularly good example of early applied geography, synthesizing information about the physical environment and local cultures to propose plans for the most appropriate way for settlement to proceed.

Scientists whose work could be described as applied geography were involved in the emergence of geography as a separate field of academic study. Among the forty-eight charter members of the Association of American Geographers, sixteen worked for the Federal government, primarily with the US Geological Survey, but also the US Census Bureau, Coast and Geodetic Survey, Department of Agriculture (USDA), Navy Hydrographic Office, Weather Bureau, and the Treasury Department's Bureau of Statistics (Dahmann 1998, pers. comm.). The list of Federal government applied geographers among the AAG charter members includes: Grove Karl Gilbert (USGS); C. Hart Merriam, director of the USDA's Biological Survey; Henry Gannett (USGS and Census Bureau); and Curtis Marbut, director of the USDA's Bureau of Soil Surveys (ibid).

The work of these individuals and others within the Federal government in the late 1800s focused on issues and problems of concern to society, ranging from public health, particularly within cities, to issues related to settlement and economic development of the arid West. Although few of those involved described themselves as "geographers," they all applied geographic concepts and techniques in describing, understanding, and seeking solutions to societal problems. Geographers as well as those of a geographical orientation adopted geographic techniques in analyzing the distribution of resources, settlement, industry, health issues and mortality, immigration, and climate (in relation to agriculture). These geographers and their colleagues developed much of the mapping and geographical infrastructure within the Federal government on which we have come to rely and from which many of the innovations at the end of the twentieth century derive.

Applied Geography, 1920–1950

Applied geography achievements within Federal and state government from the 1920s through the 1950s demonstrated that applied geography in America was increasingly broadly based. Carl Sauer's work in connection with Michigan's Land Economics Survey contributed to land-use classification and land-use planning efforts in America and abroad. Isaiah Bowman advanced applied geography as an adviser and consultant to the US government from World War I through World War II, participating in settlement of boundary disputes, promoting expeditions to and the mapping of Latin America, and serving as a consultant in international affairs.

Federal service attracted many to Washington to assist in the work with New Deal or war-related programs (Harris 1997). Harlan Barrow's work on water-resources projects between 1933 and 1941, to support the Public Works Administration, and, later to allocate Upper Rio Grande River resources among Colorado, New Mexico, and Texas, heralded a role for applied geographers as policy-makers. Geographers in planning positions served the Tennessee Valley Authority and numerous federal-local work relief projects. Gilbert White's contributions to our knowledge of water resources and flood control began with his work with the Mississippi Valley Committee, the National Resources Committee, and the National Resources Planning Board during the 1930s. He continued this work during a two-year stint within the Bureau of the Budget (predecessor of the Office of Management and Budget). His work as an applied geographer within various agencies of the federal government provided ample opportunities to translate scientific research into policy (Reuss 1993).

The concentration of geographers in Washington, DC, during the 1940s in support of the war effort resulted in a number of positive contributions to geography as a discipline (Harris 1997). In particular, the weaknesses in geographic education became apparent as geographers applied geographic concepts to specific cartographic, logistical, intelligence gathering, and political and economic topics. Ackerman (1945) noted the lack of training in the more systematic branches of geography and predominance of geographers with regional studies backgrounds. The clustering of geographers, many of whom were young professionals, also stimulated considerable discussion and an exchange of ideas regarding the application of geographic concepts. Harris (1997) notes that his collaboration with Edward Ullman on the "Nature of Cities" might never have occurred if the two had not been in Washington together during the early 1940s.

This clustering of geographers, many of whom were not members of the AAG, in Washington led directly to the formation of the American Society of Professional Geographers (ASPG) to serve the needs of professional geographers; the society's journal—the *Professional Geographer*—provided a publication outlet for applied geographers. The ASPG was incorporated into the AAG in 1948 and the *Professional Geographer* became one of the AAG's two journals. The Middle Atlantic Division of the AAG, with much of its membership based in the Washington DC, area and composed of Federal government geographers, remained as the successor to the APSG.

Applied Geography, 1950–1980

Following World War II, legislation at the federal level, the Federal Housing Act of 1949, and the Highway Act of 1956 stimulated planning activities in housing and transportation. Also, city planning became important as millions of veterans demanded housing, rural land became converted to housing tracts, shopping centers, and industrial parks, and the automobile became ubiquitous. Applied geographers found careers in burgeoning federal bureaucracies that administered legislation in the areas of housing and transportation, or in community organizations that were designed to arrest urban blight. Later, the civil rights movement of the 1960s underscored the need for basic research on the spatial aspects of minority housing and employment. Geographers contributed significantly to knowledge of these problems and to devising recommendations for solving them. The Demonstration Cities and Metropolitan Development Act of 1956 served to promote the work of regional planning agencies to ensure that transportation, land use, recreation, utilities, and other programs would be developed in comprehensive, region-wide contexts. These agencies needed geographic information to develop these programs and applied geographic specialists to help manage them. The deconcentration of population in metropolitan areas, coupled with the growth of the suburbs (the 1970 census showed for the first time that the suburban population exceeded that of central cities), the fuel crisis, and increasing awareness of environmental issues gave rise to increasingly sophisticated planning analyses. Applied geographers have had an important role in this work. Applied geographers have worked with federal and state transportation agencies to help plan the national highway system, model traffic flow volumes and seasonal transportation patterns, and devise programs to distribute billions of dollars in funding for public transportation, highway improvements, and improved air, rail, and freight services.

In addition to addressing housing, social, and transportation problems, applied geographers work on issues related to population, food, and the environment. An awareness of the human–environmental interaction has led to applied geographic research in land and water management, deforestation, erosion, desertification, and agriculture. Medical geographers have conducted applied research directed toward improving health conditions both in the US and abroad, focusing especially on ecological-epidemiological problems and issues related to the delivery of health services, particularly in sparsely populated regions. Applied rural geographers have explored the dimensions of change on the rural–urban fringe around metropolitan centers, and focused attention on the issues associated with rural development, particularly, sustainability of development, land-use change, and the ecological effects of federal, state, and local government policies and private-sector development.

By the end of the 1950s, the need for applied cartographers was apparent throughout the federal government, and in particular among defense and intelligence agencies. Three decades later, applied cartographers had a larger calling, based to some extent on a broader interpretation of the role of mapping. Today, federal agencies produce millions of maps each year, the impetus for their creation stemming often from a problem or perceived need related to spatial analysis, location or accessibility, planning, the allocation of services, or the prediction of future spatial patterns. The first comprehensive digital street map of the United States, the TIGER® (Topologically Integrated Geographic Encoding and Referencing) system, arose out of the need to prepare for the 1990 decennial census (Carbaugh and Marx 1990).

Applied Geography: Late Twentieth Century

The last twenty years have shown that applied geography can be the catalyst for solving problems in ways that possibly could not have been envisioned a generation ago. One of the most important areas in which applied geographers have contributed in the last twenty years is the area of marketing and business geography. These fields have spawned research into site selection and evaluation, sales forecasting, and real estate development. The geography of crime, a relatively new area for applied geographers, has led to efforts by multiple federal agencies, under Department of Justice leadership, to pool resources, talents, and technology to address issues such as community policing and the anticipation of where, and under what conditions, crimes are likely to occur. Emergency services planning is another area in which applied geographers have made contributions in the last several decades. The E-911 emergency response services common to many areas have been developed with the input of applied geographers. Applied geographers have participated in the local and national response to the events of September 11, 2001, and organizations such as the Federal Geographic Data Committee that represent them have led the call for greater attention to the integration and coordination of spatial information from many sources to ensure more effective responses to such events in the future.

Changes in Academia Fuel Applied Geographic Research in the 1990s

Changes in the culture and institutional setting of academia have made it possible for applied geographic research to thrive in the 1990s. These include:

- The need for increased funding in academia, and the use of strategic planning to justify the survival of academic departments; and
- The rise of research groups, institutes, and related bodies that tap the intellectual, collaborative, and financial strengths of multiple disciplines, universities, and government agencies.

Funding Decisions and Strategic Planning Benefit Applied Research

The last fifteen years have witnessed changes in the institutional climate for universities and colleges in North America. The need for funding has been led to various cost-cutting measures and strategies (I. Warde 2001). A department's decisions to hire, grant tenure, expand instructional programs, and acquire new technologies or facilities depends on its ability to support funded research. Government agencies underwrite much of this research, and government interest in solving specific problems related to programs, policies, or strategies, many of which have spatial aspects to them, makes it possible for applied geographic research to flourish and geography departments that support such research to grow.

The Role of University Research Groups

A review of geography department websites in Canada and the US reveals that many have links to research groups, centers, institutions, laboratories, or other facilities that are engaged in applied geographic and other research and that serve as the sources of funding for faculty in a variety of disciplines. Urban, environmental, and physical geography in particular are the focus of much of the research that takes place in such groups, which have programmatic and funding linkages with other universities, disciplines, and government agencies. Because they represent the pooling of faculty talents in several disciplines, or even across multiple academic institutions, these groups are well qualified to study emerging research problems in government, and take on large-scale investigative research programs funded by foundations or other organizations. Often the research involves the use of technology in areas such as image processing, remote sensing, geographic information systems, and automated cartography. The National

Aeronautical and Space Administration (NASA's) Affiliated Research Center Program, the Center for Urban and Regional Analysis of Ohio State University, the University of Texas Austin's Environmental Science Institute, the University of Buffalo's Canada-United States Trade Center, Syracuse University's Maxwell Center, the University of West Virginia's Regional Research Institute, and GEOIDE, a multidisciplinary research and development program to support geomatics in Canadian government, universities, and institutions, are illustrative of the scope and variety of these research groups.

Development of Tools and Techniques for Aiding Geographic Analysis

The growth of applied geography has been fueled by the development and exchange of geographic tools and resources that have contributed to innovation in academia, government, and the private sector. Applied geographers within the Census Bureau and the US Geological Survey were instrumental in the creation of the first nationwide digital spatial databases. The Census Bureau's TIGER system has been integral to the growth of the geographic information systems (GIS) industry. The Census Bureau's TIGER/Line® files (which are extracts of the TIGER database) form the basis for many of the on-line mapping programs available over the Internet. Remote sensing, statistical modeling, and global positioning system technology have revolutionized the study of the physical environment and made available maps and other resources that have contributed to an understanding of the complexity of human–environmental relationships.

Governments have made available a vast number of data resources that have been used to study geographic problems. These include administrative records from state governments, demographic and economic data from decennial censuses and surveys, and areal classifications, such as metropolitan area standards. The increasing use of these data sources in the legislative and policy arena has focused attention on their strengths and weaknesses, and the need to improve them. To that end, for example, in response to the need to make its programs more useful and cost effective, the Census Bureau is planning to change the way it takes a census. Subject to congressional approval, the agency will replace its once-a-decade collection of detailed social, demographic, and economic information by means of a decennial census "long form" with a continuous survey, the American Community Survey (ACS), that collects information from a small sample of the population each month and every year (see <www.census.gov/acs/www/>, accessed 15 February 2003). ACS data will be produced as annually updated single- and multi-year estimates of characteristics of geographic areas and population groups. Current data from the ACS provides a potentially powerful resource for decision-making, marketing, policy analysis, and numerous other geographic applications that take advantage of census data (Torrieri and Holland 2000). Fundamentally revised geographic area standards for metropolitan, urban, and rural areas became effective with the publication of Census 2000 data. The change in these standards had direct impact on the many programs in government and industry that use the standards as a basis for classifying land and population for transportation, environmental, and other programs, and as a basis for spatial decision-making in real estate, banking, and other businesses.

Journals, Conferences, Organizations, the Internet, and the Media

Applied geographers publish widely, although several notable journals once created expressly as outlets for their work, *Applied Geography*, and *Journal of Applied Geographic Studies*, have ceased publication. They also make presentations at AAG, American Geographical Society, and other professional conferences sponsored by organizations with ties to disciplines such as urban affairs, transportation geography, rural geography, regional science, geographic information science, sociology, demography, engineering, and environmental sciences. The Applied Geography Conference, first held in 1978, continues to hold an annual meeting. The goal of the conference, in the words of its founder, John Frazier, was to "bring together persons of diverse backgrounds to share problems, expertise, and future projects and to initiate a dialogue that would continue after the meetings" (quoted in Harvey 1997). The Applied Geography Specialty Group of the AAG has a membership of nearly 209 (as of 1 February 2003), produces a newsletter for its membership, and awards the Anderson Medal each year in recognition of highly distinguished service to the profession of geography.

Applied geographers participate in a number of geographic organizations in addition to the AAG. Notable among these are the Urban Regional Information Systems Association, the University Consortium for Geographic Information Science, and the National States Geographic Information Council. These organizations play a leading role at the state and national level as advocates for the sharing of geospatial information, and the standardization of spatial data sets.

The growth of the Internet has made geographic information more accessible, and has acted as a catalyst for applied and other geographic research. A growing number of websites make data, tutorials, lists of funding resources, links to geographic topics, and related information available. Notable sites, last accessed on 15 February 2003, include those of Statistics Canada (<www.statcan.ca>) and the University of Colorado at Boulder (<www.colorado.edu/geography/virtdept/resources/contents.htm>). Online and distance learning opportunities may further stimulate applied geographic research in the future. For example, the availability of online courses in GIS could make applied geographers out of those who might otherwise never learn about geography in a traditional classroom setting.

The media, perhaps more than any other factor, has made applied geography visible and more familiar to the public. Computer-assisted reporting and the use of GIS and maps in the analysis and presentation of information is widespread among daily newspapers. This increased use of geographical analysis and mapping tools was especially evident in articles focusing on results from the Census 2000 as choropleth maps routinely accompanied reports on demographic trends. The AAG, in recognition of these developments, instituted an AAG Media Achievement Award to recognize creative techniques and other contributions that engage non-geographers to think spatially. Many major newspapers now run series of articles on topics that involve geographic themes and topics. *The Washington Post* published a four-part article on the proposed plan to revive the Florida Everglades that featured detailed maps and descriptions of the problems and issues encountered by stakeholders in this plan (Grunwald 2002). The article highlighted the technological approaches being considered and the trade-offs in undertaking such a project. One could clearly envision after reading the article why the proposed plan, which addresses issues related to water quality, ecosystem management, economic development, and agribusiness, could generate applied geographic research for a range of clients in government and the private sector.

Themes in Applied Geographic Research

While applied geographic research takes place in many geographic subfields, several broad categories of research are notable: (1) market and location analysis; (2) medical geography; (3) land-use planning, environmental issues, and policy; and (4) geography of crime. These categories are illustrative, but by no means exhaustive, of the kinds of research and work in which most applied geographers are engaged.

Market and Location Analysis

Applied geographers in the private sector engage primarily in research and work related to market analysis—that is, the identification of potential customer bases—and location analysis—identification of potential locations for retail establishments, shopping centers, and other businesses. Although much of this work remains proprietary, and therefore, beyond notice of many geographers, papers presented at the Applied Geography Conferences and at the research meetings hosted by the International Council of Shopping Centers, have detailed insights gained by applied geographers in business and industry.

Although the familiar dictum "location, location, location" is known to many nongeographers, interest in geographical analysis has been increasing among the private sector in part because of the dissemination of GIS and its ability to facilitate spatial analysis and more easily manage spatially referenced data.

Medical Geography

Research in medical geography has taken a strongly applied approach, ever since Dr John Snow mapped the incidence of cholera in London and showed that the disease is water-borne. Since that time, medical geographers have studied the incidence of disease and mortality within various segments of the population as well as the geographic dimensions of, and constraints on, the delivery of health-care services. Health-service providers now use GIS to identify potentially at-risk populations; locate hospitals and health clinics (sometimes to maximize qualification for federal subsidies and reimbursements), and define service areas for health facilities by mapping the residential locations of populations receiving care.

Geographers, by focusing on the spatial aspects of networks and delivery systems, provide valuable insights regarding approaches to analyzing needs and problems and developing policy solutions. Pyle (1995), for example, reviews the various schools of thought affecting the way in which health-care planning programs use and analyze geographical information or bring various geographical perspectives to bear in the development of policy. Gesler and Ricketts (1992) and colleagues have contributed to greater understanding of the issues and problems facing providers and recipients of health-care services throughout rural America. Applied geographers within the National Center for Health Statistics and the Centers for Disease Control (CDC) have worked closely with colleagues in academia as well as from other professions to map epidemiological data and mortality throughout the United States (Pickle *et al.* 1996). The *Atlas of United States Mortality* and earlier cancer mortality atlases produced by the National Cancer Institute in the 1970s and 1980s demonstrated the power of maps to reveal previously unnoticed clusters of counties with high rates of cancer. Questions raised in response to patterns visible in the maps led to studies that revealed linkages "between shipyard asbestos exposure and lung cancer and snuff dipping and oral cancer" (Pickle *et al.* 1996). The CDC, in conjunction with the World Health Organization, also has facilitated medical geography research through its EpiInfo and EpiMap systems, providing systems for managing and mapping epidemiological data available free of charge to the public via the Internet.

Land-Use Planning, Environmental Issues, and Policy

This category covers a broad variety of work ranging from community planning to regional economic development studies. Applied geographers engaged in this area can be found in local, state, tribal, and federal government, non-governmental organizations, regional councils of government, and academic departments. Work may range from processing, review, and approval of subdivision development plans, preparation of general plans for local governments to regional economic development issues and the statistical categories used to produce metropolitan-level data (Mayer 1982; Morrill *et al.* 1999). Increasingly, land-use planning and regional development issues are intertwined with environmental issues. More importantly, though, an increasingly well-educated public is becoming concerned about environmental issues and the effect of increased development on the landscape and general quality of life.

Significant contributions of applied geographers to land-use planning focus on the relationship between "geography, law, and public policy," to use the subtitle from Rutherford Platt's examination of the roots of American outlooks toward uses of land as well as the effect of land-use policies on society (Platt 1996). The synthesis of concepts from a number of disciplines and professions—economics, geography, law, ecology, sociology, policy sciences—can, and often does, produce cogent analyses of the effects of land use and zoning policy and the resultant residential and commercial patterns that appear on the landscape. Although not written by a geographer, Larsen (1995) provides an excellent examination of Texas land-use policies that have resulted in the largely unregulated development of *colonias* in the Rio Grande valley. Although *colonias* meet residents' needs for low-cost housing options, *colonia* populations also exhibit high rates of water-borne and other communicable diseases because of the lack of adequate water and waste disposal infrastructure as well as high levels of social, linguistic, and cultural isolation. Applied research in the *colonias* has focused on the incidence of disease; provision of appropriate infrastructure; and the balance between regulations to ensure quality housing and the need to provide low-cost housing for local populations (GAO 1990; Holz and Davies 1993; Chapa *et al.* 1996; P. Ward 1999; Stuesse and Ward 2001). Such studies often weave in and out of the intersection between land-use policy and its far-reaching effects on society, highlighting the synthesizing nature of applied geography in the examination of causes of, and search for solutions to, social and environmental problems.

Geography of Crime

Police officers, crime analysts, and other researchers in the public and private sector have long been interested in crime analysis. The New York City Police Department, for example, has traced the use of maps back to at least 1900 (Harries 1999). Desktop mapping has revolutionized crime research and the technology of policing as fundamentally as it has changed other applied geographic disciplines. Real-time access to spatially enabled crime data is a near certain development. This should raise important concerns about the risks of arriving at premature conclusions on the basis of real-time, but possibly unconfirmed, information (Harries 1999). In 1997, the Crime Mapping Research Center at the National Institute of Justice created the Mapping and Analysis for Public Safety Program to promote research supporting the practice of criminal justice. Funded training and resources identified on the Institute of Justice website

(<www.ojp.usdoj.gov/nij/maps/training.html>, last accessed 15 February 2003) include distance learning tools and on-site courses focusing on applied geographic research and development for criminal justice applications. The website is an important contribution to the future development of applications in the geography of crime.

Future Directions

The merits of client-driven versus curiosity-driven research undoubtedly will continue to generate discussion and debate as applied geographers identify their roles within the discipline of geography and in relation to other disciplines (Gibson 1998; Wellar 1998). Yet the debate seems oddly irrelevant since applied geographic research has become mainstream to the extent that it characterizes much of the research in the discipline of geography today. Business, industry, and government seem more inclined to turn to geographers for solutions to problems. Moreover, its seems clear that the application of geographic methods in addressing national, state, provincial, local, and tribal policy issues will continue to grow in the future. In the twenty-first century, applied geographic research will help push the discipline of geography well beyond its current boundaries.

References

Ackerman, Edward A. (1945). "Geographic Training, Wartime Research, and Immediate Professional Objectives," *Annals of the Association of American Geographers*, 35/4.

Carbaugh, Larry W., and Marx, Robert W. (1990). "The TIGER System: A Census Bureau Innovation for Serving Data Analysts," *Government Information Quarterly*, 7/3.

Chapa, Jorge, Eaton, David, and Jones, Bess Harris, *et al.* (1996). *Colonia Housing and Infrastructure: Current Population and Housing Characteristics, Future Growth, and Housing, Water, and Wastewater Needs*. Lyndon B. Johnson School of Public Affairs, Policy Research Project on Colonia Housing and Infrastructure. Austin: University of Texas.

Dahmann, Donald C. (1998). Personal communication.

Frazier, John W. (1982). "Applied Geography: A Perspective," in John W. Frazier (ed.), *Applied Geography: Selected Perspectives*. Englewood Cliffs, NJ: Prentice-Hall.

Frazier, John W., Epstein, Bart J., and Schoolmaster, F. Andrew (1995). "Contributions to Applied Geography: The 1978–1994 Applied Geography Conferences," *Papers and Proceedings of Applied Geography Conferences*, 18: 1–13.

General Accounting Office, Resources, Community, and Economic Development Division (1990). *Rural Development: Problems and Progress of Colonia Subdivisions Near Mexico Border*. Washington, DC.

Gesler, Wilbert, and Ricketts, Thomas (eds.) (1992). *Health in Rural North America*. New Brunswick, NJ: Rutgers University Press.

Gibson, Lay James (1998). "The Road Less Traveled: Client-Driven Research for Geography Undergraduates," *Papers and Proceedings of the Applied Geography Conferences*, 21: 464.

Golledge, Reginald G., Loomis, Jack M., and Klatsky, Roberta L. (1997). "A New Direction for Applied Geography," *Applied Geographic Studies*, 1: 151–68.

Grunwald, Michael (2002). "The Swamp: Can $8 Billion Restore the Everglades?" *The Washington Post*, 23 June: A1.

Harries, Keith (1999). *Mapping Crime: Principle and Practice*. Washington, DC: US Department of Justice.

Harris, Chauncey D. (1997). "Geographers in the U.S. Government in Washington, DC during World War II," *Professional Geographer*, 49/2: 245–56.

Hart, John Fraser (1989). "Why Applied Geography?" Martin S. Kenzer (ed.), *Applied Geography: Issues, Questions, and Concerns*. Boston: Kluwer.

Harvey, Milton (1997). "The Editor's Note: How the Parts Fit Together," *Applied Geographic Studies*, 1: 1–6.

Holz, Robert K., and Davies, Christopher Shane (1993). "Third World Colonias: Lower Rio Grande Valley, Texas," LBJ School of Public Affairs. Working Paper No. 72 Austin: University of Texas.

Johnston, R. J. (1993). "Meet the Challenge: Make the Change," in R. J. Johnston (ed.), *The Challenge for Geography*. Cambridge, Mass.: Blackwell.

Kenzer, Martin S. (1989). "Applied Geography: Overview and Introduction," in Martin S. Kenzer (ed.), *Applied Geography: Issues, Questions, and Concerns*. Boston: Kluwer.

Larsen, Jane E. (1995), "Free Markets Deep in the Heart of Texas," *The Georgetown Law Journal*, 84/2 (December): 179–260.

Mayer, Harold M. (1982). "Geography in City and Regional Planning," in John W. Frazier (ed.), *Applied Geography: Selected Perspectives*. Englewood Cliffs, NJ: Prentice-Hall.

Morrill, Richard, Cromartie, John, and Hart, Gary (1999). "Metropolitan, Urban, and Rural Commuting Areas: Toward a Better Depiction of the United States Settlement System," *Urban Geography*, 20/8: 727–48.

Palm, Risa I., and Brazel, Anthony J. (1992). "Applications of Geographic Concepts and Methods," in Ronald F. Abler, Melvin G. Marcus, and Judy M. Olson (eds.), *Geography's Inner Worlds*. New Brunswick, NJ: Rutgers University Press.

Pickle, Linda W., Mungiole, Michael, Jones, Gretchen K., and White, Andrew A. (1996). *Atlas of United States Mortality*. Hyattsville, Md.: National Center for Health Statistics.

Platt, Rutherford H. (1996). *Land Use and Society: Geography, Law, and Public Policy*. Washington, DC: Island Press.

Pyle, Gerald F. (1995). "Some Geographical Perspectives on Policy Initiatives in Health Planning," *Croatian Medical Journal*, 36/3.

Reuss, Martin (1993). *Water Resources People and Issues: Interview with Gilbert F. White*. Fort Belvoir, Va.: Office of History, US Army Corps of Engineers.

Stuesse, Angela C., and Ward, Peter M. (eds.) (2001). *Irregular Settlement and Self-Help Housing in the United States: Memoria of a Research Workshop*. Cambridge, Mass.: The Lincoln Institute of Land Policy.

Torrieri, Nancy, and Holland, Jennifer (2000). "Census Sense, Part 3: Enabling an Annual Census." *Geospatial Solutions*, Nov.: 38–42.

Turner, B. L., II (2002). "Contested Identities: Human–Environment Geography and Disciplinary Implications in a Restructuring Academy," *Annals of the Association of American Geographers*, 92/1: 52–74.

Ward, Peter M. (1999). *Colonias and Public Policy in Texas and Mexico: Urbanization by Stealth*. Austin, Tex.: University of Texas Press.

Warde, Ibrahim (2001). "For Sale: U.S. Academic Integrity." *Le Monde Diplomatique*. March (‹http://mondediplo.com/2001/03/11academic›, last accessed 15 February 2003).

Wellar, Barry (1998). "Combining Client-Driven and Curiosity-Driven Research in Graduate Programs in Geography: Some Lessons Learned and Suggestions for Making Connections," *Papers and Proceedings of the Applied Geography Conferences*, 21: 513.

CHAPTER 35

The History of Geography

Geoffrey J. Martin

The history of geography is an untidy term that sprawls across an extended chronology and embraces an ill-defined body of thought, knowledge, and ideas.[1] Different countries have different histories of geographical thought, and not infrequently North American practitioners study one or more of these national histories. The period of investigation for some may exceed two thousand years while for others it may be decades only. There is no mainstream, but a variety of individual efforts—some oriental, some occidental, some published, others unsung—that do not of themselves come together. The historian of geography studies what other people have thought, said, and studied concerning matters geographical. Islands of knowledge form in seas of ignorance. The canvas is vast and for most of us choices must be made. Over the last thirty years it is probably true to assert that a majority of workers in this enterprise have made special studies of segments of the history of North American geography and its antecedents.

The larger purpose of this type of investigation is to understand what has gone before, to comprehend how progress is made in the advancement of thought and how such a body of knowledge in its evolution brought us to recent time. It is a form of historical inquiry cognizant of the context of times past and guarded as to the limitations imposed upon us by lack of sources.

In this country, among the academic pioneers of this genre were E. van Cleef, C. T. Conger, J. Paul Goode, C. O. Sauer, E. C. Semple, and E. L. Stevenson of whom four undertook some of their studies in Germany. Arguably, formal recognition of this branch of our field was entered into the literature by Wright (1925, 1926). Development of this body of knowledge has, however, been slow.

Occasionally the terms "history of geography" and "historical geography" have been mistaken for one another or used interchangeably. The history of geography seeks to reveal the direction that individuals, institutions, books, beliefs, and concepts have taken in the eventual construction of a discipline and profession. The history of geography relates to and intermingles with histories of other fields (most notably perhaps with geology). The story of the acquisition of geographical knowledge is long and convoluted: knowledge gained has been integrated with erroneous fact, conception, and belief encouraging Wright's later advocacy of "geosophy" (Wright 1947).

In recent times a number of geographers in the US have perceived the modern period of geography to date from 1859, the year that witnessed the birth of Alfred Hettner, the publication of Darwin's *Origin of Species*, and the deaths of Carl Ritter and Alexander von Humboldt. Some courses in the history of geography either recognized or began with this date, and were presented chronologically. Other courses occasionally dealt with the ancient period, and some the Middle Ages. Notable among practitioners who attempted that larger sweep in their course work since the 1950s were Clarence Glacken, Preston James, George Kish, Fred Lukermann, and Curtis Manchester. Other courses offered were frequently studies in method, technique, or problems, clustered in seminar format. Lack of content in a course on

[1] This essay was submitted to the editors in spring 1999.

the history of geography often resulted from a lack of specialization in the genre. Few indeed have been the academics who invested their career paths substantially in this subject matter. Academic posts have rarely been available in the subject, and students have been reluctant to specialize in a branch of learning for which there has been little demand on the campus. There have been few institutions where one could undertake a doctoral dissertation in this subject. This has meant that the field has not advanced as it might otherwise have done. Gaps in the knowledge of a national geography remain large and many. Thirty years ago R. L. Barrett could recall Davis planning to form an American Geographers Club or Association (founded as the Association of American Geographers). Currently, for purposes of professional geography, living memory takes us back only to the mid-1930s. Without preservation of archival material, written forms of the oral tradition, or accurate narrative accounts of these years, parts of a history of the disciplinal past may become essentially lost (Armstrong 1999).

Geography departments have only rarely sponsored dissertations concerning the history of geography. Indeed, over the last fifty years dissertation studies concerning the work of individual geographers have been more frequently undertaken within departments of geology, history, and education than those of geography. The mindset of our profession has bent toward the new and the relevant. Less attention has been given to the evolution of our subject as a field of study. Nevertheless much has been accomplished that warrants attention.

Numerous are the ways in which our history—the geographers, the literature, the ideas, university departments, societies, associations, etc.—can be and has been studied. Useful and ongoing surveys of studies in the history of geography and associated matters since the 1980s have been provided by approximately annual offerings in *Progress in Human Geography* by T. F. Glick (1985–90), N. Smith (1990–2), F. Driver (1994–6), and M. Bassin (1996–).

Since *c.*1970 with regard to those objects of study a commentary has arisen arguing that more attention should be given to the context in which the individual lives his or her life. Berdoulay (1981*a*) renders the case explicit and suggests more attention be given to intellectual climate, contemporary belief, and the larger historical context. D. R. Stoddart, M. Bassin, and others have written on the same theme. This is not now, nor has it been, a disputed position. However, modest differences may have arisen between historians of the subject concerning the amount and type of context to be included. Too much may conjure a surreptitious form of determinism: too little may bring the charge of inattention to this all-enveloping surround. The matters of how much of the context, and of what kind, should be invoked in a study are a function of wide reading and careful judgement.

Individuals react to their surround in different ways. Some conform and some reject part of their surround; others may tend to remove themselves from a mainstream of discourse and shared values in a way that allows them perhaps to retain more of their individuality than otherwise would be the case. The archival deposit (if available) may provide clues to the degree to which the subject geographer absorbs, and is affected by, the contextual surround. It is a matter on which historians of geography may differ in degree, but not in principle.

Nevertheless, the reluctance of some authors to visit archival deposits, to have the authors speak for themselves, is worrisome. Part of the problem is that visiting deposits many miles distant is expensive of both time and money. Locating relevant material is uncertain, and visits to multiple deposits may become necessary. The fact is that singularly few people have trodden this path in the last thirty years. And the paucity of such activity means that the existing archival holdings will be inadequately searched, and that repositories will make no effort to collect and make available archival materials for which there seems to be little scholarly demand. Since many of the existing holdings are rarely visited by geographers, the worth of a thorough contextual overlay may be compromised by an inadequate archival underpinning. Lack of archival search and data acquisition has encouraged some authors to fill gaps with narrative explanation of what seems to them might reasonably have occurred.

There has also been concern and criticism that presentist or whiggish modes of selection and arrangement have been too frequently demonstrated in texts embracing a more extended chronology. The essence of this concern is that in attempting to explain the present via disciplinary history the author is tempted to simplify what is complex and create a seamless history that fuses with or justifies a legitimate and deserving present. Needed, it is supposed, is the integration of the best research into a logical sequence and composition. Owing, however, to lacunae in the literature, and the lack of studies deriving from archival sources and careful research, the larger ideal synthesis may not be possible at present. Meanwhile compromise is made in such works as Dickinson (1969, 1976), Tinkler (1985), Martin and James (1993), and Johnston (1997). Books of "building blocks," that is, collections of essays in the history of geography that make no attempt at synthesis into a larger whole, include Dunbar (1996), Livingstone (1992), Stoddart (1986), and Wright (1966).

The most popular narrative mode in the last thirty years seems to have been the study of the individual geographer. It is within the individual that the act of creation takes place and it is the individual writ large who has made geographical science what it has become. The impulse to science is at root an individual matter. It is also part genetic, part upbringing, part opportunity, and part economic and institutional circumstance. Studies of the contributions of individual geographers ideally would come together and mesh in science to form a larger whole. Revealed are many varieties of contribution, data frequently not available elsewhere, including intellectual genesis of thought and idea, and the function of interdisciplinary exchange. *Current Geographical Publications* indicates that in the last decade alone some thirty geographers have been written about by American authors. *Geographers: Biobibliographical Studies*, first published in 1977, now provides some fifty essays by American authors, largely though not exclusively about American geographers.

The Archival Sources

While secondary sources are frequently helpful they also contain errata which are occasionally copied and recopied. It is in the correspondence, documents, and photographs of the archival holding that one will find material with which to authenticate or contradict such statements. German science has long since recognized this, and sought to overcome the problem by establishing an index to archival holdings throughout the country: the Autographen—Katalog in der Zentralkartei, Staatsbibliothek Preussischer Kulturbesitz. More than 350,000 letters maintained in university and other public holdings are cataloged, each containing name of sender and receiver, date, title, and location of collection. Letters between American and German geographers are included in the catalog, which makes it a valuable resource for study in the history of American geography. This model accomplishment stands alone, however. There are numerous other archival holdings in Europe containing material relevant to such study. In the US the National Union Catalog of Manuscript Collections (1959–) has been an essential guide. Since 1984 the Research Libraries Information Network has provided computer search facilities. In 1989 to facilitate international archival search the IGU Commission on the history of geography agreed to develop an international directory for geographical archives. Archival holdings in the history of geography have been provided for several countries and these inventories have been published in the *Newsletter* of the Commission.

Within the US over the last thirty years we have witnessed the rise of some archival holdings and the decline of others. From an unofficial holding of little size in the 1960s, the Association of American Geographers collection now extends to approximately 270 linear feet and is housed with the American Geographical Society's library collection in the Golda Meir Library, University of Wisconsin, Milwaukee. The AGS archival collection, however, extending to more than 400 linear feet, is located in its New York headquarters. It remains, arguably, the single most valuable collection of its kind, notwithstanding considerable destruction and displacement of large numbers of letters and documents in the 1970s. The National Geographic Society's holdings are extensive and maintained with meticulous care. They date back to the Society's founding in 1888 and embrace a wide variety of geographical matters.

Valuable materials have been lost when individual geographers have decided to destroy their papers. Conversely, other geographers, for example J. K. Wright, have arranged their papers in a most orderly manner and left instructions for their disposition. And, of course, with the passage of time, and especially if an agreed-upon depository is not available to a holding, materials may be dissipated and lost. Such is the case with many geographers: it is also true for much of the correspondence relating to some of the early International Geographical Congresses and the formation of the International Geographical Union. (Currently the Royal Geographical Society functions as the depository for IGU materials.) Losses of archival material have also been commonplace in the matter of university departments, societies, associations, and periodicals.

University libraries do frequently house geographers' papers and departmental records in special collections. Such is the case with the papers of J. Morse and E. Huntington at Yale, C. O. Sauer and J. B. Leighly at Berkeley, W. H. Hobbs at the University of Michigan, C. F. Marbut at the University of Missouri, D. W. Johnson and R. E. Dodge at Columbia, W. W. Atwood and others at Clark (Koelsch and Campbell 1982), and many more. Yet other collections are housed in the Library of Congress, the larger public libraries, geographical societies, and museums; some are still privately held.

Another medium in archival preservation is the film. "Talking films" were shown to some geographers in 1912 on the 250th anniversary of the Royal Society. Fifty-eight years later M. W. Dow began use of the film as an archival medium and initiated the project now known as "Geographers on Film." This series includes some 490 films of geographers—as individuals or in groups—talking of their work, experiences, and acquaintances. This series is available in VHS format and constitutes a unique link to the past for a younger generation which did not know geographers now deceased. While most of Dow's interviews have been with American geographers, A. Buttimer and T. Hägerstrand initiated an "Invitation to Dialogue" at Lund, Sweden, filming geographers from twenty countries in an attempt to investigate the sociology of knowledge. Buttimer (1983) describes the enterprise.

Prior to the invention of the adhesive postage stamp (UK 1840, USA 1847, France 1849, Bavaria 1849, followed by other German States), person-to-person communication was more common in gatherings of members of societies and other organizations. With the coming of the postage stamp, people both more and less remote from these organizations could participate in scientific or literary exchange. In the latter years of the twentieth century, ever-increasing use of the telephone, email, and fax has reduced the habit of letter-writing. This will probably exert an impact, as yet unmeasurable, on archival accumulation. And as space for archival mass becomes more difficult to acquire, the dangerous practice of "thinning" may be exercised.

The Geographers

Many are those who have made contributions to our field; fewer are those who have been recognized and written about. Historians of geography have not been much concerned with the private lives of individuals, but rather with how their contributions were conceived and developed and the longer-run significance of the work of discipline-building. Clustering individuals together in groups sometimes, for example, called "schools," may enlarge the perspective. Dunbar's (1992, 1996) biographical dictionary constitutes a helpful source in terms of chronology (dates of birth, degrees, and death).

Since the late 1960s two figures have been prominent in the literature, W. M. Davis and C. O. Sauer. Davis (1850–1934) was the subject of a doctoral dissertation by Rigdon (1933). This work was superseded by Chorley and Beckinsale (1964–91) in the remarkably comprehensive three volumes of *The History of the Study of Landforms*. These works provide intellectual ancestry of those who came before Davis, as well as information on those who were his students and followers. Beckinsale and Chorley (1981) is a biobibliography of Davis. Meanwhile in Blouet (1981), Beckinsale, Hartshorne, Martin, and Koelsch provide articles on Davis and on the New England Meteorological Society. Martin (1974*b*) writes of the Penck(s)–Davis conflict and King and Schumm (1980) contains transcribed lectures given by Davis at the University of Texas, 1926–7.

Davis's students are frequently listed in two categories —those specializing in physiography and those who studied "ontography." Curiously those who made their mark in the former have been less visited than those who studied "the life response." Indeed, very little has been written of the physiographers in the last forty years, and with the exception of D. W. Johnson and R. A. Daly, little other than obituaries and *Dictionary of American Biography* statements has been contributed; students such as L. Martin, H. E. Gregory, and R. S. Tarr, though influential in the formation of American geography, remain unstudied. Davis's students studying especially the human response to physical environments have fared better. Dissertations were written concerning A. P. Brigham (Burdick 1951), R. E. Dodge (Griffin 1952), and I. Bowman (Knadler 1958), G. Martin then wrote intellectual biographies of M. Jefferson (1968), E. Huntington (1973), and I. Bowman (1980) selected as representing three varieties of the life response theme: Jefferson adopted anthropography (not to be confused with anthropogeography), Huntington adopted (initially) physiological climatology, and Bowman—early in his career—placed response in the context of region.

Arguably Davis's most significant mentor, N. S. Shaler, stands revealed in an autobiography (1909), a dissertation (Berg 1957), a biobibliography (Koelsch 1979), an essay (Bladen 1983), and further contributions (Livingstone 1981, 1987). On the mining geologist, Raphael Pumpelly, who strode the Harvard Yard, becoming friends with Shaler and later Davis, see Champlin (1989, 1992). Koelsch (1979, 1983*b*, 1998) provided biobibliographies of Wallace W. Atwood, Robert DeC. Ward, and Charles F. Brooks, and (1983*a*) a study of the Harvard geographer-historian, Justin Winsor. LaRocco (1978) provides a life of Derwent S. Whittlesey. Smith (1987) was the first extended publication concerning the demise of geography at Harvard. Some commentaries followed with further comment from Smith (1988). Glick (1988) also wrote on the Harvard demise with special reference to the contribution of Edward Ullman and includes analysis of the work of the committee that reported on the geography question.

Curiously little attention has been given to the development of geography at Harvard's sister-institution,

Yale University. Moss (1995) was the fourth book published on the interesting figure of Jedidiah Morse. Wright (1961, repr.1966) wrote of D. C. Gilman. Martin (1973, 1980, 1988) wrote of one-time Yale geographers Huntington and Bowman, and provided a history of geography at Yale University *c*.1770–1970. Smith (1986) (and other essays) studied Bowman in his post-Yale years.

Carl Sauer (1889–1975), at odds with Davis in a number of ways, still awaits his biographer. Most of Sauer's life was spent on the west coast (Davis resided on the east coast); he developed cultural geography as a genre (Davis developed physiography); he rejected the causal notion (Davis encouraged it); he embraced German geography and geographers (Davis was in conflict with German geography *c*.1910–34); he remained essentially in North America (Davis travelled widely). There were similarities: both had remarkable intellectual capacity and fine literary style. Both were prolific letter writers, both published extensively (though Davis more so), and both had remarkable records of producing outstanding students.

Leighly (1963) edited a compilation of some of the more important writings of Carl Sauer to that time. Eighteen years later, and six years after Sauer's death, Callahan (1981) produced a selection of his essays. Together, these two publications constitute a large selection from the published work of Sauer and jointly constitute a Sauer primer.

Leighly (1976) and Parsons (1976) wrote extended obituaries of the departmental colleague they knew so well. Leighly (1978) also provided a biobibliography of Sauer. Parsons (1979) wrote of Sauer's later years. Kenzer (1986) wrote a careful and well-documented dissertation on the early years of Sauer's life. This led to his (1985*a*, *b*, 1987), the latter an edited collection of original essays resulting from a special session of the AAG annual meeting in Detroit, 1985. Similarly, a special session of the AAG held on the occasion of the centennial of Sauer's birth (Baltimore 1989) produced a series of papers currently in search of a publisher. Meanwhile further works on Sauer include Riess (1976), West (1979, 1982), Hooson (1981), Entrikin (1984), Solot (1986), and, more recently, Parsons (1996). Stoddart (1997) wrote a chapter on Sauer as geomorphologist in a work honoring the contribution of Richard J. Chorley. W. W. Speth produced a series of essays concerning historicist anthropogeography, geography at Berkeley, and Carl Sauer. Perhaps especial mention might be made of his (1981, 1987). Twelve of these essays have been gathered and published as Speth (1999).

Earlier history of the department including its founding in 1898 is developed in Dunbar (1978, 1981). The coming of Lincoln Hutchinson (1901) and Ruliff Holway (1904) have attracted less attention than Davidson, who was the first appointment to the department. More recently Parsons (1988) wrote a biobibliography of Leighly and D. H. Miller (1988) offered a memorial in the *Annals*.

Davis and Sauer have probably been the most dominant figures of the last 120 years (though arguably not the most written about—J. Morse and M. F. Maury may earn this distinction). Yet in terms of impress on our thinking, and the direction taken by the mentors' students, few would disagree. For Davis helped the birthing of American physiography in the arid lands, then provided the initial organizing concept of the discipline, arranged field trips and talks for schoolteachers and normal schoolteachers, placed his students into college-level posts, founded the Association of American Geographers, and provided our first generation of professional geographers. Sauer was beneficiary to this accomplishment. He imported the fastnesses of his thought from Germany, and more particularly, it may be, from Schlüter, and fashioned a viewpoint that we came to know as cultural geography. His largeness of vocabulary, vision, and intellect were only three of the strengths that facilitated a train of publications which shifted the locus of our thought. It is probable that these two figures will continue to demand attention; especially in the case of Sauer, for whom there is much living memory, and relating to whom some archival material yet remains unworked.

Geographers at Chicago, Clark, and Minnesota have been occasionally studied and form clusters of significance in our enterprise. Of the Chicago group Pattison (1981, 1982) wrote of Rollin Salisbury, first chair of the Geography Department, and produced his biobibliography. Koelsch (1962, 1969) edited H. H. Barrows' lectures on the historical geography of the United States, and provided analysis of that historical geography. Biobibliographies were written of R. S. Platt by R. S. Thoman, J. Paul Goode by G. Martin, and H. C. Cowles by G. F. Rogers and J. M. Robertson. Although Cowles had been appointed to the Botany Department at the University of Chicago he had a close association with the geographers, and was made a founding member of the Association of American Geographers in 1904 and president in 1910. The English-born T. G. Taylor was a member of the department from 1928 to1935, but spent more time in departments at the Universities of Sydney and Toronto. He wrote a very detailed autobiography, which remains in the National Archives, Canberra. Reduced greatly in length, and edited by Alasdair Alpin-MacGregor, it was published as *Journeyman Taylor*

(1958). Tomkins (1967) wrote of Taylor and the development of Canadian geography and Sanderson (1988) on his work as a pioneer geographer.

Since Wallace W. Atwood assumed the presidency of Clark University in 1920, it has awarded more doctoral degrees in geography than any other institution in the country. Koelsch (1987) has placed the geographic accomplishment in context. More particularly see his (1979, 1980, 1988; also 1982, 1983). Special attention has been given to Ellen C. Semple, a frequent lecturer at Clark, by Bronson (1973), Berman (1974, 1982), Bushong (1975, 1984), and James, Bladen, and Karan (1983). A number of scholars interested in the work of Semple delivered papers at Vassar College during "The Evalyn A. Clark Symposium on Excellence in Teaching: Geography and the Legacy of Ellen Churchill Semple," held in April, 1992. Unfortunately proceedings of this meeting were not published. Work on Semple has been hampered by the scattering and presumed loss of much of her correspondence.

At Minnesota in the 1920s and 1930s were "the three horsemen" brought in by D. H. Davis: R. Hartshorne, R. H. Brown, and S. N. Dicken. This was a fine nucleus around which to build a department, but Hartshorne and Dicken left for Wisconsin and Oregon respectively, and unfortunately Brown died prematurely after World War II. On Brown see especially McManis (1978), Miles (1982, 1985), and Dicken (1985), which, *en passant*, provides an alternative student view of C. O. Sauer.

The role of the female in the history of our geography is slowly being excavated. Unfortunately the history of geography in the normal school has seldom been written, and this is arguably where female geographers made their largest contribution. An exception is Berman (1988). Women dominated and maintained geography in the normal schools from the 1860s to the 1940s. A small part of this history is to be found in Whittemore (1972). Semple and M. K. Genthe, the two female founding members of the Association of American Geographers, are known to us quite differently. Semple is a well-recorded personage: Genthe is little recorded, and it seems not even the subject of an obituary in a geographical periodical. Little has been written of Gladys M. Wrigley, the last living student of H. J. Mackinder, with the exceptions of McManis (1990) and Mikesell (1986); the latter wrote about the first year of the *Geographical Review*, the year in which Wrigley became an editor. This was a woman who developed a very keen geographic sense of problem, a large knowledge, and superb literary expression. Millicent Todd Bingham, an early authority on the geography of Peru and translator of La Blache's *Principles of Human Geography*, has been written of by Berman (1980, 1987). More recently Monk (1998) wrote "The Women Were Always Welcome at Clark," part of a larger study of women in American geography.

Unwritten are the four female founders of the National Geographic Society (January 1888) and Eliza R. Scidmore, corresponding secretary for the National Geographic Society from 1899, author of eleven articles published in the magazine, two books on Alaska, and not less than five on the Orient. Scidmore remains perhaps the most published woman in the history of American geography. Clearly there is much yet to be done.

Little has been accomplished by North American historians of geography in pre-nineteenth-century geography since Glacken (1967). Ptolemy (an Ian Jackson rendered column in the *AAG Newsletter*), was visited by Marshall (1972) who provided a list of manuscript editions of Ptolemy's *Geographia*. Ptolemy, Mercator, and others were visited by Warntz and Wolff (1971). Kish (1978) covers the period from Hesiod to Humboldt. Columbus received considerable attention on the quincentennial of his Atlantic voyage. Butzer (1992) engaged the Columbian debate and provided a fine addition to the literature. Warntz (1991, 1989) investigated Bernhard Varenius and the *Geographia Generalis* and Martin (1998) wrote on geographical thinking in New England from the early 1600s to c.1920. G. M. Lewis and Allen (1981) have made significant contributions to Amerindian antecedents of American academic geography and public land policy and exploration in the West. Particular mention might be made of Allen's edited three-volume study concerning exploration in North America (1997).

Of especial significance for pre-nineteenth-century sources is Sitwell (1993), a guide to books published in English that purport to describe all the countries in the world before 1888. Also of much value, and spanning the period from the mid-eighteenth century to *c.*1984 is Dunbar (1985), which includes references in English, French, German, Italian, and Spanish, and is of service to specialist and non-specialist alike. Both these works constitute significant additions to the bibliographic literature.

The Ideas, the Anniversaries, and the Institutions

Occasionally some of the larger ideas—the moving forces—are written about, each revealing a unique history. Lewthwaite (1966) reviews notions that have been with us for centuries, and that dominated our geography in the early years of this century. Lowenthal has written several essays concerning the career and contribution of George Perkins Marsh, superseding his biography (1958). These works bring into American geography the notion originating with Marsh, that humankind makes a profound impact on its habitat—the inverse of a determining environment. Mikesell (1975) traces the history of D. S. Whittlesey's concept of sequent occupance. Martin (1974*a*) re-examines the causative notion of civilization and climate in Huntington's correspondence and book, *Civilization and Climate*. Lowenthal and Bowden (1976) edited a volume of essays in historical geosophy in honor of John Kirtland Wright. Handley (1993) has explored Wright's subjective notion, geosophy. Stoddart (1981*a*) wrote of Darwinism in American geography. Mather and Sanderson (1995, 1998) wrote about a man of ideas—C. Warren Thornthwaite—and produced a biobibliography of him.

Anniversary and jubilee occasions have also been noticed and appropriate publications have contributed to the skein of the history of geography. Whittemore (1972) observed the seventy-fifth anniversary of the *Journal of Geography*. In 1976 the fiftieth anniversary of the Society of Women Geographers won special publication. Three years later in 1979 the Association of American Geographers recognized its seventy-fifth anniversary with a book (James and Martin 1978) and a special number of the *Annals* (1979). The latter represented a wide range of viewpoints and especially revealed the gulf that had begun to yawn between the generations. The National Geographic Society celebrated its centennial in 1988, an occasion celebrated by Bryan (1987: rev. 1997). In 1997 the International Geographical Union marked its seventy-fifth anniversary, noticed in Martin (1996). Of a different genre was the special number of the *Annals* commemorating the fiftieth anniversary of the publication of R. Hartshorne's *The Nature of Geography* (Entrikin and Brunn 1989). This was a curious mix of essays, some invited, others self-invited, some appreciative, others critical of his œuvre. E. W. Miller (1993) edited a volume of papers printed to mark the fiftieth anniversary of the American Society for Professional Geographers. This included contributions by S. B. McCune, F. W. McBryde, E. W. Miller, C. D. Harris, W. W. Ristow, M. F. Burrill, and R. F. Abler.

Perhaps the institutions have been documented less adequately. Histories of university geography departments are few. Those at Yale, Dartmouth, Mt. Holyoke, Salem, Clark, and Harvard have been written in whole or part in Harmon and Rickard (1988). W. Wallace wrote a history of geography at the University of New Hampshire. Departments in the Pacific Northwest including the University of Oregon, Oregon State, Western and Central Washington Universities, Simon Fraser, Victoria, and Montana State have been documented in Datel and Dingemans (1990). The University of Wisconsin department was accounted for in 1978, on the fiftieth anniversary of its separation from geology. Twelve years later further elaboration of this history was accomplished by Olmstead (1987). Dunbar wrote the history of geography at the University of California (Berkeley and Los Angeles) from 1868 to1941 (1981, repr. 1996). Ryan (1983) wrote the history of geography at the University of Cincinnati. There are other departmental histories, usually of a brief and informal nature: the accomplishment is limited.

Three other institutions might be mentioned, each of which has facilitated advancement in the history of geography. The first of these was established in 1970 as the Archives and Association History Committee. This (standing) Committee of the AAG has sponsored a variety of sessions and workshops at the annual meetings of the Association, ensuring a place for the presentation and discussion of papers for nearly thirty years. It was to consolidate the work of this committee that the *History of Geography Newsletter*, renamed the *History of Geography Journal*, was established. The second of the three institutions mentioned above is The History of Geographic Thought Commission of the International Geographical Union. This Commission was founded in New Delhi in 1968. Since then meetings of the group have invariably included a US presence and participation, and have resulted in publications that include Blouet (1981), Hooson (1994), and Godlewska and Smith (1994). Several other works also have derived from this Commission. The third institution is the History of Geography Specialty Group of the AAG, founded and chaired by Paul Frederic in 1998.

The international tone is extended by a cluster of works that helped to reveal French geography further. They include Buttimer (1971), Dunbar (1978, 1981, 1982), and a book of translations from the works of eminent French geographers (1983). Berdoulay (1974, 1981*b*) began as a dissertation, and was revised and published in French. West (1990) translated some of the work of German geographers, with his own commentary. And Blouet (1987) brought our attention to the British geography of an earlier period with a biography of Halford Mackinder.

Concluding Remarks

In an earlier period, *c.*1880–1910, American geography imported ideas and exported students (largely to Germany) (Martin 1989). During the last thirty years we have exported ideas and imported students. Our geography is in a state of continuous change; arguably the speed of change has increased in recent years. As the number of professional geographers has increased, specializations have multiplied. Our close proximity to a very few of these only increases our tendency toward micro-studies and lessens our vision of the larger whole. In this context the history of geography may become a theoretical self-contained enterprise, a specialization in itself, no longer considered a prerequisite to geographic comprehension. Conversely, a geography devoid of its history seems to have begun to establish itself.

Our geography earlier this century had a more or less agreed-upon viewpoint. In our day several generations of geographers coexist who in concert have multiple and sometimes contentious viewpoints. Those who muddied their boots with the plane-table apparatus now mingle with the student who secures GIS computer printouts. The discipline continues to exist, but it is more loosely knit. The Balkanization of our field by tens of specialty groups has exacerbated the tendency to eclecticism. The result is a centrifugal tendency moving the discipline (whose folk and professional models are already dissonant) away from a core of subject matter and toward ways of reasoning. Objectives have become so numerous and various that a careful study of the history of our geography is necessary to explain the diaspora.

For the historians of geography the undertaking is large and the workers are few. Courses in the subject and well-qualified teachers for those courses continue to decrease. Many university departments are left without a course in the subject. Fewer students study this genre, and without posts for this subject advertised in the *AAG Newsletter* (not one in the last forty years) it is hardly surprising that novitiates in search of a university appointment avoid this specialization. Where courses in the history of geographical thought are offered, they are so various that there is little uniformity in subject matter. Some courses embrace some 2,300 years of the history of geography from Ancient Greece to the present; others dwell on the last fifty years of Anglo-American geography; others deal with philosophical or methodological issues; and yet others thread their way around a list of readings. There is no core to this enterprise, no accepted way.

The concept of the history of geography is simple; its study complex. For the history of geography is a synoptic investigation that attempts a simultaneous view of an extensive enterprise. Yet fieldwork (in the archives) is rarely practiced in the course (or by teachers of the course) and smatterings of secondary-source writings must suffice. A national history of our geography does not exist, yet this would provide a valuable common ground for student, pioneer, and quester alike. Our history is scattered in many dozens of archival deposits both here and abroad. With an insufficiency of investigative workers, many of these deposits remain unworked. Particularly difficult areas in the history of our field have been skirted or avoided. More recently, we have allowed ourselves to discuss and negotiate at length such matters as "the contextual" rather than mining the ore in the archives. Context is, of course, always important, but the facts of the matter are even more important. However, the creation and maintenance of archival deposits from the papers of noted geographers and geographical societies has been attended with some measure of success.

The identification of difficulties presents opportunity for new growth, and there remains an untapped reservoir of curiosity and unspoken appreciation for the evolution of our undertaking in many geographers. For the sacralization of science brought with it the belief that we ought to comprehend our origins and evolution, which in turn, generated the excitement of endless search. Gender studies investigate a segment of thought not previously undertaken, and though perhaps more programmatic than investigative, could exhume the unwritten history of the female contribution to American geography. M. W. Dow's *Geographers on Film* begun in 1970, now extending to 286 individual interviews on film and 226 additional holdings, is a unique repository revealing the lore and likeness of past geography and geographers. It constitutes both inspiration and teaching tool. This unique collection is soon to be placed on deposit with the NSF-sponsored National Gallery of the Spoken Word, Michigan State University. This "virtual library" will then become accessible to an ever-larger audience and may facilitate workshops in the history of geography as successfully accomplished at the AAG annual meetings in Milwaukee (1975) and Salt Lake City (1977). The opportunity to accomplish is before us.

The task of the history of geography specialty group is to maintain a forum for the ongoing discussion of our undertaking and accomplishment, and to advance scholarship that contributes to a more comprehensive understanding of the evolution of the discipline.

References

Allen, J. L. (1981). "Working the West: Public Land Policy, Exploration, and the Preacademic Evolution of American Geography," in Blouet (1981: 57–68).

—— (ed.) (1997) *North American Exploration*, i. *A New World Disclosed*; ii. *A Continent Defined*; iii. *A Continent Comprehended*. Lincoln, Nebr.: University of Nebraska Press.

Armstrong, P. H. (1999). "The Disappearing Archive." Meeting of the IGU Commission on the History of Geographic Thought, Leipzig.

Beckinsale, R. P. (1981). "W. M. Davis and American Geography: 1880–1934," in Blouet (1981: 107–22).

Beckinsale, R. P., and Chorley, R. J. (1981). "William Morris Davis, 1850–1934." *Geographers Biobibliographical Studies*, 5: 27–33.

Berdoulay, V. (1974). "The Emergence of the French School of Geography, 1870–1914." Unpublished Ph.D. thesis, University of California, Berkeley.

—— (1981*a*). "The Contextual Approach," in Stoddart (1981*b*: 8–16).

—— (1981*b*). *La Formation de l'école francaise de géographie (1870–1914)*. Paris: Bibliothèque Nationale. (Rev. edn. 1995.)

Berg, W. L. (1957). "Nathaniel S. Shaler: A Critical Study of an Earth Scientist." Unpublished doctoral dissertation, University of Washington, Seattle.

Berman, M. (1974). "Sex Discrimination and Geography: The Case of Ellen Churchill Semple." *Professional Geographer*, 26/1: 8–11.

—— (1980). *Professional Geographer*, 32/2: 199–204.

—— (1982). "Great Lady of American Geography." *Clark Now*, 12/2: 21–3.

—— (1987). "Millicent Todd Bingham." *Geographers Biobibliographical Studies*, 11: 7–12.

Bladen, W. A. (1983). "Nathaniel Southgate Shaler and Early American Geography," in W. A. Bladen and P. P. Karan (eds.), *The Evolution of Geographic Thought in America: A Kentucky Root*. Dubuque, Iowa: Kendall Hunt, 12–27.

Blouet, B. W. (ed.) (1981). *The Origins of Academic Geography in the United States*. Hamden, Conn.: Archon Books.

—— (1987). *Halford Mackinder: A Biography*. College Station, Tex.: Texas A&M University Press.

Bronson, J. C. (1973). "Ellen Semple: Contributions to the History of American Geography." Unpublished Ph.D. thesis, St Louis University.

Bryan, C. D. B. (1987). *The National Geographic Society: One Hundred Years of Adventure and Discovery*. New York: H. N. Abrams (rev. 1997).

Burdick, A. E. (1951). "The Contributions of Albert Perry Brigham to Geographic Education." Unpublished doctoral dissertation, George Peabody College for Teachers.

Bushong, A. D. (1975). "Women as Geographers: Some Thoughts of Ellen Churchill Semple." *Southeastern Geographer*, 15: 102–9.

—— (1984). "Ellen Churchill Semple, 1863–1932." *Geographers Biobibliographical Studies*, 8: 87–94.

Buttimer, A. (1971). *Society and Milieu in the French Geographic Tradition*. Association of American Geographers, Monograph Series, 6. Chicago: Rand McNally.

—— (1983). *The Practice of Geography*. London: Longman.

Butzer, K. W. (1992). "The Americans Before and After 1492: Current Geographical Research." *Annals of the Association of American Geographers*, 82/3.

Callahan, B. (1981). *Selected Essays, 1963–1975*. Berkeley, Calif.: Turtle Island Foundation.

Champlin, M. D. (1989). "Raphael Pumpelly and American Geology in the Gilded Age." Ph.D. dissertation, UCLA.

—— (1992). "Raphael Pumpelly, 1837–1923." *Geographers Biobibliographical Studies*, 14: 83–92.

Chorley, R. J., and Beckinsale, R. P. (1991). *The History of the Study of Landforms*, iii. *Historical and Regional Geomorphology, 1890–1950*. London: Routledge.

Chorley, R. J., Beckinsale, R. P., and Dunn, A. J. (1973). *The History of the Study of Landforms*, ii. *The Life and Work of William Morris Davis*. London: Methuen.

Chorley, R. J., Dunn, A. J., and Beckinsale, R. P. (1964). *The History of the Study of Landforms*, i. *Geomorphology before Davis*. London: Methuen.

Datel, R. E., and Dingemans, D. J. (eds.) (1990). *Yearbook of the Association of Pacific Coast Geographers*. Corvallis, Oreg.: Oregon State University Press, 52: 139–228.

Dicken, S. N. (1985). *The Education of a Hillbilly; Sixty Years in Six Colleges—the Memoirs of Samuel Newton Dicken*. Eugene, Oreg.: Lane County Geographical Society and Department of Geography, University of Oregon, 162.

Dickinson, R. E. (1969). *The Makers of Modern Geography*. New York: Frederick A. Praeger.

—— (1976). *The Regional Concept: The Anglo-American Leaders*. London: Routledge & Kegan Paul.

Dunbar, G. S. (1978*a*). "George Davidson, 1825–1911." *Geographers Biobibliographical Studies*, 2: 33–7.

—— (1978*b*). *Elisée Reclus: Historian of Nature*. Hamden, Conn.: Archon Books.

—— (1981*a*). "Geography in the University of California (Berkeley and Los Angeles), 1868–1941." Published by the author, printed by DeVorss, Marina del Rey, Calif. (Rev. edn. 1996.)

—— (1981*b*). "Elisée Reclus: An Anarchist in Geography," in Stoddart (1981*b*: 154–64).

—— (1983). *The History of Geography: Translation of some French & German Essays*. Malibu, Calif.: Undena Publications.

—— (1985). *The History of Modern Geography: An Annotated Bibliography of Selected Works*. New York: Garland Publishing.

—— (1992). *A Biographical Dictionary of American Geography in the Twentieth Century*. Baton Rouge, La.: Department of Geography and Anthropology, Louisiana State University.

—— (1996*a*). *The History of Geography*. Privately printed. Utica, NY: Dodge-Graphic Press.

—— (1996*b*). *A Biographical Dictionary of American Geography in the Twentieth Century*. 2nd edn. Baton Rouge, La.: Department of Geography and Anthropology, Louisiana State University.

Entrikin, J. N. (1984). "Carl O. Sauer: Philosopher in Spite of Himself." *Geographical Review*, 74: 387–408.

Entrikin, J. N., and Brunn, S. D. (eds.) (1989). *Reflections on Richard Hartshorne's* The Nature of Geography. Washington, DC: Occasional Publication of the Association of American Geographers.

Glacken, C. J. (1967). *Traces on the Rhodian Shore: Nature and Culture in Western Thought from Ancient Times to the End of the Eighteenth Century*. Berkeley and Los Angeles: University of California Press.

Glick, T. F. (1988). "Before the Revolution: Edward Ullman and the Crisis of Geography at Harvard, 1949–1950," in Harmon and Rickard (1988: 49–62).

Godlewska, A., and Smith, N. (eds.) (1994). *Geography and Empire*. The Institute of British Geographers Special Publications Series. Oxford: Blackwell.

Griffin, P. F. (1952). "The Contribution of Richard Elwood Dodge to Educational Geography." Dissertation, Columbia University.

Handley, M. (1993). "John K. Wright and Human Nature in Geography." *Geographical Review*, 83/2: 183–93.

Harmon, J. E., and Rickard, T. J. (eds.) (1988). *Geography in New England*. Special Publication of the New England/St. Lawrence Valley Geographical Society.

Hartshorne, R. (1981). "William Morris Davis—the Course of Development of his Concept of Geography," in Blouet (1981: 139–49).

Hooson, D. J. M. (1981). "Carl O. Sauer," in Blouet (1981: 165–74).

—— (ed.) (1994). *Geography and National Identity*. The Institute of British Geographers Special Publications Series. Oxford: Blackwell.

James, P. E., and Martin, G. J. (1979). *The Association of American Geographers: The First Seventy-Five Years, 1904–1979*. Association of American Geographers.

James, P. E., Bladen, W. A., and Karan, P. P. (1983). "Ellen Churchill Semple and the Development of a Research Paradigm," in W. A. Bladen and P. P. Karan (eds.), *The Evolution of Geographic Thought in America: A Kentucky Root*. Dubuque, Iowa: Kendall/Hunt.

Johnston, R. J. (1997). *Geography and Geographers: Anglo-American Human Geography since 1945*. 5th edn. London: Arnold.

Kenzer, M. S. (1985*a*). "Milieu and the 'Intellectual Landscape': Carl O. Sauer's Undergraduate Heritage." *Annals of the Association of American Geographers*, 75/2: 258–70.

—— (1985*b*). "Carl O. Sauer: Nascent Human Geographer at Northwestern." *The California Geographer*, 25: 1–11.

—— (ed.) (1987). *Carl Sauer: A Tribute*. Corvallis, Ore.: Oregon State University Press.

King, P. B., and Schumm, S. A. (eds.) (1980). *The Physical Geography (Geomorphology) of William Morris Davis*.

Kish, G. (ed.) (1978). *A Source Book in Geography*. Cambridge, Mass.: Harvard University Press.

Knadler, G. A. (1958). "Isaiah Bowman: Background of his Contribution to Thought." Ed.D. dissertation, Indiana University.

Koelsch, W. A. (ed.) (1962). *Barrows' Lectures on the Historical Geography of the United States as Given in 1933*. Research Paper, 77. Chicago: University of Chicago, Department of Geography.

—— (1969). "The Historical Geography of Harlan H. Barrows." *Annals of the Association of American Geographers*, 59: 632–51.

—— (1979). "Wallace W. Atwood, 1872–1949." *Geographers Biobibliographical Studies*, 3: 133–7.

—— (1980). "Wallace Atwood's 'Great Geographical Institute'." *Annals of the Association of American Geographers*, 70/4: 567–82.

—— (1981). "The New England Meteorological Society, 1884–1896: A Study in Professionalization," in Blouet (1981: 89–104).

—— (1983*a*). "A Profound Though Special Erudition: Justin Winsor as Historian of Discovery." *Proceedings of the American Antiquarian Society*, 93/1: 55–94.

—— (1983*b*). "Robert DeC. Ward." *Geographers Biobibliographical Studies*, 3: 13–18.

—— (1987). *Clark University: A Narrative History*, 1887–1987. Worcestes, Mass.: Clark University Press.

—— (1988). "Geography at Clark: The First Fifty Years, 1921–1971," in Harmon and Rickard (1988: 40–8).

—— (1998). "Charles F. Brooks." *Geographers Biobibliographical Studies*, 18: 10–20.

Koelsch, W. A., and Campbell, S. W. (1982). "Resources for the History of Geography in the Clark University Archives." Association of American Geographers. *History of Geography Newsletter*, 2: 32–4.

LaRocco, J. (1978). "The Life and Thought of Derwent S. Whittlesey." MS thesis, Southern Connecticut State University, New Haven, Conn.

Leighly, J. B. (1963). *Land and Life: A Selection from the Writings of Carl Ortwin Sauer*. Berkeley and Los Angeles: University of California Press.

—— (1976). "Carl Ortwin Sauer, December 24, 1889–July 18, 1975." *Annals of the Association of American Geographers*, 66: 337–48.

—— (1978). "Carl Ortwin Sauer: 1889–1975." *Geographers Biobibliographical Studies*, 2: 99–108.

Lewthwaite, G. R. (1966). "Environmentalism and Determinism: A Search for Clarification." *Annals of the Association of American Geographers*, 56: 1–23.

Livingstone, D. N. (1981). "Environment and Inheritance: Nathaniel Southgate Shaler and the American Frontier," in Blouet (1981: 123–38).

—— (1987). *Nathaniel Southgate Shaler and the Culture of American Science*. Tuscaloosa, Ala.: The University of Alabama Press.

—— (1992). *The Geographical Tradition*. Oxford: Blackwell.

Lowenthal, D. (1958). *George Perkins Marsh, Versatile Vermonter*. New York: Columbia University Press.

Lowenthal, D., and Bowden, M. J. (eds.) (1975). *Geographies of the Mind: Essays in Historical Geosophy in Honor of John Kirtland Wright*. New York: Oxford University Press.

McManis, D. (1978). "A Prism to the Past: The Historical Geography of Ralph Hall Brown." *Social Science History*, 3/1: 72–86.

—— (1990). "The Editorial Legacy of Gladys M. Wrigley." *Geographical Review*, 80/2: 169–81.

Marshall, D. W. (1972). "A List of Manuscript Editions of Ptolemy's *Geographia*." *Geography and Map Division: Special Libraries Association Bulletin*, 87: 17–38.

Martin, G. J. (1968). *Marla Jefferson: Geographer* Ypsilanti, Mich.: Eastern Michigan University Press.

—— (1972). "Robert LeMoyne Barrett, 1871–1969: Last of the Founding Members of the Association of American Geographers." *The Professional Geographer*, 24/1: 29–31.

—— (1973). *Ellsworth Huntington: His Life and Thought*. Hamden, Conn.: Archon Books.

Martin, G. J. (1974*a*). "*Civilization and Climate* Revisited." *Geography & Map Division, Special Libraries Association Bulletin*, 96: 10–17.

— (1974*b*). "A Fragment on the Penck(s)–Davis Conflict." *Geography & Map Division, Special Libraries Association Bulletin*, 98: 11–27.

— (1980). *The Life and Thought of Isaiah Bowman*. Hamden, Conn.: Archon Books.

— (1981). "Ontography and Davisian Physiography," in Blouet (1981: 279–89).

— (1988). "Geography, Geographers and Yale University, *c*.1770–1970," in Harmon and Rickard (1988: 2–9).

— (1989). "On American and German Geography circa 1850–1940." *Proceedings, New England/St. Lawrence Valley Geographical Society*, 15–25.

— (1996). "One Hundred and Twenty Five Years of Geographical Congresses and the Formation of the International Geographical Union: or, From Antwerp to The Hague." *International Geographical Union Bulletin*, 46: 5–26.

— (1998). "The Emergence and Development of Geographic Thought in New England." *Economic Geography*, Special Issue, Spring, 1–13.

Martin, G. J., and James, P. E. (1993). *All Possible Worlds: A History of Geographical Ideas*. 3rd edn. John Wiley.

Mather, J. R., and Sanderson, M. (1995). Norman, Okla.: University of Oklahoma Press.

— (1998). "Charles Warren Thornthwaite, 1899–1963." *Geographers Biobibliographical Studies*, 18: 113–29.

Mikesell, M. W. (1975). "The Rise and Decline of 'Sequent Occupance': A Chapter in the History of American Geography," in Lowenthal and Bowden (1975: 149–69).

— (1986). "Year One of the Geographical Review." *Geographical Review*, 76/1: 1–9.

Miles, L. J. (1982). "Ralph Hall Brown: Gentle Scholar of Historical Geography." Unpublished dissertation, University of Oklahoma.

— (1985). "Ralph Hall Brown, 1898–1948." *Geographers Biobibliographical Studies*, 9: 15–20.

Miller, D. H. (1988). "In Memoriam: J. B. Leighly, 1895–1986." *Annals of the Association of American Geographers*, 78/2: 347–57.

Miller, E. W. (ed.) (1993). *The American Society for Professional Geographers: Papers Printed on the Occasion of the Fiftieth Anniversary of Its Founding*. Washington, DC: Occasional Publication of the Association of American Geographers, 3.

Monk, J. (1998). "The Women Were Always Welcome at Clark." *Economic Geography*, Special Issue, Spring, 14–30.

Moss, R. J. (1995). *The Life of Jedidiah Morse: A Station of Peculiar Exposure*. Knoxville: University of Tennessee Press.

Olmstead, C. W. (1987). *Science Hall: The First Century*. Madison, Wis.: Department of Geography, University of Wisconsin, 42.

Parsons, J. J. (1976). "Carl Ortwin Sauer, 1889–1975." *Geographical Review*, 66: 83–9.

— (1979). "The Later Sauer Years." *Annals of the Association of American Geographers*, 69/1: 9–15.

— (1988). "John Leighly 1895–1986." *Geographers Biobibliographical Studies*, 12: 113–19.

— (1996). "'Mr. Sauer' and the Writers." *Geographical Review*, 86/1: 22–41.

Pattison, W. D. (1981). "Rollin Salisbury and the Establishment of Geography at the University of Chicago," in Blouet (1981: 151–63).

— (1982). "Rollin D. Salisbury, 1858–1922." *Geographers Biobibliographical Studies*, 6: 105–13.

Riess, R. O. (1977). "Some Notes on Salem, Salem Normal School, and Carl Sauer: 1914." *Proceedings of the New England/St. Lawrence Valley Geographical Society*, 6: 63–8.

Rigdon, V. E. (1933). "The Contribution of William Morris Davis to Geography in America." Dissertation, University of Nebraska.

Ristow, W. W. (1982). "Three Months in the Field." *Clark Now*, 12/1: 19–24.

— (1983). "Three Months in the Field: September 12 to December 12, 1934." *History of Geography Newsletter*, 3: 21–9.

Ryan, B. (1983). *Seventy-Five Years of Geography at the University of Cincinnati*. Cincinnati: The University of Cincinnati, Department of Geography.

Sanderson, M. (1988). *Griffith Taylor: Antarctic Scientist and Pioneer Geographer*. Ottawa: Carleton University Press.

Sitwell, O. F. G. (1993). *Four Centuries of Special Geography: An Annotated Guide to Books that Purport to Describe All the Countries in the World Published in English Before 1888, with a Critical Introduction*. Vancouver: University of British Columbia Press.

Smith, N. (1986). "Bowman's New World and the Council on Foreign Relations." *Geographical Review*, 76/4: 438–60.

— (1987). "Academic War over the Field of Geography: The Elimination of Geography at Harvard, 1947–1951." *Annals of the Association of American Geographers*, 77/ 2: 155–72.

Solot, M. (1986). "Carl Sauer and Cultural Evolution." *Annals of the Association of American Geographers*, 76/4: 508–20.

Speth, W. W. (1981). "Berkeley Geography," in Blouet (1981: 221–44).

— (1987). "Historicism: The Disciplinary World View of Carl O. Sauer," in Kenzer (1987: 11–39).

— (1999). *How It Came To Be: Carl O. Sauer, Franz Boas and the Meaning of Anthropogeography*. Ellensburg, Wash.: Ephemera Press.

Stoddart, D. R. (1981*a*). "Darwin's Influence on the Development of Geography in the United States, 1859–1914," in Blouet (1981: 265–78).

— (ed.) (1981*b*). *Geography, Ideology & Social Concern*. Oxford: Blackwell.

— (1986). *On Geography and Its History*. Oxford: Blackwell.

— (1997). "Carl Sauer: Geomorphologist," in D. R. Stoddart (ed.), *Process and Form in Geomorphology*. London: Routledge, 340–79.

Taylor, T. G. (1958). *Journeyman Taylor*. London: Robert Gale.

Tinkler, K. J. (1985). *A Short History of Geomorphology*. Totowa, NJ: Barnes & Noble.

Tomkins, G. S. (1965). "Griffith Taylor and Canadian Geography." Unpublished doctoral dissertation, University of Washington, Seattle.

Warntz, W. (1981). "*Geographia Generalis* and the Earliest Development of American Academic Geography," in Blouet (1981: 245–63).

— (1989). "Newton, The Newtonians, and the *Geographia Generalis Varenii*." *Annals of the Association of American Geographers*, 79: 165–91.

Warntz, W., and Wolff, P. (eds.) (1971). *Breakthroughs in Geography*. New York: The New American Library.

West, R. C. (1979). *Carl Sauer's Fieldwork in Latin America.* Dellplain Latin American Studies, 3. Ann Arbor, Mich.: University Microfilms International.
— (1982). *Andean Reflections: Letters from Carl O. Sauer while on a South American Trip under a Grant from the Rockefeller Foundation*, 1942. Dellplain Latin American Studies, 11. Boulder, Colo.: Westview Press.
— (ed.) (1990). *Pioneers of Modern Geography.* Geoscience & Man, 28. Baton Rouge, La.: Louisiana State University Press.
Whittemore, K. T. (1972). "Celebrating Seventy-Five Years of *The Journal of Geography*, 1897–1972." *The Journal of Geography*, 71/1: 7–18.
Wright, J. K. (1925). "The History of Geography: A Point of View." *Annals of the Association of American Geographers*, 15/4: 192–201.
— (1926). "A Plea for the History of Geography." *Isis*, 8/3: 477–91.
— (1947). "Terrae Incognitae: The Place of the Imagination in Geography." *Annals of the Association of American Geographers*, 37/1: 15.
— (1961). "Daniel Coit Gilman, Geographer and Historian." *Geographical Review*, 51: 381–99. Repr. (1966), 168–87.
— (1966). *Human Nature in Geography.* Cambridge, Mass.: Harvard University Press.

Part VI

Regional Geography

CHAPTER 36

Geography of Africa

Ezekiel Kalipeni, Joseph R. Oppong, and Benjamin Ofori-Amoah

Introduction

This chapter reviews the state of North American geographical research on Africa in the 1990s. During the 1980s research on Africa dwelt on the many crises, some real and some imagined, usually sensationalized by the media, such as the collapse of the state in Sierra Leone, Liberia, Somalia, and Rwanda and the economic shocks of structural adjustment programs. The 1990s witnessed momentous positive changes. For example, apartheid ended in South Africa and emerging democratic systems replaced dictatorial regimes in Malawi and Zambia. Persuaded that Africa had made progress on many fronts largely due to self-generated advances, some scholars began to highlight the positive new developments (Gaile and Ferguson 1996).

Due to space limitations, selecting works to include in this review has been difficult. In many instances we stayed within five cited works (first authorship) for anyone scholar to ensure focus on the most important works and to achieve a sense of balance in the works cited. Thus, research reviewed in this chapter should be treated as a sample of the variety and quality of North American geographical work on Africa. One major challenge was where to draw the boundary between "geography," "not quite geography," and "by North American authors" versus others. In these days of globalized research paradigms, geography has benefited tremendously from interchanging ideas with other social and natural science disciplines. Thus, separating North American geographic research in the 1990s from other groundbreaking works that profoundly influence the discipline of geography is difficult. For example, while the *empirical subject matter* included agriculture, health, gender, and development issues, the related *theoretical paradigm* often included representation, discourse, resistance, and indigenous development within broader frameworks influenced by the ideas of social science scholars such as Foucault (1970, 1977, 1980), Said (1978), Sen (1981, 1990), and Scott (1977, 1987). This chapter engages these debates. Building upon T. J. Bassett's (1989) review of research in the 1980s, the chapter develops a typology for the growing research on African issues and related theoretical orientations (Table 36.1).

The reviewed works fall into the three main subdisciplines of geography—human geography (by far the most dominant), physical geography now commonly referred to as earth systems science or global change studies, and geographic information systems (GIS) (Table 36.1). Within these three main subdisciplines, theoretical perspectives overlap (Watts 1993). In particular, the realization that conventional, narrowly focused disciplinary perspectives and approaches were inadequate to explain Africa's rapid and complex changes led geographers to embrace and even devise more complex and integrated interdisciplinary approaches. The most important of these transitions or developments in African geographical research during the 1990s include:

Table 36.1 *Research themes, subtopics, and theoretical perspectives of Africanist geographical research*

Research themes	Major subtopics	Theoretical orientation
Human geography Population, resources and the environment	Population growth and agrarian transformation Land in African agrarian change Resources and the environment Gender and resource contestation Resources and pastoral conflicts Food and hunger	Boserupian, political ecology, political economy, post-colonial, post-structural, feminist perspectives
Population dynamics	Fertility Mortality Migration processes	Demographic transition model, theory of Demographic response, Political economy
Development discourse	The power of development Development misconception	Post-colonial/ Postmodern/ post-structural, feminist perspectives, sustainable/ green development approaches
Policy-oriented or impact analysis studies	The human factor perspective Structural adjustment programs (SAPs) Spatial analysis	Modernization, dependency perspectives, core/periphery concept, human factor hypothesis
Urban and regional development	Industrial restructuring Informal sector studies Labor Segregation and resistance	Neo-liberal (market supremacy), political economy, political ecology, globalization
The geography of disease and health care	Disease ecology HIV/AIDS in Africa Structural adjustment and health Geography and health care	Disease ecology, political economy, location–allocation models
Global change and earth systems science	Global change (Causes and impacts) Hydrology of wetlands Landscape ecology	Global climate models, Earth systems, GIS as a tool for monitoring global change
Geographic Information Systems	Vegetation change Urban morphology	Land-use planning

- post-colonial/post-structuralist/postmodern approaches derived from the works of Said (1978), Foucault (1970, 1977, 1980), Sen (1981, 1990), and Scott (1977, 1987);
- political ecology championed by Atkinson (1991), T. J. Bassett (1993), Blaikie (1994), Blaikie and Brookfield (1987), and Bryant (1992);
- Boserupian perspectives on population and environment promoted by Tiffen *et al.* (1994);
- challenging environmental orthodoxies, particularly a reassessment of taken-for-granted ideas about the environment championed by Leach and Mearns (1996), and Fairhead and Leach (1996, 1998);
- development from below/grassroots initiatives (Taylor and Mackenzie 1992);
- recognition of the importance of indigenous knowledge;
- the impact of globalization on economic development particularly in agriculture and industrial restructuring;
- policy-oriented studies, e.g. application of spatial allocation models and the ambitious Southern African Migration Project;
- social geographies pertaining to gender and other issues;
- global environmental change research involving climatologists, geomorphologists, hydrologists, and biogeographers utilizing an integrative/systems framework.

The disciplinary subdivisions of human geography, earth systems science, and GIS guided by the above theoretical approaches form the organizational basis for the works reviewed in this chapter. Many other works do not fit neatly into the new developments but rather within previously established paradigms such as spatial modeling, cultural ecology, and political economy. Wherever possible we attempt to distinguish between continuing and new lines of scholarship.

Population, Resources, and the Environment

Research in the 1990s in this area departed from the dominant neo-Malthusian approaches of the previous decade, which usually bemoaned the combination of rapid population growth and economic and environmental decline. Indeed, many geographers remain convinced that ecological degradation is a human-induced problem, but recent works have concentrated on exploring the role of population growth on agrarian transformation, the role of land tenure in agrarian change, farmer–pastoralist conflicts, the environment, and issues of gender and resource contestation. The guiding frameworks for these recent works include the Boserupian perspective, political ecology, and political economy.

Population Growth and Agrarian Transformation

A new body of literature stresses that increasing population densities induce positive agricultural transformation. B. L. Turner *et al.*'s (1993) *Population Growth and Agricultural Change in Africa*, is of particular significance. Within the Boserupian tradition, this volume investigates the relationship between population growth in high-density areas of Africa and agricultural intensification and concludes that population growth is just one of many factors that affect agricultural change. Differences in environment, market access, social institutions, land tenure, technology, and politics make it extremely difficult to establish a straightforward relationship between population growth, environmental degradation, and agricultural intensification.

Using agricultural innovation in Kenya and evolution of maize production in Northern Nigeria as case studies, Goldman (1993) and Smith *et al.* (1994) highlight the synergistic effect of technology in a population-driven intensification process. Despite high population densities, agricultural intensification produced substantial increases in productivity and farmer welfare due to a good road system, high yielding varieties of maize, widespread adoption of fertilizer made possible in part by fertilizer subsidies, and a favorable market infrastructure. Smith *et al.* (1994) and Kull (1998) argue for quantum leap technologies in sub-Saharan Africa to accelerate sustainable agricultural intensification, particularly in areas where the preconditions exist. They note that market production rather than population pressure *per se* has driven the agricultural transformation in the study areas, and it is highly unlikely that population growth alone could have produced these changes. Consequently, these works present a forceful argument for the introduction of appropriate technologies for agricultural transformations, where conditions are favorable. However, such reliance on imported technologies, often financed by the World Bank and subsidized by the state, may be an unsustainable and possibly dangerous prescription as exemplified by World Bank-supported tube-well projects in Nigeria that have lowered the water table.

In a recent volume, T. J. Bassett (2001) examines the production and intensification of cotton in Côte d'Ivoire, one of the few long-running success stories. He argues that while the introduction of new strains of cotton in francophone West Africa was in part as a result of agronomic research by French scientists, positive change in the growth of the cotton economy in West Africa was brought about by tens of thousands of small-scale peasant farmers. In contradiction to the dominant accounts of crisis, doom, and gloom, Bassett's work shows agricultural intensification to result from the cumulative effect of decades of incremental changes in farming techniques and locally engendered social organization by peasants.

Perhaps the most influential work in the Boserupian tradition is Tiffen *et al.* (1994). In the hotly debated book, *More People, Less Erosion: Environmental Recovery in Kenya*, British and African geographers promote Boserupian ideas using Machakos District in Kenya as a case study (Tiffen and Mortimore 1994; Mortimore and Tiffen 1994). The central argument in the Machakos story is that even as population densities have increased, agropastoral productivities have increased and a degraded landscape has flourished with trees, terraces, and productive farms. The underlying assumption is that population pressure gives rise to its own solution, namely, agricultural intensification.

Other scholars have contested the Machakos story from a number of angles. For example, Murton (1999) notes that examination of the "Machakos experience" of population growth and environmental transformation at household level, shows neither a homogenous experience nor a fully unproblematic one. Data in Murton's study show how such changes in Machakos District have been accompanied by a polarization of land holdings, differential trends in agricultural productivity, and a decline in food self-sufficiency. Murton (ibid.) raises doubt about the universal relevance of the Machakos story and for that matter the Boserupian perspective as a model. Political ecologists have also criticized the Boserupian perspective, and cultural ecology analysis in general, for focusing on local dynamics while excluding relevant economic and political processes operating at broader scales.

Land in African Agrarian Change

Debates on indigenous land tenure systems versus privatization intensified during the 1990s. T. J. Bassett and Crummey's (1993) work, *Land in African Agrarian Systems* questions the traditional thought that indigenous tenure systems impede increasing agricultural productivity. The contributions in this collection challenge the notion that sweeping privatization of land will reverse declining agricultural productivity. Instead, they reveal how land access, control, and management are embedded in dynamic social, political, and economic structures that fluctuate and change over time. The works of Fairhead and Leach (1996, 1998) about indigenous production systems and forest production/generation support the Bassett/Crummey thesis that indigenous land tenure systems may not obstruct resource conservation and sound agricultural practices. Similarly, Awanyo (1998) notes that the economic behavior of cocoa farmers in Ghana is complexly linked to the politics of land tenure and cultural expectations and obligations instead of price incentives and private tenure.

Other works examine conflicts between the state and the peasantry that arise over access to land. In both Zimbabwe and South Africa, disparity between black and white land ownership and control makes the "land question" the most difficult resource issue, and its resolution, the most important task of the respective post-colonial and post-apartheid governments (Weiner and Levin 1991; Masilela and Weiner 1996; Levin and Weiner 1997; Levin 1998; Zinyama 1992). Increasing issues of class, race, regionalism, and a general breakdown of the rule of law may be evidence of the failure of the Zimbabwe government's policy on land reform. While the works on Zimbabwe look at both colonial and post-colonial times, Mackenzie (1995, 1998) examines land and gender issues in a historical context, focusing on colonial Kenya. Moore (1998, 1999) also looks at conflicts over access to environmental resources and land between state administrators and peasants in a state-administered resettlement scheme bordering Nyanga National Park in eastern Zimbabwe. These insightful analyses show that struggles over resources, shifting political alliances, and competing agendas have encountered the salient differences of gender, generation, class, education, and traditional authority.

Globalization and Agrarian Change

The impact of globalization on agrarian change in Africa is addressed under the rubric of contract farming—an arrangement between a grower and firm(s) in which non-transferable contracts specify one or more conditions of marketing and production (P. D. Little and Watts 1994). Watts (1994) compares the nature and origins of contract farming in developed and developing countries, including Kenya and Nigeria, and concludes that it illuminates the new configuration of state, capital, and small-scale commodity production, and a changing international division of labor. Thus, capital can dominate agriculture not only by expanding the frontier of capitalist enterprise but also through the establishment of elaborate networks of social regulation and control. Goodman and Watts (1997) extends our knowledge about the complexity and diversity of local and national levels of agrarian change in the context of global restructuring of the world's food and agricultural systems.

Using the Jahaly-Parchar project in Gambia, Carney (1994) also analyzes the changes that can result from the introduction of contract farming and concludes that the project produced transformations that policy-makers and project planners had never imagined. Attempts by donor agencies to entitle plots to women started a chain of conflicts that turned the household into a terrain of intense struggle, negotiation, and partial victories that have severely hindered the ability of contract farming to increase market surplus. Mather (1999) illustrates the impact of globalization on agriculture emphasizing local market power and global market strategies pursued by large national and multinational corporations. These works claim that contract farming presents a new form of Third World agriculture being pushed by global restructuring. However, contract farming is not different from the way in which cocoa, tea, coffee, and other cash crops were introduced into African economies. Only the crops are different. It is doubtful whether the impact of this new trend on African agriculture will be different from what existing cash crops have done to Africa.

Resources, the Environment, and Development

Many works utilizing the political economy, political ecology, and liberation ecology frameworks have tackled the perplexing issues of multinational corporations versus local resources, common property rights and indigenous knowledge, land for agricultural extensification versus wildlife conservation, and afforestation/ deforestation issues. For example, Watts (1996) relates the rise of home-grown, politically charged environmental movements, such as that led by the late Saro-Wiwa in Ogoniland, Nigeria, to the larger landscapes of international exploitation and political economy. Stanley (1990) elaborates the contradictions of oil exploitation in the Niger Delta and the wider socioeconomic and environmental impacts of the industry.

Jarosz (1993, 1996) takes issue with blaming population growth and shifting cultivation for deforestation in the developing world. In Madagascar, Jarosz links discourse analysis with peasant resistance and response during the colonial period to argue that the colonial cash-crop economy, rather than population growth, produced deforestation. This supports the works of Fairhead and Leach (1996, 1998) and Leach and Mearns (1996) which have challenged 100 years of received wisdom on the degradation of the African environment. Nevertheless, political ecology as an approach has been critiqued for overemphasizing the deleterious effects of state and international policies. Besides, most works employing the framework sideline the "ecology" in "political ecology" in favor of the "political." Biophysical processes are often missing in the equation. In addition, others have also criticized it for having too little politics. Perhaps collaboration with geo-scientists might better articulate and redefine the application of the political ecology framework.

Neumann (1995, 1998) and Schroeder (1995) evaluate the paradox of remedial efforts in Tanzania and Gambia to restore, preserve, and stabilize the integrity of nature through terracing, reforestation, or the creation of buffers. These studies show that despite the emphasis on "positive" incentive of profit-taking, participation, and benefit-sharing, many of the new integrated conservation and development projects in Africa have coercive elements, often constitute an expansion of state authority into remote rural areas, and do not benefit the majority. Two articles that focus on grassroots initiative in control of resources are Matzke and Nabane (1996) and Barkan *et al.* (1991). In Nigeria, Barkan *et al.* show that hometown associations serve as a civil virtue, shadow state, bulwark against state power, local growth machine, intermediary broker of linkages, and also express attachment to place. In Zimbabwe, Matzke and Nabane describe how local initiative in wildlife management has provided a new paradigm for local economic development in rural areas that lie close to national parks and wildlife preservation areas. This initiative, dubbed Communal Areas Management Program for Indigenous Resources (Campfire), rationalizes that people from wildlife producer communities should benefit from the wildlife that their land produces on a sustainable basis. The authors argue that Campfire has produced remarkable success—fencing has reduced loss of human life from animal attacks, and revenue generation and job opportunities have increased. The works of Neumann (1998), Matzke and Nabane (1996), and Barkan *et al.* (1991) have contributed greatly to policy and social impact analysis. Neumann's work distinguishes between genuine empowerment and local control from top-down initiatives that masquerade as community development, exemplified by Campfire. *Development from Within* edited by Taylor and Mackenzie (1992) provides an excellent critical appraisal of community development initiatives to preserve the environment. Using precise and historical understanding of specific places through local case studies, this volume addresses the tricky issue of mobilizing village resources by and under the control of the rural poor themselves. Several other studies have explored women's environmental initiatives (e.g. the Green Belt Movement in Kenya), and why gender matters greatly in assessing the success or failure of such bottom-up initiatives. Another important body of literature in which geographers have contributed significantly is the impacts of dams. Roder (1994), for example, examines the consequences of the Kainji Dam on the local population around Lake Kainji and shows that neither the predictions of great losses to the economy of the people, nor the expectations that they would take many decades to adjust to the lake have come true. The people have met the changes and carried on with their lives.

The lessons from these works are clear. Community empowerment gives people an opportunity to manage their resources and make decisions about their own future. While this is not a new finding, the operational evidence of this in Africa is not very common. More of these studies will help instill the hope that grassroots Africa has much untapped potential for economic development.

Gender and Resource Contestation

The rise of post-structuralist, post-colonial, postmodern, and feminist critiques of development discourse (Crush 1995; Mackenzie 1995) have spawned an interesting set of case studies that examine the complex intersections among gender, agrarian change, environmental discourse, access to resources, and indigenous knowledge (Carney 1992, 1993*a*; Schroeder 1993, 1997; Rocheleau *et al.* 1996). In these studies, the term "resources" refers to household resources in both urban and rural areas. Carney links property rights and gender conflict to environmental change and reveals repeated gender conflicts over rural resources in Gambia as male household heads concentrate land holdings to capture female labor for surplus production. Schroeder's work in The Gambia questions agroforestry as an effective means of stabilizing the environment because it does not always yield the intended results. Commoditization of tree crops and incentives to enhance the rate of tree planting ultimately led to shifting patterns of resource access and control and produced gender conflict between husbands and wives due to multiple tenure claims to land.

Rocheleau *et al.* (1996) focus on women's struggle in Kenya to gain access to natural resources such as water and wood in their double burden of household production and reproduction. These works utilize political ecology, social institutions, situated knowledges and practices, to argue that historical contexts and processes have been influential in creating environmental and social problems which, in turn, have resulted in the marginalization and suffering of women in Kenya. Other scholars have explored the role of women in development and the changing conditions of women in rural and urban Africa. Johnston-Anumonwo (1997) offers a holistic approach to the study of gender issues in Africa and notes the many significant contributions women made to the overall development of the continent despite their marginalization. Yeboah (1998) uses a Ghanaian case study and a novel geographical and historical framework to compare the economic status of the two genders and concludes that Ghanaian women are slowly closing the gender gap in economic status. Osei (1998) arrives at a similar conclusion in a study of the gender factor in rural energy systems in Africa and shows that while women dominate the rural energy system, they have not suffered any significant, negative socioeconomic impacts from being engaged in the system. However, both Yeboah's and Osei's conclusions contradict the majority of the literature which emphasizes the barriers women face in trying to get fair access to resources, and how this affects their ability both to succeed and to use the environment. Nevertheless, a common thread running though the works reviewed in this section is their insightful analysis of the role of local, national, and international forces in exacerbating the problems and burdens women face in their lives and the detailed examination of how women, households, and communities respond and adapt to these challenges.

Resources and Pastoral Conflicts

Several excellent studies within the political ecology framework have offered detailed analyses of agrarian and pastoral conflicts stressing the embeddedness of these struggles in historical context, social relations, and linkages with wider geographical and social settings (P. Little 1992; T. J. Bassett 1993; M. Turner 1993; Heasley and Delehanty 1996; Bascom 1990*a*; Johnson 1993; Campbell 1991). Bascom (1990*a*) examines the factors that are bringing refugee cattle herds to extinction and jeopardizing pastoralism as a way of life for Eritrean refugees. He attributes the demise of cattle herders to dwindling amount of rangeland and transformation of grazing rights, all of which have resulted from long structural processes.

Bencharifa and Johnson (1990) examine changes in pastoral farming in Morocco that promote and obstruct intensification. Increasing productivity, a decrease in annual and interannual fallow cycle, a shift from traditional extensive pastoral system to a much less nomadic life, and gradual abandonment of the tent as a primary residence for a permanent house and village location are all visible changes, although quite inequitably distributed. Campbell (1991) examines the impact of socioeconomic and political conditions over the past century upon the coping strategies of the Masai of the Kajiado District of Kenya and shows that both colonial and post-colonial development policies and strategies failed to recognize the complexity of society–environment interactions, which form the basis of Masai pastoralism. Campbell calls for development strategies that enhance the diversity and productivity of the Masai economy to prevent their economic and social marginalization.

P. Little (1992) looks at how the Lake Baringo area in Kenya changed from a food-surplus to a food-deficit area of famine, impoverishment, and ecological degradation. He questions conventional explanations by colonial officials and post-colonial ecologists that overstocking and the shortsightedness of pastoralists have damaged the ecological basis. His analysis goes beyond a "blame the victim" paradigm to examine the complex web of external and internal factors in a historical context. A parallel study by M. Turner (1993) in the Sahelian belt of West Africa argues that the persistent reliance by environmental analysts on carrying capacity models oversimplifies range ecology and excludes social processes from causal analysis.

T. J. Bassett (1993) presents a case study of peasant–herder conflicts among the Senufo and Fulani of northern Côte d'Ivoire in the wider context of national and international forces. The study shows that the interactive effects of land-use conflicts at three overlapping levels (local, regional, and national) continue to undermine both the expansion of Fulani livestock production and the intensification of agricultural systems in the savanna region. Heasley and Delehanty (1995) examine disputes over manure in Southwestern Niger, which reveal broad strategies for natural resource control employed by farmers and herders in a transitional and conflictual agro-pastoral economy. What needs to be emphasized is that most of these studies are about the value of the political ecology approach rather than about pastoralism *per se*.

Food and Hunger

In the 1990s food and hunger continued to occupy the attention of North American geographers but to a lesser degree than in the 1980s. These works include Bascom (1990*b*), Campbell (1990*a*, *b*, 1999), Campbell *et al.* (2000), Zinyama *et al.* (1990), and Griffith and Newman (1994). Bascom's (1990*b*) work on the Sudan shows how state intervention favors large-scale commercial farming to the detriment of peasant agriculture. A common belief that emerged in the 1980s was the widespread use of wild foods in Africa to mitigate food shortages. Zinyama *et al.* (1990) found that in Zimbabwe, wholesale dependence on wild foods during periods of food shortages was uncommon and that about 71 per cent of the people interviewed in the seven villages in the low rainfall regions still relied on maize, millet, or sorghum as their staple. Major sources for this variation were continued availability of other sources of staple foods either through government food distribution or generation of cash from other sources of rural occupation, and remittances from relatives living in the urban areas for food purchases. Another legacy of the 1980s famine research was the establishment of the famine Early Warning Systems (EWS). Campbell's (1990*a*) work focused on how to improve the accuracy of the database, and argued for using indicators based on the observed responses of people vulnerable to food shortages. Incorporating the employed range of sequential coping strategies that vary between places, age groups, gender, and social status is critical. More recent and continuing work by Campbell and his colleagues is looking at linkages between climate change, land-use change, and human welfare and policy (see e.g. Campbell 1999; Campbell *et al.* 2000).

Eliminating Hunger in Africa by Griffith and Newman (1994) was born out of dissatisfaction with the progress made after the 1980s regarding elimination of famine. The book combines new methods of data collection and analysis with a more sensitive critique that sees Africa's plight not in isolation but as part of a world order in which values and perceptions are important. Gaile's work shows that spatial analysis can be used to establish market distribution centers for the successful implementation of Kenya's food security program (Gaile 1994). Ralston *et al.* (1994) highlight the role of transportation in the distribution of food aid especially in landlocked countries in Africa. M. Little (1994) discusses the broader context of famine in Africa under the New World Order and argues that political and economic insecurity results in food insecurity. Listing the struggle for national integration, acceptance of comparative advantage and indebtedness as the main causes of political and economic insecurities in Africa, she concludes that what Africa needs is not charity but justice. Unless Western perception of Africa changes, food aid will not solve the food problem.

Other studies have used the microperspective approach by placing human systems within the context of their particular natural and social environment. Among these are household-level studies in urban Africa. For example, Drakakis-Smith (1990) and Drakakis-Smith *et al.* (1995) examine the food impacts of structural adjustment programs on the poor at the household level in Harare. Others have examined the impact of changing social and economic circumstances on household access to food by gender. Most of the studies under this rubric have been influenced by the work of Sen (1981, 1990) on entitlements, basic freedoms, social justice, and an equitable future for everyone. Other studies on the issue of food are more general and have dealt with US food aid policy (e.g. Bush 1996; Kodras 1993). While these works have provided interesting perspectives, the food and hunger issue is more complex than the issues addressed here. We think that discussion on food and hunger in Africa should not focus narrowly on distribution from outside but should also examine internal production and provision of appropriate infrastructure and mechanisms.

Indigenous Environmental Knowledge

A growing literature addresses indigenous environmental knowledge and argues that rural peoples throughout the developing world understand their environment well, particularly its possibilities for sustaining livelihoods. Consequently, natural resource management practices reflect cultural appraisals of what is possible and desirable in a given setting. Katz (1991), focusing on children's environmental knowledge in Sudan, its means of acquisition and use, is an excellent example. Similarly, Nyamweru (1996, 1998) examines indigenous environmental knowledge about preservation of sacred groves in Kenya. Fairhead and Leach (1996) also show that the islands of dense forest in the Guinea savanna were actually created by local inhabitants around their villages and are not relics of previously extensive forest cover that has been degraded by population growth. Carney (1993*b*) highlights the rice cultivation technology brought to the New World by African slaves. This corrects the notion that Africa did not contribute any technological development to the New World.

While the literature is mute on the art of indigenous mapmaking in Africa, T. J. Bassett (1998) dispels the eurocentric and pejorative view that Africans were incapable of making maps the same way Europeans did. Adopting an expanded definition of maps to include mnemonic maps, body art, the layout of villages, and the design and orientation of buildings, Bassett (ibid.) offers an intriguing overview of the range of maps existing in Africa's historical record and concludes that like those of other traditional cultures, African maps are social constructions whose form, content, and meaning vary with the intentions of their makers.

Population Dynamics

Parameters of population change such as the dynamics of fertility, mortality, and migration processes and how these relate to environmental change and resource scarcities have been relatively neglected, but Kalipeni (1995, 1996) attempts to fill this gap. Kalipeni (1995) adds a spatial dimension to the discussion of African fertility through a spatial-temporal framework for the demographic transition theory. His analysis shows that fertility levels in Africa have begun to decline and, some African countries, Zimbabwe, Botswana, Kenya, and Nigeria, for example, may be in the initial stages of an irreversible fertility transition.

The 1990s saw very little work on migration processes, and the primary focus was refugee movements from war-torn areas and migration policy in Southern Africa (Crush 1998, 1999; Crush and James 1997; Crush *et al.* 1991; Crush and Williams 1999; Hyndman 1999; Wood 1994). The most ambitious project is the Southern Africa Migration Project, a decade-long study of many aspects of migration in Southern Africa led by the Canadian geographer Jonathan Crush. This project is designed to formulate and implement new initiatives on cross-border migration in the region and promote public awareness of the role, status, and contribution of foreign immigrants of African origin in South Africa. The project has great potential to influence the important policy issue regarding what relationship should exist between post-apartheid South Africa and its neighbors.

Wood (1994) offers a framework for the causes of refugee flows and the consequences of such flows at the international and local scales. Hyndman (1999) adds a geographical dimension to the study of "new" safe spaces and discourses emerging within humanitarian circles since the end of the Cold War. She argues that while camps continue to house refugees, the meaning and value of "refugee" have changed dramatically and efforts to prevent people from crossing political borders to seek safety are increasing, producing a distinct geopolitical discourse and a new set of safe spaces. James Newman's (1995) book, *The Peopling of Africa: A Geographic Interpretation*, offers a refreshing synthesis of African pre-colonial migrations. This work uses the power of prose and narrates the population, human, and historical geography of pre-colonial Africa using concepts of space, region, and place to derive a coherent original synthesis of the richness and diversity of Africa from the beginning of humankind to the time just before European colonization.

Development Discourse

Development discourse refers to the language, words, and images used by development experts to construct the world in a way that legitimates their intervention in the name of development. This approach seeks to establish the authority of experts and their opinions that this (and not that) is the way that the world actually is and ought to be. Geographers have contributed significantly in critiquing development discourse, particularly its characteristic language of crisis and disintegration, which gives justification for intervention. This critique arises out of an old suspicion of a hidden agenda behind the introduction and abandonment of development strategies in Africa, as well as the perpetuation of development strategies and notions that are detrimental to Africa's development.

These works have a twofold mission. First, they call on African people, governments, and policy-makers to disabuse their minds about lies concerning Africa's development. Second they encourage them to critically examine development concepts, strategies, and theories before either adopting or abandoning them. While this may not be a new perspective, a number of influential interdisciplinary books have fruitfully engaged and critiqued this perspective. See for example Taylor and Mackenzie's (1992) *Development from Within*, Crush's (1995) *Power of Development*, Corbridge's (1995) *Development Studies: A Reader*, Adjibolosoo and Ofori-Amoah's (1998), *Addressing Misconceptions About Africa's Economic Development*, and Godlewska and Smith's (1994) *Geography and Empire*. Together these works offer a rich set of papers that collectively explore the language of development, and its rhetoric and meaning within different political and institutional contexts. These works utilize post-colonial, postmodern, and feminist critiques and focus on rhetoric versus performance.

An important issue raised in this literature is the persistent (mis)representation of development and of Africa itself. Most has been influenced by the seminal work of Said (1978) on Orientalism. For example, in *Geography and Empire* Rothenberg (1994) critiques *National Geographic* images and the complicity of geography as a discipline in perpetuating colonial and post-colonial myths about Africa and development. Indeed the articles by Bassett (1993, 1994, 1998) on colonial mapmaking exemplify this point. Myers (1998) shows how Lusaka and Zanzibar were enframed by the plans developed by Eric Dutton, an officer in the service of colonial hegemony.

Policy Oriented and Impact Analysis Studies

The past decade saw an impressive collection of work on policy and impact analysis. Some of these studies were prescriptive, falling within the location/allocation modeling tradition, while others denounced the deleterious impacts on African economies and livelihoods of structural adjustment programs. A new effort shifts the primary focus of development to the human factor—people's attitude and behavior—as the basic prerequisite for successful economic development.

Prescriptive Studies

Proponents of this perspective include Mehretu and Sommers (1990, 1992) and Ralston *et al.* (1994). These works remind development geographers interested in Africa that modeling using spatial analysis—neoclassical style—is still valid for Africa although structuralist development geographers have ruled out diffusion as a failure of the modernization project. Indeed, critiquing growth center strategy, Gaile (1992, 1994) and Gaile and Ngau (1995) have consistently argued that access issues that incorporate social capital, isolation effects, and institutional constraints can inform decision-making to address food security and poverty alleviation. Gaile's work in Kenya shows that spatial analytical techniques such as proximal areas, distance decay analysis, and central place concepts of threshold and range provide a better strategy for rural–urban development.

Mehretu and Sommers (1990) employ spatial analysis to explore the differences between developed and developing countries in national preferences and how competing preferences are resolved. Ralston *et al.* (1994) is in the same tradition. Wubneh (1994) interrogates the relationship between interregional disparities and political unrest in Ethiopia and attempts to analyze the impact of the policies and programs of the socialist government on interregional disparities among Ethiopia's provinces. Despite its limitations, spatial analysis has great potential to contribute meaningfully to development, but its contributions have been in theory rather than in practice. Perhaps the major problem is that geographers often write from positions far removed from the exercise of power in Africa, and with the cutbacks imposed by structural adjustment programs (SAPs), little money is available to start implementing spatial optimization projects.

A few studies (Adjibolosoo and Ofori-Amoah 1998; Ofori-Amoah 1996, 1998) have introduced a new prescriptive approach, the human factor (HF) perspective. The central message of this perspective is that Africa has a human factor problem that needs fixing before any meaningful development can take place. Human factor is defined as the spectrum of personality characteristics and dimensions of human performance that enable social, economic, and political institutions to function and remain functional over time.

Social Impact Assessment Studies

By all measures, research on structural adjustment programs (SAPs) rode the waves of policy impact studies in the 1990s. Among these are Mengisteab and Logan (1995), Ould-Mey (1996), Riddell (1992), Samatar (1993, 1994), Logan and Mengisteab (1993), Owusu (1998*a*, *b*), and Carmody (1999). With structural and post-structural theoretical orientation these works invariably begin with a good background of the rationale of the SAPs in Africa, and using selected case studies, highlight the failure of SAPs in resuscitating African economies. According to these studies, SAPs were prescribed at the beginning of the 1980s as the only solution for the internally generated economic stagnation by the International Monetary Fund (IMF) and the World Bank, the two major sponsors of SAP, together with their governmental supporters. This stagnation stems from market distortions, which in turn are due to overwhelming government intervention in the domestic economy. SAPs thus prescribe three groups of reforms: deflationary measures, which consists of removal of subsidies and reduction of public expenditures, institutional changes in the form of trade liberalization and privatization, and expenditure-switching measures in the form of currency devaluation and export promotion.

After a decade of Africa's experience with SAP, the unanimous verdict is clear—while SAPs may enhance economic growth, they do not address the pertinent development issues facing African countries. In a comprehensive survey of twenty-three African countries that adopted SAPs, Logan and Mengisteab (1993) did not find significant differences in economic performance and social welfare between weak and strong SAP reformers. In Somalia, Samatar (1993) shows that liberalization of the banana and rice economies reversed previous negative trends in production and improved profits in the industry, but nearly 75 per cent of the earnings from exports went to overseas interests, depriving Somalia of an important source of capital. In Mauritania, Ould-Mey (1996) shows that SAPs have denationalized Mauritania—the country has almost lost its capacity to conceive, design, fund, implement, monitor, and evaluate original development programs. The nerve center of decision-making has actually moved away from the state to international financial institutions and NGOs amidst accentuation of regional, ethnic, religious, linguistic, and cultural adherence loyalties. Carmody (1999, 2001) examines the impact of economic liberalization on Zimbabwe's manufacturing sector as well as political governance in the 1990s. Carmody (1999) reports that instead of reviving and strengthening the textiles, clothing, and footwear industries in Zimbabwe, SAPs contributed to the collapse of the industries. Carmody (2001) explores in greater depth World Bank and International Monetary Fund reasons behind instituting radical economic reforms, as well as how widespread deindustrialization led to disastrous political and economic consequences for the people of Zimbabwe. After presenting a strong argument that the current political crisis in Zimbabwe is intricately connected to the sudden economic structural adjustment, Carmody advocates various alternatives that are based on a better understanding of the local politico-economic context.

Even the experience of Ghana, the "star pupil" of SAP proponents, could not bear out a different verdict. Owusu (1998) shows that while SAPs led to huge export growth in the forest products sector that enabled the country to absorb the nation's annual debt interest payment, the traditional link between saw mills and domestic wood processors were broken and replaced by subcontracting to foreign firms. Furthermore, spatial integration was precluded and the program has produced severe deforestation. In another paper Owusu (2001) demonstrates the pervasive effects of the restructuring on the livelihood strategies of many people, as their established means of income generation have been disrupted. Other adverse impacts of SAPs on the psychology, politics, education, economy, and the social fabric of African countries are identified in the edited work of Mengisteab and Logan (1995). As Ould-Mey (1996) argues, SAPs were never intended to be a development strategy for Africa; their primary mission was to solve the internal economic problems of developed countries. This primary mission has unfortunately been further strengthened by two factors—the helplessness of African countries to withstand the pressure and the tendency for African governments to adopt the strategy as a metonym.

Urban and Regional Development

During the 1990s many geographers in North America, South Africa, and Europe grappled with issues of urban development, industrial restructuring, the informal sector, labor and regional development particularly related to the South African transition. Rogerson (1999*a*) edited a special issue of *GeoForum* that examines the deconstruction of apartheid's migration regime (Crush 1999), migration and environment in Southern Africa (McDonald 1999), and re-regulating the citrus industry in local and global context (Mather 1999). Other works have focused on labor for the mines and farms, the post-apartheid city, resistance through the construction of squatter camps, the informal sector, industrial restructuring and regional development, segregation, and urban protests (Watts 1997; Lemon 1995; Drakakis-Smith 1992; Rogerson 1996, 1999; Pickles and Weiner 1991; Pickles 1991; Hart 1998; Saff 1995, 1998*a*, *b*, 2001). These works highlight the momentous changes within South Africa and its positioning within Africa. Until recently, many scholars and officials have found it convenient to treat South Africa as if it was not really part of Africa—its issues were treated as unique.

Pickles (1991) for example, shows that although industrial decentralization fostered by the apartheid era stimulated increased levels of employment and production in rural areas, this was achieved at great expense to the state ultimately resulting in poor working environments for those employed, and few benefits to rural communities. Weiner and Levin (1991) warn against planning the post-apartheid agrarian landscape arguing instead for greater rural political mobilization and participation in the reconstruction process. Rogerson (1996, 1999) argues for a coherent and comprehensive national urban and regional development strategy to avoid excessive conflict between the effects of implicit and explicit spatial interventions that exacerbate regional inequalities. Hart (1998) warns that a narrow focus on the industrial sector and agrarian dispossession due to global competition are undermining redistribution and equitable development in South Africa.

While the bulk of the work on urbanization and regional change is on the South African transition, works on other parts of Africa show some interesting changes in technology and industrialization (McDade 1997; McDade and Malecki 1997; Spring and McDade 1998). Noting the pitiful performance of most African countries in industrialization and manufacturing trends, McDade concludes that a conducive macroeconomic environment, adequate supply of capital, adequate infrastructure, and a motivated, skilled labor supply are necessary for industrialization to take off in sub-Saharan Africa.

In urban geography, Sanders (1992) attacks the Eurocentric bias of the study of African urbanization in a very provocative essay and argues that while this is good for interdisciplinary reasons, it has not produced solutions to Africa's urban problems, nor shed any light on the nature of the urbanization process. Consequently, the study of African urbanization is an important development topic that requires a new debate within the realm of radical geography. On a slightly different note, Myers (1996) utilizes the example of Zanzibar's Ngambo neighborhoods, to show how toponomy and boundary-making embody a complex spatial discourse on power in the urban landscape. Stock (1995: 193–236) and Aryeetey-Attoh (1997: 182–222) provide a detailed survey of cities in sub-Saharan Africa, their origin and growth, internal structures (with reference to selected cities), followed by the causes, consequences, and policy implications of the urbanization process.

Employing the strategic concepts of place and space, two recent studies examine the evolution of pre-colonial settlements of Kano in Northern Nigeria and the Oasis Sijilmassa in Morocco (Lightfoot and Miller 1996; Nast 1994; 1996). Nast (1996) attempts to marry critical geographical notions of landscape and spatial praxis with Foucault's (1977) notion of "archaeology," to produce "spatial archaeology." Using the Kano palace of Northern Nigeria as a case study, Nast explores how slavery, gender, paternity, and motherhood were social-spatially constructed during a time of great regional change in Northern Nigeria *c.*1500. In a similar fashion, Lightfoot and Miller (1996) offer a detailed archaeological study of the rise and fall of a walled oasis in medieval Morocco, i.e. the city of Sijilmassa. Moving to the twentieth century, Cooper (1997) examines gender with reference to women's housing strategies in Maradi, Niger. The strength of this work lies in its consideration of gender through the use of multi-disciplinary methodological approaches in which oral tradition, field reconnaissance, remote sensing, historical documentation, and archaeological fieldwork are combined to produce coherent narratives of place and space. Konadu-Agyemang (1991) examines the condition and spatial organization of urban housing and the absence of squatter settlements in Ghanaian cities while Stewart (1996, 1997, 1999) examines the urban form and planning in Egypt as well as the dependent nature within the global economy of African urbanization.

The Geography of Disease and Health Care

Medical geography research on Africa during the past decade continued along the traditional lines of disease ecology and geography of health care with a clear trend toward linking health with its political and economic context.

Disease Ecology

John Hunter continued his work in disease ecology focusing on *filariasis* (elephantiasis) and *dracunculiasis* (guinea worm) in Ghana, West Africa (Hunter 1992, 1996, 1997*a*, *b*). Hunter (1997*a*) provides a breadth-taking description of the transmission and health effects of the guinea worm, using a historical, political and ecological perspective to highlight the role of both the colonial and post-colonial administrations in the process. On the effects of guinea worm control efforts in Ghana's Upper region Hunter (1997*b*) calls attention to the "walk-away syndrome" that commonly characterizes donor-host projects. After infrastructure is established, program advisers walk away with technical satisfaction while ignoring post-implementation human resource issues assuming that regional authorities possessed enough technical and financial capacity and were sufficiently prepared to maintain the program. Hunter's work on disease ecology benefits from his sustained interest and research, so that there is both a sense of long-term patterns and more recent changes and problems. Frustrated with the static nature of the disease ecology framework that dominated work on various diseases during the 1980s, some medical geographers are now calling for a major shift to more exciting ones such as political ecology (e.g. Mayer 1996).

HIV/AIDS in Africa

At the 1998 international AIDS conference in Geneva, researchers concluded unanimously that AIDS in sub-Saharan Africa was the most serious health problem in the world today. Surprisingly, Africa's alarmingly high and increasing rates of HIV-infection, AIDS mortality and number of AIDS orphans attracted few North American geographers. Initially, nongeographers such as Caldwell (1995), Rushton and Bogaert (1989), and Rushing (1995) constructed theories based on the eccentricities of African sexuality and in the case of Rushton and Bogaert outright racial determinism. Initial geography work focused on the spatial diffusion of the virus and its etiology (Shannon *et al.* 1991). Gould (1993) traces the origins of the AIDS epidemic to African monkeys and argues that initial denial by African leaders in several countries escalated its spread. The two books by Gould (1993) and Shannon *et al.* (1991), both of which are conjectural in orientation, indirectly influenced the representations of Africa as a "diseased continent" with little hard evidence about the spread of AIDS to the rest of the world. Geographers are contributing to critical reviews of these initial theories citing overgeneralization and unwillingness to look beyond simplistic approaches that focus on the peculiarities of individual sexual behavior rather than the social, economic, and political contingencies informing behavior.

Recent geography research on HIV-AIDS in Africa has usually chosen a political economy or structuralist theoretical perspective (Oppong 1998*a*; M. T. Bassett and Mhloyi 1991; Kalipeni 2000), and typically recommend education with empowerment. Good (1995) argues against intervention strategies that ignore poverty. Expecting people to abandon behaviors that bring pleasure and immediate gratification is unreasonable, especially when it brings them income or power, even while posing risks to personal and family health. Thiuri (1997) examining the threat and impact of AIDS on women in sub-Saharan Africa advocates empowerment programs to enable women to protect themselves from the disease. Oppong (1998*a*) focusing on Ghana's rare HIV-AIDS pattern (reversed Pattern Two, higher rural than urban rates, lower rates in regions of high polygamy) cautions against overgeneralization by arguing that HIV diffusion patterns probably reflect the spatial distribution and social networks of vulnerable social groups such as women compelled by economic difficulties into commercial sex work.

Within the social sciences, the few books available on AIDS in Africa are more general in their treatment (e.g. Barnett and Blaikie 1992), utilize spatial rather than social analysis (e.g. Shannon *et al.* 1991), or encompass broader geographic perspectives within critical analytical frameworks (e.g. Bond *et al.* 1997). A number of geographic studies have examined the impacts of AIDS on community and household well-being (e.g. Kesby 2000). Kesby's field data from Zimbabwe show that unequal gender relations and poor communication between men and women about sexual matters within the community and household setting play a critical role in the rapid transmission of HIV. Analysis of how to improve communication remains a crucial area for investigation. In theoretical terms the focus is changing from AIDS as a disease to social discourse and geographical analysis of African AIDS (Packard and Epstein 1991).

Structural Adjustment and Health

The health implications of structural adjustment policies in Africa was a subject of great interest (Turshen 1999; Stock 1995; Stock and Anyinam 1992; Logan 1995*b*). Stock (1995) shows that IMF-sponsored cutbacks in health-sector spending have produced large staff lay-offs and significant salary reductions, mere consulting clinics without drugs and sometimes without medical staff, and out-migration of doctors. Similarly, user fees and cost recovery programs have caused people to delay treatment or do without it, resulting in higher disease and death rates. Stock argues a solution that focuses on relieving debt repayment to enable African governments to deliver rudimentary health services. Turshen (1999) argues that recent outbreaks of new infectious diseases such as Ebola, Marburg, and AIDS demonstrate the need for the very same public health system the World Bank and IMF have systematically dismantled.

Nongeographical work has examined the health and nutrition impacts of SAPs on mothers and infants, following from Cornia and Helleiner's (1994) groundbreaking study for UNICEF. Costello (1994) argues that most of the child mortality in World Bank tables should be viewed with suspicion as such data are based on extrapolation rather than direct measurements. Directly measured estimates for Zambia, Malawi, Botswana, and Zimbabwe show an increasing trend in infant mortality starting from the 1980s through the 1990s. Other studies show a declining nutritional intake for mothers and children, which can be directly linked to structural adjustment programs (Streefland *et al.* 1995).

Geography and Health Care

Improving geographic accessibility to health services using location–allocation models was another area of research (Oppong 1996; Oppong and Hodgson 1994, 1998). Data limitations notwithstanding and contrary to critics such as Phillips (1990), these works argue that given the substantial improvements in location–allocation (LA) modeling, it is possible to achieve spatial accessibility with better locational choices and recommend such studies for prioritizing locations for siting health facilities (Oppong and Hodgson 1994, 1998). They also show how using an interaction-based LA model to locate health facilities might help to reduce the spread of HIV-AIDS in West Africa, and how better equipment and staffing of such facilities would attract more users and reduce bypassing and Intinerant Drug Vendors (Oppong and Hodgson 1998). Oppong and Hodgson also point out that lack of a spatial planning tradition, data limitations, bureaucracy, and inherent technical problems are among the major obstacles impeding acceptance of LA models in health-care planning in sub-Saharan Africa. These authors caution against reckless applications of location-allocation models. Indeed one can use the models and get "numbers" as output, but if the input data are of poor quality, the results cannot be trusted and indeed may create new problems if used as the basis for public policy decisions. A few studies call for a revival of African traditional medicine (Oppong 1998*b*; Stock 1995). Good (1991) examines how pioneer medical missions impacted the geography of health and social changes in colonial Africa and hindered the advancement of traditional medicine. Good develops a historical context and conceptual framework for investigating such topics and offers a detailed research agenda for future research on this issue.

Global Change and Earth Systems Science

Tracking developments in the field traditionally called physical geography, which researchers increasingly call earth systems science or global change, was difficult because much of the work on Africa is published, not in traditional geography journals, but in specialized interdisciplinary journals. What needs to be emphasized at the outset is that there is a tremendous amount of work on the African physical environment by North American geo-scientists and their counterparts in Africa and Britain conducting work in many countries and across several disciplines. The emerging idea of earth systems science as an integrative science has made the old-fashioned term "physical geography" obsolete, so that people who used to call themselves physical geographers are now able to cross traditional disciplinary boundaries with ease and are often working on multidisciplinary team projects. Key areas of research include global change (causes and impacts), hydrology of wetlands (mostly British research), and landscape ecology. Theoretical models include global climate models, earth systems, and remote sensing/GIS as tools for monitoring global change.

Geographers have played a key role in this largely interdisciplinary literature, particularly the causes and impacts of climate change in the Sahel, rainfall patterns, and El Niño effects in Africa (Nicholson 1999; Nicholson and Farrar 1994; Nicholson and Kim 1997; Nicholson *et al.* 1997; Lamb *et al.* 1998). Contrary to conventional wisdom, Nicholson and his co-researchers demonstrate that neither the Saharan boundary or vegetation cover in the Sahel has changed in sixteen years, and "productivity" assessed by water-use efficiency of the vegetation cover remains unchanged (Nicholson *et al.* 1998). Tarhule and Woo (1997, 1998) and Woo and Tarhule (1994), analyzing historical information on droughts, famines, and rainfall data from northern Nigeria, conclude that the most disruptive historical famines occurred when the cumulative rainfall deficit fell below 1.3 times the standard deviation of long-term mean annual rainfall. Other works that have examined climate change and its effects on drought, rainfall variability, and food production systems include Swearingen (1994), Campbell (1994), Fuller and Prince (1996), Hamly *et al.* (1998), and Ropelewski *et al.* (1993).

Sand transport, dune formation, desert landscapes, soil degradation, and river morphology studies have advanced rapidly particularly due to Nickling who has worked for almost twenty years in Mali and produced about a dozen papers on these topics (Nickling and Gillies 1993; Nickling and Wolfe 1994). Lancaster *et al.* (2000) use orbital radar images of the central Namib Desert to show the extent of relict fluvial deposits associated with former courses of the Tsondab and Kuiseb rivers. In many studies, Dyer and Torrance (1993) examined soil and food production potentials and suggest a wide range in food production potential throughout Ethiopia. European, particularly British, geographers and African geo-scientists have been active in research broadly defined as physical geography. Following Adams's pioneering work on wetlands and downstream impacts of dams (Adams 1990, 1993), many hydrologists and ecologists are studying wetland ecology/hydrology. The Cambridge group led by Grove and Adams is actively examining the multifaceted physical geography processes affecting the continent. In a recent book, Adams *et al.* (1996) detail the physical geography of Africa, its geomorphological and biogeographical aspects, and the impact of human agency on African environments at the local, regional, and continental scales.

Remote Sensing and Geographic Information Systems

North American geography is undergoing an unparalleled revolution in spatial data analysis. Personal computers, remote sensing, and global positioning systems have expanded the quantity and types of spatially referenced data about the human habitat and the physical environment. Besides, Geographic Information Systems (GIS) have eased the collection, management, and analysis of huge volumes of spatially referenced data. Thus, geographers can evaluate spatial processes and patterns and display the results and products of such analyses at the touch of a button.

GIS and remote sensing have grown as tools for studying environmental change in Africa. Nellis *et al.* (1997*a*) use Landsat thematic mapper (TM) digital data for the Southern District of Botswana to analyze variations in grazing intensity associated with rural land practices. Nellis *et al.* (1997*b*) use higher spatial resolution system, the spot hrv multispectral digital data, to analyze the urban morphology of Gaborone, the capital of Botswana. Such analysis could be used to compare the processes that structure cities in the developing and developed world. Anyamba and Eastman (1996) use the Normalized Difference Vegetation Index (NDVI) now being derived from satellite imagery to study climate trends within Southern Africa. Other studies that discuss, critique, or utilize GIS and remotely sensed images and their applicability to the African context include Fairhead and Leach (1996), T. J. Bassett and Zueli (2000), Kyem (1999*a*, *b*, 2001).

Conclusion

This review testifies to the liveliness of North American geographical research on Africa. T. J. Bassett's (1989) review of the 1980s contributions by North American geographers to Africanist research identified a number of insufficiently developed substantive areas including gaps in physical geography (geomorphology and biogeography), historical cartography, political and cultural geography, and regional geography. While these same topics continue to be under-represented as areas of research during the 1990s, T. J. Bassett's (1989) call to revitalize African regional and cultural geography seems to have been answered by the rise of several excellent and comprehensive textbooks (Stock 1995; Aryeetey-Attoh 1997; Newman 1995). Stock's and Aryeetey-Attoh's books provide a coherent thematic interpretation of the regional, cultural and development status of the continent. Newman follows the traditional regional geography approach in a powerful narrative of the many pre-colonial migrations and state formations throughout Africa. Another advance during the 1990s is the Rockefeller African Dissertation Awards program, coordinated through the University of California, Berkeley, with Michael Watts having a leading role. This program starts to address the difficulties many young African scholars face in getting solid training in methodologies appropriate for use in African settings, and funding for African dissertation research. This important step may answer the critical question of where the next generation of African and Africanist scholars will originate.

One important neglected area by human geographers is the political and electoral geography of the continent. Transitions such as democratization in quite a few countries and the collapse of organized authority in others (Liberia, Somalia, Democratic Republic of the Congo, Rwanda, etc.) have received precious little in the way of attention by geographers. With the exception of a few works such as those by Glassman and Samatar (1997), Samatar (1992), Cliffe and Bush (1994), this area is ripe for research given the rapidly changing global order. Another area of neglect is the swiftly expanding area of remote sensing and GIS. Swedish, French, and other European geographers have been very active in this area in the African context. One of the few active American groups in this regard is the Idrisi group at Clark University. Idrisi is a software package developed at Clark University in raster analytical functionality covering the full spectrum of GIS and remote sensing needs from database query, to spatial modeling, to image enhancement and classification. Although research on the physical geography processes of the continent by North American geographers continues to lag behind, a few individuals were very active in this area during the 1990s. The lag is in part a reflection of lack of specialized training in this area in American universities.

Perhaps due to overreaction against the neo-Malthusian debacle, demographic factors have been completely ignored in recent research. As Goldman (1994) notes in his critique of Little's book, the major weakness of the literature that uses political ecology as its guiding framework is its muteness on dynamics of population change and demographic factors. The human population has grown dramatically during the past thirty years throughout sub-Saharan Africa and this fact alone probably had major impacts on resource scarcities, out-migration, and waged employment, and made social conflicts more likely. In reading this body of work, however, it appears that most of the authors are constrained by the political economy and political ecology frameworks and ignore population in their analyses either as a strong reaction against the neo-Malthusian hypothesis or for fear of being branded neo-Malthusian. Yet, excluding demographic factors in these important debates may produce shortsighted policy formulations.

In conclusion, North American geographical research on Africa adopted newer methodological and theoretical perspectives during the 1990s. Several prominent geographers embraced and/or helped to advance post-colonial/ post-structuralist studies of representation and resistance and critiques of development discourse (e.g. Crush 1995; Pile and Keith 1997). Political ecology, indigenous knowledge, Boserupian perspectives on population, feminist analyses, globalization, and the dramatic shift in South African geography in the post-apartheid era emerged as important research themes. In the twenty-first century, we predict intensified use of these newer interdisciplinary approaches and the development of novel approaches and theories, while discarding old and static methodologies.

Acknowledgements

We are grateful to the University of Illinois Research Board for the funding that made this project possible. We would like to thank Stella Sambakunsi, Sosina Asfaw, and Leo Zulu for their untiring library research. We are also extremely indebted to the anonymous reviewers, particularly the one reviewer who offered very detailed constructive suggestions. Without the assistance of these individuals, it would have been extremely difficult to write such a comprehensive and inclusive chapter. However, the ideas expressed in this paper and any errors belong to the authors.

References

Adams, W. M. (1990). "Dam Construction and the Degradation of Floodplain Forest on the Turkwel River, Kenya." *Land Degradation and Rehabilitation*, 1: 189–98.

— (1993). "Indigenous Use of Wetlands and Sustainable Development in West Africa." *Geographical Journal*, 159/2: 209–18.

Adams, W. M., Goudie, A. S., and Orme, A. R. (1996). *Physical Geography of Africa*. London: Oxford University Press.

Adjibolosoo, S., and Ofori-Amoah, B. (eds.) (1998). *Addressing Misconceptions About Africa's Economic Development: Seeing Beyond the Veil*. Lewiston, NY: The Edwin Mellen Press.

Anyamba, A. J., and Eastman, R. (1996). "Interannual Variability of NDVI over Africa and its Relation to El Nino/Southern Oscillation." *International Journal of Remote Sensing*, 17/13: 2533–48.

Aryeetey-Attoh, S. (ed.) (1997). *Geography of Sub-Saharan Africa*. Upper Saddle River, NJ: Prentice Hall.

Atkinson, T. (1991). *Principles of Political Ecology*. London, Belhaven Press.

Awanyo, L. (1998). "Culture, Markets and Agricultural Production: A Comparative Study of the Investment Patterns of Migrant and Citizen Cocoa Farmers in the Western Region of Ghana." *Professional Geographer*, 50/4: 516–30.

Barkan, J. D., Mcnulty, M. L., and Ayeni, M. A. O. (1991). "Hometown Voluntary Associations, Local Development and the Emergence of Civil Society in Western Nigeria." *Journal of Modern African Studies*, 29/3: 457–80.

Barnett, T., and Blaikie, P. M. (1992). *AIDS in Africa: Its Present and Future Impact*. Chichester: Wiley.

Bascom, J. B. (1990*a*). "Border Pastoralism in Eastern Sudan." *Geographical Review*, 80/4: 416–30.

— (1990*b*). "Food, Wages, and Profits: Mechanized Schemes and the Sudanese State." *Economic Geography*, 66/2: 140–55.

Bassett, M. T., and Mhloyi, M. (1991). "Women and AIDS in Zimbabwe: The Making of an Epidemic." *International Journal of Health Services*, 21/1: 143–56.

Bassett, T. J. (1989). "Perspectives on Africa in the 1980s," in G. Gaile and C. Willmott (eds.), *Geography in America*. Columbus, Ohio: Merrill, 468–87.

— (1993). "Land Use Conflicts in Pastoral Development in Northern Côte d'Ivoire," in Thomas J. Bassett and Donald Crummey (eds.), *Land in African Agrarian Systems*. Madison, Wis.: University of Wisconsin Press, 131–54.

— (1994). "Cartography and Empire Building in Nineteenth-Century West Africa." *Geographical Review*, 84/3: 316–35.

— (1998). "Indigenous Mapmaking in Intertropical Africa," in D. Woodward and M. Lewis (eds.), *The History of Cartography, vol. ii. bk. 3: Cartography in the Traditonal African, American, Arctic, Australian, and Pacific Societies*. Chicago: University of Chicago Press, 24–48.

— (2001). *The Peasant Cotton Revolution in West Africa: Côte d'Ivoire, 1880–1995*. Cambridge: Cambridge University Press.

Bassett, T. J., and Crummey, D. (eds.) (1993). *Land in African Agrarian Systems*. Madison, Wis.: University of Wisconsin Press.

Bassett, T. J., and Zueli, K. B. (2000). "Environmental Discourses and the Ivorian Savanna." *Annals of the Association of American Geographers*, 90/1: 67–95.

Bencharifa, A., and Johnson, D. L. (1990) "Adaptation and Intensification in the Pastoral System of Morocco," in John G. Galaty and Douglas L. Johnson (eds.), *The World of Pastoralism: Herding Systems in Comparative Perspective*. New York: Guilford Press, 394–416.

Blaikie, P. (1989). "Environment and Access to Resources in Africa." *Africa*, 59/1: 18–40.

— (1994). *Political Ecology in the 1990s: An Evolving View of Nature and Society*. CASID Distinguished Speaker Series, 13. East Lansing: Center for Advanced Study of International Development, Michigan State University.

Blaikie, P., and Brookfield, H. (1987). *Land Degradation and Society*. London: Methuen.

Bond, G., Kreniske, J., Susser, I. and Vincent, J. (eds.) (1997). *AIDS in Africa and the Caribbean*. Boulder, Colo.: Westview Press.

Bryant, R. (1992). "Political Ecology: An Emerging Research Agenda in Third World Studies." *Political Geography Quarterly*, 11/1: 108–25.

Bush, R. (1996). "The Politics of Food and Starvation." *Review of African Political Economy*, 68: 169–95.

Caldwell, J. C. (1995). "Understanding the AIDS Epidemic and Reacting Sensibly to It (Editorial)." *Social Science & Medicine*, 41/3: 299–302.

Campbell, D. J. (1990*a*). "Community-Based Strategies for Coping with Food Scarcity: A Role in African Famine Early-Warning Systems." *GeoJournal*, 20/3: 231–41.

— (1990*b*). "Strategies for Coping with Severe Food Deficits in Rural Africa." *Food and Foodways*, 4/2: 143–62.

— (1991). "The Impact of Development upon Strategies for Coping with Drought among the Maasai of Kajiado District, Kenya," in *Pastoral Economies in Africa and Long-Term Responses to Drought*. Aberdeen: University of Aberdeen, 116–28.

— (1994). "The Dry Regions of Kenya," in Michael H. Glantz (ed.), *Drought Follows the Plow: Cultivating Marginal Areas*. Cambridge: Cambridge University Press, 77–90.

— (1999). "Response to Drought among Farmers and Herders in Southern Kajiado District, Kenya: A Comparison of 1972–1976 and 1994–1996." *Human Ecology*, 27/3: 377–416.

Campbell, D. J., Gichohi, H., Mwangi, A., and Chege, L. (2000). "Land Use Conflict in S. E. Kajiado District, Kenya." *Land Use Policy*, 17/4: 337–48.

Carmody, P. R. (1999). "Neoclassical Practice and the Collapse of Industry in Zimbabwe: The Cases of Textiles, Clothing and Footwear." *Economic Geography*, 74/4: 319–43.

— (2001). *Tearing the Social Fabric: Neoliberalism, Deindustrialization, and the Crisis of Governance in Zimbabwe*. Portsmouth, NH: Heinemann.

Carney, J. A. (1992). "Peasant Women and Economic Transformation in the Gambia." *Development and Change*, 2: 67–90.

— (1993*a*). "Converting the Wetlands, Engendering the Environment: The Intersection of Gender with Agrarian Change in The Gambia." *Economic Geography*, 69/4: 329–48.

— (1993*b*). "From Hands to Tutors: African Expertise in the South Carolina Rice Economy." *Agricultural History*, 67/3: 1–30.

Cliffe, L., and Bush, R. (1994). *The Transition to Independence in Namibia*. Boulder, Colo.: Lynne Rienner.

Cooper, B. M. (1997). "Gender, Movement, and History: Social and Spatial Transformations in 20th Century Maradi, Niger." *Environment and Planning. D, Society & Space*, 15/2: 195–21.

Corbridge, S. (ed.) (1995). *Development Studies: A Reader*. London: E. Arnold.

Cornia, G. A., and Helleiner, G. K. (1994). *From Adjustment to Development in Africa: Conflict, Controversy, Convergence, Consensus?* Basingstoke: Macmillan.

Costello, A. (1994) *Human Face or Human Façade? Adjustment and the Health of Mothers and Children*. London: Centre for International Development.

Crush, J. (ed.) (1995). *Power of Development*. London: Routledge.

— (ed.) (1998). *Beyond Control: Immigration and Human Rights in a Democratic South Africa*. Cape Town: Queen's University/Idasa.

— (1999). "Fortress South Africa and the Deconstruction of Apartheid's Migration Regime." *Geoforum*, 30/1: 1–11.

Crush, J., and James, W. (1997). *Crossing Boundaries: Mine Migrancy in a Democratic South*. Ottawa: IDRC.

Crush, J., and Williams, V. (eds.) (1999). *The New South Africans? The Immigration Amnesties and Their Aftermath*. Cape Town: Idasa/Queen's University.

Crush, J., Jeeves, A., and Yudelman, D. (1991). *South Africa's Labor Empire: A History of Black Migrancy to the Gold Mines*. Boulder, Colo.: Westview Press.

Drakakis-Smith, D. (1990). "Food for Thought or Thought About Food: Urban Food Distribution Systems in the Third World," in R. B. Potter and A. T. Salau (eds.), *Cities and Development*. London: Mansell, 100–20.

—— (1992). *Urban and Regional Change in Southern Africa*. London: Routledge.

Drakakis-Smith, D., Bowyer-Bower, T., and Tevera, D. (1995). "Urban Poverty and Urban Agriculture: An Overview of the Linkages in Harare." *Habitat International*, 19/2: 183.

Dyer, J. A., and Torrance, J. K. (1993). "Agroclimatic Profiles for Uniform Productivity Areas in Ethiopia." *Water International*, 18/4: 189–99.

Fairhead, J., and Leach, M. (1996). *Misreading the African Landscape: Society and Ecology in the Forest Savanna Mosaic*. New York: Cambridge University Press.

—— (1998). *Reframing Deforestation: Global Analyses and Local Realities with Studies in West Africa*. London: Routledge.

Foucault, M. (1970). *The Order of Things: An Archeology of the Human Sciences*. London: Tavistock.

—— (1977). *The Archeology of Knowledge*. London: Tavistock.

—— (1980). *Power/Knowledge*. Brighton: Harvester Press.

Fuller, D. O., and Prince, S. D. (1996). "Rainfall and Foliar Dynamics in Tropical Southern Africa: Potential Impacts of Global Climatic Change on Savanna Vegetation." *Climatic Change*, 33/1: 69–96.

Gaile, Gary L. (1992). "Improving Rural–Urban Linkages Through Small Town Market Based Development." *Third World Planning*, 14/2: 131–48.

—— (1994). "Thought for Food: Models and Policies for Food Security and Increased Agricultural Productivity," in Griffith and Newman (1994).

Gaile, G. L., and Ferguson, A. (1996). "Success in African Social Development: Some Positive Indications." *Third World Quarterly*, 17/3: 557–72.

Gaile, G. L., and Ngau, P. M. (1995). "Identifying the Underserved of Kenya: Populations without Access to Small Towns." *Regional Development Dialogue*, 16/2: 100.

Glassman, J., and Samatar, A. I. (1997). "Development Geography and the Third World State." *Progress in Human Geography*, 21/2:164–98.

Godlewska, A., and Smith, N. (eds.) (1994). *Geography and Empire*. Oxford: Blackwell.

Goldman, A. (1993). "Agricultural Innovation in Three Areas of Kenya: Neo-Boserupian Theories and Regional Characterization." *Economic Geography*, 69: 44–71.

—— (1994). "Review of the *Elusive Granary: Herder, Farmer, and State in Northern Kenya*." *Professional Geographer*, 46/1: 129–30.

Good, C. M. (1991). "Pioneer Medical Missions in Colonial Africa." *Social Science & Medicine*, 32/1: 1–11.

—— (1995). "Editorial: Incentives Can Lower the Incidence of HIV/AIDS in Africa." *Social Science & Medicine*, 40/4: 419.

Goodman, D., and Watts, M. (1997). *Globalising Food: Agrarian Questions and Global Restructuring*. London: Routledge.

Gould, P. (1993). *The Slow Plague: The Geography of the AIDS Pandemic*. Cambridge, Mass.: Blackwell.

Griffith, D. A., and Newman, J. L. (eds.) (1995). *Eliminating Hunger in Africa: Technical and Human Perspectives*. Syracuse: Syracuse University Press.

Hamly, M. E., Sebbari, R., Lamb, P. J., Ward, M. N., and Portis, D. H. (1998). "Variability of Rainfall Regimes ('Hydrological' Point of View)—Towards the Seasonal Prediction of Moroccan Precipitation and Its Implications for Water Resources Management." *IAHS Publication*, 252: 79–88.

Hart, G. (1998). "Multiple Trajectories: A Critique of Industrial Restructuring and the New Institutionalism." *Antipode*, 30/4: 333–56.

Heasley, L., and Delehanty, J. (1996). "The Politics of Manure: Resource Tenure and the Agropastoral Economy in Southwestern Niger." *Society and Natural Resources*, 9: 31–46.

Hunt, C. W. (1996). "Social Vs Biological: Theories of the Transmission of AIDS in Africa." *Social Science & Medicine*, 42/9: 1283–96.

Hunter, J. M. (1992). "Elephantiasis: A Disease of Development in North East Ghana." *Social Science & Medicine*, 35/5: 627–49.

—— (1996). "An Introduction to Guinea Worm on the Eve of its Departure: Dracunculiasis Transmission, Health Effects, Ecology and Control." *Social Science & Medicine*, 43/9: 1399–425.

—— (1997*a*). "Bore Holes and the Vanishing of Guinea Worm Disease in Ghana's Upper Region." *Social Science & Medicine*, 45/1: 71–89.

—— (1997*b*). "Geographical Patterns of Guinea Worm Infestation in Ghana: An Historical Contribution." *Social Science & Medicine*, 44/1: 103–22.

Hyndman, J. (1999). "A Post-Cold War Geography of Forced Migration in Kenya and Somalia." *Professional Geographer*, 51/1: 104–14.

Jarosz, L. (1993). "Defining and Explaining Tropical Deforestation: Shifting Cultivation and Population Growth in Colonial Madagascar (1896–1940)." *Economic Geography*, 69/4: 366–79.

—— (1996). "Defining Deforestation in Madagascar," in Richard Peet and Michael Watts (eds.), *Liberation Ecologies: Environment, Development, and Social Movements*. London: Routledge, 148–64.

Johnson, L. D. (1993). "Nomadism and Desertification in Africa and the Middle East." *GeoJournal*, 31/1: 51–66.

Johnston-Anumonwo, I. (1997). "Geography and Gender in Sub-Saharan Africa," in Aryeetey-Attoh (1997: 262–85).

Kalipeni, E. (1995). "The Fertility Transition in Africa." *Geographical Review*, 85/2: 287–301.

—— (1996). "Demographic Response to Environmental Pressure in Malawi." *Population and Environment*, 17/4: 285–308.

—— (2000). "Health and Disease in Southern Africa: A Comparative and Vulnerability Perspective." *Social Science and Medicine*, 50/7: 965–83.

Katz, C. (1991). "Sow What You Know: The Struggle for Social Reproduction in Rural Sudan." *Annals of the Association of American Geographers*, 81/3: 488–514.

Kesby, M. (2000). "Participatory Diagramming as a Means to Improve Communication About Sex in Rural Zimbabwe: A Pilot Study." *Social Science and Medicine*, 50/12: 1723–41.

Kodras, J. E. (1993). "Shifting Global Strategies of United States Foreign Food Aid, 1955–1990." *Political Geography*, 12/3: 232–46.

Konadu-Agyeman, K. (1991). "Reflections on the Absence of Squatter Settlements in West African Cities: The Case of Kumasi, Ghana." *Urban Studies*, 28/1: 139–51.

Kull, C. A. (1998). "Leimavo Revisited: Agrarian Land-Use Change in the Highlands of Madagascar." *Professional Geographer*, 50/2: 163–76.

Kyem, P. A. K. (1999*a*). "Examining the Discourse About the Transfer of GIS Technology to Traditionally Non-Western Societies." *Social Science Computer Review*, 17/1: 69–73.

—— (1999*b*). "Diagnosing User's Perception of Change in the Transfer and Adoption of Geographic Information System's Technology." *Applied Geographic Studies*, 3/2: 1–16.

—— (2001). "Embedding GIS Applications into Resource Management and Planning Activities of Local Communities: A Desirable Innovation or a Destabilizing Enterprise?" *Journal of Planning Education and Research*, 20/1: 176–86.

Lamb, P. J., Bell, M. A., Finch, J. D. (1998). "Variability of Rainfall Regimes ('Meteorological' Point of View)—Variability of Sahelian Disturbance Lines and Rainfall During 1951–1987." *IAHS Publication*, 252: 19–28.

Lancaster, N., Schaber, G. G., and Teller, J. T. (2000). "Orbital Radar Studies of Paleodrainages in the Central Namib Desert." *Remote Sensing of Environment*, 71/2: 216–25.

Leach, M., and Mearns, R. (1996). *The Lie of the Land: Challenging Received Wisdom on the African Environment*. Oxford: James Currey.

Lemon, A. (1995). *The Geography of Change in South Africa*. Chichester, NY: J. Wiley.

Levin, R. (1998). "Land Restitution, Ethnicity, and Territoriality: The Case of the Mmaboi Land Claim in South Africa's Northern Province," in E. Kalipeni and P. T. Zeleza (eds.), *Sacred Spaces and Public Quarrels: African Economic and Cultural Landscapes*. Lawrenceville, NJ: Africa World Press, 323–56.

Levin, R., and Weiner, D. (eds.) (1997). *"No More Tears . . ." Struggles for Land in Mpumalanga, South Africa*. Trenton, NJ: Africa World Press.

Lightfoot, D. R., and Miller, J. A. (1996). "Sijilmassa: The Rise and Fall of a Walled Oasis in Medieval Morocco." *Annals of the Association of American Geographers*, 86/1: 78–101.

Little, M. (1994). "Charity versus Justice: The New World Order and the Old Problem of World Hunger," in Griffith and Newman (1994: 23–38).

Little, P. (1992). *The Elusive Granary: Herder, Farmer, and State in Northern Kenya*. Cambridge: Cambridge University Press.

Little, P. D., and Watts, M. J. (eds.) (1994). *Living Under Contract: Contract Farming and Agrarian Transformation in Sub-Saharan Africa*. Madison, Wis.: University of Wisconsin Press.

Logan, B. I. (1995*a*). "Can Sub-Saharan Africa Successfully Privatize its Health Care Sector?" in Mengisteab and Logan (1995).

—— (1995*b*). "The Traditional System and Structural Transformation in Sub-Saharan Africa." *Growth and Change*, 26/4: 495–523.

Logan, B. I., and Mengisteab, K. (1993). "IMF-World Bank Adjustment and Structural Transformation in Sub-Saharan Africa." *Economic Geography*, 69: 1–24.

McDade, B. E. (1997). "Industry, Business Enterprises, and Entrepreneurship in the Development Process," in Aryeetey-Attoh (1997: 325–44).

McDade, B. E., and Malecki, E. J. (1997). "Entrepreneurial Networking: Industrial Estates in Ghana." *Tijdschrift voor Economische en Sociale Geografie (Journal of Economic and Social Geography)*, 88/3: 262–72.

McDonald, D. A. (1999). "Lest the Rhetoric Begin: Migration, Population and the Environment in Southern Africa." *Geoforum*, 30/1: 13–25.

—— (ed.) (2000). *On Borders: Perspectives on Migration in Southern Africa*. New York: St. Martin's Press.

Mackenzie, F. (1995). "Selective Silence: A Feminist Encounter with Environmental Discourse in Colonial Africa," in Jonathan Crush (1995: 100–14).

—— (1998). *Land, Ecology and Resistance in Kenya, 1880–1952*. Edinburgh: Edinburgh University Press.

Masilela, C., and Weiner, D. (1996). "Resettlement Planning in Zimbabwe and South Africa's Rural Land Reform Discourse." *Third World Planning Review*, 18/1: 23–43.

Mather, C. (1999). "Agro-commodity Chains, Market Power and Territory: Re-regulating South African Citrus Exports in the 1990s." *Geoforum*, 30/1: 61–70.

Matzke, G. E., and Nabane, N. (1996). "Outcomes of a Community Controlled Wildlife Utilization Program in a Zambezi Valley Community." *Human Ecology*, 24/1: 65–85.

Mayer, J. D. (1996). "The Political Ecology of Disease as One New Focus for Medical Geography." *Progress in Human Geography*, 20/4: 441–56.

Mehretu, A., and Sommers, L. M. (1990). "Towards Modeling National Preference Formation for Regional Development Policy: Lessons from Developed and Less Developed Countries." *Growth and Change: A Journal of Urban and Regional Policy*, 21/3: 31–47.

—— (1992). "Trade Patterns and Trends in the African-European Trading Area: Lessons for Sub-Saharan Africa from the Era of the Lome Accords 1975–1988." *Africa Development*, 17/2: 5–26.

Mengisteab, K., and Logan, B. I. (eds.) (1995). *Beyond Economic Liberalization in Africa: Structural Adjustment and the Alternatives*. London: Zed Books.

Moore, D. S. (1998). "Clear Waters and Muddied Histories: Environmental History and the Politics of Community in Zimbabwe's Eastern Highlands." *Journal of Southern African Studies*, 24/2: 377–403.

—— (1999). "The Crucible of Cultural Politics: Reworking "Development" in Zimbabwe's Eastern Highlands." *American Ethnologist*, 26/3: 654–89.

Mortimore, M., and Tiffen, M. (1994). "Population Growth and a Sustainable Environment: The Machakos Story." *Environment*, 36/8: 10.

Murton, J. (1999). "Population Growth and Poverty in Machakos District, Kenya." *Geographical Journal*, 165: 37–46.

Myers, G. A. (1996) "Naming and Placing the Other: Power and the Urban Landscape in Zanzibar." *Tijdschrift voor Economische en Sociale Geografie (Journal of Economic and Social Geography)*, 87/3: 237–46.

—— (1998). "Intellectual of Empire: Eric Dutton and Hegemony in British Africa." *Annals of the Association of American Geographers*, 88/1: 1–27.

Nast, H. J. (1994). "The Impact of British Imperialism on the Landscape of Female Slavery in the Kano Palace, Northern Nigeria." *Africa*, 64/1: 34–73.

— (1996). "Islam, Gender, and Slavery in West Africa Circa 1500: A Spatial Archaeology of the Kano Palace, Northern Nigeria." *Annals of the Association of American Geographers*, 86/1: 44–77.

Nellis, M. D., Bussing, C. E., Coleman, T. L., Nkambwe, M., and Ringrose, S. (1997*a*). "Spatial and Spectral Dimensions of Rural Lands and Grazing Systems in the Southern District of Botswana." *Geocarto International*, 12/1: 41–7.

Nellis, M. D., Bussing, C. E., Nkambwe, M., and Coleman, T. L. (1997b). "Urban Land Use and Morphology in a Developing Country Using SPOT HRV Data: Gaborone, Botswana." *Geocarto International*, 12/1: 91–5.

Neumann, R. P. (1995). "Ways of Seeing Africa: Colonial Recasting of African Society and Landscape in Serengeti National Park." *Ecumene*, 2: 149–69.

— (1998). *Imposing Wilderness: Struggles Over Livelihood and Nature Preservation in Africa*. Berkeley: University of California Press.

Newman, J. L. (1995). *The Peopling of Africa: A Geographical Interpretation*. New Haven, Conn.: Yale University Press.

Nicholson, S. E. (1999). "Historical and Modern Fluctuations of Lakes Tanganyika and Rukwa and their Relationship to Rainfall Variability." *Climatic Change*, 41/1: 53–71.

Nicholson, S. E., and Farrar, T. J. (1994). "The Influence of Soil Type on the Relationships Between NDVI, Rainfall, and Soil Moisture in Semiarid Botswana, I. NDVI Response to Rainfall." *Remote Sensing of Environment*, 50/2: 107–20.

Nicholson, S. E., and Kim, J. (1997). "The Relationship of the El Nino Southern Oscillation to African Rainfall." *International Journal of Climatology*, 17/2: 117–35.

Nicholson, S. E., Tucker, C. J., and Ba, M. B. (1998). "Desertification, Drought, and Surface Vegetation: An Example from the West African Sahel." *Bulletin of the American Meteorological Society*, 79/5: 815–29.

Nicholson, S. E., Marengo, J. A., Kim, J., Lare, A. R., and Galle, S., and Kerr, Y. H. (1997). "A Daily Resolution Evapoclimatonomy Model Applied to Surface Water Balance Calculations at the HAPEX-Sahel Supersites." *Journal of Hydrology*, 189/1–4: 946–64.

Nickling, W. G., and Gillies, J. A. 1993. "Dust Emission and Transport in Mali, West Africa." *Sedimentology*, 40/5: 859–68.

Nickling, W. G., and Wolfe, S. A. (1994). "The Morphology and Origin of Nabkhas, Region of Mopti, Mali, West Africa." *Journal of Arid Environments*, 28/1: 13–30.

Nyamweru, C. (1996). "Sacred Groves Threatened by Development." *Cultural Survival Quarterly*, Fall: 19–20.

— (1998). *Sacred Groves and Environmental Conservation*. Canton, NY: Frank P. Piskor Lecture, St. Lawrence University.

Ofori-Amoah, B. (1996). "Human Factor Development and Technological Innovation in Africa," in S. B.-S. K. Adjibolosoo (ed.), *Human Factor Engineering and the Political Economy of African Development*. Westport, Ctonn.: Praeger, 111–24.

— (1998). "The Human Factor Perspective and Development Education," in V. G. Chivaura and C. J. Mararike (eds.), *The Human Factor Approach to Development in Africa*. Harare: University of Zimbabwe Press, 34–43.

Oppong, J. R. (1996). "Accommodating the Rainy Season in Third World Location-Allocation Applications." *Socio-Economic Planning Sciences*, 30/2: 121–37.

— (1998*a*). "African Traditional Medicine: Dangerous, Anachronistics or Panacea," in S. B.-S. K. Adjibolosoo and B. Ofori-Amoah (eds.), *Addressing Misconceptions About Africa's Economic Development: Seeing Beyond the Veil*. Lewiston, NY: The Edwin Mellen Press, 97–111.

— (1998*b*) "A Vulnerability Interpretation of the Geography of HIV/AIDS in Ghana, 1986–95." *Professional Geographer*, 50/4: 437–48.

Oppong, J. R., and Hodgson, M. J. (1994). "Spatial Accessibility to Health Care Facilities in Suhum District, Ghana." *The Professional Geographer*, 46/2: 199–209.

— (1998). "An Interaction-Based Location-Allocation Model for Health Facilities to Limit the Spread of HIV-AIDS in West Africa." *Applied Geographic Studies*, 2/1: 29–41.

Osei, W. Y. (1998) "The Gender Factor in Rural Energy Systems in Africa: Some Evidence from Ghana," in Adjibolosoo and Ofori-Amoah (1998: 84–96).

Ould-Mey, M. (1996). *Global Restructuring and Peripheral States: The Carrot and the Stick in Mauritania*. London: Littlefield Adams Books.

Owusu, J. H. (1998). "Current Convenience, Desperate Deforestation: Ghana's Adjustment Program and the Forestry Sector." *Professional Geographer*, 50/4: 418–36.

— (2001). "Urban Impoverishment and Multiple Modes of Livelihood in Ghana." *The Canadian Geographer*, 45/3: 401–27.

Packard, R. M., and Epstein, P. (1991). "Epidemiologists, Social Scientists, and the Structure of Medical Research on AIDS in Africa." *Social Science and Medicine*, 33/7: 771–94.

Phillips, D. R. (1990). *Health and Health Care in the Third World*. New York: John Wiley.

Pickles, J. (1991). "Industrial Restructuring, Peripheral Industrialization, and Rural Development in South Africa." *Antipode*, 23/1: 67–91.

Pickles, J., and Weiner, D. (1991). "Rural and Regional Restructuring of Apartheid: Ideology, Development Policy and the Competition for Space." *Antipode*, 23/1: 2–32.

Pile, M., and Keith, M. (eds.) (1997). *Geographies of Resistance*. London: Routledge.

Ralston, B., Ray, J., and Harrison, G. (1994) "Issues in the Transportation of Food Aid," in Griffith and Newman (1994: 57–79).

Riddell, J. B. (1992). "Things Fall Apart Again: Structural Adjustment Programmes in Sub-Saharan Africa." *Journal of Modern African Studies*, 30/1: 53–68.

Rocheleau, D. E., Thomas-Slayter, B. P., and Wangari, E. (eds.) (1996). *Feminist Political Ecology: Global Issues and Local Experience*. London: Routledge.

Roder, W. (1994). *Human Adjustment to Kainji Reservoir in Nigeria: An Assessment of the Economic and Environment Consequences of a Major Man-made Lake in Africa*. Lanham, Md.: University Press of America.

Rogerson, C. M. (1996). "Urban Poverty and the Informal Economy in South Africa's Economic Heartland." *Environment and Urbanization*, 8/1: 167–81.

— (ed.) (1999). *South Africa: Different Geographies. Geoforum*, Special Issue, 30/1. London: Pergamon Press.

Rogerson, C. M., and McCarthy, J. J. (eds.) (1992). *Geography in a Changing South Africa: Progress and Prospects*. Cape Town: Oxford University Press.

Ropelewski, C. F., Lamb, P. J., Portis, D. H. (1993). "The Global Climate for June to August 1990: Drought Returns to Sub-Saharan West Africa and Warm Southern Oscillation Episode Conditions Develop in the Central Pacific." *Journal of Climate*, 6/11: 2188–212.

Rothenberg, T. Y. (1994). "Voyeurs of Imperialism: *The National Geographic Magazine* Before World War II," in Godlewska and Smith (1994: 155–72).

Rushing, W. A. (1995). "The Cross Cultural Perspective on AIDS in Africa," in W. A. Rushing, *The AIDS Epidemic: Social Dimensions of an Infectious Disease*. Boulder, Colo.: Westview Press, 59–90.

Rushton, J. P., and Bogaert, A. F. (1989). "Population Differences in Susceptibility to AIDS: An Evolutionary Analysis." *Social Science and Medicine*, 28/12: 1211–20.

Saff, G. (1995). "Residential Segregation in Postapartheid South Africa: What Can Be Learned from the United States Experience." *Urban Affairs Review*, 30/6: 782–808.

—— (1998*a*). *Changing Cape Town: Urban Dynamics, Policy, and Planning During the Political Transition in South Africa*. Lanham, Md.: University Press of America.

—— (1998*b*). "The Effects of Informal Settlement on Suburban Property Values in Cape Town, South Africa." *Professional Geographer*, 50/4: 449–64.

—— (2001). "Exclusionary Discourse Towards Squatters in Suburban Cape Town." *Ecumene*, 8/1: 87–107.

Said, E. (1978). *Orientalism*. London: Routledge.

Samatar, A. I. (1992). "Destruction of State and Society in Somalia: Beyond the Tribal Convention." *The Journal of Modern African Studies*, 30/4: 625–41.

—— (1993). "Structural Adjustment as Development Strategy? Bananas, Boom, and Poverty in Somalia." *Economic Geography*, 69/1: 25–43.

Sanders, R. (1992) "Eurocentric Bias in the Study of African Urbanization: A Provocation to Debate." *Antipode*, 24/3: 203–13.

Schroeder, R. A. (1993). "Shady Practice: Gender and the Political Ecology of Resource Stabilization in Gambian Garden/Orchards." *Economic Geography*, 69/4: 349–65.

—— (1995). "Contradictions Along the Commodity Road to Environmental Stabilization: Foresting Gambian Gardens." *Antipode*, 27/4: 325–42.

—— (1997). "'Re-claiming' Land in The Gambia: Gendered Property Rights and Environmental Intervention." *Annals of the Association of American Geographers*, 87/3: 487–508.

Scott, J. C. (1977). *The Moral Economy of the Peasant: Rebellion and Subsistence in Southeast Asia*. New Haven: Yale University Press.

—— (1987). *Weapons of the Weak: Everyday Forms of Peasant Resistance*. New Haven: Yale University Press.

Sen, A. (1981). *Poverty and Famine*. Oxford: Clarendon Press.

—— (1990). "Food, Economics, and Entitlements," in J. Dreze and A. Sen (eds.), *The Political Economy of Hunger*. Oxford: Clarendon Press, 35–50.

Shannon, G. W., Pyle, G. F., and Bashshur, R. (1991). *The Geography of AIDS: Origins and Course of an Epidemic*. New York: Guilford Press.

Smith, J., Barau, A. D., Goldman, A., and Mareck, J. H. (1994). "The Role of Technology in Agricultural Intensification: The Evolution of Maize Production in the Northern Guinea Savanna of Nigeria." *Economic Development and Cultural Change*, 42/3: 537–54.

Spring, A., and McDade, B. (1998). *African Entrepreneurship: Theory and Reality*. Gainesville, Fla.: University Press of Florida.

Stanley, W. R. (1990). "Socioeconomic Impact of Oil in Nigeria." *GeoJournal*, 22/1: 67–79.

Stewart, D. J. (1996). "Cities in the Desert: The Egyptian New-Town Program." *Annals of the Association of American Geographers*, 86/3: 459–81.

—— (1997). "African Urbanization: Dependent Linkages in a Global Economy." *Tijdschrift voor Economische en Sociale Geografie (Journal of Economic and Social Geography)*, 88/3: 251–62.

—— (1999). "Changing Cairo: The Political Economy of Urban Form." *International Journal of Urban and Regional Research*, 23/1: 128–47.

Stock, R. (1995). *Africa South of the Sahara: A Geographic Interpretation*. New York: Guilford Press.

Stock, R., and Anyinam, C. (1992). "National Governments and Health Policy in Africa," in T. Falola and D. Ityavar (eds.), *The Political Economy of Health in Africa*. Athens, Ohio: Ohio University Center for International Studies.

Streefland, P., Harnmeijer, J. W., and Chabot, J. (1995). "Implications of Economic and Structural Adjustment Policies on PHC in the Periphery," in J. Chabot, J. W. Harnmeijer, and P. H. Streefland (eds.), *African Primary Health Care in Times of Economic Turbulence*. Amsterdam: Royal Tropical Institute, 11–17.

Swearingen, W. (1994). "Northwest Africa," in Michael H. Glantz (ed.), *Drought Follows the Plow: Cultivating Marginal Areas*. Cambridge: Cambridge University Press, 103–16.

Tarhule, A., and Woo, M. (1997). "Towards an Interpretation of Historical Droughts in Northern Nigeria." *Climatic Change*, 37/4: 601–16.

—— (1998). "Changes in Rainfall Characteristics in Northern Nigeria." *International Journal of Climatology*, 18/11: 1261–71.

Taylor, D. R. F., and Mackenzie, F. (eds.) (1992). *Development from Within: Survival in Rural Africa*. London: Routledge.

Thiuri, P. (1997). "The Threats of AIDS to Women and Development in Sub-Saharan Africa." in E. Kalipeni, and P. Thiuri (eds.), *Issues and Perspectives on Health Care in Contemporary Sub-Saharan Africa*. Lewiston, NY: Edwin Mellen Press, 73–114.

Tiffen, M., and Mortimore, M. (1994). "Malthus Controverted: The Role of Capital and Technology in Growth and Environment Recovery in Kenya." *World Development*, 22/7: 997–1010.

Tiffen, M., Mortimore, M., and Gichuki, F. (1994). *More People, Less Erosion: Environmental Recovery in Kenya*. Chichester, NY: J. Wiley.

Turner, B. L., II, Hyden, G., and Kates, R. W. (eds.) (1993). *Population Growth and Agricultural Change in Africa*. Gainesville, Fla.: University of Florida Press.

Turner, M. (1993). "Overstocking the Range: A Critical Analysis of the Environmental Science of Sahelian Pastoralism." *Economic Geography*, 69/4: 402–21.

Turshen, M. (1999). *Privatizing Health Services in Africa*. New Brunswick, NJ: Rutgers University Press.

Watts, M. J. (1993). "The Geography of Post-Colonial Africa: Space, Place and Development in Sub-Saharan Africa." *Singapore Journal of Tropical Geography*, 14/2: 173–90.

— (1994). "Life Under Contract: Contract Farming, Agrarian Restructuring, and Flexible Accumulation," in Little and Watts (1994: 21–77).

— (1996). "Nature as Artifice and Artifact," in Richard Peet and Michael Watts (eds.), *Liberation Ecologies: Environment, Development, and Social Movements*. London: Routledge, 243–68.

— (1997). "State Violence, Local Resistance and the National Question in Nigeria," in Pile and Keith (1997).

Weiner, D., and Levin, R. (1991). "Land and Agrarian Transition in South Africa." *Antipode*, 23/1: 92–120.

Woo, M., and Tarhule, A. (1994). "Streamflow Droughts of Northern Nigerian Rivers." *Hydrological Sciences Journal*, 39/1: 19–34.

Wood, W. B. (1994). "Forced Migration: Local Conflicts and International Dilemmas." *Annals of the Association of American Geographers*, 84/4: 607–34.

Wubneh, M. (1994). "Discontinuous Development, Ethnic Collective Movements, and Regional Planning in Ethiopia." *Regional Development Dialogue*, 15/1: 120–44.

Yeboah, Ian E. A. (1998). "Geography of Gender Economic Status in Urban Sub-Saharan Africa: Ghana, 1960–1984." *The Canadian Geographer*, 42/2: 158–73.

Zinyama, L. (1992). "Local Farmer Organizations and Rural Development in Zimbabwe," in Taylor and Mackenzie (1992).

Zinyama, L. M., Matiza, T., and Campbell, D. (1990). "The Use of Wild Foods During Periods of Food Shortage in Rural Zimbabwe." *Ecology of Food and Nutrition*, 24: 251–65.

CHAPTER 37

American Ethnic Geography

Lawrence E. Estaville, Susan W. Hardwick, James P. Allen, and Ines M. Miyares

Introduction

Because the American Ethnic Geography Specialty Group was established in 1992 and was, therefore, not a part of the original *Geography in America* anthology in 1989, we think it is beneficial to present briefly the development and context of American ethnic geography into which we can place more current work. In 2000 the American Ethnic Geography Specialty Group changed its name to the Ethnic Geography Specialty Group; but because almost the whole of this report deals with the decade of the 1990s, we use the specialty group's original name throughout.

American Ethnic Geography

American ethnic geography encompasses the geographic dimensions and experiences of ethnic groups in the United States and Canada. Its roots are in cultural-historical and population geography. As such, American ethnic geography reflects the epistemologies and methodologies of human geography. Like geographers in general, most American ethnic geographers are empirical and inductive in their research.

Because ethnicity is a complex concept, scholars who research ethnicity have been troubled over the years by definitional conundrums. Although in his 1974 study Isajiw determined that most ethnic researchers never explicitly define the meaning of ethnicity, he examined twenty-seven characteristics of ethnicity to construct a definition of North American ethnicity as "an involuntary group of people who share the same culture or . . . descendants of such people who identify themselves and/or are identified by others as belonging to the same involuntary group" (ibid. 122). To Isajiw, then, a person is either born into an ethnic group and is therefore socialized as Anglo, Chinese, French, Polish, etc., or can decide at some point in her/his life which ethnic identity fits best, or other people can perceive a person's ethnicity. As underscored in the *Harvard Encyclopedia of American Ethnic Groups* (1980), these latter internal/external modes of ethnic identification have become increasingly more significant in North America. Paradoxically, in today's multiethnic American society, many ethnic groups are celebrating their heritages with renewed vigor, while, simultaneously, many people are less bound by past ethnic loyalties and have either used innovative terms of self-identification to describe their multiethnicity or simply refused to be categorized ethnically.

The growth in ethnic geography during the past three decades has paralleled the growth in other disciplines interested in ethnic studies in America. Yet, as with sociology and anthropology, a paucity of operational definitions of ethnicity characterize geography, particularly those based on subjectively chosen ethnic identities by individuals or external identification by others. Geographers and other scholars have recognized that not all traditional ethnic groups have "melted" into the larger American society and that many recent migrants hold tenaciously to their cultures. Dramatic twentieth-century shifts in the national demographics of the United States and Canada will result in an American population with no majority ethnic group in this century. Geographers who study the dynamics of America's ethnic diversity emphasize such geographical expressions as migration, settlement, landscape impress, place development, homelands, acculturation, assimilation, and territorial tensions.

The Roots of American Ethnic Geography

The early tradition of geographers investigating landscapes replete with cultural artefacts is still vigorous today as is underscored in the relatively recent literature on house types, barns, fences, and other landscape markers. However, in the 1960s geographers began to use many more government data sources in their ethnic studies, thereby producing a complementary avenue of investigation that would parallel the landscape tradition.

Catalytic to the increased usage of census and survey data in the 1960s were the quantitative revolution in geography as a whole, the increasingly critical use of computers, and, more specifically to ethnic studies, the 1964 federal Civil Rights Act and the 1965 Immigration Act, two landmark events that irreversibly forced Americans to look at themselves from unfamiliar perspectives and to form unprecedented relationships. Many geographers reacted to the government's civil rights mandates by incorporating large amounts of census data and other government survey information to investigate, from both contemporary and historical viewpoints, these new ethnic permutations and connections.

Over the years, geographers have used surnames from the US manuscript censuses and other sources to delimit spatially American ethnic groups and then to study their geographical experiences (e.g. Meigs 1941; Jordan 1966, 1969; Lemon 1966; Jakle and Wheeler 1969; Meinig 1969, 1971; Nostrand 1970, 1975, 1980; Villeneuve 1972; Estaville 1986, 1988, 1993; Haverluk 1998; Sheskin 1998*a*, *b*). The 1970 and 1980 US censuses were, however, historic turning points for ethnic groups and ethnic geography research. The 1970 census provided mother tongue data at census tract resolution, and the 1980 census specifically allowed all respondents to express their ethnic identities. Several population geographers (e.g. Roseman *et al.* 1981; Boswell 1984; Boswell and Curtis 1984, 1991; Allen and Turner 1988; and McHugh 1989) began to use these data to study and map ethnic distributions across the United States or in selected ethnic places of America, especially those in urban areas. And these places became even more diverse as immigrants from Asia, Latin America (particularly Mexico), the Middle East, Russia, and Africa added new dimensions to the traditional American ethnic landscape.

Meanwhile, as they studied ethnic groups, geographers began to experiment with the usage of personal ethnic identification as well as external societal perceptions of ethnic group identities. Nostrand (1973), for instance, framed a self-referent survey in an article titled "'Mexican American' and 'Chicano': Emerging Terms for a People Coming of Age," and, more recently, Hardwick's humanistic method explored the settlement and acculturation of new Russian immigrants in California in her book *Russian Refuge* (1993). Adopting a subjective mode has nevertheless presented serious obstacles when searching for past ethnic experiences. Yet it can be done as Lemon (1972) demonstrated in his historical analysis winnowed from contemporary accounts regarding the perceptions of European ethnic groups in eighteenth-century southeastern Pennsylvania, or as Estaville (1990) has shown by drawing on contemporary insights to explore the decline in French language usage by the Louisiana French in the nineteenth century.

American ethnic geography includes the investigation of important historical and contemporary sequences, streams, and tensions of migration, settlement, acculturation, and assimilation. During the late twentieth century, American ethnic groups were increasingly enjoying another link in such sequences—revitalization, much of which centers on new-found pride in their ethnic heritages. A salient question, however, that underlies some of this resurgence in ethnic identity of traditional American ethnic groups is how much of this enthusiasm is from the heart and how much is commercial venture—the selling of ethnic "stuff." Whatever the case may be, teaching our children about the long history of our continent's ethnic diversity is indeed another compelling goal of ethnic geographers.

Today, then, American ethnic geography has become a vital yet eclectic blend of cultural, historical, and population geographies that pursues the understanding of spatial relationships of ethnic groups in the United States and Canada and of those ethnic places that people built long ago or are now energetically creating.

The American Ethnic Geography Specialty Group

In 1991, after he noticed several papers about the ethnic geographies of the United States and Canada being given in general sessions without any specialty group affiliation at the AAG meeting in Miami, Lawrence E. Estaville organized the American Ethnic Geography Specialty Group (AEGSG); it had its initial meeting in San Diego the following year. Who were the charter members of this fledgling specialty group? Interestingly, signing the AAG petition to establish the specialty group or attending the initial meeting were several prominent cultural and historical geographers, for example, Charles S. Aiken, James M. Goodman, John Fraser Hart, Terry G. Jordan, Allen G. Noble, Richard L. Nostrand, David J. Wishart, and Wilbur Zelinksy. Geographers from other specialty groups were likewise charter members of the AEGSG, for instance: James P. Allen, Thomas D. Boswell, Ines M. Miyares, Curtis C. Roseman, Ira Sheskin, and Nancy K. Torrieri had strong affiliations with the Population Geography Specialty Group, and J. Douglas Heffington, Stephen C. Jett, Robert A. Rundstrom, and George A. Van Otten were some of the leaders of the American Indian Specialty Group (an older specialty group than the AEGSG, although a subset of the more ethnically encompassing AEGSG; we defer to this specialty group to present in this anthology information about works on the Native Americans).

The AEGSG selected Estaville (1992–4) as its first chair. Subsequent chairs have been James P. Allen (1994–6), Ines M. Miyares (1996–8), J. Douglas Heffington (1998–2000), Carlos Teixeira (2000–2), and Ira M. Sheskin (2002–present). The AEGSG membership grew from 87 in 1992 to 112 in 1998. The specialty group's newsletter, *The American Ethnic Geographer*, expanded its coverage over the years and now includes short articles and book reviews. In 1994 the specialty group honored Terry G. Jordan as its first Distinguished Scholar for his outstanding contributions to the understanding of American ethnic geography. Subsequent Distinguished Scholars have been Wilbur Zelinsky (1995), R. Cole Harris (1996), Allen G. Noble (1997), James P. Allen (1998), Michael P. Conzen (1999), Richard L. Nostrand (2000), Susan W. Hardwick (2001), and Thomas D. Boswell (2002).

Themes, Methods, and Connections

Migration and Settlement

Classic migration theories have influenced geographic studies of the migration paths and settlement patterns of various ethnic groups in North America. Ethnic geographers (e.g. Jordan 1966; Meinig 1971; Groves and Muller 1975; Ward 1971; Ostergren 1979, 1981*b*; Constantinou and Diamantides 1985; LeBlanc 1985, 1988; McKee 1985; Anderson 1987; Allen and Turner 1988, 1997; Conzen 1990; Nostrand 1992; Estaville 1993; Hardwick 1993; Hardwick and Holtgrieve 1996; Hiebert 1993; Haverluk 1997) have added well-developed spatial dimensions to these basic theories.

Among the earliest publications that laid a foundation for a geographical analysis of these fundamental themes was Jordan's *German Seed in Texas Soil* (1966), a study that was influenced by Walter Kollmorgen's work. Jordan's seminal book discussed the push-pull factors associated with German migration to Texas and provided information about conditions in their European homeland to analyze German settlement patterns in Texas. *German Seed in Texas Soil* quickly became a classic among geographers interested in analyzing historic ethnic migration and settlement patterns on the land.

Two other important works underscoring these themes include Meinig's *Southwest: Three Peoples in Geographical Change, 1600–1970* (1971) and Ward's *Cities and Immigrants: A Geography of Change in Nineteenth-Century America* (1971). Meinig focused his attention on one distinctive North American region, and Ward examined comparative settlement patterns for the entire nation. While these companion volumes concentrate on migration, economic change, and settlement at different scales and different time periods, both illustrate methods still used by geographers to investigate the migration and settlement of various ethnic groups —careful synthesizing of relevant archival material, analysis of historical census data, and detailing settlement patterns of particular ethnic groups to document changes through space and time.

Geographical analysis of African-American migration paths within the United States have been of particular interest to some ethnic geographers. Several publications (e.g. Deskins 1969, 1981; Rose 1970; Groves and Muller 1975; Aiken 1985, 1987; McHugh 1987, 1989; Cromartie and Stack 1989; Johnson and Roseman 1990; Roseman and Lee 1998) have contributed new knowledge on the migration and settlement of African Americans and have provided important data for policy-makers interested in understanding more about the complexities of the African-American geographical experience.

Other ethnic geographers have used data on migration and settlement as background for analyzing diverse topics of investigation. Anderson (1987), for example, has been concerned with uncovering the dynamics between race, racial discourse, power, and institutional practice by examining the historical settlement patterns of the large numbers of Chinese in Vancouver. In another Canadian study, Hiebert's (1993) analysis of the Jewish ethnic economy focused on the garment industry and settlement patterns within Toronto in the early twentieth century. And Sheskin (1993, 1998, 1999) has painted a broader interpretation of Jewish urban settlement in America.

Marston (1988) examined the migration and settlement of Irish immigrants in Lowell, Massachusetts as a case study about politics and ethnic identity in this industrial city, while Ostergren (1979, 1981*b*) documented the migration and settlement of Swedes in the Upper Middle West to ascertain more about cultural transfer and community building by immigrants on both sides of the Atlantic. More recently, Godfrey (1988) investigated immigrant settlement patterns in San Francisco to understand better neighborhood evolution, change, and cultural identity in this multicultural city. Jordan (1989) introduced the cultural ecological concept of preadaptation as a vehicle to explain the rapid seventeenth-century European settlement of the eastern woodlands of North America. In the same year, Constantinou (1989) interpreted the intergenerational differences in three dominant themes—language, sociocultural activities, and politics—of Greek-American ethnicity in northeastern Ohio.

Because distributions are so thoroughly geographical, it is not surprising that geographers studying aspects of ethnicity have given much consideration to the understanding of spatial patterns. However, it was not until the late 1980s that geographers could create computer-generated maps of detailed ethnic distributions. Such technological developments allowed the production of Allen and Turner's reference volume, *We the People: An Atlas of America's Ethnic Diversity* (1988). This atlas displays in full color the distributions of sixty-seven ethnic groups at the county level. More recently, Zaniewski and Rosen (1998) published a computer-generated atlas of ethnic diversity in Wisconsin.

Ethnic Landscapes

Ethnic geographers have long been interested in classifying, mapping, analyzing, and, more recently, preserving ethnic landscapes, material imprints resulting from settlement patterns—observable features that identify, define, and delimit ethnic communities. These efforts have grown directly out of Sauer's landscape tradition in his "Morphology of Landscape" (1925), Kniffen's (1965) and Glassie's (1975) work mapping and analyzing house types in the American South, Hart's *The Look of the Land* (1972), and Jackson's attention to vernacular landscapes celebrated in his highly regarded *Landscape* magazine and other publications (1970, 1984).

A representative list of features studied by ethnic geographers includes rural house types (Zelinsky 1954; Jordan 1964, 1985; Glassie 1971, 1975; Noble 1984; Jordan and Kaups 1989); cemeteries (Francaviglia 1971; Jordan 1982); field and village patterns (Francaviglia 1978; Alanen and Tishler 1980; Wood 1986; Smith 1999); barns (Francaviglia 1972; Noble 1984; Comeaux 1989; Gritzner 1990); gardens (Alanen 1990; Airriess and Clawson 1991; Airriess 1994; Helzer 1994; Miyares 1998*a*); and urban features such as yard shrines, house types, fences, religious sites, murals, and commercial establishments (Ward 1971; Lewis 1975; Conzen 1980; Curtis 1980; Arreola 1981, 1984, 1988; Ostergren 1981*a*; Anderson 1987; Alanen 1989; Hardwick 1993).

Ethnic landscapes often contain artefacts that provide evidence of ways that unique ethnic traditions have been incorporated into diverse local environments. These processes and patterns have quite often led to multiple ethnic imprints in one place resulting from different settlement histories at varying time periods. The history of ethnic settlement may also be observed on the landscape through isolated "ethnic islands," residual ethnic areas set apart from the larger and more dominant ethnic milieu around them. In 1985 Noble summarized the work on ethnic islands that had been published up to the mid-1980s. More recently, Conzen (1993) elaborated on the theme of ethnic bounded spaces and argued that ethnic islands from the same source country could be interconnected to reveal an "ethnic archipelago." This framework has been applied by Miyares (1996, 1998*a*) in studying the Hmong community, by Hardwick (1993) in her analysis of Russian religious landscapes of the North American Pacific Rim, and by McHugh *et al.* (1997) in understanding the national geography of Cubans relative to Miami.

The search for a deeper understanding of sense of place and place-making lies behind many of the more recent studies of landscape signatures by ethnic geographers. For instance, Jordan and Kaups's book, *The American Backwoods Frontier: An Ethnic and Ecological Interpretation* (1989), provides a cogent example of the search for origins and modifications of ethnic features in the landscape. Building on this emphasis on place-making and ethnic landscape features, Wood (1997) discussed the new ways Vietnamese Americans have created a distinctive and blended sense of place for themselves in the suburbs of Virginia.

Ethnic geographers have given increasingly more attention to the historic preservation of ethnic features. From the work on historical preservation in urban areas (e.g. Lewis 1975; Datel and Dingemans 1980), geographers have begun to document landscape change with an eye toward preservation of these features to maintain the integrity and visibility of the ethnic community. Ethnic markers in towns and cities such as Fredericksburg, Texas (German), Solvang, California (Danish), and San Francisco (Chinese) highlight the significance of this type of research *vis-à-vis* a growing reawakening of ethnic identities and their relationships to ethnic landscapes in small towns and urban neighborhoods. Indeed, such research has become quite useful to communities across North America that now wish to enhance their ethnic identity as it relates to economic development. Hoelscher's (1998*a*, *b*) analysis of ethnic tourism and ethnic memory elaborates on these themes.

The most comprehensive collection of papers detailing various expressions of American ethnic landscapes is Noble's book, *To Build in a New Land: Ethnic Landscapes in North America* (1992). In this volume, Noble and nineteen other contributing authors trace the development of distinctive buildings and other ethnic landscape features of many nineteenth-century immigrant groups who settled in the United States and Canada. Through maps, drawings, floor plans, photographs, and text, the book provides a tool for interpreting and comparing vernacular landscapes that serve as the underlying geography of much of rural America.

Homelands

The concept of a homeland is certainly an ancient one. The word "home" evokes extraordinarily strong emotions. The homes of groups of people are homelands, places that are inherently geographical. Until very recently, however, geographers did not explore the meaning and power of the homeland concept.

Both Carlson's book, *The Spanish-American Homeland: Four Centuries in New Mexico's Río Arriba* (1990), and Nostrand's article, "The Hispano Homeland in 1900" (1980), used the term "homeland" only to summon a people's profound sense of place without defining what a homeland means. In his award-winning book, *The Hispano Homeland* (1992), however, Nostrand began to formalize his homeland concept and proposed three basic elements for a homeland: a people, a place, and identity with place. In 1993, in collaboration with Estaville as contributing coeditors of a special issue of the *Journal of Cultural Geography* on American homelands, Nostrand expanded the basic components of a homeland to include a people, a place, a sense of place, control of place, and time. Nostrand and Estaville wrote in the introduction: "It seems axiomatic that, for a homeland to exist, there must be a resident population that occupies a place over which it has some degree of control and with which it has some kind of bond. . . . [and this] process of bonding between a people and a place takes time" (pp. 1–2).

Most of the eleven papers of the special issue of the *Journal of Cultural Geography* had their roots in AAG paper sessions at the 1991 Miami meeting and the 1992 San Diego meeting where geographers began scholarly conversations about the meaning of a homeland. Indeed, the editors of the special issue asked the authors of the eleven papers to incorporate their own definitions of a homeland into their articles. Jordan, in his paper "The Anglo-Texas Homeland" (1993: 75), proposed that:

> a homeland is a geographical area long inhabited by a self-conscious group exhibiting a strong sense of attachment to the region and exercising some measure of social, economic, and/or political control over it, while at the same time functioning as a peripheral part of a larger and more powerful independent state. . . . Its closest analogy is the nation-state, but homelanders lack and generally do not seek independence.

In his paper, Conzen (1993) follows the European model of a homeland and argues that the major elements of a homeland, a special kind of culture region to him, are indigenity, exclusivity, cultural vitality, resilience, and scale. He also introduces into the discussion the idea of ethnic substrates, the notion that, because of their wide reach across the continent, some ethnic groups, such as the English, German, and Irish, have the potential to influence many places today beyond what their relatively minor ethnic proportions would suggest.

Although Jordan contends that homelands do not have to be linked to ethnic groups as they may be traditionally conceptualized (therefore the Anglo-Texan Homeland), the papers in the special issue underscore that some ethnic groups are strongly associated with homelands in the United States. And as such, the homeland concept and its definitional discussion has become one of the principal intellectual motifs of American ethnic geography in the 1990s.

In fact, a sixteen-chapter anthology titled *Homelands: A Geography of Culture and Place across America* edited by Nostrand and Estaville (2001), expands the geographical coverage as well as the definitional debate. Moreover, in their introductory chapter, the editors juxtapose the homeland concept against Frederick Jackson Turner's frontier thesis with its essential idea of free land and Walter Prescott Webb's dry land thesis and advance the homeland concept as perhaps the most satisfying of the three frameworks for explaining the European settlement of the United States.

In this collection of papers about homelands in the United States, Conzen elaborates extensively on the necessity of homeland development being firmly fused with the building of a nation-state—the European model. In doing so, Conzen rethinks his original elements of a homeland and proposes a blueprint of three essential dimensions (each having three unique criteria of its own): identity, territoriality, and loyalty. On the other hand, Jordan-Bychkov, who, like Conzen, considers homelands as "special culture areas," agrees with Nostrand and Estaville that homelands do not have to evolve into nation-states. Further, Nostrand and Estaville suggest that Conzen's homeland typology can fit comfortably within their five basic homeland components and that there is an American homeland model that does not require an all-important political underpinning. The work in this century in developing the American homeland concept will certainly be central to the understanding of the ethnic geographies in North America.

Patterns of Assimilation

Ethnic geographers have searched for innovative geographical methods to document and analyze adjustments when ethnic groups encounter new physical environments and different peoples. Estaville (1987), for example, drew mainly from primary historical accounts and nineteenth-century manuscript census schedules to explain in a revisionist work that the thousands of Anglos who invaded Louisiana after 1803 (i.e., before the tum of the twentieth century), seriously eroded the culture of Louisiana's Cajuns. Haverluk (1998) argued that Hispanic communities have followed three models of assimilation—continuous, discontinuous, and new—that document variability in the traditional assimilation process in the US.

Using the Public Use Microdata Sample file, Allen and Turner (1996*b*) tested in Southern California the traditional model of immigrant settlement and change that had been based on the experience of European immigrants in the early twentieth century. Despite massive metropolitan expansion and the fact that many immigrants arrived with substantial skills and money, there was still a tendency for new immigrants to cluster in enclaves and for more acculturated immigrants to disperse into distant suburbs. Wong (1998) measured the prevalence of multiracial households across the United States as an indicator of varying levels of ethnic social integration. Wong found, for example, that multiracial households were most common in Hawaii (Alaska, Oklahoma, and California followed in rank). Rates of intermarriage between ethnic groups are another direct indicator of relative levels of social integration. In *The Ethnic Quilt* (1997), Allen and Turner used rates of intermarriage to measure how groups in Southern California differed in their social interconnectedness.

Hardwick and Miyares have undertaken more humanistic approaches to the study of ethnic acculturation and assimilation. Hardwick (1993) compared the acculturation experience of six different religious groups of ethnic Russian refugees who migrated to the North American Pacific Rim and concluded that the majority of the newly arriving post-Soviet Protestant groups were having a difficult time, both culturally and economically, adjusting to life in central California in the late 1990s.

Miyares (1997), using in-depth interviews and participant observation, combined a behavioral and spatial perspective to explain the acculturation of Hmong refugees in California's San Joaquin Valley. Her study resulted in a greater understanding of the ways traditional culture is maintained in the Hmong community, while, simultaneously, settlement patterns and use of space only minimally resemble traditional Hmong values.

Urban American Ethnic Geography

Urban American ethnic geography is such an integral part of urban social geography and urban sociology that attempting to distinguish between ethnic geography and these other specializations is difficult. Because so many geographers and sociologists have studied aspects of ethnicity in urban and metropolitan areas, only a few key sources can be mentioned here.

In the early 1960s an increasing awareness on the part of whites of the discrimination and injustices inflicted upon blacks led to growth in both concern for and interest in urban black communities. The availability of data by race for entire cities and small neighborhoods (census tracts) made possible geographic research on the changing distributions and intensities of segregation among racial groups (although a biologically derived concept, some researchers have used race as a marker for particular ethnic studies). Morrill's 1965 study of the black ghetto serves as a benchmark, examining the growth of the ghetto as a study in spatial diffusion. Deskins (1969), in his review of the dearth of research on the African-American community, characterized Morrill's investigation as the first to deal dynamically with patterns of racial change in neighborhoods. Subsequent to Deskins's 1969 call for action, there has been a growth in geographic literature on the evolution and nature of the black ghetto (Rose 1970; Salisbury 1971; Tata *et al.* 1975; Darden 1981; Deskins 1981; Shaffer and Smith 1986), black residential mobility (Boswell *et al.* 1998; Roseman and Knight 1974; Zonn 1980), black segregation (Darden 1995; Darden *et al.* 1997), and black return migration to, and new settlement patterns, in the American South (Aiken 1985, 1987; McHugh 1987; Cromartie and Stack 1989).

The 1965 Immigration Act resulted in a much more diverse immigrant population. As in the early decades of the twentieth century, coastal cities served as ports-of-entry, resulting in a myriad of new ethnic communities in historically diverse cities such as New York, Los Angeles, San Francisco, and Miami. Ease of communication and transportation, coupled with dispersed refugee populations (Boswell and Curtis 1984; McHugh *et al.* 1997; Miyares 1998*b*), made urban ethnic diversity or urban multiethnicity a national phenomenon (Dingemans and Datel 1995). Allen and Turner (1989) measured the relative ethnic diversity of cities across the United States and found certain types of places to be the most ethnically diverse. Because immigrants and refugees have generally chosen larger metropolitan areas as their destinations, such urban places have attracted ample research attention.

A leading topic of investigation is the relative levels of residential segregation in different places and among different groups as well as change over time. Measuring ethnic residential segregation is critical because spatial separation is considered to be a key indicator of the degree of socioeconomic separation between groups. Alternatively, researchers have measured group spatial patterns in terms of the relative ethnic diversity within different sections of a city or suburb. Studies of ethnic segregation and diversity have been conducted in such cities as Chicago (Howenstine 1996), Los Angeles (Nemeth 1991; Allen and Turner 1995, 1996*a*, *b*; Clark 1996; Thieme and Laux 1996), Miami (Sheskin 1982, 1991; Boswell 1993; Boswell and Cruz-Baez 1997), New York (Miyares and Gowen 1998), and Sacramento (Dingemans and Datel 1995), as well as in many other US cities (Darden 1995; Darden *et al.* 1997) and several Canadian cities (Bourne 1986; Teixeira 1996).

Mapping of intra-metropolitan ethnic distributions in detail shows dramatically how neighborhoods differ in ethnic composition. Bowen (1998) mapped ethnic patterns in several large metropolitan areas and made these available to the public through the Internet (<http://geogdata.csun.edu>, last accessed 25 November 2002). For greater Los Angeles, Turner and Allen's *Atlas of Population Patterns in Metropolitan Los Angeles and Orange Counties* (1991) maps neighborhood ethnic changes during the 1980s in a way that illustrates clearly the patterns associated with dynamic processes of urban change. Further, *The Ethnic Quilt* (Allen and Turner 1997) contains detailed maps of thirty-four ethnic populations in Los Angeles with historical explanations for ethnic neighborhood concentrations and change.

Traditional models of urban ethnic segregation, assimilation, and change based on European immigrant patterns indicate that poorer and more recently arrived immigrant and ethnic populations tend to cluster together, dispersing spatially only after much cultural and economic assimilation. This relationship between assimilation and spatial concentration seems also to apply to the settlement of the newer immigrants (Allen and Turner 1996*b*). However, another aspect of the model may not be as valid as scholars might expect. It has been thought that assimilated immigrants are residentially dispersed in suburbs rather than clustered in enclaves (Zelinsky and Lee 1998). However, the mapping of current ethnic distributions in Southern California shows significant ethnic group concentrations in both older and newer suburbs depending on income levels (Allen and Turner 1997).

Another urban research focus is patterns of entrepreneurship (self-employment) among different ethnic groups and the degree to which ethnic spatial concentration assists such businesses. As immigrant communities grew in number and economic strength, geographers began to study the spatial processes of ethnic enclave economies. Once the purview of sociologists, American ethnic geographers now make both theoretical (e.g. Hiebert 1993; Kaplan 1998; Li 1998) and empirical (e.g. Laux and Thieme 1996; D. O. Lee 1995; Razin and Langlois 1996; Kaplan 1997; Y. Lee *et al.* 1997; Miyares 1998*b*; Park and Kim 1998; Razin and Light 1998; Teixeira 1998; Zhou 1998) contributions to the study of immigrant and refugee economies.

A question that sits at the juncture of urban ethnic diversity and segregation and the enclave economy is the role of real estate agents as "gatekeepers," facilitating the development of enclaves through information sharing and overt and covert steering. Palm's 1985 study of ethnic steering by real estate agents in the urban housing market moved the question of housing market discrimination beyond being an issue solely of race. Recent studies (Clark 1992; Miyares and Gowen 1998; Teixeira 1995; Teixeira and Murdie 1997) show that with the growing diversity of immigrant populations, intra-ethnic preferences, and inter-ethnic prejudice can play just as important a role as race in the development of ethnic neighborhoods. Additionally, these studies punctuate that real estate agents, through controlling available information on the housing market, become powerful catalysts in the ethnic economy by helping integrate or segregate coethnics.

Teaching American Ethnic Geography

Although a few geography departments have had long histories of offering courses about ethnic geography, and although basic concepts and themes grounded in ethnic studies have been woven into introductory human and cultural geography courses for at least the past two decades, university courses focused primarily on ethnic geography, particularly about the United States and Canada, began to be taught more widely in the mid-1980s. Although there was no textbook for teaching American ethnic geography, instructors initially depended on their own reading lists of research articles and chapters by ethnic geographers and scholars in other disciplines and on their own personal experiences and those of their students. Some instructors had their students gain basic understandings of the complexity of ethnic diversity by reading more general works such as Zelinsky's classic book, *The Cultural Geography of the United States* (1973).

In the mid-1980s the National Council for Geographic Education (NCGE) assumed the lead in publishing books and other materials useful in teaching about American ethnic geography. For many years, McKee's anthology, *Ethnicity in Contemporary America: A Geographical Appraisal* (1985), was the only book about American ethnic geography that university instructors had available for their classrooms. The anthology summarizes the migration, settlement, and landscapes of specific non-European ethnic groups, but limits a geographical analysis of European groups to its final chapter that generalizes the contemporary European influence and manifestations in only urban ethnic islands in the United States.

In 1992 Noble's anthology, *To Build in a New Land*, offered university teachers another book-length survey of American ethnic geography with an emphasis on the landscape impress and material culture of some two dozen ethnic groups in North America. The book's graphic sketches of facets of everyday life of these ethnic groups is perhaps its outstanding pedagogic contribution. Conzen's *The Making of the American Landscape* (1990) also contributed a fine collection of papers, particularly his own chapter on "Ethnicity on the Land," that provide students with invaluable perspectives of the development of the American ethnic landscape. In 1988 Allen and Turner's *We the People* gave teachers an extraordinary visual way to let their students investigate the distributions of ethnic groups across the continent.

Meanwhile, introductory human/cultural geography textbooks began to include information about ethnic groups in the United States and Canada. Perhaps *The Human Mosaic: A Thematic Introduction to Cultural Geography* (Jordan-Bychkov and Domosh 1999) outlines the most discerning explanation of ethnic geography in such introductory texts. Brown (1997) underscores, however, that only four of ten popular introductory human/cultural texts he examined have chapters devoted to ethnic geography. Accompanying these general textbooks are such supplemental works as *Annual Editions: Race and Ethnic Relations* (Kronkowski 1991) that are clearly aimed at college students.

The NCGE continues its commitment to provide materials for teaching American ethnic geography. In 1997 it published *Teaching American Ethnic Geography*, the only work to date to focus specifically on teaching American ethnic geography in universities. This anthology, edited by Estaville and Rosen, has the National Geography Standards woven throughout its eighteen chapters.

Although *Teaching American Ethnic Geography* does not intend to encompass all American ethnic groups, topics range from Kate A. Berry's active learning framework for students as they link ethnicity, race, and gender with the environment, and J. Douglas Heffington's field adventure down the Blues Highway through Mississippi's Delta region that enables students to learn to interpret the landscape, to Curtis C. Roseman and J. Diego Vigil's explicit coupling of geography and history with student investigations of their geoethnic histories. Three papers juxtapose teaching about Native-American perspectives: George A. Van Otten outlines several myths that are central themes of a body of misinformation about Native Americans; Robert A. Rundstrom guides students through a list of notable literature so they can gain new understandings and connections with Indian ways; and Martha L. Henderson lets her students peer through the eyes of William Goodbear of the Winnebago Tribe to comprehend the intricacies of bioregionalism. Although these papers are pedagogic in nature, they present noteworthy philosophical, methodological, and definitional insights for teachers as well as scholars.

Most recently, some ethnic geographers (e.g. Estaville 1996; Hardwick and Brown 1998) have advocated using the World Wide Web as a vehicle to teach about American ethnic geography. The Web is indeed a large and fascinating source of information about ethnic groups, particularly those in the United States and Canada. If carefully guided through the minefields of misinformation and prejudices, students can gain a unique appreciation of the richness of American ethnic geography via the Internet. Although there are hundreds of websites containing information about American ethnic groups that are useful in the classroom, perhaps a good starting point for teachers and students, other than doing searches for particular ethnic groups, is the US Census Bureau Web site (<http://www.census.gov/>, last accessed 25 November 2002).

Future Work

Certainly the foregoing research trends will continue, some more vigorously than others. But like geographers in general, ethnic geographers have begun to use the latest geographic techniques and tools, particularly GIS and the Internet. GIS, a powerful analytical tool, can provide ethnic geographers with layers of information about regions of ethnic dominance, areas of relative ethnic pluralism, and zones of ethnic tensions and the means for probing the possible relationships that can lead to models of causation. For example, ethnic proportions can be overlaid by economic factors, educational variables, and transportation networks. Visual, statistical, and/or mathematical techniques can then analyze the data from surface to surface in an effort to discover significant relationships that may suggest possible explanations.

The enormous amount of information on the Internet gives ethnic geographers data at a variety of scales. This information is generally quantitative in nature but a large amount of qualitative material in the form of written documents, organizational charters and constitutions, place descriptions, individual opinions and insights, and surnames, for instance, is available from hundreds of websites, listservs, and bulletin boards. Moreover, geographers have begun to discuss the research possibilities regarding American ethnic groups that use the Internet to maintain or strengthen their ethnic identities and ties (Estaville 1996). This discussion has brought forth several questions that need to be addressed concerning the identification of variables to measure ethnic identity on the Internet, types and significance of sites that have information about ethnic groups, and spatial patterns and relationships of the Internet sites. Of course, the use of the Internet as a vehicle in distance education programs to teach about American ethnic geography is becoming increasingly important (e.g. Hardwick and Brown 1998) and should greatly expand in the future.

At the 1999 AAG meeting in Honolulu, geographers first began exploring in a formal sense the directions that can lead toward an applied American ethnic geography (Estaville 1999). The wide-ranging conversation in this panel session suggested research on ethnic crime, ethnic street gangs, ethnic marketing, ethnic tourism, environmental equity, and ethnic politics. Some American ethnic geographers, it seems, want to apply their knowledge to try to help solve troubling societal problems, to help communities remain economically viable, or simply to become entrepreneurs themselves. As America becomes more diverse, applied ethnic geographers will unquestionably have critical contributions to make.

Indeed, what will America's ethnic diversity look like in the twenty-first century? Models in American ethnic geography have been built on the experiences of immigrants in the nineteenth and early twentieth centuries that all assume a unidirectional migration process: people leaving their home country and going to North America where they recreate a familiar cultural milieu to facilitate the adjustment process. Changes in transportation and communications technology, particularly the Internet, will bring into question how scholars understand contemporary American ethnicity. Zelinsky and Lee (1998) proposed an alternative paradigm for understanding immigrant communities at the dawn of the new millennium—"heterolocalism." Ethnic communities without propinquity can, for instance, maintain cultural identity via the Internet and virtual economies through "e-commerce." Economic globalism brings thousands of foreign nationals to the United States and Canada who often create businessmen's enclaves without losing ties to their home country. Recent changes in naturalization laws allowing for dual citizenship have led to a growth in transnational communities in which transmigrants travel regularly between homes in two or more nations. Zelinsky and Lee challenge researchers to rethink how we study contemporary and future American ethnicity in light of heterolocalism.

Although international connections are proliferating so that many immigrants are finding themselves to be binational and bicultural, the children of immigrants, via formal education and the media, are quickly acculturating to the dominant culture. For example, even in Miami and San Diego where the Spanish language is easily accessible, studies show that by the later years of high school the second generation is more comfortable and skilled in English than in Spanish (Portes and Schauffler 1994; Rumbaut 1997). Moreover, intermarriage between ethnic groups will no doubt increase in places with greater ethnic diversity, resulting in a growing multiethnic population. In such places, the changed appearance and identities of people will not only prompt a redefinition of the meaning of "American" but will most likely also cause dramatic changes in the everyday life of these ethnically diverse places. Concomitantly, as some new immigrant groups become more economically successful than others, social and economic tensions may arise between ethnic groups. Such ethnic tensions, in fact, may well cause a return to a more segregated society based on class and ethnicity.

All these population trends suggest that American ethnic geography will become even more significant socially, economically, and politically. American ethnic geographers clearly have much work to accomplish.

Acknowledgements

We thank Thomas D. Boswell, Terry G. Jordan-Bychkov, Richard L. Nostrand, and Carlos Teixeira for their insightful comments that strengthened this chapter.

References

Aiken, Charles S. (1985). "New Settlement Patterns of Rural Blacks in the American South." *Geographical Review*, 75: 383–404.

— (1987). "Race as a Factor in Municipal Underbounding." *Annals of the Association of American Geographers*, 77: 564–79.

Airriess, Christopher A. (1994). "Vietnamese Market Gardens in New Orleans." *Geographical Review*, 84: 16–31.

Airriess, Christopher A., and Clawson, David L. (1991). "Versailles: A Vietnamese Enclave in New Orleans, Louisiana." *Journal of Cultural Geography*, 12: 1–14.

Alanen, Arnold R. (1989). "Wilderness and Landscape: Finnish Immigrant Expressions in the Upper Midwest." *Terra*, 101: 31–3.

— (1990). "Immigrant Gardens on a Mining Frontier," in *Action*, M. Francis and R. T. Hester (eds.), *The Meaning of Gardens: Idea, Place, and Action*. Cambridge: MIT Press, 160–5.

Alanen, Arnold R., and Tishler, W. H. (1980). "Finnish Farmstead Organizations in Old and New World Settings." *Journal of Cultural Geography*, 1: 66–81.

Allen, James P., and Turner, Eugene K. (1988). *We The People: An Atlas of America's Ethnic Diversity*. New York: Macmillan.

— (1989). "The Most Ethnically Diverse Urban Places in the United States." *Urban Geography*, 10: 523–39.

— (1995). "Ethnic Differentiation by Blocks within Census Tracts." *Urban Geography*, 16: 344–64.

— (1996*a*). "Ethnic Diversity and Segregation in the New Los Angeles," in Roseman, Curtis C., Laux, Hans Dieter, and Thieme, Gunther (eds.), *Ethnicity: Geographic Perspectives on Ethnic Change in Modern Cities*. Lanham, Md.: Rowman & Littlefield, 1–30.

— (1996*b*). "Spatial Patterns of Immigrant Assimilation." *Professional Geographer*, 48: 140–55.

— (1997). *The Ethnic Quilt: Population Diversity in Southern California*. Northridge, Calif.: Center for Geographical Studies, California State University, Northridge.

Anderson, Kay J. (1987). "The Idea of Chinatown: The Power of Place and Institutional Practice in the Making of a Racial Category." *Annals of the Association of American Geographers*, 77: 580–98.

Arreola, Daniel P. (1981). "Fences as Landscape Taste: Tucson's Barrios. *Journal of Cultural Geography*, 2: 96–105.

— (1984). "Mexican American Exterior Murals." *Geographical Review*, 74: 409–24.

— (1988). "Mexican American Housescapes." *Geographical Review*, 78: 299–315.

Boswell, Thomas D. (1984). "The Migration and Distribution of Cubans and Puerto Ricans Living in the United States." *Journal of Geography*, 83: 65–72.

— (1993). "The Cuban-American Homeland in Miami." *Journal of Cultural Geography*, 13: 133–48.

Boswell, Thomas D., and Cruz-Baez, Angel David (1997). "Residential Segregation by Socioeconomic Class in Metropolitan Miami: 1990." *Urban Geography*, 18: 474–96.

Boswell, Thomas D., Cruz-Baez, Angel David, and Zijlstra, Pauline (1998). "Housing Preferences and Attitudes of Blacks Toward Housing Discrimination in Metropolitan Miami." *Urban Geography*, 19: 189–210.

Boswell, Thomas D., and Curtis, James R. (1984). *The Cuban American Experience*. Totowa: Rowman & Littlefield.

— (1991). "The Hispanization of Metropolitan Miami," in Thomas D., Boswell (ed.), *South Florida, Winds of Change*. Miami: Association of American Geographers, 140–61.

Bourne, Larry S. (1986). *Canada's Ethnic Mosaic: Characteristics and Patterns of Ethnic Origin Groups in Urban Areas*. Toronto: Centre for Urban and Community Studies, University of Toronto.

Bowen, William. (1998). Electronic Map Library ‹http://geogdata.csun.edu.›, last accessed 25 November 2002.

Brown, Brock J. (1997). "Using Personal Accounts to Incorporate Ethnic Issues into Introductory College Human Geography," in Estaville and Rosen (1997: 31–6).

Carlson, Alvar W. (1990). *The Spanish-American Homeland: Four Centuries in New Mexico's Rio Arriba*. Baltimore: Johns Hopkins University Press.

Clark, William A. V. (1992). "Residential Preferences and Residential Choice in a Multi-ethnic Context." *Demography*, 29: 451–66.

— (1996). *The California Cauldron: Immigration and the Fortunes of Local Communities*. New York: Guilford Press.

Comeaux, Malcolm L. (1989). "The Cajun Barn." *Geographical Review*, 79: 47–62.

Constantinou, Stavros T. (1989). "Dominant Themes and Intergenerational Differences in Ethnicity: The Greek Americans." *Sociological Focus*, 22: 99–117.

Constantinou, Stavros T., and Diamantides, Nicholas D. (1985). "Modeling International Migration: Determinants of Emigration from Greece to the United States, 1820–1908." *Annals of the Association of American Geographers*, 75: 352–69.

Conzen, Michael P. (1980). "The Morphology of Nineteenth-century Cities in the United States," in W. Borah, J. Hardoy, and G. Stelter, (eds.), *Urbanization in the Americas: The Background in Comparative Perspective*. Ottawa: National Museum of Man, 119–41.

— (1990). *The Making of the American Landscape*. London: Unwin Hyman.

— (1993). "Culture Regions, Homelands, and Ethnic Archipelagos in the United States: Methodological Considerations." *Journal of Cultural Geography*, 23: 13–29.

— (2001). "Homelands and Cultural Space," in Nostrand and Estaville (2001).

Cromartie, John, and Stack, Carol B. (1989). "Reinterpretation of Black Return and Nonreturn Migration to the South 1975–1980." *Geographical Review*, 79: 297–310.

Curtis, James R. (1980). "Miami's Little Havana: Yard Shrines, Cult Religion and Landscape." *Journal of Cultural Geography*, 1: 1–15.

Darden, Joe T. (1981). *The Ghetto: Readings with Interpretations*. Port Washington, NY: Kennikat Press.

— (1995). "Black Residential Segregation since the 1948 Shelley v. Kraemer Decision." *Journal of Black Studies*, 25: 680–91.

Darden, Joe T., Bagakas, Joshua G., and Ji, Shun-Je (1997). "Racial Residential Segregation and the Concentration of Low- and High-Income Households in the 45 Largest U.S. Metropolitan Areas." *Journal of Developing Societies*, 13: 171–94.

Datel, Robin E., and Dingemans, Dennis. (1980). "Historic Preservation and Urban Change." *Urban Geography*, 1: 229–53.

Deskins, Donald R. (1969). "Geographical Literature on the American Negro, 1949–1968." *Professional Geographer*, 21: 145–9.

— (1981). "Morphology of a Black Ghetto." *Urban Geography*, 2: 95–114.

Dingemans, Dennis, and Datel, Robin (1995). "Urban Multiethnicity." *Geographical Review*, 84: 458–77.

Estaville, Lawrence E. (1986). "Mapping the Louisiana French." *Southeastern Geographer*, 26: 90–113.

— (1987). "Changeless Cajuns: Nineteenth-century Reality or Myth?" *Louisiana History*, 28: 117–40.

— (1988). "The Louisiana French in 1900." *Journal of Historical Geography*, 14: 342–60.

— (1990). "Louisiana-French Language in the Nineteenth Century. *Southeastern Geographer*, 30: 107–20.

— (1993). "The Louisiana French Homeland." *Journal of Cultural Geography*, 13: 31–45.

— (1996). "American Ethnic Geography on the Internet." Paper presented at the AAG meeting, Charlotte, NC.

— (1999). Toward an Applied Ethnic Geography. Panel session at the AAG meeting, Honolulu, HI.

Estaville, Lawrence E., and Rosen, Carol J. (1997). *Teaching American Ethnic Geography*. Indiana, Pa.: National Council for Geographic Education.

Francaviglia, Richard V. (1971). "The Cemetery as an Evolving Cultural Landscape." *Annals of the Association of American Geographers*, 61: 501–9.

— (1972). "Western Barns: Architectural Form and Climatic Considerations." *Yearbook of the Association of Pacific Coast Geographers*, 34: 152–60.

— (1978). *The Mormon Landscape: Existence, Creation, and Perception of a Unique Image in the American West*. New York: AMS.

Glassie, Henry H. (1971). *Pattern in the Material Folk Culture of the Eastern United States*. Philadelphia: University of Pennsylvania Press.

— (1975). *Folk Housing in Middle Virginia*. Knoxville: University of Tennessee Press.

Godfrey, Brian J. (1988). *Neighborhoods in Transition: The Making of San Francisco's Ethnic and Nonconformist Communities*. Berkeley: University of California Press.

Gritzner, Charles F. (1990). "Log Barns of Hispanic New Mexico." *Journal of Cultural Geography*, 10: 21–34.

Groves, P. A., and Muller, E. K. (1975). "The Evolution of Black Residential Areas in Late Nineteenth-century Cities." *Journal of Historical Geography*, 1: 169–91.

Hardwick, Susan W. (1993). *Russian Refuge: Religion, Migration, and Settlement on the North American Pacific Rim*. Chicago: University of Chicago Press.

Hardwick, Susan W., and Holtgrieve, Donald G. (1996). *Valley for Dreams: Life and Landscape in the Sacramento Valley*. Lanham, Md.: Rowman & Littlefield.

Hardwick, Susan W., and Brown, Brock J. (1998). "Step Up to Geography." FIPSE distance learning project focused on Latino teachers in South Texas.

Hart, John F. (1972). *The Look of the Land*. Englewood Cliffs, NJ: Prentice-Hall.

Haverluk, Terrence W. (1997). "The Changing Geography of U.S. Hispanics, 1850–1990." *Journal of Geography*, 96: 134–45.

— (1998). "Hispanic Community Types and Assimilation in Mex-America." *Professional Geographer*, 50: 465–80.

Helzer, Jennifer J. (1994). "Continuity and Change: Hmong Settlement in California's Sacramento Valley." *Journal of Cultural Geography*, 2: 51–63.

Hiebert, Daniel (1993). "Jewish Immigrants and the Garment Industry of Toronto, 1901–1931: A Study of Ethnic and Class Relations." *Annals of the Association of American Geographers*, 83: 243–71.

Hoelscher, Steven (1998*a*). *Heritage on Stage: The Invention of Ethnic Place in America's Little Switzerland*. Madison: University of Wisconsin Press.

— (1998*b*). "Tourism, Ethnic Memory, and the Other-directed Place." *Ecumene*, 5: 371–98.

Howenstine, Erick. (1996). "Ethnic Change and Segregation in Chicago," in (eds.) Curtis C. Roseman, Hans Dieter Laux, and Gunther Thieme, (eds.), *Ethnicity: Geographic Perspectives on Ethnic Change in Modern Cities*. Lanham, MD: Rowman & Littlefield, 31–50.

Isajiw, Wsevold W. (1974). "Definitions of Ethnicity." *Ethnicity*, 1: 111–24.

Jackson, J. B. (1970). *Landscape: Selected Writings of J. B. Jackson*, (ed.) E. H. Zube. Amherst, Mass.: University of Massachusetts Press.

— (1984). *Discovering the Vernacular Landscape*. New Haven: Yale University Press.

Jakle, John A., and Wheeler, James O. (1969). "The Changing Residential Structure of the Dutch Population in Kalamazoo." *Annals of the Association of American Geographers*, 59: 441–60.

Johnson, James H., Jr., and Roseman, Curtis C. (1990). "Increasing Black Outmigration from Los Angeles: The Role of Household Dynamics and Kinship Systems." *Annals of the Association of American Geographers*, 80: 205–22.

Jordan, Terry G. (1964). "German Houses in Texas." *Landscape*, 14: 24–6.

— (1966). *German Seed in Texas Soil: Immigrant Farmers in Nineteenth-century Texas*. Austin: University of Texas Press.

— (1969). "Population Origins in Texas." *Geographical Review*, 59: 83–103.

— (1982). *Texas Graveyards: A Cultural Legacy*. Austin: University of Texas Press.

— (1985). *American Log Buildings: An Old World Heritage*. Chapel Hill: University of North Carolina Press.

— (1989). "New Sweden's Role on the American Frontier: A Study in Cultural Preadaptation." *Geografiska Annaler*, 71: 71–83.

— (1993). "The Anglo-Texan Homeland." *Journal of Cultural Geography*, 13: 75–86.

Jordan, Terry G., and Kaups, Matti. (1989). *The American Backwoods Frontier: An Ethnic and Ecological Interpretation*. Baltimore: Johns Hopkins University Press.

Jordan-Bychkov, Terry G., and Domosh, Mona (1999). *The Human Mosaic: A Thematic Introduction to Cultural Geography*. New York: Longman.

Kaplan, David H. (1997). "The Creation of an Ethnic Economy: Indochinese Business Expansion in Saint Paul." *Economic Geography*, 73: 214–33.

— (1998). "The Spatial Structure of Urban Ethnic Economies." *Urban Geography*, 19: 489–502.

Kniffen, Fred (1965). "Folk Housing: Key to Diffusion." *Annals of the Association of American Geographers*, 55: 549–77.

Kronkowski, John A. (1991). *Annual Editions: Race and Ethnic Relations, 1991–1992*. Guilford, Conn.: Dushkin Publishing.

Lamme, Ary J., and McDonald, Douglas B. (1993). "The 'North Country' Amish Homeland." *Journal of Cultural Geography*, 13: 107–18.

Laux, Hans Dieter, and Thieme, Gunther (1996). "Model Minority or Scapegoat? The Experience of Korean Immigrants in Los Angeles," in Klaus Frantz (ed.), *Human Geography in North America: New Perspectives and Trends in Research*. Innsbruck: Innsbrucker Geographische Studien 26, 65–80.

LeBlanc, Robert G. (1985). "The Francophone 'Conquest' of New England: Geopolitical Conceptions and Imperial Ambition of French-Canadian Nationalists in the Nineteenth Century." *American Review of Canadian Studies*, 15: 288–310.

— (1988). "L'Émigration, Colonisation et Rapatriement: The Acadian Perspective." *Les Cahiers de la Société Historique Acadienne*, 19: 71–104.

Lee, Dong Ok (1995). "Koreatown and Korean Small Firms in Los Angeles: Locating in the Ethnic Neighborhoods." *Professional Geographer*, 47: 184–95.

Lee, Yuk, Cameron, Theresa, Schaeffer, Peter, and Schmidt, Charles G. (1997). "Ethnic Minority Small Business: A Comparative Analysis of Restaurants in Denver." *Urban Geography*, 18: 591–621.

Lemon, James T. (1966). "The Agricultural Practices of National Groups in Eighteenth-Century Southeastern Pennsylvania." *Geographical Review*, 56: 467–96.

— (1972). *The Best Poor Man's Country: A Geographical Study of Early Southeastern Pennsylvania*. Baltimore: Johns Hopkins University Press.

Lewis, Pierce F. (1975). "Common Houses, Cultural Spoor." *Landscape*, 19: 1–22.

Li, Wei (1998). "Los Angeles's Chinese Ethnoburb: From Ethnic Service Center to Global Economy Outpost." *Urban Geography*, 19: 502–18.

McHugh, Kevin E. (1987). "Black Migration Reversal in the United States." *Geographical Review*, 77: 171–82.

— (1989). "Hispanic Migration and Population Redistribution in the United States." *Professional Geographer*, 41: 429–39.

McHugh, Kevin E., Miyares, Ines M., and Skop, Emily H. (1997). "The Magnetism of Miami: Segmented Paths in Cuban Migration." *Geographical Review*, 87: 504–19.

McKee, Jesse O. (1985). *Ethnicity in Contemporary America: A Geographical Appraisal*. Dubuque: Kendall-Hunt.

Marston, Sallie A. (1988). "Neighborhood and Politics: Irish Ethnicity in Nineteenth Century Lowell, Massachusetts." *Annals of the Association of American Geographers*, 76: 414–32.

Meigs, Peveril (1941). "An Ethno-telephonic Survey of French Louisiana." *Annals of the Association of American Geographers*, 31: 243–50.

Meinig, Donald W. (1969). *Imperial Texas: An Interpretative Essay in Cultural Geography*. Austin: University of Texas Press.

— (1971). *Southwest: Three Peoples in Geographical Change, 1600–1970*. New York: Oxford University Press.

Miyares, Ines M. (1996). "To Be Hmong in America: Settlement Patterns and Early Adaptation of Hmong Refugees in the United States." in Klaus Frantz (ed.), *Human Geography in North America: New Perspectives and Trends in Research*, Innsbruck: Innsbrucker Geographische Studien, 26.

— (1997). "Changing Perceptions of Space and Place as Measures of Hmong Acculturation." *Professional Geographer*, 49: 214–23.

— (1998*a*). *The Hmong Refugee Experience in the United States: Crossing the River*. New York: Garland Publishing.

— (1998*b*). "'Little Odessa'-Brighton Beach, Brooklyn: An Examination of the Former Soviet Refugee Economy in New York City." *Urban Geography*, 19: 518–30.

Miyares, Ines M., and Gowen, Kenneth J. (1998). "Recreating Borders? The Geography of Latin Americans in New York City." *CLAG Yearbook*, 24: 31–44.
Morrill, Richard (1965). "The Negro Ghetto: Problems and Alternatives." *Geographical Review*, 55: 339–61.
Nemeth, David J. (1991). "A Case Study of Rom Gypsy Residential Mobility in the United States." *Yearbook of the Pacific Coast Geographers*, 53: 131–54.
Noble, Allen G. (1984). *Wood, Brick, and Stone: The North American Settlement Landscape*. i. *Houses*; ii. *Barns and Farm Structures*. Amherst: University of Massachusetts Press.
— (1985). "Rural Ethnic Islands," in McKee (1985: 241–57).
— (1992). *To Build in a New Land: Ethnic Landscapes in North America*. Johns Hopkins University Press.
Nostrand, Richard L. (1970). "The Hispanic-American Borderland: Delimitation of an American Culture Region. *Annals of the Association of American Geographers*, 60: 638–61.
— (1973). "'Mexican American' and 'Chicano': Emerging Terms for a People Coming of Age." *Pacific Historical Review*, 42: 389–406.
— (1975). "Mexican Americans Circa 1850." *Annals of the Association of American Geographers*, 65: 378–90.
— (1980). "The Hispano Homeland in 1900." *Annals of the Association of American Geographers*, 70: 382–96.
— (1992). *The Hispano Homeland*. Norman: University of Oklahoma Press.
Nostrand, Richard L., and Estaville, Lawrence E. (1993). "The Homeland Concept." *Journal of Cultural Geography*, 13: 1–4.
— (eds.) (2001). *Homelands: A Geography of Culture and Place across America*. Baltimore, Md.: Center for American Places, Johns Hopkins University Press.
Ostergren, Robert C. (1979). "A Community Transplanted: The Formative Experience of a Swedish Immigrant Community in the Upper Middle West." *Journal of Historical Geography*, 5: 189–212.
— (1981*a*). "The Immigrant Church as a Symbol of Community and Place on the Landscape of the American Upper Midwest." *Great Plains Quarterly*, 1: 224–38.
— (1981*b*). "Land and Family in Rural Immigrant Communities." *Annals of the Association of American Geographers*, 71: 400–11.
Palm, Risa (1985). "Ethnic Segmentation of Real Estate Agent Practice in the Urban Housing Market." *Annals of the Association of American Geographers*, 75: 58–68.
Park, Siyoung, and Kim, Kwang C. (1998). "Intrametropolitan Location of Korean Business: The Case Study of Chicago." *Urban Geography*, 19: 613–31.
Portes, Alejandro, and Schauffler, Richard (1994). "Language and the Second Generation: Bilingualism Yesterday and Today." *International Migration Review*, 28: 640–61.
Razin, Eran, and Light, Ivan H. (1998). "The Income Consequences of Ethnic Entrepreneurial Concentrations." *Urban Geography*, 19: 554–76.
Razin, Eran, and Langlois, André (1996). "Immigrant and Ethnic Entrepreneurs in Canadian and American Metropolitan Areas–A Comparative Analysis," in A. Laperrière, V. Lindström, and T. P. Seiler (eds.), *Immigration and Ethnicity in Canada*, Canadian Issues, 28. Montreal: Association of Canadian Studies, 127–44.
Rose, Harold (1970). "The Development of an Urban Subsystem: The Case of the Negro Ghetto." *Annals of the Association of American Geographers*, 60: 1–17.
Roseman, Curtis C., Sofranko, Andrew J., and Williams, James D. (1981). *Population Redistribution in the Midwest*. Ames, Ia.: Iowa State University.
Roseman, Curtis C., and Knight Prentice L., III (1974). "Residential Environment and Migration Behavior of Urban Blacks." *Professional Geographer*, 27: 160–5.
Roseman, Curtis C., and Lee, Seong Woo (1998). "Linked and Independent African American Migration from Los Angeles." *Professional Geographer*, 50: 204–14.
Rumbaut, Ruben G. (1997). "Assimilation and its Discontents: Between Rhetoric and Reality." *International Migration Review*, 31: 923–60.
Salisbury, Howard G. (1971). "The State within a State: Some Comparisons between the Urban Ghetto and the Insurgent State." *Professional Geographer*, 23: 105–12.
Sauer, Carl O. (1925). "The Morphology of Landscape." *University of California Publications in Geography*, 2: 19–53.
Shaffer, Richard, and Smith, Neil (1986). "The Gentrification of Harlem?" *Annals of the Association of American Geographers*, 76: 347–65.
Sheskin, Ira M. (1982). *Population Study of the Greater Miami Jewish Community*. Miami: Greater Miami Jewish Federation.
— (1991). "Jews in South Florida," in Thomas D. Boswell, (ed.), *South Florida, Winds of Change* Miami: Association of American Geographers, 163–80.
— (1993). "Jewish Metropolitan Homelands." *Journal of Cultural Geography*, 13: 119–32.
— (1998*a*). "The Changing Spatial Distribution of American Jews," in Harold Brodsky (ed.), *Land and community: Geography in Jewish studies*, Bethesda, Md.: University of Maryland Press, 185–221.
— (1998*b*). "A Methodology for Examining the Changing Size and Spatial Distribution of a Jewish Population: A Miami Case Study." *Shofar Studies in Jewish Geography*. Special Issue, 17: 97–116.
— (1999). *How Jewish Communities Differ: Variations in the Findings of Local Jewish Demographic Studies*. New York: North American Jewish Data Bank, City University of New York.
Smith, Jeffrey S. (1999). "Anglo Intrusion on the Old Sangre de Cristo Land Grant." *Professional Geographer*, 51: 170–83.
Tata, Robert J., Van Horn, Sharyn, and Lee, David (1975). "Defensible Space in a Housing Project: A Case Study from a South Florida Ghetto." *Professional Geographer*, 28: 297–303.
Teixeira, Carlos (1995). "Ethnicity, Housing Search, and the Role of the Real Estate Agent: A Study of Portuguese and Non-Portuguese Real Estate Agents in Toronto." *Professional Geographer*, 47: 176–83.
— (1996). "The Suburbanization of Portuguese Communities in Toronto and Montreal: From Isolation to Residential Integration?" in A. Laperrière, V. Lindström, and T. P. Seiler, (eds.), *Immigration and Ethnicity in Canada*, Canadian Issues, 28. Montreal: Association of Canadian Studies, 181–201.
— (1998). "Cultural Resources and Ethnic Entrepreneurship: A Case Study of the Portuguese Real Estate Industry in Toronto." *Canadian Geographer*, 42: 267–81.

Teixeira, Carlos, and Murdie, Robert A. (1997). "The Role of Real Estate Agents in the Residential Relocation Process: A Case Study in Suburban Toronto." *Urban Geography*, 18: 497–520.

Thernstrom, Stephan, Orlov, Ann, and Handlin, Oscar (1980). *Harvard Encyclopedia of American Ethnic Groups*. Cambridge, Mass.: Harvard University Press.

Thieme, Gunther, and Laux, Hans Dieter (1996). "Los Angeles—a Multi-ethnic Metropolis: Spatial Patterns and Socio-economic Problems as a Result of Recent Migration Processes," in Klaus Frantz (ed.), *Human Geography in North America: New Perspectives and Trends in Research*. Innsbruck: Innsbrucker Geographische Studien, 26: 81–96.

Turner, Eugene, and Allen, James P. (1991). *Atlas of Population Patterns in Metropolitan Los Angeles and Orange Counties*. Northridge, Calif.: Department of Geography, California State University, Northridge.

US Census Bureau website, ‹http://www.census.gov/›, last accessed 25 November 2002.

Villeneuve, Paul Y. (1972). "Propinquity and Social Mobility within Surname Groups in a French-Canadian Colony." *Proceedings of the Association of American Geographers*, 4: 100–3.

Ward, David (1971). *Cities and Immigrants: A Geography of Change in Nineteenth-century America*. New York: Oxford University Press.

Wong, David W. S. (1998). "Spatial Patterns of Ethnic Integration in the United States." *Professional Geographer*, 50: 13–30.

Wood, Joseph S. (1986). "The New England Village as an American Vernacular Form," in C. Wells (ed.), *Perspectives in Vernacular American Architecture*. Columbia: University of Missouri Press, 54–63.

—— (1997). "Vietnamese American Place Making in Northern Virginia." *Geographical Review*, 87: 58–72.

Zelinsky, Wilbur (1954). "The Greek Revival House in Georgia." *Journal of the Society of Architectural Historians*, 13: 9–12.

—— (1973). *The Cultural Geography of the United States*. Englewood Cliffs, NJ: Prentice-Hall.

Zelinsky, Wilbur, and Lee, Barrett A. (1998). "Heterolocalism: An Alternative Model of the Sociospatial Behavior of Immigrant Ethnic Communities." *International Journal of Population Geography*, 4: 281–98.

Zhou, Yu (1998). "How Do Places Matter? A Comparative Study of Chinese Ethnic Economies in Los Angeles and New York City." *Urban Geography*, 19: 531–54.

Zaniewski, Kazimierz J., and Rosen, Carol J. (1998). *The Atlas of Ethnic Diversity in Wisconsin*. Madison, Wis.: University of Wisconsin Press.

Zonn, Leo E. (1980). "Information Flows in Black Residential Search Behavior." *Professional Geographer*, 32: 43–50.

CHAPTER 38

American Indian Geography

Robert Rundstrom, Douglas Deur, Kate Berry, and Dick Winchell

Contemporary geographical research concerning North America's native peoples is most conspicuous for its remarkably diverse set of subjects, methods, and epistemological stances. Indeed, it would be hard to find another AAG specialty group whose members do research in as many corners of the natural and social sciences and humanities.[1] Some perspectives developed quite recently, while others emanate from a century of prior research by geographers, especially Carl Sauer and his students. We think these observations important enough to require opening our review with a description, albeit a painfully brief one, of the historical context for the current scene.

In the early twentieth century, as now, there was a great deal of cross-fertilization between anthropology and geography. Deterministic thinking associated with environmentalist theory (e.g. Hans 1925; Huntington 1919; Semple 1903) elicited many critical responses from both fields. For example, the geographer-turned-anthropologist Franz Boas and his students sought to illuminate the full complexity of Native American life, producing a vast corpus of empirical studies. Many addressed geographical topics, including Native North American place-names, environmental knowledge, and resource use. These works were frequently termed "ethnogeographies" (e.g. Barrett 1908; Boas

The American Indian Specialty Group changed its name to Indigenous Peoples Specialty Group (IPSG) in 2000, reflecting both a desire to include geographers studying and working with indigenous peoples anywhere on earth and an awareness that North American Indians and Inuit increasingly interact in globalized networks of indigenous peoples. This review was written prior to IPSG's inception, so no research on indigenous peoples outside North America is reported here. We included books, journal articles, chapters in edited volumes, and selected dissertations published by North American geographers since 1990. We tried to be complete in the first two categories, and sincerely apologize for any omissions. We inserted additional citations from anthropology, literature, and other fields only if we thought these works had been influential in the discipline or were otherwise noteworthy. We made no effort to search the so-called gray literature, the enormous body of tribal, federal, state, regional, and local government reports produced by bureaucrats and professional consultants, although much of this literature is relevant and worth reading. See Ballas (1966), Carlson (1972), Jett (1994), and Winchell *et al.* (1994) for the geographic literature on North American Indians published before 1990.

[1] The number of members in the specialty group has always been modest at best. For example, the official AAG tally for 2001 was only eighty. This is due mostly to a preference by many for devoting precious time and energies to just one specialty group, typically a much larger one associated with a major thematic or regional interest. Groups dedicated to biogeography, cultural ecology, cultural geography, ethnic geography, historical geography, and political geography have many members who study these subjects as applied to North American Indian lands and lives. Thus, the body of relevant work published since 1990 is much larger than membership in the specialty group suggests.

1934; Harrington 1916). Others attempted sweeping continental studies of regional variation based on historical and cultural processes (Kroeber 1939; Wissler 1926). The historicist critique of environmentalist theory resonated with a young geographer, Carl Sauer. Sauer (1920) long had interests in American Indian land-use practices, or "land management" in current parlance. Regular interaction with Boas's students, especially Kroeber and Lowie, coupled with independent development of their own geographical ideas, led Sauer and his students to expand their research on North American Indian cultural geography, including such subjects as settlement patterns (e.g. Sauer and Brand 1930), plant use (e.g. Carter 1945), and resources and material and oral culture (e.g. Kniffen 1939). Sauer, his large number of Ph.D. students, and his student's students, continued to define this research agenda throughout the twentieth century (e.g. Kniffen *et al.* 1987; Sauer 1971). The continued relevance of this work was signaled recently by the reissue of two classic texts in new editions (Denevan 1992*a*; Waterman 1993).

To be sure, much has changed in the way geographers study American Indians. Most noteworthy perhaps are those who use post-colonial and other social theory to unpack meanings of space and place, those who promote or criticize maps and GIS, and those who engage in advocacy on behalf of Indian economic development and land claims. Consequently, work on and with American Indians is more fully integrated into the whole of geography, as geographers with all sorts of interests and approaches are now thinking about Indians and their lands. But in what follows, the reader will also find it easy to discern direct links to Boasian-Sauerian geographies of American Indians via studies in cultural ecology, diffusion studies, and regional-historical analyses. Moreover, geographers continue to exchange ideas with anthropologists, now drawing inspiration from authors such as Keith Basso (1996) and Hugh Brody (1988, 1991).

New Work on Material Landscapes

Archaeologists and anthropologists typically have dominated the study of American Indian material landscapes using the methods of cultural ecology, a situation that in the extreme can lead to complete ignorance of geographical research on a subject like the cultural landscape (e.g. Stoffle *et al.* 1997). Following Sauer's lead, most geographically trained cultural ecologists conduct work in Latin America. Recently though, geographers have been in the vanguard of significant revisionist movements within North American cultural ecology (see Ch. 8, Cultural Ecology, for related work). Geographers have been instrumental in illuminating the sophistication of American Indian resource management and the degree to which native peoples intentionally modified their environments both prior to and after European contact. For example, the study of the fur trade and other similar networks has been overhauled by the attention paid to Indians as active trade agents, modifiers of landscapes, and manipulators of markets and geopolitics (C. Harris 1995; Marsden and Galois 1995; Tough 1990; Weissling 1992; Works 1992). Sluyter (2001) proposed a comprehensive theoretical framework for understanding native and non-native landscape production throughout the Americas as part of a system of colonial relations, and Butzer (1990) and Denevan (1992*b*) have produced syntheses of the impact of indigenous resource management for North America. In doing so, they demonstrate the importance of indigenous environmental modifications in later European resettlement. Geographers also have influenced environmental historians such as Crosby (1994), Cronon (1983), and White (1992), who discuss American Indian resource management as a foundation for studies of biotic change during the contact and historical periods. In turn, the impact on geography of the ideas of these three environmental historians has been significant.

Similarly, several geographers have identified plant cultivation practices and agricultural landforms that had been overlooked by archaeologists and anthropologists. Geographers have identified ridged agricultural fields and other evidence of intensive resource management in the upper Midwest (Brown 1991; Gartner 1997), and have expanded our understanding of human modification of environments in the American southwestern deserts (Doolittle 1992), Alaska (Highleyman 1994), and central California (Preston 1990). Anthropogenic fire regimes established by American Indians played an especially crucial role in the development of Ozark vegetation in the nineteenth century (Batek *et al.* 1999). Deur (1999, 2000*a*, 2002) has conclusively demonstrated the presence of low-intensity plant cultivation on the pre-colonial northwest coast of the US, research directly contradicting received academic wisdom on the region's agricultural history. Gritzner (1994) and Peacock (1998), working in the semi-arid interior plateau country, have shown that the peoples of this region managed plants intensively through prescribed burning, selective harvesting, and the transporting and transplanting of plant materials. Doolittle's (2002) recent book uses land-use evidence, especially field patterns and related systems, as the basis for a survey of pre-contact food production throughout North America. At 600 pages, the volume stands as the new standard reference on the systemic character of prehistoric food production across the continent.

All of these studies integrate biogeography and the analysis of settlement distributions and archaeological features, often with conventional ethnographic methods. This literature also has enhanced our understanding of such related topics as pre-contact native settlement distribution, population size, cultural complexity, and the foundations of post-contact conflicts over access to resource sites (Denevan 1992*a*, 1996; Jacobs 1996; Meinig 1993).

Another tradition associated with Sauer and his students involves sorting out the material cultural landscape by classifying artifacts, explaining their likely diffusion, and producing regional descriptions. The following are recent indications of this continuing legacy: a regional analysis of the Four Corners area (Brown 1995); a typology of Canadian Metis houses and farmsteads (Burley and Horsfall 1989); analyses of architectural form as a manifestation of cultural identity (Blake and Smith 2000; Jett 1992*b*; Porter 1992); a masterfully comprehensive volume on Indian architecture (Nabokov and Easton 1989); a comparative study of gravesites and outdoor funerary practices among Navajos, Mormons, and Zuni (Cunningham 1989); examinations of Mescalero Apache settlement and housing (Henderson 1990, 1992*c*); a diffusion study supporting the thesis that there was very early Asian contact in the pre-Christian Americas (Jett 1991); and a study of conflicting cultural landscapes at Santa Clara Pueblo by a member of that pueblo (Swentzell 1997). In another revisionist work, Jordan and Kaups (1989) argued that early New Sweden produced a successful mix of Finnish and Delaware Indian forest cultures that expanded into the Upland South and became the most successful frontier culture in North America. However, Kay (1991) was disappointed by the authors' handling of questions of backwoods relationships, intermarriage, and other social customs only from the European side.

(Re)Sources

The material elements of the biosphere normally called "resources" are also worth considering as "sources" when thinking about Indian Country because so many issues associated with the biosphere, for example, water quality and quantity, are indistinguishable from such topics as sovereignty, land dispossession and restoration, planning and development, sacredness, and even gambling and tourism. These subjects are *sources* of physical, economic, political, and spiritual livelihoods. This manipulation of language may seem thoroughly postmodern, but we are not trying to be playful. In this section, we use this ambiguity in the hope of more clearly expressing the intellectual interpenetration of these seemingly dissimilar topics.

The Biosphere

Vecsey and Venables (1980) were among the first to unite non-Indian academic and Indian perspectives in one volume dealing with environmental issues in a historical context. Since then, the evolving capacities of tribal governments as sovereigns capable of managing natural (re)sources and protecting environmental quality has increased—modestly in some cases, dramatically in others. Individual tribal members, working as attorneys or leaders for tribal governments or intertribal organizations such as the Native American Rights Fund and The Council of Energy Resource Tribes, increasingly raise their voices to characterize human–environment relations from an Indian perspective (e.g. LaDuke 1999). Geographers interested in tribal (re)source management often are reacting to contemporary protection issues raised within Indian Country, and also are examining how their research, tools, and praxis effect tribal knowledge, and use and management of natural (re)sources (Fouberg 1997; Shute and Knight 1995; Stea 1984; Wishart 1990, 1995*a*).

Water has been of special concern to geographers working on reservations in the arid and semi-arid western US. The fact that water is not only physically and economically crucial, but also an agent of social cohesion and cultural reaffirmation may be self-evident, but few human geographers seem to have recognized this where Indians are concerned. The major exception is Berry (1997*a*, 1998*a*, 1998*b*, 2000), who has published four recent articles comparing the role of water for two Paiute communities and one non-Indian community in northwestern Nevada, and the role of water in the historical geography of coastal California.

Newton (1995) has looked at response to flood hazard in the North. Endter (1987) and Berry (1993, 1998*a*) have examined the significance of cultural values in the actions, policies, and court decisions that effect water allocation for tribes. McNally (1993, 1994) and Jacobsen

(1992) have considered how water project development and marketing of water rights have been influenced by congressional actions and negotiated water rights settlements. The sort of work McGuire (1991) has done on the rhetorical analysis of water rights legislation and associated hearings and media coverage is another area where geographers could contribute more.

Significant contributions also have been made in the analysis of tribal political control over wildlife, domesticated livestock, and plant (re)sources. Silvern (1995*a*, 1995*b*, 1999) analyzed tribal control of treaty-guaranteed, off-reservation fish and wildlife through examination of the structure of conflict between hunters and fishers in Wisconsin and among six Ojibwe groups. And when it comes to pauperizing a population, the government-ordered Navajo sheep-reduction program of the 1930s is probably the most infamous example (Flanders 1998). Will Graf and others confirmed that when it comes to Indian–white relations "a little scientific knowledge [is] a dangerous commodity" (Graf 1992: 10–11). Graf finds the Navajo correct in asserting that climate variation was the cause of Colorado River sedimentation, not the number of sheep they grazed. Yet, the Bureau of Reclamation and Soil Conservation Service were strangely comfortable in recommending the reduction program when they had only incomplete data.

In jarring contrast, co-management agreements for wildlife and other biota are so common across Canada now that one writer thinks there is sufficient impetus for including them as part of the constitutional rights of First Nations (Notzke 1995). Sneed (1997) describes one such agreement in Canada's Kluane National Park in Yukon, comparing it to the apparently dismal consequences of a similar effort across the border in Wrangell-St Elias National Park in the US. However, successful cooperative co-management arrangements are not completely unknown in the US, as Waage (2001) demonstrates in a fascinating story of the salmon recovery effort in northeastern Oregon, where a Nez Perce and Euro-American alliance evolved out of concerted efforts to articulate a shared ideology and a reimagined past.

Sovereignty, Dispossession, Land Claims, and Land Restoration

Sovereignty is a subject of overwhelming importance to North American Indians made all the more potent by recent attacks on Indian sovereignty in the US Congress. A key element is the relationship of land claims and land rights to contemporary tribal powers and land-use decision-making (Fixico 1998; Gonzalez and Cook-Lynn 1999; Riggs 2000). Sovereignty is also at the center of the assertion of tribal powers and autonomy within state, provincial, and national political structures. In particular, the recognition of tribal governments as legitimate controlling agents of tribal (re)sources has put sovereignty center-stage among natural-resource geographers, economic geographers and planners, and cultural and political geographers. For example, Winchell (1992, 1996) described the position of tribal government and the concept of inherent sovereignty as the basis of tribal powers while offering models for incorporation of inherent sovereignty within contemporary comprehensive planning for tribes, and Wishart (1996) has recognized that contemporary reservations can work effectively in some cases as bulwarks against further dispossession by attacks on sovereignty. Bays and Fouberg (2002) edited a set of the most recent papers by geographers on the subject of tribal/state relations as played out in government-to-government encounters. See also Ch. 12, Political Geography, for related work.

Marks (1998) broadly presents the record and legacy of dispossession, and Bjorklund (1992) has the best single-map summary of the continent-wide geography of Indian removals during the colonial period. At a more local scale, we have Comeaux's (1991) description of how water and development economics were used to dispossess the Pima and Maricopa, Hannah's (1993, 2002) work on Oglala Lakota resistance to dispossession, Bays's (1998) book on Cherokee dispossession in relation to settlement patterns, Wishart's (1994) J. B. Jackson prize-winning book detailing the elimination of people from nineteenth-century Nebraska, and Shipek's (1988) volume on southern California where Indians were literally pushed into the rocks. The latter two are the culmination of decades of archival and, in Shipek's case, ethnographic fieldwork. Notable too is the spate of recent books on the elimination of an Indian presence in the making and maintenance of the US National Park System (Burnham 2000; Keller and Turek 1998; Spence 1999). Bertolas (1998) also examined this kind of dispossession, where a place is redefined as a "wilderness" empty of humans to separate Indians, the Cree in this case, from their own socially constructed landscapes. Bone (1992) presents the same thing for the Canadian North. In its older forms, dispossession continues in many parts of the US, as Youngbear-Tibbets (1991) and Churchill (1993) have shown. But there are newer or less commonly used methods too, including gerrymandering the Navajo reservation to reduce its representation (Phelps 1991), and contesting a tribe's right to select its own electricity provider, thereby calling the

"Indian-ness" of, in one case, the Ft. Totten Reservation, into question in a court of law (Wishart and Froehling 1996). By tracking the history of land ownership and population dynamics at Ft. Totten, geographers Wishart and Froehling were able to assist the Sioux and demonstrate another way geographers can apply their skills in the courtroom.

Political geographer Imre Sutton (1991) summarized the legal meaning of "Indian Country," an entity upheld by the US Supreme Court on a number of occasions, and one that provides a legal basis for the numerous land claims brought in efforts to regain dispossessed land. Sutton has used the *American Indian Culture And Research Journal* on three occasions, in 1988, 1991, and 2000, as an outlet for hosting "mini-symposia" on Indian land claims. The most recent one analyzes the viability of settlement acts, ongoing aboriginal land title, conflict over submerged lands, refusals of payment, resource access, and many other issues (Sutton 2000). Behnke (2002) examined the special case of Alaska, where an unusual human and legal history have made native claims to subsistence rights the specific focus of broader land and resource claims. Land claims is also the subject of a detailed study by Churchill (2000), who provides a much-needed corrective to the illusion of a beneficent and culturally neutral Indian Claims Commission dispensing, between 1946 and 1978, what those in the federal government thought would be final justice in the settlement of Indian land claims. Meanwhile, Hallock (1996) has argued for a thorough redefinition of Indian Country itself.

Canadian geographers also have been active in land-claims research. Duerden (1996) sees land claims in Yukon as a land location-allocation problem, Peters (1992) draws lessons from the Cree situation, and Usher *et al.* (1992) provide a valuable diachronic summary of tenure systems, Indian and Inuit concepts of rights, and ongoing claims in British Columbia and the North.

Although Parker (1989) and others have summarized the US situation, some US geographers appear willful in their studied disregard for Indian land rights. A 1999 issue of *Political Geography* contains a debate over the future of the Hanford Plant just across the Rattlesnake Hills from the Yakama Reservation in south central Washington. A respected political geographer argued that Yakama Indian rights to use and live on the land at Hanford, rights guaranteed by federal treaty, were those only of a "local interest group" that had not exercised these "local customs" in the 50 or so years since Hanford was built (Morrill 1999). A critic (Martin 1999) argued, to no apparent effect in the original author's rejoinder, that the Yakama should be understood legally as a nation pursuing intergovernmental relations with the federal government, not as just another interest group; that the Yakama did not merely fail to exercise their treaty rights but had actually been removed from the area that became Hanford and had long been prevented from returning; and that geographers' idea of scale is sometimes so rigid and fixed that they make mistakes when judging jurisdictional bounds in Indian Country.[2]

Out of the claims process can come land restorations, although it is highly uncommon in the US. Morehouse (1996) recounts the success of the Havasupai in regaining trust over 185,000 acres of Grand Canyon National Park. Five other tribes gained very little, but Morehouse's analysis suggests the Havasupai were more successful because they broadened their rhetorical arguments beyond the narrow concerns of the moment to include most of the other interests in the Grand Canyon involved in that negotiation. Similarly, Haberfeld (2000) recounts the remarkable story of how a small group, the 300-member Timbisha Shoshone, reacquired a non-contiguous but permanent land base of 10,000 acres in and around Death Valley National Monument. Sutton (1994) has called for the restoration of small parcels of national forest land around Los Angeles to Indians in southern California as a means of ending a history of maltreatment and confrontation over ownership.

Sacred Land

Another aspect of the colonial legacy has been the need for Indians to document their cosmology and religion as it applies to the non-Indian idea of sacred land, although the federal government did not allow "sacred land" as a legal land-claim defense until the 1970s. Blake (1999), Martin (1991), Jett (1995), and Forbes-Boyte (1997, 1999*a*, 1999*b*) explore how elements of Tohono O'odham, Navajo, Cheyenne, and Lakota spirituality have been embedded in particular places. Such works encourage the reader to consider the significance of spiritual connections to the environment in light of contemporary residential development and other land management activities, including outdoor recreation, legislative action, and the outcomes of litigation. These connections are seldom straightforward however. Jett (1992*a*) suggested there may be varying degrees of sanctity for Navajo, and Griffith (1992) explained in

[2] A 1992 issue of the same journal (then named *Political Geography Quarterly*) was devoted to a debate on the meaning of the year 1492. That debate too was cast in Eurocentric terms, even as many of the contributors agreed that Eurocentrism was especially inappropriate.

detail the multivalent possibilities of places where ideas of sacredness held by the Tohono O'odham, Yaqui, Mexican, and Anglo-American peoples intersect. Similarly, Sundstrom (1997) has attempted to set the record straight for the Black Hills using archival documents in an ethnohistorical reconstruction describing what is held sacred in the area, since when, and by whom (at least seven Indian groups). Her research framework might be especially useful in other places comparably layered with stereotypes and misunderstanding. Research on indigenous sacred landscapes reflects broader trends within geography that emphasize the role of spirituality in actively connecting people with their environments through experience, practice, and myth (Pulido 1998). See also Ch. 7, Cultural Geography, for more related work.

Popularizing sacred lands has not been without repercussion of course. For example, American Indians have been dehumanized and native cultures trivialized when non-Indian environmentalists fixate on indigenous ecological spirituality and activity while ignoring other significant aspects of their community and culture. Willems-Braun (1997) makes this point about marginalization of the Nuu-chah-nulth of British Columbia, who have been denied an active role in debates over wilderness preservation because a perspective on the environment has frequently been imposed upon them rather than asserted by them.

Economic Development and Planning

More than three decades ago, Brian Goodey (1970) called on geographers to work as researcher-advocates in support of Indian tribes and their economic development. Since that time, the nature of land use, economic development, and tourism, especially as related to gambling, have undergone rapid change in Indian Country, which has increased the demand for geographic research. Many of these research efforts have been viewed positively by tribes, and there is an increase in the number of geographers who work for tribes directly as consultants, or who have otherwise pursued research on American Indian aspects of land use and economic development.

The most widely recognized academic study of tribal economic development has been a series of extensive field-based studies of tribal businesses across the US by Cornell and his associates, who have worked closely with geographers and presented their findings at AAG annual meetings (Cornell 1992; Cornell and Gil-Swedberg 1995; Cornell and Kalt 1990, 1996). These studies examined successful economic development projects on reservations across the nation, linking both quantitative analysis of tribes, settings, activities, and economic impacts with qualitative assessment of project complexity in an effort to frame an attribute model of tribal economic development success. In Canada, Anderson and Bone (1995) undertook a similar effort. They inventoried economic development strategies among First Nations across Canada and advocated a "contingency perspective" as a theoretical framework.

Geographers remain active in case studies of specific tribal lands and economic development, focusing especially on traditional concepts of land (Dobbs 1997; Frantz 1996*a*, *b*; Henderson 1992*a*; Van Otten 1992). There is also a literature identifying difficulties in using US Census, Census Canada, Canada's Department of Indian Affairs and Northern Development (DIAND), and US Bureau of Indian Affairs (BIA) data for analysis of population and economic and community development (Armstrong 1994; Eschbach 1993; Gregory and Daly 1997; McKean *et al.* 1995; Shumway and Jackson 1995; Stephenson *et al.* 1995). However, no geographer had published a comprehensive book-length assessment of American Indian economic development in English until 1999, when Klaus Frantz's (1999) book appeared. Titled *Indian Reservations in the United States*, the volume is actually a thorough statistical treatment of economic conditions among Indians living primarily in Arizona.

Land issues have gained prominence within the urban planning literature as well. Zaferatos (1998) described the power context for community planning using a case study, while Pinkham and Ruppert (1996) edited a special issue of *Landscape and Urban Planning* on tribal planning, and Sweeney (1996) outlined tribal planning law. Pinkham's (1996) perspective on environmental and planning issues, emanating from his position as a Nez Perce tribal leader, is especially noteworthy because it is a rare contribution to the academic literature by a tribal elder, and because it helped encourage geographers and others to pursue efforts to document and support the efforts of tribal leaders.

A crucial issue in tribal planning is the need to establish a framework for government-to-government relations between tribal governments and local city, county, and state governments when work must be accomplished in a regional context. A major effort in this regard was carried out by geographer Shirley Solomon (1990, 1992) through the Northwest Renewable Resources Center (1993), which produced extensive coordination among tribal groups and local governments. Models for government-to-government relations and the roots of tribal government and sovereignty are presented in a recent guide for positive tribal/county relations produced by the Northwest Renewable Resources Center (1992).

Another important planning issue, but one only rarely studied by geographers, is health care. One study in Montana found that Indian women were twice as likely as non-Indians to have to travel outside their county to a birthing facility, and even more likely to receive poor obstetric care near home (Von Reichert *et al.* 1995). Another study on Indian Country health care found that access to health services was severely restricted on the Round Valley Reservation in northern California, not only due to the extreme distances between the reservation and comprehensive-care facilities, but also because trained medical and dental professionals could not be consistently retained (Dillinger 2002).

Gambling and Tourism

Federal legislation in the 1980s and 1990s made gambling a potentially significant component of economic development in Indian Country. Tribal gambling (or "gaming," as it is euphemistically termed) facilities, laws, and practices are examined by geographers in two edited books (Lew and Van Otten 1998; Meyer-Arendt and Hartmann 1998) and in articles by Winchell *et al.* (1997, 1998), and d'Hauteserre (1998), but Mason's (2000) in-depth study of gambling in New Mexico and Oklahoma, explaining its link to sovereignty and national politics, is the most penetrating to date. Casino gambling on Indian land has had a significant impact on gambling overall in the US, and has potential as a new source of revenue and jobs in Indian Country. Specific case studies of gambling issues and casino development found in the above works indicate generally rapid growth with, at best, a mix of positive and negative economic impacts for tribes in many regions. Issues of concern include geographic distribution of profits, persistently high regional unemployment, reorganization and rescaling of regional political economies, short-term gains versus long-term stability, and increasing intra-regional competition among many gambling locations.

Tourism on American Indian reservations, the sale of arts and crafts, and the other aspects of cross-cultural touristic encounters have long had a big impact in Indian Country, though differentially so from region to region. In the past, such activities were generally directed and controlled by non-Indians, and the bulk of revenues and proceeds went to non-Indians. The significance of cultural (re)sources in tourism has been studied by Jett (1990), while Lew (1996) and Lew and Van Otten (1998) emphasize the impacts of tourism and the increased control exercised by Indians in managing tourism facilities and activities. Woodley (1991) advocated a participatory-planning approach where consultants know all relevant cultures brought into contact at tourist visitor centers. D'Arcus (2000), perhaps the most skeptical geographer writing on Indian tourism, examined the ongoing reproduction of New Mexico's "land of enchantment" image through promotion of the tourist gaze in the city of Gallup.

Landscape, Representations, and Identities

Post-colonialism and Landscapes of the Mind

Recently, geographers have begun applying post-colonial or other social theory to American Indian studies. Anglophone post-colonial studies generally have emerged from and focus on the British colonial experience; accordingly, post-colonial American Indian research is well represented within Canadian universities, most notably the University of British Columbia. These researchers seek to deconstruct and assess the imprint of a European worldview on the lands, minds, and bodies of Indians. They draw inspiration from the Anglophone post-colonial literature (e.g. Said 1993), as well as Lefebvre, Lacan, Derrida, and Foucault. These diverse theoretical currents are found in the work of Brealey (1995), Clayton (1996), Hannah (1993), Harris (1997), Ripmeester (1995), and Sparke (1998) on colonial surveillance and control of indigenous spaces, the appropriation of such spaces through ethnocentric textual or cartographic representation, and the various Indian responses to such practices. Blomley (1996) has employed similar themes, investigating Indian blockades in British Columbia as a mechanism for tribal reappropriation of space. Given this critique, Galois (1994) has attempted to provide a revised empirical overview of historical Kwakiutl settlement that is free from colonial bias.

Explaining the mechanisms of representation and identity is fundamental in Silvern's (1995*b*) work on Ojibwe/non-Indian collisions over treaty rights in Wisconsin, Deur's (2000*b*) examination of the Makah whaling revival, Crow's (1994) admonition that geographers should spend more time in Indian communities, Schnell's (2000) study of Kiowa places, Berry's (1997*a*) comparative work on waterscapes and representation for Paiute and non-Paiute communities in northern

Nevada, and it is linked explicitly to matters of gender in Morin's (1998), Peters's (1997, 1998), and Wishart's (1995*b*, 1997, 1999) various studies of women's constructions of Indians-as-others in historic "contact zones" such as railroad depots, of the collision between definitions of "city person" and "Indian," and of other questions associated with gender and representation. In an unusual link to population geography, Jackson (1994) sifted the evidence of population decline in the US Southwest and developed the thesis that deaths were related not only to disease but to place/space destruction and reductions in cohesion, identity, and sense of place.

Still others have studied the Men's Movement's appropriation and distortion of aspects of Indian environmental and social relations (Bonnett 1996), the contribution George Catlin's early nineteenth-century paintings of Indian life made to building a national racial heritage (John 2001), how Indian photographic images were used by railroads as promotional devices to encourage tourists to visit US national parks (Wyckoff and Dilsaver 1997), the needs of contemporary Arctic environmentalists in identifying with stranded whales and stripped seals more than with the Inuit (Wenzel 1992), and how identity is materialized in particular landscapes and sense of place (Brooks 1995; Erdrich 1988; Rundstrom 1992; Silko 1987, 1995). Unpacking mental landscape representations is the focus in work on Inuit/Eskimo navigational skills (Pelly 1991; Sonnenfeld 1994). And finally, the intimate links among social structure, individual and group identity, and place embodied in place-names has long been and continues to be of special interest to geographers (Deur 1996; Fair 1997; Herman 1999; Jett 1997; Muller-Wille 1987; Rundstrom 1996*a*).

Maps and GIS

The study of Indian maps and GIS has become quite popular since 1990, and among other trends, led to a revitalized historical cartography due in part to research on Indian and Inuit maps, mapping discourse, and the interpretations made of them by explorers and contemporary geographers (Belyea 1992, 1997, 1998; Galloway 1998; Harley 1994; Lewis 1987*a*, *b*, 1991; Nabokov 1998; Pearce 1998; Rundstrom 1990, 1991, 1996*b*; Sparke 1995). This work views Indian and Inuit maps made in the past as emanating from different assumptions and discourses about the world as a home for humans and others, different especially from what Europeans and Euro-North Americans were producing. These maps actively disrupted European discourse on geography, history, and identity, and represent one side of what was, and still is to some extent, an apparently unbridgeable gap in worldviews held by colonizers and the colonized. The other side, the European and Euro-North American mapping of Indian Country, has also come under scrutiny (Brealey 1995; Cole 1993; Harley 1992; Lewis 1991; Rundstrom 1993, 1998). Geographers have questioned these maps, often finding them ideological weapons serving non-Indian interests.

Three major works were also published that vastly improved our understanding of indigenous cartography in North America, led by Malcolm Lewis's (1998*a*) long-awaited collection of scholarly essays, entitled *Cartographic Encounters: Perspectives on Native American Mapmaking and Map Use*, and his massive cartobibliographic summary for David Woodward's History of Cartography Project (Lewis 1998*b*). Finally, Mark Warhus published his own summary of Indian mapping, *Another America: Native American Maps and the History of Our Land* (Warhus 1997). All three are replete with outstanding map reproductions and new substantive interpretations.

A number of important atlases now augment American Indian studies, all of which are either totally or significantly devoted to portraying Indian geographies (Champagne-Aishihik Indian Band 1988; R. C. Harris 1987; Prucha 1990; Tanner 1987). Although they are less impressive graphically, several other atlases deserve more attention because they emanate from Indians or Inuit themselves, and represent a rising trend of "mapping back" or "counter-mapping" the colonizers. *A Zuni Atlas* (Ferguson and Hart 1985) was published because Zuni elders decided it was time to make certain protected geographical information available to the public for the first time in their effort to press a land claim. The Inuit have been especially active in counter-mapping, first with a map series containing their own place-names (Avataq Cultural Institute 1991), then with the *Nunavut Atlas* (Tungavik Federation of Nunavut 1992), an extensive compilation of wildlife patterns and other environmental data gathered from Inuit elders.

In the past seven years, a national American Indian GIS association has developed, linked to ESRI regional, national, and international conferences, and many tribes have been very active participants in remote-sensing and GIS applications and research (Bailey *et al.* 2001*b*; Bailey *et al.* 2001*a*; Winchell and Marchand 1994). Continued efforts to link the US Census Bureau's TIGER-file maps and updated map products with tribal information have also helped tribal governments initiate detailed in-house spatial-data analysis. The BIA also established a GIS arm that encourages and stockpiles tribal GIS in their own library. For example, Marozas (1991) suggested ways tribes could use GIS, with BIA assistance, in water rights litigation.

Partly in response to these developments, and partly because many Indian communities are deeply suspicious of BIA-imposed tribal governments where GIS managers are housed, a consortium of tribes in the northern plains and another in the Rio Grande corridor have closed their doors to the BIA and other federal agencies, denying access to their GIS in an attempt to protect sensitive and sacred information. These and other problems caused one geographer to raise questions about the path of GIS development in Indian Country (Rundstrom 1995). Thus, GIS has come to be seen by T. Harris and Wiener (1998) and many other geographers as a contradictory technology that both empowers and marginalizes people and communities. Harris and Wiener review the "GIS and Society" debates that emerged in geography in the 1990s, recommending "community-integrated" GIS that focus on local empowerment.

Identity and Pan-Indianism

In a 1976 interview, the Kiowa writer, N. Scott Momaday (1976), identified an ecological sensibility in Indian thought, one distinguished from that of non-Indian North Americans by the presence of what he termed a sense of "reciprocal appropriation." Then, geographers Stea and Wisner (1984) wrote of a pan-Indian ecological worldview projected outward in solidarity with other indigenous or Fourth World peoples. Since that time some geographers have examined pan-Indianness as an evolving identity complex crucial to understanding the link between place and action (Gonzalez 1998; Pulido 1996; Rundstrom 1994).

Unraveling some of the complexities of group identity, environmental consciousness, and (re)source use is a related endeavor undertaken for southwestern Indian women by Parezo *et al.* (1987) and Smith and Allen (1987). Herman's work (1995) also has yielded insights into changes in native identity, Hawaiian identity in this case, in response to colonization and corresponding shifts in environmental knowledge and language.

Research, Writing, and Teaching

Increased self-reflection in research, writing, and teaching is another recent trend influencing those working with and teaching about American Indians. Seeking appropriate contexts for interaction and research in conjunction with Indian governments and peoples, geographers have read what Indian writers have had to say to academics about intellectual property rights, citation protocols, and other matters associated with colonialism and the written word (e.g. Mihesuah 1996; Ruppert 1996). Rundstrom and Deur (1999) and Tompkins (1986) have suggested ways to translate these ideas into ethical practice.[3] Finally, teaching about Indians and Indian County has begun to receive more attention. Five papers have appeared explaining rationales for pedagogical strategies used by geographers in the classroom (Berry 1997*b*; Henderson 1992*b*, 1997; Rundstrom 1997; Van Otten 1997).

Proposals for Future Research

There is a sense among many members of the American Indian Specialty Group that geographers can and should become more active and involved in issues of importance to the residents of Indian Country. Some of the work cited here leads in that direction, but geographers can do much more: historical studies exploring continuities in land use and governance for land claims; land-use and place-name mapping and GIS; examinations of the spatial basis for self-governance and self-determination; critical approaches to the role of space and place in the social construction of Indians via public perceptions, legislative agendas, corporate intentions, and classroom teaching; continued work in deconstructing colonial legacies and post-colonial discourse, and striving toward genuine polyvocality; analysis of the health-care distribution system of the majority society and its relationship to alternative medical systems available through local cultural practices. An omission that seems particularly glaring is the subject of urbanism and urban Indians (Peters' work is the major exception), which is all the more remarkable given the much greater attention geographers gave the subject in the 1970s. Geographers might also work on reconnecting ideology and material landscape, which have largely been explored separately. In all of this, and as we believe this chapter shows, it is essential to remember that American Indian and Inuit communities throughout North America are places where research and an understanding of the meaning of sovereignty and self-determination now go hand in hand.

[3] Some geographers engaged in non-Indian research find it useful to employ Indian geographies as a metaphor in their writing (e.g. Smith 1992), but also, strangely, as a means to express disdain (Page and Walker 1994).

References

Anderson, Robert, and Bone, Robert (1995). "First Nations Economic Development: A Contingency Perspective." *Canadian Geographer*, 39: 120–30.

Armstrong, R. (1994). "Developing Social Indicators for Registered Indians Living on Reserve: the DIAND Experience." *Social Indicators Research*, 32: 235–49.

Avataq Cultural Institute (1991). *Inuit Place Name Map Series*. Inukjuak: Avataq Cultural Institute.

Bailey, K., Frohn, R., Beck, R., and Price, M. (2001*a*). "Remote Sensing Analysis of Wild Rice Production Using Landsat 7 for the Leech Lake Band of Chippewa in Minnesota." *Photogrammetric Engineering and Remote Sensing*, 67: 189–92.

Bailey, K., Beck, R., Frohn, R., Pleva, D., Plumber, D., Price, M., Krute, R., Ramos, C., and South, R. (2001*b*). "Native American Remote Sensing Distance Education Prototype (NARSDEP)." *Photogrammetric Engineering and Remote Sensing*, 67: 193–7.

Ballas, Donald (1966). "Geography and the American Indian." *Journal of Geography*, 65: 156–68.

Barrett, Samuel (1908). *Ethno-geography of the Pomo and Neighboring Indians*. Berkeley: University of California Press.

Basso, Keith (1996). *Wisdom Sits in Places: Landscape and Language Among the Western Apache*. Albuquerque: University of New Mexico Press.

Batek, M., Rebertus, A., Schroeder, W., Haithcoat, T., Compas, E., and Guyette, R. (1999). "Reconstruction of Early Nineteenth-Century Vegetation and Fire Regimes in the Missouri Ozarks." *Journal of Biogeography*, 26: 397–412.

Bays, Brad (1998). *Townsite Settlement and Dispossession in the Cherokee Nation, 1866–1907*. New York: Garland.

Bays, Brad, and Fouberg, Erin (eds.) (2002). *The Tribes and the States: Geographies of Intergovernmental Interaction*. Lanham, Md.: Rowman & Littlefield.

Behnke, Steven (2002). "Alaska's Contested Rural Landscapes and the Subsistence Claims of Alaska Natives," in K. Berry and M. Henderson (eds.), *Geographical Identities of Ethnic America: Race, Space, and Place*. Reno: University of Nevada Press, 149–73.

Belyea, Barbara (1992). "Amerindian Maps: The Explorer as Translator." *Journal of Historical Geography*, 18: 267–77.

—— (1997). "Mapping the Marias: The Interface of Native and Scientific Cartographies." *Great Plains Quarterly*, 17: 165–84.

—— (1998). "Inland Journeys, Native Maps: Hudson's Bay Company Exploration 1754–1802," in Lewis (1998*a*: 135–55).

Berry, Kate (1993). "The McCarran Water Rights Amendment of 1952: Policy Development, Interpretation, and Impact on Cross-Cultural Water Conflicts," Ph.D. thesis, University of Colorado, Boulder.

—— (1997*a*). "Of Blood and Water." *Journal of the Southwest*, 39: 79–111.

—— (1997*b*). "Projecting the Voices of Others: Issues of Representation in Teaching Race and Ethnicity." *Journal of Geography in Higher Education*, 21: 283–9.

—— (1998*a*). "Race for Water? Native Americans, Eurocentrism, and Western Water Policy," in D. Camacho (ed.), *Environmental Injustices, Political Struggles: Race, Class, and the Environment*. Durham, NC: Duke University Press, 101–24.

—— (1998*b*). "Values, Ideologies and Equity in Water Distribution: Historical Perspectives from Coastal California, United States," in R. Boelens and G. Davila (eds.), *Search for Equity: Concepts of Justice and Equity in Peasant Irrigation*. Assen: Van Gorcum, 189–97.

—— (2000). "Water Use and Cultural Conflict in 19th Century Northwestern New Spain and Mexico." *Natural Resources Journal*, 40: 759–81.

Bertolas, Randy (1998). "Cross-Cultural Environmental Perception of Wilderness." *Professional Geographer*, 50: 98–111.

Bjorklund, E. (1992). "The Namerind Continent's Euro-American Transformation, 1600–1950," in D. Janelle (ed.), *Geographical Snapshots of North America*. New York: Guilford Press, 5–10.

Blake, Kevin (1999). "Sacred and Secular Landscape Symbolism at Mount Taylor, New Mexico." *Journal of the Southwest*, 41: 487–509.

Blake, Kevin, and Smith, Jeffrey (2000). "Pueblo Mission Churches as Symbols of Permanence and Identity." *Geographical Review*, 90: 359–80.

Blomley, Nicholas (1996). "Shut the Province Down: First Nations Blockades in British Columbia, 1984–1995." *BC Studies*, 3: 5–35.

Boas, Franz (1934). *Geographical Names of the Kwakiutl Indians*. Columbia University Contributions to Anthropology, 20. New York: Columbia University Press.

Bone, Robert (1992). *The Geography of the Canadian North: Issues and Challenges*. Toronto: Oxford University Press.

Bonnett, Alastair (1996). "The New Primitives: Identity, Landscape and Cultural Appropriation in the Mythopoetic Men's Movement." *Antipode*, 28: 273–91.

Brealey, Ken (1995). "Mapping Them 'Out': Euro-Canadian Cartography and the Appropriation of the Nuxalk and Ts'ilhqot' in First Nations' Territories, 1793–1916." *Canadian Geographer*, 39: 140–56.

Brody, Hugh (1988). *Maps and Dreams: Indians and the British Columbia Frontier*. Vancouver: Douglas & McIntyre.

—— (1991). *The People's Land: Inuit, Whites, and the Eastern Arctic*. Vancouver: Douglas & McIntyre.

Brooks, Drex (1995). *Sweet Medicine: Sites of Indian Massacres, Battlefields, and Treaties* (with essay by Patricia Limerick). Albuquerque: University of New Mexico Press.

Brown, James (1991). *Aboriginal Cultural Adaptations in the Midwestern Prairies*. New York: Garland.

Brown, Kenneth (1995). *Four Corners: History, Land, and People of the Desert Southwest*. New York: HarperCollins.

Burley, David, and Horsfall, Gayel (1989). "Vernacular Houses and Farmsteads of the Canadian Metis." *Journal of Cultural Geography*, 10: 19–33.

Burnham, Philip (2000). *Indian Country, God's Country: Native Americans and the National Parks*. Washington, DC: Island Press.

Butzer, Karl (1990). "The Indian Legacy in the American Landscape," in M. Conzen (ed.), *The Making of the American Landscape*. Boston: Unwin Hyman, 27–50.

Carlson, Alvar (1972). "A Bibliography of the Geographical Literature on the American Indian, 1920–1971." *Professional Geographer*, 24: 258–63.

Carter, George (1945). *Plant Geography and Culture History in the American Southwest.* Viking Fund Publications in Anthropology, 5. New York: Wenner-Gren.

Champagne-Aishihik Indian Band (1988). *From Trail to Highway Kwaday kwatan ts'an ek'an tan kwatsin: A Highway Guide to the Places and the People of the Southwest Yukon and Alaska Panhandle.* Victoria, BC: Morriss Printing.

Churchill, Ward (1993). *Struggle for the Land: Indigenous Resistance to Genocide, Ecocide, and Expropriation in Contemporary North America.* Monroe, Me.: Common Courage Press.

— (2000). "Charades, Anyone? The Indian Claims Commission in Context." *American Indian Culture and Research Journal*, 24: 43–68.

Clayton, Daniel (1996). "Islands of Truth: Vancouver Island from Captain Cook to the Beginnings of Colonialism," Ph.D. thesis, University of British Columbia.

Cole, Daniel (1993). "One Cartographic View of American Indian Land Areas." *Cartographica*, 30: 47–54.

Comeaux, Malcolm (1991). "Creating Indian Lands: The Boundary of the Salt River Community." *Journal of Historical Geography*, 17: 241–56.

Cornell, S. (ed.) (1992). *What Can Tribes Do? Strategies and Institutions in American Indian Economic Development.* American Indian Manual and Handbook Series, 4. Los Angeles: UCLA American Indian Studies Center.

Cornell, S., and Gil-Swedberg, M. (1995). "Sociohistorical Factors in Institutional Efficacy: Economic Development in Three American Indian Cases." *Economic Development and Cultural Change*, 43: 237–55.

Cornell, S., and Kalt, J. (1990). "Pathways from Poverty: Economic Development and Institution-Building on American Indian Reservations." *American Indian Culture and Research Journal*, 14: 89–125.

— (1996). *Harvard Project on American Indian Economic Development.* Los Angeles: UCLA American Indian Studies Center.

Cronon, William (1983). *Changes in the Land: Indians, Colonists, and the Ecology of New England.* New York: Hill & Wang.

Crosby, Alfred (1994). *Germs, Seeds, and Animals: Studies in Ecological History.* Armonk, NY: M. E. Sharpe.

Crow, Dennis (1994). "My Friends in Low Places: Building Identity For Place and Community." *Environment and Planning D: Society and Space*, 12: 403–19.

Cunningham, Keith (1989). "Navajo, Mormon, Zuni Graves: Navajo, Mormon, Zuni Ways," in R. Meyer (ed.), *Cemeteries and Gravemarkers: Voices of American Culture.* Ann Arbor: UMI Research Press, 197–215.

D'Arcus, B. (2000). "The 'Eager Gaze of the Tourist' Meets 'Our Grandfathers' Guns': Producing and Contesting the Land of Enchantment in Gallup, New Mexico." *Environment and Planning D: Society and Space*, 18: 693–714.

d'Hauteserre, Anne-Marie (1998). "Foxwoods Casino Resort: An Unusual Experiment in Economic Development." *Economic Geography: Special Issue for the 1998 Annual Meeting of the AAG, Boston, Massachusetts*, 112–21.

Denevan, William (1992*a*). *The Native Population of the Americas in 1492.* 2nd edn. Madison: University of Wisconsin Press.

— (1992*b*). "The Pristine Myth: The Landscape of the Americas in 1492." *Annals of the Association of American Geographers*, 82: 369–85.

— (1996). "Carl Sauer and Native American Population Size." *Geographical Review*, 86: 385–97.

Deur, Douglas (1996). "Chinook Jargon Placenames as Points of Mutual Reference: Discourse, Environment, and Intersubjectivity in an Intercultural Toponymic Complex." *Names*, 44: 291–321.

— (1999). "Salmon, Sedentism, and Cultivation: Towards an Environmental Prehistory of the Northwest Coast," in D. Goble and P. Hirt (eds.), *Northwest Lands, Northwest Peoples: An Environmental History Anthology.* Seattle: University of Washington Press, 119–44.

— (2000*a*). "A Domesticated Landscape: Indigenous Plant Cultivation on the Northwest Coast," Ph.D. thesis, Louisiana State University.

— (2000*b*). "The Hunt for Identity: On the Contested Targets of Makah Whaling," in J. Norwine and J. Smith (eds.), *Worldview Flux: Perplexed Values Among Postmodern Peoples.* Lanham, Md.: Rowman & Littlefield, 145–73.

— (2002). "Rethinking Precolonial Plant Cultivation on the Northwest Coast of North America." *Professional Geographer*, 54: 140–57.

Dillinger, Teresa (2002). "Coping With Health Care Delivery on the Round Valley Indian Reservation," in K. Berry and M. Henderson (eds.), *Geographical Identities of Ethnic America: Race, Space, and Place.* Reno: University of Nevada Press, 201–27.

Dobbs, G. (1997). "Interpreting the Navajo Sacred Geography as a Landscape of Healing." *Pennsylvania Geographer*, 35: 136–50.

Doolittle, William (1992). "Agriculture in North America on the Eve of Contact: A Reassessment." *Annals of the Association of American Geographers*, 82: 386–401.

— (2002). *Cultivated Landscapes of Native North America.* New York: Oxford University Press.

Duerden, Frank (1996). "Land Allocation in Comprehensive Land Claim Agreements." *Applied Geography*, 16: 279–88.

Endter, Joanne (1987). "Cultural Ideologies and the Political Economy of Water in the United States West: Northern Ute Indians and Rural Mormons in the Uintah Basin," Ph.D. thesis, University of California, Irvine.

Erdrich, Louise (1988). "A Writer's Sense of Place," in M. Martone (ed.), *A Place of Sense: Essays in Search of the Midwest.* Iowa City: University of Iowa Press, 34–44.

Eschbach, K. (1993). "Changing Identification Among American Indians and Alaskan Natives." *Demography*, 30: 635–52.

Fair, Susan (1997). "Inupiat Naming and Community History: The Tapqaq and Saniniq Coasts near Shishmaref, Alaska." *Professional Geographer*, 49: 466–80.

Ferguson, T., and Hart, E. (1985). *A Zuni Atlas.* Norman: University of Oklahoma Press.

Fixico, D. (1998). *The Invasion of Indian Country in the Twentieth Century: American Capitalism and Tribal Natural Resources.* Boulder: University Press of Colorado.

Flanders, N. (1998). "Native American Sovereignty and Natural Resource Management." *Human Ecology*, 26: 425–49.

Forbes-Boyte, Kari (1997). "Indigenous People, Land, and Space: The Effects of Law on Sacred Places, The Bear Butte Example," Ph.D. thesis, University of Nebraska.

— (1999*a*). "Fools Crow Versus Gullett: A Critical Analysis of the American Indian Religious Freedom Act." *Antipode*, 31: 304–23.

—— (1999*b*). "Litigation, Mitigation, and the American Indian Religious Freedom Act: The Bear Butte Example." *Great Plains Quarterly*, 19: 23–34.

Fouberg, Erin (1997). "Tribal Territory and Tribal Sovereignty: A Study of the Cheyenne River and Lake Traverse Indian Reservations," Ph.D. thesis, University of Nebraska.

Frantz, K. (1996*a*). "Education on the Navajo Indian Reservation: Aspects of the Maintenance of Cultural Identity as Seen from a Geographical Point of View," in K. Frantz and R. Sauder (eds.), *Ethnic Persistence and Change in Europe and America*. Innsbruck: University of Innsbruck, 223–46.

—— (1996*b*). "The Salt River Indian Reservation: Land Use Conflicts and Socio-Economic Change on the Outskirts of the City of Phoenix" ("Die Salt River Indianerreservation: Land-nutzungskonflikte und sozio-okonomisher Wandel am Rande der Grossstadt Phoenix"). *Geographische Rundschau*, 48: 206–12.

—— (1999). *Indian Reservations in the United States: Territory, Sovereignty, and Socio-Economic Change*. Chicago: University of Chicago Press.

Galloway, Patricia (1998). "Debriefing Explorers: Amerindian Information in the Delisles' Mapping of the Southeast," in Lewis (1998*a*: 223–40).

Galois, R. (1994). *Kwakwiul ai kiul ai wakw Settlements, 1775–1920: A Geographical Analysis and Gazetteer*. Vancouver: University of British Columbia Press.

Gartner, William (1997). "Four Worlds without an Eden: Pre-Columbian Peoples and the Wisconsin Landscape," in R. Ostergren and T. Vale (eds.), *Wisconsin Land and Life*. Madison: University of Wisconsin Press, 331–49.

Gonzalez, A. (1998). "The (Re)articulation of American Indian Identity: Maintaining Boundaries and Regulating Access to Ethnically-Tied Resources." *American Indian Culture and Research Journal*, 22: 199–225.

Gonzalez, Mario, and Cook-Lynn, Elizabeth (1999). *The Politics of Hallowed Ground: Wounded Knee and the Struggle for Indian Sovereignty*. Urbana: University of Illinois Press.

Goodey, Brian (1970). "The Role of the Indian in North Dakota's Geography: Some Propositions." *Antipode*, 2: 11–24.

Graf, William (1992). "Science, Public Policy, and Western American Rivers." *Transactions of the Institute of British Geographers*, 17: 5–19.

Gregory, R., and Daly, A. (1997). "Welfare and Economic Progress of Indigenous Men of Australia and the U.S.: 1980–1990." *Economic Record*, 73: 101–19.

Griffith, James (1992). *Beliefs and Holy Places: A Spiritual Geography of the Pimeria Alta*. Tucson: University of Arizona Press.

Gritzner, Janet (1994). "Native-American Camas Production and Trade in the Pacific Northwest and Northern Rocky Mountains." *Journal of Cultural Geography*, 14/2: 33–50.

Haberfeld, Steven (2000). "Government-to-Government Negotiations: How the Timbisha Shoshone Got Its Land Back." *American Indian Culture and Research Journal*, 24: 127–65.

Hallock, Dorothy (1996). "Second Contact: Redefining Indian Country," in K. Frantz (ed.), *Human Geography in North America: New Perspectives and Trends in Research*. Innsbruck: Department of Geography, University of Innsbruck, 195–208.

Hannah, Matthew (1993). "Space and Social Control in the Administration of the Oglala Lakota ("Sioux'), 1871–1879." *Journal of Historical Geography*, 19: 412–32.

—— (2002). "Varieties of Oglala Lakota Resistance to Displacement and Control," in K. Berry and M. Henderson (eds.), *Geographical Identities of Ethnic America: Race, Space, and Place*. Reno: University of Nevada Press, 116–29.

Hans, W. (1925). "The American Indian and Geographic Studies." *Annals of the Association of American Geographers*, 15: 86–91.

Harley, J. (1992). "Rereading the Maps of the Columbian Encounter." *Annals of the Association of American Geographers*, 82: 522–42.

—— (1994). "New England Cartography and the Native Americans," in W. Baker *et al.* (eds.), *American Beginnings: Exploration, Culture, and Cartography in the Land of Norumbega*. Lincoln: University of Nebraska Press, 287–313.

Harrington, J. (1916). *Ethnogeography of the Tewa Indians*. Bureau of American Ethnology Report, 29. Washington DC: Smithsonian Institution.

Harris, Cole (1995). "Towards a Geography of White Power in the Cordilleran Fur Trade." *Canadian Geographer*, 39: 131–40.

—— (1997). *The Resettlement of British Columbia: Essays on Colonialism and Geographical Change*. Vancouver: University of British Columbia Press.

Harris, R. Cole (1987). *Historical Atlas of Canada*, i. *From the Beginning to 1800*. Toronto: University of Toronto Press.

Harris, Trevor, and Daniel Wiener (1998). "Empowerment, Marginalization, and 'Community-Integrated' GIS." *Cartography and GIS*, 25: 67–76.

Henderson, Martha (1990). "Settlement Patterns on the Mescalero Apache Reservation Since 1883." *Geographical Review*, 80: 226–38.

—— (1992*a*). "American Indian Reservations: Controlling Separate Space, Creating Separate Environments," in L. Dilsaver and C. Colton (eds.), *The American Environment: Interpretations of Past Geographies*. Lanham, Md.: Rowman & Littlefield, 115–34.

—— (1992*b*). "Cultural Diversity in Geography Curriculum: The Geography of American Indians." *Journal of Geography*, 91: 113–18.

—— (1992c). "Maintaining Vernacular Architecture on the Mescalero Apache Reservation." *Journal of Cultural Geography*, 13/1: 15–28.

—— (1997). "Teaching Environmental Conservation from a Native-American Perspective," in L. Estaville and C. Rosen (eds.), *Teaching American Ethnic Geography*. Indiana, Pa.: National Council for Geographic Education, 105–10.

Herman, R. (1995). "Kalai'aina: Carving the Land. Geography, Desire and Possession in the Hawaiian Islands," Ph.D. thesis, University of Hawaii.

—— (1999). "The Aloha State: Place Names and the Anti-Conquest of Hawai'i." *Annals of the Association of American Geographers*, 89: 76–102.

Highleyman, Emily (1994). "Subsistence and Culture: The Gwich'in Indians in Alaska," Ph.D. thesis, University of Wisconsin-Madison.

Huntington, E. (1919). *The Red Man's Continent: A Chronicle of Aboriginal America*. New Haven: Yale University Press.

Jackson, Robert (1994). *Indian Population Decline: The Missions of Northwestern New Spain, 1687–1840*. Albuquerque: University of New Mexico Press.

Jacobs, Wilbur (1996). *The Fatal Confrontation: Historical Studies of American Indians, Environment, and Historians*. Albuquerque: University of New Mexico Press.

Jacobsen, Judith (1992). "The Navajo Indian Irrigation Project and Quantification of Navajo *Winters* Rights." *Natural Resources Journal*, 32: 823–53.

Jett, Stephen (1990). "Culture and Tourism in the Navajo Country." *Journal of Cultural Geography*, 11/1: 85–107.

— (1991). "Further Information on the Geography of the Blowgun and its Implications for Early Transoceanic Contacts." *Annals of the Association of American Geographers*, 81: 89–102.

— (1992*a*). "An Introduction to Navajo Sacred Places." *Journal of Cultural Geography*, 13/1: 29–39.

— (1992*b*). "The Navajo in the American Southwest," in A. Noble (ed.), *To Build in a New Land: Ethnic Landscapes in North America*. Baltimore: Johns Hopkins University Press, 331–44.

— (1994). *A Bibliography of North American Geographers' Works on Native Americans North of Mexico, 1971–1991*. Lawrence, Kan.: Haskell Indian Nations University in conjunction with the American Indian Specialty Group of the Association of American Geographers.

— (1995). "Navajo Sacred Places: Management and Interpretation of Mythic History." *Public History*, 17: 29–47.

— (1997). "Place-Naming, Environment, and Perception Among the Canyon de Chelly Navajo of Arizona." *Professional Geographer*, 49: 481–93.

John, G. (2001). "Cultural Nationalism, Westward Expansion and the Production of Imperial Landscape: George Catlin's Native American West." *Ecumene*, 8: 175–203.

Jordan, Terry, and Kaups, Matti (1989). *The American Backwoods Frontier: An Ethnic and Ecological Interpretation*. Baltimore: Johns Hopkins University Press.

Kay, Jeanne (1991). "Landscapes of Women and Men: Rethinking the Regional Historical Geography of the United States and Canada." *Journal of Historical Geography*, 17: 435–52.

Keller, Robert, and Turek, Michael (1998). *American Indians and National Parks*. Tucson: University of Arizona Press.

Kniffen, Fred B. (1939). "Pomo Geography." *University of California Publications in American Archaeology and Ethnology*, 36: 353–400.

Kniffen, Fred B., Gregory, Hiram, and Stokes, George (1987). *The Historic Indian Tribes of Louisiana: From 1542 to Present*. Baton Rouge: Louisiana State University Press.

Kroeber, A. (1939). *Cultural and Natural Areas of Native North America*. Berkeley: University of California Press.

LaDuke, Winona (1999). *All Our Relations: Native Struggles for Land and Life*. Cambridge, Mass.: South End Press.

Lew, A. (1996). "Tourism Management on American Indian Lands in the U.S.A." *Tourism Management*, 17: 355–65.

Lew, A., and Van Otten, George (eds.) (1998). *Tourism and Gaming on American Indian Lands*. Elmsford, NY: Cognizant Communications.

Lewis, G. (1987*a*). "Indian Maps: Their Place in the History of Plains Cartography," in F. Luebke *et al.* (eds.), *Mapping the North American Plains*. Norman: University of Nebraska Press, 63–80.

— (1987*b*). "Misinterpretation of Amerindian Information as a Source of Error on Euro-American Maps." *Annals*, 77: 542–63.

— (1991). " 'La Grande Rivière et Fleuve de l'Ouest': The Realities and Reasons Behind a Major Mistake in the Eighteenth-Century Geography of North America." *Cartographica*, 28: 54–87.

— (ed.) (1998*a*). *Cartographic Encounters: Perspectives on Native American Mapmaking and Map Use*. Chicago: University of Chicago Press.

— (1998*b*). "Maps, Mapmaking, and Map Use by Native North Americans," in D. Woodward and G. Lewis (eds.), *The History of Cartography*, vol. ii bk. 3. *Cartography in the Traditional African, American, Arctic, Australian, and Pacific Societies*. Chicago: University of Chicago Press, 51–182.

McGuire, Thomas (1991). "Indian Water Rights Settlements: A Case Study in the Rhetoric of Implementation." *American Indian Culture and Research Journal*, 15: 139–69.

McKean, J., Taylor, R., and Liu, Wen (1995). "Inadequate Agricultural Databases for American Indians." *Society and Natural Resources*, 8: 361–6.

McNally, M. (1993). "The 1985 Fort Peck-Montana Compact: A Case Study," in T. McGuire *et al.* (eds.), *Indian Water in the New West*. Tucson: University of Arizona.

— (1994). "Water Marketing: The Case of Indian Reserved Rights." *Water Resources Bulletin*, 30A: 963–70.

Marks, Paula (1998). *American Indian Dispossession and Survival*. New York: William Morrow.

Marozas, Bryan (1991). "The Role of Geographic Information Systems in American Indian Land and Water Rights Litigation." *American Indian Culture and Research Journal*, 15: 77–93.

Marsden, Susan, and Galois, Robert (1995). "The Tsimshian, The Hudson's Bay Company, and the Geopolitics of the Northwest Coast Fur Trade, 1787–1840." *Canadian Geographer*, 39: 169–83.

Martin, Deborah (1999). "Transcending the Fixity of Jurisdictional Scale." *Political Geography*, 18: 33–8.

Martin, E. (1991). "Between Heaven and Earth." *Planning*, 57: 24–34.

Mason, W. (2000). *Indian Gaming: Tribal Sovereignty and American Politics*. Norman: University of Oklahoma Press.

Meinig, Donald (1993). *The Shaping of America: A Geographical Perspective on 500 Years of History*, ii. *Continental America, 1800–1867*. New Haven: Yale University Press.

Meyer-Arendt, Klaus, and Hartmann, Rudi (eds.) (1998). *Casino Gambling in America: Origins, Trends, and Impacts*. Elmsford, NY: Cognizant Communications.

Mihesuah, Devon (ed.) (1996). "Writing About (Writing About) American Indians." *American Indian Quarterly*, 20: 1–108.

Momaday, N. (1976). "Native American Attitudes to the Environment," in W. Capps (ed.), *Seeing with a Native Eye: Essays on Native American Religion*. New York: Harper & Row, 79–85.

Morehouse, Barbara (1996). *A Place Called Grand Canyon: Contested Geographies*. Tucson: University of Arizona Press.

Morin, Karen (1998). "British Women Travellers and the Constructions of Racial Difference Across the Nineteenth-Century American West." *Transactions of the Institute of British Geographers*, 23: 311–30.

Morrill, Richard (1999). "Inequalities of Power, Costs and Benefits Across Geographic Scales: The Future Uses of the Hanford Reservation." *Political Geography*, 18: 1–23.

Muller-Wille, Ludger (1987). *Inuttitut Nunait Atingitta Katirsutauningit Nunavimi (Kupaimmi, Kanatami)/Gazetteer of Inuit Place Names in Nunavik (Quebec, Canada)/Repertoire toponymique inuit du Nunavik (Quebec, Canada)*. Inukjuak: Avataq Cultural Institute.

Nabokov, Peter (1998). "Orientations From Their Side: Dimensions of Native American Cartographic Discourse," in G. Lewis (ed.), *Cartographic Encounters: Perspectives on Native American Mapmaking and Map Use*. Chicago: University of Chicago Press, 241–69.

Nabokov, Peter, and Easton, Robert (1989). *Native American Architecture*. New York: Oxford University Press.

Newton, John (1995). "An Assessment of Coping with Environmental Hazards in Northern Aboriginal Communities." *Canadian Geographer*, 39: 112–20.

Northwest Renewable Resources Center (1992). "A Short Course on Tribal-County Intergovernmental Coordination." Tribes and Counties Intergovernmental Cooperation Project. Seattle: Northwest Renewable Resources Center.

—— (1993). "Working Effectively at the Local Level: Tribal/County Cooperation and Coordination." Proceedings. Seattle: Northwest Renewable Resources Center.

Notzke, Claudia (1995). "A New Perspective in Aboriginal Natural Resource Management: Co-Management." *Geoforum*, 26: 187–209.

Page, Brian, and Walker, Richard (1994). "*Nature's Metropolis*: The Ghost Dance of Cristaller and Von Thunen." *Antipode*, 26: 152–62.

Parezo, Nancy, Hays, Kelley, and Slivac, Barbara (1987). "The Mind's Road: Southwestern Indian Women's Art," in V. Norwood and J. Monk (eds.), *The Desert is No Lady: Southwestern Landscapes in Women's Writing and Art*. New Haven: Yale University Press, 125–45.

Parker, Linda (1989). *Native American Estate: The Struggle Over Indian and Hawaiian Lands*. Honolulu: University of Hawaii Press.

Peacock, Sandra (1998). "Putting Down Roots: The Emergence of Wild Plant Food Production on the Canadian Plateau," Ph.D. thesis, University of Victoria.

Pearce, Margaret (1998). "Native Mapping in Southern New England Indian Deeds," in Lewis (1998*a*: 157–86).

Pelly, David (1991). "How Inuit Find their Way in the Trackless Arctic." *Canadian Geographic* (August/September): 58–64.

Peters, Evelyn (1992). "Protecting the Land Under Modern Land Claims Agreements: The Effectiveness of the Environmental Regime Negotiated by the James Bay Cree in the James Bay and Northern Quebec Agreement." *Applied Geography*, 12: 133–45.

—— (1997). "Challenging the Geographies of 'Indianness': The *Batchewana* Case." *Urban Geography*, 18: 56–61.

—— (1998). "Subversive Spaces: First Nations Women and the City." *Environment and Planning D: Society and Space*, 16: 665–85.

Phelps, Glenn (1991). "Mr. Gerry Goes to Arizona: Electoral Geography and Voting Rights in Navajo Country." *American Indian Culture and Research Journal*, 15: 63–92.

Pinkham, A. (1996). "A Traditional American Indian Perspective on Land Use Management." *Landscape and Urban Planning*, 36: 93–101.

Pinkham, A., and Ruppert, D. (1996). Introduction. *Landscape and Urban Planning*, Special Issue, 36: 91–2.

Porter, Frank, III (1992). "American Indians in the Eastern United States," in A. Noble (ed.), *To Build in a New Land: Ethnic Landscapes in North America*. Baltimore: Johns Hopkins University Press, 119–35.

Preston, William (1990). "The Tulare Lake Basin: An Aboriginal Cornucopia." *California Geographer*, 30: 1–24.

Prucha, Francis (1990). *Atlas of American Indian Affairs*. Lincoln: University of Nebraska Press.

Pulido, L. (1996). "Development of the 'People of Color' Identity in the Environmental Justice Movement of the Southwestern United States." *Socialist Review*, 26/3–4: 145–80.

—— (1998). "The Sacredness of 'Mother Earth': Spirituality, Activism and Social Justice." *Annals of the Association of American Geographers*, 88: 719–23.

Riggs, C. (2000). "American Indians, Economic Development, and Self-Determination in the 1960s." *Pacific Historical Review*, 69: 431–63.

Ripmeester, Michael (1995). "'It is Scarcely to be Believed . . .': The Mississauga Indians and the Grape Island Mission, 1826–1836." *Canadian Geographer*, 39: 157–68.

Rundstrom, Robert (1990). "A Cultural Interpretation of Inuit Map Accuracy." *Geographical Review*, 80: 156–68.

—— (1991). "Mapping, Postmodernism, Indigenous People, and the Changing Direction of North American Cartography." *Cartographica*, 28: 1–12.

—— (1992). "Mapping the Inuit Ecumene of Arctic Canada," in D. Janelle (ed.), *Geographical Snapshots of North America*. New York: Guilford Press, 11–15.

—— (1993). "The Role of Ethics, Mapping, and the Meaning of Place in Relations Between Indians and Whites in the United States." *Cartographica*, 30: 21–8.

—— (1994). "American Indian Placemaking on Alcatraz Island, 1969–1971." *American Indian Culture and Research Journal*, 18: 189–212.

—— (1995). "GIS, Indigenous Peoples, and Epistemological Diversity." *Cartography and Geographic Information Systems*, 22: 45–57.

—— (1996*a*). "An Arctic Soliloquy on Inuit Place-Names and Cross-Cultural Fieldwork." *Names*, 44: 333–58.

—— (1996*b*). "Expectations and Motives in the Exchange of Maps and Geographical Information Among Inuit and *Qallunaat* in the Nineteenth and Twentieth Centuries," in L. Turgeon *et al.* (eds.), *Cultural Transfer, America and Europe: 500 Years of Interculturation*. Quebec: Les Presses de l'Université Laval, 377–95.

—— (1997). "Teaching American-Indian Geographies," in L. Estaville and C. Rosen (eds.), *Teaching American Ethnic Geography*. Indiana, Pa.: National Council for Geographic Education, 99–104.

—— (1998). "Mapping, The White Man's Burden." *Community Property Digest*, 45 (May): 7–9.

Rundstrom, Robert, and Deur, Douglas (1999). "Reciprocal Appropriation: Toward An Ethics of Cross-Cultural Research," in J. Proctor and D. Smith (eds.), *Geography and Ethics: Journeys in a Moral Terrain*. London: Routledge, 237–50.

Ruppert, D. (1996). "Intellectual Property Rights and Environmental Planning." *Landscape and Urban Planning*, 36: 117–23.

Said, E (1993). *Culture and Imperialism*. New York: Alfred A. Knopf.

Sauer, Carl (1920). "The Geography of the Ozark Highland of Missouri." *Bulletin of the Geographical Society of Chicago*, 7. Repr. (1974), London Greenwood Press; New York: AMS Press.

—— (1971). *Sixteenth Century North America: The Land and People as Seen by the Europeans*. Berkeley: University of California Press.

Sauer, Carl, and Brand, Donald (1930). *Pueblo Sites in Southeastern Arizona*. Berkeley: University of California Press.

Schnell, Steven (2000). The Kiowa Homeland in Oklahoma. *Geographical Review*, 90: 155–76.

Semple, Ellen (1903). *American History and its Geographic Conditions*. Boston: Houghton Mifflin.

Shipek, Florence (1988). *Pushed Into the Rocks: Southern California Indian Land Tenure, 1769–1986*. Lincoln: University of Nebraska Press.

Shumway, J., and Jackson, R. (1995). "Native American Population Patterns." *Geographical Review*, 85: 185–201.

Shute, Jeremy, and Knight, David (1995). "Obtaining an Understanding of Environmental Knowledge: Wendaban Stewardship Authority." *Canadian Geographer*, 39: 101–11.

Silko, Leslie (1987). "Landscape, History, and the Pueblo Imagination," in D. Halpern (ed.), *On Nature: Nature, Landscape, and Natural History*. San Francisco: North Point Press, 83–94.

— (1995). "Interior and Exterior Landscapes: The Pueblo Migration Stories," in G. Thompson (ed.), *Landscape in America*. Austin: University of Texas Press, 155–69.

Silvern, Steven (1995*a*). "Nature, Territory and Identity in the Wisconsin Ojibwe Treaty Rights Conflict," Ph.D. thesis, University of Wisconsin-Madison.

— (1995*b*). "Nature, Territory and Identity in the Wisconsin Treaty Rights Controversy." *Ecumene*, 2: 267–92.

— (1999). "Scales of Justice: Law, American Indian Treaty Rights and the Political Construction of Scale. *Political Geography*, 18: 639–68.

Sluyter, Andrew (2001). "Colonialism and Landscape in the Americas: Material/Conceptual Transformations and Continuing Consequences." *Annals of the Association of American Geographers*, 91: 410–28.

Smith, Neil (1992). "New City, New Frontier: The Lower East Side as Wild, Wild West," in M. Sorkin (ed.), *Variations on a Theme Park: The New American City and The End of Public Space*. New York: Hill & Wang, 61–93.

Smith, Patricia, and Allen, Paula (1987). "Earthy Relationships, Carnal Knowledge: Southwestern American Indian Writers and Landscape," in V. Norwood and J. Monk (eds.), *The Desert is No Lady: Southwestern Landscapes in Women's Writing and Art*. New Haven: Yale University Press, 146–73.

Sneed, Paul (1997). "National Parklands and Northern Homelands: Toward Co-management of National Parks in Alaska and the Yukon," in S. Stevens (ed.), *Conservation Through Cultural Survival: Indigenous Peoples and Protected Areas*. Washington, DC: Island Press, 135–54.

Solomon, Shirley (1990). "Tribal-County Intergovernmental Coordination: Towards a New Regionalism." *The Western Planner*, 11/1: 18–20.

— (1992). "Tribal and County Planners Discuss Coordination." *The Western Planner*, 13/3: 14.

Sonnenfeld, Joseph (1994). "Way-Keeping, Way-Finding, and Way-Losing: Disorientation in a Complex Environment," in K. Foote *et al.* (eds.), *Re-reading Cultural Geography*. Austin: University of Texas Press, 374–86.

Sparke, Matthew (1995). "Between Demythologizing and Deconstructing the Map: Shawnadithit's New-found-land and the Alienation of Canada." *Cartographica*, 32: 1–21.

— (1998). "A Map that Roared and an Original Atlas: Canada, Cartography, and the Narration of a Nation." *Annals of the Association of American Geographers*, 88: 463–95.

Spence, Mark (1999). *Dispossessing the Wilderness: Indian Removal and the Making of the National Parks*. New York: Oxford University Press.

Stea, David (1984). "Energy Resource Development, Energy Use, and Indigenous Peoples: A Case Study of Native North Americans." *Geography Research Forum*, 7: 28–42.

Stea, David, and Wisner, Ben (1984). "Introduction—The Fourth World: A Geography of Indigenous Struggles." *Antipode*, 16/2: 3–12.

Stephenson, P., Elliott, S., Foster, L., and Harris, J. (eds.) (1995). *A Persistent Spirit: Towards Understanding Aboriginal Health in British Columbia*. Vancouver: Western Geographical Press, 31.

Stoffle, Richard, Halmo, David, and Austin, Diane (1997). "Cultural Landscapes and Traditional Cultural Properties: A Southern Paiute View of the Grand Canyon and Colorado River." *American Indian Quarterly*, 21: 229–49.

Sundstrom, Linea (1997). "The Sacred Black Hills: An Ethnohistorical Review." *Great Plains Quarterly*, 17: 185–212.

Sutton, Imre (1991). "Preface to Indian Country: Geography and Law." *American Indian Culture and Research Journal*, 15: 3–35.

— (1994). "Indian Land, White Man's Law: Southern California Revisited." *American Indian Culture and Research Journal*, 18: 265–70.

— (2000). "Not All Aboriginal Territory is Truly Irredeemable." *American Indian Culture and Research Journal*, 24: 129–62.

Sweeney, A. (1996). "Tribal Land-Use Power: A Primer for Planners." American Planning Association-Planning Advisory Service Memo.

Swentzell, Rina (1997). "Conflicting Landscape Values: The Santa Clara Pueblo and Day School," in P. Groth and T. Bressi (eds.), *Understanding Ordinary Landscapes*. New Haven: Yale University Press, 56–66.

Tanner, Helen (ed.) (1987). *Atlas of Great Lakes Indian History*. Norman: University of Oklahoma Press.

Tompkins, Jane (1986). "'Indians': Textualism, Morality, and the Problem of History," in H. Gates, Jr. (ed.), *"Race," Writing, and Difference*. Chicago: University of Chicago Press, 59–77.

Tough, Frank (1990). "Indian Economic Behavior, Exchange and Profits in Northern Manitoba During the Decline of Monopoly, 1870–1930." *Journal of Historical Geography*, 16: 385–401.

Tungavik Federation of Nunavut (1992). *Nunavut Atlas*. Edmonton: Canadian Circumpolar Institute and the Tungavik Federation of Nunavut.

Usher, Peter, Tough, Frank, and Galois, Robert (1992). "Reclaiming the Land: Aboriginal Title, Treaty Rights and Land Claims in Canada." *Applied Geography*, 12: 109–32.

Van Otten, George (1992). "Economic Development on Arizona's Native American Reservations." *Journal of Cultural Geography*, 13/1: 1–14.

— (1997). "The Native-American Experience: Teaching Facts Instead of Fiction," in L. Estaville and C. Rosen (eds.), *Teaching American Ethnic Geography*. Indiana, Pa.: National Council for Geographic Education, 91–7.

Vecsey, C., and Venables, R. (1980). *American Indian Environments: Ecological Issues in Native American History*. Syracuse: Syracuse University Press.

Von Reichert, Christiane, McBroom, William, Reed, Fred, and Wilson, Paul (1995). "Access to Health Care and Travel for Birthing: Native American-White Differentials in Montana." *Geoforum*, 26: 297–308.

Waage, S. (2001). "(Re)claiming Space and Place Through Collaborative Planning in Rural Oregon." *Political Geography*, 20: 839–57.

Warhus, Mark (1997). *Another America: Native American Maps and the History of Our Land*. New York: St. Martin's Press.

Waterman, T. (1993). *Yurok Geography*. Trinidad, Calif.: Trinidad Museum Society.

Weissling, Lee (1992). "Inuit Life in the Eastern Canadian Arctic, 1922–1942: Change as Recorded by the RCMP." *Canadian Geographer*, 35: 59–69.

Wenzel, George (1992). "Global 'Warming', the Arctic and Inuit: Some Sociocultural Implications." *Canadian Geographer*, 36: 78–9.

White, Richard (1992). *Land Use, Environment, and Social Change: The Shaping of Island County, Washington*. Seattle: University of Washington Press.

Willems-Braun, Bruce (1997). "Buried Epistemologies: The Politics of Nature in (Post)colonial British Columbia." *Annals of the Association of American Geographers*, 87: 3–31.

Winchell, Dick (1992). "Tribal Sovereignty as the Basis for Tribal Planning." *The Western Planner*, 13/3: 9–11.

— (1996). "The Consolidation of Tribal Planning in American Indian Tribal Government and Culture," in K. Frantz (ed.), *Human Geography in North America: New Perspectives and Trends in Research*. Innsbruck: Department of Geography, University of Innsbruck, 209–24.

Winchell, Dick, and Marchand, M. (1994). "Tribal Implementation of GIS." *Cultural Survival Quarterly* (Winter): 49–51.

Winchell, Dick, Devereaux, L., and Habte, A. (1997). "The Planning Impacts of Tribal Gaming in Washington." *The Western Planner*, 18/2: 10–12.

Winchell, Dick, Lounsbury, J., and Sommers, L. (1998). "Indian Gaming in the U.S.: Distribution, Significance and Trends." *Focus*, 44/4: 1–10.

Winchell, Dick, Goodman, James, Jett, Stephen, and Henderson, Martha (1994). "Geographic Research on Native Americans," in G. Gaile and C. Willmott (eds.), *Geography in America*. Columbus, Ohio.: Merrill, 239–55.

Wishart, David (1990). "Compensation for Dispossession: Payments to Indians for their Lands on the Central and Northern Great Plains in the 19th Century." *National Geographic Research*, 6/1: 94–109.

— (1994). *An Unspeakable Sadness: The Dispossession of the Nebraska Indians*. Lincoln: University of Nebraska Press.

— (1995*a*). "Evidence of Surplus Production in the Cherokee Nation Prior to Removal." *Journal of Economic History*, 55: 120–38.

— (1995*b*). "The Roles and Status of Men and Women in Nineteenth Century Omaha and Pawnee Societies: Postmodernist Uncertainties and Empirical Evidence." *American Indian Quarterly*, 19: 509–18.

— (1996). "Indian Dispossession and Land Claims: The Issue of Fairness," in K. Frantz (ed.), *Human Geography in North America: New Perspectives and Trends in Research*. Innsbruck: Department of Geography, University of Innsbruck, 181–94.

— (1997). "The Selectivity of Historical Representation." *Journal of Historical Geography*, 23: 111–18.

— (1999). "The Death of Edward McMurtry." *Great Plains Quarterly*, 19: 5–22.

Wishart, David, and Froehling, Oliver (1996). "Land Ownership, Population, and Jurisdiction: The Case of the *Devils Lake Sioux Tribe v. North Dakota Public Service Commission*." *American Indian Culture and Research Journal*, 20: 33–58.

Wissler, Clark (1926). *The Relationship of Nature to Man in Aboriginal America*. London: Oxford University Press.

Woodley, Alison (1991). "Cross-Cultural Interpretation Planning in the Canadian Arctic." *Journal of Cultural Geography*, 11/2: 45–62.

Works, Martha (1992). "Creating Trading Places on the New Mexican Frontier." *Geographical Review*, 82: 268–81.

Wyckoff, William, and Dilsaver, Lary (1997). "Promotional Imagery of Glacier National Park." *Geographical Review*, 87: 1–26.

Youngbear-Tibbetts, Holly (1991). "Without Due Process: The Alienation of Individual Trust Allotments of the White Earth Anishinaabeg." *American Indian Culture and Research Journal*, 15: 93–138.

Zaferatos, N. (1998). "Planning the Native American Tribal Community: Understanding the Basis of Power Controlling the Reservation Territory." *Journal of the American Planning Association*, 64: 395–410.

CHAPTER 39

Asian Geography

Nanda Shrestha, Martin Lewis, Shaul Cohen, and Mary McDonald

A massive continent, stretching from Turkey and the eastern shores of the Mediterranean and Red Seas to the Pacific, from the Indian Ocean to the vast desert of Mongolia right through the towering Himalayas and the plateau of Tibet, Asia is a colossal geographic collage. One can find in Asia virtually every form of landscape, both real and imagined, including James Hilton's (1933) Shangri-La, planted in the imaginative geography of Western travelers and tourists (also see Bishop 1989). As the cradle of three of the world's early civilizations, Asia is a magnificent tapestry of cultural diversity. Asia has also given birth to all the major institutional religions that are practiced today: Hinduism, Buddhism, Judaism, Christianity, Islam, and others. As such, few would deny its enormous historical significance and contributions to human progress in every respect—spiritually, materially, and intellectually. Home to some 60 per cent of the world's population, Asia is a human mosaic that is unparalleled in history (Table 39.1). So it is hardly surprising that Asia offers endless research challenges and opportunities, in virtually every field of geographical studies. With this in mind, this chapter is divided into four major sections. First, we provide a brief journalistic survey of major regional political events across Asia. This is followed by a segment on the state of Asian geography in America in the second part. Third, we discuss some of the developments, trends, and research themes in Asian geography in America during the period 1988–2000.[1] Finally, we conclude the chapter with some general remarks on the vexing question of what lies ahead for regional geography.

We explore this question not because we foresee an imminent demise of regional geography, but because some of the remarkable developments during the 1990s have definite impacts on the way we see and do regional geography. Particularly relevant to our exploratory discussion of the emerging dilemma or challenge for regional geography are: (1) the collapse of the Soviet Union, leading to the end of the Cold War, and (2) the phenomenal speed of advancements in the field of

[1] Developments in Asian geography in America prior to this period have been detailed in the chapter on Asia that Karan *et al.* prepared for the first compendium of *Geography in America* edited by Gaile and Willmott (1989). In view of the geopolitical reality of that time period, that Asia chapter made little attempt to extend its coverage to what is commonly known as Central Asia (much of which was within the domain of the former Soviet Union). In order to maintain material consistency for the sake of temporal comparisons within the regional geographical parameters laid out in the previous Asia chapter, we have decided to confine the present discussions of developments in Asian Geography in America to East, Southeast, South, and West Asia.

Table 39.1 *Selected sociodemographic characteristics of Asian countries, 1998*

Countries	Land Area	Population	Rate of Population	Infant Mortality	Population under	Urban	GNP per capita
East Asia							
China	3,600,930	1,242.5	1.0	31	26	30	750
Hong Kong	382	6.7	0.4	4	18	—	24,290
Macao	8	0.5	1.0	5	25	97	—
Japan	145,375	126.4	0.2	4	15	78	40,940
Korea, North	46,490	22.2	0.9	39	28	59	—
Korea, South	38,128	46.4	1.0	11	22	79	10,610
Mongolia	604,826	2.4	1.6	49	36	57	360
Taiwan	13,970	21.7	1.0	7	23	75	—
Southeast Asia							
Brunei	2,035	0.3	2.2	8	34	67	—
Cambodia	68,154	10.8	2.4	116	44	14	300
Indonesia	705,189	207.4	1.5	66	34	37	1,080
Laos	89,112	5.3	2.8	97	45	19	400
Malaysia	126,853	22.2	2.1	10	35	57	4,370
Myanmar	253,880	47.1	2.0	83	36	25	—
Philippines	115,124	75.3	2.3	34	38	47	1,160
Singapore	236	3.9	1.1	4	23	100	30,550
Thailand	197,255	61.1	1.1	25	27	31	2,960
Vietnam	125,672	78.5	1.2	38	40	20	290
South Asia							
Afghanistan	251,772	24.8	2.5	150	41	18	—
Bangladesh	50,260	123.4	1.8	82	43	16	260
Bhutan	18,147	0.8	3.1	71	43	15	390
India	1,147,950	988.7	1.9	72	36	26	380
Maldives	116	0.3	3.3	30	46	25	1,080
Nepal	52,819	23.7	2.2	79	43	10	210
Pakistan	297,637	141.9	2.8	91	41	28	480
Sri Lanka	24,954	18.9	1.3	17	35	22	740
Southwest Asia							
Bahrain	266	0.6	2.0	14	31	88	—
Cyprus	3,568	0.7	0.7	8	25	68	—
Gaza/W. Bank		2.9	4.0	30	48		—
Iran	631,660	64.1	1.8	35	40	61	—
Iraq	168,869	21.8	2.8	127	43	70	—
Israel	7,961	6.0	1.5	7	30	90	15,870
Jordan	34,336	4.6	2.5	34	41	78	1,650
Kuwait	6,880	1.9	2.3	10	29	100	—
Lebanon	3,950	4.1	1.6	34	34	87	2,970
Oman	82,031	2.5	3.9	27	47	72	—
Qatar	4,247	0.5	1.7	12	27	91	—
Saudi Arabia	829,996	20.2	3.1	29	42	80	—
Syria	70,958	15.6	2.8	35	45	51	1,160
Turkey	297,154	64.8	1.6	42	31	64	2,830
United Arab E.	32,278	2.7	2.2	11	30	82	—
Yemen	203,849	15.8	3.3	77	47	25	380
Asia	**10,355,277**	**3,524.86**					

Source: Population Reference Bureau (PRB) 1998: 6–7.

information and communication technologies.[2] The Cold War's end has deeply affected the configuration of regional geography. This is particularly pronounced in Asia, an intensely contested territory of many hot and cold geopolitical battles over the past fifty years and now a crucial frontier of global capitalism.

As the defunct Cold War has prompted some scholars to advance the competing theories of the borderless world (the end of history, the end of nation-states, and, hence, the end of geography) and the clash of civilizations (cultural/religious wars), questions arise about regional geography and its relevance, viability, and future directions. Of specific concern is the normative question of how geographers ought to conceptualize and analyze regional issues in view of galloping globalization and potential civilizational conflicts, when both of these phenomena defy the conventional notion of region as a contiguous spatial entity. Then there are issues about national alignments or groupings, a form of regionalization of different sovereign nations, based on such characteristics as the level of economic and technological development or particular political and economic agenda. Complicating such alignments is the tendency for them to shift from time to time. All these raise questions about what constitutes a region. What are the parameters of regional analyses? However, the regional review of Asian geographic research in America offered later in this chapter reveals that Asianist geographers have already begun to chart this vast terrain from different angles, especially in the context of East and Southeast Asia (also see Lin 2000; Yeung 1998). Certainly, there is a lot more to be done, but the foundation has been laid. Therefore, much of our attention is focused on the issue of information and communication technologies as it is inherently related to one fundamental function of regional geography: to produce, synthesize, and disseminate regional information. It is precisely this function, after all, that forms the public and popular domain of regional geography, regardless of how we view and project it. The issue revolves around the question of *speed* as related to the production and dissemination of regional data and information. The phenomenal advancements in these technologies have fundamentally altered the nature of such regional information generation and distribution, thereby significantly affecting the role of regional geography. With this premiss, our principal expectation is that this exploratory discussion will set the stage for a lively and constructive discourse on the future direction(s) of regional geography which, in large measure, determines the fate of Asian geography.

[2] While the first two issues are by-products of the sudden end of the Cold War, mass advancements of information and communication technologies are offshoots of the capitalist imperative to expand the profit frontiers—or, to apply David Harvey's logic, the technological fix to the limits to capital (although many of them had their origin in the Cold War-based military research and development).

A Glimpse of the Asian Scenes and Economic Crisis

This review covers some of the notable changes that Asia has witnessed since 1988, focusing mainly on those that have significant implications for its evolving role in the global economy and international politics and how they inform and reorient its regional geography. This survey is intended to serve two specific purposes simultaneously. First, it sets a broad geographical context for the regional focus of this chapter. Second, it indicates that almost all the issues underlying this brief survey of the Asian scenes provide an excellent foundation for geographical research.

To begin with, although Deng Xiaoping has died, his post-Mao policy of Four Modernizations continues to serve as the steering wheel of China's ever-deepening involvement with the market economy, often euphemistically dubbed "market socialism." Once inducted into the World Trade Organization, China's market economy will most likely be fully formalized. In addition, Hong Kong has been finally returned to its fold. However, suspicion about China dealing a crushing blow to its relative economic autonomy, still runs relatively rampant within this tiny territory where the practice of free market often reaches its extremes. Across the strait, Taiwan, despite its deepening economic woes, has emerged as a leading foreign investor in China, but remains defiant and vigilant of its intention. In the meantime, the election of Chen Shui-bian as the island's president has not only ended the Kuomintang's hold on power, but also added a new episode to what may be labeled a "big Taiwan question." Mr Chen has long been a fervent advocate of Taiwan's independence, and was elected despite China's vehement opposition and the Kuomintang's use of the terror card during the campaign. The question is bound to hang heavy as both Washington and Beijing keep a close watch on the realpolitik of international diplomacy.

Also long gone is North Korea's Kim Il Sung, one of the products of the Cold War that led to the Korean War and the subsequent partition of the peninsula. However, his son seems to carry on his legacy, remaining faithful to its communist ideals and defying immense pressure from the United States. So far, it is the only unyielding Asian chip in the capitalist domino play of the Cold War that the United States has waged in Asia for the past fifty years. For the US, it is a nagging gnat that refuses to be swatted away. Accustomed to the ingrained Cold War mentality—or hegemonic thinking as some might claim—the US appears determined to continue its policy until all the domino chips are brought down to comply with its global design. Further down along the peninsula, the presidency of Kim Dae Jung, once a self-proclaimed socialist who barely survived an assassination plot in the early 1980s, represents a remarkable feat of personal achievement. From a national viewpoint, his ascension to power is not only a clear rebuke of South Korea's dictatorial tradition and regional politics, but also a significant victory for democracy. In addition to carefully guiding the country through its serious economic crisis, Kim Dae Jung has seemingly made some inroad in his peninsular diplomacy to gradually nudge North and South Korea toward a zone of mutual trust and, hopefully, eventual reunification.

In Southeast Asia, Vietnam has followed the same economic path that China has charted, opening up its borders to welcome ever-expansive agents of global capital. But in Indonesia, the Asian economic crisis has swallowed its first political victim as President Suharto was forced to abdicate the power that he usurped more than thirty years ago from President Sukarno through a military coup. Furthermore, Indonesia has lost its repressive grip over East Timor after it failed to drown its liberation aspirations in the pool of East Timorese blood. And, now, Suharto's successor Habibie has been replaced by Megawati Sukarnoputri, a daughter of Indonesia's first president, Sukarno. However, her rise to presidency has been marred by public doubts about her ability effectively to govern this huge archipelago nation of more than 200 million people, the largest Muslim country in the world. A similar political development occurred in the Philippines where President Joseph Estrada was deposed on perjury charges and replaced by President Gloria Macapagal Arroya, the late president Diosdado Macapagal's daughter. Both Indonesia and Philippines continue to suffer from the grim economic outlooks. No less troubling is the growing Islamic opposition to what Indonesian protesters consider to be the blatant US bias against Arabs and Islam in its self-declared war on Afghanistan. In the Philippines, the Muslim guerrilla movement has been quite active for many years and now may spill its own wrath against US citizens and interest in the country. The economic picture of Thailand, the nation that was viewed as a buffer state between the British and French colonial empires, remains equally gloomy, although its southern neighbors, Malaysia and Singapore, have been relatively less affected by Asia's ongoing economic crisis. To its west, the rich land of Myanmar continues to be mired in militarism. By aborting the outcome of the national election, its military rulers have denied Daw Aung San Suu Kyi her electoral right to form a civilian, democratic government.

On the Indian subcontinent, what is notable is the rise of Hindu fundamentalism in India, a country that prides itself in religious pluralism and openness. It is a scenario that is no less daunting than that seen in other religions such as Christianity and Islam. This undercurrent of fundamentalism has erected the specter of religious nativism as it has brought a staunchly Hindu party to power. Under the leadership of Prime Minister Atal Bihari Vajpayee, the party has won two consecutive elections. Additionally, the nuclear age has arrived on the subcontinent. India has passed its nuclear test, prompting Pakistan to follow suit. Now that both countries have successfully joined the nuclear club, the subcontinent has been thrust into a heightened state of militarism. Zulfikar Ali Bhutto, the late prime minister of Pakistan, envisioned a day when his country would have what he referred to as its own "Islamic" bomb to match the "Christian" bomb, the "Jewish" bomb, and the "Hindu" bomb.

Ironically, however, Bhutto's dream was attained under the rule of Prime Minister Nawaz Sharif, the arch-enemy of his daughter, former Prime Minister Benazir Bhutto, the first female head of an Islamic state, who has risen to power twice and gone down both times as a fallen angel. In October 1999, Mr Sharif was overthrown in a bloodless coup in which General Pervez Musharraf seized power, thus short-circuiting, once again, any hope of lasting democracy in Pakistan. It is, however, relevant to note that President Musharraf's rule is quite shaky, especially given his open support to the Bush administration's overt decision to overthrow the Taliban government in Afghanistan.[3] Standing at the crossroads of

[3] Unless the US succeeds in achieving its goals of capturing Osama bin Laden—the man who is "Wanted, Dead or Alive" as President Bush declared—and driving the Talibans out of power and installing a pro-American government within a fairly short period of time, say a few month, the chances of President Musharraf's continued rule will be drastically diminished. The longer the US is drawn into this nasty war in Afghanistan, the greater the probability of his overthrow. Also involved in this whole regional political equation is the national interest of India, the historical and geographical enemy of Pakistan. Musharraf's recent visit to India has achieved little in terms of minimizing the enmity between the two nations.

South, Southwest, and Central Asia, Afghanistan was once regarded as the former Soviet Union's Vietnam. But now it has come under relentless and enormous strikes by the US forces because of its refusal to hand over Mr Osama bin Laden, believed to be the mastermind behind the September 11 attacks that shook up America and shattered its ingrained sense of invincibility.

The real question is not, however, who will win the war. What is at issue is the prospect of increased and widespread Islamic militancy, multi-front hit-and-run struggles to haunt the US which most Muslims perceive to be the real power behind Israeli attacks against Palestinians. In these struggles, "terror" will, of course, be the weapon of choice, for it is the most potent weapon of the weak who have little military ability to engage in conventional battles against the US, the mightiest power in the world. As the magnitude of the US-led destruction of poor and militarily weak Afghanistan, whose only reliable and biggest weapon against the West's firepower may be its hostile topography and its people's tenacity and dogged resiliency, goes way beyond any reasonable international norms of proportionality, the war will be treated by the Muslim masses across the world as a new crusade against Islam irrespective of the repeated denials by the Bush and Blair administrations. As innocent Afghanis (including children and women) are counted as collateral casualties and countless hit the refugee trails covered with clouds of dust, and as the crushing winter joins the forces of massive hunger to devour many weak and feeble refugees, along with the injured, those tragic pictures will be beamed in every corner of the world by CNN and BBC (assuming that the US and Britain do not muzzle or restrain them from showing such pictures; on October 17, Duncan Campbell of Guardian News Service reported that "The Pentagon has spent millions of dollars to prevent Western media from seeing highly accurate civilian satellite pictures of the effects of bombing in Afghanistan. The images, which are taken from Ikonos, an advanced civilian satellite launched in 1999, are better than the spy satellite pics . . . The decision to shut down access to satellite images was taken last Thursday, after reports of heavy civilian casualties from the overnight bombing." This was subsequently reported by Michael Gordon in *The New York Times*).

As those pictures are intimately linked to devastating US strikes, the public sentiment in the Muslim world will be inflamed, filled with rage and bitterness toward the US. So, even if the US manages to demolish the Taliban infrastructure and regime and liquidate bin Laden, it will not be able to defeat terrorism, especially if it fails to restrain Israel from its current course of policy/action that Prime Minister Ariel Sharon has pursued and address Palestinian rights and tragedies with sincerity and seriousness. The Talibans may lose, but they will not vanish. And anti-US terrorism will rear its head in many places and in many forms until the US is prepared to deal with the root cause of "terrorism" in the realms of policies rather than in the fields of battles.[4]

Although the future is difficult to forecast with any certainty, the likely long-range outcome of the short-sighted US military destruction of Afghanistan may prove to be one of more chaos, complication, and consternation. The country is bound to be further fragmented, geographically, politically, and ethnically. If Afghanistan emerges as the epicenter of global Islamic militancy and becomes politically unstable—both of

[4] Since this chapter was completed and submitted in late 2001, much has happened in Asia—from the Korean Peninsula to Turkey. However, instead of revising it to capture these new developments, we have decided to leave it intact, for it reflects our analysis of future national and international political outcomes at the time we prepared the chapter. The reader should have the opportunity to judge our analysis and perspectives in light of what has transpired since then. Having said this, we highlight two major developments that have wide-ranging ramifications. First, on the Afghanistan frontier, the US military might did not take much time to drive the Taliban government out of power, along with bin Laden's forces, and to install a pro-American government headed by Hamid Karzai. But the war is hardly over. The US failed to capture or kill bin Laden—he is believed to be still alive and operating in the shadows; resistance against the US occupation is on the rise. Evidently, the Karzai government has failed to gain much control outside of Kabul. The fact that American commandoes are having to take over Karzai's personal security arrangements suggests that the chances of national unity and consolidation are slim. In other words, this clearly signals that the US is bogged down in what can be described as the Afghan quagmire. Compounding the problem is the dogged preparation and pursuit by the George W. Bush administration to extend the war frontier to Iraq, even before completing the mission in Afghanistan. This Bush policy has fueled the fire of anti-Americanism all across Asia and elsewhere, including among such staunch US allies as Egypt, Jordan, Pakistan, Saudi Arabia, South Korea, Taiwan, and Turkey. If the US invasion of Iraq occurs—as it seems almost inevitable—in all likelihood it will cause immense turmoil throughout the Middle East, thus complicating the Bush administration's global war on terrorism. Iran is bound to perceive the US presence on both the eastern (Afghanistan) and western (Iraq) sides of its borders as an impending threat to its territorial sovereignty. And the US will be clearly distrusted and seen in public eyes as a new, resource-hungry imperial bully, fully bent on directly controlling the oilfields of the Middle East. Second, North Korea sees an imminent threat to its national security from the current US policy and position. By deciding to reactivate its nuclear plants, the country has openly projected its image as a nuclear power. This has led to a dangerous engagement with the US in what can be described as nuclear brinkmanship. Although we do not foresee any immediate flare-up of massive mushroom clouds in East Asia from either side—a sort of replay of the Hiroshima and Nagasaki tragedies—there is little doubt that North Korea's recent pronouncements have not only put the US on notice, but also fully rendered the "Bush policy" hypocritical, myopic, and questionable.

which are more than mere prospects—the ripple effects of the American war will reach far beyond the immediate region. Within South Asia, the ever-fragile balance of power in its regional politics will be disrupted, thereby further compounding the quagmire of Kashmir, a huge landmine for both India and Pakistan; more conflicts and more suffering will follow. If this scenario unfolds, the US will have only plowed anew the very field that the British left fractured. As the late Chinese Premier Chou En Lai once remarked about one of the most damning legacies of British colonialism, "Wherever the British went, they left a little tail behind." Once savaged by European colonialism, Asia's regional geography now appears to be inseparably fated to the omnipresent tentacles of the American global agenda which at times is designed more for domestic political consumption and expediency than for lasting international peace and invariably based on one-sided power relations rather than mutually respectful cooperation.

Even the tiny Himalayan nation of Nepal has made some headlines as it became in the early 1990s the first country in Asia to have formed a popularly elected *communist* government. Although the minority communist government has lost its power, the party is currently the main opposition party in the parliament. Excluded from it is the Maoist group that since early 1996 has been launching a nationwide people's (guerrilla) war, currently establishing its control over more than 25 per cent of the country. This is an interesting development precisely because it is taking place at a time in history when communism has suffered serious setbacks in the aftermath of the Soviet disintegration and when globalization is seemingly the order of the day. In addition, the country witnessed a horrendous royal massacre, never seen in history before, one in which every member of King Birendra's immediate royal lineage was wiped out by his own son Crown Prince Dipendra who later took his own life. Consequently, the royal lineage was passed on to Birendra's brother Gyanendra. And Sri Lanka is still deeply submerged in its ongoing ethnic civil war. As this war has been raging for many years, it has consumed many lives on both the Tamil and Sinhalese sides, the two feuding ethnic and religious factions in the country.

In Southwest Asia, although almost everything seems stalemated, much has happened since 1988. The region is a major concern for the US for two primary reasons. First, it is currently the largest source of oil. Second, some of the principal choke points of global sea lanes are located in the Middle East, thus making its regional geography vitally critical for the US and for global capitalism. If any of these points is closed, it could cause major havoc for the global shipping industry and hence for the global economy. The memory of Egypt being bombed and invaded by the Western powers, namely the British and French troops, in 1956 when President Nasser decided to nationalize the Suez Canal is still fresh in many minds. Aside from lending military support to the Israeli forces, the objective of Western invasion of Egypt was to insure that the canal, perhaps the most important artery of global shipping, stayed open.

More than thirty years after the Shah's dethronement, Iran still operates outside the American sphere of influence, with a firmly fundamentalist bent in its theocratic rule. Hopes were raised by the electoral victory of reform-minded President Mohammad Khatami about the prospect of initiating a dialogue with the US, but they have yet to materialize. In addition to the current war against Afghanistan discussed above, the Gulf War that the US orchestrated against Iraq constitutes not only a significant event, but also a powerful marker of American resolve to keep the region under its control. In the mid-1970s, President Gerald Ford openly threatened that, if necessary, the US would intervene in the Middle East militarily to keep the oil pipelines flowing without any interruption. And President (father) George Bush fulfilled Ford's promise some fifteen years later. However, the much-anticipated demise of Saddam Hussein was apparently overstated.

Significant developments of both a positive and negative nature took place in regard to the Arab–Israeli conflict. They include a peace treaty between Israel and Jordan, Israeli withdrawal from Southern Lebanon, and, certainly, the creation of the Palestine National Authority (or Palestinian homeland) with its patchwork rule over portions of the West Bank and Gaza Strip. Yet, as of this writing, negotiations are once again far from producing a peaceful resolution to the entrenched conflict and a viable state for Palestinians. One recent development that is clearly noteworthy is that President (son) Bush has, although passingly, floated the US wish to see the formation of the Palestinian State. Adding to his comment, Secretary of State Powell stated, "There's no question the continuing difficulties between the Israelis and the Palestinians are . . . a major disturbing feature that causes the kind of situation we are in to exist, where it fuels discontent" (Strobel 2001: 8A). It is not clear, however, how hard Bush will strive to create the Palestinian State, especially if the war in Afghanistan is a short affair and the Arab support for it proves to be not that critical. Moreover, acutely strained Israeli–Palestinian relations remain a giant impediment to the question of Palestinian statehood. In addition, political leadership in the region is in transition, notably with dynastic rule of various sorts providing successors in

Jordan and Syria and in Morocco. In the aftermath of the September 11, 2001 attacks in New York (World Trade Center) and Washington (Pentagon)—two powerful symbols of US domination of global finance and global politics (and military)—the regional geography of the Middle East again is being portrayed as a political quagmire and a source of terrorism. As in the past, the complexity of Southwest Asia is overwhelmed by common images that feed cultural stereotypes in which Arabs are branded as terrorists. What is most disturbing about such xenophobic tendencies is that the whole region and its people are chained to the shackles of these cultural stereotypes as they are generalized across the whole region. One spoiled apple in a basket full of apples, and the whole basket is labeled rotten.

But few of these political developments have been as geographically overarching in their scopes as the ongoing economic crisis that started in Japan, once considered the vanguard of what many proclaimed to be "The Pacific Century." The crisis soon diffused in many countries in East and Southeast Asia. As the crisis lingers, it threatens to drown the vision of capitalist globalism.[5] In January 1998, *Business Week* carried a cover story on the Asian economic crisis, clearly documenting that it not only mars many Asian countries, but also stalks the much vaunted sanctity of today's global economy. It wrote, "The US sees the next century as the age of globalization, free market and increasing prosperity. It's a beautiful vision. A devastated Asia will make a mockery of it" (*Business Week* 1998: 30). In the same article, Eisuke Sakakibara, Japan's vice-minister for international affairs, bluntly remarks: "This isn't an Asian crisis, it is a crisis of global capitalism" (p. 28). This is a sobering comment whose gravity is heightened by the fact that Japan, once thought to be rock-solid and generally immune to such periodic crisis of capitalism, remains buried in it. In short, the dragon land of Asia has been a drag on the global economy.

In November 2000, *Business Week* published a special report on the crisis of global capitalism that Mr Sakakibara spoke of three years ago. The report was not only timely in that it was published when anti-globalization protests were sprouting all over and the US economy was showing signs of teetering after a relatively long joy ride, but also surprisingly frank in its assessment of globalization's debilitating impacts on the poor and downtrodden across the globe. At any rate, the dyke that kept the global economy from being swept away was mainly anchored to the twin pillars of America's booming bull market and insatiable consumerism. Even before these signs of America's economic decline were on the horizon, some wondered loudly how long this US dyke would defend against the Asian crisis (see Greider 2000). One does not have to wonder any more. Caught in the vortex of Asia's persistent economic flu and the weight of its own oversaturation, the US economy is experiencing a recession. Given this, whether Asia's economic vitality is fundamental to the continued US economic well-being and to the global economy is a moot question. What is interesting to note here is that this is the first time all three centers in the tripolar system of global capitalism—the US, Japan, and Western Europe—are undergoing a period of serious downturn at the same time (also see British Broadcasting Service 2001). These are potentially dire times. So the questions are: how long will this global economic bust last, which of the three poles will reemerge first from the economic sinkhole and begin pulling the heavy wagon of global capitalism, and when will the Asian economic boat float again so it can lift the global economy? Answers to these questions are beyond the boundary of this chapter. Our attempt, therefore, is naturally limited to a cursory discussion of the Asian economic predicament in relation to the last question.

Long subjugated to the rapacity of European colonialism, Asia emerged as an economic powerhouse in the 1970s and 1980s. During the development decades of the 1950s and 1960s, Western economic gurus were almost universally disdainful of the so-called Asian values or what Karl Marx called the Asian mode of production. To them, these values were the root cause of Asia's sluggish economic growth and backwardness. Then came the 1970s and 1980s when, under the aegis of export-oriented industrialization and led by Japan, many East and Southeast Asian countries posted exploding rates of economic growth that few other countries could match.[6]

[5] Although globalization is often discussed as if it is a fairly recent phenomenon, something that started in the early 1970s, in reality its historical roots can be traced back to 1500 AD. It has been going on in different forms at different time periods. For example, in terms of the global process of economic domination and control by a few countries, mercantilism and colonialism are two distinct phases of globalization although they may not have been identified as such. In other words, its manifestation and dominant power may have changed from one period to another, but not necessarily its underlying process and mechanism (see Shrestha 1988).

[6] During the 1970s and 1980s, several countries in East and Southeast Asia adopted a growth policy of what is commonly known as export-oriented industrialization rather than the industrial policy of import substitution that was being pursued in South Asia and in Latin America. Export-oriented industrialization resulted in rates of economic growth that few other countries could match during those periods.

Centrally configured into US geopolitics, Taiwan and South Korea greatly benefited from the Cold War as US capital was infused, along with the US push for land reforms in both countries. For example, as a leading contractor for the US during its involvement in Vietnam, South Korea found a reliable source of capital formation which proved to be the lifeline of its industrial drive. In the wake of such growth, the very Asian, neo-Confucian values that were once berated as being Asia's inherent ills were being waved like a victory banner to explain their miraculous economic growth. There were talks of the "Pacific Century," i.e. the resurgence of Asia as a globally dominant power. In the aftershocks of the region's economic earthquake, however, these same values have once again come into question as they are suspected of having contributed to the region's economic collapse (for a detailed discussion, see *The Economist* 1998: 23–8).

Regardless of the raging debate over the economic efficacy of the Asian values and ongoing economic crisis, Asia's position in the global economy is, to repeat, unshakable as well as critical. First, despite the fact that Japan finds its economic foundation cracking, it is still the second largest economy in the world and one of the three poles of global capitalism. Second, in terms of gross national product, Asia controls almost one-third of the world's output, with China gradually emerging as a global economic player. Third, as already noted, Asia accounts for 60 per cent of the world's population; that alone is enough to secure its pivotal place in the global economy. Simply expressed, it constitutes a huge reservoir of human capital—or labor as it is conveniently called. Probably, few would dispute that, in the past thirty years, it is the vast pool of Asian labor that has served as a human machine, powering and propelling the contemporary global economy to a height that it could not reach during its colonial phase. As more and more multinational companies from Japan, the US, and Western Europe relocated or expanded their manufacturing operations to Asia, the vital role of its labor power in creating value for the global economy acquired greater importance. In terms of the spatial division of globalized manufacturing (especially garment/textile, consumer electronic, and computer-related hardware production), the gravity of Asian labor has shifted over time, from Japan to four little dragons (Hong Kong, Singapore, South Korea, and Taiwan)—and now to China and Southeast Asia. As the labor costs rise in these countries, there are signs that this gravity is slowly moving toward South Asia, a region that has for long been much maligned and neglected by the pundits and players of globalization. Endowed with an abundant supply of computer programmers, India is already emerging as an engineering Mecca for many American software companies, led by the daddy of them all: Microsoft.

Fourth, besides labor, how can one overlook its vast consumer frontier, especially given its middle-class population boom in the last decade? China alone is estimated to have 250–300 million middle-class people. Once a laughing-stock in the arena of economic growth, India was largely unscathed by the Asian economic crisis and is now experiencing a relatively steady pace of growth since it embarked on the path of liberalization in the early 1990s. It adds another 200 or so million to the global market of consumers. Other Asian countries also contribute to this large pool. Not only do these Asian consumers possess enough disposable income to fan the culture of consumerism, but they are also rife with vociferous appetites for Western/American products. To put it plainly, the success of capitalist globalism relies as much on the globalization of consumerism as it does on the globalization of production and service. Finally, Asia is not only a global manufacturing center, it has also become a primary hub of software production, the nerve center of the ongoing information and communication revolution which itself is a vital engine of today's globalization. That is, although the US is the epicenter of the information age, much of the software programming and engineering operation is outsourced to Asia.

To summarize, in the last fifty years since the end of World War II, Asia has traversed a vast space of what is routinely called development—from poverty to prosperity to current economic perplexity. Asia has certainly attracted the attention of the American business community (Krugman 1998; *Business Week* 1998, 2000; *The Economist* 1998; Montagnon 1998). The question that concerns us most is: how has Asian regional geography in America fared during this relatively rapid journey of Asia, especially over the past ten years? Before exploring this question, we first assess the state of Asian geography in America.

Asian Geography in America

As revealed by Karan *et al.* (1989), the American foundation of concerted research and field studies on Asia was laid in the early 1900s. Built on this early foundation, Asian regional geography gained its popularity in America in the 1930s and 1940s as the need for systematic knowledge about Asian countries during and after World War II became critical for strategic intelligence and defense purposes. It is, therefore, no surprise that

World War II and the subsequent Cold War (including the Korean and Vietnam wars) were instrumental in the continued growth of Asian regional geography. As the Cold War intensified following the end of World War II and as the Domino Theory was figured into the American geopolitical calculus, Asia came to occupy center stage. It was a battleground between communism and America-guided capitalist globalism, hinged on the twin goals of its ideological triumph and commercial advance (Greider 2000). This growing importance of Asia in America's geopolitical play prompted a flurry of studies on Asian countries. In short, wars were mighty good for the growth of Asian geography, and hence for the discipline as a whole.

As America became increasingly entrenched in the Cold War, the 1950s saw the publications of several geography (text) books on Asia. For instance, Cressey's *Asia's Lands and Peoples* (1951) provided the most complete descriptive reference source on the continent, thus constituting a breakthrough for Asian regional geography in America. It was followed by Spencer's *Asia, East by South* (1954), which offered a detailed portrayal of Monsoon Asia's historical and cultural geography. In his edited volume, *The Pattern of Asia*, Ginsburg (1958) discussed Asian political and economic problems from a geographical viewpoint. All this boded well for Asian geography as it continued to flourish as a subset of American geography. Among this new band of geographers was a fresh Ph.D. graduate named P. P. Karan, the first non-American among the widely published Asianist geographers, whose role in raising the profile of South Asia within Asian geography is highly notable. Almost forty years after his first book came out in 1960, Karan remains extremely active and prolific in Asian geographic research as clearly evidenced by his numerous publications. Karan is one of the very few geographers whose research coverage is as extensive as his depth of knowledge of Asia. His research expands from Southwest to East Asia, from an isolated periphery such as Bhutan or Tibet to a metropolitan country such as Japan.

One common strand of most of these studies was that they generally offered a fairly conventional view and interpretation of Asia. However, some two decades later Rhoads Murphey (1977) offered a refreshingly new approach to Asian geographic research. Taking a historical angle, Murphey's analysis focused on how colonialism structured and arranged the political and economic systems of Asia, namely India and China, from a spatial perspective. More specifically, Murphey investigated the colonial city and the treaty-port. The founding of international trading centers at port cities was aimed at transforming the traditional inward-facing economies that were dominated by profiteering European operations. According to Murphey, the colonial port dynamics played critical roles in transforming the Indian administrative and economic systems (also see Shrestha and Hartshorn 1993). The pattern of India was duplicated in China as the latter fell under Western control, following the conclusion of what is popularly known as the Opium War in the early 1840s.

Since the end of the Vietnam war, however, the fortunes of Asian regional geography in America have largely faded as indicated by steady declines in its curricular focus. No comprehensive Asian geography textbooks have been published since then. While the so-called quantitative revolution in the 1960s is often attributed as a setback for regional geography in general because of increased emphasis on topical and systematic themes (see Karan *et al.* 1989: 508), America's Vietnam debacle seems to have cast a dark shadow on Asian geography. In the meantime, a dramatic reduction in the Cold War fever stemming from the sudden disintegration of the former Soviet Union and the Berlin wall has further eroded its progress. It is generally true that geography has attracted significant attention from other disciplines and experienced a noticeable rise at public schools over the past several years (Rediscovering Geography Committee 1997). Few would deny the contribution of the Geographic Alliance project toward this recent popularity of basic geography. As the Rediscovering Geography Committee implied in its report, such seeming popularity of geography to "outsiders" has largely failed to translate into the growth of Asian regional geography—or regional geography as a whole—among the "insiders" (particularly in Ph.D. granting departments). It is perhaps no exaggeration that the past twelve years—a period which directly coincides with the collapse of communism and subsequent market triumphalism across the globe—can indeed be considered a curricular watershed for Asian geography in America.

Status of Asian Geography in Doctoral Departments

The 1994–5 *Guide to Programs in Geography* lists a total of fifty-three geography departments offering Ph.Ds in the United States (Association of American Geographers 1994). Only nineteen of these mention Asia as part of their program emphasis, and three of them have no faculty listed as Asianists and another five have only one faculty member each with some kind of Asia focus (Table 39.2).

Table 39.2 *Status of Asian geography at Ph.D. departments in the United States, 1994*

Region	Geography departments (No.)	Asia Program		Total no. of faculty	No. of faculty with Asia focus	% of total	Regional distribution of faculty with Asia focus				
		Yes	No				EA	SEA	SA	SWA	Other
Northeast	9	3	6	142	5	3.5	1	2	1	1	
South	12	3	9	204	19	9.3	5	4	3	5	2
Midwest	15	6	9	266	15	5.6	7	2	3	1	2
West	17	7	10	254	21	8.3	7	4	3	4	3
TOTAL	**53**	**19***	**34**	**866**	**60**	**6.9**	**20**	**12**	**10**	**11**	**7**

* Three departments had no faculty listed as Asianists and five other departments had only one faculty, indicating Asia focus. That is, only eleven of these departments had two or more Asianist faculty. With the exception of the University of Hawaii (six) and the University of Texas-Austin (four), none had more than three.

Source: Association of American Geographers (AAG) (1994).

Given this reality, the outlook of the Asian geography curriculum in America is hardly inspiring. It is plausible to infer from Table 39.2 that, in view of the low priority given to Asian geography, the pipeline of Ph.D. graduates with a regional focus on Asia is diminishing. With less than 7 per cent of the faculty in these Ph.D. granting departments composed of Asianists—roughly one per department on average—it would be naive to expect future growth in Asian geography. Furthermore, many of the Asianists in these departments are approaching retirement. But retiring Asianists are seldom replaced by young scholars pursuing Asian geography. Furthermore, as it is obvious from the current discussion, young Asianist geographers are a shrinking breed anyway. So the future of Asian geography in America is fuzzy. There are, however, some indications of research in Asia by geographers who are not area specialists. This is a heartening sign.

Asian Geography Research Publications in American Geographical Journals

The research frontier of Asian geography is somewhat brighter than the Ph.D. curriculum regarding Asia or the production of Ph.D. graduates with an Asia focus. In their 1989 article, Karan *et al.* included two tables, showing the number of articles on Asia published in the *Annals* and *Geographical Review* over the period 1955–87 (Karan *et al.* 1989: tables 1 and 2). Overall, 6.5 per cent of the articles published in the *Annals* and 11.4 per cent of the articles in *Geographical Review* were on Asian topics. For the period between 1981 and 1987, the respective figures were 4.1 and 16.2 per cent. For this study, we focused on the ten-year period following the previous tabulation of publications done by Karan *et al.* Our compilation of research publications in selected American geography journals during the 1988–98 period reveals a slightly better output for the same two journals (Table 39.3).

In other words, as seen from Table 39.3, *Annals* and *Geographical Review* respectively carried 6.7 and 16.3 per cent of their articles on Asian topics. For the same period, *Economic Geography* and *The Professional Geographer* had a similar share—8.7 and 8.4 per cent, respectively. *Antipode* featured the lowest number (4%). On average, 8.8 per cent of the articles published in these selected journals were focused on Asia. The *Geographical Review* has consistently been the one most inclined to publish Asian geography articles, whereas *Antipode* has published the least. The *Annals*, as the "mouthpiece" of the Association of American Geographers, has the greatest influence in the field. Unfortunately, relatively few articles on Asian geography are appearing between its covers. Will Graf (1999: 2), past President of the AAG, stated that "Of the five largest specialty groups, only one, Urban, is frequently represented in the *Annals*. This lack of representation is problematic because if the flagship journal is to fairly represent the discipline on a global basis, it ought to have a range of products from the membership." While it is difficult to determine whether this tendency is an outcome of a low rate of submission of Asian geography manuscripts, a high rate of their rejection, or some combination of the two, the trend shown by the *Annals* is hardly encouraging. Regardless of the reason(s), it is likely that authors have, to extend Graf's argument, the perception that the journal is not

Table 39.3 *Number of articles on Asia in selected American geography journals, 1988–1998*

Journals	Total no. of articles	No. of articles on Asia	%	Regional distribution of articles on Asia				
				EA	SEA	SA	SWA	Other
Annals (AAG)	298	20	6.7	12	2	3	3	
Geographical Review[a]	269	44	16.3	9	14	8	9	4
Economic Geography[b]	208	18	8.7	11	2	3	1	1
Antipode	151	6	4.0		2	3	1	
Professional Geographer	249	21	8.4	5	2	5	6	3
TOTAL	**1,175**	**109**	**8.8**	**37**	**22**	**22**	**20**	**8**

[a] The total number is for the period 1988–97.
[b] The total number includes the articles published in the special issue on the Boston meeting of the AAG.

too receptive to Asian geography papers. In fact, the AAG as a whole has given little voice to Asian geography. In the past twenty-five years or so, few members of its corps of officers have been drawn from the group of professed Asianists to represent the discipline to a larger audience. Whether true or not, the general perception is that the Association suffers from a clubbish mentality as, many times, the same group of people moves from one officership (position) to another, almost like the rotational movement of the same baseball managers from one team to another. In many respects, the exclusion of Asianists may have a direct linkage to the Association's general and historical tendency to exclude minorities from its corps of officers. Speaking broadly of the lack of minority representation, Graf (ibid.) has boldly asserted: "it is vital that we have broad and *effective* representations of the range of perspectives and experiences that minorities bring to our organization" (emphasis added; also see Rediscovering Geography Committee 1997).

Irrespective of these facts, Asian geography research remains vital and vibrant; it has done very well in terms of overall publications. In addition, Asianist geographers in America have published in various other related journals, thus clearly revealing their wider appeal and receptivity beyond the domain of geography.

Research Trends in the Subregions of Asia

Asia is simply too large to be conceptualized as a single area. Accordingly, it is conventionally divided into four subregions to provide a spatial framework for organizing scholarly activities. Specifically, the present coverage of American geographical research on Asia is focused on: East Asia, Southeast Asia, South Asia, and Southwest Asia. Although East Asia encompasses China (including Taiwan and Hong Kong), it is not included in the regional research section of this chapter for the simple reason that Chinese geography has its own separate specialty group in the AAG (see Ch. 41, Geography of China). We do, however, include here some research on China's ties with the other regions of Asia. As explained in note 1, we have also decided not to cover Central Asia in the current chapter.

East Asia

From 1988 to 2001, an energetic segment of researchers and research publications in North America addressed Korea and Japan, here taken as East Asia. The works below represent disparate but linked geographies of historical development and contemporary crises across many scales.

1. Spaces of the Past. The historical construction of East Asian regions has been illuminated by Kären Wigen, first in a 1992 article questioning the space-consciousness of histories of Japan. Her 1995 book focused on a Japanese locale, the Ina Valley, through its nineteenth-century transformation from autonomy in interregional trade to subordination in the silk economy of the Meiji Period. Wigen detailed economic interests underlying material transformations and shifts of power within and without the region. Wigen went on to examine the production of regionalist discourses in Japan's Shinano (1996, 1998, 2000) and in East Asia (1999). Analysis of once-central places marginalized by modernity continued in Yoon

Hong-Key's historical geography (1997) of a temple town, Saidaiji in Okayama Prefecture. Yoon found the old town struggling against the growing dominance of Okayama City in transport and retailing. The history of Japan's contemporary urban hierarchy also inspired Yoshio Sugiura's (1993) study of early electric power companies' choices of locations and fuels. Edwina Palmer's historical geography shed new light on Japan's early twentieth-century influenza epidemic (Rice and Palmer 1993). Siebert (2000*a*) explained why ancient province names appear in modern railway station names.

2. Cartography. The history of cartography in Korea and Japan was advanced in Harley and Woodward's comprehensive project. Their second volume included chapters by Unno (1994) and Ledyard (1994) tracing the native roots of cartography in Japan and Korea, and the impacts of European mapping practices. Nemeth (1993) interpreted the cosmography of Korean historical maps. Yonemoto (1999) read the interests of Japan's early modern state in its charted seas. She further revealed Tokugawa Japan's expansive culture of popular map-making and spatial consciousness (Yonemoto 2000). Siebert applied GIS technologies to visualize the historic urban growth of Tokyo (Siebert 2000*c*).

3. Cultural Landscape. Geographers have continued to examine East Asian places as lived and interpreted on the ground. Childs (1991) explored the landscapes of Japan's snow country to understand Kawabata's theme of the ambivalent urban soul. Latz (1992) reviewed environmental influences on aesthetics of place in Japan. Ikagawa (1993) showed "natural" landscapes of sand and pine in Western Japan to result from centuries of human disturbance of upslope soils and forests. Patchell and Hayter (1997) probed various woods prized in Japanese home construction and their multiple sources of value. Mather *et al.* (1998) insisted that Japanese landscapes remain a merger of land and life, and argued their point with many photos of 1990s Japan. Concerning the cultural landscapes of Korea, Nemeth's 1987 work on the influences of geomancy on Cheju Island stands as one of the last important contributions by North American geographers; the 1990s left the cultural geography of Korea much underexplored.

4. Cities in Japan and Korea. The themes of urbanization and comparative national experiences have continued to intrigue geographers. Japan and Korea appeared in general and comparative works about cities in East Asia. Karan and Kornhauser analyzed Japan in Costa *et al.* 1989. Kornhauser's (1991) urban survey of Japan entered its second edition. Song *et al.* (1994) surveyed rapidly urbanizing South Korea in Dutt *et al.* Latz examined farm survival at the Tokyo urban fringe in Ginsburg *et al.* (1991). Machimura offered Tokyo's imperial palace district as a space of modernity in the capital in Kim *et al.* (1997). Siebert (2000*b*) traced land use from Tokyo to suburban Kanagawa.

The decade brought a wealth of articles and books on particular cities, often noting changes in industrial structure. Such were the works of Edgington for Kansai (1990*b*), Yokohama (1991*a*), Nagoya (1992), and Osaka (2000). Machimura examined urban restructuring in Tokyo (1992), S. O. Park (1993, 1994) and Nahm (1999) did so in Seoul, and Shapira in Kitakyushu (1994). B. G. Park (1998) compared housing policies in South Korea and Singapore under conditions of urban industrial growth. The polarization of the fortunes of cities and the continuing growth of primate cities, especially through social increase of population (Liaw 1992; Masai 1994), was seen as the problematic outcome of "one point development." The dominance of both Tokyo and Seoul was reflected in geographic research. Roman Cybriwsky published several rich studies of Tokyo during the past decade. His 1988 study of Shibuya introduced his approach to distinctive nodes of transport, commerce, and consumption. Cybriwsky expanded this approach to five "epitome districts" in his first book about Tokyo (1991). Cybriwsky's 1998 book further interprets the spatial structure of the Shogun's city and previews Tokyo's futuristic megaprojects for the next millennium. This volume added to a new World Cities series, which included a volume about Seoul by Joochul Kim and Sang Chuel Choe (Kim and Choe 1997). A volume on *The Japanese City* edited by P. P. Karan and K. Stapleton gave Tokyo further treatment by Cybriwsky (1997) and Okamoto (1997). The volume included wide-ranging examinations of life in cities across Japan by several authors, including Karan (1997) and Mather (1997). Sociologists Fujita and Hill, editors of *Japanese Cities in the World Economy* (1993), contributed research on Osaka to *The Japanese City* (Fujita and Hill 1997). The 1995 earthquake in Kobe tragically demonstrated the hazards of urban life, the difficulties of reconstruction (Edgington *et al.* 1999), and the relevance of research on vulnerabilities and preparations for the next big one (Palm 1998; Wisner 1998). We thus end the 1990s with a much richer literature on the urban geography of Japan and Korea.

5. Patterns and Processes in the Space Economy. Economic geographers, including many whose regional emphasis is not East Asia, focused on Japan and Korea in the 1990s. Firms' practices, locations, and restructuring strategies have been areas of intense research and lively debate among geographers in North America, Asia,

Europe, and Australia-New Zealand. How have Japanese and Korean manufacturing firms, especially those in electrical and transport machinery industries, organized their production socially and spatially? Industrial organization, labor practices, and locational patterns were studied by Kenney and Florida (1988, 1993), Florida and Kenney (1990), Glasmeier and Sugiura (1991), Patchell (1993), and Hayashi (1994). Korean firms' industrial structure and production systems were examined by Choo (1994) and Suarez-Villa and Han (1990).

Economic growth in East Asian nations has also been seen in terms of place. Ettlinger (1991) compared components of regional competitive advantage in Japan and California. The state models within which East Asian development should be understood were discussed by Hart-Ladsberg and Burkett (1998), Douglass (1994), and Auty (1997). The specific ways states create place for industries have been studied Edgington(1994*c* and 1999), Glasmeier (1988), McDonald (1996*b*), and Markusen and Park (1993). Policies toward small business in Japan and the US were compared by Aoyama (1996, 1999) and by Aoyama and Teitz (1996). Aoyama (2000*a*) found Japanese state policy solutions for troubled small enterprises reaching their limits in the late 1990s. Women's labor in development across Asia was surveyed by Prorok *et al.* (1998).

Rural industries and rural policies of Japan also remained a focus of North American geographers, showing a countryside increasingly usurped by industrialization and by urban-biased planning measures. Latz's studies at the beginning of the decade saw agricultural land and water use surviving alongside urban and industrial land use (Latz 1989, 1991). Later studies showed farm labor and farmland absorbed by the reach of branch plants into regions such as Tohoku (McDonald 1996*a*, 1997). Communities losing labor to out-migration can no longer sustain high value fruit production (Brucklacher 1998). As primary industries throughout Japan shrank, Japan's investments in supply lines from abroad increased, as Parker (1997) demonstrated for the case of coal and McDonald (2000) for food.

The strength of the yen since 1986 has been an important factor driving the investment decisions of Japan's industries. Aoyama (1997, 2000*b*) compared firms' overseas locational strategies in theory and practice. Pressures on firms to move production abroad and employment effects in Japan were studied by Rimmer (1997) and by Edgington (1993, 1994*a*, 1997). Japanese investment in Australia was also analyzed by Edgington (1990*a*, 1991*b*). Japan's hierarchy of manufacturing networks in Asia was questioned by Edgington and Hayter (2000). The movement of banking services into the United States was traced by Hultman and McGee (1990). Capital flows into North American real estate were studied by Warf in New York (1988) and by Edgington in Canada and the US (1994*b*, 1995*a*, 1996). Edgington and Haga (1998) studied service firms' locational decisions in Pacific Rim cities. Korean business locations in the Chicago area were traced by Park and Kim (1998). Patterns of commerce within Japanese and Korean cities have been likewise restructured by information technologies (Aoyama 2001*a*, *b*; S. O. Park 2000).

Japanese auto and steel transplants abroad have captured the attention of many geographers. Japan's investments in auto and steel manufacturing in North America have been studied by Rubenstein (1988, 1992), Mair *et al.* (1988) Elhance and Chapman (1992) Florida and Kenney (1992, 1994), Kenney and Florida (1992), Jones and North (1991), Mair (1992, 1993, 1994), and Reid (1995). Sadler (1994) studied the case of Japanese auto factory location in Europe. Given the much smaller flow of foreign direct investment by North American firms into Japan, few geographers have traced investment into Japan, but see Hayter and Edgington (1997).

North–South trade within Asia and the regional structure of the Asia-Pacific economy were traced throughout the 1990's by geographers such as Graham (1993), Latz (1993), Poon and Pandit (1996), Poon (1997*a*, *b*), Poon and Thompson (1998), and Thompson and Poon (1998). Regional interdependencies were seen in the financial crisis of the late 1990s by observers such as Poon and Perry (1999), Poon and Thompson (2001), Edgington and Hayter (2001), and B. G. Park (2001).

Economic transnationalization has had social concomitants for Japan and Korea, and for their citizens abroad, often involving unforeseen boundaries and struggles. Nicola (1997) examined foreigners marrying into Japan. Experiences of immigrants and their descendants in North America were traced by Kobayashi and Jackson (1994) and Kobayashi (1996). Tyner (1998) linked the 1940s politics of eugenics to the incarceration of Japanese Americans in the US. Kwon (1990) examined Koreans' livelihood in the Bronx in terms of "opportunity structure" in their new setting.

6. *Political Concerns.* Economic and social change in East Asia has had political ramifications, both in regime change within individual countries and in bilateral relations within the East Asia region. Political dynamics on the Korean peninsula were studied in light of demographics by Fuller and Pitts (1990), regional rivalries by Dong Ok Lee and Stan Brunn (1996), and United States' interests by Pitts (1997). Relations between Japan and the US were studied by Grant and Nijman (1997) through relative levels of foreign aid to the region, and by

O'Tuathail (1992, 1993) through tendentious cooperation on defense projects. Geographers saw geoeconomic concerns replacing former geopolitical considerations in trans-Pacific diplomacy (Kodras 1993). Edgington observed new North American regional responses to trade opportunities with Japan (1995*b*). R. Grant (1993*a*, 1993*b*) and O'Loughlin and Anselin (1996) examined the boundaries of trade competition between the US and Japan. Ufkes called for new research into the politics of Asia's changing agricultural trade (1993*a*), and contributed a study of Japan–US feed and beef trade (1993*b*). East Asian fisheries policy study was aided by a new marine atlas by Morgan and Valencia (1992). Changing themes in political geography within Japan were analyzed by Yamazaki (1997) and Fukushima (1997). Japan's emergent leadership in global environmental issues met with the skepticism of Taylor (1999). The slow thaw of the cold war in Northeast Asia has introduced new possibilities for cooperation among states in the region for economic development, as in a proposed Tumen River special economic zone on the Sea of Japan studied by Marton *et al.* (1995) and by Morgan and Olson (1992). New questions as to how geographers and policy-makers should think about regions in Pacific-Asia have been posed by Murphy (1995), Watters and McGee (1997), Forbes (1997), and McGee (1997).

7. Studying East Asia. A good measure of self-critique has suffused academic geography as it apprehends East Asia in the 1990s. Asianist geographers have asked whether the structures of our discipline meet the intellectual challenges of today's Pacific Rim (Alwin 1992; Ginsburg 1993), whether geographies of Asia can escape Eurocentrism (McGee 1991), and whether the Asia inherited from past regional geographies can serve geography today (Lewis and Wigen 1997). East Asia as a subject within the secondary and undergraduate curriculum has continued to command the thoughtful attention of academic geographers (Latz and Borthwick 1992; Cybriwsky 1996; P. Grant 1998; Nemeth 1998; Latz 1996, 1999). Geographers in East Asia have also combed North America to compile useful reference works such as Hong's (1999) bibliography of Korean geography.

Southeast Asia

Over the past decade, American geographers working on Southeast Asia have pursued the traditional themes of geographical research associated with the region while simultaneously branching into new terrain. Principal among the new areas of interest are gender studies, political geography, and the study of social identity. Recent work shows an increasing concern with economic and political processes that link Southeast Asia with other parts of the world and, ultimately, with the global economy. Research on the region has also grown more reflexive, with several studies examining the fundamental geographical concepts through which Southeast Asia has been framed.

1. Environmental Concerns. The relationship between people and the natural environment has long been a central theme of geographical publications on Southeast Asia. Key areas of concern in the 1990s include deforestation, agroforestry, soil erosion, and other forms of upland environmental degradation, aquaculture and its related ecological problems, and the creation of protected areas.

Deforestation, arguably the most serious environmental problem in Southeast Asia, has been examined in depth by Kummer (1991, 1992; Kummer and Turner 1994). Through an exhaustive statistical study, Kummer has shown that forest loss in the Philippines—a process now almost completed—has been caused largely by politically protected, profit-seeking logging firms and not, as the popular imagination often sees it, by land-hungry shifting cultivators. Kummer *et al.* (1994) have examined environmental conditions on Cebu, an island often viewed as the paradigm of total ecological despoliation. Again confounding popular beliefs, Kummer and his co-workers discovered that the island's ecosystem is far more resilient than was previously believed. Several of their conclusions about Cebu are confirmed by Bensel and Remedio (1992*a*, *b*, 1995), who have examined the island's wood-fuel market in painstaking detail. They discovered small-scale rural entrepreneurs responding to market forces by reforesting significant areas—a process that is not, however, without its own environmental problems (see also Bensel and Harris 1995 and Bensel and Kummer 1996). Worrisome levels of soil erosion and other forms of environmental degradation were discovered by DuBois (1990) in his study of the neighboring island of Siquijor.

Several geographers have focused on similar environmental issues elsewhere in Southeast Asia. Hafner (1994) has investigated small-scale reforestation and agroforestry efforts in Thailand (see also Hafner and Apichatvullop 1990), and Suryanata (1994) has reported on agroforestry in Java. Detailed studies of Thai fishing and aquaculture have been conducted by Flaherty and Karnjanakesorn (1993, 1994; see also Flaherty and Vandergeest 1998). As they show, the widespread conversion of mangrove forests and other coastal lands—and now even of freshwater rice fields—to shrimp ponds has disruptive environmental consequences. Thailand's urban environmental problems, coincident with its rapid industrialization, have come under scrutiny by Hussey (1993). The difficult struggle to ensure the protection of natural ecosystems, meanwhile, has been investigated by Aiken (1994) in Malaysia, and Hafner and Apichatvullop (1990) in Thailand.

The linkages between environmental and economic issues have been the subject of a number of recent studies. Crooker and Martin (1996) and Lewis (1989, 1992) have examined highland vegetable production in northern Thailand and northern Luzon, respectively, while Surayanata (1994) has investigated highland fruit production in Java. Lewis (1992) concluded that environmental degradation in the vegetable districts of northern Luzon has been exacerbated by an indigenous system of belief and ritual that encourages high-risk economic behavior. For upland Java, Suryanata (1994) argues that certain agro-economic formations have resulted in the increasing social differentiation of poorer and wealthier cultivators, whereas others have reinforced small-scale production. Issues of social differentiation in Southeast Asian agriculture have also been studied by Hart (1989, 1991, 1992). Huke and Huke have examined agroecology at a far larger scale, surveying Southeast Asia as a whole. Their written reports, articles, and published maps document in detail the complex spatial patterning of rice, Asia's most important crop (Huke 1990; Huke and Huke 1989, 1990).

2. Economic and Urban Questions. Although rural and urban issues in Southeast Asia have usually been considered in isolation, Leinbach, in collaboration with several co-authors, has demonstrated how inextricably connected they can be. On the one hand, Indonesian peasants often depend crucially on off-farm income, including remittances from elsewhere in the country (Leinbach 1992; Leinbach and Bowen 1992; Leinbach and Watkins 1998). On the other hand, agricultural settlement schemes, particularly those associated with Indonesia's controversial transmigration program, often require close linkages to the urban economy if they are to succeed. Leinbach *et al.* have also demonstrated how labor-allocation decisions among transmigrant families often depend on the family's specific life-cycle state, a finding that builds on yet modifies the influential Chayanovian model of peasant economic behavior (Leinbach and Smith 1994; Leinbach *et al.* 1992). Migration in Indonesia and elsewhere in Southeast Asia has also been investigated by Goss and others (1992; see also Goss and Leinbach 1996; Ulack and Watkins 1991; Watkins *et al.* 1993; and Silvey 1997); other demographic issues have been examined by Ulack (1989) and Leinbach (1988).

Urban geography remains a vibrant area of Southeast Asian geographical research. Studies of urban morphology have been conducted for Bangkok (Thomson 1998*b*), Belawan, Sumatra (Airriess 1991), and Melaka (Cartier 1993). Ford (1993) has advanced a general model of Southeast Asian urban form, while McGee (1991*a*, *b*, 1995) has elucidated the new "megaurban" structure that is emerging as large cities expand (see also McGee and Greenberg 1992). Continuing his work on hill stations, Reed (1995) has outlined the growth of Dalat, Vietnam, from a small hamlet to a sizeable city. Urban poverty and slumscapes—familiar features of all Southeast Asian cities save Singapore—have been investigated by Thomson (1998*a*; 1991) and Goss (1990). According to Goss, the exploitation of Manila's poor involves both socioeconomic and spatial processes. Sicular (1989) examined one of the poorest communities in the region: urban scavengers in western Java. Historical explorations in urban geography include Cobban's (1993) work on public housing in colonial Indonesia and Doeppers' (1991, 1994, 1996) investigations of the historical development of Manila. Tyner (2000) has more recently examined contemporary Manila in the context of the global city literature, arguing for the importance of labor mobility in defining the "global city."

Transportation geography also continues to command attention. In a major study of the region's airline industry, Bowen and Leinbach (1995) argue that pragmatic governments have sought to balance the benefits and risks of deregulation through proactive policy initiatives. Airriess (1989, 1991, 1993) has focused on shipping, a particularly vital industry in Southeast Asia. He shows how containerization is changing the shipping business throughout the ASEAN region, transforming port-hinterland transportation patterns and reshaping urban form.

Broader questions of economic development in Southeast Asia have been addressed by a number of geographers, including some whose primary research sites lie in other regions of the world. Scott (1994), for example, compared the spatial economics of agglomeration in the Los Angeles jewelry industry with that of Bangkok; Murphy (1995) mapped out the patterns of economic regionalization in East and Southeast Asia; and Glassman and Samatar (1997), comparing Thailand and Botswana, examined the third-world state's role in fostering development. The latter authors concluded that elite unity in Thailand allowed the Thai state effectively to promote economic growth. Several Southeast Asianists have looked more specifically at the developmental effects of ASEAN (Hussey 1991; McGee and Greenberg 1992), direct foreign investment and trade policies (Leinbach 1995; Chang and Thomson 1994), and migration (Leinbach 1989). Leinbach (1995) concluded that direct foreign investment in Indonesia is not sufficient to ensure rapid industrialization, that Indonesia faces increasing competition in this regard from China and especially Vietnam, and that the country needs to emphasize technology and higher value-added production. Regional economics in the "growth triangle" of Singapore, the Riau Archipelago, and Johor has been the subject of an innovative dissertation by Macleod (1995). Economic history has been tackled by Doeppers (1991), who determined that the effects of the Great Depression in the Philippines were far more spatially complex than scholars had previously assumed, whereas the economic crisis of the late 1990s was the focus of an important article by Poon and Perry (1999), who similarly argued for profound spatial variability. In an important study of this crisis, Glassman (2001), arguing from a Marxian perspective, stresses the declining profitability of manufacturing and increased international competition. Focusing on labor organizing in the Philippines, Kelly (2001) has concluded that a simple antagonism between global capital and local labor does not obtain. He instead stresses the geographical complexity, informality, and fluidity of local labor-control regimes.

3. *Social Contexts and Political Questions.* The most important new developments have been in the fields of social and political geography. Several scholars have highlighted the previously neglected dimension of gender: Silvey (1997) has shown how gender influences migration patterns in an export processing zone in southern Sulaweisi; Hart, investigating Malaysia's Muda region (1992), has highlighted gender dynamics to show the limitations of both the neo-classical theory of the farm household and of Scott's notion of "everyday forms of peasant resistance" (1991); Tyner (1996*a*, *b*, 1997) has convincingly argued that Southeast Asian international labor migration must be understood in gendered terms; and Yasmeen (1997) has examined the role of women in Bangkok's public sphere through the novel concept of the "foodscape" (see also Watkins *et al.* 1993; Walker 1997). The social construction of both place and identity has also emerged as new research foci. Cartier's (1993, 1996, 1997) work in Melaka, for example, illustrates how cultural identity may be place-based, thereby allowing the formation of powerful localized social movements, yet may be simultaneously employed to advance claims about common national identity. Tyner (1997) shows how Filipino migrant entertainers have been socially constructed as "disreputable," thereby justifying their exploitation. Lewis's (1991) work in northern Luzon demonstrates how cultural identity can vary tremendously depending on social context and the spatial scale of analysis. Cartier (1998*a*) and Thomson (1993) have examined the complex patterns of identity among Southeast Asians of Chinese ancestry, with Thomson showing that the Chinese in Thailand, with the encouragement of the Thai state, are increasingly identifying themselves as "Thai."

The study of tourism in Southeast Asia has recently emerged as an active field of research. Several works have assessed the environmental and economic impact of tourist facilities in particular areas: Dearden (1991) in northern Thailand, Hussey (1989) in a Balinese village, and Lenz (1993) in Vietnam. In a series of innovative articles, Cartier (1996, 1997, 1998*b*) elucidated the role of tourism in Melaka, demonstrating how the state and developers in concert have created high-cost, environmentally damaging "ersatz leisurescapes," often sacrificing natural and historical amenities in the process (see also Chang 1999 on heritage tourism in Singapore). Generally missing, however, from these studies is in-depth research into tourism's ingrained role in the growth of prostitution and its public health consequences in the region.

Political geography in Southeast Asia has evolved in a number of different directions. National integration in Indonesia—a notoriously complex and politically charged issue—has been studied in depth by Drake (1989, 1992). Thomson (1996), questioning the political role of the Sino-Thai by analyzing electoral returns, concluded that the Chinese in Thailand do not form a distinct political group. Dissecting the political geography of ethnic rebellion in war-torn Myanmar (Burma), Thomson (1995) also determined that the unitarist policies of the Burmese government consistently undermine national integration (see also Thomson 1994 on Thailand's May 1992 democratic uprising). Kummer (1991, 1992, 1995) has uncovered the political processes underlying Philippine deforestation, and has examined the problems that this poses for researchers; in many cases, statistics have been destroyed by state functionaries eager to conceal the extent of forest destruction.

4. *The Geography of Geographical Research in Southeast Asia.* Geographical scholarship on Southeast Asia has been conducted at a wide range of spatial scales. While a number of studies have focused tightly on individual village communities or urban districts, others have targeted provinces or other sub-national regions, and a large number have taken the nation-state as their unit of analysis. Others have grappled with Southeast Asia

as a macro-region (or at least those Southeast Asian countries belonging to ASEAN). A significant departure from past efforts is the sizeable number of studies examining Southeast Asia in broader global or comparative terms (see e.g. Airriess 1991, 1993; Bowen and Leinbach 1995; Drake 1992; Glassman and Samatar 1997; Macleod 1995; Warf 1998). Kelly (1997), inverting the terms in his Philippine study, demonstrated that processes of globalization can only be understood in the context of local social and economic relations; his work also counters the common notion that globalization is both inevitable and necessary. Another innovative approach is to examine Southeast Asian peoples and influences in other regions of the world. Examples here include Tyner's (1996*a*, *b*, 1997, 1999) work on overseas Filipino entertainers and on the historical exclusion of Philippine migrants from the United States, Cartier's (1995) study of Singaporean influence on urban development in Shanghai, and Law's (2001) work on how women migrants from the Philippines create their own senses of place in foreign cities through body politics, leisure activities, and sensory experiences. Finally, scholars have begun to examine Southeast Asia metageographically, interrogating the fundamental spatial categories used to think about the region and its place in the world. The lead here was set by Southeast Asian scholars (Savage *et al.* 1993), and was subsequently picked up by Lewis and Wigen (1997). Schwartzberg's (1994*a*, *b*) brilliant work on indigenous cartography in the region has also shed light on metageographical issues. In a somewhat different register, Tuason (1999) has examined the metageographical "ideology of Empire" implicit in the *National Geographic* magazine's coverage of the Philippines.

As in the past, certain parts of Southeast Asia have been much more closely studied than others. The Cebu region (including the neighboring island of Siquijor) is arguably the single most heavily scrutinized place; other parts of the Philippines have also received much attention. Thailand, Indonesia, and Malaysia have all been the subject of numerous geographical studies. Singapore, by contrast—despite its status as the financial, telecommunications, and transportation hub of the region—has been surprisingly little studied (but see Macleod 1995; Macleod and McGee 1996; Airriess 1998; Chang 1999). More understandable is the relative invisibility of Burma, Cambodia, Laos, and Vietnam—countries that present serious obstacles for foreign researchers. Reed (1995) none the less has demonstrated the feasibility of conducting research in Vietnam, while Thomson (1995; see also Huke 1998*a*, *b*) has shown that significant work can be done on Myanmar despite the restrictions and repression of its government. Acker, meanwhile, is conducting important research on Cambodia. Only Laos remains truly *terra incognito* on the map of American geographical scholarship, the exception being the edited work by Dutt (1996), who provides a valuable and comprehensive coverage of the whole of Southeast Asia.

To conclude, relatively little work by geographers on Southeast Asia has been conducted in the idiom of postcoloniality, an idiom that has substantially transformed area studies in the humanities over the past decade (for exceptions, see Silvey 1997; Tyner 1997; Cartier 1997). But it must be recognized that the geography of Southeast Asia has become profoundly post-colonial in one respect: intellectual leadership on the subject is increasingly to be found within the region itself. Singapore, neglected as it is by American scholars, is now probably the most dynamic center of Southeast Asian geography, with the *Singapore Journal of Tropical Geography* forming its premier scholarly publication (see also Kong *et al.* 1996). One might hope that more collaborative work between US and Southeast Asian geographers will enrich our understanding of this dynamic, diverse, and—for Americans—still relatively poorly understood part of the world.

South Asia

Once a center of bold and innovative development experiments, South Asia has turned into a stepchild of Asia. In the 1950s and 1960s, such experiments included systematic policies of village development, national planning as an institutional paradigm of development (which is still practiced), the green revolution, and import substitution-based industrialization. Today, in the world dominated by material development and growth, South Asia rarely gets a mention in the American press when its attention turns to Asia. Much of the focus is almost exclusively directed toward East and Southeast Asia. It is believed that this obvious neglect of South Asia is a direct reflection of its inability to rise up the Darwinian totem pole of economic growth which is invariably measured in terms of gross national product. For instance, in the mid-1960s, India and South Korea started roughly at the same level in terms of their GNP per capita income: \$150. Today, South Korea's GNP per capita has soared to over \$10,000, whereas India's remains staggeringly low at less than \$400 (Table 39.1). As India goes, so goes the rest of South Asia. Next to sub-Saharan Africa, South Asia is routinely described by the World Bank as the poorest region in the world. It is, therefore, no surprise that poverty occupies center court in its policy and political debate as well as research agenda.

Although American geographical research on South Asia is relatively productive, the circle of geographers conducting such research is narrow. For much of its foundation, South Asian regional geography in America owes a great deal to a small group of early pioneers in the field. For example, the works by John Brush, P. P. Karan, Rhoads Murphey, Joseph Schwartzberg, and David Sopher proved to be quite influential in generating research interest in South Asian issues (Karan *et al.* 1989). The foundation that they laid was later expanded by the second generation of South Asianist geographers, among whom stand Nigel Allan, Surinder Bhardwaj, Ashok Dutt, and Allen Noble. In their continuing efforts to uplift the research profile of South Asian geography in America, Allan has done much research on the mountain rimland. His research in the area of cultural geography and ecology has spawned a new breed of geographers who can be classified as the third generation of South Asianists. Bhardwaj has been a leading geographer among those studying religions. His work has been extended by Karan (1994), Stoddard (1988), and Stoddard and Morinis (1997). Dutt's and Noble's works, on the other hand, are largely focused on urbanization and regional planning. Their numerous publications are not only notable in their breadth, but also in terms of the depth of contribution they have made to our understanding of Asian urban systems and planning. Also included in this second generation of South Asianists is Lakshman Yapa, who continues to raise challenging questions about poverty and development.

One common denominator among these first and second generations of South Asianist geographers is that they have largely taken what can be called a conventional route to geographical analysis in that their studies are relatively free of controversial positions and issues about spatial patterns and processes. Murphey and Yapa are exceptions to this general rule. Taking a stance against the diffusionist model of modernization, Yapa (1977, 1993) has consistently questioned the material and ecological efficacy of the green revolution. In an important work on the history of cartography, Edney (1997) shows how ideologies of imperialism informed the British mapping of the Indian subcontinent. Kenny (1995) documents how the British established hill stations in various parts of India not only to recreate little England in the torrid colony, but also to separate themselves from the "inferior" subjects, thus projecting an image of their racial superiority. In other words, the wall of general conformity and continuity in the geographical analysis of South Asia is now coming down. Among the new (third) generation of South Asianist geographers in America, there is little uniformity. Although they continue to walk down the same path mapped by their predecessors, there is much diversity in the views, perspectives, methodologies, and historical lenses that they deploy in their geographical studies and analyses of South Asia. With this clarification, we now discuss two major, but interrelated, themes found in American geographical research on South Asia.

1. Poverty and Development Dilemma. When independence finally came to India after years of struggle against the British, the air was filled with euphoria. Rising material expectations swept across the country. Development seemed imminent and ready to slay the dragon called poverty to free the masses from their material abyss. Jawaharlal Nehru thunderously proclaimed that India had to achieve in twenty years what the West did in 200 years. More than half a century has passed since independence, but poverty is still rampant and development remains a distant mirage. No wonder why poverty and development continue to hold the subcontinent captive as a haunting specter of its relentless past (Shrestha 1997; Yapa 1996, 1998).

In this ongoing drama of poverty, Ashok Dutt and his colleagues espouse regional planning and development as a sound economic strategy (see Noble *et al.* 1998; Pomeroy and Dutt 1998). Along this same line, they have also provided detailed studies of urban systems of South Asia (Costa *et al.* 1989; Das and Dutt 1993; Dutt 1993; Dutt *et al.* 1994; also see Dutt *et al.* 1997; Mookherjee 1994; Mookherjee and Tiwari 1996; Rasid and Odemerho 1998). It can be deduced from their studies of urban systems that cities serve as vital growth nodes, key central places from which modernization is presumed to diffuse into hinterlands, thus facilitating regional development across the nation (also see Shrestha 1990). On the other hand, Chakravorty (1993, 1994, 1995, 1996, 1999, 2000*a*, *b*) offers a different view on this issue in his extensive treatment of the role of urbanization in development. Utilizing a political economy perspective, he asserts that for urban planning and development in democratic societies with market economies to be successful, the planning process must be transparent and the goals equity oriented. This is particularly true of societies making the transition from a state-controlled to a market economy (also see Chakravorty and Gupta 1996). In addition, he shows, particularly in reference to Calcutta, how big cities tend to perpetuate social inequities. At the country level, Karan and Ishii (1994) have offered a comprehensive analysis of development and change in Nepal, a country no less beset by poverty than any of its neighbors in the region (also see Karan

1989; Karan and Ishii 1996). Zurick (1989, 1990, 1992, 1993) has discussed the roles of spatial development and traditional knowledge, along with tourism, in Nepal's rural transformation and subsequent development. Tourism and its cultural effects have also been a focus of analysis that Stanley Stevens conducted in Nepal (Stevens 1988*a*, *b*, 1993*b*; Shrestha 1998*a*). In his valuable research, Metz (1989, 1990, 1994) has, on the other hand, emphasized the various subsistence strategies of hill residents in central Nepal.

On the Pakistan side, Nigel Allan (1989, 1991, 1995) has spent much time trying to develop a geographical understanding of mountain habitat and society. His analysis of the impact of road construction as part of overall regional development efforts on mountain society has shed illuminating light on this problem. He has also cast doubt on the utility of what is now being fashionably peddled as sustainable development throughout South Asia and beyond (Allan 1998). Butz (1995) and MacDonald (1998*a*) have extended Allan's discussion of the impact of roads on mountain society to mountain portering as an economic survival strategy in northern Pakistan. While MacDonald traces its colonial roots, Butz reveals a historical conflict between commoners and royalty over the issue of portering regulations (also see MacDonald and Butz 1998). In addition, MacDonald (1996*a*) discusses indigenous labor arrangements and household security, whereas Butz examines pastoralism as a source of resource in sustaining mountain communities (also see Butz 1994; Butz and Eyles 1997; MacDonald 1996*b*, 1998*b*).

One of the central themes in the poverty and development discourse is population and its associated issues such as family planning and health-care delivery. These issues have drawn the attention of Paul's research endeavors, especially in the context of rural Bangladesh. His research ranges from family planning practices to health search and reproductive behaviors to infant mortality and AIDS (Paul 1990, 1991, 1992, 1994, 1997). Paul has also investigated flood damage to crops, a development and environmental hazard issue that has serious implications for those living in the low-lying coastal areas of Bangladesh and other Asian countries (Paul 1993, 1997, 1998). Regarding female health care in India, Tripathi's (2000) empirical findings reveal interesting dimensions of rural and urban women's health-seeking behaviors. In the case of Shrestha's research, however, much of the population- and development-related emphasis has been placed on frontier migration and settlement in the Tarai region of Nepal. Through his extensive research, he has discovered that this rural-agricultural development strategy has boomeranged. As the population gravity of Nepal shifts from the hills to the Tarai as a result of growing migration, massive land encroachment and ecopolitical conflicts between the state and landless migrant peasants surface as serious issues that have significant ramifications for the country's regional political economy (Shrestha 1989, 1990, 1998*b*; also see Conway *et al.* 2000; Shrestha and Conway 1996; Shrestha *et al.* 1999). In their investigation, Shrestha *et al.* (1993) have established that migrants' economic success at the frontier largely depends on the timing of migration as well as migrants' previous socioeconomic class background at the place of origin.

Another running theme in South Asia's poverty and development debate is the effectiveness of the green revolution in not only increasing food supply, but also advancing the cause of local and national development and hence reducing the level of poverty. Since its inception in the mid-1960s, the green revolution has received much research attention. While few question its ability to increase agricultural yields, many are doubtful of its ability to reduce poverty and achieve some sense of social parity in India and other South Asian countries (for atlases of rice and wheat production in South Asia, see Huke *et al.* 1994, 1993; Woodhead *et al.* 1994, 1993). Both Yapa (1977) and Das (1995, 1998*a*, *b*, *c*) have clearly demonstrated how the green revolution has generally failed to attain the latter goal of curtailing poverty despite a sustained and noticeable growth in total crop production. The structural flaws, according to Das, lie with the way in which the state is organized and operates in India (or, for that matter, other South Asian countries as well). Consequently, like its neighbors, the Indian state is patently incapable of alleviating poverty, for it is unable to distribute the economic benefits of the green revolution among social classes (Das 1997; Yapa 1977, 1996). This is a general sentiment that Stokke (1997) echoes in his examination of Sri Lanka's authoritarian state in the age of market liberalization. Stokke specifically contends that governance in Sri Lanka has not been the outcome of a certain economic system, but rather the result of specific political projects and accumulation strategies.

What we can discern from the works discussed above are two major conclusions that are closely entwined. First, the state has been the central and sole player in virtually every phase of development planning, projects, and strategies. This is true in every country in South Asia. While the logic of such a centralized development process or so-called "state capitalism" was firmly rooted in the belief that the state would achieve greater efficiency in the allocation of scarce capital resources and lead to both sectoral and spatial balance in national development, the outcomes have been anything by efficient and effective. Second, despite more than four decades of planned development, poverty as defined and measured by the World Bank has displayed little sign of retreat from the region's socioeconomic vista. Not surprisingly, therefore, some serious questions about official development and its failed messianic vision of uplifting the poor from poverty are being raised in many quarters. Leading the charge in this area is Yapa (1998) who, citing the case of Sri Lanka, stresses that anti-poverty measures, invariably founded on the international discourse of poverty and development, do not serve the interest of the poor. In his discursive analysis, Yapa further articulates that although Sri Lanka is often touted as an exemplary case of direct poverty alleviation because of its long history of social welfare, its official development policy fails to see that poverty is socially constructed. Because of this fundamental failure, the policy, he argues, is hardly in a position to cure poverty (also see Yapa 1996). In a similar vein, Shrestha (1999) documents how contemporary development is innately flawed as it systematically leaves behind a trail of victims, all in the name of development. He goes so far as to claim categorically that foreign aid, which now drives much of the development policy and activity in most of these countries, has to go if they are to realize the full potential of indigenous development, one that is self-sufficient and self-reliant as well as nationally sustainable. His argument is embedded in the logic that foreign aid is little more than a form of colonialism in its post-colonial garb, largely a product of the Cold War dominated foreign policy that is carefully crafted to serve Western interests rather than indigenous citizens and their communities. As such, it only engenders, and then deepens, Western dependency; it does not, it cannot, and it will not cure poverty.

2. Environmental Degradation and Ecological Breakdown. This is a theme that has gained growing popularity among South Asianist geographers. What started out as a loose theory of Himalayan degradation has now evolved into a well-entrenched field of research in and on South Asia. Once again, Karan (1991, 1989*a*, *b*) has generally led the way in this field (also see Karan and Ishii 1994; Zurick and Karan 1999). But many have pursued this topic and made significant contributions to our geographical understanding of environmental degradation and ecological breakdowns, especially in South Asia's mountain rimland. For example, Ives and Messerli (1989) have discussed at some length the various issues, revolving around the Himalayan environment and its human occupants as well as its regional political repercussions. The debate and discussion have been further extended by Allan (1995), who is now becoming increasingly wary of what he calls the "transaction costs" of doing fieldwork that is often physically demanding and time consuming.

A substantial amount of the geographic research in this field is focused on Nepal and the central Himalayan belt. Although these studies generally center around the theme of environmental degradation, one can observe much diversity in terms of specific research agendas advanced by different geographers. While Bishop (1990) and Zurick (1988) have generally investigated the human consequences of environmental degradation and ecological breakdowns in the western hills of Nepal, Brower (1991) has dwelt on the pastoral system and its impact on cultural landscapes in the eastern mountains, especially around the Everest region. Stevens (1993*a*, 1997) has elevated this cultural ecology analysis of these eastern mountain communities to an even higher plateau in his thorough book, *Claiming the High Ground* (1993). Metz (1990, 1991, 1994, 1997), on the other hand, has generally confined himself to the central hills of Nepal where his research emphasis has been placed on the hill conservation practices, subsistence system, and forest product use patterns (for a similar case in southeastern India, see Balachandran 1995).

One exception to this general hill and mountain bias in the cultural ecology studies of environmental degradation is the political ecology work of Shrestha (1990), who has mostly concentrated his research on the ecological drama being played out in the Tarai lowlands. He has shed historical light on the dynamics of eco-political battles raging in the Tarai, and recently examined the environmental degradation taking place in the Kathmandu valley (Shrestha 1998*a*; Shrestha and Conway 1996). On the India side, Paul Robbins has devoted a considerable amount of his intellectual capital exploring various aspects of political ecology and ecosystem (Robins 1998*a*, *b*, 2000*a*, *b*, 2001). One of the themes that he has diligently highlighted is the issue of common resource access and control in relation to effective resource management. Tracing the history and effects of rights to forest and pasture in Rajasthan, India, he contends that neither the central state nor the local community is a necessarily superior manager of nature (Robbins 1998*a*, 2000*b*).

To conclude, there are many other pertinent topics that demand the research attention of South Asianist geographers in America, for example, a horrendous state of urban pollution throughout the region and its profound impact on public health (see Shrestha 1998*a*; Wallach 1996). It is a pending crisis that will not only pound on South Asia's cities, but cause immense human tragedies, thus compounding its massive poverty and development efforts. Ethnic conflicts and violence as well as the consequences of market liberalization and globalization for its poor masses are also very timely and relevant themes for geographic research. One recent exception to the general lack of globalization focus in South Asian geography in America is the work of Rupal Oza (2001), who takes a critical look at the local opposition to globalization in relation to the protest by different organizations to the 1996 Miss World pageant held in Bangalore, India. At this point, however, it is unlikely that the two sets of issues discussed above will yield much ground to other research areas in the foreseeable future.

Southwest Asia

Nearly a quarter of a century ago, Michael Bonine (1976) published a query in the *Professional Geographer*, asking "Where is the Geography of the Middle East?" Despite the effort of a cadre of dedicated scholars, the same question might be asked today. A problem exists, of course, as to what constitutes research on Southwest Asia. The number of faculties listing Southwest Asia as an interest in the AAG guide can be counted on two hands, with digits to spare. This highlights both a problem of nomenclature and a challenge to geographers who (re)conceptualize the region in question. Geography textbooks dealing with the Middle East are not unanimous on this point. Fisher (1950) includes the territory from Iran in the East to Libya in the West, and from Turkey in the North to the Sudan in the South. Drysdale and Blake (1985) group the Middle East and North Africa, arguing that they are an integral unit, thus extending their boundary to Morocco in the West. Beaumont *et al.* (1988) acknowledge cultural linkages across North Africa, but bound their text with Libya's western borders. Whether or not one stops at some point between Egypt and the Atlantic, it is clear that "Southwest Asia" is not a regional appellation that holds much appeal for those who generally think of themselves as scholars of the Middle East. For that reason (and the non-inclusion of North Africa in the treatment of the Africa Specialty Group in this volume) a more inclusive approach is taken in this section. Yet, even with expanded boundaries, the question remains, where is the geography of the Middle East?

With the inclusion of North Africa, the Middle East, Southwest Asia, and specific country names as research interests, there are still fewer than forty such self-identified geographers associated with geography programs in North America. Bonine's plea to generate a larger group of scholars to study this region has not been met; indeed, a significant number of geographers interested in the Middle East are nearing retirement. What, then, is the nature of research conducted over the last decade? A number of concentrations, if not intellectual themes, can be identified during this period. Some of them have their roots in earlier geographic work on the region, such as urban morphology and sacred space/landscapes. Adaptation to arid environments continues to draw interest, and has been joined by a rapidly growing literature on water conflict. Conflict over natural resources is a subset of political dispute in general, and a number of facets of the Israeli–Palestinian conflict have drawn attention, as has the indigenous Palestinian Arab population of Israel, the West Bank, and the Gaza Strip. Physical geography has been notably lacking for the region as a whole, though with a number of localized exceptions.

1. *Cities, Territory, and Politics in the Middle East.* The Islamic city clearly continues to fascinate those who study both its morphology and meaning (Stewart 2001). Iranian cities have long been a staple of this literature, with Kheirabadi's (1991) work joining that of Hemmasi (1994), examining both form and function. It is significant that this work continues to grow alongside that of scholars in Iran, despite the impediments to research in Iran following the Islamic Revolution (Hemmasi 1992). Bonine and Ehlers (1994) have contributed a bibliography of Islamic cities that is grouped by region, going beyond the Middle East to include those parts of the world populated by Muslim majorities. Cohen (1994) addressed colonial planning as a template for green space in Jerusalem, while Stewart turned her attention to the creation of new towns in Egypt (1996*a*) and the recreation of an old one in Beirut (1996*b*), and a political economy of morphology in relation to the region's largest city, Cairo (1999). A collection edited by Bonine (1997) reflects the mainstreaming of Middle East urban interests, as captured in its title *Politics, Poverty, and Population in Middle Eastern Cities.* Unfortunately, not counting Bonine's work, geography is represented by only one contribution, that of Vasile (1997*a*), who has studied migration, politics, and religion in relation to the impoverished communities of Tunis. This study provides a link to the growing interest in the interplay between the state, unequal development, and the role of Islamist groups in the Middle East. The focus on politics in and of urban areas is also developed in relation to Jerusalem, where Emmett (1993) examined political and religious landscapes in that city, as well as a variety of "solution" scenarios (also see Emmett 1996,

1997*b*). Cohen (1993) focused on manipulation of the environment as a tool both Israelis and Palestinians employ in trying to control that city. In an examination of a similar dynamic with a slightly different set of protagonists, Emmett (1995, 1997*a*) examined the struggle between Muslims and Christians for control of the town of Nazareth, while Falah (1992, 1996*a*, 1997) pursued the topic between Jewish, Muslim, and Christian Israelis in Nazareth and other mixed towns in Israel. Territorial struggle and landscape representation in the Israeli–Palestinian conflict received considerable attention. Cohen and Kliot (1994) examined the symbolic nomenclature imposed on the landscape by Israel, while Falah (1990, 1991*a*, *b*, 1996*a*) discussed the imposition of an Israeli imprint on the land at the expense of the Palestinians living there. Cohen (2000) examined the impact of territorial fragmentation on identity, place, and land tenure in the West Bank. Parmenter (1994) describes Palestinian identity, culture, and representation in relation to the struggle against Israeli occupation. In a rare departure from the Israeli–Palestinian case in the political geographic literature, Drysdale (1992) commented on Syrian–Iraqi relations in terms of boundary, resource, and territorial disputes.

2. *Water and Arid Lands.* Focus on the political carries over into issues of water allocation, reflecting popular perception that water is a matter of war and peace in the Middle East (Kolars 1992; Drake 1997; Amery and Wolf 2000). Early in the decade geographers were engaged in assessing scarcity; recently they have turned their attention to strategies for dividing or sharing water to mitigate conflicts over usage. Kolars (1994*a*, *b*) has focused on the Euphrates from a variety of perspectives, including the generation of hydro-power in Turkey and the implications for downstream users, and the impacts of withdrawal on the ecology of the Arabian Gulf. Kolars (1995) has also examined the controversial issue of extra-basin transfers from the Jordan, and the Litani (1993), the latter pursued by Amery (1993) as well. Wolf and Dinar (1994*a*, *b*, 1997) have focused on the Jordan River and the aquifers relevant to the Israeli–Arab conflict, to develop models for markets and regional cooperation (also see Wolf and Murakami 1995). International efforts, albeit of a different nature, are also necessary to retard desertification and manage arid lands, as detailed in the work of Benchefira and Johnson (1991) and Johnson (1996). Such research continues to provide new insights into nomadism and its impact on the environment (Johnson 1993*a*, *b*), and, along with pastoralism (Benchefira and Johnson 1990), has been discussed in terms of resource management. Grazing impacts after the cessation of grazing have also been studied in Israel (Blumler 1993), though here from the physical rather than the political perspective taken by Falah. The work of Pease *et al.* (1998), Tindale *et al.* (1998), and Tchakerian *et al.* (1997) approaches arid environments from a geomorphic and eolian modeling angle, representing a small but significant exception in the lacuna of physical geography of the region's deserts.

3. *Geography in the Middle East.* While the description above is by no means exhaustive, it does capture the major elements of the work carried out in the last decade. There is, of course, a substantial corpus of research achieved by scholars in the region, including a significant amount of material published in English-language journals. Much of this work deals with local issues that would be difficult for foreign-based researchers to maintain, including research in areas that are inaccessible to foreigners. Some of it, particularly that of physical geographers in the Arab world, is carried out by those who received their training in the United States or Canada. It would be difficult to capture the breadth and depth of the work done by indigenous researchers, particularly as they publish in Arabic, Turkish, Farsi, Hebrew, English, French, and perhaps other languages as well. Fortunately, there are already summaries of some of the work being carried out in the Middle East. Gradus and Lipshitz (1996) have edited a collection of short essays by more than fifty Israeli geographers, organized in eleven thematic groups. While the contributions are compact, their bibliographies provide a key to the development of Israeli geography, in terms of both individual research programs, and the nature of the discipline in Israel as a whole. While Israeli geographers are not strangers to the pages of journals based here and in Britain, this volume gives a much fuller picture of their work. Yehiya Farhan (1996) of the University of Jordan provides a survey of geographic work in Arabic-speaking countries, tracing influences of the American, British, French, and Russian schools of geography. Farhan also provides a quantitative breakdown of the faculty and research interests at more than fifty geography departments in the Arab states, organizing them along national lines. Unfortunately, according to Farhan the majority of the Arab geographers do not belong to international professional associations, and their work is rarely published internationally, let alone in English, a point affirmed by Falah (1999). The recent genesis of *The Arab World Geographer*, an English-language journal based in the US, may help to rectify this situation. That is one of the stated goals of the journal, and, if successful, it should stimulate the interest of foreign geographers and help generate rich collaborative efforts.

4. *Towards the Future.* It is difficult to explain the paucity of research on the geography of the Middle East. Certainly the cause is not lack of interest among the general population. Indeed, whether it be the oil reserves in the region, the internecine conflicts, international conflagrations, or the ongoing religious attachments to history and territory, the English-speaking world is fed a steady diet of information and image concerning the Middle East. If there is comfort in company, anthropologists, too, have noted the relative neglect of the region in their research (Abu Lughod 1990). Particularly because of this long-standing problem, much remains to be done. Falah (1998: 1) posits the Middle East as "one of the most culturally cohesive regions on the planet." While this may be true at some levels, the complexity and diversity of the region, in both cultural and physical terms, should serve as a magnet to geographers. The Middle East is also a region of historical richness and modern ferment. It continues to interact with worlds to the East and the West, and the failure to explore its depths is a critical shortfall in furthering our understanding of the world. The Middle East should not be truncated as Southwest Asia, appended to Europe, as in Blaut's (1993) colonizer's model of the world, nor kept at undue distance stemming from a reticence to engage following Said's (1978) critique of orientalism (Abu Lughod 1990). Instead, geographers need to engage the region, starting from a new approach to what constitutes the Middle East, and from there, to all that geographers bring to the study of peoples and the physical environment.

Conclusions: Challenges for Asian/Regional Geography

Regional geography, historically the bedrock of the discipline, has arrived at a critical juncture. As noted earlier, the world has undergone a dramatic reconfiguration since the early 1990s, largely due to two overarching forces. The first catalyst behind this reconfiguration is the abrupt end of the Cold War resulting from the sudden dismantling of the Soviet Union. Gone is the map of bipolar geopolitics deeply rooted in the ideological conflicts between communism and capitalism, the map that shaped many regional contours and contortions for more than fifty years. It is now replaced by a map based on the rise of unipolar geoeconomics under the tutelage of capitalist globalism—also fashionably called "globalization." The second major development of the decade is a great leap forward in the field of information and communication technologies (ICT). In this concluding section, we first offer a short discussion of implications of the Cold War's demise, specifically in conjunction with globalization and the thesis of civilizational clash, and then proceed to our discursive exploration of how information and communication technological advancements affect regional geography as the primary producer and transmitter of regional information.

The End of the Cold War and Regional Geography

1. *Globalization and Regional Geography.* As the intensely contested bipolar space of geopolitics lost its regional configuration, unipolar capitalist globalism (notwithstanding its internal contradictions and competition) was seen as a leveler of national boundaries and barricades. Swept by globalization and expecting to capitalize on its windfalls, countries raced to adopt market liberalization as the economic mantra of the day. Global capital thus found itself free from national fetters to embark on a furious march across the world, subduing one nation after another as if they were spellbound. Then there was a global surge of ICT. As it resolutely defied both the friction of distance and national boundaries and jurisdictions, reaching almost every corner of the world with one key touch, it seemed further to fortify the emerging vision of unipolar geoeconomics: one world under global capital. Not only did this scenario lend support to the triumph of capitalism over communism, but it was also seen as a testament to the primacy of capitalist *economics* over nationalist *politics*.

Given these trends, many rushed to claim the "end of history" (Fukuyama 1993) and "the end of the nation state" (Ohmae 1995). More joined the parade. What these proclamations amounted to, in essence, was the death of *nationalism*—a prime root of international conflicts and often an impediment to free trade—and, hence, an emergence of what Ohmae (1990) called "the borderless world," a homogenous world tailor-made for global capital. For geographers, the moral of the story as told by these globalists was: no more geography. In view of the Internet's global reach and the tidal waves of e-commerce, Gary Hamel and Jeff Sampler (1998: 88), in fact, went so far as boldly to announce "the end of geography" as "E-commerce breaks every business free of its geographic moorings. No longer will geography bind a company's aspirations or the scope of its market . . . Customers, as well as producers, will escape the shackles of geography."

To be sure, if there was ever any question about the historical absurdity about such myopic claims, the terror of September 11, 2001, and the Bush administration's subsequent efforts to forge a multinational coalition in its fight against terrorism and "terrorist" nations have totally exposed their nakedness. As Fareed Zakaria (2001) aptly noted, that was "the end of the end of history" debate. Geopolitics may be dead, but history is not. And geography still matters. Globalization is not the end state of regional geography, though it has changed. In the wake of the Bush administration's Afghan war, regional geography is suddenly hot, but absent of the Cold War fervor. Nothing injects life into regional geography like international wars do. Almost overnight, Central Asia is back in the political picture, with its profile raised and roles magnified, as the US battles against Afghanistan's Taliban government. Even Russia, the Cold War enemy just until about a decade ago, is ready to lend support to the US efforts. Times have certainly changed; yet, once again, it appears that the fate, fortune, and formulation of Asian regional geography is intricately intertwined with the American global policy, plan, and play.

But there is a bigger issue. Although there is obviously little reason to give any credence to the thinking about the dissipation of history and the state, *globalization* is real and, by its very nature, geographical. Certainly, many questions are lately being raised about its national efficacy and outcomes. Yet few would dispute that globalization has altered the landscapes of national and regional political economies, e.g. regional flows of global industrial and finance capital, regional industrial location and production, cultural relations and transformations, and national alliances and realignments. One insidious side of globalization that has profound implications for regional geography and regional political economy and that has received minimal attention from Asianist (and other) geographers is the growing operation of what may be termed the "underground economy," for example, drug trafficking, human trafficking (for drudgery labor and prostitution in foreign countries), and money laundering. Not to be ignored is terrorism which has become increasingly globalized. In addition to undermining national borders and integrity, this type of globalization leads to enormous human tragedies and to the rise of cross-border banditry and lawlessness, thus threatening civil governance and security. Moreover, as globalization continues to produce winners and losers out of both people and places, even age-old poverty has grown a new face, a face that is grittier and much more complex than ever before (*Business Week* 2000; Yapa 1996; also see Soros 1997). And, as a result, regional geography has entered a new phase, one in which its frontiers have yet to be fully delineated and foci defined. Fortunately, however, we can take some comfort in the fact that many geographers—Asianists and others alike—have already laid a viable and valuable foundation for further geographical inquiries of globalization in relation to regional/local geographies and regional political economies (also see Yeung 1998).

2. The Clash of Civilizations and Regional Geography. In addition to the "end of history and the state" ideas, the defunct Cold War has spawned a competing thesis that may be characterized as "civilizational regionalism," a geographical configuration in which the boundaries of the political groupings of nations are drawn along the civilizational lines. This is what Samuel Huntington called the "clash of civilizations" in his 1993 article in *Foreign Affairs.* According to Huntington, a civilization (civilizational grouping) is a cultural entity with its foundation deeply seeped in religious identity, affinity, and affiliation, e.g. Christian (Western) or Islamic (with its geographical gravity centered in the Middle East). Apparently, the objective of his West vs. the rest thesis was two-pronged. On the one hand, he was urging Western (namely US) policy-makers to consolidate the West's newly attained unipolar global domination by preemptively fending off any challenge from the rest, Islam in particular (also see Said 2001). On the other hand, it was his bold attempt to advance the central axis of global politics in the post-Cold War world. He declared that:

> the fundamental source of conflict in this new world will not be primarily ideological or primarily economic. The great divisions among humankind and the dominating source of conflict will be cultural. Nation states will remain the most powerful actors in world affairs, but the principal conflicts of global politics will occur between nations and groups of civilizations. The clash of civilizations will dominate global politics. The fault lines between civilizations will be the battle lines of the future (Huntington 1993: 22).

Notwithstanding his seeming confusion about what is "ideological," "economic," or "cultural" and his belligerent vision of continued Western domination, not to mention his gross cultural generalization (e.g. to categorize Islam as a singular, overriding, or all-unifying cultural force regardless of national boundaries and interests), the thesis of civilizational clash does have implications for regional geography at least on two counts. First, its foundational basis defies the conventional notion of region as a contiguous geographical entity. How should a region be defined and configured?

Based on issues? For instance, can we establish a region of fear or despair? Based on geographical or some other features? Such a scenario would require new ways to configure and define regions, along with regional analyses. Additionally, if his theory were to materialize in the practical world, how does it impact globalization which demands an atmosphere of political and economic uniformity and certainty as well as global cultural convergence and similarities?

Second, one could argue that what has transpired in many Islamic nations from Indonesia to Morocco and beyond in the aftermath of the massive US bombing of Afghanistan refutes—and supports—his thesis. If one were to examine it through the prism of Islamic national governments, the thesis falls flat as many rulers have backed the US coalition, thus nullifying any notion of civilizational clash. Although, in some cases, the level of support has been somewhat tacit and low-key, in other cases, rulers have brutally suppressed public protests against US actions. Consequently, demonstrators have been seriously injured and many have been killed. And the Bush administration has gone out its way to distance itself from Huntington's theory. Referring to its war on Afghanistan and bin Laden, President Bush has claimed that this is not a clash of civilizations. This is ironic in that while the Bush administration's US-centric global policy of Western domination of the world is clearly consistent with Huntington's relatively explicit policy prescription, it wants to avoid any appearance of its subscription to the basic tenet of his theory. Nevertheless, the general public sentiment in the Muslim world, especially at the street level, does tend to accord support to the theory's principal point. It is evident that the anti-US sentiment is deeply seated among a large segment of the Muslim masses throughout the world, for they regard the American Middle Eastern policy as consistently anti-Arab and one-sided and also supportive of repressive regimes in the region. As this tendency pervades, the civilizational clash scenario deepens at least at the street level.

At any rate, whether there is any veracity to Huntington's thesis is not our primary concern. What is important are the issues posed in his article, issues that are at the core of what constitutes regional geography or how to configure it in the face of these new developments. Few would dispute that the new world is vastly different from the one defined by the Cold War. In view of the rise of ethnic animosities and atrocities in many countries, often expressed in religious terms, religions appear poised to frame the discourse and direction of national and regional politics and economics. Given this, national alignments and realignments based on religious affinity and affiliations have gained added significance in regional geography. Furthermore, where and when globalization (or economic development policy) fails to fulfil its promise, leaving countless by the wayside, religion readily moves in to fill the void, often taking on the form of fundamentalism or even nativism. As traditionalism gathers momentum, modernity comes under heavy assaults as it is invariably associated with Westernization which, in turn, is seen as a Christian bastion. Cultural relations usually turn into cultural conflicts, both internally and externally. Development and modernity as a reflection of Westernization and changing times are called into question. All these issues figure prominently in regional geography and in its reconfiguration in the new world abruptly left behind by the Cold War, a world trapped in the parallel tendencies of global cultural convergence and cultural conflicts, hopes and despairs, and mass antagonism against globalization and national alliance with global capital.

Rapid Advancements in Information and Communication Technologies and Regional Geography

To repeat, the phenomenal speed of advancements in information and communication technologies has posed a significant challenge for regional geography. It is precisely because of this challenge that we have decided to explore in this last segment a key question confronting regional geography: that of its relevancy and long-range viability. Specifically, how relevant is regional geography in this age of information, an age in which everything is defined in terms of *speed*—the speed at which regional information flows? In essence, speed is what determines one's competitive advantage in the open space of information networks. So how we respond to this challenge will play a paramount role in shaping the future of regional geography, including, of course, Asian geography.

Golledge (2000: 3) states that "Geography is a *field*. It incorporates physical, human and technical components." Nowhere is this triangular synthesis of physical, human, and technical components more explicitly demonstrated than in regional geography. Simply put, from the very beginning, regional geography has been an integral part of the core identity of what we fondly call "geography." In the contemporary context, the role of regional geography can be separated into two broad tracts. For the lack of a better word, we call the first tract *conceptual* and the second *factual-informational*. The research trends discussed in the preceding section largely exemplify what the *conceptual* tract constitutes, meaning geographers investigate issues, expound theories, propose policies, and deploy various techniques to augment the quality and content of their regional geographic inquiries. This is the domain where most regional geographers feel comfortable, where the academic surrounding tends to elevate our status in the theater of disciplinary vigor and visibility *vis-à-vis* other disciplines. And there is little dispute about its value, both from a theoretical and policy perspective. Yapa (1991), for instance, notes that among the principal problems facing most Asian countries are poverty, ecological degradation, massive rural-urban migration and the consequent growth of cities. Addressing these problems in a manner that is effective as well as locally sustainable, Yapa argues, requires a very detailed knowledge of regions. To make his case, he uses an example of energy use in a local context and illustrates why regional geography is central to the task of matching energy sources to end-uses through appropriate technology.

But when the focus is on the *factual-informational* tract, it tends to engender a sense of uneasiness for many regional geographers. Because the primary function of this tract is to generate, synthesize, and supply a wide range of detailed factual information about different places/regions, it rarely dwells in the realms of conceptual space. And, somewhat paradoxically, it is precisely this role that bestows geography with its popular identification in the public arena. This is usually perceived as a disciplinary stigma, something that could typecast regional geographic research as being shallow, thus reducing geographers' academic status amongst their colleagues outside the discipline. When the public perception or a casual discussion binds us to this role, our posture often slides into a defensive mode. As much as we herald the *conceptual* prowess of regional geography—and herald it we must—the reality is that it is the *factual-informational* arena where much of the day-to-day public consumption value of regional geography lies. This is what has historically given regional geography its secure place in geopolitics as well as geoeconomics (whether manifested in the form of mercantilism, colonialism, or globalization as it is called today) as much of the demand for detailed regional information in America comes from the military and business establishments—the two primary patrons and drivers of regional geography.

The information is used for the purpose of military intelligence and operation as well as global business expansion. So, as long as America's geostrategic and geoeconomic designs revolve around its global interests, there will be no question about the necessity of such information. In the wake of American firms' continued global march, the need for regional information is bound to soar. To apply a business term, this is a substantial "niche market" for regional geography. In fact, the number of publications containing country-level information and its clearinghouses is proliferating. For example, the Economic Intelligence Unit of the same company that publishes *The Economist* has a vast collection of such data and information, which it can make available to its customers at moment's notice. Can regional geography compete with such clearinghouses? In other words, this is where we face a stiff competition, i.e. maintaining regional geography's grip on the production and provision of regional information. This is why the question of regional geography's relevance becomes magnified and the issue of *speed* gains added urgency and significance. Whether we allow ourselves to be eventually rendered irrelevant by our failure to adapt to the speed of information and communication technologies or deploy them with a high level of creativity, energy, and vigor to strengthen Asian/regional geography, but without sacrificing the philosophical core of the discipline—this is the essence of the challenge that lies ahead. Simply put, how do we remain relevant and maintain our viability in the public domain as a producer and transmitter of place or region-specific factual knowledge while at the same time solidifying our theoretical position?

Let us illustrate this issue of speed with one example from Michael Crichton's *Congo*, a novel which in itself provides a sound knowledge of Zaire's regional geography. Crichton (1980) develops a central plot around a race among the competing industrial powers for the access and control of resources—in this case, for the Type IIb blue diamonds that are presumed to provide the next great leap in the speed of microchip technology, including the military technology such as missiles. In the novel, one of the companies engaged in this race is Texas-based ERTS. Crichton (ibid.: 62) writes:

> "It's business," Karen Ross said. "Four years ago, there were no companies like ERTS. Now there are nine around the world, and what they all sell is competitive advantage, meaning speed. Back in the sixties, a company—say an oil company—might spend months or years investigating a possible site. But that's no longer competitive; business decisions are made in weeks or days. The pace of everything has sped up [time–space convergence, as David Harvey calls it]. We are already looking to the nineteen-eighties, where we will provide answers in hours. Right now the average ERTS contract runs a little under three weeks, or five hundred hours." (parenthetical words supplied)

In the past twenty years since this novel was published, the rate of speed itself has accelerated, and is now moving much faster than one would have imagined back then. In fact, there seems to be no limit to how fast it will get. So we ask once again: can Asian/regional geography still be relevant in the age of information? To be sure, regional geography can certainly generate, process, and provide detailed information, and the basic demand for such information has not changed. What have changed, however, are both the nature of that demand and the mode of geographical information gathering and delivery. Recent developments indicate a tectonic shift in both these areas. In the past, much of the regional geography information was collected, processed, and provided by geographic field agents. As such, the information was generally comprehensive, generated at the ground level ("ground truthing" as one might call it today), systematically arranged into categories, and invariably transmitted in print forms such as books, journals, travel accounts, and reports—a form of information technology that is most suitable to the industrial age. The process was quite involved and time-consuming as extensive fieldwork had to be conducted and ground-level intelligence and information had to be compiled in meticulous detail. Naturally, therefore, such a relatively slow pace is anachronistic to the galloping speed of today's information age. In sum, we have no competitive advantage in this area, and whatever comparative advantage we enjoyed is already lost.

We live in a different world, a world that previously existed only in science fiction. This is difficult for some of us to navigate, not because we lack technical competencies, but because we still cherish the more leisurely and methodical pace of the past. We rejoiced in the challenge that came with doing fieldwork that often tested our ability to negotiate many intricate details and contours of human interactions and relationships. We enjoyed collecting information one piece at a time, carefully sifting through its texture and content before cataloging it into different forms and formats and preparing maps. But today we can hardly afford the luxury of time. Time and space are converging at a lightening pace. Microchips and satellites now rule the world, while the Internet, the prodigy of microchips and satellites, builds its own information empire, fully capable of superseding and supplanting the informational role of regional geography. All that seems to matter today is the speed at which information is gathered, processed, and made available. "Instant" is the word of this age as information is demanded instantly. And regional geography with its industrial-age speed is too sluggish to compete in this race of instantaneousness. What is even more disconcerting is that analogous to the "Death of the Salesmen" (*The Economist* 2000), regional geography may be just a few clicks away from being replaced (or even displaced) by a piece of software, often programmed and operated by a nongeographer.

Apparently then, in this age of information, we do not need field geographers to gather and process ground-level information. Today, even wars, as evidenced in Kosovo and now in Afghanistan, are often being fought over the skies and from distant locations; one does not have necessarily to engage in a field battle or combat zone, since bombers and laser- and satellite-guided missiles can deliver deadly blows with pinpoint accuracy. In such military confrontations, the type of local and regional geography information that geographers produce may not be timely enough or relevant to the generals waging wars. Sure, ground-level detailed information is still required as the US military establishment is discovering in Afghanistan where the terrain can prove to be the most treacherous enemy. Full knowledge of specific establishments, installations, or complexes with exact coordinates is required. But satellites can readily deliver it, and do so instantly, up to the hour, and with magnificent spatial details. After all, from the US perspective, wars are waged these days not for territorial conquests, but to debilitate enemies, to tame them into submission to the US plan. Given their slow pace, regional geographers are unable to offer such detailed information within the required time-frame. So it is all about speed, the speed of information acquisition and transmission. Similarly, business executives demand instant geographical information and in a format that is compressed into a page, not voluminous books and articles, since they rarely seem to have time to read extensive information.

In light of these recent trends, one may legitimately ask why anyone would pursue Asian/regional geography. We are witnessing an explosion of regional geography information and its speedy delivery, perhaps with little input and contribution from regional geographers. But where is regional geography? To repeat, the information that is now being demanded is mostly technical briefs, factual and specific, reminiscent of the kind of answer we generate to the foundational question of geography: *where*? It is being demanded in the same format and size as the microchips that generate it, process it, and then deliver it at the speed of a mouse click, all miniaturized into capsules and packaged like TV soundbites—trendy and timely, up-do-date and accessible, and, of course, easily digested. When people can readily retrieve such capsulized country-specific geographical information, including maps, from omnipresent websites such as *Lonelyplanet* (<www.lonelyplanet.com>, last accessed 28 November 2002) or some other providers—and invariably cost-free—there is hardly any need to rely on regional geography or to scour books and journals, the form of industrial-age information technology that demands time, patience, and analytical attention, the very elements that seem to be in short supply in these hectic, "not-fast-enough" times. In essence, the Internet and satellites are increasingly displacing regional geography. To regional information seekers, the only geography that counts is what they see and extract from the Internet. But what is dismaying to watch in this rapidly unfolding drama is that many Ph.D.-granting geography departments are, wittingly or unwittingly, relinquishing their roles and responsibilities to these very forces that undercut regional geography. If the wings of regional geography are clipped, is the discipline as a whole not bound to suffer?

In a response to the comments on *Rediscovering Geography*, a report by the Rediscovering Geography Committee, Wilbanks (1999: 157) charges that the reviewers' concerns are "rooted more in our internal dialogues than in our external usefulness." This is an interesting but contradictory remark, especially given that the report itself recognizes the need to revive regional geography with greater emphasis on field research in foreign areas. Our question is: how can geography continue to be *externally* useful, if its vital core is *internally* decaying? Our external shine depends on our internal strength, namely regional geography, and how we reinforce its foundation, retooling its ability to compete in terms of producing and delivering regional information to its consumers in a format that is fast and readily accessible. Yet, on the priority list, regional geography ranks low. The fact that the Committee felt compelled to call for the renewal of regional geography is a clear indication of its dire state.

How ironic that Ben Franklin's proverbial quote, "Time is money," that was uttered more than 200 years ago during the agrarian age, is now haunting geography, a discipline in which everything is measured in *distance* which is obviously time-consuming and hence costly. But this very phenomenon of *distance* that gives geography its core identity now threatens regional geography, not only because fieldwork has to be conducted in distant places that demand tremendous transaction costs, but mainly because we have not managed to close the *distance* between the time at which geographic information is generated and the time at which it is made available for public consumption. Despite the rapid pace of time–space convergence, in regional geography this distance is still too wide to overcome in order to compete effectively with the Internet and satellites.

In the final analysis, what is becoming increasingly evident is an entrenched discord between the day-to-day role and utility of regional geography in America (and its corollary, Asian geography) as a generator and transmitter of spatial information and its scientific theoretical underpinnings (also see Golledge 2000). The mounting challenge that we face as a discipline at this historical crossroads is a serious one. The question of how to cope with this challenge runs deep. From a pragmatic perspective, do we let this "niche market" go altogether, allowing it to take its own course, or do we take some technical path illuminated by such popular and pertinent tools as GIS, GPS (global positioning system), and remote sensing, trying to find a speedy channel of geographical information generation and delivery and hoping to compete effectively with the omnivorous Internet and thereby salvage whatever is left of regional geography? This path may prove to be expedient in the short run as it will help to keep regional geography solvent—but only as a technical field. Even then, we may be doing something that technical experts with little foundational knowledge of geography can easily perform (and outperform), thereby raising pertinent questions about the actual composition of the geographic community (also see Rediscovering Geography Committee 1997). Who is a geographer in other words? A city cab driver? A travel agent? A computer scientist with a vast pool of geographic information, but one who never took even one geography course and knows nothing about its foundation? A global tourist? Even more importantly, with such a technical path, devoid of any theoretical linchpins, we run the risk of sacrificing the very soul of geography that the discipline has painstakingly cultivated over the past forty years. Referring to these geographic technical tools, Yapa (1999: 152) pointedly warns that "Impressive as the new analytical capability is, I hope that we can avoid the mistakes we made in an earlier era, in the 'technique-rich, concept-poor' intellectual milieu of the so-called quantitative revolution in geography." Or do we, on the other hand, firmly hold our academic ground and stare at the prospect of becoming irrelevant, permanently sidelined by the satellites and Internet? These are difficult questions, and admittedly we have no clear answers to them. These are the questions, however, that the whole discipline needs to ponder—and ponder it with a great sense of urgency and gravity—not just those aligned with regional geography.

Acknowledgements

First, we would like to extend our profound appreciation to Drs Nigel Allan, Hussein Amery, Sanjoy Chakravorty, Dennis Conway, David Edgington, P. P. Karan, Tom Leinbach,

George Lin, Bimal Paul, and Lakshman Yapa and two reviewers for their insightful comments on the earlier draft of this chapter. Second, we would like to point out that, when we initially prepared this chapter, our coverage was limited to the 1988–98 period. In light of the change in this volume's publication schedule, we have changed our coverage period to 1988–2000. To reflect this change, we have updated the chapter to incorporate some of the new and pertinent information, including certain events that occurred in 2001, i.e. prior to its final revision in early October 2001. We would, however, like to caution that the scope of our updates is somewhat narrow largely because of the severe constraints of time.

References

General References Related to Asia

Association of American Geographers (1994). *Guide to Programs in Geography in the United States and Canada 1994–95.* Washington, DC: Association of American Geographers.

Bishop, P. (1989). *The Myth of Shangri-La: Tibet, Travel Writing and the Western Creation of Sacred Landscape.* London: Athlone Press.

British Broadcasting Service (2001). "Sharp Global Slowdown Predicted." ‹news.bbc.co.uk//hi/english/business/newsid_1608000/1608281.stm›, last accessed 19 October 2001.

Brush, J. E. (1949). "The Distribution of Religious Communities in India." *Annals of the Association of American Geographers*, 39: 81–98.

—— (1968). "Spatial Patterns of Population Distribution in Indian Cities." *Geographical Review*, 58: 362–91.

Business Week (1998). "What To Do About Asia." 26 January, 26–33.

—— (2000). "Special Report: Global Capitalism: Can It Be Made to Work Better?" 6 November, 72–100.

Campbell, D. (2001). "Pentagon Buys All Satellite War Images." Guardian News Service, 17 October, received through email, last accessed 18 October 2001.

Cox, K. (ed.) (1997). *Spaces of Globalization: Reasserting the Power of the Local.* New York: Guilford Press.

Cressey, G. B. (1951). *Asia's Lands and Peoples.* New York: McGraw-Hill.

Crichton, M. (1980). *Congo.* New York: Ballantine Books.

Economist, The (1998). "Asian Values Revisited." 25 July, 23–8.

—— (2000). "Death of the Salesmen." ‹www.economist.com›, 22 April, last accessed 27 September 2001.

Edney, M. H. (1997). *Mapping an Empire: The Geographical Construction of British India, 1765–1843.* Chicago: University of Chicago Press.

Fukuyama, F. (1992). *The End of History and the Last Man.* New York: Free Press.

Gaile, G. L., and Willmott, C. J. (eds.) (1989). *Geography in America.* Columbus, Ohio: Merrill.

Ginsburg, N. (ed.) (1958). *The Pattern of Asia.* Englewood Clifis: Prentice-Hall.

Golledge, R. (2000). "Reuniting the Field of Geography: No Core, No Periphery." *AAG Newsletter*, February.

Gordon, M. E. (2001). "Pentagon Corners Output of Satellite Images of Afghanistan." ‹www.nytimes.com/2001/10/19/international/asia/19PENT.html›, last accessed 19 October 2001.

Graf, W. (1999). "Who We Are, and Are Not." *AAG Newsletter*, March.

Greider, W. (2000). "Shopping Till We Drop." *The Nation*, 10 April.

Hamel, G., and Sampler, J. (1998). "The E-Corporation: More than Just Web-Based, it's Building a New Industrial Order." *Fortune*, 7 December, 80–92.

Hilton, J. (1933). *Lost Horizon.* London: Pan.

Huntington, S. P. (1993). "The Clash of Civilizations." *Foreign Affairs*, 72: 22–49.

Karan, P. P. (1960). *Nepal: A Physical and Cultural Geography.* Lexington: University of Kentucky Press.

—— (1989*a*). "Mountains in Crisis: Man versus Nature." *Explorers Journal*, 67: 162–8.

—— (1989*b*). "Environment and Development in Sikkim Himalaya: A Review." *Human Ecology*, 17: 257–71.

Karan, P. P., *et al.* (1989). "Asia," in Gaile and Willmott (1989: 506–45).

Kenny, J. T. (1995). "Climate, Race, and Imperial Authority: The Symbolic British Hill Stations in India." *Annals of the Association of American Geographers*, 85.

Krugman, P. (1998). "Saving Asia: It's Time to Get Radical." *Fortune*, 7 September, 74–81.

Lin, G. C. S. (2000). "State, Capital, and Space in China in an Age of Volatile Globalization." *Environment and Planning A*, 32: 455–71.

Montagnon, P. (1998). "Prognosis Dismal for Asian Flu." *Financial Times, Part 2*, 2 October, 17–20.

Murphey, R. (1977). *The Outsiders: Westerners in India and China.* Ann Arbor: University of Michigan Press.

Ohmae, K. (1990). *The Borderless World: Power and Strategy in the Interlinked Economy.* London: HarperCollins.

—— (1995). *The End of the Nation State: The Rise of Regional Economies.* London: HarperCollins.

Population Reference Bureau (1998). *1998 World Population Data Sheet.* Washington, DC: Population Reference Bureau.

Rediscovering Geography Committee (1997). *Rediscovering Geography: New Relevance for Science and Society.* Washington, DC: National Academy Press.

Said, E. W. (2001). "The Clash of Ignorance." *The Nation*, 22 October, 11–13.

Schwartzberg, J. E. (1965). "The Distribution of Selected Castes in the North Indian Plain." *Geographical Review*, 55: 477–95.

Schwartzberg, J. E., *et al.* (1978). *An Historical Atlas of South Asia.* Chicago: University of Chicago Press.

Shrestha, N. R. (1988). "Historical Evolution of the World System," in T. A. Hartshorn and J. W. Alexander, *Economic Geography*. 3rd edn. Englewood Cliffs: Prentice-Hall, 16–26.

Shrestha, N. R., and Hartshorn, T. A. (1993). "A Neocapitalist Perspective on Third World Urbanization and Economic Development," in J. Diddee and V. Rangaswamy (eds.), *Urbanisation: Trends, Perspectives and Challenges*. New Delhi: Rawat Publications, 1–38.

Sopher, D. E. (1968). "Pilgrim Circulation in Gujrat." *Geographical Review*, 58: 392–425.

— (1980). *Exploration of India: Geographical Perspectives in Society and Culture*. Ithaca: Cornell University Press.

Soros, G. (1997). "The Capitalist Threat." *The Atlantic Monthly* (Digital Edition). ‹www.theatlantic.com/issues/97feb/capital/capital.htm›, last accessed 23 August 2001.

Spencer, J. E. (1954). *Asia, East by South: A Cultural Geography*. New York: Wiley.

Strobel, W. P. (2001). "Using a Stronger Strategy: White House May Have a New Attitude toward Mideast." *The Tallahassee Democrat*, 13 October, 8A.

Wilbanks, T. J. (1999). "Response to Reviewers' Comments." *Annals of the Association of American Geographers*, 89: 157–59.

Yapa, L. (1991). "Is GIS Appropriate Technology." *International Journal of Geographic Information*, 5: 41–58.

— (1996). "What Causes Poverty? A Postmodern View." *Annals of the Association of American Geographers*, 86: 707–28.

— (1999). "Rediscovering Geography: On Speaking Truth to Power." *Annals of the Association of American Geographers*, 89: 151–75.

Yeung, H. W. C. (1998). "Capital, State and Space: Contesting the Borderless World." *Transactions of the Institute of British Geographers*, 23: 291–309.

Zakaria, F. (2001). "The End of the End of History." *Newsweek*, 24 September, 70.

East Asia

Alwin, J. (1992). "North American Geographers and the Pacific Rim: Leaders or Laggards?" *Professional Geographer*, 44: 369–76.

Aoyama, Y. (1996). "Local Economic Development or National Industrial Strategy: Recent Trends in Small Business Policy in Japan and the United States." *Review of Urban and Regional Development Studies*, 8: 1–14.

— (1997). "Integrating Business and Location: An Overview of Two Theoretical Frameworks on Multinational Firms." *Berkeley Planning Journal*, 11: 49–70.

— (1999). "Policy Interventions for Industrial Network Formation: Contrasting Historical Underpinnings of the Small Business Policy in Japan and the United States." *Small Business Economics*, 12: 217–31.

— (2000*a*). "Industrial Network Formation and Regulation: The Case of Japan's SME Policy," in E. Vatne and M. Taylor (eds.), *The Networked Firm in the Global World: Small Firms in New Environments*. Aldershot: Ashgate, 45–63.

— (2000*b*). "Keiretsu, Networks and Locations of Japanese Electronics Industry in Asia." *Environment and Planning A*, 32: 223–44.

— (2001*a*). "Information Society of the East: Technological Adoption and Electronic Commerce in Japan," in T. R. Leinbach and S. D. Brunn (eds.), *Worlds of E-Commerce: Economic, Geographical and Social Dimensions*. New York: John Wiley.

— (2001*b*). "Structural Foundations for E-Commerce Adoption: A Comparative Organization of Retail Trade Between Japan and the United States." *Urban Geography*, 22: 130–53.

Aoyama, Y., and Teitz, M. B. (1996). *Small Business Policy in Japan and the United States: A Comparative Analysis on Objectives and Outcomes*. Policy Papers in International Affairs. Berkeley, Calif.: IAS Publications.

Auty, R. M. (1997). "The East Asian Growth Model: South Korean Experience," in Watters and McGee (1997: 161–9).

Brucklacher, A. D. (1998). "Crunch for Aomori Apple Growers." *Japan Quarterly*, 45: 49–56.

Childs, I. R. W. (1991). "Japanese Perception of Nature in the Novel Snow Country." *Journal of Cultural Geography*, 11: 1–20.

Choo, S. (1994). "Production System Changes in the Korean Consumer Electronics Sector." *Growth and Change*, 25: 165–82.

Costa, F. J., Dutt, A. K., Ma, L. J., and Noble, A. G. (1989). *Urbanization in Asia: Spatial Dimensions and Policy Issues*. Honolulu: University of Hawaii Press.

Cybriwsky, R. (1988). "Shibuya Center, Tokyo." *Geographical Review*, 78: 48–61.

— (1991). *Tokyo: The Changing Profile of an Urban Giant*. Boston: G. K. Hall.

— (1996). *Japan: Nippon*. American Geographical Society's Around the World Program. Granville, Ohio: McDonald & Woodward.

— (1997). "From Castle Town to Manhattan Town with Suburbs: A Geographical Account of Tokyo's Changing Landmarks and Symbolic Landscapes," in Karan and Stapleton (1997: 56–78).

— (1998). *Tokyo: The Shogun's City at the Twenty First Century*. World Cities Series. New York: Wiley.

Douglass, M. (1994). "The 'Developmental State' and the Newly Industrialized Economies of Asia." *Environment and Planning A*, 26: 543–66.

Dutt, A. K., Costa, F. J., Aggarwal, S., and Noble, A. G. (1994). *The Asian City: Processes of Development, Characteristics, and Planning* Dordrecht: Kluwer.

Edgington, D. W. (1990*a*). *Japanese Business Down Under: Patterns of Japanese Investment in Australia*. London: Routledge.

— (1990*b*). "Managing Industrial Restructuring in the Kansai Region of Japan." *Geoforum*, 21: 1–22.

— (1991*a*). "Economic Restructuring in Yokohama: From Gateway Port to International Core City." *Asian Geographer*, 10: 62–78.

— (1991*b*). "Japanese Manufacturing Investment in Australia: Corporations, Governments and Bargaining." *Pacific Affairs*, 64: 65–86.

— (1992). "Industrial Restructuring and New Transport Infrastructure in Chukyo, Japan." *Geography*, 77: 268–70.

— (1993). "The Globalization of Japanese Manufacturing Corporations." *Growth and Change*, 24: 87–106.

— (1994*a*). "The Geography of Endaka: Industrial Transformation and Regional Employment Changes in Japan, 1986–1991." *Regional Studies*, 28: 521–35.

— (1994*b*). "The New Wave: Patterns of Japanese Foreign Investment in Canada During the 1980s." *Canadian Geographer*, 38: 28–36.

Edgington, D. W. (1994*c*). "Planning for Technology Development and Information Systems in Japanese Cities and Regions," in Shapira, Masser, and Edgington (1994: 126–54).

— (1995*a*). "Locational Preferences of Japanese Real Estate Investors in North America." *Urban Geography*, 16: 373–96.

— (1995*b*). "Trade, Investment and the New Regionalism: Cascadia and Its Economic Links with Japan." *The Canadian Journal of Regional Science*, 18: 333–56.

— (1996). "Japanese Real Estate Investment in Canadian Cities and Regions, 1985–1993." *The Canadian Geographer*, 40: 292–305.

— (1997). "The Rise of the Yen, 'Hollowing Out' and Japan's Troubled Industries," in Watters and McGee (1997: 170–89).

— (1999). "Firms, Governments and Innovation in the Chukyo Region of Japan." *Urban Studies*, 36: 305–40.

— (2000). "Osaka." *Cities*, 17: 305–18.

Edgington, D. W., and Haga, H. (1998). "Japanese Service Sector Multinationals and the Hierarchy of Pacific Rim Cities." *Asia Pacific Viewpoint*, 39: 161–78.

Edgington, D. W., and Hayter, R. (2000). "Foreign Direct Investment and the Flying Geese Model: Japanese Electronics Firms in Asia-Pacific." *Environment and Planning A*, 32: 281–304.

— (2001). "Japanese Direct Foreign Investment and the Asian Financial Crisis." Geoforum, 32: 103–20.

Edgington, D. W., Hutton, T., and Leaf, M. (1999). "The Postquake Reconstruction of Kobe: Economic, Land Use, and Housing Perspectives." *Japan Foundation Newsletter*, 27: 13–17.

Elhance, A. P., and Chapman, M. (1992). "Labor Market of a US-Japanese Automobile Joint Venture." *Growth and Change*, 23: 160–82.

Ettlinger, N. (1991). "The Roots of Competitive Advantage in California and Japan." *Annals of the Association of American Geographers*, 81: 391–407.

Florida, R., and Kenney, M. (1990). "High-Technology Restructuring in the USA and Japan." *Environment and Planning A*, 22: 233–52.

— (1992). "Restructuring in Place: Japanese Investment, Production, Organization, and the Geography of Steel." *Economic Geography*, 68: 146–73.

— (1994). "The Globalization of Japanese R&D: The Economic Geography of Japanese R&D Investment in the United States." *Economic Geography*, 70: 344–69.

Forbes, D. (1997). "Regional Integration, Internationalisation and the New Geographies of the Pacific Rim," in Watters and McGee (1997: 13–28).

Fujita, K., and Hill, R. C. (eds.) (1993). *Japanese Cities in the World Economy*. Philadelphia: Temple University Press.

— (1997). "Together and Equal: Place Stratification in Osaka," in Karan and Stapleton (1997: 106–33).

Fukushima, Y. (1997). "Political Geographers of the Past, X. Japanese Geopolitics and Its Background: What Is the Real Legacy of the Past?" *Political Geography*, 16: 407–22.

Fuller, G., and Pitts, F. R. (1990). "Youth Cohorts and Political Unrest in South Korea." *Political Geography Quarterly*, 9: 9–22.

Ginsburg, N. (1993). "Commentary on Alwin's 'North American Geographers and the Pacific Rim'." *Professional Geographer*, 45: 355–7.

Ginsburg, N., Koppel, B., and McGee, T. (eds.) (1991). *The Extended Metropolis: Settlement Transition in Asia*. Honolulu: University of Hawaii Press.

Glasmeier, A. K. (1988). "The Japanese Technopolis Programme: High-Tech Development Strategy or Industrial Policy in Disguise." *International Journal of Urban and Regional Research*, 12: 268–84.

Glasmeier, A. K., and Sugiura, N. (1991). "Japan's Manufacturing System: Small Business, Subcontracting and Regional Complex Formation." *International Journal of Urban and Regional Research*, 15: 395–414.

Graham, J. (1993). "Firm and State Strategy in a Multipolar World: The Changing Geography of Machine Tool Production and Trade," in H. Noponen, J. Graham, and A. Markusen (eds.), *Trading Industries, Trading Regions: International Trade, American Industry, and Regional Economic Development*. New York, Guilford Press, 140–74.

Grant, P. (1998). "Geography's Importance to Japan's History." *Education About Asia*, 3: 44–6.

Grant, R. (1993*a*). "Against the Grain: Agricultural Trade Policies of the US, the European Community and Japan at the GATT." *Political Geography*, 12: 247–62.

— (1993*b*). "Trading Blocs or Trading Blows? The Macroeconomic Geography of US and Japanese Trade Policies." *Environment and Planning A*, 25: 273–91.

Grant, R., and Nijman, J. (1997). "Historical Changes in United States and Japanese Foreign Aid to the Asia-Pacific Region." *Annals of the Association of American Geographers*, 87: 32–51.

Hart-Ladsberg, M., and Burkett, P. (1998). "Contradictions of Capitalist Industrialization in East Asia: A Critique of 'Flying Geese' Theories of Development." *Economic Geography*, 74: 87–110.

Hayashi, T. (1994). "Restructuring of Items Within a Product Line: A Case Study of Japanese Multinational Enterprises in the Electric Machinery Industry." *Environment and Planning A*: 509–26.

Hayter, R., and Edgington, D. W. (1997). "Cutting Against the Grain: A Case Study of MacMillan Bloedel's Japan Strategy." *Economic Geography*, 73: 187–213.

Hong, K. H. (1999). *A Bibliography of Korean Geography in Western Languages, 1945–1995*. Seoul: Hyorim.

Hultman, C. W., and McGee, L. R. (1990). "The Japanese Banking Presence in the United States and its Regional Distribution." *Growth and Change*, 21: 69–79.

Ikagawa, T. (1993). "White Sand and Blue Pine: A Nostalgic Landscape of Japan." *Landscape*, 32: 1–7.

Jones, P. N., and North, J. (1991). "Japanese Motor Industry Transplants: The West European Dimension." *Economic Geography*, 67: 105–23.

Karan, P. P. (1997). "The City in Japan," in Karan and Stapleton (1997: 12–39).

Karan, P. P., and Stapleton, K. (eds.) (1997). *The Japanese City*. Lexington: University Press of Kentucky.

Kenney, M., and Florida, R. (1988). "Beyond Mass Production: Production and the Labor Process in Japan." *Politics and Society*, 16: 121–58.

— (1992). "The Japanese Transplants: Production Organization and Regional Development." *Journal of the American Planning Association*, 58: 21–38.

—— (1993). *Beyond Mass Production: The Japanese System and its Transfer to the US*. New York: Oxford University Press.

Kim, J., and Choe, S. C. (1997). *Seoul: The Making of a Metropolis*. World City Series. New York: Wiley.

Kim, W. B., Douglass, M., and Choe, S. C., and Ho, K. C. (eds.) (1997). *Culture and the City in East Asia*. Oxford Geographical and Environmental Studies. New York: Oxford University Press.

Kobayashi, A. (1996). "Birds of Passage or Squawking Ducks? Writing Across Generations of Japanese-Canadian Literature," in R. King, J. Connell, and P. White (eds.), *Writing Across Two Worlds: Literature and Migration*. London: Routledge, 216–28.

Kobayashi, A., and Jackson, P. (1994). "Japanese Canadians and the Racialization of Labour in the British Columbia Sawmill Industry." *B. C. Studies*, 103: 33–58.

Kodras, J. E. (1993). "Shifting Geopolitical Strategies of US Foreign Food Aid, 1955–1990." *Political Geography*, 12: 232–46.

Kornhauser, D. H. (1991). *Japan, Geographical Background to Urban-Industrial Development*. 2nd edn. New York: Wiley.

Kwon, S. C. (1990). "Korean Small Businesses in the Bronx, New York: An Alternative Perspective of Occupational Adaptation of New Immigrants." *Urban Geography*, 11: 509–18.

Latz, G. (1989). *Agricultural Development in Japan: The Land Improvement District in Concept and Practice*. Geography Research Paper, 225. Chicago: University of Chicago Press.

—— (1991). "The Persistence of Agriculture in Urban Japan: An Analysis of the Tokyo Metropolitan Area," in Ginsburg, Koppel, and McGee (1991: 217–38).

—— (1992). "The Experience of Place in Japan." *Asian Art*, 5: 2–7.

—— (1993). "Regional Factors Influencing Pacific Basin Trade: The Example of Korea and United States." *Waseda Journal of Asian Studies*, 15: 1–17.

—— (1996). *Faculty Guide for The Power of Place: World Regional Geography Telecourse*. Burlington, Vt.: Annenberg/CPB Project.

—— (1999). "The Geography Discipline Network Guides: A North American Perspective." *Journal of Geography in Higher Education*, 23: 258–61.

Latz, G., and Borthwick, M. (1992). *University Student and Faculty Guides for the Pacific Century Telecourse*. 2 vols. The Annenberg/CPB Project. Boulder, Colo.: Westview Press.

Ledyard, G. (1994). "Cartography in Korea," In J. B. Harley and D. Woodward (eds.) *The History of Cartography*, vol. 2, bk. 2. *Cartography in the Traditional East and Southeast Asian Societies*. Chicago: University of Chicago Press. 235–345.

Lee, D. O., and Brunn, S. D. (1996). "Politics and Regions in Korea: An Analysis of the Recent Presidential Election." *Political Geography*, 15: 99–119.

Lewis, M. W., and Wigen, K. E. (1997). *The Myth of Continents: A Critique of Metageography*. Berkeley, Calif.: University of California Press.

Liaw, K. L. (1992). "Interprefectural Migration and its Effects on Prefectural Populations in Japan: An Analysis Based on the 1980 Census." *Canadian Geographer*, 36: 320–35.

McDonald, M. G. (1996*a*). "Farmers as Workers in Japan's Regional Economic Restructuring, 1965–1985." *Economic Geography*, 72: 49–72.

—— (1996*b*). "Farming Out Factories: Japan's Law to Promote the Introduction of Industry into Agricultural Village Areas." *Environment and Planning A*, 28: 2041–61.

—— (1997). "Agricultural Landholding in Japan: Fifty Years after Land Reform." *Geoforum*, 28: 55–78.

—— (2000). "Food Firms and Food Flows in Japan 1945–1998." *World Development*, 28: 487–512.

McGee, T. G. (1991). "Presidential Address: Eurocentrism in Geography: The Case of Asian Urbanization." *Canadian Geographer*, 35: 332–44.

—— (1997). "Globalization, Urbanization and the Emergence of Sub-global Regions, A Case Study of the Asia Pacific Region," in Watters and McGee (1997: 29–45).

Machimura, T. (1992). "The Urban Restructuring Process in Tokyo in the 1980s: Transforming Tokyo into a World City." *International Journal of Urban and Regional Research*, 16: 114–28.

—— (1997). "Building a Capital for Emperor and Enterprise: The Changing Urban Meaning of Central Tokyo," in Kim *et al.* (1997: 151–66).

Mair, A. (1992). "Just-In-Time Manufacturing and the Spatial Structure of the Automobile Industry: Lessons from Japan." *Tijdschrift voor Econ. en Soc. Geografie*, 82: 82–92.

Mair, A. (1993). "New Growth Poles? Just-In-Time Manufacturing and Local Economic Development Strategy." *Regional Studies*, 27: 207–21.

Mair, A. (1994). *Honda's Global Local Corporation*. New York: St. Martin's Press.

Mair, A., Florida, R., and Kenney, M. (1988). "The New Geography of Automobile Production: Japanese Transplants in North America." *Economic Geography*, 64: 352–73.

Markusen, A., and Park, S. O. (1993). "The State as Industrial Locator and District Builder: The Case of Changwon, South Korea." *Economic Geography*, 69: 157–81.

Marton, A., McGee, T., and Paterson, D. G. (1995). "Northeast Asian Economic Cooperation and the Tumen River Area Development Project." *Pacific Affairs*, 68: 8–33.

Masai, Y. (1994). "Metropolitization in Densely Populated Asia: The Case of Tokyo," in Dutt *et al.* (1994: 119–26).

Mather, C. (1997). "Urban Landscapes of Japan," in Karan and Stapleton (1997: 40–55).

Mather, C., Karan, P. P., and Iijima, S. (1998). *Japanese Landscapes: Where Land and Culture Merge*. Lexington: University Press of Kentucky.

Morgan, J., and Olson, H. (1992). "Chinese Access to the Sea of Japan and Integrated Economic Development in Northeast Asia." *Ocean and Coastal Management*, 17: 57–79.

Morgan, J., and Valencia, M. J. (1992). *Atlas for Marine Policy in the East Asian Seas*. Berkeley: University of California Press.

Murphy, A. B. (1995). "Economic Regionalization and Pacific Asia." *Geographical Review*, 85: 127–40.

Nahm, K. B. (1999). "Downtown Office Location Dynamics and Transformation of Central Seoul, Korea." *Geojournal*, 49: 289–99.

Nemeth, D. J. (1987). *The Architecture of Ideology: Neo-Confucian Imprinting on Cheju Island, Korea*. University of California Publications in Geography, 26. Berkeley: University of California Press.

—— (1993). "A Cross-Cultural Cosmographic Interpretation of Some Korean Geomancy Maps." *Cartographica*, 30: 85–97.

—— (1998). "Geographic Gateways to Seeing and Understanding Korea." *Education about Asia*, 3: 47–50.

Nicola, P. (1997). "International Marriage in Japan: 'Race' and 'Gender' Perspectives." *Gender, Place and Culture*, 4: 321–34.

O Tuathail, G. (1992). "Pearl Harbor Without Bombs: A Critical Geopolitics of the US-Japan FSX Debate." *Environment and Planning A*, 24: 975–94.

— (1993). "Japan as Threat; Geo-Economic Discourses on the USA-Japan Relationship in US Civil Society," in C. H. Williams (ed.), *The Political Geography of the New World Order*. New York: Belhaven Press, 181–209.

Okamoto, K. (1997). "Suburbanization of Tokyo and the Daily Lives of Suburban People," in Karan and Stapleton (1997: 79–105).

O'Loughlin, J., and Anselin, L. (1996). "Geo-Economic Competition and Trade Bloc Formation: United States, German, and Japanese Exports, 1968–1992." *Economic Geography*, 72: 131–60.

Palm, R. (1998). "Urban Earthquake Hazards: The Impact of Culture on Perceived Risk and Response in the USA and Japan." *Applied Geography*, 18: 35–46.

Park, B. G. (1998). "Where Do Tigers Sleep at Night? The State's Role in Housing Policy in South Korea and Singapore." *Economic Geography*, 74: 272–88.

— (2001). "Labor Regulation and Economic Change: A View on the Korean Economic Crisis." *Geoforum*, 32: 61–75.

Park, S. O. (1993). "Industrial Restructuring and the Spatial Division of Labor: The Case of the Seoul Metropolitan Region, the Republic of Korea." *Environment and Planning A*, 25: 81–93.

— (1994). "Industrial Restructuring in the Seoul Metropolitan Region: Major Triggers and Consequences." *Environment and Planning A*, 26: 527–41.

— (2000). "Innovation Systems, Networks, and the Knowledge-Based Economy in Korea," in J. H. Dunning (ed.), *Regions, Globalization, and the Knowledge-Based Economy*. New York: Oxford University Press.

Park, S., and Kim, K. C. (1998). "Intrametropolitan Locations of Korean Businesses: The Case of Chicago." *Urban Geography*, 19: 613–31.

Parker, P. (1997). "Canada-Japan Coal Trade: An Alternative Form of the Staple Production Model." *Canadian Geographer*, 41: 248–66.

Patchell, J. (1993). "Composing Robot Production Systems: Japan as a Flexible Manufacturing System." *Environment and Planning A*, 25: 923–44.

Patchell, J., and Hayter, R. (1997). "Japanese Precious Wood and the Paradoxes of Added Value." *Geographical Review*, 87: 375–95.

Pitts, F. R. (1997). "Uncle Sam and the Two Koreas in the Asian Century," in S. Sanders (ed.), *The US Role in the Asian Century*. Lanham, Md.: University Press of America, 339–71.

Poon, J. (1997*a*). "The Cosmopolitanization of Trade Regions: Global Trends and Implications." *Economic Geography*, 73: 288–302.

— (1997*b*). "Inter-Country Trade Patterns Between Europe and the Asia-Pacific: Regional Structure and Extra-Regional Trends." *Geografiska Annaler*, 79: 41–55.

Poon, J., and Pandit, K. (1996). "Geographic Structure of Cross-National Trade Flows and Region-States." *Regional Studies*, 30: 273–85.

Poon, J. P. H., and Perry, M. (1999). "The Asian Economic Flu: A Geography of Crisis." *Professional Geographer*, 51: 184–95.

Poon, J., and Thompson, E. R. (1998). "Foreign Direct Investment and Economic Growth: Evidence from Asia and Latin America." *Journal of Economic Development*, 23: 141–60.

— (2001). Effects of the Asian Financial Crisis on Transnational Capital. *Geoforum*, 32: 121–31.

Prorok, C. V., Chhokar, K. B., Park, S., and Cartier, C. L. (1998). *Asian Women and their Work: A Geography of Gender and Development*. Indiana, Pa.: National Council for Geographic Education.

Reid, N. (1995). "Just-In-Time Inventory Control and the Economic Integration of Japanese-Owned Manufacturing Plants with the County, State and National Economies of the United States." *Regional Studies*, 29: 345–55.

Rice, G. W., and Palmer, E. (1993). "Pandemic Influenza in Japan, 1918–19: Mortality Patterns and Official Responses." *Journal of Japanese Studies*, 19: 389–420.

Rimmer, P. (1997). "Japan's Foreign Direct Investment in the Pacific Rim, 1985–1993," in Watters and McGee (1997: 113–32).

Rubenstein, J. M. (1988). "Changing Distribution of American Motor-Vehicle-Parts Suppliers." *Geographical Review*, 18: 288–98.

— (1992). *The Changing US Auto Industry: Geographical Analysis*. London: Routledge.

Sadler, D. (1994). "The Geographies of Just-In-Time: Japanese Investment and the Automotive Components Industry in Western Europe." *Economic Geography*, 70: 41–59.

Shapira, P. (1994). "Industrial Restructuring and Economic Development Strategies in a Japanese Steel Town: The Case of Kitakyushu," in Shapira, Masser, and Edgington (1994: 155–83).

Shapira, P., Masser, I., and Edgington, D. W., (eds.) (1994). *Planning for Cities and Regions in Japan*. Liverpool: Liverpool University Press.

Siebert, L. (2000*a*). "Rail Names as Indicators of Enduring Influence of Old Provinces in Modern Japan." *Geographical Review of Japan Series B*, 73: 1–26.

— (2000*b*). "Urbanization Transition Types and Zones in Tokyo and Kanagawa Prefectures." *Geographical Review of Japan Series B*, 207–24.

— (2000*c*). "Using GIS to Document, Visualize, and Interpret Tokyo's Spatial History." *Social Science History*, 24: 537–74.

Song, N., Dutt, A. K., and Costa, F. J. (1994). "The Nature of Urbanization in South Korea," in Dutt, Costa, Aggarwal, and Noble (1994: 127–41).

Suarez-Villa, L., and Han, P. H. (1990). "The Rise of Korea's Electronics Industry: Technological Change, Growth, and Territorial Distribution." *Economic Geography*, 66: 273–92.

Sugiura, Y. (1993). "Spatial Diffusion of Japanese Electric Power Companies, 1887–1906: A Discrete Choice Modeling." *Annals of the Association of American Geographers*, 83: 641–55.

Taylor, J. (1999). "Japan's Global Environmentalism: Rhetoric and Reality." *Political Geography*, 18: 535–62.

Thompson, E. R., and Poon, J. (1998). "Determinants of Japanese, US and UK Foreign Direct Investment in East and Southeast Asia." *Journal of Asian Business*, 14: 15–39.

Tyner, J. A. (1998). "The Geopolitics of Eugenics and the Incarceration of Japanese Americans." *Antipode*, 30: 251–69.

Ufkes, F. (1993*a*). "The Globalization of Agriculture." *Political Geography*, 12: 194–7.

— (1993*b*). "Trade Liberalization, Agro-Food Politics, and the Globalization of Agriculture." *Political Geography*, 12: 215–31.

Unno, K. (1994). "Cartography in Japan," in J. B. Harley and D. Woodward (eds.), *The History of Cartography*, vol. ii, Bk. 2.

Cartography in the Traditional East and Southeast Asian Societies. Chicago: University of Chicago Press. 346–477.

Warf, B. (1988). "Japanese Investments in the New York Metropolitan Region." *Geographical Review*, 78: 257–71.

Watters, R. F., and McGee, T. G. (eds.) (1997). *Asia-Pacific: New Geographies of the Pacific Rim*. Vancouver: UBC Press.

Wigen, K. (1992). "The Geographic Imagination in Early Modern Japanese History: Retrospect and Prospect." *Journal of Asian Studies*, 51: 3–29.

—— (1995). *The Making of a Japanese Periphery, 1750–1920*. Berkeley: University of California Press.

—— (1996). "Politics and Piety in Japanese Native-Place Studies: The Rhetoric of Solidarity in Shinano." *Positions: East Asia Cultures Critique*, 4: 491–518.

—— (1998). "Constructing Shinano: The Invention of a Neo-Traditional Region," in S. Vlastos (ed.), *Mirror of Modernity: Japan's Invented Traditions*. Berkeley: University of California Press, 229–42.

—— (1999). "Culture, Power, and Place: The New Landscapes of East Asian Regionalism." *American Historical Review*, 104: 1183–201.

—— (2000). "Teaching about Home: Geography at Work in the Prewar Nagano Classroom." *Journal of Asian Studies*, 59: 550–74.

Wisner, B. (1998). "Marginality and Vulnerability: Why the Homeless of Tokyo Don't 'Count' in Disaster Preparations." *Applied Geography*, 18: 25–33.

Yamazaki, T. (1997). "Political Geography around the World, X. Political Geography in Post-War Japan: Publication Tendencies." *Political Geography*, 16: 325–42.

Yonemoto, M. (1999). "Maps and Metaphors of the 'Small Eastern Sea' in Tokugawa Japan (1603–1868)." *Geographical Review*, 89: 169–87.

—— (2000). "The 'Spatial Vernacular' in Tokugawa Maps." *Journal of Asian Studies*, 59: 647–66.

Yoon, H. K. (1997). "The Origin and Decline of a Japanese Temple Town: Saidaiji Monzenmachi." *Urban Geography*, 18: 434–50.

Southeast Asia

Aiken, S. R. (1994). "Peninsular Malaysia's Protected Area Coverage, 1903–1992: Creation, Rescission, Excision, and Intrusion." *Environmental Conservation*, 21: 49–56.

Airriess, C. A.(1989). "The Spatial Spread of Container Transport in a Developing Regional Economy: North Sumatra, Indonesia." *Transportation Research. Part A: General*, 23: 453–61.

—— (1991). "Global Economy and Port Morphology in Belawan, Indonesia." *Geographical Review*, 81: 183–96.

—— (1993). "Export-Oriented Manufacturing and Container Transport in ASEAN." *Geography: Journal of the Geographical Association*, 78: 31–43.

—— (1998). "Information Technology and the Maritime Transport Service Economy of Singapore." Unpublished Presentation at the Annual Meeting of the Association of American Geographers, Boston.

Bensel, T. G., and Harris, R. C. (1995). "Energy Policy and Economic Development in the Philippines, 1973–2000." *Journal of Energy and Development*, 20: 187–227.

Bensel, T. G., and Kummer, D. M. (1996). "Fuelwood Consumption and Deforestation in the Philippines." *Human Organization*, 55: 498–9.

Bensel, T. G., and Remedio, E. M. (1992*a*). "The Forgotten Fuel? Biomass Energy in the Philippines." *Philippine Quarterly of Culture and Society*, 20: 24–48.

—— (1992*b*). "The Woodfuel Supply System for Cebu City, Philippines: A Preliminary Analysis." *Philippine Quarterly of Culture and Society*, 20: 157–69.

—— (1995). "Residential Energy Use Patterns in Cebu City, Philippines." *Energy*, 20: 173–88.

Bowen, J. T., and Leinbach, T. R. (1995). "The State and Liberalization: The Airline Industry in the East Asian NICs." *Annals of the Association of American Geographers*, 85: 468–93.

Cartier, C. L. (1993). "Creating Historic Open Space in Melaka." *Geographical Review*, 83: 359–73.

—— (1995). "Singaporean Investment in China: Installing the Singapore Model in Sunan." *Chinese Environment and Development*, 6: 119–44.

—— (1996). "Conserving the Built Environment and Generating Heritage Tourism in Peninsular Malaysia." *Tourism Recreation Research*, 21: 45–53.

—— (1997). "The Dead, Place/Space, and Social Activism: Constructing the Nationscape in Historic Melaka." *Environment and Planning D: Society and Space*, 15: 555–86.

—— (1998*a*). "Preserving Bukit China: The Cultural Politics of Landscape Interpretation in Melaka's Chinese Cemetery," in E. Sinn (ed.), *The Last Half Century of Chinese Overseas*. Hong Kong: Hong Kong University Press, 65–77.

—— (1998*b*). "Megadevelopment in Malaysia: From Heritage Landscapes to 'Leisurescapes' in Melaka's Tourism Sector." *Singapore Journal of Tropical Geography*, 19: 151–76.

Chang, K., and Thomson, C. (1994). "Taiwanese Foreign Direct Investment and Trade with Thailand." *Singapore Journal of Tropical Geography*, 15: 112–27.

Chang, T. C. (1999). "Local Uniqueness in the Global Village: Heritage Tourism in Singapore. *Professional Geographer*, 51: 91–103.

Cobban, J. L. (1993). "Public Housing in Colonial Indonesia, 1900–1940." *Modern Asian Studies*, 27: 871–96.

Crooker, R. A., and Martin, R. N. (1996). "Marketing Distance Decay and Highland Vegetable Production in Northern Thailand." *The Geographical Bulletin*, 38: 80–91.

Dearden, P. (1991). "Tourism and Sustainable Development in Northern Thailand." *Geographical Review*, 81: 400–13.

Doeppers, D. F. (1991). "Metropolitan Manila in the Great Depressions: Crisis for Whom?" *Journal of Asian Studies*, 50: 511–35.

—— (1994). "Tracing the Decline of the Mestizo Category in Philippine Life in the Late 19th Century." *Philippine Quarterly of Culture and Society*, 22: 80–9.

—— (1996). *Gender Representation in Migration to Metropolitan Manila in the Late Nineteenth Century*. Taipei: Institute of Economics, Academia Sinica.

Drake, C. (1989). *National Integration in Indonesia: Patterns and Policies*. Honolulu: University of Hawaii Press.

—— (1992). "National Integration in China and Vietnam." *Geographical Review*, 82: 295–312.

DuBois, R. (1990). *Soil Erosion in a Coastal River Basin: A Case Study from the Philippines*. University of Chicago Geography Research Paper, 232. Chicago: University of Chicago Press.

Dutt, A. (ed.) (1996). *Southeast Asia: A Ten Nation Region*. Dordrecht: Kluwer.

Flaherty, M., and Karnjanakesorn, C. (1993). "Commercial and Subsistence Fisheries Conflicts in the Gulf of Thailand: The Case of Squid Trap Fishers." *Applied Geography*, 13: 243–58.

— (1995). "Marine Shrimp Aquaculture and Natural Resource Degradation in Thailand." *Environmental Management*, 19: 27–37.

Flaherty, M., and Vandergeest, P. (1998). "Low-Salt Shrimp Aquaculture in Thailand: Goodbye Coastline, Hello Khon Kaen!" *Environmental Management*, 22: 817–30.

Ford, L. R. (1993). "A Model of Indonesian City Structure." *Geographical Review*, 83: 374–96.

Glassman, J. (2001). "Economic Crisis in Asia: The Case of the Philippines." *Economic Geography*, 77: 122–47.

Glassman, J., and Samatar, A. I. (1997). "Development Geography and the Third-World State." *Progress in Human Geography*, 21: 164–98.

Goss, J. D. (1990). "Production and Reproduction among the Urban Poor of Metro Manila: Relations of Exploitation and Conditions of Existence." Unpublished Ph.D. dissertation, University of Kentucky, Department of Geography.

— (1992). "Research Report: Transmigration in Maluku." *Cakalele*, 3: 87–98.

Goss, J. D., and Leinbach, T. R. (1996). "Development and Differentiation: A Case Study of Transmigration Settlement in West Ceram," in D. Mearns and C. Healey (eds.), *Remaking Maluku: Social Transformation in Eastern Indonesia*. Darwin: Centre for Southeast Asian Studies, Northern Territory University, 80–94.

Hafner, J. A. (1994). "Beyond Basic Needs: Participation and Village Reforestation in Thailand." *Community Development Journal*, 30: 72–82.

Hafner, J. A., and Apichatvullop, Y. (1990). "Farming the Forest: Managing People and Trees in Reserved Forests in Thailand." *Geoforum*, 21: 331–46.

Hart, G. (1989). "Changing Mechanisms of Persistence: Reconfigurations of Petty Production in a Malaysian Rice Region." *International Labor Review*, 128: 815–29.

— (1991). "Engendering Everyday Resistance: Gender, Patronage and Production Politics in Rural Malaysia." *The Journal of Peasant Studies*, 19: 93–121.

— (1992). "Household Production Reconsidered: Gender, Labor Conflict, and Technological Change in Malaysia's Muda Region." *World Development*, 20: 809–23.

Huke, R. E. (1990). *Human Geography of Rice in Southeast Asia*. Manila: International Rice Research Institute.

— (1998*a*). *Myanmar*. New York: American Geographical Society.

— (1998*b*). "Myanmar Visited: Sumatra Revisited." *Focus*, 45: 18–25.

Huke, R. E., and Huke, E. H. (1990). "Introducing a Geographic Information System at IRRI." *IRRI Social Science Division Papers*, 90–102.

— (1989). *Human Geography of Rice in Southeast Asia*. Manila: International Rice Research Institute.

Hussey, A. (1989). "Tourism in a Balinese Village." *Geographical Review*, 79: 311–25.

— (1991). "Regional Development and Cooperation through ASEAN." *Geographical Review*, 81: 87–98.

— (1993). "Rapid Industrialization in Thailand 1986–1991." *Geographical Review*, 8: 14–28.

Kelly, P. F. (1997). "Globalization, Power, and the Politics of Scale in the Philippines." *Geoforum*, 28: 151–71.

— (2001). "The Political Economy of Local Labor Control in the Philippines." *Economic Geography*, 77: 1–22.

Kong, L., Yeoh, B., and Teo, P. (1996). "Singapore and the Experience of Place in Old Age." *Geographical Review* 86: 529–49.

Kummer, D. (1991). *Deforestation in the Postwar Philippines*. University of Chicago Geography Research Paper, 234. Chicago: University of Chicago Press.

— (1992). "Measuring Forest Decline in the Philippines: An Exercise in Historiography." *Forest and Conservation History*, 36: 185–9.

— (1995). "The Political Use of Philippine Forestry Statistics." *Crime, Law, and Social Change*, 22: 163–80.

Kummer, D., Concepcion, R., and Canizares, B. (1994). "Environmental Degradation in the Uplands of Cebu." *Geographical Review*, 84: 266–76.

Kummer, D., and Turner, B. L., II (1994). "The Human Causes of Deforestation in Southeast Asia." *Bioscience*, 44: 323–28.

Law, L. (2001). "Home Cooking: Filipino Women and Geographies of the Senses in Hong Kong." *Ecumene*, 8: 264–83.

Leinbach, T. R. (1988). "Child Survival in Indonesia." *Third World Planning Review*, 10: 255–69.

— (1989). "The Transmigration Programme in Indonesian National Development Strategy: Current Status and Future Requirement." *Habitat International*, 13: 81–93.

— (1995). "Trade Regimes, Foreign Investment and the New Liberalization Environment in Indonesia," in M. Green and R. B. McNaughton (eds.), *The Location of Foreign Direct Investment: Geographic and Business Approaches*. Aldershot: Avebury, 151–74.

— (1992). "Small Towns, Rural Linkages, and Employment." *International Regional Science Review*, 14: 317–23.

Leinbach, T. R., and Bowen, J. T. (1992). "Diversity in Peasant Economic Behavior: Transmigrant Households in South Sumatra, Indonesia." *Geographical Analysis*, 24: 335–51.

Leinbach, T. R., and Smith, A. (1994). "Off-Farm Employment, Land, and Life Cycle: Transmigrant Households in South Sumatra, Indonesia." *Economic Geography*, 70: 273–96.

Leinbach, T. R., and Watkins, J. F. (1998). "Remittances and Circulations Behavior in the Livelihood Process: Transmigrant Families in South Sumatra, Indonesia." *Economic Geography*, 74: 45–63.

Leinbach, T. R., Watkins, J. F., and Bowen, J. (1992). "Employment Behavior and the Family in Indonesian Transmigration." *Annals of the Association of American Geographers*, 82: 23–47.

Lenz, R. (1993). "On Resurrecting Tourism in Vietnam." *Focus*, 43: 1–6.

Lewis, M. W. (1989). "Commercialization and Community Life: The Geography of Exchange in a Small-Scale Society." *Annals of the Association of American Geographers*, 79: 390–410.

— (1991). "Elusive Societies: A Regional-Cartographical Approach to the Study of Human Relatedness." *Annals of the Association of American Geographers*, 81: 605–26.

— (1992*a*). *Wagering the Land: Ritual, Capital, and Environmental Degradation in the Cordillera of Northern Luzon, 1900–1986*. Berkeley: University of California Press.

— (1992*b*). "Agricultural Regions in the Philippine Cordillera." *Geographical Review*, 82: 29–42.

Lewis, M. W., and Wigen, K. (1997). *The Myth of Continents: A Critique of Metageography*. Berkeley: University of California Press.

McGee, T. G. (1991*a*). "The Emergence of *Desakota* Regions in Asia: Expanding a Hypothesis," in N. Ginsburg, B. Koppel, and T. G. McGee (eds.), *The Extended Metropolis: Settlement Transition in Asia*. Honolulu: University of Hawaii Press, 3–25.

— (1991*b*). "Southeast Asian Urbanization: Three Decades of Change." *Prisma* 51: 3–16.

— (1995). "Metrofitting the Emerging Mega-Urban Regions of ASEAN: An Overview," in T. G. McGee and I. M. Robinson (eds.), *The Mega-Urban Regions of Southeast Asia*. Vancouver: University of British Columbia Press, 3–26.

McGee, T. G., and Greenberg, C. (1992). "The Emergence of Extended Metropolitan Regions in ASEAN: Toward the Year 2000." *ASEAN Economic Bulletin*, 9: 22–44.

Macleod, S. A. (1995). "Shadows beneath the Wind: Singapore, World City and Open Region." Unpublished Ph.D. dissertation, University of British Columbia, Department of Geography.

Macleod, S., and McGee, T. G. (1996). "The Singapore-Johore-Riau Growth Triangle: An Emerging Extended Metropolitan Region," in F. Lo and Y. Yeung (eds.), *Emerging World Cities in Pacific Asia*. Tokyo: United Nations University Press, 417–64.

Murphy, A. B. (1995). "Economic Regionalization and Pacific Asia." *Geographical Review*, 85: 127–40.

Poon, J. P. H., and Perry, M. (1999). "The Asian Economic 'Flu': A Geography of Crisis." *Professional Geographer* 51: 184–95.

Reed, R. R. (1995). "From Highland Hamlet to Regional Capital: Reflections on the Colonial Origins, Urban Transformation, and Environmental Impact of Dalat," in T. Rambo, R. R. Reed, L. T. Cuc, and M. R. Digregoria (eds.), *The Challenges of Highland Development in Vietnam*. Honolulu: East-West Center Program on Environment; Hanoi: Hanoi University Center for Natural Resources and Environmental Studies; and Berkeley: University of California, Center for Southeast Asian Studies, 34–62.

Savage, V., Kong, L., and Yeoh, B. S. A. (1993). "The Human Geography of Southeast Asia: An Analysis of Post-War Development." *Singapore Journal of Tropical Geography*, 14: 229–50.

Schwartzberg, J. W. (1994*a*). "A Nineteenth Century Burmese Map Relating to Colonial Expansion in Southeast Asia." *Imago Mundi*, 46: 117–27.

— (1994*b*). "Cartography in Southeast Asia," in J. B. Harley, and D. Woodward (eds.), *The History of Cartography*, vol. ii, bk. 2. *Cartography in the Traditional East and Southeast Asian Societies*. Chicago: University of Chicago Press, 689–842.

Scott, A. J. (1994). "Variations on the Theme of Agglomeration and Growth: The Gem and Jewelry Industries in Los Angeles and Bangkok." *Geoforum*, 25: 249–63.

Sicular, D. (1989). "Scavengers and the Development of Solid Waste Management in Indonesian Cities." Unpublished Ph.D. dissertation, University of California, Berkeley, Department of Geography.

Silvey, R. (1997). "Placing the Migrant: Gender, Identity, and Development in South Sulaweisi, Indonesia." Unpublished Ph.D. dissertation, University of Washington, Department of Geography.

Suryanata, K. (1994). "Fruit Trees under Contract: Tenure and Land Use Change in Upland Java, Indonesia." *World Development*, 20: 1567–79.

Thomson, C. N. (1991). "Urban Residential Land Use Mapping in Metropolitan Bangkok Using Thematic Mapper Data." *Asian-Pacific Remote Sensing Journal*, 4: 15–24.

— (1993). "Political Identity among Chinese in Thailand." *Geographical Review*, 83: 397–409.

— (1994). "Thailand's May 1992 Uprising: Factors in the Democratic Reform." *Asian Profile*, 22: 487–500.

— (1995). "Political Stability and Minority Groups in Burma." *Geographical Review*, 85: 269–85.

— (1996). "Electoral Geography of the Sino-Thai in Thailand's National Elections, 1979–1995." *Professional Geographer*, 48: 392–404.

— (1998*a*). "Examination of the Baseline Data for Bangkok's 'Congested Communities.'" Unpublished Presentation at the Annual Meeting of the Association of American Geographers, Boston.

— (1998*b*). "Teaching Geography Using the City as a Model: The Example of Bangkok." *Education about Asia*, 3: 4–10.

Tuason, J. A. (1999). "The Ideology of Empire in *National Geographic Magazine's* Coverage of the Philippines, 1898–1908." *Geographical Review*, 89: 34–53.

Tyner, J. A. (1996*a*). "Constructions of Filipino Migrant Entertainers." *Gender, Place and Culture*, 3: 77–93.

— (1996*b*). "The Gendering of Philippine International Labor Migration." *Professional Geographer*, 48: 405–16.

— (1997). "Constructing Images, Constructing Policy: The Case of Filipino Migrant Performing Artists." *Gender, Place and Culture*, 4: 19–35.

— (1999). "The Geopolitics of Eugenics and the Exclusion of Philippine Immigrants from the United States." *Geographical Review*, 89: 54–73.

— (2000). "Global Cities and Circuits of Global Labor: The Case of Manila, Philippines." *Professional Geographer*, 52: 61–74.

Ulack, R. (1989). "The Elderly of Cebu, the Philippines." *Philippine Geographical Journal*, 33: 103–15.

Ulack, R., and Watkins, J. F. (1991). "Migration and Regional Population Aging in the Philippines." *Journal of Cross-Cultural Gerontology*, 6: 383–411.

Walker, J. (1997). "The Housewifization of Filipino Domestic Servants: The State, Gender, and Labor Export." Unpublished Presentation at the Annual Meeting of the Association of American Geographers, Fort Worth.

Warf, B. (1998). "Teaching Indonesia: A World-Systems Perspective." *Education About Asia*, 3: 17–23.

Watkins, J. F., Leinbach, T. R., and Falconer, K. (1993). "Women, Family and Work in Indonesian Transmigration." *Journal of Developing Areas*, 27: 377–99.

Yasmeen, G. (1997). "Bangkok's Foodscape: Public Eating, Gender Relations, and Urban Change." Unpublished Ph.D. dissertation, University of British Columbia, Department of Geography.

South Asia

Allan, N. J. R. (1989). "Kashgar to Islamabad: The Impact of the Karakoram Highway on Mountain Society and Habitat." *Scottish Geographical Magazine*, 105: 130–41.

Allan, N. J. R. (1991). "From Autarky to Dependency: Society and Habitat Relations in the South Asian Mountain Rimland." *Mountain Research and Development*, 11: 65–74.

—— (ed.) (1995). *Mountains at Risk: Cultural Issues in Environmental Studies*. Delhi: Manohar.

—— (1998). "A Failed Mountain Book." *Himal*, 11: 33–9.

Balachandran, C. S. (1995). "LACONEX: A Rule-Based Implementation to Advise on Plot-Level Land Conservation in Vilathikulam, Southeastern India." *Land Degradation and Rehabilitation*, 6: 265–83.

Bhardwaj, S. M. (1973). *Hindu Places of Pilgrimage in India: A Study in Cultural Geography*. Berkeley: University of California Press.

Bishop, B. C. (1990). *Karnali under Stress*. Chicago: Department of Geography Research Publications, University of Chicago.

Brower, B. (1991). *Sherpa of Khumbu: People, Livestock and Landscape*. Delhi: Oxford University Press.

Butz, D. (1994). "A Note on Crop Distribution and Micro-environmental Conditions in Holshal and Ghoshushal Villages, Pakistan." *Mountain Research and Development*, 14: 89–97.

—— (1995). "Legitimating Porter Regulation in an Indigenous Mountain Community in Northern Pakistan." *Environment and Planning D: Society and Space*, 13: 381–414.

—— (1996). "Sustaining Indigenous Communities: Symbolic and Instrumental Dimensions of Pastoral Resource Use in Shimshal, Northern Pakistan." *The Canadian Geographer*, 40: 36–53.

Butz, D., and Eyles, J. (1997). "Reconceptualizing Senses of Place: Social Relations, Ideology and Ecology." *Geografiska Annaler*, 79: 1–25.

Chakravorty, S. (1993). "The Distribution of Population and Income: Explorations Using Asian Cases." *GeoJournal*, 29: 115–24.

—— (1994). "Equity and the Big City." *Economic Geography*, 70: 1–22.

—— (1995). "Patterns of Urbanization, Urban Concentration and Income Distribution: Implications for Development." *Urban Geography*, 16: 622–42.

—— (1996). "Too Little, in the Wrong Places? The Mega City Programme, Efficiency and Equity in Indian Urbanization." *Economic and Political Weekly*, 31: 2565–72.

—— (1999). "Liberalism, Neoliberalism, and Capability Generation: Toward a Normative Basis for Planning in Developing Nations." *Journal of Planning Education and Research* 19: 77–85.

—— (2000*a*). "How Does Structural Reform Affect Regional Development? Resolving Contradictory Theory with Evidence from India." *Economic Geography* 76: 367–94.

—— (2000*b*). "From Colonial City to Global City? The Far-From-Complete Spatial Transformation of Calcutta," in P. Marcuse and R. van Kempen (eds.), *Globalizing Cities: A New Spatial Order?* Oxford: Blackwell, 56–77.

Chakravorty, S., and Gupta, G. (1996). "Let a Hundred Projects Bloom: Structural Reform and Urban Development in Calcutta." *Third World Planning Review*, 18: 415–31.

Conway, D., Bhattarai, K., and Shrestha, N. R. (2000). "Population-Environment Relations at the Forested Frontier of Nepal: Tharu and Pahari Survival Strategies in Bardiya." *Applied Geography*, 20: 221–42.

Das, R. (1995). "Poverty and Agrarian Social Structure: A Case Study of Rural India." *Dialectical Anthropology*, 20: 169–92.

—— (1998*a*). "The Geographical Unevenness of India's Green Revolution: A Political Economy Approach." *Journal of Contemporary Asia*, 29.

—— (1998*b*). "The Green Revolution, Agrarian Productivity, and Labor." *International Journal of Urban and Regional Research*, 22: 122–35.

—— (1998*c*). "The Social and Spatial Character of the Indian State." *Political Geography*, 17: 787–808.

Das, R., and Dutt, A. (1993). "Rank-Size Distribution and Primate City Characteristics in India." *GeoJournal*, 29: 125–37.

Dutt, A. (1993). "Cities of South Asia," in S. D. Brunn and J. F. Williams (eds.), *Cities of the World: World Regional Urban Development*. New York: HarperCollins, 351–87.

Dutt, A., and Geib, M. M. (1998). *Atlas of South Asia: A Geographical Analysis of Countries*. New Delhi: Oxford and IBH Publishing Co.

Dutt, A., Mitra, C., and Halder, A. (1997). "Slum Location and Cycle of Poverty: Calcutta Case." *Asian Profile*, 25: 413–25.

Dutt, A., Pomeroy, G., and Wadhwa, V. (1996). "Cultural Patterns of India," in L. R. Singh (ed.), *New Frontiers in India Geography*. Allahabad: R. N. Dubey Foundation.

Dutt, A., Costa, F. J., Aggarwal, S., and Noble, A. G. (1994). *The Asian City: Processes of Development, Characteristics, and Planning*. Dordrecht: Kluwer.

Edney, M. H. (1997). *Mapping an Empire: The Geographical Construction of British India, 1765–1843*. Chicago: University of Chicago Press.

Huke, E., Huke, R., and Woodhead, T. (1993). *Rice-Wheat Atlas of Bangladesh*. Manila: International Rice Research Institute.

—— (1994). *Rice-Wheat Atlas of Nepal*. Manila: International Rice Research Institute.

Ives, J. D., and Messerli, B. (1989). *The Himalayan Dilemma*. London: Routledge.

Karan, P. P. (1989*a*). "Mountains in Crisis: Man versus Nature." *Explorers Journal*, 67: 162–8.

—— (1989*b*). "Environment and Development in Sikkim Himalaya: A Review." *Human Ecology*, 17: 257–71.

—— (1990). *Bhutan: Environment, Culture and Development Strategy*. Delhi: Intellectual Publishing.

—— (1991). "Ecology, Environmental Conservation and Development in Bhutan Himalaya," in G. S. Rajwar (ed.), *Recent Researches in Ecology, Environment and Pollution*. Delhi: Today's and Tomorrow's Publishers, vi. 167–97.

—— (1994). "Patterns of Pilgrimage to the Sikh Shrine of Guru Gobind Singh at Patna." *Journal of Asian and African Studies*, 46–7: 275–84.

Karan, P. P., and Ishii, H. (1994). *Nepal: Development and Change in a Landlocked Himalayan Kingdom*. Tokyo: Institute for the Study of Languages and Cultures of Asia and Africa, Tokyo University of Foreign Studies.

—— (1996). *Nepal: A Himalayan Kingdom in Transition*. Tokyo: United Nations University Press.

MacDonald, K. I. (1996*a*). "Indigenous Labour Arrangements and Household Security in Northern Pakistan." *Himalayan Research Bulletin*, 16: 29–35.

—— (1996*b*). "Population Change in the Upper Braldu Valley, Baltistan, 1900–1990: All Is Not as It Seems." *Mountain Research and Development*, 16: 351–66.

— (1998*a*). "Push and Shove: Spatial History and the Construction of a Portering Economy in Northern Pakistan." *Comparative Studies in Society and History*, 40: 287–317.

— (1998*b*). "Rationality, Representation, and the Risk Mediating Characteristics of a Karakoram Mountain Farming System." *Human Ecology*, 26: 287–321.

MacDonald, K. I., and Butz, D. (1998). "Investigating Portering Relations as a Locus for Transcultural Interaction in the Karakoram Region of Northern Pakistan." *Mountain Research and Development*, 18: 333–43.

Metz, J. J. (1989). "A Framework for Classifying Subsistence Production Types of Nepal." *Human Ecology*, 17: 147–76.

— (1990). "Conservation Practices at an Upper Elevation Village of West Nepal." *Mountain Research and Development*, 10: 7–15.

— (1991). "A Reassessment of the Causes and Severity of Nepal's Environmental Crisis." *World Development*, 19: 805–20.

— (1994). "Forest Product Use at an Upper Elevation Village in Nepal." *Environmental Management*, 18: 371–90.

— (1995). "Development in Nepal: Investment in the Status Quo." *GeoJournal*, 35: 175–84.

— (1997). "Vegetation Dynamics of Several Little Disturbed Temperate Forests in East Central Nepal." *Mountain Research and Development*, 17: 333–51.

Mookherjee, D. (1994). "The Urban System and Emerging Structure: An Application of Gibbs' Measure to the Case of India," in G. O. Braun (ed.), *Managing and Marketing of Urban Development and Urban Life*. Berlin: Dietrich Reimer Verlag, 115–23.

Mookherjee, D., and Tiwari, R. C. (1996). "Perspectives on the Urbanization Process: The Indian Pattern," in N. L. Chiang, J. F. Williams, and H. L. Bednarek (eds.), *Proceedings: Fourth Asian Urbanization Conference*. East Lansing: Michigan State University, 307–18.

Noble, A. G., Costa, F. J., Dutt, A. K., and Kent, R. B. (1998). *Regional Development and Planning for the 21st Century: New Priorities, New Philosophies*. Aldershot: Ashgate.

Oza, R. (2001). "Showcasing India: Gender, Geography, and Globalization." *Signs: Journal of Women in Culture and Society*, 26: 1067–95.

Paul, B. K. (1990). "Factors Affecting Infant Mortality in Rural Bangladesh." *Rural Sociology*, 55: 522–40.

— (1991). "Family Planning Availability and Contraceptive Use in Rural Bangladesh." *Socio-Economic Planning Science*, 25: 269–82.

— (1992). "Health Search Behavior of Parents in Rural Bangladesh." *Environment and Planning A*, 24: 963–73.

— (1993). "Flood Damage to Rice Crop in Bangladesh." *Geographical Review*, 83: 150–9.

— (1994). "Aids in Asia." *Geographical Review*, 84: 367–79.

— (1995). "Farmers' Response to the Flood Action Plan of Bangladesh: An Empirical Study." *World Development*, 23: 299–309.

— (1997). "Changes in Reproductive Behavior in Bangladesh." *Geographical Review*, 87: 100–4.

— (1998). "Coping with the 1996 Tornado in Tangail, Bangladesh: An Analysis of Field Data." *Professional Geographer*, 50: 287–301.

Pomeroy, G. M., and Dutt, A. (1998). "Population Dynamics and Planning: China and India," in Noble *et al.* (1998: 81–101).

Rasid, H., and Odemerho, F. O. (1998). "Assessment of Urban Flood Problems by Residents of Slums and Squatter Settlement: Benin City, Nigeria, vs. Dhaka, Bangladesh." *Arab World Geographer*, 1: 136–54.

Robbins, P. (1998*a*). "Authority and Environment: Institutional Landscapes in Rajasthan, India." *Annals of the Association of American Geographers*, 88: 410–35.

— (1998*b*). "Nomadization in Rajasthan, India: Migration, Institutions, and Economy." *Human Ecology*, 26: 87–112.

— (2000*a*). "Pastoralism and Community in Rajasthan, India: Interrogating Categories in Arid Lands Development," in A. Agarwal and K. Sivaramakrishna (eds.), *Agrarian Environments: Resources, Representations, and Rule in India*. Durham, NC: Duke University Press, 191–215.

— (2000*b*). "The Rotten Institution: Corruption in Natural Resource Management." *Political Geography*, 19: 423–43.

— (2001). "Fixed Categories in a Portable Landscape: The Causes and Consequences of Land Cover Categorization." *Environment and Planning A*, 33: 161–79.

Shrestha, N. R. (1989). "Frontier Settlement and Landlessness among Hill Migrants in Nepal Tarai." *Annals of the Association of American Geographers*, 79: 370–89.

— (1990). *Landlessness and Migration in Nepal*. Boulder, Colo.: Westview Press.

— (1997). "A Postmodern View or Denial of Historical Integrity? The Poverty of Yapa's View of Poverty." *Annals of the Association of American Geographers*, 87: 709–16.

— (1998*a*). "Environmental Degradation of Kathmandu: Losing Shangri-La?" *Education About Asia*, 3: 11–18.

— (1998*b*). "Local Labour and Global Capital: A Historical Perspective on Nepali Migration to India." *Arab World Geographer*, 1: 60–78.

— (1999). *In the Name of Development: A Reflection on Nepal* (Nepal edn.). Kathmandu: Educational Enterprise, and Lanham, Md.: University Press of America (1997).

Shrestha, N. R., and Conway, D. (1996). "Ecopolitical Battles at the Tarai Frontier of Nepal." *International Journal of Population Geography*, 2: 313–31.

Shrestha, N. R., Velu, R., and Conway, D. (1993). "Frontier Migration and Upward Mobility: The Case of Nepal." *Economic Development and Cultural Change*, 41: 787–816.

Shrestha, N. R., Conway, D., and Bhattarai, K. (1999). "Population Pressure and Land Resources in Nepal: A Revisit, Twenty Years Later." *The Journal of Developing Areas*, 33: 245–68.

Stevens, S. (1988*a*). "Sacred and Profaned Himalayas." *Natural History*, 97: 26–35.

— (1988*b*). "Tourism and Development in Nepal." *Kroeber Anthropological Society Papers*, 67–8: 67–80.

— (1993*a*). *Claiming the High Ground: Sherpas, Subsistence, and Environmental Change in the Highest Himalaya*. Berkeley: University of California Press.

— (1993*b*). "Tourism, Change, and Continuity in the Mount Everest Region, Nepal." *Geographical Review*, 83: 410–27.

— (ed.) (1997). *Conservation through Cultural Survival*. Washington, DC: Island Press.

Stoddard, R. H. (1988). "Characteristics of Buddhist Pilgrimages in Sri Lanka," in S. M. Bhardwaj and G. Rinschede (eds.), *Pilgrimages in World Religions*. Berlin: Dietrich Reimer Verlag.

Stoddard, R. H., and A. Morinis (eds.) (1997). *Sacred Spaces, Sacred Places: The Geography of Pilgrimages*. Baton Rouge: Geoscience Publications, Department of Geography, Louisiana State University.

Stokke, K. (1997). "Authoritarianism in the Age of Market Liberalism in Sri Lanka." *Antipode*, 29: 437–55.

Tripathi, S. (2000). "Health Seeking Behavior: Q Structures of Rural and Urban Women in India with Reproductive Tract Infections and Sexually Transmitted Diseases." *Professional Geographer*, 52: 218–32.

Wallach, B. (1996). *Losing Asia*. Baltimore: Johns Hopkins University Press.

Woodhead, T., Huke, R., and Huke, E. (1993). *Rice-Wheat Atlas of Pakistan*. Manila: International Rice Research Institute.

— (1994). *Rice-Wheat Atlas of India*. Manila: International Rice Research Institute.

Yapa, L. (1977). "The Green Revolution: A Diffusion Model." *Annals of the Association of American Geographers*, 67: 350–9

— (1996). "What Causes Poverty? A Postmodern View." *Annals of the Association of American Geographers*, 86: 707–28.

— (1998). "The Poverty Discourse and the Poor in Sri Lanka." *Transactions of the Institute of British Geographers*, 23: 95–115.

Zurick, D. N. (1988). "Resource Needs and Land Stress in Rapti Zone, Nepal." *Professional Geographer*, 40: 428–44.

— (1989). "Historical Links between Settlement, Ecology, and Politics in the Mountains of West Nepal." *Human Ecology*, 17: 229–55.

— (1990). "Traditional Knowledge and Conservation as a Basis for Development in a West Nepal Village." *Mountain Research and Development*, 10: 23–33.

— (1992). "Adventure Travel and Sustainable Tourism in the Peripheral Economy of Nepal." *Annals of the Association of American Geographers*, 82: 608–28.

— (1993). "The Road to Shangri La is Paved: Spatial Development and Rural Transformation in Nepal." *South Asia Bulletin*, 1–2: 35–44.

Zurick, D. N., and Karan, P. P. (1999). *Himalaya: Life on the Edge of the World*. Baltimore: Johns Hopkins University Press.

Southwest Asia

Abu-Dawood, A. S., and Karan, P. P. (1990). *International Boundaries of Saudi Arabia*. Delhi: Galaxy Publications.

Abu Lughod, L. (1990). "Anthropology's Orient: Boundaries of Theory on the Arab World," in H. Sharabi (ed.), *Theory, Politics, and the Arab World: Critical Responses*. New York: Routledge Press.

Amery, H. (1993). "The Litani River of Lebanon." *Geographical Review*, 83: 229–37.

Amery, H., and Wolf, A. (eds.) (2000). *Water in the Middle East: A Geography of Peace*. Austin: University of Texas Press.

Beaumont, P., Blake, G., and Wagstaff, J. M. (1988). *The Middle East A Geographical Study*. New York: Halsted Press; John Wiley Books.

Benchefira, A., and Johnson, D. (1990). "Adaptation and Intensification in the Pastoral Systems of Morocco," in J. Galaty and D. Johnson, (eds.), *World of Pastoralism*. New York: Guilford Press.

— (1991). "Changing Resource Management Strategies and their Environmental Impacts in the Middle Atlas Mountains of Morocco." *Mountain Research and Development*, 11: 183–94.

Blaut, J. (1993). *The Colonizer's Model of the World: Geographical Diffusionism and Eurocentric History*. New York: Guillford Press, 394–416.

Blumler, M. (1993). "Successional Pattern and Landscape Sensitivity in the Mediterranean and Near East," in D. Thomas and R. Allison (eds.), *Landscape Sensitivity*. New York: John Wiley Publishers, 287–305.

Bonine, M. (1976). "Where is the Geography of the Middle East?" *Professional Geographer*, 28: 190–6.

— (ed.) (1997). *Population, Poverty, and Politics in Middle East Cities*. Gainesville: University of Florida Press.

Bonine, M., and Ehlers, E. (1994). "The Middle Eastern City and Islamic Urbanism: An Annotated Bibliography of Western Literature." *Bonner Geographische Abhandlungen*, 91.

Cohen, S. B., and Kliot, N. (1994). "Place-Names in Israel's Ideological Struggle over the Administered Territories." *Annals of the Association of American Geographers*, 82: 652–80.

Cohen, S. E. (1993). *The Politics of Planting: Israeli-Palestinian Competition for Control of Land in the Jerusalem Periphery*. Chicago: University of Chicago Press.

— (1994). "Greenbelts in London and Jerusalem." *Geographical Review*, 84: 74–89.

— (2000). "An Absence of Place: Expectation and Realization in the West Bank," in A. Murphy and D. Johnson, (eds.), *Cultural Encounters With the Environment: Enduring and Evolving Geographic Themes*. Lanham, Md.: Rowman & Littlefield, 283–303.

Dinar, A., and Wolf, A. (1994*a*). "International Markets for Water and the Potential for Regional Cooperation: Economic and Political Perspectives in the Western Middle East." *Economic Development and Cultural Change*, 43: 43–66.

— (1994*b*). "Economic Potential and Political Considerations of Regional Water Trade: The Western Middle East Example." *Resource and Energy Economics*, 16: 335–56.

— (1997). "Economic and Political Considerations in Regional Cooperation Models." *Agricultural and Resource Economics Review*, 26: 7–22.

Drake, C. (1997). "Water Resource Conflicts in the Middle East." *Journal of Geography*, 96: 4–12.

Drysdale, A. (1992). "Syria and Iraq—The Geopathology of a Relationship." *GeoJournal*, 28: 347–55.

Drysdale, A., and Blake, G. (1985). *The Middle East and North Africa: A Political Geography*. Oxford: Oxford University Press.

Emmett, C. (1993). "Political Manifestations in the Religious Landscape of Jerusalem." *Architecture and Behavior*, 9: 261–74.

— (1995). *Beyond the Basilica: Christians and Muslims in Nazareth*. Chicago: University of Chicago Press.

— (1996). "The Capital Cities of Jerusalem." *Geographical Review*, 86: 233–58.

— (1997*a*). "Conflicting Loyalties and Local Politics in Nazareth." *Middle East Journal*, 51: 535–53.

— (1997*b*). "The Status Quo Solution for Jerusalem." *Journal of Palestine Studies*, 27: 16–28.

Falah, G. (1990). "Arabs versus Jews in Galilee: Competition for Regional Resources." *GeoJournal*, 21: 325–36.

—— (1991*a*). "Israeli 'Judaization' Policy in Galilee." *Journal of Palestine Studies*, 20: 69–85.

—— (1991*b*). "Pre-state Jewish Colonization in Northern Palestine and its Impact on Local Bedouin Sedentarization 1914–48." *Journal of Historical Geography*, 17: 289–309.

—— (1992). "Land Fragmentation and Spatial Control in the Nazareth Metropolitan Area." *Professional Geographer*, 44: 30–44.

—— (1996*a*). "Living Together Apart: Residential Segregation in Mixed Arab-Jewish Cities in Israel." *Urban Studies*, 33.

—— (1996*b*). "The 1948 Israeli-Palestinian War and its Aftermath: The Transformation and De-signification of Palestine's Cultural Landscape." *Annals of the Association of American Geographers*, 86: 256–85.

—— (1997). "Ethnic Perceptual Differences of Housing and Neighbourhood Quality in Mixed Arab-Jewish Towns in Israel." *Environment & Planning A*, 29: 1663–74.

—— (1999). "No Longer 'Lost in the Desert:' Human Geography from Within the Arab World Since 1985, Part I." *Arab World Geographer*, 2.

Farhan, Y. (1997). "Developments in Modern Arab Geography." *Geography* (Bulletin of the Israeli Geographical Association), 14: 3–5.

Fisher, W. B. (1978). *The Middle East: A Physical, Social, and Regional Geography*. London: Methuen.

Gradus, Y., and Lipshitz, G. (eds.) (1996). *The Mosaic of Israeli Geography*. Beer Sheva: The Ben-Gurion University of the Negev Press.

Held, C. (1994). *Middle East Patterns: Places, People, and Politics*. Boulder, Colo.: Westview Press.

Hemmasi, M. (1992). "Origin and Growth of Modern Geography in Iran." *Professional Geographer*, 44: 465–9.

—— (1994). "Gender and Spatial Population Mobility in Iran." *Geoforum*, 25: 213–26.

Johnson, D. (1993*a*). "Nomadism and Desertification in Africa and the Middle East." *GeoJournal*, 31: 51–66.

—— (1993*b*). "Pastoral Nomadism and the Sustainable Use of Arid Lands." *Arid Lands Newsletter*, 33: 26–34.

—— (1996). "Development Trends and Environmental Deterioration in the Agropastoral Systems of the Central Middle Atlas, Morocco," in W. Swearingen and A. Bencerifa (eds.), *North African Environment at Risk*. Boulder, Colo.: Westview Press, 35–53.

Kheirabadi, M. (1991). *Iranian Cities: Formation and Development*. Austin: University of Texas Press.

Kolars, J. (1992). "Water Resources of the Middle East." *Canadian Journal of Development Studies*, Special Issue: 103–19.

—— (1993). *The Litani River in the Context of Middle Eastern Water Resources*. Oxford: Centre for Lebanese Studies.

—— (1994*a*). "Managing the Impact of Development: The Euphrates and Tigris Rivers and the Ecology of the Arabian Gulf—A Link in Forging Tri-Riparian Cooperation," in A. Bagis (ed.), *Water as an Element of Cooperation and Development in the Middle East*. Ankara: Hacettepe University and the Friedrich-Naumann Foundation.

—— (1994*b*). "Problems of International River Management: The Case of the Euphrates," in A. Biswas (ed.), *International Waters of the Middle East: From Euphrates-Tigris to Nile*. Oxford: Oxford University Press, 44–94.

—— (1995). "Observations Regarding Water Sharing and Management: An Intensive Analysis of the Jordan River Basin with Reference to Long-distance Transfers." *Water Resources Development*, 11: 351–75.

Parmenter, B. (1994). *Giving Voice to Stones: Place and Identity in Palestinian Literature*. Austin: University of Texas Press.

Pease, P., Tchakerian, V., and Tinsdale, N. W. (1998). "Aerosols Over the Arabian Sea: Geochemistry and Source Regions for Desert Aeolian Dust." *Journal of Arid Environments*, 39: 477–96.

Said, E. (1978). *Orientalism*. New York: Random House.

Stewart, D. (1996*a*). "Cities in the Desert: The Egyptian New-Town Program." *Annals of the Association of American Geographers*, 86: 459–80.

—— (1996*b*). "Economic Recovery and Reconstruction in Postwar Beirut." *Geographical Review*, 86: 487–504.

—— (1999). "Changing Cairo: The Political Economy of Urban Form." *International Journal of Urban and Regional Research*, 23.

—— (2001). "Middle East Urban Studies: Identity and Meaning." *Urban Geography*, 22: 175–81.

Tchakerian, V., Tinsdale, N. W., and Pease, P. (1997). "Geomorphology and Sediments of the Wahiba Sand Sea, Sultanate of Oman." *Supplementi Di Geografia Fisica e Dinamica Quaternaria*, 3: 372–3.

Tinsdale, N. W., Pease, P., and Tchakerian, V. (1998). "Aerosols Over the Arabian Sea: Transport Pathways and Concentrations of Dust and Sea Salt." *Deep Sea Research*, pt. II, *Tropical Studies in Oceanography*, 46: 1557–76.

Vasile, E. (1997*a*). "Devotion as Distinction, Piety as Power: Religious Revival and the Transformation of Space in the Illegal Settlements of Tunis," in M. Bonine (ed.), *Population, Poverty, and Politics in Middle East Cities*. Gainesville, Fla.: University of Florida Press, 113–40.

—— (1997*b*). "Re-turning Home: Transnational Movements and the Transformation of Landscape and Culture in the Marginal Communities of Tunisia." *Antipode*, 29.

Wolf, A., and Murakami, M. (1995). "Techno-Political Decision Making for Resources Development: The Jordan River Watershed." *Water Resources Development*, 11: 147–62.

CHAPTER 40

Canadian Studies

Christopher D. Merrett and Thomas Rumney

Introduction

The idea of Canadian Studies runs counter to recent economic trends that challenge the salience of the boundary between Canada and the United States. Books such as *Nine Nations of North America* (Garreau 1981) declare that the boundary is an irrelevant curiosity from a bygone era. Regional boundaries are more important than national boundaries when studying the geography of North America. Canada is viewed as nothing more than the "thirteenth Federal Reserve District" of the United States (Kindleberger 1987: 16). These statements suggest that Canada is not different from the United States, so why study geography from a Canadian perspective? One response is that the study of Canada endures because scholars persist in the idea that events (and the perception of events) in Canada differ from those elsewhere in North America.

This chapter emphasizes research conducted by members of the Canadian Studies Specialty Group (CSSG) of the Association of American Geographers (AAG) between the late 1980s to the present. However, where relevant, we also include research conducted by other AAG members, scholars from the Association for Canadian Studies in the United States (ACSUS), and in some cases, Canadian and other scholars who have published in American journals. While we intend to present a comprehensive survey, it is by no means exhaustive. Our goal is to identify some major themes and diverse ways that Canadian geography has been studied in the United States. The chapter examines research conducted from the late 1980s to the present, and is organized around six themes: (1) economic geography and free trade; (2) political geography of national identities; (3) urban and social geography; (4) Canada's regions: historical and cultural perspectives; (5) physical geography; and (6) the future of Canadian Studies.

Economic Geography and Free Trade

Differing Approaches to Economic Geography

The deregulation of the Canadian economy accelerated with the implementation of the Canada-United States Free Trade Agreement (FTA) in 1989 and the North American Free Trade Agreement (NAFTA) in 1994. Free trade influenced much of the recent research done on Canada's economic geography. Prior to the FTA, Americans paid little attention to the Canadian economy (Romey 1989). The neglect stems from the lack of controversy between Canada and the United States. The few studies of American origin were from a neoclassical, regional science perspective (e.g. MacDougall 1970).

In contrast, the "staples theory" of Harold Innis and other political economists has influenced much Canadian research (Clement and Williams 1989). This tradition is evident in texts such as *Heartland and Hinterland* (McCann 1987), and in monographs (Barnes 1996). Until recently, this tradition did not influence American research on Canada. American interest in Canadian geography increased when free trade became a contentious issue in the 1980s (Merrett 1996). With free trade as a backdrop, this section covers three themes in the economic geography of Canada: (1) Canadian foreign direct investment (FDI) and trade; (2) producer services; and (3) political economy.

FDI and Free Trade

In 1984, Brian Mulroney became Canada's prime minister by campaigning against free trade with the United States. Once in office, Mulroney negotiated the FTA with the Reagan administration. Many Canadians felt betrayed because part of the Canadian identity is based on a long-standing anti-Americanism (Lipset 1990). Engaging in free trade threatens that aspect of the Canadian identity. In order to calm Canadian nationalists, free trade supporters argued that jobs, factories, and investments would not leave Canada for lower production costs in the United States (Tremblay 1988). But researchers found that free trade might prompt Canadian firms to relocate to the United States (Harrington *et al.* 1986: 165). Furthermore, this study found that communities in upstate New York would consider using incentives to influence where Canadian FDI occurred.

More recently, American researchers examined the impact of tariffs and exchange rates on Canadian FDI and trade. While exchange rates have a modest impact on FDI, high Canadian production costs and American tariffs were larger factors prompting Canadian firms to move south (Solocha *et al.* 1989). Warf and Cox (1992, 1993) found that as tariffs drop between Canada and the United States, the geography of trade and transportation changed too. This finding is confirmed by a shift-share analysis of NAFTA's impact on American export patterns (Hayward and Erickson 1997).

A study by MacPherson and McConnell (1992) argues that while Canadian firms have an increasing presence in western New York, the FTA was not a primary factor spurring this FDI. However, these authors note that free trade might increase future Canadian FDI in western New York as Canadian firms, burdened by higher taxes and labor costs, more fully compete in a continental economy. MacPherson (1996) and Merrett (1997, 1998*b*) later confirmed that free trade has made the United States an increasingly attractive place for Canadian investment in the 1990s.

Producer Services

Commercial enterprises use producer services to support their business activities. According to Malecki (1991: 61), producer services include information processing such as software development, goods-related services such as transportation, and personal support services such as accommodations and travel. Harrington (1989) predicted that trade in producer services would increase due to the FTA, but that it could vary by region and type of producer service—research confirmed by Coffey (1995). Harrington (1992) also examined the behavior of Canadian and American banks operating within a continental economy.

MacPherson (1991) studied technical consultants in the producer services. He compared the extent to which small manufacturing firms in Toronto and Buffalo relied on outside expertise for product development, concluding that firms in Toronto have more outside linkages than comparably sized firms in Buffalo. Researchers at the Canada-United States Trade Center at SUNY Buffalo continue their analysis of firms in the Canada-US border region (MacPherson 1997; Bagchi-Sen 1999; Bagchi-Sen and MacPherson 1999). Cornish (1997*a*, *b*) extended the research on producer services by showing how software developers benefit from locating among other software developers and producer services.

The increasing attention given to the role of producer services highlights the increasingly "social character" of the capitalist economy (Gertler and DiGiovanna 1997). Instead of competing with each other, some firms are sharing proprietary information for mutual benefit. But Gertler and DiGiovanna (ibid.) question the degree to which firms can collaborate. They found that collaboration between firms is hindered by differing corporate and national cultures.

Political Economy: Converging Paradigms

The discussion so far has emphasized neoclassically conceived studies. But the "critical" perspective found in staples theory is seen in recent American studies of Canadian natural resource use and critiques of free trade. These studies disagree with neoclassical assertions that capitalist markets tend towards equilibrium

(Clement and Williams 1989: 7–8). They counter that Canadian economic development was based on the extraction and export of unprocessed natural resources, or staples (Hornsby 1989; Rigby 1990, 1991). Because extracted commodities were processed elsewhere, Canada was forced into unequal trade relations that perpetuated regional inequalities (Watkins 1989).

Recently, American scholars have drawn on this tradition to examine the connections between the Canadian economy, politics, and natural resource base (Meserve 1991; Bertolas 1995). For example, studies show how integration between the United States and Canada has restructured the Canadian meatpacking and poultry industry (Bowler 1994; Broadway 1997). With mergers and wage cuts accompanying this restructuring, Broadway (1997) questions the sustainability of rural agricultural communities and Canadian sovereignty.

Here, a convergence can be seen between American and Canadian research on Canadian economic geography. American studies now identify the role of natural resources in Canada's colonial past and neocolonial present. But instead of relying on staples theory, Willems-Braun (1997*a*, *b*) draws on the work of Said (1979) and Bhabha (1994) to develop a "post-colonial" explanation for natural resource use in British Columbia. Although not without critics (e.g. Sluyter 1997), Willems-Braun (1997*a*: 8–9) shows how the provincial government and timber firms use value-neutral discourses of "scientific management" and sustainable yields to mask relations of domination over Native Canadians, communities, and environmentalists.

Discourse analyses also provide a "critical" perspective on free trade. Free trade was promoted using "scientific" concepts such as comparative advantage to argue that everybody benefits from the FTA. But skeptics used regulation theory and critical geopolitics to show how vulnerable segments of society such as women and organized labor could suffer from free trade (Merrett 1996). But while these segments of society are vulnerable, they are not powerless. Sparke (1996, 1997) documents the vigorous response made by opponents to free trade, despite having to confront the Canadian corporate elite.

The Political Geography of National Identities

Americans have at times disparaged Canada as a bland northern extension of the United States (Garreau 1981). Yet, Canada's recent political developments are as tumultuous as any transformation in the developed world. The rise of Québécois nationalism in the 1960s led to debates over federal–provincial power-sharing, and the treatment of individuals and communities of interest under a new Constitution. This process also spawned new political parties as long-silent voices emerged to contest Canada's future. While Canadians argued over internal politics, Canada transformed its external relations by signing the FTA and NAFTA. The emergence of regional identities has fragmented the Canadian polity, eliciting research on: (1) competing nationalisms and language conflict; (2) electoral geography; (3) the geopolitics of trade and international relations; (4) land claims by Native Canadians; and (5) the portrayal of Canada in American scholarship.

When Canada became independent in 1867, it was confederated as two nations within one state. Nationalist tensions between the English-speaking majority and the French-speaking minority have existed ever since (Wynn 1987). In order to preserve their identity, French Canadians retreated into agrarian communities, shielded by the Roman Catholic Church. The Anglophone minority in Quebec dominated life in Montreal and Quebec City. Québécois nationalists recognized that their isolationism could condemn French Canadians to perpetual poverty (ibid.). After World War II, nationalists launched a Quiet Revolution to pursue prosperity and cultural autonomy in Quebec. They changed it from a rural, religious, isolated society to one based on urbanism, secularism, and economic engagement with Anglo-America. This transformed Canada's political geography because language replaced religion as the symbol of French Canadian identity.

Kaplan (1994*a*, *b*) documents how the Canadian government attempted to reconcile the contradictory visions of Canada held by English and French Canadians. But reconciliation is difficult because, increasingly, Canadians are migrating into linguistically homogeneous communities (Kaplan 1995). As linguistic groups become more segregated, demands will be made for further devolution of federal power (Kaplan 1994*a*). This will fuel nationalist sentiments in Quebec, prompting English-speaking Quebeckers to leave for less politically volatile localities (Lo and Teixeira 1998). The Canadian government attempted to mollify Québécois nationalists by making French a second official language. Some French Canadians, though, showed their continued discontent by supporting the pro-separatist Parti Québécois, Bloc Québécois, and two referenda on sovereignty.

Regions outside Quebec resent the "distinct society" status given to Quebec under proposed constitutional reforms. Regional rancor gave rise to new political parties such as the Reform Party and Bloc Québécois, transforming Canada's electoral geography (Carty and Eagles 1998; Laidman 1998). Particular attention has been paid to the 1988 federal election because the primary issue was the implementation of the FTA. Through quirks in the geography of representation, Mulroney turned a plurality of the Canadian votes into a parliamentary majority. Consequently, Mulroney could claim a mandate to implement the FTA. T. Pratt (1997) and Jenkins (1998) provide further insights into Canada's cultural mosaic and the contested nature of elections in Canada.

While Quebec separatism and Anglophone regionalism weakened Canada internally, free trade posed an external threat to Canada (Merrett 1996). This argument runs counter to the optimistic accounts portrayed in American popular press reports (Vesilind 1990). Ongoing constitutional discussions weakened the ability of Canada to maintain its east–west identity, leading to increased north–south interaction with the United States. M. Schwartz (1998) argues that Quebec separatism and free trade work synergistically to fragment Canada. Sparke (1996) examined the intersection of national identities, free trade, and gender. His analysis is important because most gender studies on Canada published in the United States have been written by Canadian geographers (e.g. Chouinard 1989; G. Pratt 1989; Rose 1989, but see Gilbert 1997).

The key point about Canada's embrace of free trade is that it marks an abrupt change in its geopolitical strategy for survival. During the 1970s, Canada adopted economic nationalist policies to diminish its reliance on the United States. This strategy failed in the 1980s as Canada became increasingly entangled in the North American economic system (Merrett 1991; Meserve 1991). Increasing economic integration also complicates the resolution of boundary disputes between the United States and Canada. It is axiomatic to say that Canada and the United States share the longest undefended border in the world. But heated disputes over boundary delineation in the Gulf of Maine or fishing rights in the Pacific Ocean reveal that boundaries and sovereignty still matter (Gray 1994; Rogers and Stewart 1997).

Continental integration prompted scholars to question the ability of Canadian and American geographers to envision worlds beyond North America. Even though Canada has free trade agreements with Chile and Mexico, Robinson and Long (1989) bemoan the declining interest by North American geographers in Latin America. Similarly, just as Canada and the United States increased their trans-Pacific linkages, Alwin (1992) questions the ability of North American geographers to provide insights into the Pacific Rim. This inability may result from North America's retreat into an isolationist trade bloc (Michalak and Gibb 1997).

But the homogenizing impact of continentalism is challenged by the Native Canadian quest for territorial sovereignty within Canada. While Elkins (1995) downplays the territorial imperative driving national identities, Native groups relentlessly fight with Federal and provincial governments for increased territorial sovereignty (Bertolas 1995; Sparke 1998). In the words of Barsh (1997: 1), the ongoing land claims by Native Canadians mean that Canada's "political geography . . . [is] still far from settled." This is confirmed by the emergence of a new territory, Nunavut, in 1999.

The peaceful cession of power to Nunavut contrasts with the violence connected to other Native Canadian land claims, including those in Quebec. Quebec nationalists assert that if they leave Canada, the territorial integrity of Quebec is non-negotiable. The Cree in Quebec riposte that they will not willingly leave Canada (Barsh 1997). This exchange underscores the long-standing mistrust between Native Canadians and the rest of Canada (Sparke 1995, 1998).

A final topic is the portrayal of Canada in American geography textbooks. Sparke (1998: 466) argues that "maps contribute to the construction of spaces that later they seem only to represent." Maps do more than just relay information. They construct perceptions that people have about places. Similarly, scholarly accounts of Canada and its geography are more than mere reportage. Maps and ideas about Canada presented to Americans shape their attitudes toward Canada. Hence, it is vital that scholars of Canadian geography pay attention to the way that Canada is portrayed in the texts that they use.

Consider a sampling of American geography textbooks. They tend either to erase the Canadian identity by combining it with regions of the United States or to ignore it completely (e.g. Garreau 1981). In *The Regional Geography of Anglo-America* (White *et al.* 1979), there is no section devoted to the development of Canada as an independent nation-state. However, there is a chapter on "The Anglo-American City" that emphasizes the uniform character of cities, regardless of their Canadian or American location.

More recently, *Regional Landscapes of the United States and Canada* (Birdsall *et al.* 1999) confirms the suspicion that Canada does not exist (despite the title) in the minds of many American geographers. The erasure of Canada might stem from its smaller population and geopolitical stature. However, this is difficult to accept when California has its own chapter, but Canada, an independent nation-state, does not (ibid). Without knowledge of Canada, students are less likely to learn how the tension between national and regional identities shapes the political and economic landscapes of North America.

Urban and Social Geography

National Differences

The spatial pattern of Canadian urban development has been compared to an archipelago, a string of isolated urban islands punctuating an unending expanse of wilderness (R. Harris 1987). The island metaphor aptly describes the paradoxical nature of Canadian society. Canadians are simultaneously urban yet isolated, dispersed along 4,000 miles of land between the Atlantic and Pacific Oceans (ibid.). Canada struggled to connect this chain of isolated settlements along an east–west axis, against the powerful north–south attraction of the United States. One nationalist justified a publicly funded Canadian Broadcasting Corporation to unite Canada, by saying it was either "the state or the United States" (Babe 1988: 27). If the government did not intervene to unify Canada, the lure of the United States would erode Canadian national identity.

Canadian cities certainly developed within a continental milieu, but were shaped by a powerful set of nationalist policies. But scholars in America, and even in Canada, who write about North American cities, have downplayed the distinctiveness of Canadian cities (C. Harris 1997). Witness the Canadian textbook on urban geography, *The North American City* (Yeates and Garner 1980). As the title suggests, this book implies that Canadian and American cities are more alike than different. Consider also the comparison of North American cities found in *The Places Rated Almanac* (Savageau and Boyer 1993). The authors rate the 343 metropolitan areas in the United States and Canada to determine "the best places to live in North America." While this rating scheme is fascinating, it is fundamentally flawed because the authors use American standards of urban livability to evaluate Canadian cities.

Many scholars now argue to the contrary that differences occur between and within North American cities based on national, ethnic, and class- and gender-based factors. In response to Yeates and Garner (1980), Goldberg and Mercer (1986) wrote *The Myth of the North American City*. They argued that Canadian cities differed from American cities across a range of social, residential, and morphological issues, prompting a wave of comparative urban ethnic and gender- and class-based studies (e.g. Ewing 1992).

Race, Gender, and Class Differences

Despite recent arguments to the contrary (e.g. Raad and Kenworthy 1998), the impact of Goldberg and Mercer (1986) cannot be denied. Sensitivity to Canadian–American differences increased as geographers moved away from the positivist approaches that predominated in the 1970s (e.g. Lemon 1996*a*). Teixeira (1995) examines the role of Portugese ethnicity in the Toronto housing market. Mitchell (1996, 1997, 1998) documents the influx of Chinese immigrants into Vancouver during the 1990s. Feminist research examines the interplay between gender, place, and identity in Canadian, American, British, and Australian cities (Gilbert 1997). Studies of comparative racial segregation find that blacks are more residentially isolated in the United States than in Canada (Fong 1994, 1996). This suggests that more racially integrated neighborhoods exist in Canada, and that during periods of economic restructuring, minorities do better in Canadian than in American cities.

Some attention has been paid to the interplay between race, and gender and welfare provision. While this topic may fall within social geography, the research on Canadian social welfare has an overtly urban focus. For example, Laws (1989) examines the local impacts of privatization on mental health care provision. In the 1990s, she considered community responses to declining welfare provision and elder care in Toronto (Laws 1994*a*, *b*). Other studies examine the gender gap and spatial inequities in caregiving for the elderly (Hallman and Joseph 1997). Gender biases are also identified in recent welfare reform efforts because single mothers comprise a large proportion of welfare recipients in the United States and Canada (Merrett 1999). Comparisons between Canadian and American cities on poverty, homelessness, and racial segregation suggest that the local social safety-net is more supportive in Canadian cities (Broadway 1989, 1995; Daly 1991). But there is speculation that Canadian cities will become more like American cities if cuts to the Canadian welfare state continue (Dear and Wolch 1993; Lemon 1996*b*).

Given its prominence in North American political discourse, surprisingly few studies have examined the geography of Canada's vaunted but crumbling health-care system. One study documenting its decline shows how the Government of Ontario closed hospital beds to cut costs (Rosenberg and James 1994). As costs declined, spatial inequalities emerged with respect to health-care access. This undermined the cherished Canadian principle of universal access to health care. Ironically, as American health-care reformers promote the Canadian model, Canadian politicians have identified the American "for-profit" model as an option for Canada.

Canada's Regions: Cultural and Historical Perspectives

Four regions stand out in the research on Canada's cultural and historical geography: Quebec, the Prairies, Atlantic Canada, and Ontario. Paradoxically, Ontario, the most populous province, receives comparatively little attention. In addition to the regional focus, historical studies on Canada's population geography and interpretations of its built environment will be discussed.

The forces that perpetuate the distinct cultures of Acadia and French Canada have been perennially popular topics (LeBlanc 1988, 1993*a*, *b*). An entire issue of the *Journal of Cultural Geography* was devoted to Quebec (Beach 1988). A study of the broad bean as a cultural icon and practical artefact shows how this legume is an important cultural identifier in the Saguenay Valley of Quebec (Gade 1994). A study of Montreal's nineteenth-century transportation system reveals the impacts of new technologies on Montreal's neighborhoods (Boone 1996, 1997).

The Prairies also comprise a substantial research emphasis. Studies on early twentieth-century settlement patterns in Saskatchewan document challenges confronted by immigrant settlers and how the Canadian government used land to provide relief during the Great Depression (Bowen 1995*a*, *b*, 1998*a*, *b*). The focus on Saskatchewan also includes Wurtele's study of community change during the inter-war years.

In Manitoba, Darlington (1996, 1998) examined the ethnic impress, particularly by Ukrainians and Eastern Europeans, on the rural landscapes of that province. Christoph Stadel has collaborated with many geographers to examine the cultural geographies of Manitoba, and the demographics of prairie settlement systems (Stadel, 1992, 1993; Stadel *et al.* 1989; Stadel and Suida, 1995).

The border separating the Prairies and Great Plains has inspired several historical studies. Widdis (1997) outlines the development of a "borderlands" region, where the boundary simultaneously integrates and differentiates Canadians and Americans living close to the 49th Parallel. Affirming the existence of the borderlands, M. Schwartz (1997) documents the cross-border contacts between Canadian and American protest movements.

Research on Atlantic Canada has focused on ethnic, social, and economic components of Cape Breton's historical geography from the early 1700s into the 1800s (Hornsby, 1989, 1992*a*, *b*). Atlantic Canada is also considered in the magisterial work of Meinig (1986). And, Konrad (1990) "rediscovers" the historical geography of the land of Norumbega.

In Ontario, Teixeira (1995, 1997) examines Portugese communities in Canada, paying attention to their locations, social systems, and neighborhoods. Historical studies have examined post-1945 settlement patterns in Peterborough (Brunger *et al.* 1991), resource evaluation in Victorian Canada, and social welfare during the nineteenth century (Anderson 1992*a*, *b*, 1993).

Research into demographics and migration has focused on regional mortality rates in Canada and the United States during the nineteenth century (Innes 1988), infant mortality in Ottawa for 1901 (Mercier 1997), and migration in the Upper Canada frontier from 1841 to 1950 (Kinquist 1992). More recently, Kaplan (1995) and Newbold (1996, 1997) studied linguistic, economic, and regional factors prompting migration within Canada. Other research on migration focuses on methodology and micro-data availability in Canada, the United States, and Britain (Amrhein and Flowerdew 1992; Fotheringham and Pellegrini 1996). Among several chapters dealing with American railroads, Vance (1995) focuses on the development of Canada's railroad system after 1840. Also, Hudson (1988), Jordan (1989), and Meinig (1989, 1995) should be mentioned for their sweeping, if tangential accounts of Canadian development in the face of American continentalism.

Studies on material landscapes include the development of housing forms over the past three centuries, as well as housing for the elderly (Ennals and Holdsworth 1998; Holdsworth and Laws 1994; Holdsworth and Simon 1993). Bloomfield (1994) documents the geography of Tim Horton's Doughnuts stores, a powerful cultural and commercial symbol on the Canadian landscape. The book, *To Build a New Land*, examines early Canadian buildings, land division, and landscape change (Noble 1992). Other studies of functional landscapes focus on the utilitarian versus symbolic functions of Canadian and American grain elevators (Carney 1995), and the "last" of our needs from a cultural landscape, the tombstone (Norris 1988).

Physical Geography

Identifying research in physical geography is a challenge for two reasons. First, disciplinary boundaries in the earth sciences are fluid, making it difficult to differentiate between geographers, geomorphologists, and geologists. Second, most members of the CSSG are not physical geographers, despite the importance of physical geography within the discipline as a whole. Notwithstanding these caveats, we identify three themes in physical geography: (1) high altitude environments; (2) climate change; and (3) water resources.

The dynamic nature of mountain landscapes makes them ideal places to study downslope processes such as avalanches and sediment flows. Researchers have compared snow avalanche paths on opposite sides of the continental divide in Alberta and Montana (Butler 1989; Butler *et al.* 1992). Other research has focused on avalanche-induced sediment flows in glacial lakes in British Columbia (Weirich 1986). Explanations for the advance or retreat of glaciers rely on both climatic variables and landscape characteristics (Sprenke *et al.* 1999; Sloan and Dyke 1998).

Geographers have studied climate change by analyzing altitudinal and latitudinal treeline shifts in Canada. In the Rocky Mountains, Reasoner and Huber (1999) examine paleoclimate changes during the Holocene. At high latitudes, geographers have used dendrochronology to show that the Canadian treeline has moved northwards due to recent temperature increases (MacDonald *et al.* 1998). However, the utility of treeline fluctuation as a measure of long-term climate change remains unclear.

Canada has abundant fresh water. But temporal and spatial variations in water availability have prompted studies into water conservation practices among Prairie farmers (Kromm 1991, 1993). As precipitation varies, so does the flow of water through river channels. Leboutillier and Waylen (1993) develop a model to predict flow frequencies for rivers in British Columbia. One of Canada's great dam builders, is of course, the beaver. This symbol of Canadian identity is examined as an agent of landscape and hydrological change (Butler and Malanson 1994).

NAFTA has prompted a convergence of political, economic, and physical geographies, spurring considerable research on boundaries and shared water resources between the United States and Canada (e.g. Rabe and Zimmerman 1995). What is surprising, given the centrality of the human–environmental relationship in geography, is the lack of research conducted by geographers in this area. For example, no geographers contributed to a special issue of the *American Review of Canadian Studies* devoted to transborder environmental issues (A. Schwartz 1997). This is an area where geographers should lend their expertise to enliven the debate over sovereignty and natural-resource use (but see Irvine *et al.* 1995; Lawrence 1995).

Reflections on the Future of Canadian Studies

The future of Canadian Studies looks bright. Geographers who want to teach about Canada from an American perspective have many texts to choose from. In addition to Birdsall *et al.* (1999), authors such as Ennals (1993), Konrad (1996), and Rumney and Beach (1993), have all published texts on the geography of Canada or Canada as a part of North America. For those conducting research, extensive bibliographies have been published that emphasize the geography of Canada (Conzen *et al.* 1993; Rumney 1996, 1999*a*, *b*). The previous sections documented, we hope, the diversity and quantity of research that forms the foundation for future studies on Canada's geography. But we return to the question raised in the opening paragraph: is the idea of Canada, and hence Canadian Studies, an outdated and irrelevant approach? Empirical studies confirm that Canadian politics and culture have a significant and continuing impact on free trade, producer services, and urban geography (Goldberg and Mercer 1986; Gertler and DiGiovanna 1997; O'Grady and Lane 1997).

The national boundary even impedes the flow of information between Canada and the United States. Research on geographic knowledge "surfaces" shows that Canadians know more about far-away places in North America than do Americans (Curtis 1998). This finding points to a broader reality—that Canadians know more about America than Americans know about Canada. We conclude that the boundary between Canada and the United States remains a salient line of demarcation between two countries, not some cartographic relic, as Garreau (1981) asserts.

The durability of Canada within North America prompts us to think about the future of Canadian Studies. We predict both continuity and change. Over a decade ago, Romey (1989) exhorted geographers to pay attention to Canada because it was undergoing profound change. Canada had just implemented free trade with the

United States. Quebec had still not signed the repatriated constitution. An influx of immigrants was changing the complexion of the Canadian cultural mosaic. In the year 2000, these processes continue to beset Canada.

In reality, the issues of Canadian–American relations, Quebec sovereignty, and immigration have affected Canada since Confederation. These are perennial issues that will provide grist for geographic research far into the future too. The expansion of NAFTA into a hemispheric free trade area remains a distinct possibility. Separatist sentiments have subsided in Quebec, but they have not gone away. A resurgent Québécois nationalism is bound to affect Canada in the future. And like the United States, Canada is a country of immigrants. Geographers will have ample opportunity in the future to study ethnic assimilation as immigrants continue to settle in Canada.

But geographers are beginning to expand their research into new areas too. Issues such as global warming were not part of public discourse a decade ago. At the beginning of the twenty-first century, wide agreement exists that global temperatures have risen. The challenge now is to understand the causes and consequences of climate change. Answers to this question may lie in northern Canada. A recent article in the *New York Times* argues that global warming will be most pronounced in sub-Arctic regions (Stevens 2000). Physical geographers in America have taken a leading role by conducting research in the sub-Arctic regions of Canada (e.g. MacDonald *et al.* 1998).

In the future, geographers need to take an even greater role explaining the complexities of human–environmental relationships in North America. The intensification of, and interconnections between, global trade, international migrations, national identities, resource scarcity, and international security will cause problems of cultural and biological survival. With Canada's vast natural resources and sparse population, it will play an increasingly important role in North American and hemispheric affairs. And geographers in the United States are well positioned to articulate what this role might be.

References

Alwin, J. (1992). "North American Geographers and the Pacific Rim: Leaders or Laggards." *Professional Geographer*, 44/4: 369–76.

Amrhein, C., and Flowerdew, R. (1992). "The Effect of Data Aggregation on a Poisson Regression Model of Canadian Migration." *Environment and Planning A*, 24: 1381–91.

Anderson, R. (1992*a*). "Making Wilderness Smile: Professional Resource Evaluation in Victorian Canada." Unpublished Ph.D. dissertation, York University.

—— (1992*b*). "The Irrepressible Stampede: Tramps in Ontario, 1870–1880." *Ontario History*, 84/1: 32–56.

—— (1993). "Garbage Disposal in the Greater Toronto Area: A Preliminary Historical Geography." *Operational Geographer*, 11/1: 7–13.

Babe, R. (1988). "Copyright and Culture." *Canadian Forum*, 67 (February): 26–9.

Bagchi-Sen, S. (1999). "The Small- and Medium-Sized Exporter's Problems: An Empirical Analysis of Canadian Manufacturers." *Regional Studies*, 33/3: 231–45.

Bagchi-Sen S., and MacPherson, A. D. (1999). "The Competitive Characteristics of Small- and Medium-Sized Manufacturing Firms in the U.S. and Canada: Empirical Evidence from a Cross-border Region." *Growth and Change*, 30/3: 315–36.

Barnes, T. (1996). "External Shocks: Regional Implications of an Open Staple Economy," in J. Britton (ed.), *Canada and the Global Economy: The Geography of Structural and Technological Change*. (Montreal: McGill-Queen's University Press, 48–68.

Barsh, R. (1997). "Aboriginal Peoples and Quebec: Competing for Legitimacy as Emergent Nations." *American Indian Culture and Research Journal*, 22/1: 1–29.

Beach, R. (1988). "A travers les frontiers: Quebec and Quebec Studies." *Journal of Cultural Geography*, 8/2: 81–94.

Bertolas, R. (1995). "Cross-Cultural Environmental Perception Toward Wilderness and Development of James Bay, Quebec." Ph.D. dissertation, SUNY Buffalo.

Bhabha, H. (1994). *The Location of Culture*. New York: Routledge.

Birdsall, S., and Florin, J. (1992). *Regional Landscapes of the United States and Canada*. 4th edn. New York: John Wiley.

Birdsall, S., Florin, J., and Price, M. (1999). *Regional Landscapes of the United States and Canada*. 5th edn. New York: John Wiley & Sons.

Bloomfield, A. (1994). "Tim Hortons: Growth of a Canadian Coffee and Doughnut Chain." *Journal of Cultural Geography*, 14/2: 1–16.

Boone, C. (1996). "The Politics of Transportation Services in Suburban Montreal: Sorting Out the 'Mile End Muddle', 1893–1909." *Urban History Review*, 24/2: 25–39.

—— (1997). "Private Initiatives to Make Flood Control Public: The St. Historical Gabriel Levee and Railway Company in Montreal, 1886–1890." *Historical Geography*, 25: 100–12.

—— (1998*a*). "Forward to a Farm: The Back-to-the-Land Movement as a Relief Initiative in Saskatchewan During the Great Depression." Ph.D. dissertation, Queen's University.

—— (1998*b*) "Little Saskatoon: An Experiment in Land Settlement During the Great Depression." *Saskatchewan History*, 50/2: 10–28.

Bowen, D. (1995*a*). "Forward to a Farm: Land Settlement as Unemployment Relief in the 1930s" *Prairie Forum*, 20/2: 207–30.

— (1995*b*). "Preserving Tradition, Confronting Progress: Social Change in a Mennonite Community, 1959–1965." *American Review of Canadian Studies*, 25/1: 53–77.

Bowler, I. (1994). "The Institutional Regulation of Uneven Development: The Case of Poultry Production in the Province of Ontario." *Transactions of the Institute of British Geographers*, 19/3: 346–58.

Bradshaw, M. (1997). *The New Global Order: A World Regional Geography*. Chicago: Brown & Benchmark.

Broadway, M. (1989). "A Comparison of Patterns of Deprivation Between Canadian and U.S. Cities." *Social Indicators Research*, 21: 531–51.

— (1995). "The Canadian Inner City 1971–91: Regeneration and Decline." *Canadian Journal of Urban Research*, 4/1: 1–20.

Broadway, M. (1997). "Alberta Bound: Canada's Beef Industry." *Geography*, 82/4: 377–9.

Brunger, A., Bowles, R., and Wurtele, S. (1991). "Patterns of Fringe Community: Post-World War II Settlement Near Peterborough, Ontario." in Kenneth B. Beesley (ed.), *Rural and Urban Fringe Studies in Canada*, North York: York University, Department of Geography, Monograph 21: 183–207.

Brunn, S., Yeates, M., and Zeigler, D. (1993). "Cities of the United States and Canada." in S. Brunn and J. Williams (eds.), *Cities of the World: World Regional Urban Development*. 2nd edn. New York: Harper Collins.

Butler, D. (1989). "Rockfall Hazard Inventory, Ram River, Mackenzie Mountains." *Canadian Geographer*, 33/3: 269–73.

Butler, D., and Malanson, G. (1994). "Beaver Landforms." *Canadian Geographer*, 38/1 (Spring): 76–9.

Butler, D., Malanson, G., and Walsh, G. (1992). "Snow-Avalanche Paths: Conduits from the Periglacial-Alpine to the Subalpine-Depositional." *Periglacial Geomorphology: Proceedings of the 22nd Annual Symposium in Geomorphology*. New York: Wiley, 185–202.

Carney, G. (1995). "Grain Elevators in the United States and Canada: Functional or Symbolic." *Material Culture*, 27/1: 1–24.

Carty, R., and Eagles, M. (1998). "The Political Ecology of Local Party Organization: The Case of Canada." *Political Geography*, 17/3: 589–609.

Chouinard, V. (1989). "Social Reproduction and Housing Alternatives: Cooperative Housing in Post-War Canada." in J. Wolch and M. Dear (eds.), *The Power of Geography: How Territory Shapes Social Life*. Boston: Unwin Hyman, 222–37.

Clement, W., and Williams, G. (1989). "Introduction", in W. Clement and G. Williams (eds.), *The New Canadian Political Economy*. Montreal: McGill-Queen's University Press, 3–15.

Coffey, W. (1995). "Producer Services Research in Canada." *Professional Geographer*, 47/1: 74–81.

Conzen, M., Rumney, T., and Wynn, G. (1993). *A Scholar's Guide to Geographical Writing on the American and Canadian Past*. Geography Research Paper, 235. Chicago: University of Chicago.

Cornish, S. (1997). "Product Innovation and the Spatial Dynamics of Market Intelligence: Does Proximity to Markets Matter?" *Economic Geography*, 73: 143–65.

— (1997). "Strategies for the Acquisition of Market Intelligence and Implications for the Transferability of Information Inputs." *Annals of the Association of American Geographers*, 87/3: 451–70.

Curtis, A. (1998). "Comparisons in the Spatial Knowledge Surfaces of Subjects from Canada and the United States." *Canadian Geographer*, 42/1: 53–61.

Daly, G. (1991). "Local Programs Designed to Address the Homelessness Crisis: A Comparative Assessment of the United States, Canada and Britain." *Urban Geography*, 12/2.

Darlington, J. (1996). "To Cluster or Disperse? One Immigrant Group's Response to the Rectangular Land Survey System of the Canadian Prairies." *Transactions, Pioneer American Society*, 19: 1–10.

— (1998). "The Ukrainian Impress on the Canadian West." in F. Incovetta, P. Draper, and R. Ventresca (eds.), *A Nation of Immigrants: Women, Workers, and Communities in Canadian History, 1840s–1960s*. Toronto: University of Toronto Press, 128–62.

Dear, M., and Wolch, J. (1993). "Homelessness." in L. Bourne and D. Ley (eds.), *The Changing Social Geography of Canadian Cities*. Montreal: McGill-Queen's University Press, 298–308.

Elkins, D. (1995). *Beyond Sovereignty: Territory and Political Economy in the Twenty-First Century*. Toronto: University of Toronto Press.

Ennals, P. (ed.) (1993). *The Canadian Maritimes: Images and Encounters*. Indiana, Pa.: National Council for Geographic Education.

Ennals, P., and Holdsworth, D. (1998). *Homeplace: The Making of the Canadian Dwelling Over Three Centuries*. Toronto: University of Toronto Press.

Ewing, G. (1992). "The Bases of Difference Between American and Canadian Cities." *Canadian Geographer*, 36/3: 266–279.

Fong, E. (1994). "Residential Proximity Among Racial Groups in U.S. and Canadian Neighborhoods." *Urban Affairs Quarterly*, 30: 285–97.

— (1996). "A Comparative Perspective on Racial Residential Segregation: American and Canadian Experiences." *Sociological Quarterly*, 37/2: 199–226.

Fotheringham, A., and Pellegrini, P. (1996). "Microdata for Migration Analysis: A Comparison of Sources in the US, Britain and Canada." *Area*, 28/3: 347–57.

Frank, A. (1979). *Dependent Accumulation and Underdevelopment*. New York: Monthly Review Press.

Frederic, P. (1995). "Maine Natural Resource-Based Manufacturing Firms: Canadian Trade and Sustainability." in C. Bryant and C. Marois (eds.), *The Sustainability of Rural Systems*. Montreal: University of Montreal Press, 331–44.

Gade, Daniel (1994). "Environment, Culture, and Diffusion: The Broad Bean in Quebec." *Cahiers de Geographie du Quebec*, 38/104: 137–50.

Garreau, J. (1981). *Nine Nations of North America*. New York: Avon.

Gertler, M., and DiGiovanna, S. (1997). "In Search of the New Social Economy: Collaborative Relations Between Users and Producers of Advanced Manufacturing Technologies." *Environment and Planning A*, 29: 1585–602.

Gilbert, M. (1997). "Feminism and Difference in Urban Geography." *Urban Geography*, 18/2: 166–79.

Goldberg, M., and Mercer, J. (1986). *The Myth of the North American City: Continentalism Challenged*. Vancouver: University of British Columbia Press.

Gray, D. (1994). "Canada's Unresolved Maritime Boundaries." *Geomatica*, 48/2: 131–44.

Hallman, B., and Joseph, A. (1997). "Exploring Distance and Caregiver Gender Effects in Eldercare: Towards a Geography of Family Caregiving." *Great Lakes Geographer*, 4/2: 15–29.

Harrington, J. (1989). "Implications of the Canada—United States Free Trade Agreement for Regional Provision of Producer Services." *Economic Geography*, 65: 314–28.

— (1992). "Determinants of Bilateral Operations of Canadian and U.S. Commercial Banks." *Environment and Planning A*, 24: 137–51.

Harrington, J., Burns, K., and Cheung, M. (1986). "Market-Oriented Foreign Investment and Regional Development: Canadian Companies in Western New York." *Economic Geography*, 62: 155–66.

Harris, C. (1997). "The Nature of Cities and Urban Geography in the Last Half Century." *Urban Geography*, 18/1: 15–35.

Harris, R. (1987). "Regionalism and the Canadian Archipelago," in L. McCann (ed.), *Heartland and Hinterland: A Geography of Canada*. 2nd Edition. Scarborough: Prentice-Hall, 532–59.

Hayward, D., and Erickson, R. (1995). "The North American Trade of U.S. States: A Comparative Analysis of Industrial Shipments." *International Regional Science Review*, 18/1: 1–31.

Holdsworth, D., and Laws, G. (1994). "Landscapes of Old Age in Coastal British Columbia." *Canadian Geographer*, 38/2: 174–81.

Holdsworth, D., and Simon, J. (1993). "Housing Form and the Use of Domestic Space." in John Miron (ed.), *House, Home and Community: Progress in Housing Canadians, 1945–1986*. Montreal: McGill-Queen's University Press, 188–202.

Holmes, J. (1988). "Industrial Restructuring in a Period of Crisis: An Analysis of the Canadian Automobile Industry." *Antipode*, 20: 19–51.

Hornsby, S. (1989). "Staple Trade, Subsistence Agriculture and Nineteenth-Century Cape Breton Island." *Annals of the Association of American Geographers*, 79/3: 411–34.

— (1992*a*). *Nineteenth-Century Cape Breton: A Historical Geography*. Montreal: McGill-Queen's University Press.

— (1992*b*), "Patterns of Scottish Emigration to Canada, 1750–1870." *Journal of Historical Geography*, 18/4: 397–416.

Hudson, J. (1988). "North American Origins of Middle Western Frontier Populations." *Annals of the Association of American Geographers*, 78/3: 395–413.

Innes, F. (1988). "Differential Regional Mortalities: Some Canadian and American Contrasts in the Nineteenth Century." *Geoscience and Man*, 25: 41–56.

Irvine K., Pratt, E., Leonard, M., Taylor, S., and McFarland, K. (1995). "Impacts of Fluctuating Water Levels and Flows to Hydropower Production on the Great Lakes: Planning for the Extremes." *Great Lakes Geographer*, 2/1: 67–85.

Jenkins, R. (1998). "Untangling the Politics of Electoral Boundaries in Canada." *American Review of Canadian Studies*, 28/4: 517–38.

Jordan, T. (1989). "Preadaptation and European Colonization in Rural North America." *Annals of the Association of American Geographers*, 79/4: 489–500.

Kaplan, D. (1994*a*). "Population and Politics in a Plural Society: The Changing Geography of Canada's Linguistic Groups." *Annals of the Association of American Geographers*, 84/1: 46–67.

— (1994*b*). "Two Nations in Search of a State: Canada's Ambivalent Spatial Identities." *Annals of the Association of American Geographers*, 84/4: 585–606.

— (1995). "Differences in Migration Determinants for Linguistic Groups in Canada." *Professional Geographer*, 47/2: 115–25.

Kindleberger, C. (1987). *International Capital Movements*. New York: Cambridge University Press.

Kinquist, C. (1992). "Migration and Madness on the Upper Canadian Frontier, 1841–1850." In Donald H. Akensen (ed.), *Canadian Papers in Rural History*. Gananoque: Langdale Press, Viii. 129–162.

Konrad, V. (1996). *Geography of Canada*. East Lansing: Michigan State University Press.

— (1990). "Observations: Rediscovering Historical Geography in the Land of Norumbega." *Canadian Geographer*, 34: 277–9.

Kromm, D. (1994). "Water Conservation in the Irrigated Prairies of Canada and the United States." *Canadian Water Resources Journal*, 18/4: 451–8.

— (1991). *On-Farm Adoption of Water Saving Practices in Southern Alberta*. Occasional Paper, 1. Lethbridge: University of Lethbridge, Water Resources Institute.

Laidman, S. (1998). "The Geography of Canadian Federal Elections: An Analysis of the 1988 and 1993 Federal Elections." Master's thesis, Vermont.

Lawrence P. (1995). "Great Lakes Shoreline Management in Ontario." *Great Lakes Geographer*, 2/2: 1–20.

Laws, G. (1989). "Privatization and Dependency on the Local Welfare State," in J. Wolch and M. Dear (eds.), *The Power of Geography: How Territory Shapes Social Life*. Boston: Unwin Hyman, 238–57.

— (1994*a*). "Community Activism Around the Built Form of Toronto's Welfare State." *Canadian Journal of Urban Research*, 3/1: 1–28.

— (1994*b*). "Aging, Contested Meanings, and the Built Environment," *Environment and Planning A*, 26/11: 1787–802.

LeBlanc, R. (1988). "A French-Canadian Education and the Persistence of la Franco-Americaine." *Journal of Cultural Geography*, 8/2: 49–64.

— (1993*a*). "The Acadian Migrations," in D. Louder and E. Waddell (eds.), *French America: Mobility, Identity and Minority Experience Across the Continent*. Baton Rouge: Louisiana State University Press, 164–90.

— (1993*b*). "The Franco-American Response to the Conscription Crisis in Canada, 1916–1918." *American Review of Canadian Studies*, 23/3: 343–72.

Leboutillier, D., and Waylen, P. (1993). "Regional Variations in Flow-Duration Curves for Rivers in British Columbia, Canada." *Physical Geography*, 14/4: 359–78.

Lemon, J. (1996). "Liberal Dreams and Nature's Limits: Great Cities of North American Since 1600." *Annals of the Association of American Geographers*, 86/4: 745–66.

Lipset, S. (1990). *Continental Divide: The Values and Institutions of the United States*. New York: Routledge.

Lo, M., and Teixeira, C. (1998). "If Quebec Goes . . . The 'Exodus' Impact." *Professional Geographer*, 50/4: 481–98.

McCann, L. (1987). "Heartland and Hinterland: A Framework for Analysis," in L. McCann (ed.), *Heartland and Hinterland: A Geography of Canada*. 2nd edn. Scarborough: Prentice-Hall, 3–37.

MacDonald, G., Szeicz, J., Claricoates, J., and Dale, K. (1998). "Response of the Central Canadian Treeline to Recent Climatic

Changes." *Annals of the Association of American Geographers*, 88/2: 183–209.

MacDougall, E. (1970). "An Analysis of Recent Changes in the Number of Farms in the North Part of Central Ontario," *The Canadian Geographer*, 14/2: 125–38.

McNaughton, R. (1992). "The Export Behavior of Small Canadian Firms." *Professional Geographer*, 44/2: 170–80.

MacPherson, A. (1991). "New Product Development Among Small Industrial Firms: A Comparative Assessment of the Role of Technical Service Linkages in Toronto and Buffalo." *Economic Geography*, 67: 136–46.

— (1996). "Shifts in Canadian Direct Investment Abroad and Foreign Direct Investment in Canada," in J. Britton (ed.), *Canada and the Global Economy: The Geography of Structural and Technological Change*. Montreal: McGill-Queens University Press, 69–83.

— (1997). "The Role of Producer Service Outsourcing in the Innovation Performance of New York State Manufacturing Firms." *Annals of the Association of American Geographers*, 87/1: 52–71.

MacPherson, A., and McConnell, J. (1992). "Recent Canadian Direct Investment in the United States: An Empirical Perspective from Western New York." *Environment and Planning A*, 24: 121–36.

Malecki, E. (1991). *Technology and Economic Development: The Dynamics of Local, Regional and National Change*. New York: Longman.

Meinig, D. (1986). *The Shaping of America. A Geographical Perspective on 500 Years of History*, i. *Atlantic America, 1492–1800*. New Haven: Yale University Press.

— (1989). "The Historical Geography Imperative." *Annals of the Association of American Geographers*, 79/1: 79–87.

— (1995). *The Shaping of America. A Geographical Perspective on 500 Years of History*, ii. *Continental America, 1800–1867*. New Haven: Yale University Press.

Melamid, A. (1994). "Geographical Record: International Trade in Natural Gas." *Geographical Review*, 84/2: 216–21.

Mercier, M. (1997). "Infant Mortality in Ottawa, 1901." Master's thesis, Carleton University.

Merrett, C. (1991). *Crossing the Border: The Canada-United States Boundary*. Borderlands Monograph Series. Orono: University of Maine Press.

Merrett, C. (1996). *Free Trade: Neither Free Nor About Trade*. Montreal: Black Rose.

— (1997). "Nation-States in Continental Markets: The Political Geography of Free Trade." *Journal of Geography*, 96/2: 105–12.

— (1998*a*). "Review of J. Britton (ed.), *Canada and the Global Economy: The Geography of Structural and Technological Change*." *Canadian Geographer*, 42/2: 205–7.

— (1998*b*). "The Rhetoric and Rationale of Free Trade: A Political Geography Perspective," in F. Davidson, J. Leib, F. Shelley, and F. Webster (eds.), *Teaching Political Geography*. Indiana, Pa.: National Council for Geographical Education, 83–90.

— (1999). "The Geography of Welfare Reform in Canada and the United States." Paper presented at the Conference of the Association for Canadian Studies in the United States, Pittsburgh, Pa. 18 November.

Meserve, P. (1991). "Boundary-Water Issues Along the Forty-Ninth Parallel," in R. Lecker (ed.), *Borderlands: Essays in Canadian-American Relations*. Toronto: ECW Press, 80–112.

Michalak, W., and Gibb, R. (1997). "Trading Blocs and Multilateralism in the World Economy." *Annals of the Association of American Geographers*, 87/2: 264–79.

Mitchell, K. (1996). "Visions of Vancouver: Ideology, Democracy and the Future of Urban Development." *Urban Geography*, 17/6: 478–501.

— (1997). "Conflicting Geographies of Democracy and the Public Sphere in Vancouver, British Columbia." *Transactions, Institute of British Geographers*, 22/2: 162–70.

— (1998). "Reworking Democracy: Contemporary Immigration and Community Politics in Vancouver's Chinatown." *Political Geography*, 17/6: 729–50.

Newbold, K. (1996). "Income, Self-Selection, and Return and Onward Interprovincial Migration in Canada." *Environment and Planning A*, 28/6: 1019–34.

— (1997). "Primary, Return, and Onward Migration in the US and Canada: Is There a Difference?" *Papers in Regional Science*, 76/2: 175–98.

Noble, A. (1992). *To Build in a New Land: Ethnic Landscapes in North America*. Baltimore: Johns Hopkins University Press.

Norris, D. (1988). "Ontario Gravestones." *Markers*, 5: 122–49.

Nissen, E. (1997). "Rating Metro Areas in the United States and Canada for the Arts and Recreation." *Journal of Regional Analysis and Policy*, 27/1: 47–54.

O'Grady, S., and Lane, H. (1997). "Culture: An Unnoticed Barrier to Canadian Retail Performance." *Journal of Retailing and Consumer Services*, 4/3: 159–70.

Pratt, G. (1989). "Incorporation Theory and the Reproduction of Community Fabric," in J. Wolch and M. Dear (eds.), *The Power of Geography: How Territory Shapes Social Life*. Boston: Unwin Hyman, 293–316.

Pratt, T. (1997). "The Multicultural Mosaic in Canada: Sovereignty, Identity, and Cultural Hegemony." Master's thesis, University of Kansas.

Raad, T., and Kenworthy, J. (1998). "The U.S. and Us: Canadian Cities Are Going the Way of Their U.S. Counterparts into Car-Dependent Sprawl." *Alternatives*, 24/1: 14–22.

Rabe, B., and Zimmerman, J. (1995). "Regime Emergence in the Great Lakes Basin." *International Environmental Affairs*, 7/4: 346–63.

Reasoner, M., and Huber, U. (1999). "Postglacial Paleoenvironments of the Upper Bow Valley, Banff National Park, Alberta, Canada." *Quarternary Science Reviews*, 18/3: 475–92.

Rigby, D. (1990). "Regional Differences in Manufacturing Performance: The Case of the Canadian Food and Beverage Industry." *Environment and Planning A*, 22: 79–100.

— (1991). "The Existence, Significance, and Persistence of Profit Rate Differentials." *Economic Geography*, 67: 210–22.

Robinson, D., and Long, B. (1989). "Trends in Latin Americanist Geography in the United States and Canada." *Professional Geographer*, 41/3: 304–14.

Rogers, R., and Stewart, C. (1997). "Prisoners of Their Histories: Canada–U.S. Conflicts in the Pacific Salmon Fishery." *American Review of Canadian Studies* 27/2: 253–69.

Romey, W. (1989). "Study of Canadian Geography," in G. Gaile and C. Willmott (eds.), *Geography in America*. Columbus: Merrill, 563–80.

Rose, D. (1989). "A Feminist Perspective of Employment Restructuring and Gentrification: The Case of Montreal," in

J. Wolch and M. Dear (eds.), *The Power of Geography: How Territory Shapes Social Life*. Boston: Unwin Hyman, 118–38.

Rosenberg, M., and James, A. (1994). "The End of the Second Most Expensive Health Care System in the World: Some Geographical Implications." *Social Science and Medicine*, 39/7: 967–81.

Rumney, T. (1999*a*). *Geography of Canada Bibliography Series*, ii. *Prairies*. Plattsburgh: Plattsburgh State University, Center for the Study of Canada.

— (1999*b*). *Geography of Canada Bibliography Series*, iii. *Atlantic Canada*. Plattsburgh: Plattsburgh State University, Center for the Study of Canada.

Rumney, T., and Beach, R. (1993). "Canadian Geography Studies in the United States," in K. Gould, J. Jockel, and W. Metcalfe (eds.), *Northern Exposures: Scholarship on Canada in the United States*. Washington: Association for Canadian Studies in the United States, 321–40.

Said, E. (1979). *Orientalism*. New York: Vintage.

Savageau, D., and Boyer, R. (1993). *Places Rated Almanac*. Rev. edn. New York: Prentice-Hall Travel.

Schwartz, A. (ed.) (1997). "Theme Issue. Red, White and Green: Canada-U.S. Environmental Relations." *The American Review of Canadian Studies*, 27/3: 321–488.

Schwartz, M. (1997). "Cross-Border Ties Among Protest Movements: The Great Plains Connection." *Great Plains Quarterly*, 17/2: 119–30.

— (1998). "NAFTA and the Fragmentation of Canada." *American Review of Canadian Studies*, 28/1–2: 11–28.

Sloan, V., and Dyke, L. D. (1998). "Decadal and Millenial Velocities of Rock Glaciers, Selwyn Mountains, Canada." *Geografisak Anneler, Series A: Physical Geography*, 80/3–4: 237–49.

Sluyter, A. (1997). "On 'Buried Epistemologies': The Politics of Nature in (Post) Colonial British Columbia." *Annals of the Association of American Geographers*, 87/4: 700–2.

Solocha, A., Soskin, M., and Krasoff, M. (1989). "Canadian Investment in Northern New York: Impact of Exchange Rates on Foreign Direct Investment." *Growth and Change*, 20/1: 1–16.

Sparke, M. (1995). "Between Demythologizing and Deconstructing the Map: Shawnadithit's New-Found-Land and the Alienation of Canada." *Cartographica*, 32/1: 1–21.

— (1996). "Negotiating National Action: Free Trade, Constitutional Debate and the Gendered Geopolitics of Canada." *Political Geography*, 15/6: 615–39.

— (1997). "Review of *Free Trade: Neither Free Nor About Trade* by Christopher Merrett." *Canadian Geographer*, 41/2: 216–17.

— (1998). "A Map that Roared and an Original Atlas: Canada, Cartography, and the Narration of Nation." *Annals of the Association of American Geographers*, 88/3: 463–95.

Sprenke, K., Miller, M., McGee, S., Adema, G., and Lang, M. (1999). "The High Plateau of the Juneau Icefield, British Columbia: Form and Dynamics." *Canadian Geographer*, 43/1: 99–103.

Stadel, C. (1992). "The Seasonal Resort of Wasagaming, Riding Mountain National Park, Manitoba," in H. John Selwood and John C. Lehr (eds.), *Reflections from the Prairies: Geographical Essays*. Winnipeg: University of Winnipeg, Department of Geography, 61–72.

— (1993). "Continuity and Change of a Non-Primate City in the Canadian Prairies: The Example of Brandon, Manitoba." *Die Erde*, 124/3: 225–36.

Stadel, C., and Suida, H. (eds.) (1995). *Theses and Issues of Canadian Geography*, bk. 28 vol. i (Salzburg: Salzburger Geographische Arbeiten.

Stadel, C., Kinnear, M., and Everitt, John C. (1989). "Recreation Homes and Hinterlands in Southwest Manitoba: The Example of Minnedos Lake Developments." *Saskatchewan Geography*, 2: 23–9.

Stevens, W. (2000). "Global Warming: The Contrarian View." *The New York Times*, 29. February (‹http://www.partners.nytimes.com/library/national/science/022900sci-environ-climate.html›, last accessed 25 February 2003).

Teixeira, C. (1995). "Ethnicity, Housing Search, and the Role of the Real Estate Agent: A Study of Portugese and Non-Portugese Real Estate Agents in Toronto." *Professional Geographer*, 47/2: 176–83.

— (1997). "The Suburbanization of Portugese Canadians in Toronto." *Great Lakes Geographer*, 4/1: 25–39.

Tremblay, R. (1988). "250 Economists Say it's Our Best Option." *Canadian Speeches*, 2 (Aug./Sept.): 5–6.

Vance, J. (1995). *The North American Railroad: Its Origin, Evolution and Geography*. Baltimore: Johns Hopkins University Press.

Vesilind, P. (1990). "Common Ground, Different Dreams: The U.S.–Canada Border." *National Geographic*, 177/2.

Warf, B., and Cox, J. C. (1992). "The U.S.-Canadian Free Trade Agreement and North American Maritime Trade." *Maritime Policy & Management*, 19/1: 19–29.

— (1993). "The U.S.-Canada Free Trade Agreement and Commodity Transportation Services Among U.S. States." *Growth and Change*, 24/3: 341–64.

Watkins, M. (1989). "The Political Economy of Growth," in W. Clement and G. Williams (eds.), *The New Canadian Political Economy*. Montreal: McGill-Queen's University Press, 17–35.

White, C., Foscue, E., and McKnight, T. (1979). *Regional Geography of Anglo-America*. 5th edn. Englewood Cliffs: Prentice-Hall.

Widdis, R. (1997). "Borderland Interaction in the International Region of the Great Plains: An Historical-Geographical Perspective." *Great Plains Research*, 7/1: 103–37.

Weirich, F. (1986). "A Study of the Nature and Incidence of Density Currents in a Shallow Glacial Lake." *Annals of the Association of American Geographers*, 76/3: 396–413.

Willems-Braun, B. (1997*a*). "Buried Epistemologies: The Politics of Nature in (Post) Colonial British Columbia." *Annals of the Association of American Geographers*, 87/1: 3–31.

— (1997*b*). "Reply: On Cultural Politics, Sauer, and the Politics of Citation." *Annals of the Association of American Geographers*, 87/4: 703–8.

Wynn, G. (1987). "Forging a Canadian Nation," in R. Mitchell and P. Groves (eds.), *North America: The Historical Geography of a Changing Continent*. Totowa: Rowman & Littlefield, 373–409.

Yeates, M., and Garner, B. (1980). *The North American City*. 3rd edn. New York: Harper & Row.

CHAPTER 41

Geography of China

C. Cindy Fan, Laurence J. C. Ma, Clifton W. Pannell, and K. C. Tan

The pioneering work of North American geographers in the 1970s and early 1980s has laid a sound foundation for research and fieldwork opportunities in China (Karan *et al.* 1989), a nation largely closed to academics and others until the late 1970s. Building on this foundation, research on China geography since the last decade or so has witnessed phenomenal achievements in mass, diversity, and intellectual depth.

Geographers have for a long time been captivated by China's sheer size. But their recent scholarship must also be understood against the backdrop of that nation's groundbreaking reforms and transformations since the late 1970s. China's decisive integration into the world economy, in conjunction with its sweeping processes of socialist transition, has fascinated geographers all over the world. As this review will show, these changes have indeed defined the priorities and research agendas of many North American geographers. Their research has been significantly facilitated and stimulated by extensive and increasing contacts with scholars in China, through long-standing collaborations, fieldwork, and international conferences especially since 1979 when exchange of ideas and visits among North American and Chinese geographers have flourished.[1] China's opening its doors has also motivated a large number of excellent Chinese students to pursue advanced training in North American institutions, who have added considerably to the human resources of China geography there.

The research output of North American China geographers is outstanding, which parallels the growth of China geography in general (Selya 1992*b*). The scope of their attention, however, is also uneven. In this review, we attempt to highlight achievements and dominant themes of inquiry since the late 1980s, as well as areas for future improvement.[2]

[1] Authors are listed alphabetically. Among the geographers on the Chinese mainland who have long collaborated with, or who have otherwise had significant impact on, North American geographers are Cai Qiangguo, Cai Yunlong, Cui Gonghao, Gu Chaolin, Hou Renzhi, Hu Zhaoliang, Liu Changming, Liu Peitong, Ning Yuemin, Shen Daoqi, Wang Ying, Wu Chuanjun, Xu Xueqiang, Yan Xiaopei, Yan Zhongmin, Yao Shimou, Ye Shunzan, Zhao Songqiao, and Zhou Yixing. At the same time, many geographers in Hong Kong and Taiwan have also had considerable impact on the work of North American geographers.

[2] This chapter focuses primarily on peer-reviewed publications and significant contributions in English by North American geographers during the period 1989 to 1999. The bibliography is selective and not intended to be exhaustive. The reader should be reminded that the Asian Geography Specialty Group has a separate chapter that deals with the remainder of Asia excluding China, and additional scholarly insights on recent geographic research on Asia may be reviewed in that chapter.

Economic and Spatial Restructuring

Recent research on China's economic geography has concentrated on two substantive foci: regional development and foreign investment-induced growth. Both are rooted in China's economic transformations and have far-reaching spatial implications.

Regional Impacts of the Reforms

An overriding question geographers ask is in what ways China's economic reforms have impacted the regional disparity in income, industrial output, investment, and processes leading to these changes. Veeck's (1991) edited volume, *The Uneven Landscape*, represents one of the earlier efforts in examining the spatially uneven results of China's reforms. In both empirical and conceptual terms, research in this subfield has significantly gained sophistication. Attempts to measure the extent and changes of regional inequality range from earlier studies comparing a handful of provinces to multi-scalar approaches encompassing interregional, interprovincial, and intraprovincial analyses (Fan 1995*a*; Wei 1999). A body of research has produced some definitive findings, namely, the regional gap between eastern or coastal China and western or interior China has widened, interprovincial inequality declined during the 1980s, the trend of intraprovincial inequality varied from one province to another, and rural-urban inequality has increased (Fan 1995*a*, *b*; Leung 1991; Lo 1989*a*, *b*; Shen 1999; Su and Veeck 1995; Wei 1998; Wei and Ma 1996; Ying 1999). Fan (1995*b*), Wei (1998), and Wei and Ma (1996) point out that the paradox of decline of interprovincial inequality amidst heightened uneven regional development during the 1980s can be explained by the rapid growth of a new coastal core/corridor, whose effect was offset by the lagging growth of old traditional industrial centers also along the coast and especially in the northeast. Continued growth of the new core/corridor is likely to bring about regional divergence in economic development in the future.

Most subprovincial studies have focused on Guangdong (Fan 1995*a*, *b*; C. M. Luk 1991) and Jiangsu (Fan 1995*a*; Ma and Fan 1994; Su and Veeck 1995), reflecting multiplier processes in scholarship, data availability, and fieldwork opportunities. Research on the Pearl River Delta is especially noteworthy (Lin 1997; Lo 1989*a*, *b*). Lin's (1997) book, *Red Capitalism in South China*, offers an in-depth spatial analysis of economic growth and transformation in that region, focusing on market forces, the role of Hong Kong, and changes in basic industrial and transport infrastructure. A special issue of *Chinese Environment and Development* (1995) is devoted to Jiangsu, including specific studies examining the widening gap between Subei (northern Jiangsu) and Sunan (southern Jiangsu). Analyses of intraprovincial inequality in Shanxi (Leung 1991), Fujian, Zhejiang, Anhui, and Hunan (Fan 1995*a*) have also been conducted. Studies on Taiwan are fewer, with some attention on changes in regional development (Selya 1993) and the industrial and service economies (Selya 1994*b*; Todd and Hsueh 1992).

Though we are still far from developing a cohesive theory of Chinese regional development, recent research has offered some powerful explanations and potential conceptual frameworks. Researchers point to shifts in political and development philosophies, especially China's abandonment of the socialist ideology of equality and adoption of comparative advantage as a guiding principle of regional development (Fan 1997; Ma and Wei 1997). Explanations have also focused on specific agents and media of development, namely, the central state, local entrepreneurs and governments, small towns, and foreign investment (see below). Fan (1995*a*) interprets their effects on regional development using the "development from above, below, and outside" framework (Shen 1999), and Wei (1999) argues for a multi-mechanism research approach.

Political-economic and/or institutional approaches have been popularized by scholars who argue that the central state remains a key determining factor of regional economic growth, through balanced development strategies (Xie and Dutt 1991), preferential policies (Fan 1995*b*), investment (Ma and Wei 1997; Shen 1999; Su and Veeck 1995; Wei 1995*b*), development plans (Leung 1991), promotion of market forces through urban centers (Tan 1990), and fiscal decentralization (Tang *et al.* 1993; Wei 1996). Recent studies have sought to emphasize forces at work below the central-state level, including local entrepreneurs and human resources (Leung 1996; Shen 1999), small towns and cities (Lin and Ma 1994; Ma and Lin 1993; Tan 1991), and rural collective and/or industrial enterprises (Fan 1995*a*; Lo 1990; Marton 1995; Veeck and Pannell 1989). More attention is needed for identifying the micro-level agents and processes that explained for changes in the Chinese spatial economy.

Foreign Investment

There is no question that foreign investment and trade have emerged as a significant factor of China's quest for rapid economic growth. Geographers are especially interested in examining the spatial distribution of foreign investment and determinants of that distribution. Some fine analytical work includes those examining the roles of state policy and incentives, urban infrastructure, accessibility, urbanization and agglomeration economies, social affinity (Gong 1995; Leung 1990, 1996), and the regional impacts of foreign trade (Fan 1992). There is overall agreement that foreign investment and trade have accelerated urban growth and exacerbated regional disparities of economic development (Fan 1992; Leung 1990; Xie and Costa 1991).

Notable in this body of literature are studies that highlight the critical roles of business practices, kinship ties, and cultural affinities in reducing transaction costs and channeling foreign investment toward selected localities (Cartier 1995; Wu 1997). Hsing (1995, 1996) has done some intriguing analyses on the provocative and sometimes sensitive topic of Taiwanese investment in mainland China. Her 1998 book, *Making Capitalism in China: The Taiwan Connection*, is an excellent example of the high-quality analytical work done by geographers on the complex political economy of foreign investment in China and its murky conflation with politics and business ties in Taiwan. Leung (1993) examines how industrial development and personal linkages interlock in a close-knit and effective manner in Hong Kong's neighbor, the Pearl River Delta, and provides a conceptual precursor to the above-cited works of Hsing and Lin. The 1998 essay of Hayter and Han investigates China's open door policy on foreign investment and its linkage to technology transfers through Hong Kong, Taiwan, and Singapore. The idea of the increasing recognition of that spatial reality termed "greater China" is implicit if not explicitly manifested here in their essay.

Processes of Urbanization and Population Movements

Research on Chinese cities and urbanization has always attracted the largest number of China geographers who, as in the previous decades, continue to produce an impressive corpus of literature (Pannell 1990). It would not be an exaggeration to claim that, outside of China, North American geographers have done more work on Chinese cities and urban development than scholars in any other social sciences discipline. Together with increasing attention on migration, which intertwines with processes of urbanization, this body of literature has shed much light on various aspects of individual cities, the urban system, and urbanization during socialist transition.

Definitions, Urban Growth, and Rural–Urban Relations

Geographers have been at the forefront of research to unravel the mystery of the shifting definitions and complex typologies of China's urban and rural populations and to ascertain their sizes. Pioneer work in the 1980s has helped social scientists examine for the first time the intricate methodological issues involved in studying China's urban system. Important earlier research on urban policy and urbanization have been reviewed and brought up to date by Chan (1992, 1994*a*, *c*), Chan and Zhang (1999), Tan (1993*a*), and L. Zhang and Zhao (1998). It should be noted that this line of research represents essential groundwork that must be accomplished and sustained before the nature of China's urbanization, either in the socialist or reform period or in the future, can be meaningfully analyzed.

A large number of individual cities have been studied from different perspectives (Chang 1998; Fung 1996; Fung *et al.* 1992; Hsu 1996), including comprehensive monographs on Hong Kong (Lo 1992), Taipei (Selya 1995), Shanghai, Tianjin, and Guangzhou (Yusuf and Wu 1997). Significant progress has been made in documenting the morphology and internal structure of Chinese cities in the border region (Gaubatz 1996, 1998) and elsewhere (Gaubatz 1995*a*, *b*, 1999; Lo 1994, 1997). On the other hand, urban planning (Xie and Costa 1993), changing urban land use and suburbanization have not attracted much attention, despite their importance for China's urban future.

An impressive body of literature has examined the impact of economic reforms on urban change. Various issues have been addressed, ranging from the role of socialist central planning in urban growth (Xie and Costa 1991), urban industrial development (Luo and Pannell 1991), to the provision of urban infrastructure services (Chan 1997). A major block of the urban literature relates to the transition of China's urban system at the national level (Chang 1989; Chang and Kim 1994; Han and Wong 1994; Hsu 1994; Pannell 1995) as well as regional level (Lo 1989*a*; Pannell and Ma 1997; Tan 1991). Issues of the relationship between city size, urban growth, and national economic development have received much attention from geographers, as have the roles of large cities (Pannell 1992; Wei 1995*a*) and small towns (Lin and Ma 1994; Ma and Fan 1994; Ma and Lin 1993; Pannell and Veeck 1989; Tan 1990, 1991, 1993*a*) and the effect of institutional factors (Fan 1999*b*) in urban and economic development. The question of city size in China's future urban development demands more serious scrutiny as large cities are already highly congested and their facilities overburdened by residents and migrants. More attention should be directed to the development of small towns as centers of rural–urban migration and economic production. Along this line, Ma and his co-authors have been developing the notion of "urbanization from below" (e.g. Ma and Fan 1994), a conceptually rich area that merits further consideration.

Many geographic studies on China's countryside, still home to two-thirds of the nation's population, are conducted in relation to processes of urbanization and industrialization. They include research on small towns cited above, rural–urban relations (Tan 1993*b*), rural industrialization (Shen 1999; Su and Veeck 1995), and rural organizational reform (Tan and Luo 1995). Notable is McGee's (1989, 1991) work on *desakota*, which examines economic development in the zones between and spanning rural and urban entities. Building on that work, Marton's (1994) and Marton and McGee's (1996) research on the lower Yangtze Basin focuses on policy formulation related to rural industrialization and the conceptualization of extended metropolitan regions. More case studies and theoretical analyses of the rural–urban interface are clearly needed to advance our understanding of this important area of research.

Migration Studies

Geographers can be proud of their achievement in analyzing the patterns of migration in China. The causes and consequences of massive rural-to-urban migration that began in the early 1980s as a result of relaxed state control of migration have been critically examined. Known sometimes as the "floating population," the broad spatial patterns of the migrants has been mapped (Ma 1996), their classifications attempted (Chan 1996; Fan 1999*a*), and the interactions between migration, economic growth, and social change explained (Fan 1996; Smith 1996). The relationship between migration and urbanization, and the important roles of the *hukou* institution in engendering multiple tracks of migration, have also received much attention, as the studies by Chang (1996), Chan (1994*a*, *b*), Chan and Zhang (1999), and Fan (1999*a*) exemplify. Aside from these established areas of inquiry, more recent concerns of Western geographic research, such as the power of place in migration and settlement formation (Ma and Xiang 1998), and the role of gender (Fan and Huang 1998), are also represented in China geographers' recent works. On the other hand, geographers in North America have paid only scant attention to migration in Taiwan (Selya 1992*a*) and have not specifically studied migration to, from, and within Hong Kong.

Food, Resources, and Environmental Challenges

Heightened attention on the relationship between reform-led economic transformations and environmental changes characterizes geographers' research on China's resources and environment. This area of research is also illustrated with new debates on the one hand and increasingly sophisticated scientific inquiries on the other.

This is an increasingly challenging and salient topic among geographers, economists, agricultural scientists, and others, including those who anticipate a looming confrontation between China's growing population, economy, and limited resource base to the extent of a crisis. The geographer who has analyzed most profoundly China's ability to deal with its population growth and growing consumption demands is Smil (1998). His 1993 book, *China's Environmental Crisis*, created almost as much discussion and controversy as the polemical and strident writings of Lester Brown. Smil's work touches on all aspects of China's environment—land, water, air, energy resources, pollution, and population growth all seen in the context of the limits to national growth and development. In response to Lester Brown's warnings, Smil's (1995*a*, *b*) multifaceted analyses conclude that given appropriate policy and central government commitment, China can feed itself during the first quarter of the twenty-first century by improving farm efficiency, reducing post-harvest waste, using better systems of raising farm animals, and consuming a greater variety of protein. At the same time Smil notes the challenge to China that results from some of its current farming practices that he asserts will not be sustainable over the long term.

Themes of production efficiencies are examined by Veeck and his colleagues (Rozelle *et al.* 1997; Veeck *et al.* 1995; Veeck and Pannell 1989), who have pioneered studies of the spatial variations in farm productivity associated with changes in the nature of cropping systems and supplementary forms for raising farm family incomes. VanderMeer and Li (1998) have studied changes in production and management in Fujian, and

Lo (1996) has examined the integrated agriculture-aquaculture system in the Pearl River Delta. Muldavin's (1996, 1997) recent studies in Heilongjiang examine policy issues and raise concerns over the impacts of agricultural reforms on the environment in rural China.

A number of special issues of *Chinese Environment and Development* focus on environmental themes, a testimony of geographers' attention on environmental implications of China's reforms. They include Xu and Tan's (1995) article which uses the agroecosystem health approach to analyze China's agriculture, Harris's (1996) work on wildlife conservation, and Chan's (1994*b*) paper on economic development. Tan's (1989) work also highlights the problem of environmental degradation due to rural industrialization. In addition, Whitney (1991*a, b*, 1992) has conducted some important studies on waste management, soil erosion, and sustainability. The 1993 book by Whitney and Luk, *Megaproject*, is a remarkable contribution by North American geographers to the important research and debates on China's Three Gorges Project.

Scientific inquiries into environmental changes in China have witnessed not only increased attention by physical geographers, but also international funding and collaborations. One example is Shiu Luk's brilliant work on soil erosion, sponsored by the Canadian government. He has led teams of North American and Chinese researchers to conduct a series of excellent studies on the laterite soils of Guangdong and the loess soils of North China (Hamilton and Luk 1993; Li *et al.* 1995; S. H. Luk 1992; S. H. Luk and Cai 1990; S. H. Luk *et al.* 1989, 1990, 1993; S. H. Luk and Woo 1997; Zhu 1990). Another example is Liu's work, which focuses on paleoenvironmental records of vegetational changes and assessments of the relative impacts of climate and human disturbance (Kremenetski *et al.* 1998; Liu and Qiu 1994; Liu *et al.* 1998).

Cultural Landscape, Historical Geography, and Tourism

Given China's long history and diverse regional cultures, it is surprising that geographers have not paid more attention to its rich historical and cultural landscapes. Knapp has written more than any other scholar in any discipline on material cultural landscape. Based on extensive fieldwork, he has amply documented and analyzed China's rural landscape and vernacular architecture, including bridges and rural house types, their internal design, construction, decorations, and symbolism (Knapp 1989, 1990, 1993, 1994, 1998*a*). Using the mode of interpretative cultural geography, he examines the notions that villages are places impregnated with meanings that can be read as texts (Knapp 1992, 1998*b*). Through literary writer Shen Congwen's essays written in the 1930s, Oakes (1995) integrates gender, regionalism, and modernity to interpret the landscape of western Hunan.

Despite the abundance of historical materials on China, geographers have done little work on historical geography, due partially to the difficulty of using historical documents. The few writings on historical urban development in the 1960s and 1970s are regrettably not joined by more recent publications. The only exception is Hsu's work on historical cartography, including a useful survey and introduction to China's historical atlases (Hsu 1997). In 1986, seven maps of the Kingdom of the Qin, the earliest extant in China dated around 300 BC, were discovered in today's Tianshui, Gansu. Hsu (1993) has expertly analyzed their content, symbolism, and cartographic execution, compared them with the well-known Han maps that she had studied earlier, and analyzed the constraints for the development of traditional Chinese cartography in the context of Chinese scientific tradition.

Tourism has received considerable attention from geographers. Notable are Lew and Yu's (1995) edited volume, *Tourism in China*, and Y. W. Zhang and Lew's (1997) paper, which sketch out the basic characteristics of tourism development in China. Geographers have also studied ethnic tourism in Xinjiang (Toops 1992, 1993, 1995) and Guizhou (Oakes 1997), and airline liberalization (Yu and Lew 1997). An excellent beginning in exploring the theme of cultural space has been made by Oakes (1992, 1993, 1997), who, using ethnic minority groups in southwest China as case studies, examines how ethnic identities are consciously created and localized in space in the wake of economic reforms and in the context of growing national and international interests in ethnic tourism. His 1998 book, *Tourism and Modernity in China*, elucidates more fully these ideas, the notion of modernity, and the relationship between tourism and cultural production.

New and Future Areas of Research

As we take stock of the achievements of North American geographers in China research, it is also clear to us that these achievements have been uneven, and that there are emerging areas and long-standing gaps in the scholarship that need more attention. Specifically, the following areas warrant higher priorities in geographers' research agenda.

The research on economic geography has generally been dominated by investigations of urban-industrial or rural-agricultural sectors, with only scant attention given to other sectors. There are some exceptions. Comtois's work (1990), including a collaboration with Rimmer (1996) which examines shipping and commercial activity through Hong Kong into China, provides a valuable reminder to the community of China geographers of the significance of transportation to the broader field of economic geography. Another exciting work is by Han and Pannell (1999), who focus on the growth and spatial variation in privatization, a vital issue in the transformation of China's industrial and service economies. These are good examples of the many diverse issues in the Chinese economy that deserve more attention by geographers.

In contrast to the accomplishments by geographers in analyzing issues of urbanization and migration in China, they have almost completely neglected fertility and mortality, two building blocks for understanding population change. There is some attention on sex ratio in Taiwan (Selya 1994*a*), but demographic studies on China are rare. This is especially problematic given the importance of the one-child policy and aging processes in determining the growth, social, and spatial dynamics of the largest country in the world. Similarly, medical geography has witnessed only sparse research. Smith (1993, 1998) and Smith and Dai (1995) have shown that the benefits of improved health care and diet since the reforms have not been evenly distributed over space, with increasing disparities between urban and rural places which exacerbate gaps inherited from pre-reform eras. Health-care, diet, and income inequalities are also evident within urban and rural areas, where the rural poor, especially women, lag most seriously behind. Lam and her associates (Lam *et al.* 1993) have studied the spatial patterns of cancer mortality in the Tai Hu (Tai Lake) region in the 1970s. But generally speaking medical geography has not been a central research area for most China geographers in North America.

Geographers have as a whole been lagging in their examination of gender and gender-related issues in China, especially when compared with non-geography China scholars. But new exciting beginnings are emerging in the 1990s, as witnessed by Cartier's (1998) work arguing for the centrality of gender in understanding the geography of China and the nature of China's reforms, Fan and Huang's (1998) study which foregrounds the agency of migrant women and their migration experiences, and Oakes's (1995) study of China's gendered landscape from a literary perspective. It is hoped that a critical mass of scholars and writings will soon come into being and that they will enable more visible contributions by geographers to this exciting subfield.

North American geographers who specialize in China's regional development and urban and economic geography have produced many studies rich in empirical and quantitative analyses. At the same time, we have found relatively little work using ethnographic and qualitative methods, although China geographers do confront ethnic issues seen embedded in some of our discourses on migration, gender, urban, tourism, cultural, and historical studies. Likewise, political, social, cultural, and historical geography, and rigorous theoretical inquiries are generally underrepresented. Given the centrality of political economy to much recent research in China studies, it is surprising that North American geographers have not developed more fully political economic and other theoretical perspectives specific for the study of this region. In addition, there is a general lack of research on the roles of space and place in China's long history and in shaping China's cultural landscape, landscape as text, geopolitics, regionalism, regional and place identities, national minorities, and ethnicity among the Han Chinese. These challenges must be met before China geographers can play more central roles in Sinology and China studies in general.

We should point out that geographers have not been very active in producing texts on China for graduate and undergraduate teaching, despite the popularity of courses on China at both levels throughout North America. One exception is Smith's 1991 text, *China: People and Places in the Land of One Billion*, which is an excellent and comprehensive survey of China with emphases on the spatial organization of production, distribution, and consumption. Our view is that geographers should be more proactive illustrating the importance of geographic perspectives in understanding the complexity of China, by not only publishing in specialized outlets but also by writing for geography and non-geography courses. Finally, North American geographers should also seriously consider reviving the efforts for a journal on China geography. We believe that the publication experience of *Chinese Environment and Development* (formerly *Chinese Geography and Environment*), which included many excellent papers during its life-span from 1986 to 1996, as well as collective commitments from China geographers, would go a long way toward establishing a unique and high-quality journal focusing on geographic research on China.

Conclusion

Viewed as a whole, the development of China geography in North America since the late 1980s has been most impressive, though the gains have been uneven and some areas remain underexplored. Urban, regional, and economic changes in the post-reform regime have almost monopolized geographers' attention, leaving only a small number of individuals working on physical, cultural, historical, and medical geographies. Given the relatively small, albeit growing, number of China geographers in North America, our assessment is that they have achieved selective excellence with new exciting work in emerging areas.

China geographers have begun to move away from a tradition that was largely empirical, to research approaches encompassing new concepts and modes of discourse, and contextually sensitive theories. Perspectives emphasizing political economy, regional cultures, social networks, and human–environment interactions, for example, have enabled the scholarship of China geographers to gain increasing recognition among non-China specialists and non-geography China experts, as witnessed by the success of China geographers publishing in disciplinary flagship journals such as *Annals of the Association of American Geographers*, and leading China journals such as *China Quarterly*. Yet the overall impact of China geography on the rich field of Sinology remains to be strengthened, and the process of intellectual maturation from a regional mode of research to one more substantive and theoretical is not yet complete. But the speed of progress in the last decade or so, and continued opportunities for cross-fertilization across the Pacific and with non-geography colleagues, have convinced us that whereas the field of China geography has borne impressive fruits in the past, the ground is so fertile that more and better fruits can be produced in the years ahead.

References

Cartier, C. (1995). "Singaporean Investment in China: Installing the Singapore Model in Sunan." *Chinese Environment and Development*, 6/1–2: 117–44.

—— (1998). "Women and Gender in Contemporary China," in C. V. Prorok and K. B. Chhokar (eds.), *Asian Women and Their Work: A Geography of Gender and Development*. Indiana, Pa.: National Council on Geographic Education, 1–7.

Chan, K. W. (1992). "Economic Growth Strategy and Urbanization Policies in China, 1949–82." *International Journal of Urban and Regional Research*, 16/2: 275–305.

—— (1994*a*). *Cities With Invisible Walls: Reinterpreting Urbanization in Post-1949 China*. Hong Kong: Oxford University Press.

—— (1994*b*). "Economic Development in the Reform Era: Environmental Implications." *Chinese Environment and Development*, 4/4: 3–24.

—— (1994*c*). "Urbanization and Rural-Urban Migration in China Since 1982: A New Baseline." *Modern China*, 20/3: 243–81.

—— (1996). "Post-Mao China: A Two-Class Urban Society in the Making." *International Journal of Urban and Regional Research*, 20/1: 134–50.

—— (1997). "Urbanization and Urban Infrastructure Services in the PRC," in C. Wong (ed.), *Financing Local Government in the People's Republic of China*. Hong Kong: Oxford University Press, 83–125.

Chan, K. W., and Zhang, L. (1999). "The *Hukou* System and Rural-Urban Migration in China: Processes and Changes." *China Quarterly*, 160: 818–55.

Chang, S. D. (1989). "The Changing Patterns of Chinese Cities, 1953–1984," in C. K. Leung *et al.* (eds.), *Resources, Environment, and Regional Development*. Hong Kong: Center of Asian Studies, University of Hong Kong, 302–31.

—— (1996). "The Floating Population: An Informal Process of Urbanization in China." *International Journal of Population Geography*, 2: 197–214.

—— (1998). "Beijing: Perspectives on Preservation, Environment, and Development." *Cities*, 15: 13–25.

Chang, S. D., and Kim, W. B. (1994). "The Economic Performance and Regional Systems of China's Cities." *Review of Urban and Regional Development Studies*, 6: 58–77.

Comtois, C. (1990). "Transport and Territorial Development in China, 1949–1985." *Modern Asian Studies*, 24: 777–818.

Comtois, C., and Rimmer, P. J. (1996). "Refocusing on China: The Case of Hong Kong's Hutchison Whampoa." *Pacific Viewpoint*, 37/1: 89–102.

Fan, C. C. (1992). "Regional Impacts of Foreign Trade in China, 1984–1989." *Growth and Change*, 23/2: 129–59.

—— (1995*a*). "Developments From Above, Below and Outside: Spatial Impacts of China's Economic Reforms in Jiangsu and Guangdong Provinces." *Chinese Environment and Development*, 6/1–2: 85–116.

—— (1995*b*). "Of Belts and Ladders: State Policy and Uneven Regional Development in Post-Mao China." *Annals of the Association of American Geographers*, 85/3: 421–49.

—— (1996). "Economic Opportunities and Internal Migration: A Case Study of Guangdong Province, China." *Professional Geographer*, 48/1: 28–45.

—— (1997). "Uneven Development and Beyond: Regional Development Theory in Post-Mao China." *International Journal of Urban and Regional Research*, 21/4: 620–39.

—— (1999*a*). "Migration in a Socialist Transitional Economy: Heterogeneity, Socioeconomic and Spatial Characteristics of Migrants in China and Guangdong Province." *International Migration Review*, 33/4: 950–83.

—— (1999*b*). "The Vertical and Horizontal Expansions of China's City System." *Urban Geography*, 20/6: 493–515.

Fan, C. C., and Huang, Y. (1998). "Waves of Rural Brides: Female Marriage Migration in China." *Annals of the Association of American Geographers*, 88/2: 227–51.

Fung, K.-I. (1996). "Satellite Towns: Development and Contributions," in Y. M. Yeung and Y. W. Sung (eds.), *Shanghai: Transformation and Modernization*. Hong Kong: Chinese University of Hong Kong Press, 321–40.

Fung, K.-I., Yan, Z. M., and Ning, Y. M. (1992). "Shanghai: China's World City," in Y. M. Yeung and X. W. Hu (eds.), *China's Coastal Cities: Catalysts for Modernization*. Honolulu: University of Hawaii Press, 124–52.

Gaubatz, P. (1995*a*). "Changing Beijing." *Geographical Review*, 85/1: 79–96.

—— (1995*b*). "Urban Transformation in Post-Mao China: Impacts of the Reform Era on China's Urban Form," in D. Davis *et al.* (eds.), *Urban Spaces in Contemporary China*. Cambridge: Cambridge University Press, 28–60.

—— (1996). *Beyond the Great Wall: Urban Form and Transformation on the Chinese Frontiers*. Stanford: Stanford University Press.

—— (1998). "Mosques and Markets: Traditional Urban Form on China's Northwestern Frontiers." *Traditional Dwellings and Settlements Review*, 9: 7–21.

—— (1999). "China's Urban Transformation: Patterns and Processes of Morphological Change in Beijing, Shanghai and Guangzhou." *Urban Studies*, 36/9: 1495–521.

Gong, H. (1995). "Spatial Patterns of Foreign Investment in China's Cities, 1980–1989." *Urban Geography*, 16/3: 198–209.

Hamilton, H., and Luk, S. H. (1993). "Nitrogen Transfers in a Rapidly Eroding Agroecosystem: Loess Plateau, China." *Journal of Environmental Quality*, 22/1: 133–40.

Han, S. S., and Pannell, C. W. (1999). "The Geography of Privatization in China, 1978–1996." *Economic Geography*, 75/3: 272–96.

Han, S. S., and Wong, S. T. (1994). "The Influence of Chinese Reform and Pre-Reform Policies on Urban Growth in the 1990s." *Urban Geography*, 15: 537–64.

Harris, R. B. (1996). "Perspectives on Wildlife Conservation in China and Taiwan." *Chinese Environment and Development*, 6/4: 3–12.

Hayter, R., and Han, S. S. (1998). "Reflections on China's Open Door Policy Towards Foreign Direct Investment." *Regional Studies*, 32/1: 1–16.

Hsing, Y. (1995). "Taiwanese Guerilla Investors in Southern China." *Review of Urban and Regional Studies*, 7: 24–34.

—— (1996). "Blood Thicker Than Water: Interpersonal Relation and Taiwanese Investment in Southern China." *Environment and Planning A*, 28/12: 2241–61.

—— (1998). *Making Capitalism in China: The Taiwan Connection*. New York: Oxford University Press.

Hsu, M. L. (1993). "The Qin Maps: A Clue to Later Chinese Cartographic Development." *Imago Mundi*, 45: 90–100.

—— (1994). "The Expansion of the Chinese Urban System, 1953–1990." *Urban Geography*, 15/6: 514–36.

—— (1996). "China's Urban Development: A Case Study of Luoyang and Guiyang." *Urban Studies*, 33/6: 895–910.

—— (1997). "An Inquiry into Early Chinese Atlases Through the Ming Dynasty," in J. A. Wolter and R. E. Grim (eds.), *Images of the World: The Atlas Through History*. Washington, DC: Library of Congress, 31–50.

Karan, P. P., Shrestha, N., Dickason, D. G., Hafner, J. A., Oshiro, K. K., and Al-Khameri, S. (1989). "Asia," in G. L. Gaile and C. J. Willmott, *Geography in America*. Columbus, Ohio: Merrill, 506–45.

Knapp, R. G. (1989). *Chinese Vernacular Architecture: House Form and Culture*. Honolulu: University of Hawaii Press.

—— (1990). *The Chinese House: Craft, Symbol, and the Folk Tradition*. Hong Kong: Oxford University Press.

—— (ed.) (1992). *Chinese Landscapes: The Village As Place*. Honolulu: University of Hawaii Press.

—— (1993). *Chinese Bridges*. Hong Kong: Oxford University Press.

—— (1994). "Popular Rural Architecture," in D. B. Wu and P. Murphy (eds.), *Handbook of Chinese Popular Culture*. Westport: Greenwood, 327–46.

—— (1998*a*). *Chinese Living Houses: Folk Beliefs, Symbols, and Household Ornamentation*. Honolulu: University of Hawaii Press.

—— (1998*b*). "Chinese Villages As Didactic Texts," in W. H. Yeh (ed.), *Landscape, Culture, and Power in Chinese Society*. Berkeley: Institute of East Asian Studies, University of California, 110–28.

Kremenetski, C. V., Liu, K., and MacDonald, G. M. (1998). "The Late Quaternary Dynamics of Pines in Northern Asia," in D. M. Richardson (ed.), *Ecology and Biogeography of Pines*. Cambridge: Cambridge University Press, 95–106.

Lam, N. S. N., Qiu, H. L., Zhao. R., and Jiang, N. (1993). "A Fractal Analysis of Cancer Mortality Patterns in China," in N. S. N. Lam and L. De Cola (eds.), *Fractals in Geography*. Englewood Cliffs, NJ: Prentice-Hall, 247–62.

Leung, C. K. (1990). "Locational Characteristics of Foreign Equity Joint Venture Investment in China, 1979–1985." *Professional Geographer*, 42/4: 403–20.

—— (1991). "Shanxi Energy Base: Planning Principles and Regional Development Implications," in G. Veeck (ed.), *The Uneven Landscape: Geographical Studies in Post-Reform China*. Baton Rouge, La.: Geoscience Publications, 125–50.

—— (1993). "Personal Contacts, Subcontracting Linkages, and Development in the Hong Kong-Zhujiang Delta Region." *Annals of the Association of American Geographers*, 83/2: 272–302.

—— (1996). "Foreign Manufacturing Investment and Regional Industrial Growth in Guangdong Province, China." *Environment and Planning A*, 28/3: 513–36.

Lew, A. A., and Yu, L. (eds.) (1995). *Tourism in China: Geographical, Political, and Economic Perspectives*. Boulder: Westview Press.

Li, G., Luk, S. H., and Cai, Q. G. (1995). "Topographic Zonation of Infiltration in the Hilly Loess Region, North China." *Hydrologic Processes*, 9/2: 227–35.

Lin, G. C. S. (1997). *Red Capitalism in South China: Growth and Development of the Pearl River Delta*. Vancouver: University of British Columbia Press.

Lin, G. C. S., and Ma, L. J. C. (1994). "The Role of Towns in Chinese Regional Development: The Case of Guangdong Province." *International Regional Science Review*, 17/1: 75–97.

Liu, K., and Qiu, H. (1994). "Late-Holocene Pollen Records of Vegetational Changes in China: Climate or Human Disturbance?" *Terrestrial, Atmospheric and Oceanic Sciences*, 5/3: 393–410.
Liu, K., Yao, Z., and Thompson, L. G. (1998). "A Pollen Record of Holocene Climatic Changes From the Dunde Ice Cap, Qinghai-Tibetan Plateau." *Geology*, 26/2: 135–8.
Lo, C. P. (1989*a*). "Population Change and Urban Development in the Pearl River Delta: Spatial Policy Implications." *Asian Geographer*, 8/1–2: 11–33.
— (1989*b*). "Recent Spatial Restructuring in Zhujiang Delta, South China: A Study of Socialist Regional Development Strategy." *Annals of the Association of American Geographers*, 79/2: 293–308.
— (1990). "The Geography of Rural Regional Inequality in Mainland China." *Transactions, Institute of British Geographers*, NS 15/4: 466–86.
— (1992). *Hong Kong*. London: Belhaven Press.
— (1994). "Economic Reforms and Socialist City Structure: A Case Study of Guangzhou, China." *Urban Geography*, 15/2: 128–49.
— (1996). "Environmental Impact on the Development of Agricultural Technology in China: The Case of the Dike-Pond System of Integrated Agriculture-Aquaculture in the Zhujiang Delta of China." *Agriculture, Ecosystems and Environment*, 60/2–3: 183–95.
— (1997). "Dispersed Spatial Development: Hong Kong's New City Form and Its Economic Implications After 1997." *Cities*, 14/5: 273–7.
Luk, C. M. (1991). "Regional Development and Open Policy: The Case of Guangdong," in G. Veeck (ed.), *The Uneven Landscape: Geographical Studies in Post-Reform China*. Baton Rouge, La.: Geoscience Publications, 151–70.
Luk, S. H. (1992). "Soil Erosion and Land Management in the Loess Plateau Region, North China." *Chinese Geography and Environment*, 3/4: 3–27.
Luk, S. H., and Cai, Q. G. (1990). "Laboratory Experiments on Crust Development and Rainsplash Erosion of Loess Soils, China." *Catena*, 17/3: 261–76.
Luk, S. H., Cai, Q. G., and Wang, G. P. (1993). "Effects of Surface Crusting and Slope Gradient on Soil and Water Losses in the Hilly Loess Region, North China." *Catena*, 24: 29–45.
Luk, S. H., Dubbin, W. E., and Mermut, A. R. (1990). "Fabric Analysis of Surface Crusts Developed Under Simulated Rainfall on Loess Soils, China." *Catena Supplement*, 17: 29–49.
Luk, S. H., Chen, H., Cai, Q. G., and Jia, Z. J. (1989). "Spatial and Temporal Variations in the Strength of Loess Soils, China." *Geodema*, 45: 303–17.
Luk, S. H., and Woo, M. K. (1997). "Soil Erosion in South China." *Catena*, 29/2: 93–221.
Luo, Y., and Pannell, C. W. (1991). "The Changing Pattern of City and Industry in Post-Reform China," in G. Veeck (ed.), *The Uneven Landscape: Geographical Studies in Post-Reform China*. Baton Rouge, La.: Geoscience Publications, 29–52.
Ma, L. J. C. (1996). "The Spatial Patterns of Interprovincial Rural-to-Urban Migration in China, 1982–1987." *Chinese Environment and Development*, 7/1–2: 73–102.
Ma, L. J. C., and Fan, M. (1994). "Urbanisation From Below: The Growth of Towns in Jiangsu, China." *Urban Studies*, 31/10: 1625–45.
Ma, L. J. C., and Lin, C. (1993). "Development of Towns in China: A Case Study of Guangdong Province." *Population and Development Review*, 19/3: 583–606.
Ma, L. J. C., and Wei, Y. (1997). "Determinants of State Investment in China, 1953–1990." *Tijdschrift Voor Economische En Sociale Geografie*, 88/3: 211–25.
Ma, L. J. C., and Xiang, B. (1998). "Native Place, Migration and the Emergence of Peasant Enclaves in Beijing." *China Quarterly*, 155: 546–81.
McGee, T. G. (1989). "Urbanisasi or Kotadesasi? Evolving Patterns of Urbanization in Asia," in F. Costa, A. K. Dutt, L. J. C. Ma, and A. G. Noble (eds.), *Urbanization in Asia: Spatial Dimensions and Policy Issues*. Honolulu: University of Hawaii Press, 93–108.
— (1991). "The Emergence of *Desakota* Regions in Asia: Expanding a Hypothesis," in N. Ginsburg, B. Koppel, and T. G. McGee (eds.), *The Extended Metropolis: Settlement Transition in Asia*. Honolulu: University of Hawaii Press, 3–25.
Marton, A. M. (1994). "Challenges for Metrofitting the Lower Yangzi Delta." *Western Geography*, 4: 62–83.
— (1995). "Mega-Urbanization in Southern Jiangsu: Enterprise Location and the Reconstitution of Local Space." *Chinese Environment and Development*, 6/1–2: 9–42.
Marton, A. M., and McGee, T. G. (1996). "Processes of Mega-Urban Development in China: The Lower Yangzi Delta and Kunshan." *Asian Geographer*, 15/1–2: 49–70.
Muldavin, J. S. S. (1996). "Impact of Reform on Environmental Sustainability in Rural China." *Journal of Contemporary Asia*, 26/3: 289–321.
— (1997). "Environmental Degradation in Heilongjiang: Policy Reform and Agrarian Dynamics in China's New Hybrid Economy." *Annals of the Association of American Geographers*, 87/4: 579–613.
Oakes, T. (1992). "Cultural Geography and Chinese Ethnic Tourism." *Journal of Cultural Geography*, 12: 3–17.
— (1993). "The Cultural Space of Modernity: Ethnic Tourism and Place Identity in China." *Environment and Planning D*, 11/1: 47–66.
— (1995). "Shen Congwen's Literary Regionalism and the Gendered Landscape of Chinese Modernity." *Geografiska Annaler B*, 77: 93–107.
— (1997). "Ethnic Tourism in Rural Guizhou: Sense of Place and the Commerce of Authenticity," in M. Picard and R. Wood (eds.), *Tourism, Ethnicity, and the State in Asian and Pacific Societies*. Honolulu: University of Hawaii Press, 35–70.
— (1998). *Tourism and Modernity in China*. London: Routledge.
Pannell, C. W. (1990). "China's Urban Geography." *Progress in Human Geography*, 14/2: 214–36.
— (1992). "The Role of Great Cities in China," in G. E. Guldin (ed.), *Urbanizing China*. New York: Greenwood Press, 11–40.
— (1995). "China's Urban Transition." *Journal of Geography*, 94: 394–403.
Pannell, C. W., and Ma, L. J. C. (1997). "Urban Transition and Interstate Relations in a Dynamic Post-Soviet Borderland: The Xinjiang Uygur Autonomous Region of China." *Post-Soviet Geography and Economics*, 38/4: 206–29.
Pannell, C. W., and Veeck, G. (1989). "Zhujiang Delta and Sunan: A Comparative Analysis of Regional Urban Systems and Their Development," in X. Q. Xu, A. Yeh, and C. E. Wen (eds.), *The*

Environment and Space Development in the Pearl River Delta. Beijing: Xinhua Publishing, 237–52.
Rozelle, S., Veeck, G., and Huang, J. (1997). "The Impact of Environmental Degradation on Grain Production in China: 1975–1990." *Economic Geography*, 73/1: 44–66.
Selya, R. M. (1992*a*). "Illegal Migration in Taiwan: A Preliminary Overview." *International Migration Review*, 26/3: 787–805.
—— (ed.) (1992*b*). *The Geography of China, 1975–1991: An Annotated Bibliography*. East Lansing: Asian Studies Center, Michigan State University.
—— (1993). "Economic Restructuring and Spatial Changes in Manufacturing in Taiwan, 1971–1986." *Geoforum*, 24/2: 115–26.
—— (1994*a*). "Abnormally Elevated Sex Ratios in the Republic of China on Taiwan: An Exploratory Review." *American Asian Review*, 12/3: 15–36.
—— (1994*b*). "Taiwan as a Service Economy." *Geoforum*, 25/3: 305–22.
—— (1995). *Taipei*. New York: John Wiley.
Shen, X. (1999). "Spatial Inequality of Rural Industrial Development in China, 1989–1994." *Journal of Rural Studies*, 15/2: 179–99.
Smil, V. (1993). *China's Environmental Crisis: An Inquiry into the Limits of National Development*. Armonk, NY: M. E. Sharpe.
—— (1995*a*). "Feeding China." *Current History*, 94: 280–4.
—— (1995*b*). "Who Will Feed China?" *China Quarterly*, 143: 801–13.
—— (1998). "China's Energy and Resource Uses: Continuity and Change." *China Quarterly*, 156: 935–51.
Smith, C. J. (1991). *China: People and Places in the Land of One Billion*. Boulder, Colo.: Westview Press.
—— (1993). "(Over)Eating Success: The Health Consequences of the Restoration of Capitalism in Rural China." *Social Science and Medicine*, 37: 761–70.
—— (1996). "Migration as an Agent of Change in Contemporary China." *Chinese Environment and Development*, 7/1–2: 14–55.
—— (1998). "Modernization and Health Care in Contemporary China." *Health and Place*, 4: 125–39.
Smith, C. J., and Dai, F. (1995). "Health, Wealth, and Inequality in the Chinese City." *Health and Place*, 1: 167–77.
Su, S.-J., and Veeck, G. (1995). "Foreign Capital Acquisition by Rural Industries in Jiangsu." *Chinese Environment and Development*, 6/1–2: 169–92.
Tan, K. C. (1989). "Small Town Development and the Environment: Hope and Reality," in C. K. Leung (ed.), *Resources, Environment and Regional Development*. Hong Kong: Centre for Asian Studies, University of Hong Kong, 332–52.
—— (1990). "China's New Spatial Approach to Economic Development." *Chinese Geography and Environment*, 2/4: 3–21.
—— (1991). "Small Towns and Regional Development in Wenzhou," in G. Veeck (ed.), *The Uneven Landscape: Geographical Studies in Post-Reform China*. Baton Rouge, La.: Geoscience Publications, 207–34.
—— (1993*a*). "China's Small Town Urbanization Program: Criticism and Adaptation." *GeoJournal*, 29/3: 155–62.
—— (1993*b*). "Rural-Urban Segregation in China." *Geography Research Forum*, 13: 71–83.
Tan, K. C., and Luo, H. (1995). "The Rural Enterprise as the Emergent Core of New Socioeconomic Organizations in the Chinese Countryside," in M. Chen, C. Comtois, and L. N. Shyu (eds.), *East Asia Perspectives*. Montreal: Canadian Asian Studies Association, 157–78.
Tang, W. S., Chu, D. K. Y., and Fan, C. C. (1993). "Economic Reform and Regional Development in China in the 21st Century," in Y. M. Yeung (ed.), *Pacific Asia in the 21st Century: Geographical and Developmental Perspectives*. Hong Kong: The Chinese University Press, 105–33.
Todd, D., and Hsueh, Y.-C. (1992). "The Industrial Complex Revisited: Petrochemicals in Taiwan." *Geoforum*, 23/1: 29–40.
Toops, S. (1992). "Tourism in Xinjiang, China." *Journal of Cultural Geography*, 12: 19–34.
—— (1993). "Xinjiang's Handicraft Industry." *Annals of Tourism Research*, 20/1: 88–106.
—— (1995). "Tourism in Xinjiang: Practice and Place," in A. A. Lew and L. Yu (eds.), *Tourism in China: Geographical, Political, and Economic Perspectives*. Boulder, Colo.: Westview Press, 179–202.
VanderMeer, C., and Li, J. (1998). "Water Distribution in Farmer Managed Rice Irrigation Systems, Fujian, China." *Geographical Review of Japan*, 71/2: 91–105.
Veeck, G. (ed.) (1991). *The Uneven Landscape: Geographical Studies in Post-Reform China*. Baton Rouge, La.: Geoscience Publications.
Veeck, G., and Pannell, C. W. (1989). "Rural Economic Restructuring and Farm Household Income in Jiangsu, People's Republic of China." *Annals of the Association of American Geographers*, 79/2: 275–92.
Veeck, G., Li, Z., and Gao, L. (1995). "Terrace Construction and Productivity on Loessal Soils in Zhongyang County, Shanxi Province, PRC." *Annals of the Association of American Geographers*, 85/3: 450–67.
Wei, Y. (1995*a*). "Large Chinese Cities: a Review and a Research Agenda." *Asian Geographer*, 14/1: 1–13.
—— (1995*b*). "Spatial and Temporal Variations of the Relationship Between State Investment and Industrial Output in China." *Tijdschrift Voor Economische En Sociale Geografie*, 86/2: 129–36.
—— (1996). "Fiscal Systems and Uneven Regional Development in China, 1978–1991." *Geoforum*, 27/3: 329–44.
—— (1998). "Regional Inequality of Industrial Output in China, 1952–1990." *Geografiska Annaler B*, 80/1: 1–15.
—— (1999). "Regional Inequality in China." *Progress in Human Geography*, 23/1: 48–58.
Wei, Y., and Ma, L. J. C. (1996). "Changing Patterns of Spatial Inequality in China, 1952–1990." *Third World Planning Review*, 18/2: 177–91.
Whitney, J. (1991*a*). "A Soil-Erosion Management Geographical Information System for North China," in P. Wilkinson and W. Found (eds.), *Resource Analysis Research in Developing Countries*. Atkinson Monograph in Geography. Toronto: York University, 75–97.
—— (1991*b*). "Waste Economy and the Dispersed Metropolis in China," in N. Ginsburg, B. Koppel, and T. G. McGee (eds.), *The Extended Metropolis: Settlement Transition in Asia*. Honolulu: University of Hawaii Press, 177–91.
—— (1992). "Asia," in R. Stren, R. White, and J. Whitney (eds.), *Sustainable Cities: Urbanization and the Environment in International Perspectives*. Boulder, Colo.: Westview Press, 229–32.

Whitney, J., and Luk, S. H. (1993). *Megaproject: A Case Study of China's Three Gorges Project*. Armonk, NY: M. E. Sharpe.

Wu, W. (1997). "Proximity and Complementarity in Hong Kong-Shenzhen Industrialization." *Asian Survey*, 37/8: 771–93.

Xie, Y., and Costa, F. J. (1991). "The Impact of Economic Reforms on the Urban Economy of the People's Republic of China." *Professional Geographer*, 43/3: 318–35.

—— (1993). "Urban Planning in Socialist China: Theory and Practice." *Cities*, 10: 103–14.

Xie, Y., and Dutt, A. K. (1991). "Spatial Disparities of Urban Socio-Economic Development in the People's Republic of China." *Geoforum*, 22/1: 55–67.

Xu, W., and Tan, K. C. (1995). "Chinese Agriculture: An Agroecosystem Health Approach." *Chinese Environment and Development*, 6/3: 3–28.

Ying, L. G. (1999). "China's Changing Regional Disparities During the Reform Period." *Economic Geography*, 75/1: 59–70.

Yu, L., and Lew, A. A. (1997). "Airline Liberalization and Development in China." *Pacific Tourism Review*, 1: 1129–36.

Yusuf, S., and Wu, W. (1997). *The Dynamics of Urban Growth in Three Chinese Cities*. New York: Oxford University Press.

Zhang, L., and Zhao, S. X. (1998). "Re-Examining China's 'Urban' Concept and the Level of Urbanization." *China Quarterly*, 154: 330–81.

Zhang, Y. W., and Lew, A. A. (1997). "The People's Republic of China: Two Decades of Tourism." *Pacific Tourism Review*, 1: 161–72.

Zhu, T. X. (1990). "Silt Reduction by Precipitation Variation and Human Activity in the Middle Reaches of the Yellow River." *Journal of Arid Land Research*, 3/3: 33–45.

CHAPTER 42

European Geography

Boian Koulov, Linda McCarthy, Anton Gosar, and Daniel Knudsen

The European Specialty Group (ESG) was founded with considerable enthusiasm in 1992. Its organization and the rapid membership increase were in response to the historic changes following the fall of the Iron Curtain, the reintegration of the European continent, and a heightened interest in the evolution of European political and economic life. The purpose of the ESG is to move beyond the Cold War legacy of East–West division of the continent and foster research, teaching, and scholarly interaction on the geography of the new Europe. The ESG also serves as a bridge between US geographers working on Europe and their counterparts in the rest of the world. Finally, the group promotes the study of Europe within the discipline of geography and facilitates the exchange of information and ideas among its members and Europeanists in other disciplines, government, and private agencies.

Research on Europe has been undertaken at a variety of spatial scales. A number of books reflect the pan-European scale (Berentsen 1993, 1997; Harris 1991, 1993*a*, *b*, 1997; Jordan 1996; McDonald 1997; Murphy 1991; Unwin 1998). The national scale also has received attention due to the continued importance of the different national contexts despite increased European integration, in conjunction with difficulties created by the lack of comparable statistical databases at a sub-national scale for the countries across Europe. Regardless of spatial scale several consistent themes have emerged. Within political geography focus is clearly on the new divisions of Europe, states–nations relationships, sub-national political transformation, the twin forces of democratization and nationalism, and ethnic conflict. Within economic geography research has centered around issues of "widening" versus "deepening" in the EU, globalization and pan-European integration, the impacts and implications of the incorporation of Central and Eastern European nations into the European economy, and the spatially uneven nature of economic change. Geographers also have been active in addressing issues of environmental damage, population, and migration.

This chapter takes a regional approach that reflects the typical focus of most research. The material is treated systematically within sections on Western, Nordic, Eastern, and Mediterranean Europe.

Western Europe

The 1990s was a time of heightened economic and political integration for Western Europe. The Single European Market was introduced, Austria, Finland, and Sweden joined the European Union (EU), and Eastern Europe reentered the capitalist space economy. In addition, processes of internationalization associated with corporate restructuring and technological innovation continued to impact this region. Governments at all levels attempted to adapt to these changes.

European Political and Economic Integration

Considerable attention has been paid to the contradictions associated with the EU's moves toward increased political and economic integration. Research on the political dimension of European integration included a focus on the EU as a political institution. Attention to the spatially uneven nature of political processes and patterns allowed unique insights into important inter-regional and international differences that contributed to conflicts over voting power in the EU Council of Ministers, such as that arising from the UK Conservative Party's fears over a loss of national sovereignty (Johnston 1994).

A significant contribution by geographers resulted from their concern to incorporate the importance of spatial scale into their analyses. Williams (1994) anticipated that the increase in EU competencies since the mid-1980s would perpetuate the democratic deficit in the decision-making process (where major elements of EU activities are not directly accountable to the European or national parliaments). He argued that the Treaty on European Union, heads of state, and the European Commission commitment to subsidiarity (where decisions are made at the lowest effective scale possible) will help remedy the democratic deficit only if appropriate decision-making scales for particular policies are also established.

There has been a concern to incorporate the importance of local context in influencing electoral patterns and processes within and between particular member states (Murphy 1990; Murphy and Hunderi-Ely 1996; O'Loughlin and Parker 1990). O'Loughlin *et al.* (1994) stressed that space and place, while neglected in the social sciences, are critical analytical concepts for understanding the electoral behavior of voters. In particular, differences in local and regional contexts resulting from variations in historical-spatial development are important variables in explaining voting choices, in addition to the social characteristics of voters that are the more commonly used explanatory indicators.

The work by geographers revealed how problems associated with increased European integration stem from economic as well as political factors. Research on the economic aspects of enhanced integration has examined the formation and operation of the European trade bloc within the global marketplace (Dicken and Öberg 1996; Warf 1992). Analyses at a European scale identified that this politics of global territorial defense is replicated at national and sub-national scales, contributing to increased competition within Europe itself (Dicken and Öberg 1996; Gertler and Schoenberger 1992; O'Loughlin and Anselin 1996).

In the area of public policy, analyses have shown how variations in economic conditions and political and cultural concerns lead to problems in establishing appropriate and effective EU-wide policies. The formulation of EU immigration policies proves difficult due to national interests that promote exclusionary policies and the retention of national control over the immigration of non-EU nationals (Huntoon 1998; Kofman 1995; Leitner 1995, 1997). Likewise, national farm politics and protectionism strongly influence EU agricultural policies (Grant 1993).

Analyses of the relationship between EU industrial policy and regional development have included examination of the feasibility of balancing the EU's economic goals with its social and environmental ones (Berentsen 1997; Duncan 1994; Murphy 1991; Sadler 1992*a*, *b*; Williams 1994). Monk and García (1996) identified that the neoliberal agenda of the EU and many Western European countries restricted the provision of social services and linked them to labor force participation. They predicted further reductions in the welfare benefits traditionally enjoyed by European women as well as increased gender-based inequalities in the labor market. Gibbs (1996) highlighted some contradictions between different EU policies. Some policies encourage environmentally sensitive development while others assist large-scale infrastructure projects in economically lagging regions. Similarly, promoting free trade within the Single European Market has stimulated a higher volume of road transportation and associated vehicle emissions.

Amid the increased integration, Western Europe has also experienced political fragmentation. There have been efforts to develop theories of nationalism and the sovereign state within the context of forces for disintegration such as the ascent of regionalism (Austin 1992; F. M. Davidson 1996, 1997; Flint 1993; Johnson 1998; Murphy 1991, 1992*a*, 1996). Emerging East–West and cross-border regional linkages have challenged the appropriateness of the traditional national scale of analysis (Ivy 1995; Murphy 1993; Pavlik 1995).

There is debate over the extent to which enhanced European integration has resulted in a "hollowing out" of the state. Some authors have suggested that the national level may no longer be the optimum scale at which to achieve economic growth and competitiveness in light of the growing relative importance of the supranational and local levels of government (Agnew 1990; Amin and Thrift 1995; F. M. Davidson 1997; Dunford and Kafkalas 1992). Others argue that the national governments continue to control a range of functions, which, however, are increasingly tied to economic processes of change and governmental relations at supranational and local scales (Jessop *et al.* 1999; Johnston and Pattie 1996).

Urban and Regional Change

On the one hand, the internationalization of the economy prompted research on the importance of the global financial market and corporate investment and restructuring for Western Europe (Foley and Watts 1996; Hallsworth 1996; Kirkham and Watts 1998; Murphy 1992*b*; Pallares-Barbera 1998; Watts 1990). Dicken (1992) concluded that automobile industry restructuring was driven as much by the threat of Japanese competition as by the Single European Market and the expansion eastwards of market and production opportunities.

On the other hand, the concurrent localization of development promoted research into local restructuring processes (Amin and Thrift 1995; Cooke *et al.* 1992; Storper 1993, 1995). At a sub-national scale of analysis, much attention has been paid to the demographic, economic, political, social, and urban morphological changes occurring within and between particular cities or regions (Berentsen 1996; Corson and Minghi 1994; Dorling and Woodward 1996; Fusch 1994; McCarthy 1990; Pattie and Johnston 1990; Wilvert 1994).

There is ongoing debate over the nature of flexible specialization and its spatial expression as regional industrial districts such as Emilia-Romagna and Baden-Württemberg (Storper and Scott 1989; Storper and Walker 1989). Martinelli and Schoenberger (1991) argued that flexibility is not confined to small-scale, non-hierarchical, integrated production complexes. Much recent work has attempted to develop a cogent explanation for the resurgence of regional economies that incorporates the temporal, spatial, and sector-specific nature of these industrial districts (Gertler 1996; Glasmeier 1991, 1995; Storper 1992, 1993, 1997; Storper and Harrison 1991).

Geographers have highlighted the spatially uneven nature of the impacts of greater economic integration for the cities and regions across Western Europe. Some expect that the centrally located, larger, more technologically innovative firms can best exploit the opportunities of European integration (Amin and Thrift 1995). There is a consensus that EU funding is inadequate to improve the competitiveness of the smaller traditional enterprises within the poorest peripherally located regions in the face of increased international competition from within and outside the EU (Dunford and Kafkalas 1992; Sadler 1992*b*).

In fact, empirical analyses investigated whether there has been convergence or divergence in prosperity within and across member states during the recently increased integration. Regional analyses identified that the gap in income levels between the richest and poorest EU regions remains substantial. Some researchers argue that this has its roots in the dominant neoliberal model of development and integration (Dunford and Kafkalas 1992). During the period of slower economic growth since the mid-1970s, declining productive investment was concentrated in the economically strongest core locations, where higher unemployment reduced opportunities for migrants from the periphery, at a time when the national governments were reducing their redistributive role. Income disparities between the EU's cities, while similarly entrenched, are higher than for the regions due to factors such as the limited EU policy and funding targeted specifically for cities (McCarthy 1998).

The Politics of Urban Economic Development

There has been a concern to incorporate the importance of the larger political economy into theories of local economic and political change (Dunford and Kafkalas 1992; Ettlinger 1994; Leitner and Sheppard 1998; McCarthy 1999; Papadopoulos 1996). Despite problems of conceptual compatibility, attempts have been made to reconstruct urban regime theory by integrating it and the regulation approach (Lauria 1997). This conceptual synthesis may provide a way of connecting the role and activities of local political agents and institutions in economic development (the concern of urban regime theorists), to changes in the national and international political economy (the focus of the regulation approach). Jessop *et al.* (1999) used the example of local business leadership in Manchester's failed bid to host the Olympic Games in 2000 to illustrate how the scope for local political agency is structured by larger economic and political contexts and relations.

Comparative analyses of the similarities and differences in the organization of local economic development efforts between the various Western European countries as well as between cities in Western Europe and the US have been undertaken in an effort to improve our conceptualizations of the politics of urban economic development (Beauregard *et al.* 1992). Much research focused on the apparent convergence on a neoliberal model in the form and organization of local economic development policy and efforts in the UK and the US. While larger factors associated with the structural transformations of late capitalism, such as deindustrialization, promoted this convergence, institutional and cultural differences between the two countries resulted in significant national variations in public policy and outcomes (Gaffikin and Warf 1993; Wood 1996).

Nordic Europe

The principal treatments of Nordic Europe remain those in standard textbooks, particularly that by Jordan (1996), Malmström (1997), and McDonald (1997). These texts focus on the role that the physical environment and resource-based industry plays in virtually all the economies of Norden and they provide a solid grounding in the geography of Nordic Europe. Other scholarship has focused on economic, political, or medical geography in the region. Research in economic geography concerning Nordic Europe principally has focused on the nature of "learning economies," while that in political geography results from both issues of "widening" versus "deepening" within the EU and the social construction of regions. Research in medical geography exists because the long history of precise medical record-keeping in the region makes the longitudinal study of disease possible.

Economic Geography

Hansen (1991) notes that industrialization in Jutland is characterized by flexible production and labor practices, networks of small firms that cooperate and compete in international competition, and a highly unionized, high-wage labor force. He observes that this form of economic organization is a classic example of flexible specialization. He explains the emergence of this particular system of production in terms of the unique economic history of the region.

Knudsen (1996) discusses Danish economic restructuring during the 1980s. He argues that the distinct cultural history of Denmark resulted in the unique form of Danish economic restructuring. Throughout much of the 1980s and early 1990s Denmark's economy was characterized by high wages and high unemployment. This can be explained in terms of cultural attitudes about economic misfortune. In Denmark, it is believed that economic misfortune is the unavoidable consequence of the vagaries of the world market.

Investigation of the socio-cultural underpinnings of the Nordic economies remains an important focus of current research. Articulated best in recent European publishing on learning economies (Lundvall 1992; Maskell *et al.* 1998), these issues have had a significant impact on the North American geographic literature. It is reasonable to expect that this literature will continue to grow rapidly within the next few years. Current industrial production in Northern Europe takes place under the highest factor costs and most stringent environmental requirements in the world. As factor costs in industry increase and environmental regulations become more stringent in other parts of the advanced capitalist world, Nordic Europe may increasingly be looked to for answers.

Political Geography

Murphy and Hunderi-Ely (1996) provide an analysis of the 1994 EU membership vote in Nordic Europe. In the 1994 referendum, Sweden and Finland joined the EU, while Norway voted not to join. Murphy and Hunderi-Ely show that support for the EU membership was highly variable across the region. Voting against membership was strongest in rural areas eligible for substantial EU subsidies, while voting for membership was strongest in urban areas that had the most to gain from cheaper consumer goods. This pattern of voting is indicative of deeper motivations that have historical and cultural undertones. In particular, in each country the "yes" campaign argued that EU membership allowed improved access to European markets, increased political representation, and enhanced peace and security. The "no" campaign in each country argued that EU membership entailed loss of sovereignty, rural decline, and tremendous economic cost. Analysis of Nordic Europe provides an exemplar of the future of the EU. The discussion concerning the EU in Nordic Europe is a prelude to a much larger discussion of "widening versus deepening" that will take place as the EU expands eastward in the first decade of the twenty-first century.

Häkli (1998) examines the historical underpinnings of Finnish administrative regions and uses the institutional production of space in Finland to create a history of the present. He argues that current administrative regions in Finland have their origin in the historically based provinces of the nation, but that the historical basis of provinces has never been the subject of adequate scrutiny. Häkli points out that only a relative few of the current administrative regions have a distinct cultural identity in Finnish history and that even those are characterized by a fuzziness in terms of boundaries. The remainder are the result of historical mapping and surveying or the imposition of central authority on the land. Indeed it is common that cultural and administrative boundaries work at cross-purposes. Furthermore, even cultural boundaries are the result of a discursive process concerning membership in a cultural group.

Medical Geography

The relatively homogeneous populations and long history of medical record collection made possible by national health-care systems has meant that Northern Europe has proven an important testing ground for geographical theory of epidemics since at least the publication of Cliff *et al.* (1981). This tradition continues in several articles by Löytönen (1991*a*, *b*, 1994) on the spread of HIV virus and AIDS in Nordic Europe, and Aase and Bentham (1994) on malignant melanoma and its relationship to ozone depletion near the poles. The latter finds a close relationship between UV exposure and malignant melanoma throughout Nordic Europe.

That Nordic Europe has received little attention in the 1990s by American geographers is surprising. The region provides an interesting exception to much of the rest of the advanced capitalist world by remaining strongly committed to democratic socialism. It also provides a startling example of how social forces structure economies. At the same time, Nordic Europe provides insight into the problems of an increasingly integrated Europe.

Eastern Europe

The dramatic changes in the post-Cold War map of Europe brought about a radical reinterpretation of the notion of Europe and European identity. The external reorientation and the internal transformation of Eastern Europe dominated the research of US geographers that specialize in Europe, as well as those focusing on the former Soviet Union (see Ch. 44, Russia, Central Eurasia, and East Europe, for further insight). The political, and, to a lesser extent, the economic aspects of the transition ranked at the top of their agenda.

The Effects of Democratization

The most comprehensive volume on the new political geography of Eastern Europe resulted from cooperation between geographers in the US and West Europe. O'Loughlin and van der Wusten (1993) produced a set of edited chapters that covers the diverse aspects of the new security situation, the social and political transformations, effected by the transition to democracy, and the new electoral geographies of Eastern Europe. In a world-systems interpretation, O'Loughlin views the decline of the Soviet economy as a conditioning factor for the change in Eastern Europe. The combination of communist legacies and the more recent ethno-territorial problems resulted in the emerging political order in Europe.

The political fragmentation, which accompanied the devolution of power in the eastern part of Europe, has been particularly striking against the background of Western Europe's increasing cohesion. The dissolution of all three federations in Eastern Europe and Western Europe's increasing cohesion and the unification of Germany were topics garnering the most interest in the 1990s. Harris (1991, 1997) was the first to comment on the effects of German reunification, including the renewed interest in the former East Prussia. Even before the split of Czechoslovakia, the Soviet Union, and Yugoslavia, Wixman (1991) examined ethnic minorities, conflicts, and nationalism across Eastern Europe, including part of the Soviet Union. He concluded that "secessionist movements of the Baltics and the potential disintegration of Yugoslavia are the most serious issues for the US and its European allies." Wixman emphasized specifically that Yugoslavia "poses a great threat to the Balkan region as a whole."

The destructive results of competing nationalisms for the nation–state relationship was further examined in the context of the Russian Federation and its "Near Abroad" (Harris 1993*a*, *b*, *c*; 1994*a*, *b*; Kaiser 1995*a*). Kaiser (1995*b*) explored socio-cultural, ethnographic, and socioeconomic factors to assess prospects for disintegration along ethno-territorial lines. Pavlinek (1995) challenged the ethno-nationalist explanations for the disintegration of Czechoslovakia. His evaluation of the influence of the different factors for the split reasoned that regional inequalities and, particularly, uneven economic development were much more consequential.

The emerging democratic political systems in Eastern Europe provided unique opportunities for study of their multi-party electoral geographies. Martis *et al.* (1992) and Koulov (1995) emphasized the importance of geographic context for understanding election results in various countries and electoral districts, as well as a number of socioeconomic variables. The new electoral geography of Eastern Europe reflects core–periphery and rural–urban cleavages, as well as age and education differences among the different geographic regions. In addition, it demonstrates the salience of external factors, such as the developments in the core of the Soviet Union-dominated geopolitical system, for the diverse forms, the pace, and substantive results of the 1989 political change in the countries of the region.

The number of states in Eastern Europe doubled in the early 1990s. Nevertheless, US geographers see few guarantees at the end of the millennium that the process of political fragmentation has exhausted itself. At the very least, two federal structures, rump Yugoslavia, and Bosnia and Herzegovina, still endure. In both cases, conflicts, including threats of secession, mar the last federations in Central and Eastern Europe with important economic and security implications for the continent as a whole.

Geo-economic Reorientation: Back to Europe

The political insecurity further increased the economic instability, which followed the disintegration of the Council for Mutual Economic assistance. US geographers addressed the reorientation of Eastern European economies towards the European Union (EU) to point out the key role of Western investment and trade. Ivy (1995) found East–West air transport linkages improved and leading the way for full economic integration of the region. Murphy (1992*a*) and Crocker and Berentsen (1998) warned against concentration of economic interaction in the capital cities and the western parts of the countries, since it reinforces existing regional inequalities within individual states. Such tendencies complicate the economic restructuring process and jeopardize the stability of these countries.

An overarching issue in US geographical research in Eastern Europe of the 1990s was how different regions cope with economic change. Berentsen (1996, 1999) analyzed the regional population dynamics in Germany in the aftermath of the collapse of the command economy to identify declining industrial cities and former district capitals, as well as regions experiencing industrial restructuring, suburban growth, or relative stabilization. His comparative assessment of the socioeconomic development of Eastern Germany for the ten-year period since the abandonment of central planning aimed at understanding the key results of the transition in this part of Europe and drawing conclusions for the future development of other "transition" economies. Crocker and Berentsen (1998) indicated that the regions with the best relative market accessibility may potentially enjoy the highest benefits from expanded trade and increasing economic integration within a future EU framework. Buckwalter (1995, 1998) put forward the possibility that the former Soviet sphere could conceivably become an economic "shatter belt" in post-Cold War Europe, since many of the policies that determine their prosperity would originate elsewhere.

Research interest in the intended and unintended consequences of the transition to market economy in Eastern Europe is likely to increase. European geographers have already taken the lead in exploring the impacts of economic transformation in individual states on their prospects of potential EU membership. Williams *et al.* (1998) found the comparisons with Southern Europe in the 1980s to be particularly instructive in terms of understanding both the formidable task of membership and the changing approaches of the EU to its enlargement.

Among the most important contributions of US geography to the newly emerged "transition" theory is the further elucidation of the "scale of research analysis" problem. Murphy (1992*b*) criticizes the preoccupation with the nation-state as a unit of analysis, since it directs attention away from often-contradictory developments at other regional scales. The volume on the geography of Southeast Europe, edited by Hall and Danta (1996), presents a promising step in the right direction. The continuing collaboration between these geographers will soon be realized in another publication on European Union expansion into Eastern Europe.

While collaborations among US and West European geographers abound, there are still relatively few joint publications with East Europeans. Berentsen (1992) was one of the first to present the works of East European geographers to the West as a guest editor of a special issue of *Tijdschrift voor Economische en Sociale Geografie* on regional problems and regional planning in Central and Eastern Europe. Most of the multi-authored publications in the 1990s involve Bulgaria, Slovakia, and Poland (e.g. Paskaleva *et al.* 1998) Significant potential exists for further academic cooperation, especially in the new East European states.

Other topics of US research on Eastern Europe include environmental, urban, social, and rural geography (e.g. Danta 1993*a*, *b*; Zaniewski 1991). These publications, however, focus mostly on individual states. Unfortunately, the physical geography of Eastern Europe has been totally ignored by US geographers as a topic of research.

Mediterranean Europe

American geographers have handled topics of Mediterranean Europe in a distinctive manner: they've disregarded areas of peace and prosperity and have focused on subjects of potential or real conflict. Impacts of environmental neglect and causes of political crises and ethnic tensions have been discussed. In this part of the chapter, we will not elaborate on the former Yugoslavia.

The inclusion of the region into the EU, with the exception of the area of former Yugoslavia and Albania, has in the 1990s transformed economies, demographic characteristics, and cultural landscape features. Major studies by American geographers are therefore related to the process of the political and economic transformation, to migration and minority questions, and to traffic- and tourism-related problems. The lasting discussions concern: European, national, and regional diversity (Agnew 1990, 1994, 1995, 1997, 1998; Agnew and Bolling 1993; Murphy 1993, 1999); the political status of Gibraltar and Cyprus, Basque and Catalan autonomy in Spain, and the regionalism of Corsica in France and "Padania" in Italy (Glassner 1996; Poulsen 1995); the pollution of the Mediterranean Sea (Diem 1997); the impact of tourism on environment and culture (Gosar 1993); and the geographic relevance of minority rights. Diverse border issues, in particular along the border of the European Union have been discussed as well (Minghi 1994*a*, *b*). Industrialization and tourism have caused serious air and general pollution problems in particular in and around major cities, such as Barcelona, Marseilles, Nice, Milan, Rome, Naples, and Athens. Ozone layers and algal bloom affect tourism and travels in general. Enormous damage is done to cultural monuments; Venice is a striking example of it (Diem 1997). The economic development is responsible for severe social problems as well. Interstate and rural–urban migrations have left several areas without the young and dynamic male population. The Pyrenees in Spain and France, the Southern Alps in France and Italy, but in particular the Mezzogiorno of Italy have become synonymous for demographic problems and, consequently, underdevelopment (Murphy 1993). Cities are employment targets for many legal and illegal immigrants from Eastern Europe (Albania, ex-Yugoslavia), Africa, and Asia. The racial (and ethnic) structure of Southern Europe's urban areas has in the 1990s changed dramatically (Leitner 1995, 1997). All the countries of the European South have a "black" or hidden economy that contributes to the gross domestic product but is not included in the official statistics.

To level out the burden of the European north–south disparity, in the 1990s four-lane highways were constructed and fast railtracks were put in place. The trans-alpine traffic has had an enormous impact on the environment and development. This is most visible along the Brenner and Mont Blanc highway and rail axis. Northern Italy's and northern Spain's core industrial areas are well connected with the rest of Europe. The Chunnel made the Alps and the Mediterranean more accessible to the British, and the fast rail connections and highways have allowed North Europeans, in particular Germans, to reach the European Sunbelt more easily. Along with the Spanish Balearic Islands, several Aegean islands in Greece and Turkey have in the 1990s become non-autochthonous retirement communities (R. Davidson 1998). The hindering effect of the war in ex-Yugoslavia has slowed down a similar development along the Adriatic.

In the literature on Southern Europe, special attention was given to the European Union's eastern borders. Here, the states of the EU connect with the most fragile part of Europe. The Shatter Belt of Europe is still trembling (Gosar and Klemencic 1994). Minority problems and the general economic situation along the borders have therefore a high priority in regional policies. Several trans-border EU programs have been put in place. Some are a continuation of the previous bilateral state agreements, such as between Italy and Slovenia, and some are new joint venture programs. Elections and, consequently, the decision-making process have enormous impacts on the geography of border regions (Minghi 1997).

Nowhere can the effect of tourism be seen better than in Southern Europe. Comprising no more than 7 per cent of the planet's landmass, the continent of Europe is more intensively visited by the world's tourists than all other continents taken together. One billion tourism trips are made annually by European residents within their own country, 300 million Europeans make holidays in another state of the continent, and 15 million tourists arrive yearly from other parts of the planet (Rafferty 1993).

In the 1990s research in tourism geography has focused on the opening of the European East and the unification of Germany. In the 1990s, Eastern Europe, particularly cities such as Prague and Budapest, opened their doors to Western tourists and business people. Also, freed from the oppression of communism, the new East European middle class increasingly can be found on holiday at Southern Europe's most prestigious destinations (Hall 1998).

Conclusion

Europe is under-researched in US geography, given its relative importance in the world politico-economic system and the wealth of development models it offers. In the 1990s, geographic research on Europe was driven primarily by the fundamental and unique transformation of the political and economic map of this region. Methodologically, this is expressed in the search for relevant case and topic-specific scales of analysis. The state scale has come under strong attack. Conversely, border areas, especially along the borders of the European Union, trans-border programs, regional policies, and the institutional production of space (e.g. the Euroregions) acquire fresh importance. A major challenge in this respect will be to better conceptualize the political and economic changes occurring at a variety of spatial scales and the interactions between these scales (see Agnew 1990, 1994, 1995, 1996, 1997, 1998; Agnew and Bolling 1993; Berentsen 1997; Harris 1991, 1993*a*, *b*, 1997; Jordan 1996; McDonald 1997; Murphy 1991; Unwin 1998).

Geographic research in the 1990s invariably integrates a number of topical areas. Political geography, however, leads both in quantity of publications and introduction of relatively new research topics, such as electoral geography of Eastern Europe. The increasing complexity of the European political map and the intra-European political and economic realignments will certainly continue to draw the attention of US geographers. Themes that are likely to persist encompass the geographic differences in the power shift away from the state scale and their causes and consequences, including the resulting territorial entities and political identities. In particular, geographers will endeavor conceptually and empirically to address the new divisions in Europe, states–nations relationships, democratization, nationalism, and ethnic conflict as political fragmentation occurs at various spatial scales at the same time as supranational integration forces continue apace.

After political geography, economic geography is next in popularity among US geographers specializing in Europe. A major focus of research is the ongoing changes to the EU both as an institution and as a union of member states. Geographers analyze empirically and theorize about the processes and implications of ongoing "deepening," involving further economic and political integration for the existing member states, as well as subsequent "widening" with the inclusion of new member countries in the future. Researchers will also assess the diverse impacts and implications of the incorporation of different East European states in the expanding European Union. Attention will continue to focus on the increasing economic integration occurring across Europe as a whole amid processes of globalization involving corporate restructuring and the spatially uneven nature of the associated urban and regional change. Moreover, researchers will attempt to study the politics of urban and regional economic development as governments at all levels attempt to respond and adapt to changes in the larger political economy.

Geographers will also be involved actively in public policy analyses addressing issues cutting across regional and state borders such as environmental damage and population redistribution, especially interstate and interregional migrations. Demographic crisis regions, which experience severe depopulation, aging, and gender imbalances, are certain to be research foci in the new millennium. European integration raises important questions about the conflicting goals of economic efficiency versus social and spatial equity at a variety of scales. The geographic analysis of the voting patterns in Eastern and Nordic Europe, for example, exhibit interesting similarities. In 1994, Norway's rural and peripheral areas voted against European Union membership. The rural and peripheral areas in Eastern Europe demonstrate similar hesitation in respect to the rapid adoption of the "free" market. Clearly, cross-regional comparisons will present new challenges for US geographers in their attempts to bridge the traditional scales of analysis that were used to form the main structure of this chapter. Environmental geography, as it relates to uneven development, expansion of trans-state infrastructure, as well as increased flows of goods and people, including tourists, is yet another enticing area in current and future US geography on Europe. The authors can confidently forecast much greater topical diversification in the near future. Nevertheless, we support the opinion that the political fragmentation should not invite intellectual segmentation within the discipline.

The pressing problems faced by Europe in the immediate future stem largely from the successes of European integration. Europe is now the largest consumer market in the world. As such, it is a major focus of global capitalism. Global capitalism, in conjunction with the growing supra-national power of the European Union has tended to alienate increasingly larger portions of everyday life from local control. Economic, social, and political structures are much less embedded than was the case a decade ago. Furthermore, the inherent uneven nature of development under capitalism has been lent a certain meanness under the neoliberal policies that swept Europe in the late 1980s and 1990s and continue today as countries struggle to enter the Euro-Zone. The result of this alienation of economic, social, and political power has been an increasing emphasis on cultural difference at local and regional levels. Not infrequently, cultural difference has sparked intolerance and separatism and degenerated into open ethnic conflict.

References

Aase, A., and Bentham, G. (1994). "The Geography of Malignant Melanoma in the Nordic Countries: The Implications of Stratospheric Ozone Depletion." *Geografiska Annaler*, 76B/2: 129–39.

Agnew, J. A. (1990). "Political Decentralization and Urban Policy in Italy: From 'State-Centered' to 'State-Society' Explanation." *Policy Studies Journal*, 18/3: 768–84.

—— (1994). "The National Versus the Contextual: The Controversy Over Measuring Electoral Change in Italy Using Goodman Flow-of-Vote Estimates." *Political Geography*, 13/3: 245–54.

—— (1995). "The Rhetoric of Regionalism: The Northern Italian League in Italian Politics, 1983–94." *Transactions, Institute of British Geographers*, 20/2: 156–72.

—— (1996). "Mapping Politics: How Context Counts in Electoral Geography." *Political Geography*, 15/2: 129–46.

—— (1997). "The Dramaturgy of Horizons: Geographical Scale in the 'Reconstruction of Italy' by the New Italian Parties, 1992–95." *Political Geography*, 16/2: 99–121.

—— (1998). "The Impossible Capital: Monumental Rome Under Liberal and Fascist Regimes, 1870–1943." *Geografiska Annaler, Series B: Human Geography*, 80/4: 229–40.

Agnew, J. A., and Bolling, K. (1993). "Two Romes or More? The Use and Abuse of the Center–Periphery Metaphor." *Urban Geography*, 14/2: 119–44.

Amin, A., and Thrift, N. (1995). *Globalization, Institutions, and Regional Development in Europe*. Oxford: Oxford University Press.

Austin, C. Murray (1992). "Geographical Perspectives of Nationalism." *History of European Ideas*, 15/4–6: 621–9.

Beauregard, R. A., Lawless, P., and Deitrick, S. (1992). "Collaborative Strategies for Reindustrialization: Sheffield and Pittsburgh." *Economic Development Quarterly*, 6/4: 418–30.

Berentsen, W. H. (1992). "Introduction to the Special Issue: Regional Problems and Regional Planning in Central and Eastern Europe." *Tijdschrift voor Economische en Sociale Geografie*, 83/5: 339–41.

—— (1993). "A Geopolitical Overview of Europe." *Social Education*, 57/4: 170–6.

—— (1996). "Regional Population Changes in Eastern Germany after Unification." *Post-Soviet Geography and Economics*, 37/10: 615–32.

—— (ed.) (1997). *Contemporary Europe: A Geographic Analysis*. 7th edn. New York: Wiley.

—— (1999). "Socioeconomic Development in Eastern Germany, 1989–1998: A Comparative Perspective." *Post-Soviet Geography and Economics*, 40/1: 27–43.

Buckwalter, D. (1995). "Spatial Inequality, Foreign Investment, and Economic Transition in Bulgaria." *The Professional Geographer*, 47/3: 288–97.

—— (1998). "Geopolitical Legacies and Central European Export Linkages: A Historical and Gravity Model Analysis of Hungary." *The Professional Geographer*, 50/1: 57–71.

Cliff, A. D., Haggett, P., Ord, J. K., and Versey, G. D. (1981). *Spatial Diffusion: An Historical Geography of Epidemics in an Island Community*. Cambridge: Cambridge University Press.

Cooke, P., Moulaert, F., Swyngedouw, E., Weinstein, O., and Wells, P. (1992). *Towards Global Localization: The Computing and Telecommunications Industries in Britain and France*. London: UCL Press.

Corson, M. W., and Minghi, J. V. (1994). "Reunification of Partitioned Nation-States: Theory Versus Reality in Vietnam and Germany." *Journal of Geography*, 93/3: 125–31.

Crocker, J. E., and Berentsen, W. H. (1998). "Accessibility Between the European Union and East-Central Europe: Analysis of Market Potential and Aggregate Travel." *Post-Soviet Geography and Economics*, 39/1: 19–44.

Danta, D. (1993*a*). "Ceausescu's Bucharest." *Geographical Review*, 83/2: 70–182.

—— (1993*b*). "Diffusion of Mesoamerican Food Complex to Southeastern Europe." *Geographical Review*, 83/2: 194–204.

Davidson, F. M. (1996). "The Rise and Fall of the SNP since 1983: Analysis of a Regional Party." *Scottish Geographical Magazine*, 112/1: 11–19.

—— (1997). "Integration and Disintegration: A Political Geography of the European Union." *Journal of Geography*, 96/2: 69–75.

Davidson, R. (1998). *Travel and Tourism in Europe*. 2nd edn. London: Longman.

Dicken, P. (1992). "Europe 1992 and Strategic Change in the International Automobile Industry." *Environment and Planning A*, 24/1: 11–31.

Dicken, P., and Öberg, S. (1996). "The Global Context: Europe in a World of Dynamic Economic and Population Change." *European Urban and Regional Studies*, 3/2: 101–20.

Diem, A. (1997). "Southern Europe," In Berentsen (1997: 383–430).

Dorling, D., and Woodward, R. (1996). "Social Polarisation 1971–91: A Micro-Geographical Analysis of Britain." *Progress in Planning*, 45/2 pt. 2: 63–122.

Duncan, S. (ed.) (1994). "The Diverse Worlds of European Patriarchy." *Environment and Planning A*, 26/8: 1174–296; 26/9: 1339–434.

Dunford, M., and Kafkalas, G. (eds.) (1992). *Cities and Regions in the New Europe: the Global–Local Interplay and Spatial Development Strategies*. London: Belhaven.

Ettlinger, N. (1994). "The Localization of Development in Comparative Perspective." *Economic Geography*, 70: 144–66.

Flint, C. (1993). "Back to Front? The Existence and Threat of Extremism in English Nationalism." *Political Geography*, 12/2: 180–4.

Foley, P. D., and Watts, H. D. (1996). "New Process Technology and the Regeneration of the Manufacturing Sector of an Urban Economy." *Urban Studies*, 33/3: 445–57.

Fusch, R. (1994). "The Piazza in Italian Urban Morphology." *Geographical Review*, 84/4: 424–38.

Gaffikin, F., and Warf, B. (1993). "Urban Policy and the Post-Keynesian State in the United Kingdom and the United States." *International Journal of Urban and Regional Research*, 17: 67–84.

Gertler, M. S. (1996). "Worlds Apart: The Changing Market Geography of the German Machinery Industry." *Small Business Economics*, 8: 87–106.

Gertler, M. S., and Schoenberger E. (1992). "Industrial Restructuring and Continental Trade Blocks: The European Community and North America." *Environment and Planning A*, 24: 2–10.

Gibbs, D. C. (1996). "European Environmental Policy: The Implications for Local Development." *Regional Studies*, 30/1: 90–2.

Glasmeier, A. (1991). "Technological Discontinuities and Flexible Production Networks: The Case of the World Watch Industry." *Research Policy*, 20: 469–85.

—— (1995). "Flexible Districts, Flexible Regions? The Institutional and Cultural Limits to Districts in an Era of Globalization and Technological Paradigm Shifts," in Amin and Thrift (1995: 118–46).

Glassner, M. I. (1996). *Political Geography*. 2nd edn. Chichester: Wiley.

Gosar, A. (1993). "International Tourism and its Impact on the Slovenian Society and Landscape." *GeoJournal*, 30/3: 339–48.

Gosar, A., and Klemencic, V. (1994). "Current Problems of Border Regions Along the Slovene-Croatian Border," in W. A. Gallusser (ed.), *Political Boundaries and Coexistence*. Bern: P. Lang, 88–94.

Grant, R. (1993). "Against the Grain: Agricultural Trade Policies of the US, the European Community and Japan at the GATT." *Political Geography*, 12/3: 247–62.

Häkli, J. (1998). "Discourse in the Production of Political Space: Decolonizing the Symbolism of Provinces in Finland." *Political Geography*, 17/3: 331–63.

Hall, D. (1998). "Tourism and Travel," in Unwin (1998: 311–29).

Hall, D., and Danta, D. (eds.) (1996). *Reconstructing the Balkans: A Geography of New Southeast Europe*. Chichester: John Wiley.

Hallsworth, A. G. (1996). "Short-Termism and Economic Restructuring in Britain." *Economic Geography*, 72/1: 23–37.

Hansen, N. (1991). "Factories in Danish Fields: How High-wage, Flexible Production Has Succeeded in Peripheral Jutland." *International Regional Science Review*, 14/2: 109–32.

Harris, C. (1991). "Unification of Germany in 1990." *Geographical Review*, 81/2: 170–82.

—— (1993*a*). "A Geographic Analysis of Non-Russian Minorities in Russia and Its Ethnic Homelands." *Post-Soviet Geography*, 34/9: 543–97.

—— (1993*b*). "Ethnic Tensions in Areas of the Russian Diaspora." *Post-Soviet Geography*, 34/4: 233–8.

—— (1993*c*). "New European Countries and Their Minorities." *Geographical Review*, 83/3: 301–20.

—— (1994*a*). "Ethnic Tensions in the Successor Republics in 1993 and Early 1994." *Post-Soviet Geography*, 35/4: 185–203.

—— (1994*b*). "Research Note on the Release of More Detailed Data from Three Soviet Censuses." *Post-Soviet Geography*, 35/1: 59–62.

—— (1997). "Europa Regional: Focus on Former East Prussia." *Post-Soviet Geography and Economics*, 38/1: 59–62.

Huntoon, L. (1998). "Immigration to Spain: Implications for a Unified European Union Immigration Policy." *International Migration Review*, 32/2: 423–50.

Ivy, R. L. (1995). "The Restructuring of Air Transport Linkages in the New Europe." *The Professional Geographer*, 47/3: 280–8.

Jessop, B., Peck, J., and Tickell, A. (1999). "Retooling the Machine: Economic Crisis, State Restructuring, and Urban Politics," in A. E. G. Jonas and D. Wilson (eds.), *The Urban Growth Machine: Critical Perspectives, Two Decades Later*. Albany, NY: State University of New York Press, 141–59.

Johnson, N. C. (1998). "Nations and Peoples," in Unwin (1998: 85–99).

Johnston, R. J. (1994). "The Conflict over Voting Power in the European Union Council of Ministers, March 1994." *Environment and Planning A*, 26: 1001–12.

Johnston, R. J., and Pattie, C. J. (1996). "Local Government in Local Governance: The 1994–95 Restructuring of Local Government in England." *International Journal of Urban and Regional Research*, 20/4: 671–96.

Jordan, T. G. (1996). *The European Cultural Area*. New York: HarperCollins.

Kaiser, R. (1995*a*). "Nationalizing the Work Force: Ethnic Restratification in the Newly Independent States." *Post-Soviet Geography*, 36/2: 87–111.

—— (1995*b*). "Prospects for the Disintegration of the Russian Federation." *Post-Soviet Geography*, 36/7: 426–35.

Kirkham, J. D., and Watts, H. D. (1998). "Multi-locational Manufacturing Organisations and Plant Closures in Urban Areas." *Urban Studies*, 35/9: 1559–75.

Knudsen, D. (1996). "Contrasting Restructuring in Denmark and the American Midwest." *Geografisk Tidsskrift*, 96/1: 81–94.

Kofman, E. (1995). "Citizenship for Some But Not for Others: Spaces of Citizenship in Contemporary Europe." *Political Geography*, 14/2: 121–37.

Koulov, B. (1995). "Geography of Electoral Preferences: The 1990 Great National Assembly Elections in Bulgaria." *Political Geography*, 14/3: 241–58.

Lauria, M. (ed.) (1997). *Reconstructing Urban Regime Theory: Regulating Urban Politics in a Global Economy*. Thousand Oaks, Calif: Sage.

Leitner, H. (1995). "International Migration and the Politics of Admission and Exclusion in Postwar Europe." *Political Geography*, 14/3: 259–78.

—— (1997). "Reconfiguring the Spatiality of Power: The Construction of a Supranational Migration Framework for the European Union." *Political Geography*, 16/2: 123–43.

Leitner, H., and Sheppard, E. (1998). "Economic Uncertainty, Inter-Urban Competition and the Efficacy of Entrepreneurialism," in T. Hall and P. Hubbard (eds.), *The Entrepreneurial City: Geographies of Politics, Regime and Representation*. Chichester: Wiley, 285–307.

Löytönen, M. (1991*a*). "The Box-Jenkins Forecast of HIV Seropositive Population in Finland, 1991–1993." *Geografiska Annaler*, 73B/2: 121–31.

— (1991*b*). "The Spatial Diffusion of Human Immunodeficiency Virus Type 1 in Finland, 1982–1997." *Annals of the Association of American Geographers*, 81/1: 127–51.

— (1994). "HIV and AIDS in the Nordic Countries." *Geografiska Annaler*, 76B/2: 117–27.

Lundvall, B. Å. (ed.) (1992). *National Systems of Innovation: Towards a Theory of Innovation and Interactive Learning*. London: Pinter.

McCarthy, L. (1990). "Evolution, Present Condition and Future Potential of the Smithfield Area of Dublin." *Irish Geography*, 23/2: 90–106.

— (1998). "Analysis of Urban Income Inequalities in an Integrating Europe: Overcoming the Absence of Comparable Units of Analysis and Economic Data." *Papers and Proceedings of the Applied Geography Conferences*, 21: 203–12.

— (1999). "The Politics of Mediating Global Economic Change: Linking the Local and the International in Western Europe." *Space & Polity*, 3/2: 217–31.

McDonald, J. (1997). *The European Scene: A Geographic Perspective*. 2nd edn. Englewood Cliffs, NJ: Prentice Hall.

Malmström, V. H. (1997). "Region Nord: the European North," in W. H. Berentsen (1997: 342–82).

Martinelli, F., and Schoenberger, E. (1991). "Oligopoly is Alive and Well: Notes for a Broader Discussion of Flexible Accumulation," in G. Benko and M. Dunford (eds.), *Industrial Change and Regional Development: the Transformation of New Industrial Spaces*. London: Belhaven, 117–33.

Martis, K. C., Kovacs, Z., Kovacs, D., and Peter, S. (1992). "The Geography of the 1990 Hungarian Parliamentary Elections." *Political Geography*, 11/3: 283–305.

Maskell, P., Eskelinen, H., Hannibalsson, I., Malmberg, A., and Vatne, E. (1998). *Competitiveness, Localized Learning and Regional Development: Specialization and Prosperity in Small Open Economies*. London: Routledge.

Minghi, V. J. (1994*a*). "European Borderlands: International Harmony, Landscape Change and New Conflict." *Eurasia: World Boundaries*, 3: 89–98.

— (1994*b*). "The Impact of Slovenian Independence on the Italo-Slovene Borderland: An Assessment of the First Three Years," in W. A. Gallusser (ed.), *Political Boundaries and Coexistence*. Berne: P. Lang, 88–94.

— (1997). "Voting and Borderland Minorities: Recent Italian Elections and the Slovene Minority in Eastern Friuli-Venezia Giulia." *GeoJournal*, 43/3: 263–71.

Monk, J., and García, M. D. (1996). "Placing Women of the European Union," in M. D. García and J. Monk (eds.), *Women of the European Union: the Politics of Work and Daily Life*. London: Routledge, 1–30.

Murphy, A. B. (1990). "Electoral Geography and the Ideology of Place: The Making of Regions in Belgian Electoral Politics," in R. J. Johnston, F. M. Shelley, and P. J. Taylor (eds.), *Developments in Electoral Geography*. London: Routledge, 227–41.

— (1991). "The Emerging Europe of the 1990s." *The Geographical Review*, 81/1: 1–17.

— (1992*a*). "Urbanism and the Diffusion of Substate Nationalist Ideas." *History of European Ideas*, 14: 639–45.

— (1992*b*). "Western Investment in East-Central Europe: Emerging Patterns and Implications for State Stability." *The Professional Geographer*, 44/3: 249–59.

— (1993). "Emerging Regional Linkages within the European Community: Challenging the Dominance of the State." *Tijdschrift voor Economische en Sociale Geografie*, 84/2: 103–18.

— (1996). "The Sovereign State System as Political-Territorial Ideal: Historical and Contemporary Considerations," in T. J. Biersteker and C. Weber (eds.), *State Sovereignty As Social Construct*. Cambridge: Cambridge University Press, 81–120.

— (1999). "Rethinking the Concept of European Identity," in G. Herb and D. Kaplan (eds.), *Nested Identities: Nationalism, Territory, and Scale*. Boulder, Colo.: Rowman & Littlefield, 53–73.

Murphy, A. B., and Hunderi-Ely, A. (1996). "The Geography of the 1994 Nordic Vote on European Union Membership." *The Professional Geographer*, 48/3: 284–97.

O'Loughlin, J., and Anselin, L. (1996). "Geo-Economic Competition and Trade Bloc Formation: United States, German, and Japanese Exports, 1968–1992." *Economic Geography*, 72/2: 131–60.

O'Loughlin, J., and Parker, A. J. (1990). "Tradition Contra Change: the Political Geography of Irish Referenda, 1937–87," in R. J. Johnston, F. M. Shelley, and P. J. Taylor (eds.), *Developments in Electoral Geography*. London: Routledge, 60–85.

O'Loughlin, J., and van der Wusten, H. (eds.) (1993). *The New Political Geography of Eastern Europe*. London: Belhaven; New York: Halsted.

O'Loughlin, J., Flint, C., and Anselin, L. (1994). "The Geography of the Nazi Vote: Context, Confession, and Class in the Reichstag Election of 1930." *Annals of the Association of American Geographers*, 84/3: 351–80.

Pallares-Barbera, M. (1998). "Changing Production Systems: The Automobile Industry in Spain." *Economic Geography*, 74/4: 344–59.

Papadopoulos, A. G. (1996). *Urban Regimes and Strategies: Building Europe's Central Executive District in Brussels*. Chicago: The University of Chicago Press.

Paskaleva, K., Shapira, P., Pickles, J., and Koulov, B. (eds.) (1998). *Bulgaria in Transition: Environmental Consequences of Economic and Political Transformation*. Aldershot: Ashgate.

Pattie, C. J., and Johnston, R. J. (1990). "One Nation or Two? The Changing Geography of Unemployment in Great Britain, 1983–1988." *The Professional Geographer*, 42/3: 288–98.

Pavlik, C. E. (1995). "Götterdämmerung or a Brave New World?" *International Regional Science Review*, 17/3: 361–6.

Pavlinek, P. (1995). "Regional Development and the Disintegration of Czechoslovakia." *Geoforum*, 26/4: 351–72.

Poulsen, T. (1995). *Nations and States—A Geographic Background to World Affairs*. Englewood Cliffs, NJ: Prentice Hall.

Rafferty M. (1993). *A Geography of World Tourism. Europe*. Englewood Cliffs, NJ: Prentice-Hall.

Sadler, D. (1992*a*). "Beyond '1992': the Evolution of European Community Policies towards the Automobile Industry." *Environment and Planning C: Government and Policy*, 10: 229–48.

— (1992*b*). "Industrial Policy of the European Union: Strategic Deficits and Regional Dilemmas." *Environment and Planning A*, 24: 1712–30.

Storper, M. (1992). "The Limits to Globalization: Technology Districts and International Trade." *Economic Geography*, 68/1: 60–93.

— (1993). "Regional 'Worlds' of Production: Learning and Innovation in the Technology Districts of France, Italy and the USA." *Regional Studies*, 27/5: 433–55.

— (1995). "The Resurgence of Regional Economies, Ten Years Later: The Region as a Nexus of Untraded Interdependencies." *European Urban and Regional Studies*, 2/3: 191–221.

— (1997). *The Regional World: Territorial Development in a Global Economy*, New York: Guilford.

Storper, M., and Harrison, B. (1991). "Flexibility, Hierarchy and Regional Development: The Changing Structure of Industrial Production Systems and Their Forms of Governance in the 1990s." *Research Policy*, 20: 407–22.

Storper, M., and Scott, A. J. (1989). "The Geographical Foundations and Social Regulation of Flexible Production Complexes," in J. Wolch and M. Dear (eds.), *The Power of Geography*. London: Unwin Hyman, 21–40.

Storper, M., and Walker, R. (1989). *The Capitalist Imperative: Territory, Technology, and Industrial Growth*. Oxford: Blackwell.

Unwin, T. (ed.) (1998). *A European Geography*. Harlow: Longman.

Warf, B. (1992). "International Finance, Europe 1992 and Euro-American Cities." *Landscape and Urban Planning*, 22: 255–67.

Watts, H. D. (1990). "Manufacturing Trends, Corporate Restructuring and Spatial Change," in D. Pinder (ed.), *Western Europe: Challenge and Change*. London: Belhaven, 56–71.

Williams, A. M. (1994). *The European Community: The Contradictions of Integration*. 2nd edn. Oxford: Blackwell.

Williams, A. M., Balaz, V., and Zajaz, S. (1998). "The EU and Central Europe: The Remaking of Economic Relationships." *Tijdschrift voor Economische en Sociale Geografie*, 89/2: 131–49.

Wilvert, C. (1994). "Spain: Europe's California." *Journal of Geography*, 93/2: 74–9.

Wixman, R. (1991). "Ethnic Nationalism in Eastern Europe," in US Central Intelligence Agency, *Eastern Europe: The Impact of Geographic Forces on a Strategic Region*. Washington, DC: US CIA, 36–47.

Wood, A. (1996). "Analysing the Politics of Local Economic Development: Making Sense of Cross-National Convergence." *Urban Studies*, 33/8: 1281–95.

Zaniewski, K. (1991). "Housing Inequalities Under Socialism: the Case of Poland." *Geoforum*, 22/1: 39–53.

CHAPTER 43

Latin American Geography

David J. Robinson, César Caviedes, and David J. Keeling

Introduction

The brief essay that constitutes this chapter demonstrates a resurgence of work on the region that bodes well for the future. A new generation of scholars is replacing those who have for many years provided leadership in a variety of subfields. Several old hands have retired, some are still publishing *con gusto* (Denevan, Siemens, Horst, Preston), others we have lost forever (Parsons, Stanislawski, West, Eden) though through their works (Denevan 1989, 1999; Pederson 1998) and their students they remain with us.

It perhaps needs to be said that this brief account of the Latin Americanist historiography of the last decade should not be viewed in isolation. Too often we are marginalized as mere "regionalists" in an age that surely lacks well-trained ones (National Research Council 1997). Our efforts, be they in historical, environmental, cultural, political, or socioeconomic also need to be seen as crucial components of each of these thematic subfields.

Natural Environments and Human Interaction

The work of North American geographers in the different fields of physical geography is being conducted under the paradigmatic premiss that the environment is a physical milieu and the place of residence and activity for humans. From that perspective, global environmental change, climatic crises, and increased pressures on biotic resources by increasing populations have been among the concerns of the scholars and politicians who, in 1992, called the Global Conference on the Environment (UNCED) in Rio de Janeiro. There, new agendas for the integrated study of humans and the biosphere were formulated.

Latin American Environmental Research within Global Change

Investigations about past and present impacts of humans on natural environments have refocused on the ways in which socioeconomic circumstances preside over environmental change (Turner 1991). The explanatory avenues and paradigmatic tenets of the natural and social sciences are now closely integrated in the treatment of humans and their environments (Turner 1997*a*). Latin America, where demographic growth and urban sprawl are testing the resilience and the limits of natural environments, and resource exploitation is exerting critical pressure on finite resources, provides a showcase for this type of analysis. With the interpretation of ecological imbalances in tropical Latin America as the outcome of human land occupancy and overexploitation of resources, the discourse about sustainable development, political ecology, cultural ecology, or human ecology has benefited from numerous Latin American case studies: today, these concepts have acquired universal application in policy-making (Turner 1997*b*).

International research coordinating bodies, such as the International Geosphere-Biosphere Program, the Intergovernmental Panel on Climate Change, the International Human Dimension Program, and the Land-Use/Cover Change project, have established as major objectives the creation of observational bases that document surface land-cover variations—using remote sensing or geographic information systems techniques—to discern how human agents and biophysical processes contribute to landscape deterioration and, eventually, propose alternatives for sustainable development and managed conservation (Geoghegan *et al.* 1998).

Their research effort has centered on the South American and Central American tropics, still the habitat of native groups and the last stands of tropical humid forests. Destructive land-use practices trigger ecological disturbances and climate fluctuations that threaten the preservation of biotic resources and lead to the extinction of traditional forms of livelihood. Biodiversity as the underlying support for native populations in the Andes and the impact of introduced crops has been critically tested by Zimmerer (1996): he concluded that these communities know how to balance the alternatives for their survival. The preservation of native forests and original vegetation stands is advocated with the rationale that their disappearance may signify an irreversible loss of gene pools from species of potential pharmaceutical or biogenetic value.

The question of human occupance and environmental challenges resulting in cultural adaptation is particularly stimulating in Latin America, where the interface between environment and culture hinges heavily on the cross-fertilization from diverse mainsprings. On the subjects of past environments, traditional economies, and forms of livelihood, the collection of essays *Culture, Form and Place* (Mathewson 1993) contains several studies in which past environments are evoked to demonstrate how foreign resource exploitation modalities reshaped the economic utilization of New World landscapes. In another collection, *Nature's Geography. New Lessons for Conservation in Developing Countries*, edited by Zimmerer and Young (1998), case studies from tropical Latin America show that nature's resilience is slowly waning as resource exploitation and ecological crises severely tax its capacity for recovery. These examples of scholarship are spearheading comparable studies in other tropical regions of the world facing similar problems, but where scientific insights receive less consideration in policy-making.

Land Supporting Capacity

Latin Americanist geographers have turned their attention to the question of land carrying capacity, particularly in the humid tropics. This concern arises from the present population increase in Latin America, and from the interest in the size of the population that once inhabited Amazonia (Denevan 1996). Empirical evidence from other tropical settings attest to declining soil fertility once the land is opened and to progressive erosion as the result of human pressures on sensitive natural environments (Binford *et al.* 1987; Brenner and Binford 1988).

Drawing on reliable soil surveys and reports of colonization schemes from localities in the Amazon basin, Weischet and Caviedes (1993) demonstrated that the actual productivity of low-status soils in the humid tropics is hampered by natural limitations that all advances in soil science have been unable to surmount, as the experiments of Pedro Sánchez in Yurimaguas, Peru, demonstrate.

What the physical properties of soils *per se* can tell us about the success of agricultural intensification in areas of difficult occupance has prompted a renewed interest in Amazonian black earths (*terras pretas*) assumed to be of anthropogenic origin (Smith 1999). Recently Woods and McCann (1999) have postulated that, although many developed in association with habitation sites, others seem to be the result of the regenerative power of these tropical soils through biotic inoculation. Latin Americanist geographers offers different insights into the question of the developmental potential of such a vast space as Amazonia (N. J. H. Smith *et al.* 1995). A different form of land use based on wetland agriculture, capable of supporting larger populations and complex social institutions in pre-Hispanic Guatemala and southern Mexico, is documented by Sluyter (1994).

In keeping with this applied use of contributions from the field of physical geography, sediment series—offering clues for determining climatic variations of sub-recent date—have been meticulously examined in northern Mexico to assess the degree of landscape deterioration caused by human occupance (Butzer and Butzer 1993), while Sluyter (1996) has investigated the ecological effects and landscape transformations caused by cattle-ranching in colonial Mexico.

Resource Inventories and Environmental Change

Resource inventories from which blueprints can be drafted for the conservation of endangered forests, wetlands, wildlife refuges, ecological preserves, native reserves, or gene-pool biotic preserves, have utilized remote sensing imagery and GIS information, especially in regions of difficult access or of unmanageable dimensions for field surveys (Mast *et al.* 1997). While the extent of disturbances in ecosystems and paleoecological environments seems to indicate heavy degradation in tropical and subtropical regions with large population concentrations (Binford *et al.* 1987), major concerns in environmental studies are the natural forests of the South American tropics (Keating 1997), relict humid forests of Central America (Herlihy 1997), retreating primary and secondary forests in Guatemala (Colchester 1991), and the few remaining pockets of natural forests in the overpopulated Dominican Republic (Brothers 1997*a*). Sambrook *et al.* (1999) link the destructive deforestation in that country to impoverished local communities which have intensified cutting practices.

The dwindling woodlands of Central America have been mapped as a first step towards their conservation and better management by native peoples (Herlihy and Leake 1997), while deforestation in the Central Andes is being surveyed to assess the degree of human encroachment there (Young 1998). In the Mayan area, Steinberg (1998) reports on efforts to adjust *milpa* to innovative agroforestry systems, and in southern Chile, remnant stands of monkey-puzzle tree (*Araucaria araucana*) are being considered in the context of the resource rights of native populations (Aegesen 1998).

Descriptive studies have advanced knowledge about the agricultural and forest resources in areas of South America where original cultivars are still an important component of the inhabitants' diet. Significant works include N. J. H. Smith *et al.* (1992, 1999), and Zimmerer's monographs on the variety of potatoes and their cultivating procedures (1991, 1998).

Even the inquiry into the peculiarities of Latin America's vegetal realm has outgrown the traditional system analysis or process description. Plant associations and fauna that have changed owing to severe climatic oscillations in the high mountains and southern latitudes of South America have been detailed by Veblen and his associates (Veblen *et al.* 1992, 1996) and by Villalba (1997), while the morphological characteristics and human disturbances of vegetation in Costa Rica have been investigated on the basis of modern pollen spectra. Alterations to *páramo* vegetation in Costa Rica due to natural and human-induced fires have been reported from prehistorical times (Horn 1993) to the present when roads were opened in these environments (Horn 1989), while Keating (1998) assessed the role of humans in disturbing paramos in Ecuador. High mountain ecology and human occupation has also been addressed by Allan *et al.* (1988), which contains insightful chapters on agricultural utilization of tropical mountain slopes, highland ecology, vertical articulation of agrarian productivity, and comparative transhumance.

Process-oriented geomorphological investigations are still pursued with an emphasis on humans as activators of erosive processes. The rate of materials removed has been estimated from sedimentation in lakes of Ecuador and Costa Rica (Wallin and Harden 1996). Soil erosion as environmental degradation has been surveyed in the Andes of Ecuador (Harden 1991) and also in water reservoirs that can result in loss of agricultural land (Harden 1996). In a more holistic approach Zimmerer (1993*b*) treats these problems including the perceptional aspects of farming communities to provide a better understanding of the implications of soil erosion in agricultural productivity and socioeconomic systems.

The Imperatives of Climate

The descriptive or synoptic climatologies of the past have been replaced by studies on climate variability and catastrophic sequels. The impact of El Niño on regional precipitation, surface hydrology, agricultural output, and social dislocation is manifested in works on rainfall forecast and run-off fluctuations gaged against Southern Oscillation or the North Atlantic Indices, sea-surface temperatures, and the Trade Winds. Waylen and Caviedes (1987, 1990) determined the effects of El Niño and La Niña on precipitation and river discharge that endanger populations and agriculture in Peru and Chile. Attention has shifted now to precipitation and run-off variability in Costa Rica (Waylen *et al.* 1996*a*, *b*). Precipitation and run-off are now assessed with a view to providing forecasting models to prevent depletion of reservoirs and avert deficits in the supply of electric energy (George *et al.* 1998). The complex interactions between the South Pacific ocean circulation and other variables affecting precipitation in Central America have been summarized by Waylen *et al.* (1998).

Droughts, the flip-side of El Niño events in Latin America, have received attention from Liverman (1990) who studied dryness and its impact on humans in central-western Mexico, and who is now focusing on the sequels of devastating hurricanes on the Pacific coast of Mexico that tend to occur in the wake of a warm El Niño cycle. Among the few biogeographers to recognize the signature of the Southern Oscillation on droughts in South America are Veblen *et al.* (1999) in their study of fires in northern Patagonia. Dixon (1991) details the effects of hurricanes on nature, populations, and human works in the Caribbean basin, especially in the Yucatan Peninsula after the 1982/3 El Niño. In the search for past hurricane occurrences surrogate historical sources have revealed changing patterns probably connected with the fluctuations of distant atmospheric factors, such as the Southern Oscillation and the Quasi-Biennial Oscillation (Caviedes 1991*b*).

Henry Diaz and Vera Markgraf (1992) have compiled the multiple sequels of El Niño. The effects of El Niño on fishing resources has been modeled with autoregressive techniques to demonstrate that natural fluctuation cycles in fish populations are more decisive than overfishing in determining fisheries collapses, such as the one of the Peruvian anchovies in 1972/3 (Caviedes and Fik 1991).

Environmental Crises and Paleoecological Reconstructions

The realization that climatic crises have left their imprints on the Latin American landscape and caused substantial environmental change has prompted searches for traces left by Quaternary temperature drops in the Cordillera Central of Hispaniola (Orvis *et al.* 1997), or studies of the impact of Pleistocene precipitation on snow-line depression and northward shifts of vegetation along Chile (Caviedes 1990) and of the vegetational sequences that led to paramos in Costa Rica (Horn 1993).

Lakes provided environments around which early human groups became established and their sediments hold imprints of ecological crises and stages of cultural development that have motivated numerous studies, such as those of the karst lakes in Guatemala (Brenner *et al.* 1990), the correspondence between human occupance and accumulation cycles in the Titicaca basin (Binford *et al.* 1992), and Costa Rican lakes (Horn and Haberyan 1993). Using palynological techniques, Kennedy and Horn (1997) prove that even the "virgin" forests of La Selva Biological Station occupy lands that were burned and farmed in prehistoric times. Paleoenvironmental findings from the Gulf of Mexico suggest that maize cultivation was practiced around 4100 BC and that eustatic sea-level rises prompted the development of wet-farming strategies among prehistoric cultivators (Sluyter 1997). On the shores of Lake Titicaca, Binford *et al.* (1997) have used lacustrine sediments and shore vegetation to discern how these indicators reflect climate change and its effects on human activities.

Contemporary Socioeconomic, Urban, and Political Themes

If the 1980s were the lost decade for Latin America then the 1990s have been a decade of restructuring and rediscovery, with geographers contributing in myriad ways to our understanding of how Latin American countries have developed and changed. Latin-Americanist geographers in North America have demonstrated a renewed level of theoretical and empirical rigor in their analyses of the region, particularly in the field, and several very promising strands of research have emerged in recent years. These research strands can be summarized into four major macro-analytical themes.

The Complexities of Globalization

The first theme addresses the response of Latin American societies to the forces and processes of globalization, privatization, and neoliberal programs. Such studies run the gamut from very general regional surveys to specific analyses of countries, internal regions, and urban centers. During the past decade, many Latin-Americanists in North America have been attracted to analyses of the development patterns and processes driven by theories of globalization and the world economy. Several regional surveys have either been updated to reflect contemporary conditions (Blouet and Blouet 1996) or are welcome additions to the literature (Caviedes and Knapp 1995; Clawson 1997; Richardson 1992), while more specific studies have attempted to examine change within specific countries (Keeling 1997) or regions (Wesche and Bruneau 1990). A continued focus on country-level or regional (i.e. the Southern Cone or Andean area) responses to globalization augurs well for a more detailed understanding of the region in the wider world (Winberry 1998). For example, recent essays on globalization and neoliberalism in the Caribbean suggest a variety of research approaches that could be applied to Latin America generally (Klak 1998). Latin American geographers particularly have done an excellent job of synthesizing the internal response of their countries to the forces of the world economy (Becker and Egler 1992; Roccatagliata 1997).

Theories of underdevelopment and spatial organization continue to offer challenges for geographic analysis (Bromley 1992), with the issues of national planning, decentralization, and regional development at the forefront of contemporary research agendas (Keeling 1998; Kent *et al.* 1998; Scarpaci 1998*a*). A promising line of research involves teasing out the subtleties in changing national and regional identities as a consequence of globalization (Price 1996), territorial loss (Elbow 1996), or regional integration and how those identities are depicted in film, music, and the media (Godfrey 1993; Roberts 1995*a*). Analyses of the North American media's coverage of the region and of major events such as the Summit of the Americas also can shed light on perceptions of how neoliberal policies are restructuring life and livelihood (Vanderbush and Klak 1996). Perhaps the most intriguing and promising avenue of research on Latin America involves understanding the diffusion and impact of technology across boundaries and how tools such as the Internet and GIS are shaping responses to local and national social problems (Froehling 1997; Salazar 1996).

Problematizing Urban Growth

Research on the second theme has refocused the analytical spotlight on urban environments by engaging a variety of methodological perspectives to explicate more clearly the contemporary global–local urban dynamic. Latin-Americanist geographers in the 1990s have reinvigorated research on the problems of urban growth and change, particularly in terms of impacts on the human condition. Lawson and Klak (1993) argued specifically for a critical and comparative approach to urban analysis suggesting that the articulation of conceptual linkages in geographic research offered great promise in policy-making arenas. In this vein, the development of a world-city paradigm, along with theories of urban responses to global economic forces, has encouraged research that looks more critically at the roles of cities beyond their immediate national hinterland (Keeling 1996; Segre *et al.* 1997; Ward 1998). Researchers are moving beyond singular case studies that show little concept development into broader and critical areas of analysis.

A renewed focus on Latin American urbanization has fostered a healthy debate over modeling the Latin American city. Ford's (1996) attempt to rethink the urban land-use model first developed in the late 1970s and widely adopted by Latin Americanists has encouraged critiques of how cities are conceptualized and structured (Crowley 1998). For example, Arreola and Curtis (1993) have adapted the model to Mexican border-towns generally, in order to explicate the structure of their urban morphology. This debate suggests that a comparative approach to urban landscapes within the context of global restructuring and modernization could further refine the model (Godfrey 1991). Researchers are beginning to build a body of empirical evidence on how urban built environments are responding to neoliberal restructuring and on whether there are significant similarities in Latin American cities to the built infrastructure of world cities in other regions.

Geographers have also made great strides in expanding research on the smaller cities and towns of Latin America (Brown *et al.* 1994; Smith 1998), and on the improvements in transport and technology that are changing the rural spatial dynamics (Rudel and Richards 1990). Urbanization in the national frontiers, particularly in the Brazilian Amazon, has attracted the most attention over the past decade (Browder and Godfrey 1997; Sternberg 1997). The northern border cities of Mexico also have attracted much attention in recent years as growth driven by *maquiladoras*, migration, and NAFTA policies continues to impact the development of these trans-frontier metropolises (Herzog 1991; MacLachlan and Aguilar 1998).

Rural to urban migration appears to have slowed in Latin America during the 1990s, yet it remains a powerful force in structuring urban environments and provides fertile ground for geographic analysis (Brown 1991). Recent empirical research from Quito examined the life cycle of the urban immigrant by drawing on the "three phases of urbanization" model developed for the Latin American city in the 1960s (Klak and Holtzclaw 1993). Many of the migration patterns mapped in the 1960s remain evident today, which suggests opportunities exist for further analysis of this very important aspect of demographic shift. Another relatively new area of urban analysis is how the media create images of the city and urban life, and what impacts television, radio, and film have on the migration process (Klak 1994).

Issues of social justice and welfare in Latin American cities also have reinvigorated urban research agendas over the past decade, driven in part by massive urban growth since the 1970s and by the major economic downturn that occurred in the 1980s. Earlier in the decade, Scarpaci (1991) spearheaded important research on urban social welfare policy in the aftermath of de-industrialization and the economic crises of the 1980s, particularly on health care in the shanty towns (Weil and Scarpaci 1992). Latin-Americanist geographers are also responding to the challenge to advance theories and models related to the urban poor, their housing conditions, and their responses to the forces of neoliberalism and globalization (Klak 1993; Cravey 1998*a*).

A Political Economy Approach to Development

The third research theme takes a political-economic approach to Latin America, with a particular focus on commodity production, national and regional development strategies, integration, and the environment. The intersection of politics and economics continued to be a key element in Latin American research during the 1990s, with issues such as the democratization of societies (Caviedes 1991*a*), national subversion and insurgency (Kent 1993), and economic development policies in the context of regional integration (Elbow 1997; Sternberg 1994) at the forefront of some exciting empirical and theoretical work. Research on industrial subcontracting (Roberts 1995*b*), on the role of industrial policy in regional development (Mutersbaugh 1997), and on the growth of maquiladoras and their impact on regional economic integration (Cravey 1998*b*), especially in terms of gender equity, has paved the way for a much more sophisticated analysis of economic change driven by neoliberal policies. Wiley (1996), for example, has examined the role of commodities such as bananas in the neoliberal political economy, developing important comparative questions about commodity relationships between Latin America, the European Union, and other banana-producing regions (Wiley 1998). Others have taken a political ecology approach to commodity production (Coomes and Barham 1997; Grossman 1998), land-use issues (Dedina 1996), colonization (Weil and Weil 1993), and the historical legacy of extractive economies (Barham and Coomes 1996).

Rural development policies and NGO activity in Latin America also spurred important research during the 1990s on the intersection of politics and economic growth (Bebbington 1997). Price (1994*a*) examined the role of NGOs in the emerging field of ecopolitics, Stadel (1994, 1997) explored the links between NGO activity in the Andes, poverty, and the mobilization of human resources, while Keese (1998) discussed land-use change and NGOs. Latin-Americanist geographers have contributed particularly to teasing out the implications of broader institutional change that can help to determine the local and regional geographies that are derived from more general processes. Another growing arena of research relates to ecotourism and the politics of tourist development, especially in environmentally sensitive regions such as the Amazon (Wesche 1996). Tourism growth poles and their contribution to national and regional development also raise many important questions related to national economic planning strategies (Coffey 1993; Meyer-Arendt 1990; Sambrook *et al.* 1994; Scarpaci 1998*b*; Smith and Andino 1996) and more research is needed in this area, particularly in terms of environmental implications (Roberts 1996). For example, the relationship between population growth, frontier development, and land degradation remains a major ecological concern in Latin America (Brothers 1997*b*; Ryder 1994; Sambrook *et al.* 1994).

The role of transport infrastructure in the policies of regional development attracted new research during the decade (Keeling 1998, 1994), suggesting that the interrelated themes of production and reproduction in national and regional economies and how transport and communication facilitate their interconnectivity could yield new insights into regional growth and change. Klak (1993) opened another promising line of political-economic research by examining state policy in the provision of housing within the conceptual frameworks of pluralism, structural dependency, and structural relative autonomy.

Contemporary Societies

Finally, the last major theme takes a strong humanistic approach to analyzing daily life in Latin America that draws generally from the sociological/structuration theories of Anthony Giddens and others. Particularly important here are studies on gender divisions, the reorganization of labor, social polarization, and the provision of public services. For example, much groundbreaking research by Latin Americanists over the past decade has occurred within the broader context of understanding the response of contemporary societies to the forces of change. Perhaps the most enlightening and theoretically challenging research focuses on the problem of gender-based discrimination inherent in the neoliberal development model currently pursued in Latin America (Hays-Mitchell 1995) and on identifying gender issues in employment (Faulkner and Lawson 1991; Lawson 1995), migration (Ellis *et al.* 1996), the informal economy, and political participation (Hays-Mitchell 1995). Cravey (1998*a*) recently provided a solid framework for analysis in this field by outlining contemporary gender and feminist theories of the state and offering suggestions for future research directions.

Research on topics as disparate as the formation of social capital (Perreault *et al.* 1998), on evaluating the geography of health in Latin America (Weil 1995), and on the spatial dynamics of migrant remittances and their impacts on recipient communities (Conway and Cohen 1998; Jokisch 1997) also augur well for developing a clearer understanding of how Latin American societies are situated at the dawn of the twenty-first century. The concept of community participation in development suggests avenues for further research (Kent and Rimarachin 1994; Santana 1996), as does the issue of religious change and its impact on indigenous societies (Steinberg 1997).

More detailed research at the household level also is needed to shed light on the responses of communities to the neoliberal industrial transition model (Cravey 1998*b*). Recent studies of environmental issues show promise and suggest that researchers are beginning to engage broader theories and methodologies (Arbona 1998; Newsome *et al.* 1997; Palm and Hodgson 1993).

Current census predictions about the demographic impact of Latin Americans in the United States suggest an opportunity for more research on the social and political implications of an increasing Hispanic presence. In this vein, several researchers have examined residential segregation, assimilation, and identity in cities such as Miami and New York (Boswell 1993; Boswell and Cruz-Báez 1997; Miyares and Gowen 1998). Others have applied Hispanic assimilation models to communities in the southwestern border states (Arreola 1993; Haverluk 1998) or have pursued the spatial dynamic of illegal immigration (Harner 1995).

Historical and Cultural Perspectives

Despite the haste of many to globalize both the economic and the ecological, or to postmodernize the landscape into texts and mere sign systems, it is salutory to note that for many Latin-Americanist geographers the past, and empirical reality as interpreted through field and archival data, still loom large. A review of the literature reveals that historical evolution and cultural contexts are increasingly seen as the most appropriate ways in which to analyze the contemporary scene, be it from either an environmental or human perspective.

Landscape Evolution

A persistent interest has centered on the changing form and functions of landscapes as well as their constituent elements. A significant set of studies used the 500th anniversary of the arrival of Columbus to interpret both landscapes and land uses in several parts of the region. For Mexico K. W. Butzer has provided a series of meticulously researched papers either alone (1990, 1992*a*, *b*, 1993), or with colleagues (Turner and Butzer 1992; Butzer and Butzer 1993) that document the pre-colonial land-use patterns and the impact that early colonial man had on them. Denevan (1992*b*) also successfully demolished the mythic "pristine" nature of the indigenous environment demonstrating that degradation, albeit at lower levels, was not unknown before the Europeans arrived. The transfer into New Spain of European cattle has been analyzed (Butzer and Butzer 1995; Hope Enfield 1998), studies that complement the impact of the "ungulate irruption" reported on by Melville (1994). It is also good to see the reverse process analyzed—the transfer of Meso-American food crops to Europe (Andrews 1993). Land grants and other legal documentation have been ingeniously used to provide clues as to the environmental conditions and land-use and settlement changes in early colonial Mexico (Sluyter, 1997) and twentieth-century Brazil (Brannstrom 1997).

Several syntheses of landscape evolution have also appeared, including Siemens's classic study of central Veracruz's wetlands over 1,500 years (1998) and his analysis of nineteenth-century Veracruz (1990), West's portrait of Sonora's personality (1993), and Pasqualetti's collection of Aschmann's landscapes (1997). Though we now have available a series of case studies of environmental impacts over relatively long time periods (Zimmerer and Young 1998), geographers have yet to match the bold outlines of macro-regional landscape change as exemplified in Dean's (1995) study of Brazilian deforestation, or the historians' overview of the colonial countryside (Hoberman and Socolow 1996).

Agricultural Systems: From Aboriginal to Traditional

A theme that has continued to provoke the attention of researchers has been the features of the agricultural past, be they irrigation and terracing systems in the Andes (Knapp 1992; Denevan, 2001) or Mexico (Doolittle 1990, 1995, 1998; 2001; Davis 1990), the dynamics of pre-Hispanic and smallholder agriculture in Ecuador (Knapp 1991, 1992, 1998), or agricultural biodiversity during Inca rule (Zimmerer 1993*a*). In each case the authors are at pains to demonstrate the skills of past agriculturalists as they adapted the natural environment to their multifarious needs, as well as emphasizing that the passage from aboriginal to the traditional of the contemporary period was one characterized by selective abandonment and adoption of farming practices (Knapp 1991). Mathewson (1992) has reminded us of the relevance of the "people without history" in our rush to judgement over vestigial agricultural practices. Denevan (1992*b*) has examined the tools used in prehistoric cultivation systems and has also provided a new model of Amazonian settlement (1996). Major advances in our knowledge of Andean terracing and irrigation have come from the Peruvian Colca valley project (Treacy 1994). Zimmerer has demonstrated the relevance of fallowing systems as well as the complexity of overlapping agri-zones (1999). He has also provided the most complete study to date of the historical ecology of the Andean valley of Paucartambo in which he meticulously documents land use, population, and cropping systems over five centuries (1996).

Population Structures and Change

The changing population of Latin America continues to attract scholarly attention. The mysterious origins of the Tarascans have been decifered (Malmstrom 1995), and the massive depopulation initiated by the introduction of European epidemic diseases analyzed, using innovative simulation modeling at the macro-level (Whitmore 1992), providing a statistical probability base for historical interpretations (Cook 1998). Refined studies at the regional level in both Meso- and Andean America (Newson 1995, 1996; Cook and Lovell 1991; Whitmore 1996; Whitmore and Williams 1997) have confirmed the significant variations in both rates and consequences of this cultural trauma. The ubiquity of colonial migration has been documented (Robinson 1990*b*), and several studies have identified the availability and potential use of specific population sources. For Central America we now have a comprehensive guide (Lovell and Lutz 1995) as well as studies of parish registers (Robinson and Doenges 1997; Horst 1991). The structure of colonial families (Robinson 1990*c*) and regional populations have provided useful insights into the colonial past, as have estimates of marriage mobility (Doenges 1996). The detailed analysis of tribute lists has provided clues as to the spatial and socioeconomic structure of the early colonial Peruvian townships (Robinson 2003). We can now better understand the relationships between political power and settlement structures in New Granada (Herrera Angel 1996) and the fate of Indian community lands in Venezuela (Samudio 1997). For the early twentieth century we have a new interpretation of the regional migration promoted by coffee cultivation in Venezuela (Price 1994*b*), and the Dawseys have provided glimpses into the experiences of one unexpected immigrant group (Dawsey and Dawsey 1995).

Colonial Systems and Structures

Significant components and processes of colonialism have received considerable attention. The development of the encomienda in Guatemala has been shown to have had significant political dimensions (Kramer 1994) as well as important links to settlement history (Kramer *et al.* 1990). We now know much more of the regional variations in mining (Craig and West 1994). Trade in key products (Mack 1998; Camille 1996), trading routes (Fifer 1994; Driever 1995; Jett 1994), and trading posts (Butzer 1997), have all received attention, demonstrating that goods moved along well-defined but mostly randomly selected trails. We still, however, lack a comprehensive study mapping the significant routeways through which the colonial system of mail and travellers passed (Robinson 1992*a*).

Colonial cartographic contributions are generally absent from the record of the last decade though art historians (Mundy 1993), literary specialists (Mignolo 1995), and historians (Gruzinski 1987) have shown what can be done with these specialized sources. The entire practice of capturing the colonial past has now been dissected and exemplified (Miller 1991).

Post-Colonial Patterns and Processes

Several authors have focused on the complexities of post-colonial Latin America. Gade (1992) has elegantly argued the case for relating identity and landscape in the post-conquest Andes. Robinson (1990*a*) has assessed the impact of French revolutionary ideas in early republican Spanish America as nations were forged by force and lofty ideals, while Lambert (1994) demonstrates the problems of center-periphery tensions in early republican Mexico. Brazil's early nineteenth-century industrial development based on salt beef products demonstrates linkages between production and slave labor (Bell 1993), and the extensive but highly differentiated patterns of *gaúcho* cattle-ranching in southern Brazil after 1850 again speak of ecological, technical, and commercial processes operating at different scales and with varying intensity (Bell 1998). Such was the significance of the diffusion of sheep that one may properly speak of "merinomania" in southern South America after 1820 (Bell 1995).

Oddities such as geophagy (Hunter *et al.* 1989; Horst 1990) have been analyzed and reported on in text and photos (Horst and Bond 1995) emphasizing the persistence of religious and cultural traditions that are also documented in Andean (Gade 1999) as well as Belizean contexts (Steinberg 1996), the latter study cogently arguing against the assumption of transferring "tradition" from one component to another. In the varied use of palms (Horst 1997), leaves (Voeks 1990), and flowers

(Parsons 1992*b*) one can witness the cultural imprint of millennia. The episodic natural disasters of both pre-Hispanic (Knapp and Mothes 1998) and modern periods (Horst 1995) have also been documented though we still know far too little of the droughts (Ouweneel 1996), volcanic eruptions, earthquakes, and hurricanes (Caviedes 1991*b*) of the past. Urban studies are also singularly few in the historical/cultural publication record of the last decade, the study of Rio de Janeiro's streetcars (Boone 1994) providing but a glimpse into the riches of the urban archives. An overview of urbanization is now available (Greenfield 1994) but the field still awaits its major author.

Geographers continue to contribute to the study of ethnic groups in Latin America, especially Native Americans. Ethnic markers and emerging ethnic identities have been identified and mapped in their historical contexts (Knapp 1990; Caviedes and Knapp 1995), as well as Latin Americans within the US (Arreola 1993; Boswell 1993; Miyares and Gowen 1998). Participatory resource mapping (geomatics) becomes important in some contexts (Herlihy 1997; Herlihy and Leake 1997) as does the role of ethnicity in sustainable development projects involving NGOs (Bebbington 1997; Bebbington and Thiele 1993).

The language of "place" in the region is seen to have been significant and ever-changing (Robinson 1992*b*), and as well as an introduction to Spanish as a language for geographical expression (Driever 1990) we also now have the luxury of a comprehensive Spanish/English geographical dictionary (Driever 1994). A few pioneers have ventured into the spaces of literary (Caviedes 1996) or cinematic (Godfrey 1993) images of place-making. However, all evidence points to the fact that among Latin Americanists historical and cultural geography is thriving as never before.

Towards the Next Millennium

The techno-economic force of globalization still needs to be explicated at the regional and local levels, and while some Latin Americanists have gone global, others have stood firm relying on grounded empirical evidence of the contexts and processes that are patently different at multi-scale levels. If, as looks likely, trade and regional integration are to be the order of the day, then we badly need studies of transportation and information systems. "Latin" America undeniably now includes significant portions of the USA.

Even though we welcome two recent atlases (Toledo Maya Council 1997; MERCOSUL 1997) we still lack a comprehensive one of the whole region (with CD-ROM) and, though we have a trilogy of textbooks (Clawson 1997; Caviedes and Knapp 1995, Blouet and Blouet 1996), new ones are already needed. Alternatively, with online Internet courses rapidly developing, perhaps we should be thinking of a database of images and primary documentation modeled on ECO (Early Canada Online 1999) that would make obsolete the very notion of a set text. Cartography appears to be a subject of little interest to most Latin-Americanist geographers (Gartner 1998) though art historians and literati have made notable contributions to the interpretations of landscapes and cultural meaning (Mundy 1996; Codding 1994; Whitehead 1998).

In closing, while we may properly celebrate the wealth of intellectual production noted above, we have to take note of, and respond effectively to, the new marginalization among Latin Americans that is evident from the impact of the most recent of the new technologies, adding to and reinforcing age-old patterns of poverty and racial/linguistic prejudice.

References

Aegesen, D. L. (1998). "Indigenous Resource Rights and Conservation of the Monkey-Puzzle Tree (*Araucaria* araucana, Araucariaceae): A Case Study from Southern Chile." *Economic Botany*, 52/2: 146–60.

Allan, Nigel J. R., Knapp, G., Stadel, C. (eds.) (1988). *Human Impact on Mountains*. Totowa, NJ: Rowman & Littlefield.

Andrews, Jean (1993). "Diffusion of Mesoamerican Food Complex to Southeastern Europe." *Geographical Review*, 83: 194–204.

Arbona, Sonia I. (1998). "Commercial Agriculture and Agrochemicals in Almolonga, Guatemala." *Geographical Review*, 88/1: 47–63.

Arreola, Daniel D. (1993). "The Texas-Mexican Homeland." *Journal of Cultural Geography*, 3/2: 61–74.

Arreola, Daniel D., and Curtis, James R. (1993). *The Mexican Border Cities: Landscape Anatomy and Place Personality*. Tucson: University of Arizona Press.

Barham, B. L., and Coomes, O. T. (1996). *Prosperity's Promise: The Amazon Rubber Boom and Distorted Economic Development*. Boulder, Colo.: Westview Press.

Bebbington, Anthony (1997)."New States, new NGOs? Crises and Transitions among Rural Development NGOs in the Andean Region." *World Development*, 25/11: 1755–65.

Bebbington, Anthony, and Thiele, Graham (1993). *NGOs and the State in Latin America: Rethinking Roles in Sustainable Agricultural Development*. London: Routledge.

Becker, Bertha K., and Egler, Claudio A. G. (1992). *Brazil: A New Regional Power in the World-Economy*. New York: Cambridge University Press.

Bell, Stephen (1993). "Early Industrialization in the South Atlantic: Political Influences on the Charqueadas of Rio Grande do Sul before 1860." *Journal of Historical Geography*, 19/4: 399–411.

—— (1995). "Aimé Bonpland and Merinomania in Southern South America." *The Americas*, 51/3: 301–23.

—— (1998). *Campanha Gaúcha: A Brazilian Ranching System, 1850–1920*. Stanford: Stanford University Press.

Binford, M. W., Brenner, M., and Engstrom, D. R. (1992). "Temporal Sedimentation Patterns in the Nearshore Littoral of Lago Huiñaimarca," in C. Dejoux, and A. Iltis (eds.), *Lake Titicaca*. Dordrecht: Kluwer, 30–9.

Binford, M. W., Brenner, M., Whitmore, T. J., Higuera-Gundy, A., and Deevey, E. S. (1987). "Ecosystems, Paleoecology and Human Disturbance in Subtropical and Tropical America." *Quaternary Science Review*, 6/1: 115–28.

Binford, M. W., Kolata, A. L., Brenner, M., Janusek, J., Seddon, M. T., Abbott, M. B., and Curtis, J. (1997). "Climate Variation and the Rise and Fall of Andean Civilization." *Quaternary Research*, 47/1: 235–48.

Blouet, Brian W., and Blouet, Olwyn M. (1996). *Latin America and the Caribbean: A Systematic and Regional Survey*. 3rd edn. New York: Wiley.

Boone, Christopher (1994). "Streetcars and Popular Protest in Rio de Janeiro: The Case of the Rio de Janeiro Tramway, Light and Power Company, 1903–1920." *CLAG Yearbook*, 20: 49–58.

Boswell, Thomas D. (1993). "Racial and Ethnic Segregation Patterns in Metropolitan Miami, Florida, 1890–1990." *Southeastern Geographer*, 33/1: 82–109.

Boswell, Thomas D., and Cruz-Baez, Angel D. (1997). "Residential Segregation by Socioeconomic Class in Metropolitan Miami: 1990." *Urban Geography*, 18/6: 474–96.

Brannstrom, Christian, (1997). "Brazilian County-Level Juridical Documents as Sources for Historical Geography: A Case Study from Western São Paulo State, Brazil." *CLAG Yearbook*, 23: 41–50.

Brenner, M., Leyden, B., and Binford, M. W. (1990). "Recent Sedimentary Histories of Shallow Lakes in the Guatemala Savannas." *Journal of Paleolimnology* 4/2: 239–52.

Bromley, Ray (1992). "Development, Underdevelopment, and National Spatial Organization." *CLAG Yearbook*, 17/18: 249–59.

Brothers, T. S. (1997*a*). "Deforestation in the Dominican Republic: a Village View." *Environmental Conservation*, 24/3: 213–23.

—— (1997*b*). "Rapid Destruction of a Lowland Tropical Forest, Los Haitises, Dominican Republic." *Ambio*, 26/8: 551–2.

Browder, John O., and Godfrey, Brian J. (1997). *Rainforest Cities: Urbanization, Development, and Globalization of the Brazilian Amazon*. New York: Columbia University Press.

Brown, Lawrence A. (1991). *Place, Migration, and Development in the Third World: An Alternative View*. New York: Routledge.

Brown, Lawrence A., Sierra , R., Digiacinto, S., and Smith, W. R. (1994). "Urban-system Evolution in Frontier Settings." *Geographical Review*, 84/3: 249–65.

Butzer, Elizabeth K. (1997). "The Roadside Inn or Venta: Origins and Early Development in New Spain." *CLAG Yearbook*, 23: 1–16.

Butzer, K. W. (1990). "Ethno-agriculture and Cultural Ecology in Mexico: Historical Vistas and Modern Implications." *CLAG Benchmark*: 139–52.

—— (1992*a*). "From Colombus to Acosta: Science, Geography and the New World." *Annals of the Association of American Geographers*, 82/3: 543–56.

—— (1992*b*). "The Americas before and after 1492: An Introduction to Current Geographical Research." *Annals of the Association of American Geographers*, 82/3: 345–68, 369–85.

—— (1993). "No Eden in the New World." *Nature*, 362/6415: 15–18.

Butzer, K. W., and Butzer, E. K. (1993). "The Sixteenth Century Environment of the Central Mexican Bajio: Archival Reconstruction from Colonial Land Grants and the Question of Spanish Ecological Impact," in Mathewson (1993: 89–124).

—— (1995). "Transfer of the Mediterranean Livestock Economy to New Spain: Adaptation and Ecological Consequences," in Turner *et al.* (1995: 151–93).

Camille, Michael A. (1996). "Historical Geography of the Belizean Logwood Trade." *CLAG Yearbook*, 22: 77–86.

Caviedes, C. N. (1990). "Rainfall Variation, Snowline Depression and Vegetational Shifts in Chile during the Pleistocene." *Climatic Change*, 16/1: 99–114.

—— (1991*a*)."Five Centuries of Hurricanes in the Caribbean." *GeoJournal*, 23: 301–10.

—— (1991*b*). *Elections in Chile: The Road to Democratization*. Boulder, Colo.: Lynne Rienner.

—— (1996). "Tangible and Mythical Places in José M. Arguedas, Gabriel García Márquez, and Pablo Neruda." *GeoJournal*, 38/1: 99–107.

Caviedes, C. N., and Fik, T. (1993). "Modeling Change in the Chilean/Peruvian Eastern Pacific Fisheries." *GeoJournal*, 30/4: 369–80.

Caviedes, César, and Knapp, Gregory (1995). *South America*. Englewood Cliffs, NJ: Prentice-Hall.

Clawson, David L. (1997). *Latin America & the Caribbean: Lands and Peoples*. New York: WCB/McGraw Hill.

Codding, Mitchell A. (1994). "Perfecting the Geography of New Spain: Alzate and the Cartographic Legacy of Sigüenza y Góngora." *Colonial Latin American Review*, 3: 185–219.

Coffey, Brian (1993). "Investment Incentives as a Means of Encouraging Tourism Development: The Case of Costa Rica." *Bulletin of Latin American Research*, 12: 83–90.

Colchester, M. (1991). "Guatemala: The Clamour for Land and the Fate of the Forests." *The Ecologist*, 21/4: 177–86.

Conway, Dennis, and Cohen, Jeffrey H. (1998). "Consequences of Migration and Remittances for Mexican Transnational Communities." *Economic Geography*, 74/1: 26–44.

Cook, N. David. (1998). *Born to Die: Disease and New World Conquest, 1492–1650*. Cambridge: Cambridge University Press.

Cook, N. David, and Lovell, George (eds.) (1991). *Secret Judgments of God*. Norman: University of Oklahoma Press.

Coomes, O. T., and Barham, B. L. (1997). "Rain Forest Extraction and Conservation in Amazonia." *The Geographical Journal*, 163/2: 180–8.

Craig, Alan K., and West, Robert C. (eds.) (1994). *In Quest of Mineral Wealth: Aboriginal and Colonial Mining and Metallurgy in*

Spanish America. Baton Rouge, La.: Geoscience Publications, Louisiana State University.
Cravey, Altha J. (1998*a*). *Women & Work in Mexico's Maquiladoras*. New York: Rowman & Littlefield.
— (1998*b*). "Engendering the Latin American State." *Progress in Human Geography*, 22/4: 523–42.
Crowley, William K. (1998). "Modeling the Latin American City." *Geographical Review*, 88/1: 127–30.
Davis, Clint (1990). "Water Control and Settlement in Colonial Mexico's First Frontier: The *bordo* System of the Eastern Bajío." *CLAG Yearbook*, 16: 73–81.
Dawsey, Cyrus B., and Dawsey, James (1995). *The Confederados: Old South Immigrants in Brazil*. Tuscaloosa: University of Alabama Press.
Dean, Warren (1995). *With Broadax and Firebrand: The Destruction of the Brazilian Atlantic Forest*. Berkeley: University of California Press.
Dedina, Serge (1996). "The Political Ecology of Transboundary Development: Land Use, Flood Control and Politics in the Tijuana River Valley." *Journal of Borderland Studies*, 10/1: 89–110.
Denevan, William M. (ed.) (1989). *Hispanic Lands and Peoples: Selected Writings of James J. Parsons*. Dellplain, 23, Boulder Colo.: Westview Press.
— (1992*a*). "The Pristine Myth: the Landscape of the Americas in 1492." *Annals of the Association of American Geographers*, 82/3: 369–85.
— (1992*b*). "Stone vs. Metal Axes: The Ambiguity of Shifting Cultivation in Prehistoric Amazonia." *Journal of the Steward Anthropological Society*, 20: 153–65.
— (1996). "A Bluff Model of Riverine Settlement in Prehistoric Amazonia." *Annals of the Association of American Geographers*, 86: 654–81.
— (1999). "James J. Parsons, 1915–1997," in P. Armstrong and Martin, Geoffrey J. (eds.) *Geographers Biobibliographical Studies*, 19: 86–101.
— (2001). *Cultivated Landscapes of Native Amazonia and the Andes*. New York: Oxford University Press.
Descola, Philippe (1994). *In the Society of Nature: A Native Ecology in Amazonia. Cambridge*: Cambridge University Press.
Diaz, H. F., and Markgraf, V. (eds.) (1992). *El Niño. Historical and Paleoclimatic Aspects of the Southern Oscillation*. Cambridge: Cambridge University Press.
Dixon, C. (1991). "Yucatan after the Wind: Human and Environmental Impact of Hurricane Gilbert in the Central and Eastern Yucatan Peninsula." *GeoJournal*, 23/4: 337–45.
Doenges, Catherine E. (1996). "Marriage Mobility in Late Colonial Celaya, Mexico." *CLAG Yearbook*, 22: 65–76.
Doolittle, William E. (1990). *Canal Irrigation in Prehistoric Mexico: The Sequence of Technological Change*. Austin: University of Texas Press.
— (1995). "Indigenous Development of Mesoamerican Irrigation." *Geographical Review*, 85: 301–23.
— (1998). "Challenging Regional Stereotypes: The Case of Northern Mexico." *Southwestern Geographer*, 2: 15–39.
— (2001). *Cultivated Landscapes of Native North America*. New York: Oxford University Press.
Driever, Steven L. (1990). "Spanish as a Language for Geographical Expression." *CLAG Yearbook*, 16: 3–14.
— (1994). *Spanish/English Dictionary of Human and Physical Geography*. London: Greenwood Press.
— (1995). "The Veracruz–Mexico City Routes in the Sixteenth Century and the Study of Pre-industrial Transport in Historical Geography." *Geografía y Desarrollo*, 6: 5–18.
Early Canadia Online (ECO). (1999). ‹http://www.canadiana.org/›, last accessed 16 September 2001.
Elbow, Gary (1996). "Territorial Loss and National Image: The Case of Ecuador." CLAG *Yearbook*, 22: 93–105.
— (1997). "Regional Cooperation in the Caribbean: The Association of Caribbean States." *Journal of Geography*, 96/1: 13–22.
Ellis, Mark, Conway, Dennis, and Bailey, Adrian J. (1996). "The Circular Migration of Puerto Rican Women: Towards a Gendered Explanation." *International Migration*, 34/1: 31–58.
Faulkner, A. H., and Lawson, Victoria (1991). "Employment versus Empowerment: A Case Study of the Nature of Women's Work in Ecuador." *Journal of Development Studies*, 27: 16–47.
Fifer, Valerie (1994). "Andes Crossing: Old Tracks and New Opportunities at the Uspallata Pass." *CLAG Yearbook*, 20: 35–48.
Ford, Larry (1996). "A New and Improved Model of Latin American City Structure." *Geographical Review*, 86/3: 437–40.
Froehling, Oliver (1997). "The Cyberspace War of Ink and Internet in Chiapas, Mexico." *Geographical Review*, 87/2: 291–307.
Gade, Daniel W. (1992). "Landscape, System, and Identity in the Post-Conquest Andes." *Annals of the Association of American Geographers*, 82: 460–77.
— (1999). *Nature and Culture in the Andes*. Madison: University of Wisconsin Press.
Gartner, William G. (1998). "Mapmaking in the Central Andes," in David Woodward and Malcolm Lewis (eds.), *The History of Cartography*. Chicago: University of Chicago Press, ii. bk. 3, 257–300.
Geoghegan, J., Pritchard, L., Ogneva-Himmelberger, Y., Chowdhury, R. R., Sanderson, S., and Turner, B. L., II (1998). "'Socializing the Pixel' and 'Pixelizing the Social' in Land-Use and Land-Cover Change," in *People and Pixels: Linking Remote Sensing and Social Science*. Washington, DC: Committee on the Human Dimensions of Global Environmental Change, National Research Council, 51–69.
George, R. K., Waylen, P. R., and S. Laporte (1998). "Interannual Variability of Annual Streamflow and the Southern Oscillation in Costa Rica." *Hydrological Sciences Journal*, 43/3: 409–23.
Godfrey, Brian J. (1991). "Modernizing the Brazilian City." *Geographical Review*, 81/1: 18–34.
— (1993). "Regional Depiction in Contemporary Film." *Geographical Review*, 83/4: 428–40.
Greenfield, Gerald M. (ed.) (1994). *Latin American Urbanization: Historical Profiles of Major Cities*. Westport, Conn.: Greenwood Press.
Grossman, Lawrence (1998). *The Political Ecology of Bananas: Contract Farming, Peasants, and Agrarian Change in the Eastern Caribbean*. Charlotte, NC: UNC Press.
Gruzinski, Serge (1987). "Colonial Indian Maps in Sixteenth-Century Mexico." *Res*, 13: 46–61.
Harden, C. P. (1991). "Andean Soil Erosion." *National Geographic Research and Exploration*, 7/2: 216–31.
— (1996). "Interrelationships between Land Abandonment and Land Degradation: A Case from the Ecuadorian Andes." *Mountain Research and Development*, 16/2: 274–80.

Harner, John P. (1995). "Continuity Amidst Change: Undocumented Mexican Migration to Arizona." *Professional Geographer*, 47/4: 399–411.

Haverluk, Terrence W. (1998). "Hispanic Community Types and Assimilation in Mex-America." *Professional Geographer*, 50/4: 465–80.

Hays-Mitchell, Maureen (1995). "Voices and Visions from the Streets: Gender Interests and Political Participation among Women Informal Traders in Latin America." *Environment and Planning D: Society and Space*, 13: 445–69.

Herlihy, P. H. (1997). "Indigenous Peoples and Biosphere Reserve Conservation in Mosquitia Rain Forest Corridor," in S. F. Stevens (ed.), *Conservation through Cultural Survival*. Washington DC: Island Press, 99–129.

Herlihy, P. H., and Leake, A. P. (1997). "Participatory Research Mapping of Indigenous Lands in the Honduran Mosquitia," in A. R. Pebley and Luis Rosero-Bixby (eds.), *Demographic Diversity and Change in the Central American Isthmus*. Santa Monica: Rand.

Herrera Angel, Marta (1996). *Poder local, población y ordenamiento territorial en la Nueva Granada, siglo XVIII*. Bogota: Archivo General de la Nación.

Herzog, Lawrence A. (1991). "Cross-national Urban Structure in the Era of Global Cities: The U.S.-Mexico Transfrontier Metropolis." *Urban Studies*, 28/4: 519–33.

—— (1999). *From Aztec to High Tech: Architecture and Landscape Across the Mexico-United States Border*. Baltimore: Johns Hopkins University Press.

Hoberman, Luisa S., and Socolow, Susan M. (1996). *The Countryside in Colonial Latin America*. Albuquerque: University of New Mexico Press.

Hope Enfield, Georgina (1998). "Lands, Livestock and the Law: Territorial Conflicts in Colonial West Central Mexico." *Colonial Latin American Review*, 7/2: 205–24.

Horn, S. P. (1992). "Holocene Fires in Costa Rica." *Biotropica*, 24/3: 354–61.

—— 1993. "Postglacial Vegetation and Fire History in the Chirripó páramo of Costa Rica." *Quaternary Research*, 40/1: 107–16.

Horn, S. P., and Haberyan, K. A. (1993). "Costa Rican Lakes." *National Geographic Research and Exploration*, 9: 86–103.

Horst, Oscar H. (1990). "Arcilla geofágica en América." *Mesoamérica*, 19: 169–76.

—— (1991). "La utilización de archivos eclesiásticos en la reconstrucción de la historia demográfica de San Juan Ostuncalco." *Mesoamérica*, 22: 211–31.

—— (1995). "1902, año de caos: el impacto político y socioeconómico de las catástrofes naturales en Guatemala." *Mesoamérica*, 30: 309–26.

—— (1997). "The Utility of Palms in the Cultural Landscape of the Dominican Republic." *Principes. Journal of the International Palm Society*, 41/1: 15–28.

Horst, Oscar, and Bond, J. (1995). *The Crucified Black Christ of Esquipulas, 1595–1995: A Photographic Record of Pilgrims and Sacred Traditions*. Guatemala: CIRMA.

Hunter, John, Horst, Oscar, and Thomas, R. N. (1989). "Religious Geophagy as a Cottage Industry: The Holy Clay Tablet of Esquipulas, Guatemala." *National Geographic Research*, 5/3: 281–95.

Jett, Stephen (1994). "Cairn Trail Shrines in Middle and South America." *CLAG Yearbook*, 20: 1–8.

Jokisch, Brad D. (1997). "From Labor Circulation to International Migration: The Case of South-Central Ecuador." *CLAG Yearbook*, 23: 63–75.

Keating, P. (1997). "Mapping Vegetation and Aanthropogenic Disturbances in Southern Ecuador with Remote Sensing Techniques: Implications for Park Management." *CLAG Yearbook*, 77–90.

—— (1998). "Effects of Anthropogenic Disturbances on Páramo Vegetation in Podocarpus National Park, Ecuador." *Physical Geography*, 19/3: 221–38.

Keeling, David J. (1994). "Regional Development and Transport Policies in Argentina: An Appraisal." *The Journal of Developing Areas*, 28/4: 487–502.

—— (1996). *Buenos Aires: Global Dreams, Local Crises*. New York: John Wiley.

—— (1997). *Contemporary Argentina: A Geographical Perspective*. Boulder, Colo.: Westview.

—— (1998). "Transportation, Regional Development, and Economic Potential in Mexico," in Allen G. Noble *et al.* (eds.), *Regional Development and Planning for the 21st Century: New Priorities, New Philosophies*. Brookfield, Vt.: Ashgate, 101–23.

Keese, James R. (1998). "International NGOs and Land-Use Change in a Southern Highland Region of Ecuador." *Human Ecology*, 26/3: 451–68.

Kennedy, L. M., and Horn, S. P. (1997). "Prehistoric Maize Cultivation at the La Selva Biological Station." *Biotropica*, 29/3: 368–70.

Kent, Robert B. (1993). "Geographical Dimensions of the Shining Path Insurgency in Peru." *Geographical Review*, 83/4: 441–54.

Kent, Robert B., and Rimarachin, Jesús C. (1994). "Rural Public Works Construction in the Andes of Northern Peru." *Third World Planning Review*, 16/4: 358–74.

Kent, Robert B., Guardia, Edgar, and Sibille, Olav K. (1998). "Decentralization, Popular Participation, and Changing Patterns of Urban and Regional Development in Bolivia," in Allen G. Noble *et al.* (eds.), *Regional Development and Planning for the 21st Century: New Priorities, New Philosophies*. Brookfield, Vt.: Ashgate, 351–64.

Klak, Thomas (1993). "Contextualizing State Housing Programs in Latin America: Evidence from Leading Housing Agencies in Brazil, Ecuador, and Jamaica." *Environment and Planning A*, 25: 653–76.

—— (1994). "Havana and Kingston: Mass Media Images and Empirical Observations of Two Caribbean Cities in Crisis." *Urban Geography*, 15/4: 318–44.

—— (ed.) (1998). *Globalization and Neoliberalism: The Caribbean Context*. New York: Rowman & Littlefield.

Klak, Thomas, and Holtzclaw, M. (1993). "The Housing, Geography, and Mobility of Latin American Urban Poor: The Prevailing Model and the Case of Quito, Ecuador." *Growth and Change*, 24/2: 247–76.

Kleinpenning, Jan. M. G. (1995). *Peopling the Purple Land: A Historical Geography of Rural Uruguay, 1500–1915*. Amsterdam: CEDLA.

Knapp, Gregory W. (1990). "Cultural and Historical Geography of the Andes." *Yearbook, CLAG*, 17/18: 165–75.

—— (1991). *Andean Ecology: Adaptive Dynamics in Ecuador*. Dellplain, 27. Boulder, Colo.: Westview Press.

—— (1992). *Riego precolonial y tradicional en la sierra norte del Ecuador*. Quito: Abya-Yala.

Knapp, Gregory W., and Mothes, Patricia (1998). "Quilota Ash and Human Settlements in the Equatorial Andes," in Patricia Mothes (ed.), *Actividad Volcánica y pueblos precolombinos en el Ecuador*. Quito: Abya-Yala, 139–55.

Kramer, Wendy (1994). *The Politics of Encomienda Distribution in Early Spanish Guatemala, 1524–1544: Dividing the Spoils*. Dellplain, 31. Boulder, Colo.: Westview Press.

Kramer, W., Lovell, George W., Lutz, C. H. (1990). "Encomienda and Settlement: Towards a Historical Geography of Early Colonial Guatemala." *CLAG Yearbook*, 16: 67–72.

Lambert, Dean P. (1994). "Regional Core-Periphery Imbalance: The Case of Guerrero, Mexico, since 1821." *CLAG Yearbook*, 20: 59–71.

Lawson, Victoria (1995). "Beyond the Form: Restructuring Gender Divisions of Labor in Quito's Garment Industry under Austerity." *Environment and Planning D: Society and Space*, 13: 414–44.

Lawson, Victoria A., and Klak, Thomas (1993). "An Argument for Critical and Comparative Research on the Urban Economic Geography of the Americas." *Environment and Planning A*, 25: 1071–84.

Liverman, D. M. (1990). "Drought Impacts in Mexico: Climate, Agriculture, Technology, and Land Tenure in Sonora and Puebla." *Annals of the Association of American Geographers*, 80: 49–73.

Lovell, W. George (1992). "'Heavy Shadows and Black Night': Disease and Depopulation in Colonial Spanish America." *Annals of the Association of American Geographers*, 82/3: 426–43.

—— (1995). *A Beauty that Hurts: Life and Death in Guatemala*. Toronto: Between the Lines.

Lovell, W. George, and Lutz, C. H. (1995). *Demography and Empire: A Guide to the Population History of Spanish Central America, 1500–1821*. Dellplain, 33. Boulder, Colo.: Westview Press.

Lutz, C. H., and Lovell, W. George (1990). "Core and Periphery in Colonial Guatemala," in Carol A. Smith (ed.), *Guatemalan Indians and the State, 1540 to 1988*. Austin: University of Texas Press, 35–51.

Mack, Taylor E. (1998). "Contraband Trade through Trujillo, Honduras, 1720s–1782." *CLAG Yearbook*, 24: 45–56.

MacLachlan, Ian, and Aguilar, Adrian G. (1998). "Maquiladora Myths: Locational and Structural Change in Mexico's Export Manufacturing Industry." *Professional Geographer*, 50/3: 315–31.

Malmström, Vincent H. (1995). "Geographical Origins of the Tarascans." *Geographical Review*, 85: 31–40.

Mast, J. N., Veblen, T. T., and Hodgson, M. E. (1997). "Tree Invasion within a Pine/Grassland Ecotone: An Approach with Historic Aerial Photography and GIS Modeling." *Forest Ecology and Management*, 93: 181–94.

Mathewson, Kent (1992). "Tropical Riverine Regions: Locating the 'People Without History.'" *Journal of the Steward Anthropological Society*, 20: 167–80.

—— (ed.) (1993). *Culture, Form and Place. Essays in Cultural and Historical Geography*. Geoscience and Man, 32. Baton Rouge, La.: Geoscience Publications.

Melville, Elinor G. K. (1994). *A Plague of Sheep: Environmental Consequences of the Conquest of Mexico*. Cambridge: Cambridge University Press.

MERCOSUL: um atlas cultural, social e econômico (1997). Rio de Janeiro: Instituto Herbert Levy.

Meyer-Arendt, Klaus J. (1990). "Patterns and Impacts of Coastal Recreation along the Gulf Coast of Mexico," in P. Fabbri (ed.), *Recreational Uses of Coastal Areas*. Dordrecht: Kluwer, 133–48.

—— (1991). "Tourism Development on the North Yucatan Coast: Human Response to Shoreline Erosion and Hurricanes." *GeoJournal*, 23: 327–36.

Meyer-Arendt, Klaus J., Sambrook, Richard A., and Kermath, Brian M. (1992). "Seaside Resorts in the Dominican Republic: A Typology." *Journal of Geography*, 91: 219–25.

Mignolo, Walter D. (1995). "Putting the Americas on the Map: Cartography and the Colonization of Space." *The Darker Side of the Renaissance: Literacy, Territoriality, and Colonization*. Ann Arbor: University of Michigan Press, 259–313.

Miller, Arthur G. (1991). "Transformations of Time and Space: Oaxaca, México, circa 1500–1700," in S. Kuchler and W. Melion (eds.), *Images of Memory: On Remembering and Representation*. Washington, DC: Smithsonion Institute Press, 141–75.

Miyares, Inés M., and Gowen, Kenneth J. (1998). "Recreating Borders? The Geography of Latin Americans in New York City." *CLAG Yearbook*, 24: 31–44.

Mundy, Barbara E. (1993). *The Mapping of New Spain: Indigenous Cartography and the Maps of the Relaciones Geográficas*. Chicago: University of Chicago Press.

Mutersbaugh, Tad (1997). "Rural Industrialization as Development Strategy: Cooperation and Conflict in Co-op, Commune and Household in Oaxaca, Mexico." *CLAG Yearbook*, 23: 91–105.

National Research Council (NRC) (1997). *Rediscovering Geography: New Relevance for Science and Society*. Washington: National Academy Press.

Newsome, Tracy, Aranguren, Freddy, and Brinkmann, Robert (1997). "Lead Contamination Adjacent to Roadways in Trujillo, Venezuela." *Professional Geographer*, 49/3: 331–41.

Newson, Linda (1995). *Life and Death in Early Colonial Ecuador*. Norman: University of Oklahoma Press.

—— (1996). "The Population of the Amazon Basin in 1492: A View from the Ecuadorian Headwaters." *Transactions of the Institute of British Geographers*, NS 21: 5–26.

Northrop, L. A., and Horn, S. P. (1996). "PreColumbian Agriculture and Forest Disturbance in Costa Rica: Paleoecological Evidence from Two Lowland Rainforest Lakes." *The Holocene*, 6/3: 289–99.

Offen, Karl (1998). "An Historical Geography of Chicle and Tunu Gum Production in Northeastern Nicaragua." *CLAG Yearbook*, 24: 57–75.

Orvis, K. H., Clark, G. M., Horn, S. P., and Kennedy, L. M. (1997). "Geomorphic Traces of Quaternary Climates in the Cordillera Central, Dominican Republic." *Mountain Research Development*, 17: 323–31.

Ouweneel, Arij (1996). *Shadows over Anáhuac: An Ecological Interpretation of the Crisis and Development in Central Mexico*. Albuquerque: University of New Mexico Press.

Palm, Risa, and Hodgson, Michael E. (1993). "Natural Hazards in Puerto Rico." *Geographical Review*, 83/3: 280–9.

Parsons, James J. (1992*a*). "Geography," in P. H. Covington (ed.), *Latin America and Caribbean Studies: A Critical Guide*. Westport: Greenwood Publishing, 267–75.

Parsons, James J. (1992*b*). "Southern Blooms: Latin America and the World of Flowers." *Queen's Quarterly: A Canadian Review*, 99: 542–61.

—— (1996*a*). "'Mr. Sauer' and the Writers." *Geographical Review* 86: 22–41.

—— (1996*b*). "Carl Sauer's Vision of an Institute for Latin American Studies." *Geographical Review*, 86: 377–84.

Pasqualetti, Martin J. (ed.) (1997). *The Evolving Landscape: Homer Aschmann's Geography*.

Pederson, Leland R. (1998). "In Memoriam: Dan Stanislawski, 1902–1997." *Annals, Association of American Geographers*, 88/4: 699–705.

Perreault, Thomas A., Bebbington, Anthony J., and Carroll, Thomas F. (1998). "Indigenous Irrigation Organizations and the Formation of Social Capital in Northern Highland Ecuador." *CLAG Yearbook*, 24: 1–16.

Potter, Robert B., and Conway, Dennis (1996). *Self-help Housing: The Poor and the State in the Caribbean*. Knoxville, Tenn.: The University of Tennessee Press.

Price Marie D. (1994*a*). "Ecopolitics and Environmental Non-governmental Organizations in Latin America." *Geographical Review*, 84/1: 42–58.

—— (1994*b*). "Hands for the Coffee: Migrants and Western Venezuela's Coffee Economy, 1870–1930." *Journal of Historical Geography*, 20: 62–80.

—— (1996). "The Venezuelan Andes and the Geographical Imagination." *Geographical Review*, 86/3: 334–55.

Relva, M. A., and Veblen, T. T. (1998). "Impacts of Introduced Large Herbivores on *Astrocedrus chilensis* Forests in Northern Patagonia, Argentina." *Forest Ecology and Management*, 108/1–2: 27–40.

Richardson, Bonham C. (1992). *The Caribbean in the Wider World, 1492–1992*. New York: Cambridge University Press.

—— (1997). *Economy and Environment in the Caribbean: Barbados and the Windwards in the Late 1800s*. Gainsville, Fla.: University Press of Florida.

Roberts, J. Timmons (1995*a*). "Subcontracting and the Omitted Social Impacts of Development Projects: Household Survival at the Carajás Mines in the Brazilian Amazon." *Economic Development and Cultural Change*, 43/4: 735–58.

—— (1995*b*). "Expansion of Television in Eastern Amazonia." *Geographical Review*, 85/1: 41–9.

—— (1996). "Global Restructuring and the Environment in Latin America," in Roberto P. Korzeniewicz and William C. Smith (eds.), *Latin America in the World Economy*. Westport, Conn.: Greenwood Press, 187–210.

Robinson, David J. (1990*a*). "Liberty, Fragile Fraternity and Inequality in Early Republican Spanish America: Assessing the Impact of French Revolutionary Ideal." *Journal of Historical Geography*, 20: 1–16.

—— (ed.) (1990*b*). *Migration in Colonial Latin America*. Cambridge: Cambridge University Press.

—— (1992*a*). *Mil leguas por América: de Lima a Caracas, 1740–1741. Diario de don Miguel de Santisteban*. Bogota: Banco de la República.

—— (1992*b*) "A linguagem e o significado de lugar na América Latina." *Revista de História*, 121: 67–110.

—— (2003). Lari Collaguas: Económia, sociedad y población, 1604–1605. Lima: Universidad Católicia del Perú.

Robinson, David J., and Doenges, Catherine E. (1997). "Fuentes parroquiales de Guatemala Colonial." *MesoAmérica*, 33: 195–214.

Roccatagliata, Juan A. (ed.) (1997). *Geografía Económica Argentina: Temas*. 2nd edn. Buenos Aires: El Ateneo.

Rudel, Thomas, and Richards, S. (1990). "Urbanization, Roads, and Rural Population Change in the Ecuadorian Andes." *Studies in Comparative International Development*, 25/3: 73–89.

Ryder, Roy (1994). "Land Evaluation for Steepland Agriculture in the Dominican Republic." *Geographical Journal*, 160/1: 74–86.

Salazar, Deborah A. (1996). *Internet and Geographic Information System Technology: Bridging the U.S.–Mexico Information Frontier*. Austin, Tex. Bureau of Economic Geology.

Sambrook, Richard A., Kermath, Brian M., and Thomas, Robert N. (1994). "Tourism Growth Poles Revisited: A Strategy for Regional Economic Development in the Dominican Republic." *CLAG Yearbook*, 20: 1–10.

Samudio, Edda (1997). "The Dissolution of Indian Community Lands in the Venezuelan Andes: The Case of La Mesa." CLAG *Yearbook*, 23: 17–26

Sambrook, Richard A., Pigozzi, Bruce W., and Thomas, Robert N. (1999). "Population Pressure, Deforestation, and Land Degradation: A Case Study from the Dominican Republic." *Professional Geographer*, 51/1: 25–40.

Santana, Deborah Berman (1996). *Kicking off the Bootstraps: Environment, Development, and Community Power in Puerto Rico*. Tucson: University of Arizona Press.

Scarpaci, Joseph L. (1991). "Primary-Care Decentralization in the Southern Cone: Shantytown Health Care as Urban Social Movement." *Annals of the Association of American Geographers*, 81/2: 103–26.

—— (1998*a*). "Reflections on Cuban Socialism and Planning in the Special Period," in Allen G. Noble *et al.* (eds.) *Regional Development and Planning for the 21st Century: New Priorities, New Philosophies*. Brookfield, Vt.: Ashgate, 225–43.

—— (1998*b*). "The Changing Face of Cuban Socialism: Tourism and Planning in the Post-Soviet Era." *CLAG Yearbook*, 24: 97–109.

Segre, Roberto; Coyula, Mario, and Scarpaci, Joseph L. (1997). *Havana: Two Faces of the Antillean Metropolis*. New York: Wiley.

Siemens, Alfred (1989). *Tierra configurada*. Mexico City: Consejo Nacional para la Cultura y las Artes.

—— (1990). *Between the Summit and the Sea: Central Veracruz in the Nineteenth Century*. Vancouver: University of British Columbia Press.

—— (1998). *A Favored Place: San Juan River Wetlands, Central Veracruz, A. D. 500 to the Present*. Austin: University of Texas Press.

Sluyter, A. (1994). "Intersive Wetland Agriculture in Mesoamerica: Space, Time and Form." *Annals of the Association of American Geographers*, 84/4: 557–84.

—— (1996). "The Ecological Origins and Consequences of Cattle Ranching in Sixteenth-Century New Spain." *The Geographical Review*, 86/2: 161–77.

—— (1997). "Regional, Holocene Records of the Human Dimension of Global Change: Sea-Level and Land-Use Change in Prehistoric Mexico." *Global and Planetary Change*, 14: 127–46.

Smith, Betty E. (1998). "Residential Land Development in Highland Ecuador," in Allen G. Noble *et al.* (eds.), *Regional Development and Planning for the 21st Century: New Priorities, New Philosophies*. Brookfield, Vt.: Ashgate, 169–86.

Smith, Betty E., and Andino, Mario (1996). "An Integrative Approach to Tourism and Sustainable Development in Sigchos, Ecuador." *International Third World Studies Journal and Review*, 8: 63–9.

Smith, Nigel J. H. (1999). *The Amazon River Forest*. New York: Oxford University Press.

Smith, Nigel J. H., Serrao, E. A., Alvim, P., and Falesi, I. C. (1995). *Amazonia. Resiliency and Dynamism of the Land and its People*. Tokyo: United Nations University Press.

Smith, Nigel J. H., Williams, J. T., Plucknett, D. L., and Talbot, J. P. (1992). *Tropical Forests and their Crops*. Ithaca, NY: Comstock Publishing Associates.

Stadel, Christoph (1994). "The Role of NGOs for the Promotion of Children in the Highlands of Bolivia," in H. G. Bohle (ed.), *World of Pain and Hunger: Geographical Perspectives on Disaster Vulnerability and Food Security*. Fort Lauderdale: Verlag Breitenbach, 187–208

—— (1997). "The Mobilization of Human Resources by Non-Governmental Organizations in the Bolivian Andes." *Mountain Research and Development*, 17/3: 213–28.

Steinberg, Michael K. (1996). "Folk House-Types as Indicators of Tradition: The Case of the Mopan Maya in Southern Belize." *CLAG Yearbook*, 22: 87–92.

—— (1997). "Death of the Dance, Cultural Change, and Religious Conversion among the Mopan Maya in Belize." *Cultural Survival Quarterly*, 21/2: 16–17.

—— (1998). "Political Ecology and Cultural Change: Impacts on Swidden-Fallow Agroforestry Practices among Mopan Maya in Southern Belize." *The Professional Geographer*, 50/4: 407–17.

Sternberg, Rolf (1994). "El cono sur en transición geopolítica, segunda parte: La geopolítica y el contexto económico." *Geopolítica*, 20/52: 42–51.

—— (1997). "Project Urban Infrastructure in Brazilian Amazonia." *Espacio y Desarrollo*, 7/9: 201–34.

Taylor, E. Mack (1998). "Contraband Trade through Trujillo, Honduras, 1720s–1782." *CLAG Yearbook*, 24: 45–56.

Toledo Maya Cultural Council (1997). *Maya Atlas: The Struggle to Preserve Maya Land in Southern Belize*. Toledo: TMCC.

Treacy, John (1994). *Las chacras de Corporaque: andenería y riego en el valle del Colca*. Lima: Instituto de Estudios Peruanos.

Turner, B. L., II (1991). "Thoughts on Linking the Physical and Human Sciences in the Study of Global Environmental Change." *Research and Exploration*, 7/2: 133–5.

—— (1997*a*). "Spiral, Bridges, and Tunnels: Engaging Human Environment Perspectives in Geography?" *Ecumene*, 4/2: 196–217.

—— (1997*b*). "The Sustainability Principle in Global Agendas: Implications for Understanding Land-Use/Cover Change." *Geographical Journal*, 163/2: 133–40.

Turner, B. L., II, Gómez, Sal, González Bernáldez, F., and Di Castri, F. (eds.) (1995). *Global Land-Use Change: A Perspective from the Columbian Encounter*. Madrid: Consejo Superior de Investigaciones Científicas.

Turner, B. L., II, and Butzer, K. W. (1992). "The Columbian Encounter and Land-Use Change." *Environment*, 43/8: 16–20, 37–44.

Vanderbush, Walter, and Klak, T. (1996) " 'Covering' Latin America: The Exclusive Discourse of the Summit of the Americas (as Viewed through the *New York Times*)." *Third World Quarterly*, 17/3: 537–56.

Veblen, T. T., Kitzberger, T., and Lara, A. (1992). "Disturbance and Forest Dynamics Along a Transect from Andean Rain Forest to Patagonian Shrubland." *Journal of Vegetation Science*, 3: 507–20.

Veblen, T. T., Donoso, C., Kitzberger, T., and Rebertus, A. J. (1996). "Ecology of Southern Chilean and Argentinean Nothofagus," in T. T. Veblen, R. S. Hill, and J. Read (eds.), *Ecology and Biogeography of Nothofagus Forests*. New Haven: Yale University Press, 293–353.

Veblen, T. T., Donoso, C., Kitzberger, T., Vollalba, R., and Donnegan, J. (1999) "Fire History in Northern Patagonia: The Role of Humans and Climatic Variation." *Ecological Monographs*, 69/1: 47–67.

Vender, JoAnn C. (1995). "Culture, Place, and School: Improving Primary Education in Rural Ecuador." *CLAG Yearbook*, 20: 107–19.

Villalba, R. (1997). "Regional Patterns of Tree Population Age Structure in Northern Patagonia: Climatic and Disturbance Influences." *The Journal of Ecology*, 85/2: 113–25.

Voeks, Robert (1990). "Sacred Leaves of Brazilian Candomblé." *Geographical Review*, 80: 118–31.

Wallin, T. R., and Harden, C. P. (1996). "Estimating Trail-Related Soil Erosion in the Humid Tropics: Jatun-Sacha, Ecuador, and La Selva, Costa Rica," *Ambio*, 25/8: 517–22.

Ward, Peter M. (1998) *Mexico City*. 2nd edn. New York: John Wiley.

Waylen, P. R., and Caviedes, C. N. (1987), "El Niño and Annual Floods in Coastal Peru," in L Mayer and D. Nash (eds.), *Catastrophic Flooding*. Boston: Allen & Unwin: 57–78.

—— (1990). "Annual and Seasonal Fluctuations of Precipitation and Streamflow in the Aconcagua River Basin." *Journal of Hydrology*, 120: 79–102.

Waylen, P. R., Caviedes, C. N., and Quesada, M. (1996*a*). "Interannual Variability of Monthly Precipitation in Costa Rica." *Journal of Climate*, 16: 1137–48.

Waylen, P. R., Quesada, M., and Caviedes, C. N. (1996*b*). "Temporal and Spatial Variability of Annual Precipitation in Costa Rica and the Southern Oscillation." *International Journal of Climatology*, 16/1: 173–93.

Waylen, P. R., Caviedes, C., Poveda, C. N., Quesada, M., and Mesa, O. (1998). "Rainfall Distribution and Regime in Costa Rica and its Response to El Niño-Southern Oscillation." *CLAG Yearbook*, 24: 75–84.

Weil, Connie (1995). "Especial: Una invitación a la geografía de la salud." *Geoistmo*, 6–7/1–2: 1–76.

Weil, Connie, and Scarpaci, Joseph L. (eds.) (1992). *Health and Health Care in Latin America During the Lost Decade: Insights for the 1990s*. Minneapolis: Prisma Institute, University of Minnesota.

Weil, Jim, and Weil, Connie (1993). *Verde es la Esperanza: Colonización, Comunidad y Coca en la Amazonia*. La Paz: Editorial Los Amigos del Libro.

Weischet, W., and Caviedes, C. N. (1993). *The Persisting Ecological Constraints of Tropical Agriculture*. London: Longman.
Wesche, Rolf (1996). "Developed Country Environmentalism and Indigenous Community Controlled Ecotourism in the Ecuadorian Amazon." *Geographische Zeitschrift*, 84/3–4: 157–68.
Wesche, Rolf, and Bruneau, Thomas (1990). *Integration and Change in Brazil's Middle Amazon*. Ottawa: University of Ottawa Press.
West, Robert Cooper (1993). *Sonora: Its Geographical Personality*. Austin: University of Texas Press.
Whitehead, Neil L. (1998). "Indigenous Cartography in Lowland South America and the Caribbean," in David Woodward and M. Lewis (eds.), *The History of Cartography*. Chicago: University of Chicago Press, ii. bk. 3, 301–26.
Whitmore, Thomas M. (1992). *Disease and Death in Early Colonial Mexico: Simulating Amerindian Depopulation*. Dellplain, 28. Boulder, Colo.: Westview Press.
—— (1996). "Population Geography of Calamity: The Sixteenth- and Seventeenth-Century Yucatán." *International Journal of Population Geography*, 2: 291–311.
Whitmore, Thomas M., and Turner, B. L., II (1992). "Landscapes of Cultivation in Mesoamerica on the Eve of Conquest." *Annals of the Association of American Geographers*, 82/3: 402–25.
Whitmore, Thomas M., and Williams, Barbara J. (1997). "Famine Risk in the Contact-Era Basin of Mexico." *Ancient Mesoamerica*, 8: 1–16.
Wiley, James (1996). "The European Union's Single Market and Latin America's Banana-Exporting Countries." *CLAG Yearbook*, 22: 31–40.
—— (1998). "The Banana Industries of Costa Rica and Dominica in a Time of Change." *Tijdschrift Voor Economische en Sociale Geografie*, 89/1: 66–81.
Winberry, John J. (1998). "Regional Economic Integration in Latin America: Strategies and the Role of Gateway Cities." *Southeastern Latin Americanist* (SECOLAS), 29: 45–56.
Woods, W. I., and McCann, J. M. (1999). "The Anthropogenic Origin and Persistence of Amazonian Dark Earths." *CLAG Yearbook*, 25: 7–14.
Woods, W. I., Wells, C. L., and Meyer, D. W. (1996). "Cautions on the Use of Soil Data for Prehistoric Reconstructions: A Belizian Example." *Papers and Proceedings of the Applied Geography Conferences*, Kansas City, 18: 209–17.
Young, Emily (1994). "The Impact of IRCA on Settlement Patterns among Mixtec Migrants in Tijuana, Mexico." *Journal of Borderland Studies*, 11/2: 1–21.
Young, K. R. (1998). "Deforestation in Landscapes with Humid Forests in the Central Andes," in Zimmerer and Young (1998: 75–99).
Zimmerer, Karl. S. (1991). "Managing Diversity in Potato and Maize Fields of the Peruvian Andes," *Journal of Ethnobiology*, 11/1: 23–49.
—— (1993*a*). "Soil Erosion and Social (Dis)courses in Cochabamba, Bolivia: Perceiving the Nature of Environmental Degradation." *Economic Geography*, 69/3: 312–28.
—— (1993*b*). "Agricultural Biodiversity and Peasant Rights to Subsistence in the Central Andes during Inca Rule." *Journal of Historical Geography*, 19/1: 15–32.
—— (1996). *Changing Fortunes. Biodiversity and Peasant Livelihood in the Peruvian Andes*. Berkeley: University of California Press.
—— (1998). "The Ecogeography of Andean Potatoes: Versatility in Farm Regions and Fields Can Aid Sustainable Development." *BioScience*, 48/6: 445–54.
—— (1999). "Overlapping Patchworks of Mountain Agriculture in Peru and Bolivia: Toward a Regional-Global Landscape Model." *Human Ecology*, 27: 135–65.
Zimmerer, K. S., and Young, K. R. (eds.) 1998. *Nature's Geography. New Lessons for Conservation in Developing Countries*. Madison: University of Wisconsin Press.

CHAPTER 44

Russian, Central Eurasian, and East European Geography

Beth Mitchneck, Robert Kaiser, James Bell, and Kurt Englemann

The end of ideology and imposed centralization has meant the return of geography, just as it signals the return, rather than 'the end' of history.

David Hooson (1994: 134)

Introduction

Over the past decade, the societies that encompass Russia, Central Eurasia, and East Europe have experienced profound and radical change. Today, the region is making uneven progress toward democratic modes of government and market-oriented economies. The fluid dynamics of change within the region make it one of the most exciting and rewarding areas of research within geography. Across Russia, Central Eurasia, and East Europe vital lessons can be learned about the contextual nature of political and economic transition. At the same time, crucial insights can be obtained into the more universal process of regional transformation and the social reconstruction of place identity. This region is a laboratory for testing the relevancy of geographic research and theory for a post-socialist world.

This chapter reviews the major changes in the practice and orientation of geographic research in the region since the collapse of state-socialism (see Ch. 39, Asian Geography, for further information on the Central Asian countries). This chapter comments on the methodological, conceptual, and topical evolution of this area-specialty over the last decade. It concludes by contemplating the possible directions of future geographic research in the region.

Changing Nature of Geographic Research in the Region

Prior to the fall of the Berlin Wall in 1989 and the dissolution of the Union of Soviet Socialist Republics in 1991, the societies of Russia, Central Eurasia, and East Europe were typically defined in political and economic terms as a unified region, known as the Soviet bloc. While national and cultural differences across the Soviet bloc were not ignored, they were treated as less significant than the uniform pattern of planned economies and communist regimes that governed the region. The region was further unified through the political and economic primacy of Moscow, where decisions were made that directly impacted the states throughout Russia, Central Eurasia, and East Europe.

In the Soviet era, geographic research in the region focused largely on strategic questions relating to the efficiency, efficacy, and future trajectory of the state-socialist model of economic and political development. The topics explored by geographers ranged from issues of agricultural production to urban structure to regional economic investment to domestic and international migration. However, analysis was hampered by the superficial and less than reliable quality of officially published data, by restricted archival access, and by prohibitions on survey- or interview-based research. In addition, official protocol constrained collaboration with colleagues in the region. Because of these limitations, geographic research in Russia, Central Eurasia, and East Europe tended toward the empirical investigation of social, economic, and environmental trends using whatever information we could discern from the available resources.

The demise of state-socialism dramatically changed the nature of geographic research in the region. Accessibility to statistical data published by state agencies has improved, while new sources of statistical information have emerged in the private and non-governmental sectors. The range of published data has expanded to include opinion polls, election returns, and customized survey results. And, perhaps most important for historical geographic research, national and local governments have opened previously restricted historical archives. Internet access to governmental and non-governmental sources in the region has also facilitated data collection.

This abundance of information has not, however, resolved the issue of data quality or even of accessibility. Research agencies in the former Soviet bloc only gradually adopt Western social scientific standards. New ethical questions also arise regarding the appropriate compensation for access to published, archived, or private information. The funding of cooperative research has been a welcome supplemental source of income for in-country specialists, who have suffered greatly from inflation and delays in the payment of salaries. But financial support for conducting research on the former Soviet bloc countries from the American side has declined, so US researchers are financially less able to take advantage of joint research opportunities. The decline of the USSR Academy of Sciences and its branch institutions has hindered access and collaboration. Additional questions of access for graduate students and for those conducting unfunded research further complicate the research process.

On the other hand, opportunities for field research have vastly expanded the scope, scale, and methodology of research in the region. Travel to distant sites means that research need not focus on the few major cities in the region. Also, researchers can now utilize survey techniques, interviews, and participant observation to analyze political and economic change. Communities, groups, and even individuals are now frequent subjects in research across the region. This has created the potential for multi-dimensional analyses that investigate the interactions and linkages between phenomena that span the local, regional, national, and international scales.

In conjunction with the numerous changes in the practice of geographic research, the demise of the Soviet bloc launched renewed interest in conceptual issues related to defining the region and comprehending the processes of regional transformation. New political and economic alignments mean that old regional definitions have less salience today. Depending on the research project, the Russian Far East can be included as part of the Pacific Rim, East Europe as part of an expanding Western Europe, Central Asia as an extension of the Middle East (see Ch. 39 for further information), or the Baltic Republics as part of Scandinavia. These reconfigurations reflect the fluidity of political and economic relations in the post-socialist world and indicate the tremendous potential for research in Russia, Central Eurasia, and East Europe to shed light on the dynamics of regional change and the consolidation of new territorially based identities.

The topics explored by geographers conducting research in the region have multiplied since the collapse of state-socialism. The question of political and economic transition in the societies of the former Soviet bloc, however, dominates the research agenda. In the following section, the breadth of geographic research in this dynamic region is reviewed.

Major Research Themes

Below, we group individual research contributions into three principal thematic categories: geography of economic transition, geography of demographic processes, and geography of political transition. Problems of environmental degradation constitute a key sub-theme within the geography of economic transition, while the relationship between national identity and territory is highlighted within the geography of political transition.

Geography of Economic Transition

Geographers analyze the spatial dimensions of economic change in the former Soviet bloc and the capacity of the region's emerging economies to adapt to the realities of the new global order. Some of the topics researched include defining the success or failure of the transition and documenting its implications for the economy, the population, and spatial interaction. As such, geographical research has focused on the issues of natural resource exploitation, trade, the natural environment, and the emerging new economic actors. Many of these topics constituted major research themes previously, but new topics, such as the role that marketization and economic relations plays in changing economic patterns at the global and interstate scales, have come to the forefront of the research agenda.

As during Soviet times, natural resource use continues to form a major research theme. Current research has changed, however, from documenting patterns of extraction and utilization to looking at questions of restructuring, the international role in national resource issues, and the importance of the sector to the financial survival of national economies. In the case of the Russian Federation, crude oil and natural gas constitute the leading export sectors of the economy and the principal means by which the Russian government services its international debt. Sagers (1996) has described the dramatic decline in oil production since the late 1980s and its eventual stabilization in the late 1990s. He has also evaluated several fossil fuel projects, including the project to send East Siberian gas to China (Sagers and Nicoud 1997), the development of offshore oil deposits near Sakhalin Island (1995), and the extraction and transportation of Caspian Sea petroleum (1994).

A new focus on international relations and natural resources has emerged in the literature that points to the importance of market relations over political ones binding together regional economies. Dienes (1993*a*) describes the replacement of Soviet-era economic links with "geoeconomic ties" to areas outside the region. Bass and Dienes (1993) address the collapse of the "secret infrastructure" of the defense industries and the political challenges they have generated, for Russia and Ukraine in particular. They lament the indiscriminate application of the export-led model of economic development as a basis for economic transition to a market economy (1998). Crocker and Berentsen (1998) apply a mathematical model of market potential and travel to determine the comparative accessibility of East-Central European countries to European Union markets. Bond and Levine (1997) examine the constraints on the Armenian copper and molybdenum industries, including political relations with Turkey and Azerbaijan. In a separate article (1993), they document how the Russian manganese industry declined to such a degree that the country has changed from being a net exporter to a net importer.

Other geographers consider the general international implications of economic transition. Bradshaw and Lynn (1994) assert that a combined application of world-systems theory and a regional geographic approach provides a more useful framework for examining the development prospects for the post-Soviet states. Lee and Bradshaw (1997) also explain the worsening prospects for Russian–South Korean bilateral trade, which has collapsed in large part due to the Asian financial crisis. Murphy (1991) addresses the problems of economic integration of Eastern Europe with the European Community (now European Union). Johansen (1995) examines factors affecting retail outlet site selection in the Baltic Republics. ZumBrunnen (1993) analyzes the problems associated with the development of new commercial structures in Russia and the various voucher and other privatization schemes.

Even as the pattern of economic interactions among the states of Russia, Central Eurasia, and East Europe unfolds, we find that the set of economic relations within national contexts continues to change. Several geographers have looked at this issue at various geographic scales. Zimine and Bradshaw (1999) examine the regional adaptation to Russia's economic crisis in Novgorod Oblast′. Liebowitz (1996) addresses Russia's center–local fiscal relations in the post-Soviet context. He emphasizes the inertia of the past (Soviet) system of ad-hoc negotiations and intergovernmental transfers, which continues to influence fiscal relations. Liebowitz argues that the specific type of federal structure is less important than the need to meet the demands of constituent republics while preserving the integrity of the Russian state. In particular, central–local economic relations remain in a fluid situation. For example, Mitchneck analyzes central–local fiscal relations in urban Russia and the new role of local governments in managing economic activities in the post-Soviet era. Mitchneck (1994, 1997*a*, *b*) concludes that local governments became the primary provider of social welfare and implementer of market reforms in the immediate post-Soviet period. She also demonstrates that these changes vary dramatically by place and level of government. Using a case study of participation in the development of local economic development strategies in Novosibirsk, Stenning (1999) demonstrates that democratization of Russian regions is limited in part because of the interaction between marketization and democratization.

Other geographers comparing regions within states have focused on the theme of the differential impact of economic reform in other contexts. Ioffe and Nefedova (1997, 2001*a*) identify two types of territories that have fared better than others during the transformation process: those experiencing rapid reform, and those where reforms have yet to be implemented. They suggest that areas of half-reform have experienced the most profound declines in terms of output, yields, and assets. Ioffe and Nefedova also observe a regional disparity in food imports and argue that this disparity has led to political divisions with clear spatial patterns: manufacturing areas of the country advocate free trade, while agricultural areas advocate protectionism.

Other analyses of the agrarian context in economic transformation focus on the differences across mode of ownership. Craumer (1994) portrays the complex spatial variability of state and collective farm reorganization and peasant farm development. His monograph, *Rural and Agricultural Development in Uzbekistan* (1995), has contributed to an understanding of the challenges of agrarian change in Central Asia in the post-Soviet period (see Ch. 39 for further information). In a more recent edited volume, Engelmann and Pavlakovic (2001) provide a set of essays whose authors assess the environmental constraints to rural development in Eurasia and the Middle East.

Large cities throughout the reglon also have been focal points in the transition to market capitalism. These urban centers serve as the destination for foreign and domestic investment, and they witness dramatic transformations in their social and physical landscapes. Moscow has emerged as a global city of sorts (Gritsai 1997). Bater, focusing on the effects of market reforms on the urban landscape, has begun to document the effects on Moscow (1994*a*, *b*, *c*, 2001; Bater *et al.* 1998). Bater notes that the gap between social classes has increased with the development of a central business district and housing privatization. Pavlovskaya (2001) teases out the changes to everyday life for residents living in the center of Moscow and traces how those changes vary by gender and class. Argenbright (1999) assesses the complex interrelationship among the transitions occurring in Moscow, the new public spaces that are being constructed, and the question of civil society, and indicates that there is by no means a clear connection between transition to a consumer-based society and democratization. Ioffe and Nefedova (2000, 2001*b*) examine the development of "suburbanization," Russian style, and find significant differences from the West in the development of the spaces between urban and rural environs. Mitchneck (1998) looks at a smaller city, Yaroslavl', and notes that its physical landscape is dramatically changing to attract the gaze of both the foreign investor and the local resident. Corruption is another issue of great importance for cities in the transition. While geographers have not directly tackled this issue, several works deal with corruption and the use of social capital as it relates to urban development (Pavlovskaya 2001; Mitchneck 2001*a*, *b*).

Environmental Degradation

The decline of the socialist bloc resulted in the acknowledgement of grave environmental degradation in Russia, Central Eurasia, and East Europe. Today, the costs of ameliorating problems associated with the central planning legacy of excessive air pollution, contaminated water supplies, or toxic waste weigh heavily on the emerging economies of the region. At the same time, the long-term degradation and depletion of agricultural and mineral resources limits the material base for economic development. *Environmental Resources and Constraints in the Former Soviet Republics* (1995), a volume edited by Pryde, provides a country-by-country overview of the legacy of Soviet-era environmental degradation and the challenges facing these states in the post-Soviet era. Pryde (with Bradley 1994) examines the geography of radioactive contamination in the former Soviet Union. They also document the expansion and financial problems of the *zapovedniki* (Russian nature reserve system) (1997). ZumBrunnen documents water and other types of degradation in Ukraine (1992) and the Russian Far North (1997). Preserving Russian nature for both commercial and conservation purposes is discussed by several authors. Backman (1999) looks at the commercial use of Siberian forest resources while Pryde (1999) evaluates Russian attempts to conserve Siberian resources through the promotion of privately operated nature reserves and nature education.

Beyond the general topic of environmental degradation, the severe problem of the Aral Sea basin of Central Asia continues to attract major research attention. Micklin documents the decline of the Aral Sea and its consequences in several publications (1991, 1996, 2000; Micklin and Williams 1997). He studies a wide range of geopolitical issues, from water resource conflict among Central Asian states to prospects of the Sea's restoration. Smith also examines real and potential conflict over water in the Central Asian region (1995) as well as ecosystem decline in the deltas of the Amu Darya and Syr Darya (1994), and soil salinization resulting from irrigation agriculture in Uzbekistan (1992). (See Ch. 39 for further information.)

Some new methods of analysis accompany research on environmental issues. For example, O'Lear (1996) examines the new technologies that may link and enable environmental movements in the former Soviet Union.

O'Lear (2001) continues to examine similar issues with a focus on Azerbaijan.

Geography of Demographic Processes

Population geography has always constituted a major theme among geographers studying the region. The enormous increase in published demographic data has allowed researchers to reexamine established themes related to the distribution and redistribution of population as well as to document and explain some of the more important demographic changes associated with the transition period.

The transition has resulted in population declines throughout the region through a combination of increased death rates, declining fertility rates, and changing migration patterns. For example, Russia is the first major industrial country to experience an absolute decline in population. Heleniak (1995, 1996) shows how the decline in fertility in Russia is spatially homogeneous, while mortality is spatially heterogeneous. Heleniak (1999) also documents the out-migration trends from the Russian North and identifies the spatial impact of this trend. Berentsen (1996) documents population decline in the former Eastern Germany. He identifies emerging factors affecting demographic behavior in the post-Soviet period, such as foreign businesses and local governments. Rowland (2001) contrasts the urban population declines of Russian and European cities in the post-Soviet period with urban population increases in post-Soviet Uzbekistan. Hanks (2000*a*) examines the spatial patterns of FDI, labor-force, and demographic change in post-Soviet Uzbekistan (see Ch. 39 for more information).

In Russia, urban population declines have appeared after more than seven decades of intense urbanization. Rowland (1994, 1995*a*, *b*, 1996*b*, 1997*b*) first traces the regional shift eastward in the location of rapidly growing urban places during the late Soviet period, and the subsequent shift to the interior central and southern regions of Russia in the post-Soviet period. In this same body of work, he then documents the important phenomenon of declining and disappearing towns and creates a typology of the spatial patterns and economic functions of these urban places. Rowland (2001) has extended this work to look at the spatial distribution of secret cities in Russia and Kazakhstan during the Soviet period.

Internal migration dynamics began to change well before the break-up of the former Soviet Union. Mitchneck and Plane (1995) began a conceptualization of the role that transition plays in affecting migration behavior. Others examine the more precise impacts of the changing migration patterns such as the spatial shift of centers of out-migration (Heleniak 1997*b*) and other demographic and migration changes (see Demko *et al.* 1997; Cole and Filatotchev 1992 for examples). Heleniak (1997*b*), in particular, emphasizes the ethnic component of migration trends.

Geography of Political Transition

An equally profound reconfiguration of the political arena has accompanied economic and demographic transition in Russia, Central Europe, and East Europe. First, profound changes in geopolitical systems at the global scale have occurred in the wake of the disintegration of the socialist bloc and the end of the Cold War. Second, formal state institutions, such as national parliaments, have been transformed from merely consultative bodies into legitimate legislative organs. Third, elected officials and representatives—at all scales—have become increasingly responsive to their constituencies. Last, new spaces of informal activism and social mobilization have opened up since the collapse of state-socialism. Geographers have investigated developments in both the formal and informal spheres of political transition in the former Soviet bloc at all these levels.

An entirely new subdiscipline has emerged in the wake of the collapse of the communist system—the electoral geography of the former Soviet Union and Eastern Europe. Elections, which were contrived and completely controlled during the Soviet era, have become the subject of extensive research. In a series of studies using *rayon* (district)-level election returns a scale of measurement that was previously unavailable to non-Soviet researchers in the post-Soviet period, Clem and Craumer (1993, 1995*a*, *b*, 1997, 1998*a*, *b*, 2000*a*, *b*) examine the spatial variation of referenda and the voting patterns of Russian parliamentary and presidential elections. By performing multivariate statistical and other analyses, Clem and Craumer (1993) show that Russian voters display pronounced spatial voting patterns, and that socioeconomic conditions play a major role in determining these patterns. Clem and Craumer (1996) also demonstrate how the reelection of President Yeltsin was achieved through an understanding of voting patterns and a concentration of campaign effort on highly populated areas. Also using a *rayon*-level GIS of Russia, Perepechko (1999) conflates the polygons into pre-Revolutionary *guberniyas* in order to explore the geographical relationships between major pre-Revolutionary and post-Soviet political parties. His probit analysis of the interactions between spatial variations in the partisanship and socioeconomic characteristics of the regions and voting yields some striking patterns exemplifying both regional change and regional continuity of political space.

This new body of work suggests that the development of stable political cleavages in Russia is relatively weak. Social and economic interests are not clearly expressed in a meaningful way in national politics. The work of Clem and Craumer, along with that of O'Loughlin *et al.* (1996) examines divisions in parliamentary elections in the Russian Federation and raises new issues about the regional basis of politics in Russia and the possibility of a stable national political culture emerging in the country. By performing multivariate statistical analyses, Clem and Craumer's work shows that distinctions exist in the electoral behavior between town and countryside—suggesting that the intra-regional variation is even greater or more significant than the interregional variation in voting. O'Loughlin *et al.* (1997) take this a step further in a study of Moscow and its electoral patterns. O'Loughlin (2001) extends his work on Russia to the Ukrainian context to look at regional issues in Ukrainian politics. Craumer and Clem (1999) have also analyzed the regional patterns in Ukrainian parliamentary elections. Mitchneck *et al.* (2002) analyze the extent to which political and economic decentralization occurred in Russia during the 1990s and suggest that the processes have been uneven at best.

The transition has also elevated the importance of studying local and regional politics. Mitchneck (2001*a*, *b*) looks at place-specific differences in the management of local politics and the perceptions of local and regional policy-makers concerning the conduct of and participation in local politics. Graham and Regulska (1997) use three case studies in Poland in an effort to understand barriers to and opportunities for the political participation of women as emerging actors in political space. Bell (1997) and Pavlovskaya (1998) also consider the evolving nature of local politics and urban social movements in Moscow.

Identity, Nation, and Territory

Shadowing research into the emerging character of both formal and informal politics in Russia, Central Eurasia, and East Europe is the question of nationalism. After decades of suppression by Communist Party authorities, nationalism has evolved into a potent ideological force in many post-socialist societies. Nationalists have demonstrated the capacity to thwart the more inclusive notions of democracy. Yet a simple relationship between nationalist movements and democracy is not apparent. The advance of indigenous peoples into positions of political power more often than not represents divisions along factional lines rather than ethno-national lines. Research on the phenomenon of nationalism and its manifestation in different contexts is vital to understanding the emerging political landscape of the region.

In *The Geography of Nationalism in Russia and the USSR*, Kaiser (1994) illustrates the need for understanding the Soviet and pre-Soviet past in order to comprehend the causal mechanisms of contemporary nationality movements. He dispels the notion that "primordial communities" existed since time immemorial and were disrupted by Soviet interference, only to reemerge in the post-Soviet context. Instead, Kaiser argues that the development of nationalism, which didn't exist prior to the nineteenth century in the pre-Russian realm, was accelerated under the Soviet system through *korenizatsiya*, increased mobilization, "reactive ethnicity," and other processes. Territorial units were not "empty containers" through which Soviet nationality policy was exercised; rather, territory became associated with feelings of homeland and fostered exclusionary tendencies. Decentralization accelerated the development of distinct national units, but conflicts have arisen in the post-Soviet period when perceived ancestral territories overlap. In other publications, Kaiser studies Russian–Kazakh relations in Kazakhstan (1995*a*), the indigenization of the workforce in the former Soviet Union (1995*b*), and the changing status of Russians in the newly independent states of the former Soviet Union (Chinn and Kaiser 1996) and the non-Russian republics of the Russian Federation (Kaiser 2000*a*, *b*, *c*).

The relationship between ethnicity, territory, and the assertion of national political rights is a crucial issue in many areas of the former Soviet bloc. In a large body of work, Fondahl (1995*a*, *b*, 1996*a*, *b*, *c*, 1997*a*, *b*) addresses the situation of indigenous groups in Siberia and the Russian North, where non-natives predominate numerically and in other ways. Fondahl outlines the contemporary goals of indigenous peoples in these areas—true multiculturalism, control over natural resources, and economic and political decision-making—and she discusses the realistic chances of achieving these goals. Fondahl notes the lack of progress in land reform legislation and true political power-sharing, and she has identified the causal factors of territorial disputes in administrative units where indigenous peoples reside. She observes that the prospects for indigenous peoples are much more favorable when they command positions of power within their eponymous territorial units.

Hanks (2000*b*) provides a comparable analysis of subaltern nationalism and devolution in post-Soviet Uzbekistan.

The relationship between national identity and territory is further complicated by the fact that numerous nations straddle international borders. Harris (1993*b*) provides a statistical overview of the Russian populations residing in the former USSR. Schwartz (1991) also examines the redistribution of minorities during the late soviet period. Since the collapse of state-socialism, significant numbers of ethnic minorities have migrated to their respective "home territories." Heleniak (1997*a*) studies the obverse of this migration process—the exodus of Russians from Transcaucasia and Central Asia—which has had a negative impact on the economies in those areas. Yet, large numbers of ethnic minorities remain in states throughout the region. Harris and Kaiser examine ethnic relations on an interstate level. Kaiser (1994*b*) examines the relationship between demographic trends among ethnic groups and interstate relations in Central Asia. Noting that the Russian government seeks to protect Russian minorities in the former Soviet Union, Harris (1994) analyzes the process through which the Russian Federation exercises control over other post-Soviet republics (see Ch. 39 for further information). Harris (1993*a*) also compares the processes of the dissolution of the Soviet Union with that of Yugoslavia. Velikonja (1994) observes that in the case of Slovenia, traditional local, regional, and national identities continue to predominate, while the supranational (Yugoslav) identity is largely restricted to recent immigrants from other Yugoslav republics.

One of the basic elements shaping contemporary manifestations of nationalism in Russia, Central Eurasia, and East Europe is the real and imagined historical and cultural geographies that underlie contemporary national formations. Using an innovative set of ethnographic methodologies, Drennon (2001) looks at the ways in which a human geography of shared space but distinctly different national identities arose in the Balkan Peninsula. Kristof (1994) examines visions of "Fatherland" and the essential nature of eastern territories in the case of Poland. Stebelsky (1994) traces the development of Ukrainian national identity from the proto-Ukrainian ethnos of the forest-steppe to the establishment of the "Movement in Support of Restructuring" (Rukh), which gained momentum during the last years before independence. Bassin (1994, 1999) describes the role played by geographers in shaping the imperial ideology of Russian nationalism as the Russian Empire expanded eastward. He also compares attitudes among Russian and American historians toward the frontier thesis, with the former displaying the influence of Marxist philosophy. Frank and Wixman (1997) identify the establishment of Soviet-era ethnically based territorial units as a major factor in the historical development of the Turkic nations of the middle Volga, although they note the contradictory nature of the effect of Soviet nationality policies on these ethnic groups.

The significance of monuments and commemorative sites in national identity formation has become increasingly apparent since the fall of communism, which ushered in a dramatic change in the monumental landscape throughout the region. The role of monuments and commemorative sites in grounding the fiction of national identity, in nationalizing history while at the same time historicizing the image of the nation, has meant that these sites have become focal points in the contestation and renegotiation of the meaning of national identity. Sidorov (2000, 2001) studies the politics of scale in nationalization processes, particularly as they relate to the resurrection of the Christ the Savior Cathedral in Moscow. Similarly, Foote *et al.* (2000) assesses the ways in which national identity is inscribed in the landscape through the renegotiation of commemorative spaces in post-socialist Hungary. Bell (1999) provides an in-depth analysis of the official uses of public spaces in Tashkent to redefine national identity in Uzbekistan.

Future Directions of Research

The region of Russia, Central Eurasia, and East Europe has fragmented and recombined since the collapse of the Berlin Wall in 1989 and the dissolution of the Union of Soviet Socialist Republics in 1991. Where once an Iron Curtain defined a coherent geographical unit of political and economic analysis, today there is an intricate mosaic of international borders, multiethnic states, and partially to fully open market economies. Area specialists are faced with the challenge of rescaling their research to illuminate developments within specific countries, or even regions within countries, without losing sight of the more general dynamics of political and economic change impacting the region as a whole. Fragmentation of the political map should not be mirrored by an equivalent intellectual fragmentation. The community of geographers that applied itself to the analysis of the Soviet bloc still has a role to play in grappling with the nature of regional transformation. As stated at the outset of the chapter, the very fluidity of the region makes it a prime research candidate for investigating the processes by which regions are reconstituted through time and space.

Within the overarching framework of regional transformation the possibilities for empirical research are both wide and varied. Broadly speaking, it seems likely that geographers will continue to analyze the nature and meaning of political and economic transition across the territories of the former Soviet bloc. One of the most exciting aspects of this research will be the ability of geographers to design and test original hypotheses regarding the universal and contextual factors that influence the movement toward democratic governments and market-oriented economies. This opens the possibility of contributing not only to the understanding of societal change in Russia, Central Eurasia and East Europe, but also to the conceptualization of political and economic transitions across the globe.

One of the keys to successful future research will be collaboration with colleagues from institutions in the region, non-area specialists in the West, and Europe area specialists who are oriented more toward Europe than Russia and Central Eurasia. Parallels also exist in Latin American and Asian transitions that can and do serve as points of comparison for research. In the first instance, collaboration among those studying this wide range of regions represents an exciting opportunity to fuse insider and outsider views of political and economic change. It also affords an opportunity to establish co-operative relationships that aid the intellectual and career development of geographers on both sides of the Atlantic. In the second instance, non-area specialists can assist in the formulation of research projects that speak to a wider audience and that test theories and explanations that have been tried in other contexts but not yet subjected to the peculiarities of post-socialist societies. In the last instance, it is imperative that comparative work within the former Soviet bloc continues. Maintaining communication with area specialists who have reoriented themselves toward the region of Europe is vital in this regard. Progress has been made on each of these fronts, and there is reason to believe that various modes of collaborative research will predominate in the future.

Reflecting the need for a strong regional perspective and for comparative analyses, the future methodologies employed by geographers studying Russia, Central Eurasia, and East Europe should encompass a range of techniques and respond to phenomena at different scales. The recent innovation of ethnographic or interview-based research promises to highlight the local effects of political and economic transition and the role of localities in the construction of new social formations in the former Soviet bloc. In addition, it offers an opportunity to place human actors within broader social relations. However, intensive, local-scale methods should not overshadow the importance of more broadly based survey and statistical research that has the capacity to illuminate representative trends or ruptures in the path to liberal capitalist societies. Across Russia, Central Eurasia, and East Europe, regionally and nationally scaled studies will be vital to understanding the structural components of societal transformation.

The rapid and dramatic pace of change in Russia, Central Eurasia, and East Europe necessitates that future research will be driven in part by the need for basic, exploratory studies that can accurately describe the empirical conditions extant in particular localities or countries. But as suggested above, there will also be a place for theoretically inspired research that tests specific hypotheses regarding political and economic transformation, contributes to a theory of transition, and links research in the region to central philosophical concerns in the discipline.

Geographers conducting research in Russia, Central Eurasia, and East Europe have already begun to demonstrate a greater sensitivity to the theoretical implications of their research. One of the central questions before area specialists is the level of theoretical abstraction that will actually enhance specific empirical studies. In many cases, theories that serve as explanatory frameworks will prove to be the most utilitarian and beneficial. However, not all future research should be limited to testing explanatory models.

In the ideological reorientation of the former Soviet bloc toward some forms of liberal, capitalist democracy, deeply profound philosophical questions arise concerning the nature of state-socialism and the prerequisites for Western-style civil society—the basis, it seems, for consolidating both market and democratic institutions. These questions have been broached outside the discipline of geography by sociologists and political scientists. However, the role of space and geography in the creation, dissolution, and aftermath of state socialism has not been rigorously theorized. Geographers may be understandably wary of reifying now defunct Soviet ideologies by creating grand narratives of "how" and "why" place and space mattered to state socialism. However, these more abstract theories ultimately apply to a very practical, and grounded, question: what was the nature of Soviet modernity? This basic question colors contemporary research across Russia, Central Eurasia, and East Europe, and its resolution offers an invaluable opportunity to deepen the understanding of political and economic change throughout this diverse region.

REFERENCES

Argenbright, R. (1999). "Remaking Moscow: New Places, New Selves." *Geographical Review*, 89/1: 1–22.

Backman, C. A. (1999). "The Siberian Forest Sector: Challenges and Prospects." *Post-Soviet Geography and Economics*, 40/6: 453–69.

Bass, I. and Dienes, L. (1993). "Defense Industry Legacies and Conversion in the Post-Soviet Realm." *Post-Soviet Geography*, 34/5: 302–17.

Bassin, M. (1994). "Russian Geographers and the 'National Mission' in the Far East," in D. Hooson (ed.), *Geography and National Identity*. Oxford: Blackwell, 112–33.

— (1999). *Imperialist Visions: Nationalist Imagination and Geographical Expansion in the Russian Far East, 1840–1865*. Cambridge: Cambridge University Press.

Bater, J. (2001). "Adjusting to Change: Privilege and Place in Post-Soviet Central Moscow." *Canadian Geographer*, 45/2: 237–51.

Bater, J. H., Amelin, V. N., and Degatyarev, A. A. (1994). "Moscow in the 1990s: Market Reform and the Central City." *Post-Soviet Geography*, 35/5: 247–66.

— (1998). "Market Reform and the Central City: Moscow Revisited." *Post-Soviet Geography and Economics*, 39/1: 1–18.

Bell, J. E. (1997). "A Place for Community?: Urban Social Movements and the Struggle Over the Space of the Public in Moscow." Ph.D. dissertation, University of Washington.

— (1999). "Redefining National Identity in Uzbekistan: Symbolic Tensions in Tashkent's Official Public Landscape." *Ecumene*, 6/2: 183–213.

Berentsen, W. H. (1996). "Regional Population Changes in Eastern Germany After Unification". *Post-Soviet Geography and Economics*, 37/10: 615–32.

Bond, A. R., and Levine, R. M. (1993). "The Manganese Shortfall in Russia." *Post-Soviet Geography*, 34/5: 293–301.

— (1997). "Development of the Copper and Molybdenum Industries and the Armenian Economy." *Post-Soviet Geography and Economics*, 38/2: 105–20.

Bradshaw, M. J., and Lynn, N. J. (1994). "After the Soviet Union: The Post-Soviet States in the World System." *Professional Geographer*, 46/4: 439–49.

Chinn, J., and Kaiser, R. (1996). *Russians as the New Minority in the Soviet Successor States*. Boulder, Colo.: Westview Press.

Clem, R. S., and Craumer, P. R. (1993). "The Geography of the April 25 (1993) Russian Referendum." *Post-Soviet Geography*, 34/8: 481–96.

— (1995*a*). "A Rayon-Level Analysis of the Russian Election and Constitutional Plebiscite of December 1993." *Post-Soviet Geography*, 36/8: 459–75.

— (1995*b*). "The Geography of the Russian 1995 Parliamentary Election: Continuity, Change, and Correlates." *Post-Soviet Geography*, 36/10: 587–616.

— (1996). "Roadmap to Victory: Boris Yel'tsin and the Russian Presidential Elections of 1996. *Post-Soviet Geography and Economics*, 37/6: 335–54.

— (1997). "Urban-Rural Voting Differences in Russian Elections, 1995–1996: A Rayon-Level Analysis." *Post-Soviet Geography and Economics*, 38/7: 379–95.

— (1998*a*). *Urban–Rural Voting Differences in Russian Elections, 1995–1996: A Rayon Level Analysis*. Washington, DC: National Council for Eurasian and East European Research.

— (1998*b*). *Regional Patterns of Voter Turnout in Russian Elections, 1993–1996*. Washington, DC: National Council for Eurasian and East European Research.

— (2000*a*). "Regional Patterns of Political Preference in Russia: The December 1999 Duma Elections." *Post-Soviet Geography and Economics*, 41/1: 1–29.

— (2000*b*). "Spatial Patterns of Political Choice in the Post-Yel'tsin Era: The Electoral Geography of Russia's 2000 Presidential Election." *Post-Soviet Geography and Economics*, 41/7: 465–82.

Cole, J. P., and Filatotchev, I. V. (1992). "Some Observations on Migration within and from the Former USSR in the 1990s." *Post-Soviet Geography*, 33/7: 432–51.

Craumer, P. R. (1994). "Regional Patterns of Agricultural Reform in Russia." *Post-Soviet Geography*, 35/6: 329–51.

— (1995). *Rural and Agricultural Development in Uzbekistan*. London and Washington, DC: Royal Institute of International Affairs, Russian and CIS Programme; Distributed by the Brookings Institution.

Craumer, P., and Clem, J. (1999). "Ukraine's Emerging Electoral Geography: A Regional Analysis of the 1998 Parliamentary Elections." *Post-Soviet Geography and Economics*, 40/1: 1–26.

Crocker, J. E., and Berentsen, W. H. (1998). "Accessibility Between the European Union and East-Central Europe: Analysis of Market Potential and Aggregate Travel." *Post-Soviet Geography and Economics*, 39/1: 19–44.

Demko, G., Pontius, S., and Zaionchkovskaya, Z. (1997). *Population Under Duress: Geodemography of Post-Soviet Russia*. Boulder, Colo.: Westview Press.

Dienes, L. (1993*a*). "Prospects for Russian Oil in the 1990s: Reserves and Costs." *Post-Soviet Geography*, 34/2: 79–110.

— (1993*b*). "Economic Geographic Relations in the Post-Soviet Republics." *Post-Soviet Geography*, 34/8: 497–529.

— (1998). *Exports from the Former USSR: Philosopher's Stone or Fool's Gold*. Papers in Russian, East European, and Central Asian Studies, 18. Seattle: The Henry M. Jackson School of International Studies, University of Washington.

Drennon, C. (2001). "Searching for Order on the Balkan Peninsula." *Geographical Review*, 91/1–2: 407–13.

Engelmann, K., and Pavlakovic, V. (eds.) (2001). *Rural Development in Eurasia and the Middle East: Land Reform, Demographic Change, and Environmental Constraints*. Seattle: Russian, East European and Central Asian Studies Center at the Henry M. Jackson School of International Studies, University of Washington, in assoc. with University of Washington Press.

Fondahl, G. (1995*a*). *Siberia's First Nations*. Washington, DC: National Council for Soviet and East European Research.

— (1995*b*). "The Status of Indigenous Peoples in the Russian North." Special Issue on The Russian North in Transition. *Post-Soviet Geography*, 36/4: 215–24.

— (1996*a*). "Contested Terrain: Changing Boundaries and Identities in Southeastern Siberia." *Post-Soviet Geography and Economics*, 37/1: 3–15.

Fondahl, G. (1996*b*). "Evenks and Land Reform in Siberia: Progress and Obstacles." *Fourth World Bulletin*, 5/102: 38–42.

— (1996*c*). *Evenks and Land Reform in Russia: Progress and Obstacles*. Washington, DC: National Council for Soviet and East European Research.

— (1997*a*). "Siberia: Assimilation and its Discontents," in I. A. Bremmer and R. Taras (eds.), *New States, New Politics: Building the Post-Soviet Nations*. Cambridge: Cambridge University Press, 190–232.

— (1997*b*). "Environmental Degradation and Indigenous Land Claims in Russia's North." in E. A. Smith and J. McCarter (eds.), *Contested Arctic: Indigenous Peoples, Industrial States, and the Circumpolar Environment*. Seattle: Russian, East European, and Central Asian Studies Center at the Henry M. Jackson School of International Studies in assoc. with University of Washington Press, 68–87.

Foote, K., Toth, A., and Arvay, A. (2000). "Hungary after 1989: Inscribing a New Past on Place." *Geographical Review*, 90/3: 301–34.

Frank, A., and Wixman, R. (1997). "The Middle Volga: Exploring the Limits of Sovereignty," in I. A. Bremmer and R. Taras (eds.), *New States, New Politics: Building the Post-Soviet Nations*. Cambridge: Cambridge University Press, 140–89.

Graham, A., and Regulska, J. (1997). "Expanding Political Space for Women in Poland: An Analysis of Three Communities." *Communist and Post-Communist Studies*, 30/1: 65–81.

Gritsai, O. (1997). "Moscow Under Globalization and Transition: Paths of Economic Restructuring." Urban Geography, 18/2: 155–65.

Hall, D. R., and Danta, D. R. (eds.) (1996). *Reconstructing the Balkans: A Geography of the New Southeast Europe*. Chichester: Wiley.

Hanks, R. (2000*a*). "Emerging Spatial Patterns of the Demographics, Labour Force and FDI in Uzbekistan". *Central Asian Survey*, 19/3–4: 348–63.

— (2000*b*). "A Separate Space?: Karakalpak Nationalism and Devolution in Post-Soviet Uzbekistan." *Europe—Asia Studies*, 52/5: 939–53.

Harris, C. D. (1993*a*). "The New Russian Minorities: A Statistical Overview." *Post-Soviet Geography*, 34/1: 1–27.

— (1993*b*). "New European Countries and their Minorities." *Geographical Review*, 83/3: 301–21.

— (1994). "Ethnic Tensions in the Successor Republics in 1993 and Early 1994." *Post-Soviet Geography*, 35/4: 185–203.

Heleniak, T. (1997*a*). "Internal Migration in Russia During the Economic Transition (1989 to End-1996)" *Post-Soviet Geography and Economics*, 38/2: 81–105.

— (1997*b*). "The Changing Nationality Composition of the Central Asian and Transcaucasian States." *Post-Soviet Geography and Economics*, 38/6: 357–78.

— (1995). "Economic Transition and Demographic Change in Russia, 1989–1995." *Post-Soviet Geography*, 36/7: 446–58.

— (1996). "Russia's Age Structure in 1996: A Research Report." *Post-Soviet Geography and Economics*, 37/6: 386–95.

Heleniak, T. (1999). "Out-Migration and Depopulation of the Russian North during the 1990's." *Post-Soviet Geography and Economics*, 40/3: 155–205.

Hooson, D. (1994). "Ex-Soviet Identities and the Return of Geography," in D. Hooson (ed.), *Geography and National Identity*. Oxford: Blackwell, 134–40.

Ioffe, G. V., and Nefedova, T. (1997). *Continuity and Change in Rural Russia: A Geographical Perspective*. Boulder, Colo.: Westview Press.

— (2000). *The Environs of Russian Cities*. Lewiston, NY: Edwin Mellen.

— (2001*a*). "Russian Agriculture and Food Processing: Vertical Cooperation and Spatial Dynamics." *Europe–Asia Studies*, 53/3: 389–418.

— (2001*b*). "Land Use Changes in the Environs of Moscow." *Area*, 33/3: 273–86.

Johansen, H. (1995). "Locating Western-Style Retail in FSU Baltic," in M. Tykkylainen (ed.), *Local and Regional Development During the 1990s Transition in Eastern Europe*. Aldershot: Avebury, 47–64.

Kaiser, R. J. (1994*a*). *The Geography of Nationalism in Russia and the USSR*. Princeton, NJ: Princeton University Press.

— (1994*b*). "Ethnic Demography and Interstate Relations in Central Asia," in Roman Szporluk (ed.), *National Identity and Ethnicity in Russia and the New States of Eurasia*. Armonk, NY: M. E. Sharpe.

— (1995). "Nationalizing the Work Force: Ethnic Restratification in the Newly Independent States." *Post-Soviet Geography*, 36/2: 87–111.

— (2000*a*). *The Geography of Nationalization and Nationalism in Post-Soviet Russia: Political Indigenization and Homeland Making in Russia's Republics*. Report One. Washington, DC: National Council for Eurasian and East European Research.

— (2000*b*). *The Geography of Nationalization and Nationalism in Post-Soviet Russia: Ethnic Restratification in Russia's Republics*. Report Two. Washington, DC: National Council for Eurasian and East European Research.

— (2000*c*). *The Geography of Nationalization and Nationalism in Post-Soviet Russia: Titular Nationalization and Interethnic Relations in Russia's Republics*. Report Three. Washington, DC: National Council for Eurasian and East European Research.

Kraus, M., and Liebowitz, R. D. (eds.) (1996). *Russia and Eastern Europe After Communism: The Search for New Political, Economic, and Security Systems*. Boulder, Colo.: Westview Press.

Kristof, L. K. D. (1994). "The Image and the Vision of the Fatherland: The Case of Poland in Comparative Perspective," in D. Hooson (ed.), *Geography and National Identity*. Oxford: Blackwell, 221–32.

Lee, C., and Bradshaw, M. J. (1997). "South Korean Economic Relations with Russia." *Post-Soviet Geography and Economics*, 38/8: 461–77.

Liebowitz, R. D. (1996). "Russia's Center-Periphery Fiscal Relations During Transition," in Michael Kraus and R. D. Liebowitz (eds.), *Russian and Eastern Europe After Communism: The Search for New Political, Economic, and Security Systems*. Boulder, Colo.: Westview Press.

Micklin, P. P. (1991*b*). "Touring the Aral: Visit to an Ecologic Disaster Zone." *Soviet Geography*, 32/2: 90–105.

— (2000). *Managing Water in Central Asia*. London: Royal Institute of International Affairs.

Micklin, P. P., and Williams, W. D. (eds.) (1996). *The Aral Sea Basin*. New York: Springer-Verlag.

Mitchneck, B. (1994). "The Changing Role of the Local Budget in Russian Cities: The Case of Yaroslavl," in Theodore Friedgut and Jeffrey Hahn (eds.), *Local Power and Post-Soviet Politics*. Armonk, NY: M. E. Sharpe.

— (1997*a*). "Restructuring Russian Urban Budgets: 1991–1995". *Europe-Asia Studies*, 49/6: 989–1015.
— (1997*b*). "The Development of Local Government in Russia," in Michael J. Bradshaw (ed.), *Geography and Transition in the Post-Soviet Republics*. London: Belhaven Press.
— (1998). "The Heritage Industry Russian Style: The Case of Yaroslavl'." *Urban Affairs Review*, 34/1: 28–51.
— (2001*a*). "The Regional Governance Context in Russia: A General Framework." *Urban Geography*, 22/4: 360–82.
— (2001*b*). "Regional Governance Regimes in Russia: The Cases of Yaroslavl and Udmurtia," in Jeffrey W. Hahn (ed.), *Regional Russia in Transition: Studies from Yaroslavl'*, Washington, DC: Woodrow Wilson Center Press.
Mitchneck, B., and Plane, D. (1995). "Migration Patterns During a Period of Political and Economic Shocks in the Former Soviet Union: A Case Study of Yaroslavl Oblast." *Professional Geographer*, 47/1: 17–30.
Mitchneck, B., Solnick, S., and Stoner-Weiss, K. (2002). "Political and Economic Decentralization in Russia: How Far, How Fast, and to What End?" in Blair A. Ruble, Jodi Koehn, and Nancy E. Popson (eds.), *Fragmented Space: Center-Region-Local in the Russian Federation*. Washington, DC: Woodrow Wilson Center Press.
Murphy, A. B. (1991). "The Emerging Europe of the 1990s." *Geographical Review*, 81/1: 1–17.
O'Lear, S. R. M. (1996). "Using Electronic Mail (E-mail) Surveys for Geographic Research: Lessons from a Survey of Russian Environmentalists." *Professional Geographer*, 48/2: 209–17.
— (2001). "Azerbaijan: Territorial Issues and Internal Challenges in Mid-2001." *Post-Soviet Geography and Economics*, 42/4: 305–12.
O'Loughlin, John (2001). "The Regional Factor in Contemporary Ukrainian Politics: Scale, Place, Space, or Bogus Effect?" *Post-Soviet Geography and Economics*, 42/1: 1–33.
O'Loughlin, J., Kolossov, V., and Vendina, O. (1997). "The Electoral Geographies of a Polarizing City: Moscow, 1993–1996." *Post-Soviet Geography and Economics*, 3810: 567–601.
O'Loughlin, J., Shin, M., and Talbot, P. (1996). "Political Geographies and Cleavages in the Russian Parliamentary Elections." *Post-Soviet Geography and Economics*, 37/6: 355–85.
Pavlovskaya, M. (1998). "Everyday Life and Social Transition: Gender, Class, & Change in the City of Moscow." Ph.D. dissertation, Clark University.
— (2001). "Privatization of the Urban Fabric: Gender and Local Geographies of Transition in Downtown Moscow." *Urban Geography*, 22/1: 4–28.
Perepechko, A. (1999). "Spatial Change and Continuity in Russia's Political Party System(s): Comparisons of the Constituent Assembly Election of 1917 and Parliamentary Election of 1995." Ph.D. dissertation, University of Washington.
Pryde, P. R., (ed.) (1995). *Environmental Resources and Constraints in the Former Soviet Republics*. Boulder, Colo.: Westview Press.
— (1999). "The Privatization of Nature Conservation in Russia." *Post-Soviet Geography and Economics*, 40/5: 383–93.
Pryde, P. R. and Bradley, D. J. (1994). "The Geography of Radioactive Contamination in the Former USSR." *Post-Soviet Geography*, 35/10: 557–93.
— (1997). "Post-Soviet Development and Status of Russian Nature Reserves." *Post-Soviet Geography and Economics*, 38/2: 63–80.
Pryde, P. R., Berentsen, W. H., and Stebelsky, I. (1989). "Soviet and East Europe," in G. L. Gaile and C. J. Willmott (eds.), *Geography in America*. Columbus, Ohio: Merrill. 546–62.
Rowland, R. H. (1994). "Declining Towns in the Former USSR." *Post-Soviet Geography*, 35/6: 352–65.
— (1995*a*). "Rapidly Growing Towns in the Former USSR and Russia, 1970–1993." *Post-Soviet Geography*, 36/3: 133–56.
— (1995*b*). "Declining Towns in Russia, 1989–1993." *Post-Soviet Geography*, 36/7: 436–45.
— (1996*a*). "Russia's Disappearing Towns: New Evidence of Urban Decline, 1979–1994." *Post-Soviet Geography and Economics*, 37/2: 63–87.
— (1996*b*). "Russia's Secret Cities." *Post-Soviet Geography and Economics*, 37/7: 426–
— (1997*a*). "Patterns of Dynamic Urban Population Growth in Russia, 1989–1996: A Research Report." *Post-Soviet Geography and Economics*, 38/3: 171–87.
— (1997*b*). *Regional Population Trends in the Former USSR, 1939–51, and the Impact of World War II*. Pittsburgh, Pa.: Center for Russian and East European Studies, University of Pittsburgh.
— (2001). "Urban Growth in Uzbekistan During the 1990s." *Post-Soviet Geography and Economics*, 42/4: 266–304.
Sagers, M. J. (1994). "The Oil Industry in the Southern Tier Former Soviet Republics." *Post-Soviet Geography*, 35/5: 267–98.
— (1995). "Prospects for Oil and Gas Development in Russia's Sakhalin Oblast." *Post-Soviet Geography*, 36/5: 274–90.
— (1996). "Russian Crude Oil Production in 1996: Conditions and Prospects." *Post-Soviet Geography and Economics*, 37/9: 523–87.
Sagers, M. J., and Kryukov, V. (1993). "The Hydrocarbon Processing Industry in West Siberia." *Post-Soviet Geography*, 34/2: 127–52.
Sagers, M. J., and Nicoud, J. (1997). "Development of the East Siberian Gas Field and Pipeline to China: A Research Communication." *Post-Soviet Geography and Economics*, 38/5: 288–96.
Schwartz, Lee (1991). "USSR Nationality Redistribution by Republic, 1979–1989: From Published Results of the 1989 All-Union Census." *Soviet Geography*, 32/4: 209–48.
Sidorov, D. (2000). "National Monumentalization and the Politics of Scale: The Resurrections of the Cathedral of Christ the Savior in Moscow." *Annals of the Association of American Geographers*, 90/3: 548–72.
— (2001). *Orthodoxy and Difference: Essays on the Geography of Russian Orthodox Churches in the 20th Century*. San Jose, Calif.: Pickwick Publications.
Smith, D. R. (1992). "Salinization in Uzbekistan (Soil Salinization as Result of Irrigation)." *Post-Soviet Geography*, 3/1: 21–33.
— (1994). "Change and Variability in Climate and Ecosystem Decline in Aral Sea Basin Deltas." *Post-Soviet Geography*, 35/3: 142–65.
— (1995). "Environmental Security and Shared Water Resources in Post-Soviet Central Asia." *Post-Soviet Geography*, 36/6: 351–70.
Stebelsky, I. (1994). "National Identity of Ukraine," in D. Hooson (ed.), *Geography and National Identity*. Oxford: Blackwell, 233–48.
— (1997). "The Toponymy of Ukraine." *Post-Soviet Geography and Economics*, 38/5: 276–87.

Stenning, A. (1999). "Marketization and Democratization in the Russian Federation: The Case of Novosibirsk." *Political Geography*, 18/5: 591–617.

Velikonja, J. (1994). "The Quest for Slovene National Identity," in D. Hooson (ed.), *Geography and National Identity*. Oxford: Blackwell, 249–56.

Zimine, D., and Bradshaw, M. (1999). "Regional Adaptation to Economic Crisis in Russia: The Case of Novgorod Oblast." *Post-Soviet Geography and Economics*, 40/5: 335–53.

ZumBrunnen, C. (1992). "Environmental Conditions," in I. S. Koropeckyj (ed.), *The Ukrainian Economy: Achievements, Problems, Challenges*. Cambridge, Mass.: Harvard Ukrainian Research Institute: distributed by Harvard University Press, 312–58.

— (1993). "Problems and Prospects for the Development of New Commercial Structures in Russia and the Russian Far East." *Annals of Japan Association of Economic Geographers*, 39/1: 50–67.

— (1997). "A Survey of Pollution Problems in the Soviet and Post-Soviet Russian North," in E. A. Smith and J. McCarter (eds.), *Contested Arctic: Indigenous Peoples, Industrial States, and the Circumpolar Environment*. Seattle: Russian, East European, and Central Asian Studies Center at the Henry M. Jackson School of International Studies in assoc. with University of Washington Press, 88–121.

Part VII

Values, Rights, and Justice

CHAPTER 45

Values, Ethics, and Justice

Audrey Kobayashi and James Proctor

Questions of ethics, values, justice, and the moral principles according to which we engage in geographical scholarship, have always been a part of geography, but for the past two decades—and perhaps even more significantly, since the events of September 11, 2001—they have become a central part of the lexicon of American and international geographical scholarship. The Values, Justice and Ethics Specialty Group (VJESG) was formed in 1997 to respond to a felt need for geographers to focus on both the ethical issues that inform our academic work, and the ways in which that work is connected to larger societal issues.[1] The concerns of the group have been less with a particular range of topics or approaches than with the ethical questions that cut across the entire discipline, on the assumption that such questions are bounded neither by subject matter nor by theoretical constraints.

The group was formed at a time when questions of *whether* geographers should be concerned about the moral, ethical implications of their work had long since been replaced with questions of *how* geographers could focus attention on these issues. Concern is with the very difficult questions that link personal commitment, or reflexivity, with larger questions of research and pedagogy. One of the best sources of evidence of the importance of such questions, and of the intellectual sophistication with which they are being asked, is the journal *Ethics, Place and Environment*, inaugurated in 1998. This group felt a need, therefore, for a geographical forum in which to explore the relationship between American geography and the world in which it operates.

While a relatively small number of geographers works in a more narrowly defined field that might be called moral philosophy (Sack 1997; Smith 1997,1998*a*, 2000), for the vast majority, ethical questions connect the academic and the personal lives of geographical practitioners, in ways that influence directly the questions they ask, the methodological and theoretical choices they make, and, perhaps most importantly, their personal relations with their research subjects and their own communities. As I. Hay (1998: 73) suggests, "the place to start that process is on our [geographers'] own professional bodies." This chapter is a very cursory attempt to describe the context in which such issues are currently being taken up by geographers and, taking our cue from David Smith (1998*b*), to raise some issues about how geographers have cared.

It would be accurate to say that ethics have always been an aspect of geography. Immanuel Kant, for example, was fundamentally concerned with developing an explicit understanding of ethics upon which his vision of geography could be layered. In the present, we are less concerned with the specifics of his moral system than with the normative grid that Kantian thought, as well as

[1] At the annual meeting of the AAG in 2001, a decision was made to merge the Values, Ethics and Justice Specialty Group with the Human Rights Specialty Group, creating a new Values, Ethics and Human Rights Specialty Group. Many if not all the concerns of the two groups are cross-cutting. In this review, however, we have focused our discussion on the emphases of scholars examining questions of ethical concern to geographers.

that of other Enlightenment thinkers, effectively placed over subsequent notions of space and rationality. Most geographers today would recognize that while most of the content of Kant's geography has long-since been replaced with new ideas, his influence on the fundamental ways in which we think about human being, and his role in shaping the values of colonial and imperial expansion that underlie today's political and economic systems, was profound. Kant provides just one example of a thinker whose work is essential to understanding today's moral landscape, as the cumulative efforts of those in a position to influence the way in which society is ordered to set the moral guidelines and boundaries within which people regulate their lives. Although, as Burch (1997) has recently pointed out, that normative grid does not create a simple epistemological divide between the higher pursuit of moral philosophy and the practical concerns of the geographer, as is commonly believed, its greatest significance lies none the less in the ways in which it has directed and infused notions of correct action and proper beliefs, in both popular and scholarly contexts.

Recent texts devoted to philosophical interpretations of geography and environmental ethics (Light and Smith 1997; Proctor and Smith 1999) reveal a rich legacy of questions that link human being and nature, and that question the nature of human being, in an ethical context. This collection emphasizes the fact that ethical understanding, while it may involve a significant amount of self-reflexivity on the part of the scholar, needs to be placed within a long and complex history of intellectual and social development that lies well beyond the scope of this essay.

In the more immediate present, our work is directly a result of the so-called "relevance debate" of the early 1970s. What began as a rather emotional response to the influence of the "quantitative revolution" initiated what is perhaps the most important period of social science theorizing in the history of our discipline. The critical approaches to understanding spatiality that are now placing geographers on the intellectual map within the wider social sciences owe a huge debt to the questions about moral values raised by people such as Anne Buttimer (1974), Yi Fu Tuan (1974), and Wilbur Zelinsky (1975). Furthermore, the early work of radical geographers such as David Harvey (1973), William Bunge (1971), and Richard Peet (1975, 1978) laid the ground for a critical morality that was to become deeply entwined with the so-called "humanist" work of those positioned under the "values" banner, although it took some time before the links began to be made explicitly (Kobayashi and Mackenzie 1989). Beginning about a decade ago, however, it was clear that both these strands, the humanist and the Marxist, had influenced a new "critical" turn in geography (Sayer 1989) that both situated the discipline within emerging debates over postmodernism (Ley 1989) and challenged geographers to expand the limits of discussion over normative issues. Geographers of the baby-boom generation, the immediate post-World War II era, grew up with the concerns of the civil rights movement, the early stages of second-wave feminism, the ecological movement, and the 1960s anti-war movement. Students during the turbulent times of the 1960s, they are the aging professors of the discipline today, and their life experiences are profoundly etched upon the moral questions that we ask now as well as upon the broader set of issues that define public policy (Harvey 1974).

From a more academic perspective, what Sayer and Storper (1997) have called the "normative turn" in social theory and social science stems, therefore, from a refutation of the notion that value-free science is possible and upon recognizing the socially constructed nature of knowledge and of moral values. The notion of value-free science became popular in the immediate post-World War II intellectual environment when scholars turned to positivist thinking as a means of removing the doubt and bias of value judgements from their work, substituting instead universal and irrefutable "truth" as a basis for knowledge. The search for universal truths, however, proved not only impossible, but also dangerous because it led to a certain blindness to the biases and values that are entwined in all attempts to achieve objective knowledge. It was that recognition, combined with their commitment to human rights, that led geographers of the 1970s and 1980s in a new direction.

The possibility of neutrality has in most of the past quarter-century of writing by geographers been debunked as not only unprovable, but also itself a normative speculation that informs the kinds of geographical knowledge that are possible (Gregory 1979; Pickles 1985), and therefore also informs the values and ideological positions through which we interpret social processes, and the languages we invoke to describe them. Debates over subjectivity and reflexivity, and the relationship and importance of personal values over structural conditions, have provided a central strand of geographical thinking throughout that period.

By the early 1990s, there developed a considerable body of literature that illustrated the ethical dilemmas inherent in epistemological choices. These are especially apparent in areas such as cartography (Harley 1991; Monmonier 1991; Rundstrom 1993) and geographic information systems (Wasowski 1991; Lake 1993; Curry 1994; Crampton 1995), as well as planning (Lake 1993;

Entrikin 1994). Similar issues have been raised with respect to the geographer's professional capacity as teacher (Havelberg 1990; Katz and Kirby 1991; Smith 1995; Merrett 2000), writer (Brunn 1989; Curry 1991), and researcher with a particular moral obligation to her or his subjects (Eyles and Smith 1988; England 1994; A. M. Hay 1995). Although physical geography has been less thoroughly infused with ethical questions, several physical geographers have raised serious questions concerning the place of the geographer in ensuring environmental sustainability (Kates 1987; Manning 1990; Cooke 1992; Reed and Slaymaker 1993), and a recent special issue of the *Annals* has been entirely devoted to this theme (1998). Experts in geographic information systems have asked questions about the ethics of data usage (Gilbert 1995), and the relationship between data manipulation and larger social issues such as the provision of water resources (Clark 1998).

These more or less introspective works represent a very important assessment of how questions of ethics and values affect the formal discipline of geography. Much more voluminous by far is the literature that addresses the ethics and values of geographical subjects empirically. Providing a conceptual framework for these moral concerns, Proctor and Smith (1999) have suggested that three dominant metaphors—space, place, and nature—dominate geographical scholarship. Recognizing that these metaphors, and the manner of their usage, need to be subjected to much deeper critical analysis than this cursory overview allows, we now turn to describing some of the recent trends in a geography of ethics.

The issue of spatial justice has motivated a whole generation of social geographers since Harvey introduced the concept in 1973, and remains a dominant geographical theme (Smith 1994). The theme of spatial justice ranges from the unequal relationships established between the developed and developing world (Corbridge 1993, 1998; Slater 1997), to territorial justice (Boyne and Powell 1991), to access to social justice through migration (Black 1996), as well as the spatial attributes of the issues—such as war and environmental disasters—that create the conditions from which refugees are fleeing. For all the variety of interpretations of spatial justice, questions of class remain fundamental (Harvey 1993), and provide a moral point of departure for geographical justice, in common with issues of "race," sexuality, ability, and other markers of difference.

For feminist geographers and the growing numbers of queer-identified geographers, the metaphor of spatial justice has probably provided the impetus for the largest body of work, concerned with issues of home, work, and childcare, transportation and other forms of access, social services, and, more recently, the moral construction of public space, an issue taken up also by those concerned with the plight of the homeless, or the need to assert rights around different forms of sexuality. Work in these areas is becoming too voluminous to cite individual examples, so we shall instead make a few remarks on the major ethical themes of recent work. Feminist geography in particular began from the ethical position of wanting to include women in a male-dominated world in which patriarchy is spatially expressed (Women and Geography Study Group 1984). Feminism in geography, and in general, is thus significant for the fact that an ethics of care and inclusion is a starting point and justification, rather than a conclusion or an object of study. The ethic extends fundamentally to developing methods of inquiry that are gender-inclusive (McDowell 1992), and that challenge the many ways in which masculine control has resulted in normative spaces (Rose 1993) and the widespread spatial domination of "heteropatriarchy" (Valentine 1993).[2]

Especially in the United States, but also in Canada and Britain, issues of spatial justice are the main concern of anti-racist geographies. As with feminism, anti-racists take as their starting point the moral goal of countering and eliminating racism. As O. Dwyer (1997) has recently shown, the preponderance of this work has been aimed at understanding and changing the growing disparities that exist in major cities, but recent issues also include geopolitics (Delaney 1993; Forrest 1995), the construction of racialized identities (Natter and Jones 1997), and racism and migration (Kobayashi 1995).

The concept of place receives continued, indeed renewed attention, both in a critical reflection on concerns over the objectification of community or of regional identities, as well as in the context of understanding how moral relationships are produced in industrial, developing, advanced capitalist, or postmodern societies. There is a huge range of work on the moral geographies of places, that includes the moral order of the city and the street (Jackson 1984; Driver 1988; Ogborn and Philo 1994; Valentine 1996), the countryside (Matless 1994), the everyday (Birdsall 1996; Johnston and Valentine 1995), institutions (Ploszajska 1994) and in some cases even the worlds of fantasy or the future, all of which provide a sense of how societies structure the kinds of normative activities that are appropriate to particular places, and what happens when normative

[2] This term refers to a normative grid in which value and power are centred on the moral system of the heterosexual white male. Social systems are built to conform to such values.

landscapes are transgressed (Cresswell 1996). Recent work on embodiment emphasizes the construction of places from a perspective of how gender, "race" (Jackson 1994; C. Dwyer 1998), ability (Butler 1998), health, age (Skelton and Valentine 1998), or sexuality (Bell *et al.* 1994; Valentine 1996) work distinctively or in tandem to influence the quality of life (Kobayashi and Peake 1994; Chouinard and Grant 1995). These geographies of the 1990s differ from the much more romanticized work of previous decades in that their focus is so strongly on injustice and exclusion (Sibley 1995). Sack (1999: 42) goes further, however, to suggest that a theoretical focus on place raises the geographer's ability to make moral judgements because "Place emphasizes our own geographical agency and draws attention to the breadth of moral concerns." Drawing strongly upon earlier work by Tuan (1989, 1991) Sack's work shows that the act of place-making is itself a moral act, thus opening up some interesting avenues for an empirical understanding of social morals.

Perhaps the most prolific area of moral geography in recent years has surrounded environmental issues. A recent collection devoted to this theme (Light and Smith 1997) shows not only that the intellectual heritage of environmental issues has influenced geographical thinking profoundly, but that contemporary philosophical questions engage the spectrum of critical thinking from Heidegger's and Mead's spatial ontology (Steelwater 1997) to the influence of the Frankfurt School in situating environmental concerns within a modernist context (Gandy 1997). Several geographers have noted the importance of cultural context in understanding environmental ethics (Gandy 1997; Simmons 1993; Booth 1997; Wescoat 1997), while others have established a strong basis for understanding the social theoretical basis for environmental understanding (O'Riordan 1981; Lewis 1992; Proctor 1995; Proctor 1998*c*), some calling for a specifically political role for the geographer (Pepper 1993; Pulido 1996*a*). Others highlight the ethical tensions between environmental issues and those surrounding questions of "race" or class (Bowen *et al.* 1995; DeLuca and Demo 2001; Pastor *et al.* 2001,) or gender (Whatmore 1997), or the differences between care for the environment and the right to a safe environment (Rogge 2001).

The emphasis on environmental issues also shows that the debate over essentialism is not closed. For many, perhaps most, understanding environmental ethics as thoroughly socially constructed and representative of dominant ideological and moral regimes has provided a major impetus for linking concerns for environmental and social justice (Harvey 1993; Low and Gleeson 1998). In this perspective the environment is viewed not as something of simple intrinsic value, but as a web of values based in human history, and human decision-making. Following the pathfinding work by the sociologist Bullard (1990), for example, geographers have paid increasing attention to issues of environmental racism to show the complex ways in which environmental ideologies are deeply racialized, politicized, and place-based (Laituri and Kirby 1994; Cutter 1995; Pulido 1996*b*, *c*; Heiman 1996; Westra and Wenz 1995). In contrast, however, some feminist geographers have been less concerned with environmental injustice than with a feminist ethics of caring through ecofeminism. Although ecofeminism has been developed primarily by nongeographers, geographers have played a significant role in recovering ecological projects from masculinist interpretations, and in turning around some of the more essentialized ecofeminist concepts in favor of a social constructionist environmental agenda (Seager 1994). Yet essentialist positions are quite widespread (though implicit) in geographical research devoted to nature conservation, where demonstration of, for instance, spatial inequity in habitat protection is deemed sufficient justification for action (Dion 2000). What is missing is the realization that value is a relational construct and does not reside solely in nature itself (nor is it simply "projected" on nature by humans, the other pole of non-relationality). In short, nature remains contested terrain to geographers precisely in the ways that its normative assumptions are grounded, whether in nature itself, or culture, or some engagement of the two.

The work of both feminist and anti-racist geographers refers consistently to the question of whether it is possible to be moral without being normative, and whether it is possible to be normative without being dominant and, thus, oppressive. Cutting a very long and complex philosophical story short, these questions in theoretical geography have centered around issues of essentialism, or the belief that there are unchangeable and essential attributes of human beings (such as, for example, qualities of gender, or "race") that determine their behavior, their abilities or their condition, and that should influence the ways in which social scientists interpret human systems. Geographers today adopt a position of social constructivism, which claims that the human condition is a result of historically produced human action rather than predetermined and immutable traits. Morality is therefore historically contingent, a product of human action rather than of universal moral values, and thoroughly dependent upon the ways in which human beings organize themselves in relationships that are powerful, spatial, and usually in some way or another involve the assertion of dominance and subordination, whether these relationships occur between human beings, between humans and animals, or between humans and nature.

Finally, a fourth broad area concerns questions of geographical activism. In its more critical forms, the normative turn has also led to attempts by geographers and other social scientists to *affect* social processes by joining social movements in hopes of forging new normative social systems. This area has two major dimensions. The first is an increasing range of writing around questions of academics as activists, a theme that has emerged repeatedly in the journals, in the sessions at this and other meetings, including a major conference on critical geography held in Vancouver in 1997. We have come a long way since the days of Bill Bunge's geographical expeditions. This literature, and the activist actions that sustain it, focus not only on a simple advocacy position for geographers (which is important in itself), but also address more complex theoretical and methodological concerns regarding situated knowledge (Katz 1992), the role of the intellectual in effecting social change (Knopp 1999), questions of relationality and identity (Larner 1995), and the ethical dilemmas involved in restructuring the relationship between researcher and subject in terms of class (Rose 1997), or in the context of developing countries (Mohan 1999). Feminist geographers in particular have addressed the question of unequal power relations (McLafferty 1995; Katz 1996) and material conditions (Gilbert 1995), as well as political credibility when scholarship and activism coincide (Kobayashi 1994).

A related concern involves asking questions of how we treat one another as geographers and fellow members of the academy, bringing questions of ethics into our classrooms, offices, even our homes. In particular, those concerned with questions of marginality have asked probing questions of the ways in which the student/teacher relationship is structured through power relations. A special issue of the *Journal of Geography in Higher Education* (1999) on teaching sexualities in the geography classroom includes a number of papers that show that teaching is a form of activism fraught with moral issues, while an issue of the *Journal of Geography* (19: 1999) is devoted to "teaching race." There is also growing concern with relations among colleagues over a range of issues from employment equity to questions around how the process of differencing creates divides according to gender, "race," sexuality, or ability. The recent piece by Gill Valentine (1998) in *Antipode* has not only shocked the discipline into an awareness of problems of harassment, but has generated a huge upwelling of intellectual support for understanding how we structure human relations according to ethical terms.

This example pushes us to note that discussion of ethical issues among geographers, and more generally, occurs increasingly in electronic form via the Internet, more informally through email list-servs and through the somewhat more formal medium of World Wide Web publication. The medium itself gives rise to ethical questions concerning how knowledge and human relationships are restructured as a result of electronic communication, especially given the space-time compression involved and the substantial changes to the meaning of "distance," and we look forward to the emergence of debate over these issues. It is also important to note, however, that new forms of communication have meant immediate and widespread engagement with ethical and political issues. For example, Shell Oil's sponsorship of the Royal Geographical Society/Institute of British Geographers represents one of the most emotional and ethically charged issues in the history of the discipline, and emerged predominantly through the listserv and website devoted to critical geography (<http://www.mailbase.ac.uk/lists/crit-geog-forum/>, now disabled). In the United States and Canada, a group of anti-racist geographers has recently developed an Anti-racist Manifesto calling on geography departments to adopt affirmative action policies in recruiting graduate students. This project too has developed largely through electronic communication. While we would not want to suggest that electronic relations have replaced, or even have the potential to replace, either face-to-face dialogue or traditional forms of publication, we do believe that the level of debate and discussion, and as a result the level of political involvement and engagement, have been increased in recent years through net talk, and that this development is directly related to larger social and ethical concerns regarding the significance of new electronic landscapes (Morley and Robins 1995).

Conclusion

We find ourselves in the somewhat ironic position of concluding that the label Values, Ethics, and Justice *Specialty* Group is something of a misnomer, for questions of professional and personal ethics cut across every specialty group, and few if any issues in human or physical geography are unconnected to values and justice. Certainly it was this recognition that led to the convergence of this group with the Human Rights Specialty Group, with the hope that by coming together we would create a stronger focus on questions of common concern. But our fundamental mission has not been to carve out a special intellectual space but, rather, to recognize that all geographies represent moral positions and locate themselves within a moral terrain, whether self-critically acknowledged or not. It is equally important that issues of ethics, values, and morality not be Balkanized as the concern of a few who decide to place one of their limited "specialty" choices on the VEJSG line when they renew their memberships each year, for it is our belief that this group has designs neither on occupying a distinctive territory of empirical investigation, nor on acting as the discipline's moral gatekeepers.

What place is there, then, for a specialty group on Values, Ethics, and Justice? We already have specialty groups devoted to feminism, sexuality, socialism, and native issues. There is a movement afoot to create a new specialty group devoted to anti-racism. Most of the members of the VEJSG also belong to some of these other groups. And the VEJSG can certainly make no claim to have identified a specific realm of geographic inquiry to call its own, no territory of difference. We would argue that there is room for the group, none the less, for two reasons. The first is the purely instrumental reason that there is sufficient interest, as shown by strong attendance at annual meeting sessions and healthy submissions to the journal *Ethics, Place, and Environment*. The second is that across the contested field that represents our discipline, with its increasing awareness of and attention to ethical questions, there is an important place for the synthetic focus on how we address moral concerns. Our task, more broadly specified, is to bring these issues to the fore and hope that in so doing we might stimulate wide debate on the ethical implications of our work as geographers. And, if recent writing across the discipline is anything to go by, that debate is ongoing, intense, highly theorized, and deeply ethical, a sign that many of us have cared very deeply.

References

Annals of the Association of American Geographers, (1998). 88/2.

Bell, D., Binnie, J., Cream, J. and Valentine, G. (1994). "All Hyped Up and No Place to Go." *Gender, Place and Culture*, 1: 31–47.

Birdsall, S. S. (1996). "Regard, Respect, and Responsibility: Sketches for a Moral Geography of the Everyday." *Annals of the Association of American Geographers*, 86: 619–29.

Black, R. (1996). "Immigration and Social Justice: Towards a Progressive European Immigration Policy?" *Transactions of the Institute of British Geographers*, 21: 64–75.

Booth, A. (1997). "Critical Questions in Environmental Philosophy," in A. Light and J. M. Smith, (eds.), *Geography and Philosophy, I: Space, Place and Environmental Ethics*. Lanham, Md.: Rowman & Littlefield, 255–74.

Bowen, W. M., Salling, M. J., Haynes, K. W. and Cyran, E. J. (1995). "Toward Environmental Justice: Spatial Equity in Ohio and Cleveland." *Annals of the Association of American Geographers*, 85: 641–63.

Boyne, G., and Powell, M. (1991). "Territorial Justice: A Review of Theory and Evidence." *Political Geography Quarterly*, 10: 263–81.

Brunn, S. D. (1988). "Issues of Social Relevance Raised by Presidents of the Association of American Geographers: The First Fifty Years." *Ethics, Place, Environment*, 1: 77–92.

— (1989). Editorial: "Ethics in Word and Deed." *Annals of the Association of American Geographers*, 79: iii–iv.

Bullard, R. D. (1990). *Dumping in Dixie: Race, Class, and Environmental Quality*. Boulder, Colo.: Westview Press.

Bunge, W. (1971). *Fitzgerald: Geography of a Revolution*. Cambridge, Mass.: Schlenkman.

Burch, R. (1997). "On the Ethical Determination of Geography: A Kantian Prolegomenon," in Light and Smith (1997: 15–47).

Butler, R. (1998). "Rehabilitating the Images of Disabled Youths," in Skelton and Valentine (1998: 83–100).

Buttimer, A. (1974). *Values in Geography*. Commission on College Geography, Resource Paper 24. Washington, DC: Association of American Geographers.

Chouinard, V., and Grant, A. (1995). "On Being Not Even Anywhere Near 'The Project': Revolutionary Ways of Putting Ourselves in the Picture." *Antipode*, 27: 137–66.

Clark, M. J. (1998). "Putting Water in Its Place: A Perspective on GIS in Hydrology and Water Management." *Hydrological Processes*, 12: 823–34.

Cooke, R. U. (1992). "Common Ground, Shared Inheritance: Research Imperatives for Environmental Geography." *Transactions of the Institute of British Geographers*, 15: 131–51.

Corbridge, S. (1993). "Marxisms, Modernities and Moralities: Development Praxis and the Claims of Distant Strangers." *Environment and Planning D: Society and Space*, 11: 449–72.

— (1998). "Development Ethics: Distance, Difference, Plausibility." *Ethics, Place and Environment*, 1: 35–54.

Crampton, J. (1995). "The Ethics of GIS." *Cartography and Geographic Information Systems*, 22: 84–9.

Cresswell, T. (1996). *In Place/Out of Place: Geography, Ideology and Transgression*. Minneapolis: University of Minnesota Press.

Curry, M. R. (1991). "On the Possibility of Ethics in Geography: Writing, Citing and the Construction of Intellectual Property." *Progress in Human Geography*, 15/2: 125–47.

— (1994). "Geographic Information Systems and the Inevitability of Ethical Inconsistency," in J. Pickles (ed.), *Ground Truth*. New York: Guilford Press.

Cutter, S. (1995). "Race, Class and Environmental Justice." *Progress in Human Geography*, 19: 107–18.

Delaney, D. (1993). "Geographies of Judgment: The Doctrine of Changed Conditions and the Geopolitics of Race." *Annals of the Association of American Geographers*, 83: 48–65.

DeLuca, K., and Demo, A. (2001). "Imagining Nature and Erasing Class and Race: Carleton Watkins, John Muir, and the Construction of Wilderness." *Environmental History*, 6: 541–60.

Dion, M. (2000). "The Moral Status of Non-Human Beings and Their Ecosystems." *Ethics, Place and Environment*, 3: 221–9.

Driver, F. (1988). "Moral Geographies: Social Science and the Urban Environment in Mid-Nineteenth Century England." *Transactions of the Institute of British Geographers*,13: 275–8.

Dwyer, C. (1998). "Contested Identities: Challenging Dominant Representations of Young British Muslim Women," in Skelton and Valentine (1998: 35–49).

Dwyer, O. (1997). "Geographical Research about African Americans: A Survey of Journals, 1911–1995." *Professional Geographer*, 49: 441–51.

England, K. V. L. (1994). "Getting Personal: Reflexivity, Positionality and Feminist Research." *Professional Geographer*, 46: 80–9.

Entrikin, J. N. (1994). "Moral Geographies: The Planner in Place." *Geography Research Forum*, 14: 113–19.

Eyles, J., and Smith, D. M. (1988). *Qualitative Methods in Human Geography*. Oxford: Polity Press.

Forrest, B. (1995). "Taming Race: The Role of Space in Voting Rights Litigation." *Urban Geography*, 16: 98–111.

Gandy, M. (1997). "Ecology, Modernity and the Intellectual Legacy of the Frankfurt School." in Light and Smith (1997: 231–54).

Gilbert, D. (1995). "Between Two Cultures: Geography, Computing and the Humanities." *Ecumene*, 2: 1–13.

Goodenough, S. (1977). *Values, Relevance and Ideology in Third World Geography*. Milton Keynes: The Open University.

Gregory, D. (1979). *Ideology, Science and Human Geography*. New York: St. Martin's Press.

Harley, J. B. (1991). "Can There Be a Cartographic Ethics?" *Cartographic Perspectives*, 10: 9–16.

Harvey, D. (1973). *Social Justice and the City*. London: Edward Arnold.

— (1974). "What Kind of Geography for What Kind of Public Policy?" *Transactions of the Institute of British Geographers*, 20: 18–24.

— (1993). "Class Relations, Social Justice and the Politics of Difference," in M. Keith and S. Pile (eds.), *Place and the Politics of Identity*. New York: Routledge, 41–66.

Havelberg, G. (1990). "Ethics as an Educational Aim in Geography Teaching." *Geographie und Schule*,12: 5–15.

Hay, A. M. (1995). "Concepts of Equity, Fairness and Justice in Geographical Studies." *Transactions of the Institute of British Geographers*, 20: 500–8.

Hay, I. (1998). "Making Moral Imaginations: Research Ethics, Pedagogy and Professional Human Geography." *Ethics, Place and Environment*, 1: 55–76.

Heiman, M. K. (1996). "Race, Waste, and Class: New Perspectives on Environmental Justice." *Antipode*, 28: 111–21.

Jackson, P. (1984). "Social Disorganization and Moral Order in the City." *Transactions of the Institute of British Geographers*, 9: 168–80.

— (1994). "Black Male: Advertising and the Cultural Politics of Masculinity." *Gender, Place and Culture*, 1: 49–60.

Johnston, L., and Valentine, G. (1995). "Wherever I Lay My Girlfriend That's My Home: The Performance of Surveillance of Lesbian Identities in Domestic Environments," in D. Bell and G. Valentine (eds.), *Mapping Desire: Geographies of Sexualities*. London: Routledge, 99–113.

Journal of Geography in Higher Education, (1999). 23/1.

Kates, R. W. (1987). "The Human Environment: The Road Not Taken, the Road Still Beckoning." *Annals of the Association of American Geographers*, 77: 525–34.

Katz, C. (1992). "All the World is Staged: Intellectuals and the Projects of Ethnography." *Environment and Planning D: Society and Space*, 10: 495–510.

— (1996). "The Expeditions of Conjurers: Ethnography, Power, and Pretense," in D. L. Wolf (ed.), *Feminist Dilemmas in Fieldwork*. Boulder, Colo.: Westview Press, 170–84.

Katz, C., and Kirby, A. (1991). "In the Nature of Things: The Environment and Everyday Life." *Transactions of the Institute of British Geographers*, 16: 259–71.

Knopp, L. (1999). "Out in Academia: The Queer Politics of One Geographer's Sexualisation." *Journal of Geography in Higher Education*, 23: 116–23.

Kobayashi, A. (1994). "Coloring the Field: Gender, 'Race' and the Politics of Fieldwork." *Professional Geographer*, 46: 73–80.

— (1995). "Challenging the National Dream: Gender Persecution and Canadian Immigration Law," in P. Fitzpatrick (ed.), *Nationalism, Racism and the Rule of Law*. Aldershot: Dartmouth, 61–74.

Kobayashi, A., and Mackenzie, S. (eds.) (1989). *Remaking Human Geography*. Boston: Unwin Hyman.

Kobayashi, A., and Peake, L. (1994). "Unnatural Discourse: 'Race' and Gender in Geography." *Gender, Place and Culture*, 1: 225–43.

Laituri, M., and Kirby, A. (1994). "Finding Fairness in America's Cities? The Search for Environmental Equity in Everyday Life." *Journal of Social Issues*, 50: 121–39.

Lake, R. W. (1993). "Planning and Applied Geography: Positivism, Ethics, and Geographic Information Systems." *Progress in Human Geography*, 17: 404–13.

Larner, W. (1995). "Theorising 'Difference' in Aotearoa, New Zealand." *Gender, Place and Culture*, 2: 177–90.

Lewis, G. M. (1992). "Metrics, Geometrics, Signs, and Language: Sources of Cartographic Miscommunication Between Native and Euro-American Cultures in North America." *Cartographica*, 30: 98–106.

Ley, D. (1989). "Fragmentation, Coherence, and Limits to Theory," in Kobayashi and Mackenzie (1989: 227–44).

Light, A., and Smith, J. M. (eds.) (1997). *Philosophy and Geography, I: Space, Place and Environmental Ethics*. Lanham, Md.: Rowman & Littlefield.

Low, N., and Gleeson, B. (1998). "Situating Justice in the Environment: The Case of BHP at the Ok Tedi Copper Mine." *Antipode*, 30: 201–26.

McDowell, L. (1992). "Doing Gender: Feminism, Feminists and Research Methods in Human Geography." *Transactions of the Institute of British Geographers*, 17: 399–416.

McLafferty, S. (1995). "Counting for Women." *Professional Geographer*, 47: 436–42.

Manning, E. W. (1990). "Presidential Address: Sustainable Development, the Challenge." *The Canadian Geographer*, 34: 290–302.

Matless, D. (1994). "Moral Geography in Broadland." *Ecumene*, 1: 125–55.

Merrett, C. D. (2000). "Teaching Social Justice: Reviving Geography's Neglected Tradition." *Journal of Geography*, 99: 207–18.

Mohan, Giles (1999). "Not So Distant, Not So Strange: The Personal and the Political in Participatory Research." *Ethics, Place and Environment*, 2: 41–54.

Monmonier, M. (1991). "Ethics and Map Design: Six Strategies for Confronting the Traditional One-Map Solution." *Cartographic Perspectives*, 10: 3–8.

Morley, D., and Robins, K. (1995). *Spaces of Identity: Global Media, Electronic Landscapes, and Cultural Boundaries*. London: Routledge.

Natter, W., and Jones, J. P. (1997). "Identity, Space and Other Uncertainties," in G. Benko and U. Strohmayer (eds.), *Space and Social Theory*. Oxford: Blackwell, 141–61.

Ogborn, M., and Philo, C. (1994). "Soldiers, Sailors and Moral Locations in Nineteenth-Century Portsmouth." *Area*, 26: 221–31.

O'Riordan, T. (1981). *Environmentalism*. London: Pion.

Pastor, M., Jr., Sadd, J. and Hipp, J. (2001). "Which Came First? Toxic Facilities, Minority Move-in, and Environmental Justice." *Journal of Urban Affairs*, 23: 1–21.

Peet, R. (1975). "Inequality and Poverty: A Marxist-Geographic Theory." *Annals of the Association of American Geographers*, 65: 564–71.

— (ed.) (1978). *Radical Geography*. London: Metheun.

Pepper, D. (1993). *Eco-socialism: From Deep Ecology to Social Justice*. London: Routledge.

Pickles, J. (1985). *Phenomenology, Science and Geography: Spatiality and the Human Sciences*. Cambridge: Cambridge University Press.

Ploszajska, T. (1994). "Moral Landscapes and Manipulated Spaces: Gender, Class and Space in Victorian Reformatory Schools." *Journal of Historical Geography*, 20: 413–29.

Proctor, J. D. (1995). "Whose Nature? The Contested Moral Terrain of Ancient Forests," in W. Cronon (ed.), *Uncommon Ground: Toward Reinventing Nature*. New York: W. E. Norton.

— (1998*a*). "Ethics in Geography: Giving Moral Form to the Geographical Imagination." *Area*, 30: 8–18.

— (1998*b*). "Expanding the Scope of Science and Ethics: A Response to Harman, Harrington and Cerveny's 'Balancing Scientific and Ethical Values in Environmental Science.'" *Annals of the Association of American Geographers*, 88: 290–296.

— (1998*c*). "Geography, Paradox and Environmental Ethics." *Progress in Human Geography*, 22: 234–55.

Proctor, J. D., and Smith, D. M. (eds.) (1999). *Geography and Ethics: Journeys in a Moral Terrain*. New York: Routledge.

Pulido, L. (1996*a*). *Environmentalism and Economic Justice: Two Chicano Struggles in the Southwest*. Tucson: University of Arizona Press.

— (1996*b*). "Multiracial Organizing among Environmental Justice Activists in Los Angeles," in M. J. Dear, H. E. Schockman, and G. Hise (eds.), *Rethinking Los Angeles*. Thousand Oaks, Calif.: Sage, 171–89.

— (1996*c*). "A Critical Review of the Methodology of Environmental Racism Research." *Antipode*, 28: 142–59.

Reed, M. G., and Slaymaker, O. (1993). "Ethics and Sustainability: A Preliminary Perspective." *Environment and Planning A*, 25: 723–39.

Rogge, M. J. (2001). "Human Rights, Human Development and the Right to a Healthy Environment: An Analytical Framework." *Canadian Journal of Development Studies*, 22: 33–50.

Rose, G. (1993). *Feminism and Geography: The Limits of Geographical Knowledge*. Cambridge: Polity.

— (1997). "Situating Knowledges: Positionality, Reflexivities and Other Tactics." *Progress in Human Geography*, 21: 305–20.

Rundstrom, R. A. (1993). "The Role of Ethics, Mapping and the Meaning of Place in Relations between Indians and Whites in the United States." *Cartographica*, 30: 21–8.

Sack, R. (1997). *Homo Geographicus: A Framework for Action, Awareness and Moral Concern*. Baltimore: Johns Hopkins University Press.

— (1999). "A Sketch of a Geographic Theory of Morality." *Annals of the Association of American Geographers*, 89/1: 26–44.

Sayer, A. (1989). "On the Dialogue between Humanism and Historical Materialism in Geography," in Kobayashi and Mackenzie (1989: 206–26).

Sayer, A., and Storper, M. (1997). "Ethics Unbound: For a Normative Turn in Social Theory. Guest Editorial." *Environment and Planning D: Society and Space*, 15: 1–17.

Seager, J. (1994). *Earth Follies*. London: Routledge.

Sibley, D. (1995). *Geographies of Exclusion: Society and Difference in the West*. London: Routledge.

Simmons, I. G. (1993). *Interpreting Nature: Cultural Constructions of the Environment*. London: Routledge.

Skelton, T., and Valentine, G. (eds.) (1998). *Cool Places: Geographies of Youth Cultures*. London: Routledge.

Slater, D. (1997). "Spatialities of Power and Postmodern Ethics: Rethinking Geopolitical Encounters." *Environment and Planning D: Society and Space*, 15: 55–72.

Smith, D. M. (1994). *Geography and Social Justice*. Oxford: Blackwell.

— (1995). "Moral Teaching in Geography." *Journal of Geography in Higher Education*, 19: 271–183.

— (1997). "Geography and Ethics: a Moral Turn?" *Progress in Human Geography*, 21: 583–90.

— (1998*a*). "Geography and Moral Philosophy: Some Common Ground." *Ethics, Place, and Environment*, 1: 7–34.

— (1998*b*). "How Far Should We Care? On the Spatial Scope of Beneficence." *Progress in Human Geography*, 22: 15–38.

— (2000). "Moral Progress in Human Geography: Transcending the Place of Good Fortune." *Progress in Human Geography*, 24: 1–18.

Steelwater, Eliza (1997). "Mead and Heidegger: Exploring the Ethics and Theory of Space, Place and the Environment," in Light and Smith (1987: 189–208).

Tuan, Y.-F. (1974). "Space and Place: A Humanistic Perspective," in C. Board *et al.* (eds.), *Progress in Geography*, 6. London: Edward Arnold, 211–52.

— (1989). *Morality and Imagination*. Madison: University of Wisconsin Press.

—— (1991). A View of Geography. *Geographical Review*, 81: 99–107.

Valentine, G. (1993). "(Hetero)sexing Space: Lesbian Perceptions and Experiences of Everyday Space." *Environment and Planning D: Society and Space*, 11: 395–413.

—— (1998). " 'Sticks and Stones May Break My Bones': A Personal Geography of Harassment." *Antipode*, 30: 305–32.

Wasowski, R. J. (1991). "Some Ethical Aspects of International Satellite Remote Sensing." *Photogrammetric Engineering and Remote Sensing*, 57: 41–8.

Wescoat, J. L., Jr. (1997). "Muslim Contributions to Geography and Environmental Ethics: The Challenges of Comparison and Pluralism," in Light and Smith (1997).

Westra, L., and Wenz, P. S. (eds.) (1995). *Faces of Environmental Racism: Confronting Issues of Global Justice*. Lanham, Md.: Rowman & Littlefield.

Whatmore, S. (1997). "Dissecting the Autonomous Self: Hybrid Cartographies for a Relational Ethics." *Environment and Planning D: Society and Space*, 15: 37–53.

Women and Geography Study Group (1984). *Geography and Gender: An Introduction to Feminist Geography*. London: Hutchinson, in assoc. with the Explorations in Feminism Collective.

Zelinsky, W. (1975). "The Demigod's Dilemma." *Annals of the Association of American Geogaphers*, 65: 123–43.

CHAPTER 46

Human Rights

Rex Honey

Scholarship addressing the geography of human rights—the geographical analysis of the ways cultures conceive of justice and understand just behavior—improves both our understanding of human rights and our understanding of geography. A full understanding of struggles over human rights requires a geographical perspective, a consideration of the contexts in which the struggles occur. Conceptualizations of human rights and the abuse of human rights do not just happen. They are the products of human action in particular cultural and environmental settings. They are place-based and socially constructed, products of processes not only tied to place but also altering the significance of place. Human rights scholarship that omits geographical background and geographical consequences misses the target because it fails to capture both the cultural struggles over what a just society is and the *milieu* of interrelated sites of injustice.

If geographical research addressing oppression does not explicitly address human rights, then virtually by definition such work does so implicitly. At its core, human rights scholarship addresses oppression. Hence, such geographical scholarship as the work of Knopp (1997) addressing gay rights and of Monk (1998) addressing women's rights fits into the scope of the geography of human rights in America.

To fulfill its potential as a scholarly discipline examining the human condition, geography needs to focus on human rights. The spotlight of geographic education's five themes—place, location, region, nature–society relations, movement—should be focused on what truly matters in people's lives, including human rights. Likewise, the study of human rights, with its vexing examination of rights and wrong, needs the nuanced sensitivity of geography, with its study of cultural and environmental contexts. To wit, *geography needs human rights and human rights needs geography.* Geographical research that has been done or is in progress explains why. Indeed, the roster of former presidents of the Association of American Geographers contains many individuals whose professional and personal lives were committed to the furtherance of human rights. Among them are such luminaries as Richard Hartshorne, Harold Rose, Gilbert White, Julian Wolpert, and Richard Morrill, each of whom struggled for the advance of human rights in his personal life as well as his scholarly work.

Nevertheless, formal emphasis on human rights is really a phenomenon of the late twentieth and twenty-first centuries, not only in geography as a discipline but among people more generally. As Murphy (1994) and Wood (1994), among others, have shown, the emergence of human rights as an important issue has in recent decades dramatically altered international law, corralling the unlimited power of a state against its people. State sovereignty had been inviolable, all-encompassing. The emergence of global human rights standards has constrained state power. At the birth of a new millennium, rulers and governors are accountable for their treatment of their subjects and citizens, and this fact is a major accomplishment in the improvement of human well-being. Furthermore, the human rights movement has done more than take the boot of the state off the neck of the oppressed. It has also focused global attention when any boot has unjustly threatened any neck, as evidenced by the work of Blomley (1994), Pred (1997), and Honey (1999*b*), among other geographers.

Global and regional institutions have joined the struggle over human rights. Some of these, including a number of organs of the United Nations, are the product of state action. Among them are the UN Commission on Human Rights and the UN Relief Works Organization, as well as a series of multi-state bodies operating at the regional scale. Others—among them such organizations as Human Rights Watch and Amnesty International—are of a very different nature. They are non-governmental organizations (NGOs) created to expose and diminish human rights violations (Drake 1994). They are significant players on the world scene, embarrassing governments and other powers involved in the systematic violation of rights.

The late twentieth century's formal emphasis on human rights, as a consequence, is matched by an ethical emphasis as scholars, jurists, pundits, and people more generally have joined the struggle over human rights. This struggle is many faceted, involving determination of what fundamental rights should be protected as well as where and how whose rights are violated. In a world with shifting bases of power as well as divergent conceptualizations of justice, this is a fascinating and challenging subject. It is a subject rich in geographical nuances and requiring careful consideration of geographical context, in the realms both of nature–society and cultural understanding. Let us examine how this is so.

Geography of the Human Rights Movement

The human rights movement emerged from a particular political geography characterized by its own conflicts and struggles. It clearly was an extension of the Enlightenment, reaching back to the principles of the American and French Revolutions and beyond (Toal 1999). Likewise, it was an extension of two twentieth-century phenomena, the Holocaust and the Cold War (Honey 1999*b*). The Holocaust stripped away resistance to constraints on sovereign power. Genocide was intolerable, full stop. The Cold War provided the geopolitical context from which the first significant success of the human rights movement emerged. That success was the Universal Declaration of Human Rights, adopted by the United Nations on 10 December 1948. By that time the Cold War was in full force. The forces of democracy and capitalism stood against the forces of communism. The Iron Curtain divided Europe, the bamboo curtain Asia. Advocates of two distinctive world-views pushed their own rights agendas (Weston 1992). At the risk of oversimplification—it is an oversimplification but one that illuminates the major shape of the debate—advocates of democracy pushed for the civil and political liberties that marked the American and French Revolutions and their descendants in political philosophy: power resided in the people who were inherently endowed with rights that the state (and others) could not abrogate. Among these were such political rights as the right to install and dismiss a government through fair elections and such civil rights as freedom of expression and thought. These "bourgeois" rights were not part of the communist rights agenda. Rather, communists emphasized economic and social rights, particularly the former. These included the right to a job as well as the right to education and health care. These so-called second-generations rights, according to this argument, had priority over civil and political rights.

At mid-century, when the human rights movement first became an influential force, the world was in upheaval. The Cold War was only the most immediately dangerous aspect of that upheaval. The age of empire was reaching its formal conclusion as the peoples living under colonial rule demanded the rights their colonial masters had taught them. Western European dominance of the world was clearly at an end with its vestiges in tropical colonies doomed. Industrial economies were giving way to post-industrial ones. Class structures were to varying degrees unstructured—or at least reconfigured. Women for numerous reasons had both their traditional security and their traditional constraints eased. In a time of uncertainty, representatives of the governments of the world adopted a series of articles in support of basic rights for all people. The list is a product of its historical geography. It is not a list that would gain approval today, but it is one widely supported, in large part because it has become instrumental in the protection of people across countries (Weston 1992). The human rights movement has provided a moral anchor amidst the global chaos.

Universality and Its Alternatives

The moral anchor of the human rights movement is contested. A major issue is the universality of human rights.

Can one find broadly if not unanimously acceptable standards of human (and state) behavior? Those answering "yes" tend to follow one of two strands of thought (Brown 1999). Natural-law arguments have tended to win the intellectual debates, but particularistic arguments have been the ones implemented in the international human rights arena (Teson 1992; Harvey 1996).

The idea of generally acceptable, basic human rights has been widely recognized. Debate continues—and doubtlessly will continue—over the specifics. The protection of civil and political rights, for example, is a broadly accepted litmus test of a just society. Anyone abrogating these rights is roundly condemned. The consensus on these rights is part of a global shift toward the norms of liberal democracy, encompassing multi-party democracy and a market economy. With regard to these first-generation rights, debate continues over the details. Which groups, for example, merit civil rights protection? Prohibitions of racial and gender discrimination are the norm in progressive societies. Extending such protection to the gay community is a much more contentious issue, championed in places with liberal orientations, condemned in more conservative places (Honey 2002).

While the debate over the precise set of first-generation rights is largely over moral issues, those over second- and third-generation rights tend to be more strategic. Promoting civil liberties generally amounts to getting governments to stop oppressing their people; the remedy is simply ending state oppression. Promoting social and economic rights is another matter altogether. Wealthy countries, often led by the United States, have been reluctant to accept global standards that they suspect may be used as levers to get them to cover the costs of implementation. Finding a globally acceptable formula for protecting environmental and survival rights—third-generation rights—is similarly difficult, because rich countries fear the price tag and poorer countries fear limits on their economic growth. An illustration is the fact that the United States Senate, following the lead of the American Bar Association, took twenty years to ratify the UN covenant on genocide because of fears over sovereignty. Another is US refusal to follow the Kyoto Accords.

The challenges to universalism are real. These challenges in general follow three arguments. Geographers have much to contribute in each case. One is that different societies operate on different cosmologies and therefore with different logics. Implicit in the human rights literature is an emphasis on this lifetime rather than an afterlife or series of lives. Denying people their rights in this lifetime is not to be justified by some presumed benefit after this life. Such an assumption is problematic in societies emphasizing different cosmologies. Requests for cultural exceptions to an international norm often follow this argument. Charlesworth suggests that such requests be studied through an interrogation of the cultural, asking what the culture really values and supports (Charlesworth 1999). Geographers are well placed to provide this interrogation, examining societies rigorously rather than reifying a form of oppression.

Fundamentalist interpretations of Scripture provide a second challenge to universalism. True debate is not possible when a believer holds that a particular translation of a religious tract is the direct word of God, whether as part of the Torah, Bible, Koran, or any other sacred source. Gay rights debates can turn on such issues. New Zealand is among the countries that prohibit discrimination against people on the basis of sexual preference. Many local communities in the United States have similar protections, though no state does. Conservative Muslims, Christians, and Jews find state support of gay rights more than inappropriate; they find it abominable. These arguments reject the idea of universal rights for gay groups as a principle. Fundamentalist interpretations quite often posit universal standards of behavior, albeit different standards for each such group. They argue against the rights delimited in the Universal Declaration or some other set of rights rather than against the concept of global standards. Today challenges to women's rights present a less-contentious but still contended issue within this second challenge to universality, but at the beginning of the twentieth century women were just beginning to gain the right to vote, let alone equality before the law. Again, through cultural interrogation, geographers have much to offer.

Examination of requests for exceptions to universal standards shows that most of these requests fall into the third category of exceptions to universality (Honey 1999*a*). Simply put, these are efforts of oppressors to maintain power. Foremost examples are the arguments dictators use that Western senses of democracy violate their culture, as in the case of Burma in the 1990s. Similarly, the Taliban justified their particularly oppressive form of patriarchy on Muslim interpretations that distort Islam. Southern Baptist requirements of wifely obedience are of a similar nature. Those geographers with an acute understanding of the cultures of a particular place are well positioned to examine the veracity of such arguments.

Students of human rights largely accept the existence of global standards of behavior, inviolable norms in a civilized world (Harvey 1996). This does not mean that human rights scholars agree on what is acceptable behavior. Rather, they recognize that the human rights standards of any time and place are socially constructed and subject to struggles, both within and between cultures.

Human Rights as Part of the Geography of America

Struggles over rights have in many ways defined the United States of America. These struggles are part of American lore and identity as well as American history, from the efforts of the Plymouth pilgrims to escape religious persecution through the Emancipation Proclamation to the Americans with Disabilities Act. These struggles have been inextricably woven into the fabric of the United States as a whole as well as in its regional differentiation and identity. In many ways these struggles are the story of expanding the franchise (and all other rights) as the definition of the people grew from white men of property to ever more encompassing conceptualizations.

All Americans—and most educated people anywhere—know the significance of the Declaration of Independence, Constitution, and Bill of Rights. Each of these amounted to a revolution in human affairs, strengthening the power of the people collectively at the expense of the rulers, particularly hereditary rulers. All Americans—and again most educated people elsewhere—know how limited the reach of these innovations was. Slavery, after all, is built into the US Constitution and defined the political geography of the United States in a way that still remains. Slavery and the racism that survived it led to the monumental struggles that have dominated domestic politics in the United States since—the Civil War, Jim Crow laws, the Civil Rights Movement, Affirmative Action. Geographers have documented these struggles, as exemplified by the works of Richard Morrill, Harold Rose, Fred Shelley, and Jeremy Brigham. Struggles over rights, both individual and collective, have been part of the definition and perception of American regions, dividing North from South but also suburb from central city, among other divisions.

Geographers have illuminated, in many studies, struggles over human rights within the United States. From David Smith's seminal work in 1973, geographers have been examining issues of social justices in the US. Yapa (1996), Shaw (1996), Crump (1997), Kodras (1997), and Mitchell (1998) provide various perspectives on the problems of poverty. Morrill (1994) and Ingalls *et al.* (1997), among others, document problems within the American electoral system. Henderson (1996), Holloway (1998), and Brigham (1999*b*) examine issues of racism. Weiss (1994) offers an intriguing atlas detailing the geographical distribution of attitudes that, *inter alia*, impinge on human rights in America. These works and others document progress in terms of well-recognized rights, including some, such as the inappropriateness of racial discrimination, that were contentious just a few decades ago. Other work addresses ongoing debates. Three important ones in the United States involve gun control, the death penalty, and gay rights (Brigham 1998*a*; Baldus *et al.* 1986; Knopp 1997).

American geographers of course research topics outside as well as within the United States. Their work, along with that of their foreign colleagues, includes a growing body of work on human rights issues. Doreen Massey (1999), Valerie Preston (1999), Janine Wiles, and Bonnie Hallman (1999) formed the North American contingent at a highly successful conference by the International Geographical Union's Commission on Gender and Geography. The geographical literature on women's rights is growing rapidly and includes significant contributions, exemplified by the work of Pain (1991) on sexual violence, Spain (1992) on gendered spaces, Kinnaird and Momsen (1993) on gender and development, McIlwaine (1996) on sex workers, and Riphenburg (1997) on women in Zimbabwe. Publications by Falah (1989), Newman (1984, 1991), Yiftachel (1995), and Yiftachel and Segal (1998) take a human rights approach to the Palestinian–Israeli conflict. Hyndman (1997), Money (1997), and Wood (1994) deal with migrants and refugees.

The story of human rights and American geography has another aspect, too, that of the role of the United States in fostering or foiling human rights elsewhere. In recent decades an interrupted series of American administrations have advocated human rights as part of foreign policy. James McCormick (1999) traces the subject throughout American history, showing how, since Carter in particular, successive administrations have worked to advance the cause of human rights outside the United States. Geographers such as Blaut (1993), Cohen (1994), and Taylor (1996) have developed similar themes, but they have shown the ways in which the United States has impeded as well as enhanced struggles for human rights elsewhere. This work makes a solid case that the government of the United States has been simultaneously the world's greatest advocate of human rights and one of its greatest threats.

Geographers can bring to the human rights debates two kinds of knowledge based on their particular research. One is knowledge of cultures: not just a knowledge reifying cultures but one based on a solid interrogation of them, gleaned from years or even decades of work. Another is a knowledge of nature–society relations. These aspects of knowledge, changing as scholars advance understanding, should be applied to the various generations of human rights. Debate over first-generation rights—the kinds codified in the Bill of Rights and the constitutions of other liberal democracies—is typically limited to instances of conflict, as illustrated with fair trials or freedom of the press, rather than the rights generally, excepting the most virulently oppressive societies, such as the Taliban. Second-generation rights involve social and economic rights—such things as the right to a job or decent medical care. These rights are more difficult because they involve doing something rather than not doing something. Who, after all, is to guarantee the job or medical care? Third-generation rights involve survival, both of cultures and the global environmental system. Clearly geographers have much to contribute to enhanced understanding of these rights.

The formal recognition of human rights on the global scale began a half-century ago. Formal human rights scholarship within geography really began within the last decade. The geographical analysis of human rights is a way for geographers, in America and elsewhere, to make truly significant statements affecting the human condition.

References

Baldus, D. C., Pulaski, C. A. Woodworth, G. (1986). "Arbitrariness and Discrimination in the Administration of the Death Penalty: A Challenge to State Supreme Courts." *Stetson Law Review*, 15/2: 135–261.

Blaut, J. (1993). *The Colonizers Model of the World*. New York: Guilford.

Blomley, N. (1992). "The Business of Mobility: Geography, Liberalism, and the Charter of Rights." *Canadian Geographer*, 36/3: 236–53.

— (1994). "Mobility, Empowerment and the Rights Movement." *Political Geography*, 10.

Brigham, J. J. (1998*a*). "The Glock Gun Spiel: A Case Study in Political Geography and Human Rights." *Focus*, 45/1: 7–10.

— (1998*b*). *The Cousin County Project: Iowa and Mississippi in the Civil Rights Movement*. Iowa City: University of Iowa.

Brown, C. (1999). "Universal Human Rights? An Analysis of the 'Human-Rights Culture' and Its Critics," in R. Patman (ed.), *Universal Human Rights*? London: Macmillan, 31–49.

Charlesworth, H. (1999). Public lecture at the University of Iowa.

Cohen, S. B. (1994). "Geopolitics in the New World Era: A New Perspective on an Old Discipline," in G. J. Demko and W. B. Wood (eds.), *Reordering the World*. Boulder, Colo.: Westview Press.

Crump, J. R. (1997). "Teaching the Political Geography of Poverty." *Journal of Geography*, 96/2: 98–104.

Drake, C. (1994). "The United Nations and NGOs: Future Roles," in G. J. Demko and W. B. Wood (eds.), *Reordering the World*. Boulder, Colo.: Westview Press.

Falah, G. (1989). "Israelization of Palestine Human Geography." *Progress in Human Geography*, 5/3: 535–50.

Hallman, B. (1999). "Making Time and Making Do: Gender, Time-Space and the Transition into Eldercare." Paper presented at the IGU Commission on Gender and Geography, Dunedin, New Zealand.

Harvey, D. (1973). *Social Justice and the City*. London: Arnold.

— (1996). *Justice, Nature and the Geography of Difference*. Cambridge: Blackwell.

Henderson, M. L. (1996). "Geography, First Peoples, and Social Justice." *The Geographical Review*, 86/2: 278–83.

Holloway, S. R. (1998). "Exploring the Neighborhood Contingency of Race Discrimination in Mortgage Lending in Columbus, Ohio." *Annals of the Association of American Geographers*, 88/2: 252–76.

Honey, R. D. (1998). "Universal Human Rights?" *New Zealand International Review*, 24/1: 18–20.

— (1999*a*). "Nested Identities in Nigeria," in G. H. Herb and D. H. Kaplan (eds.), *Nested Identities*. Lanham, Md.: Rowman & Littlefield, 175–97.

— (1999*b*). "Human Rights and Foreign Policy," in R. Patman (ed.), *Universal Human Rights*? London: Macmillan, 226–48.

— (2002). "Cultural Struggles over Human Rights", *Geoforum*, Special Issue on human rights.

Hyndman, J. (1997). "A Post-Cold War Geography of Forced Migration in Kenya and Somalia." *Professional Geographer*, 51/1: 104–14.

Ingalls, G. I., Webster, G. R. and Leib, J. L. (1997). "Fifty Years of Political Change n the South: Electing African Americans and Women to Public Office." *Southeastern Geographer*, 37/2, 140–61.

Kinnaird, V., and Momsen, J. (1993). *Different Places, Different Voices*. London: Routledge.

Knopp, L. (1997). "Sexuality and Urban Space: A Framework for Analysis," in David Bell and Gill Valentine (eds.), *Mapping Desire: Geographies of Sexuality*. London: Routledge.

Kodras, J. (1997). "The Changing Map of American Poverty in an Era of Economic Restructuring and Political Realignment." *Economic Geography*, 73/1: 67–93.

Massey, D. (1999). *Geography and Women's Studies*. Dunedin: IGU Commission on Gender and Geography.

McCormick, J. (1999). "Human Rights and the Clinton Administration: American Policy at the Dawn of a New Century," in R. Patman (ed.), *Universal Human Rights?* London: Macmillan.
McIlwaine, C. (1996). "The Negotiation Space Among Sex Workers in Cebus City, the Philippines." *Singapore Journal of Tropical Geography*, 17/2: 150–64.
Mitchell, D. (1998). "Anti-Homeless Laws and Public Space." *Urban Geography*, 19/2: 98–104.
Money, D. (1997). "No Vacancy: The Political Geography of Immigration Control in Advanced Industrial Countries," *International Organization*, 51/4: 685–720.
Monk, J. (1998). "The Women Were Always Welcome at Clark." *Economic Geography*, Special Issue (Spring): 14–30.
Morrill, R. (1984). "The Responsibility of Geography." *Annals of the Association of American Geographers*, 74/1: 1–8.
— (1994). "Electoral Geography and Gerrymandering: Space and Politics," in G. J. Demko and W. B. Wood (eds.), *Reordering the World*. Boulder, Colo.: Westview Press.
Murphy, A. B. (1994). "International Law and the Sovereign State," in G. J. Demko and W. B. Wood (eds.), *Reordering the World*. Boulder, Colo.: Westview Press.
Newman, D. (1984). "Ideological and Political Influences on Israeli Urban Colonization: The West Bank and Galilee Mountains." *Canadian Geographer*, 28/2: 142–55.
— (1991). "On Writing Involved Political Geography." *Political Geography*, 10: 195–9.
Pain, R. (1991). "Space, Sexual Violence and Social Control: Integrating Geographical and Feminist Analyses of Women's Fear of Crime." *Progress in Human Geography*, 15/4: 415–31.
Pred, A. (1997). "Somebody Else, Somewhere Else: Racisms, Racialized Spaces and the Popular Geographical Imagination in Sweden." *Antipode*, 29/4: 383–416.
Preston, V. (1999). "Flexible Bodies, Flexible Workers: Women in the Canadian Grocery Sector." Paper presented at the IGU Commission on Gender and Geography, Dunedin, New Zealand.
Riphenburg, C. (1997). "Women's Status and Cultural Expression: Changing Gender Relations and Structural Adjustment in Zimbabwe." *Africa Today*, 44/1: 33–50.
Rose, H. M. (1971). *The Black Ghetto: A Spatial Behavioral Perspective*. New York: McGraw-Hill.
Shaw, W. (1996). *The Geography of United States Poverty: Patterns of Deprivation, 1980–1990*. New York: Garland.
Shelley, F. M. (1994). "Geography, Territory and Ethnicity: Current Perspectives from Political Geography." *Urban Geography*, 45–58.
Smith, D. M. (1973). *The Geography of Social Well-being in the United States*. New York: McGraw-Hill.
— (1994). *The Geography of Social Justice*. Oxford: Blackwell.
— (1999). "Geography Community and Morality," *Environment and Planning A*, 31: 10–35.
Spain, D. (1992). *Gendered Spaces*. Chapel Hill, NC: University of North Carolina Press.
Taylor, P. J. (1996). *The Way the Modern World Works*. Chichester: John Wiley.
Teson, F. R. (1992). "International Human Rights and Cultural Relativism," in R. P. Claude and B. H. Weston (eds.), *Human Rights in the World Community*. Philadelphia: University of Pennsylvania Press, 42–51.
Toal, G. (1999). "Understanding Critical Geopolitics: Geopolitics and Risk Society." *Journal of Strategic Studies*, 22/2–3: 107–24.
Weiss, M. J. (1994). *Latitudes and Attitudes: An Atlas of American Tastes, Trends, Politics, and Passions*. Boston: Little, Brown.
Weston, B. H. (1992). "Human Rights," in R. P. Claude and B. H. Weston (eds.), *Human Rights in the World Community*. Philadelphia: University of Pennsylvania Press, 14–30.
Wood, W. B. (1994). "Crossing the Line: Geopolitics of International Migration," in G. J. Demko and W. B. Wood (eds.), *Reordering the World*. Boulder, Colo.: Westview Press.
Yapa, L. (1996). "What Causes Poverty? A Postmodern View." *Annals of the Association of American Geographers*, 86/4: 707–28.
Yiftachel, O. (1999). "Regionalism Among Palestinian-Arabs in Israel," in G. H. Herb and D. H. Kaplan (eds.), *Nested Identities*. Lanham, Md.: Rowman & Littlefield, 237–66.
Yiftachel, O., and Segal, M. D. (1998). "Jews and Druze in Israel: State Control and Ethnic Resistance." *Ethnic and Racial Studies*, 21/3: 476–506.

CHAPTER 47

Geographic Perspectives on Women

Ann M. Oberhauser, Donna Rubinoff, Karen De Bres, Susan Mains, and Cindy Pope

Introduction

Since its inception in the mid-1970s, feminist geography has significantly impacted the discipline. Open nearly any human geography textbook, review course offerings in most of the top geography programs, or examine recent publications by well-known human geographers and you will discover the influence of gender and feminist perspectives. This has not always been the case, however. In the previous volume of *Geography in America*, the chapter written for the Geographic Perspectives on Women (GPOW) specialty group noted that until recently, geography was written as if "men were representative of the species" (Gruntfest 1989: 673). Many areas of the discipline have gone beyond this approach and are beginning to recognize the role of gender in human spatial behavior. This chapter draws from the foundations of feminist geography that highlight the role of women in geographical analysis before focusing on the multiple voices and variety of perspectives that comprise this area of the discipline.

The thematic topics presented in this chapter address feminist analyses of methodology, gender and work, Third World development, cultural geography, identity and difference, and pedagogy. This discussion is by no means exhaustive, but addresses those issues that have been particularly influential during the 1990s. While feminist geographers borrow from, and contribute to, research outside the discipline, this chapter largely focuses on work undertaken by researchers more formally associated with geography. The discussion also includes work by non-American geographers due to the collaborative and international nature of feminist geography that makes it difficult to limit research to one country. The cross-fertilization of ideas and experiences is evident in the growing number of conferences, publications, institutional exchanges, and research endeavors that involve a variety of feminist scholars from multiple disciplines and nationalities.

According to the editors of this book, our charge was to write an assessment of feminist geography that is "comprehensive, current, forward-looking, and influential." The task of writing this chapter has occurred in a participatory and collaborative manner that reflects feminist projects (and produced some lively telephone conversations!). As authors, we represent a diverse group of scholars in terms of our positions in the academy, areas of specialization, and institutional affiliations. Our interest in feminist geography is shaped by our involvement and background in a variety of specializations that include cultural, political, development, economic, and medical geography. The diversity of our approaches will be evident in a manner that, it is hoped, does not detract from, but enriches our analysis.

To capture the full flavor of the subfield and incorporate the voices of other feminist geographers, we solicited input through a survey distributed on the GPOW listserv and a presentation at the 1999 annual meeting of the Association of American Geographers (AAG). The survey was administered in October of 1998 and raised questions about the following topics: how people became involved in gender geography; major theories, concepts, and/or methods that have impacted the subfield; noteworthy articles and books; the role of feminist pedagogy and other professional activities in geography; and future directions of the subfield. The survey generated responses that guided our discussions about the subfield and its contribution to geography. The diversity of respondents in regards to ideological background, nationality, and position in higher education was particularly useful in capturing a wide range of feminist perspectives.

In addition, the survey helped us articulate major themes and future directions in feminist geography, as well as providing personal stories about experiences as faculty and students in the discipline. Two themes surfaced from the surveys and are referred to in several sections of this chapter. First, nearly all the respondents mentioned the personal dimensions of their involvement in feminist geography. This came out in explanations about how they came into the subfield due to their commitment to feminism, the support gained from professional networks in the subfield (especially important where female mentors are absent in departments and institutions), and the role of feminism in their own lives. A faculty member in a prominent geography department, Vera Chouinard, remarked that she "became interested as a graduate student but personal experiences of barriers to women in academic geography and beyond really fueled that interest." Another faculty member, Altha Cravey, stated that her employment as a construction worker for ten years was instrumental in shaping her academic research on the gendering of work. These links between personal experiences and academic endeavors among feminists relate to the topics outlined in the discussions below.

A second theme in many of the surveys was the relevance of feminist research to research topics within geography. For example, several people entered this area because they discovered the importance of feminism and gender to migration, identity politics, economic restructuring, and post-structuralism. One respondent commented about her interest as an undergraduate in population geography and the importance of gender to this area of study. "It occurred to me that population was a 'woman's issue'—you had to reflect women's experiences when discussing population rather than simply crunching numbers to develop demographic models." Indeed, decades of research in feminist geography demonstrates that women, gender, and feminism are critical aspects of most areas of geographical inquiry.

This chapter is organized into eight sections that address the development of feminist thought and the dynamic and interrelated nature of the topics outlined here. In the second section, feminist methodology is examined as an area that has made major contributions to the discipline as a whole. Feminist research includes a multitude of methods that challenge conventional approaches and are based on epistemological claims that value everyday experiences and knowledge. Gender and work is examined in the third section. This was an important topic among early feminists in the discipline and continues to advance geographic analyses of labor markets, workforce diversity, and the culture of work and economic organizations. The fourth section provides an overview of how feminist geography intersects with gender and development and analyses of third world globalization. This broad and expanding area in the subfield has made important contributions to environmental studies and political ecology.

The cultural dimensions of contemporary feminist geography are explored in the fifth section. Feminism has widened the scope of cultural geography, confronting masculinist perspectives on landscape and offering a feminist understanding of culture that recognizes cultural identity and social constructions of place. The sixth section examines perhaps one of the most interdisciplinary and dynamic approaches in feminist geography, gendered identity and the politics of difference and diversity. This analysis outlines how power and knowledge are reproduced through identity, gender, representation, and space, yet stresses the need to interrogate and break down these terms. The seventh section addresses pedagogy as part of the praxis of feminist geography. Many of us are engaged in teaching as a form of praxis within institutions of higher education, but also in our communities and social networks. The discussion highlights the crucial role of feminism in our pedagogical approaches.

The final section examines the overall position of feminist geography in the discipline and speculates on the future directions of this area of study. Our analysis reinforces the need to ground our work and not lose sight of the material basis of difference and inequality in our society. The multiple and dynamic discourses in the field of feminist geography are impacting the discipline and will continue to advance our thinking about social relations and spatial processes into the twenty-first century.

Methodological Breakthroughs and Boundaries

Feminist methodology is arguably one of the most important contributions of feminism to the discipline of geography. During the 1990s, influential essays and special sections in journals, books, and papers presented at numerous national and international conferences have significantly impacted methodological debates within the discipline. Indeed, feminist research that is sensitive to power relations and their influence on the research process has become a cornerstone of feminist thought and analysis (McDowell 1992). According to Mattingly and Falconer Al-Hindi (1995), the goal of feminist geography is to increase our understanding of gender and provide the knowledge useful to the struggle for gender equity. Feminist methodology is central to generating the knowledge for both these aims.

Four traditions within feminist geographic research are outlined here in regard to their dominant theoretical approaches, methodological concerns, and selected research topics (Table 47.1). As indicated in the text, these four traditions are by no means distinct phases, but represent a general shift of ideas in feminist geographical research. This section focuses on the methodological issues that have shaped feminist research in the 1990s, specifically, the subject/object relationship, the multiple modes or techniques employed in feminist research, and the politics of research.

Early feminist geography was based on feminist empiricism and a critique of positivist research methods that focus on objectivity and the privileging of certain forms of knowledge. This work provided important critiques of sexist biases or androcentrism in geographic research (Table 47.1). Monk and Hanson (1982), for example, claimed that conventional geographic research failed to account for women's experiences in studies of migration, travel patterns, employment trends, and livelihood strategies. These claims paralleled widespread challenges to positivist methodologies that separated the object from the subject of research to ensure neutral and value-free research.

In contrast, the underlying assumptions in feminist approaches highlight subjectivity and biases inherent in the research process (McDowell 1992; Staeheli and Lawson 1995). England (1994) notes that this type of research allows the personal and partial aspect of research to surface. Over the past two decades, feminist geography has increasingly challenged the basis of geographic knowledge in efforts to highlight how the

Table 47.1 *Traditions within feminist geographic research*

	Theoretical approaches	Methodological concerns	Selected research topics
Women in geography	"Counting" women The geography of women Feminist empiricism	Mapping spatial patterns of women's activities and status Challenging positivist research	Women in the city Women and employment Women and development
Socialist feminism	Socialist Feminism Marxism Gender and Development	Historical materialism Combining theory and praxis	Relationships between patriarchy and capitalism Spatial and social structures of home and workplace Gender roles in the third world
Third-world feminism/politics of difference	Poststructuralism Postcolonialism Race theory	Discourse analysis Participatory research Life histories Politics of fieldwork	Challenge essentialist and Eurocentric forms of knowledge Gender planning and development Differences across the life course
Feminism and "new" cultural geography	Queer theory Postmodernism Psychoanalytical theory Cultural representation	Positionality and reflexivity Textual analysis Narratives Ethnography	Situated knowledge Sexuality and space The body and identity politics Imaginary and symbolic spaces

Source: Johnston *et al.* 2000; Jones, Nast, and Roberts 1997; WGSG 1997.

relationship between power and epistemology is central to methodological discussions. Nast (1994: 55), for example, examines how the fundamental difference in the sociospatial organization of women's and men's lives "has fostered ways of knowing or epistemologies that are different from those of men." Thus, understanding the social dimensions of lived experiences and their relationship to how they inform research agendas and knowledge claims are crucial to feminist methodology.

The notion of subjectivity in feminist research raises critical questions about the basic assumptions of how knowledge and power influence the research process. Feminist geography does not draw a clear line between "researcher" and "researched," recognizing that researchers themselves are part of the social structure of the project (Women and Geography Study Group 1997). This is illustrated in Gilbert's (1994, 1998) work on survival strategies of low-income women and the role of networks in finding various social and economic support systems. During her fieldwork for this project, she confronted the sometimes "messy" relationship between herself as an academic researcher and the women she interviewed.

The social and spatial situatedness of researchers and the communities they are studying also emphasizes the importance of relationality and reflexivity in fieldwork (Table 47.1). Nagar's (1997) ethnographic research on Tanzanian Asians includes fieldwork in a community that she describes as a kaleidoscope of social sites of segregated, gendered, classed, raced, and communalized spaces. She claims that people's perceptions of her "continually shaped the structure and interpretation of the narratives that were produced in the course of my work" (Nagar 1997: 208). In contrast to masculinist research that defines who participates and what questions are posed, feminist methodology emphasizes the subjective and reflexive nature of research. For numerous scholars, this involves the participants by letting them guide the questioning and help define the terms of fieldwork (Rocheleau 1995; Townsend *et al.* 1995).

Additionally, the manner in which feminist research is conducted involves multiple methods and techniques that are appropriate to the social context and purpose of the research. Feminist geographers draw from a wide variety of qualitative and quantitative methods that have enriched our understanding of the complex conceptual and empirical aspects of research. A major contribution to this discussion was the collection of articles in *The Professional Geographer* (1994, 1995) that critically addressed the complexities and politics of "fieldwork" and examined the role of quantitative methodology in feminist geographic research. While the emphasis in much of feminist geography is on qualitative methods, McLafferty (1995: 438) states that "quantitative methods are well suited to describe and probe the main aspects of women's lives, to analyze spatial association, and to document spatial and temporal inequalities." In addition to quantitative analyses, excellent examples of qualitative methods such as ethnography (Dyck 1993; Nagar 1998), life histories (Townsend *et al.* 1995), intensive interviews (Gilbert 1998; Oberhauser 1997), and visual arts (Monk and Norwood 1987) demonstrate how a variety of approaches allows us to understand the materiality and meaning of women's experiences. The goal is to highlight the complementary nature of multiple research methods to search for and understand relevant patterns and trends.

Permeating the epistemology, subjectivity, and multiple methods of feminist research discussed above is the political agenda of feminist research (see Table 47.1). An implicit goal of this approach is to uncover and challenge power relations within the research process and underlying the topic of study. For example, Oberhauser (1997) examines the multiple aspects of power relations in the domestic sphere as both a site of income generation and a "field" of research. The social location of participants has been carefully addressed by many who seek to reveal and work around the "social terrain" of the field, if not to overcome the potentially exploitative nature of researching "the other" (Nast 1994). Gibson-Graham (1994) addresses this issue in her research on women in mining communities in central Queensland, Australia. She successfully employed members of the community to undertake workshops and assist in the overall project as a means of empowering and involving the subjects in the research process. Likewise, Rocheleau's (1995) project on resource management in the Dominican Republic incorporated participatory research to uncover the gendered structure of forestry projects. The important lesson in this type of research is the need to be aware of the social contexts and consequences in which research takes place. As numerous feminist projects demonstrate, the recursive relationships between gender, class, ethnicity, age, and sexuality structure dominant power relations and therefore, political struggles and activism (Katz and Monk 1993; Momsen and Kinnaird 1993; Radcliffe and Westwood 1993).

In sum, since 1990, significant advances have been made and lively debates have taken place concerning feminist geographic methodology. This area remains a cornerstone of feminist work as more attention is paid to diversity and critical representation of the subjects of research. There remains considerable work to be done, however. Hanson (1997) articulates the need to open up new horizons in feminist geography by devising approaches that will reveal the unexpected and not affirm what one already believes. As discussed above, these approaches may involve experimenting with new methods, exploring a combination of methods, or involving other collaborators to better understand what we know and, more importantly, how we came to know it.

The Geography of Feminist Economics

Feminist geography has also contributed to our understanding of the gendered nature of economic processes. This section traces the development of feminist economic geography from its roots in political economy to contemporary themes addressing the cultural dimensions of the workplace, embodied labor, and gender and economic restructuring. The discussion draws from three themes in feminist geography that have informed spatial analyses of the economy by challenging masculinist approaches to it, illuminating alternative social categories such as gender, race, and ethnicity for analyzing economic processes, and offering a range of methodologies for doing economic research.

First, feminist theory challenges masculinist approaches that neglect the role of gender in economic processes. Examples of gender biases in conventional economic geography reflect the aggregation of workforce data with the implicit assumption that male and female work experiences are similar. Increasing attention to empirical differences in men's and women's economic activities has affected theoretical and methodological approaches that no longer ignore the implications of gender in economic behavior (Hanson and Pratt 1995).

Second, feminist geography considers multiple social relations that impact and are impacted by the economy. Contemporary feminist theory, and post-structuralism in particular, have opened up economic analyses to gender as well as race, ethnicity, age, sexuality, and other forms of power and identity in society (Gilbert 1998; Peake 1993). These social constructions are critical in examining economic processes such as the location of factories, industrial restructuring, household income-generating strategies, and other economic processes.

Finally, feminist geography has influenced methodological issues concerning what questions are asked and how research is conducted in economic geography. For example, feminist research questions male-biased assumptions about who is the primary contributor to household and indeed national economies. The prevalence of activities outside the formal waged labor force demand that we shift our approach to research to obtain a better understanding of women and gender issues in the workforce and the household (Folbre 1994; Oberhauser 1995). Qualitative research methods such as intensive interviews, participant observation, and personal narratives are commonly used in feminist research and can be usefully applied in economic analyses. The remainder of this discussion draws from these themes to outline four trends in the development of feminist economic geography. These trends do not necessarily represent separate and consecutive phases, but reflect the general development of thinking in feminist geography.

During the 1970s and early 1980s, feminist empiricism examined the spatial distribution of women's work on national and global scales as a means of "counting" women and acknowledging their economic contribution to society (Table 47.1). Early empirical analyses described the discipline as androcentric due to the marginal role and status of females in the discipline (Lee and Schultz 1982) and the neglect of women's experiences in geographic inquiry (Mazey and Lee 1983; Monk and Hanson 1982). Gender and work was an important theme in this early literature because it highlighted women's inequality in the home, the workplace, and society as a whole (Bowlby *et al.* 1989; Monk and Hanson 1982).

The second trend had its roots in political economy or Marxist geography. Some of the earliest pieces on women and work appeared during the late 1970s and early 1980s and were couched solidly within a historical materialist framework with a strong focus on gender and employment in urban areas (Hayford 1974; Mackenzie and Rose 1983). The influence of socialist feminism was important during this period as patriarchy and capitalism were used to explain how gender roles and divisions of labor contribute to social and economic inequality (Mackenzie 1989; Bowlby *et al.* 1989).

Many of the respondents to the GPOW survey noted the importance of socialist feminism and Marxism to the development of ideas and the strengthening of community within feminist geography. Vera Chouinard, for example, remarked that, "socialist feminist theory has been important in encouraging analyses of gender as one of many relations of inequality and oppression in our societies. Personally, training in Marxist geography was important in encouraging consideration of how relations such as gender are actively contested and sometimes changed in the course of struggles against oppression." Other feminists trained in geography during the 1970s and early 1980s found a sense of camaraderie among socialist geographers that is evident in their work today. Damaris Rose notes the importance of this group during her graduate education and especially the presence of certain students who have gone on to significantly influence this subfield.

> I started graduate school . . . where I met Suzanne Mackenzie as another new graduate student who had come from BC and was doing "women's studies." I couldn't for the life of me figure out what this had to do with geography! . . . Within a short while she and I and several others were involved in a reading group on Marxist theory and the city, where among other things we were trying to grapple with extending concepts such as the reproduction of labour power.[1]

These accounts demonstrate that during the 1970s and 1980s, much of feminist geography was based in socialist geography and formed part of a critical school of thought in the discipline.

The third trend in feminist economic geography has extended social categories to include other forms of power and identity such as race, ethnicity, and sexuality. This trend addressed the complexity of gender and work in diverse geographical and socioeconomic contexts and led to a reconceptualization of spatial processes that occur outside the formal boundaries of public workplaces. In particular, these approaches were helpful in opening up social categories that allowed us to deconstruct dominant notions of women and work to include globalization, informal activities, and immigrant labor. Globalization and the increasing incorporation of women in the economies of developing regions accentuate the fluid boundaries of work and home and formal and informal economic activities (Faulkner and Lawson 1991; Hays-Mitchell 1993; Lawson 1995*a*; Oberhauser 1995). This area of feminist research includes a theoretically diverse and empirically rich literature that offers critical perspectives on the contested role of gender as it is mediated by race, ethnicity, sexuality, class, and geographical contexts (Marchand and Parpart 1995; Momsen and Kinnaird 1993).

The link between women's economic strategies in an era of global restructuring and the social construction of gender identity raises questions regarding the categories of analysis and universalizing assumptions about women's labor and gender relations. In a compelling analysis of Mexican women in the *maquiladora* industry, Wright (1997) examines the importance of gender identity in the work performed by these women. Representation of the Mexican "Woman" as a docile, submissive, and tradition-bound worker stems from the dominant Western discourse of third-world women. This discourse is challenged by feminists throughout the developing world who document how women engage in strategies of resistance and struggle to empower themselves economically (Beneria and Feldman 1992; Cravey 1997; Tinker 1990). A more detailed analysis of gender and development is presented in the fifth section of this chapter. Overall, this trend has helped redefine categories and explore alternative methodologies to research the social and spatial construction of gender and work in different cultural and geographical contexts.

The fourth trend examines the gendering of work through a focus on identity, culture, and representation. Feminist discussions about economic processes shifted in the early 1990s from materialist notions of production and consumption in the capitalist economy to social and cultural dimensions of workplace dynamics and the role of gender in the organization of firms (Hanson and Pratt 1995; McDowell 1997; McDowell and Court 1994). Increased focus on these aspects of economic activities coincided with the expansion of the service-based economy and growing attention to cultural practices of work (Halford and Savage 1997). What has been labeled the "cultural turn" in human geography has influenced feminist economic geography considerably through analyses of the ways in which masculinity and femininity are constructed in the workplace. Case studies and intensive fieldwork illustrate how the embodied nature of work influences the social construction of gender divisions of labor, employment experiences, and contemporary trends in industrialized and developing economies. For example, Massey's (1995, 1997) research on social dimensions of the high-technology workplace describes the construction of masculine space that embodies the elite concept of reason. She argues that Western economies and their places of "science" are socially constructed in ways that separate mental reason on the one hand and the materialities of the world on the other (Massey 1997). This recent shift in analyses of gender and work focuses on the embedded and embodied nature of economic processes in cultural contexts and institutions. Gender is an integral aspect of this cultural order of the economy and part of the process of restructuring in ways that are unique to feminist geography.

[1] The discipline lost a vibrant member when Suzanne Mackenzie died after a long struggle with cancer in the fall of 1998. Her work has had a tremendous impact on the subfield of feminist geography, particularly in the area of urban and economic geography.

In conclusion, beginning with its roots in empiricist research on women and work, feminist perspectives on economic geography offer recommendations for emancipatory, inclusive analyses of the social and spatial dimensions of economic processes. These analyses are grounded in the claim that women, people of color, and other disempowered members of society remain marginalized in the workforce due to socially constructed norms and dominant power relations. Multiple theoretical and methodological approaches are used to identify and explore the positions and institutional context of these power relations. Additionally, comparative research and exploration of parallel experiences of workers (especially women) in diverse cultural and political economic contexts is necessary. Globalization is breaking down boundaries and requiring more sophisticated knowledge of economic processes at multiple scales. Finally, contemporary feminist analyses of economic processes reveal how culture is embedded in and constitutive of gender and work. Analyses such as these will contribute to the emancipation of workers and the formation of institutions that are sensitive to multiple social differences among workers and the diverse forms of production that comprise economic strategies.

Geographic Perspectives on Gender and Development

Over the past decade, attention to women's and gender issues in third-world development and global arenas has generated substantial empirical information and contributed to theoretical perspectives on development, globalization, and post-colonialism. Building on important historical trends outlined below, this subfield of Gender and Development (GAD) has grown and evolved in complexity, thanks to the contributions of researchers and activists from a wide range of disciplines and applied venues. At this point, much of the work in GAD has been done by feminist economists, political scientists, ecologists, anthropologists, and development theorists. Nevertheless, feminist geographers have contributed to this field in important ways that are spelled out below in analyses of political and cultural ecology, globalization and industrialization, and post-colonial/transnational studies. Because a substantial amount of feminist work on development and third-world issues has been done by nongeographers, this chapter will also incorporate their work.

Background to Gender and Development

Beginning in the 1950s and 1960s, mainstream development theory focused on modernization and Westernization as a solution to underdevelopment. At this time, development scholars and practitioners viewed women as outside the central arena of development and primarily in need of health or other welfare assistance in their roles as mothers and housewives (Moser 1993). The concept of "women in development," or WID, had its origins with the landmark publication of Ester Boserup's *Woman's Role in Economic Development* (1970). On the heels of the second wave of the feminist movement in the West, the UN named 1975–85 the Decade for Women, and sponsored the first World Conference on Women in 1975 in Mexico City. Western feminists at this conference proposed "gender equity" as a global political rallying point, but third-world feminists heavily criticized this liberal feminist perspective for its failure to recognize a more complex intermingling of oppressions including class, race, and imperial forms of domination. Nevertheless, interest in WID spawned an outpouring of Western academic research into women in the third world and a renewed attention to the principle of basic needs as a development policy.

During the early 1980s, as the neoliberal paradigm and the IMF reasserted their influence as a result of the world debt crisis, development institutions shifted their attention to women's productive and reproductive roles for their potential contribution to and subsidization of economic growth. Examples of this type of development included aid for women's agricultural schemes and support for small-scale enterprise, micro-credit, and village banking schemes (Kabeer 1994; Moser 1993). Consequently, a strong and vocal group of both northern and southern academic and activist critics had emerged to protest the implications of neoliberal policy controlled by Western development and financial institutions. Some of these critics argued that WID efforts resulted in increasing the wealth and power of elites or that gender bias in some development programs affected women and children negatively (Afshar and Dennis 1992; Beneria and Feldman 1992; Gladwin 1991; Shiva 1989; Sparr 1994). Other critics argued that gender bias would actually hinder sustainable development (Elson 1995*a*; Jacobson 1992).

Part of the critique of mainstream development and WID concerned the need to decenter Eurocentric perspectives and conceptualize alternative analytical tools for development that would integrate marginalized peoples. These critics argued for the incorporation of perspectives from third-world women who were being directly affected by hegemonic development policies, and initiated the concept of empowerment in which voice, power, and development were integrally linked. Post-colonial or alternative development perspectives were initiated by a group of feminists in the South, when DAWN (Development Alternatives with Women for a New Era) published its manual *Development, Crises, and Alternative Visions* (Sen and Grown 1987).

Some of this literature conceptualized social relations of gender in which household or personal power relations were seen as institutional locations of analysis that lowered barriers to development. This concept would serve as the basis for a radical reorientation from women toward gender relations as an analytical tool and reflect the emergence of institutions as an analytical framework within development theory. This shift signaled a theoretical reorientation away from mainstream and WID development theory to a more progressive notion of gender and development that incorporated not only gender, but also alternative perspectives, methodologies, and epistemologies. (See Kabeer (1994) and Momsen (1991) for more detailed overviews.)

Contemporary Gender and Development

During the late 1980s, at least three conceptual approaches emerged that would set the stage for significant changes in WID/GAD research and theory. These approaches include environmental sustainability, economic restructuring and globalization, and post-structural/post-colonial epistemologies. The enormous growth of multidisciplinary feminist scholarship is paralleled by the increasing complexity of the subfield over the decade.

The first approach, environmental sustainability, spawned a sizable body of work on women and the environment (see Dankelman and Davidson 1988; Shiva 1989). This work has been generally categorized under the term Gender, Environment, and Development (GED) and includes general feminist overviews and critiques of environment and development (Braidotti *et al.* 1994; Harcourt 1994; Scott 1995; van den Hombergh 1993). Seagar's *Earth Follies* (1993) is an important geographic contribution to this approach.

One aspect of this literature is ecofeminism, which links the nature and causes of patriarchy and environmental destruction (Mies and Shiva 1993; Merchant 1980; Plumwood 1992; Warren, 1994). According to ecofeminists, these two forms of oppression have similar origins. Women are also seen as integral to the solution of environmental deterioration. In contrast, materialist environmental feminism critiques essentialist relations between women and nature and analyzes the political economy, institutional, and material aspects of gender in relation to the environment (Agarwal 1992, 1994; C. Jackson 1993, 1995; Nanda 1997).

Geographers have also been instrumental in the development of feminist political ecology that addresses gendered environmental relations. This perspective extends the analyses of cultural and political ecology by incorporating both the large-scale political economic forces of power and distribution and the household scale to better understand environmental degradation (Carney 1992, 1993; Schroeder 1993, 1997; Leach 1994; Rocheleau *et al.* 1996; Sachs 1996).

Economic restructuring and globalization have been a second key influence on GAD during the last decade. This has not only led to a shift in the meaning of development in which it is no longer possible to simply understand it as a linear process from third world to first world; but as a reconfiguration into a network of global systems, spatial relationships, cultures, and identities. Specific responses to globalization have been produced on multiple academic fronts. Feminist economists, for example, critique development policy and its negative impact on local lives while challenging the gender bias of economic development and globalization. The feminist alternative highlights these invisibilities and attempts to correct mistakes of mainstream economics (see Blumberg 1991; Folbre 1993, 1994; Sparr 1994; Agarwal 1994; Bakker 1994; Elson 1995*a*, *b*). Geographers such as Hart (1991, 1992), Radcliffe and Westwood (1993, 1996), Gibson-Graham (1996), and Cravey (1998) use a political economic framework to address the gendered nature of globalization and development.

Furthermore, feminist political scientists have elaborated a critique of the linkages between geopolitical/international relations and restructuring/development. This perspective examines the influence of hegemonic political forces on the integration of the global economy and development. According to Enloe (1990) and Peterson and Runyon (1993), this process is highly gendered. Scholars have also begun to highlight women's social movements and post-colonial resistance that are responding to global processes (Afshar 1996; Waylen 1996).

Third, feminist post-structural and post-colonial epistemologies have extended these feminist political and economic analyses of globalization and development by challenging the Western, colonial, and modernist presumptions underlying the origins, nature, and production of knowledge. Building on the earlier work of DAWN, new theoretical perspectives from third-world feminists and women of color in the West (Antrobus 1996; Mohanty *et al.* 1991) as well as post-structural perspectives (Marchand and Parpart 1995) have questioned mainstream and Western feminist epistemologies.

The feminist critique of science emerged during the mid-1980s as a strong epistemological alternative to modernist thinking (Hartsock 1983; Keller 1985; Haraway 1989, 1991; Harding 1992, 1998), and has served as a foundation for the valorization of traditional and situated knowledge and methodological alternatives. Feminist geographers have been active in this arena, beginning with several edited collections that are part of the groundbreaking series *International Studies of Women and Place*, edited by Momsen and Monk. (See Momsen and Kinnaird 1993; Radcliffe and Westwood 1993; Townsend 1995; and Fenster 1999.)

The Future of Gender and Development

Clearly, the last decade has witnessed enormous changes in the structure of global economic development and environmental studies that have accompanied an explosion of research and practice around gender and deveopment. This phenomenon has begun to include the voices of those who have historically been left out of development discourses or were misrepresented as a homogenous whole. As a result of these changes, questions about the very definition of development have evolved; conversations surrounding methodologies of research have expanded; and relations of power have been decentered from the global North. These events bode well for southern participants in the process, but they leave questions about the role of American and other Western geographers. On one hand, geographers have been somewhat underrepresented in the interdisciplinary research on gender and development. On the other, the evolution of the global arena during the last decade has opened exciting new research spaces for geographers that include border zones and migration, networks and transnationalism, interrelationships between the global economy and environment, and knowledge of local places. These offer complex new arenas for investigation in which feminist geographers can make important contributions to both GAD and mainstream development and globalization theory.

Gender and Cultural Geography

Geographers who study culture employ a wide variety of theoretical approaches and are situated in a wide range of topics within human geography. This discussion explores how some of this research relates to gender by comparing the perspectives of traditional and contemporary cultural geography. The first part of this section addresses how these perspectives approach the topic of gender and culture. Since gender was not considered to be an important topic by many traditional adherents to cultural geography, one of the first topics to be considered by feminist cultural geographers was the identification of women as a group of people who are different from men. In the context of historical geography, Kay (1991) argues that while historical geographers are not necessarily prejudiced against women, many are unsure as to how to incorporate information on women into their research. Using the narratives of three frontier women, Kay presents a more balanced impression of women and men in studies of regional economies and landscape modification. In a similar vein, Mikesell (1994) admitted that cultural geography had traditionally ignored more than half the human population. Pulido (1997) also discusses the problem of representation for the historically invisible in her research on identities articulated by low-income women of color involved in environmental justice struggles.

A second way in which traditional cultural geographers and adherents to "new" cultural geography differ is based upon different ways of viewing the landscape. Material landscape studies are one of the principle foci in cultural geography (Duncan and Ley 1993; Craig 1998). These studies are seen by others as implying that there is a neutral way of viewing a landscape. Many feminist geographers who study landscapes would disagree with this implication, saying that what is presented as "neutral" is often one form of a white, middle-class, male gaze (Monk 1992; Rose 1994). As Rose (1993: 87) points out, "more recent work on landscape has begun to question the visuality of traditional cultural geography, however, as part of a wider critique of the latter's neglect of the power relations within which landscapes are embedded."

Traditional cultural geography also differs from other adherents of cultural geography in its use of dualisms or dichotomies to describe cultural spaces or behavior. Traditional cultural geographers sometimes view the natural landscape as "maternal" and consider nature and culture as a dualism (Rose 1993). But by using gender as a social and/or spatial construct, feminist geographers have embarked on several different but related courses.

Some, for example, will look at femininity or masculinity as cultural identities that are place-specific and entail different spatial behaviors.

In the 1990s, questions of politics and representation, as well as questions of dualism were addressed by feminist geographers writing about culture. Some researchers tried to spatialize the dualism already mentioned between what was seen as a masculine disembodied reason and feminine embodiment and emotion (Bondi 1992). Duncan's (1996) *BodySpace*, for example, highlights new directions in research on the role of space and place in the performance of gender and sexuality. Other studies that critically examine politics and representation as well as dualisms surrounding culture and nature are Anderson and Gale (1992), Monk (1992), Duncan and Ley (1993), and Bondi and Domosh (1998). Monk (1992: 122) has contributed to these studies through her discussions of the "invisible ideologies and social conditioning that support distinct gender roles, the power inequalities between men and women; and the differing meanings that men and women attach to their surroundings." Finally, the dualisms that define cultural aspects of gender roles are often examined in urban arenas. (For examples see Monk and Norwood 1987; Rose 1993; and Valentine 1997.) Bondi (1992), for example, discusses gender symbols in contemporary urban landscapes, while Peake (1993) looks at the ways in which the relations of race, class, gender, and sexuality are played out across urban social space.

Geographers writing about gender and culture tend to do so in a way that emphasizes the interwoven nature of gender, class, space, and place and draws from historical and contemporary case studies. Bondi and Domosh (1998) discuss the doctrine of separate spheres for men and women in both a spatial and ideological context, with particular attention paid to the changing contours of the relationship between gender divisions and distinctions between public and private spaces. In this sophisticated essay, the authors examine one of the classic themes and problems of feminist literature, the conflicts between women's use of public and private space.

As the new century unfolds, geographers are debating the way in which, or even in some cases if, the subject of gender and culture should be analyzed. Although geographers with an interest in gender are increasingly addressing cultural topics, much of this research is not always recognized, much less endorsed, by all adherents of cultural geography. The following section examines how feminist geography has advanced our understanding of contemporary cultural geography through its incorporation of gender, identity, and representation in social spaces.

Gender, Identity, and Space

Feminist geography emphasizes the need to break down terms such as difference, patriarchy, resistance, and space by exploring the ways in which power and knowledge are (re)produced through these concepts (Foord and Gregson 1986; McDowell 1986; Massey 1991; Blum and Nast 1996; Jones *et al.* 1997). This section outlines some of the key issues concerning gender, identity, and representation in the 1980s and the incorporation of identity politics in the 1990s. The discussion focuses on the growing literature addressing power/space/identity relations by analyzing important conceptual frameworks that have emerged in feminist geography, especially through the integration of post-colonial, post-structural, and gay and lesbian theories of identity and space.

Identity in Context

The intersections between identity and space have been of increasing interest to feminist geographers (Natter and Jones 1993; Massey 1994; Pile and Thrift 1995). During the 1980s, geographers in the US and beyond explored what was meant by the categories "woman," "housewife," "employee," "private/public" in an effort to interrogate the assumptions behind those terms and their geographies (Mackenzie and Rose 1983; Massey 1984; McDowell 1986; Pratt and Hanson 1995). As outlined in Table 47.1, socialist feminist geographers explored the relationship between capitalism and patriarchy in an effort to understand how both factors were important in explaining women's oppression (e.g. Foord and Gregson 1986; McDowell 1986). While radical feminists emphasized the differences between women and men, socialist feminists interrogated the differences, based not only on gender but also along class lines, particularly in terms of a gendered division of labor. More recently, feminist geographers have increasingly questioned the focus on class and gender as key modes of inquiry and have called for a broader analysis of what "difference" means. Drawing on the work of a range of theorists, feminist geographers have argued that a focus on class and gender inadequately explained important differences (and alliances) that also coexist in relation to race, ethnicity, nationality, and sexuality. In particular, feminist geographers have critiqued the idea of gendered identities and spaces as "natural" and ontologically sealed or that sociospatial positions are fixed and that a geographic truth can be found (Rose 1993).

This critique of essentialist and unified notions of gender and space also contributes to analyses of the links between identity and methodology. As outlined above, feminist methodology attempts to break down separations between the identities of the researcher and the researched in order to convey the ways in which they are intertwined (often in very specific power relations) and to illustrate the ways in which methods and theories shape the means by which knowledge, spaces, and identities are constructed (Domosh 1989; Nast 1994; Dyck 1995). Feminist geographers have built on critiques of class, gender, and patriarchy by breaking down the categories of women/men by race, gender, class, sexuality, and nationality. These critiques highlight the constantly shifting nature of these identities across categories and at different moments in different places (Bondi 1993; England 1994; Pratt and Hanson 1994).

Difference and Deconstruction

Since the late 1980s, the influence of postmodernism and post-structuralism can be clearly documented in feminist research. Postmodernism has helped shift the focus in this area from grand theory and meta-narratives to multiple positions and an increased attention toward changing identities in various spaces (Soja 1989; Butler 1990; Haraway 1991). As Vera Chouinard comments, "Postmodern debates helped fuel interest in cultural representations of women and women's lives, albeit from critical feminist perspectives primarily. Theories of the body and social difference are enriching our understanding of women's oppression and resistance, for example, drawing attention to the complex ways in which women's bodies are regulated in specific places."

A considerable amount of debate in feminist geography has also focused on the dangers of a depoliticized postmodern geography that fails fully to analyze the situated knowledges of various marginalized groups, e.g. poor women, people with disabilities, children, gays and lesbians, and women of color (Deutsche 1991; Massey 1991; Pile and Rose 1992; Gibson-Graham 1994). Feminist geographers are increasingly asking questions about what we mean by difference, how we represent fluid identities, and what options are there for feminist politics. The move towards a fuller understanding of identity and space has reflected an effort to shift away from representations of the two as static and separate, and toward an exploration of the ways in which identities and spaces are discursively (re)created. At the same time, within feminist geography there have been concerns about encouraging effective (and nuanced) feminist praxis while broadening the impact of feminist theories within the discipline. Barbara Morehouse suggests, "Thinking from the perspective of the dilemmas of postmodernism, I see a major point of contention being how to resolve the tensions between the tendency to pursue an exclusivist focus in feminist geography versus a need to place feminist geography within larger contexts, as one of many perspectives and forms of experience, without losing the power of feminist critiques."

More recently, feminist geographers have increasingly shifted their focus towards post-structuralist modes of inquiry (Mouffe 1992; Natter and Jones 1993; Jones and Moss 1995; Lawson 1995*b*; Pile and Thrift 1995). Post-structuralist research has involved a thorough analysis of the use of language and the way in which discourse becomes a key means to communicate and recreate identities and space. Representation, therefore, becomes the mediator and medium through which identities and spaces are (re)produced. In addition, feminist post-structuralists emphasize the impossibility of a unified subject, a "fixed" means of knowing, or the separation of the material and representational. Post-structuralist research does have some commonality with postmodernism in that both use deconstruction as one of their keys modes of analysis, i.e. they critique truth claims embedded within a narrative to unearth inconsistencies and contradictions that are not always explicitly represented. This work has drawn extensively from research outside the discipline, particularly in the area of cultural studies, the Frankfurt School, and social theory more generally. Post-structuralism has had considerable influence within geography in the study of identity politics, spatiality, power, and representation (P. Jackson 1991; Bondi 1992; Knopp 1992; Natter and Jones 1997; Nast 1998). This approach has also impacted studies that critically examine histories of geographic thought (Domosh 1991; Pratt 1992; Rose 1993), symbolic spaces (Martin and Kryst 1998), art/media (Monk and Norwood 1987; Zonn and Aitken 1994; Monk 1997*a*), and geographies of resistance and activism (Meono-Picado 1997; Moss 1997).

Nationalism, Identity, and Transnationalism

In addition to post-structuralism, post-colonialism has dramatically influenced feminist geography. In particular, research by women of color and third-world feminists such as Kandiyoti (1991), Mohanty *et al.* (1991), and Alexander and Mohanty (1997) challenge Eurocentrism within feminist research through analyses that address the "invisibility" of whiteness, the significance of diasporic identities, and the pervasiveness of neocolonial relations in a variety of contexts (Said 1978; Bhabha 1986; Spivak 1988; Hall 1990; hooks 1991). Postcolonial feminist geography, therefore, has broadened its focus beyond the binary division of male/female towards the social construction of gender identities/relations across a variety of contexts, and by interweaving the importance of race, ethnicity, nationality, sexual orientation, colonialism, culture, and class (e.g. Valentine 1993; Blunt 1994; Kobayashi and Peake 1994; Morin 1995; Mills 1996; Radcliffe and Westwood 1996).

This literature also examines the uneven power relations between women in different social positions by drawing on a range of cultural and social theories (including post-structuralism) that analyze multiple subjectivities and sites of gendered identities (e.g. Mohanty 1988; McDowell 1991; hooks 1991; Katz 1992). A key body of work within this area addresses the gendered nature of national identities and transnational links within gender and geography (Nash 1994; Nagar 1997) that includes a reevaluation of the discipline's imperial past, e.g. through an examination of travel writing and historiography. Feminist historical geography plays an important role in interrogating the ways in which gendered geographies have been reformulated in various contexts. In this literature, increased attention is given to the various social and physical "borders" that are created and reproduced to maintain specific gendered and racialized identities through immigration, nationalism, the impacts of colonialism, and human rights (Blunt 1994; Blunt and Rose 1994; Fenster 1998; Mains 2000*a*; Morin and Guelke 1998). Overall, post-colonialism provides insight to the means by which Western subjectivities are constructed, the uneven power relations within and beyond the academy, and the ways in which gender, ethnicity, and nationalism are reproduced through a variety of sociospatial relations.

Sexuality, Identity, and Spaces of the Body

As borders around local, national, and international identities have come under closer scrutiny, so too have spaces of the body. A growing amount of feminist geography has paid close attention to the ways in which certain bodies are marked as being different or marginal and thus associated with particular places, while others are deemed normal and neutral and in many ways omnipresent. Such representations are reflected in "places through the body" (Nast and Pile 1998), e.g. the ways in which bodies become the sites in which spatial, physical, and mental ability, gender, and racial identities are mapped out, reinscribed, and challenged (Elder 1995; Duncan 1996; Dorn 1998; Nast and Pile 1998; Mains 2000*b*). In addition, feminist geographers have paid increasing attention to the intersections of sexuality, gender, and space, by exploring the means through which sexuality is associated with the body, while at the same time being publicly monitored (Bell and Valentine 1995; Johnston 1996). The body has become an important site of analysis for feminist geography because it has been increasingly viewed as a social and political in addition to biological space (the latter has also been shown to be socially (re)constructed) (Gibson-Graham 1996; Dorn 1998; Nast and Pile 1998).

Critical analyses of masculinity and representation are increasingly included in analyses of gay and lesbian identities in urban and rural spaces (Berg 1994; Knopp 1995; Rothenberg 1995), and heterosexuality and psychoanalysis (Blum 1998; Lukinbeal and Aitken 1998; Nast 1998). Drawing on the work of Foucault and queer studies, feminist geographers have explored the relationships between sexuality and space to reveal a vast array of multiple identities and negotiations of identity and geography. For example, while exploring the ways in which discussions of capitalism and market forces have been reproduced, Gibson-Graham (1996) highlight the gendered nature of these discourses, and the usefulness of queer theories for breaking down existing hegemonic forms of analysis. Drawing on the work of Eve Sedgwick, Gibson-Graham argues that the tools for undertaking a rethinking of capitalism and economic development are already present in analyses of sexual morphology. "For queer theorists, sexual identity is not automatically derived from certain organs or practices or genders but is instead a space of transivity" which offer opportunities to subvert "monolithic representations of capitalism" (Gibson-Graham 1996: 140). Therefore, although often regarded as private, feminist geographers have shown the ways in which sexuality is frequently monitored (and disciplined) in very public relations and spaces. Nast and Pile (1998: 3) comment, "The body is both mobile and channeled, both fluid and fixed, into places," and as such, they illustrate the uniqueness and commonalities in the way we experience and reproduce space through a negotiation of bodies and places.

The Body and Medical Geography

Related to this emphasis on the body is the growing literature in feminist medical geography that examines both the metaphorical and physical challenges of the body and disease. While medical research is quite new to the field of geography, feminist medical geography is an even newer field that has been gaining momentum since the 1980s. The majority of medical geographers focusing on women's health has been concerned with reproduction, infant mortality, and child survival (Davis Lewis and Kieffer 1994). Specific examples of feminist geographers who address issues in the medical field include McLafferty and Tempalski's (1995) focus on the shifting geographic patterns of women's reproductive health in New York City and Laws' (1995) analysis of how biological bodies are socially encoded through state policies, popular culture, and academic discourses. Drawing from postmodernism, Longhurst (1994) uses the figurative Pregnant Woman to question the epistemology and ontology of geographical discourse. She problematizes the constructionist and essentialist feminist approaches to the body and aims to make questions of the body explicit in geography.

One of the most important topics that medical geographers have been concerned with is the spatial diffusion and socioeconomic characteristics of people infected with the HIV virus and AIDS. Most of the work by medical geographers has been concerned with mapping disease transmission and patterns (see Gould 1991; Shannon and Pyle 1989). While this is important to know for public health policy, feminist medical geographers have gone a step further to analyze the social production and interpretation of women's health and illness with regard to HIV/AIDS. Kearns (1996) states that geographers are in a good position to analyze the social repercussions of the epidemic because of their willingness to integrate social theory concepts such as the structure/agency debate and to engage with public health policy. Brown (1995), for example, discusses the geographies of exclusion when examining AIDS discourse and shows the importance of structures and individual agency in his case study. Kearns (1996) acknowledges the challenge for medical geographers to integrate global, national, and state-level analysis of virus distribution with the local specificity of individual life experience. This challenge has been increasingly addressed by feminist geographers who utilize qualitative methods to understand the effects of the AIDS epidemic on women (Wilton 1996; Del Casino 2001), women's vulnerability (Craddock 2000), and the variation of gendered experience internationally (Asthana 1996). Many of the geographers who have taken up issues of individual experience of HIV and HIV prevention programs for women have conducted research in Africa, Asia, and Australia. Perhaps the most notable in this field are Asthana's (1996) work with commercial sex workers in India.

While feminist research in medical geography lags behind contributions made by feminist medical anthropologists, sociologists, and cultural studies scholars, a growing number of geographers are concerned with issues of representation, the female body, and health-care policy. For example, *New Geographies of Women's Health*, edited by Dyck *et al.* (2001) brings together feminist scholars to problematize the intersections of race, ethnicity, class, and gender as they pertain to women's health. This is the first volume devoted entirely to feminist medical geography. It includes such topics as globalization and women's health, health-care access, the embodiment of health and illness, perceptions, and the role of place. In addition, medical geographers are integrating traditional geographic research methods with qualitative methods to create a more comprehensive understanding of disease and health (Barron McBride 1993; Dyck and Kearns 1995). Drawing from this integrative approach, Davis Lewis and Kieffer (1994) call for the expansion of research to include not only women's reproductive health but also the quality of life throughout the entire life cycle in a context that addresses physical, mental, social, and economic health.

Challenges for the Future: (Re)Presenting Gender and Space

In sum, feminist geographers are increasingly challenging dominant concepts of representation and space. The conceptual frameworks and methodological approaches involved in this type of critical research are complex and multiple. Important issues raised in the broad area of gender, identity, and space include the interconnection of activist and academic identities, using feminist identities and spaces to retheorize hegemonic geographies, and interrogating geographies of both the left and the right in ways that challenge the reproduction of these categories. These are not simple projects, but some of the few that offer possibilities for intriguing and provocative feminist geographies in the future.

Feminist Pedagogy and Teaching in Geography

"Democracy is an extremely important element in the way I structure all classes. I like to lay bare power in the classroom and deal with it on an ongoing basis if nothing else than to show how power works on the ground," (Pamela Moss, survey response). Understanding power relations in society and applying this knowledge to classroom teaching through feminist pedagogy is a crucial dimension of feminist geography. As demonstrated in Moss's quote, feminists not only teach about gender and unequal power relations, they apply these concepts to their pedagogical approach and interaction with students in the classroom. This section briefly reviews some of the general issues relating to teaching and pedagogy in feminist geography. Despite the diverse institutional settings and backgrounds of feminist geographers in respect to curriculum and instruction, several common concerns and issues impact our approach to the classroom and students. This discussion highlights research in this area and draws from responses to the survey distributed on the GPOW listserv in the analysis of a feminist commitment to open, inclusive styles of learning in ways that actively engage students in the learning process.

Feminist pedagogy and teaching occupies a significant proportion of our professional lives, yet until recently it has not been a major area of research in gender geography. The majority of work that has been done on this topic addresses the inclusion of gender, sexuality, and feminism in geography curricula (Monk 1985; McDowell 1997; Knopp 1999), gender and student learning in the classroom (Nairn 1997), and the use of feminist methodology for student projects (Madge 1994; Raghuram *et al.* 1998). Additionally, research on geography curricula has taken the form of multidisciplinary projects such as the National Council for Geographic Education's Finding a Way project funded by National Science Foundation. This extensive project educates secondary education teachers about gender and multicultural issues in the geography curriculum (Monk 1997; Sanders 1999). While research such as this has been useful in advancing geographic education, especially for women and minorities, a comprehensive and critical analysis of feminist pedagogy in geography is long overdue. Consequently, many feminist geographers draw from the growing literature on pedagogy in women's studies to inform their teaching methods and approach.

Two topics will be explored in this section. The first topic is a critical analysis of the institutional frameworks in which we practice feminist pedagogy and teaching. These institutions include the discipline of geography, the academic setting of universities or colleges, and geography curricula. The discussion examines the status of women, feminism, and gender studies in each of these institutions. The second topic addresses ways in which feminism is applied in the classroom. The analysis focuses on implementing inclusive, participatory approaches in feminist teaching within these different institutional contexts. The discussion includes a critical evaluation of feminist pedagogy and provides suggestions for future research.

Institutional Barriers to and Opportunities for Feminist Pedagogy

Improving the status of women in geography is a crucial factor in advancing feminist teaching and pedagogy. This discussion focuses on the range of institutional barriers to and opportunities for feminist pedagogy in geography, starting with the status of women in the discipline. Figure 47.1 represents the status of women in the Association of American Geographers since 1974. Four measures depicting the status of women include their membership in the AAG, level of education, employment in universities, and their status as students. Overall, women comprise a relatively small, but increasing proportion of geographers in all these categories. The proportion of AAG members who are women has increased significantly from 15.4 per cent in 1974 to nearly 28.9 per cent in 1997 (AAG 1998). The percentage of female geographers who are employed in universities, however, has increased slowly, from 24.4 per cent in 1981 to 31.7 per cent in 1997 compared to 49 per cent of male geographers. This is partly due to the fact that a greater proportion of females in this organization are students or work in private industry. In fact, the percentage of female students is not only increasing at a faster rate than that of female faculty, it is also growing faster than that of male students (AAG 1998).

In general, the presence of female faculty members is important to mentor students in the discipline, in addition to promoting curriculum changes and introducing more inclusive approaches to teaching. The need to improve the status of women as a means of promoting feminist issues in the discipline and academic institutions was found among several of the survey respondents. Jennifer Hyndman noted the difference between the department where she obtained her degree and her current department where she is a faculty member. She states, "I work in an interdisciplinary department where well over 50 per cent of the faculty are women and 30 per cent are people of color. It is an unusual but dynamic and comfortable mix." In contrast, the climate in departments with no women or only a few is more likely to be uncomfortable or alienating for women, as well as for gays and lesbians and minorities.

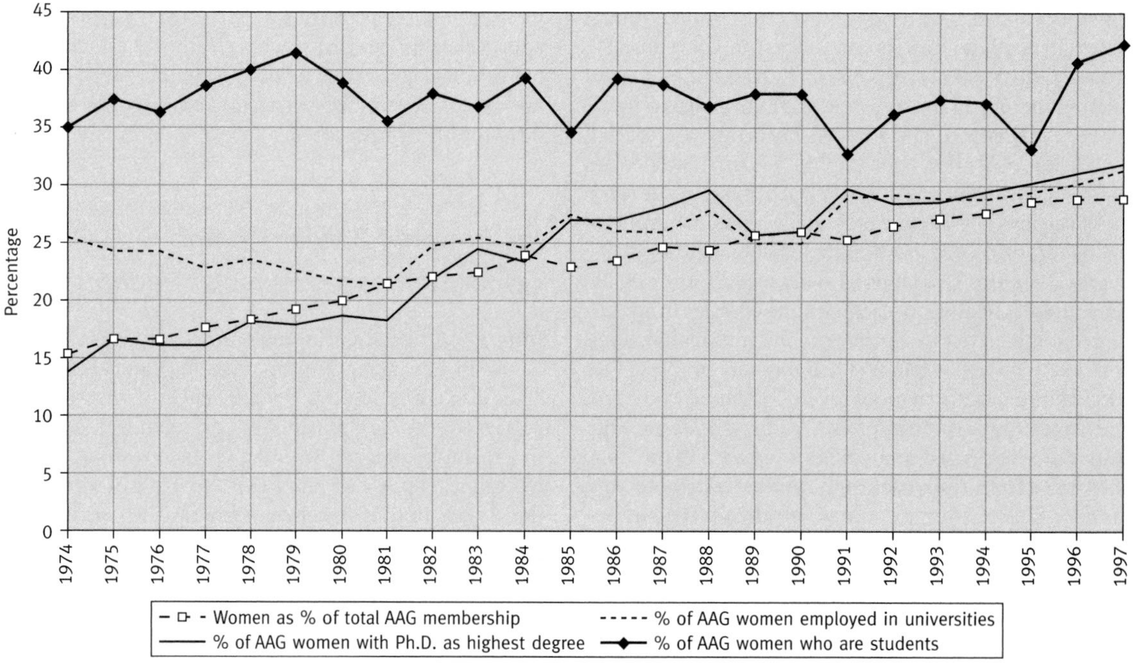

Fig. 47.1 The status of women in the AAG, 1974–97

Focusing on women, however, overlooks the potential contributions of men who encourage females to apply for grants, mentor them, or provide other means of support for their active participation in the discipline. For example, one respondent who wished to remain anonymous attributed her early interest in women in development to two male professors in her department during a time when there were few women. She credits these professors with being open-minded enough to consider a feminist pursuit when this was not at all common.

Finally, although women may be hired as faculty, as has been the case in many top departments in the country, attitudes towards women, feminism, and diversity are sometimes slow to change. The survey question about how departments or universities have changed as a result of increasing awareness of gender issues in the discipline drew this cautious response from Deborah Martin:

> There have certainly been changes in how fellowships and teaching assistantships are awarded, in recognition of the influence of the "boy's network" on past decisions and attempts to reduce gender bias. I think that there are still plenty of personal biases against certain types of gender and post-structuralist research, but the institutionalized bias against it is much reduced.

Thus, while the numbers reflect a growing presence of women, feminist approaches in teaching and research face continued resistance among certain faculty and students.

Curriculum is another important area where feminist approaches have changed the content of and material used in courses in both secondary and higher education. The Finding A Way project mentioned above introduces teaching material to secondary geography teachers that incorporates strategies to stimulate young women of diverse racial and ethnic backgrounds to study geography and improve their academic performance (Monk 1997*b*). In higher education, gender in geography is not considered a central part of the curriculum and depends very much on faculty and students in the programs. Despite the ad-hoc incorporation of gender into the curriculum, several major geography departments offer courses in feminist geography at the undergraduate if not graduate level. These courses have evolved from general overviews of gender geography to critical analyses of specific topics within the subfield. For example, courses are more frequently offered with titles such as Women in the Global Factory, Environmental Feminism, and Gender and Human Rights.

Additionally, more texts are being written for use in feminist geography courses. One series that was mentioned by several survey respondents as contributing significantly to the subfield is Routledge's *International*

Studies of Women and Place, edited by Janet Momsen and Janice Monk. Other books that are useful in gender geography courses are *Thresholds in Feminist Geography* by Jones *et al.* (1997) and *Feminist Geographies* by the Women and Geography Study Group (1997). The next decade will certainly see the options for teaching materials expand, particularly on the Internet where a wide variety of gender and feminist resources is available and will be increasingly incorporated into courses.[2]

The presence of a single or a few courses in gender geography, however, only partially fulfills the objective of feminist teaching and pedagogy. The goal of feminist geography is to incorporate gender in nearly every course since it is an integral part of political, urban, economic, and other subfields of geography. Again, this goal faces resistance by some members of the discipline. As Deborah Martin comments, "there is recognition that gender issues need to be incorporated into our curriculum, although the completeness of that enterprise is yet to be seen (not everyone has taken gender perspectives on board in their classroom; some view it as the domain and responsibility of a few profs in a few classes)."

Expanding pedagogical discussions to include sexuality and queer studies as well as gender has only recently gained attention in academic arenas. For example, a special issue of *The Journal of Geography in Higher Education* examines the implications of using queer theory in the geography classroom and in other forms of professional practice (Knopp 1999). The authors analyze how dominant cultural values enter the classroom in both positive and negative ways as social relations and institutions are examined from critical, anti-heterosexist perspectives (Elder 1999; England 1999). Additionally, students are sometimes resistant to the inclusion of gender and sexuality in courses that are not perceived as relevant to feminist analyses. For example, a faculty member who responded to the survey commented that absence among males was considerably higher than usual when she covered the topic of gender in her political geography course. In sum, although the chilly climate is improving for feminists in the institutional frameworks of the discipline, colleges, universities, and geography curricula, there is continued resistance to gender and feminist approaches in the academy.

Practicing Feminism Inside and Outside the Classroom

Feminist pedagogy is an issue that engages many of us both inside and outside the classroom. The tendency for feminists to adopt an open and inclusive approach to teaching is apparent in discussions by Nairn (1997) and Raghuram *et al.* (1998) who view the classroom as a contested site in which curriculum content and participation must be created with female students in mind. The importance of these themes is illustrated in several responses to the survey question about how current debates in gender geography and feminism influence teaching and/or pedagogical approaches. In survey responses, Carolyn Cartier states that feminism has substantially influenced her teaching "by problematizing and not abusing power relations, in being a supportive mentor, in being really accessible, and self-reflexive in how I comport myself with my students," and Kim England remarks,

> I am very committed to student-centered learning and encouraging all students to speak in my class. I try very much to be a facilitator rather than the all-knowing professor. This requires lots of advanced organizing and planning, and also requires a great deal of "tongue biting" on my part during some classes! My teaching evaluations usually indicate that the students really like this approach and appreciate the opportunity to shape their own learning. I use inclusive language and wherever possible use examples that draw on difference and diversity.
>
> I try to build self-awareness, or "consciousness raising" into my classroom, as an active-learning pedagogy that's informed by feminism. I don't want students to think that whatever they think is valid regardless of evidence, but I want their experiences to be legitimated in the classroom and to be part of critical thinking and analysis. (Deborah Martin, survey response)

These quotes underscore the ethical and political concerns that permeate feminist teaching, research, and other professional activities. Specifically, they represent a particular approach to feminist pedagogy and teaching that is sensitive to hierarchical power relations and supportive of students' direct involvement in the learning process (McDowell 1994; Monk 1996).

An important aspect of feminist pedagogy is the mentoring that occurs outside the classroom and the expectation that feminist teachers are accessible to students. Although the rewards are great, there are also personal

[2] In the 1990s, several online resources have been developed that provide information and facilitate communication among those interested in feminist geography. The Department of Geography at the University of Kentucky maintains a feminist geography Internet discussion group (GEOGFEM <geogfem@lsv.uky.edu>, last accessed February 2003). In addition, an online feminist geography bibliography is available (<http://www.emporia.edu/socsci/fembib/index.htm>, last accessed February 2003.). Finally, a GPOW website is online at <http://www.masu.nodak.edu/divisions/hssdiv/meartz/gpow/gpow.htm>, last accessed February 2003.

sacrifices involved in this type of commitment. Pamela Moss expressed concern over the personal toll that being a feminist in geography is taking on women in the discipline as they continue to battle for acceptance and recognition. Another respondent stated

> As I have increasingly become aware of, and struggled to resist, the masculinist practices within the discipline itself, I've found myself in a series of ethical dilemmas. I've found that being a feminist in geography is often a contradictory experience. For me, it was quite a surprising realization, because I had previously thought the construction industry, which I worked in for many years, was far more sexist than academia could ever be. (Altha Cravey)

In addition to the dominance of masculinist practices in academia that can be alienating, feminists are experiencing the negative impacts of the academy taking on a more pragmatic, business approach to education and research.

In conclusion, this section addresses the institutional aspects of our teaching and pedagogy that include the status of women in geography, the academic institutions where we teach and do research, and curriculum in feminist geography. While some research has looked at gender issues in education and curriculum, feminist pedagogy remains a relatively understudied area of research in feminist geography. The strategies we have developed in our teaching, however, have been informed by our research and broader feminist beliefs in addressing issues that challenge barriers to women's empowerment inside and outside academia.

Future Directions for Feminist Geographies

As can be seen in the discussion above, feminist geography has expanded tremendously during the last decade. In addition to its separate identity as a geographical subfield, its presence has grown to varying degrees within traditional specialty areas of geography such as economic, cultural, historical, urban, and development geographies, among others. Feminist geography is central to the initiation and expansion of a range of new approaches and topics such as post-colonial theory, identity/representation/space, feminist political ecology, and geographies of the body. Furthermore, it plays a central role in redefining what it means "to do" geographical research, both methodologically and analytically.

Yet, each of these categories highlights gaps and opportunities for future research and the realization of feminist goals. For example, within traditional cultural geography, and despite extensive feminist work in the "new" cultural geography, there is still a need for work that will allow feminist theory to be fully recognized, valued, and incorporated into that area. Within economic geography, continued feminist analysis and activism can perhaps contribute to the formation of emancipatory institutions within the arena of work. Within development and globalization, new global spatial forms present an enormous opportunity for geographers to influence the directions of the production of knowledge and constructions of political action, both within geography and within the cross- and multidisciplinary setting.

Within other traditional specialty areas, broader opportunities exist for feminist analysis, for example, within political geography and rural studies. Within GIS, which is one of the fastest-growing areas in geography, concern for social issues has emerged within the last decade, but without much attention by feminist geographers. Another underrepresented arena for feminist geographers is the domain of physical geography. Although there has been little cross-fertilization between human and physical geography, an important entry point for this project exists within cultural and political ecologies, which address environment/society relations, and especially within various feminist environmentalisms, including feminist political ecology.

Within feminist geography, several concerns and issues remain open for attention and debate. First, feminist geographers emphasize the need to rethink gendered geographies as meaning more than women's places, which is reflected in a gradually growing amount of research addressing spaces of masculinity, as well as an increasing emphasis on multiple, intersecting vectors of identity within feminist theory of difference. This reorientation introduces the crucial question of the prioritization of gender within the feminist frame of analysis. Another key issue that has surfaced along with the growing array of theoretical perspectives in feminist geography is the challenge of providing a "common ground" or nodal points from which constructive dialog among feminist geographers can continue. In general, both within specialty areas and across geography, there is a need for more comparative research, on an intradisciplinary level, cross-culturally, and across multiple spatial scales. A final crucial debate within feminist geography addresses the ghettoization of feminist geography. Should it maintain a separate presence, both conceptually and within its own specialty group; or should it ultimately permeate all subfields, curriculums and courses?

Within the professional arena, we have highlighted the importance of the academic institution as a venue for accomplishing broader feminist objectives, and have seen that important progress has been made in that arena. Yet, important work still remains to be accomplished around the status of women, feminism, and diversity within the institutional framework. This includes expanding the number of positions available to women and feminists; and expanding gender and feminist approaches in courses and curricula, texts and teaching materials, and pedagogy.

Looking beyond the discipline, several issues address the relationship between feminist geography, other academic disciplines, and other research collaborators. As evidenced by several respondents to the survey, feminist geography does not fully realize its potential to inform feminist studies and other fields outside the discipline. Continued extradisciplinary influence can be accomplished through judicious choice of publication venues, and feminist geographers should continue to be open to multidisciplinary collaboration. Feminist geography is diversifying methods of study and redefining boundaries between different subjects. The conceptual frameworks and methodological approaches involved in this type of critical research are complex and multiple, and offer fruitful prototypes for collaboration.

Extending beyond the academic arena, important issues raised in feminist geography, especially in methodology, pedagogy, and the broad area of gender, identity, and space, offer potential insights into ways that feminist geography can influence praxis and the political arena. These ideas include the interconnection of academic and activist identities, using feminist identities and spaces to retheorize hegemonic geographies, and interrogating geographies of both the left and the right in ways that challenge the reproduction of these categories.

None the less, it is important to consider the delicate questions of how these opportunities and investigations should be negotiated with our colleagues around the world. The growing diversity of voices and attention to a broad range of identities within gender/feminist geography intersect with important questions about developments of technology and power relations that will allow people on the margins to speak for themselves. This phenomenon raises serious questions about the long-term role of feminist geographers as the primary interpreters or channels of knowledge. Some would argue that the role of Western academics should disappear in this arena as native researchers and activists step to the fore and claim their rightful power. Alternatively, this begs the question of how this process will be integrated into the existing context, and how we as feminist researchers and activists can facilitate an expansion of knowledge and power. Our choice of research questions, our methodological techniques, and the relationships between academics, pedagogy, and activism all play an important role in how we answer these questions.

The first decade of the new millennium seems an appropriate time for feminist geographers to challenge dominant concepts of economic restructuring, cultural landscapes, gender and development, identity and space, and other important geographical concerns. Although feminist geography is not without its issues, as one survey respondent commented, "I'm reluctant to make forecasts of the future. However, I will say that as long as feminist geographers continue to address issues that make real differences in women's lives, and are willing to work towards empowerment of marginalized women inside and outside of academia, the future of feminist geography will be bright."

References

Afshar, H. (ed.) (1996). *Women and Politics in the Third World.* London: Routledge.

Afshar, H., and Dennis, C. (eds.) (1992). *Women and Adjustment Policies in the Third World.* New York: St. Martin's Press.

Agarwal, B. (1992). "The Gender and Environment Debate: Lessons From India." *Feminist Studies*, 18/1: 119–58.

— (1994). *A Field of One's Own: Gender and Land Rights in South Asia.* Cambridge: Cambridge University Press.

Alexander, M. J., and Mohanty, C. T. (eds.) (1997). *Feminist Genealogies, Colonial Legacies, Democratic Futures.* New York: Routledge.

Anderson, K., and Gale, F. (1992). *Inventing Places: Studies in Cultural Geography.* Melbourne: Longman Cheshire.

Antrobus, P. (1996). "Bringing Grassroots Women's Needs to the International Arena." *Development*, 3: 64–7.

Association of American Geographers (AAG) (1998). *Guide to Programs in Geography in the United States and Canada.* Washington, DC: AAG.

Asthana, S. (1996). "The Relevance of Place in HIV Transmission and Prevention: Geographical Perspectives on the Commercial Sex Industry in Madras," in R. A. Kearns and W. M. Gesler (eds.), *Putting Health into Place: Landscape, Identity and Well-Being.* Syracuse, NY: Syracuse University Press.

Bakker, I. (ed.) (1994). *The Strategic Silence: Gender and Economic Policy.* London: Zed Books.

Barron McBride, A. (1993). "From Gynecology to Gyn-Ecology: Developing a Practice Research Agenda for Women's Health." *Health Care for Women International*, 14/4: 315–26.

Bell, D., and Valentine, G. (eds.) (1995). *Mapping Desire*. New York: Routledge.

Beneria, L., and Feldman, S. (eds.) (1992). *Unequal Burden: Economic Crises, Persistent Poverty, and Women's Work*. Boulder: Westview Press.

Berg, L. D. (1994). "Masculinity, Place and a Binary Discourse of 'Theory' and 'Empirical Investigation' in the Human Geography of Aotearoa/New Zealand." *Gender, Place and Culture*, 1/2: 245–60.

Bhabha, H. (1986)."Remembering Fanon: Self, Psyche and the Colonial Condition." Foreword to F. Fanon, *Black Skin, White Masks*. London: Pluto Press, pp. vii–xxvi.

Blum, V. (1998). "Ladies and Gentlemen: Train Rides and Other Oedipal Stories," in Nast and Pile (1998: 263–80).

Blum, V., and Nast, H. (1996). "Where's the Difference? The Heterosexualization of Altering Henri Lefebvre and Jacques Lacan." *Society and Space*, 14: 559–80.

Blumberg, R. L. (1991). "Income Under Female vs. Male Control: Hypotheses from a Theory of Gender Stratification and Data from the Third World," in R. L. Blumberg, (ed.), *Gender, Family and Economy; The Triple Overlap*. Newbury Park, Calif.: Sage Publications.

Blunt, A. (1994). "Mapping Authorship and Authority: Reading Mary Kingsley's Landscape Descriptions," in Blunt and Rose (1994: 51–72).

Blunt, A., and Rose, G. (eds.) (1994). *Writing Women and Space: Colonial and Postcolonial Geographies*. New York: Guilford Press.

Bondi, L. (1992) "Gender Symbols and Urban Landscapes." *Progress in Human Geography*, 16/2: 157–69.

— (1993). "Locating Identity Politics," in M. Keith and S. Pile (eds.), *Place and the Politics of Identity*. New York: Routledge, 84–101.

Bondi, L., and Domosh, M. (1992). "Other Figures in Other Places: On Feminism, Postmodernism and Geography." *Environment and Planning A. Society and Space*, 10: 199–213.

— (1998). "On the Contours of Public Space: A Tale of Three Women." *Antipode*, 30/3: 270–89.

Boserup, E. (1970). *Women's Role in Economic Development*. New York: St. Martin's Press.

Bowlby, S., Lewis, J., and McDowell, L. (1989). "The Geography of Gender," in R. Peet and N. Thrift (eds.), *New Models in Geography*. London: Unwin and Hyman, iv. 157–75.

Braidotti, Rosi *et al.* (1994). *Women, the Environment and Sustainable Development*. London: Zed Books.

Brown, M. (1995). "Ironies of Distance: An Ongoing Critique of the Geographies of AIDS." *Environment and Planning D: Society and Space*,13: 159–83.

Butler, J. P. (1990). *Gender Trouble: Feminism and the Subversion of Identity*. New York : Routledge.

Carney, J. (1992). "Peasant Women and Economic Transformation in The Gambia." *Development and Change*, 23/2: 67–90.

— (1993). "Converting the Wetlands, Engendering the Environment: The Intersection of Gender with Agrarian Change in The Gambia." *Economic Geography*, 69/4: 329–47.

Craddock, S. (2000). "Disease, Social Identity, and Risk: Rethinking the Geography of AIDS." Transactions from the Institute of British Geographers, 25: 153–68.

Craig, M. (1998). *Cultural Geography*. London: Routledge.

Cravey, A. (1997). "The Politics of Reproduction: Households in the Mexican Industrial Transition." *Economic Geography*, 73/2: 166–86.

— (1998) *Women and Work in Mexico's Maquiladoras*. Lanham, Md.: Rowman & Littlefield.

Dankelman, I., and Davidson, J. (1988). *Women and Environment in the Third World*. London: Earthscan.

Davis Lewis, N., and Kieffer, E. (1994). "The Health of Women: Beyond Maternal and Child Health," in D. R. Phillips and Y. Verhasselt (eds.), *Health and Development*. London: Routledge, 122–37.

Del Casino, V. J., Jr. (2001). "Healthier Geographies: Mediating the Gaps Between the Needs of People Living with HIV and AIDS and Health Care in Chiang Mai, Thailand." *The Professional Geographer*, 53/3: 407–21.

Deutsche, R. (1991). "Boys Town." *Environment and Planning D: Society and Space*, 9: 5–30.

Domosh, M. (1989). "The New York World Building: A Method for Interpreting Landscape." *Area*, 21: 347–55.

— (1991). "Towards a Feminist Historiography of Geography." *Transactions of the Institute of British Geographers*, 16: 95–104.

Domosh, M., and Seager, J. (2001). *Putting Women in Place: Feminist Geographers Make Sense of the World*. London: Guilford Press.

Dorn, M. (1998). "Beyond Nomadism: The Travel Narratives of a Cripple," in Nast and Pile (1998: 183–206).

Duncan, N. (ed.) (1996). *Bodyspace: Destabilizing Geographies of Gender and Sexuality*. London: Routledge.

Duncan, J. S., and Ley, D. (1993). *Culture/Place/Representation*. Routledge: London.

Dyck, I. (1993). "Ethnography: A Feminist Method?" *The Canadian Geographer*, 37/1: 52–7.

— (1995). "Putting Chronic Illness 'In Place': Women Immigrants' Accounts of their Health Care." *Geoforum* 26: 247–60.

Dyck, I., and Kearns, R. (1995). "Transforming the Relations of Research: Towards Culturally Safe Geographies of Health and Healing." *Health and Place*, 1: 137–47.

Dyck, I., Davis Lewis, N., and McLafferty, S. (2001). *Geographies of Women's Health*. New York: Routledge.

Elder, G. (1995). "Of Moffies, Kaffirs and Perverts: Male Homosexuality and the Discourse of Moral Order in the Apartheid State," in Bell and Valentine (1995: 56–65).

— (1999). "'Queerying' Boundaries in the Geography Classroom." *Journal of Geography in Higher Education*, 23/1: 86–93.

Elson, D. (ed.) (1995*a*). *Male Bias in the Development Process*. Manchester: Manchester University Press.

— (1995*b*). "Gender Awareness in Modeling Structural Adjustment." *World Development*, 23/11: 1851–68.

England, K. (1994). "Getting Personal: Reflexivity, Positionality, and Feminist Research." *The Professional Geographer*, 46/1: 80–9.

— (1999). "Sexing Geography, Teaching Sexualities." *Journal of Geography in Higher Education*, 23/1: 94–101.

Enloe, C. (1990). *Bananas, Beaches and Bases: Making Feminist Sense of International Politics*. Berkeley, Calif.: University of California Press.

Faulkner, A., and Lawson, V. (1991). "Employment versus Empowerment: A Case Study of the Nature of Women's Work in Ecuador." *The Journal of Development Studies*, 27/4: 16–47

Fenster, T. (1998). "Ethnicity, Citizenship, Planning and Gender: The Case of Ethiopian Immigrant Women in Israel." *Gender, Place and Culture*, 5/2: 177–90.

—— (ed.) (1999) *Gender, Planning and Human Rights*. London: Routledge.

Folbre, N. (1993). "How Does She Know? Feminist Theories of Gender Bias in Economics." *History of Political Economy*, 25/1: 167–84.

—— (1994). *Who Pays for the Kids?: Gender and the Structures of Constraint*. London: Routledge.

Foord, J., and Gregson, N. (1986). "Patriarchy: Towards a Reconceptualisation." *Antipode* 18/2: 186–211.

Gibson-Graham, J. K. (1994). " 'Stuffed If I Know': Reflections on Post-Modern Feminist Social Research." *Gender, Place and Culture*, 2/1: 205–24.

—— (1996). *The End of Capitalism (As We Knew It): A Feminist Critique of Political Economy*. Cambridge: Blackwell.

Gilbert, M. (1994). "The Politics of Location: Doing Feminist Research at 'Home'." *The Professional Geographer*, 46/1: 90–6.

—— (1998). " 'Race', Space and Power: The Survival Strategies of Working Poor Women." *Annals of the Association of American Geographers*, 88/4: 595–621.

Gladwin, C. (ed.) (1991). *Structural Adjustment and African Women Farmers*. Gainseville, Fla.: University of Florida Press.

Gould, P. (1991). "Modelling the Geographic Spread of AIDS for Educational Intervention," in R. Ulack and W. F. Skinner (eds.), *AIDS and the Social Sciences*. Lexington, Ky.: University of Kentucky Press, 30–44.

Gruntfest, E. (1989) "Geographic Perspectives on Women," in G. Gaile and Cort Willmott (eds.), *Geography in America*. Columbus, Ohio: Merrill, 673–83.

Halford, S., and Savage, M. (1997). "Rethinking Restructuring: Embodiment, Agency and Identity in Organizational Change," in R. Lee and J. Wills (eds.), *Geographies of Economies*. London: Arnold, 108–17.

Hall, S. (1990). "Cultural Identity and Diaspora," in J. Rutherford (ed.), *Identity: Community, Culture, Difference*. London: Lawrence & Wishart, 222–37.

Hanson, S. (1997). "Introduction to Part 2: As the World Turns: New Horizons in Feminist Geographic Methodologies," in Jones, Nast, and Roberts (1997: 119–28).

Hanson, S., and Pratt, G. (1995). *Gender, Work, and Space*. London: Routledge.

Haraway, D. (1989). *Primate Visions: Gender, Race and Nature in the World of Modern Science*. New York: Routledge.

—— (1991). *Simians, Cyborgs and Women. The Re-Invention of Nature*. London: Free Association Books.

Harcourt, W. (ed.) (1994). *Feminist Perspectives on Sustainable Development*. London: Zed Books.

Harding, S. (1992). *Whose Science? Whose Knowledge*? Milton Keynes: Open University Press.

—— (1998). *Is Science Multicultural?: Postcolonialisms, Feminisms, and Epistemologies*. Bloomington: Indiana University Press.

Hart, G. (1991). "Engendering Everyday Resistance: Gender, Patronage, and Production Politics in Rural Malaysia." *Journal of Peasant Studies*, 19/1: 93.

—— (1992). "Household Production Reconsidered: Gender, Labor Conflict, and Technological Change in Malaysia's Muda Region." *World Development*, 20/6: 809–23.

Hartsock, N. (1983). "The Feminist Standpoint: Developing the Grounds for a Specifically Feminist Historical Materialism," in *Money, Sex and Power: Toward a Feminist Historical Materialism*. Boston: Northeastern University Press.

Hayford, A. (1974). "The Geography of Women: An Historical Introduction." *Antipode*, 6: 1–18.

Hays-Mitchell, M. (1993). "The Ties that Bind: Informal and Formal Sector Linkages in Streetvending: The Case of Peru's *Ambulantes*." *Environment and Planning A*, 25: 1085–102.

hooks, b. (1991). *Yearning: Race, Gender, and Cultural Politics*. Boston: South End Press, 23–31.

Jacobson, J. (1992). *Gender Bias: Roadblock to Sustainable Development*. Washington, DC: Worldwatch Institute.

Jackson, C. (1993). "Environmentalisms and Gender Interests in the Third World." *Development and Change*, 24: 6449–677.

—— (1995). "Radical Environmental Myths: A Gender Perspective." *New Left Review*, 210: 124.

Jackson, P. (1991). "The Cultural Politics of Masculinity: Towards a Social Geography." Transactions of the Institute of British Geographers, 16: 199–213.

Johnston, J. (1996). " 'Flexing Femininity': Female Body-Builders Refiguring 'The Body'." *Gender, Place and Culture*, 3/3: 327–40.

Johnston, R. J., Gregorys D., Pratt, G., Watts, M., Smith, D. M., and Johnston R. (eds.) (2000). *Dictionary of Human Geography*. Oxford: Blackwell.

Jones, J. P., and Moss, P. (1995). "Guest Editorial: Democracy, Identity, Space." *Environment and Planning D: Society and Space*, 13: 253–7.

Jones, J. P., Nast, H. J., and Roberts, S. M. (eds.) (1997). *Thresholds in Feminist Geography: Difference, Methodology, and Representation*. Oxford: Rowman & Littlefield.

Kabeer, N. (1994). *Reversed Realities*. London: Verso.

Kandiyoti, D. (1991). "Identity and Its Discontents: Women and the Nation." *Millenium: Journal of International Studies*, 20/3: 429–43.

Katz, C. (1992). "All the World is Staged: Intellectuals and the Projects of Ethnography." *Environment and Planning D: Society and Space*, 10: 495–510.

Katz, C., and Monk, J. (eds.) (1993). *Full Circles: Geographies of Women over the Life Course*. London: Routledge.

Kay, J. (1991). "Landscapes of Women and Men: Rethinking the Regional Historical Geography of the United States and Canada." *Journal of Historical Geography*, 17/4: 435–52

Kearns, R. A. (1996). "AIDS and Medical Geography: Embracing the Other?" *Progress in Human Geography* 20/1: 123–31.

Keller, E. F. (1985). *Reflections on Gender and Science*. New Haven: Yale University Press.

Knopp, L. (1992). "Sexuality and the Spatial Dynamics of Capitalism." *Environment and Planning D: Society and Space*, 10: 651–69.

—— (1995). "Sexuality and Urban Space: A Framework for Analysis," in Bell and Valentine (1995: 149–61).

—— (1999). "Queer Theory, Queer Pedagogy: New Spaces and New Challenges in Teaching Geography." *Journal of Geography in Higher Education*, 23/1: 77–9.

Kobayashi, A., and Peake, L. (1994). "Unnatural Discourse: 'Race' and Gender in Geography." *Gender, Place and Culture*, 2: 225–44.

Laws, G. (1995). "Theorizing Ageism: Lessons from Postmodernism and Feminism." *The Gerontologist*, 35/1: 112–18.

Lawson, V. (1995*a*). "Beyond the Firm: Restructuring Gender Divisions of Labor in Quito's Garment Industry under

Austerity." *Environment and Planning D: Society and Space*, 13: 415–44.

— (1995*b*). "The Politics of Difference: Examining the Quantitative/Qualitative Dualism in Post-Structuralist Feminist Research." *The Professional Geographer*, 47: 449–57.

Leach, M. (1994). *Rainforest Relations: Gender and Resource Use Among the Mende of Gola, Sierra Leone*. Washington, DC: Smithsonian Institution Press.

Lee, D., and Schultz, R. (1982). "Regional Patterns of Female Status in the United States." *The Professional Geographer*, 34: 32–41.

Longhurst, R. (1994). "The Geography Closest In—the Body . . . the Politics of Pregnability." *Australian Geographical Studies*, 32/2: 214–23.

Lukinbeal, C., and Aitken, S. C. (1998). "Sex, Violence and the Weather: Male Hysteria, Scale and the Fractal Geographies of Patriarchy," in Nast and Pile 1998: 356–80.

McDowell, L. (1986). "Beyond Patriarchy: A Class-Based Explanation of Women's Subordination." *Antipode*, 18/3: 311–21.

— (1991). "The Baby and the Bath Water: Diversity, Deconstruction and Feminist Theory in Geography." *Geoforum* 22: 123–33.

— (1992). "Doing Gender: Feminism, Feminists and Research Methods in Human Geography." *Transactions of the Institute of British Geographers*, 17/4: 399–416.

— (1994). "Polyphony and Pedagogic Authority." *Area*, 26/3: 241–8.

— (1997). *Capital Culture: Gender at Work in the City*. Oxford: Blackwell.

McDowell, L., and Court, G. (1994). "Missing Subjects: Gender, Power and Sexuality in Merchant Banking." *Economic Geography*, 70: 229–51.

Mackenzie, S. (1989). "Women in the City," in R. Peet and N. Thrift (eds.), *New Models in Geography*. Boston: Unwin Hyman, p. ii.

Mackenzie, S., and Rose, D. (1983). "Industrial Change, the Domestic Economy and Home Life," in J. Anderson, S. Duncan, and R. Hudson (eds.), *Redundant Spaces in Cities and Regions? Studies in Industrial Decline and Social Change*. London: Academic Press, 155–99.

McLafferty, S. L. (1995). "Counting for Women." *The Professional Geographer*, 47/4: 436–42.

McLafferty, S. L., and Tempalski, B. (1995). "Restructuring and Women's Reproductive Health: Implications for Low Birth-weight in New York City." *Geoforum*, 26/3: 309–23.

Madge, C. (1994). "Gendering Space: A First Year Geography Field Work Exercise." *Geography* 79/345: 330–8.

Mains, S. (2000*a*). "An Anatomy of Race and Immigration Politics in California." *International Journal of Social and Cultural Geography*, 1/2: 143–54.

— (2000*b*). "Maintaining National Identity at the Border: Scale, Masculinity, and the Policing of Immigration in Southern California," in A. Herod and M. W. Wright (eds.), *Geographies of Power: Placing Scale*. London: Blackwell, 192–214.

Marchand, M. H., and Parpart, J. L. (eds.) (1995). *Feminism/Postmodernism/Development*. London: Routledge.

Martin, A., and Kryst, S. (1998). "Encountering Mary: Ritualization and Place Contagion in Postmodernity," in Nast and Pile (eds.), *Places Through the Body*. New York: Routledge, 207–29.

Massey, D. (1984). *Spatial Divisions of Labor*. New York: Methuen.

— (1991). "Flexible Sexism." *Environment and Planning D: Society and Space*, 9: 31–57

— (1994). *Space, Place, and Gender*. Minneapolis: University of Minnesota Press.

— (1995). "Masculinity, Dualisms and High Technology." *Transactions of the Institute of British Geographers*, 20: 487–99.

— (1997). "Economic/Non-Economic," in R. Lee and J. Wills (eds.), *Geographies of Economies*. London: Arnold, 27–36.

Mattingly, D. J., and Falconer Al-Hindi, K. (1995). "Should Women Count? A Context for the Debate." *The Professional Geographer*, 47/4: 427–37.

Mazey, M. E., and Lee, D. R. (1983). *Her Space, Her Place: A Geography of Women*. Washington, DC: Association of American Geographers.

Meono-Picado, P. (1997). "Redefining the Barricades: Latina Lesbian Politics and the Creation of an Oppositional Public Sphere," in Jones, Nast, and Roberts (1997: 319–38).

Merchant, C. (1980). *The Death of Nature: Women, Ecology, and the Scientific Revolution*. New York: Harper & Row.

Mies, M., and Shiva, V. (1993). *Ecofeminism*. Halifax: Fernwood Publications.

Mikesell, M. (1993). "Afterward: New Interests, Unsolved Problems, and Persisting Tasks," in K. Foote, P. Hugill, K. Mathewson, and J. M. Smith (eds.), *Rereading Cultural Geography*. Austin: University of Texas Press.

Mills, S. (1996). "Gender and Colonial Space." *Gender, Place and Culture*, 3/2: 125–48.

Mohanty, C. T. (1988). "Under Western Eyes: Feminist Scholarship and Colonial Discourses." *Feminist Review*, 30: 65–88.

Mohanty, C. T., Russo, A., and Torres, L. (eds.) (1991). *Third World Women and the Politics of Feminism*. Bloomington: Indiana University Press.

Momsen, J. H. (1991). *Women and Development in the Third World*. London: Routledge.

Momsen, J. H., and Kinnaird, V. (eds.) (1993). *Different Places, Different Voices: Gender and Development in Africa, Asia and Latin America*. London: Routledge.

Monk, J. (1985). "Feminist Transformation: How Can It be Accomplished?" *Journal of Geography in Higher Education*, 9/1: 101–5.

— (1992). "Gender in the Landscape: Expressions of Power and Meaning," in Anderson and Gale (1992: 123–38).

— (1996). "Partial Truths: Feminist Perspectives in Ends and Means," in M. Williams (ed.), *Understanding Geographical and Environmental Education*. London: Cassell Education, 274–86.

— (1997*a*). "Introduction to Part 3: Marginal Notes on Representations," in Jones, Nast, and Roberts (eds.), *Thresholds in Feminist Geography: Difference, Methodology, Representation*. New York: Rowman & Littlefield, 241–54.

— (1997*b*). "Finding a Way." *New Zealand Journal of Geography*, 7: 7–11.

Monk, J., and Hanson, S. (1982). "On Not Excluding Half of the Human in Human Geography." *The Professional Geographer*, 34/1: 11–23.

Monk, J., and Norwood, V. (eds.) (1987). *The Desert is No Lady: Southwestern Landscapes in Women's Writings and Art*. New Haven: Yale University Press.

Monk, J., Betteridge, A. and Newgall, A. (1991). "Introduction: Reaching for Global Feminism in the Curriculum." *Women's Studies International Forum*, 14/4: 239–49.

Morin K. (1995). "A 'Female Columbus' in 1887 America: Marking New Social Territory." *Gender, Place and Culture*, 2: 191–208.

Morin, K., and Guelke, J. K. (1998). "Strategies of Representation, Relationship, and Resistance: British Women Travelers and Mormon Plural Wives, ca. 1870–1890." *Annals of the Association of American Geographers*, 88/3: 436–62.

Moser, C. O. N. (1993). *Gender Planning and Development*. London: Routledge.

Moss, P. (1997). "Places of Resistance, Spaces of Respite: Franchise Housekeepers Keeping House in the Workplace and at Home." *Gender Place and Culture*, 4/2: 179–96.

Mouffe, C. (1992). "Feminism, Citizenship, and Radical Democratic Politics," in J. Butler and J. Scott (eds.), *Feminists Theorize the Political*, New York: Routledge, 369–84.

Nairn, K. (1997) "Hearing from Quiet Students: The Politics of Silence and of Voice in Geography Classrooms," in Jones, Nast, and Roberts (1997: 93–115).

Nagar, R. (1997). "Exploring Methodological Borderlands through Oral Narratives," in Jones, Nast, and Roberts (1997: 203–24).

—— (1998). "Communal Discourses, Marriage, and the Politics of Gendered Social Boundaries among South Asians Immigrants in Tanzania." *Gender, Place and Culture*, 5/2: 117–39.

Nanda, M. (1997). "History is What Hurts: A Materialist Feminist Perspective on the Green Revolution and Its Ecofeminist Critics," in R. Hennessy and C. Ingraham (eds.), *Materialist Feminism: A Reader in Class, Difference, and Women's Lives*. London: Routledge.

Nash, C. (1994). "Remapping the Body/Land: New Cartographies of Identity, Gender, and Landscape in Ireland," in Blunt and Rose (1994: 227–50).

Nast, H. (1994). "Opening Remarks on 'In the Field'." *The Professional Geographer*, 46/1: 54–66.

—— (1998). "Unsexy Geographies." *Gender, Place and Culture*, 5/2: 191–206.

Nast, H., and Pile, S. (1998). *Places Through the Body*. New York: Routledge.

Natter, W., and Jones, J. P. (1993). "Signposts Towards a Poststructuralist Geography," in J. P. Jones, W. Natter, and T. Schatzki (eds.), *Postmodern Contentions: Epochs, Politics, Space*. New York: Guilford, 165–203.

—— (1997). "Identities, Space and Other Uncertainties," in G. Benko and U. Strohmayer (eds.), *Space and Social Theory*. Cambridge: Blackwell, 141–61.

Oberhauser, A. M. (1995). "Gender and Household Economic Strategies in Rural Appalachia." *Gender, Place and Culture*, 2/1: 51–70.

—— (1997). "The Home as 'Field': Households and Homework in Rural Appalachia," in Jones, Nast, and Roberts (1997: 165–82).

Peake, L. (1993). " 'Race' and Sexuality: Challenging the Patriarchied Structuring of Urban Social Space." *Environment and Planning D: Society and Space*, 11: 415–32.

Peterson, V. S., and Runyon, A. S. (1993). *Global Gender Issues*. Boulder: Westview Press.

Pile, S., and Rose, G. (1992). "All or Nothing? Politics and Critique in the Modernism-Postmodernism Debate." *Environment and Planning D: Society and Space*, 10: 123–36.

Pile, S., and Thrift, N. (eds.) (1995). *Mapping the Subject: Geographies of Cultural Transformation*. London: Routledge.

Plumwood, V. (1992). "Feminism and Ecofeminism: Beyond the Dualistic Assumption of Women, Men and Nature." *The Ecologist*, 22/1: 8–13.

Pratt, G. (1992). *Imperial Eyes: Travel Writing and Transculturation*. London: Routledge.

Pratt, G., and Hanson, S. (1994). "Geography and the Construction of Difference." *Gender, Place and Culture*, 1/1: 5–30.

The Professional Geographer (1994). "Women in the Field: Critical Feminist Methodologies and Theoretical Perspectives." 46/1: 54–102.

—— (1995). "Should Women Count? The Role of Quantitative Methodology in Feminist Geographic Research." 47/4: 426–66.

Pulido, L. (1997). "Community, Place and Identity," in Jones, Nast, and Roberts (1997: 11–28).

Radcliffe, S., and Westwood, S. (1993). *'Viva' Women and Popular Protest in Latin America*. London: Routledge.

—— (1996). *Remaking the Nation: Place, Identity and Politics in Latin America*. London: Routledge.

Raghuram, P., Madge, C., and Skelton, T. (1998). "Feminist Research Methodologies and Student Projects in Geography." *Journal of Geography in Higher Education*, 22/1: 35–48.

Rocheleau, D. (1995). "Maps, Numbers, Text, and Context: Mixing Methods in Feminist Political Ecology." *The Professional Geographer*, 47/4: 458–66.

Rocheleau, D., Thomas-Slayter, B. and Wangari, E. (eds.) (1996). *Feminist Political Ecology: Global Issues and Local Experiences*. London: Routledge.

Rose, G. (1993). *Feminism and Geography: The Limits of Geographical Knowledge*. Cambridge: Polity Press.

—— (1994). "The Cultural Politics of Place: Local Representation and Oppositional Discourse in Two Films." *Transactions of the Institute of British Geographers*, 19: 46–60.

Rothenberg, T. (1995). " 'And She Told Two Friends': Lesbians Creating Urban Space," in Bell and Valentine (1995: 165–81).

Sachs, C. (1996). *Gendered Fields: Rural Women, Agriculture, and Environment*. Boulder: Westview Press.

Said, E. (1978). *Orientalism*. London: Routledge.

Sanders, R. (1999). "Introducing 'White Privilege' into the Classroom: Lessons from Finding A Way." *Journal of Geography*, 98: 169–75.

Schroeder, R. (1993). "Shady Practice: Gender and the Political Ecology of Resource Stabilization in Gambian Garden/Orchards." *Economic Geography*, 69/4: 349–65.

—— (1997). " 'Re-claiming' Land in The Gambia: Gendered Property Rights and Environmental Intervention." *Annals of the Association of American Geographers*, 87/3: 487–503.

Scott, C. V. (1995). *Gender and Development*. Boulder: Lynne Reinner.

Seagar, J. (1993). *Earth Follies: Coming to Feminist Terms with the Global Environmental Crisis*. New York: Routledge.

Sen, G., and Grown, C. (1987). *Development, Crises, and Alternative Visions*. New York: Monthly Review Press.

Shannon, G. W., and Pyle, G. F. (1989). "The Origin and Diffusion of AIDS: A View from Medical Geography." *Annals of the Association of American Geographers*, 79: 1–24.

Shiva, V. (1989). *Staying Alive*. London: Zed Books.

Sparr, P. (ed.) (1994). *Mortgaging Women's Lives: Feminist Critiques of Structural Adjustment*. London: Zed Books.

Spivak, G. C. (1988). "Can the Subaltern Speak?," in C. Nelson and L. Grossberg (eds.), *Marxism and the Interpretation of Culture*. Basingstoke: Macmillan Education.

Soja, E. (1989). *Postmodern Geographies: The Reassertion of Space in Critical Social Theory*. London: Verso.

Staeheli, L., and Lawson, V. (1995). "Feminism, Praxis, and Human Geography." *Geographical Analysis*, 27/4: 321–38.

Tinker, I. (ed.) (1990). *Persistent Inequalities: Women and World Development*. New York: Oxford University Press.

Townsend, J., in collaboration with U. Arrevillage, J. Bain, S. Cancina, S. F. Frenk, S. Pacheco, and E. Perez (1995). *Women's Voices from the Rainforest*. London: Routledge.

Valentine, G. (1993). "Heterosexing Space: Lesbian Perceptions and Experiences of Everyday Spaces." *Environment and Planning D: Society and Space*, 11: 395–413.

— (1997). "Making Space: Separatism and Difference," in Jones, Nast, and Roberts (1997: 65–76).

van den Hombergh, H. (1993). *Gender, Environment and Development: A Guide to the Literature*. Utrecht: International Books.

Warren, K. J. (ed.) (1994). *Ecological Feminism*. London: Routledge.

Waylen, G. (1996). "Analyzing Women in the Politics of the Third World," in Afshar (1996).

Wilton, R. D. (1996). "Diminishing Worlds: HIV/AIDS and the Geography of Everyday Life." *Journal of Health and Place*, 2/2: 1–17.

Women and Geography Study Group of the Royal Geographical Society with the Institute of British Geographers (1997). *Feminist Geographies: Explorations in Diversity and Difference*. Harlow: Longman.

Wright, M. (1997). "Crossing the Factory Frontier: Gender, Place, and Power in the Mexican Maquiladora." *Antipode*, 29/3: 278–302.

Zonn, L., and Aitken, S. (eds.) (1994). *Place, Power, Situation, and Spectacle: A Geography of Film*. New York: Rowman & Littlefield.

CHAPTER 48

Geography of Religion and Belief Systems

Robert H. Stoddard and Carolyn V. Prorok

Spatial and environmental dimensions of religious behavior, artefacts, and attitudes are grist for the geographer's intellectual mill because spiritually motivated convictions and actions play an important role in human affairs. It is not surprising, therefore, that the geography of religion and belief systems is an important, emerging field of study.

We commence this chapter with a definition of the field, particularly as it entails distinctions that arise out of the highly personal nature that religious belief is accorded in the academy and society at large. A limited review and summary of trends in the field over the past decade follows, building on Kong's (1990) and Sopher's (1967, 1981) overviews. Although North American geographers are emphasized here, research in the geography of religion is thoroughly entwined in terms of scholars' national origin, university training, and research perspectives, thus making distinctions in nationality difficult. Moreover, geographers who do not consider themselves to be geographers of religion and numerous nongeographers also make significant contributions to this field because their work clearly incorporates both religious and geographic components in their analysis and subject matter.

We note that traditional empirical studies largely dominate the work published in the last decade. Nevertheless, humanistic research (Weightman 1996; Cooper 1997*a*; Prorok 1997; Osterrieth 1997), and the application of contemporary critical theory (Fielder 1995; Kong 1993*a*, *b*; Prorok 2000) in this field is gaining ground, particularly via recent dissertations and presentations at AAG meetings. Additional comments about future challenges and opportunities conclude the chapter.

The Focus of Study

No universally accepted definition of religion exists, as illustrated by the hundreds already published and others continually being introduced (see e.g. a separate bibliographical category devoted annually to this definitional task in *Social Compass: International Review of Sociology of Religion*). Another indication that the term "religion" lacks a single, precise definition is the continual struggle, expressed repeatedly within the American judicial system, with questions about what are truly "religious" activities. Definitions vary in their emphasis on three contrasting perspectives: (1) a transcendental divinity; (2) an immanent spirituality that permeates all of life; and (3) an ethical philosophy. Even though a formal definition of religion is seldom stated by geographers, it appears from their writings that most accept a definition close to the following: *Religion is a system of beliefs and practices that attempts to order life in terms of culturally perceived ultimate priorities.*

When the field of study is titled "geography of religion *and belief systems*," the commonality of studies becomes even less precise. In general, the transitional zone between what is regarded as religious and what is accepted as non-religious includes those beliefs and activities considered spiritual and containing a sense of "oughtness" or obligation. When persons believe they conscientiously "should" behave in a certain way (rather than just because "traditionally everyone does it this way") or they have faith that certain group action carries a synergism greater than one's own power (Bartkowski and Swearingen 1997), their actions fall within the realm regarded as "other religious-like belief systems." This more inclusive term also encompasses "civil religion," which refers to beliefs and practices expressed through informal mass adherence, often containing a strong nationalist fervor. Even though beliefs may not necessarily be organized systematically, the popular acceptance of national mythology, revered symbols, and hallowed places closely resemble characteristics of formal religions. As with other belief systems containing culturally perceived ultimate priorities, the motives for honoring patriotic icons, preserving hallowed grounds, and visiting sanctified sites originate from a sense of "shouldness."

Methodological Issues

A fundamental methodological issue concerns the epistemological stance of persons seeking to explain religious phenomena. Two views adopted here to simplify the discussion are those of Believers and Observers.

Believers are those who hold certain beliefs *on faith*. They often believe specific conditions exist because of transcendental forces, such as divine will. They affirm that the divine is manifested in sacred texts, holy places, and/or absolute ethics. From the perspective of Believers, certain research questions posed by outside observers are irrelevant—and irreverent—because such queries indicate doubt about what is already "known." The category of Believers also includes persons who, while not always believing in divine intervention, declare that certain objects (say, a national flag) and places (illustrated by the site where a national hero was martyred) become sacred. For those holding non-theistic philosophical positions (for instance, the sanctity of Mother Earth), and even for those who recognize human elements in the creation of sacred spaces (such as battlefield cemeteries), beliefs about certain behavior, nevertheless, exist as fundamental assumptions. Although non-theistic Believers may be receptive to studies about how certain sacred places function within the society, from their perspective the element of sanctity *per se* should not—and cannot—be challenged.

"Observers" refers to non-Believers, who may hold varying positions about Believers' convictions, ranging from empathy to hostility. Positions of geographic Observers, seeking to understand and explain the spatial and ecological aspects of religion and belief systems, vary. For geographers who are personally skeptics, secularists, or agnostics, the stance of Observer is that of an outsider. In contrast, for geographers studying a particular religion while personally being a Believer of that same religion, there exists a tension between trying to understand relationships and conditions objectively while also unquestionably believing some were created by the divine. For other geographers who are Believers but who seek to observe a religion or belief system different from their own, the role of outside Observer is similar though not equivalent to that of a relatively dispassionate inquirer.

Some scholars of religion declare that Observers cannot fully comprehend and thus truly "explain" a religion of which they are not a believing member because an Observer's own values and experiences distort a true understanding of the motives affecting the behavior of Believers. However, the degree to which this outsideness affects scholarship epistemologically is probably not different from other geographic studies about which researchers may hold perspectives ranging from empathy to hostility.

The contrasting views of Observers and Believers are indicated by the customary differentiation between the *geography of religion* and *religious geography*. The latter pertains to information about the geographic (that is, spatial and ecological) characteristics of religious phenomena written from the perspective of a Believer in those same phenomena (Isaac 1965). Writings about sacred geometry and/or the locations of a divinity's manifestation on Earth (Singh 1991), environmental events recorded in sacred texts (Mather and Mather 1997), or advocacy for religious sentiment in urban development (Pacione 1999) are a few examples. Although such materials may be examined by geographers of religion for the purpose of gaining insight into the motivations of Believers, they are not considered a part of the academic field being discussed here. The debate concerning Believer versus Observer epistemological positions in the field is an important one given the desire of some academic geographers not to restrict study to a particular faith system or sacred text nor to become disseminators of sectarian viewpoints (Raivo 1997; Dawson 2000).

Likewise, recent writings labeled as "spiritual geography" are not considered here as a part of the geography of religion. Although they often deal with religious feelings about the natural environment at particular places, each is representative of unique, personal experiences written from the viewpoint of a Believer (Norris 1993; Henderson 1993; Pulido 1998). In contrast, geographers of religion have focused primarily on religious behavior and artefacts of groups of people.

Another methodological issue in this field concerns the directional relationships between religion and other phenomena. One approach commences with observations about the spatial characteristics of religion and then seeks to explain those locational and ecological aspects in terms of relationships with other phenomena, which is similar to the way studies in economic, political, and physical geography are normally designed. The alternative approach begins with religious beliefs and attempts to demonstrate their impact or effect on the spatial and ecological characteristics of other phenomena. Although most branches of geography are seldom defined in this manner, the geography of religion has a long history of such "impact" studies.

To a certain extent, this discussion about the two approaches exaggerates a distinction that may be insignificant because relationships among phenomena are seldom understood well enough to specify which ones have the greatest influence on a complex of others. Since reciprocal relationships permeate most human relationships, it is difficult to sort out the degree that religious beliefs affect, and are affected by, numerous economic, political, social, and environmental phenomena.

Clusters of Emphasis

For the purposes of organization and to identify trends, we have grouped the body of literature into three clusters: studies on the impact of religion, distributional studies, and geographic studies of religious phenomena.

Studies on the Impact of Religion

Religious beliefs and behavior are so significantly and thoroughly intertwined with the lifeways of numerous communities around the world that geographers have often focused upon the impact of religion on other human endeavors. The basic message is that religious beliefs and practices influence natural environments as well as the locations and spatial characteristics of specified economic, political, cultural, or physical phenomena. The degree to which geographical characteristics of these other features are affected by religion is seldom measured, but the extent of the impact is deemed significant. Examples of such studies include the influence of religion on the distribution of urban neighborhoods and agricultural settlements (Katz 1991), the characteristics and political behavior of a region (Webster 1997), the distribution of specific plants and their uses (Voeks 1990), and settlement patterns (Abruzzi 1993). Geographers of religion have a long tradition of studying the role of religious belief and behavior on the formation of cultural landscapes. Such studies include the plurality of religions evident in urban landscapes (Weightman 1993; Numrich 1997; Ley and Martin 1993), settlement mythology (Kuhlken 1997), symbolic landscapes (Malville 1998), and urban design (Sinha 1998).

A recent trend, by geographers and other scholars, is an increase in studies on the role of religion in forming beliefs and attitudes about the environment. Typical are questions about the relationship between theology and conservation (Braden 1999), varying environmental positions of Christian groups (Curry-Roper 1990), sacred interpretations of ecophysiology by modern scientists (Cooper 1997*a*), and the merits of environmental activism (Tharan 1997).

Distributional Studies

Primary locational data on religious affiliations are difficult to obtain, especially in the United States where recent governmental censuses have not collected such information. This means that geographers studying religious affiliation may necessarily devote more attention to the acquisition of data than in many other fields of geography. One publication providing valuable distributional data originates from the Glenmary Research Center (Bradley *et al.* 1992); another, which incorporates data from numerous sources, is the atlas of religious change in America (Newman and Halvorson 2000). Of particular importance is the monumental revision of the atlas of religion in America (Gaustad and Barlow 2000), which reflects contemporary trends in cartographic technology. In addition, many atlases and encyclopedia collect and present a wide variety of religious phenomena at the national and international levels (e.g. O'Brien *et al.* 1993; Brockman 1997; Barrett *et al.* 2000; Smart 1999).

Other distributional publications are those combining locational data with suggested relationships but with limited analysis. Such is exemplified by the distribution of churches in a US county (Andrews 1990), prevalence of almshouses in Jerusalem (Shilong 1993), regional expression of poverty programs supported by religious dioceses (Pacione 1991), sociological characteristics of religion in different regions of the United States (McGuire 1991), and the distribution of Canadian Mormons in North America (Louder 1993). Also assigned to this cluster are studies in which religious affiliation is a critical element in the identification of a population, even though it is often difficult to apportion the role of religion within the mingled cultural complex of a particular group (Sheskin 1993; Morin and Guelke 1998; Hardwick 1991, 1993*a*, *b*; Tharan 1997; Emmett 1995).

Geographic Studies of Religious Phenomena

This cluster encompasses research designed to understand and explain the spatial variations in those populations, features, and activities regarded as religious. Several studies have examined factors associated with membership in religious organizations (Katz and Lehr 1991; Krindatch 1996), while other geographic questions have dealt with religious buildings and observable features (Diamond 1997; Prorok and Kimber 1997; Prorok and Hemmasi 1993; Bhardwaj and Rao 1998; Prorok 1998, 1991). Although some institutional structures (such as schools, soup kitchens, and cemeteries) associated with religious organizations may be studied geographically, they are excluded from this cluster if they are not treated primarily as phenomena of religion. A study that illustrates well a geographic analysis of religion is by Stump (2000), who examines the commonalities and political implications of fundamentalist movements in several world religions.

A topic that recently has attracted considerable academic attention concerns sacred places, which are those Earth locations deemed by Believers to be holy territory. Because the locations of most sanctified sites are identified by observable religious features and rituals, they can be analyzed geographically. In many cases, certain rituals occur only at a particular sacred site, which creates a tremendous motivation for journeys by Believers to that unique place. Thus, even though the ritualistic movements within the confines of the holy site occur at a scale not often studied by geographers, travel to these religious nodes (i.e. pilgrimages) creates considerable geographic interest.

Another characteristic of a sacred place is the manifestation of power required to control the land and its use. Because a sacred place consists of an actual areal plot (which is often clearly demarcated), it incorporates all the properties associated with territorial control of land. Frequently sacred sites are contested with accompanying issues of ownership, maintenance, and access to the site, as well as the very identification of its status as "sacred." Disputes over territory may involve more than just the religious attributes of participants because the stakes are magnified when group members believe—or are emotionally convinced—that certain territory belongs to them. That is, they base their right to specific real estate not only from a long history of attachment to the area, or the strength of internationally accepted documents of sovereignty, but also from a belief that the land has been assigned to them by a divine power (Benvenisti 2000; Friedland and Hecht 1991; Emmett 1996, 1997).

Other conflicts over ownership, maintenance, and access to land that is sacred to indigenous people may not always generate as much publicity as regional wars, but they involve some of the same fundamental issues. The desecration of Native Americans' sacred sites often occurs from intrusion by roads, mines (Jett 1992), and tourists (Price 1994). Similar conflicts occur in other regions where economic interests clash with the religious geography of indigenous groups (Fielder 1995). Controversies may also erupt over the use of land where sanctified structures have existed for long periods of time but now occupy sites coveted by economic developers or are deleteriously affected by recent development (Hobbs 1992; Kong 1993*b*; Tobin 1998).

This trend toward a heightened interest in sacred places is evidenced by the variety of studies. The general topic has been publicized by academicians (Park 1994; Cooper 1997*b*) and in popular books (Harpur 1994; Brockman 1997). Researchers have focused on the emotional experiences of Believers (Geffen 1998; Dobbs 1997; Forbes-Boyte 1998), on the concept of sacred places by specific religious groups (Bascom 1998), and on cemeteries as sanctified areas (Nakagawa 1990; Yeoh and Hui 1995; Teather 1999). Others have studied the process through which places become sacred (Nolan and Nolan 1997; Prorok 1997; Bhardwaj 1990; Berg 1998; Singh 1997), and on the geometric patterns of a religious cosmology along with the sacred sites that define it (Gutschow 1994; Singh 1994; Singh and Malville 1995; van Spengen 1998; Buffetrelle 1998; Kuhlken 1997; Grapard 1998; Malville 1998). Similarly, scholars have attempted to understand spatial behavior associated with places revered as part of civil religion (Zelinsky 1990; Sellers and Walters 1993; Foote 1997; Sherrill 1995; Azaryahu 1996; Ben-Israel 1998). In the United States, the Gettysburg Battlefield, the Vietnam War Memorial, Arlington National Cemetery, the Arizona in Pearl Harbor, and Graceland all illustrate sites sanctified primarily by the multitude of pilgrims, who possess a strong desire to visit and to experience a sense of spirituality there.

Because both the topics of sacred places and religious movement are inherently geographic, the study of pilgrimage constitutes an important emphasis by geographers of religion. It is not surprising, given the large amount of research on pilgrimage, that the relevant literature covers a variety of aspects. General essays about pilgrimage include a summary of their distinctive geographic characteristics (Stoddard and Morinis 1997*a*), a comprehensive review of and commentary on contemporary pilgrimage studies by geographers (Bhardwaj 1997), and a discussion about the differentiation between journeys motivated by religion versus tourism (Smith 1992). Pilgrimages have been classified according to the type of attraction (Nolan and Nolan 1992) and by length of journey, route configurations, and frequency of events (Stoddard 1997).

A substantial number of studies illuminate pilgrimage traditions by attempting to understand relationships among pilgrims' backgrounds, motivations for religious undertakings, size of nodal fields, and distributions of sacred sites by concentrating on a single religious tradition in one region (Din and Hadi 1997; Nolan and Nolan 1997; Rinschede 1990; Jackowski 1990; Jackowski and Smith 1992; Jackson *et al.* 1990; Hudman and Jackson 1992; Bhardwaj 1990; Cameron 1990; McDonald 1995; Zelinsky 1990). Other studies emphasize the complex of interrelationships occurring at a single pilgrimage site, such as in Mecca (Rowley 1997); at the basilica for Our Lady of Consolation in Carey, Ohio (Faiers and Prorok 1990); the shrine in Belleville, Illinois (Giuriati *et al.* 1990); the sangam at Prayag, India (Caplan 1997); the Himalayan site at Muktinath (Kaschewsky 1994); the Sikh shrine at Patna, India (Karan 1997); Lourdes (Giuriati and Lanzi 1994); Graceland in Memphis, Tennessee (Davidson *et al.* 1990), and the set of Buddhist temples located around Shikoku island (Shimazuki 1997), to name a few. Methodological approaches vary, with one group searching for general relationships, such as locational associations with major world pilgrimage sites (Stoddard 1994), the effects of scale on observed relationships (Rinschede 1997), and the empirical validity of the core-versus-periphery hypothesis (Sopher 1997; Cohen 1992). Others struggle with understanding the tensions people experience between travel and attachment to home (Osterreith 1997) and contemplate the pilgrimage experience in and of itself (Wagner 1997). An interesting pilgrimage variant is to visit a wandering ascetic rather than a fixed earth location (McCormick 1997).

Understanding religious pilgrimages (which involves more complexity than provided, for example, by economic models) requires examination from many disciplinary efforts. Thus, geographers of pilgrimage have welcomed insights by scholars in other disciplines (especially anthropology, sociology, and religious studies), who, unfortunately, have not always realized (or seem to consider) geographers as contributors or geographic literature as relevant (Crumrine and Morinis 1991; Naquin and Yu 1992; Reader and Walters 1993; Carmichael *et al.* 1994; Chidester and Linenthal 1995; Kedar and Werblowsky 1998).

Conclusion

Even though some forms of religious activity may have declined in certain regions of the world during the last century, religion continues to be a critical component of many human events. Religion is often a vital element in understanding, for example, the ethno-regional conflicts over the possession and control of specific territory, the gathering of millions of worshipers who journey to particular places, the patterns of electoral results in various democracies, and local disputes about the role of governmental regulations of public institutions and land use. Consequently, as long as geographers seek to explain the spatial behavior of humans, there will exist the need to analyze religion and similar belief systems geographically.

As an organized field of study, the geography of religion and belief systems (GORABS) is relatively young but, as judged by Zelinksy (1994: 126), showing "every sign of viability and growth". While geographic questions have been asked about religious behavior and artefacts for a long time, the first formal gathering of interested scholars did not occur until the 1965 annual meeting of the AAG (Stoddard 1990). The rise of radical and humanistic geographies in the 1960s and 1970s (Peet 1998) may have played a role in the formal emergence of this specialty inasmuch as "touchy-feely" geography became more and more acceptable in general. Even so, many geographers of religion at the time were not really interested in joining the touchy-feely side of the academy. Instead, they struggled to gain respect for research with a strong empirical basis, in part to show that peoples' religious behavior deserves the same attention and respect as their social, cultural, political, and economic behavior. Dedication to studying religion as Observers differentiated these geographers from those who might promote a particular sectarian viewpoint and thus undergirded their strong emphasis on empirical work within the field.

Calls have been made to supplement this empirical emphasis with more theoretical approaches (Sopher 1981; Levine 1986; Cooper 1992; Wilson 1993) although a recent article by Kong (2001) indicates that this is already beginning to occur. And, while a few geographers of religion have successfully combined empirical and contemporary critical ideas in their work (Stump 2000), there is still the challenge of devising theoretical frameworks applicable to people's spatial behavior that is religiously motivated (and thus contrasting with economic models). Forays into theoretical approaches among contemporary geographers of religion have relied mainly on critical theories (e.g. feminist, post-colonial, and postmodern) and philosophies (e.g. phenomenology and existentialist) that did not originate in geographic thought but which have been adopted by geographers in a number of specializations (Peet 1998). Other credible contributions to theoretical frameworks made by nongeographers (Malville 1999; Tweed 1997; Brenneman and Brenneman 1995) serve as future challenges to geographers of religion.

Because religion permeates many aspects of human behavior we predict academic attempts to understanding its spatial and environmental manifestations will continue to be vigorously pursued, particularly in the various clusters of topics as discussed above. As with other fields of geography, more sophisticated methods of observation and analysis undoubtedly will be employed in the future. Other scholars, striving to gain greater insight into the complex interrelationships among religious and other human attributes, motivations, and behavioral patterns will conduct further studies using critical theories and humanistic perspectives. And the call for creating original, relevant theories of spatio-religious behavior will increasingly challenge scholars. These endeavors will require sensitively applied skills in observation and analysis as well as true empathy while, at the same time, carefully avoiding the advocation of a particular religious doctrine. We are convinced that these goals can be achieved and geographers of religion will continue to organize and disseminate substantial new knowledge in the twenty-first century.

References

Abruzzi, W. S. (1993). *Dam That River!: Ecology and Mormon Settlement in the Little Colorado River Basin*. Lanham, Md.: University Press of America.

Andrews, A. (1990). "Religious Geography of Union County, Georgia." *Journal of Cultural Geography*, 10: 1–19.

Azaryahu, M. (1996). "The Spontaneous Formation of Memorial Space: The Case of *Kikar Rabin*, Tel Aviv." *Area*, 28: 501–13.

Barrett, D. B., Kurian, G. T., and Johnson, T. M. (2000). *World Christian Encyclopedia: A Comparative Survey of Churches and Religions AD30–AD2000*. 2nd edn. New York: Oxford University Press.

Bartkowski, J. P., and Swearingen, W. S. (1997). "God Meets Gaia in Austin, Texas: A Case Study of Environmentalism as Implicit Religion." *Review of Religious Research*, 38: 308–24.

Bascom, J. (1998). "The Religious Geography of Evangelical Christians in North America." *Pennsylvania Geographer*, 36: 148–68.

Ben-Israel, H. (1998). "Hallowed Land in the Theory and Practice of Modern Nationalism," in B. Z. Kedar and R. J. Z. Werblowsky (eds.), *Sacred Space: Shrine, City, Land*. New York: New York University Press, 78–94.

Benvenisti, M. (2000). *Sacred Landscape: The Buried History of the Holy Land Since 1948*. Berkeley: University of California Press.

Berg, E. (1998). "The Sherpa Pilgrimage to Uomi Tsho in the Context of the Worship of the Protector Deities: Ritual Practices, Local Meanings, and This-Wordly Request." *Himalayan Research Bulletin*, 18: 19–34.

Bhardwaj, S. M. (1990). "Hindu Deities and Pilgrimages in the United States," in G. Rinschede and S. M. Bhardwaj (eds.), *Pilgrimage in the United States*. Berlin: Dietrich Reimer, 211–28.

—— (1997). "Geography and Pilgrimage: A Review," in Stoddard and Morinis (1997*b*: 1–23).

Bhardwaj, S. M., and Rao, M. N. (1998). "The Temple as a Symbol of Hindu Identity in America?" *Journal of Cultural Geography*, 17: 125–43.

Braden, K. (1999). "Description of the Earth in Four Spiritual Maps." *Geography of Religions and Belief Systems*, 21/1: 1–3.

Bradley, M. B., Green, N. M., Jones, D. E., Lynn, M., and McNeil, L. (1992). *Churches and Church Membership in the United States 1990*. Atlanta: Glenmary Research Center.

Brenneman, W. L., Jr., and Brenneman, M. G. (1995). *Crossing the Circle at the Holy Wells of Ireland*. Charlottesville: University Press of Virginia.

Brockman, N. C. (1997). *Encyclopedia of Sacred Places*. Santa Barbara: ABC-CLIO.

Buffetrelle, K. (1998). "Reflections on Pilgrimages to Sacred Mountains, Lakes and Caves," in B. Z. Kedar and R. J. Z. Werblowsky (eds.), *Sacred Space: Shrine, City, Land*. New York: New York University Press, 18–34.

Cameron, C. (1990). "Pilgrims and Politics: Sikh Gurdwaras in California," in G. Rinschede and S. M. Bhardwaj (eds.), *Pilgrimage in the United States*. Berlin: Dietrich Reimer Verlag, 193–209.

Caplan, A. (1997). "The Role of Pilgrimage Priests in Perpetuating Spatial Organization within Hinduism," in Stoddard and Morinis (1997*b*: 209–33).

Carmichael, D. L., Hebert, J. L., Reeves, B., and Schanche, A. (eds.) (1994). *Sacred Sites, Sacred Places*. London: Routledge.

Chidester, D., and Linethal, E. T. (eds.) (1995). *American Sacred Space*. Bloomington: Indiana University Press.

Cohen, E. (1992). "Pilgrimage Centers: Concentric and Excentric." *Annals of Tourism Research*, 19: 33–50.

Cooper, A. (1992). "New Directions in the Geography of Religion." *Area*, 24: 123–9.

—— (1997*a*). "Sacred Interpretations of the Ecophysiology in Tropical Forest Canopies." *Pennsylvania Geographer*, 35/2: 5–31.

—— (1997*b*). *Sacred Mountains: Ancient Wisdom and Modern Meaning*. Edinburgh: Floris Books.

Crumrine, N. R., and Morinis, E. A. (eds.) (1991). *Pilgrimage in Latin America*. New York: Greenwood Press.

Curry-Roper, J. M. (1990). "Contemporary Christian Eschatologies and Their Relation to Environmental Stewardship." *Professional Geographer*, 42: 57–69.

Davidson, J. W., Hecht, A., and Whitney, H. A. (1990). "The Pilgrimage to Graceland," in G. Rinschede and S. M. Bhardwaj (eds.), *Pilgrimage in the United States*. Berlin: Dietrich Reimer, 229–52.

Dawson, A. H. (2000). "Geography, Religion and the State: A Comment on an Article by Michael Pacione." *Scottish Geographical Journal*, 116: 59–65.

Diamond, E. (1997). "Places of Worship: The Historical Geography of Religion in a Midwestern City, 1930–1960." *Pennsylvania Geographer*, 35: 45–68.

Din, A. K., and Hadi, A. S. (1997). "Muslim Pilgrimage from Malaysia," in Stoddard and Morinis (1997*b*: 161–82).

Dobbs, G. R. (1997). "Interpreting the Navajo Sacred Geography as a Landscape of Healing." *Pennsylvania Geographer*, 35: 136–50.

Emmett, C. (1995). *Beyond the Basilica: Christians and Muslims in Nazareth*, Geography Research Paper, 237. Chicago: Chicago University Press.

—— (1996). "The Capital Cities of Jerusalem." *The Geographical Review*, 86: 233–58.

—— (1997). "The Status Quo Solution for Jersusalem." *Journal of Palestine Studies*, 26: 16–28.

Faiers, G. E., and Prorok, C. V. (1990). "Pilgrimage to a 'National' American Shrine: 'Our Lady of Consolation' in Carey, Ohio," in G. Rinschede and S. M. Bhardwaj (eds.), *Pilgrimage in the United States*. Berlin: Dietrich Reimer, 137–47.

Fielder, J. (1995). "Sacred Sites and the City: Urban Aboriginality Ambivalence and Modernity," in R. Wilson and A. Dirlik (eds.), *Asia/Pacific as Space of Cultural Production*. London: Duke University Press, 101–19.

Foote, K. E. (1997). *Shadowed Ground: America's Landscapes of Violence and Tragedy*. Austin: University of Texas Press.

Forbes-Boyte, K. (1998). "It is All Sacred: Foothill Konkow Religious Perceptions of Sacred Place." *The Pennsylvania Geographer*, 36: 5–29.

Friedland, R., and Hecht, R. D. (1991). "The Politics of Sacred Space: Jerusalem's Temple Mount/al-Sharif," in Scott and Simpson-Housley (1991: 21–61).

Gaustad, E. S., and Barlow, P. L. (2000). *New Historical Atlas of Religion in America*. New York: Oxford University Press.

Geffen, J. P. (1998). "Landscapes of the Sacred: Avebury as a Case Study." *Geography of Religions & Belief Systems*, 20/2: 1, 3–4.

Giuriati, P., Myers, P. M. G., and Donach, M. E. (1990). "Pilgrims to 'Our Lady of the Snows' Belleville, Illinois in the Marian Year: 1987–1988," in G. Rinschede and S. M. Bhardwaj (eds.), *Pilgrimage in the United States*. Berlin: Dietrich Reimer, 49–192.

Giuriati, P., and Lanzi, G. (1994). "Pilgrims to Fatima as Compared to Lourdes and Medjugorje," in S. Bhardwaj, G. Rinschede, and A. Sievers (eds.), *Pilgrimage in the Old and New World*. Berlin: Dietrich Reimer, 57–79.

Grapard, A. B. (1998). "Geotyping Sacred Space: The Case of Mount Hiro in Japan," in B. Z. Kedar and R. J. Z. Werblowsky (eds.), *Sacred Space: Shrine, City, Land*. New York: New York University Press, 215–58.

Gutschow, N. (1994). "Varanasi/Benares: The Centre of Hinduism?" *Erdkunde*, 48: 94–209.

Hardwick, S. (1991). "The Impact of Religion on Ethnic Survival: Russian Old Believers in Alaska." *The California Geographer*, 31: 19–36.

—— (1993*a*). "Origin and Diffusion of Russian Baptists and Pentecostals in Russia and Ukraine." *Pennsylvania Geographer*, 31: 2–13.

—— (1993*b*). *Russian Refuge: Religion, Migration, and Settlement on the North American Pacific Rim*. Chicago: University of Chicago Press.

Harpur, J. (1994). *Atlas of Sacred Places: Meeting Points of Heaven and Earth*. New York: Henry Holt.

Henderson, M. L. (1993). "What Is Spiritual Geography?" *Geographic Review*, 83: 469–72.

Hobbs, J. J. (1992). "Sacred Space and Touristic Development At Jebel Musa (Mt. Sinai), Egypt." *Journal of Cultural Geography*, 12: 99–113.

Hudman, L. E., and Jackson, R. H. (1992). "Mormon Pilgrimage and Tourism." *Annals of Tourism Research*, 19: 107–21.

Isaac, E. (1965). "Religious Geography and the Geography of Religion," in *Man and the Earth*, University of Colorado Studies, Series in Earth Sciences, 3. Boulder: University of Colorado Press, 1–14.

Jackowski, A. (1990). "Development of Pilgrimages in Poland: Geographical—Historical Study," in G. Lallanji and D. P. Dubey (eds.), *Pilgrimage Studies: Text and Context*. Allahabad: The Society of Pilgrimage Studies, 241–50.

Jackowski, A., and Smith, V. L. (1992). "Polish Pilgrim-Tourists." *Annals of Tourism Research*, 19: 92–106.

Jackson, R. H., Rinschede, G., and Knapp, J. (1990). "Pilgrimage in the Mormon Church," in G. Rinschede and S. M. Bhardwaj (eds.), *Pilgrimage in the United States*. Berlin: Dietrich Reimer, 27–61.

Jett, S. C. (1992). "An Introduction to Navajo Sacred Places." *Journal of Cultural Geography*, 13: 29–39.

Karan, P. P. (1997). "Patterns of Pilgrimage to the Sikh Shrine of Guru Gobind Singh at Patna," in Stoddard and Morinis (1997*b*: 257–68).

Kaschewsky, R. (1994). "Muktinath—A Pilgrimage Place in the Himalayas," in S. Bhardwaj *et al.* (eds.), *Pilgrimage in the Old and New World*. Berlin: Dietrich Reimer, 139–68.

Katz, Y. (1991). "The Jewish Religion and Spatial and Communal Organization: The Implementation of Jewish Religious Law in the Building of Urban Neighborhoods and Jewish Agricultural Settlements in Palestine at the Close of the Nineteenth Century," in Scott and Simpson-Housley (1991: 3–19).

Katz, Y., and Lehr, J. C. (1991). "Jewish and Mormon Agricultural Settlement in Western Canada: A Comparative Analysis." *The Canadian Geographer*, 35: 128–42.

Kedar, B. Z., and Werblowsky, R. J. Z. (eds.) (1998). *Sacred Space: Shrine, City, Land*. New York: New York University Press.

Kong, L. (1990). "Geography and Religion: Trends and Prospects." *Progess in Human Geography*, 14: 355–71.

—— (1993*a*). "Ideological Hegemony and the Political Symbolism of Religious Buildings in Singapore." *Environment & Planning D: Society & Space*, 11: 23–46.

—— (1993*b*). "Negotiating Conceptions of 'Sacred Space': A Case Study of Religious Buildings in Singapore." *Transactions of the Institute of British Geographers*, 18: 342–58.

—— (2001). "Mapping 'New' Geographies of Religion: Politics and Poetics in Modernity," *Progress in Human Geography*, 25/2: 211–33.

Krindatch, A. (1996). *Geography of Religions in Russia*. Decatur, Ga.: Glenmary Research Center.

Kuhlken, R. (1997). "Sacred Landscapes and Settlement Mythology in the Fiji Islands." *Pennsylvania Geographer*, 35: 173–202.

Levine, G. J. (1986). "The Geography of Religion." *Transactions, Institute of British Geographers*, 11: 428–40.

Ley, D., and Martin, R. B. (1993). "Gentrification as Secularization: The Status of Religious Belief in the Post-Industrial City." *Social Compass*, 40: 217–31.

Louder, D. R. (1993). "Canadian Mormons in their North American Context: A Portrait." *Social Compass*, 40: 271–90.

McCormick, T. (1997). "The Jaina Ascetic as Manifestation of the Sacred," in Stoddard and Morinis (1997*b*: 235–56).

McDonald, D. (1995). "Changes Along the Peyote Road." *Geography of Religions and Belief Systems*, 17/2: 1–4.

McGuire, M. B. (1991). "Religion and Region: Sociological and Historical Perspective." *Journal for the Scientific Study of Religion*, 30: 544–7.

Malville, J. M. (1998). "The Symbolic Landscape of Vijayanagara." *Pennsylvania Geographer*, 36: 30–54.

—— (1999). "Complexity and Self-Organization in Pilgrimage Systems." *Proceedings, International Seminar: Pilgrimage and Complexity*. New Delhi: Indira Gandhi National Centre for the Arts.

Mather, J. R., and Mather, S. P. (1997). "Deborah—Prophetess and Possible Applied Climatologist." *Pennsylvania Geographer*, 35/2: 32–44.

Morin, K., and Guelke, J. Kay (1998). "Strategies of Representation, Relationship, and Resistance: British Women Travelers and Mormon Plural Wives, c 1870–1890." *Annals of the Association of American Geographers*, 88: 436–62.

Nakagawa, T. (1990). "Louisiana Cemeteries as Cultural Artifacts." *Geographical Review of Japan*, Series B, 63: 139–55.

Naquin, S., and Yu, C. F. (eds.) (1992). *Pilgrims and Sacred Sites in China*. Berkeley: University of California Press.

Newman, W. M., and Halvorson, P. (2000). *Atlas of Religious Change in America, 1952–1990*. 2nd edn. Atlanta, Ga.: Glenmary Research Center.

Nolan, M., and Nolan, S. (1992). "Religious Sites as Tourism Attractions in Europe." *Annals of Tourism Research*, 19: 68–78.

—— (1997). "Regional Variations in Europe's Roman Catholic Pilgrimage Traditions," in Stoddard and Morinis (1997*b*: 61–93).

Norris, K. (1993). *Dakota: A Spiritual Geography*. New York: Ticknor & Fields.

Numrich, P. D. (1997). "Recent Immigrant Religions in a Restructuring Metropolis: New Religious Landscapes in Chicago." *Journal of Cultural Geography*, 17: 55–76.

O'Brien, J., Palmer, M., and Barrett, D. B. (eds.) (1993). *The State of Religion Atlas*. New York: Simon & Schuster.

Osterrieth, A. (1997). "Pilgrimage, Travel and Existential Quest," in Stoddard and Morinis (1997*b*: 25–39).

Pacione, M. (1991). "The Church Urban Fund: A Religio-Geographical Perspective." *Area*, 23: 101–10.

—— (1999). "The Relevance of Religion for a Relevant Geography." *Scottish Geographical Journal*, 115: 117–31.

Park, C. C. (1994). *Sacred Worlds: An Introduction to Geography and Religion*. London: Routledge.

Peet, R. (1998). *Modern Geographical Thought*. Oxford: Blackwell.

Price, N. (1994). "Tourism and the Bighorn Medicine Wheel: How Multiple Use Does Not Work for Sacred Land Sites," in Carmichael *et al.* (1994: 259–64).

Prorok, C. V. (1991). "Evolution of the Hindu Temple in Trinidad." *Caribbean Geography*, 3: 73–93.

—— (1997). "Becoming a Place of Pilgrimage: An Eliadean Interpretation of the Miracle at Ambridge, Pennsylvania," in Stoddard and Morinis (1997*b*: 117–39).

—— (1998). "Dancing in the Fire: Ritually Constructing Hindu Identity in a Malaysian Landscape." *Journal of Cultural Geography*, 17: 89–114.

—— (2000). "Boundaries are Made for Crossing: The Feminized Spatiality of Puerto Rican Espiritismo in New York City." *Gender Place & Culture: A Journal of Feminist Geography*, 7: 57–80.

Prorok, C. V., and Hemmasi, M. (1993). "Muslims and their Mosques in Trinidad: A Geography of Religious Structures and Ethnic Identity." *Caribbean Geography*, 4: 28–48.

Prorok, C. V., and Kimber, C. T. (1997). "The Hindu Gardens of Trinidad: Cultural Continuity and Change in a Caribbean Landscape." *Pennsylvania Geographer*, 35: 98–135.

Pulido, L. (1998). "The Sacredness of 'Mother Earth'; Spirituality, Activism, and Social Justice." *Annals of the Association of American Geographers*, 88: 719–23.

Raivo, P. J. (1997). "Comparative Religion and Geography: Some Remarks on the Geography of Religion and Religious Geography." *Temenos*, 33: 137–49.

Reader, I., and Walter, T. (eds.) (1993). *Pilgrimage in Popular Culture*. Basingstoke: Macmillan.

Rinschede, G. (1990). "Catholic Pilgrimage Places in the United States," in G. Rinschede and S. M. Bhardwaj (eds.), *Pilgrimage in the United States*. Berlin: Dietrich Reimer, 63–135.

— (1997). "Pilgrimage Studies at Different Levels," in Stoddard and Morinis (1997*b*: 95–115).

Rowley, G. (1997). "The Pilgrimage to Mecca and the Centrality of Islam," in Stoddard and Morinis (1997*b*: 141–59).

Scott, J., and Simpson-Housley, P. (eds.) (1991). *Sacred Places and Profane Spaces: Essays in the Geographics of Judaism, Christianity, and Islam*. New York: Greenwood Press.

Sellers, R. W., and Walters, T. (1993). "From Custer to Kent State: Heroes, Martyrs and the Evolution of Popular Shrines in the USA," in Reader and Walters (1993: 179–200).

Sherrill, R. A. (1995). "American Sacred Space and the Contest of History," in Chidester and Linenthal 1995: 313–40).

Sheskin, I. M. (1993). "Jewish Metropolitan Homelands." *Journal of Cultural Geography*, 13: 19–32.

Shilong, Z. (1993). "Ashkenzai Jewish Almshouses in Jerusalem." *Journal of Cultural Geography,* 14: 5–48.

Shimazuki, H. T. (1997). "The Shikoku Pilgrimage: Essential Characteristics of a Japanese Buddhist Pilgrimage Complex," in Stoddard and Morinis (1997*b*: 269–97).

Singh, R. P. B. (1991). "Rama's Route After Banishment: A Geographic Viewpoint." *Journal of Scientific Research*, 41B: 9–46.

— (1994). "The Sacred Geometry of India's Holy City, Varanasi: Kashi as Cosmogram." *National Geographical Journal of India*, 40: 1–31.

— (1997). "Sacred Space and Pilgrimage in Hindu Society: The Case of Varanasi," in Stoddard and Morinis (1997*b*: 191–207).

Singh, R. P. B., and Malville, J. M. (1995). "Cosmic Order and Cityscape of Varanasi (Kashi): Sun Images and Cultural Astronomy." *National Geographical Journal of India*, 41: 9–88.

Sinha, A. (1998). "Design of Settlements in the Vaastu Shastras." *Journal of Cultural Geography*, 17: 27–41.

Smart, N. (ed.) (1999). *Atlas of the World's Religions*. New York: Oxford University Press.

Smith, V. L. (1992). "Introduction: The Quest in Guest." *Annals of Tourism Research*, 19: 1–17.

Sopher, D. (1967). *Geography of Religions*. Englewood Cliffs, NJ: Prentice-Hall.

— (1981). "Geography and Religion." *Progress in Human Geography*, 5: 510–24.

— (1997). "The Goal of Indian Pilgrimage: Geographical Considerations," in Stoddard and Morinis (1997*b*: 183–90).

Stoddard, R. H. (1990). "Some Comments about the History of The Geography of Religion and Belief Systems." *Geography of Religions & Belief Systems*, 12/2: 1–3.

— (1994). "Major Pilgrimage Places of the World," in S. M. Bhardwaj, G. Rinschede, and A. Sievers (eds.), *Pilgrimage in the Old and New World*. Berlin: Dietrich Reimer, 17–36.

— (1997). "Defining and Classifying Pilgrimages," in Stoddard and Morinis (1997*b*: 41–60).

Stoddard, R. H., and Morinis, A. (1997*a*). "Introduction: The Geographic Contributions to Studies of Pilgrimage," in Stoddard and Morinis (1997*b*: pp. ix–xi).

— (1997*b*). *Sacred Places, Sacred Spaces: The Geography of Pilgrimages*. Baton Rouge, La.: Geoscience Publications.

Stump, R. W. (2000). *Boundaries of Faith: Geographical Perspectives on Religious Fundamentalism*. Lanham, Md.: Rowman & Littlefield.

Teather, E. K. (1999). "High-Rise Homes for the Ancestors: Cremation in Hong Kong." *Geographical Review*, 89: 409–30.

Tharan, Z. J. (1997). "The Jewish American Environmental Movement: Stewardship, Renewal and the Greening of Diaspora Politics." *Pennsylvania Geographer*, 35: 69–97.

Tobin, D. P. (1998). "Moving a Monastery in Modern Times: The Abbey of St. Walburga, Colorado." *The Pennsylvania Geographer*, 36: 124–34.

Tweed, T. (1997). *Our Lady of the Exile: Diasporic Religion at a Cuban Catholic Shrine in Miami*. New York: Oxford University Press.

van Spengen, W. (1998). "On the Geographical and Material Contextuality of Tibetan Pilgrimage," in A. McKay (ed.), *Pilgrimage in Tibet*. Richmond: Curzon Press, 35–51.

Voeks, R. (1990). "Sacred Leaves of Brazilian Candomble." *Geographical Review*, 80: 118–31.

Wagner, P. L. (1997). "Pilgrimage: Culture and Geography," in Stoddard and Morinis (1997*b*: 299–323).

Webster, G. R. (1997). "Religion and Politics in the American South." *Pennsylvania Geographer*, 35: 173–202.

Weightman, B. A. (1993). "Changing Religious Landscapes in Los Angeles." *Journal of Cultural Geography*, 14: 1–20.

— (1996). "Sacred Landscapes and the Phenomenon of Light." *Geographical Review*, 86: 59–71.

Wilson, D. (1993). "Connecting Social Process and Space in the Geography of Religion." *Area*, 25: 75–8.

Yeoh, B. S. A., and Hui, T. B. (1995). "The Politics of Space: Changing Discourses on Chinese Burial Grounds in Post-War Singapore." *Journal of Historical Geography*, 21: 184–201.

Zelinsky, W. (1990). "Nationalistic Pilgrimages in the United States," in G. Rinschede and S. M. Bhardwaj (eds.), *Pilgrimage in the United States*. Berlin: Dietrich Reimer, 253–67.

— (1994). *Exploring the Beloved Country: Geographic Forays into American Society and Culture*. Iowa City: University of Iowa.

Index of Names

Index of Subjects